The Complete Dinosaur

LIFE OF THE PAST   James O. Farlow, editor

INDIANA UNIVERSITY PRESS   Bloomington & Indianapolis

THE COMPLETE
DINOSAUR

Edited by

M. K. Brett-Surman
Thomas R. Holtz, Jr.
James O. Farlow
Bob Walters, Art Consultant

SECOND EDITION

The editors offer special thanks to Jim Whitcraft for creating the illustrations that appear on the opening page of each chapter.

This book is a publication of

Indiana University Press
Office of Scholarly Publishing
Herman B Wells Library 350
1320 East 10th Street
Bloomington, Indiana 47405 USA

iupress.indiana.edu

The paper used in this publication meets the minimum requirements of the American National Standard for Information Sciences–Permanence of Paper for Printed Library Materials, ANSI Z39.48–1992.

Manufactured in the
United States of America

Library of Congress
Cataloging-in-Publication Data

The complete dinosaur / edited by M. K. Brett-Surman, Thomas R. Holtz Jr., and James O. Farlow ; Bob Walters, art director.–2nd ed.
   p. cm.–(Life of the past)
   Includes index.
   ISBN 978-0-253-35701-4 (cl : alk. paper)–ISBN 978-0-253-00849-7 (ebook) 1. Dinosaurs. I. Brett-Surman, M. K., [date]- II. Holtz, Thomas R., [date]- III. Farlow, James Orville [date].
   QE862.D5C697 2012
   567.9–dc23
                                    2011050297

3 4 5 6 7   22 21 20 19 18 17

This second edition is dedicated to our colleagues, and our friends:

Halszka Osmólska

John H. Ostrom

John S. McIntosh

W. A. S. Sarjeant

Edwin Colbert

Tobe Wilkins

Jim Adams

Robin Reid

Donna Engard

Thomas Jericho

You advanced our science. You made a difference.

The editors wish to thank the staff at Indiana University Press for their hard work and dedication, which allowed our "opusaurus" to be born after a ten-year gestation.

Contents

Dinosauria

In April 1842, Richard Owen coined the term *Dinosauria* in a footnote on page 103 of his *Report on British Fossil Reptiles*, and defined this new name as meaning "fearfully great, a lizard." Since that time the name has always, incorrectly, been translated as "terrible lizard." How did this etymological and aesthetic error occur? Modern dictionaries always give the meaning of *deinos* as "terrible." This is correct, if one uses the word as an adjective – but Owen used the superlative form of *deinos*, just as did Homer in the *Iliad*. A check of a Greek–English lexicon from Owen's time will confirm this (Donnegan 1832). Dinosaurs are not lizards, nor are they terrible. They are, instead, the world's most famous living superlative.

J.O.F. and M.K.B.-S.

References

Farlow, J. O., and M. K. Brett-Surman. 1997. *The Complete Dinosaur.* Bloomington: Indiana University Press.

Donnegan, J., MD. 1832. *A New Greek and English Lexicon: Principally on the Plan of the Greek and German Lexicon of Schneider.* First American edition from the second London edition, revised and enlarged by R. B. Patton. Boston: Hilliard, Gray.

Owen, R. 1842. *Report on British Fossil Reptiles.* Part II. Report of the Eleventh Meeting of the British Association for the Advancement of Science 1841: 60–204.

Contributors

C

Art Andersen Virtual Surfaces, Inc., 832 East Rand Road, Suite 16, Mt. Prospect, IL 60056 USA

J. David Archibald Department of Biology, San Diego State University, 5500 Campanile Drive, San Diego, CA 92182-4614 USA

Paul M. Barrett Department of Palaeontology, The Natural History Museum, Cromwell Road, London SW7 5BD, UK

Michael J. Benton School of Earth Sciences, Wills Memorial Building, University of Bristol, Queen's Road, Bristol, BS8 1RJ, UK

Brent H. Breithaupt Bureau of Land Management, Wyoming State Office, P.O. Box 1828, Cheyenne, WY 82003 USA

Michael K. Brett-Surman Museum Specialist for Fossil Reptiles and Amphibians, National Museum of Natural History of the Smithsonian Institution, Washington, DC 20560 USA

Emily Buchholtz Department of Biological Sciences, Wellesley College, Wellesley Hills, MA 02481-8203 USA

Elisabeth K. Burton Department of Integrative Biology and Museum of Paleontology, University of California, Berkeley, CA 94720 USA

Richard J. Butler Bayerische Staatssammlung für Paläontologie und Geologie, Richard-Wagner-Str. 10, 80333 Munich, Germany

Kenneth Carpenter Prehistoric Museum, Utah State University–Eastern, 155 E Main Street, Price, UT 84501 USA

Matthew T. Carrano Department of Paleobiology, National Museum of Natural History, Smithsonian Institution, P.O. Box 37012, MRC 121, Washington, DC 20013-7012 USA

Ralph E. Chapman New Mexico Virtualization, LLC, 102 El Morro Street, Los Alamos, NM 87544 USA

Karen Chin Department of Geological Sciences and University of Colorado Museum, UCB 265, University of Colorado at Boulder, 2200 Colorado Avenue, Boulder, CO 80309-0399 USA

Daniel J. Chure Dinosaur National Monument, Box 92, Jensen, UT 84035 USA

†Edwin H. Colbert (1905–2001) led fossil-collecting expeditions to all parts of the globe and published many books and articles about dinosaurs and other extinct vertebrates, including *The Little Dinosaurs of Ghost Ranch* and *The Great Dinosaur Hunters and Their Discoveries.*

David W. Dilkes Department of Biology and Microbiology, University of Wisconsin Oshkosh, 800 Algoma Boulevard, Oshkosh, WI 54901-8440 USA

Gregory M. Erickson Department of Biological Science, 4011 King Life Sciences, 319 Stadium Drive, Florida State University, Tallahassee, FL 32306-4295 USA

James O. Farlow Department of Geosciences, SB 242, Indiana University Purdue University Fort Wayne, 2101 Coliseum Boulevard E, Fort Wayne, IN 46805-1499 USA

Nicholas C. Fraser National Museums Scotland, Chambers Street, Edinburgh, Scotland, EH1 1JF, UK

Peter M. Galton Professor Emeritus, University of Bridgeport, Bridgeport, CT; mailing address: 315 Southern Hills Drive, Rio Vista, CA 94571-2153 USA

Nicholas R. Geist Department of Biology, Sonoma State University, 1801 E Cotati Avenue, Rohnert Park, CA 94928-3609 USA

David D. Gillette Museum of Northern Arizona, 3101 N Fort Valley Road, Flagstaff, AZ 86001 USA

Amy E. Harwell Oregon State University, Zoology Department, Corvallis, OR 97331 USA

Donald Henderson Curator of Dinosaurs, Royal Tyrrell Museum, P.O. Box 7500, Highway 838 Midland Provincial Park, Drumheller, Alberta T0J 0Y0 Canada

Douglas Henderson P.O. Box 368, 117 Makowski Lane, Whitehall, MT 59759 USA

Willem J. Hillenius Department of Biology, College of Charleston, 66 George Street, Charleston, SC 29424 USA

Casey M. Holliday Program in Integrative Anatomy, Department of Pathology and Anatomical Sciences, University of Missouri School of Medicine, University of Missouri, Columbia, MO 65212 USA

Thomas R. Holtz Jr. Department of Geology, Building 237, Room 1117, University of Maryland, College Park, MD 20472 USA

David W. E. Hone Institute of Vertebrate Palaeontology & Palaeoanthropology, Chinese Academy of Sciences, 142 Xizhimenwai Dajie, 100044 Beijing, China

John R. Horner Museum of the Rockies, Montana State University, Bozeman, MT 59717 USA

John R. Hutchinson Evolutionary Biomechanics Department, Structure and Motion Lab, Veterinary Basic Sciences, The Royal Veterinary College, Hawkshead Lane, North Mymms, Hatfield, Herts AL9 7TA, UK

Terry D. Jones Department of Biological Sciences, California State University–Stanislaus, Turlock, CA 95382 USA

James I. Kirkland State Paleontologist, Utah Geological Survey, 1594 W North Temple, Suite 3110, P.O. Box 146100, Salt Lake City, UT 84114-6100 USA

John A. Long Natural History Museum of Los Angeles County, 900 Exposition Boulevard, Los Angeles, CA 90007 USA

Peter Makovicky Department of Geology, The Field Museum, 1400 S Lake Shore Drive, Chicago, IL 60605-2496 USA

Mark Marshall Eli Lilly and Company, Lilly Corporate Center, Indianapolis, IN 46285 USA

Neffra Matthews Bureau of Land Management, National Operations Center, Denver, CO 80225-0047 USA

Kenneth J. McNamara Department of Earth Sciences, University of Cambridge, Downing Street, Cambridge CB2 3EQ, UK

Ralph E. Molnar 402 West Apache Road, Flagstaff, AZ 86001 USA

Darren Naish Ocean and Earth Science, National Oceanography Centre, University of Southampton, Southampton, SO14 3ZH, UK

Kevin Padian Department of Integrative Biology and Museum of Paleontology, University of California, Berkeley, CA 94720 USA

J. Michael Parrish Office of the Dean, College of Science, San José State University, One Washington Square, San José, CA 95192-0099 USA

Gregory S. Paul 3109 N Calvert Street, Baltimore, MD 21218 USA

Devon E. Quick Oregon State University, Zoology Department, Corvallis, OR 97331 USA

Elizabeth Rega Associate Professor of Anatomy, Western University of Health Sciences, 309 E Second Street, Pomona, CA 91766-1854 USA

†R. E. H. Reid (1924–2007) Irish paleontologist, formerly Queen's University Belfast, was one of the leading authorities on the interpretation of the internal microscopic structure of dinosaur bone.

Kristina Curry Rogers Geology Department, Macalester College, 1600 Grand Avenue, St. Paul, MN 55105 USA

John A. Ruben Department of Zoology, Oregon State University, Corvallis, OR 97331-2914 USA

†William A. S. Sarjeant (1935–2002) formerly Department of Geological Sciences, University of Saskatchewan, Saskatoon, Saskatchewan S7N 0W0 Canada

Mary Higby Schweitzer Department of Marine, Earth, and Atmospheric Sciences, Room 3135 Jordan Hall, Campus Box 8208, North Carolina State University; and North Carolina State Museum of Natural Sciences, Raleigh, NC 27695 USA

David A. E. Spalding 1105 Ogden Road, Pender Island, British Columbia V0N 2M1 Canada; professional affiliation: Department of Geological Sciences, University of Saskatchewan

Hans-Dieter Sues Department of Paleobiology, National Museum of Natural History, Smithsonian Institution, MRC 121, P. O. Box 37012, Washington, DC 20013-7012 USA

Corwin Sullivan Key Laboratory of Evolutionary Systematics of Vertebrates, Institute of Vertebrate Paleontology and Paleoanthropology, 142 Xizhimenwai Dajie, Beijing, 100044, China

François Therrien Royal Tyrrell Museum of Paleontology, Drumheller, Alberta T0J 0Y0 Canada

Bruce H. Tiffney Department of Earth Sciences, University of California, Santa Barbara, CA 93106 USA

Hugh S. Torrens Lower Mill Cottage, Furnace Lane, Madeley, Crewe CW3 9EU, UK

Jeffrey A. Wilson Museum of Paleontology and Department of Geological Sciences, University of Michigan, 1109 Geddes Road, Ann Arbor, MI 48109-1079 USA

Lawrence M. Witmer Department of Biomedical Sciences, College of Osteopathic Medicine, Life Sciences Building, Room 123, Ohio University, Athens, OH 45701 USA

Xing Xu Institute of Vertebrate Paleontology and Paleoanthropology of the Chinese Academy of Sciences, Beijing, China

Adam M. Yates School of Geosciences, University of Witwatersrand, Johannesburg, South Africa

Darla K. Zelenitsky Department of Geoscience, University of Calgary, 2500 University Drive NW, Calgary, Alberta T2N 1N4 Canada

Dinosaurs: The Earliest Discoveries

David A. E. Spalding and †William A. S. Sarjeant (1935–2002)

1

The first trackers of dinosaurs were probably other dinosaurs, as tracks have been found apparently showing carnivorous species following herbivores (Lockley 1991, 184). More recently, there is evidence that some early people, whose livelihood came partly from tracking, killing, and dismembering animals, sometimes observed and found significance in tracks, bones, and eggs of long-extinct species of no culinary value.

Traditional knowledge of large fossils has been found to persist among aboriginal peoples on several continents. Pertinent observations have been documented, but often in sources that have not generally received the attention of paleontologists until recent decades. Dinosaur trackways in situ have apparently been marked by petroglyphs and pictographs of uncertain age, so that they can be seen to have been of some significance to their finders. Other specimens have been collected in ancient times and are now found in archaeological contexts. The surviving oral and published record is widely scattered through ancient, medieval, and later literature and appears in the forms of folklore, the tales of travelers, visual records of legendary events, and oral data collected and documented by anthropologists, dinosaur researchers, and aboriginal people. Prescientific cultures have offered a variety of explanations for the remains they observed ranging from mythological to protoscientific.

In this chapter we summarize what is known of early observations of dinosaurs, in approximately chronological sequence before the rise of modern paleontology, discuss the scientific discoveries which led to the naming of the first two genera, and mention the other genera named before Owen recognized a common identity among the remains in 1841.

Simpson's classic paper (1942, 131) presents a framework for the history of fossil vertebrate discoveries. He recognizes a number of periods in North American vertebrate paleontology, of which we are here concerned with the first three. Simpson's prescientific period includes early discoveries and removal of some specimens to Europe, but "no truly scientific study . . . had been made." This period extends in North America "from the earliest times to about 1762." Simpson's protoscientific period extends from about 1762 to 1799, in which "vertebrate paleontology was not yet a true science but basic methods were being invented and sporadically applied." In the pioneer scientific period (1799–ca. 1842), Cuvier established the subject "as a true and defined science," while others adopted Cuvier's methods. While there is room for discussion of the appropriate dates of application of the periods outside North America, Simpson's structure provides a useful framework.

Other writers have extended and elaborated on Simpson's approach, paying particular attention to the early beginnings of science in the Western world. Numerous classic dinosaur texts have been pulled together in Weishampel and White (2003). Surveys of early dinosaur discoveries have been published by Buffetaut and Le Loeuff (1993), Delair and Sarjeant (1975, 2002), and Sarjeant (1987, 1997, 2003). Recent books by Adrienne Mayor (2000, 2005) and Jose Sanz (1999), and papers by Mayor (2007) and Mayor and Sarjeant (2001) have addressed discoveries of fossils in ancient civilizations and ethnographic contexts, shedding much light on the beginnings of discovery and interpretation of fossil remains.

It is now clear that many dinosaur and other fossil discoveries have been made by prescientific societies. Some fossil discoveries are commemorated in place-names (Mayor 2007). Some are found only in archaeological contexts, for which no explanations are recorded. The record of the Mediterranean and Chinese civilizations documents ancient fossil discoveries, for some of which there is a written or even visual record offering contemporary interpretations. Ethnographic data from many cultures around the world show different interpretations of vertebrate fossil remains. Some myths of monsters may have roots in fossil discoveries, what Mayor (following Dodson) calls "fossil legends" for traditional tales that specifically refer to physical evidence (Mayor 2005, xxix). In some instances, logical explanations reflecting awareness of geological change, deep time, and ancestral relationships show the development of protoscientific ideas in nonscientific cultures. Medieval societies in Europe begin with the same variety of types of explanations for fossils. Through the Renaissance, more or less fanciful explanations are offered until the emergence of truly scientific methods and explanations came in the last two centuries.

Although history shows a broad evolution of interpretations of vertebrate fossils from legendary to protoscientific to scientific, the progression of ideas appears to be linear only when the most scientific are considered. For instance, less than four decades separate "Noah's Raven" from Hitchcock's *Ornithoidichnites*, but it is likely that the traditional views continued in folk belief in the area. Even in our own day, legendary and scientific explanations may be offered of the same occurrence by different segments of society. Thus the tracks (supported by associated forgeries) at the Paluxy River site of Dinosaur State Park, Texas, are viewed by "creation scientists" as proof of the contemporaneity of dinosaurs and humans before Noah's flood (Morris 1980), while the same site is interpreted by paleontologists as showing Cretaceous sauropod and theropod tracks (Jacobs 1995). Forged human tracks from this site are documented as far back as 1939 and discussed in a context of other fossil-related puzzles by Mayor (2005, 302) as "frauds and specious legends."

Ancient Asia

Perhaps the oldest evidence of human connection with dinosaurs comes from the Gobi Desert of Mongolia, where in a site ranging in age from late Paleolithic to early Neolithic Roy Chapman Andrews found "bits of dinosaur egg shell, drilled with neat round holes—evidently used in necklaces by primitive peoples" (Andrews 1943, 238; Carpenter et al. 1994, 1).

Dragons played an important part in Chinese lore as far back as the protohistoric period; two emperors are reputedly immediate descendents of dragons, and two azure-colored dragons are reported to have presided over the birth of Confucius. Dragons have been figured in Chinese art as far back as 1100 BC (Andersson 1934).

A possible connection between Chinese dragons and the fossil remains of dinosaurs was noted as early as 1886, when British/Tasmanian geologist and folklorist Charles Gould (1834–1895) drew attention to the Chinese "dragon bones" and figured an *Iguanodon* skeleton in his gathering of dragon myths (Gould 1886, 199).

Mayor (2000, 39) documents that the *I Ching*, a compilation drawing on older traditions going back to 1000 BC, includes as a good omen "Dragons encountered in the fields" and suggests that these refer to bones plowed up during agriculture. In the second century BC, bones (possibly of dinosaurs) found during digging of a canal in northern China led to it being called the Dragon-Head Waterway (Mayor 2007). A later report of dragon bones from Wucheng (now Santain County, Sichuan province) is documented by Cheng Qu during the period of the western Jin dynasty, AD 265–317, in a work entitled *Hua Yang Goo Zhi* (Needham 1959). Although "dragon bones" are generally from fossil mammals, Dong (in Dong and Milner 1988) considered it highly probable that these particular examples represented dinosaurs.

Large fossilized bones were found near Jabalpur in what is now Madhya Pradesh in India by W. H. Sleeman and reported by G. G. Spilsbury in the 1830s. These were not described until 1868 in Hugh Falconer's posthumous memoirs or named until Richard Lydekker included some of the material in *Titanosaurus indicus* in 1877 (Barrett et al. 2008).

In the 1930s, in what was then Indochina, French geologist Josué-Heilmann Hoffet reported a dinosaur caudal vertebra from near Phalane. "Alerted by this discovery, the natives, who had often seen similar bones but had thought they came from buffaloes, told me about places where they thought such remains were to be found. . . . The bones belonged to genies, and evil would befall anyone who removed them" (Taquet 1994, 148). The "bones" proved to be only "limestones sculpted into bizarre shapes," though there were dinosaur bones in the vicinity. It was not until 1990 that new work was done in the area by Philippe Taquet, who located one of Hoffet's local assistants. The near-scientific explanation—water buffalo was the largest mammal known to the locals—proved to be firmly intermingled with folkloric traditions, for the elders remembered that "they filled forty-three baskets of bones of an ox called . . . the magnificent ox. . . . When they began to dig, there was a lightning storm . . . this was a . . . warning" (155–156). Taquet found it necessary to buy a pig from the local priest and sacrifice it before he was able to proceed with his fieldwork (156).

Classical Mediterranean

Mayor (2000, 5) notes that Cuvier was well aware of fossil finds in classical times, but that such records have largely escaped the attention of scientists (and classicists) in more recent years. She has documented many references to discoveries of large fossil bones in the ancient world, from the Mediterranean east to India and China, in areas now known for their remains of fossil vertebrates. She shows that fossil bones were well known to inhabitants

of many countries, were gathered in temples and other public places, and were often interpreted as remains of giants, monsters, legendary heroes, and other fantastic creatures.

While many of these finds are undoubtedly attributable to proboscideans, cetaceans, and other mammals, some may reflect dinosaur discoveries. A remarkable trail of documentation points to the possibility that Gobi dinosaurs may have given rise to the legendary griffin (Fig. 1.1), which became a popular motif in Greece around 700 BC (This suggestion has, however, been criticized by specialists in Gobi dinosaurs; see, for instance, Novacek 1996, 140ff.), though the same author warms slightly to the idea later (2002, 296).

Later, around 430 BC, Herodotus, in pursuit of tales of flying reptiles in Egypt, made a special trip to see "bones and spines in incalculable numbers, piled in heaps, some big and some small." Mayor (2000, 135) suggests these might have been spinosaurs, known from Egypt.

Mythical Monsters in Medieval Europe

In Europe, fossil footprints of dinosaurs and their progenitors were also linked to legends. While it is known that some dragon tales were inspired by bones of Pleistocene rhinoceros and bear (Buffetaut 1987, 13–14), it is also possible that footprints exposed in the Rhine Valley, Western Germany, may have inspired the story of the slaying of the dragon by the hero Siegfried (Kirchner 1941). When a footprint in Triassic sandstone of *Chirotherium*—the track maker was a nondinosaurian early archosaur—was incorporated into the stonework of Christ Church, Higher Bebington, Cheshire,

England, it came to be known locally as the Devil's Toenail. Many other dragon tales from Britain may ultimately derive from tracks or bones, including a report by J. Trundle (1614) of a "Strange and Monstrous Serpent or Dragon lately discovered and yet living" in Sussex. This dragon was said to be 9 feet (almost 3 m) long, and poor woodcuts of its limb bones were given.

In France, it is probably not coincidental that there is a concentration of dinosaur legends in Provence, where dinosaur eggs are abundant. One dragon at Aix is reputed to have been burst asunder by St. Margaret; another at Tarascon was first vanquished by Hercules, then (in a remarkable show of ecumenical spirit) by St. Martha. At Draguignan, the mayor has the right to have any of his godchildren christened "Drac" (Huxley 1979).

Further south, in Portugal, dinosaur tracks at Cabo Espichel are plainly visible (though not readily accessible) in the cliffs. Perched on the cliff edge is a small chapel, Capela da Memoria (memory chapel), which celebrates the legend of Nossa Senhora da Pedra da Mua (our lady of the mule tracks), commemorating the arrival of the Virgin at this location to evangelize Portugal. Inside the building, an eighteenth-century mural of painted tiles shows the event, with the Virgin riding a mule on the cliff top, escorted by angels and welcomed by residents. In the mural, the "mule's" tracks ascending the cliff are clearly shown, inadvertently providing the first illustration that certainly figures dinosaur tracks (Sanz 2000, 269, 2003, 18–19; Santos and Rodrigues 2008).

Archaeological and Ethnographic Data from the New World

Adrienne Mayor's *Fossil Legends of the First Americans* (2005) has shown unequivocally that "Native Americans observed, collected, and attempted to explain the remains of extinct . . . vertebrate species long before contact with Europeans" (297). Cuvier's documentation of early records of fossils included North American native discoveries, which helped to confirm his theory of worldwide extinctions. But earlier theorizers were at work, as surveys of myth in relation to fossils still apparent show native awareness of many earth science concepts. Douglas Wolfe (leader of the Zuni Basin Paleontology Project) is quoted as saying, "It's all there in one elegant myth, evolution, extinction, climate change, deep time, geology and fossils" (Mayor 2005, 116).

A number of dinosaur track sites are marked by pictographs or petroglyphs of generally unknown age and significance. In Paraiba, Brazil, footprints of carnivorous dinosaurs exposed on a bedding-plane surface of Lower Cretaceous sandstone are "incorporated into a design involving other symbols of unknown significance, carved beside the footprints into the rock surface" (Sarjeant 1997; G. Leonardi in Ligabue 1984). Lockley (1991, 185) cites two similar instances in Utah and also notices use of dinosaur track images on "snake priest's aprons" worn by Hopi dancers in an area where tracks are well known. Mayor records that "the dancers . . . weave these designs into their costumes because large, three-toed fossil tracks impressed in rocks were believed to have been made by the Kachina spirit who sends the rain" (2005, 142) (Fig. 1.2). A pictograph appearing to represent a tridactyl footprint appears close to *Eubrontes* tracks at Flag Point track site in Utah, and has been dated to between AD 1000 and 1200 (Mayor and Sarjeant 2001,

151). Mayor also documents track sites in Arizona known to the Navajos as "The Place with Bird Tracks" (2005, 139) and "Big Lizard Tracks" (2007, 256). A geological feature is explained by the Lakota tradition of the "Big Racetrack" in South Dakota, when in the "first sunrise of time," all the strange creatures were summoned for a great race during which the animals became buried (Mayor 2007, 258).

Mayor shows that Montana natives had collected dinosaur bones for use in hearths (2005, 273), and that Hopi and Pueblo potters sought out gastroliths to burnish pottery (157). Other kinds of fossils have been sought for building and tool making, or collected for their ornamental or magical significance.

Near-scientific interpretations have been made of dinosaur bones by aboriginal peoples. Jean-Baptiste L'Heureux, a French Canadian who lived during the early nineteenth century among the Peigan people of Alberta, Canada, recorded that bones shown to him in what is now Dinosaur Provincial Park were revered as those of "the grandfather of the buffalo" (Spalding 1999, 22) – a near-scientific interpretation, since the buffalo was the largest animal known to the First Nations of the Plains. Mayor (2005) has shown that "the grandfather" was a widespread interpretation of fossil remains among North American First Nations, being also applied for instance to bison in the Ohio Valley, and possibly elephantids from Louisiana. In nearby British Columbia, other dinosaur track sites have led to the naming by the Gitksan of a mountain called the Giant Marmot, one of whose ridges is named "Where You Find the Tracks of the Giant Marmot" (Mayor 2007, 259)

Modern residents of Brazil have interpreted dinosaur tracks as the footprints of saints (Mayor and Sarjeant 2001).

Out of Africa . . . and Australia

In Algeria, the first dinosaur tracks found in Africa were described by paleontologists Le Mesle and Péron in 1880 after being brought to their attention by a French officer (Buffetaut 1987, 180–181). Local Arabs thought they had been made by a giant bird, which Mayor and Sarjeant (2001) speculate may have been related to the Rukh (Roc) of the *Arabian Nights* (itself based on *Aepyornis* eggs from Madagascar). The scientists came to a similar conclusion and considered the tracks to be made by birds.

In Cameroon, local residents see dinosaur tracks from a different cultural perspective. Louis Jacobs (1993, 261) describes a site found in 1988 that was well known to local Muslim inhabitants: "Living in the bush as they do . . . they could certainly recognize spoor . . . even if they did not know what creatures had made the trails. Some of the more devout and imaginative among them claimed that they had seen, in amongst the tracks the knee, elbow and forehead prints of one of the genuflecting Islamic faithful praying to Mecca."

Another striking example is that reported from the Bushmen of southwestern Africa, who were quite familiar with dinosaur tracks. Paul Ellenberger, a French paleontologist fluent in their language, not only heard songs and tales about the footprints and their makers, but also found that both were depicted in their paintings. Indeed, he reported (Mossman 1990)

1.2. Zuni Kachina figure wearing costume with dinosaur track designs, ca. 1905.

Photo by Adrienne Mayor.

that the paintings of those unknown track makers were strikingly like iguanodonts, even to having forefeet of the right proportions.

The Bardi people of northwest Australia regarded dinosaur tracks as the trail of a legendary giant "emu-man," who (as the track distribution and alignment suggests) walked into the ocean and then returned. Where he sat down, his feathers stuck in the mud, as was perhaps suggested by fossil ferns found in the area. A Tjapwurong tradition in western Victoria tells of giant birds resembling the Tertiary Dromornithids, whose contemporaneity with early humans is suggested by carved footprints and cave paintings in other parts of Australia (Mayor and Sarjeant 2001).

Noah's Raven—From Folklore to Science

A slab bearing five footprints of a dinosaur in Triassic red sandstones of the Connecticut Valley was plowed up about 1802 by a Massachusetts farm boy, Pliny Moody, who described them as "three-toed like a bird's" and installed the slab as a doorstep. (The specimen is figured by Steinbock 1989, 28.) Moody's tracks were first reported in a local newspaper about 1804 and were solemnly considered to be those of the raven that, when sent out by Noah from the ark to seek land, perversely failed to return. Dr. James Deane (1801–1858) also reported a discovery in 1835 of tracks from a quarry near Greenfield, Massachusetts, which were first described as "turkey tracks made 3000 years ago" (Thulborn 1990, 38). Deane brought all of these to Reverend Edward Hitchcock's attention. Hitchcock (1793–1864) acquired the "Noah's Raven" tracks in about 1839, and 2 years later, he named the tridactyl tracks *Ornithoidichnites fulicoides* (38), roughly translatable as the "coot-like stony bird track."

Hitchcock has been much criticized because he interpreted these tracks as those of birds. However, as one of us has pointed out elsewhere (Spalding 1993, 84), the moa discoveries in New Zealand had brought knowledge of giant birds to America by this time, and in the absence of knowledge of large bipedal reptiles, Hitchcock's interpretation—of three-toed tracks at any rate—was a scientific one.

Fossils as Remains of Life

In Europe, again and again, fossils had been recognized as the remains of once-living creatures—in ancient Greece, for instance, by Xenophanes of Colophon and Xanthos of Sardis (Adams 1938, 11–12; Mayor 2000, 210); in the fifteenth century by Leonardo da Vinci; and in the sixteenth century by the Dane Niels Stensen, called Steno. But the ideas of these intellectual pioneers either remained secret (as were Leonardo's until his codified notes were first read four centuries later) or were rejected in favor of explanations that now seem as fantastic as any folkloric interpretations. Steno's views were opposed by Martin Lister of London's Royal Society, who in 1671 (p. 2282) stated categorically that fossils were "never any part of an animal." The Welsh naturalist Edward Lhuyd temporized (1699), believing them to be a product of minute spawn of animal life, carried inland by vapors arising from the ocean and growing within the rocks. Even Robert Hooke's careful demonstration of the organic nature of fossils, presented to the Royal Society of London sometime after 1668 but only published posthumously

(1705), did not convince all his contemporaries. Thus Robert Plot (Fig. 1.3), curator of Oxford's Ashmolean Museum, concluded instead (1705) that the fossil shells of invertebrates in rocks were merely Lapides sui generis, stones "formed into an Animal Mould" by "some extraordinary plastic virtues latent in the Earth" to serve as ornaments for the Earth's secret places, in the fashion that flowers adorned its surface.

Illustrations and Explanations

By the late seventeenth century, some of the basic methods of vertebrate paleontology had developed in Europe as individuals began to search for, extract, and collect specimens, which were increasingly preserved with documentation in organized collections for ongoing study. Scientific illustrations and descriptions of important specimens began to be published. While the debate about the organic origin of invertebrate fossils continued, the nature and anatomical position of fossil bones became increasingly recognized. Some attempts at identification and explanation approached our present understanding, while others continued to be less scientific.

When Plot (1677, 11) discovered and illustrated a dinosaur bone, he recognized it correctly as being "a real Bone, now petrified"—more specifically, "the lowermost part of the Thigh-Bone." From its great size—"In Compass near the capita Femoris, just two foot [0.6 m], and at the top above the Sinus . . . about 15 inches [0.45 m]"—he concluded that it "must have belonged to some greater animal than either an Ox or Horse; and if so in all probability it must have been the Bone of some Elephant, brought hither during the Government of the Romans in Britain" (12). Plot's bone came from the Middle Jurassic strata of Cornwell, near Chipping Norton, Oxfordshire. The bone is now lost, but in 1871 John Phillips identified it from the illustration as the distal end of a femur of a large megalosaur or small cetiosaur (Buffetaut 1987). It is now regarded as part of the femur of a *Megalosaurus*.

The bone was reillustrated by Richard Brookes in 1763, just after the starting point of zoological nomenclature with publication of Carl von Linné's *Systema Naturae* (1758). The illustration was captioned "Scrotum humanum." From comparison with the labeling of Brookes's other illustrations, it is evident that this was merely a descriptive appellation; he knew quite well that Plot's specimen was part of a bone. However, the name was taken very seriously by a French philosopher, Jean-Baptiste Robinet, who held the eccentric concept that fossils were attempts by Nature to reproduce in other fashions the organs of humankind. Robinet (1768) not only accepted that Plot's specimen was a scrotum, but also believed it showed the musculature of the testicles and the vestiges of a urethra (Buffetaut 1979). Though Robinet's concept was not taken seriously by other savants, it has been (somewhat mischievously) suggested that in view of its binomial format and date, *Scrotum humanum* was the earliest scientific name for a dinosaur (Halstead 1970; Delair and Sarjeant 1975). However, that proposition has now been firmly rejected by the cognoscenti of zoological taxonomy (Halstead and Sarjeant 1993).

Plot was succeeded at the Ashmolean Museum by Welsh naturalist Edward Lhuyd (1660–1709), who illustrated several fossil teeth in his

1.3. Robert Plot (1640–1696), the first illustrator of a dinosaur bone.

Lithophylacii Britannici Ichnographia (1699), which were reproduced by R. T. Gunther (1945, 140–142). Though considered by the author to be the remains of fish, two of these seem to be dinosaur teeth. Specimen 1328 closely resembles those of *Megalosaurus* described by Buckland and comes from the same location of Stonesfield, then variously spelled (Barrington 1773, 172). Specimen 1352 named *Rutellum implicatum* is surely the tooth of a cetiosaur, and it is thus the earliest record of any sauropod. It came from Caswell, now Carswell, about 5 miles (8 km) southwest of Whitney, where Thames gravels overlie Coral Rag. Unfortunately, the specimens were not among those presented to the university in 1708 and are lost.

The eighteenth century saw other documented discoveries in England of bones now known to be dinosaurian. John Woodward (1665–1728), professor at Gresham College, London, actively sought fossils, corresponding internationally, sending out collectors, and publishing a guide to collection methods (Levine 1977). At an unknown date he acquired a specimen of dinosaur limb bone from Stonesfield. If complete, the bone would have closely resembled a limb bone of a carnosaur, perhaps *Megalosaurus*. The specimen appears as number A1 in the posthumously published catalog (Woodward 1728) of his fossil collection and is still preserved in the Woodward collection of the Sedgwick Museum, University of Cambridge. It thus represents the earliest discovered dinosaur bone still known to survive in a collection. However, Woodward's theoretical works on geology ascribe large vertebrates to the "Universal Deluge."

The next recorded find was also made in Oxfordshire. Joshua Platt (1669–1763), an English dealer in curiosities, found three large vertebrae, surely of dinosaurs, at Stonesfield. In 1754 he unwisely he sent them to Peter Collinson, a Quaker merchant and botanist, for examination; Collinson did nothing with them, and their fate is uncertain. Around 1757 Platt found an enormous thighbone at Stonesfield, again probably of *Megalosaurus*. Though incomplete, it measured 2 feet 5 inches (approximately 81 cm) in length, its width across the condyle being 8 inches (22.4 cm) and across the shaft, 4 inches (11.2 cm). Platt reported this second discovery in a short note (1758) accompanied by a careful drawing by J. Mynde. The bone was listed in an unpublished catalog of Platt's collection (1773) but has since been lost (Delair and Sarjeant 2002, 188). What is surely the same bone was mentioned in a survey of Oxford's fossils published in 1757 by "A.B.," which seems to have been a pen name used by David Steuart Erskine (1742–1829), later the 11th earl of Buchan. He adds the information that it was found "three months since," which would be in December 1756 or early January 1757. A.B. indicates he "formerly met with two pieces of bone, and some vertebrae of the same kind" (1757, 122). It seems possible that the other bone passed into the collection of Smart Lethuieullier (1701–1760), whose manuscript catalog figures specimen 26, a broken rib "said to come from Stunfield in Oxfordshire." The collection, including the bone, has been lost. A.B. was also aware of large bones of "several kinds" found in a pit by William Frankcombe (1734–1767) in the Kimmeridge Clay (Late Jurassic) at Shotover before 1754, since lost.

Eighteenth-century Discoveries in the United Kingdom

Part of a scapula, almost certainly of *Megalosaurus*, again from Stonesfield, was presented by a Dr. Watson to what was then the Woodwardian Museum of Cambridge University in 1784; though it survives (specimen D.11.35c), it has never been described or illustrated.

Apart from Plot's bone, all these fossils came from Stonesfield. Lhuyd's recurved tooth and the Cambridge scapula can be assigned with reasonable confidence to *Megalosaurus*, and Woodward's limb bone fragment and Platt's thigh bone may also belong to the same genus. In the absence of a specimen or figure, the vertebrae cannot be assigned. While the size caused remark, the earliest recognition of their reptilian significance seems to come from the German geologist Abraham Gottlob Werner (1749–1817) when he alluded to them as "a ?saurian present in the compact limestone near Blenheim, Woodstock" (1774, 141). Stonesfield is close to Blenheim, and the Stonesfield Slate mined there may readily be regarded as a compact limestone. This recognition is perhaps the source of a similar reference by Thomas Maurice (1820, 1: 470), and John Whitehurst suggested the material was crocodilian (1786, 29).

Eighteenth-century Finds in France

Across the English Channel, the Normandy coast may also have yielded dinosaur bones during the eighteenth century. Bones collected from the Vaches Noires cliffs by the Abbé Dicquemare (1733–1789) and reported in 1776 may have included vertebrae and a femur of a dinosaur (Taquet 1984; Buffetaut et al. 1993); however, the descriptions were brief and unaccompanied by illustrations. Vertebrae collected by the Abbé Bachelet of Rouen from the vicinity of Honfleur, illustrated by Georges Cuvier (1808), were certainly those of a theropod dinosaur (Lennier 1887) but were misinterpreted as having belonged to crocodiles of an unusual type.

Early Nineteenth-century Discoveries in the United Kingdom

Inadequately documented finds continue into the nineteenth century. William Smith obtained from Cuckfield in Sussex in 1809 "several bones of gigantic dimensions, which, with the rest of his collection, were transferred in 1815 to the British Museum" (Phillips 1844, 63). One of these, an *Iguanodon* tibia, is illustrated by Charig (1983, 50). Mantell does not mention in his journal visiting Cuckfield until July 6, 1819 (Curwen 1940, 8), nor did he apparently meet Smith until 1833 (Dean 1999, 126). Clearly, then, Smith anticipated Mantell's discovery of *Iguanodon* bones, and informal information of Smith's discovery may have drawn Mantell's attention to the site.

Also in 1809, an elongate caudal centrum was found at Dorchester-on-Thames, Oxfordshire. It was acquired for the Woodwardian collections of the University of Cambridge, and in Seeley's 1869 catalog it is regarded as cetiosaurian and probably from the Oolite (Middle Jurassic). It survives in the Sedgwick Museum collection (no. J230 05) and was first figured in 1975 (Delair and Sarjeant 1975, 11 and fig. 3).

Before 1816, the Cretaceous strata of the Isle of Wight yielded bones –surely of dinosaurs– to the geologist Thomas Webster (1773–1844). These were mentioned in a general work on the island (Englefield 1816), and Webster later referred to them as "large saurian bones" (1824, 1829). *Iguanodon* bones are now found in the Wealden beds at that locality, so the early finds may represent discoveries of that dinosaur ahead of Mantell, though that cannot be confirmed because the specimens are lost.

The earliest documented dinosaur bones discovered in North America likewise merit only brief attention in this history. In 1787 a "large thighbone" – perhaps a hadrosaur limb – was found near Woodbury Creek, Gloucester County, New Jersey, in what are now known to be Late Cretaceous strata. It was reported to the American Philosophical Society on October 5 by Dr. Caspar Wistar (1761–1818) and Timothy Matlack and was thought to be lost, but may now have been located by Donald Baird in the Academy of Natural Sciences in Philadelphia (Weishampel and Young 1996, 58).

The second observation must have been that by William Clark. In the course of his exploratory expedition through the recently acquired Louisiana Purchase with Meriwether Lewis in 1806, Clark noted a large rib bone in a cliff on the south bank of the Yellowstone River, about 6 or 7 miles (9–11 km) below Pompey's Tower (now Pompey's Pillar) close to what would be the site of Billings, Montana. In his journal, Clark noted it as being 3 feet (0.9 m) in length, "tho' a part of the end appears to have been broken off" and about 3 inches (7.6 cm) in circumference. He obtained "several pieces of this rib: the bone is neither decayed nor petrified but very rotten" and thought it to be a bone of an immense fish (Clark, quoted in Simpson 1942, 171–172). This find was in the Late Cretaceous Hell Creek Formation and was surely a dinosaur bone. Breithaupt (1999, 60) says that the specimen and associated records are "not locatable" but considers the fossil "was most likely a poorly mineralized, Late Cretaceous dinosaur rib from the Hell Creek Formation."

Fossil bones discovered by Solomon Ellsworth Jr. during blasting for a well near East Windsor into the red sandstones of the Connecticut Valley, in contrast, were small enough to be misinterpreted as human bones (Smith 1820). These are in the Yale Peabody Museum (YPM 2125). Almost a century passed before Richard S. Lull (1915) recognized their dinosaurian character, considering them to be bones (parts of the forelimb and foot, the hind limb, and the tail of a small saurischian dinosaur) of a small coelurosaur. A more recent reexamination by Peter M. Galton (1976) indicates instead that they are bones of a prosauropod, *Anchisaurus colurus* (Weishampel and Young 1996, 58).

In the century or so succeeding Plot's discovery, then, dinosaur bones had been repeatedly found both in Europe and North America, and had almost as often been misinterpreted. It was a further find from Stonesfield, Oxfordshire, that properly launched at last the scientific study of dinosaurs.

It is not certain whether the Reverend William Buckland (1784–1856) (Fig. 1.4), reader of mineralogy at the University of Oxford, personally discovered dinosaur specimens at Stonesfield because he did not record the sources of his material. John Phillips (1800–1874) claimed later that some of them "were discovered before his [Buckland's] day" (1871, 196). In 1815 physician and geologist John Kidd (1775–1851) reported that there were the remains of "one or more quadrupeds" at Stonesfield. The stone was mined for roofing,

1.4. William Buckland (1784–1856), the first scientist to describe and name a dinosaur.

and workers descended shafts more than 40 feet (12 m) deep (Cadbury 2001, 64). It seems likely that quarrymen, other collectors, and perhaps Buckland himself discovered the material that was eventually described, but that Buckland's primary role was to describe specimens he had acquired from various sources.

His material comprised several huge teeth, recurved and with serrated edges, and a partial lower jaw with a tooth; these were early recognized to be reptilian and compared with the similar, albeit much smaller, teeth of the living monitor lizard. Cuvier saw these specimens while visiting Oxford in 1818 and later reported (1824) that they had been found several years before his visit. In a letter to Buckland written from Cuvier's laboratory in 1821, the Irish naturalist Joseph B. Pentland was already enquiring plaintively: "Will you send your Stonesfield reptile or will you publish it yourself?" (Sarjeant and Delair 1980, 262).

However, Buckland was a man of diverse concerns that ranged beyond his clerical duties and geological interests; he was a veritable polymath. With so much else to do, his progress toward publishing his discovery was not rapid. It has been suggested (Cadbury 2001, 67) that this was because Buckland was reluctant to publish material that would not help reconcile geology with the Bible. However, as McGowan (2001, 77) points out, Genesis states, "There were giants in the Earth." Buckland's delays could reasonably have been caused by more practical considerations directly related to *Megalosaurus*. Buckland seems to have been trying to gather more material, for his eventual presentation to the Geological Society was "in the hope that such persons as possess other parts of this extraordinary reptile may also transmit to the society . . . further information" (quoted in Cadbury 2001, 108). Buckland was also anticipating the cooperation of his clerical and geological colleague, the Reverend William Daniel Conybeare (1787–1857). In a letter written to Pentland on July 11, 1822, Buckland stated that Conybeare

is about to take up immediately the Stonesfield Monitor & to publish
a joint paper with me on that Animal, but we have not yet determined
through which Channel to give it publication. My great object will be that
it will be in time for Cuvier's Book. Tell me what time that will be.

As early as 1821 Conybeare had given passing mention to the "Huge Lizard" of Stonesfield (incidentally in an account of a fossil marine reptile), but their joint project did not come to fruition. Yet Buckland's discovery was already becoming quite well known.

In a general paleontological text published in 1822, James Parkinson illustrated one of the teeth and wrote:

Megalosaurus (*Megalos* great, *saurus* a lizard). An animal apparently
approaching the Monitor in its mode of dentition, and not yet described.
It is found in the calcareous slate of Stonesfield. . . . Drawings have
been made of the most essential parts of the animal, now in the
[Ashmolean] Museum of Oxford; and it is hoped a description may
shortly be given to the public. The animal must in some instances, have
attained a length of 40 feet [12 m], and stood eight feet [2.4 m] high.

On the strength of this mention, Parkinson has sometimes been credited with the authorship of the name *Megalosaurus*. However, this was a period before the rules of taxonomy had been properly formulated, let alone rigorously applied, and such preliminary sharing of information was not yet considered improper.

Another unpublished letter, this time from Buckland to Cuvier himself and written on July 9, 1823, makes it quite clear that Buckland was close to publishing his researches on the "Stonesfield Monitor":

> My Dear Baron, Herewith I send you Proof Plates of the great Animal
> of Stonesfield, to which I mean to give the name of Megalosaurus
> & which I shall publish either in 2nd part of V. [volume] 5 or
> the 1t [first] part of V. 6 of the Geological Transactions.

Cuvier was indeed becoming impatient. Pentland, again acting as his amanuensis, wrote to Buckland on February 28, 1824:

> Our friend Cuvier has this moment requested me to write to you on the subject
> of the paper which you proposed publishing on the Stonesfield reptile the
> Megalosaurus. He is now at that part of his work where he intends speaking
> of your reptile, and wishes to know if your paper has been yet published – and
> in what form? And in what work? (Sarjeant and Delair 1980, 304)

However, before this letter was sent, Buckland had at last read his paper to the Geological Society at its London headquarters on February 20, 1824. Its publication later that year constituted the earliest scientific description of a dinosaur – though, of course, that name for those reptiles had not yet been formulated.

With the hindsight of much greater knowledge and better specimens, modern paleontologists are well aware of the distinctions between *Megalosaurus* and *Iguanodon*. But both Buckland and Mantell were less sure. It seems possible that Buckland's appeal for "other parts of this extraordinary reptile" may have been directed (at least in part) at Mantell, for there is documentation to show that both of them thought that at least some of Mantell's bones belonged to *Megalosaurus*. By 1822 Mantell had teeth of carnivorous reptiles from Sussex (Cadbury 2001, 87), and in 1823 Cuvier and Buckland both regarded Mantell's *Iguanodon* teeth as belonging to a rhinoceros (99). This surely suggested to Buckland that Mantell's big bones were therefore from the same animal as those from Stonesfield, for no other large land mammals from the period were known, nor were there clear ideas of the stratigraphical distribution to be expected.

Mantell's journal (as published by Curwen 1940, 51) does not give an account of Buckland's presentation of *Megalosaurus*, which Mantell attended. But he does report on March 6, 1824, that "Professor Buckland came express from Oxford, with my friend Mr. Lyell to inspect my Tilgate fossils. I had met the Professor at a meeting of the Geological Society, about three weeks since, and shewn him some specimens of bones and vertebrae of the Megalosaurus from Tilgate Forest." Buckland clearly wanted to include this information in his paper, for only 6 days later, on March 12, as Buckland was completing his paper for publication, Mr. Warburton of the Geological Society wrote to him: "Whatever you have to say on the subject of the Stonesfield animal [i.e., *Megalosaurus*] found at Cuckfield, Sussex [Mantell's site], must be forwarded at once" (quoted in Cadbury 2001, 110). In the published version, Buckland generously acknowledged Mantell's discoveries, and he referred to the large size of "a thigh bone of another of the same species which has been discovered in the ferruginous sandstone of Tilgate Forest." (quoted in Cadbury 2001, 111). It was only with Cuvier's acknowledgement that the *Iguanodon* teeth could belong to a reptilian herbivore, and Mantell's publication of *Iguanodon*, that the distinction between the two giant reptiles became clear.

Apart from Mantell's carnivore material (which is from a later period and presumably represents a different genus), all the carnivore material known to Buckland came from the underground excavations at Stonesfield. Owen (1842, 103) suggests the possibility that this material represents a single individual, though today carnivore bone beds are also known.

Gideon Mantell and *Iguanodon*

Sussex surgeon Gideon Algernon Mantell (Fig. 1.5) also played an important role in the discovery of the dinosaurs, but it has been widely misrepresented. An oft-repeated story (e.g., in Colbert 1983, 13–15) tells how, while her husband was visiting a patient early in 1822, his wife, Mary Ann, found some fossil teeth in a pile of road metal (stone rubble used for road making); that, excited by her find, Gideon ascertained from which quarry the road metal had come, finding more teeth and bones there; and that these events marked the beginning not merely of his concern with *Iguanodon*, but also of the scientific study of dinosaurs, with Buckland only spurred into publishing results amassed later by the fear that Mantell might anticipate him.

Unfortunately, as Dean (1993, 208–211) demonstrates, this romantic story does not withstand scrutiny. In his book on *The Fossils of the South Downs*, published in May 1822, Mantell makes mention that "teeth, vertebrae, bones and other remains of an animal of the lizard tribe, of enormous magnitude [have been] discovered in the county of Sussex." Dean points out also that though Mary Ann may indeed have found *Iguanodon* teeth for him, Gideon subsequently named himself as their discoverer.

Even with such problems and uncertainties set aside, the story of Mantell's recognition of *Iguanodon* remains a complex and fascinating one. Upon displaying some of his finds at the Geological Society on June 21, 1821, Mantell aroused little interest. When Charles Lyell took one of the teeth to Paris for Cuvier (Fig. 1.6) to examine, it was dismissed as being the upper incisor of a rhinoceros, while some metatarsals were considered to belong to a species of hippopotamus (Mantell 1850, 195). The actual tooth is now in the Museum of New Zealand Te Papa Tongarewa (Yaldwyn et al. 1997). It bears two inscriptions. The first, in Mantell's writing, notes, "This was the first tooth of the Iguanodon sent to Baron Cuvier, who pronounced it to be the Incisor of Rhinoceros." However, on the other side, in what is apparently Lyell's hand, the label states, "This however was at an evening party. The next morning he told me that he was satisfied it was something quite different. Sir C.L. 4 Feb 59." It is not clear when (or even if) Lyell told Mantell of Cuvier's change of mind – the date of the note is after Mantell's death. However, Mantell sent drawings of more teeth in June 1824, and Cuvier replied that though the teeth were not carnivorous, he thought them reptilian (Cadbury 2001, 116).

This seems to have confirmed Mantell's suspicions, and in September 1824 the crucial breakthrough came when he was examining bones and teeth in the Hunterian Museum of the Royal College of Surgeons in

1.5. Gideon Algernon Mantell (1790–1852), the first scientist to describe and name a herbivorous dinosaur.

London. Another visitor on that day, Samuel Stutchbury, drew Mantell's attention to the dentition of the living iguana. Why, the Sussex teeth were merely gigantic equivalents of the iguana's teeth!

When Mantell wrote again to Cuvier, the great French scientist (who, as we have seen, had already abandoned his initial reaction) was convinced by the new interpretation, writing: "N'aurons nous pas ici un animal nouveau, un reptile herbivore?" [Have we not here a new animal, a herbivorous reptile?] (quoted by Mantell 1851, 231–232).

Mantell seems originally to have proposed calling his "new" reptile *Iguanosaurus* (Anonymous 1824), but this name was dropped in favor of the more euphonious *Iguanodon*, proposed to him by Conybeare. On February 10, 1825, almost a year after Buckland's paper, Mantell proudly reported his discovery to the Royal Society of London, and his account was published in its *Philosophical Transactions* later that year.

Megalosaurus had been interesting, but after all, gigantic reptilian carnivores were already well known; were there not living crocodiles in British India, called muggers, that attained lengths of over 30 feet (9 m)? A giant reptilian herbivore, though—why, that was indeed a wholly novel concept!

If Buckland had been less preoccupied with other matters, he might have anticipated Mantell in the description of *Iguanodon*. Adam Sedgwick (1822) noted that Buckland had discovered "cetacean" bones at Sandown Bay, Isle of Wight, before Christmas 1822; as Buckland himself later reported (1824, 392, 1829), these were in fact bones of *Iguanodon*. He had also already obtained for the Oxford museum a bone of even more gigantic size, found in fragments by the geologist Hugh Strickland (1811–1853) in a railway cutting near Enslow Bridge, Oxfordshire. Carefully pieced together by Buckland, these proved to constitute a femur 4 feet 3 inches (1.3 m) long (Strickland 1848). After being carefully cemented and bound round with wire, this bone was "long the object of admiration in the Oxford classroom for geology" (Phillips 1871, 247). Buckland also acquired or examined other bones of comparable character—among them, vertebrae from Middle Jurassic localities near Chipping Campden and near Thame, Oxfordshire, plus a "blade bone" of enormous size from the latter place, while a whole batch of fossil bones were obtained for him by William Stowe from near Buckingham. All of these, as Buckland wrote in a letter to Stowe (quoted in Phillips 1871, 245), belonged to "some yet undescribed reptile of enormous size, larger than the Iguanodon, and of which I am collecting scattered fragments into our museum, in hope ere long of being able to make of its history."

Yet Buckland never did write further on these gigantic bones. That task was left to Richard Owen, who described and named *Cetiosaurus*—the earliest sauropod dinosaur to be discovered—in an address to the Geological Society of London on June 30, 1841. This served as prelude to Owen's major account of British fossil reptiles, given to the British Association in August 1841, and published the following year. Hugh Torrens (1995) has shown that the name Dinosauria was not applied until the published version.

Buckland's Other Dinosaur Discoveries

1.6. Jean Léopold Nicolas Frédéric (called Georges), Baron Cuvier (1769–1832), the greatest anatomist of the early nineteenth century.

Further Finds in England

In the meantime, other dinosaur bones were being discovered. Saurian bones from strata exposed at Swanwich (now Swanage) Bay, Dorset, were given passing mention by the geologist William H. Fitton (1824). Vertebrae and an imperfect femur were dug up at Headford Wood Common, Sussex, in 1824 (Murchison 1826). All three discoveries were made in Wealden (Lower Cretaceous) strata. Probably all three were *Iguanodon* bones, but the specimens (if they survive) have not been identified and studied.

While Buckland was being again distracted, Mantell pursued his dinosaurs. He was accumulating further bones of *Iguanodon*, from the Wealden, including bone fragments from the cliffs about Sandown, Isle of Wight, and most notably, a slab from Maidstone, Kent. This displayed a partial skeleton (Fig. 1.7), a discovery so important that it has been facetiously styled his "mantel-piece" (Spalding 1993, 23); it allowed him to attempt a reconstruction, upon which the early restorations of that creature were to be based. Mantell also discovered the remains of an armored dinosaur – postcranial bones associated with dermal elements and armor plate – which he named *Hylaeosaurus*; this was the first ankylosaur to be described (1833). Before his death in 1852, he was to describe and name two further dinosaur genera.

Samuel Stutchbury reentered the story when, in association with S. H. Riley, he reported the earliest reptilian remains from the English Triassic – specifically, from the so-called Magnesian Conglomerate of the Bristol district (Riley and Stutchbury 1836, 1840). Three genera were recognized, two of which (*Thecodontosaurus* and *Palaeosaurus*) are accepted nowadays

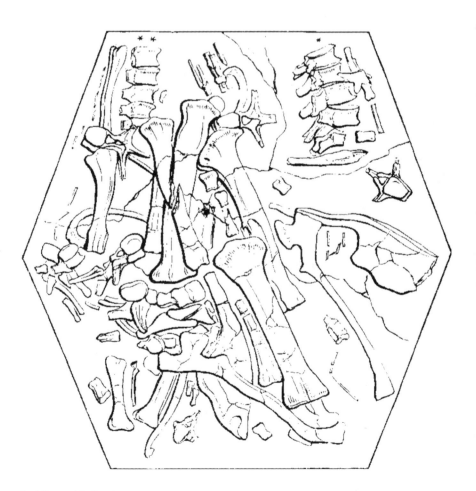

1.7. Gideon Mantell's "Mantel-piece": the first substantial associated dinosaur skeleton discovered.

as dinosaurs; the former is now considered a prosauropod, while the latter is of dubious affinity.

The first recognition of dinosaurs in France came in 1828, when A. de Caumont (1828) reported *Megalosaurus* bones from the Middle Jurassic oolite of Caen, Normandy. In the ensuing years, the paleontologist Jacques-Amand Eudes-Deslongchamps (1794–1867) painstakingly assembled bones and bone fragments from the quarries and construction sites around Caen. When he became confident that he had enough material, he published a description of the partial skeleton thus reconstituted. In tribute to the discoverer of *Megalosaurus*, he named it *Poekilopleuron bucklandi* (1838). Eudes-Deslongchamps thought the animal to have been largely marine, though well able "to rest on the shore and bask in the sun" (Buffetaut et al. 1993, 162). This was the last of the misapprehensions about dinosaurs to precede Owen's establishment of the Dinosauria—though, in justice, one must note that there have been many since!

The first finds in Germany were in Triassic red sandstones and followed close on the discoveries near Bristol. In 1837 Christian Erich Hermann von Meyer (1801–1869) described and named *Plateosaurus*, which was destined long to remain the most fully known of the dinosaurs we now call prosauropods. (More than 100 skeletons have been found in southern Germany and Switzerland.) A tooth recovered from the Jurassic strata of southern Russia by A. Zborzewski was rather unnecessarily given its own generic name, *Macrodontophion* (1834); it was probably that of a carnosaur.

Summary

Where conditions are suitable, bones, tracks, and even eggshells seem to have attracted attention from early times. For some ancient discoveries, no explanation has come down to us; other early explanations conceal a germ of truth in legend, while near-scientific explanations come from several cultures in Africa, Asia, and North America as well as Europe. The first recorded scientific discovery and most other early finds were from the Jurassic strata of England, though there were early observations also in France and North America. During almost 150 years following Robert Plot's first illustration of a dinosaur bone in 1677, these fossils were often misinterpreted in a variety of fashions—and even when rightly considered as bones, they were attributed to the remains of elephants, crocodiles, and fish. During the eighteenth century, organized collections were being made in England and France that included dinosaur bones and teeth. Marine reptiles were described first, but the earliest recognition of bones of long-extinct land creatures was made by William Buckland, in his study of the carnivorous *Megalosaurus* (1824). Gideon Mantell's researches on *Iguanodon*, first of the herbivorous dinosaurs to be identified, overlapped Buckland's work and were published only a little later (1825). Though Buckland accumulated sauropod bones also, he did not describe them; this was done eventually by Owen, in his preliminary description of *Cetiosaurus* (1841). In addition to further discoveries in England, giant reptiles recognized and named before Owen's "creation" of the dinosaur (1842) included *Plateosaurus* von Meyer

(1837), from Germany, and *Poekilopleuron* Eudes-Deslongchamps (1838), from France, plus a single tooth from Russia.

Acknowledgments

The research for the first edition of this chapter was undertaken by William Sarjeant under tenure of the Natural Science and Engineering Research Council of Canada (operating grant 8393) and with the assistance of Linda Dietz. Letters quoted on early French dinosaur finds were courteously furnished by transcription for Catherine Hustache of the Muséum National d'Histoire Naturelle, Paris, and were published with permission.

After the death of William Sarjeant in July 2002, the revision was undertaken by David Spalding, with the help of papers on aspects of the subject already published by Sarjeant and various collaborators, and new data from his own research. Helen Wong, Dr. Sarjeant's secretary, provided valuable assistance in suggesting and locating research materials and illustrations. Darren Tanke provided data from the 2008 History of Geology Group conference. Suggestions from an anonymous reviewer have been helpful.

References

"A.B." [Erskine, D. S.] 1757. Curious discoveries in making new roads. *Gentleman's Magazine* 27: 20–21.

Adams, F. D. 1938. *The Birth and Development of the Geological Sciences.* Baltimore: Williams & Wilkins. Reprint, 1954, New York: Dover Books.

Andersson, J. G. 1934. *Children of the Yellow Earth.* Reprint, 1973, Cambridge, Mass.: MIT Press.

———. 1943. *Under a Lucky Star.* New York: Viking Press.

Anonymous. 1824. Organic remains. *New Monthly Magazine* 12 (December): 575

Barrett, P. M., M. T. Carrano, and J. A. Wilson. 2008. The discovery of dinosaurs in the Lameta Beds of Central India. In Moody 2008, 32–33.

Barrington, D. 1773. Some account of a fossil lately found near Christchurch, in Hampshire. *Philosophical Transactions of the Royal Society* 48: 171–172.

Breithaupt, Brent H. 1999. The first discoveries of dinosaurs in the American West. In D. D. Gillette (ed.), *Vertebrate Paleontology in Utah,* 59–65. Miscellaneous Publication 99-1. Salt Lake City: Utah Geological Survey.

Brookes, R. 1763. *The Natural History of Waters, Earths, Stones, Fossils, and Minerals, with their Virtues, Properties, and Medicinal Uses: To Which is added, The Method in which Linnaeus has treated these Subjects.* Vol. 5. London: Newberry. 5th ed., 1772.

Buckland, W. 1824. Notice on the Megalosaurus, or great fossil lizard of Stonesfield. *Transactions of the Geological Society of London* 1 (ser. 2): 390–396.

Reprint, Weishampel and White 2003, 51–59.

———. 1829. On the discovery of the bones of the iguanodon, and other large reptiles, in the Isle of Wight and Isle of Purbeck. *Proceedings of the Geological Society of London* 1: 159–160.

Buffetaut, E. 1979. A propos du reste de dinosaurien le plus anciennement décrit: l'interprétation de J. B. Robinet (1768). *Histoire et Nature* 14: 79–84.

———. 1987. *A Short History of Vertebrate Palaeontology.* New York: Methuen.

Buffetaut, E., G. Cuny, and J. Le Loeuff. 1993. The discovery of French dinosaurs. *Modern Geology* 18: 161–182. Reprint, Sarjeant 1995, 159–180.

Cadbury, D. 2001. *The Dinosaur Hunters.* London: Fourth Estate.

Carpenter, K., K., F. Hirsch, and J. R. Horner (eds.). 1994. *Dinosaur Eggs and Babies.* Cambridge: Cambridge University Press.

Caumont, A. de. 1828. *Essai sur la topographie géognostique du département du Calvados.* Caen, France: Chalopin.

Charig, A. 1983. *A New Look at the Dinosaurs.* New York: Facts on File.

Colbert, E. H. 1983. *Dinosaurs: An Illustrated History.* Maplewood, N.J.: Hammond.

Conybeare, W. D. 1821. Notice of the discovery of a new fossil animal, forming a link between the ichthyosaurus and crocodile, together with general remarks on the osteology of the ichthyosaurus. *Transactions of the Geological Society of London* 5 (ser. 1): 559–594.

Curwen, E. C. (ed.). 1940. *The Journal of Gideon Mantell, Surgeon and Geologist.* London: Oxford University Press.

Cuvier, G. 1808. Sur des ossemens fossiles de crocodiles, et particulièrement sur ceux des environs du Havre et de Honfleur, avec des remarques sur les squelettes des sauriens de la Thuringe. *Annales du Muséum National d'Histoire Naturelle, Paris* 12: 73–110.

——. 1824. *Recherches sur les ossemens fossiles du quadrupèdes.* Rev. ed. 6 vols. Paris: Dulour et d'Ocagne.

Dean, D. R. 1993. Gideon Mantell and the discovery of *Iguanodon. Modern Geology* 18: 209–219. Reprint, Sarjeant 1995, 207–218.

——. 1999. *Gideon Mantell and the Discovery of Dinosaurs.* Cambridge: Cambridge University Press.

Delair, J. B., and W. A. S. Sarjeant. 1975. The earliest discoveries of dinosaurs. *Isis* 66: 5–25.

——. 2002. The earliest discoveries of dinosaurs: the records re-examined. *Proceedings of the Geologists Association* 113: 185–197.

Dong, Z., and A. C. Milner. 1988. *Dinosaurs from China.* London: British Museum (Natural History) China Ocean Press.

Englefield, H. C. 1816. *A description of the principal picturesque beauties, antiquities, and geological phoenomena, of the Isle of Wight . . . With additional observations on the strata of the island, and their continuation in the adjacent parts of Dorsetshire.* London: Payne and Foss.

Eudes-Deslongchamps, J. A. 1838. Mémoire sur le Poekilopleuron Bucklandii, grand saurien fossile intermédiaire entre les crocodiles et les lézards. *Mémoires de la Société Linnéenne de Normandie* (Calvados) 6: 37–146.

Fitton, W. H. 1824. Enquiries respecting the geological relations of the beds between the Chalk and the Purbeck Limestone in the south-east of England. *Annals of Philosophy* 8: 365–383.

Galton, P. 1976. Prosauropod dinosaurs (Reptilia: Saurischia) of North America. *Postilla* 169: 4.

Gould, C. 1886. *Mythical Monsters.* London: W. H. Allen & Co. Reprint, 2002, as *Dragons, Unicorns, and Sea Serpents: A Classic Study of the Evidence for their Existence.* Dover Publications.

Gunther, R. T. 1945. *Early Science in Oxford. Life and Letters of Edward Lhwyd.* Vol. 14. Oxford: The Author.

Halstead, L. B. 1970. *Scrotum humanum* Brookes 1763—the first named dinosaur.

Journal of Insignificant Research 5: 14–15.

Halstead, L. B., and W. A. S. Sarjeant. 1993. *Scrotum humanum* Brookes—the earliest name for a dinosaur? *Modern Geology,* 18: 221–224. Reprint, Sarjeant 1995, 219–222.

Hooke, R. 1705. *The posthumous works of Robert Hooke . . . containing his Cutlerian lectures, and other discourses, read at the meetings of the illustrious Royal Society.* London: Waller.

Huxley, F. 1979. *The Dragon: Nature of Spirit, Spirit of Nature.* London: Thames and Hudson.

Jacobs, L. 1993. *Quest for the African Dinosaurs: Ancient Roots of the Modern World.* New York: Villard Books.

Jacobs, L. 1995. *Lone Star Dinosaurs.* College Station: Texas A&M University Press.

Kirchner, H. 1941. Versteinerte Reptilfährten als Grundlage für ein Drachenkampf in einem Heldenlied. *Zeitschrift der Deutschen Geologischen Gesellschaft* 93: 309.

Lennier, G. 1887. Etudes paléontologiques. Description des fossiles du Cap de la Hève. *Bulletin de la Société Géologique de Normandie* 12: 17–98.

Levine, J. M. 1977. *Dr. Woodward's Shield: History, Science, and Satire in Augustan England.* Ithaca, N.Y.: Cornell University Press.

Lhuyd, E. 1699. *Lithophylacii Britannici Ichnographia, sive, lapidum aliorumque fossilium Britannicorum singulari figura insignium; quotquot hactenus vel ipse invenit vel ab amicis accepit distributio classica, scrinii sui lapidarii repertorium cum locis singulorum natalibus exhibens. Additis rariorum aliquot figuris aere incisis; cum epistolis ad clarissimos viros de quibusdam circa marina fossilia & stirpes minerales praesertim notandis. Nusquam magis erramus quam in falsis inductionibus; saepe enim ex aliquot exemplis Universale quiddam colligimus; idque perperam cum ad ea, quae excipi possunt animum non attendimus.* London: Lipsiae, Gleditsch & Weidmann.

Ligabue, G. (ed.). 1984. *Sulle Orme dei Dinosauri.* Esplorazioni e Richerche 9. Rome: Erizzo for Le Società del Gruppo ENI.

Lister, M. 1671. Fossil shells in several places of England. *Philosophical Transactions of the Royal Society of London* 6: 2281–2284.

Lockley, M. G. 1991. *Tracking Dinosaurs: A New Look at an Ancient World.* Cambridge: Cambridge University Press.

Lull, R. S. 1915. Triassic life of the Connecticut Valley. *Connecticut Geological and Natural History Survey, Bulletin* 24: 1–285. 2nd ed., *Connecticut Geological and Natural History Survey, Bulletin* 81: 1–336. 1953.

Mantell, G. A. 1822. *The Fossils of the South Downs; or, Illustrations of the Geology of Sussex.* London: Relfe.

——. 1825. Notice on the *Iguanodon,* a newly discovered fossil reptile, from the sandstone of Tilgate Forest, in Sussex. *Philosophical Transactions of the Royal Society of London* 115: 179–186. Reprint, Weishampel and White 2003, 68–71.

——. 1833. *The Geology of the South-East of England.* London: Longman.

——. 1850. *A Pictorial Atlas of Fossil Remains consisting of coloured illustrations selected from Parkinson's "Organic Remains of a Former World" and Artis's "Antediluvian Phytology."* London: Bohn.

——. 1851. *Petrifactions and their teachings; or, A hand-book to the gallery of organic remains of the British Museum.* London: Bohn.

Maurice, T. 1820. *The History of Hindostan: its arts and its science, as connected with the history of the other great empires of Asia, during the most ancient periods of the world.* 2 vols. Bulmer, London.

Mayor, A. 2000. *The First Fossil Hunters: Paleontology in Greek and Roman Times.* Princeton, N.J.: Princeton University Press.

——. 2005. *Fossil Legends of the First Americans.* Princeton, N.J.: Princeton University Press.

——. 2007. Place names describing fossils in oral traditions. In L. Piccardi and W. B. Masse (eds.), *Myth and Geology.* Geological Society, London, Special Publication 273: 245–261.

Mayor, A., and W. A. S. Sarjeant. 2001. The folklore of footprints in stone: from classical antiquity to the present. *Ichnos* 8: 143–163.

McGowan, Chris. 2001. *The Dragon Seekers: How an Extraordinary Circle of Fossilists Discovered the Dinosaurs and Paved the Way for Darwin.* Cambridge, Mass.: Perseus Publishing.

Meyer, H. von. 1837. Mitteilung an Prof. Bronn (*Plateosaurus engelhardti*). *Neues Jahrbuch für Mineralogie, Geologie und Palaeontologie* 317.

Moody, R., E. Buffetaut, D. Martill, and D. Naish (eds.). 2008. *Dinosaurs (and Other Extinct Saurians): A Historical Perspective: Abstracts Book.* London: History of Geology Group.

Morris, J. 1980. *Tracking Those Incredible*

Dinosaurs . . . And the People Who Knew Them. San Diego, Calif.: C.L.P. Publishers.

Mossman, D. J. 1990. Book review of Gillette and Lockley, *Dinosaur Tracks and Traces. Ichnos* 1: 151–153.

Murchison, R. I. 1826. Geological sketch of the north-western extremity of Sussex, and the adjoining parts of Hants. and Surrey. *Transactions of the Geological Society of London* (ser. 2) 2: 103–106.

Needham, J. 1959. *Mathematics and the Sciences of the Heavens and the Earth.* Science and Civilisation in China 3. Cambridge: Cambridge University Press.

Norman, D. B. 1993. Gideon Mantell's "Mantel-Piece"; the earliest well-preserved ornithischian dinosaur. *Modern Geology* 18: 225–246. Reprint, Sarjeant 1995, 223–243.

Novacek, M. 1996. *Dinosaurs of the Flaming Cliffs.* New York: Doubleday.

———. 2002. *Time Traveler: In Search of Dinosaurs and Ancient Mammals from Montana to Mongolia.* New York: Farrar, Straus and Giroux.

Owen, R. 1841. A description of a portion of the skeleton of the *Cetiosaurus,* a gigantic extinct saurian occurring in the oolitic formations of different parts of England. *Proceedings of the Geological Society of London* 3: 457–462.

———. 1842. Report on British fossil reptiles. *Reports of the British Association for the Advancement of Science* 11: 60–204.

Parkinson, J. 1822. *Outlines of Oryctology. An Introduction to the Study of Fossil Organic Remains, especially of Those Found in the British Strata; intended to aid the student in his enquiries respecting the nature of fossils, and their connection with the formation of the earth.* London: The Author.

Phillips, J. 1844. *Memoirs of William Smith, LL.D., author of the "Map of the Strata of England and Wales."* London: John Murray. Reprint, 2003, Bath: Bath Royal and Literary Institution.

———. 1871. *Geology of Oxford and the Valley of the Thames.* Oxford: Clarendon Press.

Platt, J. 1758. An account of the fossile thigh-bone of a large animal, dug up at Stonesfield near Woodstock, in Oxfordshire. *Philosophical Transactions of the Royal Society of London* 50: 524–527.

Plot, R. 1677. *The Natural History of Oxfordshire, Being an Essay toward the Natural History of England.* Oxford: The Author. 2nd ed., London: Brome, 1705. Reprint of extracts, Weishampel and White 2003, 9–15.

Riley, S. H., and S. Stutchbury. 1836. A description of various fossil remains of three distinct saurian animals discovered in the Magnesian Conglomerate near Bristol. *Proceedings of the Geological Society of London* 2: 397–399.

Riley, S. H., and S. Stutchbury. 1840. A description of various fossil remains of three distinct saurian animals discovered in the Magnesian Conglomerate near Bristol. *Transactions of the Geological Society of London* (ser. 2) 5: 349–357.

Robinet, J. B. 1768. *Considérations philosophiques de la gradation naturelle des formes de l'être, ou les essais de la Nature qui apprend à faire l'Homme.* Paris: Saillant.

Santos, Vanda F., and Luis A. Rodrigues. 2008. Portuguese dinosaur tracks: XVth century imagery to XXIth century reality. In Moody 2008, 63–64.

Sanz, Jose Luis. 2000. *Dinosaurios. Los senores del pasado.* Barcelona: Ediciones Martinez Roca.

———. 2003. *Mitologia de los dinosaurios.* Museo Nacional de Ciencias Naturales.

Sarjeant, W. A. S. 1987. The study of fossil vertebrate footprints: a short history and selective bibliography. In Leonardi, G. (ed.), *Glossary and Manual of Tetrapod Footprint Palaeoichnology,* 1–19. Brasilia: República Federation do Brasil, Departamento Nacional da Produço Mineral.

———. 1997. History of dinosaur discoveries, I. Early discoveries. In P. J. Currie and K. Padian (eds.), *Encyclopedia of Dinosaurs,* 340–347. New York: Academic Press.

———. 2003. Footprints before the flood: incidents in the study of fossil vertebrate tracks in 19th century Britain. Proceedings of the INHIGEO Meeting, Portugal. *Geological Resources and History* 2001: 63–86

——— (ed.). 1995. *Vertebrate Fossils and the Evolution of Scientific Concepts.* Reading: Gordon & Breach.

Sarjeant, W. A. S., and J. B. Delair. 1980. An Irish naturalist in Cuvier's laboratory. The letters of Joseph Pentland, 1820–1832. *Bulletin of the British Museum (Natural History), Historical Series* 6: 245–319.

Sedgwick, A. 1822. On the geology of the Isle of Wight. *Annals of Philosophy* (n.s.) 3: 329–335.

Seeley, H. G. 1869. *Index to the fossil remains of Aves, Ornithosauria, and Reptilia, from the Secondary System of Strata arranged in the Woodwardian Museum of the University of Cambridge.* Cambridge: Cambridge University Press.

Simpson, G. G. 1942. The beginnings of vertebrate palaeontology in North America. *Proceedings of the American Philosophical Society* 85: 130–188.

Smith, N. 1820. Fossil bones found in red sandstone. *American Journal of Science* 2: 146–147.

Spalding, D. A. E. 1993. *Dinosaur Hunters.* Toronto: Key Porter Books. Reprint, 1995, Rocklin, Calif.: Prima Books.

———. 1999. *Into the Dinosaurs' Graveyard: Canadian Digs and Discoveries.* Toronto: Doubleday Canada.

Steinbock, R. T. 1989. Ichnology of the Connecticut Valley: a vignette of American science in the early nineteenth century. In D. D. Gillette and M. G. Lockley, *Dinosaur Tracks and Traces,* 27–32. Cambridge: Cambridge University Press.

Strickland, H. E. 1848. On the geology of the Oxford and Rugby railway. *Proceedings of the Ashmolean Society* 1848: 192.

Taquet, P. 1984. Cuvier-Buckland-Mantell et les dinosaures. In E. Buffetaut, M. Mazin, and E. Salmon (eds.), *Actes du Symposium paléontologique Georges Cuvier.* Montbéliard, France: Ville de Montbéliard, 475–491.

———. 1994. *Dinosaur Impressions: Postcards from a Paleontologist.* Translation published in 1998. Trans. K. Padian. Cambridge: Cambridge University Press.

Thulborn, T. 1990. *Dinosaur Tracks.* London: Chapman & Hall.

Torrens, H. 1995. The dinosaurs and dinomania over 150 years. Reprint, Sarjeant 1995, 255–284.

Trundle, J. 1614. *True and Wonderful, a discourse relating to a Strange and Monstrous Serpent or Dragon lately discovered and yet living, to the great annoyance and diverse slaughters both of men and catell, by his strong and violent poison. In Sussex, two miles from Horsham, in a wood called St Leonard's Forest, and thirtie miles from London, this present month of August, 1614.* London.

Webster, T. 1824. Geological observations on the sea cliffs at Hastings, with some remarks on the beds immediately below the chalk. *Philosophical Magazine* 63: 455–456.

———. 1829. Observations on the strata at Hastings, in Sussex. *Transactions of the Geological Society, London* 2 (2): 30–36.

Weishampel, D. B., and L. Young. 1996. *Dinosaurs of the East Coast.* Baltimore, Md.: Johns Hopkins University Press.

Weishampel, D., and N. M. White (eds.). 2003. *The Dinosaur Papers, 1676–1906.* Washington, D.C.: Smithsonian Books.

Werner, A. G. 1774. *A Treatise on the External Characters of Fossils. : Von der ausserlichen Kennzeichen der Fossilien.*

Crusius. Leipzig, Germany. English translation published in 1805. Translated by T. Weaver. Dublin: M. N. Mahon.

Whitehurst, J. 1786. *An Inquiry into the Original State and Formation of the Earth: deduced from facts and the laws of nature*. London: Bork.

Woodward, J. 1728. *Fossils of all kinds, digested into a method suitable to their mutual relation and affinity; with the names by which they were known to the ancients, and those by which they are this day known; and not conducing to the setting forth the natural history and the main uses, of some of the most considerable of them, as also several papers tending to the further advancement of the knowledge of minerals, of the ores of metalls, and of subterranean productions*. London: the author.

Yaldwyn, J. C., J. T. Garry, and A. P. Mason. 1997. The status of Gideon Mantell's "first" iguanodon tooth in the Museum of New Zealand Te Papa Tongarewa. *Archives of Natural History* 24 (3): 397–421.

Zborzewski, A. 1834. Aperçu des recherches physiques rationelles, sur les nouvelles curiosités Podolie—Volhyniennes, et sur les rapports géologiques avec les autres localités. *Bulletin de la Société des Naturalistes de Moscou* 7: 224–254.

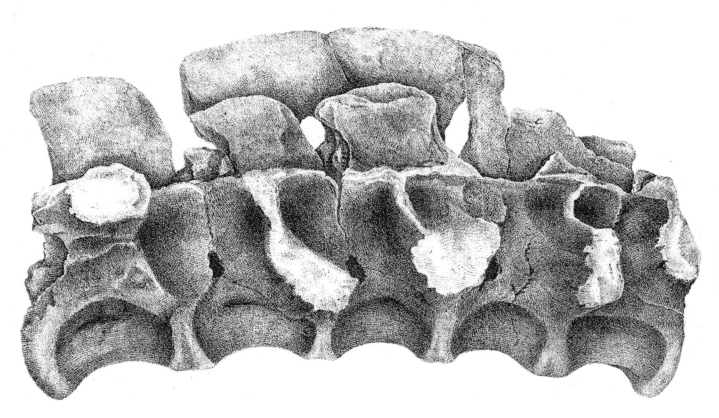

2.1. The sacral vertebrae of *Megalosaurus*
(from Buckland 1824, plate 42, fig. 1).

Politics and Paleontology: Richard Owen and the Invention of Dinosaurs

2

Hugh S. Torrens

The claim has been made that "dinosaurs are the most American animals that have ever lived" (Kirby et al. 1992, 28). While this is clearly true now, the *invention* of dinosaurs, by the English anatomist Richard Owen (1804–1892), who coined his name for these "fearfully great lizards" in April 1842, was entirely English. All our early knowledge of this new group of reptiles had come from England, and all the paleontological material that supported their invention had been found in English rocks.

The work of the great French comparative anatomist Georges Cuvier (1769–1832) was, however, crucial. He had declared in 1825 that *Plesiosaurus* (a kind of Mesozoic marine reptile) was "the most heteroclite . . . and the most monstrous [animal] that had yet been found amid the ruins of the former [i.e., fossil] world" (Buckland 1837, 1: 202). This was just after another English contender for "most monstrous" fossil animal had been described – *Megalosaurus*, revealed by the Oxford academic Reverend William Buckland (1784–1856) to the Geological Society of London, on the very same day that the first complete *Plesiosaurus* was described, in February 1824 (see Weishampel and White 2003, 51–67).

Megalosaurus was entirely based on isolated fragments of bones and associated teeth found in underground slate mines at Stonesfield and gathered together in nearby Oxford, England. The unique lower jaw of this had already been acquired by October 1797, when it passed into Sir Christopher Pegge's (1765–1822) Anatomy School collection at Christ Church (Gunther 1925, 191). It then possessed two serrated teeth, having been purchased for 10s.6d. (=£0.525). This proves how right John Phillips (1800–1874) was to later suggest that "some of the specimens in the Oxford Museum were collected before Buckland's day" (Phillips 1871, 196).

The most critical feature of this new genus of dinosaur-to-be only emerged when dinosaurs were invented nearly 20 years later. Buckland had already shown in 1824 that *Megalosaurus* possessed five vertebrae in its sacral, or hip, region (Fig. 2.1), which were all ankylosed, or fused, together (Buckland 1824).

Remarkably, Buckland already knew of two specimens that showed this feature. One, of "5 Fossil Vertebra," was then in the Geological Society's museum in London, having been presented by Henry Warburton (1784–1858) on June 21, 1817 (Moore et al. 1991, 142). The other specimen, that still preserved at Oxford and figured by Buckland (Fig. 2.1), had been acquired by an Oxford undergraduate student named Philip Barker Webb (1793–1854) from a Stonesfield well-sinker and stonemason called George Griffin. Griffin's letter dated July 14, 1814, mentions "the grate [*sic*] Back

Bone," which he had just sold to Webb (letter in Oxford University Museum of Natural History archives). Webb was later famous as a botanist and traveler and in close contact with Cuvier. So when Cuvier came to visit Oxford in 1818 he was able to see all this *Megalosaur*-to-be material. The fact that at least two sacra were already available then disposes of the recent claim that this Stonesfield material must have been "the scattered remnants of a single individual, coming to light over a period of several decades" (Delair and Sarjeant 2002, 194).

A third "monstrous" contender, *Iguanodon*, soon followed in 1825, but one with quite different teeth—teeth that indicated an herbivorous diet. These were revealed by the provincial surgeon and scientist Dr. Gideon Mantell (1790–1852), based in Lewes, Sussex. But his paper was read for him to the rival Royal Society in London (Mantell 1825; Dean 1999a). The rivalry between these two societies had started in 1808, when the long-standing president of the Royal Society, Sir Joseph Banks (1743–1820), with Humphrey Davy (1778–1829) and others, had unsuccessfully tried to control the future of the new (1807) Geological Society (Rudwick 1963). Having failed, in 1809 they resigned from it.

The work of William Smith (1769–1839), who had shown that particular rock strata in England could be identified on the basis of their fossil contents, meant that the relative antiquity of the Middle Jurassic *Megalosaurus* specimens from Stonesfield (from underground mines) could be demonstrated. The stratigraphic horizon of *Iguanodon* proved much more uncertain. However, if much stratigraphic expertise was available in England by 1820, there was little good comparative anatomy available in Britain. Much of the critical expertise in deciphering such new, and always still fragmentary, fossil vertebrate material still had to come from France, from Cuvier. Cuvier had pioneered the new science of comparative anatomy and had used it to demonstrate both that extinction was a frequent fact in the fossil record and that fossil animals could often be "reconstructed" using surprisingly limited evidence.

Cuvier visited England again in 1830, and the English anatomist Richard Owen, who was fluent in French, was chosen to show him around. Cuvier in turn invited Owen to visit Paris from July to September 1831, a visit that greatly influenced Owen in Cuvierian methods (Owen 1894, 1: 48–58). But Cuvier died in May 1832, thus creating a scientific vacuum. A power struggle now developed throughout Europe to take over Cuvier's scientific mantle.

An early contender was the founder of vertebrate paleontology in Germany, Christian Erich Hermann von Meyer (1801–1869) (Keller and Storch 2001). In 1832 he published his first attempt at a taxonomic division of the group of fossil reptiles then classed within the Order Sauria. But this was merely based on their inferred organs of locomotion, and in it *Megalosaurus* and *Iguanodon* were grouped together as "Saurians with Limbs similar to those of the heavy Land Mammals" (Meyer 1832, 201). This paper was translated into English in 1837 by Mantell's linguist-curator George F. Richardson (1796–1848) (Richardson 1837). New footnotes in the paper, provided by Mantell, his translator, or the editor, Edward Charlesworth, noted that Mantell's new 1833 genus "*Hyleosaurus* . . . probably [also] belongs to this division." Owen was certainly aware of Meyer's work when he "invented" dinosaurs in 1842, because he then noted how Meyer had assigned "no other grounds [than this heavy-footedness] for their separation from other Saurians" (Owen 1842, 103).

The decade 1822–1832 was one of political frenzy in England, with the Reform Bill of 1831 and the Reform Act of 1832. These political initiatives brought the first thorough attempts to redraw the political map of Britain and to define who should, and who should not, have the right to vote. The 1832 act was "one of the most momentous pieces of legislation in the history of modern Britain" (Evans 1983). It changed a system that distributed parliamentary representation through limited patronage to one that relied on slightly more democratically based voting. It brought an end to fears of political revolution in England. These fears had made many suspicious of all things French, particularly even in science, where pre-Darwinian theories of evolution or transformism (Laurent 1987) were seen both as highly French and equally revolutionary.

The decade 1822–1832 was equally one of major significance for British science. The deaths of two Royal Society presidents, Banks in 1820 and Davy in 1829, had brought much debate about how English science, where "the pursuit of science [still] does not constitute a distinct profession" (Babbage 1830, 10), was to be controlled and how it might best be advanced. "The Decline of Science in England" and the too-dominant role of the metropolitan Royal Society in leading English science were widely discussed. The British Association for the Advancement of Science (BAAS) had come into existence in 1831 as another rival to the Royal Society. BAAS was to be an itinerant and provincial organization, hoping to encourage British scientists and to advance its science by annual meetings at provincial centers. Such debates also helped science, and especially geology, to gain a better place in popular culture (O'Connor 2007) and soon led to the invention of the word *scientist*. Elitism also suffered, and "upward social mobility based on meritorious accomplishment" could become significantly more common thereafter (Dean 1986). This then encouraged those from lower backgrounds to aspire upward; Gideon Mantell and John Phillips provide two fine examples from geology.

In a scientific England rocked both by the passage of the Reform Act in June 1832 (and Cuvier's death in May), reforms in science were soon being promoted by BAAS. Battle lines were also drawn up in the still small world of English vertebrate paleontology. The protagonists attempting to usurp Cuvier's reputation for English science were Robert Grant (1793–1874), Gideon Mantell, and Richard Owen.

Grant was the poorly paid professor of comparative anatomy and zoology at London's new and "godless" University College. It was godless because it was open, unlike Oxford and Cambridge universities, to dissenters. It asked no religious tests and placed "no barrier to the education of any sect" (Bellot 1929, 56). It was also based in metropolitan, and thus central, London, but Grant could be progressively marginalized by Owen, both because of Grant's earlier espousal of evolutionary, and thus revolutionary, French science, and by the low prestige of his new London University chair against those at the older Oxbridge universities, who supported Owen. It did not help that Grant was both badly paid and, above all, his research badly supported. His work has recently been brought to life by Adrian Desmond,

2.2. An engraving of Richard Owen in the 1840s, holding the complete moa femur from New Zealand (from [Timbs] 1852).

who has revealed how the plaudit of the "Cuvier of this country" had been bestowed on Grant even before Cuvier's death (Desmond 1989, 98), and who was then the "English Cuvier" between then and 1835–1836 (Desmond 1989, 122, 175). Grant enters the story of dinosaurs as an early target of Owen, who believed in creation, not evolution (Desmond 1979). New light on Owen's attitudes to both Grant and to evolution has been shed by the late Stephen J. Gould (2001). But in view of all that is now known, it is clear that Grant was by 1842 merely a minor target for Owen, because Owen had largely disposed of him before Owen's invention of dinosaurs.

Gideon Mantell was the real target of Owen's dino-inventing researches. Mantell was an amateur, the son of "Mr Thomas Mantell [1750–1807], Boot-maker and a Master Gardener" at Lewes, in Sussex (*Sussex Weekly Advertiser*, July 13, 1807, 3), who was both a political radical (Whig) and a dissenter (Methodist). Lewes was a center of dissent, and between 1768 and 1774 was the home of another radical, Thomas Paine (1737–1809), the author of *The Rights of Man* of 1791, who may well have known Mantell senior there (Dean 1999a, 7–9). Mantell junior's life has been illuminated by Dennis Dean (1990, 1999a). But such a family background could only encourage confrontation with Owen's political beliefs. However, Mantell was – for Owen, even more dangerously – a mere provincial. As Robert Bakewell (1767–1843), Mantell's close friend and author of the first adequate textbook on geology in England, wrote in 1830 after a visit to Mantell's already impressive museum of fossils in Lewes, "There is a certain prejudice . . . prevalent . . . in large cities, which makes [people] unwilling to believe that persons residing in provincial towns or in the country can do anything important for science" (Bakewell 1830, 10). Mantell's reputation as a fine fossil collector and vertebrate paleontologist also extended back further than Owen's, to 1825.

In contrast, Owen was a professional comparative anatomist employed at the Royal College of Surgeons in metropolitan London. He was nearly a generation younger than Mantell. He was more energetic, and by the 1840s he was in much better health; he was also a centrally placed Christian (Anglican) believer, was better paid, and with his scientific activities, properly patronized by the liberal establishment circles of Oxbridge and the BAAS (Fig. 2.2).

Dinosaurs-to-be became weapons in the ideological wars that developed among these men. The fossils that collectors like Mantell had gathered so assiduously now inspired new debates over whether species had evolved by evolutionary transmutation, from other species, against the belief they had resulted from separate acts of divine creation. These competing theories caused tensions that built up among these three scientists. In addition, there were other tensions between mere fossil collectors, like Mantell, who gathered his own material with real assiduity, and research workers, like Owen, who did not.

When the BAAS first entered the field of vertebrate paleontology, it supported the young but foreign (Swiss) naturalist, who later made such an impact in America, Louis Agassiz (1807–1873). Agassiz was granted £210 for his work on British fossil fishes in 1835 and 1836 (1836, *Report of BAAS*, 1835 meeting, 5: xxvii). In 1837 the BAAS asked Owen to undertake a similar commission on the "Fossil Reptiles of Great Britain" and granted him £200 (perhaps near $16,000 – see Rudwick 1985, 461 – or perhaps something like $40,000 in modern money) in 1838 toward this (1838, *Report of BAAS*, 1837 meeting, 7: xvi, xix, xxiii, and 1839, *Report of BAAS*, 1838 meeting, 8: xxviii). The three-man committee that granted Owen funds helpfully included his father-in-law, William Clift (1775–1849). In 1838 the Geological Society's premier Wollaston Medal was awarded to Owen. This had earlier, in 1835, gone to Mantell, which had intensified their rivalry.

Leading lights in the BAAS soon saw it as a slight to British science that British fossil treasures had been revealed by foreigners. Owen was a rising star at the Royal College of Surgeons and St. Bartholomew's Hospital in London. The BAAS now chose to support the metropolitan, conservative Owen against the provincial, radical Mantell, who had earlier been encouraged by the BAAS's rival, the London Royal Society. When the BAAS president discussed Owen's first reptilian results in 1840, he duly evoked the name of Cuvier: "Could that eminent man view the progress which our young countryman is making towards the completion of the temple of which the French naturalist was the great architect" (1841, *Report of BAAS*, 1840 meeting, 10: xl). In BAAS's eyes, at least, a new man now wore Cuvier's mantle: Owen.

Both previous occupants of this garment, Grant and Mantell, were by these actions marginalized by the BAAS. After BAAS research support went to Owen, Grant henceforth had to make do with what little he could earn from teaching and student fees. Mantell had also previously been hailed as Cuvier's successor in England, but until he had sold his fossil collection in 1838 to the British Museum (and so to Owen), his financial plight was possibly even worse than Grant's (Dean 1999a, 164–166, 182).

Owen's first report on marine British fossil reptiles was duly read to BAAS in August 1839. It was hailed as the work "of the greatest comparative anatomist living" by William Lucas of Hitchin (Bryant and Baker 1934 1: 179). It was quickly confirmed by many others (and the BAAS president) as the work of the new "English Cuvier" (Desmond 1989, 238, 333). By autumn 1840 Owen was gathering data for his second report on the remaining fossil reptiles to the BAAS (Owen 1894, 1: 169–172). At the BAAS meeting in Glasgow in 1840, its president had announced that Owen had already "collected . . . equally numerous materials" toward this second part, and reported the "abundance of new information" that Owen had collected (1841, *Report of BAAS*, 1840 meeting, 10: xli–xlii and 443–444). Owen had been greatly helped in all this work by the sale of the Mantell collection to the British Museum in 1838 (Cleevely and Chapman 1992). Equally helpful to Owen was the offer made to him in 1840 by another provincial collector, George Bax Holmes (1803–1887), of the use of his Sussex fossil collections – rather than to the local expert Mantell (Cooper 1993).

On August 2, 1841, Owen gave his now famous lecture on the remaining British fossil reptiles to Section C of BAAS. The meeting was held that year in highly provincial Plymouth, in Devon. Neither Mantell nor Grant was present. The newspapers were divided about any interest such provincial science might have. The London *Times*, long unenthusiastic about both the BAAS and science, simply complained that Owen's lecture was of "very long detail" (August 3, 1841). Local Devonian newspapers were more proud of the events taking place in their midst. Three Plymouth newspapers reported

Owen's paper, giving an essentially similar notice to prove that syndicated reporting is nothing new (*Devonport Independent; Devonport Telegraph & Plymouth, Devonport and Stonehouse Herald*, August 7, 1841). All show that Owen ran systematically through the different groups of fossil reptiles he had by then recognized to occur in British rocks.

After describing the crocodiles, Owen next discussed the "extinct species which in their organization bear a relation to some existing at the present time such as the *Iguana*." Of these forms, only *Iguanodon* was separately noted, because a new discovery of "the best specimen [had just been made] in working a quarry near the town of Horsham, for building the church; so enormous was this animal that its claw was six times as large as the claw of an Elephant, all the rest of the body being on the same scale. From the position of the remains, it appeared as if the whole of the skeleton were there, and it was probable that the head of the animal was now under the church"! This proves that Owen still thought (as did Mantell) that *Iguanodon* had been truly gigantic in size, when he lectured at Plymouth in August 1841.

This was because Owen was still following Mantell's lead, whose largest estimate for the length of this Horsham, Sussex, specimen was that it might have even reached up to 200 feet, according to its collector (Hurst 1868, 225). But this new specimen that Owen had used at Plymouth to demonstrate its still gigantic size was not one collected by his new rival, Mantell. This Horsham specimen had been brought to scientific attention by Holmes (Cooper 1993, 205), whose allegiance was to Owen, not Mantell.

A second vital point that these Devonian reports confirm is that Owen also believed that these fossil "animal remains . . . never exhibit any indications that their forms graduated, or by any process passed into another species. They appeared to have sprung from one creative act." Owen had thus used *Iguanodon* and its allies at Plymouth to loudly and clearly proclaim his antievolutionary stance, in continued opposition to Grant (Gould 2001).

Other reports of Owen's Plymouth lecture appeared in the London *Athenaeum* (August 21, 1841, 649–650) and in the *Literary Gazette* (August 7, 1841, 509–511, and August 14, 1841, 513–519). The *Literary Gazette* even noted that Owen had actually named two new species of *Cetiosaurus* in his Plymouth address; *C. hypoolithicus* and *C. epioolithicus*. Both are hitherto unnoticed, but because Owen never described them, they must remain nomina nuda. Other notices of Owen's lecture appeared; in French (*L'Institut*, 10, no. 420, January 13, 1842, 11–13 – derived from the *Literary Gazette*), in German (*Neues Jahrbuch für Mineralogie, Geographie, Geologie, Petrographie, Jahrgang* 1842, part 2, 491–494 – from the previous French source) and in American English (*American Eclectic*, no. 2, 1842, 587–588). The *Athenaeum* specifically noted how Owen had included all three existing genera *Iguanodon*, *Megalosaurus*, and *Hylaeosaurus* merely among the "gigantic forms of terrestrial Saurians," and not in any new order. Owen certainly did not name dinosaurs at Plymouth. They still remained for him to invent.

The fullest report was that given by the *Literary Gazette*. This gives the clearest indication of what Owen actually said, and believed, at Plymouth. It confirms the following:

1. Owen thought Holmes's new Horsham *Iguanodon* was still "six times greater [in size] than the largest elephant." Clearly Owen

still regarded any dinosaurs-to-be as gigantic *and* still agreed with Gideon Mantell's existing size estimates.

2. Owen also said that "there was no gradation or passage of one form into another, but that [each] were distinct instances of Creative Power, living proofs of a divine will and the work of a divine hand, ever superintending and ruling the existence of our world" among these fossil reptiles. This view was clearly in opposition to Grant, as we have seen.

3. Owen also claimed that Mantell's name *Iguanodon* was unsuitable, as it implied a false relationship with *Iguana*. Devonian newspaper sources also note how Owen had observed that "no existing lizard differed more from the *Iguana* than did *Iguanodon* and that the name *Iguanodon* created an erroneous idea of its affinities." This was clearly a new, and pointed, dig at Mantell, who had named *Iguanodon* in 1825.

Crucially, the *Literary Gazette* report also revealed that:

4. Owen only recognized "four great families of reptiles," namely Sauria, Chelonia, Ophidia, and Bactrachia.

5. Owen had divided the relevant one of these (Sauria) into only four groups:
   i. Enaliosauria—the extinct marine reptiles.
   ii. Loricate, or crocodilian, Sauria.
   iii. Lacertians, or squamate, Sauria—Lizards.
   iv. Pterodactyls.

Note again the complete absence of any concept of "dinosaurs" in this report.

Mantell's response to the report of Owen's speech at Plymouth was rapid (1841a). Two of his three points concerned the lizardlike Sauria. First, Mantell claimed he had named *Iguanodon* only because of "the general resemblance in *external form* of [its] fossil teeth with iguana." Second, Mantell claimed that Owen's "new" identification of "the supposed teeth of *Hylaeosaurus*" had already been made 4 years before, by Mantell. Mantell's printed response had been moderate, but his true feelings of Owen were revealed to his friend Benjamin Silliman (1779–1864) in faraway New Haven, Connecticut: "It is too bad [for Owen] to censure those [like Mantell] who have cleared the way for him. I am resolved no longer tamely to submit to such injustice: and if any of my friends again serve me [thus] I will retaliate" (Spokes 1927, 133).

All the many reports of Owen's Plymouth speech make it crystal clear that he had made no mention there of any new order (or of any dinosaurs). All sources show that Owen simply grouped *Iguanodon*, *Megalosaurus*, and *Hylaeosaurus* with lizards, *within* his "Lacertian, or Squamate, division of the Saurian Order" in the Class Reptilia.

Taxonomy is an inclusive science. Any statement that *Iguanodon* and related genera were then placed in a particular division of a named order excludes any possibility that they could have then been included in any other division of this, or indeed any other, order. If one is a postgraduate, one cannot at the same time also be an undergraduate at the same institution. These are equally mutually exclusive categories, within wholly hierarchical systems. Owen simply never mentioned dinosaurs at Plymouth in 1841 because he had not yet invented them. We should pay tribute to those few Victorian journalists who so accurately reported on Owen's lecture that we can now be sure. We may wonder perhaps why today's journalists are in general so different in their attitudes to the importance of science and in reporting it. We might also ask why historians today are so averse to using such vital sources as contemporary newspapers.

After Plymouth, Owen stayed some time in the West Country, returning to London on September 11, 1841, after 47 days away (William Clift's MSS diary, Royal College of Surgeons Library, London). This included an unpleasant return by steamship from Plymouth (Austin 1991, 238, 252). He soon got back to work on the third part of his book on *Odontography* (Owen 1894, 1: 187), and on gathering data for the further report for the BAAS that he had been commissioned to undertake, on British fossil mammals.

A letter Owen wrote, on December 23, 1841, to paleontologist John Phillips shows he was then in Cambridge working on vertebrate material in that University's fine Geological Museum. This ends, "I shall return to London next week to finish the revision of the old Report [i.e., that he had delivered at Plymouth] before again starting in quest of materials for the new [report on mammals]" (Oxford University Museum, Phillips MSS, 1841/65.1).

The major spur for Owen's complete revision of this Plymouth report had come in November 1841, when Mantell's new *Iguanodon* paper (in Mantell 1841b) was published in the *Philosophical Transactions of the Royal Society of London*. Mantell was by now Owen's chief rival in reptilian studies. One hundred extra offprinted copies of both the papers that Mantell had published in this journal that year were supplied to him late in November by R. & J. E. Taylor, printers (Taylor Day Book 1839–1845, 117, St. Bride Printing History Library, London). Mantell's diary confirmed that on December 9 and 31 he sent out "copies of my papers on *Iguanodon* and Turtle from the *Philos. Trans.* . . . just published to many friends in England and to many savants in France." Owen must have received one of the first of these. Mantell's offprints were grandly titled "A Memoir on the Fossil Reptiles of the South-East of England by Dr. Mantell." At least two copies survive. William Lonsdale's (1794–1871) copy, sent him by Mantell in December 1841, is in the Brighton Public Library. The other, still in its original binding, sent out on February 1, 1842, survives in the library of the Yorkshire Museum (*Yorkshire Evening Press*, June 30, 1993). All 100 copies had been distributed by Mantell by February 3, 1842 (MSS diary at Alexander Turnbull Library, Wellington, New Zealand).

These Mantell publications also helped fan the long resentment between the Geological Society and the Royal Society in London. When Mantell continued to publish his Wealden reptile studies in the Royal Society's journal, the then president of the Geological Society, Roderick Murchison (1792–1871), commented, "Whilst I understand . . . the motive which led [Mantell] to communicate his last memoir on the *Iguanodon* to the same Society to which he had addressed his first account of that Saurian, I regret that he should not have communicated to ourselves other paleontological memoirs." Murchison sourly concluded that "so long as the Royal Society produces volumes adorned by the . . . first mathematicians, physiologists and chemists, so long will it maintain its high place, little heeding our humbler pursuits" (Murchison 1842, 653).

In this *Iguanodon* paper, Mantell had described the lower jaw of *Iguanodon* and still compared it, in words and illustrations, with that of *Iguana*. He also still wrote of the enormous, colossal magnitude of *Iguanodon*, which, from its largest known femur, he thought must have belonged to an animal with a thigh 7 feet in circumference! But more significantly Mantell had added at the end:

> I leave to those who have more leisure and ability for the task, the complete restoration of the fauna of the country of the *Iguanodon* and *Hylaeosaurus*. For this purpose I have presented to my distinguished friend, Professor Owen, drawings by M. [Joseph] Dinkel, of the principal specimens in my collection. . . . Removed from the field of my former labours, having disposed of my collection [to the British Museum], the fruit of twenty-five years assiduous research, and being engaged in the duties of an arduous profession [medicine], this memoir is, in all probability, the last contribution which it will ever be in my power to offer in this interesting department of paleontology . . . "Je termine ici mes travaux." (Mantell 1841b, 144–145)

The reason was that, just before this, in October 1841, Mantell had suffered a nearly fatal carriage accident, and then paralysis from some spinal disease. These crises, along with the departure of his wife, who had left him in 1839 (Curwen 1940, 140; Dean 1999a, 192–193), brought Mantell to his lowest ebb. These statements were surely the final stimuli that Owen needed to completely revise his earlier Plymouth report, now believing that Mantell had completely abandoned the field of these "ancient reptiles" to him.

Owen's post-Plymouth revisions were certainly major and highly significant. On January 4, 1842, his best friend, the barrister William John Broderip (1789–1859), asked Owen's opinion on a piece Broderip had just written on fossil spondylid bivalves, "wherein genera are cut down more mercilessly than you have docked the Saurians. I believe with as much justice" (Broderip to Owen, January 4, 1842, Owen MSS, Natural History Museum London, 5/111). Then on January 14, 1842, Broderip confirmed to William Buckland that Owen's "docking of the Saurians" had involved an enormous reduction in their size. Broderip now reported, "I am happy to say that I have now before me the beginnings of the end of Owen's *Report* on fossil reptiles—an *opus magnum* in itself though he has reduced the sesquipedality of some of your old friends [*Iguanodon*, etc.] 'with tails as long at St Martins Steeple'" (British Library, Add. MSS 40500 fol. 247–248). The famous steeple of St. Martin-in-the-Fields Church in London was (and still is) 192 feet high. Owen had now completely changed his opinion of the size of *Iguanodon* from what he had claimed in his Plymouth lecture—where they had been up to six times bigger than the biggest elephant (and thus up to 200 feet long)!

Murchison's presidential address to the Geological Society of London on February 18, 1842, noted that "Owen . . . will shortly lay before the world the results of his researches into the extinct Saurians of our island" (1842, 652). Owen's long-awaited second report on British fossil reptiles was finally published in London, for the BAAS, in the first fortnight of April 1842, 8

2.3. Title page of the offprint of Owen's dinosaur-inventing paper, published in April 1842–despite its 1841 date (author's collection).

months after Plymouth, in an edition of 1,500 copies, priced at 13s/6d (*Publishers Circular*, April 14, 1842, 114; but see also Torrens 1993, 274).

Only in this publication did Owen invent dinosaurs. Mantell's copy has been reprinted by Dean (1999b). The printers' records survive to confirm the complex history of the volume's gestation and its exact chronology. They also demonstrate how many changes Owen made to this particular report. It received numerous corrections in proof as Owen continually revised it (see R. & J. E. Taylor, MSS Check Book 1836–1842, fol. 145, and MSS Day Book 1839–1845, fol. 187, St. Bride Printing Library, London).

The post-Plymouth dating of some of these changes is clear from the printed text. For example, on January 23, 1842, George Bax Holmes had sent Owen details of some new *Goniopholis* (crocodile) scutes. When this discovery was published in April, Owen noted they "been discovered since the first sheets of this *Report* went to press" (Owen 1842, 194). The publisher's records for this BAAS report also survive and show that 16 sale copies of the first 50 of the full Plymouth report remained unsold by May 12, 1842 (John Murray archives, National Library of Scotland, Edinburgh). Owen had a mere 25 offprints of his paper printed for his own use (see R. & J. E. Taylor, MSS Day Book 1839–1845, f. 187), but, to add to the confusion, these all carry the incorrect date of 1841–whether to deliberately mislead is not known. The copy illustrated here (Fig. 2.3) was originally Broderip's. Buckland's copy was illustrated by Gould (2001, 186). Intriguingly, Mantell's annotated copy (Dean 1999b) is not one of these offprints but one that came from the full report.

This 1841 date has since caused some to believe that such offprints were issued as preprints, and to claim that Owen's complete printed report had been ready, and was issued, *before* his Plymouth lecture (Gardiner 1990)! This internal evidence, with the printer's and publisher's records, and the fact that these supposed preprints carry the correct pagination from the final, completed volume of 1842, all show how impossible such claims are.

Others of Owen's revisions were inspired by the recent publication of his rival Mantell's two papers (Mantell 1841b). But the most important revision was to recognize a quite new order or suborder of fossil reptiles: the Dinosauria. Owen must have had a sudden realization that allowed him to recognize this in time for the report's publication. This moment happened whenever it was that Owen discovered that the sacral vertebrae in *Iguanodon* were fused together, as they had long been known to be in *Megalosaurus*, in which this feature was well known. Such fusion, to strengthen the sacral vertebrae in such a large animal, Owen took to be "*the* adaptation of the Dinosaurs to terrestrial life." Owen had found the unique specimen, showing that this new "character . . . altogether peculiar among Reptiles" occurred also in *Iguanodon* (Owen 1842, 103). It was preserved in the museum (Fig. 2.4) of the London wine merchant and radical philanthropist William Devonshire Saull (1783–1855).

This specimen allowed Owen to infer a quite new relationship on which he entirely based this new order of Dinosauria. To quote Wittgenstein, "A relationship of similarity or . . . difference does not rest out there waiting for us to find it, but depends upon [our] *classifying* two things as similar or different. In order to say two things are similar . . . we select from the

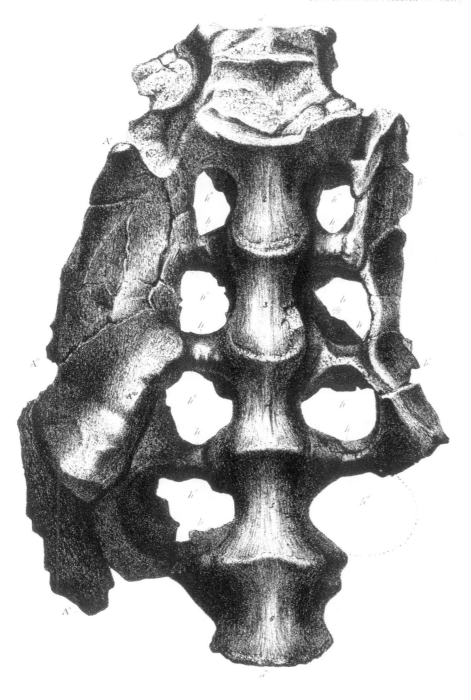

Phil Trans. MDCCCXLIX. Plate XXVI.

2.4. The sacral vertebrae of the Saull specimen of *Iguanodon* x 0.5 (from Mantell 1849, plate 26). The fusion of the vertebrae is clearly evident in this, the first illustration of the historic Saull specimen to be published. But the feud between Mantell and Owen even extended to the number of sacral vertebrae that were so fused. Mantell here saw six–numbered 1–6 downward–while Owen earlier recognized only five.

myriad of possibilities the relevant properties whereby we judge two things to be similar" (Cooper 1991, 967).

The shared fusion of the sacral vertebrae in these new dinosaurs was by far the most important of those properties. Others are listed by Owen (1842, 102–103). The role of the sacrum in *Megalosaurus* is clearly shown, as is the small numbers of bones on which that particular reconstruction was based, in Figure 2.5. The recognition of this dino-defining feature by Owen was strikingly original. It allowed him to place the two genera *Megalosaurus* and *Iguanodon* in his new order or suborder of Dinosauria. He also inferred

2.5. *Megalosaurus,* as shortened by Owen, showing internally how few bones were then available to help this reconstruction (from Owen 1854, 20).

that the third genus, *Hylaeosaurus*, belonged here too, on the basis of the very same, if limited and fragmentary, sacral evidence that he had discovered in the Mantell collection, which was now in his care (Owen 1842, 113–114).

Saull had opened his museum in the city of London in 1833. He was a radical socialist who believed in education, and his museum was freely accessible every Thursday to all, especially the working classes. His museum was clearly another of the benefits that the 1832 Reform Act had helped to bring about. Owen must have been rather chagrined, as a member of the Anglican Tory establishment, to find that this vital single specimen, on which, as Owen later carefully recorded, "the characters of the Order Dinosauria were mainly founded" (1855, 11), was preserved in the museum of a socialist radical! It alone had justified Owen's creation of a completely new order of reptiles. Furthermore, to help in Owen's war with Mantell, the specimen had come neither from Mantell's collection nor from Sussex, but from Brook, on the southern coast of the Isle of Wight, Hampshire. It survives as a truly historic object in the Natural History Museum in London (Fossil Reptilia, Reg. 37685; see Fig. 2.4). I have highlighted the importance of Brook as the dino-inventing locality in a paper on the forgotten English geologist nephew of "the father of American geology" (Torrens 2010). It was William Buckland who had first announced the importance of this locality in a paper read in 1829 but not published until 1835 (Weishampel and White 2003, 120–125).

Exactly when Owen first came across this Saull specimen is unknown, but clearly he must have realized its significance only after the Plymouth meeting. We can be sure of this because in the reports of his lecture there, such fused sacral vertebrae are only noted as present in *Megalosaurus* (*Literary Gazette*, August 7, 1841, 517). Later, when Owen claimed that he had found the Saull specimen in 1840, this was in a source that wrongly claimed that his paper had been published in 1841 (Owen 1855, 9). So Owen must have discovered this critical object between September 1841 and March 1842.

Mantell's reaction to Owen's 1842 publication is recorded in letters to Silliman and in his own annotated copy (reproduced by Dean 1999b). On April 30, 1842, Mantell noted Owen's was "an elaborate and very masterly paper," but that he, Mantell,

> had *again* [emphasis added] to regret a want of honour, and I may say justice, towards those but for whose labor and zeal, [Owen] could never have obtained the materials for his own reputation. He has in several instances behaved very disingenuously towards me; altered names which I had imposed . . . and stated many inferences as if originating with himself, when I had long since published the same. . . . I believe he would have altered *Iguanodon* and *Hylaeosaurus* had I not sent the letter of remonstrance to the *Literary Gazette.*

Silliman agreed, although he had not yet read Owen's memoir; "his treatment of you and that of Agassiz too, is unjust and dishonourable." Silliman then perceptively asked, "Have you reduced the length of the *Iguanodon*? Something in your remarks impressed me as if you had topped his tail a little." Mantell's reply, dated August 4, 1842, confirmed that "you were quite right as to your conjecture that subsequent discoveries have led me to believe that the tail of the *Iguanodon* was short and flat in a vertical direction . . . yet you will see [in Owen's BAAS report] that he has stated the probable size of the *Iguanodon* to be much shorter, and yet has taken no notice whatever that I had already announced it"—that is, that "the tail was not long," as Mantell had certainly announced (1841b, 140). "There are numerous other lamentable offences of the same kind" (Spokes 1927, 135–136).

That Owen's bold invention of dinosaurs, as a separate group of creatures, was within a period of highly political work across the whole of science has been made wonderfully clear by Desmond (1989). Although the members of Owen's new order were still large in size, they were no longer gigantic (and not just their tails). The size of both *Megalosaurus* and *Iguanodon* had been cut down (as had Mantell) to about 30 feet long rather than the previous "up to 200 feet" (Owen 1842, 142–143). Owen saw dinosaurs as four-footed and—crucially—as highly advanced reptiles, like present-day pachydermal mammals. Such reptiles, vested with advanced mammalian characters, became a nail to be sunk into the progressionists' or transmutationists' coffins because such characters clearly could only have been given to these extinct reptiles at their creation, by God. Owen was clearly in 1842 a creationist. So dinosaurs also became a final weapon of support in Owen's antievolutionary battle with Grant. His major scaling down of the size of dinosaurs, and his deadly realization of what Mantell had failed to realize, that they were not just large lizards (Fig. 2.5), were similar strikes against his now more important rival Mantell, whom he clearly thought had abandoned paleontology.

In 1845 Meyer formally named his *Pachypodes*, the group he had earlier separated as his "heavy-footed Sauria." But he had been as heavy-footed as his animals and was too late to achieve any priority over the name Dinosauria (Meyer 1845), although after Mantell's death, Thomas Huxley (1825–1895), Owen's later rival, did try to rehabilitate Meyer's name (Huxley 1870, 32–33). But unlike Owen's novel, sacral, discovery, Meyer had given no proper taxonomic reason why they should be so separated together.

Late in 1849 Mantell was awarded the Royal Society's Royal Medal for his work on dinosaurs, which he had renewed after 1841 (Dean 1999a,

230–235). The history of this award reveals how deep the rivalry between Owen and Mantell had become. Mantell's papers on fossil reptiles (Mantell 1841b) had been refused consideration for this medal in 1846, when Owen instead was awarded the medal in November 1846 for his paper on belemnites (a group of extinct mollusks related to squid). This Mantell had, with considerable justice, noted was "a tissue of blunders from beginning to end. So much for Medalism" (Curwen 1940, 212). Donovan and Crane (1992) have since shown how much justice there is in Mantell's claim. In this highly charged atmosphere, there was now a second attempt in 1849 to honor Mantell, with Owen highly active in seeking to deny Mantell any such award. Mantell noted Owen to have said that "all I [Mantell] had done was to collect fossils and get others to work them out"! (Curwen 1940, 243, 245–247). But after skillful lobbying by his friends, including Charles Lyell (1797–1875), Mantell finally received the Royal Medal on January 4, 1850 (Curwen 1940, 249), for his work on *Iguanodon* (Mantell 1849).

It was under such circumstances that Mantell wrote to Lyell later in 1850 to prove that he was still not convinced of Owen's flash of dino-inventing genius:

> In [Owen's] Report on British reptiles the only new fact in the osteology of the *Iguanodon*, stated by Owen is the construction of the sacrum which is peculiar O.[wen] *supposes* [emphasis added] to *Iguanodon*, *Megalosaurus* and *Hylaeosaurus*. . . . I am now certain there were other contemporary genera with the sacrum composed of several anchylosed vertebrae: you are aware, that after all there is nothing wonderful in this: the common crocodile &c has a sacrum formed of two anchylosed vertebrae. (American Philosophical Society, MSS Darwin/Lyell Correspondence, B/D 25, October 7, 1850)

By October 1851 Mantell had changed his mind on this, with a frankness that is praiseworthy, in view of the new vendetta that had flared up between him and Owen (Benton 1982, 124). He now realized that "the most important and novel feature in relation to the osteology of the Wealden reptiles enumerated in Professor Owen's *Reports*, was the remarkable structure of the *Sacrum* in the three extinct genera of Dinosaurians. . . . [N]o one had previously suspected that in these reptiles the pelvic arch was composed of *more than two* [emphasis added] anchylosed vertebrae" (Mantell 1851, 268).

It was this character that enabled Mantell to claim just before his death in 1852 that perhaps "there were five, if not six, genera of Wealden reptiles with a similar construction of the sacrum" (Mantell 1851, 224). Mantell had now been able to confirm that *Hyleosaurus* also possessed such fused sacral vertebrae, on the basis of new material (Mantell 1851, 325). The additional genera were:

1. *Pelorosaurus* of 1850 – his "unusually gigantic & stupendous" saurian of "enormous magnitude" (Mantell 1850), which Owen (1842, 94–100) had earlier assigned to *Cetiosaurus* (Mantell 1851, 330–332). This was Mantell's final attempt to prove to Owen that some dinosaurs had indeed been "stupendous" in size. But Mantell still had no knowledge of its sacrum (Mantell 1851, 332).
2. *Cetiosaurus* itself, which Mantell nearly recognized as "dinosaurian" because it "apparently had portions of a sacrum of the dinosaurian type," a claim he first published in 1851 (see Mantell 1854, 332).

3. The last genus that Mantell had in mind in 1851 was probably his *Regnosaurus* of 1848 (Mantell 1851, 333), a genus of which Mantell had only a portion of the lower jaw and whose vertebrae were still quite unknown.

Mantell also noted in 1851 "the spirit of self-aggrandisement and jealousy [by which Owen] had exerted [his] baneful influence over this department of [vertebrate] paleontology" (Mantell 1851, 226, and see 257, 286, etc.). Huxley duly confirmed in November 1851 that "it is astonishing with what an intense feeling of hatred Owen is regarded by the majority of his contemporaries, with Mantell as arch-hater" (Huxley 1908, 1: 136). The vendetta between the two continued unabated until Mantell's self-induced death late in 1852. But it cost Owen the presidency of the Geological Society of London in 1852, when Leonard Horner and William Hopkins decided that Edward Forbes would be a better candidate in view of Owen's bitter "antagonism to Mantell" (Dawson 1946, 85–86). Owen's obituary of Mantell provides the best evidence of the depth of his antagonism. Mantell's "want of exact scientific . . . knowledge," Owen wrote, had simply "compelled him to have recourse to those possessing it," because of course it was Owen who had "first perceived [Mantell's] error respecting [*Iguanodon's*] bulk, which had arisen out of an undue enthusiasm touching its marvellous nature" (*Literary Gazette*, November 13, 1852, 842). As the late John Wennerbom has recently pointed out, there is "more than an element of truth in this spiteful comment" (2001, 58–59).

In a final irony, the first recorded outbreak of dinomania (Torrens 1993, 279), late in 1853 and 1854, when the Crystal Palace restorations were put on display, used Owen's interpretations of the animals he had invented (Fig. 2.6).

If only Mantell had been able to react differently to the invitation he had received before Owen, from the Crystal Palace Company. Its board of directors' minutes of August 10, 1852, resolved "that a Geological Court be constructed containing a collection of full sized models of the Animals & plants of certain geological periods, and that Dr. Mantell be requested to superintend the formation of that collection" (Alexander Turnbull Library, Wellington, New Zealand, MS Papers 0083-032). Mantell's diary noted, on August 20, 1852, that "Mr [William] Thompson [Thomson (1825–1899)], Secretary of the Nat. Hist. Dept. of the New Crystal Palace called, & I found that the plan intended to be carried out as to Geology, was merely to have models of extinct animals. . . . I therefore declined the superintendence of such a scheme" (Dell 1983, 90). Mantell was then less than 3 months away from his early death, and as his diary for September 16, 1852, had already noted, "*in truth I am used up*" (Dell 1983, 91).

If only Mantell had been able to undertake these restorations, we would now see quite different visions of those wonderful fossil animals of which he had uncovered much of the evidence. These would first have been much larger than the Owen versions that so spectacularly survive in the Crystal Palace Park near London today (Doyle and Robinson 1993). We know that in 1851 Mantell was still urging that *Iguanodon* had been much bigger than Owen's "reduction" of 1842 had allowed (Mantell 1851, 312). Mantell had since discovered, in the autumn of 1849, "the most stupendous humerus of

THE SECONDARY ISLAND.

1. Mososaurus.	6. Hylæosaurus.	10, 11. Teleosaurus.	18—20. Labyrinthodo.
2, 3. Pterodactyles.	7. Megalosaurus	12—14. Ichthyosaurus.	21—22. Dikynodon.
4, 5. Iguanodon.	8, 9. Pterodactyles of Oolite.	15—17. Plesiosaurus.	

2.6. Panorama of the Extinct Animals on the Secondary Island at Crystal Park (from Anonymous 1877, 27). The dinosaurs, numbered 4, 5, 6, and 7, are in the reconstructions envisaged by Owen, not Mantell.

a terrestrial reptile ever discovered; it is 4 and a half feet in length." This became the original for his new genus *Pelorosaurus* (Mantell 1850).

Equally significantly, it is also clear than Mantell had long realized, from study of the well-preserved "Maidstone *Iguanodon*," that in this animal "the hinder extremities, in all probability, resembled the unwieldy contour of those of the Hippopotamus or Rhinoceros, and were supported by strong, short feet, protected by broad ungual phalanges: the fore feet [however] appear to have been less bulky, and adapted for seizing and pulling down the foliage and branches of trees" (Mantell 1851, 311–313). Mantell had been clearly aware that *his Iguanodon* had had a more kangaroo-like posture since at least 1841 (Mantell 1841b, 140), as opposed to the quadrupedal and more rhinocerine version proposed by Owen, which was that finally recreated for him by Benjamin Waterhouse Hawkins (1807–1894 – see Bramwell and Peck 2008) at the Crystal Palace.

Conclusions

There are many lessons for both historians and scientists in this complex story of dino invention.[1] First, Martin Rudwick's (1985, 465) statement, that "the official *Report* of each [BAAS] meeting was not published until several months later [than the meeting it purported to report] and [so] its summaries cannot be relied on as an accurate record of what was actually read at the time," is abundantly confirmed. Second, it also reveals that Owen, in so much modifying his script after it had been delivered at Plymouth, before it was printed, was simply doing what he had unjustly accused another rival, Alexander Nasmyth (died 1848), of having unfairly done just before him in a similar presentation to the BAAS (Nasmyth 1842). Furthermore, it is now clear that Owen was in the habit of secretly altering his BAAS papers after they had been delivered (Charlesworth 1846, 25–27). Owen thus might well have had a vested interest in recording false dates on his offprinted publications (as on Fig. 2.3).

Another message is the need for historians to understand something of the science of which they write the history. Dennis Dean's "partisanship of Mantell's cause" (Wennerbom 2001, 58) attempted to raise Mantell and denigrate Owen in an otherwise fine biography (Dean 1999a). Dean entitled his book *Gideon Mantell and the Discovery of Dinosaurs*. In this Mantell must become "both the primary discoverer and designer of dinosaurs" (Dean 1999a, 190–191). But Dean then had to place Owen in the historically impossible position of having had "eleven possible dinosaurs available to him," of which Owen "included three, misclassified six and omitted two." This is history with fearful dino-scale hindsight. Before April 1842, when dinosaurs were invented (and the word only slowly entered all the world's languages), there were no such things, and they can have only a prehistory. Attempts to identify "the first dinosaur bone to come to learned attention" or to be "found in North America, or probably the world" (Simpson 1942, 153, 178) are meaningless in historical terms (Desmond 1979, 224). The first dinosaur bone to be found can only be the *Iguanodon* sacrum that Owen found in Saull's museum. All other dinosaur discoveries could be recognized to be so only after this. Thus any other "first" dinosaur bones can only be such with hindsight. If one finds a wheel in an archaeological excavation, it does not prove the discovery of a motor car. The discovery of any "car" depends on the recognition of its critical features, of which wheels are only one. The fused sacral vertebrae were the unique and critical feature of Owen's diagnosis of dinosaurs.

For Mantell to have discovered "eight dinosaurs" (Dean 1999a: flyleaf, i, and 4) is even more history with hindsight. Of these eight,

1. *Megalosaurus* cannot be one. According to Delair and Sarjeant (2002, 194), this genus is known only from Middle Jurassic strata, not from the Early Cretaceous of Mantell's supposed finds.

2 and 3. Mantell's *Iguanodon* and *Hylaeosaurus* were only dinosaurs-to-be and were both first recognized as dinosaurs only by Owen.

4. *Pelorosaurus*. No one yet, as we have seen, had any knowledge of its sacrum, so it could not yet be diagnosed as dinosaurian.

5. *Cetiosaurus* is perhaps the dinosaur to which Mantell is most nearly entitled to be regarded as the discoverer, although, as we have seen, and as Dean correctly notes, Mantell still "did not quite realise it was a dinosaur" (Dean 1999a, 4).

6. The taxonomic position of *Regnosaurus*, based only on a portion of jaw, is still debated (Galton 1997, 296). It too was simply not diagnosable as dinosaurian during Mantell's life. This is exactly why Owen noted, "What evidence is there that it belongs [to the dinosaurs]?" (Dean 1999b: opposite 120 at end).

7. "Though *Hypsilophodon* would have been Mantell's seventh dinosaur, he failed to recognise it as a separate genus, having only a few vertebrae on which to base his identification" (Dean 1999a, 236).
8. "Mantell's 'second species' of *Pelorosaurus* may, however, be an entirely different dinosaur altogether. If so, it would have been his eighth" (Dean 1999a, 239).

We must conclude that Mantell properly "discovered," ignoring all hindsight, no new dinosaurs!

Elsewhere, Dean (1999a, 190, 1999b, 13) records that Owen's creation of dinosaurs "was a mistake from the beginning." This was supposedly because "it amalgamated two distinct classes of saurians having substantially different features." But here is more hindsight, because this concept was only first proposed by Henry Govier Seeley in 1887 (see Chapter 3 in this volume). Dean claimed that the "osteological definitions on which Owen based his concept have now been entirely discredited." To claim this in a book published in 1999 is equally discreditable. As Benton has recorded, the "collapse of this polyphyletic view [of the origin of dinosaurs had] came quickly and dramatically about 1984," 15 years earlier! Moreover, the once more monophyletic dinosaurs were defined by a number of synapomorphies (characters), one of which is still the presence of "three or more sacral vertebrae" (Benton 1997, 205–206) – exactly the character by which Owen had first recognized their novelty.

Note

1. Internet claims at, for example, http://dml.cmnh.org/2002Aug/msg00632.html, of my supposed plagiarism of the dino work of the American historian Pauline Carpenter Dear need to be answered. I was certainly aware of her unpublished work (from *Isis* 76: (2), 215, 1985). But I only read it after she kindly sent it to me (on February 9, 2004). That we came to such similar conclusions only proves that we have independently used the same sources.

References

Anonymous. 1877. *Crystal Palace: A Guide to the Palace and Park.* London: Dickens and Evans.

Austin, F. (ed.). 1991. *The Clift Family Correspondence, 1792–1846.* Sheffield: Centre for English Cultural Tradition and Language.

Babbage, C. 1830. *Reflections on the Decline of Science in England.* London: B. Fellowes and J. Booth.

Bakewell, R. 1830. A visit to the Mantellian Museum at Lewes. *Magazine of Natural History* 3: 9–17.

Bellot, H. H. 1929. *University College London, 1826–1926.* London: University of London Press.

Benton, M. J. 1982. Progressionism in the 1850s. *Archives of Natural History* 11: 123–136.

———. 1997. Origin and early evolution of dinosaurs. In J. O. Farlow and M. K. Brett-Surman (eds.), *The Complete Dinosaur.* Bloomington: Indiana University Press.

Bramwell, V., and R. M. Peck. 2008. *All in the Bones: A Biography of Benjamin Waterhouse Hawkins.* Philadelphia: Academy of Natural Sciences.

Bryant, G. E., and G. P. Baker. 1934. *A Quaker Journal.* 2 vols. London: Hutchinson.

Buckland, W. 1824. Notice on the *Megalosaurus* or great fossil lizard of Stonesfield. *Transactions of the Geological Society of London* (ser. 2) 1: 390–396.

———. 1837. *Geology and Mineralogy considered with reference to Natural Theology.* 2nd ed. 2 vols. London: W. Pickering.

Charlesworth, E. 1846. On the occurrence of a species of *Mosasaurus*. . . . *London Geological Journal* 1: 23–32.

Cleevely, R. J., and S. D. Chapman. 1992. The accumulation and disposal of Gideon Mantell's fossil collections. *Archives of Natural History* 19: 307–364.

Cooper, C. C. 1991. Social construction of invention through patent management. *Technology and Culture* 32: 960–998.

Cooper, J. A. 1993. George Bax Holmes (1803–1887) and his relationship with Gideon Mantell and Richard Owen. *Modern Geology* 18: 183–208.

Curwen, E. C. 1940. *The Journal of Gideon Mantell*. London: Oxford University Press.

Dawson, W. R. 1946. *The Huxley Papers*. London: Macmillan.

Dean, D. R. 1986. Review [of Rudwick 1985]. *Annals of Science* 43: 504–507.

——. 1990. A bicentenary retrospective on Gideon Algernon Mantell (1790–1852). *Journal of Geological Education* 38: 434–443.

——. 1999a. *Gideon Mantell and the Discovery of Dinosaurs*. Cambridge: Cambridge University Press.

——. 1999b. *The First "Dinosaur" Book: Richard Owen on British Fossil Reptiles (1842)*. New York: Scholars' Facsimiles.

Delair, J. B., and W. A. S. Sarjeant. 2002. The earliest discoveries of dinosaurs[-to-be]. *Proceedings of the Geologists' Association* 113: 185–197.

Dell, S. 1983. Gideon Algernon Mantell's unpublished journal, June–November 1852. *Turnbull Library Record* 16: 77–94.

Desmond, A. 1979. Designing the dinosaur. Richard Owen's response to Robert Edward Grant. *Isis* 70: 224–234.

——. 1989. *The Politics of Evolution*. Chicago: University of Chicago Press.

Donovan, D. T., and M. D. Crane. 1992. The type material of the Jurassic cephalopod *Belemnotheutis*. *Palaeontology* 35: 273–296.

Doyle, P., and E. Robinson. 1993. The Victorian "geological illustrations" of Crystal Palace Park. *Proceedings of the Geologists' Association* 104: 181–194.

Evans, E. J. 1983. *The Great Reform Act of 1832*. London: Routledge.

Galton, P. M. 1997. Stegosaurs. In J. O. Farlow and M. K. Brett-Surman (eds.), *The Complete Dinosaur*. Bloomington: Indiana University Press.

Gardiner, B. G. 1990. Clift, Darwin, Owen and the Dinosauria. *Linnean* 6: 19–27.

Gould, S. J. 2001. *The Lying Stones of Marrakech*. London: Vintage.

Gunther, A. E. 1925. *Early Science in Oxford*. Vol. 3, *The Biological Sciences*. Oxford: For the Subscribers.

Hurst, D. 1868. *Horsham, its History and Antiquities*. London: William Macintosh.

Huxley, L. 1908. *Life and Letters of Thomas Henry Huxley*. 3 vols. London: Macmillan.

Huxley, T. H. 1870. On the classification of the Dinosauria. . . . *Quarterly Journal of the Geological Society of London* 26: 32–51.

Keller, T., and G. Storch (eds.). 2001. *Hermann von Meyer, Frankfurter Bürger und Begründer der Wirbeltierpaläontologie in Deutschland*. Kleine Senckenberg-Reihe 40. Schweizerbart: Stuttgart.

Kirby, D., K. Smith, and M. Wilkin. 1992. *The New Roadside America*. New York: Fireside.

Laurent, G. 1987. *Paléontologie et Évolution en France, 1800–1860*. Paris: Editions du C. T. H. S.

Mantell, G. A. 1825. On the teeth of the Iguanodon. *Philosophical Transactions of the Royal Society* 115: 179–186.

——. 1841a. Fossil reptiles. *Literary Gazette* (August 28, 1841): 556–557.

——. 1841b. A *Memoir on the Fossil Reptiles of the South-East of England*. (An offprinted combination of two papers published in the *Philosophical Transactions of the Royal Society of London*.) London: Published privately.

——. 1849. Additional observations on the osteology of the *Iguanodon* and *Hylaeosaurus*. *Philosophical Transactions of the Royal Society* 1849: 271–305.

——. 1850. On the *Pelorosaurus*; an undescribed gigantic terrestrial reptile. *Philosophical Transactions of the Royal Society* 1850: 379–390.

——. 1851. *Petrifactions and their Teachings*. London: H. G. Bohn.

——. 1854. *Geological Excursions round the Isle of Wight*. 3rd ed. London: H. G. Bohn.

Meyer, H. von. 1832. *Palaeologica zur Geschichte der Erde und ihrer Geschöpfe*. Frankfurt: Schmerber.

——. 1845. System der fossilien Saurier. *Neues Jahrbuch für Mineralogie, Geologie und Paläontologie* 278–285.

Moore, D. T., J. C. Thackray, and D. L. Morgan. 1991. A short history of the Museum of the Geological Society of London, 1807–1911. *Bulletin of the British Museum of Natural History* (historical ser.) 19 (1): 51–160.

Murchison, R. I. 1842. Presidential address to the Geological Society of London. *Proceedings of the Geological Society of London* 3: 637–687.

Nasmyth, A. 1842. *A Letter to the Right Hon Lord Francis Egerton*. . . . London: Churchill.

O'Connor, R. 2007. *The Earth on Show*. Chicago: University of Chicago Press.

Owen, R. 1842. Report on British fossil reptiles: part II. *Report of the British Association for the Advancement of Science* 1841: 60–204.

——. 1854. *Geology and the Inhabitants of the Ancient World*. London: Crystal Palace Library.

——. 1855. *Fossil Reptilia of the Wealden Formations, Part 2*. London: Palaeontographical Society Monograph.

——. 1894. *The Life of Richard Owen*. 2 vols. London: J. Murray.

Phillips, J. 1871. *The Geology of Oxford and the Valley of the Thames*. Oxford: Clarendon Press.

Richardson, G. F. 1837. Translation of Hermann von Meyer's *On the structure of the Fossil Saurians*. *Magazine of Natural History* (n.s.) 1: 281–293, 341–353.

Rudwick, M. J. S. 1963. The foundation of the Geological Society of London. *British Journal for the History of Science* 1: 325–355.

——. 1985. *The Great Devonian Controversy*. Chicago: University of Chicago Press.

Simpson, G. G. 1942. The beginnings of vertebrate paleontology in North America. *Proceedings of the American Philosophical Society* 86: 130–188.

Spokes, S. 1927. *Gideon Algernon Mantell Surgeon and Geologist*. London: J. Bale.

Sues, H.-D. 2011 (this volume). European dinosaur hunters of the nineteenth and twentieth centuries. In M. K. Brett-Surman, Thomas R. Holtz Jr., and James O. Farlow (eds.), *The Complete Dinosaur*, 2nd ed., 45–59. Bloomington: Indiana University Press.

[Timbs, J.] 1852. *The Year Book of Facts in Science and Art*. London: Simpkin, Marshall.

Torrens, H. S. 1993. The dinosaurs and dinomania over 150 years. *Modern Geology* 18: 257–286.

——. 2010. William Perceval Hunter (1812–1878): forgotten English student of dinosaurs-to-be and Wealden rocks. In R. Moody et al. (eds.), *Dinosaurs and Other Extinct Saurians: A Historical Perspective*. Special Publication of the Geological Society of London 343:31–47.

Weishampel, D. B., and N. M. White. 2003. *The Dinosaur Papers, 1676–1906*. Washington, D.C.: Smithsonian Books.

Wennerbom, J. 2001. Gideon Mantell: bone collector or palaeontologist? *Metascience* 10: 57–59.

European Dinosaur Hunters of the Nineteenth and Twentieth Centuries

Hans-Dieter Sues

3

Although the first undisputed fossils of dinosaurs were found in England (Delair and Sarjeant 2002) and the first students of dinosaurs were active in Europe, the United States became the center for study of dinosaurs during the second half of the nineteenth century, when the combined efforts of Edward Drinker Cope, Joseph Leidy, and Othniel Charles Marsh resulted in a large number of remarkable discoveries. As a result, present-day students often overlook the important contributions made by early European researchers, even those from England. This chapter provides a brief introduction to the lives and careers of some of the principal European dinosaur researchers (some of whom did no fieldwork) and dinosaur hunters of the nineteenth and twentieth centuries. Like any review of this subject, it owes much to Colbert's (1968) classic survey of the history of research on dinosaurs. Since the publication of Colbert's book, however, numerous studies on the history of paleontology and related fields by professional historians of science (e.g., Desmond 1982) have shed much new light on the subject and often substantially revised traditional accounts.

During the nineteenth century, clergymen, medical doctors, and merchants collected most dinosaurian fossils in England and on the Continent. Unless these collectors studied the finds themselves, they would lend, donate, or sell them to the leading naturalists of the day for scientific study and formal description. In addition to drawing on this source of material, two of the foremost early European vertebrate paleontologists, Baron Georges Cuvier (1769–1832) and Sir Richard Owen (1804–1892), enjoyed great influence and patronage at the highest levels of state and did not hesitate to marshal the formidable resources of their nations to secure fossils of interest to them. European dinosaur researchers only started embarking on fieldwork of their own in the early twentieth century, presumably inspired by the spectacular successes of their American colleagues Edward Drinker Cope and Othniel Charles Marsh.

England

The Reverend William D. Fox (1813–1881) was an important collector of Early Cretaceous dinosaurian remains and other vertebrate fossils from the Isle of Wight (Blows 1983; Martill and Naish 2001). A native of Cumberland, Fox became a clergyman and moved to the Isle of Wight in 1862 as curate to the parish of Brixton (now Brighstone). He befriended the poet Alfred, Lord Tennyson, who resided at Farringford near Freshwater, only a few miles west of Brixton. When Fox first became interested in dinosaurs is not known, but he frequently corresponded with Sir Richard Owen and read the great anatomist's monographs. Owen later would describe many of the important finds made by Fox during the 1860s. Fox resigned his post in 1867 but continued to live in Brixton and collect fossils of "old dragons" in the region. In 1875, he became curate of Kingston, near Shorwell, on the Isle of Wight. Contemporaries observed that Fox was more interested in "old dragon" bones than in the spiritual well-being of his charges. In 1882, the Trustees of the British Museum (Natural History) in London purchased his important collection, which comprises over 500 specimens.

Fox also interacted with John Whitaker Hulke (1830–1895), a renowned surgeon who was best known for his work in ophthalmology and became president of the Royal College of Surgeons in 1893. A remarkably accomplished man with diverse interests, Hulke was elected to the Royal Society in 1867 for his anatomical work on the retina in vertebrates. He took up geological fieldwork as a hobby. For many years, Hulke was the only person to be granted unrestricted access to Fox's fossil collection and undertook the scientific study of some of these materials. Hulke himself also collected and prepared dinosaurian and other vertebrate fossils and published extensively on Jurassic and Cretaceous dinosaurs and other reptiles from England.

Another important collector was George Bax Holmes (1803–1887), a wealthy Quaker from Sussex (Cooper 1992). Holmes started collecting in the early 1830s. By 1840, his growing collection of vertebrate fossils, including numerous specimens of *Iguanodon*, had already attracted the attention of Sir Richard Owen. Owen first examined Holmes's collection in the summer of that year. Holmes offered his entire collection to Owen for study. He expected that Owen would describe and figure many of his fossils (Cooper 1993). Subsequently, he became disappointed when Owen did not illustrate a sufficient number of his specimens, and his pride was hurt by the fact that Owen increasingly drew on fossils from other private collections. Their relationship turned sour in the 1850s when Holmes discovered that Owen had cataloged some specimens borrowed from him into the collections of the Royal College of Surgeons. After his death, Holmes's collection was sold. It is now housed in the Booth Museum of Natural History in Brighton.

Harry Govier Seeley (1839–1909) was a leading Victorian student of dinosaurs, but he also published many important papers on other groups of reptiles and on therapsids. The son of an impoverished London artisan, Seeley (Fig. 3.1) was acutely sensitive about his humble background but advanced himself through intelligence and hard work (Desmond 1982). He first became interested in geology after attending popular lectures on natural history with his father. After a brief stint as a law student in London, Seeley entered Cambridge University but apparently never took a degree. He became an assistant to Reverend Adam Sedgwick at the Woodwardian Museum in Cambridge in 1859. The Woodwardian is home to a large collection of Late Cretaceous vertebrate fossils – especially bones of pterosaurs and dinosaurs – from the Cambridge Greensand, which soon attracted Seeley's attention and formed the subject of his early papers. In 1872, Seeley gave up his assistantship and settled in London. He became a prolific author and for a number of years undertook public lecture tours all over the British Isles on behalf of the Gilchrist Trust (Swinton 1962). In 1876, Seeley became professor of geography at King's College and professor of geography and geology at Queen's College, London. In 1881, he was appointed dean

3.1. Portrait of Harry Govier Seeley.
Courtesy of the late W. E. Swinton.

of Queen's College. Finally, in 1896, Seeley assumed the combined professorships of geology, mineralogy, and geography at King's College, which he held for the remaining years of his life.

Seeley published numerous papers on dinosaurian remains from England and continental Europe. He first proposed the fundamental split of dinosaurs into Ornithischia and Saurischia on the basis of differences in the structure of the pelvic girdle (Seeley 1887). The later years of Seeley's career were primarily devoted to the scientific study of Permian and Triassic therapsids and reptiles from the Karoo of South Africa, some of which he had collected himself during a trip to that country in 1889.

France

Although nineteenth-century French vertebrate paleontologists were primarily concerned with the study of fossil mammals, they included several prominent students of dinosaurs (Buffetaut et al. 1993). The father of vertebrate paleontology, Georges Cuvier, first reported postcranial bones of a Late Jurassic theropod dinosaur, which had been collected by an Abbé Bachelet in the vicinity of Honfleur in Normandy in the 1770s. Cuvier interpreted the bones as the remains of a crocodilian (Cuvier 1800, 1808). Allain (2001) restudied these remains and referred to the theropod *Streptospondylus altdorfensis* Meyer, 1832.

A surgeon and avid naturalist, Jacques Amand Eudes-Deslongchamps (1794–1867) reported on the first major find of dinosaurian remains in Normandy—a partial skeleton of a large theropod dinosaur from Middle Jurassic (Bathonian) strata near Caen. Many of the bones, embedded in limestone intended for use in construction, had already been removed and badly damaged by souvenir hunters by the time Eudes-Deslongchamps first learned about the existence of this find in July 1835. With much effort, he managed to salvage and reassemble a number of caudal vertebrae, ribs, gastralia, and limb bones. Eudes-Deslongchamps (1838) documented these remains in great detail, distinguishing them from Buckland's *Megalosaurus* and placing them in a separate taxon, *Poekilopleuron bucklandii*. Unfortunately, the original material of *P. bucklandii* was lost when an air raid on Caen during the Allied liberation of Normandy in 1944 destroyed the Musée de la Faculté des Sciences de Caen. However, plaster casts of a number of bones have survived in the collections of the Muséum National d'Histoire Naturelle in Paris and the Peabody Museum of Natural History at Yale University (Allain and Chure 2002).

An important early student of French dinosaurs was Henri-Emile Sauvage (1842–1917), curator at the Muséum d'Histoire Naturelle in Boulogne-sur-Mer. Between 1873 and 1914 he published a number of studies on Late Jurassic dinosaurian and other vertebrate remains from the Boulonnais region of northern France. In 1882, Sauvage also described intriguing dinosaurian bones and teeth from the Lower Cretaceous in the eastern part of the Paris basin, but most of these fossils, at that time in private hands, have since vanished (Buffetaut et al. 1993). Of this material, Allain (2005) restudied the possible allosauroid *Erectopus superbus* on the basis of a maxilla fragment, which had been rediscovered in the shop of a fossil dealer, and plaster casts of limb bones in the collections of the Muséum National d'Histoire Naturelle in Paris.

3.2. Portrait of Abbé Albert Félix de Lapparent. From http://www.geowiki.fr/.

The Upper Cretaceous strata of Provence have long been a source of well-preserved dinosaurian eggs. They have also yielded bones of a variety of dinosaurs, the first of which may have been known as early as 1840 (Buffetaut et al. 1993). These remains were first identified as dinosaurian in 1869 when Philippe Matheron (1807–1899), a geologist from Marseilles, published an account on well-preserved bones of an ornithopod, which he named *Rhabdodon priscus* and already correctly interpreted as related to *Iguanodon*. Matheron also reported on bones of a "monstrous saurian" (now considered a titanosaurian sauropod) that he named *Hypselosaurus priscus* and regarded as a crocodile-like aquatic reptile. Remarkably, he first suggested a possible association of two eggshell fragments with the skeletal remains of *Hypselosaurus*.

The most influential student of French dinosaurs during the first half of the twentieth century was Albert Félix (sometimes given with a hyphen) de Lapparent (1905–1975; Fig. 3.2). A grandson of the famous geologist Albert Auguste Cochon de Lapparent (1839–1908), he was ordained as a priest in 1929 and then studied geology (Montenat 2008). Lapparent first became interested in dinosaurs during the course of his thesis research on the sedimentary geology of Provence between Var and Durance (Taquet 2007). In 1939 he undertook the first systematic excavations at Fox-Amphoux (Var), a locality for Late Cretaceous dinosaurs and associated vertebrates first discovered in the nineteenth century (Lapparent 1947). Although his research interests partially shifted to geological and paleontological exploration in other countries (see below), Lapparent continued work on French occurrences of dinosaurian bones and tracks for the remainder of his career.

German Dinosaur Hunters

Hermann von Meyer (full name: Christian Erich Hermann von Meyer; 1801–1869; Fig. 3.3) is considered the founder of vertebrate paleontology in Germany (Keller and Storch 2001). Although he pursued a career in politics and finance, Meyer started studying fossil vertebrates (and certain invertebrates, especially crustaceans) in his spare time in 1828. Over a period spanning some three decades, he published some 300 monographs and papers, many of which were illustrated with exquisite lithographs drawn by his own hand. In 1837, Meyer announced the discovery of skeletal remains of a dinosaur from Germany, the sauropodomorph *Plateosaurus engelhardti*, from the Upper Triassic Feuerletten (now Trossingen Formation) near Nuremberg. Earlier, Meyer had first recognized the distinctiveness of dinosaurs in terms of their limb structure (Meyer 1832). However, his subsequent designation for the group, Pachypodes, never gained wide usage and subsequently lost out to Owen's (1842) more evocative Dinosauria. Meyer (1859) also described the only known skeletal remains of the enigmatic ornithischian dinosaur *Stenopelix valdensis* from the Lower Cretaceous (Berriasian) Obernkirchen Sandstone (Bückeberg Formation) near Bückeburg, northwestern Germany.

The most famous German student of dinosaurs, and a genuine dinosaur hunter, was Friedrich von Huene (1875–1969; Fig. 3.4), a scion of an old family of German nobles in the Baltic region. During an extraordinarily productive career spanning some six decades, he produced hundreds of publications, including several large monographs and books, on dinosaurs and other reptiles, therapsids, and amphibians (Reif and Lux 1987). The son of a Lutheran minister, Huene was deeply religious and initially contemplated a career in theology. His evangelical faith profoundly shaped his scientific career, and he published several works on religion and science for a general audience. Huene felt a vocation to reveal to his readers the marvels of divine creation (Turner 2009). No hardship and sacrifice was to prevent him from following this calling. Huene never sought promotion to full professor because the administrative duties associated with such a position would have taken away precious time for research.

While in his late 60s, Huene wrote his autobiography, which was published in 1944 and provides a detailed account of his life and research. Already as a young boy living in Basel, Switzerland, where his father taught at a school for preachers, he was an avid collector of fossils. Huene studied geology and biology, first at the University of Basel and then at the University of Tübingen. After receiving his doctorate for a study of Ordovician brachiopods from the Baltic region in 1898, Huene joined the faculty of the University of Tübingen. There he quickly immersed himself in research on Triassic dinosaurs and other reptiles. His extensive comparative studies of dinosaurian and other reptilian fossils in European and American museums led to a steady stream of monographs and papers, which was barely slowed by Huene's military service during World War I. Huene became an early proponent of Seeley's division of Dinosauria into Saurischia and Ornithischia.

Unlike some of his European contemporaries, Huene did not confine himself to studying fossils in museums and private collections but actively undertook fieldwork to recover new material. In 1906, bones of *Plateosaurus* were discovered near the small town of Trossingen in Württemberg (southern Germany). Excavations led by Eberhard Fraas from the Königliches Naturalien-Kabinett zu Stuttgart in 1911–1912 yielded remains of 12 skeletons of this dinosaur. Huene was eager to recover additional specimens, but the outbreak of the war and the subsequent economic collapse of Germany made it impossible for him to secure funding for his project. A lucky turn of events came in 1920 during a visit to Tübingen by the Canadian vertebrate paleontologist William Diller Matthew (1857–1930), at that time a curator at the American Museum of Natural History in New York. Matthew proposed collaboration with the American Museum, which would provide the necessary funding as well as some collecting expertise, and the resulting fossil collections would be divided between the two institutions. The scientific study of the entire material was to be assigned to Huene. During three consecutive summer seasons, from 1921 to 1923, Huene's teams collected remains of some 14 skeletons, two of which were virtually complete.

On New Year's Day 1923, Huene received a letter from the director of the museum in La Plata, Argentina, inviting him to study old and new collections of sauropod bones from the Upper Cretaceous of Patagonia. He eagerly accepted this offer and set out on a long journey. Not content with studying the fossils housed in the La Plata museum, Huene wanted to examine their geological setting, and on a fieldtrip to Patagonia, he discovered another promising site in late 1923. Later he published a monograph on the Late Cretaceous dinosaurs of Patagonia (Huene 1929), which laid the foundation for all subsequent work on Cretaceous dinosaurs from Argentina.

3.3. Portrait of Christian Erich Hermann von Meyer. From a lithograph by C. J. Allemagne (1837).

3.4. Portrait of Friedrich von Huene (early 1940s). From frontispiece to Huene (1944).

In 1924, after returning from his long trip, which also took him to South Africa, Huene received a letter from a German geologist concerning vertebrate fossils from the southern Brazilian state of Rio Grande do Sul. On Christmas Day of that year, he received a crate of Triassic vertebrate remains from that region collected by a local German doctor. Although this and subsequent shipments did not contain any dinosaurian bones, Huene was excited by this new material and secured funding for an expedition, which he undertook in 1928–1929. His effort led to the discovery of the first diverse assemblage of Triassic tetrapods from South America. Huene later identified a few bones collected by him as dinosaurian, but this identification has not been generally accepted. His success encouraged Llewellyn Ivor Price (1905–1980), a Brazilian researcher then working at the Museum of Comparative Zoology at Harvard University, to undertake further exploration around the town of Santa Maria in Rio Grande do Sul in 1936. That work resulted in the discovery of a partial skeleton that Colbert (1970) subsequently identified as one of the oldest known dinosaurs, *Staurikosaurus pricei* (see also Bittencourt and Kellner 2009).

The culmination of Huene's research on dinosaurs was the publication of a landmark study on the evolutionary history of the Saurischia, in which he reviewed all material known at that time (Huene 1932). Aside from a major study on Late Cretaceous dinosaurs from central India (Huene and Matley 1933), Huene worked mostly on other groups of tetrapods for the remainder of his career.

The Dinosaurs of Tendaguru

In 1907, Eberhard Fraas (1862–1915), a curator at the Königliches Naturalien-Kabinett zu Stuttgart (the precursor of today's Staatliches Museum für Naturkunde Stuttgart), accompanied two German businessmen on a journey to the German Protectorate East Africa (present-day Tanzania). The businessmen were interested in developing the economic potential of this region and hoped to profit from Fraas's geological expertise. Already a renowned paleontologist, Fraas (Fig. 3.5) had previously undertaken fieldwork in Egypt and German Southwest Africa (present-day Namibia) and was instrumental in building the important collection of vertebrate fossils at the Stuttgart museum (Walther 1922; Wild 1991). On the day of his departure for East Africa, he received news from a member of the Commission for the Geographic Exploration of the Protectorates that a Mr. Bernhard Sattler, an engineer with the Lindi-Schürfgesellschaft, a Hanover-based mining and exploration company, had discovered a gigantic bone weathering out of a bush path near Tendaguru Hill in the interior of the colony. Sattler had dutifully reported his unusual find to his superiors, who in turn had notified the chairman of the commission in Berlin.

On his arrival in Dar es Salaam, Fraas received an official request from the commission to follow up on Sattler's report. Thus, he was forced to organize an expedition on the spot; fortunately, the German colonial authorities provided logistic support for this venture. After conducting two other geological reconnaissance trips through the territory and into the British colonies of Uganda and Kenya, Fraas left the coastal town of Lindi for Tendaguru on August 31, 1907. He had contracted amoebic dysentery, which made travel difficult for him. A local German official, a senior military doctor, and 60 native helpers accompanied Fraas. It took a 5-day trip on foot through the coastal plain and across a densely forested high plateau to reach the region around Tendaguru Hill. Sattler met the expedition and took Fraas to the site of his discovery. The wealth of huge bones and bone fragments of dinosaurs weathering out of the ground at Sattler's site overwhelmed Fraas. Despite his poor health and a lack of adequate supplies for conducting a large excavation, Fraas initiated digs at several points in order to obtain unweathered, articulated skeletal remains. Sattler supervised the recovery and conservation of specimens for transport to Germany. Fraas's debilitating illness, compounded by malaria, finally forced him to return to Germany in late September. He never recovered his health and died in March 1915.

One of the greatest deposits of dinosaurian remains in the world had been discovered. In 1908, the year Fraas published the first paper on dinosaurian material from Tendaguru, the German authorities enacted special protection for this region. The director of the Museum für Naturkunde of the Königliche Friedrich-Wilhelm-Universität in Berlin, Wilhelm von Branca (1844–1928), was keen to follow up on Fraas's discovery. With great enthusiasm, Branca successfully set out to raise from government and private sources the large sums of money necessary to organize and undertake well-staffed expeditions for the purpose of excavating skeletons of large dinosaurs and transporting them back to Berlin for preparation, study, and exhibition (Branca 1914). Werner Janensch (1878–1969), a vertebrate paleontologist at the museum, was put in charge of the project and led three highly successful campaigns at Tendaguru between 1909 and 1911 (Hennig 1912). At the end of the 1911 field season, the research team decided that additional work remained to be done, and the geologist Hans Reck (1886–1937) led a fourth and final expedition in 1912. Maier (2003) has published a detailed account of the history of paleontological exploration at Tendaguru.

Janensch and his colleagues identified two horizons rich in dinosaurian remains at Tendaguru, both of which they determined to be Late Jurassic (Kimmeridgian) in age on the basis of associated marine invertebrate fossils. Drawing only on limited geological reconnaissance, Fraas (1908) had initially considered the dinosaur-bearing strata at Tendaguru to be Late Cretaceous in age. A dense cover of mostly shrubs and small trees at Tendaguru prevented standard quarrying operations. Large crews of workers dug numerous pits and trenches in the vicinity of Tendaguru Hill. Porters transported the crated fossils in some 5,400 four-day-long marches across forested terrain to the coastal town of Lindi for shipment by sea to Germany. Although local labor was inexpensive, the funds were quickly expended on the huge crews necessary for the formidable task at hand. During the 1909 field season, the Berlin team employed 170 native workers, 400 in 1910, and 500 each for the 1911 and 1912 seasons. The total weight of fossil material shipped to Berlin amounted to some 185 metric tons.

Conservation and preparation of the vast collections were time-consuming. According to Branca (1914), one fragile sauropod vertebra required some 450 hours of cleaning and conservation, and it took 160 hours to reconstruct a 2-m-long sauropod scapula from 80 individual fragments of bone. The outbreak of World War I in 1914 dashed plans by Janensch and his colleagues to return to East Africa. The German Protectorate East

3.5. Portrait of Eberhard Fraas. From Walther (1922).

Africa became the Tanganyika Territory of British East Africa after the war. In 1924, an expedition from the British Museum (Natural History) in London returned to Tendaguru to collect dinosaurian material for that institution. The Canadian William E. Cutler (1878–1925), who already had experience collecting dinosaurs in the badlands of Alberta, led this expedition. One member of Cutler's crew was a young Englishman named Louis S. B. Leakey (1904–1982), who would go on to become a legendary explorer of human origins. After 8 months of hard work at Tendaguru, Cutler died of malaria at Lindi. An English explorer named Frederick W. H. Migeod (1872–1952), who had experience living and traveling in Africa but no paleontological training, succeeded Cutler as expedition leader, and in 1927 was succeeded by John Parkinson (1872–1947), who later published a book about the British efforts (Parkinson 1930). The work by the British Museum teams yielded little additional dinosaurian material, which has remained largely unstudied. Between 1993 and 2000, German research teams revisited Tendaguru on several occasions for paleontological reconnaissance and to collect bulk samples of sediment for the recovery of small vertebrate remains. However, no further excavations for large specimens were undertaken. Even with modern means of transportation, Tendaguru remains a challenging location for fieldwork.

In Berlin, Janensch was entrusted with the study of much of the dinosaurian material from Tendaguru and charged with supervising the assembling and mounting for display of the skeletons of several dinosaurs, including the spectacular mount of the giant sauropod *Brachiosaurus brancai*, which was first unveiled in 1937. A series of classic monographs published by Janensch between 1925 and 1961 reflects a lifetime of careful scientific work on this remarkable collection of Late Jurassic dinosaurs. Almost miraculously, most of the Tendaguru specimens escaped the destruction of Berlin during the final months of World War II and some 45 years of neglect by the regime of the German Democratic Republic. In recent years, the skeletons of various Tendaguru dinosaurs have been conserved and reassembled during a major renovation at the Museum für Naturkunde in Berlin.

The Man Who Would Be King

One of the most remarkable figures in the history of dinosaurian paleontology – a field not lacking in colorful personalities – was Franz (Ferenc) Baron Nopcsa von Felsö-Szilvás (1877–1933).

Kubacska (1945) published a biography of and letters by Nopcsa, and Weishampel and Reif (1984) have provided a thoughtful assessment of his scientific contributions.

The last male representative of an old Hungarian noble family in Transylvania, Nopcsa (Fig. 3.6) was very much a product of the turbulent times that formed the heyday of the Austro-Hungarian Empire before World War I. A highly cultivated man with a remarkable gift for languages, Nopcsa became a paleontologist by mere chance. The discovery of Late Cretaceous dinosaurian bones by him and his sister, Ilona, in the region around Szentpéterfalva in 1895 led him to visit the famous geologist Eduard Suess (1831–1914) at the University of Vienna. Suess identified the remains as dinosaurian, but when Nopcsa pressed him for further details, Suess urged the young man to undertake a detailed study himself. Thus, Nopcsa

3.6. (Left) Portrait sketch of Franz (Ferenc) Nopcsa von Felsö-Szilvás by F. Marton (1926). (Right) Photograph of Nopcsa in Albanian clothing with rifle. From A. Kubacska (1945).

enrolled as a geology student at the University of Vienna. At the age of only 22, he presented to the Academy of Sciences in Vienna and subsequently published a detailed study of the skull of the basal hadrosaur *"Limnosaurus"* (now *Telmatosaurus*) *transsylvanicus* (Nopcsa 1899). Nopcsa continued collecting and studying the distinctive Cretaceous dinosaurs and associated fauna from Transylvania, resulting in a series of important papers. He first argued that the small size of the Transylvanian dinosaurs was related to their existence on a Late Cretaceous island archipelago; this hypothesis has received significant support from research in recent years (e.g., Stein et al. 2010). Nopcsa later published extensively on other dinosaurs, especially from England and France, as well as other groups of extinct reptiles (Lambrecht 1933). Heavily influenced by the ideas of the Belgian paleontologist Louis Dollo (see below), Nopcsa also sought to understand the paleobiology of these animals. Like his friend Friedrich von Huene, he was an early proponent of Seeley's views regarding a fundamental dichotomy in dinosaurian phylogeny. Nopcsa took a keen interest in major biological and geological issues and tried to address the difficulties of neo-Lamarckian evolutionary ideas then prevalent in Europe (Lambrecht 1933; Weishampel and Reif 1984).

However, Nopcsa's restless personality was not content with a quiet life of scholarship. Like many a young nobleman in fin-de-siècle Austria-Hungary, he sought adventure and glory. Following a vacation in Greece after completing his dissertation in 1903, Nopcsa first visited Albania, which was at that time a remote backwater of the Ottoman Empire on the Balkan Peninsula. He became obsessed with this "wild" land and its people. Nopcsa devoured the available literature on Albania and its culture and history, studied the dialects of the region, and traveled extensively through that country. In 1913, a conference of Europe's leading powers designated most of the territory populated by the Albanian people as an independent country. Austria-Hungary, worried about a power vacuum developing in the Balkans, decided to install a compliant government in this new nation. Nopcsa, politically naive, to say the least, proposed to the Imperial High Command in Vienna to stage an invasion of Albania and install himself as the pro-Austrian ruler of the new country. Nopcsa ingeniously planned to generate the necessary cash flow for the impoverished country by marrying the daughter of an American millionaire. However, his vision was not shared by the imperial government in Vienna, which instead installed Prince Wilhelm zu Wied as ruler of Albania. Prince Wilhelm's tenure there proved to be short-lived, and he and his family were forced to flee for their lives just a few months later. Nopcsa never lost his interest in Albania; he published many studies on Albanian culture and customs and left behind even more unpublished material (Robel 1966).

During World War I, Nopcsa served as an officer in the Imperial Austro-Hungarian Army. He carried out dangerous undercover missions in the heavily contested border region between Hungary and Romania and in 1916 led a force of Albanian volunteers. As a nobleman on the losing side of the conflict, Nopcsa's fortunes collapsed in the economic and political chaos that engulfed central Europe after the war ended in 1918. The family's estates in Transylvania, which had passed into Romanian control, were confiscated without indemnification. During a visit to one of his former estates, Nopcsa was ambushed and nearly killed by a mob of rebellious peasants. Finally, things seemed to improve when the Hungarian government put Nopcsa in charge of the Geological Survey in 1925. Initially full of plans for reorganizing the survey, his manner quickly put him at odds with both his superiors and subordinates, and he finally resigned in 1929. With his Albanian companion, Bajazid Elmaz Doda, Nopcsa relocated to Vienna, and the two then set out on a long tour on motorcycle through Italy until they ran out of money. Nopcsa then settled in Vienna to recover his health and resume scholarly work. Some of his research from this period was innovative and far ahead of its time, such as the application of data from bone histology to the classification of dinosaurs and other fossil tetrapods and inferences concerning global tectonics. Despondent about his poor health and faced with poverty, Nopcsa shot Bajazid before taking his own life in April 1933.

Louis Dollo: Dinosaurs and Paleobiology

Today it is rarely appreciated that paleobiology has its intellectual foundations in the work of the Belgian paleontologist Louis Dollo (1857–1931) (Abel 1928, 1931; Gould 1970). Dollo is more widely known for his studies on well-preserved skeletons of the Early Cretaceous ornithopod dinosaur *Iguanodon* that were uncovered by workers in the Saint Barbe coal mine near Bernissart in southern Belgium in 1878.

Dollo (Fig. 3.7), of Breton descent, was born and educated in Lille, France. As an engineering student, he became interested in geology and zoology and did course work in these subjects. After two years in France following completion of his studies, Dollo relocated to Belgium. In 1882, he became a junior naturalist in the Musée Royal d'Histoire Naturelle de Belgique in Brussels and commenced the study of the Bernissart specimens of *Iguanodon*, which had been excavated under the able supervision of chief technician Louis De Pauw and transported to Brussels (Casier 1960). Dollo also supervised the difficult preparation and assembling for display of several skeletons of *Iguanodon*, both as freestanding mounts and as originally preserved in the rock. The complete exhibit opened to the public in 1905. Dollo became a Belgian citizen in 1886 and worked at the museum in Brussels until his retirement in 1925. He was appointed curator at the museum in 1891 and professor of paleontology at the University of Brussels in 1909 (Van Straelen 1933).

Dollo published some 19 papers on the Bernissart iguanodons, concluding with a brief summary (Dollo 1923). Unfortunately, his growing international reputation and recognition caused the envy of his superiors and colleagues. His director and a leading foreign paleontologist conspired to prevent Dollo from conducting further research on fossil reptiles. The ban was quickly rescinded when Dollo used this involuntary break to publish a landmark study on the evolution of lungfishes (Abel 1928). Dollo's pro-German sympathies during World War I further alienated him from his Belgian colleagues (Gould 1970). His situation finally improved in later years when one of his former students, Victor Van Straelen, assumed the directorship of the Musée Royal d'Histoire Naturelle de Belgique (Abel 1931).

Dollo eschewed detailed descriptions in favor of notes, characterized by their singular format of brief, numbered statements. More than a century after its original discovery, Norman (1980) finally provided an excellent monographic study of *Iguanodon bernissartensis*, which since has been designated as the type species of the genus *Iguanodon*. In addition to his work on dinosaurs, Dollo studied many other groups of extinct and extant vertebrates, ranging from fishes to mammals (Abel 1928, 1931; Van Straelen 1933). Not content with describing and naming fossil vertebrate remains, he attempted to relate form to function in extinct animals—to develop what he called an "ethological paleontology" (Dollo 1910) and what is today referred to as paleobiology. Dollo explored many issues of dinosaurian biology, such as feeding, locomotion, and posture, and his papers laid the foundations for subsequent work in this field. Surprisingly, he showed little if any interest in the geological evidence for reconstructing the ancient environments in which these animals lived and died (Abel 1931). Dollo apparently also never undertook any fieldwork of his own.

3.7. Portrait of Louis Dollo. From frontispiece to *Dollo-Festschrift der Palaeobiologica* (1928).

European Dinosaur Hunters in North Africa

Between 1911 and 1914, the collector Richard Markgraf (1856–1916), working for Ernst Stromer von Reichenbach (1871–1952; Fig. 3.8), a vertebrate paleontologist at the Paläontologisches Museum (since 1919 Bayerische Staatssammlung für Paläontologie und historische Geologie) in Munich, amassed a collection of skeletal remains of dinosaurs and other vertebrates from Late Cretaceous (Cenomanian) strata in the El-Bahariya oasis in the Great Western Desert of Egypt. Stromer himself visited Bahariya only once in 1911, and then largely confined himself to geological observations. He was not initially interested in dinosaurs but was looking for early mammals in Africa. A native of northern Bohemia, Markgraf had worked as an itinerant musician before ending up ill and penniless in Egypt, where he set himself up as a commercial collector of fossils and other natural collectibles. Despite increasing difficulties in dealing with the British colonial authorities

3.8. Portrait of Ernst Stromer von Reichenbach (1927). From a presentation photograph.

in Egypt on the eve of World War I, Markgraf and Stromer managed to ship a considerable quantity of fossil material to Munich for preparation and study. However, much of this collection, having been inspected and inadequately repacked by the colonial authorities, was badly damaged and did not arrive in Munich until 1922. Markgraf tried to continue collecting for Stromer even after the outbreak of World War I, but the conflict severed all contacts between the two men, and Markgraf died in 1916.

Between 1914 and 1936, Stromer and a team of collaborators published a series of monographic studies on the geology of and fossils from Bahariya. Most noteworthy were Stromer's descriptions of the giant theropod dinosaurs *Spinosaurus aegyptiacus*, *Carcharodontosaurus saharicus*, and *Bahariasaurus ingens* (Stromer 1915, 1931, 1934). Unfortunately, a bombing raid on Munich by the British Royal Air Force in April 1944 destroyed all specimens of dinosaurs and other large vertebrates from Bahariya. The director of Stromer's museum, relying on political propaganda, had repeatedly refused entreaties to remove these collections for safekeeping outside the city. In 2000, an expedition from the University of Pennsylvania revisited Bahariya and recovered the first new dinosaurian material since Markgraf's pioneering efforts. Most notable among the new finds are postcranial bones of an enormous titanosaurian sauropod, which fittingly was named *Paralititan stromeri* (Smith et al. 2001). Nothdurft (2002) has provided a detailed history of research at Bahariya.

French explorers in the vast territories of North and West Africa formerly under French colonial rule reported dinosaurian fossils as early as 1905. Between 1946 and 1959, Albert F. de Lapparent from the Institut Catholique de Paris undertook nine expeditions to many regions of the Sahara Desert to explore the stratigraphy of the Mesozoic basins and occurrences of vertebrate fossils in formations collectively known as the Continental Intercalaire (Lapparent 1960). He mostly traveled accompanied by one or two guides and a few camels through this remote and challenging terrain (Taquet 2007; Montenat 2008). During his distinguished career, Lapparent also collected and studied dinosaurian skeletal remains, eggs, and tracks in Iran, Morocco, Portugal, Spain, and Svalbard.

In 1964, while prospecting for uranium ore, geologists of the French Atomic Energy Commission (CEA) found Early Cretaceous dinosaurian bones at Gadoufaoua in northeastern Niger. Jean-Paul Lehman, then director of the Institut de Paléontologie at the Muséum National d'Histoire Naturelle in Paris, sent one of his students, Philippe Taquet, to follow up on this discovery. During several expeditions between 1965 and 1973, Taquet and his colleagues systematically surveyed this remarkably fossil-rich region and collected some 25 tons of material, much of which remains unstudied. Taquet's best-known discovery was the "fin-backed" iguanodontian ornithopod *Ouranosaurus nigeriensis* (Taquet 1976). Starting in 1992, expeditions led by Paul C. Sereno (University of Chicago) have repeatedly revisited the Gadoufaoua region and recovered numerous remarkable specimens of dinosaurs.

During his distinguished career, including a term as director of the Muséum National d'Histoire Naturelle in Paris, Taquet has continued paleontological exploration in North Africa and other regions of the world (Taquet

1994, 1998). In collaboration with the Swiss geologist Michel Monbaron, he made important discoveries of new dinosaurs in Lower and Middle Jurassic strata in the High Atlas of Morocco.

The first European dinosaur hunters were, for the most part, dedicated amateurs who labored with much enthusiasm and often at great personal sacrifice. Subsequently, vertebrate paleontology, like most scientific disciplines, increasingly became the domain of academically trained professionals. Yet amateurs continue to make many significant contributions to paleontology. For example, the collector William Walker found a large manual ungual of an Early Cretaceous theropod in a clay pit near Ockley in Surrey (England) in 1983 and alerted paleontologists at the Natural History Museum in London to his discovery. A team from the museum subsequently recovered much of a skeleton of this remarkable new dinosaur, which was named *Baryonyx walkeri* in honor of its discoverer (Charig and Milner 1986). Similarly, in recent decades, quarrymen and private collectors have discovered a number of new specimens of *Archaeopteryx*, the famous transitional form between dinosaurs other than birds and birds, from the limestone quarries in the Upper Jurassic Solnhofen Formation in Bavaria (Wellnhofer 2008).

During the late nineteenth century and for much of the twentieth century, American vertebrate paleontologists dominated the study of dinosaurs in the field and laboratory. To this day, however, important new discoveries of dinosaurian remains are made across Europe, especially in France and Spain. European scientists continue to collect and study dinosaurs worldwide. Research on dinosaurs is an international undertaking, which is both its charm and its promise.

References

Abel, O. 1928. Louis Dollo. Zur Vollendung seines siebzigsten Lebensjahre. *Palaeobiologica* 1: 7–12.

———. 1931. Louis Dollo. 7. Dezember 1857–19. April 1931. *Palaeobiologica* 4: 321–344.

Allain, R. 2001. Redescription de *Streptospondylus altdorfensis*, le dinosaure théropode de Cuvier, du Jurassique de Normandie. *Geodiversitas* 23: 349–367.

———. 2005. The enigmatic theropod dinosaur *Erectopus superbus* (Sauvage, 1882) from the Lower Albian of Louppy-le-Château (Meuse, France). In K. Carpenter (ed.), *Carnivorous Dinosaurs*, 72–86. Bloomington: Indiana University Press.

Allain, R., and D. J. Chure. 2002. *Poekilopleuron bucklandii*, the theropod dinosaur from the Middle Jurassic (Bathonian) of Normandy. *Palaeontology* 45: 1107–1121.

Bittencourt, J. S., and A. W. A. Kellner. 2009. The anatomy and phylogenetic position of the Triassic dinosaur *Staurikosaurus pricei* Colbert, 1970. *Zootaxa* 2079: 1–56.

Blows, W. T. 1983. William Fox (1813–1881), a neglected dinosaur collector of the Isle of Wight. *Archives of Natural History* 11: 299–313.

Branca, W. 1914. Allgemeines über die Tendaguru-Expedition. *Archiv für Biontologie* 3: 1–13.

Buffetaut, E., G. Cuny, and J. Le Loeuff. 1993. The discovery of French dinosaurs. *Modern Geology* 18: 161–182.

Casier, E. 1960. *Les Iguanodons de Bernissart*. Bruxelles: Éditions du Patrimoine de l'Institut Royal des Sciences Naturelles de Belgique.

Charig, A. J., and A. C. Milner. 1986. *Baryonyx*, a remarkable new theropod dinosaur. *Nature* 324: 359–361.

Colbert, E. H. 1968. *Men and Dinosaurs: The Search in Field and Laboratory*. New York: E. P. Dutton.

———. 1970. A saurischian dinosaur from the Triassic of Brazil. *American Museum Novitates* 2405: 1–39.

Cooper, J. A. 1992. The life and work of George Bax Holmes (1803–1887) of Horsham, Sussex: a Quaker vertebrate fossil collector. *Archives of Natural History* 19: 379–400.

——. 1993. George Bax Holmes (1803–1887) and his relationship with Gideon Mantell and Richard Owen. *Modern Geology* 18: 183–208.

Cuvier, G. 1800. Sur une nouvelle espèce de crocodile fossile. *Bulletin de la Société Philomatique de Paris* 2: 159.

——. 1808. Sur les ossemens fossiles de crocodiles, et particulièrement sur ceux des environs du Havre et d'Honfleur, avec des remarques sur les squelettes des sauriens de la Thuringe. *Annales du Muséum d'Histoire Naturelle, Paris* 12: 73–110.

Delair, J. B., and W. A. S. Sarjeant. 2002. The earliest discoveries of dinosaurs: the records re-examined. *Proceedings of the Geologists' Association* 113: 185–197.

Desmond, A. 1982. *Archetypes and Ancestors: Paleontology in Victorian London, 1850–1875*. Chicago: University of Chicago Press.

Dollo, L. 1910. La paléontologie éthologique. *Bulletin de la Société belge de Géologie, de Paléontologie et d'Hydrologie* 23: 377–421.

——. 1923. Le centenaire des Iguanodons (1822–1922). *Philosophical Transactions of the Royal Society of London B* 212: 67–78.

Eudes-Deslongchamps, J. A. 1838. Mémoire sur le *Poekilopleuron Bucklandii*, grand saurien fossile intermédiaire entre les crocodiles et les lézards; découvert dans les carrières de la Maladrerie, près Caen, au mois de juillet 1835. *Mémoires de la Société Linnéenne de Normandie* 6: 37–146.

Fraas, E. 1908. Ostafrikanische Dinosaurier. *Palaeontographica* 55: 105–144.

Gould, S. J. 1970. Dollo on Dollo's Law: irreversibility and the status of evolutionary laws. *Journal of the History of Biology* 3: 189–212.

Hennig, E. 1912. *Am Tendaguru. Leben und Wirken einer deutschen Forschungs-Expedition zur Ausgrabung vorweltlicher Riesensaurier in Deutsch-Ostafrika*. Stuttgart: E. Schweizerbart'sche Verlagsbuchhandlung.

Huene, F. von. 1929. Los saurisquios y ornitisquios del Cretáceo Argentino. *Anales del Museo de La Plata* 3 (2): 1–196 + atlas of 44 plates.

——. 1932. Die fossile Reptil-Ordnung Saurischia, ihre Entwicklung und Geschichte. *Monographien zur Geologie und Paläontologie* (ser. 1) 4: 1–361 + atlas of 56 plates.

——. 1944. *Arbeitserinnerungen*. Halle: Kaiserlich Leopoldinisch-Carolinisch Deutsche Akademie der Naturforscher.

Huene, F. von, and C. A. Matley. 1933. The Cretaceous Saurischia and Ornithischia of the central provinces of India. *Palaeontologia Indica* (n.s.) 21(1): 1–74.

Keller, T., and G. Storch (eds.). 2001. *Hermann von Meyer. Frankfurter Bürger and Begründer der Wirbeltierpaläontologie in Deutschland*. Kleine Senckenberg-Reihe 40. Stuttgart: Schweizerbart'sche Verlagsbuchhandlung (Nägele u. Obermiller).

Kubacska, A. Tasnádi. 1945. *Franz Baron Nopcsa*. Leben und Briefe ungarischer Naturforscher 1. Budapest: Verlag des Ungarischen Naturwissenschaftlichen Museums.

Lambrecht, K. 1933. Franz Baron Nopcsa+, der Begründer der Paläophysiologie, 3. Mai 1877 bis 25. April 1933. *Palaeontologische Zeitschrift* 15: 201–222.

Lapparent, A. F. de. 1947. Les Dinosauriens du Crétacé supérieur du Midi de la France. *Mémoires de la Société géologique de France* (n.s.) 56: 1–54.

——. 1960. Les Dinosauriens du "Continental Intercalaire" du Sahara central. *Mémoires de la Société géologique de France, Nouvelle Série* 88A: 1–57.

Maier, G. 2003. *African Dinosaurs Unearthed: The Tendaguru Expeditions*. Bloomington: Indiana University Press.

Martill, D. M., and D. Naish (eds.). 2001. *Dinosaurs of the Isle of Wight*. Palaeontological Association Field Guides to Fossils 10. London: Palaeontological Association.

Matheron, P. 1869. Notice sur les reptiles fossiles des dépôts fluvio-lacustres crétacé du bassin à lignite de Fuveau. *Mémoires de l'Académie des Sciences, (Belles-)Lettres et (Beaux-)Arts de Marseille* 1868–69: 345–379.

Meyer, H. von. 1832. *Palaeologica zur Geschichte der Erde und ihrer Geschöpfe*. Frankfurt am Main: Verlag von Siegmund Schmerber.

——. 1837. Mittheilungen, an Professor Bronn gerichtet. *Neues Jahrbuch für Mineralogie, Geognosie, Geologie und Petrefakten-Kunde* 1837: 314–317.

——. 1859. *Stenopelix valdensis*, ein Reptil aus der Walden-Formation Deutschland's. *Palaeontographica* 7: 25–34.

Montenat, C. 2008. *Une famille de géologues, les Lapparent. Un siècle d'histoire & d'aventures de la géologie*. Paris: Vuibert/Société Géologique de France.

Nopcsa, F. von. 1899. Dinosaurierreste aus Siebenbürgen. (Schädel von *Limnosaurus transsylvanicus* nov. gen. et spec.) *Denkschriften der kaiserlichen Akademie der Wissenschaften Wien, mathematisch-naturwissenschaftliche Classe* 68: 555–591.

Norman, D. B. 1980. On the ornithischian dinosaur *Iguanodon bernissartensis* from the Lower Cretaceous of Bernissart (Belgium). *Mémoires de l'Institut Royal des Sciences Naturelles de Belgique* 178: 1–103.

Nothdurft, W. 2002. *The Lost Dinosaurs of Egypt*. New York: Random House.

Owen, R. 1842. Report on British fossil reptiles. Part II. *Reports of the British Association for the Advancement of Science, 11th Meeting, Plymouth, 1841*: 60–204.

Parkinson, J. 1930. *The Dinosaur in East Africa: An Account of the Giant Reptile Beds of Tendaguru, Tanganyika Territory*. London: H. F. & G. Witherby.

Reif, W.-E., and W. Lux. 1987. Evolutionstheorie und religiöses Konzept im Werk des Wirbeltierpaläontologen Friedrich Freiherr von Huene (1875–1969). Mit einer Bibliographie. *Bausteine zur Tübinger Universitätsgeschichte* 3: 91–140.

Robel, G. 1966. *Franz Baron Nopcsa und Albanien. Ein Beitrag zu Nopcsas Biographie*. Albanische Forschungen 5. Wiesbaden: Harrassowitz.

Seeley, H. G. 1887. On the classification of the fossil animals commonly named Dinosauria. *Proceedings of the Royal Society of London* 43: 165–171.

Smith, J. B., M. C. Lamanna, K. J. Lacovara, P. Dodson, J. R. Smith, J. C. Poole, R. Giegengack, and Y. Attia. 2001. A giant sauropod dinosaur from an Upper Cretaceous mangrove deposit in Egypt. *Science* 292: 1704–1706.

Stein, K., Z. Csiki, K. Curry Rogers, D. B. Weishampel, R. Redelstorff, J. L. Carballido, and P. M. Sander. 2010. Small body size and extreme cortical bone remodeling indicate phyletic dwarfism in *Magyarosaurus dacus* (Sauropoda: Titanosauria). *Proceedings of the National Academy of Sciences* 107: 9258–9263.

Stromer, E. 1915. Ergebnisse der Forschungsreisen Prof. E. Stromers in den Wüsten Ägyptens. II. Wirbeltier-Reste der Baharîje-Stufe (unterstes Cenoman). 3. Das Original des Theropoden *Spinosaurus aegyptiacus* nov. gen., nov. spec. *Abhandlungen der Königlichen Bayerischen Akademie der Wissenschaften, Mathematisch-Physikalische Klasse* 28 (3): 1–32.

——. 1931. Ergebnisse der Forschungsreisen Prof. E. Stromers in den Wüsten Ägyptens. II. Wirbeltier-Reste der

Baharîje-Stufe (unterstes Cenoman). 10. Ein Skelett-Rest von *Carcharodontosaurus* nov. gen. *Abhandlungen der Bayerischen Akademie der Wissenschaften, Mathematisch-naturwissenschaftliche Abteilung, Neue Folge* 9: 1–23.

———. 1934. Ergebnisse der Forschungsreisen Prof. E. Stromers in den Wüsten Ägyptens. II. Wirbeltier-Reste der Baharîje-Stufe (unterstes Cenoman). 13. Dinosauria. *Abhandlungen der Bayerischen Akademie der Wissenschaften, Mathematisch-naturwissenschaftliche Abteilung, Neue Folge* 22: 1–79.

Swinton, W. E. 1962. Harry Govier Seeley and the Karroo reptiles. *Bulletin of the British Museum (Natural History)* 3: 1–39.

Taquet, P. 1976. Géologie et paléontologie du gisement de Gadoufaoua (Aptien du Niger). *Cahiers de Paléontologie* 17: 1–191.

———. 1994. *L'Empreinte des Dinosaures.* Paris : Éditions Odile Jacob.

———. 1998. *Dinosaur Impressions: Postcards from a Paleontologist.* Translated by K. Padian. Cambridge: Cambridge University Press.

———. 2007. On camelback: René Chudeau (1864–1921), Conrad Kilian (1898–1950), Albert Félix de Lapparent (1905–1975) and Théodore Monod (1902–2000), four French geological travellers cross the Sahara. In P. N. Wyse Jackson (ed.), *Four Centuries of Geological Travel: The Search for Knowledge on Foot, Bicycle, Sledge and Camel,* 183–190. Geological Society of London, Special Publication 287.

Turner, S. 2009. Reverent and exemplary: 'dinosaur man' Friedrich von Huene (1875–1969). In M. Kölbl-Ebert (ed.), *Geology and Religion: A History of Harmony and Hostility,* 223–243. Geological Society of London, Special Publication 310.

Van Straelen, V. 1933. Louis Dollo (1857–1931). Notice biographique avec liste bibliographique. *Bulletin du Musée Royal d'Histoire Naturelle de Belgique* 9 (1): 1–29.

Walther, J. 1922. Eberhard Fraas. *Verhandlungen der Gesellschaft Deutscher Naturforscher und Ärzte* 87: 334–336.

Weishampel, D. B., and W.-E. Reif. 1984. The work of Franz Baron Nopcsa (1877–1933): dinosaurs, evolution and theoretical tectonics. *Jahrbuch der Geologischen Bundes-Anstalt Wien* 127: 187–203.

Wellnhofer, P. 2008. *Archaeopteryx. Der Urvogel von Solnhofen.* Munich: Verlag Dr. Friedrich Pfeil, Munich.

Wild, R. 1991. Die Ostafrika-Reise von Eberhard Fraas und die Erforschung der Dinosaurier-Fundstelle Tendaguru. *Stuttgarter Beiträge zur Naturkunde, Serie C* 30: 71–76.

North American Dinosaur Hunters

4

**†Edwin H. Colbert (1905–2001),
David D. Gillette, and Ralph E. Molnar**

Although it is likely that American Indians found dinosaurian fossils and recognized them as the remains of dead, maybe ancient, animals (Mayor 2005), the first dinosaur hunters were the unwitting discoverers of unknown animals. In 1802 a New England farm boy named Pliny Moody found some footprints in reddish-brown sandstones near his home at South Hadley, Massachusetts. Because these impressions had the appearance of large bird tracks, they were at the time popularly referred to as trackways made by "Noah's Raven." During the early years of the nineteenth century, other such tracks and trackways came to light, and they soon became objects of study by Professor Edward B. Hitchcock, the president of Amherst College. For several decades, from 1836 to 1865, Hitchcock ranged back and forth and up and down the Connecticut Valley, ferreting out footprints that were widely exposed in the Late Triassic and Early Jurassic sandstones and siltstones of the region. Many of them he collected and removed to Amherst, where in time a museum was built for their reception. In 1858 he published a large monograph entitled *Ichnology of New England*, in which he described most of the footprints as having been made by large birds.

When Hitchcock began his work, the concept of dinosaurs, established by Richard Owen in 1842 (see Torrens, Chapter 2 of this volume), was still a matter of future history. But even after Owen's pronouncement, and the pioneer work on dinosaurs in England by Owen, William Buckland, and Gideon Mantell, the possibility of the footprints having been made by dinosaurs did not enter the perception of Hitchcock. He went to his grave believing that what is now the Connecticut Valley was once inhabited by a varied population of birds, large and small. Needless to say, by midcentury the true nature of these footprints as dinosaurian was becoming evident.

In 1855, more than a decade after Owen's recognition and naming of the Dinosauria, Dr. Ferdinand V. Hayden, in charge of one of the governmental surveys of the western territories, found some teeth near the confluence of the Judith and Missouri rivers, in what is now Montana, as well as some vertebrae and a toe bone in South Dakota. These fossils were given to Dr. Joseph Leidy of Philadelphia for identification. The specimens were described by Leidy in 1856. They included several hadrosaur teeth, which he named *Trachodon mirabilis*, as well as some teeth of a theropod dinosaur, designated by Leidy as *Deinodon horridus*. In addition he named two teeth *Troodon formosus* and *Palaeoscincus costatus*, which he indicated as being "lacertilian." These were the first dinosaurs to be named from North America.

Then in 1858 two important dinosaurs were discovered, one from the East Coast, the other from badlands in terra incognita, deep in the canyons

The First Discoveries of Dinosaurs in the New World

4.1. Edward Drinker Cope (1840–1897).

of the Southwest. A partial dinosaur skeleton was discovered in a Cretaceous marl pit near Haddonfield, New Jersey, across the river from Philadelphia. Leidy was instrumental in the excavation of this skeleton, which he described in 1858 and 1859 as *Hadrosaurus foulkii*, the first dinosaur skeleton to be described in North America. Significantly, Leidy recognized the dominant bipedality of *Hadrosaurus*, thereby establishing a basic adaptation in dinosaurian evolution, quite in contrast to Owen's original concept that dinosaurs such as *Iguanodon* were rhinoceros-like quadrupedal tetrapods.

A military expedition under Captain John S. Macomb from Santa Fe, New Mexico, discovered the first sauropod dinosaur in North America in 1858. The adventurous physician Dr. John Strong Newberry was the civilian naturalist on the expedition, which, among other charges, sought the confluence of the Grand and Green rivers in southern Utah. During a difficult descent into a canyon they named Cañon Pintado, Newberry discovered fragmentary bones at the base of a vertical sandstone cliff, then cut "moki holes" to locate the bones in situ. They collected several limb and foot bones with the intention of sending them to Leidy. But Newberry's career as a naturalist was interrupted by the Civil War, and so description of his collection was delayed. Cope later described the specimen as *Dystrophaeus viaemalae* and determined its age as Triassic. The site was forgotten and details of the expedition lost for nearly a century, until a naturalist from Moab, Utah, became interested in Macomb's route and discovered the layer from which Newberry had collected the bones (Madsen 2010). Its stratigraphic position is at the base of the Morrison Formation and represents not only the earliest sauropod in North America, but also the first discovered (Gillette 1996).

O. C. Marsh and E. D. Cope: High-Stakes Paleontology

The discovery of large Mesozoic fossil bones in England and in North America, as well as the realization that such fossils were the remains of a new and previously unrecognized group of extinct reptiles, inspired two particular Americans to become deeply involved with the search for dinosaurs. Edward Drinker Cope and Othniel Charles Marsh (Figs. 4.1, 4.2) entered into this new field of paleontological endeavor with gusto – to such a degree, indeed, that soon their rivalry turned into bitter animosity, resulting in a fossil feud unparalleled in the history of paleontology.

Leidy, professor of anatomy in the medical school of the University of Pennsylvania, almost immediately found himself in the middle of a vituperative battle between the two warring paleontologists, so he soon retired from this field of research and turned his attention to other matters. He was a quiet, dignified man, who found the loud quarrel between Cope and Marsh too harsh to bear.

Cope was the son of a wealthy Quaker shipping magnate in Philadelphia. Marsh was the nephew of a wealthy businessman, George Peabody. Both men had access to ample funds with which to pursue their paleontological quests, Cope as a freelance scholar with ties to the Academy of Natural Sciences, Philadelphia, and Marsh as a member of the Yale faculty and director of the Yale Peabody Museum, posts established for him by his munificent uncle, who spent his adult life in England. Both were as independent as may be imagined, and both had very strong opinions about fossils, about each other, and about the world in general. The Cope–Marsh feud is so well known that it need not be described here (see Osborn 1931; Schuchert and Le Vene 1940; Plate 1964; Colbert 1968; Lanham 1973; Shor 1974; and the appendix of this volume for details). Suffice it to say that as a result of their rivalry, they amassed extensive collections of Mesozoic and Cenozoic fossils in western North America, including many skulls and skeletons of dinosaurs. Indeed, their animosity toward each other did have the positive result of opening to the astonished eyes of paleontologists and of the public across the globe the new and exciting world of dinosaurs and other extinct forms.

The early period of tentative exploration and research that occupied the first half of the nineteenth century, followed by the fierce Cope–Marsh rivalry of the last three decades of the century, may be regarded as a prologue to the golden age of dinosaur exploration and research in North America, from the 1890s to the 1920s. These were the years when paleontologists from the American Museum of Natural History in New York, the United States National Museum (Smithsonian Institution), the Carnegie Museum of Pittsburgh, the National Museum of Canada in Ottawa, and the Royal Ontario Museum in Toronto worked vigorously in the Upper Jurassic and Cretaceous beds of western North America. They discovered dinosaur treasures year after year and described them in numerous scientific journals. Of course there were dinosaur hunters from other museums as well, but these institutions were at the forefront of dinosaurian field exploration and laboratory research. These were years filled with the excitement of discovery and the satisfactions of lucid research, frequently followed by the appearance of excellent publications.

One may think of this golden age as having been inaugurated by the extensive excavations in the Upper Jurassic Morrison Formation at Bone Cabin, Wyoming—an area near Como Bluff that had been previously worked by Marsh and his associates and assistants, notably Samuel Wendell Williston, Arthur Lakes, and William Reed. The force behind the Bone Cabin excavations was Henry Fairfield Osborn, who had come to the American Museum in 1891, where he initiated an active program in vertebrate paleontology. Osborn was especially interested in Mesozoic mammals that might be found in the Morrison sediments, but at the same time, he realized the importance of collecting dinosaur skeletons, both for research and exhibition. The Bone Cabin program commenced in 1898, with W. D. Matthew, Walter Granger, Barnum Brown, and R. S. Lull, Albert Thomson, Peter Kaisen, and others all working together to unearth gigantic dinosaur skeletons (Fig. 4.3) as well as Jurassic mammals (in a subsidiary quarry known as Quarry Nine). The work at Bone Cabin, lasting through 6 years, began a program of dinosaur collecting that eventually would enable the American Museum to have the largest collection of these reptiles in the world—in this respect surpassing the collections at Yale and the National Museum that had been assembled by O. C. Marsh and his assistants.

Among the participants in the Bone Cabin project, the names of Barnum Brown and Richard Swann Lull are especially significant. Brown went on from Bone Cabin to devote the remainder of his long life to dinosaurs, especially those from the Cretaceous beds of Alberta, as will be told. He was

The Golden Age of Dinosaur Paleontology

4.2. Othniel Charles Marsh (1831–1899).

4.3. American Museum paleontologists at Nine Mile Quarry, Como Bluff, Wyoming, in 1899. From left to right: Richard Swann Lull, Peter Kaisen, and William Diller Matthew.

very much an individualist who preferred to labor in the field by himself, variously assisted by Peter Kaisen and George Olsen of the American Museum fossil laboratory, but usually without other collaborators. And he could be secretive about what he was up to. What he was up to generally resulted in amazing collections, properly excavated, commonly accompanied by scanty field notes consisting of locality data in illegible pencil scrawled on scraps of newspaper. He had a habit of locating fossils in one field season, burying them as a dog would bury a precious bone, and then returning to them the next year. Thus he had something on tap to ensure the success of the summer to come. This was but one of the facets of a man around whom there are numerous anecdotes and legends.

Lull devoted his life to dinosaurs, perhaps more in the laboratory than in the field. He was a dignified and imposing person, as befitted his position as a Yale professor. At the university he was known for his lucid classroom lectures on vertebrate evolution and for his prodigious efforts as director of the newly erected Peabody Museum, which replaced the old building in which Lull's predecessor, O. C. Marsh, had ruled his paleontological world with cold authority.

In 1909 the Bone Cabin excavations met their match for scientific importance, thanks to work by Earl Douglass for the Carnegie Museum, who discovered articulated dinosaur skeletons in the Morrison Formation, about 20 miles to the east of Vernal, Utah. From this discovery, the famous Carnegie Quarry was developed—a dinosaur-collecting project that extended until 1923 and resulted in the amassing of prodigious collections of Upper

Jurassic dinosaurs. Indeed, Douglass devoted the remainder of his career to the Carnegie Quarry, even to the extent of giving up his life in Pittsburgh and moving to Utah, where he established a homestead near the quarry on which he built a primitive cabin and settled in with his wife and child. He was truly a dedicated person. In Utah, across the continent from the Carnegie Museum, he was at least spared the dubious pleasure of working under the heavy thumb of W. J. Holland, the dictatorial director of the museum.

The large scope and success of the Carnegie Quarry excavation were due not only to the foresight and the dedicated efforts of Douglass, but also to unlimited financial support from the industrialist Andrew Carnegie, who enthusiastically provided funds so that (among other things) casts of a complete skeleton of *Diplodocus*, excavated from the quarry, could be produced and distributed to various museums throughout the world. And everywhere that *Diplodocus* went, W. J. Holland also went, to supervise the setting up of the plaster bones, and incidentally, to his great satisfaction, to collect honorary degrees. The Carnegie Quarry eventually was enlarged into an in situ exhibit that today forms the centerpiece of the Dinosaur National Monument.

Certainly one of the outstanding authorities on dinosaurs during those years when the American Museum and the Carnegie Museum were conducting large quarrying operations in the Morrison Formation was Charles Gilmore of the United States National Museum (Smithsonian Institution). Gilmore, a friendly, amiable person, remarkably modest about his accomplishments, not only spent long hours with his fossils, but he also had the admirable habit of writing down his observations and conclusions, thereby publishing a series of exceptionally thorough monographs on Jurassic and Cretaceous dinosaurs. During his long career at the National Museum, he conducted 16 paleontological expeditions, almost all of them in western North America. Oddly enough, however, only seven of these expeditions were devoted to the collection of dinosaurs – one in the Morrison beds of Utah, the others in the Cretaceous beds in various western states. Almost all of Gilmore's other expeditions were primarily for the purpose of collecting fossil mammals.

During those exciting years of exploration and collecting in the Morrison Formation by the American and Carnegie museums, and in the Cretaceous sediments of various western states by the National Museum, remarkably rich deposits of Cretaceous dinosaurs had also been discovered in the western Canadian provinces, particularly in Alberta. Dinosaur bones of Cretaceous age were first found in western Canada by Dr. George Dawson (Fig. 4.4), the son of Sir William Dawson, one of the giants of nineteenth-century geology. George Dawson, working as a Canadian representative of the International Boundary Commission, was a thoroughly trained geologist who had completed his graduate studies in England. He had studied under Thomas Henry Huxley – Darwin's champion – and had the perspicacity to submit fossils to Cope for verification as to their dinosaurian relationships.

It is an interesting fact that Dawson and two men who followed him as pioneer dinosaur hunters in Canada would all at first glance appear to have been ill suited for their task. Dawson was a hunchback, almost a dwarf, who in spite of his disability was a vigorous field geologist. The other two men, Joseph Burr Tyrrell and Lawrence M. Lambe, both took up field geology

4.4. George Dawson (the short man standing in the center of the photograph) and his field party at Fort McLeod, British Columbia, in 1879.

and paleontology because they suffered from ill health and decided in each case that only a vigorous outdoor life would ensure for them reasonably long lives. (The regimen was truly successful for Lambe; he lived to the ripe old age of 99.) Tyrrell discovered dinosaur bones in the valley of the Red Deer River in Alberta in 1884. Lambe, a member of the Canadian Geological Survey, made a boat trip down the Red Deer River in 1897, thereby traversing extensive exposures of Upper Cretaceous sediments, from which he collected dinosaur fossils. He studied these specimens in collaboration with Professor Osborn, and they published a joint monograph on the fossils.

Extensive explorations for Cretaceous dinosaurs in Alberta were carried on during the decade of 1910 to 1920 by two teams of paleontologists, who engaged in friendly rivalry as they discovered and collected dinosaurian treasures from the Red Deer River region. One team consisted of Barnum Brown and his assistants, Peter Kaisen and George Olsen, of the American Museum of Natural History in New York; the other group was made up of Charles H. Sternberg and his three sons, George, Charles M. (Fig. 4.5), and Levi, who were collectors working for the National Museum of Canada and for the Royal Ontario Museum.

The saga of the Sternberg family constitutes a remarkable chapter in the history of North American dinosaur collecting. The elder Sternberg and his twin brother migrated from their home in Iowa to Kansas when they were in their late teens, and there Charles became fascinated by various fossils that he discovered in the Cretaceous Dakota sandstone. In short order he was established as a field assistant to Cope, and this was the beginning of a long life devoted to paleontological collecting, a lifestyle that his sons continued with distinction.

In the years of World War I, Barnum Brown and the Sternbergs explored the Red Deer River fossil beds by floating down the river on barges

Colbert, Gillette, and Molnar

(Fig. 4.6), which served as their headquarters. With auxiliary motorboats, they ranged up and down the river, climbing up from landing places to explore the exposures. It was a successful technique, and they collected numerous Late Cretaceous dinosaur specimens (Fig. 4.7), including many articulated skeletons that today can be seen in New York, Ottawa, and Toronto. Today Red Deer River dinosaurs, as represented by original specimens and casts, can be seen in a magnificent display at the Royal Tyrrell Museum in Drumheller, Alberta, a large museum located at the edge of the Red Deer River badlands.

After the golden age of dinosaur exploration and research in North America, a time that may more properly be designated as the first golden age, there was something of a lull in the pursuit of dinosaur studies—a lull that extended for perhaps two decades, encompassing the time of World War II. Of course the exploration for dinosaurs continued through this period, as did laboratory research and publications, but during those years there was a great emphasis on mammals and their ancestors, the therapsids—the mammal-like reptiles. These tetrapods were of significance because, it was said, they were in the mainstream of evolution—from primitive vertebrates to the advanced mammals. The dinosaurs in the eyes of many paleontologists were "gee-whiz" fossils, nice to have in the exhibition halls where they could impress the public, but generally speaking off on an evolutionary sideline that left no living descendants, and therefore of lesser consequence to the evolutionist.

The older dinosaurian studies had been largely descriptive, as would be expected with such a wealth of specimens discovered and waiting to be examined. They were in essence osteological studies, aimed at the interpretation of morphological anatomy and taxonomic relationships. But after World War II, there appeared a new generation of paleontologists, exemplified by John Ostrom of the Yale Peabody Museum, who realized that the skeletons of dinosaurs revealed much more than previously had

The Dark Age and Renaissance of Dinosaur Studies

4.6. Barnum Brown's flatboat on the Red Deer River in Alberta.

been seen. Consequently there were ever more sophisticated studies of the bones themselves, such as bone structure, histology, and the implications concerning dinosaurian physiology. There were new interpretations of various bony structures that are so prevalent among the dinosaurs, and the light that such adaptations may throw upon behavior. There were studies of ontogenetic growth patterns and their significance. And there were the new and exciting discoveries of and research on dinosaurian trace fossils and productions, such as tracks and trackways, eggs, nests, and other objects in the fossil record.

There was also a change in the basic philosophy of vertebrate paleontology that percolated over from a change in the understanding of evolutionary processes, known as the modern synthesis. The notion of a linear advance from fish (or before) to man was replaced by the – ironically – earlier Darwinian notion that evolution was controlled by environmental change through natural selection rather than by some inherent tendency toward human sophistication. The evolutionary revolution was brought to the attention of many evolutionary biologists by Sir Julian Huxley, grandson of Thomas Henry, in England in the 1940s, and by G. Ledyard Stebbins, Ernst Mayr, and Theodosius Dobzhansky (who, incidentally, never acquired an advanced degree) in the United States soon after. More directly relevant to paleontology was the work of George Gaylord Simpson, who carried this change of basic attitude into vertebrate paleontology. It was this change that led to the directing of more attention to what had previously been considered evolutionary sidelines, such as dinosaurs. Furthermore, there was a renaissance of dinosaur collecting all over the world, not merely in North America, with the resultant discoveries of new dinosaurs of previously unsuspected form and variety. Indeed, our knowledge of dinosaurs has expanded at a truly remarkable rate during the past five decades. Therefore, perhaps it is appropriate to speak of an ever-expanding second golden age of dinosaur collecting and research, an age in which we are now living.

One of the first important events of postwar dinosaur fieldwork was the discovery, by Edwin H. Colbert and his American Museum field crew,

Colbert, Gillette, and Molnar

4.7. Barnum Brown collecting a skeleton of the hadrosaur *Corythosaurus* in Alberta in 1912.

of a deposit of Late Triassic dinosaurs at Ghost Ranch, New Mexico (Colbert 1995). Here, within a quarry of limited extent, were found literally hundreds of articulated skeletons and partial skeletons of the Triassic theropod *Coelophysis bauri* and other closely related archosaurs. Very large blocks containing the intricately interlaced skeletons of these dinosaurs were initially collected by Colbert's crew from the American Museum of Natural History (1947–1948), and subsequently by representatives of the Carnegie Museum, the New Mexico Museum of Natural History, the Ruth Hall Paleontology Museum at Ghost Ranch, the Museum of Northern Arizona, and Yale Peabody Museum (1983–1985). The fossils are still being prepared and studied at the several involved institutions, and at a considerable number of other paleontological laboratories to which blocks have been distributed. Newly discovered bone beds in nearby sites promise to greatly expand our knowledge of the Ghost Ranch fauna, and in turn the origin and evolution of early dinosaurs. The material from these Ghost Ranch localities gives ample opportunity for research on ontogenetic changes, sexual dimorphism, and individual variation, as well as functional and biometric studies.

In dinosaur collecting, spectacular discoveries were made in the Morrison Formation in Utah and Colorado, notably between 1927 and 1967 at the Cleveland-Lloyd Quarry in the San Rafael Swell by Ferdinand F. Hintze, Golden York, William Lee Stokes, and James Madsen and their associates of the University of Utah, and from 1972 to 1982 in the Dry Mesa Quarry along the Utah–Colorado border by James Jensen and his associates at Brigham Young University. At this latter site, various skeletal elements of prodigious size have been found, indicative of sauropod dinosaurs that seem to surpass even *Brachiosaurus* in size.

A discovery of unusual significance was that of *Deinonychus antirrhopus*, collected from the Lower Cretaceous Cloverly Formation in Montana by the Yale Peabody Museum, under the direction of John Ostrom. This discovery, and the research that has resulted from it, as well as from other discoveries of fossils related to *Deinonychus*, opened new horizons of knowledge concerning the dromaeosaurids, which have proved to be among the most aggressive and perhaps the most intelligent of predatory theropods.

It was his work on *Deinonychus* that led Ostrom to the unexpected conclusion, when he went on to study fossils of the early bird *Archaeopteryx lithographica*, that birds are descendants of dromaeosaurid theropods. Following after the change in evolutionary thinking of the "modern synthesis" came the (somewhat belated) general application of phylogenetic systematics in North America, a school proposed in the 1940s by German entomologist Willi Hennig. The view of this school was that the descendants of any taxonomic group (clade) were still members of that clade. This had the startling implication, combined with Ostrom's work showing the similarities of *Archaeopteryx* and *Deinonychus*, that the dinosaurs weren't extinct after all. They still survive–as birds.

Dale Russell has been particularly interested in, among other things, the subject of dinosaurian intelligence, as a result of his discovery of *Troodon* in the Upper Cretaceous sediments of western Canada. Indeed, Russell and his Canadian confreres, among them Philip Currie and the late William Sarjeant, have initiated vigorous new programs of dinosaur collecting in the western Canadian provinces.

At the same time, John Horner of the Museum of the Rockies in Bozeman, Montana, has carried on a spectacular program of collecting in the Upper Cretaceous beds of that state, where he has found prodigious deposits of the hadrosaur *Maiasaura peeblesorum*, consisting not only of individuals spanning an ontogenetic series from hatchling to adult, but also of almost countless numbers of eggs and nests of this dinosaur. As a result of his work, especially histological studies carried on with Armand de Ricqlès of Paris and Kevin Padian of the University of California, Berkeley, much new information has been gained not only as to ontogeny, but also as to possible behavior patterns, particularly possible parental care in this dinosaur. This work continues unabated.

Our understanding of how dinosaurs lived was fundamentally affected at the instigation of Robert T. Bakker, an undergraduate student of Ostrom's at Yale. Bakker proposed that the notion of dinosaurs as enormous cold-blooded, slow-moving, and slow-witted giant lizards was not supported by careful study of the specimens, particularly of such creatures as *Deinonychus*. Inspired by the work of, and discussions with, Ostrom, he proposed that dinosaurs were homeothermic, endothermic creatures with high metabolic rates. This conclusion was based in part on the pedal structure of *Deinonychus*, with a large spurlike claw on one of the toes. Although it is not yet clear whether this claw was used in attacking prey animals or in combat with other members of its own species–or both–it seems clear that in order

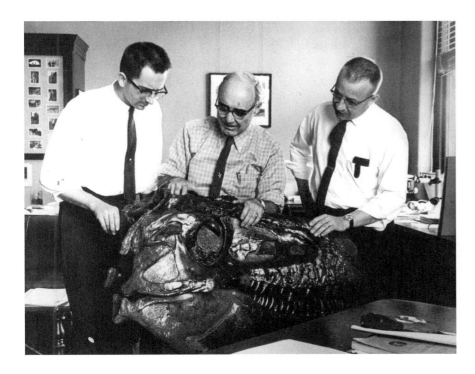

4.8. Edwin H. "Ned" Colbert in his office at the American Museum of Natural History with two of his many distinguished students, Dale Russell (left) and John Ostrom (right), who went on to become prominent dinosaur hunters. The skull is that of *Gorgosaurus libratus*.

Courtesy of the Colbert Library, Museum of Northern Arizona.

to effectively use it, the animal must have balanced on one foot – not the kind of activity that dinosaurs were believed capable of in the 1950s. Bakker's studies stimulated a new wave of fieldwork in the exploration of dinosaurs that has spanned the past four decades.

Ostrom and Bakker are two of many paleontologists who found their initial inspiration directly and indirectly from the late Edwin H. Colbert. Like Ostrom, Dale Russell was Colbert's student. Ostrom's students at Yale University who have developed distinguished careers in dinosaur paleontology include Walter Coombs, Peter Dodson, James O. Farlow, Thomas Holtz, Ralph E. Molnar, and Kevin Padian. These descendants of the Colbert heritage (Fig. 4.8) have guided students in turn and stimulated continued research through their students and associates. These paleontologists are among today's leaders in dinosaur paleontology and have participated in fieldwork that continues the tradition of research firmly rooted in collection of fossils and the study of their geological context.

Finally, it should be said that dinosaur hunters, not only in North America but also on the other continents, are today giving unprecedented attention to dinosaur tracks and trackways, with a new appreciation of such fossils as keys to dinosaurian locomotion and other behavior traits. Thus the study of footprints has come full circle since the days of Hitchcock and the Connecticut Valley tracks.

For decades the study of dinosaur tracks was a rather static, descriptive branch of paleontological research. But it was given new significance by R. T. Bird's discovery and excavation in 1940 of huge sauropod and theropod trackways of Cretaceous age along the Paluxy River in Texas (Bird 1985). Today dinosaur trackways are being vigorously studied in the field and preserved in situ on a worldwide scale. In North America dinosaur track sites have been discovered by the thousands, some containing millions of tracks. The rejuvenation of track studies is a product of this renaissance of dinosaur paleontology. Because tracks show behavior and can be preserved only under certain circumstances, they have proved to be useful in behavioral studies, habitat interpretations, locomotor studies, and census analysis. The number of dinosaur ichnologists dramatically expanded in the 1980s and discovery of dinosaur track sites continues unabated.

This survey of the work being carried on by dinosaur hunters of the modern golden age of dinosaurian research is admittedly incomplete. A comprehensive account of the activities of the new generation of dinosaur collectors after 1980 would be much too long for inclusion in the space available here. And it is too soon to evaluate the eventual influence they will have on the future of dinosaurian paleontology in, and outside of, North America. But some of the high spots of dinosaur hunting during the middle and late years of the twentieth century have been described, and perhaps they are sufficient to show that the search for dinosaurs and the resulting research are being carried on today with an intensity beyond anything achieved in the past. Dinosaurs are today a lively subject, not only in paleontological circles, but also among the general public. Indeed, one is today confronted by dinosaurs everywhere – in museums, in stores, in moving pictures, in books, and always in the field.

References

Bird, R. T. 1985. *Bones for Barnum Brown: Adventures of a Dinosaur Hunter*. Fort Worth: Texas Christian University Press.

Colbert, E. H. 1968. *Men and Dinosaurs, The Search in Field and Laboratory*. London: Evans Brothers.

Colbert, E. H. 1995. *The Little Dinosaurs of Ghost Ranch*. New York: Columbia University Press.

Gillette, D. D. 1996. Origin and early evolution of the sauropod dinosaurs of North America: the type locality and stratigraphic position of *Dystrophaeus viaemalae* Cope 1877. In A. C. Huffman, W. R. Lund, and L. H. Godwin (eds.), *Geology and Resources of the Paradox Basin*. Utah Geological Association Guidebook 25: 313–324.

Hitchcock, E., 1858. *Ichnology of New England. A Report on the Sandstone of the Connecticut Valley, Especially Its Fossil Footmarks, Made to the Government of the Commonwealth of Massachusetts*. Boston: William White.

Lanham, U. N. 1973. *The Bone Hunters*. New York: Columbia University Press.

Leidy, J. 1856. Notice of remains of extinct reptiles and fishes, discovered by Dr. F. V. Hayden in the Bad Lands of the Judith River, Nebraska Territory. *Proceedings of the Academy of Natural Sciences, Philadelphia* 72–73.

Leidy, J. 1858. [Remarks concerning *Hadrosaurus*]. *Proceedings of the Academy of Natural Sciences, Philadelphia* 215–218.

Leidy, J. 1859. *Hadrosaurus foulkii*, a new saurian from the Cretaceous of New Jersey, related to the *Iguanodon*. *American Journal of Science* 27: 266–270.

Madsen, S. K. 2010. *Exploring Desert Stone: John N. Macomb's Expedition to the Canyonlands of the Colorado*. Logan: Utah State University Press.

Mayor, A. 2005. *Fossil Legends of the First Americans*. Princeton, N.J.: Princeton University Press.

Osborn, H. F. 1931. *Cope: Master Naturalist*. Princeton, N.J.: Princeton University Press.

Owen, R. 1842. *Report on British Fossil Reptiles. Part II. Report of the British Association for the Advancement of Science*. Eleventh Meeting, Plymouth, July 1841, 60–204.

Plate, R. 1964. *The Dinosaur Hunters. Othniel C. Marsh and Edward D. Cope*. New York: David McKay.

Schuchert, C., and C. M. LeVene. 1940. *O. C. Marsh – Pioneer in Paleontology*. New Haven, Conn.: Yale University Press.

Shor, E. 1974. *The Fossil Feud between E. D. Cope and O. C. Marsh*. Detroit, Mich.: Exposition Press.

Torrens, H. S. 2011 (this volume). Politics and paleontology: Richard Owen and the invention of dinosaurs. In M. K. Brett-Surman, Thomas R. Holtz Jr., and James O. Farlow (eds.), *The Complete Dinosaur*, 2nd ed., 24–43. Bloomington: Indiana University Press.

The Search for Dinosaurs in Asia

5

Corwin Sullivan, David W. E. Hone, and Xing Xu

From 1828 to 1831, Captain (later Major-General Sir) William Henry Slee-man (Fig. 5.1) of the British army was the colonial administrator of Jubbul-pore in central India, the modern Jabalpur District. Sleeman was a capable officer who later became a relentless enemy of the cultlike fraternity of murderers that the British knew as the Thuggee, at one time holding the picturesque title of Commissioner for the Suppression of Thuggee and Dacoity. However, he also displayed a keen intellectual curiosity toward his surroundings, and his 1844 book *Rambles and Recollections of an Indian Of-ficial* is filled with colorful observations of many different aspects of Indian society. Sleeman's interest in the geology and natural history of India was apparently more casual, but in 1828 he went looking for fossils in the vicinity of his house at Jabalpur on the Narmada River. On a hill called Bara Simla, within sight of the house, he found some well-preserved petrified trees and "some fossil bones of animals" (Sleeman 1844, 127) that he did not bother to discuss in any detail.

The bones that Sleeman's amateur curiosity brought to light on Bara Simla in 1828 were the first dinosaur fossils from anywhere in Asia to un-dergo scientific study and description. In 1832 they were deposited in the Indian Museum in Calcutta (now often spelled Kolkata). They later came to the attention of Hugh Falconer, the superintendent of the Geological Survey of India. Falconer described two vertebrae found by Sleeman in his posthumous memoirs (Fig. 5.2) but did not provide a taxonomic name. It fell to Lydekker (1877) to describe Sleeman's two vertebrae, along with an additional incomplete femur collected in 1871, as the type series of the new sauropod genus and species *Titanosaurus indicus*.

Because the characters Lydekker used to diagnose his new taxon are now known to be more broadly distributed among titanosaurian sauropods, *Titanosaurus* now lacks distinctive features of its own and may be regarded as a nomen dubium (Wilson and Upchurch 2003). Nevertheless, Sleeman's specimens have their place in history as the first evidence of the titanosaurs, an important clade of Cretaceous sauropods, and as the first Asian dinosaur fossils to come to the attention of scientists. The presence of *"Titanosaurus"* in both India and South America was an early clue to the characteristic presence of titanosaurs in Gondwanan faunas of the later Mesozoic.

Beginning with the very first specimens to be studied scientifically, the dinosaurs of Asia have not only added to the catalogue of dinosaur diversity but also provided evidence for broader inferences about dinosaurs and the Mesozoic world they inhabited. The present chapter tells the story of dinosaur research on Earth's largest continent in a series of sections that are delimited on the basis of chronology rather than geography, so that the

A First Glimpse in India

5.1. Major-General Sir William Henry Sleeman (1788–1856), discoverer of the first Asian dinosaurs to undergo scientific description.

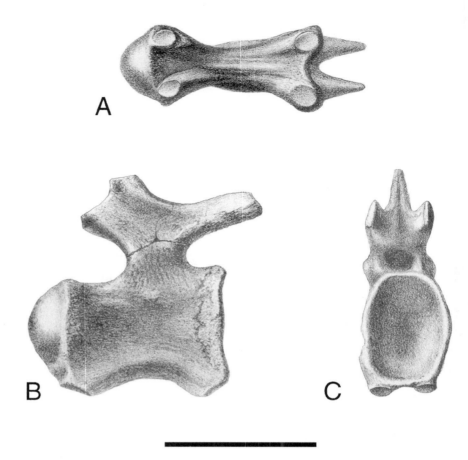

5.2. One of the indeterminate titanosaur caudal vertebrae found by Sleeman, initially described in Hugh Falconer's posthumous memoirs, and named *Titanosaurus indicus* by Lydekker (1877). A, Ventral view; B, right lateral view; C, anterior view. From Falconer (1868, pl. 34, figs. 3–5).

historical narrative passes from early to late across Asia as a whole. A final section discusses the importance of the Asian record to our current understanding of dinosaurs in general.

Remote Antiquity: The Bones of Dragons

As in other parts of the world, there are places in Asia where the bones or the traces of dinosaurs are so numerous and strikingly preserved that they can hardly have gone unnoticed by the peoples who inhabited the region for tens of thousands of years before the rise of science. Berkey and Nelson (1926) described finding in the Gobi Desert ancient beads that had been made from the fossil eggshells of birds, and occasionally of nonavian dinosaurs. A speculative but intriguing possibility is that reports emerging from the same region in the millennia before Christ may have brought to the Hellenic world descriptions of *Protoceratops*, the frilled but hornless basal neoceratopsian dinosaur that occurs there in considerable abundance. Mayor (2000) suggested that accounts of *Protoceratops* might have been the basis for Greek myths of the gryphon (Fig. 5.3), a creature that lived in the east and combined the quadrupedal body of a mammal with the head (and often the wings and talons, admittedly) of an eagle.

The bones of fossil vertebrates also have a well-known place in traditional Chinese medicine. According to Lei Hiao, writing in ad 400, a "dragon bone" (*lóng g*) would have curative properties if washed twice in hot water, ground to powder, placed in a bag with two eviscerated swallows, and then mixed with other medicines (Morgan and Lucas 2002). Equally colorful superstitions about fossils abounded in Europe until around the

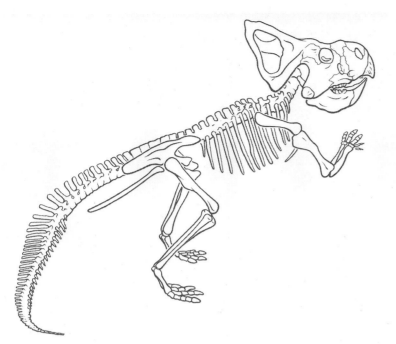

5.3. Early encounters with fossils of the basal neoceratopsian *Protoceratops* in Central Asia may have inspired Greek legends of the gryphon. Drawings by Ed Heck, coloring by Daniel Loxton.

Courtesy of Adrienne Mayor.

time of the Industrial Revolution. The medicinal use of fossils in China was indirectly beneficial to vertebrate paleontologists around the turn of the twentieth century: a few early studies of Chinese fossil mammals (e.g., Koken 1885) were based on material purchased from apothecaries. In 1885, the year of Koken's account, some 20 tons of dragon bones were apparently exported from China (Mayor 2000). However, it is possible that fossils of dinosaurs were rarely or never used in Chinese medicine; mammalian fossils are generally more abundant and obvious where they occur.

In parts of rural Thailand, there have been recent examples of dinosaur bones being kept in Buddhist temples, apparently as curiosities rather than objects of veneration (Martin et al. 1999). However, touching a dinosaur bone is said to bring luck and longevity. The bones are sometimes identified by the local people as petrified wood or as the bones of elephants. Similarly, modern villagers in Laos have been known to interpret dinosaur bones as belonging to buffaloes or oxen with supernatural properties, and Taquet (1998) was obliged to pay for the sacrifice of a pig before being guided to a fossil site. These folk attitudes to dinosaurs in Southeast Asia may be of considerable antiquity.

However, there is no evidence that premodern people anywhere in Asia took a scholarly interest in dinosaur fossils, beyond possibly identifying them as the bones of mythological creatures.

1828–1902: Prospecting in the Colonies

In the decades immediately following Sleeman's 1828 discovery on Bara Simla, only a limited amount of paleontological work on dinosaurs took place in Asia. British geologists and interested amateurs occasionally discovered dinosaur bones in central India, but virtually all of these remains were fragmentary and poorly preserved (Huene and Matley 1933). Both sauropod elements and isolated theropod teeth were represented, but these were apparently the only identifiable dinosaur specimens recovered from India in the nineteenth century. This record may seem surprisingly meager, but the known dinosaur fauna of India is somewhat scanty even at present (Sahni 2003). The colonial geologists, professional and amateur, did well considering the limitations of the fossil record.

Elsewhere in Asia, dinosaur paleontology was virtually nonexistent during the same time period. The only definite exception that has come to our attention was the discovery of dinosaur tracks along the Yagnob River, in what is now Tajikistan, by the Russian geologist G. D. Romanovsky in 1882 (Ryan 1997). Around the turn of the twentieth century, however, Asian dinosaur studies began to gather steam. Collecting and research activities gradually became much more extensive and widespread, and increasingly involved Asian paleontologists as well as Westerners.

1902–1930: The Age of Expeditions

In 1902 the study of Asian dinosaurs moved decisively beyond India. The critical moment came when Colonel Manakin of the Russian army obtained fossil bones from fishermen who had found the specimens on the banks of the Amur River (Godefroit 2006). The Amur, called in Chinese Heilongjiang (Black Dragon River), divides the Manchurian region of northeastern China from the eastern part of Asiatic Russia.

Geologists from the Russian Geological Committee in St. Petersburg visited the area in 1914 and over the winter of 1915–1916. Dinosaur bones were found in situ near the village of Jiayin on the Chinese side of the Amur, and excavations followed in the summers of 1916 and 1917. Although the finds included some theropod teeth, the most impressive discovery was a partial skeleton that became the holotype of the hadrosaur *Mandschurosaurus amurensis*. The species is now considered a nomen dubium (Bolotsky and Godefroit 2004), but the bones described by Riabinin nevertheless established the presence of dinosaurs in northeastern Asia.

It was not long before scientific explorers from the West also became involved in the search for dinosaurs in China and Mongolia. The whole region was gradually becoming more open to expeditions of all types, as transport became more efficient and contacts multiplied between Asia and the West. In 1892, for example, Vladimir A. Obruchev discovered some rhinoceros bones and teeth while crossing the Gobi (Natalín 2006). By the 1920s several Western paleontologists were employed by Chinese universities and/or the Geological Survey of China. The chief paleontologist of the survey was an American, Amadeus W. Grabau, and the Swede Johann G. Andersson and his Austrian assistant, Otto A. Zdansky, were also closely associated with the survey. Davidson Black, a Canadian anatomist and anthropologist, taught at Peking Union Medical College. Andersson, Zdansky, and Black would all play critical roles in the excavation of the Zhoukoudian locality, which yielded *Homo erectus* ("Peking Man") specimens and many other excellent fossil mammals in the 1920s and 1930s. Zdansky also collected some dinosaur bones in Shandong Province, China, in 1922 and 1923, following up initial finds of dinosaur fossils by the German mining engineer Berhagel in 1916 (Dong 1997). The material collected by Zdansky was eventually described by Wiman (1929) as the sauropod *"Helopus"* (now *Euhelopus*).

The Central Asiatic Expedition

The first spectacular discoveries of dinosaurs in Asia came as part of the Central Asiatic Expedition (CAE) of the American Museum of Natural History (AMNH) in New York. In hindsight, these impressive finds often tend to obscure the original focus of the CAE, which leaned heavily toward mammalian paleontology and even paleoanthropology. Henry Fairfield Osborn, the leading paleontologist at the AMNH and the president of the museum from 1908 to 1933, had long been interested in the potential of Asia as a new field for exploration in search of fossil mammals. His AMNH colleague William D. Matthew (1915) went further, regarding Asia as a "center of dispersal" for humans and most other primates, which he considered to have spread outward from this home region in successive waves. An early popular account of the CAE (Andrews 1926) was even entitled *On the Trail of Ancient Man,* and Osborn contributed a foreword describing Asia as nothing less than "a palaeontologic garden of Eden" (vii) and "the home of the remote ancestors of man" (ix). Curiously, Osborn also claimed in this foreword to have predicted in an earlier (1900) article that Asia would prove to be a center of mammalian evolution, even though the earlier article said no such thing.

Nevertheless, it seems clear that the driving force behind the CAE was a combination of Osborn's theoretical expectations and the enterprising zeal of a rugged AMNH field man named Roy Chapman Andrews (Fig. 5.4), who is often cited (e.g., Preston 1986) as a possible inspiration for the fictional swashbuckling archaeologist Indiana Jones. Before the beginning of the CAE in 1921, Andrews had already carried out two Asiatic expeditions on

5.4. Roy Chapman Andrews (1884–1960) surveys the landscape of central Inner Mongolia during the 1928 Central Asiatic Expedition field season. Seated is Tserin, the Mongolian who led the CAE camel train in 1928 and 1930.

Courtesy of the American Museum of Natural History.

the museum's behalf, so that what became the CAE was originally launched as the Third Asiatic Expedition of the AMNH. The first two expeditions had been zoological in nature; Andrews had traveled through parts of Asia collecting extant animals for the museum. For part of the duration of the Second Asiatic Expedition he had also acted as a civilian informant for the U.S. navy (Morgan and Lucas 2002). The two expeditions seem to have whetted Andrews's appetite for Asia in general and Mongolia in particular, and news of paleontological discoveries by Andersson and others had apparently convinced Andrews that Osborn might be willing to send him back to the region to search for fossils. Osborn is said to have accepted Andrews's proposal over lunch one day in 1920 (Preston 1986).

The Third Asiatic Expedition, as it was initially called, was not a single venture into Asia but rather a whole series of field operations carried out from 1921 to 1930 (Morgan and Lucas 2002). Although Andrews was the formal leader and principal organizer of the expedition, the chief paleontologist was Walter Granger of the AMNH. Granger, unlike Andrews, was both a patient and meticulous collector and a writer of academic papers, often in collaboration with Osborn or with the mammalian paleontologist William D. Matthew. The Third Asiatic Expedition effectively began with Granger's arrival in China in 1921. In August he joined Andersson and Zdansky in the initial excavations of the fissures at Zhoukoudian. Granger spent much of the following winter exploring and collecting in Sichuan Province, south China, where he found abundant fossil mammals.

Granger made three further winter trips to the south of China during the 1920s. However, the best-known and most impressive discoveries

5.5. Central Asiatic Expedition camel train near Tsagaan Nuur (the "white lake"), central Mongolia, 1925.

Courtesy of the American Museum of Natural History.

of the Third Asiatic Expedition, and all of the dinosaur finds, occurred during the summer forays into the Gobi that made up the CAE proper. There were five such field seasons, in 1922, 1923, 1925, 1928, and 1930, and their history has been recounted repeatedly (Andrews et al. 1932; Preston 1986; Colbert 2000; Morgan and Lucas 2002). The expedition of 1922 was essentially a prospecting venture, whose scientific personnel included the geologists Charles P. Berkey and Frederick K. Morris as well as Andrews and Granger. Travel and prospecting were carried out in automobiles, but a train of camels (Fig. 5.5) under the expert command of a Mongol named Merin moved fuel and other supplies into the Gobi along a fixed route. The cars and trucks would meet the camels, when necessary, to pick up supplies and deposit specimens.

The Gobi desert in 1922 was far from unexplored territory, even from the North American and European viewpoint. Andrews himself had visited the region in the course of the Second Asiatic Expedition. However, the vertebrate paleontology and geology of the Gobi Desert had never been extensively investigated, and it must have come as a considerable relief when Granger picked up some fragmentary mammalian fossils only a few days into the 1922 expedition. Apparently these initial finds included a rhinoceros tooth (Andrews 1926), in a curious echo of Obruchev's discovery 30 years before. The next day's prospecting turned up some dinosaur fossils. These finds were evidently made in the vicinity of Erlian (or Iren Dabasu), a settlement lying just within the Chinese-controlled region of Inner Mongolia. Beyond Erlian lay the independent country of Mongolia proper.

No doubt buoyed by this initial success, the members of the expedition moved north to Ulan Bator, which they knew as Urga, and then traveled southwest almost to the edge of the Altai Mountains. The most impressive dinosaur finds of the 1922 season came at sites encountered along a trail situated slightly north of the mountains, as the expedition moved east to return to winter headquarters in Beijing. Among these sites, the one that was ultimately most productive was known to the expedition as Shabarakh Usu or the Flaming Cliffs, so called because of the way the red rock of the cliffs seemed to glow in the setting sun. The name used in modern publications is Bayan Zag. In 1922, however, the finds at this site were modest and serendipitous. The expedition stopped one afternoon to ask directions from some local nomads, and the expedition photographer, James B. Shackelford, took advantage of the delay to investigate a nearby outcrop. He soon spotted a skull that Granger and Gregory (1923) eventually described as a new ceratopsian, *Protoceratops andrewsi*. Although ceratopsians are horned dinosaurs, both etymologically and colloquially, *Protoceratops* was unique at the time in lacking horns, even though its well-developed frill clearly indicated its ceratopsian status. Granger and Gregory correctly interpreted the new genus as representing a primitive, hitherto unknown, grade of ceratopsian phylogeny. This skull was an important discovery, but the only other really notable specimen collected at the Flaming Cliffs that afternoon was a fossil eggshell that Granger suspected belonged to an extinct bird. The expedition resumed its return journey the following morning, leaving behind a number of specimens that were too fragmentary, too heavy, or too firmly embedded in the rock to be collected. The site clearly deserved further attention, but time in the field was running short.

Thanks to "the bright observation of Wong, the Mongolian chauffeur engaged in Urga" (Osborn 1923, 2), a more impressive discovery occurred slightly earlier during the eastward journey. At a site now called Öösh, Wong apparently spotted what would prove to be the best dinosaur specimen found by the expedition in 1922, a nearly complete skeleton of an ornithischian that Osborn dubbed *Psittacosaurus*. Osborn was originally dubious about the exact affinities of the new animal, but after the skeleton had been fully prepared, he confidently declared *Psittacosaurus* to be an iguanodontid (Osborn 1924a). Subsequent research has demonstrated that *Psittacosaurus* is a ceratopsian even more primitive than *Protoceratops*, lacking not only horns but also a well-developed frill. However, Osborn's conclusions were perhaps not unreasonable considering the limited information that was available in the 1920s, and considering the absence in *Psittacosaurus* of many typical ceratopsian characteristics.

In 1923, an expanded CAE returned to both Erlian and Bayan Zag for much more systematic and extensive collecting. Critical to this effort were three expert field technicians, George Olsen, Peter Kaisen, and Albert F. Johnson, and several Chinese assistants, including Gan Chuanbao (nicknamed Buckshot!) and Liu Xigu, both of whom later spent a year at the

5.6. Walter Granger (1872–1941, foreground) excavating a dinosaur nest at the Flaming Cliffs of Bayan Zag, 1925. Granger was the chief paleontologist of the Central Asiatic Expedition.

Courtesy of the American Museum of Natural History.

AMNH learning paleontological techniques (Andrews et al. 1932). With known fossiliferous localities and a larger crew of experienced collectors to exploit them, a few weeks of collecting around Erlian yielded several important specimens, some of which were described as new taxa by Gilmore (1933). The tyrannosaurid *Alectrosaurus*, the hadrosaurs *Mandschurosaurus mongoliensis* (later renamed *Gilmoreosaurus*) and *Bactrosaurus*, and a manus and pes that became the type material of *Ornithomimus asiaticus* (later renamed *Archaeornithomimus*) were all found during this period.

A similar collecting effort at Bayan Zag in July and August produced the richest dinosaur haul of the entire CAE. As at Erlian, the lion's share of the finds was made by the American field men, though it was the Chinese taxidermist "Chih Sho-lun" who discovered the type specimen of the troodontid theropod *Saurornithoides* (Osborn 1924b). Two other new theropods, *Oviraptor* and *Velociraptor*, were also collected at Bayan Zag in 1923. *Velociraptor* is of course a dromaeosaurid, swift-moving and raptorial but considerably smaller and presumably less dangerous than the *Jurassic Park* books and films suggest. *Oviraptor* was one of the first oviraptorosaurs to be described and was by far the first to be known from more than fragmentary material.

The name *Oviraptor,* "egg stealer," was inspired by the circumstances of the taxon's discovery. Earlier in the excavation, George Olsen had found a cluster of three fossil eggs, which Granger identified as dinosaurian. Further searching revealed many more eggs, some of which were arranged in nests (Fig. 5.6). Because of the local abundance of *Protoceratops,* it was natural for the expedition members and their scientific associates to assume that the eggs belonged to this taxon. The type skull of *Oviraptor* was found in close association with a nest, which "immediately put the animal under suspicion of having been overtaken by a sandstorm in the very act of robbing the dinosaur egg nest" (Osborn 1924b, 9). Osborn acknowledged that this suspicion might prove unfounded, but it was many years before the matter could be clearly resolved (see below).

From the point of view of dinosaur paleontology, the triumphant field season of 1923 represented the peak of the CAE. The 1925 expedition was launched with a more archaeological emphasis, including as it did the Danish-born AMNH archaeologist Nels C. Nelson (see Berkey and Nelson 1926). Although there were many significant discoveries, including the beads of fossil eggshell referred to above and some fossil lizards and mammals, little dinosaur material of significance was collected. The expeditions of 1928 and 1930 were confined to the Gobi of Inner Mongolia, and most of the haul in both years consisted of Tertiary mammals.

To some extent, the focus on mammals corresponded with the theoretical underpinning of the CAE, Osborn's idea that Asia had been a center of mammalian diversification. Although the dinosaur discoveries are better known today, it is worth remembering that many of the most important CAE finds were of fossil mammals, and that these discoveries often shared at least equal billing with the dinosaurs in contemporary reports. A good *Indricotherium* skull was one of the great treasures unearthed by the first CAE of 1922, and in 1923 and 1925 Bayan Zag yielded the first reasonably intact skulls of Mesozoic mammals to be described from anywhere in the world (Simpson 1925; Gregory and Simpson 1926). Against this background, the discovery of numerous Tertiary mammals in 1928 and 1930, including a tremendously rich collection of the "shovel-tusked" elephant *Platybelodon,* can be seen as a continuation of the earlier success of the CAE.

Nevertheless, the scope of the CAE was unquestionably curtailed by political developments in the mid- to late 1920s. During this period much of China was dominated by local warlords, and early in 1926 Beijing was repeatedly bombed by the small private air force of the Manchurian warlord Zhang Zuolin. At first the bombing did little damage, and CAE members and other foreigners even amused themselves by watching ineffectual bombing runs from the roof of the Peking Hotel (Preston 1986). Gradually, however, the situation became more serious. On April 12 Andrews was caught in the open during a bombing run and had to take cover under a railway car (Andrews et al. 1932). He was unhurt, apart from burned fingers when he picked up a piece of hot shrapnel, but five Chinese were killed in the incident. Zhang finally captured Beijing in June, and continuing instability helped to keep the CAE out of the field in the summers of both 1926 and 1927.

The Gobi itself was increasingly rife with banditry, and CAE members sometimes had to reach for their guns. The archaeologist Nelson is said to have once deployed a more creative strategy, frightening off a group of bandits by the simple expedient of removing his glass eye. There was also trouble of a more official nature because the Mongolian government was close to Soviet Russia and suspicious of the CAE's intentions in exploring the Gobi. These fears may not have been unfounded, given Andrews's previous work in Asia as a civilian informant for the U.S. navy and the fact that American and British military personnel accompanied some of the CAE trips into the Gobi (Morgan and Lucas 2002). There was possibly also a degree of scientific rivalry because Russia apparently sent paleontological expeditions of its own into Mongolia during the 1920s (Kurochkin and Barsbold 2000).

On the Chinese side, the CAE had difficulty with an unofficial but seemingly influential body called the Society for the Preservation of Cultural Objects following the 1928 expedition (Andrews et al. 1932). There were further accusations of espionage, and even opium smuggling, but the society's most substantive complaint was that the CAE had "stolen China's priceless treasures" (Preston 1986, 108). In line with common practice at the time, important specimens were being taken back to the AMNH, where they remain to this day. Andrews and Osborn had also mightily annoyed the Chinese and Mongolian authorities by auctioning off a dinosaur egg for $5,000 after the 1923 expedition. This seemed to demonstrate that the objects being removed from the Gobi were financially as well as scientifically valuable.

The society briefly seized the collections made during the 1928 expedition and forced the cancellation of the planned expedition of 1929. Under the terms of an agreement with the society, the 1930 expedition eventually proceeded with the accompaniment of scientists from the Geological Survey of China. The Chinese contingent included Zhang Xiti, his young assistant Yang Zhongjian, and the famous Jesuit paleontologist Pierre Teilhard de Chardin, who was attached to the survey at the time.

Despite the relative success of the 1930 expedition, the combination of banditry and bureaucracy prevented further work by the CAE. In early 1933 Andrews was still hoping that he would be able to return to the Gobi via the puppet state of Manchukuo that the Japanese had established in northeastern China, but the plan came to nothing. Nevertheless, Granger, Andrews, and the CAE as a whole had uncovered the first really substantial Asian dinosaur fauna, as well as contributing significantly to the archaeology and mammalian paleontology of the Gobi.

Elsewhere in Asia

The CAE was responsible for the most spectacular and significant dinosaur discoveries made in Asia between 1902 and 1930, but important work was also taking place elsewhere. In India, the British geologist Charles A. Matley carried out from 1917 to 1919 the first systematic excavations ever undertaken at Sleeman's old locality of Bara Simla, and he also worked at Pisdura in 1920. With the assistance of Durgasankar Bhattacharji of the Geological Survey of India, Matley assembled a substantial collection of sauropod, theropod, and ornithischian elements, although most of the material was rather fragmentary. Many of the sauropod elements were assigned by Huene and

Matley (1933) to Lydekker's genus *Titanosaurus*, although a skull fragment and some postcranial bones from Bara Simla were assigned to the newly erected species *Antarctosaurus septentrionalis*. This taxonomic decision had biogeographical implications because *Antarctosaurus* is an Argentine sauropod genus, but referral of the Indian material to *Antarctosaurus* is probably invalid (Wilson and Upchurch 2003). Similarly, some of the specimens from Pisdura were referred to *Antarctosaurus* or to the Malagasy taxon *Laplatasaurus madagascarensis* (Huene and Matley 1933).

The most important theropod specimens collected by Matley and Bhattacharji were large skull roof fragments that were used to define the genera *Indosuchus* and *Indosaurus* (Huene and Matley 1933). Barnum Brown of the AMNH also visited Bara Simla in 1922, collecting further cranial bones and vertebrae of *Indosuchus* that were finally described by Chatterjee (1978) half a century later. Although Chatterjee regarded *Indosuchus* as a tyrannosaur and *Indosaurus* as a megalosaur, it is now clear that both taxa are actually abelisaurid ceratosaurs (Novas et al. 2004). A number of smaller theropod specimens described by Huene and Matley (1933) also show abelisauroid features. The supposed stegosaur *Lametasaurus* was based on various elements from the carnosaur bed at Bara Simla, which turned out to represent theropod pelvic and sacral material combined with sauropod and possibly crocodyliform bones (Wilson et al. 2003; Lamanna et al. 2004). Ultimately, Matley's discoveries at Bara Simla and Pisdura demonstrated the essentially Gondwanan character of the Cretaceous dinosaur fauna of India, with titanosaurids and abelisauroids as salient taxa. However, the record is by no means rich in comparison to that of the Gobi.

The American geologist George Louderback visited the Sichuan Basin in south China from 1913 to 1915 and apparently recovered some large theropod bones (Young 1935a). On a larger scale, the year 1927 marked the commencement of the Sino-Swedish Expedition (SSE), actually a multidisciplinary series of scientific ventures in northwestern China. The SSE continued until 1935 and involved a large roster of Chinese, Swedish, and other European scientists under the leadership of the Swedish aristocrat Sven Hedin. However, the focus of the SSE was more archaeological than paleontological. The most notable dinosaur finds were fragmentary remains of a sauropod from Xinjiang, *Tienshanosaurus chitaiensis* (Young 1937), and assorted fragmentary ornithischians from Inner Mongolia and Gansu that Bohlin (1953) used as the basis for several new genera and species. These taxa, including *Tienshanosaurus*, are now considered nomina dubia. However, *Tienshanosaurus* does represent the first dinosaur ever discovered in Xinjiang Autonomous Region, which eventually became a rich source of specimens. Around the same time, Yang Zhongjian published with Teilhard de Chardin a report of the first dinosaur footprint from Asia, an isolated iguanodont track from Yang's native Shaanxi Province (Teilhard de Chardin and Young 1929).

Meanwhile, dinosaur body fossils were discovered in the Kyzyl Kum Desert of Kazakhstan in the 1910s (Ryan 1997), although they were only formally described by Riabinin in the 1930s. The early Kazakh finds considered by Riabinin included fragmentary Cretaceous hadrosaurs and an isolated sauropod femur. Although the taxa he erected are now regarded as invalid (Maryańska 2000; Norman and Sues 2000), some of the hadrosaur specimens are nevertheless highly informative. This material represents the first collection of dinosaur body fossils known from Central Asia.

By 1930, then, the general nature of the Cretaceous dinosaur assemblages in the Gobi was already relatively well understood, although many crucial details remained to be filled in. Representatives of almost all of the major dinosaur groups that are known from the Gobi today had been discovered and more or less accurately interpreted (the identification of *Psittacosaurus* as an iguanodontid was, of course, a significant exception). In India, the titanosaurs and abelisaurs that constitute the most important components of the local dinosaur fauna had were already known, although the ceratosaur affinities of the abelisaurs were not understood.

1930–1946: War and Peace

The final CAE venture into the Gobi in 1930 marks the end of what might be considered the first golden age of Asian dinosaur paleontology. Until 1946 there were no further major expeditions to Asia from outside the region, and even the work of paleontologists based in Asia was somewhat curtailed during this period. The main reason was, of course, the great conflict that began as the Mukden Incident (or Manchurian Incident) of September 18, 1931, flared into full-scale war between China and Japan in 1937, and merged into World War II in 1941.

Under the circumstances, the amount of fieldwork and research on dinosaurs that took place in Asia during this period is remarkable. To some extent, the conflict was even beneficial to Japanese paleontology because control of northeastern China (the puppet state of Manchukuo, from which Roy Chapman Andrews had hoped to reenter the Gobi) gave Japanese scientists access to large tracts of fossiliferous territory. The Japanese geologist S. Satô discovered small theropod footprints in the Tuchengzi Formation of Liaoning Province (Yabe et al. 1940), and further investigations by Tokio Shikama (1942) revealed hundreds of tracks similar to the classic ichnogenus *Grallator* from the Lower Jurassic of New England. Shikama's finds represented the first extensive sample of dinosaur tracks to be discovered in China. Nagao (1936) described the hadrosaur *Nipponosaurus* from another Japanese possession, the island of Sakhalin at the mouth of the Amur River. *Nipponosaurus* was based on a rather poorly preserved subadult hadrosaur, which appears to represent a lambeosaurine (Suzuki et al. 2004). Sakhalin has since reverted to Russia.

Meanwhile, thousands of kilometers away, the French geologist Josué-Heilmann Hoffet became the first person to describe dinosaur fossils from Southeast Asia. In the course of mapping the geology of Laos, Hoffet discovered the fragmentary remains of both sauropod and ornithopod dinosaurs. He referred the sauropod to a new species, *Titanosaurus falloti*, but this must be considered yet another nomen dubium, although the specimens are clearly titanosaurian (Wilson and Upchurch 2003). They are among a relatively small number of titanosaur finds known from northern landmasses, so the Laotian titanosaur material is of definite biogeographic interest. Taquet (1998) interpreted the ornithopod material as pertaining to an iguanodontid, rather than to a hadrosaur as Hoffet believed. Hoffet himself was apparently killed by Japanese troops shortly after they invaded French Indochina in March 1945 (Taquet 1998).

5.7. Yang Zhongjian (1897–1979), known in the West as C. C. Young, the first prominent Chinese vertebrate paleontologist and the first director of the Institute of Vertebrate Paleontology and Paleoanthropology.

Courtesy of the Institute.

However, the most important work on dinosaurs being carried out in Asia at this time was undoubtedly that of Yang Zhongjian (Fig. 5.7), the talented young paleontologist who had participated in the CAE Gobi trip of 1930. Yang, who became famous in the West under the anglicized name C. C. Young, was a prolific and versatile scientist whom Sun and Zhou (1991) described as the founder of Chinese vertebrate paleontology. In many respects the designation is well deserved. After completing a bachelor of science degree in geology at Peking University in 1923, Yang began graduate studies in vertebrate paleontology at Munich University. His dissertation was published in *Palaeontologia Sinica* under the title "Fossile Nagetiere aus Nord-China" (fossil rodents from north China), and this paper (Young 1927) represents the first published work of vertebrate paleontology by a Chinese researcher. The training he received at Beijing and Munich, followed by collaboration upon his return to China with distinguished colleagues such as Teilhard de Chardin, made Yang into a skilled and perceptive paleontologist who produced thorough, well-illustrated descriptions of fossils from a wide variety of taxonomic groups. Though he published throughout his career mainly in Chinese journals, his ability to write papers in excellent English helped to make his work accessible to an international audience.

Yang was primarily interested in mammalian paleontology throughout the early part of his career, until approximately 1940. Beginning in 1928, for example, Yang took charge of excavations at the Peking Man site at Zhoukoudian, under the auspices of the newly founded Cenozoic Research Laboratory of the Geological Survey of China. However, his affiliation with the Cenozoic Research Laboratory did not prevent him from occasionally studying Mesozoic vertebrates, including dinosaurs, from sites around China. In 1934, for instance, he and his close colleague, Bian Meinian (also known as M. N. Bien or Edward M. Bien), revisited the *Euhelopus* beds at Mengyin in Shandong Province and recovered some additional bones of *Euhelopus* and other dinosaur taxa (Young 1935b). In the summer of 1936 Yang prospected in both the Triassic of Shaanxi and the Cretaceous of Sichuan with the University of California paleontologist C. L. Camp, following up Louderback's preliminary discoveries. In the course of the Sichuan expedition Yang and Camp discovered a partial sauropod skeleton that Yang later described as *Omeisaurus junghsiensis*, taking the generic name from the "famous sacred mountain Omeishan" (Young 1939, 310) lying near the locality.

Yang continued his research, although under difficult circumstances, through the height of the conflict against the Japanese. With the Japanese in control of most of coastal China, including Beijing, Yang spent most of the war years in the inland cities of China. His dinosaur work at this time focused on the Lower Jurassic Lufeng Formation of Yunnan Province, which he investigated together with Bian. Despite wartime constraints, Yang managed to complete some of his most important and lasting dinosaur studies during this time, publishing thorough descriptions of the new Lufeng prosauropods *Lufengosaurus* (Young 1941) and *Yunnanosaurus* (Young 1942). From 1944 to 1945, Yang embarked on a visit to the United States, where he pursued his research at the AMNH. Unlike Bian, who left China permanently in the 1940s to begin a new career in America, Yang returned to Beijing after the war and cemented his position as China's leading vertebrate paleontologist of the mid-twentieth century.

Following the open conflicts of the late 1930s and early 1940s, global politics exerted a subtler influence on Asian dinosaur paleontology in the next few decades. Mongolia had been a Soviet ally since the 1920s, and in 1949 Mao Zedong proclaimed the founding of the People's Republic of China. Communist governments also ultimately came to power in North Korea and some Southeast Asian countries, and even noncommunist Asian nations such as India and Malaysia were not necessarily close to the West.

As a result, scientific cooperation between the West and much of Asia was curtailed in comparison to the prewar situation. In particular, the rich fossil beds of the Gobi and many productive sites in China proper were effectively closed to Western researchers and were instead studied by Chinese and Mongolian paleontologists in frequent cooperation with colleagues from Soviet bloc countries.

Paleontology Returns to Mongolia

In 1946, the Paleontological Institute of the Russian Academy of Sciences (PIN) acted on a long-standing invitation from the government of the Mongolian People's Republic to carry out fieldwork in the Gobi. In contrast to the CAE field seasons, which had been restricted to spring and summer, the first of the Mongolian Paleontological Expeditions (MPE) began in August and remained in Mongolia until January 1947, although only two and a half months were actually spent in the field (Kurochkin and Barsbold 2000). This expedition was essentially a reconnaissance venture carried out by eight scientists and technicians, including the leader, Ivan A. Efremov, and his second-in-command, Anatole K. Rozhdestvenskii. Unlike the CAE, the MPE was entirely motorized, and instead of a supply caravan, the expedition established a fixed base at Dalanzadgad in south-central Mongolia. Working primarily from this point, the expedition visited Bayan Zag and also discovered a number of new localities in the Gobi. Among them were Late Cretaceous sites that yielded a range of dinosaurs including hadrosaurs, sauropods, theropods, and ankylosaurs.

The 1947 expedition was on a much larger scale, with over 30 personnel again led by Efremov. MPE personnel began to arrive in Mongolia in November 1947, and the winter was spent in restocking the base at Dalanzadgad. The expedition began actual fieldwork in March 1948 and remained in Mongolia until October. Collecting initially focused on a cluster of localities in the eastern Gobi, and subsequently shifted first to Bayan Zag and then to the Nemegt Valley west of Dalanzadgad. Bayan Zag and an eastern Gobi site called Bayan Shiree both yielded ankylosaur specimens, but the most spectacular discoveries were made in the Nemegt Valley (Colbert 2000; Kurochkin and Barsbold 2000). Here the expedition identified a series of important and productive Upper Cretaceous sites, including one at Altan Uul that earned the evocative nickname of Dragon's Tomb. These sites yielded specimens of the large hadrosaur *Saurolophus*,

5.8. Soviet-Chinese expedition personnel removing overburden at a site in the Alashan Desert of Inner Mongolia, 1960.

Courtesy of the Institute of Vertebrate Paleontology and Paleoanthropology.

previously known from North America, and a tyrannosaur called *Tarbosaurus* that is sometimes regarded as synonymous with the North American genus *Tyrannosaurus*. The new discoveries at Nemegt thus provided strong evidence of a close linkage between the Late Cretaceous faunas of Asia and North America.

The MPE collected further *Saurolophus* material at Nemegt during its final field season in the summer of 1949 and visited a number of additional sites. Expedition members found small dinosaurs at Bayan Zag, but they also spent considerable time at Tertiary sites that yielded a variety of vertebrates and Permian ones that yielded only plants (Kurochkin and Barsbold 2000). Lower Cretaceous sites in the west of Mongolia produced some sauropod remains.

Despite the success of all three of these field seasons, the planned expedition of 1950 never materialized, and Russian paleontologists would not return to the fossil deposits of Mongolia until the 1960s. It is fair to say that the MPE recovered fewer novel types of dinosaur than the earlier CAE, but the MPE opened up extremely important localities in the Nemegt Basin and collected many specimens of great paleobiogeographic interest.

Notwithstanding the institutional affiliations of the present authors, an important event in the history of Chinese paleontology was the founding of the Institute of Vertebrate Paleontology and Paleoanthropology (IVPP) in 1953, with Yang Zhongjian as its first director. In 1959, the IVPP and the PIN carried out a cooperative expedition to Inner Mongolia under the joint leadership of the Gobi veteran Rozhdestvenskii and the Chinese paleontologist Zhou Mingzhen (Chow Minchen). With a crew of over 60 people, the expedition collected Cretaceous dinosaurs around the old CAE stomping

5.9. Finds from the 1959 Soviet-Chinese expeditions to Inner Mongolia on display at the IVPP, 1960. The banner optimistically predicts eternal friendship between China and the Soviet Union.

Courtesy of the Institute of Vertebrate Paleontology and Paleoanthropology.

ground of Erenhot in the Chinese Gobi, and a subsequent trip in 1960 (Fig. 5.8) excavated on a similar scale in the Alashan Desert of Inner Mongolia (Currie 1997a). Significant discoveries included the huge theropod *Chilantaisaurus tashuikouensis* and the hadrosaur *Probactrosaurus*. The material was divided between the IVPP and the PIN, and some of it was displayed at the IVPP (Fig. 5.9), but many specimens apparently remain unprepared. Further joint expeditions were planned, but the project was essentially abandoned by the Russian side owing to increasing hostility between the Soviet and Chinese governments. The series of IVPP expeditions to Inner Mongolia continued until 1965, but the political climate would bar Soviet bloc paleontologists from China, although not from Mongolia, for many years.

During the 1960s, however, scientists from Russia and Poland participated in separate Gobi expeditions organized jointly with Mongolian colleagues. These cooperative ventures effectively marked the beginning of vertebrate paleontology in Mongolia itself and were extraordinarily successful. The Polish–Mongolian Paleontological Expeditions (PMPE) began with a reconnaissance trip in 1963, carried out by 11 paleontologists, including the respective Polish and Mongolian coleaders Julian Kulczycki and Naydin Dovchin (Colbert 2000). The team divided its efforts between visiting previously established localities and scouting for new ones, in both Cretaceous and Tertiary strata. In subsequent years the Mongolian contingent was led by Rinchen Barsbold; the Polish contingent was led by Kazimierz Kowalski in 1964 and by the noted paleomammalogist Zofia Kielan-Jaworowska from 1965 onward (Kielan-Jaworowska et al. 2000).

Large parties entered the field in 1964 and 1965 to carry out full-scale excavations, with the most notable dinosaur discoveries coming, predictably, from Bayan Zag and the Nemegt Basin. Bayan Zag yielded juvenile skeletons of *Protoceratops* and the ankylosaur *Pinacosaurus*, and a number of eggs and nests like those discovered by the CAE four decades before. From Nemegt localities came new specimens of *Tarbosaurus* and *Saurolophus*, the previously unknown sauropods *Nemegtosaurus* and *Opisthocoelicaudia*, and three skeletons of a new ornithomimosaur, *Gallimimus*. The most mysterious of the Nemegt finds was a pair of enormous theropod forelimbs and shoulder girdles, preserved in isolation apart from associated fragments of the axial skeleton. The specimen was christened *Deinocheirus mirificus* by Osmólska and Roniewicz (1970). *Deinocheirus* may be a primitive ornithomimosaur, but its affinities have remained essentially uncertain to the present day.

Small PMPE field parties returned to Bayan Zag to search for small vertebrate fossils every year from 1967 to 1979, underscoring the importance of Mesozoic mammal and lizard fossils to the expedition. However, large-scale quarrying in the Nemegt Valley and elsewhere resumed in the final PMPE field seasons of 1970 and 1971. The Nemegt localities, including a new and exceptionally rich site known as Khulsan, produced abundant dinosaurs including theropods, a pachycephalosaurid and an ankylosaurid. However, the most impressive single find was the so-called fighting dinosaurs, a *Protoceratops* and a *Velociraptor*, discovered by Andrzej Sulimski at the Tugrigiin Shiree site west of Bayan Zag in 1971.

Two overlapping joint Russian–Mongolian expeditions, the Soviet–Mongolian Geological Expedition and the Soviet–Mongolian Paleontological Expedition (SMPE), were launched in 1967 and 1969, respectively (Kurochkin and Barsbold 2000). Regular work continued for decades, though on a reduced scale following the decline of the Soviet Union at the end of the 1980s. Discoveries of dinosaurs and other fossil vertebrates were numerous and helped to build up a far more complete picture of the Late Cretaceous dinosaur fauna of the Gobi. Perhaps the most significant single contribution of the SMPE was the discovery of the first substantial specimens of therizinosaurian dinosaurs. Although *Therizinosaurus* itself had been named much earlier, on the basis of large ungual phalanges and a few associated fragments discovered during the MPE (Maleev 1954), the material was sufficiently undiagnostic that the affinities of the animal were largely a matter of guesswork. The title of Maleev's (1954) paper, "New Turtle-Like Reptile in Mongolia," is ample evidence of this uncertainty. The more extensive material discovered by the SMPE included incomplete skeletons of *Segnosaurus* and *Erlikosaurus*, which were immediately recognized as dinosaurs. However, the affinity of these "segnosaurs" with *Therizinosaurus*, and the identity of therizinosaurs as derived theropods, were established only much later (see Clark et al. 2004).

Elsewhere in Asia

As in the 1920s, the spectacular dinosaurs being unearthed in Mongolia in the postwar years overshadowed to some extent discoveries from elsewhere in Asia. However, the four decades between 1946 and 1986 saw such extensive progress in dinosaur paleontology across the entire continent that only some high points can be mentioned here.

Outside Mongolia, the most important development was probably the maturation of dinosaur paleontology in China as a scientific discipline. In the early stages this was primarily due to the continuing work of Yang Zhongjian, whose primary research interests were now in reptiles and non-mammalian synapsids. In 1951 he directed the excavation of the peculiar Cretaceous hadrosaur *Tsintaosaurus* at Laiyang, Shandong, and later in the 1950s he described the sauropod *Mamenchisaurus* from the Late Jurassic of Sichuan, which remains the longest dinosaur so far known from Asia (Dong 1997).

The rich Jurassic beds of Sichuan Province continued in the following decades to be exploited by Chinese paleontologists, both from the IVPP and from other institutions such as the Chongqing Municipal Museum (now called simply the Chongqing Museum). A particularly significant event was the discovery in 1976 of an incomplete sauropod skeleton, eventually named *Shunosaurus lii* (Dong et al. 1983), in Middle Jurassic strata of the Xiashaximiao Formation exposed at Dashanpu near the city of Zigong. Further excavations at the site in the following years yielded thousands of bones from sauropods, stegosaurs, theropods, and ornithopods, including some articulated skeletons of previously unknown taxa.

Significant dinosaur excavations were carried out in Xinjiang Autonomous Region and Shandong Province during the 1960s. The dinosaurs found in Xinjiang during this period were fragmentary taxa such as the theropod *Tugulusaurus* (Rauhut and Xu 2005), but they nevertheless represented the first dinosaur material discovered in the region since the Sino-Swedish expeditions. Shandong Province produced an enormous hadrosaur, *Shantungosaurus*, with an estimated length of approximately 15 m (Dong 1997).

The dinosaurs of India continued to receive attention during the post-war decades, particularly from researchers at the Geological Studies Unit of the Indian Statistical Institute in Calcutta, and some important new material was collected (summarized by Jain 1989). The Lameta beds produced occasional specimens, but perhaps the most important discovery was that of multiple individuals of the sauropod *Barapasaurus* from the Lower Jurassic Kota Formation of the Pranhita–Godavari valley. *Barapasaurus* is still among the most complete sauropods known from the Lower Jurassic (Upchurch et al. 2004). Prosauropod material was also discovered in Lower Jurassic exposures of the Dharmaram Formation in the state of Andhra Pradesh, southern India (Kutty 1969). Dinosaur eggshell specimens, including some complete eggs and even clutches from the Lameta Formation, began to be discovered in the Upper Cretaceous of India beginning in the 1980s.

This period also saw the first discoveries of dinosaur material in the Asian countries of Japan, South Korea, and Thailand. The initial Japanese discovery was an isolated, incomplete, and undiagnostic sauropod humerus found in the Lower Cretaceous strata of Iwate Prefecture in 1978 (Kobayashi et al. 2006). Further dinosaur body fossils were discovered in the following years, in addition to tracks, but the quality of these finds was uniformly rather poor (Manabe et al. 1989). Similarly, the first South Korean dinosaur find consisted of eggshell pieces discovered in 1972 in the Early Cretaceous Hasandong Formation of Hadong County, South Gyeongsang

Province (Lee and Lee 2006). Discoveries of bones and teeth followed, although these finds were generally very fragmentary. However, hundreds of ornithopod, sauropod, and theropod trackways were discovered at the Dukmyeongri locality within the Jingdong Formation, which is slightly younger than the Hasandong, during the 1980s (Lim et al. 1989).

The first Thai find probably also seemed rather unpromising when it was discovered in 1976 by Suthem Yaemniyom, a geologist based at the department of mineral resources in Bangkok. The specimen consisted of the distal end of a sauropod femur, recovered from the Early Cretaceous Sao Khua Formation near Phu Wiang. In contrast to the Japanese and South Korean discoveries, however, the Thai record of dinosaur body fossils quickly proved to be extensive. In 1978 two sauropod vertebrae were discovered in the same area, and in 1980 a humerus came to light in a Buddhist temple in a neighboring province. Thai workers followed up these initial finds in collaboration with French colleagues, and their efforts were rewarded in 1982 with the discovery of a partial sauropod skeleton from the Sao Khua. The specimen was excavated over the following 5 years and became the type of *Phuwiangosaurus* (Martin et al. 1999).

1986–Present: Asian Dinosaurs across the Turn of the Millennium

A welcome trend in the 1980s was a general warming of previously frosty relations between the West and several Asian countries, most notably China. The new atmosphere greatly increased the potential for scientific cooperation in many areas, dinosaur paleontology not least among them. Greater openness allowed Western expertise and resources to be brought to bear on the study of Asian dinosaurs to an extent not seen since the 1920s, to the benefit of the field as a whole.

The first major cooperative venture was the Canada–China Dinosaur Project (CCDP). The CCDP conducted annual fieldwork from 1986 to 1991, and the project was truly reciprocal in that Chinese and Canadian scientists hunted dinosaurs together in both countries (Dong 1993; Currie 1997b). Dong Zhiming and other leading Chinese paleontologists lent their expertise to excavations in Alberta and the Canadian arctic, while Canadian researchers such as Phillip J. Currie and Dale A. Russell helped recover important specimens from Inner Mongolia and from the Junggar Basin of Xinjiang. On the Chinese side, the potential of Xinjiang as an important source of Jurassic dinosaurs was finally realized with the discoveries of the holotypes of the theropods *Sinraptor* (Currie and Zhao 1993) and *Monolophosaurus* (Zhao and Currie 1993), and the playfully named sauropod *Mamenchisaurus sinocanadorum* (Russell and Zheng 1993). The Ordos Basin of Inner Mongolia yielded *Sinornithoides* (Russell and Dong 1993), a troodontid specimen of unprecedented completeness. Exposures of Djadokhta-equivalent strata at another Inner Mongolian site, Bayan Mandahu, produced an *Oviraptor* perched on a nest of eggs, in addition to juvenile specimens of the ankylosaur *Pinacosaurus* (Currie 1997b) and embryonic specimens of the ceratopsians *Protoceratops* and *Bagaceratops* (Dong and Currie 1993).

Beginning in 1990, the AMNH returned to Mongolia in the form of the Mongolian American Museum Expedition (MAE), a collaborative effort with Mongolian researchers (Norell 1997). Among the impressive results of this latter-day successor to the CAE was the discovery in 1993 of an entirely new Djadokhta locality, Ukhaa Tolgod. This site has continued to yield important dinosaur, lizard, chelonian, and mammal specimens up to the present. Among the many significant finds have been an oviraptorid nest containing a nearly complete embryo, and a dromaeosaur called *Tsaagan* that represents only the second member of this taxon (after the classic *Velociraptor*) ever discovered in the Djadokhta Formation (Norell et al. 2006).

Perhaps the greatest single surprise ever to emerge from the dinosaur record of Asia came to light when Ji and Ji (1996) published a brief description of a new dinosaur, *Sinosauropteryx prima*, from the Lower Cretaceous Yixian Formation of Liaoning Province in northeastern China. The presence of short, simple plumelike structures associated with the vertebral column of the specimen led Ji and Ji to describe it as an ancestral bird, but it soon became clear that the skeletal anatomy of *Sinosauropteryx* unambiguously identified it as a nonavian theropod. In other words, *Sinosauropteryx* was a feathered dinosaur, or at least a protofeathered one. Many similar specimens, from a variety of theropod clades, have since come to light.

Many other dinosaur discoveries, far too numerous to list exhaustively, have been made virtually throughout Asia since 1986. Field crews from the IVPP and other Chinese institutions have scoured the People's Republic in search of dinosaurs and have been rewarded with particularly spectacular finds in Xinjiang and Inner Mongolia in addition to the steady stream of feathered and featherless dinosaurs from Liaoning. A moderately extensive Early Cretaceous dinosaur fauna is now known from Fukui Prefecture in Japan, including the iguanodontoid *Fukuisaurus* and the relatively basal tetanuran theropod *Fukuiraptor* (Kobayashi et al. 2006). New hadrosaurs have emerged from the vicinity of the Amur River (Godefroit 2006), and further spectacular dinosaur footprint sites have come to light in South Korea (Matsukawa et al. 2006). The known complement of Thai dinosaurs has grown exponentially, and it is now clear that the Thai record is more stratigraphically complete than that of any other East Asian country apart from China (Buffetaut et al. 2006). Thailand now boasts a Triassic sauropod, *Isanosaurus*; basal eusauropods, theropods, and a single stegosaur vertebra from the Jurassic; and sauropods, theropods, and ornithischians from the Cretaceous. Particularly notable are the possible tyrannosaur *Siamotyrannus* and the possible spinosaur *Siamosaurus*, both known from the Early Cretaceous Sao Khua Formation. Although much of the material is fragmentary, the diversity of taxa is impressive and provides a basis for comparing Thai dinosaur faunas to their stratigraphic equivalents in China.

The Lameta Formation of India has also produced new dinosaurs in recent years. Wilson et al. (2003) described a new abelisaurid, *Rajasaurus narmadensis*, from associated cranial and postcranial material. This taxon is possibly the most complete theropod ever described from India. Another important discovery was the partial skeleton of a new titanosaur, initially described as *Titanosaurus colberti* (Jain and Bandyopadhyay 1997) but subsequently reassigned to the new genus *Isisaurus* (Wilson and Upchurch 2003). Major-General Sir William Henry Sleeman, comparing the new find to the great petrified vertebrae he spotted on Bara Simla one day in 1828, would no doubt have been intrigued and delighted.

Taking Stock: The Scientific Importance of Asian Dinosaurs

When considering the importance of Asian dinosaur discoveries, we might ask the question: "What would we not know were it not for these finds?" In other words, if no dinosaurs had ever been found in Asia, to what extent would the field of dinosaur paleontology be diminished? Over the last 180 years, vast numbers of localities have been discovered across the Middle East, Asiatic Russia, the Indian subcontinent, China, Japan, and beyond. Almost every Asian country can boast at least one dinosaur find, and a select few, such as Mongolia, have become central to dinosaur research and famous the world over for their fossils. Around 200 dinosaur body fossil genera are now known from Asia from well over 100 formations (data in Weishampel et al. 2004a), and there have also been an impressive number of ichnological discoveries. Major sites, such as the *Shantungosaurus* locality in Zhucheng, Shandong Province, China, continue to be discovered or at least newly excavated. The recent reopening of the *Shantungosaurus* locality has yielded thousands of bones of this giant hadrosaur, in addition to new taxa such as the horned dinosaur *Sinoceratops* and the tyrannosaur *Zhuchengtyrannus*. Other new dinosaurs from the site await description.

The Asian (and especially Mongolian and Chinese) dinosaur record is characterized by numerous sites, a large diversity of taxa, abundant specimens, and often a high quality of preservation. As a result, Asian dinosaurs are becoming increasingly important and have had an enormous impact in some areas of research in recent years as more material is uncovered. It is becoming increasingly evident that several clades of dinosaurs originated in Asia. Some are even known almost exclusively from the continent, so that without Asian sites our knowledge of dinosaur diversity would be substantially reduced.

Important Times and Places

A number of Asian geological units are world-famous for their dinosaur content (Table 5.1). Many Asian localities provide windows into slices of geological time that are poorly represented elsewhere. Asia is also the largest of Earth's continents and therefore represents an important piece of the global puzzle that is Mesozoic paleobiogeography. A notable example is the strong similarity between North American taxa and those of Eastern Asia in the Late Cretaceous. The two regions are thought to have been geographically well separated at this point, yet the similarity of their faunas strongly suggests that some species were much more widely distributed than has previously been assumed (Holtz et al. 2004).

The Middle Jurassic is perhaps the most poorly known interval of dinosaur history, with few localities worldwide, low faunal diversity, and low abundance of specimens. However, the Xiashaximiao (Lower Shaximiao) Formation of Zigong Prefecture, Sichuan Province, China, provides probably the single best Middle Jurassic faunal sample in the world. The Xiashaximiao has yielded early stegosaurs (e.g., *Gigantspinosaurus, Huayangosaurus*), theropods (e.g., *Xuanhanosaurus, Gasosaurus*), numerous sauropods (e.g., *Omeisaurus, Shunosaurus*), and ornithopods (e.g., *Agilisaurus*). All of these taxa are known from multiple, and generally complete and articulated, specimens (Peng et al. 2005). The Shangshaximiao (Upper Shaximiao) extends into the Late Jurassic and provides still more material.

Table 5.1. Several of the Most Important Dinosaur-Bearing Stratigraphic Units in Asia

Stratigraphic Unit	Country	Geological Age	Important Taxa
Nemegt Formation	Mongolia	Late Cretaceous	*Tarbosaurus, Gallimimus, Nemegtosaurus, Homalocephale, Saurolophus*
Djadokhta Formation	Mongolia	Late Cretaceous	*Velociraptor, Oviraptor, Pinacosaurus, Protoceratops*
Lameta Formation	India	Late Cretaceous	*Indosuchus, Compsosuchus, Titanosaurus*
Xinminbao Group	China	Early Cretaceous	*Sinornithoides, Probactrosaurus, Archaeoceratops*
Yixian Formation	China	Early Cretaceous	*Sinosauropteryx, Beipiaosaurus, Caudipteryx, Jeholosaurus, Liaoceratops*
Jiufotang Formation	China	Early Cretaceous	*Microraptor, Psittacosaurus*
Shaximiao Formation	China	Middle Jurassic	*Sinraptor, Mamenchisaurus, Toujiangosaurus, Gasosaurus*

The Middle–Upper Jurassic Shishugou Formation of Xinjiang Autonomous Region has produced further important taxa of exceptional quality, including *Monolophosaurus* and the basal tyrannosauroid *Guanlong* as well as the bizarre toothless and herbivorous ceratosaur *Limusaurus* and the basal alvarezsauroid *Haplocheirus*. While the Shishugou has so far yielded relatively few taxa in comparison to the Shaximiao, the potential of the former has yet to be fully explored. Much more material and many new discoveries are expected in years to come.

The Nemegt and Djadokhta formations of the Gobi were perhaps the first Asian rock units to gain worldwide recognition for their dinosaurs. Large numbers of theropods, including the exciting oviraptorosaurs and dromaeosaurs often found in dramatically lifelike poses (e.g., Fig. 5.10, an oviraptorosaur sitting on its nest; also the famous fighting dinosaurs), sparked the imagination of the general public. From the time of the CAE onward, new and bizarre species of dinosaur have been discovered in great numbers, and these finds have helped to demonstrate to the global public that dinosaur paleontology is not simply a matter of excavating further specimens of *Triceratops* in North America or *Iguanodon* in Europe.

The crowning glory of Asian dinosaur paleontology is perhaps the Jehol Biota of northeastern China, particularly western Liaoning Province (Chang et al. 2003). The Jehol Biota is known from numerous Lower Cretaceous Lagerstätten (fossil sites of exceptional quality) that occur within the Jiufotang Formation and the underlying Yixian Formation. Although the fossils are typically crushed flat, the quality of the bone is often excellent, and soft tissues are often preserved as well. This is the only assemblage of Mesozoic Lagerstätten with large numbers of dinosaurs and the only one from a terrestrial environment, and the dividends from its discovery have been spectacular. In addition to a multitude of birds and dinosaurs, there are fish, amphibians, pterosaurs, various non-dinosaurian reptiles, mammals, insects, plants, crustaceans, and more (Norell 2005). Furthermore, thanks to the quality of the preservation, almost every species is known from multiple, complete, and articulated specimens, as well as preserving soft tissues like skin, feathers, stomach contents, and even in some cases musculature. Gills are present in tadpoles, the membranous wings of pterosaurs are retained, dinosaurs are clothed in skin and feathers, and these tissues tend to be preserved in their original positions.

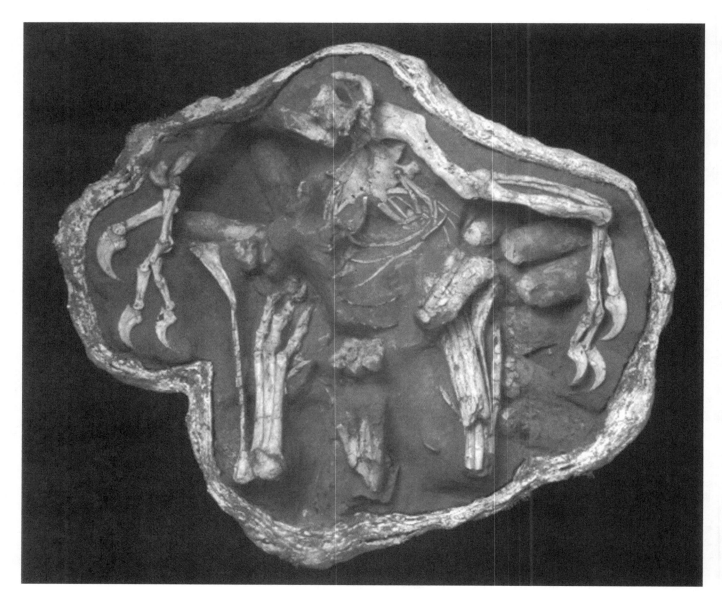

5.10. A nesting oviraptorosaur sitting on a clutch of eggs, from the Upper Cretaceous Djadokhta Formation of Ukhaa Tolgod, Mongolia.

Image courtesy of Luis Chiappe.

Thus, thanks to the Jehol, a relatively complete picture of a dinosaur-dominated ecosystem can be put together with every major component represented. This is simply quite unlike anything else known on Earth for a time so ancient (the Messel beds of the Eocene of Germany are perhaps the only reasonable comparison). Numerous new species have been discovered, and many reveal a level of detail that is almost unprecedented among dinosaur specimens from other sites. Much work still has to be completed, but already the Jehol has added at least 20 new dinosaur species to the known canon, and the pace of discovery shows little sign of slowing. An entertaining account of the early discoveries, and much of the subsequent research on the Jehol and the dinosaur–bird transition, is provided by Mark Norell in his 2005 book *Unearthing the Dragon*.

More recently, some attention has shifted to the Jehol-like Daohugou Biota, preserved within the Tiaojishan Formation and equivalent rocks of northeastern China. These strata have yielded several important new fossil vertebrates, including four feathered dinosaurs: the basal troodontid

Anchiornis (Hu et al. 2009), the scansoriopterygids *Epidendrosaurus* (Zhang et al. 2002) and *Epidexipteryx* (Zhang et al. 2008), and the enigmatic *Pedopenna*, represented only by a single feathered hind limb (Xu and Zhang 2005). The age of the beds that contain the Daohugou Biota has been controversial, but there is an emerging consensus that they are Jurassic rather than Cretaceous (Hu et al. 2009). This would imply that the Daohugou Biota significantly preceded the Jehol Biota in geological time. Furthermore, it seems likely that the Daohugou Biota is also slightly older than the Upper Jurassic German deposits that yielded the famous *Archaeopteryx*. If this conclusion is confirmed by further stratigraphic studies, *Anchiornis* and the other Daohugou dinosaurs will be established as tangible fossil evidence that some advanced feathered nonavian theropods close to the origin of birds did indeed precede *Archaeopteryx* in time, contrary to essentially spurious "temporal paradox" arguments that are sometimes advanced against the maniraptoran ancestry of birds (e.g., Feduccia et al. 2007).

Feathers

Fossil feathers are not unique to Asia, but feathered non-avian dinosaurs are. *Archaeopteryx* aside, few feathered birds are known from other regions either. While the well-preserved and numerous (both in terms of species and specimens) early birds with feathers tell us a great deal about the early evolution of birds and the origin of flight, the feathered dinosaurs are still more remarkable.

It had long been assumed that some dinosaurs probably had feathers, given the still rather dinosaurian nature of the fully feathered avian *Archaeopteryx*. However, no corroborating evidence was available until the discovery of feathered dinosaurs in China. At least five major theropod clades can be categorically demonstrated to have had feathers, protofeathers, or integumentary filaments that show the gradual acquisition and evolution of feathers in dinosaurs, from basic "dino fuzz" to fully functioning asymmetrical flight feathers. More than 10 non-avian theropods can be convincingly shown to have had feathers, including at least one example in every major clade of tetanuran theropods except for the ornithomimids. Examples include *Sinosauropteryx* among the compsognathids, *Dilong* in the tyrannosauroids, *Beipiaosaurus* among the therizinosaurs, *Shuvuuia* among the alvarezsauroids, *Caudipteryx* among the oviraptorosaurs, *Anchiornis* from the troodontids, and finally *Sinornithosaurus* from the dromaeosaurids (Xu and Norell 2006).

This diversity of feathered theropods makes it possible to trace the evolution of feathers across the phylogeny, and the fact that the feathers are often preserved in their correct positions on the body provides important additional information about the anatomical distribution and potential function of feathers in each taxon. For example, the small dromaeosaur *Microraptor gui* has flight feathers on its arms and legs and was able to glide between trees (Xu et al. 2003; Fig. 5.11). Another highly significant dinosaur specimen from the point of view of feather evolution, hailing from Asia but not from Liaoning, is a recently discovered Mongolian *Velociraptor* that exhibits papillae on the ulna corresponding to the feather-bearing quill

5.11. Holotype specimen of *Microraptor gui,* a small dromaeosaurid theropod from the Lower Cretaceous Jiufotang Formation of Liaoning Province, China. Note the elongate feathers on both the forelimbs and the hind limbs.

knobs of extant birds (Turner et al. 2007). This clear osteological correlate of feather attachment allows the presence of feathers to be inferred in the *Velociraptor* specimen, and of course its presence can also be sought in other dinosaurs that lack actual soft tissue preservation.

Finally, mention should also be made of a *Psittacosaurus* specimen, which appears to show possible filamentous integumentary structures, representing the first documented occurrence of such structures in an ornithischian dinosaur (Mayr et al. 2002). The interpretation of this specimen is controversial, but in any case it has now been superseded by the fascinating *Tianyulong*–a heterodontosaurid with similar plumelike integumentary structures on its back (Zheng et al. 2009). This perfectly illustrates the excitement still generated by Asian fossils, and the surprises that may yet lie in store.

"Rare" Families

No major dinosaur clade is known exclusively from Asia, but several are represented only poorly from other parts of the world, so that most of our knowledge of these groups is drawn from Asian material. Other Asian species are important because they are known from complete or nearly complete specimens, or because of their pivotal phylogenetic position. Most notably, maniraptoran theropods are superbly represented in the Asian fossil record and are relatively rare elsewhere, which makes Asian fossils critical to resolving questions about the origins of birds.

While the eponymous troodontid *Troodon* comes from North America and is well represented by numerous fossils, almost all other diagnosed species are from Asia. Similarly, both the therizinosaurs and alvarezsaurids are best known from Asia, with more limited material known from the Americas, and the vast majority of known oviraptorosaurs are from Asia. Not only does Asia hold the bulk of the known diversity for these clades, but it also has the greatest absolute number of specimens, including numerous complete examples of some taxa (e.g., *Caudipteryx*). Only with regard to dromaeosaurids does Asia not dominate the maniraptoran fossil record, with around half the known species found in Europe and North America. Still, the best material and most numerous specimens are Asian, and thus Asia can be considered key to our understanding of the origins and evolution of maniraptorans in general, and by extension, birds.

Among the ornithischians, basal ceratopsians are very well represented in Asia, with some nine species of *Psittacosaurus* being found in China, Mongolia, Russia, and Thailand at least. A number of more derived taxa are known from North America, but the vast majority of species and specimens are Asian, and the clade likely originated in Eastern Asia (You and Dodson 2004). Similarly, stegosaurs are well known in Asia, with roughly half the known species occurring on this continent. These include all of the most basal and earliest known stegosaurs, implying that this clade probably also originated in Asia (Maidment et al. 2008).

Among both sauropods and ankylosaurs, many basal taxa are known from Asia. Mongolia is especially rich in ankylosaurs, for instance *Shamosaurus*, whereas China and India have yielded important sauropods such as *Shunosaurus* and *Kotasaurus*, respectively. The Triassic sauropod *Isanosaurus* can also be added as a relatively recent discovery from Thailand (Buffetaut et al. 2000). While these species are not believed to represent the earliest or most primitive members of the clades in question, they do provide much information about the early evolution of these groups. Further taxa like the basal tyrannosaurs *Dilong* and *Guanlong* fill important gaps in the evolutionary histories of other clades and add greatly to our understanding.

Behavior

Paleobehavior is an incredibly ephemeral subject; almost any evidence can lend itself to numerous interpretations that must be untangled carefully. However, there are several dramatic fossils from Asia that provide convincing and relatively unambiguous indications of dinosaur behavior.

The most fascinating and frequently cited example is the famous "fighting dinosaurs," a *Velociraptor* and *Protoceratops* that were found together by the PMPE in a quite literal deadlock (Norell and Makovicky 2004). The claws of the former are buried in the torso of the latter, whose jaws are clamped around the arm of its adversary. This type of confrontation may have been a rare event rather than a regular occurrence, but it is undeniable that these particular animals were involved in genuine combat at the time of their death and rapid burial.

Another well-cited example is the vindication of *Oviraptor*, the "egg thief," named for the association of skeletons with nests of eggs, which were initially assumed to belong to *Protoceratops*. However, subsequent discoveries showed not only that the name was misapplied, but that it could not have been more inappropriate. Fossilized embryos inside the eggs showed that *Oviraptor* was the parent, not the predator, and there are now several known specimens of adult *Citipati* (a very close relative) brooding on nests

5.12. A crèche of six *Psittacosaurus* juveniles preserved together. From the Lower Cretaceous Yixian Formation of Liaoning Province, China.

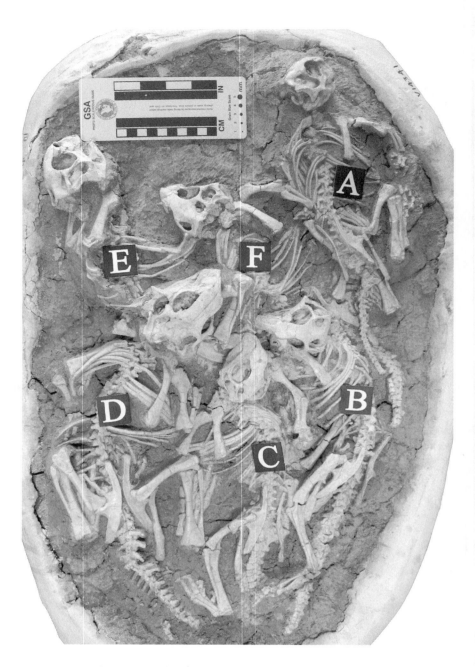

to protect and presumably incubate the eggs (Norell et al. 1995; Fig. 5.10). Oviraptorosaurs might well have eaten eggs on occasion, but presumably not from their own nests.

Nests of *Protoceratops* itself are now known, featuring hatched babies and no trace of eggshell (Weishampel et al. 2000). This suggests that at least some parental care was performed. Other nests including both eggs and hatchlings are known for the ubiquitous *Psittacosaurus*, and these specimens provide much-needed information about dinosaur social behavior. Nests with both juveniles and the presumed parent show that there was care beyond hatching provided by parents (Meng et al. 2004), and whole groups of juvenile *Psittacosaurus* individuals of varying ages show that crèches of juveniles stayed together after leaving the nest (Zhao et al. 2007; Fig. 5.12).

Development, Growth and Morphological Variation

Because some Asian dinosaurs are known from dozens or even hundreds of individuals, the Asian record can provide exceptionally valuable details about growth and development, population structure, and sexual dimorphism in many species. The most notable example is probably the incredibly abundant *Psittacosaurus*, for which hundreds of individuals are known (You and Dodson 2004).

One of the few convincingly documented cases of sexual dimorphism in dinosaurs is that of *Protoceratops* (Lambert et al. 2001), with differences between the sexes apparent in both the skull and the rest of the body. Similar morphological studies have been carried out for *Psittacosaurus*, showing the range of ornamentation and skull shape in the numerous different species. This taxon is known in such numbers and across such a broad ontogenetic range that growth rates can be calculated by bone histology (Erickson and Tumanova 2000), and every stage of life, from embryo to adult, is represented in the record! *Citipati* can also be added to the exclusive club of dinosaurs for which such extensive ontogenetic information is available

5.13. Institute of Vertebrate Paleontology and Paleoanthropology expedition members trenching around a large and recalcitrant block containing Late Cretaceous dinosaur bones. Bayan Mandahu, Inner Mongolia, China, summer 2009.

because eggs, embryos, juveniles, and adults are all known for this taxon as well (Osmólska et al. 2004). Asia therefore provides information on growth and development in both a representative saurischian and a representative ornithischian.

Some Prospects for the Future

Active dinosaur research continues in all of the countries (Fig. 5.13), and at many of the individual localities, mentioned in this review. Important new specimens continue to emerge from Liaoning and Xinjiang in China, from Mongolia, from the Sao Khua Formation of Thailand, and from many other parts of Asia. Some of the most recent discoveries have been astonishing; the feathered tyrannosaur *Dilong* from Liaoning (Xu et al. 2004), the enormous and suitably named oviraptorosaur *Gigantoraptor* from Inner Mongolia (Xu et al. 2007) and the Asain centrosaurine *Sinoceratops* (Xu et al. 2010) stand out as particular surprises. New localities in Henan Province, which was not previously thought to have been particularly rich in dinosaurs, are proving almost fantastically productive. In contrast to the classic Mesozoic formations of North America, many geologically promising parts of Asia are only now opening up to thorough paleontological prospecting. The global dinosaur community is well aware of the richness of many Asian localities and will hardly be startled by news of the next incredible find from the Nemegt Basin of Mongolia or the Lameta Formation of India. The next real surprise is likely to come from a rock stratum familiar only to regional geologists of the country in question, exposed at a site whose name Western paleontologists will initially find difficult to pronounce. We can only be confident that we will not have to wait too long.

Acknowledgments

We thank Adrienne Mayor, Ashok Sahni, Jeff Wilson, Luis Chiappe, Zhao Qi, the Institute of Vertebrate Paleontology and Paleoanthropology, and the American Museum of Natural History for providing images and/or granting permission to reproduce them. Qi Zhao was helpful in securing images from the Institute of Vertebrate Paleontology and Paleoanthropology archives. The authors are supported by the Chinese Academy of Sciences and the National Natural Science Foundation of China.

References

Andrews, R. C. 1926. *On the Trail of Ancient Man*. New York: G. P. Putnam's Sons.

Andrews, R. C., W. Granger, C. H. Pope, and N. C. Nelson. 1932. *The New Conquest of Central Asia: A Narrative of the Explorations of the Central Asiatic Expeditions in Mongolia and China, 1921–1930*. Vol. 1 of Natural History of Central Asia. New York: American Museum of Natural History.

Berkey, C. P., and N. C. Nelson. 1926. Geology and prehistoric archaeology of the Gobi Desert. *American Museum Novitates* 222: 1–16.

Benton, M. J., M. A. Shishkin, D. M. Unwin, and E. N. Kurochkin (eds.). 2000. *The Age of Dinosaurs in Russia and Mongolia*. Cambridge: Cambridge University Press.

Bohlin, B. 1953. Fossil reptiles from Mongolia and Kansu. *Sino-Swedish Expedition Publications* 37: 1–113.

Bolotsky, Y. L., and P. Godefroit. 2004. A new hadrosaurine dinosaur from the Late Cretaceous of far eastern Russia. *Journal of Vertebrate Paleontology* 24: 351–365.

Buffetaut, E., V. Suteethorn, G. Cuny, H. Tong, J. Le Loeuff, S. Khansubha, and S. Jongautchariyakul. 2000. The earliest known sauropod dinosaur. *Nature* 407: 72–74.

Buffetaut, E., V. Suteethorn, and H. Tong. 2006. Dinosaur assemblages from Thailand: a comparison with Chinese faunas. In Lu et al. 2006, 19–37.

Chang, M., P. Chen, Y. Wang, and Y. Wang (eds.). 2003. *The Jehol Biota*. Shanghai: Shanghai Scientific and Technical Publishers.

Chatterjee, S. 1978. *Indosuchus* and *Indosaurus*, Cretaceous carnosaurs from India. *Journal of Paleontology* 52: 570–580.

Clark, J. M., T. Maryańska, and R. Barsbold. 2004. Therizinosauroidea. In Weishampel et al. 2004b, 151–164.

Colbert, E. H. 2000. Asiatic dinosaur rush. In Benton et al. 2000, 211–234.

Currie, P. J. 1997a. Sino-Soviet expeditions. In Currie and Padian 1997, 1:661–662.

———. 1997b. Sino-Canadian dinosaur project. In Currie and Padian 1997, 1:661.

Currie, P. J., and K. Padian (eds.). 1997. *The Dinosaur Encyclopedia*. San Diego: Academic Press.

Currie, P. J., and X.-J. Zhao. 1993. A new carnosaur (Dinosauria, Theropoda) from the Jurassic of Xinjiang, People's Republic of China. *Canadian Journal of Earth Sciences* 30: 2037–2081.

Dong, Z. 1997. Chinese dinosaurs. In Currie and Padian 1997, 1:118–124.

Dong, Z., S. Zhou, and Y. Zhang. 1983. The dinosaurian remains from Sichuan Basin, China [in Chinese with English summary]. *Palaeontologia Sinica Series C* 23: 1–145.

Dong, Z.-M. 1993. The field activities of the Sino-Canadian Dinosaur Project in China, 1987–1990. *Canadian Journal of Earth Sciences* 30: 1997–2001.

Dong, Z.-M., and P. J. Currie. 1993. Protoceratopsian embryos from Inner Mongolia, People's Republic of China. *Canadian Journal of Earth Sciences* 30: 2248–2254.

Erickson, G. M., and T. A. Tumanova. 2000. Growth curve of *Psittacosaurus mongoliensis* Osborn (Ceratopsia: Psittacosauridae) inferred from long bone histology. *Zoological Journal of the Linnean Society* 130: 551–566.

Falconer, H. 1868. Notes on fossil remains found in the Valley of the Indus below Attock, and at Jubbulpoor. In C. Murchison (ed.), *Palaeontological memoirs and notes of the late Hugh Falconer*, vol. 1, *Fauna antiqua sivalensis*, 414–419. London: Robert Hardwicke.

Feduccia, A., L. D. Martin, and S. Tarsitano. 2007. *Archaeopteryx* 2007: quo vadis? *Auk* 124: 373–380.

Gilmore, C. W. 1933. On the dinosaurian fauna of the Iren Dabasu Formation. *Bulletin of the American Museum of Natural History* 67: 23–78.

Godefroit, P. 2006. Latest Cretaceous hadrosaurid dinosaurs from Heilongjiang Province (P.R. China) and Amur region (far eastern Russia). In Lu et al. 2006, 103–114.

Granger, W., and W. K. Gregory. 1923. *Protoceratops andrewsi*, a pre-ceratopsian dinosaur from Mongolia. *American Museum Novitates* 72: 1–9.

Gregory, W. K., and G. G. Simpson. 1926. Cretaceous mammal skulls from Mongolia. *American Museum Novitates* 225: 1–20.

Holtz, T. R., R. E. Chapman, and M. C. Lamanna. 2004. Mesozoic biogeography of Dinosauria. In Weishampel et al. 2004b, 627–642.

Hu, D., L. Hou, L. Zhang, and X. Xu. 2009. A pre-*Archaeopteryx* troodontid theropod from China with long feathers on the metatarsus. *Nature* 461: 640–643.

Huene, F. v., and C. A. Matley. 1933. Cretaceous Saurischia and Ornithischia of the central provinces of India. *Palaeontologia Indica* (n.s.) 21: 1–74.

Jain, S. L. 1989. Recent dinosaur discoveries in India, including eggshells, nests and coprolites. In D. D. Gillette and M. G. Lockley (eds.), *Dinosaur Tracks and Traces*, 99–108. Cambridge: Cambridge University Press.

Jain, S. L., and S. Bandyopadhyay. 1997. New titanosaurid (Dinosauria: Sauropoda) from the Late Cretaceous of central India. *Journal of Vertebrate Paleontology* 17: 114–136.

Ji, Q., and S. Ji. 1996. On the discovery of the earliest fossil bird in China (*Sinosauropteryx* gen. nov.) and the origin of birds [in Chinese]. *Chinese Geology* 233: 30–33.

Kielan-Jaworowska, Z., M. J. Novacek, B. A. Trofimov, and D. Dashzeveg. 2000. Mammals from the Mesozoic of Mongolia. In In Benton et al. 2000, 573–626.

Kobayashi, Y., M. Manabe, N. Ikegami, Y. Tomida, and H. Hayakawa. 2006. Dinosaurs from Japan. In Lu et al. 2006, 87–102.

Koken, E. 1885. Ueber fossile Säugethiere aus China. *Palaeontologische Abhandlungen* 3: 29–114.

Kurochkin, E. N., and R. Barsbold. 2000. The Russian–Mongolian expeditions and research in vertebrate palaeontology. In In Benton et al. 2000, 235–255.

Kutty, T. S. 1969. Some contributions to the stratigraphy of the Upper Gondwana Formations of the Pranhita-Godavari Valley, Central India. *Journal of the Geological Society of India* 10: 33–48.

Lamanna, M. C., J. B. Smith, Y. S. Attia, and P. Dodson. 2004. From dinosaurs to dyrosaurids (Crocodyliformes): removal of the post-Cenomanian (Late Cretaceous) record of Ornithischia from Africa. *Journal of Vertebrate Paleontology* 24: 764–768.

Lambert, O., P. Godefroit, H. Li, C.-Y. Shang, and Z. Dong. 2001. A new species of *Protoceratops* (Dinosauria: Neoceratopsia) from the Late Cretaceous of Inner Mongolia (P.R. China). *Bulletin de l'Institut Royal des Sciences Naturelles de Belgique, Sciences de la Terre* 71 (suppl.): 5–28.

Lee, Y.-N., and H.-J. Lee. 2006. Hasandong vertebrate fossils in South Korea. In Lu et al. 2006, 129–139.

Lim, S.-K., S.-Y. Yang, and M. G. Lockley. 1989. Large dinosaur footprint assemblages from the Cretaceous Jindong Formation of South Korea. In D. D. Gillette and M. G. Lockley (eds.), *Dinosaur Tracks and Traces*, 333–336. Cambridge: Cambridge University Press.

Lu, J. C., Y. Kobayashi, D. Huang, and Y.-N. Lee (eds.). 2006. *Papers from the 2005 Heyuan International Dinosaur Symposium*. Beijing: Geological Publishing House.

Lydekker, R. 1877. Notices of new and other Vertebrata from Indian Tertiary and Secondary rocks. *Records of the Geological Survey of India* 10: 30–43.

Maidment, S. C. R., D. B. Norman, P. M. Barrett, and P. Upchurch. 2008. Systematics and phylogeny of Stegosauria (Dinosauria: Ornithischia). *Journal of Systematic Palaeontology* 6: 367–407.

Maleev, E. A. 1954. New turtle-like reptile in Mongolia [in Russian]. *Priroda* 1954: 106–108.

Manabe, M., Y. Hasegawa, and Y. Azuma. 1989. Two new dinosaur footprints from the Early Cretaceous Tetori Group of Japan. In D. D. Gillette and M. G. Lockley (eds.), *Dinosaur Tracks and Traces*, 309–312. Cambridge: Cambridge University Press.

Martin, V., V. Suteethorn, and E. Buffetaut. 1999. Description of the type and referred material of *Phuwiangosaurus sirindhornae* Martin, Buffetaut and Suteethorn, 1994, a sauropod from the Lower Cretaceous of Thailand. *Oryctos* 2: 39–91.

Maryańska, T. 2000. Sauropods from Mongolia and the former Soviet Union. In Benton et al. 2000, 456–461.

Matsukawa, M., M. Lockley, and L. Jianjun. 2006. Cretaceous terrestrial biotas of East Asia, with special reference to dinosaur-dominated ichnofaunas: towards a synthesis. *Cretaceous Research* 27: 3–21.

Matthew, W. D. 1915. Climate and evolution. *Annals of the New York Academy of Sciences* 24: 171–416.

Mayor, A. 2000. *The First Fossil Hunters: Paleontology in Greek and Roman Times.* Princeton, N.J.: Princeton University Press.

Mayr, G., S. D. Peters, G. Plodowski, and O. Vogel. 2002. Bristle-like integumentary structures at the tail of the horned dinosaur *Psittacosaurus. Naturwissenschaften* 89: 361–365.

Meng, Q., J. Liu, D. J. Varrichio, T. Huang, and C. Gao. 2004. Parental care in an ornithischian dinosaur. *Nature* 431: 145–146.

Morgan, V. L., and S. G. Lucas. 2002. Walter Granger, 1872–1941, paleontologist. *Bulletin of the New Mexico Museum of Natural History and Science* 19: 1–58.

Nagao, T. 1936. *Nipponosaurus sachalinensis*, a new genus and species of trachodont dinosaur from Japanese Saghalien. *Journal of the Faculty of Science of Hokkaido Imperial University Series 4* 3: 185–220.

Natalín, B. A. 2006. Eduard Suess and Russian geologists. *Jahrbuch der Geologischen Bundesanstalt* 146: 217–243.

Norell, M. A. 1997. Central Asiatic expeditions. In Currie and Padian 1997, 1:100–105.

———. 2005. *Unearthing the Dragon.* New York: Pi Press.

Norell, M. A., J. M. Clark, L. M. Chiappe, and D. Dashzeveg. 1995. A nesting dinosaur. *Nature* 378: 774–776.

Norell, M. A., J. M. Clark, A. H. Turner, P. J. Makovicky, R. Barsbold, and T. Rowe. 2006. A new dromaeosaurid theropod from Ukhaa Tolgod (Ömnögov, Mongolia). *American Museum Novitates* 3545: 1–51.

Norell, M. A., and P. J. Makovicky. 2004. Dromaeosauridae. In Weishampel et al. 2004b, 196–209.

Norman, D. B., and H.-D. Sues. 2000. Ornithopods from Kazakhstan, Mongolia and Siberia. In Benton et al. 2000, 462–479.

Novas, F. E., F. L. Agnolin, and S. Bandyopadhyay. 2004. Cretaceous theropods from India: a review of specimens described by Huene and Matley (1933). *Revista del Museo Argentino de Ciencias Naturales* (n.s.) 6: 67–103.

Osborn, H. F. 1900. The geological and faunal relations of Europe and America during the Tertiary Period and the theory of the successive invasions of an African fauna. *Science* 11: 561–574.

———. 1923. Two Lower Cretaceous dinosaurs of Mongolia. *American Museum Novitates* 95: 1–10.

———. 1924a. *Psittacosaurus* and *Protiguanodon*: two Lower Cretaceous iguanodonts from Mongolia. *American Museum Novitates* 127: 1–16.

———. 1924b. Three new Theropoda, *Protoceratops* zone, Central Mongolia. *American Museum Novitates* 144: 1–12.

Osmólska, H., P. J. Currie, and R. Barsbold. 2004. Oviraptorosauria. In Weishampel et al. 2004b, 165–183.

Osmólska, H., and E. Roniewicz. 1970. Deinocheiridae, a new family of theropod dinosaurs. *Palaeontologia Polonica* 21: 5–19.

Peng, G, Y. Ye, Y. Gao, C. Shu, and S. Jiang. 2005. *Jurassic Dinosaur Faunas in Zigong* [in Chinese with English summary]. Chengdu: Sichuan People's Publishing House.

Preston, D. J. 1986. *Dinosaurs in the Attic: An Excursion into the American Museum of Natural History.* New York: St. Martin's Press.

Rauhut, O. W. M., and X. Xu. 2005. The small theropod dinosaurs *Tugulusaurus* and *Phaedrolosaurus* from the Early Cretaceous of Xinjiang, China. *Journal of Vertebrate Paleontology* 25: 107–118.

Russell, D. A., and Z.-M. Dong. 1993. A nearly complete skeleton of a new troodontid dinosaur from the Early Cretaceous of the Ordos Basin, Inner Mongolia, People's Republic of China. *Canadian Journal of Earth Sciences* 30: 2163–2173.

Russell, D. A., and Z. Zheng. 1993. A large mamenchisaurid from the Junggar Basin, Xinjiang, People's Republic of China. *Canadian Journal of Earth Sciences* 30: 2082–2095.

Ryan, M. J. 1997. Middle Asian dinosaurs. In Currie and Padian 1997, 1:442–444.

Sahni, A. 2003. Indian dinosaurs revisited. *Current Science* 85: 904–910.

Shikama, T. 1942. Footprints from Chinchou, Manchoukuo, of *Jeholosauripus*, the Eo-Mesozoic dinosaur. *Bulletin of the Central National Museum of Manchoukuo* 3: 21–31.

Simpson, G. G. 1925. A Mesozoic mammal skull from Mongolia. *American Museum Novitates* 201: 1–11.

Sleeman, W. H. 1844. *Rambles and recollections of an Indian official.* Vol. 1. London: J. Hatchard and Son.

Sun, A.-L., and M.-Z. Zhou. 1991. Professor Yang Zhong-Jian (C. C. Young)–founder of Chinese vertebrate palaeontology. In H.-Z. Wang, G.-R. Yang, and J.-Y. Yang (eds.), *Interchange of Geoscience Ideas between the East and the West: Proceedings of the XVth International Symposium of INHIGEO*, 113–116. Wuhan: China University of Geosciences Press.

Suzuki, S., D. B. Weishampel, and N. Minoura. 2004. *Nipponosaurus sachalinensis* (Dinosauria: Ornithopoda): anatomy and systematic position within Hadrosauridae. *Journal of Vertebrate Paleontology* 24: 145–164.

Taquet, P. 1998. *Dinosaur Impressions: Postcards from a Paleontologist.* Translated by K. Padian. Cambridge: Cambridge University Press.

Teilhard de Chardin, P., and C. C. Young. 1929. On some traces of vertebrate life in the Jurassic and Triassic beds of Shansi and Shensi. *Geological Society of China Bulletin* 8: 131–135.

Turner, A. H., P. J. Makovicky, and M. A. Norell. 2007. Feather quill knobs in the dinosaur *Velociraptor. Science* 317: 1721.

Upchurch, P., P. M. Barrett, and P. Dodson. 2004. Sauropoda. In Weishampel et al. 2004b, 259–322.

Weishampel, D. B., P. M. Barrett, R. A. Coria, J. Le Loeuff, X. Xu, X. Zhao, A. Sahni, E. M. P. Gomani, and C. R. Noto. 2004a. Dinosaur distribution. In Weishampel et al. 2004b, 517–606.

Weishampel, D. B., P. Dodson, and H. Osmólska (eds.). 2004b. *The Dinosauria.* 2nd ed. Berkeley: University of California Press.

Weishampel, D. B., D. E. Fastovsky, M. Watabe, R. Barsbold, and Kh. Tsogt-baatar. 2000. New embryonic and hatchling dinosaur remains from the Late Cretaceous of Mongolia. *Journal of Vertebrate Paleontology* 20 (suppl.): 78A.

Wilson, J. A., and P. Upchurch. 2003. A revision of *Titanosaurus* Lydekker (Dinosauria–Sauropoda), the first dinosaur genus with a "Gondwanan" distribution. *Journal of Systematic Palaeontology* 1: 125–160.

Wilson, J. A., P. C. Sereno, S. Srivasata, D. K. Bhatt, A. Khosla, and A. Sahni. 2003. A new abelisaurid (Dinosauria, Theropoda) from the Lameta Formation (Cretaceous, Maastrichtian) of India. *Contributions from the Museum of Paleontology, the University of Michigan* 31: 1–42.

Wiman, C. 1929. Die Kreide-dinosaurier aus Shantung. *Palaeontologia Sinica Series C* 1: 1–67.

Xu, X., and M. A. Norell. 2006. Non-avian dinosaur fossils from the Lower

Cretaceous Jehol Group of western Liaoning, China. *Geological Journal* 41: 419–437.

Xu, X., M. A. Norell, X. Kuang, X. Wang, Q. Zhao, and C. Jia. 2004. Basal tyrannosauroids from China and evidence for protofeathers in tyrannosauroids. *Nature* 431: 680–684.

Xu, X., Q. Tan, J. Wang, X. Zhao, and L. Tan. 2007. A gigantic bird-like dinosaur from the Late Cretaceous of China. *Nature* 447: 844–847.

Xu, X., and F. Zhang. 2005. A new maniraptoran dinosaur from China with long feathers on the metatarsus. *Naturwissenschaften* 92: 173–177.

Xu, X., Z. Zhou, X. Wang, X. Kuang, F. Zhang, and X. Du. 2003. Four-winged dinosaurs from China. *Nature* 421: 335–340.

Xu, X., K. Wang, X. Zhao, and D. Li. 2010. The first ceratopsid dinosaur from China and its biogeographical implications. *Chinese Science Bulletin* 55: 1631–1635.

Yabe, H., Y. Inai, and T. Shikama. 1940. Discovery of dinosaurian footprints from the Cretaceous (?) of Yangshan, Chinchou. *Proceedings of the Imperial Academy of Tokyo* 16: 560–563.

You, H., and P. Dodson. 2004. Basal ceratopsia. In Weishampel et al. 2004b, 478–493.

Young, C. C. 1927. Fossile Nagetiere aus Nord-China. *Palaeontologia Sinica Series C* 5: 1–82.

———. 1935a. On the reptilian remains of the Tzuliuching Formation (Tzekuei Series) near Chungking, Szechuan. *Bulletin of the Geological Society of China* 14: 67–72.

———. 1935b. Dinosaur remains from Mengyin, Shantung. *Bulletin of the Geological Society of China* 14: 519–533.

———. 1937. A new dinosaurian from Sinkiang. *Palaeontologia Sinica Series C* 2: 1–25.

———. 1939. On a new Sauropoda, with notes on other fragmentary reptiles from Szechuan. *Bulletin of the Geological Society of China* 19: 279–315.

———. 1941. A complete osteology of *Lufengosaurus huenei* Young (gen. et sp. nov.) from the Lufeng, Yunnan, China. *Palaeontologia Sinica Series C* 7: 1–53.

———. 1942. *Yunnanosaurus huangi* (gen. et sp. nov.), a new prosauropod from the red beds at Lufeng, Yunnan. *Bulletin of the Geological Society of China* 22: 63–104.

Zhang, F., Z. Zhou, X. Xu, and X. Wang. 2002. A juvenile coelurosaurian theropod from China indicates arboreal habits. *Naturwissenschaften* 89: 394–398.

Zhang, F., Z. Zhou, X. Xu, X. Wang, and C. Sullivan. 2008. A bizarre Jurassic maniraptoran from China with elongate ribbon-like feathers. *Nature* 455: 1105–1108.

Zhao, Q., P. M. Barrett, and D. A. Eberth. 2007. Social behaviour and mass mortality in the basal ceratopsian dinosaur *Psittacosaurus* (Early Cretaceous, People's Republic of China). *Palaeontology* 50: 1023–1029.

Zhao, X.-J., and P. J. Currie. 1993. A large crested theropod from the Jurassic of Xinjiang, People's Republic of China. *Canadian Journal of Earth Sciences* 30: 2027–2036.

Zheng, X., H. You, X. Xu, and Z. Dong. 2009. An Early Cretaceous heterodontosaurid dinosaur with filamentous integumentary structures. *Nature* 458: 333–336.

Dinosaur Hunters of the Southern Continents

Thomas R. Holtz Jr.

Most North Americans, Europeans, and Asians are understandably more familiar with the dinosaurs that inhabited these regions than they are with the forms that once lived in other parts of the world. With some rare exceptions (in particular, the Humboldt Museum in Berlin), southern dinosaurs are not exhibited in northern museums or featured in popular books written by northern authors. While many dinosaur enthusiasts in North America, for example, are familiar with forms such as *Ankylosaurus*, *Velociraptor*, and *Tyrannosaurus*, they are mostly unaware of the southern counterparts, such as *Minmi*, *Noasaurus*, and *Abelisaurus*.

However, dinosaurs of the southern continents are important to vertebrate paleontologists. On a simple level, they help to document the overall number of dinosaur species from across the globe at any given time period. Additionally, some fossils from the southern continents demonstrate unusual forms and adaptations unknown from northern dinosaurs, such as the brow horns of the South American carnivore *Carnotaurus*, the bizarre vertebrae of the Australian armored dinosaur *Minmi*, and the greatly exaggerated dorsal spines of the South American sauropod *Amargasaurus*. The recently discovered sauropod *Futalognkosaurus* is the largest dinosaur known from a relatively complete skeleton (Calvo et al. 2007), and *Puertasaurus* and *Argentinosaurus* (known from much more incomplete material) are at least as large or larger (Novas et al. 2005). It should be noted, however, that fossils of a comparably large sized specimen of the sauropod *Alamosaurus* have recently been described from New Mexico, so at least some North American dinosaurs achieved this size (Fowler and Sullivan 2011). The largest known carnivorous dinosaurs are all from the southern continents: *Spinosaurus*, *Carcharodontosaurus*, *Giganotosaurus*, and *Mapusaurus* (the first two from Africa, and second two from South America) (Therrien and Henderson 2007). Fossils from the southern continents, South America in particular, have been critical in tracing the origin and early evolution of the Dinosauria (Langer et al. 2010). Finally, as discussed in later chapters, comparison of fossils from the various continents north and south has been used to trace the history and changes in diversity of Mesozoic terrestrial ecosystems.

By the term *southern continents*, scientists mean the continents and surrounding islands of South America, Africa, Australia, Antarctica, and the Indian subcontinent. The latter may seem an unusual inclusion because it is currently part of Asia and so might appear to be part of the northern continents (North America, Europe, and the rest of Asia). However, the great triangular landmass south of the Himalayas has been associated with the truly southern lands in both geologic and human history. As is well known, all the continents of Earth formed a single landmass (Pangaea) during the beginning of the Age of Dinosaurs (see also Molnar, Chapter 39 of this volume). When Pangaea split up, it divided into two supercontinents, Laurasia and Gondwana (sometimes called Gondwanaland). Laurasia contained North America, Europe, and most of modern Asia, while Gondwana contained the rest of the landmasses. Thus the southern continents discussed in this section once formed a single southern supercontinent. Laurasia and Gondwana continued to break up during the Mesozoic and later Cenozoic eras, forming the modern continental configuration. Sometime during the Cenozoic, the landmass that is now India collided with Asia. The northern coast of India and the southern coast of Asia were crushed, crumpled, and raised up by the collision, forming the Himalayan plateau and the tallest mountains in the world.

The Gondwana lands share a similar human history. Most were colonies of the European imperial powers during the past 500 years (and some until the middle decades of the twentieth century). Vertebrate paleontology and dinosaur science historically has been predominantly a northern science, with the major research centers in the eastern part of North America and the western and central countries of Europe. During much of the past 150 years, dinosaur research in the remnants of Gondwana has been an exercise in imperial paleontology (Buffetaut 1987), with European explorers and scientists traveling to the south to obtain fossils for northern institutions (see Sues, Chapter 3 of this volume). Consequently the present section could probably be retitled "Dinosaur Hunters *in* the Southern Continents." However, since the 1960s there has been a growth in dinosaur research by scientists native to Gondwana's continents.

Many of the southern countries are poor relative to their northern counterparts. Consequently, many southern governments must necessarily spend most of their time and resources trying to stop the spread of disease, trying to provide enough food, water, and clothing for their people, and fighting in various civil and international conflicts, leaving little or no money or resources for scientific research. Illiteracy remains a great problem among southern nations. Under such conditions, dinosaur science is a luxury that many nations are unwilling to support or incapable of supporting. It is probably not coincidental that some of the best southern vertebrate paleontology comes from nations (Argentina, South Africa, Australia) that do not experience these problems.

Dinosaur Research in India

In 1871, a Mr. Medlicott found a broken bone measuring 117 cm (46 in.) in length near Jabalpur. This proved to be the femur (thighbone) of a dinosaur larger than any currently known at that time. Vertebrae of the tail of this animal were found in the same vicinity. On the basis of this material, the noted British paleontologist Richard Lydekker named the creature *Titanosaurus indicus* (1877). This fossil is notable as being the first significant dinosaur discovery of the southern continents. It comes from a family of long-necked sauropod dinosaurs (the Titanosauridae) that once dominated the Gondwanan faunas. Friedrich von Huene (see Sues, Chapter 3 of this volume) and C. A. Matley described various Indian dinosaurs during the 1920s and 1930s (Matley 1923; Huene and Matley 1933).

6.1. Sankar Chatterjee (on the ground at left) in the field.

Photograph courtesy of Texas Tech University News and Publications Photo Service.

Despite this auspicious beginning, dinosaur research never succeeded as well in India as on the other southern continents. Although dozens of papers by Indian geologists and paleontologists describe individual bones or teeth from across the subcontinent (Loyal et al. 1966; Carrano et al. 2010), relatively few nearly complete skeletons have been discovered (Wilson et al. 2003; Wilson and Upchurch 2003; Novas 2010). Most of these fossils are from the Cretaceous Period and are from titanosaur sauropods or abelisauroid theropods. Notably, a great number of dinosaur eggs are known from the central Indian Lameta Formation (Vianey-Liaud et al. 1987), presumably from the titanosaurs. Indian dinosaurs from the end of the Cretaceous are associated with the Deccan Traps, a vast complex of igneous rocks that indicates that a huge region of the subcontinent was involved in volcanic activity at the end of the Age of Dinosaurs (Pande 2002).

As an interesting side note, one paleontologist born and educated in India is now an active dinosaur researcher in the northern continents. Sankar Chatterjee (Fig. 6.1), born in Calcutta and trained at the prestigious Indian Statistical Institute, described various Mesozoic fossils of his homeland, including the primitive sauropod dinosaur *Barapasaurus* (with Jain and others). However, perhaps his greatest discoveries come from Upper Triassic sediments from central Texas. Since he became a professor of geology at Texas Tech University in 1979, Chatterjee has described many species of dinosaurs and other fossil reptiles from the Dockum Group.

Although very few specimens have been recovered so far, Pakistan has recently begun to yield dinosaur fossils of the same general clades as in India (Wilson et al. 2005; Malkani 2006).

Dinosaur Hunters in and of Africa

Much of the paleontological research in Africa has been conducted by outsiders (mostly Europeans). Some of the major discoveries and expeditions were made by European paleontologists (see Sues, Chapter 3 of this volume) and so will be touched on only briefly here. However, renewed interest in African dinosaurs by African and American paleontologists has revealed new forms and localities in this continent.

Dinosaur fossils have been found over most of Africa. Indeed, some of the first fossils from the African region were actually found on the island of Madagascar. Discovered by a doctor named Félix Salètes in the northerly region of Mevarana (sometimes spelled Maevarano), these fossils were described by the French scientist Charles Depéret in 1896. They included a new species of *Titanosaurus* and fragments of a large predatory dinosaur (*Majungasaurus*). Although various French scientists have reported new fragments over the past century, it was only in 1993 that a new paleontological expedition went to Madagascar with the express purpose of finding new Mesozoic fossils. Organized by Madagascan authorities and scientists from the State University of New York (Krause et al. 2007), these expeditions have reported the discovery of new, more complete skeletons (Sampson et al. 2007), including several species of Mesozoic birds (Forster et al. 1996). Not only have these expeditions yielded substantial insights into the paleoecology of this region at the end of the Cretaceous (Rogers et al 2007), they have also resulted in the formation of the Madagascar Ankizy Project (http://www.ankizy.org), a health and education humanitarian outreach project to benefit the children and families who live on the land that produces such wonderful fossils.

North African countries have revealed many interesting dinosaur specimens, particularly from the middle of the Cretaceous Period. German expeditions under Ernst Stromer von Reichenbach to Egypt and various French expeditions under R. Lavocat (1954), Albert-Félix de Lapparent (1960), and Philippe Taquet to western North Africa (Niger) recovered tons of fossil material. It is interesting that each of these men discovered a different dinosaur with a tall dorsal (back) fin: the predatory *Spinosaurus aegyptiacus* by Stromer, the long-necked sauropods *Rebbachisaurus garasbae* by Lavocat and *R. tamesnensis* by Lapparent, and the duckbill relative *Ouranosaurus nigeriensis* by Taquet. In addition to ongoing studies by French expeditions, North American teams have been investigating North African fossils, including Morocco (Russell 1996), Libya (Smith et al. 2010), and most especially Niger and Egypt (Rauhut and López-Arbello 2009; Remes et al. 2009; Sereno et al. 1994, 1996, 1998, 1999; Smith et al. 2001),

Few dinosaurs have been found in West Africa. However, several distinct footprint morphologies and scrappy bones and teeth of ornithopods, theropods, and sauropods are reported from the Koum Basin of Cameroon (Jacobs et al. 1996) and ongoing expeditions to Angola reveal the presence of more complete material (Mateus et al. 2011).

East African paleontology is well known for both early humans and dinosaurs. Separated by tens of millions of years but only some hundreds of kilometers are the early hominid fossils of Olduvai Gorge and Lake Turkana on the one hand, and the spectacular dinosaurs of Tendaguru Hill in Tanzania on the other. Tendaguru, as described by Sues in Chapter 3 of this volume, is the classic imperial paleontological dig, with hundreds of trained local workers under the supervision of a few German scientists. However, Tendaguru is not the only dinosaur locality in eastern Africa. In 1924 a farmer named E. C. Holt in what was then the Nyasaland Protectorate of the British Empire (now independent Malawi) discovered fossil bones. He reported these to F. Dixey, the director of the Geological Survey of the protectorate. First described in 1928 by S. H. Haughton (a South African paleontologist), this titanosaurid dinosaur is now known as *Malawisaurus dixeyi*. Long ignored by scientists, this and other fossils from the Middle Cretaceous of Malawi have received new attention as the result of a series of expeditions in the 1980s (Jacobs et al. 1990, 1993, 1996; Jacobs 1993; Gomani 2005). Under the direction of Louis L. Jacobs, paleontologist at Southern Methodist University and director of the Shuler Museum of Paleontology in Dallas, Texas, these expeditions have included other Americans such as Dale A. Winkler and William R. Downs as well as native Malawian scientists, in particular Zefe M. Kaufulu of the University of Malawi and Elizabeth M. Gomani of the Department of Antiquities of Malawi. The Malawian expeditions and the recent Madagascan studies have proven to be models for additional paleontological expeditions in the poorer countries of the world, where the resources of richer northern institutions and the expertise of both northern and local scientists can work together to uncover new information about the past.

Southern Africa is best known for its fossils older and younger than the dinosaurs. The Karoo Beds contain some of the best Permian and Triassic therapsid ("mammal-like reptiles") fossils in the world, while sedimentary rocks from the last few million years document some of the apelike ancestors and relatives of humans (*Australopithecus* and *Paranthropus*). However, southern African rocks also contain fossils from the beginning and end of the Jurassic Period. The earliest Jurassic rocks of southern Africa have provided excellent fossils of various primitive bird-hipped dinosaurs (especially *Leothosaurus* and *Heterodontosaurus*), early sauropodomorphs (*Aardonyx, Anteotronitrus, Massospondylus,* and *Vulcanodon*), and the primitive carnivores *Dracovenator* and *Megapnosaurus* (formerly "*Syntarsus*"). The latter was described and studied by Michael Raath of the Port Elizabeth Museum in Humewood, South Africa (Raath 1969, 1990). Like its North American relative *Coelophysis* (and possibly belonging to that genus), *Megapnosaurus rhodesiensis* is known from dozens of specimens. The most completely known primitive ornithischians (*Heterodontosaurus, Lesothosaurus, Stormbergia,* and *Eocursor*) are all from southern Africa. Late Jurassic fossils of southern Africa are less well known. The South African armored dinosaur *Paranthodon* and the long-necked sauropod *Algoasaurus* are preserved on in fragments, but in Zimbabwe better material of the sauropods *Brachiosaurs, Camarasaurus, Dicraeosaurus,* and *Janenschia* has been found (Raath and MacIntosh 1987). From the Early Cretaceous comes *Nqwebasaurus*, a small carnivorous dinosaur (de Klerk et al. 2000).

Of all the remains of Gondwana, South America has arguably proven the most productive in terms of important discoveries (Novas 2009). What is unusual is that active programs of dinosaur research in the South American nations are young even by southern standards, even more unusual in light of numerous paleontological discoveries in that continent over the past century and a half. Darwin discovered many unusual mammal and bird fossils during his voyage in the H. M. S. *Beagle*, which were described by his colleague (and eventual foe) Sir Richard Owen. South America, in particular Argentina, produced its own internationally respected vertebrate paleontologists during the latter part of the nineteenth century. Chief among

South American Discoveries: Keys to Dinosaur Evolution

these were the brothers Florentino and Carlos Ameghino and Francisco P. Moreno. These scientists concentrated on the strange and diverse mammals that evolved in South America during its isolation as an island continent for most of the Cenozoic. Moreno reported the presence of dinosaur fossils in 1891, but little scientific work was done with these. As discussed by Sues (Chapter 3, this volume), Friedrich von Huene worked at the Museum of La Plata, Argentina, describing many new dinosaur forms during the 1920s.

Argentine paleontologists began to turn to dinosaur science during the middle part of the twentieth century. Between 1959 and 1962, various Argentine institutions (in particular the Instituto Miguel Lillo of Tucamán) recovered fossils from the Ischigualasto Basin. Among the various Triassic fossils they discovered were the remains of a primitive meat-eating dinosaur, which Osvaldo A. Reig (1963) named *Herrerasaurus ischigualastensis*. An earlier (1936) expedition from the Museum of Comparative Anatomy of Harvard University to Brazil had uncovered a similar dinosaur, but it was not until 1970 that it was described and named *Staurikosaurus pricei* by the American Museum of Natural History's dinosaur expert Edwin H. Colbert. More recently, scientists from Argentina and the University of Chicago have discovered more complete skeletons of *Herrerasaurus* (Sereno and Novas 1992, 1993; Novas 1993; Sereno 1993) as well as the smaller *Eoraptor* (Sereno et al. 1993), and *Eodromaeus* (Martinez et al. 2011).

During the middle part of the century, expeditions to Argentina by Alfred Sherwood Romer (1894–1973) of Harvard University discovered a wealth of fossils from the dawn of the Age of Dinosaurs. Among the most unusual discoveries were the remains of the primitive relatives of dinosaurs, in particular *Lagerpeton* and *Marasuchus*. These discoveries are significant in and of themselves, but one of the greatest benefits to dinosaur research from these expeditions concerns a man rather than a fossil. A protégé of Romer during the American's southern expedition, José F. Bonaparte has become one of the most prolific paleontological discoverers and researchers of all time. With dozens of papers on the subject of Mesozoic vertebrates, Bonaparte ties for second place (with Chinese paleontologist Xu Xing) among all dinosaur specialists, living or dead, in the number of valid generic names (24) he has assigned to newly discovered dinosaurs on the basis of the compilation of Benton (2008) on data up through 2005.

Nicknamed the Master of the Mesozoic, Bonaparte has greatly expanded our knowledge of South American dinosaurs since 1970. Furthermore, he has been joined by many colleagues and students in his research, including Andrea Arcucci, Jorge Calvo, Rudolpho Coria, Martin Ezcurra, Rubén Martinez, Fernando Novas, Jamie Powell, and Leonardo Salgado, among others. In addition to the early dinosaurs (Novas 1996a, 1996b), Argentine scientists have discovered some of the best fossils from the Middle Jurassic (only poorly known in Europe, unknown in North America, and otherwise well known only from China), unusual forms from the Early Cretaceous (the horned predator *Carnotaurus* and the high-spined sauropod *Amargasaurus*), armored sauropods, and giant long-snouted dromaeosaurids (Novas et al. 2009). With his student Fernando Novas, Bonaparte (Bonaparte and Novas 1985) first recognized a group of giant flesh eaters (the Abelisauridae), to which many of the predators from other southern continents appear to belong. Within the Neuquén Basin is one of the most

6.2. R.A. (Tony) Thulborn.
Photograph courtesy of Sue Turner.

Thomas R. Holtz Jr.

complete records of Cretaceous dinosaur evolution in the world (Leanza et al. 2004).

While Argentina has produced more dinosaur fossils than the rest of South America, fossils of this group have been found in Chile (Salgado et al. 2008; Kellner et al. 2011b), Uruguay (Soto and Perea 2008), and most especially in Brazil (Kellner 1996; Kellner and Campos 2000). Known Brazilian dinosaurs include some of the oldest and most primitive (Langer et al. 2010), as well as diverse communities of Late Cretaceous species (Candeiro et al. 2006; Kellner et al. 2011a).

Dinosaurs Down Under: Discoveries in Australia and New Zealand

The Land Down Under has a low population density and lacks the extensive badlands of Mesozoic sedimentary rocks with which the Americas and Asia are blessed. Consequently, the record of dinosaurs from Australia has been less extensive than some other continents. Since the 1970s, however, Australian dinosaur paleontologists and their neighbors in New Zealand have discovered some significant material (Angolin et al. 2010).

One of the most productive dinosaur workers of modern times is Richard Anthony Thulborn (Fig. 6.2). Tony Thulborn, a vertebrate paleontologist at the Department of Zoology at the University of Queensland, has conducted research in a variety of fields of dinosaur studies. He is one of a small group of experts on dinosaur trackways, and described (with Wade and others) the Lark Quarry track site. Located in western Queensland, this locality includes many types of bipedal dinosaur tracks, allowing Thulborn to study not only the diversity of organisms there but also the variety of gaits and other aspects of locomotion that the trackmakers used (Thulborn and Wade 1984; for a recent reevaluation of these finds, see Romilio and Salisbury 2011). Additionally, Thulborn has made important discoveries in the early evolution of the bird-hipped ornithischian dinosaurs, including "fabrosaurids" of southern Africa and the primitive armored *Scelidosaurus*. Furthermore, he has written important papers on the origin of birds and on the thermal biology of dinosaurs.

Other paleontologists engaged in active dinosaur research in Australia include Ralph Molnar (for many years of the Queensland Museum, but recently retired to the American Southwest), Patricia Vickers-Rich of Monash University, Vickers-Rich's husband, Thomas Rich, of the Victoria Museum in Melbourne, and Scott Hucknull of the Queensland Museum. Of particular interest is Molnar's discovery of an unusual armored dinosaur named *Minmi paravertebrata* (Molnar 1980, 1996). This animal – known from well-preserved fossils – exhibits features from most of the major groups of armored dinosaurs as well as some unusual, unique structures.

The main focus of the dinosaurian research of Rich and Vickers-Rich lies in the discovery of an unusual fauna of dinosaurs from the mid-Cretaceous. It is not the individual dinosaurs themselves that are so unusual as they are types (small bipedal ornithischians and predatory theropods) that are known from other regions at that same time. However, the fossils from Dinosaur Cove in southern Victoria were from animals that lived within the Cretaceous Antarctic Circle (Rich et al. 1988; Rich and Rich 1989; Rich 1996; Vickers-Rich 1996). Even though dynamite was needed to get the fossils out of the hard rock, the bones found showed a diversity of organisms and a high quality of preservation, well worth the effort. These fossils, and others from the southern coast of Australia, will help scientists understand the adaptations that dinosaurs used to survive the long (but perhaps not frigid) Cretaceous polar winters. Martin (2009) has reported burrows probably made by the ornithischians, which may have allowed them to winter over.

Fossils of giant Australian dinosaurs (sauropods and theropods) had been known from fragments, but good skeletons were finally discovered and described by expeditions from the Queensland Museum and the Australian Age of Dinosaurs Museum of Natural History (Hocknull et al. 2009).

Although most of the work on dinosaurs in the Australian region has been based on that continent, new discoveries are coming to light in neighboring New Zealand. In a review by Molnar and Wiffen (1994), at least five species of dinosaur were recognized from fossils collected on the North Island. These included the fragmentary remains of sauropods, theropods, an ankylosaur, and a *Dryosaurus*-like ornithopod. Almost all of these were found by the late Joan Wiffen (1922–2009), called by some the Dragon Lady of New Zealand. These fossils hint at the potential of even more dinosaur skeletons waiting to be discovered in this island nation (Wiffen 1996; Angolin et al. 2010).

The Frozen Lands: Dinosaur Hunting in Antarctica

Antarctica is unlike any other continent on Earth. It is the only major landmass that does not have an indigenous people, and indeed lacks a native terrestrial vertebrate fauna. Much of the continent is covered beneath hundreds to thousands of meters of glacial ice.

However, these icy conditions began only relatively recently in geologic time. Throughout the Mesozoic, terrestrial animals and plants flourished in Antarctica, which was then united with South America, Africa, India, and especially Australia. Indeed, Mesozoic vertebrate fossils have been known from this southernmost land since Peter Barrett, a New Zealand geologist, discovered a fragment of an Early Triassic amphibian in 1967 (Barrett et al. 1968). It would be two decades, however, before the first Antarctic dinosaurs were found.

Fossil collecting in Antarctica is difficult, because of both the extreme cold and the lack of extensive outcrop. Dozens of previously unknown dinosaur species may be preserved as fossils in inland Antarctica, but because of the ice cap, we may never be able to reach them. Only in those few places where Mesozoic rocks break through the ice on the coastline and islands around Antarctica proper can dinosaur fossils be found.

The first Antarctic dinosaur, a Late Cretaceous armored dinosaur, was discovered in the Santa Maria Formation in 1986 by scientists of the Instituto Antartico Argentino (Gasparini et al. 1987, 1996; Olivero et al. 1991). A medium-sized bipedal herbivore was found in rocks of about the same age shortly afterward by British scientists (Milner et al. 1992). Most recently, unusual Early Jurassic species were found deeper within the Antarctic interior by American scientist William Hammer (Hammer and Hickerson 1993, 1994; Smith et al. 2007a). Among these Jurassic discoveries is a strange predatory dinosaur with a crest that is perpendicular, rather than parallel, to the long axis of the skull. This bizarre animal, named *Cryolophosaurus ellioti*, appears to be related to *Dilophosaurus* of North America and *Dracovenator* of South Africa (Smith et al. 2007b). Ongoing expeditions to these frozen wastes will continue to bring new remains to light. Should the deglaciation of the continental interior of Antarctica occur, this would have disastrous effects on human society as a result of the tremendous increase in sea level it would bring, but dinosaur paleontologists of such a world would be rewarded by exposure of fossils from formations that would otherwise be inaccessible to study.

For most of history, dinosaur research in the southern continents has been conducted by Europeans and North Americans who were after fossils to display in museums in their native lands. The past decades have seen a rise in local southern paleontologists interested in the "fearfully great reptiles." In cooperation with scientists from the north, southern paleontologists have added greatly to our understanding of the origin and history of the dinosaurs and their distributions and diversity through time, and have given us a better understanding of the features, adaptations, and habitats of these ancient creatures. In the next several decades, great new discoveries are expected by dinosaur hunters of the southern continents.

References

Angolin, F. L., M. D. Ezcurra, D. F. Pais, and S. W. Salisbury. 2010. A reappraisal of the Cretaceous non-avian dinosaur faunas from Australia and New Zealand: evidence for their Gondwanan affinities. *Journal of Systematic Palaeontology* 8 (2): 257–300.

Barrett, P. J., R. J. Braillie, and E. C. Colbert. 1968. Triassic amphibian from Antarctica. *Science* 161: 460–462.

Benton, M. J. 2008. How to find a dinosaur, and the role of synonymy in biodiversity studies. *Paleobiology* 34 (4): 516–533.

Bonaparte, J. F., and F. E. Novas. 1985 *Abelisaurus comahuensis*, n. g., n. sp., Carnosauria del Cretácico Tardio de Patagonia. *Ameghiniana* 21: 256–265.

Buffetaut, E. 1987. *A Short History of Vertebrate Palaeontology.* London: Chapman & Hall.

Carrano, M. T., J. A. Wilson, and P. M. Barrett. 2010. The history of dinosaur collecting in central India, 1828–1947. Geological Society, London, Special Publication 343: 161–173.

Calvo, J. O., J. D. Porfiri, B. J. González Riga, and A. W. A. Kellner. 2007. Anatomy of *Futalognkosaurus dukei* Calvo, Porfiri, González Riga & Kellner, 2007 (Dinosauria, Titanosauridae) from the Neuquén Group (Late Cretaceous), Patagonia, Argentina. *Arquivos do Museu Nacional, Rio de Janeiro* 65: 511–526.

Candeiro, C. R. A., F. Fanti, F. Therrien, and M. C. Lamanna. 2011. Continental fossil vertebrates from the mid-Cretaceous (Albian–Cenomanian) Alcântara Formation, Brazil, and their relationship with contemporaneous faunas from North Africa. *Journal of African Earth Sciences* 60 (3): 79–92.

Candeiro, C. R. A., A. R. Santos, T. H. Rich, T. S. Marinho, and E. C. Oliveira. 2006. Vertebrate fossils from the Adamantina Formation (Late Cretaceous), Prata paleontological district, Minas Gerais State, Brazil. *Geobios* 39 (3): 319–327.

Coria, R. A., and L. Salgado. 1995. A new giant carnivorous dinosaur from the Cretaceous of Patagonia. *Nature* 377: 224–226.

———. 1996. A basal iguanodontian (Ornithischia: Ornithopoda) from the Late Cretaceous of South America. *Journal of Vertebrate Paleontology* 16: 445–457.

De Klerk, W. J., C. A. Forster, S. D. Sampson, A. Chinsamy, and C. F. Ross. 2000. A new coelurosaurian dinosaur from the Early Cretaceous of South Africa. *Journal of Vertebrate Paleontology* 20 (2): 324–332.

Forster, C. A., L. M. Chiappe, D. W. Krause, and S. D. Sampson. 1996. The first Mesozoic avifauna from eastern Gondwana. *Journal of Vertebrate Paleontology* 16 (suppl. to 3): 34A.

Fowler, D. W., and R. M. Sullivan. 2011. The first giant titanosaurian sauropod from the Upper Cretaceous of North America. *Acta Palaeontologica Polonica* 56: 685–690.

Gasparini, Z., E. Olivero, R. Scasso, and C. Rinaldi. 1987. Un ankylosaurio (Reptilia, Ornisthischia) campaniano en el continente antartico. *Anais do X Congreso Brasileiro de Paleontologia* 1: 131–141.

Gasparini, Z., X. Pereda-Superbiola, and R. E. Molnar. 1996. New data on the ankylosaurian dinosaur from the Late Cretaceous of the Antarctic Peninsula. *Memoirs of the Queensland Museum* 39: 583–594.

Gomani, E. M. 2005. Sauropod dinosaurs from the Early Cretaceous of Malawi, Africa. *Paleontologia Electronica* 8 (1): 37 pp.

Hammer, W. R., and W. J. Hickerson. 1993. A new Jurassic dinosaur fauna from Antarctica. *Journal of Vertebrate Paleontology* 13 (suppl. to 3): 40A.

———. 1994. A crested theropod dinosaur from Antarctica. *Science* 264: 828–830.

Hocknull, S. A., M. A. White, T. R. Tischler, A. G. Cook, N. D. Calleja, T. Sloan, and D. A. Elliot. 2009. New Mid-Cretaceous (latest Albian) dinosaurs from Winton, Queensland, Australia. *PLoS ONE* 4 (7): e6190.

Huene, F. von, and C. A. Matley. 1933. The Cretaceous Saurischia and Ornithischia of the central provinces of India. *Palaeontologia Indica* (n.s.) 21 (1): 1–74.

Jacobs, L. L. 1993. *Quest for the African Dinosaurs: Ancient Roots of the Modern World*. New York: Villard Books.

Jacobs, L. L., D. A. Winkler, W. R. Downs, and E. M. Gomani. 1993. New material of an Early Cretaceous titanosaurid sauropod from Malawi. *Palaeontology* 36: 523–534.

Jacobs, L. L., D. A. Winkler, and E. M. Gomani. 1996. Cretaceous dinosaurs of Africa: examples from Cameroon and Malawi. *Memoirs of the Queensland Museum* 39: 595–610.

Jacobs, L. L., D. A. Winkler, Z. M. Kaufulu, and W. R. Downs. 1990. The dinosaur beds of northern Malawi, Africa. *National Geographic Research* 6: 196–204.

Kellner, A. W. A. 1996. Remarks on Brazilian dinosaurs. *Memoirs of the Queensland Museum* 39: 611–626.

Kellner, A. W. A., S. A. K. Azevdo, E. B. Machado, L. B. de Carvalho, and D. D. R. Henriques. 2011a. A new dinosaur (Theropoda, Spinosauridae) from the Cretaceous (Cenomanian) Alcântara Formation, Cajual Island, Brazil. *Anais da Academia Brasileira de Ciências* 83 (1): 99–108.

Kellner, A. W. A., and D. A. Campos. 2000.

Brief review of dinosaur studies and perspectives in Brazil. *Anais da Academia Brasileira de Ciências* 72 (4): 509–538.

Kellner, A. W. A., D. Rubilar-Rogers, A. Vargas, and M. Suárez. 2011b. A new titanosaur sauropod from the Atacama Desert, Chile. *Anais da Academia Brasileira de Ciências* 83 (1): 211–219.

Krause, D. W., S. D. Sampson, M. T. Carrano, and P. M. O'Connor. 2007. Overview of the history of discovery, taxonomy, phylogeny, and biogeography of *Majungasaurus crenatissimus* (Theropoda: Abelisauridae) from the Late Cretaceous of Madagascar. In S. D. Sampson and D. W. Krause (eds.), *Majungasaurus crenatissimus (Theropoda: Abelisauridae) from the Late Cretaceous of Madagascar*. Society of Vertebrate Paleontology Memoir 8: 1–20.

Langer, M. C., M. D. Ezcurra, J. S. Bittencourt, and F. E. Novas. 2010. The origin and early evolution of dinosaurs. *Biological Reviews* 85 (1): 55–110.

Leanaza, H. A., S. Apesteguía, F. E. Novas, and M. S. de la Fuente. 2004. Cretaceous terrestrial beds from the Neuquén Basin (Argentina) and their tetrapod assemblages. *Cretaceous Research* 25: 61–87.

Loyal, R. S., A. Khosla, and A. Sahni. 1996. Gondwanan dinosaurs of India: affinities and palaeobiogeography. *Memoirs of the Queensland Museum* 39: 627–638.

Malkani, M. S. 2006. Biodiversity of saurischian dinosaurs from the Latest Cretaceous of Pakistan. *Journal of Applied and Emerging Sciences* 1 (3): 108–140.

Martin, A. J. 2009. Dinosaur burrows in the Otway Group (Albian) of Victoria, Australia, and their relation to Cretaceous polar environments. *Cretaceous Research* 30 (5): 1223–1237.

Martinez, R. N., P. C. Sereno, O. A. Alcober, C. E. Colombi, P. R. Renne, I. P. Montañez, and B. S. Currie. 2011. A basal dinosaur from the dawn of the dinosaur era in southwestern Pangaea. *Science* 331: 206–210.

Mateus, O., L. L. Jacobs, A. S. Schulp, M. J. Polcyn, T. S. Tavares, A. B. Neto, M. L. Morais, and M. T. Antunes. 2011. *Angolatitan adamastor*, a new sauropod dinosaur and the first record from Angola. *Anais da Academia Brasileira de Ciências* 83 (1): 221–233.

Matley, C. A. 1923. The Cretaceous dinosaurs of the Trichinopoly district and the rocks associated with them. *Record of the Geological Society of India* 61: 337–349.

Milner, A. C., J. J. Hooker, and S. E. K. Sequeira. 1992. An ornithopod dinosaur

from the Upper Cretaceous of the Antarctic Peninsula. *Journal of Vertebrate Paleontology* 12 (suppl. to 3): 44A.

Molnar, R. E. 1980. An ankylosaur (Ornithischia: Reptilia) from the Lower Cretaceous of southern Queensland. *Memoirs of the Queensland Museum* 20: 77–87.

———. 1996. Preliminary report on a new ankylosaur from the Early Cretaceous of Queensland, Australia. *Memoirs of the Queensland Museum* 39: 653–668.

———. 2011 (this volume). Principles of biogeography. In M. K. Brett-Surman, Thomas R. Holtz Jr., and James O. Farlow (eds.), *The Complete Dinosaur*, 2nd ed., 924–957. Bloomington: Indiana University Press.

Molnar, R. E., and J. Wiffen. 1994. A Late Cretaceous polar dinosaur fauna from New Zealand. *Cretaceous Research* 15: 689–706.

Novas, F. E. 1993. New information on the systematic and postcranial skeleton of *Herrerasaurus ischigualastensis* (Theropoda: Herrerasauridae) from the Ischigualasto Formation (Upper Triassic) of Argentina. *Journal of Vertebrate Paleontology* 13: 425–450.

———. 1996a. Alvarezsauridae, Cretaceous basal birds from Patagonia and Mongolia. *Memoirs of the Queensland Museum* 39: 675–702.

———. 1996b. Dinosaur monophyly. *Journal of Vertebrate Paleontology* 16: 723–741.

———. 2009. *The Age of Dinosaurs in South America*. Bloomington: Indiana University Press.

Novas, F. E., S. Chatterjee, D. K. Rudra, and P. M. Datta. 2010. *Rahiolisaurus gujaratensis*, n. gen. n. sp., a new abelisaurid from the Late Cretaceous of India. In S. Bandyopadhyay (ed.), *New Aspects of Mesozoic Biodiversity*, 45–62. Heidelberg: Springer.

Novas, F. E., D. Pol, J. I. Canale, J. D. Porfiri, and J. O. Calvo. 2009. A bizarre Cretaceous theropod dinosaur from Patagonia and the evolution of Gondwanan dromaeosaurids. *Proceedings of the Royal Society B: Biological Sciences* 276: 1101–1107.

Novas, F. E., L. Salgado, J. Calvo, and F. Angolin. 2005. Giant titanosaur (Dinosauria, Sauropoda) from the late Cretaceous of Patagonia. *Revista del Museuo Argentino de Ciencias Naturales Bernardino Rivadavia* 71 (1): 37–41.

Olivero, E. B., Z. Gasparini, C. A. Rinaldi, and R. Scasso. 1991. First record of dinosaurs in Antarctica (Upper Cretaceous, James Ross Island): paleogeographical implications. In M. R. A. Thomson, J. A. Crane, and J. W. Thompson (eds.), *Geological Evolution of Antarctica: Proceedings of the Fifth International Symposium on Antarctic Earth Sciences*, 617–622. Cambridge: British Antarctic Survey.

Pande, K. 2002. Age and duration of the Deccan Traps, India: a review of the radiometric and paleomagnetic constraints. *Proceedings of the Indian Academy of Science (Earth and Planetary Sciences)* 111 (2): 115–123.

Raath, M. A. 1969. A new coelurosaurian dinosaur from the Forest Sandstone of Rhodesia. *Arnoldia* 4: 1–25.

———. 1990. Morphological variation in small theropods and its meaning in systematics: evidence from *Syntarsus rhodesiensis*. In P. Currie and K. Carpenter (eds.), *Dinosaur Systematics: Approaches and Perspectives*, 91–105. Cambridge: Cambridge University Press.

Raath, M. A., and J. S. McIntosh. 1987. Sauropod dinosaurs from the Central Zambezi Valley, Zimbabwe, and the age of the Kadzi Formation. *South African Journal of Geology* 90: 107–119.

Rauhut, O. W. M., and A. López-Arbarello. 2009. Considerations on the age of the Tiouaren Formation (Iullemmeden Basin, Niger, Africa): implications for Gondwanan Mesozoic terrestrial vertebrate faunas. *Palaeogeography, Palaeoclimatology, Palaeoecology* 271 (3–4): 259–267.

Remes, K., F. Ortega, I. Fierro, U. Joger, R. Kosma, J. M. M. Ferrer, O. A. Ide, and A. Maga. 2009. A new basal sauropod from the Middle Jurassic of Niger and the early evolution of Sauropoda. *PLoS ONE* 4 (9): e6924.

Rich, P. V., T. H. Rich, B. E. Wagstaff, J. McEwen Mason, C. B. Douthitt, R. T. Gregory, and E. A. Felton. 1988. Evidence for low temperatures and biologic diversity in Cretaceous high latitudes of Australia. *Science* 242: 1403–1406.

Rich, T. 1996. Significance of polar dinosaurs in Gondwana. *Memoirs of the Queensland Museum* 39: 711–717.

Rich, T. H., and P. V. Rich. 1989. Polar dinosaurs and biotas of the Early Cretaceous of southeastern Australia. *National Geographic Research* 5: 15–53.

Rogers, R. R., D. W. Krause, K. Curry Rogers, A. H. Rasoamiaramanana, and L. Rahantarisoa. 2007. Paleoenvironment and paleoecology of *Majungasaurus crenatissimus* (Theropoda: Abelisauridae) from the Late Cretaceous of Madagascar. In S. D. Sampson and D. W. Krause (eds.), *Majungasaurus crenatissimus* (Theropoda: Abelisauridae) from the Late Cretaceous of Madagascar. Society of Vertebrate Paleontology Memoir 8: 21–31.

Romilio, A., and S. W. Salisbury. 2011. A reassessment of large theropod dinosaur tracks from the mid-Cretaceous (late Albian–Cenomanian) Winton Formation of Lark Quarry, central–western Queensland, Australia: a case of mistaken identity. *Cretaceous Research* 32 (2): 135–142.

Russell, D. A. 1996. Isolated dinosaur bones from the Middle Cretaceous of the Tafilalt, Morocco. *Bulletin du Muséum National d'Histoire Naturelle, Paris* (ser. 4) 18: 349–402.

Salgado, L., R. de la Cruz, M. Suárez, M. Fernández, Z. Gasparini, S. Palma-Heldt, and M. Fanning. 2008. First Late Jurassic dinosaur bones from Chile. *Journal of Vertebrate Paleontology* 28 (2): 529–534.

Sampson, S. D., and D. W. Krause (eds.) 2007. *Majungasaurus crenatissimus* (Theropoda: Abelisauridae) from the Late Cretaceous of Madagascar. Society of Vertebrate Paleontology Memoir 8.

Sereno, P. C. 1993. The pectoral girdle and forelimb of the basal theropod *Herrerasaurus ischigualastensis*. *Journal of Vertebrate Paleontology* 13: 425–450.

Sereno, P. C., A. L. Beck, D. B. Dutheil, B. Gado, H. C. E. Larsson, G. H. Lyon, J. D. Marcot, O. W. M. Rauhut, R. W. Sadlier, C. A. Sidor, D. J. Varricchio, G. P. Wilson, and J. A. Wilson. 1998. A long-snouted predatory dinosaur from Africa and the evolution of spinosaurids. *Science* 282: 1298–1302.

Sereno, P. C., A. L. Beck, D. B. Dutheil, H. C. E. Larsson, G. H. Lyon, B. Moussa, R. W. Sadlier, C. A. Sidor, D. J. Varricchio, G. P. Wilson, and J. A. Wilson, 1999. Cretaceous sauropods from the Sahara and the uneven rate of skeletal evolution among dinosaurs. *Science* 286: 1342–1347.

Sereno, P. C., D. B. Dutheil, M. Iarochene, H. C. E. Larsson, G. H. Lyon, P. M. Magwene, C. A. Sidor, D. J. Varricchio, and J. A. Wilson. 1996. Predatory dinosaurs from the Sahara and Late Cretaceous faunal differentiation. *Science* 272: 986–991.

Sereno, P. C., C. A. Forster, R. R. Rogers, and A. M. Monetta. 1993. Primitive dinosaur skeleton from Argentina and the early evolution of Dinosauria. *Nature* 361: 64–66.

Sereno, P. C., and F. E. Novas. 1992. The complete skull and skeleton of an early dinosaur. *Science* 258: 1137–1140.

———. 1993. The skull and neck of the basal theropod *Herrerasaurus ischigualastensis*. *Journal of Vertebrate Paleontology* 13: 451–476.

Sereno, P. C., J. A. Wilson, H. C. E. Larrson, D. B. Dutheil, and H.-D. Sues. 1994. Early Cretaceous dinosaurs from the Sahara. *Science* 266: 267–271.

Smith, J. B., M. C. Lamanna, A. S. Askar, K. A. Bergig, S. O. Tshakreen, M. M. Abugares, and D. Tab Rasmussen. 2010. A large abelisauroid theropod dinosaur from the Early Cretaceous of Libya. *Journal of Paleontology* 84 (5): 927–934.

Smith, J. B., M. C. Lamanna, K. J. Lacovara, P. Dodson, J. R. Smith, J. C. Poole, R. Giegengack, and Y. Attia. 2001. A giant sauropod dinosaur from an Upper Cretaceous mangrove deposit in Egypt. *Science* 292: 1704–1706.

Smith, N. D., P. J. Makovicky, W. R. Hammer, and P. J. Currie. 2007b. Osteology of *Cryolophosaurus ellioti* (Dinosauria: Theropoda) from the Early Jurassic of Antarctica and implications for early theropod evolution. *Zoological Journal of the Linnean Society* 151: 377–421.

Smith, N. D., P. J. Makovicky, D. Pol, W. R. Hammer, and P. J. Currie. 2007a. The dinosaurs of the Early Jurassic Hanson Formation of the Central Transantarctic Mountains: phylogenetic review and synthesis. In A. K. Cooper and C. R. Raymond (eds.), *Antarctica: A Keystone in a Changing World*. USGS Open-File Report 2007-1047, Short Research Paper 003.

Soto, M., and D. Perea. 2008. A ceratosaurids (Dinosauria, Theropoda) from the Late Jurassic–Early Cretaceous of Uruguay. *Journal of Vertebrate Paleontology* 28 (2): 439–444.

Sues, H.-D. 2011 (this volume). European dinosaur hunters of the nineteenth and twentieth centuries. In M. K. Brett-Surman, Thomas R. Holtz Jr., and James O. Farlow (eds.), *The Complete Dinosaur*, 2nd ed., 45–59. Bloomington: Indiana University Press.

Therrien, F., and D. M. Henderson. 2007. My theropod is bigger than yours . . . or not: estimating body size from skull length in theropods. *Journal of Vertebrate Paleontology* 27: 108–115.

Thulborn, R. A., and M. Wade. 1984. Dinosaur trackways in the Winton Formation (mid-Cretaceous) of Queensland. *Memoirs of the Queensland Museum* 21: 213–517.

Vianey-Liaud, M., S. L. Jain, and A. Sahni. 1987. Dinosaur eggshells (Saurischia) from the Late Cretaceous Intertrappen and Lameta formations (Deccan, India). *Journal of Vertebrate Paleontology* 7: 408–424.

Vickers-Rich, P. 1996. Early Cretaceous polar tetrapods from the Great Southern Rift Valley, southeastern Australia. *Memoirs of the Queensland Museum* 39: 719–723.

Wiffen, J. 1996. Dinosaurian paleobiology: a New Zealand perspective. *Memoirs of the Queensland Museum* 39: 725–731.

Wilson, J. A., and P. Upchurch. 2003. A revision of *Titanosaurus* Lydekker (Dinosauria: Sauropoda), the first dinosaur genus with a "Gondwanan" distribution. *Journal of Systematic Palaeontology* 1 (3): 125–160.

Wilson, J. A., P. C. Sereno, S. Srivastava, D. K. Bhatt, A. Khosla, and A. Sahni. 2003. A new abelisaurid (Dinosauria, Theropoda) from the Lameta Formation (Cretaceous, Maastrichtian) of India. *Contributions from the Museum of Paleontology, The University of Michigan* 31 (1): 1–42.

Wilson, J. A., M. S. Malkani, and P. D. Gingerich. 2005. A sauropod braincase from the Pab Formation (Upper Cretaceous, Maastrichtian) of Balochistan, Pakistan. *Gondwana Geological Magazine* 8: 101–109.

2

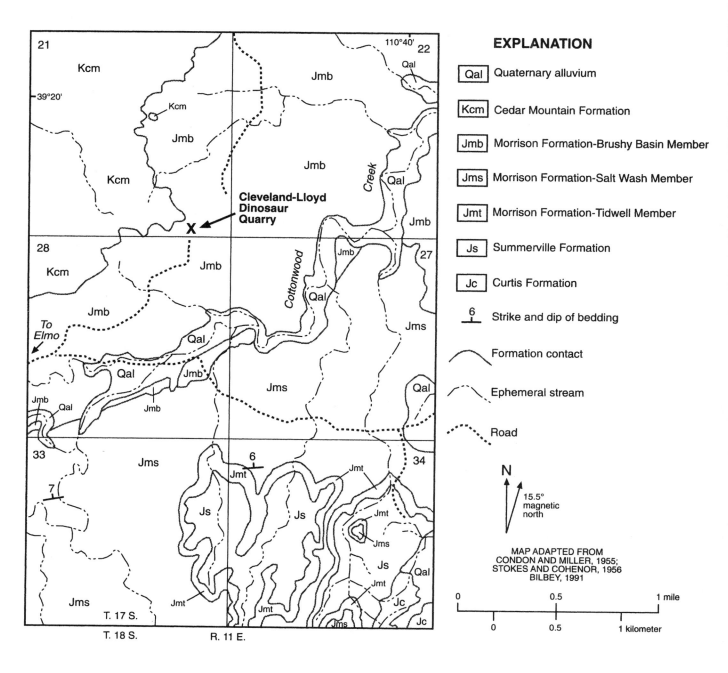

EXPLANATION

Qal	Quaternary alluvium
Kcm	Cedar Mountain Formation
Jmb	Morrison Formation-Brushy Basin Member
Jms	Morrison Formation-Salt Wash Member
Jmt	Morrison Formation-Tidwell Member
Js	Summerville Formation
Jc	Curtis Formation

6 ⊢ Strike and dip of bedding

⌢ Formation contact

Ephemeral stream

Road

N
15.5°
magnetic
north

MAP ADAPTED FROM
CONDON AND MILLER, 1955;
STOKES AND COHENOR, 1956
BILBEY, 1991

0   0.5   1 mile

0   0.5   1 kilometer

Hunting for Dinosaur Bones

David D. Gillette

7

The successful hunt for dinosaur bones requires knowledge of local geology and stratigraphy, time to conduct the search, an ability to distinguish rocks from fossils, and a little bit of luck. There are two cardinal rules of field paleontology: first, the people engaged in the work of prospecting, mapping, and excavation must have permission from the landowner or land manager; and second, those involved in field activities must respect the landscape. These practices apply equally to public lands (property owned by the federal, state, provincial, or local government), tribal lands, and private property.

Vertebrate fossils on federal land in the United States are protected by the Paleontological Resources Preservation Act of 2009. States or local governments may also have laws and regulations that protect fossils on lands they administer. Some nations have restrictive laws that prohibit export of fossils and require special permits for excavation. Paleontologists must always abide by laws that apply to the property where they are working.

For effective results, field paleontologists rely on geologic maps (Fig. 7.1) that plot the areal distribution of geologic formations (formally defined mappable units). The maps can be used to narrow the search area so that time is not wasted exploring areas where fossil bones should not be expected. Volcanic ash rarely contains fossils, while sedimentary rocks can be expected to have fossils. Searching for dinosaur bones in areas where there are igneous rocks is futile. On the other hand, not all sedimentary rocks have dinosaur bones. In general, dinosaur fossils occur where the sediments were deposited on land by lakes or streams rather than in the sea. Moreover, they are found only in rocks ranging in age from the Late Triassic to the Late Cretaceous.

For example, the Morrison Formation in the American West is an Upper Jurassic sedimentary unit consisting of mudstones, siltstones, and sandstones that can be distinguished from underlying and overlying formations. To a paleontologist who wants to search for Upper Jurassic dinosaurs, geologic maps that show the locations of Morrison Formation exposures are indispensable. Recognizing the target formation in the field requires a basic understanding of the local stratigraphy, or succession of formations. With experience, the field paleontologist can read the rock record to interpret environments of deposition and further narrow the search area to those places where fossil bone may be found.

Once the paleontologist recognizes the formation in the field, the actual search begins. A good approach is to search where other fossils have

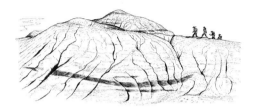

Where and How Do You Look?

7.1. Simplified geologic map in the vicinity of the Cleveland-Lloyd Dinosaur Quarry, eastern Utah. More than 10,000 bones of the Late Jurassic carnivore *Allosaurus* have been excavated from this site in the Brushy Basin Member of the Morrison Formation (Jmb). The legend is arranged in stratigraphic order, with the oldest formation (Curtis Formation of Jurassic Age, Jc) in the map area at the bottom and the youngest (Cedar Mountain Formation of Cretaceous age, Kcm, and alluvium of Quaternary age, Qal, respectively) at the top. The strike and dip symbol indicates the regional tilt of these formations as 6 degrees downward to the north.

been recovered. The abundance of fossils is usually patchy: in some places dinosaur bones are numerous, and in others the rocks may be barren. Paleontologists often visit old sites in the quest for more bones from a particular individual, or in hopes of finding bones of new dinosaurs exposed by erosion subsequent to the last excavation.

In some formations, any dinosaur bone is important and should be excavated and placed in a museum with proper field records. In others, dinosaur bones may be so abundant that discretion is called for. Priority may be established for rare or unusual species, for certain bones that are diagnostic, or for articulated skeletons. In all cases, bones discovered in the course of a search should be left undisturbed, including fragments of bone that have eroded out of the rock and been carried down slope by gravity. The locations of these fragments provide clues to the position of bones hidden the rocks. There is nothing to be gained by piling the fragments into small mounds, and much to be lost by disturbing these important clues.

A careful search around exposed bones may reveal other bones. Occasionally, dinosaur skeletons are articulated, and the exposure of one or a few bones may indicate a partial or complete skeleton. If the bones seem important, they should be photographed and documented. Photographs ideally should be taken from several angles and distances to show the bones, the setting of the site, and the overall landscape. This information helps the scientist locate the bones and make decisions concerning excavation and further study. If made by an amateur or a scientist whose specialization is not dinosaurs, the discovery should be brought to the attention of a dinosaur paleontologist. Field notes, even if crude, are indispensable for future decision making; these notes should include a sketch of the bones, a sketch of the landscape with landmarks for reference, the date, notes on how to get there, the names of the discoverers, and any other information that might be useful.

Whether conducted by amateurs or professionals, the documentation process after discovery is critical for further evaluation. Identification of bones to the species level is often difficult or impossible before excavation, but a dinosaur specialist can usually make an educated guess. Researchers seldom recognize immediately that a discovery belongs to a new species, but such a possibility should always be considered in the preliminary evaluation.

Decisions to excavate or not to excavate rest on an informal set of criteria that may change over the course of years or decades. What is recognized as an unimportant discovery by one paleontologist may prove to be critical to others. For this reason, indiscriminate collecting of dinosaur bones for personal collections and commercial gain should be discouraged.

Selecting Which Bones to Excavate

Evaluation of a discovery requires deliberation. The commitment to excavate dinosaur bones requires not only the time and expense of the field activity, but also commitment to laboratory preparation (generally five to ten times more expensive and time-consuming than the fieldwork), study, and storage in a museum. Because dinosaur bones can be large and extremely heavy, storage is expensive. For large bones especially, the space requirements are staggering.

Provided that budget and time are available, the paleontologist responsible for the fieldwork must decide whether to excavate according to the following criteria: (1) Are the bones articulated? (2) Are there unusual or important faunal and floral associations that may have scientific value? (3) Do the bones belong to a baby, juvenile, or subadult? (4) Are cranial bones likely to be found? (5) Are there paleoenvironmental or stratigraphic relations that make the dinosaur important? (6) Are there unusual features or preservation that might offer important new information? (7) Is there a real possibility that the newly discovered bones represent a new genus or species? (8) Are there any indications of cause of death, circumstances of burial, or other information that relates to the paleobiology of the dinosaurs? (9) Is there a possibility that the skeleton may serve as a display specimen in a museum exhibit?

Other questions may apply too, according to the research interests of the paleontologist and the needs of the landowner or land manager. Positive answers to any of these questions may be sufficient to justify initiating a test excavation. Conversely, if these or similar questions cannot be answered positively, the bones should not be excavated.

Before launching of a full-scale excavation that might require months or even years of fieldwork and laboratory preparation, a test excavation may be appropriate. Because dinosaur bones are usually encased in hard rock and may require considerable disturbance of the vicinity to remove, the landowner or land manager must participate in the planning of such a test excavation. The goal should be to determine the real or potential extent of bones, the depth of bones beneath the surface, and better answers to the questions posed above. Even a relatively minor disturbance should be evaluated for its potential damage to the landscape: Are rare or endangered species of plants or animals likely to be affected? Will the activities alter drainage or require modification of roads or trails? Is it safe? Will there be an effect on livestock or local wildlife? Are there any indications of archaeological materials that might be disturbed by excavation? In areas where these questions are likely to be important, especially in western North America, specialists may assist in the evaluation. If conducted on federal lands, such an evaluation may be formally prepared as an environmental impact statement as required by law. Although delays and adjustments may be frustrating, this deliberate planning is important for the wise management of an area where other concerns are equally important.

If deemed important after a preliminary test excavation, a full-scale field operation is warranted.

Mapping the Excavation

Dinosaurs are no longer trophies. Instead they are scientific specimens whose context is as important as the bones themselves. As bones are exposed, their occurrence must be mapped with care (Fig. 7.2). A grid system laid out on a regular pattern with stakes and strings is the most commonly adopted procedure. Grid intervals are typically a meter in length. For mapping in greater detail, some paleontologists use a portable meter-square frame divided by strings into decimeters (10-cm intervals). The frame can be laid over a meter square in the mapping grid to assist in the drawing of the

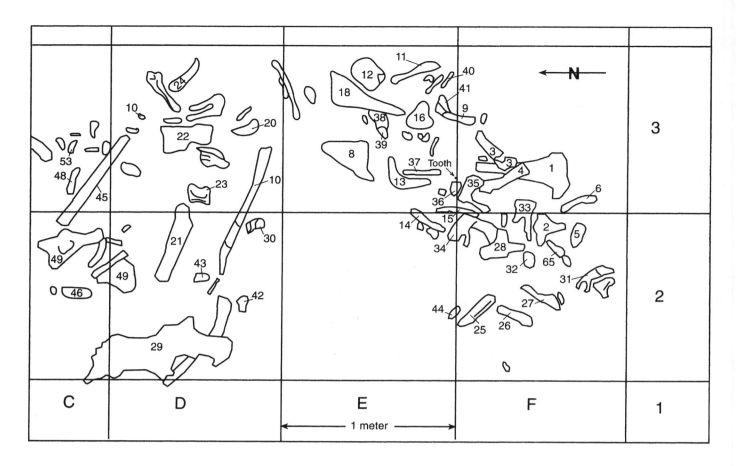

7.2. Portion of a dinosaur quarry map, from a site near Green River, Utah, in the Upper Jurassic Morrison Formation. Most of the bones are from the theropod dinosaur *Allosaurus*. Numbers correspond to specimen records with notes for reference during laboratory preparation and subsequent research. For example, bone number 29, an ilium of *Allosaurus,* is mainly in the western half of quadrant D-2.

map. This map becomes a permanent field record that must be deposited with the bones at the museum that receives the fossils. The hand-drawn map, like all other documents produced in the fieldwork, becomes archival information that gives the bones scientific context. On all maps, north–east–south–west orientations are standard and should be laid out with an accurate compass.

Careful photographic documentation often enhances the mapping but should not be used alone. Photographs should include a scale and, where possible, an exact location in the grid. All photographs should be recorded in a field notebook with the location, date, and subject as a cross-reference for labeling them later. Like the field maps, the photographic record is an essential part of the archival information concerning an excavation and should be deposited with the collection. Many paleontologists require that all participants deposit copies of their digital photos with their field notes at the conclusion of the field activity.

Paleontologists increasingly use technologically advanced survey equipment that records exact position on the earth's surface and exact elevation measured from orbiting satellites with remarkable accuracy, often with a confidence interval of a meter or less. Field equipment that uses geographical positioning system (GPS) technology should be used for all exploration and excavation records. The data generated from GPS records can be applied to geographical information system technology that produces maps and relational data bases for analysis and archival records. Computer-assisted drafting (CAD) applications may be used to store data that can be manipulated, edited, and altered for various forms of presentation.

Maps and photographs are a critical part of every excavation. In many respects they are as important as the fossils themselves, for these records are the only permanent documentation of the orientation of the skeleton, the nature of the sediments in which they occur, and their exact location for future reference. These visually oriented field records must also be supplemented by written accounts, best produced daily. Some paleontologists require all personnel on site to keep a field journal and insist that a copy of their journal be deposited with the specimens at the conclusion of the fieldwork.

The field notes should accurately records all aspects of the fieldwork, including date, personnel, operations conducted, discoveries and preliminary evaluations, sketches of field observations that can be keyed to the maps, and notes on stratigraphy and geology. The journal may also include a wide variety of other subject matter, such as weather, local flora and fauna, and memorable events of the day. Some paleontologists formalize this aspect of fieldwork by requiring completion of specially prepared forms, while others prefer an unstructured, open format for their journals. These records are important as a kind of diary for each individual, but they are invaluable documents for museum archives.

Excavation of fossil bones is tricky. Although television documentaries may give the false impression that techniques are easy and the results are immediate, in actual practice, the work of excavation requires considerable planning, time, and labor. While techniques vary from site to site and from one paleontologist to another, certain aspects are universal. These are described below to provide an overview of the procedures and materials, but the techniques cannot be learned without in-the-field training by an experienced paleontologist.

Moreover, safety is an important aspect of an excavation. Where bones are particularly large, sometimes weighing hundreds of kilograms and occasionally a ton or more, or where diggings are deep, there may be considerable danger. Blocks of rock and bones that weigh from several to as much as 10 tons are occasionally removed from quarry sites; such operations are extremely hazardous and should be attempted only by seasoned experts. Similarly, sites themselves can be hazardous, particularly if they are situated in steep terrain or near unstable topography where falling debris or landslides are likely. There are no good excuses for taking shortcuts that sacrifice the safety of the participants. Occasionally, excavations draw a crowd of spectators; their safety, too, is important and cannot be ignored.

Tools and equipment should be chosen to suit the job. Sometimes bulldozers and jackhammers are required for removal of layers of rock above the skeleton. Such extreme procedures should be adopted only with the full consent of the landowner or land manager. The principle of minimum disturbance should be adopted, especially when using heavy equipment: Move only enough rock and disturb only enough of the landscape to accomplish the goals of the excavations. More often, small power tools driven by a generator and hand tools are appropriate. For working close to bone, hand tools as small as dental picks may be required.

Initial exposure of bones ideally uncovers only their upper surfaces. The act of lifting or chiseling the rock away may have a profound impact on the bone. Fossil dinosaur bones have been confined by rocks for at least 65 million years, in total darkness, and protected from atmospheric conditions. At the time of exposure, the confining rocks are partially removed, and the bones may expand and crack. They may also undergo disturbingly rapid chemical alteration when exposed to air, especially desiccation, which causes the bones to shrink, further opening any natural cracks in them. Newly exposed bones many also change color, usually as the result of oxidation of minerals contained in the pores of the bones (reduced iron compounds altering to oxidized iron). These changes in the bones can be controlled by application of chemical hardeners that penetrate the open spaces and cracks, slowing or halting the undesirable effect.

Repairs should be made immediately, if practicable, but may have to wait for the stability and security of the laboratory. Chemical hardeners, adhesives, and glues vary in their quality and suitability. The guidance of technical personnel acquainted with excavation materials is essential. Use of an inappropriate glue or hardener may irreparably damage the bones. No single glue or adhesive is universally satisfactory because their suitability varies with weather conditions, rock type, and moisture in the bones and surrounding rock.

After the upper surfaces of the bones are exposed, stabilized, and mapped, a plan for excavation must be developed, much like the children's game of pick-up sticks. If bones are close to each other, or even one atop another, the order of their removal becomes a critical concern. Sometimes several bones can be removed together, or individual bones can be isolated and removed separately. Occasionally when a skeleton is articulated, the problem is reversed: for ease of excavation and for safety, the bones must be separated along natural cracks.

With a well-reasoned plan prepared in advance, the next step is to isolate each bone or group of bones by digging vertical trenches around their perimeter. The trenches produce a pedestal capped by the bones and supported by underlying rock. This activity must be conducted with care, because collapse of the rocks and the bones can destroy hours of work and require days of repair work in the laboratory. For best results, the bones should not be exposed on their sides or undersurface but instead should be left with adhering rock to keep natural cracks in them from expanding, causing the bones to fall apart. Any new bone surface exposed during trenching may be stabilized with chemical hardeners, if necessary. Each bone must be numbered with permanent ink, and the numbers recorded with the map records and field notes.

At some point in the trenching activity, the bone must be further stabilized by application of a jacket that makes the block rigid. This is usually burlap soaked in plaster, although fiberglass and other materials are sometimes used instead. For economy and versatility, however, the burlap-and-plaster techniques are by far the most frequently used. The plaster must not be set directly on the bone. Instead, the bones should first be covered with successive layers of damp tissue, paper towels, or strips of newspaper that separate the plaster from the bone and provide a tight cushion for transport.

7.3. The pedestal of rock that these paleontologists are removing from beneath the block is replaced by props to hold the block in place. As the undercutting proceeds, bandages of plaster and burlap are carefully applied to the sides and undersurface to lock the rock and bones of a partial skeleton of *Allosaurus* into place. Once the undercutting is completed, the block will be turned upside down to complete the application of the plaster case. The quarry floor was exposed by removing nearly a meter of overburden from above the bone level. The site is near Green River, Utah, in the Morrison Formation.

This first application of burlap and plaster "bandages" locks the bones and rock into place; it is similar to the medical practice of setting a broken bone with plaster and gauze. After the jacket sets, the pedestal must be carefully undercut (Fig. 7.3) and new bandages applied to the side and undersurface as the pedestal is narrowed. If the block is large or heavy, lumber or steel braces might be required to encase the block. These can be fixed to the top and sides of the block and should be completely encased in additional layers of plaster bandages. For huge blocks, steel banding and timbers may be required. During the undercutting stage, the block should be stabilized with props to prevent its shifting and consequent injury to workers or damage to the bone.

Eventually the undercutting activity and successive layers of bandages produce a plaster-encased block, standing on a narrow pedestal, which can be turned over without damage to the contents. Turning small blocks that weigh 10 or 20 kg is simple, but the larger the block, the more difficult the process, because there is always a danger that the contents of the block will shift and break loose from their confinement. For very large blocks, weighing hundreds of kilograms, there is also an obvious safety hazard akin to that of felling trees in the forest. The exact direction and movement of the block as it is rotated can be predicted and to some extent controlled, but often a large block pivots differently than expected. This procedure should be conducted only by an expert.

Before the block's being turned over, the outer surface must be labeled with permanent ink, indicating the field number for each bone contained inside, a north arrow for orientation, the site name and number, the date, the collector's name, and other information that will be pertinent to museum storage. This information is critical because the blocks may not be opened for laboratory preparation for weeks, months, or even years.

After the block is upside down, the small opening left by the pedestal must be covered with plaster bandages and the block entirely enclosed. The bock can now be transported without damage to the bones, although each block should be padded and locked into place to minimize jostling and bouncing. All blocks should be handled with care during transport.

The transfer of bones to the museum involves proper transfer of documentation at the same time. The fossils become the responsibility of the museum's collection manager, who is concerned with the care of the fossils. The blocks may be placed in permanent or temporary storage, where they will be secure until readied for laboratory preparation, or they may be sent directly to a technical laboratory, called the preparation laboratory, for in-house removal of the bones from the remaining rock, repairs, and stabilization. Ultimately, the bones are to be made ready ("prepared") for study, storage, or exhibit. Field notes should describe on-site treatment of bones, listing which glues, hardeners, or adhesives were applied. The field record of materials used in the excavation becomes vital to the museum preparators, who must continue the process that was begun in the field.

Technological Applications in Excavations

On the ground, the prediction of where bones are located in the subsurface is difficult or impossible. At best, the experience of the seasoned field worker is the most reliable source of information. The goal is to predict how much of the skeleton remains buried, its orientation, its depth, and its extent. If paleontologists know this information in advance of the commencement of an excavation, they can plan a complete excavation that will discover all the bones, disturb the minimum amount of rock, and complete the work in minimum time at minimum expense.

While most of the techniques discussed above have changed but little over the past century, certain new applications of modern technology hold considerable promise for improving efficiency in the field. Although every field paleontologist dreams of being able to locate bones from aerial photos or satellite images, the best these techniques can do is decipher certain aspects of the surface geology. In North America and in many parts of the world, such applications of airborne remote sensing are superfluous because geologic maps are cheaper and the interpretations have already been conducted.

On the other hand, ground-based remote sensing may be used profitably at many sites. Presently, no single technique has been developed for paleontology, but borrowed or hired equipment that has been developed for other disciplines such as hydrology or archaeology is often available. Paleontological applications have special problems, however, and the service of an experienced engineer familiar with the equipment and the theory behind its design is essential. Ideally, paleontologists need a device that can take an "X-ray" of the ground to see the bones beneath the surface. To date, this idea remains a fantasy, despite movie depictions that show entire skeletons in exquisite detail.

The most promising technique is ground-penetrating radar. From a mobile unit about the size and shape of a power lawn mower and mounted on wheels or carried on poles, radio waves are transmitted into the ground along a preestablished grid pattern. Reflections along boundary layers are recorded on the receiving device and show a continuous subsurface profile along each grid line (Fig. 7.4). The layers may be changes in rock type or

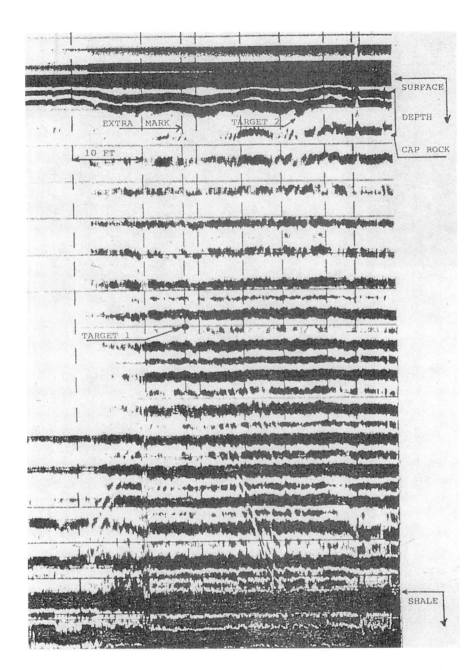

Labels within the figure: SURFACE, DEPTH, CAP ROCK, SHALE, EXTRA MARK, TARGET 2, TARGET 1, 10 FT

7.4. This record of a single traverse along a grid line for ground-penetrating radar at the *Seismosaurus* site (Gillette 1994) in the Morrison Formation in New Mexico shows two possible targets as the peaks of the parabolic disturbance patterns. Ground surface is at the top of the recording; target 1 is at a depth of 8 feet. Target 1 marked the position of the vertebral column of the skeleton *Seismosaurus hallorum* from the top of a sandstone mesa that capped the bone level 2 to 3 m below. Note that the horizontal scale is compressed.

Courtesy Sandia National Laboratory, Albuquerque, New Mexico.

composition, changes in saturation of water, natural cracks in the ground, or the boundary between rock and bone. Rather than the image of a bone being produced in the record, the indication of possible bone is a disturbance pattern in the profile. Deciphering the patterns requires experience and considerable intuition. In one excavation where the techniques were tried as experiments, such disturbance patterns represented bones with about 50% accuracy. Such applications have proven more successful in archaeological sites, where large disturbances such as buildings or caverns can be detected.

A different technique that uses seismic data also holds considerable promise for the future. Acoustic diffraction tomography, or seismic tomography, requires strategically placed vertical holes that penetrate far beneath the projected level of bones, preferably at least twice the depth. Specially designed hydrophones in a fixed array are suspended in these holes, which

are sealed with pipe and filled with water. On the surface, a high-powered 8-gauge Magnum shotgun mounted on wheels, affectionately called a betsy (Fig. 7.5), sends a lead slug into the ground at predetermined points. The shock wave that passes from that point is transmitted into the ground in all directions. The hydrophones receive the shock waves and record the exact time of arrival. Reception times along the array of hydrophones that are anomalous, either too slow or too fast, indicate the existence of something buried in the subsurface and in a direct line between the shotgun blast and the vertical array of hydrophones. Geometric calculations can locate the buried object, which the paleontologists hope is a bone or a skeleton, and through repeated tests from different points, the depth and extent of the object can be resolved. Like ground-penetrating radar, acoustic diffraction tomography requires the services of trained engineers and is so expensive that its potential lies only in large excavations.

Magnetic properties of buried bones are sometimes different from those of surrounding rocks because of their chemical composition. Measurement of this difference is the objective of proton-free precision magnetometry, which records the intensity of the earth's magnetic field on a predetermined grid at the surface of the earth. The remarkably precise instrument is mounted on a pole and has the decided advantage of mobility so that it can be carried into remote areas with ease. Variations in the intensity of the magnetic field as plotted on a grid map may indicate the existence of something other than rock beneath the surface. Fossil bones are one possibility, but other subsurface materials may also produce an anomaly.

In some places, fossil bones and fossil wood contain radioactive isotopes of uranium. These isotopes decay, and the emission of energy, or ionizing radiation, may be recorded by handheld counters. The traditional Geiger counter measures radiation of this sort, but it is not sufficiently sensitive to detect radiation emitted from buried objects. Instead, radiation can be detected from subsurface sources by scintillation counters that take measurements of radiation on a grid pattern. Unusually high counts that differ from the normal background radiation may indicate bone beneath the surface. This technique seems to be useful, if at all, only for bones that are quite shallow, no more than a half meter from the surface.

At the museum, the tedious process of preparation of the bones is the final step leading to their exhibit, study, or storage in the permanent cataloged collections. The purposes of preparation are to fully expose the bones, complete all necessary repairs, and stabilize them for curation. For dinosaur bones, this procedure is time-consuming and expensive because of the bones' size. Like other stages in discovery and excavation, laboratory preparation techniques are best learned by hands-on experience under the supervision of an experienced preparator. In general, hand tools and dental picks are appropriate for most preparation, but often microscopes and needles are required. For removal of rock far from bone, the gentle impact of pneumatic tools that operate like miniature jackhammers is quite effective.

Repairs should be made with glues and adhesives approved by the chief preparator or the collections manager. As with the chemicals applied during excavation, no single hardener or glue is universally used, and suitability varies with the originating location and nature of the fossils. In recent years, collection managers have begun to recognize that many adhesives and preservatives, though useful during preparation, actually lose their effectiveness with age and accelerate the degradation of the fossil bone. One widely used chemical, commonly called glyptal, which was popular for more than four decades, disintegrates with age and is no longer recommended. Loss of effectiveness of glues and chemical hardeners can be devastating to museum collections. In general, the guiding principle in the use of chemicals and adhesives is "least is best."

If the bones are weak, structural supports may be necessary. These are custom designed to fit the bone and may consist of plaster-and-burlap bandages that form a cradle, fiberglass, steel bands, lumber, or any other structural material that adds support and strength.

Many paleontologists have adopted the practice of retaining samples of untreated bone and rock for future reference. In some cases, these materials have yielded new fossils of microscopic size, such as pollen, spores, or the teeth of small animals, thus greatly enhancing the knowledge of the paleoenvironmental setting of an excavation site. These samples should be curated with the bones produced by the excavation.

In all cases, laboratory preparation records are important. Many preparation laboratories use forms to record which adhesives are used for each bone along with a detailed record of preparation activities on each bone. If in the future the bone needs repair or additional preparation, these treatment reports are critical if the original hardeners or adhesives are to be dissolved and replaced.

Curation and Conservation

Dinosaur bones present special curation problems because of their size. Their weight should be spread evenly through the use of pads or specially designed cradles made of plaster and burlap bandages, or foam carved to fit the bone. The bones should be stored in a clean, dark place where they are protected from changes in humidity, temperature, and decomposer organisms. Such conditions are best achieved in specially designed storage areas where temperature and humidity are constant. All bones must be labeled, and their locations and associated archives recorded in a collections catalog. They should be monitored periodically for damage to the bones or labels, and for deterioration of hardeners and glues. No one wants to pick up a dinosaur bone from a museum cabinet and have it fall apart because of decay or alteration during curation.

Curation and conservation of dinosaur bones actually begins in the field with the application of proper techniques. Museum professionals assigned the responsibility of curation are increasingly concerned with long-term effects of every action taken on a bone. Many of these personnel belong to the professional society called the Society for the Preservation of Natural History Collections. This group meets annually and publishes a wealth of technical information on conservation and curation in the journal *Collection Forum*. Membership is open to anyone interested in the subject of conservation of museum collections. Similarly, conservation research and evaluation of techniques are often reported in the journal *Curator*, published by the American Museum of Natural History. Curation standards for dinosaur bones and fossil vertebrates in general are evolving at a rapid pace. For new museums or personnel newly assigned curation responsibilities, membership in the SPNHC is a step in the right direction.

Collection Management

Besides direct responsibility for the care of the bones, collection management entails a myriad of other duties. Records associated with the collection are essential, for they are the only primary source of information concerning the excavation. Included in collection management duties are the maintenance of at least three filing systems, preferably on a computerized data management system: archives files, locality files, and specimen catalog files. These documents are among the most critical in the museum, for they give meaning to the collections. They should be managed with the utmost care, and they should always be kept up to date. In addition, archival records themselves often require special treatment and conservation: photographs may be unstable, paper may become brittle with age, and ink may fade; decay by fungus or bacteria can destroy paper.

The ultimate purposes of the museum collection are research and education. Thus the fossils should be accessible to qualified personnel for study. This aspect of collections is often overlooked in the architectural design for the collections area of the museum. Layout space, desk space, tables, good lighting, and access to electrical outlets are essential for suitable study of fossils. Microscopes and measuring devices are also important and should be made available to curators and visiting researchers. If specimens are to be displayed, their whereabouts must be recorded in the catalog and a note placed on the shelves where they are ordinarily stored. When visiting researchers need to examine an exhibit specimen, they should be allowed access to the bones, or the bones should be temporarily removed from the display. Museum visitors are seldom disturbed by a note in an exhibit stating that a specimen is removed for study; such a note indicates active use of a museum's collections.

This long road leading to the addition of another dinosaur bone or skeleton to a museum collection began with the discovery in the field. The road never ends, because the bones recovered during an excavation and deposited in a museum collection will remain available for study for as long as museums exist.

Field and laboratory techniques in paleontology have been the subject of several technical books and articles. Some are listed in the References below. Many of these were published more than a decade ago. For more current information, especially for technological applications, Web-based searches are likely to yield useful results.

Further Reading

References

Converse, H. H., Jr. 1984. *Handbook of Paleo-Preparation Techniques.* Gainesville: Florida State Museum.

Crowther, P. R., and W. A. Wimbledon (eds.). 1988. *The Use and Conservation of Paleontological Sites.* Special Papers in Paleontology 40. London: Paleontological Association.

Feldmann, R. M., R. E. Chapman, and J. T. Hannibal (eds.). 1989. *Paleotechniques.* Knoxville, Tenn., Paleontological Society Special Publication 4.

Fitzgerald, G. R. 1988. Documentation guidelines for the preparation and conservation of paleontological and geological specimens. *Collection Forum* 4 (2): 38–45.

Gillette, D. D. 1994. *Seismosaurus, the Earth Shaker.* New York: Columbia University Press.

Kummel, B., and D. Raup. 1965. *Handbook of Paleontological Techniques.* San Francisco: W. H. Freeman.

Leiggi, P., and P. May. 1994. *Vertebrate Paleontological Techniques.* Vol. 1. New York: Cambridge University Press.

Rixon, A. E. 1976. *Fossil Animal Remains: Their Preparation and Conservation.* London: Athlone Press.

8.1. The right hands (manus) of (A) a human (*Homo sapiens*) and (B) the herbivorous dinosaur *Dollodon bampingi* in anterior view. The spike in the hand of *Dollodon* is homologous to (occupies the same anatomical position as) the human thumb. The opposable digit in the hand of *Dollodon* is homologous to the human pinkie but is analogous to (has the same function as) the human thumb.

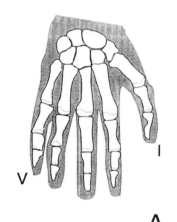

A

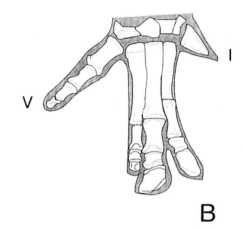

B

The Osteology of the Dinosaurs

Thomas R. Holtz Jr. and M. K. Brett-Surman

8

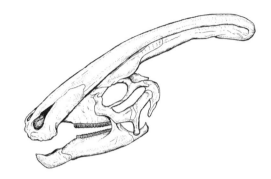

With rare but spectacular exceptions, dinosaur body fossils consist almost entirely of bones and teeth. The soft parts of the body—skin, muscles, and other organs—were destroyed by decay processes fairly quickly after death. Only bones and teeth, the hard mineralized parts of a dinosaur, are durable enough to be preserved over tens of millions of years. Except for footprints and much rarer traces such as eggs and skin impressions, fossilized skeletal material represents the only physical remains of the ancient dinosaurs. Thus the osteology (the study of bones) of dinosaurs is our main source of knowledge about these extinct animals. This chapter is intended to give a brief overview of some of the general aspects of the dinosaurian osteology; later chapters will give specifics concerning the details of particular groups of dinosaurs or aspects of their biology revealed by their bony remains.

Dinosaurs are tetrapod vertebrates—in other words, animals with bony skeletons and four limbs. All tetrapods, including amphibians, mammals, turtles, lizards, and birds, are built along the same general body plan. For example, the forelimb, or arm, of all tetrapods has one upper arm bone closest to the body, two bones below the elbow, several small bones in the wrists, and then a series of longer finger bones. (In some animals, such as snakes, the forelimbs have disappeared, but the ancestors of these animals had arms of the basic structure.) The reason all these animals share this common body plan is that all are descended from the same ancestral stock with that plan. The differences between the particular shapes of the bones arise from the same body plan having been modified, or adapted, to different uses—for example, the wings in birds or bats, the digging claws of moles, the grasping hand of a *Velociraptor*, or the pillarlike forefoot of a *Brachiosaurus*. Because of this common descent, we can recognize bones that are homologous; that is, they represent bones descended from the same original structure. Thus the upper arm bone of any tetrapod is homologous to the upper arm bone of any other tetrapod.

For an example of homology, compare the right hand of a human and the right forepaw of the plant-eating dinosaur *Dollodon* (formerly a species of *Iguanodon*; see Paul 2008) (Fig. 8.1). These hands are oriented in the same direction, with the back of the hand facing us and the palm facing away. In the dinosaur, the homologue to the thumb has been fused into a large pike. The last digit, homologous to the pinkie of a human, has evolved into an opposable finger. Opposability (the ability to place the digit on the palm) is characteristic of the thumbs of humans. Two different anatomical features in two different animals that have the same function but that are evolved from different parts of the body are called analogous. Thus the opposable digit of *Dollodon* is analogous to the human thumb, but homologous to the human pinkie.

It must be noted that in comparative anatomy, the term *homologous* was not originally used in an evolutionary context. Sir Richard Owen, who coined the term *Dinosauria* in 1842 (see Torrens, Chapter 2 of this volume), is also responsible for the use of *homology* in an anatomical context. Owen (1846, 1849) believed that for each major group of organisms, there was a single basic body plan, or blueprint, of which all species in that group are variations. This body plan, the archetype, was not considered to have ever existed in the physical universe; rather, it was a mental construct representing the simplified anatomical organization of each major group of organisms, such as vertebrates, mollusks, or insects. In this context, the pectoral fin of a trout, the wings of birds, the forelimbs of horses, and the arms of humans were considered homologous because each was a variation of the same structure in the vertebrate archetype.

Sir Charles Darwin and his primary advocate, Thomas Henry Huxley, co-opted the concept of homology into the new theory of evolution by means of natural selection. In their view, where all animals of the same body plan have a common origin, a homologous structure in two or more organisms represents variations of the same structure that was present in a real common ancestor. (See Desmond 1982 for a detailed discussion of the social and political as well as the scientific conflict between Owen and Huxley over the concepts of archetypes and ancestors.)

Anatomical Names, Directions, and Views

Because the basic bony anatomy, or skeleton, of all tetrapods is based on an ancestral body plan, all homologous bones can be given the same name. Because the anatomists who coined the various names used the classical languages Latin and Greek for scientific discourse, as did naturalists, astronomers, and other early scientists, most of these bones are named in Latin. Similarly, other structures in the skeleton (such as the socket for the eye or for the nostril) are also given Latin names.

While a single standardized nomenclature for osteological (and other anatomical) features of all tetrapods would seem a worthwhile goal (Harris 2004), in practice, there are several different similar by distinct systems (see review in Wilson 2006). In particular, human anatomy (with a set of anatomical directions based on our rather atypical posture and orientation), domestic mammals, and birds each have a formal standardized terminology. In contrast, traditional vertebrate paleontological texts (such as Williston 1925; Romer 1956; and Carroll 1988) employ a hodgepodge of terms lacking a single formal code. The authors of the present volume follow this traditional system, nicknamed the Romerian system by Wilson (2004). It is important to note, however, that a major volume in the study of dinosaurs (Weishampel et al. 2004) uses the Nomina Anatomica Avium or NAA (Baumel and Witmer 1993) for its nomenclature. We provide the NAA equivalents for Romerian terms in this chapter to help the reader translate between the two books.

Before some of the important bones and other structures in the skeleton of dinosaurs are described, the principle of anatomical direction should be discussed. In order to describe the positional relationships of one bone to another in the skeleton of an animal, a series of pairs of directions have been invented. Like *north* and *south* or *up* and *down*, these directions always have an opposite pointing the other way. Unlike *north* and *south* and *up* and *down*, however, these directions are not based on the external environment. Instead, they are internal to each organism, regardless of how it may move about in the outside world. In this way, they more closely correspond to the nautical terms *fore*, *aft*, *port*, and *starboard*.

Romerian anatomical directions are based on the standard posture of most tetrapods (that is, all forelimbs on the ground, head pointing forward, belly toward the floor and back toward the ceiling), so that the homologous directions in a human being (with only our feet on the floor, our face pointing the same direction as our belly, our belly pointed forward and our backs pointed behind us) are oriented in somewhat different external directions. However, a crawling baby is oriented in essentially the same position of most other tetrapods (the primary difference being that babies do not walk on the soles of their feet).

Anatomical directional terms can be used to describe surfaces. Just as the north face of a house is (not too surprisingly) the part of the house that faces north, so to the anterior surface of a bone is the surface on the front (anterior) face of that bone.

The first pair of these is anterior (sometimes called cranial) and posterior (sometimes called caudal). *Anterior* means "toward the tip of the snout," and *posterior* means "toward the tip of the tail." For example, the shoulders are anterior to the hips, the skull is anterior to the neck, and the nostrils are anterior to the eye sockets. Conversely, the hips are posterior to the shoulders, the neck posterior to the skull, and the eye sockets posterior to the nostrils. Because these terms are independent of the external environment, they remain the same regardless of the position an animal is in (i.e., if a cat curls up, its tail is still posterior to the skull). In the NAA, anterior is replaced by *rostral* ("of the snout") for features within the skull or *cranial* ("of the head") when discussing that direction in the rest of the body. The NAA equivalent of posterior is *caudal* ("of the tail").

A second pair of anatomical directions in both the Romerian and NAA systems is dorsal and ventral. *Dorsal* means "toward, and beyond, the spine" (or more simply "up"), while *ventral* means "toward, and beyond, the belly" (or, generally, "down"). In the skull, the teeth are ventral to the eyes, and the upper jaw is dorsal to the lower jaw. In the cranium, the lower surface is referred to as *palatal*; in the hand as *palmar*; and in the foot as *plantar*.

The next pair of directions in both the Romerian and NAA systems, *medial* and *lateral*, refer to directions relative to an imaginary plane through the center of the body, which runs from the tip of the snout through the tip of the tail, bisecting the body into a right and a left half. These two directions refer to the relative positions of bones to each other and to this imaginary midline. Medial refers to bones or structures that are closer to the midline (i.e., closer to the center), and lateral means farther away from the midline (i.e., farther out, or more right or more left). The shoulder blades are lateral to the ribs, and the spine is medial to the ribs.

A last pair is used primarily for directions within the limbs (the arms and legs) and is sometimes applied to the tail. *Proximal* means "closer to the trunk," while *distal* means "farther out from the trunk." For example, the hip is proximal to the knee, and the wrist is distal to the elbow.

Although these four pairs, anterior/posterior, dorsal/ventral, medial/lateral, and proximal/distal, are generally used to describe the relationships of bones to one another, they can also be used adverbially to describe how an anatomical structure is constructed. For example, the teeth of the upper jaw point ventrally, the snout of most animals projects anteriorly from the eyes, and the ischium (a bone of the hip) points posteroventrally (back and down) in all dinosaurs.

The names of the anatomical directions can be used to describe the particular surface of the bone illustrated in a photograph or drawing. To see the dorsal anatomical view of the skull, for example, means to see the top surface. The ventral view would be a picture of the bottom of a bone or skeleton, the anterior view the front, the posterior view the back. There is both a right lateral and a left lateral view of the skeleton, depending on whether you are viewing the right or left, respectively, of the animal. A medial view would show the surface of a bone which normally faces the midline.

In the following section, we will examine the major bones in the skeletons of dinosaurs. Drawings of various dinosaurs are used to illustrate the position and general shape of these bones. However, the Dinosauria were a very diverse group of animals, so there is considerable variation in the details of their skeletons. Elsewhere in this book you will find drawings of the osteology of the different dinosaur groups.

A note on bone itself. Bone is a composite material, containing the calcium phosphate mineral hydroxylapatite ($Ca_{10}(PO_4)_6(OH)_2$) in a matrix of the protein collagen. This combination combines the mechanical strength of the hydroxylapatite with the flexibility of collagen. Developmentally, some bones are dermal bones—that is, they initially form as (generally) flat plates in the dermis. Most of the facial and skull roof bones are dermal bones, for instance. Other bones are preformed in cartilage before ossifying (turning to bone). This developmental type is called endochondral bone, and it generally includes the long bones of the limbs and bones with complex three-dimensional shapes.

Sections of the Skeleton

The Skull

The skeleton of a dinosaur, or other tetrapod, can be divided into two main divisions: the skull, which refers to all the bones and teeth of the head, and everything else. The skull is mostly composed of many different bones that, like most of the bones of the body, are paired (i.e., there is one of those bones on the left side of the skull and one on the right side of the skull). There are also bones, which are single, and these usually lie on the midline. For example, the supraoccipital bone is right above the hole where the spinal cord enters the skull, and it is not paired. The outlines of the individual bones of the skull are recognized as sutures, where the different bones meet.

The skull itself is divided into two major sections. The upper part of the skull, including the eyes, nostrils, upper jaw, and braincase, is called the cranium (plural crania). The lower jaw is composed of the left and right mandibles.

It is sometimes easiest to recognize the different individual bones of the skull by starting out with nonbony landmarks. Landmarks are particular homologous features that are recognizable from animal to animal. Two of the best landmarks are the eye sockets and the nostril openings. The technical term for the eye socket is the orbit, while each individual nostril is called a naris (plural nares). (Note that while some refer to the nares as the "nostrils," this is not an accurate description. The nostril is a fleshy structure that is often positioned lateral to the anterior end of the naris, but in some species—including possibly many dinosaurs—the nasal passage continues for some distance in the soft tissues on the outside of the face, to emerge as a nostril (Witmer 2001).

Among the other landmarks of the skull are the teeth. Made of materials (dentine and enamel) even tougher and more durable than true bone, teeth occur only on particular bones of the skull. Among the other landmarks of the skull are the teeth. Made of material even tougher and more durable than true bone, teeth occur only on particular bones of the skull. Each tooth consists of two major parts: a crown that projects above the gum line, and a root that extends into the jawbones. The root and the interior of the crown are comprised of the material dentine (like bone, a composite of hydroxylapatite and collagen), while the crown is covered by the even tougher material enamel (which contains hydroxylapatite but no collagen). In most dinosaurs, like most other vertebrates, the entire crown is covered by enamel, but in certain advanced ornithischians, only one side of the crown is enameled. Dentine is softer than enamel, so the tooth became self-sharpening because the dentine wore away more quickly than the enamel when the teeth ground against each other. In many dinosaurs, there was no or little occlusion (that is, direct contact between teeth of the upper and lower jaws). Mostly they slid past each other in a slicing action. There are some exceptions, however. In the advanced sauropods (giant long-necked plant eaters), there is occlusion between the tips of the teeth. In ceratopsians (horned dinosaurs), the cutting surfaces of the teeth are oriented in a vertical plane that produces a scissorlike action. Because of their combination of slicing teeth and huge jaw muscles (ceratopsians may have had the strongest jaw muscles of any herbivorous animal), they have been called the first Cuisinarts. Hadrosaurs (duckbilled dinosaurs) were actually capable of chewing their food, in that their teeth came together in a grinding action.

Like crocodilians and mammals, but unlike many other tetrapods, the roots of dinosaurs are fairly long and held in sockets within the jawbone. As with most nonmammalian vertebrates, dinosaurs continued to produce new teeth within each socket throughout their life span.

In the dinosaurian upper jaw are two tooth-bearing bones. The most anterior is the premaxilla, and the posterior one (which is almost always much larger) is the maxilla. The premaxilla is ventral to the naris, and the maxilla is ventral to another opening. This opening is called the antorbital fenestra (plural fenestrae), literally "the window anterior to the orbit." This opening is found in many dinosaurs and their closest relatives. In the beaked, ornithischian (bird-hipped) dinosaurs, the antorbital fenestrae are very reduced or entirely closed over, while in the saurischian (lizard-hipped)

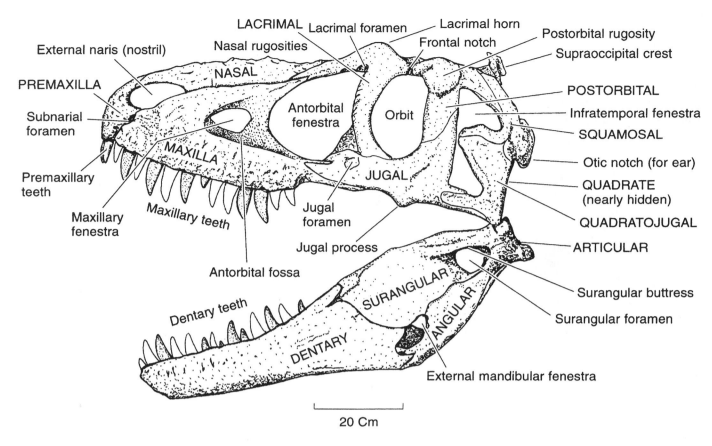

External naris (nostril)

PREMAXILLA

Subnarial foramen

Premaxillary teeth

Maxillary fenestra

LACRIMAL — Lacrimal foramen — Lacrimal horn

Nasal rugosities — Frontal notch

NASAL

Antorbital fenestra — Orbit

MAXILLA

Maxillary teeth

Jugal foramen

JUGAL

Jugal process

Antorbital fossa

Dentary teeth

SURANGULAR

DENTARY

ANGULAR

Postorbital rugosity

Supraoccipital crest

POSTORBITAL

Infratemporal fenestra

SQUAMOSAL

Otic notch (for ear)

QUADRATE (nearly hidden)

QUADRATOJUGAL

ARTICULAR

Surangular buttress

Surangular foramen

External mandibular fenestra

20 Cm

8.2. The skull of the tyrant dinosaur (tyrannosaurid) *Daspletosaurus torosus* in left lateral view, illustrating the important bones (in capital letters) and landmarks of the skull. The subnarial foramen is a character found only in the Saurischia, and the maxillary fenestra and jugal foramen are unique to certain members of the Theropoda. The nasal rugosities, frontal notch, jugal process, and surangular buttress are specializations of tyrant dinosaurs and are not found in most dinosaurs.

Illustration by Tracy L. Ford.

dinosaurs, these fenestrae are often very large. The antorbital fenestra sits in a depression, the antorbital fossa (plural fossae). In some advanced meat-eating dinosaurs, there are additional openings anterior to the antorbital fenestra. These are called the maxillary fenestra (or sometimes the accessory antorbital fenestra) and the promaxillary fenestra (which is more anterior on the inside of the antorbital fenestra). Figure 8.2 illustrates most of these landmarks on the skull of the tyrant dinosaur *Daspletosaurus*.

In some dinosaurs with a horny beak, teeth in the premaxilla and sometimes even in the maxilla are absent. When a dinosaur, or a jawbone, has no teeth, it is said to be edentulous. In the ceratopsian (frilled or horned) dinosaurs, there is an additional bone anterior to the edentulous premaxilla. This bone is called the rostral (or "snout") bone. The rostral is a single bone joining the two premaxillae.

In the rear of the skull, posterior to the orbit, lie additional openings in the skull. These are called the lateral temporal (or infratemporal) fenestra and the supratemporal fenestra. The lateral temporal fenestra is a large opening on the side of the skull, while the supratemporal fenestra is on the dorsal surface of the skull. Both are associated with the attachment of jaw muscles.

From these various landmarks, the positions of some of the other important skull bones can be determined (Figs. 8.2, 8.3). The jugal, or cheekbone, is posterior to the maxilla and ventral to the orbit. The lacrimal is a small bone between the antorbital fenestra and the orbit. The quadrate is a major bone in the rear of the skull, where the cranium articulates with the mandible (lower jaw). All dinosaurs (including birds) have a quadrate/articular jaw joint; in other words, a bone in the back of the lower jaw bone,

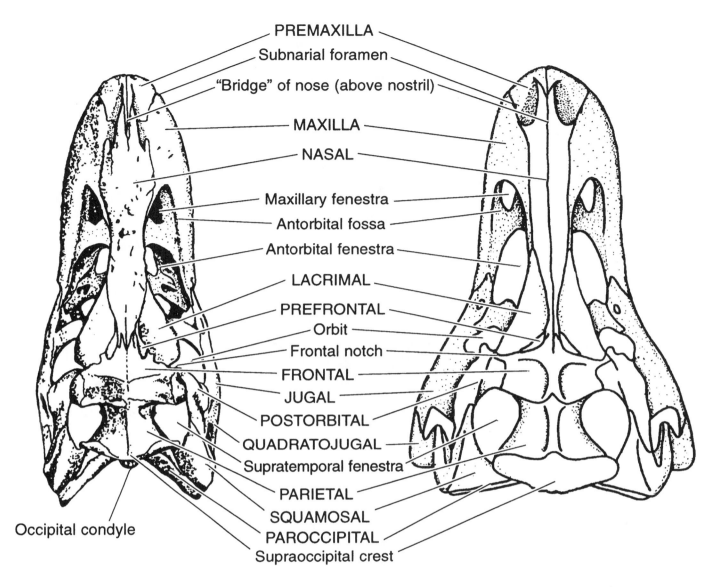

PREMAXILLA

Subnarial foramen

"Bridge" of nose (above nostril)

MAXILLA

NASAL

Maxillary fenestra

Antorbital fossa

Antorbital fenestra

LACRIMAL

PREFRONTAL

Orbit

Frontal notch

FRONTAL

JUGAL

POSTORBITAL

QUADRATOJUGAL

Supratemporal fenestra

PARIETAL

SQUAMOSAL

PAROCCIPITAL

Supraoccipital crest

Occipital condyle

8.3. The skulls of the tyrant dinosaurs *Daspletosaurus torosus* (left) and Tyrannosaurus rex (right) in dorsal view, illustrating the important paired bones (in capital letters) and landmarks on the dorsal surface of the skull.

Illustration by Tracy L. Ford.

called the articular, articulates with the quadrate bone in the skull. (Mammals have a dentary/squamosal jaw joint, meaning that the bones forming the joint in mammalian jaws are not homologous to the bones forming the joint in dinosaurian jaws.)

A series of paired bones lie along the dorsal and posterior surface of the skull (Fig. 8.3). These bones meet along the midline and so form mirror images, right and left, of each other. The most anterior are the nasals, long paired bones on the dorsal surface of the skull, posterior to the premaxilla. Posterior to the nasals are the frontals. The parietals are paired bones above the braincase on the posterior surface of the skull, posterior to the frontal. The squamosals are on the posterior surface of the skull.

There are many bones that are joined together around the brain cavity. These tightly sutured bones are collectively called the braincase and lie inside the outer skull bones listed above. Many of the bones of the braincase are fragile or were only partially ossified, and so this part of the cranial anatomy can be poorly preserved. The spinal cord exits from the brain through the foramen magnum (or "great opening") on the posterior of the skull. Beneath the foramen magnum is a structure called the occipital

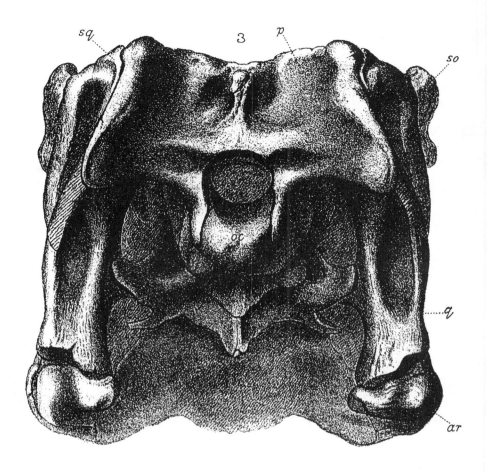

8.4. The skull of the armored dinosaur *Stegosaurus stenops* in posterior view, illustrating some important bones and landmarks of the rear of the skull. Illustration from Ostrom and McIntosh (1966). Original is a lithograph from a never-completed monograph on the Stegosauria to have been written by O. C. Marsh. Abbreviations: ar = articular; oc = occipital condyle; p = parietal; q = quadrate; sq = squamosal; so = supraorbial.

condyle, a rounded knob joint (or condyle), where the cranium articulates with the vertebral column (Fig. 8.4). In humans, other mammals, and our extinct relatives, as well as in amphibians, there are two occipital condyles (one right, one left), but in dinosaurs and other reptiles, there is only a single rounded knob directly ventral to the foramen magnum.

In dinosaurs, as in most nonmammalian vertebrates, the mandible is composed of several different bones. The tooth-bearing bone of the mandible is called the dentary, and there are several hones posterior to it, which form the connection with the cranium. In mammals, the mandible is formed exclusively by the dentary. In the Ornithischia (bird-hipped dinosaurs), there is an extra bone in front of the dentaries. Called the predentary, this bone joins the two dentaries and forms a strong beak.

The Axial Skeleton

All the bones in the skeleton except for the skull are collectively referred to as the postcranium ("posterior to the cranium") (Fig. 8.5). The postcranium can he divided into two sections, often called the axial and appendicular skeletons. The axial skeleton is the core of an animal-+-its spine, trunk, and tail (the vertebrae). The appendicular skeleton refers to the forelimbs and hind limbs, and the girdles that attach the limbs to the trunk.

The most important part of the axial skeleton is the vertebral column. This column, the backbone, is composed of many individual elements. Each of these hones is called a vertebra (plural vertebrae). A vertebra is composed of a large spool- or cylinder-shaped structure, the centrum ("body,"

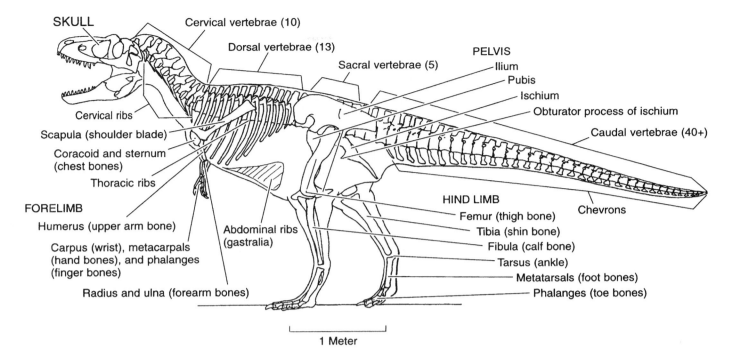

SKULL
Cervical vertebrae (10)
Dorsal vertebrae (13)
Sacral vertebrae (5)
PELVIS
Ilium
Pubis
Ischium
Obturator process of ischium
Caudal vertebrae (40+)
Cervical ribs
Scapula (shoulder blade)
Coracoid and sternum (chest bones)
Thoracic ribs
FORELIMB
Humerus (upper arm bone)
Carpus (wrist), metacarpals (hand bones), and phalanges (finger bones)
Radius and ulna (forearm bones)
Abdominal ribs (gastralia)
HIND LIMB
Femur (thigh bone)
Tibia (shin bone)
Fibula (calf bone)
Tarsus (ankle)
Metatarsals (foot bones)
Phalanges (toe bones)
Chevrons

1 Meter

plural centra) ventrally, and a neural arch dorsally. The NAA equivalent to these terms are the corpus and arch, respectively. Between the centrum and the neural arch is the neural canal, through which runs the spinal cord. On each neural arch are two sets of fingerlike projections called the zygapophyses. Directed forward (angled upward and inward) are the prezygapophyses (cranial zygapophyses in the NAA). These articulate with the postzygapophyses (caudal zygaopophyses in the NAA) facing backward and angled down and out on the vertebra in front. These zygapophyses control the amount of movement between two vertebrae. Projecting dorsally from the neural arch is a neural spine (spinose process in the NAA) to which are attached the back muscles (and which form the bumps down your spine).

The vertebral column of a dinosaur, like those of most nonmammalian tetrapods, is divided into four major segments (a mammalian column can be divided into five). These sections are the cervical (neck), dorsal (back; thoracic in the NAA terminology), sacral (hip), and caudal (tail) vertebrae (Fig. 8.5). (In mammals, the dorsals can be divided into two separate segments: a thoracic, or chest, series, which has large ribs; and a lumbar, or lower back, series, in which there are no ribs.) The sacral vertebrae are often fused into a single structure in dinosaurs, called a sacrum (plural sacra). Dinosaurs are recognized as having three or more sacral vertebrae fused together (unlike most other reptiles, such as lizards and crocodiles, which have only two). The vertebrae of each of the different sections are shaped differently, reflecting the different requirements of the various sections of the body (i.e., flexibility in the neck, strength in the hips, etc.). Figure 8.6 illustrates a vertebra of the giant herbivorous dinosaur *Apatosaurus*.

The vertebrae of some dinosaurs are quite complex, with additional ridges, prongs, and other structures (Fig. 8.7). One such structure is the hypantrum (a small anterior projection at the base of the neural spine), which articulates with the hyposphene (a small posterior projection at the base of the neural spine) of the preceding vertebra. Pleurocoels are openings along

8.5. The skeleton of the tyrant dinosaur *Daspletosaurus torosus* in left lateral view, illustrating the postcranial skeleton.

Illustration by Tracy L. Ford.

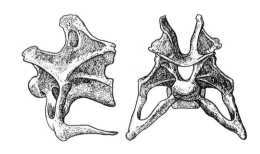

8.6. The eighth cervical (neck) vertebra of the gigantic sauropod dinosaur *Apatosaurus louisae*, in left lateral (left) and anterior (right) views.

Illustration by Tracy L. Ford, modified from Gilmore (1936).

8.7. Schematic representation of the four main types of vertebral articulations. In each case, the anterior end of the vertebra is to the left. (A) Amphiplatyan, flat on both anterior and posterior ends; (B) amphicoelous, concave on both anterior and posterior ends; (C) procoelous, concave anterior end, strongly convex posterior end; (D) opisthocoelous, strongly convex anterior end, concave posterior end.

← anterior   posterior →

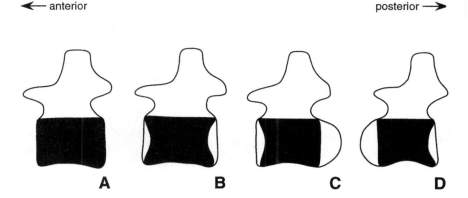

A   B   C   D

the lateral surfaces of vertebrae which open to a chamber inside the bone of the centrum and/or the neural arch. A pleurocoel can be a simple cavity, or it can be a complex system of chambers and channels. In some of the saurischian dinosaurs, there are a series of thin sheets (or lamina): Wilson (1999) provides a detailed description and terminology of these structures.

Lateral to the cervical and dorsal vertebrae are the ribs. Ribs (for some reason the only major part of the skeletal anatomy more commonly referred to by their English name than by their Latin name, costae) are long, narrow, paired bones forming a cage around the vital organs. The ribs of dinosaurs attach to the vertebrae at the bottom of the neural arch and the top of the centra by two separated projections. The ventralmost of these projections is the capitulum (plural capitula; "little head"), while the dorsalmost is the tuberculum (plural tubercula; "little lump"). Along the belly of some dinosaurs are gastralia (singular gastralium), or "belly ribs," which strengthened the ventral side of the animal and acted as a girdle to hold in the viscera ("guts"). The gastralia were also important parts of the dinosaurian respiratory system (Carrier and Farmer 2000). Ventral to the caudal vertebrae are the chevrons, structures that are something like "tail ribs" or upside-down neural arches.

The Appendicular Skeleton

The appendicular skeleton refers to the forelimbs and hind limbs, and the girdles that attach these limbs to the body. Although there is a great similarity between the structures of the forelimb and the hind limb, the pattern of the girdles is very different.

The pectoral girdle attaches the forelimb to the trunk (Fig. 8.8). The largest of the bones in the pectoral girdle is the scapula (plural scapulae), or "shoulder blade." Ventral to the scapula is the coracoid. On the posterior surface of the girdle, the region where the scapula and coracoid meet forms a circular shoulder joint (the glenoid fossa). In some dinosaurs, clavicles ("collarbones") are present, which attach the shoulder girdle to a series of fused bones along the ventral region of the chest. These fused bones formed the sternum (plural sterna), or "breastbone." In theropods (carnivorous dinosaurs), there is a furcula ("wishbone," plural furculae) instead of clavicles. In some of the sauropodomorph dinosaurs, the clavicles are not actually fused but do contact each other in a furcula-like pattern (Yates and Vasconcelos 2005).

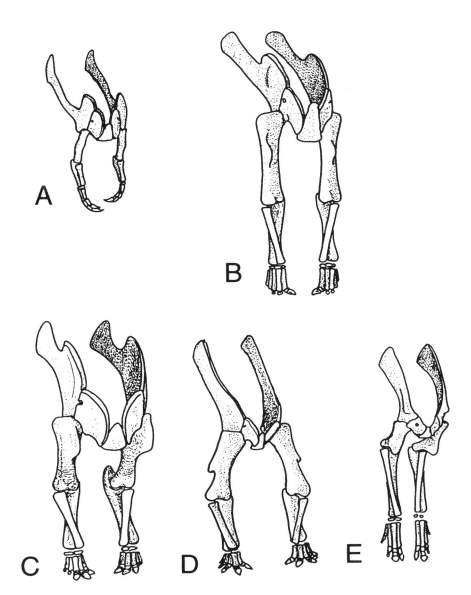

8.8. Forelimbs of the Dinosauria, in oblique right anterolateral view. (A) *Tyrannosaurus rex;* (B) *Apatosaurus louisae;* (C) *Stegosaurus stenops;* (D) *Agujaceratops mariscalensis;* (E) *Corythosaurus casuarius.* Not to scale.

Illustrations by Tracy L. Ford; modified from Osborn (1916), Gilmore (1936), Galton (1990), Lehman (1989), and Weishampel and Horner (1990).

Most of the forelimb, or arm, is made up of three bones. There is a single upper arm bone, the humerus (plural humeri), which joins the two forearm bones at the elbow. Of the two forearm bones, the ulna (plural ulnae) is the larger and more posterior, while the radius (plural radii) is generally smaller and more anterior. The many small bones of the wrist are known as the carpals. Distal to the carpals are the long bones of the palm of the hand, the metacarpals. The metacarpals are numbered in Roman numerals, from I to V, with I the most medial (inside, near the thumb) and V the most lateral (outside). The fingers are called digits, and they are numbered with the same scheme as the metacarpals (with digit I the thumb and digit V the pinkie). The individual bones of the finger are called phalanges (singular phalanx). The distalmost of the phalanges are sometimes called the unguals, and supported the horny claws or hooves. Collectively, the digits, metacarpals, and carpals form the hand, or manus (plural manus).

The hind limbs attach to the axial skeleton at the pelvic girdle (Fig. 8.9). Also known as the pelvis ("hip," plural pelves), the pelvic girdle is composed of three bones per side. The largest of these is the ilium (plural ilia), which is the dorsalmost and which connects to the sacrum. Attaching beneath

8.9. Pelvic girdles of the Dinosauria; (A–D) in left lateral view, (E) in right lateral view. Saurischian pelves: (A) *Tyrannosaurus rex*; (B) *Apatosaurus excelsus*. Ornithischian pelves: (C) *Stegosaurus stenops*; (D) *Agujaceratops mariscalensis*; (E) *Corythosaurus casuarius*. Not to scale.

Illustration by Tracy L. Ford.

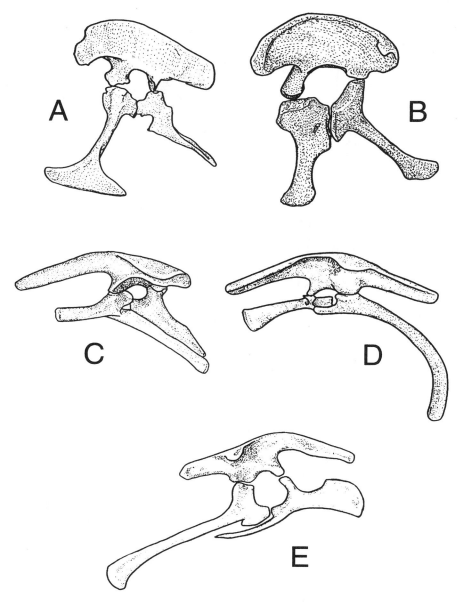

the ilium are the other two bones. The pubis (plural pubes) attaches anteriorly, and the ischium (plural ischia) attaches posteriorly. In most of the lizard-hipped or saurischian dinosaurs, the pubes point anteroventrally and the ischia posteroventrally (Fig. 8.9A–B). However, in the bird-hipped or ornithischian dinosaurs and certain lizard-hipped groups, the pubes point posteroventrally as well (Fig. 8.9C–E). Nevertheless, a pubis can always be distinguished from an ischium because the pubis attaches to the ilium anterior to the ischium. The ilium, pubis, and ischium form an open, round hole in the pelvis. Called the acetabulum (plural acetabula), this opening is the hip socket. In most tetrapods, including most mammals, turtles, lizards, and crocodiles, the acetabulum has a solid sheet of bone forming the medial wall of the socket. This condition is called a closed acetabulum. The Dinosauria, however, are specialized in having an open or perforate acetabulum, one in which there was a hole all the way through the socket, and thus no medial wall of bone.

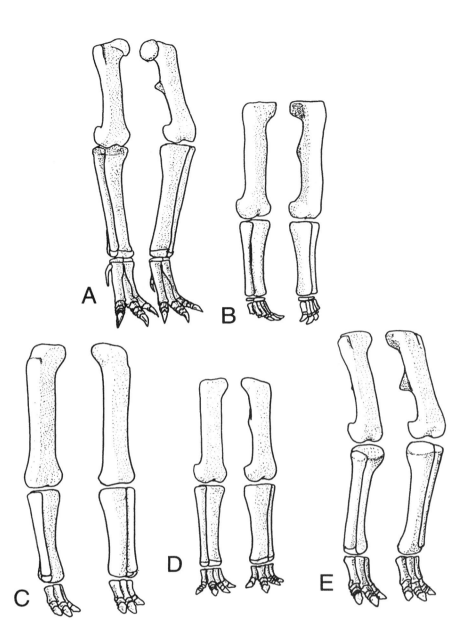

8.10. Hind limbs of the Dinosauria, in oblique right anterolateral view. (A) *Tyrannosaurus rex;* (B) *Apatosaurus louisae;* (C) *Stegosaurus stenops;* (D) *Agujaceratops mariscalensis;* (E) *Corythosaurus casuarius.* Not to scale.

Illustration by Tracy L. Ford.

The pattern of bones in the hind limb, or leg, closely matches that of the forelimb (Fig. 8.10). There is a single upper leg, or thigh, bone, the femur (plural femora). The femur joins the lower leg at the knee, but there is no well-formed kneecap in the leg of a dinosaur. There are two bones in the lower leg: the tibia ("shin bone," plural tibiae), the larger and more medial of the two, and the fibula (plural fibulae), the thinner and more lateral one. Distal to the tibia and fibula are the tarsals, the small bones of the ankle. Unlike the complex ankle region of most tetrapods, the tarsals of dinosaurs are fairly simple. Two proximal elements, the larger, more medial astragalus (plural astagali) and the smaller, more lateral calcaneum (plural calcanea), adhere to the distal ends of the tibia and fibula, respectively. The other tarsals form a row of bones adhering to the long bones of the foot. There is no pronounced heel (posterior projection) in the ankle of dinosaurs, only a roller joint between the astragalus/calcaneum and the distal tarsals. The long bones of the foot are called the metatarsals and are numbered I to V in a medial-to-lateral fashion. Like the fingers, the toes are called digits, numbered from the medialmost (I, the big toe in humans) to the lateralmost (V, the little toe in humans). Again as in the manus, each toe bone is a phalanx (plural phalanges), and the distalmost phalanges are unguals. The digits, metatarsals, and tarsals are collectively called the foot, or pes (plural pedes).

Unlike crocodiles, bears, and humans, which are plantigrade (flat-footed), dinosaurs are digitigrade. This means that dinosaurs walked on their toes, as chickens, cats, and dogs do. In order to distribute the weight of the animal and to act as a shock absorber, there was a pad of cartilage and connective tissues behind the foot. When you see the footprint of a large dinosaur, the front edges are marked by the bony claws, while the main depression is made by the nonbony pad.

Some dinosaurs also have a second set of bones in the body that arise out of the epidermis (outside skin). These bony growths, or osteoderms, form the many and varied patterns of armor seen in many dinosaurs. Most notable of these are the plates and spikes in stegosaurs and ankylosaurs. These were all anchored in the skin by connective tissue.

In order to strengthen the vertebrae in ornithischians, many of the tissues that connected the vertebrae together became filled with calcium and literally turned to bone. These are the famous ossified tendons of the bird-hipped dinosaurs, and they look somewhat like parallel strands of spaghetti. Those that occur below the tail and run across the chevrons between the caudal vertebrae are called hypaxial tendons. Those that occur above the vertebral centra and run across the neural spines are called epaxial tendons. In these dinosaurs, the base of the tail is stiff and not very mobile relative to the hips, while the tail becomes more mobile further posteriorly. In some saurischians (particularly some of the more advanced theropods), another path is followed. Instead of ossifying the tendons, the prezygapophyses of the vertebrae started to elongate and grow over several vertebrae at one time. In *Deinonychus*, these zygapophyses can cover as many as 12 vertebrae at one time. Similarly, the chevrons of *Deinonychus* were elongated to stiffen the tail. In dinosaurs such as this, the tail was most mobile anteriorly and immobile distally, the opposite condition from what is found in ornithischians.

References

Baumel, J. J., and L. M. Witmer. 1993. Osteologia. In J. J. Baumel, J. E. Breazile, H. E. Evans, and J. C. Vanden Berge (eds.), *Handbook of Avian Anatomy: Nomina Anatomica Avium*, 2nd ed., 45–132. Publications of the Nuttall Ornithological Club 23.

Carrier, D. R., and C. G. Farmer. 2000. The evolution of pelvic aspiration in archosaurs. *Paleobiology* 26: 271–293.

Carroll, R. L. 1988. *Vertebrate Paleontology and Evolution*. New York: W. H. Freeman.

Desmond, A. 1982. *Archetypes and Ancestors: Paleontology in Victorian London, 1850–1875*. Chicago: University of Chicago Press.

Galton, P. 1990. Stegosauria. In D. B. Weishampel, P. Dodson, and H. Osmólska (eds.), *The Dinosauria*, 435–455. Berkeley: University of California Press.

Gilmore, C. W. 1936. Osteology of *Apatosaurus*, with special reference to specimens in the Carnegie Museum. *Memoirs of the Carnegie Museum* 11: 175–300.

Harris, J. D. 2004. Confusing dinosaurs with mammals: tetrapod phylogenetics and anatomical terminology in the world of homology. *Anatomical Record, Part A* 281A: 1240–1246.

Lehman, T. 1989. *Chasmosaurus mariscalensis*, sp. nov., a new ceratopsian dinosaur from Texas. *Journal of Vertebrate Paleontology* 9: 137–162.

Nomina Anatomica. 1983. Baltimore, Md.: Williams and Wilkins.

Nomina Anatomica Veterinaria. 1983. Ithaca, N.Y.: World Association of Veterinary Anatomists, Cornell University Press.

Osborn, H. F. 1916. Skeletal adaptations of *Ornitholestes, Struthiomimus, Tyrannosaurus. Bulletin of the American Museum of Natural History* 35: 733–771.

Ostrom, J. H., and J. S. McIntosh. 1966. *Marsh's Dinosaurs: The Collections from Como Bluff.* New Haven, Conn.: Yale University Press.

Owen, R. 1846. Report on the archetype and homologies of the vertebrate skeleton. *Report of the British Association for the Advancement of Science, Southampton Meeting,* 169–340.

———. 1849. *On the Nature of Limbs.* London: Van Voorst.

Paul, G. S. 2008. A revised taxonomy of the iguanodont dinosaur genera and species. *Cretaceous Research* 29: 192–216.

Romer. A. S. 1956. *Osteology of the Reptiles.* Chicago, Ill.: University of Chicago Press.

Torrens, H. S. 2011 (this volume). Politics and paleontology: Richard Owen and the invention of dinosaurs. In M. K. Brett-Surman, Thomas R. Holtz Jr., and James O. Farlow (eds.), *The Complete Dinosaur,* 2nd ed., 24–43. Bloomington: Indiana University Press.

Weishampel, D. B., and J. R. Horner. 1990. Hadrosauridae. In D. B. Weishampel, P. Dodson, and H. Osmólska (eds.), *The Dinosauria,* 534–561. Berkeley: University of California Press.

Weishampel, D. B., P. Dodson, and H. Osmólska (eds.) 2004. *The Dinosauria.* 2nd ed. Berkeley: University of California Press.

Williston, S. W. 1925. *Osteology of the Reptiles.* Cambridge, Mass.: Harvard University Press.

Wilson, J. A. 1999. A nomenclature for vertebral laminae in sauropods and other saurischian dinosaurs. *Journal of Vertebrate Paleontology* 19: 639–653.

Wilson, J. A. 2006. Anatomical nomenclature of fossil vertebrates: standardized terms or *"lingua franca"? Journal of Vertebrate Paleontology* 26 (3): 511–518.

Witmer, L. M. 2001. Nostril position in dinosaurs and other vertebrates and its significance for nasal function. *Science* 293: 850–853.

Yates, A. M., and C. C. Vasconcelos. 2005. Furcula-like clavicles in the prosauropod dinosaur *Massospondylus. Journal of Vertebrate Paleontology* 25: 466–468.

9.1. How muscles control joint movement. Illustrated with a 3D digitized right hind limb of *Tyrannosaurus* (Museum of the Rockies specimen MOR 555). See Hutchinson et al. (2005), Hutchinson and Gatesy (2000), and Hutchinson and Garcia (2002) for more detailed explanation. A, Right side view with the hip, knee, ankle, and toe joints labeled (black arrows indicate joint angles), with lines representing the trunk, thigh, shank, metatarsus, and foot segments. Flexion acts to decrease the joint angles shown; extension acts to increase them. B, Front view of a body cross section (right half). The hip abduction angle is shown (lower black curved arrow). Abduction of the thigh draws it away from the body midline, in the plane of the picture (upper black curved arrow). Medial and lateral rotation act about the long axis of the femur (gray curved arrow).

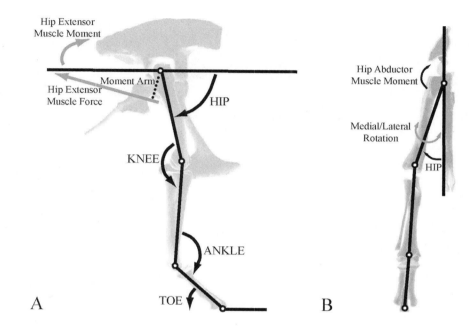

Reconstructing the Musculature of Dinosaurs

David W. Dilkes, John R. Hutchinson,
Casey M. Holliday, and Lawrence M. Witmer

9

Whenever one attempts to flesh out the skeleton of a dinosaur, some or most of the muscles must be reconstructed. However, evidence of muscular tissue is only rarely preserved in the fossil record (Kellner 1996; Briggs et al. 1997; Dal Sasso and Signore 1998). Why is the exercise of reconstructing muscles important when it seems to have so little actual evidential support? Perhaps more so than any other aspect of its soft anatomy, the addition of muscles helps to breathe life into a dinosaur and transform it from a collection of bones into a functioning organism that interacted with its environment. Images of power, agility, swiftness of movement, and complex behaviors may be conveyed to the viewer by the reconstruction of muscles. It may be tempting to view such efforts as outside the realm of science and more the concern of paleoartists and museum exhibit designers. Yet there is no sound reason to be dismissive, and there is as much valid science in placing a muscle on a skeleton as in generating a cladogram or producing any other hypothesis in historical science.

By virtue of the nature of the available data in fossils, paleontologists must use patterns of soft tissue arrangements in living species to determine whether or not a particular tissue was present in a dinosaur (or any other extinct organism, for that matter) and the most likely pattern or arrangement of that tissue. Historically, the basic approach to the reconstruction of muscles in a dinosaur has been to select one or perhaps more living models judged to be relevant because they are closely related to dinosaurs or are distinctly primitive reptiles. The muscle patterns of birds, crocodiles, and lizards have been selected most often as models for dinosaur muscles. The pattern of muscles in one of these living models was assumed to be true also for the dinosaur, and its muscles were fitted to the bones of the dinosaur.

For example, the muscles of the pelvic girdle and hind limbs of ornithischian dinosaurs were often reconstructed according to the pattern observed in living birds because both groups have a pubis that has shifted posteriorly against the ischium (Gregory 1919; Romer 1927a; Galton 1969). The rationale was that convergence in osteology suggested convergence in musculature. On the other hand, pelvic girdle and hind limb muscles of saurischian dinosaurs, especially theropods such as *Tyrannosaurus rex*, were often reconstructed according to a crocodylian model because crocodiles were viewed as archaic reptiles with a primitive pattern of muscles for archosaurs (Romer 1923c). Less consistency has been present in past choices of living models for the reconstruction of the muscles of the pectoral girdle and forelimbs of dinosaurs. Lizards and crocodiles have been used as the model for sauropod forelimb muscles (Borsuk-Bialynicka 1977), lizards alone for hadrosaur forelimb muscles (Lull and Wright 1942), and crocodiles and birds for the forelimb muscles of the ornithopod *Iguanodon* (Norman 1986). Birds alone have been used recently for restoring the forelimb muscles of *T. rex* (Carpenter and Smith 2001).

Recent studies on musculature of dinosaurs (Dilkes 2000; Hutchinson 2001a, 2001b; Carrano and Hutchinson 2002; Jasinoski et al. 2006; Langer et al. 2007) have emphasized the importance of placing any reconstructions within an appropriate phylogenetic context in order to trace out the homology of features on bones that are causally associated with a muscle, the likely presence of a muscle, and its attachment site or sites. As explained in detail by Bryant and Russell (1992) and the extant phylogenetic bracket method of Witmer (1995a), this approach generates explicit hypotheses for soft tissues and allows one to evaluate the levels of confidence in the reconstruction of each muscle. More than a single living group must be used to establish a connection between the attachment of a muscle and a bony feature, and the homologies of these bony features must be demonstrated by incorporating numerous fossil taxa. In this way, one can produce hypotheses for the reconstruction of a muscle or other soft tissue and its likely evolutionary history that are testable by both phylogeny and osteology. A method sometimes used in previous approaches—assuming a function of a muscle and then reconstructing the anatomy that best fits that function—is not favored (Lauder 1995; Witmer 1995a).

Quality and quantity of data from extant taxa and evidence of homology are also critical factors. While published accounts do provide much data on muscle arrangements, dissection checks the accuracy of these accounts and reveals individual variation within a species and variation between species so that the information is not reliant on a single taxon as an exemplar for a larger clade. Descriptions of the forelimb muscles of crocodylians highlight examples of variation in attachment sites (Meers 2003; Jasinoski et al. 2006). Recent work (Abdala and Diogo 2010) has greatly clarified homologies of muscles of the pectoral girdle and forelimb among tetrapods.

Regardless of how one attempts to reconstruct a muscle in a dinosaur, the strength of that hypothesis rests largely on the strength of the connection between a feature on the skeleton and the attachment of that muscle. Simply pointing out that a muscle was likely to have been present is merely a single step of the process; one must also determine where the muscle most likely attached to a bone. Not surprisingly, there are numerous complications for reconstruction of muscles. Muscles exist in many shapes and their attachments can vary from a well-defined tendon to a broad, ill-defined insertion onto a flat or concave surface. There exists considerable variation in the number of features that can be linked to the attachment of a muscle. Ridges, crests, projections, and trochanters are all candidates for the attachment of muscles, but not every example of these structures is necessarily an attachment site (McGowan 1979, 1982).

Furthermore, not all muscle attachments will leave evidence in the form of a discrete scar. A muscle scar is associated typically with the attachment of a tendon and is formed by the deposition of crystals of the mineral apatite along the collagen fibers of the tendon. These mineralized fibers are known as Sharpey's fibers and can appear as a region of often highly prominent bumps and pits. Muscle scars are less seldom present when a muscle has a broader or fleshy attachment (Bryant and Seymour 1990). Relative sizes of muscles are often discussed in terms of relative areas of attachment to a bone. However, there is not always a one-to-one correspondence between attachment area and the size of a muscle because muscles may attach to neighboring muscles as well as bone. If the area of attachment is well defined and modern taxa suggest that the muscle did not likely attach to other muscles, one can attempt to estimate size and strength of a muscle.

Our primary goals in this chapter are to reconstruct the most likely lines of actions of the major cephalic (head), axial, and appendicular (limb) muscles of dinosaurs, to point out problems of homology, to infer the likely functional roles of certain groups of muscles, to outline controversies of musculoskeletal function in dinosaurs, and to suggest possible future directions for research. More detailed discussions of many of these topics may be found in the references.

Muscle Actions

A muscle exerts a force (Fig. 9.1) along its line of action. The shortest distance from the line of action to the center of the joint is called the moment (or lever) arm, and the muscle force times the moment arm is the muscle moment—the rotational force (torque) applied about the joint. These four terms encompass the fundamentals of musculoskeletal function (Fig. 9.1).

By *function*, we mean how the muscles rotate (or prevent rotation of, in the opposite direction of the muscle's function) joints that they span, and hence in what direction they move (or prevent movement of) body segments. Muscular anatomy is often complex; muscles usually have moment arms about more than one functional axis of a joint, or even more than one joint. Determining muscle function accurately is thus an extremely complex exercise for living animals (e.g., Zajac 2002; Gatesy et al. 2010), and it is even more so for extinct dinosaurs. Therefore, the line of action of the muscle (and the corresponding moment arm) is often assumed to tell the main function of the muscle. This inference is often correct, but caution is warranted in accepting such inferences of function without more rigorous tests (e.g., Gatesy 1994; Lauder 1995). Indeed, muscles often follow quite curved paths (see various figures herein), not straight lines from origin to insertion, further complicating reconstruction of their anatomy and function.

We categorize muscles here according to their three major actions, which apply to any limbed animal, for convenience of description and because this is historical practice for anatomical descriptions, despite the important caveats already mentioned. Those actions (Fig. 9.1) are flexion/extension, abduction/adduction, and lateral/medial rotation. Protraction/retraction are actions often used to describe the movements of whole body segments rather than a joint and are analogous to flexion/extension.

Cephalic Musculature

Cephalic myology remains a complicated and only partially understood aspect of dinosaur biology. Although there are several different muscular systems within the head, only the jaw and neck muscles have received significant scrutiny. This section introduces different groups of head muscles known in extant archosaurs and highlights their general attachment areas. This information is then used as the basis for inferring muscular anatomy in dinosaurs. Past studies of dinosaur head myology and several controversies regarding particular muscle groups are then reviewed. Finally, some novel research avenues are presented that may address the numerous questions still plaguing this aspect of dinosaur functional morphology.

Cephalic musculature in amniotes is composed of several interrelated groups with differing functions. Muscles associated with the feeding apparatus are primarily responsible for the movement of the jaws, tongue, and throat during food capture, processing, and ingestion. They also perform secondary functions during vocalization and manipulation of objects. Other muscle groups are associated with modulation of the fleshy nostril, movement of the eye and eyelids, stabilization of the columella and eardrum, control of movements between the head and neck, and even the ruffling of feathers in the skin in birds. Experimental and modeling studies demonstrate the impact of muscles on skull shape (Bock 1974; Herring 1993; Rayfield et al. 2001; Herrel et al. 2002). Muscles occupy a large volume of the head, constraining the size, shape, or course of other structures such as arteries, eyes, and air sinuses, and the topological relationships of these structures and associated bones have been used to identify homologous soft and hard tissue structures in a number of vertebrates (Edgeworth 1936; Säve-Söderbergh 1945; Witmer 1995b, 1997; Holliday et al. 2006; Holliday and Witmer 2007; Holliday 2009). Cephalic musculature clearly plays a number of roles involved with development, architecture, and function of the skull, and thus its reconstruction is critical to inferring feeding behavior and ecology in dinosaurs.

Previous Research

Cephalic muscular anatomy of extant archosaurs and sauropsids in general is well known (for more complete descriptions and bibliographies, see Haas 1973; Schumacher 1973; Elzanowski 1987; Vanden Berge and Zweers 1993; Cleuren and de Vree 2000; Schwenk 2000; Holliday and Witmer 2007). Despite the wealth of anatomical data in extant taxa, the homologies of some of the cephalic muscle groups are only beginning to be understood in extant Archosauria (e.g., Iordansky 2000; Zusi and Livezey 2000; Tsuihiji 2005, 2010; Holliday and Witmer 2007, 2009), and evolutionary patterns of muscular anatomy have only recently begun to be explored in dinosaurs (Holliday 2009). These problems have led to disparate muscular nomenclatures among crocodylians, birds, and sauropsids in general, although Tsuihiji (2005), Holliday and Witmer (2007), Snively and Russell (2007a), Holliday (2009), and Tsuihiji (2010) made strides to bridge this gap. We will briefly introduce the common nomenclature but focus on the general attachments of cephalic muscles in extant archosaurs (birds and crocodylians). Minimally, this provides a framework to constrain some of the functional inferences of dinosaur jaw muscles; it also addresses some problems that are common in reconstructions of dinosaur head musculature.

Nonfeeding Muscles

Jaw muscles have rightly captured the attention of most workers, but they are not the only muscles within the heads of vertebrates. Although these other muscles obviously are critical to an animal's biology, their study is only in the early stages, and they consequently are not emphasized here. For example, narial muscles are found in extant crocodylians and many squamates, but not birds (Witmer 1995b). These muscles constrict or dilate the fleshy nostril in concert with narial vascular erectile tissues. Although the narial muscles rarely leave unambiguous osteological correlates that can be checked for in dinosaurs, the erectile tissues often do (Witmer 2001). Eye muscles are another group that previously has received virtually no attention from dinosaur paleontologists. Six extraocular muscles attach the bulb of the eye to the bony orbit and are responsible for controlling movements of the eyeball. Moreover, accessory muscles move the eyelids or indirectly move the orbital contents. The osteological correlates of eye muscles are subtle but generally identifiable in extant archosaurs. Certainly the foramina associated with the relevant cranial nerves (CN III, IV, VI) are often identified in descriptive works, and the surrounding bony landscape also reveals telltale muscle scars. These details, when correctly analyzed, may reveal insights into the position of the eyeball within the orbit, general orientation of the visual axis, and relative eye mobility in dinosaurs (Schmitz and Motani 2010).

Hyolingual Muscles

Hyolingual muscles include muscles that move the tongue and hyoid apparatus. Tongues are muscular organs that are key components of the feeding apparatus in many vertebrates. Lingual anatomy, however, is difficult to reconstruct because the hyobranchial elements that support the tongue are largely cartilaginous. Moreover, the hyoid elements that do ossify are only loosely connected to the skull and are often lost. The hyoid apparatuses of crocodylians and birds seem to differ somewhat, but in fact they are simplified in similar ways in comparison to other sauropsids (e.g., loss of the second ceratobranchial and ceratohyal; Schumacher 1973; Tomlinson 2000). Likewise, the tongues of extant archosaurs share relatively restricted mobility. In crocodylians the tongue is broad, fatty, and only weakly mobile, and in birds the tongue is thin, often cornified, and is generally limited to fore–aft movement. In neither archosaur group is there anything like the mobility seen in mammalian and lepidosaurian tongues (Schwenk 2000), which have a complicated intrinsic musculature that archosaurs lack.

Elements reported as hyoid bones are known from most clades of dinosaurs (e.g., ceratopsians: Colbert 1945; sauropods: Gilmore 1925; McIntosh 1990; coelophysids: Raath 1977; tyrannosaurids: Gilmore 1946; maniraptorans: Ji et al. 1998), but the morphological diversity of these elements has not been assessed systematically. In most cases, these bones are simple rods that can be interpreted as the first ceratobranchials. What little that is known about dinosaurs suggests that they resembled extant archosaurs in lacking second ceratobranchials and ceratohyals, and thus likewise may have had tongues of limited mobility. Nevertheless, Carpenter (1997) inferred highly mobile tongues in ankylosaurs on the basis of their elongate hyoids, and Brett-Surman (1997) suggested that some ornithopod tongues were capable of extending rostrally around the cheek teeth to clear the oral vestibule of food. The diversity of archosaur, sauropsid, and amniote hyobranchial functional morphology complicates inferences of the structures and their functions in dinosaurs. At this stage, it is not entirely clear just what dinosaurs could do with their tongues, but a fair statement is that we currently have little reason to believe that dinosaurs had particularly mobile tongues.

Jaw Muscles

Jaw muscles are composed of a complex assortment of pinnate and straplike muscles adapted to either generating force or increasing bite speed. They attach between different portions of the braincase, palate, and mandible. The bulk of the muscles are jaw adductors that are responsible for closing the mouth (mm. adductor mandibulae externus, pterygoideus, pseudotemporalis, adductor mandibulae posterior), whereas other muscles aid in abducting (opening) the mouth by depressing the lower jaw (e.g., m. depressor mandibulae) or move or stabilize bones of the palate (e.g., mm. levator pterygoideus, protractor pterygoideus, protractor quadratus) (Fig. 9.2).

Many broad, comparative studies described gross patterns of the jaw musculature in Sauropsida as well as Crocodylia and Aves (for reviews, see Säve-Söderbergh 1945; Schumacher 1973; Elzanowski 1987; Vanden Berge and Zweers 1993; Cleuren and de Vree 2000; Schwenk 2000; Holliday and Witmer 2007). Traditionally, jaw muscles have been identified on the basis of their topological relationships to the divisions of the trigeminal nerve and particular bones, although Rieppel (1978) showed that these criteria are not always infallible. The arrangement of archosaur jaw muscles is not the same as that in mammals, or even squamates and turtles. Holliday and Witmer (2007) found that the divergent specialization of crocodylians included a departure from the common neuromuscular topology found in birds, lizards, and turtles. The relationships of the jaw muscles to other structures in the head (e.g., blood vessels, nerves, air sinuses) appear to be quite similar in dinosaurs and other sauropsids, and fairly robust hypotheses of homology have been proposed by Holliday (2009) on the basis of morphological and transformation evidence. However, systematic comparisons of extant archosaur head development have not been adequately integrated with those of Lepidosauria (and more broadly, Sauropsida), and phylogenetically informative characters have yet to be incorporated into analyses to test these morphology-based hypotheses. Nevertheless, the wealth of anatomical and functional data on extant archosaurian jaw muscles provide adequate information to illustrate the major anatomical and functional trends of dinosaurian jaw muscles. We will group the muscles on the basis of their traditional identifications, briefly describe the major functional units of the jaw musculature in extant archosaurs, and discuss their significance in dinosaur biology.

M. CONSTRICTOR INTERNUS DORSALIS

These deep muscles include mm. levator pterygoideus, protractor pterygoideus, and protractor quadratus as well as the extraocular muscle m. depressor palpebrae inferioris. They connect the rostrolateral part of the braincase to the medial surfaces of the pterygoid and quadrate (Fig. 9.2). Powered kinesis, a largely avian and squamate phenomenon (Metzger 2002),

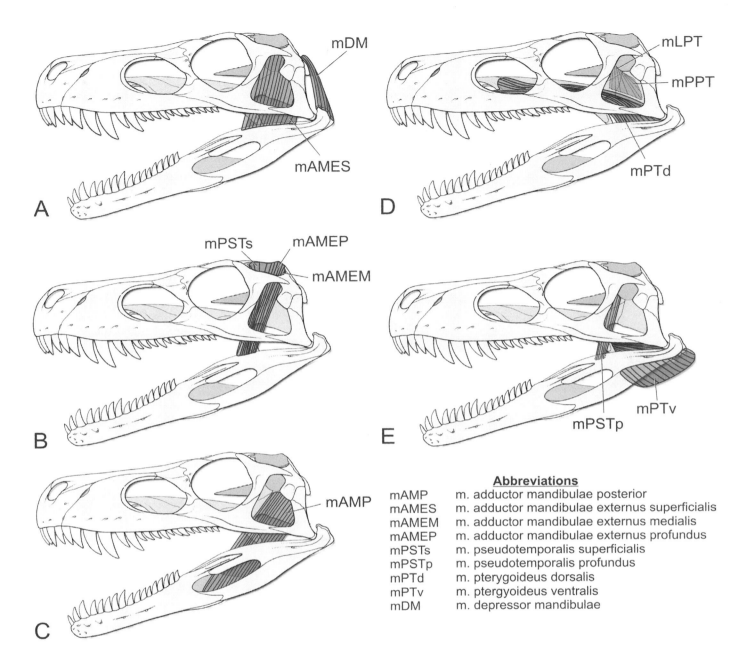

Abbreviations

mAMP	m. adductor mandibulae posterior
mAMES	m. adductor mandibulae externus superficialis
mAMEM	m. adductor mandibulae externus medialis
mAMEP	m. adductor mandibulae externus profundus
mPSTs	m. pseudotemporalis superficialis
mPSTp	m. pseudotemporalis profundus
mPTd	m. pterygoideus dorsalis
mPTv	m. ptergyoideus ventralis
mDM	m. depressor mandibulae

9.2. Generalized reconstruction of hypothesized jaw muscles in a basal dinosaur (e.g., *Herrerasaurus*) modified from Sereno and Novas (1992).

involves movement of the palatoquadrate by these muscles, which in turn abducts the rostrum independently of the braincase. Crocodylians do not have protractor muscles. In birds, the muscles attach to the interorbital septum and the ventrolateral surface of the laterosphenoid ventral to the orbital contents. They then attach to the medial and dorsomedial aspects of the pterygoid and quadrate. Weishampel (1984) identified scars for levator pterygoideus and protractor quadratus in *Hypsilophodon* and levator ptery-goideus in *Corythosaurus* and related these muscles to the pleurokinetic apparatus of ornithopods. Holliday and Witmer (2008) and Holliday (2009) identified numerous dinosaur taxa with scars for m. protractor pterygoideus and some with m. levator pterygoideus, showing that virtually all dinosaurs likely possessed some form of these muscles. The scars for the muscles are large in tyrannosaurs and hadrosaurs, and modest in size in most other taxa. However, the functional significance of these muscles remains questionable because even clearly akinetic taxa like ankylosaurs bear osteological correlates of protractor muscles.

M. ADDUCTOR MANDIBULAE INTERNUS

This muscle group is bounded medially by the braincase and laterally by the maxillary division of the trigeminal nerve (CN V_2) in sauropsids. The muscle group is always further divided into the pseudotemporalis and the pterygoideus muscles.

M. pseudotemporalis superficialis.

M. pseudotemporalis superficialis consistently attaches to the lateral surface of the laterosphenoid in extant archosaurs and also in nonavian dinosaurs (Holliday 2009). The muscle then attaches near the medial region of the coronoid process in all archosaurs. In crocodylians, the muscle bears a large sesamoid cartilage, the cartilago transiliens, and may actually join as part of a two-bellied muscle with m. intramandibularis, which attaches to the medial mandibular fossa (Holliday and Witmer 2007). The cartilago transiliens is a keystone structure in crocodylian jaw musculature, connecting several different muscles and tendons to the mandible. In ratites, M. pseudotemporalis superficialis occupies the entirety of the temporal fossa (Elzanowski 1987; Holliday and Witmer 2007) but is quite small in most other birds, where it usually attaches to the rostral lamina of the laterosphenoid rostroventromedial to the temporal fossa.

M. pseudotemporalis profundus.

M. pseudotemporalis profundus is a smaller muscle that attaches to the lateral surface of the epipterygoid and then to the medial surface of the mandible in lizards and turtles. An attachment on the epipterygoid is likely the case in many dinosaur taxa (e.g., nonavian theropods, most primitive taxa of each major clade) in that some taxa (e.g., tyrannosaurs) have fossae for the muscle on the element. However, both extant archosaur taxa have lost the epipterygoid, as have derived sauropods, ornithopods, and ceratopsians, making inferences of the muscle's presence or attachments difficult in these fossil groups (Holliday 2009). That said, crocodylians maintain a vestige of this muscle, which attaches from the ventral laterosphenoid to the medial mandibular fossa, and birds maintain large bellies of the muscle where it attaches to the orbital process of the quadrate. Thus a dinosaur without an epipterygoid may still have M. pseudotemporalis profundus, but interpreting where it attached can be challenging.

Mm. pterygoideus dorsalis and ventralis (Fig. 9.2B).

This muscle is typically partitioned into a dorsal (anterior) and ventral (posterior) portion in archosaurs, although birds often exhibit further subdivisions (e.g., Webb 1957; Witmer 1995b; Holliday and Witmer 2007). The dorsal belly runs from the dorsal surface of the palatine and pterygoids to the caudomedial aspect of the mandible, usually ventral to the jaw joint. In crocodylians, the ventral belly attaches to the caudal edge of the pterygoid and wraps around the retroarticular process to then attach on the caudolateral aspect of the mandible. This mandibular attachment is easily identified in many dinosaurs that have a smooth fossa on the caudolateral surface of the mandible. In birds, the dorsal and ventral pterygoideus muscles attach on the respective surfaces of the palatine and pterygoid and then to the caudomedial surface of the mandible, usually on the medial mandibular process.

M. ADDUCTOR MANDIBULAE EXTERNUS

M. adductor mandibulae externus lies lateral to the maxillary division of the trigeminal nerve (CN V₂) and medial to the temporal bars and skin. Morphology and specific nomenclature for this complex's various bellies is the most labile among extant archosaurs. Holliday and Witmer (2007) and Holliday (2009) provided a comprehensive review of nomenclature and found that the homologies of the bellies of mm. adductor mandibulae externus profundus, medialis, and superficialis were the best supported across archosaurs and sauropsids as a whole, although these parts themselves may develop subdivisions in particular groups, like birds.

M. adductor mandibulae externus profundus.

This muscle attaches to the lateral surface of the braincase in all archosaurs except ratites. It is the only muscle in the dorsotemporal fossa in crocodyliforms (Holliday and Witmer 2009), whereas it occupies the caudomedial portion of the fossa in dinosaurs, medial to m. adductor mandibulae externus medialis. The muscle consistently attaches to the lateral portions of the coronoid eminence or process in crocodylians, dinosaurs, and birds.

M. adductor mandibulae externus superficialis.

This muscle attaches across the upper temporal bar formed by the postorbital and squamosal and is the most superficial, or laterally positioned, jaw muscle. This parallel-fibered muscle attaches to the dorsolateral surface of the surangular, often leaving a broad, flat shelf marking its attachment. The muscle may also attach to the inner surface of the temporal fascia.

M. adductor mandibulae externus medialis.

The position of m. adductor mandibulae externus medialis is more difficult to infer precisely, although it is certain dinosaurs possessed the muscle. Correlates of the muscle can occasionally be identified on the caudal surface of the dorsotemporal fossa on the parietal or squamosal; however, most of its attachments are actually to the fascia and tendons of surrounding external muscles, making identification challenging. The muscle likely attached between the surangular and coronoid eminence in dinosaurs.

M. adductor mandibulae posterior.

M. adductor mandibulae posterior lies caudal to the maxillary nerve and medial to the mandibular nerve. It consistently attaches to the body of the quadrate in all living reptiles and also likely in dinosaurs. The muscle is pinnate in crocodylians, but is parallel fibered in birds, lizards, and turtles. The muscle rarely leaves muscle scars on the quadrate of dinosaurs, indicating that it was also likely parallel fibered. M. adductor mandibulae posterior consistently attaches to Meckel's cartilage within the caudal half of mandibular fossa, along the dorsal surface of the angular. M. adductor mandibulae posterior is the muscle that lies deep to the external mandibular fenestra in dinosaurs and other archosaurs.

M. intramandibularis.

This muscle attaches to the rostral portion of the mandibular fossa in crocodylians and some birds (e.g., *Struthio*). Dorsally, it attaches to the cartilago transiliens in crocodylians and from an intertendon shared with m. pseudotemporalis superficialis in birds. As noted above, Holliday and Witmer (2007) suggested M. intramandibularis actually may be a continuation of the pseudotemporalis muscle, rather than being a separate muscle, with the cartilago transiliens and avian intertendons being sesamoids forming within the muscle. At this time, it is difficult to identify osteological correlates of M. intramandibularis

M. intermandibularis.

This broad, thin muscle spans the floor of the mouth in living and fossil archosaurs and reptiles. Each side of the muscle attaches along the medial surface of the dentary and then converges along a midline raphe. M. intermandibularis is innervated by the mandibular division of the trigeminal nerve and is likely homologous with the mammalian m. mylohyoid.

M. depressor mandibulae.

M. depressor mandibulae is innervated by the facial nerve (CN VII) and is responsible for the powered opening (abduction) of the jaws. The muscle attaches to the caudolateral surface of the braincase, dorsolateral to the occipital condyle (Fig. 9.2). It attaches distally to the retroarticular process of the mandible. In some birds, it has several bellies, attaching to lateral and medial aspects of the process, and is often used in prying and gaping. In lizards it has an extra belly that extends out on the dorsolateral portion of the neck. Like other archosaurs, in dinosaurs, m. depressor mandibulae likely attached to the lateral portion of the paroccipital process and the retroarticular process, wrapping around the caudal edge of the external acoustic meatus and tympanic membrane.

Case Studies in Dinosaurs

Jaw muscles are one of the most studied soft tissues in dinosaurs and, along with tooth morphology (e.g., Crompton and Attridge 1986) and mechanical analysis (e.g., Molnar 1998; Rayfield et al. 2001; Henderson 2002), are common lines of evidence of feeding biology and skull function. Jaw muscles have been reconstructed in numerous fossil archosaurs, including basal archosaurs (Anderson 1936; Crompton and Attridge 1986; Walker 1990), basal dinosaurs (Crompton and Attridge 1986; Galton 1986), ankylosaurs (Haas 1969), ceratopsians (Lull 1908; Russell 1935; Haas 1955; Ostrom 1964, 1966; Dodson 1996; Sereno et al. 2010), ornithopods (Ostrom 1961; Galton 1974; Weishampel 1984; Crompton and Attridge 1986; Rybczynski et al. 2008), sauropods (Janensch 1936; Haas 1963; Zhang 1988), nonavian theropods (Adams 1919; Molnar 1973, 1998; Raath 1977; Horner and Lessem 1993; Witmer 1997; Bakker 1998; Rayfield et al. 2001; Bimber et al. 2002), and birds (Bühler et al. 1988; Witmer and Rose 1991).

Jaw muscle reconstructions in dinosaurs have ranged from assessments of attachment sites (Dollo 1884; Haas 1955; Ostrom 1966; Witmer 1997; Holliday 2009) and analyses of relative function (e.g., comparative angulation of sauropod adductor muscles; Upchurch and Barrett 2000) to measurements of volumes of sculpted muscles (Rayfield et al. 2001). All of these reconstructions prove useful and are common exercises in hypothesizing function (e.g., Walker 1990; Mazzetta et al. 1998). However, many of these studies suffer in that they prematurely hypothesize higher-order inferences such as feeding behavior and ecology before rigorous anatomical or phylogenetic tests of their reconstructions were made. In some cases, jaw muscles were reconstructed on the basis of iguanian (e.g., Haas 1955, 1969) or mammalian (e.g., Lull 1908; Adams 1919) muscular patterns. Problems such as these have led to arguments regarding the attachment site of muscles (e.g., the antorbital cavity) (Fig. 9.3) or even the presence of particular muscles (e.g., cheeks). The rest of this section will illustrate functional inferences that can be drawn from jaw muscle reconstructions and further explore some of these controversies.

Antorbital Cavity

Misunderstanding systemic interactions in the head can lead to dubious hypotheses of muscular attachment and subsequently errant functional interpretations of bony structures. Moreover, such reconstructions of muscles, or other soft tissues for that matter, can significantly affect functional hypotheses of not only the muscles, but also other neighboring systems, such as pneumatic or sensory structures. For example, the role of the antorbital cavity has been postulated to be the attachment site of the dorsal pterygoideus muscle (e.g., Adams 1919; Galton 1974; Bakker 1986, 1998; Horner and Lessem 1993; Molnar 1998, 2008) (Fig. 9.3). The muscle was hypothesized to attach on the lateral margin of the antorbital cavity, then course medially into the antorbital fossa and over the dorsal surface of the palate, then attach to the medial surface of the mandible (Fig. 9.3A). This attachment site was then used to infer enormous bite forces and predatory behavior, particularly in tyrannosaurs (Adams 1919; Molnar 1998). However, Witmer (1995b, 1997) determined on the basis of the consistent topological relationship of facial structures across Sauropsida that the antorbital cavity largely housed a paranasal air sinus. Witmer (1997) did suggest that the dorsal pterygoideus was among the contents of the antorbital cavity, but that it was restricted to the floor of the cavity and did not attach to the margins of the antorbital fossa.

Whether or not the antorbital cavity was filled by muscle has a major impact on inferences of head function. If true, it would mean that the dorsal

9.3. Examples of conflicting cephalic muscle attachments in dinosaurs. A, Dinosaurian (e.g., *Allosaurus*) pterygoid muscle attaches to antorbital cavity (adapted from Molnar 1998) or palate (adapted from Witmer 1997). B, Ceratopsian (e.g., *Triceratops*) temporal muscle attaches across entire frill (adapted from Ostrom 1964) or along margin of dorsotemporal fenestra (adapted from Dodson 1996).

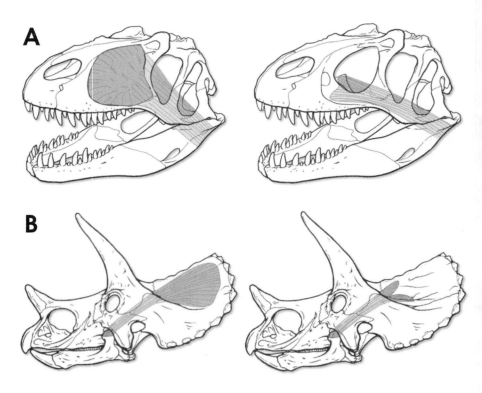

pterygoideus was, in some cases (e.g., *Coelophysis*; Bakker 1986), by far the largest jaw muscle, which in turn affects estimates of maximum bite force and muscle moment (Sinclair and Alexander 1987; Molnar 1998) and of force transmission through the skull (Rayfield et al. 2001). Likewise, such a large muscle would constrain the morphology of neighboring tissues such as air sinuses, the eyes and tongue, and blood vessels. On the other hand, a more modest dorsal pterygoideus, as indicated by available data (Witmer 1997; Holliday 2009), has far different implications for all of the above considerations. Thus, the two different dorsal pterygoideus morphologies not only imply significantly different adaptations to feeding, but also affect functional hypotheses of respiration and sight, and likely all other tissues they neighbor.

Ceratopsian Frills

Another example of ambiguous muscle attachment relates to the frill of ceratopsians. Expanded frills of many neoceratopsians were originally proposed to serve as attachment sites of enlarged jaw musculature (Lull 1908; Haas 1955; Ostrom 1964, 1966) (Fig. 9.3). According to this model, the largely pinnate musculature fanned out across the entire dorsal surface of the frill and then coursed ventrally through the dorsotemporal fenestra to attach on the coronoid process (Fig. 9.3B). However, Dodson (1996) expressed reservation about this attachment hypothesis, supporting a more conservative muscle attachment around the margin of the dorsotemporal fenestra. Moreover, the horns of other ceratopsians or the teeth of predators commonly damaged the frills (Farlow and Holtz 2002; Farke et al. 2009). The bone texture of the frill indicates more of a vascular and keratinous surface than muscular surface (Horner and Marshall 2002; Holliday 2009). Again, these two different reconstructions have different effects on skull function in that a muscular frill would require a much longer and more massive muscle, with consequent implications for mechanics. However, if the frill is not a site for muscular attachment, then other functional explanations must be sought, such as defensive weaponry (Sternberg 1929; Bakker 1986), thermoregulation (Farlow et al. 1976; Wheeler 1978), sexual selection (Farlow and Dodson 1975; Sampson et al. 1997; Sampson 2001; Knell and Sampson 2011), or species recognition (Padian and Horner 2011a, 2011b). Thus, resolution of muscle attachments is necessary for not only justifiable inferences of feeding function, but also behavior and physiology.

Cheeks

The presence of cheek muscles in ornithischian dinosaurs has long been debated (Hatcher et al. 1907; Lull 1908; Russell 1935; Lull and Wright 1942; Haas 1955; Ostrom 1961, 1964; Galton 1973; Witmer 1995a; Papp and Witmer 1998; Papp 2000; Barrett 2001; Knoll 2008). The medially inset dentition and the corresponding bony recesses of many ornithischian skulls have been postulated to be the osteological correlates of a cheek muscle, and associated foramina would transmit the relevant motor nerve bundles. Not only did Galton (1973) use these bony data to support the reconstruction of muscular cheeks, but he further hypothesized that all ornithischian dinosaurs were highly efficient foragers that must have had muscular cheeks to keep food from falling out of their mouths. Furthermore, he additionally postulated that this feature of the feeding apparatus was the key innovation for the adaptive radiation of Ornithischia and subsequent competitive displacement of prosauropod dinosaurs.

However, the evolutionary origin of this cheek muscle is contentious. No living sauropsid has a homolog of the mammalian cheek muscle (buccinator; Witmer 1995a), requiring that ornithischians evolved a "new" cheek muscle. Furthermore, the commonly noted osteological correlates of ornithischian cheeks, buccal emarginations, do not correspond to those of mammals (the only known taxon with cheeks) in that mammalian teeth are not medially inset. Moreover, the nerve foramina mentioned above do not transmit motor nerves in any vertebrate but rather sensory nerves (and also vessels). Papp (2000) suggested the novel jaw morphology of ornithischians is the presence of a more extensive keratinous rhamphotheca. Sereno et al. (2010) hypothesized a novel muscle, the pseudomasseter, as a cheek muscle in *Psittacosaurus* but provided no evidence of the evolutionary transformation leading to this new feature. Interestingly, two specimens of nodosaurid ankylosaur (*Panoplosaurus*, *Edmontonia*) preserve a large osteoderm residing within their buccal emarginations (Lambe 1919), revealing that at least some ornithischians minimally had integument in this region. A novel, dinosaurian cheeklike structure – muscular or not – could have existed, but given the great disparity between the bony structures in mammals and ornithischians, it would have functioned differently from the mammalian system. Ironically, the best evidence for cheeks in ornithischians, the nodosaurid example above, is coupled with some of the clearest evidence that enhanced mastication was unlikely to have been the selective force in that ankylosaurs have tiny, vestigial teeth with little evidence of significant wear. Regardless, the presence, composition, and evolution of dinosaur cheeks are challenging hypotheses to test but are critical for adequate understanding of the evolution of herbivory in dinosaurs.

Powered Cranial Kinesis

Hypotheses of skull function can often suffer when muscular systems are not included in the analysis. For instance, the role of protractor muscles in the functioning of cranial kinetic systems is poorly understood in extant reptiles and often completely ignored in dinosaurs. Since the original work of Versluys (1910), cranial kinesis has been the subject of many anatomical, functional, and experimental studies in Sauropsida and has formed the foundation for studies of kinesis in dinosaurs. Although there have been many studies of squamate kinesis (for reviews, see Metzger 2002; Holliday and Witmer 2008), only a few in vivo analyses have focused on extant archosaur taxa (Zweers 1974; Van den Hueval 1992), although there have been numerous postmortem, biomechanical comparisons of avian kinetic apparatuses (for a review, see Zusi 1993). Kinetic systems have been hypothesized and investigated in some extinct archosaurs including basal taxa (Walker 1990; Sereno and Novas 1992), ornithopods (Norman 1984; Weishampel 1984; Rybczynski et al. 2008), theropods (Bakker 1986; McClelland 1990; Mazzetta et al. 1998; Larsson 2008), and early birds (Gingerich 1973; Bühler et al. 1988; Zusi and Warheit 1992; Zweers et al. 1997; Chiappe et al. 1999; Degrange et al. 2010).

Studies of extant archosaur cranial kinesis are problematic in that most functional conclusions have been reached through manipulation of dead specimens or superficial observations of skeletal morphology, all of which may possibly exaggerate functional hypotheses. Little is known about the physiological and behavioral activities of jaw muscles controlling kinesis in sauropsids in general. Do protractor muscles actively power kinesis, as they do in geckos (Herrel et al. 2000), or do they simply stabilize movements within an otherwise passive system (Holliday and Witmer 2008)? Many hypotheses of cranial kinesis in dinosaurs overlook the role of muscles completely and mainly hypothesize a passive system necessary for swallowing larger prey items (e.g., Bakker 1986). Supporting hypotheses of pleurokinesis, Weishampel (1984) identified scars for protractor muscles on the braincases of ornithischians, and Holliday and Witmer (2008) found that protractor muscle scars were actually ubiquitous among dinosaurs and could be quite large in tyrannosaurs and hadrosaurs. So the presence of protractor muscles is not debated; however, their function remains contentious. By means of several objective criteria, Holliday and Witmer (2008) found that most dinosaurs lack the necessary permissive kinematic linkages in the temporal bars and palate that would actually allow movement to occur. Moreover, no nonavian dinosaurs have been found to have a hinge joint on their skull roof. In ankylosaurs, patent or open palatobasal joints are maintained but the pterygoid bones are fused together. Patent otic (quadrate–squamosal) joints are common among dinosaurs. Even crocodylians, which have every bone in their skull sutured, have immobile synovial joints on the laterosphenoid and quadrate. These and other data indicate that these joints are likely growth plates only secondarily adapted for cranial kinesis and that the protractor muscles may serve a postural rather than powering role (Holliday and Witmer 2008). Although some nonavian dinosaurs may ultimately be shown to have been kinetic, to date, no analyses have met the criteria. Regardless, cranial kinesis, its adaptive significance, and its true distribution among dinosaurs and other reptiles remains a poorly understood phenomenon.

Solutions

How can we properly assess the site of attachment, the function of a muscle, or even the feeding biology of dinosaurs if so much of it may hinge on lost soft tissues? The case studies above illustrate how problems with phylogeny, structural misinterpretation, or premature functional inference lead to conflicting ideas about behavior. What are some possible solutions that will allow paleontologists to more accurately describe the anatomy, function, and evolution of dinosaurian feeding ecology?

Determining phylogenetic congruence of muscular attachment sites or the osteological correlates of muscles allow hypotheses to be tested on the presence and location of particular muscles in the skull. Studying crocodylians and birds (the extant phylogenetic bracket of dinosaurs; Witmer 1995a, 1995b, 1997) allows determination of the causal association of cephalic muscles and the bony scars they leave. Tracking these correlates through dinosaurs as a congruence test of homology illuminates phylogenetic transformation series of various characteristics. By means of these correlates, along with the use of extant data based on the topology of nerves, arteries, air sacs, and muscles, we can more accurately identify where homologous muscles attach on the skull. Some correlates are easily identifiable and commonly used, such as the coronoid process, the dorsotemporal fenestra, or the retroarticular process. However, others are more subtle and more difficult to interpret, such as the smaller scars on the braincase or the mandible.

Using larger bony characteristics to gauge muscle functional anatomy, however, may prove useful. The size or volume of structures such as the adductor chamber, the dorsotemporal fenestra, the pterygoid, or the orbital process of the quadrate may be associated with the cross-sectional area or pinnation angle of the attaching muscles. These anatomical parameters are often used to estimate force potential of the muscle (Sinclair and Alexander 1987; Rayfield et al. 2001). Mandibular attachment sites such as the retroarticular process or coronoid process may be used to measure mechanical advantage of particular muscles that can be compared across taxa. For instance, how might muscular functional inferences differ based on the skull morphologies of *Albertosaurus* and the more robust *Tyrannosaurus*? Are there ecological patterns in jaw muscle functional anatomy that can be found by comparing trophic guilds of dinosaurs (Dollo 1884), or among different types of herbivorous dinosaurs (Upchurch and Barrett 2000)? Finally, incorporating these anatomical inferences into modeling analyses, such as finite element analysis (Rayfield et al. 2001; Daniel and McHenry 2001; Porro 2008), animation (Rybczynski et al. 2008), or multibody modeling (Curtis et al. 2010), may help further elucidate the interaction of muscular forces and skull shape.

Finally, tooth wear has been a commonly used indicator of diet and feeding biology (Norman and Weishampel 1985; Weishampel and Norman 1989; Fiorillo 1998; Barrett 2000) and has been used to infer feeding mechanics and possible jaw muscle function (Weishampel 1984; Williams et al. 2009). With better understanding of muscle attachment sites and functional anatomy, paleontologists may be able to better integrate tooth wear patterns with jaw movement patterns, thus further elucidating feeding biology and cephalic function in dinosaurs.

Axial Musculature

The axial muscles of the cervical (neck), trunk, and caudal (tail) regions are grossly similar, but there are some regional specializations that we will discuss. In each region, the axial muscles are divided into two major groups (Fig. 9.4): the epaxial musculature, surrounding the neural spines on the dorsal surfaces of the vertebrae, and the hypaxial musculature, below the neural spines, on the ventral surfaces of the vertebrae. The hypaxial portion generally was larger than the epaxial portion, as it surrounded the cavities and organs of the neck, trunk, and tail. It had a larger number of specialized subdivisions. Little attention has been given to axial muscles in dinosaurs, most of it in the last decade or so, so our treatment will be brief.

Cervical (Neck) Muscles

Cervical muscles had been studied the least of all axial muscles among dinosaurs until recently. Fortunately, some of their homologies are fairly well resolved (Tsuihiji 2005, 2007, 2010; Snively and Russell 2007a, 2007b), and initial reconstructions of their associated ligaments have been proposed (Tsuihiji 2004). They were similar to other axial muscle groups; like the trunk hypaxial muscles, their hypaxial portion had ribs (cervical ribs; Fig. 9.4A) embedded in them. The cervical muscles extended from the head back to the pectoral girdle, and in dinosaurs they predominantly served to support and move the mobile head and neck. Exactly how the cervical muscles moved or supported the S-shaped neck of many dinosaurs, the scapula, or the elongate necks of sauropods remains poorly understood, but in general they should have acted like other axial muscles. Snively and Russell (2007b, 2007c) used biomechanical modeling to estimate how neck muscles might have powered feeding in tyrannosaurs and other large theropods, finding evidence for functional disparity that would have complemented suspected variation in feeding styles demonstrated by cranial anatomy. Schwarz-Wings et al. (2007) reconstructed muscles and pneumatic structures in sauropod necks that differed greatly from some previous reconstructions such as that by Tsuihiji (2004). There is ample opportunity for more progress in this area, which is relevant to controversies about sauropod neck posture and mobility.

Trunk Muscles

The typical dinosaur had a deep, narrow trunk with a large rib cage for the attachment of hypaxial muscles, whereas the epaxial muscles were limited to a small space on top of the vertebrae (Fig. 9.4B). Longitudinally, these

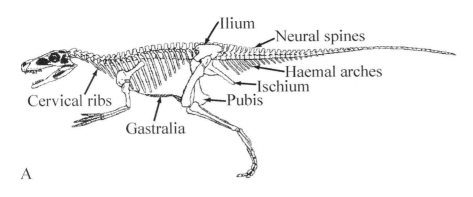

Ilium
Neural spines
Haemal arches
Ischium
Pubis
Cervical ribs
Gastralia

A

9.4. A representative dinosaur, *Herrerasaurus* (modified from Sereno and Novas 1992), in left side view showing noteworthy skeletal features (A) and the rough boundaries of axial muscle groups (B).

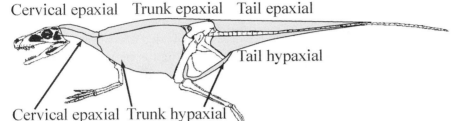

Cervical epaxial   Trunk epaxial   Tail epaxial
Tail hypaxial
Cervical epaxial   Trunk hypaxial

B

muscles spanned the space between the cervical and the pelvic or caudal regions of the axial skeleton.

EPAXIAL TRUNK MUSCLES

The epaxial muscles helped to stiffen the trunk or extend its intervertebral joints, bowing the trunk upward to resist the sag imposed by gravity. It seems that most dinosaurs had little flexibility of the joints in this region, especially laterally, so these muscles presumably served a mainly supportive role rather than actuating much motion, unlike in more basal tetrapods. Schwarz-Wings (2010) produced a novel reconstruction of thoracic epaxial muscles in select sauropods, whereas Organ (2006a) focused a similar effort on archosaurs with specific reference to ornithopods and their remarkable ossified tendons (Organ 2006b).

HYPAXIAL TRUNK MUSCLES

At least three major hypaxial muscle groups (abdominal obliques, transverse abdominals, and rectus abdominis) were likely present in dinosaurs (Dilkes 2000). The hypaxial muscles covered the rib cage and ran between the pectoral and pelvic girdles. Together, these muscles could have acted to compress or expand the trunk during ventilation and other activities and to stiffen the rib cage and abdomen. More aspects of this system are considered further below in relation to ventilatory function.

Caudal (Tail) Muscles

The transverse processes (Fig. 9.5A) of the vertebrae lay in the plane dividing the hypaxial and epaxial regions (Fig. 9.4B). Gatesy and Dial (1996) and Gatesy (2001) investigated how some of the tail muscles were co-opted

9.5. Epaxial (lower arrow) and hypaxial (upper arrow) ossified tendons (ot) in the proximal (A) and distal (B) tail of an ornithopod (*Tenontosaurus,* modified from Winkler et al. 1997), viewed from the left side. Double-headed arrows are for 10-cm scale.

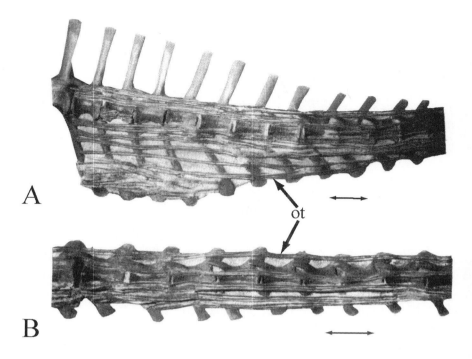

for use in flight in birds. Arbour (2009) and Arbour and Snively (2009) as well as Carpenter et al. (2005) used reconstructions of tail musculature in ankylosaurs and stegosaurs to reconstruct the mechanics of striking with weaponized tails in these taxa.

EPAXIAL TAIL MUSCLES

The epaxial tail muscles extended from the back of the ilium (sometimes marked by a roughened area on the back corner) along the length of the tail, thinning distally. They thus formed an almost continuous longitudinal muscle mass from head to tail, unlike the hypaxial muscles, which were interrupted by the pelvis and hind limb. In dinosaurs with narrow pelves that enclosed the sacrum, such as some coelurosaurian dinosaurs (including birds), this connection was reduced, providing increased decoupling of the tail musculature from the rest of the body (Gatesy and Dial 1996; Hutchinson 2001a). In conjunction with these muscular changes, mobility of the tail was restricted to the proximal end by bony supports that stiffened the tail distally (Gatesy 2001). Many dinosaurs evolved tall neural spines for attachment of these epaxial muscles, which would have helped to stiffen the tail, preventing it from drooping ventrally.

HYPAXIAL TAIL MUSCLES

The hypaxial muscles extended ventrally from the posterior pelvis and intervening soft tissues to the distal tail (Dilkes 2000). An attachment point for part of the hypaxial muscle origin can be seen on the swollen end of the ischium (the ischial boot; Fig. 9.6A) of some dinosaurs. Fibers of these muscles are visible in some well-preserved specimens such as a mummified ornithopod (Brown 1916) and perhaps in a young theropod specimen, *Scipionyx* (Dal Sasso and Signore 1998). Parts of this hypaxial muscle group presumably acted to stiffen the tail or flex it up or down.

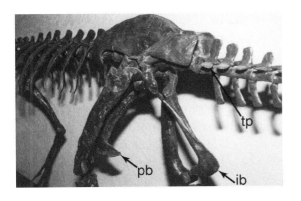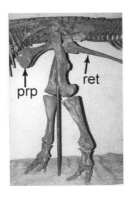

9.6. Osteological correlates of axial muscle attachments in dinosaurs. A, Ornithomimid theropod *Struthiomimus* (display cast at the American Museum of Natural History; reversed), in left oblique rear view showing the pubic boot (pb), the ischial boot (ib), and the transverse processes (tp). B, Ornithopod dinosaur *Edmontosaurus* (display cast at the Denver Museum of Natural History), in left side view showing the prepubic process (prp) and retroversion of the pubis (ret).

Problems and Solutions for the Axial Musculature of Dinosaurs

Although axial musculature has not been a focus of much dinosaur research until recently, its anatomy has bearing on several paleobiological issues, such as the orientation of the axial column and its function during ventilation and locomotion. As with the cephalic musculature, more work on the relationship between muscular and bony form and function in living animals is badly needed in order to better understand such things in dinosaurs.

Some studies (e.g., Ruben et al. 1997, 1999) have suggested that a fourth, deeper hypaxial trunk muscle group, M. diaphragmaticus, extended from the liver back to the pubis in many dinosaurs. The existence of a diaphragmatic muscle in dinosaurs, however, remains contentious (e.g., Ruben et al. 1997, 1999; Carrier and Farmer 2000; Paul 2002). It is difficult to test this hypothesis because many other abdominal and pelvic muscles attach to the same areas of the pelvis and there are thus no unambiguous osteological correlates for an M. diaphragmaticus (Romer 1923a; Hutchinson 2001a). Likewise, the few fossils that some claim to show a diaphragmatic-like hepatic piston anatomy can be otherwise interpreted (Paul 2002). Furthermore, it seems that the diaphragmatic muscles of crocodylians evolved separately from any in other archosaurs (Carrier and Farmer 2000), making reconstructions of such nonhomologous muscles in dinosaurs less convincing. In any case, the derived pelvic and thoracic morphology of ornithischians and maniraptoran theropods (including birds) shows that if a diaphragmatic muscle were present in some dinosaurs, it was lost in other clades (e.g., Codd et al. 2008).

In many dinosaurs, the more ventral parts of the hypaxial muscles had gastralia (Fig. 9.4A), or belly ribs, embedded in them, anchoring the muscles and adding support to the belly. Ornithischian dinosaurs, possibly some sauropods, and some theropods (including later birds) lost these gastralia. Claessens (1997) proposed that the gastralia functioned in ventilation as a levered framework and that some hypaxial muscles fanned out during inspiration, expanding the trunk cavity. Dinosaurs that lost the gastralia presumably had other mechanisms for ventilation, such as movements of the rib cage, sternum, or pelvis (Carrier and Farmer 2000; Codd et al. 2008). With or without gastralia or diaphragmatic muscles, the functions of axial muscles during ventilation in dinosaurs have recently enjoyed a surge of research attention.

One of the most striking specializations of the tissues in the axial column of dinosaurs is the development of ossified tendons (Fig. 9.5) running crosswise between vertebrae in most ornithischian dinosaurs except stegosaurs (e.g., Brown 1916; Galton 1990; Organ 2006a, 2006b). Whether these tendons were tendons of axial musculature, nonmuscular ligaments, or something else was previously unresolved (Reid 1996), but Organ (2006a) has made a convincing case that they are ossifications of the tendons within a layer of M. transversospinalis (an epaxial muscle group). Regardless, it is obvious that these structures consist of true bone. In living animals, similar ossifications develop around a collagenous tissue precursor, although cartilage is sometimes involved (Reid 1991, 1996). Certainly these ossifications added rigidity to the vertebral column, resisting sagging or lateral flexion (Organ 2006a, 2006b). Some maniraptoran dinosaurs also had bony uncinate processes lying between ribs that likely performed an analogous stiffening role, especially during ventilation (Paul 2002), although their function even in living animals remains somewhat mysterious (but see Codd et al. 2008). These puzzles could be solved by more research on how similar features develop and function in living animals, their detailed structure and evolution in dinosaurs, and functional models of dinosaurs.

One obvious attachment region for many hypaxial trunk muscles, apart from the ribs and sternum (breastbone), was the end of the pubic bones of the pelvis, which had a knoblike or roughened region (the pubic boot; Fig. 9.6A) in many dinosaurs. Along with forming a large attachment area for hypaxial trunk muscles, some of which may have extended beyond the pubis to attach to a similar boot on the ischium (Carrier and Farmer 2000; Hutchinson 2001a), the pubic boot may have supported dinosaurs when they sat down (Charig 1972; Paul 1988). A forward projection of the upper end of the pubis (Fig. 9.6B), the prepubic process (or anterior ramus), evolved in ornithischians mainly as an abdominal muscle attachment (Romer 1927a; Charig 1972; Dilkes 2000). Carrier and Farmer (2000) proposed that the prepubic process anchored an iliopubic muscle that came down from a process on the ilium and was involved in a mobile pelvic mechanism for expiration. There is no example of such a muscle in living archosaurs, and other epaxial and hind limb muscles attach to this region of the ilium, so this inference seems unlikely. The prepubis is unique to ornithischians, and its unique function or functions may remain elusive. Likewise, which muscles attached to which areas of the pubis, or how large they were, is difficult to precisely infer because the bony surfaces are generally smooth, and the muscles vary widely in living animals. In general, this difficulty with circumscribing muscle areas is a major problem for the study of dinosaur muscle functions. Some studies have partly circumvented this obstacle (e.g., Gatesy 1990; Hutchinson and Garcia 2002).

The orientation of the vertebral column in dinosaurs remains uncertain in many cases, as the bones themselves only provide so much information. In particular, the typical angle of the body to the horizontal (i.e., pitch) is controversial; reconstructions range from more upright as in humans to more horizontal as in birds (e.g., Tarsitano 1983; Paul 1988), and it probably varied with anatomy and gait. Musculoskeletal models could shed some light on this, but because living animals vary the pitch of their vertebral column during different activities, so should have dinosaurs. Likewise, neck orientations of sauropods have been controversial (Stevens and Parrish 1999, 2005; Dzemski and Christian 2007; Taylor et al. 2009), but neck musculature has only begun to be integrated into analyses of this problem (e.g., Christian 2010; Christian and Dzemski 2007; Schwarz-Wings et al. 2010).

Muscles in the appendages (limbs and fins) of vertebrates differentiate during early stages of development into masses that are positioned dorsally and ventrally along the skeletal core of the appendage. This basic division of limb muscles persists in the fins of adult fish where each muscle mass in a fin has a relatively simple functional role: the dorsal mass elevates (lifts) the fin and ventral mass depresses (lowers) the fin. In contrast to the musculature of a fin, the limbs of terrestrial vertebrates (tetrapods) have more complex functional roles of support of the body and the elevation, protraction, and retraction of the limbs. As a consequence, distinction between dorsal and ventral muscle masses is less clear in limbs of adult tetrapods than in fins because each mass splits into numerous smaller muscles. Only in the more distally located extensor and flexor muscles can one still find evidence of this basic dorsal and ventral division. Despite the loss of a dorsal–ventral separation during ontogeny, limb muscles of modern tetrapods are arranged customarily in dorsal and ventral groups. It is presumably safe to say that this dorsal–ventral separation was primitive for dinosaurs and was inherited by birds (Romer 1927b).

Forelimb Musculature

Organization of the muscles of the forelimb of dinosaurs was probably the same as in other tetrapods. Unlike the pelvic girdle with its sacral ribs that attached the girdle directly to the vertebrae, the pectoral girdle of dinosaurs was held against the rib cage in a muscular sling consisting of muscles from the axial and appendicular skeletons. The only other connection to the vertebrae was between the pectoral girdle and the cartilaginous sternal ribs and sternum and, if present, sternal plates. For a quadrupedal dinosaur, this arrangement meant that the trunk was suspended between the pair of forelimbs. The majority of the muscles that form this muscular sling consist of axial muscles. Two of these axial muscles are M. trapezius and M. rhomboideus, which originate from the connective tissue sheet associated with the cervical and thoracic vertebrae and attach to the dorsal and dorsomedial edges of scapula and cartilaginous suprascapula. A third axial muscle, M. serratus ventralis, originates from the lateral surface of the rib cage and attaches to the inner side of the scapula below the attachments of M. trapezius and M. rhomboideus. Last, M. pectoralis of the ventral group of limb muscles extends from the sternum to the inner side of the humerus.

Muscles of the forelimb of dinosaurs are grouped according to the basic dorsal and ventral subdivision of tetrapod limb muscles. The muscle classification of Romer (1922, 1944) will be followed, but modified slightly to emphasize the actions of extension/flexion and adduction/abduction at joints and protraction/retraction of the limb. There have been far fewer attempts to reconstruct the forelimb muscles of dinosaurs in comparison to the pelvic girdle, but several different groups of dinosaurs consisting of obligatory bipedal, obligatory quadrupedal, and facultatively quadrupedal

Appendicular (Limb) Muscles

9.7. Reconstruction of lines of action of major muscles of the forelimb from their origin to insertion in the early dinosaur *Herrerasaurus ischigualastensis* (modified from Sereno 1993). A, Lateral view of forelimb showing lines of action of axial muscles that attach to the scapula and those dorsal muscles of the arm that are located along lateral side of forelimb. B, Medial view of forelimb showing lines of action those dorsal muscles of the arm that are situated along medial side of forelimb and the flexors of the forearm. C, Lateral view of forelimb showing lines of action of ventral muscles of the arm and extensors of the forearm. Numbers 1 to 5 refer to digits 1 to 5.

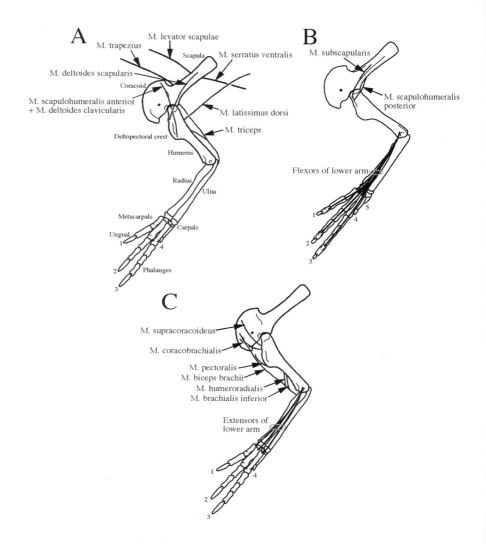

taxa are represented in the literature. Detailed accounts of forelimb muscles of dinosaurs include hadrosaurs (Dilkes 2000; Lull and Wright 1942), sauropods (Borsuk-Bialynicka 1977; Langer et al. 2007), theropods (Carpenter and Smith 2001; Nicholls and Russell 1985; Jasinoski et al. 2006), ankylosaurs (Coombs 1978), and *Iguanodon* (Norman 1986).

Dorsal Muscles of the Upper Arm

This complex consists of five major groups: M. latissimus dorsi, M. deltoideus (scapularis and clavicularis), M. subcoracoscapularis (M. subscapularis and M. subcoracoideus), M. scapulohumeralis (anterior and posterior), and M. triceps brachii (Dilkes 2000; Jasinoski et al. 2006; Figs. 9.7A, B).

SUPERFICIAL MUSCLES OF UPPER ARM

The fan-shaped M. latissimus dorsi reached from its broad origin along the connective tissue of the neural spines of the thoracic vertebrae to cover most of the other muscles of the upper arm and narrow to a tendon that inserted on the dorsal surface of the humerus. A prominent muscle scar may accompany insertion of this muscle onto the humerus, as in *Iguanodon bernissartensis* (Norman 1980), hadrosaurs (e.g., *Maiasaura peeblesorum*;

Dilkes 2000), ceratopsians (e.g., *Torosaurus*; Johnson and Ostrom 1995), and sauropods (e.g., *Opisthocoelicaudia skarznskii*; Borsuk-Bialynicka 1977). The shape of this muscle scar may also vary ontogenetically, as demonstrated for *Maiasaura* (Dilkes 2001). The action of M. latissimus dorsi would have been to abduct and retract the humerus about the shoulder joint. M. teres major is a division of M. latissimus dorsi that originates from the lateral surface of the scapula in crocodylians and turtles. Its presence in dinosaurs is therefore equivocal, but it has been reconstructed in those instances (Coombs 1978; Norman 1986) where crocodylians were the chosen model.

DEEP MUSCLES OF UPPER ARM

Two other prominent muscles from the shoulder girdle, M. deltoideus scapularis and M. deltoideus clavicularis (and its division, M. scapulohumeralis anterior), likely attached along the dorsal surface of the deltopectoral crest of the humerus. Their lines of action suggest that this pair of muscles had similar functions to that of M. latissimus dorsi. In view of the more laterally placed insertion of the two deltoideus muscles onto the humerus in comparison to M. latissimus dorsi, these muscles may have also abducted the humerus during retraction of the limb. Given the possible closeness of the insertion of M. deltoideus scapularis to the shoulder joint, contraction of this muscle may have also helped to stabilize the joint. One of the deeper extensor muscles, M. scapulohumeralis posterior, was likely covered entirely by M. latissimus dorsi and M. deltoideus scapularis and inserted next to the shoulder joint. Along with M. deltoideus scapularis, M. scapulohumeralis posterior may have acted to stabilize the shoulder joint. As discussed by Carpenter (2002) for theropods, the relative degree of protraction–retraction and adduction–abduction may depend on the configuration of the glenoid and shape of the head of the humerus.

At least one division of M. subcoracoscapularis (likely M. subscapularis) was present in dinosaurs. Presence of M. subcoracoideus is equivocal, but it has been restored in dromaeosaurids as originating from the medial side of the coracoid and merging with the tendon of M. subscapularis (Jasinoski et al. 2006). Origin of M. subscapularis was most likely along the medial side of the scapula, and it inserted in close proximity to M. scapulohumeralis posterior. Action of M. subscapularis was probably to adduct and retract the humerus and possibly stabilize the shoulder joint.

EXTENSORS OF THE FOREARM

Only a single forearm extensor, M. triceps brachii, will be considered. The number of divisions of this muscle is extremely variable among tetrapods, and while it is probable that at least one division originated from the scapula and a second part originated from the shaft of the humerus in dinosaurs, additional divisions may have been present. A right humerus of *Tyrannosaurus rex* from the individual known popularly as Sue has a pathological region opposite the deltopectoral crest that has been interpreted as the result of partial tearing of a medial head of M. triceps humeralis (Carpenter and Smith 2001). However, this interpretation has been questioned; the feature has been attributed alternatively to a traumatic event that affected the right side of the individual (Brochu 2003). Regardless of the actual number of divisions of the triceps muscles in dinosaurs, all probably inserted onto the proximal end of the ulna (the olecranon process) because this insertion is true for all living tetrapods. The primary action of the triceps muscle was the extension of the forearm about the elbow.

Ventral Muscles of the Upper Arm

FLEXORS/ADDUCTORS OF THE UPPER ARM

Three prominent muscles of the ventral muscle division, M. supracoracoideus, M. pectoralis, and M. coracobrachialis, likely had important functional roles of adduction and flexion of the humerus (Fig. 9.7C). M. pectoralis is one of the muscles that forms the muscular sling to suspend the trunk in a quadruped. Its likely line of action was primarily in a mediolateral direction to provide significant adduction; however, given the possible broad nature of its origin from the sternum, some fibers that attached along the anterior margin of the sternum may have helped to protract the humerus, while other fibers that attached more posteriorly provided some retraction. Fibers of the pectoralis probably attached along the edge of ossified sternal plates present in a variety of dinosaurs such as ceratopsians and stegosaurs, as well as many ornithopods, sauropodomorphs, and theropods. In those theropods such as *Tyrannosaurus rex* that lack sternal plates, the pectoralis most likely originated from any cartilaginous sternal elements and ribs. Protraction of the humerus was a likely component of the action of M. supracoracoideus because its origin from the coracoid was anterior to its probable insertion onto the ventral surface of the deltopectoral crest of the humerus proximal to the insertion of M. pectoralis.

FLEXORS OF THE FOREARM

Tetrapods have three major muscles that serve to flex the forearm: M. biceps brachii, M. brachialis inferior, and M. humeroradialis. Origin of M. biceps brachii includes the coracoid. A second origin from the humerus is equivocal, with no muscle scar on the humeri of hadrosaurs that could be attributed to this muscle (Dilkes 2000), but a scar on the humerus of eumaniraptoran theropods might be evidence for this additional origin (Jasinoski et al. 2006). These three muscles would have acted as powerful flexors of the forearm in an antagonistic manner to the action of M. triceps brachii. A tubercle on the coracoid of theropods referred to customarily as the biceps tubercle or occasionally as the acrocoracoid (Russell and Dong 1993) is often interpreted as the origin of M. biceps brachii. In other dinosaurs, such as hadrosaurs and sauropods, there is a prominent projection along the anterior margin of the coracoid that was the likely attachment point for the biceps muscle. Not all paleontologists accept the standard interpretation of the theropod biceps tubercle. Soon after Ostrom (1974) identified a protuberance on the coracoid of the theropod *Deinonychus antirrhopus* as the biceps tubercle and homologous to the biceps tubercle of *Archaeopteryx* and the crocodylomorph *Sphenosuchus*, Walker (1977) argued that this hypothesis of homology was wrong. He identified this process on the coracoid of *Deinonychus* as a likely tendinous attachment for M. coracobrachialis brevis. Brochu (2003) and Carpenter and Smith (2001) followed the customary interpretation of the tubercle for the biceps muscle, but Carpenter (2002) offered a different interpretation. He considered the tubercle to be a product of the actions of M. coracobrachialis and M. supracoracoideus that attached to neighboring regions of the coracoid. More recently, Jasinoski et al. (2006) argue for attachment of M. biceps brachii to the biceps tubercle, noting that according to their restoration, M. supracoracoideus originated more ventrally on the coracoid than restored by Carpenter (2002) and thus was too far from the biceps tubercle. The controversy over the homology of this tubercle on the coracoids of different dinosaurs and its connection to the presence of muscles is an example of the "hole or doughnut" problem (Witmer 1995a), where tension from the action of a muscle may be a direct cause for the existence of a projection (i.e., the doughnut) or the projection is the indirect result of surrounding soft tissues such as other muscles (i.e., the hole; in this case, the tubercle). It serves as a useful illustration of the potential conflicting interpretations of deceptively simple osteological features and the necessity of looking at alternative soft tissue explanations for these structures.

According to the pattern in modern tetrapods, the other pair of flexor muscles of the forearm had extensive origins from the shaft of the humerus. All three flexor muscles most likely attached onto the proximal end of the radius and perhaps also the ulna, although doubtfully as far distally as inferred by Johnson and Ostrom (1995) for *Torosaurus*. Often prominent and tightly constrained muscle scars are associated with the insertion of these muscles in a variety of dinosaurs such as hadrosaurs (Dilkes 2000), *Velociraptor* (Norell and Makovicky 1999; Jasinoski et al. 2006), and *Tyrannosaurus* (Carpenter and Smith 2001). A distinctive biceps tuberosity that was the likely insertion point for the biceps muscle is present on the radius of the small theropod *Microraptor gui* (Xu et al. 2003) and a therizinosauroid (Zhang et al. 2001).

A humerus of the sauropod *Camarasaurus* has a pathology that has been interpreted similarly as most likely due to trauma and infection at the attachment site of M. brachialis inferior and M. brachioradialis (=M. humeroradialis) (McWhinney et al. 2001). As a result of this injury and the accompanying restriction of blood supply and compression of nerves, the individual almost certainly walked with a limp.

DORSAL AND VENTRAL MUSCLES OF THE FOREARM

Only a few of the numerous muscles that act to extend and flex the forearm will be considered (Fig. 9.7B, C). Extensors are part of the dorsal muscle mass and are identified as those that originate from the lateral epicondyle (ectepicondyle) of the humerus. Flexors are those that are part of the ventral muscle mass and originate from the medial epicondyle (entepicondyle) of the humerus. In addition to extension or flexion of the forearm about the elbow, some of these muscles may have also acted to extend, flex, and abduct the hand. Major extensor muscles include M. extensor carpi radialis,

M. extensor carpi ulnaris, and M. extensor digitorum. Those of the flexor complex include M. flexor carpi radialis, M. flexor carpi ulnaris, and M. flexor digitorum longus.

Insertion of the extensor muscles of the forearm was likely onto the base of the metacarpals (bones in the palm) (Dilkes 2000; Carpenter and Smith 2001) or possibly a combination of the carpals (M. extensor carpi radialis and M. extensor carpi ulnaris) and metacarpals (M. extensor digitorum). Similarly, the flexor carpi radialis and ulnaris muscles likely attached to the metacarpals, while M. flexor digitorum longus attached to the base of the last bone (ungual) of each finger.

INTRINSIC MUSCLES OF THE HAND

No attempt will be made to reconstruct the complex series of muscles of the individual fingers in view of their uncertain homologies with muscles in other tetrapods, although progress has been made recently (Diogo et al. 2009; Abdala and Diogo 2010), and in view of the absence of clear osteological features that would correlate the presence of a specific muscle. It is highly probable that the hands of predatory dinosaurs with long fingers had a variety of small muscles that allowed for complex actions of grasping and manipulating prey. In contrast, these muscles may have been reduced in number or size in obligatory quadrupedal dinosaurs such as larger ceratopsians and sauropodomorphs, where fingers were largely enclosed in soft tissue and carried a significant portion of the animal's weight.

Forelimb Musculature of Bipedal and Quadrupedal Dinosaurs: Controversies and Possible Directions

Given the considerable range of morphologies of the forelimbs of different dinosaurs, their dramatically different functional roles in strictly bipedal and quadrupedal dinosaurs, the possible multitude of functions in those that were likely facultatively bipedal (e.g., many prosauropods, hadrosaurs, and basal iguanodontians), and possible ontogenetic changes in the function of the forelimb (e.g., *Maiasaura*; Dilkes 2001), it is highly problematic whether one can look at any specific muscle and ascribe a specific set of functions. Rather, there were likely a series of possible actions, the relative importance of which remains to be determined for most dinosaurs. Explicitly biomechanical studies such as that of Carpenter and Smith (2001) show how one can proceed with an isolated muscle, but the true challenge is to model the complexity of interactions of the forelimb muscles, incorporate more experimental data from living tetrapods, and ultimately place these efforts within a phylogenetic context.

The relative size and shape of the scapulae and coracoids vary considerably among theropods. The scapula of early bipedal theropods such as *Herrerasaurus* is relatively long and narrow (Sereno 1993). Among other theropods, the scapula may be expanded dorsally, and this expansion may be related to expansion of dorsal muscles of the forelimb. Numerous muscle scars on the scapula of *Tyrannosaurus* (Brochu 2003) may be evidence of this spread of dorsal muscles. Proximal expansion along the contact with the coracoid is also present in many theropods. Size of the coracoid also varies relative to the scapula in theropods, such as the especially large coracoid in *Deinonychus*. Carpenter (2002) has proposed that these differences in the sizes of the coracoids are correlated with differences in relative sizes of protractor, retractor, and adductor muscles. Differences in relative sizes of these functional groups of muscles may indicate differing abilities among theropods to grasp, hold, and manipulate prey. For example, those theropods with enlarged coracoids such as *Deinonychus* may have had relatively larger muscles for protraction and adduction that suggests an adaptation for reaching out and grasping prey. Naturally, any interpretations of the relative functional importance of a muscle in a dinosaur depend on whether one can accurately estimate the size of a muscle that likely had a broad, fleshy attachment.

Among facultative quadrupeds such as many basal sauropodomorphs and large ornithopods such as *Iguanodon* and hadrosaurs, there is a similar expansion of the distal end of the scapula, although the absence of clearly identifiable muscle scars makes any functional interpretations difficult. There is a dramatic expansion of the scapula next to the coracoid in obligate quadrupeds such as stegosaurs and sauropods. This expansion likely displaced the attachments for M. scapulohumeralis anterior and M. supracoracoideus anterior to the glenoid. Ankylosaurs have greatly enlarged coracoids relative to the scapula that may have similarly shifted anteriorly the attachments for M. supracoracoideus, M. biceps brachii, and M. coracobrachialis (Coombs 1978). A lateral scapular ridge called the pseudoacromion process on nodosaurid ankylosaurs was the probable attachment site for M. scapulohumeralis anterior. In an interesting example of convergent evolution, this ridge with the pseudoacromion is analogous to the mammalian scapular spine and its acromion. The posterior shift of the pseudoacromion has apparently provided an increased area for attachment of M. supracoracoideus relative to ankylosaurid ankylosaurs. Coombs (1978) interpreted this apparent enlargement of the supracoracoideus as an adaptation to increase the power of this muscle for locomotion and possibly for digging.

Proper orientation of the scapulocoracoid against the rib cage continues to be controversial in theropods and sauropods (Jasinoski et al. 2006; Schwarz et al. 2007). Depending on the orientation, muscles may attach to the scapulocoracoid in different locations with resultant differences in their actions, and different muscles may form the muscular sling to support the trunk in a quadruped. Functional interpretations of muscles are clearly influenced by the position of the scapulocoracoid. A more horizontal position of the scapulocoracoid in dromaeosaurids was considered evidence that the two parts of M. rhomboideus inserted onto the anterodorsal margin of the scapula, as in birds (Jasinoski et al. 2006). Consequently, this muscle became a protractor of the scapula and replaced M. levator scapulae, the presence of which is equivocal in theropods and absent in birds. Rather than develop a single model of muscles attaching to the pectoral girdle of sauropods on the basis of phylogeny, Schwarz et al. (2007) offered separate interpretations on the basis of both crocodylian and avian models. In each model, different muscles formed the muscular sling.

One area of continuing controversy in the study of quadrupedal dinosaurs is the forelimb posture of the ceratopsians. For nearly a century, the debate has centered on whether ceratopsians had a more sprawled stance, similar to that of lizards, or a more erect, mammal-like posture whereby the forelimbs were protracted and retracted along an anterior–posterior arc. Work has explored the evidence for these two contrasting interpretations and whether forelimb posture was intermediate (e.g., Johnson and Ostrom 1995; Dodson 1996; Dodson and Farlow 1997; Paul and Christiansen 2000; Thompson and Holmes 2007; Rega et al. 2010). Musculature of the forelimb, specifically those muscles that attached to the deltopectoral crest of the humerus, has played an interesting part in this ongoing debate. Johnson and Ostrom (1995) argued that the large deltopectoral crest has a more reasonable mechanical role if the humerus is oriented in a primarily horizontal direction so that the primary action of the pectoralis muscle is adduction. A more vertical position of the humerus would introduce significant twisting of the humerus by the pectoralis rather than adduction. Furthermore, the size of the deltopectoral crest was thought to be more consistent with a sprawling tetrapod, where a presumably enlarged pectoralis would generate significant force for support of the front of the ceratopsian with its heavy skull. Those who favor a more erect posture for the forelimbs (e.g., Paul and Christiansen 2000) have argued that size of the deltopectoral crest is poorly correlated with a sprawling posture. For this debate to progress further, and to determine how the forelimb of ceratopsians functioned, future studies need to combine a reconstruction of forelimb muscles within a clear phylogenetic framework to show the stages in the evolution of the forelimb osteology and musculature of the larger ceratopsians, and with biomechanical data from living taxa that examine both joint morphology and interactions of muscles. The use of trackway data to match prints to articulated forelimb skeletons and the study of pathologies would constrain the range of postures (Thompson and Holmes 2007; Rega et al. 2010).

Hind Limb Musculature

Romer's work (1923a, 1923b, 1923c, 1927a, 1927b) remains the classic treatment of the thigh muscles in particular, but a more explicit phylogenetic perspective and new fossil discoveries have revealed much more about the anatomy and history of dinosaurian hind limb muscles. Detailed discussions of these muscles are provided by Dilkes (2000), Carrano and Hutchinson (2002), Hutchinson (2001a, 2001b, 2002), and references therein. Figure 9.8 shows several examples of osteological correlates of hind limb muscles, whereas Figures 9.9 and 9.10 show the major muscle groups.

Dorsal Thigh Muscles

TRICEPS FEMORIS

As its name describes, the triceps femoris group had three parts centered on the femur. All three converged to form their primary tendon of insertion, the knee extensor tendon. Dinosaurs, like extant birds, had a large crest on the front of the tibia, the tibial crest (Fig. 9.8A), that provided extra leverage and area for extensor tendon insertion. Unlike living birds, nonavian dinosaurs did not have a kneecap or patella as a mobile bony attachment associated with their extensor tendon. This bone in extant birds, mammals, and other animals such as lizards increases the moment arm of these muscles about the knee. The primary action of the triceps femoris group was to extend the knee joint, so it was especially important for bipedal dinosaurs that supported all their weight on their hind limbs.

DEEP DORSAL THIGH MUSCLES

The homologies of these muscles remain contentious (for more information, see Romer 1923a, 1923b, 1923c, 1927a, 1927b; Walker 1977; Rowe 1986; Gatesy 1994; Carrano 2000; Hutchinson 2001b). They originated from the space in front of and slightly above the hip joint, mostly from the ilium. Their insertions were only on the proximal end of the femur on a series of trochanters that give valuable clues to how these muscles evolved within the Dinosauria (Hutchinson 2001b). As in living animals, the dinosaurian deep dorsal muscles acted variably to flex, abduct, and medially rotate the hip joint. Some parts balanced the body on one leg at a time during bipedal movement (Hutchinson and Gatesy 2000). Evidence for this abductor function includes the lesser (or anterior) trochanter (not homologous with lesser trochanter of mammals), which is an easily recognizable deep dorsal thigh muscle insertion (Fig. 9.8B). The lesser trochanter was ancestrally present

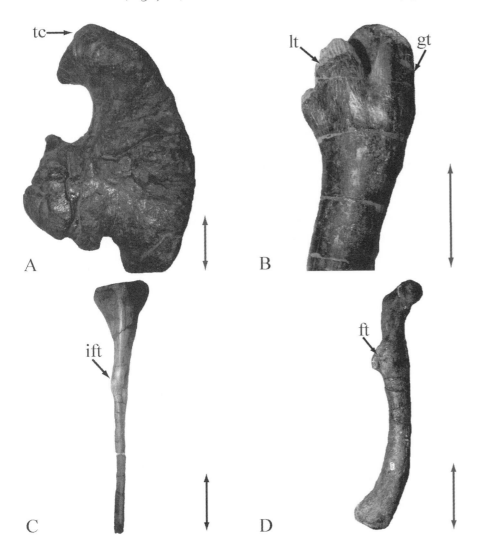

9.8. Hind limb muscle attachments in dinosaurs. Double-headed arrows are for 10-cm scale. A, Top view of a left tibia of the large theropod *Allosaurus* (American Museum of Natural History specimen AMNH 680), showing the tibial crest (tc). B, Side view of a right femur (reversed) of *Allosaurus* (Yale Peabody Museum specimen YPM 4944), showing the lesser (lt) and greater (gt) trochanters. C, Side view of a right fibula (reversed) of a tyrannosaur (*Alectrosaurus*, specimen AMNH 6554), showing the fibular tubercle (ift). D, Inner (medial) view of a right femur (reversed) of the basal dinosaur *Herrerasaurus* (Museo de Ciencias Naturales, Universidad Nacional de San Juan specimen PVSJ 373), showing the fourth trochanter (ft).

in all dinosaurs but evolved along different paths after their Late Triassic divergence (Carrano 2000; Dilkes 2000).

M. ILIOFIBULARIS

Certainly this was a fairly large muscle in most dinosaurs, as in birds, judging from the large space on the ilium behind the hip joint for its origin. It tapered to a stout tendon near the knee joint and then attached to a large tubercle (the fibular tubercle; Fig. 9.8C) or scar on the side of the fibula, which is readily visible in many dinosaur skeletons. In theropod dinosaurs, its tendon seemed to have created stresses in the tibia and fibula that induced formation of a large fibular crest that bound the two bones more tightly together (Müller and Streicher 1989). M. iliofibularis was a hip extensor, but it also could flex the knee joint, opposing the action of the triceps femoris muscles.

Ventral Thigh Muscles

FLEXOR CRURIS

The ancestral condition among reptiles is to have up to seven heads of this muscle group in a complex arrangement around the pelvis, all tapering to tendons that insert around the top of the tibia. The dinosaurian flexor cruris muscles remain a mystery because in living animals they leave few scars on bones, sometimes originating only from ligaments or other muscles. Moreover, their muscle homologies remain problematic, and they are variable. Yet we can reconstruct at least two parts with confidence in dinosaurs (Fig. 9.10), one (the external head, M. flexor tibialis externus) coming from the posterior end of the ilium and another (the internal head, M. flexor tibialis internus "3") coming from a large scar near the base of the ischium, behind the hip joint. As hamstring muscles like M. iliofibularis, these muscles could extend the hip and flex the knee. Dinosaurs continued an archosaurian trend of reducing these muscles, which were also presumably important adductors in early sprawling reptiles but of less utility to the more erect dinosaurs (Romer 1923a). Eventually birds had only two main muscles, although these muscles were still fairly large. The flexor cruris muscles evolved within theropod dinosaurs to become an integral part of the crouched limb design of birds (Gatesy 1990; Carrano and Hutchinson 2002).

ADDUCTORS

Crocodiles and birds have two heads of this muscle group, unlike other reptiles, which only have one, so we can presume that dinosaurs also had two adductors (Fig. 9.10; Mm. adductores femorii 1 + 2). These muscles originated from the dorsal and ventral borders of the ischium (Dilkes 2000; Hutchinson 2001a) and inserted on the posterior distal end of the femur. They probably changed relatively little in dinosaurs, but in contrast to what their name implies, they may have had a larger role in hip extension than adduction (Hutchinson and Gatesy 2000). Dinosaurs ancestrally had an erect posture, which reduced the requirement for adductor muscle moments during support of the body, and their long ischia transformed the adductors into extensors by moving the line of action of the adductors behind, rather than below, the hip joint.

LATERAL ROTATORS

Two major muscle groups, Mm. puboischiofemorales externi (three heads) and M. ischiotrochantericus, were ancestrally present in dinosaurs (Fig. 9.9). The first group had a large set of origins from wide apronlike expanses of bone on the pubis and ischium, and the second group originated from the inside surface of the ischium, coming to lie on the opposite side of the ischium sometime late in bird evolution (Hutchinson 2001a). Both groups sent tendons to insert on scars on the outer side of the proximal femur (the greater trochanter; Fig. 9.8B; Dilkes 2000; Hutchinson and Gatesy 2000; Hutchinson 2001b). They seem to have acted mainly in opposition to the deep dorsal thigh muscles, able to laterally rotate (and adduct; Romer 1923a; Charig 1972) the femur. The pubic muscle heads also had some capacity to flex the hip in dinosaurs that had pubic bones pointed forward, whereas the ischial heads extended the hip. In dinosaurs that had retroverted pubes (Fig. 9.6B), the line of action was drawn just below and behind the hip joint, eliminating their hip flexor moment arms and leaving them mainly as lateral rotators of the femur (and adductors of the lower leg). Hutchinson and Gatesy (2000) proposed that the lateral rotation function of these muscles was important for nonavian dinosaur locomotion (as in birds) to spin the femur about its long axis while the leg was airborne, and as a consequence of this lateral rotation, to draw the lower leg inward (i.e., adduct it), positioning the foot closer to the body midline for the next step.

CAUDOFEMORALIS MUSCLES

Unlike many mammals, most reptiles (including birds and other dinosaurs) have two main components of this group (Fig. 9.10), which play a crucial role in locomotion. The first part, M. caudofemoralis brevis, has its origin revealed by a large fossa (bony shelf) on the lower edge of the ilium behind the hip. The second part, M. caudofemoralis longus, had a much larger origin in most dinosaurs, spanning part of the length of the tail beneath the more superficial hypaxial tail muscles. These two large muscles converged on the fourth trochanter, a large crest on the back of the femur (Fig. 9.8D) that is distinct in most nonavian dinosaurs. Reid (1991) described areas of heavy bone remodeling in this area, corroborating the attachment of a large muscle. The caudofemoral muscles were presumably the chief agents of hip extension (femoral retraction) in most nonavian dinosaurs (Gatesy 1990, 2001), providing much force and leverage to propel the body forward during locomotion. In basal theropod dinosaurs, the tail began shortening and stiffening distally, reducing the area available for the origins of these muscles, although they were still large even in coelurosaurs such as tyrannosaurs (Persons and Currie 2011). In concert, the fourth trochanter became smaller during theropod evolution, and the role of the caudofemoral muscles in femoral retraction was likewise reduced. As a result, birds use femoral

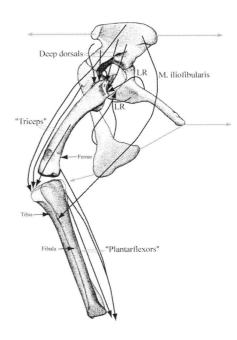

9.9. Major hind limb muscle paths, from origin to insertion, in the early dinosaur *Herrerasaurus* (modified from Novas 1993). Left hind limb in side view. Gray arrows indicate the general paths of trunk and caudal musculature. Gray ellipses surround major hind limb muscle groups. LR = lateral rotator muscles. For simplicity, the foot bones are not shown in these reconstructions.

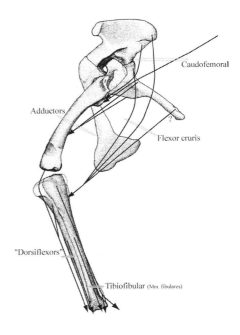

9.10. Major hind limb muscle paths, from origin to insertion, in the early dinosaur *Herrerasaurus* (modified from Novas 1993). Left hind limb in side view ? indicates another possible (but ambiguously present) flexor cruris muscle head. Other details as in Figure 9.9.

retraction less to produce their long, crouched strides during walking. They instead rely more on their large hamstring (and plantarflexor) muscles to move the knee joint (and ankle and toes) during walking (Gatesy 1990). Indeed, many birds have lost part or all of these caudofemoral muscles. Gatesy and Dial (1996; see also Gatesy 2001) proposed another novel idea: this reduction of tail-based musculature decoupled the tail from the hind limb, enabling birds to use their tails as an independent locomotor module during flight.

Lower Leg Muscles

The muscles of the lower leg (past the knee) have been less well studied in dinosaurs (see Dilkes 2000; Hutchinson 2002), but some general conclusions may be drawn. The tiny intrinsic muscles of the feet will not be considered here.

PLANTARFLEXORS

Dinosaurs had two main sets of plantarflexor muscles (Fig. 9.9) that developed mainly from a ventral embryonic mass. The first, superficial, group (M. gastrocnemius) only extended (plantarflexed) the ankle joint and had little or no action at the toes. M. gastrocnemius had at least two heads, a lateral one from the end of the femur and a medial one from the top of the tibia. Their tendons joined to form an Achilles tendon that inserted on the back of the metatarsus, often well marked by scars. The second, deeper, group (the digital flexors) was less effective at ankle extension but sent tendons to the toes that enabled it to flex (clench) all of them. The digital flexor group had many subdivided heads (Hutchinson 2002). These muscle heads sent many independent tendons to the toes, allowing for some differential control of the toes in conjunction with the intrinsic foot muscles.

DORSIFLEXORS

Two main muscles of dorsal developmental origin are noteworthy (Fig. 9.10): M. extensor digitorum longus, which occupied much of the space on the front of the tibia and split into single tendons that went to scarred regions on each toe bone, and M. tibialis anterior, which came down from the top of the tibia to insert on tubercles on the upper ends of the metatarsal bones. Homology of these muscles among living animals is not completely resolved (Hutchinson 2002). These muscles acted to flex the ankle and extend the toe joints, primarily when the foot was off the ground during locomotion.

TIBIOFIBULAR MUSCLES

The remainder of the muscles of the lower leg can be lumped into a set of odd muscles (Fig. 9.10) that were vestiges of muscles from earlier ancestors of dinosaurs. These muscles act to move the fibula with respect to the tibia (i.e., rotate one about the other's long axis) in living lizards, turtles, and crocodiles, which have fairly mobile tibiae and fibulae. In dinosaurs, the tibia and fibula were much less free, or even unable, to move relative to each other, so these muscles must have been less crucial.

Problems and Solutions for the Hind Limb Musculature of Dinosaurs

The single dinosaur receiving the most attention regarding muscle anatomy has long been the large coelurosaur *Tyrannosaurus* (Romer 1923a, 1923c; Walker 1977; Carrano and Hutchinson 2002). *Tyrannosaurus* has often been used as an exemplar of the saurischian condition (e.g., Romer 1923a, 1923c), despite its specialized morphology and phylogenetic distance from basal saurischians. A more expansive phylogenetic investigation (Carrano and Hutchinson 2002; also see Brochu 2003) contradicted some of Romer's and others' conclusions, showing that *Tyrannosaurus* was not an ideal model for the ancestral saurischian condition.

Less attention has been given to the hind limb muscles of sauropodomorphs than to other dinosaurs, perhaps because muscle scars are less evident on many sauropod fossils, despite their large size. Unfortunately, the few such studies published, such as Romer (1923c), Borsuk-Bialynicka (1977), or Christian et al. (1996), over rely on lizard or crocodylian models, eschewing potentially informative data from birds. Conversely, Cooper (1981) studied excellent muscle scarring in the basal sauropodomorph *Massospondylus* but applied mainly an avian model for his reconstruction. Sauropodomorphs deserve more research on their musculoskeletal form and function.

What is sometimes labeled as the saurischian condition is mostly plesiomorphic for dinosauromorphs. It was ornithischian dinosaurs whose pelvic muscle anatomy seems to have diverged most greatly in the Triassic. Ornithischian dinosaurs have long been characterized by their superficially birdlike pelvis. Romer's (1927a, 1927b) investigation of chick development for reconstructing ornithischian myology remains a respected classic, but Dilkes (2000) showed how a phylogenetic perspective reveals important differences that studies using bird models missed.

As expected, dinosaur hind limb muscles emphasize the actions expected in an erect biped (Romer 1923a; Charig 1972; Hutchinson 2002), even ancestrally (e.g., Figs. 9.4, 9.8, 9.9, 9.10). Muscles that ancestrally adducted/abducted or medially/laterally rotated the limbs (especially those distal to the hip joint) were reduced or restructured, shifting muscle moment arms in favor of flexors and extensors. The capacity of muscles to control the three-dimensional position of the foot and other distal limb joints became concentrated at the hip joint, which remained a ball-and-socket joint, while the lower leg joints became increasingly more hingelike in dinosaurs, even fusing together in some lineages.

Yet so far, muscles provide only a few key details about dinosaur locomotion; this dynamic area of research has only begun to move beyond classical functional anatomy. (See Romer 1923a, 1923b, 1923c; Charig 1972; Walker 1977; Tarsitano 1983; and references therein for classical examples.) However, the work of Gatesy (e.g., 1990, 2001) is the new gold standard for integrating myological data into locomotor studies of dinosaurs, combining experimental data from living animals with anatomical data from dinosaurs into a phylogenetic context. Carrano (2000) and Hutchinson and Gatesy (2000) used similar approaches, particularly with a focus on how the unique stance and gait of extant birds evolved (reviewed by Hutchinson and Allen 2009). But many questions about stance and gait proposed by classical works remain unresolved. No study has yet shown how all of the muscles of a dinosaur hind limb worked together during locomotion, or established the precise speeds that dinosaur muscles enabled. Although this is a major goal of research on dinosaur musculature, it is a daunting problem that years of anatomical, functional, and computer modeling work may not conclusively resolve, even with trackway information, considering the many unknowns and complexities.

The locomotion of large dinosaurs remains a particularly contentious issue. Hutchinson and Garcia (2002) used a simple muscle reconstruction and biomechanical model of *Tyrannosaurus* to show that its leg muscles were not large enough to enable fast running (above 5–11 m/s or 11–25 mph). Subsequent studies have elaborated on and generally supported this modeling approach (Hutchinson 2004a, 2004b; Hutchinson et al. 2005, 2007; Sellers and Manning 2007; Gatesy et al. 2009; reviewed by Hutchinson and Allen 2009). Bakker (1986), Paul and Christiansen (2000), and others have argued that some bulky processes on ceratopsid hind limbs offered large space for muscle attachment, and thus enabled trotting or even galloping in all ceratopsians regardless of body size. Many details of such studies remain controversial (e.g., Johnson and Ostrom 1995; Dodson and Farlow 1997).

These studies of large animals have not demonstrated that the limb muscles were large enough to enable fast running; they merely observed that the attachments seem large in comparison with living analogs. Functional analogies with rhinoceroses, elephants, birds, or long-legged artiodactyls are all limited in their ability to test functional hypotheses about dinosaur locomotion. They often overlook dissimilarities in search of similarities, rendering qualitative or vague rather than quantitative or specific results (Hutchinson and Allen 2009). Assumptions are made about form–function relationships in extant animals that are not necessarily accurate (Lauder 1995). Biomechanical studies such as the approach taken by Hutchinson and Garcia (2002) could quantitatively test whether any impressive morphologies in dinosaurs corresponded to extreme locomotor performance.

Conclusions

Recent studies on the reconstruction of muscles and other soft tissues in dinosaurs have emphasized the need to go beyond simply looking at one or perhaps two living models. Instead, attention must be paid to multiple living models within a clearly articulated phylogeny, including fossil data to polarize character states and test homologies, in order to explicitly demonstrate the relative support for the reconstruction of each soft tissue feature. More attention must be given to the details of soft tissue anatomy in a diversity of extant species and the nature of any correlation between a feature on a bone and any soft tissue structures. Unexpected conclusions may be a result (e.g., Witmer 2001). These reconstructions should also examine the evolutionary and developmental history of the features on the skeleton that are correlated with soft tissues in living species so that the homology of these skeletal features can be traced and evolutionary changes to the soft tissues revealed. As a vital step in any discussion of musculoskeletal function in a dinosaur, muscles need to be reconstructed as accurately as possible using all available data. Yet at the same time, the ambiguities of these reconstructions must also be explicitly considered. Experimental data from living species on how muscles and other soft tissues operate together must also be incorporated into discussions of form and function in dinosaurs. The studies that will have the most enduring influence will be those that hold the highest feasible standards for the methods and evidence used.

References

Abdala, V., and R. Diogo. 2010. Comparative anatomy, homologies and evolution of the pectoral and forelimb musculature of tetrapods with special attention to extant limbed amphibians and reptiles. *Journal of Anatomy* 217: 536–573.

Adams, L. A. 1919. A memoir on the phylogeny of the jaw muscles in recent and fossil vertebrates. *Annals of the New York Academy of Science* 28: 51–166.

Anderson, H. T. 1936. The jaw musculature of the phytosaur, *Machaeroprosopus*. *Journal of Morphology* 59: 549–587.

Arbour, V. M. 2009. Estimating impact forces of tail club strikes by ankylosaurid dinosaurs. *PLoS ONE* 4(8): e6738.

Arbour, V. M., and E. Snively. 2009. Finite element analyses of ankylosaurid dinosaur tail club impacts. *Anatomical Record* 292: 1412–1426.

Bakker, R. T. 1986. *Dinosaur Heresies*. New York: William Morrow.

———. 1998. Brontosaur killers: late Jurassic allosaurids as sabre-tooth cat analogues. *Gaia* 15: 145–158.

Barrett, P. M. 2000. Prosauropod dinosaurs and iguanas: speculations on the diets of extinct reptiles. In H.-D. Sues (ed.), *Evolution of Herbivory in Terrestrial Vertebrates: Perspectives from the Fossil Record*, 42–78. Cambridge: Cambridge University Press.

———. 2001. Tooth wear and possible jaw action of *Scelidosaurus harrisonii* Owen and a review of feeding mechanisms in other thyreophoran dinosaurs. In K. Carpenter (ed.), *The Armored Dinosaurs*, 25–52. Bloomington: Indiana University Press.

Bimber, O., S. M. Gatesy, L. M. Witmer, et al. 2002. Merging fossil specimens with computer-generated information. *IEEE Computer* 35: 25–30.

Bock, W. J. 1974. The avian skeletomuscular system. In D. S. Farner and J. R. King. (eds.), *Avian Biology*, 4:119–257. New York: Academic Press.

Borsuk-Bialynicka, M. 1977. A new camarasaurid sauropod *Opisthocoelicaudia skarzynskii* gen. n., sp. n. from the Upper Cretaceous of Mongolia. *Palaeontologia Polonica* 37: 5–37.

Brett-Surman, M. K. 1997. Ornithopods. In J. O. Farlow and M. K. Brett-Surman (eds.), *The Complete Dinosaur*, 330–346. Bloomington: Indiana University Press.

Briggs, D. E. G., P. R. Wilby, B. P. Pérez-Moreno, J. L. Sanz, and M. Fregenal-Martínez. 1997. The mineralization of dinosaur soft tissue in the Lower Cretaceous of Las Hoyas, Spain. *Journal of the Geological Society, London* 154: 587–588.

Brochu, C. A. 2003. Osteology of *Tyrannosaurus rex*: insights from a nearly complete skeleton and high-resolution computed tomographic analysis of the skull. *Society of Vertebrate Paleontology Memoir* 7: 1–138.

Brown, B. 1916. *Corythosaurus casuarius*: skeleton, musculature and epidermis. *Bulletin of the American Museum of Natural History* 35: 709–716.

Bryant, H. N., and A. P. Russell. 1992. The role of phylogenetic analysis in the inference of unpreserved attributes of extinct taxa. *Philosophical Transactions of the Royal Society of London, Series B* 337: 405–418.

Bryant, H. N., and K. L. Seymour. 1990. Observations and comments on the reliability of muscle reconstruction in fossil vertebrates. *Journal of Morphology* 206: 109–117.

Bühler, P., L. D. Martin, and L. M. Witmer. 1988. Cranial kinesis in the Late Cretaceous birds *Hesperornis* and *Parahesperornis*. *Auk* 105: 111–122.

Carpenter, K. 1997. Ankylosaurs. In J. O. Farlow and M. K. Brett-Surman (eds.), *The Complete Dinosaur*, 307–329. Bloomington: Indiana University Press.

———. 2002. Forelimb biomechanics of nonavian theropod dinosaurs in predation. *Senckenbergiana lethaea* 83: 59–76.

Carpenter, K., F. Sanders, L. A. McWhinney, and L. Wood. 2005. Evidence for predator–prey relationships: examples for *Allosaurus* and *Stegosaurus*. In K. Carpenter (ed.), *The Carnivorous Dinosaurs*, 325–350. Bloomington: Indiana University Press.

Carpenter, K., and M. Smith. 2001. Forelimb osteology and biomechanics of *Tyrannosaurus rex*. In D. Tanke and K. Carpenter (eds.), *Mesozoic Vertebrate Life*, 90–116. Bloomington: Indiana University Press.

Carrano, M.T. 2000. Convergence and the evolution of dinosaur locomotion. *Paleobiology* 26: 489–512.

Carrano, M. T., and J. R. Hutchinson. 2002. Pelvic and hindlimb musculature of *Tyrannosaurus rex* (Dinosauria: Theropoda). *Journal of Morphology* 253: 207–228.

Carrier, D. R., and C. G. Farmer. 2000. The evolution of pelvic aspiration in archosaurs. *Paleobiology* 26: 271–293.

Charig, A. 1972. The evolution of the archosaur pelvis and hind-limb: an explanation in functional terms. In K. A. Joysey and T. S. Kemp (eds.), *Studies in Vertebrate Evolution*, 121–155. Edinburgh: Oliver & Boyd.

Chiappe, L. M., J. Shu'an, J. Qiang, and M. A. Norell. 1999. Anatomy and systematics of the Confuciusornithidae (Theropoda: Aves) from the late Mesozoic of Northeastern China. *Bulletin of the American Museum of Natural History* 242.

Christian, A. 2010. Some sauropods raised their necks – evidence for high browsing in *Euhelopus zdanskyi*. *Biology Letters* 6: 823–825.

Christian, A., and G. Dzemski. 2007. Reconstruction of the cervical skeleton posture of *Brachiosaurus branchai* Janensch, 1914 by an analysis of the intervertebral stress along the neck and a comparison with the results of different approaches. *Fossil Record* 10: 38–49.

Christian, A., D. Koberg, and H. Preuschoft. 1996. Shape of the pelvis and posture of the hindlimbs in *Plateosaurus*. *Paläontologische Zeitschrift* 70: 591–601.

Claessens, L. 1997. Gastralia. In P. Currie and K. Padian (eds.), *Encyclopedia of Dinosaurs*, 269–270. New York: Academic Press.

Cleuren, J., and F. De Vree. 2000. Feeding in crocodilians. In K. Schwenk (ed.), *Feeding: Form, Function, and Evolution in Tetrapod Vertebrates*, 337–358. New York: Academic Press.

Codd, J. R., P. L. Manning, M. A. Norell, and S. F. Perry. 2008. Avian-like breathing mechanics in maniraptoran dinosaurs. *Proceedings of the Royal Society B, Biological Sciences* 275: 157–161.

Colbert, E. H. 1945. The hyoid bones in *Protoceratops* and in *Psittacosaurus*. *American Museum Novitates* 1301: 1–10.

Coombs, W. P., Jr. 1978. Forelimb muscles of the Ankylosauria (Reptilia, Ornithischia). *Journal of Paleontology* 52: 642–657.

Cooper, M. R. 1981. The prosauropod dinosaur *Massospondylus carinatus* Owen from Zimbabwe: its biology, mode of life and phylogenetic significance. *Occasional Papers of the National Museums and Monuments of Rhodesia B, Natural Sciences* 6: 689–840.

Crompton, A. W., and J. Attridge. 1986. Masticatory apparatus of the large herbivores during Late Triassic and Early Jurassic times. In K. Padian (ed.), *The Beginning of the Age of Dinosaurs: Faunal Change across the Triassic–Jurassic Boundary*, 223–236. Cambridge: Cambridge University Press.

Curtis, N., M. E. H. Jones, S. E. Evans, J. Shi, P. O'Higgins, and M. J. Fagan. 2010. Predicting muscle activation patterns from motion and anatomy: modeling the skull of *Sphenodon* (Diapsida: Rynchocephalia). *Journal of Royal Society Interface* 7: 153–160.

Dal Sasso, C., and M. Signore. 1998. Exceptional soft-tissue preservation in a theropod dinosaur from Italy. *Nature* 392: 383–387.

Daniel, W. J. T., and C. McHenry. 2001. Bite force to skull stress correlation – modeling the skull of *Alligator mississippiensis*. In G. C. Grigg, F. Seebacher, and C. E. Franklin (eds.), *Crocodilian Biology and Evolution*, 135–143. Chipping Norton: Surrey Beatty and Sons.

Degrange, F. J., C. P. Tambussi, K. Moreno, L. M. Witmer, and S. Wroe. 2010. Mechanical analysis of feeding behavior in the extinct "terror bird" *Andalgalornis steulleti* (Gruiformes: Phorusrhacidae). *PLoS ONE* 5 (8): e11856.

Dilkes, D. W. 2000. Appendicular myology of the hadrosaurian dinosaur *Maiasaura peeblesorum* from the Late Cretaceous (Campanian) of Montana. *Transactions of the Royal Society of Edinburgh: Earth Sciences* 90: 87–125.

———. 2001. An ontogenetic perspective on locomotion in the Late Cretaceous dinosaur *Maiasaura peeblesorum* (Ornithischia: Hadrosauridae). *Canadian Journal of Earth Sciences* 38: 2105–1227.

Diogo, R., V. Abdala, M. A. Aziz, N. Lonergan, and B. A. Wood. 2009. From fish to modern humans – comparative anatomy, homologies and evolution of the pectoral girdle and forelimb musculature. *Journal of Anatomy* 214: 694–716.

Dodson, P. 1996. *The Horned Dinosaurs: A Natural History*. Princeton, N.J.: Princeton University Press.

Dodson, P., and J. O. Farlow. 1997. The forelimb carriage of ceratopsid dinosaurs. In D. L. Wolberg, E. Stump, and G. D. Rosenberg (eds.), *Dinofest International: Proceedings of a Symposium Held at Arizona State University*, 393–398. Philadelphia, Pa.: Academy of Natural Sciences.

Dollo, L. 1884. Cinquième note sur les dinosauriens de Bernissart. *Bulletin du Musée Royal d'Histoire Naturelle de Belgique* 3: 129–146.

Dzemski, G., and A. Christian. 2007. Flexibility along the neck of the ostrich (*Struthio camelus*) and consequences for the reconstruction of dinosaurs with extreme neck length. *Journal of Morphology* 268: 701–714.

Edgeworth, F. H. 1936. *The Cranial Muscles of Vertebrates*. Cambridge: Cambridge University Press.

Elzanowski, A. 1987. Cranial and eyelid muscles and ligaments of the tinamous (Aves: Tinamiformes). *Zoologische Jahrbuch für Anatomie* 116: 63–118.

Farke, A. A., E. D. S. Wolff, and D. H. Tanke. 2009. Evidence of combat in *Triceratops*. *PLoS ONE* 4 (1): e4252.

Farlow, J. O., and P. Dodson. 1975. The behavior significance of frill and horn morphology in ceratopsian dinosaurs. *Evolution* 29: 353–361.

Farlow, J. O., C. V. Thompson, and D. E. Rosner. 1978. Plates of the dinosaur *Stegosaurus*: forced convection heat loss fins? *Science* 192: 1123–1125.

Farlow, J. O., and T. R. Holtz Jr. 2002. The fossil record of predation in dinosaurs. *Paleontological Society Papers* 8: 251–266.

Fiorillo, A. R. 1998. Dental microwear patterns of the sauropod dinosaurs *Camarasaurus* and *Diplodocus*: evidence for resource partitioning in the late Jurassic of North America. *Historical Biology* 13: 1–16.

Galton, P. M. 1969. The pelvic musculature of the dinosaur *Hypsilophodon* (Reptilia: Ornithischia). *Postilla* 131: 1–64.

———. 1973. The cheeks of ornithischian dinosaurs. *Lethaia*. 6: 67–89.

———. 1974. The ornithischian dinosaur *Hypsilophodon* from the Wealden of the Isle of Wight. *Bulletin of the British Museum of Natural History, Geological Series* 25: 1–152.

———. 1986. Herbivorous adaptations of Late

Triassic and Early Jurassic dinosaurs. In K. Padian (ed.), *The Beginning of the Age of Dinosaurs: Faunal Change across the Triassic–Jurassic Boundary*, 203–222. Cambridge: Cambridge University Press.

———. 1990. Stegosauria. In D. B. Weishampel, P. Dodson, and H. Osmólska (eds.), *The Dinosauria*, 435–455. Berkeley: University of California Press.

Gatesy, S. M. 1990. Caudofemoral musculature and the evolution of theropod locomotion. *Paleobiology* 16: 170–186.

———. 1994. Neuromuscular diversity in archosaur deep dorsal thigh muscles. *Brain Behavior Evolution* 43: 1–14.

———. 2001. The evolutionary history of the theropod caudal locomotor module. In J. Gauthier and L. F. Gall (eds.), *New Perspectives on the Origin and Early Evolution of Birds: Proceedings of the International Symposium in Honor of John H. Ostrom*, 333–345. New Haven, Conn.: Peabody Museum of Natural History, Yale University.

Gatesy, S. M., M. Baeker, and J. R. Hutchinson. 2009. Constraint-based exclusion of limb poses for reconstructing theropod dinosaur locomotion. *Journal of Vertebrate Paleontology* 29: 535–544.

Gatesy, S. M., D. B. Baier, F. A. Jenkins, and K. P. Dial. 2010. Scientific rotoscoping: a morphology-based method of 3-D motion analysis and visualization. *Journal of Experimental Zoology* 313A: 244–261.

———. 1996. Locomotor modules and the evolution of avian flight. *Evolution* 50: 331–340.

Gilmore, C. W. 1925. A nearly complete articulated skeleton of *Camarasaurus*, a saurischian dinosaur from the Dinosaur National Monument. *Memoir of the Carnegie Museum* 10: 347–384.

———. 1946. A new carnivorous dinosaur from the Lance Formation of Montana. *Smithsonian Miscellaneous Collections* 106: 1–19.

Gingerich, P. D. 1973. Skull of *Hesperornis* and early evolution of birds. *Nature* 243: 70–73.

Gregory, W. K. 1919. The pelvis of dinosaurs: a study of the relations between muscular stresses and skeletal forms. *Copeia* 69: 18–20.

Haas, G. 1955. The jaw musculature in *Protoceratops* and in other ceratopsians. *American Museum Novitates* 1729: 1–24.

———. 1963. A proposed reconstruction of the jaw musculature of *Diplodocus*. *Annals of the Carnegie Museum* 36: 139–157.

———. 1969. On the jaw musculature of *Ankylosaurus*. *American Museum Novitates* 2399: 1–11.

———. 1973. Muscles of the jaws and associated structures in the Rhynchocephalia and Squamata. In C. Gans and T. S. Parsons (eds.), *Biology of the Reptilia*, vol. 4, *Morphology D*, 285–490. New York: Academic Press.

Hatcher, J. B., O. C. Marsh, and R. S. Lull. 1907. The Ceratopsia. *U.S. Geological Survey Monographs* 49: 1–300.

Henderson, D. M. 2002. The eyes have it: the sizes, shapes, and orientations of theropod orbits as indicators of skull strength and bite force. *Journal of Vertebrate Paleontology* 22: 766–778.

Herrel, A., P. Aerts, and F. De Vree. 2000. Cranial kinesis in geckoes: functional implications. *Journal of Experimental Biology* 203: 1415–1423.

Herrel, A., J. C. O'Reilly, and A. M. Richmond. 2002. Evolution of bite performance in turtles. *Journal of Evolutionary Biology* 15: 1083–1094.

Herring, S. W. 1993. Epigenetic and functional influences on skull growth. In J. Hanken and B. K. Hall (eds.), *The Skull*, vol. 1, *Development*, 153–206. Chicago: University of Chicago Press.

Holliday, C. M. 2009. New insights into dinosaur jaw muscle anatomy. *Anatomical Record* 292: 1246–1265

Holliday, C. M., R. C. Ridgely, A. Balanoff, and L. M. Witmer. 2006. Report on a novel vascular device of the Caribbean flamingo (*Phoenicopterus ruber*) using CT scanning and digital visualization techniques. *Anatomical Record* 288 (10): 1031–1041.

Holliday, C. M., and L. M. Witmer. 2007. Archosaur adductor chamber evolution: integration of musculoskeletal and topological criteria in jaw muscle homology. *Journal of Morphology* 268: 457–484.

———. 2008. Cranial kinesis in dinosaurs: intracranial joints, protractor muscles, and their significance for cranial evolution and function in diapsids. *Journal of Vertebrate Paleontology* 28 (4): 1073–1088.

———. 2009. The epipterygoid of crocodyliforms and its significance for the evolution of the orbitotemporal region of eusuchians. *Journal of Vertebrate Paleontology* 29 (3): 715–733.

Horner, J. R., and D. Lessem. 1993. *The Complete T. rex*. New York: Simon and Schuster.

Horner, J. R., and C. Marshall. 2002. Keratinous covered dinosaur skulls. *Journal of Vertebrate Paleontology* 22 (3): 31A.

Hutchinson, J. R. 2001a. The evolution of pelvic osteology and soft tissues on the line to extant birds (Neornithes). *Zoological Journal of the Linnean Society* 131: 123–168.

———. 2001b. The evolution of femoral osteology and soft tissues on the line to extant birds (Neornithes). *Zoological Journal of the Linnean Society* 131: 169–197.

———. 2002. The evolution of hindlimb tendons and muscles on the line to crown-group birds. *Comparative Biochemistry and Physiology* A 133: 1051–1086.

———. 2004a. Biomechanical modeling and sensitivity analysis of bipedal running ability. I. Extant taxa. *Journal of Morphology* 262: 421–440.

———. 2004b. Biomechanical modeling and sensitivity analysis of bipedal running ability. II. Extinct taxa. *Journal of Morphology* 262: 441–461.

Hutchinson, J. R., and V. Allen. 2009. The evolutionary continuum of limb function from early theropods to birds. *Naturwissenschaften* 96: 423–448.

Hutchinson, J. R., F. C. Anderson, S. Blemker, and S. L. Delp. 2005. Analysis of hindlimb muscle moment arms in *Tyrannosaurus rex* using a three-dimensional musculoskeletal computer model. *Paleobiology* 31: 676–701.

Hutchinson, J. R., and M. Garcia. 2002. *Tyrannosaurus* was not a fast runner. *Nature* 415: 1018–1021.

Hutchinson, J. R., and S. M. Gatesy. 2000. Adductors, abductors, and the evolution of archosaur locomotion. *Paleobiology* 26: 734–751.

Hutchinson, J. R., V. Ng-Thow-Hing, and F. C. Anderson. 2007. A 3D interactive method for estimating body segmental parameters in animals: application to the turning and running performance of *Tyrannosaurus rex*. *Journal of Theoretical Biology* 246: 660–680.

Iordansky, N. N. 2000. Jaw muscle of the crocodiles: structure, synonymy, and some implications on homology and functions. *Russian Journal of Herpetology* 7: 41–50.

Janensch, W. 1936. Die Schädel der Sauropoden *Brachiosaurus*, *Barosaurus*, und *Dicraeosaurus* aus den Tendaguruschichten Deutsch-Ostafrikas. *Palaeontographica* 7: 147–298.

Jasinoski, S. C., A. P. Russell, and P. J. Currie. 2006. An integrative phylogenetic and extrapolatory approach to the reconstruction of dromaeosaur (Theropoda: Eumaniraptora) shoulder musculature. *Zoological Journal of the Linnean Society* 146: 301–344.

Ji, Q., P. J. Currie, M. A. Norell, and S.-A. Ji. 1998. Two feathered dinosaurs from northeastern China. *Nature* 393: 753–761.

Johnson, R. E., and J. H. Ostrom. 1995. The forelimb of *Torosaurus* and an analysis of the posture and gait of ceratopsian dinosaurs. In J. J. Thomason (ed.), *Functional Morphology in Vertebrate Paleontology*, 205–218. New York: Cambridge University Press.

Kellner, A. W. A. 1996. Fossilized theropod soft tissue. *Nature* 379: 32.

Knell, R. J., and S. D. Sampson. 2011. Bizarre structures in dinosaurs: species recognition or sexual selection? A response to Padian and Horner. *Journal of Zoology* 283: 18–22.

Knoll, F. 2008. Buccal soft anatomy in *Lesothosaurus* (Dinosauria: Ornthischia). *Neues Jahrbuch für Geologie und Paläontologie-Abhandlungen* 248: 355–364.

Lambe, L. M. 1919. Description of a new genus and species (*Panoplosaurus mirus*) of an armoured dinosaur from the Belly River Beds of Alberta. *Transactions of the Royal Society of Canada* (ser. 3) 13: 39–50.

Langer, M. C., M. A. G. França, and S. Gabriel. 2007. The pectoral girdle and forelimb anatomy of the stem-sauropodomorph *Saturnalia tupiniquim* (Upper Triassic, Brazil). *Special Papers in Palaeontology* 77: 113–137.

Larsson, H. C. E. 2008. Palatal kinesis of *Tyrannosaurus rex*. In P. Larsen and K. Carpenter (eds.), *Tyrannosaurus rex: The Tyrant King*, 245–254. Bloomington: Indiana University Press.

Lauder, G. V. 1995. On the inference of function from structure. In J. J. Thomason (ed.), *Functional Morphology in Vertebrate Paleontology*, 1–18. Cambridge: Cambridge University Press.

Lull, R. S. 1908. The cranial musculature and the origin of the frill in the ceratopsian dinosaurs. *American Journal of Science* 25: 387–399.

Lull, R. S., and N. E. Wright. 1942. Hadrosaurian dinosaurs of North America. *Geological Society of America Special Paper* 40: 1–242.

Mazzetta, G. V., R. A. Fariña, and S. F. Vizcaíno. 1998. On the palaeobiology of the South American horned theropod *Carnotaurus sastrei* Bonaparte. *Gaia* 15: 185–192.

McClelland, B. K. 1990. Anatomy and cranial kinesis of the *Allosaurus* skull. Master's thesis, University of Utah, Salt Lake City.

McGowan, C. 1979. The hind limb musculature of the Brown Kiwi, *Apteryx australis mantelli*. *Journal of Morphology* 160: 33–73.

———. 1982. The wing musculature of the Brown Kiwi *Apteryx australis mantelli*

and its bearing on ratite affinities. *Journal of Zoology* 197: 173–219.

McIntosh, J. S. 1990. Sauropoda. In D. B. Weishampel, P. Dodson, and H. Osmólska (eds.), *The Dinosauria*, 345–401. Berkeley: University of California Press.

McWhinney, L., K. Carpenter, and B. Rothschild. 2001. Dinosaurian humeral periostitis: a case of a juxacortical lesion in the fossil record. In D. Tanke and K. Carpenter (eds.), *Mesozoic Vertebrate Life*, 364–377. Bloomington: Indiana University Press.

Meers, M. B. 2003. Crocodylian forelimb musculature and its relevance to Archosauria. *Anatomical Record Part A* 247A: 891–916.

Metzger, K. 2002. Cranial kinesis in lepidosaurs: skull in motion. In P. Aerts, K. D'Août, A. Herrel, and R. Van Damme (eds.), *Topics in Functional and Ecological Vertebrate Morphology*, 15–46. Maastricht: Shaker Publishing.

Molnar, R. E. 1973. The cranial morphology and mechanics of *Tyrannosaurus rex* (Reptilia: Saurischia). Ph.D. dissertation, University of California, Los Angeles.

———. 1998. Mechanical factors in the design of the skull of *Tyrannosaurus rex* (Osborn, 1905). *Gaia* 15: 193–218.

———. 2008. Reconstruction of jaw musculature of *Tyrannosaurus rex*. In P. Larson and K. Carpenter (eds.), *Tyrannosaurus rex*, 254–281. Bloomington: Indiana University Press.

Müller, G. B., and J. Streicher. 1989. Ontogeny of the syndesmosis tibiofibularis and the evolution of the bird hindlimb: a caenogenetic feature triggers phenotypic novelty. *Anatomy and Embryology* 179: 327–339.

Nicholls, E. L., and A. P. Russell. 1985. Structure and function of the pectoral girdle and forelimb of *Struthiomimus altus* (Theropoda: Ornithomimidae). *Palaeontology* 28: 643–677.

Norell, M. A., and P. J. Makovicky. 1999. Important features of the dromaeosaurid skeleton II: information from newly collected specimens of *Velociraptor mongoliensis*. *American Museum Novitates* 3282: 1–45.

Norman, D. B. 1980. On the ornithischian dinosaur *Iguanodon bernissartensis* from the Lower Cretaceous of Bernissart (Belgium). *Institut Royal des Sciences Naturelles de Belgique Mémoire* 178: 1–103.

———. 1984. On the cranial morphology and evolution of ornithopod dinosaurs. *Symposiums of the Zoological Society of London* 52: 521–547.

———. 1986. On the anatomy of *Iguanodon*

atherfieldensis (Ornithischia: Ornithopoda). *Bulletin de l'Institut Royal des Sciences Naturelles de Belgique* 56: 281–372.

Norman, D. B., and D. B. Weishampel. 1985. Ornithopod feeding mechanisms: their bearing on the evolution of herbivory. *American Naturalist* 126: 151–164.

Novas, F.E. 1993. New information on the systematics and postcranial skeleton of *Herrerasaurus ischigualastensis* (Theropoda: Herrerasauridae) from the Ischigualasto Formation (Upper Triassic) of Argentina. *Journal of Vertebrate Paleontology* 13: 400–423.

Organ, C. L. 2006a. Thoracic epaxial muscles in living archosaurs and ornithopod dinosaurs. *Anatomical Record* 288: 783–793.

———. 2006b. Biomechanics of ossified tendons in ornithopod dinosaurs *Paleobiology* 32: 652–665.

Ostrom, J. H. 1961. Cranial morphology of the hadrosaurian dinosaurs of North America. *Bulletin of the American Museum of Natural History* 122: 39–186.

———. 1964. A functional analysis of jaw mechanics in the dinosaur *Triceratops*. *Postilla* 88: 1–35.

———. 1966. Functional morphology and evolution of the ceratopsian dinosaurs. *Evolution* 20: 290–308.

———. 1974. The pectoral girdle and forelimb function of *Deinonychus* (Reptilia: Saurischia): a correction. *Postilla* 165: 1–11.

Padian, K., and J. R. Horner. 2011a. The evolution of "bizarre structures" in dinosaurs: biomechanics, sexual selection, social selection, or species recognition? *Journal of Zoology* 283: 3–17.

———. 2011b. The definition of sexual selection and its implications for dinosaurian biology. *Journal of Zoology* 283: 23–27.

Papp, M. J. 2000. A critical appraisal of buccal soft-tissue anatomy in ornithischian dinosaurs. Master's thesis. Ohio University, Athens.

Papp, M. J., and L. M. Witmer. 1998. Cheeks, beaks, and freaks: a critical appraisal of buccal soft-tissue anatomy in ornithischian dinosaurs. *Journal of Vertebrate Paleontology* 18 (suppl. to 3): 76A.

Paul, G. S. 1988. *Predatory Dinosaurs of the World*. New York: Simon and Schuster.

———. 2002. *Dinosaurs of the Air*. Baltimore, Md.: Johns Hopkins University Press.

Paul, G. S., and P. Christiansen. 2000. Forelimb posture in neoceratopsian dinosaurs: implications for gait and locomotion. *Paleobiology* 26: 450–465.

Persons, W. S., and P. J. Currie. 2011. The tail of *Tyrannosaurus*: reassessing the size and locomotive importance of the *M. caudofemoralis* in non-avian theropods. *Anatomical Record* 294: 119–131.

Porro, L. 2008. Accuracy in finite element modeling of extinct taxa: sensitivity analyses in *Heterodontosaurus tucki*. *Journal of Vertebrate Paleontology* 28: 128A.

Raath, M. A. 1977. The anatomy of the Triassic theropod *Syntarsus rhodesiensis* (Saurischia: Podokesauridae) and a consideration of its biology. Ph.D. dissertation, Rhodes University, Grahamstown, South Africa.

Rayfield, E., D. Norman, C. Horner, J. Horner, P. Smith, J. Thomason, and P. Upchurch. 2001. Cranial design and function in a large theropod dinosaur. *Nature* 409: 1033–1037.

Reid, R. E. H. 1991. The histology of dinosaurian bone, and its possible bearing on dinosaurian physiology. *Symposium of the Zoological Society of London* 52: 629–663.

———. 1996. Bone histology of the Cleveland-Lloyd dinosaurs and of dinosaurs in general, part I. Introduction: introduction to bone tissues. *Brigham Young University Geology Studies* 41: 25–71.

Rega, E., R. Holmes, and A. Tirabasso. 2007. Habitual locomotor behavior inferred from naual pathology in two Late Cretaceous chasmosaurine dinosaurs, *Chasmosaurus irvinensis* (CMN 31357) and *Chasmosaurus belli* (ROM 843). In M. J. Ryan, B. J. Chinnery-Allgeier, and D. A. Eberth (eds.), *New Perspectives on Horned Dinosaurs*, 340–354. Bloomington: Indiana University Press.

Rieppel, O. 1978. Streptostyly and muscle function in lizards. *Experientia* 34: 776–777.

Romer, A. S. 1922. The locomotor apparatus of certain primitive and mammal-like reptiles. *Bulletin of the American Museum of Natural History* 46: 517–606.

———. 1923a. The ilium in dinosaurs and birds. *Bulletin of the American Museum of Natural History* 48: 141–145.

———. 1923b. Crocodilian pelvic muscles and their avian and reptilian homologues. *Bulletin of the American Museum of Natural History* 48: 533–552.

———. 1923c. The pelvic musculature of saurischian dinosaurs. *Bulletin of the American Museum of Natural History* 48: 605–617.

———. 1927a. The pelvic musculature of ornithischian dinosaurs. *Acta Zoologica* 8: 225–275.

———. 1927b. The development of the thigh musculature of the chick. *Journal of Morphology* 43: 347–385.

———. 1944. The development of the tetrapod limb musculature – the shoulder girdle of *Lacerta*. *Journal of Morphology* 74: 1–41.

Rowe, T. 1986. Homology and evolution of the deep dorsal thigh musculature in birds and other Reptilia. *Journal of Morphology* 189: 327–346.

Ruben, J. A., C. Dal Sasso, N. R. Geist, W. J. Hillenius, T. D. Jones, and M. Signore. 1999. Pulmonary function and metabolic physiology of theropod dinosaurs. *Science* 283: 514–516.

Ruben, J. A., T. D. Jones, N. R. Geist, and W. J. Hillenius. 1997. Lung structure and ventilation in theropod dinosaurs and early birds. *Science* 278: 1267–1270.

Russell, D. A., and Z.-M. Dong. 1993. A nearly complete skeleton of a new troodontid dinosaur from the Early Cretaceous of the Ordos Basin, Inner Mongolia, People's Republic of China. *Canadian Journal of Earth Sciences* 30: 2163–2173.

Russell, L. S. 1935. Musculature and function in the Ceratopsia. *Bulletin of the National Museum of Canada* 77: 39–48.

Rybczynski, N., A. Tirabasso, P. Bloskie, R. Cuthbertson, and C. M. Holliday. 2008. A 3D cranial animation of *Edmontosaurus* (Hadrosauridae) for testing feeding hypotheses. *Paleontologica Electronica* 11(2): 1–14.

Sampson, S. D. 2001. Speculations on the socioecology of ceratopsid dinosaurs (Ornithischia: Neoceratopsia). In D. Tanke and K. Carpenter (eds.), *Mesozoic Vertebrate Life*, 263–276. Bloomington: Indiana University Press.

Sampson, S. D., M. J. Ryan, and D. H. Tanke. 1997. Craniofacial ontogeny in centrosaurine dinosaurs (Ornithischia: Ceratopsidae): taxonomic and behavioral implications. *Zoological Journal of the Linnean Society* 221: 293–337.

Säve-Söderbergh, G. 1945. Notes on the trigeminal musculature in non-mammalian tetrapods. *Nova Acta Regiae Societatis Scientiarum Upsaliensis* (ser. 4) 13: 1–50.

Schumacher, G. H. 1973. The head muscles and hyolaryngeal skeleton of turtles and crocodilians. In C. Gans (ed.), *Biology of the Reptilia*, vol. 4, *Morphology D*, 101–200. London: Academic Press.

Schwarz, D., E. Frey, and C. A. Meyer. 2007. Novel reconstruction of the orientation of the pectoral girdle in sauropods. *Anatomical Record* 290: 32–47.

Schwarz-Wings, D. 2009. Reconstruction of the thoracic epaxial musculature of diplodocid and dicraeosaurid sauropods. *Journal of Vertebrate Paleontology* 29: 517–534.

Schwarz-Wings, D., E. Frey, and C. A. Meyer. 2007. Pneumaticity and soft-tissue reconstructions in the neck of diplodocid and dicraeosaurid sauropods. *Acta Palaeontologica Polonica* 52: 167–188.

Schwarz-Wings, D., C. A. Meyer, E. Frey, H. R. Manz-Steiner, and R. Schumacher. 2010. Mechanical implications of pneumatic neck vertebrae in sauropod dinosaurs. *Proceedings of the Royal Society B: Biological Sciences* 277: 11–17.

Schwenk, K. 2000. Feeding in lepidosaurs. In K. Schwenk (ed.), *Feeding: Form, Function, and Evolution in Tetrapod Vertebrates*, 175–291. New York: Academic Press.

Schmitz, L., and R. Motani. 2010. Morphological differences between eyeballs of nocturnal and diurnal amniotes revisited from optical perspectives of visual environments. *Vision Research* 50: 936–943.

Sellers, W. I., and P. L. Manning. 2007. Estimating dinosaur maximum running speeds using evolutionary robotics. *Proceedings of the Royal Society B* 274: 2711–2716.

Sereno, P. C. 1993. The pectoral girdle and forelimb of the basal theropod *Herrerasaurus ischigualastensis*. *Journal of Vertebrate Paleontology* 13: 425–450.

Sereno, P. C., and F. E. Novas. 1992. The complete skull and skeleton of an early dinosaur. *Science* 258: 1137–1140.

Sereno, P. C., Z. Xijin, and T. Lin. 2010. A new psittacosaur from Inner Mongolia and the parrot-like structure and function of the psittacosaur skull. *Proceedings of the Royal Society B* 277 (1679): 199–209.

Sinclair, A. G., and R. M. Alexander. 1987. Estimated forces exerted by the jaw muscles of some reptiles. *Journal of Zoology, London* 213: 107–115.

Snively, E., and A. P. Russell. 2007a. Functional morphology of neck musculature in Tyrannosauridae (Dinosaura, Theropoda) as determined via a hierarchical inferential approach. *Zoological Journal of the Linnean Society* 151: 759–808.

———. 2007b. Functional variation of neck muscles and their relation to feeding style in Tyrannosauridae and other large theropod dinosaurs. *Anatomical Record* 290: 934–957.

Snively, E., and A. P. Russell. 2007c. Craniocervical feeding dynamics of *Tyrannosaurus rex*. *Paleobiology* 33: 610–638.

Sternberg, C. M. 1929. A new species of horned dinosaur from the Upper Cretaceous of Alberta. *Bulletin of the National Museum of Canada* 54: 34–37.

Stevens, K. A., and J. M. Parrish. 1999. Neck posture and feeding habits of two Jurassic sauropod dinosaurs. *Science* 284: 798–800.

———. 2005. Digital reconstructions of sauropod dinosaurs and implications for feeding. In K. Curry Rogers and J. A. Wilson (eds.), *The Sauropods: Evolution and Paleobiology*, 178–200. Berkeley: University of California Press.

Tarsitano, S. 1983. Stance and gait in theropod dinosaurs. *Acta Paleontologica Polonica* 28: 251–264.

Taylor, M. P., M. J. Wedel, and D. Naish. 2009. Head and neck posture in sauropod dinosaurs inferred from extant animals. *Acta Palaeontologica Polonica* 54: 213–220.

Thompson, S., and R. Holmes. 2007. Forelimb stance and step cycle in *Chasmosaurus irvinensis* (Dinosauria: Neoceratopsia). *Palaeontologia Electronica* 10: 1–17.

Tomlinson, C. A. 2000. Feeding in paleognathous birds. In K. Schwenk (ed.), *Feeding: Form, Function, and Evolution in Tetrapod Vertebrates*, 359–394. New York: Academic Press.

Tsuihiji, T. 2004. The ligament system in the neck of *Rhea americana* and its implication for the bifurcated neural spines of sauropod dinosaurs. *Journal of Vertebrate Paleontology* 24: 165–172.

———. 2005. Homologies of the *Transversospinalis* muscles in the anterior presacral region of Sauria (crown Diapsida). *Journal of Morphology* 263: 151–178.

———. 2007. Homologies of the *Longissimus, Iliocostalis*, and hypaxial muscles in the anterior presacral region of extant Diapsida. *Journal of Morphology* 268: 986–1020.

———. 2010. Reconstructions of the axial muscle insertions in the occipital region of dinosaurs: evaluations of past hypotheses on Marginocephalia and Tyrannosauridae using the extant phylogenetic bracket approach. *Anatomical Record* 293: 1360–1386.

Upchurch, P., and P. M. Barrett. 2000. The evolution of sauropod feeding mechanisms. In H.-D. Sues (ed.), *Evolution of Herbivory in Terrestrial Vertebrates: Perspectives from the Fossil Record*, 79–122. Cambridge: Cambridge University Press.

Van den Hueval, W. F. 1992. Kinetics of the skull in the chicken (*Gallus gallus domesticus*). *Netherlands Journal of Zoology* 42: 561–582.

Vanden Berge, J. C., and G. A. Zweers. 1993. Myologia. In J. J. Baumel (ed.), *Handbook of Avian Anatomy: Nomina Anatomica Avium*, 2nd ed., 189–247.

Cambridge: Nuttall Ornithological Society.

Versluys, J. 1910. Sterptostylie bei Dinosaureirn, nebst Bemerkungen über die Verwandtschaft der Vögel und Dinosaurier. *Zoologischer Jarbusch Anatomie* 30: 175–260.

Walker, A. D. 1977. Evolution of the pelvis in birds and dinosaurs. In S. M. Andrews, R. S. Miles, and A. D. Walker (eds.), *Problems in Vertebrate Evolution, Linnean Society Symposium* 43: 319–358. New York: Academic Press.

Walker, A. D. 1990. A revision of *Sphenosuchus acutus* Haughton, a crocodylomorph reptile from the Elliot Formation (late Triassic or early Jurassic) of South Africa. *Philosophical Transaction of the Royal Society of London B* 330: 1–120.

Webb, M. 1957. The ontogeny of the cranial bones, cranial peripheral and cranial parasympathetic nerves, together with a study of the visceral muscles of *Struthio. Acta Zoologica* 38: 81–203.

Weishampel, D. B. 1984. Evolution of jaw mechanisms in ornithopod dinosaurs. *Advances in Anatomy, Embryology, and Cell Biology* 87: 1–110.

Weishampel, D. B., and D. B. Norman. 1989. Vertebrate herbivory in the Mesozoic: jaws, plants, and evolutionary metrics. *Geological Society of America Special Paper* 238: 87–100.

Wheeler, P. E. 1978. Elaborate CNS cooling structures in large dinosaurs. *Nature* 275: 441–443.

Williams, V. S., P. M Barrett, and M. A. Purnell. 2009. Quantitative analysis of dental microwear in hadrosaurid dinosaurs, and the implications for hypotheses of jaw mechanics and feeding. *Proceedings of the National Academy of Sciences of the United States of America* 106 (27): 11194–11199.

Winkler, D. A., P. A. Murry, and L. L. Jacobs. 1997. A new species of *Tenontosaurus* (Dinosauria: Ornithopoda) from the Early Cretaceous of Texas. *Journal of Vertebrate Paleontology* 17: 330–348.

Witmer, L. M. 1995a. The extant phylogenetic bracket and the importance of reconstructing soft tissues in fossils. In J. Thomason (ed.), *Functional Morphology in Vertebrate Paleontology* 19–33. New York: Cambridge University Press.

———. 1995b. Homology of facial structures in extant archosaurs (birds and crocodilians), with special reference to paranasal pneumaticity and nasal conchae. *Journal of Morphology* 225: 269–327.

———. 1997. The evolution of the antorbital cavity of archosaurs: a study in soft-tissue

reconstruction in the fossil record with an analysis of the function of pneumaticity. *Memoirs of the Society of Vertebrate Paleontology, Journal of Vertebrate Paleontology* 17 (suppl. to 1): 1–73.

———. 2001. Nostril position in dinosaurs and other vertebrates and its significance for nasal function. *Science* 293: 850–853.

Witmer, L. M., and K. D. Rose. 1991. Biomechanics of the jaw apparatus of the gigantic Eocene bird *Diatryma*: implications for diet and mode of life. *Paleobiology* 17: 95–120.

Xu, X., Z. Zhou, X. Wang, X. Kuang, F. Zhang, and X. Du. 2003. Four-winged dinosaurs from China. *Nature* 421: 335–340.

Zajac, F. E. 2002. Understanding muscle coordination of the human leg with dynamical simulations. *Journal of Biomechanics* 35: 1011–1018.

Zhang, X.-H., X. Xu, P. Sereno, X-W. Kuang, and L. Tan. 2001. A long-necked therizionosauroid dinosaur from the Upper Cretaceous Iran Dabasu Formation of Nei Mongol, People's Republic of China. *Vertebrata PalAsiatica* 39: 282–290.

Zhang, Y. 1988. *The Middle Jurassic Dinosaur Fauna from Dashanpu, Zigong, Sichuan.* Vol. 1, *Sauropod Dinosaurs: Shunosaurus* [in Chinese with English summary]. Chengdu, China: Sichuan Publishing House of Science and Technology.

Zusi, R. L. 1993. Patterns of diversity in the avian skull. In J. Hanken and B. K. Hall (eds.), *The Skull*, 2:391–437. Chicago: University of Chicago Press.

Zusi, R. L., and B. C. Livezey. 2000. Homology and phylogenetic implications of some enigmatic cranial features in galliform and anseriform birds. *Annals of the Carnegie Museum* 69: 157–193.

Zusi, R. L., and L. I. Warheit 1992. On the evolution of intraramal mandibular joints in pseudodontorns (Aves: Odontopterygia). In K. Campbell (ed.), *Papers in Avian Paleontology Honoring Pierce Brodkorb*. Science Series, Natural History Museum of Los Angeles County 36: 351–360.

Zweers, G. A. 1974. Structure, movement, and myography of the feeding apparatus of the mallard (*Anas platyrhynchos* L.): a study of functional anatomy. *Netherlands Journal of Zoology* 24: 323–467.

Zweers, G. A., J. C. Vanden Berge, and H. Berkhoudt. 1997. Evolutionary patterns of avian trophic diversification. *ZACS Zoology* 100: 25–57.

Dinosaur Paleoneurology

Emily Buchholtz

10

Dinosaur paleoneurology began in 1871 with the report of a "large reptilian skull from Brooke, Isle of Wight, probably dinosaurian, and referable to the genus *Iguanodon*." In that report, J. W. Hulke (1871, 199) described the internal surface of a partial braincase (Fig. 10.1), inferring the structure of the brain and cranial nerves that it once housed. Beginning in the early 1880s, H. G. Seeley, O. C. Marsh, and their followers described endocasts made from braincases and sacra in the rapidly expanding collections of dinosaurs in Europe and North America. Growing fascination with dinosaurs was enhanced by what appeared to be incredible discrepancies between large body size and small brain size. In the Victorian intellectual atmosphere in which vertebrates were ranked by their similarity to humans, dinosaurs were compared unfavorably to living reptiles and mammals. Remnants of this "scala natura" view of diversity are preserved in the common usage of the term *dinosaurian* to refer to ideas that are hopelessly outdated and to brain capacities that are dangerously small.

Four scientists stand out among the many who have made contributions to dinosaur paleoneurology: O. C. Marsh, T. Edinger, H. J. Jerison, and J. A. Hopson. O. C. Marsh (1831–1899) was among the first to collect and produce endocasts; he also was the first to codify trends in brain evolution over time with what he called his general law of brain growth (1874).

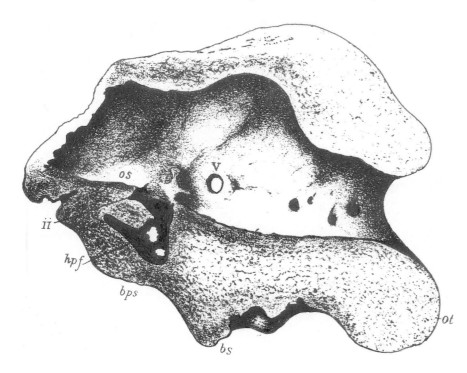

10.1. Internal view of the braincase of *Iguanodon,* slightly modified from the image published by J. W. Hulke in 1871. bps = basipresphenoid; bs = basisphenoid; hpf = hypophyseal fossa; os = orbitosphenoid; ot = occipital tuberosity; ii, V = foramina for cranial nerves. Hulke did not provide a scale.

Edinger (1892–1967) integrated anatomical (biological) and stratigraphic (geological) information, allowing her to date the neurological innovations she documented across many vertebrate groups. She challenged Marsh's law and documented the widely dispersed literature (1929, 1975) in the rapidly growing subdiscipline of paleoneurology. H. J. Jerison (born 1925) revolutionized the study of brain size and intelligence in vertebrates with documentation of the allometric relationship between brain size and body size. His encephalization quotient (1969, 1973) allows estimation of the behavioral sophistication of animals of different body sizes, taxonomic affiliations, and geological time periods. J. A. Hopson (born 1935) meticulously described known dinosaur endocasts, expanding the list of neural features identifiable by documentation of the relationship between brains and endocasts in living reptiles (1977, 1979). He used encephalization quotients and indicators of dominant sensory systems to predict behavior and lifestyle in different dinosaur subgroups.

Modern attempts to understand dinosaur neuroanatomy are still hampered by stereotyped ideas of dinosaur behavior and evolution, by the paucity of endocasts, and by the few neural features that can be interpreted from endocast surface anatomy and size (Northcutt 2001). Nevertheless, recent advances of discovery and technology continue to expand the data set available, and improving phylogenetic reconstruction provides an evolutionary framework within which to evaluate dinosaur neural structures and to infer their correlates in behavior and intelligence.

Source Materials for Dinosaur Paleoneurology

The central nervous system of vertebrates is supported and protected by the bony braincase and vertebral column. These encasing hard tissues are the major source of information about the soft nervous tissues of dinosaurs, which inevitably decayed soon after death. Endocasts are internal molds of the braincase or vertebral neural canal but do not exactly duplicate the external features and size of the brain or spinal cord they once housed. In *Die fossilen Gehirne* (Fossil Brains), the first major treatise of paleoneurology, Edinger (1929) emphasized that not only neural tissues, but also support and vascular tissues are housed within neural cavities. Most significant of these are the supporting (meningial) membranes and fluids, and the circulatory vessels and sinuses. As a result, the brain itself represents a variable percentage of endocast volume, with estimates ranging from 50% to 95%.

The first endocasts described were natural, the result of braincase infilling and subsequent erosion of the cranial bones. These have been supplemented by artificial endocasts made by cleaning and then filling braincases with a variety of casting materials. Computed tomography (CT) now also allows digital "preparation" of an endocast without any removal of the infilling matrix or casting. CT endocasts have been generated for theropods (Brochu 2000; Sanders and Smith 2005), sauropods (Sereno et al. 2007), psittacosaurs (Zhou et al. 2007), and the early bird *Archaeopteryx* (Dominguez Alonzo et al. 2004), among others. Spiral CT scanning was used by Rogers (1998, 1999) to explore the possibility that a natural endocast of *Allosaurus fragilis* preserved both internal and surficial features. Somewhat surprisingly, this procedure generated images of structures that appear to correspond to brain regions and to connective and vascular tissues. Computers may also be used to generate very accurate estimates of endocast volume and of cortical surface areas. High-field magnetic resonance imaging easily distinguishes between neural and nonneural tissues in living reptiles (Anderson et al. 2000) and may be able to provide standards for interpreting the degree of fill in fossil crania.

Despite the increase in number and type of endocasts, their interpretation is still largely restricted to inferences from size and surface anatomy. As a result, much of our understanding of dinosaur brains still depends on traditional comparative methods that use living taxa as references. Cladistic reconstruction of phylogenetic relationships continues to improve the identification of appropriate reference sets for comparative inferences (Witmer 1995).

Basic Guide to the External Anatomy of the Dinosaur Brain

Vertebrate neural anatomy is conservative: the patterns of sensory and motor signal transmission and the location of processing sites for different sensory modalities are preserved across all vertebrates. This consistency—and details of neural anatomy in the living avian and crocodylian relatives of dinosaurs—provides the context within which dinosaur neuroanatomy is understood.

Regions of ventricular expansion divide the vertebrate brain topographically into the forebrain (prosencephalon), midbrain (mesencephalon), and hind brain (rhombencephalon). Each of these subunits contributes to the ventral brainstem, and each is also associated with one of the dorsal laminated integration areas (cerebrum, optic tectum, cerebellum) and therefore with one of the three major special senses (smell, sight, hearing). These brain areas can often be recognized on the external surface of the brain and are therefore visible on some dinosaur endocasts (Fig. 10.2).

Forebrain (Olfactory Bulbs and Lobes, Cerebrum, Diencephalon)

OLFACTORY SYSTEM

The forebrain is primitively associated with chemosensation and the sense of smell. Action potentials generated by chemosensory neurons in the olfactory epithelium of the nasal cavities project to the olfactory bulb and then to the lateral (olfactory) cortex. Both olfactory bulbs and cortex are frequently distinguishable on the anterior and ventral aspects of dinosaur endocasts, and olfaction appears to have been a dominant sense in many dinosaurs. The vomeronasal system is not known in either crocodilians or birds, but large vomeronasal bulbs were identified dorsal to the olfactory bulbs on an endocast of *Tarbosaurus* by Saveliev and Alifanov (2007).

CEREBRUM

The primitive function of the cerebrum was the processing of olfactory information, and this function is retained in all vertebrates. Additionally, however, the cerebral cortex serves to integrate the information of many different sensory systems, and its elaboration is therefore associated with neural processing and intelligence. The area of the cortex involved in this

A

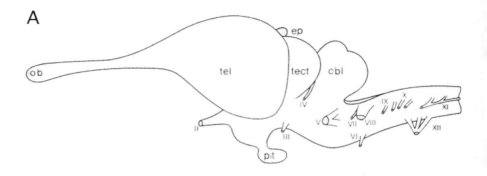

B

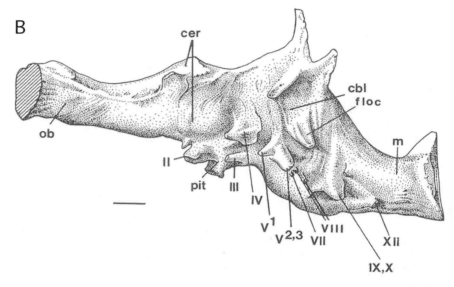

C

I	Olfactory	VII	Facial
II	Optic	VIII	Octaval
III	Oculomotor	IX	Glossopharyngeal
IV	Trochlear	X	Vagus
V	Trigeminal	XI	Spinal accessory
VI	Abducens	XII	Hypoglossal

elaboration is strikingly different in sauropsids and mammals (Fig. 10.3). Birds and reptiles possess a hypertrophied and centrally located dorsal ventricular ridge. Its expansion is particularly marked in birds. In contrast, it is the dorsal pallium that is extensively thickened and enlarged as the isocortex in mammals. When significantly enlarged, the cortex may overlap the midbrain or even the cerebellum, limiting their dorsal exposure, with the result that these structures may not be visible on endocasts.

In a study of mammal brains, Northcutt and Kaas (1995) concluded that expansion of the cortex must have occurred multiple times in different mammalian orders, producing superficially similar but nonhomologous patterns of cortical enlargement. Edinger (1948) made a similar hypothesis on the basis of her observations of cortical expansion in the endocasts of fossil horses. It seems likely that the expansion of the dorsal ventricular ridge in dinosaurs must also have undergone multiple independent expansions, most notably in ornithopods and small theropods.

Medial pallium
Dorsal pallium
Lateral pallium

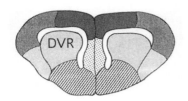

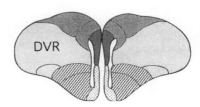

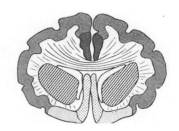

10.3. Differential expansion of pallial areas in living reptiles (left), birds (center), and mammals (right). Note that the lateral pallium is dominant in birds and reptiles, and the dorsal pallium in mammals. DVR = dorsal ventricular ridge. Adapted from Kardong (2006, fig. 16.43).

DIENCEPHALON

Only the dorsal and ventral surfaces of the diencephalon are visible on endocasts. The dorsal surface, the epithalamus, may give rise to as many as four dorsal evaginations. Primitively, this complex was probably a light sensitive organ, and it retains this and related circadian roles in some vertebrates today. The posterior pineal gland (epiphysis) has an endocrine function related to control of serotonin and melatonin levels. The anterior parietal or parapineal organ may form a photoreceptor organ ("third eye"), sometimes with a lens, that is exposed to the surface at an opening within or between the parietal bones in many living reptiles. No definitive opening for a parietal organ is known in dinosaurs, but impressions of parietal organs on the inner surface of skull roofing bones have been reported in theropods and ornithopods (Russell 1972; Rich and Rich 1989; Kundrát 2007).

Nuclei of the ventrally located hypothalamus are involved in the regulation of homeostasis and are connected via tracts and/or hormones to the pituitary gland. The anterior pituitary (adenopophysis) produces and/or stores a wide range of hormones including growth hormone, a key determinant of body size.. In many animals, the pituitary gland is supported in a discrete bony capsule, allowing its size to be estimated independently of the rest of the brain. The large optic nerves (cranial nerve II) from the eye intersect at the optic chiasma on the ventral surface of the diencephalon and are occasionally visible on endocasts. They continue as the optic tracts on the lateral surface of the diencephalon, en route to sites in the diencephalon and optic tectum.

Midbrain (Optic Tectum)

The midbrain is primitively associated with vision. Its roof or tectum is laminate in cellular organization, consisting of topographically arranged cells that receive and integrate input from optic/somatosensory and vestibular systems. Enlargements of the tectum are called optic lobes in birds and reptiles. The optic tectum is only rarely visible on dinosaur endocasts. In some cases, a venous sinus may have overlapped the midbrain, preventing its surface exposure. In others, disparity in size between the brain and the braincase may have prevented impression of tectal shapes on the internal surface of the braincase. Laterally displaced optic lobes are particularly large and distinctive in pterosaurs (Witmer et al. 2003) and small coelurosaurian dinosaurs (Kundrát 2007), groups that appear to have used vision as their dominant sensory modality.

Hind Brain (Medulla, Cerebellum,
Balance and Hearing Structures)

MEDULLA

The ventral portion of the hind brain, the medulla, is a major route of ascending and descending neural traffic between the spinal cord and the brain. It is also the site of reflex centers for the control of basal functions (respiration, heartbeat) and of the roots of cranial nerves V–XII. Of these, the trigeminal (V, with distinct ophthalmic, maxillary and mandibular subdivisions) and facial (VII) are noteworthy as large general nerves that innervate the jaw and face musculature. The paths by which these nerves exited from the braincase are often visible on endocasts and have phylogenetic significance (Currie 1997).

CEREBELLUM

The cerebellum, or "little brain," is a dorsal elaboration of the hind brain and serves as an integration site for information controlling balance, posture, muscle tone, and movement. Cerebellar subdivisions are rarely identifiable on endocasts, although Kundrát (2007) reported possible cerebellar folia in the oviraptorid *Conchoraptor*. The lateral cerebellar flocculus, associated with the vestibulo-ocular reflex, can be recognized adjacent to or even within the membranous labyrinth in animals with high demands for coordination and balance (e.g., pterosaurs, Witmer et al. 2003).

BALANCE (VESTIBULAR) AND HEARING (ACOUSTIC) STRUCTURES

The fragile tissues of the organs of equilibrium and hearing, the membranous labyrinth and sacculus, are typically encased in bone immediately lateral to the brain cavity at the point of entry of the octaval nerve (VIII). The membranous labyrinth responds to changes in head orientation and is of large relative dimension in vertebrates (such as birds) with high balance and control requirements. In almost all vertebrates, the labyrinth is composed of three semicircular canals (anterior, posterior, lateral) oriented at right angles to each other. The lateral semicircular canal is oriented horizontally in living vertebrates (de Beer 1947; Blanks et al. 1972; Erichsen et al. 1989), providing a method of predicting habitual position of the head in extinct animals (Hopson 1979; Witmer et al. 2003). Endocasts of the osseous labyrinth are rare, but it can also be visualized with CT scans (e.g., Larsson 2001; Sanders and Smith 2005). CT images of osseous labyrinths were used by Sereno et al. (2007) to show the radical repositioning of the head that accompanied extreme adaptations for herbivory in diplodocid sauropods (Plate 3).

Spinal Cord

The spinal cord is the posterior continuation of the central nervous system; its nerves innervate sequential dermatomes or segments of the body. In reptiles, the cord and the column are isosegmental—that is, the cord that supplies a given body segment lies within the neural canal at the same segmental level. The enlargements in cord cross section at pectoral and pelvic levels reflect increased neural supply to the forelimbs and hind limbs. Similar enlargements in the neural canal may be used to predict limb innervation and the segmental location of the limbs along the column of dinosaurs (Giffin 1995).

Vascular Tissues

The brain is richly supplied with blood, and the impressions of its arteries and veins (valleculae) can be identified on the dorsal surface of dinosaur endocasts when the brain fit tightly in the braincase. More commonly, large and irregularly shaped dorsal and ventral vascular sinuses in the dura mater obscure neural anatomy, particularly from the midbrain level posteriorly. The internal carotid artery is frequently visible at its ventral entry to the pituitary fossa, and the middle cerebral vein at the level of the cerebellum. Sedlmayr (2001) recently presented a preliminary description of a dual endocranial venous system that is unique to archosaurs. The close contact between these sinuses and arteries suggest a function for these vessels in temperature regulation of the brain.

Brain Size and Intelligence in Dinosaurs

Dinosaur brain size has been a subject of enduring interest and controversy. Large brain size relative to body size is associated with enhanced information processing and intelligence. Disagreements exist with regard to the optimal methods of estimating dinosaur brain and body size, and of calculating the allometric relationships between them.

Trends in the brain size of extinct animals were addressed first by Lartet (1868), and soon after by Othniel Marsh (1874). Marsh codified his observations of mammalian brain size in his famous laws of brain growth, quoted below from an 1876 presentation:

> The conclusions reached may be briefly stated as follows: *First*, all tertiary mammals had small brains; *second*, there was a gradual increase in the size of the brain during this period; *third*, this increase was mainly confined to the cerebral hemispheres, or higher portion of the brain; *fourth*, in some groups, the convolutions of the brain have gradually become more complicated; *fifth*, in some, the cerebellum and olfactory lobes have even diminished in size. (61)

Marsh later extended his observations to dinosaurs (Marsh 1879b, 358):

> The Dinosaurs from our Western Jurassic follow the same law, and had brain cavities vastly smaller than existing reptiles. Many other facts point in the same direction, and indicate that the general law will hold good for all extinct vertebrates.

Marsh's presentation notably lacks any reference to the size of the animal, now recognized as a primary determinant of brain size. Edinger (1962) harshly criticized his omission, documenting the (relatively) small brains of large mammals. Relationships of brain mass to body mass in living vertebrates have now been described by many workers, among them Snell (1891), Dubois (1897), Jerison (1969, 1973), and Platel (1976, 1979). A common feature of these relationships is an allometric exponent, typically near two thirds, indicating that brain size does not increase as fast as body size:

$$E \text{ (brain mass)} = k \text{ (} y \text{ intercept)} \times P \text{ (body mass)}^{2/3}$$

Regressions for all vertebrates have approximately the same slope, but values of k differ by taxon. For animals of the same body mass, k is approximately 10 times greater in birds and mammals than it is in fish, amphibians, or reptiles, indicating a 10-fold difference in brain mass. Allometric regressions provide the means of predicting an expected brain mass for a vertebrate of known taxon and body mass.

Accurate estimation of relative brain size in an extinct animal requires preservation of both cranial and postcranial remains, preferably from the same individual or at least individuals of approximately the same body size. This ideal is rarely met in dinosaurs. Body size is frequently estimated by scaling up models whose volume may be measured by water displacement. Such models are subject to the artistic interpretation of the modeler. The relationship between long bone circumference and weight in mammals and birds documented by Anderson et al. (1985) offers the possibility of accurate estimates from much more restricted material. Radinsky (1967) presented a regression of brain mass on foramen magnum area in mammals, proposing that medullar (and thus foramen magnum) cross section is directly related to body size. This method has the advantage of eliminating the need for postcranial material, but unfortunately it has not been duplicated in reptile studies.

Water displacement is a common method of volume estimation for endocranial casts as well. Jerison (1969, 1973) presented an alternative graphic integration method that allows accurate volume estimates from lateral and dorsal views of an endocast. Hurlburt and Lewis (1999) successfully predicted brain and even lobe sizes of living birds using elliptic sinusoids as models, a method that could be applied to at least some dinosaurs. CT scans now also allow computer-generated estimates of virtual endocast volumes. None of these models directly addresses the unresolved issue of percentage fill of the braincase. A general estimate of 50% fill almost certainly ignores differences between individuals of different life stages (percentage fill appears to decrease with age), body size (percentage fill probably decreases as body size increases), and taxonomic affiliation (percentage fill is probably higher in species with greater behavioral complexity). Imprints of vascular or neural structures on the interior surface of the skull roof (valleculae) are commonly accepted as indicators of greater fill (Osmólska 2004; Evans 2005).

In a classic paper, Jerison (1969) charted body and brain mass of living vertebrates. He used polygons to map the area of the regression represented by living reptiles (Fig. 10.4). Extending the lines of his polygon, he was then able to predict the chart area into which brain masses of dinosaurs would fall if they followed the same allometric relationship as living reptiles. His results allowed him to reject the prediction of Marsh (1879b, 358) that dinosaurs had "brain cavities vastly smaller than any existing reptiles." This holds true for even the very large dinosaurs, whose small brains caused such consternation historically. These data predict that the average dinosaur was

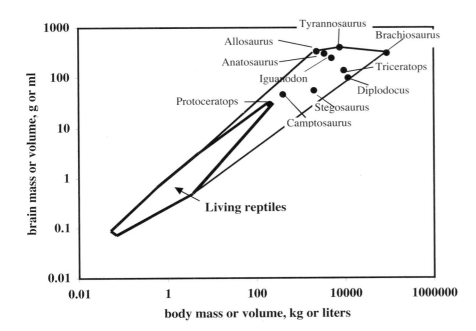

10.4. The relationship between body weight and brain weight in living reptiles and dinosaurs, redrawn from Jerison (1969, fig. 4).

no more or less intelligent than living reptiles of smaller body size. There is thus no reason to expect that dinosaurs could "barely eat and sleep and muddle through life" or that brain size, per se, was "an important reason" for dinosaur extinction, as has been suggested in many popular works (e.g., Andrews 1953, 140). Variations in relative brain size among dinosaur taxa are included in the discussion of individual groups below.

Behavioral Implications of Dinosaur Endocast Anatomy by Taxonomic Group

Many dinosaur endocasts preserve enough anatomical detail to allow estimation of total brain size. The encephalization quotient (EQ) devised by Jerison (1973) is used in the comparison of brain size in animals of different body size. The regression of brain mass on body mass described above allows prediction of an expected brain volume for a reptile of a given body size. The EQ is the ratio of observed brain volume to this expected value. Interpretation of the metric typically assumes that relatively larger brain sizes (>1.0) are correlated with enhanced information processing (intelligence) and behavioral complexity. A relationship between total brain size and behavioral complexity exists in living birds (Madden 2001) and in lizards (Platel 1976, 1979), and was similarly proposed for dinosaurs by Hopson (1977). Because reptiles in general have smaller brains for a given body size then do birds or mammals, it is reasonable to infer that dinosaur behavior was more stereotyped and less plastic than that of these groups.

The relative size of the major regions of sensory processing are also apparent on some endocasts, allowing the identification of dominant sensory modalities of extinct species. This approach is supported by correlations between the size of particular brain regions and the complexity and/or type of behavior in living birds (Bang and Cobb 1968; Canady et al. 1984; Healy and Krebs 1992; Lefebvre et al. 1997; Garamszegi et al. 2002).

Hopson (1977, 1979) published both the major extant summary of dinosaur endocast anatomy and an interpretation of dinosaur intelligence and behavior by subgroup. He used endocasts of recent crocodilians and of the aetosaurid *Desmatosuchus*, originally described by Case (1921), to predict

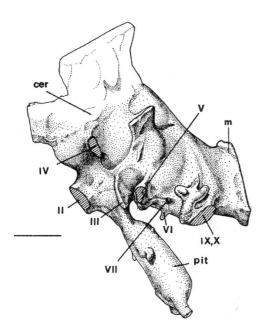

10.5. Lateral view of the endocast of the sauropod *Diplodocus longus,* adapted from Hopson (1979, fig. 16). Note the lack of anterior (olfactory) structures and the steep flexure between the midbrain and medulla. cer = cerebrum; m = medulla; pit = pituitary; II, III, IV, V, VI, VII, IX, X = corresponding cranial nerve roots. Scale bar = 2 cm.

ancestral brain anatomy in archosaurs. The *Desmatosuchus* endocast is narrow and elongate with prominent cephalic and pontine flexures. Its large olfactory bulbs suggest that the animal had a well-developed sense of smell, as do many living reptiles. The small cerebral hemispheres lack surficial anatomy, indicating incomplete braincase fill, suggesting modest intelligence. There is no indication of the midbrain. Living alligators have an EQ slightly below 1.0, and it is reasonable to assume that *Desmatosuchus* did as well.

Sauropods

Sauropod endocasts can be recognized on the basis of their anterior/posterior compression, transverse expansion, and steep flexures (Fig. 10.5). The short and deep shape may reflect packaging requirements in the small sauropod head (Hopson 1979). Incomplete ossification of the sauropod braincase often results in anterior truncation of many endocasts and poor representation of olfactory lobes. Dural sinuses also often obscure endocast detail (but see Sereno et al. 2007). The pituitary fossa is relatively large and has been used as an argument for pituitary gigantism in sauropods.

Because sauropods were the largest dinosaurs, the allometric relationship between body and brain size predicts that they should possess the smallest brain:body ratio. Janensch (1938) estimated this ratio to be 1:200,000 in *Brachiosaurus*. Hopson (1979) argued that surficial detail justifies an assumption of complete braincase fill in sauropods. Despite this concession and the relatively small brain predicted by allometric analysis, sauropod endocasts still produce EQs well below that of the average living reptile (Hopson 1979; Sereno et al. 2007). These values support other indicators of a slow, herbivorous lifestyle, very modest information processing, and passive defense strategies. Using the anatomy of elephants as a comparison, Knoll et al. (2006) used the relatively small size of the facial nerve orifice in *Diplodocus* to argue against the hypothesis (Bakker 1971) that it possessed a muscular trunk or proboscis.

Theropods

Brain anatomy in theropods is relatively well known and falls into two distinct anatomical groups, suggesting different dominant sensory modalities and lifestyles. Allosaurids and large coelurosaurs retain primitive endocast shape and distinct cerebral and pontine flexures (Fig. 10.2). The optic tectum is obscured and must have been small. The cerebral hemispheres are not separable and show only modest convexity. The prediction of extremely large olfactory bulbs in *Tyrannosaurus rex* (Brochu 2000) has been scaled downward, but it still seems likely that smell was a dominant sense (Rogers 1998, 1999; Stokstad 2005; Saveliev and Alifanov 2007). Rogers (1998, 1999) described the vestibular apparatus of *Allosaurus fragilis*, which more closely resembles that of crocodylians than that of lizards, turtles, or birds. He used semicircular canal orientation to suggest that *Allosaurus* held its head at or very slightly (≤10 degrees) inclined to the horizontal. Burish et al. (2004) included endocranial casts of *Tyrannosaurus*, *Allosaurus*, and *Carcharodontosaurus* in an analysis of relative telencephalic (~cerebral) size and behavior in birds. Their data indicate values for theropods at or below the range of living reptiles, and predict low social complexity.

Coelurosaurs

Endocasts of small coelurosaurs (Russell 1969, 1972; Colbert and Russell 1969; Kundrát 2007) display a strikingly different anatomical pattern (Fig. 10.6). They retain details of brain anatomy and roofing bone sutures on their surfaces, suggesting that the brain filled the braincase nearly completely. Brain flexures are minimal and olfactory bulbs are small, indicating that smell was not a dominant sense. Cerebral hemispheres are separable, convex, and expanded laterally and/or posteriorly (Kundrát 2007), suggesting an active intelligence. The large optic lobes are visible either dorsally or displaced laterally by the large cerebrum, as in living birds. Russell (1969) associated the large optic lobes with large eyes and binocular vision, and it is likely that sight was the dominant sense. Kundrát (2007) described an expanded cerebellum with presumptive cerebellar folia among the avian-like characters of the oviraptorid theropod *Conchoraptor*, inferring excellent balance and coordination.

Encephalization quotients of small coelurosaurs vary with predictions of body mass and percentage of braincase fill, but even conservatively, they are far higher than those of any other dinosaur group, overlapping those of living birds (Hopson 1977; Kundrát 2007). Larsson et al. (2000) estimated cerebral volumes by superimposing ellipsoids on endocasts with surficial indications of cerebral extent. Their data suggest at least three stages of increase of relative cerebral size to total brain size over a period of only 40 million years: of coelurosaurs over allosaurs, of *Archaeopteryx* over coelurosaurs, and of ornithurine birds over *Archaeopteryx*. The high encephalization values of small coelurosaurs indicate an active, complex, and social lifestyle that agrees well with their frequent interpretation as pack hunters.

Stegosaurs

Stegosaur endocasts are notorious for their relatively small size, infamously said to be "no larger than a walnut" despite body size "bigger than an elephant" (Andrews 1953, 83). Despite such fame, brain anatomy in stegosaurs is unfortunately poorly known, and some of the specimens first described by Marsh (1880) are now recognized as composites. Proportions and anatomy of endocasts of *Stegosaurus* vary little from those of ancestral archosaurs, with an elongate shape, large olfactory lobes, and extremely narrow cerebral hemispheres. Lack of surface detail suggests that the brain did not fill the braincase. EQ estimates are below 0.6 (Hopson 1977), agreeing well with predictions of a slow herbivorous lifestyle. Sacral enlargements (see below) almost certainly did not house a "sacral brain."

Ankylosaurs

Ankylosaurs retain ancestral archosaur endocast anatomy, with parallel forebrain and medullar regions and only modest cerebral expansion. Details of structure suggest divergent sensory specializations in the two subfamilies.

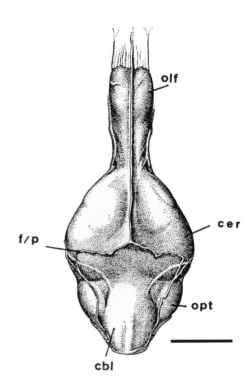

10.6. Dorsal view of the finely detailed endocast of *Troodon formosus* (adapted from Hopson 1979, fig. 13). cer = cerebrum; cbl = cerebellum; f/p = frontoparietal suture; olf = olfactory bulb; opt = optic lobe. Scale bar = 2 cm.

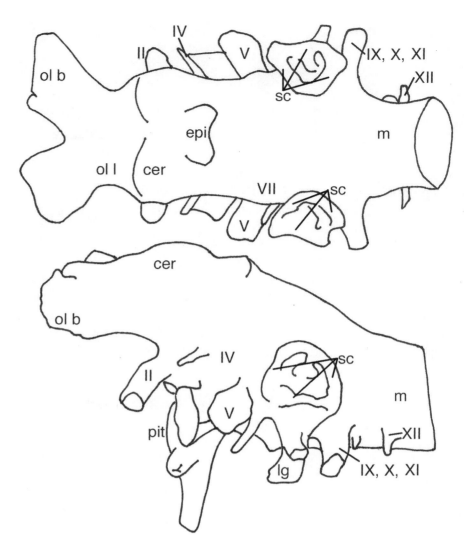

Endocast surface details are preserved more clearly in nodosaurids (*Struthiosaurus*) than in ankylosaurids (*Euoplocephalus*), suggesting more complete fill. The braincase of the nodosaurid *Silvisaurus* described by Eaton (1960) has impressions of both a rounded cerebrum and the optic tectum; this is highly unusual and suggests the importance of vision. In contrast, the *Euoplocephalus* cerebrum (Coombs 1978) (Fig. 10.7) is narrow and terminates in large divergent olfactory bulbs. These bulbs may reflect the lateral placement of the nares in the broad muzzle and/or an enlarged olfactory epithelium in the complex air passages of ankylosaurids (Coombs 1978; Carpenter 1997). Hopson's (1979) calculation of an EQ for *Euoplocephalus* of 0.52 is consistent with ankylosaur herbivory and a largely passive lifestyle, although the presence of a terminal tail club is powerful evidence for an active defense strategy.

Ornithopods

The endocasts of ornithopods are distinguished by their lack of a pontine flexure and a modest cephalic flexure. The resulting near linear profile is demonstrated clearly in endocasts of the genera *Iguanodon* (Fig. 10.8) and *Kritosaurus* (Ostrom 1961; Hopson 1979). Subdued flexure may reflect the

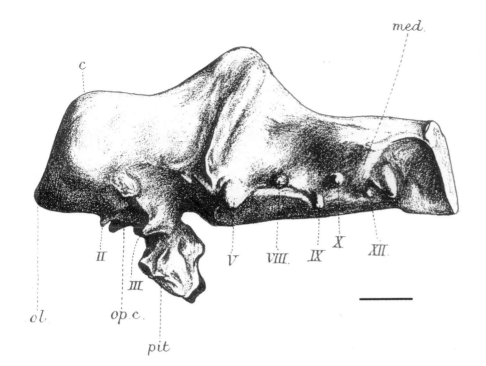

10.8. Endocast of the ornithopod *Iguanodon*, slightly modified from the image published by Andrews (1897, plate XVI). c. = cerebrum; med. = medulla; ol. = olfactory bulb; op. c. = optic chiasma; pit. = pituitary; II, III, V, VIII, IX, X, XII = corresponding cranial nerve roots. Scale bar = 2 cm.

bipedal posture of ornithopods, which may be inferred from the differential size of the fore- and hind limbs and by the structure of the tail. Ornithopod endocasts also exhibit a large and convex forebrain, although the cerebral hemispheres are not separable. The recent description of valleculae on the inner surface of the skull of *Edmontosaurus* (Jerison et al. 2001) and of an unnamed lambeosaurine hadrosaur (Evans 2005) indicate relatively complete braincase fill, consistent with the interpretation of an active and flexible lifestyle inferred from limb structure and estimates of brain size (Hopson 1979). Neuroanatomy is also helpful in limiting the possible interpretations of the function of cranial crests in lambeosaurine hadrosaurs (Evans 2005). The distribution of the lambeosaurine olfactory system is outside the nasal passages, eliminating the possibility that olfaction is a significant factor in their hypertrophy.

The endocast of a juvenile hypsilophodontid *Leaellynasaura amicagraphica* recently found in Australia (Rich and Rich 1988, 1989) shows unusual fidelity of the brain surface (Fig. 10.9). It exhibits large and clearly separated cerebral hemispheres, well-defined and laterally placed optic lobes, and a parietal body, all clear indicators of extensive braincase fill. The authors suggest a possible correlation between the implied visual acuity of the animal and extended periods of winter darkness at the high latitudes of its discovery. Their hypothesis is supported by recent work by Garamszegi et al. (2002), who report coevolution of brain size and eye size in response to nocturnal activity in birds.

Marginocephalians

The endocasts of ceratopsian marginocephalians have few distinguishing features, retaining primitive archosaur anatomy. The forebrain and the medulla are horizontal, and the lack of surficial anatomy suggests incomplete fill of the braincase. There is no clear distinction been cerebral, midbrain,

10.9. Natural endocast of a juvenile hypsilophodontid, *Leaellynasaura amicagraphica*, showing fine details of dorsal anatomy. cer = cerebrum; op. l. = optic lobe; p. b. = parietal body. Length of entire skull is 52 mm.

Photograph provided courtesy of Thomas H. Rich and Patricia Vickers-Rich.

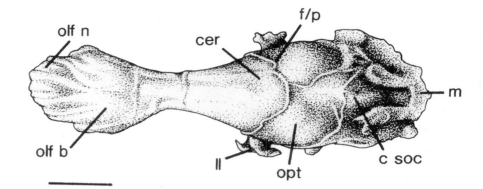

10.10. Dorsal view of the endocast of a pachycephalosaur, *Pachycephalosaurus wyomingensis,* from Giffin (1989, fig. 1A). c soc = cartilage pit of supraoccipital; cer = cerebrum; f/p = frontoparietal suture; m = medulla; olf b = olfactory bulb; olf n = olfactory nerve; opt = optic lobe; II = optic nerve. Scale bar = 2 cm.

Courtesy of the Society of Vertebrate Paleontology.

and cerebellar regions. Endocasts of *Psittacosaurus lujiatunensis* suggest that smell was the dominant sense (Zhou et al. 2007). Brown and Schlaikjer (1940) compared endocasts of the small-bodied ceratopsian *Protoceratops* with those of larger ceratopsids, noting the proportionately larger size of the brain in the former and of the pituitary in the latter. Although many authors (e.g., Farlow and Dodson 1975) have postulated that the large frills of ceratopsians were used in agonistic or display behaviors, EQ values well below 1.0 (Hopson 1977) suggest largely stereotyped behavior.

The robust domes of pachycephalosaur marginocephalians have ensured preservation of numerous specimens from which endocasts may be made. Large olfactory bulbs are connected to the brain by short, divergent olfactory tracts. The forebrain is only modestly expanded, but rounded cerebral shapes in some taxa (e.g., *Stegoceras*, *Pachycephalosaurus*) suggest close contact between brain and the roofing bones of the dome (Fig. 10.10), as do recently described vascular imprints on the endocranium (Evans 2005). The cerebellum and optic tectum are only rarely distinguishable. The occiput is rotated ventrally in derived taxa, possibly a result of head use in agonistic or display behavior (Giffin 1989; Goodwin et al. 1998).

Endocrine Control of Gigantism in Dinosaurs

The pituitary body is housed in a discrete fossa at the base of the braincase, allowing estimation of its volume separate from that of the brain proper. Long before discovery of pituitary growth hormone (GH), Nopcsa (1917) noted that the pituitary fossa represented a larger portion of the endocast in large dinosaurs than in small ones. Reasoning from known results of pituitary disturbances in man, he suggested that an enlarged pituitary gland was the cause of giant size in some dinosaurs.

After the discovery of GH, Janensch (1935, 1938) made quantitative measurements of total endocast and pituitary volume. He documented pituitary bodies that were 4.5% (*Brachiosaurus*), 9.5% (*Barosaurus*), and an amazing 9.9% (*Dicraeosaurus*) of total endocast volume. Despite these values, he rejected Nopcsa's argument for a relationship between pituitary volume and gigantism because the ratio of pituitary to body size was still smaller than that of living domestic fowl. Edinger later (1942a, 1942b) revisited this debate, supporting Nopcsa's hypothesis that phylogenetic increase in body size was accompanied by increase in pituitary body size, but not total brain size. She supported her argument with parallel examples from ceratopsian and sauropod dinosaurs, and from dinornithid and aepyornithid birds.

Nopcsa (1917) also suggested that an enlarged pituitary as a cause of sauropod extinction, on the basis of sterility of humans with pituitary dysfunction. Edinger (1942a) considered and rejected his hypothesis on the basis of the concurrent extinction of dinosaurs of many body sizes and on the basis of differences between pathological and normal enlargement of the pituitary gland.

The spinal cord, like the brain, is protected by bone, lying within the neural canal of the vertebral column. Although the absolute size of the cord is unknowable, relative increases at limb levels over interlimb levels can be used to predict the density of the innervation to, and therefore the use of, the limbs. In general, the size of spinal enlargements in dinosaurs corresponds well with that of living lepidosaurs, crocodylians, and birds (Giffin 1995). Dinosaurs with forelimbs that appear to have had manipulative function (*Deinonychus*, *Saurornitholestes*) have much larger brachial level spinal enlargements than do those with limbs of reduced size or missing digits (*Tyrannosaurus*, *Carnotaurus*). Similarly, sacral level spinal enlargements of many dinosaurs (e.g., *Dysalotosaurus*, *Plateosaurus*, *Triceratops*, *Thecospondylus*) fall in the range of living reptiles.

Othniel Marsh (1879a, 87) noted that the sauropod *Morosaurus* (= *Camarasaurus*) had a massively enlarged sacral neural canal, writing that

> a striking feature of this sacrum is seen in the large size of the neural canal, which, strange to say, is here two or three times the diameter of the brain cavity. This is a most suggestive fact, and without parallel in known vertebrates.

Sacral canals with large bladderlike intervertebral enlargements were later described (Janensch 1939) from other sauropods (*Dicraeosaurus*, *Barosaurus*, *Apatasaurus*) and less irregularly shaped ones from stegosaurs (*Stegosaurus*, *Kentrosaurus*) (Marsh 1881; Hennig 1915) (Fig. 10.11). The exaggerated enlargements appear to be restricted to these two groups. Sauropods and stegosaurs are isolated phylogenetically, but they share large size, graviportal build, herbivorous diet, and low encephalization quotients.

Early interpretations of sacral enlargements as indicators of the presence of a second sacral "brain" may have origin in Marsh's (1881) description of the sacral cavity of *Stegosaurus* as a posterior braincase. The addition was thought necessary for innervation and coordination of the hind limbs and/or tail, something the small brain presumably could not manage alone. This interpretation was repeated frequently in the popular literature and even inspired poetry. Yet details of enlargement size and position do not support this interpretation (Giffin 1991). Other possible interpretations include the presence of pneumatic sacs (Janensch 1939, 1947) or of neural support or nutritional tissues (Krause 1881; Halstead 1969). Of the latter, the most likely is a large avian-style glycogen body, located in the roof plate of neural segments centered at the approximate level of the sacral plexus (Giffin 1991). Recently, Necker et al. (2000) described segmentally organized semicircular canal–like structures associated with accessory lobes in the sacral spinal cord of the pigeon. He reported expansion of neural canal size associated with these structures, for which he presents behavioral evidence of a

Neural Canal Enlargements and the "Sacral Brain" Debate

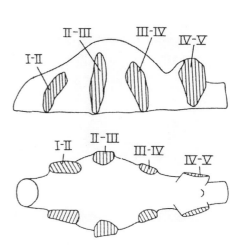

10.11. Lateral (above) and dorsal views of the sacral endocast of the stegosaur *Kentrosaurus aethiopicus*, adapted from Janensch (1939, fig. 4). Roman numerals identify intervertebral spaces between adjacent sacral vertebrae. The cast is approximately 28 cm in length.

function in equilibrium. The possibility of a dinosaur analog is interesting, although tenuous because of the phylogenetic distance between birds and either stegosaurs or sauropods.

The Future of Dinosaur Paleoneurology

Scientific understanding of dinosaur paleoneurology has increased greatly since Hulke's first descriptive work on the endocranium of *Iguanodon*. Among the factors enhancing paleoneurological work are an increase in the number and taxonomic breadth of specimens available for study, the improving comparative context among living animals provided by the rapidly expanding field of vertebrate neuroscience, reconstruction of the history of neural innovations using robust phylogenetic hypotheses, and the advent of noninvasive visualization methods that do not require mechanical preparation. Because all of these causative factors are active areas of discovery and research, it seems certain that our understanding of dinosaur neuroanatomy and its cognitive and behavioral implications will continue to expand as well.

References

Anderson, C. L., G. W. Kabalka, D. G. Layne, J. P. Dyke, and G. M. Burghardt. 2000. Noninvasive high field brain imaging of the garter snake. *Copeia* 2000: 265–269.

Anderson, J. F., A. Hall-Martin, and D. A. Russell. 1985. Long-bone circumference and weight in mammals, birds and dinosaurs. *Journal of Zoology* 207: 53–61.

Andrews, C. W. 1897. Note on a cast of the brain-cavity of *Iguanodon*. *Annals and Magazine of Natural History* 19: 585–591.

Andrews, R. C. 1953. *All About Dinosaurs*. New York: Random House.

Bakker, R. T. 1971. Ecology of the brontosaurs. *Nature* 229: 172–174.

Bang, B. G., and S. Cobb. 1968. The size of the olfactory bulb in 108 species of birds. *Auk* 85: 55–61.

Blanks, R. H. I., I. S. Curthoys, and C. H. Markham. 1972. Planar relationships of semicircular canals in the cat. *American Journal of Physiology* 223: 55–62.

Brochu, C. A. 2000. A digitally-rendered endocast for *Tyrannosaurus rex*. *Journal of Vertebrate Paleontology* 20 (1): 1–6.

Brown, B., and E. M. Schlaikjer. 1940. The structure and relationships of *Protoceratops*. *Annals of the New York Academy of Science* 40: 133–265.

Burish, M. J., H. Y. Kueh, and S. S. Wang. 2004. Brain architecture and social complexity in modern and ancient birds. *Brain Behavior and Evolution* 63: 107–124.

Canady, R. A., D. E. Kroodsma, and F. Nottebohm. 1984. Population differences in complexity of a learned skill are correlated with the brain space involved. *Proceedings of the National Academy of Sciences of the United States of America* 81: 6232–6234.

Carpenter, K. 1997. Ankylosauria. In P. J. Currie and K. Padian (eds.), *Encyclopedia of Dinosaurs*, 16–20. San Diego: Academic Press.

Case, E. C. 1921. On an endocranial cast from a reptile, *Desmatosuchus spurensis*, from the Upper Triassic of Western Texas. *Journal of Comparative Neurology* 33: 133–147.

Colbert, E. H., and D. A. Russell. 1969. The small Cretaceous dinosaur *Dromaeosaurus*. *American Museum Novitates* 2380: 1–49.

Coombs, W. P., Jr. 1978. An endocranial cast of *Euoplocephalus* (Reptilia, Ornithischia). *Palaeontographica* A 161: 176–182.

Currie, P. J. 1997. Braincase anatomy. In P. J. Currie and K. Padian (eds.), *Encyclopedia of Dinosaurs*, 81–85. Academic Press: San Diego.

de Beer, G. R. 1947. How animals hold their heads. *Proceedings of the Linnean Society of London* 159: 125–139.

Dominguez Alonzo, P., A. C. Milner, R. A. Kechum, M. J. Cookson and T. B. Rowe. 2004. The avian nature of the brain and inner ear of *Archaeopteryx*. *Nature* 430: 666–669.

Dubois, E. 1897. Sur le rapport du poids de l'encéphale avec la grandeur du corps chez mammifères. *Bulletins de la Société d'anthropologie de Paris* 8: 337–376.

Eaton, T. H. 1960. A new armored dinosaur from the Cretaceous of Kansas. *University of Kansas Paleontological Contributions, Vertebrata* 8: 1–21.

Edinger, T. 1929. Die fossilen Gehirne. *Ergebnisse der Anatomie und Entwicklungsgeschichte* 28: 1–249.

——. 1942a. The pituitary body in giant animals, fossil and living: a survey and a suggestion. *Quarterly Review of Biology* 17: 31–45.

——. 1942b. L'encéphale des Aegyptornithes. *Bulletin de l'Academie Malgache* 24: 25–50.

——. 1948. Evolution of the horse brain. *Memoirs of the Geological Society of America* 25: 1–177.

——. 1962. Anthropocentric misconceptions in paleoneurology. *Proceedings of the Rudolf Virchow Medical Society* 19: 56–107.

——. 1975. Paleoneurology 1804–1966, an annotated bibliography. *Advances in Anatomy, Embryology and Cell Biology* 49: 1–258.

Erichsen, J. T., W. Hodos, C. Evinger, B. B. Bessette, and S. J. Phillips. 1989. Head orientation in pigeons: postural, locomotor and visual determinants. *Brain Behavior and Evolution* 33: 268–278.

Evans, D. C. 2005. New evidence on brain–endocranial cavity relationships in ornithischian dinosaurs. *Acta Palaeontologica Polonica* 50 (3): 617–622.

Farlow, J. O. and P. Dodson. 1975. The behavioral significance of frill and horn morphology in ceratopsian dinosaurs. *Evolution* 29: 353–361.

Garamszegi, L. A., A. P. Møller, and J. Erritzøe. 2002. Coevolving avian eye size and brain size in relation to prey capture and nocturnality. *Proceedings of the Royal Society of London B* 269: 961–967.

Giffin, E. B. 1989. Pachycephalosaur paleoneurology (Archosauria: Ornithischia). *Journal of Vertebrate Paleontology* 9: 67–77.

——. 1991. Endosacral enlargements in dinosaurs. *Modern Geology* 16: 101–112.

——. 1995. Postcranial paleoneurology of the Diapsida. *Journal of Zoology, London* 235: 389–410.

Goodwin, M. B., E. A. Buchholtz, and R. E. Johnson. 1998. Cranial anatomy and diagnosis of *Stygimoloch spinifer* (Ornithischia: Pachycephalosauria) with comments on cranial display structures in agonistic behavior. *Journal of Vertebrate Paleontology* 18: 363–375.

Halstead, L. B. 1969. *The Pattern of Vertebrate Evolution*. Edinburgh: Oliver and Boyd.

Healy, S. D., and J. R. Krebs. 1992. Food storing and the hippocampus in corvids: amount and size are correlated. *Proceedings of the Royal Society of London B* 248: 241–245.

Hennig, E. 1915. *Kentrosaurus aethiopicus*, der Stegosauride des Tendaguru. *Sitzungsberichte der Gesellschaft Naturforschender Freunde zu Berlin* 1915: 219–247.

Hopson, J. A. 1977. Relative brain size and behavior in archosaurian reptiles. *Annual Review of Ecology and Systematics* 8: 429–448.

——. 1979. Paleoneurology. In C. Gans, R. G. Northcutt, and P. Ulinski (eds.), *Biology of the Reptilia*, 9:39–146. San Francisco: Academic Press.

Hulke, J. W. 1871. Note on a large reptilian skull from Brooke, Isle of Wight, probably dinosaurian, and referable to the genus *Iguanodon*. *Quarterly Journal of the Geological Society of London* 27: 199–206.

Hurlburt, G. R., and A. Lewis. 1999. Brain part volume estimation from bird endocasts. *Journal of Vertebrate Paleontology* 19: 54A.

Janensch, W. 1935. Die Schädel der Sauropoden *Brachiosaurus, Barosaurus* und *Dicraeosaurus* aus den Tendaguruschichten Deutsch-Ostafrikas. *Palaeontographica, Supplement* 7: 145–298.

——. 1938. Gestalt und Größe von *Brachiosaurus* und anderen riesenwüchsigen Sauropoden. *Biologe* 7: 130–134.

——. 1939. Der sakrale Neuralkanal einiger Sauropoden und anderer Dinosaurier. *Palaeontologische Zeitschrift* 21: 171–193.

——. 1947. Pneumatizität bei Wirbeln von Sauropoden und anderen Saurischien. *Palaeontographica Supplement* 7: 1–25.

Jerison, H. J. 1969. Brain evolution and dinosaur brains. *American Naturalist* 103: 575–588.

——. 1973. *Evolution of the Brain and Intelligence*. New York: Academic Press.

Jerison, H. J., J. R. Horner, and C. Horner. 2001. Dinosaur forebrains. *Journal of Vertebrate Paleontology* 21 (suppl. to 3): 65A.

Kardong, K. V. 2006. *Vertebrates: Comparative Anatomy, Function, Evolution*. 4th ed. Boston: McGraw Hill.

Knoll, F., P. M. Galton, and R. López-Antoñanzas. 2006. Paleoneurological evidence against a proboscis in the sauropod dinosaur *Diplodocus*. *Geobios* 39: 215–221.

Krause, W. 1881. Zum Sacral Gehirn der Stegosaurier. *Biologisches Centralblatt* 1: 461.

Kundrát, M. 2007. Avian-like attributes of a virtual brain model of the oviraptorid theropod *Conchoraptor gracilis*. *Naturwissenschaften* 94: 499–504.

Larsson, H. C. E. 2001. The endocranial anatomy of *Carcharodontosaurus saharicus* (Theropoda: Allosauroidea) and its implications for theropod brain evolution. In D. Tanke and K. Carpenter (eds.), *Mesozoic Vertebrate Life*, 19–33. Bloomington: Indiana University Press.

Larsson, H. C. E., P. C. Sereno, and J. A. Wilson. 2000. Forebrain enlargement among nonavian theropod dinosaurs. *Journal of Vertebrate Paleontology* 20: 615–618.

Lartet, E. 1868. De quelques cas de progression organique vérifiables dans la succession des temps géologiques sur des mammifères de même famille et de même genre. *Comptes Rendus Academie de Sciences* 66: 1119–1122.

Lefebvre, L., P. Whittle, E. Lascaris, and A. Finkelstein. 1997. Feeding innovations and forebrain size in birds. *Animal Behavior* 53: 549–560.

Madden, J. 2001. Sex, bowers and brains. *Proceedings of the Royal Society of London B* 268: 833–838.

Marsh, O. C. 1874. Small size of the brain in Tertiary mammals. *American Journal of Science* (ser. 3) 8: 66–67.

——. 1876. Recent discoveries of extinct animals. *American Journal of Science* (ser. 3) 12: 59–61.

——. 1879a. Principal characters of American Jurassic dinosaurs. Part II. *American Journal of Science* (ser. 3) 17: 86–92.

——. 1879b. History and methods of palaeontological discovery. *American Journal of Science* (ser. 3) 18: 323–359.

——. 1880. Principal characters of American Jurassic dinosaurs Part III. *American Journal of Science* (ser. 3) 19: 253–259.

——. 1881. Principal characters of American Jurassic dinosaurs Part IV. Spinal cord, pelvis, and limbs of *Steogsaurus*. *American Journal of Science* (ser. 3) 21: 167–170.

Necker, R., A. Janßen, and T. Beissenhirtz. 2000. Behavioral evidence of the role of lumbosacral anatomical specializations in pigeons in maintaining balance during terrestrial locomotion. *Journal of Comparative Physiology A* 186: 409–412.

Nopcsa, F. v. 1917. Über Dinosaurier. 2. Die Riesenformen unter den Dinosauriern. *Centralblatt für Mineralogie* 1917: 332–351.

Northcutt, R. G. 2001. Changing views of brain evolution. *Brain Research Bulletin* 55 (6): 663–674.

Northcutt, R. G., and J. H. Kaas. 1995. The emergence and evolution of mammalian neocortex. *Trends in Neuroscience* 18: 373–379.

Osmólska, H. 2004. Evidence on relation of brain to endocranial cavity in oviraptorid dinosaurs. *Acta Palaeontologica Polonica* 49 (2): 321–324.

Ostrom, J. H. 1961. Cranial morphology of the hadrosaurian dinosaurs of North America. *Bulletin of the American Museum of Natural History* 122: 33–186.

Platel, R. 1976. Analyse volumétrique comparée des principales subdivisions encéphaliques chez les reptiles sauriens. *Journal für Hirnforschung* 17: 513–537.

———. 1979. Brain weight–body weight relationships. In C. Gans, R. G. Northcutt, and P. Ulinski (eds.), *Biology of the Reptilia*, 9:147–171. San Francisco: Academic Press.

Radinsky, L. B. 1967. Relative brain size: a new measure. *Science* 155: 836–838.

Rich, T. H., and P. V. Rich 1988. A juvenile dinosaur brain from Australia. *National Geographic Research* 4 (2): 148.

———. 1989. Polar dinosaurs and biotas of the Early Cretaceous of Southeastern Australia. *National Geographic Research* 5 (1): 15–53.

Rogers, S. W. 1998. Exploring dinosaur neuropaleobiology: computed tomography scanning and analysis of an *Allosaurus fragilis* endocast. *Neuron* 21: 673–679.

———. 1999. *Allosaurus*, crocodiles, and birds: evolutionary clues from spiral computed tomography of an endocast. *Anatomical Record* 257: 162–173.

Russell, D. A. 1969. A new specimen of *Stenonychosaurus* from the Oldman Formation (Cretaceous) of Alberta. *Canadian Journal of Earth Science* 6: 595–612.

———. 1972. Ostrich dinosaurs from the Late Cretaceous of Western Canada. *Canadian Journal of Earth Sciences* 9: 375–402.

Saveliev, S. V., and V. R. Alifanov. 2007. A new study of the brain of the predatory dinosaur *Tarbosaurus bataar* (Theropoda, Tyrannosauridae). *Paleontological Journal* 41: 281–289.

Sanders, R. K. and D. K. Smith. 2005. The endocranium of the theropod dinosaur *Ceratosaurus* studied with computed tomography. *Acta Palaeontologica Polonica* 50 (3): 601–616.

Sedlmayr, J. C. 2001. Encephalic blood vessels and evolution of the archosaur braincase [abstract]. *American Zoologist* 41 (6): 1584.

Sereno, P. C., J. A. Wilson, L. M. Witmer, J. A. Whitlock, A. Maga, O. Ide, and T. A. Rowe. 2007. Structural extremes in a Cretaceous dinosaur *PLoS One* 2 (11): e1230.

Snell, O. 1891. Die Abhängigkeit des Hirngewichtes von dem Körpergewicht und den geistigen Fähigkeiten. *Archiv für Psychiatrie und Nervenkrankheiten* 23: 436–446.

Stokstad, E. 2005. *Tyrannosaurus rex* gets sensitive. *Science* 310: 966–967.

ten Donkelaar, H. J. and R. Nieuwenhuys. 1979. The brainstem. In C. Gans, R. G. Northcutt, and P. Ulinski (eds.), *Biology of the Reptilia*, 9:133–200. San Francisco: Academic Press.

Witmer, L. M. 1995. The extant phylogenetic bracket and the importance of reconstructing soft tissues in fossils. In J. J. Thomason (ed.), *Functional Morphology in Vertebrate Paleontology*, 19–33. New York: Cambridge University Press.

Witmer, L. W., S. Chatterjee, J. Franzosa, and T. Rowe. 2003. Neuroanatomy of flying reptiles and implications for flight, posture and behavior. *Nature* 425: 950–953.

Zhou, C.-F., K.-Q. Gao, R. C. Fox and X.-K. Du. 2007. Endocranial morphology of psittacosaurs (Dinosauria: Ceratopsia) based on CT scans of new fossils from the Lower Cretaceous, China. *Palaeoworld* 16 (4): 285–293.

The Taxonomy and Systematics of the Dinosaurs

11

Thomas R. Holtz Jr. and M. K. Brett-Surman

Taxonomy, the naming of names, is the scientific practice and study of labeling and ordering like groups of organisms. It should not be confused with systematics, the scientific study of the diversity of organisms within and among clades (genetically related groups of organisms). Both help us to understand the world of organisms, but each practice helps us in a different way: systematics, to understand relationships among organisms, and taxonomy, to give internationally standardized names to organisms and groups of organisms in order to increase the efficiency of communication among researchers.

All languages have common names for different plants and animals. The main problem is that all languages have *different* names for the same plants and animals. This was not a problem until natural historians started to catalog and study the floras and faunas from around the world. It was realized by Western Europeans that the animals and plants of India, eastern Asia, the Pacific Islands, and particularly the New World of North and South America had great economic value as sources of medicine, spices, food, and furs. Whoever was the first to find and identify new plants and animals of economic importance in these regions would have the best access to these resources. Thus it became important to identify and classify these organisms.

There was initially a great deal of confusion as the great exploring and colonizing European countries each used their own names for the plants and animals they were discussing in the scientific literature. The only compromise that pleased all concerned was that every organism would be given a formal, official name based on Latin or Greek, the language of the most highly educated Europeans and of the Catholic Church. In the seventeenth century, Caspar Bauhin (1623) and John Ray (1686–1704) invented the precursors of the later binomial (two-name) system. They introduced the concept of genus and species. It was not until the eighteenth century that these names were organized into a hierarchy of divisions (kingdom, class, order, family, genus, species) by Carl Linné (formally known as Carolus Linnaeus), a natural historian and botanist in mid-1700s Sweden, and his successors (Linné 1758).

The basic principle of the Linnaean taxonomy is the nested hierarchy, in which each group is nested in a series of larger and larger, and thus more inclusive, groups. Each group is a taxon (plural taxa), a named group of organisms. Living taxa are recognized by their unique combinations of anatomical characteristics—bones, skin, hair/feathers/scales, physiology, DNA sequences, reproductive features, and so on. Extinct vertebrate taxa can be defined only by their bones and teeth.

What's in a Name? Taxonomy

209

Taxa are named in Latin or Latinized forms of other languages. Among other languages, Greek is the most common in taxonomy, but any other language (including English, Mongolian, Sanskrit, and the invented languages of J. R. R. Tolkien) will do, as long as it has Latinized endings. All taxa, of whatever level, must be in Latinized form. Taxonomic names can be named after anything the discoverer decides, commonly including the following:

- Features of the anatomy: Mammalia, for the mammary glands.
- General appearance: *Anatotitan*, "titanic duck," for a duckbilled dinosaur.
- Behavior (alleged or otherwise): *Tyrannosaurus rex*, the "king tyrant lizard."
- Name of their discoverer or other significant individual: *Lambeosaurus*, discovered by L. Lambe.
- The location from which it was found: *Albertaceratops*, a horned dinosaur first discovered in Alberta, Canada.

The most basic taxonomic levels, to which every living organism can be assigned, are the genus (plural *genera*) and species. The rules of nomenclature (the official naming of taxa) are based on species. Species names are listed only along with the genus name, never by themselves. Species are, literally, more specific words than genera: species refer to a smaller total number of organisms than do genera. The international rules for naming a new species are governed by the International Code of Zoological Nomenclature.

The exact biological or philosophical boundary where one species ends and another begins (or, more practically, whether a given specimen is assignable to a particular species) is the subject of much debate among biologists and paleontologists. Different criteria are used to define species by different scientists. Some, for example, use the degree of similarity or difference in the genetic code of a newly found organism in comparison with a cataloged species. Others define the boundary of a species on the basis of evolutionary divergence or lineage splitting: all individuals that had a more recent common ancestor with a cataloged specimen than with another cataloged specimen are included in the first specimen's species.

One way many biologists determine whether two or more individuals are members of the same species is to observe the results of mating between individuals. If under natural conditions they mate and produce offspring that themselves can produce offspring, then the original two individuals are members of the same species. If the two individuals cannot mate or do not produce living offspring, or the offspring they produce are sterile, then the two original specimens are not members of the same species. Of course, this is an impossible test with fossil individuals! For dinosaur studies, an individual is assigned to a species only if it shows a high degree of physical similarity in many parts of the skeletal anatomy to others thought to be in that species. The determination of which characteristics of the skeleton to use in classification is somewhat subjective.

A genus is defined as a group of one or more closely related species. Genus names are often listed by themselves. (Most people know dinosaurs only by their generic, not their specific, names; for example, people say

"*Triceratops*," not "*Triceratops horridus*.") Genera are, literally, more generic than species: genera refer to a larger number of individuals than do species. The Linnaean binomials are thus the reverse of the European style of naming individual people in which the personal name comes first and the surname last. If we were to write our names in the form of a Linnaean binomial, we would be called *Holtz thomas* and *Brett-Surman michael*. The following are some examples of the scientific names of modern animal species: humans, *Homo sapiens* ("thinking person"–Linnaeus was an optimist), abbreviated *H. sapiens*; cats, *Felis cattus* ("cat cat"); dogs, *Canis familiaris* ("familiar dog"); moose, *Alces alces* ("elk elk"); and the American alligator, *Alligator mississippiensis* ("Mississippi alligator"). Dinosaur species include *Tyrannosaurus rex* ("king tyrant lizard"), *Apatosaurus excelsus* ("surpassingly deceptive lizard"), *Triceratops horridus* ("roughened three-horned face"), and *Iguanacolossus fortis* ("mighty iguana colossus").

Linnaean species names are written *Genus species*, always in italics (or underlined in the case of handwriting or typescript). The species name is abbreviated *G. species*. Although it once was common practice, the trivial name is never capitalized: for example, *Tyrannosaurus Rex* (a usage seen in many popular books) is incorrect; the proper form is *Tyrannosaurus rex*. It is never proper taxonomic grammar to use a trivial nomen by itself (for the example above, *rex* or *Rex* is never correct; only *Tyrannosaurus rex* or the abbreviation *T. rex*).

Type Specimens, Priority, Synonymy, and Validity

Linnaean taxonomy is based on the idea of type specimens. A type is the actual individual specimen that is first given the name. Types are the name holders only; they are not sacred objects that represent what a species should look like. They are simply the first reference specimen to carry a new name. This specimen must be deposited in an accredited institution where it is available for study; it must be cataloged (for example, the type specimen for the duckbilled dinosaur *Edmontosaurus annectens*, in the collections of the National Museum of Natural History [Smithsonian Institution], is cataloged under number USNM 2414); and it must be described in the scientific literature where its name is presented. Additional specimens are assigned to a species (or genus or other taxon) on the basis of how closely a taxonomist believes the new specimen is related to a type. If the specimen is very similar, showing all the features, it is probably the same taxon. However, if the specimen does not show any features that disqualify it but shows no features that are definitely distinctive of the taxon, it can be questionably assigned to the taxon. If the specimen shows new features, it may be a new taxon. For example, William Buckland's first dinosaur was not like any other known reptile, so he considered it a new genus and named it *Megalosaurus*. Gideon Mantell's first dinosaur was likewise not identical to any known reptile, so he considered it a new genus, and his specimens were made the type for *Iguanodon*.

Each species has a type specimen. Each genus, in turn, has a type species—the first species to be given that generic name. Not all types are complete skeletons. Most fossil vertebrate types are incomplete skeletal material. So it is not uncommon for two (or more) names to be proposed for what later turns out to be the same genus, or even the same species.

When this happens, the oldest valid name (figured by date of its publication) has priority and is the official name of that taxon. The younger names are considered junior synonyms and are not used. For example, the name *Troodon formosus* was given to a dinosaur tooth named by Joseph Leidy in 1856. Much later, in 1932, Charles M. Sternberg named a fragmentary skeleton of a small dinosaur *Stenonychosaurus inequalis*. More complete skeletons found after the 1960s showed that the tooth called *Troodon* and the fragments called *Stenonychosaurus inequalis* belonged to the same species. Because the former was named 76 years earlier, it had priority, and the small birdlike dinosaur is properly called *Troodon formosus*.

Many factors might result in variation between individuals, which should be considered in deciding if two specimens belong of the same genus or same species. In nearly all species, there are changes in shape as they organism grows up; this is known as ontogenetic variation. In many, sexual variation exists between males and females. If a species exists over a wide region, there are often regional variations between different subpopulations. Perhaps most common of all are individual variations: except for identical twins and the like, no two individuals have exactly the same combination of genes, and thus they are all different from each other to some degree. A paleontologist needs to consider all these factors as possibilities before concluding that two specimens are actually different from each other because they represent different species in the same genus, or even different genera.

If a paleontologist concludes that two previously named species are just ontogenetic, sexual, regional, or individual variants of the same species, then he or she must use the older of the valid names for this taxon. For example, Scannella and Horner (2010) consider the specimens of the giant horned dinosaur previously called *Torosaurus* to be just the adult form of the better-known *Triceratops*. Because *Triceratops* was named earlier, this would be the proper name for the genus.

Although some specimens may be named new species on valid grounds at the time, later discoveries can show that these fossils are not distinct. The names based on these types are then considered invalid. Only those specimens with distinct features can be valid types. For example, in 1928 S. H. Haughton referred some dinosaur bones from Africa to a new species, *Gigantosaurus dixeyi*. The type material of *Gigantosaurus* (G. *megalonyx* from England, named in 1869 by H. Seeley) turned out to have no characters distinct from other dinosaur species or genera. "*Gigantosaurus*" *dixeyi* was left without a proper generic name until Jacobs and colleagues transferred the species to a new genus *Malawisaurus* in 1993.

Family and Family Group Names

Biologists have long recognized that animals can be grouped together hierarchically not only into species and genera, but also into larger and larger groups. For example, lions (*Panthera leo*) and tigers (*Pathera tigris*) can be grouped with domestic cats (*Felis cattus*) because of many similarities, including their retractable claws. Cats can be grouped with bears and dogs because of their specialized cheek teeth. Cats, dogs, and bears can be grouped with horses, humans, and whales because they all give milk. And so on, throughout the living world.

Each of these groupings can be considered a taxon. For most taxa larger than the genus, there are no specific rules for names, other than the requirement that the names be in Latin or a Latinized form of other languages. For example, Richard Owen used three known genera (*Megalosaurus*, *Iguanodon*, and *Hylaeosaurus*) that together were so distinct from all others that he gave them their own name, Dinosauria (Latinized Greek for "fearfully great lizards").

Unlike species names (which are always of the form *Genus species*), taxa from the genus and higher taxonomic levels have one-word names only: Felidae, Carnivora, Mammalia. Unlike the species or genus names, taxa higher than the genus are never italicized. Other than that, there are few rules for most higher taxonomic names. One special type of taxon that does have special rules of nomenclature is the family (and the related subfamily and superfamily). A family is an assemblage of closely related genera, such as the cats (great and small), the dogs (from foxes to timber wolves), or the ostrich dinosaurs. Each family has a type genus (just as genera have type species and species have type specimens) from which that family gets its name. The family name comes from the name of the type genus (above, *Felis*, *Canis*, and *Ornithomimus*, respectively), modifying the ending according to Latin rules (generally dropping the *-is* or *-us*), and adding the suffix *-idae* (Latin for "of the family of"). Thus, the cat family is Felidae, the dog family Canidae, and the ostrich dinosaur family Ornithomimidae (remember that because a family is larger and more inclusive than a genus or species, the name is not italicized). When families are spoken of informally, their names are used in the lowercase, and the *-idae* ending becomes *-id*: above, felid, canid, and ornithomimid (see Table 11.1).

Table 11.1. Traditional Family Group Suffixes

Rank	Formal Suffix	Vernacular Suffix	Examples (Formal, Vernacular)
Superfamily	-oidea	-oid	Hadrosauroidea, hadrosauroid
Family	-idae	-id	Hadrosauridae, hadrosaurid
Subfamily	-inae	-ine	Hadrosaurinae, hadrosaurine

Traditionally, each genus belonged to a family, even if that genus was the only member of the family. However, some scientists now use families only when two or more genera are grouped together. Sometimes a family has so many genera that it becomes important to recognize groups within the family that contain more than one genus. A new taxon, the subfamily, is used for these smaller divisions. Subfamily names are formed by taking a type genus (just as in family names) and adding *-inae* instead of *-idae* to the shortened genus name. For example, Ceratopsidae, the family of horned dinosaurs, contains more than 30 distinct genera. Those closer to *Chasmosaurus*, with long frills and longer beaks, are grouped into the subfamily Chasmosaurinae, while those closer to *Centrosaurus*, with shorter frills and deeper beaks, are grouped into the subfamily Centrosaurinae. Some taxonomists split subfamilies into even smaller divisions (such as tribes, subtribes, and supergenera), but this practice is not yet common in dinosaur taxonomy.

On the other hand, sometimes taxonomists want to recognize a taxon that includes a family and other closely related families or genera. The most common way to do this is to name a superfamily. Superfamilies are formed by taking a type family and changing the *-idae* ending to *-oidea*. For example, paleontologists recognize that the Allosauridae and Sinraptoridae, two families of carnivorous dinosaurs, are closely related. They are grouped together into the superfamily Allosauroidea.

Systematics

Systematics is the scientific study of the diversity of organisms within and among clades (genetically related groups of organisms). Systematics is related to taxonomy in that the former is the practice of identifying evolutionarily significant groups of organisms, while the latter is the practice of naming those evolutionarily significant groups. Traditionally, there have been two different methods of systematics employed by vertebrate paleontologists, evolutionary systematics (sometimes called evolutionary taxonomy or gradistics), and phylogenetic systematics (often called cladistics). Because the methods vary between these types of systematics, the taxonomy associated with each also varies.

Evolutionary Systematics: Grades

Evolutionary systematics ("gradistics") is an eclectic system of classification based on morphological similarity and the Linnaean taxonomic hierarchy. Groups of organisms are recognized by their physical resemblances. In order to be considered valid, all members of gradistic taxon must have a common ancestor that was also considered a member of that taxon (for example, the common ancestor of all lizards must be a lizard and the common ancestor of all dinosaurs, a dinosaur). However, unlike cladistics, a gradistic taxon can exclude descendant groups if the descendants share a great number of anatomical advances not shared by any other member of the larger group. In other words, only those organisms of the same grade of development are included together in a taxon, while descendents of a higher grade of development are excluded from this taxon. For example, snakes all lack legs and eyelids and have many specialized characters not shared by their ancestors (which were lizards), so the snakes (Ophidia) are excluded from the lizards (Lacertilia) under gradistic systematics. Similarly, under gradistics, birds (Aves) are excluded from their reptilian ancestors because birds possess many specialized features (toothless beaks, wishbones, feathers, warm-bloodedness, and many more) not found in turtles, lizards, snakes, crocodiles, and the like. Some taxonomists under gradistics even allow groups of animals of the same grade of organization not descended from the same ancestor to be placed in their own taxon. Although this practice was never common, some still use this extreme version of gradistics.

Under evolutionary taxonomy, all taxa are assigned a Linnaean rank. The standard Linnaean ranks are phylum, class, order, family, genus, and species. Linnaean taxonomy is a system of nested hierarchies; in practical terms, this means that each phylum contains one or more classes, each class one or more orders, and so on. As commonly used, each species in an evolutionary taxonomy must belong to a genus, family, order, class, and

Table 11.2. Linnaean Taxonomic Ranks and the Systematics of the Duckbilled Dinosaur Species *Anatotitan copei*

Traditional Rank	Additional Subordinal Rank
Phylum Chordata	Phylum Chordata
Class Reptilia	Class Reptilia
Order Ornithischia	Order Ornithischia
	Parvorder Genasauria
	Nanorder Cerapoda
	Hyporder Euornithopoda
	Suborder Ornithopoda
	Infraorder Inguanodontia
	Gigafamily Dryomorpha
	Megafamily Anklopollexia
	Grandfamily Styracosterna
	Hyperfamily Iguanodontia
	Superfamily Hadrosauroidea
Family Hadrosauridae	Family Hadrosauridae
	Subfamily Hadrosaurinae
	Tribe Edmontosaurini
Genus *Anatotitan*	Genus *Anatotitan*
Species *Anatotitan copei*	Species *Anatotitan copei*

Note: See text for discussion. Note that most of these ranks are no longer used by either gradistics or cladistics. Traditional ranks are those required by standard evolutionary systematics; additional subordinal ranks are ranks below order used to more precisely describe the systematic position of the species. Parvorder Genasauria to Superfamily Hadrosauroidea from Sereno (1986); Family Hadrosauridae to Species *Anatotitan copei* from Brett-Surman (1988).

phylum, even if that species is the only known representative of each of those higher taxa (a case of redundant taxonomic names).

It has long been recognized that there are subgroups that are intermediate between Linnaean ranks, so various prefixes (*super-, sub-, infra-,* and many others) have been used for these intermediate ranks (superclass, subfamily, infraorder, for example). Also, some taxonomists have added additional ranks (division, cohort, etc.) that are intercalated between previous ranks. However, this practice ultimately resulted in a bewildering number of ranks and subranks, as shown in Table 11.2. Because the number of intercalated subranks increased to the point of being unmanageable, there has been a trend to abandon the rank concept by both gradistics and cladistics.

The gradistic system of taxonomy has been useful, in various incarnations, since the 1700s. Much of our understanding of major living groups comes from research under the evolutionary system of systematics. A taxon in gradistic systematics must have a common ancestor but may exclude one or more groups of descendants. For example, the superclass Tetrapoda ("four-footed ones") has long been considered to be composed of four classes: Amphibia (cold-blooded tetrapods that have no scales and must reproduce in water), Reptilia (cold-blooded tetrapods that have scales and lay their eggs on land), Aves (warm-blooded tetrapods that have scales and feathers and lay their eggs on land), and Mammalia (warm-blooded tetrapods that have hair, give milk, and either lay eggs on land or have internal eggs). Almost all cultures recognized these classes, especially birds and mammals. However, evolutionary biologists soon recognized that reptiles are descendants of extinct amphibians (as "amphibians" were traditionally conceived by Linnaean taxonomists), and that birds and mammals are descendants of different groups of "reptiles" (although mammalian ancestors are now not considered members of Reptilia).

Until the 1970s, this was the system most widely used. It is a system designed to provide a taxonomy in which like groups are placed into the same hierarchical level (such as families), and to provide evolutionary statements that are built into the scheme. One can look at an evolutionary classification and see which groups are most closely related, their level of organization (body plans, or grades) as reflected in their Linnaean ranking (e.g., orders within a class were assumed to be evolutionary equals), and sometimes a degree of anatomical complexity. Evolutionary taxonomy recognizes grades of evolution that do not reflect total genetic relationships. A good example of a grade taxon is the reptilian order "Thecodontia." This is a group of archosaurian reptiles that were all placed in the same order because they had thecodont (socket-tooth) dentition. Because they were more "advanced" than earlier reptiles and more "primitive" than the dinosaurs, birds, pterosaurs, and crocodiles, they were put in their own group, which reflected their level (grade) of evolution rather than their particular relationships to other archosaurs. However, thecodonts were not characterized by any unique features. Instead, they had features shared with all other archosaurs, but at the same time they lacked the specializations of the more advanced forms (dinosaurs, birds, pterosaurs, and crocodiles).

Phylogenetic Systematics: Clades

Evolutionary taxonomy (gradistics) has been correctly criticized for being too subjective and trying to put too much information into a system that tries to pigeonhole everything into a classification scheme that is a rigid artificial abstraction. To meet the need for a more objective system, one close to the reality of evolution, with or without all the Linnaean ranks, in the 1950s entomologist Willi Hennig invented what is known as phylogenetic systematics, also known as cladistics (Hennig 1950, 1966). Cladistics has now replaced evolutionary systematics as the method used by most vertebrate paleontologists.

A method was needed to show the closeness of ancestry between two groups because independent evidence shows that most species arise as splitting events (when two or more parts of a population of organisms, separated by some sort of barrier, follow different evolutionary pathways as a result of natural selection and genetic drift). For any three taxa, two must share a more recent common ancestor than the third. For example, the horned dinosaurs *Triceratops* and *Chasmosaurus* are both more closely related to each other than either is to a *Centrosaurus*. This can be shown graphically (Fig. 11.1) by a cladogram, which shows that *Triceratops* and *Chasmosaurus* are joined with each other above the level at which *Triceratops*, *Chasmosaurus*, and *Centrosaurus* are all joined.

The point where two (or more) lines in a cladogram join is called a node. A node is recognized as a taxon itself–specifically the taxon containing all taxa that join at that node. In the cladogram in Figure 11.1, the node at the bottom of the cladogram (the base) represents the taxon Ceratopsidae.

The taxon that shares a splitting event with another taxon is called the sister taxon or sister group. In the cladogram in Figure 11.1, *Chasmosaurus* is

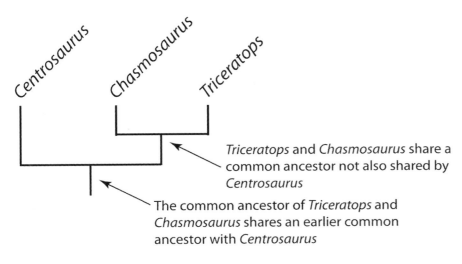

Triceratops and *Chasmosaurus* share a common ancestor not also shared by *Centrosaurus*

The common ancestor of *Triceratops* and *Chasmosaurus* shares an earlier common ancestor with *Centrosaurus*

the sister taxon to *Triceratops*, *Centrosaurus* is the sister taxon to *Chasmosaurus* plus *Triceratops*, and, conversely, the group *Chasmosaurus* plus *Triceratops* is the sister taxon to *Centrosaurus*. In general practice, when scientists refer to the sister group to a particular taxon, they mean the closest that is known to science, and not just the closest on a very simplified cladogram.

From a cladogram, we can recognize three types of groups (Fig. 11.2). Monophyletic ("single branch") groups are composed of a single ancestor and all of its descendants. Mammals have long been recognized as a monophyletic taxon. Paraphyletic ("nearly a branch") groups are the grades of evolutionary taxonomy: a single ancestor, but not all descendants. Lizards are paraphyletic if snakes are excluded from lizards. Similarly, reptiles are paraphyletic if birds are excluded from the Reptilia. Polyphyletic ("multiple branches") groups have multiple ancestors, which have long been regarded as invalid by taxonomists. Grouping mammals and birds together without including reptiles (as a grade; see below) is a polyphyletic grouping because mammals had a separate origin within the reptiles from that of the birds.

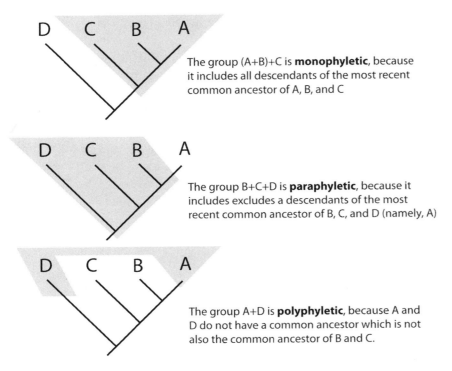

The group (A+B)+C is **monophyletic**, because it includes all descendants of the most recent common ancestor of A, B, and C

The group B+C+D is **paraphyletic**, because it includes excludes a descendants of the most recent common ancestor of B, C, and D (namely, A)

The group A+D is **polyphyletic**, because A and D do not have a common ancestor which is not also the common ancestor of B and C.

11.2. A hypothetical cladogram portraying possible relationships among four groups of organisms. From top to bottom: monophyletic (an ancestor and all of its descendants), paraphyletic (an ancestor but not all of its descendants), and polyphyletic (no immediate common ancestor). Traditional gradistic taxonomy accepts the use of both paraphyletic and monophyletic groups, but cladistics requires that all taxa be monophyletic.

Monophyletic groups are called clades ("branches"). Phylogenetic systematics seeks to find the relationships among taxa to form clades. Because our interest is in monophyletic groups, it is important to use only smaller monophyletic groups while conducting a phylogenetic analysis.

(A parenthetical note: The terms *evolutionary systematics* and *phylogenetic systematics* can be confusing at times. Scientists who use the methodology of evolutionary systematics are interested in determining phylogenies—that is, evolutionary trees that depict ancestor–descendant relationships. Workers who use phylogenetic systematics are interested in the recency of common ancestry and the interrelationships of clades, without making any statements about which taxa were ancestral to other taxa. This approach accepts biological evolution as the sole reason for the existence of the branching pattern of life. The terms *gradistics* and *cladistics* more accurately reflect the procedures used by these different schools of thought: gradists sometimes accept the use of paraphyletic grades of organisms in their taxonomies, while cladists accept only the use of monophyletic clades.)

Phylogenetic Analysis

Phylogenetic analyses are the various methods used to determine the interrelationships of a group of organisms. For example, we might wish to examine the relationships of the long-necked herbivorous dinosaur group Sauropodomorpha. In particular, we wish to resolve whether the somewhat more primitive sauropodomorphs of the Late Triassic and Early Jurassic form a paraphyletic series to the Jurassic and Cretaceous Sauropoda, or whether these genera form their own monophyletic group, the Prosauropoda (Fig. 11.3). In the first case, some of the nonsauropod sauropodomorphs (for example, *Massospondylus*) are more closely related to true sauropods than are other basal sauropodomorphs (for example, *Plateosaurus*). In the second case, the Prosauropoda as a whole forms the sister group to Sauropoda.

Phylogenetic analyses are thus searches for clades. How is this search conducted?

Many biologists use genetic or other biomolecular similarities between organisms to search for clades. For fossil groups (because the genes have all decayed over millions of years), the tool for finding clades is the search for shared derived characters (synapomorphies). First, scientists examine the characters of organisms, that is their physical features (shape of the bones and their relationship to one another, presence or absence of rare structures, etc.). Then they look at how these characters are distributed among various taxa, both within and outside the groups they wish to study. They determine which characters are found in taxa both inside and outside the group of interest, which characters are found only within the group of interest but among all members of that group, and which characters occur only within subsets of the group of interest.

Characters that are found in all members of the group of interest (and possibly outside the group) are called primitive characters, and they presumably were present in some common ancestor of all of the creatures in which they are now found. For example, five fingers are primitive for mammals, and hair is primitive for a primate. Therefore, the presence of five fingers

Primitive (basal) Sauropodomorpha

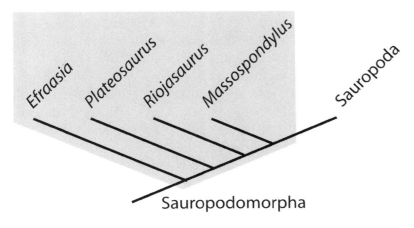

11.3. Two possible cladograms for the sauropodomorph dinosaurs of the Late Triassic and Early Jurassic (see also Yates, Chapter 21 of this volume). A, Some of the early branches of the sauropodomorphs (e.g., *Massospondylus*) share a more recent common ancestor with the sauropods than they do with more primitive prosauropods (e.g., *Plateosaurus*). In this scheme, Sauropoda are the direct descendants of the basal sauropodomorphs, and the similarities between *Massospondylus* and sauropods represent shared derived characters. B, The basal sauropodomorphs are all more closely related to each other than any is to the sauropods, making a monophyletic group (clade) Prosauropoda. In this scheme, Sauropoda is the sister taxon to Prosauropoda, and the similarities between *Massospondylus* and sauropods represent convergence.

Monophyletic Prosauropoda

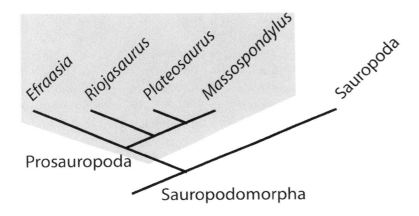

cannot help us determine which mammals are most closely related to each other, nor can the presence of hair help us understand the cladistic relationships within primates. Primitive characters are considered primitive homologies (see Holtz and Brett-Surman, Chapter 8 of this volume).

In contrast, characters that are found in only a few groups are probably shared derived characters, which evolved in some relatively recent common ancestor. For example, having five fingers is a shared derived character of the tetrapods when compared to all other vertebrates (including fishes), and hair is a shared derived character of mammals among tetrapods. Consequently, the tetrapods share a common ancestor that lived more recently than the common ancestor of all vertebrates, and mammals share a common ancestor that lived more recently than the common ancestor of all tetrapods. We can therefore use the presence of a five-fingered hand to distinguish tetrapods from other vertebrates, and hair to distinguish mammals from other tetrapods. Derived characters are advanced homologies. Shared primitive characters do not help resolve a cladogram, while shared derived characters do.

Unique derived characters (those found only within a single group), while important for understanding the biology of animals, do not help resolve the cladistic relationships among taxa. For example, because feathers are unique to birds among modern tetrapods, they do not help us to recognize which other group of tetrapods is the sister group to birds.

Convergences are a special sort of character. Because of similar functions or behaviors, two or more groups of organisms can independently acquire similar features. Although at first these resemblances might seem to be shared derived characters, additional evidence shows that they are convergent. For example, it might at first appear that the upright posture of some mammals and dinosaurs is a shared derived character of a mammal–dinosaur group. However, the dinosaurian skull, vertebrae, limbs, tail, and indeed most of the rest of the skeleton share more derived characters with other reptiles than with mammals. Thus the upright posture in mammals and dinosaurs is convergent.

Definition and Diagnosis

As might be imagined, there is a difference between the definition and diagnosis of different groups in the gradistic and cladistic systems of taxonomy. In the former, definition and diagnosis are for all intents and purposes identical, and are character based. In cladistics, definitions are based on taxa, and diagnoses are recognized by characters.

Under gradistics, the definition (the meaning of a taxon name) and the diagnosis (the way in which that taxon is recognized) are essentially the same. Taxa are defined by their characters (derived or primitive), so that a gradistic Reptilia could be defined as all amniotes (animals that reproduce by means of a specialized shelled egg, or derivatives of that style of reproduction) that have scales but lack feathers, fur, or warm-bloodedness. The diagnosis of the gradistic Reptilia would then be the presence of an amniotic egg and scales, and the lack of feathers, fur, and warm-bloodedness.

Similarly, under gradistics, dinosaurs would be defined as the superorder (or class, or subclass, etc.) Dinosauria, that group of archosaurian reptiles with upright limbs, three or more sacral vertebrae, and an open hip socket. The superorder Dinosauria consists of the two orders Saurischia and Ornithischia. The Ornithischia would be considered to have had an ancestor among the Saurischia because the saurischian skeleton is the less "advanced" of the two. This information could be presented as an evolutionary tree showing when each group originated and from which clade each group arose.

Under cladistics, the definition of a taxon is based on the relationships of two or more taxa (Sereno 2005). De Queiroz and Gauthier (1990, 1992, 1994) recognized two main kinds of phylogenetic definitions, branch based (formerly called "stem based") and node based. A third form, derived character based, is unstable (Padian and May 1993; Bryant 1994; Holtz 1996; Sereno 2005), in that the character used to diagnose the clade may be found to have evolved independently more than once. On the other hand, branch-based and node-based taxon definitions will always represent natural clades because all organisms share common ancestry to one degree or another.

Branch-based definitions are of the form "taxon X and all organisms sharing a more recent common ancestor with taxon X than with taxon Y"

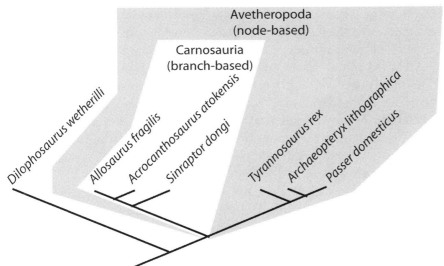

11.4. A cladogram of some carnivorous dinosaurs, showing the two main types of phylogenetic taxon definitions. Carnosauria is a branch-based taxon (*Allosaurus fragilis* and all theropods closer to *Allosaurus* than to *Passer domesticus*). Avetheropoda is a node-based taxon (all descendants of the most recent common ancestor of *Allosaurus fragilis* and *Passer domesticus*).

(Fig. 11.4A). For example, Saurischia is defined as *Megalosaurus bucklandii* and all taxa sharing a more recent common ancestor with *Megalosaurus bucklandii* than with *Iguanodon bernissartensis*. Node-based definitions are of the form "the most recent common ancestor of taxon X and taxon Y, and all descendents of that common ancestor." For example, Dinosauria is defined as the most recent common ancestor of *Megalosaurus bucklandii* and *Iguanodon bernissartensis*, and all of that ancestor's descendents.

Considering still broader, more inclusive groups, the category Reptilia is now considered a stem-based taxon: the lizard *Lacerta agilis*, the crocodile *Crocodylus niloticus*, and all taxa sharing a more recent common ancestor with these species than with *Homo sapiens* (Modesto and Anderson 2004). As a consequence, Reptilia contains turtles, lepidosaurs (lizards, including snakes, and the tuatara), and archosaurs (crocodiles and birds and their extinct relatives), as well as their extinct kin. Thus Aves (the birds) is part of the larger monophyletic Reptilia. Mammals, however, are not part of this clade, because our ancestors diverged from the common ancestor of all reptiles (as now defined) before the turtle–lepidosaur–archosaur divergence (Fig. 11.5). Thus under cladistics, the ancestors of mammals by definition were not reptiles (i.e., were not part of the clade Reptilia). (Some ornithologists take extreme exception to this demotion.)

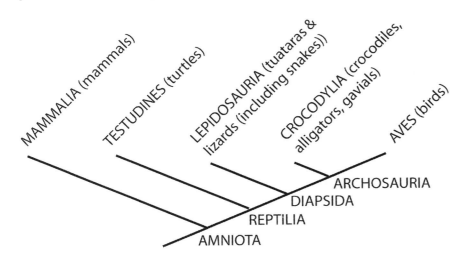

11.5. Cladogram of the living Amniota. Birds (Aves) are part of the monophyletic taxon Reptilia, but mammals (Mammalia) are not.

Diagnosis of taxa under the phylogenetic taxonomic system follows definition. After the distribution of shared derived character within a cladogram is determined, those shared derived characters that unite taxa into a stem-based or node-based taxon are used as the diagnosis of that taxon.

This phylogenetic system of nomenclature has its critics (Benton 2000) and its supporters (Bryant and Cantino 2002). Recently the International Society for Phylogenetic Nomenclature has developed an International Code of Phylogenetic Nomenclature, or PhyloCode (Cantino and de Queiroz 2010); whether working systemacists adopt this code and its specific regulations remains to be seen.

In the chapters that follow, the various contributing authors discuss dinosaurian relationships from both the gradistic and cladistic schools of systematics. It is informative to compare and contrast the conclusions the authors draw about the ancestors and ancestral characteristics of the dinosaur groups they study on the basis of the different methods they use to understand the evolutionary relationships of these wonderful ancient animals.

References

Bauhin, C. 1623. *Pinax Theatri Botanici.* Basel.

Benton, M. 2000. Stems, nodes, crown clades, and rank-free lists: is Linnaeus dead? *Biological Reviews* 75: 633–648.

Brett-Surman, M. K. 1988. Revision of the Hadrosauridae (Reptilia: Ornithischia) and their evolution during the Campanian and Maastrichtian. Ph.D. dissertation, George Washington University.

Bryant, H. N. 1994. Comments on the phylogenetic definition of taxon names and conventions regarding the naming of crown clades. *Systematic Biology* 43: 124–130.

Bryant, H. N., and P. D. Cantino. 2002. A review of criticisms of phylogenetic nomenclature: is taxonomic freedom the fundamental issue? *Biological Reviews* 77: 39–55.

Cantino, P. D., and K. de Queiroz. 2010. International code of phylogenetic nomenclature, version 4c. http://www.ohio .edu/phylocode/preface.html.

De Queiroz, K., and J. Gauthier. 1990. Phylogeny as a central principle in taxonomy: Phylogenetic definitions of taxon names. *Systematic Zoology* 39: 307–322.

———. 1992. Phylogenetic taxonomy. *Annual Review of Ecology and Systematics* 23: 449–480.

———. 1994. Toward a phylogenetic system of biological nomenclature. *Trends in Ecology and Evolution* 9: 27–31.

Haughton, S. H. 1928. On some reptilian remains from the dinosaur beds of Nyassaland. *Transactions of the Royal Society of South Africa* 16: 67–75.

Hennig, W. 1950. *Grundzüge einer Theorie der phylogenetischen Systematik.* Berlin: Deutscher Zentralverlag.

Hennig, W. 1966. *Phylogenetic Systematics.* Urbana: University of Illinois Press.

Holtz, T. R., Jr. 1996. Phylogenetic taxonomy of the Coelurosauria (Dinosauria: Theropoda). *Journal of Paleontology* 70: 536–538.

Holtz, T. R., and M. K. Brett-Surman. 2011 (this volume). The osteology of the dinosaurs. In M. K. Brett-Surman, Thomas R. Holtz Jr., and James O. Farlow (eds.), *The Complete Dinosaur,* 2nd ed., 134–149. Bloomington: Indiana University Press.

Jacobs, L. L., D. A. Winkler, W. R. Downs, and E. M. Gomani. 1993. New material of an Early Cretaceous titanosaurid sauropod from Malawi. *Palaeontology* 36: 523–534.

Leidy, J. 1856. Notices of remains of extinct reptiles and fishes, discovered by Dr. F. V. Hayden in the Bad Lands of the Judith River, Nebraska Territories. *Proceedings of the Academy of Natural Science, Philadelphia* 8: 72–73.

Linné, C. 1758. *Systema Natura per Regina Tria Naturae, Secundum Classes, Ordines, Gernera, Species cum Characterisbus, Differentiis, Synonymis, Locis. Edition decimal, reformata, Tomus I: Regnum Animalia.* Laurentii Salvii, Holmiae.

Modesto, S. P., and J. S. Anderson. 2004. The phylogenetic definition of Reptilia. *Systematic Biology* 53: 815–821.

Padian, K., and C. May. 1993. The earliest dinosaurs. In S. G. Lucas and M. Morales (eds.), *The Nonmarine Triassic*, 379–380. New Mexico Museum of Natural History and Science Bulletin 3.

Ray, J. 1686–1704. *Historia plantarum*. 3 vols. London: S. Smith and B. Waldorf.

Scannella, J. B., and J. R. Horner. 2010. *Torosaurus* Marsh, 1891, is *Triceratops* Marsh, 1889 (Ceratopsidae: Chasmosaurinae): synonymy through ontogeny. *Journal of Vertebrate Paleontology* 30: 1157–1168.

Sereno, P. C. 1986. Phylogeny of the bird-hipped dinosaurs (Order Ornithischia). *National Geographic Research* 2: 234–256.

———. 2005. The logical basis of phylogenetic taxonomy. *Systematic Biology* 54: 595–619.

Sternberg, C. M. 1932. Two new theropod dinosaurs from the Belly River Formation of Alberta. *Canadian Field-Naturalist* 46: 99–105.

Yates, A. 2011 (this volume). Basal Sauropodomorpha: the "prosauropods." In M. K. Brett-Surman, Thomas R. Holtz Jr., and James O. Farlow (eds.), *The Complete Dinosaur*, 2nd ed., 424–443. Bloomington: Indiana University Press.

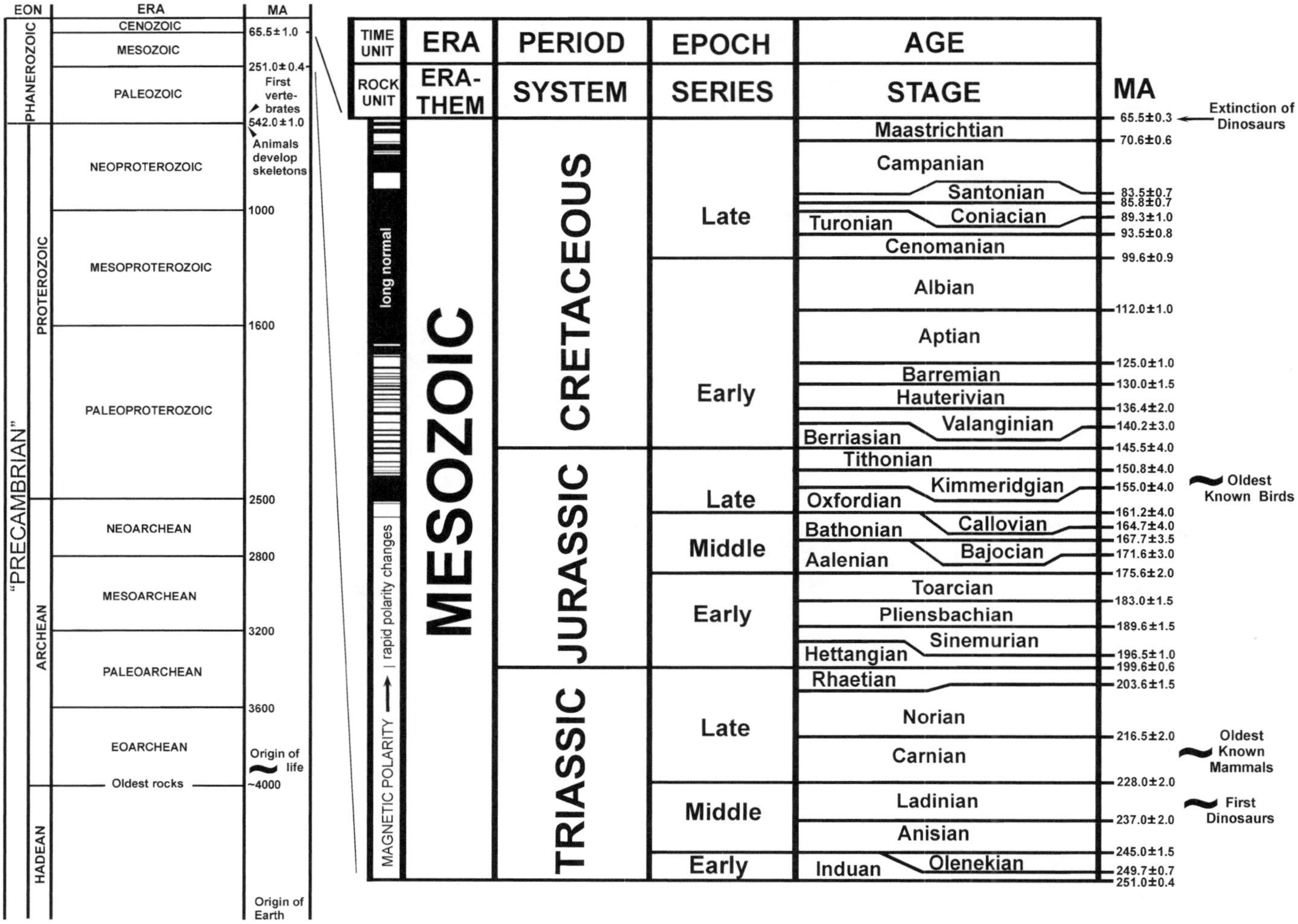

TIME UNIT	ERA	PERIOD	EPOCH	AGE		MA	
ROCK UNIT	ERA-THEM	SYSTEM	SERIES	STAGE			

long normal	**MESOZOIC**	**CRETACEOUS**	Late	Maastrichtian		65.5±0.3	← Extinction of Dinosaurs
						70.6±0.6	
				Campanian			
					Santonian	83.5±0.7	
						85.8±0.7	
				Turonian	Coniacian	89.3±1.0	
						93.5±0.8	
				Cenomanian		99.6±0.9	
			Early	Albian		112.0±1.0	
				Aptian		125.0±1.0	
				Barremian		130.0±1.5	
				Hauterivian		136.4±2.0	
					Valanginian	140.2±3.0	
				Berriasian		145.5±4.0	
		JURASSIC	Late	Tithonian		150.8±4.0	
				Oxfordian	Kimmeridgian	155.0±4.0	～ Oldest Known Birds
						161.2±4.0	
			Middle	Bathonian	Callovian	164.7±4.0	
						167.7±3.5	
				Aalenian	Bajocian	171.6±3.0	
						175.6±2.0	
			Early	Toarcian		183.0±1.5	
rapid polarity changes				Pliensbachian		189.6±1.5	
				Hettangian	Sinemurian	196.5±1.0	
						199.6±0.6	
		TRIASSIC	Late	Rhaetian		203.6±1.5	
				Norian		216.5±2.0	Oldest Known Mammals
MAGNETIC POLARITY →				Carnian		228.0±2.0	～ First Dinosaurs
			Middle	Ladinian		237.0±2.0	
				Anisian		245.0±1.5	
			Early	Induan	Olenekian	249.7±0.7	
						251.0±0.4	

Left column chart:

EON	ERA	MA	
PHANEROZOIC	CENOZOIC	65.5±1.0	
	MESOZOIC	251.0±0.4	
	PALEOZOIC	First verte-brates	
		542.0±1.0	
"PRECAMBRIAN" / PROTEROZOIC	NEOPROTEROZOIC	Animals develop skeletons	
		1000	
	MESOPROTEROZOIC	1600	
	PALEOPROTEROZOIC		
		2500	
ARCHEAN	NEOARCHEAN	2800	
	MESOARCHEAN	3200	
	PALEOARCHEAN	3600	
	EOARCHEAN	Origin of life	
	Oldest rocks	~4000	
HADEAN		Origin of Earth	

12.1. Geological timescale of the Earth with emphasis on the Mesozoic. Modified after Gradstein et al. (2004) and International Commission on Stratigraphy (2006), with magnetostratigraphy after the 1999 GSA Timescale.

Dinosaurs and Geologic Time

12

James I. Kirkland and James O. Farlow

The histories of paleontology and geology are intertwined because developments in both sciences have been dependent on each other since their beginnings (Albritton 1986; Berry 1987). Geology gives paleontology context, and therefore geological discoveries have been critical in developing our understanding of dinosaurs. Nowhere is this more evident than in understanding the distribution of dinosaurs through geologic time.

In 1699, Danish geologist Nicholaus Steno established some basic principles by which the relative timing of geological events could be determined. These are (1) the principle of superposition, in which each rock layer was laid down one upon another, (2) the principle of original horizontality, in which rock layers were originally laid down horizontally and were inclined only if secondarily tilted, and (3) the principle of original continuity, in which rock layers were assumed to be continuous over great distances. Together with Scottish geologist Charles Lyell's principle of crosscutting relationships, where any feature cutting across another must have formed after the layer it crosses, these four principles gave geology its initial foundation as a science (Dott and Prothero 2003; Stanley 2009).

Archbishop Usher's 1654 pronouncement that the genealogy recorded within the Bible indicated that the Earth was created at 9:00 AM on October 26, 4004 BC, reflected a literal view of Genesis. This restricted, 6,000-year chronology strengthened a catastrophic view of Earth history to account for the development of Earth's geological features in such a short amount of time. This view of the Earth laid the grounds for developing the overall framework for geologic time, where divisions were made at distinct breaks in the geologic and fossil records. Thus, the geologic time intervals we continue to use today are variable in their duration and are based on abrupt changes in the European rock record caused by gaps in the rock record and, in a few cases, mass extinctions.

In 1795, an alternative view of geologic time was proposed by Scottish geologist James Hutton, who recognized that the processes shaping the Earth today are the same processes responsible for forming older geological features on the Earth. This principle of uniformitarianism led Hutton to propose, "No vestige of a beginning, no prospect of an end." From this conceptual framework developed by Steno, Lyell, and Hutton, the subdivisions of a relative geologic timescale were gradually established over the next 50 years.

Geologists divide Earth history into extremely long time intervals known as eons, and eons are in turn subdivided into eras, eras into periods, periods into epochs, and epochs into ages. Chronostratigraphic rock units are the rocks deposited during these intervals of geologic time. Thus, rocks

The Relative Framework of Geologic Time

deposited during an eon are referred to as an eonothem, those during an era are an erathem, those during a period are a system, those during an epoch are a series, and those during an age are a stage. The names used for both geologic time units and chronostratigraphic rock units are the same and are derived from names used in developing a relative geologic timescale in Europe during the eighteenth and early part of the nineteenth centuries.

The most recent geologic timescale (Gradstein et al. 2004) divides the Earth's history into four eons (Fig. 12.1). Earth history prior to the oldest rocks preserved at the Earth's surface is referred to the Hadean. The Hadean can be visualized by gazing up at the moon, which still preserves the scars of meteor impacts that occurred during part of the Hadean Eon. The oldest rocks on earth date back to about 4 billion (4,000 million) years and mark the beginning of the Archean Eon. The earliest fossils, representing bacteria or similar organisms, are found in rocks of just slightly younger age. The Archean Eon ended 2.5 billion (2,500 million) years ago, followed by the Proterozoic Eon. The Proterozoic Eon ended at about 542 million years ago. During this long interval of time, living things became gradually more diverse and complex, with the first true animals appearing toward the end of the Proterozoic Eon.

The first appearance of animals with skeletal hard parts marks the beginning of the Phanerozoic Eon (time of evident life), which began at the close of the Proterozoic and continues to the present. Eons prior to the Phanerozoic are commonly lumped together as Precambrian as in the time before the Cambrian Period at the beginning of the Phanerozoic. The Phanerozoic is divided into three eras, which in ascending order are the Paleozoic (old life), the Mesozoic (intermediate life), and our own era, the Cenozoic (recent life).

Nonavian dinosaurs are restricted to the Mesozoic (Fig. 12.1). The Mesozoic is divided into three periods. Dinosaurs originated during the middle of the Triassic Period, which refers to the threefold division of rocks of this age in Germany and is divided into three epochs, early, middle, and late. The medial Jurassic Period was named for marine rocks in the Jura Mountains between France and Switzerland and is also divided into early, middle, and late epochs. The terminal Cretaceous Period was named for extensive chalk (the Latin word for chalk is *creta*) cliffs along the English Channel between Great Britain and France and, although it is the longest Mesozoic period, has only been divided into early and late epochs. Whereas periods and epochs span tens of millions of years, the ages/stages into which they are divided only span millions of years in most cases. Although ages/stages are sometimes divided into early, middle, and late subages/substages, ages/stages form the basic division of geologic time in the Mesozoic. It is important to note that the term *stage* is commonly misused in place of *age* in the geologic literature.

Although various subdivisions of geologic time have been proposed for regions outside Europe, such as Gulfian for the Early Cretaceous Epoch and Montanian for the Late Cretaceous Epoch in North American, in recent years there has been a general consensus to develop uniform international nomenclature primarily on the basis of the original European terms (Rudwick 1976; Albritton 1986; Berry 1987; Gradstein et al. 2004; U.S.

Geological Survey Geologic Names Committee 2007). The subdivisions of the Mesozoic are mostly based on European nomenclature.

Because the various intervals of geologic time were originally based on rock sequences, geologists frequently use Triassic, Jurassic, and Cretaceous or terms such as the Carnian, Oxfordian, or Campanian in two senses. One refers to the rock intervals themselves, and the other refers to the time intervals represented by those rocks. If we discuss the Carnian Stage of the Upper Triassic, for example, we are using the term to refer to the rocks deposited during the Carnian Age of the Late Triassic; Upper Triassic rocks were deposited above Lower and Middle Triassic rocks. On the other hand, the Late Triassic refers to the time span after the Early and Middle Triassic and before the Early Jurassic. Thus "Lower" in the stratigraphic sense corresponds to "Early" in the temporal sense, and "Upper" to "Late." "Middle," conveniently, is used in either the stratigraphic or the temporal sense.

Adjectives such as *early, middle,* and *late,* in addition to their use as modifiers of names of formally defined intervals of geologic time, are also used to describe informal stretches of geologic history. For example, one can talk informally about the "later Mesozoic," "middle Cretaceous," or "earliest Jurassic." These are not official terms of the geologic past but can be useful when referring to geological or evolutionary events. Note that *later, middle,* and *earliest* are not capitalized when one uses them informally. Figure 12.1 may seem to contain a daunting number of names, but the use of this terminology enables paleontologists to pinpoint when particular dinosaurs lived as finely as the resolution of the stratigraphic record permits. The reader will encounter these terms repeatedly throughout this book; we encourage readers to return to Figure 12.1 to help keep these chronological terms straight. Be aware, though, that some contributors to this book use slightly different subdivisions of Mesozoic time than those outlined in Figure 12.1.

An Absolute Geologic Timescale

Through the nineteenth century, quantitative methods were difficult to apply to dating the age of the Earth. Lord Kelvin's 1897 estimated age of the Earth as 24–40 million years, a value based on the hypothetical cooling of the Earth from an initial molten state, was the most widely accepted age estimate for many years. In fact, the timescale in use by the American Museum of Natural History in the 1920s reflects this view of the world, with an estimate of approximately 12–13 million years for the age of the *Protoceratops* beds in Mongolia (Andrews 1926). The discovery of radioactive decay not only provided a mechanism for heating the interior of the Earth, thus greatly expanding its potential age beyond Lord Kelvin's estimate, but also permitted the development of radiometric dating, a basis for putting specific dates on the geologic timescale. The decay of uranium to lead dates the oldest rocks on the Earth at 3.96 billion years old and the age of the Earth at about 4.6 billion years, on the basis of dates obtained from meteorites and lunar rock samples (Dalrymple 1994).

Various naturally occurring radioactive isotopes are suitable for radiometric dating in the Mesozoic, including uranium 235, uranium 238, thorium 232, rubidium 87, and potassium 40, which have half-lives of hundreds to thousands of millions of years. Carbon 14, with a half-life of just 5,750

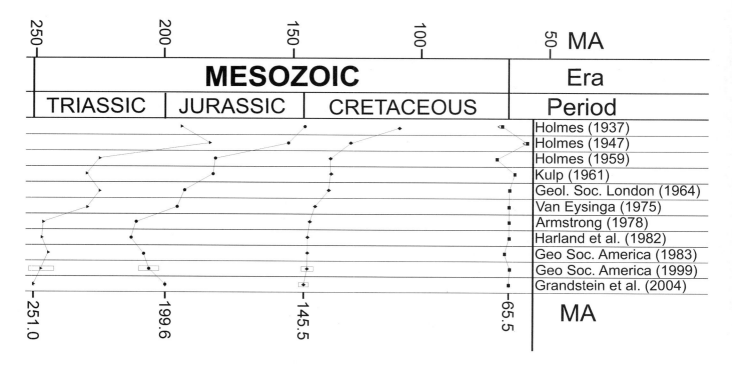

12.2. Comparison of proposed dates for period boundaries of Mesozoic. Boxes indicate error.

years, is too short lived to have any utility in dating Mesozoic rocks. In order for these radioactive isotopes to be useful in dating, they must be incorporated into stable minerals at the time the rock is formed, from which point they begin to decay. For these reasons, volcanic stratiform rocks such as lava flows and ash falls have the greatest potential for providing high-resolution dates of the stratigraphic sequences in which they are preserved. Some volcanic minerals, such as zircon, are very stable and may be redeposited millions to hundreds of millions of years after they first crystallized.

The resolution of radiometric dating has improved over the past 100 years, with the determination of dates becoming more consistent (Fig. 12.2). Some methods, such as potassium/argon dating of glauconite and fission-track dating, have essentially been abandoned. Argon[40]/argon[39] (potassium-to-argon decay series) dating of single sanidine (feldspar) crystals and uranium/lead dating of zircon crystals are currently considered to be the most accurate means of dating volcanic ash falls. Additionally, because uranium may be incorporated into calcium carbonate in ancient soils and springs, uranium/lead dating of radiogenic carbonates is being applied to terrestrially deposited rock sequences (Rasbury et al. 1998; Ludvigson et al. 2010).

A new method of using laser ablation of zircon crystals to provide dates, at a much reduced cost compared to other methods, permits random zircons from sandstones to be dated to determine the age of the youngest zircons and thus a maximum age for the rock. With coeval volcanic activity, dating sandstones in this way can provide maximum dates within a few million years for rocks that previously had been considered impossible to date (Gehrels 2000; Dickenson and Gehrels 2003, 2009). Because zircon is extremely durable, this method provides a spectrum of dates that may be used to trace ancient river systems by preserving a record of the ages of the zircons distributed in different age rocks throughout the drainage basin of the river (e.g., Prokopiev et al. 2008). This methodology has also been applied to Mesozoic aeolian rocks (Dickenson and Gehrels 2009). While uncertainties and errors in the geologic timescale still exist, improvements in methodologies continue to improve accuracy and precision (Gradstein and Ogg 2007).

Certain intervals of the geologic timescale are still poorly constrained, in part as a result of an absence of recognized datable materials. For some of these time intervals, the dates are estimated by interpolating from known dates. For the Middle Jurassic, the uncertainties of the dates are longer than for the stages themselves (Fig. 12.1). A recent revision of the age of the late Carnian Age of the Late Triassic Epoch indicates that as originally defined, the subsequent Norian Age is more than 20 million years long (Furin et al. 2006). Thus, the Norian is currently the longest age of the Mesozoic, if not the entire Phanerozoic. The next version of the global geological timescale must make allowances for this significant change.

A fundamental flaw in developing the divisions of geologic time is derived from the initial divisions being based on perceived breaks in the fossil record. For many, if not most, of the sites where these divisions were originally defined, the boundaries are marked by unconformities, such that the entire time interval in question is not represented at the type locality. This is particularly true with regard to the stages/ages. For example, the Upper Cretaceous Campanian Stage in its type area in France only represents a portion of the Campanian Stage as it has been used elsewhere (Hancock et al. 1996). Thus, in recent years, instead of attempting to correlate to the initial sites where the ages of the geologic timescale were initially defined, sites are now being selected across the globe, where the lower boundary of each stage, and in some cases substage, is ideally developed. At these sites, the rocks form a continuous sequence across the boundary, and there are multiple means of correlation present within the stratigraphic sequence. Relative to dinosaurian paleontology, it is unfortunate that marine rocks consistently possess the required properties, whereas terrestrial rocks do not.

These new global boundary stratotype sections and points (GSSPs) are approved by the International Commission on Stratigraphy and ratified by the International Union of Geological Sciences (Gradstein et al. 2004). Many of the GSSPs have had a "golden" spike mounted at the stratotype boundary sections with the stage or system engraved on the placard. The base of the Cenozoic Era (top of Cretaceous) was one of the first GSSPs, defined in 1991 at the bottom of the iridium-bearing boundary clay at El Kef, Tunisia (Molina et al. 2006). The GSSPs for the base and top of a particular stage may commonly be designated on different sides of the planet, but in some cases, the GSSPs for a stage are in relatively close proximity. For example, the placement of the base of the Upper Cretaceous, Cenomanian Stage, was ratified in 2002 at 36 m below the top of the Marnes Bleues Formation, Mont Risans, Haute-Alpes, southeastern France (Kennedy et al. 2004), whereas the placement of the base of the overlying Turonian Stage was ratified in 2003 at the base of Bed 86, Rock Canyon Anticline, west of Pueblo, Colorado (Kennedy et al. 2005). The placement of the base of the Lower Jurassic Sinemurian Stage was ratified at 0.9 m above the base of Bed 145, East Quantoxhead, Wastchet, West Somerset, southwestern England (Bloos and Page 2002), whereas the base of the overlying Pliensbachian

Stage was ratified in 2005 at the placement of the base of Bed 73b, Wine Haven section, Robin Hood's Bay, Yorkshire, England (Meister et al. 2006) (Fig. 12.1). As of this writing, many of the GSSPs for the Mesozoic have yet to be ratified, although most potential sites have been identified. Continuously updated data on the International Geological Timescale and the GSSPs are posted to the International Commission on Stratigraphy's Web site (http://www.stratigraphy.org). Although perhaps confusing to the nongeologist, the establishment of this system of GSSPs promises to bring stability to the global correlation of geological and biological events.

Lithostratigraphy

Because the divisions of the geologic timescale are now defined by single points on the Earth's surface and radiometric dates are only possible at a few sites, integrating geological and paleontological histories from sites across the planet must be accomplished by other means of correlation. Advances in the correlation of rocks regionally and globally are critical to our further understanding of the Earth's geological and biological history.

The physical correlation of rock layers, or lithostratigraphy, is limited by the lateral continuity of the rocks across their area of deposition and their physical distinctiveness in the vertical succession. As environments of deposition shift geographically through time, the resulting rock layers are actually diachronous; that is, the rock layers cross time lines laterally (Fig. 12.3). For example, as sea level rises, the beach environment and its sands shift inland. It may take millions of years as shoreline processes shift inland and deposit a sheet of coastal sand, which forms sandstone, across hundreds of miles or kilometers. For this reason, most rock layers are useful for time control in only small geographic areas. Exceptions are rocks formed by short-term events such as volcanic ash layers or impact ejecta blankets. Sequences of sedimentary rocks with the same characteristics deposited under the same environmental conditions may be repeated in the same area under fluctuating environmental conditions, leading to further confusion in correlating rocks, particularly in areas having sporadic exposures of outcrop. Additionally, the rates at which sedimentary rock have accumulated may vary considerably in different areas. Various data collected from stratigraphic sequences may be plotted against stratigraphic thickness or against time (Fig. 12.4).

The basic unit of lithostratigraphy is the formation, which is a mapable package of rock that has consistent internal characteristics and distinct upper, lower, and lateral boundaries. A formation may be divided into members, and when highly distinctive, an individual bed may be named. Several formations in sequence may be united together as a group or, rarely, a supergroup (Boggs 1995). These terms are capitalized as part of formal lithostratigraphic unit names for a specific geographic type area with clearly defined measured type section. Thus, the type specimen of *Diabloceratops eatoni* (UMNH VP 16699) was found 51.72 m above the base of the middle mudstone member of the Wahweap Formation at Utah Museum of Natural History locality 1092 (Utah locality 42Ka800V) in the south-central Kaiparowits Plateau, Grand Staircase–Escalante National Monument, Kane County, south-central Utah (Kirkland and DeBlieux 2009). Having precise

12.3. Simplistic model of general diachroneity of lithostratigraphic units with rising and falling sea level compared to chronostratigraphic markers and the ranges of several hypothetical organisms versus time and space.

geographic data tied to described and measured stratigraphic sections and unique specimens curated into public collections is critical to constraining temporal and paleoenvironmental data such as dinosaur occurrences, paleosols, rock descriptions, and radiometric dates. In addition to providing important contextual data, the site may then be reexamined in the future as needed. A principal means of testing geological and paleontological theories is other scientists independently reexamining important localities and the specimens on which these theories are based.

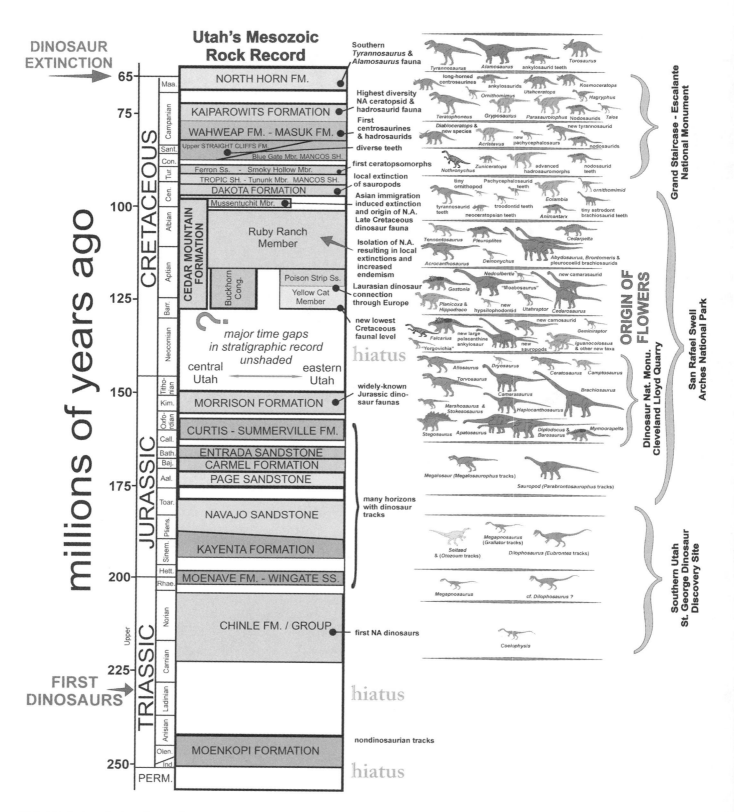

DINOSAUR EXTINCTION

FIRST DINOSAURS

millions of years ago

Utah's Mesozoic Rock Record

Southern *Tyrannosaurus* & *Alamosaurus* fauna

Highest diversity NA ceratopsid & hadrosaurid fauna

First centrosaurines & hadrosaurids

diverse teeth

first ceratopsomorphs

local extinction of sauropods

Asian immigration induced extinction and origin of N.A. Late Cretaceous dinosaur fauna

Isolation of N.A. resulting in local extinctions and increased endemism

Laurasian dinosaur connection through Europe

new lowest Cretaceous faunal level

widely-known Jurassic dinosaur faunas

many horizons with dinosaur tracks

first NA dinosaurs

nondinosaurian tracks

hiatus

hiatus

hiatus

CRETACEOUS

Maa.

Campanian

Sant.
Con.
Tur.
Cen.
Albian
Aptian
Barr.
Neocomian

JURASSIC

Titho-nian
Kim.
Oxfo-rdian
Call.
Bath.
Baj.
Aal.
Toar.
Pliens.
Sinem.
Hett.
Rhae.

TRIASSIC

Upper
Norian
Carnian
Ladinian
Anisian
Olen.
Ind.

PERM.

65
75
100
125
150
175
200
225
250

NORTH HORN FM.

KAIPAROWITS FORMATION

WAHWEAP FM. - MASUK FM.

Upper STRAIGHT CLIFFS FM.

Blue Gate Mbr. MANCOS SH.

Ferron Ss. - Smoky Hollow Mbr.

TROPIC SH. - Tununk Mbr. MANCOS SH.

DAKOTA FORMATION

Mussentuchit Mbr.

CEDAR MOUNTAIN FORMATION

Ruby Ranch Member

Buckhorn Cong.

Poison Strip Ss.

Yellow Cat Member

major time gaps in stratigraphic record unshaded

central Utah ⟷ eastern Utah

MORRISON FORMATION

CURTIS - SUMMERVILLE FM.

ENTRADA SANDSTONE

CARMEL FORMATION

PAGE SANDSTONE

NAVAJO SANDSTONE

KAYENTA FORMATION

MOENAVE FM. - WINGATE SS.

CHINLE FM. / GROUP

MOENKOPI FORMATION

Grand Staircase - Escalante National Monument

San Rafael Swell / Arches National Park

ORIGIN OF FLOWERS

Dinosaur Nat. Monu. Cleveland Lloyd Quarry

Southern Utah St. George Dinosaur Discovery Site

12.4. Distribution of select dinosaur fauna through Mesozoic time on the Colorado Plateau based on an idealized stratigraphic sequence in Utah plotted against a linear timescale.

Modified after Kirkland and Madsen (2007).

Sequence Stratigraphy

Globally, sea level has risen and fallen throughout Earth's history, and given a tectonically stable coastline, it has been suggested that eustatic sea level high and low stands can be correlated from continent to continent (Vail et al. 1977; Hallam 1992). However, there is a great deal of overprinting by tectonics and climate. As sea level rises, sediments are trapped upslope resulting in condensation and erosion downslope; as sea level falls, erosional surfaces form by downcutting and extend downslope. Thus, breaks in the geological record bound the sequences of strata deposited during intervals of sea level rise and fall, as opposed to the simplistic model shown in Figure 12.3. The study of these stratigraphic sequences, or system tracts, has been driven by the search for oil and gas, which has had a profound effect on the interpretation of sedimentary rocks deposited along the margins of the continents (Emery and Meyers 1996; Miall 1997).

The effects of sea level changes are recorded well inland from the shoreline, and thus a sequence stratigraphic approach is important in order to understand dinosaur-bearing rock sequences (Shanley and McCabe 1994; Miall 2006). Additionally, within isolated continental basins, tectonic and climatic changes may be reflected by sedimentary breaks, changes in the sedimentary architecture, and water-level changes in lacustrine (lake) systems, which result in basin-wide sedimentary sequences bounded by erosional surfaces, laterally extensive amalgamated river-deposited sandstones and conglomerates, or marker paleosol (fossil soils) horizons (Kraus 1999). Even within a formation or member, if two dinosaurs do not occur in the same sequence, they are generally separated by a significant stratigraphic break.

Biostratigraphy

The recognition by William "Strata" Smith at the beginning of the nineteenth century that a vertical sequence of different fossil species was consistently repeated from region to region facilitated his mapping of sedimentary rocks over a wide region of Britain (Winchester 2001). The correlation of rocks based on the vertical distribution of fossil species and assemblages of fossil species is called biostratigraphy. Biostratigraphy works extremely well for dating rocks because evolutionary theory indicates that a species only exists on Earth between its initial origin and its final extinction, and evolutionary processes are such that no species reappears after it has gone extinct (e.g., Prothero 2007). Superficial complications in lithostratigraphic correlations resulting from the diachronicity of individual rock layers and the repetition of similar rock types in time can be resolved because each fossil species is unique and had a limited temporal distribution. The distribution of fossils was fundamental to the development of the relative geologic timescale (Berry 1987). Biostratigraphy has been one of the major tools in geology for 200 years, and much of the primary data on which the history of life has been based resulted from collecting fossil data to solve geological problems.

The basic unit in biostratigraphy for a rock unit is the zone, with the corresponding time unit being the biochron. A range zone is the body of rock encompassing the known stratigraphic and geographic distribution of a taxon (International Commission on Stratigraphy n.d., http://www.stratigraphy.org/bio.htm). A biochron is the total temporal interval of a taxon from its origin to its extinction. Thus, like in lithostratigraphy, biostratigraphic units are not time-parallel units.

Other types of zones have broader utility. A concurrent range zone is the body of strata encompassing the overlap of two range zones. Composite range zones are the primary basis for most current high-resolution biostratigraphic correlations. The boundaries of these zones are defined on the lowermost and uppermost occurrences of multiple taxa (e.g., Kauffman et al. 1993; Gradstein et al. 2004). Zones are named for a taxon that characterizes them. For example, the base of the Upper Cretaceous Turonian Stage is placed at the base of Bed 86, Rock Canyon anticline, Pueblo, Colorado, which preserves the lowest examples of the ammonites, *Watinoceras devonensis*, *W. praecursor*, *W. depressum*, and *Quitmaniceras reaseri*, and the inoceramid bivalve *Mytiloides puebloensis* of the *Watinoceras devonensis* Zone, which is adjacent to the lowest occurrences of the planktonic foraminifer *Praeglobtruncana helvetica* and 28 cm above the first occurrence of the inoceramid bivalve *Mytiloides hattini* and the last occurrence of *Inoceramus pictus* (Kauffman et al. 1993; Kennedy et al. 2000, 2005).

The distribution of a species in space and time is based on biological factors of an organism's life history. A new taxon first appears at one geographic location and over some time interval becomes more numerous and expands its geographic range to some maximum extent; subsequently the taxon's geographic range becomes more restricted until the taxon goes extinct at some other specific geographic location (Fig. 12.3). The distribution of a fossil in space and time is also based on its dispersal characteristics after death and its potential to be preserved as a fossil. The best fossils for biostratigraphy are referred to as index fossils. These are geographically wide-ranging species that have limited temporal distributions. Index fossils have several properties:

- · They evolve rapidly.
- · They disperse rapidly over wide geographic areas.
- · They are not restricted to only one or a few habitats.
- · They are abundant, which often relates to being low on the food chain.
- · They are prone to extinction.
- · They have hard parts that are likely to fossilize.
- · They have distinct morphologic features that permit confident identification.

Within marine environments, invertebrate fossils are most often used in biostratigraphic correlation. Microfossils of planktonic organisms are particularly useful because they are often extremely abundant. Furthermore, planktonic taxa include organisms whose fossilizable parts vary considerably in their chemical composition: aragonite (pteropods), calcite (coccoliths, planktonic foraminifera), silica (diatoms, radiolarians), phosphate (conodonts, icthyoliths), and complex organic molecules (dinoflagellates, acritarchs). Therefore, at least one microfossil type will be preserved under almost all chemical environments developed during the formation of marine strata. Within the Mesozoic, the shells of ammonites are particularly useful; these fossils are abundant and morphologically distinctive (both as whole shells and internal molds). Ammonites had long planktonic larval stages allowing for widespread dispersal by marine currents, and their gas-filled shells had a high potential for postmortem drift. Consequently, many ammonite species have near-global distributions (e.g., Kennedy and Cobban 1976). Ammonites evolved so rapidly that the resolution of correlations based on them exceeded the error of the best radiometric dates until only a few years ago (e.g., Obradovich 1993). For the Cretaceous, inoceramid bivalves are extremely useful in long-range biostratigraphic correlation because inoceramids were abundant in a variety of shallow to deep marine environments, evolved rapidly, are readily identified under a variety of preservational regimes, and apparently had long planktonic larval stages. Historic taxonomic ambiguity has limited their widespread utility until recently (e.g., Kauffman et al. 1993; Kennedy et al. 2000). Because of the usefulness of ammonites, inoceramids, and other invertebrates, global biostratigraphic correlations of marine strata are extraordinarily accurate throughout the Mesozoic.

Within terrestrial environments, pollen, spores, and freshwater microfossils have had the most utility in biostratigraphic correlations. Palynology (the study of pollen and spores) provides one of geology's best tools for correlating the superior marine biostratigraphic record with terrestrial rocks because pollen and spores are readily transported far and wide by wind and water. However, the tough organic remains of palynomorphs are readily reworked, are not preserved in rocks that are oxidized or are alkaline, and, while some taxa have relatively time-parallel first and last occurrences, many others are markedly diachronous with latitude or elevation.

Significant contributions to correlating terrestrial rocks have been made using fossil vertebrates. Fossil mammals have been particularly useful because they are identifiable on the basis of their teeth. Continental land-mammal ages are a well-respected tool for biostratigraphic correlation. They are composite range zones named for particular geological areas where they are represented (Matthew 1915; Woodburne 2004). Mammals work well for correlations in the Cenozoic, but because Mesozoic mammals were small and therefore more restricted in their habitats, dinosaurs are beginning to have greater importance in the correlation of terrestrial Mesozoic strata (Jerzykiewicz and Russell 1991; Kirkland et al. 1998). Because dinosaurs are large, they theoretically should have dispersed rapidly across a variety of habitats. Ideally, the most robust biostratigraphic systems are based on multiple taxonomic groups, which maximizes the likelihood that suitable fossil material will be recovered (Kauffman et al. 1993; Eaton and Kirkland 2003). Land vertebrate ages are beginning to receive more respect as a potentially important tool in correlating terrestrial strata during the Mesozoic. Dinosaurs are limited in their biostratigraphic utility because they are rare, and at least a skull, if not an entire skeleton, is required to accurately identify them. Dinosaurs are nevertheless potentially important for terrestrial biostratigraphy because they evolved fairly rapidly, and not many genera of dinosaur are documented as ranging over more than a few million years (Wang and Dodson 2006). A notable exception is the Asian ceratopsian genus *Psittacosaurus*, which spans about the last 30 million years of the Lower Cretaceous (Lucas 2006). Thus, the occurrence of a dinosaur taxon at two different sites suggests a relatively robust correlation.

Magnetostratigraphy

With the recognition that the polarity of Earth's magnetic field reverses in a periodic but not cyclic manner, geologists began to use the residual record of magnetism preserved in rocks as a means of correlation. Initially, volcanic rocks were used in paleomagnetic studies because a strong magnetic signal is preserved when certain minerals align their magnetic domains with the Earth's magnetic field as the lava cools. Fine-grained sedimentary rocks also record the paleomagnetic polarity under which they were deposited because tiny magnetic grains align themselves with the Earth's magnetic field as they settle through the water column or form as part of cements lithifying the rock. For any relatively long interval, the pattern of the paleomagnetic record is unique, given a relatively constant rate of deposition. The paleomagnetic record for about the last 175 million years is relatively well established, as are several isolated intervals further back in time (Fig. 12.1). Magnetostratigraphy has a great advantage in being a global record and has been used to great effect in precise dating and correlation of terrestrial strata (e.g., Sankey and Gose 2001; Muttoni et al. 2004; Kent and Olsen 2008).

In order to date sedimentary rocks using magnetostratigraphy, many oriented rock samples are taken through the stratigraphic section to get a record of the pattern of normal and reversed magnetic polarity over a long interval of time. These data need to be calibrated using biostratigraphic and other chronostratigraphic methods and then compared with the known paleomagnetic record. The paleomagnetic record in the strata can be overprinted by thermal or chemical alteration. Additionally, there is an interval of normal polarity approximately 40 million years long during the middle of the Cretaceous Period in which magnetostratigraphy is essentially useless. This is known as the Cretaceous Long Normal (Fig. 12.1).

Cyclic and Event Stratigraphy

To some degree, uniformitarianism, while providing the critical key to developing the science of geology, put some blinders on geological thought in that many geologists equated it with gradualism—the idea that processes shaping the Earth are limited to those that occur on an average pleasant day. Derek Ager (1981) noted that much of the stratigraphic record resulted from short-term catastrophic events, such as hurricanes, floods, and landslides. He noted that events that are rare on a day-to-day basis are sure to occur over geologic time. This view of catastrophic uniformitarionism provides a basis for finer correlations of strata than are possible under a gradualistic view.

Stratigraphic events include both erosional and depositional events. Erosional events include the formation of laterally extensive storm deflation surfaces across beach profiles, or scours within river channels formed during flood events. Such erosive features are of limited utility beyond interpreting local depositional histories. Depositional events include local to regional features such as crevasse splays or sheet-flood deposits, turbidites or subaquatic debris flows, and tempestites or storm deposits (e.g., Einsele et al. 1991). Paleosols typically develop at the top of crevasse splays and provide an important visual clue for making correlations within floodplain sequences and are useful in constructing local geological histories because they are often repeated stratigraphically in the same environmental setting (e.g., Kraus 1999; Kirkland 2006). Note that when there is deposition on the floodplain, there is erosion within the adjoining river system; therefore, dinosaurs preserved in channel sandstones and those preserved in the adjoining finer-grained floodplain strata are unlikely to have lived at the same time.

Other geologic events have broader applications. Volcanic ash falls may be deposited over large regions in a variety of environments and have provided isochronous (time parallel) surfaces on which correlations based on other methodologies may be tested (e.g., Hattin 1985; Kauffman 1988; Laurin and Sageman 2007). Large impact events such as at the end of the Cretaceous have the potential to leave a global isochronous record.

Climatic cyclicity resulting from the cycles in Earth's orbital parameters has been recognized as controlling glacial episodes within our current ice age. These Milankovitch cycles have been termed the "pacemakers of the ice ages" (Hays et al. 1976). Variations in solar input in the northern and southern hemisphere are controlled by the precessional (wobble) and obliquity (tilt) cycles of the Earth's spin axis, and variations of solar input for the entire Earth are caused by fluctuations in the eccentricity of Earth's orbit (House and Gale 1996; Olsen and Whiteside 2008). The recognition of Milankovitch orbital parameters as the forcing mechanism for marine and terrestrial cyclic strata has provided a powerful tool for calibrating the geological timescale, because datable strata rarely occur in the immediate proximity of chronostratigraphic unit boundaries (Aziz et al. 2008; Barron et al. 1985; Sageman et al. 1998, 2006; D'Argenio et al. 2004; Olsen et al. 2005; Gradstein and Ogg 2007; Locklair and Sageman 2008). Within Upper Triassic lacustrine sequences, varves (thin sediment couplets formed from annual variations in sediment input and organic productivity) are also being used to calibrate Milankovitch cyclicity. Thus, research into cyclic stratigraphy is, in turn, providing an important method for fine tuning the long-term history of Earth's orbital parameters (Olsen and Whiteside 2008).

Although the stacking of fining-upward sequences such as tempestites and turbidites superficially resembles interbedded lithologies formed by orbital forcing, they are not; instead, they are formed by repeated random events such as storms and floods. Additionally, some repeated sedimentary sequences are autocyclic—that is, they repeat as a result of feedback mechanisms acting on the system. Delta lobe switching and river channel migration are autocyclic; both are controlled by sediment accumulation reaching a tipping point, resulting in a shift in the locus of sedimentation (delta lobe or channel), which eventually can return as sediment accumulation exceeds its tipping point laterally.

Chemostratigraphy

A new area of research that has considerable promise for intercontinental correlation is in the area of chemostratigraphy using stable isotopes of carbon. The stable isotopes of carbon are ^{13}C and ^{12}C. The proportions of ^{13}C to ^{12}C in Earth's atmosphere are reflected by their proportions in both marine carbonates and in buried organic carbon. The much more abundant ^{12}C is preferentially taken up by photosynthesis. Without polar icecaps, Mesozoic deep ocean water masses were sourced from warm, saline, poorly oxygenated, equatorial water, as opposed to modern deep water masses, which are sourced from cold, well-oxygenated, polar water. Periodically, during intervals of high planktonic productivity, the decomposition of organic material as it drifts down to the seafloor would consume all the oxygen in the water column making the world's deep oceans anoxic. During each global anoxic events, vast amounts of organic carbon were buried on the seafloor, resulting in a significant positive shift in the proportions of ^{13}C to ^{12}C in Earth's atmosphere that is mirrored in buried carbon globally (e.g., Pratt 1985; Weissert et al. 1998). Several global anoxic events in the middle Cretaceous have caused global positive shifts in the proportions of ^{13}C that are a particularly useful tool for chronostratigraphic correlations during the long Cretaceous interval of normal polarity (Kennedy et al. 2000; Gradstein et al. 2004).

Relative to dinosaur paleontology, the atmospheric record of stable carbon isotopes is also well recorded in terrestrially buried organic carbon (Hasegawa 1997; Grocke 1998, 2002; Heimhofer et al. 2003). A new area of investigation is the examination of stable isotopes from primary carbonate in paleosol carbonate nodules (Ludvigson et al. 2003, 2004, 2007, 2010; Baines et al. 2003; Koch et al. 2003). This research is significant for correlations in dinosaur paleontology because dinosaur bones are often preserved in environments with abundant carbonate paleosol nodules, whereas the alkalinity of the sediments in such environments precludes the preservation of pollen and spores. Because these stable isotope records of carbon reflect a global atmospheric signal, they represent a means of global correlation.

Basinal and subsurface correlations can be made by using gamma radiation as a proxy for the distribution of uranium and thorium in a stratigraphic section. Closely spaced readings of gamma radiation are plotted against their stratigraphic occurrence. Because gamma radiation is one of the tools petroleum geologists use to correlate rocks in the subsurface from well data, this methodology provides the opportunity for comparing subsurface information with data collected in outcrops (Zelt 1985; Leckie et al. 1997).

Dating and Correlating Dinosaur-Bearing Strata

With such a diversity of dating and correlation methodologies, one would think it would be simple to date dinosaur-bearing strata. In some cases, it is relatively easy, such as when radiometrically datable material is in intimate association with dinosaur fossils, or when dinosaur remains are found in marine strata. It is also relatively straightforward to determine the age of dinosaur localities in areas where there is interfingering of marine and terrestrial dinosaur-bearing strata, such as in the southwestern United States (Blakey and Ranney 2008). In other regions, such as in central Asia, where there is a mosaic of tectonically controlled isolated basins without any marine connections, dating dinosaur bearing strata is considerably more problematic.

Both central Asia and the southwestern United States preserve exceptionally complete dinosaur records, but the extraordinary richness of correlation tools in the southwestern United States permits a precise integration of its dinosaur record with the geologic timescale (Fig. 12.4). This provides opportunities to test new methodologies for correlation of terrestrial strata, which can then be applied to improving the temporal resolution in other regions of the world such as central Asia.

High-resolution event stratigraphic methodologies were developed to establish detailed reference sections for use in marine stratigraphy through trenching or coring the strata to collect all potential data that may be used for chronostratigraphic correlation (e.g., Kauffman 1988). These methodologies may also be applied to developing a robust framework of correlation for terrestrial strata (e.g., Olsen et al. 2003). It is important to apply as many lines of evidence as possible for use in correlation, so that various unrelated data may be used as a cross check for errors or biases.

The prospects for greater temporal resolution of Earth's dinosaur faunas are certain as these dating and correlation methodologies are more widely applied. Furthermore, it is nearly assured that new methods yet to be developed will add new dating and correlation tools for this purpose. Currently, it has been suggested that for portions of the Cenozoic, the resolution of the geologic timescale is only limited by chaos theory (Gradstein and Ogg 2007). The future of precise resolution of dinosaurian history is extremely hopeful.

Paleontology is a historical science. Consider studying world history from a book where the pages and chapters are mixed up. In understanding history, it is critical to know the sequence of events. It might be possible to develop a relative history of cause and result, but deeper meaning of what has gone on would be impossible to discern. Let us look at a couple examples where high-resolution temporal correlations may enhance dinosaur paleontology.

Importance of Knowing the Relative Ages of Dinosaur Faunas

Paleogeographical reconstructions can benefit from both robust phylogenetic hypotheses and precise temporal control. Much of the geographic history of the Mesozoic revolves around the breakup of the supercontinent Pangaea and the resulting increase in provinciality of terrestrial faunas. In the Northern Hemisphere, the opening of the Atlantic Ocean and the subsequent formation of the Alaskan land bridge, along with rising and falling sea levels, are the primary controls of the distribution of dinosaurs during the late Mesozoic.

Radiometric dates in the Mussentuchit Member of the Cedar Mountain Formation of central Utah fix the first appearance of dinosaur faunas characteristic of the Late Cretaceous at the base of the Upper Cretaceous (Fig. 12.4). Because many of these taxa are related to Early Cretaceous taxa known in central Asia, it has been hypothesized that the terrestrial Mussentuchian vertebrate fauna represents the first North American fauna with strong biogeographic ties to those of Asia and dates the advent of the Alaskan land bridge and the draining of Dawson or Midwestern Strait (Jeletzky 1970, 1984) of western Canada at the end of the Early Cretaceous (Cifelli et al. 1997; Kirkland et al. 1998).

Another example is derived from the discovery of the basal therizinosaur dinosaur *Falcarius* in the basal strata of the Cedar Mountain Formation in rocks that predate the development of the Alaskan land bridge (Fig. 12.4). Because most therizinosaurs are known from Asia, it had been assumed that their presence in North America in the Upper Cretaceous was the result of their direct migration into North America from Asia. However, the discovery of *Falcarius* in rocks essentially of the same age as those preserving the oldest therizinosaurs in Asia demonstrates that therizinosaurs must have been widely distributed across the Northern Hemisphere during the Early Cretaceous because Europe was the only connection between North America and Asia at that time (Kirkland et al. 2005). The proof of this hypothesis would be the future discovery of a therizinosaur in the Lower Cretaceous of Europe.

Ghost Lineages

If the remains of a dinosaur are found in rocks that formed during the Turonian Age of the Cretaceous Period, it follows that that taxon of dinosaur lived during that time interval. A more difficult question is how one should interpret the absence of a fossil of a particular time in rocks of a given age.

A classic problem along these lines has to do with evolutionary origins of birds such as *Archaeopteryx*. Phylogenetic analysis strongly supports that the closest relatives of birds are among the feathered, maniraptoran theropod dinosaurs (e.g., Kirkland et al. 2005; Turner et al. 2007). However, most maniraptorans occur in rocks younger than the Upper Jurassic in which *Archaeopteryx* occurs. A few paleontologist have taken this (along with other lines of evidence) to mean that theropods are unlikely to have been the ancestors of birds and that similarities between the groups represent evolutionary convergence rather than a close relationship (Hou et al. 1996; Feduccia 1996).

Such a conclusion assumes that our knowledge of when various dinosaurs lived is fairly complete; it takes the geological occurrence of dinosaur taxa at close to face value. On the other hand, if dromaeosaurs or other maniraptorans really are the closest relatives (the sister taxon) of birds, as indicated by most phylogenetic studies, then the theropods in question must have originated earlier than the Late Jurassic when *Archaeopteryx* existed. Such hypothetical temporal range extension of groups of organisms prior to their earliest known fossil occurrences are known as ghost lineages (e.g., Norell and Novacek 1992a, 1992b; Weishampel and Heinrich 1992; Norell 1993; Benton and Storrs 1994, 1996; Weishampel 1996). If a particular taxon split off earlier than its presently known first occurrence in the fossil record, then fossil hunting in rocks where its lineage is predicted to occur could eventually reveal that the group did in fact exist earlier as predicted. Stratigraphic congruence between geological data and phylogenetic hypotheses happens when there are few, short ghost lineages, which indicates that there is good temporal support for the phylogenetic hypothesis. There is relatively good congruence within most dinosaur lineages. The many recent discoveries of tiny maniraptoran dinosaurs at the feathered dinosaur sites in the middle Lower Cretaceous Yixian Formation, Liaoning, China, suggests that there may be many tiny, fragile maniraptoran dinosaurs that would rarely be preserved as fossils (e.g., Chan et al. 2004). The recent discovery of a derived maniraptoran theropod in the Upper Jurassic Morrison Formation (Hartman et al. 2005) does much to minimize these ghost lineages, and the subsequent discovery of an upper Middle Jurassic maniraptoran in China (Xu et al. 2009) eliminates many of them.

Another example is the recently developed phylogenetic hypothesis that the presence of an accessory antorbital fenestra in the Campanian *Bagaceratops* and *Magnirostris* among Asian ceratopsians indicates that

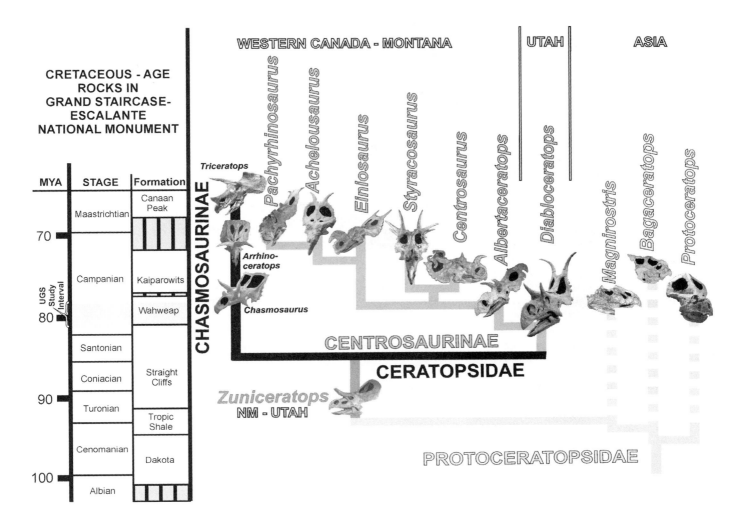

CRETACEOUS - AGE ROCKS IN GRAND STAIRCASE-ESCALANTE NATIONAL MONUMENT

MYA	STAGE	Formation
	Maastrichtian	Canaan Peak
70		
	Campanian	Kaiparowits
80		Wahweap
	Santonian	Straight Cliffs
	Coniacian	
90	Turonian	
		Tropic Shale
	Cenomanian	Dakota
100		
	Albian	

UGS Study Interval

WESTERN CANADA - MONTANA   UTAH   ASIA

Triceratops — *Pachyrhinosaurus* — *Achelousaurus* — *Einiosaurus* — *Styracosaurus* — *Centrosaurus* — *Albertaceratops* — *Diabloceratops* — *Magnirostris* — *Bagaceratops* — *Protoceratops*

Arrhino-ceratops

Chasmosaurus

CHASMOSAURINAE

CENTROSAURINAE

CERATOPSIDAE

Zuniceratops
NM - UTAH

PROTOCERATOPSIDAE

they are the closest sister taxa to the North American middle Turonian horned ceratopsian *Zuniceratops*, because this feature is also recognized in *Zuniceratops* and in later primitive ceratopsids (Kirkland and DeBlieux 2009). This phylogenetic hypothesis results in a ghost lineage spanning the approximately 10 million years needed to provide a potential ancestor for both *Bagaceratops* and *Zuniceratops* (Fig. 12.5).

The different contributors to this book differ in how much weight they are willing to give to the earliest known temporal occurrences of particular kinds of dinosaurs, as opposed to phylogenetic predictions of when these groups might have originated. The reader should keep this in mind in interpreting the authors' hypotheses of relationships among different kinds of dinosaurs. However, if a suspect fossil results in a phylogenetic hypothesis that leads to numerous significant extensions of ghost lineages, it should be considered suspect.

12.5. Phylogenetic hypothesis of the relationships among the crown Ceratopsia. Ghost lineages for Campanian Asian protoceratopsids are indicated with dashes.

Modified after Kirkland and DeBlieux (2009).

Albritton, C. C., Jr. 1986. *The Abyss of Time: Unraveling the Mystery of the Earth's Age.* Los Angeles: Jeremy P. Tarcher.

Ager, D. V. 1981. *The Nature of the Stratigraphic Record.* New York: John Wiley and Sons.

Andrews, R. C. 1926. *On the Trail of Ancient Man.* New York: G. P. Putnam's Sons.

References

Armstrong, R. L. 1978. Pre-Cenozoic Phanerozoic time scale – computer file of critical dates and consequences of new and in progress decay-constant revision. In G. V. Cohee, M. F. Glaessner, and H. D. Hedberg (eds.), *Contributions to the Geological Timescale: Papers Given at the Geological Time Scale Symposium 106.6, 25th IGC Sydney, Australia, August 1976*. American Association of Petroleum Geologists Studies in Geology 6: 73–91.

Aziz, H. A., F. J. Hilgen, G. M. van Luijk, A. Sluijs, M. J. Kraus, J. M. Pares, and P. D. Gingerich. 2008. Astronomical climate control on paleosols stacking patterns in the upper Paleocene–lower Eocene Willwood Formation, Bighorn Basin, Wyoming. *Geology* 36: 531–534.

Baines, S. R. D., Norris, R. M. Corfield, G. J. Bowen, P. D. Gingerich, and P. L. Koch. 2003. Marine–terrestrial linkages at the Paleocene–Eocene boundary. In S. L. Wing, P. D. Gingerich, B. Schmitz, and E. Thomas (eds.), *Causes and Consequences of Globally Warm Climates in the Early Paleogene*. Geological Society of America Special Paper 369: 1–9.

Barron, E. J., M. A. Arthur, and E. G. Kauffman. 1985. Cretaceous rhythmic bedding sequences: a plausible link between orbital variations and climate. *Earth and Planetary Sciences Letters* 72: 327–340.

Benton, M. J., and G. W. Storrs. 1994. Testing the quality of the fossil record: paleontological knowledge is improving. *Geology* 22: 111–114.

——. 1996. Diversity in the past: comparing cladistic phylogenies and stratigraphy. In M. E. Hochberg, J. Clobert, and R. Barbault (eds.), *Aspects of the Genesis and Maintenance of Biological Diversity*, 19–40. Oxford: Oxford University Press.

Berry, W. 1987. *Growth of a Prehistoric Time Scale: Based on Organic Evolution*. Rev. ed. Palo Alto, Calif.: Blackwell Scientific Publications.

Blakey, R., and W. Ranney. 2008. *Ancient Landscapes of the Colorado Plateau*. Grand Canyon, Ariz.: Grand Canyon Association.

Bloos, G., and K. N. Page. 2002. Global stratotype section and point for base of the Sinemurian Stage (Lower Jurassic). *Episodes* 25: 22–28.

Boggs, S., Jr. 1995. *Principles of Sedimentology and Stratigraphy*. New York: Freeman.

Chan, M., P. Chen, Y. Wang, and Y. Wang (eds.). 2004. *The Jehol Biota: The Emergence of Feathered Dinosaurs, Beaked Birds, and Flowering Plants*. Shanghai: Shanghai Scientific and Technical Publishers.

Cifelli, R. L., J. I. Kirkland, A. Weil, A. R. Deinos, and B. J. Kowallis. 1997. High-precision $^{40}Ar/^{39}Ar$ geochronology and the advent of North America's Late Cretaceous terrestrial fauna. *Proceedings of the National Academy of Sciences of the United States of America* 94: 11163–11167.

Dalrymple, G. B. 1994. *The Age of the Earth*. Stanford: Stanford University Press.

D'Argenio, B., A. G. Fischer, S. I. Premoli, H. Weissert, and V. Ferreri (eds.). 2004. *Cyclostratigraphy. An Essay of Approaches and Case Histories*. Society for Sedimentary Geology Special Publication 81.

Dickenson, W. R., and G. E. Gehrels. 2003. U-Pb ages of detrital zircons from Permian and Jurassic eolian sandstones of the Colorado Plateau, USA: paleogeographic implications. *Sedimentary Geology* 163: 29–66.

——. 2009. U-Pb ages of detrital zircons in Jurassic eolian and associated sandstones of the Colorado Plateau: evidence for transcontinental dispersal and intraregional recycling of sediment. *Geological Society of America Bulletin* 121: 408–433.

Dott, R. H., Jr., and D. Prothero. 2003. *Evolution of the Earth*. 7th ed. New York. McGraw-Hill.

Eaton, J. G., and J. I. Kirkland. 2003. Nonmarine Cretaceous vertebrates of the Western Interior Basin. In P. J. Harries (ed.), *High-Resolution Approaches in Stratigraphic Paleontology*, 263–313. Topics in Geobiology 21. Boston: Kluwer Academic Publishers.

Einsele, G., W. Ricken, and A. Seilacher (eds.). 1991. *Cycles and Events in Stratigraphy*. Berlin: Springer-Verlag.

Emery, D., and K. J. Myers. 1996. *Sequence Stratigraphy*. Palo Alto, Calif.: Blackwell Science.

Feduccia, A. 1996. *The Origin and Evolution of Birds*. New Haven, Conn.: Yale University Press.

Furin, S., N. Preto, M. Rigo, G. Roghi, P. Gianolla, J. I. Crowley, and S. A. Bowring. 2006. High-precision U-Pb zircon age from the Triassic of Italy: implications for the Triassic time scale and the Carnian origin of calcareous nannoplankton and dinosaurs. *Geology* 34: 1009–1012.

Gehrels, G. E. 2000. Introduction to detrital zircon studies of Paleozoic and Triassic strata in western Nevada and northern California. In M. J. Soregham and G. E. Gehrels (eds.), *Paleozoic and Triassic Paleogeography and Tectonics of Western Nevada and Northern California*. Geological Society of America Special Paper 347: 1–17.

Gradstein, F. M., and J. G. Ogg, 2007. Geologic time scale 2004 – why, how, and where next! *Lethaia* 37 (2): 175–181.

Gradstein, F. M., J. G. Ogg, and A. G. Smith (eds.). 2004. *A Geologic Time Scale, 2004*. Cambridge: Cambridge University Press.

Grocke, D. R. 1998. Carbon isotope analyses of fossil plants as a chemostratigraphic and palaeoenvironmental tool. *Lethaia* 31: 1–13.

——. 2002. The carbon isotope composition of ancient CO_2 based on higher-plant organic matter. *Philosophical Transactions of the Royal Society of London* (ser. A) 360: 633–658.

GSA [Geological Society of America]. 1999. GSA timescale. *Geological Society of America*. http://www.geosociety.org/science/timescale/timescl.pdf.

Hallam, A. 1992. *Phanerozoic Sea-Level Changes*. New York: Columbia University Press.

Hancock, J. M., A. S. Gale, S. Gardin, et al. 1996. The Campanian Stage. *Bulletin de l'Institut Royal des Dciences Naturelles de Belgique, Sciences de la Terre* 63: 133–148.

Harland, W. B., A. V. Cox, P. G. Llewellyn, et al. 1982. *A Geological Time Scale*. Cambridge: Cambridge University Press.

Hartman, S., D. Lovelace, and W. Wahl. 2005. Phylogenetic assessment of a maniraptoran from the Morrison Formation. *Journal of Vertebrate Paleontology* 25 (suppl. to 3): 67A–68A.

Hasegawa, T. 1997. Cenomanian–Turonian carbon-isotope events recorded in terrestrial organic matter from northern Japan. *Palaeogeography, Palaeoclimatology, Palaeoecology* 130: 251–273.

Hattin, D. E. 1985. Distribution and significance of widespread, time parallel pelagic limestone beds in the Greenhorn Limestone (Upper Cretaceous) of the central Great Plains and southern Rocky Mountains. In L. M. Pratt, E. G. Kauffman, and F. B. Zelt (eds.), *Fine-Grained Deposits and Biofacies of the Cretaceous Western Interior Seaway: Evidence of Cyclic Sedimentary Processes*. Society of Economic Paleontologists and Mineralogists Fieldtrip Guidebook 4: 28–37.

Hays, J. D., J. Imbrie, and N. J. Shackleton. 1976. Variations in the Earth's orbit: pacemakers of the ice ages. *Science* 194: 1131–1133.

Heimhofer, U., P. A. Hochuli, S. Burla, N. Andersen, and H. Weissert. 2003.

Terrestrial carbon-isotope records from coastal deposits (Algarve, Portugal): a tool for chemostratigraphic correlation on an intrabasinal and global scale. *Terra Nova* 15: 8–13.

Holmes, A. 1937. *The Age of the Earth*. London: Nelson.

——. 1947. *The Construction of a Geological Time Scale*. Transactions of the Geological Society of London, Special Report 8.

——. 1959. A revised geological timescale. *Transactions of the Geological Society of Glasgow* 21: 183–216.

Hou, H., L. D. Martin, Z. Zhou, and A. Feduccia. 1996. Early adaptive radiation of birds: evidence from fossils in northeastern China. *Science* 274: 1164–1167.

House, M. R., and A. G. Gale (eds.). 1996. *Orbital Forcing Timescales and Cyclostratigraphy*. The Geological Society (London) Special Publication 85.

International Commission on Stratigraphy. 2006. International stratigraphic chart. *International Union of Geological Sciences*. http://stratigraphy.org/cheu.pdf.

Jeletzky, J. A. 1970. Cretaceous macrofossils: biochronology chapter, pt. V. In R. W. Douglas (ed.), *Geological Survey of Canada Paper* 67-72: 1–66.

——. 1984. Jurassic–Cretaceous boundary beds of western and Arctic Canada and the problem of the Tithonian–Berrimian stages in the Boreal Realm. In G. E. G. Westermann (ed.), *Jurassic-Cretaceous Biochronology and Paleogeography of North America*. Geological Association of Canada Special Paper 27: 175–255.

Jerzykiewicz, T., and D. A. Russell. 1991. Late Mesozoic stratigraphy and vertebrates of the Gobi Basin. *Cretaceous Research* 12: 345–377.

Kauffman, E. G. 1988. Concepts and methods of high-resolution event stratigraphy. *Annual Review of Earth and Planetary Sciences* 16: 605–654.

Kauffman, E. G., B. B. Sageman, J. I. Kirkland, W. P. Elder, P. J. Harries, and T. Villamil. 1993. Molluscan biostratigraphy of the Western Interior Cretaceous Basin, North America. In W. G. E. Caldwell and E. G. Kauffman (eds.), *Evolution of the Western Interior Basin*. Canadian Association of Geologists Special Paper 39: 397–434.

Kennedy, W. J., and W. A. Cobban. 1976. Aspects of ammonite biology, biogeography, and biostratigraphy. Palaeontological Association Special Papers in Palaeontology 17: 1–94.

Kennedy, W. J., A. S. Gale, J. A. Lees, and M. Caron. 2004. The global boundary stratotype section and point (GSSP) for the base of the Cenomanian Stage, Mont Risou, Haute-Alpes, France. *Episodes* 27: 21–32.

Kennedy, W. J., I. Walaszczyk, and W. A. Cobban. 2000. Pueblo Colorado, USA, candidate global boundary stratotype section and point for the base of the Turonian Stage of the Cretaceous and for the Middle Turonian substage, with a revision of the Inoceramidae (Bivalvia). *Acta Geologica Polonica* 50 (3): 295–334.

——. 2005. The global boundary stratotype section and point for the base of the Turonian Stage of the Cretaceous: Pueblo, Colorado, USA. *Episodes* 28: 93–104.

Kent, D. V., and P. E. Olsen. 2008. Early Jurassic magnetostratigraphy and paleolatitudes from the Hartford continental rift basin (eastern North America): testing for polarity bias and abrupt polar wander in association with the central Atlantic magmatic province. *Journal of Geophysical Research* 113, B06105.

Kirkland, J. I. 2006. Fruita Paleontological Area (Upper Jurassic, Morrison Formation), western Colorado: an example of terrestrial taphofacies analysis. In J. R. Foster and S. G. Lucas (eds.), *Paleontology and Geology of the Upper Jurassic Morrison Formation*. New Mexico Museum of Natural History and Science Bulletin 36: 67–95.

Kirkland, J. I., and D. D. DeBlieux. 2009. Centrosaurine ceratopsids from the middle Campanian Wahweap Formation, Grand Staircase–Escalante National Monument, southern Utah. In M. J. Ryan, B. J. Chinnery-Allgeier, D. A. Eberth, and P. E. Ralrick (eds.), *New Perspectives on Horned Dinosaurs*, 117–140. Bloomington: Indiana University Press.

Kirkland, J. I., S. G. Lucas, and J. W. Estep. 1998. Cretaceous dinosaurs of the Colorado Plateau. In S. G. Lucas, J. I. Kirkland, and J. W. Estep (eds.), *Lower to Middle Cretaceous Non-Marine Cretaceous Faunas*. New Mexico Museum of Natural History and Science Bulletin 14: 67–89.

Kirkland, J. I., and S. K. Madsen. 2007. The Lower Cretaceous Cedar Mountain Formation, eastern Utah: the view up an always interesting learning curve. In W. R. Lund (ed.), *Field Guide to Geological Excursions in Southern Utah*. Geological Society of America Rocky Mountain Section 2007 Annual Meeting, Grand Junction Geological Society. Utah Geological Association Publication 35: 1–108.

Kirkland, J. I., L. E. Zanno, S. D. Sampson, J. M. Clark, and D. D. DeBlieux. 2005. A primitive therizinosauroid dinosaur from the Early Cretaceous of Utah. *Nature* 435 (7038): 84–87.

Koch, P. L., W. C. Clyde, R. P. Hepple, M. L. Fogel, S. L. Wing, and J. C. Zachos. 2003. Carbon and oxygen isotope records from paleosols spanning the Paleocene–Eocene boundary, Bighorn Basin, Wyoming. In S. L. Wing, P. D. Gingrich, B. Schmitz, and E. Thomas (eds.), *Causes and Consequences of Globally Warm Climates in the Early Paleogene*. Geological Society of America Special Paper 369: 49–64.

Kraus, M. J. 1999. Paleosols in clastic sedimentary rocks: their geologic applications. *Earth Science Reviews* 47: 41–70.

Kulp, J. L. 1961. Geological time scale. *Science* 133: 1105–1114.

Laurin, J., B., and B. Sageman. 2007. Cenomanian–Turonian coastal record in SW Utah, USA: orbital-scale transgressive–regressive events during oceanic anoxic event II. *Journal of Sedimentary Research* 77: 731–756.

Leckie, R. M., J. I. Kirkland, and W. P. Elder. 1997. Stratigraphic framework and correlation of a principle reference section of the Mancos Shale (Upper Cretaceous), Mesa Verde, Colorado. In O. J. Anderson, B. S. Kues, and S. G. Lucas (eds.), *Mesozoic Geology and Paleontology of the Four-Corners Region*. New Mexico Geological Society Guidebook 1997: 163–216.

Locklair, R. E., and B. B. Sageman. 2008. Cyclostratigraphy of the Upper Cretaceous Niobrara Formation, Western Interior, USA: a Coniacian–Santonian orbital timescale. *Earth and Planetary Science Letters* 269: 540–553.

Lucas, S. G. 2006. The *Psittacosaurus* biochron. Early Cretaceous of Asia. *Cretaceous Research* 27: 189–198.

Ludvigson, G. A., L. A. González, R. M. Joeckel, A. Al-Suwaidi, J. I. Kirkland, and S. Madsen. 2007. Carbonate carbon isotope excursions in the Ruby Ranch Member of the Cedar Mountain Formation and their terrestrial paleoclimatic impact. *Geological Society of America Abstracts with Programs* 39 (5): 33.

Ludvigson, G. A., L. A. González, R. M. Joeckel, S. J. Carpenter, and J. I. Kirkland. 2004. Stable isotopic records of the terrestrial paleoclimatic consequences of mid-Cretaceous (Aptian–Albian) oceanic anoxic events [abstract 0584]. *American Geophysical Union, Fall Meeting Program*, 41A: 302.

Ludvigson, G. A., L. A. González, J. I.

Kirkland, and R. M. Joeckel. 2003. Terrestrial carbonate records of the carbon isotope excursions associated with mid-Cretaceous (Aptian–Albian) oceanic anoxic events. *Geological Society of America Abstracts with Programs* 35 (7): 289.

Ludvigson, G. A., R. M. Joeckel, L. A. González, E. I. Gulbranson, E. T. Rasbury, G. J. Hunt, J. I., Kirkland, and S. Madsen. 2010. Correlation of Aptian–Albian carbon isotope excursions in continental strata of Cretaceous Foreland Basin of eastern Utah. *Journal of Sedimentary Research* 80: 955–974.

Matthew, W. D. 1915. Methods of correlation by fossil vertebrates. *Geological Society of America Bulletin* 27: 515–524.

Meister, C., M. Aberhan, J. Blau, J.-L. Dommergues, S. Feist-Burkhardt, E. A. Hailwood, M. Hart, S. P. Hesselbo, M. W. Hounslow, M. Hylton, N. Morton, K. Page, and G. D. Price. 2006. The global boundary stratotype section and point (GSSP) for the base of the Pliensbachian Stage (Lower Jurassic), Wine Haven, Yorkshire, UK. *Episodes* 29: 93–106.

Miall, A. D. 1997. *The Geology of Stratigraphic Sequences.* Berlin: Springer-Verlag.

———. 2006. *The Geology of Fluvial Deposits: Sedimentary Facies, Basin Analysis, and Petroleum Geology.* Berlin: Springer-Verlag.

Molina, E., L. Alegret, I. Arenillas, J. A. Arz, N. Gallala, J. Hardenbol, K. von Salis, E. Steurbaut, N. Vandenberghe, and D. Zaghib-Turki. 2006. The global boundary stratotype section and point for the base of the Danian Stage (Paleocene, Paleogene, "Tertiary," Cenozoic) at El Kef, Tunisia – original definition and revision. *Episodes* 29: 263–273.

Muttoni, G., D. V. Kent, P. E. Olsen, P. D. Stefano, W. Lowrie, S. M. Bernasconi, and F. M. Hernandez. 2004. Tethyan magnetostratigraphy from Pizzo Mondello (Sicily) and correlation to the Late Triassic Newark astrochronological polarity time scale. *Geological Society of America Bulletin* 116: 1043–1058.

Norell, M. A. 1993. Tree-based approaches to understanding history: comments on rank, rules, and quality of the fossil record. *American Journal of Science* 293A: 407–417.

Norell, M. A., and M. J. Novacek. 1992a. Congruence between superpositional and phylogenetic patterns: comparing cladistic patterns with fossil records. *Cladistics* 8: 319–337.

———. 1992b. The fossil record and evolution: comparing cladistic and paleontologic evidence for vertebrate history. *Science* 255: 1690–1693.

Obradovich, J. D. 1993. A Cretaceous time scale. In W. G. E. Caldwell and E. G. Kauffman (eds.), *Evolution of the Western Interior Basin.* Geological Association of Canada Special Paper 39: 379–396.

Olsen, P. E., D. V. Kent, M. Et-Touhami, and J. H. Puffer. 2003. Cyclo-, magneto-, and bio-stratigraphic constraints on the duration of the CAMP event and its relationship to the Triassic–Jurassic boundary. In W. E. Hames, J. G. McHone, P. R. Renne, and C. Ruppel (eds.), *The Central Atlantic Magmatic Province: Insights from Fragments of Pangea.* Geophysical Monograph Series 136: 7–32.

Olsen, P. E., and J. H. Whiteside. 2008. Pre-Quaternary Milankovitch cycles and climate variability. In V. Gornitz (ed.), *Encyclopedia of Paleoclimatology and Ancient Environments*, 826–835. New York: Springer.

Olsen, P. E., J. H. Whiteside, P. M. LeTourneau, and P. Huber. 2005. Jurassic cyclostratigraphy and paleontology of the Hartford basin. In B. J. Skinner and A. R. Philpotts (eds.), *97th New England Intercollegiate Geological Conference*, A4-1–A4-51. New Haven, Conn.: Department of Geology and Geophysics, Yale University.

Pratt, L. M. 1985. Isotopic studies of organic matter and carbonate in rocks of the Greenhorn Marine Cycle. In L. M. Pratt, E. G. Kauffman, and F. B. Zelt (eds.), *Fine-Grained Deposits and Biofacies of the Cretaceous Western Interior Seaway: Evidence of Cyclic Sedimentary Processes.* Society of Economic Paleontologists and Mineralogists Fieldtrip Guidebook 4: 38–48.

Prokopiev, A. V., J. Toro, E. L. Miller, and G. E. Gehrels. 2008. The paleo–Lena River – 200 m.y. of transcontinental zircon transport in Siberia. *Geology* 36: 699–702.

Prothero, D. R. 2007. *Evolution: What the Fossils Say and Why It Matters:* New York: Columbia University Press.

Rasbury, E. T., G. N. Hanson, W. J. Meyers, W. E. Holt, R. H. Goldstein, and A. H. Saller. 1998. U-Pb dates of paleosols; constraints on late Paleozoic cycle durations and boundary ages. *Geology* 26: 403–406.

Rudwick, M. J. S. 1976. *The Meaning of Fossils: Episodes in the History of Paleontology.* 2nd ed. New York: Watson Academic Publications.

Sageman, B. B., S. R. Meyers, and M. A. Arthur. 2006. Orbital timescale and new C-isotope record for Cenomanian–Turonian boundary stratatype. *Geology* 34: 125–128.

Sageman, B. B., J. Rich, M. A. Arthur, W. E. Dean, C. E. Savarda, and T. J. Bralower. 1998. Multiple Milankovitch cycles in the Bridge Creek Limestone (Cenomanian–Turonian), Western Interior Basin. *SEPM Concepts in Sedimentology and Paleontology* 6: 153–171.

Sankey, J. T., and W. A. Gose. 2001. Late Cretaceous mammals and magnetostratigraphy, Big Bend, Texas. *Occasional Papers of the Museum of Natural Science* (Louisiana State University) 77: 1–16.

Shanley, K. W., and P. J. McCabe. 1994. Perspectives on the sequence stratigraphy of continental strata. *American Association of Petroleum Geologists Bulletin* 78: 544–568.

Stanley, S. M. 2009. *Earth Systems History.* New York: W. H. Freeman.

Turner, A. H., D. Pol, J. A. Clarke, G. M. Erickson, and M. A. Norell. 2007. A basal dromaeosaurid and size evolution preceding avian flight. *Science* 317: 1378–1381.

U.S. Geological Survey Geologic Names Committee. 2007. Divisions of geologic time – major chronostratigraphic and geochronologic units. Fact sheet. *U.S. Geological Survey.* http://pubs.usgs.gov/fs/2007/3015/.

Vail, P. R., R. M. Mitchum Jr., and S. Thompson III. 1977. Seismic stratigraphy and global changes in sea level, part 4: global cycles of relative changes of sea level. In C. E. Payton (ed.), *Seismic Stratigraphy.* American Association of Petroleum Geologists Memoir 26: 83–97.

Van Eysinga, F. W. B. (comp.). 1975. *Geological Timetable.* 3rd ed. Amsterdam: Elsevier.

Wang, S. C., and P. Dodson. 2006. Estimating the diversity of dinosaurs. *Proceedings of the National Academy of Sciences of the United States of America* 103 (37): 13601–13605.

Weishampel, D. B. 1996. Fossils, phylogeny, and discovery: a cladistic study of the history of tree topologies and ghost lineage durations. *Journal of Paleontology* 16: 191–197.

Weishampel, D. B., and R. E. Heinrich. 1992. Systematics of Hypsilophodontidae and basal Iguanodontia (Dinosauria: Ornithopoda). *Historical Biology* 6: 159–184.

Weissert, H., A. Lini, K. B. Föllmi, and O. Kuhn. 1998. Correlation of Early Cretaceous carbon-isotope stratigraphy and platform drowning events: a possible link? *Palaeoecology, Paaleogeography, Palaeoclimatology* 137: 189–203.

Winchester, S. 2001. *The Map that Changed*

the World: William Smith and the Birth of Modern Geology. New York: Harper Collins.

Woodburne, M. O. (ed.). 2004. *Late Cretaceous and Cenozoic Mammals of North America – Biostratigraphy and Geochronology.* New York: Columbia University Press.

Xu, X., Q. Zhao, M. Norell, C. Sullivan, D. Hone, G. Erickson, X. Wang, F. Han, and Y. Guo. 2009. A new feathered maniraptoran dinosaur fossil that fills a morphological gap in avian origin. *Chinese Science Bulletin* 54: 430–435.

Zelt, F. G. 1985. Paleoceanographic events and lithologic/geochemical facies of the Greenhorn Marine Cycle (Upper Cretaceous) examined using natural gamma-ray spectrometry. In L. M. Pratt, E. G. Kauffman, and F. B. Zelt (eds.), *Fine-Grained Deposits and Biofacies of the Cretaceous Western Interior Seaway: Evidence of Cyclic Sedimentary Processes.* Society of Economic Paleontologists and Mineralogists Fieldtrip Guidebook 4: 49–59.

13.1. Paleontologist in the field using handheld GPS (see Breithaupt et al. 2001, 2004; Matthews and Breithaupt 2001; Matthews et al. 2006). Image copyright Brent Breithaupt and Neffra Matthews.

Technology and the Study of Dinosaurs

13

Ralph E. Chapman, Art Andersen,
Brent H. Breithaupt, and Neffra A. Matthews

It would seem to be the ultimate oxymoron. What in the world could technology do with dinosaurs? Although we cannot totally exclude the possibility that some dinosaurs used primitive technology—some avian dinosaurs are known to have used tools in their foraging behavior (e.g., Millikan and Bowman 1967)—in this chapter we will review the many ways that technology greatly improves and expands what paleontologists and related professionals can do with dinosaurs. This version of the chapter differs tremendously from that of the first edition (Chapman 1997) because the application of technology to the study of dinosaurs has exploded over the past decade and a half. Gone are discussions of analytical methods in favor of a detailed discussion of the technology and how it has been applied to dinosaurs.

The importance of the application of technology by paleontologists to dinosaurs and their fossils cannot be overemphasized. Technology can improve, sometimes greatly, our capabilities to locate fossils, excavate and prepare them better, maintain collections of them more efficiently and safely, study them in much more detail and with many more possibilities, and present them to the public in a more effective and exciting manner. In short, technology has allowed dinosaur paleontology to expand in ways never before seen.

Paleontologists and their associates interact with dinosaurs in five ways: finding them in the field, preparing them in the laboratory after they have been removed from the field, including them in museum collections, studying them within the vast range of possible research topics, and sharing them with students and the public during outreach. We will review how technology has affected—and probably will affect—each of these elements. The application of recent technology has affected each phase of dinosaur work in a different way and to a different degree.

In the Field

Finding dinosaur fossils in the field is of the utmost importance because without them, we would have nothing to study, keep in collections, or present to the public. Consequently, finding them in the field is of the highest priority to dinosaur paleontology. Until relatively recently, paleontologists used the same basic technology for doing this necessary fieldwork that they have been using for more than a century—in fact, that technology is still important for the work they do. Mostly with maps and compass, and materials for extraction, the paleontologists would go into the field, find fossils, do his or her best at locating the find on the available maps, and then take whatever notes possible on the geological context within which the fossils exist. Then he or she would remove them and take them to the preparation laboratory. This approach evolved slowly as the maps and compasses improved, vehicles made it easier to keep track of where the paleontologist was on the ground, and airplanes allowed for aerial reconnaissance. However, the last 30 years has seen the introduction of new technology that is totally changing how the paleontologist interacts with the field and finds fossils.

Prospecting has traditionally started in the laboratory through the study of geological maps for potential field areas. The paleontologist uses these maps to locate the formations that can represent the proper environments that are of the right age, that are exposed at the surface, and that have a reasonable geological structure and landscape morphology to provide potential outcrops for prospecting. This is still incredibly important, but the paleontologist now has even more help during this part of the prospecting phase. This comes in the form of geographic information systems (GIS). These are complex databases that provide access to vast quantities of stored information on geological context, along with data for environmental, landscape, and biological parameters. The variables involved can then be studied, and the combinations that provide the most potential for identifying productive quarries for dinosaurs can be inferred for areas not yet explored. GIS laboratories have been established at most major universities, and most paleontologists may now access them, if they are at all motivated. In its more detailed form, GIS can be used to predict many aspects of fossil abundance in well-known formations, as was done by Oheim and Hall (2005) for the Upper Cretaceous Two Medicine Formation (see also Matthews et al. 2006).

The possibilities are expanded further through input from earth-imaging satellites and the power of remote sensing–that is, obtaining data about an object or landscape without actually being in contact with it. Here, satellite photographs, multiband digital imagery, multispectral and hyperspectral imagery, radar, lidar, ifsar, sonar, and other approaches provide data for the landscape, with resolutions down to less than a meter. Variations in these data can provide accurate input about surface and even subsurface geology. This technology has mostly been used for mineral exploration, geohydrology, and other economic disciplines, but the potential for its use in exploration for vertebrate paleontology is considerable (see, for example, Lillesand et al. 2004; Matthews et al. 2006; Kaye and Cavigelli 2005).

Another useful development is the assembly of digital libraries of aerial photographs for some private areas, but mostly the public lands governed by the Bureau of Land Management and the United States Forest Services, as well as other federal and state agencies. These images and those funded by private industry are now widely available through streaming Web map services. These services make imagery available to many people. Important measurements can be made with these photographs through the science and art of photogrammetry (Mikhail et al. 2001; Matthews et al. 2006).

This digitization of these resources traditionally used by paleontologists, combined with the new data generated by satellite imagery and GIS, has made the first phase of prospecting for fossils much more powerful. Combine this with the expansion of paleontologists into countries and regions not previously accessible, and the result is the unprecedented increase in productivity we see today in finding new dinosaur material. This should not only continue in the near future, but should increase even further as these new resources are more widely applied.

Can technology assist in the second phase of prospecting for dinosaurs—actually looking at the ground and finding the fossils? Yes, although much of this potential has yet to be realized at this time. This includes locating possible areas identified in the first phase, keeping track of where you are in the field in real time, mapping the quarry/area in detail, including bone and/or footprint orientations, and keeping track of the fossils themselves.

Once you know where you want to go to prospect for fossils, you have to actually go there and know when you've gotten there. Further, when you run into fossils, you need to know where you are located with a high level of accuracy. This has frequently been a significant source of error, and in the past, when dealing with paper maps, it was almost inevitably difficult to do well and required a lot of hard work. We still must work with maps in some format—electronic maps are now commonly used in the field—but global positioning systems (GPS) have helped ease the work and reduce errors considerably (Fig. 13.1). This was pioneered by the crew of paleontologists from the American Museum of Natural History when they returned to the Mongolian Gobi Desert in 1990 (McKenna 1992). As they traveled around Mongolia trying to identify old quarries from the days of Roy Chapman Andrews, as well as trying to find new dinosaur fossils, there was little to help the expedition identify where they were among the monotonous dunes. GPS helped them save time and find a lot more dinosaurs. GPS use coded data from an overlapping network of earth-orbiting navigational satellites. When a sufficient number of these satellites is available, the position of the receiver can be located. The accuracy is dependent on the precision of the GPS's clock. For most current systems, the accuracy will be within a meter or so of the actual position—far better than was possible with more conventional systems, and all located in a modestly priced handheld device (see Owings 2005).

Once you have an accurate location of a site, there are many other pieces of data that should be taken in the field, if possible. Of importance is the ability to collect accurate information and to ensure that that information stays available and associated with the fossils. The promulgation of handheld and other transportable computers allows notes and other information to be recorded directly into electronic media, with easy backup both on site and through communications. This is incredibly important because many of the mistakes made through the years for field areas have occurred through the transcription of field notes. Further, field notebooks, still an essential part of the process, have been lost, stolen, or damaged by fire or water. It makes more sense to send the information directly back to the lab as a backup. More than one field season has been ruined by the loss of the information taken in the field. One of us (R.E.C.) knows of two field notebooks that have been stolen (the thieves did not know what they were taking) from other paleontologists, ruining much research, causing worry, wasting time, and costing thousands of dollars to redo the work.

Another potential problem area in the field is with the packaging of specimens with the proper identification data. All too often, bones or other fossils end up without the proper connection to their locality. With portable printing devices becoming more common, these data can be printed on site

13.2. Paleontologists in the field using electronic distance measurement to study positions of footprints at the Middle Jurassic Red Gulch Dinosaur Tracksite in Wyoming (see Breithaupt et al. 2001, 2004; Matthews and Breithaupt 2001; Matthews et al. 2006).

Images copyright Brent Breithaupt and Neffra Matthews.

13.3. Use of an unmanned airborne vehicle in the field to provide low-altitude aerial photographs (see Breithaupt et al. 2001, 2004; Matthews and Breithaupt 2001; Matthews et al. 2006).

Image copyright Brent Breithaupt and Neffra Matthews.

and included with the wrapped fossils, even, if necessary, in a tiny encoded form for small items.

One of the aspects of collecting that separates most professional paleontologists from many amateurs is the auxiliary information taken along with the collected fossil. These data can be essential in the proper interpretation of the context of the fossil – how it got to the point of being collected – and this can provide important clues toward many research problems. Extracting these data has always taken a lot of time and effort, but this has also been ameliorated with technology that allows quarries and their included fossils to be mapped with a submillimeter-level precision only dreamed of 20 years ago. These electronic distance measurement systems range widely in cost, but reasonably priced approaches, including leasing equipment, are available (e.g., Breithaupt et al. 2001; Evans et al. 2005; Fig. 13.2). Such equipment permits the location of even the most delicate fossils (Nave and Matney 2005; see also Kaye 2007) to be documented. For some applications, low-altitude aerial photography with devices such as small blimps or other unmanned airborne vehicles can assist (Breithaupt et al. 2001; Adams and Breithaupt 2003; Fig. 13.3). The results can include not only documentation of the general context of the quarry, but also the three-dimensional position of each bone, artifact, or feature within the quarry, including orientation data for elongate structures (Fig. 13.4). Innovations in computer software that make multiview pixel matching possible are available to a wider segment of the population at low, or in some cases no, cost. Combined with digital cameras that have improved sensors, dense and accurate point clouds can be produced for an overlapping pair of photographs (Leberl et al. 2010). These and other advances in technology make close-range photogrammetry an affordable option for 3-D documentation of both field and museum subjects (Matthews and Breithaupt 2009). For a detailed discussion of procedures useful for documenting structures like footprints in the field, see the review by Breithaupt et al. (2004) and the technical note by Matthews (2008). The two stand-alone technologies of close-range photogrammetry and ground-based lidar have formed a most efficient partnership for the documentation and 3-D visualization of fossil footprint sites (Breithaupt et al. 2004; Bates et al. 2009b).

New white-light, laser, or other related area scanners can provide a three-dimensional digitization of a whole quarry (or building, or other geological feature) at an incredibly high resolution, depending on the distance of the scanner from the various parts of the feature. These data, along with that from EDMs and other related devices, can be used independently to extract information, or they can all then be coordinated within a GIS context to maximize the use of different data types simultaneously.

Technology can also provide help in deducing what might be in the ground under exposed materials. This was best illustrated in the Michael Crichton (1990) book *Jurassic Park* and the movie based on it. There, a procedure known as computer-assisted sonic tomography (CAST) was used to do a field sonogram on a buried skeleton of *Velociraptor*. It worked marvelously, and a whole skeletal outline could be seen. In the real world, however, such technologies do not work quite as well. Many people have been trying to apply technologies such as ground-penetrating radar to clarify underground structure before excavation proceeds (e.g., Main and Fiorillo

Chapman, Andersen, Breithaupt, and Matthews

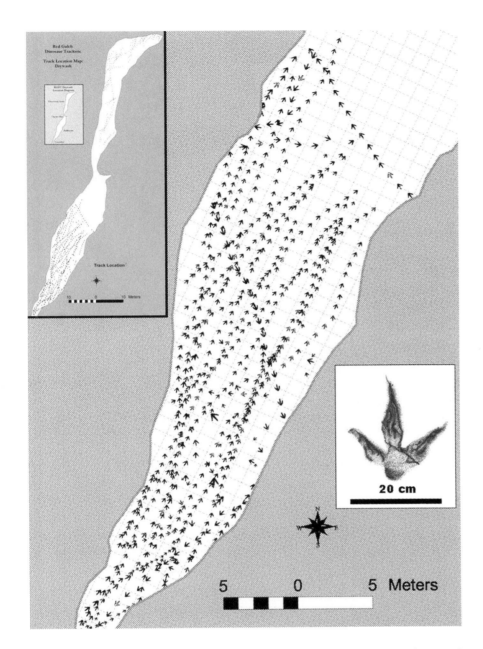

13.4. Electronic distance measurement–produced plot of the Ballroom at the Red Gulch Dinosaur Tracksite in Wyoming showing the position of various Middle Jurassic dinosaur footprints (see Breithaupt et al. 2001, 2004; Matthews and Breithaupt 2001; Matthews et al. 2006).

Image copyright Brent Breithaupt and Neffra Matthews.

2002; Slotnick 2001; Jefferson 1989; Main and Hammon 2003; Gardner and Taylor 1994). This is an approach that tends to work better with archeological and paleoanthropological sites where unconsolidated sediment is involved (e.g., Jorstad and Clark 1995). The most extensive attempt at using this technology so far has been the relatively unsuccessful but innovative work associated with the excavation of the skeleton of *Seismosaurus* (Witten et al. 1992; Gillette 1994), where many scanning technologies were applied. The effectiveness of any method of this type will depend on what it measures and how different the dinosaur bones are from the surrounding matrix relative to this parameter. If the bones act like a surrounding limestone, for example, then it will be difficult to discriminate between the bones and the limestone with almost any scanner. In other circumstances, especially with less consolidated sandstones or mudstones, the bones should be easier to extract virtually. Clearly, however, this technology will develop further and will soon start enabling paleontologists to visualize before they excavate.

13.5. Microscribe articulating arm digitizer. Precision is about 0.25–0.3 mm. The stylus tip is placed on the point on the fossil being studied, and the three-dimensional coordinates of that position are sent to a spreadsheet package. Distances can be calculated between points by three-dimensional Euclidean distance (the hypotenuse of the three-dimensional triangle).

One promising technology to locate fossil bones comes from NASA and their use of scanning Raman laser spectroscopy to identify rocks on Mars. Tom Kaye (2007) described how this technology could be used to identify fossil bones by laser scanning the face of a cliff or hill. The hydroxyl apatite in the bones will fluoresce, allowing for their quick identification.

When it comes to the actual extraction of dinosaur fossils from the outcrop, the relevant technological advances have been more on the mechanical side. Basically, we have more powerful devices useful for extracting fossils faster with less damage. Various vehicles help remove jackets from the field with less damage than was possible in the horse-and-wagon days. It is enticing to think of what megacollectors like John Bell Hatcher and Charles Sternberg would have accomplished with a large air compressor and helicopter access.

In the Preparation Laboratory

The fossil preparation laboratory should be the perfect environment for adapting new technology to dinosaurs. However, the development and use of advanced technology in the paleontology preparation laboratory has proceeded more slowly than in the other areas. Certainly, the mechanical technology used to do the actual preparation work has improved markedly over the past decades as air abrasive and minijackhammer systems have evolved tremendously, increasing the throughput of specimens. However, a surprisingly small amount of experimentation has occurred with scanning and related technologies. Basic works include Zangerl (1965), Hooper (1965a, 1965b), Rolfe (1965a, 1965b), Chapman (1989), Zangerl and Schultze

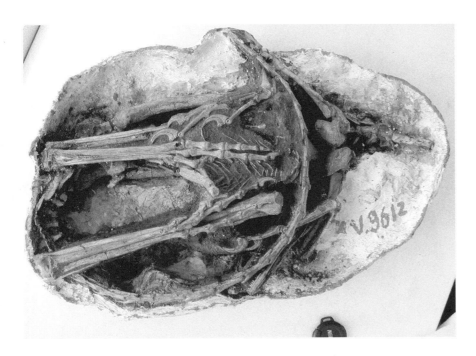

(1989), Plotnick and Harris (1989), Gillette (1994), Witten et al. (1992), Clark and Morrison (1994), and Harbersetzer (1994). More recent applications of advanced x-ray and ultraviolet light applications useful for preparation work include the study by Czerkas and Ji (2002b) on the Early Cretaceous bird *Omnivoropteryx sinousaorum*, Czerkas and Xu's (2002) study of the Early Cretaceous bird *Archaeovolans repatriatus*, and Czerkas and Ji's (2002a) work on the Late Jurassic(?) pterosaur *Pterorhynchus wellnhoferi*. We will discuss more advanced technological options available for preparation.

Once a jacket has entered the laboratory, it is important to keep track of the position of the various elements contained in it at the start and during the preparation process. This can be done in a cumbersome way, but there are relatively inexpensive articulating arm digitizers available that can provide the three-dimensional coordinates of materials within the jacket area as they are exposed (Fig. 13.5). This is a miniature and more mechanical version of the quarry mapping discussed for the field, but with relatively little effort, the positions of specimens and features can be documented with an accuracy of 0.5 mm or better. If the digitizer must be moved in and out of the area during the long preparation process, the placement and digitization of three artificial and permanent landmarks will help reorient the data each time.

One of the biggest problems encountered during the preparation process is that we do not know what is inside the jacket until it is exposed with the pneumatic and other tools that are part of the preparer's standard toolbox. Computed tomographic (CT) scanning, to be discussed in greater detail below, can be helpful here (Cavigelli et al. 2001), but relatively little use has been made of it within this context. The pioneering work on this is the study by Conroy and Vannier (1984) on fossil skulls. CT has great potential for preparation, and a test project on a skeleton of the small troodontid dinosaur *Sinornithoides youngi* (Russell and Dong 1993; IVPP V9612) revealed just how useful and powerful such an approach can be (Andersen et al. unpublished data; Figs. 13.6–13.8). That original skeleton was prepared as

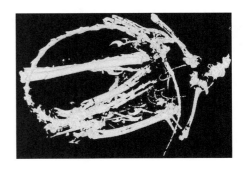

13.7. CT scanning of skeleton of *Sinornithoides youngi* (Andersen et al. unpublished data). Figure shows only extracted virtual bones using CT scan; the matrix has been removed digitally.

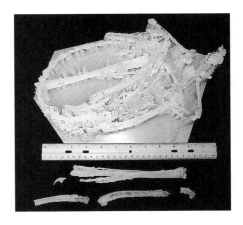

13.8. Rapid prototyping (three-dimensional printing) of the bones from the skeleton of *Sinornithoides youngi* (Russell and Dong 1993; see Andersen et al. unpublished data). Image shows complete skeleton produced with sterolithography in the background, including support struts under the "bones." Foreground shows individual "bones" removed and separated.

13.9. Digital photography setup using copy stand and direct feed into connected laptop computer (in background). Camera would mount on the top of the copy stand.

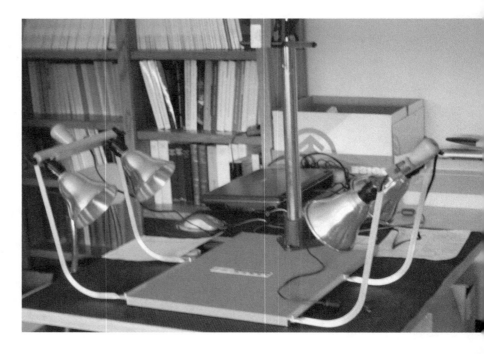

a platform or relief-style mount of an almost complete dinosaur. The mount was used to maximize the study and display of the skeleton while protecting its fragile bones. The problem is that, typical of this style of mount, many of the bones are not exposed and useful for study, and many parts of those that are exposed are not visible as well. Scanning can solve these problems.

The skeleton (Fig. 13.6) was scanned with a relatively old CT scanner, which limited the quality of the result but showed just how successful we will be in the future doing this kind of work with better scanning systems. The whole relief mount was CT scanned at the direction of Phil Currie (then of the Royal Tyrrell Museum of Palaeontology) and Steve Godfrey (Calvert Marine Museum), and the resulting data file was then provided to Andersen and Chapman for this work. Using high-level software, and taking advantage of the fact that the bones and matrix scanned differently, Andersen and Chapman were able to extract the whole skeleton virtually (Fig. 13.7). This allowed it to be viewed from all directions, and the resulting skeleton was printed out three-dimensionally, a process called rapid prototyping (Fig. 13.8). After doing this, the individual bones could be separated out—all without any further preparation work done on the specimen. The only problem with this project was the coarse initial resolution of the CT scanner. A higher-resolution scanner would provide "bones" even more usable for most research and outreach projects.

Scanning of yet-to-be-prepared bones or even prepared bones can also provide a nondestructive approach to the duplication of fossil material via three-dimensional printing technology. Conventional casting can be a destructive process, even when done well by an experienced professional. Scanning a bone is essentially nondestructive (the scanning technician must be careful with the specimen, of course) and produces a virtual copy of the bone or other element. Once edited and perfect, this can then be printed out using rapid prototyping technology at any scale, or the molding elements can be directly produced, which can then be used to make

13.10. Miniature bones generated by rapid prototyping after CT scanning of the *Tyrannosaurus rex* skeleton often referred to as Stan. Skeleton owned by the Black Hills Institute of Geological Research.

Image copyright Peter Larson and the BHIGR.

many replicas of the specimen (Jabo et al. 1999; Chapman and Andersen 1999; Chapman et al. 1999, 2003; Andersen et al. 2003). A great option here is the ability to produce copies of the bones out of durable materials such as aluminum (Deck et al. 2007) that maximize their use for education and outreach applications.

Technology and Collections

Collections work is another area of high-potential improvement that may use technology that has yet to reach its potential. The first improvements are in obvious areas. As computers have gotten faster and data storage has become cheaper, the size and usability of collections databases has vastly improved. Now you can search even the largest collections databases in less than a second and extract all the data available for a specimen. The bottleneck is with data entry, which has been sped up somewhat, thanks to technology, but is still the slowest part of the process.

An important ongoing development is the increase in our ability to associate images—both two-dimensional and three-dimensional—with the rest of the collections data. The more useful data associated with specimens in a collections database, the better. Incorporating images allows paleontologists and others from around the world to be able to browse through collections

13.11. Photogrammetric recording of an individual theropod footprint from Middle Jurassic Red Gulch Dinosaur Tracksite of Wyoming (see Breithaupt et al. 2001, 2004; Matthews and Breithaupt 2001; Matthews et al. 2006). Left, Basic setup for recording photogrammetric images in the field. Right, three different versions of the resulting model of the theropod footprint.

Images copyright Brent Breithaupt and Neffra Matthews.

and determine what specimens are of interest to a particular project. Digital photography (Fig. 13.9) mostly reduces two-dimensional photographic costs to personnel time, so comprehensive documentation of major collections at a reasonable price is possible in a way never before realized. This will make it harder and harder for important specimens to hide undiscovered in collections drawers as more and diverse experts view the specimens without having to travel thousands of miles to get to the museum they are housed in.

All paleontological specimens deteriorate, even in museums, although, it is hoped, at a relatively slow pace in most cases. The deterioration process can accelerate as a result of a number of climatic factors (such as humidity, temperature, and their variation), as well as the inevitable problems encountered during interaction with humans. To maximize the utility of museum collections, important specimens should be virtualized to capture as much of the specimen data for posterity as possible. That way, if a specimen should be lost or catastrophically damaged, the virtual file will preserve much of the important information. This scanning can include CT scanning or surface scanning—to be discussed in detail below—and should be done at the highest resolution possible (Chapman and Andersen 1999; Andersen and Chapman 1999; Deck et al. 2004).

It is a major problem that the very interaction of paleontologists, technicians, and the public with real specimens usually causes them some damage. Specimens that have been measured frequently with calipers, for example, can develop grooves from the contact of the jaws of the calipers with the specimen. This grooving inevitably makes the subsequent measurements inaccurate. Virtualizing specimens can allow many of the measurements to be taken in virtual space and at no cost to the quality of the specimen and at no decrease in the accuracy of the measurements (it can be quite the opposite because many measurements are difficult to take with calipers). Articulating arm point digitizers (Fig. 13.5) can make even more mechanical measurements less costly to the specimens because only the tip of a stylus comes in contact with a position on the specimen. Further, the number of measurements taken using this method over a specified time period can be five or 10 times that done with calipers, at a comparable resolution, and measurements such as angles can be calculated simultaneously. Virtualizing specimens allows extremely large specimens to be measured more easily and accurately, and virtual specimens can be sent to colleagues thousands of miles away in a few seconds (Figs. 13.7 and 13.10).

Technology can also increase the collections we can extract from the field and increase those available in museums. Many fossils such as dinosaur footprints are difficult or impossible to remove from where they are found. As an example, individual footprints of dinosaurs from an Early Jurassic site in Culpeper, Virginia, that were taken to the Smithsonian Institution's National Museum of Natural History came with associated rock that weighed many hundreds of pounds and occupied volumes of at least 10 cubic feet—for a single track! Clearly, a better way of archiving tracks from many sites is needed. Outdoor surface scanners can be used to capture whole track sites. These can be reproduced by means of rapid prototyping as small but accurate models that can become part of museum collections. These scanners (see Arakawa et al. 2002; Azuma et al. 2002), and technology such as photogrammetry (Matthews and Breithaupt 2001; Breithaupt et al. 2001, 2004; Matthews et al. 2006; Matthews 2008; Fig. 13.11), moiré photography (Ishigaki and Fujisaki 1989; Lim et al. 1989), or other optical approaches (Goto et al. 1996; Gatesy et al. 2005) can capture the individual tracks in great detail. These can be kept as part of virtual collections but can also be duplicated by rapid prototyping to provide physical models in the collections area.

Virtual museum catalogs or collections of fossil footprints or other vertebrate fossils can also be made available via the Web. These virtual collections can be especially effective when created by photogrammetry, where the surface model and image texture are unified in the production process, or through the use of surface scanning and secondary image texture matching (see below). File formats, such as 3-D PDF, provide for easy viewing and manipulation of these 3-D models without additional software viewers; even more options are possible with high-end software packages.

Research Potentials

The potential for technology to help improve and expand our research into dinosaurs is almost endless. We will not discuss the many types of high-level data analysis now possible. Because of the awesome increase in speed and storage seen in computers in the past decade or two, our ability to analyze the shapes (morphometrics) and relationships (phylogenetic analysis) of dinosaurs has expanded tremendously over this period. Technology has allowed us to analyze aspects of dinosaur biology and taphonomy not previously available to paleontologists. This includes the study of isotopic concentrations in bone and other materials (e.g., Barrick et al. 1996, 1997; Barrick and Showers 1994, 1995; Goodwin 2002) and the study of paleobiomolecules and paleo-DNA pioneered by Mary Schweitzer, summarized in Chapman et al. (2002) and Schweitzer and Marshall (this volume). These are specialized fields that require significant background to understand, and we suggest visiting the cited references and those cited therein. We will concentrate on the technological applications that allow us to reconstruct dinosaurs and study how they worked as machines and animals. Technology has become essential here.

The rise of digital photography (Fig. 13.9) has been a tremendous boon for paleontologists, as it has automated and facilitated many of the basic steps for documenting sites and specimens that could take up long hours

13.12. Virtualization of dinosaur skeleton with an articulating arm digitizer by Ray Wilhite. Left, Wilhite digitizing skeletal elements from the *Tyrannosaurus rex* referred to as Stan (owned by the BHIGR). Right, Resultant model of the skeleton. See also Wilhite (2003).

Images courtesy of Ray Wilhite and Virtual Surfaces, Inc., and copyright Ray Wilhite.

even 20 years ago. Digital single lens reflex (SLR) cameras are now available at a reasonable price with interchangeable lenses and high-resolution, full-size (35 mm) sensors. This means a paleontologist can take many pictures in the field and preview them for acceptability, which can save expensive trips back into the field for more documentation. In the laboratory, specimen photographs are easily taken and edited, typically saving 75% or more of the time it used to take (Chapman 2003; Matthews et al. 2006).

Technology has a long history of application to dinosaur studies through the use of microscopes and thin-section work. The quality of such devices has continued to improve, and combined with digital imaging, these devices have made it much easier to study bone (e.g., Horner et al. 2000; Chinsamy et al. 1998), enamel (e.g., Stokosa 2005), and eggshell microstructure (e.g., Varricchio et al. 2002), as well as paleopathology (e.g., Rothschild and Tanke 2005; Rothschild et al. 1997). Diligent application of the scanning electron microscope especially allows paleontologists to take a close look at even the finest microstructure (e.g., Wegweiser et al. 2004).

At the heart of the application of technology to research is, again, the virtual capture of the bones themselves, which allows them to be studied individually, as composite and articulating groups, or as whole animals, and with muscles and skin attached. It all starts with the individual bones, however. The potentials for the use of this technology for research into dinosaur footprints are also spectacular (e.g., Gatesy et al. 1999) and are discussed in the chapter by Farlow et al. (this volume).

How do you capture the bone? You do it with a three-dimensional scanning device. There are three main options here. You can use the three-dimensional point digitizers (e.g., Fig. 13.5) mentioned above. Here, you digitize the bone or footprint as a series of manually selected points—the more, the better. This was done effectively for footprints by Farlow (1991, 1993; Farlow and Chapman 1997) and Rasskin-Gutman et al. (1997), but with a bone of any size, it will take a few thousand points to derive a decent but still more stylized representation of the bone. It takes a lot of work and a lot of time but can be useful in some cases. Wilhite (2003; Fig. 13.12) has used this approach in an innovative way to study how sauropod limbs work. All in all, however, it is best to let technology speed up the process and make much more accurate virtual bone models.

That leaves CT scanning and related approaches, as well as surface scanners for the highest-quality products. Both have their strengths and weaknesses, and we are lucky to have both available to us.

CT scanning (Figs. 13.7 and 13.13) is now a relatively old technology that has evolved greatly over the past 10 years. Even most medical scanners encountered on a day-to-day basis have resolutions that would have been greatly envied 10 years ago. Essentially, CT scanning is a three-dimensional

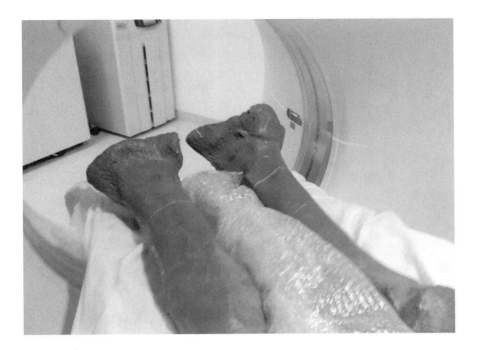

13.13. CT scanning of dinosaur bone. Dinosaur Bone in CT device awaiting scanning. Note hole and belt width that restrict maximum size of specimen being scanned.

Image copyright the Smithsonian Institution.

x-ray done through an object slice by slice. A source generates the x-ray, and a collector takes in the signal for processing.

For each slice, an x-ray source is either rotated around the object with a collector on the opposite side also rotating, or there are multiple sources and collectors distributed around the object. The different absorption rates of the materials around, inside, and on the surface of the object will be seen as different levels in the signal received by a collector. Bone will typically look different from muscle or matrix, for example. Each slice is like a poker chip of whatever thickness you select to scan at. The more slices, the more time and memory involved, but also the thinner the poker chips and the more accurate the resulting models.

Typical medical CT systems will do a maximum of about 0.5-mm slices. Micro-CT systems can do slices down to 7 μm in thicknesses, and other systems can easily do even thinner slices. These slices are then exported as individual but associated files, which then must be merged with specialized software into three-dimensional solid models. This can be a complex process: equivalent structures on each slice must be associated and used to build each element in the final product, which includes the internal data. For example, if doing a CT of a human leg, the femur will be on the inside. To assemble that whole bone, the part of each cross section that is the femur must be identified in each slice, and these connected up to make the virtual femur.

Medical scanners can be relatively weak because the object being scanned is sometimes alive and it is usually beneficial if it stays that way. Industrial CT scanners, adapted for nonliving objects, can use three or four times as much power (a motor will not die if overexposed) and can get much finer resolution.

The advantage to CT systems is that they are available almost everywhere, and most medical centers enjoy scanning objects for paleontologists from time to time. The process of model building has been nicely automated for most systems, which may involve the use of high-level independent

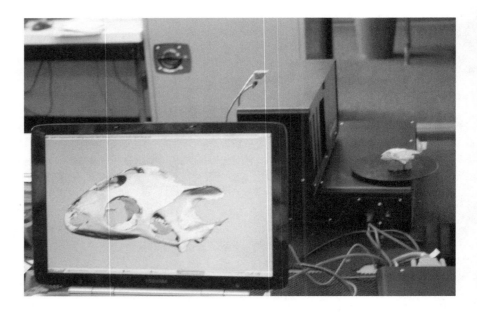

13.14. Surface scanning of turtle skull in the Idaho Virtualization Laboratory. Scanning was done by a Cyberware M-15 laser surface scanner. The scanner is virtualizing the skull in the background, and the resulting model is being shown in the foreground on the computer monitor. Resolution is up to approximately 0.1 mm.

software, and you get internal information for the fossils most of the time. The disadvantages are that resolution levels can be relatively coarse; they have a size limit for what they can scan (i.e., the specimen must fit through the hole in the scanner); and the systems are not generally transportable: they take up lots of room and usually require a lead-lined area for safety. The exception is micro-CT systems, which are high resolution and fully self-contained. They do have a size limit in terms of what they can scan, but they do it at amazing resolution. Industrial CT scanning devices can vary in size and are powerful; they also can be quite high-resolution devices. All CT systems cost $100,000 or more, so most paleontologists have to either rent time on a system or get that time donated. The results are worth it, as evidenced by the high-quality studies now being published, including Brochu's (2003) detailed analysis of the skeleton of the *Tyrannosaurus rex* skeleton often referred to as Sue, the study of Rowe et al. (1999) of *Alligator mississippiensis*, and many others (e.g., Wedel 2003; Tykoski et al. 2002; Franzosa and Rowe 2005; see the massive output of the Witmer Laboratory at Ohio University).

Other scanning technologies may also apply to fossils, but there has been relatively little exploration done using these different technologies. For example, Schwarz et al. (2005) were quite successful applying neutron tomography to examine the internal structure of diplodocid bones. Other technologies, such as magnetic resonance imaging, have yet to be applied seriously and may not be useful because they specialize in soft tissue morphology and not hard parts, but to our knowledge, this remains to be tested.

Surface scanning is a different animal, and many different methods are used to accomplish it (Fig. 13.14). In some cases, lasers are used, but alternative optical methods are applied in others. The disadvantage of surface scanners is that they can be difficult to gain access to, they do not reveal internal structure, and they typically take more user time and interaction. The advantage is that they are generally of higher resolution than many CT devices and can produce much higher quality scans for the surfaces they virtualize. Further, you can often associate and overlay photographic data right on top of the surface data, a process called texture mapping, so

Chapman, Andersen, Breithaupt, and Matthews

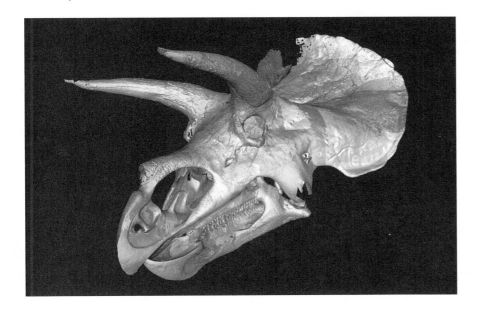

13.15. Virtualization of the Smithsonian Institution's *Triceratops* skeleton. This is the model generated with a white-light surface scanner. Different colors represent different individual passes by the scanner. The final model is a composite of all these passes merged together. Final model consisted of over 20 million points.

Image copyright the Smithsonian Institution.

13.16. Rapid prototyping of humerus of the Smithsonian Institution's *Triceratops*. This prototyping is done by using successively smaller drill bits to cut out high-resolution, full-sized left humerus using mirror-imaged model of the original right humerus.

Image copyright the Smithsonian Institution.

the resulting models look just like the original object. We will discuss this approach using a surface laser scanner, although other surface scanning methods will use their own technologies.

Laser surface scanning is a deceptively simple process (Fig. 13.14). There are various methods used for laser scanning, but the most common method is to use a detector mounted to the laser projector. The projected laser beam strikes the target surface, and the detector records and calculates the target distance through triangulation. The scanners can scan, record, and calculate the hundreds of thousands of distances in less than a second. It takes a high-precision device to accomplish this. Many scanners can give a resolution of 0.1 mm or better. In both, the specimen is rotated in front of the laser source to give readings across the whole three-dimensional shape.

Finally, white-light scanners and photography-based technologies are available for capturing the morphology of specimens in the field or laboratory—something relatively unexplored except by Matthews and Breithaupt (2001) and Breithaupt et al. (2001, 2004) for dinosaur footprints at the Middle Jurassic Red Gulch Dinosaur Tracksite in Wyoming (Figs. 13.1–13.4, 13.11). They captured the morphology of individual footprints by photogrammetry (but see also Ishigaki and Fujisaki 1989; Lim et al. 1989; Goto et al. 1996; Gatesy et al. 2005). A recent proliferation of optical scanners intended for outdoor use, as well as a decline in their price, will provide even more options for capturing whole sites and specimens that cannot be removed from the field. The products and applications of white-light scanners overlap with those of surface laser scanners. The result of both processes is a three-dimensional virtual model of the original. These virtual models will then open a panorama of research possibilities when studied individually and in combination.

Of particular interest is the ability to replicate the structures, or parts of them, with rapid prototyping technology (Fig. 13.8). The most extensive application of rapid prototyping technology to paleontology was the Smithsonian Institution's *Triceratops* project. During this work, this technology was used in a wide variety of contexts and utilized a number of different rapid prototyping methods (Figs. 13.15–13.18).

13.17. Rapid prototyping of the skull of the Smithsonian Institution's *Triceratops*. This enlarged skull (approximately 7 feet long) was reproduced using a model based on the original 6-foot-long skull. The reproduction was done using more than 30 blocks of stereolithography that were then subsequently fused together. The prototype was painted blue to protect the final product.

Image copyright the Smithsonian Institution.

With rapid prototyping, the output can be at any scale; very small bones can become large enough to work with, and very large bones—say, the forelimb bones of a *Triceratops*—can be brought down to a size where a researcher can study how they articulate as easily as with the skeleton of a modern dog (Chapman et al. 2001a, 2001b). The technology for this can include mechanical approaches that cut the shape from a solid block of material, or other approaches that build the shape layer by layer. Milling systems drill out the object from a block of material and use progressively smaller bits to heighten the detail until the final product looks surprisingly good, especially with a bit of sanding (Fig. 13.16). Another system cuts out each layer from individual sheets of paper and fuses the results into a hard and accurate replica. Finally, the most common methods, such as stereolithography, build up the final model by solidifying ultrathin layers of polymers or other substances; one method even uses a sugar–water mixture (Andersen et al. 2001; Burns 1993).

A lot can be learned by studying bones individually, and virtual models can lead to spectacular results, including improving our means of restoring soft part morphology. Starting with the groundbreaking study by Conroy and Vannier (1984), as well as pioneering work by Lawrence Witmer, Stephen Gatesy, and others, CT scans, along with computer modeling and animation software, have totally changed the ways we look at dinosaurs, especially the restoration of their heads (e.g., Whitlock 2001; Witmer 1995, 1997, 2001; Bimber et al. 2002; Gatesy et al. 1999; Figs. 13.19, 13.20). A pioneering facility in the application of CT systems, especially ultra-high-resolution micro-CT work, is the DigiMorph laboratory at the University of Texas, Austin, where many significant studies and publications have vastly improved our understanding of the morphology of many fossil and recent forms (e.g., Rowe et al. 1995; Fig. 13.22). A specialized area is the scanning of dinosaur braincases in situ with CT scanners and using them to reconstruct the brains of the animals by filling in the spaces (e.g., Larsson et al. 2000; Brochu 2000; Dominguez Alonso et al. 2004; Witmer 2004; Franzosa and Rowe 2005; Fig. 13.7D). Scanning technology is even useful for detecting forgeries such as the supposed specimen of *Archaeoraptor*, which turned out to be a chimera

13.18. Bronze model of the skull of the Smithsonian Institution's *Triceratops*. Sculpture is now located at the Constitution Avenue entrance into the National Museum of Natural History. The sculpture was produced from a mold taken from the large prototype skull. An alternative option was simply to produce the mold elements directly using rapid prototyping, thus avoiding the molding phase altogether.

Image copyright the Smithsonian Institution.

Chapman, Andersen, Breithaupt, and Matthews

13.19. Modeling detailed dinosaur anatomy with CT-scanned data as a base. Extraction of the brain of a *Tyrannosaurus rex* by Lawrence Witmer.

Image provided by and copyright Lawrence Witmer (see Witmer 1995, 1997, 2001).

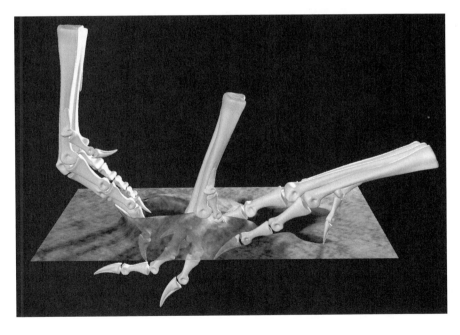

13.20. Modeling of the formation of a footprint by a theropod dinosaur. The models were used to follow the whole process of track formation, from foot insertion to extraction by the animal, and the subsequent effects on the footprint shape (see Gatesy et al. 1999; Bimber et al. 2002).

Image courtesy of and copyright Stephen M. Gatesy.

(Rowe et al. 2001). Finally, applying finite-element analysis to these virtual models can show a researcher how stress patterns will promulgate across a structure as force is applied to it at selected positions (e.g., Rayfield et al. 2001; Fig. 13.21; see also Farke et al. 2010). This provides clues as to what the animal could and could not do it real life.

The possibilities get even better as multiple models are combined in studies of function in dinosaurs. Here, legs, cranial complexes, and even

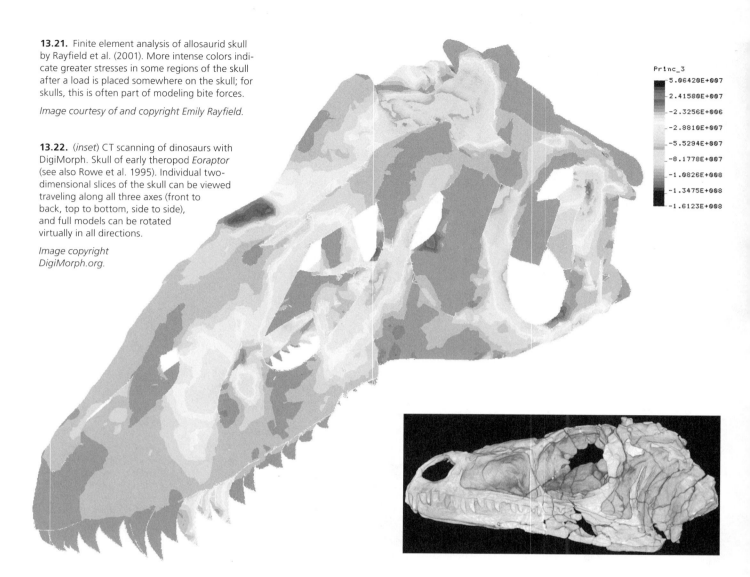

13.21. Finite element analysis of allosaurid skull by Rayfield et al. (2001). More intense colors indicate greater stresses in some regions of the skull after a load is placed somewhere on the skull; for skulls, this is often part of modeling bite forces.

Image courtesy of and copyright Emily Rayfield.

13.22. (*inset*) CT scanning of dinosaurs with DigiMorph. Skull of early theropod *Eoraptor* (see also Rowe et al. 1995). Individual two-dimensional slices of the skull can be viewed traveling along all three axes (front to back, top to bottom, side to side), and full models can be rotated virtually in all directions.

Image copyright DigiMorph.org.

```
Princ_3
 5.06420E+007
 2.41580E+007
-2.3256E+006
-2.8810E+007
-5.5294E+007
-8.1778E+007
-1.0826E+008
-1.3475E+008
-1.6123E+008
```

studying how dinosaurs actually worked. The last 5 years have seen a great explosion in the scanning and virtualization of dinosaur skeletons, mostly for functional studies.

As computers improve further in speed and storage, the effectiveness and accuracy of the animations will improve as well, and we will get a progressively better picture of the animals behind the dinosaur skeletons. Eventually, the most difficult movements to model, such as chewing in the incredibly kinetic skulls of hadrosaurs, will even be approachable as more than just schematic diagrams (e.g., Weishampel 1984). In the meantime, innovative approaches, such as the Dinomorph program developed by Kent Stevens (Stevens 2002; Stevens and Parrish 1999, 2005; Fig. 13.23), which uses more stylized virtual bones—the important articulation surfaces are done with much more accuracy—allows a detailed analysis of the potential positions of articulating bones. This program led to the disturbing but well-supported possibility that sauropod necks tend to be much more horizontal and limited in movement than most researchers previously thought (Stevens and Parrish 1999), although much work still needs to be done regarding this question.

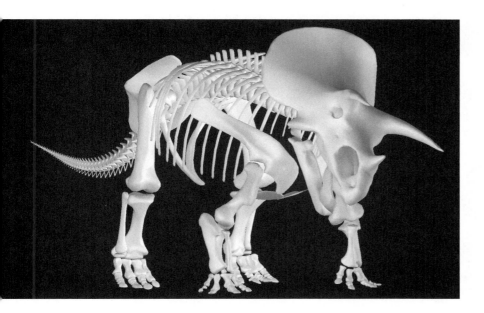

13.23. Articulated skeleton of *Triceratops* using innovative program Dinomorph (see also Stevens 1992; Stevens and Parrish 1999). These virtual bones are more detailed in key areas where bones articulate and are less detailed in other areas.

Images courtesy of and copyright Kent Stevens.

Education and Outreach

Education and outreach remain another area of tremendous potential for applying technology to dinosaurs, with perhaps the greatest potential impact. Traditionally, outreach includes formal classroom education and other, less formal venues such as museums, and informal, and often distance, learning.

Museums want to exhibit dinosaur skeletons in a way that excites and energizes their visitors. However, even though it is still commonly done, mounting real bones is no longer a viable option because the process is destructive to the bones themselves in both the short and long term (Moltenbrey 2001; Hutchinson 1999). Most dinosaur species are only known from one or fewer articulated skeletons, and most dinosaur skeletons are far from complete. Even when multiple skeletons are available, they will often be from animals of different sizes, growth stages, and sexes. Mounting a full skeleton can consist of assembling a group of bones that often do not go together all that well.

Technology can come to the rescue here and make generating an accurate, complete, and dynamic skeleton possible. We have already discussed the *Sinornithoides* project, which uses the relief mount (Andersen et al. unpublished data; Figs. 13.6–13.8). The problem with many bones, especially in mounts like this, is that they can be compressed during the process of fossilization. Software can allow paleontologists to take scans of the bones, then reinflate and undistort the result to get new ones.

There are other options for using the bones available to fill in for those bones that are not. Skeletons are bilaterally symmetrical. This fact can be used to help make new versions of many missing bones. Once virtualized, a left humerus, for example, can be mirror imaged to produce a right humerus. Even parts of bones from the axis (midline) of the animal can be filled in if the part from the other side of the bone is there. Here you cut out a good section, mirror image it, and fuse it into the missing area. For serial elements like vertebrae in a series (e.g., dorsals), which vary mostly by size, one vertebra can be replicated at the new size. This works for bones coming from different-sized individuals; they can be scaled up or down to

13.24. Mounted miniature *Triceratops* skeleton at the Smithsonian Institution's National Museum of Natural History. This is the original mount posture modeled using one-sixth-scale bones generated by rapidly prototyping the original bones.

Image copyright the Smithsonian Institution.

fit the skeleton—as long as the element does not change shape appreciably with the size difference (i.e., allometry). The result is a skeleton that is not the same as an individual animal but an informed composite that is a good representative of the species.

The best example of this is the recent *Triceratops* project at the Smithsonian Institution's National Museum of Natural History (Chapman et al. 1999, 2001a, 2001b; Andersen et al. 1999, 2001; Hutchinson 1999; Jabo et al. 1999; Moltenbrey 2001). Here, the mounted skeleton—the first ever mounted horned dinosaur—was falling apart because the bones were disintegrating from the stresses of being displayed for almost 100 years. The mount was to be replaced with a cast skeleton. This gave the museum a chance to fix other problems the mount had, such as hadrosaur back feet (*Triceratops* feet were unknown when the skeleton was originally mounted) and mismatched bones. The original mount was a composite from many individuals collected for O. C. Marsh by John Bell Hatcher and his crew. The work done to fix the mount and improve it used technology extensively and probably used more and different rapid prototyping approaches than any other project in any context.

First, the skull was too small (Fig. 13.17)—about a foot too short in length—for the rest of the skeleton, although of a reasonable shape. This mount is the reference source for many reconstructions of *Triceratops*, especially in the popular literature, which ended up with heads that were too small. There were two options to correct this. First, there was a skull of

the appropriate size, but only one side was available. This could have been scanned, mirror imaged, and fused into a complete skull. The other choice was just to scale up the original an extra foot in length. The latter was done, and a complex prototype (Fig. 13.17) was generated in more than 30 separate blocks that were then fused together into the completed skull. This was then molded and cast to produce a skull for the new mount.

A few other bones were from a smaller animal than was desirable, or were just not of sufficient quality. The better-quality bones were taken from the other side of the animal. These were virtualized and mirror imaged to produce the needed element (e.g., the right humerus was used to generate a left one). These were prototyped at full size using a milling method, and these new "bones" were then molded and cast to make the needed bones (Fig. 13.16).

Finally, once a whole skeleton was available, it was prototyped at one-sixth scale to produce a miniature skeleton (Fig. 13.24). This was used in two ways. First, each miniature prototype was molded and cast, then used to produce a miniature replica of the original mount in its original posture. This display was the first mount of a horned dinosaur, and this, along with digitizing an array of points on that original, helped preserve what the original mount looked like for historical reasons. Second, determining the correct posture with the original bones is incredibly difficult because they are so big. This effort was facilitated with the miniatures because it was then possible to work each articulation manually—still the best method to feel the way the bones should go together.

The result of all this work was a new mount (Fig. 13.25) in a better and, presumably, more accurate posture, a replaced mount of cast material, a miniature replica of the old mount (Fig. 13.24), a series of original bones, and even a prototype in display cases. These vitrines can protect the bones much better than if they were in a mount. Technology improved paleontology in many ways in just this one project. The new skull was also used to help produce a bronze cast of the skull for placement outside the museum entrance (Fig. 13.18).

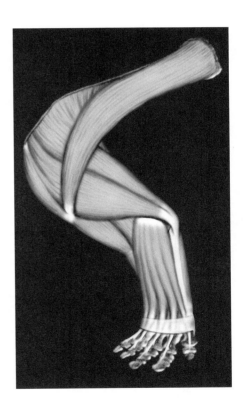

13.26. Addition of virtual muscles on the leg of *Triceratops* by Walters et al. (2000, 2001). Original model is the virtual *Triceratops* skeleton based on the Smithsonian Institution's new mount. Rods show possible pathways for musculature.

Image copyright the Smithsonian Institution.

In formal classroom situations, technology can help extend the time students have with skeletal material, an essential part of learning. The great interest young people have in dinosaurs should be enhanced by as much contact with the fossils as possible. It would be expensive and impractical to outfit standard classrooms with casts of dinosaurs, especially the large ones, but students can have tremendous access to these bones in virtual form. Reduced forms of these files can also be used to allow students to try developing animations of walking dinosaurs themselves. Similarly, distance and informal learning about dinosaurs can be tremendously improved through access to these fossil bones through the Web.

Finally, artistic reconstruction of dinosaurs can be made much more accurate starting with virtual dinosaur skeletons. Sculptors can use small, accurate skeletal reconstructions as a basis for their work, and two-dimensional artists can pose virtual skeletons in whatever positions they want before the process of adding muscle and skin to the animals produces a fully fleshed-out reconstruction (Walters et al. 2000, 2001; Fig. 13.26).

Summary

We are in the middle of a revolution. Technology is vastly changing every aspect of the way we interact with dinosaurs. Fieldwork will be easier and more productive, and we will end up with far more accurate positional data. Preparation work will be done much more efficiently, greatly enhancing the number of specimens prepared and providing us all with more dinosaur material in less time. Collections will be used a lot more and yet not be damaged as much. The data contained in them will be archived in case of damage or loss. As more specimens are virtualized and made available on the Web, researchers will be able to do their work without having to scrounge for thousands of dollars to cover the time and expense of traveling from museum to museum to examine and measure specimens. The work being done will vastly improve what we know about dinosaurs as animals. Finally, the way we present dinosaurs to students and others will be progressively more dynamic, maintaining and increasing the interest of the general public in dinosaurs and in science in general.

Acknowledgments

There are a number of individuals we would like to thank for their contribution to our thoughts in the long term. First, we would like to thank Jim Farlow, Mike Brett-Surman, and Tom Holtz for administrative support and patience. Linda Deck has been a sounding board on these topics for over two decades, as have Bob Walters and Tess Kissinger. The pioneering work of Kent Stevens, Dave Weishampel, Tim Rowe, Jack Horner, Scott Sampson, Larry Witmer, Mary Schweitzer, and others has been an inspiration. R.E.C. would also like to thank the staff of the Idaho Virtualization Laboratory—Christian Petersen, Robert Schlader, and Nicholas Clement—for hours of discussion.

References

Adams, T. L., and B. H. Breithaupt. 2003. Documentation of Middle Jurassic dinosaur tracks at the Yellow Brick Road Dinosaur Tracksite, Bighorn Basin, Wyoming. *Wyoming Geo-notes* 76: 28–32.

Andersen, A., and R. E. Chapman. 1999. Conserving the data in natural history collections: approaches for automatic scanning and digitizing. In *Program and Abstracts, Annual Meeting*, 19. Washington, D.C.: Society for the Preservation of Natural History Collections.

Andersen, A., R. E. Chapman, and L. T. Deck. 2010. Sue goes virtual. *Computer Graphics World* 33 (7).

Andersen, A., R. E. Chapman, J. Dickman, and K. Hand. 2001. Using rapid prototyping technology in vertebrate paleontology. *Journal of Vertebrate Paleontology* 21 (suppl. to 3): 28A.

Andersen, A. F., R. E. Chapman, K. Kenny, and H. C. E. Larsson. 1999. Animation of 3-D digital data: the walking *Triceratops*. *Journal of Vertebrate Paleontology* 19 (suppl. to 3): 29A.

Andersen, A., R. E. Chapman, and B. Wilcox. 2003. Three-dimensional techniques for capturing and building virtual models of complex objects for use in scientific and industrial applications, data archiving, and the entertainment industry. In A. J. Woods, M. T. Bolas, J. O. Merritt, and S. A. Benton (eds.), *Stereoscopic Displays and Virtual Reality Systems X*, 461–468. Proceedings of SPIE–IS&T Electronic Imaging, SPIE (International Society for Optical Engineering), vol. 5006.

Arakawa, Y., Y. Azuma, A. Kano, T. Tanijiri, and T. Miyamoto. 2002. A new technique to illustrate and analyze dinosaur and bird footprints using 3-D digitizer. *Memoir of the Fukui Prefectural Dinosaur Museum* 1: 7–18.

Azuma, Y., Y. Arakawa, Y. Tomida, and P. J. Currie. 2002. Early Cretaceous bird tracks from the Tetori Group, Fukui Prefecture, Japan. *Memoir of the Fukui Prefectural Dinosaur Museum* 1: 1–6.

Barrick, R. E., A. G. Fischer, and W. J. Showers. 1996. Comparison of thermoregulation of four ornithischian dinosaurs and a varanid lizard from the Cretaceous Two Medicine Formation: evidence from oxygen isotopes. *Palaios* 11: 295–305.

Barrick, R. E., and W. J. Showers. 1994. Thermophysiology of *Tyrannosaurus rex*. *Science* 265: 222–224.

———. 1995. Oxygen isotope variability in juvenile dinosaurs (*Hypacrosaurus*): evidence for thermoregulation. *Paleobiology* 21: 450–459.

Barrick, R. E., M. K. Stoskopf, and W. J. Showers. 1997. Oxygen isotopes in dinosaur bone. In J. O. Farlow and M. K. Brett-Surman (eds.), *The Complete Dinosaur*, 474–490. Bloomington: Indiana University Press.

Bates, K. T., P. L. Falkingham, B. H. Breithaupt, D. Hodgetts, W. I. Sellers, and P. L. Manning. 2009a. How big was "Big Al"? Quantifying the effect of soft tissue and osteological unknowns on mass predictions for Allosaurus (Dinosauria: Theropoda). *Paleontological Electronica* 12 (3): 14.

Bates, K. T., P. L. Falkingham, D. Hodgetts, J. O. Farlow, B. H. Breithaupt, M. O. O'Brien, N. Matthews, W. I. Sellers, and L. P. Manning. 2009b. Digital imaging and public engagement in paleontology. *Geology Today* 25 (4): 134–139.

Bimber, O., S. M. Gatesy, L. M. Witmer, R. Raskar, and L. M. Encarnacao. 2002. Merging fossil specimens with computer-generated information. *IEEE Computer* 35 (9): 25–30.

Breithaupt, B., N. A. Matthews, and T. Noble. 2004. An integrated approach to three-dimensional data collection at dinosaur tracksites in the Rocky Mountain West. *Ichnos* 11 (1/2): 11–26.

Breithaupt, B., E. H. Southwell, T. L. Adams, and N. A. Matthews. 2001. Innovative documentation methodologies in the study of the most extensive dinosaur tracksite in Wyoming. In V. L. Santucci and L. McClelland (eds.), *Proceedings of the 6th Fossil Resources Conference*, 113–122. Technical Report NPS/NRGRD/GRDTR-01/01. Washington, D.C.: U.S. Department of the Interior, National Parks Service, Geological Resources Division.

Brochu, C. A. 2000. A digitally rendered endocast for *Tyrannosaurus rex*. *Journal of Vertebrate Paleontology* 20 (1): 1–6.

———. 2003. Osteology of *Tyrannosaurus rex*: insights from a nearly complete skeleton and high-resolution computed tomographic analysis of the skull. *Journal of Vertebrate Paleontology* 22 (suppl. to 4): 138.

Burns, M. 1993. *Automated Fabrication: Improving Productivity in Manufacturing*. Englewood Cliffs, N.J.: Prentice-Hall.

Cavigelli, J. P., R. Ritchie, and B. H. Breithaupt. 2001. The use of CT scans as an aid to fossil preparation. *Journal of Vertebrate Paleontology* 21 (3): 39A.

Chapman, R. E. 1989. Computer assembly of serial sections. In R. M. Feldmann, R. E. Chapman, and J. T. Hannibal (eds.), *Paleotechniques*. Paleontological Society Special Publication 4: 157–164.

———. 1997. Technology and the study of dinosaurs. In J. O. Farlow and M. K. Brett-Surman (eds.), *The Complete Dinosaur*, 112–135. Bloomington: Indiana University Press.

———. 2003. Digital measurement and capture of images. In E. R. S. Hodges (ed.), *The Guild Handbook of Scientific Illustration*, 2nd ed., 40–43. New York: John Wiley & Sons.

Chapman, R. E., and A. Andersen. 1999. Conserving the data in natural history collections: using technology to minimize wear and damage on specimens during research. *Program and Abstracts, Annual Meeting*, 24. Washington, D.C.: Society for the Preservation of Natural History Collections.

Chapman, R. E., A. F. Andersen, and S. J. Jabo. 1999. Construction of the virtual *Triceratops*: procedures, results, and potentials. *Journal of Vertebrate Paleontology* 19 (suppl. to 3): 37A.

Chapman, R. E., A. Andersen, and B. Wilcox. 2003. Studying extinct animals using three-dimensional visualization, scanning and prototyping. In A. J. Woods, M. T. Bolas, J. O. Merritt, and S. A. Benton (eds.), *Stereoscopic Displays and Virtual Reality Systems X*, 469–477. Proceedings of SPIE–IS&T Electronic Imaging, SPIE (International Society for Optical Engineering), vol. 5006.

Chapman, R. E., R. Johnson, and K. A. Stevens. 2001a. The posture of *Triceratops*: insight from three-dimensional modeling, scanning, and prototyping. *Journal of Morphology* 248 (3): 215.

Chapman, R. E., N. A. Matthews, M. H. Schweitzer, and C. C. Horner. 2002. Applying 21st century technology to very old animals. In J. G. Scotchmoor, D. A. Springer, B. H. Breithaupt, and T. Fiorillo (eds.), *Dinosaurs: The Science behind the Stories*, 137–144. Society of Vertebrate Paleontology, Paleontological Society. Alexandria, Va.: American Geological Institute.

Chapman, R. E., R. A. Snyder, S. Jabo, and A. Andersen. 2001b. On a new posture for the horned dinosaur *Triceratops*. *Journal of Vertebrate Paleontology* 21 (suppl. to 3): 39A–40A.

Chinsamy, A., T. Rich, and P. Vickers-Rich. 1998. Polar dinosaur bone histology. *Journal of Vertebrate Paleontology* 18 (2): 385–390.

Clark, S., and I. Morrison. 1994. Methods and use of CT scan and x-ray. In P. Leiggi and P. May (eds.), *Vertebrate Paleontological Techniques*, 1: 323–329. New York: Cambridge University Press.

Conroy, G. C., and M. W. Vannier. 1984. Noninvasive three-dimensional computer imaging of matrix-filled fossil skulls by high-resolution computed tomography. *Science* 226: 1236–1239.

Crichton, M. 1990. *Jurassic Park*. New York: Ballantine.

Czerkas, S. A. and X. Xu. 2002. A new toothed bird from China. In S. J. Czerkas (ed.), *Feathered Dinosaurs and the Origin of Flight*, 1: 43–61. *Dinosaur Museum Journal* 1. Blanding, Utah: Dinosaur Museum.

Czerkas, S. A., and Q. Ji. 2002a. A new rhamphorhynchoid with a headcrest and complex integumentary structures. In S. J. Czerkas (ed.), *Feathered Dinosaurs and the Origin of Flight*, 1:15–41. *Dinosaur Museum Journal* 1. Blanding, Utah: Dinosaur Museum.

———. 2002b. A preliminary report on an omnivorous volant bird from northeast China. In S. J. Czerkas (ed.), *Feathered Dinosaurs and the Origin of Flight*, 1: 127–35. *Dinosaur Museum Journal* 1. Blanding, Utah: Dinosaur Museum.

Deck, L. T., R. E. Chapman, and A. Andersen. 2004. Conservation of vertebrate fossils and collections data using virtualization. *Journal of Vertebrate Paleontology* 24 (suppl. to 3): 51A.

Deck, L. T., R. Schlader, N. Clement, W. Gibbs, and R. E. Chapman. 2007. Especially durable prototypes of fossil specimens and replicas for use in public programming. *Journal of Vertebrate Paleontology* 27 (suppl. to 3): 67A.

Dominguez Alonso, P., A. C. Milner, R. A. Ketcham, M. J. Cookson, and T. B. Rowe. 2004. The avian nature of the brain and inner ear of *Archaeopteryx*. *Nature* 430: 666–669.

Evans, T., K. Poole, J. Smith, and S. Novak. 2005. A simple, cheap, non-invasive, and fast quarry mapping system with comparisons to common quarry mapping techniques. *Journal of Vertebrate Paleontology* 25 (suppl. to 3): 55A.

Farke, A. F., R. E. Chapman, and A. Andersen. 2010. Modeling structural properties of the frill of *Triceratops*. In M. J. Ryan, B. J. Chinnery-Allgeier, and D. A. Eberth (eds.), *New Perspectives on the Horned Dinosaurs*, 264–270. Bloomington: Indiana University Press.

Farlow, J. O. 1991. *On the Tracks of Dinosaurs. A Study of Dinosaur Footprints.* New York: Franklin Watts.

———. 1993. *The Dinosaurs of Dinosaur Valley State Park*. Austin: Texas Parks and Wildlife Department.

Farlow, J. O., and R. E. Chapman. 1997. The scientific study of dinosaur footprints. In J. O. Farlow and M. K. Brett-Surman (eds.), *The Complete Dinosaur*, 519–553. Bloomington: Indiana University Press.

Farlow, J. O., R. E. Chapman, B. Breithaupt, and N. Matthews. 2011 (this volume). The scientific study of dinosaur footprints. In M. K. Brett-Surman, Thomas R. Holtz Jr., and James O. Farlow (eds.), *The Complete Dinosaur*, 2nd ed., 712–459. Bloomington: Indiana University Press.

Franzosa, J., and T. Rowe. 2005. Cranial endocast of the Cretaceous theropod dinosaur *Acrocanthosaurus atokensis*. *Journal of Vertebrate Paleontology* 25 (4): 859–864.

Gardner, S. P., and L. H. Taylor. 1994. Ground penetrating radar survey of Bone-Cabin Quarry. In *Forty-Fourth Annual Field Conference*, 39–60. Wyoming Geological Association.

Gatesy, S. M., K. M. Middleton, F. A. Jenkins Jr., and N. H. Shubin. 1999. Three-dimensional preservation of foot movements in Triassic theropod dinosaurs. *Nature* 399: 141–144.

Gatesy, S. M., N. H. Shubin, and F. A. Jenkins Jr. 2005. Anaglyph stereo imaging of dinosaur track morphology and microtopography. *Palaeontologia Electronica* 8 (1) [10A]: 1–10.

Gillette, D. D. 1994. *Seismosaurus: The Earth Shaker*. New York: Columbia University Press.

Goodwin, M. B. 2002. Stable isotopes and dinosaur endothermy effects to the burial environment on hadrosaur biogeochemistry. *Journal of Vertebrate Paleontology* 22 (3): 59A

Goto, M., M. Araki, and S. Tomono. 1996. A method for representation of dinosaur footprints using 3-D photographs. *Bulletin of the Toyama Science Museum* 19: 1–7.

Harbersetzer, J. 1994. Radiography of fossils. In P. Leiggi and P. May (eds.), *Vertebrate Paleontological Techniques*, 1: 329–339. New York: Cambridge University Press.

Hooper, K. 1965a. X-ray microscopy in morphological studies of fossils. In B. Kummel and D. Raup (eds.), *Handbook of Paleontological Techniques*, 320–326. San Francisco: Freeman.

———. 1965b. Analytical techniques applied to fossils and shells. In B. Kummel and D. Raup (eds.), *Handbook of Paleontological Techniques*, 360–387. San Francisco: Freeman.

Horner, J. R., A. de Ricqlès, and K. Padian. 2000. Long bone histology of the hadrosaurid dinosaur *Maiasaura peeblesorum*: growth dynamics and physiology based on an ontogenetic series of skeletal ele

ments. *Journal of Vertebrate Paleontology* 20 (1): 115–129.

Hutchinson, H. 1999. Relief for weary bones. *Mechanical Engineering* 121 (6): 54–57.

Ishigaki, S., and T. Fujisaki. 1989. Three dimensional representation of *Eubrontes* by the method of moiré typography. In D. D. Gillette and M. G. Lockley (eds.), *Dinosaur Tracks and Traces*, 421–425. Cambridge: Cambridge University Press.

Jabo, S. J., P. A. Kroehler, A. F. Andersen, and R. E. Chapman. 1999. The use of three-dimensional computer imaging and scale-model prototypes in the mounting of a cast of *Triceratops*. *Journal of Vertebrate Paleontology* 9 (suppl. to 3): 54A.

Jefferson, G. T. 1989. Digitized sonic location and computer imaging of Rancho La Brea specimens from the Page Museum salvage. *Current Research in the Pleistocene* 6: 45–47.

Jorstad, T., and J. Clark. 1995. Mapping human origins on an ancient African landscape *Professional Surveyor* 15 (4): 10–12.

Kaye, T. G. 2007. Detection of micro-fossils using laser-stimulated fluorescence. *Proceedings of the 13th Annual Symposium in Paleontology and Geology*. Casper, Wyo.

Kaye, T. G., and J. P. Cavigelli. 2005. Remote detection of fossils using infrared spectroscopy. *Journal of Vertebrate Paleontology* 25 (suppl. to 3): 71B.

Larsson, H. C. E., P. C. Sereno, and J. A. Wilson. 2000. Forebrain enlargement among nonavian theropod dinosaurs. *Journal of Vertebrate Paleontology* 20 (3): 615–618.

Leberl, F., A. Irschara, T. Pock, P. Meixner, M. Gruber, S. Scholz, and A. Wiechert. 2010. Point clouds: lidar versus 3D vision. *Photogrammetric Engineering and Remote Sensing* 76 (10): 1123–1134.

Lillesand, T. M., R. W. Kiefer, and J. W. Chipman. 2004. *Remote Sensing and Image Interpretation*. 5th ed. Hoboken, N.J.: Wiley.

Lim, S.-K., S. Young, and M. G. Lockley. 1989. Large dinosaur footprint assemblages from the Cretaceous Jindong Formation of Korea. In D. D. Gillette and M. G. Lockley (eds.), *Dinosaur Tracks and Traces*, 333–336. Cambridge: Cambridge University Press.

Main, D. J., and A. R. Fiorillo. 2002. Results of the ground penetrating radar mapping techniques at a dinosaur quarry in Big Bend National Park. *Geological Society of America, Abstracts with Programs* 34 (6): 539–540.

Main, D. J., and W. S. Hammon III. 2003.

The application of ground penetrating radar as a mapping technique at vertebrate fossil excavations in the Cretaceous of Texas. *Cretaceous Research* 24: 335–345.

Matthews, N. A. 2008. Resource documentation, preservation, and interpretation: aerial and close-range photogrammetric technology in the Bureau of Land Management. Technical Note 428. *U.S. Department of the Interior, Bureau of Land Management, Heritage Resources.* http://www.blm.gov/wo/st/en/prog/more/CRM/publications.html.

Matthews, N. A., and B. H. Breithaupt. 2001. Close-range photogrammetric experiments at Dinosaur Ridge. *Mountain Geologist* 38 (3): 147–153.

———. 2009. Close-range photogrammetric technology for paleontological resource documentation, preservation, and interpretation. In S. E. Foss, J. L. Cavin, T. Brown, J. I. Kirkland, and V. L. Santucci (eds.), 2009 *Proceedings of the Eighth Conference on Fossil Resources,* 94–96. Salt Lake City, Utah: Bureau of Land Management, Utah State Office.

Matthews, N. A., T. A. Noble, and B. H. Breithaupt. 2006. The application of photogrammetry, remote sensing and geographic information systems (GIS) to fossil resource management. In S. G. Lucas, J. A. Spielmann, P. M. Hestor, J. P. Kenworthy, and V. L. Santucci (eds.), *Fossils from Federal Lands,* 119–131. New Mexico Museum of Natural History and Science Bulletin 34.

McKenna, P. C. 1992. GPS in the Gobi: dinosaurs among the dunes. GPS *World* 3 (6): 20–26.

Mikhail, E. M., J. S. Bethel, and J. C. McGlone. 2001. *Introduction to Modern Photogrammetry.* Hoboken, N.J.: Wiley.

Millikan, G. C., and R. I. Bowman. 1967. Observations on Galapagos tool-using finches in captivity. *Living Bird* 6: 23–41.

Moltenbrey, K. 2001. No bones about it. *Computer Graphics World* 24 (2): 24–30.

Nave, J., and S. Matney. 2005. Reflectorless versus retro-reflector EDM for data collection at the Gray Fossil site: a statistical analysis for precision. *Journal of Vertebrate Paleontology* 25 (suppl. to 3): 95A.

Oheim, K., and J. Hall. 2005. Finding fossils: a GIS-based suitability analysis of the Two Medicine Formation in central Montana. *Journal of Vertebrate Paleontology* 25 (suppl. to 3): 97A–98A.

Owings, R. 2005. GPS *Mapping.* Fort Bragg, Calif.: Ten Mile Press.

Plotnick, R. E., and G. Harris. 1989. X-ray analysis using energy dispersive and wavelength dispersive spectroscopy. In R.

M. Feldmann, R. E. Chapman, and J. T. Hannibal (eds.), *Paleotechniques.* Paleontological Society Special Publication 4: 179–185.

Rasskin-Gutman, D., G. Hunt, R. E. Chapman, J. L. Sanz, and J. J. Moratalla. 1997. The shapes of tridactyl dinosaur footprints: procedures, problems and potentials. In D. L. Wolberg, E. Stump, and G. D. Rosenberg (eds.), *Dinofest International,* 377–383. Philadelphia, Pa.: Academy of Natural Sciences.

Rayfield, E. J., D. B. Norman, C. C. Horner, J. R. Horner, P. M. Smith, and J. J. Smith. 2001. Cranial design and function in a large theropod dinosaur. *Nature* 409: 1033–1037.

Rolfe, W. D. I. 1965a. Uses of infrared rays. In B. Kummel and D. Raup (eds.), *Handbook of Paleontological Techniques,* 344–350. San Francisco: Freeman.

———. 1965b. Uses of ultraviolet rays. In B. Kummel and D. Raup (eds.), *Handbook of Paleontological Techniques,* 350–360. San Francisco: Freeman.

Rothschild, B. M., and D. Tanke. 2005. Theropod paleopathology. State-of-the-art review. In K. Carpenter (ed.), *The Carnivorous Dinosaurs,* 351–365. Bloomington: Indiana University Press.

Rothschild, B. M., D. Tanke, and K. Carpenter. 1997. *Tyrannosaurus* suffered from gout. *Nature* 387: 357.

Rowe, T., C. A. Brochu, and K. Kishi. 1999. Cranial morphology of *Alligator mississippiensis* and phylogeny of the Alligatoroidea. *Journal of Vertebrate Paleontology* 19 (suppl. to 2): 1–100.

Rowe, T., W. Carlson, and W. Bottorff. 1995. *Thrinaxodon: Digital Atlas of the Skull.* 2nd ed., CD-ROM. Austin: University of Texas Press.

Rowe, T., R. Ketcham, C. Denison, M. Colbert, X. Xu, and P. J. Currie. 2001. The *Archaeoraptor* forgery. *Nature* 410: 539–540.

Russell, D. A., and Z. Dong. 1993. A nearly complete skeleton of a new troodontid dinosaur from the Early Cretaceous of the Ordos Basin, Inner Mongolia, People's Republic of China. *Canadian Journal of Earth Sciences* 30: 2163–2173.

Schwarz, D., P. Vontobel, E. H. Lehmann, C. A. Meyer, and G. Bongartz. 2005. Neutron tomography of internal structures of vertebrate remains: a comparison with x-ray computed tomography. *Palaeontologia Electronica* 8 (2) [30A]: 11 pp.

Schweitzer, M. H., and M. Marshall. 2011 (this volume). Claws, scales, beaks, and feathers: molecular traces in the fossil record. In M. K. Brett-Surman, Thomas R.

Holtz Jr., and James O. Farlow (eds.), *The Complete Dinosaur,* 2nd ed., 273–284. Bloomington: Indiana University Press.

Shay, D., and J. Duncan. 1993. *The Making of Jurassic Park.* New York: Ballantine Books.

Slotnick, R. S. 2001. Fossil hunting, by radar. *American Scientist* 89 (1): 26–27.

Stevens, K. A. 2002. Dinomorph: parametric modeling of skeletal structure. *Senckenbergiana Lethaea* 82 (1): 23–34.

Stevens, K. A., and J. M. Parrish. 1999. Neck posture and feeding habits of two Jurassic sauropod dinosaurs. *Science* 284: 798–800.

———. 2005. Digital reconstructions of sauropod dinosaurs and implications for feeding. In K. C. Rogers and J. A. Wilson (eds.), *The Sauropods: Evolution and Paleobiology,* 178–200. Berkeley: University of California Press.

Stokosa, K. 2005. Enamel microstructure variation within the Theropoda. In K. Carpenter (ed.), *The Carnivorous Dinosaurs,* 163–178. Bloomington: Indiana University Press.

Tykoski, R. S., T. B. Rowe, R. A. Ketcham, and M. W. Colbert. 2002. *Calsoyasuchus valliceps,* a new crocodyliform from the Early Jurassic Kayenta Formation of Arizona. *Journal of Vertebrate Paleontology* 22 (3): 593–611.

Varricchio, D. J., J. R. Horner, and F. D. Jackson. 2002. Embryos and eggs for the Cretaceous theropod dinosaur *Troodon formosus. Journal of Vertebrate Paleontology* 22 (3): 564–576.

Walters, R., R. E. Chapman, and R. A. Snyder. 2001. Fleshing-out *Triceratops*: adding muscle and skin to the virtual *Triceratops. Journal of Vertebrate Paleontology* 21 (suppl. to 3): 111A.

Walters, R. F., R. E. Chapman, R. A. Snyder, and B. J. Mohn. 2000. Using virtual skeletons as a basis for reconstructing fossil vertebrates. *Journal of Vertebrate Paleontology* 20 (suppl. to 3): 76A.

Wedel, M. J. 2003. The evolution of vertebral pneumaticity in sauropod dinosaurs. *Journal of Vertebrate Paleontology* 23 (2): 344–357.

Wegweiser, M. D., B. H. Breithaupt, N. A. Matthews, J. W. Sheffield, and E. S. Skinner. 2004. Paleoenvironmental and diagenetic constraints on Late Cretaceous dinosaur skin from western North America. *Sedimentary Record* 2 (1): 4–8.

Weishampel, D. B. 1984. Evolution of jaw mechanisms in ornithopod dinosaurs. *Advances in Anatomy, Embryology and Cell Biology* 87: 1–110.

Whitlock, K. 2001. Saving face. *Perspectives* (Ohio State University) 5 (11): 14–21.

Wilhite, R. 2003. Digitizing large fossil skeletal elements for three-dimensional applications. *Palaeontologia Electronica* 5 (1): 10 pp.

Witmer, L. M. 1995. Homology of facial structures in extant archosaurs (birds and crocodilians), with special reference to paranasal pneumaticity and nasal conchae. *Journal of Morphology* 225: 269–327.

——. 1997. The evolution of the antorbital cavity of archosaurs: a study in soft-tissue reconstruction in the fossil record with an analysis of the function of pneumaticity. *Memoirs of the Society of Vertebrate Paleontology, Journal of Vertebrate Paleontology* 17 (suppl. to 1): 1–73.

——. 2001. Nostril position in dinosaurs and other vertebrates and its significance for nasal function. *Science* 293: 850–853.

——. 2004. Inside the oldest bird brain. *Nature* 430: 619–620.

Witten, A., D. D. Gillette, J. Sypniewski, and W. C. King. 1992. Geophysical diffraction tomography at a dinosaur site. *Geophysics* 57: 187–195.

Zangerl, R. 1965. Radiographic techniques. In B. Kummel and D. Raup (eds.), *Handbook of Paleontological Techniques*, 305–320. San Francisco: Freeman.

Zangerl, R., and H.-P. Schultze. 1989. X-radiographic techniques and applications. In B. Kummel and D. Raup (eds.) *Handbook of Paleontological Techniques*, 165–177. San Francisco: Freeman.

Claws, Scales, Beaks, and Feathers: Molecular Traces in the Fossil Record

14

Mary Higby Schweitzer and Mark Marshall

Introduction

Over the last few decades, discoveries of fossil representatives of extinct taxa have increased at an exponential rate, and these new finds have added greatly to our understanding of evolutionary branches on the tree of life, when subjected to phylogenetic analyses based on morphological features preserved in the bones. The field of molecular biology has likewise advanced exponentially as scientists have begun to uncover the relationships between DNA, proteins, and other biomolecules, and how these molecules relate to both life processes and evolutionary history. The genomes of many organisms have been sequenced, and protein sequences and proteomes obtained by mass spectrometry are also known. By comparing the sequences of bases in DNA or amino acids in proteins derived from various taxa, we can predict phylogenetic relationships based on the number of changes in known genes (Fig. 14.1). Melding the fields of paleontology and molecular biology holds great promise for increasing our understanding of the history of life on this planet.

Proteins are the products of genes, and they are the means by which genetic instructions are carried out. Because they are encoded by the strands of DNA that each organism carries within its cells, proteins reflect genetic distances between taxa in the same manner as DNA. These protein distances can be exploited by the immune system of any animal. For example, if a host rabbit is injected with a purified bird protein, the rabbit will produce antibodies that will recognize the foreign protein and bind to it. The antibody binds to its target in a specific lock-and-key configuration (Fig. 14.2). Antibodies recognize the molecular conformation (essentially,

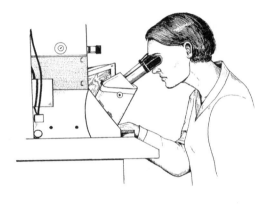

A

Turkey	1	L I D L P T P S N I S A W W N F G S L L A V C L I T Q I L T G L L
Human	1	F I D L P T P S N I S A W W N F G S L L G T C L I L Q I T T G L F
Alligator	1	L I D L P T P S N I S A W W N F G S L L G L T L L I Q I L T G F F
Iguana	1	F I D L P A P S N I S A W W N F G S L L G L C L I M Q T L T G L F

Turkey	34	L A M H Y T A D T T L A F S S V A Y T C R N V Q Y G W L L H N L H
Human	34	L A M H Y S P D A S T A F S S I A H I T R D V N Y G W I I R Y L H
Alligator	34	L M M H F S S S D T L A F S S V S Y T S R E V W F G W L I R N L H
Iguana	34	L A M H Y T A D I S S A F S S V A H I C R D V Q H G W L I R N L H

B

Turkey	1	TTA ATC GAC CTC CCA ACC CCA TCC AAC ATC TCC GCT
Human	1	TTc ATC GAC CTC CCc ACC CCA TCC AAC ATC TCC GCa
Alligator	1	cTA ATt GAC CTa CCA ACa CCc TCa AAC ATC TCC GCT
Iguana	1	TTc ATC GAC CTg CCc gCC CCc TCC AAC ATC TCt GCa

Turkey	37	TGA TGA AAC TTC GGC TCC CTA CTA GCA GTA TGC CTC
Human	37	TGA TGA AAC TTC GGC TCa CTA CTA GaC Gcc Tca acC
Alligator	37	TGA TGA AAC TTt GGa TCa CTA CTA Ggc cTA acC CTa
Iguana	37	TGA TGA AAC TTC GGC TCa CTA CTA GcA cTt TGC CTa

14.1. A comparison of protein (a) and DNA (b) sequences of the *cytochrome B* gene and its expressed protein, derived from various taxa. As can be seen, this is a highly conserved gene with little overall variability, despite the phylogenetic distances that separate the taxa in question. The protein sequences for the same taxa likewise show a high degree of overall similarity.

three-dimensional shape, charge, and hydrophobicity) of epitopes, which are small regions of the whole protein molecule. In general, the evolutionary distance between organisms, reflected by variations in protein primary structure, dictates the final three-dimensional structure of the protein. The greater the evolutionary distance between the host animal and the animal from which the protein is derived, the greater the immune response. Conversely, if the antibodies thus produced are allowed to react with proteins from different sources, they will react most intensely, with the one closest in either primary sequence or three-dimensional conformation to the protein used to generate the antibodies. Usually conformation in three dimensions is a direct product of primary structure. Therefore, an immunological approach can give a good first approximation of the identity of molecules that may be present in a tissue, as well as a reasonable estimate (coupled with other methods) for phylogenetic placement.

Molecular Longevity

Biomolecules are divided into many classes, and within these classes, rates of degradation vary widely. Some molecules, such as some hormone proteins or messenger RNA, have short life spans, degrading in minutes within the living organism. Other molecules, including lipids, some modified polysaccharides such as chitin or lignin, and structural proteins such as collagen or keratin, must be extremely durable in order to carry out their intended function. Chitin and various lipids have been identified in fossils millions of years old (Stankiewicz et al. 1997, 1998).

Collagen is the primary protein in bone, making up 88% or more of the organic matrix (Miller 1984). The structure of collagen resembles a tightly wound three-stranded rope, with each strand bound to the others via interactions between its constituent amino acids (Fig. 14.3). In addition, when bone begins to calcify, the crystals of bone mineral hydroxyapatite $(Ca_{10}(PO_4)_6(OH)_2)$ align along the length of the fibrils of forming collagen, and in some cases even completely encase the protein molecule (Weiner et al. 1989; Glimcher et al. 1990; Salamon et al. 2005; Sykes et al. 2000). Because collagen has a high affinity for apatite mineral, this interaction serves to provide further resistance to degradation. Finally, collagen is not very water soluble, and only after degradation has begun and it is broken into small fragments can it be carried away by infiltrating water. All of these factors combine to make collagen a durable protein likely to persist for long periods of time. Its presence has been identified in fossils as old as 80 Ma (Asara et al. 2007a, 2007b; Schweitzer et al. 2007, 2009, and references therein).

Our increasing understanding of the durability and degradation processes of biomolecules, the enhanced sensitivity and specificity of analytical techniques, and our greater knowledge of the relationships between DNA, proteins, and evolution have led to efforts to extract biomolecules from fossils of extinct animals. Such molecules or molecular fragments lead to a new awareness of physiology and genetics from these organisms that are unable to be deduced by morphology alone.

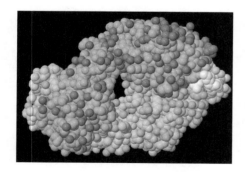

14.2. Three-dimensional space-filling model of an antibody bound to its ligand (Tormo et al. 1994). This binding is very specific and is a result of the interaction between the three-dimensional shapes of the antibody binding pocket and the ligand or epitope to which it binds. This three-dimensional shape is the result of the primary structure (amino acid sequence) of the antibody protein, which is in turn encoded by the base sequence of DNA.

Fossils from various ages have been tested for biomolecular content, with varying degrees of success. The advent of polymerase chain reaction (PCR) technologies (Mullis et al. 1986) has been applied to obtain DNA sequence data from fossils (DeSalle 1992; Cano et al. 1993; Woodward et al. 1994; Cooper 1994; Krings et al. 1997; Lindqvist et al. 2010; Paabo et al. 2004 and references therein), although DNA reported to have been recovered from fossils many millions of years old has been met with skepticism (Hedges and Schweitzer 1995; Austen et al. 1997; Cooper and Poinar 2000). PCR uses enzymes to make many copies of rare template molecules. While this makes it ideal for the study of rare, ancient DNA molecules, its extreme sensitivity means that DNA from a contaminating living source is much more likely to be amplified than the degraded and fragmentary molecules of truly ancient DNA. Efforts to verify sequence data from specimens similar to those used in the early reports for very old fossils have been unsuccessful, and the validity of these works has been called into question (Austen et al. 1997; Cooper and Poinar 2000).

When working with ancient proteins, however, chemical amplification is not a factor. Proteins that are vertebrate specific can be distinguished from microbial or other potential contaminants both by sequence data and by antibody reactivity. Applying multiple analytical techniques can also rule out contamination as a source of signal. For example, antibodies against a particular protein, such as collagen, can be applied to ancient samples to test for cross-reactivity. If positive, extracts of the same fossil material can be used to make antibodies as described above. These antibodies can then tested against a battery of purified proteins from living animals, and when cross-reactivity occurs, it is indicative of the source protein in the fossil tissues that antibodies were generated against. When these experiments are coupled with analytical tests on the bone extracts, such as mass spectroscopy, the presence of protein fragments or protein-derived compounds is independently supported.

The first efforts to test fossils for amino acid/protein content were carried out by Abelson (1956), who was able to identify amino acids in extracts of Ordovician invertebrate shells. Scientists have since identified the presence of collagen (Tuross et al. 1980; Rowley et al. 1986; Grupe 1995; Schweitzer et al. 2002), osteocalcin (Ulrich et al. 1987; Muyzer et al. 1992; Nielsen-Marsh et al. 2002), albumin (Prager et al. 1980; Tuross 1989), hemoglobin (Loy and Wood 1989; Loy et al. 1990; Schweitzer et al. 1997), and keratin (Gillespie 1970; Schweitzer et al. 1999a, 1999b) from fossils of various ages, from Mesozoic to Pleistocene.

Despite these reports, the controversy that surrounds the reports of endogenous molecules (Lindahl 1992; Austen et al. 1997) is based on the elucidation of molecular kinetics of degradation of macromolecules (Collins et al. 1999), on laboratory experiments that attempt to simulate the aging process by exposing solubilized molecules to artificially extreme conditions (Lindahl 1992), and on the difficulties of releasing the molecules of interest from heterogeneous organic complexes formed through cross-linking during degradation (Poinar et al. 1998).

Molecular Signals in the Fossil Record

14.3. Space-filling model showing the three-dimensional structure of the collagen molecule. This molecule consists of two identical alpha-one chains and one slightly different alpha-two chain. The overall structure of the molecule is highly constrained because the frequent twists in the molecule as the chains wrap around each other require that the smallest amino acid, glycine, be present at each turn. Therefore, collagen is easily recognized by an overall amino acid content that is one third glycine, and a primary sequence of the pattern repeat GLY-X-Y, where X and Y are quite often the amino acids proline or lysine.

Preservation Potential

As mentioned above, the chemical characteristics of organic molecules such as hydrophobicity or charge play an important role in predicting molecular survival and are often closely related to the role of the protein. Proteins that have a role in structure or support, such as collagen and keratin, survive both in living tissues and in fossil specimens far longer than other proteins, such as hormones or enzymes.

In addition to structure, environmental and geological factors may contribute to the potential for recovery of molecular fragments from the fossil record. If a fossil is buried rapidly or is buried in sediments where exposure to water, oxygen, or both is limited, degradative processes are slowed and biomolecular life may be extended. If molecules have a mineral surface on which to adsorb, they are stabilized and thus protected. Finally, if clays are present in the surrounding sediments, degradative enzymes are inactivated, also contributing to preservation (Butterfield 1990).

Keratin Proteins

The skin, the first barrier between our internal and external environments, must resist abrasion, form an impenetrable defense against microbial invasion, and maintain the delicate water balance needed for life chemistry. Keratin, the dominant protein in skin and skin-derived tissues, is vital to these functions, and as such exhibits great durability. Because keratin incorporates many hydrophobic amino acids for its role in maintaining water balance, it is also insoluble in water, and it resists the hydrolytic damage that is often the starting point for degradation. Structures that were originally keratinous, such as feathers or claw sheath material, are second only to bone and teeth in frequency of occurrence (Logan et al. 1991, reviewed in Schweitzer 2011). Because these tissues in life are not normally stabilized through biomineralization, as are bone and tooth proteins, their persistence in the fossil record implies that their constituent proteins were sufficiently resistant to persist until mineralization occurred.

Keratin is actually the product of many genes, all derived from a common precursor. Furthermore, there are two main types of keratins that are phylogenetically distinct. Alpha keratins are produced by all vertebrate epidermal cells, and an antiserum raised against avian or reptilian alpha keratins will cross-react with human skin or nail. However, beta keratin proteins are unique to reptiles and birds. The genes that encode the beta keratins arose sometime after the divergence of the mammals and reptiles (Fig. 14.4); thus, this protein is only expressed in skin and skin-derived structures of reptiles and birds. Antibodies against avian beta keratins will react to crocodile skin but will not react to any mammal tissues.

The skin, scales, claw sheaths, beaks, and other structures of birds and reptiles consist of both alpha and beta keratins. Although the proteins that make up these various structures differ from one another in sequence, size, protein filament diameter, and electrophoretic properties (Brush 1980; Gregg et al. 1983; Wyld and Brush 1983; Shames and Sawyer 1987; Shames et al. 1989), they are similar enough in protein structure that antibodies generated against a tissue composed of both alpha and beta keratin in birds will cross-react with similar tissues in crocodiles.

Virtually all epidermal structures of birds and reptiles, except feathers, consist of both alpha and beta keratins. Only beta keratin proteins exist

14.4. A, Alignment of amino acid sequences of various avian keratinous structures, compared with human hair alpha keratin as the outgroup. B, Phylogenetic tree representing the key events in the origin, diversification, and distribution of the keratin gene families among vertebrates. Because the beta keratin proteins are only possessed by modern archosaurs and squamates, the origin of this gene family must have occurred after the divergence of the mammalian lineages. While scales, beaks, skin, and claws of members of these taxa consist of both alpha and beta keratins, among extant taxa, only feathers consist solely of beta keratins. The earliest point at which epidermal structures consisting of only beta keratin *could* have occurred, therefore, was after the divergence of birds and crocodiles. Whether pterosaurs and dinosaurs possessed such beta-keratin-only structures is a matter of speculation; fossil evidence shows that some dinosaur lineages had feathers that are at least morphologically identical to those of modern birds. The beta keratin proteins that are unique to extant feathers demonstrate a deletion when compared to beta keratins of scales, beaks, and other epidermal structures. This deletion may have occurred before the labeled point on this diagram but did occur at least at this point.

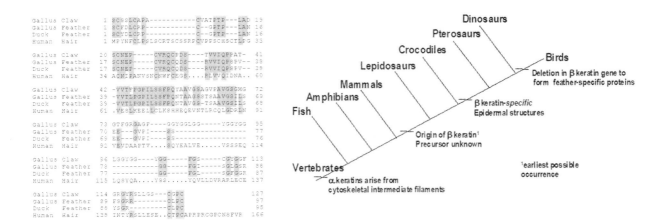

in the feathers of postembryonic birds (Knapp personal communication). Therefore, beta keratin antibodies will react with feathers, while alpha keratin antibodies will not.

Rahonavis ostromi

In 1995, a team of scientists (including the authors of this chapter) from the State University of New York, Stony Brook, and the Universite d'Antananarivo recovered a spectacular specimen of a new, primitive bird from the Cretaceous Maevarano Formation of Madagascar (Forster et al. 1998a, 1998b). During preparation of the sickle-shaped left ungual phalanx, we noted that fibrous material covered part of the regions of the bone. This material was not seen in association with any other element of the skeleton, nor was it observed in the depositional sediments. Because of the appearance and location of this material, we hypothesized that it represented the remains of the keratinous claw sheath that in life would have overlain the ungual.

The depositional setting of this specimen favored preservation of molecular compounds, although even the best environment does not guarantee such preservation. The bird was preserved in articulation, indicating burial rapid enough to slow microbial decay. Furthermore, analyses of the sediments reveals a semiarid environment, and the particles of clay mixed with quartz probably limited exposure to what water did pass through the system. Additionally, there has been virtually no cycling or other alterations of the immature sediments since the Cretaceous (R. Rogers personal communication), indicating that the specimen was never exposed to excessive heat or pressure.

The fibrous material from the ungual of the bird showed that it was morphologically similar to the keratinous sheath covering the ungual of an extant emu, in that it exhibited the same fibrous and ridged pattern as the emu claw (Schweitzer et al. 1999a) when studied by electron microscopy. However, unlike the emu, the ancient claw material was overlain with a thin coating of mineral. Subsequent elemental analyses showed that high levels of calcium and phosphorous were localized to the claw sheath material. Because the surrounding sediment were not phosphatic (R. Rogers personal communication), bone apatite is the likely source of this mineral coating.

Examples of Keratin in the Fossil Record

14.5. Illustration of the principles of in situ immunohistochemistry. 1, Tissue is first embedded in a polymer, preferably one that is water soluble, which will allow penetration of the antibody preparations, and sectioned or ground to appropriate thinness. Epitopes within the tissue are then exposed by etching, if necessary. 2, Specific primary antiserum is then applied to the tissues, and the antibodies bind to the epitopes to which they are specific. 3, Excess sera are removed through multiple washes, and a secondary antibody that recognizes the bound primary antibodies is applied and allowed to incubate with the tissues. 4, After multiple washes, a fluorescent-labeled conjugate is applied, which recognizes the biotin complexed to the secondary antibodies. 5, The fluorescent label is then detected by appropriate laser stimulation.

1. Prepare embedded thin section.

2. Bind antigen-specific rabbit antibodies.

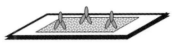

3. Bind biotin-coupled anti-rabbit antibodies.

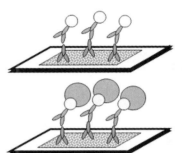

4. Bind FITC-conjugated avidin.

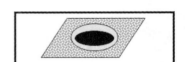

5. Detect complexes by confocal microscopy.

We embedded both the emu and Rahonavis claw samples in a polymer resin and collected 0.5-μm sections. No grinding was required, which is rather unusual for fossil material and indicates that there was little secondary mineralization. Grains of mineral will break the glass knives when they strike it, and indeed, for this reason, we were unable to section the sediments surrounding this material in a similar manner. When we visualized the sections under transmitted light, the fibrous character was confirmed, and the fibers were shown to run somewhat parallel to one another.

We applied antisera against either alpha or beta keratin proteins to the sections of claw material, as detailed in Figure 14.5. As controls, we also used sera from rabbits that had not been immunized with keratin (normal sera) and therefore should not react with keratin proteins in tissues. To show that reactivity was specific for keratin and was not just a generalized immune response, we also applied antisera human adrenocorticotropin hormone, a common, short-lived signaling protein. Finally, we incubated the beta keratin antiserum with minced feathers or with extracts of feathers. This causes competition for the binding site in the antibodies, thus blocking their ability to bind to the same components in the claw tissues. All of these antisera were classified as primary because they contained the antibodies that would directly bind to the tissues.

After incubation, sections were washed to remove unbound antibodies, then incubated with a second antisera that would recognize the primary antibodies bound to the sample. Finally, a fluorescent label, which fluoresces brightly when stimulated by laser light, was bound to the secondary antibodies. If the primary antibodies had not bound to tissue, none of the other antibodies would bind either, and fluorescence would not occur. This procedure builds an antibody sandwich that will take a small signal and amplify it so that it can be detected (Fig. 14.6).

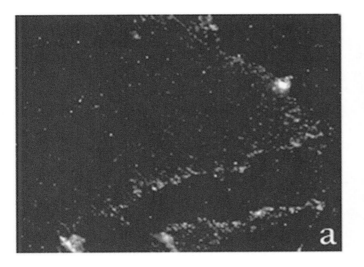

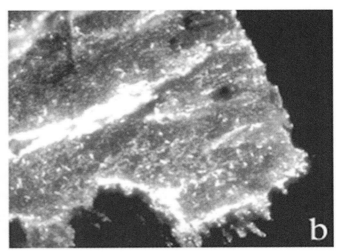

Our results clearly showed that both alpha and beta keratin antibodies bound specifically to *Rahonavis* claw material, just as with sections of modern emu claw sheath, demonstrating that alpha and beta keratin proteins were present in both claw samples. In negative controls, neither normal nor nonrelevant sera gave signal above background under identical test and data-gathering conditions, indicating that signal was specific for keratin. The results showed that antibodies recognized components in the ancient claw tissues that had molecular features in common with extant claw, and consistent with keratin proteins.

Mass spectroscopy data showed amino acid and peptide fragments consistent with proteinaceous material, supporting the immunological data. The specificity of the immunological reactions and the identification of peptides in the tissues both support the hypothesis that beta keratin protein fragments are preserved in these samples.

Shuvuuia deserti

Rahonavis is not the only Mesozoic fossil associated with originally keratinous material that reacts with antibodies against keratins. In 1993, the

14.6. Example of immunohistochemical results on sections of *Rahonavis ostromi* claw sections. A, A 0.5-μm section of claw, exposed to sera from a rabbit that had not been immunized against beta keratin. Lack of fluorescent signal indicates no antibody bound to tissues. B, A 0.5-μm section of claw, exposed to rabbit sera containing antibodies against beta keratin, a protein not produced by mammals. Primary sera were applied at 1:50 dilutions, and data were integrated over half a second under identical conditions. C, Light micrograph of section in (B) showing fibrous nature of tissue.

American Museum of Natural History recovered an exceptionally well-preserved articulated skeleton of *Shuvuuia deserti*, a new species of alvarezsaurid (Dashzeveg et al. 1995). Small white fibers were noted in the sediments surrounding the skull and cervical region of this skeleton, some in clumps and others oriented parallel to one another. The location and orientation of these fibers suggested the possibility that they were remnants of a feathery body covering.

Although it was difficult to physically separate the delicate fibers from the surrounding sediment, we were able to collect them using ethanol to wet the sediments. At first examination, the fibers were not obviously similar to most extant bird feathers in that they did not branch. However, when compared to morphologically specialized feathers that also do not branch, such as those removed from the head of a turkey vulture, the similarities between samples were apparent (Schweitzer et al. 1999b).

We obtained thin sections of these fibers as described above. Because fungal hyphae or plant roots might also have a somewhat similar gross appearance to these fibers, we used standard chemical tests to eliminate the possibility that these were sources of the fibers. Fiber sections were exposed to cellulase, an enzyme that digests the cellulose in plant cell walls. While the fibers were not altered, plant material used as a positive control was completely digested. Likewise, a stain used to identify fungal contaminants was ineffective in staining the fibers from *Shuvuuia*. More importantly, sections of the fibers revealed a hollow core—a feature common to feathers but not to plants or fungi.

We incubated sections of the *Shuvuuia* fibers using the same keratin antisera and controls described above. Like the *Rahonavis* material, the fibers and extant feathers showed strong reactivity with the beta keratin antisera. However, unlike the claw samples, minimal staining was observed with the alpha keratin antiserum in both duck feather and *Shuvuuia* fibers, consistent with the fact that feathers consist almost exclusively of beta keratin. As before, incubation with beta keratin antisera that had been blocked to binding by extant feather extracts resulted in greatly reduced fluorescent signal, supporting the specificity of antibody response.

In addition, 95-nm sections of the fibers were visualized with transmission electron microscopy. Electron-dense mineral surrounded pockets containing filaments approximately 3–6 nm in diameter, consistent with the known diameter of beta keratin protein filaments.

Finally, mass spectroscopy again revealed the presence of amino acid fragments in the interior of these fibers. Although we cannot determine that the amino acids we detect are of ancient origin or that the amino acids are wholly endogenous to the sample, it does verify the presence of protein building blocks within the samples of *Shuvuuia* fibers. A complete discussion of these data is available in Schweitzer et al. (1999a, 1999b).

Discussion

All of the evidence presented here demonstrates that beta and alpha keratin antisera cross-react with components in structures in the fossil record that were originally keratinous. While the possibility exists that the results of any one of these tests can be explained by other processes not requiring the preservation of protein components, when taken together, the data are consistent with the preservation of actual protein fragments in the original tissues. The facts that mammals do not produce beta keratin and that beta keratin antibodies do not react with human tissues are further evidence that the original structures remain and are not derived from contamination.

These data also demonstrate the utility of immunological methods, in combination with other analytical techniques, to confirm the identity of enigmatic fossil structures. For example, while the feathers of *Archaeopteryx* are morphologically indistinguishable from those of modern birds, extant feathers have a wide variety of shapes, depending on location and function (Lucas and Stettenheim 1972). When considering the evolution of these complex structures, the a priori assumption is that their precursors were undoubtedly simpler in morphology, and we may not be able to recognize these simpler precursors through morphology alone. Molecular methodologies may help to identify structures in the fossil record that are related to extant feathers at the molecular level, though morphologically dissimilar.

Molecular methods may also be used to detect phylogenetic signal. With adequate controls, variability in strength of antibody binding may provide a rough estimate of relationships among taxa. And if present in sufficient quantities, it may eventually be possible to unravel sequence data contained in fossil specimens, allowing more accurate and detailed phylogenies to be constructed.

While these are all exciting and important possibilities, current data are extremely limited. The idea that endogenous molecules can be preserved over geological time periods is still controversial. Bench experiments and understanding of the processes of molecular degradation indicate that it is unreasonable for even small regions of proteins to remain intact. However, these experiments subject biomolecular compounds to unrealistically extreme conditions that are not likely to occur in nature, and therefore they may not be the best models for estimating the interactions that occur during the diagenetic history of a fossil. Much more work needs to be done on many more specimens to begin to obtain a clearer picture of the potential for molecular preservation in the fossil record.

Acknowledgments

This research used a variety of analytical techniques and required the input and advice of numerous experts. In particular, we would like to acknowledge Dr. Loren Knapp for his contribution of the polyclonal avian keratin antisera and for his advice and suggestions; Dr. Recep Avci of the ICAL facility of Montana State University, who contributed countless hours in the generation and interpretation of the mass spectroscopy results, and Nancy Equall for her imaging assistance; D. Krause, Cathy Forster, Mark Norell, and Luis Chiappe for contribution of specimens for analyses and for their unfailing support; C. Paden and J. Watt for immunochemical advice and laboratory facilities and E. Lamm and J. R. Horner for some of the histological preparations; and S. Pincus for advice, counsel, and support of all kinds. This research was funded in part by generous contributions from Dr. M. Marshall and Dr. G. Ellis, and by National Science Foundation grant EAR-97-53187.

Note

This article was written for the second edition of *The Complete Dinosaur* in 2003. During the time since I wrote the first version of this chapter, the field has made major advances in understanding factors involved in both the preservation and recovery of original biomolecules in fossil material of all ages. Fossils have been identified that exhibit exceptional preservation down to the microscopic level, and using new technologies capable of high-resolution, extremely sensitive analyses have been employed to determine whether preservation extends to the molecular level. Multiple investigators have employed a plethora of methods to test the extent of preservation and to compare the data recovered from fossils with similar data from extant taxa to identify modification and alteration.

In this chapter, my colleague Mark Marshall and I present some of the data supporting the preservation of keratin proteins associated with claw and other epidermal tissues recovered from the Cretaceous, related to peer-reviewed papers we published in 1999. Since the time of this writing, we have identified tissues and cells recovered from demineralized dinosaur material that are still soft, flexible, and transparent, and we have sequenced collagen protein from two dinosaurs—sequences that retain phylogenetic information.

The data we gathered for the preservation of keratin in two Cretaceous specimens, reported here, was compelling but does not stand alone. Establishing the preservation of molecules beyond theoretical limits of preservation requires cross-disciplinary evidence and adequate controls; it also requires an understanding of what these data can and cannot indicate about the specimens from which they were obtained. There are limits to all methods, and it is important that when going beyond conventional wisdom, all methods support the same conclusion. Sequence data are the holy grail for this kind of information, but because sequences generated by mass spectrometry are, at some level, subject to interpretation, they likewise cannot stand alone. We did not have the technology in the late 1990s to sequence keratin proteins from such small samples, but the multiple methods we used all supported the hypothesis that keratin was preserved in these epidermal structures derived from Cretaceous fossils.

Some aspects of this chapter are outdated in light of new data and new understanding, but the methods and approach presented here are still valid. For further and more recent information, the reader is referred to the papers below and references therein. Many other investigators have presented data from similar studies.

References

Asara, J. M., J. S. Garavelli, D. A. Slatter, M. H. Schweitzer, L. M. Freimark, M. Phillips, and L. C. Cantley. 2007a. Interpreting sequences from mastodon and *Tyrannosaurus rex*. *Science* 317: 1324–1325.

Asara, J. M., M. H. Schweitzer, M. P. Phillips, L. M. Freimark, and L. C. Cantley. 2007b. Protein sequences from mastodon (*Mammut americanum*) and dinosaur (*Tyrannosaurus rex*) revealed by mass spectrometry. *Science* 316: 280–285.

Asara, J. M., and M. H. Schweitzer. 2008a. Response to comment on "Protein sequences from mastodon and *Tyrannosaurus rex* revealed by mass spectrometry." *Science* 319: 33d.

Asara, J. M., M. H. Schweitzer, L. C. Cantley, J. S. Cottrell. 2008b. Response to comment on "Protein sequences from mastodon and *Tyrannosaurus rex* revealed by mass spectrometry." *Science* 321: 1040c.

Abelson, P. H. 1956. Paleobiochemistry. *Scientific American* 195: 83–92.

Austen, J. J., A. J. Ross, A. B. Smith, R. A. Fortey, and R. H. Thomas. 1997. Problems of reproducibility – does geologically ancient DNA survive in amber preserved insects? *Proceedings of the Royal Society of London B* 264: 467–474.

Butterfield, N. J. 1990. Organic preservation of non-mineralizing organisms and the taphonomy of the Burgess Shale. *Paleobiology* 16 (3): 272–286.

Brush, A. H. 1980. Chemical heterogeneity in keratin proteins of avian epidermal structures: possible relations to structure and function. In R. I. C. Spearman and P. A. Riley (eds.), *The Skin of Vertebrates*, 87–111.

Cano, R. J., H. N. Poinar, N. J. Pieniazek, A. Acra, and G. O. Poinar. 1993. Amplification and sequencing of DNA from a 120–135-million-year old weevil. *Nature* 363: 536–538.

Cooper, A. 1994. Ancient DNA sequences reveal unsuspected phylogenetic relationships within New Zealand wrens (Acanthisittidae). *Experientia* 50 (6): 558–563.

Cooper, A., and H. N. Poinar. 2000. Ancient DNA: do it right or not at all. *Science* 289 (5482): 1139.

Chiappe, L. M., M. A. Norell, and J. M. Clark. 1998. The skull of a relative of the stem-group bird, *Mononykus*. *Nature* 392: 275–278.

Collins, M. J., E. R. Waite, and A. C. van Duin. 1999. Predicting protein decomposition: the case of aspartic-acid racemization kinetics. *Philosophical Transactions of the Royal Society of London B* 354 (1379): 51–64.

Dashzeveg, D., M. J. Novacek, M. A. Norell, J. M. Clark, L. M. Chiappe, A. Davidson, M. C. McKenna, L. Dingus, C. Swisher, and A. Perle. 1995. Unusual preservation in a new vertebrate assemblage from the Late Cretaceous of Mongolia. *Nature* 374: 446–449.

DeSalle, R., J. Gatesy, W. Wheeler, and D. Grimaldi. 1992. DNA sequences from a fossil termite in Oligo-Miocene amber and their phylogenetic implications. *Science* 257: 1933–1936.

Forster, C. A., S. D. Sampson, L. M. Chiappe, D. W. Krause. 1998a. The theropod ancestry of birds: new evidence from the Late Cretaceous of Madagascar. *Science* 279: 1915–1919.

——. 1998b. Genus correction. *Science* 280: 185.

Gillespie, J. M. 1970. Mammoth hair: stability of alpha keratin structure

and constituent proteins. *Science* 170: 1100–1101.

Glimcher, M. J., L. Cohen-Solala, D. Kossiva, and A. de Ricqles. 1990. Biochemical analyses of fossil enamel and dentin. *Paleobiology* 16 (2): 219–232.

Gregg, K., S. D. Wilton, and G. E. Rogers. 1983. Avian keratin genes: organisation and evolutionary inter-relationships. In P. Nagley, A. W. Linnane, W. J. Peacock, and J. A. Pateman (eds.), *Manipulation and Expression of Genes in Eukaryotes*, 65–71.

Grupe, G. 1995. Preservation of collagen in bone from dry sandy soil. *Journal of Archaeological Science* 22: 193–199.

Hedges, S. B., and M. H. Schweitzer. 1995. Detecting dinosaur DNA. *Science* 268 (5214): 1191–1192.

Hongo, C., K. Noguchi, K. Okuyama, Y. Tanaka, and N. Nishino. 2005. Repetitive interactions observed in the crystal structure of a collagen-model peptide, [(Pro-Pro-Gly)9]3. *Journal of Biochemistry (Tokyo)* 138: 135–144.

Huq, N. L., A. Tseng, and G. E. Chapman. 1990. Partial amino acid sequence of osteocalcin from an extinct species of ratite bird. *Biochemistry International* 21 (3): 491–496.

Krings, M., A. Stone, R. W. Schmitz, H. Krainitzki, M. Stoneking, and S. Paabo. 1997. Neandertal DNA sequences and the origin of modern humans. *Cell* 90 (1): 19–30.

Lindahl, T. 1992. Instability and decay of the primary structure of DNA. *Nature* 362: 709–715.

Lindqvist, C., S. C. Schuster, Y. Sun, S. L. Talbot, J. Qi, A. Ratan, L. P. Tomsho, L. Kasson, E. Zeyl, J. Aars, W. Miller, O. Ingolfsson, L. Bachmann, and O. Wiig. 2010. Complete mitochondrial genome of a Pleistocene jawbone unveils the origin of polar bear. *Proceedings of the National Academy of Sciences of the United States of America* 107: 6118–6123.

Logan, G. A., M. J. Collins, and G. Eglinton. 1991. Preservation of organic biomolecules. In P. A. Allison and D. E. G. Briggs (eds.), *Taphonomy: Releasing the Data Locked in the Fossil Record*, 1–24. New York: Plenum Press.

Loy, T. H., R. Jones, D. E. Nelson, B. Meehan, J. Vogel, J. Southon, and R. Cosgrove. 1990. Accelerator radiocarbon dating of human blood proteins from late Pliestocene art sites in Australia. *Antiquity* 64: 110–116.

Loy, T. H., and A. R. Wood. 1989. Blood residue analysis at Cayonu Tepesii,

Turkey. *Journal of Field Archaeology* 16: 451–460.

Lucas, A. M., and P. R. Stettenheim. 1972. *Avian Anatomy? The Integument, Part I.* Agricultural Handbook 362. Washington, D.C.: U.S. Department of Agriculture.

Miller, A. 1984. Collagen: the organic matrix of bone. *Philosophical Transactions of the Royal Society of London B* 304: 455–477.

Mullis, K. B., F. Faloona, S. J. Scharf, R. K. Saiki, G. T. Horn, and H. A. Erlich. 1986. Specific enzymatic amplification of DNA in vitro: the polymerase chain reaction. *Cold Spring Harbor Symposia on Quantitative Biology* 51: 263–273.

Muyzer, G., P. Sandberg, M. H. J. Knapen, C. Vermeer, M. J. Collins, and P. Westbroek. 1992. Preservation of the bone protein osteocalcin in dinosaurs. *Geology* 20: 871–874.

Nemethy, G., K. D. Gibson, K. A. Palmer, N. Yoon, G. Paterline, A. Zagari, S. Rumsey, and H. A. Scheraga. 1992. Parameters in polypeptides 10. Improved geometrical parameters and non-bonded interactions for use in the ECEPP/3 algorithm, with application to proline-containing peptides. *Journal of Physical Chemistry* 96: 6472.

Nielsen-Marsh, C. M., P. H. Ostrom, H. Ghandi, B. Shapiro, A. Cooper, P. V. Hauschka, and M. J. Collins. 2002. Sequence preservation of osteocalcin protein and mitochondrial DNA in bison bones older than >55 ka. *Geology* 30 (1): 1099–1102.

Organ, C. L., M. H. Schweitzer, W. Zheng, L. M. Freimark, L. C. Cantley, and J. M. Asara. 2008. Molecular phylogenetics of mastodon and *Tyrannosaurus rex*. *Science* 320: 499.

Paabo, S., H. N. Poinar, D. Serre, V. Jaenicke-Despres, J. Hebler, N. Rohland, M. Kuch, J. Krause, L. Vigilant, and M. Hofreiter. 2004. Genetic analyses from ancient DNA. *Annual Reviews of Genetics* 38: 645–679.

Poinar, H. N., M. Hofreiter, W. G. Spaulding, P. S. Martin, B. A. Stankiewicz, H. Bland, R. P. Evershed, G. Possnert, and S. Paabo. 1998. Molecular coproscopy: dung and diet of the extinct ground sloth *Nothrotheriops shastensis*. *Science* 281 (5375): 402–406.

Prager, E. M., A. C. Wilson, J. M. Lowenstein, and V. M. Sarich. 1980. Mammoth albumin. *Science* 209: 287–289.

Qiang, J., P. J. Currie, M. A. Norell, J. Shu-An. 1998. Two feathered dinosaurs from northeastern China. *Nature* 393: 753–761.

Rowley, M. J., P. V. Rich, T. H. Rich, and I. R. Mackay. 1986. Immunoreactive collagen in avian and mammalian fossils. *Naturwissenschaften* 73: 620–623.

Salamon, M., N. Tuross, B. Arensburg, and S. Weiner. 2005. Relatively well preserved DNA is preserved in the crystal aggregates of fossil bones. *Proceedings of the National Academy of Sciences of the United States of America* 102 (30): 13783–13788.

Schweitzer, M. H., R. Avci, T. Collier, and M. Goodwin. 2008. Microscopic, chemical and molecular methods for examining fossil specimens. *Palevol* 7: 159–184.

Schweitzer, M. H., C. L. Hill, J. M. Asara, W. S. Lane, and S. H. Pincus. 2002. Identification of immunoreactive material in mammoth fossils. *Journal of Molecular Evolution* 55: 696–705.

Schweitzer, M. H., M. Marshall, K. Carron, D. S. Bohle, S. Busse, E. Arnold, C. Johnson, and J. R. Starkey. 1997. Heme compounds in dinosaur trabecular tissues. *Proceedings of the National Academy of Sciences of the United States of America* 94: 6291–6296.

Schweitzer, M. H., Z. Suo, R. Avci, J. M. Asara, M. A. Allen, F. Teran Arce, and J. R. Horner. 2007. Analyses of soft tissue from *Tyrannosaurus rex* suggest the presence of protein. *Science* 316: 277–280.

Schweitzer, M. H., J. A. Watt, R. Avci, C. A. Forster, D. W. Krause, L. Knapp, R. Rogers, I. Beech, and M. Marshall. 1999b. Keratin-specific immunoreactivity in claw sheath material from a late Cretaceous bird from Madagascar. *Journal of Vertebrate Paleontology* 19 (4): 712–722.

Schweitzer, M. H., J. A. Watt, R. Avci, L. Knapp, L. Chiappe, M. Norell, and M. Marshall. 1999a. Beta-keratin specific immunological reactivity in feather-like structures of the Cretaceous Alvarezsaurid, *Shuvuuia deserti. Journal of Experimental Zoology, Part B, Molecular and Developmental Evolution* 285: 146–157.

Schweitzer, M. H., W. Zheng, C. L. Organ, R. Avci, Z. Suo, L. M. Freimark, V. S. Lebleu, M. B. Duncan, M. G. Vander Heiden, J. M. Neveu, W. S. Lane, J. S. Cottrell, J. R. Horner, L. C. Cantley, R. Kalluri, and J. M. Asara. 2009. Biomolecular characterization and protein sequences of the Campanian hadrosaur *Brachylophosaurus canadensis. Science* 324: 626–629.

Schweitzer, M. H. 2011. Soft tissue preservation in terrestrial Mesozoic vertebrates. *Annual Review of Earth and Planetary Sciences* 39: 187–216.

Shames, R. B., L. W. Knapp, W. E. Carver, L. D. Washington, and R. H. Sawyer. 1989. Keratinization of the outer surface of the avian scutate scale: interrelationship of alpha and beta keratin filaments in a cornifying tissue. *Cell and Tissue Research* 257: 85–92.

Shames, R. B, and R. H. Sawyer. 1987. Expression of beta-keratin genes during development of avian skin appendages. In A. A. Moscona and A. Monroy (eds.), *The Molecular and Developmental Biology of Keratins*, 235–253.

Stankiewicz, B. A., D. E. G. Briggs, R. P. Evershed, M. B. Flannery, and M. Wuttke. 1997. Preservation of chitin in 25-million year old fossils. *Science* 276: 1541–1543.

Stankiewicz, B. A., H. N. Poinar, D. E. G. Briggs, R. P. Evershed, and G. O. Poinar Jr. 1998. Chemical preservation of plants and insects in natural resins. *Proceedings of the Royal Society of London B* 265: 641–647.

Sykes, G. A., M. J. Collins, and D. I. Walton. 2000. The significance of a geochemically isolated intracrystalline organic fraction within biominerals. *Organic Geochemistry* 23: 1059–1065.

Tormo, J., D. Blaas, N. R. Parry, D. Rowlands, D. Stuart, and I. Fita. 1994. Crystal structure of a human rhinovirus neutralizing antibody complexed with peptide derived from viral capsid protein VP2. *EMBO Journal* 13: 2247.

Tuross, N. 1989. Albumin preservation in the Taima-Taima mastodon skeleton. *Applied Geochemistry* 4: 255–259.

Tuross, N., D. R. Eyre, M. E. Holtrop, M. J. Glimcher, and P. E. Hare. 1980. Collagen in fossil bones. In P. E. Hare (ed.), *Biogeochemistry of Amino Acids*, 53–63. New York: J. Wiley & Sons.

Ulrich, M. M., W. R. Perizonius, C. F. Spoor, P. Sandberg, and C. Vermeer. 1987. Extraction of osteocalcin from fossil bones and teeth. *Biochemical and Biophysical Research Communications* 149: 712–719.

Weiner, S., W. Traub, H. Elster, and M. J. DeNiro. 1989. The molecular structure of bone and its relation to diagenesis. *Applied Geochemistry* 4: 231–232.

Woodward, S. R., N. J. Weyand, and M. Bunnell. 1994. DNA sequence from Cretaceous period bone fragments. *Science* 266 (5188): 1229–1232.

Wyld, J., and A. H. Brush. 1983. Keratin diversity in the reptilian epidermis. *Journal of Experimental Zoology* 255: 387–396.

Dinosaurs as Museum Exhibits

15

Kenneth Carpenter

Ever since the first dinosaur skeleton was mounted for exhibition in 1868 at the Philadelphia Academy of Natural Sciences, the public has been fascinated by these extinct animals. Despite that public interest, natural history museums were slow to seize on that curiosity as a way of enticing visitors. Not until steel magnate and philanthropist Andrew Carnegie donated casts (replicas) of the sauropod skeleton *Diplodocus carnegii* to the principal museums of South America and Europe did museums come to realize the draw of dinosaurs as a means of increasing attendance and hence revenue. Dinosaurs are now such an integral part of most natural history museums that the public has come to expect the association.

15.1. *Tyrannosaurus rex* at the American Museum of Natural History reflects changing ideas about dinosaur posture. A, As originally mounted in 1915. B, As remounted in 1995.

A, Courtesy Department of Library Services, American Museum of Natural History, photographed by A. E. Anderson.

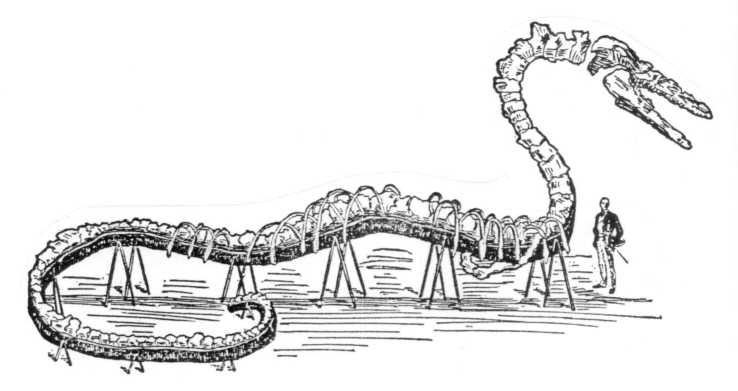

15.2. One of the earliest mounted skeletons, the archaeocete whale *Basilosaurus,* as part of a traveling exhibit. The mounting technique used wood and steel, permitting the skeleton to be dismantled and moved from location to location. From Lucas (1902).

15.3. The chaotic workshop of Waterhouse Hawkins around 1868 (A) showing the partially mounted skeletons of *Hadrosaurus foulkii* (a) and *Dryptosaurus aquilunguis* (b), mounted pelvis (backward) and femur of a cast of *Hadrosaurus* (c); the specimen that would become the holotype of *Ornithotarsus immanis* (d) used as a guide to reconstructing the ankle region of *Hadrosaurus;* left-side of a mold of *Hadrosaurus* (e), a life-size mold for the Paleozoic Museum to be built in Central Park, New York City. Reference skeletons include an ostrich (f) and emu (g). Note the use of steel pipes and rods to support the bones, the same techniques used today. Missing parts are reconstructed (light-colored parts). In 1984 a cast of the bones of *Hadrosaurus* was used to reconstruct a more contemporary posture (B), but this has since been dismantled.

3A, courtesy of the Academy of Natural Sciences of Philadelphia Library.

Dinosaur Skeletons in the Public Eye

Today we are in the midst of a new golden age of dinosaur studies, which began in the 1970s. With advances in communication, there has been a steady stream of popular dinosaur books, movies, and television documentaries. Many financially strapped museums have turned to dinosaurs to draw the visitors by expanding or renovating dinosaur galleries, or bringing in temporary dinosaur exhibitions. Meanwhile, museum gift shops offer stuffed dinosaurs, dinosaur erasers, wooden dinosaur skeleton kits, dinosaur cookie cutters, and anything else remotely dinosaurian to the public as souvenirs.

A more serious approach by some museums has been to present to the public what dinosaur paleontologists have learned. Gone is the view that *Tyrannosaurus* was a lumbering giant that stood on its hind legs, tail dragging on the ground, and head high in the clouds (as another bipedal dinosaur is depicted in Fig. 15.1A). Instead, *Tyrannosaurus* is seen as a moderately swift predator, using its tail to counterbalance a horizontal body (Fig. 15.1B). The public has responded favorably to these new exhibits (meaning that attendance is up), and as a result, increasing numbers of museums have mounted or remounted dinosaur skeletons. For example, the American Museum of Natural History saw a 20% increase in attendance after its renovated dinosaur hall opened in 1995, and the Field Museum in Chicago at 67% increase in 1999 (Anonymous 2002).

Almost as soon as fossilized vertebrate skeletons were found, attempts were made to assemble them for the public. The earliest skeletal mount was not a dinosaur but was probably that of the giant extinct ground sloth, *Megatherium,* mounted in 1788 at the Museo Nacional de Ciencias Naturales, in Madrid, Spain (Boyd 1958). This was followed in 1806 by that of a mastodon displayed at Charles Peale's museum in Philadelphia, Pennsylvania. The

legs of the skeleton may be seen in Peale's 1822 painting *The Artist in His Museum* (Alexander 1983, fig. 3). The techniques used in the mounts were unfortunately not recorded. This exhibit of a long-extinct animal skeleton had a tremendous impact on the public in the United States, and it became clear to entrepreneurs that there was money to be made with similar displays. Soon traveling exhibitions that featured extinct animals appeared in many cities, both in the United States and in Europe. For a price, the public could see these fossilized skeletons. Competition for the public's money resulted in some shady one-upmanship. One entrepreneur claimed to have the largest mastodon skeleton in the world. Actually, its exaggerated length of 10 m (32 feet) and height of 4.5 m (15 feet) were due to several partial skeletons being used to make a composite skeleton (Simpson 1942). Yet another entrepreneur boasted of having the skeleton of a sea monster, which was actually a composite archaeocete whale with an exaggerated length of 35 m (114 feet) (Kellogg 1936). This mount was rather crude, using boards and metal bars to hold the skeleton together (Fig. 15.2). Nevertheless, this innovation made it easier to disassemble the skeleton and move it from city to city for repeated exhibition.

It was only natural that once fossil skeletons were displayed, someone would mount a dinosaur skeleton. In 1868 Waterhouse Hawkins, a well-known sculptor from England, was asked to recreate various prehistoric animals for a museum in New York City. Hawkins spent time at the Academy of Natural Sciences of Philadelphia, where some of the more important fossil specimens of the time were curated. Among these were the skeletons of the dinosaurs *Hadrosaurus foulkii* and *Dryptosaurus aquilunguis*. In order to get various body proportions correct for his sculptures, Hawkins approached Joseph Leidy, who was in charge of the collections, about molding and casting these skeletons, and he offered to mount the real bones as skeletons for exhibition at the academy. Hawkins convinced Leidy that mounting the skeletons would increase visitor attendance. Neither skeleton was complete (Fig. 15.3), forcing Hawkins to rely a great deal on his own imagination and discussions with Leidy to reconstruct the missing bones. Hawkins utilized many of the techniques that are still used today: vertebrae were drilled and strung onto a steel rod like beads, limb bones were held to the supporting armature with steel bands, and missing bones were modeled as mirror images of their preserved counterparts or were modeled from the best analogous living animal (Fig. 15.3A).

The tripodal pose used by Hawkins for the mounts was based on Leidy's observation that "the enormous disproportion between the fore and hind parts of the skeleton of *Hadrosaurus* has led me to suspect that this great herbivorous Lizard sustained itself in a semierect position on the huge hinder extremities and tail" (Leidy 1865, 97). The reconstructed mammalian-like characters used in the mount, including the seven cervicals (neck vertebrae) and the shape of the scapula (shoulder blade), reflect the influence of the kangaroo skeleton. The public saw the final result of Hawkins's hard work in 1868, and as Hawkins had predicted, visitor attendance at the academy soared. While preparing the *Hadrosaurus* skeleton for display, Hawkins also made plaster of paris casts of the skeletons (Fig. 15.3A). One set of casts was to be used in the museum in New York City, while others were distributed to the Smithsonian Institution in Washington, D.C., to Princeton University,

and to the Field Museum of Natural History in Chicago. Once assembled, these were some of the first casts of a dinosaur skeleton ever displayed.

Many of the mounting techniques used by Hawkins were improved during the late 1870s and early 1880s by Louis Dollo at the Musee Royal d'Histoire Naturelle in Brussels. Dollo and his assistants mounted a group

15.4. One of the skeletons of *Iguanodon* as found in the coal mine at Bernissart, Belgium. These were some of the first completely articulated dinosaur skeletons found. Such discoveries allow the bones in disarticulated dinosaurs to be placed in their correct anatomical position. From Anonymous (1897).

15.5. Dollo's workshop showing an *Iguanodon* skeleton being mounted (a). Note the life-size drawing used as a guide. After positioning, the bones were temporarily suspended using cables (right side) from scaffolding before being permanently attached to a steel armature. Another skeleton awaits mounting (b). The skeletons of an ostrich and kangaroo stand just in front of the dinosaur's knees (c). Note the influence that Hawkins and Leidy had with the tripodal stance. From Anonymous (1897).

of *Iguanodon* skeletons that had been collected from a coal mine at Bernissart in Belgium. Finding so many of the skeletons with all of their bones articulated was a big bonus for Dollo (Fig. 15.4). There could be no question about where each bone belonged or their shape. As a result, there was much less guesswork than Hawkins faced. As an aid for mounting, life-size drawings of the skeletons were prepared showing where each bone fit (Fig. 15.5). Individual bones or sections of bone were suspended from wooden scaffolding in positions they would assume in the drawings, and they were attached by metal bands to an iron armature. Dollo used the skeletons of an emu and a kangaroo as references in the mounting to ensure that the bones were articulated as anatomically accurately as possible. Today, the skeletons remain the oldest dinosaur mounts (Fig. 15.6).

The use of comparative anatomy in the mounting of dinosaur skeletons became even more important when the first sauropod skeleton was mounted in 1905. In preparation for mounting this skeleton, William Matthew and Walter Granger of the American Museum of Natural History dissected several modern reptiles in order to better understand how muscles and joints worked (Matthew 1905). The muscle scars on the reptile limb bones were matched with those on sauropod limbs. Matthew and Granger then used strips of paper to connect the origin and insertion points of the sauropod muscles. The bones were adjusted so as not to violate probable muscle movement. Once the bones were properly positioned, they were fastened permanently to a steel armature. The use of dissections with comparative and functional anatomy remains important in the mounting of dinosaur skeletons today. This is especially true now that newer dinosaur mounts attempt to breathe life into the fossil bones. Skeletons are postured so as to appear dynamic, inviting the viewer to imagine the living animal walking, running, or fighting.

Let us look briefly at why knowledge of comparative and functional anatomy is so important in the mounting of dinosaur skeletons. The vertebrate skeleton is the framework for the body. It has many functions, among them providing attachment sites for muscles. Muscles work by pulling, not by pushing. They do this through shortening and thickening of muscle fibers. Thus, a muscle connecting opposite sides of a joint will pull the two bones toward each other with the joint acting as the hinge (Fig. 15.7). How much motion is possible for the bones depends on the type of joint, as well as the surrounding soft tissue. Motion is actually less than what the joints might suggest because of the restriction by connective tissue, muscle, and skin (Fig. 15.8). Cartilage and connective tissue at a joint leaves a scar around the rim of the joint surface (Fig. 15.9). The scar delineates the maximum amount of motion possible. To move the limbs beyond the limits imposed

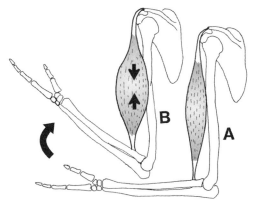

15.7. To understand how limbs in dinosaur skeletons moved, it is important to understand how muscles and joints work in living animals. To move the lower arm from its position in (A), the biceps muscle contracts, pulling the arm toward it (B). The elbow is a simple hinge joint allowing movement only in one plane.

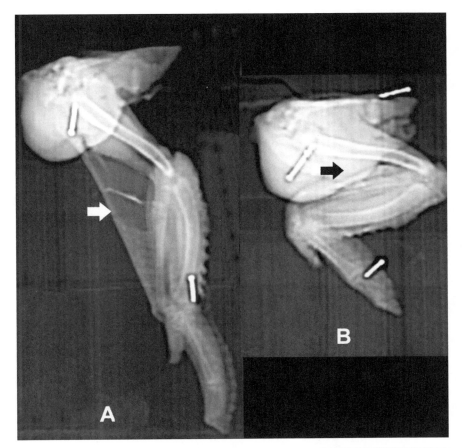

15.8. The influence of soft tissue is usually not appreciated in mounting of skeletons. X-rays showing the arm (wing) of a chicken extended (A) and flexed (B). Note how the skin and ligaments in (A) (arrow) prevent the arm from actually extending further than the joint alone would suggest, and how the mass of the muscles prevents full flexion in (B).

Courtesy of Steven White, Kaiser Permanente.

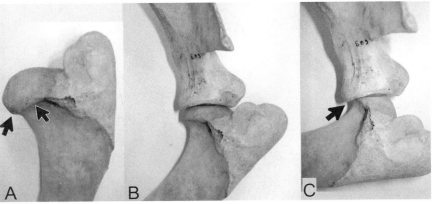

15.9. Example of the limitations of a joint in a modern deer. A, Lower edge for the cartilage cap on the upper arm bone, or humerus, denoted by arrows. B, Neutral position of the humerus in the glenoid (arm socket). Note that the rim shown in (A) is outside the socket. C, Disarticulated arm in the glenoid. Note that the rim is now within the glenoid. This would imply tearing of the ligaments that bind the bones together, as well as rupture of the fluid-filled synovial capsule.

15.10. An excruciatingly painful disarticulated shoulder in a sauropod (*Camarasaurus*) mount. Note the abnormal gap at arrow and inclusion of the cartilage rim denoted by a dart into the glenoid (shoulder socket). The cause for this ailment is a too vertically mounted scapula (shoulder blade). Articulated skeletons have repeatedly shown that the scapula is almost parallel to the dorsal vertebrae.

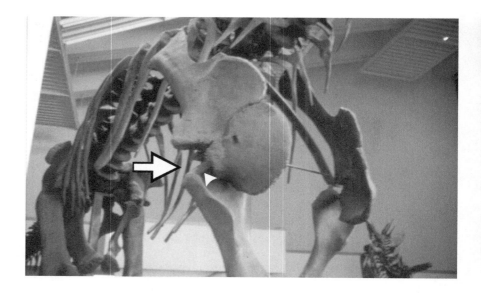

by the scars would, in life, indicate a damaged joint. In a dinosaur skeleton, a damaged joint would be implied if, for example, the cartilage scars around the head of the humerus (upper arm bone) were incorporated into the glenoid (shoulder socket) (Fig. 15.10). Various studies have shown that dinosaur

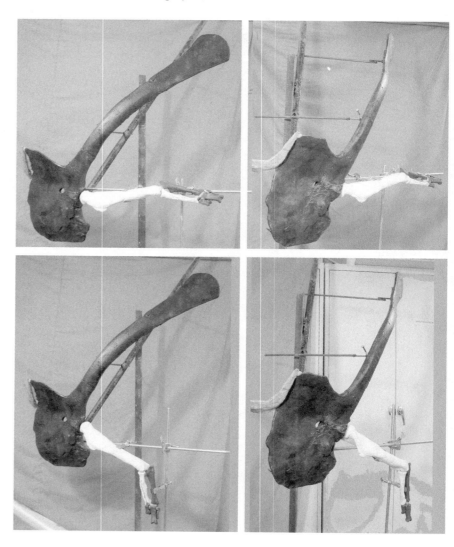

15.11. Actualistic study of limb motion in *Tyrannosaurus* using casts in side and front views. Upper row, Lower arm fully extended and retracted. Lower row, Arm protracted and lower arm flexed. These positions represent the maximum ranges of motion without violating join morphology. Actual range of motion was less as a result of the influence of soft tissue.

15.12. A more dynamic pose for a sauropod skeleton can result in the tail high in the air, as seen in this specimen at the Denver Museum of Nature & Science.

limb motion is not always what might be expected. For example, few theropods can move their forearms forward enough to reach their mouths (Fig. 15.11; Carpenter 2002; Senter 2006)

The stance of mounted bipedal dinosaur skeletons has changed over the years because the living dinosaur is no longer thought to have used its tail as a prop in kangaroo fashion (Fig. 15.1A). Instead, the tail was carried in the air as a counterbalance to the body over the hind legs (Fig. 15.1B). The tail moved slightly from side to side to maintain balance alternately over the leg during each step. This change in what we think the tail was for is based on the straight position of the back and tail in dinosaurs found as articulated skeletons (e.g., Fig. 15.4). The result of this new interpretation can result in some rather surprising mounts, such as the *Diplodocus* skeleton with the tail high in the air at the Denver Museum of Nature & Science (Fig. 15.12).

The work to extract dinosaur bones from the ground, clean them of the encasing rock, restore the missing parts, and mount the skeleton for display can take years. At least 7 years were needed for the *Apatosaurus* skeleton at the American Museum of Natural History in New York City when it

Mounting Skeletons

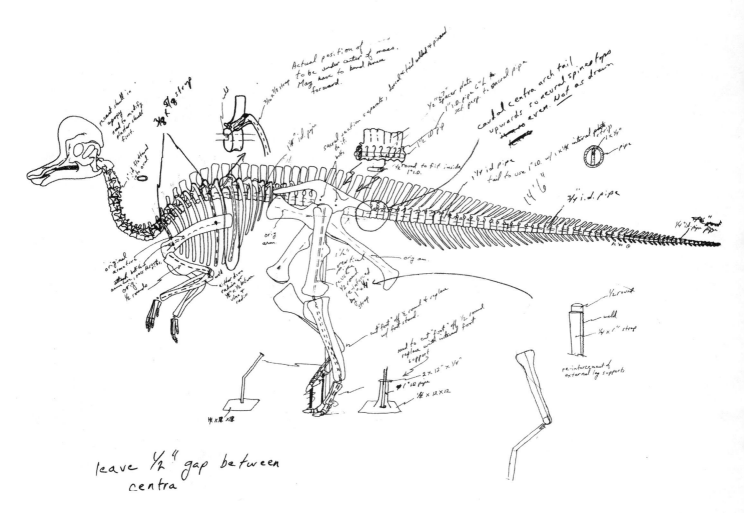

15.13. A sketch of the dinosaur *Corythosaurus* showing the proposed position of the limbs and tail. Notes were also made on the sketch for how the armature was to be made. Much of the armature was internal or on the back side of bones to minimize their visibility (dashed lines). Also shown (facing page) is the final mount. During mounting, it became necessary to add a vertical pipe to support the front of the body rather than cable the skeleton from the ceiling, as was originally planned. Other changes from the original plan include turning the neck and head, making the arms straighter, and adding acrylic rods on the back and tail to replicate ossified tendons. Displayed at the Philadelphia Academy of Natural Sciences.

first went on display in 1905 (Matthew 1905). It is because of the time and expense involved in preparing skeletons for exhibition that some museums have begun mounting plastic casts of skeletons. This has generated a controversy that does not have a satisfactory solution. On the one hand, mounting plastic skeletons does allow museums to have skeletons relatively cheaply, and lightweight mounts can be posed in active, lifelike positions more easily. But on the other hand, the museum visitor often wants to see real fossils because museums are often the only place they may see them. Other considerations in the real versus cast debate include whether the desired specimen for the exhibit is in the museum's collection, or, failing that, the availability and cost of obtaining and preparing the required specimen versus the cost of obtaining a cast. Finally, the exhibit schedule is a concern: is there enough time to put a real skeleton on exhibit? In the end, the decision is made on the basis of numerous factors, not all of them scientific. Some museums with real bone skeletons, such as the Smithsonian Institution's National Museum of Natural History, have begun to replace their real bone skeletons with replicas for conservation reasons.

Many older museums have also begun a program of dismantling and remounting their dinosaur skeletons to reflect the new information paleontologists have about how the animals stood or moved (Fig. 15.1). Remounting skeletons is often part of a larger program to renovate dinosaur exhibits. Computer stations with touch screens now dot the exhibit hall, and with

a touch, anyone can call up information about a particular dinosaur. Remounting a skeleton is often cheaper than preparing a new specimen or purchasing and mounting a cast. The results often make the skeleton look so different that it seems to the visitor that there is a new skeleton on exhibit.

Assembling skeletons involves more than fastening bones to a metal frame. A considerable amount of planning is necessary to ensure that the mount is a good one. Sketches or scale models are prepared showing the finished mount in various views. Such views allow for changes to be made in the posture before assembly makes this impossible or too costly to alter. Sketches and models also show whether the allotted exhibit space is adequate, not cramping the skeleton. A sketch of the skeleton can also become a blueprint, allowing the steps and materials used in the mounting process to be thought out in detail (Fig. 15.13). In planning for the mount, it is important to know if the animal was habitually bipedal, walking only on the hind legs, or quadrupedal, walking on all four limbs. Mounting a bipedal dinosaur can be difficult because the forelimbs cannot be used to help support the skeleton. Instead, steel cables from the ceiling can be used, or the skeleton can rest atop vertical steel pipe supports (Fig. 15.14A).

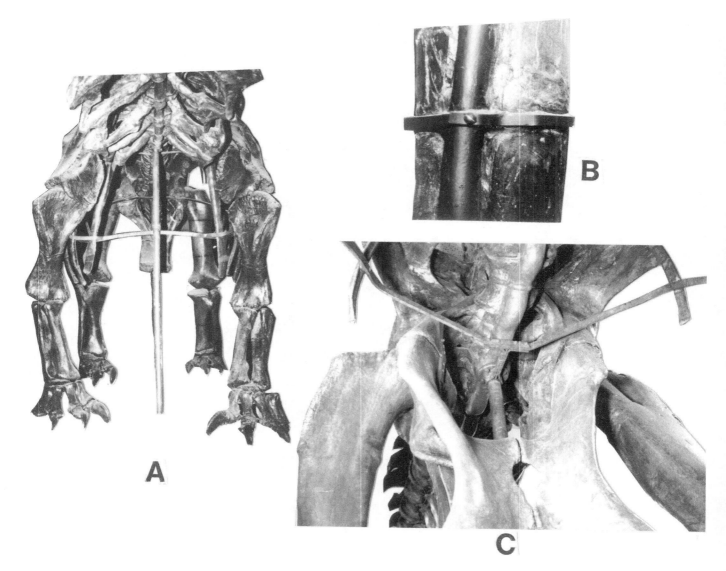

15.14. The weight of heavy dinosaur skeletons can be supported by vertical steel pipes (A). Bones may be clamped to the armature, which has been bent to conform to the bone contour (B). Custom armature made at a foundry supports the vertebral column in this sauropod skeleton.

It is rare for a dinosaur skeleton chosen to be mounted to have all of its bones preserved because erosion, which uncovers most fossils, destroys what it uncovers. Missing bones can be replaced with casts or actual bones from another individual. Sometimes the missing parts can be sculpted from wood, plaster of Paris, papier-mâché, epoxy putty, or ceramic clay fired in an oven (Figs. 15.3 and 15.15). Sometimes two or more partial or fragmentary skeletons can be combined to make a single skeleton. This has been done, for example, with the *Triceratops* skeleton at the American Museum of Natural History.

The tools and equipment needed for mounting a skeleton can be obtained from a hardware store; rarely are specialized tools used. The common tools include hammers, saws, pliers, wrenches, nuts and bolts, electric drills, a welder, a bench vise, paints or wood stain, and artist brushes of various sizes.

Bones are attached in some way to a support armature usually made of steel (Fig. 15.14, 15.15). Because fossilized bone is brittle, the weight of each bone (except perhaps the smallest, lightest ones) is borne by the armature (Fig. 15.15), a technique used by Dollo in mounting *Iguanodon* (Fig. 15.6).

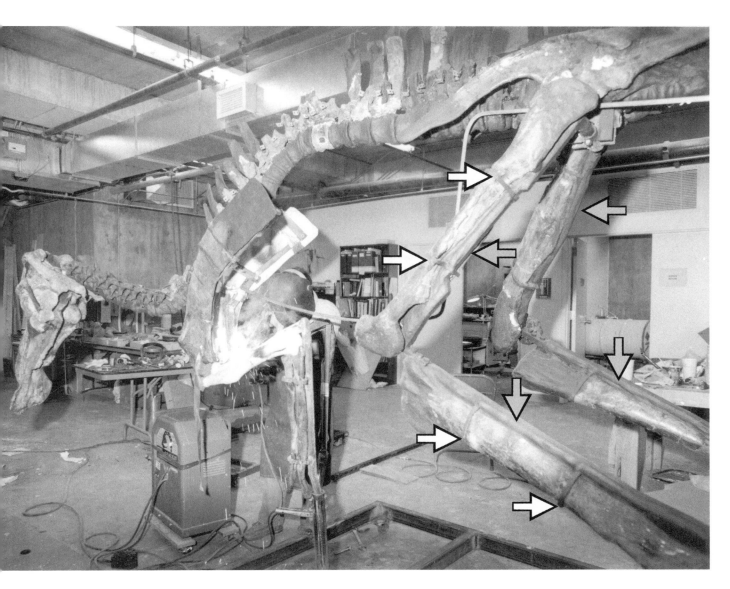

15.15. Mounting of a skeleton of *Maiasaura*. Damaged and missing parts were reconstructed and painted a darker color than the actual fossil so that it would be apparent what was real and what was not. Because the skeleton is on all four limbs, the support armature extends down all four legs (gray arrows), with metal bands welded around the bones (white arrows). Duct tape was used to hold the bones to the armature until the bands could be welded. The base of this mount is a steel frame.

This armature may be custom formed to the surface of the bones or placed inside casts or bones that have been drilled out (Fig. 15.16). If the armature is outside, it is placed on the underside or back side of bones so that it is least noticeable to the visitor. The steel selected must be of a diameter large enough to support the bone or skeleton without being too obvious. The armature can be welded (Fig. 15.14) or bolted together. The bones are then attached with steel straps, pins, or bolts (Fig. 15.15B). For vertebrae, the support steel can be strung through the cast, or a hole can be drilled through the center of the bones and then strung like beads on a steel pipe prebent to the shape of the vertebral column (Fig. 15.16). A custom-cast steel armature can also be made to support the vertebrae from the underside (Fig. 15.14C). However, this practice is now rare because it requires a nearby steel foundry, and most of the foundries in the United States have closed.

Another method for displaying dinosaur skeletons is the panel mount. The skeleton may be left embedded in the rock in which it was found or mounted so that only one side of the skeleton is visible (Fig. 15.17), especially if one side of the skeleton has been damaged through erosion. The damaged

15.16. Mounting a cast of *Tyrannosaurus* was easy because the cast was hollow, allowing the bones to be strung on an internally placed armature, then foamed into place.

15.17. Panel mount of *Corythosaurus* at the Royal Tyrrell Museum of Palaeontology in which only one side of a skeleton is visible. Such mounts are used when one side of the skeleton is damaged due to erosion or too much of one side missing. Light-colored bones are reconstructed.

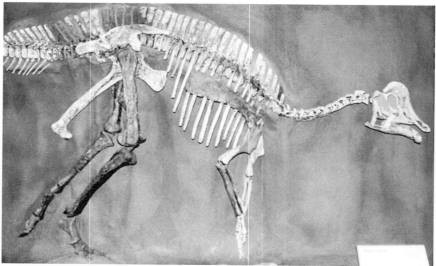

15.18. Part of a temporary exhibit of Chinese dinosaurs on display in the Cityplaza shopping mall, Taikoo Shing, Hong Kong, in 2005. Such exhibits open for a few months and then travel on.

side is embedded in plaster of Paris simulated to resemble rock, while the bones of the uneroded side are cleaned for display.

Once a freestanding mount has been completed, the base must be finished in some way to make it visually appealing. Usually a barrier is installed to protect the fossil from souvenir collectors (a sad fact of the modern museum; glass barriers are visible in Figs. 15.1B and 15.6). Finally, labels must be prepared giving the name of the dinosaur and some information about the animal. Once all of this is completed, the skeleton is ready for its public debut (Fig. 15.1B; note signage at feet).

Other Dinosaur Exhibits

There are several types of dinosaur exhibits other than permanent exhibits of skeletons in museums. Temporary or traveling exhibitions have gained popularity in recent years. Such exhibits tend to remain at one place for a few months before being shipped to another. One advantage of such exhibits is that they expose the visitor to a greater variety of dinosaur skeletons than would otherwise be possible. For example, several exhibitions of Chinese dinosaurs have traveled around North America, Japan, and Europe (Fig. 15.18). Most traveling dinosaur exhibits are designed to make it easy for the skeletons to be assembled for display. Skeletons are often modular, with pins or bolts holding segments of armature with bones already attached to them. Both cast and real bone skeletons are used in these traveling exhibitions.

Another type of dinosaur exhibit uses restorations of the animals as they might have appeared in life. The first attempts were the fanciful imagination of Waterhouse Hawkins and Richard Owen (who gave Dinosauria its name). Hawkins created life-size restorations of several extinct animals, including the dinosaurs *Iguanodon*, *Megalosaurus*, and *Hylaeosaurus* (Fig. 15.19). These sculptures were displayed in 1853 at the Crystal Palace, one of the first world's fairs, and are the first example of cooperation between

Within the image: PUBLISHED DEC 7 1854 BY G. BAXTER PROPRIETOR & PATENTEE LONDON

15.19. Illustration of life-size dinosaurs by Waterhouse Hawkins on the grounds of the Crystal Palace (background). Among the dinosaurs (left to right) are *Megalosaurus, Hylaeosaurus,* and two *Iguanodons.* Note the people for scale. First published in 1854.

15.20. Life-size diorama of two *Stygimoloch* pachycephalosaurs fighting. The diorama is a useful method of showing dinosaurs in their natural habitat while also conveying some aspect of their behavior. Denver Museum of Nature & Science.

scientist and artist. These restorations can still be seen in Hyde Park, London. A variation of this approach attempts to reconstruct the dinosaur in what scientists believe was its natural environment in a scene called a diorama (Fig. 15.20). Since the exhibition at the Crystal Palace, life restorations of dinosaurs have appeared at several world's fairs, including those in St. Louis in 1904 and New York City in 1964 (Fig. 15.21). Richard Lull attempted another version of the life restoration of a dinosaur at the Peabody Museum at Yale University, using a technique pioneered by the Denver Museum of Natural History in 1918. Lull mounted a skeleton of the dinosaur *Centrosaurus* inside a partial life restoration of the animal. For accuracy, he used actual skin impressions of ceratopsians to replicate the skin surface. The result can still be seen at the Peabody Museum. On the right side, the fleshed-out animal is seen, while on the left side, a mounted skeleton inside a partial body shell is visible.

For a brief time in the 1990s, robotic or animatronic dinosaurs were popular. Pioneered by the Disney Company in the 1960s, robotic dinosaurs attempted to make dinosaurs come alive. Many of the early attempts only vaguely looked like the dinosaurs they were meant to represent. Most looked like overweight, misproportioned monsters with jerky head motions accompanied by loud hissing of their pneumatic motors. Later versions had the input of dinosaur paleontologists, and the results were thought to more closely resemble what we think dinosaurs actually looked like.

Many museums use these animatronic dinosaurs in temporary exhibits to increase museum attendance. Several attempts of developing a walking robotic dinosaur have not yet been successful but could usher in a new wave of animatronics in museums.

The most recent dinosaur exhibition innovation is the virtual museum. The use of the Internet has broken dinosaurs out of their tradition museum setting. Even so, many museums now host Web sites to highlight their exhibits, such as the Smithsonian Institution, and even small museums like the Dakota Dinosaur Museum in Dickinson, North Dakota. Virtual museums, as they have come to be called, provide a Web-based learning environment. One of the foremost examples of an educational site is hosted by the Museum of Paleontology at the University of California. Some sites provide a virtual expedition to a dinosaur excavation, such as the Virtual Dinosaur Dig hosted by the Smithsonian Institution. A list of some of these virtual Web sites is presented in Table 15.1. Another source of Web-based information includes dinosaur databases. Some of these are quite extensive, especially the professional Paleobiology Database. A few representative examples of these databases are also presented in Table 15.1. For a discussion of dinosaur-related topics, there is the Dinosaur Mailing List, a mass e-mail-based system of communication, as well as several blogs. Of course, a lot of information can be found using a search engine, such as Google.

15.21. Life-size reconstruction of *Tyrannosaurus* at the 1964 World's Fair in New York City.

Courtesy of Don Glut.

Table 15.1. Online Dinosaur Resources (does not imply an endorsement of content)

Description	URL
Virtual dinosaur Web sites	
Dinosaur exhibit tour (Smithsonian)	http://paleobiology.si.edu/dinosaurs/interactives/tour/main.html
Dinosaur dig (Smithsonian)	http://paleobiology.si.edu/dinosaurs/interactives/dig/dinodig.html
Exhibit tour (Dakota Dinosaur Museum)	http://www.realnd.com/dakotadinoindex.htm
Exhibit and collection tours (Dinosaur National Monument)	http://www.nps.gov/history/museum/exhibits/dino/
Educational (Museum of Paleontology, University of California)	http://www.ucmp.berkeley.edu/exhibits/index.php
Exhibit (American Museum of Natural History)	http://www.amnh.org/exhibitions/permanent/fossilhalls/virtualtours/
Dinosaur databases, blogs, and the like	
Dinosauria	http://www.dinosauria.com/
Dinobase (University of Bristol)	http://dinobase.gly.bris.ac.uk/
Dino Directory (Natural History Museum, London)	http://internt.nhm.ac.uk/jdsml/nature-online/dino-directory/
The Paleobiology Database (more than dinosaurs)	http://paleobackup.nceas.ucsb.edu/cgi-bin/bridge.pl
Ask a Biologist	http://www.askabiologist.org.uk/
Tetrapod Zoology (Darren Naish's blog)	http://scienceblogs.com/tetrapodzoology/
Sauropod Vertebra Picture of the Week (joint blog by Darren Naish, Matt Wedel, & Mike Taylor)	http://svpow.wordpress.com/
Dinochick blogs (Rebecca Foster's blog)	http://paleochick.blogspot.com/
Dinosaur Mailing List	http://dml.cmnh.org/

The Future

It is difficult to know what changes will occur with museum exhibits of dinosaurs. If the past is any guide, dinosaur skeletons will always be popular. Armatures will probably be less noticeable with the use of carbon fiber and epoxy resins in place of steel. New casting materials may be stronger and have a more fossillike appearance. Finally, walking robotic dinosaurs may appear, stalking the halls of the museum. There may even be genetically altered birds that express dinosaurian characters currently repressed. These might be the closest we will ever have to dinosaurs in the traditional sense.

References

Alexander, E. 1983. *Museum Masters; Their Museums and Their Influence.* Nashville, Tenn.: American Association for State and Local History.

Anonymous. 1897. *Guide dans Les Collections: Bernissart et Les Iguanodons.* Brussels: Musee Royal D'Histoire Naturelle.

Boyd, J. P. 1958. The *Megalonyx*, the *Megatherium*, and Thomas Jefferson's lapse of memory. *Proceedings of the American Philosophical Society* 102 (5): 420–435.

Carpenter, K. 2002. Forelimb biomechanics of nonavian theropod dinosaurs in predation. *Senckenbergiana Lethaea* 82: 59–76.

Gangewere, R. J. 2002. Carnegie's dinosaurs: a world treasure. http://www.carnegiemuseums.org/cmag/bk_issue/2002/mayjun/cmnh1.htm.

Kellogg, R. 1936. A review of the Archaeoceti. *Carnegie Institute of Washington Publication* 482: 1–366.

Leidy, J. 1865. Memoir on the extinct reptiles of the Cretaceous formations of the United States. *Smithsonian Contributions to Knowledge* 14: 1–135.

Lucas, F. 1902. *Animals of the Past.* New York: McClure, Philips.

Matthew, W. D. 1905. The mounted skeleton of *Brontosaurus*. *American Museum Journal* 5 (2): 64–70.

Senter, P. 2006. Comparison of forelimb function between *Deinonychus* and *Bambiraptor* (Theropoda: Dromaeosauridae). *Journal of Vertebrate Paleontology* 26: 897–906.

Simpson, G. 1942. The beginnings of vertebrate paleontology in North America. *Proceedings of the American Philosophical Society* 86: 130–188.

16.1. A, Pencil study of a grove in Deer Creek, East Fork, Sequoia National Forest. B, *Starvation Creek Grove,* a nature study based on field sketches made by the author in the Sequoia National Forest. The site is situated along a small tributary of Deer Creek in the southern Sierra Nevada north of Lake Isabella, southeast of the Tule River Indian Reservation and near the headwaters of the Kern River.

Restoring Dinosaurs as Living Animals

Douglas Henderson

16

The illustration of dinosaurs as living animals is an interpretive work of imagination based on scientific inquiry. Such paleoillustration depicts, in views and scenes, the living appearance of ancient life, presented in a form borrowed from our direct experience and familiarity with the natural world as we see it today. The interpretive role of paleontological art has put the work of both paleontologists and artists before a wide audience.

Paleontological artists may be scientists or amateurs—anyone who shares an interest in the arts and the earth sciences. Paleoillustration incorporates a traditional approach to art, a development and use of style, medium, and approach to subject unique to each artist. It requires some introduction to the sciences and an appreciation for the limitations in producing complete

images from the fossil record. The work that results from a collaborative effort between artist and scientist turns paleontology and related earth sciences into a rich and modern form of storytelling.

Depicting dinosaurs as living animals has little part to play in the actual science of paleontology. Scientists are mostly concerned with seeing their observations and interpretations expressed and disseminated in objectively written publications intended primarily for the scientific community. While technical drawing of fossil material or simple diagrams and reconstructions may serve this purpose, fully rendered images of great beasts passing in silhouette against western sunsets and the like generally do not. These more romantic works—which nonetheless tell truths—are akin to the arts of theater and literature. However, as Emerson wrote in his poem "Wood Notes I" (1899) of the poet's expression of a personal, subjective experience with the natural world: "What he knows, nobody wants." When the concern of science is to be empirical and documentary, imaginative imagery is regarded as the wrong language.

On the other hand, for many people, the vision of dinosaurs as living animals has validity as art for its own sake. In addition, paleontological art has considerable value for writers and editors of books and magazines, museum curators, paleontologists, and others concerned with interpreting and presenting the science of paleontology to the general public.

The Reasoning, Research, and Procedures of Paleoillustration

The illustration of dinosaurs, or any other aspect of earth history, has a speculative component. This is the result of several factors, including the incomplete nature of the fossil record, varying interpretations based on fossil material, and our inability to observe the specific behavior and natural history of dinosaurs in life. The dinosaur artist must often contend with many unknowns, many possible scenarios, and only a few sure inferences beyond the fossil material itself.

Factors other than science can influence illustrations, such as the expectations and interpretations of the editors, curators, or others who commission illustrations. In addition, the artist may be influenced by long-standing assumptions about the nature of dinosaurs that retain unquestioned acceptance through long-term repetition in the work of other artists and scientists alike.

The scientific artist consequently cannot be expected to render a scene that is literally true. Paleoillustrations, like the science they reflect, are just ideas. They can honor both the reasoned view of science and the unexplored realms that science suggests to the mind's eye.

The imaginative contribution of artists is very much determined by the knowledge, experience, observation, and slant that they bring to their work. An important basis of paleoillustration is effective drawing skills, including a basic understanding of perspective, a sense of composition, some command of a medium, and practice at life drawing. Drawing in the field, a disciplined form of observation, is a learning process. Our subjective experience of the natural world can be captured in drawings, field sketches, and nature studies. In the process, some knowledge can be gained of nature and the natural composition of landforms, river courses, stands of trees—a certain wonderful dishevelment of natural systems (Fig. 16.1). This familiarity with the modern natural world becomes a guide and a fine gauge of normalcy when representing the prehistoric past.

Paleoillustration allows for a flexible approach in representing the science of paleontology. Artists have been called on to summarize the broad evolutionary and anatomical diversity of dinosaurs over the entire Mesozoic Era, but they have also depicted the diversity of plants and animals that lived at specific times and places. Illustration has taken the form of murals, scenes of condensed knowledge, showing a whole fauna and flora in a single image. Alternatively, restorations have been cast as a series of natural scenes patterned according to some pace by which nature is experienced, one thing or one herd at a time. Images may be dominated by dinosaurs, or they may show dinosaurs as only part of a larger physical landscape or ecological community. Science finds any of these approaches valid. The artist, for reasons of esthetics or some appreciation of the story told by scientific data, may find one approach more desirable than others.

Paleoillustration requires some familiarity with the science of paleontology. To begin with, one needs a broad understanding of the earth's ancient geography, climate, flora, and fauna, and the changes in these over time, apart from information about the specific animals to be featured in a scene. Such general background information is available in textbooks and books written for the general public (such as Russell 1977, 1989).

The scientific literature, though technical and not written for the layperson, is a valuable source of information about particular organisms. However, its availability beyond university libraries may be limited. Correspondence and discussion with willing paleontologists provide the best opportunities for instruction and guidance. In addition, paleontologists generally collect an extensive library of data related to their studies, which can become a resource available to the artist.

Visiting museums and private collections provides an opportunity to see, sketch, and photograph all manner of fossil material. This can include full-size, articulated fossil skeletons of dinosaurs, permitting their features to be studied from many points of view. Exhibits in many museums now include life-size, lifelike sculptures of dinosaurs as well.

Visiting zoos allows the observation of living animals, which may share with dinosaurs some elements of overall body size or anatomy, or which may be close or distant relatives of dinosaurs. Such animals might include large mammals, ground birds, and some reptiles.

Paleoillustration can also require an overview of paleobotany. Textbooks and fossil plant field guides discuss the evolution and classification of plants, and indicate those forms that composed Mesozoic floras. The scientific literature is descriptive and technical, but it is often unsatisfactory to the artist interested in reconstructing the living appearance of ancient plants. A living plant consists of many anatomical modules (roots, stems, branches, leaves, reproductive organs) that commonly separate before fossilization. Consequently, it is often hard to determine what an entire ancient plant looked like.

However, the fossil record suggests many phylogenetic relationships between Mesozoic plants and living ones. Plants related to those characteristic of Mesozoic floras, such as ferns, tree ferns, cycads, various conifers, and primitive flowering plants (such as magnolia and dogwood), can be found

in greenhouses and botanical gardens. Many hardwood trees found in the eastern and southeastern United States have leaves remarkably similar to fossil leaves of Cretaceous age. For artists, this similarity logically suggests the possibility that the living appearance of some modern conifers and hardwood trees presents forms and patterns common to ancient floras (Fig. 16.2). Unfortunately, paleobotanists caution that some of these similarities may be more apparent than real.

Another resource is the work of other scientific artists, especially those who have undertaken a study of dinosaur anatomy, presented reasonable theories of reconstruction, and produced illustrations that depict the living appearance of dinosaurs. The work of Robert Bakker, Mark Hallet, and especially Greg Paul offers convincing restorations, revealing accurate skeletal dimensions, ranges of limb motion, sizes and attachments of muscles, and interpretations of postures and gaits. A familiarity with this work aids in restoring dinosaurs from technical drawings and new or unfamiliar fossil material.

16.2. A study based on Black Mountain Grove, Sequoia National Forest.

A finished illustration (Fig. 16.3) can be used to describe the steps and considerations involved in producing an image, and the nature of the fossil information on which it is based. The intent of the scene is to show a large group of *Coelophysis*, a 10-foot carnivorous dinosaur, passing through a Triassic forest in an upland environment. The scene is based on fossils collected from the Chinle Formation, a unit of sedimentary deposits left by rivers that crossed the southwestern United States some 225 million years ago.

Reconstruction of a Late Triassic Scene

16.3. Restoration of a large group of the small carnivorous dinosaur *Coelophysis* in the environment of Late Triassic New Mexico.

In 1947, paleontologist Edwin Colbert discovered a fossil bone bed at Ghost Ranch, New Mexico (Colbert 1995). It was composed almost entirely of many articulated and disarticulated *Coelophysis* skeletons, including both adults and juveniles. The large concentration of animals suggested that *Coelophysis* lived or associated in large groups. (It should be noted, however, that some paleontologists think that the Ghost Ranch theropods constitute two different kinds of dinosaur; Sullivan 1994; Sullivan et al. 1996.)

Another goal of the scene is to represent some of the plants of the Chinle flora, including two kinds of ferns, a tree fern, a horsetail, a cycad, a streamside shrub called *Sanmiguelia*, and a large conifer tree, *Araucarioxylon*. Fossil remains of these plants are often found in what during the Triassic were lowland floodplains. *Araucarioxylon* is known from an abundance of fossil logs, the best examples of which are found in Petrified Forest National Park in Arizona. During the Late Triassic, the logs were the remains of trees with their bark and limbs stripped away, the result of stream transport during periodic flood events (Ash 1986; Long and Houk 1988; Vince Santucci personal communication). The illustration that was prepared represents a forested upland setting where the fossil logs might have originated.

Many details about *Coelophysis*, such as its habits, preferred environments, prey, skin patterns, and color, are unknown. What can be inferred from the fossil record, however, affords a wealth of imagery with which to work.

The fine preservation of many *Coelophysis* specimens allows the artist to restore the dinosaur to something close to its living appearance. *Coelophysis* was a bipedal, slightly built, fast-running hunter with small, grasping arms and small, sharp teeth suited only for handling prey smaller than itself. *Coelophysis* shared the Late Triassic American Southwest with a great many other reptiles. It lived amid a flora comprising a rich assemblage of plants. *Coelophysis* may have been common among the variety of upland, forest, river, and lowland floodplain landscapes and habitats available to it. The illustration touches on only a few of these ideas.

The illustration's preliminary outline sketches of *Coelophysis* were based on Colbert's (1989) skeletal reconstruction. The restoration of its living appearance follows the interpretations of Greg Paul. The wedge-shaped skull, lacking the complex of facial muscles of mammals, needs only a cover of skin to give it a living appearance. The neck of the animal was slender and held in an S curve at rest. There was a slight arch to the spine. The chest narrowed from side to side at the hip region, which was no wider than the connective tissue that spanned the pubis and ischium, the ends of which defined the lower outline of the body. The tail was moderately flexible, like a lizard's. The musculature and general appearance of the neck, legs, and feet were birdlike.

The reconstruction of the Chinle flora in the illustration is based on publications by paleobotanists and discussions with paleontologists, as well as my own observations. The Chinle flora is known from fossil plant material found in localities across the Southwest. In addition to fossil logs, stems, roots, and a few cones, there is a diversity of leaf impressions. All of this indicates that the Chinle flora included many kinds of plants—many more

16.4. Refined preliminary outlines of the small carnivorous dinosaur *Coelophysis,* showing the appearance of the animal from different views. Most of the work of planning the restoration is done in this step of the project.

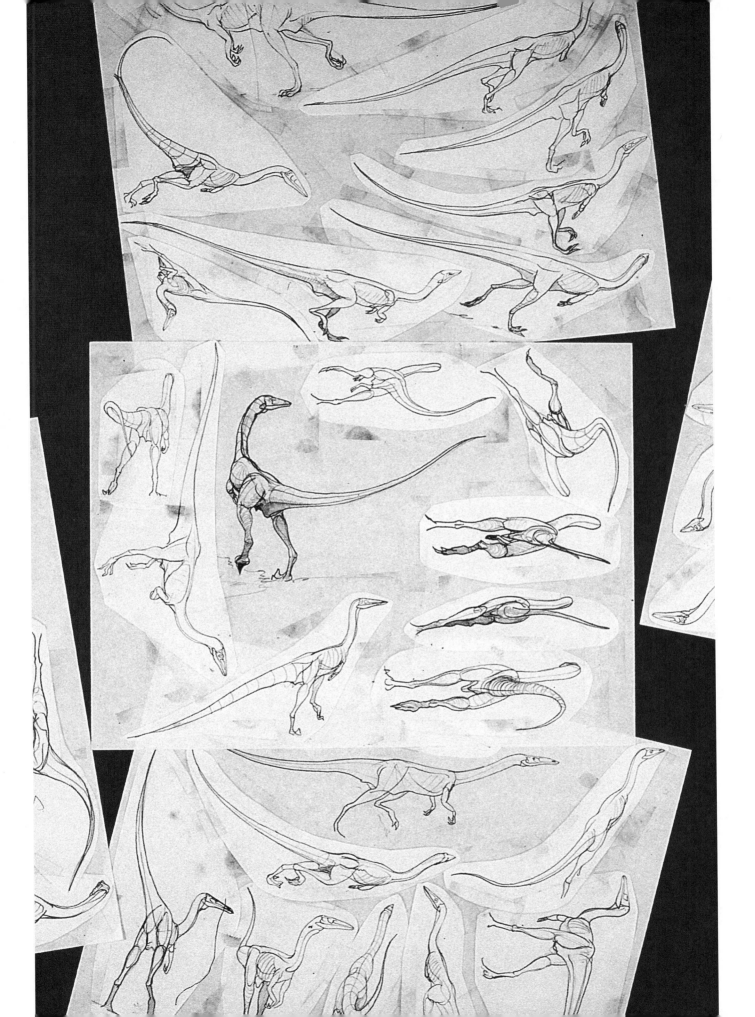

16.5. Test drawings (A) and a refined composition (B) for *Coelophysis*.

than are included in the illustration. Some of these fossil plants, especially the ferns and some unique shrubs, can be accurately reconstructed. Others are represented only by isolated leaves or other structures, and the whole plant is unknown.

In illustrations, partially known material can be represented by showing the plant structure that is known or by adopting some living appearance based on a relationship or similarity to living plants. In the *Coelophysis* scene, the tern fern *Itopsidema*, known only by its fossil trunk, is reconstructed both as a trunk only and with vague foliage appropriate for a living tree fern.

The conifer tree *Araucarioxylon* is one of three species of trees known only from large and often well-preserved fossil logs. Other conifer fossils include a half-dozen or more types of fossil foliage of small size, superficially resembling springs of juniper and cryptomeria, and one large type resembling fir twigs. None of the foliage can be associated with the logs.

The reconstruction of *Araucarioxylon* can be based on observations of the fossil logs, which have many prominent structures; these include traces of roots, swollen bases, wood surface patterns, long, tapering trunks, and a variety of limb scars. The limb scars are spaced in varying patterns; often these scar patterns are sparse, but in some specimens, they extend

along nearly the length of the trunk. The scars show limbs both large and small, which invariably exited the trunk at an upward and often steep angle. These features suggest that *Araucarioxylon* may have resembled a giant, long-trunked Utah juniper or the tall Monterey cypress that forests a good potion of Golden Gate Park in San Francisco.

Once the purpose of an illustration has been decided on and the design of its various elements considered, the next step is to bring the separate components together into a single scene. Rough sketches of an overall scene are drawn, perhaps many times, to establish some satisfactory composition. Once a composition or design is determined, it becomes a guide to the further preliminary work of refining characters, and also a reference while working on the illustration itself. If the illustration is one requested by a publisher or curator, then the preliminary work may be subject to review and approval by editors, writers, and consultants.

Refining characters can be the most time-consuming portion of illustration, requiring that each individual animal be properly outlined in different poses, angles of view, and size relative to distance across the scene. Artists may refer to skeletal drawings, photographs, or their familiarity with dinosaur proportions. Some artists construct, or refer to, small models and sculptures. In the case of the *Coelophysis* illustration, an army of individual dinosaurs was drawn (Fig. 16.4), and the figures were cut out, shuffled around, and taped into place around outlined tree trunks sketched on a sheet of paper. This refined the composition into a carefully choreographed outline.

Such an outline can then be transferred to a new surface of paper (or canvas, or whatever surface is needed for the particular project) on which the illustration will be drawn or painted. Over several days, weeks, or even months, the work is slowly built up with line and shading to a finished image (Fig. 16.5). Paleoillustration is a problem-solving process to which each artist will bring her or his own solutions.

The illustration of dinosaurs is an interpretive tool in service of both science and art. It presents views of ancient life that broaden the appeal and understanding of paleontology beyond the circle of scientists. In addition, paleoillustration allows the individual artist to explore the realm of ideas and legitimate fantasy suggested in the objective interpretation of the fossil record.

The work of paleoillustration is unfinished. (See Plate 26 for another rendering of *Coelophysis*.) A search through the sources of paleontological information, from discoveries in the field to the scientific literature, both old and new, reveals a wealth of data that have not been represented in illustration. The collaborative effort of art and science has much left to tell.

References

Ash, S. 1986. *Petrified Forest: The Story behind the Scenery.* Rev. ed. Petrified Forest, Ariz.: Petrified Forest Museum Association.

Colbert, E. H. 1989. The Triassic dinosaur *Coelophysis.* Museum of Northern Arizona Bulletin 57.

———. 1995. *The Little Dinosaurs of Ghost Ranch.* New York: Columbia University Press.

Long, R. A., and R. Houk. 1988. *Dawn of the Dinosaurs: The Triassic in Petrified Forest.* Petrified Forest, Ariz.: Petrified Forest Museum Association.

Russell, D. A. 1977. *A Vanished World: The Dinosaurs of Western Canada.* Natural History Series 4. Ottawa, Ontario: National Museum of Natural Sciences, National Museums of Canada.

———. 1989. *An Odyssey in Time: The Dinosaurs of North America.* Toronto: University of Toronto Press.

Sullivan, R. M. 1994. Topotypic material of *Coelophysis bauri* Cope and the *Coelophysis–Rioarribasaurus–Syntarsus* problem. *Journal of Vertebrate Paleontology* 14 (suppl. to 3): 48A.

Sullivan, R. M., S. G. Lucas, A. Heckert, and A. P. Hunt. 1996. The type locality of *Coelophysis,* a Late Triassic dinosaur from north-central New Mexico (USA). *Paläontolgische Zeitschrift* 70: 245–255.

The Clades of Dinosaurs

3

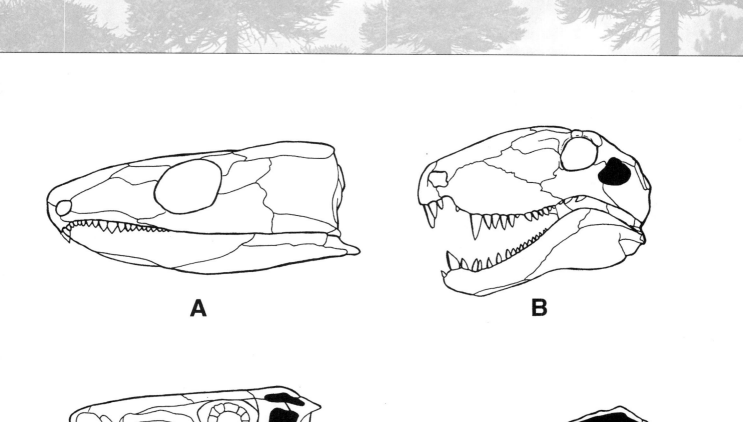

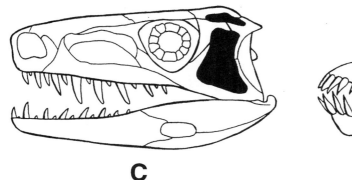

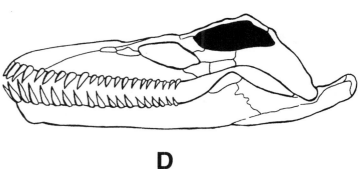

17.1. The four major types of amniote skulls, based on openings in the temporal region. A, Primitive pattern, retained in modern turtles (anapsid). No temporal openings in the cheek region. B, Synapsid pattern (mammals, mammal-like reptiles), with a single temporal opening below the postorbital/squamosal suture. C, Diapsid pattern (lizards, snakes, archosaurs, etc.) with two temporal openings separated by the postorbital/squamosal suture. D, Euryapsid pattern (many marine reptiles). This is a modified version of the diapsid pattern, where the lower temporal opening becomes open ventrally and is lost entirely in many groups. A similar pattern is seen in many lizards.

Evolution of the Archosaurs

17

J. Michael Parrish

Most readers will be familiar with groups such as dinosaurs, pterosaurs, and crocodiles, but the larger group to which all of these organisms belong, the Archosauria, is more obscure. Archosauria was initially erected by Cope (1869) to include dinosaurs, crocodilians, and all their presumed common ancestors. It has been slightly redefined by modern systematists to include the last common ancestor of the two extant groups of archosaurs—the crocodilians and the birds—and all of the descendants of that common ancestor. This is the sense in which I will use the name here.

The amniotes (the evolutionary group containing reptiles, mammals, and birds) have historically been differentiated on the basis of the arrangement of openings in the cheek region of the skull behind the orbit (Fig. 17.1). The pattern that is seen in fishes and amphibians, and that is primitive for the amniotes, is a solid cheek, without any openings. This pattern, termed anapsid, is also seen in early amniotes like captorhinids and pariesaurs, and is retained today in turtles, although some studies suggest that turtles may have acquired this condition secondarily (DeBraga and Rieppel 1997).

Very early in the evolution of the amniotes, certainly by the middle of the Carboniferous (300 My), a lineage of organisms emerged that had a single opening located low on the cheek region (Figs. 17.1, 17.2). This group, the synapsids, represents a large evolutionary line that included two groups of mammal-like reptiles, the pelycosaurs and the therapsids, as well as the mammals themselves, which appeared at about the same time that the dinosaurs did, in the later part of the Triassic Period.

A bit later in the Carboniferous, another major group of amniotes emerged; these animals had two openings in the cheek region. This group, the diapsids, includes lizards, snakes, crocodiles, birds, and a number of extinct groups. Several groups of extinct amniotes, principally the marine reptile groups Ichthyosauria and Sauropterygia (plesiosaurs and relatives) were formerly thought to form a fourth group, the Euryapsida, on the basis of their possession of an opening high on the cheek region. The extent to which these euryapsids are closely related to one another is still being debated (e.g., Rieppel 1993), but the present consensus is that the euryapsid pattern is simply a modification of the diapsid arrangement, in which the lower opening became open ventrally and was subsequently lost. Most recently, the name Euryapsida has been applied to a group comprising the Sauropterygia plus the turtlelike placodonts (Rieppel 1993).

Even though they represent important evolutionary features, the functional significance of these cranial openings remains obscure. However,

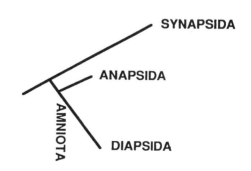

17.2. Cladogram showing the relationships of the major amniote groups.

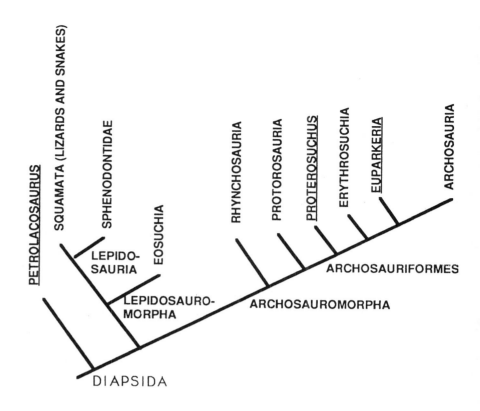

17.3. Phylogeny of the Diapsida, modified from Gauthier (1984).

they appear to correlate with sites of attachment for the major jaw muscles and with sites of possible high stress concentration within the skull.

The earliest diapsid is an animal called *Petrolacosaurus*, known from the Late Carboniferous of Kansas (Reisz 1981). By the beginning of the Permian (285 My), the diapsids had split into two divergent lines, one leading to lizards and snakes, the other leading to the archosaurs (Fig. 17.3). The Lepidosauromorpha includes a number of groups that are superficially lizardlike, but that lack key features shared by later members of the group (Gauthier et al. 1988). The Lepidosauria are a subset of the lepidosauromorph that appeared by the late Permian that includes all of the modern lepidosauromorphs: lizards, snakes, and the tuatara *Sphenodon*. These groups share distinctive patterns of bone ossification, limb structure, and ear morphology.

The other major group of diapsids is the Archosauromorpha (Gauthier 1984). Primitive members of this group are a diverse group of tetrapods, including the beaked, herbivorous rhynchosaurs, the protorosaurs (mostly strange, long-necked aquatic forms such as *Tanystropheus*), and the trilophosaurs (a group of terrestrial plant eaters with sturdy skulls and distinctive three-cusped teeth). These reptiles share a distinctive tarsal (ankle) pattern that will be discussed below.

Near the end of the Permian Period, another line of archosauromorphs appears, the Archosauriformes. This group was historically identified as the Archosauria and includes the archosaurs as now recognized, as well as a number of more basal groups (Gauthier 1984). The archosauriforms are distinguished by a number of features, notably the presence of an antorbital fenestra (an opening on the side of the snout between the orbit and the external naris) and of the laterosphenoid bone in the braincase (Clark et al. 1993).

At the end of the Carboniferous and in the Early Permian, most of the large terrestrial vertebrates were basal synapsids that are often grouped together as a paraphyletic group known as pelycosaurs. These forms diminished in abundance near the middle of the Permian, to be succeeded by a variety of therapsids, a group that arose from one group of carnivorous pelycosaurs, the Sphenacodontidae. Archosauriforms appeared near the end of the Permian but did not become dominant land vertebrates until about the Middle Triassic, as the abundance of therapsids steadily decreased. For the last half of the Triassic, the dominant terrestrial vertebrates were various types of early archosaurs, with the most familiar archosaur groups, the dinosaurs, pterosaurs, and crocodiles, making their appearance in the last third of the period. It was only near the end of the Triassic that the dinosaurs became relatively abundant and assumed the position as the dominant land vertebrates that they held until the end of the Mesozoic Era.

Much has been written about the evolutionary dynamics that drove this succession of dominant vertebrates. A strictly progressive interpretation (e.g., Colbert 1973) was favored until sometime in the 1970s, the idea being that each group that succeeded another was competitively superior to its predecessors. Thus archosaurs were considered to have outcompeted the therapsids, the dinosaurs to have outcompeted the other archosaurs, and so on. Benton (1983) offered a more opportunistic interpretation, pointing out that the extinctions of major groups generally preceded the radiation of the forms that later succeeded them. This left open the question of what caused the extinctions, which have generally been linked to environmental changes or sometimes to catastrophic events such as impacts of extraterrestrial objects (e.g., Olsen et al. 2002).

One way in which the dinosaurs were historically considered superior to their predecessors involved their mode of locomotion. The earliest tetrapods, such as *Ichthyostega* from the Devonian Period, used their limbs as little more than points of connection with the ground; most of the muscular force for forward movement was still supplied by wavelike, side-to-side movement of the trunk, a pattern inherited from their fish ancestors (Coates and Clack 1990). The pattern present in early amniotes, and retained in extant forms such as lizards and turtles, is somewhat similar. The proximal parts of the limbs project horizontally from the body, forming right angles with the distal parts of the limbs, which project downward at the elbow or knee to terminate in the forefeet and hind feet. Such animals exhibit a sprawling gait, with movement within the limbs contributing less to forward movement of the body than sideways movements of the trunk, although some forward momentum is imparted by rotation of the proximal parts of the limbs around their long axes (Charig 1972; Brinkman 1980).

Crocodilians and other nondinosaurian archosauriforms were long considered to be inferior to dinosaurs in their locomotor capabilities. In an influential paper, Charig (1972) depicted crocodilians as exhibiting a semierect gait, with the capability of bringing the limbs close to the body in a nearly erect "high walk" in addition to walking with a typical sprawling gait. Several studies of limb movement in early archosauriforms (Parrish 1986), pterosaurs (Padian 1983), and crocodylomorphs (Walker 1970; Crush 1984; Parrish 1987; but see Gatesy 1991 for a slightly different interpretation of modern crocodilian locomotion) suggest that all but the earliest archosauriforms had what appear to be fully erect gaits, with the limbs aligned within a vertical plane parallel to the body's vertical axis of symmetry through the body's midline. With this arrangement, which humans share with other mammals, birds, and dinosaurs, the limbs move within a single plane, with only flexion and extension occurring at the elbow/knee and wrist/ankle joints (Brinkman 1980; Parrish 1986). In conjunction with such an erect stance, the trunk is usually stiffened to prevent side to side movement, and almost all of the muscular effort involved in flexion and extension of the limb segments translates into forward movement. Other studies of the kinematics of crocodilian locomotion (Reilly and Elias 1998) and of the breathing dynamics of crocodilians (Carrier 1987; Seymour et al. 2004) independently support the notion that crocodilians independently developed their variable, semierect gait from an erect ancestor.

If dinosaurs and most of their archosaurian predecessors and contemporaries had erect stances, then posture alone cannot be cited as a competitive advantage dinosaurs held with respect to their relatives. Dinosaurs and other erect archosaurs do differ markedly in their limb proportions, as well as in the posture of their feet (Parrish 1986; Chatterjee 1985). In dinosaurs, the tibia and fibula are generally longer than the femur, whereas the opposite condition pertains in most nondinosaurian archosaurs. Dinosaurs also held the metatarsus off the ground in what is known as a digitigrade stance, whereas the other archosaurs mostly retained the primitive arrangement, a plantigrade stance, where the entire foot lands on the ground during normal movement.

Early History of the Archosauriforms

The earliest archosauriform is *Archosaurus*, known from fragmentary material from the Late Permian of central Russia (Tatarinov 1960). Even more fragmentary material from the Late Permian of southern Africa may belong to a similar taxon (Parrington 1956). Although these fossils are incomplete, they seem to represent animals similar to *Proterosuchus*, the next archosauriform to appear in the fossil record. *Proterosuchus* (Fig. 17.4) is quite well known. It is represented by complete skeletons from South Africa and China, and by more fragmentary material from Argentina and India. This reptile is superficially similar to modern crocodilians in size, body proportions, and inferred ecological habits. *Proterosuchus* appears to have retained the sprawling gait of its earlier amniote ancestors, and its long, low skull and relatively homogeneous, conelike teeth suggest that this animal may have favored small vertebrates such as fish rather than larger items as its predominant diet. Initially, a number of different genera and species of earliest Triassic archosauriforms were named from South Africa, where these animals are relatively abundant. However, subsequent study by several

17.4. Skeletal reconstruction of *Proterosuchus* (after Paul in Parrish 1986).

workers (Cruickshank 1972; Clark et al. 1993; Welman and Flemming 1993) has shown that the other genera, *Chasmatosaurus* and *Elaphrosuchus*, were based on specimens of *Proterosuchus* in which the skull was crushed dorsoventrally such that the snout drooped downward and the quadrate bone at the back of the skull was angled posteriorly. Two other early Triassic forms, *Kalisuchus* (Thulborn 1979) and *Tasmaniosaurus* (Camp and Banks 1978; Thulborn 1986), are known from Australia and do appear to be distinct from *Proterosuchus*, although neither taxon is represented by very complete material.

In the latter part of the Early Triassic, two other groups of archosauriforms appeared. The first of these, the Erythrosuchidae, were a group of large terrestrial carnivores that persisted into the Middle Triassic and are represented by at least seven genera (Parrish 1992). The erythrosuchids were the first group of archosauriforms to exhibit a distinctive type of skull, which is tall and compressed mediolaterally, with large, recurved teeth (Fig. 17.5). Erythrosuchids, some of which had skulls over a meter in length, appear to have been the dominant terrestrial predators of their day. Although information about their limb morphology, and particularly the structure of their feet, is relatively scanty, well-preserved material of *Vjushkovia triplocostata* (PIAS 951) from southern Russia suggests an animal with a much more erect stance than was present in *Proterosuchus* or earlier amniotes, even though a great deal of rotation of the proximal limb elements probably took place during locomotion. The most primitive erythrosuchids, *Garjainia* from Russia and *Fugusuchus* from China, were also the earliest to appear. Both retain a relatively narrow snout and skulls that are lower in profile than those of the slightly later *Erythrosuchus* and *Vjushkovia*.

Another important early archosauriform is *Euparkeria* (Fig. 17.6), known from the late Early Triassic of South Africa (Ewer 1965). *Euparkeria* is known from 10 specimens from a single locality in the Aliwal North region of South Africa. This was a much smaller animal than other early archosauriforms; the largest specimen would be less than a meter in total length. It is the first archosauriform that can be clearly shown to have had

17.6. Skeletal reconstruction of *Euparkeria* (after Paul in Parrish 1986).

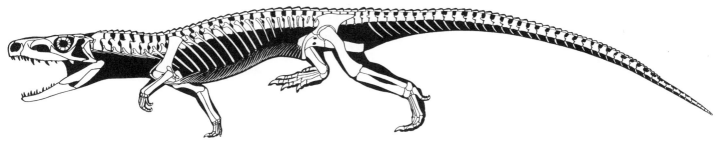

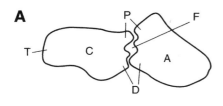

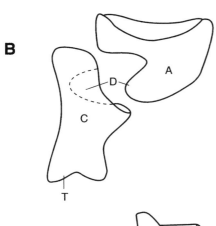

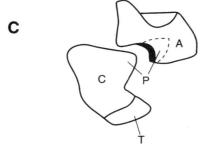

17.7. Schematic of the major types of archosauriform ankles (after Parrish 1986). A, Primitive archosauriform pattern with two pairs of articular facets. B, Crocodile-normal pattern, where the distal of the two facet pairs is modified into a rotary ankle joint. C, Crocodile-reverse pattern, where the proximal of the two facet pairs is modified into a rotary ankle joint. P = proximal facet pair; D = distal facet pair; F = perforating foramen; A = astragalus; C = calcaneum; T = calcaneal tuber.

armor, a feature that is retained in some form by most of the subsequent archosauriforms. Although *Euparkeria* lacks some archosaurian features that place it just outside of the group, it probably represents the best fossil example we have of what the common ancestor of dinosaurs and crocodilians might have looked like. Often restored in a very dinosaurian bipedal pose, *Euparkeria* was more likely (based on limb proportions) to have been predominantly quadrupedal.

The Archosauria in the strict sense originated by the middle of the Triassic. Several features characterize this group, including a new foramen through which the internal carotid artery took a different path into the braincase than that seen in other diapsids. Archosaurs are represented by two different lineages: one, dubbed the *Ornithodira* by Gauthier (1984), led to dinosaurs and birds; the other, named the *Crurotarsi* by Sereno (1991), led to crocodilians.

Historically, one of the features considered critical for elucidating archosaurian phylogeny has been the structure of the ankle (Fig. 17.7). Archosauromorphs, including early archosauriforms like *Proterosuchus*, are united by a distinctive ankle pattern where the two proximal ankle bones, the astragalus and calcaneum, are connected by two pairs of ball-and-socket joints (Brinkman 1981); the more proximal of these joints consists of a socket on the astragalus and a ball on the calcaneum, whereas the distal pair has a socket on the calcaneum and a ball on the astragalus. In the crurotarsans, the ventral of these joints becomes movable, such that the main joint between the foot and the lower leg is between the astragalus and calcaneum, rather than between those two bones and the distal tarsals. This arrangement was dubbed "crocodile-normal" by Chatterjee (1982) because it is the pattern present in the extant Crocodylia. In the Ornithosuchidae, the more proximal of the two sets of facets became elaborated, such that a mobile joint developed with the ball on the calcaneum and the socket on the astragalus. This arrangement was termed "crocodile-reverse" by Chatterjee (1982). In the Ornithodira, the functional ankle joint is of the primitive type, between the proximal and distal tarsals, although the details of the articulation between the two bones most resembles the pattern seen in the Ornithosuchidae (Chatterjee 1982; Bonaparte 1982).

Different assumptions about the evolutionary implications of these structures led to two divergent evolutionary interpretations. The first, held by Thulborn (1980) and Sereno (1991), is that the Ornithodira have retained the primitive, fixed ankle, and that the Crurotarsi (including Ornithosuchidae) comprise a separate evolutionary lineage (Fig. 17.8; Sereno and Arcucci 1990), which has developed a movable joint between the calcaneum and astragalus. The second view, espoused by Gauthier (1984) and Parrish (1986), called for the ornithosuchids being united with the Ornithodira in a lineage, the Ornithosuchia, which has the crocodile-reverse ankle as a derived character, and whose last common ancestor with the Crocodylotarsi had the primitive, two-faceted type of ankle joint. Another study (Parrish 1993) found the two phylogenetic arrangements almost equally likely, and others, by Juul (1994), Dyke (1998), and Brusatte et al. (2010), followed Sereno and Arcucci (1990) in nesting the crocodile-reverse Ornithosuchidae within the Crurotarsi. Dyke (1998) suggested that ankle morphology is not particularly useful in elucidating archosaur phylogeny.

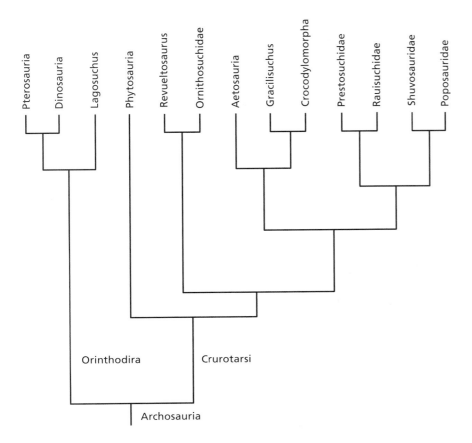

The earliest members of the Ornithodira are the lagosuchids, small, long-legged archosaurs from the Middle Triassic of Argentina (Romer 1971, 1972; Bonaparte 1975a; Sereno and Arcucci 1993, 1994). Although the two presently known lagosuchids, *Marasuchus* and *Lagerpeton*, are poorly known, they had pelvic structures, limb proportions, and reduced ankles that clearly link them with the dinosaurs. *Lagerpeton* had a peculiar hind limb, with short pelvic elements and a foot with elongated lateral digits that may have been utilized for perching.

The other group that appears to belong within the Ornithodira comprises the pterosaurs, or flying reptiles, the earliest well-known forms of which are from the Late Triassic of Italy. A link between the pterosaurs and the other ornithodirans may be provided by a peculiar animal, *Scleromochlus*, from the Late Triassic of Scotland, which has skull structure and limb proportions similar to those of pterosaurs and an ankle structure similar to those of lagosuchids (Gauthier 1984; Benton 1999), although not all workers endorse a close relationship between *Scleromochlus* and pterosaurs (Sereno 1991). Alternative phylogenies proposed by Wild (1984) and Bennett (1996) place the pterosaurs outside the archosaurs, with Wild suggesting that they have closer evolutionary relationships with early lepidosauromorphs and Bennett placing them just outside of Archosauria. These scenarios are in conflict with the considerable character evidence amassed by Gauthier (1984), Padian (1984), and Sereno (1991) in support of a dinosaur/pterosaur clade. The early pterosaurs are clearly flying reptiles, but they are considerably different from their later relatives: they are relatively small and retain long tails. *Eudimorphodon*, one of the earliest well-known pterosaurs, also has complex, multicuspid teeth.

17.9. Skeletal reconstruction of *Rutiodon*, modified from Camp (1930).

Dinosaurs appeared in the fossil record by the beginning of the Late Triassic. Three-toed, dinosaur-like footprints are known as early as the end of the Early Triassic, but these could have also been made by close relatives to the dinosaurs like *Marasuchus*.

In the Middle and Late Triassic, most described archosaurs belonged to the Crocodylotarsi, a confusing and, until recently, little-known group (Benton and Clark 1988; Parrish 1993). The most primitive crocodylotarsans are represented by a group that is not abundant until the Late Triassic, during which they are among the most common of archosaur fossils. The Parasuchia, or phytosaurs (Fig. 17.9), were crocodile-like animals with long, narrow snouts, dorsally placed nostrils, and flattened skulls (Camp 1930; Chatterjee 1978). Their limb structure and the sedimentary settings in which their remains were found suggest that these animals, like modern crocodilians, were amphibious to aquatic. They represented a wide variety of ecological types, from small, presumably mostly piscivorous, forms like *Mystriosuchus* with rodlike snouts and conical teeth to giant, predatory forms like *Nicrosaurus* with 1.5-m-long skulls, large teeth, and relatively broad snouts.

One confusing aspect of crurotarsan evolution lies in the fact that several distinct groups had the high, narrow skull pattern seen earlier in erythrosuchids and shared with most meat-eating dinosaurs (Gauthier 1984; Parrish 1993; Gower 2000; Gower and Nesbitt 2006). These groups have been grouped differently by different workers over the years. The earliest of these groups, the Prestosuchidae (Fig. 17.10), are known from South America and Europe in the Middle Triassic, and they included gigantic forms like the Brazilian genus *Prestosuchus* (Huene 1942; Barberena 1978). These seem to have been mostly erect, but they do not have any limb adaptations for rapid movement. The Rauisuchidae, which appeared in Brazil at the same time the Prestosuchidae, have a series of derived modifications of the skull, as well as a mediolaterally compressed foot with a reduced lateral digit and specialized ankle joint that indicates the capability for more rapid movement (Bonaparte 1984). Another group, the Poposauridae, appeared

17.10. Skeletal reconstruction of the prestosuchid *Saurosuchus* (after Paul in Parrish 1986).

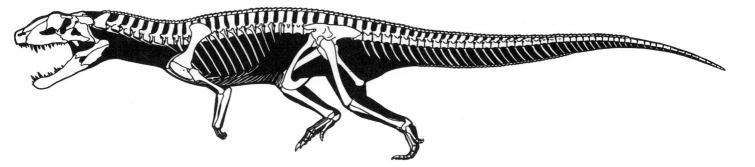

17.11. The poposaurid *Postosuchus* (after Paul in Parrish 1986).

in the Middle Triassic (Chatterjee 1985). Poposaurs (Fig. 17.11) superficially resemble other Rauisuchidae, but they have cranial specializations, including an eustachian tube system, that support their placement as the closest relatives of the crocodylomorphs (Parrish 1987, 1993; Benton and Clark 1988). More recently, a taxon dubbed the Shuvosauridae has emerged; it is based on two late Triassic taxa from the southwestern United States, *Shuvosaurus* and *Effigia*, that are both strongly convergent on Cretaceous theropod dinosaurs (Nesbitt and Norell 2006; Nesbitt 2007). Despite many phylogenetic analyses, the relative positions of these carnivorous taxa remain a matter of dispute, largely because a number of incomplete, and in some cases chimeric, taxa have hampered their phylogenetic analysis.

Members of a fourth group of terrestrial predators, the Ornithosuchidae, also have skull profiles and body sizes similar to the other three groups (Bonaparte 1975b). They also appear to have been capable of relatively rapid movement, but they are the only one of these large predator groups that exhibits the crocodile-reverse tarsal pattern. Because all of these groups overlap with one another stratigraphically, their classification has been confusing. Their ecological roles were probably relatively similar. It seems likely that they differed mainly in the type of prey they fed upon, with groups like the poposaurs probably specializing in more agile prey than, for example, the prestosuchids.

Before the Late Triassic, virtually all archosauriforms were carnivores. One possible exception was *Lotosaurus*, an enigmatic rauisuchian from the Middle Triassic of southern China that lacked teeth entirely, but instead appears to have had a curving, turtlelike beak (Zhang 1975; Parrish 1993). *Lotosaurus* also had elongated neural spines along the trunk, which suggest that it may have had a fleshy sail like those of some pelycosaurs and dinosaurs. In the Late Triassic, the first group of clearly herbivorous archosaurs appeared, the Aetosauria (Walker 1961; Parrish 1994). Aetosaurs (Fig. 17.12) also had turtlelike beaks anteriorly, and most had simple, conical teeth in the cheek region. The most primitive aetosaurs, like *Aetosaurus* from Germany, had pointed snouts, but more derived forms, like *Stagonolepis* and *Desmatosuchus*, had flattened, upturned snouts that appear to have been an adaptation for rooting up vegetation. Other apparent digging adaptations are found in their powerful limbs and broad, elongate digits. The aetosaurs possessed a complete coat of protective dermal armor that, in some derived forms like *Typothorax*, forms a broad oval carapace similar to a turtle's shell (Heckert et al. 2010). Another herbivorous crurotarsan is *Revueltosaurus* (Parker et al. 2005). Originally known from teeth attributed to an herbivorous dinosaur,

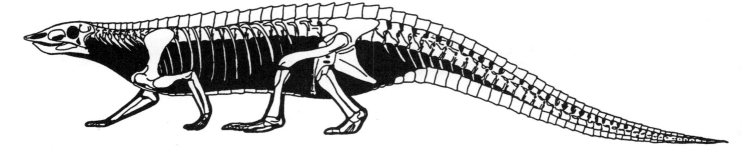

17.12. Skeletal reconstruction of *Stagonolepis*, modified from Walker (1961).

this animal is now known to be an armored crurotarsan that is most closely related to either the Aetosauria (Parker et al. 2005) or the Ornithosuchidae (Brusatte et al. 2010).

The Crocodylomorpha, including crocodilians and a number of their fossil relatives, appeared in the Late Triassic (Walker 1970; Clark and Sues 2002). Crocodylomorphs can be distinguished by a number of specializations of the skull, including the development of extensive pneumatic cavities and a quadrate bone that tilts anteriorly and develops extensive sutures with other bones at the back of the skull. Other specializations include an elongate wrist and specialized shoulder and pelvic girdles. The earliest crocodylomorphs were a group of forms often united as the Sphenosuchia, an apparently monophyletic group that originated in the early part of the Late Triassic and persisted until the Late Jurassic. Sphenosuchians (Fig. 17.13) had most of the distinctive crocodylomorph cranial features but also had narrow, elongate limbs and mediolaterally compressed feet. The sphenosuchians, as well as other early crocodylomorphs like the Protosuchidae, appear to have been erect, fully terrestrial animals and some, such as the appropriately named *Terrestrisuchus*, were probably agile, rapid runners.

Thus the modern view we have of crocodilians being sluggish, mostly aquatic predators did not hold for most of their Triassic and Jurassic ancestors. It was in the Early Jurassic that the first aquatic crocodylomorphs appeared, but these belonged to lines like the Metriorhynchidae and Teleosauridae, which developed paddlelike appendages from their limbs. The main line of crocodilians did not assume their modern amphibious habits until well into the Mesozoic. The unique crocodilian locomotor pattern, instead of being an intermediate semierect stage between sprawling and erect forms, appears to be a modification of the erect stance seen in early crocodylomorphs; it allows modern crocodilians to use their limbs as laterally projecting swimming appendages in addition to the erect high walk

17.13. The early crocodylomorph *Pseudhesperosuchus* (after Paul in Parrish 1986).

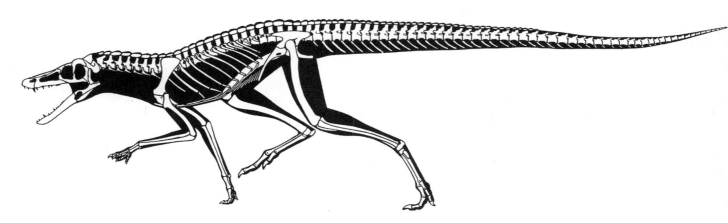

that they employ when traveling overland (Brinkman 1980; Parrish 1987; Reilley and Elias 1998).

By the end of the Triassic Period, all of the archosauriforms other than dinosaurs, pterosaurs, and crocodylomorphs had disappeared. Groups underwent steady replacement during that period, and the final decline in diversity near the Triassic–Jurassic boundary does correlate roughly with a dramatic increase in the abundance of dinosaur fossils, notably those of small theropods like *Coelophysis* and prosauropods like *Plateosaurus* (Sander 1992).

References

Barberena, M. C. 1978. A huge thecodont skull from the Triassic of Brazil. *Pesquisas* 7: 111–129.

Bennett, S. C. 1996. The phylogenetic position of the Pterosauria within the Archosauromorpha. *Zoological Journal of the Linnean Society* 118: 261–309.

Benton, M. J. 1983. Dinosaur success in the Triassic: a noncompetitive ecological model. *Quarterly Review of Biology* 58: 29–55.

———. 1999. *Scleromochlus taylori* and the origin of dinosaurs and pterosaurs. *Philosophical Transactions of the Royal. Society of London B* 354: 1423–1446.

Benton, M. J., and J. M. Clark. 1988. Archosaur phylogeny and the relationshipsof the Crocodylia. In M. J. Benton (ed.), *Phylogeny and Classification of Amniotes.* Systematics Association Special Volume 35A: 295–338.

Bonaparte, J. F. 1975a. Nuevos materiales de *Lagosuchus tamalpayensis* Romer (Thecodontia, Pseudosuchia) y su significado en el origen de los Saurischia. *Acta Geologica Lilloana* 13: 5–90.

———. 1975b. The family Ornithosuchidae (Archosauria: Thecodontia). *Colloque International Centre National de la Recherche Scientifique* 218: 48–501.

———. 1982. Classification of the Thecodontia. *Geóbios Mémoir Spéciale* 6: 99–112.

———. 1984. Locomotion in rauisuchid thecodonts. *Journal of Vertebrate Paleontology* 3: 210–218.

Brinkman, D. 1980. The hindlimb step cycle of *Caiman sclerops* and the mechanics of the crocodilian tarsus and metatarsus. *Canadian Journal of Zoology* 58: 2187–2200.

———. 1981. The origin of the crocodiloid tarsi and the interrelationships of thecodontian archosaurs. *Breviora* 464: 1–22.

Brusatte, S. L., M. J. Benton, J. B. Desojo, and M. C. Langer. 2010. The higher level phylogeny of Archosauria (Reptilia: Diapsida). *Journal of Systematic Palaeontology* 8: 3–47.

Camp, C. L. 1930. A study of the phytosaurs. *Memoirs of the University of California* 10: 1–161.

Camp, C. L., and M. R. Banks. 1978. A proterosuchian reptile from the Early Triassic of Tasmania. *Alcheringa* 2: 143–158.

Carrier, D. R. 1987. The evolution of locomotor stamina in tetrapods: circumventing a mechanical constraint. *Paleobiology* 13: 326–341.

Charig, A. J. 1972. The evolution of the archosaur pelvis and hindlimb: an explanation in functional terms. In K. A. Joysey and T. S. Kemp (eds.), *Studies in Vertebrate Evolution*, 121–155. Edinburgh: Oliver and Boyd.

Chatterjee, S. 1978. A primitive parasuchid (phytosaur) from the Upper Triassic Maleri Formation of India. *Paleontology* 21: 83–127.

———. 1982. Phylogeny and classification of the thecodontian reptiles. *Nature* 295: 317–320.

———. 1985. *Postosuchus*, a new thecodontian reptile from the Triassic of Texas and the origin of tyrannosaurs. *Philosophical Transactions of the Royal Society of London B* 309: 395–460.

Clark, J. M., J. A. Gauthier, J. Welman, and J. M. Parrish. 1993. The laterosphenoid bone of early archosaurs. *Journal of Vertebrate Paleontology* 13: 48–57.

Clark, J. M., and H.-D. Sues. 2002. Two new basal crocodylomorph archosaurs from the Lower Jurassic and the monophyly of the Sphenosuchia. *Zoological Journal of the Linnean Society* 135: 77–95.

Coates, M. I., and J. A. Clack. 1990. Polydactyly and the earliest known tetrapod limbs. *Nature* 347: 66–69.

Colbert, E. H. 1973. *Wandering Lands and Animals.* New York: E. P. Dutton.

Cope, E. D. 1869. Synopsis of the extinct Batrachia, Reptilia, and Aves of North America. *Transactions of the American Philosophical Society* 14: 1–252.

Cruickshank, A. R. I. 1972. The proterosuchian thecodonts. In K. A. Joysey and T. S. Kemp (eds.), *Studies in Vertebrate Evolution*, 89–119. Edinburgh: Oliver and Boyd.

Crush, P. 1984. A late Upper Triassic sphenosuchid crocodile from Wales. *Palaeontology* 27: 133–157.

DeBraga, M., and O. Rieppel. 1997. Reptile phylogeny and the interrelationships of turtles. *Zoological Journal of the Linnean Society* 120: 281–354.

Dubiel, R. F., J. T. Parrish, J. M. Parrish, and S. C. Good. 1991. The Pangean megamonsoon – evidence from the Upper Triassic Chinle Formation, Colorado Plateau. *Palaios* 6: 347–370.

Dyke, G. J. 1998. Does archosaur phylogeny hinge on the ankle joint? *Journal of Vertebrate Paleontology* 18: 558–562.

Ewer, R. F. 1965. The anatomy of the thecodont reptile *Euparkeria capensis* Broom. *Philosophical Transactions of the Royal Society of London B* 248: 379–435.

Gatesy, S. M. 1991. Hind limb movements of the American alligator (*Alligator mississippiensis*) and postural grades. *Journal of Zoology* 224: 577–588.

Gauthier, J. A. 1984. A cladistic analysis of the higher systematic categories of the Diapsida. Ph.D. dissertation, University of California, Berkeley.

Gauthier, J. A., R. Estes, and K. de Queiroz 1988. A phylogentic analysis of Lepidosauromorpha. In R. Estes and G. Pregill (eds.), *Phylogenetic Analysis of the Lizard Families*, 15–98. Stanford: Stanford University Press.

Gower, D. J. 2000. Rauisuchian archosaurs: an overview. *Neues Jahrbuch für Geologie und Paläontologie, Abhandlungen* 218: 447–488.

Gower, D. J., and S. J. Nesbitt. 2006. The braincase of Arizonasaurus babbitti – further evidence of the non-monophyly of Rauisuchia. *Journal of Vertebrate Paleontology* 26: 79–87.

Heckert, A. B., S. G. Lucas, L. F. Rinehart, M. D. Celesky, J. A. Speilmann, and A. P. Hunt. 2010. Articulated skeletons of the aetosaur *Typothorax coccinarum* Cope (Archosauria: Stagonolepididae) from the Upper Triassic Bull Canyon Formation (Revueltian, early-mid Norian), eastern New Mexico, USA. *Journal of Vertebrate Paleontology* 30: 619–642).

Huene, F. 1942. *Die fossilen Reptilien des Südamerikanischen Gondwanalandes.* Munich: C. H. Beck.

Juul, L. 1994. The phylogeny of basal archosaurs. *Paleontologica Africana* 31: 1–38.

Nesbitt, S. J. 2007. The anatomy of *Effigia okeeffeae* (Archosauria, Suchia), theropod-like convergence, and the distribution of related taxa. *Bulletin of the American Museum of Natural History* 302: 1–84.

Nesbitt, S. J., and M. A. Norell. 2006. Extreme convergence in the body plan of an early suchian (Archosauria) and ornithomimid dinosaurs (Theropoda). *Proceedings of the Royal Society of London B* 273: 1045–1048.

Padian, K. 1983. A functional analysis of flying and walking in pterosaurs. *Paleobiology* 9: 218–239.

———. 1984. The origin of pterosaurs. In W.-E. Reif and F. Westphal (eds.), *Third Symposium on Terrestrial Mesozoic Ecosystems, Short Papers*, 163–168. Tübingen: Attempto Verlag.

Parker, W. G., R. B. Irmis, S. J. Nesbitt, J. W. Martz, and L. S. Browne. 2005. The Late Triassic pseudosuchian *Revueltosaurus callenderi* and its implications for the diversity of early ornithischian dinosaurs. *Proceedings of the Royal Society of London B* 272: 973–979.

Parrington, F. R. 1956. A problematic reptile from the Upper Permian. *Annals and Magazine of Natural History* 12: 333–336.

Parrish, J. M. 1986. Locomotor evolution in the hindlimb and pelvis of the Thecodontia (Reptilia: Archosauria). *Hunteria* 1 (2): 1–35.

———. 1987. The origin of crocodilian locomotion. *Paleobiology* 13: 396–414.

———. 1992. Phylogeny of the Erythrosuchidae. *Journal of Vertebrate Paleontology* 12: 93–102.

———. 1993. Phylogeny of the Crocodylotarsi and a consideration of archosaurian and crurotarsan monophyly. *Journal of Vertebrate Paleontology* 13: 287–308.

———. 1994. Cranial osteology of *Longosuchus meadei* and a consideration of the phylogeny of the Aetosauria. *Journal of Vertebrate Paleontology* 14: 196–209.

Olsen, P. E., D. V. Kent, H.-D. Sues, C. Koeberl, H. Huber, A. Montanari, E. C. Rainforth, S. J. Fowell, M. J. Szajna, and B. W. Hartline. 2002. Ascent of dinosaurs linked to an iridium anomaly at the Triassic–Jurassic boundary. *Science* 296: 1305–1307.

Reilly, S. M., and J. A. Elias. 1998. Locomotion in *Alligator mississippiensis*: kinematic effects of speed and posture and their relevance to the sprawling-to-erect paradigm. *Journal of Experimental Zoology* 201: 2559–2574.

Reisz, R. R. 1981. A diapsid reptile from the

Pennsylvanian of Kansas. *Special Publications of the Museum of Natural History, University of Kansas* 7: 1–174.

Rieppel, O. 1993. Euryapsid relationships: a preliminary analysis. *Neues Jahrbuch für Geologie und Paläontologie, Abhandlungen* 188: 241–264.

Romer, A. S. 1971. The Chañares (Argentina) Triassic reptile fauna. X. Two new but incompletely known long-limbed pseudosuchians. *Breviora* 378: 1–10.

———. 1972. The Chañares (Argentina) Triassic reptile fauna. XV. Further remains of the thecodonts Lagosuchus and Lagerpeton. *Breviora* 394: 1–7.

Sander, M. 1992. The Norian *Plateosaurus* bonebeds of central Europe and their taphonomy. *Palaeogeography, Palaeoclimatology, Palaeoecology* 93: 255–299.

Sereno, P. C. 1991. Basal archosaurs: phylogenetic relationships and functional implications. Society of Vertebrate Paleontology Memoir 2. *Journal of Vertebrate Paleontology* 11 (suppl. 4): 1–53.

Sereno, P. C., and A. B. Arcucci. 1990. The monophyly of crurotarsal archosaurs and the origin of bird and crocodile ankle joints. *Neues Jahrbuch Geologie und Paläontologie, Abhandlungen* 180: 21–52.

———. 1993. Dinosaur precursors from the Middle Triassic of Argentina: *Lagerpeton chanarensis*. *Journal of Vertebrate Paleontology* 13: 385–399.

———. 1994. Dinosaur precursors from the Middle Triassic of Argentina: *Marasuchus lilloensis*, gen. nov. *Journal of Vertebrate Paleontology* 14: 53–73.

Seymour, R. S., C. L. Bennett-Stamper, S. D. Johnston, D. R. Carrier, and G. C. Grigg. 2004. Evidence for endothermic ancestors of crocodiles at the stem of archosaur evolution. *Physiological and Biochemical Zoology* 77: 1051–1067

Tatarinov, L. P. 1960. Otkrytie psevdizukhii v verkhnie Permi SSSR. *Paleontologicheskii Zhurnal* 1960: 74–80.

Thulborn, R. A. 1979. A proterosuchian thecodont from the Rewan Formation of Australia. *Memoirs of the Queensland Museum* 19: 331–355.

———. 1980. The ankle joints of archosaurs. *Alcheringa* 4: 141–161.

———. 1986. The Australian Triassic reptile *Tasmaniosaurus triassicus* (Thecodontia, Proterosuchia). *Journal of Vertebrate Paleontology* 6: 123–142.

Walker, A. D. 1961. Triassic reptiles from the Elgin area: *Stagonolepis, Dasygnathus*, and their allies. *Philosophical Transactions Royal Society of London B* 244: 103–204.

———. 1970. A revision of the Jurassic crocodile *Hallopus*, with remarks on the classification of crocodiles. *Philosophical Transactions Royal Society of London B* 257: 323–372.

———. 1990. A revision of *Sphenosuchus acutus* Haughton, a crocodylomorph reptile from the Eliot Formation (Late Triassic or Early Jurassic) of South Africa. *Philosophical Transactions Royal Society of London B* 330: 1–120.

Welman, J., and A. Flemming. 1993. Statistical analysis of skulls of Triassic proterosuchids (Reptilia, Archosauromorpha) from South Africa. *Paleontologia Africana* 30: 113–123.

Wild, R. 1984. Flugsaurier aus den Obertrias von Italien. *Naturwissenschaften* 71: 1–11.

Zhang, F. 1975. A new thecodont, *Lotosaurus*, from the Middle Triassic of Hunan [in Chinese with English summary]. *Vertebrata Palasiatica* 13: 144–148.

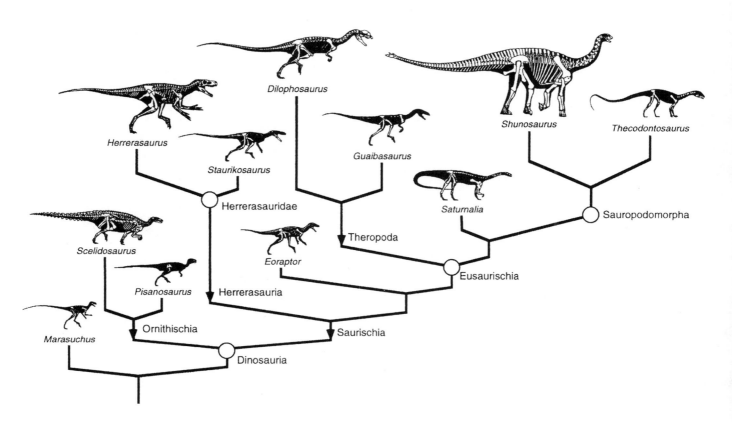

18.1. Cladogram showing suggested relationships of the basal dinosaurs.

Courtesy of Max Langer.

Origin and Early Evolution of Dinosaurs

Michael J. Benton

18

The dinosaurs arose in the Triassic, and probably during the Early to Middle Triassic. They entered a world far different from the typical Age of Dinosaurs scenes, a world in which the dominant herbivores were synapsids (dicynodonts and chiniquodontids) and rhynchosaurs, and carnivores were cynodonts and basal archosaurs of various kinds, previously called thecodontians (see Chapter 17, this volume). Into this world came the dinosaurs, initially small bipedal carnivores. They rose to dominance at some point during the second half of the Triassic. Certainly, by the end of the Triassic Period, dinosaurs were abundant and reasonably diverse, and all the major lineages had emerged and diversified.

Since 1980, paleontologists' views have changed dramatically, and new specimens and new methods have revolutionized our understanding of the origin of the dinosaurs. Most important has been the widespread use of cladistics as the key tool in disentangling the tree of life (see Chapter 11, this volume). Second, new, high-precision methods of dating the rocks give a much firmer timescale of events. There are still debates, however, about the relative importance of different groups of land vertebrates through the Triassic and the kinds of ecological processes that might have been involved in the rise and initial expansion of the clade Dinosauria.

Three key topics will be explored here: phylogeny (defining what is a dinosaur, spurious early records, the first dinosaurs), geology (dating the rocks), and models (how evolutionary radiations happen).

Definition of the Dinosauria

Richard Owen assumed in 1842 that his new group, the Dinosauria, was a real group, a monophyletic group, or a clade, in modern parlance (see Chapter 2, this volume). In other words, he assumed that the Dinosauria had a single ancestor, and that the group included all the descendants of that ancestor. This view was commonly held during most of the nineteenth century, but it was shaken by Harry Seeley's demonstration in 1887 that there were two major dinosaur groups, the Saurischia and Ornithischia, distinguished by the nature of their pelvic arrangements (see Chapter 3, this volume).

Perhaps, thought Seeley, the Saurischia and Ornithischia were distinct evolutionary branches that had arisen from separate ancestors. Seeley was merely espousing a commonly held view at the time called the persistence of types, a view promoted especially by Thomas Henry Huxley. The idea was that large changes in form could not happen readily in evolution, and the fossil record showed how major groups retained their main characteristics for long periods of time. Hence, paleontologists had to look for very long periods of initial evolution that led up to each major group, and often these initial spans of evolution were missing from the fossil record.

Seeley's view dominated during most of the twentieth century, and many dinosaur paleontologists made it even more complex. Not only had the Saurischia and Ornithischia evolved from separate ancestors, but so too had some of the subdivisions within those two groups, probably the two saurischian groups, the Theropoda and Sauropodomorpha, and possibly even some of the main ornithischian groups known in 1900, the Ornithopoda, Ceratopsia, Stegosauria, and Ankylosauria. In the end, the dinosaurs became merely an assemblage of large extinct reptiles of the Mesozoic that shared little in common. Hence, dinosaurs were seen as a polyphyletic group, deriving from two, three, or more sources among the basal archosaurs (reviewed in Benton 2004).

The collapse of the polyphyletic view came quickly and dramatically in about 1984 (Fig. 18.1). This had been presaged in short papers by Bakker and Galton (1974) and Bonaparte (1976), who saw many unique characters shared by both saurischian and ornithischian dinosaurs. Some brief papers published in 1984 were followed by more substantial accounts (Gauthier 1986; Sereno 1986; Novas 1989, 1994; Benton 1990); all applied strictly cladistic approaches to the data, and they independently agreed strongly that the Dinosauria of Owen (1842) is a monophyletic group, defined by a number of synapomorphies, including the following:

- Possible absence of a postfrontal bone (Fig. 18.2B).
- Jugal branching backward into two distinct processes to contact the quadratojugal (Fig. 18.2B).
- Temporal musculature extending anteriorly onto the skull roof, marked by a distinct fossa on the frontal bone.
- Epipophyses (additional facets on the postzygapophyses) on the cervical vertebrae.
- Elongate deltopectoral crest on the humerus (Fig. 18.2A).
- Possible three or fewer phalanges in the fourth finger of the hand (Fig. 18.2C).
- Fully open acetabulum (Figs. 18.2A, 18.3A).
- Possible brevis shelf on the ventral surface of the postacetabular part of the ilium.
- Asymmetrical fourth trochanter with a steeply angled distal margin on the femur.
- Articulation facet for the fibula occupying less than 30% of the transverse width of the astragalus.

Three of these 10 characters are listed as "possible" because incomplete knowledge of basal dinosauromorphs such as *Marasuchus* and *Eucoelophysis* makes their placement uncertain (Benton 1999, 2004; Langer and Benton 2006; Brusatte et al. 2010; Langer et al. 2010; Nesbitt 2011).

Many other dinosaurian features arose lower down the cladogram in close outgroups of Dinosauria, such as *Marasuchus*, *Silesaurus*, *Eucoelophysis*, and Pterosauria (see Parrish, Chapter 17 in this volume). These include several characters that were given by Gauthier (1986), Benton and Clark

18.2. The early dinosaur *Herrerasaurus ischigualastensis* from the Ischigualasto Formation of Argentina. A, Skeleton in side view; B, skull in side view; C, left hand viewed from above; D, right foot viewed from above. A–C, based on Sereno (1994); D, based on Novas (1994).

Michael J. Benton

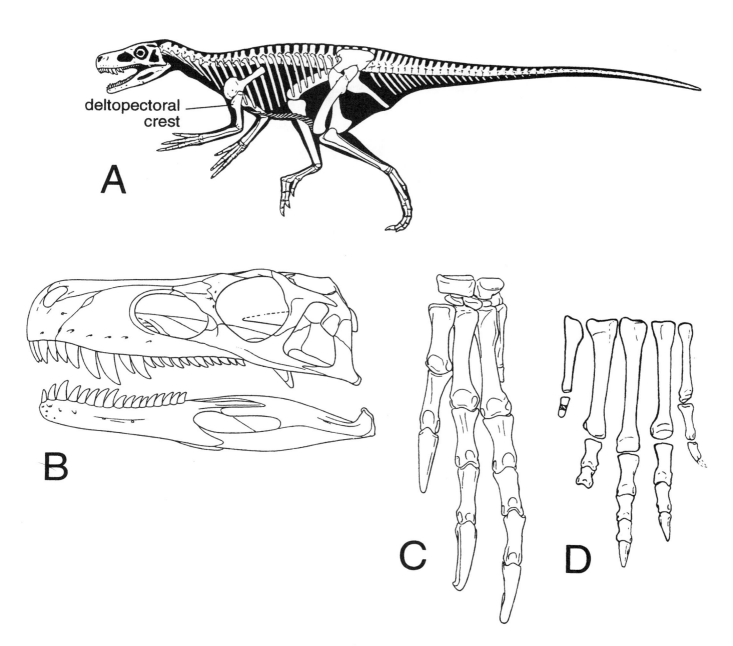

deltopectoral crest

A

B

C

D

(1988), and Novas (1996) as dinosaurian synapomorphies: elongate vomers, an elongate scapula, a symmetrical hand, three or more sacral vertebrae, and the ascending astragalar process and attachment site on the anterior face of the tibia. These have been dropped because they are seen in other basal archosaurs or they are not convincingly present in all dinosaurs—for example, *Herrerasaurus* has two sacral vertebrae, whereas ornithosuchids, *Scleromochlus*, *Silesaurus*, and Pterosauria apparently have three.

Over the years, many claims have been made about the date of the first dinosaurs. Earliest supposed records have been based on isolated bones, groups of bones, and footprints from Middle Triassic, Lower Triassic, and even Permian rocks. With a clear cladistic definition of the Dinosauria, it should now be possible to weed these out.

Many supposed early dinosaurs were described on the basis of isolated vertebrae, skull elements, and limb bones from the Triassic of Germany.

What Is the Oldest Dinosaur?

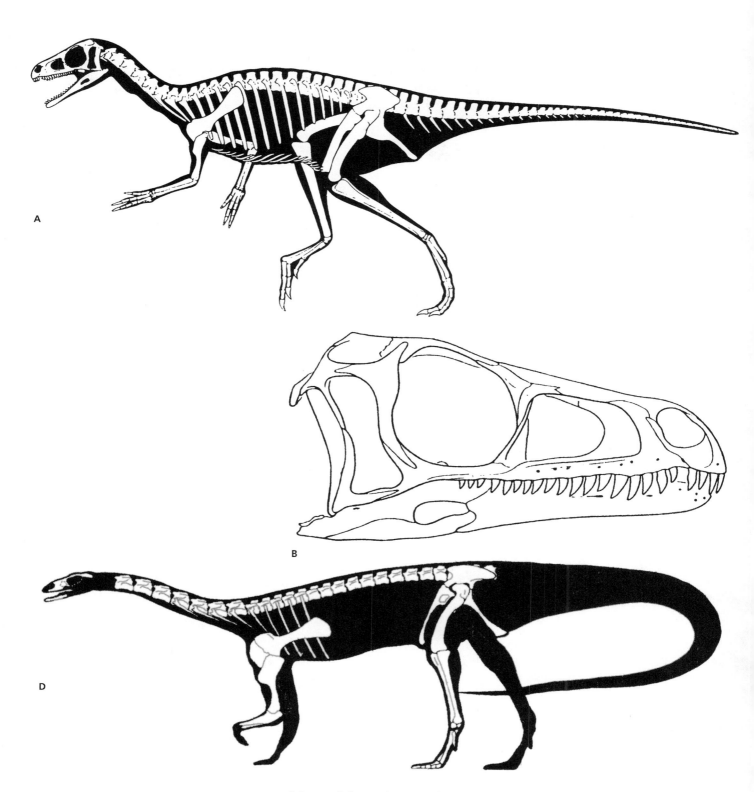

18.3. Early dinosaurs. A, B, Skeleton and skull of *Eoraptor lunensis* from the Ischigualasto Formation of Argentina, both in side view. C, D, Restored skeleton and hind limb of *Saturnalia tupinquim,* both in side view; the hind limb shows key muscles in backswing and forward positions of the stride. E, F, Partial skeleton and lower jaw of *Pisanosaurus mertii* from the Ischigualasto Formation of Argentina. A, B, Based on Sereno et al. (1993); C, D, based on Langer (2003); E, F, based on Bonaparte (1976).

Many of these elements have turned out to belong to prolacertiforms, to belong to rauisuchid archosaurs, or to be indeterminate (Benton 1986a).

Unusually early records of dinosaurs based on footprints extend back into the Middle and Lower Triassic, and even the Permian. These early finds have all been based on three-toed footprints, with no indication of a palm print, a good indication that the trace was made by a dinosaur standing up on its toes (the digitigrade posture), since most other Permo-Triassic tetrapods left plantigrade four- and five-toed prints. However, some of these

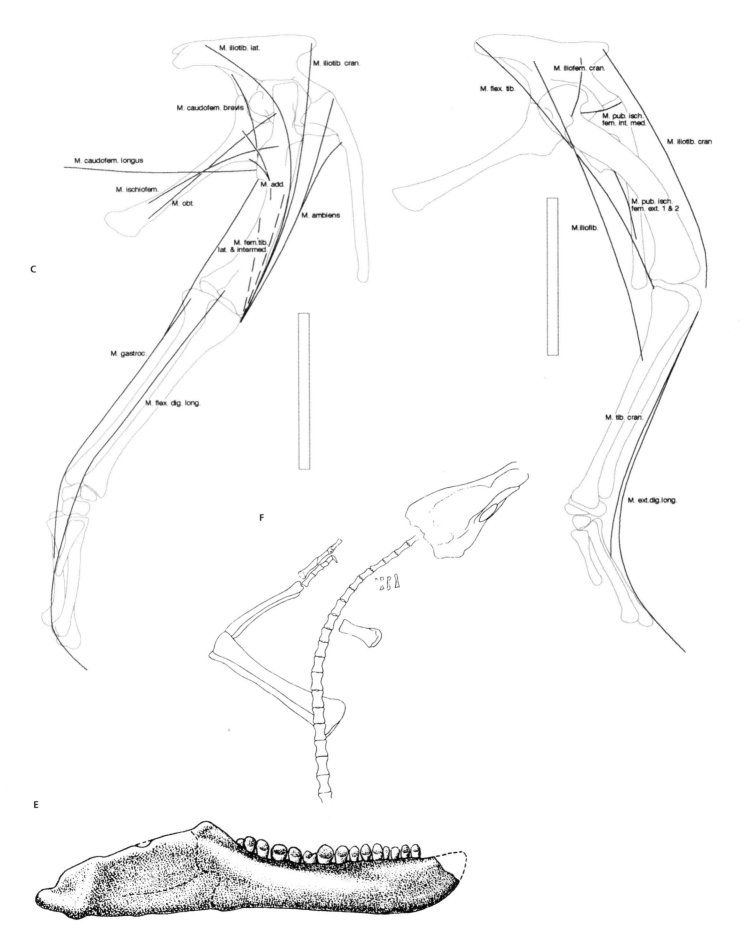

C

F

E

early records have turned out to be either broken fragments of five-toed prints, invertebrate tracks (king crabs leave tiny "three-toed" impressions), or inorganic sedimentary structures (King and Benton 1996). Newer reports, however, identify convincing trackways of well-preserved three-toed digitigrade footprints from the Lower and Middle Triassic of Europe (Brusatte et al. 2011), and it is yet to be clarified whether these were made by dinosaurs or, more likely, by basal dinosauromorphs that had already adopted the upright digitigrade posture before Dinosauria emerged.

The report of the silesaurid dinosauromorph *Asilisaurus* in the late Anisian of Tanzania (Nesbitt et al. 2010) shows a much earlier origin of Dinosauromorpha than had been assumed previously (*Marasuchus* is Ladinian; *Silesaurus* and *Eucoelophysis* are Carnian and Norian, respectively), and suggests that true Dinosauria ought to extend back to that time as well, although convincing fossils to fill that gap have yet to be found.

The First Dinosaurs

The first unquestionable dinosaurs are all Carnian in age, and late Carnian at that (Fig. 18.4). Although rare elements in their faunas, late Carnian dinosaurs are now known from many parts of the world, and these are detailed here. Most of the specimens are incomplete, and they will be summarized briefly first, followed by a fuller account of the superb South American early dinosaurs.

The only possible Carnian dinosaur from Europe is *Saltopus* from the Lossiemouth Sandstone Formation of Elgin, Scotland, but the single specimen is equivocal, and it is most probably a nondinosaurian ornithodiran (Benton and Walker 2011). From Africa comes *Azendohsaurus* from the Argana Formation of Morocco, based on a tooth (Gauffre 1993), but with such limited material that its identity is open to question. Some small prosauropod jaws from Madagascar (Flynn et al. 1999) have been presented as representing possibly the oldest known dinosaur. The remains are clearly dinosaurian; dating of the fossiliferous unit is uncertain but is likely Carnian. *Alwalkeria* from the Maleri Formation of India, based on a partial skull and skeleton, appears to be a small theropod (Chatterjee 1987).

A number of dinosaurs have been reported from the Carnian of North America, but most are based on isolated elements. Many have been misidentified; most are likely from basal archosaurs or the basal diapsid *Trilophosaurus*, or are simply nondiagnostic (Irmis et al. 2007; Nesbitt et al. 2007). Named forms include the supposed theropods *Camposaurus*, *Caseosaurus*, and *Eucoelophysis* and the ornithischian *Tecovasaurus*, which were all probably basal dinosauromorphs.

The late Carnian Santa Maria and Caturrita formations of Brazil, and the Ischigualasto Formation of Argentina have been much more productive than all the other units of the same age elsewhere in the world. This, as well as the fact that *Marasuchus* and *Lagerpeton*, close outgroups of the Dinosauria, are exclusively South American, suggests that the dinosaurs perhaps arose in that continent (Langer et al. 2010).

The Santa Maria Formation is the source of *Staurikosaurus pricei*, *Saturnalia tupinquim*, and *Teyuwasu barbarenai*, and the Caturrita Formation of *Guaibasaurus candelariensis*. The Ischigualasto Formation is even better known as the source of *Herrerasaurus ischigualastensis*, *Eoraptor lunensis*,

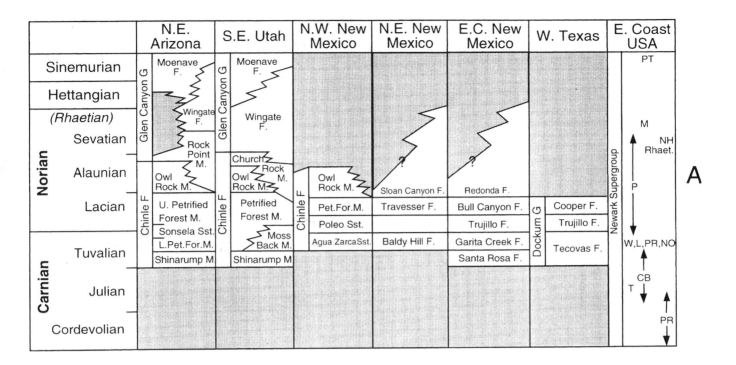

Eodromaeus murphi, and *Pisanosaurus mertii*. These Carnian dinosaurs were moderately sized animals, all lightweight bipeds, mostly 1–2 m long.

Staurikosaurus and *Herrerasaurus* are members of the Family Herrerasauridae. *Herrerasaurus* (Fig. 18.2) is known in some detail (Novas 1989, 1992, 1994; Sereno and Novas 1992, 1994; Sereno 1994) from 11 specimens, including some partial skeletons. These show a slender lightweight biped (Fig. 18.2) ranging in length from 3 to 6 m. The skull (Fig. 18.2B) is narrow and low. There is a sliding joint in each lower jaw, which allowed the jaws to flex and grasp struggling prey. The neck is slender. The forelimbs are less

18.4. Stratigraphy of vertebrate-bearing sequences in the Late Triassic and earliest Jurassic of North America (A) and various parts of Gondwanaland (B). The dates are based largely on comparisons of tetrapods with the German sequence by Olsen et al. (1987, 1990), Benton (1994a, 1994b), and others. CB = Cow Branch Formation; L = Lockatong Formation; M = McCoy Brook Formation; NH = New Haven Arkose; NO = New Oxford Formation; P = Passaic Formation; PR = Pekin Formation; PT = Portland Formation; T = Turkey Branch Formation; W = Wolfville Formation. Modified from Benton (1994a).

than half the length of the hind limbs, and the hand is elongated. Digits IV and V of the hand (Fig. 18.4C) are reduced, and the long penultimate phalanges of the hand indicate that it was adapted for grasping. There are two sacral vertebrae, one less than the normal dinosaurian condition, and the acetabulum is perforate. The femur has an inturned subrectangular head to fit into the pelvic bowl, and the tibia bears a cnemial crest. The foot (Fig. 18.2D) is digitigrade (the animal stands high on its toes), the calcaneum is reduced in size, and the astragalus bears an ascending process on the front of the tibia. Two other species from the Ischigualasto Formation, *Ischisaurus cattoi* and *Frenguellisaurus ischigualastensis*, are probably synonyms of *Herrerasaurus ischigualastensis*. *Staurikosaurus* is similar, but it differs from *Herrerasaurus* in the shape of the ilium and the distal outline of the tibia. *Eoraptor* is another small carnivorous dinosaur (Fig. 18.3A, B), but only 1 m long (Sereno et al. 1993). It has a lower snout than *Herrerasaurus*, there is no intramandibular joint, and the hand is shorter. Otherwise, *Eoraptor* differs from other basal dinosaurs in being generally more primitive rather than in possessing a great many unique features. Finally, *Eodromaeus murphi* has been described as a basal theropod (Martínez et al. 2011).

Panphagia protos, *Saturnalia tupinquim*, and *Teyuwasu barbarenai* were all described as basal sauropodomorphs. *Panphagia* (Martínez and Alcober 2009) is based on a single partial skeleton of an animal originally about 1.3 m long. It appears to be the most basal sauropodomorph so far described, sharing many features with *Saturnalia*, but also sharing hollow bones, sublanceolate teeth, and overall proportions with *Eoraptor*.

Saturnalia is known from three skeletons, only partly described so far (Langer et al. 1999; Langer 2003; Langer and Benton 2006). *Saturnalia* has small, leaf-shaped, serrated-edged teeth, characteristic of prosauropods. There are at least two sacral vertebrae, and possibly a third, caudosacral, element. The pelvic elements are similar to other early dinosaurs: a short ilium, with an extended anterior process and a short posterior process, and a largely closed acetabulum. The pubis and ischium are long and slender, and the pubes meet in the midline in a deep pubic apron. The femur is sigmoid, with an inturned head, and the tibia is straight. The ankle is dinosaurian, and there are five digits on the foot, the fifth most reduced (Fig. 18.3C, D). *Teyuwasu* has been described only briefly on the basis of a dinosaurian femur and tibia (Kischlat 1999) and is probably not a validly established taxon.

Guaibasaurus candelariensis was described (Bonaparte et al. 1999, 2007) on the basis of some vertebrae, a scapulocoracoid, and hip and hind limb elements as a basal saurischian ancestral to the sauropodomorphs, although Langer (2000) suggested it might be a basal theropod. Two sacral vertebrae are preserved, but there was probably at least a third. The scapula is slender. The acetabulum is only partially open, and the hind limb elements are similar to those of *Herrerasaurus*.

Pisanosaurus is another tiny dinosaur (Bonaparte 1976) known from incomplete remains (Fig. 18.3E, F). The lower jaw (Fig. 18.3F) indicates that this specimen is an ornithischian dinosaur: the teeth are broad and diamond shaped, with a spatulate outer face, and there is a prominent shelf along the outer face of the lower jaw that marks the bottom of a soft cheek. The hind limb and foot appear dinosaurian, with a reduced calcaneum, an ascending process on the astragalus, and a functionally three-toed digitigrade foot.

Finally, *Eocursor parvus* is the most complete Triassic ornithischian yet described, based on a selection of skeletal elements, including skull fragments, spinal elements, pelvis, long leg bones, and unusually large grasping hands. Together, these indicate an animal about 1 m long (Butler et al. 2007; Butler 2010). The skull, poorly known in *Pisanosaurus*, is similar to Early Jurassic basal ornithischians in *Eocursor*, with a slightly inset tooth row (implying possible cheeks) and low, triangular teeth.

The systematics of the early dinosaurs has been disputed. For example, *Herrerasaurus* and *Staurikosaurus* were formerly regarded as primitive forms that were neither saurischians nor ornithischians. However, Sereno and Novas (1992, 1994), Sereno et al. (1993), Novas (1994), and Sereno (1994) argued that the Herrerasauridae and *Eoraptor* are basal theropods (Fig. 18.1), not least because they share the intramandibular joint (Fig. 18.4A). However, many other characters of theropods are absent. A new cladistic analysis of basal dinosaurs by Langer (2004; Langer and Benton 2006; Langer et al. 2010) places *Eoraptor* and Herrerasauridae as basal saurischians, successive outgroups to the clade compirisng Theropoda and Sauroporomorpha). In a further migration of *Eoraptor*, Martínez et al. (2011) class it as a sauropodomorph!

The Saurischia may be diagnosed by a number of characters of the skull, vertebrae, and limbs (Gauthier 1986; Sereno et al. 1993), including the following: the jugal overlaps the lacrimal; the thumb is robust; the bases of metacarpals III and IV lie on the palmar surfaces of manual digits III and IV, respectively; and the calcaneal proximal articular face is concave. The shape of the saurischian pelvis (Fig. 18.2A) is not useful diagnostically because this pattern is primitive and shared with the ancestors of the dinosaurs, and indeed with most other reptiles.

The Saurischia may include Herrerasauridae and *Eoraptor* as basal taxa, and then the two major clades, Theropoda and Sauropodomorpha. *Guaibasaurus* turns out to be the basal theropod, and *Panphagia* is the basal sauropodomorph, lying immediately below *Saturnalia* (Langer et al. 1999; Langer and Benton 2006; Brusatte et al. 2010; Langer et al. 2010).

The Ornithischia has long been recognized as a monophyletic group; ornithischians are diagnosed by the presence of triangular teeth, with the largest tooth in the middle of the tooth row. In addition, there is a coronoid process behind the tooth row in the lower jaw (Fig. 18.3F) and a reduced external/mandibular fenestra (cf. Fig. 18.3B, a saurischian, and 18.3F, an ornithischian). The classic ornithischian trademark, a predentary bone at the front of the lower jaw, is not seen in *Pisanosaurus* (Fig. 18.3F) or *Eocursor*, in both cases because of poor preservation, but both are widely accepted as basal ornithischians.

Relationships of the First Dinosaurs: Saurischia and Ornithischia

In early accounts of the history of vertebrate life on land during the Triassic (e.g., Colbert 1958; Romer 1970), the stratigraphic schemes were little

Stratigraphy

more refined than lower, middle, and upper. Even in later accounts (e.g., Bonaparte 1982; Benton 1983; Charig 1984), there was little improvement, and most of the stratigraphic assignments of tetrapod-bearing rock units were based on comparisons of the vertebrates themselves. It is little use in attempts to study patterns of evolution to sequence faunas in terms of the nature of the faunas themselves!

The standard stratigraphic scheme for the Triassic rocks is based on ammonoids (Tozer 1974, 1979) and hence can only be applied to marine rocks. There is an independent palynological scheme, based on pollen and spores, for dating continental Triassic rocks (e.g., Visscher and Brugman 1981; Roghi et al. 2009; Smith 2009), but this scheme is not always reliable. Continental Triassic rocks have been correlated here and there with the marine scheme by using isolated tetrapod finds and other crossover evidence, and this has led to a more independent dating scheme for Triassic terrestrial tetrapod faunas (Ochev and Shishkin 1989; Benton 1991, 1994a, 1994b; Brusatte et al. 2010). Radiometric dating of volcanic layers associated with the Ischigualasto Formation (Rogers et al. 1993) gave a date of 227 million years ago (mya), which is in good agreement with biostratigraphic assumptions. This has been revised to a range of 231.4 and 225.9 mya, based on dating multiple volcanic ash beds (Martínez et al. 2011). Current work involving the establishment of further radiometric dates (e.g., Furin et al. 2006), as well as magnetostratigraphy (determining the reversed and normal magnetization of rocks through time; e.g., Muttoni et al. 2004) are providing firmer correlations between terrestrial and marine rocks (Brusatte et al. 2010; Langer et al. 2010), and these suggest that many vertebrate-bearing units formerly assigned to the late Carnian now move up to the early Norian, and the Norian stage extends in duration to some 20 million years. The correlations of major tetrapod faunas of the Late Triassic are shown in Figure 18.4.

Ecological Models for the Origin of the Dinosaurs

A major faunal turnover took place on land during the Late Triassic. Various long-established groups, sometimes termed paleotetrapods (synapsids, "thecodontians," temnospondyl amphibians, rhynchosaurs, prolacertiforms, procolophonids), were replaced by new reptilian types, sometimes termed neotetrapods (turtles, crocodilians, dinosaurs, pterosaurs, lepidosaurs, mammals). It had long been assumed that this replacement was a long, drawn-out affair involving competition (Bonaparte 1982; Charig 1984), with the dinosaurs leading the way in driving out the synapsids, rhynchosaurs, and "thecodontians." The success of the dinosaurs was explained by their superior adaptations, such as their upright posture, their initial bipedalism, their speed and intelligence, or their posited endothermy.

I have opposed this assumption of long-term competitive replacement (Benton 1983, 1986a, 1991, 1994a, 1994b, 2004). First evidence came from quantitative studies of tetrapod faunas through the Triassic (Benton 1983), which did not show a long-term decline of paleotetrapod groups and a matching rise of neotetrapod groups. Indeed, new groups generally did not supplant previously existing groups. The study revealed that there was a dramatic changeover from late Carnian and early Norian faunas that contain rare dinosaurs, to those in the mid- and late Norian (Fig. 18.4), where dinosaurs dominate.

Later studies (Benton 1986a, 1991, 1994b) clarified this changeover. Redating suggested that this event happened at the Carnian–Norian boundary, then dated at about 220 mya. Further redating (e.g., Muttoni et al. 2004; Brusatte et al. 2010) has moved the Carnian–Norian boundary down to 227–228 mya, but the turnover event that affected vertebrates remains at about 220 mya, now somewhere between the early and late Norian, as originally indicated using older conventions by Benton (1983).

During this terrestrial extinction event, the kannemeyeriid dicynodonts, chiniquodontids, traversodontids, and rhynchosaurs all died out or diminished sharply in diversity and abundance (single, isolated survivors of some of these groups in the later Norian hardly constitute continuing dominance!). These four families had made up 40–80% of all the late Carnian and early Norian faunas, representing the dominant medium- and large-size herbivores worldwide. Other groups that disappeared at this time were temnospondyl amphibians (Mastodonsauridae, Trematosauridae), archosauromorphs (Prolacertidae), basal archosaurs (Proterochampsidae, Scleromochlidae), and dinosaurs (Herrerasauridae, Pisanosauridae). Hence, 10 of the 24 late Carnian families died out (a loss of 42%), and continental tetrapod faunas were dramatically depleted in terms of diversity and abundance.

A detailed study of the Ischigualasto Formation by Rogers et al. (1993), subsequently updated by Martínez et al. (2011), confirmed the view that there was no long-term ecological replacement of paleotetrapods by neotetrapods because members of both assemblages coexist without evidence of a decline of the former and a rise of the latter. Dinosaurs appear early in the Ischigualasto sequence, but they never increase above a diversity of three or four species, nor a percentage representation of 6% of all specimens collected.

There were other extinctions during this time among marine organisms (foraminifera, ammonoids, bivalves, bryozoans, conodonts, coral reefs, echinoids, and crinoids; Benton 1986b; Hallam and Wignall 1997). At this time, there was a worldwide series of climatic changes from humid to arid (Simms and Ruffell 1990). These may have been triggered by events associated with the beginning of rifting of the supercontinent Pangaea. Major floral changes occurred too, with the disappearance of the *Dicroidium* floras of southern continents, and the spread worldwide of northern conifer-dominated floras. Perhaps the drying climates favored conifers over seed ferns such as *Dicroidium*, and perhaps the dominant Carnian herbivores were unable to adapt to new kinds of vegetation, and died out.

The origin and early expansion of dinosaurs has been assessed by comparisons of diversity and disparity change (Brusatte et al. 2008a, 2008b). Disparity is the morphological diversity of a group, representing body design, diet, and lifestyle. When dinosaurs and crurotarsans in the Late Triassic were compared, it emerged that crurotarsans were twice as disparate as dinosaurs (Fig. 18.6). In other words, the crurotarsans were experimenting with many more different ways to make a living, more so than the dinosaurs. These results show that during the first 30 million years of their history, dinosaurs were living alongside and sharing niches with a group (the crurotarsans) that was evolving at the same pace and exploring twice the number of different lifestyles. This pattern held across the entire 30 million years that dinosaurs and crurotarsans lived side by side. Thus, there is no evidence that dinosaurs were doing anything better than crurotarsans. Both groups were doing quite well, but it actually was the crurotarsans that were doing better in one key trait (disparity), as well as being more diverse and abundant.

First Radiation of the Dinosaurs

Dinosaurs diversified to a limited extent during the late Carnian and early Norian, and increasingly after the early Norian extinction event. In the middle and late Norian, for the first time, mass accumulations of dinosaur skeletons are found – for example, the famous death assemblage of several hundred individuals of the theropod *Coelophysis* at Ghost Ranch, New Mexico, in the Upper Petrified Forest Member of the Chinle Formation. For the first time too, dinosaurs became relatively diverse, and they began to exhibit that feature for which the group is famous: large size. Specimens of *Plateosaurus* from the Stubensandstein and Knollenmergel of Germany reached lengths of 6–8 m. Overall, dinosaurs had switched from being minor players in the late Carnian and early Norian, at faunal abundances of less than 6%, to being the dominant land reptiles, with abundances of 25–60% in the middle and late Norian.

The end-Triassic mass extinction event, 200 mya, had major effects on life in the sea and on land (Hallam and Wignall 1997). The end-Triassic event has been explained as the result of an impact, analogous to the Cretaceous–Tertiary impact that caused the final extinction of Dinosauria, by Olsen et al. (1990, 2002) and Bice et al. (1992). A major impact crater, the Manicouagan structure in Quebec, was identified as the smoking gun for a catastrophic extraterrestrial impact at the Triassic–Jurassic boundary (Olsen et al. 1990). Shocked quartz was found at a Tr-J boundary section in Italy (Bice et al. 1992), as well as an iridium anomaly and a fern spike in the eastern United States (Olsen et al. 2002), all three classic evidence for the Cretaceous–Tertiary impact. However, none of these indicators of impact at the Triassic–Jurassic boundary is particularly convincing. First, the Manicouagan impact structure was redated (Hodych and Dunning 1992) away from the Tr-J boundary (200 mya) with an age of 214 mya (Fig. 18.5). Second, the nature of the lamellae in the shocked quartz was not adequate to rule out other explanations, such as a volcanic source for the material (Bice et al. 1992). Third, the iridium anomaly and fern spike are modest in comparison with Cretaceous–Tertiary examples (Olsen et al. 2002).

18.5. Timescale of major events in the Late Triassic, showing the time line of the early dinosaurs from the Ischigualasto Formation of Argentina, through the current date of the Manicouagan impact and the two mass extinctions.

Period	Stage	Boundary date (Ma)	Events
Jurassic	Hettangian		
Late Triassic	Norian	202	Triassic-Jurassic mass extinction
			214-Ma Manicouagan impact
		220	Carnian-Norian mass extinction
	Carnian		228-Ma Ischigualasto dinosaurs
		230	

Note: Ma = million years ago.

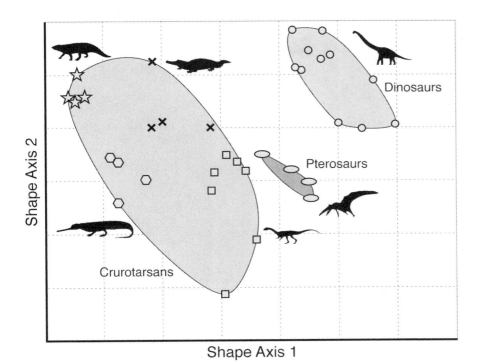

18.6. The morphospace occupied by crurotarsans was much larger than that occupied by dinosaurs or pterosaurs during the Carnian and Norian (230–200 mya), the time when dinosaurs were supposedly outcompeting the crurotarsans. Note that the extremely small morphospace occupied by the Late Triassic pterosaurs reflects their uniform range of form. (Redrafted from Brusatte et al. 2008a.)

Current evidence suggests that the end-Triassic event was geologically sudden (Deenen et al. 2010) and probably closely linked to massive volcanic eruptions of the Central Atlantic Magmatic Province (Whiteside et al. 2010), which produced a few degrees of global warming (Hallam and Wignall 1997; McElwain et al. 1999; Whiteside et al. 2010). Global warming led to stagnation of the oceans with anoxia on the seabed, and a calcification crisis occurred as surfaced water became more acid. On land, the volcanic gases, mixed with atmospheric water, fell as acid rain, presumably killing plants in huge quantities, and so damaging the bases of food chains.

The global expansion of dinosaurs was evidently a three-step phenomenon, with splitting of the main lineages (Ornithischia, Theropoda, Sauropodomorpha) in the Carnian, expansion of dinosaurian diversity and abundance from two or three species and up to 6% of individuals in late Carnian and early Norian faunas to four or five species and 25–60% of faunal composition in the middle and late Norian, and finally radiation of further theropod and sauropodomorphs, as well as ornithischians in the earliest Jurassic.

Bakker, R. T., and P. M. Galton. 1974. Dinosaur monophyly and a new class of vertebrates. *Nature* 248: 168–172.

Benton, M. J. 1983. Dinosaur success in the Triassic: a noncompetitive ecological model. *Quarterly Review of Biology* 58: 29–55.

———. 1986a. The Late Triassic tetrapod extinction events. In K. Padian (ed.), *The Beginning of the Age of Dinosaurs: Faunal Change across the Triassic–Jurassic Boundary*, 303–320. Cambridge: Cambridge University Press.

———. 1986b. More than one event in the Late Triassic mass extinction. *Nature* 321: 857–861.

———. 1990. Origin and interrelationships of dinosaurs. In D. B. Weishampel, P. Dodson, and H. Osmólska (eds.), *The Dinosauria*, 11–30. Berkeley: University of California Press.

———. 1991. What really happened in the Late Triassic? *Historical Biology* 5: 263–278.

References

——. 1994a. Late Triassic terrestrial vertebrate extinctions: stratigraphic aspects and the record of the Germanic Basin. *Paleontologia Lombarda* (n.s.) 2: 19–38.

——. 1994b. Late Triassic to Middle Jurassic extinctions among tetrapods: testing the pattern. In N. C. Fraser and H.-D. Sues (eds.), *In the Shadow of the Dinosaurs: Triassic and Jurassic Tetrapod Faunas*, 366–397. Cambridge: Cambridge University Press.

——. 1999. *Scleromochlus taylori* and the origin of dinosaurs and pterosaurs. *Philosophical Transactions of the Royal Society B* 354: 1423–1446.

——. 2004. Origin and interrelationships of dinosaurs. In D. B. Weishampel, P. Dodson, and H. Osmólska (eds.), *The Dinosauria*, 2nd ed., 7–19. Berkeley: University of California Press.

Benton, M. J., and J. M. Clark. 1988. Archosaur phylogeny and the relationships of the Crocodylia. In M. J. Benton (ed.), *The Phylogeny and Classification of the Tetrapods, Vol. 1: Amphibians and Reptiles*, 295–338. Oxford: Clarendon Press.

Benton, M. J., and A. D. Walker. 2011. *Saltopus*, a dinosauriform from the Upper Triassic of Scotland. *Earth and Environmental Science Transactions of the Royal Society of Edinburgh.*

Bice, D. M., C. R. Newton, S. McCauley, P. W. Reiners, and C. A. McRoberts. 1992. Shocked quartz at the Triassic–Jurassic boundary in Italy. *Science* 255: 443–446.

Bonaparte, J. F. 1976. *Pisanosaurus mertii* Casamiquela and the origin of the Ornithischia. *Journal of Paleontology* 50: 808–820.

——. 1982. Faunal replacement in the Triassic of South America. *Journal of Vertebrate Paleontology* 21: 362–371.

Bonaparte, J. F., G. Brea, C. L. Schultz, and A. G. Martinelli. 2007. A new specimen of *Guaibasaurus candelariensis* (basal Saurischia) from the Late Triassic Caturrita Formation of southern Brazil. *Historical Biology* 19: 73–82.

Bonaparte, J. F., J. Ferigolo, and A. M. Ribeiro. 1999. A new Early Late Triassic saurischian dinosaur from Rio Grande do Sul State, Brazil. In Y. Tomida, T. H. Rich, and P. Vickers-Rich (eds.), *Proceedings of the Second Gondwanan Dinosaur Symposium* 1: 89–109.

Brusatte, S. L., M. J. Benton, M. Ruta, and G. T. Lloyd. 2008a. Superiority, competition, and opportunism in the evolutionary radiation of dinosaurs. *Science* 321: 1485–1488.

——. 2008b. The first 50 mya of dinosaur evolution: macroevolutionary pattern and morphological disparity. *Biology Letters* 4: 733–736.

Brusatte, S. L., S. J. Nesbitt, R. B. Irmis, R. J. Butler, M. J. Benton, and M. A. Norell. 2010. The origin and early radiation of dinosaurs. *Earth-Science Reviews* 101: 68–100.

Brusatte, S. L., G.Niedźwiedzki, and R. J. Butler. 2011. Footprints pull origin and diversification of dinosaur stem-lineage deep into Early Triassic. *Proceedings of the Royal Society of London, Series B* 278: 1107–1113.

Butler, R. J. 2010. The anatomy of the basal ornithischian dinosaur *Eocursor parvus* from the lower Elliot Formation (Late Triassic) of South Africa. *Zoological Journal of the Linnean Society* 16: 648–684..

Butler, R. J., R. M. H. Smith, and D. B. Norman. 2007. A primitive ornithischian dinosaur from the Late Triassic of South Africa, and the early evolution and diversification of Ornithischia. *Proceedings of the Royal Society B* 274: 2041–2046.

Charig, A. J. 1984. Competition between therapsids and archosaurs during the Triassic period: a review and synthesis of current theories. *Symposia of the Zoological Society of London* 52: 597–628.

Chatterjee, S. K. 1987. A new theropod dinosaur from India with remarks on the Gondwana–Laurasia connection in the Late Triassic. *Geophysics Monographs* 41: 183–189.

Colbert, E. H. 1958. Tetrapod extinctions at the end of the Triassic. *Proceedings of the National Academy of Sciences of the United States of America* 44: 973–977.

Deenen, M. H. L., M. Ruhl, N. R. Bonis, W. Krijgsman, W. M. Kuerschner, M. Reitsma, and M. J. van Bergman. 2010. A new chronology for the end-Triassic mass extinction. *Earth and Planetary Science Letters* 291: 113–125.

Flynn, J. J., J. M. Parrish, B. Rakotosamimanana, W. F. Simpson, R. L. Whatley, and A. R. Wyss. 1999. A Triassic fauna from Madagascar, including early dinosaurs. *Science* 286: 763–765.

Furin, S., N. Preto, M. Rigo, G. Roghi, P. Gianolla, J. L. Crowley, and S. A. Bowring. 2006. High-precision U-Pb zircon age from the Triassic of Italy: implications for the Triassic time scale and the Carnian origin of calcareous nannoplankton and dinosaurs. *Geology* 34: 1009–1012.

Gauffre, F.-X. 1993. The prosauropod dinosaur *Azendohsaurus laaroussii* from the Upper Triassic of Morocco. *Palaeontology* 36: 897–908.

Gauthier, J. A. 1986. Saurischian monophyly and the origin of birds. *Memoirs of the California Academy of Sciences* 8: 1–55.

Hallam, A., and P. B. Wignall. 1997. *Mass Extinctions and Their Aftermath*. Oxford: Oxford University Press.

Hodych, J. P., and G. R. Dunning. 1992. Did the Manicouagan impact trigger end-of-Triassic mass extinction? *Geology* 20: 51–54.

Holtz, T. R., Jr., and M. K. Brett-Surman. 2011 (this volume). The taxonomy and systematics of the dinosaurs. In M. K. Brett-Surman, Thomas R. Holtz Jr., and James O. Farlow (eds.), *The Complete Dinosaur*, 2nd ed., 209–223. Bloomington: Indiana University Press.

Irmis, R. B., W. G. Parker, S. J. Nesbitt, and J. Liu. 2007. Early ornithischian dinosaurs: the Triassic record. *Historical Biology* 19: 3–22.

King, M. J., and Benton, M. J. 1996. Dinosaurs in the Early and Mid Triassic? The footprint evidence from Britain. *Palaeogeography, Palaeoclimatology, Palaeoecology* 122: 213–225

Kischlat, E.-E. 1999. A new dinosaurian "rescued" from the Brazilian Triassic: *Teyuwasu barbarenai*, new taxon. *Paleontologia em Destaque, Boletim Informativo da Sociedade Brasileira de Paleontologia* 14 (26): 58.

Langer, M. C. 2000. Early dinosaur evolution: new evidence from South Brazil. *Abstracts of the 5th European Workshop on Vertebrate Palaeontology* 1: 43–44.

——. 2003. The pelvic and hind limb morphology of the stem-sauropodomorph *Saturnalia tupinquim* (Late Triassic, Brazil). *PaleoBios* 23 (2): 1–30.

——. 2004. Basal Saurischia. In D. B. Weishampel, P. Dodson, and H. Osmólska (eds.), *The Dinosauria*, 2nd ed., 25–46. Berkeley: University of California Press.

Langer, M. C., F. Abdala, M. Richter, and M. J. Benton. 1999. A sauropodomorph dinosaur from the Upper Triassic (Carnian) of southern Brazil. *Comptes Rendus de l'Académie des Sciences, Paris* 329: 511–517.

Langer, M. C., and M. J. Benton. 2006. Early dinosaurs: a phylogenetic study. *Journal of Systematic Palaeontology* 4: 309–358.

Langer, M. C., M. D. Ezcurra, J. S. Bittencourt, and F. E. Novas. 2010. The origin and early evolution of dinosaurs. *Biological Reviews* 85: 55–110.

Martínez, R. N., and O. A. Alcober. 2009. A basal sauropodomorph (Dinosauria: Saurischia) from the Ischigualasto Formation (Triassic, Carnian) and the early evolution of Sauropodomorpha. *PLoS One* 4 (2): 1–12.

Martínez, R. N., P. C. Sereno, O. A. Alcober, C. E. Colombi, P. R. Renne, I. P. Montañez, and B. S. Currie. 2011. A basal dinosaur from the dawn of the dinosaur era in southwestern Pangaea. *Science* 331: 206–210.

McElwain, J. C., D. J. Beerling, and F. I. Woodward. 1999. Fossil plants and global warming at the Triassic–Jurassic boundary. *Science* 285: 1386–1390.

Muttoni, G., D. V. Kent, P. E. Olsen, P. Di Stefano, W. Lowrie, S. M. Bernasconi, and F. Martín Hernández. 2004. Tethyan magnetostratigraphy from Pizzo Mondello (Sicily) and correlation to the Late Triassic Newark astrochronological polarity time scale. *Geological Society of America Bulletin* 116: 1043–1058.

Nesbitt, S. J. 2011. The early evolution of archosaurs: Relationships and the origin of major clades. *Bulletin of the American Museum of Natural History* 352: 1–292.

Nesbitt, S. J., R. B. Irmis, and W. G. Parker. 2007. A critical re-evaluation of the Late Triassic dinosaur taxa of North America. *Journal of Systematic Palaeontology* 5: 209–243.

Nesbitt, S. J., C. A. Sidor, R. B. Irmis, K. D. Angielczyk, and R. M. H. Smith. 2010. Ecologically distinct dinosaurian sister group shows early diversification of Ornithodira. *Nature* 464: 95–98.

Novas, F. E. 1989. The tibia and tarsus in Herrerasauridae (Dinosauria, incertae sedis) and the origin and evolution of the dinosaurian tarsus. *Journal of Paleontology* 63: 677–690.

———. 1992. Phylogenetic relationships of the basal dinosaurs, the Herrerasauridae. *Palaeontology* 35: 51–62.

———. 1994. New information on the systematics and postcranial skeleton of *Herrerasaurus ischigualastensis* (Theropoda: Herrerasauridae) from the Ischigualasto Formation (Upper Triassic) of Argentina. *Journal of Vertebrate Paleontology* 13: 400–423.

———. 1996. Dinosaur monophyly. *Journal of Vertebrate Paleontology* 16: 723–741.

Ochev, V. G., and M. A. Shishkin. 1989. On the principles of global correlation of the continental Triassic on the tetrapods.

Acta Palaeontologica Polonica 34: 149–173.

Olsen, P. E., S. J. Fowell, and B. Cornet. 1990. The Triassic/Jurassic boundary in continental rocks of eastern North America: a progress report. *Geological Society of America Special Paper* 247: 585–593.

Olsen, P. E., D. V. Kent, H.-D. Sues, C. Koeberl, H. Huber, A. Montanari, E. C. Rainforth, S. J. Fowell, M. J., Szajna, and B. W. Hartline. 2002. Ascent of dinosaurs linked to an iridium anomaly at the Triassic–Jurassic boundary. *Science* 296: 1305–1307.

Owen, R. 1842. Report on British fossil reptiles. Part II. *Report of the British Association for the Advancement of Science 1841* (1842): 60–204.

Parrish, J. M. 2011 (this volume). Evolution of the archosaurs. In M. K. Brett-Surman, Thomas R. Holtz Jr., and James O. Farlow (eds.), *The Complete Dinosaur*, 2nd ed., 316–329. Bloomington: Indiana University Press.

Rogers, R. R., C. C. Swisher III, P. C. Sereno, C. A. Forster, and A. M. Monetta. 1993. The Ischigualasto tetrapod assemblage (Late Triassic) and ^{40}Ar/^{39}Ar calibration of dinosaur origins. *Science* 260: 794–797.

Roghi, G., P. Gianola, L. Minarelli, C. Pilatil, and N. Preto. 2009. Palynological correlation of Carnian humid pulses throughout western Tethys. *Palaeogeography, Palaeoclimatology, Palaeoecology* 290: 89–106.

Romer, A. S. 1970. The Triassic faunal succession and the Gondwanaland problem. In *Gondwana Stratigraphy: IUGS Symposium Buenos Aires, 1967*, 375–400. Paris: UNESCO.

Seeley, H. G. 1887. On the classification of the fossil animals commonly called Dinosauria. *Proceedings of the Royal Society* 43: 165–171.

Sereno, P. C. 1986. Phylogeny of the bird-hipped dinosaurs (Order Ornithischia). *National Geographic Research* 2: 234–256.

———. 1994. The pectoral girdle and fore-limb of the basal theropod *Herrerasaurus ischigualastensis*. *Journal of Vertebrate Paleontology* 13: 425–450.

Sereno, P. C., C. A. Forster, R. R. Rogers, and A. M. Monetta. 1993. Primitive dinosaur skeleton from Argentina and the

early evolution of Dinosauria. *Nature* 361: 64–66.

Sereno, P. C., and F. E. Novas. 1992. The complete skull and skeleton of an early dinosaur. *Science* 258: 1137–1140.

———. 1994. The skull and neck of the basal theropod *Herrerasaurus ischigualastensis*. *Journal of Vertebrate Paleontology* 13: 451–476.

Simms, M. J., and A. H. Ruffell. 1990. Climatic and biotic change in the late Triassic. *Journal of the Geological Society of London* 147: 321–327.

Smith, D. G. 2009. Late Triassic palynology and the definition of the lower boundary of the Rhaetian standard age/stage. *Geological Magazine* 114: 153–156.

Sues, H.-D. 2011 (this volume). European dinosaur hunters of the nineteenth and twentieth centuries. In M. K. Brett-Surman, Thomas R. Holtz Jr., and James O. Farlow (eds.), *The Complete Dinosaur*, 2nd ed., 45–59. Bloomington: Indiana University Press.

Torrens, H. S. 2011 (this volume). Politics and paleontology: Richard Owen and the invention of dinosaurs. In M. K. Brett-Surman, Thomas R. Holtz Jr., and James O. Farlow (eds.), *The Complete Dinosaur*, 2nd ed., 24–43. Bloomington: Indiana University Press.

Tozer, E. T. 1974. Definitions and limits of Triassic stages and substages: suggestions prompted by comparisons between North America and the Alpine–Mediterranean region. *Schriftenreihe der Erdwissenschaftlichen Kommissionen, Österreichische Akademie der Wissenschaften* 2 195–206.

———. 1979. Latest Triassic ammonoid faunas and biochronology, western Canada. *Paper of the Geological Society of Canada* 79: 127–135.

Visscher, H., and W. A. Brugman. 1981. Ranges of selected palynomorphs in the Alpine Triassic of Europe. *Reviews in Palaeobotany and Palynology* 34: 115–128.

Whiteside, J. H., P. E. Olsen, T. Eglinton, M. E. Brookfield, and R. N. Sambrotto. 2010. Compound-specific carbon isotopes from Earth's largest flood basalt eruptions directly linked to the end-Triassic mass extinction. *Proceedings of the National Academy of Sciences of the United States of America* 107: 6721–6725.

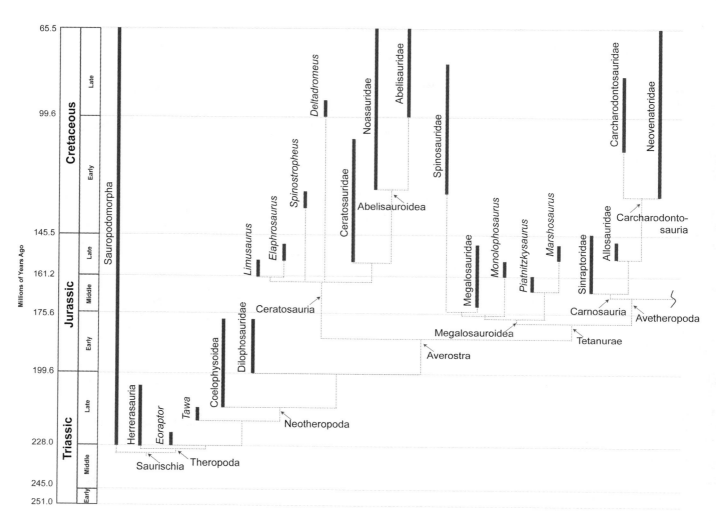

19.1. Phylogeny of the basal theropods, based primarily on Benson (2009), Benson et al. (2010), Brusatte and Sereno (2008), Langer and Benton (2006), Nesbitt et al. (2009), Smith et al. (2007, 2008), and Xu et al. (2009a). Timescale from Gradstein et al. (2004). Solid lines = known range from oldest to youngest specimens; dashed lines = inferred branching relationships between groups. Continued on Figure 19.2.

Theropods

19

Thomas R. Holtz Jr.

The Theropoda ("beast-footed ones") are a clade of saurischian dinosaurs commonly referred to as the carnivorous dinosaurs. It is certainly true that the majority (perhaps all) of the dinosaurs with a taste for flesh were theropods, and indeed, from the beginning of the Jurassic Period (199.6 Ma) until the end of the Cretaceous (65.5 Ma), these were the dominant terrestrial predators on every continent (Figs. 19.1, 19.2). However, this group produced a series of omnivorous and herbivorous forms, just as the placental group Carnivora has produced omnivorous raccoons and herbivorous pandas. Theropods famously include the largest terrestrial carnivores in Earth's history: taxa such as *Spinosaurus*, *Giganotosaurus*, and *Tyrannosaurus* dwarf both their rauisuchian predecessors and their mammalian successors. Yet many theropods were quite small: *Epidendrosaurus*, *Microraptor*, *Mei*, and *Hesperonychus* are among the smallest dinosaurs of the Mesozoic. Theropods represent the most successful group of dinosaurs; they alone survived the catastrophe at the end of the Cretaceous in the form of that most specialized group of feathered theropods: Aves (the birds), about which see Naish (this volume).

19.2. Phylogeny of the coelurosaurian theropods, based primarily on Holtz et al. (2004), Li et al. (2009), Longrich and Currie (2009a, 2009b), Mayr et al. (2005), Senter (2007), Zanno et al. (2009), and Zhang et al. (2008). Timescale from Gradstein et al. (2004). Solid lines = known range from oldest to youngest specimens; dashed lines = inferred branching relationships between groups. Question marks show uncertainty of phylogenetic position of *Archaeopteryx* (Mayr et al. 2005). Date for *Eshanosaurus* is uncertain; may be as old as Early Jurassic or as late as Early Cretaceous (Barrett 2009).

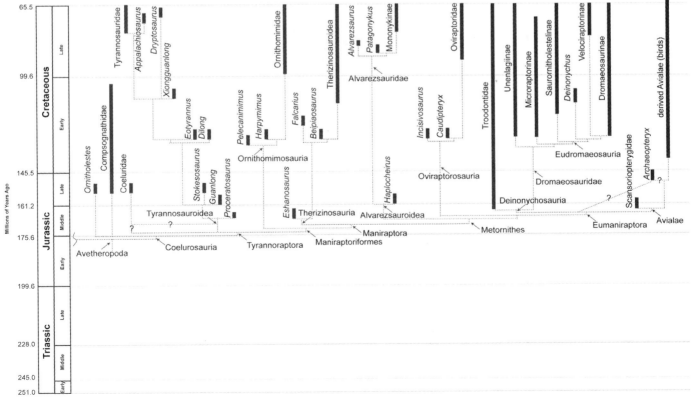

Theropods are the only major clade of dinosaur to retain the ancestral dinosaurian habit of obligate bipedality throughout their entire history. In other clades, larger and more specialized forms often reverted to partial (iguanodontian ornithopods) or obligate (derived sauropodomorphs, derived thyreophorans, derived marginocephalians) quadrupedality in response to increasing body size, development of heavy armor, enlarged frilled skulls, and so forth. Even the largest theropods—rivaling modern African elephants in mass—were strict bipeds.

Most theropods retained the ziphodont (bladelike) teeth typical of other archosauriforms (such as *Euparkeria*, ornithosuchids, rauisuchians, and erythrosuchians), and indeed of many other carnivorous vertebrates (Farlow et al. 1991). Additionally, typical theropods retained the long grasping hands shared by primitive members of the other dinosaur groups (heterodontosaurids and *Eocursor* among the ornithischians; early sauropodomorphs). In particular, the manual claws of theropods were recurved, tapering to sharp points; this presumably allowed them to more effectively clutch onto prey items.

The skull bones of theropods show numerous chambers and openings associated with a complex of air sacs (pneumatic diverticula), especially in the maxillae, nasals, orbital region, palate, and (especially in derived forms) throughout the braincase (Witmer 1997). Postcranially, theropods had hollow limb bones and vertebrae (in the cervical region in primitive theropods, and more extensively in derived forms). The pattern of postcranial pneumaticity strongly suggests the presence of a birdlike flow-through lung, which was developed relatively early in theropod history (O'Connor and Claessens 2005), or perhaps even earlier in dinosaurian history (Wedel 2006).

Origin of Theropods

19.3. Skeletal reconstructions of basal theropods. A, Late Triassic herrerasaur *Herrerasaurus ischigualastensis,* body length approximately 3.9 m; B, Late Triassic basal theropod *Tawa hallae,* body length approximately 2.0 m; C, Late Triassic coelophysoid *Coelophysis bauri,* body length approximately 2.7 m; D, Early Jurassic dilophosaurid *Dilophosaurus wetherilli,* body length approximately 6 m.

*A, C, D, Illustrations by G. S. Paul;
B, from Nesbitt et al. (2009).*

Theropoda has been defined (as a branch-based taxon) as *Passer domesticus* (the European house sparrow) and all taxa sharing a more recent common ancestor with it than with *Cetiosaurus oxoniensis* (Holtz and Osmólska 2004). However, a revision (D. Naish et al. personal communication) has suggested replacing *Passer domesticus* with *Allosaurus fragilis.* Consequently, the origin of theropods and sauropodomorphs occurred as a single divergence. Because there is debate about whether the earliest carnivorous dinosaurs were truly theropods (see below), the oldest unquestionably theropod fossils are present from the Norian Age (middle Late Triassic) onward (Nesbitt et al. 2007). The existence of the Carnian Age (early Late Triassic) sauropodomorphs *Panphagia* and *Saturnalia* indicates that the split between Theropoda and its sister group Sauropodomorpha had occurred by the early Late Triassic, and at least some studies have suggested that the Carnian carnivorous dinosaurs are true theropods. Three-toed footprints from Middle Triassic of Argentina (Marsicano et al. 2007) might represent extremely early theropods, but they might alternatively be from a nondinosaur instead.

There are definite Carnian Age carnivorous dinosaurs in the form of *Eoraptor* of Argentina and Herrerasauria (including *Herrerasaurus* of Argentina and *Staurikosaurus* of Brazil) (Fig. 19.3A). These are considered basal members of Theropoda by some workers (Rauhut 2003; Sereno 2007; Nesbitt et al. 2009). However, other recent studies place them as

Thomas R. Holtz Jr.

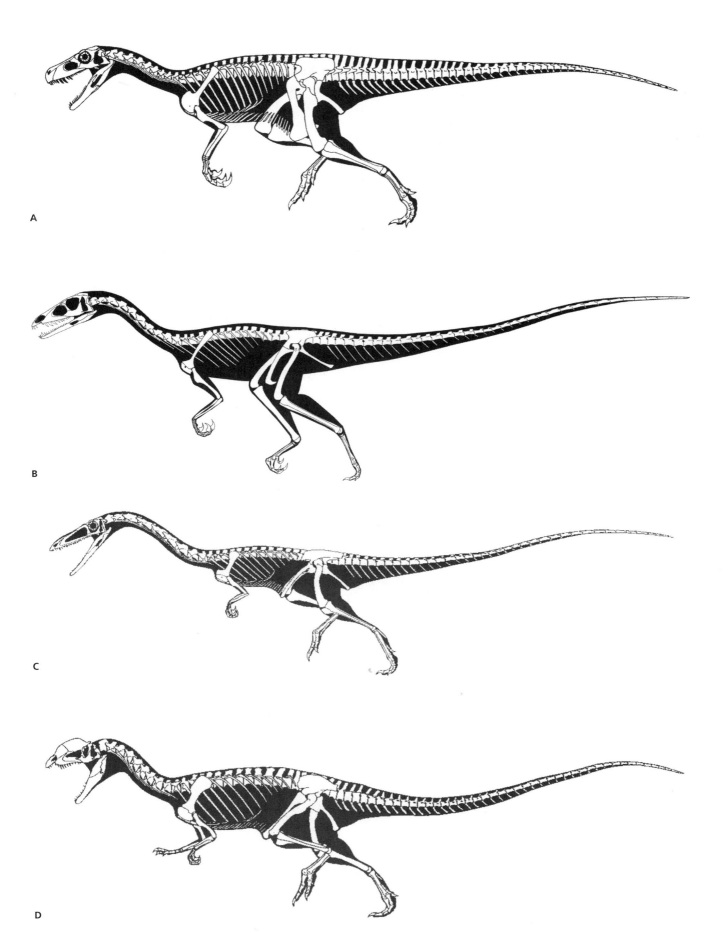

A

B

C

D

basal saurischians that diverged before the theropod–sauropodomorph split (Langer 2004; Langer and Benton 2006), in which case the similarities between these Triassic forms and true theropods, such as ziphodont teeth and grasping claws, are either the result of the retention of the primitive meat-eating habit or of convergence resulting from their predatory adaptations.

Another possible early theropod is *Guaibasaurus* of the Late Triassic of Brazil (Langer and Benton 2006). Like later theropods, it had long, slender pubes (Bonaparte et al. 2007). However, the skeleton of this dinosaur is incompletely known, so additional material may show that it is not part of Theropoda proper. Indeed, an analysis by Ezcurra (2010) has placed *Guaibasaurus* in a clade of primitive sauropodomorphs rather than as a theropod.

The uncertainty over the evolutionary positions of *Eoraptor* and Herrerasauria renders Theropoda difficult to diagnose. Features that seem to unite these Triassic taxa and later theropods include the following:

· Prezygapophyses in the distal caudals extending one quarter or more the length of the previous caudal, strengthening the rear portion of the tail.
· Humerus shorter than 60% the length of the femur.
· Proximal ends of the metacarpals abutting each other but not overlapping.
· Shaft of metacarpal IV much more slender than those of metacarpals I–III.
· Manual digit V lacking phalanges.

The recently discovered *Tawa*–a 2-m-long primitive theropod from the Norian of New Mexico (Fig. 19.3B)–represents a genus intermediate in form between basal carnivorous dinosaurs such as herrersaurs and *Eoraptor* and the definite theropods (the neotheropods) (Nesbitt et al. 2009). This is a more gracile form than *Herrerasaurus*. Initial phylogenetic analyses including this genus place it in a Theropoda that contains Herrerasauria as its earliest branch (Fig. 19.1). Like the more derived theropods, it has pneumatic depressions (pleurocoels) in its anterior cervical cervicals and a manus 40% or more the length of the humerus.

Neotheropoda, the more derived theropods (comprising the common ancestor of *Coelophysis bauri* and *Passer domesticus*, and all of its descendants), share the following features:

· The lacrimal extends onto the top of the skull (lower in herrerasaurs, *Eoraptor*, and sauropodomorphs; unknown in *Guaibasaurus*).
· An intramandibular joint is present consisting of a hinge between the tooth-bearing dentary and the postdentary bones (absent in sauropodomorphs, present in herrerasaurs, arguably present in *Eoraptor* [Sereno 2007], unknown in *Guaibasaurus*, not present in *Tawa*). This joint may possibly have functioned as a shock absorber for holding struggling prey.
· Clavicles are fused into a single bone, the furcula (unknown in herrerasaurs, *Eoraptor*, *Guaibasaurus*, and *Tawa*). Yates and Vasconcelos (2005) have shown that basal sauropodomorphs have clavicles

that are separate bones but meet at the midline. Furculae—the wishbones of modern birds—may have functioned in nonavian theropods as springs or shock absorbers, perhaps to deal with stresses on the forelimbs and shoulders from struggling prey.

· The medialmost distal carpal is significantly larger than the other distal carpals.

· Metacarpal and manual digit V have been lost, so only four or fewer fingers were present. (Metacarpal V is present in sauropodomorphs, herrerasaurs, *Eoraptor*, and *Tawa*; the hand of *Guaibasaurus* had at least four complete metacarpals; Bonaparte et al. 2007. However, it is uncertain whether metacarpal V was lost in *Guaibasaurus* or simply not preserved.)

· There are five or more sacral vertebrae (three or fewer in basal sauropodomorphs, *Eoraptor*, herrerasaurs, *Guaibasaurus*, and *Tawa*).

· Pedal digits I and V are reduced, making the foot functionally tridactyl (three-toed), in which metatarsal V lacks a toe, while metatarsal I is reduced with a small digit I. (The primitive state is found in basal sauropodomorphs, *Eoraptor*, herrerasaurs, *Guaibasaurus*, and *Tawa*.)

Neotheropods become an important group of predators in the Late Triassic and are the dominant group of terrestrial carnivores throughout the entire Jurassic and Cretaceous.

There are two major clades of advanced neotheropods, Ceratosauria and Tetanurae. Additionally, there are two primitive branches, the Coelophysoidea and Dilophosauridae (Fig. 19.3). In late twentieth-century (Gauthier 1986; Holtz 1994; Sereno 1999) and some early twenty-first-century (Ezcurra and Cuny 2007) studies, dilophosaurids were considered a type of coelophysoid. Additionally, this whole grouping was traditionally found to be closer to Ceratosauria than to tetanurines. However, most newer analyses that place Ceratosauria and Tetanurae in a clade (Averostra sensu Ezcurra and Cuny 2007: *Ceratosaurus nasicornis*, *Allosaurus fragilis*, their most recent common ancestor, and all of its descendants) exclusive of coelophysoids, and follows recent work suggesting that dilophosaurids were closer to averostrans than to Coelophysoidea proper (Rauhut 2003; Smith et al. 2007; Xu et al. 2009a). Furthermore, the analysis of Nesbitt et al. (2009) found that even the Coelophysoidea and Dilophosauridae as used in this chapter were paraphyletic, with coelophysoids being the basalmost phase of neotheropod evolution and the dilophosaurids the next most derived.

Early Theropods: Coelophysoids, Dilophosaurids, Ceratosaurs, Spinosauroids, and Carnosaurs

When theropods first appeared, much larger archosaurs such as ornithosuchids and rauisuchians were the top predators (Parrish this volume). Triassic neotheropods tended to be fairly small, such as 1-m-long *Procompsognathus* to 3-m-long *Coelophysis*. These lightly built early forms mostly fall phylogenetically within the Coelophysoidea (Carrano and Sampson 2004; Ezcurra and Cuny 2007; Smith et al. 2007; but see Nesbitt et al. 2009 for an alternative hypothesis.)

Coelophysoids, characterized by elongate cervical centra and shortened ischia, were the first major theropod radiation. They appear (best studied in the form of North American *Coelophysis bauri*) in the Rhaetian (late Late Triassic), when they are midsize carnivores (2–4 m long). They are among the first groups of dinosaurs to show up in North America (Nesbitt et al. 2007), where their footprints are one of the most common fossils.

Until recently, the coelophysoids and dilophosaurids were thought to form a clade united by a kink between the maxilla and the premaxilla (often with a corresponding large dentary fang below); this may have served as a spot to hold onto narrow prey (modern crocodilians with a similar snout pattern do the same). However, more complete sampling of the taxa and characters shows that this is a trait shared by all early neotheropods and lost in averostrans. In other words, coelophysoids in the old (1990s) sense of the term were the paraphyletic ancestors of the averostrans (Rauhut 2003; Smith et al. 2007; Nesbitt et al. 2009).

True coelophysoids tended to be long and slender, with slender skulls (Fig. 19.3C). The first coelophysoids were midsize (2–4 m long) forms such as *Coelophysis* (first known from the Late Triassic of the American Southwest, but also found in the Early Jurassic of southern Africa under the synonyms "*Syntarsus*" (preoccupied by an insect) and "*Megapnosaurus*"). However, there were small (~1–2 m long) coelophysoids, such as the Late Triassic *Procompsognathus* of Europe and the Early Jurassic *Segisaurus* of the American Southwest.

Dilophosaurids were larger, typically 4–6 m long (Fig. 19.3D). They are definitely present in the Early Jurassic in the form of *Dracovenator* of South Africa, *Dilophosaurus* of the American Southwest, "*Dilophosaurus*" *sinensis* (which does not appear to belong to the genus *Dilophosaurus*) of China, and *Cryolophosaurus* of Antarctica. The latter was long thought to be a primitive tetanurines, until a more complete study of this animal by Smith et al. (2007) revealed traits shared with *Dilophosaurus* and its kin. *Berberosaurus* of Early Jurassic North Africa was initially considered an early ceratosaurs (Allain et al. 2007), but recent work places it too as a dilophosaurid (Xu et al. 2009a). Dilophosaurids thus seem to represent a global radiation, the first large dinosaur predators, and the first time dinosaurs were the top (apex) predators in their environment, because the big predatory rauisuchians that dominated the Triassic were extinct. Most dilophosaurids have crests on their skulls, but because some coelophysoids and some primitive tetanurines have similar crests, this may have been a trait common to basal neotheropods. (Smith et al. 2007 did not use Charig and Milner's 1990 name, Dilophosauridae, to describe their newly discovered clade, but this appropriate term might be defined as "*Dilophosaurus wetherilli* and all taxa sharing a more recent common ancestor with it than with *Coelophysis bauri*, *Ceratosaurus nasicornis*, and *Allosaurus fragilis*.") There are a few other large (4–6 m long) primitive neotheropods of the Late Triassic (*Gojirasaurus* of the American Southwest; *Zupaysaurus* of Argentina) that might be giant coelophysoids, early dilophosaurids, or intermediate between the two.

Dilophosaurids share several derived features with later theropods (averostrans), including a promaxillary fenestra (an extra skull opening between the naris and the antorbital fenestra, but argued to be present in herrerasaurs; Sereno 2007); a lacrimal fenestra (pneumatic opening into the upper corner of the lacrimal); and a reduced total number of maxillary teeth. Although common in the Late Triassic and Early Jurassic, no coelophysoids or dilophosaurids are known after the end of the Early Jurassic. They seem to have been completely replaced by the averostrans.

Averostra (comprising Ceratosauria and Tetanurae; Fig. 19.1) has sometimes been called Neotheropoda (e.g., Padian et al. 1999; Holtz and Osmólska 2004); however, this chapter follows the usage of Rauhut (2003), Ezcurra and Cuny (2007), and Nesbitt et al. (2009) to include with Neotheropoda the more inclusive group that also contains Coelophysoidea and Dilophosauridae. Averostrans can be recognized by asymmetrical premaxillary teeth (not simple cones), straplike scapulae, and expanded anterior ends of the ilia, with hooklike blades.

The basal members of both Ceratosauria and Tetanurae typically have mediolaterally narrow, dorsoventrally deep skulls, sometimes called hatchet heads or, more technically, the oreinirostral condition. Functional analyses (Busbey 1995; Rayfield 2005) suggest that this skull pattern is effective for striking hard against a victim and slicing it up. However, it is not particularly strong in torsion (shaking back and forth), and so these dinosaurs probably did not latch onto their prey with their jaws for extended struggles (Holtz 2002).

Ceratosauria (defined as *Ceratosaurus nasicornis* and all taxa sharing a more recent common ancestor with it than with *Passer domesticus*) includes various basal forms, "elaphrosaurs," ceratosaurids, noasaurids, and abelisaurids (Carrano and Sampson 2007) (Fig. 19.4). Traditionally (Gauthier 1986; Holtz 1994; Sereno 1999) the coelophysoids and dilophosaurids were found to be part of this assemblage. If Early Jurassic *Berberosaurus* (Allain et al. 2007) is a ceratosaur rather than a dilophosaurid, it is the oldest known member of this clade; otherwise, known ceratosaurs are from the early Late Jurassic (*Limusaurus*) until the end of the Cretaceous. They began as a rare part of the theropod community, but in the Late Cretaceous they dominate most of the world, particularly the southern continents and Europe. Ceratosaurs have dorsal parapophyses that project far laterally, six or more sacral vertebrae, deep coracoids, reduced muscle scars and joint surfaces on the humerus, and reduced manual phalanges that suggested that they had relatively weak, and in some cases possibly even functionless, hands (Xu et al. 2009a).

Primitive ceratosaurs include the only partially known slender mid–Early Cretaceous *Spinostropheus* of Africa, and giant *Deltadromeus* of late Early/early Late Cretaceous of northern Africa (Carrano and Sampson 2007). Much more completely known is the slender *Elaphrosaurus* of Late Jurassic eastern Africa and western North America, and the recently discovered *Limusaurus* of the early Late Jurassic China (Xu et al. 2009a). These latter two, which seem to form a clade of "elaphrosaurs" (to which *Spinostropheus* might also belong) are known from relatively complete skeletons, and in the case of *Limusaurus* a skull—the only one known for a primitive ceratosaur. In the case of the latter animal, the skull is toothless, and indeed closely resembles those of their distant ornithomimid kin or the even more distantly related shuvosaurid crurotarsans such as *Effigia* and *Shuvosaurus* (Nesbitt 2007). The toothless beak of *Limusaurus* strongly suggests a diet other than strictly meat, and it is thus the oldest currently known example of a noncarnivorous theropod (and the only outside of Coelurosauria). Whether the toothless beak was also found in the as-yet-undiscovered skull of *Elaphrosaurus* and other primitive ceratosaurs remains to be seen.

Another clade of early ceratosaurs—and a group that most assuredly is not toothless—are the predatory Ceratosauridae. This clade of early ceratosaurs is best known from the 6–8-m-long *Ceratosaurus* of the Late Jurassic in western North America and Europe (Fig. 19.4A), but new specimens from the Early Cretaceous of Europe and South America (Rauhut 2004; Soto and Perea 2008) are increasing our knowledge of this group.

The Abelisauroidea of the Cretaceous were the most speciose of all ceratosaur groups. Abelisauroids as a whole are characterized by enlarged external mandibular fenestrae, arched sacra, rounded humeral heads, and doubled lateral and medial grooves on the pedal unguals. Abelisauroidea includes two major divisions. The first of these are the small, slender Noasauridae, ranging in size from tiny (<1 m long) *Ligabuenio* and *Velocisaurus* of South America and *Laevisuchus* of India to 2–m-long *Genusaurus* of Early Cretaceous Europe, *Noasaurus* of Late Cretaceous South America, and *Masiakasaurus* of Late Cretaceous Madagascar. Of these dinosaurs, only *Masiakasaurus* is known in substantial detail (Carrano et al. 2002) (Fig. 19.4B).

The sister group to Noasauridae is Abelisauridae, a clade that includes the top predators of South America, India, Madagascar, and Europe during the Late Cretaceous, and possibly of the unsampled or poorly sampled communities of Late Cretaceous continental Africa, Australasia, and Antarctica. The first abelisaurids (such as late Early Cretaceous *Kryptops* and early Late Cretaceous *Rugops* of Africa and *Ekrixinatosaurus* of South America) were minor predators compared to their titanic spinosaurid and carcharodontosaurid neighbors. With the extinction of those two tetanurine groups, however, abelisaurids seem to have come into their own. Abelisaurids are further specialized from other abelisauroids by rugose (horny) texture of the bones of the face (this supported some sort of softer tissue, but whether keratin or some thickened skin is not certain); short, rounded snouts; thickened skull roofs; squat, thickened teeth; forearms highly reduced—the ulnae and radii are practically no more than carpals; and relatively short and stocky hind limbs (Fig. 19.4C). The particulars of their forelimbs show that they were useless in grappling; their tough skulls and stout teeth suggest that they may have used their skulls to hold onto prey with their jaws in order to kill it. Notable Late Cretaceous abelisaurids include *Rajasaurus* and *Indosaurus* of India; *Majungasaurus* (formerly "*Majungatholus*"; see Sampson and Krause 2007 for an extensive review of this dinosaur) of Madagascar; and *Abelisaurus*, *Aucasaurus*, *Carnotaurus*, and *Skorpiovenator* of South America. Abelisaurids range all the way until the end of the Cretaceous.

The remaining theropods form the Tetanurae ("stiff tails"): *Passer domesticus* and all taxa sharing a more recent common ancestor with it than with *Ceratosaurus nasicornis*. Tetanurines (some prefer the form "tetanurans") are specialized from other theropods in possessing teeth restricted to the front of the jaws, proportionately larger hands, and interlocking tail vertebrae in at least the distal half of the tail (which may have served as dynamic stabilizers). Basal tetanurines tended to be large (5–8 m long) oreinirostral carnivores. There are three major clades within Tetanurae:

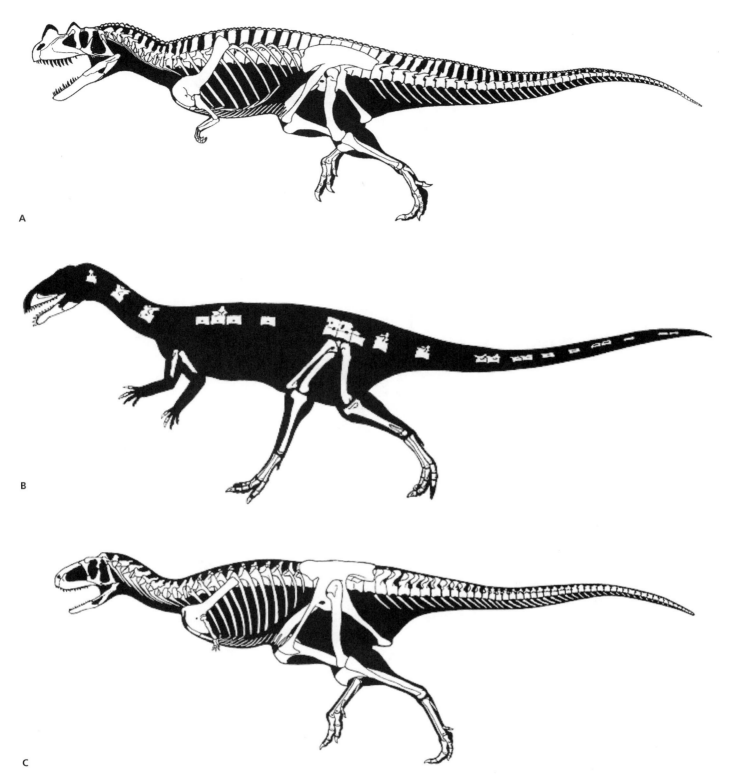

A

B

C

Megalosauroidea, Carnosauria, and Coelurosauria. The latter two together form the clade Avetheropoda. *Piatnitzkysaurus* and *Condorraptor* of Middle Jurassic Argentina (Smith et al. 2007) and *Xuanhanosaurus* and "*Szechuanosaurus*" *zigongensis* (Holtz et al. 2004) appear in some analyses to be tetanurines that diverged from the others before the spinosauroid–avetheropod split; however, analyses by Benson (2009) and Benson et al. (2010) support these animals as primitive members of Megalosauroidea.

19.4. Skeletal reconstructions of ceratosaurs. A, Late Jurassic ceratosaurid *Ceratosaurus nasicornis*, body length approximately 5.7 m; B, Late Cretaceous noasaurid *Masiakasaurus knopfleri*, body length approximately 1.8 m; C, Late Cretaceous abelisaurid *Aucasaurus garridoi*, body length approximately 5.1 m.

A, Illustration by G. S. Paul; B, from Carrano et al. (2002); C, illustration by by S. Hartman.

A

B

19.5. Skeletal reconstructions of megalosauroids. A, Early Late Jurassic basal megalosauroid *Monolophosaurus jiangi,* body length of preserved skeleton approximately 3.2 m; B, Early Cretaceous spinosaurid *Suchomimus tenerensis,* body length approximately 12 m.

Illustrations by S. Hartman.

One of the first major clades of tetanurines is the Megalosauroidea (Fig. 19.5). Traditionally called Spinosauroidea (e.g., Sereno 1999; Holtz et al. 2004), Benson (2009) and Benson et al. (2010) argue for the use of this older name. Megalosauroids share elongate skulls and powerful forelimbs. While some analyses (Sereno 1999; Holtz et al. 2004; Benson 2009; Benson et al. 2010) found a division between a primitive mostly Jurassic clade (Megalosauridae) and a derived Cretaceous spinosaurid group, others (Smith et al. 2007, 2008) find that taxa such as *Duriavenator, Eustreptospondylus, Streptospondylus, Magnosaurus,* and *Dubreuillosaurus* of Middle Jurassic Europe, massive *Torvosaurus* of Late Jurassic North America and Europe, and Early Cretaceous *Afrovenator* of northern Africa form a paraphyletic series relative to the strictly Cretaceous Spinosauridae. *Monolophosaurus* of the Middle Jurassic of China (Fig. 19.5A) might represent the sister group to the carnosaur–coelurosaur group Avetheropoda (Smith et al. 2007) or may be a primitive megalosauroid (Benson 2009; Benson et al. 2010). *Lourinhanosaurus* of the Late Jurassic of Portugal may be a megalosaurid (Allain 2005; Mateus et al. 2006), but it might instead be a sinraptorid carnosaur (Holtz et al. 2004; Brusatte and Sereno 2008; Benson et al. 2010).

The derived megalosauroid clade Spinosauridae (Fig. 19.5B) is distinctive and phylogenetically well supported. The spinosaurids are a group of mid-Cretaceous (late Early to middle Late Cretaceous) giant (8–14 m long) theropods characterized by elongate crocodile-like snouts with conical teeth

(Rayfield et al. 2007) and tall dorsal neural spines (which in the eponymous *Spinosaurus* formed a sail along the back.) Additionally, spinosaurids share with at least some of the basal spinosauroids an enormously enlarged thumb claw. The adaptations of the crocodile-like spinosaurid jaws and teeth, as well as their gut contents, suggest that they added large fish as well as dinosaurs to their diet (Holtz 2002; Rayfield et al. 2007). All spinosaurids have been discovered in environments in which large fish are common. Spinosaurids include *Baryonyx* of Early Cretaceous Europe; *Suchomimus* (which may simply be a species of *Baryonyx*) of the Early Cretaceous of northern Africa; *Irritator* of the Early Cretaceous of Brazil; and giant (14 m long) *Spinosaurus* of the early Late Cretaceous of northern Africa. *Spinosaurus* is one contender for the largest known theropod of all time (Dal Sasso et al. 2005; Therrien and Henderson 2007). The youngest known spinosaurid, and indeed the youngest known megalosauroid, is an unnamed taxon known from a single tooth from the middle Late Cretaceous of China (Hone et al. 2010).

Avetheropods share the following transformations from the ancestral tetanurine condition: a maxillary fenestra (an additional opening in the maxilla posterior to the promaxillary fenestra but anterior to the antorbital fenestra); extremely complex air sac chambers in vertebrae (convergently evolved in abelisaurids and in some sauropods); extreme reduction of manual digit IV (and independently in the advanced carnosaurs and coelurosaurs loss of metacarpal IV). (Note that the traditional counting of theropod digits is used here; Xu et al. 2009a offer an alternative interpretation.) Avetheropods generally fall in two clades, Carnosauria and Coelurosauria. Before the 1990s, these terms were often used as synonyms for "big theropod" and "little theropod," respectively. Thus larger coelophysoids, ceratosaurs, megalosauroids, and tyrannosaurid coelurosaurs were considered carnosaurs, while small coelophysoids were included with the coelurosaurs. Since the rise of cladistic studies, however, these names are restricted to two branches of the derived tetanurines.

Carnosauria was the dominant group of large theropods from the Middle Jurassic through the Early Cretaceous (Fig. 19.6). (Note that some authors [Padian et al. 1999; Holtz et al. 2004] use Carnosauria for the entire clade of *Allosaurus fragilis* and all taxa sharing a more recent common ancestor with it than with *Passer domesticus*; others [Sereno 1999; Brusatte and Sereno 2008] abandon this term and refer to this clade as Allosauroidea, a name the former authors restrict to a major subgroup of carnosaurs.) Carnosaurs are best known in the form of Late Jurassic North American and European *Allosaurus*. Carnosaurs are characterized by extra openings on the maxillae and nasals, large naris size, and various other features. Definite carnosaurs include the Sinraptoridae of the Middle and Late Jurassic of China (such as *Sinraptor* and *Yangchuanosaurus*) and Europe (*Poekilopleuron*, *Metriacanthosaurus*, and *Lourinhanosaurus*) (Naish and Martill 2007; Benson 2009; Benson et al. 2010) and the Allosauridae of Late Jurassic North America and Europe, including *Allosaurus* (19.6A) and giant 13-m-long *Saurophaganax*, currently the largest known Jurassic theropod, although specimens of *Torvosaurus* may reach nearly this size.

A comprehensive analysis by Benson et al. (2010) indicates that the Cretaceous members of Carnosauria form a clade, Carcharodontosauria. This group includes two major branches. One of these, Carcharodontosauridae,

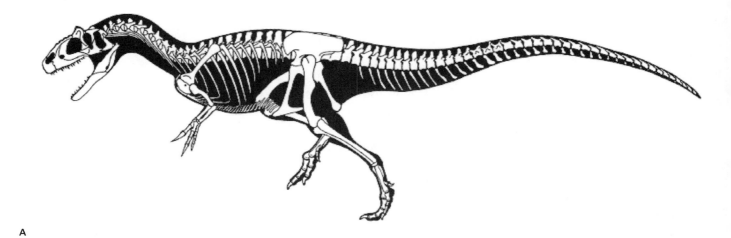

A

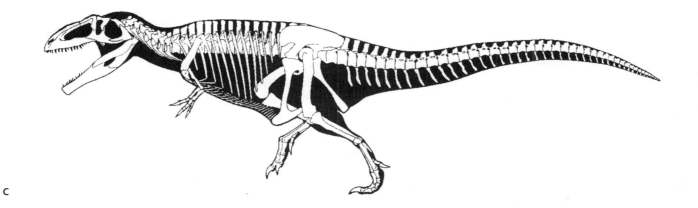

B

C

19.6. Skeletal reconstructions of carno-saurs. A, Subadult Late Jurassic allosaurid *Allosaurus fragilis*, body length approximately 7 m; B, Early Cretaceous carcharodontosaurid *Acrocanthosaurus atokensis,* body length ap-proximately 11.5 m; C, early Late Cretaceous carcharodontosaurid *Giganotosaurus caro-linii,* body length approximately 13 m.

A, B, Illustrations by S. Hartman;
C, illustration by G. S. Paul.

includes medium-size (8–10 m long) *Eocarcharia* of late Early Cretaceous northern Africa and early Late Cretaceous *Shaochilong* of China (Brusatte et al. 2009a, 2010), and giant (12–13 m long) *Acrocanthosaurus* of the late Early Cretaceous of North America (Fig. 19.6B). The most specialized carcharodontosaurids are those of the late Early Cretaceous and early Late Cretaceous of South America and Africa. These include *Tyrannotitan,* *Giganotosaurus,* and *Mapusaurus* of the late Early Cretaceous of South

America and *Carcharodontosaurus* of the early Late Cretaceous of Africa. These are among the largest theropods known: in particular, *Mapusaurus* and *Giganotosaurus* (Fig. 19.6C) exceeded *Tyrannosaurus rex* in size and rivaled the largest *Spinosaurus* specimens in mass, although the latter was probably longer, given the relatively long snout and neck of spinosaurids (Dal Sasso et al. 2005; Therrien and Henderson 2007).

The sister taxon to Carcharodontosauridae is Neovenatoridae (Benson et al. 2010). This clade includes the primitive *Neovenator* of the Early Cretaceous of Europe (Brusatte et al. 2008) and giant *Chilantaisaurus* of late Early or early Late Cretaceous Asia (Benson and Xu 2008), as well as a geographically widespread clade Megaraptora. The latter is a group of relatively gracile predatory dinosaurs with enlarged thumb claws, including *Fukuiraptor* of Early Cretaceous Japan and *Siamotyrannus* of the same age in Thailand (Holtz et al. 2004; Brusatte and Sereno 2008), Australian *Australovenator* (Hocknull et al. 2009), and middle Late Cretaceous South American *Aerosteon* (Sereno et al. 2008) and *Megaraptor*. The latter was originally thought to be a possible coelurosaur (Novas 1998), but newer analyses suggested megalosauroid (Smith et al. 2008) or carnosaur (Smith et al. 2007) affinities. Benson et al. (2010) recognized the union of the various megaraptorans within Neovenatoridae. In the later Late Cretaceous, the role of top predator was relinquished by the carcharodontosaurs and assumed by the abelisaurids in much of the world, and by tyrannosaurids in North America and Asia (Brusatte et al. 2009a, 2009b). However, the megaraptoran *Orkoraptor* (Novas et al. 2008a; Benson et al. 2010) persisted until the late Maastrichtian, making it the youngest known carnosaur.

The Coelurosauria ("*Coelurus* lizards") first appear in the Middle Jurassic (Holtz 2000; Rauhut 2003) or even earlier (Barrett 2009) and have persisted ever since (Fig. 19.2). They were ancestrally carnivorous, but various coelurosaurian subclades evolved numerous omnivorous and herbivorous lineages. The coelurosaurs differ from other theropods by possessing enlarged brains, at least twice the size of those of other theropods of the same body size (Larsson et al. 2000); a long, slender, tridactyl manus; boat-shaped (more specifically, kayak-shaped) chevrons, making the distal part of the tail particularly slender; and long, narrow metatarsals. Additionally, primitive coelurosaurs possessed a partial body covering of protofeathers—simple, apparently hollow downlike tufts (Ji and Ji 1996; Prum and Brush 2002; Xu et al. 2009b).

The oldest known coelurosaur is *Proceratosaurus* of the Middle Jurassic of England (Holtz 2000; Rauhut 2003; Naish and Martill 2007); recent study demonstrates this taxon to be a primitive tyrannosauroid (Rauhut et al. 2010). Only the skull is known at present. Other basal coelurosaurs include a trio of Late Jurassic, western North American forms: *Ornitholestes*, *Coelurus*, and *Tanycolagreus* (Carpenter et al. 2005a, 2005b). It may be that one or more of these basal forms are actually basal tyrannosauroids (Senter 2007). These and other early coelurosaurs were relatively small (2–4 m), slender animals with skulls full of small, bladelike, serrated teeth. Their narrow grasping hands suggest they adapted to catching small prey; their light build, slender limbs, and narrow dynamic stabilizing tail, all

Primitive Coelurosaurs: Compsognathids and Tyrannosauroids

useful both in chasing prey and avoiding predators, suggest relatively agile animals.

An important group of small primitive theropods is the Compsognathidae (Fig. 19.7A). This group ranges from the 1-m-long *Compsognathus* of the Late Jurassic of Europe and *Sinosauropteryx* of the Early Cretaceous of China to *Sinocalliopteryx* of the Early Cretaceous of China, at 2.5 m long. Compsognathids are also known from Late Jurassic (Göhlich and Chiappe 2006) and Early Cretaceous Europe (Naish and Martill 2007) and South America (Naish et al. 2004); they represented a minor radiation of small-bodied dinosaurs. Gut contents show that they ate lizards and small mammals (Currie and Chen 2001; Hurum et al. 2006). Their primitive and unspecialized anatomy has made these a difficult group of theropods to place phylogenetically: they have been considered basal to tyrannosauroids and maniraptoriforms (Sereno 1999; Holtz et al. 2004), primitive maniraptorans (Hwang et al. 2004, among others), or primitive tyrannosauroids (Zanno et al. 2009).

The most long-lived and ecologically significant group of primitive coelurosaurs was Tyrannosauroidea, the "tyrant dinosaurs" (Fig. 19.7B, C). Best known from the later Late Cretaceous Asia and North American Tyrannosauridae (Holtz 2001; Currie et al. 2003), recent discoveries reveal a long history of tyrant dinosaurs going back to the Middle Jurassic (Rauhut et al. 2010; Averianov et al. 2010). Basal tyrannosauroid specializations include fused nasals allowing for a stronger bite (Snively et al. 2006), incisiform premaxillary teeth with a U-shaped basal cross section that are much smaller than the maxillary teeth; and slender metacarpal III. *Guanlong* of the Middle–Late Jurassic boundary of China is a 3 m or longer crested basal tyrannosauroid (Xu et al. 2006). Like other early coelurosaurs, the arms were fairly long. Other Jurassic tyrannosauroids include poorly known *Stokesosaurus* and *Aviatyrannis* of North American and Europe. Little (1.5 m long) *Dilong* of the Early Cretaceous of China was the first tyrannosauroid found with protofeathers (Xu et al. 2004). Larger is *Eotyrannus* of Europe (Hutt et al. 2001), with an adult size of more than 4.5 m (Fig. 19.7B). It was dwarfed by other theropods in its community: the carnosaur *Neovenator* and the spinosaurid *Baryonyx*. Contemporaneous with *Dilong* and *Eotyrannus*, but phylogenetically closer to the derived tyrannosauroids, is 3-m-long *Raptorex* (Sereno et al. 2009). This tyrannosauroid had the highly reduced forelimb length, elongate hind limbs, and specialized metatarsus of the later members of the clade.

Tyrannosauroids increased in size again, with 5-m-long *Xiongguanlong* of later Early Cretaceous China (Li et al. 2009), 6-m or longer *Dryptosaurus* of the Late Cretaceous of eastern North America, comparable-size *Alectrosaurus* of Asia, *Appalachiosaurus* of eastern North America, *Bishtahieversor* of the American Southwest (Carr and Williamson 2010), and the Tyrannosauridae proper. As with *Raptorex*, the forelimbs were extremely reduced in length in at least *Dryptosaurus* and *Alectrosaurus*; the arms of the others are not presently known or described. In *Dryptosaurus*, the arm is short but has a large claw. In these theropods, the distal hind limbs (tibia, metatarsi) are elongated, an indication of cursorial (running) ability. Additionally, these taxa share with *Raptorex* an arctometatarsus (pinched metatarsus: Holtz 1995), proportionately slender metatarsals with metatarsal

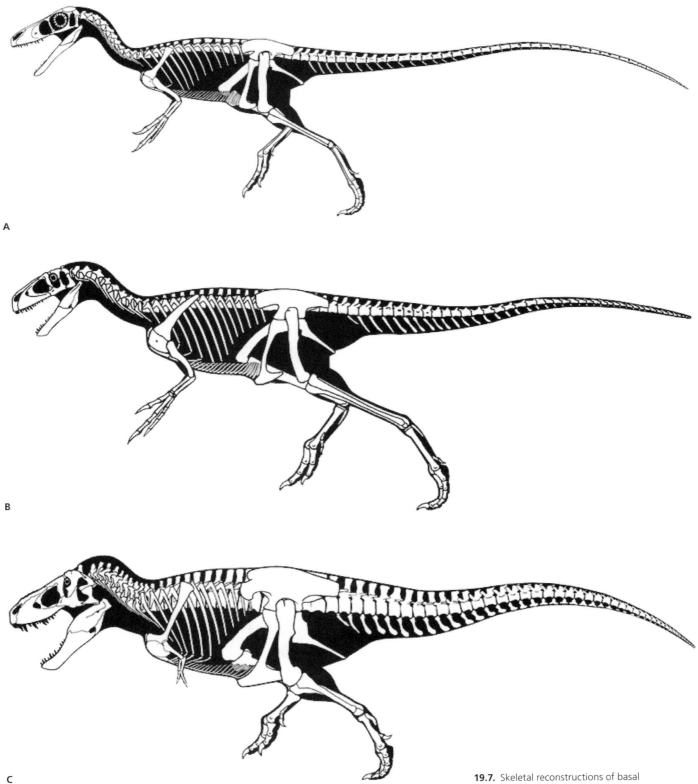

A

B

C

III pinched out proximally between II and IV. This adaptation is associated with greater strength and greater turning ability (Henderson and Snively 2004; Snively et al. 2004). It is found in other coelurosaurs that have cursorial limb proportions, and thus it is likely an adaptation for greater speed than other dinosaurs of the same body size (Holtz 1995; Snively et al. 2004).

19.7. Skeletal reconstructions of basal coelurosaurs. A, Early Cretaceous compsognathid *Huaxiagnathus orientalis,* body length approximately 1.7 m; B, Early Cretaceous basal tyrannosauroid *Eotyrannus lengi,* body length approximately 3.5 m; C, Late Cretaceous tyrannosaurid *Tyrannosaurus rex,* body length approximately 13 m.

Illustrations by S. Hartman.

Tyrannosauridae are one of the last groups of large-bodied theropods to evolve, showing up only in the last 20 million years or so of the Late Cretaceous of North America and Asia (Carr et al. 2005). Although for most of their history tyrannosauroids were minor predators in their habitats, tyrannosaurids were by far the largest flesh eaters in their environments. The most complete skeleton of the small tyrannosaurid *Alioramus* was approximately 4.5 m long, but this was only a subadult (Brusatte et al. 2009b). Most tyrannosaurid species reached at least 10 m, and at least one genus (*Tyrannosaurus*) reached up to 13 m (Fig. 19.7C). Tyrannosaurids were specialized relative to their ancestors by possessing proportionately large skulls, thickened maxillary and dentary teeth with deep roots (not the simple bladelike teeth of most other theropods; such teeth would have been useful for grabbing and holding onto prey and smashing through bone, rather than just slicing up meat; Holtz 2002; Snively et al. 2006); wide skulls allowing for forward-facing eyes (giving them potentially binocular vision; Stevens 2006) and increased neck muscle attachment (Snively and Russell 2007a, 2007b); and extremely short forelimbs in which manual digit III is lost, making a didactyl (two-fingered) hand. (Because no complete hand is known from the tyrannosauroids most closely related to Tyrannosauridae, it is quite possible that the didactyl hand evolved earlier, perhaps even as early as *Raptorrex*.)

Tyrannosaurids include the relatively slender *Albertosaurus* and *Gorgosaurus* of western North America; long-skulled *Alioramus* of Asia; more heavily built *Daspletosaurus* of western North America and *Tarbosaurus* of Asia; and giant 13-m-long, 6-ton *Tyrannosaurus* of western North America. Tyrannosaurids seem to have relied solely on their jaws to kill their food (Holtz 2002). Their long legs meant that they were faster than their potential prey (hadrosaurids, ceratopsids), although adults of the 2 ton or greater size range may not have been fast runners (Farlow et al. 1995; Hutchinson and Garcia 2002). (Juvenile tyrannosaurids, though, would have been among the fastest dinosaurs; Hutchinson 2004). At least some tyrannosaurids have been found in groups of different ages, possibly even family associations (Currie 2000).

Noncarnivorous Theropods: Ornithomimosaurs, Therizinosaurs, Oviraptorosaurs, Alvarezsauroids

The remaining coelurosaurs (Maniraptoriformes) all have brains that are twice again as large or larger (based on skull size) as the more basal coelurosaurs and a reduction of skull size relative to body size. If the phylogeny of Zanno et al. (2009) or something similar is correct (that is, that compsognathids are basal to maniraptoriforms and that ornithomimosaurs, therizinosaurs, alvarezsauroids, and oviraptorosaurs form a paraphyletic series relative to eumaniraptorans), than Maniraptoriformes is also characterized by a reduction in tooth size, an increase in tooth count, and a loss of the ancestral ziphodont tooth form. In this case, it appears that maniraptoriforms were ancestrally not primarily carnivorous and instead shifted to a more omnivorous or herbivorous lifestyle. In this case, the predatory status of some deinonychosaurs and birds would be a reversal to the more general condition of the theropods.

The basalmost lineage of the maniraptoriforms is the Ornithomimosauria, the ostrich dinosaurs. Ornithomimosaurs differ from the ancestral

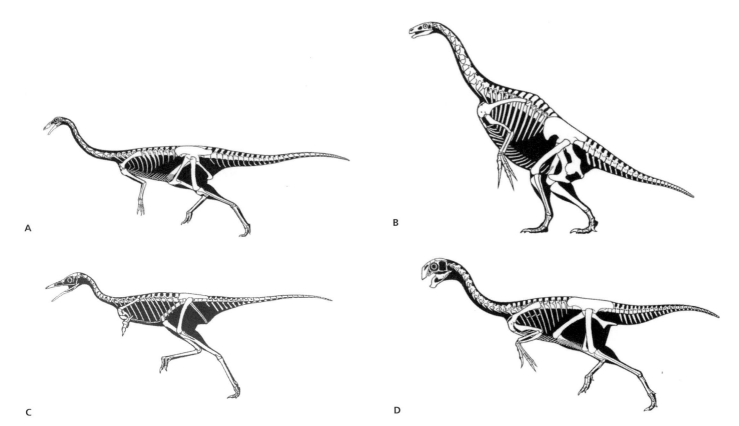

A

B

C

D

19.8. Skeletal reconstructions of nonhypercarnivorous maniraptoriforms. A, Late Cretaceous ornithomimid *Gallimimus bullatus,* body length approximately 5.5 m; B, Late Cretaceous therizinosauroid *Nothronychus mckinleyi,* body length approximately 4.5 m; C, Late Cretaceous alvarezsaurid *Shuvuuia deserti,* body length approximately 0.6 m; D, Late Cretaceous oviraptorid *Khaan mckennai,* body length approximate 1.2 m.

A, B, D, Illustrations by S. Hartman;
C, illustration by G. S. Paul.

state by reduced skull size, reduced tooth size, elongate neck, and a hooking and clamping hand in which all metacarpals the same length (except in the primitive *Harpymimus* of the Early Cretaceous of Asia) (Fig. 19.8A). Their adaptations suggest a move away from predation toward a more omnivorous or even herbivorous lifestyle.

Primitive ornithomimosaurs are known from the Early Cretaceous of Europe (*Pelecanimimus*) and Asia (*Harpymimus, Shenzhousaurus*) and the Late Cretaceous of Asia (*Garudimimus, Sinornithomimus*). The latter two share with the Late Cretaceous derived group Ornithomimidae skulls with toothless beaks, and *Sinornithomimus* shares with Ornithomimidae an arctometatarsus (convergently evolved with Tyrannosauridae and its kin). These dinosaurs are typically considered among the most cursorial of all theropods (Farlow et al. 2000). Ornithomimidae includes Late Cretaceous western North American (*Struthiomimus, Ornithomimus*) and Asian (*Gallimimus, Anserimimus*) taxa. Giant *Deinocheirus,* known only from its arms and a few isolated bones, may be a *Tyrannosaurus*-size ornithomimid (or more primitive ornithomimosaur; Senter 2007). At least some ornithomimosaurs lived in herds (Kobayashi and Lü 2003; Varricchio et al. 2008).

The remaining theropods form the clade Maniraptora ("hand grabbers"). Maniraptorans are the most diverse clade of dinosaurs. None retains a basal theropod form; indeed, few retain the ancestral carnivorous condition. The major groups within Maniraptora include the Therizinosauria, Alvarezsauroidea, Oviraptorosauria, and Eumaniraptora.

Maniraptorans show numerous specializations, notably another increase in relative brain size; elongated forelimbs; large, bony sternum for attachment of the muscles that pull the arms inward; and the semilunate carpal (a pulley-shaped block of wrist bones that allowed greater folding motion while sacrificing motion in any other plane). If a phylogeny similar to those of Sereno (1999), Holtz et al. (2004), or Hwang et al. (2004) is correct and oviraptorosaurs and therizinosaurs share a most recent common ancestor with each other than with any other group, then the following traits are also diagnostic of Maniraptora: laterally directed shoulder joints; honest-to-goodness true feathers on at least the arms and tail; and direct brooding on nests of eggs (Varricchio and Jackson 2004), although this trait might have been present in more basal coelurosaurs. However, should Zanno et al. (2009) be correct, these latter traits seem to be restricted to a subgroup of Maniraptora comprising oviraptorosaurs and eumaniraptorans (as shown in Fig. 19.2).

One possible, but problematic, shared derived feature of Maniraptora is a backward-pointing pubis. Most coelurosaurs (and saurischians in generally) have a vertically oriented or anteriorly oriented pubis. In therizinosauroids, alvarezsaurids, the basal troodontid *Sinovenator*, dromaeosaurids, *Archaeopteryx*, and other avialians the pubis points backward; in the basal therizinosaur *Falcarius*, the basal alvarezsauroid *Haplocheirus*, oviraptorosaurs, and troodontids other than *Sinovenator*, it points vertically or anteriorly. So it is difficult to say which condition is found in the most recent common ancestor of all maniraptorans. Changes in the muscle attachments in the hind limbs of maniraptorans (Hutchinson and Gatesy 2000) shows a switch from the femur-and-tail power stroke found in other dinosaurs (inherited from the early diapsids) to one where the flexion of the knee is more important.

The oldest maniraptorans are some possibly Middle Jurassic eumaniraptorans (Zhang et al. 2002; Xu and Zhang 2005), and this clade is definitely present by the earliest Late Jurassic (Hu et al. 2009). A dentary from the Early Jurassic of China is considered by some authors to be a therizinosaur (Xu et al. 2001; Barrett 2009): however, it might simply be a derived sauropodomorph (Rauhut 2003). The dating of this specimen is questionable, and while an Early Jurassic age is possible, it might be from the Middle Jurassic, or even as young as the Early Cretaceous (Barrett 2009).

Therizinosauria ("scythe reptiles") have been considered sauropodomorphs and late surviving proto-ornithischians (Paul 1984), but are in fact coelurosaurian theropods (Clark et al. 1994). Before the 1990s, they were often called the segnosaurs. Similar to the ornithomimosaurs, this group is characterized by small skulls and elongated necks, but the rest of the skeleton demonstrates that they are maniraptoran (Fig. 19.8B). This group is definitely known only from the Cretaceous and only from Asia and North America at present. The basalmost form is Early Cretaceous *Falcarius* of western North America (Kirkland et al. 2005; Li et al. 2007; Zanno et al. 2009). It retains a relatively elongate metatarsus and a vertically oriented pubis. The derived therizinosaurs form the clade Therizinosauroidea and are characterized by shortened metatarsi in which all four toes touch the ground, and backward-pointing pubes. In this case, like the ornithischians, this is almost certainly to accommodate a large gut for digesting plants.

Therizinosaurs seem to have been primarily, if not strictly, herbivores. Their stumpy feet and short legs show them to have been among the slowest theropods. To defend themselves (and possibly to help them feed), they had huge claws. They ranged from bear-size taxa such as *Erlikosaurus* to *Tyrannosaurus*-size *Therizinosaurus* with 1-m-long claws. If *Therizinosaurus* retained the feathers of its ancestors, it was likely the largest feathered animal known.

Therizinosauria and Oviraptorosauria may together form a clade Oviraptoriformes, united by the shared presence of leaf-shaped teeth with large denticles, indicating that plants formed a major part of their diet. There are additional skull, vertebral, and limb characters that support a monophyletic Oviraptoriformes; on the other hand, some characteristics support a group of Oviraptorosauria plus Eumaniraptora exclusive of therizinosaurs (Zanno et al. 2009).

Alvarezsauridae is a highly specialized group of Late Cretaceous theropods. Alvarezsaurids are known from South and North America, Europe, and Asia (Martinelli and Vera 2007). They have numerous birdlike features and were once thought to have been specialized flightless birds (Chiappe et al. 1996). Most alvarezsaurids range in size of chickens to turkeys, but the recently discovered *Kol* of Mongolia represents a larger, rhea-size genus (Turner et al. 2009). The newly discovered earliest Late Jurassic *Haplocheirus* of China (Choiniere et al. 2010) represents a basal member of the lineage leading to Alvarezsauridae; collectively, *Haplocheirus* and the later clade form the group Alvarezsauroidea.

Alvarezsaurids have small, beaky skulls with tiny teeth and bizarrely powerful arms with a huge thumb claw and exceedingly small digits II and III (Fig. 19.8C). The hand of *Haplocheirus* was less specialized, but it was still characterized by a much larger thumb than typical theropods. While *Haplocheirus* has a primitive pelvic condition, alvarezsaurids have a backward-pointing pubis. Unlike the therizinosauroid and ornithischian situation, this backward position of the pubis is more likely associated with changes in the locomotory muscles toward knee-driven power from the ancestral tail-and-femur-driven power than for increased gut space.

Only a little is known of *Alvarezsaurus* itself; somewhat more is known for the more derived *Achillesaurus* and *Patagonykus* (both of South America). The highly derived Parvicursorinae (known as Mononykinae to some researchers), in contrast, is known from many excellent specimens. The best studied are the Asian taxa *Mononykus*, *Parvicursor*, *Shuvuuia*, and *Xixianykus* (Xu et al. 2010). More poorly known are North American *Albertonykus*. Like tyrannosaurids and ornithomimids, parvicursornies have an arctometatarsus. The parvicursorines show numerous cursorial adaptations, but these were almost certainly defensive. They seem to have been insectivores (Longrich and Currie 2009a), and their forelimbs may have been used to batter into ant and termite nests (Senter 2005). Alvarezsauroids have been found from deserts to well-watered environments.

Another odd batch of maniraptorans is the Oviraptorosauria. This group is characterized by short, boxy skulls. The basalmost oviraptorosaurs are toothy *Incisivosaurus*, *Protarchaeopteryx*, and *Caudipteryx*, all from the Early Cretaceous of China (Senter 2007). Other early branches of the oviraptorosaurs are *Microvenator* of the Early Cretaceous of western North America and specialized *Avimimus* (which had an arctometatarsus) of the

Late Cretaceous of Asia. Trace fossil evidence suggests that the latter seems to have dwelt in large flocks.

These basal branches of Oviraptorosauria are relatively small, about the size of chickens to turkeys. The more derived Oviraptoridae contain forms that range from turkey to human to tyrannosaur size. Oviraptorids are derived by loss of all teeth as well as other cranial specializations (Fig. 19.8D). Some oviraptorids had an arctometatarsus and relatively short arms; others had short metatarsi and relatively short feet. Oviraptorids are limited to the Late Cretaceous of Asia and North America, including Asian *Oviraptor*, crested *Citipati* and *Rinchenia*, crestless *Khaan* and North American *Hagryphus*, arctometatarsus-footed *Chirostenotes*, and a similar, larger, crested but not yet named form from the end of the Cretaceous. By far the largest oviraptorosaur, though, is the recently discovered *Gigantoraptor* of Asia; it is as large as an *Albertosaurus* or other smaller tyrannosaurid (Xu et al. 2007).

Interpreting the life habits of oviraptorosaurs is confusing. While the ancestral ones seem to be convincingly herbivorous (Xu et al. 2002), there are lizards in the gut contents of some oviraptorids; perhaps they were omnivorous. Many oviraptorosaurs have been found in desert environments, but others are found in forested regions.

Deinonychosaurs: The Raptors

Eumaniraptora ("true maniraptorans"), the last remaining group of theropods, comprises two primary branches, Deinonychosauria and Avialae (Fig. 19.9). A deinonychosaur might be present in the Middle Jurassic (Ji et al. 2005), although the dating of the unit in question is not certain. *Archaeopteryx*, *Anchiornis* (Hu et al. 2009), and an unnamed Morrison Formation troodontid (Hartman et al. 2005) demonstrate that both clades of eumaniraptorans are definitely present by Late Jurassic. The actual shared derived characters that unite the branches of Eumaniraptora at present as a whole are a bit problematic as a result of the uncertainty of the position of *Archaeopteryx* (Mayr et al. 2005) and the discovery that basal deinonychosaurs are far more *Archaeopteryx*-like than are the classic deinonychosaurs (Hwang et al. 2002; Makovicky et al. 2005). Some probable transformations at the base of Eumaniraptora include small body size (basal members of each clade are crow size; Turner et al. 2007); quite long arms, often as long or longer than the hind limbs, with proportionately large hands; tail mobile near the base but stiff the rest of the length; a distally placed metatarsal I so that pedal digit I touches the ground (or branch); backward-facing pubis, present in dromaeosaurids, avialians, and the basal troodontid *Sinovenator*; and long leg feathers of the same design as the long arm and tail feathers.

One group of possible Middle Jurassic (although it might be Late Jurassic, or even earliest Cretaceous!) eumaniraptorans are the tiny Scansoriopterygidae (Zhang et al. 2008). These are some of the smallest nonavialian dinosaurs (pigeon size), although at least two of the known specimens are not fully grown (Fig. 19.9C). Another contemporary is *Pedopenna*. Because both the scansoriopterygids and *Pedopenna* are quite small and both have the distally placed pedal digit I, they may have spent some time up in the trees. The skull of *Pedopenna* is not known, but that of the scansoriopterygids suggests that it might have been an insect eater. At least some analyses

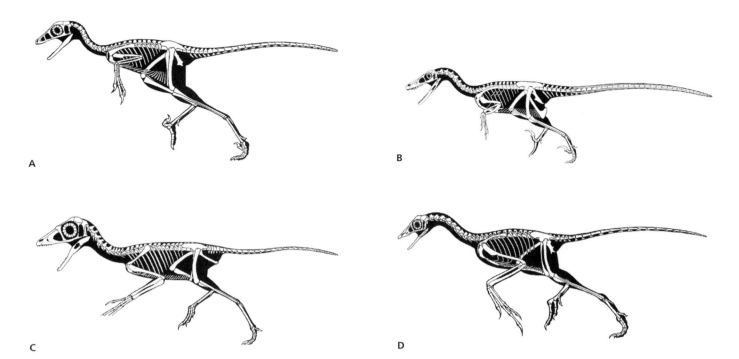

A   B

C   D

(Zhang et al. 2008) indicate that these tiny dinosaurs are the basalmost members of Avialae.

The remaining eumaniraptorans belong to either Deinonychosauria or to Avialae. However, there are several eumaniraptorans whose exact placement is uncertain. The best-studied and longest-known eumaniraptoran of the Late Jurassic is *Archaeopteryx* (Fig. 19.9D). This taxon came from the Late Jurassic of Europe. Isolated bones from western North America of the same age might be from a similar taxon (Jensen and Padian 1988). Some paleontologists regard all the German specimens as a single genus and species (*A. lithographica*); others divide them up into several genera (*Archaeornis, Jurapteryx, Wellnhoferia,* as well as *Archaeopteryx*; see review by Mayr et al. 2007). While typically considered a basal member of Avialae (see Naish, this volume), some recent studies find that it branched off before the deinonychosaur–avialian split (Mayr et al. 2005)—a hypothesis previously suggested by Paul (1988, 2002)—or as a basal member of Deinonychosauria (Xu et al. 2011). The first and second specimens of this theropod were the first fossil skeletons with feathers ever discovered (in the 1860s and 1870s). Indeed, these were the first good maniraptoran fossils of the Mesozoic ever found. Consequently, *Archaeopteryx* has spent most of the last century and a half being considered a bird. But because feathers have been found in other theropods and the skeletons of basal deinonychosaurs have been discovered, the birdiness of *Archaeopteryx* has decreased.

There remain a few features that *Archaeopteryx* shares with Avialae (modern birds and the theropods closer to birds than to deinonychosaurs), but a new specimen also shows some features more similar to deinonychosaurs than to avialians (Mayr et al. 2005, 2007). So at present, the phylogenetic position of this little crow-size dinosaur remains uncertain. Also uncertain is whether this dinosaur was capable of flight (Senter 2006).

Deinonychosauria ("*Deinonychus* lizards") may have first appeared in the Middle Jurassic (possibly; definitely by the earliest Late Jurassic) and

19.9. Skeletal reconstructions of eumaniraptorans. A, Middle Jurassic? troodontid *Jinfengopteryx elegans,* body length approximately 0.55 m; B, Late Cretaceous dromaeosaurid *Velociraptor mongoliensis,* body length approximately 1.6 m; C, juvenile Middle Jurassic? scansoriopterygid *Epidendrosaurus ninchengensis,* body length approximately 0.25 m; D, basal avialian (or basal eumaniraptoran) *Archaeopteryx siemensii,* body length approximately 0.4 m.

Illustrations by S. Hartman.

died out at the end of the Cretaceous. They are known from all over the world, and in habitats from deserts to coasts. Deinonychosaurs possess a sickle-shaped claw on the retractable pedal digit II (comparable to the retractable claws of cats). (Note that a slight development of the sickle claw and the retractable pedal digit II also show up in *Archaeopteryx*; Mayr et al. 2005, 2007.)

Deinonychosaurs divide into two clades (at least in our present understanding): Dromaeosauridae and Troodontidae. The basal members of both groups are only about crow size (similar in size to *Archaeopteryx*, *Pedopenna*, and basal avialians).

The unquestionably carnivorous dromaeosaurids are known throughout the Cretaceous and from North and South America, Europe, Asia, Africa, and Madagascar. Basal dromaeosaurs (Unenlagiinae and Microraptorinae; Makovicky et al. 2005; Longrich and Currie 2009b) include some crow- to turkey-size taxa: *Rahonavis* of the Late Cretaceous of Madagascar (and initially thought to be a bird); *Microraptor* of the Early Cretaceous of China (the first eumaniraptoran for which the tail feathers were known); tiny *Hesperonychus* of Late Cretaceous Alberta; long-snouted *Buitreraptor* of Late Cretaceous Argentina; and others. However, at 5 m long, the unenlagiine *Austroraptor* (Novas et al. 2008b) is one of the largest eumaniraptorans of all time. This Late Cretaceous Argentine form had a long snout with numerous small teeth and short forelimbs—highly different from the typical dromaeosaurid condition and suggesting that it had moved into a different form of predation than its ancestors.

The derived groups of Dromaeosauridae (Saurornitholestinae, Velociraptorinae, *Deinonychus*, and Dromaeosaurinae; collectively the Eudromaeosauria; Longrich and Currie 2009b) were coyote to grizzly bear size (Fig. 19.9B). Well-studied examples include coyote-size *Velociraptor* of the Late Cretaceous deserts of Mongolia; wolf-size *Deinonychus* of the Early Cretaceous of western North America; lion-size *Achillobator* of mid–Late Cretaceous Asia; and grizzly bear–size *Utahraptor* of the Early Cretaceous of western North America, with *Austroraptor* the largest known eumaniraptoran.

Dromaeosaurid caudals were more tightly interlocked than in typical theropods, and in eudromaeosaurs, extensions from the chevrons and neural arches grew extremely long (Ostrom 1969). The tail was thus an extreme dynamic stabilizer. Some basal dromaeosaurids had elongate metatarsi (indeed, they have a primitive form of the arctometatarsus, dubbed a "subarctometatarsus" by White 2009), but the majority had relatively short stout metatarsi and tibiae. This suggests that they had sacrificed speed, perhaps for agility, the better to turn quickly while pursuing prey or escaping predators, especially with the help of the stiffened tail. Despite certain popular books and movies to the contrary, the dromaeosaurids show no signs of being speed specialists.

While the small primitive forms may have eaten small prey (skewering it with the sickle claw in the manner of modern secretary birds, perhaps), the larger forms were predators of dinosaurs. The "fighting dinosaurs" specimen of *Velociraptor* shows it in combat with *Protoceratops*, the hands used to grasp the head of the herbivore while the sickle claw was ripping into the throat, similar to the attacks used by large cats (Carpenter 2000). Some (controversial) evidence suggests that *Deinonychus* may have attacked the much larger iguanodontian *Tenontosaurus* in groups (packs or mobs) (Maxwell and Ostrom 1995; but see Roach and Brinkman 2007). Note that the sickle claw may have also been used to climb, either up trees, or up the sides of victims (Manning et al. 2006)!

In the Early Cretaceous, dromaeosaurids were major midsize predators, and in the deserts of Late Cretaceous Asia, they were among the largest carnivorous dinosaurs present. However, with the rise of the tyrannosaurids, large-bodied dromaeosaurids disappeared in the Northern Hemisphere, and the Late Cretaceous dromaeosaurids of Asia and North America were fairly small animals.

The other clade of deinonychosaurs – Troodontidae – were crow- to deer-size theropods. The oldest troodontid may be *Jinfengopteryx* of China, described originally as a bird but now recognized as a troodontid (Turner et al. 2007) (Fig. 19.9A). It might be from the Middle Jurassic, but because the rocks it was found in are not well dated, it could be as young as earliest Cretaceous. Like *Archaeopteryx*, *Jinfengopteryx* had long tail feathers all along its tail. *Anchiornis* is an earliest Late Jurassic Chinese troodontid (Hu et al. 2009) or possibly a close relative of *Archaeopteryx* (Xu et al. 2011). A Late Jurassic troodontid is known from western North America, but it has not yet been named or fully described (Hartman et al. 2005). The last troodontids are from the end of the Cretaceous.

Troodontids differ from other eumaniraptorans by an increase in the number of teeth (although this might simply be a retention of the ancestral maniraptoriform condition; Zanno et al. 2009), reduced arm length, and increased distal hind limb length. Tiny *Sinovenator* of the Early Cretaceous of China has a backward-pointing pubis and lacks an arctometatarsus; later troodontids show a reversal to a forward-pointing pubis and have an arctometatarsus. Troodontids have relatively lightly built snouts (Norell et al. 2009), suggesting that they did not tackle large prey, particularly because their forelimbs were fairly short and lightly built as well. The largest and most derived troodontids are not particularly large – turkey to deer size – and had forward-facing eyes and leaf-shaped teeth with large denticles on the back. At least some of these derived troodontid taxa may have had some plants in their diet (Holtz et al. 2000), although microwear abrasion patterns on the teeth of these theropod strongly suggest that they ate at least some meat (Fiorillo 2008)

The more completely known troodontids include *Mei* and *Sinornithoides* of the Early Cretaceous of Asia (and both known from specimens buried in a sleeping position!), *Saurornithoides* and *Zanabazar* of the Late Cretaceous of Asia (Norell et al. 2009), and *Troodon* (previously called "*Stenonychosaurus*") of the Late Cretaceous of western North America, largest of the Troodontidae.

The large eyes of troodontids suggest to some that they may have been nocturnal. Possibly related to that, troodontids are by far the most common dinosaur in polar Late Cretaceous North American fossil sites, where light levels would have been low during the winter, although they were no more common than dromaeosaurids in the rest of their range (Fiorillo 2008).

Theropod Senses and Feeding

Although uncovering the sensory capacity of extinct animals poses some great difficulty (indeed, it is difficult to gauge the sensory ability of many living animals), there has been research to assess how well theropods could see, hear, and smell. Understanding these attributes may help us to understand the world as they perceived it, including their ability to detect potential food and predators.

Stevens (2006) assessed the possibility that some theropod dinosaurs had the ability to achieve stereopsis (increased depth perception due to increased overlap between the range of vision of both eyes relative to the ancestral form). He found that the more basal theropods studied had relatively little overlapping vision, giving them wide visual scope around the landscape but less ability to focus directly in front of them. In contrast, the coelurosaurs studied (tyrannosaurids and deinonychosaurs) had considerable potential for stereopsis, which might indicate an increased ability to deliver a focused bite on a food object (living or dead). Farlow (1994) demonstrated that tyrannosaurids (and by implication other large-bodied theropods including carnosaurs and spinosaurids) would be able to see over quite large distances as a result of the elevation of their eyes. This would be advantageous for early detection of potential food, foes, or mates.

Gleich et al. (2005) estimated the hearing capacity of the carnosaur *Allosaurus* on the basis of measurements of the preserved dimensions of the inner ear and comparison of these attributes with those of living archosaurs for whom the hearing sensitivity had been tested. They found that *Allosaurus* would its best frequency of hearing below 1.5 kHz and highest frequency of hearing of about 3 kHz, similar to values for an emu (*Dromaius*; 1.5 and 3.8 kHz, respectively) but considerably below that of humans (4 and 17.6 kHz, respectively: Heffner 2004).

The olfactory ability of theropods has been estimated by Zelenitsky et al. (2009) by comparing the relative size of the olfactory bulb as preserved in endocasts to the rest of the brain. Within theropod dinosaurs, they found that most lineages (including ceratosaurs, carnosaurs, basal tyrannosauroids, troodontids, and basal avialians) had olfactory bulbs that scaled along the same curve, but that two sets of theropods were markedly off this curve. The first of these represented noncarnivorous (or at least nonhypercarnivorous) theropods such as ornithomimosaurs and oviraptorosaurs, which plotted significantly below other theropods and thus likely had greatly reduced olfactory ability. In contrast, dromaeosaurids and tyrannosaurids had significantly larger than expected olfactory bulbs, and thus may have had an increased sensitivity to smell compared to even closely related taxa. Presumably dromaeosaurids and tyrannosaurids would have had an increased olfactory component to their food-finding behaviors than other theropods, although of course olfaction also has other, not mutually exclusive functions in modern carnivorous animals, such as smell-based displays, including the marking of territory.

Farlow and Holtz (2002) and Holtz (2002) reviewed general patterns of theropod feeding. Like the Jabberwock of Lewis Carroll, the theropod predatory armament consisted of "jaws that bite" and "claws that catch." Each of these was modified in different ways among the Theropoda. The ancestral theropods (and herrerasaurs) combined bladelike serrated teeth (ancestral for Archosauria) and grasping hands (a dinosaur-shared derived character) with the intramandibular joint and powerful claws. In addition, neotheropods at least had the furcula as a possible aid for holding onto prey. Coelophysoids and dilophosaurids retained these ancestral traits and evolved the premaxillary–maxillary kink to hold onto victims. Ceratosaurs reduced their manual grasp, and thus would have presumably been restricted to jaw-based attacks; this is taken to an extreme in the stumpy-armed, round-skulled abelisaurids. Tetanurines, in contrast, increased the size of the hands and the power of the arms. Within the primitive tetanurines, the spinosaurids evolved numerous additional adaptations associated with the ability to capture large fish (Rayfield et al. 2007). Hone and Rauhut (2010) have suggested that theropods in general may have preferentially preyed upon and consumed whole juvenile dinosaurs, resulting in the apparent low proportion of juvenile dinosaur fossils relative to the expected level given clutch size.

In many environments, several different theropods shared the same habitat. In some cases, they may have partitioned the resources by body size (although the juveniles would still overlap). But in the case of the spinosaurids, there seems to have been evolution of the ability to access meat that other theropods couldn't: fish. Similarly, spinosaurids could travel more easily from lake to lake and also capture food more easily on land than the giant crocodyliforms that were their main competitor for fish.

Basal coelurosaurs show the ancestral theropod predatory adaptations: bladelike serrated teeth, intramandibular joint, grasping hands, powerful claws, and furculae. Additionally, the enlarged brains, narrow metatarsi, and narrow tail would have aided coelurosaurs in predation because they helped with agility. Basal tyrannosauroids inherited these basal coelurosaur conditions, with modifications in their feeding adaptations (incisor-like scraping U-shaped premaxillary teeth and fused nasals). Tyrannosauroidea underwent correlated progression in selection for increased jaw and tooth power (larger skull, deeply rooted thickened teeth, massive jaw and neck muscles) and decreased arm power (reduced arm length and loss of manual digit III)–a shift from forelimb plus jaw attacks to jaw-only attacks (Snively et al. 2006). Biomechanical studies of theropod jaws show that tyrannosaurids had estimated bite forces three to five times more forceful than those of carcharodontosaurids of comparable body size (for a comprehensive review and study, see Therrien et al. 2005).

Basal coelurosaurs represented the minor predators of many Jurassic and Early Cretaceous environments. From these forms, the evolution of non-meat-eating specializations allowed for a diversification of maniraptoriform coelurosaurs into niches previously unoccupied by theropods. For example, in the well-sampled faunas of Late Cretaceous Asia, small nonpredatory coelurosaurs are common, while small ornithopods–relatively common in the rest of the Jurassic and Cretaceous–are absent. It is not yet known if the earlier appearance of nonpredatory ceratosaurs (as seen by *Limusaurus*) resulted in a similar radiation in the Late Jurassic or Early Cretaceous.

Among the carnivorous theropods, there seems to be a historical shift from a dilophosaurid-dominated Early Jurassic to a long phase (Middle Jurassic through early Late Cretaceous) in which megalosauroids and carnosaurs are the apex predators of their faunas. Tyrannosauroids and ceratosaurs spend much of this phase as medium-size predators, but with the disappearance of megalosauroids and most carnosaurs prior to the last 20 million years of the Cretaceous, these previously smaller-bodied taxa evolved into the giant Tyrannosauridae and Abelisauridae, apex predators in their respective regions.

Theropod Locomotion

In general, basal theropods retain the ancestral dinosaurian striding habit. Overall, theropods tended to be more slender than similar-size herbivorous dinosaurs, and consequently they were most likely faster, more agile, or both (Farlow et al. 2000). This agility was aided in the tetanurines by evolution of the tightly locked distal tail as a dynamic stabilizer

In some ceratosaurs (*Elaphrosaurus*, *Limusaurus*, and some noasaurids), there is evolution of cursorial (fast-running) limb proportions; this evolved many times independently in coelurosaurs (Holtz 1995; Farlow et al. 2000). Advanced tyrannosauroids (especially tyrannosaurids), ornithomimosaurs (especially ornithomimids), some oviraptorosaurs, parvicursorine alvarez-sauroids, some basal dromaeosaurids, and advanced troodontids each independently evolved more elongated hind limbs and the arctometatarsus for increased cursoriality (Holtz 1995; Snively et al. 2004). In the case of the tyrannosauroids, this enhanced running ability was associated with predation (chasing down prey), while in the others it may have been more defensive. Perhaps not coincidentally, these cursorial noncarnivorous coelurosaurs co-occur with tyrannosauroids.

In contrast, the limbs of at least some abelisaurids are proportionately short and stocky, suggesting that they were slower than other theropods of their body size. However, if sauropods were their main prey, they might not need to be particularly fast. Similarly, therizinosauroids show shifts away from their more cursorial ancestral behavior toward slower motion, as seen by the proportionately short metatarsi and stocky legs. Because these were nonpredatory forms, they would not have to pursue their food (plants).

Changes in the pelvic and hind limb muscles from basal theropods up through the maniraptorans into modern-style birds shows a shift from the ancestral dinosaur striding pattern to a knee-driven form of walking (Hutchinson 2002; Hutchinson and Gatesy 2001). Small eumaniraptorans are within a size range in which tree climbing may have been a potential habit, and their distally placed pedal digit I may have helped to allow them to better clutch branches. Even juvenile members of larger eumaniraptorans may have taken advantage of a scansorial (tree climbing) habit even if their adult phase was more arboreal, just as today the Komodo dragon *Varanus komodoensis* transitions from a scansorial to an arboreal habit as juveniles to terrestrial adults (Imansyah et al. 2008).

Deinonychosaurs, basal avialans, and possibly other basal maniraptorans such as oviraptorosaurs were likely able to take advantage of a form of locomotion that lizards, and indeed all other nonmaniraptorans, would have been forbidden: wing-assisted incline running (Dial 2003; Bundle and Dial 2003; Tobalske and Dial 2007; Dial et al. 2008). In this form of locomotion, feathered arms of birds (from the hatchling stage onward) are flapped anteriorly and posteriorly, generating thrust down through the feet. (This is opposed to the normal flight stroke, in which the wings move more dorsoventrally, generating thrust forward.) By this motion, modern birds can generate sufficient traction against a substrate, allowing them to run up vertical (or even greater than vertical!) surfaces such as tree trunks. Experimental evidence shows that even a partial development of arm feathers gives a bird sufficient thrust to run up steep surfaces (Dial 2003). Because nonavialian maniraptorans such as deinonychosaurs and oviraptorosaurs have the same set of adaptations (both skeletal and feather) that would allow this form of wing beat, perhaps this behavior was employed to allow small members of these clades to escape the jaws and claws of their larger relatives in order to hide in the branches of trees. Similarly, the presence of a somewhat winglike feathered forelimb (as found in oviraptorosaurs and deinonychosaurs) would allow for controlled flapping descent out of trees (Dial et al. 2008).

Although theropods outside of birds do not show particular swimming adaptations, there are sets of theropod tracks that demonstrate that they were capable of such locomotion (Ezquerra et al. 2007).

Size Evolution in the Theropoda: Gigantism and Miniaturization

Early theropods (e.g., *Guaibasaurus*, coelophysoids) are in the 2–4-m range, and consequently are not much larger than the most primitive members of the other dinosaur branches (e.g., *Eoraptor*, *Saturnalia*, *Pisanosaurus*, *Eocursor*). Dilophosaurids and basal averostrans grew larger, to about 6–7 m long. This size increase occurs after the Triassic/Jurassic extinctions and the loss of most of the potential competitors among large-bodied predators. (Small predators such as sphenosuchian crocodylomorphs and eutriconodont mammals remained as potential rivals for juvenile and small-bodied adult theropods in the Jurassic and Cretaceous.) In many of the averostran groups (basal ceratosaurs, ceratosaurids, abelisaurids, basal megalosauroids, spinosaurids, carnosaurs, tyrannosauroids, ornithomimosaurs, therizinosaurs, oviraptorosaurs), there are at examples of gigantism well beyond this 6–7-m size. Giants greater than 5 tons are known to have arisen independently in Spinosauridae, Carcharodontosauridae, and Tyrannosauridae.

Miniaturization also occurs, however. Early coelurosaurs (at 2–3 m or so) were smaller than their 6–7-m length outgroups (basal carnosaurs, basal spinosauroids, and basal ceratosaurs). Some coelophysoids, noasaurids, compsognathids, oviraptorosaurs, alvarezsauroids, and eumaniraptorans all produced forms with adult body size lengths of less than the 2–3-m ancestral length. Eumaniraptorans in particular show a great decrease in body size (Turner et al. 2007).

Of course, the most dramatic miniaturization within the theropods – indeed, within all of Dinosauria – occurs within Avialae. The average avialian is considerably smaller than the average member of any other clade of dinosaur. For more on this issue, consult Naish (Chapter 20, this volume).

References

Allain, R. 2005. The postcranial anatomy of the megalosaur *Dubreuillosaurus valesdunensis* (Dinosauria: Theropoda) from the Middle Jurassic of Normandy, France. *Journal of Vertebrate Paleontology* 25: 850–858.

Allain, R., R. Tykoski, N. Aquesbi, N.-E. Jalil, M. Monbaron, D. Russell, and P. Taquet. 2007. An abelisauroid (Dinosauria: Theropoda) from the Early Jurassic of the High Atlas Mountains, Morocco, and the radiation of ceratosaurs. *Journal of Vertebrate Paleontology* 27: 610–624.

Averianov, A. O., S. A. Krasnolutskii, and S. V. Ivantsov. 2010. A new basal coelurosaur (Dinosauria: Theropoda) from the Middle Jurassic of Siberia. *Proceedings of the Zoological Institute* 314: 42–57.

Barrett, P. M. 2009. The affinities of the enigmatic dinosaur *Eshanosaurus deguchiianus* from the Early Jurassic of Yunnan Province, People's Republic of China. *Palaeontology* 52: 681–688.

Benson, R. B. J. 2009. A description of *Megalosaurus bucklandii* (Dinosauria: Theropoda) from the Bathonian of the UK and the relationships of Middle Jurassic theropods. *Zoological Journal of the Linnean Society* 158: 882–935.

Benson, R. B. J., M. T. Carrano, and S. L. Brusatte. 2010. A new clade of archaic large-bodied predatory dinosaurs (Theropoda: Allosauroidea) that survived to the latest Maastrichtian. *Naturwissenschaften* 97: 71–78.

Benson, R. B. J., and X. Xu. 2008. The anatomy and systematic position of the theropod dinosaur *Chilantaisaurus tashuikouensis* Hu, 1964 from the Early Cretaceous of Alashan, People's Republic of China. *Geological Magazine* 145: 778–789.

Bonaparte, J. F., G. Brea, C. L. Schultz, and A. G. Martinelli. 2007. A new specimen of *Guaibasaurus candelariensis* (basal Saurischia) from the Late Triassic Caturrita Formation of southern Brazil. *Historical Biology* 19: 73–82.

Brusatte, S. L., R. B. J. Benson, and S. Hutt. 2008. The osteology of *Neovenator salerii* (Dinosauria: Theropoda) from the Wealden Group (Barremian) of the Isle of Wight. Palaeontographical Society Monograph 631: 1–75.

Brusatte, S. L., R. B. J. Benson, D. J. Chure, X. Xu, C. Sullivan, and D. E. W. Hone. 2009a. The first definitive carcharodontosaurid (Dinosauria: Theropoda) from Asia and the delayed ascent of tyrannosaurids. *Naturwissenschaften* 96: 1051–1058.

Brusatte, S. L., T. D. Carr, G. M. Erickson, G. S. Bever, and M. A. Norell. 2009b. A long-snouted, multihorned tyrannosaurid from the Late Cretaceous of Mongolia. *Proceedings of the National Academy of Sciences of the United States of America* 106: 17261–17266.

Brusatte, S. L., D. J. Chure, R. B. J. Benson, and X. Xu. 2010. The osteology of *Shaochilong maortuensis*, a carcharodontosaurid (Dinosauria: Theropoda) from the Late Cretaceous of Asia. *Zootaxa* 2334: 1–46.

Brusatte, S. L., and P. C. Sereno. 2008. Phylogeny of Allosauroidea (Dinosauria: Theropoda): comparative analysis and resolution. *Journal of Systematic Palaeontology* 6: 155–182.

Bundle, M. W., and K. P. Dial. 2003. Mechanisms of wing-assisted incline running (WAIR). *Journal of Experimental Biology* 206: 4553–4564.

Busbey, A. B., III. 1995. The structural consequences of skull flattening in crocodilians. In J. Thomason (ed.), *Functional Morphology in Vertebrate Paleontology*, 173–192. Cambridge: Cambridge University Press.

Carpenter, K. 2000. Evidence of predatory behavior by carnivorous dinosaurs. *Gaia* 15: 135–144.

Carpenter, K., C. Miles, and K. Cloward. 2005a. New small theropod from the Upper Jurassic Morrison Formation of Wyoming. In K. Carpenter (ed.), *The Carnivorous Dinosaurs*, 23–48. Bloomington: Indiana University Press.

Carpenter, K., C. Miles, J. H. Ostrom, and K. Cloward. 2005b. Redescription of the small maniraptoran theropods *Ornitholestes* and *Coelurus* from the Upper Jurassic Morrison Formation of Wyoming. In K. Carpenter (ed.), *The Carnivorous Dinosaurs*, 49–71. Bloomington: Indiana University Press.

Carr, T. D., and T. E. Williamson. 2010. *Bistahieversor sealeyi*, gen. et sp. nov., a new tyrannosauroid from New Mexico and the origin of deep snouts in Tyrannosauroidea. *Journal of Vertebrate Paleontology* 30: 1–16.

Carr, T. D., T. E. Williamson, and D. B. Schwimmer. 2005. A new genus and species of tyrannosauroid from the Late Cretaceous (Middle Campanian) Demopolis Formation of Alabama. *Journal of Vertebrate Paleontology* 25: 119–143.

Carrano, M. T., and S. D. Sampson. 2004. A review of coelophysoids (Dinosauria: Theropoda) from the Early Jurassic of Europe, with comments on the late history of the Coelophysoidea. *Neues Jahrbuch für Geologie und Paläontologie Monatshefte* 2004: 537–558.

——. 2007. The phylogeny of Ceratosauria (Dinosauria: Theropoda). *Journal of Systematic Palaeontology* 6: 183–236.

Carrano, M. T., S. D. Sampson, and C. A. Forster. 2002. The osteology of *Masiakasaurus knopfleri*, a small abelisauroid (Dinosauria: Theropoda) from the Late Cretaceous of Madagascar. *Journal of Vertebrate Paleontology* 22: 510–534.

Charig, A. J., and A. C. Milner. 1990. The systematic position of *Baryonyx walkeri*,

in the light of Gauthier's reclassification of the Theropoda. In K. Carpenter and P. J. Currie (eds.), *Dinosaur Systematics: Approaches and Perspectives*, 127–140. Cambridge: Cambridge University Press.

Chiappe, L. M., M. A. Norell, and J. M. Clark. 1996. Phylogenetic position of *Mononykus* (Aves: Alvarezsauridae) from the Late Cretaceous of the Gobi Desert. *Memoirs of the Queensland Museum* 39: 557–582.

Choiniere, J. N., X. Xu, J. M. Clark, C. A. Forster, Y. Guo, and F. Han. 2010. A basal alvarezsauroid theropod from the early Late Jurassic of Xinjiang, China. *Science* 327: 571–574.

Clark, J. M., A. Perle, and M. A. Norell. 1994. The skull of *Erlicosaurus andrewsi*, a Late Cretaceous "segnosaur" (Theropoda: Therizinosauroidea) from Mongolia. *American Museum Novitates* 3115: 1–39.

Currie, P. J. 2000. Possible evidence of gregarious behavior in tyrannosaurids. *Gaia* 15: 271–277.

Currie, P. J., and P.-J. Chen. 2001. Anatomy of *Sinosauropteryx prima* from Liaoning, northeastern China. *Canadian Journal of Earth Sciences* 38: 1705–1727.

Currie, P. J., J. H. Hurum, and K. Sabath. 2003. Skull structure and evolution in tyrannosaurid dinosaurs. *Acta Palaeontologica Polonica* 48: 227–234.

Dal Sasso, C., S. Maganuco, E. Buffetaut, and M. A. Mendez. 2005. New information on the skull of the enigmatic theropod *Spinosaurus*, with remarks on its size and affinities. *Journal of Vertebrate Paleontology* 25: 888–896.

Dial, K. P. 2003. Wing-assisted incline running and the evolution of flight. *Science* 299: 402–404.

Dial, K. P., B. E. Jackson, and P. Segre. 2008. A fundamental avian wing-stroke provides a new perspective on the evolution of flight. *Nature* 451: 985–989.

Ezcurra, M. 2010. A new early dinosaur (Saurischia: Sauropodomorpha) from the Late Triassic of Argentina: a reassessment of dinosaur origin and phylogeny. *Journal of Systematic Palaeontology* 8: 371–425.

Ezcurra, M. D., and G. Cuny. 2007. The coelophysoid *Lophostropheus airelensis*, gen. nov.: a review of the systematic of "*Liliensternus*" *airelensis* from the Triassic–Jurassic outcrops of Normandy (France). *Journal of Veretbrate Paleontology* 27: 73–86.

Ezquerra, R., S. Doublet, L. Costeur, P. M. Galton, and F. Pérez-Lorente. 2007. Were non-avian theropod dinosaurs able to swim? Supportive evidence from an Early

Cretaceous trackway, Cameros Basin (La Rioja, Spain). *Geology* 35: 507–510.

Farlow, J. O. 1994. Speculations about the carrion-locating ability of tyrannosaurs. *Historical Biology* 7: 159–165.

Farlow, J. O., D. L. Brinkman, W. L. Abler, and P. J. Currie. 1991. Size, shape, and serration density of theropod dinosaur lateral teeth. *Modern Geology* 16: 161–198.

Farlow, J. O., S. M. Gatesy, T. R. Holtz Jr., J. R. Hutchinson, and J. M. Robinson. 2000. Theropod locomotion. *American Zoologist* 40: 640–663.

Farlow, J. O., and T. R. Holtz Jr. 2002. The fossil record of predation in dinosaurs. *Paleontological Society Papers* 8: 251–265.

Farlow, J. O., M. B. Smith, and J. M. Robinson. 1995. Body mass, bone "strength indicator," and cursorial potential in *Tyrannosaurus rex. Journal of Vertebrate Paleontology* 15: 713–725.

Fiorillo, A. R. 2008. On the occurrence of exceptionally large teeth of *Troodon* (Dinosauria: Saurischia) from the Late Cretaceous of Alaska. *Palaios* 23: 322–328.

Gauthier, J. A. 1986. Saurischian monophyly and the origin of birds. In K. Padian (ed.), *The Origin of Birds and the Evolution of Flight. Memoirs of the California Academy of Science* 8: 1–55.

Gleich, O., R. J. Dooling, and G. A. Manley. 2005. Audiogram, body mass, and basilar papilla length: correlations in birds and predictions for extinct archosaurs. *Naturwissenschaften* 92: 595–598.

Göhlich, U. B., and L. M. Chiappe. 2006. A new carnivorous dinosaur from the Late Jurassic Solnhofen archipelago. *Nature* 440: 329–332.

Gradstein, F., J. Ogg, and A. Smith (eds.). 2004. *A Geologic Time Scale, 2004.* Cambridge: Cambridge University Press.

Hartman, S., D. Lovelace, and W. Wahl. 2005. Phylogenetic assessment of a maniraptoran from the Morrison Formation. *Journal of Vertebrate Paleontology* 25 (suppl. to 3) 67A–68A.

Heffner, R. S. 2004. Primate hearing from a mammalian perspective. *Anatomical Record A* 281A: 1111–1122.

Henderson, D. M., and E. Snively. 2004. *Tyrannosaurus* en pointe: allometry minimized rotational inertia of large carnivorous dinosaurs. *Proceedings of the Royal Society B: Biological Letters* 271 (suppl. 3): S57–S60.

Hocknull, S. A., M. A. White, T. R. Tischler, A. G. Cook, N. D. Calleja, T. Sloan, and D. A. Elliot. 2009. New Mid-Cretaceous (latest Albian) dinosaurs from Winton, Queensland, Australia. *PLoS ONE* 4 (7): e6190.

Holtz, T. R., Jr. 1994. The phylogenetic position of Tyrannosauridae: implication for theropod systematics. *Journal of Paleontology* 68: 1100–1117.

———. 1995. The arctometatarsalian pes, an unusual structure of the metatarsus of Cretaceous Theropoda (Dinosauria: Saurischia). *Journal of Vertebrate Paleontology* 14: 480–519.

———. 2000. A new phylogeny of the carnivorous dinosaurs. *Gaia* 15: 5–61.

———. 2001. The phylogeny and taxonomy of the Tyrannosauridae. In D. Tanke and K. Carpenter (eds.), *Mesozoic Vertebrate Life: New Research Inspired by the Paleontology of Philip J. Currie*, 64–83. Bloomington: Indiana University Press.

———. 2002. Theropod predation: evidence and ecomorphology. In P. H. Kelly, M. Koweleski, and T. A. Hansen (eds.), *Predator–Prey Interactions in the Fossil Record. Topics in Geobiology* 17: 325–340.

Holtz, T. R., Jr., D. L. Brinkman, and C. L. Chandler. 2000. Denticle morphometrics and a possibly omnivorous feeding habit for the theropod dinosaur *Troodon. Gaia* 15: 159–166.

Holtz, T. R., Jr., R. E. Molnar, and P. J. Currie. 2004. Basal Tetanurae. In D. B. Weishampel, P. Dodson, and H. Osmólska (eds.), *The Dinosauria*, 2nd ed., 71–110. Berkeley: University of California Press.

Holtz, T. R., Jr., and H. Osmólska. 2004. Saurischia. In D. B. Weishampel, P. Dodson, and H. Osmólska (eds.), *The Dinosauria*, 2nd ed., 21–24. Berkeley: University of California Press.

Hone, D. W. E., and O. W. M. Rauhut. 2010. Feeding behaviour and bone utilization by theropod dinosaurs. *Lethaia* 43: 232–244.

Hone, D. W. E., X. Xu, and D.-Y. Wang. 2010. A probable baryonychine (Theropoda: Spinosauridae) tooth from the Upper Cretaceous of Henan Province, China. *Vertebrata PalAsiatica* 48: 19–26.

Hu, D., L. Hou, L. Zhang, and X. Xu. 2009. A pre-*Archaeopteryx* troodontid theropod from China with long feathers on the metatarsus. *Nature* 461: 640–643.

Hurum, J. H., Z.-X. Luo, and Z. Kielan-Jaworowska. 2006. Were mammals originally venomous? *Acta Palaeontologica Polonica* 51: 1–11.

Hutchinson, J. R. 2002. The evolution of hindlimb tendons and muscles on the line to crown-group birds. *Comparative Biochemistry and Physiology Part A: Molecular and Integrative Physiology* 133: 1051–1086.

———. 2004. Biomechanical modeling and sensitivity analysis of bipedal running. II. Extinct taxa. *Journal of Morphology* 262: 441–461.

Hutchinson, J. R., and M. Garcia. 2002. *Tyrannosaurus* was not a fast runner. *Nature* 415: 1018–1021.

Hutchinson, J. R., and S. M. Gatesy. 2000. Adductors, abductors, and the evolution of archosaur locomotion. *Paleobiology* 26: 734–751.

Hutt, S., D. Naish, D. M. Martill, M. J. Barker, and P. Newbery. 2001. A preliminary account of a new tyrannosauroid theropod from the Wessex Formation (Early Cretaceous) of southern England. *Cretaceous Research* 22: 227–242.

Hwang, S. H., M. A. Norell, Q. Ji, and K. Gao. 2002. New specimens of *Microraptor zhaoianus* (Theropoda: Dromaeosauridae) from Northeastern China. *American Museum Novitates* 3381: 1–44.

———. 2004. A large compsognathids from the Early Cretaceous Yixian Formation of China. *Journal of Systematic Palaeontology* 2: 13–30.

Imansyah, M. J., T. S. Jessop, C. Ciofi, and Z. Akbar. 2008. Ontogenetic differences in the spatial ecology of immature Komodo dragons. *Journal of Zoology* 274: 107–115.

Jensen, J. A., and K. Padian. 1988. Small pterosaurs and dinosaurs from the Uncompahgre fauna, Brushy Basin Member, Morrison Formation: ?Tithonian, Late Jurassic, western Colorado. *Journal of Paleontology* 63: 364–373.

Ji, Q., and S. A. Ji. 1996. On the discovery of the earliest bird fossil in China and the origin of birds [in Chinese]. *Chinese Geology* 233: 30–33.

Ji, Q., S. A. Ji, J. Lü, H. You, W. Chen, Y. Liu, and Y. Liu. 2005. First avialian bird from China. *Geological Bulletin of China* 24: 197–210.

Kirkland, J. I., L. E. Zanno, S. D. Sampson, J. M. Clark, and D. D. Deblieux. 2005. A primitive therizinosauroid dinosaur from the Early Cretaceous of Utah. *Nature* 435: 84–85.

Kobayashi, Y., and J.-C. Lü. 2003. A new ornithomimid dinosaur with gregarious habits from the Late Cretaceous of China. *Acta Palaeontologica Polonica* 48: 235–259.

Langer, M. C. 2004. Basal Saurischia. In D. B. Weishampel, P. Dodson, and H. Osmólska (eds.), *The Dinosauria*, 2nd ed., 25–46. Berkeley: University of California Press.

Langer, M. C., and M. J. Benton. 2006. Early dinosaurs: a phylogenetic study.

Journal of Systematic Palaeontology 4: 309–358.

Larsson, H. C. E., P. C. Sereno, and J. A. Wilson. 2000. Forebrain enlargement among nonavian theropod dinosaurs. *Journal of Vertebrate Paleontology* 20: 615–618.

Li, D., M. A. Norell, K.-Q. Gao, N. D. Smith, and P. J. Makovicky. 2009. A longirostrine tyrannosauroid from the Early Cretaceous of China. *Proceedings of the Royal Society B: Biological Sciences* 277: 183–190.

Li, D., C. Peng, H. You, M. C. Lamanna, J. D. Harris, K. J. Lacovara, and Zhang J. 2007. A large therizinosauroid (Dinosauria: Theropoda) from the Early Cretaceous of Northwestern China. *Acta Geologica Sinica* 81: 539–549.

Longrich, N. R., and P. J. Currie. 2009a. *Albertonykus borealis*, a new alvarezsaur (Dinosauria: Theropoda) from the Early Maastrichtian of Alberta, Canada: implications for the systematics and ecology of the Alvarezsauridae. *Cretaceous Research* 30: 239–252.

———. 2009b. A microraptorine (Dinosauria–Dromaeosauridae) from the Late Cretaceous of North America. *Proceedings of the National Academy of Sciences of the United States of America* 106: 5002–5007.

Makovicky, P. J., S. Apesteguía, and F. L. Agnolín. 2005. The earliest dromaeosaurid theropod from South America. *Nature* 437: 1007–1011.

Manning, P. L., D. Payne, J. Pennicott, P. M. Barrett, and R. A. Ennos. 2006. Dinosaur killer claws or climbing crampons? *Biology Letters* 2: 110–112.

Marsicano, C. A., N. S. Domnanovich, and A. C. Mancuso. 2007. Dinosaur origins: evidence from the footprint record. *Historical Biology* 19: 83–91.

Martinelli, A. G., and E. I. Vera. 2007. *Achillesaurus manazzonei*, a new alvarezsaurid theropod (Dinosauria) from the Late Cretaceous Bajo de la Carpa Formation, Río Negro Province, Argentina. *Zootaxa* 1582: 1–17.

Mateus, O., A. Walen, and M. T. Antunes. 2006. The large theropod fauna of the Lourinhã Formation (Portugal) and its similarity to the Morrison Formation, with a description of a new species of *Allosaurus*. *New Mexico Museum of Natural History and Science Bulletin* 36: 123–129.

Maxwell, W. D., and J. H. Ostrom. 1995. Taphonomy and paleobiological implications of *Tenontosaurus–Deinonychus*

associations. *Journal of Vertebrate Paleontology* 15: 707–712.

Mayr, G., B. Pohl, S. Hartman, and D. S. Peters. 2007. The tenth skeletal specimen of *Archaeopteryx*. *Zoological Journal of the Linnean Society* 149: 97–116.

Mayr, G., B. Pohl, and D. S. Peters. 2005. A well-preserved *Archaeopteryx* specimen with theropod features. *Science* 310: 1483–1486.

Naish, D. 2011 (this volume). Birds. In M. K. Brett-Surman, Thomas R. Holtz Jr., and James O. Farlow (eds.), *The Complete Dinosaur*, 2nd ed., 379–423. Bloomington: Indiana University Press.

Naish, D., and D. M. Martill. 2007. Dinosaurs of Great Britain and the role of the Geological Society of London in their discovery: basal Dinosauria and Saurischia. *Journal of the Geological Society* 164: 493–510.

Naish, D., D. M. Martill, and E. Frey. 2004. Ecology, systematics, and biogeographical relationships of dinosaurs, including a new theropod, from the Santana Formation (?Albian, Early Cretaceous) of Brazil. *Historical Biology* 16: 57–70.

Nesbitt, S. J. 2007. The anatomy of *Effigia okeeffeae* (Archosauria, Suchia), theropod-like convergence, and the distribution of related taxa. *Bulletin of the American Museum of Natural History* 302: 1–84.

Nesbitt, S. J., R. B. Irmis, and W. G. Parker. 2007. A critical re-evaluation of the Late Triassic dinosaur taxa of North America. *Journal of Systematic Palaeontology* 5: 209–243.

Nesbitt, S. J., N. D. Smith, R. B. Irmis, A. H. Turner, A. Downs, and M. A. Norell. 2009. A complete skeleton of a Late Triassic saurischian and the early evolution of dinosaurs. *Science* 326: 1530–1533.

Norell, M. A., P. J. Makovicky, G. S. Bever, A. M. Balanoff, J. M. Clark, R. Barsbold, and T. Rowe. 2009. A review of the Mongolian Cretaceous dinosaur *Sinornithoides* (Troodontidae: Theropoda). *American Museum Novitates* 3654: 1–63.

Novas, F. E. 1998. *Megaraptor namunhuaiquii*, gen. et sp. nov., a large-clawed Late Cretaceous theropod from Patagonia. *Journal of Vertebrate Paleontology* 18: 4–9.

Novas, F. E., M. D. Ezcurra, and A. Lecuona. 2008a. *Orkoraptor burkei* nov. gen et sp., a large theropod from the Maastrichtian Pari Aike Formation, Southern Patagonia, Argentina. *Cretaceous Research* 29: 468–480.

Novas, F. E., D. Pol, J. I. Canale, J. D. Porfiri, and J. O. Calvo. 2008b. A bizarre Cretaceous theropod dinosaur from Patagonia and the evolution of Gondwanan dromaeosaurids. *Proceedings of the Royal Society B: Biological Sciences* 276: 1101–1107.

O'Connor, P. M., and L. P. A. M. Claessens. 2005. Basic avian pulmonary design and flow-through ventilation in nonavian theropod dinosaurs. *Nature* 436: 253–256.

Ostrom, J. H. 1969. Osteology of *Deinonychus antirrhopus*, an unusal theropod from the Lower Cretaceous of Montana. *Bulletin of the Peabody Museum of Natural History* 30: 1–165.

Padian, K., J. R. Hutchinson, and T. R. Holtz Jr. 1999. Phylogenetic definitions and nomenclature of the major taxonomic categories of the carnivorous dinosaurs (Theropoda). *Journal of Vertebrate Paleontology* 19: 69–80.

Parrish, J. M. 2011 (this volume). Evolution of the archosaurs. In M. K. Brett-Surman, Thomas R. Holtz Jr., and James O. Farlow (eds.), *The Complete Dinosaur*, 2nd ed., 316–329. Bloomington: Indiana University Press.

Paul, G. S. 1984. The segnosaurian dinosaurs: relics of the prosauropod–ornithischian transition? *Journal of Vertebrate Paleontology* 4: 507–515.

———. 1988. *Predatory Dinosaurs of the World*. New York: Simon and Schuster.

———. 2002. *Dinosaurs of the Air: The Evolution and Loss of Flight in Dinosaurs and Birds*. Baltimore, Md.: Johns Hopkins University Press.

Prum, R. O., and A. H. Brush. 2002. The evolutionary origin and diversification of feathers. *Quarterly Review of Biology* 77: 261–295.

Rauhut, O. W. M. 2003. The interrelationships and evolution of basal theropod dinosaurs. *Special Papers in Palaeontology* 69: 1–213.

———. 2004. Provenance and anatomy of *Genyodectes serus*, a large-toothed ceratosaur (Dinosauria: Theropoda) from Patagonia. *Journal of Vertebrate Paleontology* 24: 894–902.

Rauhut, O. W. M., A. C. Milner, and S. Moore-Fay. 2010. Cranial osteology and phylogenetic position of the theropod dinosaur *Proceratosaurus* from Middle Jurassic of England. *Zoological Journal of the Linnean Society* 158: 155–195.

Rayfield, E. J. 2005. Aspects of comparative cranial mechanics in the theropod dinosaurs *Coelophysis*, *Allosaurus*, and

Tyrannosaurus. Zoological Journal of the Linnean Society 144: 309–316.

Rayfield, E. J., A. C. Milner, V. B. Xuan, and P. G. Young. 2007. Functional morphology of spinosaur "crocodile-mimic" dinosaurs. *Journal of Vertebrate Paleontology* 27: 892–901.

Roach, B. T., and D. L. Brinkman. 2007. A reevaluation of cooperative pack hunting and gregariousness in *Deinonychus antirrhopus* and other nonavian theropod dinosaurs. *Bulletin of the Peabody Museum of Natural History* 48: 103–138.

Sampson, S. D., and D. W. Krause (eds.). 2007. *Majungasaurus crenatissimus* (Theropoda: Abelisauridae) from the Late Cretaceous of Madagascar. *Memoir of the Society of Vertebrate Paleontology* 8: 1–184.

Senter, P. 2005. Function in the stunted forelimbs of *Mononykus olecranus* (Theropoda), a dinosaurian anteater. *Paleobiology* 31: 373–381.

——. 2006. Scapular orientation in theropods and basal birds, and the origins of flapping flight. *Acta Palaeontologica Polonica* 51: 305–313.

——. 2007. A new look at the phylogeny of Coelurosauria (Dinosauria: Theropoda). *Journal of Systematic Palaeontology* 5: 439–463.

Sereno, P. C. 1999. The evolution of dinosaurs. *Science* 284: 2137–2147.

——. 2007. The phylogenetic relationships of early dinosaurs: a comparative report. *Historical Biology* 19: 145–155.

Sereno, P. C., R. N. Martinez, J. A. Wilson, D. J. Varricchio, O. A. Alcober, and H. C. E. Larsson. 2008. Evidence for avian intrathoracic air sacs in a new predator dinosaur from Argentina. *PLoS ONE* 3 (9): e3303.

Sereno, P. C., P. Tan, S. L. Brusatte, H. J. Kreigstein, X. Zhao, and K. Cloward. 2009. Tyrannosaurid skeletal design first evolved at small body size. *Science* 326: 418–422.

Smith, N. D., P. J. Makovicky, F. L. Agnolin, M. D. Ezcurra, D. F. Pais, and S. W. Salisbury. 2008. A *Megaraptor*-like theropod (Dinosauria: Tetanurae) in Australia: support for faunal exchange across eastern and western Gondwana in the Mid-Cretaceous. *Proceedings of the Royal Society B: Biological Sciences* 275: 2085–2093.

Smith, N. D., P. J. Makovicky, W. R. Hammer, and P. J. Currie. 2007. Osteology of *Cryolophosaurus ellioti* (Dinosauria: Theropoda) from the Early Jurassic of Antarctica and implications for early

theropod evolution. *Zoological Journal of thevLinnean Society* 151: 377–421.

Snively, E., D. M. Henderson, and D. S. Phillips. 2006. Fused and vaulted nasals of tyrannosaurid dinosaurs: implications for cranial strength and feeding mechanics. *Acta Palaeontologica Polonica* 51: 435–454.

Snively, E., and A. P. Russell. 2007a. Functional variation of neck muscles and their relation to feeding style in Tyrannosauridae and other large theropod dinosaurs. *Anatomical Record* 290: 934–957.

——. 2007b. Craniocervical feeding dynamics of *Tyrannosaurus rex*. *Paleobiology* 33: 610–638.

Snively, E., A. P. Russell, and G. L. Powell. 2004. Evolutionary morphology of the coelurosaurian arctometatarsus: descriptive, morphometric and phylogenetic approaches. *Zoological Journal of the Linnean Society* 142: 525–553.

Soto, M., and D. Perea. 2008. A ceratosaurid (Dinosauria, Theropoda) from the Late Jurassic–Early Cretaceous of Uruguay. *Journal of Vertebrate Paleontology* 28: 439–444.

Stevens, K. A. 2006. Binocular vision in theropod dinosaurs. *Journal of Vertebrate Paleontology* 26: 321–330.

Therrien, F., and D. M. Henderson. 2007. My theropod is bigger than yours . . . or not: estimating body size from skull length in theropods. *Journal of Vertebrate Paleontology* 27: 108–115.

Therrien, F., D. M. Henderson, and C. B. Ruff. 2005. Bite me: biomechanical models of theropod mandibles and implications for feeding behavior. In K. Carpenter (ed.), *The Carnivorous Dinosaurs*, 179–237. Bloomington: Indiana University Press.

Tobalske, B. W., and K. P. Dial. 2007. Aerodynamics of wing-assisted incline running in birds. *Journal of Experimental Biology* 210: 1742–1751.

Turner, A. H., S. J. Nesbitt, and M. A. Norell. 2009. A large alvarezsaurid from the Late Cretaceous of Mongolia. *American Museum Novitates* 3648: 1–14.

Turner, A., D. Pol, J. Clarke, G. Erickson, and M. A. Norell. 2007. A basal dromaeosaurid and size evolution preceding avian flight. *Science* 317: 1378–1381.

Varricchio, D. J., and F. Jackson. 2004. Two eggs sunny-side up: reproductive physiology in the dinosaur *Troodon formosus*. In P. J. Currie, E. B. Koppelhus, M. A. Shugar, and J. L. Wright (eds.), *Feathered Dragons: Studies on the Transition from*

Dinosaurs to Birds, 215–233. Bloomington: Indiana University Press.

Varricchio, D. J., P. C. Sereno, X. Zhao, L. Tan, J. A. Wilson, and G. H. Lyon. 2008. Mud-trapped herd captures evidence of distinctive dinosaur sociality. *Acta Palaeontologica Polonica* 53: 567–578.

Wedel, M. J. 2006. Origin of postcranial skeletal pneumaticity in dinosaurs. *Integrative Zoology* 2: 80–85.

White, M. A. 2009. The subarctometatarsus: intermediate metatarsus architecture demonstrating the evolution of the arctometatarsus and advanced agility in theropod dinosaurs. *Alcheringa* 33: 1–21.

Witmer, L. M. 1997. The evolution of the Antorbital cavity of archosaurs: a study in soft-tissue reconstruction in the fossil record with an analysis of the function of pneumaticity. *Memoir of the Society of Vertebrate Paleontology* 3: 1–73.

Xu, X., Y.-N. Cheng, X.-L. Wang, and C.-H. Chang. 2002. An unusual oviraptorosaurian dinosaur from China. *Nature* 419: 291–293.

Xu, X., J. M. Clark, J. Mo, J. Choiniere, C. A. Forster, G. M. Erickson, D. W. E. Hone, C. Sullivan, D. A. Ebert, S. Nesbitt, Q. Zhao, R. Hernandez, C. Jia, F. Han, and Y. Guo. 2009a. A Jurassic ceratosaur from China helps clarify avian digital homologies. *Nature* 459: 940–944.

Xu, X., J. M. Clark, C. A. Forster, M. A. Norell, G. M. Erickson, D. A. Eberth, C. Jia, and Q. Zhao. 2006. A basal tyrannosauroid dinosaur from the Late Jurassic of China. *Nature* 439: 715–718.

Xu, X., M. A. Norell, X. Kuang, X. Wang, Q. Zhao, and C. Jia. 2004. Basal tyrannosauroids from China and evidence for protofeathers in tyrannosauroids. *Nature* 431: 680–684.

Xu, X., Q. Tan, J. Wang, X. Zhao, and L. Tan. 2007. A gigantic bird-like dinosaur from the Late Cretaceous of China. *Nature* 447: 844–847.

Xu, X., D.-Y. Wang, C. Sullivan, D. W. E. Hone, F. L. Han, R. H. Yan, and F. M. Du. 2010. A basal parvicursorine (Theropoda: Alvarezsauridae) from the Upper Cretaceous of China. *Zootaxa* 2413: 1–19.

Xu, X., and F. Zhang. 2005. A new maniraptoran dinosaur from China with feathers on the metatarsus. *Naturwissenschaften* 92: 173–177.

Xu, X., X. Zhao, and J. M. Clark. 2001. A new therizinosaurs from the Lower Jurassic lower Lufeng Formation of Yunnan, China. *Journal of Vertebrate Paleontology* 21: 477–483.

Xu, X., X. Zheng, and H. You. 2009b. New

feather type in a nonavian theropod and the early evolution of feathers. *Proceedings of the National Academy of Sciences of the United States of America* 106: 832–834.

Yates, A. M., and C. C. Vasconcelos. 2005. Furcula-like clavicles in the prosauropod dinosaur *Massospondylus*. *Journal of Vertebrate Paleontology* 25: 466–468.

Zanno, L. E., D. D. Gillette, L. B. Albright, and A. L. Titus. 2009. A new North American therizinosauroid and the role of herbivory in "predatory" dinosaur evolution. *Proceedings of the Royal Society B: Biological Sciences* 276: 3505–3511.

Zelenitsky, D. K., F. Therrien, and Y. Kobayashi. 2009. Olfactory acuity in theropods: palaeobiological and evolutionary implications. *Proceedings of the Royal Society B: Biological Sciences* 276: 667–673.

Zhang, F., Z. Zhou, X. Xu, and X. Wang. 2002. A juvenile coelurosaurian theropod from China indicates arboreal habits. *Naturwissenschaften* 89: 394–398.

Zhang, F., Z. Zhou, X. Xu, X. Wang, and C. Sullivan. 2008. A bizarre Jurassic maniraptoran from China with elongate ribbon-like feathers. *Nature* 455: 1105–1108.

Birds

Darren Naish

20

Birds are one of the most conspicuous, familiar, and speciose of vertebrate groups, containing about 10,000 extant species that occur worldwide in most habitats. If the group termed birds is understood to include *Archaeopteryx*, then their fossil record extends back to the Tithonian stage of the Late Jurassic, and their success means that the prominence of dinosaurs in the global ecosystem was not restricted to the Mesozoic. In fact, it can be argued that dinosaurs were more successful during the Cenozoic than they were before.

While the term birds is conventionally applied to the maniraptoran clade that includes *Archaeopteryx* and extant birds, authors disagree as to which technical name should be attached to this clade. The term Aves has typically been used, but many traits considered diagnostic for Aves by neontologists are restricted to subsets of the *Archaeopteryx* plus extant bird group, rendering incorrect many statements about Aves—for example, that members possess a pygostyle, or lack teeth. Gauthier (1986) therefore used Aves for the crown group only and coined Avialae for the branch-based clade that includes modern birds and all species closer to them than to deinonychosaurs. As a result of this decision, the term Aves is used differently by different authors, whereas Avialae is maximally inclusive (Fig. 20.1). The bird branch of Maniraptora is referred to here as Avialae, whereas the avialian crown is termed Neornithes. Confusing things further is that some phylogenetic studies find *Archaeopteryx* to be outside of Avialae: this could be interpreted to mean that *Archaeopteryx* is not a bird.

The skeletal anatomy of *Archaeopteryx* and of other Mesozoic birds and birdlike maniraptorans (particularly dromaeosaurids) has demonstrated to

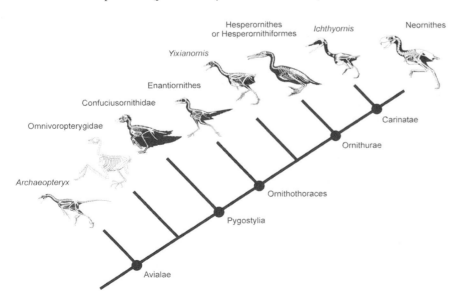

20.1. This simplified cladogram shows the relationships that have been inferred between most of the Mesozoic bird groups. *Archaeopteryx* is shown here as being outside of the clade that includes all other members of Avialae, but some studies find that the scansoriopterygids may be even more basal yet still within Avialae. A few other areas of controversy exist. Omnivoropterygids are shown here as being outside of Pygostylia, but a position closer to neornithines than are confuciusornithids has been suggested by some. The term Ornithurae is applied here to the clade that includes Hesperornithes, *Ichthyornis,* and Neornithes. As discussed in the text, some authors prefer to use the name Ornithurae for the clade that includes all birds closer to neornithines than to *Archaeopteryx*. Key trends within Avialae include the reduction of the tail, the fusion of the bones of the hand, and the enlargement of the arms. The oldest members of the clade—like *Archaeopteryx*—were extremely similar to the deinonychosaurs that were their closest relatives outside of Avialae.

the satisfaction of most paleontologists and zoologists that birds are theropod dinosaurs. Indeed, the fossil record shows that nonavialian theropods and birds grade into one another such that deinonychosaurs and other nonavialian maniraptorans are strikingly similar to basal birds, and vice versa. When birds first evolved, they were merely one among several lineages of small, feathered maniraptoran. However, alternative views have been proposed over the years (Witmer 1991).

It was well known by the 1800s that birds were essentially glorified reptiles or feathered saurians. However, a poor fossil record meant that possible avian relatives were unknown, and authors disagreed as to whether bird ancestors were arboreal or terrestrial. The first skeletal specimen of *Archaeopteryx* (the London specimen), discovered in 1861, was not immediately used as evidence for bird origins. Huxley (1870) drew attention to the strong similarity present between dinosaurs and birds and was most impressed with the birdlike features of the recently discovered *Hypsilophodon* and *Compsognathus*. His views were actually vague: he regarded dinosaurs as closer to the avian condition than other reptiles, but he did not state that birds had evolved from *among* the dinosaurs. In fact, he seems to have thought that birds and dinosaurs shared the same Paleozoic ancestor. After Huxley, Marsh (1877) proposed the evolution of birds from among dinosaurs, but he later favored the notion of a common ancestry for the two.

During the late 1800s and the first few decades of the twentieth century, a dinosaurian origin of birds was considered favorably by many workers, though possible links with pterosaurs and hypothetical early archosaurs were preferred by some. Between the 1900s and 1920s, Franz Nopcsa (1907, 1923) suggested the derivation of birds from terrestrial, cursorial, dinosaurlike reptiles (though not dinosaurs proper), while Othenio Abel (1911) argued that birds and dinosaurs shared an arboreal ancestor. Robert Broom's 1913 description of *Euparkeria* was influential, as here was an animal that seemed generalized enough to be a common ancestor to birds, dinosaurs, and other archosaurs (Broom 1913).

It is often implied that a dinosaurian origin for birds was universally favored after Huxley and before Heilmann. In fact, nebulous views on ancestry and a poor understanding of theropod anatomy meant that birds and dinosaurs were more typically imagined to have shared the same ancestral stock.

Gerhard Heilmann's *The Origin of Birds*

One of the most important twentieth-century contributions on bird origins was Gerhard Heilmann's *The Origin of Birds*, published in English in 1926. Heilmann was a Danish artist and graphic designer who became interested in bird origins in 1912 and published a series of articles on the subject between 1913 and 1916. His work was mostly ignored and even ridiculed by Danish zoologists (Ries 2007), but it was influential elsewhere. Heilmann's success may, in part, have been because non-Danish scientists assumed that he was a professional scientist.

Heilmann concluded that theropods were indisputably birdlike. However, in his view, they could not have given rise to birds because they lacked clavicles (Heilmann accepted Dollo's law—the idea that a lost biological structure could not be evolved again). He was unaware that Osborn had described clavicles in *Oviraptor* just a few years earlier (though Osborn had misidentified them as interclavicles). With theropods barred from avian ancestry, Heilmann settled on the so-called pseudosuchian thecodonts as the most likely bird ancestors. "Thecodontia" was an artificial assemblage of archosauriforms and archosaurs (today recognized as a grade and not a clade) that included *Euparkeria, Aetosaurus*, and so on, and pseudosuchians were regarded as a subset of this group. What has been called the thecodont hypothesis or pseudosuchian hypothesis then became textbook dogma for the next several decades, and any birdlike features seen in theropods were dismissed as the result of convergent evolution. Heilmann (1926) was in fact so convincing that research on bird origins all but ground to a halt.

However, a few opinions on avian origins and affinities were expressed in the following decades. During the 1930s, J. E. V. Boas argued that birds evolved from theropods closely related to *Ornitholestes* and *Compsognathus* (Boas 1930), and Nils Holmgren stated likewise during the 1950s (Holmgren 1955). It has been argued that Percy Lowe, whose papers on bird evolution were published during the 1920s, 1930s, and 1940s, supported a dinosaurian origin for birds. Lowe actually regarded dinosaurs and birds as among the descendants of a Triassic "eosuchian" stock, in which case any strong similarities between birds and dinosaurs represented parallelisms (Lowe 1935).

Many shared characters (behavioral, soft tissue, and skeletal) show that birds are more closely related to crocodilians than to other living reptiles. Close examination of the crocodilian inner ear and braincase led Alick Walker (1972) to propose that crocodilians were more closely related to birds than were dinosaurs. However, the characters concerned are widely distributed within archosaurs, and Walker later abandoned his hypothesis.

John Ostrom and the Theropod Hypothesis

A major event in our understanding of bird origins occurred when John Ostrom (1973) drew attention to the many characters shared by *Archaeopteryx* and such coelurosaurs as his recently described *Deinonychus*. Ostrom noted that *Archaeopteryx* resembled coelurosaurs in vertebral count, in the anatomy of its long, tridactyl forelimbs and mesotarsal, digitigrade hind limbs, and in scapulocoracoid and pelvic shape. Ostrom argued that "direct inheritance from a small coelurosaurian ancestor" (1973, 136) best explained these similarities. The hypothesis that birds were the direct descendants of theropods was now firmly on the agenda.

Ostrom interpreted *Archaeopteryx* as a bipedal runner, similar in predatory behavior to its nonavialian relatives but equipped with long forelimb feathers that were perhaps used as insect nets. Ostrom (1974) proposed that enlargement of the wing surface area, coupled with the development of an increasingly powerful flapping stroke used in swatting prey, allowed proto-birds to take flight in pursuit of prey.

Ostrom's proposal of a theropod origin for birds was bolstered during the 1980s when Jacques Gauthier (1986) examined the affinities of *Archaeopteryx* and other theropods and archosaurs within a cladistic analysis. Gauthier's analysis was hugely influential and formed the basis for virtually all later classifications of archosaurs. Gauthier refined Ostrom's hypothesis by showing that a specific group of coelurosaurs (deinonychosaurs) shared particularly long arms, semilunate carpal bones, and other characters with

birds, and he named Maniraptora for the clade that included deinonychosaurs and birds. Further support for Ostrom's hypothesis was provided by
Gauthier and Padian (1985). They argued that the anatomical components
and range of motion present in the avialian flight stroke were already present
in the predatory stroke of maniraptorans. The insect net hypothesis had by
now been abandoned.

Since Ostrom's work and Gauthier and Padian's early papers, further
discoveries have provided support for the theropod hypothesis. Feathers
have been reported in numerous maniraptorans outside of Avialae (including in the oviraptorosaurs *Caudipteryx* and *Protarchaeopteryx*, the dromaeosaurids *Sinornithosaurus* and *Microraptor*, the troodontid-like deinonychosaurs *Jinfengopteryx* and *Anchiornis*, and *Xiaotingia*), additional characters
shared by early birds and nonavialian maniraptorans have been identified,
and it has been realized that *Archaeopteryx* was more dromaeosaur-like
than even Ostrom had thought. Consequently, the dinosaurian heritage of
birds is extremely robust and widely accepted. Many scientists who study
living birds are largely unconcerned with the deep origins of their favorite
animals, and they remain noncommittal or ambivalent on the subject of
bird origins. However, others argue that the dinosaurian ancestry of birds is
of great interest and relevance to ornithologists because it means that many
questions about the origin of avian physiology, nesting biology, locomotor
behavior, and so on can finally be tested (Prum 2002; Kaiser 2007).

The "Birds Are Not Dinosaurs" Movement

The theropod hypothesis is not universally accepted. Some ornithologists
and paleontologists argue that theropods cannot be ancestral to birds because they do not conform, in anatomy or lifestyle, to the true bird ancestor
as imagined by these researchers. Supposedly, theropods are too large and
too specialized for terrestrial cursoriality to give rise to birds, possess anatomical characters that bar them from avialian ancestry, and appear too late
in the Mesozoic record to be ancestral to *Archaeopteryx* (e.g., Martin 1983;
Feduccia 1996, 2002). These authors argue that a number of peculiar Triassic reptiles—they include *Megalancosaurus*, *Cosesaurus*, and *Longisquama*
(Fig. 20.2) and have been dubbed avimorph thecodonts—might represent
the real closest relatives of birds. None of these taxa is at all birdlike and all
clearly belong elsewhere in reptile phylogeny.

The objections proposed by these workers have never been a problem
for the theropod hypothesis and are naive attempts to falsify a well-supported hypothesis. While the Jurassic record of small theropods is poor,
numerous fossils (among the best are those of the Middle and Late Jurassic
maniraptorans *Anchiornis* and *Xiaotingia*) show that deinonychosaurs and
other maniraptorans were present before the Tithonian. Furthermore, nonavialian maniraptorans were not all large (some deinonychosaurs and other
maniraptorans were similar in size to, or smaller than, *Archaeopteryx*), nor
do claims that they were fundamentally distinct from basal birds withstand
scrutiny.

It has been argued that the neornithine hand represents digits II–IV and
is therefore different from the maniraptoran hand, which is usually taken to
represent digits I–III. It seems peculiar to argue that a single character, or

even a complex of related characters, can trump a list of tens or hundreds of others. The strong character evidence nesting birds within coelurosaurs must mean either that the proposed II–IV formula for the avialian hand is wrong, or that an unusual embryological event—a so-called frameshift—occurred in theropod evolution. If such a frameshift did occur, true digit I was lost and true digit II became digit I. However, evidence from Hox genes indicates that the condensation axis for the first embryonic hand digit in birds receives a Hox signal normally associated with digit I (Vargas and Fallon 2005). This shows that the avialian hand likely does represent digits I–III.

Overall, the "birds are not dinosaurs" movement can be dismissed as naive because it hinges on the idea that we should make predictions about the avian ancestor before looking at fossils or phylogeny. Evolutionary hypotheses should be formulated on phylogenies, not vice versa.

Because vaned feathers are known for deinonychosaurs and oviraptorosaurs, Feduccia and colleagues have more recently argued that feathered maniraptorans are secondarily flightless members of Avialae, thereby renouncing decades of argumentation in which they stated that deinonychosaurs have no close relationship with birds (Feduccia 2002; Martin 2004; Feduccia et al. 2007). This idea of secondary flightlessness is not wholly objectionable, and some supporters of the theropod hypothesis have argued similarly (Paul 2002), but it is not supported by large analyses that incorporate good sampling of characters and taxa. Furthermore, what makes the hypothesis of Feduccia et al. untenable is their corollary that this feathered maniraptoran clade has no close affinity with the rest of Dinosauria. It is difficult to take this seriously, given that nonavialian maniraptorans have obvious affinities with nonmaniraptoran coelurosaurs, which in turn have affinities with noncoelurosaurian theropods, and so on.

A cladistic analysis that found no strong support for the theropod affinities of birds (James and Pourtless 2009) is extremely flawed. This study excluded all characters where homology has been questioned and excluded nontheropodan dinosaurs. Furthermore, the phylogenies generated by James and Pourtless (2009) are unresolved polytomies, so it cannot be said that this study failed to find support for the theropod hypothesis in particular; it actually failed to find support for *any* hypothesis! While some details of maniraptoran phylogenies may prove incorrect, the "birds are not theropods" movement is based on erroneous argumentation and fails to account for the data as well as the theropod hypothesis does.

Because birds are bipedal animals that descend from bipedal dinosaurs, it has been suggested that bird flight originated on the ground among cursorial, terrestrial reptiles. Samuel Williston is often credited with the first mention of this proposal (Williston 1879). However, because it is easier to imagine that flight could only begin once bird ancestors were jumping or gliding from a height, and because many birds are small animals with climbing or perching adaptations, the hypothesis that bird flight began in small climbing reptiles has also been popular. These two hypotheses have usually been regarded as competitors, with a ground-up model being favored by supporters of the dinosaurian origin for birds and a trees-down

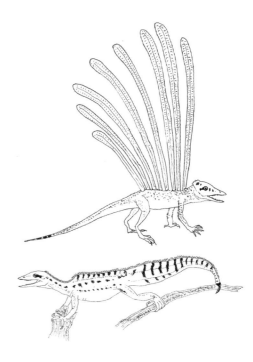

20.2. The peculiar Late Triassic reptiles *Longisquama insignis* (from Kirghizia) and *Megalancosaurus preonensis* (from Italy) have been suggested by some authors to be close to avialian ancestry. Both animals are profoundly different from birds and do not even belong to Archosauria. Sauropodomorphs and ornithischians share more characters with birds than these animals do.

The Origins of Avialian Flight

model being favored by those who argue that the affinities of birds must lie elsewhere.

A trees-down model has also been adopted by some because it has been argued that feathers must have evolved in an aerodynamic context. Supporters of this view argue that complex, vaned feathers appeared before simpler feathers did (Feduccia 1996). However, fossil theropods such as *Sinosauropteryx* show that simple, quill-like structures evolved long before vaned feathers. Furthermore, large, vaned feathers are known from non-avialian maniraptorans (such as *Caudipteryx*) that lack any features related to climbing, gliding, or flying. So far as we can tell from the phylogeny, feathers did not, therefore, evolve for flight, and it can be inferred from their phylogenetic distribution that even complex feathers were first associated with other functions. The fact that the most basal taxa with vaned feathers seem to have these feathers on the hands and tail alone might indicate that vaned feathers were initially involved in display.

The assumption that a trees-down model excludes theropods from avialian ancestry is flawed given that the nonavialian maniraptorans closest to birds are small forms likely capable of at least some climbing and perching. Dinosaur workers have mostly been skeptical of the idea that early birds, and the nonavialian maniraptorans closest to birds, might have been tree climbers, and studies on claw geometry indicate that most Mesozoic birds were not strongly arboreal or scansorial (Glen and Bennett 2007). Nevertheless, belief in a simple dichotomy between the ground-up and trees-down models is misleading because the anatomy of small deinonychosaurs and archaeopterygids indicates that they were neither arboreal nor terrestrial specialists; rather, their leg and toe proportions, their sharp, curved claws, and their enlarged, distally placed halluces indicate that they were generalists able to both climb and run (Elzanowski 2001; Hopson 2001; Paul 2002).

The juveniles of some birds use their half-grown wings to boost their escape run up trees and slopes, a behavior known as wing-assisted incline running (WAIR). This raises the possibility that small, feathered maniraptorans—surely at risk from larger theropods and other predators—could utilize WAIR as an escape tactic, even with wings that seem only half developed compared to those of birds (Dial 2003). The use of heights as refugia may then have encouraged the evolution of gliding and flapping flight and of larger wing feathers. The evolution of long arms and hands, of long feathers initially used in display, and of a flexible wrist joint that allowed the hand to be partially folded toward the ulna may therefore have combined to preadapt bird ancestors for flight. This scenario is satisfying because it agrees with the idea that birdlike maniraptorans and basal birds were generalists able to climb as well as run and forage on the ground.

Avialian Anatomy

Because Avialae encompasses taxa as disparate as the long-tailed, toothed archaeopterygids of the Jurassic and such extant neornithines as swifts, owls, and finches, it is difficult to review the anatomy of extant forms without making constant recourse to the substantially different, more typically maniraptoran morphology of Mesozoic forms. Mesozoic nonavialian theropods are covered elsewhere in this book (see Holtz, Chapter 19 of this volume), so the review of bird anatomy provided here focuses on neornithines alone.

The bird body is clearly divided into distinct anatomical modules that operate independently: a head–neck module, a pectoral module that involves the wing muscles and wings, a pelvic module that corresponds to the pelvis and hind limbs, and a caudal module. Despite their specialization for flight, most neornithines are terrestrial foragers, and only a minority feed on the wing. Enlarged legs where the bones are unusually thick are seen in many cursorial birds (some of which have reduced wings), and enlarged wings and reduced legs and feet are common in birds that forage or feed in flight. The bird skeleton is pneumatic (the bones are occupied by air sacs that are connected to the lungs), though skeletal pneumaticity is reduced or absent in diving birds. The walls of bird bones are particularly strong and thin. This has often been suggested to be a weight-saving adaptation, but bird bones are similar in mass or even heavier than those of equivalent-size mammals.

Most of the bones of the neornithine cranium are fused in adults (Fig. 20.3). Various sheetlike or rodlike regions, some of which are poorly mineralized, allow flexibility at the base of the bill, in the palate, and sometimes within the rami of the jaws. The premaxillae are typically large, with long nasal processes that form the dorsal midline, and maxillary and palatal process making up much of the rest of the rostrum. An interobital septum composed of the mesethmoid separates the eyeballs along the midline of the braincase. The quadrate is mobile, with heads that fit into sockets on the squamosal. A fingerlike structure – the orbital process – is directed inward and forward, while the pterygoid process on the medial side of the quadrate's base articulates with the pterygoid. The zygomatic bar (composed of the quadratojugal and jugal) articulates with a socket on the side of the quadrate (sometimes being fused with it) and connects anteriorly with the maxilla. Most birds possess a reduced antorbital fossa, and the lacrimals are often reduced or absent (Fig. 20.3).

Neornithine palatal structure is highly variable. Some birds (like pigeons, waders, and woodpeckers) have an open palatal structure where thin, delicate shelves from the premaxillae and maxillae fail to make contact along the midline. Others (like ibises, kingfishers, and hornbills) have an extensive bony palate formed from coossified premaxillae, maxillae, and palatines. The two main neornithine groups – paleognaths and neognaths – differ in palatal structure (Fig. 20.4). In paleognaths, large, laterally projecting basipterygoid processes are present, there is a broad connection between the pterygoids and palatines, and the vomers are long and reach posteriorly to brace the palatines. In neognaths, the basipterygoid processes are small or absent (though they are large, oval facets in waterfowl and gamebirds), the contact zone between the rodlike pterygoids and palatines is small and hingelike, and the vomers do not contact the palatines. The paleognath palate is reinforced and relatively stiff, but movement of the palatines and pterygoids in the neognath palate helps the rostrum to flex up and down when the jaws are opened. It is unclear, however, whether neognath skulls are necessarily more flexible than those of paleognaths because of these differences, and flexible zones in the paleognath rostrum mean that their skulls can still be kinetic.

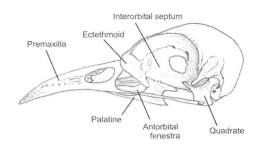

20.3. The skull of a corvid passerine. Though different names are sometimes used for some of the bones, this is clearly recognizable as a modified theropod dinosaur skull. As is the case in many passerines, the lacrimal is small or absent and a large, inflated ectethmoid takes its place. The quadrate moves anteroposteriorly thanks to the two convex heads that articulate with sockets on the squamosal. The palatines and entire rostrum are joined to the quadrate and are also mobile. The sclerotic rings are not shown.

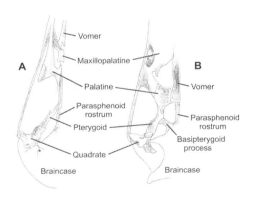

20.4. Slightly schematic representations of (A) the neognath palate and (B) the paleognath palate. These diagrams show the right side of the posterior region of the palate, with anterior at the top. Note the prominent basipterygoid process in the paleognath (versus its absence in the neognath), the simple, hingelike articulation between the pterygoid and palatine in the neognath (versus the firm union in the paleognath), and the restriction of the neognath vomer to the anterior part of the palate (versus its posterior component in the paleognath).

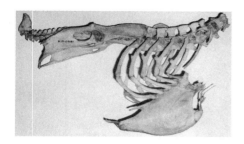

20.5. Partial skeleton of a seriema. Note the hypapophyses on the ventral surfaces of at least some of the dorsal vertebrae, the uncinate processes on the ribs, and the large, boat-shaped sternum. Mobile joints between the sternum and sternal ribs mean that dorsoventral movement of the sternum helps ventilate the body's air sacs. However, some flightless birds have a proportionally small sternum yet still have air sacs, so this is not the whole story.

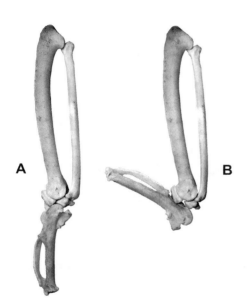

20.6. The folding action of the neornithine wrist, as illustrated by a domestic turkey. A, Wrist in unfolded position, with the long axes of the carpometacarpus and radius–ulna being essentially in line. B, The trochleated semilunate part of the carpometacarpus and shape of the radiale allow rotation in the wrist such that the carpometacarpus comes closer to the ulna. This wrist action allows the remiges to be folded away when not in use but is also used in the flight stroke.

Images courtesy of Corwin Sullivan, used with permission.

Neornithine cervical vertebrae are heterocoelous (that is, they possess saddle-shaped articular surfaces) and number between 13 and 25. Short cervical ribs that do not extend for the full length of the centrum are present, and the fingerlike dens process projects from the axis, through the short atlas, to the occipital condyle. The dorsal region consists of about 12 vertebrae (some of which are sometimes fused into a notarium). Ventral projections termed hypapophyses descend from the centra (Fig. 20.5) and are large, forked structures in some birds and are fused to form a ventral keel in others. Seven ribs articulate with the dorsal vertebrae, five of which are connected to a cartilaginous sternal rib that contacts the sternum. Ossified uncinate processes project from the posterior margin of some of the ribs in most (though not all) neornithines (Fig. 20.5). Often thought to help brace the rib cage and resist torque incurred during flight, they are attached to muscles that allow birds to move their ribs, and hence ventilate their air sacs, when sternal movement is restricted (Codd et al. 2005).

Several dorsal vertebrae (typically five or six) have been captured by the pelvis to form a synsacrum, as have a number (typically five or six) of what were originally caudal vertebrae. The synsacrum includes between 11 and 23 vertebrae, with the highest number being present in ratites. Four to seven free caudals are present, whereas the more distal caudals are fused into the laterally compressed pygostyle. The free caudals have large transverse processes, and large tail muscles attach to these and to the posterior end of the pelvis. Some of these muscles surround the rectricial bulbs: two structures that encase the roots of the large tail feathers termed rectrices. Together, these bulbs and the associated muscles allow birds to fan their rectrices, and to raise, lower, rotate, and laterally deviate the tail.

The humerus is one of the largest, most robust bones in the neornithine skeleton. It has a globular head, large deltopectoral and bicipital crests, and a series of pneumatic openings located on and around these structures. The radius and ulna are typically long and slender, and the larger, thicker ulna is usually gently curved along its length. Quill nodes (bony insertion points for the secondary feathers) are present on the ulna, though these are absent in tinamous and flamingos.

In adult neornithines, three metacarpals—usually termed the alular, minor, and major—are fused with the semilunate carpal to form the carpometacarpus. The semilunate carpal forms a grooved trochlea that articulates with a wedge-shaped radiale in the wrist. A sliding articulation between the groove and the radiale allows the carpometacarpus to be folded up close to the ulna (Fig. 20.6), thereby allowing the remiges to be folded away when not in use. A few flattened phalanges represent the abbreviated digits. The thumblike alula is mobile and articulates with a facet on the carpometacarpus. Manual claws, often thought to be restricted to Mesozoic taxa and the hoatzin *Opisthocomus hoazin*, are widespread in neornithines and occur on the alula at least in some paleognaths, waterfowl, gamebirds, waders, and others. Various birds possess spurs, clubs, or knobs on their carpometacarpi that they use as weapons.

The coracoid is typically a long-shafted bone (it is short and squat in some birds) with a transversely expanded base; this fits into a groove on the anterolateral part of the sternum. The hooked acromion process on the bladelike scapula combines with the dorsal ends of the coracoid and furcula to form the triosseal canal. The tendon of the supracoracoideus muscle passes through this canal to attach to the head of the humerus; the muscle itself is attached to the sternal keel, and by pulling downward, it raises the wing. The furcula is usually V shaped or U shaped; however, it is reduced to thin, splintlike clavicles or even lacking entirely in some parrots, owls, passerines, and others. The sternum is usually large and boat shaped, and it has a ventral keel (Fig. 20.5). Sternal rib facets line part of the sternum's lateral margins, and the bone's posterior margin is often deeply incised or perforated by openings.

The neornithine pelvis is typically wide, and the ventral tips of the pubes and ischia are well separated. The transverse processes of the sacral vertebrae are usually expanded and fused with the ilium, forming a large dorsal shield that roofs the abdominal region. The pelvis is opisthopubic, with the ischium and pubis extending posteriorly in parallel. In some neornithines, the posterior end of the ischium does not contact the ilium, but the two are usually fused and enclose the ilioischiadic foramen (Fig. 20.7). A large facet that articulates with the greater trochanter of the femur—the antitrochanter—is located above and behind the acetabulum.

Neornithines are conservative in femoral morphology. A large greater trochanter forms a tall crest on the lateral side of the proximal end. A wide channel—the rotular groove or patellar sulcus—separates the two distal condyles on the anterior surface, and a deep concavity—the popliteal fossa—excavates the posterior surface of the distal end. The tibia is fused with the astragalus and calcaneum to form the tibiotarsus. In most neornithines, two flangelike cnemial crests project anterolaterally from the tibiotarsus and a transverse bar of bone—the supratendinal bridge—spans the distal end, just proximal to the distal condyles. The neornithine fibula is reduced, and its bony part fails to reach the tarsus.

Metatarsals II–IV are fused, forming the tarsometatarsus. A posteriorly projecting process, the hypotarsus, is located at its proximal end; it may be perforated by one or more canals that enclose tendons for the toes. A small metatarsal I articulates with the hallux. Most birds possess anisodactyl feet; this is where the hallux is posteriorly directed and hence opposes digits II–IV. Kiwi have an unreversed hallux, but this is a reversal based on the presence of a reversed hallux in other birds. The hallux has been lost in various cursorial birds, and digit II is also absent in ostriches.

Several neornithine lineages possess more than one opposable toe. In quetzals and other trogons, digits I and II oppose III and IV, a condition termed heterodactyly, while digits I and IV oppose II and III in woodpeckers, parrots, and cuckoos, a condition termed zygodactyly. A fossil trackway shows that zygodactyl feet were present in at least one Mesozoic bird lineage. In swifts, mousebirds, pygmy parrots, and a few giant woodpeckers, all four toes point anteriorly, a condition termed pamprodactyly. Digit IV can be rotated to point anteriorly or posteriorly in some birds, such as ospreys and owls. This condition is termed semizygodactyly.

Despite their many skeletal peculiarities, it is of course the soft tissue structures of birds that we recognize most immediately. In life, the avian body is mostly obscured by feathers, and these organs form what is effectively an environmental suit that protects and insulates the fragile animal within and allows it to fly. In most birds, feathers are arranged in tracts

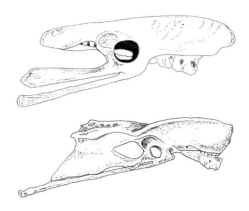

20.7. The pelvic girdles of (top) a paleognath (the moa *Emeus*) and (bottom) a neognath (the turkey *Meleagris*). Note how the neognath has fused the ischium and posterior part of the ilium, forming the ilioischiadic foramen. To those who know the anatomy of nonavialian maniraptorans, the paleognath pelvis should look familiar.

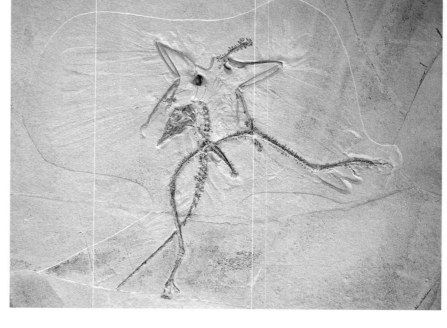

20.8. The beautifully preserved, articulated Thermopolis specimen of *Archaeopteryx*. This specimen confirms that *Archaeopteryx* is highly similar to dromaeosaurids and other maniraptorans in skull, pelvic and hind limb anatomy, and indeed recently discovered dromaeosaurids, troodontids and other maniraptorans are extremely similar to *Archaeopteryx*. In the foot, the hallux is not fully reversed and the second toe is hyperextendible. Well preserved feather impressions surround the skeleton. Ten specimens assigned to *Archaeopteryx* are currently known, but exactly how many species they represent remains controversial.

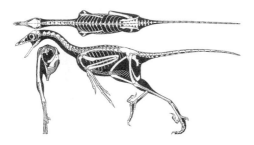

20.9. The skeleton of *Archaeopteryx lithographica* (specifically, the Berlin specimen) restored. The estimated wing size of *Archaeopteryx* overlaps with that of modern flying birds, suggesting reasonable flight abilities. However, doubt continues, and it may have been limited to gliding. The shape of the articular surfaces in the second toe shows that this digit could be hyperextended, as shown here.

Illustration by G. S. Paul, used with permission.

termed pterylae. The normal feathers of the body and limbs are the contour or vaned feathers. Some small birds, like hummingbirds, possess only about 940 of these, whereas over 25,000 have been counted on swans. Fluffy down feathers underlie contour feathers on parts of the body and provide increased insulation. About 25 long contour feathers, termed remiges, form the main flight surface of each wing, and the similarly large retrices (typically numbering 12) form the fanlike tail. The remiges can be divided into primaries, secondaries, and tertials; these are partially covered on both the dorsal and ventral surfaces of the wing by coverts.

The trachea in birds is long and modified for sound production in some groups. It is absurdly long and contains extra loops in certain cranes, swans, birds-of-paradise, and others. Deep within the chest, it branches into the two bronchi, each of which connects to the lungs. The syrinx—the main organ of sound production—is located just ahead of this divergence. It is a complicated chamber surrounded by the syringeal muscles, and its detailed anatomy varies considerably. Despite the importance of the syrinx, the larynx is not irrelevant and moves anteriorly and posteriorly, as well as dorsally and ventrally, during vocalization. This changes the size and shape of the mouth and throat cavity and seems to play a key role in how birds control the noises they produce.

The Mesozoic Avialian Radiation

Archaeopteryx from the German Solnhofen Limestone is the most famous Mesozoic bird (Figs. 20.8, 20.9, and 20.11), and some specimens are extremely well preserved (Elzanowski 2002; Mayr et al. 2007; Fig. 20.8; see also Plate 4). *Archaeopteryx* has a sloping snout; relatively long, pointed premaxillae; and a small number of short, peglike, widely spaced teeth (there are four in each premaxilla, eight to nine in each maxilla, and 11 to 12 in each dentary). The presence or otherwise of an ossified postorbital bar is controversial; a complete postorbital bar is known for some other Mesozoic birds, like confuciusornithids and some enantiornithines. The arms and legs of *Archaeopteryx* are about equal in length. The hallux is not fully reversed (though it is larger and more distally placed than that of nonavialian maniraptorans), and the specimen at the Wyoming Dinosaur Museum in Thermopolis, Wyoming (See Fig. 20.8; Plate 4), demonstrates that the second toe was at least slightly hyperextendable. The pubis is near vertical rather than opisthopubic. Contour feathers covered the neck, body, and tail; 12 asymmetrically vaned primaries were probably present, as were secondaries and coverts (tertials were probably absent). Sixteen or 17 pairs of symmetrical feathers emerged from the 21 or 22 vertebrae of the tail. Contour feathers were also present on the legs as far distally as the tibiae.

The taxonomic status of the 10 specimens conventionally included in *Archaeopteryx* remains the subject of debate. The large Solnhofen specimen differs in hand and foot characters from the others and has been named *Wellnhoferia grandis*, while some authors maintain that *A. bavarica* and *A. siemensii* are distinct from *A. lithographica*. Teeth identified as belonging to a close relative of *Archaeopteryx* have been reported from the Kimmeridgian of Guimarota in Portugal, but their sigmoidal shape and serrated carinae means that some workers doubt their proposed archaeopterygid identity.

Archaeopterygids have almost always been regarded as the most basal bird group. However, *Archaeopteryx* has been recovered as part of Deinonychosauria, and hence not within Avialae, in some studies (Xu et al. 2011). If this position is correct, *Archaeopteryx* does not tell us as much about early bird anatomy and biology as long thought, and it should not be discussed in the present chapter alongside indisputable avialians. However, it would be inaccurate to argue that the phylogenetic pattern in this part of the theropod tree is settled yet, and other research groups still find archaeopterygids to be within Avialae. It has also been suggested that the Middle Jurassic scansoriopterygids from China (Fig. 20.11) might be avialians, and perhaps further from neornithines than is *Archaeopteryx* (Zhang et al. 2008). Scansoriopterygids are small, short-skulled maniraptorans with a propubic pelvis. *Epidendrosaurus* at least had a long third finger that might have been used as a foraging or climbing tool, and its distally placed hallux suggests a climbing ability. Scansoriopterygids remain enigmatic, and their proposed position at the base of Avialae needs further testing. They recall oviraptorosaurs in some respects and could be early members of that lineage.

A number of long-tailed Early Cretaceous birds have been described from China. These are *Jeholornis*, *Jixiangornis*, *Shenzhouraptor*, and *Dalianraptor*, the first three of which are possibly synonymous. *Jeholornis* possesses a short skull with robust jaws, a relatively short hallux, 27 caudal vertebrae, and a distally placed rectricial fan. The upper jaws are toothless, but three small, conical teeth are present at the tip of the left dentary. The stomach of one specimen contains over 50 seeds, so presumably this bird was herbivorous or omnivorous. Its forelimb proportions suggest that it could fly, but the shorter forelimbs of *Dalianraptor* indicate flightlessness. It is suspected that these taxa are closer to modern birds than are archaeopterygids.

Zhongornis from the Yixian Formation—described on the basis of the tiny skeleton of a juvenile—possesses 13-14 differentiated caudal vertebrae and suggests that a reduction in their number preceded the evolution of the pygostyle. *Zhongornis* is also unusual in possessing just three phalanges in

20.10. *Archaeopteryx* was first discovered in 1861, but excellent and informative specimens have also been discovered in recent decades. Two of those are depicted here: the Munich specimen at left (discovered in 1991 and originally described as the new species *A. bavarica*), and the Eichstatt specimen (discovered in the 1950s and once suggested to represent the distinct taxon *Jurapteryx recurva*).

20.11. Life restoration of the scansoriopterygid *Epidendrosaurus ninchengensis* from the Middle Jurassic of China. All scansoriopterygids were tiny (less than 20 cm long) though none are fully grown. Their bodies were covered with branched feathers that look something like closed, flattened flowes, but there is as yet no evidence for enlarged forelimb feathers.

Image courtesy of John Conway, used with permission.

20.12. Skeletal reconstruction of *Sapeornis chaoyangensis* from the Early Cretaceous of China. *Sapeornis* was large for a Mesozoic bird, with a wingspan of perhaps 1.5 m. Its long wings suggest that it was adept at soaring. The presence of gastroliths, combined with its short snout and peglike teeth, suggest herbivory of some form.

20.13. Reconstruction of the confuciusornithid *Confuciusornis sanctus*. Ribbonlike tail feathers present in some specimens suggest sexual dimorphism. It is uncertain how *Confuciusornis* made a living. Its toe proportions suggest that it was a capable percher and probably did not spend much time on the ground. These factors, combined with its superficially kingfisher-like bill, suggest that it grabbed aquatic prey before returning to a perch. Indeed, stomach contents show that at least some confuciusornithids ate fish. From Chiappe et al. (1999).

20.14. Reconstruction of the enantiornithine *Iberomesornis romerali* from the Early Cretaceous of Spain. *Iberomesornis* is one of several enantiornithines that lack the characters that unite the members of the large enantiornithine clade Euenantiornithes. *Iberomesornis* was small (body length approximately 80 mm). From Chatterjee (1997).

The Enantiornithines

its third finger. While it has been suggested that *Zhongornis* represents an important taxon intermediate between *Jeholornis* and confuciusornithids (Gao et al. 2008), the juvenile status of the type specimen has resulted in skepticism about its proposed phylogenetic position.

Sapeornis (Fig. 20.12) is another Early Cretaceous Chinese bird with a short, deep skull, though, unlike *Jeholornis* and *Zhongornis*, it is short-tailed, with only six or seven free caudals and a pygostyle. It is also unusual in apparently lacking an ossified sternum and in possessing particularly elongate arms (more than 1.5 times longer than the legs), fusion of the distal carpals and proximal ends of the metacarpals, a reduced manual formula of 2–3–2, and procumbent premaxillary teeth. The extremely similar *Omnivoropteryx*, *Didactylornis* and *Shenshiornis* are almost certainly close relatives (or even junior synonyms) of *Sapeornis*, and all should be grouped together as the omnivoropterygids. Also appearing to belong to the same region of the tree is *Zhongjianornis*. This bird was toothless and has been interpreted as the most basal of all edentulous birds. *Jeholornis* and its possible kin, omnivoropterygids, and *Zhongjianornis* are all relatively large compared to most other Cretaceous birds.

A new Mesozoic bird clade was recognized in the 1990s; while represented by only a few taxa, the more than 1,000 collected specimens mean that the anatomy and paleobiology of these birds are comparatively well known (Chiappe et al. 1999). These are the confuciusornithids of Early Cretaceous China and Korea. Key confuciusornithid characters include toothlessness, a hypertrophied pollex claw, a small ungual on manual digit II, and a V-shaped posterior border to the sternum (Fig. 20.13). The arm is longer than the leg, the hallux is fully reversed, and the primary feathers are particularly long (about 3.5 times as long as the hand skeleton). The distal 10 or so tail vertebrae are fused to form a pygostyle, with four or five free vertebrae being present in addition. Rectrices are absent, and only a narrow tuft of unspecialized feathers surrounds the pygostyle. However, some specimens of *Confuciusornis* possess two ribbonlike tail feathers, each of which exceeds the entire length of the rest of the animal.

The enormous numbers of *Confuciusornis* discovered in the Jiufotang Formation suggest that these birds were gregarious. Confuciusornithids were suggested by some workers to be allied to enantiornithines, but the details of their anatomy better support a position intermediate between archaeopterygids and enantiornithines. The name Pygostylia is used for the clade that includes confuciusornithids and modern birds.

A large, morphologically diverse pygostylian clade occurred virtually worldwide during the Cretaceous but failed to survive beyond it. These are the enantiornithines, sometimes called "opposite birds." Cyril Walker (1981) displayed remarkable prescience in identifying this clade of archaic birds as phylogenetically intermediate between *Archaeopteryx* and neornithines. Enantiornithines were initially named as such because the articulation between the coracoid and scapula seems opposite relative to that of modern birds: in enantiornithines, a groove on the ventral surface of the scapula forms the more important part of the triosseal canal, whereas in neornithines it is the dorsal end of the coracoid (specifically, the procoracoid process) that serves this role. Fusion in the enantiornithine tarsometatarsus begins proximally, rather than distally as it does in neornithines, and this also contributed to the idea of enantiornithines as opposite birds. Currently accepted synapomorphies for the group include a third metacarpal that projects further distally than does the second, a particularly large posterior trochanter on the femur, and a narrow fourth metatarsal.

While initially assumed to have poor flight capabilities, the discovery of *Neuquenornis* from Argentina showed that proficient flight abilities were normal in the group. This was confirmed by such additional discoveries as *Sinornis* from the Early Cretaceous of China and *Iberomesornis* (Fig. 20.14) and *Concornis* from the Early Cretaceous of of Spain. A large, fully reversed hallux (almost as long as digit II) shows that many enantiornithines were capable perchers. *Eoalulavis* and *Protopteryx* demonstrate that an alula was present within the group, and the longipterygid *Shanweiniao* demonstrates that at least some enantiornithines possessed a rectricial fan (O'Connor et al. 2009).

Some enantiornithines fed on aquatic prey such as crustaceans, but members of the group were diverse in ecology and morphology and probably overlapped with ornithurines and their relatives in terms of lifestyle and ecology. Some enantiornithines (such as *Gobipteryx*) are toothless and may have been seed eaters, while amber preserved in the stomach region of *Enantiophoenix* from Lebanon suggests that it was a sap feeder (Chiappe 2007). Elongate, streamerlike rectrices have been reported for *Protopteryx*, *Dapingfangornis*, *Paraprotopteryx*, and *Bohaiornis* and may have been display structures. Longipterygids—the most basal enantiornithine clade according to some studies (O'Connor et al. 2009; Ji et al. 2011)—possess a long rostrum (equal to or exceeding about 60% the total length of the skull) with a specialized dentition (large, recurved teeth are present at the jaw tips). Slender, elongate tibiotarsi and tarsometatarsi in *Lectavis* suggest wading habits. *Yungavolucris* has a broad, strongly convex articular surface on the distal end of its second metatarsal and laterally divergent ends on its third and fourth metatarsals, so it obviously had weird feet that were perhaps specialized for swimming. Finally, the limb proportions of *Elsornis* from Mongolia indicate poor flight ability or even flightlessness (Fig. 20.15). Early enantiornithines were similar in size to finches and thrushes, whereas some Late Cretaceous forms had wingspans of a meter or more.

For all their diversity, enantiornithines did not survive beyond the Maastrichtian. It remains mysterious why neornithines (of several lineages) survived the end of the Cretaceous while enantiornithines did not.

Several Cretaceous birds share characters with neornithines that are not present in enantiornithines and more basal birds, including distally fused second and third metacarpals, a curved scapula, and completely fused metatarsals II–IV. These taxa have been united in Ornithuromorpha. Chicken-size *Patagopteryx* from the Late Cretaceous of Argentina is near the base of this clade. Its simplified pectoral girdle and short wings show that it was flightless, and its stout pelvis, legs, and feet indicate life as a

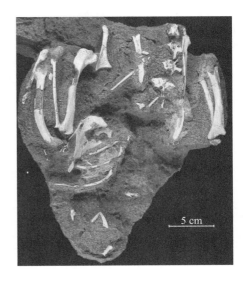

20.15. The partial skeleton of the Late Cretaceous enantiornithine *Elsornis keni* from the Djadokhta Formation of the Gobi Desert. The long-shafted coracoid, long humerus, straight ulna, curved radius, and boat-shaped sternum—all visible here—are much like those of modern birds in general shape. The wing proportions of *Elsornis* overlap with those of ratites, suggesting near-absent flight abilities.

Photo courtesy of Gareth Dyke, used with permission.

20.16. Reconstruction of *Yixianornis grabaui* from the Jiufotang Formation of China. Preserved feathers show that it had broad wing tips and a long tail, and was therefore well adapted for maneuvering in a forested habitat. Illustration by G. S. Paul, from Clarke et al. (2006).

Basal Ornithuromorphans

terrestrial walker. The gracile-limbed *Hollanda* from the Late Cretaceous of Mongolia also seems to have been markedly terrestrial; it was probably a roadrunner-like cursor.

Other taxa known from this region of the tree—they include *Hongshanornis*, *Ambiortus*, *Apsaravis*, *Yixianornis* (Fig. 20.16), and *Yanornis*—are generally thrush-size, volant land birds. At least some combined narrow, toothed jaws, gastralia, and a pubic symphysis (all primitive characters within birds) with fairly modern wing and tail morphologies. All are from the Cretaceous of Asia.

Basal Ornithurines

Within Ornithuromorpha, neornithines form a clade with the marine birds of the Late Cretaceous. The name Ornithurae is often used for this clade, though this usage contradicts Gauthier's (1986) suggestion that it should be used for all birds closer to modern species than to *Archaeopteryx*. Ornithurae is thus another problematic term, as it is used for different clades by different authors. It is here used for the clade that includes *Hesperornis* and kin, *Ichthyornis*, and Neornithes.

The hesperornithines or hesperornithiforms are toothed, foot-propelled diving birds from the Cretaceous. The best known hesperornithine, the North American *Hesperornis* (Figs. 20.17A, 20.18), reached large size (1.5 m in total length) and had a strongly reduced wing; only a rodlike humerus remains. A large patella projects from the proximal end of the tibiotarsus, the tarsometatarsus is transversely compressed, and the fourth toe is the longest. The hesperornithine foot skeleton is similar to that of grebes, and these Cretaceous birds probably had lobed toes and swam with outward-facing feet, like grebes. Flightlessness has been regarded as characteristic for hesperornithines, but early forms (like *Enaliornis* and *Pasquiaornis*) may have been volant.

Ichthyornis (Fig. 20.17B) is one of the most familiar of Mesozoic birds, a fact we owe to its early discovery (it was named in 1872) and its frequent depiction in artwork. It was a volant seabird with laterally compressed, unserrated teeth, and apart from its teeth, it probably looked gull-like. Claims that a predentary bone is present at the tip of the lower jaw are not correct. Numerous Cretaceous taxa have been united with *Ichthyornis* in an "Ichthyornithiformes." Clarke (2004) showed that none of these was really close kin of *Ichthyornis*, and that those valid taxa once named as *Ichthyornis* species in fact deserved new generic monikers (*Guildavis*, *Iaceornis*, and *Austinornis*). The first two are stem ornithurines, while *Austinornis* may be a stem galliform (Clarke 2004).

Another aquatic Cretaceous bird appears to be a stem ornithurine: *Gansus* from the Early Cretaceous of China. This tern-size bird (known from hundreds of specimens) possessed pelvic and leg characters indicative of foot-propelled diving, but its large sternal keel, long wing bones, and preserved remiges show that it was a capable flier.

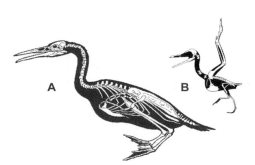

20.17. The hesperornithine *Hesperornis regalis* (A) shown approximately to scale with *Ichthyornis dispar* (B). Both species inhabited the Central American seaway during the Late Cretaceous (though related species occurred elsewhere in the world). *Hesperornis* is shown here with webbed toes, but it may in fact have had lobed toes like those of grebes. From Chatterjee (1997) and Clarke (2004).

The Rise of Modern Birds, the Neornithines

Because *Ichthyornis*, Hesperornithes, and *Gansus* seem to be the closest relatives of neornithines, it seems that neornithines originated in aquatic habitats. Many neornithine lineages then became specialized for life in terrestrial environments, however, so it does not seem that aquatic life explains how and why neornithines survived across the Cretaceous–Paleogene boundary.

Key characters demonstrating the monophyly of Neornithes include an absence of teeth, fusion of the symphysis between the two halves of the lower jaw, and the presence of three articular facets on the ventral end of the quadrate. Neornithines also tend to have a complex hypotarsus, lack gastralia, and have a more pneumatic ear region than earlier birds. It is sometimes assumed that some or all of these characters made neornithines better fliers or more adaptable feeders than nonneornithine birds. However, it should always be remembered that birds were widespread and diverse for more than 70 million years prior to the appearance of the first neornithine.

Fragmentary fossils show that neornithines had appeared by the Maastrichtian. *Vegavis* is securely dated as Late Cretaceous and confidently placed within Anseriformes (waterfowl), and fossils identified as gaviiforms (loons) are also known from this time. Several other alleged Late Cretaceous neornithines are less securely identified, including a possible lithornithid paleognath, the possible stem gamebirds *Palintropus* and *Austinornis*, the possible "pelecaniform" *Torotix*, possible tube-nosed seabirds and waders, an as-yet-undescribed penguin-like bird from Chatham Island, and a mandible fragment allegedly from a parrot (Hope 2002; Mayr 2009). Even if most or all of these records are erroneous, the Cretaceous waterfowl and loons demonstrate that paleognaths, gamebirds, and members of various other neognath clades were present at this time. This shows that the neornithine radiation was well underway before the end of the Cretaceous and that the rise of neornithines was not dependent on the end-Cretaceous extinction event. This is generally supported by molecular analyses, but it seems that nearly all Cretaceous (and Paleocene) members of extant lineages were stem taxa, and not members of the crown groups.

Why did so many neornithine lineages survive into the Cenozoic while other avialian lineages apparently did not? It is possible that neornithines were more resistant to ecological disturbance, were more mobile, or were faster at growing or reproducing than nonneornithines. However, the poor Paleocene bird record means that the apparent extinction of nonneornithines might not be as real as it appears: *Qinornis* from the early Paleocene of China has been suggested to represent a nonneornithine lineage (Mayr 2009), hinting at the possibility that some groups lingered on and were not inferior to neornithines in the survival stakes. On the other hand, it is also conceivable that some nonneornithine groups disappeared before the end of the Maastrichtian.

20.18. The hesperornithine *Hesperornis regalis* chasing prey. Feathers preserved on a specimen of *Parahesperornis* indicate that these birds had long, shaggy feathers like ratites. This seems surprising, but the thick fur coats of many swimming mammals function well as aquatic insulation.

Image courtesy of John Conway, used with permission.

Assembling the Neornithine Tree

Assembling the phylogeny of neornithine birds has been described as one of the greatest challenges facing modern phylogenetics. The majority of nonpasserine families are monophyletic, a situation that contrasts strongly with the status of passerine families. But because most nonpasserine groups appear highly distinct relative to one another, finding broader affinities has been difficult. Efforts to reconstruct neornithine phylogeny began with Huxley (1867) and Fürbringer (1888). Hans Gadow's (1893) system was particularly influential: he listed ratites first, arranged grebes, penguins, tube-nosed seabirds, herons, storks, waterfowl, and raptors in one major assemblage, and within a second assemblage, he listed parrots, rollers, woodpeckers, and passerines. By the middle decades of the twentieth century, the literature on neornithine systematics mostly concentrated on tweaking this system, with the main approach being a phenetic one (where groups were united or divided on the basis of overall similarity, rather than on the presence of shared derived characters). Even today, both popular and technical works on birds list them in the approximate order proposed by Gadow.

Phylogenetic approaches to neornithine phylogeny were first employed during the 1970s, most notably by Joel Cracraft (1972). During the 1980s and 1990s, Charles Sibley and colleagues used DNA–DNA hybridization to analyze the neornithine tree. Sibley and Ahlquist (1990) produced a mostly resolved phylogeny, dubbed the Tapestry, for over 1,100 species. They supported a divergence between Eoaves (ratites and tinamous) and Neoaves, and broke the latter down into six groups: Galloanserae (gamebirds and waterfowl), Turnicae (buttonquails), Picae (woodpeckers and kin), Coraciae (hornbills, trogons, rollers, and kin), Coliae (mousebirds), and Passerae (everything else, from cuckoos, parrots, and pigeons to cranes, waders, raptors, herons, pelicans, and passerines). While Sibley and Ahlquist's suggestions did much to inspire new work, their conclusions were not always supported by their data (Harshman 1994, 2007).

Since the late 1990s, phylogenetic studies have used both molecular and morphological data to analyze neornithine affinities (e.g., Mayr and Clarke 2003; Mayr et al. 2003; Cracraft et al. 2004; Ericson et al. 2006; Livezey and Zusi 2007; Hackett et al. 2008). These studies agree on several areas. Paleognaths and neognaths are sister groups. Within Neognathae, a sister group relationship between Galloanserae and all other neognaths is supported (Fig. 20.19). The "all other neognaths" clade is usually termed Neoaves (note that this usage differs from that favored by Sibley and Ahlquist), though the name Plethornithes is preferred by some.

Within Neoaves, a series of waterbird taxa usually cluster close to the root and might be close relatives. In addition to tube-nosed seabirds, penguins, pelicans, storks, and similar birds, this section of the tree might include gruiforms and cuckoos (Fig. 20.19). Charadriiformes—the waders,

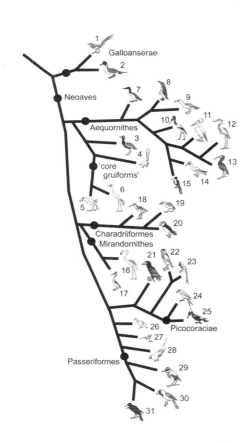

20.19. Highly simplified approximate consensus cladogram of Neognathae, incorporating data from Harshman (2007), Hackett et al. (2008), and other sources. Because many of the relationships shown here are relatively new, names do not yet exist for many of the proposed clades. Some of the relationships shown here are contentious: Ericson et al. (2003) and Hackett et al. (2008), for example, found Aequornithes to form a clade with bustards, cuckoos, and core gruiforms, and they also recovered a clade that included raptors, owls, mousebirds, woodpeckers, and rollers. Pigeons and Strisores (nightbirds, swifts, and hummingbirds) are shown here as being close to parrots and passerines, but they are recovered in some analyses as distantly related to these groups and actually outside the clade that includes Aequornithes and higher landbirds. 1, Galliformes (gamebirds or fowl); 2, Anseriformes (waterfowl or wildfowl); 3, Otididae (bustards); 4, Cuculiformes (cuckoos); 5, Rallidae (rails); 6, Gruidae (cranes); 7, Gaviiformes (loons or divers); 8, Sphenisciformes (penguins); 9, Procellariiformes (tube-nosed seabirds); 10, Ciconiidae (storks); 11, Pelecanidae (pelicans); 12, Ardeidae (herons); 13, Threskiornithidae (ibises and spoonbills); 14, Suloidea (gannets, cormorants, and anhingas); 15, Fregatidae (frigatebirds); 16, Phoenicopteriformes (flamingos); 17, Podicipediformes (grebes); 18, Scolopaci (sandpipers and kin); 19, Turnicidae (buttonquails or hemipodes); 20, Lari (auks, gulls, and kin); 21, Accipitriformes (raptors); 22, Strigiformes (owls); 23, Coliiformes (mousebirds); 24, Coracii (rollers and kin); 25, Piciformes (woodpeckers and kin); 26, Columbiformes (pigeons); 27, Strisores (nightbirds, swifts and hummingbirds); 28, Psittaciformes (parrots); 29, Suboscines (pittas, broadbills, ovenbirds, and kin); 30, Passerida (finches, thrushes, warblers, and kin); 31, Corvoidea (crows and kin).

gulls, and allies—may also be part of this assemblage. Birds of prey, owls, swifts, and nightjars then form a series of landbird lineages that are successively closer to a higher landbird group (sometimed called Anomalogonatae) that typically includes kingfishers, woodpeckers, mousebirds, parrots, and passerines. This approximate structure—which is not far off the arrangement favored by Gadow—is probably correct because it has been repeatedly recovered in molecular, morphological, and combined data sets.

Some analyses have reported the division of Neoaves into two clades termed Metaves and Coronaves (Fain and Houde 2004; Ericson et al. 2006; Hackett et al. 2008) (though note that the exact membership of these groups differs between the respective studies). Metaves includes frogmouths, pigeons, tropicbirds, flamingos, swifts, hummingbirds, mesites, kagus, grebes, and sandgrouse, while Coronaves includes the remaining neoavians. Many of the taxa included in Metaves have historically been difficult to classify. Other studies have yet to repeat the recovery of Metaves or have rejected its monophyly (Morgan-Richards et al. 2008): its existence is only supported by data from a single gene.

A major morphological analysis of all birds, incorporating data from over 2,900 characters (Livezey and Zusi 2007), is conservative compared to other recently proposed phylogenies. Loons form a clade with grebes, and flamingos with storks, for example, while the traditional Caprimulgiformes, Pelecaniformes, Coraciiformes, and Falconiformes were recovered as monophyletic (in contrast, other recent studies reject the monophyly of these groups). Livezey and Zusi (2007) produced an enormous morphological character set, but it has been argued that their conclusions were erroneous due to a low signal-to-noise ratio (Mayr 2007).

Among extant neornithines, paleognaths and various galloanseraen clades have a mostly Southern Hemisphere distribution, and most clades within the traditional Gruiformes and Caprimulgiformes are denizens of the southern continents too. This is also true for passerines and some other clades. On the basis of this evidence, Cracraft (2001) argued that the early history of neornithines occurred almost entirely in the Cretaceous of Gondwana. If true, this could mean that neornithines partially escaped the effects of the end-Cretaceous Chicxulub event, as these might have been less severe in the Southern Hemisphere. However, many supposed southern groups were present in the Northern Hemisphere during the Paleogene, so a strong southern signal for early neornithines is not as clear as Cracraft implied (Mayr 2005a, 2009).

20.20. Skull and neck of a lithornithid specimen (possibly belonging to *Lithornis vulturinus*) from the Early Eocene Fur Formation of Denmark. The articulated hand skeleton is preserved lying underneath the neck.

Photo courtesy of Bent Lindow, used with permission.

Paleognaths: Ratites, Tinamous, and Lithornithids

Because indisputable neognaths were present in the Late Cretaceous, paleognaths must have been present at this time as well. Confirmed Mesozoic paleognaths have yet to be reported (though some contenders have been suggested), and the oldest definite members of the clade are Paleocene. These are the lithornithids, a group of volant paleognaths from the Paleocene and Eocene of North America and Europe (Fig. 20.20). A large hallux and curved claws suggest perching habits, and a slender bill probably capable of flexibility at its tip implies that lithornithids probed in soft sediments for invertebrate prey.

Ratites (as traditionally understood) are flightless, ground-living birds that have reduced wings where the humerus is longer than the ulna. The ratite sternum is as broad as it is long and lacks a keel. Vestigial features (including the presence of a pygostyle in ostriches and the differentiation of primary and secondary remiges in rheas) have usually been interpreted as evidence for derivation from volant ancestors, though this is moot given the many volant birds basal to ratites on the cladogram. It has often been assumed that ratite flightlessness is ancient and that their modern distribution reflects an original Gondwanan distribution. The realization that the volant lithornithids are probably paleognaths led to the suggestion that stem ratites used overwater dispersal early in their history (Houde 1986). Phylogenies that nest the flighted tinamous within ratites provide support for this model.

Numerous different views have been expressed on the relationships between the different ratites (e.g., Bledsoe 1988; Lee et al. 1997; Cooper et al. 2001; Livezey and Zusi 2007; Harshman et al. 2008; Bourdon et al. 2009). Emus (Dromaiidae) and cassowaries (Casuariidae) are close relatives and are grouped together as the Casuariiformes, but beyond this, most relationships are in flux. Some studies find kiwi (Apterygidae) or rheas (Rheidae) to be close to casuariiforms; a sister-group relationship between kiwi and moa has often been suggested but remains controversial.

Ostriches were present in Africa from the Early Miocene onward, but were also present from the Middle Miocene to the Pleistocene in eastern Europe and Asia. They persisted into the Holocene in China and Mongolia. In general, Miocene ostriches were small compared to the extant *Struthio camelus*, while the Pliocene and Pleistocene forms were similar in size, or even much larger, than *S. camelus*. It remains disputed whether ostriches arose in Asia (possibly India) and later migrated to Africa, or vice versa. Ostriches are cursors of savannas, semideserts, and deserts and combine phenomenal leg musculature (accounting for about a third of the animal's mass) with long tarsometatarsi and didactyl feet. In some fossil species, the fourth toe was even smaller than it is in living ostriches, making it seem that these birds were evolving toward complete dependence on the third digit.

Rheas are entirely South American, and fossils from Argentina and Brazil show that they have been in existence since the Paleocene. Some phylogenies recover Rheidae as one of the youngest paleognath clades (Harshman et al. 2008; Bourdon et al. 2009); their presence in the Paleocene would mean that most paleognath clades had arisen by this time.

Palaeotis from the Middle Eocene of Germany has sometimes been identified as a stem ostrich or rhea. It was a cursorial, flightless bird about 1 m tall with a narrow pelvis and narrow beak. It almost certainly does have affinities with crown ratites, but its position within Palaeognathae has yet to be determined with confidence. *Remiornis* from the Paleocene of France may be a close relative (Mayr 2009).

The most diverse ratites are the moa of New Zealand. Two moa clades are recognized, Emeidae and Dinornithidae, but the number of species considered valid (approximately nine to 14) is in flux because molecular work has revealed additional lineages within some morphologically defined species and has identified profound sexual dimorphism in the group. Some gigantic "species" (like *Dinornis giganteus*) have turned out to be

20.21. A selection of modern and recently extinct paleognaths. A, Double-wattled or Southern cassowary *Casuarius casuarius*. The largest of the three (or four) extant cassowaries, it occurs in northeastern Australia and on New Guinea and several of the surrounding islands. B, Single-wattled cassowary *C. unappendiculatus* of New Guinea and surrounding islands. C, Brown kiwi *Apteryx* (the species-level taxonomy of *Apteryx* is currently unresolved). All kiwi are endemic to New Zealand. D, Little spotted kiwi *A. owenii*, the smallest kiwi. E, The robust-legged moa *Pachyornis elephantopus*. Its massive proportions and stout, thick-boned legs make it one of the heaviest moa, with large individuals weighing 150 kg or more. F, Head of the moa *Euryapteryx geranoides*. G, The recently extinct elephant bird *Aepyornis maximus* of Madagascar.

the females of much smaller species (like *D. stuthoides*) (Bunce et al. 2003; Huynen et al. 2003). In such species, the females were about 280% the weight and 150% the height of the largest males. Moa were herbivores of forests and montane grasslands; bill shape, gizzard contents, and coprolites show that some favored twigs and other fibrous plant fragments, while others chose leaves and fruit (Worthy and Holdaway 2002). Huge olfactory chambers indicate a good sense of smell. Moa are arguably the most flightless of all birds, lacking the wing entirely and even the glenoid on the scapulocoracoid. *Pachyornis* and *Euryapterx* have remarkably short, stout legs (Fig. 20.21E). Lines of arrested growth (see Reid, Chapter 31 of this volume) are present in moa bones and show that they matured far more slowly than other neornithines. Some species took almost a decade to reach skeletal maturity (Turvey et al. 2005). Elongate, looped tracheae indicate that some species could generate loud, resonating calls. Aside from a few Miocene eggshell fragments, moa have no pre-Pleistocene fossil record.

The Australasian emus and cassowaries form a clade, and *Emuarius* from the Oligocene of Australia may be a stem emu. Cassowaries and crown emus are known from the Pliocene and Pleistocene. Cassowaries are notable for their various peculiarities compared to other ratites. These include brightly colored wattles and naked, carunculated neck skin, a particularly long, spikelike second toe claw (Fig. 20.22), long, stiff quills on the wings, and a bony casque sheathed with keratin. The casque may be used as a foliage deflector when the bird runs through dense bush, a visual signal of age and/or sex, a digging tool, and—most remarkably—as a microphone that helps the bird detect infrasonic sounds (Mack and Jones 2003).

Kiwi are among the most bizarre of neornithines (Fig. 20.21C, D) and combine gracile, monodactyl forelimbs and hairlike feathers with a reliance on olfaction and a tactile bill. Their production of an enormous egg that is about 20% of adult mass is well known. Rather than representing a leftover from large-bodied ancestors, it seems that this represents an extreme example of reliance on a single egg (the production of an egg this large is not unique; storm petrel eggs can be 28% of adult weight). Kiwi are endemic to New Zealand, and nothing is known of their history.

The Madagascan elephant birds (*Aepyornis* and *Mullerornis*) have been recovered by some as close to ostriches and rheas, and as members of a kiwi plus casuariiform clade by others. *Aepyornis* (Fig. 20.21G) exceeded 300 kg and was alive as recently as the seventeenth century. The wings and sternum are tiny, the legs are massive, and the beak is shallow and pointed. Alleged aepyornithid bones and eggshells have been reported from the Eocene, Oligocene, Miocene, and Pliocene of Egypt, Algeria, Turkey, and the Canary Islands, but all of these records are probably erroneous. As many as 15 *Aepyornis* species have been named, but all may be synonymous.

Until recently, the South American tinamous have been regarded as the ratite sister group. However, some studies indicate that tinamous are deeply nested within Ratitae (Hackett et al. 2008; Harshman et al. 2008) and may even be the sister group to moa (Phillips et al. 2009). If correct, this means that the flightlessness seen in the different ratite clades arose convergently at least four times. This would help explain why ratites look so different in such things as wing length and pelvis shape.

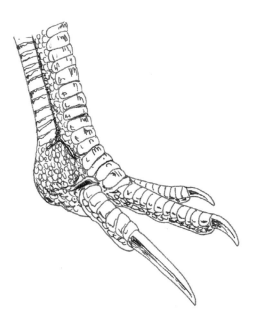

20.22. The left foot of a double-wattled cassowary. The long, pointed claw on digit II (the hallux is absent) is used as a weapon, deployed when the birds leap and kick. The claw is presumably used as a raking weapon, and not in stabbing. People have reportedly been killed by cassowaries, but only after provoking or even attacking the birds. The only verified death was that of Philip McClean in 1926. He hit the bird in an effort to kill it, tripped while running away, and was then stabbed in the neck.

Tinamous are gamebird–like birds of the grasslands, forests, and montane habitats of South and Central America. They can fly, though they do so reluctantly and are clumsy and prone to collisions. A number of forest-dwelling tinamous with plain or barred plumage are traditionally grouped together as the subfamily Tinaminae. A clade of open-habitat tinamous—the steppe tinamous or Nothurinae—is characterized by more complex plumage patterns and by the presence of posteriorly located nostrils. It seems that the invasion of open habitats occurred once within tinamou phylogeny (Bertelli and Chiappe 2005). Early Miocene fossils from Argentina represent close relatives of nothurine tinamous. They agree with paleoenvironmental data that suggest that the invasion of grasslands by tinamous happened during the Early Miocene.

Waterfowl and Gamebirds: the Galloanserae

Morphological and molecular data indicates that waterfowl or wildfowl (Anseriformes) and gamebirds or fowl (Galliformes) are sister groups and form the clade Galloanserae (or Galloanseres). Most of the characters that link them are in the palate, quadrate, and lower jaw. They include a long, mediolaterally compressed retroarticular process and large, oval facets for the pterygoids. Postcranially, modern waterfowl and gamebirds are quite different, but the gallinuloidid galliforms of the Eocene are anseriform-like in coracoid, scapula, and carpometacarpus morphology. Screamers (Anhimidae; Fig. 20.23A) are anseriforms, yet look superficially like galliforms. They have short, hooked bills and long, unwebbed toes. Because screamers are outside the clade that includes all duckbilled anseriforms (Anatoidea), this similarity may well be plesiomorphic. Modern screamers are South American, but North American and European kinds are known from the Eocene.

Anatoid anseriforms are known from the Late Cretaceous, so screamers and gamebirds must also have been present at this time. Within Anatoidea, the Australian magpie goose *Anseranas semipalmata* appears archaic, and most authors classify it outside of Anatidae (the clade that includes ducks, swans, and geese) and within its own group, Anseranatidae. *Anseranas* is long-legged with hardly any toe webbing, and unlike anatids, it does not molt all of its wing feathers at once. *Anatalavis* of the Paleocene of New Jersey and the Eocene of England might be a magpie goose (the erection of the new name *Nettapterornis* for the English species *A. oxfordi* is not widely accepted), but (unlike *Anseranas*) it was probably a filter feeder. If both screamers and magpie geese inhabited the Northern Hemisphere during the Paleogene, the Southern Hemisphere distribution of the extant taxa does not necessarily indicate a Southern Hemisphere origin for these birds.

Presbyornithids are closer to anatids than screamers or magpie geese and are best known from the Eocene of North America (Upper Cretaceous records are also known). This distribution is also not in agreement with a Southern Hemisphere origin for anatoids. The best known presbyornithid, *Presbyornis*, combines an anatid-like head with long legs. It was probably a wading filter feeder. Trackways show that it had fully webbed toes, and mass death assemblages demonstrate gregarious habits.

Anatids are the most diverse and successful waterfowl and include the whistling ducks, swans, geese, shelducks, dabbling ducks, diving ducks,

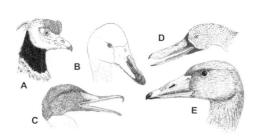

20.23. A selection of modern anseriforms. A, Northern or black-necked screamer *Chauna chavaria* of Colombia and Venezuela. Screamers feed mostly on water plants (they also eat grasses and seeds). B, Whooper swan *Cygnus cygnus*. Swans (Cygnini) and geese (Anserini) form the anatid clade Anserinae and have a fossil record extending back to the Oligocene. C, Common merganser or goosander *Mergus merganser*, a Northern Hemisphere merganser. Mergansers are part of the seaduck clade (Mergini) that also includes eiders and scoters. D, Southern or red shoveler *Anas platalea* of southern South America. E, Tundra bean goose *Anser serrirostris* of Russia and Siberia. Shovelers (D) are dabbling ducks (Anatini). Note the diversity of rhamphothecal serrations and lamellae. In geese (like E), the toothlike lamellae form an area known as the grinning patch.

mergansers, and stifftails (Fig. 20.24). The key to their success might be their unique double-piston suction pump. The tongue acts as a piston and is stored adjacent to the upper jaw rather than the lower; water is drawn in at the bill tip and then expelled outward via comblike lamellae. While originally used in shallow-water filter feeding, the components of this system have been modified for terrestrial grazing and browsing, fish seizing, and durophagy. Flightlessness has evolved repeatedly in waterfowl, most spectacularly in the recently extinct Hawaiian moa-nalos. These gooselike birds had a strongly reduced wing and pectoral girdle and turtlelike jaws. Despite their size, molecular work indicates that they are close relatives of dabbling ducks. Equally remarkable is another recently extinct Hawaiian anatid, the probably flightless *Talpanas*. Braincase characters and small orbits indicate that it was nocturnal and supremely specialized for tactile foraging with its bill, thereby convergently resembling kiwi.

It now appears that the enormous, flightless mihirungs (Dromornithidae) from Australia are anseriforms, or close relatives of them, and they are strikingly anseriform-like in many details of skull anatomy. Rounded beak tips, flattened crushing platforms at the backs of the jaws, and the presence of gastroliths indicate that mihirungs were herbivores (Murray and Vickers-Rich 2004). The oldest mihirungs are from the Late Oligocene or Early Miocene (though a possible Eocene record is known), and the youngest member of the group (*Genyornis*) survived to the Holocene. The largest mihirung (*Dromornis*) exceeded 400 kg.

The gastornithids from the Paleocene and Eocene of North America and Eurasia—often known as diatrymids after *Diatryma*, a junior synonym of *Gastornis*—also seem to be close to anseriforms. The massive, thick-boned gastornithid rostrum (Fig. 20.25) has usually been interpreted as that of a predator, but herbivory and nut cracking have also been suggested. An analysis of bite force suggests that *Gastornis* could bite through bones, and hence that it might have supplemented its diet with scavenging (Witmer and Rose 1991). However, circumstantial evidence—their galloanseraen affinities, the presence of undoubted herbivory in mihirungs, and the fact that they inhabited lush forests where fruits, seeds, and edible vegetation were abundant—suggests that gastornithids were most likely herbivorous.

It has been proposed that the bony-toothed birds (Pelagornithidae) are the sister group to Anseriformes, and that both form a clade that can be called Odontoanserae (Bourdon 2005, 2011). Further study is required before this suggestion can be accepted. More typically, pelagornithids have been regarded as close to tube-nosed seabirds or to "pelecaniforms."

Gamebirds or fowl are highly familiar, mostly as the result of the domestication of certain pheasants, guineafowl, quails, and turkeys. There are lineages that inhabit tropical forests, temperate woodlands, deserts, high mountains, and cold tundras. In general, gamebirds are strong-legged, ground-foraging birds with rotund bodies, short, rounded wings, and high wing loadings (the ratio of wing area to body mass). Megapodes (Megapodiidae) use rotting vegetation, warm soil, and even volcanically heated sand to build large or enormous nest mounds (the biggest are 5 m in diameter) in which they deposit their eggs (though some species construct burrows). Megapode hatchlings are superprecocial—that is, they are able to forage immediately after hatching and fly a few days later.

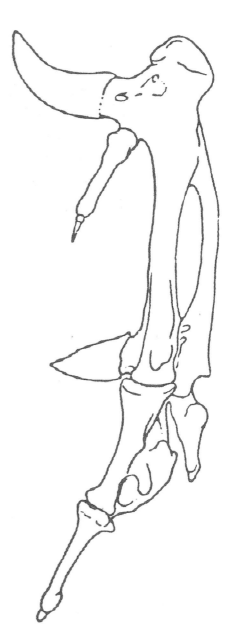

20.24. Carpometacarpus of the screamer (anhimid) *Chauna chavaria*. The remarkable daggerlike spurs are used in intraspecific combat and may even break off in fights and remain lodged in the soft tissues of the opponent. Many waterfowl are pugnacious birds, and clubs and spurs on the carpometacarpus have evolved several times within the group.

20.25. Skull and neck of the gastornithid *Gastornis*, now thought to include the species previously included in *Diatryma*. Several *Gastornis* species (differing in size, the gracility of their limbs, and proportional length of their toe bones) are known from the Eocene of Europe and North America, and Paleocene remains may also belong to the taxon. *Zhongyuanus* from the Early Eocene of China seems to be a close relative.

Neoaves: the Waterbird Assemblage

Extant galliforms range in size from less than 100 g (some quails) to over 10 kg (some turkeys). Some fossil galliforms were much larger: the flightless megapodes *Sylviornis* of New Caledonia and the Isle of Pines and *Megavitiornis* of Fiji may have weighed as much as 40 kg. Gamebirds are mostly generalist herbivores, though many will eat small animals. Several distinctive gamebird characters are seen in the furcula and sternum and seem to be related to the evolution of a large crop that takes up space originally occupied by the anterior part of the sternal keel.

Two taxa that might be stem galliforms—*Palintropus* and *Austinornis*—have been identified from the Cretaceous, but otherwise the earliest records of the clade are from the Eocene. These include the gallinuloidids of Europe and North America and the European paraortygids and quercymegapodiids (the latter are also known from Brazil). All of these taxa seem to be stem galliforms. Among other characters, they possess a primitive, cup-shaped facet for the scapula on the coracoid (in crown galliforms, the facet is flat). Crown galliforms are definitely known from the Oligocene of Australia and Europe, but a few possible Eocene records are known.

Recent phylogenies indicate that megapodes are the sister taxon to all other crown galliforms, and that curassows, chachalacas, and guans (Cracidae), guineafowl (Numididae), and New World quails (Odontophoridae) are successively closer to Phasianidae, a large clade that includes pheasants and pheasantlike taxa, grouse, turkeys, quails, and peafowl (Ksepka 2009). Megapodes and cracids have sometimes been regarded as each other's closest relatives, and even to warrant distinction as their own order (Craciformes) separate from other galliforms, while grouse and turkeys have traditionally been regarded as outside of Phasianidae.

Molecular data show that Neognathae diverged early in its history into Galloanserae and Neoaves. A ventral keel on the palatine, the absence of basipterygoid processes, and the absence of a tubercle just below the acetabulum may be neoavian synapomorphies (Mayr and Clarke 2003), but one of the best-known neoavian characters is the absence of the phallus. In contrast to most members of Palaeognathae and Galloanserae, neoavian males inseminate females via a brief cloacal kiss. It is usually thought that the loss of the phallus represents a weight-saving adaptation, but this is questionable because the weight loss concerned would have been minimal. Furthermore, members of several neoavian lineages (including various parrots and passerines) have evolved nonhomologous intromittent structures, and these are sometimes surprisingly large.

As discussed earlier, the various waterbird groups cluster at the base of Neoaves. Most of these birds share marked furrows (termed nasolabial grooves) along the sides of the beak and also long nostrils, and they are also united in genetic analyses. Loons, penguins, tube-nosed seabirds, storks and stork-like forms, herons and gannets and kin all seem to form a clade (Fig. 20.19), named Aequornithes by Mayr (2010a). Waterbirds with fully webbed (so-called totipalmate) feet and gular pouches have been classified together as the Pelecaniformes. These are the tropicbirds (Phaethontidae), frigatebirds (Fregatidae), pelicans (Pelecanidae), gannets and boobies (Sulidae), cormorants (Phalacrocoracidae), and anhingas (Anhingidae). Molecular

20.26. Several gamebird lineages have evolved spectacular plumage and other display structures. The male great argus *Argusianus argus* shown at left (displaying to a female) has raised his enormous ocellated wing and tail feathers. At right, the male domestic turkey *Meleagris gallopavo* is displaying his snood (the dangling object drooping over the bill), carunculated head and neck skin, and metallic plumage.

studies have failed to support pelecaniform monophyly: pelicans group with the African shoebill *Balaeniceps rex* and hammerkop *Scopus umbretta*, and perhaps with ibises and herons too (Hackett et al. 2008). At least some morphological studies, however, still find pelicans to be close kin of frigatebirds and Suloidea (the gannet, cormorant, and anhinga clade) (Smith 2010): the name Steganopodes has been used for the pelican-frigatebird-suloid clade. Frigatebirds and Suloidea form a clade characterized by a short tarsometatarsus and a pectinate claw on digit III. *Masillastega* from the Middle Eocene of Germany is gannetlike and indicates that most divergences within the frigatebird–suloid clade had occurred by this time, and this is further confirmed by the Eocene frigatebird *Limnofregata*. *Protoplotus* from the Paleocene or Eocene of Sumatra might be an early member of the frigatebird–suloid clade (Mayr 2009).

Frigatebirds (Fregatidae) are long-winged aerial specialists. They exhibit the lowest wing loadings of any bird, and together with swifts, they are the only birds known to sleep during flight. They have short legs and lack waterproofed plumage, and are reluctant to swim or dive. Prey are plucked from the water during flight, but frigatebirds are also pirates of other seabirds. The exploitation of dispersed prey correlates with low provisioning rates of juveniles, and this probably explains why parental care in frigatebirds is so extended: they take about a year to raise a chick. In turn, their reproductive rate is low and their life spans are long. *Limnofregata* lacks the long wings and various fusions in the pectoral region of modern frigatebirds and was probably more gull-like in ecology.

Anhingas–darters or snakebirds–are swimming predators that use a rapid darting mechanism in the neck and a daggerlike, serrated bill to spear fish and other prey. They have notably long tail feathers, the outermost of which have a peculiar corrugated texture. Particularly large anhingas inhabited South America during the Miocene. Of these, it has been suggested that *Meganhinga* was flightless (though this looks doubtful based on its limb proportions) while *Macranhinga* was a cormorant-like swimmer and diver, and not a surface skulker like extant anhingas. Shared characters of the cervical vertebrae, pelvis and tibiotarsus indicate that anhingas are the sister group to cormorants, and both are united in the suloid clade Phalacrocoracoidea (Mayr 2009; Smith 2010).

Cormorants are widespread birds of lakes, rivers, and coasts that chase fish underwater by way of foot-propelled swimming. Their earliest appearance is in the Late Eocene of England. Gannets and boobies are plunge divers that lack external nostrils and possess large pectoral air sacs. The

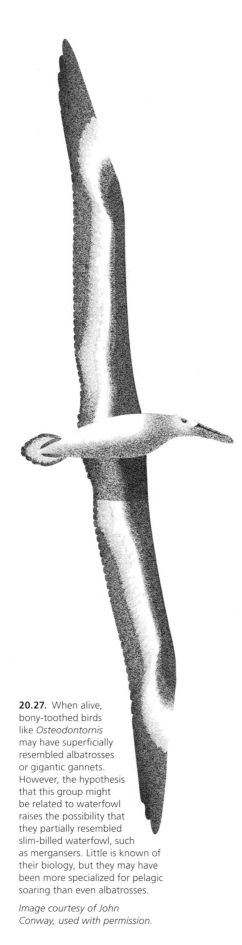

20.27. When alive, bony-toothed birds like *Osteodontornis* may have superficially resembled albatrosses or gigantic gannets. However, the hypothesis that this group might be related to waterfowl raises the possibility that they partially resembled slim-billed waterfowl, such as mergansers. Little is known of their biology, but they may have been more specialized for pelagic soaring than even albatrosses.

Image courtesy of John Conway, used with permission.

largest species (like *Morus bassanus*) launch themselves at the water at 24 m s^{-1} and can dive to depths of 34 m. *Rhamphastosula* from the Pliocene of Peru is unlike modern sulids in having a deep, curved bill.

Tropicbirds are a small group of superficially gull-like plunge divers. Their affinities are uncertain: a close relationship with tube-nosed seabirds and inclusion in Metaves have both been proposed. The prophaethontids of the Paleocene and Eocene of Eurasia, North America, and Morocco were similar to tropicbirds but perhaps with better swimming and diving abilities.

Plotopterids are flightless, wing-propelled diving birds known from the Eocene to Miocene of the northern Pacific. They resemble penguins in wing form and in their shortened tarsometatarsi. The smallest taxa are similar in size to medium-size cormorants, while the largest (*Copepteryx*) is about 1.8 m long. Plotopterids were originally suggested to be convergent with penguins and to have emerged from among Suloidea. Mayr (2005b) proposed that plotopterids and penguins might be sister groups but Smith (2010) argued that characters of the plotoperid coracoid, humerus, pelvis and hind limb better support a position within Suloidea and close to Phalacrocoracoidea.

Among the most remarkable of fossil birds are the bony-toothed birds, pseudotoothed birds, or pelagornithids (Fig. 20.27), sometimes given their own order, Odontopterygiformes. Their fossils have been discovered virtually worldwide and range in age from Late Paleocene to Pliocene. Pelagornithids were pelagic soarers, superficially resembling albatrosses but with toothlike projections lining their jaws. They also possess bony furrows along the length of the rostrum, and their thick jugal bars, intramandibular joints, and absence of a dentary symphysis imply the handling and swallowing of large prey (presumably fish and/or squid). While some species were similar in size to large albatrosses, others (like *Osteodontornis* from the Miocene of the United States) have wingspans of 5–6 m.

Pelagornithid taxonomy is confused, and revision is needed (Bourdon 2011). Their higher-level relationships are also unresolved. A superficial similarity with tube-nosed seabirds and some "pelecaniforms" (particularly sulids) has meant that pelagornithids have often been considered close relatives of one or both groups. However, no convincing character data have been presented that might support either possible affinity, and it seems that any similarities between pelagornithids and tube-nosed seabirds or sulids are convergent. Bourdon (2005, 2011) found character support for a sister-group relationship between pelagornithids and waterfowl. The ventral surface of the braincase is similar in both groups, and they also share characters of the radius, ulna, carpometacarpus, and tarsometatarsus. Pelagornithids also recall anseriforms in the presence of bony pseudoteeth (seen elsewhere only in moa-nalos) and in details of brain anatomy, and a few other features also suggest placement outside Neoaves. Nevertheless, further study is needed before we can be confident about the affinities of these spectacular birds.

Tube-nosed seabirds (Procellariiformes) are one of the most important seabird groups, with about 110 species occurring worldwide. The majority inhabit the Southern Hemisphere. Most are pelagic birds that eat squid, fish, and plankton and nest on islands or in burrows. They are unquestionably monophyletic and share tubular nostrils and a hallux composed of only a single phalanx. The bill is covered with a number of plates and has a terminal hook. Of the four extant groups, albatrosses (Diomedeidae) are large, stiff-winged soarers (Fig. 20.28); diving-petrels (Pelecanoididae) are wing-propelled divers that superficially resemble auks; storm-petrels (Hydrobatidae) are passerine-size planktivores that usually pick prey from the sea surface, while petrels, shearwaters, and fulmars (Procellariidae) include a diverse array of piscivores, planktivores, and scavengers.

Diving-petrels and true petrels appear to be sister groups, though studies differ as to whether albatrosses or storm-petrels are closest to this clade. Albatrosses are among the most specialized of birds for soaring and possess a locking joint in the shoulder as well as an accessory ossification in the elbow region that helps keep the propatagium stiff.

Tube-nosed seabirds have a poor early fossil record. *Tytthostonyx* from the Paleocene of New Jersey was described as a tube-nose, various Paleocene remains from Europe and Asia have also been referred to the group, and possible albatrosses have been reported from the Eocene. However, most of these records are doubtful or require confirmation. Definite albatrosses, storm-petrels, diving-petrels, and true petrels are known from the Miocene. The Oligocene–Miocene diomedeoidids are known from numerous remains from Europe and Iran. They appear to be long winged like petrels, but like some storm-petrels, they have large, broad feet and blunt claws. Various characters suggest placement outside the procellariiform crown.

Storks (Ciconiidae) are well known for their long legs and necks and massive, pointed bills. *Eociconia* from the Middle Eocene of China may be an early member of the group, but a more securely identified early taxon is *Palaeoephippiorhynchus* from the Early Oligocene of Egypt. Storks practice a variety of foraging techniques both on land and in water: some (such as the *Ciconia* species) snap up frogs, worms, insects, and other prey in wetlands, others (like the *Mycteria* species) have sophisticated touch receptors in their beak tissue and grab prey in turbid or muddy pools, while others (the marabous or adjutants) are terrestrial predators and scavengers. Some fossil marabous were gigantic animals 1.8 m tall (Fig. 20.29).

Herons (Ardeidae) are long-legged waders with flexible necks. The daggerlike bill is hurled forward at great speed and driven through the body of the prey. Herons are highly adaptable, and some species eat frogs, snakes, birds, and mammals in addition to fish. Some even use bait (such as pieces of bread) to catch fish. A broad bill is present in the tropical American boat-billed heron *Cochlearius cochlearius*.

A general similarity with storks and ibises has led to the inclusion of herons within Ciconiiformes, but the narrow feather tracts and presence of powder-down patches (areas where the feather barbs break down to form a fine dust used in preening and waterproofing) have been used to link herons to gruiforms such as mesites or the members of Eurypygae (Olson 1978, 1985). Some studies find herons to be close to the shoebill–pelican clade (Ericson et al. 2006; Hackett et al. 2008), a discovery that suggests an African origin for herons. *Proardea* from Quercy in France has often been regarded as the oldest heron, but its age is uncertain, and an Early Oligocene specimen from Egypt might be older.

Ibises and spoonbills (Threskiornithidae) are also long-legged and long-billed, but they differ from storks and herons in having schizorhinal nostrils (where the nostrils are long and slitlike, and extend posterior to

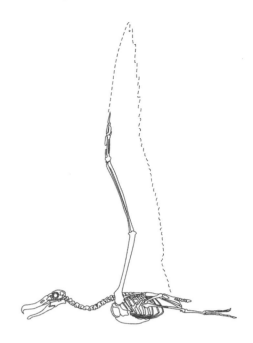

20.28. Skeleton of the wandering albatross *Diomedea exulans*, the longest-winged extant flying bird. It can have a wingspan of 3.7 m and weigh as much as 13 kg (some swans and bustards are heavier, reaching 20 kg or slightly more). Satellite tracking shows that some individuals can cover 15,000 km in a single foraging trip.

Image courtesy of Mark Witton, used with permission.

20.29. The gigantic Pleistocene marabou stork *Leptoptilos robustus* from Flores, shown to scale with the extinct hominid *Homo floresiensis*. This bird was about 1.8 m tall and had reduced flight abilities. Its strong adaptation to terrestrial life was probably related to an abundance of small prey (like rodents and lizards) and an absence of large mammalian predators.

Image by Inge van Noortwijk and provided by Hanneke Meijer, used with permission.

20.30. Spoonbills (Plataleinae) and ibises (Threskiornithinae) are sister groups within the waterbird clade Threskiornithidae. The White spoonbill *Platalea leucorodia* is an Old World species of Eurasia and Africa. Spoonbill jaws are dorsoventrally flattened and possess numerous sensory corpuscles at their tips.

20.31. Head of a shoebill *Balaeniceps rex*, also known as the whalebill or whale-headed stork. The enormous, hook-tipped bill is used to grab fish, turtles, and snakes, and local anecdotes tell of them eating antelope calves and carrion. A large shoebill stands 1.4 m tall and has a wingspan of 2.6 m.

the nasofrontal hinge) and a flexible bill that has sensitive pits at its tip. These features are shorebirdlike, and some authors have suggested that threskiornithids might be allied with Charadriiformes and/or Gruiformes (Olson 1978). It is more likely that they are close relatives of herons and the shoebill–pelican clade (Ericson et al. 2006; Hackett et al. 2008; Mayr 2010a). The Eocene threskiornithid *Rhynchaeites* differs from ibises and spoonbills (Fig. 20.30) in having shorter legs and a less sensitive bill. It indicates that the long legs and shorebirdlike bill of modern taxa were not present in the earliest members of this clade. Flightless ibises evolved on Hawaii and Jamaica but became extinct within recent centuries.

Shoebills (Fig. 20.31), hammerkops, and pelicans share a hooked tip to the rostrum and the presence of numerous small foramina on the internal surface of the sternum. Shoebills and hammerkops are long-legged birds that grab fish, frogs, and other prey from shallow pools; pelicans are short-legged and grab prey while swimming, though two species are plunge divers. An Early Oligocene shoebill is known (*Goliathia* from Egypt), while the oldest pelican is from the Early Oligocene of Europe and appears fully modern, hinting at an unknown, Southern Hemisphere history.

Loons or divers (Gaviiformes) are represented by five extant species that occur in Northern Hemisphere lakes and along seacoasts (depending on the time of year). They are heavy-bodied diving birds with daggerlike beaks, a long, narrow pelvis, and muscular legs that are placed well back on the body. A large triangular cnemial crest extends anterior to the knee, forming a large projection similar to that seen in hesperornithines and grebes, the tarsometatarsus is laterally compressed, and the long toes are webbed. Two Cretaceous loons are known: *Neogaeornis* from Chile and *Polarornis* from Antarctica (both may be synonymous). Cenozoic loons are known from Europe and North America. While loons have often been grouped with grebes, the two are different in virtually all respects, and many workers have argued that the similarities represent convergences to foot-propelled diving. A position close to tube-nosed seabirds, penguins, storks, and/or pelicans has been supported in most recent studies.

Penguins (Sphenisciformes) have an excellent fossil record that may begin in the Upper Cretaceous with an as yet unnamed stem taxon from Chatham Island. *Waimanu* from the Paleocene of New Zealand was already a flightless, wing-propelled swimmer, but it differs from modern penguins in having a long, slender bill and a proportionally longer tarsometatarsus. Daggerlike bills, sometimes likened to those of loons, are widespread in fossil penguins and indicate that they speared fish and were not planktivores like most modern penguins. Large size was common in the penguins of the Paleocene, Eocene, and Oligocene, with many being larger than the emperor penguin *Aptenodytes forsteri*. The biggest (*Anthropornis*) perhaps had a standing height of about 1.7 m. Penguin diversity was high during the Paleogene, and the post-Oligocene disappearance of giant forms may be related to the rise of pinnipeds or odontocete whales. Recent discoveries have shown that penguins aquired many of their distinctive specializations during the warm phase of the Paleogene and that, unlike living species, some fossil penguins possessed reddish-brown feather patterns (Ksepka and Ando 2011). Penguins might be close to tube-nosed seabirds or to the frigatebird–suloid clade (Fig. 20.19).

The monophyly or otherwise of a group that includes cranes, rails and their supposed allies – the Gruiformes – has been one of the most controversial subjects in neornithine systematics. Analyses suggest monophyly of a core gruiform clade that includes rails (Rallidae), finfoots (Heliornithidae), trumpeters (Psophiidae), limpkins (Aramidae), and cranes (Gruidae). Of the other taxa usually included in Gruiformes, kagus (Rhynochetidae) and sunbitterns (Eurypygidae) seem to be sister groups, forming the clade Eurypygae. Eurypygae is part of Metaves in some analyses (Fain and Houde 2004; Hackett et al. 2008), as are mesites or roatelos (Mesitornithidae). Bustards (Otididae) are not core gruiforms, and their affinities remain uncertain. The South American seriemas (Cariamidae) lack characters that link them to core gruiforms and might be part of the clade that includes parrots and passerines (Mayr and Clarke 2003; Ericson et al. 2006; Hackett et al. 2008). This suggests that a number of fossil groups usually regarded as close to seriemas (and grouped together with them in Cariamae) also belong in this region of the tree.

Rails are a successful group. There are over 140 living species; they occur worldwide and include the swamphens, wood-rails, flufftails, true rails, crakes, waterhens, and coots. Rallidae of tradition may be paraphyletic with respect to finfoots (Hackett et al. 2008). Rails have a narrow sternum and pelvis (resulting in a laterally compressed body) and are generally stout-legged birds with short wings and high wing loadings. They are good at colonizing islands, and several large, flightless island endemics have evolved. Examples include New Zealand's takahe *Porphyrio hochstetteri* (Fig. 20.32) and the ibislike *Aphanapteryx* and *Diaphorapteryx* of the Mascarenes and New Zealand, respectively. A predisposition to flightlessness has made rails prone to extinction at the hands of humans and introduced rats and cats. Various alleged rallids have been reported from the Eocene of Europe, Asia, and North America, but virtually all lack characters allowing definite classification. *Belgirallus* from the Early Oligocene of Belgium and Germany is the best-known early rail. *Amitabha* from the Eocene of Wyoming (originally described as a gamebird) was identified by Ksepka (2009) as a possible rallid sister taxon.

Finfoots are grebelike swimming birds of the African, Asian, and New World tropics. A Miocene fossil from the United States suggests that finfoots invaded North America from Asia and later migrated into South America. A South American finfoot, the sungrebe *Heliornis fulica*, possesses paired pouches under its wings that are used to carry the young.

The recently extinct, flightless adzebills (*Aptornis*) from New Zealand might be part of Eurypygae (Livezey and Zusi 2007), or closer to rails (Houde et al. 1997). *Aptornis* is notable for its robust, decurved bill; superficially, it may have resembled a gigantic flightless rail with strongly reduced wings. It was probably a predator of invertebrates and small vertebrates.

Several raillike birds from the Paleocene, Eocene, and Oligocene of North America and Europe are grouped together as the Messel rails or messelornithids. Hundreds of specimens are known. Several authors have regarded Messel rails as close relatives of kagus and sunbitterns, but a close relationship with the rail–finfoot clade was favored by Mayr (2009). Thanks to the soft tissue preserved in association with one specimen, messel rails are usually imagined to possess a fleshy head crest, similar to a chicken's comb.

20.32. Skull of a takahe *Porphyrio hochstetteri*, a large, flightless rail from New Zealand's South Island. The two takahe species (the longer-legged North Island *P. mantelli* is extinct) were once grouped together as *Notornis* but do not seem to be close relatives. Modern takahe inhabit cool upland grasslands and mostly eat tussock grass, but fossils show that they were previously common in lowland habitats. Their restriction to uplands may be due to human hunting and habitat change.

However, Mayr (2009) argued that the crest did not belong to the bird and was just a fortuitously positioned piece of organic matter.

Limpkins are long-legged, long-billed waders, similar in many respects to cranes and forming their sister group in most analyses (Sibley and Ahlquist 1990 proposed a close link with the sungrebe, but this is now thought to be incorrect). Possible fossil limpkins are known from the Oligocene; the extant species (*Aramus guarauna* of the American tropics) is a predator of apple snails.

Fragmentary fossils from the Eocene and Oligocene of Europe, Asia, Africa, and possibly North America appear to be early cranes. These are thought by some authors to be close relatives or members of *Balearica*, the African crowned cranes, and it has also been suggested that crowned cranes occurred across North America and Asia during the Miocene (Feduccia 1996). Gruine cranes diversified on the steppes and marshes of Eurasia and North America. Flightless cranes evolved on Cuba and Bermuda.

A number of cranelike birds known from the Eocene to Pliocene of Europe and Asia are grouped together within Eogruidae (here taken to include Ergilornithidae following Clarke et al. 2005). At least some eogruids were didactyl, with a strongly reduced or absent trochlea for digit II. They were cursorial and perhaps had reduced or absent flight abilities. Didactyly and a few other features once led some workers to propose that eogruids were ancestral to ostriches. The poorly known geranoidids from the Eocene of North America may be close relatives of eogruids.

As discussed above, the seriemas and their fossil relatives—grouped together as the Cariamae—are almost certainly not close relatives of core gruiforms. The proposal that seriemas might be long-legged relatives of the hoatzin (Olson 1985)—generally mostly dismissed as bizarre and speculative—has been supported in some recent analyses (e.g., Mayr and Clarke 2003; Clarke et al. 2005). Seriemas are often likened to secretarybirds, and like them are ground-hunting predators of small vertebrates and insects. Most fossil members of Cariamae, in particular idiornithids and bathornithids, seem to have been similar. Idiornithids are best known from the Eocene and Oligocene of Europe but have also been reported from the Paleocene of Brazil, while bathornithids are from the Eocene and Oligocene of North America. The best known bathornithid, *Bathornis*, was (under its junior synonym *Neocathartes*) long depicted as a sort of long-legged cursorial vulture.

Among the most spectacular of fossil birds are the phorusrhacids or terror birds (Fig. 20.33). Several phorusrhacid clades have been identified (Alvarenga and Höfling 2003). Patagornithines and phorusrhacines were large, but the members of some of the other clades were gracile, seriema-like birds. Little is known about phorusrhacid paleobiology, but it is assumed that they were cursorial predators of mammals and other vertebrates, perhaps using carefully placed strikes with the relatively rigid, narrow beak and perhaps mostly exploiting small prey, despite their often large size (Degrange et al. 2010). Tremendously strong leg bones in the gracile mesembriornithines of the Miocene and Pliocene suggest that they were able to stun or kill prey with strong kicks.

The gigantic *Brontornis* (supposedly the best-known member of the phorusrhacid clade Brontornithinae) was argued by Agnolin (2007) to be

20.33. A, Reconstructed skull of the Miocene phorusrhacid *Phorusrhacos longissimus* from Argentina. The discovery of *Kelenken guillermoi*, also from the Miocene of Argentina, shows that some giant phorusrhacids had shallower skulls than this. B, In life, phorusrhacids like this *Andalgalornis steuletti* probably resembled gigantic seriemas, but with more formidable beaks. The idea that they possessed mobile clawed fingers is erroneous: the alula was mobile, but this is normal for birds. The largest species (like *K. guillermoi*) were over 2 m tall and exceeded 160 kg.

B courtesy of John Conway, used with permission.

only convergently similar to phorusrhacids, and, on the basis of characters in its quadrate, mandible, and leg bones, to actually be an anseriform. Alvarenga et al. (2011) have since argued that the characters used to support this proposal are either erroneous or present in undoubted phorusrhacids, and that *Brontornis* can be confidently assigned to Phorusrhacidae.

Shorebirds and their kin, the Charadriiformes, are a globally distributed group of over 350 species. Most frequent freshwater and marine habitats (Fig. 20.34), though some (including pratincoles and coursers, buttonquails, and various gulls) occur in dry terrestrial places. Various Cretaceous neognaths—grouped together as the graculavids and included among the so-called transitional shorebirds—have been regarded as charadriiforms by some authors, but they lack characters that might support such an assignment. While some studies support a basal position within Neoaves for charadriiforms, others find them to be closer to the higher landbirds (Hackett et al. 2008). If *Morsoravis sedilis* from Denmark has been correctly identified (Bertelli et al. 2010), stem charadriiforms are known from the Early Eocene.

Charadriiform phylogeny has been controversial, but most studies find the crown group to consist of three clades. Charadrii contains the Antarctic sheathbills (Chionididae), thick-knees (Burhinidae), stilts and avocets (Recurvirostridae), oystercatchers (Haematopodidae), and plovers (Charadriidae). Lari (containing gulls, terns, auks and coursers) and Scolopaci (containing sandpipers, jacanas and seedsnipes) may (Baker et al. 2007) or may not (Mayr 2011a) be closer to one another than to Charadrii. The fact that auks (Alcidae) are known from the Late Eocene shows that these three clades had appeared by this time. Pratincoles and coursers (Glareolidae) are known from the Early Oligocene. Gulls (Laridae) and terns (Sternidae) are definitely known from the Miocene onward, with only fragmentary fossils reported from the Oligocene.

Charadriiforms encompass great diversity and include many peculiar, specialized taxa. Skimmers (Rynchopidae) are close kin of gulls and terns that possess a bladelike lower jaw that is trawled through the water during flight (Fig. 20.34D). They possess vertical, slitlike pupils and a suite of cranial and cervical adaptations for skim feeding. Skuas (Stercorariidae) are gull-like kleptoparasites or predators of other seabirds. Auks have traditionally been characterized as miniature flying penguins destined for flightlessness, but they are actually specialized for extremely rapid flight (up to 130 km h^{-1} in the marbled murrelet *Brachyramphus marmoratus*) that allows them to exploit predictable patches of energy-rich marine prey. Puffins and auklets are flamboyantly ornamented, and razorbills and other species can dive to depths of 140 m or more. Large, flightless auks evolved in both the Pacific (the mancallines of the Miocene, Pliocene, and Pleistocene) and Atlantic. The great auk *Pinguinus impennis* (Fig. 20.34C)—the most recent flightless auk—was hunted to extinction in about 1850.

Buttonquails or hemipodes (Turnicidae) have often been linked with rails, cranes, and allies within Gruiformes but seem nested within Charadriiformes and may be close to the auk–gull clade. These superficially quail-like birds inhabit Old World and Australasian tropical grasslands, and they prefer not to fly. Stem turnicids (*Turnipax* and *Cerestenia*) are known from

Shorebirds and Kin, and Grebes and Flamingos

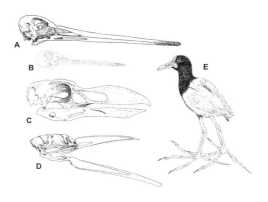

20.34. Extreme morphologies among Charadriiformes. Skulls of (A) the scolopacids *Limosa* (godwits) and (B) *Calidris mauri* (Western sandpiper), (C) the recently extinct alcid *Pinguinus impennis* (great auk), and (D) the rynchopid *Rynchops* (skimmers). (E) Wattled jacana or lily-trotter *Jacana jacana*, displaying the enormously long toes of this charadriiform group.

A, B, and D courtesy of Mark Witton; used with permission.

20.35. The flamingo skull (this belongs to *Phoenicopterus ruber*) is unique and unmistakeable. In life, the deep, downcurved rostrum houses an enlarged tongue used to pump water in and out of the mouth across filtering lamellae. *Phoenicopterus* has coarse lamellae compared to some other flamingos and hence feeds on larger prey, typically small crustaceans and mollusks. Taxa with more closely spaced lamellae feed on blue-green algae.

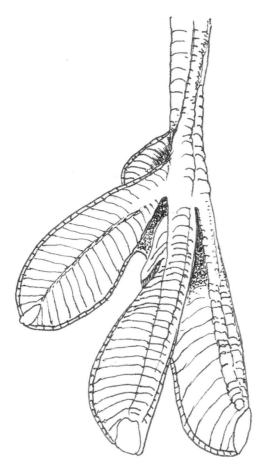

20.36. The lobed foot of a grebe. Grebe feet are thrust sideways – rather than backward – during swimming. During the power stroke, the toes separate and are used to generate lift. When the foot is drawn back to the body during the recovery stroke, the toes close tightly together to form a streamlined blade.

the Early Oligocene. *Turnipax*, at least, differs from crown turnicids in possessing a hallux. Scolopacidae includes the long-billed godwits, dowitchers, curlews, snipes, and others (Fig. 20.33A, B). Flexible bill tips allow them to grasp prey even when the bill is immersed in sediment, and pressure-sensitive organs in the bill (termed Herbst corpuscles) allow the birds to detect pressure gradients, and hence hard, concealed objects, within sediment.

Among the most bizarre of charadriiforms are the tropical jacanas or lily-trotters (Jacanidae) (Fig. 20.34E). They possess incredibly long, slender toes and use them to walk on floating vegetation. Some jacanas possess anteriorly positioned nostrils and hide underwater while using their bills as snorkels. The oldest fossil jacanas (*Nupharanassa* and *Janipes*) are from the Late Eocene and Early Oligocene. Jacanas appear to be the sister group to the painted snipes (Rostratulidae).

Flamingos (Phoenicopteriformes) are unmistakeable long-legged filter feeders of the American, African, and Asian tropics, though three species inhabit cool or temperate habitats in South America. Fossil species occurred in Australia, Europe, and the United States between the Miocene and Pleistocene. Flamingo pecularities are well documented. They form gigantic feeding and breeding aggregations, build columnar mud nests, and possess deep, downcurved bills where an enormous spiny tongue pumps water through filtering lamellae in the upper jaw (Fig. 20.35).

Historically, flamingos have been allied with waterfowl, storks, and stilts. The proposal that flamingos and grebes might be close relatives (Van Tuinen et al. 2001) first seemed bizarre but has since been supported by other studies (it has, however, been harshly criticized by some workers: see Livezey 2010). The name Mirandornithes has been proposed for the grebe plus flamingo clade. Mirandornithines might be the sister group to Charadriiformes (Morgan-Richards et al. 2008) or to mesites (Mayr 2010). Characters shared with the members of Eurypygae hint at a relationship with this clade as well.

A group of Oligocene and Miocene waterbirds, the palaelodids (sometimes called swimming flamingos), have been imagined as stem flamingos (and even reconstructed as long-legged and pink by some artists). Mayr (2009) noted that the palaelodid hind limb is grebelike in some respects (the tarsometatarsus is shorter and more mediolaterally compressed than that of crown flamingos, and the hypotarsus partially encloses the canals for the digital flexor tendons, for example), though the deep mandible suggests an enlarged tongue used in flamingo-like filter feeding. It is possible, therefore, that stem flamingos like palaelodids were swimming birds that, like grebes, used their feet for propulsion. Crown flamingos would represent a specialized, shallow-water, wading mirandornithine clade if this interpretation is correct. However, *Juncitarsus* – which is long-legged and possesses a straight beak, yet appears to be a stem mirandornithine – conflicts with this scenario.

Grebes (Podicipediformes) are known as fossils from the Miocene onward, but even the oldest forms are little different from extant species. Their stiff toe lobes (Fig. 20.36) were assumed to increase drag during the leg's power stroke. However, the feet are moved perpendicular to the bird's direction of movement, and it now seems that the toes act as a multislotted hydrofoil (Johansson and Norberg 2001). Two South American grebes are or were flightless (one became extinct in 1989). A peculiar feature of grebes is that they deliberately swallow their own feathers (feathers may compose 50% of their stomach contents by weight), and even feed them to their chicks. This may be to protect the stomach from sharp fish bones, though this hypothesis still awaits testing.

Several neornithine clades exhibit a strongly hooked beak and (usually) raptorial talons (Fig. 20.37), and they use these organs in predation or scavenging. These are the birds of prey, or raptors. The monophyly or otherwise of this assemblage, traditionally united as the Falconiformes, is a long-standing controversy. They do share some characters (such as paired canals on the pedal unguals), but in other respects, the similarities are superficial, and falcons at least have not grouped with other raptors in recent molecular phylogenies. The similarity that owls (Strigiformes) have with diurnal raptors has often been dismissed as superficial, and affinities with caprimulgiforms and parrots have been supported by some. However, owls do group with diurnal raptors in some morphological analyses (Mayr and Clarke 2003; Mayr et al. 2003; Livezey and Zusi 2007).

Accipitrids—kites, harriers, hawks, eagles, Old World vultures, and so on—are known from the Late Eocene onward, and the distinctive pedal unguals of ospreys (Pandionidae) show that these birds were present as early as the Eocene too. Some analyses find accipitrids, ospreys, and secretarybirds (Sagittariidae) to form a clade. Accipitrids include the most spectacular of neornithine predators. Even some living eagle species (like the golden eagle *Aquila chrysaetos*) can kill mammals more than 16 times their own weight, and the extinct Haast's eagle *Hieraaetus moorei* of New Zealand routinely dispatched moa that weighed more than 100 kg (Fig. 20.38).

Secretarybirds are represented today by the African species *Sagittarius serpentarius*. It is a long-legged predator of snakes, lizards, insects, and other animals, and kills them with powerful kicks. Fossil secretarybirds similar to *Sagittarius* (but smaller) are known from the Oligocene of France. A Miocene accipitrid from North America (*Apatosagittarius*) was strikingly similar to secretarybirds in hind limb morphology and probably hunted in a similar manner.

New World vultures (Cathartidae) are dedicated scavengers (though they sometimes kill live animals, and fruit and other plant material is occasionally eaten by some species). They have naked heads, perforated nostrils, and weakly curved claws. Some have an excellent sense of smell. The hypothesis that New World vultures might be closer to storks than to other raptors was proposed in the 1960s but was based on an erroneous interpretation of genetic data, and good evidence links cathartids to other raptors. Two alleged New World vultures are known from the Eocene of Europe (*Diatropornis* and *Parasarcoramphus*). However, the limited comparisons carried out so far, combined with a few notable differences from modern New World vultures, suggest that these fossils might instead be outside the clade that includes New World vultures and teratorns (Mayr 2009).

Teratorns (Teratornithidae) are another spectacular group of extinct neornithines: they are uniquely American, and famous for their sometimes enormous wingspans. The largest teratorn—and largest flying bird—is

Birds of Prey

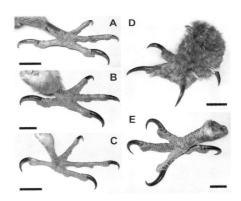

20.37. The feet of assorted predatory birds. A, Northern goshawk *Accipiter gentilis,* an accipitrid that occurs across North America and Eurasia. Note the hypertrophied talons on digits I and II. B, Red-tailed hawk *Buteo jamaicensis,* an accipitrid of North and Central America. C, Peregrine *Falco peregrinus,* a falconid with a near-global distribution. Falconids may not be close relatives of other birds of prey, and their digit I and II claws are not as enlarged as those of accipitrids. D, Great gray owl *Strix nebulosa,* a strigid of northern North America and Eurasia. Owl claws are long and weakly curved, but the toes are short and exert considerable constriction on prey. E, Osprey *Pandion haliaetus,* a cosmopolitan fish predator with horny spicules on the undersides of its toes. Unlike other raptors, its claws are all similar in size, and digit IV can be directed posteriorly. Scale bars = 20 mm.

Images courtesy of Denver Fowler; used with permission.

20.38. The giant, recently extinct eagle *Hieraaetus moorei* shown attacking a moa. Damaged moa bones (particularly pelvic girdles) show that this New Zealand eagle was a regular moa killer. With an estimated mass of 13 kg for the largest female, it is the largest known eagle. Originally classified in its own genus (*Harpagornis*), genetic data shows that it is a close relative of the small (approximately 1 kg) species conventionally included in *Hieraaetus* (but now suggested by some to be part of *Aquila*).

Argentavis from the Late Miocene of Argentina. It was 1.5 m tall when standing, had a wingspan of 6–8 m, and may have weighed over 70 kg. Though often imagined as soaring, vulturelike scavengers, some data suggest that teratorns were adept walkers that foraged for small vertebrates. Teratorns resemble New World vultures in tarsometatarsal characters and in possessing a fused prefrontal, ectethmoid, and frontal. The oldest teratorn, *Taubatornis*, is from the Late Oligocene or Early Miocene of Brazil.

Owls have an excellent fossil record extending back to the Paleocene. The phylogenetic relationships of fossil owls await investigation, but it seems likely that *Berruornis* and *Ogygoptynx* from the Paleocene and the protostrigids of the Eocene and Oligocene, at least, are outside the crown group. Owls are nocturnal hunters with enormous tubular eyes enclosed within turretlike sclerotic rings, asymmetrically positioned ears (one is usually higher up than the other), fringed feathers that permit near-silent flight, and powerful, semizygodactyl feet that are used to kill rodents and other prey (Fig. 20.37D).

Strisores: Swifts, Hummingbirds, and Nightbirds

A large group of landbirds that mostly possess short bills, wide gapes, and short hind limbs seem to be close relatives. They include the swifts and hummingbirds (Apodiformes) and the various nightbirds conventionally grouped together in Caprimulgiformes: the nightjars (Caprimulgidae), potoos (Nyctibiidae), frogmouths (Podargidae), owlet-nightjars (Aegothelidae; Fig. 20.39), and oilbirds (Steatornithidae). Mayr (2010b) found all of these groups to form a clade termed Strisores. Several studies published since 2001 have agreed that owlet-nightjars are closer to apodiforms than to other caprimulgiforms (e.g., Mayr et al. 2003; Cracraft et al. 2004; Hackett et al. 2008; Mayr 2010b), and the names Apodimorphae and Daedalornithes have both been used for the owlet-nightjar plus apodiform clade. The Early Eocene European apodimorph *Eocypselus* appears morphologically intermediate between owlet-nightjars and apodiforms like the aegialornithids. Some of the remaining caprimulgiforms—nightjars and potoos—have been united with apodimorphs in the clade Cypselomorphae (Mayr 2002). Several other fossil and living short-billed landbirds, including frogmouths and oilbirds, may be allies of this group.

The large number of nocturnal lineages within Strisores suggests that night flying was their ancestral condition, and it has been speculated that their radiation might be linked to the rise of certain Paleogene moth groups (Mayr 2009). If nocturnality was primitive for Strisores, apodiforms are secondarily diurnal. It remains uncertain whether Strisores is close to the higher landbird clade (as found by Livezey and Zusi 2007 and shown in Fig. 20.19), or only distantly related to it and in fact outside a clade that includes Aequornithes and higher landbirds (Ericson et al. 2006; Hackett et al. 2008).

The somewhat frogmouthlike fluvioviridavids from the Eocene of Europe and North America might be members of Strisores, and it also seems

20.39. Head of a large owlet-nightjar *Aegotheles insignis* from New Guinea. Owlet-nightjars inhabit Australasia from the Maluku Islands to New Caledonia and New Zealand; New Guinea is the center of their distribution. Their eyes are directed forward in owllike fashion and sensitive bristles surround the bill. Some species hawk after prey, while others pursue prey on the ground. A large, long-legged species from New Zealand (*A. novaezealandiae*) had reduced flight abilities.

that the superficially swiftlike *Protocypselomorphus* from Messel is part of this clade and perhaps close to Cypselomorphae. The archaeotrogonids (originally regarded as trogons) might also be members of Cypselomorphae. Archaeotrogonids are best known from the Oligocene of Quercy, though Eocene British specimens are also known. At least some archaeotrogonids possess a spike on the carpometacarpus that was perhaps used in combat or defense. One species (*Archaeotrogon venustus*) is remarkable for apparently having a geological range of 15 million years.

Potoos are South and Central American nightbirds that possess particularly short tarsometatarsi and an abbreviated bill. They are best known for their habit of resting in a vertical posture on a tree stump or fence post, hoping to disguise themselves as a piece of wood. Some species can see when their eyelids are closed, thanks to two or three notches in the upper eyelid. Early potoos are known from the Eocene of Europe (Fig. 20.40). Morphological and molecular data indicate a close relationship between potoos and nightjars; both share wide palatine bones and a conelike bony process in the orbit.

Frogmouths prey on insects and small vertebrates and may hunt by pouncing from a perch. They occur in Australasia as well as tropical Asia, though the Asian species are regarded as distinct (as Batrachostomidae) by some authors. Again, Eocene fossils show a Paleogene distribution in the Northern Hemisphere. The fossil species have narrower bills and longer legs than the living ones.

Within Cypselomorphae, a close affinity between swifts (Hemiprocnidae and Apodidae) and hummingbirds (Trochilidae) within the order Apodiformes is well established, despite their different anatomy and lifestyle. Several additional groups from the Paleogene of Europe and North America belong to this clade, the best known of which are the aegialornithids. Apodids are known from the Early Eocene onward, and already these birds strongly resembled modern swifts (Fig. 20.41). Swifts are uniquely adapted for life on the wing: some species mate and sleep on the wing, and may only land to breed. The *Collocalia* species famously make their cup-shaped nests from dried saliva. In the swift foot, digits I and II can oppose III and IV in pincerlike fashion, a peculiar arrangement elsewhere seen in chameleons and climbing marsupials. Two swift genera (*Collocalia* and *Aerodramus*) generate clicking noises that they use in echolocation.

Modern hummingbirds are exclusively American birds specialized for hovering and for consuming a diet of nectar (though they also eat insects). They include the smallest bird: the Cuban bee hummingbird *Mellisuga helenae*. Pronounced sexual dimorphism in bill size and shape, and differential niche utilization (where the sexes exploit different resources) are known in hummingbirds, while the sword-billed hummingbird *Ensifera ensifera* has a bill longer than the rest of its body. While catching insects, hummingbirds exhibit a unique form of mandibular kinesis where the end part of the lower jaw bends downward while simultaneously bowing outward.

Crown hummingbirds seem to have diversified during the Miocene when lowland tropical ancestors gave rise to the hummingbird clades of the Andes, West Indies, and temperate regions of Central and North America (Bleiweiss 1998). The jungornithids of Eocene Eurasia appear to be stem hummingbirds and share limb bone characters with crown hummingbirds.

20.40. The Middle Eocene potoo *Paraprefica kelleri* from Messel. Like modern potoos, it has enormous palatines and short tarsometatarsi. The feathering also shows that *Paraprefica* was similar in appearance to exant potoos, though with longer wings. Scale bar = 10 mm.

Photo provided by Gerald Mayr; used with permission.

20.41. The skeleton (with preserved feathering) of the Middle Eocene swift *Scaniacypselus szarskii* from Messel. This bird was clearly similar overall to modern swifts, though differed from them in having only a modest forking of the tail feathers.

Photo provided by Gerald Mayr, used with permission.

Unlike crown hummingbirds, jungornithids appear to have been adapted for gliding, and their short bills suggest that they caught flying insects in swiftlike fashion. Feathers preserved in one taxon (*Parargornis*) suggest that it would have resembled an owlet-nightjar when alive. Stem hummingbirds more closely related to crown hummingbirds (namely, *Eurotrochilus*) inhabited Europe during the Oligocene. *Eurotrochilus* possesses specializations for hovering flight, and its long, slender bill suggests nectarivorous habits.

Oilbirds convergently resemble some swifts in practicing echolocation. The only extant oilbird, *Steatornis caripensis*, inhabits northern South America and feeds on oil palm and laurel fruits. As is the case for various other neornithine lineages, a current Southern Hemisphere distribution does not demonstrate a Southern Hemisphere origin, as fossil oilbirds (*Prefica*) are known from the Eocene of North America. *Prefica* already seems to have been a fruit eater.

Oilbirds share several characters with trogons (Trogoniformes), including a bulbous footplate on the stapes and a particularly wide pelvis, and the two have been identified as sister groups (Mayr 2003). Trogons had previously been linked with kingfishers or mousebirds, or regarded as close to the entire coraciiform assemblage.

Trogons are brightly colored, short-billed, insectivorous or frugivorous birds of the African, Asian, and American tropics. The males are decorated with metallic-hued feathers, bare facial skin patches, facial crests, and long tail feathers (up to 70 cm in the resplendent quetzal *Pharomachrus mocinno*). Several competing phylogenies have been proposed for the approximately 40 extant taxa, but it appears most likely that the African apalodermines are outside of a clade that includes the Indomalayan harpactines and neotropical trogonines (Espinosa de los Monteros 1998; Johansson and Ericson 2004). Fossil trogons are known from the Eocene and Oligocene of Europe. They are similar to extant forms, but a few differences (such as narrower, longer skulls) suggest placement outside the crown group. *Primotrogon* from the Oligocene of France inhabited an arid environment.

Pigeons, Cuckoos, and Mousebirds

Doves and pigeons (Columbidae) are a globally distributed group of over 310 species. Distinctive characters include the presence of a cere (a soft, fleshy swelling around the nostrils), loosely rooted feathers, and the production of a milklike substance from the crop that is fed to the young. Their early fossil record is poor, with the oldest specimen being from the Late Oligocene of Australia. Ground living evolved several times within columbids. Members of Gourinae (the clade that includes the crowned pigeons *Goura* of New Guinea and the tooth-billed pigeons *Didunculus* of Samoa, Fiji, and Tonga) have repeatedly evolved large, flightless forms, the best known of which are the dodo *Raphus cucullatus* of Mauritius (which became extinct some time around 1690) and solitaire *Pezophaps solitaria* of Rogriguez (which was extinct by the 1760s). The dodo and solitaire both exhibited strong sexual dimorphism: male solitaires weighed nearly 40% more than females. Solitaires possessed rounded bony clubs on their carpometacarpi that were used in combat. Dodolike *Natunaornis* inhabited Fiji, but the white dodo of Réunion is almost certainly based on sightings of the extinct ibis *Threskiornis solitarius*. *Raphus* and *Pezophaps* are deeply nested within

Columbidae, and evidence does not support their distinction in a separate Raphidae.

It seems that columbids are closely related to sandgrouse (Pteroclidae), an Old World clade of arid-habitat birds that possess a dense plumage and high-aspect wings. They are best known for the ability of males to soak up water with their absorbent belly feathers. All but one species use this as a way of transporting water to their chicks. Fossil sandgrouse are known from the Oligocene of France. Phylogenies that find pigeons close to parrots (Livezey and Zusi 2007) result in the position shown in Fig. 20.19. However, other studies indicate that pigeons and sandgrouse are perhaps outside a clade that includes Aequornithes and higher landbirds (Ericson et al. 2006; Hackett et al. 2008).

Cuckoos (Cuculiformes) are cosmopolitan zygodactyl landbirds; most are long-tailed predators of insects (particularly caterpillars) and small vertebrates. Members of several lineages are terrestrial, and some (like the coucals) are almost pheasantlike. Of the 140 or so species, about 50 are truly parasitic. Affinities with various gruiforms (like kagus and mesites) have been proposed (Mayr and Ericson 2004; Hackett et al. 2008), but cuckoos also share some characters (such as bony recesses on the dorsal surface of the pelvis) with seriemas, hoatzins, and turacos. Fossil cuckoos are rare and poorly known. One of the best early specimens is *Eocuculus* from the Late Eocene of Colorado. Its tarsometatarsi suggest that it was not fully zygodactyl.

Cuckoos have frequently been linked with turacos (Musophagidae). Both groups are unusual among birds in practicing extensive oral processing of food items (large invertebrates and fruits, respectively) thanks to platforms on their jaw edges and/or crenellated palatal crests and bill edges (Korzun et al. 2003). Modern turacos inhabit African forests and grasslands. They are arboreal, frugivorous birds, all of which possess tall, feathery head crests. Oligocene and Miocene turacos have been reported from Egypt, Germany, and France, but the Oligocene German records were argued by Mayr (2009) to represent idiornithids.

The South American hoatzin (pronounced "what-seen") has also often been linked with cuckoos and turacos (it also uses crenellated jaw margins to chew its food), though a position close to seriemas or within Metaves has also been suggested. Despite its frequent abuse in the popular literature as an archaic, Mesozoic-grade bird, it is deeply nested within neognaths and is bizarrely specialized, not archaic. Hoatzins are unique among birds in practicing foregut fermentation: they digest leaves in an enlarged esophagus and crop. The large size of these organs has resulted in a gamebird–like reduction in the size of the sternal keel. A fossil hoatzin (*Hoazinoides*) is known from the Miocene of Colombia. A superficially hoatzinlike bird from the Eocene Green River Formation – *Foro* – might be close to hoatzins or turacos, but requires more detailed study.

Mousebirds or colies (Coliiforms) are presently restricted to sub-Saharan Africa. They are herbivorous, arboreal birds with long tails, scruffy plumage, and a habit of clambering and creeping about among branches (Fig. 20.42). Fossils show that stem mousebirds inhabited North America during the Eocene and Oligocene, and Europe between the Eocene and Miocene. Some were quite different from living mousebirds. *Chascacocolius*, for example, has a long, conical bill and massive retroarticular processes (Fig. 20.42). It

20.42. Extant mousebirds like the speckled mousebird *Colius striatus*, shown here, are short-billed climbing birds that eat fruits, seeds, and nectar. Some fossil coliiforms were very different. *Chascacocolius cacirostris* from the Middle Eocene of Germany, shown at top, had a conical rostrum and enormous retroarticular processes. These features make its skull look like that of an icterid passerine (like a grackle) and suggest that it fed by gaping.

convergently resembles icterid passerines and presumably foraged by gaping (that is, by prying upon soil or bark by forcibly opening its beak). A relationship between mousebirds and owls was weakly supported by Hackett et al. (2008) and the affinities of this group remain uncertain.

Higher Landbirds: Coraciiforms, Piciforms, Falcons, and Parrots

In the traditional classification used by Wetmore and others, an assemblage of landbird taxa was grouped together as the Coraciiformes. These include the cosmopolitan kingfishers (Alcedinidae), the todies (Todidae) of the West Indies, the tropical American motmots (Momotidae), the bee-eaters (Meropidae) and rollers (Coraciidae) of the Old World and Australasia, the Madagascan ground-rollers (Brachypteraciidae) and cuckoo-roller (*Leptosomus*), the Afro-Asian hornbills (Bucerotidae), the hoopoes (Upupidae) of Africa and Eurasia, and the African woodhoopoes (Phoeniculidae). Recent analyses have found Coraciiformes of tradition to be paraphyletic with respect to Piciformes: Mayr (2010) proposed the name Picocoraciae for the clade that includes the traditional Coraciiformes as well as Piciformes.

Shared characters, including the presence of syndactyl feet (where the second and third toes are fused at their bases), indicate that kingfishers, bee-eaters, todies, and motmots form a clade that can be called Alcediniformes. *Quasisyndactylus* from Messel seems to be an early alcediniform, and other Paleogene fossils indicate European origin for the lineages within this clade. Todies are known from the Oligocene of the United States but also from the Oligocene and Eocene of France and Germany, and an Oligocene fossil from Switzerland (*Protornis*) has been identified as a motmot. Phylogenies indicate that most of bee-eater evolution occurred in Africa, though the group may have originated in Asia.

Kingfishers (argued by some ornithologists to warrant separation into three families: Alcedinidae, Halcyonidae, and Cerylidae) have short, rounded wings, short legs, and (typically) a long, straight, proportionally large bill. Typical kingfishers are generalist predators that hunt all manner of small creatures from a perch; fish-eating plunge divers are in the minority. Kingfishers are most diverse in Southeast Asia and New Guinea. A Miocene halcyonid is known from Australia, and several Eocene and Oligocene fossils from North America and Europe are claimed to have kingfisher affinities.

Ground-rollers and rollers form a clade that can be called Coracii. These birds have relatively large heads and stout, slightly hooked bills that they use to grab insects and small vertebrates; superficially, they recall falcons or giant swifts when in flight. Extant rollers inhabit Eurasia, Africa, and Australasia. Eocene taxa show that members of Coracii formerly inhabited North America as well as Europe (Clarke et al. 2009). Some of these birds (originally described as representing a new piciform group called Primobucconidae) were far smaller than extant rollers, and stomach contents show that some were seed eaters and not predators like extant rollers.

The Madagascan cuckoo-roller or courol *Leptosomus discolor* was often considered close to ground-rollers and rollers in the past. However, molecular and morphological evidence shows that it is definitely not a roller, though its precise affinities remain to be determined. A position close to cuckoos, to Strisores, or close to the base of a clade that includes trogons and

Picocoraciae have all been suggested. *Plesiocathartes* from the Eocene and Oligocene of Europe and Eocene of North America—originally described as a New World vulture—seems to be a stem leptosomid, and is so similar overall to *Leptosomus* that Mayr (2008a) regarded the latter as a possible living fossil. The modern restriction of leptosomids to Madagascar may be relictual following their extinction elsewhere.

Hornbills occur in tropical Africa and Asia, and are mostly forest-dwelling omnivores with huge, downcurved bills. They often possess hollow bony casques (the casque of one species is solid and is used in noisy aerial jousting) and are famous for their habit of sealing up the female inside the nesting chamber. Ground hornbills (Fig. 20.43) are carnivorous, striding grassland birds that do not practice the sealing-up behavior. This is currently the only hornbill group represented in the fossil record, with species known from the Miocene of northern Africa and Europe.

Hoopoes and woodhoopoes share long, slender bills, a long hallux, and a peculiar undulating margin to the third metacarpal. They form the clade Upupiformes. Hoopoes mostly forage on the ground, while woodhoopoes cling to tree trunks in woodpecker-like fashion and probe for insects. Fossil woodhoopoes are known from the Miocene of Germany and France, and *Messelirrisor* from the Eocene of Germany seems to be a stem upupiform. It is a tiny, thin-billed bird with perching adaptations.

Over 350 species of scansorial and arboreal insectivores, frugivores, and omnivores are grouped together as the Piciformes. They possess zygodactyl feet, a unique arrangement of foot tendons, and a distinctive flange on one of the finger bones. The tropical American toucans (Ramphastidae) are well known for their gigantic, strikingly colored bills, honeyguides (Indicatoridae) eat the wax produced by social insects, jacamars (Galbulidae) are long-billed predators of flying insects, and woodpeckers (Picidae) are excavators of wood that cling and climb on tree trunks. It has sometimes been argued that jacamars and puffbirds (Bucconidae), both of the neotropics and grouped together as the Galbulae, are distinct from other piciforms (grouped together as Pici) and actually closer to rollers, but recent studies have confirmed the monophyly of traditional Piciformes.

Barbets (Capitonidae) are paraphyletic with respect to toucans (meaning that the barbet–toucan clade should be termed Ramphastidae, as this name is older), and some or all New World barbets are more closely related to toucans than they are to Old World barbets (Prum 1988). Some authors recognize four distinct barbet groups: African barbets (Lybiidae), Asian barbets (Megalaimidae), American true barbets (Capitonidae sensu stricto) and toucan-barbets (Semnornithidae), though the monophyly of most of these groups is doubtful.

Woodpeckers, the most diverse piciforms, are mostly insectivorous, but some regularly visit flowers and eat ripe fruits. How they cope with the rapid, repetitive pecking that they practice (involving decelerations of 600–1500g on impact) remains uncertain, but the small size and orientation of the brain within the cranium (its longest axis is dorsoventral) may help. Woodpeckers possess an extremely long, tentacle-like tongue that, when stored, loops around the brain and is rooted near the nostril. Wrynecks (the most basal crown woodpeckers) possess a tongue of this sort but lack the features associated with wood excavation.

20.43. A, The casqued skull of a Northern ground hornbill *Bucorvus abyssinicus*. B, Northern ground hornbill in living condition. Ground hornbills are black plumaged but have naked red or blue facial skin that signals maturity and condition. They can reach 4 kg and have a wingspan of 2 m. *Euroceros* from the Miocene of Bulgaria had smaller wings than living ground hornbills and was probably more strongly adapted for terrestrial walking.

A courtesy of Mark Witton; used with permission.

The gracilitarsids from the Paleocene and Eocene of Europe, North America, and South America, and the sylphornithids from the Eocene of Europe, might be close to the ancestry of piciforms. All are tiny, gracile-legged birds that probably resembled slender-billed passerines. A few European fossils from the Early Oligocene seem to be early members of Pici. The German Oligocene taxon *Rupelramphastoides* resembles honeyguides in skull shape and toucans in leg proportions. Miocene barbets and woodpeckers are known from Europe and North America.

Falcons (Falconidae) are usually classified together with other diurnal raptors, but anatomical differences have led many to doubt this. Some phylogenies find falcons to be close to parrots and passerines (Hackett et al. 2008). If this is accurate, it represents an outstanding case of convergent evolution. Most of the 60 or so extant falcons are South American, so the crown clade at least may have its origins here. Fragments from the Eocene and Oligocene of Europe and Antarctica have been identified as falcons but are too incomplete for positive identification. Definite falcons like *Badiostes* and *Pediohierax* are known from the Middle Miocene of South and North America. Some falcons are swift, aerial predators of other birds: the peregrine *Falco peregrinus* can reach speeds of 70 m s^{-1} in a downward stoop. Others (like the laughing falcon *Herpetotheres cachinnans*) are short-winged forest birds, while others (like caracaras) are scavengers and invertebrate eaters that may spend a lot of time foraging on the ground. A large extinct caracara from Jamaica had strongly reduced flight abilities.

Parrots (Psittaciformes) are zygodactyl birds with a unique deep beak where the ventral maxillary margin is sigmoidally curved, the nostrils are small, and the lower jaw is shorter than the upper. They also possess an enlarged deltopectoral crest, a proportionally long ulna, and long, gracile foot claws. Parrots are morphologically homogenous, but the more than 370 species inhabit tropical forests, grasslands, and deserts worldwide; some species inhabit temperate climes. Phylogenies indicate that the parrots of New Zealand (such as the kea *Nestor notabilis* and flightless kakapo *Strigops habroptilus*) and the Australasian cockatoos are the most basal extant taxa (Wright et al. 2008); this has supported the conclusion that crown parrots originated in Gondwana, perhaps before its fragmentation. However, dispersal across oceanic barriers was apparently important in the history of the group, with independent dispersal events explaining the affinities discovered between widely dispersed parrot taxa (Schweizer et al. 2010). Furthermore, the diversity of parrots in Miocene Europe (all of which are probably close to, but not part of, the crown group) implies Northern Hemisphere origin and diversification, with subsequent dispersal to the southern continents.

Several zygodactyl Eocene birds from Europe and North America seem to be stem psittaciforms. They have most often been termed pseudasturids, but because *Halcyornis* from the London Clay of the Isle of Sheppey (the earliest named fossil bird) is referable to this group, their correct name is Halcyornithidae. Unlike crown parrots, halcyornithids have elongate, slender humeri and superficially mousebirdlike skulls. The Eocene messelasturids also now seem to be stem psittaciforms and the sister group to Halcyornithidae (Mayr 2011b). Messelasturids combine a deep, hooked bill with large supraorbital processes and semizygodactyl feet with (in *Messelastur*) raptor-like unguals. These features led to previous proposals that mess-elasturids are close relatives of owls and/or diurnal raptors. Their raptorial features suggest carnivorous habits. Additional stem parrots—*Vastanavis*, *Quercypsitta* and *Psittacopes*—are known from the Eocene of Europe and India. Stem parrots were evidently abundant and diverse in the Northern Hemisphere during the Eocene.

A group of small arboreal birds with zygodactyl feet—the zygodactylids (now known to include Primoscenidae)—is known from the Eocene and Oligocene of Europe and North America. They may be the sister group to Passeriformes (Mayr 2008b). Two particularly long central rectrices are present in the tail, and two smaller ones are present on either side. Zygo-dactylids seem to include both arboreal and terrestrial species. Fruit seeds are preserved as stomach contents in *Primozygodactylus*, while a sharply pointed bill in *Zygodactylus* suggests insectivorous habits.

The Passerine Radiation

Passerines, also known as Passeriformes, songbirds, or perching birds, in-clude about 59% of all extant bird species. As such, they are the most suc-cessful dinosaurian clade, and more than half of any review of neornithine diversity should really be devoted to this one group alone. For reasons of space, the discussion of passerines provided here is brief. Passerines are mostly small terrestrial, insectivorous or graminivorous birds, and the as-sumption that they are morphologically conservative and (apart from larks and swallows) all but impossible to tell apart skeletally long obscured efforts to determine their phylogeny. The majority of passerines belong to the two great clades Suboscines (also called Tyranni) and Oscines, both of which are united as Eupasseres.

Suggestions that passerines diverged early within Neognathae are now thought to be the result of long-branch attraction, and most studies find passerines to be one of the youngest major neognath clades. It seems that many passerine families are not monophyletic; this seems to be the case for many genera too. As a result, the taxonomy of some sections of the passerine tree is rather chaotic.

While passerines are unquestionably monophyletic, it has proved dif-ficult to find synapomorphies, or to find characters that might somehow explain passerine success. A particularly long hallux and a large hallux claw characterize the group; it might have allowed them to be better at perch-ing and foraging either arboreally or terrestrially than other small birds. Interestingly, particularly large claws are also seen in falcons and parrots, raising the possibility that this is a shared character: Hackett et al. (2008) found passerines, parrots, and falcons to form a clade that also includes seriemas. While ecologically diverse, passerines have been conservative in morphology, and most are finch shaped or warbler shaped and less than 40 g in mass. Lyrebirds (Menuridae) can exceed 1 m in total length (though much of this is tail) and weigh over 1 kg, while the largest corvid (the north-ern raven *Corvus corax*) can reach 1.7 kg. Flightlessness is present in two passerine clades: in the New Zealand wrens (where both *Traversia lyalli* and *Dendroscansor decurvirostris* were flightless), and in buntings (where the long-legged bunting *Emberiza alcoveri* of the Canary Islands was flightless). All three species are extinct as a result of human intervention.

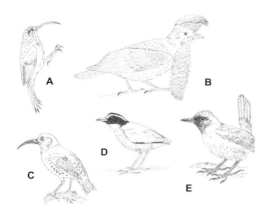

20.44. A selection of suboscine passerines. A, Red-billed scythebill *Campylorhamphus trochilirostris,* a dendrocolaptine furnariid that occurs from Panama south to Argentina. The long bill is used for probing into bark and among epiphytes. Dendrocolaptines, or woodcreepers, were long regarded as a distinct family but appear nested within Furnariidae. B, Amazonian umbrellabird *Cephalopterus ornatus,* a cotingid. One of three umbrellabird species, it is the largest South American passerine. C, Wattled false sunbird *Neodrepanis coruscans* of Madagascar, a philepittid suboscine. D, Fairy pitta *Pitta nympha* of tropical Asia, a pittid suboscine. Pittas are the largest Old World suboscine group (containing 32 species) and mostly occur in Asia and Australia (two occur in Africa). They are all short-tailed terrestrial foragers. E, Black-throated huet-huet *Pteroptochos tarnii,* a rhinocryptid (tapaculo) of Argentina and Chile. Tapaculos are terrestrial foragers with poor flight abilities, so their former presence in Cuba and the Isle of Youth is surprising because it requires overwater dispersal.

20.45. Ovenbirds (furnariids) are so named because some species create clay nests that resemble old-fashioned ovens. Some species produce stick nests, others manufacture domed nests made of moss, while others use natural cavities or burrows, or create simple, cuplike nests. A predisposed ability to vary nest architecture might explain why ovenbirds have been so successful.

Passerine biogeography indicates an origin and early diversification within the Southern Hemisphere. New Zealand wrens (Acanthisittidae) are the most basal passerines, suboscines are mostly South American, and most taxa originally included in the oscine group Corvida are Australasian. The oldest fossil passerines are from the Eocene of Australia. Passerines are absent from the north before the Early Oligocene, and during the Eocene, the role of small perching birds in North America and Europe at least was occupied by mousebirds, rollers (the "primobucconids"), and hoopoe relatives (the messelirrisorids). Fossil passerines from the Oligocene and Miocene of Germany and France appear to be outside of Eupasseres (Manegold et al. 2004), suggesting that basal passerines had reached the Northern Hemisphere before the Miocene. Suboscines (such as broadbills) and oscines (like the corvid *Miocorvus*) were present in the Northern Hemisphere by the Miocene.

Molecular work shows that New Zealand wrens are the sister group to Eupasseres. Suboscines includes the Old World broadbills (Eurylaimidae), asities (Philepittidae), and pittas (Pittidae)—all of which are grouped together as Eurylaimides—and the diverse American clade Tyrannides (Fig. 20.44). Eurylaimides is not exclusive to the Old World, as the sapayoa *Sapayoa aenigma* of northern South America (once known as the broad-billed manakin) is a close relative of African and Indo-Malayan broadbills.

Tyrannides includes tyrant flycatchers, cotingas, and manakins (grouped together as Tyrannida; they possess the simple haplophone syrinx), and the ovenbirds, woodcreepers, and antbirds (grouped together as Furnariida; they possess the complex tracheophone syrinx). Cotingas (Cotingidae) and manakins (Pipridae) are brightly colored neotropical suboscines, many of which exhibit remarkable display adaptations (Fig. 20.44B). Manakins make whirring and clicking noises by vibrating modified wing feathers. Woodcreepers (Dendrocolaptinae) are convergently similar to woodpeckers and possess stiffened rectrices, partially fused toes, legs specialized for vertical climbing, and a suite of cranial features that allow them to pry and probe into wood. Ovenbirds (Fig. 20.44) incorporate 55 genera and have been described as the most diverse neornithine family in terms of natural history and ecology. They include species that strongly resemble oscines from elsewhere in the world: there are ovenbirds that resemble thrushes, dippers, larks, thrashers, sylviid warblers, and creepers.

Sibley and Ahlquist (1990) divided oscines into two clades: Corvida (an ancestrally Australasian group, including all crowlike oscines) and Passerida (Fig. 20.46). It now seems that Corvida is paraphyletic with respect to Passerida, with the lyrebird–scrub-bird clade (Menuroidea), the bowerbird–Australian treecreeper clade, and the pardalote–fairywren–honeyeater clade (Meliphagoidea), and others, being outside of the clade that includes core corvoids and Passerida. Core corvoids include cuckooshrikes, woodswallows, vireos, orioles, vangas, drongos, birds-of-paradise, corvids, and shrikes. All outgroups to the core corvoid plus Passerida clade are Australasian, so an Australasian origin for both Oscines and Passerida is well supported (Barker et al. 2004).

Within Passerida, Sibley and Ahlquist (1990) proposed the existence of three major groups: Muscicapoidea (including dippers, thrushes, chats,

Old World flycatchers, starlings, and mynas), Sylvioidea (including tits, wrens, swallows, bulbuls, babblers, and sylviid warblers), and Passeroidea (including larks, pipits, weavers, finches, sparrows, and icterids). Subsequent studies have found some of these relationships to be incorrect and have recovered a more complex topology within Passerida. The rockjumper (*Chaetops*) and rockfowl (*Picathartes*) may represent one of the most ancient lineages within the clade, Sylvioidea may be paraphyletic with respect to Passeroidea, and waxwings and a titmouse–penduline tit clade have proved hard to place (Ericson and Johansson 2003; Voelker and Spellman 2003). Nuthatches, treecreepers, wrens, and gnatcatchers – grouped together as Certhioidea – may represent a distinct lineage closer to Muscicapoidea and Passeroidea than to Sylvioidea (Cracraft et al. 2004). The African *Hyliota* species, the Australian robins (Petroicidae), and the stenostirids of South Africa and the Indo-Malayan region also seem to be distinct lineages within Passerida (Fuchs et al. 2006, 2009). Larks are not passeroids but are part of the same sylvioid clade as swallows and the extremely confusing assemblage that includes bulbuls, sylviids, white-eyes, and babblers.

Ancestrally, passeridans were probably thin-billed insectivores, but a large number of interesting specializations have evolved within the clade. Conical, reinforced jaws have evolved repeatedly in seed-eating groups, with the most impressive rostra (such as those of hawfinches) enabling the birds to exert forces of several hundred newtons. Brushlike tongues used in nectar feeding have evolved several times, most notably in the hummingbird-like sunbirds. The African oxpeckers *Buphagus* (conventionally, but possibly incorrectly, classified as starlings) perch on large mammals and feed on skin, blood, and mucus in addition to ectoparasites, and nuthatches and treecreepers are expert vertical climbers. Passeridans have even adapted to aquatic life: dippers are the only diving birds restricted entirely to freshwater. Because rockjumpers, rockfowl, *Hyliota*, stenostirids, and various other archaic members of Passerida are African, this clade's early evolution may have occurred on this continent: some authors have even suggested that stem passeridans flew directly from Australia to Africa.

We are currently at an interesting stage in terms of our knowledge on bird diversity and evolution. Large combined morphological and molecular data sets are finally allowing ornithologists to piece together the neornithine tree, and the belief that views on neornithine phylogeny are in chaos are no longer true. In fact, many studies find the same relationships, and rough consensus now exists for some parts of the tree (Cracraft et al. 2004; Harshman 2007; Mayr 2010a).

However, views on the species-level diversity of extant birds are currently in flux as controversies over species and subspecies concepts mean that many taxa traditionally recognized as subspecies have been raised to species level. It is sometimes suggested that wholesale application of the phylogenetic species concept might result in a doubling of the currently accepted count of between 9,000 and 10,000 extant bird species, but this is not really accurate (Zink 1996). Nevertheless, taxonomic inflation and its implications for conservation priority has become a major area of discussion among ornithologists and conservationists.

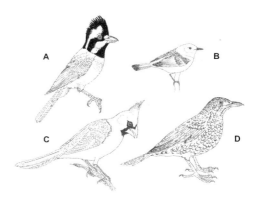

20.46. A selection of oscine passerines. A, Crested shriketit *Falcunculus frontatus*, an Australian pachycephalid core corvoid. B, Prothonotary warbler *Protonotaria citrea*, a parulid passeroid passeridan that breeds in North America and winters in Central or South America or the Caribbean. Parulids, or New World warblers, are part of the enormous nine-primaried oscine group (it contains about 1,000 species). Parulidae as conventionally conceived is nonmonophyletic. C, Northern cardinal *Cardinalis cardinalis*, a North American emberizid passeroid passeridan. Emberizids are also nine-primaried oscines and may be close kin of parulids. D, White's thrush *Zoothera dauma*, an Asian turdid muscicapoid passeridan. Thrushes are Old World passerines that seem to have crossed the Atlantic to colonize the Americas at least five times during the Miocene and Pliocene.

The Future of Neornithine Diversity

While our knowledge of avian history is improving, the future of the group looks bleak in some respects. About 12% of extant species are globally threatened, and about 180 species face imminent extinction. Habitat destruction is the greatest threat to birds, affecting 86% of globally threatened species. However, there are increasing indications that human-driven climate change is having a negative impact on how birds time their migration and breeding, and population declines are resulting. Data indicates that distributional and life cycle changes occuring in birds are not happening fast enough to keep track with climate change (Thomas 2011). Many groups currently affected by the threat of extinction contain specialized species with small ranges (examples include shearwaters, rails, and parrots) or are low in diversity and might be considered biogeographic relicts (examples include kiwi, megapodes, cranes, kagus, mesites, and ground-rollers). Such species encapsulate a substantial proportion of neornithine diversity, preserving morphological and molecular characters not present elsewhere. Their extinction will result in a substantial loss of evolutionary history (von Euler 2001) and homogenization of the global avifauna (Lockwood et al. 2000).

References

Abel, O. 1911. Die Vorfahren der Vögel und ihre Lebensweise. *Verhandlungen der Zoologisch-Botanischen Gesellschaft in Wien* 61: 144–191.

Agnolin, F. 2007. *Brontornis burmeisteri* Moreno & Mercerat, un Anseriformes (Aves) gigante del Mioceno Medio de Patagonia, Argentina. *Revista del Museo Argentino de Ciencias Naturales* (n.s.) 9: 15–25.

Alvarenga, H. M. F., L. Chiappe, and S. Bertelli. 2011. Phorusrhacids: the terror birds. In G. Dyke and G. Kaiser (eds.), *Living Dinosaurs: the Evolutionary History of Modern Birds*, 187–208. Chichester, UK: John Wiley and Sons.

Alvarenga, H. M. F., and E. Höfling. 2003. Systematic revision of the Phorusrhacidae (Aves: Ralliformes). *Papéis Avulsos de Zoologia, Museu de Zoologia da Universidade de São Paulo* 43: 55–91.

Baker, A. J., S. L. Pereira, and T. A. Paton. 2007. Phylogenetic relationships and divergence times of Charadriiformes genera: multigene evidence for the Cretaceous origin of at least 14 clades of shorebirds. *Biology Letters* 3: 205–209.

Barker, F. K., A. Cibois, P. Schikler, J. Feinstein, and J. Cracraft. 2004. Phylogeny and diversification of the largest avian radiation. *Proceedings of the National Academy of Sciences of the United States of America* 101: 11040–11045.

Bertelli, S., and L. M. Chiappe. 2005. Earliest tinamous (Aves: Palaeognathae) from the Miocene of Argentina and their phylogenetic position. *Contributions in Science, Natural History Museum of Los Angeles County* 502: 1–20.

Bertelli, S., B. E. K. Lindow, D. J. Dyke, and L. M. Chiappe. 2010. A well-preserved 'charadriiform-like' fossil bird from the Early Eocene Fur Formation of Denmark. *Palaeontology* 53: 507–531.

Bledsoe, A. H. 1988. A phylogenetic analysis of postcranial skeletal characters of the ratite birds. *Annals of Carnegie Museum* 57: 73–90.

Bleiweiss, R. 1998. Origin of hummingbird faunas. *Biological Journal of the Linnean Society* 65: 77–97.

Boas, J. E. V. 1930. Über das Verhältnis der Dinosaurier zu den Vögeln. *Morphologisches Jahrbuch* 64: 223–247.

Bourdon, E. 2005. Osteological evidence for sister group relationships between pseudo-toothed birds (Aves: Odontopterygiformes) and waterfowls (Anseriformes). *Naturwissenschaften* 92: 586–591.

———. 2011. The pseudo-toothed birds (Aves, Odontopterygiformes) and their bearing on the early evolution of modern birds. In G. Dyke and G. Kaiser (eds.), *Living Dinosaurs: the Evolutionary History of Modern Birds*, 209–234. Chichester, UK: John Wiley and Sons.

Bourdon, E., A. de Ricqlès, and J. Cubo. 2009. A new transantarctic relationship: morphological evidence for a Rheidae-Dromaiidae-Casuariidae clade (Aves, Palaeognathae, Ratitae). *Zoological Journal of the Linnean Society* 156: 641–663.

Broom, R. 1913. On the South-African pseudosuchian *Euparkeria* and allied genera. *Proceedings of the Zoological Society of London* 83: 619–633.

Bunce, M., T. H. Worthy, T. Ford, W. Hoppitt, E. Willerslev, A. Drummond, and A. Cooper. 2003. Extreme reversed sexual size dimorphism in the extinct New Zealand moa *Dinornis*. *Nature* 425: 172–175.

Chatterjee, S. 1997. *The Rise of Birds*. Baltimore, Md.: Johns Hopkins University Press.

Chiappe, L. M. 2007. *Glorified Dinosaurs: The Origin and Early Evolution of Birds*. London, UK: John Wiley and Sons.

Chiappe, L. M., S. A. Ji, Q. Ji, and M. A. Norell. 1999. Anatomy and systematics of the Confuciusornithidae (Theropoda: Aves) from the late Mesozoic of northeastern China. *Bulletin of the American Museum of Natural History* 242: 1–89.

Clarke, J. A. 2004. Morphology, phylogenetic taxonomy, and systematics of *Ichthyornis* and *Apatornis* (Avialae: Ornithurae). *Bulletin of the American Museum of Natural History* 286: 1–179.

Clarke, J. A., D. T. Ksepka, N. A. Smith, and M. A. Norell. 2009. Combined phylogenetic analysis of a new North American fossil species confirms widespread Eocene distribution for stem rollers (Aves, Coracii). *Zoological Journal of the Linnean Society* 157: 586–611.

Clarke, J. A., M. A. Norell, and D. Dashzeveg. 2005. New avian remains from the Eocene of Mongolia and the phylogenetic position of the Eogruidae (Aves, Gruoidea). *American Museum Novitates* 3494: 1–17.

Clarke, J. A., Z. Zhou, and F. Zhang. 2006. Insight into the evolution of avian flight from a new clade of Early Cretaceous ornithurines from China and the morphology of *Yixianornis grabaui*. *Journal of Anatomy* 208: 287–308.

Codd, J. R., D. F. Boggs, S. F. Perry, and D. R. Carrier. 2005. Activity of three muscles associated with the uncinate processes of the giant Canada goose *Branta canadensis maximus*. *Journal of Experimental Biology* 208: 849–857.

Cooper, A., C. Lalueza-Fox, S. Anderson, A. Rambaut, J. Austin, and R. Ward. 2001. Complete mitochondrial genome sequences of two extinct moas clarify ratite evolution. *Nature* 409: 704–707.

Cracraft, J. 1972. The relationships of the higher taxa of birds: problems in phylogenetic reasoning. *Condor* 74: 379–392.

———. 2001. Avian evolution, Gondwana biogeography and the Cretaceous–Tertiary mass extinction event. *Proceedings of the Royal Society of London B* 268: 459–469.

Cracraft, J., F. K. Barker, M. Braun, J. Harshman, G. J. Dyke, J. Feinstein, S. Stanley, A. Cibois, P. Schikler, P. Beresford, J. García-Moreno, M. D. Sorenson, T. Yuri, and D. P. Mindell. 2004. Phylogenetic relationships among modern birds (Neornithes): towards an avian tree of life. In J. Cracraft and M. Donoghue (eds.), *Assembling the Tree of Life*, 468–489. Oxford: Oxford University Press.

Degrange, F. J., C. P. Tambussi, K. Moreno, L. M. Witmer, and S. Wroe. 2010. Mechanical analysis of feeding behavior in the extinct "terror bird" *Andalgalornis steulleti* (Gruiformes: Phorusrhacidae). *PLoS ONE* 5 (8): e11856.

Dial, K. P. 2003. Wing-assisted incline running and the evolution of flight. *Science* 299: 402–404.

Elzanowski, A. 2001. The life style of *Archaeopteryx* (Aves). *Asociación Paleontológica Argentina. Publicación Especial* 7: 91–99.

———. 2002. Archaeopterygidae (Upper Jurassic of Germany). In L. M. Chiappe and L. M. Witmer (eds.), *Mesozoic Birds: Above the Heads of Dinosaurs*, 129–159. Berkeley: University of California Press.

Ericson, P. G. P., C. L. Anderson, T. Britton, A. Elzanowski, U. S. Johansson, M. Källersjö, J. I. Ohlson, T. J. Parsons, D. Zuccon, and G. Mayr. 2006. Diversification of Neoaves: integration of molecular sequence data and fossils. *Biology Letters* 22: 543–547.

Ericson, P. G. P., and U. S. Johansson. 2003. Phylogeny of Passerida (Aves: Passeriformes) based on nuclear and mitochondrial sequence data. *Molecular Phylogenetics and Evolution* 29: 126–138.

Espinosa de los Monteros, A. 1998. Phylogenetic relationships among the trogons. *Auk* 115: 937–954.

Fain, M. G., and P. Houde. 2004. Parallel radiations in the primary clades of birds. *Evolution* 58: 2558–2573.

Feduccia, A. 1996. *The Origin and Evolution of Birds*. New Haven, Conn.: Yale University Press.

———. 2002. Birds are dinosaurs: simple answer to a complex problem. *Auk* 119: 1187–1201.

Feduccia, A., L. D. Martin, and S. Tarsitano. 2007. *Archaeopteryx* 2007: quo vadis? *Auk* 124: 373–380.

Fuchs, J., J. Fjeldså, R. C. K. Bowie, G. Voelker, and E. Pasquet. 2006. The African warbler genus *Hyliota* as a lost lineage in the oscine songbird tree: molecular support for an African origin of the Passerida. *Molecular Phylogenetics and Evolution* 39: 186–197.

Fuchs, J., E. Pasquet, A. Couloux, J. Fjeldså, and R. C. K. Bowie. 2009. A new Indo-Malayan member of the Stenostiridae (Aves: Passeriformes) revealed by multilocus sequence data: biogeographical implications for a morphologically diverse clade of flycatchers. *Molecular Phylogenetics and Evolution* 53: 384–393.

Fürbringer, M. 1888. *Untersuchungen zur Morphologie und Systematik der Vögel*. 2 vols. Amsterdam: von Holkema.

Gadow, H. 1893. Vögel. II. – Systematischer Theil. In H. G. Bronn (ed.), *Klassen und Ordnungen des Thier-Reichs*, vol. 6. Leipzig: C. F. Winter.

Gao, C., L. M. Chiappe, Q. Meng, J. K. O'Connor, X. Wang, X. Cheng, and J. Liu. 2008. A new basal lineage of Early Cretaceous and its implications birds from China on the evolution of the avian tail. *Palaeontology* 51: 775–791.

Gauthier, J. 1986. Saurischian monophyly and the origin of birds. *Memoirs of the California Academy of Science* 8: 1–55.

Gauthier, J., and K. Padian. 1985. Phylogenetic, functional, and aerodynamic analyses of the origin of birds and their flight. In M. K. Hecht, J. H. Ostrom, G. Viohl, and P. Wellnhofer (eds.), *The Beginnings of Birds – Proceedings of the International Archaeopteryx Conference, Eichstätt*, 185–197. Eichstätt, Germany: Freunde des Jura-Museums Eichstätt.

Glen, C. L., and M. B. Bennett. 2007. Foraging modes of Mesozoic birds and non-avian theropods. *Current Biology* 17: 911–912.

Hackett, S. J., R. T. Kimball, S. Reddy, R. C. K. Bowie, E. L. Braun, M. J. Braun, J. L. Cjojnowski, W. A. Cox, K.-L. Han, J. Harshman, C. J. Huddleston, B. Marks, K. J. Miglia, W. S. Moore, F. H. Sheldon, D. W. Steadman, C. C. Witt, and T. Yuri. 2008. A phylogenomic study of birds reveals their evolutionary history. *Science* 320: 1763–1768.

Harshman, J. 1994. Reweaving the Tapestry: what can we learn from Sibley and Ahlquist (1990)? *Auk* 111: 377–388.

———. 2007. Classification and phylogeny of birds. In B. G. M. Jamieson (ed.), *Reproductive Biology and Phylogeny of Birds*, 1–35. Enfield, N.H.: Science Publishers.

Harshman, J., E. Braun, M. Braun, C. Huddleston, R. Bowie, J. Chojnowski, S. Hackett, K. Han, R. Kimball, B. Marks, K. Miglia, W. Moore, S. Reddy, F. Sheldon, D. Steadman, S. Steppan, C. Witt, and T. Yuri. 2008. Phylogenomic evidence for multiple losses of flight in ratite birds. *Proceedings of the National Academy of Sciences of the United States of America* 105: 13462–13467.

Heilmann, G. 1926. *The Origin of Birds.* London: H. F. & G. Witherby.

Holmgren, N. 1955. Studies on the phylogeny of birds. *Acta Zoologica* 36: 243–328.

Holtz, T. R., Jr. 2011 (this volume). Theropods. In M. K. Brett-Surman, Thomas R. Holtz Jr., and James O. Farlow (eds.), *The Complete Dinosaur*, 2nd ed., 346–378. Bloomington: Indiana University Press.

Hope, S. 2002. The Mesozoic radiation of Neornithes. In L. M. Chiappe and L. M. Witmer (eds.), *Mesozoic Birds: Above the Heads of Dinosaurs*, 339–388. Berkeley: University of California Press.

Hopson, J. A. 2001. Ecomorphology of avian and nonavian theropod phalangeal proportions: implications for the arboreal versus terrestrial origin of bird flight. In J. Gauthier and L. F. Gall (eds.), *New Perspectives on the Origin and Early Evolution of Birds: Proceedings of the International Symposium in Honor of John H. Ostrom*, 211–235. New Haven, Conn.: Peabody Museum of Natural History, Yale University.

Houde, P. 1986. Ostrich ancestors found in the Northern Hemisphere suggest new hypothesis of ratite origins. *Nature* 324: 563–565.

Houde, P., A. Cooper, E. Leslie, A. E. Strand, and G. A. Montaño. 1997. Phylogeny and evolution of 12S rDNA in Gruiformes (Aves). In D. P. Mindell (ed.), *Avian Molecular Evolution and Systematics*, 121–158. New York: Academic Press.

Huxley, T. H. 1867. On the classification of birds; and on the taxonomic value of the modification of certain of the cranial bones observable in that class. *Proceedings of the Zoological Society of London* 1867: 415–472.

——. 1870. Further evidence of the affinity between the dinosaurian reptiles and birds. *Quarterly Journal of the Geological Society, London* 26: 12–31.

Huynen, L., C. D. Millar, R. P. Scofield, and D. M. Lambert. 2003. Nuclear DNA sequences detect species limits in ancient moa. *Nature* 425: 175–178.

James, F. C., and J. A. Pourtless. 2009. Cladistics and the origins of birds: a review and two new analyses. *Ornithological Monographs* 66: 1–78.

Ji, S.-A., J. Atterholt, J. K. O'Connor, M. C. Lamanna, J. D. Harris, D.-Q. Li, H.-L. You, and P. Dodson. 2011. A new, three-dimensionally preserved enantiornithine bird (Aves: Ornithothoraces) from Gansu Province, north-western China.

Zoological Journal of the Linnean Society 162: 201–219.

Johansson, L. C., and U. M. L. Norberg. 2001. Lift-based paddling in diving grebe. *Journal of Experimental Biology* 204: 1687–1696.

Johansson, U. S., and P. G. P. Ericson. 2004. A re-evaluation of basal phylogenetic relationships within trogons (Aves: Trogonidae) based on nuclear DNA sequences. *Journal of Zoological Systematics* 43: 166–173.

Kaiser, G. W. 2007. *The Inner Bird: Anatomy and Evolution.* Vancouver: UBC Press.

Korzun, L. P., C. Erard, J.-P. Gasc, and F. J. Dzerzhinsky. 2003. Biomechanical features of the bill and jaw apparatus of cuckoos, turacos and the hoatzin in relation to food acquisition and processing. *Ostrich* 74: 48–57.

Ksepka, D. T. 2009. Broken gears in the avian molecular clock: new phylogenetic analyses support stem galliform status for *Gallinuloides wyomingensis* and rallid affinities for *Amitabha urbsinterdictensis*. *Cladistics* 25: 173–197.

Ksepka, D. T., and T. Ando 2011. Penguins past, present, and future: trends in the evolution of the Sphenisciformes. In G. Dyke and G. Kaiser (eds.), *Living Dinosaurs: the Evolutionary History of Modern Birds*, 155–186. Chichester, UK: John Wiley and Sons.

Lee, K., J. Feinstein, and J. Cracraft. 1997. The phylogeny of ratite birds: resolving conflicts between molecular and morphological data sets. In D. P. Mindell (ed.), *Avian Molecular Evolution and Systematics*, 173–211. New York: Academic Press.

Livezey, B. C. 2010. Grebes and flamingos: standards of evidence, adjudication of disputes, and societal politics in avian systematics. *Cladistics* 26: 1–11.

Livezey, B. C., and R. L. Zusi. 2007. Higher-order phylogeny of modern birds (Theropoda, Aves: Neornithes) based on comparative anatomy. II. Analysis and discussion. *Zoological Journal of the Linnean Society* 149: 1–95.

Lockwood, J. L., T. M. Brooks, and M. L. McKinney. 2000. Taxonomic homogenization of the global avifauna. *Animal Conservation* 3: 27–35.

Lowe, P. R. 1935. On the relationship of Struthiones to the dinosaurs and to the rest of the avian class, with special reference to the position of *Archaeopteryx*. *Ibis* 5: 398–432.

Mack, A. L., and J. Jones. 2003. Low-frequency vocalizations by cassowaries (*Casuarius* spp.). *Auk* 120: 1062–1068.

Manegold, A., G. Mayr, and C. Mourer-Chauviré. 2004. Miocene songbirds and the composition of the European passeriform avifauna. *Auk* 121: 1155–1160.

Martin, L. D. 1983. The origin of birds and of avian flight. In R. F. Johnston (ed.), *Current Ornithology*, 1: 105–129. New York: Plenum Press.

——. 2004. A basal archosaurian origin for birds. *Acta Zoologica Sinica* 50: 978–990.

Marsh, O. C. 1877. Introduction and succession of vertebrate life in America. *Proceedings of the American Association of the Advancement of Science* 1877: 211–258.

Mayr, G. 2002. Osteological evidence for paraphyly of the avian order Caprimulgiformes (nightjars and allies). *Journal of Ornithology* 143: 82–97.

——. 2003. On the phylogenetic relationships of trogons (Aves, Trogonidae). *Journal of Avian Biology* 34: 81–88.

——. 2005a. The Paleogene fossil record of birds in Europe. *Biological Reviews* 80: 1–28.

——. 2005b. Tertiary plotopterids (Aves, Plotopteridae) and a novel hypothesis on the phylogenetic relationships of penguins (Spheniscidae). *Journal of Zoological Systematics and Evolutionary Research* 43: 61–71.

——. 2007. Avian higher-level phylogeny: well-supported clades and what we can learn from a phylogenetic analysis of 2954 morphological characters. *Journal of Zoological Systematics and Evolutionary Research* 46: 63–72.

——. 2008a. The Madagascan "Cuckoo-roller" (Aves: Leptosomidae) is not a roller – notes on the phylogenetic affinities and evolutionary history of a "living fossil." *Acta Ornithologica* 43: 226–230.

——. 2008b. Phylogenetic affinities of the enigmatic avian taxon *Zygodactylus* based on new material from the early Oligocene of France. *Journal of Systematic Palaeontology* 6: 333–334.

——. 2009. *Paleogene Fossil Birds.* Berlin: Springer.

——. 2010a. Metaves, Mirandornithes, Strisores and other novelties – a critical review of the higher-level phylogeny of neornithine birds. *Journal of Zoological and Systematic Evolutionary Research* 49: 58–76.

——. 2010b. Phylogenetic relationships of the paraphyletic 'caprimulgiform' birds (nightjars and allies). *Journal of Zoological Systematics and Evolutionary Research* 48: 126–137.

———. 2011a. The phylogeny of charadriiform birds (shorebirds and allies) – reassessing the conflict between morphology and molecules. *Zoological Journal of the Linnean Society* 161: 916–934.

———. 2011b. Well-preserved new skeleton of the Middle Eocene *Messelastur* substantiates sister group relationship between Messelasturidae and Halcyornithidae (Aves, ?Pan-Psittaciformes). *Journal of Systematic Palaeontology* 9: 159–171.

Mayr, G., and J. Clarke. 2003. The deep divergences of neornithine birds: a phylogenetic analysis of morphological characters. *Cladistics* 19: 527–553.

Mayr, G., and P. G. P. Ericson. 2004. Evidence for a sister group relationship between the Madagascan mesites (Mesitornithidae) and the cuckoos (Cuculidae). *Senckenbergiana biologica* 84: 1–17.

Mayr, G., A. Manegold, and U. S. Johansson. 2003. Monophyletic groups within "higher land birds" – comparison of morphological and molecular data. *Journal of Zoological and Systematic Evolutionary Research* 41: 233–248.

Mayr, G., B. Pohl, S. Hartman, and D. S. Peters. 2007. The tenth skeletal specimen of *Archaeopteryx*. *Zoological Journal of the Linnean Society* 149: 97–116.

Morgan-Richards, M., S. A.Trewick, A., Bartosch-Härlid, O. Kardailsky, M. J. Phillips, P. A. McLenachan, and D. Penny. 2008. Bird evolution: testing the Metaves clade with six new mitochondrial genomes. *BMC Evolutionary Biology* 8: 20.

Murray, P. F., and P. Vickers-Rich. 2004. *Magnificent Mihirungs: The Colossal Flightless Birds of the Australian Dreamtime.* Bloomington: Indiana University Press.

Nopcsa, F. 1907. Ideas on the origin of flight. *Proceedings of the Zoological Society of London* 1907(15): 223–236.

———. 1923. On the origin of flight in birds. *Proceedings of the Zoological Society of London* 1923(31): 463–477.

O'Connor, J. K., X. Wang, L. M. Chiappe, C. Gao, Q. Meng, X. Cheng, and J. Liu. 2009. Phylogenetic support for a specialized clade of Cretaceous enantiornithine birds with information from a new species. *Journal of Vertebrate Paleontology* 29: 188–204.

Olson, S. L. 1978. Multiple origins of the Ciconiiformes. *Proceedings of the Colonial Waterbird Group* 1978: 165–170.

———. 1985. The fossil record of birds. In D. S. Farner, J. R. King, and K. C. Parkes (eds.), *Avian Biology,* 8: 79–238. Orlando, Fla.: Academic Press.

Ostrom, J. H. 1973. The ancestry of birds. *Nature* 242: 136.

———. 1974. *Archaeopteryx* and the origin of flight. *Quarterly Review of Biology* 49: 27–47.

Paul, G. S. 2002. *Dinosaurs of the Air: the Evolution and Loss of Flight in Dinosaurs and Birds.* Baltimore, Md.: Johns Hopkins University Press.

Phillips, M., G. Gibb, E. Crimp, and D. Penny. 2009. Tinamous and moa flock together: mitochondrial genome sequence analysis reveals independent losses of flight among ratites. *Systematic Biology* 59: 90–107.

Prum, R. O. 1988. Phylogenetic interrelationships of the barbets (Aves: Capitonidae) and toucans (Aves: Ramphastidae) based on morphology with comparisons to DNA–DNA hybridization. *Zoological Journal of the Linnean Society* 92: 313–343.

———. 2002. Why ornithologists should care about the theropod origin of birds. *Auk* 119: 1–17.

Reid, R. E. H. 2011 (this volume). How dinosaurs grew. In M. K. Brett-Surman, Thomas R. Holtz Jr., and James O. Farlow (eds.), *The Complete Dinosaur,* 2nd ed., 621–635. Bloomington: Indiana University Press.

Ries, C. J. 2007. Creating the Proavis: bird origins in the art and science of Gerhard Heilmann 1913–1926. *Archives of Natural History* 34: 1–19.

Schweizer, M., O. Seehausen, M. Güntert, and S. T. Hertwig. 2010. The evolutionary diversification of parrots supports a taxon pulse model with multiple trans-oceanic dispersal events and local radiations. *Molecular Phylogenetics and Evolution* 54: 984–994.

Sibley, C. G., and J. A. Ahlquist. 1990. *Phylogeny and Classification of Birds.* New Haven, Conn.: Yale University Press.

Smith, N. D. 2010. Phylogenetic analysis of Pelecaniformes (Aves) based on osteological data: implications for waterbird phylogeny and fossil calibration studies. *PLoS ONE* 5 (10): e13354.

Thomas, G. H. 2011. The state of the world's birds and the future of avian diversity. In G. Dyke and G. Kaiser (eds.), *Living Dinosaurs: the Evolutionary History of Modern Birds,* 381–403. Chichester, UK: John Wiley and Sons.

Turvey, S. T., O. R. Green, and R. N. Holdaway. 2005. Cortical growth marks reveal extended juvenile development in New Zealand moa. *Nature* 435: 940–943.

Van Tuinen, M., D. B. Butvill, J. A. W. Kirsch, and S. B. Hedges. 2001. Convergence and divergence in the evolution of aquatic birds. *Proceedings of the Royal Society of London B* 268: 1345–1350.

Vargas, A. O., and J. F. Fallon. 2005. Birds have dinosaur wings: the molecular evidence. *Journal of Experimental Zoology (Molecular and Developmental Evolution)* 304B, 86–90.

Voelker, G., and G. M. Spellman. 2003. Nuclear and mitochondrial DNA evidence of polyphyly in the avian subfamily Muscicapoidea. *Molecular Phylogenetics and Evolution* 30: 386–394.

von Euler, F. 2001. Selective extinction and rapid loss of evolutionary history in the bird fauna. *Proceedings of the Royal Society of London B* 268: 127–130.

Walker, A. D. 1972. New light on the origin of birds and crocodiles. *Nature* 237: 257–263.

Walker, C. A. 1981. New subclass of birds from the Cretaceous of South America. *Nature* 292: 51–53.

Williston, S. W. 1879. Are birds derived from dinosaurs? *Kansas City Review of Science* 3: 357–360.

Witmer, L. M. 1991. Perspectives on avian origins. In H.-P. Schultze and L. Trueb (eds.), *Origins of the Higher Groups of Tetrapods: Controversy and Consensus,* 427–466. Ithaca, N.Y.: Cornell University Press.

Witmer, L. M., and K. D. Rose. 1991. Biomechanics of the jaw apparatus of the gigantic Eocene bird *Diatryma:* implications for diet and mode of life. *Paleobiology* 17: 95–120.

Worthy, T. H., and R. N. Holdaway. 2002. *The Lost World of the Moa.* Bloomington: Indiana University Press.

Wright, T. F., E. E. Schirtzinger, T. Matsumoto, J. R. Eberhard, G. R. Graves, J. J. Sanchez, S. Capelli, H. Müller, J. Scharpegge, G. K. Chambers, and R. C. Fleischer. 2008. A multilocus molecular phylogeny of the parrots (Psittaciformes): support for a Gondwanan origin during the Cretaceous. *Molecular Biology and Evolution* 25: 2141–2156.

Xu, X., H. You, K. Du, and F. Han. 2011. An *Archaeopteryx*-like theropod from China and the origin of Avialae. *Nature* 475: 465–470.

Zhang, F., Z. Zhou, X. Xu, X. Wang, and C. Sullivan. 2008. A bizarre Jurassic maniraptoran from China with elongate ribbon-like feathers. *Nature* 455: 1105–1108.

Zink, R. M. 1996. Bird species diversity. *Nature* 381: 566.

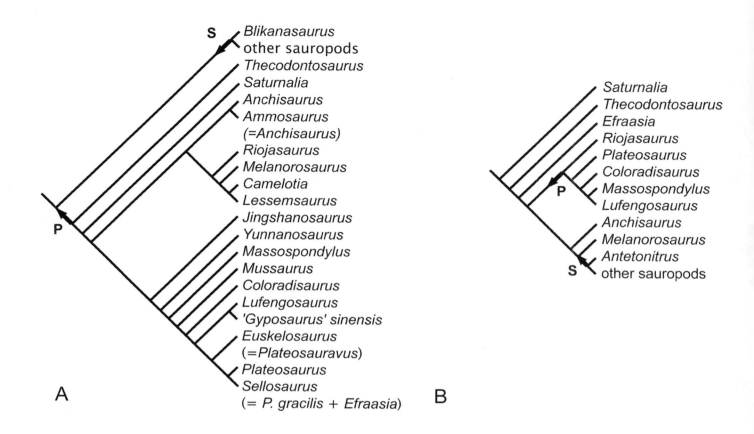

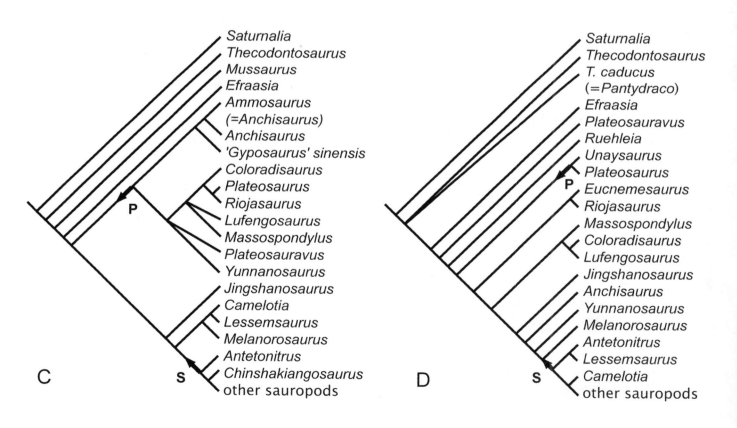

Basal Sauropodomorpha: The "Prosauropods"

Adam M. Yates

21

Very early in the evolution of dinosaurs a branch of saurischians, known as the Sauropodomorpha, split away from the predatory Theropoda and began to specialize toward herbivory. The earliest sauropodomorphs that we know had already acquired small heads, moderately elongate necks, and leaf-shaped tooth crowns. These adaptations indicate that the ability to prey upon large vertebrates of similar body size had been abandoned, although some degree of omnivory was probably retained at this stage. Large body size soon followed, and by the middle of the Norian Stage (approximately 220 mya) sauropodomorphs had become the Earth's dominant large terrestrial herbivores. Prior to the Norian the largest terrestrial herbivores were the stocky stahleckeriid dicynodonts that were unable to browse on vegetation over 1.5 m above the ground. Early sauropodomorphs, with their long necks, were the first herbivorous vertebrates that were capable of browsing among the trees. The sauropodomorph trend toward large-bodied herbivores culminated in the eusauropods and their close relatives, which may well have been in existence before the end of the Triassic Period. The early branching pattern of the Sauropodomorpha has always been a vexatious issue, in particular the precise relationship between the large columnar-limbed sauropods and the more basal sauropodomorphs. A consequence of this uncertainty is that the taxonomy of basal sauropodomorphs has remained in a state of flux.

History of Classification

The classification of basal sauropodomorphs has had an unusually convoluted history, which is still far from settled. Early dinosaur studies placed the basal sauropodomorphs among the carnivorous theropods (e.g., Marsh 1881). In 1920 Friedrich von Huene invented the taxon Prosauropoda to encompass the early saurischians with long necks and small skulls. It is now clear that not all basal sauropodomorphs form an exclusive clade, and I will use the informal term *basal sauropodomorph* rather than Prosauropoda to refer to all nonsauropod sauropodomorphs.

Huene can also be credited with first recognizing that the derived similarities between his prosauropods and sauropods indicate that they are more closely related to each other than to other dinosaur groups (Huene 1914). He coined the taxon Sauropodomorpha to encompass both sauropods and basal sauropodomorphs (Huene 1932). However, confusion with theropods persisted, and Huene abandoned Sauropodomorpha in his last work on saurischian classification, placing Prosauropoda, Sauropoda, and "Carnosauria" (large robust theropods that are no longer thought to form a clade) together in a more inclusive taxon he called Pachypodosauria (Huene

21.1. Cladograms depicting the three major classes of results from recent phylogenetic analyses. (A) "Grand prosauropod monophyly" from Galton and Upchurch (2004). (B, C) Two versions of "Core prosauropod monophyly" from (B) Yates and Kitching (2003) and (C) Upchurch et al. (2007a). (D) "Extreme prosauropod paraphyly" from Yates (2007). P denotes the clade that conforms to the currently accepted definition of Prosauropoda. Note that in (D) it refers to a clade of low diversity that does not warrant recognition as a major unit of dinosaur diversity. S denotes the clade that conforms to the definition of Sauropoda used here.

425

1948). But even though it was not reflected in his classification, Huene had not abandoned his hypothesis of genealogical relationship between basal sauropodomorphs and sauropods.

Underpinning the continued confusion between basal sauropodomorphs and theropods lay the so-called teratosaurids. These were thought to be dinosaurs with a postcranial skeleton identical to that of other basal sauropodomorphs but with a large robust skull armed with large carnosaur-like teeth. Thus it was thought that later "carnosaurs" had evolved from basal sauropodomorphs via the teratosaurids. It transpired that teratosaurids were chimaeras made of the postcrania of genuine basal sauropodomorphs that had been mixed with shed teeth or isolated skull parts of large carnivorous archosaurs, mostly rauisuchians (Benton 1986). Nevertheless, the continued confusion resulted in Romer's 1956 classification where Prosauropoda was relegated to an infraorder within the suborder Theropoda while Sauropoda was elevated to the status of a suborder, on a par with Theropoda.

Charig et al. (1965) revived Sauropodomorpha as a taxon, proposing that Prosauropoda and Sauropoda were two clades descended from a common quadrupedal pseudosuchian ancestor. Charig et al. recognized three families within Prosauropoda: Thecodontosauridae, Plateosauridae, and Melanorosauridae. This was little more than a size-based classification, with Thecodontosauridae encompassing the small, gracile species, Plateosauridae the medium-sized species, and Melanorosauridae the large robust species. No derived features were proposed to unite the members of any of these families.

Since Charig et al. it has been traditional to divide Sauropodomorpha into two large groups: Prosauropoda and Sauropoda. The three-part familial classification of Prosauropoda also remained in place, although Galton (1990) split off some of the more autapomorphic genera into additional monotypic families (Anchisauridae, Massospondylidae, Yunnanosauridae, and Blikanasauridae).

Prior to the advent of cladistic studies and the demand for strictly monophyletic classifications, the Prosauropoda–Sauropoda division was acceptable, even though many researchers had suggested that Sauropoda evolved from a branch of the Prosauropoda, with the melanorosaurids as the most likely candidates. Neverthless Prosauropoda survived the advent of strictly monophyletic classifications because most of the early cladistic studies of sauropodomorph interrelationships recovered a monophyletic Prosauropoda. However, it has become clear that there is no clade that encompasses all basal sauropodomorphs to the exclusion of traditional sauropods. Consequently the use of Prosauropoda has had to be modified, and in some situations it may not be a useful term at all.

Recent Cladistic Analyses and Classification

Cladistic analyses of basal sauropodomorph relationships have produced variable results. These can be broadly classed into three main categories: grand prosauropod monophyly(a term coined by Upchurch et al. 2005), core prosauropod monophyly, and extreme prosauropod paraphyly (Fig. 21.1). Grand prosauropod monophyly refers to those hypotheses where all known sauropodomorphs fall into two monophyletic clades: Prosauropoda and Sauropoda. In these hypotheses Prosauropoda is diverse and includes all sauropodomorphs that are not members of the derived columnar-limbed sauropod clade. Studies that support this arrangement include Sereno (1999, although the taxon sampling is limited and the results of this study are congruent with some versions of core prosauropod monophyly), Benton et al. (2000), and Galton and Upchurch (2004).

Sereno (1998) also provided the first phylogenetic definitions for clades within Sauropodomorpha. Prosauropoda was defined as the clade containing all sauropodomorphs that share a more recent common ancestor with *Plateosaurus* than with the late Cretaceous eusauropod *Saltasaurus*. Conversely, Sauropoda was defined as being all sauropodomorphs more closely related to *Saltasaurus* than to *Plateosaurus*. Within Prosauropoda the taxon Plateosauria was defined as the clade descending from the most recent common ancestor of *Plateosaurus* and *Massospondylus*. Subsequently Galton and Upchurch (2004) erected Anchisauria for the clade of *Anchisaurus*, *Melanorosaurus*, and all descendants of their most recent common ancestor. In their topology, Anchisauria lay outside of the Plateosauria.

Core prosauropod monophyly refers to those topologies where a subset of traditional prosauropods (minimally plateosaurids, riojasaurids, and massospondylids) form a clade (which should be called Prosauropoda), while others (e.g., *Saturnalia*, *Thecodontosaurus*, *Pantydraco*, and *Efraasia*) lie outside the Prosauropod–Sauropod clade; yet others (e.g. *Jingshanosaurus* and *Melanorosaurus*) are basal members of the sauropod clade. Studies supporting core prosauropod monophyly include Yates and Kitching (2003), Yates (2004) and Upchurch et al. (2007a).

Extreme prosauropod paraphyly describes those analyses where the traditional prosauropods, even core prosauropods are broken into a paraphyletic array of low-diversity clades on the stem leading to Eusauropoda (derived sauropods). In these topologies many traditional prosauropods are transferred to the base of the Sauropoda by virtue of the relatively basal position of *Plateosaurus* within this array. Examples of analyses that have found this sort of topology include Yates (2003a, 2007) and Pol and Powell (2005).

Part of the reason that core prosauropod monophyly has broken up in the most recent analyses is the inclusion of new data from *Melanorosaurus*, *Jingshanosaurus*, *Mussaurus*, and *Aardonyx*. These genera includes a mix of sauropod-like features with a number of characters that were thought to diagnose core prosauropods, thus diminishing the support for core prosauropod monophyly. In cases of extreme prosauropod paraphyly the name Prosauropoda strictly applies to just *Plateosaurus* and its closest relatives (which may include only *Unaysaurus* from Brazil).

It is the opinion of some authors (Yates 2007; Sereno 2007) that the expansion of Sauropoda well beyond its traditional boundaries has rendered Sereno's initial definition of Sauropoda undesirable. Accordingly they have proposed new definitions for Sauropoda that maintain more traditional contents for this clade when applied to strongly paraphyletic topologies. The position adopted here is that

Sauropoda includes all taxa more closely related to *Saltasaurus* than to *Melanorosaurus*, the most derived "traditional prosauropod" (Yates 2007; Taylor et al. in press).

Hypothesis testing using the comprehensive matrix of Yates (2007) indicates that we can confidently reject "grand prosauropod monophyly." However, the difference between core prosauropod monophyly and total prosauropod paraphyly is statistically insignificant so we must view both hypotheses as viable options (Yates 2007). Similar results were obtained with a slightly smaller matrix (Upchurch et al. 2007a). In this case, however, the difference between grand prosauropod monophyly and other topologies was large but not statistically significant.

The Sauropodomorph Stem: "Thecodontosaurids"

Basal Sauropodomorph Diversity and Anatomy

Traditionally the basal-most sauropodomorphs have been grouped into the family Thecodontosauridae along with other small, gracile species. However, there is no evidence that they form a clade even when advanced gracile forms such as *Anchisaurus* are removed. A recently recognized assemblage of the very earliest sauropodomorphs, named Guaibasauridae (Ezcurra 2010), might form a monophyletic radiation that would be the sister group of all other sauropodomorphs. The monophyly of this radiation is by no means settled, and it is possible that the nominate genus, *Guaibasaurus*, is in fact a basal theropod (Langer and Benton 2006; Yates 2007; Bittencourt Rodrigues 2010). Part of the problem is that we are dealing with dinosaurs so close to the common ancestor of sauropodomorphs and theropods that they had not yet acquired many of the future distinctive characteristics of either lineage. Couple this lack of distinctiveness with incomplete fossil remains, and it becomes difficult to place such taxa conclusively. Nevertheless, the guaibasaurid assemblage does contain some incontrovertible sauropodomorphs.

Saturnalia (Fig. 21.2B), from the late Carnian of Brazil, is the most completely known of these. It was a small, gracile bipedal form measuring no more than 2 m in length and standing no more than 45 cm tall at the hips. It had a reduced skull (that measured less than half the length of the femur), short but powerful forelimbs, and long hind limbs with elongate distal segments indicating the ability to run swiftly. Derived characters that it shares with other sauropodomorphs include the reduced size of the skull, leaf-shaped teeth, and a transversely expanded distal end of the humerus. Unlike other sauropodomorphs it retained on its teeth fine serrations set at right angles to the long axis of the tooth, a partially closed acetabulum of the pelvis, a well-developed shelflike process associated with the lesser trochanter of the femur (also seen in a range of early dinosaurs and basal dinosauriforms), and a slender proximal end of the fifth metatarsal (Langer and Benton 2006).

Chromogisaurus is a very similar form from coeval strata further south in Argentina. Both *Saturnalia* and *Chromogisaurus* share an exceptionally enlarged and heavily scarred olecranon process projecting from the proximal end of the ulna (Ezcurra 2010). This implies that the forearm was unusually powerful in extension (straightening of the elbow), which may well have been an adaptation to some kind of digging behaviour.

More closely related to all other sauropodomorphs are *Thecodontosaurus* and *Pantydraco*. These small forms, from the Triassic fissure fills of southwestern England and Wales (Fig. 21.2A), show significant advances

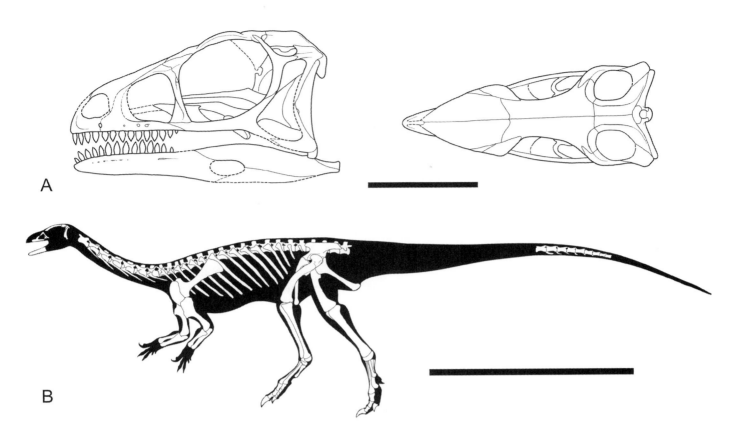

21.2. Anatomy of the basal-most sauropodomorphs. (A) Skull of *Pantydraco* in lateral and dorsal views. (B) Skeleton of *Saturnalia*. Scale bars equal 20 mm in (A) and 500 mm in (B). (A) redrawn from Yates (2003 a), (B) redrawn from Langer and Benton (2006).

including an elongated, ridgelike lesser trochanter without any trochanteric shelf and coarse, upwardly angled serrations on the teeth (Yates 2003a). *Efraasia* from the mid-Norian of Germany is distinctly more advanced than the previously mentioned genera. It shows many of the advanced features seen in the clade of *Plateosaurus* + Sauropoda (a clade that I will here refer to as Plateosauria) such as large body size (adult *Efraasia* reached approximately 6 m in length), an elongated neck, and a curious construction of the hand whereby the metacarpal of the enlarged thumb was inset into the wrist. In the latter character the proximal end of the thumb metacarpal articulates against the side of the second distal carpal, while the first distal carpal forms a cap over both these bones. Nevertheless the hand remained elongate (about 60% of the humerus + radius) as it is in *Thecodontosaurus* and the immediate sauropodomorph outgroups (Theropoda, *Eoraptor*, Herrerasauridae) and the prefrontal remained small, without the caudal extension seen in plateosaurians proper. For many years *Efraasia* was thought to be a junior synonym of *Sellosaurus*, but it has since been shown to be distinctly different from the type specimen of that genus, and has now been placed in the genus *Plateosaurus* (Yates 2003b).

Core Prosauropods: Plateosaurids, Riojasaurids, and Massospondylids

Although it is debatable whether these sauropodomorphs form a clade or not, they do form a compact group that displays little morphological disparity. In topologies of extreme basal sauropodomorph paraphyly, the core prosauropods form an assemblage at the base of the Plateosauria (all descendants of the common ancestor of *Plateosaurus* and *Massospondylus*), whereas in core prosauropod monophyly they form the Plateosauria in its entirety.

The skulls of core prosauropods have deeper snouts than those of thecodontosaurids due to the enlarged bony nostrils, which have a diameter that exceeds half the diameter of the orbit (Fig. 21.3B). Another significant feature is the lowering of the jaw joint to a point below the level of the dentary tooth row.

Sereno (1997) has argued that core prosauropods possessed a small horny beak at the tips of their jaws. This was based on the observation that the first dentary tooth is slightly inset from the tip of the mandible, that there is a pair of ridges at the front of the premaxilla that defines a raised platform, and that the area is invested with large nutritive foramina. However, the raised platform is readily visible only in *Riojasaurus*. In other genera such as *Plateosaurus* and *Massospondylus* there is only the faintest of ridges as the lateral surface of the premaxilla curves toward the midline symphysis at the tip of the snout.

A concentration of nutritive foramina can be found on the premaxilla and at the rostral end of the dentary in several other saurischians, such as herrerasaurids and coelophysoid theropods, which show no sign of ever having borne a beak. Consequently a rostral concentration of nutritive foramina is not necessarily correlated with the presence of a beak. Well-preserved premaxillae of *Massospondylus* show that the tip of the snout was invested with numerous very small foramina or pits (Fig. 21.4), indicating that there was probably a dense patch of collagen fibers in the area. These fibers probably bound the skin tightly to the bone in this area in much the same way that skin is tightly bound to the bone along the entire margins of the jaws of crocodilians. Thus most core prosauropods probably had a small patch of tough, tightly adherent skin at the tips of the jaws while *Riojasaurus* may have developed a proper beak.

The basic body plan of the postcranial skeleton includes large body size ranging from 4 to 9 m in length, robust build, a long neck, and a strong disparity in length between the fore and hind limbs (Fig. 21.3A). The long neck was inherited from an earlier common ancestor shared with *Efraasia* and is produced by the elongation of the anterior cervical vertebrae. Neck elongation is carried to an extreme in massospondylids, where these vertebrae are four times longer than the width of the anterior centrum face. A growth series of *Massospondylus* shows that hatchlings had relatively short necks and that full elongation was not achieved until adult size was reached.

In most core prosauropods the length of the forelimb (humerus + radius) is just over half the length of the hind limb (femur + tibia), although it is as long as 70% in *Riojasaurus*. Nevertheless, the large deltopectoral crest and broad distal humerus indicates that the forelimb was very powerful. The short, broad, grasping manus bore a very robust thumb with a very large trenchant claw. Indeed, in massospondylids the thumb was the longest digit in the hand. In all core prosauropods the phalanx that bore the thumb claw is strongly twisted about its axis so that the claw faced outward when extended. There is no doubt that it would have formed a formidable weapon, but it is unknown whether it was used for defense against predators or intraspecific aggression.

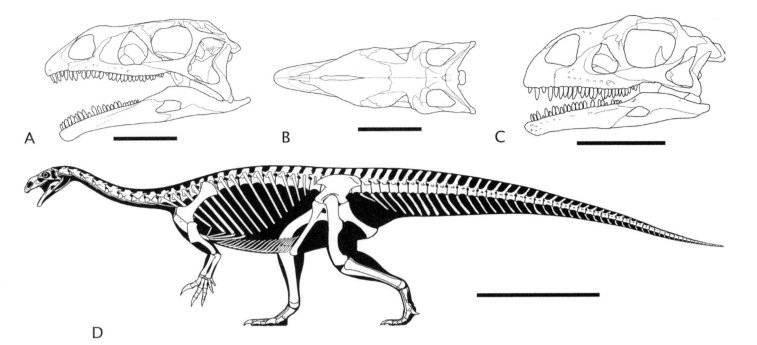

A

B

C

D

The anterior lobe of the ilium in the pelvis is a small triangular process that does not protrude beyond the anterior end of the pubic peduncle of the ilium. This indicates that the anterior ilium supported relatively less muscle mass than in many other dinosaurs, suggesting that the stride length may have been relatively shorter than most other dinosaurs' (Carrano 2000). Nevertheless, the femur had a flexed shaft and a proximally located fourth trochanter, indicating that core prosauropods did not progress with the slow, stiff-legged gait that eusauropods used. The foot retained a compact, nonspreading metatarsus that bore four well-developed, clawed toes. Although the first toe was the shortest, it bore the largest claw. The fifth toe was reduced to a single nubbin of bone borne on a short metatarsal with a slender, splintlike distal end.

It has commonly been argued that such a reduced fifth toe barred basal sauropodomorphs from being ancestors of the more derived sauropods. However, the fifth toe of eusauropods also consists of a single nubbin-like phalanx borne on a metatarsal, and differs from the fifth toe of basal

21.3. Anatomy of the basal plateosaurians. (A) Skull of *Plateosaurus* in lateral view. (B) Skull of *Plateosaurus* in dorsal view. (C) Skull of *Massospondylus* in lateral view. (D) Skeleton of *Plateosaurus*. Scale bars equal 100 mm in (A), (B), (C) and 1 m in (D).

(D) used courtesy of Scott Hartman.

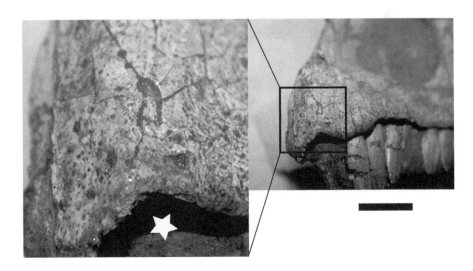

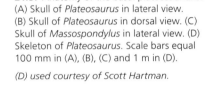

21.4. The premaxillae of *Massospondylus* in anterolateral view with an enlargement of the rostral tip. Note the fine foramina on the tip. The white star marks the empty first alveolus. Scale bar equals 10 mm.

sauropodomorphs only in that the metatarsal is more robust. It is true that metatarsal five of eusauropods is of a similar length to the other metatarsals, whereas it is significantly shorter in basal sauropodomorphs, but it must be remembered that the other metatarsals of sauropods have been shortened to produce the short, broad, elephant-like foot of sauropods. Thus it does not require the reevolution or even reenlargment of lost or reduced structures to derive the sauropod fifth toe from that of a basal sauropodomorph.

Near Sauropod Sauropodomorphs: Anchisaurians

There is a growing consensus that a number of the more advanced basal sauropodomorphs are indeed more closely related to sauropods than any of the core prosauropods are. This group minimally includes *Jingshanosaurus*, *Mussaurus*, *Aardonyx*, *Melanorosaurus*, and *Camelotia* but probably also includes *Anchisaurus* (sometimes called *Ammosaurus*, see Sereno 2007). Others, such as *Blikanasaurus*, *Chinshakiangosaurus*, *Lessemsaurus*, and *Antetonitrus* are probably true members of Sauropoda, albeit very basal ones.

Anchisaurus has traditionally been regarded as one of the more basal members of Sauropodomorpha, or Prosauropoda, based largely on its small size and gracile build. However, when the anatomical characteristics of this species are examined more closely it is found to share much in common with basal sauropods. Some of these derived characteristics include wrinkled tooth enamel, dorsally displaced parasphenoid process of the braincase, a large U-shaped fossa developed between the posteroventral tubers of the braincase, reduction in the extent of the antorbital fossa, a reduced and dorsally displaced quadrate foramen, and the relatively distal position of the fourth trochanter on the femur. The small size of *Anchisaurus* appears to be a reversal of an otherwise marked trend toward increasing size in the lineage leading to Eusauropoda. The clade Anchisauria (all descendants of the common ancestor of *Anchisaurus* and *Melanorosaurus*) is a useful name for separating these near sauropods from core prosauropods in cases of extreme basal sauropodomorph paraphyly.

As well as including one of the smallest sauropodomorphs, the near-sauropod assemblage includes the largest sauropodomorphs outside the Sauropoda. *Jingshanosaurus* from the lower Lufeng Series (Early Jurassic) of China reached lengths of more than 8.5 m, and with a femur that is 86 cm or more in length, it approached eusauropods in size. It was found to be the sister group of Anchisauria in my most recent analysis, but it shares many of the derived characters seen in anchisaurians and in future studies may prove to be a part of this group. Its well-preserved skull gives us our best example of the cranial anatomy of sauropodomorphs close to the base of Anchisauria. It is quite short and deep, with orbits that are slightly pinched ventrally and with a lower temporal fenestra that extends underneath the orbit, indicating that the skull is partially foreshortened in a sauropod-like manner. Nevertheless, it remained functionally similar to the skulls of core prosauropods, with a narrow, pointed snout (fig. 21.5A, contra Zhang and Yang 1994), no precise tooth-on-tooth contact, a shallow anterior end of the dentary, and an elongate retroarticular process.

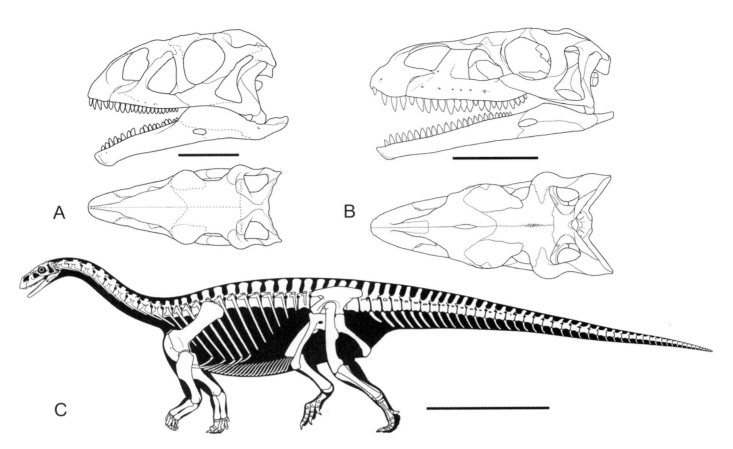

A

B

C

The postcranial anatomy remains very similar to that of core prosauropods, with relatively short forelimbs, grasping hands, a flexed femoral shaft, a proximally located fourth trochanter, and an elongate, compact metatarsus.

The newly described *Aardonyx* from the Early Jurassic of South Africa is almost as large and, given the juvenile nature of the only known specimens, probably reached an even greater size in adulthood. It shows several significant advances toward the sauropod condition compared to *Anchisaurus* and *Jingshanosaurus* (Yates et al. 2010). For example, the outer rims of its tooth sockets were raised into tall strips of bone, called lateral plates that braced the bases of the teeth against outwardly directed forces. Such forces may have been generated when the animal tugged and tore mouthfuls of foliage from plants (Upchurch et al. 2007b). This suggests the beginning of sauropod-style bulk-browsing in the advanced anchisaurians. Other sauropod-like features include a short, broad hind foot, with the most robust elements skewed toward the inner side of the foot (a condition known as entaxony), and a distally located fourth trochanter, which indicates that *Aardonyx* was slowing down and tending toward graviportalism. Nevertheless, its forelimb remained short and lacked the ability to pronate its hand; thus it probably still progressed with a bipedal gait (Yates et al. 2010).

The skull of *Melanorosaurus* is superficially more plateosaurid-like than those of other anchisaurians, but this is largely due to its secondarily elongated snout. Other details of its skull indicate that it has a closer relationship with eusauropods than most other basal sauropodomorphs do. These characters include the derived skull characters seen in other anchisaurians

21.5. Anatomy of the basal anchisaurians. (A) Skull of *Jingshanosaurus* in lateral and dorsal views. (B) Skull of *Melanorosaurus* in lateral and dorsal views. (C) Skeleton of *Melanorosaurus*. Scale bars equal 100 mm in (A), (B) and 1 m in (C).

(C) used courtesy of Scott Hartman.

mentioned above as well as an inflection in the profile of the snout at the base of dorsal process of the premaxilla, a broad ventral process of the squamosal, and an anteriorly directed, straplike maxillary process of the palatine. These features match the eusauropod-like nature of its postcranium. Derived postcranial characteristics that *Melanorosaurus* shares with Eusauropoda include a sacrum that is composed of at least four vertebrae, a long forelimb, and an ulna with a large proximal anterolateral process creating a deep radial fossa. The last two characters indicate that the animal had switched to a largely quadrupedal gait. The former character partially equalizes the difference in stride length between the fore and hind limbs, while the latter character shifts the radius into a more anteromedial position relative to the ulna, thus allowing the hand to be pronated so that the palm faces posteriorly and the hand swings in the direction of travel (Bonnan and Yates 2007).

Diet

Typically basal sauropodomorphs are regarded as herbivores. It is true that even the most basal known taxa lack the adaptations for tackling large vertebrate prey that are seen in herrerasaurids and theropods. However, some degree of omnivory, or even insectivory, especially in the case of the most basal forms, cannot be easily dismissed. The case for basal sauropodomorph herbivory is based on anatomical adaptations and paleoecological considerations. The anatomical evidence rests largely on their teeth, which are leaf shaped with expanded bases, are not posteriorly recurved, and have coarse upwardly angled marginal serrations. The expanded bases reduce the gaps between teeth, or even eliminate them, thus producing a completely continuous, or at least nearly continuous, cutting edge. The angled, coarse serrations are more suited to shredding plant material than the fine serrations set at right angles to the tooth margin that are seen in carnivorous archosaurs. The teeth of basal sauropodomorphs are remarkably similar to those of iguanine lizards, especially the genus *Iguana*, which are among the most herbivorous of modern lizards. A similar suite of adaptations can be seen in the teeth of other archosaurs that are thought to be herbivorous, most notably Ornithischia and Therizinosauroidea. However, Barrett (2000) pointed out that iguanines are better regarded as facultative omnivores as opposed to strict herbivores. This is demonstrated by the observations that wild iguanas include small vertebrate prey in their diet and captive iguanas require some meat for normal growth. The pointed, finely serrated premaxillary teeth of iguanas may be used to deal with prey. Basal sauropodomorphs also display narrow, sharply pointed, and finely serrated premaxillary teeth, indicating that they may have also included small vertebrate prey in their diet.

Other anatomical evidence that has a bearing on diet relates to the position of the jaw joint and the possible presence of fleshy cheeks. In modern mammalian herbivores the jaw joint tends to be raised above the level of the tooth rows, so that when the jaws are closed the teeth tend to come together at the same time. In contrast, carnivores have a jaw joint that is level with the tooth rows so that the jaws have a scissorlike action, with the point of contact between the upper and lower tooth rows moving from back to front during jaw closure. Plateosaurians have a jaw joint that lies below the dentary tooth row, but the effect is the same as the raised jaw joint seen in mammalian herbivores (Galton 1984). Interestingly, the jaws of thecodontosaurids (Fig. 21.2B) lack the depressed joint and would appear to have been closed in a carnivore-like manner.

Cheeks would have helped to hold plant matter in the mouth during oral processing. Evidence for their presence in both prosauropods and ornithischians comes from the lateral ridge on the dentary and unusually large neurovascular foramina of the dentary and maxilla (Galton 1984). The dentary ridge may have supported a cheek while the enlarged foramina were a mark of increased blood supply to the soft tissues of the proposed cheeks. Although both members of the extant phylogenetic bracket of dinosaurs (crocodilians and birds) lack cheeks, or the necessary soft tissue structures to serve as the progenitors of cheeks (Papp and Witmer 1998), it must be remembered that they both have highly modified jaws. Crocodilians have derived lipless mouths edged with tightly adherent cornified skin, whereas birds have greatly enlarged premaxillae that bear a horny rhamphotheca, while the maxilla is highly reduced. The maxilla is the upper jaw bone that is presumed to have borne the cheeks in basal sauropodomorphs and ornithischians. Ironically, sauropods–which were more highly specialized herbivores than the basal sauropodomorphs–may have abandoned their cheeks to allow for wider gapes to enable bulk-browsing (Upchurch et al. 2007b).

Further evidence for basal sauropodomorph herbivory can be gleaned from their high abundance. When present, plateosaurian sauropodomorphs –from core prosauropods to basal anchisaurians–are often the most abundant vertebrate fossils. Indeed, *Plateosaurus*, *Lufengosaurus*, and *Massospondylus* each account for more than 70% of vertebrate fossils from their respective formations. Similarly, the moderately diverse fauna of basal sauropodomorphs from the Triassic of southern Africa (lower Elliot Formation) accounts for more than 95% of vertebrate fossils from that unit. This dominance indicates that basal sauropodomorphs were probably the most abundant large vertebrates in their habitat. Basic ecological theory would suggest that such a large component of the fauna would have to sit at the base of the food web as primary consumers.

Countering the arguments for herbivory is the lack of an efficient masticatory apparatus capable of effectively breaking down large quantities of tough fodder. The skull and jaws of basal sauropodomorphs are weakly constructed, and the teeth slide pass each other without anything more than occasional and incidental tooth-on-tooth contact. Thus food could only be coarsely chopped and shredded. Galton (1984) suggested that a gastric mill could have triturated the food, and indeed, gastroliths are known from basal sauropodomorphs (e.g., *Efraasia* and *Massospondylus*). However, only one small *Massospondylus* skeleton (femur length approximately 210 mm, body mass approximately 50 kg) from Zimbabwe is preserved with enough gastroliths (approximately 1% of body mass) to form an effective gastric mill (Raath 1974). Furthermore, the semipolished to polished nature of all known basal sauropodomorph gastroliths is inconsistent with their use as gastric mills, which in modern animals coarsely abrade and roughen gastroliths (Wings and Sander 2007).

It is probable that the earliest sauropodomorphs had quite catholic diets. *Saturnalia* had rather small, triangular teeth that lacked coarse, upwardly angled serrations. These were clearly not efficient at processing vegetation, nor were they effective in dealing with flesh, but they would have been ideal for piercing insect and other arthropod cuticles as well as for coarsely shredding soft vegetable matter such as fleshy reproductive structures and new shoots. The powerful forelimb may have been an excellent adaptation for stripping bark, opening rotten logs, turning stones, and digging while foraging for small prey. Vegetation probably formed the bulk of the diet of larger sauropodomorphs, such as *Efraasia* and the plateosaurians. Indeed, these appear to have been the dominant browsers in most terrestrial habitats around the world from the mid-Norian to the early Toarcian.

Large body size, with commensurately large guts and long passage times, probably helped plateosaurians and their relatives deal with the inadequacies of their teeth and jaws. Nevertheless, it is possible that small amounts of meat were also included in their diets, up to the advanced, heavy-bodied anchisaurians such as *Aardonyx*, *Melanorosaurus*, and Sauropoda, which all show adaptations toward bulk browsing and obligate herbivory.

Locomotion

The basic stance of basal sauropodomorphs has proven to be a controversial issue with the full gamut of postures being advocated in the literature—from bipedal through facultatively bipedal to quadrupedal. Part of the difficulty lies in the intermediate nature of the skeletal proportions of basal sauropodomorphs. The length of the forelimb relative to the hind limb can be used as a rough guide as to the gait of an animal, although there were some dinosaurs with relatively short forelimbs that were undoubted obligate quadrupeds (e.g., stegosaurs), and there were yet others with relatively long forelimbs that were undoubted obligate bipeds (e.g., ornithomimosaurs and most maniraptorans).

The ratios displayed by adult basal sauropodomorphs fall between typical bipeds and typical quadrupeds (Cooper 1981). Similarly, the ratio of the hind limb to trunk length falls between those of obligate quadrupeds and bipeds (Van Heerden 1997). This has led some to propose that they were similarly intermediate in life with the ability to adopt either stance as need dictated. Arguing against such facultative quadrupedalism are the ranges of motion that the bony joints of the forelimb of core prosauropods could allow. The radius is straight and lacks a rotary joint with the ulna, and the humerus could not be elevated into a subhorizontal position; thus there does not seem to be any anatomical way that the radius could cross the ulna and pronate the hand. Therefore flexing the hands, with their medially facing palms, would not produce a posterior propulsive force, severely hampering any attempt at quadrupedal locomotion (Bonnan and Senter 2007). Furthermore, it appears that it was impossible for the humerus to be brought forward from a subvertical position further hindering quadrupedal locomotion. More support for obligate bipedality in core prosauropods comes from a study of a virtual articulated skeleton of *Plateosaurus* that was created by scanning the actual bones of a single skeleton. These model bones could then be articulated and manipulated within cyberspace (Mallison 2010). It was found that quadrupedal models of this digital skeleton performed poorly due to the disparity in fore- and hind-limb lengths and the limited range of motion allowed by the forelimb (Mallison 2010). One objection that can be raised against these range-of-motion studies is that the intervening joint cartilage is missing and might significantly change the range of possible motions. Nevertheless, a recent study has shown that although the cartilage caps of the limb bones of extant archosaurs (both crocodilians and birds) are quite thick, the underlying calcified cartilage (which is fossilized) accurately reflects the joint shape at least in load-bearing adult limb bones (Bonnan et al. 2010).

Perhaps the best evidence that most basal sauropodomorphs were bipedal comes from the footprint record. The most convincing candidate for a basal sauropodomorph track is the ichnogenus *Otozoum* (Rainforth 2003). Although many *Otozoum* tracks are known, only one shows traces of the hands, and even these seem to have been made when the animal stopped and briefly went down onto all fours while remaining stationary (Fig. 21.6A). These hand traces are interesting for they show that the large thumb and claw were held clear of the ground, leaving no trace, and that the hand was not pronated, that is, the hand was turned out with the palm facing in toward the midline as was predicted from the skeletal morphology (Bonnan and Senter 2007).

Quadrupedalism seems to have evolved in advanced anchisaurs (*Melanorosaurus* + Sauropoda), which show large forelimbs (forelimb/hind limb ratio of *Melanorosaurus* is close to 75%) and modifications of the radius and ulna to allow pronation of the hand (Bonnan and Yates 2007).

Another aspect of gait is the degree to which the metatarsus was held off the ground. Usually basal sauropodomorphs are portrayed as digitigrades that held their ankles off the ground with a nearly erect metatarsus. However, *Otozoum* tracks show a clear trace of a fifth toe pad (Fig. 21.6B). The only way the short fifth toe could have made contact with the ground was if the metatarsus was held at a low angle to the ground, in a subplantigrade position like that of their probable descendants the

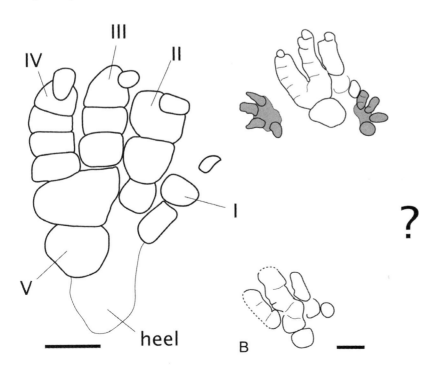

21.6. Footprints (ichnogenus *Otozoum*) that can be attributed to basal sauropodomorphs. (A) Typical *Otozoum* track showing the impressions of digits I–V. (B) *Otozoum* trackway showing manus impressions in grey. Question mark denotes the inferred position of a missing right pes print. Scale bars equal 100 mm. Redrawn from Rainforth (2003).

sauropods. A subplantigrade foot is also supported by the morphology of the tarsal-metatarsal joint and the metatarsal-phalangeal joints (Sullivan et al. 2003)

Reproduction and Growth

The eggs of basal sauropodomorphs are the oldest known for Dinosauria—indeed, they are the oldest of any amniote. Eggs are known for the Late Triassic *Mussaurus* of Argentina (Bonaparte and Vince 1979) and the slightly younger Early Jurassic *Massospondylus* from South Africa (Reisz et al. 2005). *Massospondylus* laid large clutches of subspherical eggs about 6 cm in diameter. The total number of eggs in a clutch is not yet known but a block containing part of a nest preserves six eggs. The preserved shells are exceptionally thin (0.2–0.3 mm thick), even taking into account the loss of the inner mammillary layer. Grine and Kitching (1987) reported that the microstructure of the eggshell resembled those of crocodilians. Zelenitsky and Modesto (2002) argued that the thinness of the shell, the poor definition of the shell units, and the blocky nature of the calcite crystals of the shell indicate that diagenetic recrystalisation has destroyed the original microstructure. Nevertheless, it is intriguing that the eggshells of *Mussaurus* are also extremely thin. Perhaps this represents the basal condition for Sauropodomorpha (or even Dinosauria), and later sauropods evolved thicker eggshell independently of other dinosaurs.

The *Massospondylus* eggs from South Africa also contain embryonic skeletons (Fig. 21.7). Unlike the embryonic titanosaurs from Auca Mahuevo (Salgado et al. 2005), these were well ossified late-term embryos close to hatching (one of the embryos appears to have died in the act of hatching, while one of the eggs in the clutch had hatched). They reveal many important details of the growth of this dinosaur. Their skulls are relatively large and held on short necks comprised of blocky, equant vertebrae. This is the very opposite of the condition seen in adult sauropodomorphs. During growth, the skull experiences strong negative allometry while the neck undergoes strong positive allometry. This is almost certainly a generality for sauropodomorphs and can also be seen in the very young juveniles of *Mussaurus*. These also display large skulls and short necks made of blocky vertebrae. Having such large front-heavy skeletons meant that the hatchlings had to have been quadrupedal. Indeed, the forelimbs of the embryonic *Massospondylus* are close to the hind limbs in length. Subsequent growth shows a negatively allometric growth of the forelimb, so that by adulthood they are only just over half the length of the hind limb. It is unknown at what stage of growth these dinosaurs would have switched from quadrupedalism to bipedalism (Reisz et al. 2005).

The *Massospondylus* embryos revealed one last surprise: they were edentulous. Reisz and his coauthors suggested that the embryos were so close to hatching it is unlikely that hatchlings had erupted teeth. They further speculated that an edentulous hatchling was capable of feeding itself and suggested that *Massospondylus* exhibited parental care. Although the evidence is weak, it is supported by the observation that the posthatching juveniles of *Mussaurus* were found in varying sizes in a single small area along with unhatched eggs and eggshells. This is suggestive of a nest in which the juveniles remained during their early stages of growth.

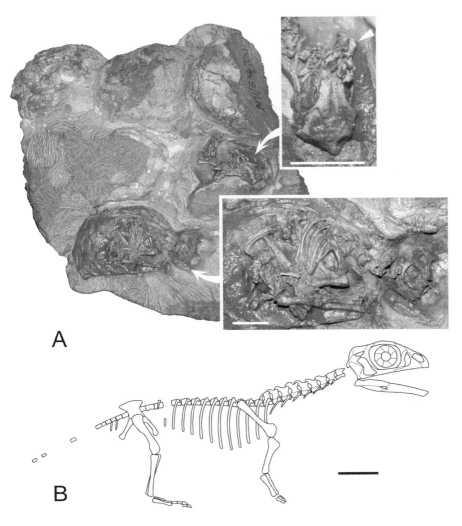

A

B

21.7. Embryos of *Massospondylus*. (A) Block of fossilized eggs with enlargements of the opened eggs showing a skull in dorsal view (top) and a skeleton in lateral view (bottom). (B) Reconstruction of the embryonic skeleton. (B) redrawn from Reisz et al. (2005). The white triangle points to the projecting edentulous tips of the dentaries. Scale bars equal 10 mm.

Distribution in Time and Space

The earliest sauropodomorphs appear at the same time as the earliest-known dinosaurs in the Carnian of western Gondwana. *Chromogisaurus* and *Panphagia* from the Ischigualasto Formation of Argentina and *Saturnalia* from the Santa Maria Formation of Brazil are the only adequately represented and named species from this time interval. A femoral fragment from the coeval Pebbly Arkose Formation of Zimbabwe, and an undescribed guaibasaurid in India indicate that similar dinosaurs may have extended across much of Gondwana. Other records of Carnian sauropodomorphs are not reliable. *Azendohsaurus* (Morocco and Madagascar) was once thought to be a Carnian sauropodomorph but is now known to be a non-dinosaurian archosauromorph (Flynn et al. 2010), while a supposed sauropodomorph tooth from the Carnian of West Texas (Harris et al. 2002) is simply insufficient evidence on which to establish a sauropodomorph identity.

By the mid-Norian Stage sauropodomorphs achieved a near-global distribution, with abundant specimens being found in South America, Africa, India, Europe, and Greenland. It was during the Norian that Plateosauria radiated to produce Plateosauridae, Riojasauridae, Massospondylidae, and Anchisauria. The former two clades were apparently short-lived and did not survive the end of the Triassic.

Strangely, there are no definite body fossils of basal sauropodomorphs from the Triassic of western North America despite the rich, well-sampled

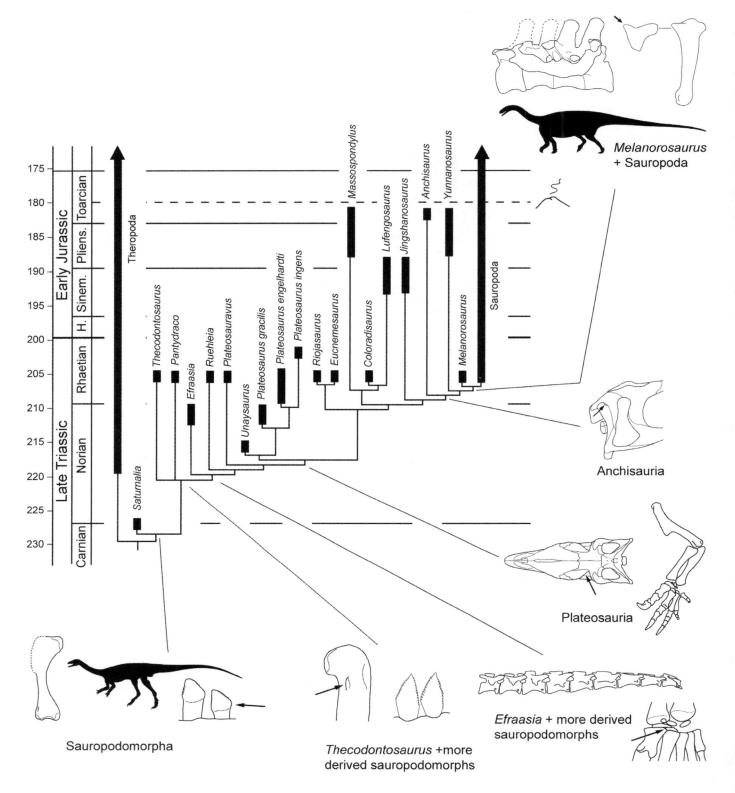

Triassic vertebrate assemblages from this region (Nesbitt et al. 2007; Rowe et al. 2010). Previous records from the Triassic of western North America include a premaxilla that probably belongs to a silesaurid dinosauriform (Irmis et al. 2007) and some postcranial fragments (Long and Murry 1995) that do not closely resemble sauropodomorphs, nor show any definite sauropodomorph synapomorphies. Basal sauropodomorph footprints (ichnogenus *Pseudotetrasauropus*) have been recorded from the Triassic of North America but as noted by Thulborn (2006), it is difficult to distinguish basal sauropodomorph tracks from other archosaurs such as poposaurids and large crocodylomorphs with reduced fifth toes on the basis of hind foot–only tracks. Indeed, many of the so-called basal sauropodomorph tracks from the Triassic of North America do not bear a close resemblance to *Otozoum*, which are definitive basal sauropodomorph tracks from the Early Jurassic (Rainforth 2003).

The oldest plausible sauropodomorph fossils from North America are footprints of advanced, obligatory quadrupedal sauropods (ichnogenus *Tetrasauropus*) from high in the Chinle Group (latest Norian or Rhaetian). However, even these have had their identity questioned (Nesbitt et al. 2007). Definitive basal sauropodomorph fossils do not appear in the North American record until the Early Jurassic (Rowe et al. 2010). Early Jurassic basal sauropodomorphs (mostly belonging to Massospondylidae and Anchisauria) are also known from South America, Africa, Antarctica, Asia, and India but are apparently absent from Europe. Most of these fossils are preserved in continental red beds and aeolian sandstones that indicate they preferred semiarid to desert environments. This probably explains their absence from the Early Jurassic of Europe, which at the time was flooded by an epicontinental sea. This sea broke the land into a small series of islands, which probably enjoyed a moist, maritime climate.

Despite their virtual worldwide success for a period of 45 million years, it appears that all surviving lineages of basal sauropodomorph dinosaurs (and their coelophysoid predators) went extinct sometime during the Toarcian, the last stage of the Early Jurassic (Allain and Aquesbi 2008). Unfortunately, dinosaur assemblages from this stage and the earliest stage of the Middle Jurassic (Aalenian) are poorly sampled, making it difficult to understand the nature of this extinction. It is intriguing that this turnover in dinosaur faunas coincides with the eruption of the Karoo-Ferrar Traps (Fig. 21.8). This event spewed more than 2.5 million cubic km of lava over southern Africa and Antarctica over a brief period 183 million years ago, during the early Toarcian (Wignall 2001).

Other similar continental flood basalt events are also coincident with major disruptions of the Earth's biota, including the catastrophic Permo-Triassic extinction event. The Karoo-Ferrar Traps have been linked to a small but global extinction event among benthic marine invertebrates (Rampino and Stothers 1988), but no similar effect on terrestrial vertebrate faunas has been detected to date. However, any effect that this event may have wrought on terrestrial vertebrate faunas is likely to be masked by the abysmal terrestrial vertebrate record from the late Toarcian and the succeeding Aalenian Stage.

21.8. Evolution of basal Sauropodomorpha according to the cladistic analysis of Yates (2007). Dashed line near the end of the Early Jurassic indicates the eruption of the Karoo-Ferrar Igneous Province. Selected synapomorphies and major morphological innovations are shown for several clades. Sauropodomorpha: Transverse width of the distal end of the humerus greater than 33% of the total humeral length; skull less than 50% of the length of the femur; tooth crowns are unrecurved and have expanded bases. *Thecodontosaurus* + more derived sauropodomorphs: lesser trochanter is a low, elongate ridge; teeth with coarse, upwardly angled serrations. *Efraasia* + more derived sauropodomorphs: elongated cervical vertebrae; metacarpal one inset into carpus. Plateosauria: elongated posterior process of the prefrontal; manus reduced to less than 45% of the humerus + radius. Sauropoda: reduced, triangular antorbital fossa shorter than orbit. Anchisauria: quadrate foramen reduced and dorsally placed. *Melanorosaurus* + more derived Sauropoda: Forelimbs more than 70% of hind limbs, with habitual quadrupedal posture; at least four sacral vertebrae; anterolateral process on the proximal end of the ulna.

References

Allain, R., and Aquesbi, N. 2008. Anatomy and phylogenetic relationships of *Tazoudasaurus naimi* (Dinosauria, Sauropoda) from the late Early Jurassic of Morocco. *Geodiversitas* 30: 345–424.

Barrett, P. M. 2000. Prosauropod dinosaurs and iguanas: speculations on the diets of extinct reptiles. In H.-D. Sues (ed.), *Evolution of Herbivory in Terrestrial Vertebrates: Perspectives from the Fossil Record*, 42–78. Cambridge: Cambridge University Press.

Benton, M. J. 1986. The late Triassic reptile *Teratosaurus*: a rauisuchian not a dinosaur. *Palaeontology* 29: 293–301.

Benton, M. J., L. Juul, G. W. Storrs, and P. M. Galton. 2000. Anatomy and systematics of the prosauropod dinosaur *Thecodontosaurus antiquus* from the Upper Triassic of southwest England. *Journal of Vertebrate Paleontology* 20: 77–108.

Bittencourt Rodrigues, J. S. 2010. Revisão filogenética dos dinossauriformes basais: implicações para a origem dos dinossauros. Ph.D. dissertation, Universidade de São Paulo, São Paulo.

Bonaparte, J. F., and M. Vince. 1979. El hallazgo del primer nido de dinosaurios triásicos (Saurischia, Prosauropoda) Triásico superior de Patagonia, Argentina. *Ameghiniana* 16: 173–182.

Bonnan, M. F., J. L. Sandrik, T. Nishiwaki, D. R. Wilhite, R. M. Elsey, and C. Vittore. 2010. Calcified cartilage shape in archosaur long bones reflects overlying joint shape in stress-bearing elements: implications for nonavian dinosaur locomotion. *Anatomical Record* 293: 2044–2055.

Bonnan, M. F. and Senter, P. 2007. Were the basal sauropodomorph dinosaurs *Plateosaurus* and *Massospondylus* habitual quadrupeds? In P. M. Barrett and D. J. Batten (eds.), *Evolution and Palaeobiology of Early Sauropodomorph Dinosaurs. Special Papers in Palaeontology* 77: 139–156.

Bonnan, M. F. and Yates, A. M. 2007. A new description of the forelimb of the basal sauropodomorph *Melanorosaurus*: implications for the evolution of pronation, manus shape and quadrupedalism in sauropod dinosaurs. In P. M. Barrett and D. J. Batten (eds.), *Evolution and Palaeobiology of Early Sauropodomorph Dinosaurs. Special Papers in Palaeontology* 77: 157–168.

Carrano, M. T. 2000. Homoplasy and the evolution of dinosaur locomotion. *Paleobiology* 26: 489–512

Charig, A. J., J. Attridge, and A. W. Crompton. 1965. On the origin of the sauropods and the classification of the Saurischia. *Proceedings of the Linnean Society* 176: 197–221.

Cooper, M. R. 1981. The prosauropod dinosaur *Massospondylus carinatus* Owen from Zimbabwe: its biology, mode of life and phylogenetic significance. *Occasional Papers of the National Museum and Monuments of Rhodesia* (ser. B) 6: 689–840.

Ezcurra, M. D. 2010. A new early dinosaur (Saurischia: Sauropodomorpha) from the Late Triassic of Argentina: a reassessment of dinosaur origin and phylogeny. *Journal of Systematic Palaeontology* 8: 371–425.

Flynn, J. J., S. J. Nesbitt, J. M. Parrish, L. Ranivoharimanana, and A. R. Wyss. 2010. A new species of *Azendohsaurus* (Diapsida: Archosauromorpha) from the Triassic Isalo Group of southwestern Madagascar: cranium and mandible. *Palaeontology* 53: 669–688.

Galton, P. M. 1984. Diet of prosauropod dinosaurs from the late Triassic and early Jurassic. *Lethaia* 18: 105-123.

———. 1990. Basal Sauropodomorpha–Prosauropoda. In D. B. Weishampel, P. Dodson, and H. Osmólska (eds.), *The Dinosauria*, 320–344. Berkeley: University of California Press.

Galton, P. M., and P. Upchurch. 2004. Prosauropoda. In D. B. Weishampel, P. Dodson, and H. Osmólska (eds.), *The Dinosauria*, 2nd ed., 232–258. Berkeley: University of California Press.

Grine, F. E., and J. W. Kitching. 1987. Scanning electron microscopy of early dinosaur egg shell structure: a comparison with other rigid sauropsid eggs. *Scanning Microscience* 1: 615–630.

Harris, S. K., A. B. Heckert, S. G. Lucas, and A. P. Hunt. 2002. The oldest North American prosauropod from the Upper Triassic Tecovas Formation of the Chinle Group (Adamanian; latest Carnian), west Texas. In A. B. Heckert and S. G. Lucas (eds.), *Upper Triassic Stratigraphy and Paleontology. New Mexico Museum of Natural History and Science, Bulletin* 20: 249–252.

Huene, F. von. 1914. Nachtrage zu meinen früheren Beschreibungen triassischer Saurischia. *Geologie und Palaeontologie Abhandlung* 12: 69–82.

———. 1920. Bemerkungen zur systematic und stammesgeschichte einiger reptilien. *Zeitschrift für Inducktive Abstammungs- und Vererbungslehre* 24: 162–166.

———. 1932. Die fossile Reptile-Ordnung Saurischia ihre Entwicklung und Geschichte. *Monographie für Geologie und Palaeontologie* (ser. 1) 4, pts. 1 and 2: 1–361.

———. 1948. Short review of the lower tetrapods. In A. L. Du Toit (ed.), *Robert Broom Commemorative Volume*. Royal Society of South Africa, Special Publication: 65–106.

Irmis, R. B., W. G. Parker, S. J. Nesbitt, and L. Jun. 2007. Early Ornithischian dinosaurs: the Triassic record. *Historical Biology* 19: 3–22.

Langer, M. C., and M. J. Benton. 2006.

Early dinosaurs: a phylogenetic study. *Journal of Systematic Palaeontology* 4: 309–358.

Long, R. A., and P. A. Murry. 1995. Late Triassic (Carnian and Norian) tetrapods from the south-western United States. *New Mexico Museum of Natural History and Science, Bulletin* 4: 1–254.

Mallison, H. 2010. The digital *Plateosaurus* I: body mass, mass distribution and posture assessed using CAD and CAE on a digitally mounted complete skeleton. *Palaeontologia Electronica* 13 (2): 8A. http://palaeo-electronica.org/2010_2/198/index.html.

Marsh, O. C. 1881. Principal characters of American Jurassic dinosaurs. Pt. 5. *American Journal of Science* (ser. 3) 21: 417–423.

Nesbitt, S. J., R. B. Irmis, and W. G. Parker. 2007. A critical re-evaluation of the Late Triassic dinosaur taxa of North America. *Journal of Systematic Palaeontology* 5: 209–243.

Papp, M. J., and L. M. Witmer. 1998. Cheeks, beaks, or freaks: a critical appraisal of buccal soft-tissue anatomy in ornithischian dinosaurs. *Journal of Vertebrate Paleontology* 18 (suppl. to no. 3): 69A.

Pol, D., and J. Powell. 2005. New information on *Lessemsaurus sauropoides* (Dinosauria, Sauropodomorpha) from the Late Triassic of Argentina. *Journal of Vertebrate Paleontology* 25 (suppl. to no. 3): 100A.

Raath, M. A. 1974. Fossil vertebrate studies in Rhodesia: further evidence of gastroliths in prosauropod dinosaurs. *Arnoldia* 7 (5): 1–7.

Rainforth, E. C. 2003. Revision and re-evaluation of the early Jurassic dinosaurian ichnogenus *Otozoum*. *Palaeontology* 46: 803–838.

Rampino, M. R., and R. B. Stothers. 1988. Flood basalt volcanism during the past 250 million years. *Science* 241: 663–668.

Reisz, R. R., S. Diane, H.-D. Sues, D. C. Evans, and M. A. Raath. 2005. Embryos of an Early Jurassic prosauropod dinosaur and their evolutionary significance. *Science* 309: 761–764.

Romer, A. S. 1956. *Osteology of the Reptiles*. Chicago: University of Chicago Press.

Rowe, T. B., H.-D. Sues, and R. R. Reisz. 2010. Dispersal and diversity in the earliest North American sauropodomorph dinosaurs, with a description of a new taxon. *Proceedings of the Royal Society* B, doi:10.1098/rspb.2010.1867.

Salgado, L., R. A. Coria, and L. M. Chiappe. 2005. Osteology of the sauropod embryos from the Upper Cretaceous of Argentina. *Acta Paleontologia Polonica* 50: 79–92.

Sereno, P. C. 1997. The origin and evolution of dinosaurs. *Annual Review of Earth and Planetary Sciences* 25: 435–489.

———. 1998. A rationale for phylogenetic definitions, with application to the higher-level phylogeny of Dinosauria. *Neues Jahrbuch für Geologie und Paläontologie Abhandlungen* 210: 41–83.

———. 1999. The evolution of dinosaurs. *Science* 284: 2137–2147.

———. 2007. Basal Sauropodomorpha: historical and recent hypotheses, with comments on *Ammosaurus major* (Marsh 1889). In P. M. Barrett and D. J. Batten (eds.), *Evolution and Palaeobiology of Early Sauropodomorph Dinosaurs. Special Papers in Palaeontology* 77: 261–289.

Sullivan, C., F. A. Jenkins, S. M. Gatesy, and N. H. Shubin. 2003. A functional assessment of hind foot posture in the prosauropod dinosaur *Plateosaurus*. *Journal of Vertebrate Paleontology* 23 (suppl. to number 3): 102A.

Taylor, M. P., P. Upchurch, A. M. Yates, M. J. Wedel, and D. Naish, D. In press. Sauropoda. In K. de Queiroz, P. D. Cantino, and J. A. Gauthier (eds.), *Phylonyms: A Companion to the PhyloCode*. Berkeley: University of California Press.

Thulborn, T. 2006. On the tracks of the earliest dinosaurs: implications for the hypothesis of dinosaurian monophyly. *Alcheringa* 30: 273–311.

Upchurch, P., P. M. Barrett, and P. M. Galton. 2005. The phylogenetic relationships of basal sauropodomorphs: implications for the origins of sauropods. *Journal of Vertebrate Paleontology* 25 (suppl. to number 3): 126A.

———. 2007a. A phylogenetic analysis of basal sauropodomorph relationships: implications for the origin of sauropod dinosaurs. In P. M. Barrett and D. J. Batten (eds.), *Evolution and Palaeobiology of Early Sauropodomorph Dinosaurs. Special Papers in Palaeontology* 77: 57–90.

Upchurch, P., Barrett, P. M., X.-J. Zhao, and X. Xu. 2007b. A re-evaluation of *Chinshakiangosaurus chunghoensis* Ye *vide* Dong 1992 (Dinosauria, Sauropodomorpha): implications for cranial evolution in basal sauropod dinosaurs. *Geological Magazine* 144: 247–262.

Van Heerden, J. 1997. Prosauropods. In J. O. Farlow and M. K. Brett-Surman (eds.), *The Complete Dinosaur*, 242–263. Bloomington: Indiana University Press.

Wignall, P. B. 2001. Large igneous provinces and mass extinctions. *Earth-Science Reviews* 53: 1–33.

Wings, O., and P. M. Sander. 2007. No gastric mill in sauropod dinosaurs: new evidence from analysis of gastrolith mass and function in ostriches. *Proceedings of the Royal Society* B 274: 635–640.

Yates, A. M. 2003a. A new species of the primitive dinosaur, *Thecodontosaurus* (Saurischia: Sauropodomorpha) and its implications for the systematics of early dinosaurs. *Journal of Systematic Palaeontology* 1: 1–42.

———. 2003b. The species taxonomy of the sauropodomorph dinosaurs from the Löwenstein Formation (Norian, Late Triassic) of Germany. *Palaeontology* 46: 317–337.

———. 2004. *Anchisaurus polyzelus* (Hitchcock): the smallest known sauropod dinosaur and the evolution of gigantism amongst sauropodomorph dinosaurs. *Postilla* 230: 1–58.

———. 2007. The first complete skull of the Triassic dinosaur *Melanorosaurus* Haughton (Sauropodomorpha: Anchisauria). In P. M. Barrett and D. J. Batten (eds.), *Evolution and Palaeobiology of Early Sauropodomorph Dinosaurs. Special Papers in Palaeontology* 77: 9–55.

Yates, A. M., M. F. Bonnan, J. Neveling, A. Chinsamy, and M. G. Blackbeard. 2010. A new transitional sauropodomorph dinosaur from the Early Jurassic of South Africa and the evolution of sauropod feeding and quadrupedalism. *Proceedings of the Royal Society* B 277: 787–794.

Yates, A. M., and J. W. Kitching. 2003. The earliest known sauropod dinosaur and the first steps towards sauropod locomotion. *Proceedings of the Royal Society of London* B 270: 1753–1758.

Zelenitsky, D. K., and S. P. Modesto. 2002. Re-evaluation of the eggshell structure of eggs containing dinosaur embryos from the Lower Jurassic of South Africa. *South African Journal of Science* 98: 407–408.

Zhang, Y., and Z. Yang. 1994. *A New Complete Osteology of Prosauropoda in Lufeng Basin, Yunnan, China: Jingshanosaurus*. Kunming: Yunnan Publishing House of Science and Technology [in Chinese with English summary].

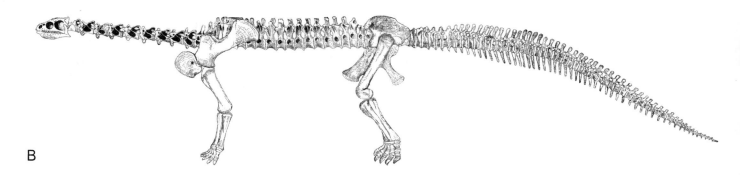

A

B

22.1. Early skeletal reconstructions of sauropod dinosaurs. A, O. C. Marsh's skeletal reconstruction of *Brontosaurus*. B, E. D. Cope's skeletal reconstruction of *Camarasaurus*.

Sauropoda

22

Jeffrey A. Wilson and Kristina Curry Rogers

Despite the appeal of *Tyrannosaurus*, the ubiquity of *Triceratops*, or the stateliness of *Stegosaurus*, sauropods are *the* iconic dinosaurs. Subject of pop songs, advertising campaigns, movies, and cartoons, the idea of a sauropod is readily conveyed with relatively little effort. This quality stems from the obvious and recognizable body plan of sauropods, as well as their large body size. These qualities, together with aspects of the sauropod fossil record and their systematic history, have conspired to create the notion that sauropods are "monolithic" animals. Paradoxically, despite their supposed monolithic nature, sauropods comprise 18%, or nearly one-fifth, of the 661 recognized dinosaur species (Weishampel et al. 2004). In this chapter, we'll explore the sauropod body plan, describe its variations, and, we hope, dispel the myth that it led to evolutionary stagnation and eventual replacement by more "advanced" herbivorous dinosaurs.

The name Sauropoda (Greek for "lizard foot") was coined by O. C. Marsh in 1878 to refer to the "general character of the feet" of sauropods, which, like lizards, possess five functional digits. Given the rarity of feet in the sauropod fossil record and the distinctiveness of more commonly preserved vertebral bones of the sauropod skeleton, we might wonder why Marsh chose this name. More importantly, why did Marsh ignore earlier, more evocative names? Sir Richard Owen, who described the first sauropod, coined "Opisthocoelia" ("back hollow") in 1859 to refer to the distinctive ball-and-socket articulations of trunk and neck vertebrae. H. G. Seeley created "Cetiosauria" ("whale lizards") in 1874 to call to mind the first sauropod, *Cetiosaurus*, known for its whalelike proportions. Although it is undeniable that Marsh and his chief rival, E. D. Cope, were driven to coin new taxonomic names at any cost (see Chapter 4 of this volume), Sauropoda was not a new name for a new taxon – it was a new name for an *old* taxon. A better explanation may be that Marsh's Sauropoda was a new name for a new idea. Marsh was interested in classification of dinosaurs, and the pre-existing names for sauropods conveyed ideas about their classification that Marsh disagreed with, requiring a fresh start. For example, Owen placed Opisthocoelia as a carnivorous subgroup of marine Crocodilia that bear little resemblance to the terrestrial Dinosauria (a name he coined in 1842). It was not until 1869 that sauropods were first classified as dinosaurs. Although Cetiosauria invoked the first named sauropod, *Cetiosaurus*, Seeley considered it to be one of only two dinosaurian orders, implying a membership that included more than just sauropods. Moreover, Seeley did not regard dinosaurs as a single evolutionary lineage. In contrast, Marsh considered Dinosauria to be a natural group and regarded sauropods as one of its three orders. Thus, Marsh's Sauropoda represented a new taxonomic idea that gained currency in popular and scientific literature. Apart from a few early holdouts such as E. S. Riggs, the scientific community adopted Sauropoda as the term of choice for the reptilian behemoths that are the subject of this chapter.

History of Discoveries

The first sauropod bones were reported by John Kingdon in 1825 to a meeting of the Geological Society of London as "certain bones of very large size, appearing to have belonged to a whale or crocodile." The great French comparative anatomist Georges Cuvier is reported to have examined the remains, which he "pronounced to be cetaceous" (Buckland 1841, 96). Kingdon's material was later given to the Oxford museum, where it was studied by Sir Richard Owen, a disciple of Cuvier and an excellent comparative anatomist in his own right (see Torrens, Chapter 2 of this volume). In 1841, Owen described Kingdon's collection as well as other bones from Oxfordshire. In that initial paper, Owen identified numerous reptilian characteristics, such as the absence of epiphyses (growth plates) on caudal vertebrae, but could identify no cetacean characteristics, other than large body size. Based on its paradoxical nature, Owen established a new genus for the collection of bones: *Cetiosaurus*, the "whale lizard."

Owen concluded his initial description with this assessment: "the vertebræ, as well as the bones of the extremities, prove its marine habits . . . the surpassing bulk and strength of the *Cetiosaurus* were probably assigned to it with carnivorous habits, that it might keep in check the Crocodilians and Plesiosauri" (1841, 462). He regarded *Cetiosaurus* as a member of a new group of crocodiles, Opisthocoelia, on the basis of ball-and-socket vertebral articulations in which the ball was placed anteriorly and the socket posteriorly, which is opposite the procoelous (= front hollow) condition found in most crocodiles. Owen's assessment was based on limited anatomical evidence—isolated vertebrae and limb bones—and with John Phillips's discovery of abundant *Cetiosaurus* bones in Oxfordshire, a new picture of sauropods emerged. Thomas Huxley examined this "splendid series of remains" before the publication of Phillips's (1871) monograph and was the first to place *Cetiosaurus* within Dinosauria. Huxley (1869) allied *Cetiosaurus* with the bipedal Iguanodontidae, which is surprising given its massive forelimb bones. Huxley's interpretation of dinosaurs as bipedal reflected his view that birds evolved from dinosaurs, and contrasted sharply with Owen's vision of dinosaurs as mammal-like quadrupeds.

Phillips (1871, 294) was the first to interpret *Cetiosaurus* as a plant-eating dinosaur. He also hypothesized that its limb bones were "suited for walking," but he was anticipated by Mantell (1850) who came to the same conclusion on the basis of the shape and structure of an upper arm bone. Phillips could not rule out the possibility that *Cetiosaurus* was amphibious, however, concluding that it was a "marsh-loving or river-side animal" nourished by "vegetable food which abounded in the vicinity, and was not obliged to contend with megalosaurus [sic] for a scanty supply of more stimulating diet." Using the lengths of limb bones scaled to proportions of living reptiles, Phillips estimated that *Cetiosaurus* measured 64 feet long, which agrees well with current estimates.

These early interpretations of *Cetiosaurus* and other dinosaurs (e.g., *Iguanodon*, *Hylaeosaurus*, *Megalosaurus*) were based on somewhat limited samples. Better assessments of sauropod paleobiology and classification emerged with discovery of abundant, articulated sauropod skeletons in western North America and eastern Africa during the late nineteenth and early twentieth centuries. Othniel C. Marsh and Edward D. Cope described many new and well-represented dinosaurs from the Morrison Formation of the western United States, which now include the sauropods *Apatosaurus* (= *Brontosaurus*), *Amphicoelias*, *Barosaurus*, *Camarasaurus*, *Diplodocus*, *Haplocanthosaurus*, and the recently named *Suuwassea*. Importantly, Cope and Marsh's fieldwork led to the discovery of the first complete sauropod skull (*Diplodocus*; Marsh 1884), skeletal reconstructions of *Apatosaurus* (Marsh 1883; Fig. 22.1A) and *Camarasaurus* (Cope in Osborn and Mook 1921: pl. 82; Fig. 22.1B), and the first museum mount of a complete sauropod skeleton (*Diplodocus*; Anonymous 1905). German exploration in coeval deposits in East Africa (present-day Tanzania) from 1907 to 1913 produced sauropod material rivaling that from North America. Werner Janensch and others led field crews at Tendaguru, where they collected more than 235,000 kg of fossils (Maier 2003, 83) that represented new genera and species such as *Dicraeosaurus* and *Gigantosaurus* (now *Janenschia*), as well as species related to the North American genera *Brachiosaurus* and *Barosaurus* (e.g., Janensch 1914, 1929a, 1929b, 1935–36, 1950, 1961). These discoveries led to some of the earliest descriptions of the sauropod body plan and to basic classification. In addition, however, the wealth of bones found in the Morrison Formation and at Tendaguru contributed to the erroneous notion that the apex of sauropod diversity occurred in the Late Jurassic, after which they were replaced by more specialized ornithischian herbivores.

In the years between the two world wars, Friedrich von Huene (see Sues, Chapter 3 of this volume) published seminal monographs describing important new Cretaceous sauropods from the southern (Gondwanan) landmasses of South America and India. Between 1923 and 1926, Huene traveled to La Plata, Argentina, to study newly collected specimens from Patagonia, as well as materials described by Richard Lydekker (*Titanosaurus*, *Argyrosaurus*, *Microcoelus*) and material collected earlier by Wichmann (1916). Huene (1929) described new sauropod genera based on the new bones (*Antarctosaurus*, "*Campylodon*" [now *Campylodoniscus*]), and new taxa based on revision of Lydekker's work (*Laplatasaurus araukanicus*, *Titanosaurus robustus*). Later, Huene jointly studied C. A. Matley's collection of dinosaurs from central India (Huene and Matley 1933). Matley's excellent collection included new theropod and sauropod genera that had previously been recognized in South America (*Antarctosaurus*) and in Madagascar and South America (*Titanosaurus*, *Laplatasaurus*). The South American, Madagascan, and Indian sauropod discoveries led to the idea of a "remarkable community of type which undoubtedly exists between the faunas of southern continents of the world" (Lydekker 1893, 3).

Subsequent sauropod discoveries built on the temporal and geographical framework established by the pioneers Owen, Phillips, Marsh, Cope, Janensch, and Huene. A wealth of information on pre-Morrison and pre-Tendaguru sauropods came from discoveries in Lower and Middle Jurassic beds of China by C. C. Young (1935, 1939, 1941, 1947, 1954, 1958) and Dong

Zhiming (e.g., 1992) of the IVPP in Beijing, and most recently by regional museums such as in Sichuan (e.g., Ye et al. 2001). Lower Jurassic sauropods were also discovered by field parties of the Indian Statistical Institute and Geological Survey of India in a series of expeditions in Andra Pradhesh beginning in the 1960s that produced a wealth of sauropod remains (e.g., Jain et al. 1979; Bandyopadhyay 1999; Yadagiri 2001; Bandyopadhyay et al. 2010). Even more ancient sauropods were discovered in Lower Jurassic deposits of southern Africa in the early 1960s, extending the temporal range of sauropods into that of their presumed prosauropod progenitors (Charig et al. 1965; Raath 1972). More recent exploration into Upper Triassic beds of southern Africa has yielded what may be the most primitive known sauropods (Galton and van Heerden 1985, 1998; Yates and Kitching 2003; Yates et al. 2011).

Cretaceous sauropods have been discovered from the southern landmasses such as South America (e.g., Salgado and Bonaparte 2007; Novas 2009), India (Jain and Bandyopadhyay 1997), Madagascar (Curry Rogers and Forster 2001), and Africa (Jacobs et al. 1993; Gomani 2005; Sereno et al. 1999; Smith et al. 2001). These discoveries have dramatically improved understanding of the anatomy of Cretaceous sauropods, and more fully elucidated differences between species and major lineages. Well preserved Cretaceous sauropods have been discovered from northern landmasses – where they had long been thought extinct – including Asia (Wiman 1929; Bohlin 1953; Nowinski 1971; Borsuk-Bialynicka 1977; Martin et al. 1994; Suteethorn et al. 1995; Allain et al. 1999), Europe (Le Loeuff 1995; Jianu and Weishampel 1999), and North America (Gilmore 1946; Tidwell et al. 1999, 2001; Wedel et al. 2000).

Prior to the Dinosaur Renaissance inaugurated by Ostrom's (1969) description of *Deinonychus*, sauropods were discovered at the rate of approximately one per year. Since that time, however, sauropod discoveries have steadily increased. This relatively sudden surge in sauropod discoveries is more striking when we consider that 25 sauropod species (20% of all) were named in the first four years following the first edition of *The Complete Dinosaur*, published in 1997.

Sauropod Anatomy

Sauropod dinosaurs were walking backbones, as anyone who has collected a skeleton in the field can attest. Most sauropods have approximately 80 vertebrae, but diplodocids such as *Diplodocus* and *Apatosaurus* may have as many as 110 vertebrae. Add to these the paired free ribs associated with the dorsal vertebrae in front of the pelvis and the chevrons associated with the vertebrae following the pelvis, and there may be as many as 165 axial elements in a sauropod skeleton! Consequently, axial bones are the most common elements found at sauropod sites, and extremities such as the skull, hands, and feet are comparably rare.

Skull

Sauropods have relatively small skulls compared to their overall body size. In *Brachiosaurus*, for instance, the skull accounts for only 1/200th of total

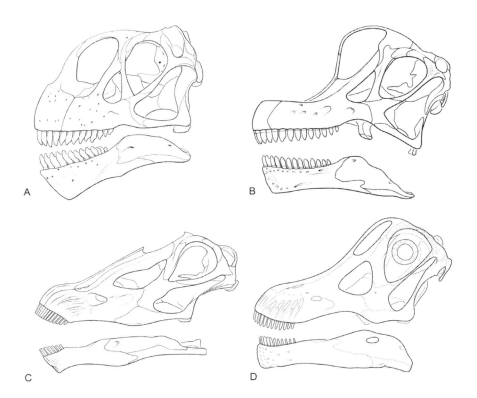

22.2. Skulls of the sauropod dinosaurs *Camarasaurus* (A), *Brachiosaurus* (B), *Diplodocus* (C), and *Nemegtosaurus* (D). A, B, and D are macronarians, C is a diplodocoid. A–C are modified from Wilson and Sereno (1998, Figs. 6A, 7A, 8A); D is from Wilson (2005, 16A). Not to scale.

body volume (Gunga et al. 2008). As we will discuss later, our estimates of sauropod metabolism suggest that they were able to ingest and digest massive amounts of vegetation, which requires the cranial and dental equipment to find and select and crop foodstuffs before swallowing. Below we discuss the basic architecture of the sauropod skull (Fig. 22.2), with attention to features that relate to food processing.

All sauropods share skull features that distinguish them from their closest relatives, prosauropods (whether monophyletic or paraphyletic). There are five large, paired openings in the sauropod skull: the external naris (bony nostril), antorbital fenestra (in front of the eye socket), orbit (eye socket), and the upper and lower temporal fenestrae (sites of attachment for jaw-closing muscles). In prosauropods, all but the upper temporal fenestra are evenly spaced and fall on an anteroposterior line. In sauropods, these openings are jumbled together and fall on a zigzag line as a result of two important skull distortions–reduction of the infraorbital region and retraction of the external nares. A second important feature shared by all sauropods is enlargement of the tooth-bearing bones. The premaxilla, maxilla, and dentary are all noticeably enlarged dorsoventrally and accommodate large teeth (or a larger number of smaller teeth) that are packed into the jaws. Most sauropods have three or four replacement teeth at each tooth position, but some diplodocoids pack between six and eight replacement teeth. Despite the size and increased packing of teeth, sauropods have fewer tooth positions than their closest relatives. Most have 15 or fewer dentary tooth positions, but basal sauropods have more than 20, and the derived diplodocoid *Nigersaurus* has more than 30. Sauropods are also distinguished from their close relatives by the shape of the tooth rows (i.e., jaw rami), which are broadly arched and U shaped when viewed from above. Some diplodocoids have tooth rows that are squared across the front so that most teeth are oriented transversely. In most sauropods, teeth are closely arranged along the tooth row, overlapping in taxa with broad tooth crowns. All sauropods have tooth rows of subequal length, which allows for teeth to meet in a one-to-one or one-to-two manner. Occlusion results in well-developed wear facets that are apical in the former condition (one-to-one occlusion) and V shaped in the latter condition (one-to-two occlusion). Precise occlusion appears to characterize sauropods primitively, although this was lost in some diplodocids (e.g., *Diplodocus*). All sauropods have characteristic "wrinkled" enamel that can be recognized on isolated crowns. Textured enamel increases the surface area of the tooth, which may increase the tooth's resistance to wear. All but the most basal sauropods lack the marginal denticles present in prosauropods. Some basal sauropods maintain denticles on only the mesial (leading) edge of the tooth. Although sauropod tooth form has been traditionally dichotomized into broad- and narrow-crowned forms (see "Sauropod Phylogeny" below), there is a diversity of tooth shapes within the group (see Chure et al. 2010).

In many sauropods, a well-defined snout extends forward of the external nares. In animals such as *Camarasaurus* and *Brachiosaurus* the snout is sharply demarcated from the rest of the skull, whereas in others, such as *Diplodocus* or *Nemegtosaurus*, it less obvious. The antorbital region just behind the tooth row is reduced in sauropods, which lack a fossa surrounding the opening. A small preantorbital opening is present in all but the most primitive sauropods. Additional maxillary fenestrae pierce the maxilla in many sauropods. The external naris is enlarged compared to that of sauropod outgroups, and in some species it is the largest skull opening (i.e., macronarians). A well-defined fossa usually surrounds the external naris. The orbit has an acute ventral margin, and the anteroventral corner of the lateral temporal fenestra usually extends below it. Although the bone on the roof of the eye socket is roughened, sauropods apparently lacked much of the orbital ornamentation present in other dinosaurs such as theropods. The supratemporal fenestrae are reduced in size, elliptical, and oriented transversely, and they are secondarily closed in rebbachisaurid sauropods (e.g., *Nigersaurus*, *Limaysaurus*). The skull roof itself is fairly short and flat in sauropods; the supratemporal fenestrae are usually visible in lateral view.

The occiput is flat transversely and is usually taller than deep. Paroccipital processes flank the foramen magnum, which leads into the brain cavity. Sauropods have relatively small brains for their body size, and the thinking part of the brain (the cerebrum) is also relatively small. Along with stegosaurs, sauropods have the smallest Encephalization Quotient among dinosaurs (see Chapter 9 of this volume). The sauropod braincase is not particularly noteworthy except for the basipterygoid processes that connect it to the palate. In most sauropods these are small, and the palate extends nearly the length the skull. In diplodocoids, however, the basipterygoid processes are narrow, elongate, and forwardly directed. The palate is shifted forward along with the quadrate, which forms the jaw joint. As a consequence the diplodocoid skull is tipped posteriorly, implying a reorientation of jaw closing muscles (see "Sauropod Paleobiology" below). Titanosaur sauropods are characterized by a second braincase-to-palate connection, via the basal tubera and the quadrate. The quadrate in most sauropods bears a fossa on its posterior surface that may be enlarged transversely in some sauropods.

Vertebral Column

The vertebral column is perhaps the most diagnostic region of the sauropod skeleton. There are four main regions of the vertebral column – the neck, trunk, sacrum, and tail, which contain variable numbers of variably shaped and sized components.

Each vertebra is constructed from a centrum and an overlying neural arch, which together form the canal through which the spinal cord passes. The neurocentral suture joining these two vertebral components is coalesced in skeletally mature animals, but the order in which vertebrae fuse (and variations therein) has been little explored. Ribs attach to the vertebral column via two costal articulations, the diapophysis and the parapophysis. In the tail, specialized chevron bones mirror the neural arches below the centrum and enclose tail vessels and nerves. They are usually not fused to the centrum.

All sauropods have an elongate neck relative to their outgroups, but no two sauropods have the same length neck (Fig. 22.3). The neck is braced at one end by the shoulders and terminates at the skull, which contains feeding and sensory apparatuses. The extensive variation in the overall length of the neck and the number and shape of constituent neck vertebrae has important effects on the feeding and sensory volumes available to individual sauropod species. More than a dozen independent changes in neck length occurred in sauropod evolution via one or more of three mechanisms: elongation, duplication, and incorporation. Elongation results from allometric vertebral growth and affects the relative length of vertebrae. Giraffe necks, for example, contain the same number of vertebrae as do human necks, but each has elongate proportions. Duplication (intercalation) occurs when new vertebrae are added to the axial column due to extended production of somites during embryogenesis, as occurs in snakes. Incorporation (transposition) involves shifting the boundary between vertebral regions, in which overall number of vertebrae does not change, but the number allotted to each region does. Although elongation, duplication, and incorporation all

22.3. Modern skeletal reconstructions of the basal sauropods *Shunosaurus* and *Gongxianosaurus*, the diplodocoids *Dicraeosaurus* and *Apatosaurus*, and the macronarians *Opisthocoelicaudia*, *Brachiosaurus*, and *Camarasaurus*. Approximately to scale. Modified from Wilson and Sereno (1998, fold-out), Wilson (2002, 1), and Wilson (2005, 1.5).

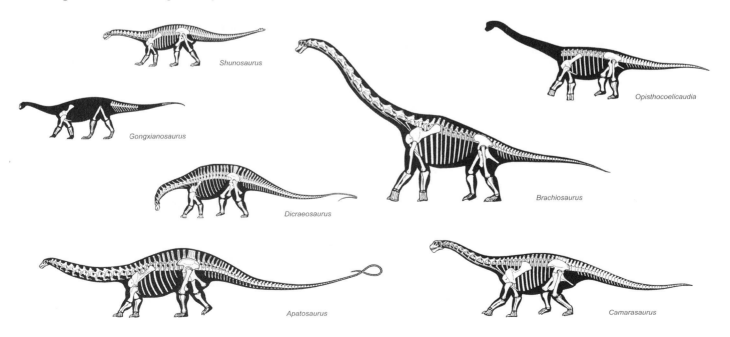

serve to elongate the neck, their possible impact on neck function (e.g., mobility), singly or in combination, has not yet been explored.

Although many vertebrate lineages modify the absolute length and proportions of their vertebral columns, there are relatively few such examples among the fully terrestrial vertebrates that support their body weight on limbs. In the two lineages where this tendency is widespread, sauropods and birds, some or all vertebrae are pneumatized. In living birds, epithelial outpocketings of the lungs invade the developing vertebral centrum and neural arch to form internal cavities and external fossae that are filled with air. Although the soft-tissue epithelia responsible for pneumatic excavations do not fossilize, the hollows and depressions they create in bone are preserved and serve as indirect evidence of pneumatization. Sauropod vertebrae (Fig. 22.4A, B) resemble those of modern birds because they have internal cavities and a complex pattern of strutlike vertebral laminae (Wilson 1999) that bound external fossae (Wilson et al. 2011). Although the greatest pneumaticity is found in sauropod neck and trunk vertebrae, in some species the sacral or even caudal vertebrae may also be pneumatized. The primary effect of pneumatization is overall reduction in the weight of the vertebral column, and thereby the load borne by the limbs. An important secondary effect, however, is that pneumatization influences the three-dimensional distribution of vertebral bone. The resultant internal and external architectures created by pneumatic cavities are oriented along lines of stress and resist loads imposed on the centrum and neural arch. For example, paired excavations in the lateral walls of the centrum in some sauropods leave an I-beam structure that is oriented to resist dorsoventral bending forces. Vertebral laminae likewise are oriented along principal directions of stress experienced by the neural spine and rib articulations. All sauropod neck vertebrae contact one another via ball-and-socket joints in which the front end of the vertebra is convex and the back half is concave. These opisthocoelous (= "back hollow") vertebrae are also present in the trunk.

Ribs are fused to each neck vertebra prior to full maturity. As in theropods and prosauropods, the neck ribs of sauropods are aligned parallel to the axis of the vertebral column. The neck ribs of many sauropods are pneumatized, particularly in the region that connects to the centrum. In most sauropods the ribs are elongate and overlapping, but in diplodocoids the ribs do not extend beyond their respective vertebra. In *Mamenchisaurus*, for example, neck ribs reach an astonishing length of 4 m and underlap the succeeding three elongate cervical vertebrae—only the dentary bone of the blue whale is longer!

The number and length of vertebrae in the trunk are reduced in sauropod evolution, often in exchange for an increase in neck or sacral vertebrae. The trunk consists of vertebrae and ribs that form a supporting arch between the fore and hind limbs and house the viscera. Gastralia (abdominal ribs) are present in some sauropods (e.g., *Jobaria*) and may have had a cartilaginous connection to the rib cage. In contrast to neck ribs, trunk ribs are oriented perpendicular to the vertebral column and are often planklike, especially in the shoulder region where they support the shoulder girdle. Neural spines of trunk vertebrae are generally shortest near the neck and gradually increase in height toward the hips. Transverse processes consist

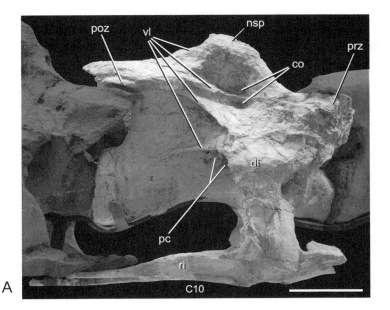

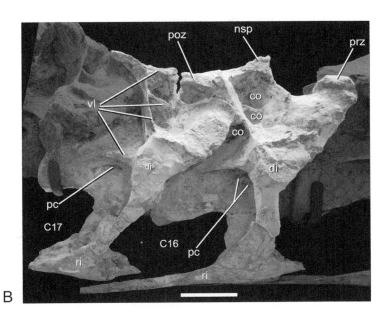

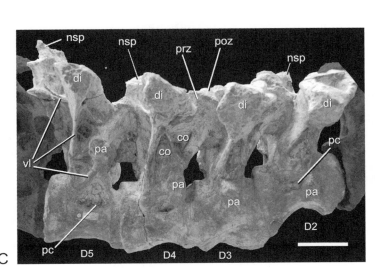

22.4. Sauropod vertebral elements. Cervical vertebra 10 (A), cervical vertebrae 16–17 (B), and dorsal vertebrae 2–5 (C) of the macronarian sauropod *Euhelopus zdanskyi.* Note the presence of numerous struts of bone, called vertebral laminae, that delimit pneumatic spaces. Scale bars equal 10 cm. Abbreviations: c, cervical vertebra; co, coel; d, dorsal vertebra; di, diapophysis; nsp, neural spine; pa, parapophysis; pc, pleurocoel; prz, prezygapophysis; poz, postzygapophysis; ri, rib; vl, vertebral laminae. Modified from Wilson and Upchurch (2009, Figs. 10, 12, 16).

of the two rib attachments (diapophysis and parapophysis) and extend at nearly right angles from the vertical axis of each dorsal vertebra. Like the neck vertebrae, trunk vertebrae are pneumatized (Fig. 22.4C).

Trunk vertebrae contact each other via three articulations: the centrum, zygapophyses, and hyposphene-hypantrum. The trunk centra closest to the shoulders are always opisthocoelous in sauropods, even in early-appearing forms such as *Tazoudasaurus*. In most sauropods, trunk centra contain lateral excavations (pleurocoels) that die out toward the hips. However, in certain taxa (e.g., *Brachytrachelopan*; Rauhut et al. 2005), these excavations have been secondarily lost. Adjacent neural arches of the trunk articulate at the zygapophyses as well as at hyposphene-hypantrum articulations (Cope 1878), which characterize all saurischian dinosaurs. Several sauropod lineages independently evolved forked neural spines in the posterior cervical and anterior dorsal vertebrae. Forked neural spines may have enclosed an enlarged nuchal ligament and provided a greater surface area for attachment. There is a tendency within some sauropods to enhance mobility within the trunk by modifying the articulations between trunk vertebrae. The ball-and-socket central articulations extend to the hips in all macronarians, and hyposphene-hypantrum articulations are lost in some members of the macronarian subgroup Titanosauria.

Although other dinosaurs contain fused bones in the neck (e.g., ceratopsians) or tail (e.g,. ankylosaurs, stegosaurs), for most sauropods the sacrum contains the only naturally fused vertebrae in the skeleton. Some sauropods, however, independently evolved tail clubs (e.g., *Shunosaurus, Mamenchisaurus*) or pathologically fused tail vertebrae at the point where the transverse processes disappear (this is a common pathology in *Diplodocus*). The sacrum is the only point in the skeleton where the limbs have a direct, bony connection to the vertebral column, and thus serves the important function of transmitting locomotor forces generated by the hind limbs to the rest of the body. For this reason, early in ontogeny the component sacral vertebrae fuse, which involves coalescence of the central articulation, the neural arch articulations (zygapophyses, hyposphene-hypantrum), and neural spine. The number of sacral vertebrae increased during sauropod evolution. Sauropod outgroups typically possessed three coalesced sacral vertebrae (e.g., prosauropods), whereas the earliest, most primitive sauropods (e.g., *Shunosaurus*) possessed four, the familiar Late Jurassic sauropods (e.g, *Diplodocus, Apatosaurus, Camarasaurus, Brachiosaurus*) possessed five, and the predominantly Cretaceous titanosaurs (e.g., *Isisaurus*) possessed six or occasionally seven (*Neuquensaurus*). The timing of fusion and morphology of sauropod sacral vertebrae suggest that the additional elements were acquired by incorporating vertebrae from the trunk and tail, rather than by developing new vertebrae. The fourth-incorporated sacral vertebra comes from the tail, whereas the fifth- and sixth-incorporated sacral vertebrae came from the trunk. In addition to the centrum and neural arches, the sauropod sacrum is characterized by fusion of sacral ribs into a sacricostal yoke, which adds lateral strength to the sacrum and contributes to the roof of the hip socket (acetabulum).

Sauropod tails usually contain between 50 and 60 vertebrae, but there are extremely long-tailed sauropods (e.g., *Diplodocus, Apatosaurus*) that have more than 80 tail vertebrae, as well as extremely short-tailed forms (e.g., *Opisthocoelicaudia*) that have as few as 35. Sauropod tails are typically quite deep and broad near the pelvis, due to the elongate neural spines atop the centrum and free chevrons below it. The neural spines are typically thickened and rugose due to their heavy ligamentous connection to the sacrum. The transverse processes (serially homologous to the ribs of the trunk region) serve as attachment sites for the femoral retractor muscles (m. caudofemoralis) and are responsible for the breadth of the tail. The transverse processes usually disappear by the fifteenth caudal vertebrae, beyond which the tail becomes shallower and narrower. In this midcaudal region, neural spines and zygapophyseal articulations become gradually reduced. The distal portion of the tail, which consists solely of caudal centra and rudimentary chevrons, is occasionally specialized in sauropods. Specializations include the "whiplash" tails present some diplodocoids and tail clubs present in some basal sauropods. The whiplash tail is composed of supernumerary, elongate (length/height ratio = 4–9.5), biconvex tail vertebrae whose function has been interpreted as either defensive (Holland 1915) or noisemaking (Myhrvold and Currie 1997) adaptations. Although biconvex vertebrae are known in a variety of sauropod species, only diplodocoids are known to possess the whiplash. Tail clubs are known in only two sauropods thus far. *Shunosaurus* has three to five enlarged distalmost caudal vertebrae that are fused into a club, whereas the tail club of *Mamenchisaurus* is shaped like a cock's comb and includes three relatively narrow caudal vertebrae. Sauropod chevrons are modified into elongate, sledlike structures in many basal sauropods and in diplodocoids.

Girdles and Limbs

The extreme body size of sauropods can be expected to have had a profound effect on the shape and proportions of the limb skeleton, which function in trunk support and locomotion (Fig. 22.5). Body mass is a volumetric measure ($L \times W \times H \propto L^3$), but the strength of bones supporting body weight is proportional to their combined cross-sectional area ($U\text{-}\sigma L \times W \propto L^2$). The disproportionate increase in body size relative to its bony support most strongly affects organisms at the extremes of body size distribution, such as sauropods.

Several modifications related to shape, proportion, and posture characterize most or all sauropod limb elements. Weight-bearing elements usually have a straight shaft that is oriented approximately perpendicular to the articular surfaces at their ends. In addition, all limb articular surfaces have a coarse, rugose texture that indicates the presence of a thick cartilage cap. The thickness of this missing articular cartilage can be estimated from articulated pelves and hind limbs, in which the volumes of the acetabulum and femoral head differ considerably. Proportional changes in early sauropod evolution include an overall lengthening of the forelimb, particularly the distal limb elements, as well as a relative shortening of distal hind-limb elements. These shape changes and proportional changes are related to the acquisition of an obligately quadrupedal, columnar limb posture, in which bending moment experienced by each element is minimized.

The *pectoral girdle* consists of three paired bones (scapula, coracoid, sternal plate) that are occasionally joined by a fourth (clavicle). The scapula

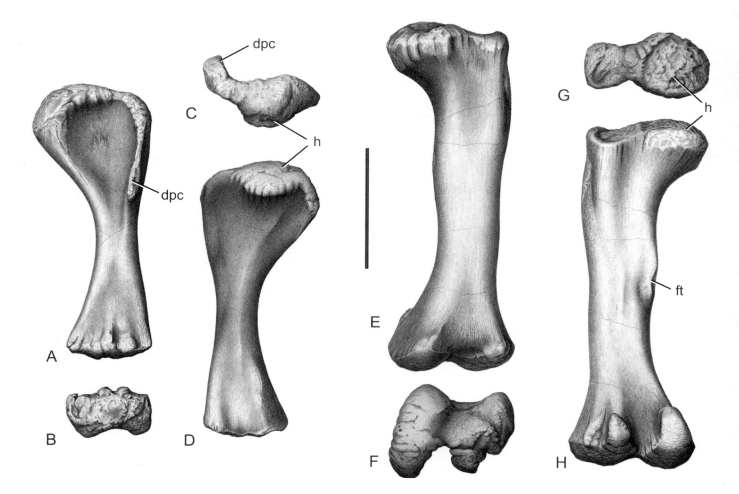

22.5. Sauropod limb bones. Humerus (A–D) and femur (E–H) of *Camarasaurus.* in anterior (A, E), distal (B, F), proximal (C, G), and posterior (D, H) views. Scale bar equals 10 cm. Abbreviations: dpc, deltopectoral crest; ft, fourth trochanter; h, head.

and coracoid fuse at or near skeletal maturity and form the concave articular surface (glenoid) for the humerus. The scapula is broadest in its acromial region near the coracoid connection and narrows abruptly into a blade, which takes a variety of shapes in sauropods. The coracoid is usually ovoid or D shaped, but in titanosaurs it is elongate and quadrangular. The paired scapulocoracoids approach one another on the midline, but there is no bony symphysis between them. From this near-midline position, the scapulocoracoids extend backward onto the trunk ribs. Clavicles are known for some sauropods, but they have been never found in articulation. In *Jobaria* (Sereno et al. 1999), the clavicles are elongate, blade-shaped structures with asymmetrical ends. The clavicles probably contacted one another at the midline and extended laterally to contact the upper surface of the coracoids, as appears to be the case in prosauropods (Yates and Vasconcelos 2005). Sternal plates contact one another along the midline but do not have a well-marked bony connection to the scapulocoracoid, suggesting that their contact was cartilaginous. In most sauropods, sternal plates are relatively small and triangular or oval in shape, but in titanosaurs the sternal plates are enlarged and crescentic.

The forelimb comprises the humerus, radius and ulna, carpals, metacarpals, and phalanges. The humerus is always the longest bone in the forelimb, and in *Brachiosaurus* the longest limb element. The humerus contacts the shoulder girdle via the well-marked and rounded humeral head that articulates with the glenoid surface formed by the scapula and coracoid. The humerus is straight-shafted and has an ovoid cross section, which is more resistant to bending along its longer, transverse, axis. The prominent deltopectoral crest serves as attachment point for muscles that protract the arm (m. deltopectoralis). The humerus contacts the radius and ulna at the lateral and medial sides, respectively, of its relatively flat distal condyles. The radius is a straight, simple bone that is usually slightly shorter than the ulna. The ulna is triradiate in proximal view and lacks the prominent olecranon process present in other dinosaurs (titanosaurs, however, reacquire the olecranon process). The number of carpal bones varies within sauropods. Primitive species retain at least three, but most post–Late Jurassic sauropods retain two or fewer. Carpal bones are blocklike, and their rugose surfaces indicate they were completely enveloped in cartilage. Some titanosaurs (e.g., *Alamosaurus, Opisthocoelicaudia*) lack ossified carpal bones, which were presumably retained as cartilaginous elements.

Sauropods have five metacarpals that are elongate and subequal in length. They range from 30% to 50% the length of the radius. In all but the most primitive sauropods, the metacarpals are vertically oriented and arranged into a tight arch of ~270°. This digitigrade forefoot contacted the substrate at the metacarpal-phalangeal joint. Sauropods reduced the total number of toe bones (phalanges) and retain only a single claw, that on the first digit (pollex). Trackways indicate that sauropods did not have distinct fingers extending from the flesh of the forefoot and that the manus was rotated outward relative to the line of travel.

The pelvic girdle consists of three paired bones (ilium, pubis, ischium) that together circumscribe the acetabulum. The ilium is the single dorsal element and forms the upper border of the acetabulum. It is fused to the sacrum via an extensive contact with the 4–6 sacral centra and ribs. In some sauropods, it may even be partially pneumatized as a result of that contact (e.g., *Euhelopus*). The ilium is semicircular and contacts the ventral pelvic elements at two peduncles that border the acetabulum. The pubic peduncle is D shaped in cross section and extends away from the body of the ilium. The ischial peduncle, in contrast, is relatively small and is flush against the body of the ilium. The iliac blade extends in front of and behind the acetabulum via its preacetabular and postacetabular processes, respectively. The shape, size, and orientation of the preacetabular process is variable within sauropods. In most sauropods it terminates at a ventrally directed point, but in *Brachiosaurus* and titanosaurs it is semicircular. The preacetabular blade typically extends slightly laterally from the body axis, but in titanosaurs it extends nearly perpendicular to the body axis. The pubis is usually much larger than the ischium (except in titanosaurs) and is pierced by a conspicuous obturator foramen. The pubis contacts the ilium proximally at its iliac peduncle and the ischium posteriorly via its ischial peduncle. Below these contacts, the pubis extends downward and slightly forward, contacting its opposite on the midline. The foot of the pubis is heavy and forms the insertion for the abdominal muscles that originate on the ribs and sternum. The ischium contacts the ilium dorsally at its iliac peduncle and the pubis anteriorly at the pubic peduncle. The ischium extends beneath the tail and has an extensive midline contact with its opposite.

The hind limb comprises the femur, tibia, fibula, tarsus, metacarpals, and phalanges. The femur is the longest limb bone in all sauropods except *Brachiosaurus*. The medially projecting femoral head serves as the contact with the acetabulum. It is ovoid in shape and bears a strongly rugose bone texture. The greater trochanter, upon which muscles that protract or rotate the femur (mm. puboischiofemoralis externi) insert, is level or nearly level with the femoral head. The femur has a broad, elliptical shaft that bears a low ridge, the fourth trochanter, marking the insertion of the caudofemoralis muscle. Like the humerus, the femoral shaft is transversely ovoid in cross section, a geometry that resists loading on the medially projecting femoral head. The distal condyles of the femur, which articulate with the lower leg bones, are distinct and strongly convex, and extend from the ventral surface of the femur around to its posterior surface. The tibia is usually about 60% the length of the femur, and the fibula is slightly longer. The tibia is always the heavier of the two lower leg bones and presumably received a proportionately greater load. A prominent cnemial crest extends anterolaterally from the proximal surface of the tibia, just below the knee. The cnemial crest is the attachment point for the triceps femoris (thigh muscles that extend the knee joint, including m. ambiens, mm. femorotibiales, mm. iliotibiales; Carrano and Hutchinson 2002). The shaft of the tibia is usually subcircular to subtriangular in cross section, with its long axis oriented transversely. The fibula is less than half the breadth of the tibia and is only slightly expanded proximally and distally. A conspicuous lateral trochanter is present on the outer surface near midshaft, forming the insertion for muscles that flex the knee (m. iliofibularis). Distally, two proximal tarsal bones meet the tibia and fibula. The medially positioned astragalus is broader than is the laterally positioned calcaneum, which is not ossified in some sauropods. Like the tibia, the astragalus likely received a greater proportion of the load. Distal tarsal elements are not ossified in any sauropod except *Blikanasaurus*, which retains distal tarsals 4 and 5. It

22.6. Sauropod body armor. Osteoderm from a Late Cretaceous titanosaur from India. From D'Emic et al. (2009, 1).

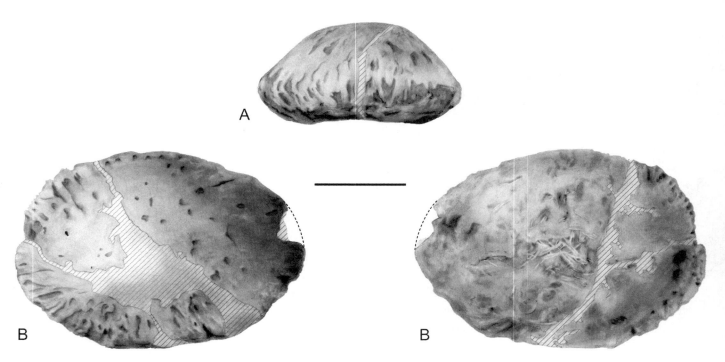

A

B

B

is not known whether distal tarsals were retained as cartilaginous elements, or if they were incorporated into one or both of the proximal tarsal bones in other sauropods. There are five metatarsals in the sauropod pes, the longest of which (metatarsal III) is approximately 25% the length of the tibia. The metatarsals are much shorter and more robust than are the metacarpals, and they were held in a near-horizontal or semidigitigrade orientation. As with the lower leg bones and the tarsal bones, the medial metatarsals are the most robust and presumably bore more load than the more gracile lateral elements, as shown in sauropod footprints. Three to four large claws are present at the ends of the inner digits of the foot; in most sauropods they are directed laterally relative to the main axis of the foot. Sauropod footprints indicate the presence of a relatively large heel pad that supported the bony skeleton of the foot.

Osteoderms in sauropods are restricted to titanosaurs (Fig. 22.6). Many of these enigmatic dermal elements have been discovered, but relatively few have been found in association with skeletons (D'Emic et al. 2009). Titanosaur osteoderms range from tiny millimeter-scale ossicles (e.g., *Saltasaurus*) to half-meter-scale elements (e.g., *Ampelosaurus*). Several functional hypotheses have been proposed for titanosaur osteoderms (e.g., defense, calcium metabolism, bracing of the axial column), but rigorous tests of these hypotheses await more complete discoveries (Salgado 2003).

Sauropod Phylogeny

Systematists are reaching consensus on many aspects of the evolutionary relationships of Sauropoda. This emerging view of sauropod descent has unfolded over the course of more than a century, beginning with traditional studies of Marsh, Janensch, and others, and continuing through to the modern cladistic analyses.

Traditional Studies

When Marsh (1878) coined the suborder Sauropoda, it included only a single family, Atlantosauridae. Several of the features Marsh (1878, 412) listed in that initial diagnosis of Sauropoda are now well-corroborated synapomorphies for the group or for more exclusive sauropod subgroups that were not identified at the time of Marsh's writing (e.g., columnar, quadrupedal posture). Marsh invented new families to accommodate the increasing sauropod diversity revealed by new discoveries worldwide (e.g., Atlantosauridae, Morosauridae, Diplodocidae, Pleurocoelidae, Titanosauridae). The formal familial diagnoses for these groups (Marsh 1884, 1895) also recognized features currently considered synapomorphies for sauropod subclades. These diagnoses, however, did not resolve how these groups were interrelated; Marsh's ranked classifications did not function as hypotheses of evolutionary descent. Janensch (1929a) was the first to produce a classification of Sauropoda that employed higher-level groupings. He recognized two principal sauropod subgroups, one with broad, laterally facing nares and spatulate tooth crowns, and the other with elevated, dorsally facing nares and narrow tooth crowns. Janensch named these two families Bothrosauropodidae and Homalosauropodidae, and recognized three and four subfamilies within each, respectively. Huene (1956) followed this dichotomous scheme, raising

Janensch's subfamilies to familial rank and Janensch's families to "family-group" rank. In contrast to that of Marsh, Janensch's classification could be interpreted as an evolutionary hypothesis that involved divergence between two lineages that differ in tooth morphology.

A dichotomous scheme for higher-level classification of sauropods based on tooth form and narial position became widely accepted, despite nomenclatural differences (Brachiosauridae versus Titanosauridae: Romer 1956, 1966; Camarasauridae versus Atlantosauridae: Steel 1970). Other traditional classifications of sauropods, however, follow Marsh in recognizing taxa of equivalent rank (usually families) with no higher-level hierarchical information (e.g., McIntosh 1990). Bonaparte (1986) also utilized serially ranked families, but he regarded Late Jurassic and younger sauropod families ("Neosauropoda") as advanced relative to older forms ("Eosauropoda").

Numerical methods for assessing phylogenetic relationships in sauropod dinosaurs were first introduced by Gauthier (1986) in his analysis of saurischian dinosaurs. His character choice reflected those cited by previous authors (e.g., Romer 1956; Steel 1970) and his topology consequently conformed to the traditional dichotomy. Since then, numerous cladistic analyses focusing on Sauropoda or its subgroups have appeared (Russell and Zheng 1994; Calvo and Salgado 1995; Upchurch 1995, 1998; Salgado et al. 1997; Wilson and Sereno 1998; Sanz et al. 1999; Curry Rogers 2001; Curry Rogers and Forster 2001; Wilson 2002; González Riga 2003; Upchurch et al. 2004; Harris 2006).

Cladistic Hypotheses

The main topological disagreement among early cladistic analyses of Sauropoda centered on the relationships of broad- and narrow-crowned sauropods. Upchurch (1995) presented the first large-scale cladistic analysis of sauropods, in which he proposed a slightly modified version of the traditional dichotomy that resolved broad tooth crowns as a primitive feature and narrow tooth crowns as a uniquely derived feature characterizing *Diplodocus*-like taxa (i.e., Diplodocoidea) and titanosaurs. Salgado et al. (1997) were the first to depart from this traditional dichotomy by providing character evidence linking narrow-crowned titanosaurs to the broad-crowned *Brachiosaurus*, rather than to the other narrow-crowned group (Diplodocoidea). This result was corroborated by Wilson and Sereno (1998). In a subsequent analysis, Upchurch (1998) produced a topology that agreed in many ways with those of Salgado et al. (1997) and Wilson and Sereno (1998) but also explored the relationships of genera not treated by either. These three analyses agree on several topological points, including the separation of early appearing genera (e.g., *Vulcanodon, Shunosaurus, Barapasaurus, Omeisaurus*) from a derived clade called Neosauropoda (Bonaparte 1986), identification of the two constituent neosauropod lineages Diplodocoidea (e.g., *Apatosaurus*) and Macronaria (e.g., *Camarasaurus*), and the positioning of the titanosaur lineage within Macronaria (Fig. 22.7).

Despite points of agreement, other topological differences persist. The most significant of these centers on the phylogenetic affinities of two different groups of Asian sauropods: the Chinese "euhelopodids" (*Shunosaurus, Omeisaurus, Mamenchisaurus, Euhelopus*) and the Mongolian

22.7. Recent phylogenetic hypotheses of sauropod interrelationships based on cladistic analysis. A, Wilson and Sereno (1998); B, Upchurch (1998); C, Wilson (2002); D, Upchurch et al. (2004).

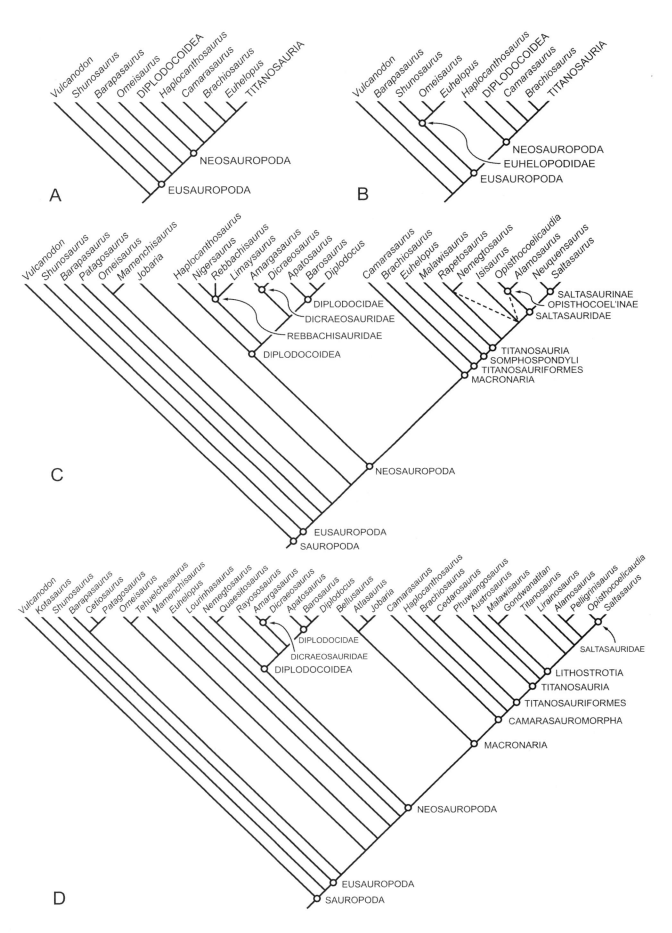

nemegtosaurids (*Nemegtosaurus*, *Quaesitosaurus*). Upchurch (1995) proposed "Euhelopodidae" as a clade that evolved while China was geographically isolated from Europe from Middle Jurassic until Early Cretaceous times (Russell 1993; Luo 1999; Barrett et al. 2002; Upchurch et al. 2002; Zhou et al. 2003). According to the results of Upchurch (1998), "Euhelopodidae" evolved independently of its sister-taxon Neosauropoda, but was eventually replaced by it during the Cretaceous (Upchurch 1995, 1998). In contrast, Wilson and Sereno (1998) suggested that Chinese sauropods are polyphyletic, with *Omeisaurus* occupying the sister-taxon to Neosauropoda (as in Upchurch 1995, 1998), but *Shunosaurus* positioned basally and *Euhelopus* positioned apically. This result was corroborated by Wilson (2002), whose analysis resolved some of Upchurch's (1998) "euhelopodid" characters as supporting the monophyly of *Omeisaurus* and *Mamenchisaurus* (Omeisauridae). Templeton tests (e.g., Larson 1994) show that "euhelopodid" paraphyly could not be statistically rejected by the matrix of Upchurch (1998), but "euhelopodid" monophyly could be rejected by the matrix of Wilson (2002). This issue has recently been resolved by the opposing camps—Wilson and Upchurch (2009) jointly restudied the remains of *Euhelopus* and concluded that it is a close relative of Titanosauria.

A second area of disagreement involves the relationships of the isolated skulls of the sauropods *Nemegtosaurus* and *Quaesitosaurus* from the Late Cretaceous of Mongolia. These slender-crowned taxa were originally described as *Dicraeosaurus*-like (Nowinski 1971), a designation consistent with the presumed diplodocid affinities of the Late Jurassic Chinese sauropod *Mamenchisaurus* (McIntosh 1990), as well as the traditional division of sauropods into narrow-crowned and broad-crowned groups. More recently, cladistic analyses have produced new hypotheses of relationships for *Nemegtosaurus* and *Quaesitosaurus*, suggesting they are monophyletic sister-taxon of diplodocoids (Yu 1993; Upchurch 1998, 1999; Upchurch et al. 2002), basal members of a clade including diplodocoids and titanosaurs (Upchurch 1995), or nested within titanosaurs (Salgado and Calvo 1997; Curry Rogers and Forster 2001; Wilson 2002; Wilson 2005a, b; Curry Rogers 2005). Although the weight of the evidence is in favor of titanosaur affinities for *Nemegtosaurus* and *Quaesitosaurus*, convergences with diplodocoids are noteworthy (Upchurch 1999; Curry Rogers and Forster 2001, 2004; see below).

In addition to areas of disagreement, there are unresolved areas resulting from lack of information. Two such areas involve the origin of sauropods and the diversification of their latest-surviving lineage, Titanosauria. Sauropods were thought to be absent from Triassic rocks, in contrast to other saurischians (prosauropods, theropods), which are found in lowermost Upper Triassic horizons. Recent discoveries of Triassic sauropod body fossils (e.g., *Isanosaurus* Buffetaut et al. 2000) and ichnofossils (see below) have provided the first opportunity to resolve sauropod origins, but additional field and museum research are needed. Renewed interest in titanosaurs, whose interrelationships remain unresolved, has been fueled by descriptions of many new discoveries in the field (e.g., Curry Rogers 2005). These include the first titanosaur with associated cranial and postcranial remains (*Rapetosaurus* Curry Rogers and Forster 2001, 2004), the first embryonic titanosaur remains

(Chiappe et al. 1998, 2001), and nearly complete associated or articulated postcranial skeletons from South America (*Mendozasaurus* González Riga 2003; *Epachthosaurus* Martínez et al. 2004; *Gondwanatitan* Kellner and Azevedo 1999), Asia (*Phuwiangosaurus* Martin et al. 1994; *Tangvayosaurus* Allain et al. 1999), India (*Isisaurus* Jain and Bandyopadhyay 1997), Europe (*Lirainosaurus* Sanz et al. 1999; *Ampelosaurus* Le Loeuff 1995, 2003), and Africa (*Malawisaurus* Jacobs et al. 1993; Gomani 2005; *Paralititan* Smith et al. 2001). Several analyses have investigated titanosaur phylogeny (e.g., Sanz et al. 1999; González Riga 2003; Curry Rogers 2005; Calvo et al. 2007), and there are several points of agreement among them. These preliminary analyses are the first step toward establishing a framework for titanosaur evolutionary history, but at least a dozen valid titanosaur genera have yet to be accommodated by a phylogenetic analysis, in addition to the many undescribed specimens uncovered in recent years.

Sauropod Paleobiology

Body size is the most recognizable and memorable characteristic of sauropods and prompted the long-standing popular view of them as dimwitted, overgrown icons of extinction. The extreme sizes attained by adult sauropods can be expected to have influenced all aspects of their biology, and it is no surprise that most questions on sauropod life history, locomotion, biomechanics, and herbivory have been framed around the improbability of living life at large body size. Walter Coombs (1975, 1978) and Robert Bakker (1968, 1971a, b; 1986) reinterpreted sauropods as dynamic, terrestrial vertebrates that fed in a tripodal stance, used their tails as weapons, and generated their own body heat via high food consumption capabilities. Although appealing, Coombs's and Bakker's supercharged sauropods represent an extreme view of sauropod capabilities. Our current view of sauropod lifestyles is filtered through the lens of their superficially "monolithic" stature, limited modern analogues, and an incompletely known fossil record that documents approximately 20 genera that are known from more than 90% of the skeleton (Upchurch et al. 2004). Major evolutionary transitions remain poorly resolved, including the origin of sauropods from their sauropodomorph ancestors, the timing of divergence of the major neosauropod lineages, and the Early Cretaceous diversification of rebbachisaurids and titanosaurs. Other pressing questions are not necessarily dependent on discoveries of new and better fossils, but rather on our ability to interpret those already collected. Although we have made inroads, particularly with the combined data sources of bone histology, eggs, nests, and ichnofossils as described below, sauropod life history stands out as a complex issue knotting physiology, development, ecology, and behavior. Basic questions remain, including the pattern of body size change within the phylogenetic, temporal, and geographic distribution of sauropods, not to mention the extreme ontogenetic body size variation that characterizes the group. In addition, the masticatory forces producing wear facets on teeth, the relative roles of oral and gastric maceration of plant material, and even the composition of sauropod diets all remain poorly understood. More challenging still are questions of sauropod thermal biology and energetics, reproductive behavior, and social interactions (Clauss 2011; Ganse et al. 2011).

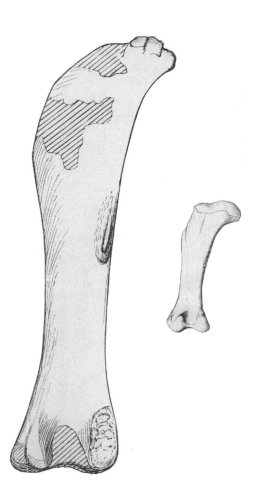

22.8. Size range in adult sauropod dinosaurs. Sauropods exhibit a wide range of adult body size, as exemplified by the femora of these two contemporaneous Argentine titanosaur sauropods, the dwarf *Neuquensaurus australis* (length, 70 cm) and the giant *Antarctosaurus giganteus* (length, 231 cm). From Huene 1929, pl. 20, 36).

Sauropods hatched from eggs, emerging as hatchlings measuring approximately 0.5 m and weighing less than 10 kg (e.g., Powell 1992; Britt and Naylor 1994; Chiappe et al. 1998, 2001). As adults, the largest sauropods are estimated to have attained body lengths of 30 m long and body weights between 40 and 70 metric tons or more (Colbert 1962; Seebacher 2001; Peczkis 1995), an upper bound reached independently within multiple sauropod lineages (Diplodocoidea: "*Seismosaurus*"; Macronaria: *Brachiosaurus*, *Argentinosaurus*). Sauropoda also includes uniquely small taxa and includes one of the few phylogenetic decreases in body size among dinosaurs (Carrano 2005; Hone et al. 2005). The smallest sauropods (e.g., *Magyarosaurus*, *Saltasaurus*) may have weighed between 1.5 and 3 metric tons and measured only 10 m long (e.g., Curry Rogers and Erickson 2005; Sander et al. 2006; Benton et al. 2010; Stein et al. 2010). Although most sauropods were gigantic, all sauropods passed through a phase in their life history (of unknown duration) when they were not yet giants, and several sauropods never grew to be gigantic as adults (Fig. 22.8). Nevertheless, in his 1991 *Natural History* article "Lifestyles of the Huge and Famous," Peter Dodson queried: "Gigantism was obviously part of their [sauropods'] formula for success. But did they become too big for their own good?" This is where we must start.

Sauropods would seem to be exemplary representatives of Cope's Rule, the tendency for lineages to increase in body size over time (Cope 1887, 1896; Benton 2002; Hone and Benton 2005; Hone et al. 2005). In order for Cope's Rule to operate, body size increase must convey advantages that outweigh its disadvantages (Peters 1983; Calder 1984; Schmidt-Nielsen 1984; Dunham et al. 1989; Spotila et al. 1991; Brown 1995; Hone and Benton 2005). Benefits of size increase might include: increased defense against predators; greater range of food sources and acceptable foods; increased success in mating and intraspecific competition; increased success in interspecific competition; extended longevity; increased intelligence (with increased brain size); the potential for thermal inertia (at very large size); survival through lean times; and resistance to climatic extremes and variation. Problems associated with increased body size might include: increased developmental time; increased resource requirements; susceptibility to extinction (due to slower generation times and slower evolutionary rates); and lower fecundity with a switch from r-selection to K-selection; and development of higher parental investment in fewer offspring (Schmidt-Nielsen 1984; McKinney 1997; Benton 2002; Hone et al. 2005; Hone and Benton 2005).

The evolutionary history of sauropod body size has only recently received detailed treatment (e.g., Yates 2004; Hone et al. 2005; Carrano 2006; Clauss 2011; Rauhut et al. 2011). These authors recognized that early sauropod evolution is characterized by an increase in body size. In fact, over the first 40 million years of sauropod evolution, body size doubled. The pattern within Neosauropoda is more complicated, with diplodocoids and macronarians both documenting positive and negative divergences from average neosauropod body size. The largest members of these clades are similar in size (diplodocoids–*Apatosaurus*; macronarians–*Brachiosaurus*, *Argyrosaurus*, *Argentinosaurus*; all weigh ~50 tons), but the smallest members are not. Macronarian taxa (e.g., *Europasaurus* <5 metric tons, Sander et al. 2006; *Neuquensaurus*, *Saltasaurus*, *Magyarosaurus* ~1.5–3 metric tons) reach significantly smaller adult body sizes than the smallest diplodocoids (*Dicraeosaurus*, *Amargasaurus* ~5–10 metric tons) (Carrano 2005). The most dramatic episode of sauropod body size change occurred in the last 30 million years of the Mesozoic, when titanosaurs evolved a tenfold difference in adult size between the smallest and largest genera (Carrano 2005; Hone et al. 2005).

Surprisingly, most changes in the upper limit for sauropod body size increases occur early in the phylogenetic history of Sauropoda; few occurred after the Upper Jurassic neosauropod radiation. Body size decreases, in contrast, occurred later in sauropod history, particularly among derived macronarians, which are the only sauropod clade to document a significant size decrease (Carrano 2005). Carrano's investigation of this pattern suggests that sauropod body size evolution is characterized by "passive" macroevolutionary processes, as indicated by: (1) an expanded range of sauropod sizes over time (and through the phylogeny), with loss of smaller taxa as larger taxa appear, and (2) weak positive correlation between ancestor-descendant changes and ancestral size. Carrano (2005) noted that both the upper and lower size limits for sauropods (50 and 1–3 tons, respectively) are considerably greater than those of other dinosaur lineages. Carrano (2005, 2006) and Hone et al. (2005) suggested that upon reaching an upper limit for body size (7.8 m, Hone et al. 2005; 50 tons, Carrano 2005), lineages tend to decrease in body size.

Sauropod Growth Rates

The growth strategies that permitted sauropods to attain such massive proportions have been difficult to pinpoint. Early workers extrapolated from reptilian growth rates, and postulated that even relatively small sauropods might have taken until age 60 to reach sexual maturity and over a century to attain their adult sizes (e.g., Case 1978b; Calder 1984). A different perspective on sauropod life history emerged in the 1960s, when Ricqlès (1968a,b) and others began to utilize bone histology to reveal life history. The internal microstructure of growing bone allows us to choose among appropriate modern analogues for an overview of growth patterns, which allows quantification of growth rates at both bone and somatic levels. Surprisingly, histological data indicate that sauropods exhibited elevated growth rates through most of ontogeny (Rimblot-Baly et al. 1995; Curry 1999; Sander 2000, 2003; Erickson et al. 2001; Curry Rogers and Erickson 2005; Sander and Clauss 2008; Sander et al. 2011). Histology, particularly when combined with scaling principles such as developmental mass extrapolation, allows us to investigate important life history questions. Did sauropods grow indeterminately? Did sauropods grow at constant rates during ontogeny, or did they experience regular cycles of relative variation in growth rates? How long did it take sauropods to reach their adult size? Did all sauropods exhibit the same basic growth strategy? What implications do sauropod growth rates have on our understanding of the evolutionary history of sauropod body size?

Most studies agree that sauropods commonly deposited highly vascularized fibrolamellar bone throughout most of ontogeny (but see Chapter 31 of this volume and Reid 1981, 1984, 1990, 1996 for other interpretations). Late in ontogeny, bone vasculature decreases dramatically and bone matrices are parallel to lamellar fibers and often punctuated by lines of arrested growth, which indicate periodicity in active growth. Some sauropod bones also appear to record ontogenetic cyclicity in rapid growth rates via cycles or "polish lines" (e.g., Sauropoda indet. Reid 1981; *Lapparentosaurus*, Ricqlès 1983; Rimblot-Baly et al. 1995; *Apatosaurus*, Curry 1999; *Janenschia*, Sander 2000; Sander and Tückmantel 2003). Sauropod bone histology (Fig. 22.9), including polish lines or growth cycles, presents a different pattern from that in extant reptiles. Histological organization indicates that sauropod bone tissue and bone growth patterns are more qualitatively similar to those of modern birds and mammals. These results suggest that prolonged ontogeny was not required for the attainment of gigantic body sizes in sauropods. Instead, sauropods attained large adult body size via increased posthatching growth rates (e.g., Rimblot-Baly et al. 1995; Curry 1999; Sander 2000; Erickson et al. 2001; Sander and Tückmantel 2003; Klein et al. 2008; Sander et al. 2011). Counts of cortical stratification marks support this hypothesis, and indicate that some sauropods might have grown to half-size by the age of 5 years and to their adult sizes between the ages of 10 and 30 (e.g., Curry 1999; Sander 2000; Sander and Tückmantel 2003; Curry Rogers and Erickson 2005; Sander et al. 2011; Dumont et al. 2011).

Bone histology, long-bone growth, and developmental mass all support ontogenetic growth strategies on par with some of the largest living vertebrates rather than the decades-long or even century-long sauropod ontogenies previously assumed. Erickson et al. (2001) analyzed dinosaurian growth trajectories and demonstrated that sigmoidal equations accurately describe growth data for all dinosaurs (Erickson et al. 2001; Curry Rogers and Erickson 2005). All members of Dinosauria grow at rates 2–56 times faster than any living reptile. As a clade, dinosaurs did not grow at rates intermediate to those of reptiles, birds, and mammals, nor at rates equivalent to those of living altricial birds. Instead, the results indicate that members of Dinosauria exhibit unique growth trajectories, with growth rates that may be slower than, faster than, or equal to typical mammalian/avian rates. Interestingly, the growth rates of dinosaurs depend on the size of the taxon of interest: the larger the dinosaur, the more rapid the overall body growth rate. The regression equation Erickson et al. (2001) established for Dinosauria included six taxa spanning the phylogenetic, size, and temporal range for the clade. It allowed quantified predictions of dinosaurian growth rates beyond the range of included taxa (Erickson et al. 2001). Sauropod growth rates approach those known for extant whales, which are absolutely and relatively some of the fastest-growing eutherians. For example, a fully grown *Apatosaurus* grew at ~14,460 g/day, compared with 20,700 g/day for a 30,000 kg gray whale (*Eschrichtius robustus*, Case 1978a). Although this rapid growth rate may seem astounding, it is less than half the highest known growth rate observed: the extant blue whale (*Balaenoptera musculus*) grows at a rate of 66,000 g/day (Case 1978a). In fact, even the enormous sauropod *Argentinosaurus*, whose adult body weight has been estimated to be 100,000 kg (Seebacher 2001) or 73,000 kg (Mazzetta et al. 2004), is still

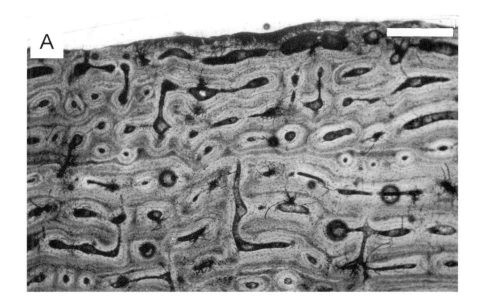

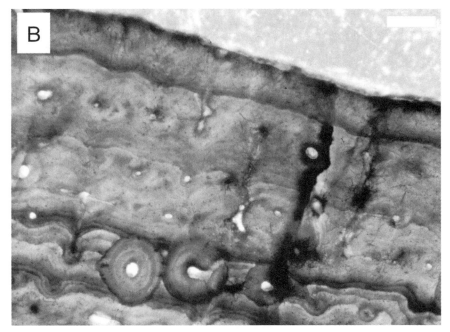

22.9. *Apatosaurus* exponential and stationary stages of growth. A, Cross section of juvenile *Apatosaurus* radius (50% adult size). Periosteal surface (toward top) documents fast-growing laminar, fibrolamellar bone in the external cortex. B, Cross section of adult *Apatosaurus* radius (90% adult size). Periosteal surface (toward top) is punctuated by the first peripheral lines of arrested growth (LAG) in the external cortex. These LAG indicate the onset of the stationary phase of growth. Scale bars equal 100μm.

predicted to have lower growth rates (55,638 g/day) than *B. musculus* (Case 1978a) (Fig. 22.10). To compare – at age five, an African elephant is only ~1 metric ton, but at the same age, *Apatosaurus* might have been closer to 20 metric tons.

Reproduction

Although other vertebrates attain comparable adult body sizes at similar growth rates, sauropods may be the only giants whose young hatch from eggs. Other large-bodied vertebrates bore live young, including chondrichthyans (sharks; Dulvy and Reynolds 1997), mammals (whales; Clapham et al. 1999), and marine reptiles such as plesiosaurs (Cheng et al. 2004), ichthyosaurs (Böttcher 1990; Maxwell and Caldwell 2003), and mosasaurs

22.10. *Apatosaurus* growth rates. A, Ontogenetic growth curve for *Apatosaurus*. The largest animal in the sample is among the largest specimens known. The growth curve for *Apatosaurus* is similar to that known in living vertebrates, and includes early ontogeny followed by an exponential stage, and finally a stationary stage. B, Comparison of exponential stage growth rates in *Apatosaurus* (A), *Maiasaura* (Ma), *Massospondylus* (Ms), *Syntarsus*, (Sy), *Psittacosaurus* (P), and *Shuvuuia* (Sh) with typical values for extant vertebrates. Standardized comparisons are made using contrasts between animals of comparable adult mass. Growth rates for included dinosaurs are as follows: *Shuvuuia* (3.4 g/day), *Psittacosaurus* (12.5 g/day), *Syntarsus* (23.9 g/day), *Massospondylus* (90.3 g/day), *Maiasaura* (2,793 g/day), and *Apatosaurus* (14,460 g/day). Modified from Erickson et al. (2001).

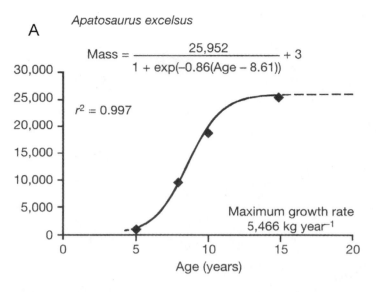

A

Apatosaurus excelsus

$$Mass = \frac{25{,}952}{1 + \exp(-0.86(Age - 8.61))} + 3$$

$r^2 = 0.997$

Maximum growth rate
5,466 kg year^{-1}

Age (years)

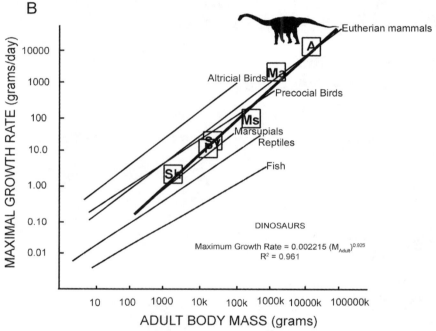

B

MAXIMAL GROWTH RATE (grams/day)

Eutherian mammals

Altricial Birds

Precocial Birds

Marsupials

Reptiles

Fish

DINOSAURS

Maximum Growth Rate = 0.002215 $(M_{Adult})^{0.925}$
$R^2 = 0.961$

ADULT BODY MASS (grams)

(Caldwell and Lee 2001). The ontogenetic variation in body size exhibited by sauropods exceeds that for any other vertebrate lineage (<1 m hatchlings, >30 m adults).

Fossil localities from the Jurassic of the western United States and Portugal and the Cretaceous of France, Spain, India, Romania, and Argentina contain clutches of subspherical eggs characterized by relatively thick eggshell comprising a single structural layer of calcite with clearly separated shell units and a tuberculate surface ornamentation (Sahni et al. 1990; Powell 1992; Vianey-Liaud et al. 1990; Mohabey 1996; Calvo et al. 1997). Traditionally, these eggs have been classified within the Megaloolithidae category of eggshell parataxonomy ("ootaxonomy," see Chapter 30 of this volume). Because of their large size and association with sauropod bones, megaoolithid eggs have often been assigned to Sauropoda (Zhao 1979; Mikhailov 1991, 1997), even before the smoking gun of diagnostic sauropod embryonic material inside was reported in 1998 (see below). Such isolated occurrences and speculative assignment to Sauropoda nevertheless provided fodder for interpretation of sauropod nest construction, egg-laying behavior, and physiology. Nest construction is most commonly inferred from clutch geometry rather than surrounding lithological structure (Dughi and Sirugue 1966; Kérourio 1981; Williams et al. 1984; Faccio 1990; Sahni et al. 1990; Powell 1992; Sanz et al. 1995; Mohabey 1996; Calvo et al. 1997). Egg-laying behavior was suggested for *Hypselosaurus* on the basis of the clutch configuration and limb anatomy. Cousin et al. (1990) suggested that arcs comprising 15–20 eggs in a radius of 1.3–1.7 m reflected the turning radius of a crouching, egg-laying female *Hypselosaurus*. Gregarious nesting and/or site fidelity have also been hypothesized for sauropods from unidentified eggs (Sanz et al. 1995; Figueroa and Powell 2000; López-Martínez 2000; Mohabey 2000), usually on the basis of abundant fragmented eggshell and clutch distribution. Taxonomically ambiguous eggs have also been utilized in studies of sauropod physiology (Case 1978b; Erben and Wedepohl 1979; Bakker 1986; Paul 1990). For example, Paul (1990) determined lifetime reproductive potential as a function of egg size, and estimated sauropod hatchling and adult body mass, whereas Case (1978b) extrapolated from egg size to determine the age of sauropods at sexual maturity. However, without diagnosable embryonic sauropods associated with eggs and coupled with detailed stratigraphic and taphonomic analyses, many of these assumptions are unwarranted.

The first concrete evidence for sauropod reproductive biology occurs in the form of *in ovo* titanosaur embryos preserved in megaoolithid eggs, discovered in nesting structures from Auca Mahuevo, Argentina (Chiappe et al. 1998, 2000, 2001, 2004, 2005; Chiappe and Dingus 2001). Six stratigraphically distinct egg layers contain hundreds of egg clutches, some preserving nest architecture. The clutches are contained in large, subcircular to subelliptical to kidney-shaped depressions in sandstone, although the depression and interstitial spaces between the eggs are filled by mudstone (Chiappe et al. 2004, 2005). These well-preserved nesting traces provided evidence that unlike extant reptiles, titanosaurs laid eggs in excavated depressions without burying them (Chiappe et al. 2004; Sander et al. 2008), although the nests may have been lined and covered with vegetation (Chiappe et al. 2005) as suggested for other megaoolithid egg sites (Kérourio 1981; Cousin et al. 1990). Each clutch at Auca Mahuevo contains 20 to 40 eggs (Chiappe et al. 2004). The eggs are spherical to subspherical, approximately 13–15 cm in diameter (average volume ~800 cm^3), and bear a tubercular surface ornamentation comprising single, rounded nodes. The eggshell consists of a single structural layer of calcite – approximately 1.3 mm thick in well-preserved samples. Gregarious behavior is suggested by the high concentration of identical eggs (11 eggs/m^2) in six stratigraphic horizons, which indicates that a single titanosaur species exhibited site fidelity at least six times (Chiappe et al. 2005). Although nest attendance by titanosaurs may be inferred by phylogenetic bracketing (all living archosaurs attend their nests), adult size and proximity between clutches suggests little or no parental care, a conclusion supported by the apparent lack of trampling at Auca Mahuevo, where most eggs show minimal crushing.

The only other association between sauropod eggs and hatchlings comes from the latest Cretaceous of India, where a partial clutch of titanosaur eggs and the front quarters of a hatchling were found in association with the 3.5-meter-long snake *Sanajeh indicus* (Wilson et al. 2010). The snake was coiled around a recently hatched egg and adjacent to the partial hatchling. Other associations between this snake and sauropod hatchlings at the same site indicate that the snake frequented the nesting ground and probably specialized in preying upon sauropod hatchlings.

In summary, sauropods lack the prolonged development times predicted for large-bodied reptilian vertebrates, instead growing rapidly to adult size in a similar fashion to modern mammals. Unlike mammals, however, sauropods were not K-strategists in terms of reproduction and fecundity. The few records we have suggest sauropods were r-strategists (like other reptiles) that made large nests with a potential for many rapidly growing offspring, a strategy that implies little parental investment. Sauropods seem to be reproducing like reptiles but growing like mammals and birds.

Fossilized Tracks and Trackways

Struggling to find adequate descriptors for sauropod size "ponderous," "behemoth," "enormous," "stupendous," and "massive" being a few — many early paleontologists assumed that sauropods could not support their body weight on land (e.g., Owen 1875; Osborn 1899; Hatcher 1901). Although their anatomy was trumpeted as a "marvel of construction . . . a mechanical triumph for great size, lightness, and strength" (Osborn 1899, 213), most early life reconstructions depicted sauropods up to their necks in ancient swamps. Alternatively, early workers such as Mantell (1850), Phillips (1871), Cope (1877, 1878), and Marsh (1877) initially postulated that the sauropods were fully terrestrial, feeding on the vegetation of mountain forests. In their eyes, sauropods may have been the largest animals capable of terrestrial locomotion. Despite these early pronouncements, Cope and Marsh eventually adopted Owen's opinion and got sauropods "into the swim" (Coombs 1975, 2).

Perhaps most surprisingly, sauropod footprints — which we might think of as evidence of terrestriality — were initially interpreted as evidence for amphibious habits (Bird 1941, 1944). In particular, the absence of a tail drag suggested a buoyant force held aloft the tail, which Bird and others assumed would drag on the ground (we now interpret the tail as held aloft in all dinosaurs). Modern interpretations of sauropods and their trackways all point to a fully terrestrial lifestyle for sauropods (Bakker 1971a, b; Coombs 1975; Farlow 1992; Lockley et al. 1994). In addition to providing evidence of the terrestrial lifestyles of sauropods, footprints provide temporal, spatial, and character data that inform hypotheses of sauropod distribution and locomotor evolution. Spatiotemporal distributions based on sauropod body fossils and ichnofossils often overlap, but ichnofossils have had particular impact on our understanding of the early distributional history of sauropods (Lockley et al. 1994; Wilson 2005c), where ichnofossils comprise a significant proportion of the Carnian-Norian sauropod fossil record (e.g., Lockley et al. 2001; Marsicano and Barredo 2004). In addition to providing a spatiotemporal record of sauropod distributions, ichnofossils have also proven useful in illuminating locomotor posture, particularly in those earliest sauropods for which body fossils often lack manual and pedal data. For example, the first sauropod with manual and pedal data is *Shunosaurus*, from the Middle Jurassic (Zhang 1988). *Shunosaurus* exhibits the highly specialized manus and pes characteristic of later sauropods, leaving the nature of the evolution of this structure difficult to discern. Ichnofossils from the Late Triassic and Early Jurassic may shed light on the timing and pattern associated with the evolution of the secondarily quadrupedal locomotor strategy of Sauropoda, and more recent ichnofossils document limb postural distinctions in more derived sauropods that are borne out in appendicular skeletal anatomy (Carrano and Wilson 2001; Wilson 2005c). Tracks are also proving useful for extending the geographic ranges of sauropods where body fossils are lacking. Recent reports of sauropod tracks from the Cretaceous of Alaska (McCrea et al. 2005) confirm the presence of sauropods at high latitudes.

Ichnological data in a phylogenetic and anatomical context can provide insight the morphological and functional diversity observed in the secondarily quadrupedal sauropods (Carrano 2005). Many aspects of sauropod appendicular morphology are associated with supporting a large body in terrestrial environments, in part because these same features are also found in other large terrestrial forms such as other dinosaurs and in large terrestrial mammals. Other aspects of limb morphology, however, appear to be unique to sauropods. An example of this is the "wide-gauge" locomotion of certain sauropods, a feature that was first recognized in trackways. Sauropod tracks, like other dinosaur trackways, exhibit variation in several parameters. Gauge width is one parameter across which there is considerable variation. Whereas most sauropod trackways record a trackmaker walking with its limbs close to the trackway midline, others record trackmakers with a much broader stance. Although there are intermediate trackway gauges, most sauropod trackways are referred to as "narrow-gauge" or "wide-gauge" trackways, respectively (Farlow 1992; Fig. 22.11). Interestingly, a subgroup of sauropods possesses appendicular characteristics that implicate them as the wide-gauge trackmakers (Wilson and Carrano 1999; Fig. 22.12). Some wide-gauge trackmakers have shortened tails, flaring preacetabular processes, and reduced ossification of the carpus, manus, and tarsus, features that may relate to occasional bipedal or tripodal rearing (e.g., Borsuk-Bialynicka 1977; Wilson and Carrano 1999).

Sauropod trackways and footprints also have potential to provide information on aspects of sauropod lifestyles not likely to be recorded in body

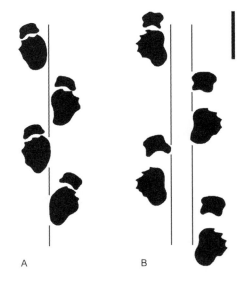

22.11. Sauropod trackway types. A, narrow-gauge trackway from the Upper Jurassic of Portugal; B, wide-gauge trackway from the Lower Cretaceous of Texas. Scale bar equals 1 m. Modified from Wilson and Carrano (1999, 1).

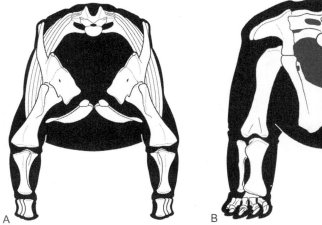
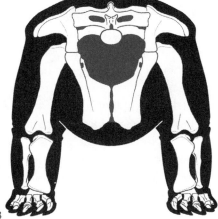

22.12. Appendicular anatomy of the wide-gauge trackmaker *Opisthocoelicaudia*. A, trunk, shoulder girdle, and forelimb; B, sacrum, pelvic girdle, and hindlimb. From Wilson (2005, 1.14).

fossils, such as behavior (e.g., Bird 1944; Ostrom 1972; Lockley et al. 1986; Farlow 1987a; Carrano and Wilson 2001; Wright 2005), soft tissues (Gatesy et al. 1999), and speed (Alexander 1976; Thulborn 1982, 1990). All sauropod trackways document an upright, quadrupedal stance and parasagittal gait. All known sauropod trackways show slow walking speeds, in keeping with their graviportal anatomy. Well-preserved sauropod trackways show depth variations within footprints, which may indicate the sequence of weight distribution in the sauropod step cycle (Wilson and Carrano 1999). A few sauropod tracksites preserve a number of parallel trackways, potentially indicating gregariousness (Bird 1944; Lockley et al. 1986; Farlow et al. 1989). Multiple sauropod trackways in the same direction on a single bedding plane are preserved in numbers up to 25. There is currently no consistent ichnological evidence for strategic herding behavior in sauropods (e.g., protecting young with adults on the outside of a group, Lockley 1991).

Herbivory

Recent analyses of sauropod feeding have focused primarily on tooth morphology and microwear (Fiorillo 1998; Christiansen 2000; Upchurch and Barrett 2000; Whitlock 2011), jaw muscle reconstruction and cranial mechanics (Calvo 1994; Barrett and Upchurch 1994; Upchurch and Barrett 2000), bioenergetics (Farlow 1987b; Dodson 1990), neck biomechanics (Stevens and Parrish 1999, 2005; Christian 2010), and isotopic analysis (Tütken 2011). These studies have clarified significant distinctions among sauropod feeding mechanisms. Basal macronarians (i.e., excluding titanosaurs) had broad, spatulate teeth that appear to have been optimal for biting off resistant vegetation, whereas the teeth of diplodocids and titanosaurs were more peglike in shape, and may have been more adept at cropping, stripping, or clipping vegetation near the ground. Most studies agree that sauropods engaged in little to no oral processing and that some form of postoral processing was needed. More recent discoveries indicate that the simple dichotomy (diplodocoid versus macronarian) may not capture the diversity of sauropod dentition. For example, the rebacchisaurid *Nigersaurus* evolved a transverse dental battery supported by high rates of tooth replacement (Sereno et al. 2007). This derived sauropod lived in the context of the evolution of angiosperms, and, in some cases, alongside euornithopods, raising the question of whether the evolution of angiosperms prompted the evolution of complex dental batteries, or whether sauropods and ornithischians competed for food resources. Sereno and Wilson (2005) compared the dental battery of *Nigersaurus* to those of euornithopods and neoceratopsians, and concluded that differences in structure and in the timing of acquisition suggest independent causes for the evolution of dental batteries.

As mentioned above, sauropods probably relied on postoral processing to break down plant food into usable energy. The most commonly cited solution to the problem of sauropod food processing is the presence of gastroliths in the stomach. These polished stones are found in a number of sauropod-bearing strata and have long been hypothesized to be analogous to those used in some modern vertebrates to help grind up food and aid in digestion (Stokes 1944; Farlow 1987b). Gillette (1991) reported the association of a number of gastroliths in the gut region of a skeleton of *Seismosaurus*, but no detailed taphonomic analysis was performed to rule out alternative sources for the stones. Wings (2003, 2004) and Wings and Sander (2007) rejected the notion of sauropod gastroliths on the basis of (1) experimental analysis of gizzard stone processing and decay in extant ostriches; (2) lack of association with skeletons; and (3) taphonomy and sedimentology/stratigraphy of "gastrolith"-producing localities. It is more likely that sauropods, like other large-bodied herbivorous animals, relied on microbial fermentation of plant food in their elongated digestive systems (e.g., Sander and Clauss 2008; Hummel and Clauss 2011).

The notion that sauropods reared up on their hind limbs to feed high in the treetops emerged with the first discoveries of their skeletons. Hatcher (1901) postulated that the "ski-shaped" chevrons of *Diplodocus* protected the underside of its tail while it acted as the "third leg" during tripodal feeding at the tops of trees. This idea is common today, despite the fact that few have investigated the biomechanical and physiological consequences of this posture. Of course, it would have been necessary for sauropods to rear up to their hind legs occasionally during life for successful mating. Only one sauropod group – Titanosauriformes – appears to exhibit anatomical features associated with tripodality in modern vertebrates. These characteristics, including a laterally flared ilium, the medially deflected head of the femur, and wide-gauge stance, as well as restricted adult body size in some derived titanosaurians (e.g., *Saltasaurus*) may indicate a propensity for more regular tripodal postures (Wilson and Carrano 1999).

Estimates of maximum feeding height for sauropods are typically based on assumptions regarding neck posture (McIntosh et al. 1997; Upchurch and Barrett 2000; Stevens and Parrish 1999, 2005; Christian 2010; Christian and Dzemski 2011). Some estimate maximum browsing height as the sum of shoulder height and neck length and assume a completely vertical neck (McIntosh et al. 1997; Upchurch and Barrett 2000), whereas others have outlined feeding envelopes using a method for three-dimensional reconstructions of sauropod skeletons (Stevens and Parrish 2005). In contrast to earlier reconstructions of giraffelike or tripodal sauropods, Stevens and Parrish (2005) inferred that most sauropods were medium to low browsers. In their view, sauropod feeding is constrained not only by the dentition, but also by vertebral flexibility and forage availability and abundance.

Additionally, it seems that even within classes of dentition, feeding behaviors may have varied between taxa (Barrett and Upchurch, 2005; Whitlock 2011). Microwear features on the teeth of *Diplodocus* and *Dicraeosaurus*, representatives of two closely related clades of narrow-crowned sauropods, appear to indicate that these animals selected foods at different feeding heights, and in potentially different ways; this may also be reflected in snout shape (Whitlock 2011; Fig. 22.13).

The relationship between sauropod feeding strategies and Mesozoic floral evolution is also of interest. At least some paleobotanists (Wing and Tiffney 1987) and vertebrate paleontologists (Bakker 1978) have suggested that "clear cutting" of Jurassic and Early Cretaceous forests by sauropod herds may have been instrumental in creating ecological conditions that favored the origin of flowering plants. Stevens and Parrish (2005) concluded that only *Brachiosaurus* and perhaps *Camarasaurus* would have been able to graze on tall gymnosperms, whereas Christian (2010) added *Euhelopus* to

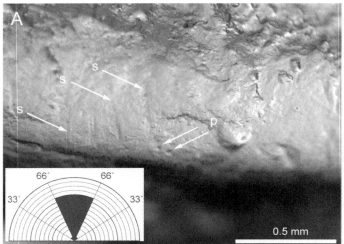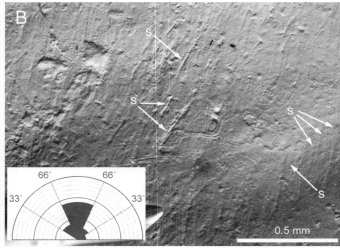

22.13. Microwear features on the teeth of the diplodocoid sauropods *Diplodocus* (A) and *Dicraeosaurus* (B). Rose diagrams indicate orientation of scratches (labiolingual axis is vertical). The relative abundance of pits (p) and the unimodal distribution of scratches (s) suggests a low-height browsing behavior in *Diplodocus,* whereas the lack of large pits and the broader distribution of scratch orientations in *Dicraeosaurus* suggests a reliance on higher browse. Modified from Whitlock (2007).

the list of potentially high-browsing sauropods. Other sauropods may have spent far more time grazing at or near ground level. Lycopods and ferns were abundant in Mesozoic terrestrial ecosystems, and might have served as a rapidly growing, renewable food source for the fast-growing sauropod eating machines (Krasilov 1981; Stevens and Parrish 2005; Gee 2011). The recent discovery of grass phytoliths in titanosaur coprolites from India indicates that grass was also an established part of at least some Cretaceous ecosystems, and that titanosaurs may have exploited a feeding envelope that included ample forage at their feet and above their heads (Prasad et al. 2005).

References

Alexander, R. M. 1976. Estimates of speeds of dinosaurs. *Nature* 261: 129–130.

Allain, R., P. Taquet, B. Battail, J. Dejax, P. Richir, M. Véran, F. Limon-Duparcmeur, R. Vacant, O. Mateus, P. Sayarath, B. Khenthavong, and S. Phouyavong. 1999. Un nouveau genere de dinosaure sauropode de la formation des Grés supérieurs (Aptien-Albien) du Laos. *Comptes Rendus de l'Academie des Sciences Paris, Sciences de la Terre et des Planétes* 329: 609–616.

Anonymous. 1905. The presentation of a reproduction of *Diplodocus carnegiei* to the trustees of the British Museum. *Annals of the Carnegie Museum* 3: 443–452.

Bakker, R. T. 1968. The superiority of dinosaurs. *Discovery* 3: 11–22.

——. 1971a. The ecology of brontosaurs. *Nature* 229: 172–174.

——. 1971b. Dinosaur physiology and the origin of mammals. *Evolution* 25: 636–658.

——. 1978. Dinosaur feeding behavior and the origin of flowering plants. *Nature* 274: 661–663.

——. 1986. *Dinosaur Heresies.* New York: William Morrow.

Bandyopadhyay, S. 1999. Gondwana vertebrate fauans. *PINSA* 65: 285–313.

Bandyopadhyay, S., D. D. Gillette, S. Ray, and D. P. Sengupta. 2010. Osteology of *Barapasaurus tagorei* (Dinosauria: Sauropoda) from the Early Jurassic of India. *Palaeontology* 53: 533–569.

Barrett, P. M., Y. Hasegawa, M. Manabe, S. Isaji, and H. Matsuoka. 2002. Sauropod dinosaurs from the Lower Cretaceous of eastern Asia: taxonomic and biogeographical implications. *Palaeontology* 45: 1197–1217.

Barrett, P. M., and P. Upchurch. 1994. Feeding mechanisms of *Diplodocus.* *Gaia* 10: 195–204.

——. 2005. Sauropodomorph diversity through time: macroevolutionary and paleoecological implications. In K. A. Curry Rogers and J. A. Wilson (eds.), *The Sauropods: Evolution and Paleobiology,* 125–156. Berkeley: University of California Press.

Benton, M. J. 2002. Cope's Rule. In M. Pagel (ed.), *Encyclopedia of Evolution,* 185–186. New York: Oxford University Press.

Benton, M. J., Z. Csiki, D. Grigorescu, R. Redelstorff, P. M. Sander, K. Stein, and D. B. Weishampel. 2010. Dinosaurs and the island rule: the dwarfed dinosaurs from Hateg Island. *Palaeogeography, Palaeoclimatology, Palaeoecology* 293: 438–454.

Bird, R. T. 1941. A dinosaur walks into the museum. *Natural History* 47: 74–81.

———. 1944. Did *Brontosaurus* ever walk on land? *Natural History* 53: 61–67.

Bohlin, B. 1953. Fossil reptiles from Mongolia and Kansu. *Sino-Swedish Expedition Publication* 37: 1–113.

Bonaparte, J. F. 1986. The early radiation and phylogenetic relationships of sauropod dinosaurs, based on vertebral anatomy. In K. Padian (ed.), *The Beginning of the Age of Dinosaurs*, 247–258. Cambridge: Cambridge University Press.

———. 1996. *Dinosaurios de America del Sur.* Buenos Aires: Impreso en Artes Gráficas Sagitario Iturri.

Borsuk-Bialynicka, M. 1977. A new camarasaurid sauropod *Opisthocoelicaudia skarzynskii* gen. n., sp. n. from the Upper Cretaceous of Mongolia. *Palaeontologia Polonica* 37: 5–64.

Böttcher, R. 1990. Neue Erkenntnisse über die Fortpflanzungsbiologie der Ichthyosaurier (Reptilia). *Stuttgarter Beiträge Naturkunde* (ser. B) 164: 13–51.

Britt, B. B., and B. G. Naylor. 1994. An embryonic *Camarasaurus* (Dinosauria, Sauropoda) from the Upper Jurassic Morrison Formation (Dry Mesa Quarry, Colorado). In K. Carpenter, K. F. Hirsch, and J. R. Horner (eds.), *Dinosaur Eggs and Babies*, 256–264. New York: Cambridge University Press.

Brown, J. H. 1995. *Macroecology.* Chicago: University of Chicago Press.

Buckland, W. 1841. *Geology and Mineralogy Considered with Reference to Natural Theology*, 2nd ed. Philadelphia: Lea & Blanchard.

Buffetaut, E., V. Suteethorn, G. Cuny, H. Tong, J. Le Loeuff, S. Khansubha, and S. Jongautchariyakul. 2000. The earliest known sauropod dinosaur. *Nature* 407: 72–74.

Calder, W. A. III. 1984. *Size, Function, and Life History.* Cambridge, Mass.: Harvard University Press.

Caldwell, M. W., and M. S. Y. Lee. 2001. Live birth in Cretaceous marine lizards (mosasaurids). *Proceedings of the Royal Society of London* B 268: 2397–2401.

Calvo, J. O. 1994. Jaw mechanics in sauropod dinosaurs. *Gaia* 10: 183–194.

Calvo, J. O., S. Engelland, S. E. Heredia, and L. Salgado. 1997. First record of

dinosaur eggshells (?Sauropoda-Megaloolithidae) from Neuquén, Patagonia, Argentina. *Gaia* 14: 23–32.

Calvo, J. O., J. Porfiri, B. J. González Riga, and A. W. A. Kellner. 2007. A new Cretaceous terrestrial ecosystem from Gondwana with the description of a new sauropod dinosaur. *Anais da Academia Brasileira de Ciências* 79: 529–541.

Calvo, J. O., and L. Salgado. 1995. *Rebbachisaurus tessonei* sp. nov. a new Sauropoda from the Albian-Cenomanian of Argentina; new evidence on the origin of Diplodocidae. *Gaia* 11: 13–33.

Carrano, M. T. 2005. The evolution of sauropod locomotion: morphological diversity of a secondarily quadrupedal radiation. In K. A. Curry Rogers and J. A. Wilson (eds.), *The Sauropods: Evolution and Paleobiology*, 229–251. Berkeley: University of California Press.

———. 2006. Body-size evolution in the Dinosauria. In M. T. Carrano, T. J. Gaudin, R. W. Blob, and J. R. Wible (eds.), *Amniote Paleobiology: Perspectives on the Evolution of Mammals, Birds, and Reptiles*, 225–268. Chicago: University of Chicago Press.

Carrano, M. T., and J. R. Hutchinson. 2002. Pelvic and hindlimb musculature of *Tyrannosaurus rex* (Dinosauria: Theropoda). *Journal of Morphology* 253: 207–228.

Carrano, M. T., and J. A. Wilson. 2001. Taxon distributions and the tetrapod track record. *Paleobiology* 27: 564–582.

Case, T. J. 1978a. On the evolution and adaptive significance of postnatal growth rates in the terrestrial vertebrates. *Quarterly Review of Biology* 53: 243–282.

———. 1978b. Speculations on the growth rate and reproduction of some dinosaurs. *Paleobiology* 4: 320–328.

Charig, A., J. Attridge, and A. W. Crompton. 1965. On the origins of the sauropods and the classification of the Saurischia. *Proceedings of the Linnaean Society of London* 176: 197–221.

Cheng, Y.-N., X.-C. Wu, and Q. Ji. 2004. Triassic marine reptiles gave birth to live young. *Nature* 432: 383–386.

Chiappe, L. M., R. A. Coria, L. Dingus, F. Jackson, A. Chinsamy, and M. Fox. 1998. Sauropod dinosaur embryos from the Late Cretaceous of Patagonia. *Nature* 396: 258–261.

Chiappe, L. M., and L. Dingus. 2001. *Walking on Eggs.* New York: Scribner.

Chiappe, L. M., L. Dingus, F. Jackson, G. Grellet-Tinner, R. Aspinall, J. Clarke, R. A. Coria, A. Garrido, and D. Loope. 2000. Sauropod eggs and embryos from the Upper Cretaceous of Patagonia. In

A. M. Bravo and T. Reyes (eds.), *1st International Symposium of Dinosaur Eggs and Embryos, Isona, Spain*: 23–29.

Chiappe, L. M., F. Jackson, R. A. Coria, and L. Dingus. 2005. Nesting titanosaurs from Auca Mahuevo and adjacent sites: understanding sauropod reproductive behavior and embryonic development. In K. A. Curry Rogers and J. A. Wilson (eds.), *The Sauropods: Evolution and Paleobiology*, 285–302. Berkeley: University of California Press.

Chiappe, L. M., L. Salgado, and R. A. Coria. 2001. Embryonic skulls of titanosaur sauropod dinosaurs. *Science* 293: 2444–2446.

Chiappe, L. M., J. G. Schmitt, F. Jackson, A. Garrido, L. Dingus, and G. Grellet-Tinner. 2004. Nest structure for sauropods: Sedimentary criteria for recognition of dinosaur nesting traces. *Palaios* 19: 89–95.

Christian, A. 2010. Some sauropods raised their necks: evidence for high browsing in *Euhelopus zdanskyi. Biology Letters* 6: 823–825.

Christian, A., and G. Dzemski. 2011. Neck posture in sauropods. In N. Klein, K. Remes, C. T. Gee, and P. M. Sander (eds.), *Biology of the Sauropod Dinosaurs: Understanding the Life of Giants*, 251–260. Life of the Past (series ed. J. O. Farlow). Bloomington: Indiana University Press.

Christiansen, P. 2000. Feeding mechanisms of the sauropod dinosaurs *Brachiosaurus, Camarasaurus, Diplodocus*, and *Dicraeosaurus. Historical Biology* 14: 137–152.

Chure, D. J., B. B. Britt, J. A. Whitlock, and J. A. Wilson. 2010. First complete sauropod dinosaur skull from the Cretaceous of the Americas and the evolution of sauropod dentition. *Naturwissenschaften* 97: 379–391.

Clapham, P. J., S. E. Wetmore, T. D. Smith, and J. G. Mead. 1999. Length at birth and at independence in humpback whales. *Journal of Cetacean Research and Management* 1: 141–146.

Clauss, M. 2011. Sauropod biology and the evolution of gigantism: what do we know? In N. Klein, K. Remes, C. T. Gee, and P. M. Sander (eds.), *Biology of the Sauropod Dinosaurs: Understanding the Life of Giants*, 3–7. Life of the Past (series ed. J. O. Farlow). Bloomington: Indiana University Press.

Colbert, E. H. 1962. The weights of dinosaurs. *American Museum Novitates* 2076: 1–16.

Coombs, W. P., Jr. 1975. Sauropod habits and habitats. *Palaeogeography,*

Palaeoclimatology, Palaeoecology 17: 1–33.

——. 1978. Theoretical aspects of cursorial adaptations in dinosaurs. *Quarterly Review of Biology* 53: 393–418.

Cope, E. D. 1877. On a gigantic saurian from the Dakota epoch of Colorado. *Paleontology Bulletin* 25: 5–10.

——. 1878. A new species of *Amphicoelias*. *American Naturalist* 12: 563–564.

——. 1887. *The Origin of the Fittest*. New York: Appleton Press.

——. 1896. *The Primary Factors of Organic Evolution*. New York: Open Court Publishing.

Cousin, R., G. Bréton, R. Fournier, and J. Watté. 1990. Dinosaur egglaying and nesting in France. In K. Carpenter, K. F. Hirsch, and J. R. Horner (eds.), *Dinosaur Eggs and Babies*, 56–74. New York: Cambridge University Press.

Curry, K. A. 1999. Ontogenetic histology of *Apatosaurus* (Dinosauria: Sauropoda): new insights on growth rates and longevity. *Journal of Vertebrate Paleontology* 19: 654–665.

Curry Rogers, K. A. 2001. The evolutionary history of the Titanosauria. Ph. D. dissertation, State University of New York, Stony Brook.

——. 2005. Titanosauria: a phylogenetic overview. In K. A. Curry Rogers and J. A. Wilson (eds.), *The Sauropods: Evolution and Paleobiology*, 50–103. Berkeley: University of California Press.

Curry Rogers, K. A., and G. M. Erickson. 2005. Sauropod histology: microscopic views on the lives of giants. In K. A. Curry Rogers and J. A. Wilson (eds.), *The Sauropods: Evolution and Paleobiology*, 303–326. Berkeley: University of California Press.

Curry Rogers, K. A., and C. A. Forster. 2001. The last of the dinosaur titans: a new sauropod from Madagascar. *Nature* 412: 530–534.

——. 2004. The skull of *Rapetosaurus krausei* (Sauropoda: Titanosauria) from the Late Cretaceous of Madagascar. *Journal of Vertebrate Paleontology* 24: 121–144.

D'Emic, M. D., J. A. Wilson, and S. Chatterjee. 2009. A definitive titanosaur (Dinosauria: Sauropoda) osteoderm from the Upper Cretaceous of India and the titanosaur osteoderm record. *Journal of Vertebrate Paleontology* 29: 165–177.

Dodson, P. 1990. Sauropod paleoecology. In D. B. Weishampel, P. Dodson, and H. Osmólska (eds.), *The Dinosauria*, 402–407. Berkeley: University of California Press.

——. 1991. Lifestyles of the huge and famous. *Natural History* 100: 30–34.

Dong, Z. 1992. *Dinosaurian Faunas of China*. Beijing: China Ocean Press.

Dughi, R., and F. Sirugue. 1966. Sur la fossilization des oeufs de dinosaures. *Comptes rendus des séance de l'Académie des Sciences* 262: 2330–2332.

Dulvy, N. K., and J. D. Reynolds. 1997. Evolutionary transitions among egg-laying, live-bearing and maternal inputs in sharks and rays. *Proceedings of the Royal Society of London B* 264: 1309–1315.

Dumont, M., A. Borbély, A. Kostka, P. M. Sander, and A. Kaysser-Pyzalla. 2011. Characterization of sauropod bone structure. In N. Klein, K. Remes, C. T. Gee, and P. M. Sander (eds.), *Biology of the Sauropod Dinosaurs: Understanding the Life of Giants*, 150–170. Life of the Past (series ed. J. O. Farlow). Bloomington: Indiana University Press.

Dunham, A. E., K. L. Overall, W. P. Porter, and C. A. Forster. 1989. Implications of the ecological energetics and biophysical and developmental constraints for life history variation in dinosaurs. In J. O. Farlow (ed.), *Paleobiology of the Dinosaurs*. Geological Society of America, Boulder, Colo., Special Paper 238: 1–19.

Erben, H, J. Hoefs, and K. H. Wedepohl. 1979. Paleobiological and isotopic studies of eggshells from a declining dinosaur species. *Paleobiology* 5: 380–414.

Erickson, G. M., K. Curry Rogers, and S. Yerby. 2001. Dinosaurian growth patterns and rapid avian growth rates. *Nature* 412: 429–433.

Faccio, G. 1990. Dinosaurian eggs from the Upper Cretaceous of Uruguay. In K. Carpenter, K. F. Hirsch, and J. R. Horner (eds.), *Dinosaur Eggs and Babies*, 47–55. New York: Cambridge University Press.

Farlow, J. O. 1987a. *Lower Cretaceous Dinosaur Tracks, Paluxy River Valley, Texas*. Waco, Tex.: Baylor University.

——. 1987b. Speculations about the diet and digestive physiology of herbivorous dinosaurs. *Paleobiology* 13: 60–72.

——. 1992. Sauropod tracks and trackmakers: integrating the ichnological and skeletal records. *Zubía* 10: 89–138.

Farlow, J. O., J. G. Pittman, and J. M. Hawthorne. 1989. *Brontopodus birdi*, Lower Cretaceous sauropod footprints from the US Gulf coastal plain. In D. D. Gillette and M. G. Lockley (eds.), *Dinosaur Tracks and Traces*, 371–394. New York: Cambridge University Press.

Figueroa, C., and J. E. Powell. 2000. Structure and permeability properties of thick dinosaur eggshells from the Upper Cretaceous of Patagonia, Argentina. Results of a computer simulation analysis. In A. M.

Bravo and T. Reyes (eds.), *1st International Symposium of Dinosaur Eggs and Embryos, Isona, Spain*: 51–60.

Fiorillo, A. R. 1998. Dental microwear patterns of sauropod dinosaurs *Camarasaurus* and *Diplodocus*: evidence for resource partitioning in the Late Jurassic of North America. *Historical Biology* 13: 1–16.

Galton, P. M., and J. van Heerden. 1985. Partial hindlimb of *Blikanasaurus cromptoni* n. gen. and n. sp., representing a new family of prosauropod dinosaurs from the Upper Triassic of South Africa. *Géobios* 18: 509–516.

——. 1998. Anatomy of the prosauropod dinosaur *Blikanasaurus cromptoni* (Upper Triassic, South Africa), with notes on the other tetrapods from the Lower Eliote Formation. *Paläontologische Zeitschrift* 72: 163–177.

Ganse, B., A. Stahn, S. Stoinski, T. Suthau, and H.-C. Gunga. 2011. Body mass estimation, thermoregulation, and cardiovascular physiology of large sauropods. In N. Klein, K. Remes, C. T. Gee, and P. M. Sander (eds.), *Biology of the Sauropod Dinosaurs: Understanding the Life of Giants*, 105–115. Life of the Past (series ed. J. O. Farlow). Bloomington: Indiana University Press.

Gatesy, S. M., K. M. Middleton, F. A. Jenkins, Jr., and N. H. Shubin. 1999. Three-dimensional preservation of foot movements in Triassic theropod dinosaurs. *Nature* 399: 141–144.

Gauthier, J. 1986. Saurischian monophyly and the origin of birds. In K. Padian, (ed.), *The Origin of Birds and the Evolution of Flight*. Memoirs of the California Academy of Sciences 8: 1–55.

Gee, C. T. 2011. Dietary options for the sauropod dinosaurs from an integrated botanical and paleobotanical perspective. In N. Klein, K. Remes, C. T. Gee, and P. M. Sander (eds.), *Biology of the Sauropod Dinosaurs: Understanding the Life of Giants*, 34–56. Life of the Past (series ed. J. O. Farlow). Bloomington: Indiana University Press.

Gillette, D. D. 1991. *Seismosaurus halli* gen. et sp. nov., a new sauropod dinosaur from the Morrison Formation (Upper Jurassic/Lower Cretaceous) of New Mexico, USA. *Journal of Vertebrate Paleontology* 11: 4176–433.

Gilmore, C. W. 1946. Reptilian fauna of the North Horn Formation of central Utah. United States Geological Survey Professional Paper 210-C: 1–52.

Gomani, E. 2005. Sauropod dinosaurs from the Early Cretaceous of Malawi,

Africa. *Palaeontologia Electronica* 8 (1): 27A. http://palaeo-electronica.org/paleo/2005_1/gomani27/issue1_05.htm.

González Riga, B. J. 2003. A new titanosaur (Dinosauria, Sauropoda) from the Upper Cretaceous of Mendoza Province, Argentina. *Ameghiniana* 40: 155–172.

Gunga, H.-C., T. Suthau, A. Bellman, S. Stoinski, A. Friedrich, T. Trippel, K. Kirsch, and O. Hellwich. 2008. A new body mass estimation of *Brachiosaurus brancai* Janensch, 1914 mounted and exhibited at the Museum of Natural History (Berlin, Germany). *Fossil Record* 11: 33–38.

Harris, J. D. 2006. The significance of *Suuwassea emilieae* (Dinosauria: Sauropoda) for flagellicaudatan intrarelationships and evolution. *Journal of Systematic Palaeontology* 4: 185–198.

Hatcher, J. B. 1901. *Diplodocus* (Marsh): its osteology, taxonomy, and probable habits, with a restoration of the skeleton. *Memoirs of the Carnegie Museum* 1: 1–63.

Holland, W. 1915. Heads and tails. A few foot notes relating to the structure of the sauropod dinosaurs. *Annals of the Carnegie Museum* 9: 273–278.

Hone, D. W. E., and M. J. Benton. 2005. The evolution of large size: how does Cope's Rule work? *Trends in Ecology and Evolution* 20: 4–6.

Hone, D. W. E., T. M. Keesey, D. Pisani, and A. Purvis. 2005. Macroevolutionary trends in the Dinosauria: Cope's rule. *Journal of Evolutionary Biology* 18: 587–595.

Huene, F. von. 1929. Los Saurisquios y Ornithisquios del Cretaceo Argentino. *Anales del Museo La Plata* 2: 1–196.

———. 1956. *Paläontologie und Phylogenie der Niederen Tetrapoden*. Jena, Germany: Gustav Fischer Verlag.

Huene, F. von, and C. A. Matley. 1933. The Cretaceous Saurischia and Ornithischia of the central provinces of India. *Memoirs of the Geological Survey of India* 21: 1–74.

Hummel, J., and M. Clauss. 2011. Sauropod feeding and digestive physiology. In N. Klein, K. Remes, C. T. Gee, and P. M. Sander (eds.), *Biology of the Sauropod Dinosaurs: Understanding the Life of Giants*, 11–33. Life of the Past (series ed. J. O. Farlow). Bloomington: Indiana University Press.

Huxley, T. H. 1869. On the classification of the Dinosauria, with observations on the Dinosauria of the Trias. *Quarterly Review of the Geological Society of London* 26: 32–51.

Jacobs, L. L., D. A. Winkler, W. R. Downs, and E. M. Gomani. 1993. New material of an Early Cretaceous titanosaurid sauropod dinosaur from Malawi. *Palaeontology* 36: 523–534.

Jain, S. L., and S. Bandyopadhyay. 1997. New titanosaurid (Dinosauria: Sauropoda) from the Late Cretaceous of central India. *Journal of Vertebrate Paleontology* 17: 114–136.

Jain, S. L., T. S. Kutty, T. K. Roy-Chowdhury, and S. Chatterjee. 1979. The sauropod dinosaur from the Lower Jurassic Kota Formation of India. *Proceedings of the Royal Society of London B* 188: 221–228.

Janensch, W. 1914. Ubersicht uber die Wirbeltierfauna der Tendaguru-Schichten, nebst einer kurzen Charakterisierung der neu aufgefuhrten Arten von Sauropoden. *Archive für Biontologie* 3: 81–110.

———. 1929a. Material und Formengehalt der Sauropoden in der Ausbeute der Tendaguru Expedition. *Palaeontographica* (suppl. 7) 2: 1–34.

———. 1929b. Die Wirbelsäule der Gattung *Dicraeosaurus hausemanni*. *Palaeontographica* (suppl. 7) 3: 39–133.

———. 1935–1936. Die Schädel der Sauropoden *Brachiosaurus*, *Barosaurus* und *Dicraeosaurus* aus den Tendaguruschichten Deutsch-Ostafrikas. *Palaeontographica* (suppl. 7) 2: 147–298.

———. 1950. Die Wirbelsäule von *Brachiosaurus brancai*. *Palaeontographica* (suppl. 7) 3: 27–93.

———. 1961. Die Gliedmaszen und Gliedmaszengürtel der Sauropoden der Tendaguru-Schichten. *Palaeontographica* (suppl. 7) 3: 177–235.

Jianu, C.-M., and D. B. Weishampel. 1999. The smallest of the largest: a new look at possible dwarfing in sauropod dinosaurs. *Geologie en Mijnbouw* 78: 335–343.

Kellner, A. W. A., and S. A. K. de Azevedo. 1999. A new sauropod dinosaur (Titanosauria) from the Late Cretaceous of Brazil. *Proceedings of the Second Gondwanan Dinosaur Symposium* 15: 111–142.

Kérourio, P. 1981. Nouvelles observations sur le mode de nidification et de ponte chez les dinosauries du Cretace terminal du Midi de la France. *Compte Rendu Sommaire Societe geologique de la France* 1: 25–28.

Kingdon, J. 1825. On the finding of large bones in the Oolite quarries about a mile from Chipping Northon near Chappel House. *London Edinburgh Dublin Philosophical Magazine* 66: 64. Reprint, *Annals of Philosophy* (n.s.) 10: 229.

Klein, N., P. M. Sander, and V. Suteethorn. 2009. Bone histology and its implications for the life history and growth of the Early Cretaceous titanosaur *Phuwiangosaurus sirindhornae*. *Geological Society of London Special Publication* 315: 217–228.

Krasilov, V. A. 1981. Changes of Mesozoic vegetation and the extinction of the dinosaurs. *Paleogeography, Paleoclimatology, Palaeoecology* 34: 207–224.

Larson, A. 1994. The comparison of morphological and molecular data in phylogenetic systematics. In B. S. B. Schierwater, G. P. Wagner, and R. DeSalle (eds.), *Molecular Biology and Evolution: Approaches and Applications*, 371–390. Basel: Birkhauser Verlag.

Le Loeuff, J. 1995. *Ampelosaurus atacis* (nov. gen., nov. sp.), un nouveau Titanosauridae (Dinosauria, Sauropoda) du Crétacé superieur de la Haute Vallee de L'Aude (France). *Comptes Rendus de L'Academie des Sciences de Paris* (ser. 2a) 321: 693–699.

———. 2003. The first articulated titanosaurid skeleton in Europe. In C. A. Meyer (ed.), *First European Association of Vertebrate Paleontologists Meeting*. Natural History Museum, Basel.

Lockley, M. G. 1991. *Tracking Dinosaurs*. New York: Cambridge University Press.

Lockley, M. G., J. O. Farlow, and C. A. Meyer. 1994. *Brontopodus* and *Parabrontopodus* ichnogen nov. and the significance of wide and narrow gauge sauropod trackways. *Gaia: revista de geosciencias* (Museu Nacional de Historia Natural, Lisbon, Portugal) 10: 126–34.

Lockley, M. G., K. Houck, and N. K. Prince. 1986. North America's largest dinosaur tracksite: implications for Morrison Formation paleoecology. *Geological Society of America Bulletin* 57: 1163–1176.

Lockley, M. G., J. L. Wright, A. P. Hunt, and S. G. Lucas. 2001. The Late Triassic sauropod track record comes into focus: old legacies and new paradigms. *New Mexico Geological Society Guidebook, 52nd Field Conference, Geology of the Llano Estacado*: 181–190.

López-Martínez, N. 2000. Eggshell sites from the Creatceous-Tertiary transition in south-central Pyrenees (Spain). In A. M. Bravo and T. Reyes (eds.), *1st International Symposium of Dinosaur Eggs and Embryos, Isona, Spain*: 95–115.

Luo, Z. 1999. A refugium for relicts. *Nature* 400: 23–25.

Lydekker, R. 1893. Contributions to the study of the fossil vertebrates of Argentina. I. The dinosaurs of Patagonia. *Anales Museo de La Plata Sec. Paleontologia* 2: 1–14.

Maier, G. 2003. *African Dinosaurs Unearthed*. Bloomington: Indiana University Press.

Mantell, G. A. 1850. On the *Pelorosaurus*; an undescribed gigantic terrestrial reptile, whose remains are associated with those of the *Iguanodon* and other saurians in the strata of the Tilgate Forest, in Sussex. *Philosophical Transactions of the Royal Society of London* 140: 391–392.

Mariscano, C. A., and S. P. Barredo. 2004. A Triassic tetrapod footprint assemblage from southern South America: palaeobiogeographical and evolutionary implications. *Palaeogeography, Palaeoclimatology, Palaeoecology* 203: 313–335.

Marsh, O. C. 1877. Notice of a new and gigantic dinosaur. *American Journal of Science* (ser. 3) 14: 87–88.

——. 1878. Principal characters of American Jurassic dinosaurs. Pt. I. *American Journal of Science* (ser. 3) 16: 411–416.

——. 1883. Principal characters of American Jurassic dinosaurs. Part VI. Restoration of *Brontosaurus*. *American Journal of Science* (ser. 3) 26: 81–86.

——. 1884. Principal characters of American Jurassic dinosaurs. Part VII. On the Diplodocidae, a new family of the Sauropoda. *American Journal of Science* (ser. 3) 27: 161–168.

——. 1895. On the affinities and classification of the dinosaurian reptiles. *American Journal of Science* (ser. 3) 50: 413–423.

Martin, V., E. Buffetaut, and V. Suteethorn. 1994. A new genus of sauropod dinosaur from the Sao Khua Formation (Late Jurassic or Early Cretaceous) of northeastern Thailand. *Comptes Rendus de l'Academie des Sciences Paris* (ser. 2) 319: 1085–1092.

Martínez, R. D., O. Giménez, J. Rodríguez, M. Luna, and M. C. Lamanna. 2004. An articulated specimen of the basal titanosaurian (Dinosauria: Sauropoda) *Epachthosaurus sciuttoi* from the early Late Cretaceous Bajo Barreal Formation of Chubut Province, Argentina. *Journal of Vertebrate Paleontology* 24: 107–120.

Maxwell, E. E., and M. W. Caldwell. 2003. First record of live birth in Cretaceous ichthyosaurs: closing an 80 million year gap. *Proceedings of the Royal Society of London B* 03b10108.S: 1–4.

Mazzetta, G. V., P. Christiansen, and R. A. Farina. 2004. Giants and bizarres: body size of some southern South American Cretaceous dinosaurs. *Historical Biology* 16: 71–83.

McCrea, R., P. Currie, and S. Pemberton. 2005. Canada's largest dinosaurs: ichnological evidence of the northernmost record of sauropods in North America. *Journal of Vertebrate Paleontology* 25: 91A.

McIntosh, J. S. 1990. Sauropoda. In D. B. Weishampel, P. Dodson, and H. Osmólska (eds.), *The Dinosauria*, 345–401. Berkeley: University of California Press.

McIntosh, J. S., M. K. Brett-Surman, and J. O. Farlow. 1997. Sauropods. In J. O. Farlow and M. K. Brett-Surman (eds.), *The Complete Dinosaur*, 264–290. Bloomington: Indiana University Press.

McKinney, M. L. 1997. Extinction vulnerability and selectivity: combining ecological and paleontological views. *Annual Review of Ecology and Systematics* 28: 495–516.

Mikhailov, K. E. 1991. Classification of fossil eggshells of amniotic vertebrates. *Acta Paleontologica Polonica* 36: 193–238.

——. 1997. Fossil and recent eggshell in amniotic vertebrates: fine structure, comparative morphology and classification. *Special Papers in Paleontology* 56: 1–80.

Mohabey, D. M. 1996. A new oospecies, *Megaloolithus matleyi*, from the Lameta Formation (Upper Cretaceous) of Chandrapur district, Maharashtra, India, and general remarks on the palaeoenvironment and nesting behavior of dinosaurs. *Cretaceous Research* 17: 183–196.

——. 2000. Indian Upper Cretaceous (Maestrichtian) dinosaur eggs: their parataxonomy and implication in understanding the nesting behavior. In A. M. Bravo and T. Reyes (eds.), *1st International Symposium of Dinosaur Eggs and Embryos, Isona, Spain*: 95–115.

Myhrvold, N. P., and P. J. Currie. 1997. Supersonic sauropods? Tail dynamics in the diplodocids. *Paleobiology* 23: 393–409.

Novas, F. E. 2009. *The Age of Dinosaurs in South America*. Bloomington: Indiana University Press.

Nowinski, A. 1971. *Nemegtosaurus mongoliensis* n. gen. n. sp. (Sauropoda) from the uppermost Cretaceous of Mongolia. *Acta Palaeontologica Polonica* 25: 57–81.

Osborn, H. F. 1899. A skeleton of *Diplodocus*. *Memoirs of the American Museum of Natural History* 1: 191–214.

Osborn, H. F., and C. C. Mook. 1921. *Camarasaurus, Amphicoelias*, and other sauropods of Cope. *Memoirs of the American Museum of Natural History* 3: 247–387.

Ostrom, J. H. 1969. Osteology of *Deinonychus antirrhopus*, an unusual theropod from the Lower Cretaceous of Montana. *Bulletin of the Yale Peabody Museum of Natural History* 30: 1–165.

——. 1972. Were some dinosaurs gregarious? *Palaeogeography, Palaeoclimatology, Palaeoecology* 11: 287–301.

Owen, R. 1841. A description of a portion of the skeleton of *Cetiosaurus*, a gigantic extinct saurian occurring in the Oolitic Formation of different parts of England. *Proceedings of the Geological Society of London* 3: 457–462.

——. 1842. Report on British fossil reptiles. Part II. Reptiles. *Report of the British Association for the Advancement of Science* 1841: 60–204.

——. 1875. Monographs on the British fossil Reptilia of the Mesozoic formations. Part II. (Genera *Bothriospondylus, Cetiosaurus, Omosaurus*). *Palaeontological Society Monographs* 29: 15–93.

Paul, G. 1990. Dinosaur reproduction in the fast lane: implications for size, success, and extension. In K. Carpenter, K. F. Hirsch, and J. R. Horner (eds.), *Dinosaur Eggs and Babies*, 244–255. New York: Cambridge University Press.

Peczkis, J. 1995. Implications of body-mass estimates for dinosaurs. *Journal of Vertebrate Paleontology* 14: 520–533.

Perry, S. F., A. Christian, T. Breuer, N. Pajor, and J. R. Codd. 2009. Implications of an avian-style respiratory system for gigantism in sauropod dinosaurs. *Journal of Experimental Zoology Part A: Ecological Genetics and Physiology* 311A: 600–610.

Peters, R. H. 1983. *The Ecological Implications of Body Size*. Cambridge: Cambridge University Press.

Phillips, J. 1871. *Geology of Oxford and the Valley of the Thames*. Oxford: Clarendon Press.

Powell, J. 1992. Hallazgo de huevos asignables a dinosaurios titanosáuridos (Saurischia, Sauropoda) de la Provincia de Río Negro, Argentina. *Acta Zoologica Lilloana* 41: 381–389.

Prasad, V., C. A. E. Strömberg, H. Alimohammadian, and A. Sahni. 2005. Dinosaur coprolites and the early evolution of grasses and grazers. *Science* 310: 1177–1180.

Raath, M. 1972. Fossil vertebrate studies in Rhodesia: a new dinosaur (Reptilia, Saurischia) from the near the Trias-Jurassic boundary. *Arnoldia* 30: 1–37.

Rauhut, O. W. M., R. Fechner, K. Remes, and K. Reis. 2011. How to get big in the Mesozoic: the evolution of the sauropodomorph body plan. In N. Klein, K. Remes, C. T. Gee, and P. M. Sander (eds.), *Biology of the Sauropod Dinosaurs: Understanding the Life of Giants*, 119–149. Life of the Past (series ed. J. O. Farlow). Bloomington: Indiana University Press.

Rauhut, O. W. M., K. Remes, R. Fechner, G. Cladera, and P. Puerta. 2005.

Discovery of a short-necked sauropod dinosaur from the Late Jurassic period of Patagonia. *Nature* 435: 670–672.

Reid, R. E. H. 1981. Lamellar-zonal bone with zones and annuli in the pelvis of a sauropod dinosaur. *Nature* 292: 49–51.

——. 1984. The histology of dinosaurian bone, and its possible bearing on dinosaurian physiology. In M. W. J. Ferguson (ed.), *The Structure, Development and Evolution of Reptiles*, 629–663. Zoological Journal of London Symposia 52. London: Academic Press.

——. 1990. Zonal "growth rings" in dinosaurs. *Modern Geology* 15: 19–48.

——. 1996. Bone Histology of the Cleveland-Lloyd dinosaurs and of dinosaurs in general. Part I. Introduction to bone tissues. *Brigham Young University Geology Studies* 41: 25–71.

Ricqlès, A. de. 1968a. Quelques observations paléohistologiques sur les dinosaurien sauropode *Bothriospondylus*. *Annales of the University of Madagascar* 6: 157–209.

——. 1968b. Recherches paléohistologiques sur les os longs des tétrapodes. I–Origine du tissu osseux plexiforme des dinosauriens sauropodes. *Annales de Paléontologie* 54: 133–145.

——. 1983. Cyclical growth in the long limb bones of a sauropod dinosaur. *Acta Paleontologica Polonica* 28: 225–232.

Rimblot-Baly, F., A de Ricqlès, and l. Zylberberg. 1995. Analyse paléohistologique d'une série de croissance partielle chez *Lapparentosaurus madagascariensis* (Jurassique Moyen): Essai sur la dynamique de croissance d'un dinosaure sauropode. *Annales de Paléontologie* 81: 49–86.

Romer, A. S. 1956. *Osteology of the Reptiles.* Chicago: University of Chicago Press.

——. 1966. *Vertebrate Paleontology.* Chicago: University of Chicago Press.

Russell, D. A. 1993. The role of central Asia in dinosaurian biogeography. *Canadian Journal of Earth Science* 30: 2002–2012.

Russell, D. A., and Zheng, Z. 1994. A large mamenchisaurid from the Junggar Basin, Xinjiang, People's Republic of China. *Canadian Journal of Earth Sciences* 30: 2082–2095.

Sahni, A. S. K., Tandon, A. Jolly, S. Bajpai, A. Sood, and S. Srinivasan. 1990. Upper Cretaceous dinosaur eggs and nesting sites from the Deccan volcano-sedimentary province of peninsular India. In K. Carpenter, K. F. Hirsch, and J. R. Horner (eds.), *Dinosaur Eggs and Babies*, 204–226. New York: Cambridge University Press.

Salgado, L. 2003. Considerations on the bony plates assigned to titanosaurs (Dinosauria, Sauropoda). *Ameghiniana* 40: 441–456.

Salgado, L., and J. F. Bonaparte. 2007. Sauropodomorpha. In Z. Gasparini, L. Salgado, and R. A. Coria (eds.), *Patagonian Mesozoic Reptiles*, 188–228. Bloomington: Indiana University Press.

Salgado, L., and J. O. Calvo. 1997. Evolution of titanosaurid sauropods II: The cranial evidence. *Ameghiniana* 34: 33–48.

Salgado, L., R. A. Coria, and J. O. Calvo. 1997. Evolution of titanosaurid sauropods I: phylogenetic analysis based on the postcranial evidence. *Ameghiniana* 34: 3–32.

Sander, P. M. 2000. Longbone histology of the Tendaguru sauropods: implications for growth and biology. *Paleobiology* 26: 466–488.

——. 2003. Long and girdle bone histology in sauropod dinosaurs: methods of study and implications for growth, life history, taxonomy, and evolution. *Journal of Vertebrate Paleontology* 23: 93A.

Sander, P. M., and M. Clauss. 2008. Sauropod gigantism. *Science* 322: 200–201.

Sander, P. M., N. Klein, K. Stein, and O. Wings. 2011. Sauropod bone histology and its implications for sauropod biology. In N. Klein, K. Remes, C. T. Gee, and P. M. Sander (eds.), *Biology of the Sauropod Dinosaurs: Understanding the Life of Giants*, 276–302. Life of the Past (series ed. J. O. Farlow). Bloomington: Indiana University Press.

Sander, P. M., O. Mateus, T. Laven, and N. Knötsche. 2006. Bone histology indicates insular dwarfism in a new Late Jurassic sauropod dinosaur. *Nature* 441: 739–741.

Sander, P. M., C. Peitz, F. D. Jackson, and L. M. Chiappe. 2008. Upper Cretaceous titanosaur nesting sites and their implications for sauropod dinosaur reproductive biology. *Palaeontographica* A 284: 69–107.

Sander, P. M., and C. Tückmantel. 2003. Bone lamina thickness, bone apposition rates, and age estimates in sauropod humeri and femora. *Paläontologische Zeitschrift* 77: 139–150.

Sanz, J. L., Moratalla, J. J., Díaz-Molina, M., López-Martínez, N., Kälin, O., and M. Vianey-Liaud. 1995. Dinosaur nests at the sea shore. *Nature* 376: 731–732.

Sanz, J. L., J. E. Powell, J. Le Loueff, R. Martinez, and X. P. Suberbiola. 1999. Sauropod remains from the upper Cretaceous of Laño (northcentral Spain). Titanosaur phylogenetic relationships. In H. Astiba, J. C. Corral, X. Murelaga, X. Oue-Extebarria, and X. Pereda-Suberbiola (eds.), *Geology and Palaeontology of the Upper Cretaceous Vertebrate-Bearing Beds of the Laño Quarry (Basque-Cantabrrian Region, Iberian Peninsula). Estudios del Museo de Ciencias Naturales de Alava* 14 (num. espec. 1): 235–255.

Schmidt-Nielsen, K. 1984. *Scaling: Why Is Animal Size So Important?* New York: Cambridge University Press.

Seebacher, F. 2001. A new method to calculate allometric length-mass relationships of dinosaurs. *Journal of Vertebrate Paleontology* 21: 51–60.

Sereno, P. C., A. L. Beck, D. B. Dutheil, H. C. E. Larsson, G. H. Lyon, B. Moussa, R. W. Sadleir, C. A. Sidor, D. J. Varricchio, G. P. Wilson, and J. A. Wilson. 1999. Cretaceous sauropods from the Sahara and the uneven rate of skeletal evolution among dinosaurs. *Science* 286: 1342–1347.

Sereno, P. C., and J. A. Wilson. 2005. Structure and evolution of a sauropod tooth battery. In K. A. Curry Rogers and J. A. Wilson (eds.), *The Sauropods: Evolution and Paleobiology*, 157–177. Berkeley: University of California Press.

Sereno, P. C., J. A. Wilson, L. M. Witmer, J. A. Whitlock, A. Maga, O. Ide, and T. A. Rowe. 2007. Structural Extremes in a Cretaceous Dinosaur. *PLoS ONE* 2 (11): e1230.

Smith, J. B., M. C. Lamanna, K. J. Lacovara, P. Dodson, J. R. Smith, J. C. Poole, R. Giengengack, and Y. Attia. 2001. A giant sauropod dinosaur from an Upper Cretaceous mangrove deposit in Egypt. *Science* 292: 1704–1706.

Spotila, J. R., M. P. O'Connor, P. Dodson, and F. V. Paladino. 1991. Hot and cold running dinosaurs: body size, metabolism, and migration. *Modern Geology* 16: 203–227.

Steel, R. 1970. Saurischia. *Handbuch der Paläoherpetologie* 13: 1–88.

Stein, K., Z. Csiki, K. Curry Rogers, D. B. Weishampel, R. Redelstorff, J. L. Carballido, and P. M. Sander. 2010. Small body size and extreme cortical bone remodeling indicate phyletic dwarfism in *Magyarosaurus dacus* (Sauropoda: Titanosauria). *Proceedings of the National Academy of Sciences USA* 107: 9258–9263.

Stevens, K. A., and J. M. Parrish. 1999. Neck posture and feeding habits of two Jurassic sauropod dinosaurs. *Science* 284: 798–800.

——. 2005. Digital reconstructions of sauropod dinosaurs and implications for feeding. In K. A. Curry Rogers and J. A. Wilson (eds.), *The Sauropods: Evolution and Paleobiology*, 178–200. Berkeley: University of California Press.

Stokes, W. L. 1944. Jurassic dinosaurs from Emery County, Utah. *Proceedings of the Utah Academy of Science, Arts, and Letters* 21: 11.

Suteethorn, V., E. Buffetaut, V. Martin, Y. Chaimanee, H. Tong, and S. Triamwichanon. 1995. Thai dinosaurs: an updated review. In A. Sun and Y. Wang, Y. (eds.), *Sixth Symposium on Mesozoic Terrestrial Ecosystems and Biota, Short Papers,* 133–136. Beijing: China Ocean Press..

Thulborn, R. A. 1982. Speeds and gaits of dinosaurs. *Palaeogeography, Palaeoclimatology, Palaeoecology* 38: 227–256.

———. 1990. *Dinosaur Tracks.* New York: Cambridge University Press.

Tidwell, V. K., K. Carpenter, and W. Brooks. 1999. New sauropod from the Lower Cretaceous of Utah, USA. *Oryctos* 2: 21–37.

Tidwell, V. K., K. Carpenter, and S. Meyer. 2001. New titanosauriform (Sauropoda) from the Poison Strip Member of the Cedar Mountain Formation (Lower Cretaceous), Utah. In D. Tanke, and K. Carpenter (eds.). *Mesozoic Vertebrate Life,* 139–165. Bloomington: Indiana University Press.

Tütken, T. 2011. The diet of sauropod dinosaurs: implications of carbon isotope analysis on teeth, bones, and plants. In N. Klein, K. Remes, C. T. Gee, and P. M. Sander (eds.), *Biology of the Sauropod Dinosaurs: Understanding the Life of Giants,* 57–79. Life of the Past (series ed. J. O. Farlow). Bloomington: Indiana University Press.

Upchurch, P. 1995. The evolutionary history of sauropod dinosaurs. *Philosophical Transactions of the Royal Society of London* B 349: 365–390.

———. 1998. The phylogenetic relationships of sauropod dinosaurs. *Zoological Journal of the Linnean Society* 124: 43–103.

———. 1999. The phylogenetic relationships of the Nemegtosauridae (Saurischia, Sauropoda). *Journal of Vertebrate Paleontology* 19: 106–125.

Upchurch, P., and P. M. Barrett. 2000. The evolution of sauropod feeding mechanisms. In H. D. Sues (ed.), *Evolution of Herbivory in Terrestrial Vertebrates,* 79–122. Cambridge: Cambridge University Press.

Upchurch, P., P. M. Barrett, and P. Dodson. 2004. Sauropoda. In D. B. Weishampel, P. Dodson, and H. Osmólska (eds.), *The Dinosauria,* 2nd ed., 259–324. Berkeley: University of California Press.

Upchurch, P., C. A. Hunn, and D. B. Norman. 2002. An analysis of dinosaurian biogeography: evidence for the existence of vicariance and dispersal patterns caused by geological events. *Proceedings of the Royal Society of London* B 269: 613–621.

Vianey-Liaud, M., Mallan, P., Buscail, O., and C. Montgelard. 1990. Review of French dinosaur eggshells: morphology, structure, mineral, and organic composition. In K. Carpenter, K. F. Hirsch, and J. R. Horner (eds.), *Dinosaur Eggs and Babies,* 151–183. New York: Cambridge University Press.

Wedel, M. J., R. L. Cifelli, and R. K. Sanders. 2000. Osteology, paleobiology, and relationships of the sauropod dinosaur *Sauroposeidon. Acta Palaeontologica Polonica* 45: 343–388.

Weishampel, D. B., P. M. Barrett, R. A. Coria, J. Le Loeuff, X. Xing, A. Sahni, E. M. Gomani, and C. R. Noto. 2004. Dinosaurian distributions. In D. B. Weishampel, P. Dodson, and H. Osmólska (eds.), *The Dinosauria,* 2nd ed., 517–606. Berkeley: University of California Press.

Whitlock, J. A. 2011. Inferences of diplodocoid (Sauropoda: Dinosauria) feeding behavior from snout shape and microwear analyses. *PLoS ONE* 6: e18304.

Wichmann, R. 1916. The beds with dinosaurs: on the southern coast of the rio Negro, in from of General Roca. *Physis* 11: 258–262.

Williams, D. L. G., Seymour, R. S., and P. Kérourio. 1984. Structure of fossil dinosaur eggshell from the Aix Basin, France. *Palaeogeography, Palaeoclimatology, Palaeoecology* 45: 23–37.

Wilson, J. A. 1999. A nomenclature for vertebral laminae in sauropods and other saurischian dinosaurs. *Journal of Vertebrate Paleontology* 19: 639–653.

———. 2002. Sauropod dinosaur phylogeny: critique and cladistic analysis. *Zoological Journal of the Linnean Society* 136: 217–276.

———. 2005a. Redescription of the Mongolian sauropod *Nemegtosaurus mongoliensis* Nowinski (Dinosauria: Saurischia) and comments on Late Cretaceous sauropod diversity. *Journal of Systematic Palaeontology* 3: 283–318.

———. 2005b. Overview of sauropod phylogeny and evolution. In K. A. Curry Rogers and J. A. Wilson (eds.), *The Sauropods: Evolution and Paleobiology,* 15–49. Berkeley: University of California Press.

———. 2005c. Integrating ichnofossil and body fossil records to estimate locomotor posture and spatiotemporal distribution of early sauropod dinosaurs: a stratocladistic approach. *Paleobiology* 31: 400–423.

Wilson, J. A., and M. T. Carrano. 1999. Titanosaurs and the origin of "wide gauge" trackways: a biomechanical and systematic perspective on sauropod locomotion. *Paleobiology* 25: 252–267.

Wilson, J. A., M. D. D'Emic, T. Ikejiri, E. M. Moacdieh, and J. A. Whitlock. 2011. A nomenclature for vertebral fossae in sauropods and other saurischian dinosaurs: *PLoS ONE* 6: e17114.

Wilson, J. A., D. M. Mohabey, S. E. Peters, and J. J. Head. 2010. Predation upon hatchling dinosaurs by a new snake from the Late Cretaceous of India: *PLoS Biology* 8: e1000322.

Wilson, J. A., and P. C. Sereno. 1998. Early evolution and higher-level phylogeny of sauropod dinosaurs. *Society of Vertebrate Paleontology Memoir* 5: 1–68 (suppl. *to Journal of Vertebrate Paleontology* 18).

Wilson, J. A., and P. Upchurch. 2009. Redescription and reassessment of the phylogenetic affinities of *Euhelopus zdanskyi* (Dinosauria: Sauropoda) from the Early Cretaceous of China. *Journal of Systematic Palaeontology* 7: 1–41.

Wiman, C. 1929. Die Kriede-dinosaurier aus Shantung. *Palaeontologica Sinica* (ser. C) 6: 1–67.

Wing, S. L., and B. H. Tiffney. 1987. The reciprocal interaction of angiosperm evolution and tetrapod herbivory. *Review of Paleobotany and Palynology* 50: 179–210.

Wings, O. 2003. Observations on the release of gastroliths from ostrich chick carcasses in terrestrial and aquatic environments. *Journal of Taphonomy* 1: 97–103.

———. 2004. Identification, distribution and function of gastroliths in dinosaurs and extant birds with emphasis on ostriches (*Struthio camelus*). Ph. D. dissertation, Rheinischen Friedrich-Wilhelms-Universität Bonn.

Wings, O., and P. M. Sander. 2007. No gastric mill in sauropod dinosaurs: new evidence from analysis of gastrolith mass and function in ostriches. *Proceedings of the Royal Society of London* B 274: 635–640.

Wright, J. L. 2005. Steps in understanding sauropod biology: the importance of sauropod tracks. In K. A. Curry Rogers and J. A. Wilson (eds.), *The Sauropods: Evolution and Paleobiology,* 252–284. Berkeley: University of California Press.

Yadagiri, P. 2001. The osteology of *Kotasaurus yamanpalliensis,* a sauropod from the Early Jurassic Kota Formation of India. *Journal of Vertebrate Paleontology* 21: 242–252.

———. 2004. *Anchisaurus polyzelus* (Hitchcock): the smallest known sauropod dinosaur and the evolution of gigantism among sauropodomorph dinosaurs. *Postilla* 230: 1–58.

Yates, A. M., and J. W. Kitching. 2003. The earliest known sauropod dinosaur and the first steps towards sauropod evolution. *Proceedings of the Royal Society of London B* 270: 1753–1758.

Yates, A. M., and C. C. Vasconcelos. 2005. Furcula-like clavicles in the prosauropod dinosaur Massospondylus. *Journal of Vertebrate Paleontology* 25: 466–468.

Yates, A. M., M. F. Bonnan, and J. Neveling. 2011. A new basal sauropodomorph dinosaur from the Early Jurassic of South Africa. *Journal of Vertebrate Paleontology* 31: 610–625.

Ye, Y., Ouyang, H., and Fu Q.-M. 2001. [New material of *Mamenchisaurus hochuanensis* from Zigong, Sichuan]. *Vertebrae PalAsiatica* 39: 266–271. [In Chinese.]

Young, C. C. 1935. Dinosaurian remains from Mengyin, Shantung. *Bulletin of the Geological Survey of China* 14: 519–533.

——. 1939. On the new Sauropoda, with notes on other fragmentary reptiles from Szechuan. *Bulletin of the Geological Survey of China* 19: 279–315.

——. 1941. A complete osteology of *Lufengosaurus huenei* Young (gen. et sp. nov.). *Palaeontologia Sinica* © 7: 1–53.

——. 1947. On *Lufengosaurus magnus* (sp. nov.) and additional finds of *Lufengosaurus huenei* Young. *Palaeontologia Sinica* © 12: 1–53.

——. 1954. On a new sauropod from Yiping, Szechuan, China. *Acta Palaeontologica Sinica* 2: 355–369.

——. 1958. New sauropods from China. *Vertebrate PalAsiatica* 2: 1–28.

Yu, C. 1993. The skull of *Diplodocus* and the phylogeny of Diplodocidae. Ph.D. Dissertation, University of Chicago.

Zhang, Y. 1988. *[The Middle Jurassic dinosaur fauna from Dashanpu, Zigong, Sichuan. Sauropod dinosaurs. Vol. 3, Shunosaurus].* Chengdu: Sichuan Science and Technology Publishing House. [In Chinese with English summary.]

Zhao, Z. 1979. Advances in the study of fossil dinosaur eggs in our country. In *Mesozoic and Cenozoic Red Beds of South China; Selected Papers from the Field Conference on the South China Cretaceous-Early Tertiary Red Beds*, 330–340. Beijing: Science Press.

Zhou, Z., P. M. Barrett, and J. Hilton. 2003. An exceptionally preserved Lower Cretaceous ecosystem. *Nature* 421: 807–814.

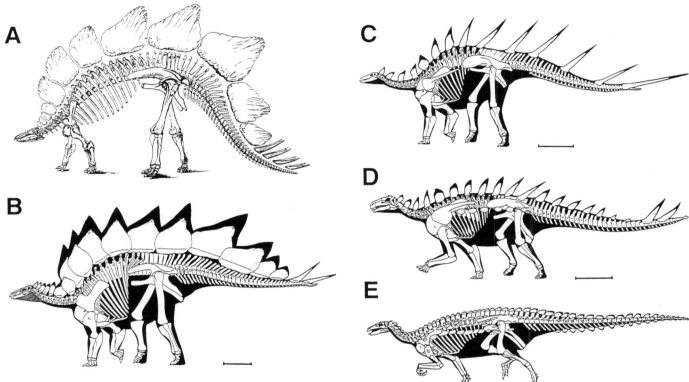

23.1. Skeletal reconstructions of stegosaurs (A–D) and *Scelidosaurus* (E). (A–B) Stegosaurid *Stegosaurus* from the Upper Jurassic of the western United States: (A) First published reconstruction of a stegosaur, by Othniel Charles Marsh in 1891, composite based on specimens of *Stegosaurus ungulatus* (YPM 1853, 1854, 1858) for vertebrae, girdles, limbs, and four pairs of tail spines plus *S. stenops* (USNM 4934) for skull and dermal armor (of the seventeen plates, four anterior plates were paired but not indicated as such on the reconstruction [Marsh 1896], and plate 17 was omitted to make room for the two extra pairs of tail spines in *S. ungulatus* [Czerkas 1987]). (B) *Stegosaurus stenops,* mostly based on holotype USNM 4934, plus USNM 4714 for last half of tail and a tail spine. (C) Stegosaurid *Kentrosaurus* from Upper Jurassic of Tanzania, East Africa, based on mounted skeleton in MFNB (see Janensch 1925). (D) Huayangosaurid *Huayangosaurus* from Middle Jurassic of the People's Republic of China, based on ZDM T7001 (see Zhou 1984) with parasacral spine after Sereno and Dong 1992. (E) The basal thyreophoran *Scelidosaurus* from Lower Jurassic of England, from Paul 1987. Abbreviations for museums: AMNH = American Museum of Natural History, New York, N.Y.; CM = Carnegie Museum of Natural History, Pittsburgh, Penn.; MFNB: Museum für Naturkunde - Leibniz Institute for Research on Evolution and Biodiversity at the Humboldt University Berlin, Germany; USNM = United States National Museum, Washington, D.C; YPM = Peabody Museum of Natural History, Yale University, New Haven, Conn.; ZDM = Zigong Dinosaur Museum, Sichuan, People's Republic. Scale lines represent 50 cm.

Figures B–E kindly supplied by Gregory S. Paul, who retains the copyright.

Stegosaurs

Peter M. Galton

23

The Stegosauria, or plated dinosaurs, are medium-sized to large (up to 9 m in total body length), quadrupedal ornithischians, the most diagnostic character of which is the two rows of upright dermal plates and/or spines, one on each side of the midline, extending from the neck region to the end of the tail (Fig. 23.1A–D). Their remains are known definitely from the Middle Jurassic to the Early Cretaceous and on all continents except Antarctica (Olshevsky and Ford 1993; Galton and Upchurch 2004; Maidment et al. 2008; Novas 2009; Maidment 2010; Pereda Suberbiola et al. in press).

Othniel Charles Marsh of Yale College erected the Stegosauria in 1877 for a new order of large extinct reptiles from the Upper Jurassic of Morrison near Denver, Colorado, USA. He considered that the back of what he then thought was a new aquatic reptile, *Stegosaurus armatus* (Greek *stege*, "roof," *saurus*, "reptile"; Latin *armatus*, "armed"), was covered by large osteoderms, or dermal plates, one of which measured more than a meter in length. These large sheets of bone were thought to be completely embedded in the skin and comparable to those of *Protostega*, a large aquatic turtle from the Late Cretaceous of Kansas (Cope 1871). Marsh (1877, 513) noted that the bones are "embedded in so hard a matrix that considerable time and labor will be required for a full description." Fortunately for O. C. Marsh, his workers in the field located quarry sites that yielded better material at Garden Park near Cañon City, Colorado, and Como Bluff in southeastern Wyoming (Ostrom and McIntosh 1999), in much softer rocks of what would become known as the Morrison Formation. Unfortunately for science, most of the original material from Morrison at the Yale Peabody Museum of Natural History is still in rock, and only a small part of it has been prepared (see Carpenter and Galton 2001; Galton 2010).

Marsh (1880) illustrated the limbs and plates plus dermal spines of *Stegosaurus ungulatus* and reduced the Stegosauria to a suborder within the Dinosauria of Richard Owen. He recognized that the tail plates were held erect, as were even the largest asymmetrical ones over the back that were arranged in a row, or perhaps rows, on either side of the midline. Marsh noted that the dermal spine might be associated with the forefoot. In this he followed Owen (1875), who, in his description of the reasonably complete, but headless, skeleton of *Omosaurus armatus* from the Upper Jurassic of England, described a dermal spine found near the radius and ulna (forearm bones) that was compared with the wrist spine of the ornithopod dinosaur *Iguanodon* (Owen 1872). Marsh (1880) included *Omosaurus* in his new family, the Stegosauridae. The discovery of an almost complete skeleton

Othniel Charles Marsh and *Stegosaurus*

of *Stegosaurus* in 1887, with the plates and spines preserved in articulation, showed that all plates were held vertically and that the spines belonged to the tail (Marsh 1887). The first skeletal reconstruction of *Stegosaurus* showed the plates in a single row along the midline plus four pairs of tail spines (Fig. 23.1A; Marsh, 1891, 1896—see below and Czerkas 1987 for more details on restorations of plates subsequent to Marsh).

Changing Concepts of the Armored Dinosaurs

The first subdivision of the Dinosauria of Owen (1842) was made by Thomas Henry Huxley (1870), who referred all forms with dermal armor to the Scelidosauridae. This included *Scelidosaurus*, a reasonably complete skeleton from the Lower Jurassic of England (Fig. 23.1E; Owen 1861, 1863). When Harry Govier Seeley (1887) split the Dinosauria into the Saurischia and Ornithischia, the Scelidosauridae went to the latter. Marsh (1889, 1896) referred all the quadrupedal ornithischians with dermal armor (along with forms with similar tooth morphology) to the Stegosauria. On the basis of differences in the pelvic girdle, Alfred Sherwood Romer (1927) first made a case for the separation of the suborder Ankylosauria ("fused or joined-together reptiles"), erected without definition by Henry Fairfield Osborn (1923), from the suborder Stegosauria. Romer limited Stegosauria to Stegosauridae plus Scelidosauridae (for *Scelidosaurus* plus a few poorly known forms). This restricted use of the Stegosauria for just the plated dinosaurs was followed by most paleontologists in the United States, Canada, and England. However, other workers followed the original usage of Marsh (1889, 1896) with a more inclusive Stegosauria (e.g., Hennig [1925] and Lapparent and Lavocat [1955] as Stegosauroidea; Nopcsa [1915, 1928] as Thyreophora, the "shield bearers," in a version that included the Ceratopsia, or horned dinosaurs). The suborder Ankylosauria was not widely accepted until Romer (1956) provided a lengthy diagnosis.

As a result of several cladistic analyses in the 1980s, the Thyreophora of Franz Baron Nopcsa von Felsö-Szilvás (1915) was reinstated by David B. Norman (1984) in the sense of Friedrich Freiherr von Huene (1956; i.e., without the Ceratopsia or horned dinosaurs). In the cladistic analysis of Maidment et al. (2008) (Fig. 23.4), Thyreophora includes stegosaurs and ankylosaurs plus their basal relatives from the Lower Jurassic, *Scutellosaurus* from the United States (Colbert 1981), *Emausaurus* from Germany (Haubold 1990), and *Scelidosaurus* (Fig. 23.4). According to Maidment et al. (2008, also Maidment 2010), characters uniting thyreophorans (Figs. 23.1, 23.2) include medially inset maxillary tooth row and adentary tooth row that is sinuous in lateral view, cortical remodeling of some bones of the skull, an ilium with a horizontal enlargement laterally and the presence of dermal armor (Maidment et al. 2008; Maidment 2010).Characters linking Eurypoda (= Stegosauria + Ankylosauria; Sereno 1986) are primitive for Stegosauria. These include the quadrate lacking a lateral sheet (Fig. 23.2A, C, E), no premaxillary teeth (Fig. 23.2A; reversed in *Huayangosaurus*, Fig. 23.2C), maxillary teeth with a prominent, ringlike cingulum, first neural arch fused to intercentrum, prezygapophyses (anterior articular processes) fused ventrally on some dorsal vertebrae (Fig. 23.3C), fused scapula and coracoid (shoulder bones), and unguals (last bone of digit) that are hoof-shaped (semi-circular in dorsal view, dorsoventrally compressed; Fig. 23.3A, B) (Maidment 2010).

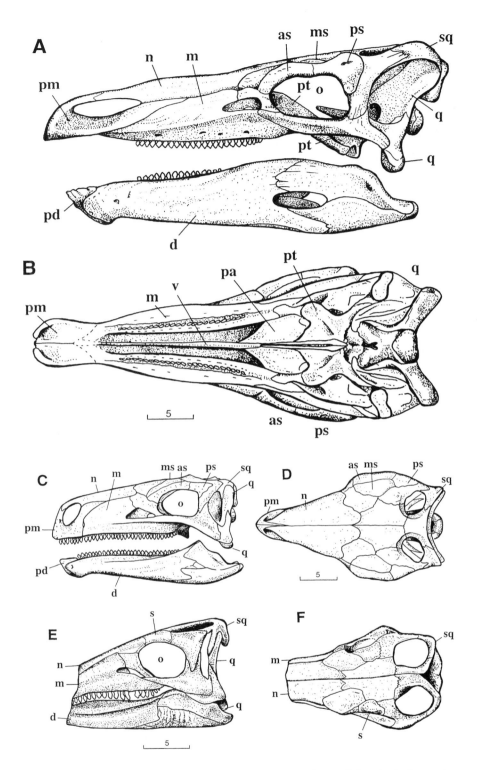

23.2. Skulls of stegosaurs and *Scelidosaurus* in left lateral view (A, C, E), ventral view (B, without lower jaw) and dorsal view (D, F). (A, B) Stegosaurid *Stegosaurus* from the Upper Jurassic of Colorado, after Sereno and Dong 1992. (C, D) Huayangosaurid *Huayangosaurus* from the Middle Jurassic of the People's Republic of China, after Sereno and Dong 1992. (E, F) Basal thyreophoran *Scelidosaurus* from the Lower Jurassic of England, after Coombs et al. 1990. Abbreviations: as = anterior supraorbital (or palpebral); d = dentary; m = maxilla; ms = medial supraorbital (or palpebral); n = nasal; o = orbit; pa = palatine; pd = predentary; pm = premaxilla; ps = posterior supraorbital (or palpebral); pt = pterygoid; q = quadrate; s = supraorbital (palpebral); sq = squamosal; v = vomer. Scale lines represent 5 cm.

The term *armored dinosaur* refers to a member of the Ankylosauria. The Stegosauria, or *plated dinosaurs*, are now restricted to medium-sized to large (up to 9 m in total body length), quadrupedal ornithischians, the most diagnostic character of which is the two rows of parasagittal plates and/or spines extending from the neck region to the end of the tail (Fig. 23.1A–D; Olshevsky and Ford 1993; Galton and Upchurch 2004; Maidment et al. 2008; Maidment 2010). The cladogram (Fig. 23.4) shows the interrelationships within the Stegosauria after two unstable taxa, *Gigantspinosaurus*

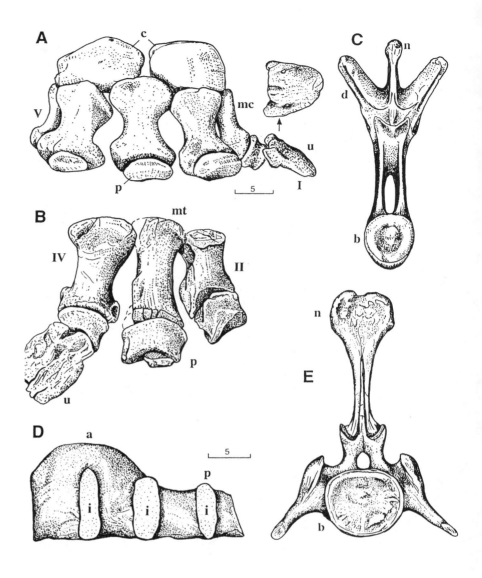

23.3. Anatomy of *Stegosaurus* from the Late Jurassic of western United States. (A) Right fore-foot and proximal carpals (wrist bones) in anterior view plus ungual I in dorsal view (unguals II and III not preserved). (B) Incomplete right hind foot in anterior view (ungual on IV probably pertains to II, other unguals not preserved). (C) Middorsal vertebra in anterior view. (D) Sacral endocast in left lateral view. (E) Anterior caudal vertebra in anterior view. A, B after Gilmore 1914, C–E after Marsh 1896. Abbreviations: a = anterior enlargement; b = body or centrum; c = carpals; d = diapophysis; i = cast of intervertebral foramen; mc = metacarpals; mt = metatarsals; n = neural spine; p = posterior enlargement; u = ungual phalanx; I–V, digits I–V. Scale lines represent 5 cm.

23.4. Cladogram of the Stegosauria and their closest ornithischian relatives. Strict consensus of five most parsimonious trees obtained from a branch-and-bound search. *Jiangjunosaurus* not included. Abbreviations: Huay = Huayangosauridae; Ornit = Ornithischia; *S. armatus* = *Stegosaurus stenops*, *S. homheni* = *Wuerhosaurus homheni*; *S. mjosi* = *Hesperosaurus mjosi*; Ste = Stegosauridae; Steg = Stegosaurinae; Thyreo = Thyreophora; * = position of labeled node; adapted from Maidment et al. 2008.

and *Jiangjunosaurus* (see below), are pruned from the data set (for details see Maidment 2010).

The group may be represented in the Rhaetian (Upper Triassic) of England by two large femoral shafts (original length of bones ~850–1100 mm) that are columnar in anterior and side views, wider transversely than anteroposteriorly, with the outer layer of compact bone not thick as in sauropod dinosaurs (Galton 2005; identity disputed by Maidment et al. 2008). A small manus print with four toes (Hill et al. 1966, pl. 15, fig. 5; Thulborn 1990, fig. 3.39a), which came from the Lower or Middle Jurassic coal measures of Queensland, could be from a stegosaur, a group not represented in Australia by skeletal remains. However, it equally well might have originated from an early ankylosaur (Thulborn 1990, 209), a group represented by skeletons of *Minmi* from the Early Cretaceous of Queensland (see Chapter 24 of this volume), so it is Thyreophora indet. Trackways of *Deltapodus brodricki* from the Middle Jurassic (Aalenian) of England are the earliest definite footprint record of the Stegosauria (see Whyte and Romano 1994, 2001). The base of a dermal spine and a few other pieces of bone from the Middle Jurassic (Bajocian) of England were identified as possibly stegosaurian by Benton and Spencer (1995; Thyreophora indet. Maidment et al. 2008). The earliest definite skeletal record of the group consists of fragmentary remains from the Middle Jurassic (Lower Bathonian) of England (Galton and Powell 1983). However, almost complete skeletons with skulls of *Huayangosaurus* (Figs 23.1D, 23.2C, D) are known from the Bathonian-Callovian of the People's Republic of China (Zhou 1984; Sereno and Dong 1992; Maidment et al. 2006).

Stegosaurs are best represented in the Upper Jurassic, with articulated skeletons (but mostly lacking skulls) known for many of the genera. Stegosaurs from Europe, and England in particular (Galton 1985), were usually discovered in commercial quarries involved in the production of bricks before the widespread use of large machinery. This stegosaur material was originally included in the genus *Omosaurus* Owen (1875) but the name was already occupied, having been applied to a crocodile by Joseph Leidy (1856). Therefore, the specimens from the Kimmeridge Clay were renamed *Dacentrurus* by Frederick August Lucas (1902). The genus is also known from France, Portugal, and Spain (Galton 1991; Maidment et al. 2008; Maidment 2010). *Lexovisaurus* Hoffstetter (1957) was applied to *Omosaurus durobrivensis* Hulke (1887) for the material from the older Lower Oxford Clay (upper Middle Jurassic) that represents a separate genus. However, the holotype of *L. durobrivensis* is a nomen dubium, being based on nondiagnostic material, so *Stegosaurus priscus* Nopcsa (1911) from the same formation was made the type species of *Loricatosaurus* (Maidment et al. 2008), a partial skeleton of which is also known from France (Galton 1990). Octavio Mateus et al. (2009) described a new stegosaur based on a partial skeleton from Portugal, *Miragaia*, a close relative of *Dacentrurus* with 17 cervical vertebrae, more than any other nonavian archosaur and matched only by a few Chinese sauropod dinosaurs (*Mamenchisaurus*, *Omeisaurus*, *Euhelopus*).

The remains of *Kentrosaurus* (Fig. 23.1C) from East Africa were excavated by hand labor under the direction of Werner Janensch and Edwin

Hennig of Berlin from 1909 to 1912 in the torrid climate at Tendaguru, in what is now Tanzania (Hennig 1925; Janensch 1925; Galton 1982, 1988; Heinrich 1999; Mallison 2011a).

Although the fossil record of stegosaurs for most of Asia is very fragmentary (Averianov et al. 2007; Maidment 2010), that of the People's Republic of China is the longest and most diverse (see Dong 1990, 1992; Olshevsky and Ford 1993; Galton and Upchurch 2004; Maidment et al. 2008; Maidment 2010). A partial skeleton of *Chialingosaurus* was described by Chung Chien Young (1959) from the Shangshaximiao Formation of Sichuan and, thanks to the efforts of Zhimin Dong, additional parts of this skeleton plus skeletons of two additional genera, *Chungkingosaurus* and *Tuojiangosaurus*, are also known (Dong et al. 1983; Peng et al. 2005). A recent taxonomic revision recognizes three valid genera: *Chungkingosaurus*, *Tuojiangosaurus*, and *Gigantspinosaurus* (Maidment and Wei 2006). *Gigantspinosaurus* was overlooked for 14 years by Western workers because the original unillustrated description by Hui Ouyang in 1992 was in an obscure publication; this genus is based on a nearly complete skeleton with an enormous parascapular spine that is over twice the length of the scapula (Peng et al. 2005, photos of mounted skeleton figs. 135, 147; Maidment et al. 2008). Chenghai Jia et al. (2007) described *Jiangjunosaurus*, which is based on a partial skull and neck with dermal plates from the Shinhugou Formation of Xiangjiang.

In addition to the well-known *Stegosaurus*, a partial skeleton of which was recently described from the Upper Jurassic of Portugal (Escaso et al. 2007 as S. *armatus*), a fairly complete disarticulated skull and skeleton without limbs from the base of the Morrison Formation of Wyoming was described by Ken Carpenter et al. (2001) as *Hesperosaurus mjosi*. However, Susannah Maidment et al. (2008) recognized only two species of *Stegosaurus* from North America, S. *mjosi* and the more derived S. *armatus* (which included the other Morrison species regarded as valid, i.e., those based on diagnosable material). However, based on differences in the dermal armor of the distal part of the tail (see below), Galton (2010) recognized four Morrison species of *Stegosaurus*, with S. *stenops* as the best represented (Fig. 23.1B; Galton 2011). In addition, based on three additional referred skeletons of *Hesperosaurus* ("Lily," "Moritz," "Victoria," see Siber and Möckli 2009), Carpenter (2010) provided a skeletal reconstruction and a list of diagnostic characters for *Hesperosaurus mjosi* with the contrasting condition for *Stegosaurus stenops*.

Stegosaurs are rare in the Cretaceous (Galton 1981; Galton and Upchurch 2004; Pereda Suberbiola et al. 2003; Maidment 2010), but, surprisingly, the earliest description of a stegosaur is that of *Craterosaurus* from the Lower Cretaceous of southern England. *Craterosaurus* was described by Harry Govier Seeley (1874) as a braincase, presumably dinosaurian, but, as Franz Baron Nopcsa von Felsö-Szilvás (1912) showed, it is actually part of the neural arch of a stegosaurian dorsal vertebra. This vertebra is Stegosauria indet., but numerous recently discovered remains from Spain are identified as *Dacentrurus* sp. or Dacentrurinae (Pereda Suberbiola et al. 2003; Cobos et al. 2010; Company et al. 2010).

Jose Bonaparte (1996) described a few vertebrae, including an almost complete cervical that he figured plus a piece of dermal armor, from the Lower Cretaceous of Argentina, as representing a small stegosaur. It was also identified as a stegosaur by Coria and Cambiaso (2007) and Novas (2009), who provided photos of the vertebra and of a dermal plate, but Maidment et al. (2008) considered the remains to represent an indeterminate ornithischian. However, it is a stegosaur (Pereda Suberbiola et al. in press) as indicated by the presence of a sculptured anterior supraorbital over the eye in the skull, the proportionally large cross section of the neural canal of the neck vertebra, the anterior height and width of which are half the height of the anterior face of the centrum (body) as in some cervicals of *Kentrosaurus* and *Stegosaurus* (Hennig 1925; Ostrom and McIntosh 1999), and the form of the dermal plate, which resembles those of *Miragaia* (Mateus et al. 2009).

Stegosaurs are also present in the Lower Cretaceous of South Africa (*Paranthodon*, front end of skull; Galton and Coombs 1981) and Asia (Dong 1990, 1992; Averianov et al. 2007; Maidment et al. 2008; Maidment 2010). A partial skeleton from the People's Republic of China, that of *Wuerhosaurus homheni* Dong (1973, 1990; also *W. ordosensis* from Mongolia [Dong 1993]), is regarded by Susannah Maidment et al. (2008) as representing a valid species of *Stegosaurus*, so this genus may have persisted in China after becoming extinct in North America. However, Carpenter (2010) has questioned the validity of this referral.

A supposed large five-toed stegosaur manus print from the Lower Cretaceous of Broome, Western Australia, was reidentified as the manus print of a sauropod dinosaur by Page (1998). A small five-toed manus print with incomplete pes prints from Broome (Thulborn 1997; Long 1998, 130, color photo of manus) could be from a stegosaur, a group not represented in Australia by skeletal remains. Thulborn (1997, 8) noted that these prints "do not correspond at all closely with what is known of the hand and foot structure in ankylosaurs." He identified them as thyreophoran (?stegosaur). However, given the close similarities to the manus and pes prints of *Deltapodus brodricki* Whyte and Romano (1994), a Middle Jurassic footprint taxon from England that was subsequently referred to the Stegosauria by Whyte and Romano (2001), these Broome prints probably represent a stegosaur.

Stegosaurs may have survived into the Upper Cretaceous in southern India. A partial skeleton from the Coniacian (mid-Upper Cretaceous) was described by P. Yadagiri and K. Ayyasami (1979), and some of the bones of the skull were reidentified by Galton (1981). Sankar Chatterjee and Dhiraj Rudra (1996, 518) noted that the holotype shows nothing relating to the stegosaurian skull and plates of the original description. They concluded that this taxon is a nomen dubium, being based on weathered limb and girdle elements that may belong to plesiosaurs, the remains of which they located in the holotype quarry. However, this important material needs to be independently reexamined and reassessed because a small tooth (Yadagiri and Ayyasami 1979, fig. 4, pl. 1, figs. 3a, b) is similar to that of *Kentrosaurus*, and the 10 photographed bones are not plesiosaurian. Stegosaur bones were also reported from the latest Cretaceous (Maastrichtian) of India (Yadagiri and Ayyasami 1979, 529). However, to date only a single plate, roughly symmetrical and subtriangular in side view (~17.4 cm wide close to base, ~19 cm high), has been illustrated (Yadagiri and Ayyasami 1978). In addition, Mateus et al. (2011) reidentified a "sauropod" manus print from the latest Maestrichtian of India (Mohabey 1986) as a pes print like those of *Deltapodus*, a Middle and Upper Jurassic footprint taxon from Europe (and

Upper Jurassic, USA) that has been referred to the Stegosauria (Whyte and Romano 2001; Milàn and Chiappe 2009; Cobos et al. 2010).

Skull

Compared to the skull of *Huayangosaurus* (from the Middle Jurassic), that of *Stegosaurus* (and also probably the rest of stegosaurids) is shallower, and above the orbits the skull roof is narrower (relative to skull length; Fig. 23.1A–D). As in other ornithischians, the tips of the anterior bones of the skull, the upper premaxillae and the lower predentary, were covered by rhamphothecae (horny sheaths) that were used to crop vegetation. In *Huayangosaurus* each premaxilla has seven teeth, but this bone lacks teeth in the other stegosaurs for which skulls are known, and the rhamphothecae were proportionally larger (Fig. 23.2A–C). Compared to teeth, a horny sheath is especially suitable as a cropping structure. It always has a continuous edge that can be self-sharpening; it can have a cutting edge and/or a flat crushing area; and it is rapidly replaced as it is worn. In the lower jaw, the first dentary tooth is next to the predentary in *Gigantspinosaurus* and *Huayangosaurus* (Fig. 23.2C), whereas in *Jiangjunosaurus*, *Kentrosaurus*, and *Stegosaurus* (Fig. 23.2A) there is an edentulous gap or diastema (also on the upper jaw in *Miragaia* and *Stegosaurus*, Fig. 23.2A, B), comparable to that present in many herbivorous mammals. The cheek tooth crowns are simple in form, with a prominent encircling cingulum (swelling) close to the root, and show a few small facets produced by tooth-to-food wear.

Ornithischian dinosaurs have been commonly restored with cheeks (1908–1940 and from 1973 on, see below) or without cheeks (1940–1973). As in other ornithischians, *Huayangosaurus* has a space external to the maxillary (upper) and dentary (lower) tooth rows (also preserved in *Gigantspinosaurus*) that is roofed by a prominent horizontal ridge on the maxilla and floored by the massive dentary (Fig. 23.2C). This ridge was one area of attachment (there was a corresponding ridge on dentary) for a structure that, although it was not close against the teeth, performed some of the functions of the cheeks of mammals. This cheeklike structure bordered a vestibule or space that would have received any chewed food that ended up outside the lower teeth. Chewing in ornithischians was occasionally interrupted so that the long, narrow-based, and slender tongue could scoop up residual food from the vestibule and return it to the mouth to be chewed again or swallowed. Our cheeks are very close to our cheek teeth and usually keep food between them. However, sometimes we have to use the tongue to retrieve food (e.g., peanut butter) from the outside of the teeth and gums because it cannot be kept between the teeth by the efforts of the buccinator muscle. Alternatively, it has been suggested that the horny beaks or rhamphothecae at the front of the ornithischian mouth may have been followed either by typically reptilian lips (Czerkas 1999), or by a continuation of the beak along the sides of the jaws (Papp and Witmer 1998; Czerkas 1999).

Cheeks were originally suggested for the ceratopsian dinosaur *Triceratops* in 1903 by Richard Swann Lull, who restored cheeks with a buccinator muscle for this genus in 1908. Barnum Brown and Erich Maren Schlaikjer in 1940, and especially Georg Haas in 1955, argued against cheeks in ornithischians because the buccinator, a facial muscle innervated by a branch of the seventh cranial nerve, is characteristic of mammals and is not present in any living reptile or bird. However, this argument does not preclude ornithischians, such as stegosaurs, from having a cheeklike structure that lacked a buccinator muscle (Galton 1973; but see Papp and Witmer 1998). In addition, Gregory Paul (2010, 25–26 with fig.) points out that the California Condor (*Gymnogyps californianus*) has a large, fleshy, elastic cheeklike structure (not present in the Black Vulture, *Coragyps atratus*) that starts superiorly about halfway along the upper jaw and attaches at about the same level on the mandible (also present in the Andean or South American Condor, photograph in Hoyo et al. 1994, 37). R. S. Lull (1908) and Loris Shano Russell (1935) restored the cheek muscles with vertically oriented fibers, but, if present in ornithischians such as stegosaurs, then they were probably pinnate, as in many mammals, with fibers derived from the M. levator anguli oris passing obliquely backward from the maxilla and dentary to the body of the cheek (Galton 1973; see also Tyson 1977). In *Stegosaurus* (also *Jungjinosaurus*, *Kentrosaurus*, *Tuojiangosaurus*) the cheek was supported inferiorly by a prominent dorsolateral edge on the dentary, separated from the teeth more medially by a shelflike area, that becomes a dorsal lamina more posteriorly so the dorsal margins of the tooth alveoli (sockets) are not visible in lateral view (Fig. 23.2A).

A deep median keel along the length of the palate (formed by vomers, palatines, and pterygoids; Fig. 23.2B) probably provided support for a soft secondary palate that was supported laterally by the maxillae. This palate separated the more dorsal nasal passages from the oral cavity so stegosaurs could continue to breathe while chewing food. The lower jaw articulates with the quadrate, the upper end of which is fused to the squamosal in adult individuals of *Stegosaurus* (Fig. 23.2A). The skull formed a solid box, and, because of the fixed quadrate and the anteroposterior tightness of the jaw joint, the jaw-closing part of the chewing cycle involved an upward movement of the teeth of the lower jaw against those of the upper jaw without the complexities present in the Heterodontosauridae and Ornithopoda (see Chapter 26 of this volume). From a study of 3-D models of the skull, Miriam Reichel (2010) concluded that *Stegosaurus* fed on leaves and smaller branches (less than 12 mm diameter) of the herbaceous short-lived plants that were common in the Morrison Formation (Parrish et al. 2004).

Vertebrae

The dorsal vertebrae of stegosaurs are tall. The increase in height is in the region between the centra and the diapophyses (transverse processes that carry ribs) (Fig. 23.3C), so the height of the body cavity is increased, which would provide more room for viscera and, in addition, for subvertebral muscles that stabilized and flexed the back. In other dinosaurs with tall dorsals, the increased height is due to elongation of the neural spines, which would provide more room for the spinalis muscles that stabilized and extended the back and/or to provide support for a saillike structure. The neural canal is larger in posterior cervical vertebrae to accommodate the brachial enlargement, a part of the spinal cord involved with the brachial nerve plexus that supplied nerves to the muscles of the massive forelimbs (Lull 1917). Some mid- and posterior dorsals appear to have a dilated neural canal in *Stegosaurus* and especially in *Kentrosaurus*. However, the size of the canal is considerably reduced by a thin transverse septum of bone that bridges the upper part of the opening. Consequently, the increase in height of the pedicel region of the dorsals was not to provide room for a larger spinal cord. The diapophyses (transverse processes) are directed upward by as much as 50–60° from the horizontal in middorsal vertebrae (Fig. 23.3C); this angle is about 25–40° in anterior and posterior dorsals. The increased angle would have provided better support for the overlying osteoderms, the parasagittal orientation of which would have concentrated the weight along the region adjacent to the ends of the diapophyses. This region is also supported by the shafts of the ribs, which, in the middorsal region, have a T-shaped cross section with the flat surface on the outside. The rib of the first sacral vertebra either is on the end of its diapophysis or forms an antero-posteriorly thin vertical extra sacral rib in *Chungkingosaurus, Dacentrurus, Kentrosaurus,* and *Stegosaurus* (Galton 1982, 1991; Galton and Upchurch 2004). However, these sacral morphs, which probably represent a sexual dimorphism, occur together only in the same quarry, with three sacra of each type, in *Kentrosaurus aethiopicus,* in Quarry St (Galton 1982). A similar sacral dimorphism, which is correlated with a few other dimorphisms in the pelvic girdle, occurs in the basal ornithopod dinosaur *Hypsilophodon foxii* (Lower Cretaceous, England; Galton 1974, in press). The morphs with the extra sacral rib may represent males (Galton 1974) or females (Galton 1999, 26). A geometric morphometric analysis of the femora of *Kentrosaurus aethiopicus* by Holly Barden and Susannah Maidment (2011) demonstrated a statistically significant shape difference of the proximal end in anterior view that, because it can be identified only in larger femora and is then independent of size, probably represents a sexual dimorphism. The robust morph, which is twice as common (14 versus 7), proximally has a more distinct head and a pronounced and better-developed greater trochanter (posterolateral corner of bone), and distally, the condyles are larger compared to the slender morph. The mounted skeleton of *Kentrosaurus aethiopicus* (Fig. 23.1C) has a sacrum with four sacral ribs (Hennig 1925, fig. 23b). Bones from the same lectotype individual from Quarry St (Heinrich 2011a) include articulated series of dorsal and caudal vertebrae, an ulna, ilia, and femora, neither of which was included in the femoral analysis of Barden and Maidment (2011), and both femora are the robust morph (see right in anterior view, Galton 1982, pl. 6, fig. 1; H. Mallison, MFNB, personal communication, for museum abbreviations see caption to Fig. 23.1). In the theropod dinosaur *Tyrannosaurus rex* there is also a sexual femoral dimorphism (Larson 2008), with the robust morph identified as female based on pelvic dimensions and fused caudal vertebrae. Analyses for medullary bone in this same series of femora showed a significant clustering with the robust forms (Schweitzer et al. 2005). This indicates that the robust morph was probably female because medullary bone is formed by female birds as a calcium reservoir to aid in egg shell formation (Dacke et al. 1993). Consequently, in *Kentrosaurus aethiopicus* the four sacral rib morph was probably female, because it is associated with the robust femoral morph, but analysis for medullary bone in the associated femora is needed to confirm this.

In all stegosaurids except *Tuojiangosaurus,* the diapophyses and the T-shaped sacral ribs fuse together to form an almost solid dorsal plate that extends from the base of the neural spines to the flat dorsal surface of the ilium (upper hip bone), with which it is continuous. This plate strengthened the sacrum, and appears to be associated with the proportionally long femur (Fig. 23.1B–D). This elevated the hip joint well above the shoulder joint so that the femur supported much of the body weight.

The neural canal of the sacrum is extremely dilated to form an endosacral enlargement that is extremely large anteriorly, with a smaller enlargement posteriorly (Fig. 23.3D). O. C. Marsh (1881, 168), who also illustrated the endocranial cast (Marsh 1880, 1881; see Galton 2001), described the sacral enlargement as a "posterior brain case," which explains the origin of the popular misconception of *Stegosaurus* having "two sets of brains, one in his head, the usual place, and the other at his spinal base" (from poem on subject by Bert L. Taylor, 1930s and 1940s columnist for *Chicago Tribune*; for rest see Fastovsky and Weishampel 2005, 116). However, this enlargement did not represent a "sacral brain" (Edinger 1961). Some of this space (anterior enlargement) was probably occupied by an enlarged spinal cord associated with the increased size of the sacral plexus that sent nerves to the large vertically held, weight-supporting hind limbs. In addition, the posterior enlargement probably housed nerves that innervated the caudofemoralis muscle, which functioned to pull the hind limb backward during walking, as well as in lateral movements of the tail with its osteoderms (Wiedersheim 1881; Lull 1910a, b, 1917; for details see Giffin 1991). The rest of the space was probably occupied by a glycogen body, similar to that seen in birds (Krause 1881), the function of which is not clear (Giffin 1991).

The posterior surface of the last centrum of the sacrum is markedly concave transversely, whereas the adjacent surface of the first caudal is markedly convex; both surfaces are vertically straight. The shape of this joint would have restricted vertical movements and facilitated lateral movements of the tail, but the latter movement was restricted by the postacetabular process of the ilium and the overlying dermal plate. The ends of the neural spines of the anterior caudal vertebrae are transversely expanded (Fig. 23.3E), so the ends are broader in anterior view than in lateral view. The centra or bodies of the posterior caudal vertebrae are almost square in shape (Fig. 23.1A–D), rather than elongate as in most other dinosaurs. These modifications of the tail aided in supporting the weight of the osteoderms of this region.

Ossified tendons are associated with the dorsal neural spines in *Hesperosaurus mjosi* (also *Huayangosaurus*, one tendon figured by Maidment et al. 2006), but none with the caudals. Many of the dorsal and sacral neural spines are also rugose because of partial ossification of the interspinous and supraspinous ligaments (Carpenter et al. 2001). Paul Sereno and Zhimin Dong (1992) suggested that the increased height of the pedicels of the dorsal vertebrae may have provided additional attachment surface to maintain trunk rigidity in the absence of ossified tendons.

The lower part of the scapula forms a broad plate. The humerus (upper arm bone) is short but massive, with expanded ends (Fig. 23.1A–D). Consequently, there was plenty of room for the attachment of powerful shoulder and pectoral muscles to support the body weight anteriorly. The forefoot is quite elephant-like, being short and relatively inflexible (Fig. 23.3A). In adults

Girdles and Limbs

there are two large, blocklike proximal carpals (wrist bones) separated by a flat intercarpal articulation; no distal carpals are preserved. Five short and robust metacarpals bear short digits, two of which terminate in a hooflike ungual (Fig. 23.3A). The phalangeal formula is 2-2-1-1-?1 in *Hesperosaurus* (Siber and Möckli 2009, 36, fig.). The formula 2-2-1-1-1, with a reduced crescent-shaped proximal phalanx on digits III, IV and V, occurs in a described but unfigured manus of *Stegosaurus* (AMNH 650; A. Watanabe, Chicago, pers. comm. contra Watanabe 2008). Phil Senter (2010) demonstrated that the usual depiction of stegosaurian metacarpals as slanted and spreading out distally (Fig. 23.3A) is incorrect. Instead they were posed vertically and arranged in a compact semitubular configuration, as in ankylosaurs and sauropod dinosaurs (Senter 2011). Senter (2010) noted that confirmation of this pose is provided by the manus print of the stegosaur *Deltapodus*, in which the imprint of the distal metacarpus formed a semicircle (see Whyte and Romano 2001; Milàn and Chiappe 2009; Cobos at al. 2010).

The ilium has a prominent horizontal supraacetabular flange, which overhangs the acetabulum (hip joint). The pubis (anteroventral hip bone, Fig. 23.1A–D) has a relatively long anterior process (at least 40% of length of posterior process), an oval-shaped, laterally directed acetabular surface (Sereno and Dong 1992), and a robust long posterior process. The femur is slender in side view but broad in anterior view, with a straight shaft of nearly uniform width, and the fourth trochanter (process on posterior surface for caudofemoralis muscles) is reduced or indistinct (Fig. 23.1A–D). In all stegosaurs except *Gigantspinosaurus* and *Huayangosaurus*, the femur is long compared to the humerus (and to the tibia, Fig. 23.1A–D), which, with the stocky proportions of the fore and hind feet (Fig. 23.3A, B), indicates a graviportal (elephantine) mode of locomotion. The pes of *Stegosaurus* (Fig. 23.3B) is relatively symmetrical about digit III, a result of the loss of digit I and the great reduction of metatarsal V, which is very small (Gilmore 1914, fig. 53 shown inverted; cf. Galton 2001, fig. 5.10S). The phalangeal formula is X-2-2-2-0 (CM 41132; Watanabe 2008) and the three robust, weight-bearing central digits (II–IV) each bear a hooflike ungual.

Integument and Dermal Armor

Integument or skin impressions have been reported by Li Xing et al. (2008) for *Gigantspinosaurus* and by Nicholai Christiansen and Emanuel Tschopp (2010) for *Hesperosaurus*. In both, the skin on the sides of the body has a ground pattern of small polygonal scales that in some places surround a larger oval tubercle to form a rosettelike pattern. In *Hesperosaurus* the skin covering a dorsal plate lacks any scalelike texture. Instead it is smooth with long and parallel, shallow grooves that is interpreted as the impression of a keratinous covering. Octavio Mateus et al. (2011) describe a plantar pes impression of *Deltapodus*. The skin texture consists of closely packed subrounded tubercles 4–8 mm in diameter (with 4–5 mm scales laterally) forming an even cover except for the smooth hooflike distal parts of the digits.

Stegosaurus has small (diameter 4–27 mm), hexagonal ossicles arranged in a rosette pattern in the throat region (Fig. 23.1B; Gilmore 1914, pl. 22, figs. 2, 3; Carpenter 1998, fig. 6). In *Huayangosaurus* there are low-keeled, lateral osteoderms, similar to those of *Scelidosaurus* (Fig. 23.1D, E), that are lost in other stegosaurs (Fig. 23.1B, C), plus two rows above the vertebral column, the only ones present in other stegosaurs, which are angled upward and slightly outward. Viewed from the side, the osteoderms of most stegosaurs except *Hesperosaurus* and *Stegosaurus* (see next paragraph) form a series that grade anteriorly from short, erect plates to longer, posterodorsally angled spines (Fig. 23.1C, D). In addition, all of the dorsal and more proximal tail plates and spines are paired, so there are a left and a right representative of each type. In all stegosaurs, the series ends with a proportionally narrow pair of spines that may extend beyond the last tail vertebra (Fig. 23.1A–D).

Articulated skeletons of *Gigantspinosaurus*, *Huayangosaurus* (Fig. 23.1D), and *Tuojiangosaurus* have a spike with a large flat sheetlike base preserved adjacent to the scapula. This spine is large (longer than the humerus) in the French partial skeleton of *Loricatosaurus* (Galton 1990) and enormous (over twice the length of the scapula) in *Gigantspinosaurus* (Peng et al. 2005; Maidment et al. 2008); the parascapular spine is secondarily lost in *Hesperosaurus* and *Stegosaurus* (Fig. 23.1B). In *Kentrosaurus* the Berlin mounted skeleton has a similar spike with a large wide subcircular base that has been restored lateral to the scapula (Fig. 23.1C). However, there is a matching concavity on the ilium against which the base of this spike fits closely, so, for this reason, Heinrich Mallison (2011a, fig. 1, tab. 1) has catalogued this spike as a parasacral, its position as originally restored (Hennig 1925; Janensch 1925).

In the articulated skeleton of the holotype of *Stegosaurus stenops* (USNM 4934, Fig. 23.1A), the dermal armor consists of a series of 17 erect and thin plates of varying sizes, with the dorsal plates being taller vertically than long anteroposteriorly. These plates extend along most of the tail with two terminal pairs of tail spines (Gilmore 1914). The 17 plates have been reconstructed in different patterns (see Czerkas 1987): as a single median row (Fig. 23.1A; Marsh 1891, 1896; Ostrom and McIntosh 1999, pl. 65; Czerkas 1987, with plates 6 to 13 angled slightly outward from the midline so there is a small amount of overlap between adjacent plates), or as two parasagittal rows with the plates arranged either in pairs (Lull, 1910a, b; Paul, 1987) or as a staggered (alternating) series (Fig. 23.1B; Gilmore 1914, 1918; Farlow et al. 1976; Bakker 1986; Paul 1992; Ford 1997; Carpenter 1998, 2010; Galton 2010). No two plates have exactly the same shape or size in USNM 4934, in which three of the 17 plates are preserved in an alternating pattern (Gilmore 1914) (Fig. 23.1B). The same alternating plate arrangement also occurs in another specimen from Garden City, Colorado (Paul 1992; Carpenter and Small 1993; Carpenter 1998, fig. 2; photograph Carpenter 2007, fig. 2) and in two others (Carpenter 1998). However, in the holotype of *S. ungulatus* (YPM 1853) there is at least one pair of larger plates that are the same size and shape (Ostrom and McIntosh 1999, pl. 59, fig.1, pl. 60; Galton 2010, figs. 3n, r). Consequently, some or all of the plates may have been arranged in pairs in this species (Paul 1987; Galton 2010, fig. 2b), in which four pairs of tail spines were described (Fig. 23.1A; Marsh 1891, 1896; Ostrom and McIntosh 1999, pls. 55, 56). Ken Carpenter and Peter M. Galton (2001) suggested that these may represent two sets of spikes from different individuals, but as no duplicate bones were excavated from the type quarry, the question remains unresolved. In this connection it is interesting that Zhiming Dong et al.

(1983, 126, fig. 102) described an articulated distal tail referred to *Chungkingosaurus* with four pairs of tail spikes, the most anterior of which were lost due to weathering.

The middorsal to anterior caudal plates of *Hesperosaurus mjosi* are oval shaped and vertically low, whereas in *Stegosaurus stenops* these plates are tall and triangular (Carpenter 2010, fig. 7). For *Stegosaurus*, Galton (2010, figs. 2–4) considered that there are possibly four separate valid species based on the dermal armor of the distal part of the tail, namely, *S. ungulatus* with three pairs of small flat dermal spines immediately anterior to the tail spikes (see Ostrom and McIntosh 1999, pls. 59, 60; Galton 2010, figs. 2b, 3a-e, k-o, r), *S. stenops* with three larger thin alternating plates in this position (Fig. 23.1B; see Gilmore 1914, fig. 58, pl. 24, figs. 1–4; Galton 2010, fig. 4), *S. sulcatus* with a very large-based spike (see Gilmore 1914, fig. 65, pl. 18; Ostrom and McIntosh 1999, pl. 58) that was possibly placed on the shoulder or more distally near the sacrum, and *S. longispinus* with two pairs of extremely elongate tail spikes with subequal bases (see Gilmore 1914, fig. 66).

Taphonomy and Paleoecology

The European record for the Jurassic provides no information on the habitat of stegosaurs because it consists of single isolated carcasses that drifted downstream and disintegrated to a varying degree before being deposited in marine sediments. In the Lower Cretaceous of Teruel, Spain, stegosaurian bones and trackways coexisted along with ornithopods and sauropods in the same formation, which represents wetlands of a restricted tidal environment (Cobos et al. 2010).

The Tendaguru fauna of Tanzania occurs in nearshore deposits, which came from land with a diverse conifer flora and a subtropical to tropical climate characterized by seasonal rainfall alternating with a pronounced dry season (Aberhan et al. 2002). A relatively minor element of the fauna was *Kentrosaurus*, of which the medium-sized individuals were favored during preservation (Russell et al. 1980). In the main Tendaguru stegosaur quarry there was an enormous concentration of partly articulated and partly sorted remains (Hennig 1925), whereas in another quarry, in which six individuals represent more than 95% of the dinosaur finds, the bones were completely disarticulated, with a predominance of hand and foot bones plus a few limb bones and vertebrae including sacra (Heinrich 1999).

A detailed study of the taphonomy of Morrison dinosaurs was performed by Peter Dodson et al. (1980), who showed that the common genera are broadly distributed in a variety of different sediments. *Stegosaurus* occurs more frequently in channel sands that represent a concentration of bones from animals that probably spent much of their lives in the floodplain areas. However, *Stegosaurus* may have inhabited areas farther from sources of water, so it was probably somewhat separated ecologically from sauropods. Morrison dinosaur carcasses typically decomposed in open, dry areas, or spent a considerable time in channels prior to deposition. Consequently, *Stegosaurus* occurs only occasionally as an articulated skeleton; usually it is part of an accumulation of 20 to 60 skeletons of other dinosaurs with only moderate to low degrees of articulation. At the Carnegie Quarry in Dinosaur National Monument, Utah, the time of formation of the quarry assemblage ranged from a few months to only a few years, thereby recording ecological time rather than evolutionary time (Fiorillo 1994). Almost complete specimens of *Stegosaurus* from quarries at Garden Park, Colorado, occur in assemblages that resulted from attritional and noncatastrophic mass mortality resulting from drought (Carpenter 1998; Evanoff and Carpenter 1998). Quarry 13 at Como Bluff, Wyoming, is unique because it consists of a concentration of *Stegosaurus* and *Camptosaurus* (an ornithopod dinosaur) remains with only a few remains of the sauropod dinosaur *Camarasaurus* (Ostrom and McIntosh 1999).

Several skeletons of stegosaurs from Sichuan, China, include skulls, and, because the degree of articulation of the skeletons is greater than those from the Morrison Formation, the carcasses were buried more rapidly in Sichuan. *Huayangosaurus* occurs in sandstones of the Xiashaximiao Formation that were deposited in a lakeshore, shallow-bank environment under low-energy conditions (Xia et al. 1984).

Stegosaurs were graviportal, or elephant-like, in their locomotion (Coombs 1978), although T. Karbek (2002) has argued that *Stegosaurus* was an agile, cursorial biped, a mode of locomotion that the large subvertical plates would seem to render very unlikely. An occasional more upright bipedal pose was probably possible, to graze on high shrubs and the low canopy of trees up to ~3.6 m. It was suggested that the body was supported by a tripod formed by the hind limbs and tail (Alexander 1985; Bakker 1986). However, based on a computer-aided design analysis of *Kentrosaurus aethiopicus*, Heinrich Mallison (2010) concluded that bipedality was more likely than using a tripod for support. In *Miragaia* the vertical feeding range was increased by lengthening the neck, produced by the probable lengthening of the centra and increasing of the number of cervicals to 17 by "cervicalization" of the anterior dorsal vertebrae (Mateus et al. 2009, ? number for dorsals). Lengthening of the neck also occurred in *Hesperosaurus*, with 13 cervicals and 13 dorsals (Carpenter 2010), and probably also in *Wuerhosaurus ordosensis* with 11 dorsals (Dong 1993, ? number for cervicals). The corresponding counts are 9 and 16 for *Huayangosaurus* (Zhou 1984) and 10 and 17 for *Stegosaurus stenops* (Carpenter 2010).

Ragna Redelstorff and Martin Sander (2009) concluded that *Stegosaurus* grew more slowly than non-thyreophoran dinosaurs, a result of the common occurrence of parallel-fibered bone and poor vascularization in the long bones of this genus compared to the highly vascularized fibrolamellar bone seen in the other dinosaurian groups. Based on histological studies, Shoji Hayashi et al. (2009) showed for *Stegosaurus* that there was a differential growth pattern, with the osteoderms maintaining a faster growth rate than the body elements and that this continued after maturity of the body.

Grant Hulburt (Toronto, pers. comm.) calculated an endocast volume of 64.2 ml for *Stegosaurus armatus* (CM 106) by Graphic Double Integration (Jerison 1973; Hurlburt 1999) using the illustrations in Galton (2001). This contrasts with the lower estimate of 45 ml based on the endocast used by James A. Hopson (1980) based on a braincase (USNM 4934) that is distorted by horizontal shear (see Galton 2001). Assuming that the brain occupies 37% of the endocast volume, as in adult alligators (Hurlburt and Waldorf 2002), CM 106 had an REQ (Reptile Encephalization Quotient, Hurlburt 1996) of

Paleobiology and Behavior

0.39 or 0.48, which correspond to body mass estimates of 3100 kg and 2000 kg respectively (Alexander 1985; Colbert 1962; Hurlburt pers. comm.). This overlaps the R E Q range (0.40–2.40) of extant (nonavian) reptiles (Hurlburt 1996). Among the dinosaurs analyzed by Hurlburt (1996), only the sauropod *Diplodocus* has a lower R E Q. The smaller stegosaur *Kentrosaurus* has an R E Q of 0.63, exceeding that of the horned dinosaur *Triceratops*, and the sauropods *Brachiosaurus* and *Diplodocus*. The ankylosaur *Euoplocephalus* has an R E Q of 0.66 (Hurlburt pers. comm.). The low value of the R E Q for stegosaurs and ankylosaurs was attributed by Hopson (1980) to their reliance on defensive armor and tail weapons, rather than speedy flight, to cope with predators.

The overall pattern of the plates and spines (Fig. 23.1B–D) is characteristic for each species, so it was probably important for the recognition of other individuals of the same species and for sexual displays (Carpenter 1998). L. S. Davitashvili (1961) suggested that this was probably the original function of the erect osteoderms, especially as the armor of all stegosaurs is ideally arranged for maximum effect during a lateral display (Spassov 1982).

It has been suggested that the plates of *Stegosaurus* were normally held horizontally, to provide a flank defense with the largest plates over the vulnerable hind limb, and that the plates could be suddenly erected by muscles to startle and deter an attacker, or to ward off attack from above (Hotton 1963; Bakker 1986). However, studies of the histology of the plates of *Stegosaurus* by Vivian de Buffrénil et al. (1986) suggest that this kind of plate mobility, which would require the evolution of new transverse back muscles to pull the plates upright, is very unlikely. Surface markings on the basal third of the plate indicate that it was embedded symmetrically in the thick, tough skin. Furthermore, the plates of *Stegosaurus* are unlikely to have functioned as armor because they do not consist of thick compact bone like the terminal pairs of tail spikes.

Apart from any use in sexual display, the plates could have functioned in temperature regulation. In an alternating arrangement, the plates could have worked well as a forced convection fin to dissipate heat, and possibly as heat absorbers from solar radiation (Farlow et al. 1976; Buffrénil et al. 1986). The plates would have formed scaffolding for the support of a richly vascularized skin, which would have acted as an efficient heat exchange structure. A heat-absorbing role for the plates would be useful if *Stegosaurus* was an ectotherm ("cold-blooded"), whereas heat loss by radiation or forced convection would help if *Stegosaurus* was ectothermic or to any degree endothermic ("warm-blooded"; Buffrénil et al. 1986). These conclusions would also probably apply to similarly large, thin dorsal plates of *Loricatosaurus*, *Hesperosaurus*, and *Wuerhosaurus*, but in other stegosaurs display was probably the main function of the plates.

Russell P. Main et al. (2005) question the thermoregulatory function of the vertical plates of *Stegosaurus*, which are not present in most stegosaurs, and favor a display and recognition function analogous to the varied cranial characters developed in other ornithischian groups. Their comparative histological studies show that the mostly thin parasagittal vertical plates and spikes of *Stegosaurus* grew mainly by basal osteogenesis, with some lateral periosteal deposition, and extensive internal remodeling, so the plates "evolved" by hypertrophic growth of the parasagittal keel of the basal thyreophoran scute. Internal "pipes" and external grooves are often present in broad, flat bones, including the cranial frills of ceratopsian dinosaurs, as well as the horns of artiodactyls, structures that do not seem to be primarily for thermoregulation. They conclude that these vascular and histological features are best regarded as constructional artifacts, a reflection of the processes and modes of bone growth. Surface vascular features on the plates probably also provided the blood supply to a keratinous covering, so, like the tail spikes, they were probably covered by horn. This was suggested for *Stegosaurus* by Charles W. Gilmore (1914; also Horner and Marshall 2002), and demonstrated for *Hesperosaurus* by Christiansen and Tschopp (2010). James Farlow et al. (2010) showed that alligator osteoderms, which are used as armor and for heat exchange with the external environment, as suggested by infrared thermographic imaging of basking caimans, had a similar internal vascularity to the plates of *Stegosaurus*. In the latter multiple large openings in the plate base lead to a longitudinally oriented vestibule from which originate multiple branching "pipes" that are best developed in the lower half of the plate. The pipes communicate with cancellous regions, some of which would have been vascular, with connections to vascular pits and grooves on the plate surface. Farlow et al. (2010) noted that thyreophoran osteoderms presumably had multiple functions as in living crocodiles. However, they considered that the potential for thermoregulation was greatest in *Stegosaurus* because of the large size and the thinness of most of the plate, the dorsal vertical orientation and alternating arrangement, and the extensive vascularity inside and outside.

The horn-covered terminal tail spines of stegosaurs would have been formidable weapons. Robert T. Bakker (1986) suggested that the loss of ossified tendons in stegosaurs was correlated with increased flexibility of the tail. He visualized *Stegosaurus* using the strong shoulder muscles to pivot its body on the very tall hind limbs (also Lull 1910b), the main weight supporters, as it arched and twisted its tail so the spines were driven into the body of an attacker. Ken Carpenter (1998) noted that the tail of *Stegosaurus* was carried high in the air parallel to the ground, a result of the superiorly facing posterior end surface of the last sacral centrum and the wedge-shaped centrum of the first caudal, with the expanded, bifid neural spines providing additional attachment surface for the supraspinous and interspinous ligaments that helped to maintain the horizontal posture without muscular effort. The horizontal posture was also aided by the bases of the large plates, which effectively locked segments of vertebrae together. This was also true for lateral motion of the tail, but because the small amount of motion between vertebrae was cumulative, the posterolaterally directed terminal tail spikes acted as a weapon that could be projected a little more than perpendicular to the long axis of the body (see also Ford 1997). Using multiple simulation methodologies, Heinrich Mallison (2011b) demonstrated that the tail of *Kentrosaurus* (Fig. 23.1C) was a dangerous weapon, which would have inflicted skin and flesh wounds with continuous rapid motion and broken bones with aimed whiplash blows.

Shoji Hayashi et al. (2008) showed in *Dacentrurus* and *Stegosaurus* that the tail spikes, unlike the plates and throat dermal ossicles, have thick compact bone with a medullary cavity, whereas those of ankylosaurs have a unique structure of supporting collagen fibers, indicating that the two groups independently used different strategies to develop defensive weapons.

Charles W. Gilmore (1914) observed that the anterior pair of the two pairs of tail spikes of *Stegosaurus* were apparently more deeply embedded by their bases in thick skin and more prone to injury and fracture than the posterior pair. Lorrie McWhinney et al. (2001) found that ~10% (5 out of 51) of tail spines examined showed broken tips with trauma-induced bone fractures and remodeled bone growth (indicating survival after injury); two of these also had posttraumatic chronic osteomyelitis (evidence of an infection), and one exhibited either a contiguous spread of the disease or a separate injury with secondary infection. The remodeling implies that the spikes were broken well before death and that the horny sheath probably protruded less than 1 cm beyond the bony tip; otherwise the horny sheath would have broken off rather than the bony spike. They concluded that the dermal tail spikes of *Stegosaurus* were primarily used in active defensive and offensive posturing in interspecific and intraspecific combat. Ken Carpenter et al. (2005) showed mathematically (see also Mallison 2011b for *Kentrosaurus*) that life-threatening wounds could be inflicted by *Stegosaurus* tail spikes on an adult of the carnivorous theropod dinosaur *Allosaurus*. They illustrated an anterior caudal vertebra of *Allosaurus* with a partly healed wound that was congruent with what a *Stegosaurus* tail-spike puncture would be expected to cause, and a *Stegosaurus* cervical plate with a bite pattern that matched that of an *Allosaurus* mouth.

References

Aberhan, M., R. Bussert, W.-D. Heinrich, E. Schrank, S. Schultka, B. Sames, J. Kriwet, and S. Kapilima. 2002. Palaeoecology and depositional environments of the Tendaguru Beds (Late Jurassic to Early Cretaceous, Tanzania). *Mitteilungen der Museum für Naturkunde Berlin Geowissenschaften* 5: 19–44.

Alexander, R. McN. 1985. Mechanics of posture and gait of some large dinosaurs. *Zoological Journal of the Linnean Society* 83: 1–25.

Averianov, A. O., A. A. Bakirov, and T. Martin. 2007. First definitive stegosaur from the Middle Jurassic of Kyrgyzstan. *Paläontologische Zeitschrift* 81: 440–446.

Bakker, R. T. 1986. *The Dinosaur Heresies: New Theories Unlocking the Mystery of the Dinosaurs and Their Extinction.* New York: William Morrow.

Barden, H. E., and S. C. R. Maidment. 2011. Evidence for sexual dimorphism in the stegosaurian dinosaur *Kentrosaurus aethiopicus* from the Upper Jurassic of Tanzania. *Journal of Vertebrate Paleontology* 31: 641–651.

Benton, M. J., and P. S. Spencer. 1995. Fossil Reptiles of Great Britain. *Geological Conservation Review Series* 10 (4). London: Chapman and Hall.

Bonaparte, J. 1996. Cretaceous tetrapods of Argentina. *Münchner Geowissenschaftliche Abhandlung* 30: 73–130A.

Brown, B., and E. M. Schlaikjer. 1940. The structure and relationships of *Protoceratops. Annals of the New York Academy of Science* 40: 133–266.

Carpenter, K. 1998. Armor of *Stegosaurus stenops*, and the taphonomic history of a new specimen from Garden Park, Colorado. *Modern Geology* 22: 127–144.

——. 2007. How to make a fossil. Pt. 1–Fossilizing bone. *Journal of Paleontological Sciences* 1: 1–10.

——. 2010. Species concept in North American stegosaurs. *Swiss Journal of Geosciences* 103: 155–162.

——. 2011 (this volume). Ankylosaurs. In M. K. Brett-Surman, Thomas R. Holtz Jr., and James O. Farlow (eds.), *The Complete Dinosaur*, 2nd ed., 505–525. Bloomington: Indiana University Press.

Carpenter, K., and P. M. Galton. 2001. Othniel Charles Marsh and the myth of the eight-spiked *Stegosaurus*. In K. Carpenter (ed.), *The Armored Dinosaurs*, 76–102. Bloomington: Indiana University Press.

Carpenter, K., C. A. Miles, and K. Cloward. 2001. New primitive stegosaur from the Morrison Formation, Wyoming. In K. Carpenter (ed.), *The Armored Dinosaurs*, 55–75. Bloomington: Indiana University Press.

Carpenter, K., F. Sanders, L. A. McWhinney, and L. Wood. 2005. Evidence for predator-prey relationships. Examples for *Allosaurus* and *Stegosaurus*. In K. Carpenter (ed.), *The Carnivorous Dinosaurs*, 325–350. Bloomington: Indiana University Press.

Carpenter, K., and B. Small. 1993. New evidence for plate arrangement in *Stegosaurus stenops*. *Journal of Vertebrate Paleontology* 13 (suppl. to 3): 28A–29A.

Chatterjee, S., and D. K. Rudra. 1996. KT events in India: impact, rifting, volcanism and dinosaur extinction. *Memoirs of the Queensland Museum* 39: 489–532.

Christiansen, N. A., and E. Tschopp. 2010. Exceptional stegosaur integument impressions from the Upper Jurassic Morrison Formation of Wyoming. *Swiss Journal of Geosciences* 103: 163–171.

Cobos, A., R. Royo-Torres, L. Luque, L. Alcalá, and L. Mampel. 2010. An Iberian stegosaurs paradise: the Villar del Arzobispo Formation (Tithonian-Berriasian) in Teruel (Spain). *Palaeogeography, Palaeoclimatology, Palaeoecology* 293: 223–236.

Colbert, E. H. 1962. The weights of dinosaurs. *American Museum Novitates* 2076: 1–16.

——. 1981. A primitive ornithischian dinosaur from the Kayenta Formation of Arizona. *Museum of Northern Arizona Bulletin* 53: 1–61.

Company, J., X. Pereda Suberbiola, and J. I. Ruiz-Omenaca. 2010. New stegosaurian (Ornithischia, Thyreophora) remains from Jurassic–Cretaceous transition beds of Valencia province (Southwestern Iberian Range, Spain). *Journal of Iberian Geology* 36: 243–252.

Coombs, W. P., Jr. 1978. Theoretical aspects of cursorial adaptations in dinosaurs. *Quarterly Review of Biology* 53: 393–418.

Coombs, W. P., Jr., D. B. Weishampel, and L. M. Witmer. 1990. Basal Thyreophora. In D. B. Weishampel, P. Dodson, and H. Osmólska (eds.), *The Dinosauria*, 1st hardback ed., 427–434. Berkeley: University of California Press.

Cope, E. D. 1871. On the fossil reptiles and fishes of the Cretaceous rocks of Kansas. United States Geological and Geographical Survey of the Territories. F. V. Hayden, United States Geologist.

Annual Report for 1870 2 (=5) [Wyoming, etc.]: 385–424.

Coria, R. A., and A. V. Cambiaso. 2007. Ornithischia. In Z. Gasparini, L. Salgado, and R. A. Coria (eds.), *Patagonian Mesozoic Reptiles*, 167–187. Bloomington: Indiana University Press.

Czerkas, S. A. 1987. A reevaluation of the plate arrangement on *Stegosaurus stenops*. In S. J. Czerkas and E. C. Olson (eds.), *Dinosaurs Past and Present*, 2: 83–99. Seattle: University of Washington Press.

——. 1999. The beaked jaw of stegosaurs and their implications for other ornithischians. In D. D. Gillette (ed.), *Vertebrate Paleontology in Utah*. Utah Geological Survey Miscellaneous Publication 99–1: 143–150.

Dacke, C. G. S., S. Arkle, D. J. Cook, I. M. Wormstone, S. Jones, M. Zaidi, and Z. A. Bascal. 1993. Medullary bone and avian calcium regulation. *Journal of Experimental Biology* 184: 63–88.

Davitashvili, L. 1961. *The Theory of Sexual Selection*. Moscow: Izdatel'stvo Akademia Nauk SSSR. [In Russian.]

de Buffrénil, V., J. O. Farlow, and A. de Ricqlès. 1986. Growth and function of *Stegosaurus* plates. Evidence from bone histology. *Paleobiology* 12: 459–473.

Dodson, P., A. K. Behrensmeyer, R. T. Bakker, and J. S. McIntosh. 1980. Taphonomy and paleoecology of the Upper Jurassic Morrison Formation. *Paleobiology* 6: 208–232.

Dong, Z. 1973. Dinosaurs from Wuerho. *Memoirs of the Institute of Vertebrate Paleontology and Paleoanthropology, Academic Sinica* 11: 45–52. [In Chinese.]

——. 1990. Stegosaurs of Asia. In K. Carpenter and P. J. Currie (eds.), *Dinosaur Systematics: Approaches and Perspectives*, 255–268. Cambridge: Cambridge University Press.

——. 1992. *Dinosaurian Faunas of China*. Beijing: China Ocean Press; New York: Springer-Verlag Press.

Dong, Z.-M. 1993. A new species of stegosaur (Dinosauria) from the Ordos Basin, Inner Mongolia, People's Republic of China. *Canadian Journal of Earth Sciences* 30: 2174–2176.

Dong, Z. M., S. W. Zhou, and Y. H. Chang. 1983. The dinosaur remains from Sichuan Basin, China. *Palaeontologica Sinica* 162 © 23: 1–166. [In Chinese with English summary.]

Edinger, T. 1961. Anthropocentric misconceptions in paleoneurology. *Proceedings of the Rudolf Virchow Medical Society of New York* 19: 56–107.

Escaso, F., F. Ortega, P. Dantas, E. Malafaia, N. L. Pimentel, X. Pereda-Suberbiola, J. L. Sanz, J. C. Kullberg, M. C. Kullberg, and F. Barriga. 2007. New evidence of shared dinosaur across Upper Jurassic Proto-North Atlantic: "*Stegosaurus*" from Portugal. *Naturwissenschaften* 94 (5): 367–374. doi: 10.1007/s00114-006-0209-8.

Evanoff, E., and K. Carpenter. 1998. History, sedimentology, and taphonomy of Felch Quarry 1 and associated sandbodies, Garden Park, Colorado. *Modern Geology* 22: 145–169.

Farlow, J. O., S. Hayashi, and G. J. Tattersall. 2010. Internal vascularity of the dermal plates of *Stegosaurus* (Ornithischia, Thyreophora). *Swiss Journal of Geosciences* 103: 173–185.

Farlow, J. O., C. V. Thompson, and D. E. Rosner. 1976. Plates of the dinosaur *Stegosaurus*: forced convection heat loss fins? *Science* 192: 1123–1125.

Fastovsky, D. E., and D. B. Weishampel. 2005. *The Evolution and Extinction of the Dinosaurs*. 2nd ed. Cambridge: Cambridge University Press.

Fiorillo, A. R. 1994. Time resolution at Carnegie Quarry (Morrison Formation: Dinosaur National Monument, Utah): implications for dinosaur physiology. *Contributions in Geology* (University of Wyoming) 30: 149–156.

Ford, T. L. 1997. How to draw dinosaurs. *Stegosaurus*. *Prehistoric Times* 22: 22–23.

Galton, P. M. 1973. The cheeks of ornithischian dinosaurs. *Lethaia* 6: 67–89.

——. 1974. The ornithischian dinosaur *Hypsilophodon* from the Wealden of the Isle of Wight. *Bulletin of the British Museum of Natural History: Geology* 25: 1–152c.

——. 1981. *Craterosaurus pottonensis* Seeley, a stegosaurian dinosaur from the Lower Cretaceous of England, and a review of Cretaceous stegosaurs. *Neues Jahrbuch für Geologie und Palaontologie, Abhandlungen* 161: 28–46.

——. 1982. The postcranial anatomy of stegosaurian dinosaur *Kentrosaurus* from the Upper Jurassic of Tanzania, East Africa. *Geologica et Palaeontologica* 15: 139–160.

——. 1985. British plated dinosaurs (Ornithischia, Stegosauria). *Journal of Vertebrate Paleontology* 5: 211–254.

——. 1988. Skull bones and endocranial casts of stegosaurian dinosaur *Kentrosaurus* Hennig, 1915 from Upper Jurassic of Tanzania, East Africa. *Geologica et Palaeontologica* 22: 123–143.

——. 1990. A partial skeleton of the

stegosaurian dinosaur *Lexovisaurus* from the uppermost Lower Callovian (Middle Jurassic) of Normandy, France. *Geologica et Palaeontologica* 24: 185–199.

———. 1991. Postcranial remains of the stegosaurian dinosaur *Dacentrurus* from the Upper Jurassic of France and Portugal. *Geologica et Palaeontologica* 254: 299–327.

———. 1999. Sex, sacra and *Sellosaurus gracilis* (Saurischia, Sauropodomorpha) – or why the character "two sacral vertebrae" is plesiomorphic for Dinosauria. *Neues Jahrbuch für Geologie und Paläontologie, Abhandlungen* 213: 19–55.

———. 2001. Endocranial casts of *Stegosaurus* (Upper Jurassic, western USA): Marsh's originals compared to complete undistorted ones. In K. Carpenter (ed.), *The Armored Dinosaurs*, 103–129. Bloomington: Indiana University Press.

———. 2005. Bones of large dinosaurs (Prosauropoda and Stegosauria) from the Rhaetic Bone Bed (Upper Triassic) of Aust Cliff, southwest England. *Revue de Paléobiologie* 24: 51–74.

———. 2010. Species of plated dinosaur *Stegosaurus* (Morrison Formation, Late Jurassic) of western USA: new type species designation needed. *Swiss Journal of Geosciences* 103: 187–198.

———. 2011. Case 3536. *Stegosaurus* Marsh, 1877 (Dinosauria, Ornithischia): proposed replacement of the type species with *Stegosaurus stenops* Marsh, 1887. *Bulletin of Zoological Nomenclature* 68 (2).

———. In press. *Hypsilophodon foxii* and other smaller bipedal ornithischian dinosaurs from the Lower Cretaceous of southern England. In P. Godefroit and O. Lambert (eds.), *Bernissart: Dinosaurs in Depth.* Bloomington: Indiana University Press.

Galton, P. M., and W. A. Coombs Jr. 1981. *Paranthodon africanus* (Broom), a stegosaurian dinosaur from the Lower Cretaceous of South Africa. *Géobios* 14: 299–309.

Galton, P. M., and H. P. Powell. 1983. Stegosaurian dinosaurs from the Bathonian (Middle Jurassic) of England, the earliest record of the Stegosauridae. *Géobios* 16: 219–229.

Galton, P. M., and P. Upchurch. 2004. Stegosauria. In D. B. Weishampel, P. Dodson, and H. Osmólska (eds.), *The Dinosauria*, 2nd ed., 343–362. Berkeley: University of California Press.

Giffin, E. B. 1991. Endosacral enlargements in dinosaurs. *Modern Geology* 16: 101–112.

Gilmore, C. W. 1914. Osteology of the armored dinosaurs in the United States National Museum, with special reference to the genus *Stegosaurus*. *Bulletin of the United States National Museum* 89: 1–136.

———. 1918. A newly mounted skeleton of the armored dinosaur *Stegosaurus stenops* in the United States National Museum. *Proceedings of the United States National Museum* 54: 383–396.

Haas, G. 1955. The jaw musculature in *Protoceratops* and in other ceratopsians. *American Museum Novitates* 1729: 1–24.

Haubold, H. 1990. Ein neuer Dinosaurier (Ornithischia, Thyreophora) aus dem unteren Jura des Nördlichen Mitteleuropa. *Revue de Paléobiologie* 9: 149–177.

Hayashi, S., K. Carpenter, and D. Suzuki. 2009. Different growth patterns between the skeleton and osteoderms of *Stegosaurus* (Ornithischia: Thyreophora). *Journal of Vertebrate Paleontology* 29: 123–131.

Hayashi, S., C. Kenneth ([sic], K. Carpenter), M. Watabe, O. Mateus, and R. Barsbold. 2008. Defensive weapons of thyreophoran dinosaurs: histological comparisons and structural differences in spikes and clubs of ankylosaurs and *Stegosaurus*. *Journal of Vertebrate Paleontology* 28 (suppl. to no. 3): 89–90A.

Heinrich, W.-D. 1999. The taphonomy of dinosaurs from the Upper Jurassic of Tendaguru (Tanzania) based on field sketches of the German Tendaguru Expedition (1909–1913). *Mitteilungen der Museum für Naturkunde Berlin Geowissenschaften* 2: 25–61.

Hennig, E. 1925. *Kentrurosaurus aethiopicus*, die Stegosaurier-funde von Tendaguru, Deutsch-Ostafrika. *Palaeontographica* (suppl. 7) 1 (1): 103–254.

Hill, D., G. Playford, and J. T. Woods. 1966. *Jurassic Fossils of Queensland*. Brisbane, Australia: Queensland Palaeontographical Society.

Hoffstetter, R. 1957. Quelques observations sur les Stégosaurinés. *Bulletin de Museum National d'Histoire Naturelle de Paris* 29 (2): 537–547.

Hopson, J. A. 1980. Relative brain size in dinosaurs – implications for dinosaurian endothermy. In R. D. K. Thomas, and E. C. Olson (eds.), *A Cold Look at the Warm-Blooded Dinosaurs*, 287–310. Selected Symposium 28, American Association for the Advancement of Science. Boulder: Westview Press.

Horner, J. R., and C. Marshall. 2002. Keratinous covered dinosaur skulls. *Journal of Vertebrate Paleontology* 22 (suppl. to no. 3): 67A.

Hotton, N., III. 1963. *Dinosaurs*. New York: Pyramid Publications.

Hoyo, J., A. Elliott, and J. Sargatal. 1994. *Handbook of the Birds of the World*. Vol. 2: *New World Vulture to Guineafowl*. Barcelona: Lynx Edicions.

Huene, F. von. 1956. *Paläontologie und Phylogenie der Niederen Tetrapoden*. Jena, Germany: Fischer.

Hurlburt, G. R. 1996. Relative brain size in Recent and fossil amniotes: determination and interpretation. Ph.D. dissertation, University of Toronto.

———. 1999. Comparison of body mass estimation techniques, using recent reptiles and the pelycosaur *Edaphosaurus boanerges*. *Journal of Vertebrate Paleontology* 19: 338–350.

Hurlburt, G. R., and L. Waldorf. 2002. Endocast volume and brain mass in a size series of alligators. *Journal of Vertebrate Paleontology* 23 (suppl. to 3): 69A.

Hulke, J. W. 1887. Note on some dinosaurian remains in the collection of A. Leeds, Esq., of Eyebury, Northamptonshire. *Quarterly Journal of the Geological Society of London* 43: 695–702.

Huxley, T. H. 1870. On the classification of the Dinosauria with observations on the Dinosauria of the Trias. *Quarterly Journal of the Geological Society of London* 26: 31–50.

Janensch, W. 1925. Ein aufgestelltes Skelett des Stegosauriers *Kentrurosaurus aethiopicus* E. Hennig aus den Tendaguru-Schichten Deutsch-Ostafrikas. *Palaeontographica* (suppl. 7) 1 (1): 257–276.

Jerison, H. J. 1973. *Evolution of Brain and Intelligence*. New York: Academic Press.

Jia, C., C. A. Foster, X. Xu, and J. M. Clark. 2007. The first stegosaur (Dinosauria, Ornithischia) from the Upper Jurassic Shishugou Formation of Xianjiang, China. *Acta Geologica Sinica* 81: 351–356.

Karbek, T. R. 2002. The case for *Stegosaurus* as an agile, cursorial biped. *Journal of Vertebrate Paleontology* 22 (suppl. to 3): 73A.

Krause, W. 1881. Zum Sacralhirn der Stegosaurier. *Biologisches Zentralblatt* 1: 461.

Lapparent, A. F. de, and R. Lavocat. 1955. Dinosauriens. In J. Piveteau (ed.), *Traité de Paléontologie* 5: 785–962. Paris: Masson et Cie.

Larson, P. L. 2008. Variation and sexual dimorphism in *Tyrannosaurus rex*. In K. Carpenter and P. L. Larson (eds.), Tyrannosaurus rex: *The Tyrant King*, 103–128. Bloomington: Indiana University Press.

Leidy, J. 1856. Notice of extinct animals discovered by Prof. E. Emmons. *Proceedings* of the Academy of Natural Sciences of Philadelphia 8: 255–256.

Long, J. A. 1998. Dinosaurs of Australia and New Zealand and Other Animals of the Mesozoic Era. Cambridge, Mass.: Harvard University Press.

Lucas, F. A. 1902. Paleontological notes: the generic name Omosaurus. Science 19: 435.

Lull, R. S. 1903. Skull of Triceratops serratus. Bulletin of the American Museum of Natural History 19: 685–695.

———. 1908. The cranial musculature and the origin of the frill in the ceratopsian dinosaurs. American Journal of Science (series 4) 25: 387–399.

———. 1910a. The armor of Stegosaurus. American Journal of Science (series 4) 29: 201–210.

———. 1910b. Stegosaurus ungulatus Marsh, recently mounted at the Peabody Museum of Yale University. American Journal of Science (series 4) 30: 361–376.

———. 1917. On the functions of the "sacral brain" in dinosaurs. American Journal of Science (series 4) 44: 471–477.

Maidment, S. C. R. 2010. Stegosauria: a historical review of the body fossil record and phylogenetic relationships. Swiss Journal of Geosciences 103: 199–210.

Maidment, S. C. R., D. B. Norman, P. M. Barrett, and P. Upchurch. 2008. Systematics and phylogeny of Stegosauria (Dinosauria: Ornithischia). Journal of Systematic Palaeontology 6: 367–407. doi: 10.1017/S1477201908002459

Maidment, S. C. R., and G. Wei. 2006. A review of the Late Jurassic stegosaurs (Dinosauria, Stegosauria) from the People's Republic of China. Geological Magazine 143: 621–634.

Maidment, S. C. R., G. Wei, and D. B. Norman. 2006. Re-description of the postcranial skeleton of the Middle Jurassic stegosaur Huayangosaurus taibaii. Journal of Vertebrate Paleontology 26: 944–956.

Main, R. P., A. de Ricqles, J. R. Horner, and K. Padian. 2005. The evolution and function of thyreophoran dinosaur scutes: implications for plate function in stegosaurs. Paleobiology 31: 291–314.

Mallison, H. 2010. CAD assessment of the posture and range of motion of Kentrosaurus aethiopicus Hennig 1915. Swiss Journal of Geosciences 103: 211–233.

———. 2011a. The real lectotype of Kentrosaurus aethiopicus Hennig 1915. Neues Jahrbuch für Geologie und Paläontologie 259: 197–206.

———. 2011b. Defence capabilities of Kentrosaurus aethiopicus Hennig 1915.

Palaeontologia Electronica 14 (2): 1–25. palaeo-electronica.org/2011_2/255/index.html.Marsh, O. C. 1877. New order of extinct Reptilia (Stegosauria) from the Jurassic of the Rocky Mountains. American Journal of Science (series 3) 14: 513–514.

———. 1880. Principal characters of American Jurassic dinosaurs. Pt. 3. American Journal of Science (series 3) 19: 253–259.

———. 1881. Principal characters of American Jurassic dinosaurs. Pt 4. Spinal cord, pelvis and limbs of Stegosaurus. American Journal of Science (series 3) 21: 167–170.

———. 1887. Principal characters of American Jurassic dinosaurs. Pt. 9. The skull and dermal armor of Stegosaurus. American Journal of Science (series 3) 34: 413–417.

———. 1889. Comparison of the principal forms of the Dinosauria of Europe and America. American Journal of Science (series 3) 37: 323–331.

———. 1891. Restoration of Stegosaurus. American Journal of Science (series 3) 42: 179–181.

———. 1896. Dinosaurs of North America. United States Geological Survey, 16th Annual Report 1894–95: 133–244.

Mateus, O., S. C. R. Maidment, and N. A. Christiansen. 2009. A new long-necked "sauropod-mimic" stegosaur and the evolution of the plated dinosaurs. Proceedings of the Royal Society B 276: 1815–1821.

Mateus, O., J. Milàn, M. Romano, and M. A. Whyte. 2011. New finds of stegosaur tracks from the Upper Jurassic Lourinhã Formation, Portugal. Acta Palaeontologica Polonica: 651–658. doi:10.4202/app.2009.0055.

McWhinney, B. M. Rothschild, and K. Carpenter. 2001. Posttraumatic chronic osteomyelitis in Stegosaurus dermal plates. In K. Carpenter (ed.), The Armored Dinosaurs, 141–156. Bloomington: Indiana University Press.

Milàn, J., and L. M. Chiappe. 2009. First American record of the Jurassic ichnospecies Deltapodus brodkricki and a review of the fossil record of stegosaurian footprints. Journal of Geology 117: 343–348.

Mohabey, D. M. 1986. Note on dinosaur foot print from Kheda District, Gujarat. Journal of the Geological Society of India 27: 456–459.

Nopcsa, F. 1911. Notes on British dinosaurs. Pt. 4. Stegosaurus priscus, sp. nov. Geological Magazine (series 5) 8: 109–115, 145–153.

———. 1912. Notes on British dinosaurs. Pt.

5. Craterosaurus (Seeley). Geological Magazine (series 5) 9: 481–484.

———. 1915. Die Dinosaurier der siebenburgischen Landesteile Ungarns. Mittheilungen aus dem Jahrbuch der Ungarischen geologischen Reichsanst 23: 1–26.

———. 1928. The genera of reptiles. Palaeobiologica 1: 163–188.

Norman, D. B. 1984. A systematic reappraisal of the reptile order Ornithischia. In W.E. Reif, and F. Westphal (eds.), Third Symposium on Mesozoic Terrestrial Ecosystems, Short Papers, 157–162. Tübingen: Attempto Verlag.

Novas, F. E. 2009. The Age of Dinosaurs in South America. Bloomington: Indiana University Press.

Olshevsky, G., and T. Ford. 1993. The origin and evolution of the stegosaurs. Gakken Mook 4: 65–103. [In Japanese.]

Osborn, H. F. 1923. Two Lower Cretaceous dinosaurs from Mongolia. American Museum Novitates 95: 1–10.

Ostrom, J. H., and J. S. McIntosh. 1999. Marsh's Dinosaurs: The Collections from Como Bluff, 2nd ed. New Haven, Conn.: Yale University Press.

Ouyang, H. 1992. Discovery of Gigantspinosaurus sichuanensis and its scapular spine orientation. Abstracts and Summaries for Youth Academic Symposium on New Discoveries and Ideas in Stratigraphic Paleontology, December 1992: 47–49. [In Chinese.]

Owen, R. 1842. Report on British fossil reptiles. Pt. 2. Annual Report of the British Association for the Advancement of Science 11: 60–204.

———. 1861. A monograph of the fossil Reptilia of the Lias formations. 1. Scelidosaurus harrisonii. Palaeontographical Society Monographs 13: 1–14.

———. 1863. A monograph of the fossil Reptilia of the Lias formations. 2. Scelidosaurus harrisonii Owen of the Lower Lias. Palaeontographical Society Monographs 14: 1–26.

———. 1872. Monograph on the fossil Reptilia of the Wealden and Purbeck formations. Supplement 4. Dinosauria (Iguanodon). Palaeontographical Society Monographs 25: 1–15.

———. 1875. Monographs of the fossil Reptilia of the Mesozoic Formations (pts. 2 and 3) (genera Bothriospondylus, Cetiosaurus, Omosaurus). Palaeontographical Society Monographs 29: 15–94.

Page, D. 1998. Stegosaur tracks and the persistence of facies: the Lower Cretaceous of Western Australia. Geology Today 14: 75–77.

Papp, M. J., and L. Witmer. 1998. Cheeks,

beaks, or freaks: a critical appraisal of buccal soft-tissue anatomy in ornithischian dinosaurs. *Journal of Vertebrate Paleontology* 18 (suppl. to 3): 69A.

Parrish, J. T., F. Peterson, and C. E. Turner. 2004. Jurassic "savannah"–plant taphonomy and climate of the Morrison Formation (Upper Jurassic, western USA). *Sedimentary Geology* 167: 137–162.

Paul, G. S. 1987. The science and art of restoring the life appearance of dinosaurs and their relatives. In S. J. Czerkas and E. C. Olson (eds.), *Dinosaurs Past and Present*, 2: 4–49. Seattle: University of Washington Press.

——. 1992. The arrangement of plates in the first complete *Stegosaurus*, from Garden Park. *Tracks in Time* (Garden Park Paleontological Society) 3 (1): 1–2.

——. 2010. *The Princeton Field Guide to Dinosaurs*. Princeton, N.J.: Princeton University Press.

Peng, G., Y. Gao, C.-K Shu, and S. Jiang. 2005. *Jurassic Dinosaur Faunas in Zigong*. Zigong: Sichuan Scientific and Technological Publishing House. [In Chinese with English summary.]

Pereda Suberbiola, X., P. M. Galton, H. Mallison, and F. Novas. In press. The record of a plated dinosaur (Ornithischia, Stegosauria) from the Early Cretaceous of Argentina, South America: a re-evaluation. *Journal of South American Earth Sciences*.

Pereda Suberbiola, X., P. M. Galton, F. Torcida, P. Huerta, L. A. Izzquierdo, D. Montero, G. Perez, and V. Urien. 2003. First stegosaurian dinosaur remains from the Early Cretaceous of Burgos (Spain), and a review of Cretaceous stegosaurs. *Revue de Espanola Paleontologie* 18: 143–150.

Redelstorff, R., and P. M. Sander. 2009. Long and girdle bone histology of *Stegosaurus*: Implications for growth and life history. *Journal of Vertebrate Paleontology* 29: 1087–1099.

Reichel, M. 2010. A model for the bite mechanics in the herbivorous dinosaur *Stegosaurus* (Ornithischia, Stegosauridae). *Swiss Journal of Geosciences* 103: 235–240.

Romer, A. S. 1927. The pelvic musculature of ornithischian dinosaurs. *Acta Zoologica* 8: 225–275.

——. 1956. *Osteology of the Reptiles*. Chicago: University of Chicago Press.

Russell, D., P. Béland, and J. S. McIntosh. 1980. Paleoecology of the dinosaurs of Tendaguru (Tanzania). *Mémoires de la Société Géologiques de France* (n. s.) 1980 (139): 169–175.

Russell, L. S. 1935. Musculature and function in the Ceratopsia. *Bulletin of the National Museum of Canada* 77: 39–48.

Schweitzer, M. H., J. L. Wittmeyer, and J. R. Horner. 2005. Gender-specific reproductive tissue in Ratites and *Tyrannosaurus rex*. Science 308: 1456–1460.

Seeley, H. G. 1874. On the base of a large lacertian cranium from the Potton Sands, presumably dinosaurian. *Quarterly Journal of the Geological Society of London* 30: 690–692.

——. 1887. On the classification of the fossil animals commonly called Dinosauria. *Proceedings of the Royal Society of London* 43: 165–171.

Senter, P. 2010. Evidence of a sauropod-like metacarpal configuration in stegosaurian dinosaurs. *Acta Palaeontologica Polonica* 55: 427–432.

——. 2011. Evidence of a sauropod-like metacarpal configuration in ankylosaurian dinosaurs. *Acta Palaeontologica Polonica* 56: 221–224.Sereno, P. C. 1986. Phylogeny of the bird-hipped dinosaurs (Order Ornithischia). *National Geographic Research* 2: 234–256.

Sereno, P. C., and Z. Dong. 1992. The skull of the basal stegosaur *Huayangosaurus taibaii* and a cladistic analysis of Stegosauria. *Journal of Vertebrate Paleontology* 12: 318–343.

Siber, H. J., and U. Möckli. 2009. *The Stegosaurs of the Sauriermuseum Aathal*. Aathal, Switzerland: Sauriermuseum Aathal.

Spassov, N. B. 1982. The bizarre dorsal plates of *Stegosaurus*: ethological approach. *Comptes Rendus de l'Academie Bulgare des Sciences* 35: 367–370.

Thulborn, T. 1990. *Dinosaur Tracks*. London: Chapman and Hall.

——. 1997. Dinosaur tracks of the Broome Sandstone. *Conference on Australiasian Vertebrate Evolution, Palaeontology and Systematics. Notes for Field Excursion:* 28–29 June 1997.

Tyson, H. 1977. Functional craniology of the Ceratopsia (Reptilia: Ornithischia) with special reference to *Eoceratops*. M.Sc. dissertation, Department of Zoology, University of Alberta, Edmonton.

Watanabe, A. 2008. *Stegosaurus*: hands, feet, and footprints. *Journal of Vertebrate Paleontology* 28 (suppl. to 3): 158A.

Whyte, M. A., and M. Romano. 1994. Probable sauropod footprints from the Middle Jurassic of Yorkshire, England. *Gaia* 10: 15–26.

——. 2001. Probable stegosaurian dinosaur tracks from the Saltwick Formation (Middle Jurassic) of Yorkshire, England. *Proceedings of the Geological Association* 112: 45–54.

Wiedersheim, R. 1881. Zur Palaeontologie Nord-Amerikas. *Biologisches Zentralblatt* 1: 359–372.

Xia, W., X. Li, and Z. Yi. 1984. The burial environment of dinosaur fauna in Lower Shaximiao Formation of Middle Jurassic at Dashanpu, Zigong, Sichuan. *Journal of Chengdu College of Geology* 2: 46–59. [In Chinese with English summary.]

Xing, I. D., G. Z. Peng, and C. K. Shu. 2008. Stegosaurian skin impressions from the Upper Jurassic Shangshaximiao Formation, Zigong, Sichuan, China: a new observation. *Geological Bulletin of China* 27: 1049–1053. [In Chinese with English abstract.]

Yadagiri, P., and K. Ayyasami. 1978. New dinosaurian remains. *Geological Survey of India News* 9 (5): 4.

——. 1979. A new stegosaurian dinosaur from Upper Cretaceous sediments of south India. *Journal of the Geological Society of India* 20: 521–530.

Young, C. C. 1959. On a new Stegosauria from Szechuan, China. *Vertebrata PalAsiatica* 3: 1–8.

Zhou, S. W. 1984. *The Middle Jurassic Dinosaurian Fauna from Dashanpu, Zigong, Sichuan*. Vol. 2: *Stegosaurs*. Chongquing: Sichuan Scientific and Technical Publishing House. [In Chinese with English summary.]

Ankylosaurs

Kenneth Carpenter

24

Ankylosaurs were short-limbed, four-legged, armor-plated dinosaurs, with a long, wide body (Fig. 24.1). It is a body encased in armor or osteoderms that characterizes the group (see also Plates 27–29). The most prominent osteoderms consisted of keeled or flat plates of bone embedded in the skin (hence the term "osteoderm"–bone skin). These osteoderms are sometimes supplemented with spines or spikes on the body and tail. In at least one group, there was also a bone club on the end of the tail. At one time it was thought that the rough texture of the skulls was due to armor fusing to the bone surface. However, recent studies have shown that in most (but not all) cases, the texture is due to remodeling of the skull surface by the overlying skin or scales (Vickaryous et al. 2001; Carpenter et al. 2001).

The taxonomy of ankylosaurs remains confusing because so many of the species are named from fragmentary material. The Ankylosauria can be divided into three families, the Polacanthidae (informally called polacanthids), the Nodosauridae (or nodosaurids), and the Ankylosauridae (or

24.1. Top and side view of *Saichania* showing the characteristic features of ankylosaurs, including the wide body and extensive covering of body armor, or osteoderms, embedded in the skin.

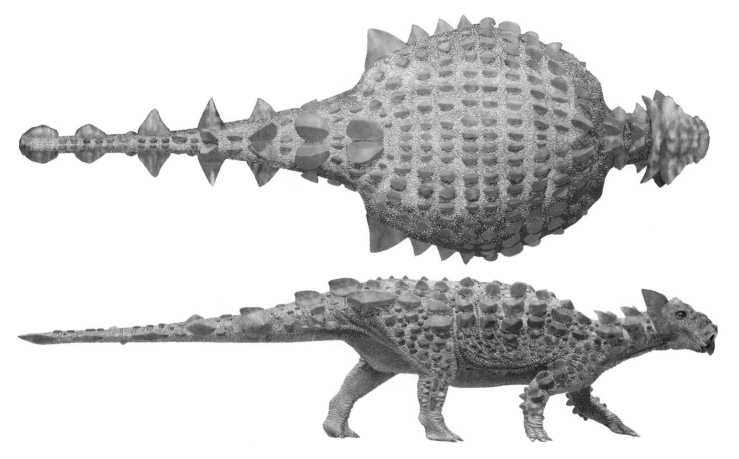

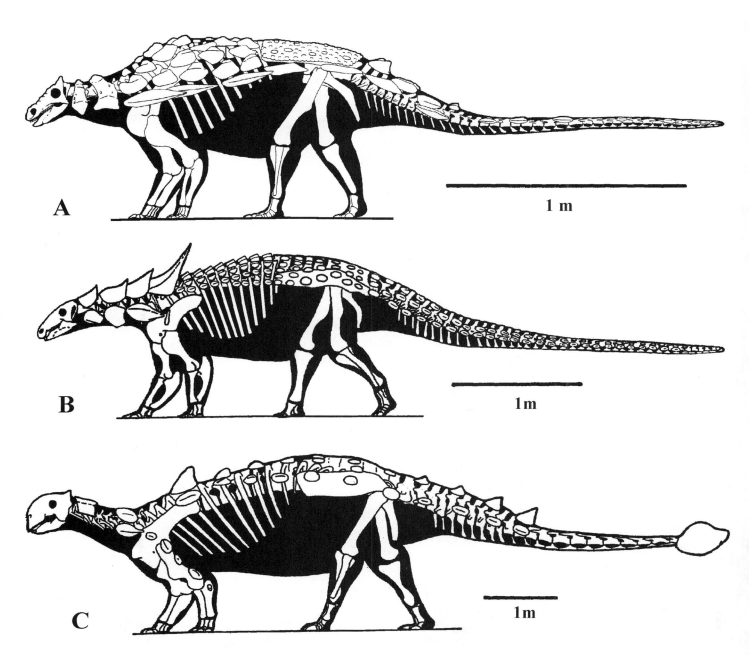

24.2. Representatives of the three families of ankylosaurs: A, Polacanthidae (*Gargoyleosaurus*); B, Nodosauridae (*Sauropelta*); C, Ankylosauridae (*Euoplocephalus*). Scales in m.

ankylosaurids), which is subdivided into the subfamilies Shamosaurinae (shamosaurines) and Ankylosaurinae (ankylosaurines). This threefold family division is not universally accepted (e.g., Vickaryous et al. 2004) despite features in the skull, scapula, and osteoderms that unify the various species into the three families (Fig 24.2).

Skull

Skeletal Features

The ankylosaur skull is greatly modified as compared to that of other dinosaurs. Whereas in most dinosaurs the skull is deeper than wide, in ankylosaurs, this is reversed (Fig. 24.3) due to lateral expansion or bulging of the maxillary bones above the tooth row (more on this below). As a general rule of thumb, the skulls of polacanthids are triangular, being as wide, or nearly as wide, as long in a manner similar to ankylosaurids (Fig. 24.3A, C).

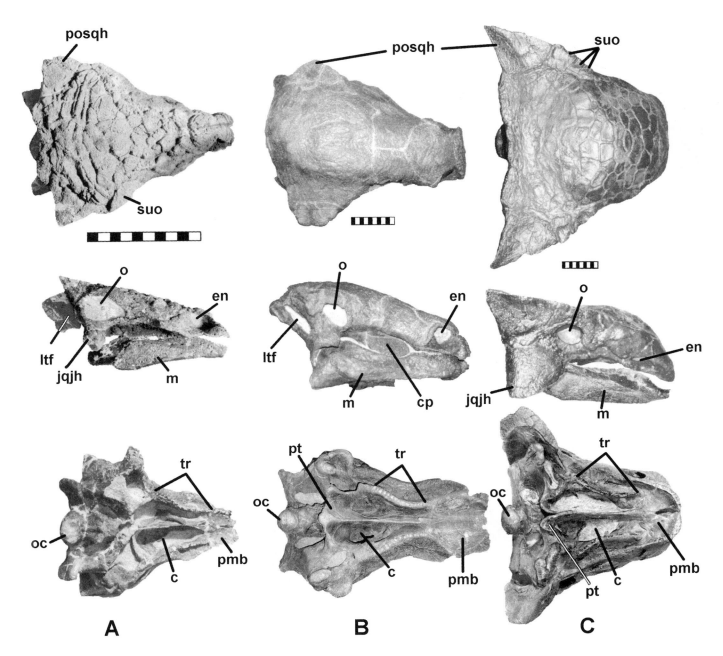

24.3. Representative skulls of the three families in top, side, and bottom views: A, Polacanthidae (*Gargoyleosaurus*); B, Nodosauridae (*Edmontonia*); C, Ankylosauridae (*Ankylosaurus*). Abbreviations: c−choana; cp−cheek plate; en−external nares; jqjh−jugal-quadratojugal horn; ltf−lateral temporal fenestra; m−mandible; o−orbit; oc−occipital condyle; pmb−premaxillary beak; posqh−postorbital-squamosal horn; pt−pterygoid (damaged in A); suo−supraorbital; tr−tooth row. Scales in cm.

In contrast, the skulls of nodosaurids (Fig. 24.3B) are proportionally longer and less wide (although still wider than typical in dinosaurs).

The surface of the skull in many ankylosaurs is rough, or knobby. At one time it was thought that this texture was due to thin sheets of osteoderms fusing to the skull surface. But recent histological studies, where thin sections are examined microscopically, have shown that in many examples the texturing is the result of remodeling of the skull surface by skin, with each bump and knob being the site of a former scale (Carpenter et al. 2001; Vickaryous et al. 2001). The scale pattern is especially noticeable on the skull of *Edmontonia*, where the textured surface consists of large, slightly raised regions that are symmetrically arranged (see Fig. 24.3B). In contrast, the scale pattern on polacanthid and ankylosaurid skulls is typically smaller, more numerous, and asymmetrical (Fig. 24.3A, C). Ankylosaur skulls also typically have horns of sorts projecting from the rear corners. These structures

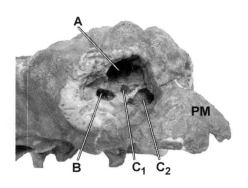

24.4. Detail of the external nares in *Saichania* adjacent to the premaxilla (PM). The nasal vestibule located inside the outer opening (external nares) is pierced by several openings: A, narial opening into the nasal chamber; B, lateral opening into the maxillary sinus; C_1 and C_2 open into the premaxillary sinuses. Almost all ankylosaurs have openings A and B.

are actually outgrowths of the postorbital and/or squamosal (upper horn), and of the jugal and/or quadratojugal (lower horn). This remodeling and outgrowths from the skull bones have obliterated most of the cranial sutures in adults. In juveniles, such as of *Pinacosaurus*, or subadults, such as of *Cedarpelta*, the cranial sutures remain distinct, and these are important for determining skull bone homologies.

Of the five major skull openings typically seen in the dinosaurs (external nares, antorbital fenestra, supratemporal fenestra, and lateral temporal fenestra), ankylosaurs have only three: external nares, orbit and lateral temporal fenestra (Fig. 24.3). The loss of the antorbital and supratemporal fenestrae occurs by expansion of the adjacent bones closing the openings (a dimple may mark the antorbital fenestra in some ankylosaurids). The external nares for the nostrils remain in their primitive position in polacanthids and nodosaurids (near the front and facing outward); in contrast, it is more variable in ankylosaurids, where it may face forward in some species (e.g., *Pinacosaurus*) or to the sides in others (e.g., *Ankylosaurus*). Within the external nares there is usually a cupped cavity (nasal vestibule) that is pierced by one or more openings (Fig. 24.4), of which the uppermost one usually opens into the nasal chamber. In *Euoplocephalus*, the air passage continues from the nasal chamber in a very convoluted, looped airway in the skull; it is slightly less convoluted in *Panoplosaurus* before opening into the roof of the mouth (Witmer and Ridgley 2008) . In most ankylosaurs there is at least one other opening, which is just within the lateral edge of the nasal vestibule. This opening usually connects with the large maxillary sinus, which causes the maxillary bone to bulge outward above the tooth row; thus the tooth row is inset from the sides of the face. In ankylosaurids, there may be additional openings into other sinuses, especially in the premaxilla. In *Ankylosaurus*, the premaxillary sinus is huge and has crowded the external nares around to the side of the face.

The orbits in ankylosaurs tend to be circular or oval, and are often walled internally along the rear margin so as to separate the jaw-closing adductor muscles from the eyeball. The dorsal or upper rim of the orbit is composed of two or three bones called supraorbitals. In some species, such as *Pinacosaurus*, these supraorbitals assume sharp, ridgelike structures over the orbit. In *Euoplocephalus* a curved disk of bone was embedded in the eyelid to protect the eyeball. The lateral temporal fenestra is visible when the skull is viewed laterally in polacanthids and nodosaurids, but is visible only in rear view in ankylosaurids. Why this change occurs only in ankylosaurids is not understood.

The looped air passage seen in *Euoplocephalus* and other ankylosaurs is a puzzle. Why such a complex configuration developed is not known, although several hypotheses have been proposed. One idea is that the looped passage increased the surface area for olfactory tissue without elongating the snout, and that therefore these animals had an especially keen sense of smell. However, the olfactory lobes and forebrain region of the brain are not especially well developed, casting doubt on this hypothesis (Coombs 1978b). Another hypothesis is that the increased surface area provided more tissue surface to warm and moisten the air. While that might be necessary for ankylosaurs living in the desert, *Euoplocephalus* lived in a hot, humid coastal environment. One other suggestion is that the enlarged chamber functioned as a resonating chamber to amplify sound (Maryańska 1977). However, animals tend to use vocal chords (or syrinx in birds) to make sounds. The most recent idea is that they indirectly reduce the mass, hence the weight, of the head (Witmer and Ridgley 2008). At present no single hypothesis explains the looped airway, especially since it is less elaborate in the ankylosaurid *Pinacosaurus* (Witmer, unpublished CT data). The airway in polacanthids and most nodosaurids is straight as in other dinosaurs.

The underside of the ankylosaur skull (Fig. 24.3) has a scoop-shaped beak at the front of the mouth. The edge of the beak has a sharp rim around the sides and front. In life, it was probably covered with a keratinous beak much like a turtle's. Ankylosaurids and some nodosaurids, such as *Edmontonia*, tend to have a broad beak, suggesting that they were general grazers, cropping low plants. Other nodosaurids and polacanthids have a proportionally narrower beak, suggesting that they were more selective browsers, cropping specific plants or plant parts. In primitive members of all three families there are also premaxillary teeth along the edges of the beak, but these are lost in more advanced members of the families. Near the center of the palate is a pair of large, elliptical openings, the palatal vacuities. These are separated by a vertical sheet of bone, which in some ankylosaurs extends to the underside of the skull roof, completely dividing the nasal chamber in half. The vacuities are where the internal nares open into the throat. The tooth rows are typically hourglass shaped, being wider (farther apart) at the front and rear. This has required a modification of the lower jaws so that the teeth can occlude. Typically this involves sloping the jaws inward.

The braincase is a robust structure that is partially underlapped by the pterygoids (overlapped when the skull is viewed upside down). The extent that the pterygoids underlap the braincase is highly variable, with no discernable pattern in the three families. The cavity for the brain reveals the shape and relative sizes of the brain hemispheres and nerves (Fig. 24.5). A cast of the brain cavity (called an endocast) approximates the size and shape of the brain, although the brain does not fill the cavity completely (Fig. 24.5B). The endocast (Fig. 24.5B) shows that the forebrain region (mostly the cerebrum) was rather small, including the olfactory lobes, so that even with a large nostril, the sense of smell was not very acute. In the midbrain region, the optical nerves are not very prominent, suggesting that vision was not very good either (like the rhinoceros today). The pituitary region is large for the size of the endocast, but that is true for most dinosaurs regardless of size. The hindbrain (cerebellum and medulla oblongata) functions to control balance, motor coordination, and body organs (heart, lung, liver, intestine, etc.). Compared to living reptiles, ankylosaurs had a brain about half the size that would be expected.

The unworn teeth of the ankylosaurs are shaped like tiny hands with the fingers close together (Fig. 24.6), which basically resembles the teeth of primitive ornithischians, such as *Lethososaurus*. As the teeth wear, the cusps, or denticles, are worn flat. The teeth are small compared to skull size, and those of ankylosaurids and polacanthids are typically smaller than those of the nodosaurids. Near the base of the tooth crown, just above the roots, there may be a shelf, or cingulum. This cingulum is more common among nodosaurids than ankylosaurids. However, even among nodosaurids, the cingulum is variable or absent even in the same tooth row. In ankylosaurids

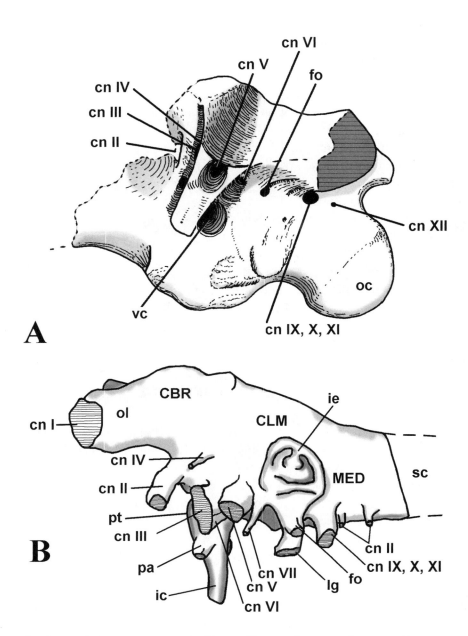

24.5. A, lateral view of a braincase (nodosaurid *Sauropelta*) showing the major foramina. B, endocast of ankylosaurid *Euoplocephalus* showing major structures (after Coombs 1978c). Abbreviations: CBR–cerebrum; CLM–cerebellum; cn I–XII–cranial nerves I–XII; fo–fenestra ovalis; ic–internal carotid; ie–inner ear (semicircular canals); lg–lagena; MED–medulla; oc–occipital condyle; ol–olfactory lobe; pa–palatine artery; pt–pituitary; sc–spinal chord; vc–vidian canal.

the base of the crown is typically swollen, and a cingulum is sometimes present. The teeth of polacanthids are the most primitive among ankylosaurs, lacking a swollen crown or a cingulum. In primitive members of all three families, there are conical teeth along the inner side of the premaxillary beak that are appropriately called premaxillary teeth. These teeth were independently lost in more derived forms of all three families. For example, premaxillary teeth are present in the primitive polacanthid *Gargoyleosaurus*, the primitive nodosaurid *Silvisaurus*, and the primitive ankylosaurid *Cedarpelta*. But they are absent in the advanced polacanthid *Gastonia*, the advanced nodosaurid *Edmontonia*, and the advanced ankylosaurid *Ankylosaurus*. Why this loss occurred in separate lineages is unknown, especially since the timing of the loss is so variable.

The shape of the teeth indicates that ankylosaurs were herbivorous, because the teeth are ill-suited for slicing flesh. Wear patterns of the teeth show tooth-to-tooth wear along the tops of the crowns in polacanthids and nodosaurids, and wear along the faces of the crowns in ankylosaurids. The

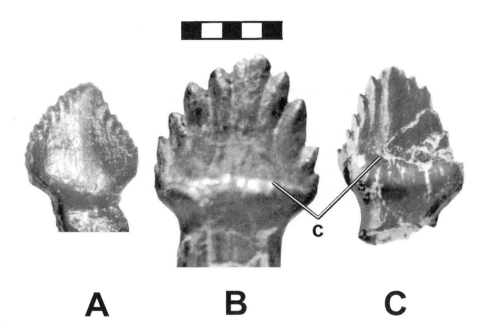

24.6. Comparison of ankylosaur teeth. A, polacanthid (*Gastonia*); B, nodosaurid (*Edmontonia*); C, ankylosaurid (*Ankylosaurus*). Abbreviation: C, cingulum. Scale in mm.

chewing motion of *Euoplocephalus*, the only ankylosaur whose jaw motion has been studied, was apparently two-phased: the jaw closes on the food, and then is pulled backward, grinding or shredding the plant material between the teeth (Rybczynski and Vickaryous 2001). This is a more complex method of chewing than pulping the plant material between the teeth with a simple up–down motion. How extensive such complex motion was among the different ankylosaurs has yet to be studied.

Postcrania

The vertebral column is divided into seven or eight cervical vertebrae in the neck, about 12 to 18 dorsal vertebrae in the back, and three or four sacral vertebrae in the pelvis (Fig. 24.7A). The sacral vertebrae are fused together with the last two to six dorsal vertebrae, forming a structure called a synsacrum. The tail has about 20 to 24 caudal vertebrae in ankylosaurids and up to 40 in nodosaurids (the count is unknown for polacanthids). The cervical vertebrae are often as wide as long, resulting in a short neck to support the large, heavy skull (Fig. 24.7B). In contrast, the dorsal vertebrae are typically longer (Fig 24.7C), producing a long body. Furthermore, the diapophyses, where the

24.7. A, vertebral column of *Saichania* is representative of most ankylosaurs (except for the tail club and handle). Fusion of the last dorsal vertebrae and sacral vertebrae form the synsacrum. Front and side views of representative ankylosaur vertebrae (*Ankylosaurus*): B, neck or cervical; C, middorsal; D, anterior caudal with fused chevron (V-shaped bone beneath the tail). *Ankylosaurus* tail club in side and top views made up of fused osteoderms. Scales for B–E in cm.

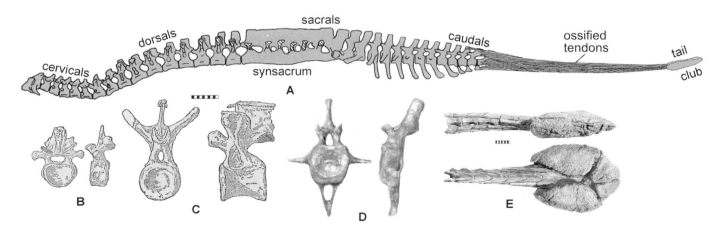

ribs attach to the vertebrae, are angled upward between 30 and 50 degrees. This high angle makes the ribs arc outward, producing a barrel chest. The last four or more ribs are often immovably fused to their vertebrae. This fusion is apparently related to the extremely wide pelvis (more on this below).

The caudal vertebrae are elongated, except near the pelvis, where they are much wider than long (Fig. 24.7D). In advanced ankylosaurids the caudals in the last half of the tail are modified to form a "handle" for a large bone club on the end of the tail. These are further strengthened with ossified tendons (Fig. 24.7A). The club is formed by the fusion of several large osteoderms to the last few tail vertebrae (Fig. 24.7E). In nodosaurids, polacanthids, and possibly shamosaurine ankylosaurids, the caudals are not modified to carry a bone club. Instead, the tail was long and slender.

The front and rear limbs of ankylosaurs are proportionally short (Fig. 24.2), especially in the ankylosaurids (Coombs 1978a, b, 1979), making the body long and low (see Plates 27–29). The coracoid and scapula (shoulder blade) are another major area separating the three families (Fig. 24.8). In polacanthids, the coracoid is a disk-shaped plate that is pierced by the coracoid foramen, whereas it is typically an elongated rectangle in nodosaurids, and a short, vertical rectangle or square in ankylosaurids. The scapula is an elongate structure that is curved so as to conform to the shape of the rib cage. The upper and lower edges of the scapular blade may diverge or may be parallel to each other, depending on the species. For example, in

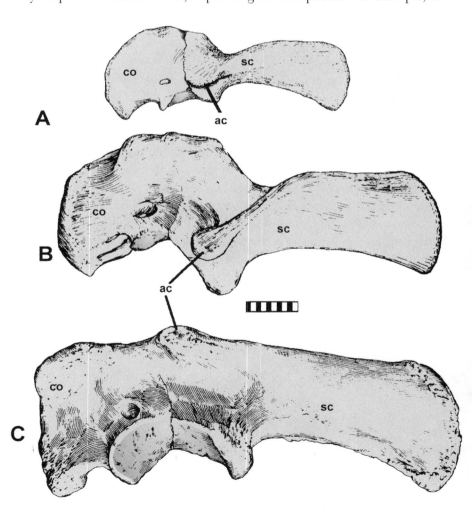

24.8. Representative left shoulder blades of ankylosaurs. A, polacanthid *Gastonia;* B, nodosaurid *Sauropelta;* C, ankylosaurid *Euoplocephalus.* Abbreviations: ac–acromion process, co–coracoid, sc–scapula. Scale in cm.

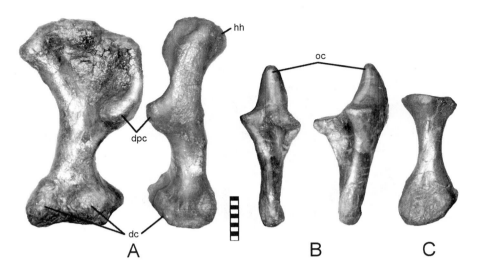

A   B   C

24.9. Representative left forelimb bones (*Euoplocephalus*) showing their characteristic stockiness. A, humerus (upper arm bone) in front and side views; B, ulna in front and side views; C, radius in side view. Abbreviations: dc – distal condyles (forms part of the elbow joint); dpc – deltopectoral crest (attachment site for many upper arm muscles); hh – humeral head (fits into the arm socket); oc – olecranon (attachment site for muscles straightening the lower arm, which is important for locomotion). Scale in cm.

Gastonia burgei, the edges diverge toward the end of the blade, but in a new, as yet unnamed, species, the edges are parallel. There is a flange or knob on the outer side of the scapula, above the shoulder socket, called an acromion process; it serves as the attachment site of the scapulohumeralis anterior muscle. In polacanthids, this flange is rather broad and projects at an angle downward and outward. In contrast, it is a ridge ending in a knob in nodosaurids, and in ankylosaurids is either a short flange or a swelling along the top edge of the scapula. The position of the knob varies among the different species of nodosaurids, making the shoulder blade important in defining different nodosaurid species. The cuplike glenoid, or shoulder socket, is shared by both the coracoid and scapula.

The upper arm bone, or humerus, is short and stout in ankylosaurs (Fig. 24.9A, B). There is a very prominent flange of bone on the upper part of the humerus called the deltopectoral crest. Although all dinosaurs have this crest, in ankylosaurs (and other quadrupedal dinosaurs) it is especially well developed. The crest was the attachment site for several major muscles used in locomotion. Typically, the humerus of nodosaurids is proportionally more slender for its length than for ankylosaurids and polacanthids. The humeral head, which fits into the glenoid, is located more on the backside of the humerus than the front side. In contrast, the condyles to which the lower arm bones articulate are more developed on the front side of the humerus than on the back side. These features tell us that the forearms were held not straight, elephantine-like, but flexed (see Fig. 24.2). The lower arm consists of a large, robust ulna (Fig. 24.9C, D) and a smaller radius (Fig. 24.9E). At the elbow, the ulna has a very tall olecranon, to which the extensor muscles were attached. The forefeet are not very well known in ankylosaurs because the bones are frequently lost prior to burial. Those that are known are short and broad. There are five digits in the forefeet of primitive ankylosaur species and four in advanced species. The toes end in wide, flat hooves.

The pelvis of ankylosaurs is considerably modified from that seen in other ornithischians. The upper pelvic bone, the ilium, is very large and is oriented almost horizontally, encompassing a large gut (Fig. 24.10). Muscle scars on the underside of the ilium show that the muscles for moving the hind leg were large and powerful. The pubis, located below the ilium, is reduced to a small block or blade of bone, or may be missing altogether. In

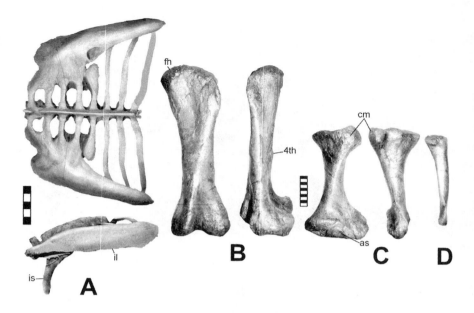

24.10. Representative hind limb and pelvis (*Euoplocephalus*). A, pelvis in top and side views. Note how much the ilium has rotated into a nearly horizontal position and is angled outward to accommodate a large belly. B, left femur, or thigh bone, in front and side views; C, left tibia, or shin bone, in front and side views; D, fibula in side view. Abbreviations: 4th—fourth trochanter (attachment for the caudofemoralis muscle, a major retractor); as—astragalus; cm—cnemial crest (attachment site of extensor muscles for the lower leg); fh—femoral head; il—ilium; is—ischium. Scales in cm.

contrast, the ischium remains prominent and projects downward in a manner more typical to saurischian dinosaurs. The ischium is rather straight shafted in ankylosaurids, but is bowed or kinked at midshaft in nodosaurids and polacanthids. The acetabulum, or hip socket, is a cuplike depression formed by the ilium and ischium, with little or no contribution by the pubis. In most dinosaurs, the acetabulum is open, allowing the femur head to extend partially through. In contrast, the acetabulum is closed in nodosaurids and ankylosaurids, and the femur head more of a ball, so that it forms a ball-and-socket joint with the acetabulum. However, the acetabulum is incompletely closed in polacanthids, especially the primitive *Mymoorapelta*; it remains a slit in the more advanced *Gastonia*. Most likely these openings were closed cartilage, making a cuplike socket.

The femur (thighbone) is pillarlike, being straight and massive (Fig. 24.10C, D) to carry the enormous weight of the ankylosaur body. The femoral head is a bulbous mass set at an angle to the shaft, rather than a cylinder set at a right angle as in bipedal dinosaurs. The fourth trochanter, where the major leg retractor, the caudofemoralis, attaches, is a roughened scar, rather than the hooklike structure seen in small bipedal dinosaurs, such as *Hypsilophodon*. The femur rests atop two short but massive lower leg bones, the tibia and fibula (Fig. 24.10E–G). The tibia is twisted along its length so that the knee portion is at right angles to the foot end. The insertion area for extensor muscles is a prominent crest, called the cnemial crest, at the front of the knee joint. The ankle consists of the astragalus and calcaneum, but as individuals mature, these fuse to the ends of the tibia and fibula respectively. As in all dinosaurs, the ankle is a simple hinge, but why this fusion occurred in ankylosaurs is unknown; it is not size related, as similar fusion occurs in *Stegosaurus*. The hind feet are short and stout, but as with the forefeet, the number of digits of the hind feet is not very well known. Nodosaurids appear to have four digits, whereas advanced ankylosaurids usually have three (the count is unknown in polacanthids).

The most distinctive feature about the ankylosaurs as dinosaurs is their armor, which is composed of flat or keeled osteoderms of different sizes and shapes, as well as elongate spines and plates. This larger armor is surrounded

A

B

C

24.11. Osteoderms, or armor of ankylosaurs in top view. A, *Gargoyleosaurus* (reconstructed from armor found in life position) showing the characteristic triangular spiny plates of polacanthids along the neck and sides of the body. Also, the armor over the pelvis is fused together. B, *Edmontonia* showing articulated armor as found, showing low-keeled neck plates and spikes along the side of the body. C, *Euoplocephalus* armor as found, including bands of thin-walled, conelike osteoderms arranged in transverse rows across the back and tail. Roughly to scale.

by a mosaic of smaller osteoderms, including irregular marble- to pea-sized osteoderms that may also extensively cover the arms, throat, and belly regions. Several specimens have been found with the armor preserved in natural or life position (Fig. 24.11B, C), allowing for fairly accurate reconstructions (Fig. 24.1).

Armor

Polacanthid armor consists of at least two rows of plates on the neck, oval, keeled plates arranged in alternating rows along the back, transverse parallel rows on the tail, and flattened, broad-based triangular plates or spines projecting along the sides of the neck, body, and tail (Fig. 24.12A). Armor on the pelvis is typically fused together as seen in *Gargoyleosaurus*, *Gastonia*, and *Polacanthus*. Those spines along the sides of the neck are characterized

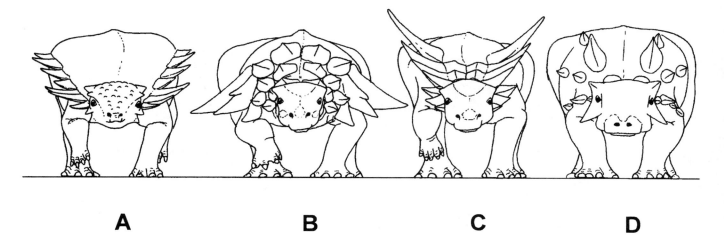

| A | B | C | D |

24.12. Front views of various ankylosaurs emphasizing the major armor (A, polacanthid *Gargoyleosaurus;* B, nodosaurid *Edmontonia;* C, nodosaurid *Sauropelta;* D, ankylosaurid *Euoplocephalus*). Spikes or spines are especially prominent in the nodosaurids, suggesting that display may have been important in this group. Not to scale.

by a long groove along their back edge whose function is apparently to accommodate the front edge of the following spine. In contrast, the armor of nodosaurids typically consists of three rows of large, platelike osteoderms on the neck, but is otherwise highly variable and may include long spines or spikes projecting from the neck and shoulders. For example, in *Edmontonia* there is a pair of forward projecting spines at the base of the neck (Fig. 24.11B, 24.12B), whereas in *Sauropelta* there are two rows of spines, with the largest spines projecting upward and outward from the neck (Fig. 24.12C; see Plate 29). Nodosaurids also typically have thick triangular plates projecting from the sides of the body, and parallel rows of keeled oval plates on the back. The armor on the hips is rarely fused, and consists of low-keeled to flat-profiled oval or circular plates surrounded by smaller osteoderms, usually arranged in a rosette pattern.

In ankylosaurids there are usually two rows of neck plates and thin-walled triangular spines or plates projecting from the sides of the neck, body, and tail. The back is usually covered by oval keeled plates arranged in parallel (*Euoplocephalus*) or both transversely parallel and alternating rows (*Saichania*). Armor on the pelvis usually is a continuation of the armor covering the back or a variation of that armor. Ankylosaurine ankylosaurids have a club formed by enlarged fused plates at the end of the tail, whereas it appears that shamosaurine ankylosaurids do not. A growth series for the ankylosaurid *Pinacosaurus* shows that the body armor was nearly absent in the very youngest specimens known (about 1 m long), being restricted to the bands of neck armor. The tail club does not appear until later in life, when the individual is over half grown (about 2.5 m). A fossilized ankylosaurid club is very heavy as a result of the minerals that fill the very porous internal structure. In life, organic tissue filled these spaces, making the club strong and resilient.

Histological studies of ankylosaur osteoderms show them to be distinct at the family level (Scheyer and Sander 2004). *Scelidosaurus* and polacanthid osteoderms have a thickened, uniform outer layer (cortex) completely surrounding the inner latticelike (trabecular) bone. Nodosaurid osteoderms have a complex three-dimensional orthogonal structural fibers at 45° angle to each other in the cortex, which overlies a thickened zone of trabecular bone. Finally, ankylosaurids have thin osteoderms that also have reinforcing structural fibers in the cortex, although these are not arranged as neatly as in nodosaurids.

Table 24.1. Chronological Distribution of Ankylosaurs. Note the uneven occurrence due to the spottiness of the fossil record.

			Polacanthidae	Ankylosauridae	Nodosauridae	Of Uncertain Family Position
Cretaceous	Late	Maastrichtian		Ankylosaurus Nodocephalosaurus Tarchia	Edmontonia Struthiosaurus	
		Campanian		Euoplocephalus Pinacosaurus Saichania	Aletopelta Antarctopelta* Edmontonia Panoplosaurus Struthiosaurus	
		Santonian		Talarurus Tsagantegia	Hungarosaurus Niobrarasaurus	
		Coniacian		Bissektipelta	Niobrarasaurus	
		Turonian				
		Cenomanian		Zhongyuanosaurus Crichtonsaurus	Animantarx Nodosaurus Silvisaurus Stegopelta	
	Early	Albian		Cedarpelta Gobisaurus Shamosaurus	Anoplosaurus Pawpawsaurus Peloroplites Prioconodon Sauropelta Texasetes	Minmi
		Aptian			Sauropelta	Minmi
		Barremian		Hoplitosaurus Gastonia Polacanthus		Liaoningosaurus
		Hauterivian				
		Valanginian	Hylaeosaurus			
		Berriasian	Hylaeosaurus			
Jurassic	Late	Tithonian				
		Kimmeridgian	Dracopelta Gargoyleosaurus Mymoorapelta			
		Oxfordian				
	Middle	Callovian				Sarcolestes
		Bathonian				
		Bajocian				
		Aalenian				
	Early	Toarcian				
		Pliensbachian				
		Sinemurian				Scelidosaurus
		Hettangian				
Triassic						

* Probably not valid since the material is not diagnostic.

Origin and Evolution

The origin of the ankylosaurs is not well understood because so few specimens are known from the Early and Middle Jurassic, when ankylosaurs apparently first appeared and diversified (see Table 24.1). Several cladistic studies have shown that ankylosaurs and stegosaurs, which together are called the thyreophorans, share a common ancestry (e.g., Norman et al. 2004), which includes *Scutellosaurus* from the Lower Jurassic of Arizona (USA) and *Emausaurus* from the Lower Jurassic of Germany. However, there is disagreement about where the split occurred, with some studies (e.g., Norman et al. 2004) placing *Scelidosaurus* from the Lower Jurassic of England (Table 24.1) as basal to both stegosaurs and ankylosaurs, along with

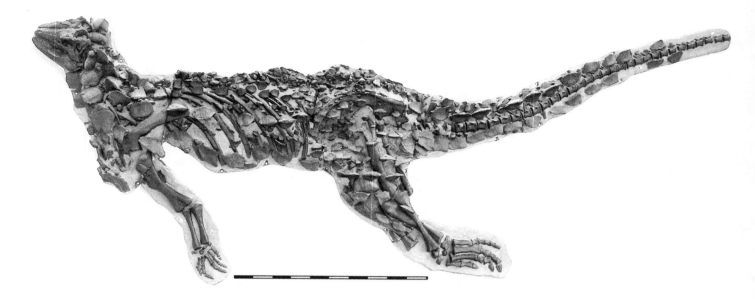

24.13. The most primitive ankylosaur, *Scelidosaurus* from the Early Jurassic of England. Cast on display at the Bristol City Museum and Art Gallery, Bristol, England. Scale length 1 m.

24.14. Hypothetical relationship of ankylosaurs. Specimens known only by skulls are not included because a skull does not represent the entire animal. Results based on 23 taxa and 73 characters organized using NEXUS Data Editor (Page 2001a), analyzed with PAUP* (Swofford 2002), trees analyzed with COMPONENT (Page 1993), and created with TreeView (Page 2001b). Although *Liaoningosaurus* is known from two skeletons, both are very immature (posthatchlings) with many of the characters needed for proper analysis not yet ossified. Therefore, it was excluded from the analysis. A, Strict consensus tree for heuristic search rooted at *Lesothosaurus* resulting in 25 trees, tree length = 191, consistency index 0.54, homoplasy index 0.45, retention index 0.76. B, Edited, and preferred, tree, because the shamosaurines in A would essentially be a separate family, rather than a subfamily, and because *Gargoyleosaurus* and *Gastonia* are recognized as separate from the ankylosaurids, but closer to each other. Support for two subfamilies of ankylosaurids has been shown by Vickaryous et al. (2004) as predicted by Tumanova (1983). Although B is not yet supported by phylogenetic analysis, I predict it will be in time. Abbreviations: A–Ankylosauria; AN–Ankylosauridae; ANK–Ankylosaurinae; NO–Nodosauridae; P–Polacanthidae; S–Shamosaurinae; T–Thyreophora.

Scutellosaurus, and other studies placing *Scelidosaurus* as the most primitive ankylosaur (e.g., Carpenter 2001; Figure 24.14).

The skull of *Scelidosaurus* shows some remodeling of the surface as seen in ankylosaurs, although this not so extensive as to give the appearance of armor. The skull still retains the narrow and tall profile such as seen in non-thyreophoran ornithischians. In addition, both the antorbital and supratemporal fenestrae, features seen in most dinosaurs, are open. The pelvis is also primitive, and in side view, it very much resembles that of primitive ornithischians, such as *Lesothosaurus*. However, in top view the ilium shows the initial stages becoming horizontal as seen in later ankylosaurs. The neck, body, and tail are extensively covered with rows of osteoderms as in ankylosaurs as well (Fig. 24.13).

Ankylosaurs from the Middle Jurassic are represented by fragmentary remains, and little is known of their relationships to later ankylosaurs. Most are so fragmentary, consisting of single bones or teeth, that they cannot be differentiated from other ankylosaurs and are not considered valid species (e.g., *Cryptodraco* is known only from a femur). Late Jurassic ankylosaurs, all polacanthids, are better known (Table 24.1), especially *Gargoyleosaurus* from the Morrison Formation. It is represented by a large portion of the skeleton with the armor preserved in life position, including the characteristic grooved neck spines of the polacanthids. A recently discovered pelvis may belong to this genus and is intermediate between that of *Scelidosaurus* and later polacanthids. Two other polacanthids known from this time are *Mymoorapelta* from the western United States and *Dracopelta* from Portugal.

The record for Early Cretaceous ankylosaurs is better known, with the first appearances of the nodosaurids and ankylosaurids, and the last appearance of polacanthids. These last polacanthids include *Polacanthus* and *Hylaeosaurus* from Europe and *Gastonia* and *Hoplitosaurus* from the western United States. Polacanthids apparently become extinct near the end of the Barremian, about the time the first nodosaurids appear, as well as the first flowering plants; a cause and effect among these has yet to be firmly established, however. *Gastonia* is the best known polacanthid, with tens of individuals of different sizes known from bone beds in the Cedar Mountain Formation.

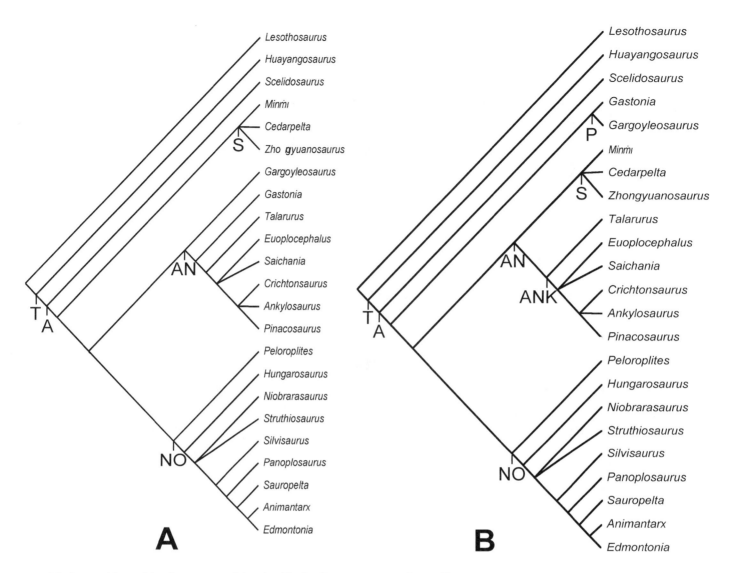

A

B

Nodosaurids suddenly appeared in the Early Cretaceous and rapidly diversified (Table 24.1). One of the earliest, *Sauropelta*, has all the hallmarks of nodosaurids already well developed, indicating even more primitive nodosaurids have yet to be found. Nodosaurids, as well as ankylosaurids, underwent a diversity and numerical decline prior to the end of the Cretaceous (Carpenter and Breithaupt 1986). Understanding the evolution of nodosaurids is complicated because most nodosaurids are known from one or two partial specimens (e.g., *Hungarosaurus*). *Edmontonia* is an exception, being represented from several skulls and partial skeletons, including some with armor preserved in life position (24.11B). In North America, nodosaurids include *Edmontonia*, *Niobrarasaurus*, *Nodosaurus*, *Panoplosaurus*, *Silvisaurus*, and *Stegopelta*. In Europe, nodosaurids are represented by *Hungarosaurus* and three species of *Struthiosaurus*. Nodosaurids have only recently been found in Asia (Tumanova et al. 2003), South America (Coria and Salgado 2001), and Antarctica (Salgado and Gasparini 2006), but the specimens are fragmentary and rare.

The first ankylosaurids all belong to the shamosaurine subfamily, which is characterized by a boxy skull, lack of horns on the upper rear (squamosal horn) corners of the skull, tapering or narrow muzzle, and apparently

24.15. Shoulder spikes of the nodosaurid *Edmontonia* as an antipredator device against the predator *Gorgosaurus*.

no tail club. The most primitive shamosaurine, *Cedarpelta*, retains teeth along the sides of the premaxilla. The shamosaurines reached their greatest diversity during the Albian, and then apparently went extinct by the end of the Cenomanian, shortly after the first appearance of the ankylosaurine subfamily, which is represented by *Crichtonsaurus*. Ankylosaurine ankylosaurids are characterized by horns in the upper (squamosal horn) and lower (jugal-quadratojugal horn) corners of the skull, usually a broad beak (moderately wide in the most primitive species, *Crichtonsaurus*), and a tail club. During the rest of their evolution, ankylosaurines greatly diversified across Asia and North America, where they were restricted. *Pinacosaurus*, *Saichania*, *Talarurus*, and *Tarchia* are known from Asia. Most of these taxa are known from skulls, and some by nearly complete skeletons with armor preserved in situ. North American ankylosaurids include *Ankylosaurus* and *Euoplocephalus*, both known from skulls and partial skeletons (Carpenter 1982; Penkalski 2001)

One other ankylosaur, *Minmi*, known from the Lower Cretaceous of Australia, is enigmatic. Although represented by two specimens, including a nearly complete skeleton (Molnar 1996), its relationship to other ankylosaurs is controversial. In some studies (e.g., Vickaryous et al. 2004), it is a primitive ankylosaur outside the three families. However, it does share with shamosaurine ankylosaurids a skull with the similar boxy profile, absence of upper cranial (squamosal) horns, moderately developed lower cranial (jugal-quadratojugal) horns, and narrow beak. The only known skull is of a young animal, as indicated by the open cranial sutures and very small, thin osteoderms resting on the skull surface.

The distribution of ankylosaurs suggests that they first appeared in Europe during the Lower Jurassic from a primitive thyreophoran related to *Scelidosaurus*. Polacanthid ankylosaurs appeared at least by the late Middle Jurassic, probably earlier, and spread from Europe to North America, where they arrived by the Late Jurassic. Nodosaurids appear in North America during the late Early Cretaceous before they appear in Europe. However, the oldest nodosaurid, *Sauropelta*, is rather advanced and indicates that considerable amount of evolution had preceded it. Unfortunately, it is not known whether nodosaurids originated in Europe or North America. Regardless, they eventually spread to South America and Antarctica by the Late Cretaceous. The most primitive shamosaurine ankylosaurid, *Cedarpelta*, suggests the subfamily originated in North America during the late Early Cretaceous and from there spread to Asia. The oldest ankylosaurine ankylosaurid, *Crichtonsaurus*, has a primitive skull, suggesting this subfamily may have originated in Asia in the early Late Cretaceous, and then migrated to North America. Ankylosaurids apparently never migrated to Europe or the Southern Hemisphere.

Ankylosaurs are characterized by a low, wide body that is encased in armor embedded in the skin (Fig. 24.1). The armor was probably formed within the skin in a manner similar to that of crocodiles and armored lizards, that is, by ossification of cartilage nodules (Moss 1969). In *Pinacosaurus*, it is known that ossification began from the neck and progressed toward the tail, but whether this was the standard order of armor formation among ankylosaurs is unknown. Traditionally ankylosaur armor was thought to have provided protection against large predators. Among nodosaurids, the forward-projecting neck spines of *Edmontonia* could certainly have been used as a weapon. The largest spine is braced against the base of the neck, and the most obvious purpose of such a spine is as an antipredator device, used to keep predators at bay (Fig. 24.15). In other ankylosaurs, the tail might have been used for defense, especially if a tail club was present (Coombs 1979), as in the ankylosaurids (Fig. 24.16; Plate 28).

Many other nodosaurids, such as *Sauropelta*, have spines that project upward, suggesting that not all spines were used against predators (Fig. 24.12C; Plate 29). Such spines are certainly noticeable, and may have been similar to the "showy" structures used by living animals in behavioral interactions, such as the horns in antelopes. Showy structures are important among many living animals for sexual and agonistic display. Sexual display is used to attract a potential mate, whereas agonistic display is used to drive away a rival. Agonistic behavior can be divided into threat and intimidation (Leuthold 1977). Threat behavior involves the prominent display of a weapon toward the opponent, thereby signaling a willingness to fight.

Biology and Behavior

24.16. Use of the tail club as an offensive weapon by *Euoplocephalus*.

Intimidation, on the other hand, is a form of "psychological warfare" in which the displaying animal tries to make itself appear larger, thus more dangerous to fight with. While the spines in *Sauropelta* emphasize height, those of *Edmontonia* project outward, making the animal seem more squat and massive than it really is (Fig. 24.12B). Polacanthids, ankylosaurids, and the nodosaurid *Panoplosaurus* lack large projecting neck spines, suggesting that frontal intimidation display may not have been important in these animals (Fig. 22.12A, D). Instead, side intimidation display may have been used, as inferred from the behavior of modern hornless (i.e., primitive) ungulates. In this type of intimidation, the animals probably stood parallel to each other with much hissing, growling, or whatever sounds ankylosaurs made.

It is also possible that ankylosaurs could have "blushed," so that the armor had a pink tint as a result of the infusion of blood under the horny covering of the armor. The vascular grooves covering the surface of most ankylosaur armor indicate that a rich supply of blood was present. Imagine two-ton ankylosaurs with pinkish armor bellowing and growling at each other! *Edmontonia* may have had an additional use for the shoulder spine other than to keep predators at bay. The spine is bifurcated, and this structure may be analogous to the tines on a deer antler. In deer, the antlers are engaged by their tines during intraspecific fighting. This keeps the antlers from slipping past one another and possibly injuring the combatants. It

24.17. Use of the bifurcated neck spike in shoving contest of strength by two male *Edmontonia* during the breeding season.

is possible that the bifurcated spine of *Edmontonia* may have functioned like an antler tine, allowing the animals to engage in shoving matches for dominance during the breeding season without serious injury (Fig. 24.17).

One final role that the body armor may have played is thermoregulation, or body temperature control (Blows 2001). The large surface area of the armor over the broad body would have provided ample opportunity for excess body heat to be dissipated, thus preventing overheating. Such a role has been documented for armor in crocodiles (Seidel 1979) and may have been crucial for ankylosaurs living in desert environments, like *Pinacosaurus*.

With short legs, ankylosaurs were built low to the ground and were probably restricted to feeding on plants or plant parts 2 meters or less from the ground. Ankylosaurs apparently had a keratinous covering on the beak for cropping the vegetation. The type of plants ankylosaurs ate depended on where the animals lived and on the shape of the muzzle (Carpenter 1982). Most of the ankylosaurids from Asia, such as *Saichania*, lived in arid or semiarid environments where the plants were adapted to water stress. Such plants tend to have a waxy coating on their leaves to retard evaporation, and this can make digestion more difficult. On the other hand, many ankylosaurids from North America lived in well-watered coastal environments, where vegetation was lush and lacked these protective structures. Interestingly, however, the teeth of the Asian and North American ankylosaurids show no major differences in size or structure as might be expected because of the different plant diet. There are, however, noticeable differences between the teeth of nodosaurids and ankylosaurids, which were contemporaneous in the Late Cretaceous of North America. The differences (Fig. 24.6) in the teeth show that they probably avoided competition for food. In addition, the shape of the muzzles differs between the two families, showing that how food was gathered also differed. The broad muzzle of the ankylosaurid *Euoplocephalus* suggests that this animal was a generalized feeder, cropping low plants with its wide beak (Carpenter 1982). On the other hand, its contemporary, the nodosaurid *Edmontonia*, had a narrower muzzle, so it could selectively crop vegetation. This partitioning of food resources enabled the ankylosaurs to cohabit the Late Cretaceous coastal plain.

Ankylosaurs apparently had a very mobile tongue, because of the large size of the throat bones (hyoids) that lay at the base of the tongue. Having a mobile tongue would allow the ankylosaurs to push and roll a wad of vegetation around in the mouth while chewing. Processing food in this manner requires some way to keep the food in the mouth, hence the need for cheeks. Most likely the cheeks were fleshy, not muscular as in mammals. Cheeks can be inferred from cheek teeth being inset from the sides of the face and by the presence of cheek osteoderms in *Panoplosaurus* and *Edmontonia*. Stomach content has been reported for *Minmi* and consists of vascular tissue fragments, seed-bearing organs, seeds, and possibly sporangia (Molnar and Clifford 2001). Because cellulose, which forms the cell walls of plants, is difficult to digest, nutrients can be extracted only by microbial fermentation. In living mammals, this may occur in special chambers of the stomach, as in the cow (foregut fermentation), or in the intestines (hindgut fermentation), as in the horse. Herbivorous lizards use hindgut fermentation (McBee and McBee 1982) as well because this method is easier to evolve. Ankylosaurs most likely utilized hindgut, or intestinal, fermentation as well. The very broad, horizontal hips suggest a very large or long rear gut (DiCroce et al. 2005). Fusion of the rearmost sets of ribs to their respective dorsal vertebrae probably functioned to make the rear portion of the body rigid as an accommodation for the enlarged gut. Most likely ankylosaurs were walking methane machines.

In summary, the armored ankylosaurs were a very successful group of ornithischian dinosaurs. They apparently originated in Europe during the Early Jurassic and waddled their way to North America and Asia, then on to South America and Antarctica before the end of the Cretaceous. Along the way, they adapted to a variety of environments and adjusted their diets accordingly. The evolution of ankylosaurs is marked by a progressive widening of the hindgut, as seen by the widening of the pelvis. This change may have been the result of changing diet and the need for a longer fermentation time.

References

Blows, W. T. 2001. Dermal armor of the polacanthine dinosaurs. In K. Carpenter (ed.), *The Armored Dinosaurs*, 363–385. Bloomington: Indiana University Press.

Carpenter, K. 1982. Skeletal and dermal armor reconstruction of *Euoplocephalus tutus* (Ornithischia: Ankylosauridae) from the Late Cretaceous Oldman Formation of Alberta. *Canadian Journal of Earth Sciences* 19: 689–697.

——. 2001. Phylogenetic analysis of the Ankylosauria. In K. Carpenter (ed.), *The Armored Dinosaurs*, 455–483. Bloomington: Indiana University Press.

Carpenter, K., and B. Breithaupt. 1986. Latest Cretaceous occurrence of nodosaurid ankylosaurs (Dinosauria, Ornithischia) in western North America and the gradual extinction of the dinosaurs. *Journal of Vertebrate Paleontology* 6: 251–257.

Carpenter, K., J. I. Kirkland, D. Burge, and J. Bird. 2001. Disarticulated skull of a new primitive ankylosaur from the Lower Cretaceous of eastern Utah. In K. Carpenter (ed.), *The Armored Dinosaurs*, 211–238. Bloomington: Indiana University Press.

Coombs, W. P. [Jr.]. 1978a. The families of the ornithischian Order Ankylosauria. *Palaeontology* 21: 143–170.

——. 1978b. Forelimb muscles of the Ankylosauria (Reptilia, Ornithischia). *Journal of Paleontology* 52: 642–658.

——. 1978c. An endocranial cast of Euoplocephalus (Reptilia: Ornithischia). *Palaeontographica* (Abt. A), 161: 176–182.

——. 1979. Osteology and myology of the hindlimb in the Ankylosauria (Reptilia, Ornithischia). *Journal of Paleontology* 53: 666–684.

Coria, R. A., and L. Salgado. 2001. South American ankylosaurs. In K. Carpenter (ed.), *The Armored Dinosaurs*, 159–168. Bloomington: Indiana University Press.

DiCroce, T., K. Carpenter, and B. Kinneer. 2005. Reconstruction of the pelvic and hind limb musculature in the ankylosaur *Gastonia*. *Journal of Vertebrate Paleontology* 25 (suppl. to 3): 51A.

Leuthold, W. 1977. *African Ungulates*. Berlin: Springer-Verlag.

Maryańska, T. 1977. Ankylosauridae (Dinosauria) from Mongolia. *Paleontologia Polonica* 37: 85–151.

McBee, R. H., and V. H. McBee. 1982. The hindgut fermentation in the green iguana, *Iguana iguana*. In G. M. Burghardt (ed.), *Iguanas of the World: Their Behavior, Ecology and Conservation*, 77–83. New York: Noyes Publications.

Molnar, R. E. 1996. Preliminary report a new ankylosaur from the Early Cretaceous of Queensland, Australia. *Memoirs of the Queensland Museum* 39: 653–668.

Molnar, R. E., and H. T. Clifford. 2001. An ankylosaurian cololite from the Lower Cretaceous of Queensland, Australia. In K. Carpenter (ed.), *The Armored Dinosaurs*, 399–412. Bloomington: Indiana University Press.

Moss, M. L. 1969. Comparative histology of dermal sclerifications in reptiles. *Acata Anatomica* 73: 510–533.

Norman, D. B., L. M. Witmer, and D. B. Weishampel. 2004. Basal Thyreophora. In D. B. Weishampel, P. Dodson, and H. Osmolska (eds.), *The Dinosauria*, 335–342. Berkeley: University of California Press.

Page, R. D. M. 1993. COMPONENT. http://taxonomy.zoology.gla.ac.uk/rod/cpw.html.

——. 2001a. NEXUS Data Editor. http://taxonomy.zoology.gla.ac.uk/rod/NDE/nde.html.

——. 2001b. TreeView (WIN 32). http://taxonomy.zoology.gla.ac.uk/rod/treeview.html.

Penkalski, P. 2001. Variation in specimens referred to *Euoplocephalus tutus*. In K. Carpenter (ed.), *The Armored Dinosaurs*, 261–297. Bloomington: Indiana University Press.

Rybczynski, N. and M. K. Vickaryous. 2001. Evidence of complex jaw movement in the Late Cretaceous ankylosaurid, *Euoplocephalus tutus* (Dinosauria: Thyreophora). In K. Carpenter (ed.), *The Armored Dinosaurs*, 299–317. Bloomington: Indiana University Press.

Salgado, L., and Z. Gasparini. 2006. Reappraisal of an ankylosaurian dinosaur from the Upper Cretaceous of James Ross Island (Antarctica). *Geodiversitas* 28: 119–135.

Scheyer, T. M., and Sander, P. M. 2004. Histology of ankylosaur osteoderms: implications for systematics and function. *Journal of Vertebrate Paleontology* 24: 874–893.

Seidel, M. R. 1979. The osteoderms of the American alligator and their functional significance. *Herpetologica* 35: 375–380.

Swofford, D. L. 2002. PAUP*. Phylogenetic Analysis Using Parsimony (*and Other Methods). Version 4. Sinauer Associates, Sunderland, Mass.

Tumanova, T. A. 1983. The first ankylosaurs from the Lower Cretaceous of Mongolia. *Transactions of the Soviet-Mongolian Palaeontological Expeditions* 24: 110–120.

Tumanova, T. A., V. R. Alifanov, and Y. L. Bolotsky. 2003. First find ankylosaur remains in Russia. *Priroda* 3: 69–70.

Vickaryous, M. K., and A. P. Russell. 2003. A redescription of the skull of *Euoplocephalus tutus* (Archosauria: Ornithischia): a foundation for comparative and systematic studies of ankylosaurian dinosaurs. *Zoological Journal of the Linnean Society*, 137: 157–186.

Vickaryous, M. K., A. P. Russell, and P. J. Currie. 2001. The cranial ornamentation of ankylosaurs (Ornithischia: Thyreophora): re-appraisal of developmental hypothesis. In K. Carpenter (ed.), *The Armored Dinosaurs*, 318–340. Bloomington: Indiana University Press.

Vickaryous, M. K., T. Maryańska, D. Weishampel. 2004. Ankylosauria. In D. B. Weishampel, P. Dodson, and H. Osmolska (eds.), *The Dinosauria*, 363–392. Berkeley: University of California Press.

Witmer, L. M., and R. C. Ridgely. 2008. The paranasal air sinuses of predatory and armored dinosaurs (Archosauria: Theropoda and Ankylosauria) and their contribution to cephalic structure. *Anatomical Record: Advances in Integrative Anatomy and Evolutionary Biology* 291: 1362–1388.

25.1. Dorsal views of the skulls of *Homalocephale* (A) and *Prenocephale* (B). Lateral view of the skull of *Stegoceras* (C) and *Prenocephale* (D). Note the primitively large supratemporal fenestrae in the flat-headed taxon *Homalocephale* (A). In advanced dome-headed forms such as *Prenocephale,* the dome extends across the whole skull roof and roofs over the supratemporal fenestrae (B). *Stegoceras* photograph courtesy of Alan Lindoe. Specimens not to scale.

A

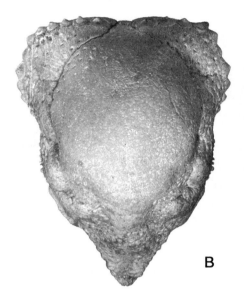

B

C

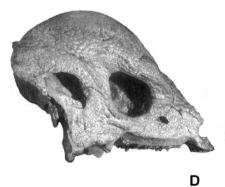

D

Marginocephalia

25

Peter Makovicky

The Marginocephalia, or margin-headed dinosaur group, comprises the thick-skulled pachycephalosaurs and the frill-bearing ceratopsians. Marginocephalians have a predominantly Cretaceous fossil record, although the earliest species of ceratopsians are now known from the Late Jurassic (Xu et al. 2006; Zhao et al. 1999; Zhao et al. 2006). Until recently, Marginocephalia was thought to constitute the sister group to all Ornithopoda (Sereno 1986, 1999), a group dating back to the Early Jurassic, which would imply a long, undiscovered history for Marginocephalia. Recently this view has been challenged by Butler et al. (2008), who found marginocephalians to nest inside the ornithopod radiation, which significantly shortens the implied missing stratigraphic range between marginocephalians and their closest relatives. It is noteworthy that Late Jurassic *Yinlong downsi* (Xu et al. 2006), the earliest member of the Ceratopsia, displays a mosaic of

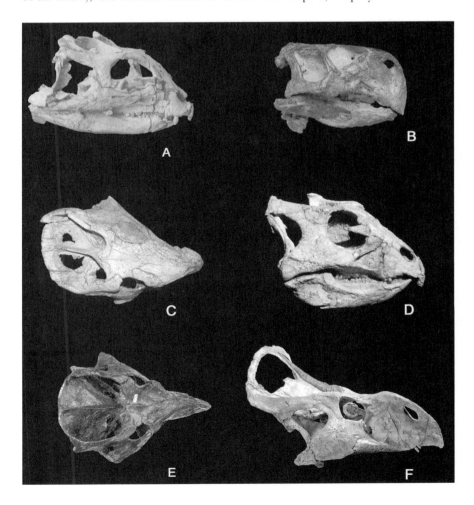

25.2. A gallery of basal ceratopsian taxa: (A) *Yinlong* in lateral view; (B) *Psittacosaurus* in lateral view; (C) *Liaoceratops* in dorsal view; (D) *Archaeoceratops* in lateral view; (E) *Leptoceratops* in dorsal view; (F) *Protoceratops* in lateral view. (B, E, F) photographed by M. Ellison, other photos by the author. Not to scale.

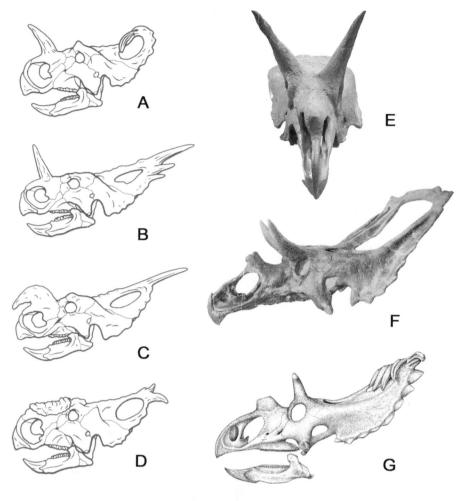

25.3. A gallery of ceratopsid taxa: (A–D) The centrosaurine genera *Centrosaurus*, *Styracosaurus*, *Einiosaurus*, and *Pachyrhinosaurus* in lateral view, displaying the diverse nasal horn forms and frill spike configurations found in this group. (E–G) The chasmosaurine genera *Triceratops* and *Pentaceratops*, and *Kosmoceratops* in anterior and lateral views respectively. Chasmosaurines differ from centrosaurines in having a longer snout, pockets in the narial region, enlarged brow horns, and long frills. *Triceratops* is anomalous in having a secondarily shortened frill without fenestrae. (A–D) taken from Sampson et al. (1997); (F) taken from Lull (1933); (G) drawing by Lukas Panzarin, courtesy of Mark Loewen. Specimens not to scale.

traits that were previously considered diagnostic of either ceratopsians or marginocephalians, suggesting that *Yinlong* is temporally close to their last common ancestor.

Although Cretaceous members of each clade are superficially very different from one another due to long separate evolutionary histories, ceratopsians and pachycephalosaurs are united by a number of derived skeletal features. Chief among these is a backward extension of the rearmost skull bones (squamosals and parietals) to form a shelf or expanded margin at the back of the skull. In pachycephalosaurs, the frill is relatively short and shelflike, and most of this margin is composed of the two outer bones (squamosals) (Fig. 25.1A). It is adorned with tubercles or spikes along the edges and corners in most pachycephalosaur species. The skull margin shows the greatest development in derived ceratopsians, where it forms a prominent frill or shield extending from the back of the skull above the neck (Figs. 25.3D–F), sporting an array of diagnostic spikes or bumps in ceratopsids. In contrast to pachycephalosaurs, the ceratopsian frill margin is comprised mainly by the parietal in all Cretaceous ceratopsians. The basal ceratopsian *Yinlong* possesses a pachycephalosaur-like condition of the frill (Fig. 25.2A), however, thus muting some earlier concerns that the different makeup of the frill in the two groups could reflect separate origins of the structure (Galton and Sues 1983). Other evolutionary novelties that unite marginocephalians are exclusion of the premaxillae from the internal nares and pubic bones that are not fused to the rest of the hip joint (Sereno 1986, 1999). Some of these features, such as the unfused pubes, also occur in ornithischians outside of Marginocephalia, however.

Pachycephalosaurs range in size from the minute and possibly juvenile *Wannanosaurus*, with an estimated skull length of less than 10 cm (4 inches) to the large *Pachycephalosaurus* with a 65-cm (2-foot) skull. In general, pachycephalosaurs display a primitive, bipedal ornithischian body plan. Conversely, their skulls are highly modified and have vertically thickened skull-roof bones that form either a wide, flat table or a convex dome in more derived species. This dome is formed by the frontal and parietal bones that roof the brain cavity, and can be up to 8 inches thick in the largest pachycephalosaurs. In highly domed species, the bones around the edge of the orbit also fuse and contribute to the structure of the dome. The skull roof is always far thicker than the depth of the brain cavity, even in flat-headed species, and its possible function has been the focus of much paleobiological study (see below). In most pachycephalosaurs the outer surface of the facial bones and part of the skull roof are ornamented with small tubercles, although the dome of advanced species is smooth or only slightly pitted.

Not surprisingly, much of the pachycephalosaur fossil record consists of isolated skullcaps, which preserve more readily than other parts of the skeleton due to their unusual thickness. Prior to the discovery of more complete materials, isolated pachycephalosaur skullcaps were interpreted as a variety of structures such as dinosaur kneecaps (Sereno 1989), and they were referred to different ornithischian clades, including stegosaurs and ceratopsians (see, e.g., Hatcher et al. 1907). Even now, the majority of named genera and species are based solely on skullcap specimens (Sullivan 2003, 2006; Longrich et al. 2010), and most of the characters used in pachycephalosaur systematics refer to this limited part of their anatomy. An unresolved debate persists regarding the validity of some of these species, because development of the dome appears to be an age-related trait (Goodwin and Horner 2004; Sullivan 2007).

A feature associated with the formation of the thickened skull roof is the progressive reduction in size of the supratemporal fenestrae. Whereas these openings are still present and large in most flat-headed pachycephalosaurs such as *Goyocephale* and *Dracorex*, they are very reduced in some dome-headed taxa such as *Stegoceras* (Fig. 25.1C) and are completely roofed over in the fully domed genera such as *Prenocephale* (Fig 25.1D) and *Pachycephalosaurus* (Maryańska and Osmólska 1974; Sereno 2000).

Pachycephalosaurs are unique among ornithischians in having an ossified interior wall to the eye socket (Fig. 25.1C), a feature that is otherwise present only in birds among dinosaurs (Maryańska and Osmólska 1974). A number of new bones evolved in pachycephalosaurs to form this wall, which would have firmly secured the eyeball and may have served as an adaptation to counteract the shock encountered during possible head-butting behaviors (Maryańska and Osmólska 1974).

The lower jaws of pachycephalosaurs are slender and have only small insertions for the major jaw muscles, indicating weak chewing capabilities. The external surface of the dentary as well as external bones at the back of

Pachycephalosauria–Cretaceous Boneheads

the jaw are ornamented with small pointed tubercles in the parts underlying the skin in the North American genus *Stegoceras* (Gilmore 1924), but are not ornamented in the Mongolian *Tylocephale* (Maryańska and Osmólska 1974).

The front (premaxillary) teeth are pointed and recurved with large denticles in *Prenocephale* and some other taxa, superficially resembling the teeth of the carnivorous Troodontidae, a similarity that caused considerable taxonomic confusion between the two groups for the first half of the twentieth century (Maryańska and Osmólska 1974). Indeed, many paleontologists considered the groups to be synonymous (e.g., Brown and Schlaikjer, [1943]) until troodontid theropod anatomy was better understood (Currie 1987). The cheek teeth of pachycephalosaurs resemble those of primitive ornithischians in being leaflike with large simple denticles along both margins. They are large in *Tylocephale*, but relatively small in other species. These size differences along with disparate wear patterns have been interpreted as evidence for differing diets among the species (Maryańska and Osmólska 1974). A kink in the upper jaw of some pachycephalosaurs, such as *Prenocephale*, is similar to the concave depression that accommodates a tusklike tooth on the lower jaw in *Heterodontosaurus*. Notably, an enlarged tusk is present near the tips of both the upper and lower jaws of *Goyocephale* (Maryańska 1990). *Dracorex* is unusual in lacking premaxillary teeth and having a rounded margin to the tip of the snout, which may have supported a keratinous pad against which the lower jaws would bite, such as is observed in manatees (Bakker et al. 2006).

Postcranial skeletal remains are rare, and no specimen has been found with a complete postcranium. Substantial postcranial remains are known for only *Stegoceras*, *Goyocephale*, *Homalocephale*, and *Stenopelix* although the pachycephalosaur affinities of the later have been questioned (Butler and Sullivan 2009). A few postcranial bones are known for *Wannanosaurus* and *Prenocephale* (Maryańska 1990). Postcranial features that unite pachycephalosaurs include unusually elongate ribs in the hip region and near the base of the tail, causing the hips to be very wide. Unlike the condition in other dinosaurs, the ischium virtually encircles the bottom half of the hip socket, thereby excluding the pubis (Maryańska and Osmólska 1974). This enlarged hip-socket section of the ischium is braced internally by two or three sacral ribs, another evolutionary novelty that is characteristic of pachycephalosaurs. In taxa that preserve the hip bone (ilium), this element is unusual in having a medial prong in dorsal view. The vertebral column of pachycephalosaurs also bears derived traits among ornithischians. Locking ridges and grooves occur on the zygapophyseal articular surfaces of the presacral vertebrae (Gilmore 1924; Maryańska 1990). These locking ridges occur in the neck vertebrae of *Dracorex*, and coincide with reduced mobility at the skull–neck joint relative to other dinosaurs, presumably representing adaptations for intraspecies agonistic behaviors using the thickened skull roof (Bakker et al. 2006). Some basal neoceratopsians have such ridges on the dorsal vertebrae (Brown and Schlaikjer 1943), but they are variably present among and even within species and may have had a separate evolutionary origin within that clade (Makovicky 2001, 2002). An intricate cylindrical basket of ossified tendons surrounds and braces the vertebrae in the outer half of the tail in *Stegoceras* and *Homalocephale*, and this feature was probably present in most other pachycephalosaurs. The forelimbs, which are known only in *Stegoceras*, are relatively short and weak. The hind limbs are stout, but proportionately short, and together with the remarkable width of the hips indicate that these animals were rather stocky and not highly cursorial.

Stenopelix from the Early Cretaceous of Germany has been considered the earliest and most primitive pachycephalosaur by some workers (Sereno 1989). It is known from a single articulated specimen that lacks the skull, but exhibits purported derived traits characteristic of pachycephalosaurs in its hip region. A number of authors (Sues and Galton 1987; Sullivan 2006) have questioned the validity and distribution of these traits, however, and *Stenopelix* is regarded as an indeterminate ornithischian by Sullivan (2006). Butler and Sullivan (2009) reevaluated the evidence for pachycephalosaurid affinities in a recent comprehensive review of *Stenopelix*, concluding that none of the evidence unambiguously supports such a referral and that the taxon is best regarded as an indeterminate marginocephalian. The roughly contemporaneous *Yaverlandia* from the Isle of Wight is represented by an isolated skullcap. Although only weakly domed, it has a thickened skull roof, suggesting that pachycephalosaurs had evolved this derived feature by the time the clade makes its appearance in the fossil record. Nevertheless, doubts have been voiced regarding the pachycephalosaurian identity of this specimen, and it is regarded as a possible theropod by Naish (as cited in Sullivan [2006]).

All other confirmed pachycephalosaur species are geographically restricted to Asia and North America and derive from Late Cretaceous sediments (Maryańska 1990). *Wannanosaurus* from China and *Homalocephale* and *Goyocephale* from Mongolia are "flat-headed" taxa, whereas the Mongolian *Prenocephale* and all described North American forms with the exception of *Dracorex* have domed skull roofs. *Stegoceras* was until recently considered the most common North American taxon and is known from many Campanian (85–72 mya) localities in Montana, Alberta, and New Mexico. *Stegoceras validum* is known from a complete skull and partial skeleton, but virtually all other specimens referred to this genus are skullcaps. Extensive variation is observed in overall size, the degree of skull-roof doming, and the development of a projected margin at the back of the skull between these specimens. This range of variation has been attributed to individual variation, sexual dimorphism, and taxonomic differences by various authors. Although a consensus has not been reached regarding taxonomic identification, the assemblage of specimens previously attributed to *Stegoceras* is currently viewed as encompassing at least three or four species and at least two to three genera. The most recent revision of the *Stegoceras* material (Sullivan 2003, 2006), has reallocated a number of *Stegoceras* species into two new genera, *Colepiocephale* and *Hanssuesia*, as well as to the Mongolian genus *Prenocephale* (Fig. 25.1D). Under this scheme, the genus *Stegoceras* is retained for the original holotype material (Gilmore 1924) and a few referred specimens. By contrast, Williamson and Carr (2002) erected the new genus *Sphaerotholus* for some of the specimens that Sullivan referred to *Prenocephale*, while Williamson and Carr retained several species of *Stegoceras*. Emphases on different characters and disagreements regarding the scope and importance of ontogenetic and individual variation in skullcap morphology underlie these taxonomic debates. Other North

American pachycephalosaurs are the large-bodied Maastrichtian genera *Pachycephalosaurus*, *Stygimoloch*, and *Dracorex*, which are characterized by bearing clusters of spikes on the edge of the shelflike skull margin. Whereas the spikes are relatively small and evenly distributed in *Pachycephalosaurus*, they are prominent and concentrated in nodes in *Stygimoloch* and *Dracorex*. *Pachycephalosaurus* and *Dracorex* also bear a tract of spikes along the nose.

Pachycephalosaur systematics is currently in a state of flux because of the competing classifications for specimens discussed above. Older systematic studies of pachycephalosaurs lumped the flat-headed species into one family and the domed taxa into another (Maryańska and Osmólska 1974; Sues and Galton 1987). All recent phylogenetic hypotheses (Sereno 2000; Williamson and Carr 2002; Sullivan 2003) agree, however, that the flat-headed taxa are basal within Pachycephalosauria and the domed genera form a monophyletic group dubbed the Pachycephalosauridae (Sereno 1999), which evolved from a flat-headed ancestor. As mentioned above, *Stenopelix* is currently regarded as the most primitive taxon in several recent phylogenies, but the lack of overlap in preserved skeletal parts with many other pachycephalosaurs makes this assignment tenuous. Despite disagreements between authors regarding validity of certain taxa and their phylogenetic positions, there is a remarkable degree of consensus between recent phylogenetic analyses regarding the interrelationships of the taxa represented by more than skullcap material. The basal branches in the pachycephalosaurian evolutionary tree (Fig. 25.4) are occupied by the flat-headed genera *Wannanosaurus*, *Goyocephale*, and *Homalocephale* (Fig. 25.1A) from Central Asia. Despite being relatively basal, all of these taxa appear late in the fossil record of the group, postdating the appearance of the more derived dome-headed taxa (Sullivan 2006). Most authors now agree that the dome-headed species constitute a single, natural (monophyletic) group, with *Stegoceras* as its basal taxon, and the name Pachycephalosauridae has been applied to this group. Other dome-headed species are more derived than *Stegoceras* in the degree of doming and in the progressive closure of the supratemporal fenestrae and are united within the clade Pachycephalosaurinae of Sereno (1999) (Fig. 25.4). In all recent phylogenetic hypotheses (Sereno 2000; Williamson and Carr 2002; Sullivan 2003) *Prenocephale* and *Tylocephale* are closely related genera, united by the linear rows of ornamental bumps extending along some of the skull bones in the temporal region (Fig. 25.1D). Together, these two genera form the sister taxon to a clade comprising the North American genera *Stygimoloch*, *Pachycephalosaurus*, and presumably *Dracorex*. *Stygimoloch* and *Pachycephalosaurus* accentuate the parietosquamosal shelf with a system of spikes. *Dracorex* displays homologous nodes of spikes on the squamosal corners, and likely represents a member of this lineage, but the absence of a dome and the widely open supratemporal fenestrae in this taxon complicate the distribution of characters within Pachycephalosauria (Sullivan 2006). These character inconsistencies have been interpreted as due to ontogenetic changes in skull doming between juveniles and adults (Horner and Goodwin 2009).

Differences between these hypotheses relate to disputed specimens referred to either *Stegoceras*, *Sphaerotholus* (Williamson and Carr 2002), *Prenocephale*, *Colepiocephale*, and *Hanssuesia* (Sullivan 2003), or

Ornatotholus (Sereno 2000). All of these disputed taxa are based solely on skullcap material and are subject to varying opinions regarding the limits of intraspecific (hereunder ontogenetic) versus interspecific variation. For example, in Sereno's (2000) cladistic analysis *Ornatotholus* falls outside of Pachycephalosauridae in an unresolved trichotomy with *Homalocephale*, and three or more additional steps are required to move it inside the latter clade. Nevertheless, both Williamson and Carr (2002) and Sullivan (2003) consider this taxon to represent a juvenile *Stegoceras* and interpret the differences in the degree of skull doming and supratemporal fenestra closure between it and adult specimens of *Stegoceras* to be ontogenetic variation. The contrast between the disagreements in defining pachycephalosaurid taxa known from skullcaps only and the consensus on taxonomy and phylogeny of taxa known from more complete remains indicates an overreliance on relatively subtle traits related to skull doming and supratemporal closure. Recent considerations and research on pachycephalosaurid ontogeny (Horner and Goodwin 2009; Sullivan 2007) have cast doubts on the taxonomic utility of many such cranial features and the validity of taxa based on them. Sullivan (2007) has even suggested that flat-headed taxa may have evolved from domed taxa through a heterochronic process, a pattern that may be more consistent with stratigraphy, and now supported to some degree by the phylogenetic analysis of Longrich et al. (2010), though the latter study ascribes phylogenetic value to some of the transformations regarded as ontogenetic by Sullivan (2007).

The rarity of pachycephalosaurian remains makes paleobiological reconstructions difficult, and as a consequence most paleobiological work has focused on the possible function of the thickened skullcaps that constitute the bulk of the pachycephalosaurid fossil record. Colbert (1955) was the first to suggest that the thickened skull roof may have been used in intraspecific combat. This idea was later expanded on by Galton (1971), who argued that male domed pachycephalosaurs used their heads as battering rams when competing for females, in analogous fashion to head-butting behaviors of bighorn sheep and other bovids. The head-butting scenario has been questioned by some, as it would have required two individuals to precisely impact the smooth and convex domes against each other, and could have led to serious cervical fractures. Nevertheless, the neck of *Dracorex* displays derived traits that have been interpreted as stiffening the joint between the skull and neck (Bakker et al. 2006), and extant bovids with head-butting behaviors exhibit many complex behaviors related to proper alignment of their horns during intraspecific agonistic encounters. Longrich et al. (2010) note that the highly convex horn-bases of musk-oxen and Cape buffalo present no hindrance to interspecific agonistic behaviors. Some authors have favored a scenario that involves flank-butting behavior instead of head butting (Goodwin et al. 1998). An alternative hypothesis for the function of the domes interprets them as thermoregulatory devices. Under this hypothesis, the large columnar vacuities seen in cross sections of some skullcaps are proposed as a reservoir for blood, possibly serving as a heat-exchange mechanism (Reid 1996).

A recent histological analysis of several pachycephalosaur domes at different growth stages shows that the large vacuities seen in some skullcaps

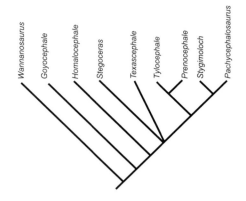

25.4. Consensus backbone of relationships among pachycephalosaurid genera common to several recent hypotheses (Sereno, 2000; Williamson and Carr 2002; Sullivan 2003). *Wannanosaurus, Homalocephale,* and *Goyocephale* are flat-headed genera, whereas the more derived species all possess some degree of doming of the skull-roof bones. Disputed taxa such as *Sphaerotholus, Colepiocephale,* and *Hanssuesia* are not shown here. *Texascephale* is shown here as a basal dome-headed taxon, but the only analysis to include it (Longrich et al. 2010) disagrees with this general consensus tree.

are a transient feature of juveniles and become infilled with bone tissue in adults (Goodwin and Horner 2004), indicating that they are an ontogenetic feature rather than an adaptation for heat exchange. Another finding of this study is the presence of large tracts of Sharpey's fibers, a bone tissue associated with attachment of skin or horn, on the outer surface of the domes. This implies that some nontrivial soft tissue structure extended beyond the dome, and depending on its shape, might have interfered with a headbutting type of combat. Perhaps in addition to, or instead of, being used for intraspecific combat, the thickened skullcaps, which are often decorated with a variety of spikes or small tubercles, may have served as visual display structures for species recognition and mate attraction (Goodwin and Horner 2004). Similar functions have been proposed for the frills and horns of ceratopsian dinosaurs.

Ceratopsia – Horns and Frills

Ceratopsians constitute the more diverse clade within Marginocephalia, and their remains are also far more abundant than those of pachycephalosaurs. Species such as *Protoceratops* and *Psittacosaurus* are the most abundant dinosaur taxa in their respective environments and therefore the subject of diverse paleobiological studies. The first remains of a ceratopsian were discovered in 1856 and consisted of a single tooth referred to the now disused taxon *Monoclonius*. By 1890, more complete remains of these animals were found, and their general appearance was known (Hatcher et al. 1907). All of the specimens collected in the early days of ceratopsian research belong to the Ceratopsidae, the most derived and diverse group within the Ceratopsia, comprising all the large-bodied species with characteristic facial horns and large frills. Not until the discovery of *Protoceratops* by the Third Central Asiatic Expedition of the American Museum of Natural History in 1922 were more basal, non-ceratopsid members of the group recognized, although specimens of the primitive genera *Leptoceratops* and *Montanoceratops* had already been discovered in 1914 and 1916, respectively. The Third Central Asiatic Expedition also discovered the remains of the primitive ceratopsian *Psittacosaurus*, but its ceratopsian affinities were not properly recognized until several decades later (Maryańska and Osmólska 1975; Sereno 1986). In recent decades a wealth of new primitive species have been discovered in Central Asia (Maryańska and Osmólska 1975; Nessov et al. 1989; Makovicky and Norell 2006), China (Dong and Azuma 1997; Zhao et al. 1999; Lambert et al. 2001; You and Dong 2003; You et al. 2003, 2005, 2010; Xu et al. 2002, 2006, 2010; Zhao et al. 2006; Zhou et al. 2006; Sereno et al. 2007, 2010) and North America (Chinnery 2004; Chinnery and Horner 2007; Wolfe and Kirkland 1998), bolstering our knowledge of the earliest chapters of ceratopsian evolution.

The principal diagnostic character of ceratopsians is the rostral bone at the tip of the upper jaws. This bone, which forms the upper part of a cropping beak, is an evolutionary novelty not present in other dinosaurs. Other features that diagnose most ceratopsians are the expansion of the skull margin into a frill overhanging the neck and the presence of outwardly bowed cheekbones, which make the skull distinctly triangular when viewed from above. Ceratopsian diversity is discussed here in light of the cladogram in Figure 25.5.

Yinlong and Chaoyangsauridae–the Earliest Marginocephalians

Marginocephalia suffered from a very long gap in the fossil record (Sereno 1997), until this was reduced by the discovery of the Late Jurassic *Yinlong downsi* (Xu et al. 2006) from the Late Jurassic of Xinjiang. Besides its stratigraphic importance, this taxon (Fig. 25.2A) has provided further evidence for the relationship between ceratopsians and marginocephalians, which was otherwise weak because long separate evolutionary histories for these lineages overprinted some of the shared derived characters uniting them. *Yinlong* is a small animal, somewhat more than a meter in length, with a wide deep skull characterized by enormous temporal fenestrae. A small apical rostral bone is present in *Yinlong*, flanked on either side by a pair of enlarged, pointed premaxillary teeth with convex labial and flat lingual faces as in some other basal ceratopsians. In some other features, however, *Yinlong* resembles pachycephalosaurs. These include tracts of rugose ornamentation along the cheeks and temporal regions of the skull, and a parietosquamosal shelf mainly expressed through lateral and caudal expansion of the squamosals. Unlike the case in other ceratopsians, the parietal is not expanded to participate in a frill.

Chaoyangsaurus youngi (Zhao et al. 1999) and *Xuanhuaceratops niei* (Zhao et al. 2006), from the uppermost Jurassic or possibly the earliest Cretaceous of China, represent the next rung of the ceratopsian evolutionary tree. As in *Yinlong*, some of the facial bones are rugose, but these taxa share outwardly arched cheek bones with other neoceratopsian more derived than *Yinlong*. All other ceratopsians belong to one of two major groups, the Psittacosauridae or the Neoceratopsia.

Psittacosaurids–Parrot-Beaked Dinosaurs

The genus *Psittacosaurus*, with 13 named species, represents the first major radiation of ceratopsians in the fossil record (Fig. 25.2B, 25.5). All discoveries of *Psittacosaurus* are restricted to the Early Cretaceous of Asia (Sereno 2000), with all but three of the named species from China, while *P. mongoliensis* and *P. sibiricus* are known from Mongolia and Russia, respectively. The fragmentary *P. sattarayaki* from Thailand is the only possible psittacosaur found outside the Central Asian region thus far, although Sereno (2000) considered this fragmentary material undiagnostic. Psittacosaurids are all relatively small and did not exceed 2 m in body length. They have moderately large but short skulls with a deep snout and wide cheekbones that end in a small point or horn (Fig. 25.2B), a synapomorphy shared with neoceratopsians, to the exclusion of *Yinlong* and chaoyangsaurids. The beak is rounded in psittacosaurs, and premaxillary teeth are absent. In a number of species, including *P. mongoliensis*, *P. sibiricus*, *P. lujiatunensis*, and *P. meileyingensis*, the lower jaw bears a deep flange along its lower edge, which is squared off in some taxa but not others. The cheek teeth have wide, pinnate crowns with only a faint primary ridge and low-angled wear facets. Psittacosaurids have relatively short forelimbs compared to their hind limbs and are thought to have had primarily bipedal locomotion. The coracoid, which forms the lower half of the shoulder girdle, bears an enlarged tuber for the upper arm musculature, and the hand is unique among ornithischians in

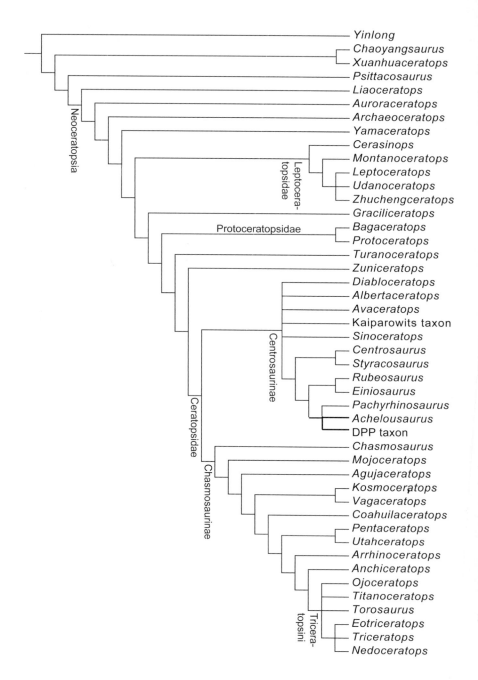

25.5. Phylogenetic relationships among ceratopsians combining results from Makovicky (2010) for basal neoceratopsians with those of Sampson and Loewen (2010) for ceratopsids. The major suprageneric taxa are listed next to the relevant nodes. Several recently named, such as *Medusaceratops,* are not shown, because they have not been placed into comprehensive phylogenetic analyses.

being asymmetric through the reduction of the fourth and fifth digits. The rest of the skeleton is relatively primitive in its features, which are generally similar to those of basal ornithischians. Complete skeletons of several *Psittacosaurus* species have been discovered with masses of relatively large gastroliths (stomach stones) preserved inside their rib cages, implying the presence of a strong "stomach mill" for postoral maceration of their food. Coupled with potential strong bite forces at the beak, the gastroliths provide evidence that these animals fed on tough vegetable matter (Sereno et al. 2010).

Psittacosaur systematics is largely based on subtle differences in a handful of cranial and dental characters, many of which have been shown to vary ontogenetically and individually. A purported new genus of psittacosaurid, *Hongshanosaurus,* was recently described from a young juvenile specimen

(You et al. 2003). Many of the characters used to diagnose it are features that are known to change with growth, however, and the distinction of this new taxon is not supported (Sereno 2010). Several other named historical species were based on juvenile material or undiagnostic traits, and these have been synonymized with more established taxa (Sereno 1990).

New insights on several important aspects of psittacosaur paleobiology have been gained recently. A pioneering life history study shows that *P. mongoliensis* grew relatively faster than modern reptiles, but more slowly than larger-bodied dinosaurs and mammals. *P. mongoliensis* did not attain somatic maturity for nine years or more (Erickson and Tumanova 2000; Erickson et al. 2001, 2009) in spite of its modest size. A recent discovery of an adult skeleton in undisturbed association with 34 posthatchling juveniles is suggestive of a high degree of parental care, perhaps within a nest structure (Meng et al. 2004). Six co-occurring juveniles of *Psittacosaurus* cf. *Lujiatunesis* from the Yixian Formation may belong to two separate cohorts, indicating crèche-forming behaviors in this taxon (Qi et al. 2007), a finding supported by histological analysis of another such sample of juveniles of mixed age (Erickson et al. 2009). The plentitude of *Psittacosaurus* specimens at some localities allows for population biological studies. Based on the most comprehensive histological sampling of a dinosaur species to date, Erickson et al. (2009) determined that *Psittacosaurus lujiatunensis* followed a Type B1 survivorship pattern, which is also observed in the tyrannosaurid *Albertosaurus* (Erickson et al. 2006), and living species of ground birds and mammals. An initial high mortality in neonates and young animals is followed by relatively low mortality between the ages of two and eight. Mortality is then observed to increase dramatically in older subadult and adult individuals, probably coincident with the attainment of sexual maturity (Erickson et al. 2009). Another remarkable discovery from Liaoning is a well-preserved *Psittacosaurus* with soft-tissue preservation. A bizarre, brushlike arrangement of quill-like structures extends along the tail of this specimen (Mayr et al. 2002). The individual quills or bristles appear to be cylindrical and possibly hollow, and have been interpreted as having a display function. The same specimen preserves large tracts of scaly skin with patterns of large and small scales, which appear to retain traces of the animal's color pattern (Lingham-Soliar and Plodowski 2010). The distribution of scales with color traces suggests that this specimen had a cryptic pattern of light and dark tracts along the trunk and limbs, and also displayed countershading with a paler ventrum (Lingham-Soliar and Plodowski 2010).

Neoceratopsia–Shearing Jaws, Large Frills, and Gigantic Skulls

The Neoceratopsia represent the most diverse radiation of ceratopsian dinosaurs, and include all taxa with a large, backward-expanded frill, including such well-known species as *Protoceratops* and *Triceratops*. Their fossil record was until recently restricted to the Late Cretaceous and therefore disjunct from that of psittacosaurids. The discovery of the cat-sized *Liaoceratops* (Xu et al. 2002) (Fig. 25.2C) in Liaoning province, China, has significantly extended their fossil record into the Early Cretaceous, and *Liaoceratops* is a contemporary of *Psittacosaurus*, so the two clades have coeval appearances in the fossil record as predicted by their relationship to one another. *Liaoceratops* is smaller than *Psittacosaurus*, and it is possible that differences in body size and dental morphology and beak shape may reflect ecological niche partitioning within the Yixian Formation paleoenvironment. *Liaoceratops* already displays some of the derived evolutionary traits that characterize the neoceratopsians. It has an enlarged frill extending over the rear of the skull, which is composed mostly of the midline parietal with smaller contributions on each side by the squamosal bones (Fig. 25.2C). The beak is pointed and narrow, unlike the beaks of psittacosaurs and *Chaoyangsaurus*. The teeth of *Liaoceratops* are enameled on only one side and are arranged in more ordered series than in psittacosaurs, so that the tooth crowns in each jaw line up to form a single cutting edge that shears against the teeth in the opposing jaw. Another neoceratopsian trait is the backward extension of the cheekbones so that the cheek horns are placed well behind the eye socket, in contrast to their suborbital position in psittacosaurs. As in other neoceratopsians, the joint between the skull and neck is formed by a spherical condyle in *Liaoceratops*. Surprisingly, *Liaoceratops* retains the jaw flange seen ancestrally in psittacosaurs, but this feature is otherwise absent among neoceratopsians (Xu et al. 2002).

Auroraceratops (You et al. 2005) and *Archaeoceratops* (Dong and Azuma 1997) (Fig. 25.2D), from the Early Cretaceous of China, and *Yamaceratops* (Makovicky and Norell 2006) from the late Early Cretaceous of Mongolia, occupy the next branches up the ceratopsian tree (Fig. 25.5). All are larger than *Liaoceratops* and are more derived in a number of features, especially in the cheek region. In *Yamaceratops* and all more advanced neoceratopsians the cheek horn is exaggerated by a novel ossification called the epijugal. The epijugal is crescent shaped in *Archaeoceratops* and some other primitive species, but it becomes conical in more advanced neoceratopsians such as *Protoceratops*. All three species retain three conical teeth in the premaxilla, which are reduced to two teeth or entirely lost in the more advanced species. These teeth do not oppose any teeth on the lower jaw, but rather would have projected over the horny cover of the lower jaw beak, with which they seem to have occluded, as evident from wear facets (Makovicky and Norell 2006). Well-preserved specimens of *Liaoceratops* and *Archaeoceratops* exhibit small frill fenestrae, demonstrating that these structures were ancestrally present in the clade.

Leptoceratopsidae

The leptoceratopsids, a small group of medium- to large-bodied species, form the next branch in the ceratopsian tree (Fig. 25.5). Although they are relatively primitive in the spectrum of ceratopsian evolution, leptoceratopsids are known only from the latter half of the Late Cretaceous, and their early history remains to be discovered. Leptoceratopsids have short, deep skulls with massive mandibles (Fig 25.2E). Their lower-jaw teeth are unique in having a bulge on their inner, enamel-free face, which gets worn into a shelf through occlusion with the upper teeth. Teeth exhibiting this characteristic occlusion pattern have been reported from the Campanian of southern Sweden (Lindgren et al. 2007), indicating that leptoceratopsids are one of only two ceratopsian clades purported to have a distribution extending beyond Asia and North America. The lower jaws of *Leptoceratops* (Sternberg 1951) and *Prenoceratops* (Chinnery 2004) from North America, *Udanoceratops* (Kurzanov 1992) from Mongolia, and *Zhuchengceratops* from China (Xu et al. 2010) are deeply curved along the bottom edge. Leptoceratopsids are considerably larger than Early Cretaceous neoceratopsians: *Leptoceratops* and *Zhuchengceratops* adults are about 2 m long, and the giant *Udanoceratops* was probably close to 4 m in length. Unusually for neoceratopsians, the frill is not fenestrated in *Leptoceratops* (Sternberg 1951) and *Prenoceratops* (Chinnery 2004), but remains fenestrated in the more basal taxon *Cerasinops*, which may have been bipedal according to Chinnery and Horner (2007).

Coronosauria

The remaining neoceratopsians are grouped within the Coronosauria (Fig. 25.5), a group that contains the well-known forms *Protoceratops* (Fig. 25.2F) and the ceratopsids (Fig. 25.3) such as *Triceratops* and *Styracosaurus*. Coronosaurs are named for their expanded frills, which, unlike those of more primitive forms, fan out sideways and are perforated by large frill openings (except in *Triceratops*). A small radiation of primitive coronosaurs from Late Cretaceous deposits of the Gobi desert is known as the Protoceratopsidae (Sereno 2000); it includes the genera *Protoceratops* (Fig. 25.2F) and the smaller *Bagaceratops* and related forms (You and Dong 2003). Both of these taxa are known from numerous specimens, including very complete growth series, ranging from tiny hatchlings with skulls of less than 3 cm in length to 2.5–3 m adults with skulls close to a meter in length (Brown and Schlaikjer 1940). Another hallmark of most coronosaurs is the presence of a horn core on the nose slightly behind or directly above the nostrils. An incipient arching of the nasals is observed in *Protoceratops* adult individuals, whereas a prominent trapezoidal horn core is seen in *Bagaceratops* (Maryańska and Osmólska 1975) and the possibly congeneric *Magnirostris* (Makovicky and Norell 2006), with some degree of fusion between the nasals observed in the horn cores of even small juveniles. *Protoceratops* is extremely abundant at several Gobi desert localities, allowing for studies of related population biology of this taxon. Dodson (1976) demonstrated dimorphism in several cranial traits in large and presumably adult specimens of *Protoceratops* from the Bayn Zag (Flaming Cliffs) locality, although recent reanalysis of this data suggests that this signal is subtle and affected by how missing measurements are analytically treated (Chapman et al. 2008), prompting Padian and Horner (2011) to consider it as insufficient distinction to qualify as sexual dimorphism. The famous "fighting dinosaurs" – in which a specimen of *Protoceratops* is preserved in an agonistic interaction with a large specimen of *Velociraptor* – represents an exceptional case of presumed predator and prey interaction in the dinosaurian record.

Recently, *Ajkaceratops*, a species from the Santonian of Hungary based on an isolated snouts and several predentary bones, has been interpreted as a close relative of *Bagaceratops*, making it only the second ceratopsian find from Europe (Ösi et al. 2010). The taxon is referred to Ceratopsia based on the presence of a rugose upper beak, interpreted as a rostral, and allied with *Bagaceratops* through the shared derived character of having an accessory fenestra along the maxillary-premaxillary suture. The reported rostral is unusual, however, in that it is completely fused to the premaxillae such that the borders of the element cannot be discerned, and a cross section of the internarial bar fails to reveal a distinct rostral component along the midline, unlike the case in psittacosaurs and neoceratopsians. Furthermore, the collected specimens lack basal neoceratopsian synapomorphies such as a keeled and caudally bifurcated predentary. More material is required to test the affinities of *Ajkaceratops*.

Ceratopsidae

The ceratopsids make up the bulk of ceratopsian diversity, with more than 35 species, almost all from the Late Cretaceous of western North America (Sampson and Loewen 2010), although the recently discovered *Sinoceratops* from the Campanian of China (Xu et al. 2010) establishes the presence of this clade in Asia. *Turanoceratops* from Turonian deposits of Uzbekistan has been posited as an Asian ceratopsid (Sues and Averianov 2009a, b), but is regarded as sister to that clade by others (Farke et al. 2009b). Ceratopsids are all large animals, ranging from rhino- to elephant-size and beyond, with giants such as *Triceratops* estimated to weigh as much as 8,000 kg. Their skulls are enormous and measure in excess of 2 m in some taxa such as *Pentaceratops* and *Titanoceratops* (Fig. 25.3F), making them the largest skulls for any known terrestrial animal. A long list of evolutionary novelties and modifications separates the ceratopsids from the less-derived Central Asian coronosaurs, but the sheep-sized *Zuniceratops* (Wolfe and Kirkland 1998) from the mid-Cretaceous of Utah is a transitional form, which bridges that character gap to some extent, as does *Turanoceratops*. Perhaps the most prominent diagnostic traits of ceratopsids are the changes to the visual display structures of the skull. *Zuniceratops* and all ceratopsids bear paired horn cores, one above each eye, and their frills are greatly expanded. The frill margin of all ceratopsids (but apparently not *Zuniceratops*) bears a variable number of epoccipital bones, forming either an undulating edge or a series of hooks and spikes. The nostrils are hugely expanded and bear a large midline lamina anteriorly, which is excavated by small pockets in the chasmosaurine subfamily (Dodson et al. 2004; Forster 1990). Ceratopsids also perfected the shearing action of their jaws through major modifications to the jawbones and especially dentition (Ostrom 1966). The number of tooth positions in each jaw increased from fewer than 20 in more primitive neoceratopsians to typically 25–35 in ceratopsids. This increase in tooth positions was augmented by adding extra rows of replacement teeth at each tooth position, so that each tooth position carries at least two replacement teeth in addition to the erupted, functional one. In order to accommodate the additional replacement teeth, the tooth roots are bifurcated and constricted from side to side. The coronoid process, which marks the insertion of the main jaw muscles, is greatly enlarged and projected upward. At the tip of the jaws, the rostral and predentary bones that form the beak evolved a sharp cutting edge.

The large body sizes of ceratopsids also led to modifications in the post-cranial skeleton. The hips are wide and elongate, and are supported by more than 10 sacral ribs on each side. The muscle-bearing outer face of the ilium

Plate 1. The Geological Time Scale. Major events in the history of Earth. Ma = millions of years ago.

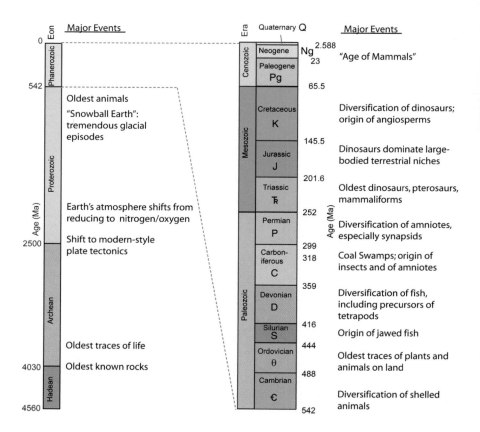

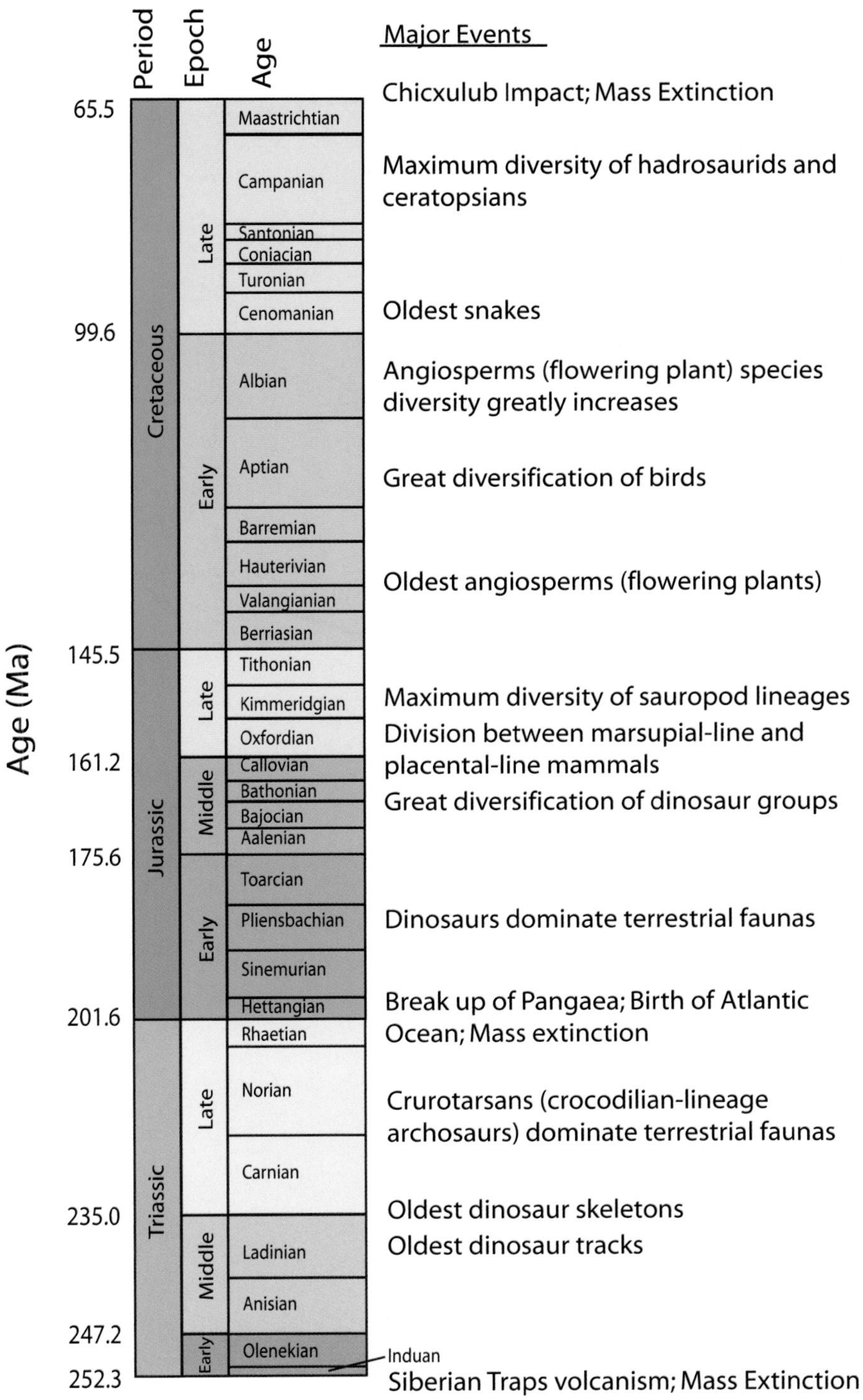

Plate 2. Major events during the Mesozoic Era, the "Age of Dinosaurs." Ma = millions of years ago.

Age (Ma)

| Period | Epoch | Age | Major Events |

Period | Epoch | Age

Major Events

- 65.5 — Maastrichtian — Chicxulub Impact; Mass Extinction
- Campanian — Maximum diversity of hadrosaurids and ceratopsians
- Santonian
- Coniacian
- Turonian
- Cenomanian — Oldest snakes — 99.6
- Albian — Angiosperms (flowering plant) species diversity greatly increases
- Aptian — Great diversification of birds
- Barremian
- Hauterivian — Oldest angiosperms (flowering plants)
- Valangianian
- Berriasian
- 145.5 — Tithonian
- Kimmeridgian — Maximum diversity of sauropod lineages
- Oxfordian — Division between marsupial-line and placental-line mammals
- 161.2 — Callovian
- Bathonian — Great diversification of dinosaur groups
- Bajocian
- Aalenian
- 175.6 — Toarcian
- Pliensbachian — Dinosaurs dominate terrestrial faunas
- Sinemurian
- Hettangian — Break up of Pangaea; Birth of Atlantic Ocean; Mass extinction — 201.6
- Rhaetian
- Norian — Crurotarsans (crocodilian-lineage archosaurs) dominate terrestrial faunas
- Carnian
- 235.0 — Oldest dinosaur skeletons / Oldest dinosaur tracks
- Ladinian
- Anisian
- 247.2 — Olenekian — Induan
- 252.3 — Siberian Traps volcanism; Mass Extinction

Cretaceous — Late / Early

Jurassic — Late / Middle / Early

Triassic — Late / Middle / Early

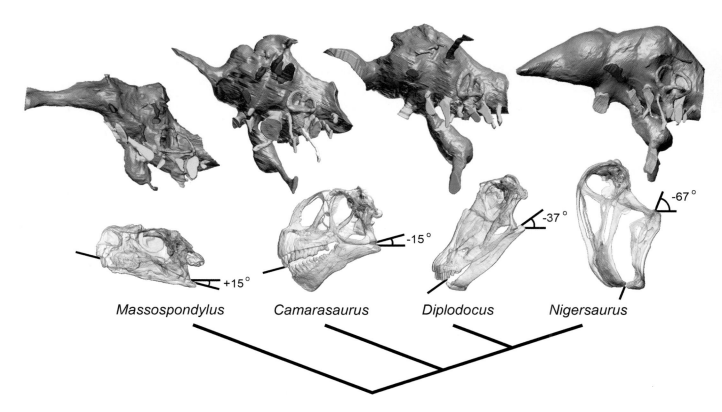

Massospondylus   *Camarasaurus*   *Diplodocus*   *Nigersaurus*

+15°   -15°   -37°   -67°

Plate 3. CT images of the osseous labyrinth of diplodocid sauropods indicate that a radical reorientation of the skull accompanied dental and skull adaptations for extreme herbivory. The skull and dural sinuses are represented in blue, nerve openings in yellow, the osseous labyrinth in pink, and the path of the internal carotid artery in red. From Sereno et al. 2007.

Plate 4. The beautifully preserved, articulated Thermopolis specimen of *Archaeopteryx*. This specimen confirms that *Archaeopteryx* was highly similar to dromaeosaurids and other maniraptorans in skull, pelvis, and hind limb anatomy, and indeed, recently discovered dromaeosaurids, troodontids, and other maniraptorans are extremely similar to *Archaeopteryx*. In the foot, the hallux was not fully reversed and the second toe was hyperextendible. Well-preserved feather impressions surround the skeleton. Ten specimens assigned to *Archaeopteryx* are currently known, but exactly how many species they represent remains controversial.

Plate 5. Simple fibrolamellar bone in the shaft of a turtle scapula. Late Cretaceous, Montana. Maximum diameter 16.5 millimeters. Collector D. Maxwell. The age of the turtle at the time of death is estimated as at least 17-19 years. Similar tissues were figured by Enlow (1969: Figs. 8, 12, 13, 15) from *Pseudemys, Crocodylus,* and *Alligator.*

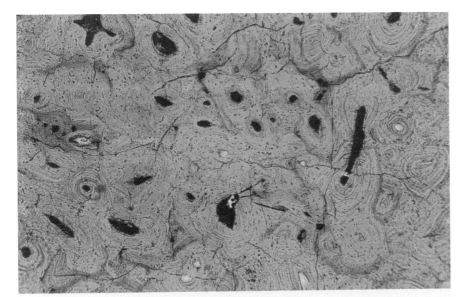

Plate 6. Dense Haversian bone in the shaft of a crocodilian femur. Bridger Formation, Eocene, Wyoming. Maximum diameter 27 millimeters. Uncatalogued Utah Division of State History specimen. Periosteal bone around the medullary cavity is extensively replaced by Haversian tissue, developed as dense Haversian bone on the left hand side as seen here. This tissue was said by de Ricqlès (1980: 124) to be now formed only in endotherms.

Plate 7. *Velociraptor* turns a corner.
Mixed media. © 2011 Scott Hartman.
www.skeletaldrawing.com.

Plate 8. *Gigantoraptor erlianensis*. Digital media.
© 2009 Jaime Chirinos. www.zooartistica.com.

Plate 9. *Epidendrosaurus* ("Tree Snacks").
Colored pencil and watercolor. © John
Bindon. www.bindonart.com.

Plate 10. *Lesothosaurus*. Mixed media. © Juan Jose Castellano. www.juanjocastellano.com.

Plate 11. The Jehol Biota. Colored pencil. ©
Jason Brougham. www.jasonbrougham.com.

Plate 12. *Alioramus.* Mixed media. ©
Fabio Pastori. www.fabiopastori.it.

Plate 13. *Caudipteryx.* Digital media. © Alain
Bénéteau. http://dustdevil.deviantart.com.

Plate 14. Hell Creek Mural (detail, *Triceratops*).
Mixed media. © 2008 Bob Walters and
Tess Kissinger. www.dinoart.com.

Plate 15. *Spinosaurus* and Giant Sturgeon.
Acrylic on board. © 2010 James McKinnon.

Plate 16. *Tianyulong.* Digital media. © Zhao Chuang and Xing Lida. http://xinglida.net.

Plate 17. *Leaellynasaura* ("Twilight
Feeding"). Colored pencils and watercolor.
© John Bindon. www.bindonart.com.

Plate 18. *Anchiornis*. Digital media. ©
Julius Csotonyi. www.csotonyi.com.

Plate 19. *Psittacosaurus*. Mixed media. © Juan
Jose Castellano. www.juanjocastellano.com.

Plate 20. Morrison Mural (detail, *Stegosaurus*).
Mixed media. © 2007 Bob Walters and
Tess Kissinger. www.dinoart.com.

Plate 21. *Koreaceratops*. Digital media.
© Julius Csotonyi. www.csotonyi.com.

Plate 22. *Struthiosaurus transsilvanica.*
Acrylics. © David Weishampel. www.
hopkinsmedicine.org/fae/DBW.htm.

Plate 23. *Sinornithosaurus* ("Twilight"). Mixed
media. © Fabio Pastori. www.fabiopastori.it.

Plate 24. *Confuciusornis.* Digital media. © Alain
Bénéteau. http://dustdevil.deviantart.com.

Plate 25. *Sinosauropteryx.* Digital media. ©
Zhao Chuang and Xing Lida. http://xinglida.net.

Plate 26. Restoration of a large group
of the small carnivorous dinosaur
Coelophysis. © Douglas Henderson.

Plate 27. A herd of the polacanthid *Gastonia* migrating during the dry season. The dry landscape is dominated by cycadeoids and some angiosperm trees. Early Cretaceous of eastern Utah. © Kenneth Carpenter.

Plate 28. The ankylosaurid *Ankylosaurus* in a floodplain forest. The landscape is composed of deciduous angiosperms, with a low ground cover of herbaceous plants that may have been a major part of the ankylosaurs diet. Late Cretaceous of southern Alberta. © Kenneth Carpenter.

Plate 29. *Sauropelta,* a nodosaurid, walks across a fern prairie. Although flowering plants had appeared by this time (note trees), grasses had not yet appeared. Early Cretaceous of southern Montana. © Kenneth Carpenter.

Plate 30. The Arundel Formation. Mixed
media. © Mary Parrish. parrishm@si.edu.

Plate 31. "The Double Death." Two
Carcharodontosaurus saharicus hoist a juvenile
Paralititan stromeri. Digital media. © Robert
Nicholls. bob.nicholls@paleocreations.com.

Plate 32. "Reaper in Paradise." Three *Albertosaurus sarcophagus* (one already dead) attacked by two *Deinosuchus riograndensis.* Acrylics. © Robert Nicholls. bob.nicholls@paleocreations.com.

is everted, and the ischium has become strongly recurved. The forelimbs, and especially the forefeet, are enlarged, and the processes of the vertebrae in the neck and back are elongated relative to their smaller relatives. In contrast, ceratopsid tails are relatively short and without the elongate neural spines seen in most non-ceratopsid neoceratopsians.

Two major divisions of the Ceratopsidae coexisted throughout most of the Campanian and Maastrichtian. The centrosaurines, also known as short-frilled ceratopsids, are a radiation of seven or more genera, most of which have a long nasal horn, short orbital horns, and a shorter frill that bears different combinations of horns or spikes in various genera (Fig. 25.3A–D). Many are spectacular, such as *Styracosaurus* (Fig. 25.3B) with an array of spikes radiating from the edge of the frill, or *Einiosaurus* (Fig. 25.3C) with a forward-curved nasal horn and two long spikes extending backward from the edge of the frill. In the derived centrosaurines *Achelousaurus* and *Pachyrhinosaurus* (Fig. 25.3D) the horn cores have evolved into large, rough bosses of bone (Sampson 1995), which may have supported a keratinous horn or some other similar structure. The recent discovery of basal centrosaurines such as *Albertaceratops* (Ryan 2007) and *Diabloceratops* (Kirkland and Deblieux 2010) reveals that centrosaurines evolved from ancestors with long postorbital horn cores. These basal taxa also lack the prominent nasal horn of centrosaurines such as *Centrosaurus* and *Styracosaurus*, and instead exhibit a small expansion of the nasal bones with a distinct epinasal horn core, a trait previously considered diagnostic of Chasmosaurinae. *Sinoceratops* was interpreted as a basal centrosaurine by Xu et al. (2010), but exhibits a more derived complement of an enlarged nasal horn and reduced postorbital horns as observed in derived centrosaurine taxa from Alberta and Montana.

Several centrosaurine species, including *Centrosaurus*, *Pachyrhinosaurus*, and *Einiosaurus*, are known from bone beds representing catastrophic mass-death occurrences of juvenile through adult animals (Dodson and Currie 1990). Beyond providing direct evidence that these animals congregated at least seasonally, such bone beds yield qualitative insights on the growth and development of these animals. Juvenile centrosaurines of different genera were very similar in appearance, and the species-specific shape of facial horns and complement of frill spikes appeared only late in development, as these animals approached or attained adulthood (Sampson et al. 1997). This observation has important implications for the validity of several centrosaurine taxa that are based on small, and presumably juvenile, skeletons, such as *Brachyceratops*. Further investigation of the relationships between frill and horn features and development are warranted, before the validity of such taxa can be established (Sampson et al. 1997).

Until recently, almost all centrosaurine taxa appear to be geographically constrained to the area covered by present-day Montana and Alberta (Dodson and Currie 1990), although the youngest centrosaurine in the fossil record, *Pachyrhinosaurus*, appears to have extended northward to the North Coast of Alaska. Recent discoveries of centrosaurines in the southwestern United States (Kirkland and Deblieux 2010; Heckert et al. 2003; Sampson and Loewen 2007) now indicate a wider geographic range for this clade. Rapid turnover between centrosaurine genera through the fossil record has been noted in the Two Medicine Formation, where it has been linked to habitat-size fluctuations related to sea-level changes in the Cretaceous epicontinental seaway (Horner et al. 1992), and also in contemporaneous strata of Dinosaur Provincial Park (Ryan et al. 2006). It is clear that most centrosaurine genera were short-lived (i.e., restricted to narrow stratigraphic horizons) and that evolution of cranial display characters occurred relatively rapidly (Sampson 1995) and in nonlinear fashion (Ryan 2007). Discovery of new taxa from within geographic and chronostratigraphic ranges that have been sampled intensely over the last century (McDonald and Horner 2010; Ryan et al. 2010) suggest a greater standing diversity of centrosaurines than previously recognized.

Chasmosaurines, which constitute the other major division of ceratopsids, have relatively longer frills and more elongate snouts than centrosaurines. In all chasmosaurines, the squamosals are long and taper toward the edge of the frill, and they extend high onto the frill margin. Their short nasal horn and long orbital horns were traditionally considered diagnostic of the clade (Lull 1933), but the discovery of basal centrosaurines with this pattern demonstrates that it is actually primitive for ceratopsids (Ryan 2007; Kirkland and Deblieux 2010). The expanded depression surrounding the nostril is a diagnostic trait of chasmosaurine species, however. Chasmosaurine frills differ widely in shape, but most taxa lack the prominent frill spikes seen in centrosaurines (Fig. 25.3E, F). Recently discovered species from Utah (Sampson et al. 2010) diverge from this pattern, however, and exhibit elaborate episquamosal and epiparietal bones. In *Kosmoceratops* (Fig. 25.3G), the relatively short frill is crowned by a rostrally oriented parapet of triangular accessory ossifications. In some taxa, such as *Chasmosaurus*, the back edge of the frill is straight or only slightly indented, whereas others, including *Triceratops* and *Pentaceratops*, have rounded frills (Fig. 25.3E, F).

Basal chasmosaurines derive from the northern part of the Western Interior Basin, but much of the subsequent Campanian evolution of the clade occurred within the southern part of the basin, with only one instance of cross-basin dispersal. This biogeographic pattern has been interpreted as the result of temporary vicariance between northern and southern parts of the basin during the early to mid-Campanian. This period of regional endemism is followed by dispersal of a clade that culminates in the Triceratopsini from the southern to the northern part of the basin (Sampson et al. 2010; Longrich 2010).

In contrast to the centrosaurines, chasmosaurine mass-death assemblages are rare and have been recorded only for *Agujaceratops*. In both *Triceratops* (Horner and Goodwin 2006) and *Agujaceratops* (Lehman 1990), horn orientation and frill features as well as skull shape have been found to vary with development. Lehman (1990) interpreted in differences in postorbital horn orientation of *Agujaceratops* as sexual dimorphism, but given that such changes are known to occur with ontogeny in other chasmosaurines, and considering the limited sample size, this claim is difficult to support (Padian and Horner 2011). Chasmosaurines from the northern Plains region (Alberta, Montana, Wyoming) belong to two temporally distinct lineages. As mentioned above, two species of *Chasmosaurus*, *C. belli* and *C. russelli*, rapidly succeed one another in the sediments of the Dinosaur Park Formation of Alberta (Dodson et al. 2004). Monophyly of the Maastrichtian taxa *Triceratops*, *Eotriceratops*, and *Torosaurus* is disputed. Sampson and

Loewen (2010) and Longrich et al. (2010) recover these forms as a clade more closely related to the southern chasmosaurine radiation than to *Chasmosaurus*, whereas Ryan (2007) finds these taxa to represent a basal grade of chasmosaurines, a pattern that appears less consistent with stratigraphy by incurring long ghost ranges for both *Triceratops* and *Torosaurus*. *Triceratops*, the best-known chasmosaurine, appears to have a secondarily foreshortened frill without frill openings, suggestive of heterochronic evolution in this lineage (Fig. 25.3E), whereas its sister taxa *Eotriceratops* and *Torosaurus* do possess such fenestrae. Skulls of *Triceratops* are relatively abundant, and historical taxonomic practices led to the erection of myriad distinct species, most of which comprised a single skull. More recent studies that take into account factors such as individual variation through rigorous quantitative methods have reduced this diversity to one, or at most two, species (Forster 1996). Scannella and Horner (2010) have taken this a step further, and conclude that *Triceratops* and *Torosaurus gladius* are synonymous with the latter representing ontogenetically older individuals of the former, in which the delayed development of frill fenestrae is manifest. Evidence for this hypothesis relies on relative staging of specimens using cranial traits including some of the features inferred to vary ontogenetically, and remains to be tested by independent histological data.

Paleobiology

Most paleobiological studies of ceratopsians have centered on the diverse ceratopsids, which are comparatively better known than other members of the group, and have focused on the function of the horns and frill. Ceratopsid horn cores were traditionally interpreted as defensive structures, but their diversity in shape and form parallels that in modern bovids, in which the horns are used primarily in interspecific combat and mate competition. This interpretation of their function is bolstered by the fact that the full expression of the species-specific horn and frill-edge characters occurs late in development, presumably as these animals attained sexual maturity (Sampson et al. 1997). If defense against predators were the main function of these structures, one would assume, they would be more prominent early in ontogeny and selection would favor a fairly limited range of morphologies, rather than the disparity that is actually encountered. Farke (2004) modeled combat between *Triceratops* scale models and concluded that horn configuration and size would not have hindered interspecific agonistic behaviors. A follow-up study on the distribution of pathologies in ceratopsid skulls found that the peripheral bones identified as most likely to be injured during combat by Farke's (2004) model experiment show a significantly greater number of injuries than would be expected at random (Farke et al. 2009a). The diversity of horn and frill morphologies coupled with stasis in functional traits related to feeding and locomotion, plus the obvious parallels to bovid horns, has led several authors to interpret these structures as being sexually selected traits (Farlow and Dodson 1975; Sampson et al. 1997; Sampson and Loewen 2010; McDonald and Horner 2010). Evidence for sexual dimorphism in frill and horn structures is weak or absent, however, leading Padian and Horner (2011) to suggest that these structures better fulfill the criteria of species recognition characters than other competing hypotheses based on sexual (Sampson 1999) or social (Hieronymus et al. 2009) selection. This conclusion rests in part on strict adherence to Darwin's original definition of criteria defining sexual selection (as opposed to more modern usage) and lack of statistical power needed to choose between population-level hypotheses.

Frill function has also been studied from a biomechanical perspective, as the frill is formed by the skull bones that anchor the large masticatory muscles of diapsids. Older works (Haas 1955; Ostrom 1964) interpreted the frill as providing an enlarged area for the jaw musculature and thus an increased power stroke for lower-jaw motion. With the predominant interpretation of the frill as a visual display structure involved in species recognition, the biomechanical hypothesis has fallen out of favor recently. Nevertheless, the relatively short frill of *Liaoceratops* shows clear signs of muscular insertion along the robust posterior edge (Fig. 25.2C), and at least in this basal form, the frill clearly did serve as a platform for enlarged chewing muscles. In a number of protoceratopsid specimens muscle insertions are evident along the caudal edges of the parietal fenestrae adjacent to the midline keel, showing that muscles did span across the fenestrae in basal neoceratopsians. This was likely not the case for ceratopsids, in which the jaw adductors appear to have been restricted to the base of the frill (Dodson et al. 2004). The frills of *Auroraceratops* specimens show similarities in shape with *Liaoceratops*, and these persist even in *Leptoceratops* (Fig. 25.2E), disappearing only in the wide-frilled coronosaurs. Therefore, a primary function of the frill as an enlarged base for jaw muscles early in ceratopsian evolution may have been secondarily coopted into visual display functions later in the history of the clade (Makovicky and Norell 2006).

Another ongoing debate regards forelimb posture and locomotion in ceratopsids. Psittacosaurids have long been thought to be facultative quadrupeds, a conclusion challenged by Senter (2007) who found, through cast manipulations, that the forelimb could not be brought under the body in a weight-bearing pose. His study did confirm that quadrupedality was possible in neoceratopsians, however. Older museum mounts of ceratopsid skeletons show the animals with bent elbows that point outward, a stance dictated by the direction of the articular surfaces between the upper and lower arm bones (Johnson and Ostrom 1995). Other reconstructions have posited the skeletons with more erect forelimbs that would have allowed for faster locomotion, perhaps even galloping (Bakker 1986), but these are at odds with bone architecture as they require full extension of the forelimb, which is precluded by the massive olecranon process. Trackways interpreted as belonging to ceratopsids (Lockley and Hunt 1995) show closely spaced forefoot prints that do not support the sprawling stance of older mounts. Paul and Christiansen (2000) argue that with reconfiguration of the rib cages and pectoral girdles of the old mounts, the forelimbs can be brought under the body into a position in line with the trackway evidence, but differing from both the sprawling and fully erect postures. A recent modeling approach by Thompson and Holmes (2007) concludes a similar forelimb posture with a partially everted elbow, but found a much lower degree of humeral excursion that would have precluded running gaits. Fujiwara (2009) presented detailed information on the articulated forelimbs of a well-preserved *Triceratops*, which indicates that the distal end of the forearm and the manus had an L-shaped configuration such that the first and second digits point forward, with the elbow bending in a fore–aft plane.

Very little is known about other aspects of ceratopsian biology. Data are emerging with regard to life history patterns. Growth studies on *Protoceratops* (Makovicky et al. 2007) indicate that this taxon grew both faster (at peak growth) and longer than *Psittacosaurus*. Age and growth patterns in ceratopsids have been the topic of intense debate. Based on histology, some authors have inferred extremely high growth rates, with adulthood being reached in under a decade (Reizner and Horner 2006; Lee 2007), while Lehman (2007) coupled demographics from an *Agujaceratops* assemblage with growth rates for extant ectothermic reptiles to propose that this chasmosaurine did not reach full size until approximately 80 years of age. Recently, Erickson and Druckenmiller used growth lines from an adult-sized limb bone of an Alaskan specimen of *Pachyrhinosaurus* to determine that it was roughly 20 years old at the time of death, implying growth rates and longevity in keeping with other dinosaurs of comparable body mass. Their data does, however, corroborate previous conclusions derived from demographics of mass-death assemblages, that linear growth was extremely rapid during the earliest years of development.

A clear trend in jaw and dental evolution toward ever larger and more refined shearing action is obvious among neoceratopsians, but so far no stomach contents have been reported from ceratopsian fossils, and most species from Central Asia occur in rocks that rarely preserved plant fossils. Likewise, little is known about the reproductive biology of the group. Although the first reported dinosaur eggs discovered in the Gobi by the Central Asiatic Expeditions in 1923 were referred to *Protoceratops* because of the abundance of its remains, they have subsequently been shown to derive from maniraptoran theropods (Norell et al. 1994). Surprisingly, in spite of the numerous dinosaur eggs collected by the many expeditions to Mongolia and bordering areas since the late 1940s, none are definitely referable to *Protoceratops* or other ceratopsians, and the same holds true for the more depauperate North American paleological record. Only recently has an isolated ceratopsian egg been identified from the early Late Cretaceous Javkhlant Formation of Mongolia (Balanoff et al. 2008). This egg was identified through digital preparation of CT scans of embryonic remains within it. The egg is ellipsoid and symmetrical in shape as is typical of crocodylians and some archosaur outgroups, but eggshell microstructure reveals three eggshell layers as in birds. As this egg was found isolated, it does not provide insights on nesting habits of ceratopsians, but the discovery of a buried nest of neonate *Protoceratops* (Fastovsky et al. 1997) and the *Psittacosaurus* family find (Meng et al. 2004) suggest that ceratopsians laid large clutches of eggs and used nesting structures.

Bakker, R. T. 1986. *The Dinosaur Heresies*. London: Penguin Books.

Bakker, R. T., P. Larson, V. Porter, S. Salisbury, and R. M. Sullivan. 2006. *Dracorex hogwartsia*, n. gen, n. sp., a spiked, flat-headed pachycephalosaurid dinosaur from the Hell Creek Formation of South Dakota. *Bulletin of the New Mexico Museum of Natural History and Science* 35: 331–345.

References

Balanoff, A. M., M. A. Norell, G. Grellet-Tinner, and M. R. Lewin. 2008. Digital preparation of a probable neoceratopsian preserved within an egg, with comments on microstructural anatomy of ornithischian eggshells. *Naturwissenschaften* 95 (6): 493–500.

Brown, B., and E. M. Schlaikjer. 1940. The structure and relationships of *Protoceratops*. *Annals of the New York Academy of Sciences* 40: 133–266.

———. 1943. A study of the troödont dinosaurs, with the description of a new genus and four new species. *Bulletin of the American Museum of Natural History* 82: 115–150.

Butler, R. J., and R. M. Sullivan. 2009. The phylogenetic position of the ornithischian dinosaur *Stenopelix valdensis* from the Lower Cretaceous of Germany and the early fossil record of Pachycephalosauria. *Acta Palaeontologica Polonica* 41(1): 21–34.

Butler, R. J., P. Upchurch, and D. B. Norman. 2008. The phylogeny of the ornithischian dinosaurs. *Journal of Systematic Palaeontology* 6: 1–40.

Chapman, R., R. Sadleir, P. Dodson, and P. Makovicky. 2008. Handling missing data in paleontological matrices: approaches for exploratory multivariate analyses in morphometrics. *Journal of Vertebrate Paleontology* 28 (suppl. to 3): 62A.

Chinnery, B. 2004. Description of *Prenoceratops pieganensis* gen. et sp nov (Dinosauria: Neoceratopsia) from the Two Medicine Formation of Montana. *Journal of Vertebrate Paleontology* 24 (3): 572–590.

Chinnery, B. J., and J. R. Horner. 2007. A new neoceratopsian dinosaur linking North American and Asian taxa. *Journal of Vertebrate Paleontology* 27 (3): 625–641.

Colbert, E. H. 1955. *Evolution of the Vertebrates*. New York: Wiley. 479.

Currie, P. J. 1987. Bird-like characteristics of the jaws and teeth of troodontid theropods (Dinosauria, Saurischia). *Journal of Vertebrate Paleontology* 7 (1): 72–81.

Dodson, P. 1976. Quantitative aspects of relative growth and sexual dimorphism in *Protoceratops*. *Journal of Paleontology* 50: 929–940.

Dodson, P., and P. J. Currie. 1990. Neoceratopsia. In P. Dodson, H. Osmólska, and D. B. Weishampel (eds.), *The Dinosauria*, 593–618. Berkeley: University of California Press.

Dodson, P., C. A. Forster, and S. D. Sampson. 2004. Ceratopsidae. In D. B. Weishampel, P. Dodson, and H. Osmólska (eds.), *The Dinosauria*, 2nd ed.,

494–516. Berkeley: University of California Press.

Dong, Z.-M., and Y. Azuma. 1997. On a primitive neoceratopsian from the Early Cretaceous of China. In Z.-M. Dong (ed.), *Sino-Japanese Silk Road Dinosaur Expedition*, 68–89. Beijing: China Ocean Press.

Erickson, G. M. and P. S. Druckenmiller. 2011. Longevity and growth rate estimates for a polar dinosaur: a *Pachyrhinosaurus* (Dinosauria, Neoceratopsia) specimen from the North Slope of Alaska showing a complete developmental record. *Historical Biology*. doi: 10.1080/08912963.2010.546856.

Erickson, G. M., P. J. Currie, B. D. Inouye, and A. A. Winn. 2006. Tyrannosaur life tables: An example of nonavian dinosaur population biology. *Science* 313: 213–217.

Erickson, G. M., P. J. Makovicky, B. D. Inouye, C. F. Zhou, and K. Q. Gao. 2009. A life table for *Psittacosaurus lujiatunensis*: initial insights into ornithischian dinosaur population biology: *Anatomical Record: Advances in Integrative Anatomy and Evolutionary Biology* 292:1514–1521.

Erickson, G. M., K. C. Rogers, and S. A. Yerby. 2001. Dinosaurian growth patterns and rapid avian growth rates. *Nature* 412 (6845): 429–433.

Erickson, G. M., and T. A. Tumanova. 2000. Growth curve of *Psittacosaurus mongoliensis* Osborn (Ceratopsia: Psittacosauridae) inferred from long bone histology. *Zoological Journal of the Linnean Society* 130 (4): 551–566.

Farke, A. A. 2004. Horn use in *Triceratops* (Dinosauria: Ceratopsidae): testing behavioral hypotheses using scale models. *Palaeontologia Electronica* 7 (1): 1–10. http://palaeo-electronica.org/paleo/2004_3/horn/issue1_04.htm.

Farke, A. A., S. D. Sampson, C. A. Forster and M. A. Loewen. 2009a. *Turanoceratops tardabilis*-sister taxon, but not a ceratopsid. *Naturwissenschaften* 96 (7): 869–870.

Farke, A. A., E. D. S. Wolff, and D. H. Tanke. 2009b. Evidence of combat in *Triceratops*. PLoS ONE 4 (1): e4252. doi:10.1371/journal.pone.0004252.

Farlow, J. O., and P. Dodson. 1975. The behavioral significance of frill and horn morphology in ceratopsian dinosaurs. *Evolution* 29 (2): 353–361.

Fastovsky, D. E., D. Badamgarav, H. Ishimoto, M. Watabe, and D. B. Weishampel. 1997. The paleoenvironments of Tugrikin-Shireh (Gobi Desert, Mongolia) and aspects of the taphonomy and paleoecology of *Protoceratops*

(Dinosauria: Ornithishichia). *Palaios* 12 (1): 59–70.

Forster, C. A. 1990. The cranial morphology and systematics of *Triceratops*, with a preliminary analysis of ceratopsian phylogeny. Ph.D. dissertation, University of Pennsylvania, Pa.

———. 1996. Species resolution in *Triceratops*: cladistic and morphometric approaches. *Journal of Vertebrate Paleontology* 16 (2): 259–270.

Fujiwara, S.-I. 2009. A reevaluation of the manus structure in *Triceratops* (Ceratopsia: Ceratopsidae). *Journal of Vertebrate Paleontology* 29: 1136–1147.

Galton, P. M. 1971. A primitive dome-headed dinosaur (Ornithischia: Pachycephalosauridae) from the Lower Cretaceous of England, and the function of the dome in pachycephalosaurids. *Journal of Paleontology* 45: 40–47. *Canadian Journal of Earth Sciences*

Gilmore, C. 1924. On *Troodon validus*, an orthopodous dinosaur from the Belly River Cretaceous of Alberta, Canada. *University of Alberta Bulletin* 1: 1–41.

Goodwin, M. B., E. A. Buchholtz, and R. E. Johnson. 1998. Cranial anatomy and diagnosis of *Stygimoloch spinifer* (Ornithischia: Pachycephalosauria) with comments on cranial display structures in agonistic behavior. *Journal of Vertebrate Paleontology* 18 (2): 363–375.

Goodwin, M. B., and J. R. Horner. 2004. Cranial histology of pachycephalosaurs (Ornithischia: Marginocephalia) reveals transitory structures inconsistent with head-butting behavior. *Paleobiology* 30 (2): 253–267.

Haas, G. 1955. The jaw musculature in *Protoceratops* and other ceratopsians. *American Museum Novitates* 1729: 1–24.

Hatcher, J. B., O. C. Marsh, and R. S. Lull. 1907. *The Ceratopsia*. U. S. Geological Survey Monograph 49. Washington, D.C.: Government Printing Office.

Heckert, A. B., S. G. Lucas, and S. E. Krzyzanowski. 2003. Vertebrate fauna of the late Campanian (Judithian) Fort Crittenden Formation, and the age of Cretaceous vertebrate faunas of southeastern Arizona (USA). *Neues Jahrbuch für Geologie und Paläontologie-Abhandlungen* 227 (3): 343–364.

Hieronymus, T. L., L. M. Witmer, D. H. Tanke, and P. J. Currie. 2009. The facial integument of centrosaurine ceratopsids: morphological and histological correlates of novel skin structures. *Anatomical Record* 292: 1370–1396.

Horner, J. R., and M. B. Goodwin. 2006. Major cranial changes during *Triceratops* ontogeny. *Proceedings of the Royal Society B* 273: 2757–2761.

———. 2009. Extreme cranial ontogeny in the Upper Cretaceous dinosaur Pachycephalosaurus. *PLoS ONE* 4 (10): e7626. doi:10.1371/journal.pone.0007626.g003.

Horner, J. R., D. J. Varricchio, and M. B. Goodwin. 1992. Marine transgressions and the evolution of Cretaceous dinosaurs. *Nature* 358 (6381): 59–61.

Johnson, R. E., and J. H. Ostrom. 1995. The forelimb of *Torosaurus* and an analysis of the posture and gait of ceratopsian dinosaurs. In J. J. Thomason (ed.), *Functional Morphology in Vertebrate Paleontology*, 205–218. Cambridge: Cambridge University Press.

Kirkland, J. I., and D. D. Deblieux. 2010. New basal centrosaurine ceratopsian skulls from the Wahweap Formation (Middle Campanian), Grand Staircase Escalante National Monument, southern Utah. In M. J. Ryan, B. J. Chinnery-Allgeier, and E. D. A. (eds.), *New Perspectives on Horned Dinosaurs: The Royal Tyrell Museum Ceratopsian Symposium*, 117–140. Bloomington: Indiana University Press.

Kurzanov, S. 1992. A gigantic protoceratopsid from the Upper Cretaceous of Mongolia. *Paleontological Journal* 24: 85–91.

Lambert, O., P. Godefroit, H. Li, C.-Y. Shang, and Z.-M. Dong. 2001. A new species of *Protoceratops* (Dinosauria, Neoceratopsia) from the Late Cretaceous of Inner Mongolia (P.R. China). *Bulletin de l'Institut Royal des Sciences Naturelles de Belgique* 71 (suppl.): 5–28.

Lee, A. H. 2007. How *Centrosaurus* (and other ceratopsians) grew to large size. In D. R. Braman (ed.), *Ceratopsian Symposium : Short Papers, Abstracts, and Programs*, 105–106. Drumheller, Alberta, Canada: Royal Tyrrell Museum of Palaeontology.

Lehman, T. M. 1990. The ceratopsian subfamily Chasmosaurinae: sexual dimorphism and systematics. In P. J. Currie and K. Carpenter (eds.), *Dinosaur Systematics: Perspectives and Approaches*, 211–229. Cambridge: Cambridge University Press.

———. 2007. Growth and population structure in the horned dinosaur *Chasmosaurus*. In K. Carpenter (ed.), *Horns and Beaks*, 259–317. Bloomington: Indiana University Press.

Lindgren, J., P. J. Currie, M. Siverson, J. Rees, P. Cederstrom, and F. Lindgren. 2007. The first neoceratopsian dinosaur remains from Europe. *Palaeontology* 50: 929–937.

Lingham-Soliar, T., and G. Plodowski. 2010. The integument of Psittacosaurus from Liaoning Province, China: taphonomy, epidermal patterns and a color of a ceratopsian dinosaur. *Naturwissenschaften* 97: 479–486.

Lockley, M. G., and A. P. Hunt. 1995. Ceratopsid tracks and associated ichnofauna from the Laramie formation (Upper Cretaceous: Maastrichtian) of Colorado. *Journal of Vertebrate Paleontology* 15 (3): 592–614.

Longrich, N. R. 2010. *Titanoceratops ouranos*, a giant horned dinosaur from the late Campanian of New Mexico. *Cretaceous Research* 32: 264–276.

Longrich, N. R., J. T. Sankey, and D. Tanke. 2010. *Texascephale langstoni*, a new pachycephalosaurid (Dinosauria: Ornithischia) from the upper Campanian Aguja Formation, southern Texas, USA. *Cretaceous Research* 31: 274–284.

Lull, R. S. 1933. *A revision of the Ceratopsia or horned dinosaurs*. Memoirs of the Peabody Museum of Natural History, 3:pt. 3. New Haven, Conn.: Peabody Museum of Natural History.

Makovicky, P. J. 2001. A *Montanoceratops cerorhynchus* (Dinosauria: Ceratopsia) braincase from the Horseshoe Canyon Formation of Alberta. In D. Tanke and K. Carpenter (eds.), *Mesozoic Vertebrate Life*, 243–262. Bloomington: Indiana University Press.

———. 2002. Taxonomic revision and phylogenetic relationships of basal Neoceratopsia (Dinosauria: Ornithischia). Ph.D. dissertation, Columbia University, New York City.

———. 2010. A redescription of the *Montanoceratops cerorhynchus* holotype, with a review of referred material. In M. J. Ryan, B. Chinnery-Allgeier, D. A. Eberth (eds.), *New Perspectives on Horned Dinosaurs*, 68–82. Bloomington: Indiana University Press.

Makovicky P. J., G. M. Erickson, R. W. Sadleir, and P. Dodson. 2007. Life history of *Protoceratops andrewsi* from Bayn Zag, Mongolia. *Journal of Vertebrate Paleontology* 27 (suppl. to 3): 109A.

Makovicky P. J., and M. A. Norell. 2006. *Yamaceratops dorngobiensis*, a new primitive ceratopsian (Dinosauria: Ornithischia) from the Cretaceous of Mongolia. *American Museum Novitates* 3530: 1–42.

Maryańska, T. 1990. Pachycephalosauria. In D. B. Weishampel, P. Dodson, and H. Osmólska (eds.), *The Dinosauria*, 564–577. Berkeley: University of California Press.

Maryańska, T., and H. Osmólska. 1974.

Pachycephalosauria, a new suborder of ornithischian dinosaurs. *Palaeontologica Polonica* 32: 45–102.

———. 1975. Protoceratopsidae (Dinosauria) of Asia. *Palaeontologia Polonica* 33: 133–181.

Mayr, G., D. S. Peters, G. Plodowski, and O. Vogel. 2002. Bristle-like integumentary structures at the tail of the horned dinosaur *Psittacosaurus*. *Naturwissenschaften* 89 (8): 361–365.

Meng, Q. J., J. Y. Liu, D. J. Varricchio, T. Huang, and C. L. Gao. 2004. Parental care in an ornithischian dinosaur. *Nature* 431 (7005): 145–146.

McDonald, A. T., and J. R. Horner. 2010. New material of "Styracosaurus" ovatus from the Two Medicine Formation of Montana. In M. J. Ryan, B. Chinnery-Allgeier, and D. A. Eberth (eds.), *New Perspectives on Horned Dinosaurs*, 156–168. Bloomington: Indiana University Press.

Nessov, L. A., L. F. Kaznyshkina, and G. O. Cherepanov. 1989. Ceratopsian dinosaurs and crocodiles of the middle Mesozoic of Asia. In T. N. Bogdanova and L. I. Kozhatsky (eds.), *Theoretical and Applied Aspects of Modern Paleontology*, 142–149. Leningrad: Nauka.

Norell, M. A., J. M. Clark, D. Demberelyin, B. Rinchen, L. M. Chiappe, A. R. Davidson, M. C. McKenna, P. Altangerel, and M. J. Novacek. 1994. A theropod dinosaur embryo and the affinities of the Flaming Cliffs dinosaur eggs. *Science* 266 (5186): 779–782.

Ösi, A., R. J. Butler, and D. B. Weishampel. 2010. A Late Cretaceous ceratopsian dinosaur from Europe with Asian affinities. *Nature* 465: 466–468.

Ostrom, J. H. 1964. A functional analysis of the jaw mechanics in the dinosaur *Triceratops*. *Postilla* 88: 1–35.

———. 1966. Functional morphology and evolution of the ceratopsian dinosaurs. *Evolution* 20: 290–308.

Padian, K., and J. R. Horner. 2011. The evolution of "bizarre structures" in dinosaurs: biomechanics, sexual selection, social selection or species recognition? *Journal of Zoology* 283 (1): 3–17.

Paul, G. S., and P. Christiansen. 2000. Forelimb posture in neoceratopsian dinosaurs: implications for gait and locomotion. *Paleobiology* 26 (3): 450–465.

Qi, Z., P. M. Barrett, and D. A. Eberth. 2007. Social behaviour and mass mortality in the basal ceratopsian dinosaur *Psittacosaurus* (Early Cretaceous, People's Republic of China). *Palaeontology* 50: 1023–1029.

Reid, R. E. H. 1996. Bone histology of the Cleveland-Lloyd dinosaurs and of dinosaurs in general. Pt. 1. Introduction: Introduction to bone tissues. *Brigham Young University Geology Studies* 41: 25–71.

Reizner, J., and J. R. Horner. 2006. An ontogenetic series of the ceratopsid dinosaur *Einiosaurus procurvicornis* as determined by long bone histology. *Journal of Vertebrate Paleontology* 26 (supp. to 3): 114A.

Ryan, M. J. 2007. A new basal centrosaurine ceratopsid from the Oldman Formation, southeastern Alberta. *Journal of Paleontology* 81 (2): 376–396.

Ryan, M. J., D. B. Brinkman, D. A. Eberth, P. J. Currie, and D. H. Tanke. 2006. A new *Pachyrhinosaurus*-like ceratopsian from the upper Dinosaur Park Formation (Late Campanian) of southern Alberta, Canada. *Journal of Vertebrate Paleontology* 26 (suppl. to 3): 117A.

Ryan, M. J., A. P. Russell, and S. Hartman. 2010. A new chasmosaurine ceratopsid from the Judith River Formation, Montana. *New Perspectives on Horned Dinosaurs: The Royal Tyrell Museum Ceratopsian Symposium*, 181–188. bloomington:

Sampson, S. D. 1995. Two new horned dinosaurs from the Upper Cretaceous Two Medicine Formation of Montana, with a phylogenetic analysis of the Centrosaurinae (Ornithischia: Ceratopsidae). *Journal of Vertebrate Paleontology* 15 (4): 743–760.

Sampson, S. D. 1999. Sex and destiny: the role of mating signals in speciation and macroevolution. *Historical Biology* 13 (2–3): 173–197.

Sampson, S. D., and M. A. Loewen. 2007. New information on the diversity, stratigraphic distribution, biogeography, and evolution of ceratopsid dinosaurs. In D. R. Braman, *Ceratopsian Symposium: Short Papers, Abstracts, and Programs*, 125–133. Drumheller: Royal Tyrrell Museum of Palaeontology.

Sampson, S. D., and M. A. Loewen. 2010. Unraveling a radiation: a review of the diversity, stratigraphic distribution, biogeography, and evolution of horned dinosaurs (Ornithischia: Ceratopsidae.). In M. J. Ryan, B. J. Chinnery-Allgeier, and D. A. Eberth (eds.), *New Perspectives on Horned Dinosaurs: The Royal Tyrrell Museum Ceratopsian Symposium*, 405–427. Bloomington: Indiana University Press.

Sampson, S. D., M. A. Loewen, A. A. Farke, E. M. Roberts, C. A. Forster, J. A. Smith, and A. L. Titus. 2010. New horned dinosaurs from Utah provide evidence for intracontinental dinosaur endemism.

PloS ONE 5 (9): e12292. doi:10.1371/journal.pone.0012292.

Sampson, S. D., M. J. Ryan and D. H. Tanke. 1997. Craniofacial ontogeny in centrosaurine dinosaurs (Ornithischia: Ceratopsidae): taxonomic and behavioral implications. *Zoological Journal of the Linnean Society* 121 (3): 293–337.

Scannella, J. B., and J. R. Horner. 2010. Torosaurus Marsh, 1891, is Triceratops Marsh, 1889 (Ceratopsidae: Chasmosaurinae): synonymy through ontogeny. *Journal of Vertebrate Paleontology* 30: 1157–1168.

Senter, P. 2007. Analysis of forelimb function in ceratopsians. *Journal of Zoology* 273: 305–314.

Sereno, P. C. 1986. Phylogeny of the bird-hipped dinosaurs (order Ornithischia). *National Geographic Research* 2: 234–256.

———. 1989. Pachycephalosaurs and ceratopsians (Ornithischia: Marginocephalia). In S. J. Culver (ed.), *The Age of Dinosaurs, 12th Annual Short Course of the Paleontological Society*, 71–79. Paleontological Society Papers 10. Knoxville: University of Tennessee Press.

———. 1990. Psittacosauridae. In B. Weishampel David, P. Dodson, and H. Osmolska (eds.), *The Dinosauria*, 579–592. Berkeley: University of California Press.

———. 1997. The origin and evolution of dinosaurs. *Annual Review of Earth and Planetary Sciences* 25: 435–489.

———. 1999. The evolution of dinosaurs. *Science* 284 (5423): 2137–2147.

———. 2000. The fossil record, systematics and evolution of pachycephalosaurs and ceratopsians from Asia. In M. Benton, M. Shishkin, D. Unwin, and E. Kurochkin (eds.), *The Age of Dinosaurs in Russia and Mongolia*, 480–516. New York: Cambridge University Press.

———. 2010. Taxonomy, cranial morphology, and relationships of parrot-beaked dinosaurs. *New Perspectives on Horned Dinosaurs: The Royal Tyrell Museum Ceratopsian Symposium*, 21–58. Bloomington:

Sereno, P. C., X. J. Zhao, L. Brown, and T. Lin. 2007. New psittacosaurid highlights skull enlargement in horned dinosaurs. *Acta Palaeontologica Polonica* 52 (2): 275–284.

Sereno, P. C., X. J. Zhao, and L. Tan. 2010 A new psittacosaur from Inner Mongolia and the parrot-like structure and function of the psittacosaur skull. *Proceedings of the Royal Society* B 277(1679): 199–209.

Sternberg, C. M. 1951. Complete skeleton of *Leptoceratops gracilis* Brown from the Upper Edmonton Member on Red Deer

River, Alberta. *National Museum of Canada Bulletin, Annual Report (1949–50)* 123: 225–255.

Sues, H. D., and A. Averianov. 2009a. Phylogenetic position of Turanoceratops (Dinosauria: Ceratopsia). *Naturwissenschaften* 96 (7): 871–872.

———. 2009b. Turanoceratops tardabilis- the first ceratopsid dinosaur from Asia. *Naturwissenschaften* 96 (5): 645–652.

Sues, H.-D., and P. M. Galton. 1987. Anatomy and classification of the North American Pachycephalosauria (Dinosauria: Ornithischia). *Palaeontographica Abt.* B 198: 1–40.

Sullivan, R. M. 2003. Revision of the dinosaur *Stegoceras* Lambe (Ornithischia, Pachycephalosauridae). *Journal of Vertebrate Paleontology* 23 (1): 181–207.

———. 2006. A taxonomic review of the Pachycephalosauridae (Dinosauria: Ornithischia). *Bulletin of the New Mexico Museum of Natural History and Science* 35: 347–366.

———. 2007. Doming, heterochrony, and paedomorphosis in the Pachycephalosauridae (Ornithischia: Dinosauria): taxonomic and phylogenetic implications. *Journal of Vertebrate Paleontology* 27 (3): 154A–155A.

Thompson, S., and R. Holmes. 2007. Forelimb stance and step cycle in *Chasmosaurus irvinensis* (Dinosauria: Neoceratopsia). *Palaeontologia Electronica* 10 (1): 5A: 17.

Williamson, T. E., and T. D. Carr. 2002. A new genus of derived pachycephalosaurian from western North America. *Journal of Vertebrate Paleontology* 22: 779–801.

Wolfe, D. G., and J. I. Kirkland. 1998. *Zuniceratops christopheri* n. gen & n. sp., a ceratopsian dinosaur from the Moreno Hill Formation (Cretaceous, Turonian) of west-central New Mexico. *Bulletin of the New Mexico Museum of Natural History and Science* 14: 303–317.

Xu, X., C. A. Forster, J. M. Clark, and J. Mo. 2006. A basal ceratopsian with transitional features from the Late Jurassic of northwestern China. *Proceedings of the Royal Society* B 273: 2135–2140.

Xu, X., P. J. Makovicky, X. L. Wang, M. A. Norell, and H. L. You. 2002. A ceratopsian dinosaur from China and the early evolution of Ceratopsia. *Nature* 416 (6878): 314–317.

You, H. L., and Z. M. Dong. 2003. A new protoceratopsid (Dinosauria: Neoceratopsia) from the Late Cretaceous of Inner Mongolia, China. *Acta Geologica Sinica* (English ed.) 77 (3): 299–303.

You, H. L., D. Q. Li, Q. Ji, M. C. Lamanna, and P. Dodson. 2005. On a new genus of basal neoceratopsian dinosaur from the Early Cretaceous of Gansu Province, China. *Acta Geologica Sinica-English Edition* 79 (5): 593–597.

You, H. L., K. Tanoue, and P. Dodson. 2010. A new species of *Archaeceratops* (Dinosauria, Neoceratopsia) from the Early Cretaceous of the Mazongshan Area, Northwestern China. In M. J. Ryan, B. J. Chinnery-Allgeier, and D. A. Eberth (eds.), *New Perspectives on Horned Dinosaurs: The Royal Tyrell Museum Ceratopsian Symposium*, 59–67. Bloomington: Indiana University Press.

You, H. L., X. Xu, and X. L. Wang. 2003. A new genus of Psittacosauridae (Dinosauria: Ornithopoda) and the origin and early evolution of Marginocephalian dinosaurs. *Acta Geologica Sinica* (English ed.) 77 (1): 15–20.

Zhao, X., Z. Cheng, and X. Xing. 1999. The earliest Ceratopsian from the Tuchengzi Formation of Liaoning, China. *Journal of Paleontology* 19 (4): 681–691.

Zhao, X., Z. Cheng, X. Xing, et al. 2006. A new ceratopsian from the Upper Jurassic Houcheng Formation of Hebei, China. *Acta Geologica Sinica* (English ed.) 80 (4): 467–73.

Zhou, C. F., K. Q. Gao, C. F. Fox, and S. H. Chen. 2006. A new species of *Psittacosaurus* (Dinosauria: Ceratopsia) from the Early Cretaceous Yixian Formation, Liaoning, China. *Palaeoworld* 15: 100–114.

26.1. Simplified phylogeny of Ornithischia focusing on the relationships of ornithopods, based upon Norman (2004) and Butler et al. (2008). Taxa formerly considered as ornithopods (*Lesothosaurus, Agilisaurus, Hexinlusaurus, Othnielosaurus*) are marked with an asterisk. Note the probable paraphyly of "Hypsilophodontidae" (e.g., *Agilisaurus, Hexinlusaurus, Othnielosaurus, Hypsilophodon,* and *Thescelosaurus*) and "Iguanodontidae" (e.g., *Iguanodon* and *Ouranosaurus*). The clade Heterodontosauridae is shown in three alternative positions: (1) as the most basal well-known ornithischians (Butler et al. 2007, 2008); (2) as the sister taxon of Marginocephalia (e.g., Xu et al. 2006); (3) as the most basal radiation of ornithopods (Sereno 1986).

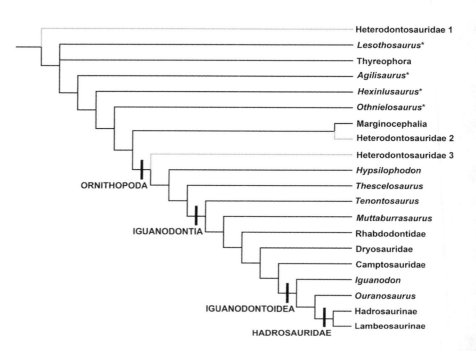

Ornithopods

Richard J. Butler and Paul M. Barrett

26

The Ornithopoda ("bird feet"), commonly known as ornithopods, were a hugely successful and diverse clade of bipedal herbivorous ornithischian dinosaurs known from the Middle Jurassic to the end of the Cretaceous. Their remains have been found on every continent, including Antarctica. Ornithopods generally retained a rather conservative morphology—they lacked the extensive armor present in stegosaurs and ankylosaurs, the thickened skull roof of pachycephalosaurs, and the horns and frills of ceratopsians—although some derived taxa (hadrosaurs) sported elaborate cranial crests. Unique joints within the skull and complex tooth batteries allowed sophisticated mastication (chewing), making them the most specialized of all dinosaurian herbivores. Their diverse range of body sizes and cranial morphology often allowed several different ornithopod species to co-occur in the same ecosystem, occupying different niches and, presumably, eating different plants. In many Cretaceous faunas, particularly those of Laurasia, they were the most successful of the herbivorous dinosaurs, in terms of both species richness and abundance.

History of Study

An ornithopod, *Iguanodon*, was the second dinosaur to be named when it was described by Gideon Mantell in 1825. Based upon the limited remains available to him, Mantell reconstructed *Iguanodon* as a giant quadrupedal herbivorous lizard, similar to a scaled-up version of the living iguana. In 1842, Richard Owen identified *Iguanodon* (along with the theropod *Megalosaurus* and the ankylosaur *Hylaeosaurus*) as a founding member of his new group, the Dinosauria.

In the 1850s, Joseph Leidy described *Iguanodon*-like animals from New Jersey (USA) as *Hadrosaurus*. Leidy reconstructed *Hadrosaurus* with a rather kangaroo-like posture, rather than the lizardlike or elephantine reconstructions proposed (on the basis of less-complete material) by Mantell and Owen. Louis Dollo proposed a similar reconstruction for *Iguanodon* in 1883, based upon the amazing discovery of over 30 mostly complete skeletons in a coal mine at Bernissart, Belgium, in 1878.

Many other major discoveries were made during the late nineteenth century, including the European genera *Hypsilophodon* and *Rhabdodon* (both named in 1869, by T. H. Huxley and P. Matheron, respectively) and the North American genera *Camptosaurus* and *Dryosaurus* (both named by O. C. Marsh: the former in 1885 and the latter in 1894). In 1881, Marsh recognized and named Ornithopoda as a distinct clade of ornithischians, basing its name on the supposed similarities between ornithopod and bird feet.

The early twentieth century witnessed the discovery of numerous hadrosaurs in Canada (including *Saurolophus*, *Hypacrosaurus*, and *Corythosaurus*), and the first Asian (e.g., *Bactrosaurus*, *Tanius*, and *Tsintaosaurus*) and African (*Dysalotosaurus*) ornithopods. Discovery of new ornithopod species continued throughout the twentieth and early twenty-first centuries, with the first descriptions of new taxa and specimens from Australia, South America, and Antarctica, as well as further new finds from other regions.

As more new ornithischian taxa were described, Ornithopoda became a taxonomic wastebasket into which all bipedal, morphologically conservative ornithischians were placed. Therefore, taxa as diverse as "fabrosaurs," early armored dinosaurs, heterodontosaurids, pachycephalosaurs, and psittacosaurs were included within Ornithopoda, despite the fact that bipedalism was actually an ancestral character shared with other dinosaurs. Many of the other characters used to unite this disparate group were also ancestral for Ornithischia and of little systematic value. This practice ended with the application of cladistic methodology to ornithischian phylogeny during the 1980s (Norman 1984a; Sereno 1984, 1986; Cooper 1985; Maryańska and Osmólska 1985). These studies restricted Ornithopoda to a natural (monophyletic) grouping of taxa. Pachycephalosaurs were pulled out and placed into their own group, which was united with Ceratopsia (including Psittacosauridae) in the larger clade Marginocephalia. Likewise, "fabrosaurs" and early armored dinosaurs such as *Scutellosaurus* were also removed from Ornithopoda, and positioned more basally within Ornithischia. Discussion continues regarding the ornithopod affinities of several other taxa, including heterodontosaurids (e.g., Butler 2005; Xu et al. 2006; Butler et al. 2008).

Classification

Ornithopoda includes all ornithischian dinosaurs more closely related to the hadrosaur *Edmontosaurus* than to the ceratopsian *Triceratops* (Norman et al. 2004). The anatomical features that define Ornithopoda are uncertain at present, and represent an area for future work. However, most ornithopods possess some or all of the following characters: cranial kinesis (pleurokinesis); premaxillae and jaw joints that are ventrally offset with respect to the maxillary tooth row; a closed or highly reduced external mandibular fenestra; and a tablike obturator process on the ischium.

A current ornithischian phylogeny is shown in Figure 26.1. The most primitive ornithopods are the "hypsilophodontids," small-bodied bipedal, cursorial taxa (most 1–3 m long) known from every continent and including such famous genera as *Hypsilophodon* (Fig. 26.2) and *Thescelosaurus* (Fig. 26.3A). The earliest genera (*Othnielosaurus* and *Yandusaurus*) are known from the Late Jurassic, although the presence of an iguanodontian in the Middle Jurassic (see below) suggests that hypsilophodontids must have originated somewhat earlier. Hypsilophodontids survived up until the end of the Cretaceous (e.g., *Parksosaurus* and *Thescelosaurus*). These animals were previously thought to represent a monophyletic clade of small ornithopods, the Hypsilophodontidae (e.g., Sereno 1986; Weishampel and Heinrich 1992). However, most recent work suggests that the hypsilophodontids do not form a natural group and that some genera (e.g., *Thescelosaurus* and *Parksosaurus*) are more closely related to iguanodontians than to other hypsilophodontids (e.g., Scheetz 1999; Buchholz 2002; Weishampel et al.

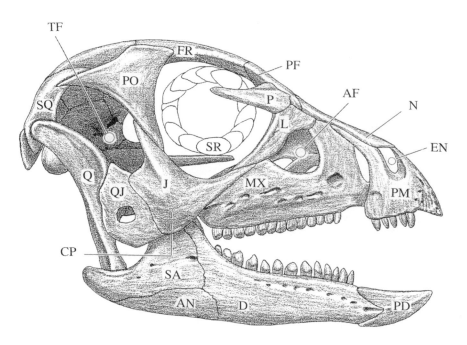

26.2. *Hypsilophodon* skull. Abbreviations as follows: AF = antorbital fenestra; AN = angular; CP = coronoid process; D = dentary; EN = external naris; FR = frontal; J = jugal; L = lachrymal; MX = maxilla; N = nasal; P = palpebral; PD = predentary; PF = prefrontal; PM = premaxilla; PO = postorbital; Q = quadrate; QJ = quadratojugal; SA = surangular; SQ = squamosal; SR = sclerotic ring; TF = lower temporal fenestra. The skull of this animal is approximately 12 cm long.

This and other drawings in this chapter by Gregory S. Paul, who retains the copyright.

2003; Butler 2005; Butler et al. 2008). Many genera (e.g., *Hypsilophodon*, *Orodromeus*, and *Thescelosaurus*) are known from multiple specimens and were highly abundant, at least locally.

Iguanodontia contains all ornithopods more closely related to the hadrosaur *Edmontosaurus* than to the hypsilophodontid *Thescelosaurus* (Norman 2004), and is characterized by the following features: premaxilla is transversely expanded and lacks teeth; dentary is deep with parallel dorsal

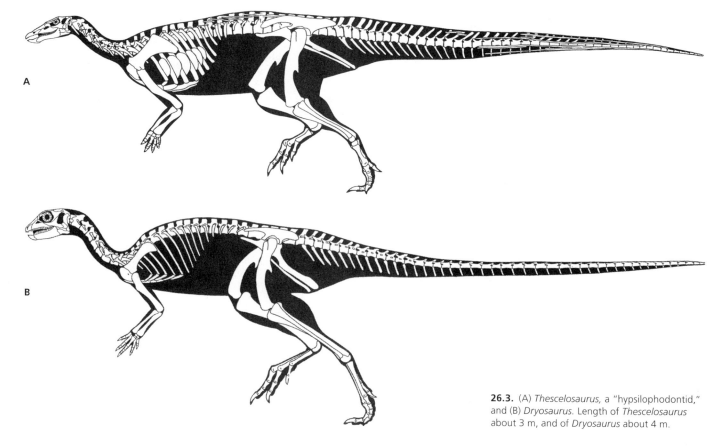

26.3. (A) *Thescelosaurus*, a "hypsilophodontid," and (B) *Dryosaurus*. Length of *Thescelosaurus* about 3 m, and of *Dryosaurus* about 4 m.

26.4. Flesh restorations of (A) *Camptosaurus* (left) and *Dryosaurus* (right) and (B) *Muttaburrasaurus*.

and ventral margins; one phalanx is lost from digit III of the hand; and anterior process of the pubis is deep and transversely compressed (Norman 2004). Included within Iguanodontia are *Tenontosaurus*, *Muttaburrasaurus*, rhabdodontids, dryosaurids, camptosaurids, "iguanodontids," and hadrosauroids. The earliest evidence for iguanodontians is a single femur (known as *Callovosaurus leedsi*) from the Middle Jurassic of England (Ruiz-Omeñaca et al. 2007), representing a dryosaurid. Well-preserved iguanodontian material is first known from the Late Jurassic of North America, Europe, and Africa (camptosaurids and dryosaurids).

The earliest-known iguanodontian clade was the Dryosauridae, comprising small (2–4 m in length), lightly built, fast-running bipeds, with elongate legs, narrow feet, and short arms (Figs. 26.3B, 26.4A). Dryosaurids are known from the Middle Jurassic of Europe (*Callovosaurus*, see above), the Late Jurassic of North America (*Dryosaurus*) and Africa (*Dysalotosaurus*), and the Early Cretaceous of Europe and Africa (*Valdosaurus*, *Elrhazosaurus*, *Kangnasaurus*: Galton 2009; McDonald et al. 2010).

Camptosaurs are known from the Late Jurassic of North America (*Camptosaurus dispar* and *Camptosaurus aphanoecetes*), England (*Cumnoria prestwichii*), and Portugal (*Draconyx*). It has also been suggested that they were present in the basal Cretaceous of England ("*Camptosaurus*" *hoggii*, recently referred to the genus *Owenodon*: Galton 2009). Recent phylogenetic analyses, however, have failed to support camptosaurid monophyly (McDonald et al. 2010), suggesting instead that they represent a grade of morphologically similar taxa. They were the first large, heavily built ornithopods (Figs. 26.4A, 26.5A), reaching up to 7 m in length, with relatively longer arms than hypsilophodontids and dryosaurs, and noticeably elongated muzzles, presumably to increase the amount of food taken and processed per bite. Their body proportions suggest that they may have been the first ornithopods to utilize both bipedal and quadrupedal gaits. The forelimb of *Camptosaurus* included a heavily ossified, blocklike wrist, and a subconical ungual on the first digit. In these features camptosaurs resembled later ornithopods such as *Iguanodon*. Unlike later ornithopods, camptosaurs retained a rather primitive pelvic and hind-limb arrangement, with a deep ilium, elongate pubic shaft, two phalanges present on the first digit of the foot, and rather pointed claws.

Tenontosaurus (Fig. 26.5B) is an enigmatic genus that was initially classified as a hypsilophodontid, but is now widely believed to be a basal iguanodontian (Forster 1990; Winkler et al. 1997). It was widespread in the Early Cretaceous of North America, with abundant and well-preserved skeletons known from Montana, Wyoming, and Texas. It reached about 7–8 m in length and had a very high skull and an unusually long neck and tail. *Tenontosaurus* is famous as the prey of *Deinonychus* (Maxwell and Ostrom 1995).

26.5. (A) *Camptosaurus* and (B) *Tenontosaurus*. Length of both taxa about 7 m.

Another problematic and unusual iguanodontian is *Muttaburrasaurus* (Fig. 26.4B) from the Early Cretaceous of Australia, which reached about 7 m in length, and was characterized by a bulbous inflated snout.

Rhabdodontids are known only from the Late Cretaceous of Europe, and were medium-sized ornithopods (3–5 m long) characterized by a number of features, including the presence of more than 12 vertical ridges on the dentary teeth, strong twisting of the anterior process of the ilium, and a femur that is strongly bowed in anterior view. *Zalmoxes robustus*, one of the best-known taxa, had a rather rotund appearance and was probably incapable of fast locomotion. It has been proposed that *Zalmoxes* is a paedomorphic dwarf, a hypothesis also proposed for other dinosaurs from the Late Cretaceous of Romania (Weishampel et al. 2003; Benton et al. 2010).

Within Iguanodontia, taxa more derived than camptosaurs are included within the Iguanodontoidea, which includes the hadrosaurs (see below) and an assemblage of more basal forms. A number of these taxa, including the sail-backed *Ouranosaurus* (Fig. 26.6B), were previously included with *Iguanodon* in the Iguanodontidae, although it now appears that Iguanodontidae is paraphyletic (e.g., Norman 2004; McDonald et al. 2010). *Iguanodon* (Figs. 26.6A, 26.7A) is a very well known taxon (Norman 2004), present in the Early Cretaceous of Europe and possibly Asia. *Iguanodon* reached large sizes (up to 10 m), utilized quadrupedal (as well as bipedal) locomotion, and

26.6. (A) *Iguanodon* and (B) *Ouranosaurus*. Length of *Iguanodon* about 9 m, and of *Ouranosaurus* about 7 m.

A

B

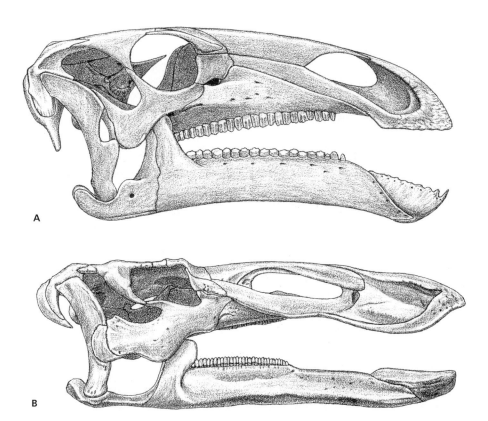

A

B

had highly unusual hands, with the bones of the thumb fused into a large triangular spike, which could presumably have been used as a defensive weapon. Recent work suggests that many of the large number of species and specimens historically referred to *Iguanodon* actually pertain to several different genera (Paul 2008; McDonald et al. 2010; Norman 2010; Carpenter and Ishida 2010): the taxonomy of this important genus is currently in the process of a major revision.

Hadrosauroidea includes taxa more closely related to *Edmontosaurus* than *Iguanodon*, and includes the important clade Hadrosauridae, consisting of the subclades Hadrosaurinae (including taxa such as *Shantungosaurus* and *Edmontosaurus*: Figs. 26.7B, 26.8A–C) and Lambeosaurinae (including taxa such as *Parasaurolophus*: Fig. 26.8D), as well as several more basal taxa such as *Telmatosaurus* from the Late Cretaceous of Romania. Hadrosaurids are exceptionally well known and are represented by many complete skeletons, as well as remains of eggs, embryos, and hatchlings; skin impressions; and coprolites. They were large (7–12 m long, average adult body weight 3 tons, largest taxa ranging up to 16 tons) animals that may have used both bipedal and quadrupedal postures. Although best known from North America, they had a near-worldwide distribution, and have been reported from Europe, Asia, South America, and Antarctica (although not from Africa, India, or Australia, to date). The tooth rows of hadrosaurs were highly modified into interlocking dental batteries made up of as many as 60 closely packed tooth families, with each individual tooth family containing 3–5 teeth. In lambeosaurs the skull was heavily modified to form a hollow supracranial crest that may have had a function in intraspecific communication.

Problematic "Ornithopods"

As discussed above, Ornithopoda has previously been used as a taxonomic wastebasket, into which almost all bipedal ornithischians have been placed at one time or another. During the last 25 years many workers have attempted to restrict the content of Ornithopoda, and define it as a monophyletic clade (e.g., Norman 1984a; Sereno 1986; Butler et al. 2008). For instance, psittacosaurs and pachycephalosaurs have been removed from Ornithopoda, and are dealt with elsewhere in this volume. Several other groups of small bipedal ornithischians that were considered ornithopods in the past are discussed briefly here. Although some of these taxa (e.g., the "fabrosaurs") are almost certainly not ornithopods, the ornithopod affinities of others (e.g., heterodontosaurids) remain contentious.

A number of problematic Triassic ornithischians have been considered as ornithopods. *Pisanosaurus*, from the Late Triassic of Argentina, is a small bipedal herbivore, known only from a maxilla, dentary, and some incomplete postcranial material. It was originally described as an ornithopod, with some authors (Bonaparte 1976) suggesting that it might belong to a clade of putative ornithopods, the heterodontosaurids (see below). More recent work has reassessed the position of *Pisanosaurus*, and it is now believed to be a very primitive ornithischian, not an ornithopod (Sereno 1991). *Technosaurus*, based upon fragmentary material from the Late Triassic of the United States, has also been considered an ornithopod, although recent work indicates that this material possesses no unequivocal ornithischian synapomorphies, and it is currently considered an archosauriform reptile of uncertain affinities (Nesbitt et al. 2007).

Lesothosaurus (often incorrectly referred to as *Fabrosaurus*) is a very small (1 m in length) bipedal ornithischian with strongly reduced forelimbs known from the Early Jurassic upper Elliot Formation of South Africa and Lesotho (Fig. 26.9A). It has frequently been referred to the family Fabrosauridae (Galton 1978). The "fabrosaurs" were believed to represent a primitive ancestral stock from which the other ornithopod groups were derived. Amongst other primitive features, fabrosaurs supposedly lacked the cheeks present in all other ornithischians. However, recent work (Sereno 1986, 1991; Butler 2005; Butler et al. 2007, 2008) has demonstrated that Fabrosauridae is not a monophyletic clade. *Lesothosaurus* cannot be included within Ornithopoda, but is a much more primitive ornithischian that may be close in morphology to the common ancestor of stegosaurs, ankylosaurs, ornithopods, ceratopsians, and pachycephalosaurs. A recent ornithischian phylogeny recovered *Lesothosaurus* as the most basal armored dinosaur (thyreophoran), but support for this position is weak and further work is required to assess its phylogenetic position (Butler et al. 2008). *Stormbergia* is a larger (about 2–3 m in length) fabrosaur from the upper Elliot Formation known, as yet, from postcranial remains only (Butler 2005). A tab-shaped process (the obturator process) is present on the ischium of the pelvis of *Stormbergia*, a character that has generally been considered diagnostic of Ornithopoda. However, nearly all other anatomical features are similar to *Lesothosaurus*, suggesting that *Stormbergia* is not an ornithopod but a more primitive ornithischian.

Heterodontosaurids were small (about 1 m or less in length) ornithischians with a number of unique features, including caninelike teeth and relatively long arms with large hands and strongly recurved claws. Most

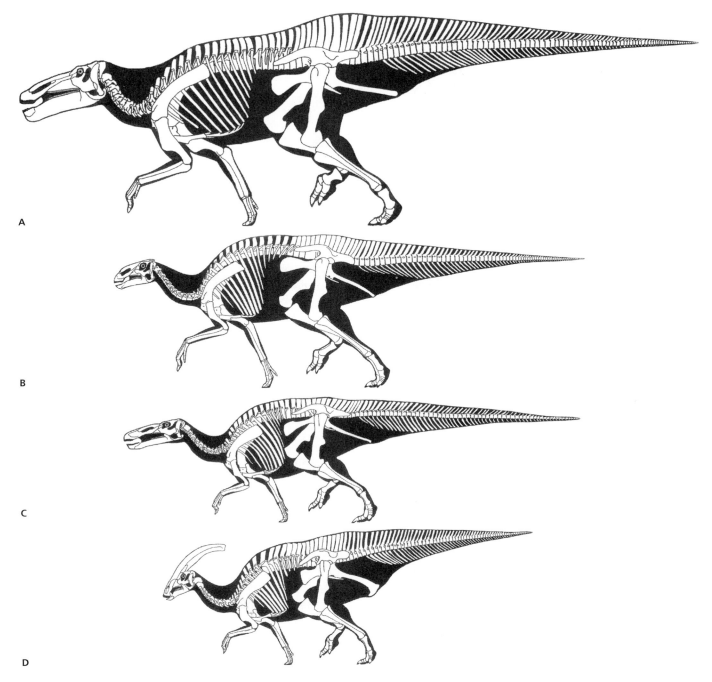

A

B

C

D

heterodontosaurid taxa (including the genera *Heterodontosaurus*, *Abrictosaurus*, and *Lycorhinus*) are known from the upper Elliot Formation of South Africa and Lesotho; however, heterodontosaurids are also known from the Late Triassic of Argentina (making them some of the earliest-known ornithischians), the Early and Late Jurassic of North America (Butler et al. 2010), and the Early Cretaceous of England (Norman and Barrett 2002). The problematic Triassic taxon *Pisanosaurus* (see above) may also be a heterodontosaurid. Recently a new species, *Tianyulong confuciusi*, was reported from the famous Jehol Group (Early Cretaceous) of Liaoning Province, China (Zheng et al. 2009). The exceptionally preserved holotype specimen of *Tianyulong* demonstrates the presence of long, singular and unbranched filamentous integumentary structures below the neck and

26.8. (A) *Shantungosaurus,* (B) *Brachylophosaurus,* (C) *Edmontosaurus* (*"Anatotitan"*), and (D) *Parasaurolophus.* Length of *Shantungosaurus* about 17 m, of *Brachylophosaurus* and *Edmontosaurus* about 12 m, and of *Parasaurolophus* about 9 m.

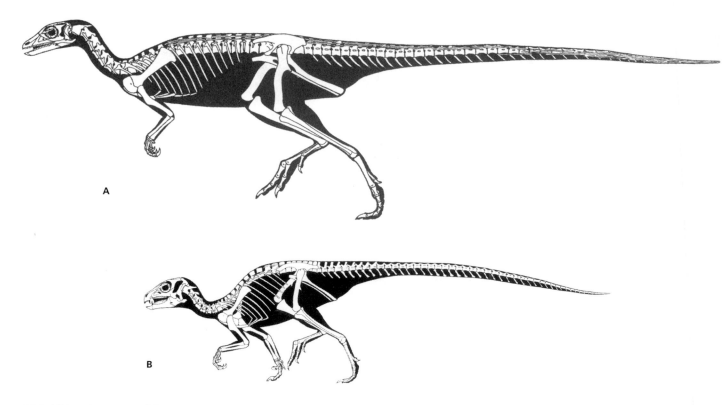

26.9. (A) *Lesothosaurus,* a "fabrosaurid," and (B) *Heterodontosaurus.* Length of each animal about 1 m.

above the back and tail. These integumentary structures in *Tianylong* have been homologized with feathers seen in birds and some nonavian theropods, and have been referred to as *protofeathers,* with the implication that the presence of filamentous integumentary structures could be a primitive condition for dinosaurs. Further work, and more specimens, is required to accurately assess the potential homology of these structures with feathers.

The phylogenetic position of heterodontosaurids is extremely problematic. Nevertheless, most workers have suggested that heterodontosaurids are the most primitive ornithopods (e.g., Sereno 1986; Norman et al. 2004), with the consequent implication that Ornithopoda originated during the Late Triassic. More recently, two alternative hypotheses have been proposed. Many paleontologists have now suggested that the heterodontosaurids were the sister clade to Marginocephalia (ceratopsians and pachycephalosaurs: e.g., Xu et al. 2006). In contrast, it is possible that heterodontosaurids represent some of the most primitive ornithischians, a position that would fit well with their early appearance in the fossil record (Butler et al. 2007, 2008; Zheng et al. 2009).

Two small ornithischians from the Middle Jurassic of China, *Agilisaurus* and *Hexinlusaurus* (previously known as *Yandusaurus multidens*) and *Othnielosaurus* from the Late Jurassic of the United States have also traditionally been included within Ornithopoda as hypsilophodontids. However, new evidence (Barrett et al. 2005; Butler 2005; Butler et al. 2008) suggests they may be more basal ornithischians.

Geographic Distribution

Ornithopods are known from virtually every corner of the globe, although no specimens have yet been described from India or Madagascar (Weishampel et al. 2004). Defining a center of origin for Ornithopoda is impossible at present, given the uncertainties regarding the taxa that are included within the clade and the scarcity and known distributions of basal taxa. The earliest definite ornithopod specimen is from the Middle Jurassic of Europe (*Callovosaurus*). By the Late Jurassic ornithopods were relatively widespread, being present in both Laurasia and Gondwanaland. Dryosaurids were present in both geographical areas at this time. Camptosaurids also appears to have been relatively widespread at this time, being present in North America and Europe. The absence of Late Jurassic ornithopods on other continents (e.g., South America) may simply reflect collection biases (e.g., lack of exploration or genuine scarcity of Late Jurassic terrestrial deposits in those areas).

By the Early Cretaceous, ornithopods had a near-cosmopolitan distribution, with small "hypsilophodontid" ornithopods and larger iguanodontians present in Europe, North America, Asia, Africa, and Australia. In general, non-hadrosaurid iguanodontians disappeared during the Late Cretaceous, with the exception of the European rhabdodontids (which may represent a relict population) and several taxa from the Late Cretaceous of South America and Antarctica. By contrast, small hypsilophodontid ornithopods remained diverse and widespread. Hadrosaurids became widely distributed during the Late Cretaceous, with the most-abundant and best-preserved material present in North America. Their remains are still unknown from Africa, India, and Australia. Hadrosaurids lived at high latitudes, with material known from the North Slope of Alaska and the Yukon Territory, Canada – areas with paleolatitudes of 70–85° N during the Late Cretaceous. The area of origin for and subsequent biogeographical history of hadrosaurs remains an area of active study (e.g. Prieto-Márquez 2010).

Paleobiology

Basal ornithischians, such as *Lesothosaurus* and *Hexinlusaurus*, had relatively simple jaw mechanisms. The skulls were akinetic, the jaws operated in a simple scissorlike fashion, and the teeth did not grind against each other (occlude) in a precise manner. Dentitions were relatively simple, with up to six premaxillary teeth (which were somewhat recurved) and a variable number of leaf-shaped cheek teeth whose margins were lined with coarse serrations (denticles). Such teeth are ideally suited to puncturing and slicing through the plant material that made up the bulk of their diets (Galton 1986), though it is possible that some primitive ornithischians were omnivorous (Barrett 2000). The snout and mandible were tipped with self-sharpening horny beaks, similar to those of living turtles, which would also have been used in food gathering.

The feeding apparatus of ornithopods exhibits several important modifications to the basal ornithischian *bauplan* (Norman and Weishampel 1991). Although basal ornithopods retain premaxillary teeth, they become reduced in number. The enamel on the maxillary and dentary teeth becomes asymmetrically distributed, so that it is much thicker on the external (labial) side of the upper teeth and the internal (lingual) surface of the lower teeth. In addition, the geometry of the tooth rows allowed the upper teeth to come into contact with their opposite numbers in the mandible during jaw closure, resulting in a precise occlusion. The teeth were able to trap and slice plant food more effectively as a consequence. Moreover, development of occlusion allowed more extensive grinding and chewing of food prior to swallowing, increasing digestive efficiency and permitting subsistence on tough, fibrous fodder. The teeth became differentially worn during this process, due to the asymmetrical distribution of the tooth enamel. As a result, the teeth were self-sharpening: this constant wear produced a sharp leading (cutting) edge on the more thickly enameled side of the tooth. Jaw joints are ventrally offset, so that the tooth rows would have met in parallel as the jaws were closed, resulting in a more even distribution of bite force. A remarkable form of cranial kinesis, termed *pleurokinesis*, also appeared (Norman 1984b; Weishampel 1984: but see Rybczynski et al. 2008 for a critique of this hypothesis). Essentially, the skull became divided into three functional units: the central skull roof and braincase; the maxilla and associated cheek and palatal bones; and the lower jaws. A distinctive hinge-joint developed between the immobile skull roof and the maxilla/cheek unit. This joint allowed the maxilla/cheek unit to rotate slightly relative to the skull roof. As the mandible was raised at the onset of the chewing cycle, the upper and lower teeth would be pushed past each other: the upward force of the lower jaw would cause the maxilla/cheek unit to rotate outward around its hinge with the skull roof. This resulted in a strong grinding action as the teeth moved past each other. Pleurokinesis, a unique feature of ornithopods, allowed chewing at a level of efficiency only otherwise achieved by mammalian herbivores (Norman and Weishampel 1985). Finally, the upper and lower tooth rows were inset from the sides of the skull. The resulting space was probably bounded by fleshy cheeks that would have been useful in preventing food from falling out of the mouth during mastication (Galton 1973). Rudimentary cheeks, restricted to the back of the mouth, may also have been present in basal ornithischians.

Although the pleurokinetic system was present in hypsilophodontids, it became even more sophisticated in iguanodontians, reaching its greatest development in hadrosaurs (Norman 1984b; Weishampel 1984). The snout lengthened to accommodate larger numbers of teeth; the antorbital fenestrae became almost vestigial (possibly to consolidate the longer snout); and the snout broadened, permitting faster assimilation rates. Premaxillary teeth were lost in iguanodontians, but additional functional teeth were added to the maxilla and dentary, culminating in the appearance of complex dental batteries in hadrosaurs. Although it has been suggested that the appearance of flowering plants (angiosperms) drove the evolution of complex jaw mechanisms in ornithopods (Bakker 1978; Weishampel and Norman 1989), evidence for coevolutionary interactions between these clades is weak (Barrett and Willis 2001; Butler et al. 2009).

The feeding mechanisms of heterodontosaurids have received much attention (Thulborn 1978; Hopson 1980; Weishampel 1984; Crompton and Attridge 1986), but there is currently no consensus regarding the jaw mechanisms of these animals. They possess many of the features present in ornithopods (horny beak, asymmetrically distributed tooth enamel, occlusion, ventrally offset jaw joint, and cheeks), but were not pleurokinetic.

In other respects, ornithopod skulls are relatively conservative. However, the nasal apparatus of iguanodontians was often elaborate, with large external nares and complex internal divisions (Norman 1998; Witmer 2001), though the functions of these structures have yet to be explored in detail. Lambeosaurine and hadrosaurine hadrosaurs possess a variety of cranial crests (Hopson 1975). These crests are likely to have served as intra- and interspecific visual signals, allowing species recognition and/or social display functions (Hopson 1975). Hadrosaurine crests (such as that of *Saurolophus*) are solid structures, whereas those of lambeosaurines (such as *Parasaurolophus* and *Corythosaurus*) are hollow. The latter are composed of the expanded premaxillae and nasals, which contain complex passages that are connected to the nasal cavity. It has been suggested that these hollow crests acted as resonators that amplified the calls of these animals, permitting acoustic communication (Weishampel 1981). Morphometric studies of lambeosaurine skulls suggest that the crests were sexually dimorphic and changed shape during ontogeny, supporting suggestions of a social function (Dodson 1975).

Basal ornithischians and ornithopods were bipedal, as shown by their long hind limbs, short trunk regions, long, counterbalancing tails, and relatively short forelimbs (Galton 1970; Coombs 1978). The backbone was held horizontally, and the axial skeleton was strengthened by the presence of a lattice of ossified tendons that extended along the length of the spine. All small ornithopods retain bipedal posture. Iguanodontians were also bipedal; their larger body size and relatively longer forelimbs suggest that they may have moved in a quadrupedal posture, at least occasionally (Norman 1980). Additional evidence for quadrupedality comes from the structure of the iguanodontian hand, whose central digits were robust, capable of weight bearing, and ended in hooflike unguals (Norman 1980), and from trackways that exhibit manus impressions (Lockley and Wright 2001). The slender hind limbs and high tibia:femur length ratios of the basal ornithischians and smaller ornithopods (e.g., *Lesothosaurus*, heterodontosaurids, hypsilophodontids, and dryosaurids) suggest that these animals may have been fast cursorial forms. Increased body size, stockier hind limbs, and greater reliance on quadrupedal locomotion indicate that more derived ornithopods (camptosaurs, iguanodontids, and hadrosaurs) were less fleet of foot (Coombs 1978).

Ornithopod hands fulfilled a variety of functions (Fig. 26.10). As mentioned previously, the hands of iguanodontians were capable of weight bearing, but they were also capable of playing additional roles in food gathering and defense (Norman 1980). For example, the terminal phalanx on digit I in *Iguanodon* and *Probactrosaurus* was modified into a conical thumb-spike that has been interpreted as a "stilletto-like" weapon to deter attacks by large predatory dinosaurs. Digit V was also highly specialized in *Iguanodon* and was capable of prehensile movements that may have been useful in food gathering (Norman 1980). The hands of basal ornithopods, heterodontosaurids, and *Lesothosaurus* were not capable of load-bearing, but seem to have had a grasping or scooping function that was probably employed during food gathering (Thulborn 1972; Galton 1974; Santa-Luca 1980; Norman et al. 2004).

Evidence from trackways, bone beds, and communal nesting sites suggests that many ornithopods, particularly hadrosaurs and some other iguanodontians, were highly social animals that lived in multigenerational herds. One bone bed in Montana (USA) is estimated to contain the remains of up to 10,000 individuals of the hadrosaur *Maiasaura* (Schmitt et al. 1998). This evidence is consistent with the development of the complex visual

26.10. Ornithopod hands, showing the variety of structure and function. (A) The heterodontosaurid *Heterodontosaurus,* showing the complete hand (right) and a close-up on digit I (left). Note that the articulation between metacarpal I and its first phalanx resulted in a medial orientation for the digit, which could have influenced grasping ability. (B) The camptosaurid *Camptosaurus,* showing the complete hand (right) and details of digit I (left). (C) The iguanodontian *Iguanodon.* Note the hooflike unguals of digits II–IV and the columnar construction of these central digits. Digit I is modified into the famous "thumb-spike," and digit V appears to have been very flexible. (D) The hadrosaurid *Edmontosaurus* ("*Anatotitan*"). Digit I is lost. Digits II–IV are modified for weight bearing, but digit V remains prehensile. Not to scale.

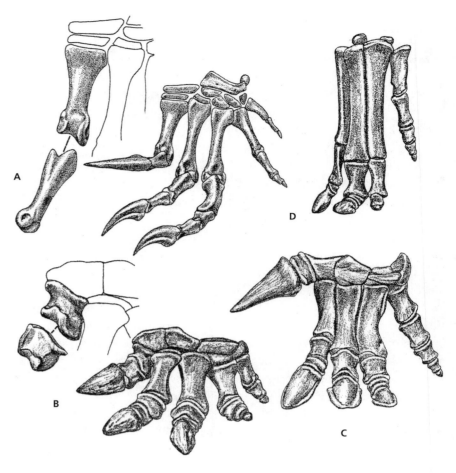

and acoustic signaling apparatus seen in these animals, which must have been a prerequisite for group living. Numerous eggs, embryos, and nests of *Maiasaura* have been excavated at a communal nest site named "Egg Mountain" in Montana (Horner and Makela 1979; Horner 1994). Growth series of many ornithopods are also known (including *Iguanodon, Dryosaurus,* and *Hypacrosaurus* [Carpenter 1994; Horner and Currie 1994]). *Maiasaura* nests were similar to those of living megapode birds, and there is evidence that the sites were used recurrently (Horner 1994, 2000). They consisted of a mound up to 1 m across, with a bowllike hollow on the upper surface that housed the clutch. Vegetation preserved in the nests suggests that the eggs may have been incubated by the decomposition of plant matter and/or that parents brought food to the nests (Horner 2000). The limb bones of hatchlings were poorly developed, adding weight to the suggestion that they were nest-bound and dependent on high levels of parental care. Growth rates of juvenile *Maiasaura* were prodigious: a 50-cm hatchling could grow to a length of 3 m in just one to two years (Horner et al. 2000). Unfortunately, little is currently known regarding nesting in other ornithopod groups or basal ornithischians.

References

Bakker, R. T. 1978. Dinosaur feeding behavior and the origin of flowering plants. *Nature* 274: 661–663.

Barrett, P. M. 2000. Prosauropod dinosaurs and iguanas: speculations on the diets of extinct reptiles. In H.-D. Sues (ed.), *Evolution of Herbivory in Terrestrial Vertebrates: Perspectives from the Fossil Record,* 42–78. Cambridge: Cambridge University Press.

Barrett, P. M., R. J. Butler, and F. Knoll. 2005. Small-bodied ornithischian dinosaurs from the Middle Jurassic of Sichuan, China. *Journal of Vertebrate Paleontology* 25: 823–834.

Barrett, P. M., and K. J. Willis. 2001. Did dinosaurs invent flowers? Dinosaur-angiosperm coevolution revisited. *Biological Reviews* 76: 411–447.

Benton, M. J., Z. Csiki, D. Grigorescu, R. Redlstorff, P. M. Sander, K. Stein, and D. B. Weishampel. 2010. Dinosaurs and the island rule: the dwarfed dinosaurs from Hateg Island. *Palaeogeography, Palaeoclimatology, Palaeoecology* 293: 438–454.

Bonaparte, J. F. 1976. *Pisanosaurus mertii* Casamiquela and the origin of the Ornithischia. *Journal of Paleontology* 50: 808–820.

Buchholz, P. W. 2002. Phylogeny and biogeography of basal Ornithischia. In D. E. Brown. (ed.), *The Mesozoic in Wyoming*, 18–34. Casper, Wyo.: Tate Geological Museum.

Butler, R. J. 2005. The "fabrosaurid" ornithischian dinosaurs of the Upper Elliot Formation (Lower Jurassic) of South Africa and Lesotho. *Zoological Journal of the Linnean Society* 145: 175–218.

Butler, R. J., P. M. Barrett, P. Kenrick, and M. G. Penn. 2009. Diversity patterns among herbivorous dinosaurs and plants during the Cretaceous: implications for hypotheses of dinosaur/angiosperm co-evolution. *Journal of Evolutionary Biology* 22: 446–459.

Butler, R. J., P. M. Galton, L. B. Porro, L. M. Chiappe, D. M. Henderson, and G. M. Erickson. 2010. Lower limits of ornithischian dinosaur body size inferred from a new Upper Jurassic heterodontosaurid from North America. *Proceedings of the Royal Society B* 227: 375–381.

Butler, R. J., R. M. H. Smith, and D. B. Norman. 2007. A primitive ornithischian dinosaur from the Late Triassic of South Africa, and the early evolution and diversification of Ornithischia. *Proceedings of the Royal Society B* 274: 2041–2046.

Butler, R. J., P. Upchurch, and D. B. Norman. 2008. The phylogeny of ornithischian dinosaurs. *Journal of Systematic Palaeontology* 6: 1–40.

Carpenter, K. 1994. Baby *Dryosaurus* from the Upper Jurassic Morrison Formation of Dinosaur National Monument. In K. Carpenter, K. F. Hirsch, and J. R. Horner (eds.), *Dinosaur Eggs and Babies*, 288–297. Cambridge: Cambridge University Press.

Carpenter, K., and Y. Ishida. 2010. Early and "Middle" Cretaceous iguanodonts in

time and space. *Journal of Iberian Geology* 36: 145–164.

Coombs, W. P. 1978. Theoretical aspects of cursorial adaptations in dinosaurs. *Quarterly Review of Biology* 393–418.

Cooper, M. R. 1985. A revision of the ornithischian dinosaur *Kangnasaurus coetzeei* Haughton, with a classification of the Ornithischia. *Annals of the South African Museum* 95: 281–317.

Crompton, A. W., and J. Attridge. 1986. Masticatory apparatus of the larger herbivores during Late Triassic and Early Jurassic times. In K. Padian (ed.), *The Beginning of the Age of Dinosaurs*, 223–236. Cambridge: Cambridge University Press.

Dodson, P. 1975. Taxonomic implications of relative growth in lambeosaurine dinosaurs. *Systematic Zoology* 24: 37–54.

Forster, C. A. 1990. The postcranial skeleton of the ornithopod dinosaur *Tenontosaurus tilletti*. *Journal of Vertebrate Paleontology* 10: 273–294.

Galton, P. M. 1970. The posture of hadrosaurian dinosaurs. *Journal of Paleontology* 44: 464–473.

——. 1973. The cheeks of ornithischian dinosaurs. *Lethaia* 6: 67–89.

——. 1974. The ornithischian dinosaur *Hypsilophodon* from the Wealden of the Isle of Wight. *Bulletin of the British Museum of Natural History: Geology* 25: 1–152.

——. 1978. Fabrosauridae, the basal family of ornithischian dinosaurs (Reptilia: Ornithopoda). *Paläontologische Zeitschrift* 52: 138–159.

——. 1986. Herbivorous adaptations of Late Triassic and Early Jurassic dinosaurs. In K. Padian (ed.), *The Beginning of the Age of Dinosaurs*, 203–221. Cambridge: Cambridge University Press.

——. 2009. Notes on Neocomian (Lower Cretaceous) ornithopod dinosaurs from England–*Hypsilophodon*, *Valdosaurus*, "*Camptosaurus*," "*Iguanodon*"–and referred specimens from Romania and elsewhere. *Review de Paléobiologie* 28: 211–273.

Hopson, J. A. 1975. The evolution of cranial display structures in hadrosaurian dinosaurs. *Paleobiology* 1: 21–43.

——. 1980. Tooth function and replacement in early Mesozoic ornithischian dinosaurs: implications for aestivation. *Lethaia* 13: 93–105.

Horner, J. R. 1994. Comparative taphonomy of some dinosaur and extant bird colonial nesting grounds. In K. Carpenter, K. F. Hirsch, and J. R. Horner (eds.), *Dinosaur Eggs and Babies*, 116–123. Cambridge: Cambridge University Press.

——. 2000. Dinosaur reproduction and parenting. *Annual Review of Earth and Planetary Sciences* 28: 19–45.

Horner, J. R., and P. J. Currie. 1994. Embryonic and neonatal morphology and ontogeny of a new species of *Hypacrosaurus* (Ornithischia: Lambeosauridae) from Montana and Alberta. In K. Carpenter, K. F. Hirsch, and J. R. Horner (eds.), *Dinosaur Eggs and Babies*, 312–336. Cambridge: Cambridge University Press.

Horner, J. R., and R. Makela. 1979. Nest of juveniles provides evidence of family structure among dinosaurs. *Nature* 282: 296–298.

Horner, J. R., A. de Ricqlès, and K. Padian. 2000. The bone histology of the hadrosaurid dinosaur *Maiasaura peeblesorum*: growth dynamics and physiology based upon an ontogenetic series of skeletal elements. *Journal of Vertebrate Paleontology* 20: 109–123.

Lockley. M. G., and J. L. Wright. 2001. Trackways of large quadrupedal ornithopods from the Cretaceous: a review. In D. H. Tanke, K. Carpenter, and M. W. Skrepnick (eds.), *Mesozoic Vertebrate Life*, 428–442. Bloomington: Indiana University Press.

Maryańska, T., and H. Osmólska. 1985. On ornithischian phylogeny. *Acta Palaeontologica Polonica* 30: 137–150.

Maxwell, W. D., and J. H. Ostrom. 1995. Taphonomy and paleobiological implications of *Tenontosaurus-Deinonychus* associations. *Journal of Vertebrate Paleontology* 15: 707–712.

McDonald, A. T., P. M. Barrett, and S. D. Chapman. 2010. A new basal iguanodont (Dinosauria: Ornithischia) from the Wealden (Lower Cretaceous) of England. *Zootaxa* 2569: 1–43.

Nesbitt, S. J., R. B. Irmis, and W. G. Parker. 2007. A critical re-evaluation of the Late Triassic dinosaur taxa of North America. *Journal of Systematic Palaeontology* 5: 209–243.

Norman, D. B. 1980. On the ornithischian dinosaur *Iguanodon bernissartensis* of Bernissart (Belgium). *Mémoire, Institut Royal des Sciences Naturelles de Belgique* 178: 1–104.

——. 1984a. A systematic reappraisal of the reptile order Ornithischia. In W.-E. Reif and F. Westphal (eds.), *Third symposium on Mesozoic terrestrial ecosystems, short papers*, 157–162. Tübingen: Attempto Verlag.

——. 1984b. On the cranial morphology and evolution of ornithopod dinosaurs. *Symposium of the Zoological Society of London* 52: 521–547.

———. 1998. On Asian ornithopods (Dinosauria: Ornithischia). 3. A new species of iguanodontid dinosaur. *Zoological Journal of the Linnean Society* 122: 291–348.

———. 2004. Basal Iguanodontia. In D. B. Weishampel, P. Dodson, and H. Osmólska (eds.), *The Dinosauria*, 2nd ed., 413–437. Berkeley: University of California Press.

———. 2010. A taxonomy of iguanodontians (Dinosauria: Ornithopoda) from the lower Wealden Group (Cretaceous: Valanginian) of southern England. *Zootaxa* 2489: 47–66.

Norman, D. B., and P. M. Barrett. 2002. Ornithischian dinosaurs from the Lower Cretaceous (Berriasian) of England. In A. R. Milner and D. J. Batten (eds.), *Life and Environments in Purbeck Times*, 161–189. Special Papers in Palaeontology 68. London : Palaeontological Association.

Norman, D. B., and D. B. Weishampel. 1985. Ornithopod feeding mechanisms: their bearing on the origin of herbivory. *American Naturalist* 126: 151–164.

Norman, D. B., H.-D. Sues, L. M. Witmer, and R. A. Coria. 2004. Basal Ornithopoda. In D. B. Weishampel, P. Dodson, and H. Osmólska (eds.), *The Dinosauria*, 2nd ed., 393–412. Berkeley: University of California Press.

Norman, D. B., and D. B. Weishampel. 1991. Feeding mechanisms in some small herbivorous dinosaurs: patterns and processes. In J. M. V. Rayner and R. J. Wootton (eds.), *Biomechanics and Evolution*, 161–181. Cambridge: Cambridge University Press.

Paul, G. S. 2008. A revised taxonomy of the iguanodont dinosaur genera and species. *Cretaceous Research* 29: 192–216.

Prieto-Márquez, A. 2010. Global historical biogeography of hadrosaurid dinosaurs. *Zoological Journal of the Linnean Society* 159: 503–525.

Ruiz-Omeñaca, J. I., X. Pereda Suberbiola, and P. M. Galton. 2007. *Callovosaurus leedsi*, the earliest dryosaurid dinosaur (Ornithischia: Euornithopoda) from the Middle Jurassic of England. In K. Carpenter (ed.), *Horns and Beaks: Ceratopsian and Ornithopod Dinosaurs*, 3–16. Bloomington: Indiana University Press.

Rybczynski, N., A. Tirabasso, P. Bloskie, R. Cuthbertson, and C. Holliday. 2008. A three-dimensional animation model of *Edmontosaurus* (Hadrosauridae) for testing chewing hypotheses. *Palaeontologica Electronica* 11 (2): 26A. http://palaeo-electronica.org/2008_2/132/index.html.

Scheetz, R. D. 1999. Osteology of *Orodromeus makelai* and the phylogeny of basal ornithopod dinosaurs. Ph.D. dissertation, University of Montana.

Schmitt, J. G., J. R. Horner, R. R. Laws, and F. Jackson. 1998. Debris-flow deposition of a hadrosaur-bearing bone bed, Upper Cretaceous Two Medicine Formation, northwest Montana. *Journal of Vertebrate Paleontology* 18 (suppl. to 3): 76A.

Sereno, P. C. 1984. The phylogeny of Ornithischia: a reappraisal. In W.-E. Reif and F. Westphal (eds.), *Third Symposium on Mesozoic Terrestrial Ecosystems, Short Papers*, 219–226. Tübingen: Attempto Verlag.

———. 1986. Phylogeny of the bird-hipped dinosaurs (order Ornithischia). *National Geographic Research* 2: 234–256.

———. 1991. *Lesothosaurus*, "fabrosaurids," and the early evolution of Ornithischia. *Journal of Vertebrate Paleontology* 11: 168–197.

Thulborn, R. A. 1972. The postcranial skeleton of the Triassic ornithischian dinosaur *Fabrosaurus australis*. *Palaeontology* 15: 29–60.

———. 1978. Aestivation among ornithopod dinosaurs of the African Triassic. *Lethaia* 11: 185–198.

Weishampel, D. B. 1981. Acoustic analyses of potential vocalization in lambeosaurine dinosaurs (Reptilia: Ornithischia). *Paleobiology* 7: 252–261.

———. 1984. Evolution of jaw mechanisms in ornithopod dinosaurs. *Advances in Anatomy, Embryology and Cell Biology* 87: 1–110.

Weishampel, D. B., P. M. Barrett, R. A. Coria, J. Le Loeuff, X. Xu, X.-J. Zhao, A. Sahni, E. M. P. Gomani, and C. R. Noto. 2004. Dinosaur distribution. In D. B. Weishampel, P. Dodson, and H. Osmólska (eds.), *The Dinosauria*, 2nd ed., 517–606. Berkeley: University of California Press.

Weishampel, D. B., and R. E. Heinrich. 1992. Systematics of Hypsilophodontidae and basal Iguanodontia (Dinosauria: Ornithopoda). *Historical Biology* 6: 159–184.

Weishampel, D. B., C.-M. Jianu, Z. Csiki, and D. B. Norman. 2003. Osteology and phylogeny of *Zalmoxes* (N. G.), an unusual euornithopod dinosaur from the latest Cretaceous of Romania. *Journal of Systematic Palaeontology* 1: 65–123.

Weishampel, D. B., and D. B. Norman. 1989. Vertebrate herbivory in the Mesozoic: jaws, plants, and evolutionary metrics. In J. O. Farlow (ed.), *Paleobiology of the Dinosaurs*, 87–100. Special Paper 238. Boulder, Colo.: Geological Society of America.

Winkler, D. A., P. A. Murry, and L. L. Jacobs. 1997. A new species of *Tenontosaurus* (Dinosauria: Ornithopoda) from the Early Cretaceous of Texas. *Journal of Vertebrate Paleontology* 17: 330–348.

Witmer, L. M. 2001. Nostril position in dinosaurs and other vertebrates and its significance for nasal function. *Science* 293: 850–853.

Xu, X., C. A. Forster, J. M. Clark, and J.-Y. Mo. 2006. A basal ceratopsian with transitional features from the Late Jurassic of northwestern China. *Proceedings of the Royal Society B* 273: 2135–2140.

Zheng, X.-T., You, H.-L., Xu, X. & Dong, Z.-M. 2009. An Early Cretaceous heterodontosaurid dinosaur with filamentous integumentary structures. *Nature* 458: 333–336.

4

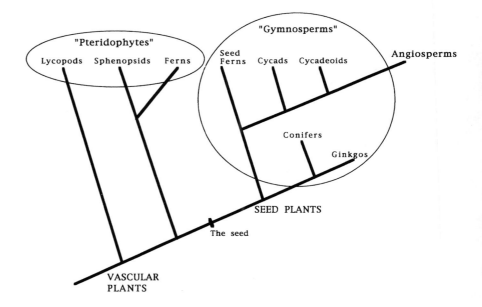

27.1. General relationships of the major groups of vascular land plants. The branching diagram shows the ancestor–descendent relationships of the groups, while the two circled units indicate the paraphyletic groups Pteridophytes and Gymnosperms. After Doyle and Donoghue 1986; see also Doyle and Donoghue 1992.

Land Plants as a Source of Food and Environment in the Age of Dinosaurs

27

Bruce H. Tiffney

Plants in a book on dinosaurs? Not as out of place as you might think. Plants are autotrophs (self-feeders), organisms that are able to capture the sun's energy directly. By contrast, dinosaurs, like all animals, are heterotrophs (other feeders), organisms that have to feed on other organisms in order to live. Since plants lie at the base of the food chain, they have had an immense influence on the evolution of both herbivores and carnivores in Earth history. The size of the available plants, their rate of growth, their ability to recover from damage, the rate at which they reproduce, their abundance in the environment, and the digestibility of their leaves and reproductive organs all combine to influence the amount of energy that herbivores can draw from them. As these features of plants change through evolutionary time, so also will the nature of the herbivore and carnivore communities dependent upon them.

Additionally, plants are important to animals in that they define the environment within which animals live. By example, a forest forms a barrier to large animals and is difficult for them to pass through. In contrast, smaller animals perceive the forest as a three-dimensional habitat, and have the option to move vertically as well as horizontally within it. Conversely, the two-dimensional surface of an open "grassland" allows free motion of large animals but limits the options for small animals. Only by burrowing can smaller animals create a three-dimensional environment in open country. This is a common solution for mammals, but among the dinosaurs, only a few birds were small enough to effectively explore it.

The interaction of plants and animals is not a one-way street. Herbivores have had an important influence on plant evolution by their choice of food, the volume that they consume, and the frequency and duration of their feeding. Excessive herbivory can destroy environments and place selective pressure on plants, possibly even leading to the extinction of existing lineages or the evolution of new ones.

In this chapter we will survey (1) the major groups of plants available to dinosaurs as food, and (2) the possible effects of this interaction between plants and dinosaurs on both groups. In looking at this interaction, we will view plants from the perspective of an herbivorous dinosaur, examining those features of plants that influence their quality as food. Similarly, we will look at herbivorous dinosaurs as plant-consuming machines. Readers should be aware of two features. First, this chapter is brief, and presents generalized information for which specific exceptions are often known to exist. Second, the study of the evolutionary interaction of herbivores and the plants they fed upon in the fossil record is in its infancy. Thus, many of the interpretations presented in the second portion of this chapter are hypotheses open to test and refutation.

Groups of Mesozoic Plants

Dinosaurs radiated into a world already populated by two major kinds of land plants, the pteridophytes and the gymnosperms, and were joined by a third kind, the angiosperms, in the mid-Cretaceous (Figs. 27.1 and 27.5). All three groupings are vascular plants, that is, they possess conducting tissue that transports water and food products within the plant body. This is in contrast to the algae (seaweeds), which are almost entirely aquatic, or to the bryophytes (mosses and associates), which are terrestrial but are restricted in their size and significance by the lack of vascular tissue. We will look at the characteristics of the three vascular plant groups in turn, and then at their distribution in space and time during the Mesozoic. We will not discuss terrestrial algae or bryophytes, as they did not form an important part of the diet of herbivorous dinosaurs.

What is the source of the following information? Where fossil plants are similar to living ones, we can infer the biology of the fossil from that of the living counterpart, recognizing that the extant plant cannot be a perfect proxy for the fossil. However, in many cases there are no close living relatives, and inferences must be made from the circumstantial evidence provided by the anatomy and morphology of the fossil, and their parallels in plants of the present day. Further insight may be gained from the depositional situation within which the fossil was found. Such evidence allows us to erect a model of Mesozoic ecosystems, but with the knowledge that this model will surely evolve with new information and understanding. Those wishing further information on Mesozoic plants, or guides to their identification and ecology, are referred to Stewart and Rothwell (1993), Taylor and Taylor (1993), and Willis and McElwain (2002).

Pteridophytes

This term, *pteridophyte*, is used to circumscribe several groups of primitive vascular plants. The pteridophytes are paraphyletic; that is, the group embraces some, but not all, of the descendents of a common ancestor. In this case, the pteridophytes all evolved from the first vascular plant, but the group does not include the other descendents of that first land plant, the seed plants. Thus, the pteridophytes are defined by the common character of their aquatic mode of reproduction, rather than by their ancestor–descendent relationships (Fig. 27.1). Of the pteridophyte clades, the most significant to our story are the ferns and their less commonly observed allies, the horsetails (sphenopsids) and the club mosses (Lycopsids): all are depicted in Figure 27.2. These three groups share a common history, originating in the Devonian. In all three groups, the commonly observed plant is diploid; that is, it possesses two copies of each individual chromosome within each cell. Within specialized structures on this diploid plant, meiosis takes place, resulting in the division of the paired chromosomes into two groups, each containing one copy of each chromosome. This is referred to as the haploid condition. These haploid chromosome groups are encased in a protective structure to form spores, which are released to be blown about in the atmosphere. These spores settle from the air, and if they land in a moist area, germinate to yield a tiny, photosynthetic, haploid plant, the gametophyte. The gametophyte bears the haploid egg and sperm. Reminiscent of its algal forbearers, fertilization occurs when sperm is released from one gametophyte and literally swims in available dew or rainwater to an egg-bearing gametophyte. The fusion of the haploid egg and sperm yields a diploid zygote, which grows into the visible plant (e.g., a fern) familiar to the average viewer. This is an amphibious life cycle, and it restricts pteridophytes in the main to areas where moisture is available. Fertilization can occur only where free water allows the sperm to swim, and the new diploid plant can grow only where fertilization has taken place. Thus, pteridophytes are uncommon in year-round or strongly seasonally dry environments.

Pteridophytes are generally fairly low growing herbaceous plants without well-developed aerial stems. The most common exceptions to this generalization are tree ferns and, in limited areas during the Triassic, a small arborescent lycopod that was the last survivor of the great late Paleozoic tree lycopods. Both possessed an unbranched aerial stem with a single growing point. These treelike forms excepted, pteridophytes almost universally possess an underground stem or rhizome that, by branching, creates many growing points with the subsequent appearance of many new above-ground individuals from one original subterranean rhizome. This offsets the limitations of the sexual reproductive system, as a single successful sexual event can literally result in the growth of a field of plants from subsequent vegetative reproduction. Further, vegetative growth confers the ability to regrow quickly after damage. The loss of existing emergent leaves is countered by the growth of new leaves from the rhizome, which remains safely beneath the ground. Additionally, with sufficient available water, pteridophytes tend to grow fairly rapidly. The foliage of most pteridophytes is relatively succulent and lacks strong mechanical defenses (tough bark, spines, etc.). However, several living pteridophytes possess chemicals that are carcinogenic in mammals, or are known to inhibit digestion, and past pteridophytes may have had similar effects upon dinosaurs.

Summing the above from the dinosaurian herbivore's point of view, the non-treelike pteridophytes offered a fast-growing, renewable resource that could be grazed without the death of the whole plant. However, these advantages were offset by the pteridophytic requirement for moisture for reproduction. Given the widespread nature of arid and seasonally arid environments in the Mesozoic, pteridophytes may not have been generally available.

27.2. Some examples of Mesozoic Pteridophytes. (A) A rhizomatous fern. (B) A sphenopsid (*Equisetum*). (C) A lycopsid. An adult brachiosaur foot is provided for relative scale.

Gymnosperms

The gymnosperms (naked seeds) include the primitive groups of seed plants. Seed plants are probably a monophyletic clade (a group that includes an ancestor and all its genetic descendents), as it is generally thought that the seed evolved only once in the Devonian. As the name is commonly applied, the gymnosperms are paraphyletic as they include several clades that have a common ancestor (the first seed plant), but exclude the most advanced of seed plants, the angiosperms (Fig. 27.1). However, recent molecular evidence has been used to suggest that gymnosperms are monophyletic (e.g.,

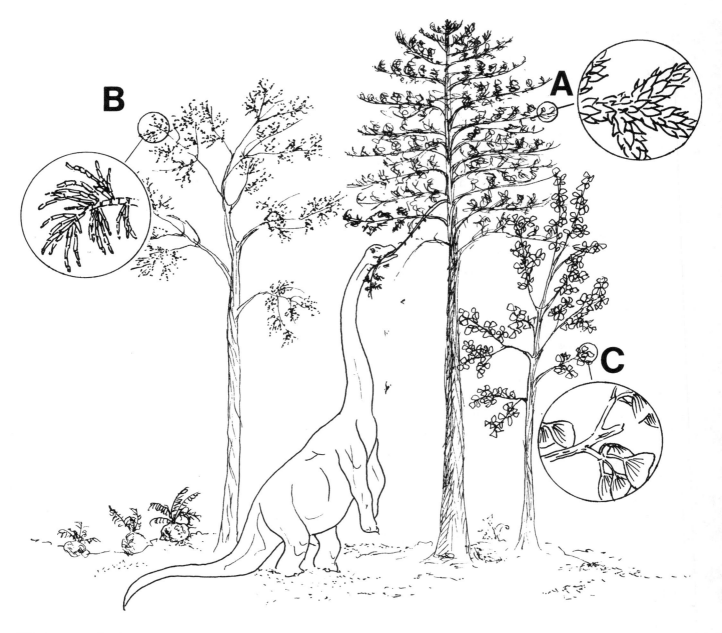

27.3. Some examples of Mesozoic gymnosperm trees. (A) *Araucaria;* a conifer of the family Araucariaceae. (B) *Pseudofrenelopsis;* a conifer of the family Cheirolepidiaceae (after Alvin 1983). (C) Ginkgo, the maidenhair tree, of the family Ginkgoaceae. A hungry adult brachiosaur is provided for relative scale. Note: while all of these taxa lived in the Mesozoic, they did not all live at the same time.

Chaw et al. 2000), and the matter is currently under debate (e.g., Rydin et al. 2002; Burleigh and Mathews 2004)

Whether monophyletic or paraphyletic, the gymnosperms are defined by the common character of possessing a naked seed, as contrasted to the angiosperms, which have a seed borne within a fruit. Several different and distinct branches of seed plants are embraced within the gymnosperms.

The most common gymnosperms in the present day and probably in the Mesozoic are the conifers (Fig. 27.3). These are the "pine trees" and their allies, including the genus *Araucaria* and its relatives, frequently depicted in dinosaur reconstructions. However, other conifers were also important in the Mesozoic. Some of the more important families include the closely related Cupressaceae (the cypress family) and Taxodiaceae (the bald cypress family), and the Podocarpaceae (the podocarps, today almost entirely limited to the Southern Hemisphere, but potentially an important plant in wetter areas of the Mesozoic). Conifers range in size from small shrubs to large trees, trees being the most common.

A second major group of gymnosperms is the cycadophytes, including the cycads and the cycadeoids (Fig. 27.4). The former group survives in the tropics and subtropics today, while the latter is extinct. Cycads and cycadeoids include both erect forms with narrow, sometimes branching, palmlike trunks, and lower forms with barrellike or elongate-hemispherical trunks. In both cases, the trunks or branches were capped with a rosette of divided leaves. While cycads and cycadeoids looked similar, they had very different modes of reproduction. While the pollen and ovules of cycadeoids were borne on a common axis on a single plant, cycads possessed pollen- and seed-bearing structures borne on separate plants.

Several other gymnosperm clades were clearly important in Mesozoic systems, but are extinct or of very limited importance in the present day and thus difficult to evaluate. These include seed ferns (small to large extinct plants that possessed fernlike foliage but that bore seeds), gnetophytes (survived by a small group of plants including *Ephedra*, commonly called Mormon tea), Czekanowskiales (an extinct conifer-like group of trees with unusual reproductive characters), and *Ginkgo* (the "maidenhair tree" of many modern city streets; see Fig. 27.3), among others.

These plants are united by the common evolutionary novelty of the seed. The seed is to the evolution of terrestrial plants what the amniotic egg (the mode of reproduction in the amniotes, the reptiles + synapsids) was to the evolution of tetrapods. In both cases this was a major innovation that freed the group from dependence on standing water for sexual reproduction. In the amniotes, fertilization was internal within the egg-bearing parent, and the resulting embryo was surrounded by a protective membrane (the amnion). The embryo was either brought to term within the mother (as in mammals) or encased in a shell containing nutrients and deposited outside the mother to mature and hatch (as in living reptiles). In plants, this advance involved the retention of the egg on the parent plant, thereby avoiding the stage of the dissemination of spores leading to the growth of a free-living, egg-bearing gametophyte. The sperm is introduced to the egg by the male spore or pollen grain, which is borne to the egg-bearing structure by air currents. The pollen grain germinates in moisture provided by the egg-bearing parent, releasing a sperm, which swims to the egg in this moisture. At an early stage, the resulting embryo, plus its store of nutritive tissue and moisture provided by the parent plant, is enclosed in a protective covering to form the seed. The seed may be dispersed from the parent plant by wind, water, or an animal, germinating in a suitable environment at some distance from the parent plant. This set of adaptations results in freedom from the necessity of standing water at the time of fertilization, and to a lesser degree, at the time of establishment of the young seedling. While primitive gymnosperms (e.g., pteridosperms) may have maintained some dependence on water for reproduction, the dominant Mesozoic gymnosperms (e.g., cycadophytes, conifers, and others) were able to colonize relatively dry sites. This led to the appearance of a widespread land flora.

How do gymnosperms fare as herbivore food? By present knowledge, all gymnosperms of the Mesozoic and modern day have been and are trees or shrubs, with branching trunks bearing the growing points above ground. There are no "herbaceous" forms in which rhizomes would allow rapid colonization by vegetative growth, and only a very few genera in

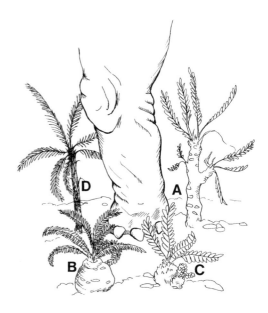

27.4. Some examples of common smaller Mesozoic gymnosperms. (A) The Cycadeoid *Williamsonia*. (B) The Cycadeoid *Cycadeoidea*. (C) A generalized low-growing cycad. (D) *Leptocycas*, a Late Triassic cycad (after Delevoryas and Hope 1971). An adult brachiosaur leg is provided for relative scale. Note: while all of these taxa lived in the Mesozoic, they did not all live at the same time.

which other forms of vegetative reproduction are present (but see Rothwell et al. 2000 for an exception). Thus, gymnosperms are dependent upon the seed for the establishment of new individuals, and are generally not able to respond to external damage by regrowing from underground buds. While some living gymnosperms (e.g., some species of pine) grow quite quickly, others (e.g., cycads, some other conifers) are slow to grow and respond to damage. From both living and fossil evidence, many gymnosperms possess thick bark and tough, resistant, often spinose foliage, generally with abundant thick-walled cells that are hard to break down and digest. Further, extant gymnosperm foliage is often rich in indigestible chemicals and resins. There are exceptions to these generalizations. Cycadophyte foliage from the Jurassic Yorkshire Delta of England is thick, but is not interpreted as tough or resistant (P. Crane personal communication 1994). Similarly, *Ginkgo*, a deciduous tree common at higher latitudes in the Mesozoic, possessed soft foliage without great amounts of resin. An extinct group of conifers, the Cheirolepidiaceae, appear to have grown in great numbers in the lower paleolatitudes in the mid and late Mesozoic, and to have borne quantities of quite succulent foliage, although we have no sense of their chemistry or rate of growth.

These general characteristics of substantial height, thick bark, high resin content, and tough foliage are often correlated in the present day with plants that grow in dry environments and/or in environments frequented by fire. Drought and fire may have been common in the early and mid Mesozoic, ensuring the dominance of gymnosperms through this time, and thus helping to determine the nature of the available food for herbivores. Pteridophytes with subterranean rhizomes, particularly ferns, might also be favored by burning in the dry season of a seasonally moist climate.

From a dinosaurian herbivore's point of view, gymnosperms were probably the most significant source of food simply because of their wide dissemination across the land. However, the generally resistant nature of Mesozoic gymnosperm foliage and its rich chemical content probably dictated that large quantities of foliage had to be consumed in order to obtain sufficient nutrition. Further, the presumed generally slow rates of Mesozoic gymnosperm growth and the lack of underground rhizomes suggest that relatively long intervals had to pass between one herbivore visit and the next to any single plant or stand (Bond 1989; Midgley and Bond 1991).

Angiosperms

The angiosperms, or flowering plants, are the most recent major group of vascular plants to evolve. There are some 250,000 to 300,000 species of angiosperms in the present day, contrasted to some 1,000 species of gymnosperms and 10–12,000 species of pteridophytes. The angiosperms first appeared in the mid-Cretaceous, and they diversified to dominate the world flora by the end of the Cretaceous (Niklas et al. 1985; Lidgard and Crane 1990; Wing and Boucher 1998; Friis et al. 2005). While the angiosperms are almost certainly a monophyletic group (that is, derived from a single common ancestor), their origins are obscure. It is difficult to define them on the basis of any particular character. In common discussion, they are the "flowering plants" in reference to their possession of flowers. This is often associated with insect pollination, although the latter character is also seen in some gymnosperms. For the present case, it is best to consider them as an "advanced" group of seed plants, as reflected by their position nested within the gymnosperms (Fig. 27.1).

The earliest angiosperms were apparently small shrubs or possibly herbs of wet, shaded areas (Feild et al. 2004), but they quickly diversified into a wide range of ecological niches and growth forms, generally characterized by a "weedy" ecology of fast growth and disturbance tolerance. Important to our interest in dinosaur fodder, the angiosperms possess modifications of the life cycle and vegetative body that allow them, on the whole, to reproduce and grow more quickly than gymnosperms (Stebbins 1981; Bond 1989; Midgley and Bond 1991). Further, many but not all angiosperms possess either underground rhizomes or an ability to sprout from the roots, allowing them to colonize large areas without sexual reproduction and to recover from grazing. While the group is quite variable, on the whole it would be safe to describe them as possessing more succulent foliage with fewer indigestible chemicals than gymnosperms.

In some respects, the angiosperms were an answer to an herbivorous dinosaur's prayer. Because of the seed habit, they grew in a wide range of terrestrial sites, and were thus almost ubiquitously available. Because of their rapid life cycle, rapid growth rate, and ability for vegetative reproduction, they were relatively tolerant of intensive herbivory and could regenerate rapidly following cropping. All of these features suggest that relative to gymnosperms, angiosperms could support larger numbers of herbivores. Also, angiosperms possess a wider range of growth forms and a greater diversity of habitat ecologies than gymnosperms. As a result, more of the world became vegetated after the appearance of the angiosperms, altering the three-dimensional environment in which dinosaurs lived.

The Succession of Paleofloras through Time

The land flora has not been homogeneous through time. During the late Paleozoic, the interplay of continental positions and global climate created many sites for the growth of moisture-loving pteridophytes, and they dominated terrestrial vegetation. Commencing in the later Pennsylvanian, terrestrial climates generally became dryer and more seasonal (termed *continental climates*) due to the formation of the supercontinent of Pangaea. This placed the reproductive cycle of existing pteridophytes under increasing stress, leading to a decline in pteridophyte diversity and a rise in that of seed plants (DiMichele et al. 2001). This process continued through the latest Paleozoic, leading to the appearance of a gymnosperm-dominated world in the Triassic (Fig. 27.5; Niklas et al. 1985), although the timing and severity of this transition varied across the planet (Rees 2002).

The Triassic gymnosperms included conifers, seed ferns, early cycadophytes, and various less well understood groups. These clades diversified through the Jurassic into the Cretaceous. The seed ferns became less and less important during the middle Mesozoic and disappeared in the Cretaceous, while the cycadophytes diversified and became increasingly

27.5. The changing contribution of various major groups of vascular plants (labeled) to the Mesozoic and Cenozoic flora, contrasted with the range of sizes of pre-dinosaurian herbivores (grey bars), dinosaurian herbivores (black bars), and mammalian herbivores (striped bars). Note that mammals did not exhibit signs of strict herbivory until the latest Cretaceous and are hence not shown for earlier periods. The vertical scale for herbivore mass is logarithmic in grams; 10^6 = one metric ton. The horizontal timescale is not linear but diagrammatic. Perm = Permian; Lo Tr = lower Triassic; Up Tr = upper Triassic; Lo J = lower Jurassic; Up J = upper Jurassic; Lo K = lower Cretaceous; Up K = upper Cretaceous; Pe = Paleocene; Eo = Eocene; Oligo = Oligocene; Mio = Miocene; Plio = Pliocene; Q = Quaternary. Dinosaur data largely from Norman 1985 and redrawn from Tiffney 1989. Plant patterns redrawn schematically after Niklas et al. 1985 and Lidgard and Crane 1990.

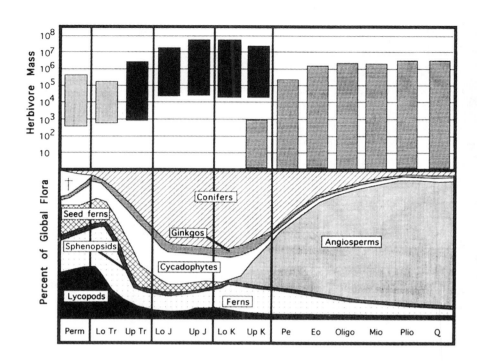

important in the Jurassic and then declined in the Cretaceous. The conifers similarly diversified through the Jurassic into the early Cretaceous and began to decline in diversity in the later Cretaceous (Niklas et al. 1985; Lidgard and Crane 1990). In the Late Triassic and especially the Jurassic, several modern lineages of ferns evolved, paralleling the breakup of Pangaea and the resulting reduction in the global area dominated by continental climates. *Equisetum* continued to be common in moist areas and may account for some herbivores displaying strong tooth wear (Whatley personal communication 2006).

The rise of the angiosperms in the later Cretaceous depended on both vegetative and reproductive features. The relatively more rapid reproductive cycle and rate of growth of angiosperms may have given them an edge over gymnosperms, particularly in the ability to colonize disturbed sites (Bond 1989). As a result, gymnosperms became increasingly restricted to marginal sites where physical features (low levels of sunlight, low temperatures, poor nutrients) put angiosperms at a greater disadvantage. Further, angiosperms are often pollinated by insects (Crepet et al. 1991), and their seeds frequently dispersed by vertebrates (Tiffney 2004). Both these features could influence rates of gene flow, and thus of speciation. This has led to the hypothesis that angiosperms have a greater ability for speciation and thus for evolutionary flexibility than gymnosperms because of their relationships with animals (Stebbins 1981), although this hypothesis has been questioned (Midgley and Bond 1991), and much remains to be learned (Bawa 1995).

Paleophytogeography

The various kinds of plants were not distributed homogeneously across the globe in the Mesozoic (see Meyen 1987, 313–323 for a brief summary; Vakhrameev 1991, for a detailed treatment; and Rees et al. 2004, for the Late Jurassic). Latitudinal climatic gradients and continental positions shifted

Bruce H. Tiffney

through the Mesozoic, influencing the distribution of the different kinds of plants and the vegetation that they created. The synthesis of patterns in Mesozoic vegetation on a global level is still in its early stages, and different sources vary in the details. Further, our knowledge of these fossil floras is far better for present Northern Hemisphere land masses than for those of the present Southern Hemisphere. Thus future research could demonstrate many of the following generalizations to be wrong.

The whole of the Mesozoic is marked by the pattern of the poles being relatively temperate and moist in comparison to the equatorial zone, which was generally warmer and dryer, tending to strong aridity at some times. Often the equatorial zone was dominated by drought-adapted conifers and cycadophytes, while communities toward the poles were increasingly dominated by moisture-requiring conifers, other gymnosperms (e.g., ginkgos, Czekanowskiales) and pteridophytes. The arid equatorial belt gradually developed during the Late Triassic through Middle Jurassic and carried on with little change through the later Cretaceous. While some authors suggest that the arid belt expanded dramatically in the later Jurassic and then contracted in the Early Cretaceous (Hallam 1984, 1993; Vakhrameev 1991), Ziegler et al. (1993) suggest that this is a result of the Northern Hemisphere continents "drifting" through arid belts, rather than a function of actual global climatic change. The angiosperms first appeared in arid locales in the equatorial belt in the later portions of the Early Cretaceous, and quickly spread to the poles by the early Late Cretaceous, invading existing conifer communities and displacing cycadophytes, pteridophytes, and lesser-known groups in the process (Saward 1992; Spicer et al. 1993).

We tend to envision the past in terms of what we are familiar with in the present. However, in doing so, we must be careful not to turn the past into the present. Thus, while many reconstructions of the Triassic–Cretaceous Earth depict a planet as densely and continuously vegetated as the present-day Earth, alternative hypotheses should be entertained. The combination of continental configuration, global climate, and available plant types in the Mesozoic suggests that there was no vegetation unit equivalent to the modern "tropical rain forest" (Saward 1992; Ziegler et al. 1993). Rather, most of the continents in the lower and middle latitudes possessed discontinuous vegetation adapted to seasonal drought. Herbivores in equatorial and subequatorial regions would encounter adequate fodder along seashores and riverine lowlands, but might find inland and upland settings had more patchy vegetation, scattered in an otherwise arid and perhaps vegetation-poor environment. At the higher latitudes, vegetation probably became more continuous and productive with the increasing availability of moisture. This vegetation was probably deciduous at the poles (Saward 1992; Spicer et al. 1993; Ziegler et al. 1993). It has been informally suggested that dinosaur distribution may have paralleled this pattern, as the greatest diversity of dinosaurs has been collected in the Mesozoic mid-latitudes of the Northern and Southern hemispheres (Hallam 1993, 296; Parrish 2003), although some of largest dinosaurs are associated with the most water-stressed areas (Rees et al. 2004). It is also possible that the perceived distribution of dinosaurs could also reflect taphonomic (Rees et al. 2004) or collection bias, rather than all the places in which they originally lived.

Dinosaurs and Plants

Much as we inferred the biology of Mesozoic plants, we have to infer the biology of Mesozoic herbivores from a combination of the morphology and geological context of the fossils and the biology of living vertebrate herbivores, both reptilian and mammalian. From living organisms, we can deduce two broad generalizations about herbivores.

First, the efficiency of food use depends on its digestibility. Plant food may be broadly categorized as resistant to digestion (much foliage, bark) or fairly digestible (fruits, seeds, some starch-rich roots). In order to gain energy from the latter source, herbivores need only crack the protective outer layer and subject the contents to mild chewing. However, to digest leaves, herbivores need either to mechanically break down the walls of the individual leaf cells to expose the cell contents or to retain the leaves for prolonged times in the gut, thereby exposing them to bacterial fermentation that is capable of breaking down the cell walls. Some dinosaurs apparently ingested stones (gastroliths), which were retained in the digestive tract (Stokes 1987). These may have functioned to aid digestion by one of two means. Classically, the muscular activity of the stomach is supposed to have caused the stones to bounce against each other, thereby crushing the trapped plant material. Alternatively, the stones may have served to mix up the contents of the stomach, ensuring more complete digestion (Gillette 1994). It is also possible the tough plant material was broken within a gizzardlike structure, or even within a muscular crop (Farlow 1987; Gillette 1994), although evidence for these soft parts has not yet been found in the fossil record.

Second, food quality and herbivore size are related. Large living herbivores require large amounts of food. Since high-quality food items (fruit, seeds) are widely scattered in the environment, large herbivores cannot afford to seek them out, and thus tend to consume great quantities of easily obtainable but low-quality food (Mellett 1982; Farlow 1987). This generates a self-reinforcing association between large size and low-quality fodder. A greater food intake requires a larger stomach. Further, fermentation requires a long residence time in the stomach, necessitating a larger stomach to hold fodder in various stages of digestion. Both factors dictate a larger whole animal. Such large herbivores become "whole-plant predators", consuming substantial portions of the plant, and exerting a very strong influence on the vegetation they inhabit. By example, elephants in African game reserves can turn forests into grasslands by their need for fodder.

In contrast, small living herbivores do not require as relatively great a volume of food, and can afford to seek out higher-quality food items, such as fruits and seeds, that are scattered in the environment. As a result, small herbivores tend to be more specialized in their food preferences, and to be plant predators at the level of individual plant organs. This creates new possibilities for specific kinds of plant–animal interactions, including pollination or fruit and seed dispersal.

These principles of herbivore–plant interaction are central to what distinguishes the age of dinosaurs from the subsequent age of mammals, as both the size of the herbivores and the nature of the available plant food change from the Mesozoic to the Cenozoic (Fig. 27.5).

Late Triassic

The Triassic spanned a major transition from a seed fern–pteridophyte flora to a conifer-cycadophyte flora, particularly in the equatorial region and southerly higher latitudes. The Early Triassic vegetation included many forms with relatively soft leaves, borne on plants of a variety of heights, from herbs through shrubs and trees. The conifer-cycadophyte vegetation that spread in the Late Triassic was dominated by plants with tougher and often spinose foliage, which was generally rich in chemical defenses. In the case of many conifers, this foliage was borne on trees. The probable reason for this change in vegetation was the aforementioned spread of continental climates and the rise in the frequency of fire accompanying the formation of Pangaea.

This floristic change appears to parallel a transition in herbivorous tetrapods (Benton 1983). The dominant herbivores in the Early and Mid Triassic were therapsids, which tended to be relatively small (up to pig size) and to feed within a few feet of the ground (Zavada and Mentis 1992). By contrast, while the early dinosaurian communities of the Late Triassic included a few small ornithischian herbivores, they also included very large, high-feeding, prosauropod herbivores like *Plateosaurus* and *Melanorosaurus*, setting the stage for the herbivore communities of the Jurassic and Early Cretaceous (Galton 1985). It is possible that the large size attained by these early saurischian herbivores was a direct result of the poor quality of the newly dominant vegetation.

Jurassic–Mid-Cretaceous

While there were many species of "small" dinosaurian herbivores (a few hundred kilograms in size) from the Triassic through the Cretaceous, two points are of note. First, the smallest dinosaurian herbivores were still perhaps two orders of magnitude larger than the smallest vertebrate herbivores among living mammals and birds. Second, while these "small" dinosaurian herbivores existed, Mesozoic terrestrial ecosystems were very strongly influenced, if not dominated, by the large dinosaurian herbivores. This is particularly true of the whole Earth in the Jurassic and Early Cretaceous and, after smaller Ornithischia radiated in the later Cretaceous of the Northern Hemisphere, of the Late Cretaceous in the Southern Hemisphere.

Classically, these large herbivores are envisioned as having fed at least to 8–10 m off the ground, and possibly to 15 m if they were able to rear back on their hind legs and balance on their tails as Bakker (1978, 1986) has suggested, although this may have been uncommon behavior (Dodson 1990a). This feeding height suggests that large sauropods commonly browsed on tree foliage, which in this case would be almost entirely provided by conifers. Recent biomechanical models (Stevens and Parrish 1999; Parrish 2003) have questioned this high-feeding model in diplodocids, based on the position of the skull on the vertebral column, and Dodson (1990a) has additionally noted the difficulty of maintaining a sufficient supply of blood to a consistently elevated head. These observations suggest the counterintuitive hypothesis that at least some sauropods stood in one place and grazed a wide range of lower vegetation as "boom feeders." Indeed, some workers have suggested these sauropods fed on vast "fern prairies" (Coe et al. 1987). While pteridophytes were common in moist areas in the Mesozoic (Krassilov 1981; Saward 1992; Spicer et al. 1993), it seems unlikely that they grew in vast stands in the arid interiors of Mesozoic continents like grasses do in the present day, as the term *prairie* would imply. Further, the radiation of low-feeding Ornithischia *after* the origin of low-growing angiosperms (see below) is circumstantial evidence that low-level fodder was not a significant resource before the appearance of the angiosperms.

Thus, the interpretation of low browsing in sauropods stands in apparent contradiction to the interpretation that most of the year-round fodder would be available at height, not at ground level. Clearly, we need to know more about the plants, the herbivores, or both, to make sense of this discrepancy. Further, we need to remember that the past is not the same as the present, and that large areas of the pre-angiosperm world may have been without vegetation.

These large sauropod herbivores generally lacked the ability to chew fodder, as they commonly possessed peglike teeth that did not occlude. It seems likely that the teeth were used as "rakes," with the animal closing its mouth around a branch and pulling its head back, peeling foliage off as it went. The combination of the poor quality of most available gymnosperm foliage, particularly in low and mid paleolatitudes, its high content of indigestible chemicals, and the inability of these herbivores to chew effectively leads to the hypothesis that most breakdown of the plant material took place in the gut. The great size of these herbivores would have allowed for a large intake and long residence time of relatively indigestible food. This would permit microbial fermentation to break down tough cell walls to allow the extraction of sufficient energy to support the organism (Farlow 1987; but note that Ghosh et al. [2003] contend that geochemical signals in titanosaur coprolites are not compatible with bacterial fermentation). Digestion could also be aided by mechanical breakdown, possibly in a gizzard or muscular crop, or in the stomach through abrasion by gastroliths. The large intake, long passage time, and possible microbial fermentation permitted by large body size might also have allowed the dinosaurs to largely negate the effects of digestion-inhibiting chemicals (Farlow 1987). Presumably, the greater diversity and density of high-latitude plant communities (Saward 1992; Ziegler et al. 1993; Spicer et al. 1993) provided a better source of fodder, and one prediction from the foregoing hypothesis is that the largest herbivorous dinosaurs from the high latitudes should have been somewhat smaller than their lower-latitude counterparts.

If the herbivores became quite large as a result of their interactions with plants, it is not surprising that the coeval carnivores evolved large size in order to deal with the herbivores. Thus, it may be possible to ascribe the size of large dinosaur predators to the nature of the Jurassic plant community.

While the focus has been on the ecologically important large herbivores, the smaller ones (e.g., some prosauropods, *Stegosaurus*, and several smaller Ornithischia) must have fed closer to the ground. Many of these would have to have contended with the same poor food quality of tough conifer and cycadophyte foliage, but others could have fed on seed ferns, pteridophytes, and other gymnosperms with softer foliage, and thus have not faced quite the same requirements of maintaining a digestive volume. Some may even have fed on the large seeds of cycads and some conifers (Weishampel 1984; Tiffney 2004). It is possible that the youngsters of sauropod herbivores may have fed on higher-quality food until their requirements for food volume forced them to seek lower-quality fodder, although if dinosaurs possessed a lower metabolism than mammals and birds, this pattern may not have been as pronounced as it is in mammals and birds (Farlow 1987).

The sheer size of the dominant sauropod herbivores, coupled with the relatively (compared to the present day) scattered nature and low food quality of the Jurassic vegetation, at least beyond the higher latitudes (Rees et al. 2004), raises some interesting questions. How many individual herbivorous dinosaurs were there at any one time in the Jurassic? As many as living mammal herbivores in the present? In the modern day, it is observed that the larger the size of a species of animal, the fewer the total number of individuals (Peters 1983). It is also recognized that the poorer the quality of the available food, the smaller the number of individual organisms that can be supported. The great size of many Jurassic herbivores, the relatively poor quality of the available food, its restricted distribution, and its low diversity suggests three predictions. First, that we might expect a lower number of species of giant herbivores relative to the number of species of present-day large herbivores. Second, that we might expect relatively few individuals to be present within each sauropod species. Third, that individuals within a species might have to migrate over very large areas in order to obtain sufficient fodder to live.

There is circumstantial evidence to support these hypotheses, or at least to suggest that the very nature of the Late Triassic–mid-Cretaceous terrestrial ecosystem was quite different from that of the ecosystems of the Tertiary and the present day. First, the taxonomic diversity of dinosaurs is remarkably low compared to that of fossil mammals. Dodson (1990b) estimated that there were between 900 and 1,200 genera and about 1,100 to 1,500 species of dinosaurs in the 160 million years of dinosaur dominance, or about 7–9 species per million years. This is approximately paralleled by the generic estimate of Fastovsky et al. 2004, although Wang and Dodson 2004, and Dodson and Wang 2004, raise the estimate to about 1,900 genera, which (based upon similar reasoning) would yield approximately 2,375 species over 160 million years, or about 15 species per million years.

By contrast there are about 3,000 genera of fossil mammals reported from the Tertiary. Assuming about four species per genus of living mammals (Nowak 1991), this suggests that there were between 3,000 (most conservatively) and 12,000 species during this 65-million-year period, or between 46 and 185 species per million years. Even allowing for various preservational biases, these figures suggest that dinosaurs had a very much lower specific diversity per unit time than would be expected by comparison with mammals. Dinosaur trackway evidence in North America suggests both that the large sauropod herbivores may have migrated long distances (e.g., Dodson 1990a) and that the individual herds were not very large. This is weak evidence for a low number of individuals in a species, but stands in sharp contrast to clear evidence for the large numbers of individuals in some herds from the later Cretaceous (see below).

However, these speculations rest on two (if not more) basic assumptions. First, what is the physiology (or is it physiologies) of the dinosaurs? Were the herbivorous sauropods endotherms, inertial homeotherms, or ectotherms? If they were ectotherms, they may have had lower food requirements than do living mammals, which are endotherms. If they were inertial homeotherms, we have the difficulty that no large, terrestrial, inertial homeotherms exist in the present day upon which we can base an estimate of necessary food intake. Second, what is the effect of size on caloric requirements for an organism the size of a large sauropod? From studies of living animals, it has been demonstrated that the basal metabolic rate decreases in a linear manner with increasing size, all the way up to an elephant (Peters 1983; Schmidt-Nielsen 1984). However, we do not have any living examples that are as large as the average herbivorous sauropod to check if this relationship continues to be linear at a size of 10 times that of an elephant. However, even if sauropod herbivores did not possess a physiology comparable to that of living mammals, it is still probable that they would have been a formidable force of deforestation in the Mesozoic.

Late Cretaceous

The nature of the herbivore–plant relationship changed dramatically with the appearance and diversification of the angiosperms. Indeed, Bakker (1986) and Wing and Tiffney (1987) observed that the disturbance provided by dinosaur herbivory may well have selected for the appearance of a disturbance-tolerant seed plant with rapid growth and abilities for vegetative reproduction—in other words, an angiosperm. By contrast, Sereno (1997) and Barrett and Willis (2001) conclude that there is no spatiotemporal correlation between the radiation of angiosperms and patterns of change in dinosaurs. Nonetheless, by the Late Cretaceous, the angiosperms provided a potential food source that grew in a wider range of habitats, in a wider range of heights and shapes, and that possessed less mechanical and chemical impediments to digestion than did the gymnosperms and ferns of the earlier Mesozoic. In short, the radiation of angiosperms resulted in far more energetically rewarding vegetation for a herbivore to prey upon. A recent paper (Prasad et al. 2005) describes specialized epidermal cells attributable to grasses ("phytoliths") in titanosaurid coprolites, raising the vision of dinosaurs (in this case sauropods) feeding on grass. While the presence of grass phytoliths in the coprolites is fascinating evidence for presence of grass taxa in the Maastrichtian, their small numbers suggests that grass was not an important part of the diet of these animals. Indeed, phytoliths are famously distributed by air currents (e.g., Romero et al. 2003) and may well have been deposited on the leaves of the other, larger plants that dominated these coprolites.

The exact nature of this vegetation is still being discovered. It is clear that the angiosperms became the most taxonomically diverse plants on the Earth's surface by the Late Cretaceous (Lidgard and Crane 1990) and that they occurred in plant communities from the equator to the poles (Saward

1992; Spicer et al. 1993). However, it is not clear how rapidly they actually came to dominate the global vegetation physically; it is possible that this high diversity of types was present in low individual numbers, possibly forming largely herbaceous vegetation at least in some areas of the globe (Wing et al. 1993; Johnson 1998).

The Late Cretaceous also witnesses a major radiation of the ornithischians, including the hadrosaurs and ceratopsians. The parallel between the radiation of the Ornithischia and angiosperms suggests the hypothesis that the angiosperms underwrote the ornithischian radiation (although some doubt this; cf. Insole and Hutt 1994; Barrett and Willis 2001; but see Leckey et al. 2003). Several lines of evidence circumstantially support this hypothesis. First, the new ornithischian dinosaurs fed at a lower level than the preceding sauropod herbivores (Bakker 1978), befitting the herbaceous, shrubby, or small-tree stature of many Cretaceous angiosperms. Second, these were relatively smaller herbivores (1–10 tons) than the preceding sauropods (10–50 tons), suggesting a more digestible food source requiring a less-prolonged period of fermentation. Third, in a case where approximately contemporaneous faunas occurred in North America (Lehman 1987), areas dominated by conifers supported a fauna rich in herbivorous sauropods (*Alamosaurus*), while areas dominated by angiosperms supported an ornithischian herbivore fauna (Leckey et al. 2003), suggesting an association of herbivore and food type. Indeed Wheeler and Lehman (2000) report associated floras and faunas from the Big Bend area of Texas, where more open floras with gymnosperms (and admittedly, some angiosperm trees) are associated with *Alamosaurus*, while more closed, angiosperm-rich floras are associated with hadrosaurs.

Finally, two changes took place in dinosaur diversity. Dodson (1990b) observed that almost 50% of the described species of dinosaurs come from the last 20 million years of their existence, while Fastovsky et al. (2004) reported that 44% of all known terrestrial dinosaur genera occurred in the Late Cretaceous (99.6 to 65.5 mya). These observations throw the numbers of pre–Late Cretaceous dinosaur species into even more stark contrast to the numbers of mammals in the Tertiary. Further, mass-kill sites with large numbers of individuals of one species of dinosaur are observed only in the Late Cretaceous. In the most dramatic case this involved a herd of up to 10,000 individuals of the duckbill herbivore *Maiasaura* (Weishampel and Horner 1990); other examples exist (e.g., Currie and Dodson 1984). While these sites could represent long-term accumulations (Eberth and Currie 2005), it is noteworthy that these appear only in the later Cretaceous, suggesting a rise in numbers of source animals.

The appearance of angiosperms was not the sole factor in this radiation; evolutionary changes within the Ornithischia also played a role. Many of the hadrosaurs had teeth that occluded, and ceratopsians had shearing teeth, providing the ability to break down plant material prior to swallowing it (Norman and Weishampel 1985; Weishampel and Norman 1989). This increased the efficiency of digestion and the energetic reward per unit of food consumed, favoring the greater success of these herbivores, and possibly their smaller size. However, similar to the case with the saurischian herbivores, some questions remain open. For example, was the physiology of these ornithischian herbivores the same as the physiology of the large sauropod herbivores, or is part of the change in herbivore diversity and size due to an unrecognized change in physiology?

The appearance of angiosperms may have also affected the environment of the dinosaurs in other ways. The Jurassic vegetation of the Earth probably included large areas in the low and mid latitudes where trees were widely spaced, or occurred in patches, forming open vegetation. Indeed, many well-known Jurassic dinosaur localities are associated with open vegetation (Krassilov 1981; Saward 1992; Ziegler et al. 1993). This parallels the modern day, where large herbivores are associated with open environments. We know that the advent of angiosperms created a more diverse flora in the Cretaceous (Lupia et al. 1999; Schneider et al. 2004; Friis et al. 2005), but what of the physical structure of the plant communities they formed? The nature of this transition is an active area of research. Wolfe and Upchurch (1987) suggest that Cretaceous angiosperm communities were "open woodlands," forming a more closed and widespread community than created by the preceding gymnosperms. Wheeler and Lehman (2000) suggest the association of duckbills with smaller wood pieces (and thus possibly a more closed community) in a coastal setting, and the association of *Alamosaurus* with larger axes (to 1 m) that were spaced 12–13 m apart, suggesting an open forest. By contrast, Wing et al. (1993) and Johnson (1998) describe herbaceous-dominated communities in the Late Cretaceous. This suggests the possibility that low-growing communities of ferns and angiosperms dominated at least some moister areas of Cretaceous western North America. If the radiation of angiosperms did change the three-dimensional environment of the Late Cretaceous, what effect did this change have upon (a) the sauropod–ornithischian transition, (b) the evolution of the Ornithischia in the Late Cretaceous, and (c) the Cretaceous–Tertiary extinction of the terrestrial dinosaurs? Regarding the last question, the development of increasingly closed vegetation may have fragmented dinosaur populations, rendering them more susceptible to extinction.

The early Tertiary saw the first appearance of a dense angiosperm forest vegetation of modern aspect (Tiffney 1984). The early Tertiary also witnessed a massive radiation of birds and mammals. These herbivores were tiny in comparison to their dinosaurian predecessors (Fig. 27.5), and they fitted within the three-dimensional structure of the newly evolved angiosperm forests rather than around its margins. Their small size suited them to feed on the organs of angiosperms, thus setting up a very different herbivore–plant interaction than prevailed in the age of dinosaurs. In fact, the most basic difference between the terrestrial ecosystems of the Mesozoic and those of the Cenozoic and the present is that of the scale of plant–animal interaction. The Mesozoic involved open vegetation and large herbivores that generally acted as "plant predators," preying on whole organisms. The early Cenozoic introduced closed vegetation housing far smaller herbivores that preyed on plant organs. Whole-plant predation reappeared only with the evolution of more open angiosperm communities in the later Tertiary.

Summary

The foregoing description of the plants of the Mesozoic summarizes generally accepted information. However, the narrative of the interactions and interrelationships of Mesozoic plants and dinosaurs is far more speculative.

Why? Twenty years ago paleobotanists and vertebrate paleontologists tended to ignore the question of how these two important elements of the Mesozoic terrestrial ecosystem interacted. This is no longer so, but the field remains young. Many patterns have been recognized, and some hypotheses have been put forward to explain these observations. However, the hypotheses require testing, modification, and retesting, and more patterns await observation. There is much here to entertain and frustrate the paleontologists of the future!

What is clear from these initial observations is that the dynamic interactions between plants and herbivores of the Late Triassic through mid-Cretaceous were different from those of the Late Cretaceous, and that both were vastly different from the herbivore–plant interactions of the birds, mammals, and angiosperms in the Tertiary and present. Students should not come away thinking that the age of dinosaurs was "just like today," only with dinosaurs instead of birds and mammals. Rather, they should view the age of dinosaurs as a unique ecosystem that functioned in its own manner, and that offers an interesting look into one of the alternative worlds that have existed on our planet (Tiffney 1992).

Acknowledgments

I thank Prof. James Farlow (Indiana University Purdue University at Fort Wayne) for inviting this photosynthetic contribution to an otherwise vertebrate-dominated party, and Dr. Scott L. Wing (Smithsonian Institution) for his assistance in developing many of the ideas put forth here. The drawings were provided by Robin M. Gowen. I thank the anonymous reviewers for their critical commentary and suggestions, which have improved this paper, even as I release them from any responsibility for its content of heretical ideas.

References

Alvin, K. L. 1983. Reconstruction of a Lower Cretaceous conifer. *Botanical Journal of the Linnean Society* 86: 169–176.

Bakker, R. K. 1978. Dinosaur feeding behavior and the origin of flowering plants. *Nature* 274: 661–663.

———. 1986. *The Dinosaur Heresies*. New York: Morrow.

Barrett, P. M., and K. J. Willis. 2001. Did dinosaurs invent flowers? Dinosaur-angiosperm coevolution revisited. *Biological Reviews* 76: 411–447.

Bawa, K. S. 1995. Pollination, seed dispersal and diversification of the angiosperms. *Trends in Ecology and Evolution* 10: 311–312.

Benton, M. J. 1983. Dinosaur success in the Triassic: a noncompetitive ecological model. *Quarterly Review of Biology,* 58: 29–55.

Bond, W. J. 1989. The tortoise and the hare: ecology of angiosperm dominance and gymnosperm persistence. *Biological Journal of the Linnean Society,* 36: 227–249.

Burleigh, J. G., and S. Mathews. 2004. Phylogenetic signal in nucleotide data from seed plants: implications for resolving the seed plant tree of life. *American Journal of Botany* 91: 1599–1613.

Chaw, S. M., C. L. Parkinson, Y. C. Chen, T. M. Vincent, and J. D. Palmer. 2000. Seed plant phylogeny inferred from all three plant genomes: monophyly of extant gymnosperms and origin of Gnetales from conifers. *Proceedings of the National Academy of Sciences* (USA) 97 (8): 4086–4091.

Coe, M. J., D. L. Dilcher, J. O. Farlow, D. M. Jarzen, and D. A. Russell. 1987. Dinosaurs and land plants. In E. M. Friis, W. G. Chaloner, and P. R. Crane (eds.), *The Origins of Angiosperms and Their Biological Consequences*, 225–258. Cambridge: Cambridge University Press.

Crepet, W. L., E. M. Friis, and K. C. Nixon. 1991. Fossil evidence for the evolution of biotic pollination. *Philosophical Transactions of the Royal Society of London B* 333 (1267): 187–195.

Currie, P. J., and P. Dodson. 1984. Mass death of a herd of ceratopsian dinosaurs. In W.-E. Reif and F. Westphal (eds.), *Third Symposium on Mesozoic Terrestrial Ecosystems, Short Papers*, 61–66. Tübingen: Attempo Verlag.

Delevoryas, T., and R. C. Hope. 1971. A new Triassic cycad and its phyletic implications. *Postilla* 150: 1–14.

DiMichele, W. A., H. W. Pfefferkorn, and R. A. Gastaldo. 2001. Response of Late Carboniferous and Early Permian plant communities to climate change. *Annual Review of Earth and Planetary Sciences* 29: 461–487.

Dodson, P. 1990a. Sauropod paleoecology. In D. B. Weishampel, P. Dodson, and H. Osmólska (eds.), *The Dinosauria*, 402–407. Berkeley: University of California Press.

———. 1990b. Counting dinosaurs: how many kinds were there? *Proceedings of the National Academy of Sciences* (USA) 87: 7608–7612.

Dodson, P., and S. C. Wang. 2004. Counting dinosaurs 2004: how many kinds were there? *Journal of Vertebrate Paleontology* (suppl.) 24 (3): 52–53.

Doyle, J. A., and M. J. Donoghue. 1986. Seed plant phylogeny and the origin of angiosperms: an experimental cladistic approach. *Botanical Review* (Lancaster) 52: 321–431.

———. 1992. Fossils and seed plant phylogeny reanalyzed. *Brittonia* 44: 89–106.

Eberth D. A., and P. J. Currie. 2005. Ceratopsian bonebeds: occurrence, origins and significance. In P. J. Currie, and E. B. Koppelhus (eds.), *Dinosaur Provincial Park: A Spectacular Ancient Ecosystem Revealed*, 501–536. Bloomington: Indiana University Press.

Farlow, J. O. 1987. Speculations about the diet and digestive physiology of herbivorous dinosaurs. *Paleobiology* 13: 60–72.

Fastovsky, D. E., Y. Huang, J. Hsu, J. Martin-McNaughton, P. M. Sheehan, and D. B. Weishampel. 2004. Shape of Mesozoic dinosaur richness. *Geology* 32: 877–880.

Feild, T. S., N. C. Arens, J. A. Doyle, T. E. Dawson, and M. J. Donoghue. 2004. Dark and disturbed: a new image of early angiosperm ecology. *Paleobiology* 30: 82–107.

Friis, E. M., K. R. Pedersen, and P. R. Crane, P. 2005. When Earth started blooming: insights from the fossil record. *Current Opinion in Plant Biology* 8: 5–12.

Galton, P. M. 1985. Diet of prosauropod dinosaurs from the Late Triassic and Early Jurassic. *Lethaia* 18: 105–123.

Ghosh, P., S. K. Bhattacharya, A. Sahni, R. K. Kar, D. M. Mohabey, and K. Ambwani. 2003. Dinosaur coprolites from the Late Cretaceous (Maastrichtian) Lameta Formation of India: isotopic and other markers suggesting a C_3 plant diet. *Cretaceous Research* 24: 743–750.

Gillette, D. D. 1994. *Seismosaurus: The Earth Shaker*. New York: Columbia University Press.

Hallam, A. 1984. Continental humid and arid zones during the Jurassic and Cretaceous. *Palaeogeography, Palaeoclimatology, Palaeoecology* 47: 195–223.

———. 1993. Jurassic climates as inferred from the sedimentary and fossil record. *Philosophical Transactions of the Royal Society of London B* 341: 287–296.

Insole, A. N., and S. Hutt. 1994. The palaeoecology of the dinosaurs of the Wessex Formation (Wealden Group, Early Cretaceous), Isle of Wight, Southern England. *Zoological Journal of the Linnean Society.* 112: 197–215.

Johnson, K. 1998. Late Cretaceous (early Maastrichtian) herbaceous vegetation from the Almond Formation, Rock Springs Uplift, Wyoming. *Geological Society of America Abstracts with Programs* 30 (7): 36.

Krassilov, V. A. 1981. Changes of Mesozoic vegetation and the extinction of dinosaurs. *Palaeogeography, Palaeoclimatology, Palaeoecology* 34: 207–224.

Leckey, E. H., B. H. Tiffney, and S. Sweeney. 2003. Cretaceous plant-animal interactions; can distribution patterns for herbivorous dinosaurs and plants be used to test for co-evolution? *Geological Society of America Abstracts with Programs* 35 (6): 503–504.

Lehman, T. M. 1987. Late Maastrichtian paleoenvironments and dinosaur biogeography in the western interior of North America. *Palaeogeography, Palaeoclimatology, Palaeoecology* 60: 189–217.

Lidgard, S., and P. Crane. 1990. Angiosperm diversification and Cretaceous floristic trends: a comparison of palynofloras and leaf macrofloras. *Paleobiology* 16: 77–93.

Lupia, R., S. Lidgard, and P. R. Crane. 1999. Comparing palynological abundance and diversity: implications for biotic replacement during the Cretaceous angiosperm radiation. *Paleobiology* 25: 305–340.

Mellett, J. S. 1982. Body size, diet, and scaling factors in large carnivores and herbivores. *Proceedings of the Third North American Paleontological Convention* 2: 371–376.

Meyen, S. V. 1987. *Fundamentals of Palaeobotany*. London: Chapman and Hall.

Midgley, J. J., and W. J. Bond. 1991. Ecological aspects of the rise of the angiosperms: a challenge to the reproductive superiority hypothesis. *Biological Journal of the Linnean Society* 44: 81–92.

Niklas, K. J., B. H. Tiffney, and A. H. Knoll. 1985. Patterns in vascular land plant diversification: an analysis at the species level. In J. W. Valentine (ed.), *Phanerozoic Diversity Patterns: Profiles in Macroevolution*, 97–128. Princeton, N.J.: Princeton University Press.

Norman, D. 1985. *The Illustrated Encyclopedia of Dinosaurs*. New York: Crescent Books.

Norman, D. B., and D. B. Weishampel. 1985. Ornithopod feeding mechanisms: their bearing on the evolution of herbivory. *American Naturalist* 126: 151–164.

Nowak, R. M. 1991. *Walker's Mammals of the World*. Baltimore: Johns Hopkins University Press.

Parrish, J. M. 2003. Mapping ecomorphs onto sauropod phylogeny. *Journal of Vertebrate Paleontology* 20: 85A–86A.

Peters, R. H. 1983. *The Ecological Implications of Body Size*. Cambridge: Cambridge University Press.

Prasad, V., C. A. E. Strömberg, H. Alimohammadian, and A. Sahni. 2005. Dinosaur coprolites and the early evolution of grasses and grazers. *Science* 310: 1177–1180.

Rees, P. M. 2002. Land plant diversity and the end-Permian mass extinction. *Geology* 30: 827–830.

Rees, P. M., C. R. Noto, M. Parrish, and J. T. Parris. 2004. Late Jurassic climates, vegetation and dinosaur distributions. *Journal of Geology* 112: 643–653.

Romero, O. E., L. Dupont, U. Wyputta, S. Jahns, and G. Wefer. 2003. Temporal variability of fluxes of eolian-transported freshwater diatoms, phytoliths, and pollen grains off Cape Blanc as reflection of land-atmosphere-ocean interactions in northwest Africa. *Journal of Geophysical Research – Oceans* 108: C5, 3153. doi:10.1029/2000JC000375.

Rothwell, G. W., L. Grauvogel-Stamm, and G. Mapes. 2000. An herbaceous fossil conifer: gymnospermous ruderals in the evolution of Mesozoic vegetation. *Palaeogeography, Palaeoclimatology, Palaeoecology* 156: 139–145.

Rydin, C., M. Kallersjo, and E. M. Friis. 2002. Seed plant relationships and the systematic position of Gnetales based on nuclear and chloroplast DNA: conflicting data, rooting problems, and the monophyly of conifers. *International Journal of Plant Sciences* 163: 197–214.

Saward, S. A. 1992. A global view of Cretaceous vegetation patterns. In P. J. McCabe and J. T. Parrish (eds.), *Controls on the Distribution and Quality of Cretaceous Coals*, 17–35. Special Paper 267. Boulder, Colo. : Geological Society of America.

Schmidt-Nielsen, K. 1984. *Scaling: Why Is Animal Size So Important?* Cambridge: Cambridge University Press.

Schneider, H., E. Schuettpelz, K. M. Pryer, R. Cranfill, S. Magalllon, and R. Lupia. 2004. Ferns diversified in the shadow of the angiosperms. *Nature* 428: 553–557.

Sereno, P. C. 1997. The origin and evolution of dinosaurs. *Annual Review of Earth and Planetary Sciences* 25: 435–489.

Spicer, R. A., P. M. Rees, and J. L. Chapman. 1993. Cretaceous phytogeography and climate signals. *Philosophical Transactions of the Royal Society of London B* 341: 277–286.

Stebbins, G. L. 1981. Why are there so many species of flowering plants? *Bioscience* 31: 573–577.

Stevens, K. A., and J. M. Parrish. 1999. Neck posture and feeding habits of two Jurassic dinosaurs. *Science* 284: 798–800

Stewart, W. N., and G. W. Rothwell. 1993. *Paleobotany and the Evolution of Plants*. Cambridge: Cambridge University Press.

Stokes, W. L. 1987. Dinosaur gastroliths revisited. *Journal of Paleontology* 61: 1242–1246.

Taylor, T. N., and E. L. Taylor. 1993. *The Biology and Evolution of Fossil Plants*. Englewood Cliffs, N. J.: Prentice Hall.

Tiffney, B. H. 1984. Seed size, dispersal syndromes, and the rise of the angiosperms: evidence and hypothesis. *Annals of the Missouri Botanical Garden* 71: 551–576.

——. 1989. Plant life in the age of dinosaurs. In K. Padian and D. J. Chure (eds.), *The Age of Dinosaurs*, 35–47. Short Courses in Paleontology 2. Knoxville, Tenn.: Paleontological Society.

——. 1992. The role of vertebrate herbivory in the evolution of land plants. *Palaeobotanist* 41: 87–97.

——. 2004. Vertebrate dispersal of seed plants through time. *Annual Review of Ecology and Systematics* 35: 1–29.

Vakhrameev, V. A. 1991. *Jurassic and Cretaceous Floras and the Climates of the Earth*. Cambridge: Cambridge University Press.

Wang, S. C., and P. Dodson. 2004. Estimating Absolute Diversity: How Many Dinosaur Genera Were There? *Geological Society of America Abstracts with Programs* 36 (5): 456.

Weishampel, D. B. 1984. Interactions between Mesozoic plants and vertebrates: fructifications and seed predation. *Neues Jahrbuch für Geologie und Paläontologie, Abh.* 167: 224–250.

Weishampel, D. B., and J. R. Horner. 1990. Hadrosauridae. In D. B. Weishampel, P. Dodson, and H. Osmólska (eds.), *The Dinosauria*, 534–561. Berkeley: University of California Press.

Weishampel, D. B., and D. B. Norman. 1989. Vertebrate herbivory in the Mesozoic: jaws, plants and evolutionary metrics. In J. O. Farlow (ed.), *Paleobiology of the Dinosaurs*, 87–100. Special Paper, 238. Denver: Geological Society of America.

Wheeler, E. A., and T. M. Lehman. 2000. Late Cretaceous woody dicots from the Aguja and Jevelina Formations, Big Bend National Park, Texas, USA. *International Association of Wood Anatomists Journal* 21: 83–120.

Willis, K. J., and J. C. McElwain. 2002. *The Evolution of Plants*. Oxford: Oxford University Press.

Wing, S. L., and L. D. Boucher. 1998. Ecological aspects of the Cretaceous flowering plant radiation. *Annual Review of Earth and Planetary Sciences* 26: 379–421.

Wing, S. L., L. J. Hickey, and C. J. Swisher. 1993. Implications of an exceptional fossil flora for Late Cretaceous vegetation. *Nature* 363: 342–344.

Wing, S. L., and B. H. Tiffney. 1987. The reciprocal interaction of angiosperm evolution and tetrapod herbivory. *Review of Palaeobotany and Palynology* 50: 179–210.

Wolfe, J. A., and G. R. Upchurch, Jr. 1987. North American nonmarine climates and vegetation during the Late Cretaceous. *Palaeogeography, Palaeoclimatology, Palaeoecology* 61: 33–77.

Zavada, M. S., and M. T. Mentis. 1992. Plant-animal interaction: the effect of Permian megaherbivores on the glossopterid flora. *American Midland Naturalist* 127: 1–12.

Ziegler, A. M., J. M. Parrish, Y. Jiping, E. D. Gyllenhaal, D. B. Rowley, J. T. Parrish, N. Shangyou, A. Bekker, and M. L. Hulver. 1993. Early Mesozoic phytogeography and climate. *Philosophical Transactions of the Royal Society of London B* 341: 297–305.

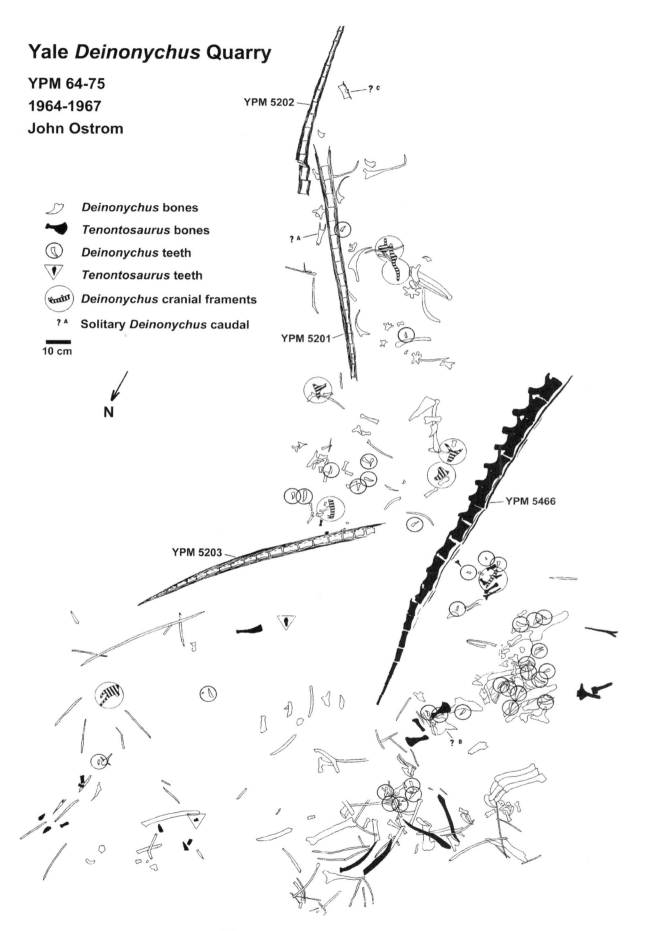

Yale *Deinonychus* Quarry

YPM 64-75
1964-1967
John Ostrom

Deinonychus bones
Tenontosaurus bones
Deinonychus teeth
Tenontosaurus teeth
Deinonychus cranial framents
? A Solitary *Deinonychus* caudal

10 cm

N

YPM 5202
? C
? A
YPM 5201
YPM 5203
YPM 5466
? B

What Did Dinosaurs Eat: Coprolites and Other Direct Evidence of Dinosaur Diets

28

Karen Chin

What did the Mesozoic dinosaurs really eat? This question has spawned numerous hypotheses from scientists, dinosaur enthusiasts, and fantasy writers. Speculations about dinosaur diets are frequently based on indirect evidence such as surveys of available food (i.e., contemporaneous organisms) and theories about foraging abilities inferred from functional morphology. Such analyses are important tools that suggest generalized dinosaur feeding strategies. Even so, indirect evidence cannot tell us more specifically which available foods were actually eaten. Did dinosaurs really feast on ferns, munch on mammals, and eat each other? We will never completely understand dinosaur food habits, but scrutiny of the fossil record has revealed a number of fortuitous traces of dinosaur feeding activities. These clues are usually rare and often controversial, but they provide paleobiological information that can help us better understand dinosaurs and their interactions with other organisms.

In order to look for direct evidence of dinosaur diets, we can consider all stages of feeding behavior, including search, capture, ingestion, digestion, and defecation (Bishop 1975). Although these activities often leave little preservable evidence, animals spend a substantial proportion of their time seeking food, so it is not surprising that some traces of feeding activity have been preserved. Clues have been gleaned from a variety of trace fossils (fossils that indicate the activity of organisms) and from distinctive assemblages of skeletal material. These disparate sources of fossil evidence provide multiple perspectives on dinosaur feeding habits.

Trackways as Evidence of the Search for Food

The act of looking for food might seem to be untraceable, but dinosaurs occasionally left tracks that suggest that they were actively seeking dinner. At a famous Early Cretaceous site along the Paluxy River in Texas, tracks from one or more theropods appear to follow several sauropod trackways. It is apparent that the sauropods preceded the theropods because the theropod footprints are superimposed on top of several of the sauropod tracks. Furthermore, sedimentological evidence suggests that this track layer was deposited in a single event; thus, it is possible that the tracks on this surface were laid down within a relatively short time frame (J. O. Farlow personal communication). Early dinosaur tracker Roland T. Bird suggested that the dovetailing trackways indicate that the theropods were hot on the heels of the sauropod herd (Farlow 1987), and more recent analysis supports that at least one theropod closely followed one of the sauropods (J. O. Farlow personal communication). This fortuitous sequence of tracks thus appears

28.1. Map of the Yale *Deinonychus* quarry in Montana (YPM 64–75) showing partial skeletal remains of both *Deinonychus* and *Tenontosaurus* and more than 35 *Deinonychus* teeth scattered about the site. The abundance of shed teeth is consistent with an increased incidence of tooth loss during feeding. Figure published in the *Journal of Vertebrate Paleontology* (Maxwell and Ostrom 1995).

to provide rare evidence of theropod hunting behavior. Hunting activity is also suggested by another set of Upper Cretaceous theropod/sauropod trackways in Bolivia, where several theropod trackways parallel and overlap prints made by a group of sauropods (Leonardi 1984; Lockley 1991).

Another possible theropod hunting scenario is presented by a Cretaceous trackway assemblage in Australia. This site contains thousands of tracks and has been interpreted as a dinosaur "stampede" triggered by a single large theropod stalking a mixed group of small coelurosaurs and ornithopods (Thulborn and Wade 1979). Although the large theropod footprints do not actually follow the smaller dinosaur tracks, the long stride lengths and parallel trackways of the more than 100 small dinosaurs suggest that they were fleeing a significant threat.

Surprisingly, fossil tracks may also provide information about the foraging behavior of herbivorous dinosaurs. A set of intriguing Cretaceous footprints in the roof of a Utah coal mine were found clustered around fossil tree trunks that were preserved in growth position. The tracks are oriented toward the tree trunks and suggest the shuffling steps of browsing hadrosaurs (Parker and Rowley 1989; L. Parker personal communication).

Taphonomic Assemblages That Indicate Predator/Prey Interactions

Predator/prey interactions can occasionally be inferred from taphonomic associations of different organisms in exceptional fossil assemblages. Concentrations of theropod teeth among the bones of other animals are particularly telling. Dinosaur teeth were continually shed as new ones grew in, so we should expect to find them in feeding areas where vigorous biting accelerated tooth loss. Probable theropod feeding sites are indicated by the discovery of numerous theropod teeth with sauropod skeletons in the Upper Jurassic of Thailand (Buffetaut and Suteethorn 1989) and Utah (Jennings and Hasiotis 2006).

Even more compelling evidence for carnivory was found in the Lower Cretaceous of Montana, where 15 different sites were found to have *Deinonychus* teeth associated with *Tenontosaurus* bones (Maxwell and Ostrom 1995). The frequent co-occurrence of these elements and the dearth of *Deinonychus* teeth in the vicinity of bones from other possible prey animals suggest that the herbivorous *Tenontosaurus* was the preferred prey of *Deinonychus*. At one exceptional locality, over 35 *Deinonychus* teeth and skeletal elements from four *Deinonychus* individuals were found with the partial remains of one *Tenontosaurus* (Fig. 28.1). The bones were found in fine overbank deposits and could not have been transported by fluvial processes. Thus the assemblage has been interpreted as the scavenged remains of a struggle between a large *Tenontosaurus* and a pack of the much smaller *Deinonychus*. The presence of both *Deinonychus* and *Tenontosaurus* bones at the site suggests that not only the prey animal but some members of the attacking *Deinonychus* pack were killed during the struggle and were subsequently consumed (Ostrom 1990; Maxwell and Ostrom 1995). A variation on this interpretation is that several *Deinonychus* died as a result of agonistic encounters as they vied to feed on the *Tenontosaurus* carcass (Roach and Brinkman 2007).

A spectacular Upper Cretaceous find from the Gobi Desert reveals the articulated skeleton of a carnivorous *Velociraptor* entangled with an herbivorous *Protoceratops* (Kielan-Jaworowska and Barsbold 1972). The relative positions of the two dinosaurs suggest that they were engaged in a struggle when they died, with the theropod's clawed feet extending into the *Protoceratops*'s throat and belly. Although this association has often been cited as an example of fighting dinosaurs, one report disputed that view and suggested that the *Velociraptor* was simply feeding on a dead or dying animal (Osmólska 1993). This scenario portrays the *Velociraptor* as a scavenger that died of unknown causes while feeding. Subsequent interpretations (Unwin et al. 1995; Carpenter 1998), however, argue that the taphonomic evidence supports the original predator/prey fight encounter. Particularly telling is the fact that the theropod's arm is firmly locked in the herbivore's jaws—a position that implies active defense. One study suggests that the struggling dinosaurs died simultaneously in a massive sandstorm (Unwin et al. 1998), whereas another proposes that the *Protoceratops* died of its wounds and the *Velociraptor* was trapped by its ensnared arm and the weight of the *Protoceratops*. These different interpretations paint different scenarios of the event recorded by this remarkable Mongolian assemblage, but all conclude that the *Velociraptor* fully intended to dine on the *Protoceratops*.

These skeletal associations tell us much about interactions between different dinosaurs because both predator and prey organisms have been identified. Fossil assemblages suggesting clear examples of predatory behavior are very rare, however, and must be carefully scrutinized so that inadvertent associations of fossil bones are not misinterpreted.

Tooth-Damaged Bone

It is reasonable to expect to find physical evidence of theropod feeding activity preserved on bones—if the bones themselves were not ingested. Most conspicuous are the very rare examples of dinosaur teeth physically stuck in the bones of their prey. In Montana, a tyrannosaurid tooth was found embedded in a *Hypacrosaurus* fibula (J. R. Horner personal communication). Further north, a *Saurornitholestes* tooth was discovered in the tibia of an azhdarchid pterosaur in Alberta. The size of the pterosaur tibia indicates a wingspan of around 6 m, whereas the *Saurornitholestes* itself was probably less than 2 m long. The significant size difference between predator and prey suggests that the small theropod opportunistically scavenged the pterosaur carcass postmortem (Currie and Jacobsen 1995). Another pterosaur bone from the Early Cretaceous of Brazil was also found with an embedded dinosaur tooth; in this case a spinosaurid had bitten into a cervical vertebra (Buffetaut et al. 2001).

Tooth-marked dinosaur bone is less rare than theropod teeth embedded in bone, but such bite marks are still relatively uncommon (e.g., Fiorillo 1991; Chure et al. 1998; Jacobsen 1998). One study of various bone assemblages found the incidence of tooth marks on dinosaur bones (0–4%) to be significantly lower than that of marks found on bones from ancient and modern communities with large mammalian carnivores (13.1–37.5%; Fiorillo 1991). Another survey of a thousand dinosaur bones from the Judith River Formation observed somewhat higher percentages of tooth-marked dinosaur bone: 14% of hadrosaurid bones, 5% of ceratopsid bone, and 2% of tyrannosaurid bones (Jacobsen 1998). The smaller percentages of tooth marks on dinosaur bones in these studies relative to those observed on mammal bones may reflect differences in carcass utilization patterns (Hunt 1987; Fiorillo 1991) or taphonomic biases (Erickson and Olson 1996).

Tooth-damaged dinosaur bone can be recognized by distinctive markings such as grooves or punctures (Fiorillo 1991; Erickson and Olson 1996; Fowler and Sullivan 2006; Hone and Watabe 2010). Although some damage may have been inflicted during nonlethal intraspecific fights (Tanke and Currie 1998; Bell and Currie 2010), most bite marks probably reflect feeding activity. Identification of tooth-marked, unhealed bone can tell us that a particular species of dinosaur was eaten, but generally does not indicate whether the prey was killed for food or opportunistically scavenged. In some cases, however, it may be possible to associate different tooth marks with specific predator activities based on the types and location of damage. Multiple bite marks on the ends of dinosaur limb bones, for example, are more likely to represent feeding traces than assault wounds (Hunt et al. 1994b; Longrich et al. 2010).

It can be difficult to determine the identity of the animals responsible for causing fossil tooth marks, because many Mesozoic vertebrates (crocodilians, dinosaurs, etc.) were capable of inflicting generalized tooth damage on bone. Well-preserved bite marks, however, may retain evidence of tooth

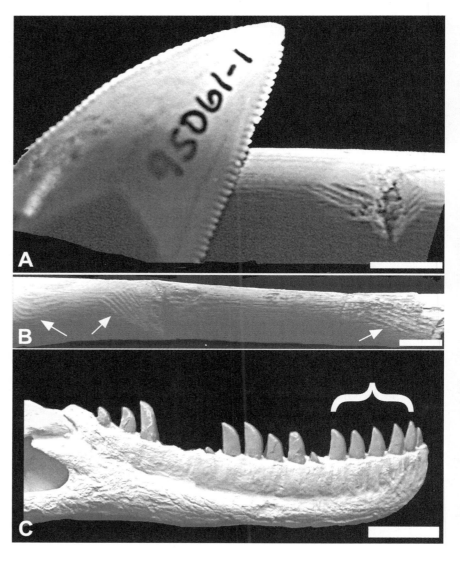

28.2. Tooth-marked bone that reflects the tooth and jaw morphology of *Majungatholus*. (A) Image of *Majungatholus* tooth with tooth marks in a fossil *Majungatholus* rib (UA 8678), showing the equivalence in denticle size and spacing between the tooth and the denticle drag marks on the bone. Scale bar = 5 mm. (B) Same rib bone as in A, with three sets of tooth marks (arrows) showing striations caused by tooth denticles. Scale bar = 5 mm. (C) Right dentary of *Majungatholus* (FMNH PR2100), showing an even pattern of eruption (bracket) and intertooth spacing that corresponds to parallel tooth marks found on other bones in the assemblage. This evidence indicates that *Majungatholus* engaged in cannibalism. Scale bar = 5 cm. Figure originally published in *Nature* (Rogers et al. 2003).

and jaw morphology that provides good information about the identity of the biter. For instance, characteristics of a predator jaw can be inferred from the spatial distribution of parallel score marks caused by teeth. In an early study, the spacing of teeth in an *Allosaurus* jaw was found to match patterns of scoring found on bones of an *Apatosaurus* (Matthew 1908). The morphology and dimensions of tooth serrations are also informative. Examination of tooth marks in the jaw of the small theropod *Saurornitholestes* indicated that they were probably made by a subadult tyrannosaurid (Jacobsen 2001). In another case, characteristic denticle drag marks and parallel tooth scorings revealed biting activity by the Late Cretaceous Madagascan abelisaurid, *Majungatholus* (Fig. 28.2).

Tooth size and morphology are even more diagnostic. Dental putty pushed into a puncture mark in a bone from the Hell Creek Formation of Montana provided a positive cast of the tooth that bit into a *Triceratops* pelvis. The size and shape of this cast indicate that *Tyrannosaurus* was the perpetrator, and additional comparable tooth marks reveal that *T. rex* fed on *Edmontosaurus* as well (Erickson and Olson 1996). Casts of puncture marks have also been used to identify *Deinonychus* tooth marks in *Tenontosaurus* bones (Gignac et al. 2010). A combination of different tooth-mark features can be used to deduce the identity of the theropods that bit into bones (e.g., Gignac et al. 2010; Hone and Watabe 2010). This information is useful because it provides evidence for specific predator (or scavenger)/prey interactions. Interestingly, tooth marks in bone have revealed cases of theropod cannibalism. The *Majungatholus* tooth marks found on numerous bones indicate that this theropod not only fed on the sauropod *Rapetosaurus* but also on members of its own species (Rogers et al. 2003). Similarly, inferred *Tyrannosaurus* bite marks on *Tyrannosaurus* bones suggest feeding activity rather than nonlethal intraspecific interactions (Longrich at al. 2010).

Stomach Contents: Evidence of Food Already Ingested

If a dinosaur died with a full belly, its partially digested last meals might have been preserved in the gut region. Such fortuitous preservation would have required fossilization of an articulated specimen that was undisturbed by erosion or scavenging. Furthermore, careful collection and documentation of the entire assemblage is necessary in order to make the case that organic matter in the gut region of the carcass is not allocthonous debris that was transported in. Several cases of possible herbivore stomach contents have been reported. Conifer needles, twigs, and seeds found in the body cavity of an *Edmontosaurus* from Wyoming have been interpreted as ingested fodder (Kräusel 1922; Weigelt 1989 [1927]). Concentrated twigs and other plant fragments were also found in the gut region of a *Corythosaurus* mummy from Alberta that had skin impressions covering most of the thoracic region (Currie et al. 1995). In both of these cases, the anatomical setting and the fragmented nature of the included plant debris are suggestive of gut contents, but it is also possible that fluvial mechanisms were responsible for deposition of the plant material (Abel 1922; Currie et al. 1995).

The origin of organic plant material found in the abdominal area of a brachylophosaurid hadrosaur from the Judith River Formation of Montana is less equivocal. The absence of significant scavenging, integrity of the carcass, and pronounced compositional differences between the materials inside and outside of the carcass suggest that the comminuted leaves in the gut region represent the dinosaur's final snacks (Tweet et al. 2008). An Early Cretaceous *Minmi* ankylosaur from Australia presents an even stronger case of herbivorous dinosaur gut contents. Because this dinosaur was preserved in marine sediments, it is highly unlikely that the plant material in the gut was hydrodynamically washed into the carcass. The calcareous mass within the body cavity of the ankylosaur includes a variety of comminuted plant tissues including probable angiosperm organs (Molnar and Clifford 2000; Molnar and Clifford 2001).

But some dinosaurs ingested inorganic materials as well. Deposits of siliceous grains found within the ribcages of twelve Chinese ornithomimid dinosaurs have been interpreted as gastrolith masses. These inorganic gut contents suggest that the toothless ornithomimids were herbivorous, because extant herbivorous birds commonly ingest grit as a digestive aid (Kobayashi et al. 1999).

Reports of theropod gut contents are more common, but can still incite controversy. The discovery of bones attributed to *Coelophysis* within the thoracic cavities of two articulated skeletons of the same species prompted an interpretation of cannibalism (Colbert 1989). However, the fact that the included elements are up to two-thirds the size of the same bones in the enclosing skeletons provoked skepticism of this conclusion (Gay 2002). This debate took an unexpected turn after a more thorough analysis. It turns out that only one of these *Coelophysis* skeletons contains bones constrained within the gut cavity, and the morphology and histology of the included bones indicate that the *Coelophysis* ingested a crocodylomorph (Nesbitt et al. 2006)!

Other reported theropod stomach contents are less contentious and indicate that dinosaurs ingested a variety of vertebrates. The holotype specimen of a tiny *Compsognathus* from the Solnhofen limestone was found to have a lizard in its belly (Nopcsa 1903; Ostrom 1978). The partial lizard skeleton is clearly sandwiched within the ribcage of the *Compsognathus* (Fig. 28.3). Interestingly, another *Compsognathus* specimen from France (Peyer 2006) and a *Sinosauropteryx* compsognathid dinosaur from China (Chen et al. 1998), also contain lizard bones in their gut cavities. These finds suggest that lizards provided a significant food source for these smaller, nonavian dinosaurs. Symmetrodont mammal bones recovered from the gut region of another *Sinosauropteryx* specimen reveal that this small dinosaur preyed on our furry ancestors as well (Currie and Chen 2001). And some dinosaurs apparently had a taste for fish; the gut region of a Lower Cretaceous *Baryonyx* dinosaur from England contained etched fish scales and teeth (from *Lepidotes*), in addition to subadult *Iguanodon* bones (Charig and Milner 1997).

As noted above, it can be difficult to verify whether residues within a thoracic cavity truly represent ingested food, and it can be even more challenging to identify gut contents that have been dislodged from a carcass. Stokes (1964) suggested that a fossilized mass of fragmented plant debris found near scattered sauropod bones represented its spilled gut contents, but this interpretation is poorly supported (Currie et al. 1995; Molnar and Clifford 2000). A better case for dissociated digestive residues from carnivorous animals can be made if skeletal elements appear to be acid-etched.

28.3. A partial skeleton of a lizard evident in the gut region of the holotype specimen of *Compsognathus longipes* (Bayerische Staatssammlung für Paläontologie und Historische Geologie specimen number AS I 563). The lizard bones are clearly enclosed by the ribs of *Compsognathus*, indicating that the dinosaur had ingested the lizard. Franz Nopcsa (1903) recognized the lizard in the gut and drew this illustration of the included bones. John Ostrom later identified the lizard as *Bavarisaurus*.

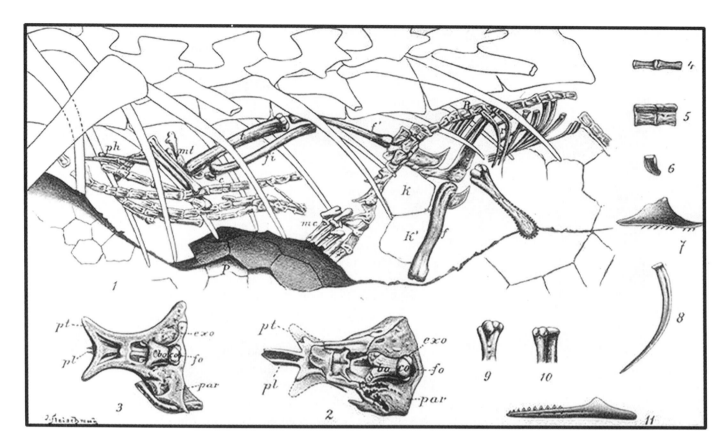

Significantly etched juvenile hadrosaur bones from the Upper Cretaceous of Montana were found near unetched tyrannosaurid bones. The contrasting surficial textures of the bones suggest that the hadrosaur bones had been subjected to gastric digestion (Varricchio 2001). Indeed, it may be that isolated gastrically etched bones will be easier to find than large articulated theropod carcasses with bona fide stomach contents.

Coprolites: The End Result of Feeding Activities

Reconstructions of the lives of dinosaurs often conveniently overlook the fact that dinosaurs produced fecal material. Lots of it. Studies of extant wild animal diets often rely on fecal analyses because many dietary components are still identifiable after passage through an animal's gut. It thus makes sense to look to dinosaur coprolites (fossil feces) for the end view of dinosaur food habits. Although a relatively small number of coprolites have been unequivocally attributed to dinosaurs, the number of reported specimens is growing.

Preservational biases help account for the paucity of known dinosaur coprolites, since the fossilization potential of fecal material is dependent on diet and on the environment of deposition. Although plant-eating dinosaurs and other terrestrial herbivores undoubtedly outnumbered their meat-eating counterparts, theropod coprolites are much more common: the relatively high phosphorus content of a carnivorous diet appears to facilitate mineralization (Bradley 1946; Chin et al. 2003). Depositional conditions are also important; most reported coprolites were produced by aquatic organisms that lived in environments subject to rapid sedimentation. In contrast, fecal matter deposited on land is less likely to be preserved because it is vulnerable to decomposition, desiccation, trampling, erosion, and coprophagy (consumption of feces). Thus the fortuitous preservation of dinosaur feces likely required rapid burial.

Recognizing possible dinosaur coprolites can be problematic, especially since many vertebrates produce similarly shaped feces. Spiral coprolites can be attributed to primitive fish (e.g., sharks) that had spiral intestinal valves, but most other fossil feces lack taxonomically distinct morphologies (Hunt et al. 1994a). Fortunately, large fecal size can be a potentially diagnostic characteristic, because large dinosaurs would have generated sizable fecal volumes, whereas small animals cannot produce large dung masses. On the other hand, large fecal masses are easily fragmented, and quantities of small, pelletoid feces can be excreted by large animals. As such, dinosaur feces may appear as broken or irregularly shaped masses that do not have typical fecal morphologies. In the absence of diagnostic shapes, other coprolite characteristics are particularly important, including chemical composition, comminuted organic contents, and depositional context.

Because indisputable dinosaur coprolites are relatively rare, the conspicuous absence of large quantities of dinosaur feces in Mesozoic sediments has led to claims that questionable material is dinosaur dung. There is speculation, for example, that numerous suspiciously shaped nodules are poorly preserved dinosaur coprolites, even though such rocks contain no organic inclusions that provide positive evidence of a fecal origin. In other cases, definite coprolites have been ascribed to dinosaurs despite the fact that the evidence for a dinosaurian origin is weak (Thulborn 1991).

Fossilized feces from herbivorous dinosaurs are particularly rare, but large, bona fide coprolites containing abundant plant material can usually be plausibly attributed to dinosaurs, because most other Mesozoic herbivores were too small to have generated much fecal material in one defecation "event." An unusual cluster of flattened coprolites from Jurassic sediments in Yorkshire, England, illustrates this (Hill 1976). The individual carbonaceous traces scarcely resemble animal droppings, but it is apparent that the collective deposit represents an assemblage of over 250 small (roughly 1 cm diameter) pelletoid feces similar to the pellet groups produced by extant deer. The coprolites contain large quantities of cycadeiod leaf cuticle, indicating that the source animal was a terrestrial herbivore. The dimensions of the flattened traces suggest that the original total fecal volume would have been approximately 130 cm^3; thus the mass can be reasonably attributed to a dinosaur.

Other reported dinosaur coprolites are permineralized and retain a three-dimensional morphology. Several nodular plant-filled specimens from the Upper Jurassic Morrison Formation exhibit exquisite preservation of conifer, cycadophyte, and fern organs. These 5–15.5 cm diameter specimens lack obvious fecal shapes, but the contents and sedimentological context strongly suggest that they are coprolites (Chin and Kirkland 1998). Another nodular mass of plant material from the Cretaceous Aguha Formation of Texas has also been interpreted as a coprolite from a plant-eating dinosaur. Bark and wood fragments have been identified in this 15 × 9 × 12 cm specimen (Baghai-Riding and DiBenedetto 2001).

Isolated coprolites are certainly informative, but assemblages of multiple specimens provide better perspectives on ancient diets. Numerous deposits

28.4. Herbivorous dinosaur coprolite from the Two Medicine Formation of Montana. (A) Specimen is predominantly composed of conifer wood fragments that appear as dark linear inclusions. Large coprolites like this often have atypical shapes because they have been broken or deformed by trampling, coprophagy, and/or weathering (MOR 771/BU-93-2). Scale at base is 11 cm long. (B) Thin section of coprolite from the same site showing fragments of conifer wood tissue and disaggregated wood cells (MOR 771/BU-89-2-Ha. Figure 5A by Karen Chin. Figure 5B courtesy SEPM (Society for Sedimentary Geology; Chin 2007).

of large dinosaur coprolites from the Upper Cretaceous Two Medicine Formation of Montana have revealed that herbivorous dinosaurs sometimes consumed surprisingly woody meals. The atypical blocky coprolites (Fig. 28.4A) lack familiar fecal morphologies, but the presence of dung beetle burrows in and around the specimens helped establish their fecal origin (Chin and Gill 1996). The large size (up to 7 l in volume), fibrous content, and paleontological context suggest that the coprolites were produced by *Maiasaura* hadrosaurs. Furthermore, characteristics of the highly concentrated (13–85%) wood fragments in the coprolites (Fig. 28.4B) suggest that the fecal producers intentionally consumed rotting wood. Decaying wood would have provided an assortment of nutritious foods, including cellulose within the decomposed wood, fungi, other microbes, and detritivorous invertebrates. Woody coprolites in three different horizons indicate that this (likely seasonal) habit of wood ingestion spanned at least 6 million years (Chin 2007)

The Late Cretaceous Lameta Formation of India also yields numerous coprolites. Many of these specimens have been attributed to titanosaurs based on large size (up to 10 cm in diameter) and plant matter content. A wide variety of inclusions have been documented in the specimens, including gymnosperm and angiosperm tissues, fungal spores, diatoms, and insect parts (e.g., Ghosh et al. 2003; Mohabey 2005). One particularly surprising discovery was the occurrence of grass phytoliths in some of the coprolites. This find predates previous fossil evidence for grasses and suggests that dinosaurs were among the earliest herbivores to exploit grasses (Prasad et al. 2005).

Although most large, plant-filled Mesozoic coprolites can be reasonably ascribed to dinosaurs, Mesozoic carnivore coprolites are more difficult to interpret, because midsized fecal masses could have been produced by a variety of organisms, including large fish, crocodilians, turtles, and moderately sized theropods. At the present time, only large Mesozoic coprolites have been credibly attributed to theropods. Irregular bone-packed masses from the Morrison Formation appear to be fossilized feces generated by sizable Jurassic theropods (Stone et al. 2000; Chin and Bishop 2007). Similarly, size, contents, and stratigraphic context indicate that a large (~2.5 l) specimen from the Upper Cretaceous Frenchman Formation of Saskatchewan was probably produced by *Tyrannosaurus*. Numerous comminuted dinosaurian bone fragments are present in the ground mass of this specimen (Chin et al. 1998).

Another tyrannosaurid coprolite contains fossilized soft tissue, as well as bone. This very large (~6 l) specimen from the Dinosaur Park Formation in Alberta (Fig. 28.5A) has undergone relatively little diagenetic alteration. Amazingly, three-dimensional impressions of muscle tissues from the ingested victim are evident inside the fecal mass (Figs. 28.5A, B). The

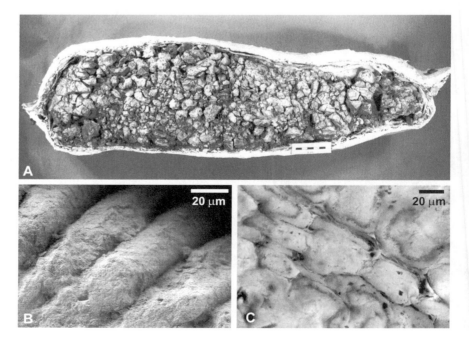

28.5. Large theropod coprolite (TMP 98.102.7) from the Dinosaur Park Formation of Alberta, Canada. (A) Specimen is estimated to be approximately 6 l in volume, and is attributed to a Campanian tyrannosaurid. Scale is 5 cm long. (B) Scanning electron micrograph of fossilized muscle cells preserved as three-dimensional casts in this coprolite. (C) Thin section showing a cross section through muscle cells in the same coprolite. The dark borders around the cells are the diagenetically altered remains of connective tissues.

Images courtesy SEPM (Society for Sedimentary Geology; Chin et al. 2003).

extraordinary preservation of "fossil meat" within this specimen indicates a relatively short gut residence time coupled with rapid lithification of the fecal mass after it was deposited (Chin et al. 2003).

* * *

Despite the transitory nature of diet, direct evidence of dinosaur feeding activity has been gleaned from a surprising variety of fossil sources. Exceptional trackways, taphonomic assemblages, tooth marks, gut contents, and coprolites provide different perspectives on the feeding activities of nonavian dinosaurs. Less typical approaches such as analyses of tooth microwear (e.g., Fiorillo 1998; Williams et al. 2009) and the isotopic composition of skeletal elements (e.g., Fricke and Pearson 2008) offer additional clues. Much of this fossil evidence is subject to different interpretations, but multiple lines of evidence can help resolve some of the ambiguities.

Information on dinosaur diet can be very general, because the taxonomic identities of the eater or the eaten are often uncertain. Yet occasionally the fossil record reveals specific foods that one particular dinosaur consumed at one particular time. Such cases provide snapshots of trophic interactions in the Mesozoic. Some observations help confirm previous speculations about dinosaur herbivory or predator/prey interactions, while others reveal unanticipated feeding behavior such as cannibalism or feeding on decaying wood. Individual examples may not be representative of feeding patterns, but combining direct evidence of dinosaur feeding activities with anatomical analyses (e.g., Barrett and Rayfield 2006), physiological and paleoenvironmental constraints (e.g., Farlow et al. 2010), and a knowledge of contemporaneous organisms (e.g., Tiffney, Chapter 27 of this volume) will improve our understanding of dinosaur food habits and shed light on the roles of dinosaurs in Mesozoic ecosystems.

References

Abel, O. 1922. Diskussion zu den Vorträgen R. Kräusel und F. Versluys. *Paläontologische Zeitschrift* 4: 87.

Baghai-Riding, N. L., and J. N. DiBenedetto. 2001. An unusual dinosaur coprolite from the Campanian Aguha Formation, Texas. *Gulf Coast Association of Geological Societies Transactions* 51: 10–20.

Barrett, P. M., and E. J. Rayfield. 2006. Ecological and evolutionary implications of dinosaur feeding behavior. *TRENDS in Ecology and Evolution* 21: 217–224.

Bell, P. R., and P. J. Currie. 2010. A tyrannosaur jaw bitten by a confamilial: scavenging or fatal agonism? *Lethaia* 43: 278–281.

Bishop, G. A. 1975. Traces of predation. In R.W. Frey (ed.), *The Study of Trace Fossils*, 261–281. New York: Springer-Verlag.

Bradley, W. H. 1946. Coprolites from the Bridger Formation of Wyoming: their composition and microorganisms. *American Journal of Science* 244: 215–239.

Buffetaut, E., D. Martill, and F. Escuillié. 2004. Pterosaurs as part of a spinosaur diet. *Nature* 430: 33.

Buffetaut, E., and V. Suteethorn. 1989. A sauropod skeleton associated with theropod teeth in the Upper Jurassic of Thailand: remarks on the taphonomic and palaeoecological significance of such associations. *Palaeogeography, Palaeoclimatology, Palaeoecology* 73: 77–83.

Carpenter, K. 1998. Evidence of predatory behavior by carnivorous dinosaurs. *Gaia* 15: 135–144.

Charig, A. J., and A. C. Milner. 1997. *Baryonyx walkeri*, a fish-eating dinosaur from the Wealden of Surrey. *Bulletin of the Natural History Museum of London (Geology)* 53: 11–70.

Chen, P.-J., Z.-M Dong, and S.-N. Zhen. 1998. An exceptionally well-preserved theropod dinosaur from the Yixian Formation of China. *Nature* 391: 147–152.

Chin, K. 2007. The paleobiological implications of herbivorous dinosaur coprolites from the Upper Cretaceous Two Medicine Formation of Montana: why eat wood? *Palaios* 22: 554–566.

Chin, K., and J. Bishop. 2007. Exploited twice: bored bone in a theropod coprolite from the Jurassic Morrison Formation of Utah, USA. In R. G. Bromley, L. A. Buatois, M. G. Mángano, J. F. Genise, and R. N. Melchor, (eds.), *Sediment-Organism Interactions: A Multifaceted Ichnology*. SEPM Special Publications 88: 377–385.

Chin, K., D. A. Eberth, M. H. Schweitzer, T. A. Rando, W. J. Sloboda, and J. R. Horner. 2003. Remarkable preservation of undigested muscle tissue within a Late Cretaceous tyrannosaurid coprolite from Alberta, Canada. *Palaios* 18: 286–294.

Chin, K., and B. D. Gill. 1996. Dinosaurs, dung beetles, and conifers: participants in a Cretaceous food web. *Palaios* 11: 280–285.

Chin, K., and J. I. Kirkland. 1998. Probable herbivore coprolites from the Upper Jurassic Mygatt-Moore Quarry, Western Colorado. *Modern Geology* 23: 249–275.

Chin, K., T. T. Tokaryk, G. M. Erickson, and L. C. Calk. 1998. A king-sized theropod coprolite. *Nature* 393: 680–682.

Chure, D. J., A. R. Fiorillo, and A. Jacobsen. 1998. Prey bone utilization by predatory dinosaurs in the Late Jurassic of North America with comments on prey bone use by dinosaurs throughout the Mesozoic. *Gaia* 15: 227–232.

Colbert, E. H. 1989. The Triassic dinosaur *Coelophysis*. *Bulletin of the Museum of Northern Arizona* 57: 1–160.

Currie, P. J., and P.-J. Chen. 2001. Anatomy of *Sinosauropteryx prima* from Liaoning, northeastern China. *Canadian Journal of Earth Sciences* 38: 1705–1727.

Currie, P. J., and A. R. Jacobsen. 1995. An azhdarchid pterosaur eaten by a velociraptorine theropod. *Canadian Journal of Earth Sciences* 32: 922–925.

Currie, P. J., E. B. Koppelhus, and A. F. Muhammad. 1995. Stomach contents of a hadrosaur from the Dinosaur Park Formation (Campanian, Upper Cretaceous) of Alberta, Canada. In A. Sun and Y. Wang (eds.), *Sixth Symposium on Mesozoic Terrestrial Ecosystems and Biota, Short Papers*, 111–114. Beijing: China Ocean Press.

Erickson, G. M., and K. H. Olson. 1996. Bite marks attributable to *Tyrannosaurus rex*: preliminary description and implications. *Journal of Vertebrate Paleontology* 16: 175–178.

Farlow, J. O., I. D. Coroian, and J. R. Foster. 2010. Giants on the landscape: modeling the abundance of megaherbivorous dinosaurs of the Morrison Formation (Late Jurassic, western USA). *Historical Biology* 22 (4): 403–429.

Farlow, J. O. 1987. *Lower Cretaceous Dinosaur Tracks, Paluxy River Valley, Texas*. Waco, Tex.: South Central Geological Society of America, Baylor University.

Fiorillo, A. R. 1991. Prey bone utilization by predatory dinosaurs. *Palaeogeography, Palaeoclimatology, Palaeoecology* 88: 157–166.

Fiorillo, A. R. 1998. Dental microwear patterns of the sauropod dinosaurs *Camarasaurus* and *Diplodocus*: evidence for resource partitioning in the Late Jurassic of North America. *Historical Biology* 13: 1–16.

Fowler, D. W., and R. M. Sullivan. 2006. A ceratopsid pelvis with toothmarks from the Upper Cretaceous Kirtland Formation, New Mexico: evidence of Late Campanian tyrannosaurid feeding behavior. In S. G. Lucas and R. M. Sullivan (eds.), *Late Cretaceous Vertebrates from the Western Interior. New Mexico Museum of Natural History and Science Bulletin* 35: 127–130.

Fricke, H. C., and Pearson, D. A. 2008. Stable isotope evidence for changes in dietary niche partitioning among hadrosaurian and ceratopsian dinosaurs of the Hell Creek Formation, North Dakota. *Paleobiology* 34: 534–552.

Gay, R. J. 2002. The myth of cannibalism in *Coelophysis bauri*. *Journal of Vertebrate Paleontology Abstracts of Papers* 22 (suppl. to 3): 57A.

Gignac, P. M., P. J. Makovicky, G. M. Erickson, and R. P. Walsh. 2010. A description of *Deinonychus antirrhopus* bite marks and estimates of bite force using tooth indentation simulations. *Journal of Vertebrate Paleontology* 30: 1169–1177.

Ghosh, P., S. K. Bhattacharya, A. Sahni, R. K. Kar, D. M. Mohabey, and K. Ambwani. 2003. Dinosaur coprolites from the Late Cretaceous (Maastrichtian) Lameta Formation of India: isotopic and other markers suggesting a C$_3$ diet. *Cretaceous Research* 24: 743–750.

Jennings, D. S., and S. T. Hasiotis. 2006. Taphonomic analysis of a dinosaur feeding site using geographic information systems (GIS), Morrison Formation, Southern Bighorn Basin, Wyoming, USA. *Palaios* 21: 480–492.

Hill, C. R. 1976. Coprolites of *Ptiliophyllum* cuticles from the Middle Jurassic of North Yorkshire. *Bulletin of the British Museum of Natural History: Geology* 27: 289–294.

Hone, D. W. E., and M. Watabe. 2010. New information on scavenging and selective feeding behaviour of tyrannosaurids. *Acta Palaeontologica Polonica* 55: 627–634.

Hunt, A. P. 1987. Phanerozoic trends in nonmarine taphonomy: implications for Mesozoic vertebrate taphonomy and paleoecology. *South Central Section, Geological Society of America Abstracts with Programs* 19: 171.

Hunt, A. P., K. Chin, and M. G. Lockley. 1994a. The palaeobiology of vertebrate coprolites. In S. K. Donovan (ed.), *The Palaeobiology of Trace Fossils*, 221–240. New York: Wiley.

Hunt, A. P., C. S. Meyer, M. G. Lockley, and S. G. Lucas. 1994b. Archaeology, toothmarks and sauropod dinosaur taphonomy. *Gaia* 10: 225–231.

Jacobsen, A. R. 1998. Feeding behaviour of carnivorous dinosaurs as determined by tooth marks on dinosaur bones. *Historical Biology* 13: 17–26.

———. 2001. Tooth-marked small theropod bone: an extremely rare trace. In D. H. Tanke and K. Carpenter (eds.), *Mesozoic Vertebrate Life*, 58–63. Bloomington: Indiana University Press.

Kielan-Jaworowska, Z., and R. Barsbold. 1972. Narrative of the Polish-Mongolian palaeontological expeditions 1967–1971. *Palaeontologia Polonica* 27: 5–13.

Kobayashi, Y., J.-C. Lu, Z.-M. Dong, R. Barsbold, Y. Azuma, and Y. Tomida. 1999. Herbivorous diet in an ornithomimid dinosaur. *Nature* 402: 480–481.

Kräusel, R. 1922. Die Nahrung von *Trachodon*. *Palaeontologische Zeitschrift* 4: 80.

Leonardi, G. 1984. Le impronte fossili di Dinosauri. In J. F. Bonaparte, E. H. Colbert, P. J. Currie, A. de Ricqlès, Z. Kielan-Jaworowska, G. Leonardi, N. Morello, and P. Taquet (eds.), *Sulle Orme dei Dinosauri*, 165–186. Venice: Erizzo Editrice.

Lockley, M. G. 1991. *Tracking Dinosaurs*. Cambridge: Cambridge University Press.

Longrich, N. R., J. R. Horner, G. M. Erickson, and P. J. Currie. 2010. Cannibalism in *Tyrannosaurus rex*. *PLoS One* 5: 1–6.

Matthew, W. D. 1908. *Allosaurus*, a

carnivorous dinosaur, and its prey. *American Museum Journal* 8: 2–5.

Maxwell, W. D., and J. H. Ostrom. 1995. Taphonomic and paleobiological implications of *Tenontosaurus-Deinonychus* associations. *Journal of Vertebrate Paleontology* 15: 707–712.

Mohabey, D. M. 2005. Later Cretaceous (Maastrichtian) nests, eggs, and dung mass (coprolites) of sauropods (titanosaurs) from India. In V. Tidwell and K. Carpenter (eds.), *Thunder-Lizards the Sauropodomorph Dinosaurs*, 466–489. Bloomington: Indiana University Press.

Molnar, R. E., and H. T. Clifford. 2000. Gut contents of a small ankylosaur. *Journal of Vertebrate Paleontology* 20: 188–190.

———. 2001. An ankylosaurian cololite from the Lower Cretaceous of Queensland, Australia. In K. Carpenter (ed.), *The Armored Dinosaurs*, 399–412. Bloomington: Indiana University Press.

Nesbitt, S. J., A. H. Turner, G. M. Erickson, and M. A. Norell. 2006. Prey choice and cannibalistic behaviour in the theropod *Coelophysis*. *Biology Letters* 2: 611–614.

Nopcsa, Baron F. 1903. Neues über *Compsognathus*. *Neues Jahrbuch fur Mineralogie, Geologie und Palaeontologie (Stuttgart)* 16: 476–494.

Osmólska, H. 1993. Were the Mongolian "fighting dinosaurs" really fighting? *Revue de Paléobiologie* spéc. 7: 161–162.

Ostrom, J. H. 1978. The osteology of *Compsognathus longipes*. Wagner. *Zitteliana* 4: 73–118.

———. 1990. Dromaeosauridae. In D. B. Weishampel, P. Dodson, and H. Osmólska (eds.), *The Dinosauria*, 269–279. Berkeley: University of California Press.

Parker, L. R., and R. L. Rowley Jr. 1989. Dinosaur footprints from a coal mine in East-Central Utah. In D. D. Gillette and M. G. Lockley (eds.), *Dinosaur Tracks and Traces*, 361–366. Cambridge: Cambridge University Press.

Peyer, K. 2006. A reconsideration of *Compsognathus* from the Upper Tithonian of Canjuers, southeastern France. *Journal of Vertebrate Paleontology* 24: 879–896.

Prasad, V., A. E. Strömberg, H. Alimohammadian, and A. Sahni. 2005. Dinosaur coprolites and the early evolution of grasses and grazers. *Science* 310: 1177–1180.

Roach, B. T., and D. L. Brinkman. 2007. A reevaluation of cooperative pack hunting and gregariousness in *Deinonychus antirrhopus* and other nonavian theropod dinosaurs. *Bulletin of the Peabody Museum of Natural History* 48: 103–138.

Rogers, R. R., D. W. Krause, and K. C. Rogers. 2003. Cannibalism in the Madagascan dinosaur *Majungatholus atopus*. *Nature* 422: 515–518.

Stokes, W. L. 1964. Fossilized stomach contents of a sauropod dinosaur. *Nature* 143: 576–577.

Stone, D. D., E. L. Crisp, and J. R. Bishop. 2000. A large meat-eating dinosaur coprolite from the Jurassic Morrison Formation of Utah. *Geological Society of America Abstracts with Programs* 32: 220.

Tanke, D. H., and P. J. Currie. 1998. Head-biting behavior in theropod dinosaurs: paleopathological evidence. *Gaia* 15: 167–184.

Tiffney, B. H. 2011 (this volume). Land Plants as a Source of Food and Environment in the Age of Dinosaurs. In M. K. Brett-Surman, Thomas R. Holtz Jr., and James O. Farlow (eds.), *The Complete Dinosaur*, 2nd ed., 568–587. Bloomington: Indiana University Press.

Thulborn, R. A. 1991. Morphology, preservation and palaeobiological significance of dinosaur coprolites. *Palaeogeography, Palaeoclimatology, Palaeoecology* 83: 341–366.

Thulborn, R. A., and M. Wade. 1979. Dinosaur stampede in the Cretaceous of Queensland. *Lethaia* 12: 275–279.

Tweet, J. S., K. Chin, D. R. Braman, and N. L. Murphy. 2008. Probable gut contents within a specimen of *Brachylophosaurus canadensis* (Dinosauria: Hadrosauridae) from the Upper Cretaceous Judith River Formation of Montana. *Palaios* 23: 624–635.

Unwin, D. M., A. Perle, and C. Trueman. 1995. *Protoceratops* and *Velociraptor* preserved in association: evidence for predatory behaviour in dromaeosaurid dinosaurs? *Journal of Vertebrate Paleontology* 15 (suppl. to 3): 57A–58A.

Varricchio, D. J. 2001. Gut contents from a Cretaceous tyrannosaurid: implications for theropod digestive tracts. *Journal of Paleontology* 75: 401–406.

Weigelt, J. 1989. *Recent Vertebrate Carcasses and their Paleobiological Implications*. Chicago: University of Chicago Press. [Translation by J. Schaefer, of *Rezente Wirbeltierleichen und ihre paläobiologische Bedeutung*. Leipzig: Verlag von Max Weg, 1927.]

Williams, V. S., P. M. Barrett, and M. A. Purnell. 2009. Quantitative analysis of dental microwear in hadrosaurid dinosaurs, and the implications for hypotheses of jaw mechanics and feeding. *Proceedings of the National Academy of Sciences* 106: 11194–11199.

29.1. Urogenital system of gnathostome vertebrates. Left, generalized embryo (undifferentiated). Center, diagrams of generalized definitive
gnathostome male and female urogenital systems.
Right, generalized nonmammalian amniote
(sauropsid) urogenital systems. The archinephric
duct becomes the ductus deferens and the
paramesonephric duct becomes the oviduct.
Abbreviations: A, archinephric (Wolffian, mesonephric) duct; C, cloaca; G, gonad; K, kidney; P,
paramesonephric (Müllerian) duct. Dashed lines and/
or lighter colors represent regressed structures.

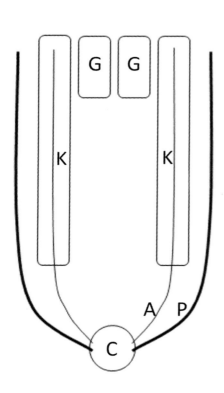

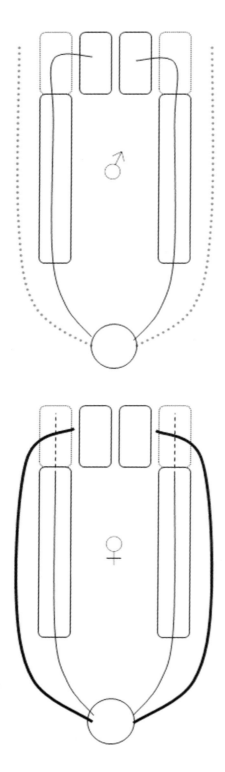

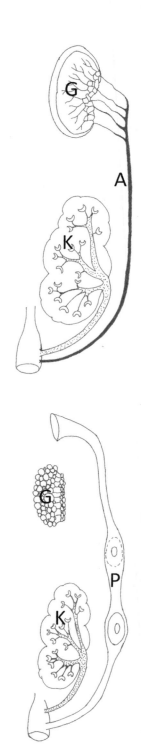

Reproductive Biology of Dinosaurs

29

Terry D. Jones and Nicholas R. Geist

Discoveries of the nests and eggs of a variety of dinosaur taxa, combined with recently described fossils of baby dinosaurs, have captured the imagination of the public and stimulated professional interest in the reproductive biology of dinosaurs (a.k.a. nonavian dinosaurs). These fossils offer tantalizing glimpses into the biology of dinosaurs and bring them to life in a way that individual bones can't. As is often the case with dinosaur biology, this area is contentious. Filling the gaps that remain in our understanding of dinosaur reproduction and, at least partially, reconciling the differences in opinion that emerge requires supplementing the intrinsic limitations of the fossil record by applying knowledge of the biology of the living relatives of dinosaurs. Some researchers, pointing to the close relationship between birds and dinosaurs, have interpreted the behavior of many dinosaurs to have been much like that of living birds. Conversely, others discern more conservative crocodilian or "reptilian" patterns of behavior from the same fossils.

Phylogenetic modeling based on the biology of extant archosaurs (crocodilians and birds) notwithstanding, there is no single key to unlock the details of dinosaur reproduction, and in many cases this methodology falls short of a definitive explanation. In a number of instances the fossils seem to indicate that the reproductive patterns of dinosaurs may have been neither particularly crocodilian nor birdlike, but uniquely dinosaurian. This realization should not be too surprising considering the long evolutionary history and great diversity of the dinosaurs as well as the unique selective pressures of the Mesozoic environment. Here we utilize both neontological and paleontological data to reconstruct the reproductive anatomy, physiology, development, and behavior of dinosaurs.

Anatomy and Development of Reproductive Tracts

Development of the reproductive tract in extant archosaurs (crocodilians and birds) follows the generalized pattern of gnathostomes (jawed vertebrates), in which two complete sets of ducts associated with the primordial gonads form within the dorsal body wall at an early, undifferentiated developmental stage: the archinephric (Wolffian, or mesonephric) and the paramesonephric (Müllerian) ducts (Austin 1989; Feguson 1985; Kardong 2006; Lofts and Murton 1973) (Fig. 29.1). The sperm ducts (ductus deferens) of male vertebrates develop from the archinephric ducts; the paramesonephric ducts degenerate. At sexual maturity, gametes (sperm) are conveyed via the ductus deferens from the testes into the cloaca, a common chamber

Reproductive Anatomy and Physiology of the Dinosaurs

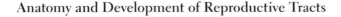

that also receives feces and urine (Austin 1989; Feguson 1985; Lofts and Murton 1973.

Conversely, in females the archinephric ducts are reduced and the paired oviducts develop from the paramesonephric ducts (Austin 1989; Feguson 1985; Kardong 2006; Lofts and Murton 1973). Although development of the reproductive tract is fairly conservative in vertebrates, there are differences in the adult female reproductive tracts of crocodilians and birds. Female crocodilians, like most gnathostomes, possess paired ovaries and oviducts with eggs passing into the cloaca (Ferguson 1985). However, in the birds the right side of the female reproductive tract tends to be vestigial, leaving a single functional ovary and oviduct (Clinton and Haines 1999; Lofts and Murton 1973). This asymmetrical reproductive system represents a derived condition, results in reduced body mass, and is probably an adaptation for flight and/or production of few, relatively large eggs (Witschi 1935).

The development of the male dinosaurian reproductive system, like that of extant archosaurs, likely exhibited the basic gnathostome pattern of paired gonads and sperm ducts. Regardless of the differences in the reproductive systems of adult female crocodilians and birds, fossil evidence makes determining the anatomy of female dinosaurs fairly straightforward. The apparent paired pattern of egg deposition in the theropod dinosaurs *Troodon* and *Oviraptor* implies the presence of paired female reproductive tracts (Clark et al. 1999; Dong and Currie 1998; Norell et al. 1995; Varricchio et al. 1997). Additionally, a recently described fossil of an oviraptorosaur with two eggs retained within the pelvic region provides direct evidence that confirms the primitive pattern of paired oviducts (Sato et al. 2005). One specimen of *Sinosauropteryx* was also described as having had two eggs fossilized within the oviduct (Chen et al. 1998). However, given the ventral location of these structures, their identification as eggs is questionable. Regardless, given the embryological similarity of the reproductive tract and the fossil evidence, it is reasonable to infer the presence of paired ovaries and oviducts for all dinosaurs.

Reproductive Functional Morphology

Numerous aspects of the physiology and many behaviors associated with reproductive function (e.g., mode of egg/shell production, reproductive mode, egg deposition pattern, sex determination, etc.) vary in living amniotes, often differing dramatically even within closely related species. Although the inherent diversity of dinosaurs makes deciphering reproductive physiology challenging, we can apply our understanding of living amniotes to arrive at some general conclusions regarding certain features of dinosaurian reproductive function.

Non-archosaurian egg-laying (oviparous) amniotes possess a "multipurpose," undifferentiated uterus, in which the eggshell membranes and outer shell are deposited sequentially (Palmer and Guillette 1992). In these taxa, eggs are relatively stationary until fully formed (Palmer 1990; Palmer and Guillette 1992). These oviparous species usually produce an egg with a highly organic, leathery shell primarily composed of fibrous proteins (Packard and DeMarco 1992). Such eggs are generally poorly allotted with water and rely on osmotic transfer through the relatively porous shell, requiring that a clutch of eggs must be laid in a carefully selected, slightly moist setting for proper development (Packard and Packard 1988).

Extant archosaurs are unique among oviparous amniotes in that they possess a regionally differentiated "assembly-line" uterus in which distinct regions produce eggshell membranes and a relatively impermeable and durable shell characterized by a unique calcareous structure (Bakst 1998; Girling 2002; Palmer and Guillette 1992). In contrast to non-archosaurian eggs, the eggs of archosaurs are well provisioned with water in the albumen (white) through the process of "plumping" (Board and Sparks 1991; Girling 2002; Packard and Packard 1988; Palmer and Guillette 1992). The archosaurian mode of egg production requires constant movement through the reproductive tract. Notably, there is a difference between crocodilians and birds in the timing of egg production. Crocodilians rapidly produce a large clutch (20–100+) of relatively small eggs that they deposit en masse (Ferguson 1985), whereas birds produce a smaller clutch of larger eggs sequentially on a daily basis (Johnson 2000).

The morphological and functional differences in these strategies have significant implications on reproductive mode. Viviparity (live birth) has evolved hundreds of times in nearly all gnathostome taxa, with the apparent exception of archosaurs (Blackburn 1998; Shine 1988, 1991). The non-archosaurian pattern of egg production (i.e., stationary) allows for extensive intrauterine egg retention, the requisite first stage in the evolution of viviparity (Girling 2002; Shine 1988; Thompson et al. 2004; Tinkle and Gibbons 1977). In contrast, the archosaurian assembly-line uterus precludes egg retention (Palmer and Guillette 1992), thus probably canalizing archosaurs to obligate oviparity. In fact, if archosaurian eggs stall within the reproductive tract, development is arrested, and nonviable eggs may result (Packard et al. 1977). While the archosaurian mode of oviparity enables the rapid continuous production of many desiccation-resistant, durable eggs, it apparently has precluded the evolution of viviparity.

Since it is unlikely that this unique assembly-line pattern of egg production evolved independently in birds and crocodilians, it is reasonable to infer that this strategy originated in their common ancestor and was present in dinosaurs as well. This phylogenetic evidence, along with the well-documented record of fossil dinosaur eggs with shell structure similar to living archosaurs (Carpenter 1999), makes it nearly certain that dinosaurs also employed obligate oviparity. Interestingly, within dinosaurs there appears to have been a range of egg-laying behaviors, from the crocodile-like mass laying strategy (e.g., sauropods, hadrosaurs) to an avian-style sequential pattern (e.g., *Troodon*, other theropods) (Carpenter 1999; Dong and Currie 1998; Varricchio et al. 1997).

External Genitalia and Copulation

External genitalia of female crocodilians and birds, as is common among vertebrates, are rather unembellished, consisting primary of the vent—the opening into the cloaca (Kardong 2006; Lofts and Murton 1973). Unlike the stereotypical morphology of the male reproductive tract of vertebrates and genitalia of females, male external genitalia have evolved independently

Reproductive Behavior
of Dinosaurs

multiple times and exhibit diverse morphology (Kardong 2006; Kelly 2002; King 1981), making their reconstruction in fossil taxa more speculative. In amniotes, fertilization is internal, and sperm is usually introduced into the female reproductive tract via an erectile intromittent organ that is everted from the vent. The presence of an intromittent organ in crocodilians and basal extant birds (e.g., tinamous, curassows, ratites, and anseriforms) may be a primitive trait; its absence in most birds appears to be a derived character reflecting a secondary loss in these taxa (Kelly 2002; King 1981). Though we are unlikely to ever encounter a fossilized dinosaur phallus, the primitive archosaurian (i.e., crocodilian) condition probably provides an appropriate model for male dinosaur genitalia.

Copulation in extant terrestrial amniote taxa in which males possess a phallus requires its close association with the cloacal opening of the female. In many extant taxa this occurs with the male mounting the female from behind. However, given the size, armament, accoutrements, and the like of many dinosaurs, such a feat may have required great finesse and compromise, if not flexibility. It is likely that most dinosaurs utilized the tail to bring the cloacae into close association, thereby allowing intromission.

Nests and Egg Deposition

Eggs of extant archosaurs are deposited in a nest that serves as a site for incubation (maintenance of a relatively constant temperature) in a protected environment. In crocodilians, incubation is generally from the nest environment (e.g., fermentation of nest materials and/or solar heat). Nests of extant crocodilians are constructed by females and are composed of holes or mounds of vegetation and/or sand into which a large clutch of eggs is deposited (Booth and Thompson 1991; Ferguson 1985). There is no pattern to egg placement, and during oviposition sand may enter between or atop the eggs (Coombs 1989; Ferguson 1985). Following oviposition, the nest is covered. Female crocodilians typically tend and/or guard their nests, but the amount of time spent attending the nest is highly variable (Carpenter 1999; Coombs 1989; Cott 1971; Deeming 1991; Drent 1972; Ferguson 1985; Shine 1988; Vleck and Hoyt 1991). Significantly, once laid, crocodilian eggs are not manipulated or turned, as postovipositional egg movement results in reduced hatching rates (Coombs 1989; Deeming 1991).

Birds are the only extant archosaurs that employ direct application of body heat (contact incubation) to their eggs. Eggs of birds are typically laid in the open, also with no apparent pattern, and are tended to by the parents. To maintain relatively constant egg temperature, parents utilize direct contact with the eggs and must rearrange them within the nest (Deeming 1991, 2002). As the eggs are moved, they may also be turned. Unlike in crocodilians, failure to turn most avian eggs decreases their viability, and jostling increases hatching rates (Deeming 1991, 2002).

All known dinosaur nests were located on the ground, and the eggs often appear to have been randomly deposited in simple pits (Carpenter 1999). In other cases the eggs were arranged in linear rows or circular patterns and/or deposited in multiple tiers in relatively shallow nests (Carpenter 1999; Clark et al. 1999; Dong and Currie 1998; Norell et al. 1995; Varricchio et al. 1997). The regular arrangement and/or layering of eggs implies that

29.2. *Oviraptor* nesting (modified from Deeming 2002).

they were not moved after being deposited (contra Norell et al. 1995; contra Varricchio et al. 1997). Furthermore, the relatively high porosity/gas conductance values from the shells of dinosaur eggs (Deeming 2002; Rahn 1991; Sabath 1991) and presence of vegetation in association with dinosaur eggs (Horner 1982; Horner and Makela 1979) are consistent with eggs in covered nests of extant archosaurs. Significantly, covering the nests likely precluded postdepositional movement.

Partial or complete burial and the apparent lack of postlaying manipulation of dinosaur eggs implies that their eggs were incubated by environmental sources rather than by direct parental contact. However, it has been suggested that some dinosaurs (e.g., *Oviraptor* and *Troodon*) may have used contact incubation (Clark et al. 1999; Dong and Currie 1998; Norell 1995). The primary evidence supporting this is the close association of adults with the nest. For example, several specimens of *Oviraptor* were discovered in association with eggs in nests. The eggs of complete *Oviraptor* nests are spirally arranged in two or more layers, and the adult is positioned with its trunk in the center of the circle rather than directly atop and in contact with the eggs (Figure 29.2) (Clark et al. 1999; Dong and Currie 1998; Norell 1995). Therefore, rather than being evidence of contact incubation, the association of the adult and nest is indicative of nest attendance (Deeming 2002; Ruben et al. 2003). Nest attendance by many or all dinosaurs is not particularly surprising in view of similar behaviors in living archosaurs. Furthermore, high eggshell-porosity values in these *Oviraptor* eggs are similar to those of crocodilian eggs (2–5 times avian eggshell porosity values) and are indicative of buried eggs (Deeming 2002), making contact incubation extremely unlikely.

Hatchling Maturity and Parental Care

With few exceptions, extant archosaurs provide a high degree of parental care to their offspring prior to and following hatching that includes, but is not limited to, nest protection and opening, egg opening, and hatchling transport, care, and protection (Shine 1988; Starck 1989; Starck and Ricklefs 1998). However, parental care is not linked with degree of hatchling development (Starck and Ricklefs 1998). Hatchlings of crocodilians are all precocial (i.e., mobile and able to leave the nest at or near the time of hatching). Bird hatchlings exhibit a range of developmental maturity from precociality in more basal taxa (e.g., ratites, tinamous, galliforms, anseriforms, etc.) to altriciality (nest-bound and dependant on parental care) in many derived taxa (Starck 1989, 1996; Starck and Ricklefs 1998). Given that crocodilians and basal birds have precocial young, altriciality exhibited by songbirds and the like is probably the derived condition (Starck and Ricklefs 1998).

The phylogentic inference is that dinosaur young, like those of crocodilians and basal birds, would also have been precocial (Feduccia et al. 1996; Geist and Jones 1996a,b; Norell and Clark 1996). This conclusion is supported by fossil evidence of a highly precocial Early Cretaceous enantiornithine bird embryo from China that possessed a well-ossified skeleton and "well-grown feather sheets" (Zhou and Zhang 2004).

However, on the basis of bone histology it has been suggested that the young of some dinosaurs may have been altricial because their long bones were thought to have been poorly ossified (Horner 2000; Horner et al. 2000; Horner et al. 2001; Horner and Weishampel 1988). This view is inconsistent with data from developmental patterns of extant juvenile archosaurs and is not supported by paleontological data (Geist and Jones 1996a,b; Ruben et al. 2003; Varricchio et al. 1997). In fact, it has been demonstrated that it is not possible to determine the extent of maturity of hatchlings according to the degree or pattern of ossification of their skeletons (Starck 1996). Therefore, the most appropriate model for dinosaurian juvenile behavior is the ancestral condition exhibited by crocodilians and/or basal extant birds. In addition, since nest attendance and parental care are the rule for archosaurs, it is likely that these were part of dinosaurian behaviors as well. This conclusion is supported by the association of hatchling and adult *Protoceratops* (discussed in Coombs 1989).

Conclusions

Given the variety of reproductive behaviors characteristic of extant archosaurs and the diversity of dinosaurs, it is difficult to model dinosaurian reproductive biology exclusively on either crocodilians or birds. Instead, it appears that dinosaurs may have exhibited features of both of these taxa as well as some that are unique. However, there are some generalities that can be drawn from paleontological and neontological data.

Dinosaurs, like other amniotes, employed internal fertilization, and their eggs developed within paired reproductive tracts. Unlike other egg-laying amniotes, archosaurs—including dinosaurs—evolved a unique "assembly-line" strategy of egg development that enabled them to rapidly produce clutches of water-rich, desiccation-resistant eggs. A functional consequence of the evolution of this strategy seems to have been that archosaurs were canalized into obligate oviparity. Significantly, various dinosaur groups appear to have used either an avian- or a crocodilian-type egg-laying strategy—sequential egg laying or en masse clutch production, respectively.

Eggshell microstructure and nest structure indicate that dinosaur nests were most likely covered by sand or vegetation. As is typical of extant archosaurs, at least some dinosaurs exhibited extensive parental care, coupled with nest attendance. This high degree of parental behavior is not indicative of developmental immaturity of hatchlings. To the contrary, fossil and phylogenetic evidence strongly supports a precocial, rather than altricial, pattern of hatchling development.

We thank J. W. Gibbons and an anonymous reviewer for their comments and suggestions and Samantha Winegarner for the *Oviraptor* illustration.

Acknowledgments

References

Austin, H. B. 1989. Müllerian-duct regression in the American alligator (*Alligator mississippiensis*): its morphology and testicular induction. *Journal of Experimental Zoology* 251: 329–338.

Bakst, M. R. 1998. Structure of the avian oviduct with emphasis on sperm storage in poultry. *Journal of Experimental Zoology* 282: 618–626.

Blackburn, D. G. 1998. Structure, function, and evolution of the oviducts of squamate reptiles, with special reference to viviparity and placentation. *Journal of Experimental Zoology* 282: 560–617.

Board, R. G., and N. H. C. Sparks. 1991. Shell structure and formation in avian eggs. In D. C. Deeming et al. (eds.), *Egg Incubation: Its Effects on Embryonic Development in Birds and Reptiles*, 71–86. Cambridge: Cambridge University Press.

Booth, D. T., and M. B. Thompson. 1991. A comparison of reptilian eggs with those of megapode birds. In D. C. Deeming et al. (eds.), *Egg Incubation: Its Effects on Embryonic Development in Birds and Reptiles*, 325–344. Cambridge: Cambridge University Press.

Carpenter, K. 1999. *Eggs, Nests, and Baby Dinosaurs*. Bloomington: Indiana University Press.

Chen, P., Z. Dong, and S. Zhen. 1998. An exceptionally well-preserved theropod dinosaur from the Yixian Formation of China. *Nature* 391: 147–152.

Clark, J. M., M. A. Norell, and L. M. Chiappe. 1999. An oviraptorid skeleton from the Late Cretaceous of Ukhaa Tolgod, Mongolia, preserved in an avian-like brooding position over an oviraptorid nest. *American Museum Novitates* 3265: 1–35.

Clinton, M., and L. C. Haines. 1999. An overview of factors influencing sex determination and gonadal development in birds. *Cellular and Molecular Life Sciences* 55: 876–886.

Coombs, W. P. 1989. Modern analogs for dinosaur nesting and parental behavior. In J. O. Farlow (ed.), *Paleobiology of the Dinosaurs*, 21–53. Boulder, Colo.: Books on Demand.

Cott, H. B. 1971. Parental care in the Crocodilia, with special reference to *Crocodylus niloticus*. *IUCN Publications* (n.s.) 32: 166–180.

Deeming, D. C. 1991. Reasons for the dichotomy in egg turning in birds and reptiles. In D. C. Deeming et al. (eds.), *Egg Incubation: Its Effects on Embryonic Development in Birds and Reptiles*, 307–323. Cambridge: Cambridge University Press.

——. 2002. Importance and evolution of incubation in avian reproduction. In D. C. Deeming (ed.), *Avian Incubation: Behaviour, Environment and Evolution*, 1–7. Oxford: Oxford University Press.

Dong, Z.-M., and P. J. Currie. 1998. On the discovery of an oviraptorid skeleton on a nest of eggs at Bayan Mandahu, Inner Mongolia, People's Republic of China. *Canadian Journal of Earth Science* 33: 631–636.

Drent, R. 1972. The natural history of incubation. In D. S. Farner (ed.), *Breeding Biology of Birds*, 262–311. Washington, D.C.: National Academy of Sciences.

Feduccia, A., L. D. Martin, and J. E. Simmons. 1996. Nesting dinosaur. *Science* 272: 1571.

Ferguson, M. W. J. 1985. Reproductive biology and embryology of the Crocodilians. In C. Gans et al. (eds.), *Biology of the Reptilia*, 14:331–491. New York: John Wiley and Sons.

Geist, N. R., and T. D. Jones. 1996a. Dinosaurs and their youth: response to Norell and Clark. *Science* 272: 1571.

——. 1996b. Juvenile skeletal structure and the reproductive habits of dinosaurs. *Science* 272: 712–714.

Girling, J. E. 2002. The reptilian oviduct. *Journal of Experimental Zoology* 293: 141–170.

Horner, J. R. 1982. Evidence of colonial nesting and site fidelity among ornithischian dinosaurs. *Nature* 297: 675–676.

——. 2000. Dinosaur reproduction and parenting. *Annual Review of Earth Planetary Sciences* 28: 19–45.

Horner, J. R., A. J. de Ricqles, and K. Padian. 2000. Long bone histology of the hadrosaurid dinosaur *Maiasaura peeblesorum*: growth dynamics and physiology based on an ontogenetic series of skeletal elements. *Journal of Vertebrate Paleontology* 20: 115–129.

——. 2001. Comparative osteohistology of some embryonic and perinatal archosaurs: developmental and behavioral implications for dinosaurs. *Paleobiology* 27: 39–58.

Horner, J. R., and R. Makela. 1979. Nest of juveniles provides evidence of family structure among dinosaurs. *Nature* 282: 296–298.

Horner, J. R., and D. B. Weishampel. 1988. A comparative embryological study of two ornithischian dinosaurs. *Nature* 332: 256–257.

Johnson, A. L. 2000. Reproduction in the female. In G. C. Wittow (ed.), *Sturkie's Avian Physiology*, 5th ed., 569–591. New York: Academic Press.

Kardong, K. V. 2006. *Vertebrates: Comparative Anatomy, Function, Evolution.* New York: McGraw-Hill.

Kelly, D. A. 2002. The functional morphology of penile erection: tissue designs for increasing and maintaining stiffness. *Integrative and Comparative Biology* 42: 216–221.

King, A. S. 1981. Phallus. In A. S. King et al. (eds.), *Form and Function in Birds*, 2:107–147. London: Academic Press.

Lofts, B., and R. K. Murton. 1973.

Reproduction in birds. In D. S. Farner et al. (eds.), *Avian Biology*, 3:1–107. New York: Academic Press.

Norell, M. A. 1995. Origins of the feathered nest. *Natural History* 104: 58–61.

Norell, M. A. and J. M. Clark. 1996. Dinosaurs and their youth. *Science* 273: 165–168.

Norell, M. A., J. M. Clark, L. M. Chiappe, and D. Dashzeveg. 1995. A nesting dinosaur. *Nature* 378: 774–776.

Packard, G. C., and M. J. Packard. 1988. The physiological ecology of reptilian eggs and embryos. In C. Gans et al. (eds.), *Biology of the Reptilia*, 16:523–605. New York: Alan R. Liss.

Packard, G. C., C. R. Tracy, and J. J. Roth. 1977. The physiological ecology of reptilian eggs and embryos, and the evolution of viviparity within the class Reptilia. *Biology Review* 52: 71–105.

Packard, M. J., and V. G. DeMarco. 1992. Eggshell structure and formation in eggs of oviparous reptiles. In D. C. Deeming et al. (eds.), *Egg Incubation: Its Effects on Embryonic Development in Birds and Reptiles*, 53–69. Cambridge: Cambridge University Press.

Palmer, B. D. 1990. Functional morphology and biochemistry of reptilian oviducts and eggs: implications for the evolution of reproductive modes in tetrapod vertebrates. Ph.D. dissertation, University of Florida, Gainesville.

Palmer, B. D., and L. J. Guillette Jr. 1992. Alligators provide evidence for the evolution of an archosaurian mode of oviparity. *Biology of Reproduction* 46: 39–47.

Rahn, H. 1991. Why birds lay eggs. In D. C. Deeming et al. (eds.), *Egg Incubation: Its Effects on Embryonic Development in Birds and Reptiles*, 345–360. Cambridge: Cambridge University Press.

Ruben, J. A., T. D. Jones, and N. R. Geist. 2003. Respiratory and reproductive paleophysiology of dinosaurs and early birds. *Physiological and Biochemical Zoology* 76: 141–164.

Sabath, K. 1991. Upper Cretaceous amniotic eggs from the Gobi desert. *Acta Palaeontologica Polonica* 36: 151–192.

Sato, T., Y. Cheng, X. Wu, D. K. Zelenitsky, and Y. Hsiao. 2005. A pair of shelled eggs inside a female dinosaur. *Science* 308: 375.

Shine, R. 1988. Parental care in reptiles. In C. Gans et al. (eds.), *Biology of the Reptilia*, 16: 275–329. New York: Alan R. Liss.

——. 1991. Influences of incubation requirements on the evolution of viviparity. In D. C. Deeming et al. (eds.), *Egg Incubation: Its Effects on Embryonic Development in Birds and Reptiles*, 361–369. Cambridge: Cambridge University Press.

Starck, J. M. 1989. Zeitmuster der ontogenesen bei nestflüchtenden und nesthockenden vögeln. *Courier Forschungsinstut Senkenberg* 114: 1–319.

——. 1996. Comparative morphology and cytokinetics of skeletal growth in hatchlings of altricial and precocial birds. *Zoologischer Anzeiger* 235: 53–75.

Starck, J. M., and R. E. Ricklefs. 1998. Patterns of development: the altricial-precocial spectrum. In J. M. Starck et al. (eds.), *Avian Growth and Development*, 8: 3–30. New York: Oxford University Press.

Thompson, M. B., S. M. Adams, and J. F. Herbert. 2004. Placental function in lizards. *International Congress Series* 1275: 218–255.

Tinkle, D. W., and J. W. Gibbons. 1977. The distribution and evolution of viviparity in reptiles. *Miscellaneous Publications of the Museum of Zoology, University of Michigan* 154: 1–55.

Varricchio, D. J., F. Jackson, J. J. Borlowski, and J. R. Horner. 1997. Nest and egg clutches of the dinosaur *Troodon formosus* and the evolution of avian reproductive traits. *Nature* 385: 247–250.

Witschi, E. 1935. Origin of asymmetry in the reproductive system of birds. *American Journal of Anatomy* 56: 119–141.

30.1. Photographs of dinosaur skeletons associated with their eggs. (A) "Brooding" oviraptorid from Mongolia collected by the AMNH, plan view. Adult skeleton, lacking the skull and tail, is preserved on top of its eggs, visible along the right side, in a birdlike brooding position. (B) Two eggs preserved inside the pelvis of an oviraptorosaur from southeastern China, right lateral view. The eggs are laterally compressed, which indicates that they were preserved while still inside the female. Eggs associated with "brooding" oviraptorids (A) are dorsoventrally compressed. All scale bars = 10 cm.

Photo courtesy of X.-C. Wu,
Canadian Museum of Nature.

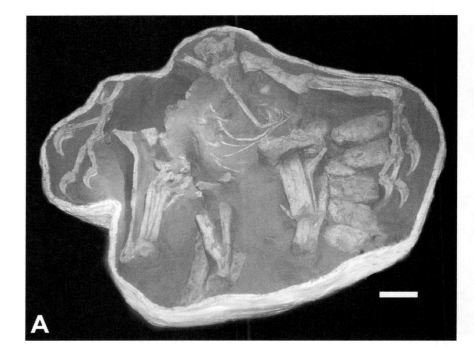

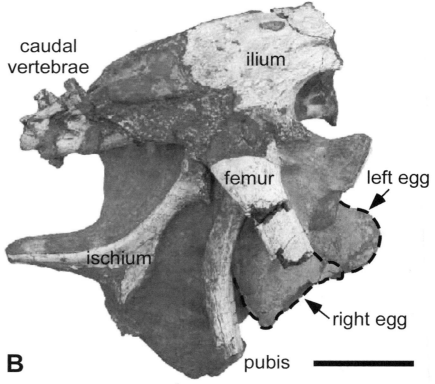

Dinosaur Eggs

Darla K. Zelenitsky, John R. Horner, and François Therrien

30

Early Discoveries

The earliest discoveries of dinosaur eggshells were made thousands of years ago during Paleolithic or Neolithic times (Carpenter et al. 1994). Evidence of these prehistoric discoveries comes from archeological sites in the Gobi Desert of Mongolia, sites that have yielded dinosaur eggshell fragments that had been cut and carved into pieces of jewelry (Andrews 1932; Pauc and Buffetaut 1998). Although the Stone Age discoverers may have recognized the fossil eggshells as unusual or unique objects, their identity and connection to dinosaurs awaited pioneering paleontologists of the nineteenth and twentieth centuries.

Dinosaur eggshells were first collected for scientific study by a priest and self-taught paleontologist named Jean-Jacques Pouech in southern France during the 1850s (Pouech 1859). He identified these specimens as fossil eggshells of extinct giant birds, probably because dinosaurs were still poorly known at the time. Subsequent comparison of the microstructure of Pouech's eggshells to that of modern eggshells, however, led to the conclusion that they had been laid by a large reptile (Gervais 1877). At the time, the egg-layer was hypothesized to have been *Hypselosaurus*, which was then thought to be a large crocodile (Matheron 1869), but is now known to be a titanosaur. Shortly thereafter, Pouech's enigmatic eggshells were forgotten, not to be recalled for several decades until the discovery of dinosaur eggs in a remote part of the world (Straelen and Denaeyer 1923).

The first fossil eggs ascribed to dinosaurs were found in 1923 during the Central Asiatic Expedition of the American Museum of Natural History to the Mongolian Gobi Desert (Andrews 1932). The size of the eggs, as well as their occurrence in Cretaceous rocks containing abundant dinosaur remains, led to the conclusion that they had a dinosaurian origin. The popularization of these spectacular finds by the American Museum of Natural History paved the way for future discoveries of dinosaur eggs and eggshells worldwide.

Distribution

Dinosaur egg remains are currently known from more than 200 localities distributed over five continents (Carpenter and Alf 1994), only a few of which are known from pre-Cretaceous times. The oldest known dinosaur eggs are those of prosauropods from the Early Jurassic of South Africa (Kitching 1979; Reisz et al. 2005). The only other known Jurassic dinosaur eggs are those of theropods from the Late Jurassic of the United States and Portugal (Hirsch 1994; Mateus et al. 1997; Mateus et al. 2001). Eggs of sauropods, ornithopods, and stegosaurs—dinosaurs well represented by skeletal remains from the Jurassic—have yet to be identified from this period.

Overall, the diversity of known Jurassic eggs is unexpectedly low considering the relatively high diversity of dinosaurs from this time (Hirsch 1994).

The vast majority of dinosaur eggs have been recovered from terrestrial sediments of Cretaceous age, with the majority of specimens found in Upper Cretaceous deposits. Rocks of this age have yielded impressive examples of dinosaur nesting sites and of eggs associated with skeletal remains. Exquisitely preserved adult theropod skeletons sitting atop egg clutches, as well as embryonic skeletons curled within eggs, have been found in ancient eolian (windblown) deposits of China and Mongolia (Norell et al. 1994, 1995; Dong and Currie 1996). Spectacular hadrosaur and sauropod nesting sites from North America and Argentina, respectively, have been found in ancient floodplain deposits, sometimes spanning areas of over 1 km^2 (Horner 1982; Horner and Currie 1994; Chiappe et al. 1998). Although eggs of several different dinosaur taxa are known from Upper Cretaceous rocks, eggs of some well-known dinosaurs from this time, such as tyrannosaurids, ornithomimosaurs, ankylosaurs, and ceratopsids, have yet to be identified.

Parentage

Until recently, the parentage of the various types of dinosaurs eggs remained uncertain. Prior to discoveries of eggs found in close association with skeletal remains, the parentage of some of these eggs had been hypothesized, often incorrectly, based on the predominant dinosaur species found in the same rock layer or locality. In the past 20 years, however, a number of discoveries of eggs closely associated with embryos or adults has allowed for the accurate determination of the parentage of several types of dinosaur eggs.

Nonavian Theropods

Although nonavian theropods comprised a relatively small percentage of overall dinosaur populations, this group is well represented in terms of specimens found closely associated with eggs. To date, the parentage of nonavian theropod eggs has been determined on the basis of eggs associated with skeletal remains from several major groups, including oviraptorosaurs, therizinosaurs, troodontids, and allosauroids.

OVIRAPTOROSAURS

Oviraptorosaurs were the first dinosaurs found in close association with eggs. In 1923, the skeleton of an adult *Oviraptor* was found in close proximity to an egg clutch in Upper Cretaceous rocks of the Gobi Desert of Mongolia (Osborn 1924). At that time, however, the *Oviraptor* (meaning "egg stealer") was assumed to have been preserved in the act of raiding a nest of *Protoceratops*, a more common dinosaur in the rock formation. It was not until 1992 that these presumed *Protoceratops* eggs were suspected of actually belonging to nonavian theropods because of the birdlike structure of the eggshell (Mikhailov 1992). Recent discoveries have since confirmed that these eggs actually belonged to oviraptorosaurs, and that the specimen found in 1923 was preserved in association with its own eggs. The first of these discoveries

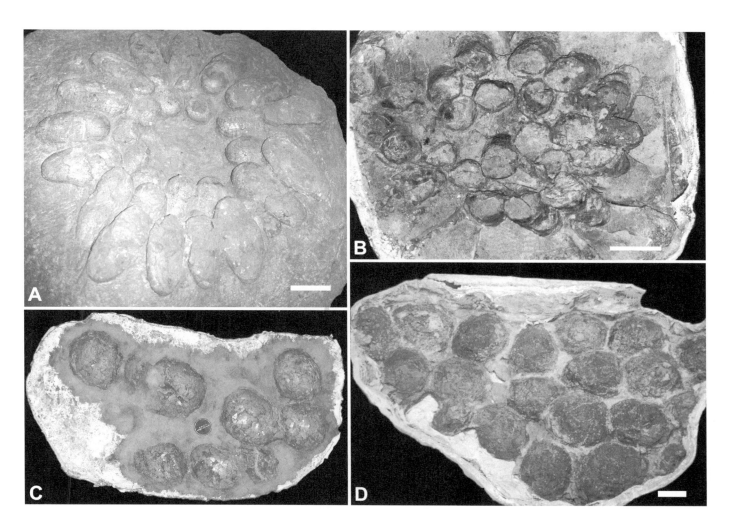

occurred in 1994, when an egg containing the curled-up embryonic skeleton of an oviraptorid was described (Norell et al. 1994). Shortly thereafter, two skeletons of adult oviraptorids were found sitting atop their egg clutches in a manner similar to brooding birds (Fig. 30.1A; Norell et al. 1995; Dong and Currie 1996). Then in 2005, a gravid oviraptorosaur preserving two eggs inside its pelvis was discovered, revealing that these dinosaurs laid two eggs at a time (Fig. 30.1B; Sato et al. 2005). These discoveries have shown that oviraptorosaur eggs, although varying greatly in size, are all elongate (twice as long as wide), are asymmetrical (i.e., pointed at one end), display a ridged ornamentation on the outer surface, and have an internal eggshell structure similar to that of modern birds. Oviraptorosaur clutches consist of multilayered rings of radially oriented, often paired eggs, with their long axes positioned subhorizontally to horizontally (Fig. 30.2A).

30.2. Photographs of dinosaur egg clutches. (A) Oviraptorid nest from China showing stacked, tiered rings of eggs, oblique view. (B) Troodontid clutch from Montana with eggs positioned subvertically in a subcircular cluster, plan view. Long axes of the eggs are directed into the page, and egg "tops" are missing. (C) Putative titanosaur sauropod clutch from France, plan view. (D) Lambeosaurine clutch from Montana, plan view. All scale bars = 10 cm.

THERIZINOSAURS

Other eggs from Upper Cretaceous rocks of China have recently been attributed to therizinosaurs, based on the description of *in ovo* embryonic remains (Kundrát et al. 2008). Little is known about therizinosaur eggs or clutches, other than that the eggs are subspherical with diameters of 7 cm by 9 cm. Therizinosaur eggs are the only known theropod eggs that are nearly spherical in shape; all other known nonavian theropod eggs are elongate.

TROODONTIDS

Eggs and embryos from Upper Cretaceous rocks of Montana, initially assigned to an ornithischian (Horner and Weishampel 1988), have since been recognized as belonging to troodontid theropods (Horner and Weishampel 1996). The eggs are elongate, 12–15 cm in length, and asymmetrical, with a smooth outer surface and an internal eggshell structure comparable to that of modern bird eggs. The egg clutch is in the form of a circular cluster with eggs oriented subvertically and arranged in pairs (Fig. 30.2B). The eggs are half-buried in sediment with the pointed ends oriented downward (Fig. 30.2B). Skeletal remains of an adult *Troodon* atop a partial clutch of these eggs suggest that this dinosaur may have sat on its nest to brood, like oviraptorids (Varricchio et al. 1997, 2002).

ALLOSAUROIDS

Eggs containing embryos from the Upper Jurassic of Portugal (Mateus et al. 1997) have been ascribed to the allosauroid *Lourinhanosaurus* (Mateus et al. 2001), and are the oldest known theropod eggs. The eggs are about 11 cm in length, and somewhat elongate, with a smooth outer surface and an internal eggshell structure comparable to that of birds. These eggs were described from a specimen consisting of an unordered cluster of approximately 100 eggs, which may represent a communal clutch laid by several females or a fluvially reworked accumulation of several separate clutches (Mateus et al. 1997).

Sauropodomorphs

Prosauropod eggs have been indentified from Lower Jurassic deposits of South Africa, based on the description of *in ovo* embryonic remains of *Massospondylus* (Kitching 1979; Reisz et al. 2005). The subspherical eggs, grouped in a cluster of six, have very thin eggshells (Zelenitsky and Modesto 2002), and are among the smallest (5.5 cm in diameter) dinosaur eggs known to contain embryos. These embryos are thought to have required parental care upon hatching, based on their body proportions and a lack of well-developed teeth (Reisz et al. 2005).

Sauropod eggs have been identified only from Upper Cretaceous deposits, which is surprising considering that these dinosaurs were at their peak diversity during the Late Jurassic. In the late 1990s, extensive nesting sites containing thousands of sauropod clutches, some with *in ovo* embryonic remains, were discovered in Argentina (Chiappe et al. 1998, 2001). These eggs, ascribed to titanosaurs based on associated embryonic remains, are spherical to subspherical in shape, are 14–16 cm in diameter, and have an outer eggshell surface ornamented with small bumps. Clutches occur in the form of unorganized, single- or multilayered clusters containing up to 35 eggs (Chiappe et al. 2004). Similar eggs from the Upper Cretaceous of France (including Pouech's enigmatic eggshells), India, and Uruguay have also been assigned to titanosaurs (Fig. 30.2C), based on their size and their occurrence in rocks preserving titanosaur skeletal remains (Carpenter et al. 1994).

Ornithischians

The only known ornithischian eggs are attributed to hadrosaurs (Horner 1999). The first hadrosaur eggs found associated with embryonic remains were those of the hadrosaurine *Maiasaura peeblesorum*, from the Upper Cretaceous of Montana (Horner and Makela 1979; Horner 1999). The eggs are spherical in shape, about 9 cm in diameter, and ornamented with a pattern of netlike ridges on the outer surface. The internal eggshell structure is unique to the Hadrosauria among dinosaur eggs of known parentage, and is without modern analogue. The clutch consists of a stacked, double-layered cluster of 16 eggs.

Other hadrosaur eggs and embryos, assigned to the lambeosaurine *Hypacrosaurus stebingeri*, are known from Upper Cretaceous rocks of Alberta and Montana (Horner and Currie 1994; Horner 1999). These eggs are also spherical, but their diameter is about twice (18–20 cm) that of *Maiasaura* eggs. The largest clutch of *Hypacrosaurus* is oval-shaped and contains 22 eggs arranged in a single layer (Fig. 30.2D). Unfortunately, the eggshell of *Hypacrosaurus* is too poorly preserved to reveal any diagnostic features of lambeosaur eggshell (Horner and Currie 1994).

Despite the discovery of thousands of dinosaur eggs, many aspects of dinosaur nesting may never be known because they are simply not preserved in the fossil record. For example, questions that relate to how nests were constructed and the duration of incubation may never be answered for most extinct dinosaurs. Some aspects of nesting, however, can be inferred from fossilized remains, in conjunction with what is known about reproductive traits of birds and crocodiles, the closet living relatives of extinct dinosaurs.

Eggs and Nests

Dinosaur eggs, like modern bird eggs, vary in shape (Fig. 30.3) and in spatial arrangement within the clutch (Fig. 30.4). The shapes of dinosaur eggs range from spherical (e.g., sauropods, hadrosaurs) to elongate (e.g., oviraptorosaurs, troodontids). The arrangement of the eggs in a clutch also varies, often as a function of the structure of the nest, in that eggs can be laid in unordered clusters (e.g., sauropods, hadrosaurs), circular clusters (e.g., troodontids), or near-perfect rings (e.g., oviraptorosaurs) (Figs. 30.2, 30.4).

Traces of dinosaur nests have occasionally been recognized in the rocks, allowing for the reconstruction of the shape of the nest. The nests vary in form from circular to kidney-shaped excavations with peripheral rims (e.g., sauropods; Chiappe et al. 2004), rimmed depressions much larger than the clutch itself (e.g., troodontids; Varricchio et al. 1997, 1999), bowllike structures (e.g., psittacosaurs; Meng et al. 2004), and mounds (Zelenitsky and Therrien 2008b). The nature of the nest structure and the arrangement

Nests and Nesting

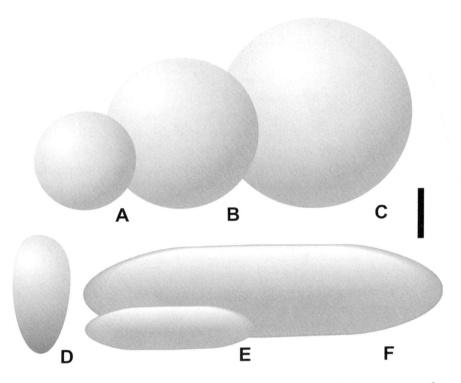

30.3. Schematic of shapes and relative sizes of various dinosaur eggs. (A) Hadrosaurine, (B) Titanosaurid, (C) Lambeosaurine, (D) Troodontid, (E) Oviraptorid, (F) Large oviraptorosaur. Scale bar = 5 cm.

of the eggs within the nest are likely related to the nesting habits (e.g., mode of incubation) of the egg-layers.

Nesting Behaviors

Dinosaurs, like modern birds and crocodiles, probably used different modes of incubation to hatch their young. Based on known egg–skeletal associations, it is likely that some dinosaurs incubated their eggs in open nests like most birds. For example, the discovery of adult oviraptorids on top of their egg clutches suggests that these dinosaurs brooded their eggs in a manner similar to birds (Fig. 30.1A; Norell et al. 1995; Dong and Currie 1996; Clark et al. 1999). Although oviraptorid clutches are laid in two or three superposed ringlike layers of eggs, the eggs of the different layers are tiered so that the inner extremity of each egg is exposed at the center of the clutch (Fig. 30.2A). This clutch geometry would have allowed at least partial contact of all eggs with the overlying adult. Similarly, in troodontid nests, the upper ends of the eggs were exposed above the substrate, allowing for direct contact with a brooding adult (Fig. 30.2B; Varricchio et al. 1997, 2002). Considering the posture of the overlying adults and the disposition of the eggs

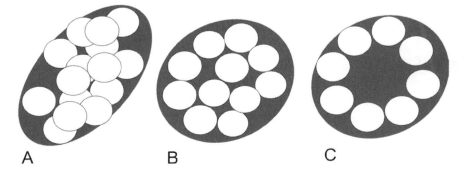

30.4. Schematic of egg arrangements in various dinosaur clutches (modified from Zelenitsky and Therrien 2008a). (A) Eggs are stacked in an unordered cluster. (B) Eggs are arranged in a circular cluster. (C) Eggs are arranged in a ring.

Zelenitsky, Horner, and Therrien

in these clutches, these theropods probably provided at least some heat for incubation of their eggs. Additionally, their egg clutches may have been surrounded peripherally with vegetation to help conceal or insulate the eggs, as such materials are used for nesting by both living crocodiles and birds.

Some dinosaurs may not have maintained such close contact with their eggs during incubation. Clutches consisting of stacked layers of eggs laid in excavations in the ground, such as those of sauropods (Chiappe et al. 2004) and hadrosaurs (Horner 1999), would have been less conducive to direct brooding by an adult. Some titanosaur clutches from Argentina appear to have been incubated in open nests (Chiappe et al. 2004; Jackson et al. 2008), whereas others from France and Spain were reported to be covered with nesting material (Seymour 1979; Deeming 2006). Other sauropod nests have been found in proximity to the preserved traces of coeval hydrothermal environments, suggesting the use of natural geothermal activity for the purposes of incubation (Grellet-Tinner and Fiorelli 2010). Thus, there appears to have been a great diversity of incubation methods among nonbrooding dinosaurs, even within a given clade.

References

Andrews, R. C. 1932. *The New Conquest of Central Asia.* New York: American Museum of Natural History.

Carpenter, K., and K. Alf. 1994. Global distribution of dinosaur eggs, nests, and babies. In K. Carpenter, K. F. Hirsch, and J. Horner (eds.), *Dinosaur Eggs and Babies,* 15–30. Cambridge: Cambridge University Press.

Carpenter, K., K. F. Hirsch, and J. R. Horner. 1994. Introduction. In K. Carpenter, K. F. Hirsch, and J. Horner (eds.), *Dinosaur Eggs and Babies,* 1–11. Cambridge: Cambridge University Press.

Chiappe L. M, R. A. Coria, L. Dingus, F. Jackson, A. Chinsamy, and M. Fox. 1998. Sauropod dinosaur embryos from the Late Cretaceous of Patagonia. *Nature* 396: 258–61.

Chiappe, L. M., L. Salgado, and R. A. Coria. 2001. Embryonic skulls of titanosaur sauropod dinosaurs. *Science* 293: 2444–2446.

Chiappe, L. M., J. G. Schmitt, F. D. Jackson, A. Garrido, L. Dingus, and G. Grellet-Tinner. 2004. Nest structure for sauropods: sedimentary criteria for recognition of dinosaur nesting traces. *Palaios* 19: 89–95.

Clark, J. M., M. A. Norell, and L. M. Chiappe. 1999. An oviraptorid skeleton from the Late Cretaceous of Ukhaa Tolgod, Mongolia, preserved in an avian-like brooding position over an oviraptorid nest. *American Museum Novitates,* 3265, 1–35.

Deeming, D. C. 2006. Allometric anaylsis of the ultrastructural morphology of dinosaur eggshells. *Palaeontology.* 49: 171–186.

Dong, Z-M., and P. J. Currie 1996. On the discovery of an oviraptorid skeleton on a nest of eggs at Bayan Mandahu, Inner Mongolia, People's Republic of China. *Canadian Journal of Earth Sciences* 33: 631–36.

Gervais, P. 1877. De la structure des coquilles calcaires des oeufs et des caractéristiques que l'on peut en tirer. *Comptes rendus des séances de l'Académie des Sciences* 84: 159–165.

Grellet-Tinner, G., and L. Fiorelli. 2010. A new Argentinean nesting site showing neosauropod dinosaur reproduction in a Cretaceous hydrothermal environment. *Nature Communications* 1: 32. doi:10.1038/ncomms1031.

Hirsch, K. F. 1994. Jurassic eggshell from the Western Interior. In K. Carpenter, K. F. Hirsch, and J. R. Horner (eds.), *Dinosaur Eggs and Babies,* 137–150. Cambridge: Cambridge University Press.

Horner, J. R. 1982. Evidence of colonial nesting and "site fidelity" among ornithischian dinosaurs. *Nature* 297: 675–76.

———. 1999. Egg clutches and embryos of two hadrosaurian dinosaurs. *Journal of Vertebrate Paleontology* 19: 607–611.

Horner, J. R., and P. J. Currie. 1994. Embryonic and neonatal morphology and ontogeny of a new species of hadrosaur (Ornithischia, Lambeosauridae) from Montana and Alberta. In K. Carpenter, K. F. Hirsch, and J. Horner (eds.), *Dinosaur Eggs and Babies*, 312–336. Cambridge: Cambridge University Press.

Horner, J. R., and R. Makela. 1979. Evidence of colony nesting and "site fidelity" among ornithischian dinosaurs. *Nature* 297: 675–676.

Horner J. R., and D. B. Weishampel. 1988. A comparative embryological study of two ornithischian dinosaurs. *Nature* 332: 256–57.

——. 1996. Correction to: A comparative embryological study of two ornithischian dinosaurs. *Nature* 383: 103.

Jackson, F. D., D. J. Varricchio, R. A. Jackson, B.Vila, and L. M. Chiappe. 2008. Comparison of water vapor conductance in a titanosaur egg from the Upper Cretaceous of Argentina and a *Megaloolithus siruguei* egg from Spain. *Paleobiology* 34: 229–246.

Kitching, J. W. 1979. Preliminary report on a clutch of six dinosaurian eggs from the Upper Triassic Elliot Formation, Northern Orange Free State. *Paleontolographica Africana* 22: 41–45.

Kundrát, M., A. R. I. Cruickshank, T. W. Manning, and J. Nudds. 2008. Embryos of therizinosauroid theropods from the Upper Cretaceous of China: diagnosis and analysis of ossification patterns. *Acta Zoologica* 89: 231–251.

Mateus, I., H., M. T. Antunes, O. Mateus, P. Taquet, V. Ribeiro, and G. Manuppella. 1997. Couvée, œufs et embryons d'un dinosaure théropode du Jurassique supérieur de Lourinha (Portugal). *Les Comptes rendus de l'Académie des sciences, Earth and Planetary Science* 325: 71–78.

Mateus, O., M. T. Antunes, and P. Taquet. 2001. Dinosaur ontogeny: the case of *Lourinhanosaurus* (Late Jurassic, Portugal). *Journal of Vertebrate Paleontology* 21: 78A.

Matheron, P. 1869. Notice sur les reptiles fossiles des dépôts fluvio-lacustres crétacés du bassin à lignite de Fuveau. *Bulletin de la Société Géologique de France* (ser. 2) 26: 781.

Meng, Q., J. Liu, D. J. Varricchio, T. Huang, and C. Gao. 2004. Parental care in an ornithischian dinosaur. *Nature* 431: 145–146.

Mikhailov, K. E. 1992. The microstructure of avian and dinosaurian eggshells: phylogenetic implications. In K. Campbell (ed.), *Contributions in Science: Papers in Avian Paleontology Honoring Pierce Brodkorb*, 361–373. Los Angeles: Los Angeles County Museum of Natural History.

Norell, M. A., J. M. Clark, L. M. Chiappe, and D. Dashzeveg. 1995. A nesting dinosaur. *Nature* 378: 774–776.

Norell, M. A., J. M. Clark, D. Demberelyin, B. Rhinchen, L. M. Chiappe, A. R. Davidson, M. C. McKenna, A. Perle, and M. J. Novacek. 1994. A theropod dinosaur embryo and the affinities of the Flaming Cliffs dinosaur eggs. *Science* 266: 779–82.

Osborn, H. F. 1924. Three new theropoda, *Protoceratops* zone, central Mongolia. *American Museum Novitates* 144: 1–12.

Pauc, P., and E. Buffetaut. 1998. Reproduction de perles circulaires en coquilles d'oeufs de dinosaures. *Oryctos* 1 : 137–146.

Pouech, H. 1859. Mémoire sur les terrains tertiaires de l'Ariège rapportés à une coupe transversale menée du Fossat à Aillières, passant par le Mas d'Azil, et projetée sur le méridien de ce lieu. *Bulletin de la Société Géologique de France* 16: 381–411.

Reisz, R. R., D. Scott, H.-D. Sues, D. C. Evans, and M. A. Raath. 2005. Embryos of an Early Jurassic prosauropod dinosaur and their evolutionary significance. *Science* 309: 761–764.

Sato, T., Y.-N. Cheng, X.-C. Wu, D. K. Zelenitsky, and Y.-F Hsiao. 2005. A pair of shelled eggs inside a female dinosaur. *Science* 308: 375.

Seymour, R. S. 1979. Dinosaur eggs: gas conductance through the shell, water loss during incubation and clutch size. *Paleobiology* 5: 1–11.

Straelen, V. van, and M. E. Denaeyer. 1923. Sur les oeufs fossiles du Crétacé superieur de Rognac en Provence. *Bulletin de l'Académie Royale Belge, Section Science* 9: 14–26.

Varricchio, D. J., J. R. Horner, and F. D. Jackson. 2002. Embryos and eggs for the Cretaceous theropod dinosaur *Troodon formosus*. *Journal of Vertebrate Paleontology* 22: 564–576.

Varricchio, D. J., F. Jackson, J. J. Borkowski, and J. R. Horner. 1997. Nest and egg clutches of the dinosaur *Troodon formosus* and the evolution of avian reproductive traits. *Nature* 385: 247–250.

Varricchio, D. J., F. Jackson, and J. R. Horner. 1999. A nesting trace with eggs for the Cretaceous dinosaur *Troodon formosus*. *Journal of Vertebrate Paleontology* 19: 91–100.

Zelenitsky, D. K., and S. P. Modesto. 2002. Re-evaluation of the eggshell structure of eggs containing dinosaur embryos from the Lower Jurassic of South Africa. *South African Journal of Science* 98: 407–408.

Zelenitsky, D. K., and F. Therrien. 2008a. Phylogenetic analysis of reproductive traits of maniraptoran theropods and its implications for egg parataxonomy. *Palaeontology* 51: 807–816.

——. 2008b. Unique maniraptoran egg clutch from the Upper Cretaceous Two Medicine Formation of Montana reveals theropod nesting behaviour. *Palaeontology* 51: 1253–1259.

Robin Reid: An Influential Figure in the Development of Modern Osteohistological Analyses for the Study of Dinosaurian Biology

GREGORY M. ERICKSON

More has been learned about dinosaur biology in the last fifteen years than was gleaned during the entire previous century and a half. This renaissance has been fueled by the application of methodologies borrowed from other disciplines to the study of dinosaur paleontology. Of these, osteohistological techniques have provided more insights than any others. Advances in our understanding of growth rates, longevity, lifespan, physiology, development, giantism, dwarfism, locomotion, display, fighting structures, diet, bite forces, parental care, tooth replacement rates, injuries, sex, timing of sexual and somatic maturity, systematics, tissue preservation, and population biology—just to name a few—have been made possible through studies of skeletal microstructure. With such prodigious advancement, the discipline is growing at a breakneck speed as young researchers flock to the field to participate in what can best be described as a scientific gold rush.

Those of us in the field are indebted to the handful of researchers who some thirty years ago helped establish the foundation for today's cutting-edge investigations. Among the most notable, and perhaps least known to newcomers, was the late Robert Edward Hay Reid (better known as Robin Reid to friends and colleagues), senior lecturer from Queen's University in Belfast, Ireland. Reid had the foresight to recognize the untapped paleobiological potential of dinosaur osteohistology and made some of the key findings on which nearly all contemporary histological research depends. Furthermore, he inspired and shared his insights with the next generation of paleohistologists including myself, Anusuya Chinsamy (University of Cape Town), and David Varricchio (Montana State University).

So who was Robin Reid? Reid was an accomplished invertebrate paleontologist who studied the classification and taxonomy of Mesozoic glass sponges and the stratigraphy of Northeastern Ireland for the majority of his professional career. However, late in his tenure, he became intrigued by the debates surrounding dinosaurian physiology. He deduced that bone histology held the clues to this long-standing mystery. This led him to undertake an unorthodox—yet remarkably successful—interdisciplinary shift in focus to dinosaur paleobiology. To ready himself for his second career, Reid mastered the comparative neontological literature on bone development and microstructure. He also studied what little was known about dinosaur bone histology. Such investigations date back to the mid-nineteenth century but were for the most part descriptive in nature, and their biological import had largely gone unrealized. He then began amassing his own study specimens and put his already seasoned skills in petrographic slide preparation and microscopy to use. Between 1984 and 2007 he wrote nearly a dozen scientific works on his dinosaur findings. These include the present peer-reviewed chapters—his final professional publications.

Having had the pleasure of reading Reid's papers at the time they were published, and seeing his career in dinosaurian paleontology unfold, I can attest to his significant influence on the field that helped shape our current views on dinosaurian biology.

Reid's more notable accomplishments include the following:

- *He was the first to discover that growth lines are pervasive in dinosaur bones.* This led to broad-scale reanalysis of how dinosaurs grew and unequivocally showed that most dinosaurs had disruptive growth, more like living reptiles. Prior to this discovery, the dogma was that dinosaurs grew rapidly and continually, just like most living birds and mammals. Hence they were likely endothermic. Current views of dinosaur growth and physiology have been retooled to account for Reid's important revelation.
- *Reid was among the first scientists to deduce that dinosaurian growth lines could be used to determine the age of these animals.* This provided some of the first quantifications of just how fast dinosaurs actually grew. Current analyses of dinosaur growth patterns, which have revolutionized studies of dinosaur life history and developmental evolution, can all be traced to his pioneering work.
- *He broadly surveyed the bones of dinosaurian taxa to reveal a much more remarkable histological diversity among taxa and within individuals than was previously realized.* This led to the recognition that generalizations about dinosaurian histology were overly simplified. As a result, rigorous multi-individual analyses are common today in which individual, developmental, interelemental, scaling, and phylogenetic influences are all controlled for *prior* to making developmental, physiological, or systematic inferences.
- *Reid was the first to link histology with the minority intermediate physiology hypothesis (i.e., dinosaurs did not have a physiology like living birds and mammals or like living reptiles).* He questioned uniformitarian approaches that sought perfect anatomical/physiological matches among living animals and divided the field. Reid's opinion is advocated by many paleontologists and remains at the forefront of current testing.
- *Finally, Reid was instrumental in questioning generalizations about reptilian bone histology that were used for comparison with dinosaur bones.* At the time, reptilian bones were generalized as being nearly avascular, with tightly packed bone fibers, and very much unlike dinosaur bone. Reid, on the other hand, through his own investigations and critical examinations of the literature, believed that reptilian bones were actually more complex, and sometimes quite dinosaur-like. His findings have stimulated current researchers not

just to study dinosaur bone histology, but to partake in extensive integrative studies of living vertebrate bone development.

Reid retired from university life in 1991 and passed away in 2007 at the age of 83. Just before his death he wrote "How Dinosaurs Grew," and "Intermediate Dinosaurs: The Case Updated" for this new edition of *The Complete Dinosaur*. These are updated versions of chapters that he wrote for the first edition of this title, published in 1997. In the latter chapter he presented his final thoughts on the nature of dinosaur physiology for future researchers to ponder and test. Notably he wrote this article in a manner that retells the sequence of his discoveries and the context in which they were made. It provides intriguing insights into a pivotal period in dinosaur paleontology, when histological analyses shifted focus from a historically descriptive subdiscipline to its current form as an integrative, question-driven pursuit.

That said, I strongly urge professional scientists, interested lay public, and, in particular, newcomers to the field, to read his final work. Whether it is a scientific masterpiece, only time will tell. Regardless, this final chapter in Reid's stellar career will bring an appreciation for his pioneering work and a greater understanding of how modern studies of osteohistology came to be.

Because nobody has ever been able to watch a nonavian dinosaur grow, or ever will, any essay on dinosaurian growth must be a study in inference. Luckily, two types of bone that formed as the animals grew tell us parts of the story; and because the same general growth processes are now common to all tetrapods, we can use information from modern forms to fill some of the gaps. But there is still a great deal we don't know, and may never know.

Logically, a study of growth should start with embryos and work through to the end of growth in adults. In recent years, various dinosaur embryos have been found (e.g., Norell et al. 1994), but few details have yet been published. So here we need to start by looking at how bones arise in living animals.

Early Development and Further Growth

First, we can assume that most bones in a dinosaur's skeleton, apart from some in the skull, were at first formed from cartilage before ossification began. In its simplest form, cartilage is a tissue in which live cells called chondrocytes occur scattered within a gellike matrix formed from the proteoglycan chondroitin sulphate. There is usually also some content of collagen fibers. Cartilage thus resembles bone in being a cellular tissue; but it differs in that cells in it are able to divide, and to generate new matrix, thus allowing it to grow interstitially. These abilities are the basis of one of the two main processes by which bones grow.

Next we can look at how a dinosaurian limb bone would develop, as a model for all bones with cartilage prototypes (Fig. 31.1). At first it would be just a rod of cartilage (a), with an external coating of a fibrous tissue called

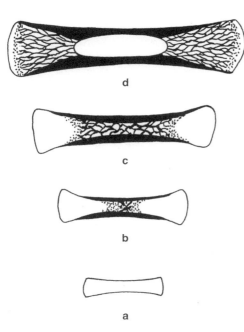

31.1. Four stages in the early development of a dinosaurian limb bone, which will be hollow in the adult, reconstructed from what is seen in modern bones. (a) The bone first appears as a simple cartilage rod, coated by an external perichondrium. (b) Tissue coating the sides (not the ends) of the bone has become the periosteum, and has started to lay down periosteal bone (solid black). Cartilage in the central part of the shaft has calcified (dotted shading), and is starting to be replaced by endochondral bone (reticulating black). (c) The central part of the shaft is now occupied by trabecular endochondral bone, with marrow-filled interspaces; and zones of calcification, under which this bone was formed, are moving out toward the terminal surfaces. (d) Uncalcified cartilage is now restricted to parts underlying the terminal surfaces, as a layer of articular cartilage, and is underlain by zones of calcification, with growing endochondral bone extending up to them. The radiating pattern of trabeculae is a characteristic of this tissue from this stage onward. A medullary (marrow) cavity has been formed, and is expanding longitudinally and radially.

the perichondrium, containing chondroblast cells, which were responsible for forming the cartilage. Early growth would be simply by formation of more cartilage at the sides and the ends, but two new processes would then start in the central parts. Bone-forming cells called osteoblasts would appear in the external perichondrium, and start to coat the cartilage surface with bone, spreading out toward the two ends but not over them (b). In modern species, the external generative tissue after this change is called the periosteum, and bone formed by it is called periosteal. About the same time, a second process would begin with calcification of the cartilage at the center of the shaft, and invasion of this calcified cartilage by tissue of perichondral (or periosteal) origin, containing cartilage-resorbing cells termed chondroclasts (b). Osteoblasts appearing in this tissue, now called the endosteum, would next start lining cavities excavated by the chondroclasts with bone described as endochondral bone. This threefold process would then spread toward both ends of the developing bone (c), extending into the terminal parts (epiphyses; Fig. 31.2), but not as far as the terminal surfaces. These would remain cartilaginous and become the articular cartilages (d).

From this stage, further growth of the bone would proceed by two processes: growth in thickness, by accretion of more periosteal bone along the sides, and growth in length, by continued formation of new cartilage at the ends, and its progressive replacement by endochondral bone. In dinosaurs, periosteal bone formed after early life is dense (compact) bone; but endochondral bone, as in other tetrapods, is typically cancellous, and built from bony struts termed *trabeculae* with spaces called *cancelli* between them. In life, these would be lined by the endosteum, and otherwise filled with marrow tissues, including (in some bones) the haemopoietic tissue in which blood cells (erythrocytes, granular leukocytes, and thrombocytes) are formed. In addition, the bone would be subject to two further processes, not concerned with growth directly, but essential to its proper development. These are remodeling, in which bone formed earlier is resorbed, and internal reconstruction, in which it is replaced by new tissues. Bone resorption, involved in both, is carried out by cells called osteoclasts, which appear in the periosteum and endosteum, so that resorption can be either external or internal.

There are two main cases here. First, because bone cannot grow interstitially, a limb bone can grow in length only by addition of bone at its ends. But because the ends also become expanded to form articulatory structures, the cylindrical shaft (diaphysis) cannot be lengthened by simple terminal growth alone, and is extended by external bone resorption in the parts (metaphyses) just short of the ends until the right shape is reached (Fig. 31.3). Periosteal bone formation then takes over. Bone resorption is also involved when bones have to change their curvature during growth—for example, when a braincase is expanding. Enlow (1962) gives a good account of these processes. Second, as a limb bone grows, its marrow can only expand radially through resorption of bone that surrounds it. In hollow limb bones, seen in theropods and some ornithischians, endosteal osteoclasts would first produce a marrow (medullary) cavity in the central part of the shaft (Fig. 31.1d), and this would then expand both longitudinally and radially. In bones that lack a marrow cavity, as in sauropods, marrow expansion would occur through outward spread of a secondary cancellous (spongy) bone

Parts   **Characteristic processes**

epiphysis   growth in length

metaphysis   external bone remodeling

diaphysis   growth in thickness

31.2. Limb bone parts and processes. At left: how the parts of a limb bone are named. Top to bottom: articulating terminal part: epiphysis; expanded subterminal part: metaphysis; shaft: diaphysis. At right: processes characteristic of these regions in dinosaurian (not mammalian) limb bones. Top to bottom: growth in length by formation of new endochondral bone; external remodeling by resorption of bone underlying the covering periosteum; growth in thickness by accretion of periosteal bone. When the ends of bones have complex shapes, the boundary between areas of bone resorption and periosteal accretion is often less regular than shown here; but these processes are still characteristic of the indicated regions. Mammalian limb bones differ in that growth in length takes place at an internal growth plate, extending transversely between the epiphysis and metaphysis (see Figure 31.6).

tissue, with marrow-filled interspaces. Internal bone reconstruction can also occur in other contexts, but these are the two important here.

Endochondral bone has only one known pattern in dinosaurs. If you look at the seemingly articular surfaces of a well-preserved limb bone, or those of the centrum or zygapophyses of a vertebra, you will find that they are formed by a dense tissue, which may be featureless or show numerous closely spaced round pores (Figs. 31.4c, d). This is not bone but calcified cartilage, which once formed part of a growth zone underlying the soft articular cartilage. Its nature may not be apparent in surface views but is readily seen in sections, in which it appears as a tissue resembling plastic foam (Fig. 31.4b), containing numerous densely packed small rounded cavities that once contained chondrocytes. In life (Fig. 31.5), the true articular surface would be formed by uncalcified cartilage, with a coating of perichondral and synovial tissues, and new cartilage would be added at the surface during growth. Some distance under it, however, would be a zone in which chondrocytes underwent enlargement (hypertrophy) and division until they became packed together in columns or without regular order. Deeper still would be a zone in which this hypertrophic cartilage was calcified, and under this a zone in which endochondral bone was replacing it. Thus, the seeming articular surfaces seen in the fossils (Figs. 31.4c, d)

Endochondral Bone and Epiphyses

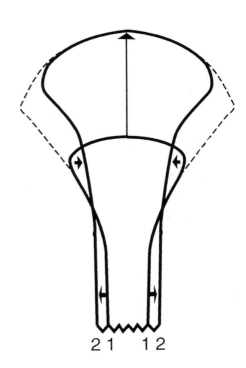

31.3. External bone remodeling: how external bone resorption allows limb bones to maintain their proper shapes during growth, and what would happen if it did not. Here overlapping solid outlines (1, 2) show the form of part of a limb bone at two successive stages of its growth. Because bone tissues cannot expand once they are formed, growth in length (vertical arrow) depends on the formation of new endochondral bone, and growth in thickness (horizontal arrows, pointing outward) depends on the accretion of new periosteal bone. But those processes alone would lead only to progressive expansion of the terminal parts (see broken outlines), without lengthening of the shaft (diaphysis); and this result is avoided by bone being resorbed (horizontal arrows, pointing inward) in parts below the growing terminal (epiphyseal) surface, allowing the shaft to be lengthened and the bone to maintain its proper shape.

31.4. Endochondral bone and calcified cartilage. (a) Both tissues as seen at the top end of an *Allosaurus* fibula. The upper half of the figure shows calcified cartilage containing spaces excavated by advancing marrow processes, with early endochondral bone (dark tissue) in places. The lower half shows trabecular endochondral bone. These tissues correspond with those represented in zones e and f of Figure 31.5. (b) Enlarged view of the calcified cartilage, in an area near the top right in (a) (arrow), showing the rounded form of the spaces (chondrocyte lacunae, chondroplasts) once occupied by cartilage cells (chondrocytes). (c) The "articular" surface of an *Ornithomimus* metatarsal, in which small pores pierce a sheet of unabraded calcified cartilage in a manner suggesting the presence of cartilage canals like those of birds. (d) The "articular" surface of an *Iguanodon* limb bone, in which abrasion has exposed features normally hidden under it. Here the seeming pores are cross sections of tubules excavated by advancing marrow processes (cf. 31.5). Marrow osteoblasts have lined them with endochondral bone, seen as dark rings around the "pores."

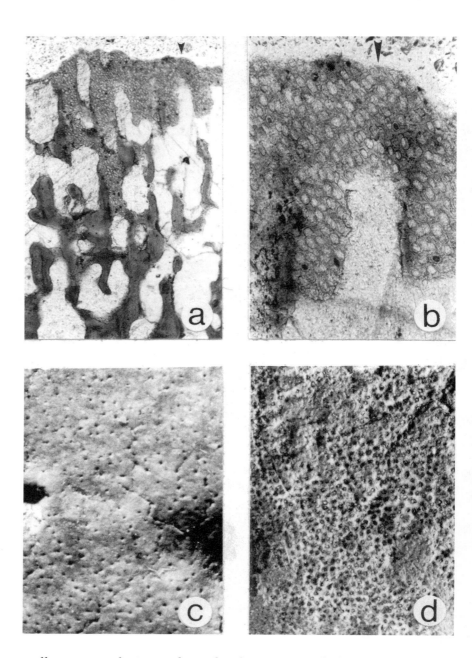

really represent the internal interface between uncalcified cartilage, which is lost, and the underlying calcified tissue. This style of longitudinal growth is now seen in turtles and crocodiles, from which the missing details have been taken here, and is also the primitive pattern for tetrapods in general (Haines, 1938, 1942).

Details of the growth zone have been little studied, but some information is available (e.g. Reid, 1984, 1996). The zone of calcified cartilage varies in thickness, from a few cells to many cells thick; while the cartilage–marrow contact ranges generally from more or less flat, with bone trabeculae simply abutting it, to deeply sculptured, with early endochondral bone extending into the zone of calcified cartilage. The endochondral trabeculae may be formed from bone only, or may contain cores or "islands" of calcified cartilage for some distance away from the growth zone. These were later destroyed by remodeling of the trabeculae. "Articular" surfaces can show continuous cartilage only, or more or less numerous small pores that

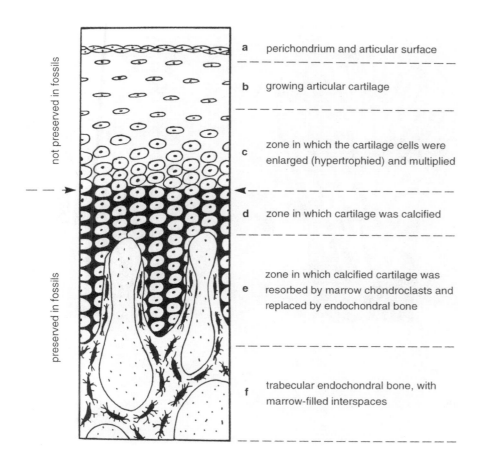

a   perichondrium and articular surface

b   growing articular cartilage

c   zone in which the cartilage cells were enlarged (hypertrophied) and multiplied

d   zone in which cartilage was calcified

e   zone in which calcified cartilage was resorbed by marrow chondroclasts and replaced by endochondral bone

f   trabecular endochondral bone, with marrow-filled interspaces

not preserved in fossils

preserved in fossils

31.5. How dinosaurian limb bones grew in length: reconstructed section through the "growth plate" in the terminal part (epiphysis: see Figures 31.2 and 31.6a) of a growing bone. (a) Perichondrium, under which new cartilage was formed, coating the true articular surface. (b) Growing articular cartilage, formed progressively under the perichondrium. Flattening of cells here is inferred from modern examples. (c) Zone in which the cartilage cells (chondrocytes) became enlarged (hypertrophied) and multiplied, as seen when the cells became arranged in vertical columns. In modern examples, such columns are formed by repeated division of a cell at the top of each column. (d) Zone in which the cartilage matrix became calcified (black shading). The interface (arrowed) at the top of this zone forms the "articular" surfaces of fossil bones. (e) Zone in which calcified cartilage was resorbed by special cells (chondroclasts) in intruding marrow processes, and the first endochondral bone was formed on the walls of the cavities they excavated. (f) Trabecular endochondral bone, with marrow-filled interspaces (dotted shading). Note that the bone cells (osteocytes: solid black) differ from the chondrocytes in having branching processes.

seem to be of two different kinds. In some theropods (e.g., *Ornithomimus*: Fig. 31.4c) they form simple round perforations in unabraded cartilage, and then suggest that the lost soft epiphysial cartilage contained cartilage canals like those now seen in birds. These contain blood vessels that enter the epiphyses laterally and have branches that run through the zones of hypertrophy and calcification (Fig. 31.5c, d) into the metaphyseal marrow (see Haines 1942, 286–287, and figs. 5E and 10). In contrast, in, for example, *Iguanodon* pores are seen only when "articular" surfaces have been abraded, and so mark the positions of tubules excavated in the calcified cartilage by advancing marrow processes (cf. Fig. 31.5). They may be outlined by cartilage only, and can then have a markedly scalloped outline due to cutting into spaces once occupied by chondrocytes, or have a thin lining of endochondral bone laid down on the walls of the tubules (Fig. 31.4d). A similar condition is known from young crocodiles (Haines 1938, fig. 8). In a third case, pores with scalloped outlines may occur in apparently unabraded surfaces, suggesting that marrow processes advanced beyond the zone of calcification.

Here we need to note a difference between dinosaurs and various other tetrapods, including mammals (Fig. 31.6). In dinosaurs, the processes leading to the replacement of cartilage with endochondral bone took place under the articular surfaces, as in crocodiles (Fig. 31.6a); but in mammals, lepidosaurs (lizards plus *Sphenodon*), and some amphibians, the epiphyses themselves develop centers of calcification, which may ossify as in mammals (Fig. 31.6b). Growth in length then takes place at a separate internal growth plate, in the form of a sheet of cartilage extending transversely

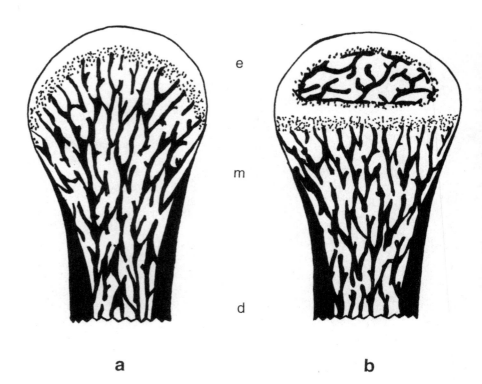

31.6. How dinosaurs (a) differed from mammals (b) in the way in which their bones grew in length. In dinosaurs (a), the zones of hypertrophic and calcified cartilage (dotted shading) formed a "growth plate" that underlay the articular cartilage (clear, at top) directly, and had endochondral trabeculae (black, reticulating) extending up to it. The growth plate thus followed the form of the articular surface; and the seeming articular surfaces of dinosaurian bones follow the form of the interface between uncalcified and calcified cartilage. This style of growth in length is now seen in turtles and crocodiles, as well as in plesiosaurs, ichthyosaurs, and all tetrapods older than Jurassic. In mammals (b), in contrast, the epiphyses (top, e) develop secondary centers of calcification, which then ossify by endochondral replacement; and growth in length by formation of endochondral bone takes place under an internal growth plate, extending transversely between the epiphysis (e) and the metaphysis (m); d = diaphysis. At maturity, this growth plate disappears as the epiphyses fuse, preventing any further growth in length; whereas, dinosaurs could potentially have grown throughout their lives, like turtles and crocodiles.

a   b

between the calcified or ossified epiphysis and the expanded metaphysis at the end of the shaft; while cartilage extending across the end of the bone has a purely articular function. In mammals, moreover, the ossified epiphyses fuse with the metaphyses when growth ceases at adult size, and no further growth in length is then possible. Because of this, and because we know the ages at which different epiphyses fuse, their condition can be used to identify adults and even determine ages.

We can finish here with a puzzle. Most of the limb bones of birds show the same style of growth in length as those of dinosaurs, and it is tempting to see this as simply an inheritance from dinosaur ancestors. But is it? In 1931, it was shown by Landauer that any of the limb bones may form calcified epiphyses in a condition described as chondrodystrophy; and as Haines (1938) saw, this must mean that the mechanism for their production exists in all these bones, although it is normally not used. Or, in genetic terms, the genes for their production must exist, but are normally "switched off." And if that is so, then the usual subarticular style of growth must be secondary, whereas nothing is known to suggest that it is anything but primitive in nonavian dinosaurs. Perhaps future work on early birds will throw light on this conundrum.

Periosteal Bone

Having seen how bones grew in length during increase in stature, we can now look at how they grew in thickness. In outline (Figs. 31.7, 31.8), periosteal growth was often continuous, as in mammals and birds, but sometimes periodic, as in reptiles, with some dinosaurs following both styles in different parts of their skeletons (Fig. 38.14). Because of a lack of systematic sampling, we do not know how common the latter condition was. Variations in growth rates can also be detected.

Continuous periosteal growth in dinosaurs seems commonly to have been rapid because the tissue produced is of a type seen when modern forms grow quickly. This tissue (Fig. 31.9), called fibro-lamellar bone (de Ricqlès 1974), is formed initially as finely cancellous bone, and then is compacted by internal deposition of more bone to form structures called primary osteons. The initial cancellous framework is built from a fast-growing tissue called woven bone, in which collagen fiber bundles are arranged without order; and its cancellous form allows a given volume of bone to produce a higher rate of radial growth than it would if laid down in compact form. Dinosaurs with bone of this type seem to have grown at rates comparable with those of large fast-growing mammals, in which the same type occurs, and could do so up to brachiosaur sizes. Some, however, have similar-looking bone, in which use of crossed polarizers shows the periosteal bone to have a layered structure, implying slower growth than true woven bone, or the whole of the bone to be formed from such tissue, without osteons, implying slower growth still (Fig. 31.10c, d). But what the actual rates were often cannot be measured, because of the absence of a time scale.

In contrast, the periosteal bone of some dinosaurs (Figs. 31.7b, c, 31.8b) shows conspicuous "growth rings," or zones, like those seen in the bones of modern reptiles, which result from the periodic slowing or cessation of growth. Slowed growth is shown by bands of dense tissue known as annuli, and arrested growth by features called resting lines (or arrest lines, or LAGS [lines of arrested growth]), marking where the surface was during growth pauses (see also Figs. 38.6b, c). Bone in the zones may be fibro-lamellar bone, implying rapid growth while it was forming, or of some slower-growing type, including finely lamellated tissues, which are usually formed slowly.

Bone with growth rings should potentially be able to yield data on ages and growth rates, if the rings are assumed to be annual as in crocodilians; but the interpretation of growth rings involves problems. The main one is that, even when present, growth rings often record only the last part of a dinosaur's growth, because most of the periosteal bone produced earlier was resorbed or was replaced by secondary tissues as the marrow expanded. But if enough is left, it may then be possible to estimate how many rings have

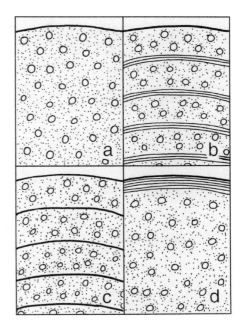

31.7. How different styles of growth are recorded in periosteal bone formed as bones grew in thickness. (a) Bone formed without interruptions or textural variations, implying continuous growth at a constant rate, with no sign of growth slowing before it ceased. This style of growth is the commonest in dinosaurian limb bones. (b) Bone divided into major growth rings, or zones, by thin bands of lamellated tissue termed *annuli*, implying a regular alternation of periods of normal growth (the zones) and slowed growth (the annuli). (c) Bone divided into growth rings by resting lines, which mark positions of the periosteal surface during periods in which growth was interrupted. (d) Bone like that seen in panel A coated externally by a layer of lamellated tissue, implying a later switch from active growth to slow accretion. This pattern is rare in dinosaurs, although known from several sauropods and small theropods. Circular features in the figures represent channels (vascular canals) containing capillary blood vessels and other connective tissues in life.

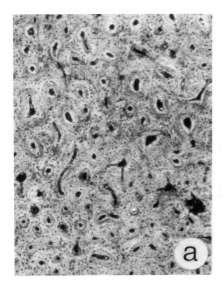
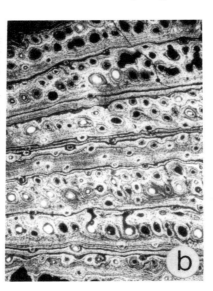

31.8. The contrasting appearances of periosteal bone when formed continuously (a) and discontinuously (b), as seen in samples from ribs of *Tyrannosaurus* (a) and a large early Cretaceous theropod (b). Bone seen in (a) shows no structural interruptions apart from those caused by vascular canals, now infilled with a dark mineral, which once contained enclosed blood vessels and other connective tissue. That in panel b, in contrast, is repeatedly interrupted by circumferential (transverse) resting lines, which divide it into growth rings termed *zones*, and mark successive positions of the periosteal surface during a series of growth pauses. Some of the resting lines are double, due to growth resumption sometimes aborting, and some are accompanied by thin layers of lamellated tissue, termed *annuli* and marking periods of slow accretion.

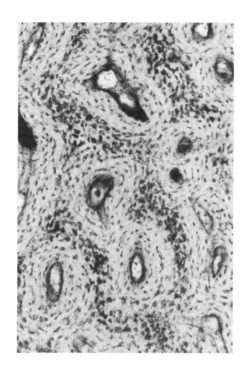

31.9. Fibro-lamellar bone from a limb bone of the sauropod *Cetiosaurus*. In this figure, the large rounded spaces are vascular canals, which contained blood capillaries, and the small dark bodies are mineral-filled spaces (lacunae) that once contained bone cells (osteocytes). The tissue was formed initially as finely cancellous (spongy) bone, represented by the dark tracts in which the osteocyte lacunae are largest, and then compacted by internal deposition of more bone that grew inward toward enclosed blood vessels. This bone formed structures called primary osteons, in which the osteocyte lacunae are smaller than in the initial framework and show a rough concentric arrangement around the vascular canals.

been lost, by extrapolation from the thicknesses of those which remain. By this method, used by Ferguson et al. (1982) to "age" alligators, a Bajocian sauropod (see Fig. 38.11) took 28 or 29 years to reach roughly the size of an adult modern elephant weighing 5 to 6 tons (Reid 1987, 1990); while the iguanodont *Rhabdodon* reached the size of a riding horse in about 16 years (Reid 1990). David Varricchio (1993) suggested 3 to 5 years for the small theropod Troodon to reach 50 kg in weight, although he warned that the resting lines relied on may not have been annual. Subsequent study of *Troodon* growth (Erickson et al. 2007) indicated that Varricchio's earlier age estimates were much too low. Another small theropod, *Saurornitholestes*, seems to have taken 6 or 7 years to reach a size comparable to that of *Troodon* (Reid 1993). Studies by Erickson et al. (2004) and Horner and Padian (2004) both show *Tyrannosaurus* as reaching full size in less than 20 years, and sometimes living as long as 28 years. Until recently no ages could be assigned to forms that grew continuously; but Martin Sander (2000) has now shown that their bone may contain a hidden zonation, in the form of small cyclical variations in bone hardness, which can be seen in polished sections using bright-field illumination. Taking these "polish lines" as annual, he interprets an example of the sauropod *Janenschia* as having ceased to grow actively at 26, and died at 38 (also see Sander et al. 2004, 2006; Klein and Sander 2008).

Knowing the ages of dinosaurs then raises the prospect of being able to quantify their growth rates, in terms of either weight or bone accretion rates; but both approaches involve problems. First, different ways of estimating weight can lead to markedly different estimates. One method is to make a model of the dinosaur concerned, determine its volume by immersing it in water, scaling up, and then estimating weight from the average density of animal tissues. An alternative method is based on the relationship between weight and the diameters of major limb bones (femora, tibiae) in modern animals. By the first method Colbert (1962) obtained an estimated weight of about 80 tonnes for the giant sauropod *Brachiosaurus*; but the estimate obtained by Anderson et al. (1985), who devised the second method, was 29 tonnes. So, did *Brachiosaurus* weigh as much as 16 5-tonne elephants, or a tonne less than only six of them: in which case, its weight would have increased at little more than a third of the rate implied by Colbert's estimate? Second, bone accretion rates can be assessed with any certainty only in material showing "polish lines," as by Sander and Tückmantel (2003), or when zones are marked only by slight cyclical variations in the density of the periosteal matrix, as in the alligator bone shown in Fig. 38.7 below, or by vascular cyclicity as in Curry's *Apatosaurus* figures (1999, figures 3a, 5a). But if annuli or resting lines (LAGs) are present, all that can be measured is the thickness of annual increments, because we do not know the lengths of the periods of active and slowed or interrupted growth. As with nonsequences in sediments, the periods represented by annuli or LAGs could be longer than the periods of active bone formation; and, while the annual rate of active accretion cannot be less than the thickness of the zones, it can potentially be several times greater.

These problems have recently led to new attempts to find solutions to them, along two lines. First, in the weight field, a mathematical technique devised by Erickson, Curry Rogers, and Yerby (2001) shows weight

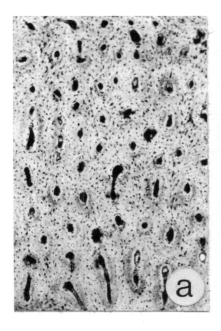

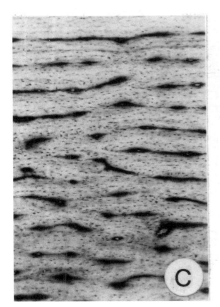

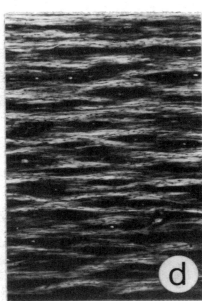

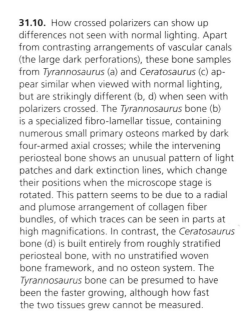

31.10. How crossed polarizers can show up differences not seen with normal lighting. Apart from contrasting arrangements of vascular canals (the large dark perforations), these bone samples from *Tyrannosaurus* (a) and *Ceratosaurus* (c) appear similar when viewed with normal lighting, but are strikingly different (b, d) when seen with polarizers crossed. The *Tyrannosaurus* bone (b) is a specialized fibro-lamellar tissue, containing numerous small primary osteons marked by dark four-armed axial crosses; while the intervening periosteal bone shows an unusual pattern of light patches and dark extinction lines, which change their positions when the microscope stage is rotated. This pattern seems to be due to a radial and plumose arrangement of collagen fiber bundles, of which traces can be seen in parts at high magnifications. In contrast, the *Ceratosaurus* bone (d) is built entirely from roughly stratified periosteal bone, with no unstratified woven bone framework, and no osteon system. The *Tyrannosaurus* bone can be presumed to have been the faster growing, although how fast the two tissues grew cannot be measured.

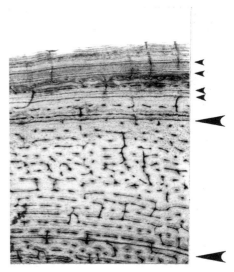

31.11. Apparent determinate growth in a small Cretaceous theropod, *Saurornitholestes*. This figure shows the outer parts of a bone built mainly from a highly vascular tissue, divided into major growth rings, or zones, by resting lines (LAGs), marking pauses in growth. The last of these zones is seen between the large arrows at right. This is then followed by bone showing several closely spaced resting lines (small arrows), and this in turn by almost avascular bone showing fine circumferential lamellation. Active growth thus seems to have been followed by a switch to slow accretion, as can happen when modern endotherms stop growing at a maximum size.

in representative dinosaurs (*Shuvuuia, Psittacosaurus, Massospondylus, Apatosaurus, Psittacosaurus, Maiasaura*) as increasing in a sigmoidal manner, at rates ranging from those of marsupials to those of precocial birds. Data used were weight estimates based on long-bone diameters, and ages obtained by counting growth rings. Their results are consistent with what would be expected from the usual nature of dinosaurian periosteal bone; but reexamination of their samples for polish lines could yield greater ages and lower rates. Second, some workers see the key to obtaining accretion rates as the notion that all similar tissues must be formed at similar rates, which they term *Amprino's Rule*. If this is so, dinosaurian rates should be obtainable by matching dinosaurian tissues with modern ones whose rates of accretion can be measured. This idea is potentially valid; but their "rule" is really only an intuitive hypothesis, which has never been adequately tested. It is well established that different basic bone tissues grow at different rates, with lamellar finely bundled bone growing slowest, and woven-fibered coarsely bundled bone fastest (cf. Pritchard 1956, 12–14); but are they formed at the same rates in all tetrapods, or at different rates in different ones, then providing only evidence of local relative growth rates? Some experimental work has been done recently in attempts to solve this problem, but results so far are conflicting. Using mallards, Castanet et al. (1996) find that different bones have different histological characters correlating with different growth rates, and de Margerie et al. (2002) describe variations in fibro-lamellar bone that correlate with growth rates; but the latter also found that different types of fibro-lamellar bone can grow at similar rates. Starck and Chinsamy (2002), using Japanese quail, instead report, "We found that a single type of primary bone varies significantly in rates of growth in response to environmental conditions. Ranging between 10–50 μm per day, rates of growth overlap with the full range of bone deposition rates that were previously associated with different patterns of bone histology" (232, abstract), and conclude (244) that "our results also nullify the actualistic hypothesis used by Horner et al. (2000) that similar bone tissue types suggest similar growth rates." De Margerie et al. (2004) also note that similar tissues can grow at different rates in different bones. Whether Amprino's Rule can tell us anything useful about dinosaurs is thus debatable.

Returning to ages assessed from LAGs, there is also a problem in that some such features have no connections with annual cyclicity. There are two distinct cases. First, adverse events in the life of any animal that normally grows continuously can interrupt growth, producing resting lines, or annuli, that are simply individual features only, with no bearing on its age. Second, in some dinosaurs sporadic occurrences of such features appear to mark occasional manifestations of a zonal cyclicity that was normally suppressed. This applies, for instance, to the occasional LAGs found by Sander (2000) in a *Janenschia* bone (see above), and to a case in which I found a small patch of typical zonal bone within a seeming "zone" in an *Allosaurus* radius (Reid 1990, 29; Fig. 38.10 this volume). Mistaking such noncyclical features for zonal ones would yield gross underestimates of age, and correspondingly exaggerated growth rates. The patch of zonal bone just noted, for instance, implies that the two LAGs that bound it were formed four years apart. In both cases, however, they can be recognized by being present or absent in different examples of the same bone, or by showing no constant pattern in different examples, as in the Cleveland-Lloyd allosaurs (Reid 1996, 34).

Interestingly, the timing of sexual maturity in the life history of dinosaurs can be determined from patterns of ontogenetic growth recorded by bone. Unlike birds, in which sexual maturity is reached after the adult body mass is attained, nonavian dinosaurs seem, like typical modern squamates and crocodilians, to have become sexually mature well before the final adult size was reached (Erickson et al. 2007).

Last, the end of growth in dinosaurs appears to have followed several different patterns. Many bones that grew continuously show no signs of growth having ceased or even slowed before death, so that their owners seem to have grown throughout their lives unless thought to have died without reaching skeletal maturity (in, e.g., *Allosaurus*: Reid 1996; Bybee et al. 2005). De Ricqlès (1980) saw growth throughout life, as in modern reptiles, as a characteristic of dinosaurs; and their lack of separate epiphyseal ossifications (Fig. 31.6) would have allowed it (but it is worth noting that even in nonavian reptiles there is an asymptotic body size). Some such bones, however, show reduced vascularity after early life (in, e.g., *Massospondylus*: Chinsamy 1993), suggesting that growth then slowed, although it continued until death. Alternatively, bones in which much of growth was continuous may show a late change to forming zonal bone, in which the zones may also become thinner progressively, showing growth as slowing correspondingly. Bone with wholly zonal periosteal bone may show similar progressive slowing, or not show it (e.g., Reid 1990, fig. 24: ornithopod *Rhabdodon*, femur, and figure 1, indet. sauropod, pubis). Finally, there are cases in which bones show a sudden late change from active growth to formation of a thin external covering of dense lamellated tissue, which presumably formed slowly. This condition was first figured from the giant sauropod *Brachiosaurus* (Gross 1934, figs. 15, 17), and has since been found in three small theropods (*Syntarsus*: Chinsamy 1990; *Troodon*: Varricchio 1993; *Saurornitholestes*: Reid 1993; Fig. 31.11 in this volume) and five more sauropods (*Apatosaurus*: Curry 1999; *Barosaurus, Janenschia*, and two *Dicraeosaurus* species: Sander 2000). What proportions of their life spans such tissue represents seems to vary. In *Janenschia* Sander (2000) saw it as representing 12 years of a 38-year life; but a *Saurornitholestes* seems to have lived longer after active growth ceased (12 years) than beforehand (6–7 years: Reid 1993). As was seen by Chinsamy (1990), such tissues resemble the accretionary bone that is sometimes formed in mammals after active growth has ceased, and suggests determinate growth like that of modern endotherms; but the actively formed bone of *Syntarsus* and *Saurornitholestes* is zonal, so that mammal-like and reptilelike features occur together. These variations in how growth ended add yet another element to the problem of trying to understand dinosaurian physiology.

Summary

The main things that bone tells us about how dinosaurs grew can be summed up as follows:

1. As in all tetrapods, the limb bones of dinosaurs grew in length by formation of new endochondral bone, and in thickness by growth of periosteal bone. Remodeling by external bone resorption allowed the shapes of bones to be maintained or adjusted during growth.
2. In limb bones endochondral bone was formed under the articular cartilages, as in crocodiles, and not under separate epiphyseal ossifications as in lizards and mammals.
3. In many dinosaurs the structure of epiphyses is similar to that found in crocodiles, and then may follow a pattern that was primitive in archosaurs; but ornithomimids and nonavian maniraptors (e.g., dromaeosaurs) can show features suggesting the presence of cartilage canals like those of birds.
4. The periosteal bone of dinosaurs was commonly a fast-growing fibro-lamellar tissue, formed continuously as in mammals; but it could also be a slower-growing tissue, or show cyclical (zonal)

interruptions like those seen in modern reptiles. Some show both styles of growth in different bones, or a change from continuous (azonal) growth to a cyclical pattern as bones grew.

5. Many dinosaurs appear to have grown throughout their lives, although some of these show evidence of growth slowing before death. In contrast, some show evidence of determinate growth like that of mammals and birds.

References

Anderson, J. F., A. Hall-Martin, and D. A. Russell. 1985. Long-bone circumference and weight in mammals, birds and dinosaurs. *Journal of Zoology, London* A 207: 53–61.

Bybee, P. J., A. H. Lee, and E.-T. Lamm. 2005. Sizing the Jurassic theropod dinosaur *Allosaurus*: assessing growth strategy and evolution of ontogenetic scaling of limbs. *Journal of Morphology* 267: 347–359.

Castanet, J., A. Grandin, A. Abourachid, and A. de Ricqlès. 1996. Expression de la dynamique de croissance dans la structure de l'os périostique chez *Anas platyrhynchos*. *Comptes Rendu de l'Académie de Science Paris* 319: 301–308.

Chinsamy, A. 1990. Physiological implications of the bone histology of *Syntarsus rhodesiensis* (Saurischia: Theropoda). *Palaeontologia africana* 27: 77–82.

———. 1993. Bone histology and growth trajectory of the prosauropod dinosaur *Massospondylus carinatus* (Owen). *Modern Geology* 18: 319–329.

Colbert, E. H. 1962. The weights of dinosaurs. *American Museum Novitates* 2076: 1–16.

Curry, K. A. 1999. Ontogenetic histology of Apatosaurus (Dinosauria, Sauropoda): new insights on growth rates and longevity. *Journal of Vertebrate Paleontology* 19: 654–665.

de Margerie, E., J. Cubo, and J. Castanet. 2002. Bone typology and growth rates: testing and quantifying "Amprino's Rule" in the mallard (*Anas platyrhynchos*). *Comptes Rendu Biologie* 325: 221–230.

de Margerie, E., J.-P. Robin, D. Verrier, J. Cubo, R. Groscoles, and J. Castenet. 2004. Assessing a relationship between bone microstructure and growth rate: a fluorescent labeling study in the king penguin chick (*Aptenodytes patagonicus*). *Journal of Experimental Biology* 207: 869–879.

de Ricqlès, A. J. 1974. Evolution of endothermy: histological evidence. *Evolutionary Theory* 1: 51–80.

———. 1980. Tissue structure of dinosaur bone: functional significance and possible relation to dinosaur physiology. In R. d. K. Thomas and E. C. Olson (eds.), *A Cold Look at the Warm Blooded Dinosaurs*, 103–139. American Association for the Advancement of Science Selected Symposium 28. Boulder, Colo.: Westview Press.

Enlow, D. H. 1962. A study of the postnatal growth and remodeling of bone. *American Journal of Anatomy* 110: 79–102.

Erickson, G. M., K. Curry-Rogers, D. J. Varricchio, M. A. Norell, and X. Xu. 2007. Growth patterns in brooding dinosaurs reveal the timing of sexual maturity in non-avian dinosaurs and genesis of the avian condition. *Biology Letters* 3: 558–561.

Erickson, G. M., K. Curry Rogers, and S. A. Yerby. 2001. Dinosaurian growth patterns and rapid avian growth rates. *Nature* 412: 429–433.

Erickson, G.M., P.J. Makovicky, P.J. Currie, M.A. Norell, S.A. Yerby, and C.A. Brochu. 2004. Gigantism and comparative life-history parameters of tyrannosaurid dinosaurs. *Nature* 430: 772-775.

Ferguson, M. W. J.; L. S. Honig; P. Bringas Jr.; and H. C. Slavkin. 1982. In vivo and in vitro development of first branchial arch derivatives in *Alligator mississipiensis*. In A. D. Dixon and B. Sarnat (eds.), *Factors and Mechanisms Influencing Bone Growth*, 275–296. New York: Alan R. Liss.

Gross, W. 1934. Die Typen des mikroskopischen Knochenbaues bei fossilen Stegocephalen und Reptilien. *Zeitschrift für Anatomie* 103: 731–764.

Haines, R. W. 1938. The primitive form of the epiphysis in the long bones of tetrapods. *Journal of Anatomy* 72: 323–343.

———. 1942. The evolution of epiphyses and endochondral bone. *Biological Reviews* 17: 267–292.

Horner, J. R., and K. Padian. Age and growth dynamics of *Tyrannosaurus rex*. *Proceedings of the Royal Society B* 271: 1875–1880.

Klein, N., and P. M. Sander. 2008. Ontogenetic changes in the long bone histology of sauropod dinosaurs. *Paleobiology* 34: 247–263.

Landauer, W. 1931. Untersuchungen über der Krüperkuhn. II. Morphologie und Histologie des Skelets, inbesondere de Skelets der langen Extremitätenknochen. *Zeitschrift für mikroskopische-anatomische Forschung* 25: 115–141.

Norell, M. A., J. M. Clark, D. Dashzevegg, R. Barsbold, L. M. Chiappe, A. R. Davidson, M. C. McKenna, A. Perle, and M. J. Novacek. 1994. A theropod dinosaur embryo and the affinities of the Flaming Cliffs dinosaur eggs. *Science* 266: 779–782.

Pritchard, J. J. 1956. General anatomy and physiology of bone. In G. H. Bourne (ed.), *The Biochemistry and Physiology of Bone*, 1–25. New York: Academic Press.

Reid, R. E. H. 1984. The histology of dinosaurian bone, and its possible bearing on dinosaurian physiology. In M. W. J. Ferguson (ed.), *The Structure, Development and Evolution of Reptiles*, 629–663. Orlando, Fla.: Academic Press.

——. 1987. Bone and dinosaurian "endothermy." *Modern Geology* 11: 133–154.

——. 1990. Zonal "growth rings" in dinosaurs. *Modern Geology* 15: 19–48.

——. 1993. Apparent zonation and slowed late growth in a small Cretaceous theropod. *Modern Geology* 18: 391–406.

——. 1996. Bone Histology of the Cleveland-Lloyd dinosaurs, and of dinosaurs in general. Part I: Introduction: Introduction to bone tissues. *Brigham Young University Geology Studies* 41: 25–71.

Sander, P. M. 2000. Longbone histology of the Tendaguru sauropods: implications for growth and biology. *Paleobiology* 26: 466–488.

Sander, P. M., N. Klein, E. Buffetaut, G. Cuny, V. Suteethorn, and J. Le Loeuff. 2004. Adaptive radiation in sauropod dinosaurs: bone histology indicates rapid evolution of giant body size through acceleration. *Organisms, Diversity, and Evolution* 4: 165–173.

Sander, P. M., O. Mateus, T. Laven, and N. Knötschke. 2006. Bone histology indicates insular dwarfism in a new Late Jurassic sauropod dinosaur. *Nature* 441: 739–741.

Sander, P. M., and C. Tückmantel. 2003. Bone lamina thickness, bone apposition rates, and age estimates in sauropod humeri and femora. *Paläontologisches Zeitschrift* 77: 161–172.

Starck, J. M., and A. Chinsamy. 2002. Bone microstructure and developmental plasticity in birds and other dinosaurs. *Journal of Morphology* 254: 232–246.

Varricchio, D. J. 1993. Bone microstructure of the Upper Cretaceous theropod dinosaur Troodon formosus. *Journal of Vertebrate Paleontology* 13: 99–104.

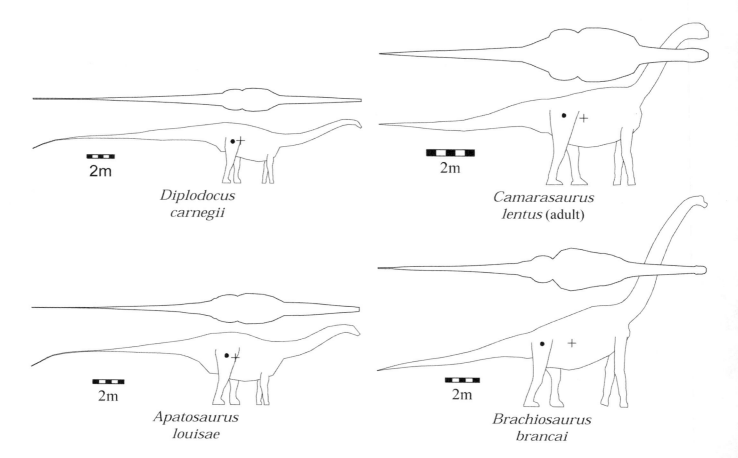

Diplodocus
carnegii

Camarasaurus
lentus (adult)

Apatosaurus
louisae

Brachiosaurus
brancai

32.1. On the left are two diplodocids with centers of mass (CM, '+') that lie immediately in front of their hips, and necks that show a gentle upward arc. On the right are two macronarians with centers of mass that are more anterior and necks that are much more erect. Differences in body shape make it difficult to use terms such as *height* and *length* to describe animals as unusual as dinosaurs.

Engineering a Dinosaur

Donald Henderson

32

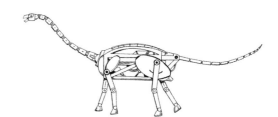

Although they may sometimes appear to be creatures of fantasy, dinosaurs existed as physical entities in a physical world. They were subject to external factors such as gravity, air and water pressures, and temperature changes, as well as internal factors such as blood and gas pressures, muscle tensions, and the stresses and strains that occur in bones, tendons, and ligaments. The material properties of an animal's tissues, and the internal and external physical forces that they are subjected to, determine the limits for what is possible for the organism in terms of growth, body support, movement, feeding methods, and life history. The study of organisms as physical systems, subject to mechanical constraints, is the field of biomechanics.

Many of the factors that zoologists consider when studying the biomechanics of living animals are based on living soft tissues, for example, the sizes and relative proportions of the muscles, the proportions of the different muscle fiber types in different muscles, and forces that these muscles can produce. The fossil record has left us, with rare exceptions, just the bones of dinosaurs, but it is from these bones that we must make inferences as to what the dinosaurs were like as living, functioning organisms. Fortunately, potential sources of information on the now-vanished soft tissues of dinosaurs do exist. Dinosaurs, birds, and crocodilians are all members of a larger group of animals, the Archosauria, and with what we know of the skeletons and anatomy of crocodiles and birds we can make plausible inferences about the anatomy of extinct dinosaurs. Another important source of information comes from studying the biomechanics and ecology of large, extant, terrestrial mammals. The skeletons of elephants, giraffes, and rhinos show many features in common with those of dinosaurs, and these similarities are interpreted to be examples of convergent evolution – the biomechanical challenges of being a 5-tonne animal are similar for both mammals and dinosaurs. Despite the long period of independent evolution of the mammal and dinosaurs, they must both respond to the same physical constraints.

This chapter presents an overview of various aspects of the biomechanics of dinosaurs. After a brief discussion of the importance of body mass as it relates to dinosaurs, the chapter follows a tour of the skeletons and bodies of various dinosaurs by progressing from the tail to the head.

Dinosaurian Body Mass

The immense size of many dinosaurs, as the largest land animals of all time, is a distinguishing feature to the group. Size can be measured in several ways: height, length, width, and weight. As linear measures, the first three of these quantities are statements about the dimensions of the animal in a single direction and ignore the other two dimensions. Quoting just a single linear dimension has the potential to be vague or misleading when describing animals as unusual as dinosaurs. The height of dinosaur – the distance measured from the top of the head from the ground – is a subjective quantity, as it would depend on how the animal was restored. As an example, the Late Jurassic sauropods *Diplodocus carnegii* and *Brachiosaurus brancai* both had total lengths of about 24m (Fig. 32.1). However, the neck of *Diplodocus* is traditionally restored with a very gentle upward arc, which results in the head being at the same level as tips of the sacral neural spines, giving the animal a "height" of about 4 m. In contrast, the neck of *Brachiosaurus* is usually restored as being subvertical, which results in the animal having a "height" of 13 m. Similarly, although *Diplodocus* and *Brachiosaurus* have equal lengths, the former is estimated to have weighed about 11 tonnes, while the latter is estimated to have weighed almost 26 tonnes (Henderson 2006). Just quoting the length of an animal says nothing about how wide or deep it is, and ignores differences in shape. Given these objections to using linear dimensions to describe the sizes of dinosaurs, the preferred way of indicating the size of a dinosaur, or any other animal, is to use body mass.

An animal's weight provides a measure of how much force the body exerts on the ground, and is independent of body shape. The weight of any object, animate or not, is the product of its mass multiplied by the acceleration that it experiences. The mass of any object is a measure of the amount of material used in its construction, and is distinct from its weight. At the surface of the Earth, all objects are constantly subjected to at least one source of acceleration – that due to the planet's gravity of 9.81 m/s^2. The product of an object's mass with this common gravitational acceleration provides a common reference force (weight) for the comparison of masses. The terms *mass* and *weight* are often used interchangeably, with the units of measure being the units of mass (kilograms), with the implicit assumption that true weight comes from multiplying this unit of measure by 9.81 m/s^2. As will be discussed in following sections, the mass (weight) of a dinosaur, or the weights of parts of its anatomy, has important implications for how the whole animal would have functioned as a living organism. The very high masses of dinosaurs, many tens of tonnes in some cases (Gunga et al. 1995; Henderson 2006; Pecksis 1994), undoubtedly exerted a strong influence on their daily lives, and what they would have been capable of biomechanically.

The importance of body mass for inferring aspects of the biology of dinosaurs, or any other animal, living or extinct, should not be underestimated. The following is a partial list of ecological and physiological parameters that are all strongly influenced by body mass: metabolic rate; the amount of food required; how many offspring/eggs are produced in a breeding season; the size of the neonate/hatchling; the time required to reach sexually maturity; life span; the number of individuals required to form a viable, self-sustaining population; the time required for an animal's body to heat up or cool down; the potential to jump; and the maximum speed of locomotion. This correspondence between body mass and ecophysiological parameters is known as biological scaling, and the interested reader is referred to Schmidt-Nielsen (1984) and Calder (1984) for comprehensive reviews of this subject.

Given the importance of mass for inferring the paleobiology of dinosaurs and other extinct animals, a variety of methods for estimating body mass in extinct animals have appeared over the last 100 years. Immersion methods rely on the construction of scaled, physical models that are then immersed in water (or sand – Gregory [1905]), with the volume of fluid displaced by the immersed model being measured (Colbert 1962). With the known scale of the model, and an assumed density for the tissues of the animal the model represents, the volume of displaced water is used to compute a mass for the full-sized animal. An extension to the immersion method was used by Alexander (1985a): he incrementally immersed his models and recorded the individual volumes displaced by segments in order to determine the distribution of body mass along the length of the model. The resulting mass distribution data was necessary for further biomechanical studies related to bending and sagging of the body (Alexander 1985a) (see "Tails," below). While most studies used scale models, Hurlburt (1999) used a life-sized model of the early, herbivorous synapsid *Edaphosaurus* for part of his mass estimation process, although he took direct measurements from the model rather than immersing it. In a similar vein, Gunga et al. (1999) measured mounted skeletons of *Brachiosaurus brancai* and *Dicraeosaurus hansemanni* directly using photogrammetric methods. These dimensional data were then used to compute volumes of the various body parts by approximating them with regular solids such as spherical caps, cylinders, and truncated cones. The computed volumes were then multiplied by appropriate densities to get mass estimates for the body parts.

Since the early 1990s, as more computing power has become available at lower cost, there has been increasing use of digital data and more mathematically sophisticated techniques to estimate body mass. The digital/mathematical model studies still make use of the same basic ideas of volumes and scaling factors. Seebacher (2001) fitted polynomial equations to outlines of body shapes of dinosaurs, and then computed the volume as solids of revolution with an applied correction in recognition of the fact that dinosaurs are not circular cylinders. Bates et al. (2009) used laser scanning (LIDAR) of several mounted dinosaur skeletons in museums to capture the all the details of the mount. The skeletal geometry data was used as a foundation for restoring soft tissues (muscles, internal organs, etc.) using 3-D computer modeling techniques to quickly and easily generate smooth surfaces. The resulting volumes were then processed to compute body segment masses, centers of mass, and moments of inertia of body parts.

No single method, digital or physical, is better, although it is easier to change the parameters of a digital model than it is to reshape a physical model. If two or more methods are used to provide body mass estimates for a dinosaur, and the two methods give results that are not that different from each other, this provides a means to bracket the range of mass estimates for the dinosaur of interest. Conversely, two or more restorations of the same animal can be "weighed" with a single method as was done by Henderson (1999) with two different *Tyrannosaurus rex* restorations – those of Paul (1988) and Smith (in Farlow et al. 1995).

Tails

With very few exceptions, the trackways of dinosaurs do not record the impressions of tails being dragged along the ground (Thulborn 1990). Given the now discarded idea that all dinosaurs dragged their tails, the sauropods and their massive tails would have been the ideal example of tail-dragging dinosaurs. The extreme rarity of sauropod trackways showing tail drags was easily explained by the old-timers, who interpreted sauropods as semi-aquatic animals – the water obviously buoyed up the tail, and it therefore left no drag mark (Lockley 1989). With the modern interpretation of sauropods, and other dinosaurs, as fully terrestrial animals, it is now clear that their tails were normally carried clear of the ground. If this is so, how did they do it?

Various horizontally projecting or flexed portions of an animal's body can be held either passively with ligaments or actively by the continual firing of muscle cells to maintain tension and resist the sagging or bending. Both of these strategies are employed in tail support, but passive ligamentous support seems to be the most developed in dinosaurs. Given the immense sizes of many dinosaurs and the high metabolic cost of using just muscle to maintain the positions of extended regions of their more massive bodies, this is a reasonable expectation. Basic to all vertebrates are ligaments that span the gaps between the neural spines, termed the *interspinous ligaments*, and the low extensibility of these ligaments will act to resist sagging of the tail caused by the action of gravity continually pulling downward on it. The interspinous ligaments are assisted by another set that runs along the very tops of the neural spines – the supraspinous ligaments.

The principal effect to be achieved by the combination of the neural spines and their connecting ligaments is to resist the turning force applied to the tail by gravity. The magnitude of this gravitational turning force (torque) is given as the product of the weight of the tail and the distance of the center of mass of the tail from its point of support at the posterior end of the sacrum. This distance represents the lever, or moment, arm of the gravitational force acting on the tail. For the tail to be held continually clear of the ground, the turning force applied to the tail by gravity has to be countered by turning forces of equal strength, but opposite direction, applied by tensile forces within the plane of the spine itself. The product of the tension (force) in the inter- and supraspinous ligaments and their associated neural spines represents this countering torque. Any torque will be a maximum when both the turning force and its lever arm are as large as possible, and the largest gravitational torques acting on the tail will be those experienced at the rear of the hips where the amount of tail projecting posteriorly will be at its maximum.

A distinguishing feature of large dinosaurs is that the heights of their neural spines reach their maximum values in the area of the sacrum and the proximal part of the tail, and this is especially noticeable in larger forms such as *Diplodocus*, *Apatosaurus*, and *Tyrannosaurus*. The torques provided by the combination of maximally tall neural spines and associated tensions transmitted by interspinous ligaments will match and counter the torques acting to drag the tail down. The decrease in the heights of the neural spines as one progresses posteriorly reflects the gradual reduction in the magnitudes of the tail-drooping torques that need to be countered. The association of tall sacral neural spines with extremely long tails observed in the diplodocid sauropods (see "Trunk," below) such as *Apatosaurus* and *Diplodocus* is not seen in relatively short tailed macronarians such as *Camarasaurus* and *Brachiosaurus*. The correlation of short tails and low sacral neural spines in this latter group lends support to the inference that tall sacral neural spines are a biomechanical optimization in the long-tailed diplodocids.

In addition to the interspinous and supraspinous ligamentous support seen in the saurischians, the ornithischians developed an additional form of support—lattices of ossified tendons. When a tendon experiences high enough tensile stress, there is a tendency for it to ossify, and this effect can be seen in living birds such as turkeys in the form of ossified tendons that transmit forces to the toes from the digital flexor muscles. The epaxial tendons of ornithopods, those that lie above the horizontal midline of the vertebral centra, form a trellislike structure that is best developed over the sacrum, where the tensions acting to pull the body downward on either side of the sacrum would be maximal. The amount of ossified tendon developed appears to correspond with the size of the animal. Large ornithopods such as *Edmontosaurus* (10 tonnes) and *Lambeosaurus* (2 tonnes) have extensive, almost matlike, meshes of ossified tendons, while much smaller ornithopods such as *Lesothosaurus* (2 kg) and *Hypsilophodon* (16 kg) have much more limited amounts of ossified tendon, but still situated in the vicinity of the sacrum (Galton 1974). A difficulty with making correlations between body size and the occurrence of ossified tendons in fossilized animals is taphonomic effect. The ossified tendons would have been attached to the bones by either unossified tendon or small slips of muscle at their proximal ends (Norman 1980, 1986), and the processes of decay and preburial transport may have resulted in the loss of the tendons in many specimens. An added complication is the final taphonomic effect of fossil preparation! In their zeal to get to the fossilized bones, preparators may just burrow past the tendons embedded in the matrix immediately adjacent to the bones without distinguishing the tendons from the matrix (personal observation).

Although they don't normally produce extensive amounts of it, incipient ossification of tendons or ligaments can be seen in saurischians. A good example of this is seen in the proximal caudal region of the *Diplodocus longus* on display at the Smithsonian Institution (Washington, D.C.) via preserved sections of its supraspinous ligament. Another example is seen in the very large *Tyrannosaurus rex* ("Sue") at the Field Museum (Chicago), where in the region of the sacrum and anterior tail, small, delicate fingers of bone arise from the posterior and anterior edges of the neural spines, indicating the beginnings of ossification of the interspinous ligaments in this exceptionally large individual. It is telling that these examples of the beginnings of tendon ossification in animals that don't normally produce them occur in regions of the body where stresses related to sagging of the body are predicted to be at their maximum values.

A peculiar analogue of this ligamentous support strategy is seen in small, agile carnivorous dromaeosaurs, such as *Deinonychus*. The distal four-fifths of the tails of these animals are sheathed in bony rods that appear to be long, interwoven tendons. However, closer inspection reveals that the "tendons" are actually a combination of elongated postzygapophyses that extend posteriorly almost to the posterior end of the next vertebra, and fantastically elongated prezygapophyses that reach forward over 7–9 of the preceding vertebrae (Ostrom 1969). This pattern of elongation is repeated by the chevrons in the ventral portion of the tail. No saurischian has evolved systems of true epaxial tendons such as those seen in the ornithischians, but the greatly modified zygapophyses and chevrons of the dromeosaurs are a functional analogue. A very similar system of zygapophyseal "tendons" evolved independently in rhamphorhynchid pterosaurs (Wellnhofer 1990).

The ventral half of the dinosaurian tail served a different purpose from that of the upper half. Its proximal region was the origination point for the main muscle that acted to pull the leg backward – the caudifemoralis longus (Gatesy 1990). Associated with the pulling action on the leg, there would also be a reciprocal pulling action on the tail. If there were not the epaxial muscles and tendons of the upper half of the tail to resist the downward pull of the caudifemoralis, the effectiveness of this muscle would be greatly reduced due to the deformation and bending of the tail itself.

More than acting as just a base for the limb retractor muscles, the tail would also act as a dynamic stabilizer for an animal that had to balance its body on just two legs. This property of dynamic stability arises from the physical principle known as the conservation of angular momentum (CAM) (Halliday et al. 1993). The angular momentum of any pivoting or rotating object is calculated as the product of its mass, the distance of its center of mass from the point of support, and the speed of rotation. An animal flicking its tail is a form of rotation, and the flicked portion of the tail possesses angular momentum. As applied to the entire body of a bipedal dinosaur, the CAM principle implies that if the animal flicks its tail to one side, the front half of the body has to rotate a bit to the same side to counteract the change in angular momentum caused by the moving tail. Similarly, if the tail is flicked upward, the anterior half of the body also has to rise as well. The amount of sideways or up–down motion of the anterior portion of the body will never be as great as that of the tail, as the combined mass of the head, neck, and trunk is many times that of the tail. One can imagine the tail of a standing, bipedal dinosaur tail gently, but frequently, flicking from side to side and up and down to steady the animal as movements of its head and arms moved the animal's center of mass and altered the overall angular momentum of the body. Controlled changes in the angular momentum of regions of the body can be used for more than just balance, and two extreme examples of the exploitation of angular momentum changes in dinosaurs can be seen in dromeosaurs and some thyreophorans.

The zygapophyseal and chevron reinforced tail of *Deinonychus*, and the specialization of its proximal caudal vertebrae for great flexibility, led to the proposition that the tail of this animal could function as a very efficient means of controlling and changing the orientation of the body (Ostrom 1969). The extreme stiffening of the tail suggests that it was adapted to resist bending when rapidly accelerated by the muscles at its base. The tail of *Deinonychus* was a slender, lightly built structure, so the muscles at its base could change the speed and direction of the tail very quickly. A rapidly pivoting tail would have a very high angular momentum, and rapid changes in the angular momentum of the tail would reappear as rapid changes in the orientation of the front half of the body due to CAM. With this sort of adaptation, it is easy to imagine *Deinonychus* as a rapid, agile predator that could turn very quickly (Fig 32.2).

Deinonychus maximized the angular momentum of its tail by maximizing its rotational speed. Stegosaurs, ankylosaurs, and the sauropod *Shunosaurus* appear to have used heavy spines and clubs as weapons, and

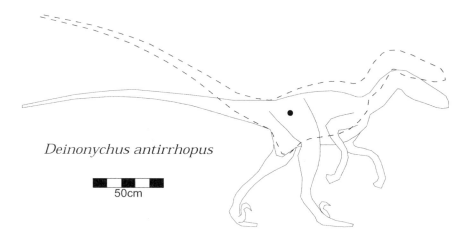

Deinonychus antirrhopus

50cm

maximized the angular momentum associated with their tail weapons by placing the added mass out on the tips of their tails (Arbour and Snively 2009). Being hit by a club or spike weapon involves an exchange of energy (Alexander et al. 1997). Upon collision between the weapon and the target, the kinetic energy of the moving club or spike is transferred very rapidly to the target, and the more kinetic energy that can be delivered to the target, the greater the damage inflicted. With masses of bone (ankylosaurs) or large spikes (stegosaurs) placed as far from the body as possible, a modest angular velocity as measured at the base of the tail becomes magnified by the long distance to the tip of tail, and maximizes the angular momentum imparted by a small amount of muscular effort. The hip and trunk regions of ankylosaurs and stegosaurs were massive relative to their tails, and would have possessed high rotation inertias. With their large rotational inertias, those portions of the body anterior to the hips would have responded only slightly to the rapid changes in angular momentum of the swinging tail, thus providing a stable base for the actions of the tail and its weapons.

Hind Limbs

Dinosaurs were primitively bipedal (Benton 1990; Gauthier 1986), as indicated by the retention of robust, long hind limbs in all dinosaurs, and almost all dinosaurs had hind limbs longer than their forelimbs. Even animals as thoroughly quadrupedal and front-heavy as *Triceratops* and the various ankylosaurs had longer hind limbs than forelimbs. A rare exception would be the Late Jurassic sauropod *Brachiosaurus* ("arm reptile"), which had arms slightly longer than its hind limbs. This retention of long hind limbs is more than just a case phylogenetic inertia, as dinosaurs, like most tetrapods, were basically "rear-wheel drive" (Ahlberg and Milner 1994). The main muscle that acted to retract the hind limb, and propel a dinosaurian body forward – the caudifemoralis longus (CFL) – was the largest muscle in the body of any dinosaur. The site of origination (base) for this muscle covered the anterior chevrons that lie below the centra, the sides of the anterior centra themselves, and the undersides of the transverse processes of these centra. The CFL's point of insertion (attachment) was on the posteromedial side of the femur, at a bony prominence termed the *fourth trochanter*. A similar arrangement of origin and insertion is seen in living crocodilians (Gatesy 1990). With the foot firmly planted on the ground, the action of the CFL pulling backward on the leg results in the leg acting a bit like the spoke of a wheel rotating about its axle. In this case the "axle" of the hind limb is the base of the foot, with the body being pulled up, forward, and over the supporting foot (Christian et al. 1996).

With its spokelike action, the hind limb is acting as a form of lever, and it is instructive to consider exactly where the CFL attaches to the leg to see what the mechanical advantages are for different points of attachment. Levers can be grouped into two main types – those that maximize the output force and those that maximize the distance that the distal end of the lever moves (Vogel 1998, 178–182). The in-lever of a force-maximizing lever is relatively longer than that of a distance-maximizing lever. The in-lever for the CFL is the distance between the point of insertion of the muscle on the femur and the hip socket. With the distance from the hip socket to the foot representing the out-lever, the mechanical advantage of the system is determined by dividing the out-lever by the in-lever, and the larger the in-lever the greater the mechanical force advantage. In animals that would require rapid acceleration, such as predators chasing prey or prey needing to escape the predators, the ideal would be for a given muscle contraction to result in a rapid movement of the limbs, and a correspondingly rapid acceleration of the body. This can be achieved by minimizing the in-lever. The disadvantage of this system is that the rapid acceleration of the foot and body is obtained at the expense of output force. This system works best for small-bodied animals, and inspection of the femora of a small predatory dinosaur such as *Coelophysis* (Colbert 1989), or small orthipods such as *Hypsilophodon* and *Heterodontosaurus* show that their fourth trochanters are no more than about one-quarter of the way down the back of their femora. In complete contrast are the insertion points of massive, quadrupedal forms such as the ankylosaurs *Edmontonia* and *Ankylosaurus*. In these heavy, slow-moving animals output force is the preferred optimization, and the point of insertion of the CFL is not less than halfway down the femur, thus improving the ratio of the output force to input force.

Determining whether an animal has the potential for running and jumping or is limited to walking depends on the strength of its bones. The weight of an animal exerts forces on its bones, and the strength of a bone is a measure of its ability to resist these forces and not break. The response of the bones to the forces that they are subject to depends on their size and shape, and this makes it difficult to compare the effects of forces of similar magnitudes on the bones of different animals. Rather than deal with forces directly, it is better to consider the stresses that act on bones. The stress within any solid is computed as the applied force divided by the area over which it acts. For a force coaxial with the long axis of the bone, the area to be considered will be that of the transverse area of the bone that lies perpendicular to the bone's long axis. For limb bones there are two main types of bone strength to consider: compressive and bending (Alexander 1985b). Compressive strength is a measure of the force directed along the axis of the bone that it can resist before failing by being crushed. Bending strength is a measure of the resistance of a limb bone to a force directed perpendicular to its long axis. Using the basic principles of engineering, Alexander (1985a) determined what sorts of forces the limb bones of dinosaurs could resist. He found that the estimated compressive strengths of all dinosaur bones were more than adequate to bear the weight of the animal, but that their bending strengths were the limiting factor for determining what sorts of activities a dinosaur could engage in.

The African elephant is the largest land living animal that we see today, and we can observe that elephants don't jump, and at their fastest they can only briefly manage a sort of shuffling, fast walk. It is the relatively low bending strengths of the elephant's main limb bones (femur, tibia, humerus) that set the limits on the animal's athletic ability. Computed strength indicators for the femur and tibia of an African elephant, 11 and 9 respectively, and similar calculations for the same bones of the Late Jurassic sauropod *Apatosaurus*, 9 and 6, are not that different (Alexander 1991). This suggests the *possibility* that an *Apatosaurus* could be almost as agile as the elephant. It should be noted that the body mass estimate used by Alexander for *Apatosaurus*, 34 tonnes, is much higher than more recent estimates of 17.5 (Paul 1997) and 16.4 (Henderson 2006) tonnes, and using this lower mass would give a higher strength for the bones. However, the elephant body form is quite compact when compared to that of the sauropod, and the dynamics of the extremely long neck and tail of *Apatosaurus* would most likely complicate matters. The implied similarities between the gait of the elephant and the sauropod are supported by the many preserved trackways interpreted to have been made by sauropods (Farlow 1992). These trackways show that sauropods took relatively short strides, and their typical mode of locomotion would have been a slow walk. To a first approximation the bending forces that a limb experiences during locomotion are proportional to the angle that the limb makes with the vertical. The short stride lengths of sauropods meant that their limb bones were not exposed to large bending stresses. It is for this reason that elephants do not gallop or jump; we can infer that sauropods were similarly limited.

The compressive strength of a bone depends on the total area of the bone that is subject to the applied force, while its bending strength depends not just on the area but also how that area is arranged about a reference line – the neutral axis of bending (Wainwright et al. 1976). The midshaft, cross-sectional shapes of the femora of many types of dinosaurs – stegosaurs, ankylosaurs, large ceratopsians, large hadrosaurs, sauropods – are elliptical (Wilson and Carrano 1999), with the long axis of the ellipse oriented perpendicular to the sagittal plane of the body (Fig 32.3). A bone that has an elliptically shaped cross section will have its greatest resistance to bending in the plane that contains the long axis of ellipse. Within the plane of the ellipse, the more bone that is situated away from the neutral axis, the greater the resistance to bending, and for the above-mentioned dinosaurs this implies that their femora were adapted to resist mediolateral forces more than anteroposterior one. Within the sauropods, Wilson and Carrano found that the degree of ellipticity of the femora of titanosaurs was greater than that of other sauropods. From an analysis of sauropod hips and knees and the shapes of their femora, these authors concluded that titanosaurs would have had their hind limbs more laterally positioned than in other sauropods. This meant that titanosaur limbs would make a larger angle with the vertical, and the bending stress acting in a plane perpendicular to the body axis would have been higher in their femora than in other sauropods. The increased mediolateral widths of their femora can be seen as a biomechanical

32.3. Demonstration of the problems of isometric increase in size of an animal's anatomy. The smaller limb is a normal-sized *Brachiosaurus* limb, and the larger limb shows the same limb with its linear dimensions increased by 150%. The mass of the second limb has increased by a factor of 3.58 (1.5³), but the stress developed in the femur has increased by 150%. The cross-sectional area of the femur has increased by only 225%, not enough to match the increased load it has to bear.

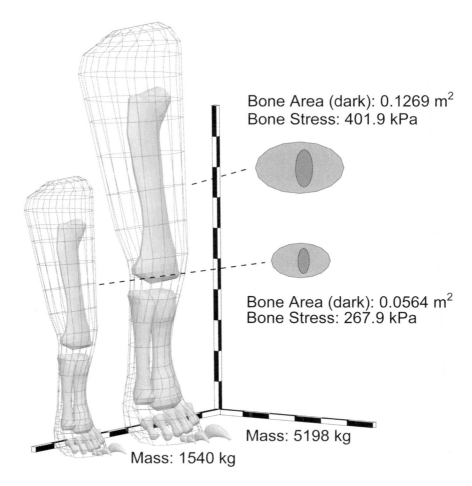

Bone Area (dark): 0.1269 m²
Bone Stress: 401.9 kPa

Bone Area (dark): 0.0564 m²
Bone Stress: 267.9 kPa

Mass: 5198 kg

Mass: 1540 kg

response to provide the added strength to enable titanosaurids to maintain the slightly more abducted limb posture.

In addition to the problems related to bending stresses in the bones, holding limbs in a bent pose takes muscular effort, and this muscular effort goes toward resisting the tendency of a flexed limb to collapse. This tendency to collapse arises because of gravity pulling the weight of the animal's body down onto all parts of the limb. Any bone that is not aligned with the vertical force of gravity will experience a turning force that is proportional to the length of the bone and to the angle that it makes with the either the hip socket or the other limbs bones that follow or precede it (Alexander 1985b). The more flexed the limb, the greater the turning forces that have to be resisted. For small animals, with their low body weights, it doesn't take much effort to hold the flexed pose, but because body mass increases faster than muscle strength when body size is increased isometrically, it would take impossibly large amounts of muscle and energy for a large animal, such as a *Tyrannosaurus rex*, to hold a flexed pose (Hutchinson and Garcia 2002). Elephants keep their limbs very erect, with the long bones vertically aligned, and have very short, stubby toes. These features characterize a type of stance and body support referred to as graviportal (Hildebrand 1982). By strategically aligning their bones so that the turning moments acting on the bones are minimized, large animals can avoid the need for unreasonably large leg and arm muscles. The repeated evolution of extreme body size in different clades of dinosaurs provides many examples of the trend toward increasing alignment of the limb bones. Within the ornithopods this trend can be observed with smaller forms, such as the Early Jurassic *Heterodonto-saurus* and the Early Cretaceous *Hypsilophodon*, being restored with their slender limb bones in flexed poses, but massive, Late Cretaceous hadrosaurs such as *Saurolophus*, *Edmontosaurus* and *Shantungosaurus*, which weighed in the range of 5–10 tonnes, having limbs best restored as erect columns.

Small animals tend to be more agile than large ones, and this is partly a result of their flexed limbs. A flexed limb can be quickly extended to propel an animal forward or sideways. In contrast, large animals, with their more erect limbs, have much less agility. A quantitative analysis of the muscles required to resist turning moments about the joints in the hind limbs of a running chicken was used to infer how much muscle would have been required in a cursorial, 6,000 kg *Tyrannosaurus rex* (Hutchinson and Garcia 2002). The results demonstrated that an adult *Tyrannosaurus* would have needed muscles on each leg equal to 40% of body mass, but estimates of the amount of leg muscles on the legs of a *Tyrannosaurus* are between 7 and 10% (Carrano and Hutchinson 2002). This analysis demonstrates the impossibility of a large bipedal animal such as *Tyrannosaurus rex* using a more flexed stance, and it also prevents the animal from running as fast as has been proposed in the past (Bakker 1986; Paul 1988).

An interesting biomechanical feature seen in the toes and claws of some of the smaller, predatory dinosaurs—the dromeosaurs, troodontids, and some of their closest relatives—is the hypertrophied claw on the second toe. This large claw would have been operated by a stout flexor tendon that attached to its base. Unlike the manual claws, however, the ratio of in-lever to out-lever of these pedal claws is 5:1, 25% greater than the 4:1 observed in hand claws. This higher ratio for the pedal claws suggests that speed of deployment was more important than force. It has been hypothesized that the hypertrophied pedal claws acted as a unique killing apparatus (Ostrom 1969), with the claws being quickly deployed and the prey being kicked and slashed with the claws. The force of penetration for these claws wouldn't necessarily come from the digital flexor muscles, but from the muscles of the hind limb itself. This would correspond with the inference that the speed of initial rotation of the claw was more important than power, as the power of penetration would come from larger muscles farther up the leg. Further evidence that speed was the preferred optimization is that the arc of motion swept out by these claws would be enhanced by coordinated rotations of the more proximal phalanges that supported the claw (Ostrom 1969). Given the other evidence that these dinosaurs were agile animals (see "Tails"), it is easy to imagine them leaping up on their prey, and using their body weight to drive the killing claws into the prey. The weight of the attacker would act to draw the claws down the sides of the prey to leave a very traumatic wound (Paul 1988, 367). As a result of a detailed finite element analysis of the enlarged pedal claw of the Early Cretaceous dromaeosaurid *Deinonychus*, Manning et al. (2009) suggested an additional function for the claw—that of a sort of climbing hook. It was shown that the claw alone was more than strong enough to support the weight of the animal, and could have also assisted in climbing.

A novel biomechanical mechanism has been observed in the lower portion of the hind limbs of some theropods—ornithomimids, tyrannosaurids, troodontids, elmisaurids, and avimimids. These taxa are characterized by a constriction of the proximal halves of their third metatarsals, and this state has been termed the *arctometatarsalian* condition (Holtz 1994). It is hypothesized that the third metatarsal, being wedged between metatarsals II and III, and held in place by elastic ligaments, acted as sort of shock absorber. When the foot first makes contact with the substrate, and is abruptly loaded with the weight of the animal, the broad ligaments anchoring metatarsal III to its neighbors would be stretched, accumulate potential energy, and distribute the force of impact across all three metatarsals (Snively and Russell 2002). This absorption of the energy of impact would lessen the chance of bone breakage when the animal was running at high speed and the forces of impact could be up to three times the weight of the animal (Alexander et al. 1979).

Trunk

Over the past 100-plus years it has been proposed several times that some sauropods could have reared up onto their hind legs in reaching for high-growing forage (Osborn 1899; Riggs 1904; Coombs 1975; Bakker 1978), or during some sort of ritual combat between conspecifics (Bakker 1986). Dodson (1990) cautioned that such behaviors may have been used only in times of stress such as periods of low food availability or during competition for mates, although he noted that the requirement for reproduction most likely involved the ability of one of the parties to stand on just its hind limbs. Three biomechanical aspects of an animal's body will determine if a bipedal stance is possible: (1) the ability of just the single pair of limbs to bear all the weight of the body; (2) the position of the center of mass of the body relative to the feet; and (3) the ability of the axial musculature and skeleton to lift the anterior body in the first place. As we have seen above, the forces normally experienced by the limb bones of an animal during locomotion can easily equal twice the body weight, so this is not a concern for a dinosaur that adopts a bipedal stance, as the bones are more than strong enough (Alexander 1985a).

The more crucial component to consider is the location of the center of mass of the body during the bipedal stance. A quadrupedal sauropod will have the greatest stability when its center of mass is somewhere above the stability region defined by the large quadrilateral, with vertices represented by the two hands and the two feet (Henderson 2006). For the sauropod to be able to stand bipedally, it is necessary that the animal tilt the front half of its body up in order to get the CM over the supporting hind feet. When the body has been tipped up enough, the CM will be positioned midway between the soles of the hind feet, and the animal will be balancing on just two legs. However, this two-legged stance is decidedly unstable, because the originally two-dimensional, quadrilateral stability region has collapsed down to just a one-dimensional, imaginary line joining the two feet. Our sauropod would be continually oscillating its CM back and forth across this line as it tried to maintain balance. For such a massive and elongate animal this does not seem like such a good idea. A much better scheme would be for tripodal support. A distinguishing feature of the diplodocid sauropods (*Diplodocus, Apatosaurus, Barosaurus*) and some cetiosaurids from the Early and Middle Jurassic of China (McIntosh 1990) is the peculiar, wedge-shaped chevrons that underlie the middle 50% of their tails (Hatcher 1901) (in fact the name *Diplodocus* means "double beam" and refers to the fact that this chevron form is best developed in this genus). If the tail was used as a "fifth leg" to form the third vertex of a triangular region of support (two hind feet plus the tail), these wedge-shaped chevrons would have done a good job of strengthening those ventral portions of the tail that would be in contact with ground. With this triangular configuration, the tail could also take some fraction of the animal's weight (Alexander 1985a).

Could a diplodocid, or early cetiosaurids, rear up into this tripodal stance? The amount of force that the muscles in the back would have to exert to lift the front half of the body up depends not just on the magnitude of the mass of the body anterior to the hips, but also the position of the center of mass of this region. A mass situated forward of the hips, when pulled on by gravity, has a tendency to tilt downward, and the magnitude of this effect is proportional to how far the body's CM is from its hips. This distance represents the moment arm of the anterior body mass relative to the hips, and the product of the mass multiplied by its moment arm is termed the moment of the mass. In a quadrupedal stance the forelimbs will act to resist this downward tilt, but if the animal wants to adopt a bipedal stance it becomes the task of the dorsal muscles not just to counter this downward tilt, but to supersede it in order to lift the anterior portion upward. The lifting of the anterior half of the body will be most strenuous at the beginning, because the CM of this region will be farthest from the hips. With increasing elevation of the front half of the body, the CM is moved closer to the hips (as measured in the horizontal direction), and the body mass moment to be countered diminishes. The concept of moment is also important in determining the ability of axial muscles to lift the front half of the body. The force of the epaxial muscles would work in conjunction with the moment arm provided by the neural spines of the posterior dorsal vertebrae and those of sacral and anterior caudal vertebrae. The very tall neural spines in this region of the axial skeleton of a diplodocid would provide a mechanical advantage to the epaxial muscles.

The centers of mass of the diplodocids are closer to their hips than in other sauropods, a result of their slender necks and long, heavy tails displacing the CM posteriorly (Alexander 1985a; Henderson 2006). The neural spines in the region of the hips are especially tall in diplodocids when compared to those of other sauropods. The combination of relatively small anterior body-mass moments, tall neural spines, and the presence of modified chevrons in diplodocids strongly suggests that these dinosaurs were capable of raising themselves up into a bipedal stance (Fig. 32.4). Another group of well-represented sauropods are the macronarians, with *Brachiosaurus* and *Camarasaurus* being among the best-known members. These animals are distinguished by relatively short tails; low dorsal, sacral, and caudal neural spines; and CMs that are situated almost midway between the shoulders and hips (Alexander 1985a; Henderson 2006) (Fig. 32.4). These animals also lack the modified chevrons that would have aided in strengthening the ventral portions of their tails. Despite the images in certain motion pictures (i.e., *Brachiosaurus* in *Jurassic Park* [1993]), it does not appear that macronarians would have adopted bipedal stance, except possibly for reproductive functions.

Even with the above arguments for and against facultative bipedalism in different sauropod groups based on differences in their skeletons,

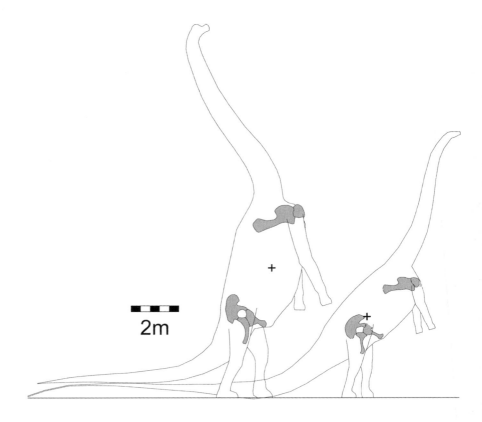

32.4. *Diplodocus* (right), with its posteriorly located center of mass (CM), is able to locate its CM between its hind feet and maintain a bipedal pose. It also has the potential to use its tail as a "third leg" for added stability. *Brachiosaurus* (left), with its more anteriorly located CM, is unable to position its CM between its feet—even when the neck is inclined backward in an attempt to balance. For this and other reasons it is unlikely that *Brachiosaurus* could adopt the stance seen in the film *Jurassic Park* (1993).

2m

these interpretations must be taken as tentative. Both African and Indian elephants are capable of getting themselves into a bipedal stance, and these animals possess neither long tails, nor tall sacral neural spines, nor centers of mass close to their hips (Henderson 2006). A study of the osteology of an animal is just the beginning in determining what an animal is capable of functionally.

A trend observed in several clades of herbivorous dinosaurs is for large forms to evolve very wide trunks. This is especially noticeable in heavily built sauropods such as *Camarasaurus* and *Opisthocoelicaudia*, and ankylosaurs such as *Euoplocephalus* and *Ankylosaurus*. The region of the spine between the shoulder and hip girdles of any quadrupedal tetrapod can be viewed as a beam, with the digestive tract and other organs being suspended below (Christian and Preuschoft 1996). A large, centrally placed load pulling down on a beam will cause the beam to sag at its middle, but the amount of sag increases exponentially with the length of the beam. As herbivorous dinosaurs grew larger and larger, they may have been encountering problems of increasing spinal sagging when the volume of the trunk increased to house a larger digestive tract. Rather than just lengthening the trunk, and then having to heavily reinforce the spine to resist sagging, it may have been better to just lengthen the trunk by a relatively small amount, but increase the side-to-side span of the ribs. This way trunk volume could increase and the sagging of the spine would be minimized.

With the overwhelming evidence that birds are flying theropod dinosaurs (Witmer and Chiappe 2002), aspects of avian anatomy and behavior have been increasingly assigned to extinct dinosaurs. One of the most intriguing is the documentation in the fossil record of the evolution of the system of air sacs seen in the spinal columns of living birds and trunk regions of extinct saurischian dinosaurs (Wedel 2008). Air sacs along the neck are inferred to have appeared no later than the Late Triassic in theropods (Sereno et al. 2008), and more extensive systems of air sacs are clearly well developed in the early bird *Archaeopteryx* by the Late Jurassic (Christiansen and Bonde 2000). Sauropods evolved their system of axial and abdominal air sacs in parallel to that of the theropods (Wedel 2003b). As well as having important implications for the physiology of extinct saurischians, it may also have a bearing on their locomotion. The presence of large air chambers within the body of a sauropod would have the effect of lowering its mean body density. To a good approximation the mean density of any animal is the same as that of water—1,000 gm/l. With internal air sacs equal to 15% of the total body volume, similar to that seen in birds (Proctor and Lynch 1997), the density of the more derived sauropods such as *Diplodocus* or *Brachiosaurus* will be less than of water. This suggests that they would be extremely buoyant when immersed in water (Henderson 2003). An earlier finding of Kermack (1951) showed that the water pressure acting on the chest of a fully immersed sauropod would collapse the animal's lung, and demonstrated that the old idea of immersed aquatic sauropods was not plausible. The buoyancy modeling study has shown that it would have been impossible for a sauropod to even sink to such a depth. Additionally, the computational models of the macronarian sauropods *Brachiosaurus* and *Camarasaurus* showed that they floated with their hands deeper than their feet, suggesting that while floating they could have made the manus-only trackways attributed to sauropods (Lockley and Rice 1990). In contrast, similar models of the diplodocoid sauropods *Diplodocus* and *Apatosaurus* showed them floating with feet deeper than the hands, making it unlikely that could have made the manus-only trackways (Henderson 2003).

Forelimbs

Relatively small, gracile forelimbs are characteristic of the early, primitive members of the main dinosaur clades—*Eoraptor* for theropods, *Thecodontosaurus* for sauropodomorphs, *Scutellosaurus* for thyreophorans, *Lesothosaurus* for ornithopods, and *Psittacosaurus* for ceratopsians. With the exception of the theropods, all other clades of dinosaurs evolved some form of quadrupedalism as their members evolved increased body sizes during the Mesozoic, and all those dinosaur groups that reverted to a four-footed stance were herbivores. Plant material is more difficult to digest than animal tissue, and herbivores are characterized by larger and more complex digestive systems (Farlow 1987). The digestive tract is contained in the region anterior to the hips, and with the increasing mass of their gut it would have become increasingly difficult for large herbivorous dinosaurs to maintain balance on just two legs. The forelimbs can be viewed as analogues of the front wheel of a wheelbarrow—they just acted to hold up the front of the body, and stop it from plowing into the ground and seriously impeding forward progression.

With the evolution of large body size in plant-eating dinosaurs, the transition between strictly bipedal gaits and quadrupedal ones can be observed both in their skeletons and in the trackways left by these animals, and is best seen in the evolution of the ornithopods. All the early ornithischians, animals such as the 1 m long Early Jurassic *Lesothosaurus* from South Africa were relatively small animals, were habitually bipedal, and weighed much less than 10 kg. During the Cretaceous, ornithischians, via the ornithopods, evolved some large to very large species – beginning with *Iguanodon bernisartensis* (2,640 kg) from the Early Cretaceous of western Europe (Norman 1980), and ending with *Edmontosaurus regalis* (10,250 kg) from the Late Cretaceous of North America and *Shantungosaurus giganteus* from the Late Cretaceous of China (25,000 kg) (unpublished results). The forelimbs of these animals became longer and much more robust, and it is apparent that the large ornithopods spent most of their time in a quadrupedal stance (Brett-Surman 1997). A very telling trackway from the Early Cretaceous of England shows the footprints A N D handprints made by a slowly walking *Iguanodon atherfieldensis* (Norman 1986). The interesting feature is that the patterns of handprint impressions are not as regular as those of a strictly quadrupedal animal, where one would expect a handprint for every footprint. It appears that this animal was delicately balanced, and most of its weight was carried on its hind limbs. The need for contact between the substrate and the hands was minimal, as evidenced by their infrequent record of contact (Wright 1999).

This phylogenetic trend from bipedality toward quadrupedality in ornithopods is also documented within the ontogeny of at least one genus of hadrosaur – *Maiasaura peeblesorum* from the Late Cretaceous of Montana. Using the abundant remains of juveniles, subadults, and adults of *Maiasaura*, Dilkes (2001) studied the osteology of the forelimb and identified a series of changes associated with growth: increasing robustness of adult humeri, allometric changes in postural muscles of the forelimb, increases in the lever arms of limb protractor muscles, increasingly closer association of the metacarpals, and changes in metacarpal cross-sectional shape to resist bending. All these observed changes can be interpreted as adaptation to cope with the increased loads that the forelimbs would have had to bear as the animal adopted a quadrupedal stance.

Among the most-specialized forelimbs of any dinosaur must be those of the sauropodomorphs. These dinosaurs show trends to reduce the phalanges and claws of digits II through V that began in the prosauropods (Galton and Upchurch 2004), with the reduction of the phalanges of digits IV and V. Simultaneously, the thumb claw became greatly enlarged, and the phalanges supporting it became very robust. *Plateosaurus* from the Late Triassic of Germany is known from many good skeletons (van Heerden 1997), and the joint at the base of its thumb claw enabled the thumb claw of this animal to be hyperflexed, perhaps to keep it clear of the substrate when the animal walked on the ground. In the more derived sauropods there was a marked trend to reduce manual digits II through V (Upchurch 1994). In the large forms such as *Brachiosaurus* and the diplodocids the various phalanges of the manus have been reduced to single little nubbins of bone. Long digits are not the best way for a large animal to transmit its great weight to the ground, and just as long limb bones are subject to bending stresses when weight forces act at a high angle to the bones, long digits would be subject to hyperextension and dislocation when loaded with the great weights of these animals. Again like the larger limb bones (such as the femur or humerus), the metacarpals of sauropods were held almost vertical in order to minimize the bending stresses that they would be subject to, and were arranged in a semicircular palisade reminiscent of a Greek column (Christiansen 1997). Bone is strongest when loaded in compression, and with the weight force borne by the forelimbs directed along the axes of the metacarpals these bones would be best able to support the animal without suffering mechanical fatigue and failing catastrophically. A final forelimb-strengthening feature of sauropods is the great reduction in the mobility of the wrist (Christiansen 1997), as indicated by the reduction of the set of carpals to just two, sometimes three, flattish bones (Bonnan 2003). Given the large loads experienced by the forelimbs, a mobile wrist could be a liability, and limiting the range of mobility can be viewed as a way of strengthening the limb.

There has been a long-running debate about the possible function of the large thumb claws of many sauropodomorphs. A common suggestion is for defense, and in life the prominent ungual would have been enlarged with the keratinous sheath that would have covered it (Van Heerden 1997) and enhanced its sharpness and deterrent effect. In the relatively smaller forms such as *Plateosaurus* this is a reasonable suggestion, although Galton and Upchurch (2004) suggested that the combination of the large claw and the flexible first digit to which it is attached would have had a grasping function.

For the very large sauropods there are two problems with the use of the thumb claw as a weapon. The first is that the claw is directed posteromedially, and the hand and lower arm would have had to be rotated by 90 degrees about its long axis to bring the claw into an anteromedial orientation for use (Upchurch 1994). The ability of the wrist and elbow to perform this rotation is doubtful. The second problem is the immense size of the forelimb of the sauropods – up to 4 m long and weighing almost 1,000 kg in something like *Brachiosaurus* (Henderson 2006) (Fig. 32.4). The forelimbs of all sauropods were specifically adapted for supporting the great weights of these animals (Christiansen 1997), and the normal "mode of operation" would have been to slowly swing the limb back and forth in a parasagittal plane. Raising the forelimbs and then rapidly swinging them mediolaterally at some adversary not only would take a large amount of muscular effort, but also would put the animal at risk of toppling over (Upchurch 1994). The center of mass of *Brachiosaurus* is very close to the midpoint of the trunk, and balance is carefully managed (Henderson 2006). With a large, heavy forelimb held out to the side or in the front of the body and moving, the normal state of balance could be jeopardized.

A more plausible suggestion for the function of the manus claws has been for use in anchoring the forefoot to a tree trunk. This would be done when a sauropod raised itself up onto its hind limbs to reach higher-growing foliage. By resting its forelimbs on the trunk of the tree on which it was feeding, the animal would be much more stable (Upchurch 1994).

A long-running controversy has been over the posture of the forelimbs of ceratopsians. Although some early workers suggested erect, mammal-like forelimb posture (Marsh 1891; Hatcher et al. 1907), the forelimbs of animals such as *Triceratops* have been typically restored with the humerus subhorizontal and projecting laterally from the body (e.g., Gilmore 1905; Osborn 1933). This forelimb stance required that the elbows be flexed at about 90 degrees, and the animal would have had a very sprawled, lizardlike posture. Later authors such as Bakker (1986) and Paul (1987) proposed that the forelimbs were held erect like those of mammals and that the forelimbs moved back and forth, again like a mammal's. The hind limbs of ceratopsians were held erect, and there has been little controversy on this aspect of their anatomy (Paul and Christiansen 2000), but the erect hind limbs would appear to be at odds with sprawling forelimbs (Bakker 1986). The strengths of the bones of adult *Triceratops* are very similar to values calculated for living rhinos (Alexander 1991), and it was suggested that *Triceratops* and other ceratopsians could have been as agile as living rhinos, which are capable of a slow gallop. Other aspects of ceratopsian anatomy also suggest that they were the Cretaceous ecological analogue of rhinos as well—heavily built, slow-moving, low-grazers—so why not endow them with a similar locomotion capability?

The discovery of a trackway made by a very large ceratopsian from the Maastrichtian of Colorado (Lockley and Hunt 1995) may have resolved the question of ceratopsian forelimb posture. This trackway shows that the manus (hand) prints were not directly in line with the pes (foot) prints as seen in a rhinoceros, nor were they far outside the pes prints as they would be if the animal had the fully sprawled posture. Instead the manus prints were just to the sides and slightly ahead of the footprints. A detailed analysis of the shoulder and elbow joints of *Triceratops* by Paul and Christiansen (2000) reveals that proper articulation of the humerus with the shoulder, and of the radius and ulna at the elbow, results in the hands being in just the positions recorded by the Maastrichtian trackways. Furthermore, the elbows are in an intermediate position, neither far out to the sides, nor directly under the body. This revised orientation for the forelimb of large ceratopsians has the limb in flexed stance, unlike the more columnar, erect stance seen in living elephants and the much larger sauropods. The humerus of *Triceratops* and other large ceratopsians is a very robust bone, and well able to resist the weight of the body acting on it in this bent pose (see "Hind Limbs" section above for a discussion of bone strength).

Another long-running debate concerning dinosaurian forelimbs has been about the possible function, or not, of the reduced forelimbs of tyrannosaurids. Early theropods all had substantial, strong-built arms with well developed claws, for example, *Coelophysis* and *Herrerasaurus*. In life, the unguals would have been covered with keratinous sheaths that would have provided larger, sharper tips to the bony cores, and it is widely accepted that the forelimbs of early theropods were used for prey capture. Additionally, the long scapulae and broad coracoids of theropods would have provided large areas for the origination of muscles that acted to powerfully protract and retract the arms. Given the prominence of strongly clawed forelimbs in most theropods, the reduced forelimbs of tyrannosaurids are unusual. Paul (1988) interpreted them as vestigial, and in a hypothetical scenario where dinosaurs did not go extinct at the end of the Cretaceous he restores tyrannosaurs as having lost their forelimbs altogether Paul (2000).

Other authors feel that tyrannosaur limbs were not useless appendages and on their way to evolutionary oblivion. Horner and Lessem (1993) noted that the attachment scar for the biceps on the ulna of *Tyrannosaurus rex* was a big as an American dime, and from that they conclude that this muscle was exerting a powerful pull on the lower arm. Carpenter and Smith (2001) also predicted a very muscular arm for *Tyrannosaurus*. Both of these studies highlight the strong humerus of *Tyrannosaurus*, which would have been able to resist high bending and compressive stresses. Carpenter and Smith argue that the strong forelimbs of *Tyrannosaurus* were used to restrain struggling prey. However, with just two, relatively small, claws on the ends of short arms, tyrannosaurs do not seem to have been as well equipped to grapple with prey via their arms. Another group of large theropods, the allosaurids from the Late Jurassic, produced some very large individuals (e.g., "Big Al") that approached the size of large tyrannosaurs. All the allosaurs retained large forelimbs with three stout claws, and the combination of longer arms and heavy claws makes prey-grasping a reasonable functional interpretation in this case.

Newman (1970) suggested a completely different function for tyrannosaur arms—that of stabilizers for when the animal was lifting itself up after having rested with its belly and head stretched out along the ground. As the hind limbs pushed the body up, the forelimbs would have braced the body from sliding forward during the lifting action. The stresses acting on the arm bones in this scenario would have been mainly compressive, and the short, stocky nature of the bones suggests that they were well able to resist such forces. Carpenter and Smith (2001) also noted that the geometry of the cross section of the humerus indicated that it was well designed to resist sudden impulse forces. Such forces would arise if the powerful hind limbs resulted in the body being rapidly jerked upward at the beginning of the lifting phase as envisaged by Newman (1970). These ideas about forelimb function were investigated using a sophisticated 3-D computer model of *Tyrannosaurus rex* by Stevens et al. (2008).

The extreme forelimb reduction seen in the derived tyrannosaurs is not unique to them. A similar but independent evolution of large predatory dinosaurs with diminutive forelimbs took place during the Cretaceous on the land masses that once formed the southern supercontinent Gondwana. The abelisaurs, a group of derived ceratosaurs more closely related to better-known forms such as *Coelophysis* (Late Triassic) (Tykoski and Rowe 2004) than to tyrannosaurs, are characterized by greatly reduced forelimbs. Similar to the tyrannosaurs, the abelisaurs also show extremely robust pectoral girdles with broad coracoids and long, wide scapulae. Clearly these reduced arms were not vestigial structures. Unlike the derived tyrannosaur condition, the abelisaurs retained more than two digits in their hands, for example, *Carnotaurus* (Bonaparte 1985) and *Aucasaurus* (Coria et al. 2002), both from Campanian/Maastrichtian of Argentina.

Neck

If the immense size of many dinosaurs is the first aspect of the group that attracts attention, then the long necks of many forms must surely be the next most conspicuous feature. The earliest known saurischian dinosaurs—theropods such as *Eoraptor* and *Herrerasaurus* and early prosauropodomorphs such as *Anchisaurus* and *Saturnalia*—all had necks that were noticeably longer than those of other archosaurs from the same time period of Middle and Late Triassic—rauisuchians, phytosaurs, aetosaurs, and rhynchosaurs. The necks of all the various dinosaur groups experienced evolutionary changes to some degree, but it was within the saurischians that the most dramatic changes took place with the extreme lengthening of the necks of sauropods.

No other group of terrestrial vertebrates has evolved necks as distinctive as those of sauropods. The most basal sauropods had 10 cervicals (Upchurch 2004), while the Late Jurassic *Mamenchisaurus* from China had 19 cervicals forming a neck that may have been 14 m long! Not only did sauropods have a higher number of cervicals, but the cervical vertebrae themselves became greatly elongated. This is especially noticeable in the Early Cretaceous brachiosaurid *Sauroposeidon*, where the exposed length of the eighth cervical is 1.25 m (Wedel et al. 2000). As bizarre as the long necks of sauropods appear, they must have had a selective advantage and been beneficial to the group, as long necks were a consistent feature of sauropods right up to the end of the Mesozoic. The challenge for paleontologists is to understand how such extreme necks could have been supported and moved, and how the other parts of the body coped with the long neck. Giraffes are often cited as ecological analogues of sauropods, but with just the standard mammalian allocation of seven cervical vertebrae, and necks that aren't much longer than their trunks, the giraffes don't even come close to the conditions seen in most sauropods. However, giraffes provide our only insight to some aspects of the biomechanics of living with a long neck.

An intriguing aspect of sauropod necks is that it is a long way to pump blood between the heart and the head if the neck is held vertically. The blood in the carotid arteries of the neck can be viewed as columns of liquid, and with the neck of an adult giraffe being about 2–2.5 m long, this column of liquid exerts a substantial downward pressure at its base. An adult giraffe has a heart about as big as a basketball, and this heart has to provide enough force not only to lift the column of blood up to the head, but also to overcome the friction associated with moving a viscous fluid such as blood through a long, narrow tube. Now imagine the forces involved in moving blood up and along the carotid arteries in something like *Mamenchisaurus* with its 14 m (?) neck. It is the problem of very high pressure at the base of the columns of blood in the carotid arteries that troubles many paleontologists and physiologists when they see sauropods restored with their necks held vertically (e.g., Bakker 1986; Paul 1997).

Lizards, snakes, and turtles normally have their heads just slightly above the levels of their hearts, and with their simple three-chambered hearts (two atria and a single, common ventricle) they do not have to contend with the problems of high blood pressure associated with greatly elevated heads and necks. Typical blood pressures of extant reptiles are between 20 and 70 mm Hg (Seymour 1976), and in these animals the circuit that carries blood to and from the lungs (pulmonary circuit) operates at the same pressure as the circuit that carries blood to the rest of the body (systemic circuit). Sauropods would be expected to have had very high systemic blood pressures if they held their heads high up–508 mm Hg in the case of *Brachiosaurus*, with a 6.5 m neck (Seymour 1976)–and they would experience leakage of the fluid component of the blood out of the circulatory system and into the lungs (pulmonary edema) if this high blood pressure was acting at the lung surfaces. Mammals have high systemic blood pressure, but have separated the systemic and pulmonary circuits with a four-chambered heart (two atria and two ventricles). This allows the pulmonary circuit to operate at a much lower pressure, and avoids the fluid leakage problem. If sauropods did have high systemic pressures, they must have had a four-chambered heart and separate pulmonary and systemic circuits. This inference is plausible given that the two groups of living archosaurs, crocodilians and birds (avian theropods), both have four-chambered hearts.

The above discussion of high systemic blood pressures rests on the assumption that sauropods normally held their necks in an erect, swan-like fashion. The blood pressure problems could be avoided if sauropods maintained their necks horizontally, as the head would be no more than half a meter above the level of the heart in something like *Apatosaurus* (Hohnke, 1973). By employing computational geometry to constrain the degree of separation of pre- and postzygapophyses of the cervical vertebrae, it has been proposed that the diplodocids *Apatosaurus* and *Diplodocus* were limited to keeping their necks subhorizontal by having the zygapophyses aligned with each other, the so-called osteological neutral pose (Stevens and Parrish 1999). However, a new problem arises with this neck orientation, as the tensile forces in the epaxial portions of the cervical and dorsal regions of the back would become very large, and there would be equally large compressive forces between the cervical centra to be resisted as well (Alexander 1985a). Another objection to the use of the osteological neutral pose is that living amniotes do not habitually have their necks in this pose (Taylor et al. 2009). These latter authors showed that mammals, turtles, lizards and snakes, crocodilians and birds, all held their necks erect.

As was seen with the tails of sauropods, a horizontally held sauropod's neck would have experienced large forces acting to bend it downward. Within the proximal third of the tail of an animal such as *Apatosaurus* the centra are massive, squat disks that were well able to resist both the great compressive forces associated with the droop-resisting tensions in the upper half of the tail and the leg retraction forces coming from the pair of left and right caudifemoralis muscles (see "Hind Limbs"). A sauropod tail did not have to be very flexible, and the massive, amphiplatyan proximal centra of *Apatosaurus* are perfectly suited to dealing with these forces. However, it is a very different story with the cervical vertebrae. As observed for giraffes, it is generally assumed that the long necks of sauropods were associated with feeding, and the long necks may have made facilitated food gathering in a forested environment (Wedel et al. 2000). Energetically, it also makes sense for an animal weighing many tonnes to move just the neck to reach all the foliage within range of the body, rather than shifting the whole body by a few steps every few minutes. To enable a sauropod to reach all the foliage in its neighborhood the neck ought to be very flexible and able to bend sideways as well as up and down (Martin 1987). This requires not only well-developed articulation between the centra, but low neural spines that will not interfere with each other when the neck is strongly flexed. Unfortunately, it appears that a weighty neck is incompatible with the requirements of flexibility, as large amounts of motion-constricting muscle and/or tendon and ligament would be required to support and brace a heavy neck.

Sauropods appear to have overcome the problem of combining neck support with flexibility by greatly reducing the amount of bone required for the construction of their necks via the development of "excavations" on the insides and outsides of their cervical and dorsal vertebrae. This has resulted in their cervical vertebrae being constructed of thin sheets and laminae of bone (Britt 1993), in contrast to the more typical forms of centra as solid cylinders and neural arches as trusses of solid rods and plates of bone. Early sauropods such as the cetiosaurids exhibit relatively modest reductions in the amount of bone used to form their cervicals (Martin 1987), but the more derived diplodocids and macronarians show extreme reductions in the amount of cervical bone, and it is in these latter groups that the necks reach the extreme lengths. *Sauroposeidon*, an Early Cretaceous brachiosaurid from Oklahoma, has extremely thin bony laminae comprising its cervicals, and it has a neck that is between 25 and 33% longer than that of *Brachiosaurus* (Wedel et al. 2000). It could be argued that the ability to reduce the amount of material used in the construction of sauropod vertebrae was a "key innovation" (Liem 1973) that led to the evolution of extreme neck forms. Incidentally, this reduction of bone in the cervical and dorsal vertebrae correlates with the likelihood that more derived sauropods had extensive pneumatization of the axial skeleton similar to that seen in birds (Reid 1997; Wedel 2003b). The resorption of bone associated with the growth and development of the system of pneumatic air sacs (Witmer 1997) would have assisted with the reduction of the mass of the cervical vertebrae.

The neural spines of the cervical and anterior dorsal vertebrae of diplodocid and camarasaurid sauropods are distinctive in that they are divided down their midlines into left and right processes. It has been proposed that a strong ligament ran down the center of this groove between the bifurcated neural spines (Alexander 1985a). With the neck of a *Diplodocus* held horizontally and subject to the downward pull and bending associated with the force of gravity, Alexander calculated what the tensile stress would have been in this central, epaxial ligament. He assumed that the ligament filled the neural spine groove, which would have given it a cross-sectional area of about $0.09 m^2$, and found the stress would have been 800 kPa. Collagen is a common connective tissue in vertebrates, but has a very high strength, far higher than would be necessary to be used in supporting the neck of *Diplodocus*, so Alexander proposed that elastin would be a better material to use for the construction of an epaxial, suspensory ligament. Prominent elastin ligaments (nuchal ligaments) are found in the necks of many grazing mammals (Hildebrand 1982), and the elastic recoil of the tendon assists with raising an animal's head when it has finished grazing or drinking and has relaxed the muscles that had actively flexed the head and neck downward. Such a passive neck-raising mechanism would have been appropriate for long-necked sauropods.

Final points to consider about the necks of sauropods are the length of the trachea and the mechanisms for moving air in and out of the lungs. Animals such as *Barosaurus*, *Mamenchisaurus*, and *Sauroposeidon*, with estimated neck lengths of 9, 14(?), and 12 m, respectively, would have had large amounts of "dead" air in their tracheas. This stagnant, oxygen-depleted, CO_2-laden air would dilute the quantity of fresh air that was taken with each inspiration and reduce the amount of CO_2 that could be exhaled (Daniels and Pratt 1992). There would also be energetic cost of pushing this dead air back and forth within the trachea, but not getting any value from it. To investigate the minimum requirements for respiration in *Mamenchisaurus*, Daniels and Pratt (1992) developed a mathematical model of the flow of air through the trachea of this dinosaur that incorporated the length and radius of the trachea, the density and viscosity of air, and the possibilities that the animal had a low, reptilian metabolism or a high, mammalian one. They determined that with its very long, narrow trachea, it would be most efficient (least mechanical effort) for *Mamenchisaurus* to have a birdlike lung and a system of air sacs to efficiently move the air in and out of the body, and to extract as much oxygen as possible from a given inhalation.

The head of any vertebrate is arguably the most complex part of its body. The head is a fusion of three distinct embryonic units: chondrocranium (the set of bones that enclose and support the brain and sense organs), splanchnocranium (forming the jaws, hyoid apparatus, and gills), and dermatocranium (the set of bones that cover the skull and jaws). From a phylogenetic point of view the complexity of the skull of any vertebrate provides a wealth of detail for unraveling the ancestry and relationships of the animal. From a biomechanical point of view the skull is a challenge waiting to be tackled. The convoluted shapes of the skull bones, their three-dimensional relationships to one another, and the complexity of the forces produced by the various muscles that lie inside and outside the skull make it very difficult to resolve all the forces and their biomechanical effects in all but the simplest cases. However, progress has been, and is being, made, on several fronts.

Alexander (1985a) established that viewing some portion of a dinosaur's skeleton as either a beam loaded in compression or subject to a laterally directed bending force enables us to calculate the strength of the beamlike skeletal element and to infer what sorts of loads the skeletal element could resist before breaking. To a good approximation, the well-spaced teeth of a carnivorous dinosaur such as a *Tyrannosaurus* or an *Allosaurus* can be treated as cantilevered beams with their roots set solidly in the edge of the jaw, while their crowns are exposed and subject to various forces. With data on the mediolateral (ML) and anteroposterior (AP) tooth crown dimensions, and crown heights from a wide range of theropod dinosaurs and other animals, it was found that the cross-sectional shapes of theropod teeth were not circular, but elliptical, with their long axes aligned anteroposteriorly (Farlow et al. 1991). Put succinctly, "The business end of a carnivorous dinosaur was its mouth" (Farlow et al. 1991), and when a theropod's mouth impacted a prey item the teeth would experience forces that would tend to bend them either sideways or forward/backward. Just as was seen in the shape of the cross sections of the hind-limb bones of titanosaurs (see "Hind Limbs" above), the preferred alignment of the cross-sectional axes of the teeth indicate that they were better able to resist forces in the AP direction. This seems reasonable, because after biting into a prey animal, the next task of the teeth can be either to stop the prey escaping forward from the mouth or to enable the predator to pull backward with its neck muscles to pull chunks of flesh from the victim (Rayfield et al. 2001).

The teeth of any animal do not function in isolation, and a tooth's performance is not just limited to its own internal strength. Any force that does not break off a tooth crown will be transmitted to its root, and that root in turn transmits the force to the bones of the skull that enclose the root. It follows that the teeth of any predaceous dinosaur ought to be well anchored. From an analysis of the shapes and sizes of the maxillae of predatory dinosaurs, and the largest teeth that they bore, it was found that there is a strong correlation between the amount of bone enclosing the base of the teeth and the size and strength of the teeth (Henderson and Weishampel 2002). Further support for the idea that larger, stronger teeth need more bony support comes from the fact that this trend to reinforce the maxilla evolved independently at least twice in two different clades of macrocarnivorous theropods (Ceratosauria and Tetanura), as well as in other nondinosaurian, carnivorous archosaurs (rauisuchians and crocodylomorphs) (Henderson and Weishampel 2002).

It is not just the strength of the cranium that matters when biting; the lower jaws also have to sustain high forces. An analysis of the bending and torsional strengths of the mandibles of extant lizards and crocodilians and extinct dinosaurs found that strength profiles differed between the specimens examined, but that they could be grouped into different categories, and inferences could be made about feeding styles and bite force (Therrien et al. 2005). For example, the strength profile of the largest living lizards – the Komodo dragon (*Varanus komodoensis*) – was similar to that of the abelisaurids *Majungasaurus* and *Carnotaurus*, while tyrannosaurids were unique in their exceptional mandibular adaptations to resist twisting of the anterior of the lower jaws.

Following an initial suggestion by Chure (1998) that the sizes of the orbits of theropods may be useful for inferring the ecology of these dinosaurs, a more comprehensive investigation found that there exists a relationship between the sizes of the orbits and strengths of the skulls that hosted them (Henderson 2002). The skull strengths were obtained for a series of skulls by treating the cranium as a beam that experienced an upward force applied on the underside of the muzzle at the position of the longest tooth. The actual strength parameters were calculated using three-dimensional digital data that enabled cross sections of the skull to be determined at the positions of the orbits, and to determine the area of the orbits as well. When the sizes of the orbits were plotted against skull strength there was a clear trend associating relatively small orbits with strong skulls. *Tyrannosaurus rex* had not only the smallest orbits (expressed as a percentage of the total lateral area of the skull), but also the strongest skull (Fig. 32.5a). The strength of the *T. rex* skull was twice that of the second strongest skull – that of the very large, Early Cretaceous allosauroid *Acrocanthrosaurus atokensis*. A surprising result was that the 100 cm long skull of the Early Cretaceous spinosaur *Suchomimus teneresis* was actually weaker than the skull of smaller animals such as *Herrerasaurus* and *Dromaeosaurus*. This effect appears to be due to the muzzle region of *Suchomimus* being very elongate relative to the posterior portion of the skull, and the lever action of this long snout would provide a large mechanical advantage to any force that impinged on the snout (Fig. 32.5b). It would seem intuitively obvious that a skull composed of a large amount of bone would be stronger than one with less bone, and with the digital data for the skulls, and the shapes and sizes of the other openings in the skulls (external nares, antorbital fenestra, temporal fenestrae), it was possible to check this hypothesis. When the total bony area of the lateral side of the theropod skulls was plotted against strengths computed for the digital skulls, there was a strong positive correlation between the two (Henderson 2002).

A thorough study that looked at the stresses throughout an entire dinosaur skull was performed by Rayfield et al. (2001). The form of the entire skull and mandible of a large specimen of *Allosaurus fragilis* ("Big Al") was captured using serial computed tomography, and this provided a finite element model of 146,398 elements. With some assumptions about the

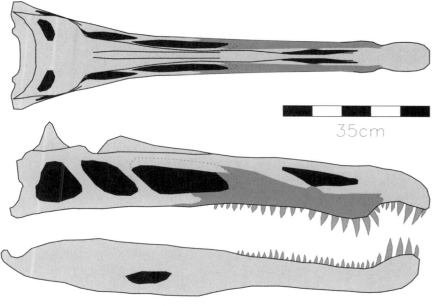

Suchomimus tenerensis

35cm

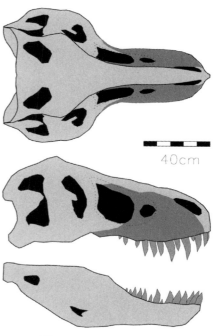

Tyrannosaurus rex

40cm

mechanical properties of the bones and teeth (e.g., elastic constants, shearing resistance, densities, etc.) it was possible to computationally apply an external stress and see how the digital skull of *Allosaurus* responded. Based on the available space for jaw adductor (closing) muscles, it was found that *Allosaurus* did not have a particularly strong bite. In contrast, it was found that the bony elements of the skull, together with their geometrical arrangement, made for a very strong skull – much higher than that required to resist the forces produced by the biting muscles. This "overengineered" skull led to the inference that *Allosaurus* may have had a slash-and-tear mode of attacking prey, similar to what is seen in living varanid lizards (Rayfield et al. 2001), and that the skull had evolved to deal with the high, brief stresses associated with the impact between predator and prey.

The observed correlations between tooth strengths of theropods and the bones and skulls that they are associated with can also be seen in non-carnivorous dinosaurs. The Late Jurassic sauropod *Camarasaurus* has very stout teeth that exhibit clear wear facets produced by tooth-to-tooth and tooth-to-food contact (Calvo 1994). Analysis of the microwear patterns on their teeth revealed long scratches, but minor amounts of pitting, and suggests that that these dinosaurs were feeding on tough, but not hard, vegetation (Calvo 1994). This animal exhibits a robustly constructed skull and a deep, thick maxilla, features indicating that this animal had a skull strong enough to apply sufficient bite force to deal with resistant food. The skull of *Camarasaurus* is very different from that of diplodocids such as *Apatosaurus* and *Diplodocus*, whose skulls are elongate and appear to be lightly built. The teeth of these latter two genera are slender and pencil-like forms that were clearly not involved in biting down on resistant materials (Barrett and Upchurch 1994). It also appears that the preferred lower jaw motion for diplodocids was a fore-and-aft, propalineal motion, which would result in the upper and lower dentitions acting to snip plant tissues from their attachments (Calvo 1994).

32.5. "Rex Rules!" The wide temporal region and broad muzzle of *Tyrannosaurus* give this animal an exceptionally strong skull. The long, low, slender snout of *Suchomimus* makes for a much weaker skull. The robust teeth of *Tyrannosaurus* are associated with a very deep, thick maxilla, and are indicative of a powerful bite. The relatively small teeth of *Suchomimus* are rooted in a slender, thin maxilla, and it is highly unlikely that this animal was capable of biting with great force. The slender, weak skulls seen in spinosaurids such as *Suchomimus* contradict aspects of the plot of the film *Jurassic Park III* (2001).

An important confirmation of the modeling studies that predict great strength for the skull of the *Tyrannosaurus* comes from the discovery and interpretation of a series of bite marks found on the pelvis of a *Triceratops* specimen from Montana (Erickson et al. 1996). Casts were made of the dental puncture marks in the ceratopsian pelvis, and this led to the identification of the puncture-making teeth as those of a *Tyrannosaurus*. With the observation that the bones of all large vertebrates have very similar mechanical properties, experiments on the amount of force required to make similar-size bite marks in the pelvic bones of extant cattle showed that somewhere between 6,410 and 13,400 N of force were required to replicate the 11.5 mm puncture marks seen in the *Triceratops* pelvis. For comparison, most living carnivores exert much lower forces: Labrador dogs: 550 N; wolves: 1,412; and lions: 4,168. A notable exception among living predators is the American alligator, which can exert up to 13,300 N of bite force (Erickson et al. 1996).

The above discussions of skull mechanics have all been related to feeding, but an unusual occurrence of exceptional skull strength relates to the thickened heads in a group of incompletely known marginocephalians – the pachycephalosaurs (Maryańska et al. 2004). The name of the group refers to the greatly thickened mass of bone covering the brain and front part of the face, and the general consensus appears to be that this thickening is somehow related to a form of ritual, intraspecific combat between individuals (Galton 1971; Sues 1978). The amount of added bone is remarkable: a *Pachycephalosaurus* skull 62 cm long had a capping thickness of 22 cm of bone (Sues 1978). Galton (1971) suggested that male pachycephalosaurs would have butted heads in a manner similar to that seen in bighorn sheep. A pair of bighorn rams will approach and collide with each other at speeds estimated to be 21 km/hr (Kitchener 1988), and using this modern example of head-butting, Alexander (1997) provided rough estimates of the forces involved, which indicate that the kinetic energies of impacting pachycephalosaurs could have been absorbed by a combination of dorsal muscle and bone. Aspects of the postcranial skeleton support the inference that pachycephalosaurs were actively colliding with each other. These features include: a prominent buttress on the posterodorsal rim of the acetabulum to reinforce the hip socket when the body would decelerate to an abrupt stop against the head of the femur upon impact; forward inclination of the region of the braincase where it connected to the spine so as to better align the force of impact occurring on the top of the skull with the spinal column itself to minimize any force components that would tend to bend the neck; and tongue-and-groove contacts between vertebrae to resist twisting forces associated with the impact (Maryańska et al. 2004).

Conclusion

There is more to the study of dinosaurs than just describing their bones and then determining who was related to whom. The immense size of many dinosaurs demonstrates that the variety of large terrestrial animals that we see today are not representative of the true potential of the vertebrate body in terms of large body size. With the quantitative techniques of biomechanics, studies of how dinosaurs supported, moved, and used their bodies provide us with a broader perspective of what vertebrates in general are capable of as physical organisms. Biomechanical studies of dinosaurs, supported by what is known of the biomechanics of living vertebrates, enable us to attach a degree of quantitative rigor to our inferences about the paleobiology of dinosaurs.

References

Ahlberg, P. E., and A. R. Milner. 1994. The origin and early diversification of tetrapods. *Nature* 368: 507–514.

Alexander, R. McN. 1985a. Mechanics of posture and gait of some large dinosaurs. *Zoological Journal of the Linnean Society* 83: 1–25.

———. 1985b. Body support, scaling, and allometry. In M. Hildebrand, D. M. Bramble, K. F. Liem, and D. B. Wake (eds.), *Functional Vertebrate Morphology*, 26–37. Cambridge, Mass.: Harvard University Press.

———. 1991. How dinosaurs ran. *Scientific American* 264: 62–68.

———. 1997. Engineering a dinosaur. In J. O. Farlow and M. K. Brett-Surman (eds.), *The Complete Dinosaur*, 414–425. Bloomington: Indiana University Press.

Alexander, R. M., R. A. Fariña, S. F. Vizcaino. 1997. Tail blow energy and carapace fractures in a large glyptodont (Mammalia, Xenarthra). *Zoological Journal of the Linnean Society* 126: 41–49.

Alexander, R. M., G. M. O. Maloiy, R. Njau, and A. S. Jayes. 1979. Mechanics of running of the ostrich. *Journal of Zoology* 187: 169–178.

Arbour, V. M., and E. Snively. 2009. Finite element analyses of ankylosaurid dinosaur tail club impacts. *Anatomical Record* 292 (9): 1412–1426.

Bakker, R. T. 1978. Dinosaur feeding behaviour and the origin of flowering plants. *Nature* 274: 661–663.

———. 1986. *The Dinosaur Heresies: New Theories Unlocking the Mystery of the Dinosaurs and Their Extinction*. New York: William Morrow.

Barrett, P. M., and P. Upchurch. 1994. Feeding mechanisms of *Diplodocus*. *Gaia* 10: 195–203.

Bates, K. T., P. L. Manning, D. Hodgetts, and W. I. Sellers. 2009. Estimating mass properties of dinosaurs using laser imaging and 3D computer modeling. *PLoS ONE* 4 (2): e4532.

Benton, M. J. 1990. Origin and interrelationships of Dinosaurs. In D. B. Weishampel, P. Dodson, and H. Osmólska (eds.), *The Dinosauria*, 11–30. Berkeley: University of California Press.

Bonaparte, J. F. 1985. A horned Cretaceous carnosaur from Patagonia. *National Geographic Research* 1: 150–152.

Bonnan, M. F. 2003. The evolution of manus shape in sauropod dinosaurs: implications for functional morphology, forelimb orientation, and phylogeny. *Journal of Vertebrate Paleontology* 23 (2): 595–613.

Brett-Surman, M. K. 1997. Ornithopods. In J. O. Farlow and M. K. Brett-Surman (eds.), *The Complete Dinosaur*, 330–346. Bloomington: Indiana University Press.

Britt, B. B. 1993. Pneumatic postcranial bones in dinosaurs and other archosaurs. Ph.D. thesis, University of Calgary.

Calder, W. A. 1984. *Size, Function, and Life History*. Cambridge, Mass.: Harvard University Press.

Calvo, J. O. 1994. Jaw mechanics in sauropod dinosaurs. *Gaia* 10: 183–193.

Carpenter, K., and M. Smith. 2001. Forelimb osteology and biomechanics of *Tyrannosaurus rex*. In D. H. Tanke and K. Carpenter (eds.), *Mesozoic Vertebrate Life*, 90–116. Bloomington: Indiana University Press.

Carrano, M. T., and J. R. Hutchinson. 2002. The pelvic and hind limb musculature of *Tyrannosaurus rex* (Dinosauria: Theropoda). *Journal of Morphology* 253: 207–228.

Christian, A., D. Koberg, and H. Preuschoft. 1996. Shape of the pelvis and posture of the hindlimbs in *Plateosaurus*. *Paläontologische Zietschrift* 70: 591–601.

Christian, A., and H. Preuschoft. 1996. Deducing the body posture of extinct large vertebrates from the shape of the vertebral column. *Palaeontology* 39 (4): 801–812.

Christiansen, P. 1997. Locomotion in sauropod dinosaurs. *Gaia* 14: 45–75.

Christiansen, P., and N. Bonde. 2000. Axial and appendicular pneumaticity in *Archaeopteryx*. *Proceedings of the Royal Society of London B* 267: 2501–2505.

Chure, D. J. 1998. On the orbit of theropod dinosaurs. In B. P. Perez-Moreno, T. Holtz, Jr., J. L. Sanz, and J. Moratalla (eds.), *Aspects of Theropod Paleobiology*. *Gaia* 15: 233–240.

Colbert, E. H. 1962. The weights of dinosaurs. *American Museum Novitates* 2076: 1–16.

——. 1989. The Triassic dinosaur *Coelophysis*. *Bulletin of the Museum of Northern Arizona* 53: 1–174.

Coombs, W. P., Jr. 1975. Sauropod habits and habitats. *Palaeogeography, Palaeoclimatology, Palaeoecology* 17: 1–33.

Coria, R. A., L. M. Chiappe, and L. Dingus. 2002. A new close relative of *Carnotaurus sastrei* Bonaparte 1985 (Theropoda: Abelisauridae) from the Late Cretaceous of Patagonia. *Journal of Vertebrate Paleontology* 22: 460–465.

Daniels, C. B., and J. Pratt. 1992. Breathing in long necked dinosaurs: did the sauropods have bird lungs? *Comparative Biochemistry and Physiology* 101A (1): 43–46.

Dilkes, D. 2001. An ontogenetic perspective on locomotion in the Late Cretaceous dinosaur *Maiasaura peeblesorum* (Ornithischia: Hadrosauridae). *Canadian Journal of Earth Sciences* 38: 1205–1227.

Dodson, P. 1990. Sauropod paleoecology. In D. B. Weishampel, P. Dodson, and H. Osmólska (eds.), *The Dinosauria*, 402–407. Berkeley: University of California Press.

Erickson, G. M., S. D. Van Kirk, J. Su, M. E. Levenston, W. E. Caler, and D. R. Carter. 1996. Bite-force estimation for *Tyrannosaurs rex* from tooth-marked bones. *Nature* 382: 706–708.

Farlow, J. O. 1987. Speculations about the diet and digestive physiology of herbivorous dinosaurs. *Paleobiology* 13 (1): 60–72.

——. 1992. Sauropod tracks and trackmakers: integrating the ichnological and skeletal records. *Zubia* 10: 89–138.

Farlow, J. O., D. L. Brinkman, W. L. Abler, and P. J. Currie. 1991. Size, shape, and serration density of theropod dinosaur lateral teeth. *Modern Geology* 16: 161–198.

Farlow, J. O., M. B. Smith, and J. M. Robinson. 1995. Body mass, bone "strength indicator," and cursorial potential of *Tyrannosaurus rex*. *Journal of Vertebrate Paleontology* 15 (4): 713–725.

Galton, P. M. 1971. A primitive dome-headed dinosaur (Ornithischia: Pachycephalosauridae) from the Lower Cretaceous of England, and the function of the dome in pachycephalosaurids. *Journal of Paleontology* 45: 40–47.

——. 1974. The ornithischian dinosaur *Hypsilophodon* from the Wealden of the Isle of Wight. *Bulletin of the British Museum of Natural History (Geology)* 25: 1–152.

Galton, P. M., and P. Upchurch. 2004. Prosauropoda. In D. B. Weishampel, P. Dodson, and H. Osmólska (eds.), *The Dinosauria*, 2nd ed., 232–258. Berkeley: University of California Press.

Gatesy, S. M. 1990. Caudofemoral musculature and the evolution of theropod locomotion. *Paleobiology* 16: 170–16.

Gauthier, J. A.1986. Saurischian monophyly and the origin of birds. *Memoirs of the California Academy of Sciences* 8: 1–55.

Gilmore, C. W. 1905. The mounted skeleton of *Triceratops prorsus*. *Proceedings of the United States National Museum* 29: 433–435.

Gregory, W. K. 1905. The weight of the Brontosaurus. *Science* (n.s.) 22 (566): 572.

Gunga, H.-Chr., K. A. Kirsch, F. Baartz, and L. Röcker. 1995. New data on the dimensions of *Brachiosaurus brancai* and their physiological implications. *Naturwisseschaften* 82: 190–192.

Gunga, H.-Chr., K. A. Kirsch, J. Rittweger, L. Röcker, A. Clarke, J. Albertz, A. Wiedemann, S. Mokry, T. Suthau, A. Wehr, W.-D. Heinrich, and H.-P. Schultze. 1999. Body size and body volume distribution in two sauropods from the Upper Jurassic of Tendaguru (Tanzania). *Mitteilungen aus dem Museum für Naturkunde Berliner Geowissenschaflichte* Reihe 2: 91–102.

Halliday, D., R. Resnick, and J. Walker. 1993. *Fundamental of Physics*. New York: John Wiley & Sons.

Hatcher, J. B. 1901. *Diplodocus* (Marsh): its osteology, taxonomy, and probable habits, with a restoration of the skeleton. *Memoirs of the Carnegie Museum* 1: 1–63.

Hatcher, J. B., O. C. Marsh, and R. S. Lull. 1907. The Ceratopsia. *United States Geological Survey Monographs* 49: 1–300.

Henderson, D. M. 1999. Estimating the masses and centers of mass of extinct animals by 3–D mathematical slicing. *Paleobiology* 25: 88–106.

——. 2002. The eyes have it: the sizes, shapes, and orientations of theropod orbits as indicators of skull strength and bite force. *Journal of Vertebrate Paleontology* 22 (4): 766–778.

——. 2003. Tipsy punters: sauropod dinosaur pneumaticity, buoyancy and aquatic habits. *Proceedings of the Royal Society of London B* (suppl.) 271: 180–183

——. 2006. Burly gaits: centers of mass, stability, and the trackways of sauropod dinosaurs. *Journal of Vertebrate Paleontology* 26 (4): 907–921.

Henderson, D. M., and D. B. Weishampel. 2002. Convergent evolution of maxilla-dental-complex among carnivorous archosaurs. *Senckenbergiana lethaea* 82 (1): 77–92.

Hildebrand, M. 1982. *Analysis of vertebrate structure*. 2nd ed. New York: John Wiley & Sons.

Holtz, T. R. 1994. The arctometatarsalian pes, an unusual structure of the metatarsus of Cretaceous Theropoda. *Journal of Vertebrate Paleontology* 70: 480–519.

Horner, J. R., and D. Lessem. 1993. *The complete T. rex*. New York: Simon and Schuster.

Hohnke, L. A. 1973. Haemodynamics in the Sauropoda. *Nature* 244: 309–310.

Hurlburt, G. 1999. Comparison of body mass estimation techniques, using recent reptiles and the pelycosaur *Edaphosaurus boanerges*. *Journal of Vertebrate Paleontology* 19 (2): 338–350.

Hutchinson, J. R., and M. Garcia. 2002. *Tyrannosaurus* was not a fast runner. *Nature* 415: 1018–1021.

Kermack, K. A. 1951. A note on the habits of sauropods. *Annals and Magazine of Natural History* 4: 830–832.

Kitchener, A. 1988. An analysis of the forces of fighting of the blackbuck (*Antilope cervicapra*) and the bighorn sheep (*Ovis canadensis*) and the mechanical design of the horns of bovids. *Journal of Zoology* 214: 1–20.

Liem, K. F. 1973. Evolutionary strategies and morphological innovation: cichlid pharyngeal jaws. *Systematic Zoology* 22: 425–441.

Lockley, M. G. 1989. Dinosaur trackways. In S. J. Czerkas and E. C. Olson (eds.), *Dinosaurs Past and Present*, 1: 81–95. Seattle: University of Washington Press.

Lockley, M. G., and A. P. Hunt. 1995. Ceratopsid tracks and associated ichnofauna from the Laramie Formation (Upper Cretaceous: Maastrichtian) of Colorado. *Journal of Vertebrate Paleontology* 15: 592–614.

Lockley, M. G., and A. Rice. 1990. Did "Brontosaurus" ever swim out to sea?: evidence from brontosaur and other dinosaur footprints. *Ichnos* 1: 81–90.

Manning, P. L., L. Margetts, M. R., Johnson, P. J. Withers, W. I. Sellers, P. L. Falkingham, P. M. Mummery, P. M. Barrett, and D. R. Raymont. 2009. Biomechanics of dromaeosaurid dinosaur claws: applications of X-Ray microtomography, nanoindentation, and finite element analysis. *Anatomical Record* 292: 1397–1405.

Marsh, O. C. 1891. Restoration of *Triceratops*. *American Journal of Science* (ser. 3), 41: 339–342.

Martin, J. 1987. Mobility and feeding of *Cetiosaurus* (Saurischia: Sauropoda)–why the long neck? In P. J. Currie and E. H. Koster (eds.), *4th Symposium Mesozoic Terrestrial Ecosystems, Short Papers*, 154–159. Drumheller, Alberta: Tyrrell Museum of Paleontology.

Maryańska, T., R. E. Chapman, and D. B. Weishampel. 2004. Pachycephalosauria. In D. B. Weishampel, P. Dodson, and H. Osmólska (eds.), *The Dinosauria*, 2nd ed., 464–477. Berkeley: University of California Press.

Newman, B. H. 1970. Stance and gait in the flesh-eating dinosaur *Tyrannosaurus*. *Biological Journal of the Linnean Society* 2: 119–123.

Norman, D. B. 1980. On the ornithischian dinosaur *Iguanodon bernissartensis* from the Lower Cretaceous of Bernissart (Belgium). *Mémoir, Institut Royal des Sciences Naturelles de Belgique* 178: 1–103.

———. 1986. On the anatomy of *Iguanodon atherfieldensis* (Ornithischia: Ornithopoda). *Bulletin de l'Institut Royal des Sciences Naturelles de Belgique: Sciences de la Terre* 56: 281–372.

Osborn, H. F. 1899. A skeleton of *Diplodocus*. *Memoirs of the American Museum of Natural History* 1: 191–214.

———. 1933. Mounted skeleton of *Triceratops elatus*. *American Museum Novitates* 654: 1–14.

Ostrom, J. H. 1969. *Osteology of Deinonychus antirrhopus, an unusual theropod from the Lower Cretaceous of Montana*. Peabody Museum of Natural History Bulletin 30. New Haven, Conn.: Peabody Museum of Natural History, Yale University.

Paul, G. S. 1987. The science and art of restoring the life appearance of dinosaurs and their relatives. In S. J. Czerkas and E. C. Olson (eds.), *Dinosaurs Past and Present*, 2: 5–39. Seattle: University of Washington Press.

———. 1988. *Predatory Dinosaurs of the World*. New York: Simon and Schuster.

———. 1997. Dinosaur models: the good, the bad, and using them to estimate the mass of dinosaurs. In D. L. Wolberg, E. Stump, and G. D. Rosenberg (eds.), *DINOfest International: Proceedings of a Symposium Held at Arizona State University*, 129–154. Philadelphia: Academy of Natural Sciences.

———. 2000. The Yucatan impact and related matters. In G. S. Paul (ed.), *The Scientific American Book of Dinosaurs*, 381–390. New York: St. Martin's Press.

Paul, G. S., and P. Christiansen. 2000. Forelimb posture in neoceratopsian dinosaurs: implications for gait and locomotion. *Paleobiology* 26: 450–465.

Pecksis, J. 1994. Implications of body-mass estimates for dinosaurs. *Journal of Vertebrate Paleontology* 14: 520–533.

Proctor, N. S., and P. J. Lynch. 1993. *Manual of Ornithology: Avian Structure and Function*. New Haven, Conn.: Yale University Press.

Rayfield, E. J., D. B. Norman, J. R. Horner, P. M. Smith, J. J. Thomason, and P. Upchurch. 2001. Cranial design and function in a large theropod dinosaur. *Nature* 409: 1033–1037.

Reid, R. E. H. 1997. Dinosaurian physiology: the case for "intermediate" dinosaurs. In J. O. Farlow and M. K. Brett-Surman (eds.), *The Complete Dinosaur*, 449–473. Bloomington: Indiana University Press.

Riggs, E. S. 1904. Structure and relationships of opisthocoelian dinosaurs. Pt. II. The Brachiosauridae. *Publications of the Field Columbian Museum (Geological Series)* 2: 229–248.

Schmidt-Nielsen, K. 1984. *Scaling: Why Is Animal Size So Important?* Cambridge: Cambridge University Press.

Seebacher, F. 2001. A new method to calculate allometric length-mass relationships of dinosaurs. *Journal of Vertebrate Paleontology* 21: 51–60.

Sereno, P. C., R. N. Martinez, G. P. Wilson, D. J. Varrichio, O. A. Alcober, and H. C. E. Larsson. 2008. Evidence for avian intrathoracic air sacs in a new predatory dinosaur from Argentina. *PLoS ONE* 3 (9): e3303.

Seymour, R. S. 1976. Dinosaurs, endothermy and blood pressure. *Nature* 262: 207–208.

Snively, E., and A. P. Russell. 2002. The tyrannosaurid metatarsus: bone strain and inferred ligament function. *Senckenbergiana lethaea* 82: 35–42.

Stevens, K. A., P. Larson, E. D. Wills, and A. Anderson. 2008. Rex, sit: digital modeling of *Tyrannosaurus rex* at rest. In P. Larson and K. Carpenter (eds.), *Tyrannosaurus Rex, the Tyrant King*, 193–204. Bloomington: Indiana University Press.

Stevens, K. A., and J. M. Parrish. 1999. Neck posture and the feeding habits of two Jurassic sauropod dinosaurs. *Science* 284: 798–800.

Sues, H.-D. 1978. Functional morphology of the dome in pachycephalosaurid dinosaurs. *Neues Jahrbuch für Geologie und Paläontologie Monatshefte* 1978 (8): 459–472.

Taylor, M. P., M. J. Wedel, and D. Naish. 2009. Head and neck posture in sauropod dinosaurs inferred from extant animals. *Acta Palaeontologica Polonica* 54 (2): 213–220.

Therrien, F., D. M. Henderson, and C. B. Ruff. 2005. Bite me: biomechanical models of theropod mandibles and implications for feeding behavior. In K. Carpenter (ed.), *The Carnivorous Dinosaurs*, 179–237. Bloomington: Indiana University Press.

Thulborn, T. 1990. *Dinosaur Tracks*. London: Chapman and Hall.

Upchurch, P. 1994. Manus claw function in sauropod dinosaurs. *Gaia* 10: 161–171.

———. 2004. Sauropoda. In D. B. Weishampel, P. Dodson, and H. Osmólska (eds.), *The Dinosauria*, 2nd ed., 259–322. Berkeley: University of California Press.

Van Heerden, J. 1997. Prosauropods. In J. O. Farlow and M. K. Brett-Surman (eds.), *The Complete Dinosaur*, 242–263. Bloomington: Indiana University Press.

Vogel, S. 1998. *Cats' Paws and Catapults: Mechanical Worlds of Nature and People*. New York: W. W. Norton.

Wainwright, S. A., W. D. Biggs, J. D. Currey, and J. M. Gosline. 1976. *Mechanical Design in Organisms*. New York: John Wiley & Sons.

Wedel, M. J. 2003a. The evolution of vertebral pneumaticity in sauropod dinosaurs. *Journal of Vertebrate Paleontology* 23 (2): 344–357.

———. 2003b. Vertebral pneumaticity, air sacs, and the physiology of sauropod dinosaurs. *Paleobiology* 29 (2): 243–255.

———. 2008. Evidence for bird-like air sacs in saurischian dinosaurs. *Journal of Experimental Zoology* 311A: 1–18.

Wedel, M. J., R. L. Cifelli, and R. K. Sanders. 2000. Osteology, paleobiology, and relationships of the sauropod dinosaur *Sauroposeidon*. *Acta Palaeontologica Polonica* 45 (4): 343–388.

Wellnhofer, P. 1990. *The Illustrated Encyclopedia of Pterosaurs*. London: Salamander Books.

Wilson, J. A., and M. T. Carrano. 1999. Titanosaurs and the origin of "wide-gauge" trackways: a biomechanical and systematic perspective on sauropod locomotion. *Paleobiology* 25: 252–267.

Witmer, L. M. 1997. The evolution of the antorbital cavity in archosaurs. A study in soft-tissue reconstruction in the fossil record with an analysis of the function of pneumaticity. *Society of Vertebrate Paleontology Memoir* 3: 1–73.

Witmer, L. M. and Chiappe, L. M. 2002. *Mesozoic Birds: Above the Heads of Dinosaurs*. Berkeley: University of California Press.

Wright, J. L. 1999. Ichnological evidence for the use of the forelimb in iguanodontid locomotion. *Special Papers in Palaeontology* 60: 209–219.

33.1. Healing fracture in a turkey vulture radius and ulna. R.W. Shufeldt used comparative material from his own research of extant bird injuries, like this one, to assess injuries in tertiary bird bones brought to him by E. D. Cope.

Figure from Shufeldt 1893.

Disease in Dinosaurs

Elizabeth Rega

33

Dinosaur disease is an intriguing subject for the scholar and public alike, resulting from the combination of often bizarre deformity with already charismatic mega fauna. Throw in the "csi" thrill of solving an ancient cold case, and the result is a heady mix, but one which runs the risk of substituting headlines for scientific rigor. To paraphrase the famed fictional sleuth Sherlock Holmes, the scientific approach should be to define the plausible range of possibility while excluding the impossible and arguing against the improbable . While attempting to avoid pedantism, this approach to paleopathology will characterize the review contained in this chapter.

Observations of ancient disease are as old as the study of fossils themselves. Dramatic examples of bone pathology attracted attention from the earliest days of paleontology (Esper 1774), with descriptions and diagnoses of abnormal bone scattered throughout both the scholarly and popular literature. Conclusions about the effect of disease on dinosaurian behavior, diet, locomotion, and the even the role of pathology in extinction are ideally based on complete skeletons, but more commonly conclusions are made from isolated specimens of diseased bone. Determining the pathological requires a thorough knowledge of the normal. Because recognition of normal variation necessitates a larger sample than just one bone or individual, deformity due to disease has been cited as grounds to challenge the establishment of new taxa displaying unexpected features. No less a scientist than pathologist and physician Rudolf Virchow famously (and mistakenly) attributed the morphology of the original Neanderthal skeleton to the ravages of arthritis, premature suture fusion, and rickets on a modern human (Rosen 1977; Virchow 1872). More recently, the protracted debate over the novel morphology of the Liang Bua hominids from the island of Flores hinged on whether the LB1 cranium manifests normal features of a new species or pathological ones of *Homo sapiens* (Argue et al. 2006; Brown et al. 2004; Gordon et al. 2008; Martin et al. 2006).

The term *paleopathology* has its origins in the study of extinct nonhuman vertebrates. Virchow has been identified as the "first genuine paleopathologist" (Klebs 1917; Scarani 2003, 95), although several earlier researchers documented pathologies from paleontological and archaeological contexts (Esper 1774; Eudes-Deslongchamps 1838; von Walther 1825; see Cook and Powell 2006). Ruffer claims credit for the term's creation (1914). He was identified as its originator by Klebs (1917) and Moodie (1918a), and modern sources (Weiss 2000) continue to perpetuate this error.

A (Brief) History of Paleopathology in Paleontology

In fact, Aufderheide and Rodriguez Martin (1998) attribute the first use of the term to physician and amateur ornithologist R. W. Shufeldt, who used it to describe the study of "all diseased or pathological conditions found fossilized in the remains of extinct or fossil animals" (Shufeldt 1893, 679). Shufeldt's expertise was sought by none other than E. D. Cope in the latter's search for evidence of predation-related injuries in a large sample of Pliocene bird skeletons. Shufeldt relied less on his experience as a physician and more on his avocational study of avian injuries in assessing the pathologies from this sample, which he found to be surprisingly few in number. He refers to his observations on fracture healing in extant birds to explain ancient fracture, with a clear emphasis on matching the progression and process of injury and avian healing (Fig. 33.1). Shufeldt makes explicit the uniformitarian approach to ancient skeletal pathology: "Animals that lived during the past ages of the world, and now long extinct, must have suffered, it would seem, from many injuries quite similar to those now sustained by their descendants of the present epoch" (1893, 679).

As with its geological counterpart, uniformitarianism in paleopathology is predicated on the assumption that past disease processes are identical to those observable in the present day. In a manner later to be characteristic of most paleopathological studies, Shufeldt proceeds to the next level of inference, proposing mechanisms for injury and a hypothesis of wing-fighting behavior that he admits has no parallels in his extant comparative sample. He concludes his brief study by outlining alternative scenarios for injury and underscores the speculative nature of his behavioral conclusions.

Paleopathology Originated with Animals

Although the discipline of paleopathology has come to be dominated by inquiries into ancient human disease, early texts regarded as seminal to the field focused on pathologies of extinct animals. Pleistocene fauna were the subject of the very earliest paleopathological reports as cave excavation gained prominence (Esper 1774; Virchow 1895; von Walther 1825).

Pathology of permineralized bone first attracted attention in the nineteenth century. Eudes-Deslongchamps (1838) noted vertebral fusion and phalangeal bone overgrowths in his description of *Poekilopleuron*. Williston (1898) described a mosasaur vertebra from the cretaceous of Kansas showing an anomalous bone lump diagnosed as a benign bone tumor. Ruffer's 1921 landmark *Studies in the Paleopathology of Egypt* is known for its detailed descriptions of soft-tissue pathologies recovered from mummified human tissue, but also featured a dramatic example of permineralized vertebral pathology in a Miocene crocodile. Ruffer arrives at the essentially descriptive diagnosis in common parlance by contemporary physicians—*spondylitis deformans*, literally "deformed inflamed spine"—for the two fused vertebrae, which are beautifully photographed and described in several high-quality plates. Although the precision of this diagnosis can be challenged based on modern diagnostic refinements, these illustrations set a high standard for documentation and description in the discipline that is rarely surpassed today.

Disease in the Dinosaurs

Documentation of dinosaurian pathology continued the trend of brief notations embedded in lengthy descriptive monographs. Hatcher (1901) documented fused caudal vertebrae in two individuals in his description of the genus *Diplodocus* (Fig. 33.2). In his redescription of Marsh's *Triceratops serratus* (= *horridus*), Hatcher noted a perforation affecting the posterior right frontal and postfrontal, deeming it "probably pathologic" as it was lacking on the opposite side (1907, 124). Such lesions have subsequently proved to be relatively common among ceratopsids (Farke et al. 2009). In the same monograph, he notes the presence in the same individual of a marked lower jaw protuberance surmounted by multiple perforations, though he does not explicitly identify it as pathological. Riggs (1903) described an enlargement in the shaft of a right fifth *Apatosaurus* rib (FMNH P25112) and interpreted it as a healed fracture callus; the adjacent sixth rib manifests breakage that Riggs declared a premortem fracture failing to heal.

Gilmore contributed numerous brief accounts of diseased dinosaur bone. His 1915 *Allosaurus* monograph mentioned but did not illustrate a striking scapular pathology. His statement exemplifies the principal concern of contemporary paleontologists in establishing the normal rather than describing the pathological: "The left scapula was injured in life and the subsequent healing produced great deformation of the bone. This pathological condition caused a widening of the blade that would have been entirely misleading as to its true form had not the opposite scapula been present" (Gilmore 1915, 504).

Brevity of pathological documentation in these monographs is consistent with an emphasis on normal characters in establishing new taxa. Consequently, even remarkable dinosaur abnormalities occasionally fell through the cracks. In spite of the magnitude (8.6 × 4.4 cm) of the erosive bone lesion in the dorsal aspect of the right ilium of the holotype of

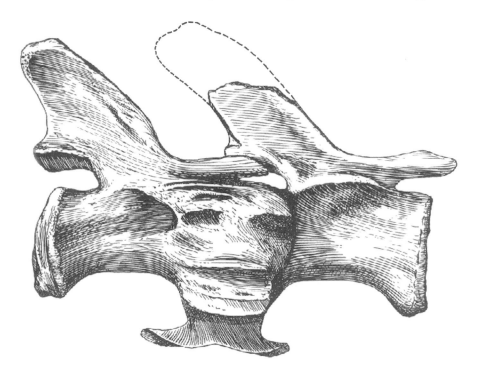

33.2. Right view, *Diplodocus* caudal vertebrae 17–18 (CM 94). Hatcher identified these as "coössified (pathologically) by their centra." Holland (1906) identified these vertebrae as 20–21. Caudals 24 and 25 are fused as well in this specimen.

Figure after Hatcher 1901, 36.

33.3. Close-up of massive erosive lesion affecting right *Camptosaurus brownii* ilium (USNM 4282). Gilmore (1909) did not mention this lesion in his initial description of the species.

Author's photo.

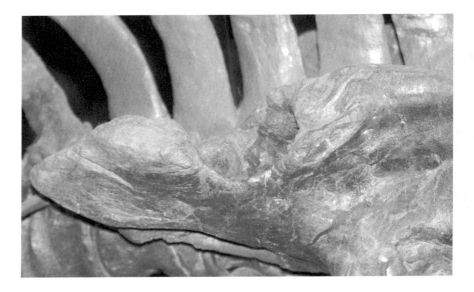

Camptosaurus browni (USNM 4282; Fig. 33.3), Gilmore makes no mention of it in his original species description (1909), waiting until his 1912 account of the mounted specimen to take note. There he describes the hip pathology and an enlarged spinous process on a proximal caudal vertebra. He attributes both injuries to well-healed trauma:

> While the wound in the ilium must have been an exceedingly painful one at the time of infliction, it in no way utterly disabled the animal, at least to the extent of leading to its death, as all the broken margins of bone have healed. Although these injuries may have been inflicted by some of its large carnivorous contemporaries, the position of the wounds suggests the idea that this was a female who might have received the injuries during copulation. (Gilmore 1912, 690)

This marks the first published instance—though not the last—that copulation is invoked to explain dinosaurian skeletal injury!

Gilmore provides slightly more detailed description and illustrations in his later monographs, devoting two full photographic plates to pathological ceratopsian bone (Gilmore 1919), including a truncated and resorbed right horn core of *Triceratops elatus* (= *horridus* USNM 4708) interpreted by Gilmore as a well-healed horn fracture. A right scapula (USNM 8013) manifests a "large bony hornlike growth" on the internal aspect of unknown origin, which presented "a serious handicap to the movement of the shoulder blade" (Gilmore, 1919, 102).

Diseased dinosaur bone was therefore a minor point of interest imbedded in descriptive monographs, with little emphasis on diagnosis beyond the anecdotal. For practice to change, Williston himself awakened the interest of a young paleontologist-in-training in Chicago named Roy Moodie.

Moodie and the Professionalization of Paleopathology

Despite confusion as to the term's originator, there is widespread agreement that the title of founder of the discipline of paleontological paleopathology belongs to Roy L. Moodie, who was responsible not only for editing and publishing Ruffer's research after his untimely death, but popularizing paleopathology with landmark publications in the scientific and popular press, featuring vertebrates from the Paleozoic through the Quaternary (Moodie 1916a, b; 1918a, b, c; 1921; 1923a, b; 1926). Moodie—unlike other prior and contemporary paleopathologists not a physician, but an anatomically trained paleontologist—was a great cataloger of ancient disease. He was provided with pathological dinosaur specimens by his mentor Williston and many other paleontologists of the day and authored the first paper devoted solely to analysis of dinosaur pathology (Moodie 1916a). His more detailed subsequent description of fused dorsal vertebrae attributed to *Apatosaurus* supplied to him by Williston employed both gross and microscopic observation (Moodie 1918b, 1923a, b).

Like Ruffer's, Moodie's diagnoses reflect contemporary clinical knowledge and terminology; these have not all stood the test of time. However, Moodie was an early adopter of techniques that have recently proved vital in characterizing dinosaurian biology. Moodie routinely included illustrations of microscopic views of thin sections of bone to investigate the disease process at the tissue level. He employed radiography (x-rays) and explicitly referenced extant disease states and processes when looking for a cause of the observed diseased bone. Although uniformitarian (as well as anecdotal and occasionally maddeningly sloppy) in his approach, Moodie anticipates concerns over pathogen evolution when he states, "Doubtless many of the diseases from which ancient animals suffered are now extinct. Their results, however, as seen in the fossilized bones, closely parallel the pathological anatomy of recent times" (Moodie 1918b, 280).

Moodie also produced landmark synthetic chronological summaries of pathology, employing his own observations and citing the publications of others, in an attempt to trace the evolution of disease through time in paleontological and medical publications (Moodie 1918a, b; 1923 a, b). According to Cook and Powell (2006), Moodie was concerned in his synthetic summaries with refuting Henry Fairfield Osborn's hypothesis that disease was a great factor in the extinction of species. He states: "A study of the paleontological evidences [*sic*] of disease, as seen in the bone lesions, does not help us much yet in an appreciation of what part disease may have played in extinction. The part may have been great, but this is a hypothetical assumption, based purely on analogy" (Moodie 1918b, 281).

Because of subsequent phylogenetic and chronological revision, Moodie's synthetic works have stood the test of time rather poorly, but no contemporary scholar was more thoroughly versed in the paleopathology of extinct animals.

The uniformitarian approach exemplified by Moodie and his predecessors remains the heart and soul of paleontological paleopathology, where description and matching of skeletal observations with extant disease manifestations form the bulk of published studies. However, paleopathological "diagnosis" is necessarily an approximate activity. Diagnosis is important to the physician and to the veterinarian, who require a precise diagnosis to cure disease. But paleopathologists are not attempting to "cure" fossils—instead, they are occupied with looking for clues to past behavior and events.

Diagnosing Dinosaur Disease

33.4. The profound modification of bone shape cause by disturbance of the periosteal membrane is illustrated by the left fibula of *Tyrannosaurus rex* FMNH PR 2081 "Sue." The massive diaphyseal hyperostosis is most likely due to chronic infection of the bone and surrounding periosteum.

Photo courtesy of C. Brochu and the FMNH.

The Problem with Paleo-Diagnosis

Clinicians and pathologists have the advantage of patient history, behavioral and physiological symptoms, pathogen culture, and immunochemical methods in arriving at a diagnosis in their living and recently dead patients. Most of these opportunities are decidedly lacking in ancient bone, although remarkable advances in understanding the evolution of disease have been accomplished utilizing molecular techniques (Bathurst and Barta 2004; Braun et al. 1998; Taylor et al. 2007). These DNA-based methods have not yet been shown to be possible with permineralized tissue; however, Straight et al. (2009) have used oxygen isotopes to document temperature elevation in healed sections of permineralized dinosaurian bone.

As permineralized bone constitutes the overwhelming majority of preserved dinosaur tissue and very few diseases presently identifiable actually leave any mark whatsoever on bones, it is not possible to directly document the entire range of diseases affecting dinosaurs. Thus, large swathes of the range of disease affecting dinosaurs—including most causes of death—are simply invisible. We are left with diseases that do make their presence known in the skeleton.

Describing Bone Disease

Paleopathological conclusions are based on description and careful comparison with extant and extinct examples of skeletal disease. Bone may appear static, but it is a living tissue that is continuously resorbed and remodeled, even in adults. Pathology is assessed in ancient skeletons by observing the characteristics of the abnormal bone—its texture, type of response, pattern across the skeleton, and microstructure. Challenges to bone or to the membranes covering the outer and inner surfaces can cause rapid bone alteration, which can radically affect the shape of skeletal elements (Fig. 33.4). Bony response most often constitutes the sole evidence of pathology visible from fossilized bones. This includes proliferation, dense reactive bone on the margins of resorption, increased fine or coarse porosity, enlarged holes for pus drainage, and cavities for suppurative or reactive tissue masses. The following factors are typically considered when examining diseased dinosaur bone:

· Number and pattern of affected areas
· Shape of pathological lesion (single, multiple, fused)
· Type of response (proliferative or resorptive or mixed)
· Location in the bone tissue itself (near the marrow cavity, if present, under the outer bone membrane, within the dense or spongy bone tissue)
· Location on the whole bone relative to its growth and development (shaft, near the growth plate, on the joint surface, under cartilage—diaphyseal, metaphyseal, epiphyseal, subchondral, respectively)
· Appearance of lesion margins (porous, sharp, or smooth)
· Texture of newly deposited bone or of resorbed bone
· Bone density

· Presence or absence of:
    Sclerotic (hard dense smooth) bone
    Vascular markings
    Drainage holes (cloacae)
    Spaces or chambers

Two-dimensional radiographs and three-dimensional computerized tomographic (CT) scans are extremely useful in assisting visualization of abnormal bone and revealing features in the interior of the bone (Fig. 33.5); Magnetic resonance imaging (MRI) is used much less frequently, as fossils lack typical MRI contrast generators and permineralized vertebrate bone images poorly, but has shown surprising utility in evaluating pathology in some invertebrate fossils (Mietchen et al., 2005).

Bone markers of disease are sadly rarely specific enough to be regarded as "pathognomic"–meaning that one identifiable type of lesion is caused by one and only one cause (Kelley and Eisenberg 1987; Salo et al. 1994). Avoiding misdiagnosis is dependent on many factors. Process-based rather than diagnostic approaches were suggested a century ago by Wood-Jones, as "there is nothing to be gained by an etiological nomenclature as long as we have so little definite knowledge about the causes" (as quoted in Klebs 1917, 262). Recently, other researchers have echoed this call (Farke et al. 2009; Rega et al. 2003; Rega 2008; Wolff and Varricchio 2005).

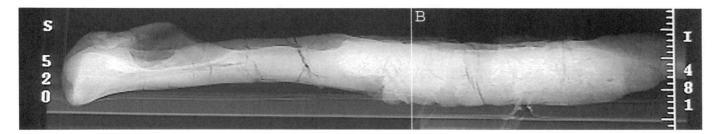

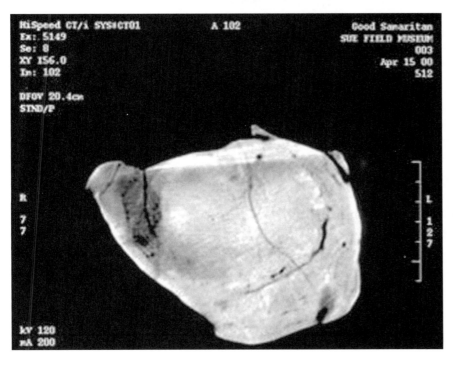

33.5. Computerized tomographic scans through the left fibula of FMNH PR 2081 "Sue." A shows a longitudinal scan with location of B indicated by dashed line, B. B depicts a cross-sectional computed axial tomography (CAT) scan at the fibular midshaft. No evidence of fracture was detected on the scan. The outer shell of the involucrum surrounds the internal dead bone known as a sequestrum The thin black crack is postmortem damage.

CAT scan courtesy of M. Carrano.

Paucity of standardization of descriptive and diagnostic terminology compounds issues of comparison between published sources (Ubelacker 2003; Capasso 2005), and conclusions regarding disease prevalence in the past remain dependent on a consistency between researchers that has not yet been achieved. Therefore, a specific paleo-diagnosis applied with undue confidence can obscure more than it reveals.

What Is an Appropriate Model for Dinosaur Disease?

When confronted with abnormal bone, paleopathologists create a "differential diagnosis" by comparing features of the pathology in question with characteristics of different bone pathologies derived from living populations. This practice relies on the uniformitarian assumption that the bone response to injury or infection in dinosaurs is similar to that of living animals and even people. Because of the vast wealth of available data, many of the comparisons in dinosaurian paleopathology are drawn from the human medical literature. However, comparative immunology provides ample grounds for caution in this regard, because "the character, quality, and progression of inflammatory reactions in non-mammalian vertebrates . . . differ from the mammalian paradigm more markedly than any other aspect of general pathology" (Therio 2004, 1).

Phylogenetic Bracketing Is Necessary but Not Sufficient

This difficulty can be partly addressed by creating a careful comparison of dinosaur bone pathology with known pathology of taxa bracketing dinosaurs, namely birds and crocodilians. Matching actual bone dynamics of dinosaurs by means of histology to that of extant models extends the approach. While extant developmental and metabolic parallels exist for aspects of dinosaurian bone, it is clear that there is no one perfect animal model for dinosaurs, which varied from each other in their size, growth dynamics, and likely also metabolism to a great extent (Starck and Chinsamy 2002; Chinsamy-Turan 2005; Padian et al. 2001, 2004). Histological diversity reflects growth rates, not phylogeny (Amprino's Rule).

Ornithosuchian bone (*sensu* Padian et al. 2004), which is characteristic of large dinosaurs and most pterosaurs, has many similarities to avian and mammalian bone in tissue type and growth rates. Haversian remodeling—where bone is resorbed, then redeposited in circumferential layers around a blood vessel—is common in many birds, most mammals and all large dinosaurs (Chinsamy-Turan 2005; Kimmel et al. 1998) and not generally found in living reptiles. Recent data suggest that crocodilians may have somewhat more limited comparative value than birds, as their bone dynamics differ from those of the large dinosaurs (Padian et al. 2004), except for small, slow-growing dinosaurs, such as *Scutellosaursus*, which may be best compared with slower-growing extant crocodiles.

But avian models cannot be employed indiscriminately. Unlike dinosaurs, bird cortical bone generally lacks lines of cyclical arrested growth (LAGs), which are a general phenomenon in all nonavian tetrapods, both ectothermic and endothermic (Lee and Werning 2008). The lack of bird LAGs is due to growth-period truncation (Turvey et al. 2005). Of all birds documented, only a single extinct order of ratites (Moa Aves: Dinornithiformes) manifest annual growth marks. Other factors render uncritical reliance on avian models problematic. Bird bone structure varies greatly between species by size and habitus (Chinsamy-Turan 2005). Extreme thinning of the cortex—not found in dinosaurs—is a prominent feature of flying (volant) birds.

Because of similarities of bone dynamics, mammalian models are not categorically inapplicable to large dinosaurs. Neither should mammalian analogies be used a priori or naively, in isolation from other phylogenetically appropriate evidence (Rega and Brochu 2001a; Rega et al. 2003; Rega 2008; Wolff and Varricchio 2005). Due to growth dynamic, the ornithosuchian bone of large dinosaurs presents structural similarities to that of larger mammals, especially rapid subperiosteal bone deposition and extensive Haversian remodeling. Some skeletal microanatomy (particularly tendon) is apomorphic for Vertebrata, while in its ossified form, dinosaur tendon has proven to be structurally derived, different from all extant forms (Organ and Adams 2005) and without extant parallel.

Evolution and Sampling

Even with careful selection of an appropriate model and observation of microscopic bone dynamics, infectious disease presents a particular difficulty, because pathogens, with their short intergenerational time, are subject to rapid evolution, rendering it highly unlikely that a modern infectious disease is identical to that in the past. Human pathogens have evolved significantly during the short span of human evolutionary history (Buikstra 1981; Ewald 1994). Interspecies transmission of significant infectious agents (de Graff et al. 2008) complicates the picture. Dinosaur pathogens certainly evolved in their behavior and manifestation in the 60–100 million years of evolutionary time separating dinosaurs and birds, and it is overwhelmingly likely that many infectious diseases in the present have no exact parallel in the Mesozoic and the reverse.

Compounding the difficulty of assessing ancient disease is the serendipity of vertebrate fossil preservation. Rigorous consideration of ancient disease must be undertaken at the level of the individual and the population (Buikstra and Cook 1980; Wright and Yoder 2003). The incompleteness of the fossil record limits conclusions about disease in dinosaurs. Differential preservation and excavation means that unbiased samples of true biological populations are extremely rare, and such samples are necessary for robust conclusions about disease distribution and prevalence.

A Way Forward?

While all these considerations may at first glance seem discouraging, we can be more confident making some conclusions than others. Without committing diagnostically to a specific infectious or metabolic agent, the sequence of events leading to the observed lesions and bone response can often be "read" by means of careful description, especially when histological sections are available. There is a hierarchy of inference in paleopathological diagnosis (Fig. 33.6). Trauma, bone breakage, fracture repair, degenerative alteration, and chronic bone changes lend themselves to more secure conclusions about the interplay of these responses with the lives of the dinosaurs than diagnoses of specific infectious diseases, acute conditions, metabolic abnormalities, or dietary deficiency. And studies of complete specimens, where the pattern of disease in the entire skeleton can be assessed, yield more secure conclusions than those based on individual isolated bones.

A useful comprehensive review of all published literature on dinosaur paleopathology is outside the scope of this chapter. The utility of such a work is also dubious, as variation in descriptive terminology, methodology, and diagnostic criteria in the literature renders comparability of results at this juncture problematic and meta-analysis largely untenable. And given that the vast majority of dinosaur pathologies documented are found on isolated bones, not whole individuals, realistic examination of skeletal pattern and prevalence is not possible for most taxa. Instead, this chapter predominately highlights cases drawn from complete individuals or aggregate studies of closely related taxa where conclusions are more secure. These cases have been chosen on the basis of their power to illustrate considerations of paleopathological methodology as well as interesting hypotheses about dinosaurian biology.

Dead Bones Do Tell Tales

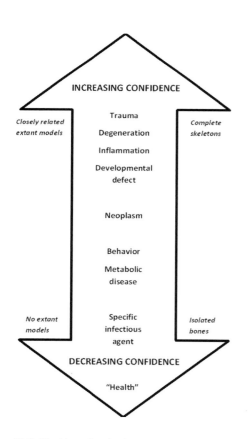

33.6. The hierarchy of paleo-pathological inference.

MOR 693 *Allosaurus fragilis* provides an ideal opportunity to examine a variety of disease processes in a highly charismatic creature. An exceptionally well preserved specimen housed at the Museum of the Rockies, it exhibits a total of nineteen bony abnormalities. Because of its 95% completeness and preservation (and not because of its size), it received the moniker "Big Al"; this subadult individual is not in fact particularly large for *Allosaurus* (Breithaupt 2001). In addition, other pathological *Allosaurus* bones from the Cleveland-Lloyd quarry provide an extensive, though disarticulated, comparative sample. Big Al embodies an introduction to paleopathology, because its pathologies (and pseudopathologies) are common in dinosaurs, the degree of skeletal completeness and size of comparative sample is close to ideal, and the quality of published analysis is high.

Way of Life and Manner of Death

Paleontological detective work tells the tale of this remarkable animal's life and death (Breithaupt 2001). The completeness and degree of bone articulation show that the carcass was not transported far postmortem. It was recovered in the classic opisthotonic "death pose" with the head contracted over the tail. Previously attributed to the desiccation and contraction of spinal tendons and ligaments (Laws et al. 1996), this posture is suggested by Faux and Padian (2007) to be more closely indicative of "death throes" based on experimental evidence and survey of affected forms, a view advanced by Moodie (1918c) and later rejected by many researchers. Faux and Padian link this manifestation to death from factors affecting the central nervous system: nutrition, asphyxiation, viral infection, or toxins. Interestingly, as this posture is observed only in ornithodiran reptiles, mammals, and marsupials, it may be correlated with high basal metabolic rates. Although clues regarding cause or manner of death remain extraordinarily rare in the fossil record, in this case Big Al's death posture might be narrowed to the causes listed above.

Big Al's pathologies were described extensively by Hanna (2002); these affect the vertebrae, ribs, gastralia, a proximal tail chevron, ilium, manual and pedal phalanges, and metatarsals. These pathologies fall into four major categories: bony proliferation, abnormal bone spurs or nodules, smooth depressions at muscle attachment sites, and overall bone texture and shape change combining bone resorption with deposition.

Distinguishing Pseudopathology and Postmortem Breakage

Several of Big Al's bones manifested multiple pockmarklike erosions. In living bone, erosive lesions are termed *lytic*, and these erosions indeed resembled lytic skeletal pathology with no healing. Bone response, such as healing, shows that an animal was alive when the injury occurred. But these erosions are entirely destructive, lacking any evidence of reactive bone formation.

True pathology must be distinguished from the postmortem diagenetic changes which can cause "pseudopathology." Insect boring, root marks, water erosion, and even partial digestion can mimic lytic skeletal lesions. Ground pressure can even change the shape of bones (Petersen et al. 1972) and impersonate bone mineralization deficiencies (osteomalacia). Even proliferative pathology can be mimicked by mineral or sediment adhesion. The presence of bone response definitively distinguishes true trauma from pseudopathology (Ortner 2003); in this case response is lacking. After careful comparison with modern dermestid beetle damage to bones, Big Al's lesions were attributed to beetle burrows created on the bones during postmortem exposure but before the skeleton was buried by a series of flooding events (Laws et al. 1996; Hasiotis et al. 1999; Fig. 33.7).

33.7. Pseudopathology. Erosion caused by beetle damage postmortem mimics osteolytic pathology in MOR 693, *Allosaurus fragilis*.

Photo courtesy of R. Hanna, with permission of the Museum of the Rockies.

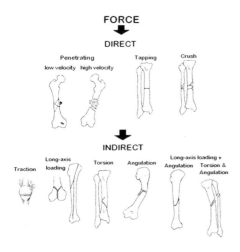

33.8. Patterns of bone breakage resulting from different types of force application.

Dinosaurian Fracture Repair

Pattern Distinguishes Postmortem Damage from Trauma

Bone damage can occur just before death (perimortem), and injuries acquired at this time would not show healing. As dinosaur bones are almost always recovered in a fragmentary state, a perimortem fracture can be mistaken for postmortem breakage. It is not, however, impossible to arrive at a conclusion of perimortem fracture absent healing, just more difficult. The shape of broken bone surfaces and pattern of breakage on those bones lacking healing can yield clues regarding whether the breakage occurred while the bone was fresh or "green" (Ortner 2003). These breaks will have a different shape (splintered or spiral), with clear indicators of the direction of force applied (Fig. 33.8). This differs from the pattern of postmortem breakage in dry or fossilized bone. Using the pattern of breakage to distinguish perimortem fracture and its causes from taphonomic breakage has been of critical importance in the analysis of the fracture and cause of death in human skeletons (Turner and Turner 1998; Rega et al. 1999).

In the Cleveland-Lloyd *Allosaurus* sample, for example, two greenstick breaks without healing in ribs are reported (UUVP 2252 and UUVP 1847: Petersen et al. 1972). As the name implies, these breaks look like broken willow sticks; they indicate that fracture occurred while the bone was very fresh, possibly just before death. Other researchers have asserted that these breaks are postmortem damage, however (Tanke and Rothschild 2002).

Healing Indicates Life after Injury

Big Al does contain numerous genuine pathologies, which are easily differentiated from pseudopathology because they all exhibit robust healing. It is clear that Big Al mounted a significant defense over a period of time, and none of these injuries were likely the direct cause of death. The presence of bony reaction is the most indicative sign of genuine premortem trauma.

Big Al has no obvious unhealed fractures, but does manifest multiple healed ones. These are indicated by bony calluses—lumps of bone proliferation—typically seen surrounding a fracture during the final stages of the normal healing sequence. In Big Al, these are macroscopically identical to the stage of fracture healing called the *hard callus* in mammals and birds (Table 33.1).

The calluses cluster on several adjacent right ribs (3rd–5th, Fig. 33.9A, B), as well as on the fifth left gastralia and chevron 2 of the tail. Postmortem damage to the callus on the fifth and sixth ribs shows the underlying broken rib surrounded by a thick layer of porous new bone. The original rib outer surface is still preserved inside the callus, because this bone died as a result of the rapid severing of its connection with the circulatory system and had not yet been resorbed by the animal. The rapid elevation of the periosteum—the membrane covering the outside of living bone—during the initial fracture response causes both bone death and new enveloping bone deposition (Tully 2002). During normal fracture healing, this old bone would gradually be remodeled by resorption and blood vessel reinvasion, and, after many years, dense, new bone with newly formed blood vessel

Table 33.1. Healing Response to Fracture in Birds and Mammals. Based on consideration of dinosaur physiology determined from microstructure studies (Padian et al. 2004), this is likely the most appropriate model of large dinosaurian response.

Time

Traumatic event with disruption of blood vessels

Lifting of soft-tissue membranes covering inside and outside of bone – periostium & endostium

Inflammatory reaction & blood clot (hematoma) formation

Soft callus formation
   Vascular & cellular invasion of fibragen clot
   Replacement with disorganized collagen & cartilage islands
   Opportunity for pseudarthrosis formation

Hard callus of woven bone replacement

Remodeling
   Resorption by bone destroying cells (osteoclasts)
   Haversian bone formation by bone-forming cells (osteoblasts)

canals (Haversian bone) would be created. However, Big Al likely died before the final stage of fracture healing could occur. The right fourth rib also shows a malalignment typical of fractured bone that is not set during repair. These observations all support Hanna's (2002) contention that these pathologies are best explained by fracture repair.

Hanna (2002) reports that the right manual phalanx II-1 of Big Al also displays evidence of healed fracture. It is covered with substantial rugose bone growth, which extends to the distal joint surface (Fig. 33.10). Scanning electron microscopy of a cross-sectional thin section reveals an oblique

33.9. Fourth and fifth dorsal ribs of MOR 693 "Big Al," showing hard callus. The fourth rib shows evidence of misalignment during healing. The fifth rib (B) shows a putative cloaca for pus drainage, indicating that bone infection may have been a complication of this fracture healing. However, liquid pus may not have been produced in dinosaurs, as it is not the characteristic response in birds.

Photos courtesy of R. Hanna.

5 cm

A

5 cm

B

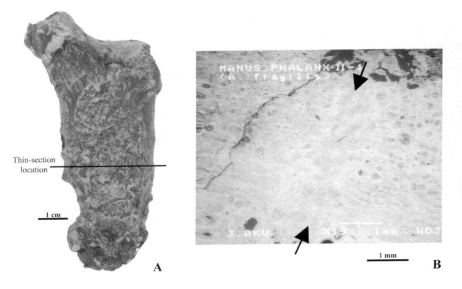

33.10. Manual phalanx of *Allosaurus fragilis* MOR 693 "Big Al." A, lateral view of right manual phalanx II-1 showing rugose growth that surrounds the cortex. In addition to being fractured longitudinally, this element shows evidence of periosteal disturbance. B, scanning electron microscope view of cross section through cortical bone of manus phalanx II-1 shows fibrolamellar bone (opaque area between arrows) that unites an oblique-longitudinal fracture through zonal lamellar bone of the cortex. Figure courtesy of R. Hanna, from Hanna 2002.

Thin-section location

1 cm

MANUS PHALANX-II-1 (A. fragilis)

3.0KV   X19   1mm   WD3

1 mm

A   B

longitudinal healed fracture line through the cortex that likely resulted from twisting (torsion) of the digit .

Broken Bones and Dinosaurian Hard Times

Are broken bones typical for *Allosaurus fragilis* or was Big Al just exceptionally unlucky? Fractures are fairly common among the *Allosaurus fragilis* specimens from the Cleveland-Lloyd dinosaur collection. This collection–82% of which is subadult (Gates 2005)–is the largest sample of *Allosaurus* specimens in the world, consisting of 70 partial skeletons, 12,000 individual bones, and isolated dinosaur eggs. Approximately 2% of bones in the collection exhibit some sort of pathology (Petersen et al. 1972). Of these pathologies, approximately one-third of them are fractures in various stages of healing (Hanna 2002), surrounded by prominent hard calluses. Five isolated ribs from the collection show similar manifestations to the healed fractures in Big Al (Hanna 2002; Madsen 1976).

In fact, fractures in various stages of healing constitute the most common type of pathology encountered in dinosaurs (Tanke and Rothschild 1997). The most commonly reported fractures occur in ribs, tail elements, fibulae, and phalanges. Methods of data collection, inconsistent reporting, and disparate sample sizes currently rule out rigorous synthesis of fracture prevalence data, but some provisional observations are possible.

Fracture Rates and Behavior

Did *Allosaurus* live an especially rough life? *Allosaurus* fractures are reported in ribs, caudal vertebrae, gastralia, a scapula, a humerus, a radius, and a fibula (Petersen et al. 1972; Molnar 2001). It is most tempting to attribute the prevalence of fractures to injuries sustained during activity associated with a predatory lifestyle. But *Allosaurus* is also very well represented in the fossil record. The large sample of both complete skeletons and disarticulated remains may well introduce representational bias relative to less completely known taxa. In a similar vein, the relationship between amounts of sedimentary rock available for sampling has already been noted to skew apparent extinction rates if the uneven sampling is left uncorrected (Peters and Foote 2002). Geologically more recent taxa are also generally better represented than the older, and may bias any findings of disease prevalence in favor of more recent taxa (Molnar and Farlow 1990).

Similarly abundant taxa should be used to compare fracture frequency. Remarkably, the abundant taxon *Coelophysis* is entirely lacking in observed fractures reported in the literature (Schwartz and Gillette 1994), while *Allosaurus* and the later tyrannosaurids appear to be very frequently affected by traumatic pathology. In the latter, this most frequently affects ribs, caudal vertebrae, neural spines, and gastralia (Molnar 2001). *Gorgosaurus libratus* manifests a surprising number of fibular lesions attributed to healing fracture (Molnar 2001). Tanke and Currie (1998) report fibular fractures in 10–15% of Albertan tyrannosaurids. Sadly, the basal small-bodied *Coelophysis*, while well represented, is not the ideal comparative. Size may be exerting as much an effect as predatory activity, with the larger taxa manifesting more traumatic pathologies than their smaller counterparts due to the more catastrophic effects of accidents on animals of large mass (Farlow et al. 1995).

Traumatic injuries are by no means limited to carnivores. Large numbers of healing fractures are also encountered in Late Cretaceous hadrosaurids, where caudal vertebrae in particular, as well as broken ossified tendons, neural spines (Fig. 33.11), ribs, and gastralia are the most commonly affected elements (Rega et al. 2002; Tanke and Rothschild 2002). *Iguanodon* also manifests fractures, particularly of the ilium (Tanke and Rothschild 2002), but not in the numbers seen in the late Cretaceous hadrosaurids. Rib fractures are also well known from late Cretaceous ceratopsians, especially *Pachyrhinosaurus*, which may be related to flank-butting behavior (Farlow and Dodson 1975).

33.11. *Edmontosaurus* tail RAM 7150 manifesting series of hypertrophic neural spines. Arrows indicate affected areas.

Author's photo.

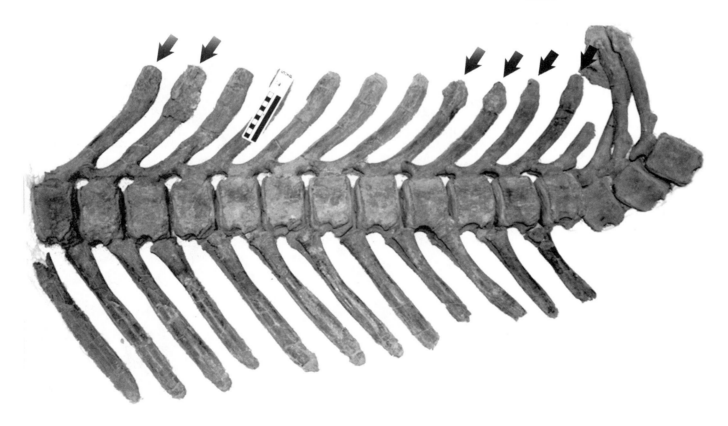

Callus Alone Does Not a Fracture Make

A proliferative lesion may have multiple causes, and healed fracture callus may be only one of them. The diagnosis of healed rib fracture advanced by Riggs (1903) provides an example of this difficulty. He assumed that the bony lump on the right fifth rib of *Apatosaurus* (FMNH P 25112) was a healed fracture callus. An adjacent sixth rib manifests breakage that Riggs declared a premortem fracture failing to heal. However, the pattern of breakage in this rib resembles postmortem breakage rather than the type occurring when the bone is green or fresh. The fifth rib may indeed have a healed fracture callus. However, the callus form is not what would be expected from frank breakage after a blow or a fall. The bulk of the callus formation appears on the posterior and interior surface of the rib, with the lateral surface unmarked. In the case of a fall, the tensile stresses would have been predominantly lateral – not medial – and the callus would manifest as the mirror image to its actual form (Fig. 33.12). In addition, a posterior nodule arising from the lump resembles an osteo-cartilagenous growth that may be a benign bone tumor. Without a radiographic to verify breakage under the lump, a number of other possibilities must be considered: elevation of the periosteum from trauma that did not break the bone, reaction to lung infection, benign bone tumor, and especially stress fracture from microstructure damage resulting from bending the rib posterior-laterally, not medially.

Stress Microfracture

Stress fractures, which result from damage to bone microstructure without frank external breakage, have been reported in an isolated ornithomimid metatarsal (Sullivan et al. 2000) and an isolated ceratopsian phalanx unattributed to species (Rothschild 1988). Diagnosis of stress fracture in the latter example benefited from sectioning along the sagittal plane, which

33.12. Possible healed fracture callus on right fifth rib of *Apatosaurus* (FMNH P 25112); courtesy of Bill Simpson and Field Museum of Natural History, Chicago. This was diagnosed by Riggs (1903) as a healed fracture. This is not the classic appearance of fracture callus, and other causes should be considered. The anterior lesion could be focal periostosis of stress fracture, and the posterior protuberance an exostosis.

demonstrated discontinuity of underlying bone material (Rothschild and Martin 1993).

Stress fractures can arise from overuse or from bone insufficiency (Hardie et al. 1998). Healing of these injuries can manifest in mammals as a bump or surface discontinuity, the result of reactive bone forming on the outer and inner bone surfaces (Salter 1999; Stover et al. 1992). This "focal periosteal callus" has been observed in domestic birds as well, with cases of substantial periosteal callus formation in reaction to increased stress on weight-bearing lower limbs of abnormally heavy turkeys (Crespo et al. 1999). The underlying cause of the surface reaction is discontinuity of underlying bone material, which is best detected using MRI in living animals (Court-Brown 2003). Caution is highly warranted when identifying subtle pathologies such as stress fractures at the gross level without supplemental histological or imaging data, however. Absent microscopic verification of bone disruption, radiography (x-ray) has successfully demonstrated discontinuity of bone mineral underlying a surface bump on mosasaur fossils (Schulp et al. 2004), adding evidence to the hypothesis of stress fracture in this case.

Conversely, a lesion that appears at the gross level like a typical frank fracture callus can prove in fact to be a stress fracture when the bone tissue is examined microscopically (Rega et al. 2004). In a series of neural spines from the early Permian synapsid *Dimetrodon gigahomogenes*, gross examination revealed several swellings in the proximal spines that resembled well-healed fracture calluses completely surrounding the shaft. Histological thin sections through the fibrolamellar bone revealed striking evidence of torsionally induced slippage between lamella, resulting in significant remodeling of shaft deformation within one to two seasons of the injury.

The precise nature of dinosaurian reaction to stress fracture has yet to be investigated at the tissue and element level to confirm that dinosaur stress fractures manifest grossly in the same manner as human ones. Dinosaur stress fractures have heretofore been identified mainly by gross lesions deemed "characteristic and unequivocal" (Rothschild and Molnar 2005; Rothschild et al. 2001, 334). Caution is warranted, as this is unverified by systematic histological or radiographic study, and stress fractures may develop in lamellar and fibrolamellar bone very differently than in nonlamellar bone (Rega et al. 2004). Because of this fundamental difficulty, hypotheses that stresses from prey capture are responsible for greater rates of these lesions in the phalanges in *Allosaurus* (Rothschild et al. 2001) should be subjected to more rigorous testing. Conversely, their absence at the gross level should not rule out stress microfracture without independent verification.

Microstructure and Fracture Healing

Straight et al. (2009) utilize oxygen isotope data to suggest that a healing fracture callus in a *Hadrosaurus* neural spine reflects an elevated temperature associated with healing, indicating capacity for localized fever or elevated temperature due to increased vascularity. Their thin-section data demonstrate dense Haversian remodeling of the callus (to an even greater extent than that seen in birds), with the original cortical bone surfaces still identifiable inside the callus (Fig. 33.13). In spite of the thermal elevation

33.13. Thin cross section through healing callus of THA hadrosaur neural spine at 37× magnification, width 6.2 mm. Photomicrograph illustrates contact between the callus and original cortex. The callus, radially oriented along the upper edge of the photo, is thin at this position. The original cortex is composed of dense secondary Haversian bone tissue. A few of the newer Haversian osteons cut into the callus, which also has several primary vascular channels (Straight et al. 2009).

Courtesy of W. Straight.

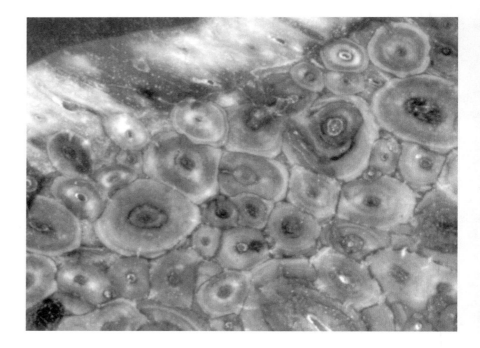

and dense Haversian remodeling, Straight et al. suggest that multiple large lacunae in this fracture callus may contain cartilage islands more reminiscent of reptilian fracture repair cartilage.

Morphologically similar abnormally enlarged and porous neural spinous processes from an *Edmontosaurus* tail RAM 7150 (Rega et al. 2002) tell a slightly different tale of bone dynamics. Gross examination suggests either healed fracture or submembrane bone deposition. In microscopic cross section, multiple episodes of resorption, deposition, and recutting of bone can be read. No fracture plane is apparent, and the section lacks any cortical boundary. If lifting of the bone membrane and subsequent bone deposition or fracture were cause of the macroscopic "swelling," their traces have been completely obliterated by remodeling. Large lacunae containing as yet underdetermined tissue are also present in the section (Fig. 33.14). It is distinctly possible that hadrosaurian neural spines are affected by an as

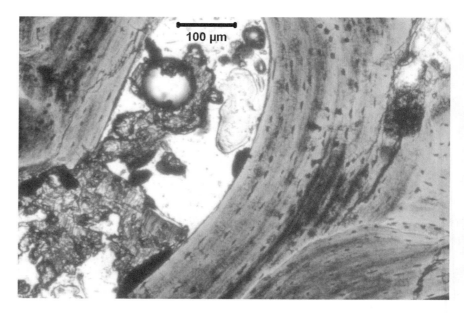

33.14. Thin section through hypertrophic neural spine of *Edmontosaurus* tail (RAM 7150).

Author's photo.

yet uncharacterized condition causing blood vessel invasion and intense remodeling.

The dense cortical bones of hadrosaurine limbs may heal differently than the highly spongy elements, such as neural spines and centra. The microstructure of a massive fracture callus on the third and fourth metacarpals of an articulated *Hadrosaurus* manus was examined by Moodie (1926), who also reported large lacunae. These spaces were circular in cross section and lined with lamellae of bone containing numerous osteocyte lacunae, which is more indicative of blood vessels inhabiting these spaces. There was no indication of mineralized cartilage. These differences between sections may be due both to different elapsed time of healing and to properties of healing varying in different anatomical locations. Systematic study of this phenomenon at the histological level is clearly warranted, in both dense cortical and cancellous elements.

Fracture Provides More Clues to Dinosaur Bone Biology

Observations on fracture healing provide further evidence that the bone dynamic of dinosaurs was closer to that of birds and mammals than to that of reptiles. Straight et al. (2006) report that fracture repair in dinosaurs "produced very different structures at significantly faster rates" than in crocodiles. In their comparative study between shaft fracture in a hadrosaur and a modern alligator, the differences include callus form, nature and resorption of cortical boundary, cancellous infilling, and localized temperature elevation caused by "aggressive cellular activity" (Straight et al. 2006, 46). In each case, the observations of dinosaur fracture were closer to those of avian/mammalian fracture healing.

Most avian fractures are repaired by means of secondary healing characterized by the aforementioned sequence of callus formation, mineralization, ossification, and Haversian remodeling. Only with surgical fixation of the bone is primary healing via remodeling without callus formation possible. Callus in the endosteal cavity typically provides the major and early support for avian fracture healing, together with surrounding subperiosteal callus. In mammals the latter forms the major healing support. In reptiles, by contrast, the bone adjacent to the fracture site develops a cartilaginous collar, which gradually expands to cover the fracture site; this mechanism is subject to more frequent failure to unite than in birds and is far slower (Mitchell 2002).

Air sacs remain a major clinical consideration in the fracture healing of pneumatized bird bone, relevant to the highly pneumatized vertebrae of saurischian dinosaurs and pterosaurs. Aspiration pneumonia, air sacculitis, or asphyxiation can result from fluid entering the air sac in the proximal fracture segment (Wissman 2006). Perimortem fractures involving the vertebrae of saurischian dinosaurs and pterosaurs may have had adverse affects on the air sacs that might have caused immediate death, therefore precluding healing. To date, no consideration of vertebral pathology in dinosaurs has been related to air-sac pathology.

Extreme cortical thinning and air-sac presence in long bones of avian species renders the dinosaur match to birds inexact in non-pneumatized dinosaur long bone. Avian air sacs are even more extensive than in dinosaurs, invading limb bones and limb girdles as well as axial elements. Nonpneumatic ratite and large mammalian limb bone may therefore prove to be a more appropriate model for dinosaurian limb fracture.

Bone Breakage, Infection, and Trauma in a *Tyrannosaurus rex* Named "Sue"

An exceptionally complete specimen of *Tyrannosaurus rex*, FMNH PR 2018 – known as "Sue" – also displays prominent bone proliferation resembling healing calluses on three adjacent right ribs, P 13–15. Postmortem breakage of the proliferative bone on P 14 reveals the smooth original cortical surface surrounded by an elevated, more porous shell, the hallmark of periosteal reactive bone, a feature also seen in the rib fractures of "Big Al." These calluses and the observable fracture plane lie along the same approximate horizontal axis, and the relative degree of healing is similar. The presence of a pseudojoint (pseudarthrosis) on P 15 demonstrates that this rib must have been subject to movement during the soft callus stage, disrupting the bone healing. Rib movement may therefore have accompanied normal breathing (Rega and Brochu 2001b), providing additional support for proposed models of theropod breathing (Carrier and Farmer 2000; Claessens 1997; Fig. 33.15). Contrary to many reports, the shiny material located within this callus is not an imbedded tooth, but fragments of cortex, with no evidence of dental enamel (Brochu 2003; Fig 33.16). With their blood supply cut off by the injury, these fragments (sequestra) paradoxically retain the appearance of healthy shiny bone, when they are in fact dead and largely beyond the reach of the scavenging immune cells contributing to the healing response.

Other pathologies that affect the right humerus and coracoids of Sue may be a consequence of a single traumatic event such as a fall or a blow affecting the right side of the body. These consist of pits and spurs resembling avulsion injuries caused by forced muscle and tendon pull (Rega and Brochu 2001a; Carpenter and Smith 2001). Given the similar degree of healing, it is likely that injuries occurred together.

Infection without Fracture

The left fibula of Sue manifests florid proliferative bone growth affecting the entire circumference of two-thirds of the shaft (Fig. 33.4). Healed fracture has been identified as the cause by some (Tanke and Rothschild 2002), however, serial cross-sectional CAT scans taken of the entire left fibula fail to support this view (Rega and Brochu 2001a; Brochu 2003). The scans manifest a continuous original outer bone surface (cortex); by contrast, cortical malalignment is the typical result of an untreated fracture in the wild (Stocker 2005). There is no evidence of subluxation or torsion, where the ends of the bone might have slid past each other. This could be due to the "splinting" effect of the tibia in the case of a fracture, but recourse to splinting does not explain the total lack of horizontal displacement demonstrated by the aligned cortex deep to the outer envelope of bone. Instead, the CAT scans of Sue's fibula show classic features of subperiosteal osteomyelitic infection, which include the remnant of dead cortex (sequestrum) surrounded

33.15. Pseudarthrosis in left thoracic rib in FMNH PR 2081 "Sue."

33.16. Sequestrum (not a tooth) present in left thoracic rib in FMNH PR 2081.

by an envelope of new bone (involucrum) present on over two-thirds of the shaft (Fig 33.5).

The surface features of healing fracture callus and subperiosteal bone deposition caused by inflammatory bone infection (pyogenic osteomyelitis) are broadly similar, as they are both the result of the same process, namely the elevation of the periosteal membrane. However, infection often results in a coarser surface, with coarser pores and externally visible openings of channels. In Sue's fibula, these openings can be seen to penetrate the involucrum, reaching the sequestrum. These are morphologically similar to pus-drainage channels called cloacae (Latin: sewer) which characterize mammalian response. This observation in Sue is complicated by the clinical observation that pus does not characterize infections in birds and reptiles. Rather these clades manifest a solid substance known as a caseating granuloma (Montali 1988; Therio 2004) which effects are different than the lytic effects of pus. Why this healed bone morphology would closely resemble a more mammalian process remains an area for future inquiry.

It is true that "a dog can have both fleas and ticks"–meaning that a fracture that is open to the environment as a result of skin and soft-tissue damage is also highly susceptible to infection. As in birds, the distal portion of the leg below in *T. rex* likely had very little soft tissue covering the bone. Avian fractures in these areas are often open and broken into multiple bone fragments (comminuted), and osteomyelitis is frequently the result of inoculation of the underlying periosteum (Wissman 2006).

Osteomyelitis without fracture also commonly occurs on the elements lying directly underneath the skin surface unprotected by muscle mass, and frequently localizes in the metaphyseal bone of young animals and the vertebrae of adults (Johnson 1994). The lower leg (tarsometatarsus) is most commonly affected in birds (Wyers et al. 1991), evoking Sue's fibular injury. Bacteria and fungi often enter the bone via trauma to the skin and underlying periosteum in the involved bone, but infection can also result from blood-borne (hematogenous) spread from a distant site. Osteomyelitic infection becomes systematic less frequently in birds than in mammals

Table 33.2. Common Causes of Osteomyelitis and Infectious Synovitis

Pathogen	Host	Reference
Bacteria		
Staphylococcus aureus	birds as well as mammals	(Stocker 2005)
Salmonella sp.	squamates	(Ramsey et al. 2002
Escherichia coli	birds	(Clark et al. 1991)
Streptococcus sp.		
Mycobacterium sp.	mammals, birds	(Kelley and Eisenberg 1987)
Actinomyces (flora)	turkeys	(Johnson 1994)
	kangaroos	(Rega 2008)
	mammals	(Resnick 1995c)
Fungi		
Blastomyces dermatididis	birds	(Johnson 1994; Ramis et al. 1998)
Coccidioides sp.	mammals	(Resnick 1995c)
Sporothrix schenckii		
Trichophyton		
Microsporum		
Aspergillis sp.	wild and domestic birds	(Stocker 2005; Bauck 1994)
	reptiles	(Heatley et al. 2001)
	mammals	(Bajracharya et al. 2006; Sherman et al. 2006)
Protozoan		
Trichomonas sp.	birds	(Stocker 2005)
Hepatozoa	mammals*	(Kocan et al. 2000)

* Although hepatozoans are prevalent in reptiles and lissamphibians, they do not typically cause symptomatic disease as they do in mammals, where the genus is well known for causing hepatozoonosis and osteomyelitis.

(Stocker 2005), being contained to the source rather than spreading to other body regions via the bloodstream. Table 33.2 lists common causes of osteomyelitis and joint infection in extant animals.

In sum, the lack of fracture plane and morphology of the proliferation weigh against a diagnosis of fracture and toward osteomyelitis in Sue. The bony manifestations of infection are limited to the fibular shaft, sparing the joint surfaces and surrounding bone, arguing for a robust response and delimitation of infection. Osteomyelitis does not seem to cause systemic (bodywide) infection in birds as seen in mammals with open wounds (Stocker 2005), and the limitation of the disease in Sue and other theropods may be due to conserved similarities in the immune systems of avian and nonavian dinosaurs.

Infection and Injury in Theropod Dinosaurs

Osteomyelitis is well documented but, surprisingly, not especially common in dinosaurs. However, *Tyrannosauridae* and *Allosauridae* may be exceptions (Molnar 2001). These families possess many specimens with potentially infectious pathology. *Tyrannosaurus rex* is becoming increasingly well known for skeletal pathology attributable to trauma or infection or both. Strangely, humeral injuries appear to be relatively common for *Tyrannosaurus rex*, a total of no fewer than seven avulsion injuries or infections being known (Larson personal communication 2008).

Big Al manifests a markedly abnormal pedal phalanx III-1 with rugose surface morphology and marked expansion of the proximal two-thirds of the shaft, extending into the proximal joint surface (Hanna 2002). As this cauliflower-shaped deformity appears to lack pus-drainage holes (cloacae), the enlargement and surface texture may also indicate a benign bone tumor (see "Do Dinosaurs have a Propensity for Cartilage Defects?" in this chapter). The right fifth metatarsal in Big Al also manifests significant shaft expansion and rugose bone on the lateral aspect of the shaft, which Hanna also attributed to osteomyelitic infection. Rothschild and Tanke (2005) assert that the latter is a misdiagnosis. However, two penetrating lesions that do appear to be cloacae (Hanna 2002) argue in favor of the infectious hypotheses for this digit (Fig. 33.17).

Bone Infections in Other Dinosaurs

Osteomyelitis is more commonly noted in carnivorous forms than in other dinosaurs. *Allosaurus* manual and pedal elements are particularly affected, though oversampling of this genus should temper conclusions concerning the meaning of the relative frequency (Petersen et al. 1972; Madsen 1976; Molnar 2001). However, herbivores are also affected. A complete specimen of the basal ceratopsid *Psittacosaurus* was found to be affected by chronic osteomyelitis of the fibula that did not spread to other bones (Lü et al. 2007). Although the osteomyelitis was attributed to tuberculosis, the proliferative, as opposed to purely destructive, nature of the lesions renders this causative agent less likely than other pathogens causing osteomyelitis (Kelley and Eisenberg 1987), unless the natural ecology of the bacillus was drastically altered in dinosaurs.

A 106-million-year-old hypsilophodontid (NMV P186047) from Australia manifested classic signs of chronic osteomyelitis in the left tibia, which affected the length and shape as it grew (Gross et al. 1993). A partial right pelvis and complete right hind limb allow comparison; the left tibia is shorter by 27 mm; radiographic examination shows a growth arrest line in the distal tibia 15 mm from the distal end. The deformation and degree of healing show that this was a long-standing condition that affected the animal during growth and from which at least a partial recovery was made, perhaps with some degree of functional deficit in life. The lack of systemic infection continues to be reminiscent of the osteomyelitic ecology in modern birds (Montali 1988; Stocker 2005). Pathology attributed to osteomyelitis or infectious periostitis in ornithischians is reported in a ceratopsian scapula (Rothschild and Martin 1993) and an isolated *Triceratops* coronoid (Rothschild 1997).

Several cases of dinosaurian osteomyelitis are proposed as secondary to primary trauma (Marshall et al. 1998; McWhinney et al. 1998; Molnar 2001). The fractured right manus phalanx II-1 in Big Al also manifests changes in the callus and joint surface that are suggestive of secondary osteomyelitic infection (Hanna 2002). Additional cases attributed to osteomyelitis can be found in an isolated hadrosaur tibia (CMN 41201; Tanke and Rothschild 2002) and perhaps also the aforementioned hadrosaur metacarpals (Moodie 1926).

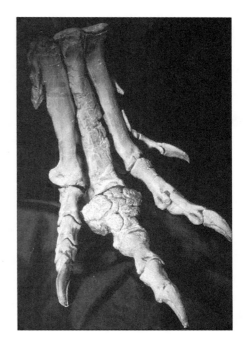

33.17. Right foot of MOR 693 "Big Al," demonstrating potentially osteomyelitic right proximal phalanx III. Other possible causes for this deformity include benign bone tumor, such as osteochondroma, and peripheral bone proliferation in response to lung disease, as seen in pulmonary hypertrophic osteoarthropathy.

Photo courtesy of MOR.

Dinosaur Arthritis

Spondyloarthropathy can have many causes, but true degenerative arthritis appears to be very rare among dinosaurs (Rothschild and Martin 2002). Inflammation (synovitis) and pathogen invasion of the joint capsule can alter joint morphology, as can the wear and tear of joint movement when articular cartilage is damaged. Defects in small areas of bone underlying joints—resulting in pockmarks on subchondral bone—have been identified as affecting hadrosaurs particularly (Rothschild and Tanke 2007). However, the typical signs of cartilage degeneration encountered in mammals, such as subchondral cysts, bone on bone rubbing in a joint (eburnation), and joint margin osteophytes are virtually never encountered in dinosaurs.

This finding is consistent with the avian process, where infectious arthritis is common, but purely degenerative arthritis does not merit a mention (Olson and Kerr 1966; van der Heide 1977). The large cartilaginous cap surrounding the epiphyses of birds and presumably dinosaurs does not seem to be subject to the same degree of mechanical degeneration as the thin layer of hyaline cartilage in the mammalian diarthrodial joint, though this has not yet been conclusively demonstrated.

Clues to Repetitive Motion

Degenerative joint damage—when it does occur in dinosaurs—can offer a unique perspective on habitual behavior or repetitive motion. Two mature chasmosaurs from the Late Cretaceous of Alberta, Canada, show bony changes of joint surfaces that are ultimately due to degenerative changes in the articular surfaces and joint capsules of the first digit, the latter due in part to repetitive trauma (Rega et al. 2010). The holotype of *Chasmosaurus irvinensis* (CMN 41357) has a complete articulated right manus that includes a first digit that is markedly abnormal in shape and orientation (Fig. 33.18). The external and joint surfaces of distal metacarpal I and the proximal phalanx are highly roughened and rugose, with pitting and ridges disrupting

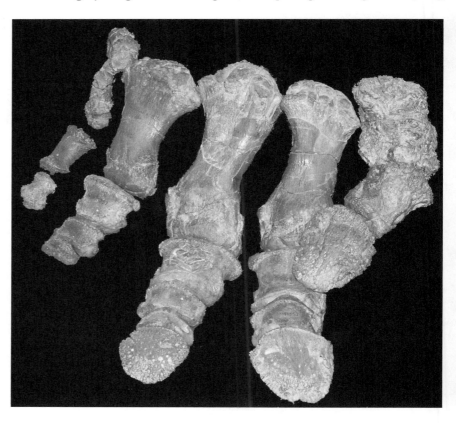

33.18. Right manus of *Chasmosaurus irvinensis* (CMN 41357). B. Close-up of metacarpal I and proximal phalanx. Angulation of thumb due to degenerative changes superimposed on infection or developmental defect.

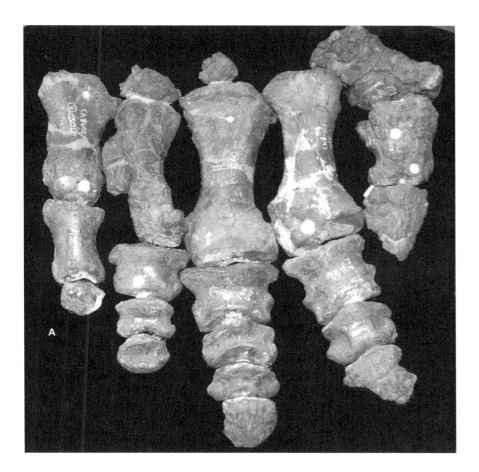

33.19. The right (A) and left (B) manus of *Chasmosaurus belli* (ROM 843), showing joint angulation of the first metatarsal phalangeal joint. Metacarpal IV on the right manus also manifests a bony excresence perhaps due to osteochondroma, a type of benign bone tumor, or to multiple hereditary exotoses.

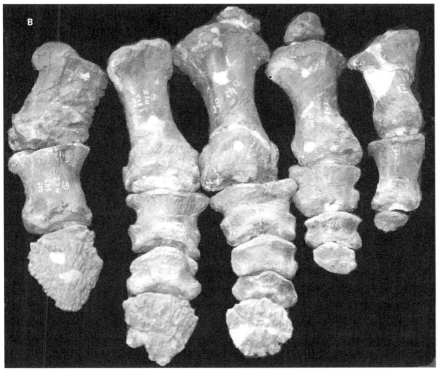

the normally smooth surfaces. Strikingly similar abnormal morphology, also involving the first digit from complete hands, is exhibited by both the right and left manus of *Chasmosaurus belli* (ROM 843), sister taxon to *C. irvinensis* (Fig. 33.19). The metacarpo-phalangeal joints in these dinosaur thumbs are angled 20–36 degrees from normal, and the joint surfaces are highly disrupted. Moreover, the bony joint surfaces fit tightly together, indicating that the presumably extensive cartilage caps covering the bone ends had degenerated.

Freedom from the Tyranny of Specific Diagnosis

Infection, trauma, and/or a growth defect (such as dyschondroplasia) are potential proximate causes on which the secondary joint degeneration accumulated through walking was superimposed. Pathological changes consistent with osteomyelitis do appear on the cortical surface of the angulated thumb metacarpals and phalanges. Bacteria and environmental fungi not typically pathological can cause severe infectious synovitis and osteomyelitis when introduced through traumatic inoculation (Destino et al. 2006). Infection can spread from joint to the subchondral bone or the reverse, as is common in cattle (Verschooten et al. 2000). Cartilage defects causing periarticular pits and joint deformity are common in some birds and may indicate similar genetic tendencies in some dinosaurs.

Whatever the ultimate cause (developmental, infectious, traumatic), the important evidence for functional inference is contained in the degenerative changes that resulted in marked joint angulation. These joints recorded habitual stress, reflecting a locomotor behavior in the large chasmosaurines incorporating manual pronation during the step cycle (Rega et al. 2010). Recourse to process-based description of disease in this case allows freedom from the tyranny of specific diagnosis.

Dinosaur Gout?

Gouty arthritis has been diagnosed in very few dinosaur cases, based on observations of spherical erosion with sharp margins and minimal bone reaction. Rothschild et al. (1997) diagnosed gout from bubblelike lesions present on casts of the right second and third metacarpals of "Sue" FMNH PR 2081, as well as on an isolated unspeciated tyrannosaur hand bone from Dinosaur Provincial Park. They attribute the cause to a diet high in meat protein, as gout is ultimately caused by excessive accumulation of the nitrogen-metabolism waste product uric acid in the body. Gouty arthritis is caused by deposition of its crystal form – monosodium urate – which creates erosive lesions at joint margins.

The paleopathologist must give careful consideration to phylogenetic comparisons when characterizing pathological bone response. The form of gout affecting bone surfaces of joints is found primarily in mammals and far less frequently in reptiles. In the crocodilian manifestation, bony gout is known but not common (Appelby and Siller 1960). Moreover, "gout should not be identified with the concept of human gout, but rather deals in changes more comparable to urate storage as observed in birds" (Frank 1965, 220, author's translation).

Although gout is quite common in some birds, the crystal deposition is virtually always on the guts, including heart, abdominal viscera, and membranes, not in the joints (Crespo and Shivaprasad 2003; Klasing and Austic 2003). A single case of gout was diagnosed in a museum specimen of a ratite on morphological grounds, but lacks clinical verification and evidence of crystal deposition (Rothschild and Rühli 2007). The avian condition results universally from kidney malfunction, dehydration, or poison affecting kidney function. Diet per se is not invoked.

Given the predominantly visceral form of gout in birds and crocodilians, gout prevalence may be in fact severely underestimated in dinosaurs, where guts are generally not preserved! The specificity of the lesion form and its lack of bone response may also be open to question. Human studies suggest that periosteal reactive bone can be in present in cases of articular gout (Liote and Ea 2006) and that the condition is not necessarily purely erosive. Moreover, the rare cases of articular gout lesions forming in some birds may be in fact subcutaneous, not within the joint capsule at all (Beach 1962). A wider differential diagnosis, including infectious agents, cysts, and cartilage inclusions (dyschondroplasia), should continue to be considered for lytic lesions adjacent to joints.

Vertebral Fusion and Joint Disease

Co-ossification is one of the most frequently reported phenomena in dinosaur vertebrae. In some cases, the fusion is regular and affects all individuals of a species. Multiple fused distal caudal vertebrae, forming a "tail club," are regarded as normal for several sauropod taxa, including *Shunosaurus* and *Mamenchisaurus hochuanensis*. Ossification of interspinous tendons is, of course, a normal characteristic of several ornithischian taxa, as well as extant birds, likely due to alteration of a regulatory genetic locus. The purpose, if any, of these normal fusions has been debated. Organ assessed the homologies and functional significance of ossified tendon in birds and other archosaurs. He cautioned that adaptive function of fusion is unclear, especially given the strong genetic component to tendon ossification and "its seemingly random phylogenetic and anatomical patterns of occurrence" (2006, 791).

Pathological spinal fusion—usually limited to two adjacent vertebral centra—is widespread in sauropods, especially *Apatosaurus*, *Diplodocus*, and *Camarasaurus*. The general category of fusion masks an underlying heterogeneity, with at least three degrees of fusion affecting the centra and intercentra. Prolific ossification is typical of some, such as the florid "pachyostotic mass" found in from *Vulcanodon* (QG24), which Raath interprets as "arthritic" fusion of caudal vertebrae 3 and 4 and associated intercentra (1972, 12). Moodie's famous fused vertebrae (*Apatosaurus* sp.) from the frontispiece of his landmark *Paleopathology* volume belongs to this category of florid proliferation, said to resemble an oak gall in degree of deformity (Moodie 1918a, 1923b; Fig. 33.20)

Some fusion is less apparently prolific. *Diplodocus carnegii* (Hatcher 1901, CM 94; Fig. 33.2) manifests this more moderate degree of co-ossification, with some accommodation or molding of proliferative bone around longitudinal tail muscles. *Diplodocus longus* (USNM 10865) belongs to this

33.20. Moodie's famous fused *Apatosaurus* vertebrae.

From Moodie 1918a.

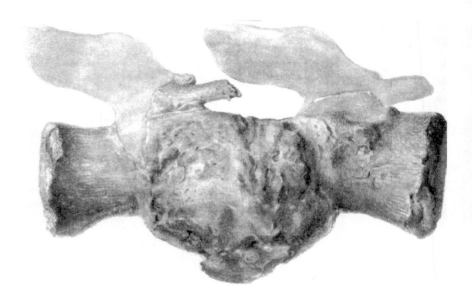

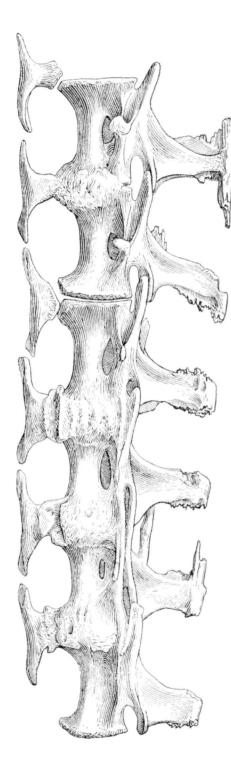

33.21. Fused caudal vertebrae in *Diplodocus longus* USNM 10865, from Gilmore 1932. Gilmore identified these as Ca 15–16 and 17–20. However, they are correctly identified as Ca 16–17 and 18–21.

category as well, with fusion affecting caudal vertebrae 16–17 and 18–21 (Gilmore 1932; Fig 33.21). Finally, fusion of centra with minimal osseous proliferation is manifest in basal sauropods, including the dorsal vertebrae of *Mamenchisaurus hochuanensis* (CCG V 20401), *Eomamenchisaurus yuanmouensis* (CXMVZA 165), and *Mamenchisaurus youngi* and the caudal vertebrae of *Cetiosauriscus stewarti* (BMNH R.3078). Minimal proliferation is also observed in the cervical vertebrae of *Apatosaurus louisae* (CM 3018).

These different forms may well have different causes. Sauropod vertebral fusion is commonly attributed to the consequences of trauma, as tail centra are frequently affected (Blumberg and Sokoloff 1961; Hatcher 1901; Holland 1906; Moodie 1916a, 1918b). But fusion occurs not just in tails but also in dorsal and cervical vertebrae of sauropods, such as seen in *Camarasaurus supremus* (AMNH 5761), *Giraffatitan brancai* (HM SII), and *Apatosaurus louisae* (CM 3018). Gilmore (1932, 14) cited "senility" as causal in a fused section of *Diplodocus longus* tail, noting the additional presence of abnormally ossified dorsal interspinous ligaments spanning Ca 18–19. He notes that the affected vertebra would not have contacted the ground in normal posture and views a traumatic etiology as unlikely. Infection (Moodie 1916a) and tumors (Moodie 1923b) have also been evoked as causal.

Pathology or Functional Modification?

Several researchers have suggested that fusion serves a functional purpose. Rothschild and Berman (1991) examined 18 sauropod tails and noted fusion of adjacent vertebral centra in eight specimens. Half the cases (N=5) of fusion were documented in *Apatosaurus* and *Diplodocus*, where caudals 17–18 and 22–23 were the most commonly fused pairs. Fusion of more than two vertebrae (four or five) or two pairs occurring in separate tail regions occurred only in *Diplodocus*, where three out of the six specimens studied were affected by multiple fusions (CM 94; USNM 10865; Fig. 33.21; AMNH 655). The authors use CT data to suggest ossification of longitudinal ligaments and attribute this condition to DISH (diffuse idiopathic skeletal hyperostosis—a diagnostic term meaning "broad skeletal overgrowth without a clear cause"). They regard this condition as functional, rather than pathological, and deem the numbers suggestive of sexual dimorphism.

DISH is known clinically from aging humans (Resnick and Niwayama 1995b) and less frequently from other mammals (Morgan and Stavenborn 1991), where it is considered a pathological phenomenon of unknown cause, by definition affecting at least four contiguous vertebrae. DISH is unknown in reptiles, where vertebral proliferation and fusion is overwhelmingly caused by bacterial infection (Isaza et al. 2000). Birds commonly fuse dorsal, sacral, and caudal vertebrae into the notarium and synsacrum, but DISH has not been documented in birds.

Rothschild and Berman expand the accepted human clinical definition of DISH by applying it to cases with fewer than four fused contiguous vertebrae. They suggest copulation as the driving mechanism. Whereas Gilmore (1912) invoked mating behavior as the cause of dinosaur injury, Rothschild and Berman suggest an adaptive copulatory rationale: "A major physical barrier to successful mating would be the thick powerful tail of the female. Perhaps, therefore, the fused caudal vertebrae occurred only in females and represent an adaptation permitting the upward and sideways arching of the tail to facilitate copulation" (Rothschild and Berman 1991, 35).

Whiplash Tails in Sauropods

Myhrvold and Currie (1997) explore the same phenomenon, attributing caudal fusion to a sexually dimorphic adaptation of the diplodocid tail as a supersonic sound-generating bullwhip. They base their argument on geometric relationships of vertebral dimensions, which show a high degree of regularity in diplodocids reminiscent of a bullwhip. In these taxa, centrum length increases to a maximum between the 18th and 25th caudal vertebrae, the region most frequently affected by fusion. Myhrvold and Currie characterize this zone as the transition between the stiff muscular base and the flexible whiplike section. They propose "injury consistent with overextension of the joint on a plane parallel to the ground, as might be achieved while whipping the tail" (Myhrvold and Currie 1997, 396) and link vertebral fusion in this region to injury or to an adaptive response. Physiologically, proliferation of bone at joint margins is likely driven by "local joint stresses and abnormal mechanical loading" as well as a physiological and genetic propensity to manifest proliferative bone response and connective tissue ossification (Dequeker and Luyten 2008, 9).

This intriguing hypothesis serves the scientific purpose of generating additional expectations that have yet to be reconciled. *Camarasaurus* certainly lacks a terminal whiplash, yet the pattern of caudal co-ossification is similar in *Camarasaurus*, with no fewer than three specimens (CM 312, UUVP 4317, and AMNH 5761) affected by fusion. Any fusion mechanism that depends on the whiplash is insufficient because it explains only some of the cases. Conversely, *Mamenchisaurus* possesses the same pattern of vertebral metrics, but neither does it have the long whiplike tail nor do the two affected specimens exhibit fusion in this "transitional region." Finally, Myhrvold and Currie point out that while diplodocid terminal tails ought to be injured by crush fractures, no examples are currently known. Clearly further consideration of the phenomenon is required.

Vertebral Fusion in Other Dinosaurs

Co-ossification has been documented in other dinosaurian taxa, including *Tyrannosaurus rex* (FMNH PR2081; Rega and Brochu 2001a), which manifests caudal fusion similar to the intermediate type noted in sauropods, including "molds" of longitudinal tail muscle. *Camptosaurus* (UUVP 5409; Peterson et al. 1972) and *Allosaurus* (multiple specimens, Peterson et al. 1972) are also affected by fusion. Peterson et al. hedged their causal bets in the *Allosaurus* cases by attributing the extensive bony overgrowth in several specimens (UUVP 3773, 5256, and 3811) to neoplasm (tumor) caused by trauma, a condition rarely encountered in the veterinary literature.

Not all fusion is created equal, and causality will ultimately hinge on precise characterization of ossified soft tissue structures, as well as alteration of apophyseal and zygapophyseal joints. The diagnosis *spondylitis deformans* (Goldthwait 1899), as cited in older paleontological literature, was a catchall category for vertebral fusion. Modern medical and veterinary literature remains sadly inconsistent, but the term *spondylosis deformans* is currently the most popular and accurate term for bone-spur production associated with deformation of the annulus fibrosis and longitudinal ligaments (Resnick and Niwayama 1995a) without an apparent inflammatory cause.

Of Cancer and Cartilage

Despite popular media attention (Whitfield 2003), evidence of cancer is relatively rare in dinosaurs. Abnormal tissue growth—neoplasm—can be slow growing and nonfatal (benign) or fast moving, aggressive, and fatal (malignant). The vast majority of neoplastic lesions documented in dinosaurs fall into the benign category. Skeletal evidence of such tumors can present as bone proliferation or as erosive cavities, which would have been filled with bone-destroying soft-tissue tumors in life.

Moodie entertained several different causes—infection, fracture, and tumor—for his famous fused *Apatosaurus* vertebrae (Moodie 1918a) before finally settling on a diagnosis of hemangioma (Moodie 1918b, 1923a, 1923b). Hemangioma is a proliferative benign neoplasm of blood vessels causing the bone deformity. Moodie cited the high vascularity of the bone assessed with microscopic thin sections in support of this conclusion. However, the florid hyperostosis of this lesion is less than satisfactory for hemangioma, and earlier hypotheses cannot presently be discarded. This iconic specimen cries out to be revisited; sadly, its location is not currently known.

Rothschild et al. (1998) claim evidence for hemangioma from a large fragment of dinosaur bone, possibly a centrum, from the Upper Jurassic Morrison Formation of Utah (TMP94.191.1). The specimen is not attributable to species. They base their conclusions on the presence of anteroposterior thick trabecular struts, separated by relatively clear zones. Macroscopic and radiological evidence was interpreted as classic signs of hemangioma.

In the same publication as above, Moodie (1918a) identified another benign proliferative tumor in the vertebra of a cretaceous mosasaur—the first osteoma described in paleontological literature, although this diagnosis has been challenged by subsequent researchers (Rothschild and Martin 1993). Osteoma is a benign focal hyperostosis, resembling a hard, dense button in many cases. It has been reported frequently in mammals and fish, much less frequently in birds (Capasso 2005). Osteomas have been reported in many taxa—*Pachyrhinosaurus*, an unspecified Cretaceous hadrosaur from Alberta (Norman and Milner 1989), and an *Apatosaurus* sp. rib fragment (Capasso 2005).

Lesions interpreted as malignant are extremely rare. A tumor of osteo-cartilagenous origin called a chondrosarcoma has been reported to affect a humerus from the Cleveland-Lloyd Quarry assignable to *Allosaurus* or *Torvosaurus* (Newman 1991). A *Camptosaurus* vertebral tumor is reported in the popular literature (Anonymous 2006). Both specimens await scholarly publication.

Osteochondroma and Errors of Cartilage Proliferation

The beautifully complete holotype of *Chasmosaurus irvinensis* (CMN 41357) is affected by a massive cauliflower-like proliferation fusing cervical vertebrae 5–9. Cervical ribs 5 and 6 are also incorporated into the mass, which is bilateral but asymmetrical, the left side being more severely affected. The fusion caused the neck to be locked twisted to the right and arched dorsally. The obliteration of cartilaginous disc spaces appears to have compressed the neck as well (Fig. 33.22).

The growth strongly resembles a condition termed *osteochondroma* (Rega and Holmes 2006). Osteochondroma is a relatively common benign bone tumor, typically with cauliflower-like appearance and cartilage cap, which can undergo malignant degeneration (Green et al. 1999). It is thought to be part of a spectrum of cartilage proliferation errors largely affecting the growth plate, such as dyschondroplasia and multiple hereditary cartilaginous exostoses.

Several lesions affecting the sister taxon *C. belli* (ROM 843) suggest that this animal too was affected by this spectrum of defects. This animal is affected by marked discrete osseous excrescences on the left metacarpal IV (Fig. 33.23) and the left eighth and ninth thoracic ribs (McGowan 1991). While the rib growth has been attributed to bony proliferation subsequent

3cm

33.22. Right cervical region of *Chasmosaurus irvinensis* (CMN 41357).

33.23. Right manus of *Chasmosaurus belli* (ROM 843). Arrow indicates osteochondroma.

to rib fracture (McGowan 1991), the truncated nature of rib 8, together with the rounded morphology of both the eighth rib and the manual exostosis, is strongly suggestive of osteochondroma/multiple hereditary cartilaginous exostoses. The latter is due to a well-documented autosomal dominant mutation (Schmale et al. 1994; Shupe et al. 1981). Multiple exostoses can also occur from mechanical disruption or ingestion of excessive vitamin A (Lynch et al. 2002).The propensity for these defects may be inherent in the genus (Rega and Holmes 2006).

Comparative Perspective

The rarity of cancer in the fossil record is in keeping with its rarity in wild extant animals. Skeletal tumors are very rare in birds, except for captive psittaciform birds (Weissengruber and Loupal 1999). Osteochondroma, osteosarcoma, chondrosarcoma, and unspecified hyperostosis rank among the tumors that do affect avian bones. Crocodilians have the lowest rates of neoplasm in all captive reptiles, with "pseudotumors" caused by fibriscess around localized infections and granulomatas caused by localized fungal infections, both external and internal (Huchzermeyer 2003). Genuine bone neoplasm is exceedingly rare, with chondrosarcoma and fibrosarcoma among cases cited. Malignant osteosarcoma is occasionally seen in snakes and lizards; when present it is usually in vertebrae. Benign chondrosarcoma is more common, especially in Colubrida, where it arises from vertebral articulations (Garner et al. 2005). But chondrosarcoma is rarely diagnosed in lizards; when seen it affects leg cartilage.

Differential diagnosis necessitates consideration of other noncancerous growths, including bone spurs. Bone spurs are very common findings in dinosaurs and can be differentiated based on their position; in a joint capsule attachment (marginal capsular osteophyte), in a ligament (syndesmophyte), as an ossifying area of muscle attachment (enthospathy), or forming a tumor (neoplasm). The shape of the growths on ROM 843 makes the former unlikely, and the form/location make it unlikely to be either a syndesmophyte or an enthesophyte (Mann and Hunt 2005, 183).

Do Dinosaurs Have a Propensity for Cartilage Defects?

Periarticular pits and joint deformity may indicate genetic tendency toward cartilage defects in some dinosaurs. Rothschild and Tanke (2007) have diagnosed lesions in hadrosaurids as osteochondrosis, a defect of cartilage inclusion in the bone around joints. Another cartilage developmental or benign neoplastic defect, known as dyschondroplasia, can account for the some of the cystic changes observed in the aforementioned deformed *Chasmosaur* thumbs and resulting bone and joint deformity.

Dyschondroplasia is a relatively common and usually benign developmental defect characterized by abnormal masses of avascular cartilage below the growth plate (Crespo and Shivaprasad 2003). This has been associated with the widespread occurrence of osteomyelitis deformity and lameness in the legs in turkeys (Wyers et al. 1991). Dyschondroplasia and the resultant condition osteochondrosis—which result from errors in growth plate endochondral bone formation and cartilage development—are common conditions in birds, including terrestrial ratites (Tully and Shane 1996). Dyschondroplasia is extremely common in some varieties of domestic fowl, such as broiler chickens, ducks, and turkeys, affecting up to 30% of flocks (Crespo and Shivaprasad 2003).

The most rewarding part of paleopathological analysis is when the evidence of pathology informs or addresses hypotheses about dinosaurian lifeways. One of the more intriguing debates concerns the evidence for biting and goring, relating to both predation and defense as well as to violent intraspecific dominance interactions.

Biting, Horn Goring, and Other Forms of Violence

Tooth Marks and Predation

Tooth-marked bone in the fossil record is likely more common than publications represent, with disarticulated bones more commonly marked than whole specimens (Currie and Jacobsen 1995). Bone marked by the teeth of predators is nearly twice as common among hadrosaurs as in armored and presumably less vulnerable ceratopsians (Erickson 2000).

Although the animals inflicting tooth marks can be identified only in rare instances, sometimes a smoking gun is found in the form of an actual tooth. A broken tooth tip embedded in an azhdarchid pterosaur from the Upper Cretaceous of Dinosaur Provincial Park implicates the theropod *Saurornitholestes langstoni* as the biter (Currie and Jacobsen 1995). According to Tanke and Currie (1998), a Dinosaur Park specimen (TMP 85.62.1) of an unidentified tyrannosaurid left dentary manifests an unhealed wound with a tip of a tyrannosaurid tooth embedded in it.

More commonly, a diagnosis of bite marks is made based on position and form of injuries. Healed damage to the neural spines of the 13th–17th caudal vertebrae in *Edmontosaurus annectens* (DMNH 1943) is attributed by Carpenter to a bite taken from the tail (1998). The spine of caudal vertebra 15 is missing its distal third, ending in a U-shaped truncation with an expansion interpreted as healed osteomyelitis. Spines 13, 14, and 16 are said to manifest pits interpreted as tooth marks lining up with the truncation to form a *T. rex* mouth-shaped "bite." Together with 17, all affected spines show a malalignment or "kink" most clearly visible from posterior. As the right preacetabular process in this animal displays an older healed lesion – possibly a healed fracture – limping might have made the animal vulnerable to predation.

The height at which the damage occurs and the shape of the lesion suggest a *Tyrannosaurus rex* as the biter (Carpenter 1998). No tooth or tooth fragment was apparently recovered from the affected area, contrary to rumor. The ossified tail tendons in DMNH 1943 are entirely sculpted, and no damage or lack thereof can be inferred from their condition.

Healing Complicates the Diagnosis of Bite Marks

Healing may have been good for this particular *Edmontosaurus* in life. When it comes to conclusively identifying biting, however, it complicates matters. The story of a bitten tail is intriguing. The more mundane fact is that the pathology described above could have other causes, including neural spine fracture and subsequent bone infection from an open wound. The truncated neural spine 15 could then result from a failure of bone union – pseudarthrosis – with pre- or postmortem loss of the distal portion. The pits interpreted as bite marks are sufficiently healed to make their identification far from conclusive. Except for those injuries with embedded teeth, well-healed bite marks, especially when infection is involved, will always be open to other interpretations. Unhealed tooth marks therefore have more utility in conclusively identifying both the biting and the biter.

Erickson and Olson (1996) demonstrate that several characteristic attributes of unhealed lesions on a *Triceratops* sp. pelvis (MOR 799) retain enough markers to more confidently identify them as *T. rex* bite marks. These include markers of "puncture and pull behavior," where biting deeply with great force was followed by drawing the mouth through the flesh and bone, leaving long gashes One triceratops pelvis showed evidence of nearly 80 gouges and punctures. Many punctures were eye shaped in cross section and spaced 4 inches apart. Complete lack of healing meant that matching microscopic evidence of striations with the serrated carinae of *T. rex* teeth was possible.

Dinosaur bite marks are not limited to prey. Tanke and Currie (1998) present several cases of theropods where intraspecific biting is hypothesized as a part of a behavioral repertoire of face biting for dominance behavior. The holotype of *Sinraptor dongi* (IVPP 10600) displays multiple healing wounds of the maxilla and surangular that they maintain are healing bite wounds. Several show a teardrop shape indicative of puncture and pull, and the spacing of the wounds on the surangular matches expectations for theropod tooth marks.

Not All Oral Holes are Bite Marks

Multiple perforations concentrated on the surangular and dentary elements of many mature *Tyrannosaurus rex* specimens (FMNH PR 2018, MOR 1125, MOR 008, AMNH 5027) have been identified as bite marks in the popular press and elsewhere. Several researchers have challenged this assertion (Rega and Brochu 2001; Brochu 2003; Wolff and Varricchio 2005) as the lesions do not conform to expectations generated by the biting hypothesis. The spacing is not regular, each perforation passes through the bone at right angles, the perforations are often in different stages of healing, the position of these lesions on the surangular and posterior dentary is one of the least accessible places on the skull for biting, and there is no teardrop shape to indicate the puncture and pull demonstrated on other bite mark examples. Instead, some of the lesions appear to originate in multiple fenestrations that coalesce into a single lesion (Figs. 33.24 and 33.25). The margins of the bone lesions on the affected *T-rex* specimens are primarily characterized by dense, hard reactive bone (called sclerotic bone). In addition, multiple lesions on the same animal appear to be in different stages of healing, some very strongly sclerotic, others thinner and less fully healed. All of these are hallmarks of a chronic disease of somewhat long standing.

In terms of the age profile of affected dinosaurs, the ubiquity of these lesions in mature theropods and their increasing prevalence with age may be

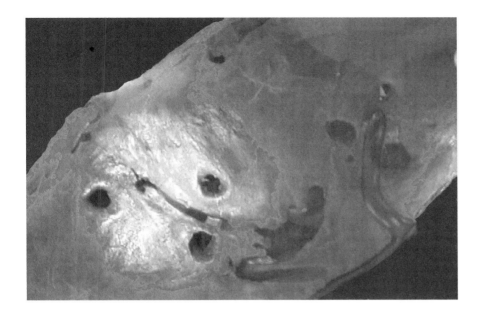

attributable to an accumulative and "typical" chronic infection. In a survey of 56 tyrannosaurids, Wolff and Varricchio (2005) found over one-fourth had some defect of the surangular or posterior dentary, which they attribute to a variety of potential causes, many of them chronic. Because these lesions affect primarily the mature *T-rex*, the disease has parallels to dental caries, in that lesions accumulate in older individuals.

In reading the bone response characteristic of these mysterious holes, the "smoking gun" for which we should be searching will be something chronic; a pathogen of oral flora affecting the oral mucosa and bone that does not kill quickly, but allows sufficient time for lesions to accumulate with advanced age. Diagnostic caution should therefore be exercised. Narrowing the range of what type of pathogen could cause these lesions is necessary, but is not the same as identifying the specific causative organism. With infection, the modern scientific approach would be based upon identifying a similar observed process and characterizing the attributes of the "unknown" ancient pathogen in light of known modern pathogen behavior. A broad differential diagnosis remains necessary, including developmental defects, bone infarct and cyst, neoplasm, and bacterial and

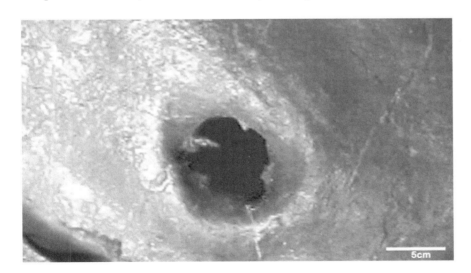

33.25. Detail of a perforating lesion on the left jaw of *Tyrannosaurus rex* FMNH PR 2081 "Sue." Multiple holes appear to have coalesced to form a larger lesion.

fungal infections. Chronic infection by bacteria or fungi tops the list, based on their prevalence in wildlife (Rega and Brochu 2001b).

Pathogens in the Past

When dealing with paleo-infection, drawing conclusions regarding specific pathogens becomes extremely difficult, due to evolution of pathogens themselves from Mesozoic to modern times. Evolution has been documented in the present to alter the nature of the host-pathogen relationship and response (Gill and Mock 1985). Thus, attribution of a disease to a specific pathogen is an area where uniformitarian assumptions are least well supported. This is further complicated by the paucity of comparative clinical data, as the veterinary literature on infection emphasizes domesticated animals, while certain kinds of pathogens, including protozoans, fungi, and fungal-like disease, are more prevalent in wild animal populations. These remain relatively poorly characterized in the literature, especially in documenting their effects on bone.

There is abundant evidence in the literature that some bacteria and multiple fungi are destructive to bone (conditions including but not limited to coccidiodomycosis, blastomycosis, aspergillosis, cryptococcosis, sporotrichosis—see Resnick 1995c). One modern pathogen that creates lesions similar to those observed and fits the life-history profile of the affected dinosaurs is actinomycosis, a fungal-like bacterial infection seen in wild and domestic animal populations, which is part of normal mouth flora in many mammalian species. Actinomycosis fits many, but not all, of the model criteria, the biggest objection being that it affects mammals and marsupials, but is rarely seen in birds.

Wolff et al. (2009) propose that the protozoan *Trichomonas gallinae* provides the best match for the observed lesions. While trichomoniasis has many attributes of the model proposed (osteolytic, intra oral), several features make the conclusion that "these animals died as a direct result of this disease, most likely through starvation" far less supportable by evidence. (Wolff 2009,1). And finally, there is the issue of scale – something which chokes a raptor or a pigeon would have to "scale up" in a *T. rex*; the relative size of this type of lesion is completely different, (i.e., much larger) in small bird throats.

For example, the observed sharp margins with little reactive bone shown by the radiographs of trichomonas-infected birds are dissimilar to the reactive bone seen in the affected *T. rex* specimens. Trichomoniasis can moreover be very rapidly fatal in birds (14 days or less) although with a less virulent milder form, affected individuals can recover. Indeed, if a trichomonas-like protozoan is the culprit, this suggests that trichomonas was less acute in its non-avian dinosaur form in the late Cretaceous.

This case underscores the difficulty with comparative models in paleopathology. Although birds are indeed the only extant dinosaurs, they are not a perfect model system, and several aspects of their bone biology are very different from extinct dinosaurs. Birds lack lines of arrested growth that are very common in dinosaurs (indeed birds are virtually unique in lacking them). Bird bone is thinned because of the demands for flight and expansion of pneumatic sacs, making the appearance of bone lesions somewhat different and curiously, perhaps making mammalian bone response to infection a more apt comparison. Bird mouths are also very different from dinosaurs (thin bones, no teeth!).

Many Causes of Ceratopsian Face Lesions

Tanke and Farke (2007) come to similar conclusions about some smooth-rimmed frill perforations that are extremely common in chasmosaurine and centrosaurine cranial material. These lesions are widespread, occurring in no fewer than six ceratopsid genera. Lack of bony indicators of infection or trauma challenged the notion that these perforations are the result of horn injuries. Instead, they suggest that the perforations are the result of chronic cutaneous infections and the horn resorption a normal effect of advanced age.

In contrast, Farke et al. (2009) find statistically suggestive evidence for different patterns of horn injury in *Triceratops* as compared to *Centrosaurus*. They employ a G-test of independence to compare the incidence rates between the two genera of injuries attributable at the gross level to infection or trauma, including healed fracture. As *Centrosaurus* has only a single nasal horn (and no massive brow horns), the authors posit that it would be most likely to show horn-combat-related differences from *Triceratops*. Indeed,

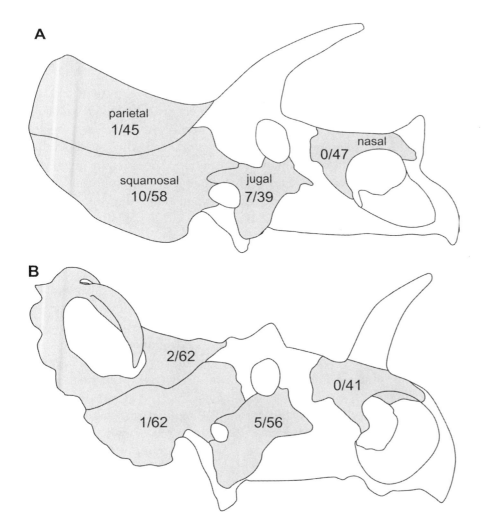

33.26. Diagram of *Centrosaurus* and *Triceratops* frequency of injury.

Figure courtesy of A. Farke.

a significantly higher number of *Triceratops* skulls showed lesions on the squamosal bone (Fig. 33.26). This reflects their expectations of the location of supraorbital horn damage due to "horn locking." No other cranial element demonstrated a significant difference.

Lower pathology rates in *Centrosaurus* may indicate visual rather than physical use of cranial ornamentation in this genus, or a form of combat focused on the body rather than the head. The authors do not claim to be reconstructing the exact nature of injury for each individual, but to be demonstrating a difference in overall pattern. The high frequency of jugal lesions in both genera invites further study. The orientation of the fracture planes on the squamosals remains enigmatic and less readily explainable by horn damage.

Dinosaurs and "Paleo-Health"

Ultimately it should be remembered that only a small percentage of dinosaur bone appears diseased. Dinosaurs have been declared both healthy (Rothschild 1997, 426) and highly diseased based on evidence from pathology. However, a pronouncement concerning past health based on frequency of diseased bone is exceedingly difficult to defend even under ideal conditions—those being a complete skeletal sample, drawn without bias from a true biological population (Ortner 2003). It is scientifically insupportable to make these conclusions with fragmentary skeletal remains widely separated in time, space, and depositional environment, as is true for most dinosaurs.

Why Are Healthy Bones Not Necessarily from Healthy Dinosaurs?

The osteological paradox was formulated by Wood et al. in their seminal 1992 paper. The principal difficulty in assessment of health lies with skeletal invisibility of most disease, particularly acute infectious disease, a major cause of death. Moreover, people and animals differ in their susceptibility to disease, with some individuals being frail and some robust under different conditions. Even in diseases that do affect bones, the very frailest individuals in a population will typically expire before showing any bony manifestations of disease, because bone generally takes a longer period of time to respond to systemic insult than do soft-tissue systems. The most acute illnesses (such as smallpox, anthrax, and plague, to name a few) are fast moving and kill the sick in days or even hours (Ewald 1994). Therefore, a fossil recovered without bony lesions may indeed have been healthy. Or it may actually have been so sick that it died immediately from the illness, leaving clean bones. The very sick and the healthy thus are osteologically indistinguishable from each other. It is the middle group of affected dinosaurs who were sick enough to manifest bony disease but healthy enough to mount a sufficient immune response to survive for a time that we are investigating when looking at "paleopathology."

Were theropods such as Big Al and Sue particularly sickly? As many of the pathologies discussed thus far resolve very slowly, one would a priori expect that there would be a greater number of traumatic and degenerative pathologies present with increasing age of the individual. Complete specimens, where available, seem to bear out this trend. Interestingly, the age-specific mortality for four assemblages of North American tyrannosaurs—*Albertosaurus*, *Tyrannosaurus*, *Gorgosaurus*, and *Daspleto-saurus*—is characterized by relatively high juvenile survivorship and increased mortality at midlife and near the maximum life span (Erickson et al. 2006), favoring the accumulation of many pathological traces. This may in part also account for the large number of pathologies observed in tyrannosaurids. These individuals, and indeed the vast majority of dinosaurs with pathological bones, belong to this middle group of animals, those healthy enough to accumulate a lifetime of wounds.

If appropriate animal models are to inform paleopathological thinking, diagnoses that are common to archosaurs should receive consideration. One example that has eluded the literature is the diagnosis of "calcified granuloma." This condition, or its healed consequences, is entirely missing from accounts of dinosaur paleopathology. Granulomas (a tumorlike mass of inflammatory tissue containing immune-system cells, prone to calcification and chronic persistence) seem to represent the most common avian and reptilian response to infection and tissue death; in contrast, granulomas are very uncommon in mammals. While extant mammals mount an aggressive inflammatory response to infection, birds manifest a pus-oozing, or suppurative, reaction far less frequently (Montali 1988). The creamy liquid pus characteristic of mammals is not typical of either birds or reptiles (Montali 1988; Therio 1994). Instead, a cheeselike mass that can be peeled away from underlying tissue is seen. Therefore, expectations of pus drainage in the case of chronic bone infection in dinosaurs may need to be examined.

The most promising avenue for future dinosaurian paleopathology lies with careful reading of the biological signal contained in the bone tissue itself , with critical reference to extant models and their applicability and with thoughtful analysis of patterns of disease related to age-at-death, size, and morphology manifested by individuals and groups of individuals. It is to be hoped that scientifically supportable evidence regarding dinosaur disease can be ultimately more interesting than a lurid headline.

Conclusions and Future Directions

Institutional Abbreviations

CMN	Canadian Museum of Nature, Ottawa Canada
CM	Carnegie Museum, Pittsburgh, Pennsylvania USA
FMNH	Field Museum of Natural History, Chicago, IL, USA
IVPP	Institute of Vertebrate Paleontology and Paleoanthropology, Beijing, China
NMVP	National Museum of Victoria, Department of Vertebrate Paleontology, Melbourne, Victoria, Australia
USNM	Smithsonian National Museum of Natural History, Washington, DC, USA
RAM	Raymond Alf Museum, Webb School, Claremont, CA USA
ROM	Royal Ontario Museum, Toronto, Canada

Acknowledgments

Thanks to Jim Farlow and Mike Brett-Surman for inviting my participation and for their patience. I benefited from the input of many scholars, including but not limited to Chris Brochu, Matt Carrano, Andy Farke, Becca Hanna,

Jason Head, Jack Horner, Rob Holmes, Peter Larsen, Pete Makovicky, Bill Simpson, Will Straight, Kevin Padian, and Ewan Wolff. They facilitated my access to material, provided stimulating discussion, and in many cases supplied figures as credited. I am grateful to Jane Buikstra and Della Cook for their comments on the manuscript, and to Della for sharing her in-press manuscript on Roy Moodie. Thanks to author Judith Horan, who contributed advice and editorial changes aimed at writing for the nonspecialist. Thanks as well to the bone crew at Western University—Amy Chew, Vicki Wedel, and Matt Wedel—and to Western University of Health Sciences avian veterinarians Teresa Morishita and Miguel Saggese for review, discussion, and truly helpful comments. I am grateful to the Western University Public Affairs and Publications departments, especially Rick Clapper and Jeff Malet for their invaluable assistance with figure 33.1. Without the efforts of Western U's excellent library research staff (Lurlene Moreno, Rudy Barraras, Kim Hoang, and Jennifer Cassidy) this paper could not have been written, while Jeff Keating provided good-humored proofreading on typically short notice. Laura Rush provided editorial advice and alcohol as needed. Nancy Lightfoot at Indiana University Press was a delight to work with and I am grateful for her patience and good humor. Thanks also to an anonymous reviewer for suggestions. As always thanks to Stuart, Darwin, and Owen Sumida.

Finally, the entire discipline of paleontological paleopathology owes a multitude of thanks to Darren Tanke—the R. L. Moodie of his age—for his tireless efforts in documenting and bringing to light the pathology of dinosaurs.

References

Appelby, E. C., and W. G. Siller. 1960. Some cases of gout in reptiles. *Journal of Pathology and Bacteriology* 80: 427–430.

Anonymous. 2006. Dinosaur tumors studies for human cancer clues. *Live Science.* http://www. livescience.com/ health/060403_dino_med.html.

Argue, D., D. Donlon, C. Groves, and R. J. Wright. 2006. Homo floresiensis: microcephalic, pygmoid, Australopithecus, or Homo? *Journal of Human Evolution* Oct; 51 (4): 360–374.

Aufderheide, A. C., and C. Rodriguez-Martin. 1998. *The Cambridge Encyclopedia of Human Paleopathology.* Cambridge: Cambridge University Press.

Bajracharya, S., M. Jayaram, M. P. Singh, and G. K. Singh. 2006. Fungal osteomyelitis of the tibia and fibula with bony ankylosis of ankle, metatarsal and tars metatarsal joints—a rare presentation. *Nigerian Journal of Orthopaedics and Trauma* 5 (2): 56–57.

Bathurst, R. R., and J. L. Barta. 2004. Molecular evidence of tuberculosis induced hypertrophic osteopathy in a 16th century Iroquoian dog. *International Journal of Osteoarchaeology* 10: 447–450.

Bauck, L. 1994. Mycoses. In B. W. Ritchie, G. L. Harrison, and L. H. Harrison (eds.), *Avian Medicine: Principles and Application,* 997–1006. Lake Worth, Fla.: Wingers Publishing.

Beach, J. E. 1962. Diseases of Budgerigars and other cage birds. A survey of post-mortem findings, Part. II. *Veterinary Record* 74: 63–68.

Blumberg, B. S., and L. Sokoloff. 1961. Coalescence of caudal vertebrae in the giant dinosaur *Diplodocus. Arthritis and Rheumatism* 4: 592–601.

Braun, M., D. C. Cook, and S. Pfeiffer. 1998. DNA from Mycobacterium tuberculosis complex identified in North American pre-Columbian human skeletal remains. *Journal of Archaeological Science* 25: 271–277.

Breithaupt, B. H. 2001. The case of "Big Al" the *Allosaurus:* a study in paleodetective partnerships. In V. L. Santucci and L. McClelland (eds.), *Proceedings of the 6th Fossil Resources Conference,* 95–106. U.S. Department of the Interior, National Parks Service, Geological Resources Division, Technical Report NPS/NRGRD/GRDTR-01/01.

Brochu, C. A. 2003. Osteology of *Tyrannosaurus rex*: insights from a nearly complete skeleton and high-resolution computed tomographic analysis of the skull. *Society of Vertebrate Paleontology Memoir* 7: 1–138.

Brown, P., T. Sutikna, M. J. Morwood, R. P. Soejono, Jatmiko, E. Wayhu Saptomo, and R. Awe Due. 2004. A new small-bodied hominid from the Late Pleistocene of Flores, Indonesia. *Nature* 431: 1055–1061.

Buikstra, J. E. 1981. *Prehistoric Tuberculosis in the Americas.* Archaeological Program Scientific Papers 5. Evanston, Ill.: Northwestern University.

Buikstra, J. E., and D. C. Cook. 1980. Palaeopathology: an American account. *Annual Review of Anthropology* 9: 433–476.

Capasso, L. L. 2005 Antiquity of cancer. *International Journal of Cancer* 113: 2–13.

Carpenter, K. 1998. Evidence of predatory behavior by theropod dinosaurs. *Gaia* 15: 135–144.

Carpenter, K., and M. Smith. 2001. Forelimb osteology and biomechanics of *Tyrannosaurus rex*. In D. H. Tanke and K. Carpenter (eds.), *Mesozoic Vertebrate Life*, 90–116. Bloomington: Indiana University Press.

Carrier, D. R., and C. G. Farmer. 2000. The integration of ventilation and locomotion in Archosaurs. *American Zoologist* 2000 40 (1): 87–100.

Chinsamy-Turan, A. 2005. *The Microstructure of Dinosaur Bone.* Baltimore: Johns Hopkins University Press.

Claessens, L. 1997. Gastralia. In P. J. Currie and K. Padian (eds.), *Encyclopedia of Dinosaurs*, 269–270. San Diego: Academic Press.

Clark, S. R., H. J. Barnes, A. A. Bickford, R. P. Chin, and R. Droul. 1991. Relationship of osteomyelitis and associated soft-tissue lesions with green liver discoloration in tom turkeys. *Avian Diseases* 35: 139–146.

Cook, D. C., and M. L. Powell. 2006. The evolution of American paleopathology. In *Bioarchaeology: The Contextual Analysis of Human Remains*, J. E. Buikstra and L. A. Beck (eds.), 281–322. San Diego: Academic Press.

Court-Brown, C. M. 2003. Intramedullary nailing of open tibial fractures. *Current Orthopaedics* 17 (3): 161–244.

Crespo, R., and H. Shivaprasad. 2003. Developmental, metabolic and other noninfectious disorders. In Y. M. Saif, H. J. Barnes, J. R. Glisson, A. M. Fadly, L. R. McDougald, and D. E. Swayne (eds.), *Diseases of Poultry*. 11th ed. Ames: Iowa State University Press.

Crespo, R., S. M. Stover, R. Droual, R. P. Chin, and H. L. Shivaprasad. 1999. Femoral fractures in a young male turkey breeder flock. *Avian Diseases* 43: 150–154.

Cummings, C., and E. Rega. 2007. A case of dyschondrosteosis in an Anglo-Saxon skeleton. *International Journal of Osteoarchaeology* 18 (4): 431–437. doi: 10.1002/oa.948.

Currie, P. J., and A. R. Jacobsen. 1995. An azhdarchid pterosaur eaten by a velociraptorine theropod. *Canadian Journal of Earth Science* 32 (7): 922–925.

Dequeker, J., and F. P. Luyten. 2008. The history of osteoarthritis-osteoarthrosis. *Annals of Rheumatic Disease* 67: 5–10.

Destino, L., D. A Sutton, A. L Helon, P. L. Havens, J. G. Thometz, R. E. Willoughby Jr., and M. J. Chusid. 2006. Severe osteomyelitis caused by *Myceliophthora thermophila* after a pitchfork injury. *Annals of Clinical Microbiology and Antimicrobials* 5: 21. doi: 10.1186/1476-0711-5.

Erickson, G. M., P. J. Currie, B. D. Inouye, and A. A. Winn. 2006. Tyrannosaur life tables: an example of nonavian dinosaur population biology. *Science* 313 (5784): 213–217.

Erickson, G. M., and K. H. Olson.1996. Bite marks attributable to tyrannosaurus rex: Preliminary description and implications. *Journal of Vertebrate Paleontology*. 16 (1): 175-178.

Esper, J. F. 1774. *Ausfürliche Nachtrichten von Neuentdeckten Zoolithen Unbekannter Vierfüssiger Thiere.* [Descriptive News of Newly Discovered Fossilized Unknown Quadrupeds.] Nuremberg: Erben. (Facsimile copy 1978. Wiesbaden: Guido Pressler Verlag.)

Eudes-Deslongchamps, J. A. 1838. Mémoire sur le *Poekilopleuron bucklandii*, grand saurien fossile, intermédiaire entre les crocodiles et les lézards. *Mémoire de la Société Linnéenne de Normandie* 6: 37–146.

Ewald, P., 1994. *Evolution of Infectious Disease.* Oxford: Oxford University Press.

Farke, A. A., E. D. S. Wolff, and D. H. Tanke. 2009. Evidence of combat in *Triceratops*. PLoS ONE 4 (1): e4252. doi:10.1371/journal.pone.0004252.

Faux, C. M., and K. Padian. 2007. The opisthotonic posture of vertebrate skeletons: postmortem contraction or death throes? *Paleobiology* 33: 201–226.

Farlow, J. O., and P. Dodson 1975. The behavioral significance of frill and horn morphology in Ceratopsian dinosaurs. *Evolution* 29 (2): 353–361.

Farlow, J. O., M. B. Smith, and J. M. Robinson. 1995. Body mass, "strength indicator," and cursorial potential of *Tyrannosaurus rex. Journal of Vertebrate Paleontology* 15 (4): 713–725.

Frank, W. 1965. Gelenk and visceralgicht bei panzerechsen (*Tomistoma schegelii* and *Gavialis gangeticus* Reptilia, Crocodylia). *Acta Tropica* 22 (3): 217–234.

Garner, M. M., S. M. Hernandez-Divers, and J. T. Raymond. 2005. Reptile neoplasia: a retrospective case study of submissions to a specialty diagnostic service. *Veterinary Clinics of North America: Exotic Animal Practice* 7: 653–671.

Gates, T. A. 2005. Late Jurassic Cleveland-Lloyd Dinosaur Quarry as a drought-induced assemblage. *Palaios* 20 (4): 363–375.

Gill, D. E., and B. A. Mock. 1985. Ecological and evolutionary dynamics of parasites: the case of *Trypanosoma diemyctyli* in the red spotted newt *Notopthalmus viridescens*. In D. Rollinson and R. M. Anderson (eds.), *Ecology of Host Parasite Interaction*, 157–183. London: Academic Press.

Gilmore, C. W. 1909. Osteology of the Jurassic reptile *Camptosaurus*, with a revision of the species of the genus, and a description of two new species. *Proceedings of the United States National Museum* 36: 197–332, and plates 6–20.

———. 1912. The mounted skeletons of *Camptosaurus* in the United States National Museum. *Proceedings of the United States National Museum* 41 (1878): 687–696, and plates 56–61.

———. 1915. On the forelimb of *Allosaurus fragilis. Proceedings of U. S. National Museum* 49 (2120): 501–513.

———. 1919. A new restoration of Triceratops, with notes on the osteology of the genus. *Proceedings of the United States National Museum* 55 (2260): 97–112.

———. 1932. On a newly mounted skeleton of Diplodocus in the United States National Museum. *Proceedings of the United States National Museum* 81 (2941) (18): 1–21.

Goldthwait, J. E. 1899. Osteo-Arthritis of the Spine: Spondylitis Deformans. *Boston Medical and Surgical Journal* 141: 128–132.

Gordon, A. D., L. Nevell, and B. Wood. 2008. The *Homo floresiensis* cranium (LB1): size, scaling, and early Homo affinities. *Proceedings of the National Academy of Sciences of the United States of America* 105 (12): 4650–4655.

de Graaf, M., A. D. M. E. Osterhaus, R. A. M. Fouchier, and E. C. Holmes. 2008. Evolutionary dynamics of human and avian metapneumoviruses. *Journal of General Virology* 89: 2933–2942.

Green, E. M., W. M. Adams, and H. Stein-berg. 1999. Malignant transformation of solitary spinal osteochondroma in two mature dogs. *Veterinary Radiology and Ultrasound* 40 (6): 634–637.

Gross, J. D., T. H. Rich, and P. Vickers-Rich. 1993. Chronic osteomyelitis in a Hypsilophodontid dinosaur in the early Cretaceous, Polar Australia. *National Geographic Research and Exploration* 9 (3): 286–293.

Hanna, R. R. 2002. Multiple injury and infection in a sub-adult theropod dinosaur (*Allosaurus fragilis*) with comparisons to *Allosaur* pathology in the Cleveland-Lloyd Dinosaur Quarry collection. *Journal of Vertebrate Paleontology* 22 (1): 76–90.

Hardie, E. M., O. Ramirez 3rd, E. M. Clary, J. N. Kornegay, M. T. Correa, R. A. Feimster, and E. R. Robertson. 1998. Abnormalities of the thoracic bellows: stress fractures of the ribs and hiatal hernia. *Journal of Veterinary Internal Medicine* 12 (4): 279–287.

Hasiotis, S. T., A. R. Fiorillo, and R. R. Hanna. 1999. Preliminary report on borings in Jurassic dinosaur bones: evidence for invertebrate-vertebrate interactions. In D. D. Gillette (ed.), *Vertebrate Paleontology in Utah*. Utah Geological Survey, Miscellaneous Publication 99–1, 193–200.

Hatcher, J. B. 1901. *Diplodocus* (Marsh): its osteology taxonomy and probable habits, with a restoration of the skeleton. *Memoirs of the Carnegie Museum* 1 (1): 64.

———. 1907. *The Ceratopsia.* Ed. R. S. Lull. U. S. Geological Survey Monograph 44: 3–155.

Heatley, J. J., M. A. Mitchell, J. Williams, J. A. Smith, and T. N. Tully Jr. 2001. Fungal periodontal osteomyelitis in a chameleon, *Furcifer pardalis*. *Journal of Herpetological Medical Surgery* 11 (4): 7–12.

Holland, W. J. 1906. The osteology of *Diplodocus* Marsh. *Memoirs of the Carnegie Museum* 2: 225–264.

Huchzermeyer, F. W. 2003. *Crocodiles: Biology, Husbandry and Diseases.* Cambridge, Mass.: CABI Publishing.

Isaza, R., M. Garner, and E. Jacobson. 2000. Proliferative osteoarthritis and osteoarthrosis in 15 snakes. *Journal of Zoo and Wildlife Medicine* 31 (1): 20–27.

Johnson, K. A. 1994. Osteomyelitis. In S. J. Birchard and R. G. Sherding (eds.), *Saunders Manual of Small Animal Practice.* 1091–1095. Philadelphia: W. B. Saunders.

Gittis, S. Miller, L. Carey, and A. Raynor (eds.), *The DinoFest Symposium,* 29–30. Academy of Natural Sciences, Philadelphia, Penn., April 17–19..

Kelley, M. A., and L. Eisenberg. 1987. Blastomycosis and tuberculosis in early American Indians: a biocultural view. *Midcontinental Journal of Archaeology* 12: 89–116.

Klasing, K. C., and R. E. Austic. 2003. Nutritional diseases. In Y. M. Saif, H. J. Barnes, J. R. Glisson, A. M. Fadly, L. R. McDougald, and D. E. Swayne (eds.), *Diseases of Poultry,* 11th ed.: 1027–1053. Ames: Iowa State University Press.

Klebs, A. C. 1917. Paleopathology. *Bulletin of the Johns Hopkins Hospital* 28: 261–266.

Kimmel, D. B., E. L. Moran, and E. R. Bogach. 1998. Animal models of osteopenia or osteoporosis. In Y. H. An and R. J. Friedman (eds.), *Animal Models in Orthopaedic Research,* 279–308. Boca Raton, Fla.: CRC Press.

Kocan, A. A., C. A. Cummings, R. J. Panciera, J. S. Matthew, S. A. Ewing, and R. W. Barker. 2000. Naturally occurring and experimentally transmitted *Hepatozoan americanum* in coyotes from Oklahoma. *Journal of Wildlife Disease* 36 (1): 149–153.

Laws, R. R., S. T. Hasiotis, A. R. Fiorillo, D. J. Chure, B. H. Breithaupt, and J. R. Horner. 1996. The demise of a Jurassic Morrison dinosaur after death: three cheers for the dermestid beetle. *Geological Society of America Abstracts with Programs* 28 (7): A-299.

Lee, A. H., and S. Werning. 2008. Sexual maturity in growing dinosaurs does not fit reptilian growth models. *Proceedings of the National Academy of Science* 105 (2): 582–587.

Liote, F., and H. Ea. 2006. Gout: An update on some pathogenic and clinical aspects. *Rheumatic Disease Clinics of North America* 32: 295–311.

Lü, J., Y. Kobayashi, Y. N. Lee, and Q. Ji. 2007. A new *Psittacosaurus* (Dinosauria: Ceratopsia) specimen from the Yixian Formation of western Liaoning, China: the first pathological psittacosaurid. *Cretaceous Research* 28 (2): 272–276.

Lynch, M., H. McCracken, and R. Slocombe. 2002. Hyperostotic bone disease in red pandas (*Ailurus fulgens*). *Journal of Zoo and Wildlife Medicine* 33 (3): 263–271.

Madsen, J. H. 1976. Allosaurus fragilis: a revised osteology. *Utah Geological and Mineral Survey Bulletin* 109: 1–163.

Mann, R. W., and D. R. Hunt. 2005. *Photographic Regional Atlas of Bone Disease.* St. Louis: Charles C. Thomas.

Marshall, C., D. Brinkman, R. Lau, and K. Bowman. 1998. Fracture and osteomyelitis in PII of the second pedal digit of *Deinonychus antirrhopus* (Ostrom 1976)

an Early Cretaceous "raptor" dinosaur. Abstract. *Palaeontology Newsletter:* 39:16. (The Paleontological Association, 42nd Annual Meeting, Abstracts and Posters, December 16–19, 1998. University of Portsmouth.)

Martin, R. D., A. M. Maclarnon, J. L. Phillips, W. B. Dobyns. 2006. Flores hominid: new species or microcephalic dwarf? *Anatomical Record Part A. Discoveries in Molecular, Cellular, and Evolutionary Biology* 288 (11): 1123–45.

McGowan, C. 1991. *Dinosaurs, Spitfires, and Sea Dragons.* Cambridge, Mass.: Harvard University Press.

McWhinney, L. A., B. M. Rothschild, and K. Carpenter. 1998. Post-traumatic chronic osteomyelitis in *Stegosaurus* dermal spikes. (Abstract). *Journal of Vertebrate Paleontology* 18 (3): 62A.

Mietchen, D., H. Keupp, B. Manz, and F. Volke. 2005. Non-invasive diagnostics in fossils: magnetic resonance imaging of pathological belemnites. *Biogeosciences* 2: 133–140.

Mitchell, M. A. 2002. Diagnosis and management of reptile orthopedic injuries. *Veterinary Clinics of North America: Exotic Animal Practice* 5: 97–114.

Molnar, R. E. 2001. Theropod paleopathology: a literature survey. In D. Tanke and K. Carpenter (eds.), *Mesozoic Vertebrate Life,* 337–363. Bloomington: Indiana University Press.

Molnar, R. E., and J. O. Farlow 1990. Carnosaur paleobiology. In D. D. Weishampel, P. Dodson, and H. Osmólska (eds.), *The Dinosauria,* 210–224. Berkeley: University of California Press.

Montali, R. J. 1988. Comparative pathology of inflammation in the higher vertebrates (reptiles, birds, mammals). *Journal of Comparative Pathology* 99: 1–26.

Moodie, R. L. 1916a. Two caudal vertebrae of a sauropodus dinosaur exhibiting a pathological lesion. *American Journal of Science* Series 4 (41): 530–531.

———. 1916b. Mesozoic pathology and bacteriology. *Science* 43: 425–426.

———. 1918a. Studies in paleopathology: I. General considerations of the evidences of pathological conditions found among the fossil animals. *Annals of Medical History* 1 (4): 374–393.

———. 1918b. Paleontological evidences of the antiquity of disease. *Scientific Monthly* 7 (3): 265–281.

———. 1918c. Studies in paleopathology: III. Opisthotonus and allied phenomena among fossil vertebrates. *American Naturalist* 52: 384–394.

———. 1921. Status of our knowledge of

Mesozoic pathology. *Bulletin of the Geological Society of America* 32: 321–326.

——. 1923a. *The Antiquity of Disease.* Chicago: University of Chicago Press.

——. 1923b. *Paleopathology. An Introduction to the Study of Ancient Evidences of Disease.* Urbana, Ill.: University of Illinois Press.

——. 1926. Studies in paleopathology: II. Excess callus following fractures of the forefoot. *Annals of Medical History* 8: 73–77.

Morgan, J. P., and M. Stavenborn. 1991. Disseminated idiopathic skeletal hyperostosis (DISH) in a dog. *Veterinary Radiology and Ultrasound* 32 (2): 65–70.

Myhrvold, N. P., and P. J. Currie. 1997. Supersonic sauropods? Tail dynamics in the diplodocids. *Paleobiology* 23 (4): 393–409.

Newman, M. E. 1991. Fossil tumor creates interest in prehistoric cancers *Journal of the National Cancer Institute* 83 (18): 1288–1289.

Norman, D. B., and A. Milner. 1989. *Dinosaurs.* Toronto: Stoddart.

Olson, N. O., and K. M. Kerr. 1966. Some characteristics of an avian arthritis viral agent. *Avian Diseases* 10 (4): 470–476.

Organ, C. L., and J. Adams. 2005. The histology of ossified tendons. *Journal of Vertebrate Paleontology* 25 (3): 602–613.

Ortner, D. J. 2003. *Identification of Pathological Conditions in Human Skeletal Remains.* San Diego: Academic Press.

Ostrom, J. H. 1976. On a new specimen of the lower Cretaceous theropod dinosaur *Deinonychus antirrhopus. Breviora* 439: 1–21.

Padian, K., A. de Ricqlès, and J. R. Horner. 2001. Dinosaurian growth rates and bird origins. *Nature* 412: 405–408.

Padian, K., J. R. Horner, and A. de Ricqlès. 2004. Growth in small dinosaurs and pterosaurs: the evolution of archosaurian growth strategies. *Journal of Vertebrate Paleontology* 24 (3): 555–571.

Peters, S. E., and M. Foote. 2002. Determinants of extinction in the fossil record. *Nature* 416 (28): 420–424.

Petersen, K., J. I. Isakon, and J. H. Madsen Jr. 1972. Preliminary study of paleopathologies in the Cleveland-Lloyd dinosaur collection. *Utah Academy Proceedings* 48: 44–47.

Raath, M. A. 1972. Fossil vertebrate studies in Rhodesia: A new dinosaur (Reptilia: Saurischia) from near the Trias-Jurassic [*sic*] Boundary. In *Arnoldia, National Museums of Rhodesia* 30 (5): 1–37.

Ramis, A., J. Fernandez-Moran, X. Gibert, and H. Fernandez-Bellon. 1998.

Dermatophytosis in a Hyacinth Macaw (*Anodorhynchus hyacinthinus*): a case report. *Proceedings of International Virtual Conferences in Veterinary Medicine: Diseases of Psittacine Birds.* http://www.vetuga.edu/vpp/ivcvm/1998/ramis/index.php.

Ramsey, E. C., G. B. Daniel, B. W. Tryon, J. I. Merryman, P. J. Morris, and D. A. Bemis. 2002. Osteomyelitis associated with *Salmonella enterica* SS arizonae in a colony of ridgenose rattlesnakes (*Crotalis willardi*). *Journal of Zoo and Wildlife Medicine* 33 (4): 301–310.

Rega, E. A. 2008. Analysis of ancient pathology and the search for an appropriate animal model. *Proceedings of the American College of Veterinary Pathologists.* American College of Veterinary Pathologists and American Society for Veterinary Clinical Pathology, Madison, Wis., USA. International Veterinary Information Service. http://www.ivis.org/proceedings/acvp/2008.

Rega, E. A., and C. Brochu. 2001a. Paleopathology of a mature *Tyrannosaurus rex. Journal of Vertebrate Paleontology* 21 (suppl. to 3): 92A.

——. 2001b. Evidence of thoracic breathing in Tyrannosaurus rex, based on recent anatomical and pathological evidence from "Sue." *Journal of Morphology* 248 (3): 274.

Rega, E. A., R. R. Hanna, and E. D. S. Wolff. 2003. A new tool for paleopathological analysis: development of a description-based classification system for pathological bones. *Journal of Vertebrate Paleontology* 23 (suppl. to 3): 89.

Rega, E. A., and R. Holmes. 2006. Manual pathology indicative of locomotor behavior in two chasmosaurine ceratopsid dinosaurs. *Journal of Vertebrate Paleontology* 26 (suppl. to 3): 114A.

Rega, E. A., R. Holmes, and A. Tirabasso. 2010. Habitual locomotor behavior inferred from manual pathology in two Late Cretaceous chasmosaurine ceratopsid dinosaurs, *Chasmosaurus irvinensis* (CMN 41357) and *Chasmosaurus belli* (ROM 843). In M. J. Ryan, B. Chinnery-Allgeier, and D. A. Eberth (eds.), *New Perspectives on Horned Dinosaurs*, 340–354. Bloomington: Indiana University Press.

Rega, E. A., J. Michaelson, and L. TenEyck. 1999..Analysis of commingled skeletal remains from a pre-contact cave site in the Jackson's Bay Cave series, Jamaica. *American Journal of Physical Anthropology* (suppl. 28): 230.

Rega, E., A. K. Noriega, S. Sumida, and A. Lee. 2004. Histological analysis of traumatic injury to multiple neural spines of an associated skeleton of *Dimetrodon*: implications for healing response, dorsal sail morphology and age-at-death in a Lower Permian synapsid. *Integrative and Comparative Biology* 44 (6): 628.

Rega, E. A., N. K. Wideman, and C. A. Brochu. 2002. Paleopathology of amniote specimens from the late Paleozoic and Mesozoic of North America: comparison of gross morphological and histological analyses. *Journal of Vertebrate Paleontology* 22 (suppl. to 3): 98.

Resnick, D., and G. Niwayama. 1995a. Degenerative disease of the spine. In D. Resnick (ed.), *Diagnosis of Bone and Joint Disorders*, 3rd ed., 3:1372–1462. Philadelphia: W. B. Saunders.

——. 1995b. Diffuse idiopathic skeletal hyperostosis (DISH): ankylosis hyperostosis and Forestier and Rotes-Querol. In D. Resnick (ed.), *Diagnosis of Bone and Joint Disorders*, 3rd ed., 3:1463–1495. Philadelphia: W. B. Saunders.

——. 1995c. Osteomyelitis, septic arthritis and soft tissue infection: organisms In D. Resnick (ed.), *Diagnosis of Bone and Joint Disorders*, 3rd ed., 4:2448–2558. Philadelphia: W. B. Saunders.

Riggs, E. S. 1903. Structure and relationship of the opisthocoelian dinosaurs. Part I: *Apatosaurus* Marsh. *Field Columbian Museum Publication* (Geological ser.) 2, no. 4: 165–196, and plates XLVII–LIII.

Rosen, G. 1977. Rudolf Virchow and Neanderthal man. *American Journal of Surgical Pathology* 1: 183–187.

Rothschild, B. M. 1988. Stress fracture in a ceratopsian phalanx. *Journal of Paleontology* 62: 302–303.

——. 1997. Dinosaurian paleopathology. In J. Farlow and M. K. Brett-Surman (eds.), *The Complete Dinosaur*, 426–448. Bloomington: Indiana University Press.

Rothschild, B. M., and L. Martin.1993. *Paleopathology.* Montclair, N.J.: Telford Press.

Rothschild, B. M., and R. E. Molnar. 2005. Sauropod stress fractures as clues to activity. In V. Tidwell and K. Carpenter (eds.), *Thunder-Lizards: The Sauropodomorph Dinosaurs*, 381–394. Bloomington: Indiana University Press.

Rothschild, B. M., and D. S. Berman. 1991. Fusion of Caudal Vertebrae in Late Jurassic Sauropods. *Journal of Vertebrate Paleontology* 11 (1): 29–36.

Rothschild, B. M., and F. R. Rühli. 2007. Comparative frequency of osseous macroscopic pathology and first report of gout in captive and wild-caught ratites. *Journal of Veterinary Medicine* (ser. A) 54 (5): 265–269.

Rothschild, B. M., and D. H. Tanke. 2005. Theropod paleopathology. In K. Carpenter (ed.), *The Carnivorous Dinosaurs*, 351–366. Bloomington: Indiana University Press.

———. 2007. Osteochondrosis in Late Cretaceous Hadrosauria. In K. Carpenter (ed.), *Horns and Beaks: Ceratopsian and Ornithopod Dinosaurs*, 171–183. Bloomington: Indiana University Press.

Rothschild, B. M, D. H. Tanke, and K. Carpenter. 1997. Tyrannosaurs suffered from gout. *Nature* 387: 357.

Rothschild, B. M., D. H. Tanke, and T. L. Ford. 2001. Theropod stress fractures and tendon avulsions as a clue to activity. In D. H. Tanke and K. Carpenter (eds.), *Mesozoic Vertebrate Life*, 331–336. Bloomington: Indiana University Press.

Rothschild B. M., D. Tanke, I. Hershkovitz. 1998. Mesozoic neoplasia: origins of hemangioma in the Jurassic age. *Lancet* 351: 1862.

Ruffer, M. A. 1914. Studies in the paleopathology of Egypt. *Journal of Pathology and Bacteriology* 18: 149–162, plates I–X.

———. 1921. *Studies in the Paleopathology of Egypt.* Ed. R. L. Moodie. Chicago: University of Chicago Press.

Saif, Y. M., H. J. Barnes, J. R. Glisson, A. M. Fadly, L. R. McDougald, and D. E. Swayne, eds. 2003. *Diseases of Poultry.* 11th ed. Ames: Iowa State University Press

Salo, W. L., A. C. Aufderheide, J. E. Buikstra, and T. A. Holcomb. 1994. Identification of *Mycobacterium tuberculosis* DNA in a pre-Columbian Peruvian mummy. *Proceedings of the National Academy of Sciences* 91: 2091–2094.

Salter, R. B. 1999. *Textbook of Disorders and Injuries of the Musculoskeletal System.* 3rd ed., Hagerstown, Md.: Lippincott Williams & Wilkins.

Scarini, P. 2003. Rudolf Virchow (1821–1902). *Virchow's Archiv* 442 (2): 95–98.

Schmale, G. A., E. U. Conrad III, and W. H. Raskind. 1994. The natural history of hereditary multiple exostoses. *Journal of Bone and Joint Surgery* (American version) 76 (7): 986–92.

Schulp, A. S., G. H. I. M. Walenkamp, P. A. M. Hofman, B. M. Rothschild, and J. W. M. Jagt. 2004. Rib fracture in *Prognathodon saturator* (Mosasauridae, Late Cretaceous). *Netherlands Journal of Geosciences* [*Geologie en Mijnbouw*] 83 (4): 251–254.

Schwartz, H. L., and D. D. Gillette. 1994.

Geology and taphonomy of the *Coelophysis* Quarry, Upper Triassic Chinle Formation, Ghost Ranch, New Mexico. *Journal of Paleontology* 68 (5): 1118–1130.

Shufeldt, R. W. 1893. Notes on paleopathology. *Popular Science Monthly* 42 (March): 679–684.

Shupe, J. L., N. C. Leone, E. J. Gardner, and A. E. Olson. 1981. Hereditary multiple exostoses in horses. *American Journal of Pathology* 104 (3): 285–288.

Starck, J. M., and A. Chinsamy. 2002. Bone microstructure and developmental plasticity in birds and other dinosaurs. *Journal of Morphology* 254: 232–246.

Stocker, L. 2005. *Practical Wildlife Care.* 2nd ed. Boston: Blackwell.

Stover, S. M., B. J. Johnson, B. M. Daft, D. H. Read, M. Anderson, B. C. Barr, H. Kinde, J. Moore, J. Stoltz, A. A. Ardans, and R. R. Pool. 1992. An association between complete and incomplete stress fractures of the humerus in racehorses. *Equine Veterinary Journal* 24: 260–63.

Straight, W. H., G. L. Davis, C. W. Skinner, A, Haimes, B. L. McClennan, and D. H. Tanke. 2009. Bone lesions in Hadrosaurs: Computed tomographic imaging as a guide for paleohistologic and stable-isotopic analysis. *Journal of Vertebrate Paleontology* 29 (2): 315–325.

Straight, W. H., H. C. W. Skinner, B. L. McClennan, A. Haims, L. Davis, and G. Olack. 2006. Comparison of bone tissues around healing fractures in hadrosaur and alligator. (Annual meeting abstract). *Geological Society of America: Abstracts with programs* 38 (7): 46. http://gsa.confex.com/gsa/2006AM/finalprogram/abstract_113591.htm.

Sullivan, R. M., D. H. Tanke, and B. M. Rothschild. 2000. An impact fracture in an ornithomimid (Ornithomimosauria: Dinosauria) metatarsal from the Upper Cretaceous (Late Campanian) of New Mexico. In S. G. Lucas and A. B. Heckert (eds.), *Dinosaurs of New Mexico*, 109–111. Bulletin 17. Albuquerque: New Mexico Museum of Natural History and Science.

Tanke, D. H., and P. Currie 1998. Head-biting behavior in theropod dinosaurs: paleopathological evidence. *Gaia* 15: 167–184

Tanke, D. H., and A. A. Farke. 2007. Bone resorption, bone lesions, and extra cranial fenestrae in ceratopsid dinosaurs: a preliminary assessment. In K. Carpenter (ed.), *Horns and Beaks: Ceratopsian and Ornithopod Dinosaurs*, 319–347. Bloomington: Indiana University Press.

Tanke, D. H., and B. M. Rothschild. 1997.

Paleopathology. In P. J. Currie and K. Padian (eds.), *Encyclopedia of Dinosaurs*, 525–529. San Diego: Academic Press.

———, eds. 2002. *Dinosores: An Annotated Bibliography of Dinosaur Paleopathology and Related Topics – 1838–1999.* Bulletin 20. Albuquerque: New Mexico Museum of Natural History and Science.

Taylor, G. M., E. Murphy, R. Hopkins, P. Rutland, and Y. Chistov. 2007. First report of *Mycobacterium bovis* DNA in human remains from the Iron Age. *Microbiology* 153: 1243–1249. doi 10.1099/mic.0.2006/002154-0.

Therio, K. A. 2004. Comparative inflammatory responses of non-mammalian vertebrates: Robbins and Coltran for the birds. *55th Annual Meeting of the American College of Veterinary Pathologists ACVP and 39th Annual Meeting of the American Society of Clinical Pathology ASVCP*, Orlando, Fla., USA, 1225–1104 [*sic*]. International Veterinary Information Service. http://www.ivis.org.

Tully, T. N. 2002. Basic avian bone growth and healing. *Veterinary Clinics of North America: Exotic Animal Practice* 5: 23–30.

Tully, T. N., and S. M. Shane. 1996. *Ratite Management, Medicine, and Surgery.* Malabar, Fla.: Krieger Publishing Company.

Turner, C. G., II, and J. Turner. 1998. *Man Corn: Cannibalism and Violence in the Prehistoric American Southwest.* Salt Lake City: University of Utah Press.

Turvey, S. T., O. R. Green, and R. N. Holdaway. 2005. Cortical growth marks reveal extended juvenile development in New Zealand moa. *Nature* 435: 940–943.

Ubelacker, D. H. 2003. Anthropological perspectives on the study on ancient disease. In C. Greenblatt and M. Spigelman (eds.), *Emerging Pathogens: The Archaeology, Ecology and Evolution of Infectious Disease*, 93–102. New York: Oxford University Press.

van der Heide, L. 1977. Viral arthritis/tenosynovitis: a review. *Avian Pathology* 6: 271–284.

Verschooten, F., D. Vermieren, and L. Devriesse. 2000. Bone infection in the bovine appendicular skeleton: a clinical, radiographic and experimental study. *Veterinary Radiology and Ultrasound* 41 (2): 250–260.

Virchow, R. 1872. Untersuchung des Neaderthal-Schadels. *Zeitschrift für Ethnology* 4: 157–165.

———. 1895. Knochen vom Höhlenbären mit krankhaften Veränderungen [Bones from cave bears with changes of disease]. *Zeitschrift für Ethnology* 27: 706–708.

von Walther, F. 1825. Über das Altertum der Knochenkrankheiten. *Journal der Chirugie und Augenheilkunde* VIII, 1).

Weiss, L. 2000. Observations on the antiquity of cancer and metastasis. *Cancer and Metastasis Reviews* 19: 193–204.

Weissengruber, G., and G. Loupal. 1999. Osteochondroma of the tracheal wall in a Fischer's lovebird (*Agapornis fischeri*, Reichenow 1887). *Avian Diseases* 43: 155–159.

Whitfield, J. 2003. Dinosaurs got cancer. *News@Nature* (20 Oct). doi: 10.1038/news031020-2, News.

Williston, S. W. 1898. Mosasaurs. *Bulletin of University Kansas Geological Survey* IV.

Wissman, M. A. 2006. Avian orthopedics. *Exotic Pet Vet . Net.* http://www.exotic-petvet.net/avian/orthopedic.html.

Wolff, E. D. S., and D. J. Varricchio. 2005. Zoological paleopathology and the case of the tyrannosaurus jaw: integrating phylogeny and the study of ancient diseases. (Annual meeting abstract). *Geological Society of America.* *http://gsa.confex.com/gsa/2005AM/finalprogram/abstract_95518.htm*.

Wolff, E. D. S., S. W. Salisbury, J. R. Horner, and D. J. Varricchio. 2009. Common avian infection plagued the tyrant dinosaurs. PLoS ONE 4 (9): e7288. doi:10.1371/journal.pone.

Wood, J. D., G. R. Milner, H. C. Harpending, and K. M. Weiss. 1992. The osteological paradox. *Current Anthropology* 33: 343–370.

Wright, L. E., and C. J. Yoder. 2003. Recent progress in bioarchaeology: approaches to the osteological paradox. *Journal of Archaeological Research* 11 (1): 43–70.

Wyers, M., Y. Cherel, and G. Plassiart. 1991. Late Clinical Expression of Lameness Related to Associated Osteomyelitis and Tibial Dyschondroplasia in Male Breeding Turkeys. *Avian Diseases*, 35 (2): 408-414.

34.1. Early Jurassic (East Berlin Formation) dinosaur footprints, Dinosaur State Park, Rocky Hill, Connecticut. The footmarks are about 30 to 40 cm long. The makers of these prints are here regarded as theropods (Farlow and Galton 2003), but Weems (1987, 1992, 2003) argues that large early Mesozoic dinosaur footprints such as these were made by prosauropods. Similar three-toed footmarks, both large and small, occur at many tracksites in early Mesozoic rocks of eastern North America. Prints like these were studied by the pioneering ichnologist Edward Hitchcock (1858, 1865).

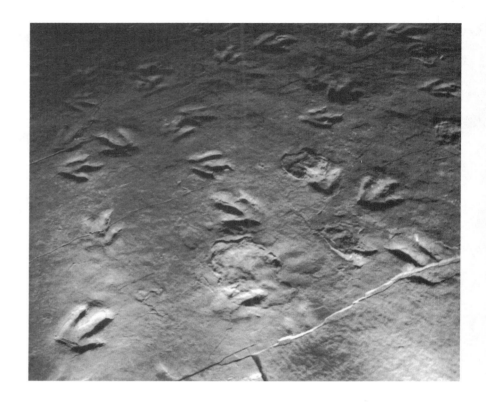

The Scientific Study of Dinosaur Footprints

34

James O. Farlow, Ralph E. Chapman,
Brent Breithaupt, and Neffra Matthews

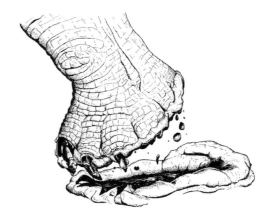

Perhaps the most vivid impression of a dinosaur as a living creature comes not from seeing a mounted skeleton, but instead from the examination of a well-preserved trackway. Looking at footprints made, one after the other, by a dinosaur going about its business millions of years ago gives one an almost palpable sense of the trackmaker as a real animal as opposed to some analytical construct created from the ruminations and speculations of an imaginative paleontologist. For one of us (Chapman), walking along the trackway and in the footprints of a single animal, probably a dilophosaur, for more than 100 yards in a quarry in Culpeper, Virginia (Lower Jurassic; see Weems 1987, 1992) was one of the thrills of a long career. Visiting the famous Lower Cretaceous Paluxy River Valley dinosaur tracksites (Farlow 1987, 1993; Jasinski 2009) is like being a tourist in one of the great cathedrals of Europe – an elegant and awe-inspiring relic of a fantastic past.

Edward Hitchcock (1858, 1865) must have experienced similar feelings. He was an early-nineteenth-century American geologist who conducted pioneering studies of the fossil footprints of the Connecticut Valley of the eastern United States (Colbert 1997), whose scientific work influenced the thinking of such literary figures of his day as Herman Melville, Emily Dickinson, Henry David Thoreau, and Henry Wadsworth Longfellow (Dean 1969). In his monumental description of this early Mesozoic footprint fauna (Fig. 34.1), the *Ichnology of New England* of 1858, Hitchcock wrote:

> What a wonderful menagerie! Who would believe that such a register lay buried in the strata? . . . At first men supposed that the strange and gigantic races which I had described, were mere creatures of imagination, like the Gorgons and Chimeras of the ancient poets. But now that hundreds of their footprints, as fresh and distinct as if yesterday impressed upon the mud, arrest the attention of the sceptic on the ample slabs of our cabinets, he might as reasonably doubt his own corporeal existence as that of these enormous and peculiar races. (190)

It is ironic that current thinking suggests that dinosaurs and other prehistoric animals may have been the actual inspiration for the very mythological creatures referred to by Hitchcock (Mayor 2000, 2005).

Dinosaur footprints provide a source of information about dinosaurs that is at least partly independent of evidence gained from skeletal material. This chapter describes how dinosaur footprints are studied, what type of information we can extract from them, how we identify the kinds of dinosaurs that made fossilized tracks, how and why dinosaur footprints are given scientific names of their own, and what such trace fossils tell us about how their makers walked and ran and where they lived.

Before we begin, however, a brief note about terminology is in order. In everyday English, the word *footprint* has an unambiguous meaning.

The word *track*, however, can refer either to an individual footprint or to a sequence of footprints made by the same animal. To avoid confusion, we will use the terms *footprint*, *print*, and *footmark* here as synonyms to refer to a mark made by an individual foot. The word *trackway* refers to a sequence of footprints made by the same animal. Finally, the term *track* can be used more generically, when it is not necessary to specify whether a single footprint or a trackway is being discussed.

How Dinosaur Footprints Are Studied: Fieldwork

One of the most important tasks in describing a dinosaur footprint site is to record the positions of the footprints relative to each other. Footprints made by the same animal need to be recognized, and the direction of travel of each animal at the site recorded. The best way to preserve such information is to make a map of the site (Fig. 34.2).

This can be done in many different ways. One simple method is to lay a square grid on the tracksite surface, with subdivisions of the grid made at fixed intervals. A similar grid is drawn—obviously at a much smaller scale!—in a field notebook; arithmetic graph paper also works well for this. The individual footprints can then be sketched in the notebook at the proper distance from, and in their orientation relative to, intersections in the grid, and at the proper scale to provide information about their size. A more elaborate version of this method is to photograph each square on a square grid drawn on the tracksite, prepare a photomosaic of the entire track surface, and then trace the position of each print from the photomosaic onto a sheet of paper or plastic. An alternative way of making a map is to measure the distance of each footprint from some fixed landmark on the tracksite, along with the compass bearing from the landmark to the print. The size and compass heading of each footprint are also measured. These measurements can then be used to draw the position, orientation, and size of each track on a map.

Mapping also can be facilitated greatly these days by technology (Chapman et al., Chapter 13 this volume). Global positioning systems (GPS) can locate the latitude and longitude of a trackway's position globally with a resolution down to less than a meter. Blimps and other unmanned airborne vehicles (UAVs) can photograph whole sites from an altitude that will provide an overall view unavailable in other ways. Electronic distance measurement devices, most often used by surveyors, can locate the position of tracks and other landmarks at a site and relative to a known landmark with submillimeter resolution. Three-dimensional scanners can capture whole tracksites in three dimensions (Bates et al. 2008, 2009), and these data can then be used with rapid-prototyping technology to produce physical models of the whole site reduced to a small and accurate scale. Individual tracks can be captured morphologically with submillimeter resolution using various types of scanners (e.g., Arakawa et al. 2002; Azuma et al. 2002), using moiré photography (Ishigaki and Fujisaki 1989; Lim et al. 1989), or photogrammetrically (Matthews and Breithaupt 2001; Goto et al. 1996). Finally, all these data can be merged within a geographical information system that will allow researchers and others to explore the tracksites virtually (Breithaupt et al. 2001; Matthews and Breithaupt et al. 2001; Breithaupt et al. 2004). This is especially important because most tracksites are ephemeral—they are available only

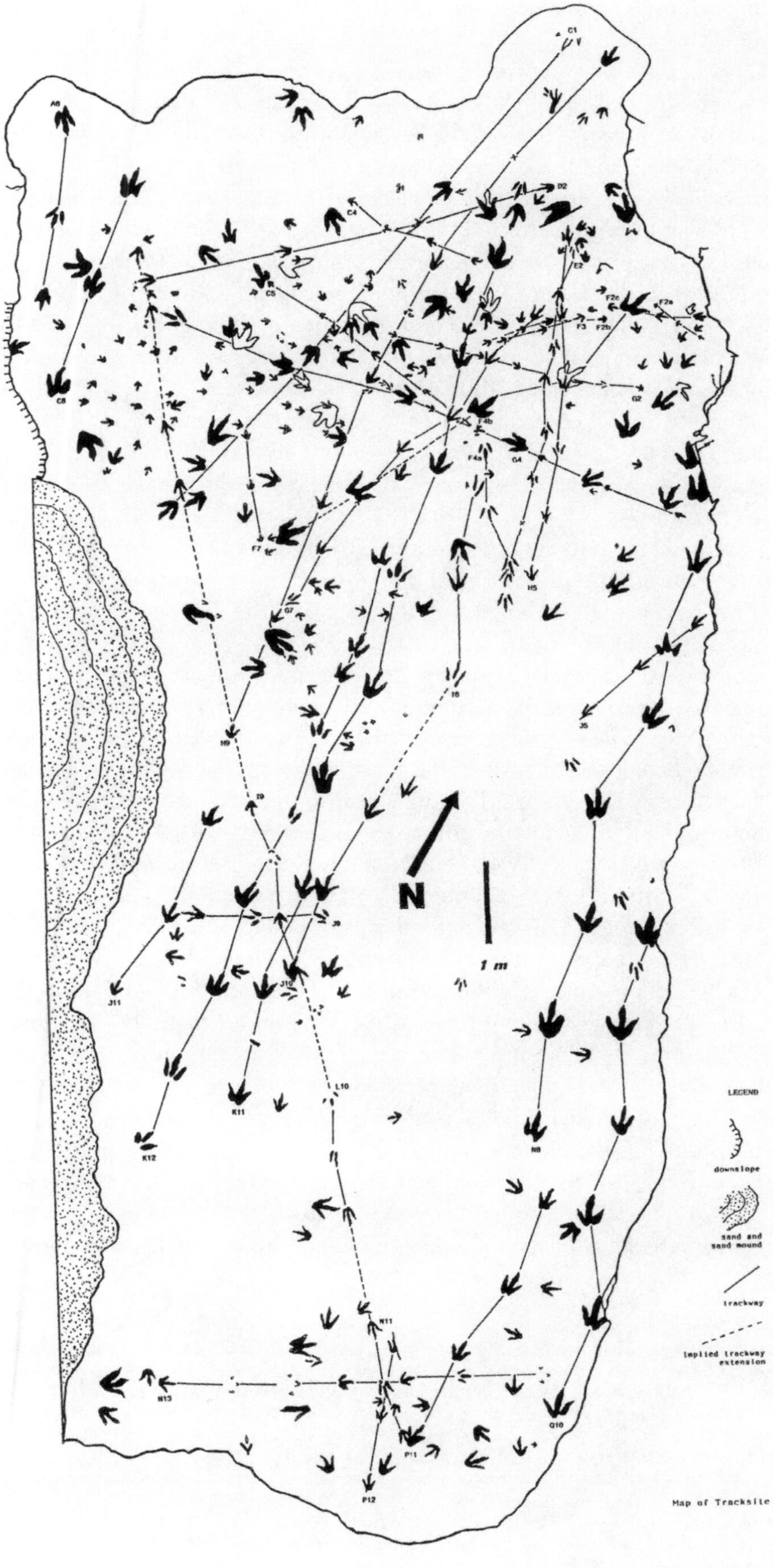

34.2. Map of an Early Jurassic dinosaur footprint site from the Moenave Formation, northeastern Arizona. The linear scale is 1 m long.

Reproduced from Irby (1996a).

LEGEND

downslope

sand and
sand mound

trackway

implied trackway
extension

Map of Tracksite

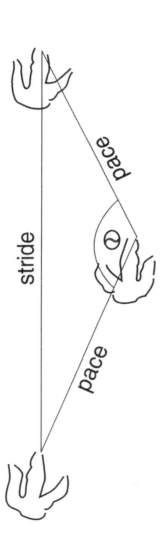

for a relatively short amount of time before a commercial quarry must continue their work, destroying the layer with the tracks (as happened to the Culpeper tracksite mentioned above), or just plain weathering does the job. Most footprints are difficult to remove and place in museum collections, and this technology provides the best opportunity to add the data provided by the tracks and tracksites into permanent collections. See illustrations included in Chapman et al. (Chapter 13 of this volume).

Using the data extracted from the field, some ichnologists (scientists who study footprints) make relatively simple, diagrammatic maps; other, more artistically talented ichnologists prepare maps in which the features of each individual footprint are beautifully rendered. There will always be an artistic component to summary figures of tracksites because photographic or scanned images can be difficult to extract data from quickly. Regardless of how the map is constructed, as much photographic and other documentation of tracksites should be taken as conditions allow. These can be used to check the accuracy of the map, and also provide a permanent record of the tracksite.

For detailed studies of the movements of trackmakers, additional measurements are useful (Fig. 34.3). The distance between successive prints of the animal's left and right feet (the pace) and the distance between two successive footprints of the same foot (the stride) obviously tell us something about how the animal was moving; the longer the animal's steps, the more quickly it was moving. The extent to which left and right footprints of the animal are aligned, or widely separated from one another, may indicate how wide-bodied or narrow-bodied the trackmaker was, and whether it had an erect or sprawling carriage. The angle made by the longest dimension of the footprint with respect to the animal's direction of travel indicates whether the trackmaker walked with its feet angling outward or pointed straight ahead, or with its toes turned inward.

In order for the makers of footprints to be identified as accurately as possible, the shapes of individual prints must be recorded as faithfully as possible. Measurements of the overall size of each footprint, of the lengths of the toes (and of segments of each toe), and of the angles between toes are often made. This can be done either in the field, from the footmarks themselves, or later in the laboratory, from casts or photographs of the prints. Photographs of individual footprints are best taken from directly overhead, and not obliquely; photographing footprints from an angle distorts their shape (Fig. 34.4). Another confounding factor is the lighting; a footmark can look quite different at dusk, when the sun's light is at a low angle to the tracksite, than it does at noon, when the sun is directly overhead; footprints

34.3. Typical measurements made on dinosaur trackways. The pace is the distance between two successive footprints made by the opposite feet, and the stride is the distance between two successive marks of the same foot. The pace angulation Θ measures how nearly prints of the left and right foot are in a single line; the pace angulation is 180 degrees if the animal puts its left and right feet directly in front of each other as it walks. The pace angulation also serves as an indirect indication of how wide the trackway is, and therefore may provide information about how broad the trackmaker's body was or whether it walked in an erect or sprawling fashion. Other measurements commonly made include the length of each separate toe mark, the angles between toes, and the angle made between individual footprints and the trackmaker's overall direction of travel. For quadrupedal dinosaurs, the position of prints of the front and hind feet can be used to estimate the distance between the trackmaker's shoulders and its hips. For more information about footprint and trackway measurements, see Leonardi (1987).

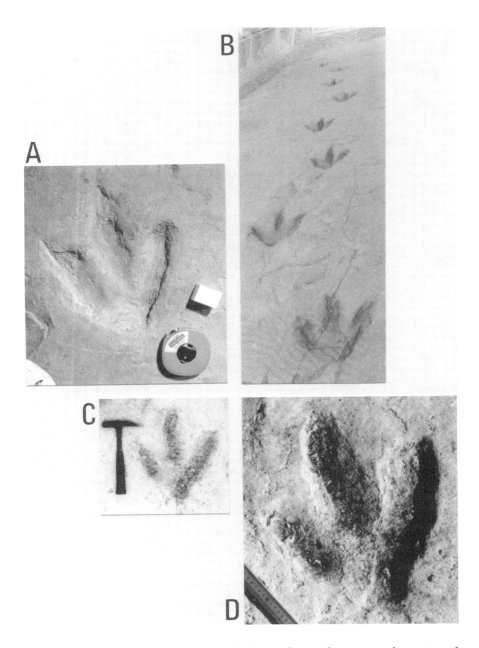

34.4. Different ways of photographing dinosaur footprints in the field. (A) An Early Jurassic theropod print (East Berlin Formation), Dinosaur State Park, Rocky Hill, Connecticut. Footmark length about 35 cm. The print is viewed from directly overhead, to provide as accurate an impression of its shape as possible. The tape indicates the size of the footmark, and the shadow of the little cardboard box shows the direction of lighting in the photograph. (B) Oblique view of an Early Cretaceous theropod trackway (Enciso Group), Los Cayos, La Rioja, Spain. The prints are about 45 cm long. Such an oblique view shows the arrangement of the footprints in the trackway, but it greatly distorts the shape of individual prints. (C) Overhead view of an Early Cretaceous theropod footmark (Fort Terrett Formation), F6 Ranch site, Kimble County, Texas, as seen when the sun is directly overhead. The footprint is about 37 cm long. (D) Overhead view of the same footprint seen in C, but here photographed at dusk, under low-angle illumination.

that are clearly evident at dawn or dusk can almost disappear when viewed at noon. As mentioned above, the morphology of footprints can also be captured directly using scanners or photogrammetry (Breithaupt et al. 2001, 2004; Matthews and Breithaupt 2001); examples will be discussed later in this chapter.

Whenever possible, it is a good idea to make as many casts of the footprints to take back to the lab for further study as is practical. The medium used for casting tracks depends on the preservation of the footprints, and also on the ichnologist's budget. Some fossil tracks are preserved as depressions in the surface across which the trackmaker walked; others are preserved as natural casts from sediments that later filled the footprints. If the footmarks are simple depressions, with no undercuts (where toes or edges of the track extend farther into the rock, below the surface of the track-bearing layer, than they do at the surface), a simple, quick, and cheap way to make a cast is to grease the rock surface with a separator (such as petroleum jelly), and then pour a batter of plaster of Paris into it. Once the plaster cures, you have your cast.

Unfortunately, many footprints in the field have been ruined when amateurs made plaster casts without knowing what they were doing. If no proper separator was used, or if plaster filled undercuts in the footprint, it may be next to impossible to remove the cast without damaging the original track. Making hard casts of footprints in the field is best left to professionals.

When there are undercuts, it is better to make a latex rubber cast of the footprint. This involves putting down thin coats of liquid or paste latex on the print, one after another, allowing time for each coat to dry before the next is added. After several such layers of latex have cured, strips of burlap or cheesecloth are pressed into a new coat of latex while it is still wet, and then still more coats of latex are painted onto the cloth. The cheesecloth or burlap makes the cast resistant to tears, and also gives it a bit of rigidity so that it maintains the shape of the track more. Working with latex in this manner can obviously be rather time consuming. Once several layers of latex have been applied to the footprint and allowed to cure, the cast can be peeled away from the rock surface. It should then be put back against the print, and a rigid backing of fiberglass, burlap, or cheesecloth dipped in plaster is made for the cast, so that it will retain its shape and not collapse under its own weight. Alternatively, the latex cast can be filled with expandable foam, or even with a mushy mixture of latex and tissue paper that will later harden. Making a support for a latex cast is especially important for casts of very large footprints, because latex eventually degrades. It is consequently a good idea to make permanent copies of latex casts of footprints in some other medium.

When footmarks are preserved as natural casts, they also can be copied, although it is often a trickier procedure. If, for example, natural casts of footprints project from the undersurface of a rock ledge, it is pretty hard to make plaster copies of them, although latex approaches may work in some cases.

Latex and plaster are not the only suitable materials for copying footprints in the field or lab, but they are relatively cheap and are readily available. Other media, such as silicone rubber, can also be used, but these are often fairly expensive.

An alternative approach is to capture the track using photogrammetry (Breithaupt et al. 2001, 2004) or a scanner (Bates et al. 2008, 2009; Chapman et al., Chapter 13 of this volume), to build a high-resolution virtual copy of the track, and then print out a copy using rapid-prototyping technology (Andersen et al. 2001; Burns 1993), which encompasses a broad range of different approaches for either carving out or building up a shape using a computer model as a starting point. This can be expensive, however, and the more mechanical approaches can be done far less expensively. There is the option, however, of producing replicas at a smaller and less expensive size. Scanning and photogrammetry will be very successful with most tracks, but those with significant overhangs may be captured in reduced form.

Sometimes ichnologists even collect the actual footprints themselves, although this can be a labor-intensive and time-consuming enterprise. Probably the most ambitious project of this kind was the collection of some 40 tons of dinosaur footprint–containing rock of Early Cretaceous age from the bed of the Paluxy River near Glen Rose, Texas, by Roland T. Bird of the

American Museum of Natural History (Bird 1985). Trackways of a sauropod and large theropod were removed in pieces, and later reassembled as an exhibit in the museum.

Moving most real footprints requires a massive effort, however. As an example, one of the individual footprints from the Early Jurassic Culpeper, Virginia, site that was removed to the Smithsonian's National Museum of Natural History came with associated rock that weighed many hundreds of pounds, and occupied a volume of at least 10 cubic feet. Museum storage would disappear very quickly if many of these tracks were collected.

Laboratory Work

Fieldwork is enjoyable, but it is only the start. The real work begins when maps, measurements, photographs, and casts are brought back to the lab for more detailed study. Casts of footprints can be photographed, scanned three-dimensionally, and illustrated under more controlled conditions in the lab than in the field (Fig. 34.5). Once all this information is gathered, then the real work of research into the tracks and trackways can proceed in the laboratory, but only after quality fieldwork has been done to support this work. This work proceeds over a wide variety of research topics, some of which we will discuss below.

The first characteristic of footprints we will discuss is shape. It is probably not surprising that different dinosaurs have different footprint shapes, but it might be surprising that there are many other factors that affect the shape of a track. First, there are nonbiological factors such as the substrate in which the animal steps. This can greatly affect many shape factors depending on the amount of water held by the sediment and the overall cohesiveness of the sediment, and whether the foot merely indents the sediment surface or punches through it (Avanzini 1998; Gatesy et al. 1999; Nadon 2001; Bimber et al. 2002; Gatesy 2003; Milàn et al. 2004, 2006; Graversen et al. 2007; Marty et al. 2009; Jackson et al. 2010). If the foot sinks deeply enough, the footprint it makes changes as the animal moves forward over the foot and finally lifts it out of the sediment, in the process destroying or altering the shape impressed when the foot initially touched down.

The same animal may therefore make different-looking tracks when walking in different substrates and/or in different ways (such track variation was the inspiration for spurious claims of dinosaur footprints occurring with human footprints; Kuban, 1989a, b). Next, there is the inevitable variation between the two feet of the animal (your two feet are not identical), and between different animals, and finally for the same animal in different life stages. Also, the same animal makes different tracks while exhibiting different behaviors. The deformation that happens during the fossilization process can also greatly affect shape. Finally, one other factor that will also affect shape will surprise most lay readers. When a footprint is impressed into a sediment layer, sometimes a number of footprints are made simultaneously in the layers under the surface (see Lockley 1991, 1997). Such undertracks have subtle to not-so-subtle shape differences from the surface track (Allen 1989, 1997; Manning 2004; Farlow et al. 2006; Milàn and Bromley 2006, 2008; Jackson et al. 2009; Marty et al. 2009), and it is often difficult to determine whether you are studying a surface track or undertrack (unless scalation patterns are preserved, indicating soft tissues and therefore a

A

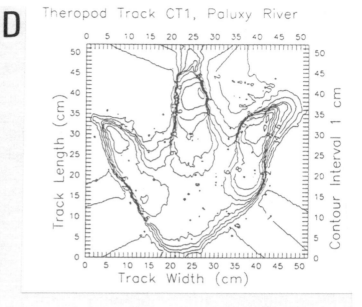

D

Theropod Track CT1, Paluxy River

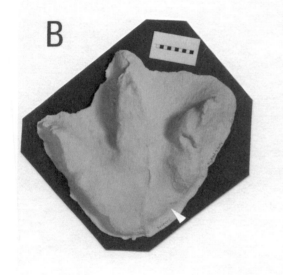

B

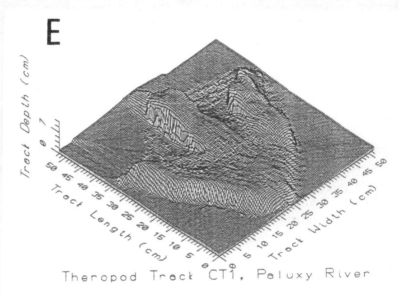

E

Theropod Track CT1, Paluxy River

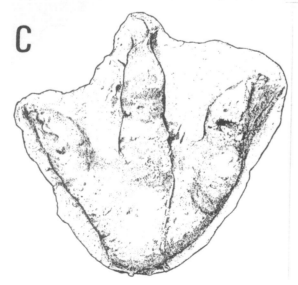

C

F

Theropod Track CT1, Paluxy River

surface track). All in all, there is a lot involved in the analysis of footprint shape.

Consequently, very powerful and rigorous methods are needed to study the tracks and how they vary and how you can tell differences among the various tracks and trackmakers. These can include relatively simple metrics such as the angle between the toes and size (Fig. 34.6), studied as histograms or in pairs. Better is to study these measurements simultaneously, using powerful multivariate statistical methods such as principal components analysis (PCA), discriminant analysis, and cluster analysis (Moratalla et al. 1988; Farlow et al. 2006). This allows all variables measured to contribute simultaneously and increases the ability to discriminate among the different types of tracks and study their variation in shape. More powerful still are methods that capture the whole track and compare the shapes geometrically, using the landmark points such as the tips of the toes or base of the track, or the whole outline simultaneously (Rasskin-Gutman et al. 1997). A combination of all these approaches allows the research to maximize the information extracted from the shape. Even then, some dinosaur tracks, even those made by species that were not related, can look very similar, especially given all the factors discussed above that affect the shape.

So what do you study with shape? You can do lots of different things. A good first step is to look at variation in track shape for a single animal, including between the left and right foot and along individual trackways (Farlow et al. 2006). This provides a baseline of variation for comparison across different animals and comparison among taxa. Rigorous morphometric analyses allow track shape to be studied in a way to help determine relationships between track shape and lithology, stride length, and the action of the animal (i.e., what the animal is doing at the time). Much work needs to be done in this area, but the tools for capturing the shapes and studying them are fully available now.

Footprints can be used as a proxy for the size of the animal that made them; specifically either leg length or height at hip can be estimated from the length of the footprint (Alexander 1976; Lockley et al. 1983; Sanz et al. 1985; Thulborn 1990; Henderson 2003). The conversion from print length to leg length is simple, but complications may arise from the way the footprints are preserved (Manning 2008).

The speed of the animal making the trackway can also be estimated, given a series of assumptions. For example, the greater the speed, the farther apart the footprints will be. You can confirm this next time you are making footprints on a beach—the faster you go, the farther apart each step will be. Also, the faster you go, the more your footprints will fall toward the midline between the left and right footprints; the trackway will get progressively thinner as speed increases. Scientists use these observations and the

34.5. Different ways of illustrating the shape of a dinosaur footprint. (A) Photograph of a print of the right foot of a large carnivorous dinosaur in situ in the dolostone bed of the Paluxy River at Dinosaur Valley State Park, Glen Rose, Texas. The footmark is about 50 cm long. (B) Photograph of a cast of the same footprint (cast made by Peggy Maceo). Because the cast was made by filling the actual print with the casting medium, the cast reverses the topography of the real footprint. Note the "notch" along the inner edge of the footmark (on the right of the print in this cast, *arrow*); this indentation occurs because the inner digit did not impress over its entire length. (C) Artist's drawing of a cast of the same footprint (drawing by Jim Whitcraft). (D) Computer-drawn topographic map of the same footprint cast. The surface of the cast was read into a computer file using a three-dimensional digitizer, after which software converted the raw data into the form shown here. (E) and (F) Wire-frame diagrams showing two perspective views of the footprint. Each of the methods of illustrating the footprint has its desirable features, and also its shortcomings. Photographs and computer images are more objective portraits of the footmark than drawings are, but may not emphasize features that the ichnologist considers important. Drawings can be prepared in such a way as to emphasize features not obvious in photographs or computer images, but may not as accurately portray the overall shape of the print. Whenever possible, it is desirable to illustrate a footprint by more than one method.

34.6. Measurements and shape analysis for typical tridactyl dinosaur footprints. Other types of footprints will have equivalent approaches. (A) Classic measurements taken from a footprint used in studies such as Moratalla et al. (1988). Measurements include two angles, labeled a and b; three lengths from the base point to the three toe tips; and three base toe widths and total width and length. (B) Geometric data taken for shape analysis such as discussed by Rasskin-Gutman et al. (1997). The data include six homologous landmarks—base point, the tips of the three toes, and the two interdigital points (hypexes) between II and III and between III and IV. In addition, two landmarks are calculated to complete the left side base of toe II and right side base of toe IV. Finally, 18 evenly spaced points are calculated between these eight points, providing the whole outline of the foot in great detail. (C) Results of shape analysis between two footprints, one from an ornithopod and one from a theropod. Arrows show shape change from ornithopod shape to theropod shape. Arrows show the footprint of the theropod is thinner and longer.

Redrawn from Rasskin-Gutman et al. (1997).

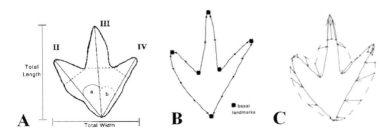

relationship between hip height and footprint length to estimate the speed the dinosaur was going when it made a trackway. This relationship is mathematical, and the equation used depends on the assumptions being made.

When trackways are long, especially when they include data from multiple animals, detailed study can demonstrate behavior among dinosaurs in a way unapproachable from any other source. An example is the famous Lower Cretaceous Paluxy River tracksite that apparently shows a theropod, presumably *Acrocanthosaurus*, following (possibly even attacking?) a sauropod, presumably *Paluxysaurus* (Bird 1985; Farlow 1987; Thomas and Farlow 1997). Detailed analysis of other trackway sites has led to discussion of possible herding behavior (e.g., Ostrom 1972; Currie 1983; Lockley et al. 1983, 1986, 2002a, b; Farlow 1987; Lockley and Matsukawa 1999; Day et al. 2004; Farlow et al. 2006; Li et al. 2008; Schulp et al. 2008).

If the footprints from a new site are significantly different from those that have previously been named in the scientific literature, it is necessary to prepare a formal description of the new footprints, and to give them their own name. Frequently, too, an ichnologist may find it necessary to restudy previously described footprints and tracksites, and to rename previously described trace fossils in some way. In any case, the final result of field and lab work is the preparation of a report for publication that describes the tracksite and its footprints, so that scientists can add this information to the overall database about dinosaurs.

Linking the Footprints and the Bones: Identifying the Makers

One of the most important tasks an ichnologist has is to try to identify that animal that made the footprints (or other traces) that he or she is investigating. This turns out to be incredibly difficult, as footprints and bones are seldom found in any association; the conditions that favor the fossilization of one seem to not favor that of the other. Occasionally an invertebrate is found at the end of a trace, or a burrowing animal in the bottom of its burrow, but this never seems to happen with dinosaurs. Further, for many dinosaurs, the morphology of the foot is sufficiently similar to that of others as to make discrimination difficult. It is not impossible, however, and we will discuss in some detail approaches intended to help make this connection. We will start, however, with the grander question of how historically paleontologists figured out that the footprints were even from dinosaurs.

Identifying the Makers of Dinosaur Footprints: The Remarkable Birds of Professor Hitchcock

At the time that Edward Hitchcock described the Early Jurassic dinosaur footprints of the Connecticut Valley, he had no reason to suppose that his trackmakers had been dinosaurs. He knew that animals such as *Iguanodon* had existed during the Mesozoic Era (Hitchcock 1858, 175), but his image of what dinosaurs had looked like undoubtedly reflected Richard Owen's interpretation of dinosaurs as huge, four-footed (quadrupedal) animals (see Torrens, Chapter 2 of this volume). The three-toed footmarks of the Connecticut Valley had obviously been made by two-legged creatures. Hitchcock concluded that most of these three-toed prints could have been made only by big, flightless birds, like living emus, ostriches, rheas, or cassowaries.

Hitchcock's confidence in his interpretation of the nature of the Connecticut Valley trackmakers was bolstered (Hitchcock 1858, 76–79, 178) by one of Owen's scientific triumphs: Owen's recognition that New Zealand had once been the home of several species of large flightless birds, the moa (Gruber 1987). The former existence of moa (the word *moa* is used as both a singular and a plural), and of the elephant bird on Madagascar, combined with the continued existence of modern flightless birds, seemed ample reason for attributing the Connecticut Valley footprints to similar creatures. Ironically enough, Hitchcock's studies of the New England footprints had a minor reciprocal influence on Owen's research, for Owen named some of his moa species after specific names of Connecticut Valley footprints (Anderson 1989).

Hitchcock was not the only pioneering American geologist to think that the Mesozoic world had been populated by huge extinct birds. Ferdinand Hayden, who organized early geological surveys across the American West, similarly believed that huge three-toed fossil footprints that he found in Wyoming in 1868 had been made by colossal birds (Deibert and Breithaupt 2006).

Hitchcock's reconstruction of a vanished world populated by a diversity of huge birds gripped the imagination of his literary contemporaries. In an essay from *The Encantadas*, Herman Melville described himself as being like "an antiquary of a geologist, studying the bird-tracks and ciphers upon the exhumed slates trod by incredible creatures whose very ghosts are now defunct." In a poem titled "To the Driving Cloud," Henry Wadsworth Longfellow (whose father-in-law helped fund Hitchcock's work; Dean 1969) imagined an ancient race of Native Americans who "once by the margin of rivers stalked those birds unknown, that have left us only their footprints."

Even so, Hitchcock had lingering doubts about the avian nature of some of the Connecticut Valley trackmakers. One of Hitchcock's footprint taxa, which he called the *Gigantitherium* (subsequently known as *Gigandipus* [Lull 1953], but presently recognized as a variant of *Eubrontes* [Rainforth 2007]), combined birdlike footprints with a most un-birdlike tail drag mark (Fig. 34.7). Hitchcock speculated (1858, 93, 179–180) that the *Gigantitherium*-maker had been a huge, bipedal, birdlike lizard or amphibian: "if a biped, its body must have had somewhat the form of a bird, in order to keep it properly balanced. . . . how very strange must have been the appearance of a lizard, or batrachian [amphibian in modern terms], with feet and body like those of a bird, yet dragging a veritable tail!" (180). Another of Hitchcock's footprint taxa, *Anomoepus*, combined a three-toed birdlike foot with a long hind-foot "heel" and a five-toed forefoot that reminded Hitchcock very strongly of the forefeet of marsupial (pouched) mammals such as kangaroos. Hitchcock found *Anomoepus* very hard to interpret: "If a phrase so compound could be used as should imply a participation in the characters of marsupials, birds, batrachians, and even lizards, it would better than any other express my present convictions; for the longer I study these tracks, the stronger my impressions become that some of these ancient animals possessed characters now more exclusively belonging to two, three, or even four [vertebrate] classes" (1858, 60).

To be fair to Hitchcock's thinking, and to see it in its proper historic context, it must be emphasized that he did not interpret the mixture of

34.7. Footprints described by Edward Hitchcock under the name *Gigantitherium* (now assigned to the ichnogenus *Eubrontes;* Rainforth 2007). Note the groove running the length of the slab, from footprint to footprint, interpreted as a tail drag mark. The tracks are about 40 to 45 cm long.

Photograph used by permission of Beneski Museum of Natural History at Amherst College.

characters of different vertebrate classes in his footprints in an evolution-ary framework; Hitchcock remained committed throughout his life to the concept of special creation (Guralnick 1972; Lawrence 1972).

Ironically, although Hitchcock was not an evolutionist, the discovery of *Archaeopteryx*, the earliest known bird, gave him considerable satisfaction. Here was a bird of very "low grade" whose skeleton showed the very features that Hitchcock had identified, and found troubling, in the Connecticut Valley footprints. In a posthumous *Supplement* to his *Ichnology* of 1858, Hitchcock concluded that the *Anomoepus*-maker must have been a "low grade" bird like *Archaeopteryx*, and "if we can presume that the *Anomoepus* was a bird, it lends strong confirmation to another still more important conclusion, which is that all the fourteen species of thick-toed bipeds which I have described in the Ichnology, and in this paper, were birds" (Hitchcock 1865, 32).

In 1858, the same year that Hitchcock's *Ichnology* was published, Joseph Leidy of the Academy of Natural Sciences of Philadelphia described the skeleton of an ornithopod dinosaur that he named *Hadrosaurus*. Noting that the hind limbs of this animal were considerably longer than its forelimbs, Leidy suggested that *Hadrosaurus* and other dinosaurs might have been bipeds (Glassman et al. 1993), an idea that had actually been anticipated by Gideon Mantell for *Iguanodon* (see Torrens, this volume, Chapter 2). This conclusion was amply confirmed in the decades that followed, with the discovery of a diversity of bipedal theropods, ornithopods, and other dinosaurs.

It didn't take long for naturalists to realize the implications of the existence of bipedal dinosaurs for the interpretation of the Connecticut Valley three-toed footprints. In 1867, a note in the *Proceedings of the Academy of Natural Sciences of Philadelphia* reported that Edward Drinker Cope "gave an account of the extinct reptiles which approached the birds" in their structure (Anonymous 1867: 234); Cope concluded that "the most bird-like of the tracks of the Connecticut sandstone" had been made by dinosaurs. The following year, in a spirited discussion of the implications of *Archaeopteryx* and the small theropod dinosaur *Compsognathus* for the correctness of the theory of evolution in general, and the evolutionary derivation of birds from dinosaurs in particular, Thomas Henry Huxley noted that the Connecticut Valley footprints proved that "at the commencement of the Mesozoic epoch, bipedal animals existed which had the feet of birds, and walked in the same erect or semi-erect fashion. These bipeds were either birds or reptiles, or more probably both" (Huxley 1868, 365). The implication was that many of these trackmakers might have been dinosaurs, rather than birds in the strict sense. In 1877, Othniel Charles Marsh of Yale University went even further, asserting that all of the Connecticut Valley three-toed footprints had been made by dinosaurs rather than birds.

Hitchcock had come very close to deducing the body form of bipedal dinosaurs from his study of the early Mesozoic footprints of New England. He had called the trackmakers birds and not dinosaurs, but given the little that was known about dinosaurs when he was studying his tracks, "bird" was as accurate an identification of the trackmakers as was possible for him to have made. For modern interpretations of the systematics and affinities of the New England Early Jurassic dinosaurian ichnofauna, see Olsen et al. (1998), Farlow and Galton (2003), Olsen and Rainforth (2003), Rainforth (2003, 2007), and Smith and Farlow (2003).

Identifying the Trackmakers

Although modern paleontologists have a much clearer understanding of what dinosaurs were like than Hitchcock did, deciding exactly what kind of dinosaur was responsible for a particular footprint can still be tricky. Except for those rare instances where dinosaur "mummies" preserve replicas of desiccated soft tissues of the foot (such as the *Corythosaurus* described by Brown [1916]), we do not know what the soft parts of the feet of dinosaur taxa—which are described on the basis of skeletal material—looked like when those animals were alive. Consequently, when we say that a given footprint is a theropod footmark, for example, what we are really saying is that the print has a shape consistent with what we think would have been made by the soft tissues around the bones of a theropod's foot when pressed into sediment. This conclusion is based on what we know about the skeletal structure of the theropod foot and the relationship between the foot skeleton and its enveloping soft tissues in modern vertebrates, particularly crocodilians and birds, the closest living relatives of dinosaurs.

Some aspects of trackmaker identification are easier than others. One important determination in identifying the maker of a trackway is whether the animal walked quadrupedally or bipedally. Some dinosaurs, including

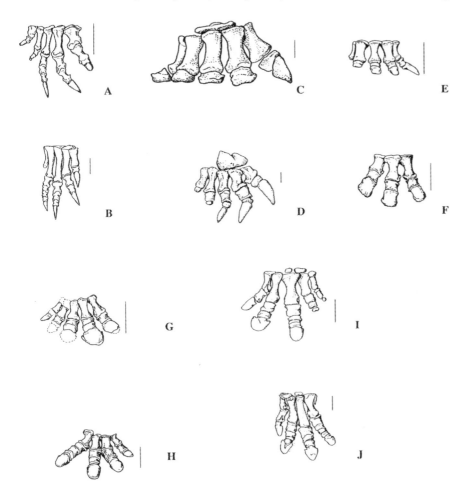

34.8. Manus (forefoot) and pes (hind foot) skeletons of quadrupedal dinosaurs. (A) and (B) Right manus and pes of the prosauropod *Plateosaurus*. (C) and (D) Right manus and pes of the sauropod *Apatosaurus*. (E) and (F) Right manus and pes of the stegosaur *Kentrosaurus*. (G) and (H) Right manus and pes of the ankylosaur *Talarurus*. (I) and (J) Left manus and pes of the ceratopsian *Centrosaurus*. Scale bar is 10 cm long.

Figures redrawn from Lull (1933); Gilmore (1936); Galton (1990); and Thulborn (1990).

34.9. Trackway of a large sauropod in the dolomitic bed (Glen Rose Formation) of the Paluxy River, Blue Hole Ballroom, Dinosaur Valley State Park, Glen Rose, Texas. A meter stick (about 39 inches long) in the middle of the trackway provides the scale. The dinosaur was moving from left to right in this view. Only hind-foot tracks are recorded, presumably because the sauropod placed its hind feet right on top of prints previously made by its forefeet. Hind-foot prints of the left and right side are well separated, making this a wide-gauge trackway.

most theropods and pachycephalosaurs, were habitual bipeds, while other dinosaurs, such as sauropods, stegosaurs, ankylosaurs, and ceratopsians, probably walked on all fours nearly all the time. Ornithopods and pro-sauropods may have used both quadrupedal and bipedal gaits with some regularity.

The shape and size of the individual footprints in a trail are also important. The various groups of quadrupedal dinosaurs differed among themselves in the size of the forefoot compared with that of the hind foot, and also in the number and shape of toes on each foot (Fig. 34.8). The many kinds of four-footed dinosaurs also differed among themselves in overall adult body size. A well-preserved trackway consisting of very large, rather elephant-like footprints, in which the hind-foot footmarks are noticeably bigger than those of the forefoot, is most likely to have been made by a sauropod. In contrast, a trackway consisting of prints of more modest size, in which the forefoot impressions have five toes and the hind-foot footmarks four, could very well have been made by an ankylosaur (McCrea and Currie 1998; Lockley et al. 1999; Petti et al. 2010). Some trackways of medium-sized quadrupedal dinosaurs characterized by tridactyl hind-foot prints are likely to have made by stegosaurs (Whyte and Romano 1994; Lockley and Hunt 1998; Milàn and Chiappe 2009; Belvedere and Mietto 2010; Cobos et al. 2010).

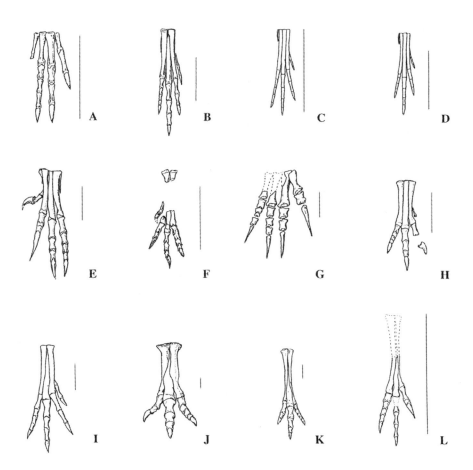

34.10. Manus and pes skeletons of theropod dinosaurs. (A) and (B) Right manus and pes of the coelophysoid *Coelophysis*. (C) Right pes of the small coelurosaur *Compsognathus*. (D) Right pes of the coelophysoid *Procompsognathus*. (E) Left pes of the dromaeosaur *Deinonychus*. (F) Left pes of the basal bird(?) *Mononykus*. (G) Right pes of the therizinosaur *Erlikosaurus*. (H) Right pes of the oviraptorosaur *Ingenia*. (I) Right pes of the caenagnathid *Chirostenotes*. (J) Left pes of the tyrannosaur *Tarbosaurus*. (K) Left pes of the ornithomimosaur *Struthiomimus*. (L) Right pes of the troodont *Borogovia*. Scale bar is 10 cm long.

Figures redrawn from Ostrom (1969, 1978); Maleev (1974); Colbert (1989); Barsbold and Maryańska (1990); Barsbold et al. (1990); Barsbold and Osmólska (1990); Currie (1990); Osmólska and Barsbold (1990); and Perle et al. (1994).

In addition to counting toes, and comparing the relative sizes of forefoot and hind-foot tracks, ichnologists also consider the spacing of footprints in trackways of quadrupedal dinosaurs—the extent to which footprints (particularly those of the hind foot) of the left side are separated from those of the right side. By analogy with railroads, quadrupedal dinosaur trackways are designated narrow-gauge if the left and right hind-foot prints are very close together or even intersect the trackway midline, and wide-gauge if they are well separated (Fig. 34.9), or middle-gauge if somewhere between those extremes (Farlow 1992; Lockley et al. 1994a, 2002a, b; Wilson and Carrano 1999; Romano et al. 2007; Day et al. 2004; Wright 2005; Schulp et al. 2008; Gonzáles Riga and Orlando Calvo 2009; Moratalla 2009; Santos et al. 2009; Marty et al. 2010). Most ichnologists think that trackway gauge has some value in discriminating among different kinds of quadrupedal dinosaur trackmakers, although it may be affected by footprint preservation or trackmaker behavior (e.g., lateral body movements; Carpenter 2009). Wilson and Carrano (1999) suggested that wide-gauge trackways might be particularly characteristic of titanosaurian sauropods.

Unfortunately, footprints of quadrupedal dinosaurs are often poorly preserved, and sometimes can be recognized as prints only because they are arranged in an obvious trackway pattern. Sometimes interpreting the major group to which the trackmaker belonged is little more than an educated guess. Furthermore, even if one is fairly certain that a particular trackway was made by, say, a sauropod, it is usually impossible to say what kind of sauropod—known from skeletal material—the trackmaker was. Indeed, in

34.11. Manus and pes skeletons of ornithopods and other potentially bipedal ornithischians. (A) and (B) Left manus and pes of the basal ornithischian *Lesothosaurus (Fabrosaurus)*. (C) and (D) Left manus and pes of the basal thyreophoran *Scutellosaurus*. (E) and (F) Right manus and pes of the basal ornithopod *Heterodontosaurus*. (G) and (H) Left manus and pes of the small euornithopod *Hypsilophodon*. (I) and (J) Left manus and right pes of the basal iguanodontian *Tenontosaurus*. (K) and (L) Right manus and pes of the basal iguanodontian *Camptosaurus*. (M) and (N) Left manus ("thumb" not shown) and pes of the iguanodontoid *Iguanodon*. (O) and (P) Right manus and pes of the hadrosaur *Edmontosaurus*. Scale bar 2.5 cm in A, B, C, E, and G, 10 cm in all other cases.

Figures redrawn from Gilmore (1909); Thulborn (1972); Galton (1974); Santa Luca (1980); Colbert (1981); Norman (1986); Forster (1990); and Thulborn (1990).

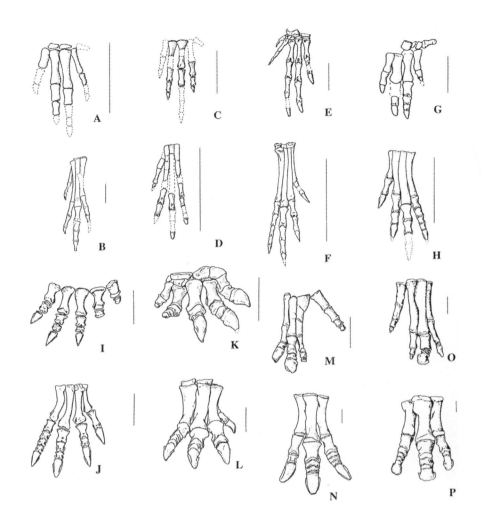

many cases the actual trackmaker may have belonged to a species whose skeletal remains have not yet been discovered.

The same problems apply to the identification of three-toed (tridactyl) prints of bipedal dinosaurs (Figs. 34.10, 34.11). Most (but probably not all; Harris et al. 1996) theropods made three-toed prints during normal locomotion. Many bipedal ornithischians were likewise tridactyl, but some forms had a long first toe that might or might not generally have touched the ground. Bipedally walking prosauropods may have made a four-toed footprint (Lockley 1991). However, Weems (1987, 1992, 2003) suggested that some of the larger three-toed tracks of the early Mesozoic (such as those shown in Fig. 34.1) were made by prosauropods, either specialized tridactyl forms whose skeletal remains have yet to be found, or *Plateosaurus*-like prosauropods that walked with the inner toe of the hind foot held above the ground. Are there any criteria that could serve to distinguish among tridactyl footprints of theropods, bipedal ornithischians, and prosauropods (if any three-toed prosauropod footmarks exist)?

Well-preserved three-toed prints often have conspicuous swellings, or digital nodes, at places along the lengths of the toe marks. At least the larger of these nodes, particularly those nearest the bases of the toes, often correspond to joints between adjacent foot and toe bones, or between toe bones (Heilmann 1927; Peabody 1948; Baird 1957; Thulborn 1990; Fig. 34.12). This

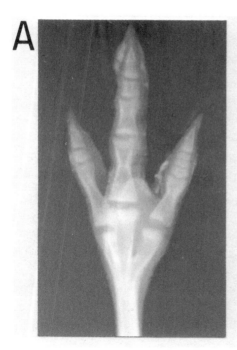

makes it possible to estimate the lengths of individual toe bones (formally known as phalanges) of the foot skeleton of a trackmaker from the pattern of the digital nodes, although this should be done only with caution (Smith and Farlow, 2003). The pattern of digital nodes does not always reflect the pattern of the foot skeleton with perfect fidelity (notice the outer toes of the rhea foot shown in Fig. 34.12). Once the foot skeleton is reconstructed from the digital node pattern, this reconstruction can be compared with real foot skeletons of three-toed bipedal dinosaurs.

It is therefore interesting to look at the proportions of foot skeletons of various kinds of dinosaurs that may have made tridactyl footprints, to see if there are any differences in foot skeletal structure among groups of potential trackmakers that might be discernible in footprints with well-preserved digital nodes. A few such distinctive features do show up (Figs. 34.13, 34.14; also see Farlow et al. 2006). Large theropods tend to have relatively longer toe bones in the middle portion of each toe, compared with the bones at either end of the toe, and relatively narrower toe bones, than do large ornithopods. Theropods also tend to have relatively shorter toe claws than do ornithischians (or prosauropods; cf. Farlow and Lockley, 1993).

Some subgroups within the major bipedal dinosaur categories can be distinguished from one another on the basis of foot proportions. Large theropods and ornithomimosaurs plot in distinct regions of the morphological space in a principal components analysis (Fig. 34.13)—a multivariate statistical method that allows researchers to study many measurements simultaneously (see Chapman 1997; Chapman et al. 1981 for discussions of this analytical technique). Hadrosaurids and *Iguanodon* similarly plot well away from other groups, because of the exaggeratedly short middle phalanges of their toes, and their relatively broad toes. *Iguanodon* and *Tenontosaurus* plot higher along the Factor 3 axis than do hadrosaurids, because of their disproportionately long toe claws.

34.12. (A) X-ray of the right foot of a young greater rhea (*Rhea americana*), a flightless bird from South America. Joints between toe bones (phalanges) are clearly visible. In the middle toe (digit III), swellings (digital nodes or pads) in the soft tissues of the toe correspond to joints between phalanges. This correspondence is less obvious in the toes of the outer digits (II and IV), which have rather short phalanges, all of which are incorporated in the same digital pad of each toe. (B) Natural cast of beautifully preserved small theropod footprints (*Anchisauripus;* presently known as *Eubrontes*–see Rainforth 2007). Notice the well-developed nodes in the footprints' toe marks.

Photograph used by permission of Beneski Museum of Natural History at Amherst College.

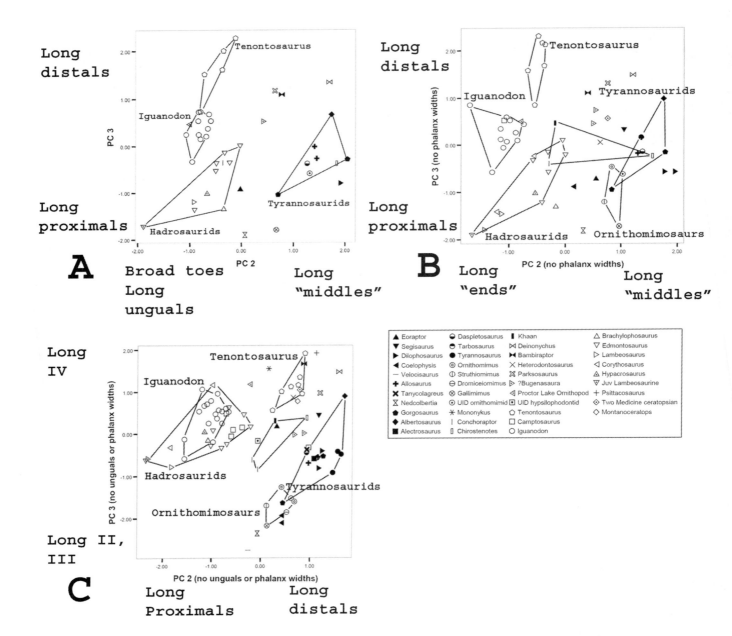

34.13. Principal components analyses (PCA) of foot skeletons of potential makers of three-toed dinosaur tracks. The first factor, related to the overall size of each specimen, has been removed. Three different versions of the PCA are done. In part A, data for lengths of all of the bones in digits II–IV are used, as well as measurements of the distal width of the second phalanx of each of the three toes. Consequently the sample size of specimens is rather small, because a foot skeleton must have a complete set of measurable toe bones in order to be included in the analysis–and many dinosaur specimens just aren't that complete. In part B, all of the phalanx (toe bone) lengths are employed in the analysis, but the three phalanx widths are excluded. This increases the number of specimens in the analysis. In part C, phalanx widths and also the lengths of the claw-bearing bones (unguals) are excluded, further increasing the sample size, but at the cost of losing information about foot shape. In analysis A, component 2 makes a contrast between specimens with long unguals and broad toes (left side of graph), on the one hand, and specimens with long phalanges from the middle region of each toe, and narrow toes, on the other (right side of graph). Component 3 makes a contrast between specimens with long proximal (closest to the leg) phalanges (bottom of vertical axis of graph) and specimens with long distal phalanges (farthest from the leg). Individual genera of dinosaurs are identified in the legend, and polygons are drawn around groups of points. The tyrannosaurid polygon is clearly separated from the hadrosaurid, *Iguanodon*, and *Tenontosaurus* polygons along component 2, and the three ornithopod polygons separate along component 3. Note, though, that the tyrannosaurid polygon encloses points for *Allosaurus*. Analysis B now has enough additional foot specimens to allow polygons to be drawn around points for ornithomimosaurs and oviraptorosaurs (the latter not labeled on the graph). Component 3 is much the same as in analysis A, but component 2 is now a contrast between feet with long phalanges from the middle regions of toes (right side of graph) and feet with long phalanges at the proximal and distal ends of toes. There is still pretty good separation among some major groups of dinosaurs, but allosaurs continue to fall among tyrannosaurid points. In analysis C component 2 is a contrast between feet with long distal vs. proximal phalanges, while component 3 is a contrast between feet with long digits II and III on the one hand, and digit IV on the other. Tyrannosaurids continue to separate from the big herbivores (but not *Allosaurus* or *Dilophosaurus*). *Tenontosaurus*, oviraptorosaurs (again not labeled), and ornithomimosaurs similarly form distinct groups, but points for *Iguanodon* and hadrosaurids can no longer be told apart. In all three versions of the PCA, there is good separation between medium-sized and large ornithopods as opposed to theropods, but not necessarily among different groups within large theropods or ornithopods. Various genera of small theropods and ornithischians fall in regions of the graphs between medium-sized and large ornithopods and theropods. Some small theropods are as much or more like small ornithischians than other theropods, and some small ornithischians as much or more like small theropods than other ornithischians.

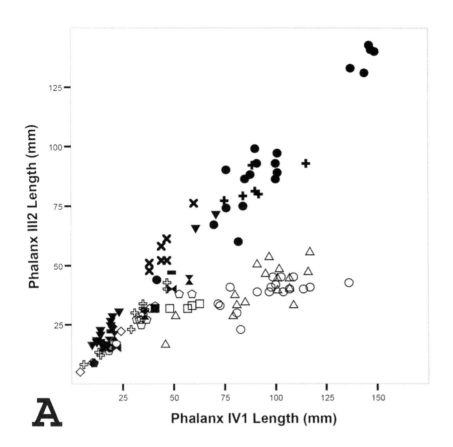

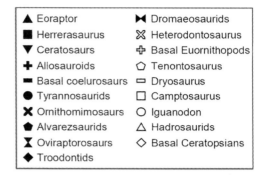

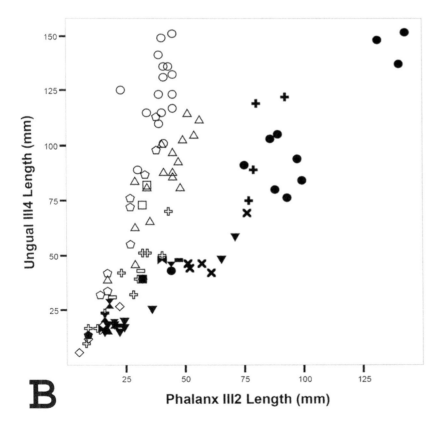

34.14. Relative lengths of selected phalanges in potential makers of tridactyl dinosaur footprints. These graphs show some of the same results as were seen in the PCAs (Figure 34.13), but because only two bones need to be preserved in a foot for it to be plotted, the sample size of feet is much larger than for the PCAs. Solid symbols are for theropods, and open symbols represent different kinds of ornithischians. (A) Plotting the length of phalanx III2 against that of phalanx IV1 shows the contrast between feet with a relatively long phalanx from the middle region of a toe (in this case phalanx III2), and feet with a relatively long proximal phalanx (IV1). (B) Separation of feet with a relatively long ungual (III4) as opposed to a relatively long phalanx from the middle region of the toe (III2). As in the PCAs, large theropods are very different from large ornithischians in both graphs, but smaller ornithischians and theropods are much harder to tell apart.

However, there is overlap among groups. The graphs suggest that one would probably not be able to distinguish footprints of tyrannosaurids from those made by *Allosaurus* and its kin. The region of morphological space occupied by small ornithopods overlaps or is adjacent to that of basal ceratopsians and several small theropods. It might therefore be difficult to discriminate among footprints of some of these dinosaurs on the basis of the proportions of the three main toes of the hind foot alone (although dromaeosaurids would probably not ordinarily leave impressions of the long claw of the second digit, and just such two-toed footprints have been found; Li et al. 2008; Kim et al. 2008). One would have to hope for the presence of additional criteria, such as the occasional impressions of digit I of the foot, or tracks made by the hand, to identify trails made by members of some of these dinosaur groups.

In some cases (notably hadrosaurids), foot skeletons of members of the same genus plot as far away from each other as they do from the foot skeletons of other genera in the group. Even when it is possible to assign a footprint to a particular group of bipedal dinosaurs on the basis of toe-bone proportions estimated from the print, it may not be possible to be certain about which genus within the group was the trackmaker.

If tridactyl prosauropod footprints exist, such prints would more likely be mistaken for the tracks of tridactyl ornithischians than for tracks made by theropods (Figs. 34.8, 34.10, 34.11). In fact, the hind foot of *Plateosaurus* is remarkably similar to that of *Tenontosaurus* (Figs. 34.8B, 34.11J), which suggests that confusing tracks made by bipedal prosauropods with those made by (for now hypothetical) early Mesozoic large, bipedal ornithischians could in theory be a real problem, regardless of the number of toes on the hind foot! Interestingly enough, one of Hitchcock's footprint types, *Otozoum*, was made by a four-toed biped that has been interpreted as a prosauropod (Lockley 1991; Farlow 1992; Rainforth 2003), an ornithopod (Thulborn 1990), and a thyreophoran ornithischian (Gierlinski 1995).

It must be emphasized that the preceding comments about the possibility of identifying the makers of tridactyl dinosaur footprints from foot proportions apply only to those instances where footprints are so well preserved that they clearly show digital pads, from which the relative lengths of individual toe bones can be estimated. Quite often three-toed footmarks do not show distinct toe pads. Are there any criteria by which the makers of tridactyl footprints can be identified on the basis of gross print shape?

One set of potential characters involves comparisons of the lengths of the three toe marks (often made in descriptions of dinosaur footprints; cf. Moratalla et al. 1988, 1992; Demathieu 1990; Gierlinski 1988, 1994; Casanovas Cladellas et al. 1993; Gierlinski and Ahlberg 1994; Romero Molina et al. 2003). Many tridactyl footprints show a conspicuous indentation along the inner margin of the footprint, near the rear of the print (Fig. 34.5; cf. Thulborn 1990), which can be useful for identifying whether a footmark was made by the left or the right foot of the dinosaur. This indentation exists because the inner toe, digit II, did not touch the ground over its entire length (Farlow et al. 2000). The back end of the impression of digit II was usually made by the joint between the first and second phalanges of that toe. In like manner, the most proximal indication of the middle toe of the foot (digit III) also represents the joint between phalanges 1 and 2. In contrast,

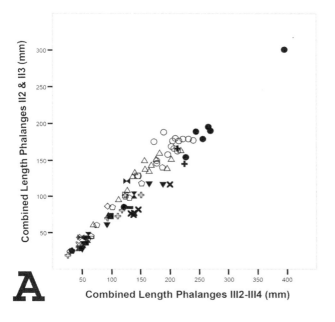

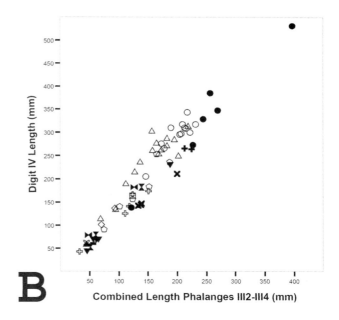

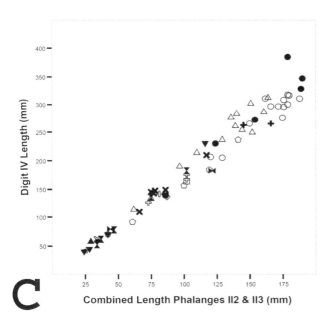

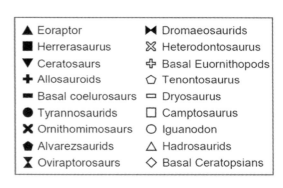

digit IV usually did impress over its entire length, and so the back end of its impression, and often the back of the footprint itself, was made by the joint between the first phalanx of the toe and the fourth metatarsal bone of the foot.

Consequently the toe marks of digits II and III will generally consist of the combined lengths of phalanges 2–3 and 2–4, respectively, while the toe mark of digit IV will be made by the combined lengths of all five phalanges of that toe. Comparing the relative lengths of the three main toes of the foot (Fig. 34.15) results in some separation of large theropods from big ornithopods, but there is a lot of overlap among groups.

Large theropods tend to have relatively narrower phalanges than do big ornithopods (Figs. 34.10, 34.11, 34.13), and so a big, three-toed print with long, skinny toe marks (Fig. 34.5) was more likely made by a theropod than an ornithopod, while a big footmark with three relatively short, thick toe

34.15. Relative lengths of the portions of the three digits (II2–3, III2–4, and IV1–5) likely to have touched the ground during normal walking. Medium-sized and large theropods tend to have a relatively long digit III, compared with medium-sized and large ornithopods. However, there is little difference between small theropods and small ornithischians in any toe-length comparison.

34.16. Computer-drawn topographic map of a large ornithopod footmark in the collection of the Royal Tyrrell Museum of Palaeontology, Alberta, Canada. Note the relatively short, thick toe marks; compare this shape with that of the large theropod print shown in Figure 34.5.

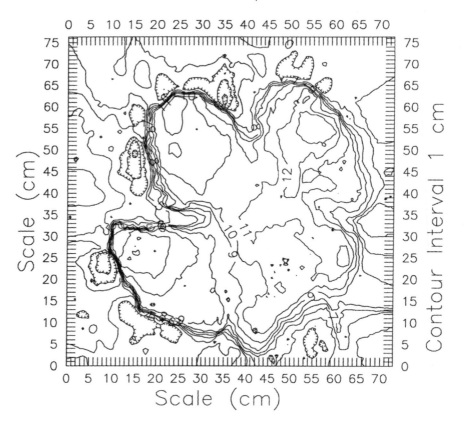

TMP87.76.6 Ornithopod Track

marks (Fig. 34.16) probably had an ornithopod maker. However, many of the smaller bipedal ornithischians had relatively long, narrow toes (Figs. 34.11, 34.13), and so toe-mark slenderness might not be very helpful in distinguishing prints of small theropods from those of small ornithischians. It is not surprising, for example, that the maker of an early Mesozoic footprint known as *Atreipus* has been identified as an ornithischian and as a theropod by different paleontologists (Thulborn 1990).

Many tridactyl dinosaurs, both theropods and ornithischians, have an outer toe (digit IV) that is conspicuously narrower than digits II and III (Figs. 34.10, 34.11). However, the troodontid *Borogovia* and the early bird(?) *Mononykus* have a very stout digit IV and a relatively slim central toe (digit III). Such differences in digit thickness might be recorded in very well-preserved footprints.

Ichnologists have proposed additional criteria for identifying the makers of tridactyl dinosaur footprints that are less readily related to the skeletal structure of the foot (Thulborn 1990, 1994; Lockley 1991; Weems 1992). These include such features as the overall width/length ratios of footprints, the angles between toe marks, the angles made between footprint long axes and the trackmaker's direction of travel, and trackway pace angulations (Fig. 34.3). Such parameters may well be useful, but probably do not permit as confident identification of the trackmaker as do parameters more directly related to skeletal structures.

The job of identifying dinosaurian trackmakers can be made somewhat easier by a consideration of the stratigraphic occurrences of dinosaur

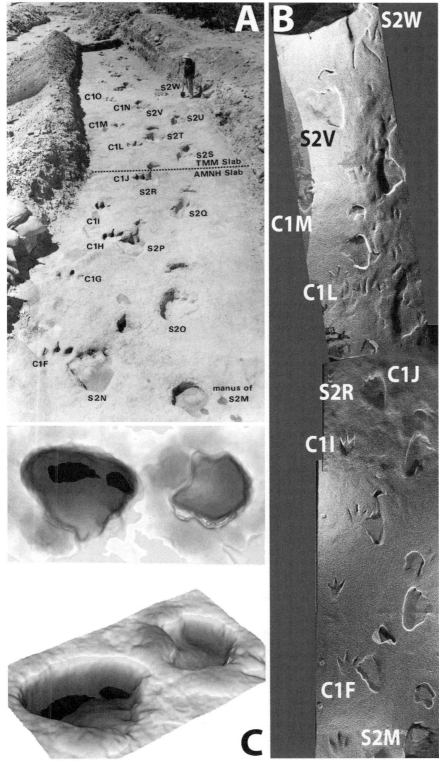

34.17. Sauropod trackway from the bed of the Paluxy River, Dinosaur Valley State Park, Glen Rose, Texas. Tracks of this kind are known by the name *Brontopodus*. (A) Photograph of the trackway taken by R. T. Bird of the American Museum of Natural History (New York) in 1940 (Bird 1985). The trackway was assigned the label S2, with individual hind-foot prints assigned letter labels. Superimposed on many of the sauropod prints are tridactyl footmarks of a big theropod (trackway C1) that may have been following the sauropod. The portion of the two trackways in the foreground was collected in pieces by Bird and reassembled at the American Museum. Parts of the two trackways beyond the dotted line were collected and reassembled at the Texas Memorial Museum (Austin). (B) Computer image of the combined American Museum and Texas Memorial Museum trackway slabs. (C) Three-dimensional computer images of right manus (forefoot) and pes (hind foot) set S2M in overhead (top) and oblique (bottom) views; manus print to the right. The two footprints are shaded from light (shallow) to deep (dark) to indicate how deeply different parts of them are impressed.

Images in B and C created by Karl Bates and Peter Falkingham. Images used by permission of American Museum of Natural History, specimen AMNH FARB 3065.

footprints. For example, the Early Cretaceous Glen Rose Formation of Texas has yielded beautifully preserved sauropod footprints (Figs. 34.9, 34.17). We cannot say for sure what kind of sauropod made those tracks. However, the fact that skeletons of a titanosauriform known as *Paluxysaurus* (formerly considered to belong to the genus *Pleurocoelus*; Rose [2007]) have been found in rocks of the same age, in the same region, and that what is known of the foot skeleton of *Paluxysaurus* is consistent with the shape of the footprints, suggests that *Paluxysaurus* could well have been the Glen Rose Formation's sauropod footprint maker. We cannot be completely certain of this, however, because a second sauropod, *Sauroposeidon*, is also known from rocks of the same age in Oklahoma (Wedel et al. 2000). The foot of *Sauroposeidon* is unknown, though, so we don't know how good a candidate for the Glen Rose sauropod trackmaker this dinosaur is. But its occurrence in rocks of the right time and about the right place complicates our assignment of the tracks to a trackmaker.

Similarly, large theropod footprints from the Glen Rose Formation are quite likely to have been made by a big allosaurid, *Acrocanthosaurus*, whose skeletal remains are found in rocks of the appropriate age from the same region (Farlow 2001; Farlow et al. 2006). Using the same reasoning, enormous tridactyl footprints with relatively narrow toe marks and/or claw marks from latest Cretaceous sediments of New Mexico and Montana could well have been made by the gigantic Maastrichtian theropod *Tyrannosaurus* (Lockley and Hunt 1994; Manning et al. 2008). Again, these assignments cannot be considered certain-just good prospects.

We can also make inferences about what dinosaurs are unlikely to have been the trackmakers of any particular stratigraphic unit. Because skeletal remains of tyrannosaurids are known only from rocks of Late Cretaceous age, it is unlikely that any of the Early Jurassic large theropod footprints of the Connecticut Valley were made by those particular large theropods. Tyrannosaurids are not presently known from Late Cretaceous rocks of South America, where abelisaurids were the common later Mesozoic large theropods; it is therefore more likely that Late Cretaceous large theropod footprints from South America were made by abelisaurids rather than by tyrannosaurids. For similar reasons we can be skeptical that any Late Cretaceous footprints of quadrupedal dinosaurs were made by prosauropods (a group that is presently known only from the Triassic and Jurassic) walking on all fours. These are, of course, statements only of what is likely, and not of what is known to be true with complete certainty. It is possible that someone will someday find the skeletal remains of an Early Jurassic tyrannosaurid or a Cretaceous prosauropod–but it seems very unlikely. It is somewhat less improbable that remains of a tyrannosaurid will someday turn up in South America in rocks of Late Cretaceous age.

Sometimes, however, stratigraphic considerations make the identification of trackmakers more complicated. Three-toed footprints similar to typical theropod footmarks have been found in Triassic rocks that are older than the most ancient skeletal fossils of carnivorous dinosaurs (Haubold 1984; Demathieu 1989; Arcucci et al. 1995–but see King and Benton 1996). Either theropods evolved rather earlier than we presently think, or these footprints were made by non-dinosaurian archosaurs with very theropod-like feet and gaits. If the latter is true, it raises the question of whether some Late Triassic or even Early Jurassic footprints that we presently regard as having been made by theropods could instead have been made by archosaurs that were related to, but were not, dinosaurs. Footprints indicate that the broader group to which dinosaurs belong (dinosauromorphs) originated fairly early in the Triassic (Brusatte et al. 2010). Even more confusing, there seem to have been Late Triassic striding, bipedal archosaurs that were more closely related to crocodilians than to dinosaurs, and yet had a very dinosaur-like, functionally tridactyl or nearly so, hind foot (Gauthier et al. 2011).

Identifying the makers of dinosaur tracks in any particular stratigraphic unit will always involve an element of uncertainty. We will probably never be able to reconstruct the composition of dinosaur faunas on the basis of dinosaur footprint assemblages alone with as much confidence as is possible for well-preserved skeletal assemblages. On the other hand, for stratigraphic units with few skeletal remains, footprints will remain our only window on those otherwise lost dinosaur communities.

Naming Dinosaur Footprints

Hitchcock's *Ichnology* of 1858 was preceded by several shorter articles about the Connecticut Valley tracks. In some of these earlier works, Hitchcock coined names that he applied specifically to the footprints themselves. By 1865, however, Hitchcock had concluded that names such as *Gigantitherium* and *Anomoepus* were suitable tags not just for the footprints, but also for the creatures that made them. This belief reflects the confidence that early-nineteenth-century naturalists had in the predictive powers of comparative anatomy. The great French anatomist Georges Cuvier had correctly predicted, from the scant skeletal evidence initially visible at the surface of the enclosing rock, that a certain fossil mammal would turn out to be a marsupial once its bones were more fully exposed (Rudwick 1976, 113–116). Richard Owen had similarly been able to deduce the former existence of moa as huge flightless birds on the basis of a fragmentary femur. Hitchcock therefore felt confident in asserting,

> The grounds on which I propose to name and describe the animals that made the fossil footmarks, are derived from comparative anatomy and zoology. These sciences show a mathematically exact relation to exist, not only between different classes and families of animals, but between different parts of the same animals. . . . Why then, in view of such facts, should we not name and describe an unknown animal, though nothing but its tracks remain from which to judge of its nature? For in truth a track . . . scarcely differs from the foot petrified; and if Cuvier's principle be true, it will generally give us a tolerably correct idea of the other parts of the body. Why can we not construct the whole animal from this petrified foot, as well as the anatomist can from a single bone belonging to some other part of the frame? (1858: 23–24).

The unfortunate answer to Hitchcock's rhetorical question is contained in the discussion of the previous section of this chapter. Although it may be possible to get a "tolerably correct idea" of the kind of animal responsible for a particular footprint, this may seldom be accurate enough to allow us to say exactly which kind of beast made the track. We would probably not be able to distinguish footprints of one species and perhaps even genus of hadrosaurid, or tyrannosaurid, from another.

Ichnologists have therefore reverted to Hitchcock's initial approach, in which scientific names are created to describe distinctive kinds of footprints themselves, rather than the animals that made them. Footprints are named as ichnogenera and ichnospecies, to distinguish them from genera and species named, in the case of fossil vertebrates such as dinosaurs, on the basis of skeletal material (cf. Baird 1957; Demathieu 1970; Haubold 1984; Sarjeant 1989, 1990; Thulborn 1990; Lockley 1991; Leonardi 1987; Sarjeant and Langston 1994). There are, however, differences of opinion as to the criteria by which the footprints of dinosaurs and other fossil vertebrates should be named.

Vertebrate footprints are only one kind of trace fossil; other vertebrate trace fossils include such things as eggs, coprolites, and bite marks (see Chin 1997; Hirsch and Zelenitsky 1997; Zelenitsky et al., Chapter 30 of this volume). Furthermore, vertebrates are not the only kinds of organisms that left traces of their activities in ancient sediments; a diversity of invertebrates, plants, and even microorganisms did so as well. Bottom-living (benthic) invertebrates, both marine and freshwater, create marks on or in sediments (or even hard rock) as they crawl, bore, burrow, and feed on the floors of seas and lakes. Frequently ichnologists cannot identify even the phylum of the animal responsible for a particular invertebrate trace fossil. On the other hand, the kinds of animal activities reflected in invertebrate trace fossils that occur in sedimentary rocks can provide very useful information about the environmental conditions under which the sediments accumulated (Osgood 1987; Maples and West 1992). Consequently ichnologists who study invertebrate traces name these sedimentary structures on the basis of the kind of tracemaker behavior recorded in the trace, and not on the basis of the generally unknowable taxonomic affinities of the tracemaker (Sarjeant and Kennedy 1973; Basan 1979; Magwood 1992; Lockley 2007). A marine invertebrate can make one kind of trace while sitting motionless on the sea floor, and an entirely different trace while crawling about. Following the procedure of ichnologists who study invertebrate traces, such different traces of the same animal are given different names.

Some ichnologists, most notably the late W. A. S. Sarjeant (e.g., Sarjeant 1990; Sarjeant and Langston 1994) advocated using the same approach when it comes to recognizing and naming vertebrate ichnotaxa. Footprints of a given characteristic shape should be given a name regardless of whether they can be correlated with a particular category of trackmaker. Furthermore, if footprints made by the same kind of animal differ in shape under different circumstances (such as when the beast is running, as opposed to walking), those different footmarks made by the same kind of animal should be given different names.

Most ichnologists agree that vertebrate footprints that have a characteristic morphology indicative of a distinctive foot structure should be named, regardless of whether the taxonomic affinities of their maker can be ascertained. The controversy arises over whether differences in footprint shape caused by differences in behavior should also be used as a basis for naming footprints, as is done with invertebrate traces.

We are skeptical about the usefulness of doing this for two reasons. First of all, even though, as already discussed, there can be uncertainty as to the maker of a particular vertebrate track, this level of uncertainty is at a much lower taxonomic level for vertebrates than for most benthic invertebrates (Schult and Farlow 1992; Lockley 2007). Being unable to tell footmarks of theropods of one family from those of another family, or even theropod prints from ornithopod tracks, is a rather different thing from being unable to distinguish burrows made by "worms" of one phylum from those of a different phylum. Tracks of dinosaurs and other vertebrates will usually preserve a greater amount of information about the taxonomic affinities of their makers than invertebrate traces will.

Secondly, differences in benthic invertebrate traces, reflecting as they do differences in the behavior of the tracemakers, preserve useful information about ecological conditions (such as the distribution of food in sediments) of the environments in which the traces were made (Osgood 1987). In contrast, differences in the shape of footprints of a particular kind of vertebrate caused by differences in the creature's behavior, or differences in the way the prints were formed and preserved, will seldom reveal anything particularly interesting about the environments (apart from substrate conditions) in which the tracks were made (Schult and Farlow 1992). Whether a dinosaur was walking, running, hopping, sitting, or limping may tell us nifty things about what the animal was up to, but these will not reflect anything ecologically significant about the environment in which the animal was doing those things (a possible exception is with swimming traces versus walking traces—Ezquerra et al. 2007). We therefore don't see anything useful to be gained from naming vertebrate ichnotaxa in exactly the same way that invertebrate trace fossils are named.

All of this leads to a basic question: What are ichnotaxa for? Or, what is the goal, and what is to be gained, by giving names to vertebrate trace fossils?

As noted by Lockley (1991, 1997), in some stratigraphic units there is a better record of dinosaurs and other Mesozoic vertebrates in the form of footprints than in body fossils. The early Mesozoic rocks of eastern North America (such as the rocks of the Connecticut Valley that Hitchcock studied; cf. Olsen et al. 1998; Olsen and Rainforth 2003; Rainforth 2003, 2007) are a good example.

Suppose that we had studied such a rock unit that was rich in dinosaur footprints but had few or no dinosaur bones (e.g., the Middle Jurassic Red Gulch and associated sites from Wyoming; Breithaupt et al. 2001; Kvale et al. 2001). Suppose, too, that we wanted to compare the dinosaur fauna inferred from the footprints of our rock unit with dinosaur faunas inferred from footprints, or known from skeletons, from some other geographic area or stratigraphic interval. If in naming ichnotaxa we had used footprint shape differences created by differences in trackmaker behavior, and not just track shape differences that reflected differences in foot structure of the trackmakers, then a list of ichnotaxa from our rock unit might show very little correspondence to the kinds of animals, and the diversity of animal taxa, that had actually inhabited the region. We would not be able to use the ichnotaxa to say very much about whether the dinosaur fauna represented by our footprint assemblage was similar to, or different from, dinosaur faunas of other places and times.

Many ichnologists therefore prefer an approach to naming vertebrate footprints (cf. Baird 1957; Olsen and Galton 1984; Farlow et al. 1989; Lockley et al. 1994a; Lockley and Hunt 1995a; Olsen and Rainforth 2003; Lockley 2007; Rainforth 2007) that is somewhere in the middle of the positions represented by Hitchcock's view of 1858 and the approach adopted by students of invertebrate traces. Footprints should be given names only when they display a distinctive shape that is unlikely to be an artifact of their formation, a shape that to some extent reflects the skeletal structure of the trackmaker. Furthermore, the foot structure inferred from the footprint should be different from the foot structure inferred from previously named ichnotaxa. The names given to footprints are not the same as those given to the creatures that made them, however. For example, the name *Brontopodus* was assigned to sauropod tracks from the Glen Rose Formation that very likely (but not certainly) were made by *Paluxysaurus* (Farlow et al. 1989).

This approach has the advantage that a list of vertebrate ichnotaxa will bear a closer relationship to the fauna of animals that made the tracks than will a list compiled on the basis of the approach used by students of invertebrate traces. There will be a correlation between skeletal taxa and ichnotaxa at some taxonomic level. What that level will be probably will vary from case to case. As noted by Olsen and Galton (1984, 94), "The ichnogenus *Grallator* [another of Hitchcock's footprint names] could have been made by any of the conservative members of the . . . Theropoda from any part of the Mesozoic. The ichnogenus *Anomoepus* could have been made by cursorial members of the families Fabrosauridae, Hypsilophodontidae, some Iguanodontidae, Psittacosauridae, and some Leptoceratopsidae."

Inspection of Figures 34.13–34.15 gives some support to that view. The presently available data for tridactyl dinosaur hind-foot proportions suggest that one ichnotaxon (perhaps an ichnogenus) could be justified for footprints made by *Tenontosaurus*, should such tracks ever be discovered. Another ichnotaxon might well incorporate all tracks of hadrosaurids. Yet another ichnotaxon could be created for tyrannosaurid (and other large theropod?) footprints, and still other ichnotaxa erected for tracks made by ornithomimids and ceratosaurs. An ichnotaxon that would likely include all the shapes of tracks made by *Hypsilophodon* and its close relatives would probably also have to include prints made by *Heterodontosaurus* and psittacosaurs, and possibly other small bipedal ornithischians as well. Thus a list of ichnotaxa from a rock unit named on the basis of the approach advocated in this chapter will probably underestimate the diversity of animals in the living fauna, but it will also be as realistic a reflection of the number and relationships of animal taxa in that fauna as can be obtained from footprint evidence.

Unfortunately, Figures 34.13–34.15 also suggest a more disturbing possibility: that a particular ichnotaxon might have to include prints made by unrelated groups of dinosaurs. Note once more the overlap between small ornithischians and some small theropods in the graphs. Although this concern is raised by comparisons of foot skeletons, and not footprints, there is another reason for thinking that we are dealing with more than a hypothetical problem.

Over the course of the Cenozoic Era, numerous groups of ground-living (often flightless) birds evolved in many parts of the world (Feduccia 1996). These include ratites (ostriches, rheas, cassowaries, emus, kiwis, moa, and elephant birds); tinamous; gastornithiforms (*Diatryma* and its kin); gruiforms (bathornithids, seriemas, phorusrhacoids, Messel rails, eogruids, ergilornithids, the adzebill, bustards, cranes, and rails); large flightless pigeonlike birds (the dodo and the solitaire); big flightless ducks and geese; galliforms (pheasants, turkeys, megapodes, and their kin); secretary birds and similar birds of prey; and mihirungs (large to gigantic flightless birds

34.18. Casts of footprints of living and extinct species of large ground-living birds. *Clockwise from the upper left-hand corner:* Right footprint of a young emu (*Dromaius novaehollandiae*); right footprint of an adult emu; left footprint of a cassowary (*Casuarius casuarius* – note the enormous claw mark on the inner toe, digit II [on the left side of the cast in this view]; the same digit in the adjacent emu footprint is on the right side of the cast); right footprint of an adult ostrich (*Struthio camelus* – note the absence of the digit II, which would have been to the right of the large toe mark for digit III in this view); right(?) footprint of a mihirung (dromornithid) from Tasmania (collection of the Queen Victoria Museum and Art Gallery, Launceston, Tasmania); right(?) footprint of a moa from New Zealand (Museum of New Zealand Te Papa Tongarewa S.24314); left footprint of a lesser (or Darwin's) rhea (*Rhea* [or *Pterocnemia*] *pennata*); right footprint of a kori bustard (*Ardeotis kori*). Although the emu is more closely related to the cassowary than to the other birds, its footprint is as much or more like that of the kori bustard. Similarly, the rhea, emu, cassowary, and moa may be more closely related to each other and to the ostrich than to the kori bustard or the mihirung, but one would be hard-pressed to see this in comparisons of footprint shapes. This suggests that footprints belonging to the same ichnotaxon will not necessarily be made by animals that are closely related, and that animal species that are close relatives might make footprints that would be placed in different ichnotaxa.

of uncertain relationships from the Australian region), to name some of the more impressive forms. The evolutionary relationships of these various groups of ground birds are not completely understood (Cooper et al. 1992; Cooper and Penny 1997; Feduccia 1996; Cooper et al. 2001; Haddrath and Baker 2001; Worthy and Holdaway 2002; Zelenitsky and Modesto 2003; Murray and Vickers-Rich 2004; Ericson et al. 2006; Livezey and Zusi 2007; Hackett et al. 2008).

That being the case, comparing footprints made by different kinds of ground-living birds (Fig. 34.18) raises some interesting questions. If the ratites do constitute a monophyletic group, it is intriguing to note that the two-toed ostrich makes a print very different from that of other ratites, a footmark that in gross appearance might resemble prints made by certain extinct gruiform birds, the similarly two-toed ergilornithids, more closely than it does the footprints of other ratites. Emus and cassowaries are universally believed to be close relatives, but the enormous claw on digit II of the cassowary foot makes cassowary prints differ more in gross appearance from emu footprints than the latter do from prints of rheas, moa, mihirungs, and bustards.

Footprints of two taxa of trackmakers can resemble each other for one of the following reasons: (1) both taxa retain the primitive foot shape of a common ancestor, a primitive foot shape that they also share with other taxa; (2) both taxa have evolved a similar foot shape that they share with no other taxa because they are part of a monophyletic group (cf. Holtz and Brett-Surman, Chapter 11 of this volume); (3) the two taxa are not close relatives, but have independently evolved a similar foot shape because of functional similarities associated with a comparable mode of life; (4) one taxon retains the primitive foot form, and the other taxon has reverted to that foot shape from a more derived condition, perhaps because it has returned to the way of life of its ancestors.

Which of these is the correct explanation for similarities in footprint form probably varies from case to case, whether we are talking about dinosaur or bird footprints. The key point, however, is that most systematists now contend that natural phylogenetic groups should be recognized only on the basis of shared derived features (see Holtz and Brett-Surman, Chapter 11 of this volume). Consequently there will be congruence between groupings of taxa on the basis of foot shapes and groupings of taxa in monophyletic clades only if foot shapes consistently change at the same time as do the characters used to define monophyletic groups (characters that could include, in the case of dinosaurs, the rest of the skeleton as well as the foot). Figures 34.13–34.15 and 34.18 suggest that this is frequently not the case.

Footprint taxa are usually distinguished from one another, or considered to be similar to each other, on the basis of overall differences or similarities in shape. Whether similarities in print shape are due to the second reason given above (the only reason that would involve monophyletic groups; Olsen 1995; Wilson and Carrano 1999; Carrano and Wilson 2001; Wright 2005) or to one of the other three explanations cannot always (or perhaps even often) be determined.

That being the case, there will inevitably be an unavoidable lack of resolution in correlations between dinosaur skeletal and footprint taxa (although we once again emphasize that this will not be as great a problem as in invertebrate ichnotaxonomy). This will significantly affect the resolution we can hope for in reconstructing the composition of ecological communities, making biostratigraphic correlations, or interpreting biogeographic patterns on the basis of dinosaur ichnotaxa.

For example, Lockley (1997) suggested that the presence of the same theropod ichnotaxa in Eurasia and North America indicates the presence of land connections between these landmasses throughout the Jurassic Period. Similarly, footprints similar to the classic Early Jurassic Connecticut Valley ornithischian ichnogenus *Anomoepus* (Olsen and Rainforth 2007) also occur in Late Jurassic rocks of northern Spain (Lockley et al. 2008). We do not question the idea that it was possible for terrestrial vertebrates to move relatively easily from one landmass to another at that time. However, the presence of the same ichnogenus on two or more continents will provide convincing evidence for intercontinental connections only if footprints from the different regions are placed in the same ichnogenus because similarities in foot shape paralleled shared derived characters that would indicate membership of the trackmakers from the different regions in a monophyletic clade. In the present examples, we would have to be confident that the theropods and small ornithischians in North America responsible for the relevant ichnogenera were more closely related to the

Eurasian makers of the same ichnogenera than either group was to other kinds of large theropods or small ornithischians.

Whether footprint shapes can yield this much taxonomic resolution is arguable. As we have already seen, closely related birds and dinosaurs do not necessarily have similar foot shapes, while distantly related bird and dinosaur taxa can have foot shapes hard to tell apart. On a brighter note, though, if footprint assemblages from different regions share many of the same ichnotaxa (cf. Olsen and Galton 1984), the chances that all of these similarities are misleading should be greatly reduced.

Regardless of just how much useful taxonomic information can be extracted from well-preserved footprints, we suspect that many – perhaps even most – dinosaur footprints do not preserve enough information about the skeletal foot structures of their makers to warrant their being given names. We doubt, for example, that many of the tridactyl dinosaur ichnotaxa that have been named over the years are really different enough from each other to be very useful proxies for the kinds of animals that made them, apart from such generalized identifications as "theropod footprint" or even "bipedal dinosaur print." We think it best to err on the side of caution. Giving a formal name to a vertebrate trace fossil creates the impression that there is something anatomically distinctive about that fossil. We advocate creating such names only in those (rare?) instances where this really is true.

Footprints and Dinosaur Locomotion

Much of our understanding about what dinosaurs were like as living animals is based on interpretations of how dinosaur skeletons function as living machines (see Alexander 1997; Henderson, Chapter 32 of this volume). One aspect of such interpretations has to do with how dinosaurs walked and ran. Reconstructions are made of the kinds of movements possible at joints between bones (based on the shapes of the surfaces of those bones). The postures of living dinosaurs are inferred from the way that limb bones articulate with each other and with the shoulder or hip girdles. Interpretations of how strenuous the activities of dinosaurs were can be made from calculations of the mechanical strengths of the bones involved.

As useful as such functional morphological studies are, it is nice to have an independent source of information for testing hypotheses about dinosaur biomechanics. Dinosaur trackways provide such an independent line of evidence.

In the early 1900s, a bitter controversy arose over the body carriage of sauropod dinosaurs (Desmond 1976). For some time, American paleontologists, following the lead of O. C. Marsh, had reconstructed the limbs of sauropods as held in an erect, elephant-like fashion. When casts of the skeleton and models of the life appearance of *Diplodocus* were prepared by the Carnegie Museum of Natural History in Pittsburgh at the behest of the museum's patron, Andrew Carnegie, and distributed to museums around the world, several paleontologists objected to the erect posture given the dinosaur. It was argued that *Diplodocus* instead was more likely to have adopted a sprawling carriage, with its belly close to the ground – as living crocodiles were said to do.

This anatomical revisionism prompted vigorous counterattacks from W. J. Holland of the Carnegie Museum, among others. It was noted that giving *Diplodocus* and other sauropods a sprawling posture would have required major dislocations of joints between limb bones, and would have positioned the deep (from top to bottom) body in such a way that a sprawling sauropod would have gouged a trench in the ground with its belly as it moved!

The case for erect sauropods was further strengthened by the discovery of a nearly complete skeleton of a juvenile *Camarasaurus*. "The articulated right hind limb furnishes indisputable evidence in favor of those who have supported the view that these animals walked in an upright quadrupedal attitude, and it should quiet for all time those who advocate a crawling, lizard-like posture for the sauropod dinosaurs" (Gilmore 1925, 349).

Support for Gilmore's view came from the discovery of the sauropod trackways by R. T. Bird of the American Museum of Natural History (Fig. 34.17), in rocks of the Glen Rose Formation exposed in the bed of the Paluxy River in north-central Texas. The distance between left and right footprints in these trackways was much too narrow for the trackmakers to have had anything other than an erect posture, with their limbs directly beneath their bodies. The trackway evidence thus provided the clinching evidence in a controversy that had previously been argued entirely on the basis of skeletal anatomy.

Although sauropod trackways settled one controversy about locomotion of the big plant eaters, they created another. Bird (1944, 1985) described another Glen Rose Formation sauropod trackway, this one from south-central Texas, that was characterized by the absence of hind-foot tracks over much of its length. Thinking it unlikely that sauropods were capable of doing handstands while walking, Bird reasonably concluded that his trackway had been made by a "swimming" sauropod that was pulling itself along by its forelimbs, while its hindquarters were floating along behind. "Manus-dominated" and "manus-only" sauropod trackways subsequently turned up at other dinosaur tracksites around the world (e.g., Pascual Arribas and Sanz Perez (2000); Day et al. 2004; Vila et al. 2005; Ishigaki and Matsumoto 2009; Santos et al. 2009).

Lockley and Rice (1990), however, proposed an alternative hypothesis for trackways of this kind. They suggested that the absence or near absence of hind-foot (pes) prints was an artifact generated by observation of sauropod trackways consisting of undertracks. If the proportion of the dinosaur's weight pushing down on the forefoot (manus), relative to the area of the manus, was less than the proportion of the dinosaur's weight pushing down on the pes, relative to the area of the pes, then there might be deeper downward deformation of underlying sediment layers beneath impressions of the manus than of the pes. Consequently sauropod trackways preserved as undertracks might only record manus tracks, or if pes tracks did occur, they might be fainter or otherwise more poorly preserved than manus prints.

Vila et al. (2005) described a sauropod tracksite in the Pyrenees Mountains of the Catalonian region of Spain that had both "regular" (containing both manus and pes prints) and manus-only trackways. Had their manus-only trackway reflected unusual behavior (such as swimming) on the part of the sauropod, the authors reasoned, this should be reflected in differences in manus print spacing between the manus-only and a regular trackway. Instead, the spacing of manus prints in the manus-only trail was similar to that of the regular sauropod trackway. Vila et al. (2005) reasonably concluded from this that the absence of pes prints in their manus-only trackway is probably due to preservation of the footprints as undertracks.

Whether this conclusion can be generalized to all manus-only sauropod trackways remains uncertain. "True" (not undertrack) manus and pes prints in sauropod trackways from the Glen Rose Formation of Texas are usually impressed to about the same depth, or, if there is a difference in depth, the pes prints are a bit deeper (Fig. 34.17C). This is just the opposite of what the undertrack hypothesis for manus-only sauropod trackways calls for.

Henderson (2004) modeled the buoyancy of different kinds of sauropod dinosaurs, and showed that forms with relatively long forelimbs, like *Brachiosaurus* and *Camarasaurus*, would likely have floated with their forelimbs hanging down farther than the rest of their bodies. Consequently it would have been possible for such sauropods to have pulled themselves along the bottom using their front feet, exactly as Bird hypothesized. So maybe some manus-only sauropod trackways really were made by "swimming"—or better, wading (Wilson and Fisher 2003)—dinosaurs after all!

Trackways have figured in another recent controversy about dinosaur locomotion, this one involving the forelimb carriage of ceratopsid dinosaurs. Most paleontologists of the early twentieth century reconstructed *Triceratops* and its kin with fully erect hind limbs, but with forelimbs held in a sprawling fashion (Dodson 1996; Dodson and Farlow 1997). This interpretation was challenged by R. T. Bakker (1986 and references therein), who argued that ceratopsids moved in a rhinoceros-like fashion, their forelimbs held just as vertically as their hind limbs.

The discovery of what were probably ceratopsid trackways (named *Ceratopsipes* by Lockley and Hunt [1995a]) added an ichnological dimension to the controversy. Although the width across left and right forefoot (manus) impressions was greater than the width across left and right hind-foot (pes) prints, Lockley and Hunt concluded that this was no more than would be expected in other quadrupedal dinosaurs that had an erect forelimb carriage. Lockley and Hunt concluded that the *Ceratopsipes*-maker held its forelimbs in either a semierect or (more likely, in their opinion) erect fashion. Paul (1991) likewise argued that the *Ceratopsipes* trackway pattern was most consistent with an erect forelimb carriage.

Although agreeing with Paul (1991) and Lockley and Hunt (1995a) that *Ceratopsipes* does not suggest a sprawling forelimb carriage in ceratopsids, Dodson and Farlow (1997) found it harder to say whether the forelimbs were held in an erect or semierect fashion. They noted problems in estimating the body size of the *Ceratopsipes*-maker, and in reconstructing the width across the ceratopsid shoulders as compared with the width across the hips, all of which would influence the interpretation of forelimb carriage from the trackway pattern, but ultimately suggested that a less than erect carriage was the best solution. Paul and Christiansen (2000) argued for a more nearly, but not completely, erect carriage for ceratopsid forelimbs. Thompson and Holmes (2007) used a specimen of the ceratopsid *Chasmosaurus* with a complete articulated forelimb to create a computer model of the dinosaur's forelimb step cycle. They concluded that the movement of the ceratopsid forelimb had no exact modern equivalent. It was best described as semierect, with an unusual rotation of the upper arm that would have put pressure on the inner side of the forefoot as it pushed the animal forward. Rega et al. (2010) described pathologies of the finger bones that were consistent with the stresses imposed by such motion.

In primitive land vertebrates, the humeri and femora project outward from the body, and there is a sharp downward bend of the limb at the elbow or knee (Bakker 1971; Charig 1972). This puts footprints of the left and right side well apart from each other. As already described, an indirect way (but not the only way; see Leonardi 1987) of describing this in quantitative terms is by the pace angulation (also called the "step angle") of a trackway (Fig. 34.3). Trackways of sprawling amphibians and reptiles often have pace angulations less than 100 degrees (Peabody 1948; Haubold 1984; Padian and Olsen 1984a; Lucas and Heckert 1995; Kubo and Ozaki 2009).

Crocodilians walking on land generally do not use a sprawling lizard-like gait, but rather move in what is called a "high walk," with the limbs held in a more nearly vertical, semierect fashion (Cott 1961; Bakker 1971; Charig 1972; Webb and Manolis 1989; Parrish 1997; Carpenter 2009; Kubo and Ozaki 2009; Farlow and Elsey 2010; Kubo 2010; Milàn and Hedegaard 2010). Ironically, crocodilians do not usually walk on land with the sprawling gait that was attributed to them by advocates of sprawling sauropods! Dinosaurs and certain other archosaurs had a fully erect gait, at least for the hind limb (Bakker 1971; Charig 1972). Trackways of semierect and fully erect quadrupedal reptiles generally have pace angulations of at least 100 degrees (Peabody 1948; Demathieu 1970; Padian and Olsen 1984b; Kubo and Ozaki 2009), sometimes reaching 180 degrees; trackways of quadrupedal dinosaurs have pace angulations between 100 and 120 degrees (Farlow et al. 1989; Thulborn 1990).

Of course, the pace angulation of a trackway depends on more than how the trackmaker holds its forearms and thighs during walking. A wide-bodied, short-legged quadruped will make a trail with a lower pace angulation than a narrow-bodied, long-legged quadruped will, even if both of them have an erect stance. The trackways of two erect animals with legs of equal length may show different pace angulations if the legs of one are directed straight downward from the shoulder and hip, but the legs of the other animal slant slightly inward, as well as downward, from the shoulder and hip. Finally, the pace angulation of the trail of a running animal can be greater than that of the same animal moving more slowly.

Trackways of bipedal dinosaurs have higher pace angulations (sometimes reaching 180 degrees, with tracks of the left and right foot falling on a straight line) than do the trackways of quadrupedal dinosaurs (Farlow 1987; Thulborn 1990; Mossman et al. 2003). A quadruped has four feet supporting its body, so the animal has a broad base of support as it walks. It isn't very likely to tip over sideways, or to fall forward or backward. Not so a biped. Consider a typical theropod, with its head and body on the one side and its tail on the other, both balanced over the beast's hips (Fig. 34.19). With only two feet constituting its base of support, our theropod constantly risks falling on its side, face, or rump (or arse, if one prefers).

Of course, if its body were wide compared with the length of its legs, with its left and right feet far apart, our theropod would be fairly stable from side to side. However, the dinosaur's head, torso, and tail would extend well

34.19. Restoration of the huge theropod *Tyrannosaurus* in its normal walking pose, as suggested by the typical pattern of theropod dinosaur trackways. The dinosaur puts one foot down nearly in front of the other, and the tail is carried well off the ground.

Drawing by Jim Whitcraft, based on a model by Matt Smith.

to the front and rear of the creature's center of balance; even standing still would involve a rather precarious balancing act to keep from tilting forward or backward. If that weren't difficult enough, with every step the animal would diminish its stability of support in the very direction in which it was already risking a fall.

On the other hand, if – like real theropods – our biped is relatively narrow-bodied and long-legged, its long legs will act to check falls of its body in the forward direction; the animal will be as stable as is possible during forward motion. It will now be unstable in the sideways direction, but it partly mitigates this risk by having a narrow body, so that its weight remains directly in line above its feet. It can also avoid moving sideways any more than is absolutely necessary! Substrate conditions may also affect the way that the animal places its feet on the ground (Mossman et al. 2003). All of these considerations are reflected in the way theropods and other bipedally walking dinosaurs moved, with the result that they usually made rather narrow trackways, with high pace angulations. See Henderson (Chapter 32 of this volume) for a detailed discussion of the biomechanics of locomotion in dinosaurs.

Countless old monster movies to the contrary notwithstanding, dinosaurs usually did not walk with their tails dragging on the ground, as lizards and crocodilians often do. Trackways of both bipedal and quadrupedal dinosaurs seldom show tail drag marks (Hitchcock's *Gigantitherium* [Figure 34.7] is a notable exception). As one would infer from skeletal functional morphology (Molnar and Farlow 1990), trails confirm that theropods moved like animated seesaws, the weight of their tails counterbalancing that of their heads and bodies over their hips (Fig. 34.19). Quadrupedal dinosaurs did not need their tails for balance, but trackways suggest that even long-tailed quadrupedals (like some sauropods) rarely allowed their tails to drag passively across the ground.

From observations of the movements of a variety of living animals (humans, horses, jirds, elephants, and ostriches), the British zoologist R. McNeil Alexander (1976) devised a formula that analyzed the length of an animal's legs (its hip height), its stride length, and its speed of movement in terms of a dimensionless parameter known as the Froude number (Alexander 1997). Alexander's formula could be recast in a form from which an animal's speed could be estimated from its trackway:

$$\text{speed (meters/second)} = 0.25 \times \text{gravitational constant}^{0.5} \times \text{stride length}^{1.67} \times \text{hip height}^{-1.17}$$

To estimate the speed of a dinosaur from its trackway, one first measures the stride directly from the trail. The hip height obviously cannot be obtained from a trackway. However, Alexander noted that, based on skeletal reconstructions, the hip height of dinosaurs was roughly four times the length of the portion of the foot that touched the ground; one could therefore estimate a trackmaker's hip height from the length of its footprint. Thulborn (1990) went on to provide more elaborate equations for estimating hip heights from footmark lengths for individual dinosaur groups, but Henderson (2003) argued that in many cases Alexander's simpler calculation worked as well or better for estimating hip height.

Many paleontologists have applied Alexander's equation to estimating dinosaur speeds (Thulborn 1990). Unsurprisingly, most speed estimates suggest that the dinosaur trackmakers were walking, at speeds between 2 and 10 km per hour. Like living animals, dinosaurs probably didn't go tearing across landscapes any more than they had to, but 5 to 10 km per hour will get you a fair distance if you keep at it long enough. It is therefore not unreasonable to speculate that some dinosaurs could have migrated over fairly long distances (Hotton 1980; Lockley 1997). Whether they actually did so is another matter (cf. Sampson and Loewen 2010 for ceratopsids).

Some trackways seem to record running on the part of their makers (Thulborn 1990; Irby 1996b). In such trails, the stride length is roughly three or more times the trackmaker's estimated hip height (Thulborn 1990). The most impressive collection of trackways of apparently running dinosaurs at any one place is the Lark Quarry, a middle Cretaceous site in western Queensland, Australia (Thulborn and Wade 1984; Romilio and Salisbury 2010). At this locality, numerous small bipedal dinosaurs dashed across a mudflat (Fig. 34.20). The estimated speeds of these little dinosaurs were about 12 to 16 km per hour, which may not seem that fast. However, because the trackmakers were small animals, this was probably about as fast as their short legs could carry them.

Some of the fastest speeds estimated for dinosaurs are based on trackways of medium-sized theropods from a Cretaceous site in Texas (Farlow 1981). Most of the trackmakers at this site were walking, but three of the theropods seem to have been running (Fig. 34.21), with estimated speeds of 30 to 40 km per hour. A comparable speed has been estimated for the

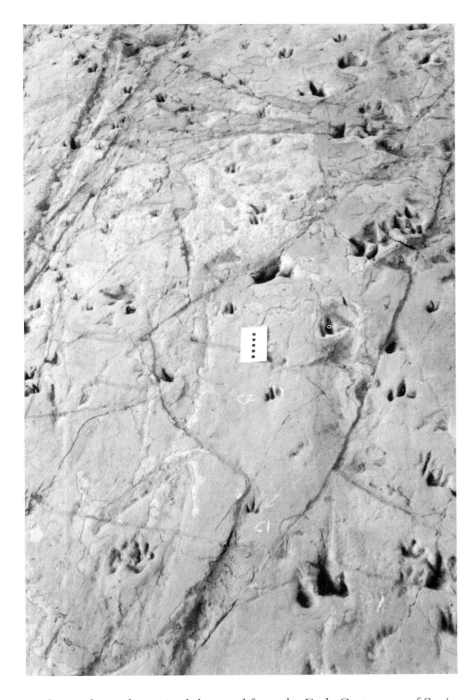

34.20. Footprints of small bipedal dinosaurs at the Lark Quarry (Winton Formation, Mid-Cretaceous), Queensland, Australia. The trackmakers were moving away from the viewer. The long stride lengths of the numerous little dinosaurs at this tracksite suggest that the trackmakers were running flat out.

trackway of a medium-sized theropod from the Early Cretaceous of Spain (Viera and Torres 1995).

To date, trackways that appear to have been made by running dinosaurs are limited to those made by bipedal forms of small to medium size. There are as yet no trackways that suggest rapid running on the part of large dinosaurs, quadrupeds or bipeds. Consequently, interpretations of the peak running abilities of big dinosaurs are presently restricted to speculation based on the functional morphology and limb strengths of dinosaur skeletons (Coombs 1978; Bakker 1986; Paul 1988; Alexander 1989, 1997; Thulborn 1990; Farlow et al. 1995; Hutchinson and Garcia 2002). Some paleontologists think that even large dinosaurs such as *Triceratops* and *Tyrannosaurus* were capable of speeds as fast as those estimated from the

34.21. Three footprints in a trackway of a running theropod dinosaur, F6 Ranch site, Kimble County, Texas. The middle footprint in this sequence (next to the meter stick) is shown in Figures 34.4C and D.

trackways of smaller dinosaurs at the above-mentioned tracksite in Texas (an interpretation that figured prominently in the movie *Jurassic Park*!). Other scientists doubt that large dinosaurs could move that quickly. As in the case of sauropod postures, trackway evidence could provide the telling evidence in this controversy. It would take only one trackway of a galloping *Triceratops* or a sprinting *Tyrannosaurus* to prove that these dinosaurs were capable of very fast locomotion. On the other hand, if—after enough tracksites are studied—no trails made by rapidly moving big dinosaurs ever turn up, it will be hard to reject the conclusion that large dinosaurs either were incapable of running, or did so very rarely.

Unfortunately, there is a caveat that must be added to considerations of dinosaur speed based on fossil trackways. Estimates of the hip height of dinosaurian trackmakers are usually calculated based on footprint lengths. But Manning (2008) showed that if measurements of print length are made on transmitted tracks rather than on the footprints actually made on the substrate surface, the measured print length may be so far off from the true footprint length as to cause speed estimates to be off by as much as tenfold. As with other aspects of dinosaur ichnology, considerable care must be taken in interpreting speed estimates based on trackways.

Extant flightless birds are able to swim, as the famous English naturalist Charles Darwin described for rheas (what he called ostriches) in South America:

> It is not generally known that ostriches readily take to the water. Mr. King informs me that at the Bay of San Blas, and at Port Valdes in Patagonia, he saw these birds swimming several times from island to island. They ran into the water both when driven down to a point, and likewise of their own account when not frightened; the distance crossed was about two hundred yards. When swimming, very little of their bodies appear above the water; their necks are extended a little forward, and their progress is slow. On two occasions I saw some ostriches swimming across the Santa Cruz river, where its course was about four hundred yards wide, and the stream rapid. (Darwin 1860, 91)

If big, flightless birds can swim, then why not bipedal, nonavian dinosaurs? Fossil trackways attributed to swimming dinosaurs have been described. However—as in the case of manus-only sauropod trackways—for some of them there is controversy over whether the prints in question really indicate swimming, as opposed to being artifacts of preservation (Coombs 1980; Farlow and Galton 2003; Ezquerra et al. 2007), and sometimes there is uncertainty over whether the prints were made by crocodiles as opposed to dinosaurs (Ezquerra et al. 2007).

The best candidates for trackways made by a swimming theropod dinosaur come from an Early Jurassic site in southern Utah, and an Early Cretaceous site in northern Spain (Milner et al. 2006; Ezquerra et al. 2007). The Spanish trackway consists of several large prints (as much as 60 cm long) composed of two or three roughly parallel grooves in the rock, sometimes with little mounds of former sediment at their posterior ends (Fig. 34.22). Left footprints are nearly parallel to the animal's direction of travel, while right footprints have their front ends turned sharply inward with respect to the direction of travel. Both the size of the prints, and their shape, suggest that the trackmaker is likely to have been a theropod.

The Jurassic swimming dinosaur tracks from Utah are equally spectacular, and similarly consist of three parallel grooves in tracksite surfaces. Prints of both large and small theropods that appear to have been swimming occur at this site, along with beautifully preserved footprints of normally walking theropods.

Several authors (e.g., Thulborn 1990; Lockley 1991, 1997, 2007; Farlow 1992; Schult and Farlow 1992; Lockley et al. 1994b, 2004; Dalla Vecchia et al. 2000, 2001; Kvale et al. 2001; Huh et al. 2003; Wright 2005; Farlow et al. 2006; Hunt and Lucas 2007; Marty 2008; Petti et al. 2008; Moratalla et al. 2009) have discussed the paleoenvironmental contexts in which dinosaur footprints were preserved, and the use of footprint assemblages in reconstructing dinosaur faunas. Certain associations of dinosaur footprint types with each other and with particular sedimentary settings (such as sauropod and theropod footprints with shallow-water carbonate depositional environments; Lockley 1991, 2007; Lockley et al. 1994b; Dalla Vecchia et al. 2000, 2001; Moreno and Pino 2002; Marty et al. 2003; Hunt and Lucas 2007; Petti et al. 2008) are recurrent features of the dinosaur footprint record. Controversy exists over the extent to which such associations between dinosaur footprint types and sedimentary facies are truly indicative of habitat preferences, as opposed to being artifacts of where and how footprints tend

Footprints and Dinosaur Paleoecology

34.22. Trackway attributed to a swimming theropod dinosaur, La Virgen del Campo tracksite, Early Cretaceous, La Rioja Province, Spain. (A) Oblique view of the trackway; the animal was moving away from the viewer. Two meters length of tape are exposed. (B) Overhead view of the right footprint in the foreground of panel A. The scale bar is marked off in centimeters.

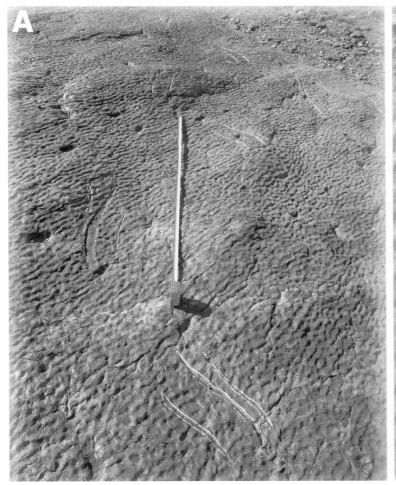 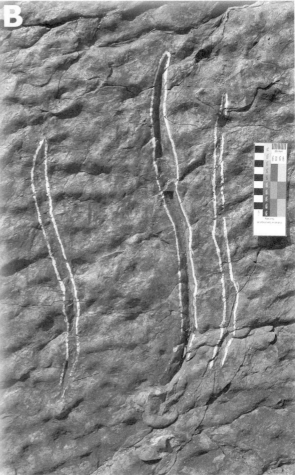

to be preserved (Lockley 1991; Farlow 1992; Lockley et al. 1994b; Meyer and Pittman 1994; McIntosh et al. 1997; Wright 2005), how the associations should be named and classified (Hunt and Lucas 2007; Lockley 2007), and the extent to which habitat preferences deduced from the ichnological record match habitat preferences suggested by the skeletal record (cf. Farlow 1992; Schult and Farlow 1992; Wright 1995; Mannion and Upchurch 2010).

Because of the large body sizes of many dinosaurian trackmakers, as well as the predatory nature of typical theropods, some authors have argued that it is unlikely that big dinosaurs were restricted to the shoreline habitats in which their tracks are commonly preserved (Farlow 1992, 2001; Meyer and Pittman 1994; cf. Dalla Vecchia et al. 2000). In the case of large theropod and sauropod trackmakers of the Glen Rose Formation, skeletal fossils of the putative trackmakers are known from more inland clastic facies (Stovall and Langston 1950; Harris 1998; Currie and Carpenter 2000; Rose 2007). On the other hand, the skeletal record of at least some large dinosaurs (e.g., ceratopsids; Eberth 2010) suggests a surprising degree of habitat specificity for such big animals, and so the track record may indeed preserve more information about the habitats frequented by dinosaurs than one would guess on the basis of what is known about the ecology of modern large mammals.

Dinosaur tracks obviously serve as proxies for their makers. Consequently the kinds of dinosaur trace fossils in footprint assemblages provide a picture of the composition of dinosaur paleocommunities that complements that provided by skeletal assemblages (Lockley 1991; Schult and Farlow 1992; Lockley et al. 1995b; Lockley et al. 2008), even if, as already discussed, the portrait painted by footprints lacks the taxonomic resolution provided by skeletal assemblages. Furthermore, because a given animal could conceivably make a huge number of footprints over its lifetime, but donate only a single skeleton to the fossil record, rare constituents of dinosaur faunas may be more likely to be recorded by footprints than by bones.

While tracks may potentially provide a qualitatively more complete record of the kinds of dinosaurs that lived in an area than the skeletal fauna does, the skeletal record may yield a quantitatively more accurate census of the members of a dinosaur community than the track record, because footprints record the activities of animals, and not the animals themselves. It is not unusual for dinosaur track assemblages to be dominated by footprints of large theropods (cf. Leonardi 1989; Schult and Farlow 1992), while in most dinosaur skeletal faunas there are considerably more specimens of large herbivores than of big meat eaters, as one would expect in a typical large-vertebrate community. The excess of large carnivore trackways may reflect greater activity on the part of the great hunters, who presumably had to range farther to find a meal than did big plant eaters.

References

Alexander, R. McN. 1976. Estimates of speeds of dinosaurs. *Nature* 261: 129–130.
——. 1989. *Dynamics of Dinosaurs and Other Extinct Giants*. New York: Columbia University Press.
——. 1997. Engineering a dinosaur. In J. O. Farlow and M. K. Brett-Surman (eds.), *The Complete Dinosaur*, 414–425. Bloomington: Indiana University Press.
Allen, J. R. L. 1989. Fossil vertebrate tracks and indenter mechanics. *Journal of the Geological Society of London* 146: 600–602.

——. 1997. Subfossil mammalian tracks (Flandrian) in the Severn Estuary, S. W. Britain: mechanics of formation, preservation and distribution. *Philosophical Transactions of the Royal Society London* B 352: 481–518.

Andersen, A., R. E. Chapman, J. Dickman, and K. Hand. 2001. Using rapid prototyping technology in vertebrate paleontology. *Journal of Vertebrate Paleontology* 21 (suppl. to 3): 28A.

Anderson, A. 1989. *Prodigious Birds: Moas and Moa-Hunting in Prehistoric New Zealand.* Cambridge: Cambridge University Press.

Anonymous. 1867. Untitled report of Professor Cope's "account of the extinct reptiles which approached the birds." *Proceedings of the Academy of Natural Sciences of Philadelphia* 1867: 234–235.

Arakawa, Y., Y. Azuma, A. Kano, T. Tanijiri, and T. Miyamoto. 2002. A new technique to illustrate and analyze dinosaur and bird footprints using 3-D digitizer. *Memoir of the Fukui Prefectural Dinosaur Museum* 1: 7–18.

Arcucci, A. B., C. A. Forster, F. Abdala, C. L. May, and C. A. Marsicano. 1995. "Theropod" tracks from the Los Rastros Formation (Middle Triassic), La Rioja Province, Argentina. *Journal of Vertebrate Paleontology* 15 (suppl. to 3): 16A.

Avanzini, M. 1998. Anatomy of a footprint: bioturbation as a key to understanding dinosaur walking dynamics. *Ichnos* 6: 129–139.

Azuma, Y., Y. Arakawa, Y. Tomida, and P. J. Currie. 2002. Early Cretaceous bird tracks from the Tetori Group, Fukui Prefecture, Japan. *Memoir of the Fukui Prefectural Dinosaur Museum* 1: 1–6.

Baird, D. 1957. Triassic reptile footprint faunules from Milford, New Jersey. *Bulletin of the Museum of Comparative Zoology* (Harvard University) 117: 449–520.

Bakker, R. T. 1971. Dinosaur physiology and the origin of mammals. *Evolution* 25: 636–658.

——. 1975. Experimental and fossil evidence for the evolution of tetrapod bioenergetics. In D. M. Gates and R. B. Schmerl (eds.), *Perspectives in Biophysical Ecology*, 365–399. Berlin: Springer Verlag.

——. 1986. *The Dinosaur Heresies: New Theories Unlocking the Mystery of the Dinosaurs and Their Extinction.* New York: William Morrow.

Barsbold, R., and T. Maryańska. 1990. Segnosauria. In D. B. Weishampel, P. Dodson, and H. Osmólska (eds.), *The Dinosauria*, 408–415. Berkeley: University of California Press.

Barsbold, R., T. Maryańska, and H. Osmólska. 1990. Oviraptorosauria. In D. B. Weishampel, P. Dodson, and H. Osmólska (eds.), *The Dinosauria*, 249–258. Berkeley: University of California Press.

Barsbold, R., and H. Osmólska. 1990. Ornithomimosauria. In D. B. Weishampel, P. Dodson, and H. Osmólska (eds.), *The Dinosauria*, 225–244. Berkeley: University of California Press.

Basan, P. B. 1979. Trace fossil nomenclature: the developing picture. *Palaeogeography, Palaeoclimatology, Palaeoecology* 28: 143–167.

Bates, K. T., P. L. Falkingham, D. Hodgetts, J. O. Farlow, B. H. Breithaupt, M. O'Brien, N. Matthews, W. I. Sellers, and P. L. Manning. 2009. Digital imaging and public engagement in paleontology. *Geology Today* 25: 134–139.

Bates, K. T., F. Rarity, P. L. Manning, D. Hodgetts, B. Vila, O. Oms, Á. Galobart, and R. L. Gawthorpe. 2008. High-resolution LiDAR and photogrammetric survey of the Fumanya dinosaur tracksites (Catalonia): implications for the conservation and interpretation of geological heritage sites. *Journal of the Geological Society* 165: 115–127.

Belvedere, M., and P. Mietto. 2010. First evidence of stegosaurian *Deltapodus* footprints in North Africa (Iouaridène Formation, Upper Jurassic, Morocco). *Paleontology* 53: 233–240.

Bimber, O., S. M. Gatesy, L. M. Witmer, R. Raskar, and L. M. Encarnacao. 2002. Merging fossil specimens with computer-generated information. *IEEE Computer* 35 (9): 25–30.

Bird, R. T. 1944. Did *Brontosaurus* ever walk on Land? *Natural History* 53: 61–67.

——. 1985. *Bones for Barnum Brown: Adventures of a Dinosaur Hunter.* Fort Worth: Texas Christian University Press.

Breithaupt, B. H., N. A. Matthews, and T. A. Noble. 2004. An integrated approach to three-dimensional data collection at dinosaur tracksites in the Rocky Mountain West. *Ichnos* 11: 11–26.

Breithaupt, B., E. H. Southwell, T. L. Adams, and N. A. Matthews. 2001. Innovative documentation methodologies in the study of the most extensive dinosaur tracksite in Wyoming. In V. L. Santucci and L. McClelland (eds.), *Proceedings of the 6th Fossil Resources Conference*, 113–122. U.S. Department of the Interior, National Parks Service, Geological Resources Division, Technical Report NPS/NRGRD/GRDTR-01/01.

Brown, B. 1916. *Corythosaurus casuarius:* skeleton, musculature, and epidermis. *Bulletin of the American Museum of Natural History* 35: 709–716.

Brusatte, S. L., G. Niedźwiedzki, and R. J. Butler. 2010. Footprints pull origin and diversification of dinosaur stem lineage deep into Early Triassic. *Proceedings of the Royal Society* B. doi:10.1098/rspb.2010.1746.

Burns, M. 1993. *Automated Fabrication: Improving Productivity in Manufacturing.* Englewood Cliffs, N.J.: Prentice-Hall.

Carpenter, K. 2009. Role of lateral body bending in crocodylian track making. *Ichnos* 16: 202–207.

Carrano, M. T., and J. A. Wilson. 2001. Taxon distributions and the tetrapod track record. *Paleobiology* 27: 564–582.

Casanovas Cladellas, M. L., R. Ezquerra Miguel, A. Fernández Ortega, F. Pérez-Lorente, J. V. Santafé Llopis, and F. Torcida Fernández. 1993. Icnitas de dinosaurios. Yacimientos de Navalsaz, Las Mortajeras, Peñaportillo, Malvaciervo y la Era del Peladillo 2 (La Rioja, España). *Zubía* 5: 9–133.

Chapman, R. E. 1997. Technology and the study of dinosaurs. In J. O. Farlow and M. K. Brett-Surman (eds.), *The Complete Dinosaur*, 112–135. Bloomington: Indiana University Press.

Chapman, R. E., A. Andersen, B. Breithaupt, and N. Matthews. 2012 (this volume). Technology and the study of dinosaurs. In M. K. Brett-Surman, T. H. Holtz, and J. O. Farlow (eds.), *The Complete Dinosaur*, 2nd ed., 246–272. Bloomington: Indiana University Press.

Chapman, R. E., P. M. Galton, W. P. Wall, and J. J. Sepkoski Jr. 1981. A morphometric study of the cranium of the pachycephalosaurid dinosaur *Stegoceras*. *Journal of Paleontology* 55 (3): 608–618.

Charig, A. J. 1972. The evolution of the archosaur pelvis and hindlimb: an explanation in functional terms. In K. A. Joysey and T. S. Kemp (eds.), *Studies in Vertebrate Evolution*, 121–155. New York: Winchester Press.

Chin, K. 1997. What did dinosaurs eat? Coprolites and other direct evidence of dinosaur diets. In J. O. Farlow and M. K. Brett-Surman (eds.), *The Complete Dinosaur*, 371–382. Bloomington: Indiana University Press.

Cobos, A., R. Royo-Torres, L. Luque, L. Alcalá, and L. Mampel. 2010. An Iberian stegosaurs' paradise: the Villar del Arzopispo Formation (Tithonian-Berriasian) in Teruel (Spain). *Palaeogeography, Palaeoclimatology, Palaeoecology* 293: 223–236.

Colbert, E. H. 1981. *A Primitive Ornithischian Dinosaur from the Kayenta Formation of Arizona*. Bulletin 53. Flagstaff: Museum of Northern Arizona Press.

———. 1989. *The Triassic Dinosaur Coelophysis*. Bulletin 57. Flagstaff: Museum of Northern Arizona Press.

———. 1997. North American dinosaur hunters. In J. O. Farlow and M. K. Brett-Surman (eds.), *The Complete Dinosaur*, 24–33. Bloomington: Indiana University Press.

Coombs, W. P., Jr. 1978. Theoretical aspects of cursorial adaptations in dinosaurs. *Quarterly Review of Biology* 53: 393–418.

———. 1980. Swimming ability of carnivorous dinosaurs. *Science* 207: 1198–1200.

Cooper, A., C. Lalueza-Fox, S. Anderson, A. Rambaut, J. Austin, and R. Ward. 2001. Complete mitochondrial genome sequences of two extinct moas clarify ratite evolution. *Nature* 409: 704–707.

Cooper, A., C. Mourer-Chauviré, G. K. Chambers, A. von Haeseler, A. C. Wilson, and S. Pääbo. 1992. Independent origins of the New Zealand moas and kiwis. *Proceedings of the National Academy of Sciences U.S.A.* 89: 8741–8744.

Cooper, A., and D. Penny. 1997. Mass survival of birds across the Cretaceous-Tertiary boundary: molecular evidence. *Science* 275: 1109–1113.

Cott, H. B. 1961. Scientific results of an inquiry into the ecology and economic status of the Nile crocodile (*Crocodilus niloticus*) in Uganda and northern Rhodesia. *Transactions of the Zoological Society of London* 29: 211–357.

Currie, P. J. 1983. Hadrosaur trackways from the Lower Cretaceous of Canada. *Acta Palaeontologica Polonica* 28: 63–73.

———. 1990. Elmisauridae. In D. B. Weishampel, P. Dodson, and H. Osmólska (eds.), *The Dinosauria*, 245–248. Berkeley: University of California Press.

Currie, P. J., and K. Carpenter. 2000. A new specimen of *Acrocanthosaurus atokensis* (Theropoda, Dinosauria) from the Lower Cretaceous Antlers Formation (Lower Cretaceous, Aptian) of Oklahoma, USA. *Geodiversitas* 22: 207–246.

Dalla Vecchia, F. M., A. Tarlao, G. Tunis, and S. Venturini. 2000. New dinosaur track sites in the Albian (Early Cretaceous) of the Istrian Peninsula (Croatia). *Memorie di Scienze Geologiche* 52: 193–292.

Dalla Vecchia, F. M., G. Tunis, S. Venturini, and A. Tarlao. 2001. Dinosaur track sites in the upper Cenomanian (Late Cretaceous) of Istrian Peninsula (Croatia). *Bolletino della Società Italiana* 40: 25–53.

Darwin, C. 1860. *The Voyage of the Beagle*. New York: Doubleday, 1962.

Day, J. J., D. B. Norman, A. S. Gale, P. Upchurch, and H. P. Powell. 2004. A Middle Jurassic dinosaur trackway site from Oxfordshire, UK. *Palaeontology* 47: 319–348.

Dean, D. R. 1969. Hitchcock's dinosaur tracks. *American Quarterly* 21: 639–644.

Deibert, J. E., and B. H. Breithaupt. 2006. Ferdinand Hayden's 1868 "huge bird tracks" in the Upper Cretaceous Almond Formation: field evidence for the first dinosaur fossil discovered in Wyoming. In S. G. Lucas and R. M. Sullivan (eds.), *Late Cretaceous Vertebrates from the Western Interior*, 69–78. Bulletin 35. Albuquerque: New Mexico Museum of Natural History and Science.

Demathieu, G. R. 1970. *Les Empreintes de Pas de Vertébrés du Trias de la Bordure Nord-Est du Massif Central*. Paris: Centre National de la Recherche Scientifique.

———. 1989. Appearance of the first dinosaur tracks in the French Middle Triassic and their probable significance. In D. D. Gillette and M. G. Lockley (eds.), *Dinosaur Tracks and Traces*, 201–207. Cambridge: Cambridge University Press.

———. 1990. Problems in discrimination of tridactyl dinosaur footprints, exemplified by the Hettangian trackways, the Causses, France. *Ichnos* 1: 97–110.

Desmond, A. 1976. *The Hot-Blooded Dinosaurs: A Revolution in Palaeontology*. New York: Dial Press.

Dodson, P. 1996. *The Horned Dinosaurs*. Princeton, N.J.: Princeton University Press.

Dodson, P., and J. O. Farlow. 1997. The forelimb carriage of ceratopsid dinosaurs. In D. L Wolberg, E. Stump, and G. D. Rosenberg (eds.), *Dinofest International*, 393–398. Philadelphia, Pa..: Academy of Natural Sciences.

Eberth, D. A. 2010. A review of ceratopsian paleoenvironmental associations and taphonomy. In M. J. Ryan, B. J. Chinnery-Allgeier, and D. A. Eberth (eds.), *New Perspectives on Horned Dinosaurs*, 428–446. Bloomington: Indiana University Press.

Ericson, P. G. P. P., C. L. Anderson, T. Britton, A. Elzanowski, U. S. Johansson, M. Källersjö, J. I. Ohlson, T. J. Parsons, D. Zuccon, and G. Mayr. 2006. Diversification of Neoaves: integration of molecular sequence data and fossils. *Biology Letters* 2: 543–547.

Ezquerra, R., S. Doublet, L. Costeur, P. M. Galton, and F. Pérez-Lorente. 2007. Were non-avian theropod dinosaurs able to swim? Supportive evidence from an Early Cretaceous trackway, Cameros Basin (La Rioja, Spain). *Geology* 35: 507–510.

Farlow, J. O. 1981. Estimates of dinosaur speeds from a new trackway site in Texas. *Nature* 294: 747–748.

———. 1987. *Lower Cretaceous Dinosaur Tracks, Paluxy River Valley, Texas*. Waco, Tex.: South Central Section, Geological Society of America and Baylor University.

———. 1992. Sauropod tracks and trackmakers: integrating the ichnological and skeletal records. *Zubía* 10: 89–138.

———. 1993. *The Dinosaurs of Dinosaur Valley State Park*. Austin: Texas Parks and Wildlife Press.

———. 2001. *Acrocanthosaurus* and the maker of Comanchean large-theropod footprints. In D. H. Tanke and K. Carpenter (eds.), *Mesozoic Vertebrate Life*, 409–427. Bloomington: Indiana University Press.

Farlow, J. O., and R. M. Elsey. 2010. Footprints and trackways of the American alligator, Rockefeller Wildlife Refuge, Louisiana. In J. Milàn, S. G. Lucas, M. G. Lockley, and J. A. Spielman (eds.), *Crocodyle Tracks and Traces*, 31–39. Bulletin 51. Albuquerque: New Mexico Museum of Natural History and Science.

Farlow, J. O., and P. M. Galton. 2003. Dinosaur trackways of Dinosaur State Park, Rocky Hill, Connecticut. In P. M. Letourneau and P. E. Olsen (eds.), *The Great Rift Valleys of Pangea: Sedimentology, Stratigraphy, and Paleontology*, 2: 248–272. New York: Columbia University Press.

Farlow, J. O., S. M. Gatesy, T. R. Holtz, and J. M. Robinson. 2000. Theropod locomotion. *American Zoologist* 40: 640–663.

Farlow, J. O., W. Langston Jr., E. E. Deschner, R. Solis, W. Ward, B. L. Kirkland, S. Hovorka, T. L. Reece, and J. Whitcraft. 2006. *Texas Giants: Dinosaurs of the Heritage Museum of the Texas Hill Country*. Canyon Lake: Heritage Museum of the Texas Hill Country.

Farlow, J. O., and M. G. Lockley. 1993. An osteometric approach to the identification of the makers of early Mesozoic tridactyl dinosaur footprints. In S. G. Lucas and M. Morales (eds.), *The Nonmarine Triassic*, 123–131. Bulletin 3. Albuquerque: New Mexico Museum of Natural History and Science.

Farlow, J. O., J. G. Pittman, and J. M. Hawthorne. 1989. *Brontopodus birdi*, Lower Cretaceous sauropod footprints from the U.S. Gulf Coastal Plain. In D. D. Gillette and M. G. Lockley (eds.), *Dinosaur Tracks and Traces*, 371–394. Cambridge: Cambridge University Press.

Farlow, J. O., M. B. Smith, and J. M. Robinson. 1995. Body mass, bone "strength indicator," and cursorial potential of *Tyrannosaurus rex*. *Journal of Vertebrate Paleontology* 15: 713–725.

Feduccia, A. 1996. *The Origin and Evolution of Birds*. New Haven, Conn.: Yale University Press.

Forster, C. A. 1990. The postcranial skeleton of the ornithopod dinosaur *Tenontosaurus tilletti*. *Journal of Vertebrate Paleontology* 10: 273–294.

Galton, P. M. 1974. The ornithischian dinosaur *Hypsilophodon* from the Wealden of the Isle of Wight. *Bulletin of the British Museum of Natural History: Geology* 25: 1–152c.

———. 1990. Basal Sauropodomorpha-Prosauropoda. In D. B. Weishampel, P. Dodson, and H. Osmólska (eds.), *The Dinosauria*, 320–344. Berkeley: University of California Press.

Gatesy, S. M. 2003. Direct and indirect features: what sediment did a dinosaur touch? *Ichnos* 10: 91–98.

Gatesy, S. M., K. M. Middleton, F. A. Jenkins Jr., and N. H. Shubin. 1999. Three-dimensional preservation of foot movements in Triassic theropod dinosaurs. *Nature* 399: 141–144.

Gauthier, J. A., S. J. Nesbitt, E. R. Schachner, G. S. Bever, and W. G. Joyce. 2011. The bipedal stem crocodilian *Poposaurus gracilis*: inferring function in fossils and innovation in archosaurian locomotion. *Bulletin of the Peabody Museum of Natural History* 52: 107–126.

Gierlinski, G. 1988. New dinosaur ichnotaxa from the Early Jurassic of the Holy Cross Mountains, Poland. *Palaeogeography, Palaeoclimatology, Palaeoecology* 85: 137–148.

———. 1994. Early Jurassic theropod tracks with the metatarsal impressions. *Przeglad Geologiczny* 42: 280–284.

———. 1995. Thyreophoran affinity of *Otozoum* tracks. *Przeglad Geologiczny* 43: 123–125.

Gierlinski, G., and A. Ahlberg. 1994. Late Triassic and Early Jurassic dinosaur footprints in the Höganäs Formation of southern Sweden. *Ichnos* 3: 99–105.

Gilmore, C. W. 1909. Osteology of the Jurassic reptile *Camptosaurus*, with a revision of the species of the genus, and descriptions of two new species. *Proceedings of the U.S. National Museum* 36: 197–332.

———. 1925. A nearly complete articulated skeleton of *Camarasaurus*, a saurischian dinosaur from the Dinosaur National Monument, Utah. *Memoirs of the Carnegie Museum* 10: 347–384.

———. 1936. Osteology of *Apatosaurus*, with special reference to specimens in the Carnegie Museum. *Memoirs of the Carnegie Museum* 11: 175–298.

Glassman, S., E. A. Bolt Jr., and E. E. Spamer. 1993. Joseph Leidy and the "Great Inventory of Nature." *Proceedings of the Academy of Natural Sciences of Philadelphia* 144: 1–19.

Gonzáles Riga, B. J., and J. Orlando Calvo. 2009. A new wide-gauge sauropod track site from the Late Cretaceous of Mendoza, Neuquén Basin, Argentina. *Palaeontology* 52 (3): 631–640.

Goto, M., M. Araki, and S. Tomono. 1996. A method for representation of dinosaur footprints using 3–D photographs. *Bulletin of the Toyama Science Museum* 19: 1–7.

Graversen, O., J. Milàn, and D. B. Loope. 2007. Dinosaur tectonics: a structural analysis of theropod undertracks with a reconstruction of theropod walking dynamics. *Journal of Geology* 115: 641–654.

Gruber, J. W. 1987. From myth to reality: the case of the moa. *Archives of Natural History* 14: 339–352.

Guralnick, S. M. 1972. Geology and religion before Darwin: the case of Edward Hitchcock, theologian and geologist (1793–1864). *Isis* 63: 529–543.

Hackett, S. J., R. T. Kimball, S. Reddy, R. C. K. Bowie, E. L. Braun, M. J. Braun, J. L. Chojnowski, W. A. Cox, K.-L. Han, J. Harshman, C. J. Huddleston, B. D. Marks, K. J. Miglia, W. S. Moore, F. H. Sheldon, D. W. Steadman, C. C. Witt, and T. Yuri. 2008. A phylogenomic study of birds reveals their evolutionary history. *Science* 320: 1763–1767.

Haddrath, O., and A. J. Baker. 2001. Complete mitochondrial DNA genome sequences of extinct birds: ratite phylogenetics and the vicariance biogeography hypothesis. *Proceedings of the Royal Society of London B* 268: 939–945.

Harris, J. D. 1998. *A Reanalysis of Acrocanthosaurus atokensis, Its Phylogenetic Status, and Paleobiogeographic Status, Based on a New Specimen from Texas*. Bulletin 13. Albuquerque: New Mexico Museum of Natural History and Science.

Harris, J. D., K. R. Johnson, J. Hicks, and L. Tauxe. 1996. Four-toed theropod footprints and a paleomagnetic age from the Whetsone Falls Member of the Harbell Formation (Upper Cretaceous: Maastrichtian), northwestern Wyoming. *Cretaceous Research* 17: 381–401.

Haubold, H. 1984. *Saurierfährten*. Wittenberg Lutherstadt, Germany: A. Ziemsen.

Heilmann, G. 1927. *The Origin of Birds*.

New York: D. Appleton. Reprint, 1972, New York: Dover.

Henderson, D. M. 2003. Footprints, trackways, and hip heights of bipedal dinosaurs—testing hip height predictions with computer models. *Ichnos* 10: 99–114.

———. 2004. Tipsy punters: sauropod dinosaur pneumaticity, buoyancy and aquatic habits. *Proceedings of the Royal Society of London B* (suppl.) 271: S180–S183.

———. 2012 (this volume). Engineering a dinosaur. In M. K. Brett-Surman, T. R. Holtz, and J. O. Farlow (eds.), *The Complete Dinosaur*, 2nd ed., 636–665. Bloomington: Indiana University Press.

Hirsch, K. F., and D. K. Zelenitsky. 1997. Dinosaur eggs. In J. O. Farlow and M. K. Brett-Surman (eds.), *The Complete Dinosaur*, 394–402. Bloomington: Indiana University Press.

Hitchcock, E. 1858. *Ichnology of New England: A Report of the Sandstone of the Connecticut Valley, Especially Its Fossil Footmarks*. Boston: William White. Reprint, 1974, New York: Arno Press.

———. 1865. *Supplement to the Ichnology of New England*. Boston: Wright and Potter.

Holtz, T. R., Jr., and M. K. Brett-Surman. 2012 (this volume). The taxonomy and systematics of the dinosaurs. In M. K. Brett-Surman, Thomas R. Holtz Jr., and James O. Farlow (eds.), *The Complete Dinosaur*, 2nd ed., 209–223. Bloomington: Indiana University Press.

Hotton, N., III. 1980. An alternative to dinosaur endothermy: the happy wanderers. In R. D. K. Thomas and E. C. Olson (eds.), *A Cold Look at the Warm-Blooded Dinosaurs*, 311–350. American Association for the Advancement of Science Selected Symposium 28. Boulder, Colo.: Westview Press.

Huh, M., K. G. Hwang, I. S. Paik, C. H. Chung, and B. S. Kim. 2003. Dinosaur tracks from the Cretaceous of South Korea: distribution, occurrences and paleobiological significance. *Island Arc* 2003: 132–144.

Hunt, A. P., and S. G. Lucas. 2007. Tetrapod ichnofacies: a new paradigm. *Ichnos* 14: 59–68.

Hutchinson, J. R., and M. Garcia. 2002. *Tyrannosaurus* on the run. *Nature* 415: 1018–1021.

Huxley, T. H. 1868. On the animals which are most nearly intermediate between birds and reptiles. *Geological Magazine* 5: 357–365.

Irby, G. V. 1996a. Paleoichnology of the Cameron Dinosaur Tracksite, Lower Jurassic Moenave Formation, northeastern Arizona. In M. Morales (ed.), *The*

Continental Jurassic, 147–166. Bulletin 60. Flagstaff: Museum of Northern Arizona.

———. 1996b. Paleoichnological evidence for running dinosaurs worldwide. In M. Morales (ed.), *The Continental Jurassic*, 109–112. Bulletin 60. Flagstaff: Museum of Northern Arizona.

Ishigaki, S., and T. Fujisaki. 1989. Three dimensional representation of *Eubrontes* by the method of moiré typography. In D. D. Gillette and M. G. Lockley, (eds.), *Dinosaur Tracks and Traces*, 421–425. Cambridge: Cambridge University Press.

Ishigaki, S., and Y. Matsumoto. 2009. Re-examination of manus-only and manus-dominated sauropod trackways from Morocco. *Geological Quarterly* 53: 441–448.

Jackson, S. J., M. A. Whyte, and M. Romano. 2009. Laboratory-controlled simulations of dinosaur footprints in sand: a key to understanding vertebrate track formation and preservation. *Palaios* 24: 222–238.

———. 2010. Range of experimental dinosaur (*Hypsilophodon foxii*) footprints due to variation in sand consistency: how wet was the track? *Ichnos* 17: 197–214.

Jasinski, L. E. 2009. *Dinosaur Highway: A History of Dinosaur Valley State Park*. Fort Worth: Texas Christian University Press.

Kim, J. Y., K. S. Kim, M. G. Lockley, S. Y. Yang, S. J. Seo, H. I. Choi, and J. D. Lim. 2008. New didactyl dinosaur footprints (*Dromaeosauripus hamanensis* ichnogen. et ichnosp. nov.) from the Early Cretaceous Haman Formation, south coast of Korea. *Palaeogeography, Palaeoclimatology, Palaeoecology* 262: 72–78.

King, M. J., and M. J. Benton. 1996. Dinosaurs in the Early and Mid-Triassic? The footprint evidence from Britain. *Palaeogeography, Palaeoclimatology, Palaeoecology* 122: 213–225.

Kuban, G. J. 1989a. Elongate dinosaur tracks. In D. D. Gillette and M. G. Lockley (eds.), *Dinosaur Tracks and Traces*, 57–72. Cambridge: Cambridge University Press.

———. 1989b. Color distinctions and other curious features of dinosaur tracks near Glen Rose, Texas. In D. D. Gillette and M. G. Lockley (eds.), *Dinosaur Tracks and Traces*, 427–440. Cambridge: Cambridge University Press.

Kubo, T. 2010. Variation in modern crocodylian limb kinematics and its effect on trackways. In J. Milàn, S. G. Lucas, M. G. Lockley, and J. A. Spielman (eds.), *Crocodyle Tracks and Traces*, 51–53. Bulletin 51. Albuquerque: New Mexico Museum of Natural History and Science,

Kubo, T., and M. Ozaki. 2009. Does pace angulation correlate with limb posture? *Palaeogeography, Palaeoclimatology, Palaeoecology* 275: 54–58.

Kvale, E. P., G. D. Johnson, D. L. Mickelson, K. Keller, L. C. Furer, and A. W. Archer. 2001. Middle Jurassic (Bajocian and Bathonian) dinosaur megatracksites, Bighorn Basin, Wyoming. *Palaios* 16: 233–254.

Lawrence, P. J. 1972. Edward Hitchcock: the Christian geologist. *Proceedings of the American Philosophical Society* 116: 21–34.

Leonardi, G. (ed.). 1987. *Glossary and Manual of Tetrapod Footprint Palaeoichnology*. Brazilia: Brazilian Department of Mines and Energy.

———. 1989. Inventory and statistics of the South American dinosaurian ichnofauna and its paleobiological interpretation. In D. D. Gillette and M. G. Lockley (eds.), *Dinosaur Tracks and Traces*, 165–178. Cambridge: Cambridge University Press.

Li, R., M. G. Lockley, P. J. Makovicky, M. Matsukawa, M. A. Norell, J. D. Harris, and M. Liu. 2008. Behavioral and faunal implications of Early Cretaceous deinonychosaur trackways from China. *Naturwissenschaften* 95: 185–191.

Lim, S.-K., S. Young, and M. G. Lockley. 1989. Large dinosaur footprint assemblages from the Cretaceous Jindong Formation of Korea. In D. D. Gillette and M. G. Lockley (eds.), *Dinosaur Tracks and Traces*, 333–336. Cambridge: Cambridge University Press.

Livezey, B. C., and R. L. Zusi. 2007. Higher-order phylogeny of modern birds (Theropoda, Aves: Neornithes) based on comparative anatomy. II. Analysis and discussion. *Zoological Journal of the Linnean Society* 149: 1–95.

Lockley, M. G. 1991. *Tracking Dinosaurs: A New Look at an Ancient World*. Cambridge: Cambridge University Press.

———. 1997. The paleoecological and paleoenvironmental utility of dinosaur tracks. In J. O. Farlow and M. K. Brett-Surman (eds.), *The Complete Dinosaur*, 554–578. Bloomington: Indiana University Press.

———. 2007. A tale of two ichnologies: the different goals and potentials of invertebrate and vertebrate (tetrapod) ichnotaxonomy and how they relate to ichnofacies analysis. *Ichnos* 14: 39–57.

Lockley, M. G., J. O. Farlow, and C. A. Meyer. 1994a. *Brontopodus* and *Parabrontopodus* ichnogen. nov. and the significance of wide- and narrow-gauge sauropod trackways. *Gaia* 10: 135–145.

Lockley, M. G., J. C. Garcia-Ramos, L. Piñuela, and M. Avanzini. 2008. A review of vertebrate track assemblages from the Late Jurassic of Asturias, Spain with comparative notes on coeval ichnofaunas from the western USA: implications for faunal diversity in siliciclastic facies assemblages. *Oryctos* 8: 53–70.

Lockley, M. G., K. J. Houck, and N. K. Prince. 1986. North America's largest dinosaur trackway site: implications for Morrison paleoecology. *Bulletin of the Geological Society of America* 97: 1163–1176.

Lockley, M. G., and A. P. Hunt. 1994. A track of the giant theropod dinosaur *Tyrannosaurus* from close to the Cretaceous/Tertiary boundary, northern New Mexico. *Ichnos* 3: 1–6.

———. 1995a. Ceratopsid tracks and associated ichnofauna from the Laramie Formation (Upper Cretaceous: Maastrichtian) of Colorado. *Journal of Vertebrate Paleontology* 15: 592–614.

———. 1995b. *Dinosaur Tracks and Other Fossil Footprints of the Western United States*. New York: Columbia University Press.

———. 1998. A probable stegosaur track from the Morrison Formation of Utah. *Modern Geology* 23: 331–342.

Lockley, M. G., A. P. Hunt, and C. Meyer. 1994b. Vertebrate tracks and the ichnofacies concept: implications for paleoecology and palichnostratigraphy. In S. Donovan (ed.), *The Paleobiology of Trace Fossils*, 241–268. New York: John Wiley and Sons.

Lockley, M. G., J. I. Kirkland, F. L. De-Courten, B. B. Britt, and S. T. Hasiotis. 1999. Dinosaur tracks from the Cedar Mountain Formation of eastern Utah: a preliminary report. In D. D. Gillette (ed.), *Vertebrate Paleontology in Utah*, 253–257. Salt Lake City: Utah Geological Survey.

Lockley, M. G., and M. Matsukawa. 1999. Some observations on trackway evidence for gregarious behavior among small bipedal dinosaurs. *Palaeogeography, Palaeoclimatology, Palaeoecology* 150: 25–31.

Lockley, M. G., and A. Rice. 1990. Did *Brontosaurus* ever swim out to sea? Evidence from brontosaur and other dinosaur footprints. *Ichnos* 1: 81–90.

Lockley, M. G., A. S. Schulp, C. A. Meyer, G. Leonardi, and D. K. Mamani. 2002a. Titanosaurid trackways from the Upper Cretaceous of Bolivia: evidence for large manus, wide-gauge locomotion and gregarious behavior. *Cretaceous Research* 23: 383–400.

Lockley, M. G., J. Wright, D. White, M. Matsukawa, J. Li, L. Feng, and H. Li. 2002b. The first sauropod trackways from China. *Cretaceous Research* 23: 363–381.

Lockley, M. G., J. L. Wright, and D. Thies. 2004. Some observations on the dinosaur tracks at Münchehagen (Lower Cretaceous), Germany. *Ichnos* 11: 261–274.

Lockley, M., G. B. H. Young, and K. Carpenter. 1983. Hadrosaur locomotion and herding behavior: evidence from footprints in the Mesaverde Formation, Grand Mesa coal field, Colorado. *Mountain Geologist* 20: 5–14.

Lucas, S. G., and A. B. Heckert (eds.). 1995. *Early Permian Footprints and Facies.* Bulletin 6. Albuquerque: New Mexico Museum of Natural History and Science.

Lull, R. S. 1933. A revision of the Ceratopsia or horned dinosaurs. *Memoirs of the Peabody Museum of Natural History* 3: 1–135.

———. 1953. *Triassic Life of the Connecticut Valley.* Bulletin 81. Hartford: Connecticut State Geological and Natural History Survey.

Magwood, J. P. A. 1992. Ichnotaxonomy: a burrow by any other name . . . ? In C. G. Maples and R. R. West (eds.), *Trace Fossils*, 15–33. Paleontological Society Short Course 5. Knoxville: University of Tennessee.

Maleev, E. A. 1974. [Gigantic carnosaurs of the family Tyrannosauridae]. *Sovm. Sov.-Mong. Paleontol. Eksped.* Trudy 1: 132–191. [In Russian].

Manning, P. L. 2004. A new approach to the analysis and interpretation of tracks: examples from the dinosauria. *Geological Society of London Special Publications* 228: 93–123.

———. 2008. *T. rex* speed trap. In P. Larson and K. Carpenter (eds.), *Tyrannosaurus rex: The Tyrant King*, 205–231. Bloomington: Indiana University Press.

Manning, P. L., C. Ott, and P. L. Falkingham. 2008. A probable tyrannosaurid track from the Hell Creek Formation (Upper Cretaceous), Montana, United States. *Palaios* 23: 645–647.

Mannion, P. D., and P. Upchurch. 2010. A quantitative analysis of environmental associations in sauropod dinosaurs. *Paleobiology* 36: 253–282.

Maples, C. G., and R. R. West (eds.). 1992. *Trace Fossils.* Paleontological Society Short Courses 5. Knoxville: University of Tennessee.

Marty, D. 2008. Sedimentology, taphonomy, and ichnology of Late Jurassic dinosaur tracks from the Jura carbonate platform (Chevenez-Combe Ronde tracksite, NW Switzerland): insights into the tidal-flat palaeoenvironment and dinosaur diversity, locomotion, and palaeoecology. *GeoFocus* (University of Fribourg, Switzerland) 21: 113–236.

Marty, D., M. Belvedere, C. A. Meyer, P. Mietto, G. Paratte, C. Lovis, and B. Thüring. 2010. Comparative analysis of Late Jurassic sauropod trackways from the Jura Mountains (NW Switzerland) and the central High Atlas Mountains (Morocco): implications for sauropod ichnotaxonomy. *Historical Biology* 22: 109–133.

Marty, D., W. A. Hug, A. Iberg, L. Cavin, C. A. Meyer, and M. G. Lockley. 2003. Preliminary report on the Courtedoux dinosaur tracksite from the Kimmeridgian of Switzerland. *Ichnos* 10: 209–219.

Marty, D., A. Strasser, and C. A. Meyer. 2009. Formation and taphonomy of human footprints in microbial mats of present-day tidal-flat environments: implications for the study of fossil footprints. *Ichnos* 16: 127–142.

Marsh, O. C. 1877. Introduction and succession of vertebrate life in America. *American Journal of Science* (ser. 3) 14: 337–378.

Matthews, N. A., and B. H. Breithaupt. 2001. Close-range photogrammetric experiments at Dinosaur Ridge. *Mountain Geologist* 38 (3): 147–153.

Mayor, A. 2000. *The First Fossil Hunters: Paleontology in Greece and Roman Times.* Princeton, N.J.: Princeton University Press.

———. 2005. *Fossil Legends of the First Americans.* Princeton, N.J.: Princeton University Press.

McCrea, R. T., and P. J. Currie. 1998. A preliminary report on dinosaur tracksites in the Lower Cretaceous (Albian) Gates Formation near Grand Cache, Alberta. In S. G. Lucas, J. I. Kirkland, and J. W. Estep (eds.), *Lower and Middle Cretaceous Terrestrial Ecosystems*, 155–162. Bulletin 14. Albuquerque: New Mexico Museum of Natural History and Science.

McIntosh, J. S., M. K. Brett-Surman, and J. O. Farlow. 1997. Sauropods. In J. O. Farlow and M. K. Brett-Surman (eds.), *The Complete Dinosaur*, 264–290. Bloomington: Indiana University Press.

Meyer, C. A., and J. G. Pittman. 1994. A comparison between the *Brontopodus* ichnofacies of Portugal, Switzerland, and Texas. *Gaia* 10: 125–133.

Milàn, J., M. Avanzini, L. B. Clemmensen, J. C. García-Ramos, and L. Piñuela. 2006. Theropod foot movement recorded by Late Triassic, Early Jurassic and Late Jurassic fossil footprints. In J. D. Harris (ed.), *The Triassic-Jurassic Terrestrial Transition*, 352–364. Bulletin 37. Albuquerque: New Mexico Museum of Natural History and Science.

Milàn, J., and R. G. Bromley. 2006. True tracks, undertracks and eroded tracks, experimental work with tetrapod tracks in laboratory and field. *Palaeogeography, Palaeoclimatology, Palaeoecology* 231: 253–264.

———. 2008. The impact of sediment consistency on track and undertrack morphology: experiments with emu tracks in layered cement. *Ichnos* 15: 18–24.

Milàn, J., and L. M. Chiappe. 2009. First American record of the Jurassic ichnospecies *Deltapodus brodricki* and a review of the fossil record of stegosaurian footprints. *Journal of Geology* 117: 343–348.

Milàn, J., L. B. Clemmensen, and N. Bonde. 2004. Vertical sections through dinosaur tracks (Late Triassic lake deposits, East Greenland)–undertracks and other subsurface deformation structures revealed. *Lethaia* 37: 285–296.

Milàn, J., and R. Hedegaard. 2010. Interspecific variation in tracks and trackways from extant crocodylians. In J. Milàn, S. G. Lucas, M. G. Lockley, and J. A. Spielman (eds.), *Crocodyle Tracks and Traces*, 15–29. Bulletin 51. Albuquerque: New Mexico Museum of Natural History and Science.

Milner, A. R. C., M. G. Lockley, and J. I. Kirkland. 2006. A large collection of well-preserved theropod dinosaur swim tracks from the Lower Jurassic Moenave Formation, St. George, Utah. In J. D. Harris, S. G. Lucas, J. A. Spielmann, M. G. Lockley, A. R. C. Milner, and J. I. Kirkland (eds.), *The Triassic-Jurassic Terrestrial Transition*, 315–328. Bulletin 37. Albuquerque: New Mexico Museum of Natural History and Science.

Molnar, R. E., and J. O. Farlow. 1990. Carnosaur paleobiology. In D. B. Weishampel, P. Dodson, and H. Osmólska (eds.), *The Dinosauria*, 210–224. Berkeley: University of California Press.

Moratalla, J. J. 2009. Sauropod tracks of the Cameros Basin (Spain): identification, trackway patterns and changes over the Jurassic-Cretaceous. *Geobios* 42: 797–811.

Moratalla, J. J., J. L. Sanz, and S. Jimenez. 1988. Multivariate analysis on Lower Cretaceous dinosaur footprints: discrimination between ornithopods and theropods. *Geobios* 21 (4): 395–408.

Moratalla, J. J., J. L. Sanz, S. Jiménez, and M. G. Lockley. 1992. A quadrupedal ornithopod trackway from the Lower Cretaceous of La Rioja (Spain): inferences on gait and hand structure. *Journal of Vertebrate Paleontology* 12: 150–157.

Moreno, K., and M. Pino. 2002. Huellas de dinosaurios en la Formación Baños del Flaco (Titoniano-Jurásico Superior), VI Región, Chile: paleoetología y paleoambiente. *Revista Geológica de Chile* 29: 151–165.

Mossman, D. J., R. Brüning, and H. P. Powell. 2003. Anatomy of a Jurassic theropod trackway from Ardley, Oxfordshire, U.K. *Ichnos* 10: 195–207.

Murray, P. F., and P. Vickers-Rich. 2004. *Magnificent Mihirungs: The Colossal Flightless Birds of the Australian Dreamtime.* Bloomington: Indiana University Press.

Nadon, G. C. 2001. The impact of sedimentology on vertebrate track studies. In D. H. Tanke and K. Carpenter (eds.), *Mesozoic Vertebrate Life*, 395–407. Bloomington: Indiana University Press.

Norman, D. B. 1986. On the anatomy of *Iguanodon atherfieldensis* (Ornithischia: Ornithopoda). *Bulletin de l'Institut Royal des Sciences Naturelles de Belgique* 56: 281–372.

Olsen, P. E. 1995. A new approach for recognizing track makers. *Geological Society of America Abstracts with Program* 27 (1): 72.

Olsen, P. E., and P. M. Galton. 1984. A review of the reptile and amphibian assemblages from the Stormberg of southern Africa, with special emphasis on the footprints and the age of the Stormberg. *Palaeontologia africana* 25: 87–110.

Olsen, P. E., and E. C. Rainforth. 2003. The Early Jurassic Ornithischian dinosaurian ichnogenus *Anomoepus*. In P. M. Letourneau and P. E. Olsen (eds.), *The Great Rift Valleys of Pangea: Sedimentology, Stratigraphy, and Paleontology*, 2: 314–368. New York: Columbia University Press.

Olsen, P. E., J. B. Smith, and N. G. MacDonald. 1998. Type material of the type species of the classic theropod footprint genera *Eubrontes, Anchisauripus*, and *Grallator* (Early Jurassic, Hartford and Deerfield basins, Connecticut and Massachusetts, U.S.A.). *Journal of Vertebrate Paleontology* 18: 586–601.

Osgood, R. G., Jr. 1987. Trace fossils. In R. S. Boardman, A. H. Cheetham, and A. J. Rowell (eds.), *Fossil Invertebrates*, 663–674. Palo Alto, Calif.: Blackwell Scientific Publications.

Osmólska, H., and R. Barsbold. 1990. Troodontidae. In D. B. Weishampel, P. Dodson, and H. Osmólska (eds.), *The Dinosauria*, 259–268. Berkeley: University of California Press.

Ostrom, J. H. 1969. *Osteology of Deinonychus antirrhopus, an Unusual Theropod from the Lower Cretaceous of Montana.* Bulletin 30. New Haven, Conn.: Peabody Museum of Natural History, Yale University.

———. 1972. Were some dinosaurs gregarious? *Palaeogeography, Palaeoclimatology, Palaeoecology* 11: 287–301.

Ostrom, J.H. 1978. The osteology of *Compsognathus longipes* Wagner. *Zitteliana* 4: 73–118.

Padian, K., and P. E. Olsen. 1984a. Footprints of the Komodo monitor and the trackways of fossil reptiles. *Copeia* 1984: 662–671.

———. 1984b. The fossil trackway *Pteraichnus*: not pterosaurian, but crocodilian. *Journal of Paleontology* 58: 178–184.

Parrish, J. M. 1997. Evolution of the archosaurs. In J. O. Farlow and M. K. Brett-Surman (eds.), *The Complete Dinosaur*, 191–203. Bloomington: Indiana University Press.

Pascual Arribas, C., and E. Sanz Pérez. 2000. Icnitas de dinosaurios en Valdelavilla (Soria, España). *Estudios Geologicos* 56: 41–61.

Paul, G. S. 1988. *Predatory Dinosaurs of the World: A Complete Illustrated Guide.* New York: Simon and Schuster.

———. 1991. The many myths, some old, some new, of dinosaurology. *Modern Geology* 16: 69–99.

Paul, G. S., and P. Christiansen. 2000. Forelimb posture in neoceratopsian dinosaurs: implications for gait and locomotion. *Paleobiology* 26: 450–465.

Peabody, F. E. 1948. Reptile and amphibian trackways from the Lower Triassic Moenkopi Formation of Arizona and Utah. *Bulletin of the Department of Geological Sciences* (University of California, Berkeley) 27: 295–468.

Perle, A., L. M. Chiappe, R. Barsbold, J. M. Clark, and M. A. Norell. 1994. Skeletal morphology of *Mononykus olecranus* (Theropoda: Avialae) from the Late Cretaceous of Mongolia. *American Museum Novitates* 3105: 1–29.

Petti, F. M., S. D'Orazi Porchetti, M. A. Conti, U. Nicosia, G. Perugini, and E. Sacchi. 2008. Theropod and sauropod footprints in the Early Cretaceous (Aptian) Apennic Carbonate Platform (Esperia, Central Italy): a further constraint on the palaeogeography of the Central-Mediterranean area. *Studi Trentini di Scienze Naturali, Acta Geologica* 83: 323–334.

Petti, F. M., S. D'Orazi Porchetti, E. Sacchi, and U. Nicosia. 2010. A new purported ankylosaur trackway in the Lower Cretaceous (lower Aptian) shallow-marine carbonate deposits of Puglia, southern Italy. *Cretaceous Research* 31: 546–552.

Rainforth, E. C. 2003. Revision and re-evaluation of the Early Jurassic dinosaurian ichnogenus *Otozoum*. *Palaeontology* 46: 803–838.

———. 2007. Ichnotaxonomic updates from the Newark Supergroup. In E. C. Rainforth (ed.), *Contributions to the Paleontology of New Jersey (II): Field Guide and Proceedings.* Geological Association of New Jersey, XXIV Annual Conference and Field Trip, East Stroudsburg University, East Stroudsburg, Pa.

Rasskin-Gutman, D., G. Hunt, R. E. Chapman, J. L. Sanz, and J. J. Moratalla. 1997. The shapes of tridactyl dinosaur footprints: procedures, problems and potentials. In D. L. Wolberg, E. Stump, and G .D. Rosenberg (eds.), *Dinofest International*, 377–383. Philadelphia, Pa.: Academy of Natural Sciences.

Rega, E., R. Holmes, and A. Tirabasso. 2010. Habitual locomotor behavior inferred from manual pathology in two Late Cretaceous chasmosaurine ceratopsid dinosaurs, *Chasmosaurus irvinensis* (CMN 41357) and *Chasmosaurus belli* (ROM 843). In M. J. Ryan, B. J. Chinnery-Allgeier, and D. A. Eberth (eds.), *New Perspectives on Horned Dinosaurs*, 340–354. Bloomington: Indiana University Press.

Romano, M., M. A. Whyte, and S. J. Jackson. 2007. Trackway ratio: a new look at trackway gauge in the analysis of quadrupedal dinosaur trackways and its implications for ichnotaxonomy. *Ichnos* 14: 257–270.

Romero Molina, M. M., F. Pérez-Lorente, and P. Rivas Carrera. 2003. Análisis de la parataxonomía con las huellas de dinosaurio. In F. Pérez-Lorente (ed.), *Dinosaurios y otros Reptiles Mesozoicos en España*, 13–32. Logroño, Spain: Gobierno de La Rioja, Universidad de La Rioja.

Romilio, A., and S. W. Salisbury. 2010. A reassessment of large theropod dinosaur tracks from the mid-Cretaceous (late Albian-Cenomanian) Winton Formation of Lark Quarry, central-western Queensland, Australia: a case for mistaken identity. *Cretaceous Research* 32: 135–142.

Rose, P. J. 2007. A new titanosauriform sauropod (Dinosauria: Saurischia) from the Early Cretaceous of central Texas and its phylogenetic relationships. *Palaeontologia Electronica* 10 (2): 8A.

http://palaeo-electronica.org
/paleo/2007_2/00063/index.html.

Rudwick, M. J. S. 1976. *The Meaning of Fossils: Episodes in the History of Palaeontology.* 2nd ed. New York: Neale Watson Academic Publication.

Sampson, S. D., and M. A. Loewen. 2010. Unraveling a radiation: a review of the diversity, stratigraphic distribution, biogeography, and evolution of horned dinosaurs (Ornithischia: Ceratopsidae). In M. J. Ryan, B. J. Chinnery-Allgeier, and D. A. Eberth (eds.), *New Perspectives on Horned Dinosaurs,* 405–427. Bloomington: Indiana University Press.

Santa Luca, A. P. 1980. The postcranial skeleton of *Heterodontosaurus tucki* (Reptilia, Ornithischia) from the Stormberg of South Africa. *Annals of the South African Museum* 79: 159–211.

Santos, V. F., J. J. Moratalla, and R. Royo-Torres. 2009. New sauropod trackways from the Middle Jurassic of Portugal. *Acta Palaeontologica Polonica* 54: 409–422.

Sanz, J.-L., J. J. Moratalla, and M. L. Casanovas. 1985. Traza icnologia de un dinosaurio Iguanodontido en el Cretacico inferior de Cornago (La Rioja, Espana). *Geologicos* 41: 85–91.

Sarjeant, W. A. S. 1989. "Ten paleoichnological commandments": a standardized procedure for the description of fossil vertebrate footprints. In D. D. Gillette and M. G. Lockley (eds.), *Dinosaur Tracks and Traces,* 369–370. Cambridge: Cambridge University Press.

———. 1990. A name for the trace of an act: approaches to the nomenclature and classification of fossil vertebrate footprints. In K. Carpenter and P. J. Currie (eds.), *Dinosaur Systematics: Approaches and Perspectives,* 299–307. Cambridge: Cambridge University Press.

Sarjeant, W. A. S., and W. J. Kennedy. 1973. Proposal of a code for the nomenclature of trace-fossils. *Canadian Journal of Earth Sciences* 10: 460–475.

Sarjeant, W. A. S., and W. Langston Jr. 1994. *Vertebrate Footprints and Invertebrate Traces from the Chadronian (Late Eocene) of Trans-Pecos Texas.* Bulletin 36. Austin: Texas Memorial Museum.

Schulp, A. S., M. Al-Wosabi, and N. J. Stevens. 2008. First dinosaur tracks from the Arabian Peninsula. *PLoS ONE* 3 (5): 3223. doi:10.1371/journal.pone.0002243.

Schult, M. F., and J. O. Farlow. 1992. Vertebrate trace fossils. In C. G. Maples and R. R. West (eds.), *Trace Fossils,* 34–63. Paleontological Society Short Course 5. Knoxville: University of Tennessee.

Smith, J. B., and J. O. Farlow. 2003. Osteometric approaches to trackmaker assignment for the Newark Supergroup Ichnogenera *Grallator, Anchisauripus,* and *Eubrontes.* In P. M. Letourneau and P. E. Olsen (eds.), *The Great Rift Valleys of Pangea: Sedimentology, Stratigraphy, and Paleontology,* 2: 273–292. New York: Columbia University Press.

Stovall, J. W., and W. Langston Jr. 1950. *Acrocanthosaurus atokensis,* a new genus and species of Lower Cretaceous Theropoda from Oklahoma. *American Midland Naturalist* 43: 696–728.

Thomas, D. A., and J. O. Farlow. 1997. Tracking a dinosaur attack. *Scientific American* (December): 48–53.

Thompson, S., and R. Holmes. 2007. Forelimb stance and step cycle in *Chasmosaurus irvinensis* (Dinosauria: Neoceratopsia). *Palaeontologia Electronica* 10 (1): 5A. http://palaeo-electronica.org/2007_1/step/intro.htm.

Thulborn, R. A. 1972. The post-cranial skeleton of the Triassic ornithischian dinosaur *Fabrosaurus australis. Palaeontology* 15: 29–60.

———. 1990. *Dinosaur Tracks.* London: Chapman and Hall.

———. 1994. Ornithopod dinosaur tracks from the Lower Jurassic of Queensland. *Alcheringa* 18: 247–258.

Thulborn, R. A., and M. Wade. 1984. Dinosaur trackways in the Winton Formation (mid-Cretaceous) of Queensland. *Memoirs of the Queensland Museum* 21: 413–517.

Torrens, H. S. 2012 (this volume). Politics and paleontology: Richard Owen and the invention of the dinosaurs. In M. K. Brett-Surman, T. R. Holtz, and J. O. Farlow (eds.), *The Complete Dinosaur,* 2nd ed., 24–43. Bloomington: Indiana University Press.

Viera, L. I., and J. A. Torres. 1995. Análisis comparativo sobre dos rastros de Dinosaurios Theropodos: Forma de marcha y velocidad. *Munibe* 47: 53–56.

Vila, B., O. Oms, and À. Galobart. 2005. Manus-only titanosaurid trackway from Fumanya (Maastrichtian, Pyrenees): further evidence for an underprint origin. *Lethaia* 38: 211–218.

Webb, G., and C. Manolis. 1989. *Australian Crocodiles: A Natural History.* Chatswood, New South Wales: Reed Books.

Wedel, M. J., R. L. Cifelli, and R. K. Sanders. 2000. *Sauroposeidon proteles,* a new sauropod from the Early Cretaceous of Oklahoma. *Journal of Vertebrate Paleontology* 20: 109–114.

Weems, R. E. 1987. A Late Triassic footprint fauna from the Culpeper Basin, northern Virginia (U.S.A.). *Transactions of the American Philosophical Society* 77: 1–79.

———. 1992. A re-evaluation of the taxonomy of Newark Supergroup saurischian dinosaur tracks, using extensive statistical data from a recently exposed tracksite near Culpeper, Virginia. In P. C. Sweet (ed.), *Proceedings of the 26th Forum on the Geology of Industrial Minerals, May 14–18, 1990,* 113–127. Charlotte: Commonwealth of Virginia, Division of Mineral Resources.

———. 2003. *Plateosaurus* foot structure suggests a single trackmaker for *Eubrontes* and *Gigandipus* footprints. In P. M. Letourneau and P. E. Olsen (eds.), *The Great Rift Valleys of Pangea: Sedimentology, Stratigraphy, and Paleontology,* 2: 293–313. New York: Columbia University Press.

Whyte, M. A., and M. Romano. 1994. Probable stegosaurian dinosaur tracks from the Saltwick Formation (Middle Jurassic) of Yorkshire. *Proceedings of the Geologists' Association* 112: 45–54.

Wilson, J. A., and M. T. Carrano. 1999. Titanosaurs and the origin of "wide-gauge" trackways: a biomechanical and systematic perspective on sauropod locomotion. *Paleobiology* 25: 252–267.

Wilson, J., and D. Fisher. 2003. Are manus-only sauropod trackways evidence of swimming, sinking, or wading? *Journal of Vertebrate Paleontology* 23 (3): 111A.

Worthy, T. H., and R. N. Holdaway. 2002. *The Lost World of the Moa: Prehistoric Life of New Zealand.* Bloomington: Indiana University Press.

Wright, J. L. 2005. Steps in understanding sauropod biology; the importance of sauropod tracks. In K. A. Curry Rogers and J. A. Wilson (eds.), *The Sauropods: Evolution and Paleobiology,* 252–284. Berkeley: University of California Press.

Zelenitsky, D. K., J. R. Horner, and François Therrien. 2012 (this volume). Dinosaur eggs. In M. K. Brett-Surman, Thomas R. Holtz Jr., and James O. Farlow (eds.), *The Complete Dinosaur,* 2nd ed., 612–620. Bloomington: Indiana University Press.

Zelenitsky, D. K., and S. P. Modesto. 2003. New information on the eggshell of ratites (Aves) and its phylogenetic implications. *Canadian Journal of Zoology* 81: 962–970.

35.1. Diagrammatic graphs illustrating the six heterochronic processes. Upper graph shows effect of changing growth rate, with increase producing acceleration and reduction neoteny. Middle graph shows effect of changing time of onset of growth; earlier onset is predisplacement, and later onset is postdisplacement. Lower graph shows effect of varying the time of growth offset, later offset being hypermorphosis, earlier offset, progenesis. For all graphs the upper line produces peramorphosis in the descendant, the lower graph paedomorphosis.

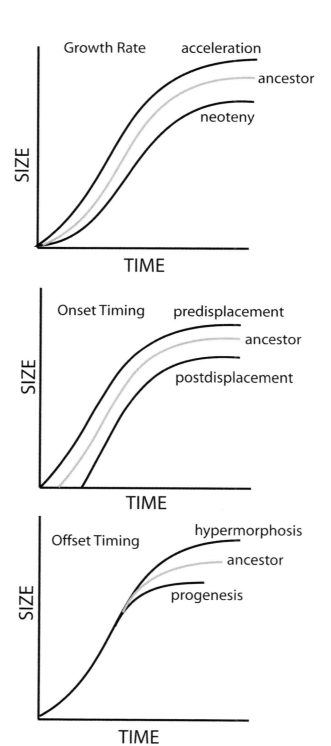

The Role of Heterochrony in Dinosaur Evolution

35

Kenneth J. McNamara and John A. Long

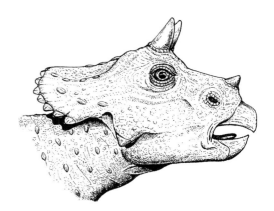

Heterochrony can be defined as "change to the timing and rate of development between individuals and between species" (Alberch et al. 1979; McKinney and McNamara 1991). Every single organism has an ontogeny – this is its life history – from the time of conception until its death. Certainly, among animals there would be none that remains morphologically static throughout its ontogeny. Not only do animals increase in size as they grow, but as they develop, from an initial egg, through larval or embryonic stages, to juvenile and then adult stages, they change in shape. You the reader, for instance, look very different now from what you looked like when you were born. This is because, like most other animals, including dinosaurs, many of your body parts have changed in relative shape during your ontogeny. Not only that, but most organisms (although maybe not all dinosaurs) have a finite period of growth, which is usually terminated by the onset of sexual maturity. At this stage growth (both in size and shape change) slows down markedly, or stops. Within populations, different individuals grow at slightly different rates, and for slightly different durations. The same is true for many species, but the differences are more pronounced. Moreover, different parts of a single individual may grow for different lengths of time and at different relative rates. In many vertebrates, the rate of growth of the head, for instance, is relatively greater than growth of the trunk or limbs *in utero*, whereas after birth these postcranial parts undergo a relatively greater amount of growth. Consequently any slight changes to the rate or duration of growth of different parts of the body during ontogeny can have a marked effect on what the adult looks like. Such changes, however slight, to the rate and duration of growth of all or part of an organism compared with its ancestor are, in our view, what define heterochrony.

Until the last quarter of the twentieth century, evolution was thought to involve just classical Darwinian natural selection and genetics, with the genes providing the mutations on which natural selection operates. However, since the publication of Steve Gould's *Ontogeny and Phylogeny* (Gould 1977), the importance of changes in development to evolution has slowly become widely recognized. Much of the pioneering work in this field was carried out by paleontologists in the 1980s who concentrated on macroevolutionary changes brought about by heterochrony (see McKinney 1988; McKinney and McNamara 1991; McNamara 1997). The 1990s saw an increasing number of biologists recognize the importance of studying developmental change in evolution. This field has become known (Hall 1992) as evolutionary developmental biology (usually abbreviated, somewhat unfortunately, to "Evo-Devo"). The realization that developmental change plays an important role in evolution (a fact recognized by a number of

761

nineteenth-century biologists—see Gould 1977) means that it can be regarded as being the third part of an evolutionary triangle. At one corner sits natural selection; at the second is genetics, while developmental change is at the third. Remove any of the three and the triangle is lost and evolution does not occur.

The growth of any organism is controlled by its genetic makeup. The genes determine, to a large degree, its morphological, physiological, and behavioral characteristics—in other words what it looks like, and how it functions and behaves. It is becoming widely recognized that so-called developmental genes, especially those that regulate embryonic or larval development, play a major role in evolution (Richardson and Brakefield 2003). One particular group of genes for instance, known as Hox genes, control the timing of expression of growth factors that essentially determine at what time and at what place a particular morphological structure starts to develop and for how long it grows. Consequently, any tinkering with the timing of expression of these controlling genes, or the period for which they can influence growth, will result in a descendant that can look markedly different from its immediate ancestor.

Such slight changes happen all the time within populations, so that variations between individuals in the duration and rate of growth of the entire organism, or maybe just some particular traits, result in morphological differences between individuals. In populations of a species this is called intraspecific variation. These variants, induced by various heterochronic processes, are then subjected to natural selection. Because heterochrony plays such an important part in providing much of the raw material on which natural selection works, heterochrony consequently plays a very significant role in evolution (McKinney and McNamara 1991; McNamara 1995, 1997). It can operate within species at this level of intraspecific variation; between sexes (McNamara 1995); between species; in the generation of evolutionary trends (McNamara 1982, 1990); and in the evolution of major morphological novelties, such as the evolution of birds from dinosaurs (McNamara 1997; McNamara and McKinney 2005). In recent years it has been recognized that dinosaurs are no different from other organisms, in that heterochrony played a major role in their evolution (Weishampel and Horner 1994; Long and McNamara 1995, 1997a, b).

Classifying Heterochrony

There are two basic types of heterochrony. When a descendant organism (either a geologically younger species or a descendant individual within a population) is compared with an ancestor, it can show either "less" or "more" growth. This can apply to the organism as a whole or to any particular morphological feature. *Paedomorphosis* (literally meaning "child-shape") is the name given to the type of heterochrony where there is less growth during ontogeny in a descendant form, compared with its ancestor. It was given this name because the descendant adult resembles the juvenile condition of the ancestor in some way. In the other type of heterochrony, the descendant form undergoes more development, and is said to show *peramorphosis* (literally "beyond-shape"). Often, though by all means not always, paedomorphic forms may be smaller than their ancestor, while peramorphic forms will be larger. Paedomorphosis and peramorphosis are descriptive terms that describe the appearance of the descendant morphology. In themselves they are *not* evolutionary processes—they are the morphological (and in some cases behavioral) end result.

The most commonly used model to describe the different heterochronic processes that produce paedomorphosis and peramorphosis is that proposed by Alberch et al. (1979). Their model simplified and quantified the terminology of the heterochronic processes (see also McNamara 1986). Although there have been recent attempts to introduce many more terms to describe the heterochronic processes (e.g., Webster and Zelditch 2005), most authors still use the Alberch et al. model. This recognizes that paedomorphosis and peramorphosis can each be produced by three quite different, but complementary, processes (Fig. 35.1). Paedomorphic morphologies are produced if the period of growth of the descendant form is prematurely stopped (*progenesis*), if the rate of growth is less in the descendant than in the ancestor (*neoteny*), or if onset of growth is delayed (*postdisplacement*). Progenesis will affect the whole organism if premature cessation of growth is caused by the earlier onset of sexual maturity. However, it may also involve only certain morphological features. Neoteny and postdisplacement usually affect only certain morphological features, and not the entire organism.

Peramorphic morphologies will evolve if the period of growth in the descendant is extended (*hypermorphosis*), if the growth rate is increased in the descendant (*acceleration*), or if onset of growth occurs earlier in the descendant than in the ancestor (*predisplacement*). Like progenesis, hypermorphosis can affect the whole organism if there is a delay in the onset of sexual maturity, as fast juvenile growth rates will persist for a longer period. Alternatively, hypermorphosis can target just certain traits. Acceleration and predisplacement affect only certain features, not the entire organism.

The terminology of heterochrony can be used both to describe the appearance of meristic characters during ontogeny (in other words, discrete structures formed during ontogeny, such as the number of vertebrae or digits) or it can be applied to the subsequent changes in shape of these structures. These are called *mitotic* and *growth* heterochrony, respectively (McKinney and McNamara 1991). In many animals, mitotic heterochrony, induced especially by pre- and postdisplacement, can be very important very early in development. This is because variations can occur between ancestors and descendants in the timing of onset of development of major morphological features. Neoteny and acceleration (that is, reduced and accelerated growth rates, respectively) will be particularly common during later ontogenetic development. Progenesis and hypermorphosis often occur at late ontogenetic stages, because they reflect variations in the time of offset of growth caused by differential occurrence of sexual maturity. This will be less important for those animals that had continuous growth. It has been suggested (see Reid, Chapter 31 of this volume) that some dinosaurs may have experienced continuous (indeterminate) growth throughout their lives. However, the majority are thought to have had determinate growth (Erickson et al. 2001; Chinsamy 1990). Even in those forms that might have had indeterminate growth, the rate of growth will still probably have been greatly reduced at onset of maturity even if it did not stop altogether.

As animals grow they not only increase in size, but also change in shape, often quite appreciably. The relationship between size and shape, known as *allometry*, arises from differential growth rates between different parts of the body or in different axes on the same structure. Thus during growth a bone may become relatively longer and thinner, because growth occurs at a faster rate along one axis than another. If, however, the relative size and shape of a structure remains the same during ontogeny relative to the animal's overall body size, growth is said to be *isometric*. In reality few, if any, organisms are known to grow isometrically (Klingenberg 1998), although some individual skeletal elements might. Usually a particular structure, such as a bone, will change shape and size relative to the size and shape of the whole organism, during ontogeny. If the bone increases in relative size, growth is said to occur by *positive allometry*. Conversely, if there is a relative decrease in size, growth is said to show *negative allometry*. Because heterochrony involves changes not only in time, but also in shape and size, there is a close relationship between allometry and heterochrony. The effect of changing growth rates (acceleration and neoteny) is to cause allometric changes. Extensions or contractions of the period of growth, that is, hypermorphosis or progenesis, have the affect of exacerbating or contracting the effects of allometric changes. Consequently, those organisms that undergo pronounced allometric change during growth are more likely to generate very different descendant adult morphologies if rates or durations of growth have changed.

Ontogenetic Growth in Dinosaurs

So, what is known about the growth and development of dinosaurs? The last decade or so has seen a great increase in our knowledge, based in part on the study of embryonic material discovered in dinosaur eggs (e.g., Carpenter et al. 1994; Horner and Weishampel 1988; Reisz et al. 2005). Thus for a number of groups we have a reasonably good understanding of the allometric changes that occur during ontogeny. From this it is possible to formulate suggestions as to the impact and type of heterochrony that occurred during the evolution of many groups of dinosaurs. Moreover, studies of the microstructure of dinosaur bone have provided a lot of information on growth rates and times of maturation (Chinsamy 1990, 2005; Erickson et al. 2001, 2004, 2007; Sander and Klein 2005). Such data are especially important when trying to assess which particular type of heterochronic process was operating. Without such actual growth-rate data it is not possible with fossil material to be sure whether, for instance, a paedomorphic trait arose by progenesis (where there was just a reduction in the time of growth) or neoteny (where there was an actual reduction in the rate of growth).

In the case of theropod dinosaurs, studies of bone microstructure indicate that growth rates, as with most dinosaurs, were very fast (Chinsamy 1990; Varricchio 1993; Erickson et al. 2004), with growth rates in *Tyrannosaurus rex* about 2 kg per day during its actively growing period (Erickson et al. 2004, 773). Analysis of the bone microstructure of the small theropod *Troodon* revealed that it passed through three distinct growth phases during its ontogeny (Varricchio 1993). However, more recent reconstructions of theropod growth curves suggest that up to five distinct growth phases could be identified: juvenile, adolescent, subadult, young adults, and senescent adults. Growth was fastest during the subadult stage, prior to the onset of maturity (Erickson et al. 2004). Moreover, all the studied theropods had determinate growth. In other words, after reaching sexual maturity they stopped growing. However, a study by Sander and Klein (2005) on the prosauropod *Plateosaurus* has suggested that in this dinosaur at least, there appears to have been a large degree of variation within the single species in the time of onset of maturity and cessation of growth. This has significance for trying to understand aspects of the life histories of these animals (see below). Bone microstructure studies (e.g., Erickson et al. 2001, 2004; Lee and Werning 2008) have shown the occurrence of periods of arrested growth, perhaps indicating the effect of seasonality on the growth rate of dinosaurs—greater growth taking place presumably during periods when food was more abundant. On the basis of such data the little maniraptoriform theropod *Shuvuuia* is thought to have reached maturity in about three years when it weighed just under 2 kg (Erickson et al. 2001). By contrast, a very large theropod, such as *T. rex*, is thought to have reached maturity at 18.5 years, when it would have weighed about 5,000 kg (Erickson et al. 2004). Counts of arrested growth in theropods led Lee and Werning (2008) to calculate that ages of reproductive maturity for *Tenontosaurus*, *Allosaurus*, and *Tyrannosaurus* were 8, 10, and 15 years, respectively.

Studies of juvenile material of the theropods *Gorgosaurus* (Russell 1970; Madsen 1976; Paul 1988), *Syntarsus* (Raath 1990), and *Tyrannosaurus* (Carpenter 1992; Molnar 1990) have allowed estimates to be made of the ontogenetic changes that occurred in these theropods (Fig. 35.2). These include, in particular, the change in the skull and jaw, from being gracile and elongate in juveniles to more massive and foreshortened in adults; a reduction in the size of the orbits; increase in size of the postorbital bar late in ontogeny (see Long and McNamara 1997a; Fig. 35.2); closure and fusion of some cranial sutures, especially in the lower jaw; increase in the number of serrations on the teeth (Currie et al. 1990); relative reduction in length of hind limb elements, but increase in their width; thickening of the metatarsals, which became more robust; relative lengthening of the presacral column and cervical neural spines; relative shortening of the tail; increase in size of the pubic boot; and relative increase in size of the pelvis overall. The astragalus also shows a significant change in shape during ontogeny (Welles and Long 1974).

Ontogenetic changes in higher ornithopods have been described in *Dryosaurus* (Carpenter 1994), *Maiasaura* and other hadrosaurs (Dodson 1975; Carpenter et al. 1994; Guenther 2009; Horner and Currie 1994; Weishampel 1981), and hypsilophodontids, such as *Orodromeus* (Horner and Weishampel 1988). The general trends in ontogenetic development in this group (Fig. 35.3) include: relative reduction in orbit size; reduction in size of palpebral bones relative to the orbit; increase in size of premaxilliary and nasal bones, producing the characteristic hadrosaur "bill"; elongation of snout in large species; development of prefrontal brow in some forms, such as *Maiasaura* and *Prosaurolophus*; extra teeth added to upper and lower jaws; reduction in size of neural canal relative to vertebra size; increase in length of neural spines.

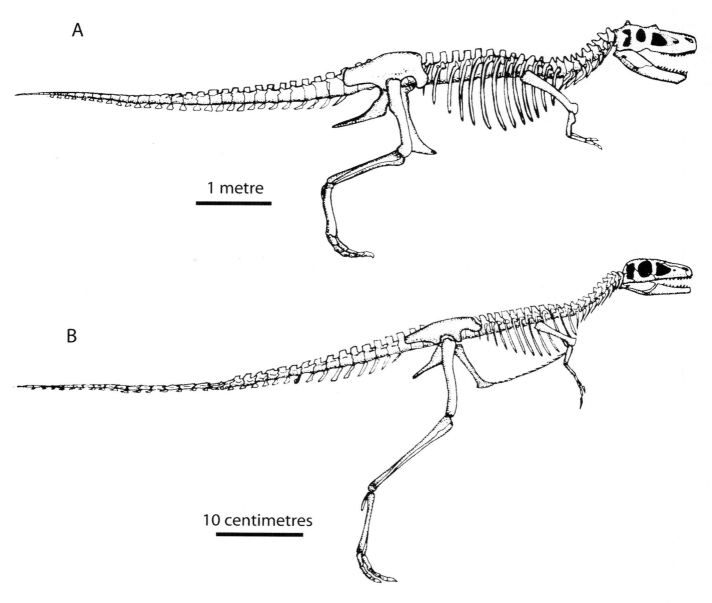

35.2. Comparison of adult (*top*) and juvenile (*bottom*) skeletons of *Gorgosaurus libratus,* scaled to same size. Adapted from Long and McNamara (1995, 8.1).

A

1 metre

B

10 centimetres

In ceratopsians the nature of ontogenetic change is well known in primitive neoceratopsians, such as *Bagaceratops* (Maryańska and Osmólska 1975; Chapman 1990) and *Protoceratops* (Carpenter et al. 1994). Much less is known of ontogenetic variation in the chasmosaurines and centrosaurines, although more recently Horner (2005) and Scannella and Horner (2010) have presented new data on the growth of *Triceratops* suggesting that *Torosaurus* may be a mature morphological variant of *Triceratops*. In neoceratopsians the main changes that occurred are decrease in orbit size; increase in snout length; increase in frill length initially, followed by a shortening; widening of the frill; widening of the jugal and quadrate areas; and, in *Protoceratops*, development of the nasal horn.

Within the sauropodomorphs, plateosaurid prosauropods show marked differences between juveniles and adults in skull morphology (Long and McNamara 1997a, b) (Fig. 35.4A, B). During growth there was a pronounced increase in snout length, with a resultant increase in the maxilla anteriorly and increase in number of teeth in the jaws. There was more posterior

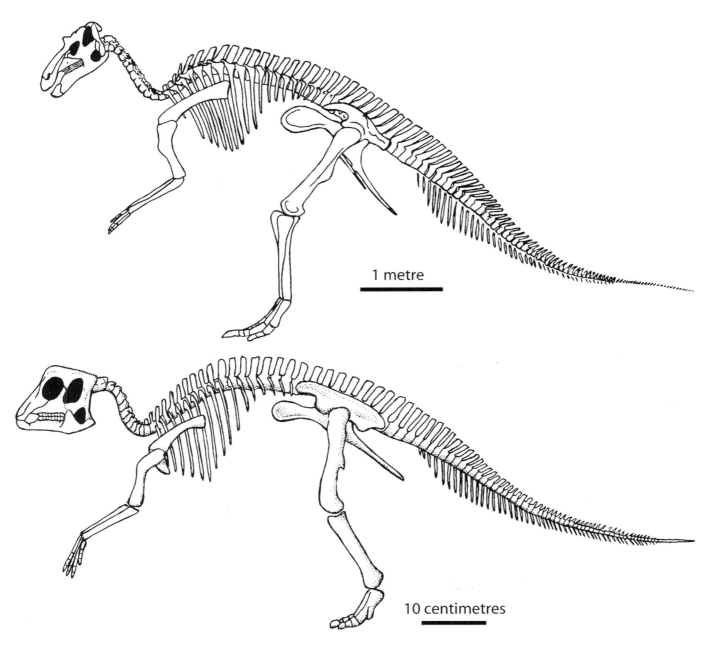

1 metre

10 centimetres

35.3. Comparison of adult (*top*) and juvenile (*bottom*) skeletons of *Maiasaura peeblesorum,* scaled to same size. Adapted from Long and McNamara (1995, 8.5).

development of the quadrate and enlargement of the ventral divisions of the temporal fenestra. A study of embryonic material of the early Jurassic prosauropod *Massospondylus* (Reisz et al. 2005) shows that this genus underwent pronounced ontogenetic changes, in both cranial and postcranial elements, the heterochronic significance of which is discussed below. In the head, the skull underwent positive allometry as the maxilla, premaxilla, and nasal bones became relatively longer. Like other dinosaurs, the embryonic orbit is relatively huge, occupying almost 40% of the total skull length. This reduced appreciably during growth. Postcranially, the tibia and the dorsal vertebrae show isometric growth, whereas the cervical vertebra and ribs underwent positive allometry. In the forelimbs, the humerus and ulna both show negative allometry as *Massospondylus* grew from its embryonic snout-vent length just prior to hatching of about 8 cm, to an adult length of up to 5 m (Cooper 1981; Reisz et al. 2005).

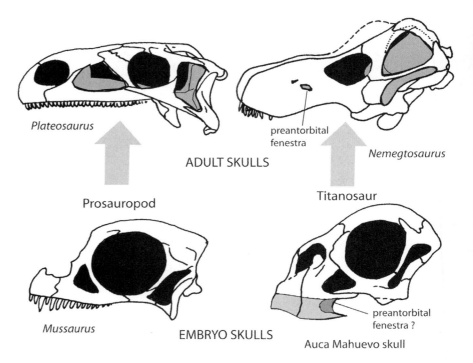

35.4. Ontogenetic evolution of sauropodomorph skull. Prosauropod skulls are based on an embryo of *Mussaurus* and adult of *Plateosaurus*. Titanosaur skulls based on embryo from Auca Mahuevo, Patagonia, and adult of *Nemegtosaurus*. Drawings of skulls based on Salgado et al. (2005, 6).

Dramatic changes occurred during the ontogeny of sauropods, especially in the skull (Fig. 35.4 C, D). These include relatively great reduction in the size of the frontals and parietals, which also migrated dorsally to the posterodorsal area of the orbit; ventral reduction in size and constriction of the orbit to a tear shape; enlargement of the rostrum; posterior expansion of the maxilla, making a connection with the quadratojugal; expansion of external nares, which migrated posterodorsally (Salgado et al. 2005).

A very small, complete skeleton of the nodosaurid ankylosaur *Liaoningosaurus* has been described from the Early Cretaceous of Liaoning, China (Xu et al. 2001). Only 34 cm long, it is quite turtlelike in possessing a large ventral plate. Juvenile features include fewer large maxillary teeth, compared with adult ankylosaurs; laterally facing postacetabular process of the ilium; presence of a pubic peduncle; distinct fingerlike lesser trochanter; tibia as long as femur; and claw-shaped manual and pedal unguals. Xu et al. (2001) have suggested that shifts in timing of ontogenetic events involving these characters has played a significant role in ankylosaur evolution.

Identifying Heterochrony in the Fossil Record

In order to identify heterochrony in the fossil record it is necessary to have some degree of ontogenetic information on both ancestral and presumed descendant forms. Whereas many ontogenetic changes will be species specific, there are many general changes that occur within particular orders and even across all dinosaurs (the decrease in size of the orbit is one such example). Thus even if the ontogeny of the descendant and presumed ancestral form are not known in detail, a number of inferences concerning likely occurrence of heterochrony can still be made. Two of the factors involved in heterochrony, namely shape and size, are always available with fossil material. Consequently, it is often possible to assess whether a particular species is either peramorphic or paedomorphic, compared with other members of the same genus, family, or order. In many cases, although it can be recognized that heterochrony has occurred in the evolution of particular species, it may be that only certain traits can be identified as being peramorphic or paedomorphic.

Elucidating which process or processes have caused the heterochrony can be more difficult when dealing with fossil material. Among fossil groups dinosaurs show the most promise of our being able to ascertain likely heterochronic processes. A number of studies have been made describing growth rates and time of onset of sexual maturity and cessation or reduction in growth in dinosaurs (see Chinsamy 2005). For instance, studies of bone microstructure in dinosaurs, such as *Troodon* (Varricchio 1993), *Syntarsus* (Chinsamy 1990, 1992), *Massospondylus* (Chinsamy 1992, 1993), *Timimus* (Constantine et al. 1998), *Maiasaura* (Horner et al. 2000), *Dryosaurus* (Chinsamy 1995), *Apatosaurus* (Curry 1999), and *Psittacosaurus* (Erickson and Tumanova 2000) show that periodic, possibly seasonal, growth reflected in growth lines in the bone, allow "real"-time information to be gleaned from such fossil material. With such information, it is possible to make more accurate inferences of the type of particular heterochronic processes in extinct species.

To date, however, little of this type of information has been assessed to interpret heterochronic processes in dinosaurs, due largely to the rather small data base. More typically paleontological studies use size as a proxy for time – that is, if one species is smaller than another, it grew for a shorter period of time. Consequently, assumptions are made that, for example, a descendant that is smaller than its ancestor, and shows paedomorphic morphological characteristics, ceased growing at an earlier age and therefore arose by progenesis. So, for instance, smaller late Cretaceous tyrannosaurs, like *Maleevosaurus*, which were "only" 5 m long, possess certain features that can be interpreted as being paedomorphic (see below) and are smaller than their presumed ancestors. Paedomorphic features can be attained by neoteny or progenesis. Either process could explain paedomorphosis in these smaller tyrannosaurs. However, studies in other groups of animals suggest that neoteny will tend to target only certain structures, not the entire organism. While progenesis is more probable, without actual growth-rate data it is not possible to determine the precise process. Thus care needs to be taken in assigning heterochronic processes to fossil material when age data is lacking (Jones 1988).

In theory, paedomorphosis and peramorphosis should occur with roughly equal frequencies (Gould 1977). However, studies of heterochrony in the fossil record suggest that this may not always be so. Studies of heterochrony in amphibians suggest a preponderance of paedomorphosis (McNamara 1988). This may be related to the large cell size of amphibians, reducing the rate of cellular division (McNamara 1997). However, the few reviews of heterochrony in dinosaurs (Long and McNamara 1995, 1997a, b) indicated that peramorphosis may have been more frequent, and, in particular, a major contributing factor to the evolution of very large body size.

Heterochrony and the Evolution of Large Body Size

Evolutionary trends toward increased body size have been so often recognized in the fossil record that they have been codified as Cope's Rule (McKinney 1990). It has been argued that peramorphosis is the prime factor in generating increases in body size. This can be by either hypermorphosis or acceleration (McNamara 1997). In their reviews of heterochrony in dinosaur evolution, Long and McNamara (1995, 1997a, b) argued that the trends of increases in body size seen in sauropods, ceratopsians, theropods, and ornithopods were driven by peramorphic processes. Not only did body size increase, but the morphological features became correspondingly more complex, suggestive of peramorphosis. Since these reviews were published, more data has come to light that supports this view, although a number of examples of paedomorphosis have since also been recognized.

Studies of the microstructure of dinosaur bone indicate that many juveniles experienced very rapid growth rates (Padian et al. 2001; Sander and Klein 2005). This would suggest that acceleration alone was the process responsible for the attainment of large peramorphic features. However, studies of patterns of theropod growth support the argument of Long and McNamara (1997a, 116) that the very large body sizes of some dinosaurs may have been achieved by a combination of both acceleration and hypermorphosis. In their study of tyrannosaurid theropods, Erickson et al. (2004) compared growth-rate data for four genera: *Albertosaurus*, *Gorgosaurus*, *Daspletosaurus*, and *Tyrannosaurus*. The first two genera are the smallest, with *Daspletosaurus* being a little larger, but *Tyrannosaurus* being appreciably larger. Was the size increase caused by faster growth rates, longer period of fast growth, or both? On the basis of bone histology, Erickson and his colleagues calculated that *Tyrannosaurus* reached a body size of up to 5.5 tonnes. Its fastest period of growth, when it was adding an impressive 2 kg per day, persisted for about 4 years, until it began to slow down as it reached somatic maturity at about 18.5 years.

The other, relatively smaller, genera (reaching a maximum body size in *Gorgosaurus* and *Albertosaurus* of about 1 tonne, and in *Daspletosaurus* of about 1.5 tonnes), experienced maximum growth rates that also persisted for up to about 4 years, but that were much less than in *T. rex*, being about ⅓ to ½ kg per day. These genera also attained maturity at a younger age than *T. rex*, at between 14 and 16 years. The theropod *Gigantoraptor* had an accelerated growth rate that was greater than most other theropods, including *Gorgosaurus* and *Albertosaurus*. This 8-meter-long, 1.5-tonne oviraptosaurian reached maturity and this body size after only about 7 years (Xu et al. 2007).

Thus, not only was *T. rex* growing much faster (acceleration) than other tyrannosaurids, but it delayed the offset of this faster growth to a later age (hypermorphosis) (Fig. 35.5). The same trend is seen when all tyrannosaurids are compared with older theropods. Other, geologically older small theropods, such as *Syntarsus*, became mature at a much younger age (about 4 years) and had maximum growth rate of just 9 kg per year (Chinsamy 1990). The combination of acceleration and hypermorphosis explains the extreme peramorphic features of *T. rex*, including its massive skull and huge hind limbs. Why it has such small forelimbs, though, is discussed below, for not every morphological trait was peramorphic. It is noteworthy that the evolution of gigantism in other reptiles, notably crocodiles and lizards, was attained only by hypermorphosis (Erickson et al. 2003).

In terms of allometric changes in tyrannosaurids, the relatively very large head, especially in later, larger forms is indicative of evolution by

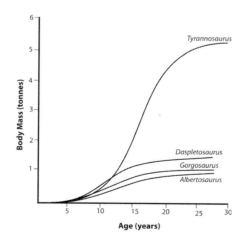

35.5. Logistic growth curves for four tyrannosaurid genera, illustrating the acceleration in growth of *Tyrannosaurus*, compared with other tyrannosaurids. Based on Erickson et al. (2004, 2).

peramorphosis, juveniles and phylogenetically earlier adults possessing a more gracile, elongate skull. This change occurred because of the greater positive allometric growth in a dorsoventral direction (and relative foreshortening), causing the evolution of a massive skull (Fig. 35.2). The similar effect of extended growth and acceleration of positive allometry also occurred in the evolution of the massive hind limbs. Selection in these large tyrannosaurids was targeting not only large body size, but also the large hind limbs that were able to support the huge body weight.

Peramorphosis is also evident in tyrannosaurids in the exclusion of the frontals from the orbits by the lachrymal bone, due to a greater amount of growth of this bone. Other peramorphic features are the increase in supraoccipital crests on the head and the development of lachrymal rugosity. Postcranially there is a dominant peramorphic trend in the evolution of the pelvis, especially the pubis. It is known that hatchling *Albertosaurus* have a long, slender pubic bone and weakly developed boot. During ontogeny, however, this structure increases relatively in size, undergoing positive allometry, the pubic bone thickening and the boot increasing appreciably in size. In the oldest tyrannosaurids, the Late Jurassic *Guanlong* (Xu et al. 2006) and the Early Cretaceous *Siamotyrannus*, the pubic bone is slender, even in the largest adults (Buffetaut et al. 1996). In its slender shape and weak development it is reminiscent of the form of the pubic bone in juvenile *Albertosaurus*. The boot is weakly developed in *Siamotyrannus* and in juvenile *Gorgosaurus*, but already very prominent in *Guanlong* (Xu et al. 2006, fig. 2e). In advanced, very large tyrannosaurids, such as *T. rex* and *T. bataar*, the pubic bone is massive and the boot is very strongly expanded, with the anterior part becoming larger than the posterior (Fig. 35.6). It has been suggested that the pubic boot was the attachment site for muscles used in the system of hepatic piston–diaphragm mechanism breathing proposed for tyrannosaurids (Ruben et al. 1997, 1999). Only by developing a peramorphically larger and more complex muscle attachment site and thickened pubic bone would the larger tyrannosaurs, such as *Tyrannosaurus*, been able to breathe effectively and maintain such a large body size and active, bipedal posture.

Some of the larger ornithopods, such as *Tenontosaurus*, which grew up to 7.5 m in length, also show evidence of evolution by peramorphosis (Long and McNamara 1995, 1997a, b). This genus was regarded as a sister taxon to *Dryosaurus*, *Camptosaurus*, and iguanodontids by Weishampel and Heinrich (1992). By comparison with the ontogenetic changes seen in *Dryosaurus*, it can be argued that a feature such as the long snout in *Tenontosaurus* is a peramorphic structure, forming from positive allometric growth of the nasal and frontal bones. Moreover, *Tenontosaurus* possesses peramorphic features including a relatively straight quadrate bone, development of a retroarticular process, and relatively small orbits that, like the long snout, are peramorphic features (Fig. 35.7).

Despite the paucity of juvenile material of sauropods, there is evidence to indicate that the very large body size, massive limbs, and elongate neck produced by the massive increase in relative size of the cervical vertebrae arose by peramorphosis (Fig. 35.8). Growth patterns of *Apatosaurus* show that it reached maturity when aged about 12 years, and grew at a maximum rate of a phenomenal 5.5 tonnes per year (Curry 1999; Erickson et al. 2001).

35.6. Ontogenetic changes in the pelvic region of *Gorgosaurus* compared with early tyrannosaurids *Guanlong* and *Siamotyrannus* and the advanced *Tyrannosaurus* and *Tarbosaurus*. Based, in part, on Long and McNamara (1997a, 3) and on Xu et al. (2006, 2e).

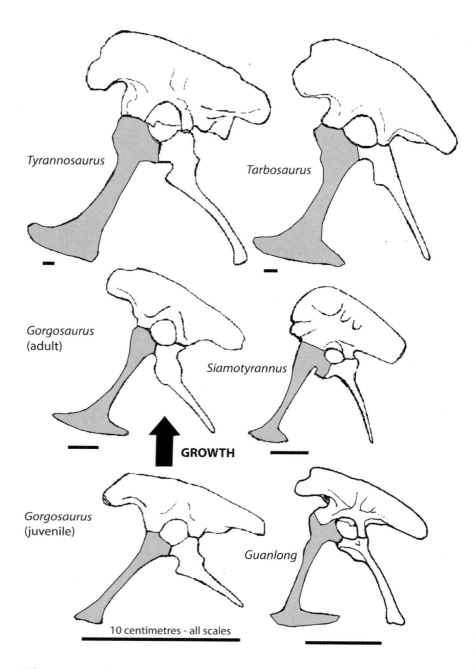

Tyrannosaurus

Tarbosaurus

Gorgosaurus (adult)

Siamotyrannus

GROWTH

Gorgosaurus (juvenile)

Guanlong

10 centimetres - all scales

This is equivalent to over 15 kg per day! Most of this rapid growth occurred between the ages of about 6 and 10 years. The prosauropod, *Massospondylus*, by contrast, while achieving maturity at a similar or even slightly later age than *Apatosaurus*, had a maximum growth rate of a mere 34.6 kg per year (Chinsamy 1993; Erickson et al. 2001). This suggests that, in contrast to theropods, the peramorphic features in sauropods arose from acceleration alone.

Juvenile material of the Early Cretaceous *Phuwiangosaurus* shows morphological features that otherwise occur in the adults of older taxa, supporting the notion of peramorphic evolution in later sauropods. Thus, while the centra and pleurocoels of the juvenile vertebrae of *Phuwiangosaurus* are very simple, they become more complex in adults. The vertebral morphology of older adult sauropods from the Middle Jurassic is like that of

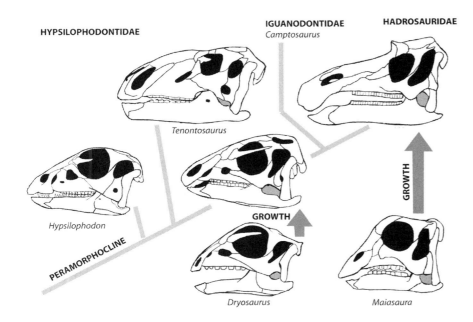

HYPSILOPHODONTIDAE

IGUANODONTIDAE
Camptosaurus

HADROSAURIDAE

Tenontosaurus

Hypsilophodon

GROWTH

GROWTH

PERAMORPHOCLINE

Dryosaurus

Maiasaura

35.7. Peramorphosis in euornithopod evolution. Based on Long and McNamara (1995, 8.4).

the juvenile *Phuwiangosaurus*. Likewise, the femora of juveniles of *Phuwiangosaurus* are more elongate than in adults, a condition found in earlier adult sauropods.

A study of sauropod embryos from the Late Cretaceous of Patagonia (Salgado et al. 2005) has shown that some of the features present in the embryos of these very derived forms compare with adult characters of some neosauropods, in particular the morphology of the temporal region. The

35.8. Comparison of a prosauropod skull and a range of sauropod skulls, to illustrate the peramorphic development of the maxilla (stippled). The tremendous size increase in sauropodomorphs from the Late Triassic through to the Late Jurassic is another peramorphic feature. Based on Long and McNamara (1997a, 10).

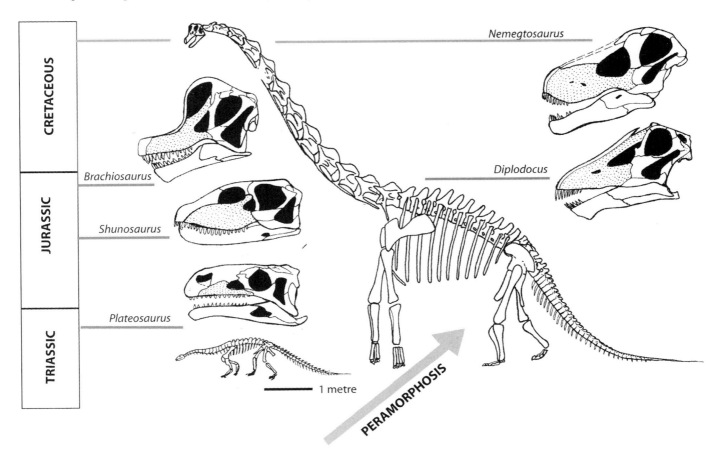

CRETACEOUS

JURASSIC

TRIASSIC

Nemegtosaurus

Brachiosaurus

Diplodocus

Shunosaurus

Plateosaurus

1 metre

PERAMORPHOSIS

embryonic skulls possess a mosaic of features that suggest that different parts of the skull evolved at different rates during ontogeny, with some characters being peramorphic, and others paedomorphic (see below).

Paedomorphosis

Within a group that otherwise seems to be dominated by peramorphosis, the theropod genus *Nanotyrannus* has been suggested as an example of an essentially paedomorphic theropod, its small size (body length of "only" 5 m) suggesting that it evolved by progenesis. Paedomorphic features include the more slender snout, wide, rounded orbit, and lack of a postorbital bar (Long and McNamara 1997a), all of which are seen in the oldest and most plesiomorphic tyrannosaurid, *Guanlong* (Xu et al. 2006). However, some authors have suggested that *Nanotyrannus* is an immature tyrannosaurid, perhaps belonging with *Tyrannosaurus* (Carr 1999; Holtz 2001). This would therefore nullify any suggestions as to its paedomorphic origins. Even if *Nanotyrannus* is considered to be an immature tyrannosaurid, Currie (2003) and Currie et al. (2003) have pointed out that its high number of maxillary teeth preclude it from being considered conspecific or even congeneric with *T. rex*. However, the degree to which *Nanotyrannus* represents a paedomorphic tyrannosaurid awaits the discovery of adult material assignable to this taxon.

Although, as discussed above, sauropod evolution was dominated by the operation of peramorphic processes, resulting in the evolution of very large body size, examples of paedomorphosis that produced so-called dwarf forms have been described. The Late Cretaceous titanosaurid *Magyarosaurus dacus* from Transylvania is much smaller than other sauropods of similar age elsewhere in the world. This small size has been attributed to paedomorphosis (Jianu and Weishampel 1999). On the basis of comparison of the humeri of *M. dacus* with ontogenetic material of other sauropods, this so-called dwarf sauropod appears to be most similar to subadults of other taxa, in both shape and size. Similarly, the euornithopod *Zalmoxes robustus* from the same horizon and locality as *Magyarosaurus* has also been interpreted as a paedomorphic dwarf (Weishampel et al. 2003). It had been suggested by Nopcsa (1914) that the Hateg Basin from where *Magyarosaurus* and *Zalmoxes* came was insular and possibly associated with an island, suggesting the possibility of them being island dwarf forms (Jianu and Weishampel 1999; Weishampel et al. 2003), much like the "dwarf" (i.e., paedomorphic) elephants that lived on Corfu (Boekschoten and Sondaar 1972) and the paedomorphic Pleistocene hippos of Sumatra (Roth 1992).

Paedomorphosis has more often been recorded in dinosaurs targeting specific morphological traits. For instance, it has been reported in another dinosaur from the Hateg Basin, the ornithopod *Telmatosaurus*, in the form of miniaturized maxillary dentition (Weishampel et al. 1993). Some features in later hadrosaurs have been interpreted as paedomorphic. Although possessing predominantly peramorphic traits, *Tenontosaurus* possesses short palpebral bones, and relatively large and elongate antorbital fenestra, features present in juveniles of *Dryosaurus* (Long and McNamara 1997a). Some features of titanosaurian skull ontogeny seem to have evolved by paedomorphosis, such as the loss of the squamosal–quadratojugal contact in some adult sauropods (Salgado et al. 2005, 87). Paedomorphosis in some postcranial elements in prosauropods may have played a crucial role in the evolution of sauropods.

Loss of premaxillary teeth in both stegosaurids and ankylosaurids occurred by paedomorphosis. These teeth are present in adults of the most primitive stegosaurid *Huayangosaurus* and in the basal thyreophoran *Scutellosaurus* but are absent in later forms. Interestingly, this same loss occurred in the evolution of birds. Similarly, the loss of premaxillary teeth during euornithopod evolution also occurred by paedomorphosis. Other paedomorphic traits occur in the postcranial skeleton of *Tenontosaurus*. Unlike hypsilophodontids, digit III on the manus was paedomorphically reduced. Gaston et al. (2003) have suggested, on the basis of a study of theropod trackways, that reduction in digit number in some dinosaurs also occurred by paedomorphosis. Digit I appears late in developmental sequences in birds and reptiles. Its absence or reduction in theropods, as deduced from trackways, can therefore be interpreted as having occurred by paedomorphosis.

Heterochrony and Developmental Tradeoffs

There is a tendency to describe particular species as either peramorphic or paedomorphic. In reality, most organisms are a cocktail of traits that evolved by both peramorphic and paedomorphic processes, such as in the case outlined above in the skull of sauropod embryos, where some features can be interpreted as peramorphic and others as paedomorphic (Salgado et al. 2005). Such a mixture of peramorphic and paedomorphic features is known as *dissociated heterochrony*. It has been suggested that peramorphic and paedomorphic traits are linked, such that the paedomorphic traits may be developmental tradeoffs for the peramorphic features (McNamara 1997). A frequent consequence of peramorphosis involving size increase is that it is often associated with paedomorphic trends in some traits. Involving, as it does, increase in body mass, evolutionary trends toward increased body size mean that there has to be a greater input of energy to enable the organism to attain the larger body and the accompanying more complex morphological features, than in its ancestor. As a result there can be tradeoffs, whereby selection pressure will favor some structures more than others. This phenomenon has also been recognized in hominid evolution by Aiello and Wheeler (1995), who call it the "expensive tissue hypothesis."

In tyrannosaurids there have been countless arguments as to why *T. rex* possessed such reduced forelimbs and a manus with only two digits. These are extreme paedomorphic features, resembling the state of the forelimb that was likely to have existed in the early embryonic development of ancestral tyrannosaurids. The recently discovered more primitive tyrannosauroid, *Dilong paradoxus* from the Early Cretaceous of Liaoning, China, although very small, being only 1.6 m in length (Xu et al. 2004), had relatively long arms, as did the oldest tyrannosaurid, *Guanlong* (Xu et al. 2006). The reduced forelimb of *T. rex* is very unlikely to have had any adaptive significance, but may have been a developmental tradeoff for the evolution of its huge body size, particularly the hind limbs and head. The peramorphic trend of increased body size, in combination with increased complexity and size of the skull and hind limbs, is offset by a paedomorphic reduction in the forelimbs and digits. *Dilong*, as well as possessing protofeathers, had a more gracile skull, like juveniles of later tyrannosaurids, and more slender hind limbs. A similar example of a possible developmental tradeoff is the presence of a single, large digit in the manus of *Mononykus*, the other digits having been traded for the single digit. The reduction and modification of digits and forelimb size reduction is a common trend that evolved in parallel in a number of tyrannosaurids and other derived maniraptorans, such as abelisaurids. Similar dissociated heterochrony involving possible developmental tradeoffs is also evident in the ornithopod *Tenontosaurus*. In addition to the peramorphic features discussed above, this genus shows evidence of paedomorphosis in the reduction of the digits and forelimbs.

Heterochrony and Evolutionary Novelties

Heterochrony has frequently been proposed as the agent for the evolution of major evolutionary novelties (McKinney and McNamara 1991; Jablonski 2000; McNamara and McKinney 2005). For instance, as early as the 1920s Walter Garstang suggested that paedomorphosis resulted in the evolution of vertebrates from a tunicate-like larva. Gould (1977) argued that of all the different types of heterochrony, progenesis was potentially the most important in producing evolutionary novelties. He suggested that this involved not just the pronounced morphological disparity between ancestor and descendant that can arise from a much earlier offset of growth, but that selection was also targeting the very life-history strategy of precocious maturation (see below). Thus natural selection may often target life history strategies rather than morphology itself.

However, it is likely that other heterochronic processes, such as neoteny and hypermorphosis, also play an important role in macroevolution. For instance, a study of the largest-ever terrestrial lizard, the Pleistocene varanid *Varanus* (*Megalania*) *prisca*, has demonstrated that it evolved by hypermorphosis (Erickson et al. 2003), a concept supporting a recent taxonomic revision that suggests it was just another large species of *Varanus* (Molnar 2004, 107). This dominant Australian carnivore was more than twice the length of any living varanid and achieved its enormous size by delaying the time of onset of maturity by 3 to 4 years. As discussed above, a combination of hypermorphosis and acceleration was probably the key to the evolution of extremely large body size and the relatively massive head and hind limbs in later tyrannosauroids.

Many groups of vertebrates, including cetaceans, ichthyosaurs, and plesiosaurs, show hyperphalangy (the production of numerous finger bones). Analysis of the growth of the dolphin embryo flipper (Richardson and Oelschläger 2002) has shown that there is a developmental basis for this phenomenon that is probably linked to hypermorphic delays in cessation of phalange production during development. Conversely, limb reduction—such as in the forelimbs of *T. rex*—could have arisen from early cessation of production of phalanges.

Variation in patterns of ontogenetic development of individual digits characterizes different groups of tetrapods. For instance, in the early embryogenesis of birds five digits start to form (see Galis et al. 2003 for review), even though as adults they have only three digits, two having failed to develop. Experiments carried out on the digits of chickens and ostriches indicates that evolutionary digit reduction in bird wings has been due to "arrested development" of digit I (i.e., progenesis of this digit), followed by its subsequent degeneration. Such may also have been the case in those theropods with reduced digit number.

Thus those theropods with reduced digit I can be interpreted as having undergone progenesis very early in development. The skeletal elements of the digit just stopped growing much earlier than in the ancestral form. This has relevance to the arguments concerning bird evolution from dinosaurs. Forelimb digits in birds have long been thought to consist of II–IV. However, in theropod dinosaurs, the presumed ancestors of birds, they are I–III. Some have argued that this apparently seemingly fundamental difference nullifies the view that birds evolved from dinosaurs (Feduccia 2003). However, it can also be argued that there is no reason why birds could not have evolved from an ancestor with digits I–III (McNamara and McKinney 2005). Heterochrony targeting specific digits provides an alternative explanation. To derive the avian II–IV pattern from the theropod I–III one, only two basic heterochronic changes would have had to have occurred: early progenesis of digit I, combined with hypermorphosis in digit IV. However, recent work by Xu et al. (2009) has found evidence for the derivation of the standard theropod digit pattern in intermediate early fossil forms of birds. Likewise, Tamura et al. (2011) have shown that wing digits in modern birds are the same as in theropod dinosaurs, namely I–III. Studies of Hox gene expression by Vargas and Fallon (2005) similarly found a pattern that identified the anteriormost digit of the bird wing as digit I, in accordance with the hypothesis that these digits are I, II, and III, as in theropod dinosaurs.

Heterochrony played an important role in the evolution of birds from dinosaurs in other ways. There have been suggestions that feathers were present not only on some small adult dinosaurs but also on the juveniles of larger dinosaurs, but not on the adults. While this is largely speculative, if it was the case, then the small-bodied dinosaurs, such as the feathered *Sinosauropteryx* and *Caudipteryx*, can be regarded as paedomorphic dinosaurs. Likewise the retention of feathers in birds can be viewed as being by paedomorphosis. Other paedomorphic features of early birds include reduction in some of the digits; the small body size; retention of a number of unfused bones; gracile, elongate skull; large orbit; relatively large brain case; tooth shape; and subsequent tooth loss. However, early "birds," such as *Archaeopteryx*, possess greatly enlarged forelimbs, indicating the operation of peramorphic processes targeting these particular skeletal elements.

Paedomorphosis in limb development may have been important in another major evolutionary event–the evolution of sauropods from prosauropods (Bonaparte and Vince 1979). In their description of embryonic material of the prosauropod *Massospondylus carinatus*, Reisz et al. (2005) argue that the hatchlings were facultatively quadrupedal, in contrast to the adults, which were at least facultatively bipedal. They base this on the size of the forelimbs, which are relatively much longer in the embryo than in the adults, being of similar length to the hind limbs (Fig. 35.9). This is the state found in adults of later sauropods. Another feature of the embryonic prosauropod is its long, horizontally held neck, in contrast to the situation

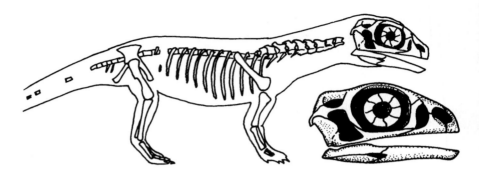

35.9. Reconstructions of the embryo of the Early Jurassic prosauropod *Massospondylus carinatus* showing the relatively long forelimbs and long, horizontally held neck, features found in adults of later sauropods. Based on Reisz et al. (2005, 3).

found in adults. This is another trait of adult sauropods, suggesting that it too may have a paedomorphic origin.

Heterochrony and Evolutionary Trends

Heterochrony is also important in the generation of evolutionary trends (McNamara 1982, 1990). This link between evolutionary trends and heterochrony arises because evolutionary trends are by definition unidirectional, in the same way as ontogenetic trajectories. For evolutionary trends to develop, the intrinsic factor of heterochrony must be combined with extrinsic factors of natural selection. Selection of either progressively more paedomorphic or more peramorphic traits often takes place along an environmental gradient in the aquatic environment, from, for example, deep to shallow water, or from coarse- to fine-grained sediments. In the case of heterochronically generated evolutionary trends in dinosaurs situated in the terrestrial environment, the agents of selection are harder to quantify.

An evolutionary trend from ancestors to descendants that show increasingly more paedomorphic characters is known as a *paedomorphocline*. If the trend produces increasingly more peramorphic descendants it is called a *peramorphocline* (McNamara 1982). Collectively these are called *heterochronoclines* (McKinney and McNamara 1991). The driving force behind heterochronoclines is often competition or predation pressure. When a heterochronocline develops by competition, the persistence of the ancestral form constrains selection in a single direction, away from the ancestral species. The resultant phylogenetic pattern will be one of cladogenesis, with branching species. However, selection induced by predation pressure will produce an anagenetic heterochronocline, whereby ancestral species are replaced by the descendant form. While a number of such anagenetic trends have been described in invertebrate groups (McNamara 1997), none have, to date, been described in dinosaurs. However, examples of cladogenetic evolutionary trends fuelled by heterochrony have been documented in neoceratopsians. The earliest and most "primitive" ceratopsian, the Early Cretaceous *Psittacosaurus*, shows a relatively small degree of morphological change during ontogeny. The most prominent changes are the relative reduction in the size of the orbit (an ontogenetic trend seen in most, if not all, dinosaurs) and the dorsal expansion of the anterior of the skull, producing a beaklike structure. In later neoceratopsians, such as *Bagaceratops*, *Breviceratops*, and *Protoceratops*, there are a number of heterochronic changes in the skull that can be interpreted as progressively more peramorphic (Fig. 35.10). There is a general increase in morphological complexity and development of frills

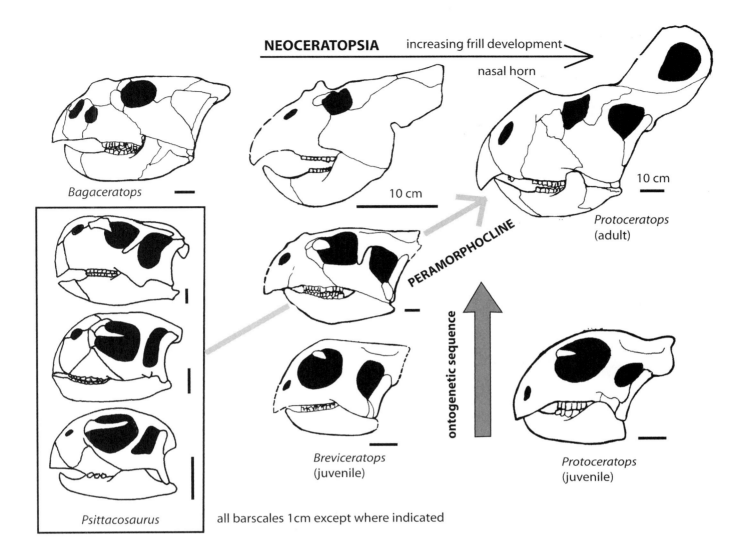

NEOCERATOPSIA increasing frill development

nasal horn

Bagaceratops

10 cm

PERAMORPHOCLINE

ontogenetic sequence

Protoceratops
(adult)

10 cm

Psittacosaurus

all barscales 1cm except where indicated

Breviceratops
(juvenile)

Protoceratops
(juvenile)

35.10. Peramorphosis in neoceratopsian evolution. Based on Long and McNamara (1995, 8.6).

and horn structures on the skull in later adults. This is particularly evident in some advanced neoceratopsians, such as *Centrosaurus, Einiosaurus,* and *Pachyrhinosaurus* (Fig. 35.11). These morphological trends are accompanied by trends of increased body size. While the juveniles of *Breviceratops* and *Protoceratops* are very similar to each other, later forms show progressively greater degrees of morphological change during ontogeny, producing a peramorphocline.

There are less pronounced morphological changes postcranially. The main trends involve flattening of the digits in the later, larger forms, to accommodate the greater body weight, and increase in size of the olecranon process of the ulna in chasmosaurines. In chasmosaurines most of the postcranial skeletal elements grow isometrically (Lehman 1990). In some large ceratopsians, such as *Triceratops,* broadening and thickening of the limbs to support the large body probably rose by peramorphosis.

There exists a close relationship between life history strategies and heterochrony. Life history strategies include a number of factors, notably size at birth; growth rates; age at maturation; body size at maturity; the number,

Heterochrony and Life History Strategies

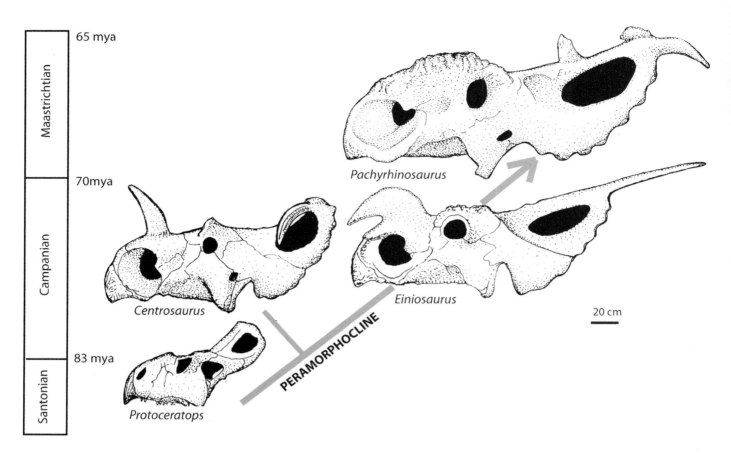

35.11. Peramorphocline in skull evolution in advanced neoceratopsians. Based on Long and McNamara (1997a, 8).

size, and sex of offspring; and length of life. Many of these factors are determined by heterochrony. There have been a number of attempts over the years to categorize life history traits, but these have met with only limited success. The most well known is the "*r-K* continuum." This is a descriptor of both environments and the life history traits of the organisms that inhabit these environments. Although the concept of the *r-K* continuum is an oversimplification of life history strategies, it does seem to have applicability at higher taxonomic levels.

So what is the *r-K* continuum? The "*r*-selected" end of the continuum incorporates unpredictable, often ephemeral environments. Consequently, selection pressure for organisms living in such environments targets those that mature rapidly, have short life spans, and have small body size. These are all produced by progenesis. These *r*-selected organisms typically produce large numbers of offspring. To date no dinosaurs have been unequivocally identified as being *r*-selected. However, one group that may fit this category are the small, feathered dinosaurs that occur in the Early Cretaceous at Liaoning. At the other extreme of the continuum are "*K*-selected" animals. These occupy unchanging, or at least predictable, environments. Organisms inhabiting such environments are characterized by having a relatively delayed onset of reproduction, long life span, and consequent large body size. These are all features characteristic of hypermorphosis. Another feature of such organisms is that they produce few, large offspring.

It has been suggested (McKinney and McNamara 1991) that as well as being generated by progenesis, *r*-selected characteristics may also be produced by acceleration. Likewise, many *K*-selected organisms appear to show some traits that evolved by neoteny, as well as hypermorphosis. However, the evidence that the evolution of *T. rex* occurred by a combination of both acceleration and hypermorphosis indicates that this model may be overly simplistic. Other factors, particularly predation pressure, are likely to have played a significant role. Given the likelihood that the prey of *T. rex* was as large as, if not larger than, itself, then selection would have favored acceleration of growth rates to attain the massive body size as quickly as possible during development. This need to rapidly attain a large body size may be interpreted as a type of selection pressure, perhaps instigated by environmental stress. Hypermorphosis, which seems to have played a relatively minor role in the evolution of *T. rex*, would have been the icing on the evolutionary cake. Similarly, it could be argued that the acceleration in growth rates that occurred in later sauropods is also indicative of a stressful environment, the stress again having been induced by potential predation pressure—the rapid attainment of a large body size to avoid being predated would have been a strong selection pressure.

With the great increase in research on bone microstructure allowing life history parameters such as growth rate and maturation time to analyzed for many groups of dinosaurs, other aspects of their life histories and physiology are able to be better understood. While much of the research is providing data on growth rates and maturation time that allow predictions of heterochrony to be proposed, research by Sander and Klein (2005) has highlighted the need to analyze as many specimens of a single species as possible. In their analysis of the prosauropod *Plateosaurus engelhardti* from the Late Triassic of central Europe they found a high level of developmental plasticity, especially in the rate of growth, with some individuals being fully grown with a body length of 4.5 m, but others grew to 10 m. They found a poor correlation between size and age, highlighting the problems in assuming age from size when trying to interpret heterochronic processes from fossil material. Environmental factors are thought to have accounted for the wide variations in growth rate. Sander and Klein point out that in modern amniotes, such strong developmental plasticity correlates with low metabolic rate and behavioral thermoregulation, suggesting that *Plateosaurus* may well have been an ectotherm.

Acknowledgments

We wish to thank Brian Choo for reading the manuscript and offering helpful suggestions for its improvement. Marcus Good is thanked for his assistance with the provision of literature.

References

Aiello, L. C., and Wheeler, P. 1995. The expensive-tissue hypothesis. *Current Anthropology* 36: 199–221.

Alberch, P., S. J. Gould, G. F. Oster, and D. B. Wake. 1979. Size and shape in ontogeny and phylogeny. *Paleobiology* 5: 296–317.

Boekschoten, G. J., and P. Y. Sondaar. 1972. On the fossil mammals of Cyprus. *Proceedings Koninklijke Nederlandse Akademie van Wetenschappen* B75: 306–338.

Bonaparte, J. F., and M. Vince. 1979. El hallazgo del primer nido de dinosaurios triásicos (Saurischia, Prosauropoda), Triásico Superiorde Patagonia, Argentina. *Ameghiniana* 16: 173–182.

Buffetaut, E., V. Suteethorn, and H. Tong. 1996. The earliest known tyrannosaur from the Lower Cretaceous of Thailand. *Nature* 381: 689–691.

Carpenter, K. 1992. Tyrannosaurids (Dinosauria) of Asia and North America. In N. Mateer and Chen Pei-ji (eds.), *Aspects of Nonmarine Cretaceous Geology*, 250–268. Beijing: China Ocean Press.

Carpenter, K. 1994. Baby *Dryosaurus* from the Upper Jurassic Morrison Formation of Dinosaur National Monument. In K. Carpenter, K. F. Hirsch, and J. R. Horner (eds.), *Dinosaur Eggs and Babies*, 288–297. Cambridge: Cambridge University Press.

Carpenter, K., K. F. Hirsch, and J. R. Horner (eds.). 1994. *Dinosaur Eggs and Babies*. Cambridge: Cambridge University Press.

Carr, T. D. 1999. Craniofacial ontogeny in Tyrannosauridae (Dinosauria, Coelurosauria). *Journal of Vertebrate Palaeontology* 19: 497–520.

Chapman, R. E. 1990. Shape analysis in the study of dinosaur morphology. In K. Carpenter and P. J. Currie (eds.), *Dinosaur Systematics: Approaches and Perspectives*, 21–42. Cambridge: Cambridge University Press.

Chinsamy, A. 1990. Physiological implications of the bone histology of *Syntarsus rhodesiensis* (Saurischia: Theropoda). *Palaeontologica Africana* 27: 77–82.

———. 1992. Ontogenetic growth of the dinosaurs *Massospondylus carinatus* and *Syntarsus rhodesiensis*. *Journal of Vertebrate Paleontology* 12: 23A.

———. 1993. Bone histology and growth trajectory of the prosauropod *Massospondylus carinatus* Owen. *Modern Geology* 18: 319–329.

———. 1995. Ontogenetic change in the bone histology of the Late Jurassic ornithopod *Dryosaurus lettowvorbecki*. *Journal of Vertebrate Paleontology* 15: 96–104.

———. 2005. *The Microstructure of Dinosaur Bone: Deciphering Biology with Fine-Scale Techniques*. Baltimore: Johns Hopkins University Press.

Constantine, A., A. Chinsamy, T. H. Rich, and P. Vickers-Rich. 1998. Periglacial environments and polar dinosaurs. *South African Journal of Science* 94: 137–141.

Cooper, M. R. 1981. The prosauropod dinosaur *Massospondylus carinatus* Owen from Zimbabwe: its biology, mode of life and phylogenetic significance. *Occasional Papers of the National Museum and Monuments* (Rhodesia) B 6: 689–840.

Currie, P. J. 2003. Cranial anatomy of tyrannosaurid dinosaurs from the Late Cretaceous of Alberta, Canada. *Acta Palaeontologica Polonica* 48: 191–226.

Currie, P. J., J. H. Hurum, and K. Sabath. 2003. Skull structure and evolution in tyrannosaurid dinosaurs. *Acta Palaeontologica Polonica* 48: 227–234.

Currie, P. J., J. K. Rigby, and R. E. Sloan. 1990. Theropod teeth from the Judith River Formation of southern Alberta, Canada. In K. Carpenter and P. J. Currie (eds.), *Dinosaur Systematics: Approaches and Perspectives*, 107–125. Cambridge: Cambridge University Press.

Curry, K. A. 1999. Ontogenetic histology of *Apatosaurus* (Dinosauria: Sauropoida): new insights on growth rates and longevity. *Journal of Vertebrate Paleontology* 19: 654–665.

Dodson, P. 1975. Taxonomic implications of relative growth in lambeosaurine hadrosaurs. *Systematic Zoology* 24: 37–54.

Erickson, G. M., K. Curry Rogers, D. J. Varricchio, M. A. Norell, and X. Xu, 2007. Growth patterns in brooding dinosaurs reveals the timing of sexual maturity in non-avian dinosaurs and genesis of the avian condition. *Biology Letters* 3: 558–561.

Erickson, G. M., K. Curry Rogers, and S. A. Yerby. 2001. Dinosaurian growth patterns and rapid avian growth rates. *Nature* 412: 429–433.

Erickson, G. M., A. de Ricqlès, V. de Buffrénil, R. E. Molnar, and M. K. Bayless. 2003. Vermiform bones and the evolution of gigantism in *Megalani* – how a reptilian fox became a lion. *Journal of Vertebrate Paleontology* 23: 966–970.

Erickson, G. M., P. J. Makovicky, P. J. Currie, M. A. Norell, S. A. Yerby, and C. A. Brochu. 2004. Gigantism and comparative life-history parameters of tyrannosaurid dinosaurs. *Nature* 430: 772–775.

Erickson, G. M., and T. A. Tumanova. 2000. Growth curve of *Psittacosaurus mongoliensis* Osborn (Ceratopsia: Psittacosauridae) inferred from long bone histology. *Zoological Journal of the Linnean Society of London* 130: 551–566.

Feduccia, A. 2003. Bird origins: problem solved, but the debate continues. *Trends in Ecology and Evolution* 18: 9–10.

Galis, F., M. Kundrát, and B. Sinervo. 2003. An old controversy solved: bird embryos have five fingers. *Trends in Ecology and Evolution* 18: 7–9.

Gaston, R., M. G. Lockley, S. G. Lucas, and A. P. Hunt. 2003. *Grallator*-dominated fossil footprint assemblages and associated enigmatic footprints from the Chinle Group (Early Triassic), Gateway Area, Colorado. *Ichnos* 10: 153–163.

Gould, S. J. 1977. *Ontogeny and Phylogeny*. Cambridge, Mass.: Belknap Press of Harvard University Press.

Guenther, M. E. 2009. Influence of sequence heterochrony on hadrosaurid dinosaur postcranial development. *Anatomical Record* 292: 1427–1441.

Hall, B. K. 1992. *Evolutionary Developmental Biology*. London: Chapman and Hall.

Holtz, T. R. 2001. The phylogeny and taxonomy of the Tyrannosauridae. In D. H. Tanke and K. Carpenter (eds.), *Mesozoic Vertebrate Life*, 64–83. Bloomington: Indiana University Press.

Horner, J. R. 2005. A new *Triceratops* growth series. *Journal of Vertebrate Paleontology* (suppl. to 3): 71A.

Horner, J. R. and P. J. Currie. 1994. Embryonic and neonatal morphology and ontogeny of a new species of *Hypacrosaurus* (Ornithischia, Lambeosauridae) from Montana and Alberta. In K. Carpenter, K. F. Hirsch, and J. R. Horner (eds.), *Dinosaur Eggs and Babies*, 312–336. Cambridge: Cambridge University Press.

Horner, J. R., A. de Ricqlès, and K. Padian. 2000. Long bone histology of the hadrosaurid dinosaur *Maiasaura peeblesorum*: growth dynamics and physiology based on an ontogenetic series of skeletal elements. *Journal of Vertebrate Paleontology* 20: 115–129.

Horner, J. R., and D. B. Weishampel. 1988. A comparative embryological study of two ornithischian dinosaurs. *Nature* 332: 256–257.

Jablonski, D. 2000. Micro- and macroevolution: scale and hierarchy in evolutionary biology and paleobiology. In D. H. Erwin and S. L. Wing (eds.), *Deep Time: Paleobiology's Perspective*, 15–52. Supplement to *Paleobiology* 26 (4). Lawrence, Kans.: Paleontological Society.

Jianu, C-M., and D. B. Weishampel. 1999. The smallest of the largest: a new look at possible dwarfing in sauropod dinosaurs. *Geologie en Mijnbouw* 78: 335–343.

Jones, D. S. 1988. Sclerochronology and the size versus age problem. In M. L. McKinney (ed.), *Heterochrony in Evolution: A Multidisciplinary Approach*, 93–108. New York: Plenum.

Klingenberg, C. P. 1998. Heterochrony and allometry: the analysis of evolutionary change in ontogeny. *Biological Reviews* 73: 79–123.

Lee, A. H., and S. Werning. 2008. Sexual maturity in growing dinosaurs does not fit reptilian growth models. *Proceedings of the National Academy of Sciences* 105: 582–587.

Lehman, T. H. 1990. The ceratopsian subfamily Chasmosaurinae: sexual dimorphism and systematics. In K. Carpenter and P. J. Currie (eds.), *Dinosaur Systematics: Approaches and Perspectives*. Cambridge: Cambridge University Press.

Long, J. A., and K. J. McNamara. 1995. Heterochrony in dinosaur evolution. In K. J. McNamara (ed.), *Evolutionary Change and Heterochrony*, 151–168. Chichester: John Wiley and Sons.

——. 1997a. Heterochrony: the key to dinosaur evolution. In D. L. Wolberg, E. Stump, and G. D. Rosenberg (eds.), *Dinofest International*, 113–123. Philadelphia: Academy of Natural Sciences.

——. 1997b. Heterochrony. In P. J. Currie and K. Padian (eds.), *The Encyclopedia of Dinosaurs*, 311–317. San Diego: Academic Press.

Madsen, J. H. 1976. *Allosaurus fragilis*: a revised osteology. *Bulletin of the Utah Geological and Mining Survey* 109: 1–163.

Maryańska, T., and H. Osmólska. 1975. Proceratopsidae (Dinosauria) of Asia. *Palaeontologica Polonica* 33: 45–102.

McKinney, M. L. (ed.) 1988. *Heterochrony in Evolution: A Multidisciplinary Approach*. New York: Plenum.

——. 1990. Trends in body size evolution. In K. J. McNamara (ed.), *Evolutionary Trends*, 75–118. London: Belhaven.

McKinney, M. L., and K. J. McNamara. 1991. *Heterochrony: The Evolution of Ontogeny*. New York: Plenum.

McNamara, K. J. 1982. Heterochrony and phylogenetic trends. *Paleobiology* 8: 130–142.

——. 1986. A guide to the nomenclature of heterochrony. *Journal of Paleontology* 57: 461–473.

——. 1988. The abundance of heterochrony in the fossil record. In M. L. McKinney (ed.), *Heterochrony in Evolution: A Multidisciplinary Approach*, 287–325. New York: Plenum.

—— (ed.). 1990. *Evolutionary Trends*. London: Belhaven.

—— (ed.). 1995. *Evolutionary Change and Heterochrony*. Chichester: John Wiley and Sons.

——. 1997. *Shapes of Time: The Evolution of Growth and Development*. Baltimore: Johns Hopkins University Press.

McNamara, K. J., and M. L. McKinney. 2005. Heterochrony, disparity, and macroevolution. *Paleobiology* 31 (2) (suppl.): 17–26.

Molnar, R. E. 1990. Variation in theory and in theropods. In K. Carpenter and P. J. Currie (eds.), *Dinosaur Systematics:*

Approaches and Perspectives, 71–79. Cambridge: Cambridge University Press.

——. 2004. *Dragons in the Dust: The Paleobiology of the Giant Monitor Lizard Megalania*. Bloomington: Indiana University Press.

Nopcsa, F. 1914. Über das Vorkommen der Dinosaurier in Siebenbürgen. *Verhandlungen der Zoologischen und Botanischen gessellschaft* 54: 12–14.

Padian, K., A. J. de Ricqlès, and J. R. Horner. 2001. Dinosaurian growth rates and bird origins. *Nature* 412: 405–408.

Paul, G. S. 1988. *Predatory Dinosaurs of the World*. New York: Simon and Schuster.

Raath, M. A. 1990. Morphological variation in small theropods and its meaning in systematics: evidence from *Syntarsus rhodesiensis*. In K. Carpenter and P. J. Currie (eds.), *Dinosaur Systematics: Approaches and Perspectives*, 91–105. Cambridge: Cambridge University Press.

Reid, R. E. H. 2012 (this volume). How dinosaurs grew. In M. K. Brett-Surman, Thomas R. Holtz Jr., and James O. Farlow (eds.), *The Complete Dinosaur*, 2nd ed., 621–635. Bloomington: Indiana University Press.

Reisz, R. R., D. Scott, H.-D. Sues, D. C. Evans, and M. A. Raath. 2005. Embryos of an early Jurassic prosauropod dinosaur and their evolutionary significance. *Nature* 309: 761–764.

Richardson, M. K., and P. M. Brakefield. 2003. Hotspots for evolution. *Nature* 424: 894–895.

Richardson, M. K., and H. H. A. Oelschläger. 2002. Time, patttern, and heterochrony: a study of hyperphalangy in the dolphin embryo flipper. *Evolution and Development* 4: 435–444.

Roth, V. L. 1992. Inferences from allometry and fossils: dwarfing of elephants on islands. *Oxford Surveys in Evolutionary Biology* 8: 259–288.

Ruben, J. A., C. Dal Sasso, N. R. Geist, W. J. Hillenius, T. D. Jones, and M. Signore. 1999. Pulmonary function and metabolic physiology of theropod dinosaurs. *Science* 283: 514–516.

Ruben, J. A., T. D. Jones, N. R. Geist, and W. J. Hillenius. 1997. Lung structure and ventilation in theropod dinosaurs and early birds. *Science* 278:1267–1270.

Russell, D. A. 1970. Tyrannosaurs from the Late Cretaceous of western Canada. *Palaeontological Publications of the National Museum of Natural Science* 1: 1–34.

Salgado, L., R. A. Coria, and L. M. Chiappe. 2005. Osteology of the sauropod embryos from the Upper Cretaceous of

Patagonia. *Acta Palaeontologica Polonica* 50: 79–92.

Sander, P. M., and N. Klein. 2005. Developmental plasticity in the life history of a prosauropod dinosaur. *Science* 310: 1800–1802.

Scannella, J. B., and J. R. Horner. 2010. *Torosaurus* Marsh, 1891, is *Triceratops* Marsh, 1889 (Ceratopsidae: Chasmosaurinae): synonymy through ontogeny. *Journal of Vertebrate Paleontology* 30: 1157–1168.

Tamura, K., N. Nomura, R. Seki, S. Yonei-Tamura, and H. Yokoyama. 2011. Embryological evidence identifies wing digits in birds as digits 1, 2, and 3. *Science* 331: 753–757.

Vargas, A. O., and J. F. Fallon. 2005. Birds have dinosaur wings: the molecular evidence. *Journal of Experimental Zoology* 304B: 85–90.

Varricchio, D. J. 1993. Bone microstructure of the Upper Cretaceous theropod *Troodon formosus*. *Journal of Vertebrate Paleontology* 13: 99–104.

Webster, M., and M. Zelditch. 2005. Evolutionary modifications of ontogeny: heterochrony and beyond. *Paleobiology* 31: 354–373.

Weishampel, D. B. 1981. The nasal cavity of lambeosaurine hadrisaurods (Reptilia: Ornithischia): comparative anatomy and homologies. *Journal of Paleontology* 55: 1046–1058.

Weishampel, D. B., and R. E. Heinrich. 1992. Systematics of the Hypsilophodontidae and basal Iguanodontia (Dinosauria: Ornithopoda). *Historical Biology* 6: 159–184.

Weishampel, D. B., and J. R. Horner. 1994. Life history syndromes, heterochrony, and the evolution of the Dinosauria. In K. Carpenter, K. F. Hirsch, and J. R. Horner (eds.), *Dinosaur Eggs and Babies*. Cambridge: Cambridge University Press.

Weishampel, D. B., C.-M. Jianu, Z. Csiki, and D. B. Norman. 2003. Osteology and phylogeny of *Zalmoxes* (n.g.), an unusual euornithopod dinosaur from the latest Cretaceous of Romania. *Journal of Systematic Palaeontology* 1: 65–123.

Weishampel, D. B., D. B. Norman, and D. Grigorescu. 1993. *Telmatosaurus transsylvanicus* from the Late Cretaceous of Romania: the most basal hadrosaurid. *Palaeontology* 36: 361–385.

Welles, S. P., and R. A. Long. 1974. The tarsus of theropod dinosaurs. *Annals of the South African Museum* 64: 191–218.

Xu, X., J. M. Clark, C. Forster, M. A. Norell, D. A. Eberth, J. Chengkai, and

Q. Zhao. 2006. A basal tyrannosauroid dinosaur from the Late Jurassic of China. *Nature* 439: 715–718.

Xu, X., J. M. Clark, J. Mo, J. Choiniere, C. A. Forster, G. M. Erickson, D. W. E. Hone, C. Sullivan, D. A. Eberth, S. Nesbitt, Q. Zhao, R. Hernandez, C. Jia, F. Han, and Y. Guo. 2009. A Jurassic ceratosaur from China helps clarify avian digital homologies. *Nature* 459: 940–944.

Xu, X., M. A. Norell, X. Kuang, X. Wang, Q. Zhao, and C. Jia. 2004. Basal tyrannosauroids from China and evidence of protofeathers in tyrannosauroids. *Nature* 431: 680–684.

Xu, X., Q. Tanm, J. Wang, X. Zhao, and L. Tan. 2007. A gigantic bird-like dinosaur from the Late Cretaceous of China. *Nature* 447: 844–847.

Xu, X., X.-L. Wang, and H.-L. You. 2001. A juvenile ankylosaur from China. *Naturwissenschaften* 88: 297–300.

Metabolic Physiology of Dinosaurs and Early Birds

36

John A. Ruben, Terry D. Jones, Nicholas R. Geist, Willem J. Hillenius, Amy E. Harwell, and Devon E. Quick

There is far more to dinosaurs than just their impressive appearance in museum displays and fearsome demeanor in movies and documentaries. In many respects, they were probably the most successful of all terrestrial vertebrates. The 150+ million year (Late Triassic to Late Cretaceous) cosmopolitan reign of dinosaurs over the terrestrial environment far exceeds the duration and magnitude of dominance by any other group of tetrapods, including mammals and their ancestors. Moreover, dinosaurian diversity was spectacular; their variety encompassed some of the most specialized herbivores and carnivores ever to have existed. Little wonder that so much energy has been devoted to the biology of these long-extinct animals.

Few aspects of dinosaur biology have engendered more debate than attempts to unravel their thermoregulatory and metabolic biology. The notion that dinosaurs, like living birds and mammals, were "warm-blooded," or endothermic, provides an alluring scenario consistent with interpretations of these animals as having led particularly active, interesting lives. The more traditional model of "cold-blooded," or ectothermic dinosaurs, is often wrongly associated with cold, dull-witted beasts leading slothful, sedentary lives. However, accurate interpretation of the biology of any group of animals necessitates that only carefully evaluated, objective evidence be considered. Unfortunately, this has not always been the case. Rather,

> the strongest impression gained from reading the literature of the dinosaur [metabolic] physiology controversy is that some of the participants have behaved more like politicians or attorneys than scientists, passionately coming to dogmatic conclusions via arguments based on questionable assumptions and/or data subject to other interpretations (Farlow 1990).

In addition, misguided peer-pressure has influenced many researchers to interpret morphology, physiology, or behavior of extinct taxa "according to the cladogram" (Dodson 2000; Feduccia 1999).

As an alternative, it may often be more revealing to examine living taxa for functionally-linked, specialized (apomorphic) structures and then identify these same derived attributes in related, extinct taxa. Thus, rather than starting with the a priori assumption that dinosaurs were endothermic, or that they possessed other apomorphic attributes like those of modern birds, we suggest it preferable to assume the presence of the primitive amniote condition (e.g., ectothermy, bi-phasic breathing) and then attempt to falsify that assumption by identifying in dinosaurs the derived morphological features known in living taxa to be causally- or functionally-linked to endothermy or air sac breathing (see also Paladino et al. 1997 or Seebacher 2003).

Here we present recent evidence for thermoregulatory and respiratory biology in dinosaurs and the earliest birds. In both cases, we have considered the shared presence, or absence, of diagnostic, derived morphological features in extinct and related extant forms to be the best indicators of respiratory and thermoregulatory patterns in taxa known only from fossils.

Metabolic and Thermoregulatory Characteristics of Living and Extinct Amniotes

Variation in metabolic rate, especially during periods of rest or routine activity, comprises the core physiological difference between ectotherms (e.g., reptiles) and endotherms (birds, mammals). Endotherms routinely have much higher rates of aerobiosis, or cellular oxygen consumption: at rest, mammalian and avian metabolic rates are typically about 5–15 times greater than those of reptiles of the same body mass and temperature (Bennett 1982, 1991; Bennett and Dalzell 1973; Bennett and Dawson 1976; Else and Hulbert 1981; Schmidt-Nielsen 1984, 1990). In the field, the metabolic rates of mammals and birds typically exceed reptilian rates by about 20 times (Nagy 1987). Furthermore, the increased aerobic capacities of endotherms allow them to sustain routine activity levels well beyond the capacity of most ectotherms. With some noteworthy exceptions, ectotherms such as reptiles are capable of bursts of intense exercise, but generally fatigue rapidly as a result of lactic acid accumulation generated by their reliance on nonsustainable, anaerobic metabolism for all activities beyond relatively slow movements. In contrast, endotherms are able to sustain even relatively high levels of activity for extended periods. This enables these animals to forage widely and to migrate over extensive distances (Bennett and Ruben 1979; Ruben 1995).

Endothermy is one of the major evolutionary developments of vertebrates and is among the most significant features that distinguish living birds and mammals from reptiles, amphibians, and fish. Endothermy, which evolved independently in birds and mammals (Kemp 1988), provides distinct physiological and ecological benefits and may be largely responsible for the present success of birds and mammals in a wide range of aquatic and terrestrial environments (Ruben 1995). Elevated rates of lung ventilation, oxygen consumption, and internal heat production (*via* aerobic metabolism), hallmarks of endothermy, enable birds and mammals to maintain thermal stability over a wide range of ambient temperatures. As a result, endotherms are able to thrive in environments with cold or highly variable thermal conditions and in many nocturnal niches generally unavailable to ectothermic vertebrates.

Thermoregulatory strategies in extant endothermic and ectothermic tetrapods also contrast sharply. During periods of either routine or accelerated levels of activity in endotherms, internal heat production rates are usually sufficient to maintain a constant body temperature over a wide range of ambient temperatures. Alternately, without access to a substantive external heat source (e.g., the sun), ectotherms, especially those of temperate latitudes, are generally unable to achieve and maintain optimal body temperatures. In such cases, Q_{10} effects often cause ectotherms to appear sluggish, even in moderate temperatures. These contrasting thermoregulatory attributes formed much of the basis for the early reconstructions of cold-blooded, brutish, slow-moving dinosaurs, but they were also the impetus for notions of endotherm superiority and the popular appeal of hot-blooded dinosaurs (e.g., Bakker 1968, 1975, 1980).

In reality, low resting metabolic rates typical of extant ectotherms hardly preclude their frequent maintenance of high body temperatures and/ or homeothermy during periods of activity. Given normal field conditions, many extant reptiles maintain marked variation between internal and external temperatures (Besson and Cree 2010, Pearson 1954; Schmidt-Nielsen 1990). Even small lizards of temperate latitudes are often sufficiently adept at solar basking that they sustain diurnal body temperatures that overlap, or in some cases exceed, those of many endotherms (Avery 1982; Greenberg 1980). In addition, some particularly large ectotherms (e.g., Komodo dragons [*Varanus*]) living in warm, equable climates remain virtually homeothermic for long periods in their natural environments. In such cases, these animals achieve inertial homeothermy by virtue of their minimal surface area-to-volume ratios, and correspondingly low heat loss rates (Frair et al. 1972; McNab and Auffenberg 1976; Paladino et al. 1990; Seebacher et al. 1999; Standora et al. 1984).

Given the generally warm, equable climates of the Mesozoic Era, behavioral thermoregulation and thermal inertia were likely to have enabled dinosaurs, whether ecto- or endothermic, to have maintained relatively high and stable body temperatures for extended periods of time. Mathematical models indicate that medium to large dinosaurs (>500 kg; some sauropods probably exceeded 30,000 kg) were likely to have been inertial homeotherms, relatively unaffected by diurnal temperature fluctuations (O'Connor and Dodson 1999; Paladino et al. 1997; Seebacher 2003; Spotila et al. 1991). In such cases, even small fluctuations in body temperatures might only have occurred on a weekly or monthly scale. As in many extant reptiles, small dinosaurs (<75 kg) might easily have utilized behavioral thermoregulation to achieve high body temperatures and homeothermy during periods of diurnal activity. Interestingly, at high latitudes during the Late Cretaceous Era, some dinosaurs might have resorted to hibernation or migration to escape seasonal cooling. These models indicate that dinosaurs were quite capable of maintaining high and stable body temperatures regardless of their metabolic status and the dynamic skeletal structures of many dinosaurs strongly suggests that they possessed bird- or mammal-like capacity for at least burst activity. Thus, even if fully ectothermic, had theropod dinosaurs possessed aerobic metabolic capacities and predatory habits equivalent to those of some large, modern tropical latitude lizards (e.g., *Varanus*), they may well have maintained large home ranges, actively pursued and killed large prey, and defended themselves fiercely (Bennett 1973).

Another variable that affects resting or routine metabolic rate in all organisms is body mass. Because of their greater mass, large ectotherms and endotherms maintain higher total metabolic rates than do small ectotherms and endotherms, respectively. However, because of the allometric relationship between animal mass and metabolic rate (where total metabolic rate scales at about mass$^{0.75}$), each gram of tissue in a small animal actually sustains a higher rate of oxygen consumption than does each gram of tissue in a larger animal. Thus, while a resting elephant obviously maintains a much higher rate of total calorie production than does a resting mouse, each gram of mouse has an oxygen consumption rate considerably greater than does each gram of elephant. The physiological mechanisms and selective factors governing the allometry of metabolism have long puzzled biologists (Darveau et al. 2002; Randall et al. 2002; Schmidt-Nielsen 1990), but the relationship between body mass and metabolic rate is well documented for all vertebrate and invertebrate classes (Hemmingsen 1960). Similar allometry almost surely occurred in dinosaurs as well (Reid 1997). However, by itself, this metabolic scaling has no implication for interpreting the ectothermic or endothermic status of dinosaurs (Reid 1997). As a result of their lower mass, small ectothermic dinosaurs undoubtedly would have maintained considerably higher mass-specific metabolic rates than would larger dinosaurs. Therefore, all other factors being equal, smaller ectothermic dinosaurs were no closer to achieving avian metabolic rates than were large dinosaurs. Similarly, large extant birds and mammals are not less endothermic than are small ones, and small extant reptiles are no closer to an endothermic metabolic status than are larger species.

Considering the significance of endothermy to the biology of extant vertebrates, it is hardly surprising that the evolution of endothermy has received so much attention (Bennett and Huey 1990; Bennett and Ruben 1979; Hopson 1973). In the past few decades, there has been considerable speculation about the possible presence of endothermy, or the lack of it, in extinct sauropsid lineages, such as pterosaurs and dinosaurs, which dominated the terrestrial and aerial environments for most of the Mesozoic Era (Farlow 1990; Padian 1983).

Until very recently, endothermy has been virtually impossible to demonstrate clearly in extinct forms. Endothermy is almost exclusively an attribute of the soft anatomy, and detecting its presence in fossilized remains is difficult as soft tissue leaves a poor or nonexistent fossil record. To support high oxygen consumption levels, endotherms possess profound structural and functional modifications that facilitate oxygen uptake, transport, and delivery. Both mammals and birds have greatly elevated rates of lung ventilation, fully separated pulmonary and systemic circulatory systems, and expanded cardiac output. They also have greatly increased blood volume and blood oxygen-carrying capacities, as well as high tissue aerobic enzymatic activities (Ruben 1995). Unfortunately, these key features of endothermic physiology are unlikely to have ever been preserved in fossils—mammalian, avian, or otherwise. Consequently, previous hypotheses concerning possible endothermy in a variety of extinct vertebrates, especially dinosaurs, have relied primarily on supposed correlations of metabolic rate with a variety of weakly supported criteria (including, but not limited to, predator-to-prey ratios, fossilized trackways, and correlations with avian or mammalian posture) (Bakker 1971, 1980, 1986; Padian 1983) Close scrutiny has revealed that virtually all of these correlations are, at best, equivocal (Bennett and Dalzell 1973; Farlow et al. 1995).

Recently, much of the thought regarding dinosaur metabolic physiology has often centered on the assumed relationship between growth rates and bone histology. Two histological types of compact bone—lamellarzonal and fibrolamellar—have been recognized in extant vertebrates, differing qualitatively in their fibril organization and degree of vascularization. Lamellarzonal bone, the primary bone of extant amphibians and most reptiles, has been associated with ectothermy. Here, poorly vascularized compact bone is deposited in relatively few primary osteons, principally by periosteal deposition, and has a layered appearance, within which incremental growth lines are often recognized (Reid 1997).

Fibrolamellar bone of many birds, mammals, and dinosaurs, is well vascularized and most of the bony matrix is deposited in abundant primary osteons that produce a fibrous, woven appearance (Reid 1997). Fibrolamellar bone is often held to be correlated with high growth rates that require rapid deposition of calcium salts. Such rapid growth is supposedly possible only in systems with the elevated metabolic rates associated with endothermy. Thus, the primary correlation is between growth rate and bone structure. Accordingly, it has been widely accepted that growth rates of extant endotherms in the wild are about an order of magnitude greater than in ectotherms. Given the widespread occurrence of fibrolamellar bone in dinosaurs, their growth is often assumed to have been rapid, as in birds and mammals. According to this scenario, like mammals and birds, dinosaurs must also have been endothermic, or nearly so (Chinsamy and Dodson 1995; Padian and Horner 2002; Reid 1997). However, this conclusion is inconsistent with a variety of paleontological and biological data.

For example, fibrolamellar bone is known to be present in the skeleton of extant, rapidly growing turtles, crocodilians, and lizards (Chinsamy and Dodson 1995; Reid 1997), which suggests that a high basal metabolic rate is not a prerequisite for such bone deposition (contra Horner, Padian et al. 2001; Padian and Horner 2002). Moreover, long bones in numerous dinosaurian genera have regions of both fibrolamellar and lamellarzonal histology (Chinsamy and Dodson 1995; Reid 1997). In addition, during the early stages of development when relative growth rates are highest in virtually all tetrapods, fibrolamellar bone is often formed, but, paradoxically, this is also the stage when particularly fast-growing altricial avian hatchlings are poikilothermic ectotherms (Dietz and Drent 1997; Olson 1992; Ricklefs 1979; Thomas et al. 1993; Visser and Ricklefs 1993).

In any case, attempts to assess growth rates in extinct archosaurs using bone histology as an indicator of bone depositional rates (Horner, Padian et al. 2001) are probably futile. Starck and Chinsamy (2002) have demonstrated that, under experimentally varied environmental and/or nutritional regimes, bone growth rate variation is sufficiently large "that extrapolations of average bone deposition rates from extant birds to fossil dinosaurs (e.g. Horner et al. 2000) are premature and inaccurate." Clearly, since bone depositional rates have not been linked to actual oxygen consumption rates and basal metabolic rates, it is inappropriate to directly infer aspects of metabolic physiology (e.g., endothermy or ectothermy) from bone texture and organization (Chinsamy et al. 1994; Chinsamy and Hillenius 2005).

Bone histology notwithstanding, there is also reason to question the presumed magnitude of variation in growth rates between endotherms and nonavian sauropsids, especially crocodilians, the closest living relatives of birds and dinosaurs. In the most frequently cited comparative study, regressions of maximally sustained growth rates for amniotes scaled positively with increasing adult body mass (slope < 0.7), but reptile y-intercept elevations ("a" values) were reportedly only about 10% those of endotherms (Case 1978). However, criteria for calculating these regressions were not equivalent—endotherm adult weight approximated mass at sexual maturity and mass at a similar stage in the ectotherm life cycle would seem appropriate to facilitate construction of regressions on an equal-footing basis. Nevertheless, American alligator (*Alligator mississippiensis*) adult weight was plotted at 160 kg, a value far in excess of the species' actual 30 kg mass at sexual maturity. In addition, growth rate for the alligator was listed at 28 g/day, rather than the more accurate 42 g/day (Case, personal communication). If the corrected daily growth increment, as well as the more appropriate 30 kg mature mass, is assumed, growth rate for the American alligator is actually about fourfold that of marsupials and approximates growth rates in many placental mammals (Ruben 1995). In this context, it is especially significant that growth rates in some alligators are virtually indistinguishable from estimated growth rates for the dromaeosaurid theropod dinosaur *Troodon* (Ruben 1995). Additionally, Chinsamy and Dodson (1995) evaluated growth rate in three genera of dinosaurs and found no broad pattern of elevated growth rates.

In another scenario, relative quantities of fossilized bone oxygen isotope (O^{16}:O^{18}) were purported to demonstrate relatively little *in vivo* variation between extremity and deep body temperature in some large dinosaurs (e.g., *Tyrannosaurus*) (Barrick and Showers 1994). This was assumed to signify that these large dinosaurs were endothermic because living endotherms, unlike ectotherms, were presumed to maintain relatively uniform extremity versus core temperatures. However, there are abundant data demonstrating that many birds and mammals often maintain extremity temperatures well below deep-body, or core, temperatures (Ruben 1995). Additionally, fossil bone oxygen isotope ratios may be strongly influenced by ground water temperatures (Kolodny et al. 1996). In any case, the conclusions reached by Barrick and Showers (1994) are not statistically supported by their own data (Ruxton 2000). Thus, bone oxygen isotope ratios in dinosaurs are likely to reveal little, if any, definitive information about dinosaur metabolic physiology.

Fisher et al. (2000) describe evidence of a fossilized four-chambered heart with a fully partitioned ventricle in an onithischian dinosaur, *Thescelosaurus*. These authors assert that, as in birds, the fossilized heart lacked a foramen of Panizza between the ventricles and had a single aortic trunk. They also suggest that these attributes are consistent with a high, perhaps endotherm-like, metabolic rate in this dinosaur. Unfortunately, several anatomical errors and questionable assumptions cast serious doubts on this interpretation.

Contrary to Fisher et al. (2000), the ventricles of extant crocodilians and birds are fully separated; the foramen of Panizza is located not between the ventricles, but at the base of the aortic trunks (Franklin et al. 2000; Goodrich 1930; White 1968, 1976). A fully partitioned, four-chambered heart is thus present in both crocodilians and birds; its presence among dinosaurs was not only likely but affords little inference regarding metabolic rates in these animals. The presence of a single aortic trunk (rather than the paired aortic trunks of extant reptilians) might be more diagnostic. However, it is clear that the cardiovascular complex of this specimen is incomplete since neither the pulmonary arteries nor the carotids are preserved (Fisher et al. 2000). Thus, the presence in life of only a single aortic trunk cannot be substantiated. Significantly, the left side of the specimen (where the missing aortic trunk would most likely have been) is also absent from the fossil and, lastly, it still remains unclear if the specimen is actually a fossilized heart (Rowe et al. 2001). This specimen seems to provide little, if any, reliable insight into the metabolic physiology of dinosaurs.

Recently, filiform or featherlike integumentary structures have been described in a variety of small theropod specimens from Early Cretaceous deposits in China. These have been interpreted by some as evidence that theropod dinosaurs possessed an insulative covering, suggestive of endo-thermic homeothermy. However, assertions of feathered dinosaurs have been strongly challenged (Feduccia et al. 2005). In any case, the presence of even a fully developed set of feathers in dinosaurs, or in the Mesozoic avian ancestors of extant birds, need not necessarily signal the presence of endothermy or even an approach to it. Like modern reptiles, some living birds utilize behavioral thermoregulation to absorb ambient heat across feathered skin. During nocturnal periods of low ambient temperatures, body temperature in the roadrunner (*Geococcyx californianus*) declines by ~4°C. After sunrise, the roadrunner exposes poorly feathered parts of its body to solar radiation and warms ectothermically to normal body tem-perature (Ohmart and Lasiewski 1971). A number of other fully-feathered extant birds can also readily absorb and use incident radiant solar energy (e.g. Hamilton and Heppner 1966; Lustick et al. 1970). Similarly, feathered ectothermic theropods, which are thought to have lived in a warm sunny climate, might easily have had similar behavioral thermoregulatory ca-pacity. Indeed, even a fully feathered *Archaeopteryx*, whether ectothermic or endothermic, could easily have achieved homeothermy. Consequently, the appearance of plumage or plumagelike covering in theropods or early birds need not be linked to any particular pattern of metabolic physiology. Because feather antecedents are likely to have appeared initially in small, gliding archosaurs from the middle Triassic Period (220 My) (Fig. 36.1) (Geist and Feduccia 2000; Jones et al. 2001; Jones, Ruben et al. 2000), the

36.1. Nonavian feathers from the Middle Triassic *Longisquama*, a small archosaur that predated the earliest dinosaurs. As in modern feathers, and unlike scales, these structures were hollow (Jones et al. 2000; Jones et al. 2001), and since they are wider distally than proximally, must have developed from a follicle. Assertions elsewhere that these structures were merely elongate, bladelike scales (Prum 2001) are falsified by the overlapped position of some individual barb elements. *Longisquama* was probably an accomplished glider, but the function of feathers, or featherlike structures, in dinosaurs remains unclear. Scale bar = 5 mm.

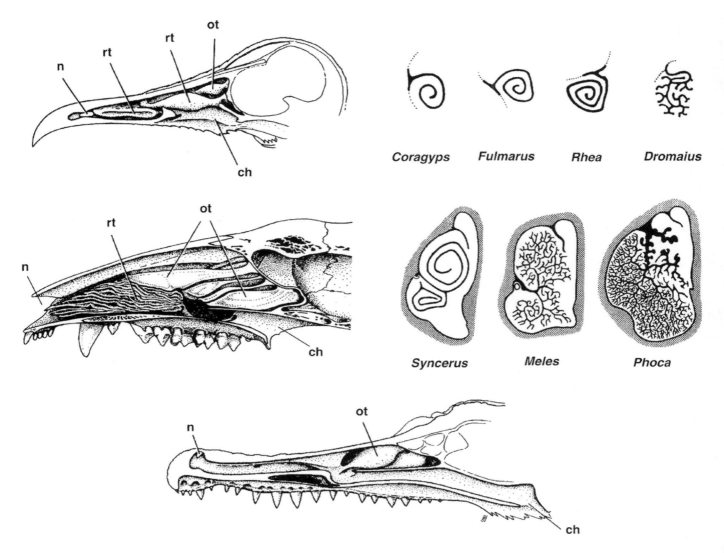

Coragyps   Fulmarus   Rhea   Dromaius

Syncerus   Meles   Phoca

36.2. Nasal passage anatomy in modern birds (top left), mammals (middle left), and crocodilians (bottom); representative cross-sections through respiratory turbinates in living birds (upper right) and mammals (lower right). Abbreviations: n, nostrils; rt, respiratory turbinate; ot, olfactory turbinate; ch, choana or internal nares. Modified from Hillenius (1994).

insulatory role and conferred endothermy by feathers or featherlike integumentary structures in dinosaurs is dubious.

Perhaps more to the point, virtually all of these arguments are based predominantly on apparent similarities to the mammalian or avian metabolic condition, without a clear functional correlation to endothermic processes per se. However, respiratory turbinates (described below) are essential to, and have a tight functional correlation with, maintenance of high rates of lung ventilation and metabolism in virtually all terrestrial mammals and birds. It has also been discovered that respiratory turbinates and, by extension, elevated lung ventilation and metabolic rates occurred in at least two groups of Permo-Triassic mammal-like reptiles (therapsids) (Hillenius 1992, 1994). Consequently, the respiratory turbinates represent the first direct morphological indicator of endothermy that can be observed in the fossil record.

Respiratory Turbinates and their Relationship to Endothermy in Mammals and Birds

Turbinate bones, or cartilages, are scroll- or bafflelike elements located in the nasal cavity of all amniotes. Olfactory turbinates (lateral sphenoids, naso- or ethmoturbinates) are generally located in blind cul de sacs immediately adjacent to the main respiratory airway and are lined with sensory (olfactory) epithelia that contain the primary receptors for the sense of smell; they occur ubiquitously in all amniotes and have no association with the maintenance of endothermy (Hillenius 1992, 1994). Respiratory turbinates (or maxilloturbinates) are mucous membrane-lined structures that protrude directly into the main nasal airway (i.e., the nasal passage proper) of all nostril-breathing terrestrial birds and mammals (Geist 2000; Hillenius 1994; Ruben 1996) (Fig. 36.2).

Only the respiratory turbinates have a strong functional association with endothermy. As mentioned previously, in both mammals and birds, endothermy is tightly linked to high levels of oxygen consumption and elevated rates of lung ventilation. Respiratory turbinates create an intermittent countercurrent exchange of respiratory heat and water between respired air and the moist, epithelial linings of the turbinates (Fig. 36.3). As cool external air is inhaled, it absorbs heat and moisture from the turbinate linings before it reaches the lungs. This not only prevents desiccation of the lungs and a potential decrease in core temperature, but it also cools the respiratory epithelia and creates a thermal gradient along the turbinates. During exhalation, this process is reversed—warm, fully saturated air from the lungs is cooled as it passes back over the respiratory turbinates. Consequently, the exhaled air becomes supersaturated as it cools and the excess water vapor condenses on the respiratory turbinate surfaces, where it can be reclaimed and recycled. Over time, a substantial amount of water and heat can thus be conserved, rather than lost to the environment (Fig. 36.3). In the absence

INHALATION

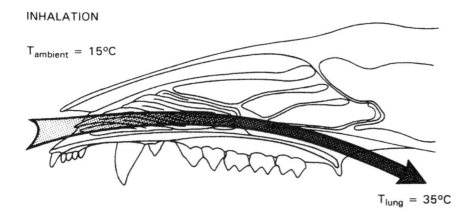

$T_{ambient} = 15°C$

$T_{lung} = 35°C$

EXHALATION

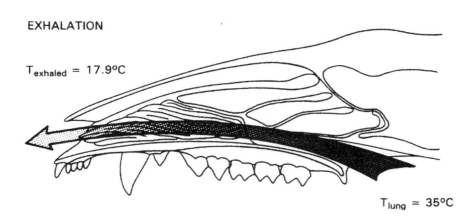

$T_{exhaled} = 17.9°C$

$T_{lung} = 35°C$

36.3. The counter-current heat exchange mechanism of the respiratory turbinates. During inhalation, air is warmed to body temperature and saturated with water vapor as it passes over the respiratory turbinates. As a result, turbinate surfaces are cooled by evaporative heat loss. Upon exhalation, the process is reversed as warm, saturated air from the lungs passes over cool turbinate surfaces. Consequently, exhaled air becomes supersaturated with water vapor and excess moisture condenses on, and warms, the surfaces of the turbinates.

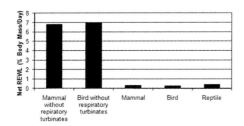

36.4. Daily net respiratory evaporative water loss (REWL) rates (i.e., REWL–metabolic water production) for a free-living 1-kg mammal, reptile, and bird, and probable REWL for a free-living mammal and bird lacking the use of respiratory turbinates (i.e., with a reptilelike nasal anatomy and net REWL rates/cm3 of O2 consumed). Without the water conserving function of respiratory turbinates, daily water flux rates in mammals and birds would be out of balance by about 30%. (Modified from Geist 2000; Hillenius 1992).

of respiratory turbinates, continuously high rates of oxidative metabolism and endothermy might well be unsustainable insofar as respiratory water and heat loss rates would frequently exceed tolerable levels, even in species of nondesert environments (Fig. 36.4) (Geist 2000; Hillenius 1994; Ruben 1996). Additionally, it has long been suggested that the ubiquitous occurrence of vascular shunts between respiratory turbinates and the brain indicates that these turbinates are also utilized as brain coolers. This would be especially critical during periods of elevated ambient temperatures or during periods of extended activity typical of many birds and mammals, when rates of internal heat production would be highest (Baker 1982; Bernstein et al. 1984).

Respiratory turbinates are present in all extant nostril-breathing terrestrial birds and mammals (Hillenius 1994; Ruben et al. 1996). The extent and complexity of the nasal cavity of birds vary with bill shape, but, in general, the avian nasal passage is elongate with three successive cartilaginous or (occasionally) ossified turbinates (Bang 1971). The anterior turbinate is often relatively simple, but the others, particularly the middle turbinates, are usually highly developed as prominent scrolls with multiple turns. Sensory (olfactory) epithelium is restricted to the posterior turbinate. Like mammalian olfactory turbinates, this structure is situated outside the main respiratory air stream, often in a separate chamber. Embryological and anatomical studies indicate that only the posterior turbinate is homologous to those of reptiles; the anterior and middle turbinates have evolved independently in birds (Witmer 1995).

The anterior and middle turbinates of birds, like the respiratory turbinates of mammals, are situated directly in the nasal passage and are covered primarily with respiratory epithelium. These turbinates are well positioned to modify bulk respired air. Previously published observations demonstrate that the respiratory turbinates of birds function as well as, or superior to, those of mammals for the recovery of water vapor contained in exhaled air (Geist 2000). Consequently, in birds these structures probably represent an adaptation to high lung ventilation rates and endothermy, fully analogous to respiratory turbinates of mammals.

Neither mammalian nor avian respiratory turbinates have analogs or homologs among living reptiles or amphibians (Witmer 1995). In living reptiles, one to three simple nasal turbinates are typically located in the posterodorsal portion of the nasal cavity and, like those of mammals and birds, are exclusively olfactory in function (Fig. 36.2). There are no structures in the nasal cavity of any extant ectotherm specifically adapted for the recovery of respiratory water vapor, nor are they as likely to be needed. Reptilian lung ventilation rates are apparently sufficiently low that respiratory water loss rates seldom create significant problems, even for desert species.

Maintenance of an analogous exchange mechanism in any portion of the respiratory tree other than the nasal cavity (i.e., the trachea) would be untenable. While the presence of such a system in the trachea would interfere with the stability of brain temperatures due to the proximity to the brain-bound carotid arteries, such exchange sites in the body cavity would necessarily preclude deep-body homeothermy (Jones and Ruben 2001). The fact that deep nasopharyngeal temperature is equivalent to core body temperature in extant mammals (Ingelstedt 1956; Jackson and Schmidt-Nielsen

Ruben, Jones, Geist, Hillenius, Harwell, and Quick

1964; Proctor et al. 1977) and birds (Geist 2000) confirms that little or no heat exchange takes place in the trachea. The widespread presence of respiratory turbinates among extant mammals and birds also indicates that these structures are likely a plesiomorphic attribute for each of these groups (Geist 2000; Hillenius 1992, 1994); the rare cases of turbinate reduction or absence among these taxa clearly represent secondary, specialized developments.

To summarize, physiological data denote that independent selection for endothermy in birds, mammals, and/or their ancestors was, by necessity, tightly associated with the convergent evolution of respiratory turbinates in these taxa. In the absence of these structures, unacceptably high rates of respiratory water and heat loss and/or central nervous system overheating would probably have posed chronic obstacles to maintenance of elevated rates of bulk lung ventilation or high, sustained levels of activity consistent with endothermy. Although independently derived in birds (Witmer 1995), these structures are remarkably similar to their mammalian analogs in morphology, function, and association with high lung ventilation rates and endothermy. Consequently, as in the therapsid–mammal lineage, the occurrence or absence of these structures provides a potential road map for revealing patterns of lung ventilation rate and metabolism in early birds and their close relatives, the dinosaurs.

The Functional Significance of Respiratory Turbinates on the Metabolic Status of Dinosaurs and Early Birds

Several factors complicate the study of the evolutionary history of respiratory turbinates in birds and, potentially, in their relatives, the dinosaurs. Although they may be ossified in many extant taxa, respiratory turbinates often remain cartilaginous and lack bony points of contact in the nasal passage of birds, thus greatly decreasing the chances for direct detection of their presence or absence in extinct taxa. Nevertheless, the presence of respiratory turbinates in extant endotherms is inevitably associated with marked expansion of the cross-sectional area of the nasal cavity proper (Ruben et al. 1996) (Fig. 36.5). Significantly, in some therapsid lineages (the immediate ancestors of mammals), relative nasal passage diameter increases steadily over time and eventually attains mammalian/avian nasal passage cross-sectional dimensions (Fig. 36.5) in the very mammal-like *Thrinaxodon*. Increased nasal passage cross-sectional area in endotherms probably serves to accommodate elevated lung ventilation rates and to provide increased rostral volume to house the respiratory turbinates.

The application of computed tomography (CT) scans to paleontological specimens has greatly facilitated noninvasive study of fine details of the nasal region in fossilized specimens, especially those which have been incompletely prepared. In the theropods *Nanotyrannus* and *Ornithomimus* (Fig. 36.6), CT scans clearly demonstrate that in life these animals were unlikely to have possessed respiratory turbinates: they are absent from the fossils and nasal passage cross-sectional dimensions are identical to those in extant ectotherms (Figs. 36.5 and 36.6). This condition is strikingly similar to the nasal region of many extant reptiles (e.g., *Crocodylus*, Figs. 36.2 and 36.5).

Similarly, the ornithischian hadrosaur *Hypacrosaurus* had a narrow, ectotherm-like nasal passage (Figs. 36.5 and 36.6) and, as in the nasal passages

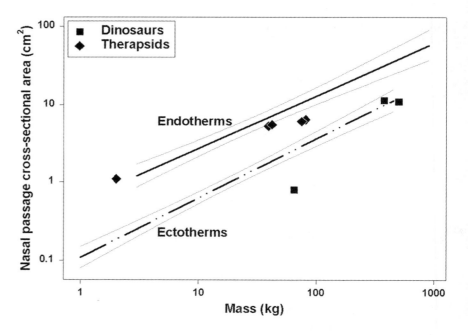

36.5. The relationship of nasal passage cross-sectional area to body mass (M) in extant endotherms (birds and mammals; cross-sectional area equals $0.57\,M^{0.68}$) and ectotherms (lizards and crocodilians; cross-sectional area equals $0.11\,M^{0.76}$). Also plotted are three genera of Late Cretaceous dinosaurs (■) and five genera of Permo-Triassic therapsids (◆) (values for dinosaurs and therapsids were not included in regression calculations). *Ornithomimus* mass = 65 kg; *Hypacrosaurus* mass = 375 kg; *Nanotyrannus* mass = 500 kg. For specimen numbers and more detailed information, the reader is referred to Ruben et al. 1996.

of theropod dinosaurs, was unlikely to have contained respiratory turbinates (Ruben et al. 1996). The nasal passage of lambeosaurines, including that of *Hypacrosaurus*, consisted of a complex of diverticula and curved passages (Weishampel 1981), but these appear to relate to sound production (Weishampel 1981, 1997) and possibly to olfactory functions (Horner, Horner, and Weishampel 2001).

Extensive modifications of the narial region, resulting in occasionally marked expansion of the nostrils (narial fossae) do occur, however, in some other ornithischians (e.g., ceratopsids, hadrosaurine ornithopods) and in some sauropod dinosaurs (e.g., brachiosaurs) (Witmer 1999). The function of narial expansion in these dinosaurs remains unclear and each case appears to be an independently derived specialization, absent in the smaller, basal members of each taxon (e.g. Dodson and Currie 1990; Forster 1990; Upchurch 1995; Weishampel et al. 1993). Thus, without specific compelling evidence to the contrary, there is no particular reason to suspect that the expanded narial regions in these particular dinosaurs are associated with elevated resting metabolic and lung ventilation rates and not sexual dimorphism.

Unfortunately, accurate quantification of nasal passage cross-sectional area can only be achieved in three-dimensionally preserved skulls and, therefore, this parameter offers little insight into the metabolic status of many archosaurians, especially early birds. However, pneumatization of the skull—in particular, the morphology of the paranasal (accessory) sinuses ("AC" in Fig. 36.6)—makes it possible to confidently infer nasal passage dimensions and, consequently, the presence or absence of respiratory turbinates in many less well-preserved specimens. In advanced theropods (tetanurans) and in archaeornithine birds (e.g., *Archaeopteryx, Confuciusornis*), the maxillary and/or promaxillary fenestrae—apertures in the rostral portion of the antorbital fossa—always open into an expansive maxillary antrum and promaxillary sinus, respectively, and are not part of the nasal passage (Witmer 1997) (Fig. 36.7). The ceilings of one or both of these

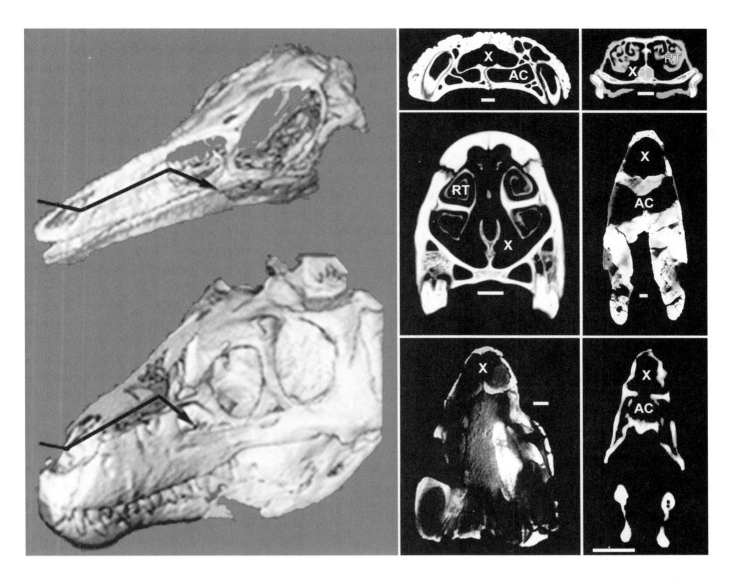

36.6. Left: left lateral CT-scans of the skulls of the theropod dinosaurs *Ornithomimus* (above) and *Nanotyrannus* (below). Arrows trace the path of airflow through the nasal passage. Right: cross-sectional CT scans of the nasal passages in (upper left) crocodile, (upper right) ostrich, (middle left) bighorn sheep, (middle and lower right) the theropods *Nanotyrannus* and *Ornithomimus*, respectively, and (lower left) the ornithischian dinosaur *Hypacrosaurus*. Respiratory turbinates are housed within voluminous nasal passages in birds and mammals. As in living reptiles, relatively narrow nasal passages in dinosaurs are probably consistent with the absence of respiratory turbinates. Scale bar = 1 cm. Abbreviations: AC, accessory (paranasal) sinus or cavity; RT, respiratory turbinate; X, mid-region of the nasal passage.

sinuses form much of the floor of the nasal passage. Conversely, in modern birds, these sinuses have been pushed caudally and their fenestrae are obliterated (Witmer 1997). This results, at least in part, from the expansion of the nasal passage to accommodate respiratory turbinates (Witmer 1997). Therefore, the occurrence of either of these fenestrae signals the likely presence of extensive paranasal sinuses that would have restricted the volume of the nasal passage in many theropods and in early birds. Accordingly, it is highly unlikely that respiratory turbinates occurred in either the earliest birds or their theropod relatives.

Thus, in the avian lineage, the appearance of birdlike characteristics, feathers, and probably powered flight, were likely to have preceded the development of endothermy. As in dinosaurs, the absence of respiratory turbinates in the earliest birds is inconsistent with their having attained an endothermic metabolic status. Significantly, other osteological features (e.g., the presence of annuli, or lines of seasonally arrested growth, in long-bone sections) of ancient birds are also consistent with this scenario (Chinsamy et al. 1994). Analysis of various attributes of the physiology and anatomy of extant birds also suggest that avian flight may have evolved prior to the

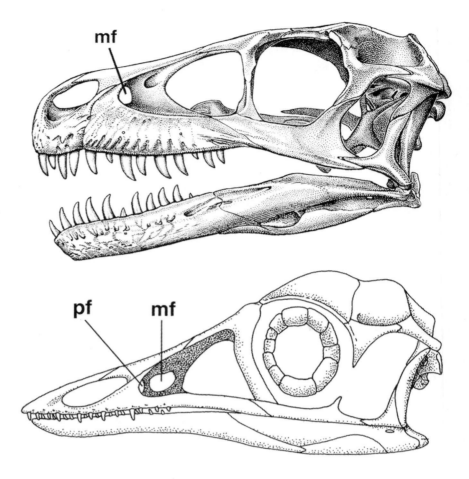

36.7. The promaxillary (pf) and maxillary (mf) fenestrae of the theropod dinosaur *Dromaeosaurus* (top) and the earliest known bird, *Archaeopteryx* (below). Both fenestrae open into paranasal sinuses that are causally linked to narrow nasal passages and the likely absence of nasal respiratory turbinates. Modified from Currie (1997) and Chatterjee (1997), respectively.

origin of avian endothermy (Randolph 1994). This contrasts sharply with respiratory evolution in the therapsid–mammal lineage, where endothermic metabolic rates were likely to have developed prior to the appearance of the earliest definitive mammal (Hillenius 1992; Ruben and Jones 2000).

Finally, we emphasize that reconstructing dinosaurs and early birds as ectothermic does not mean that these animals were necessarily vulnerable to temperature fluctuations, especially considering the mild climates of most of the Mesozoic Era (even higher latitude, seasonally cooler regions may have been warmer and more equable than thought previously [Royer et al. 2002]). Although smaller forms (< 75 kg) may still have been somewhat affected by diurnal temperature fluctuations, moderate to large forms (> 500 kg) would be nearly unaffected by such changes (i.e., they would effectively have been homeothermic), and would respond only to fluctuations on a weekly to monthly scale. Moreover, given the metabolic capacities of many modern reptiles during burst-level activity, it is likely that early birds, even though ectothermic, were quite capable of uninterrupted powered flight for distances up to one kilometer (Ruben 1991).

Further Evidence of Activity Capacity

Although evidence of the presence or absence of respiratory turbinates provides some insight into the resting or routine metabolic and lung ventilation rates in both extant and extinct taxa, these data do not necessarily serve as precise indicators of metabolic capacities during periods of exercise. However, paleontological and neontological evidence of lung morphology

Ruben, Jones, Geist, Hillenius, Harwell, and Quick

and ventilatory mechanisms in theropod dinosaurs and early birds facilitates hypotheses regarding their activity capacities.

Extant Amniote Lung Morphology and Ventilatory Mechanisms

Amniote lungs are of two morphologically and ontogenetically distinct types, each of which is derivable from hypothetical ancestral simple saclike lungs. Extant theropsids (mammals) have alveolar lungs; extant sauropsids (i.e., chelonians, lepidosaurs, rhychocephalians, crocodilians, and birds) possess septate lungs (Perry 1989; Ruben et al. 1997).

Alveolar lungs are composed of millions of highly vascularized, spherical alveoli in which ventilatory airflow is bi-directional. During inhalation, increased pleural cavity volume and decreased pleural cavity pressure results in the expansion of the alveoli. Exhalation is accomplished, at least in part, by elastic recoil of the alveoli. The unique morphology of this lung, and especially of the alveoli, removes the necessity of high volumes of supporting parenchymal tissues and allows nearly all of the lung parenchyma to function actively in gas exchange (Perry 1983, 1989). These attributes, combined with a thin blood-gas barrier, provide the alveolar lung with a high anatomical diffusion factor (ADF, mass-specific ratio of vascularized pulmonary respiratory surface area to pulmonary blood-gas barrier thickness [Duncker 1989; Perry 1983, 1992]), an attribute essential for maintenance of high rates of oxygen consumption during extended periods of intensive activity.

The general lung morphology of extant nonavian sauropsid amniotes ("reptiles") is distinct from the alveolar lungs of mammals. The generalized sauropsid septate lung (a unicameral lung) is functionally analogous to a single, oversized mammalian alveolus in which septa (vascularized ingrowths) penetrate medially from the perimeter, forming respiratory units (i.e., trabeculae, ediculae, or faveoli depending on their depth), and are the principle sites of gas exchange (Perry 1983).

Variations from this generalized sauropsid septate lung morphology range from homogeneous to heterogeneous distribution of parenchyma, from one to many chambers, from dorsally attached to unattached, and from possessing no diverticulae to exhibiting many, elaborate diverticulae (Perry 1983; Perry and Duncker 1980). As in the mammalian lung, ventilatory airflow in the reptilian lung is bi-directional. However, unlike alveoli, the respiratory units of the reptilian septate lung contribute little to air convection during ventilation. Additionally, the amount of effective parenchymal tissue (parenchymal tissue volume/respiratory surface area, an indicator of the amount of nonrespiratory, supportive tissues) of the reptilian lung is significantly greater than that of the mammalian lung (Perry 1989). The result is a low relative overall ADF in reptiles (Perry 1983). To compensate, the ventral region of the lung in some nonavian sauropsids is often relatively poorly vascularized and functions largely to assist in ventilation of dorsal, more vascularized portions of the lung (Perry 1983). Although maximal oxygen consumption rates (VO2 max) in some varanid lizards are significantly higher than those of other reptiles, no extant reptile is capable of achieving maximal aerobic respiratory exchange rates greater than about 15 to 20% of those of typical endotherms (Hicks and Farmer 1998, 1999; Ruben et al.

1998). However, in a best-case scenario based on hypothetical improvements of the circulatory system and optimized pulmonary diffusion capacity, the reptilian septate lung might be capable of attaining about 50–60% of these rates and thus might overlap with some of the less-active mammals (Hicks and Farmer 1998, 1999; Ruben et al. 1998).

Birds, like all sauropsids, also possess septate lungs, but they circumvented inherent constraints on respiratory gas exchange rates of the reptilian septate lung and have a particularly high ADF (Perry 1983). Especially high rates of lung ventilation and gas exchange are made possible by the modification of the ventral, nonvascularized chambers into a series of extensive, highly compliant air sacs (Maina and Africa 2000). These air sacs extend into the visceral cavity and aid in a specialized cross-current ventilation of the dorsal vascularized parabronchi during both inhalation and exhalation (Maina and Africa 2000).

The parabronchial lung in modern birds is securely attached to the vertebral column. In some birds – particularly those with notaria – there are distinct, inverted T-shaped hypopophyses that serve as additional sites of attachment. Diverticulae from the air sacs invade and pneumatize portions of the skeleton. Pneumatization of the avian skeleton, with the exception of the long bones of the hindlimbs in a small subset of birds, is limited to the axial skeleton and forelimbs and results from invasion by the anterior (cervical and clavicular) air sacs, but is not linked to respiratory function or specific lung morphology (Duncker 1979; McLelland 1989a; Scheid and Piiper 1989).

Mechanisms for powering lung ventilation vary among extant amniotes. Lizards and snakes lack complete transverse subdivision of the body cavity and rely primarily on lateral expansion and contraction of the ribcage (costal ventilation) to alter pleuroperitoneal cavity pressure. Partial separation of the body cavity is possible and, when present, results from either the presence of an incomplete post-pulmonary septum or, in some cases (e.g., macroteiids), by an incomplete post-hepatic septum, but not both (Duncker 1979). To some extent, mammals and crocodilians use ribcage movements to ventilate the lungs, but they also rely on diaphragm-assisted lung ventilation.

In mammals, the diaphragm consists of an airtight, transversely oriented, muscularized septum that completely subdivides the visceral cavity into anterior thoracic and posterior abdominal regions (Fig. 36.8). Muscular contraction of the diaphragm increases the volume of the pleural cavity, which reduces pleural cavity pressure, resulting in filling of the lungs.

In crocodilians, an airtight, transversely-oriented diaphragm (composed of the postpulmonary and posthepatic septa [Duncker 1979; Goodrich 1930]) also completely subdivides the visceral cavity into anterior pleural-pericardial (thoracic) and posterior abdominal regions (Fig. 36.8). Unlike the mammalian diaphragm, the crocodilian diaphragm is nonmuscular and adheres to the surface of the liver. The posterior and ventrolateral aspects of the liver capsule serve as the insertion for the diaphragmaticus muscles. These muscles consist primarily of the large, ventral portion that takes origin largely from the last pair of gastralia; a lateral portion of the muscle originates from the proximal shaft of the pubis and the small, preacetabular portion of the ischium (personal observation) (Fig. 36.9). Contraction of the

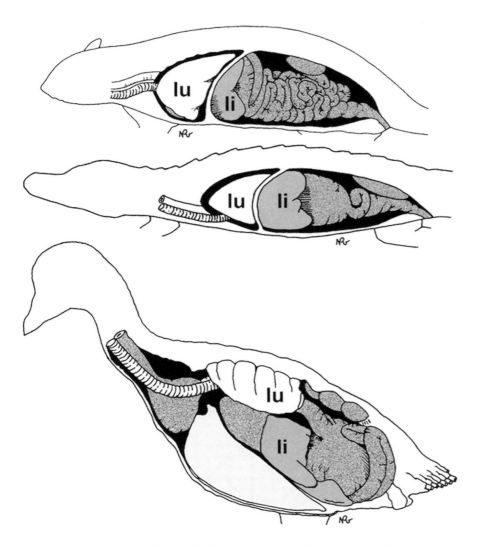

36.8. Body cavity partitioning correlates with lung ventilation mechanisms in extant amniotes. Only mammals (top) and crocodilians (middle) use active diaphragmatic lung ventilation. This necessitates a complete, transverse, fore-aft, separation of the anterior, thoracic cavity from the abdominal cavity. Birds, which are exclusive rib breathers, exhibit no distinct antero-posterior separation of the body cavity. Abbreviations: li, liver; lu, lung.

diaphragmaticus muscles pulls the liver posteriorly in a pistonlike manner, resulting in decreased pleural cavity pressure and filling of the lungs (Gans and Clark 1976).

The tri-radiate pelvis of extant crocodilians, with its stout, rodlike pubic rami, is ideally suited to provide support for the robust posterior gastralia, the primary site of origin for the ventral portion of the diaphragmaticus muscle (Figs. 36.9 and 36.10). However, it is important to note that the elongate, distinctly theropod-like pubes of early (Triassic) crocodylomorphs (e.g., *Terrestrisuchus* [Crush 1984]) probably represent the pleisiomorphic pelvic

36.9. The hepatic-piston lung ventilatory mechanism of crocodilians. Extensive ventral diaphragmatic muscles take origin primarily from the last pair of gastralia. More delicate lateral diaphragmatic muscles originate from the preacetabular processes of the ischium and the shafts of the pubes. Abbreviations: dm, diaphragmatic muscles; li, liver; lu, lung; p, pubis. Illustration by R. Jones.

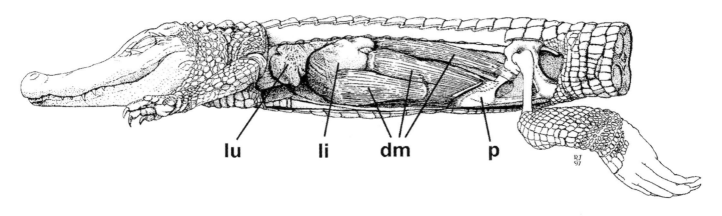

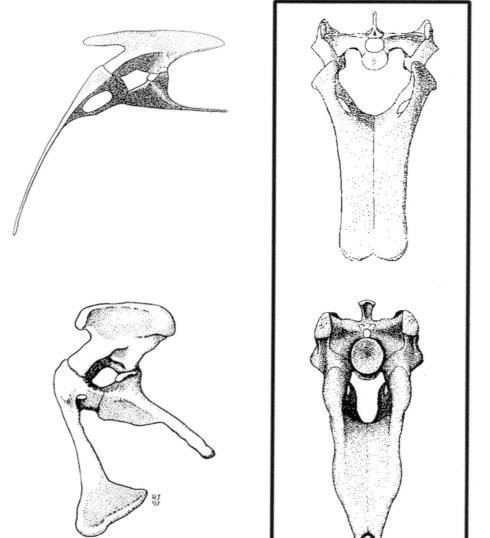

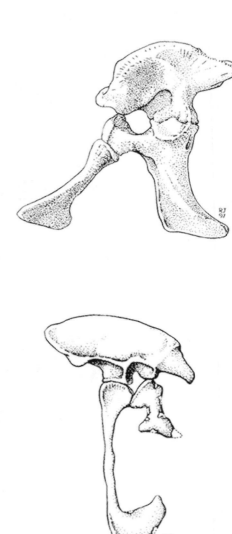

36.10. Left lateral and anterior views of the pelves of crocodylomorphs and theropod dinosaurs. Crocodylomorphs *Terrestrisuchus* (upper left, upper center) and *Alligator* (upper right); theropods *Herrerasaurus* (lower left, lower center) and *Unenlagia* (lower right). Note the marked similarity between theropod and crocodilian pubes. Modified from Ruben et al. (1997), Crush (1984) and Novas (1993). Illustrations for *Alligator*, *Herrerasaurus*, and *Unenlagia* by R. Jones.

morphology for these taxa (Fig. 36.10). Similarly to mammals, crocodilian posterior ribs are reduced (or even lost) to allow lateral expansion of the viscera when the liver is pulled caudally during inhalation (Hengst 1998).

In most tetrapods, rectus abdominus musculature functions to support the abdominal viscera. However, in crocodilians much of the rectus abdominus appears to have contributed to formation of diaphragmatic muscles (Carrier and Farmer 2000a). Consequently, in crocodilians, the gastralia, as well as passively aiding in lung ventilation by maintaining the volume of the body cavity, may play a supportive role.

Birds, like lizards, lack a mammal- or crocodilian-like diaphragm and rely principally on costal-powered lung ventilation. However, unlike reptilian ribs, those of birds possess unique intercostal and sternocostal joints that allow sagittal rotation of the sternum and shoulder girdle. Significantly, the distal end of each sternal rib is expanded transversely and forms a robust hinge joint with the thickened anterolateral border of the sternum. These modifications facilitate ventilation of abdominal air sacs as costal action results in a dorso-ventral rotation of the posterior end of the enlarged sternum, which in turn contracts and expands the air sacs (Brackenbury 1987; Fedde 1987; King 1966; Schmidt-Nielsen 1971). The highly derived avian lung-air sac system, which permeates the entire pleuroperitoneal cavity, is dependent on the aforementioned skeletal features (Duncker 1972, 1974, 1989) and precludes the distinct transverse separation of avian body cavity that is typical of diaphragm-breathing tetrapods.

Given their relationship to living reptiles and birds, extinct sauropsids – including dinosaurs – probably possessed septate lungs, although they were unlikely to have possessed avian-style, flow-through lungs. The dinosaurian ribcage-pectoral girdle complex lacks indications of avianlike thoracic musculoskeletal capacity for inhalatory filling of abdominal air sacs (see above). Although ossified sternal ribs are documented for the sauropod *Apatosaurus* (Marsh 1883; McIntosh 1990) and several derived theropods, including oviraptorids (e.g. Clark et al. 1999), ornithomimids (e.g., *Pelecanimimus*, personal observation), and dromaeosaurids (e.g. Norell and Makovicky 1999; Ostrom 1969), in each case the sternal ribs lack the unique morphology of the avian sternocostal joints. More importantly, mere sternocostal articulations alone cannot be considered diagnostic for any particular lung morphology or ventilatory mechanism – such articulations between ribs and sternae occur in most amniotes, including lepidosaurs, crocodilians, and mammals (e.g. Goodrich 1930; Romer 1956).

The sternal plates are known from a variety of dinosaur taxa, e.g., ankylosaurs (Coombs and Maryanska 1990), ornithopods (Coombs and Maryanska 1990; Forster 1990; Norman 1980), ceratopsians (Sereno and Chao 1988), sauropodomorphs (McIntosh 1990; Young 1947), as well as oviraptorids (Barsbold 1983; Clark et al. 1999), ornithomimids (Perez-Moreno and Sanz 1999), and dromaeosaurids (Barsbold 1983; Burnham et al. 2000; Maryanska et al. 2002; Norell and Makovicky 1997; Xu et al. 1999). But all are relatively short and in each case lack the thickened lateral border and the transversely oriented costal articulations that characterize the sterna of all extant birds. The sternal plates of the immature dromaeosaur *Bambiraptor* are relatively long, but its lateral edges, including the regions where the sternal ribs articulated, are thin (Burnham et al. 2000).

The presence of avianlike uncinate processes, posterodorsal projections on the thoracic ribs in oviraptorids (Clark et al. 1999) and in some dromaeosaurids (Norell and Makovicky 1999), is sometimes mentioned in the context of avian-style lung ventilation (e.g. Paul 2001; Perry 2001). However, avian uncinate processes appear to function primarily in strengthening the ribcage and stabilizing the shoulder musculature (Bellairs and Jenkin 1960; Hildebrand and Goslow 2001; King and King 1979). They have no direct role in sternocostal ventilation of the avian lungs. Uncinate processes are absent in screamers (Anseriformes: Anhimidae) and may be reduced in other birds (Bellairs and Jenkin 1960; Brooke and Birkhead 1991). Furthermore, cartilaginous or ossified uncinate processes are also known from crocodilians (Duncker 1978; Frey 1988; Hofstetter and Gasc 1969), *Sphenodon* (Hofstetter and Gasc 1969; Wettstein 1931), and even several

Early bird and dinosaur lung morphology and ventilatory mechanisms

labyrinthodont amphibians, such as *Eryops* (Gregory 1951). Thus, the presence of uncinate processes does not necessarily signal any particular suite of respiratory adaptations.

Similarly, birdlike vertebral pneumatization in a variety of saurischian dinosaurs has led some to infer the presence of an avian-style, flow-through lung air sac system in theropods (Bakker 1986; Britt et al. 1998; O'Connor and Claessens 2005; Perry 2001; Perry and Reuter 1999; Reid 1997). In particular, pneumatization of sacral vertebrae in a few theropods (e.g., *Allosaurus*, *Baryonyx*) has recently been interpreted as proof for the existence of fully functional, birdlike abdominal air sacs consistent with high-performance, avian-style, flow-through lungs in those taxa (O'Connor and Claessens 2005). Such assumptions should be regarded with skepticism. Sauropsid lung systems actually exist in a continuum of variation with respect to parenchymal heterogeneity (Duncker 1979; Perry 1983, 1989, 2001; Perry and Duncker 1980). Modern bird lungs form only one end of that continuum. For example, substantial nonrespiratory dilatations are known from turtles, snakes, chameleons, and monitor lizards, among others (Duncker 1979; Perry and Duncker 1980). The ventrolateral, ventromedial, and terminal portions of the crocodilian lung are also described as saclike (Perry 1988). While these portions of the crocodilian lung are not completely avascular, they are nevertheless substantially less densely vascularized than the craniodorsal portions of that lung. Therefore, the presence of nonvascularized dilatations (or diverticula, or air sacs) in theropods is not automatically an indication of an avianlike parabronchial respiratory system.

The latter is an important point, since skeletal pneumaticity is not limited to birds. Cranial pneumaticity occurs in modern crocodilians and the antorbital fenestra, recognized by Witmer (1995; 1997) as a manifestation of that pneumaticity, is a synapomorphy of archosauriformes (cf. Gauthier et al. 1988). Postcranial pneumaticity also appears to date back to basal archosaurs (Gower 2001). Consequently, there is no compulsory correlation between skeletal pneumaticity, even postcranial pneumaticity, and the avian parabronchial mode of respiration. Wedel (2003) also cautions against concluding that a particular pattern of pneumatization necessarily corresponds with a particular pattern of nonrespiratory dilatations (see especially his figure 4). Thus, even if the sacral pneumatopores in theropods derive from a posterior parenchymal dilatation, this is still no requisite sign of a birdlike arrangement (note that in virtually all nonavian sauropsids the nonrespiratory dilatations are posteriorly situated)—there are simply too many variables and too many possible alternative scenarios to justify that conclusion.

There is also another fundamental difficulty with scenarios for avian-style, air sac lungs in theropods—the absence of any skeletal mechanism for resistance of inhalatory paradoxical movements of the posterior non-vascularized air sacs. For example, in extant birds the synsacrum and the slender, posteriorly directed pubic bones surround and are fused to the lateral aspects of the abdominal air sacs. This arrangement prevents air sac collapse which would otherwise occur during the negative pressure, inhalatory phase of avian lung ventilation (Sapp 2004). In this context, the broad, open flanks and relatively unmodified pelves of theropods are inconsistent with the presence of extensive posterior nonvascularized air sacs that would have been required for the function of an avian style lung air sac system (Quick and Ruben 2009). Modern sauropsids with the best-developed non-vascularized pulmonary dilatations, such as snakes and chameleons, have a well-developed ribcage that encases the abdominal cavity to a much greater extent than that of any theropod.

It has been suggested that theropod dinosaurs might have mobilized their well-developed gastralia, or belly ribs, to have generated negative abdominal pressure during inhalation, thereby facilitating expansion of posterior, birdlike air sacs (Carrier and Farmer 2000a, 2000b; Claessens 2004). However, this is unlikely because: (1) crocodilians and rhynchocephalians possess well-developed gastralia, but these perform no known role in active generation of either negative or positive abdominal pressure during lung ventilation; (2) in extant crocodilians (and other living reptiles), neither of the muscles hypothesized to have caused active expansion of theropod gastralia (i.e., ischiopubis and/or ischiotruncus muscles) insert on sites appropriate to have generated such movements (*via* posterior rotation of medial gastralia elements); (3) gastralia are restricted to the ventral abdominal wall and theropods possessed deep, laterally open, lumbar flank regions. Consequently, inefficient, paradoxical medial movements of the lateral abdominal wall would likely have been unavoidable during the postulated gastralia-powered, negative abdominal pressure cycles requisite for expansion of posterior air sacs (Quick and Ruben 2009).

The presence of pneumatized vertebrae in dinosaurs signifies the likely presence of nonrespiratory diverticuli, but cannot be regarded as indicative of an extensive, avian-style air sac system. Consequently, there is no compelling evidence that any dinosaur possessed thoracic skeletomuscular modifications to ventilate a birdlike lung air sac system. It is possible that a less highly derived, proto air sac lung may have existed in some dinosaurs (Perry 1992, 2001), but, at present, there is no indication of the presence thereof, and any such reconstruction must be considered speculative (Perry and Reuter 1999).

Hepatic-piston ventilation of relatively unmodified septate lungs may have afforded theropods, early crocodylomorphs, and, perhaps, pterosaurs, the low maintenance cost of ectothermy combined with enhanced endurance similar to that in some modern endotherms. This strategy might have functioned optimally in the relatively mild, equable climatic regimes of most of the Mesozoic Era, where chronic maintenance of ectothermic homeothermy would have been possible (as in the *Varanus komodoensis* [McNab and Auffenberg 1976; Spotila et al. 1991]).

The Early Cretaceous compsognathid *Sinosauropteryx* (Chen et al. 1998) retains, along with ocular and integumentary tissues, preserved traces of much of the contents of the visceral cavity. The cavity exhibits complete thoracic-abdominal separation, delimited by a transversely oriented subdivision coincident with the anterior surface of the liver (Fig. 36.11). Additionally, theropods possessed reduced posterior dorsal (lumbar) ribs, well-developed gastralia, and a triradiate pelvis similar to that in crocodilians (Fig. 36.10). These are consistent with the hypothesis that theropod dinosaurs, like modern crocodiles (and, especially, early crocodilians), probably

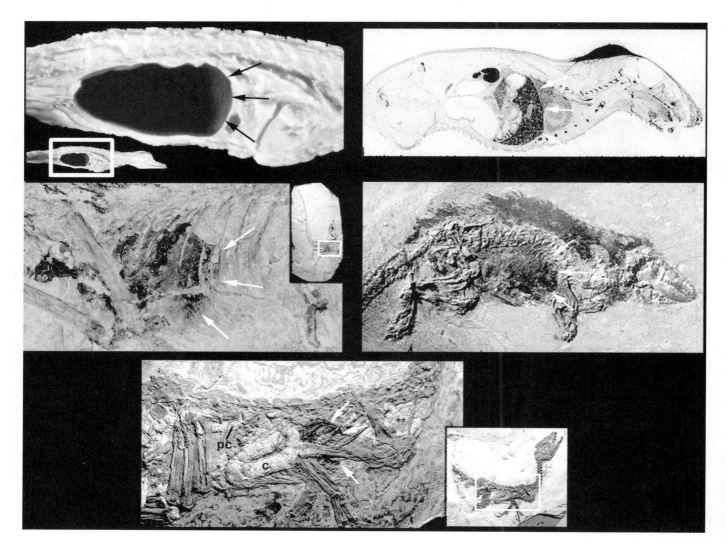

36.11. Similar fore-aft body cavity partitioning in *Alligator* (*upper left*), the theropod dinosaurs *Sinosauropteryx* (*middle left*) and *Scipionyx* (bottom), and in the rat (*top right*) and the Eocene mammal *Pholidocercus* (*middle right*) (from von Koenigswald et al. 1988). In each case, arrows delineate the complete vertical separation of the anterior thoracic and posterior abdominal cavities from one another. Insets show entire specimen. See also Figure 36.8. Abbreviations: c, colon; pc, posterior colon.

possessed a bellowslike septate lung and that the lung was ventilated, at least in part, by a hepatic-piston diaphragm powered by diaphragmatic muscles that extended between the pubes and liver. Significantly, fossilized preservation of abdominal contents whose general arrangement is consistent with diaphragm breathing is not unique to *Sinosauropteryx*. Similar transverse segregation of the viscera is seen in *Pholidocercus*, a mammal from the Eocene Messel formation of Germany (von Koenigswald et al. 1988) that undoubtedly utilized a diaphragm for lung ventilation (Figure 36.11).

Incredibly, a juvenile maniraptoran theropod dinosaur with exquisitely preserved soft tissue was recently described. As well as having a remarkably preserved skeleton, *Scipionyx samniticus* retains fossilized skeletal muscles, trachea, large intestines, and liver in situ (Dal Sasso and Signore 1998a, 1998b; Ruben et al. 1999). A section of the trachea is preserved in the posterior cervical region, immediately anterior to the scapulocoracoid complex (Dal Sasso and Signore 1998a). Like the trachea of crocodilians, *Scipionyx*'s trachea in this region is situated well ventral to the vertebral column. In contrast, the avian posterior cervical trachea, except in specialized, long-necked birds (e.g., swans), is usually positioned dorsally and adjacent to the vertebral column, thereby facilitating entry of the trachea into the dorsally attached parabronchi (McLelland 1989b).

In visible light, the liver of *Scipionyx* appears as a small hematitic halo restricted to the ventral margin of the anterior portion of the visceral cavity (Dal Sasso and Signore 1998a, 1998b). However, under ultraviolet illumination the liver fluoresces as a suboval, indigo-colored mass that extends from the vertebral column to the ventral body wall (Ruben et al. 1999) (Fig. 36.11). As in extant taxa that utilize a diaphragm for ventilation, *Scipionyx*'s liver is situated ahead of the large intestine and fills the anterior-most portion of the abdominal cavity (Ruben et al. 1999). Furthermore, as in crocodilians, mammals, and the theropod *Sinosauropteryx*, the anterior border of the liver in *Scipionyx* is transversely oriented and completely subdivides the visceral cavity into anterior thoracic and posterior abdominal regions (Ruben et al. 1999). In the theropods *Scipionyx* and *Sinosauropteryx*, and the mammal *Pholidocercus*, the pleural cavity appears empty because delicate lung tissues were not fossilized (Fig. 36.11).

Portions of the large intestine and trachea of *Scipionyx* are visible and appear to have been preserved in situ (Dal Sasso and Signore 1998a, 1998b; Ruben et al. 1999). Notably, the posterior colon, or colorectal intestine, is situated far dorsally, at about the same level as the vertebrae in the lumbar-sacral region. This condition is comparable to the position of the colon in living taxa such as crocodilians and mammals (Fig. 36.12) (Ruben et al. 1999). In contrast, the colon (or rectum) of birds is invariably suspended by the dorsal mesentery (mesocolon) so that it is situated in the mid-abdominal cavity, some distance from the roof of the cavity (Fig. 36.12) (Duncker 1979). This mid-abdominal suspension of avian large (and small) intestines is necessary to accommodate the dorsally and medially attached abdominal air sacs (which extend caudally from the dorsally attached parabronchi).

Computed tomography (CT) of the alimentary canals in extant birds and crocodilians affirm the general positions of the large and small intestines relative to the dorsal body wall. However, further similarities between *Alligator* and *Scipionyx* are also evident. Not only is the overall organization of the abdominal and thoracic cavities remarkably comparable in *Scipionyx* and *Alligator* (Fig. 36.13), but also the specific orientation of individual intestinal elements is strikingly similar. Specifically, in both taxa: (1) the large,

A

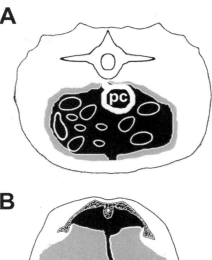

B

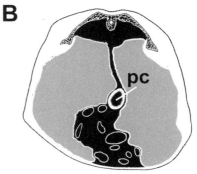

36.12. The relation of the posterior colon (pc) to the roof of the abdominal cavity in (A) crocodilians and (B) birds. Areas in gray represent the coelomic cavity. Unlike crocodilians and the theropod *Scipionyx*, the posterior colon of birds is situated ventrally, some distance from the roof of the visceral cavity, in order to accommodate abdominal air sacs. After Duncker (1979). Abbreviation: pc, posterior colon.

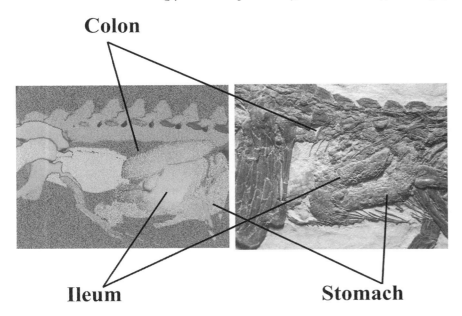

36.13. Three dimensional CT scan reconstruction of *Alligator* abdominal digestive anatomy (left) compared to preserved anatomy in *Scipionyx* (right). In both images, head is to the right and tail to the left. The overall organization of the visceral elements is strikingly similar in not only the position of the stomach, ileum, and colon, but also in the specific orientations of these intestinal elements.

curved stomach is situated immediately caudal to the liver; (2) the small intestine loops ventrally and then dorsally away from the stomach until it approximates the narrow junction with the dorsally-positioned colon at the ileo-cecal valve; (3) while maintaining a near-linear orientation as it courses along the dorsal body wall and then caudally through the pubic canal, the colon opens into the cloaca (Fig. 36.13). It is unlikely that such detailed similarity between these taxa is merely the result of post-mortem distortion. We conclude that Scipionyx's alimentary and respiratory morphology – and perhaps its physiology – was like that in extant crocodilians.

Suggestions that the obvious fore-aft subdivision of the visceral cavities in Sinosauropteryx and Scipionyx are merely the result of fortuitous cracks or post-mortem organ displacement (Paul 2001) strain credulity. Cracks that do exist in the holotype Sinosauropteryx fossil clearly do not compromise the diagnostic leading edge of the preserved abdominal remains (personal observation; Chinsamy and Hillenius 2005). Furthermore, a more recently described specimen shows the same distinct visceral geography as described here (Lingham-Soliar et al 2007). With regard to Scipionyx, there is little question that the skeleton, various muscles, and intestines have been preserved in situ and therefore, there is no logical reason to assume that the structure and placement of the liver is not positioned as it was in life.

A report of an incomplete theropod fossil (Martill et al. 2000) that includes a short section of colorectum with a supposedly birdlike, mid-abdominal location should be regarded with skepticism. The element in question courses postero-dorsally toward the pubic canal at a sharply oblique angle (ca. 45°) to the vertebral column; it then abruptly flexes ventrally as it passes through the puboischiac canal approximately parallel to the vertebral column. We are aware of no extant bird, or crocodilian, that exhibits colonic anatomy wherein the posterior, colorectal region is thrown into similarly sharp flexures that would almost certainly obstruct smooth movement of fecal matter toward the cloaca. Significantly, the avian colorectum is almost invariably a straight tube that extends through the mid-abdominal cavity in a course parallel to the vertebral column. Compared to extant archosaurians, the location of the colorectum in this specimen almost certainly reflects post-mortem ventral displacement of soft tissues.

The axial skeletal features and distinct visceral organization of Scipionyx and Sinosauropteryx are consistent with the presence of a hepatic-piston, diaphragm-ventilation mechanism. Accordingly, it is unlikely that avian-style, abdominal air sacs were present in Scipionyx or Sinosauropteryx and, by extension, in other theropod dinosaurs. Since abdominal air sacs are of fundamental importance to the function of the lung in extant birds (Duncker 1971), their likely absence is another indication that an avian style, flow-through, air sac lung was absent in theropod dinosaurs. Thus, the presence of diaphragm-assisted lung ventilation in theropods might indicate that, although these dinosaurs maintained ectotherm-like routine metabolic rates, they were, nevertheless, uniquely capable of sustaining active oxygen consumption rates and activity levels beyond those of the most active living reptiles.

Objections to the hepatic-piston model of theropod lung ventilation fall principally into two categories: (1) that there was insufficient space on the theropod pubis to have accommodated diaphragmaticus musculature; (2)

that theropods were unlikely to have been diaphragm ventilators because their pelvis, with its fixed pubes, differed from the mobile pubes of living crocodilians. These criticisms are poorly supported by both theropod and crocodilian anatomy and ventilatory physiology in living crocodilians. Hutchinson (2001) considers it unlikely that theropods had a crocodilian-like diaphragmaticus muscle because puboischiofemoralis muscles supposedly occupied the formerly accepted origin of the diaphragmaticus on the antero-posterior surfaces of the fused pubic rami, or pubic apron. However, in crocodilians, the pubic rami are not major sites of origin for the diaphragmaticus musculature. Instead, the large, ventral portion of the diaphragmaticus attaches primarily to the last pair of gastralia. Some slips of this muscle may originate directly from the pubis, but these attach to the craniolateral edge of the distal part of the pubis, ventrally adjacent to the sites of attachment of the puboischiofemoralis muscles. Other slips of the diaphragmaticus insert more proximally on the pubis and the preacetabular region of the ischium (Carrier and Farmer 2000a, 2000b; Farmer and Carrier 2000a, 2000b; Ruben et al. 1997). Elsewhere, Hutchinson (2001) asserts that the lateral surface of the distal pubis and the pubic boot of theropods served mainly for abdominal muscles, including the rectus abdominus, thereby leaving no room for insertion of the diaphragmaticus. However, insofar as the crocodilian diaphragmaticus muscles are likely derivatives of the rectus abdominus (Carrier and Farmer 2000a), there would seem to be little, if any, conflict between Hutchinson's (2001) reconstruction of theropod pelvic musculature and the presence of a diaphragmaticus muscular system in these animals. Interestingly, the pubis of some theropods exhibits marked posterior retroversion (e.g., in *Herrerasaurus* and some dromaeosaurid theropods). In such cases, the result might have been enhanced lung tidal volume as a consequence of elongation of the diaphragmaticus muscle and additional posterior travel of the liver during hepatic-piston ventilation.

In another instance, the mobile pubes of modern crocodilians were asserted to be essential for operation of a hepatic-piston diaphragm (Carrier and Farmer 2000a, 2000b; Farmer and Carrier 2000a, 2000b). Supposedly, avoidance of greatly elevated intra-abdominal pressure otherwise associated with posterior movement of the liver during inhalation is made possible only by concomitant, backward rotation of the pubes. However, experimental data broadly falsify these assertions—alligators with surgically fixed pubes exhibit no alteration in intra-abdominal pressures during exercise-induced periods of enhanced lung ventilation (Harwell et al. 2002). Instead, posterior, inhalatory rotation of the liver and lung tidal volume are reduced by about 15% in these animals. It is likely that the mobile pubis of alligators serves to enhance tidal volume (via supplemental, backward movement of the liver during inhalation) and is probably a secondary specialization to compensate for reduced tidal volume capacity in aquatic, dorso-ventrally flattened crocodilians. As noted above, early crocodylomorphs (e.g., *Terrestrisuchus*) were terrestrial, not aquatic, and possessed deeper bodies and pelvic anatomy markedly like that of theropods rather than modern crocodilians (Frey 1988).

These observations are further reinforced by a recently described, superbly preserved specimen of the bipedal stem crocodilian *Poposaurus gracilis*. This crocodilian possessed a strikingly theropod-like pelvis, including

36.14. Pelvic and tail skeleton and suprapubic musculature of modern perching birds (e.g., pigeon) and *Archaeopteryx*. In both extant and extinct arboreal birds, marked projection of the distal pubis posterior to the ilium and ischium is, and probably was, associated with suprapubic musculature rotation of the pelvis and tail. Such movements facilitate ventilation of the nonvascularized air sacs during arboreal roosting. This condition of the pubis is broadly inconsistent with the morphology of the thero-pod pelvis and with hepatic-piston diaphragmatic breathing. Anterior is to the left of the figures.

Illustration by R. Jones.

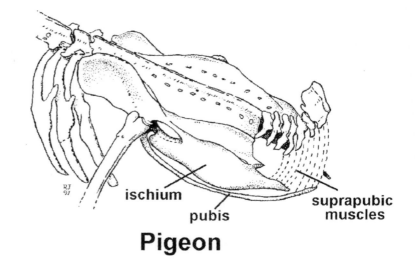

Pigeon

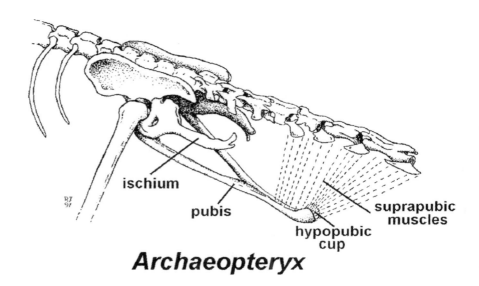

Archaeopteryx

an elongate, immobile pubis complete with a "pubic foot," virtually identical to those in many theropod dinosaurs (Gauthier et al. 2011).

Interestingly, hepatic-piston diaphragm lung ventilation may have been a primitive (plesiomorphic) attribute for the archosaurs. In addition to crocodilians and theropod dinosaurs, pterosaurs probably utilized a diaphragm in lung ventilation as well. In many advanced pterosaurs, the presence of a virtually immovable, fixed ribcage as well as a crocodile-like pubis and well-developed set of gastralia are fully consistent with diaphragm breathing (Carrier and Farmer 2000a; Jones and Ruben 2001).

Like their dinosaurian relatives, *Archaeopteryx* and early birds lacked the thoracic skeletal modifications consistent with the ability to have ventilated a fully modern avian-style lung and yet, appear not to have utilized hepatic-piston diaphragmatic lung ventilation. As in modern arboreal birds, we interpret the pubis in early birds to have been nearly horizontal and, more importantly, to have extended well posterior to the ilium and ischium (based on the arrangement of the best-preserved examples of *Archaeopteryx* (i.e., the Berlin and Solnhofen specimens) and known enantiornithine birds

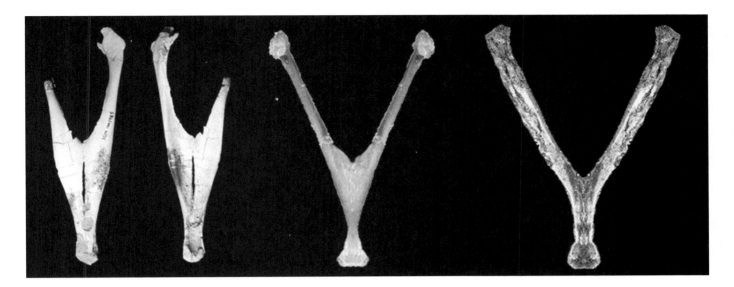

(personal observation; Martin 1991) (Fig. 36.14). Importantly, *Archaeopteryx* and some early birds (e.g., *Confuciusornis*) apparently possessed laterally expanded, dorsally concave hypopubic cups at the distal end of their pubes (Fig. 36.15) (Jones and Ruben 2001; Ruben et al. 1997). The similarity of the pubic morphology of modern and early birds indicates that this hypopubic cup may have served as the site of origin for suprapubic musculature and is inconsistent with hepatic-piston diaphragmatic lung ventilation. Consequently, it is reasonable to conclude that when roosting in trees, early birds probably also utilized suprapubic and infrapubic musculature to assist in ventilation of incipient nonvascularized posterior air sacs (Fig. 36.14).

Unlike most early birds, *Archaeopteryx* has been interpreted as adapted for a terrestrial rather than an arboreal existence (Ostrom 1991; Peters and Görgner 1992). The severely opisthopubic pelvis and the presence of the hypopubic cup of *Archaeopteryx*, as in enantiornithines, signal that early birds, including *Archaeopteryx*, were probably adapted for a substantially arboreal existence. Hence, the pelvis of *Archaeopteryx* and other early birds may evince the likely long arboreal history of their ancestry.

Given the undoubted relationship between dinosaurs and birds, how might an avian-style, air sac lung have evolved from a diaphragm-ventilating ancestor? We have previously argued that such a transition was unlikely since development of extensive posterior, nonvascularized air sacs might have compromised function of the vertically-oriented diaphragm. However, we have reconsidered this position. We propose that in early birds (e.g., *Archaeopteryx*), as in many extant reptiles, tidal volume may have been supplemented by development of incipient, nonvascularized air sacs. These might have expanded posteriorly from anterior parenchyma resulting in posterior rotation of the dorsal aspect of the post-pulmonary septum. Continued posterior migration of the dorsal aspect of the septum would have been sufficient to accommodate development of extensive posterior, nonvascularized air sacs, which were ultimately linked to the development of the avian flow-through lung and a horizontal, postpulmonary septum. Furthermore, we suggest that selection for abandonment of diaphragm-assisted breathing in early birds might have been causally linked to selection for flight

36.15. The pubes of theropods and birds. Anterior and posterior view of the pubes of *Velociraptor* (left) and posterodorsal view of the pubes of *Archaeopteryx* (center) and *Confuciusornis* (right). Only the pubes of early birds (e.g., *Archaeopteryx* and *Confuciusornis*) possess a concave, laterally expanded hypopubic cup at their distal ends. From Jones and Ruben (2001).

apparatus-assisted lung ventilation in early birds (Boggs 2002; Jenkins et al. 1988). Conversely, the distinct independence of lung ventilation from the chiropteran flight apparatus may have resulted in selection for maintenance of diaphragm-assisted lung ventilation in early bats (Lancaster et al. 1995).

Finally, claims elsewhere in this volume of putative falsification of our conclusions that theropods lacked respiratory turbinates (see Paul) amount to little more than a series of statistically unsupported anecdotes.

Summary

Utilization of comparative physiology and anatomy has provided exciting insight into the biology of many long-extinct taxa. Surprisingly, solid inferential evidence (e.g., the absence of respiratory turbinates) indicates that dinosaurs and early birds were likely to have maintained ectotherm or ectotherm-like resting, or routine, metabolic and lung ventilation rates. Nevertheless, some may have possessed the capacity for sustained activity that might have approached that in endotherms.

Based on the physiology of most extant ectotherms, in which costal ventilation well supports active rates of oxygen consumption, a specialized diaphragm to supplement ventilation in theropod dinosaurs seems superfluous. However, a variety of data suggest that expansion of lung ventilatory capacity in dinosaurs might have allowed multicameral, septate lungs to have achieved rates of O_2–CO_2 exchange that might have approached those of a few mammals with relatively low aerobic scopes (Hicks and Farmer 1998; Ruben et al. 1998). Thus, the presence of diaphragm-assisted lung ventilation in theropods might indicate that, although these dinosaurs maintained ectotherm-like routine metabolic rates, they were, nevertheless, uniquely capable of sustaining active oxygen consumption rates and activity levels beyond those of the most active living reptiles.

This hypothesized pattern of metabolic physiology in theropods may seem inconsistent with the presence of a hepatic-piston diaphragm in extant crocodilians, none of which appears to have particularly enhanced capacity for oxygen consumption during exercise (Bennett et al. 1985). However, relatively low aerobic capacity in recent crocodilians, all of which are secondarily aquatic, might not represent the ancestral condition. Because early (Triassic) crocodylomorphs (e.g., *Protosuchus* and *Terrestrisuchus*) appear to have been fully terrestrial, cursorial, and relatively active (Colbert and Mook 1951; Crush 1984), they, like theropods, might also have had enhanced aerobic capacities.

Such a scenario may have afforded theropods, early crocodylomorphs, and, perhaps, pterosaurs, the low maintenance cost of ectothermy combined with enhanced endurance similar to that in some modern endotherms. This strategy might have functioned optimally in the relatively mild, equable climatic regimes of most of the Mesozoic Era, where chronic maintenance of ectothermic homeothermy would have been possible.

Early birds, like other Mesozoic archosaurs, were probably ectothermic, but lacked the skeletal modifications indicative of hepatic-piston lung ventilation. Their lungs, while possibly including diverticula (incipient air sacs), were probably ventilated by costal expansion and contraction and were unlikely to have been capable of the activity capacity of modern birds.

References

Avery, R. A. 1982. Field studies of body temperatures and thermoregulation. In C. Gans, and H. F. Pough (eds.), *Biology of the Reptilia*, vol. 12, 93–166. London: Academic Press.

Baker, M. A. 1982. Brain cooling in endotherms in heat and exercise. *Annual Review of Physiology* 44: 85–96.

Bakker, R. T. 1968. The superiority of dinosaurs. *Discovery* 3: 11–22.

———. 1971. Dinosaur physiology and the origin of mammals. *Evolution* 25: 636–658.

———. 1975. Dinosaur renaissance. *Scientific American* 232: 58–78.

———. 1980. Dinosaur heresy–dinosaur renaissance. In R. D. K. Thomas, and E. C. Olson (eds.), *A Cold Look at Warm Blooded Dinosaurs*, 351–462. Boulder, Colo.: Westview Press.

———. 1986. *The Dinosaur Heresies*. New York: Morrow and Co.

Bang, B. 1971. Functional anatomy of the olfactory system in 23 orders of birds. *Acta Anatomica (Basel)* 79: 1–71.

Barrick, R. E., and W. J. Showers. 1994. Thermophysiology of *Tyrannosaurus rex*: evidence from oxygen isotopes. *Science* 265: 222–224.

Barsbold, R. 1983. Carnivorous dinosaurs from the Cretaceous of Mongolia. *The Joint Soviet-Mongolian Palaeontological Expedition* 19: 1–117.

Bellairs, A. d. A., and C. R. Jenkin. 1960. The skeleton of birds. In A. J. Marshall (ed.), *Biology and Comparative Physiology of Birds*, 241–300. New York: Academic Press.

Bennett, A. F. 1973. Blood physiology and oxygen transport during activity in two lizards, Varanus gouldii and Sauromalus hispidus. *Comparative Biochemistry and Physiology* 46A: 673–690.

———. 1982. The energetics of reptilian activity. In C. Gans, and H. F. Pough (eds.), *Biology of the Reptilia*, vol. 13, 155–199. New York: Academic Press.

———. 1991. The evolution of activity capacity. *Journal of Experimental Biology* 160: 1–23.

Bennett, A. F., and B. Dalzell. 1973. Dinosaur physiology: a critique. *Evolution* 27: 170–174.

Bennett, A. F., and W. R. Dawson. 1976. Metabolism. In C. Gans, and W. R. Dawson (eds.), *Biology of the Reptilia*, vol. 5, 127–223. New York: Academic Press.

Bennett, A. F., and R. B. Huey. 1990. Studying the evolution of physiological performance. In D. Futuyma, and J. Antonovics (eds.), *Oxford Surveys in Evolutionary Biology*, vol. 7, 251–284. Oxford: Oxford University Press.

Bennett, A. F., and J. A. Ruben. 1979. Endothermy and activity in vertebrates. *Science* 206: 649–654.

Bennett, A. F., R. S. Seymour, and G. J. W. Webb. 1985. Mass-dependence of anaerobic metabolism and acid-base disturbance during activity in the saltwater crocodile, *Crocodylus porosus*. *Journal of Experimental Biology* 118: 161–171.

Bernstein, M. H., H. L. Duran and B. Pinshow. 1984. Extrapulmonary gas exchange enhances brain oxygen in pigeons. *Science* 226: 564–566.

Besson, A. and A. Cree. 2010. A cold-adapted reptile becomes a more effective thermoregulator in a thermally challenging environment. *Oecologia* 163: 571–581.

Boggs, D. F. 2002. Interactions between locomotion and ventilation in tetrapods. *Comparative Biochemistry and Physiology Part A* 133: 269–288.

Brackenbury, J. H. 1987. Ventilation of the lung-air sac system. In T. J. Sellers (ed.), *Bird Respiration*, 36–69. Boca Raton, Fla: CRC Press.

Britt, B. B., P. J. Makovicky, J. Gauthier, and N. Bonde. 1998. Postcranial pneumatization in *Archaeopteryx*. *Nature* 395: 374–376.

Brooke, M. and T. R. Birkhead. 1991. *The Cambridge Encyclopedia of Ornithology*. Cambridge: Cambridge University Press.

Burnham, D. A., K. L. Derstler, P. J. Currie, R. T. Bakker, Z.-h. Zhou, and J. H. Ostrom. 2000. Remarkable new birdlike dinosaur (Theropoda: Maniraptora) from the Upper Cretaceous of Montana. *University of Kansas Paleontological Contributions* 13: 1–1

Carrier, D. R., and C. G. Farmer. 2000a. The evolution of pelvic aspiration in archosaurs. *Paleobiology* 26: 271–293.

———. 2000b. The integration of ventilation and locomotion in archosaurs. *American Zoologist* 40: 87–100.

Case, T. J. 1978. On the evolution and adaptive significance of postnatal growth rates in terrestrial vertebrates. *Quarterly Review of Biology* 53: 243–282.

Chatterjee, S. 1997. The beginnings of avian flight. In *Dinofest International: Proceedings of a Symposium Sponsored by Arizona State University*, 311–335. Philadelphia: Academy of Natural Sciences.

Chen, P.-j., Z.-m. Dong, and S.-n. Zhen. 1998. An exceptionally well-preserved theropod dinosaur from the Yixian Formation of China. *Nature* 391: 147–152.

Chinsamy, A., L. M. Chiappe, and P. Dodson. 1994. Growth rings in Mesozoic birds. *Nature* 368: 196–197.

Chinsamy, A., and P. Dodson. 1995. Inside a dinosaur bone. *American Scientist* 83: 174–180.

Chinsamy, A. and W. J. Hillenius. 2005. Physiology of non-avian dinosaurs. In D. B. Weishampel, P. Dodson, and H. Osmólska (eds.), *The Dinosauria*, Berkeley: University of California Press.

Claessens, L. P. A. M. 2004. Dinosuar Gastralia: Origin, Morphology, and Function *Journal of Vertebrate Paleontology* 24: 89–106.

Clark, J. M., M. A. Norell, and L. M. Chiappe. 1999. An oviraptorid skeleton from the Late Cretaceous of Ukhaa Tolgod, Mongolia, preserved in an avian-like brooding position over an oviraptorid nest. *American Museum Novitates* 3265: 1–35.

Colbert, E. H., and C. C. Mook. 1951. The ancestral crocodilian *Protosuchus*. *Bulletin of the American Museum of Natural History* 97: 143–182.

Coombs, W. P., Jr., and T. Maryańska. 1990. Ankylosauria. In D. B. Weishampel, P. Dodson, and H. Osmólska (eds.), *The Dinosauria*, 456–483. Berkeley: University of California Press.

Crush, P. J. 1984. A Late Upper Triassic Sphenosuchid crocodilian from Wales. *Palaeontology* 27: 131–157.

———. 1997. Theropoda. In P. J. Currie and K. Padian (eds.), *Encyclopedia of Dinosaurs*, 731–737. San Diego: Academic Press.

Dal Sasso, C. and M. Signore. 1998a. Exceptional soft-tissue preservation in a theropod dinosaur from Italy. *Nature* 392: 383–387.

———. 1998b. In situ preservation in *Scipionyx*. In J. W. M. Jagt, P. H. Lambers, E. W. A. Mulder, and A. S. Schulp (eds.), *Third European Workshop of Vertebrate Paleontology-Maastricht*, 23. Maastricht: Naturhistorisches Museum.

Darveau, C.-A., R. K. Suarez, R. D. Andrews, and P. W. Hochachka. 2002. Allometric cascade as a unifying principle of body mass effects on metabolism. *Nature* 417: 166–170.

Dietz, M. W. and R. H. Drent. 1997. Effect of growth rate and body mass on resting metabolic rate in galliform chicks. *Physiological Zoology* 70: 493–501.

Dodson, P. 2000. Origin of birds: the final solution? *American Zoologist* 40: 504–512.

Dodson, P., and P. J. Currie. 1990. Neoceratopsia. In D. B. Weishampel, P. Dodson, and H. Osmolska (eds.), *The Dinosauria*, 593–618. Los Angeles: University of California Press.

Duncker, H.-R. 1971. The lung air sac system of birds. *Advances in Anatomy, Embryology, and Cell Biology* 45: 1–171.

———. 1972. Structure of avian lungs. *Respiration Physiology* 14: 44–63.

———. 1974. Structure of the avian respiratory tract. *Respiration Physiology* 22: 1–19.

———. 1978. General morphological principles of amniote lungs. In J. Piiper (ed.), *Respiratory Function in Birds, Adult and Embryonic*, 2–15. Berlin: Springer-Verlag.

———. 1979. Coelomic cavities. In A. S. King and J. McLelland (eds.), *Form and Function in Birds*, vol. 1, 39–67. New York: Academic Press.

———. 1989. Structural and functional integration across the reptile-bird transition: locomotor and respiratory systems. In D. H. Wake, and G. Roth (eds.), *Complex Organismal Functions: Integration and Evolution in Vertebrates*, 147–169. New York: Wiley and Sons.

Else, P. L., and A. J. Hulbert. 1981. Comparison of the "mammal machine" and the "reptile machine": energy production. *American Journal of Physiology* 240: R3–R9.

Farlow, J. O. 1990. Dinosaur energetics and thermal biology. In D. B. Weishampel, P. Dodson, and H. Osmólska (eds.), *The Dinosauria*, 43–55. Berkeley: University of California Press.

Farlow, J. O., P. Dodson, and A. Chinsamy. 1995. Dinosaur biology. *Annual Review of Ecology and Systematics* 26: 445–471.

Farmer, C. G. and D. R. Carrier. 2000a. Pelvic aspiration in the American alligator (*Alligator mississippiensis*). *Journal of Experimental Biology* 203: 1679–1687.

———. 2000b. Ventilation and gas exchange during treadmill locomotion in the American alligator (*Alligator mississippiensis*). *Journal of Experimental Biology* 203: 1671–1678.

Fedde, M. R. 1987. Respiratory muscles. In T. J. Sellers (ed.), *Bird Respiration*, vol. 1, 3–37. Boca Raton, Fla.: CRC Press.

Feduccia, A. 1999. 1,2,3 = 2,3,4: Accommodating the cladogram. *Proceedings of the National Academy of Sciences* 96: 4740–4742.

Feduccia, A., T. Lingham-Soliar, and J. R. Hinchliffe. 2005. Do feathered dinosaurs exist? Testing the hypothesis

on neontological and plaeontological evidence. *Journal of Morphology* 266: 125–166.

Fisher, P. E., D. A. Russell, M. K. Stoskopf, R. E. Barrick, M. Hammer, and A. A. Kuzmitz. 2000. Cardiovascular evidence for an intermediate or higher metabolic state in an ornithischian dinosaur. *Science* 288: 503–505.

Forster, C. A. 1990. The postcranial skeleton of the ornithopod dinosaur *Tenontosaurus tilletti*. *Journal of Vertebrate Paleontology* 10: 273–294.

Frair, W., R. G. Ackman, and Mrosovsky. 1972. Body temperature of *Dermochelys coriacea*: warm turtle from cold water. *Science* 177: 791–793.

Franklin, C., F. Seebacher, G. C. Crigg, and M. Axelsson. 2000. At the crocodilian heart of the matter. *Science* 289: 1687–1688.

Frey, E. 1988. The carrying system of crocodilians – a biomechanical and phylogenetic analysis. *Stuttgarter Beiträge zur Naturkunde, Serie A (Biologie)* 426: 1–60.

Gans, C. and B. Clark. 1976. Studies on the ventilation of *Caiman crocodilus* (Reptilia: Crocodilia). *Respiration Physiology* 26: 285–301.

Gauthier, J., A. G. Kluge, and T. Rowe. 1988. Amniote phylogeny and the importance of fossils. *Cladistics* 4: 105–209.

Gauthier, J., S. J. Nesbitt, E. R. Schachner, G. S. Bever, and W. G. Joyce. 2011. The bipedal stem crocodilian *Poposaurus gracilis*: inferring function in fossils and innovation in archosaur locomotion. *Bulletin of the Peabody Museum of Natural History* 52: 107–126.

Geist, N. R. 2000. Nasal respiratory turbinate function in birds. *Physiological and Biochemical Zoology* 73: 581–589.

Geist, N. R., and A. Feduccia. 2000. Gravity-defying behaviors: Identifying models for protoaves. *American Zoologist* 40: 664–675.

Goodrich, E. S. 1930. *Studies on the Structure and Development of Vertebrates*. London: MacMillan.

Gower, D. J. 2001. Possible postcranial pneumaticity in the last common ancestor of birds and crocodilians: evidence from *Erythrosuchus* and other Mesozoic archosaurs. *Naturwissenschaften* 88: 119–122.

Greenberg, N. 1980. Physiological and behavioral thermoregulation in living reptiles. In R. D. K. Thomas, and E. C. Olson (eds.), *A Cold Look at the Warm Blooded Dinosaurs*, 141–166. Boulder, Colo.: Westview Press.

Gregory, W. K. 1951. *Evolution Emerging.* New York: Macmillan.

Hamilton, W. J., and F. Heppner. 1966. Radiant solar energy and the function of black homeotherm pigmentation: an hypothesis. *Science* 155: 196–197.

Harwell, A., D. Van Leer, J. A. Ruben, and T. D. Jones. 2002. New evidence for hepatic-piston breathing in theropods. *Journal of Vertebrate Paleontology* 22: 3S

Hemmingsen, A. M. 1960. Energy metabolism as related to body size and respiratory surfaces, and its evolution. *Reports of the Steno Memorial Hospital* 9: 7–110.

Hengst, R. 1998. Ventilation and gas exchange in theropod dinosaurs. *Science* 281: 47–48.

Hicks, J. W., and C. G. Farmer. 1998. Ventilation and gas exchange in theropod dinosaurs. *Science* 281: 47–48.

———. 1999. Gas exchange potential in reptilian lungs: implications for the dinosaur-avian connection. *Respiration Physiology* 117: 73–83.

Hildebrand, M., and G. E. Goslow, Jr. 2001. *Analysis of Vertebrate Structure.* New York: Wiley.

Hillenius, W. J. 1992. The evolution of nasal turbinates and mammalian endothermy. *Paleobiology* 18: 17–29.

———. 1994. Turbinates in therapsids: evidence for Late Permian origins of mammalian endothermy. *Evolution* 48: 207–229.

Hofstetter, R., and J.-P. Gasc. 1969. Vertebrae and ribs of modern reptiles. In C. Gans, A. d. A. Bellairs, and T. S. Parsons (eds.), *Biology of the Reptilia*, vol. 1: Morphology A, 201–310. London: Academic Press.

Hopson, J. A. 1973. Endothermy, small size and the origin of mammalian reproduction. *American Naturalist* 107: 446–452.

Horner, C. C., J. R. Horner, and D. B. Weishampel. 2001. Comparative internal cranial morphology of some hadrosaurian dinosaurs using computerized tomographic X-ray analysis and rapid prototyping. *Journal of Vertebrate Paleontology* 21: 64A.

Horner, J. R., A. J. de Ricqlès, and K. Padian. 2000. Long bone histology of the hadrosaurid dinosaur *Maiasaura peeblesorum*: growth dynamics and physiology based on an ontogenetic series of skeletal elements. *Journal of Vertebrate Paleontology* 20: 115–129.

Horner, J. R., K. Padian, and A. J. de Ricqlès. 2001. Comparative osteohistology of some embryonic and perinatal archosaurs: developmental and behavioral

implications for dinosaurs. *Paleobiology* 27: 39–58.

Hutchinson, J. R. 2001. The evolution of pelvic osteology and soft tissues on the line to extant birds (Neorithes). *Zoological Journal of the Linnean Society* 131: 123–168.

Ingelstedt, S. 1956. Studies on the conditioning of air in the respiratory tract. *Acta Oto-Larngologica* 131: 1–79.

Jackson, D. C., and K. Schmidt-Nielsen. 1964. Countercurrent heat exchange in the respiratory passages. *Proceedings of the National Academy of Sciences* 51: 1192–1197.

Jenkins, F. A., K. P. Dial, and G. E. Goslow. 1988. A cineradiographic analysis of bird flight: the wishbone in starlings is a spring. *Science* 241: 1495–1498.

Jones, T. D., J. O. Farlow, J. A. Ruben, D. M. Henderson, and W. J. Hillenius. 2000. Cursoriality in bipedal archosaurs. *Nature* 406: 716–718.

Jones, T. D., and J. A. Ruben. 2001. Respiratory structure and function in theropod dinosaurs and some related taxa. In J. Gauthier, and L. F. Gall (eds.), *Proceedings of the International Symposium in Honor of John H. Ostrom (February 13–14, 1999): New perspectives on the origins and evolution of birds.*, 443–461. New Haven: Yale University Press.

Jones, T. D., J. A. Ruben, P. F. A. Maderson, and L. D. Martin. 2001. *Longisquama* fossil and feather morphology. *Science* 291: 1899–1902.

Jones, T. D., J. A. Ruben, L. D. Martin, E. N. Kurochkin, A. Feduccia, P. F. A. Maderson, W. J. Hillenius, N. R. Geist, and V. Alifanov. 2000. Nonavian feathers in a Late Triassic Archosaur. *Science* 288: 2202–2205.

Kemp, T. S. 1988. Haeomothermia or Archosauria?: The interrelationships of mammals, birds, and crocodiles. *Biological Journal of the Linnean Society* 92: 67–104.

King, A. S. 1966. Structural and functional aspects of the avian lung and air sacs. *International Review of General and Experimental Zoology* 2: 171–267.

King, A. S., and D. Z. King. 1979. Avian Morphology. In A. S. King and MacLelland (eds.), *Form and Function in Birds*, 1–38. London: Academic Press.

Kolodny, Y., B. Lutz, M. Sander, and W. A. Clemens. 1996. Dinosaur bones: fossils or pseudomorphs? The pitfalls of physiology reconstruction from apatitic fossils. *Palaeogeography, Palaeoclimatology, and Palaeoecology* 126: 161–167.

Lancaster, W. C., O. W. Henson, Jr., and A. W. Keating. 1995. Respiratory muscle activity in relation to vocalization in flying bats. *The Journal of Experimental Biology* 198: 175–191.

Lingham-Soliar, T., A. Feduccia, and X. Wang. 2007. A new Chinese specimen indicates that "protofeathers" in the Early Cretaceous theropod dinosaur *Sinosauropteryx* are degraded collagen fibers. *Proceedings of the Royal Society B: Biological Sciences* 274: 1823–1829.

Lustick, S., S. Talbot, and E. Fox. 1970. Absorption of radiant energy in red-winged blackbirds (*Agelaius phoeniceus*). *Condor* 72: 471–473.

Maina, J. N., and M. Africa. 2000. Inspiratory aerodynamic valving in the avian lung: functional morphology of the extrapulmonary primary bronchus. *Journal of Experimental Biology* 203: 2865–2876.

Marsh, O. C. 1883. Principal characters of American Jurassic dinosaurs. Pt. VI. Restoration of *Brontosaurus*. *American Journal of Science (series 3)* 26: 81–85.

Martill, D. M., E. Frey, H.-D. Sues, and A. R. I. Cruickshank. 2000. Skeletal remains of a small theropod dinosaur with associated soft structures from the Lower Cretaceous Santana Formation of northeastern Brazil. *Canadian Journal of Earth Sciences* 37: 891–900.

Martin, L. D. 1991. Mesozoic birds and the origin of birds. In H.-P. Schultze, and L. Trueb (eds.), *Origins of the Higher Groups of Tetrapods: Controversy and Consensus*, 485–540. Ithaca, N.Y.: Comstock Publishing Associates.

Maryańska, T., H. Osmólska, and M. Wolsan. 2002. Avialan status for Oviraptorosauria. *Acta Palaeontologica Polonica* 47: 97–116.

McIntosh, J. S. 1990. Sauropoda. In D. B. Weishampel, P. Dodson, and H. Osmólska (eds.), *The Dinosauria*, 345–401. Berkeley: University of California Press.

McLelland, J. 1989. Anatomy of lungs and air sacs. In A. S. King, and J. McLelland (eds.), *Form and Function in Birds*, vol. 4, 221–279. London: Academic Press.

———. 1989. Larynx and trachea. In A. S. King and J. McLelland (eds.), *Form and Function in Birds*, vol. 4, 69–100. New York: Academic Press.

McNab, B. K., and W. A. Auffenberg. 1976. The effect of large body size on temperature regulation of the Komodo dragon, *Varanus komodoensis*. *Comparative Biochemistry and Physiology* 55A: 345–350.

Nagy, K. A. 1987. Field metabolic rates and food requirement scaling in mammals and birds. *Ecological Monographs* 57: 111–128.

Norell, M. A., and P. J. Makovicky. 1997. Important features of the dromaeosaur skeleton: information from a new specimen. *American Museum Novitates* 3215: 1–28.

———. 1999. Important features of the dromaeosaur skeleton II: information from newly collected specimens of *Velociraptor mongoliensis*. *American Museum Novitates* 3282: 1–44.

Norman, D. B. 1980. On the ornithischian dinosaur *Iguanodon bernissartensis* from the Lower Cretaceous of Bernissart (Belgium). *Institut Royal des Sciences Naturelles de Belgique. Memoirs* 178: 1–103.

Novas, F. E. 1993. New information on the systematics and postcranial skeleton of *Herrerasaurus ischigualastensis* (Theropoda: Herrerasauridae) from the Ischigualasto formation (Upper Triassic) of Argentina. *Journal of Vertebrate Paleontology* 13: 400–423.

O'Connor, M. P., and P. Dodson. 1999. Biophysical constraints on the thermal ecology of dinosaurs. *Paleobiology* 25: 341–368.

O'Connor, P. M., and L. P. A. M. Claessens. 2005. Basic avian pulmonary design and flow-through ventilation in non-avian theropod dinosaurs. *Nature* 436: 253–256.

Ohmart, R. D., and R. C. Lasiewski. 1971. Roadrunners: energy conservation by hypothermia and absorption of sunlight. *Science* 172: 67–69.

Olson, J. M. 1992. Growth, the development of endothermy, and the allocation of energy in red-winged blackbirds (*Agelaius pheoniceus*) during the nestling period. *Physiological Zoology* 65: 125–152.

Ostrom, J. H. 1969. A new theropod dinosaur from the Lower Cretaceous of Montana. *Postilla* 128: 1–17.

———. 1991. The question of the origin of birds. In H.-P. Schultze, and L. Trueb (eds.), *Origins of the higher Groups of Tetrapods: Controversy and Consensus*, 467–484. Ithaca, N.Y.: Comstock Publishing Associates.

Padian, K. 1983. A functional analysis of flying and walking in pterosaurs. *Paleobiology* 9: 218–239.

Padian, K., and J. R. Horner. 2002. Typology versus transformation in the origin of birds. *Trends in Ecology and Evolution* 17: 120–124.

Paladino, F. V., M. P. O'Connor, and J. R. Spotila. 1990. Metabolism of leatherback turtles, gigantothermy, and thermoregulation of dinosaurs. *Nature* 344: 858–860.

Paladino, F. V., J. R. Spotila, and P. Dodson. 1997. A blueprint for giants: modeling the physiology of large dinosaurs. In J. O. Farlow and M. K. Brett-Surman (eds.), *The Complete Dinosaur*, 491–504. Bloomington, Ind.: Indiana University Press.

Paul, G. S. 2001. Were the respiratory complexes of predatory dinosaurs like crocodilians or birds? In J. Gauthier and L. F. Gall (eds.), *Proceedings of the International Symposium in Honor of John H. Ostrom (February 13–14, 1999): New perspectives on the Origins and Evolution of Birds*, 463–482. New Haven, Conn.: Yale Peabody Museum of Natural History, Yale University.

Pearson, O. P. 1954. Habits of the lizard *Liolaemus multiformis multiformis* at high latitudes in southern Peru. *Copeia* 1954: 111–116.

Perez-Moreno, B. P., and J. L. Sanz. 1999. Theropod breathing mechanism: the osteological evidence. *Riv. Mus. civ. Sc. Nat. "E. Caffi" Bergamo* 20: 121–122.

Perry, S. F. 1983. Reptilian lungs: functional anatomy and evolution. *Advances in Anatomy, Embryology, and Cell Biology* 79: 1–83.

———. 1988. Functional morphology of the lungs of the Nile crocodile, *Crocodylus niloticus*: non-respiratory parameters. *Journal of Experimental Biology* 134: 99–117.

———. 1989. Mainstreams in the evolution of vertebrate respiratory structures. In A. S. King and J. McLelland (eds.), *Form and Function in Birds*, vol. 4, 1–67. London: Academic Press.

———. 1992. Gas exchange strategies in reptiles and the origin of the avian lung. In S. C. Wood, R. E. Weber, A. R. Hargens, and R. W. Millard (eds.), *Physiological Adaptations in Vertebrates. Respiration, Circulation, and Metabolism*, 149–167. New York: Marcel Dekker.

———. 2001. Functional morphology of the reptilian and avian respiratory systems and its implications for theropod dinosaurs. In J. Gauthier and L. F. Gall (eds.), *Proceedings of the International Symposium in Honor of John H. Ostrom (February 13–14, 1999): New Perspectives on the Origins and Evolution of Birds*, 429–441. New Haven, Conn.: Yale Peabody Museum of Natural History, Yale University.

Perry, S. F., and H.-R. Duncker. 1980. Interrelationships of static mechanical factors and anatomical structure in lung evolution. *Journal of Comparative Physiology, Series B* 138: 321–334.

Perry, S. F., and C. Reuter. 1999.

Hypothetical lung structure of *Brachiosaurus* (Dinosauria: Sauropoda) based on functional constraints. *Mitteilungen aus den Museum für Naturkunde in Berlin, Geowissenschaftliche Reihe* 2: 75–79.

Peters, D. S., and E. Görgner. 1992. A comparative study on the claws of *Archaeopteryx*. In K. Campbell (ed.), *Proceedings of the Second International Symposium of Avian Paleontology*, 29–37. Los Angeles: L. A. Museum of Natural History Press.

Proctor, D. F., I. Andersen, and G. R. Lundqvist. 1977. Human nasal mucosal function at controlled temperatures. *Respiration Physiology* 30: 109–124.

Prum, R. O. 2001. *Longisquama* fossil and feather morphology. *Science* 291: 1899–1900.

Quick, D. E. and J. A. Ruben. 2009. Cardio-Pulmonary anatomy in theropod dinosaurs: implications from extant archosaurs. *Journal of Morphology* 270: 1232–1246.

Randall, D., W. W. Burggren, and K. French. 2002. *Eckert's Animal Physiology: Mechanisms and Adaptations*. N.Y.: W. H. Freeman.

Randolph, S. E. 1994. The relative timing of the origin of flight and endothermy: evidence from the comparative biology of birds and mammals. *Zoological Journal of the Linnean Society* 112: 389–397.

Reid, R. E. H. 1997. Dinosaurian physiology: the case for "intermediate" physiology. In J. O. Farlow and M. K. Brett-Surman (eds.), *The Complete Dinosaur*, 449–473. Bloomington: Indiana University Press.

Ricklefs, R. E. 1979. Adaptation, constraint, and compromise in avian post-natal development. *Biological Reviews of the Cambridge Philosophical Society* 54: 269–290.

Romer, A. S. 1956. *Osteology of the Reptiles*. Chicago: University of Chicago Press.

Rowe, T., E. F. McBride, and P. C. Sereno. 2001. Dinosaur with a heart of stone. *Science* 291: 783.

Royer, D. L., C.P. Osborne, and D. J. Beerling. 2002. High CO_2 increases the freezing sensitivity of plants: implications for paleoclimatic reconstructions from fossil floras. *Geology* 30: 963–966.

Ruben, J. A. 1991. Reptilian physiology and the flight capacity of *Archaeopteryx*. *Evolution* 45: 1–17.

———. 1995. The evolution of endothermy in mammals and birds: from physiology to fossils. *Annual Review of Physiology* 57: 69–95.

———. 1996. Evolution of endothermy in mammals, birds and their ancestors. In

I. A. Johnston, and A. F. Bennett (eds.), *Animals and Temperature: Phenotypic and Evolutionary Adaptation*, 347–376. Cambridge: Cambridge University Press.

Ruben, J. A., C. Dal Sasso, N. R. Geist, W. J. Hillenius, T. D. Jones, and M. Signore. 1999. Pulmonary function and metabolic physiology of theropod dinosaurs. *Science* 283: 514516.

Ruben, J. A., W. J. Hillenius, N. R. Geist, A. Leitch, T. D. Jones, P. J. Currie, J. R. Horner, and G. Espe, III. 1996. The metabolic status of some Late Cretaceous dinosaurs. *Science* 273: 1204–1207.

Ruben, J. A., and T. D. Jones. 2000. Selective factors for the origin of fur and feathers. *American Zoologist* 40: 585–596.

Ruben, J. A., T. D. Jones, N. R. Geist, and W. J. Hillenius. 1997. Lung structure and ventilation in theropod dinosaurs and early birds. *Science* 278: 1267–1270.

———. 1998. Ventilation and gas exchange in theropod dinosaurs. *Science* 281: 47–48.

Ruxton, G. D. 2000. Statistical power analysis: application to an investigation of dinosaur thermal physiology. *Journal of Zoology* 252: 239–241.

Scheid, P., and J. Piiper. 1989. Respiratory mechanics and air flow in birds. In A. S. King and J. McLelland (eds.), *Form and Function in Birds*, vol. 4, 369–388. New York: Academic Press.

Schmidt-Nielsen, K. 1971. How birds breathe. *Scientific American* 225: 72–79.

———. 1984. *Scaling*. Cambridge: Cambridge University Press.

———. 1990. *Animal Physiology: Adaptation and Environment*. Cambridge: Cambridge University Press.

Seebacher, F. 2003. Dinosaur body temperatures: the occurrence of endothermy and ectothermy. *Paleobiology* 29: 105–122.

Seebacher, F., G. C. Grigg, and L. Beard. 1999. Crocodiles as dinosaurs: behavioural thermoregulation in very large ectotherms leads to high and stable body temperatures. *Journal of Experimental Biology* 202: 77–86.

Sereno, P. C., and S. Chao. 1988. *Psittacosaurus meileyingensis* (Ornithischia: Ceratopsia), a new psittacosaur from the Lower Cretaceous of northeastern China. *Journal of Vertebrate Paleontology* 8: 353–365.

Spotila, J. R., M. P. O'Connor, P. Dodson, and F. V. Paladino. 1991. Hot and cold running dinosaurs: body size, metabolism and migration. *Modern Geology* 16: 203–227.

Standora, E. A., J. R. Spotila, J. A. Keinath, and C. R. Shoop. 1984. Body temperatures, diving cycles, and movement in the subadult leatherback turtle, *Dermochelys coriacea*. *Herpetologica* 40: 169–176.

Starck, J. M., and A. Chinsamy. 2002. Bone microstructure and developmental plasticity in birds and other dinosaurs. *Journal of Morphology* 254: 232–246.

Thomas, D. W., C. Bosque, and A. Arends. 1993. Development of thermoregulation and energetics of nestling oilbirds (*Steatornis caripensis*). *Physiological Zoology* 66: 322–348.

Upchurch, P. 1995. The evolutionary history of sauropod dinosaurs. *Philosophical Transactions of the Royal Society of London, series B* 349: 365–390.

Visser, G. H., and R. E. Ricklefs. 1993. Development of temperature regulation in shorebirds. *Physiological Zoology* 66: 771–792.

von Koenigswald, W., G. Storch, and G. Richter. 1988. Ursprüngliche « Insectenfresser, » extravagante Igel und Langfinger. In Schaal and W. Zeigler (eds.), *Messel: ein Schaufenster in die Geschichte der Erde und des Lebens*, 159–177. Frankfurt: Waldemar Kramer.

Wedel, M. J. 2003. Vertebral pneumaticity, air sacs, and the physiology of sauropod dinosaurs. *Paleobiology* 29: 243–255.

Weishampel, D. B. 1981. Acoustic analysis of potential vocalization in lambeosaurine dinosaurs (Reptilia: Ornithischia). *Paleobiology* 7: 252–261.

———. 1981. The nasal cavity of lambeosaurine hadrosaurids (Reptilia: Ornithischia). *Journal of Paleontology* 55: 1046–1057.

Weishampel, D. B. 1997. Dinosaurian cacophony. *BioScience* 47: 150–159.

Weishampel, D. B., D. B. Norman, and D. Grigorescu. 1993. *Telmatosaurus transsylvaticus* from the Late Cretaceous of Romania: the most basal hadrosaurid dinosaur. *Paleontology* 36: 361–385.

Wettstein, O. 1931. Rhynchocephalia. In W. Kükenthal, and T. Krumbach (eds.), *Handbuch der Zoologie*, 1–235.

White, F. N. 1968. Functional anatomy of the heart of reptiles. *American Zoologist* 8: 211–219.

———. 1976. Circulation. In C. Gans and W. R. Dawson (eds.), *Biology of the Reptilia*, vol. 5, 275–334. N. Y.: Academic Press.

Witmer, L. M. 1995. Homology of facial structures in extant archosaurs (birds and crocodilians), with special reference to paranasal pneumaticity and nasal conchae. *Journal of Morphology* 225: 269–327.

———. 1997. The evolution of the antorbital cavity of archosaurs: a study in soft-tissue reconstruction in the fossil record with an analysis of the function of pneumaticity. *Journal of Vertebrate Paleontology* 17: 1–73.

———. 1999. Nasal conchae and blood supply in some dinosaurs: physiological implications. *Journal of Vertebrate Paleontology* 19: 85A.

Xu, X., Z.-L. Tang, and X.-L. Wang. 1999. A therizinosauroid dinosaur with integumentary structures from China. *Nature* 399: 350–354.

Young, C.-C. 1947. On *Lufengosaurus magnus* (sp. nov.) and additional finds of *Lufengosaurus heunei* Young. *Palaeontologica Sinica* 12: 1–53.

37.1. The metabolic grades based upon the current data on mass specific standard metabolic rates in terrestrial tetrapods used in this study.

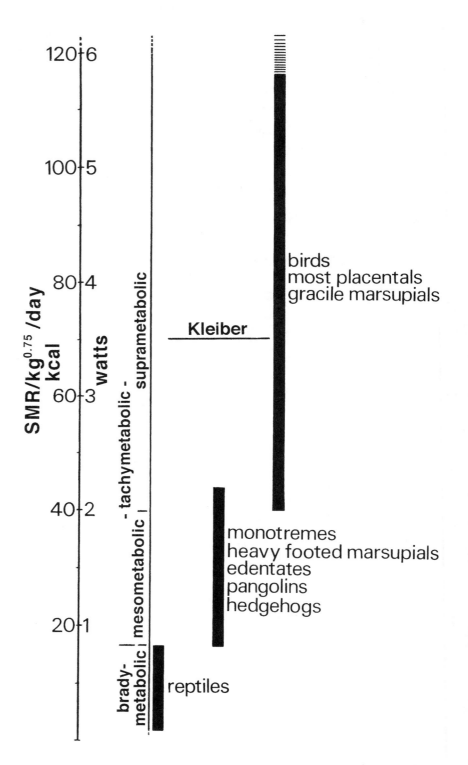

Evidence for Avian-Mammalian Aerobic Capacity and Thermoregulation in Mesozoic Dinosaurs

37

Gregory S. Paul

During the first two thirds of the twentieth century the energetic status of dinosaurs and basal birds was not particularly controversial, it being widely thought that the former's metabolics and thermoregulation were generally reptilian and the latter's avian. This consensus was shaken in the late 1960s and 1970s by arguments that dinosaurs had much the same metabolic and thermal system found in their avian descendants. Counter arguments portray dinosaurs and even early birds as thermo-energetically reptilian. Others propose that dinosaurs and basal birds were metabolic intermediates or that they exhibited a wide variety of energy and thermal levels and controls.

The assessment of Parsons (2001) – that research on the issue of dinosaur energetics has been relatively moribund since the 1978 AAAS symposium and its volume on the subject (Thomas and Olson 1980) – is far from correct. The quantity of data and research in the last third of the twentieth century and into the twenty-first century has expanded manyfold with results contradicting dinosaurs and basal birds as erect legged reptiles. Instead, the bulk of the available data indicates that dinosaurs of all types and sizes were adapted to consume and burn oxygen at rates above those exhibited in reptiles, allowing them to achieve levels of sustained activity not possible in lizards and crocodilians. Within this paradigm considerable variation almost certainly existed. Early dinosaurs possessed an intermediate anatomy that indicates the presence of a similarly intermediate, extinct thermo-energetics. Other dinosaurs had aerobic power systems most similar to those seen in mammals and birds with considerable divergence in exercise and thermoregulatory capacity. The presence of this oxygen-rich system allowed dinosaurs to do things nonmarine reptiles cannot do but which mammals and birds can: to grow rapidly (often to gigantic sizes), to achieve high levels of activity over long periods of time (e.g., migrating over long distances, living in organized groups, and caring for the young), and to fly.

The energetic systems of organisms are complex and variable even within specific groups, such as mammals, and cannot be properly defined by such simple terms as ectothermic versus endothermic (Fig. 37.1, 37.2). Below is a brief summary based in part on Paul (2002), which also defines most of the terminology. (For the purposes of this examination, resting metabolic rates can be considered roughly equivalent to slightly lower standard or basal rates.)

Insects have very low resting metabolic rates, like those of most fish and all amphibians and reptiles, so they are bradymetabolic. In many insects the maximum aerobic capacity which provides sustained motive power is also

Energetics of Living Animals

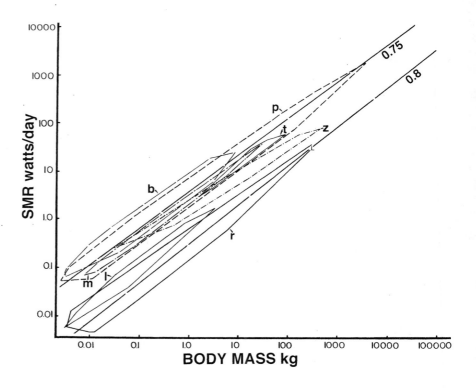

37.2. Comparison of standard metabolic rates of tetrapods, mostly continental; l, lizards at 37° C; r, reptiles at 24–30° C; L, leatherback turtles (Lutcavage et al., 1990; vertical line indicates errant data from Paladino et al. 1990); b, most birds; t–ratites; z, monotremes, heavy footed marsupials, edentates, pangolins, hedgehogs, manatees; m, most marsupials, p, most placentals. Curves are the 0.75 Kleiber for placental mammals and 0.8 for reptiles (5.5 kcal/kg$^{0.8}$/day at 24–30° C, r = 0.9840, n = 135, this and other plotted data sets available upon request) (Paladino et al. 1997).

very low, being bradyaerobic. Insects with low minimal and maximal metabolism have correspondingly low energy budgets, a combination labeled bradyenergetic. With such low levels of heat production, most insects are ectotherms dependent upon external sources for body heat. However, larger flying insects can achieve very high tachyaerobic levels, similar to those seen in mammals and birds, and are active endotherms that produce the majority of body heat internally (Heinrich 1993). The maximal/minimal aerobic ratios of insects can be extremely high, in the hundreds, because they have dispersed tracheal respiratory systems. Combining low resting metabolic rates with an elevated aerobic capacity boosts total energy budgets; such insects are heteroenergetic. Insects lack the ability to generate high levels of anaerobic power.

Vertebrates differ from insects in key regards. They have a centralized respiro-circulatory system and, if the aerobic capacity of the muscles is elevated above reptilian levels, they need more oxygen and nutrients both at rest and when exercising, which requires that the capacity and work level of the central organs be elevated, which in turn increases the organs' use of energy and boosts resting metabolism (Else and Hulbert 1985; Priede 1985; Brill 1986;, Paul and Leahy 1994; Paul 1998, 2002;, Graham and Dickson 2004). As a result, aerobic scopes are usually only about 10 and do not exceed a few dozen fold. Many vertebrates can generate high levels of anaerobic powered activity. However, intense anaerobiosis can be sustained only briefly, results in profound fatigue and in extreme cases death, and requires more food to generate a given level of power. Most fish are bradymetabolic and bradyaerobic so they are bradyenergetic, but some sharks and tuna are at least somewhat tachyaerobic and some are tachymetabolic (Brill 1986; Brill and Bushnell 1991, 2001; Graham and Dickson 2004; Bernal et al. 2005).

All amphibians and reptiles share broadly similar energetics (being consistently bradymetabolic, bradyaerobic, and bradyenergetic) with energy budgets being low. Even the most aerobically capable lizards, such as certain monitors and teiids, as well as some marine turtles, achieve maximal aerobic exercise levels barely matching the lowest seen in mammals and only a fifth as high as those typical of birds and mammals (Bickler and Anderson 1986; Thompson and Withers 2000; Clemente et al. 2009), so reptiles can sustain only low levels of activity on land. Many reptiles partly compensate with very high, tachyanaerobic power production that allows them to achieve correspondingly intense but brief periods of activity. Ectothermic reptiles are poikilotherms whose body temperatures fluctuate in tune with their surroundings—having a high body temperature in the absence of warm air requires basking in the sun. Preferred body temperatures range from 12°C for cool adapted reptiles to 41°C for hot climate examples. Energetically speaking, reptiles are basically bradyenergetic fish with legs.

The energetics of birds and mammals are far more diverse. The resting metabolic rates of some mammals, such as echidnas, tenrecs, pangolins, golden moles, and manatees, are just above or even within the maximum reptilian level. Their intermediate resting metabolisms are mesometabolic. Most resting placentals are around 10 times more energetic than reptiles, although some carnivores and camels are considerably less so. All of these are suprametabolic. Living birds are suprametabolic, although ratites and raptors are often marginally so. The aerobic capacity of mammals and birds is similarly variable. All are at least marginally tachyaerobic with those in the lower tachyaerobic zone being mesoaerobic. Most are supraaerobic. Tachymetabolic and tachyaerobic birds and mammals have high energy budgets—this combination is tachyenergetic. The latter can be divided into medium energy mesoenergetic, not seen in any bird, and the more standard supraenergetic with the high power levels seen in most mammals and in birds. The improved ability to use oxygen to produce power results in the ability to sustain high levels of activity. High levels of internal heat production make birds and mammals endotherms, except for the bizarre naked mole rats, which are ectotherms, and for manatees, which are marginally endothermic. Some mesometabolic mammals depend upon basking to help raise body temperatures (Geiser et al. 2002). Because they rely on both external and internal heat they are mesothermic. Most mammals, as well as all birds, are endotherms. Many such mammals are strong homeotherms whose healthy body temperatures are always very stable. Many placentals use heat producing brown fat to help achieve this stability. Other mammals and birds are heterothermic to varying degrees on a daily and/or seasonal basis. A few mammals are poikilothermic within limits (Schweitzer and Marshall 2001, McKechnie and Lovegrove 2002, Dausmann et al. 2004). Giraffes, for example, do not thermoregulate as long as body temperature is between 34 and 40°C (Langman and Maloiy 1989). Normal body temperatures are around 30°C in monotremes, 34–36°C in marsupials and edentates, 36–38°C in most placentals, 38–42°C in most birds, with 38–40°C in flightless examples. Note that sometimes mammals and birds can be more cold-blooded than reptiles, and sometimes the latter more warm-blooded than the former.

The absence of a dramatic metabolic gap between living bradyenergetic, ectothermic reptiles on the one hand, and tachyenergetic, endothermic mammals on the other, means that it was not possible for extinct animals to have had minimal metabolic rates that are both "well below those of modern endotherms, although still higher than in any modern reptiles" (contra Reid 1990). If a dinosaur had a resting metabolic rate above that seen in living reptiles then it was already into the mammalian zone. Even so, not all possible metabolic types exist today. It is possible that at least some dinosaurs exhibited such extinct energetics. Such speculative energetics must follow certain rules. For example, vertebrates probably cannot have the extremely high aerobic scopes seen in many insects. Because anaerobiosis and aerobiosis are mutually antagonistic, it is not possible to maximize the performance of both in the same muscles. Whether a tetrapod can combine a reptilian resting metabolism with an avian-mammalian aerobic exercise capacity, like large flying insects, is at best highly questionable because of the lack of living examples. But, if such vertebrates existed, their energy budgets were elevated enough that they would have generated most of their heat internally and would have been bradymetabolic, heteroenergetic endotherms—like similarly tachyaerobic insects, they were not ectotherms.

Power for Living on Land and Flying in the Air

Living in the water is easy. Bodies are buoyed up, and the cost of moving a given distance can be very low. The more hydrodynamically streamlined the swimmer is the better its efficiency. As a result, streamlined swimmers can move at high cruising speeds for long periods of time with little effort (Hill and Wyse 1989).

Living on land is much harder. An animal must constantly work against 1 G, and the cost of locomotion is three to 12 times higher than swimming the same distance. Because amphibians and reptiles are stuck with the low aerobic capacity of fish, they are forced to either move slowly for long periods or move rapidly on anaerobic power for brief periods and then rest to recover from the ensuing fatigue. Claims that terrestrial bradyaerobes can be quite active are misleading. While an ora can dash at a modest 14 km/h for a couple of minutes, it cannot sustain 5 km/h for hours on end the way walking birds and mammals can do. Even notoriously lazy lions are far more active in all activity parameters than the most energetic monitors (compare observations and data in Schaller 1972 to Auffenberg 1981, 1994).

In order to enjoy the high levels of sustained activity seen in energy efficient swimmers, energy inefficient land walkers must adopt a correspondingly energy inefficient tachyaerobic exercise capacity. The need to strongly elevate aerobic power in high performance land animals is the aerobic exercise capacity hypothesis. This hypothesis encompasses the aerobic capacity hypothesis, which argues that the limited maximal/minimal aerobic ratios of vertebrates forces tachyaerobes to boost the resting metabolism to tachymetabolic levels (Bennett and Ruben 1979; Bennett 1991; Paul and Leahy 1994; Hayes and Garland 1995; Paul 1998, 2002; Hayes 2010). The aerobic exercise capacity hypothesis also applies to flying vertebrates—the per unit time power requirements of flapping flight can be met only with boosted rates of oxygen consumption (even though flight is energy efficient per unit distance). The aerobic capacity hypothesis does not necessarily deny that other possible advantages of tachyenergy also boost survival potential enough to contribute to its evolution.

One can think of anaerobic power as the rough equivalent of rockets which briefly produce extreme power before it is quickly expended without taking in oxygen from the environment. Aerobic power is more akin to an air breathing jet engine that cannot produce as much burst power as a rocket, but can sustain high levels of energy for long periods. Although bradyaerobic and tachyaerobic land animals can both be active and athletic, they do so in very different ways. Bradyaerobes are much more limited in what they can do—they can be sprinters but not marathoners. Tachyaerobes can be either or both.

Because high body temperatures associated with warm habitats do not raise aerobic exercise capacity above the reptilian maximum, bradyaerobic reptiles living in warm Mesozoic climates were no more able to overcome their severe limitations than are modern land reptiles, so the applicability of the aerobic exercise capacity hypothesis remains constant over time, latitude, and climate. That a number of tachyaerobic mammals and birds are heterothermic, to a greater or lesser degree, is in line with aerobic power rather than homeothermic thermoregulation being the primary, albeit not the sole, selective force behind the elevation of metabolic rates. Although the aerobic exercise capacity hypothesis in principle applies to aquatic regimes, the low cost of locomotion may reduce the importance of sustained power relative to the need to thermoregulate in cold waters (Block et al. 1993).

Power versus Efficiency

A common theme of those who doubt that dinosaurs were highly energetic proposes that they should not have consumed large amounts of energy because doing so is inefficient and unnecessary for great archosaurs living in a toasty world where high rates of internal thermogenesis were not necessary to remain warm. But Priede (1985) notes that if energy efficiency really has been the primary goal of natural selection, then

> there should be a progressive reduction in energy expenditure by animals together with a general increase in food intake. This is clearly absurd since, taken to its logical conclusion, it would be suggest [sic] an evolutionary progression towards sessile animals with minimal locomotor energy expenditure. Evolution has in fact proceeded from sessile forms towards more active forms of life.

Some 10,000 species of birds and mammals are very energy inefficient. Even in the warm tropics mammals are the dominant tetrapods small and especially large.

The basic reason energy inefficiency can be advantageous is simple. Greater levels of power production increase the ability to sustain high levels of activity, which increases speed of sustained movement and range, which in turn increases the ability to acquire more resources that can be dedicated to reproduction. Greater levels of power production are inherently energy inefficient, but so are cars and aircraft compared to horses and bicycles as means of human transport. Indeed, the primary factor behind the performance of leading edge life-forms has been increased power, and this is the basis of the aerobic exercise capacity hypothesis.

The advantages of high aerobic power do not mean that all animals should be energy inefficient, as the multitudes of reptiles and other low metabolic rate creatures show. What it does mean is that a priori assumptions as to what animals should be like are of little use in restoring paleometabolics. Nor is it logical to presume that extinct vertebrates whose energetics are subject to dispute should be presumed to be bradyenergetic unless demonstrated otherwise, or tachyenergetic either.

Size and Metabolism

It is generally thought that metabolic rates scale close to the ¾s power (Savage et al. 2004), but this hypothesis has been challenged in favor of a scaling value as low as 2/3s, or as high as 1 for at least some metabolic factors over short segments (Dodds et al. 2001; Packard and Birchard 2008; Glazier 2009; Dodds 2010; Kolokotrones et al. 2010; also see Christiansen 2004), or no consistent scaling (White et al. 2019) – a sign that metabolic scaling knowledge remains somewhat immature. A number of researchers propose that there is a convergence between the metabolic rates of giant mammals and reptiles. In this view, the tenfold average difference between small reptiles and mammals dwindles or disappears among giants, resulting in gigantothermy in which the energetics of very large animals are similar regardless of whether they are reptiles, mammals, or dinosaurs. This hypothesis stemmed in part from falsely high metabolic measurements of leatherback sea turtles (Spotila et al. 1991 versus Lutcavage et al. 1990; Paul 1994a; Paul and Leahy 1994; Ruben 1995), the resting metabolism of elephants being many times above the level expected in reptiles of their size (Paladino et al. 1997), and the fact that a regression of the limited mass range of reptiles available does not support a significant convergence (Fig. 37.2). The order of magnitude difference appears to exist at even more gigantic dimensions (Paul 1998; Ruben et al. 2003) – the cellular differences that distinguish the basic metabolic grades are present and the poikilothermy (Carey et al. 1971) and normally sluggish activity of the biggest sharks indicate their energy production is a small fraction of that of tachyenergetic, homeothermic whales. The competing hypotheses cannot be fully tested because there is no data from a larger number of very large examples from the two groups – giraffes, rhinos, hippos, and whales on the tachymetabolic side, big oras, crocodilians, tortoises, and gigantic fish on the bradymetabolic.

Using bulk as insulation, reptilian giants can maintain fairly stable body temperatures throughout the day, allowing them to exploit their maximum aerobic exercise capacity around the clock. Seebacher (2003) joins others (Spotilla et al. 1991; O'Conner and Dodson 1999; Ruben et al. 2003) in making the basic mistake of assuming that such inertial homeothermy goes a long way to reduce the need for any large land animal to be tachyenergetic in order to be highly active 24 hours a day. But animals are not heat pumps. They are internal combustion engines whose motive power stems entirely from the oxidization of food (even anaerobic power ultimately being oxygen based). So bradyaerobic inertial homeotherms suffer around the clock from the same low and slow sustained activity levels that afflict all land reptiles. Also, homeothermy provided by mass results in body temperatures markedly below those achieved by high metabolic rates, even in warm climes, and higher aerobic capacity requires higher temperatures.

There is also debate over the relationship between size and anaerobic capacity. Small tachyaerobes are quickly exhausted by intense anaerobiosis.

Larger reptiles appear to sustain high levels of anaerobiosis for longer periods (Bennett et al. 1985), but this is not well documented because the actual activity level of the crocodilians was not measured and their subsequent mortality rates were high. Energy inefficient anaerobiosis cannot be used by land giants as a substitute for aerobiosis to sustain high levels of activity.

Supplementary heat generated by fermenting gut flora is always very limited because the great majority of energy that a herbivore can extract from the fodder has to be absorbed and burned by the animal that went to the effort to consume it.

Procedures

This chapter emphasizes the power systems of tetrapods over the manner in which they regulate body temperatures or their resting metabolic rates. Attempts to estimate the resting metabolism of extinct forms, for example, are somewhat misplaced in that they focus on when the animal was withdrawn from its habitat, rather than actively participating in it. In land tetrapods, the most important energy requirement is to generate power to do work and be active. Other metabolic and thermoregulatory adaptations are secondary results of, or requirements for, operating at a given power level in a given habitat.

This analysis is not a search for a Rosetta stone character with which one can reliably diagnose and restore the energetics of entire groups. Animals are intricate machines made of complicated interacting systems in which each serves multiple functions, precluding simplistic conclusions as to their metabolic implications. Paleoenergetics is instead a study of the preponderance of the evidence, based on as many attributes as possible. This study summarizes the wide body of evidence for high rates of aerobic power production and related issues in dinosaurs rather than examining a few problems in depth. Nor is a particular metabolism presumed to have an advantage over the others unless demonstrated otherwise. In particular it is not assumed that dinosaurs had reptilian energetics either because this was the traditional view or because it is the primitive condition and more derived conditions automatically have the burden of evidence upon them. The traditional or derived state of a system is irrelevant to whether or not it was present. Dinosaurs were an exotic group whose energetics are unmeasured by direct observation, so it is assumed that all possible systems are potentially and equally applicable until demonstrated otherwise. It is neither assumed that dinosaurs had a wide variety of metabolics nor were uniform in this regard; again, the evidence is the only determining factor.

Cruising is sustained movement over many minutes, hours, or days, and can only be powered aerobically (and is not to be confused with fast running). If a bradyaerobic tetrapod tries to use anaerobiosis to power continuously high walking speeds it will become severely fatigued and may even kill itself. If the same creature tries to regularly power persistently high walking speeds by moving fast only for short distances, resting, and then continuing for another brief period of fast walking, it will still suffer chronic fatigue and toxic illness, as well as elevated food consumption. It is the indefinitely

Cruising, Migrating and Being Social and Parental

sustainable aerobic exercise capacity, which is somewhat less than the maximum aerobic capacity, that sets the maximum cruising speed, especially during migrations. Also important is that moving a given distance costs about the same regardless of speed or limb design, posture, or number (Fedak and Seeherman 1979; Hill and Wyse 1989; Alexander 1992; Langman et al. 1995), so a slow walking, flat footed, sprawling monitor burns about as much oxygen to walk a kilometer as does a trotting, erect limbed, digitigrade wolf. Even a big snake uses the same amount of energy to move the same distance. Therefore, an animal with a reptilian anaerobic and aerobic exercise capacity will not be able to sustain high walking speeds, even if its legs are long and erect (contra Farlow 1990). Nor does large size provide an escape from the problem (contra Thulborn 1992). If cost of locomotion scales to the 0.72 power (Fedak and Seeherman 1979) and sustained aerobic locomotory power to 0.67, then increasing size degrades sustainable speed a modest amount; if the latter factor remains the same and cost scales to 0.68 (Langman et al. 1995), then size makes little difference: even if the latter factor is 0.68 while available power scales to 0.75, the differential of only 0.7 allows a 10 tonne animal to sustain a speed just half as high as one 100 times smaller. Pontzer et al. (2009) conclude that larger size actually increases the difficulty of sustaining aerobic locomotion.

Migrations on land are such arduous journeys that they demand "the total physiological attention" of tetrapods which must move rapidly in order to minimize travel time and exposure to danger (Meier and Fivizzani 1980). A high aerobic exercise capacity is therefore required to migrate on land (Paul 1988, 1991, 1994a; Pough et al. 2005; Bell and Snively 2008). Spotila et al. (1991) calculated that an energy efficient bradyaerobic land animal can migrate thousands of kilometers at 50 km/day and actually outrange less efficient tachyaerobes. But oras, perhaps the most mobile living land lizards, move at most only 10 kilometers in a day and on average move less than 2 kilometers, and Bengal monitors are less vagile (Auffenberg 1981, 1994). Big predatory mammals move much further in a day, up to 80 kilometers (Kruuk 1972; Schaller 1972). Migrating ungulates move 20–60 km/day (Nowak 1999). The pace projected in the Spotila et al. model, 3–4 km/h assuming a reasonable rest period each day, is far above that sustainable by reptiles. Spotila et al. failed to appreciate that aerobic power is more important than energy efficiency when it comes to migrating on land.

Extensive interacting between individuals, and especially moving in organized groups, is another form of activity that requires sustained effort at high levels of performance. This includes parenting, a form of social activity that leads to the common complaint of parents that their children are running them ragged. On land these actions should require a high aerobic capacity because an animal must have enough energy to spare for extended activity aside from foraging and breeding.

Reptiles, including the most aerobically capable and the largest monitors, are limited to cruising speeds of only 0.1–2.0 km/h whether on treadmills (Bickler and Anderson 1986; Thompson and Withers 1997) or in the field (Fig. 37.3). Cruising speeds of about 5 km/h cited in the literature for oras (including Auffenberg 1981) would represent an extraordinary performance for a walking reptile, but no documentation has been provided.

Such speeds are out of line with the short daily movements of oras. Videos of normally walking oras do not support such high sustained velocities, and Auffenberg (1994) cites much slower normal walking speeds for the large Bengal monitor. Eighty-three percent of the 12 dozen fossil reptile trackways representing animals from a few grams to dozens of kilograms also record estimated speeds of 0.1 to 2.0 km/h. Only one is over 3 km/h. Some reptiles walk for many hours each day, but they do so slowly (Bennett 1983). Faster walking and running speeds are always limited in duration. If the aerobic speed performance of the fastest cruising large varanids is extended to the 1 tonne size class, even the most optimistic projected maximum indefinitely sustainable cruising speed is only ~3 km/h. It is not surprising that no land reptile migrates or lives in organized mobile groups (Bennett 1983). The limited examples of reptilian social organization and parenting are unimpressive compared to the levels common among birds and mammals, both in the degree of interactions that occur and in being restricted to high density populations that only move short distances (Shine 1988; Grenard 1991; Duffield and Bull 2002). That bradyaerobic fish and sea reptiles migrate long distances at high cruising speeds (Spotila et al. 1991; Paladino et al. 1997), and/or live in schools, is irrelevant since swimming is so much more energy efficient than walking. Although leatherbacks can indefinitely sustain a swimming speed of 3–5 km/h, on land the sixfold decrease in locomotory efficiency would result in a land speed of only 0.5–0.8 km/h. Migration and high degrees of socialization are evidence for elevated aerobic capacity not because of their sophistication – after all schools of bradyaerobic fish swim across entire oceans – but because of the high energy cost of these activities on land.

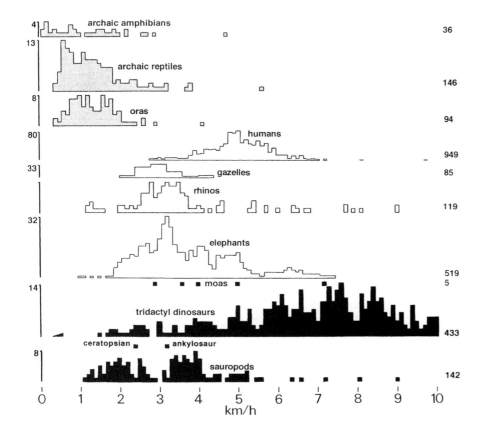

37.3. Cruising speeds in tetrapods, fossil examples estimated from trackways, living examples from motion pictures or direct observation (Paul and Leahy 1994; Paul 2000, 2002). Triangle on left is a short trackway in which decreasing stride length indicates decreasing speed. Number of trackways at a given speed on the left; sample sizes on the right. Trackway dimensions for 56 Korean sauropods, mostly juvenile, where provided, by Lockley (pers. comm.).

Tachyaerobic mammals rarely move at speeds below 2–3 km/h, easily cruise over 3 km/h, and sometimes cruise at 10 km/h or more for many hours or even days (Fig. 37.3, Paul 2002). One hundred kilogram to 6 tonne elephants normally move at 1–7 km/h, exhibiting a peak at 3–4 km/h with slowest speeds by isolated adults and all juveniles in herds. Two hundred gram ground squirrels dash about at a mean velocity of 10 km/h (Hoyt and Kenagy 1988; Kenagy and Hoyt 1989). Humans can cruise on level ground at 5 km/h for 20 hours straight aside from brief rest stops (author's experience) while big monitors manage only a third that speed over a few hours. At all sizes, tachyaerobes enjoy marked cruising speed superiority over bradyaerobes (contra Thulborn 1992). It follows that some land mammals migrate long distances, caribou, gnu, and polar bears being premiere examples. Many land mammals move in organized herds or packs, and intensive parenting is the norm.

Estimating sustained speeds from trackways is unavoidably imprecise (Manning 2008), but the consistently low, reptilian speeds recorded by basal tetrapod trackways support the basic veracity of the method as long as only gross comparisions are being made. Speeds were estimated for over 500 trackways made by a variety of dinosaurs over a 50,000 fold mass range spanning all three periods of the Mesozoic (Fig. 37.3; following procedures in Paul 2002). At all sizes, the great majority of trackways show long, avian/ mammalian-like stride lengths indicative of mammal-like speeds. Ninety-two percent were made at estimated speeds exceeding ~2 km/h and 85% at over 3 ~km/h. Many of these low speeds probably represent the minus end of estimate errors. The sole individual below 1 km/h was slowing down because strides lengths were decreasing. Only if abnormally low stride frequencies and/or inaccurately high hip-height/foot ratios are assumed can the great majority of estimated speeds of the long stride trackways be driven down to the level of shorter striding reptiles. The power generation needed to sustain the peak estimated speed for tridactyl dinosaurs, mostly theropods and ornithopods of 1 kg to 5 tonnes, ~6–8 km/h, is three to six times the aerobic exercise capacity that can be generated by reptilian aerobiosis. Contrary to Thulburn's (1992) opinion that it is size that gives dinosaurs a sustained speed advantage, even small dinosaurs outpaced reptiles of the same size or even larger. The observed pattern is fully and most compatible with theropods and ornithopods having enjoyed tachyaerobic exercise energetics matching those of similarly fast walking birds, carnivores, ungulates, and humans. With no examples under 1 km/h, the speed range of sauropods from 250 kg (none smaller are available) to 50 tonnes as estimated is similar to that observed in elephants, even more so when the highest and lowest speeds are tightened towards the mean to adjust for estimating error spread, which decreases the numbers below 2 and above 7 km/h. The sauropod's estimated peak ~3–4 km/h velocity is above even the optimistically predicted aerobic performance for a reptilian giant – most slow sauropod trackways were made by juveniles moving independently from adults, which almost always moved faster than 3 km/h. This pattern suggests adults were not less aerobically energetic than juveniles and perhaps were the opposite (see sections on circulation and bone isotopes). On the other hand, herding juveniles, which weighed at least a tonne, walked as fast as the adults – the famed Paluxy herd was doing about 3.5 km/h.

Because sauropod cruising velocities match those of elephants rather than reptiles, the first's aerobic energetics rate as fully and more compatible with those of modestly tachyaerobic elephants than with those of bradyaerobic reptiles. Contrary to Dodson's (1991) suggestion that sauropods lived in the same slow lane as bradyaerobic reptiles, they cruised in the same speed lane as tachyaerobic elephants, while theropods and ornithopods walked as fast as ground birds and ungulates. The large data set shows that dinosaur cruising speeds were nonreptilian, placing all the dinosaurs considered in the avian-mammalian speed range and requiring sustained aerobic power output beyond that which could be generated by low endurance reptiles, often by large multiples.

There is widespread agreement, based in part on trackway evidence as well as on single species fossil aggregations and other evidence, that a number of dinosaurs migrated, and/or moved in organized groups, and in some cases cared for the young (Paul 1988, 1994a, c; Russell 1989; Dodson 1991; Spotila et al. 1991; Horner and Dobb 1997; Varricchio et al. 1999, 2008; Horner 2000; Schweitzer and Marshall 2001; Meng et al. 2004; Lehman 2007; Bell and Snively 2008. Carpenter [1999] agrees on dinosaur socialization but questions parenting, as do Paul [2008] for giant theropods, and Paul [1998] and Sander and Clauss [2008] for sauropods). If so, then elevated aerobic exercise capacity was required to achieve the high levels of sustained activity on land needed to be strongly social.

The assertion by Ruben et al. (1997, also see 2003) that "if dinosaurs possessed aerobic metabolic capacities and predatory habits equivalent to those of modern tropical-latitude lizards, they may well have maintained large home ranges" is misleading. Oras and other monitor home ranges cover just a few square kilometers or less, those of similar sized predatory mammals are tens to hundreds of times larger (or more), and big herbivorous mammal home lands can cover many thousands of square kilometers (Kruuk 1972; Schaller 1972; Auffenberg 1981, 1994; Nowak 1999). Slow cruising reptilian dinosaurs could have maintained territories only a small fraction of the sizes typical of tachyaerobes. Because trackways show that dinosaurs actually walked as fast as ground birds and mammals, they could have consistently walked fast and had reptilian aerobic exercise capacity and energy budgets only if they powered almost every step partly anaerobically and if they moved no further than, and for much shorter periods of time than, reptiles. If bradyaerobic dinosaurs cruised fast, but did so only briefly, and then spent the rest of the day recovering from their chronic use of anaerobiosis, they would be less active per total time each day than many modern reptiles. Only tachyaerobic exercise capacity easily and fully explains the high cruising speeds, long range, and social/parental interactions observed in and postulated for dinosaurs (Paul 1988, 1998, 2002; Paul and Leahy 1994). This conclusion is in agreement with those of Pontzer et al. (2009, as well as Bakker 1972) utilizing similar locomotory parameters, with larger dinosaurs especially hard pressed to locomote as bradyaerobes.

The difference in cruising speeds between bradyaerobes and tachyaerobes helps reveal a causal link between limb posture and aerobic exercise

The Causal Link between Limb Posture and Aerobic Exercise

capacity. Sprawling and quadrupedalism are well suited for slow cruising bradyaerobes. The wide gauge and multiple ground contacts provide the stable platform needed by animals making short, low frequency strides. Sprawling legs can also run the high speed bursts powered by brief anaerobiosis by assuming a more dynamic, partly erect gait (Fieler and Jayne 1998), and then set the body down for the much needed recovery. Belly crawling is also good for chilled reptiles that must reach a sunny site to bask before they can fully exploit what exercise potential they have. Crocodilians and chameleons have semi-erect legs, but they can adopt a sprawling posture or use the narrow gauge posture to walk slowly along branches.

The higher aerobic powered speeds and stamina of walking tachyaerobes mean they do not benefit from the slow speed and belly resting functions of sprawling limbs. Instead, they need erect legs whose strong fore-and-aft pendulum effect promotes the longer strides and higher step frequencies that they are capable of powering over long periods. This is a physical action rather than an energetic effect, because locomotory energy efficiency remains largely the same regardless of posture. Indeed, erect legs favor long strides to the point that walking below ~2 km/h is physically awkward, tall body carriage inhibits belly resting, and the instability inherent to erect carriage and a narrow gauge further discourage slow speeds in favor of a faster pace in which dynamic balance keeps the animal from falling (Alexander 1992). Because obligatory bipedalism is inherently unstable, it also favors high dynamic walking speeds. A fully bipedal and/or erect limbed reptile would find itself tending to walk too fast to avoid the excesses of anaerobic power and would be vulnerable to falling over when chilled, yet would not be able to fully exploit the potential of the legs for a high, sustained walking pace. In this view, the evolution of erect legs probably forces aerobic capacity to be elevated above, and anaerobic capacity to be reduced below, reptilian levels.

Carrier (1987) and Carrier and Farmer (2000b) link erect and bipedal gaits with boosted aerobic exercise capacity via another argument. In salamanders and lizards, the side-to-side flexion of the dorsal columns associated with their nonerect legs hinders breathing when running, probably to the point that suprareptilian rates of oxygen uptake are prevented. Birds and mammals have escaped this constraint by adopting bipedal and/or erect gaits that do not involve lateral spinal flexion.

Limb bone morphology and trackways show that the hindlimbs of protodinosaurs, dinosaurs, and early birds were erect and could not adopt a sprawling posture, the only possible exception being small basal sinornithosaur dromaeosaurs which used the hindlegs as second wings (explaining why many of the specimens are preserved splayed out and femoral heads are more rounded than normal for dinosaurs). Dinosaur vertebral series were not suitable for strong lateral flexion. Many ornithischians and prosauropods, as well as all theropods and early birds, were strongly or entirely bipedal. Feduccia's (1996) argument that dinosaurs evolved erect limbs in order to support their enormous bodies is falsified by the fact that the dinosaurian erect gait first evolved in small forms—the erect gait may have aided gigantism but did not evolve because of it. Neither erect legs nor habitual bipedalism is compatible with reptilian energetics; both are fully in accord with dinosaurs and early birds having been tachyaerobic (Ostrom 1970).

The different aerobic performance of bradyaerobes and tachyaerobes is expressed in their musculature. Reptiles' low capacity respiro-cardiac systems cannot oxygenate large, tachyaerobic muscles; the avian-mammalian system can. Nor do reptiles need large leg muscles in order to run fast, because their bradyaerobic, tachyanaerobic leg muscles can produce about twice as much anaerobic burst power per unit mass as those of mammals and birds. The latter have to compensate for their less anaerobically capable muscles by having larger muscles. At a given mass the skeletal and leg muscles of reptiles appear to be significantly smaller than those of birds and mammals of similar mass (Ruben 1991; Paul 2002). Note that if an animal's leg muscles were as large as a mammal's and had the mass specific anaerobic power typical of reptiles, its sprint speed would exceed that of any mammal and be too extreme to be practical.

The dorsopelvis (ilium in reptiles, dinosaurs, and birds; ilium plus ischium in mammals) supports the large set of proximal locomotory muscles (Fig. 37.4). In reptiles, the ilium is short so the muscle attachment area is narrow, this being all that is needed to support a low volume of bradyaerobic, tachyanaerobic thigh muscles. There is a progressive correlation between dorsopelvis length and aerobic capacity in that mesoaerobic monotremes and insectivores tend to have somewhat longer mesoschian pelves than less energetic brevischian reptiles. In supraaerobic birds and mammals, the dorsopelvis is a more elongated plate that anchors a large set of tachyaerobic thigh muscles. A comparison of dorsopelvis length/mass shows that this structure is consistently longer, by an average of two, in longoschian mammals and birds than in brevischian diapsid reptiles. A quarter or more

Muscle Mass and Pelvic Dimensions

37.4. The relationship of dorsopelvis length to body mass in land tetrapods: r, lizards and crocodilians; C, juvenile crocodilian *Sarcosuchus;* L, semi-bipedal lizard; M, giant monitor *Megalania;* b, birds; z, monotremes and tenrecs; bm; bipedal mammals; qm, most quadrupedal mammals; pd, brevischian protodinosaurs and basal dinosaurs; bd, longoschian bipedal dinosaurs; qd, longoschian quadrupedal dinosaurs except armored ornithischians. The ⅓ slope line indicates isometric scaling, shifting longoschian dinosaurs to the reptilian level would require increasing body mass estimates two to tenfold.

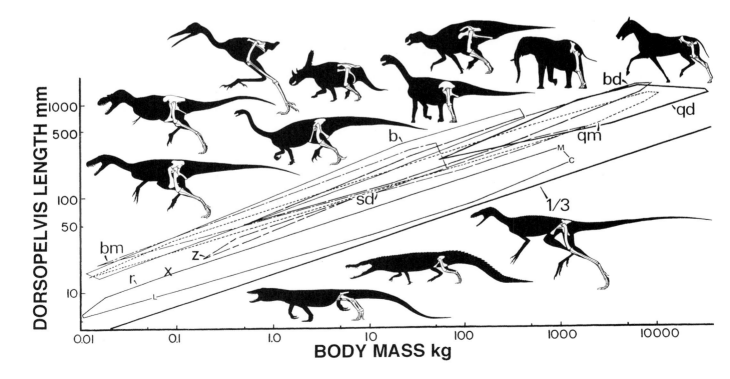

of a ratite's mass is leg muscles and to run at top speed requires that each kilogram of muscle generate over 200 W, which is achievable by tachyaerobic muscles. If there were only half as much leg muscle as per a monitor, then each kilogram would have to put out twice as much wattage, which only tachyanaerobic muscle fibers can do.

Within given groups hip dimensions tend to remain isometric, so a significant size effect is absent – even big monitors have much shorter ilia than birds and mammals of similar size. Dorsopelvic expansion is not necessary for erect or bipedal gaits; the ilia of the most bipedal modern reptiles are, and some bipedal, erect limbed dinosaurs were, much shorter than those of birds and bipedal mammals. Nor is dorsopelvis expansion a function of severe tail reduction because the latter is large in most big-hipped dinosaurs. Dorsopelvis elongation occurs in heavily armored paraeisaurs, stegosaurs, ankylosaurs, and edentates, as well as in leaping frogs whose very long and slender anterior ilia do not support thigh muscles. Elongation of the legs can also increase the volume of the limb musculature, other factors being equal.

Protodinosaurs, basal theropods, and prosauropods are interesting in that their reptile-shaped ilia were only modestly elongated compared to those of reptiles (Fig. 37.4). The pelves were therefore mesoschian and most similar to those of monotremes and insectivores in this respect. Such short pelves were not able to support a thigh musculature as broad and large as those of birds. On the other hand, the modest expansion of the ilia, combined with major elongation of the limbs, suggests that limb muscle volume was higher than in semi-bipedal reptiles. If so, then it is improbable that the leg muscles were either tachyanaerobic like reptiles, or dramatically less so like those of birds and mammals; an intermediate condition is probable. The corresponding boost in aerobic capacity should have limited anaerobic power generation leading to the loss of reptilian tachyanaerobiosis. An avian-mammalian aerobic capacity is falsified, but reptilian values are not supported either. Some form of marginal tachyaerobiosis, in the lower mesoaerobic range, is probable.

The great majority of dinosaurs had long, nonreptilian ilia (Fig. 37.4) whose length plots in the avian-mammalian range. The ilia of all known avepod (tridactyl theropods sensu Paul 2002) dinosaurs were birdlike plates in size and gross form and became especially large in the gracile tyrannosaurs and ornithomimids. Ceratopsids had exceptionally long pelves. The sauropods' shorter ilia reflect their low top speeds, although their deep pelves were larger than it may seem. Some artists, such as Knight, illogically applied narrow, reptilelike leg muscles to longoschian dinosaurs (Paul 1996). Broad, bird- and mammal-like thighs are restored by most technical researchers. The long length of most dinosaur legs also indicates that their leg muscles were larger than those of reptiles. It is very improbable that these large muscles favored anaerobic over aerobic power generation, because the resulting sprint speeds would have been as fantastical as they were impractical. Consider tyrannosaurs, whose combination of nonlocomotory mass reduction – including pneumatic skulls and skeletons, air sac filled trunks, short digestive tracts, atrophied arms, and reduced distal tails – along with deep proximal tails, very large pelves for expanded hindlimb musculature, and long legs, indicate that up to and perhaps more than a third of tyrannosaurs consisted of locomotory muscles (Paul 2008; Persons and Currie 2011). If so, then running at even ratitelike speeds required each kilogram of the leg muscle of *Tyrannosaurus* to generate just 40 watts; reptilian level tachyanerobiosis was simply not needed. Boosted aerobic exercise capacity easily and fully explains the long ilia of dinosaurs. Bradyaerobic, tachyanaerobic reptilian muscle energetics are therefore falsified in favor of higher mesoaerobic and/or supraaerobic levels (in general agreement with Pontzer et al. 2009).

Powered Flight

Ruben (1991) suggested that *Archaeopteryx* was a bradymetabolic and bradyaerobic ectotherm, an argument continued by Ruben et al. (1997). If so, and if birds are the descendants of derived predatory avepods, then the latter should not have had high metabolic rates either unless major metabolic reversals are invoked. Ruben assumed that the urvogel could power fly, but that its flight muscles were too small to generate the needed power unless they were tachyanaerobic in the reptilian manner. The same argument can be applied to basal flying dromaeosaurs, although their large sternal plates imply a better developed set of flight muscles. In any case, this hypothesis has since been refuted by the realization that the flight muscles of some birds are as tachyanaerobic as are the muscles of reptiles (Askew et al. 2001), being able to produce power bursts over 400 W/kg. Also, the flight muscles of some flight-capable birds are rather small, not being significantly larger than the arm muscles of ground bound mammals, and make up the same portion of body mass as restored for basal birds (Paul 2002).

The basic argument that basal powered fliers must be bradyaerobic and tachyanaerobic due to the problem of overly small flight muscles is clearly false because bats evolved from tachyaerobic insectivores. What is true is that—except for tiny insects to whom air is so thick that flight is extremely energy efficient—only supraaerobic insects, bats, and birds can produce the high levels of power per unit time needed to flap through modern skies. While restoring flying dino-birds as highly aerobic involves no undue speculation, restoring them as having reptilian energetics does.

Noses and Trachea

Respiratory conchae (RC) are thin sheetlike structures covered with a veneer of respiratory tissues (Fig. 37.5). RC are always set in the nasal airway through which the main airflow passes. It has been argued (see Ruben et al., chapter 36 this volume) that the dimensions of nasal airways and RC offer a means to restore the energetics of extinct amniotes, one so powerful that it may constitute a Rosetta stone, or bellwether indicator, for diagnosing paleometabolics. Ruben et al. observe that RC are absent in bradymetabolic reptiles and are well developed in tachymetabolic birds and mammals. They further observe that RC minimize the loss of water and/or body heat that occurs during the extensive breathing associated with high resting metabolic rates. The combination of causal links and empirical correlation led to the hypothesis that animals that lack RC are bradymetabolic, while those that have well-developed RC are tachymetabolic—the hypothesis is specific to resting metabolism since most animals breath through the

37.5. Variation of nasal airway (solid black) and respiratory concha (stippled: divided into anterior and middle conchae in birds, primarily maxilloturbinal in mammals) size and configuration in birds and mammals (olfactory conchae vertical solid lines), x is pre-orbital sinus. A, gull; B, black vulture; C kite; D, tropic bird; E, cormorant; F, kiwi; G, albatross. H, echidna; I, baboon; J, dog. Not to same scale except for F and H, in which bar equals 10 and 20 mm for transverse and sagittal sections respectively.

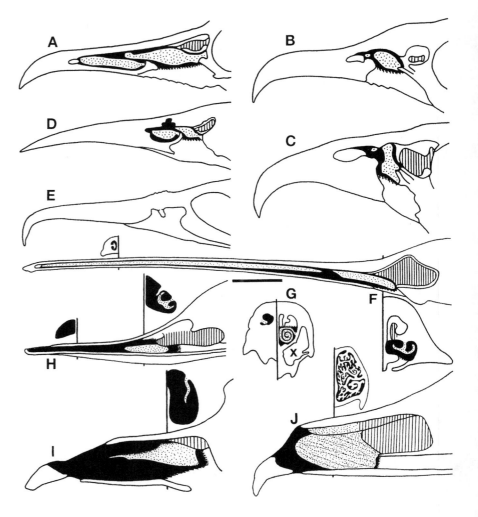

mouth when exercising. It was further argued that the cross-sectional area of the nasal airway is larger in birds and mammals than in reptiles. They explained this pattern as resulting from the need for the nasal airway to be broad in order to accommodate the high rates of air flow associated with a high rate of metabolism.

Some birds and mammals lack well-developed RC (Fig. 37.5). In birds, the anterior conchae are sometimes much less well developed than the middle conchae; in many cases neither is very large. The simple anterior conchae of kiwis and albatross are especially notable. RC are poorly developed in some raptorial birds and the nostrils of California condors are so small (Hoyos et al. 1994, 37) that they may be mouth breathers even when resting. In some marine birds, the nostrils are almost completely or entirely closed off. This is true of emperor penguins which continuously brood their egg up to four months in a habitat where freezing water is thermally expensive to ingest (Groscolas 1990). The respiratory water and heat loss experienced by obligatory oral breathers is not known. The simple RC of penguins with open nostrils keep respiratory water loss low (Murrish 1973). Water conservation is as effective in pigeons as it is in birds with better developed RC (Withers and Williams 1990, Geist 2000). RC are poorly developed in spiny anteaters, bats, elephants, saiga antelope, elephants, and most primates, including long-snouted baboons that live in arid climates.

Elephants may use the long nasal passages in their probosci to help compensate for their poorly developed RC (Paul 2002). This leads to the important point that any narrow airway from the nostrils to the lungs has the potential to condition air before it is expired. Owerkowicz and Crompton (2001) observed that, although the surface area of RC scales constant relative to metabolic rate in birds and mammals, it is three times lower in the former than in the latter despite the similar metabolisms of the two groups. Trachea can generate the temperature gradients needed to conserve water and heat. The surface areas of bird trachea are over three times larger than those of mammals, so the combined RC-trachea is comparable in both groups. Birds and mammals with poorly developed RC have additional tracheal modifications including elongation (sometimes in the form of loops in the sternal region) and subdivision.

Comparing the dimensions of nasal airways (NA) requires large samples, and in some birds both the narrow anterior and broader posterior sections must be measured (Figs. 37.6, 37.7). Cross-sectional areas of nasal airways vary up to twentyfold within birds at a given body mass. Of particular interest are ratites, most of all the recent and living island ratites, which tend to have narrower NA than other birds. The broadest section of moa and kiwi NA plots between other birds and reptiles. Because kiwis have such extremely slender beaks, their anterior nasal airway falls well below the reptile line. Roadrunners are also interesting because, although their RC are well developed and highly ossified, the internal nares are very narrow slits as little as 0.7 mm wide. The anterior nasal passages of long-nosed insectivorous spiny and giant anteaters mimic kiwis in being so slender they are in the reptile range. The anterior nasal passages of elephants are just 40 mm. The hypothesis that tachyaerobic tetrapods must breath through broad NA is falsified.

RC have never been reported in fossil birds. Taken at face value, this would indicate that they were bradymetabolic until modern times. A basic problem with using conchae to restore paleometabolics is their poor preservability: RC are almost always entirely thin sheets of cartilage in birds and, in some cases, are anchored only on cartilage sidewalls of the nasal airway. Their presence in dinosaurs must be inferred from the form and dimensions of the nasal airway, yet there is no consistent relationship between these factors. The nasal airways of birds are highly variable in form and no set of morphological indicators has been widely accepted. Even if they had been, it is not clear how they would apply to dinosaurs, whose narial passages are dramatically different from those of birds (Witmer 1995). Using nasal dimensions also appears to be unreliable. While most RC are contained in broad, long NA, this is not always the case. The anterior nasal airways of albatross and kiwis contain simple anterior conchae even though they are as slender as, or more slender than, those of reptiles. Ergo, there is currently no means by which to establish or refute the presence of RC in fossil tetrapods.

Contrary to some misleading restorations (Paul 2002), the nasal airways of *Archaeopteryx* and most predatory dinosaurs were not short and simple like those of lizards, but were long and L-shaped (Fig. 37.6). In most predatory dinosaurs, the anterior nasal tubes were relatively broader than those of kiwis, the posterior section was potentially two to four times broader

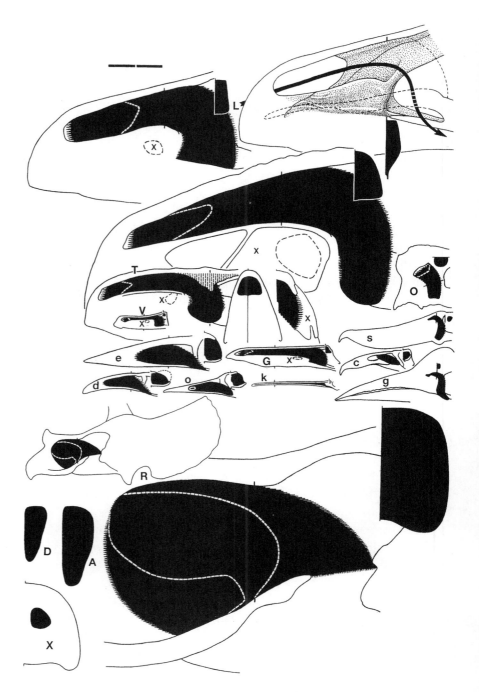

(actual dimensions cannot be restored due to lack of bony enclosure and the unknown dimensions of surrounding sinuses). This was a birdlike arrangement, and the anterior section could have contained a simple RC, while the broader posterior nasal airway was a potential site for complex middle conchae set directly above the internal nares like they are in some birds. The failure to consider the posterior section of the NA by Ruben and company causes them to underestimate the potential space for RC in theropods and basal birds. Ruben et al. (2003) point to the expansion of rostral sinuses at the expense of the NA in theropods and basal birds as evidence against well-developed RC, but this is contradicted by the empirical data in most cases so these dinosaurs and early birds appear to have diverted large NA around the sinuses. In any case, the RC may have been augmented by enlarged trachea,

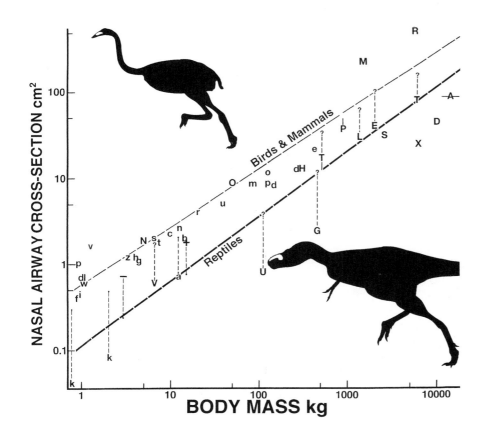

37.7. The relation of nostril (elephant only) or nasal airway (all others) cross-sectional area to body mass. L, Allosaurus; T, juvenile Tyrannosaurus; G, Gallimimus; U, Ornithomimus; V, Velociraptor; A, Apatosaurus; D, Diplodocus; O, Oviraptor; N, Ingenia; H, Hypacrosaurus, adult and juvenile; S, Parasaurolophus; P, Panoplosaurus; E, Euplocephalus; M, Centrosaurus; R, Triceratops; e, elephant bird; d, m, p, d (over 100 kg), moas; o, ostrich; r, rhea; u, emu; a, albatross; b, bustard; c, condor; d, duck (~1 kg); f, falcon; g, ground hornbill; h, heron; i, prairie chicken; k, kiwi; l, gull; n, swan; p, pheasant; s, shoebill stork; t, turkey; v, turkey vulture; w, hawk; z, goose; X, African elephant; -, echidna; +, giant anteater. Both slender anterior and broader posterior (vertical dashed lines) passage cross-sections are plotted for kiwis, albatross, and most theropods (question marks indicate approximate estimates for posterior sections). Same scale figures show that the potential space for respiratory conchae (solid white) was similar in the small headed elephant bird and a big headed juvenile Tyrannosaurus of similar mass.

especially in the long-necked ornithomimids whose slender beaks left little space for the NA. The vertical and potentially capacious nasal passages of deep-beaked oviraptors is similar to that of birds with big, deep beaks whose external nares are set so posteriorly that the NA is subvertical.

The narrow external nares of diplodocid sauropods are no more indicative of a limited aerobic metabolism than are the similarly constricted nostrils of elephants (Figs. 37.6, 37.7; contra Hengst et al. [1996]) (whose comparison of the equally large nostrils of horses is inappropriate since the latter breath entirely though the nose even when galloping). Instead, the sauropods' enlarged nasal airways were compatible with the presence of well-developed RC (Witmer and Sampson 1999), and their long necks contained exceptionally long trachea. The very long anterior nasal airways of lambeosaurine hadrosaurs are proportionally as slender as those of kiwis, and may have contained a simple RC. The greatly enlarged anterior nasal passages of hadrosaurines are compatible with the presence of well-developed RC (Witmer and Sampson 1999). The latter researchers also observed the presence of enlarged nasal airway and bony evidence for RC in ankylosaurids. The preserved nasal airways of the big headed ceratopsids are large enough to park a Buick within them. Very large anterior conchae could have been set on the cartilage spanning their enormous external nares, and the well-developed and complex midline rostral septum may have helped divert airflow to the conchae. Taken as a whole, preserved dinosaur NA cross-section area varies up to forty fold at a given mass. The arbitrary dismissal of the often-capacious NA of ornithischians as evidence for the presence of well-developed RC by Ruben et al. (2003) indicates that they do not consider their own methodology reliable when it produces results contradictory to their concept of dinosaur energetics. The apparent preservation

37.8. Same scale figures of *Giraffatitan* posed in maximum vertical reach and female sperm whale of ~30 tonnes, bar equals 3 m. The whale's 7 m long anterior airway, small lungs and normal high-pressure, 150 kg heart are shown. The sauropod is restored with a 10 m esophagus leading to a crop, 11 m trachea leading to small, birdlike lungs, and a super-high-pressure, 1 tonne heart that leaves just enough room for the lungs; the posterior cervicals as preserved are each pitched upwards 10 degrees. Center inset shows 10 degree tilt upwards between *Camarasaurus* fused posterior cervicals, right side reversed. To upper right is the elongated trachea, lung, and sternum of the trumpeter swan.

of respiratory conchae in the anterior nasal tube of a tyrannosaurid skull (Witmer and Ridgley 2010) is the most direct evidence to date that at least some dinosaurs had RC.

Long necked semi and fully aquatic reptiles such as tanystropheids show that long trachea are compatible with bradyenergy. Daniels and Pratt (1992) concluded that sauropod trachea were too long for them to have had the high rates of respiration associated with tachyenergy. This is incorrect because the costs of respiration are a small proportion of a tetrapod's energy budget regardless of trachea length or metabolic level, a fact verified by sperm whales and birds with looped trachea (Fig. 37.8) which combine very long airways with high levels of oxygen consumption; the saurischians' air sacs may have assisted in this task (Paul 1998, Wedel 2003, 2009; Perry et al. 2009). Respiratory conchae, nasal airways, and trachea are not bellwether or Rosetta stone paleometabolic indicators for dinosaurs (Seymour et al. 2008).

Heart and Lungs

Reptile respiro-circulatory systems can support maximum aerobic capacities only moderately higher than the resting metabolisms of birds and mammals (Seymour et al. 2008). Reptile lungs are often very large, but this is because they are too simple to be efficient. Even the semi-four-chambered crocodilian heart has a limited capacity and pressure. So does the unusual crocodilian pelvovisceral-pump lung system, although this may reflect the slow metabolism of living members of the group. Increasing respiratory capacity above reptilian levels requires upgrading internal lung complexity, raising lung ventilation values, and boosting cardiac output (Paul 1998, 2002; Seymour et al. 2008). The powerful, high-pressure hearts of birds and mammals have only one systemic aorta and are completely four chambered. In mammals, elevated lung capacity is accomplished with large, dead-end lungs via extreme internal complexity that greatly increases surface area/volume ratios, an effective muscular diaphragm, and vertical flexion of the dorsal trunk that enhances lung ventilation when galloping or bounding. Although the lungs of birds are less internally intricate than those of mammals, they are much more complex than those of reptiles (Perry 1989; Paul 2002). Breathing during locomotion is unhindered because the trunk and lungs are both semi-rigid. The small lungs are very efficient because they are ventilated mainly by large air sacs, with most but not necessarily all airflow being in one direction.

The simple standard lungs of reptiles are associated with a correspondingly simple trunk that is long and flexible with simple rib-vertebrae articulations. The much more complex crocodilian respiratory tract is contained within a trunk that, although still long and flexible, is more sophisticated and features uniquely elongated transverse processes that provide a smooth ceiling along which the deep liver and other organs can slide back and forth; a well-developed lumbar region; broad, mobile pubes; and a degree of unidirectional airflow (Claessens 2009a; Sanders et al. 2010). In basal crocodilians, these skeletal adaptations are less well developed, but the same may be true of their pelvovisceral pumps, which may not have appeared in the earliest terrestrial examples. This uncertainty contributes to debates over whether the unusual pelvovisceral pump evolved to oxygenate high metabolic rates in basal running crocodilians, or as part of a buoyancy control system as the group became increasingly aquatic, or initially for the former and adapted for the latter (Paul 2002; Uriona and Farmer 2008). As a result the paleometabolic diagnostic value of the crocodilian lung is uncertain.

In birds, the trunk is short and rigid and the lungs are firmly anchored within a deeply corrugated ribcage ceiling. Some air sacs invade the trunk skeleton; those along the wall of the body are ventilated by a combination of highly mobile, elongated posterior ribs; long sternal plates hinged on ossified sternal ribs; and uncinate processes that are ossified (Paul 2002). However, some of these features are also part of the flight apparatus. Active precocial juveniles, various ratites and other flightless forms, and even some fliers lack one or all major features such as large sterna, ossified sterna, ossified sternal ribs, ossified uncinates, or large abdominal air sacs (Paul 2002; Claessens 2009b). A lack of the entire, fully derived, respiratory complex present in modern adult flying birds is therefore not contrary to the presence of any degree of avian respiration, it only discredits the presence of the most extreme capacity system needed to maximize flight performance. As usual in evolution, the development of avian respiration and the corresponding anatomical adaptations was a progressive affair. The speculation by Quick and Ruben (2009) that abdominal air sacs are prevented from collapsing by the thighs when, as is true in some flightless birds, sufficiently long posterior ribs are absent is questionable. Due to a paucity of anatomical documentation, it is not yet certain that air sacs actually fill most of the gap between the ribcage and ventral pelvis when there is a large space between them, and it is difficult to see how thighs that pump fore and aft up to 70 degrees when birds run can offer stable support to air sacs in any case.

The subvertical diaphragm that ventilates the lungs of mammals is anchored on a rapidly descending edge of the ribcage immediately in front of a well-developed lumbar region.

Because predatory dinosaurs were ancestral to birds, they are expected to show evidence of a pre-avian air sac driven respiratory system, and this is the case. Some moderately derived theropods possess appendicular pneumatic structures that are strongly indicative of the presence of ventilatory air sacs (Sereno et al. 2008). As theropods evolved, the vertebrae and ribs became increasingly pneumatic, the trunk shorter and more rigid, and the posterior ribs became longer and developed more mobile hinge articulations (Fig. 37.9). These progressive changes toward the avian condition indicate that air sacs were present and expanding in size and that the posterior ribs and gastralia operated the posterior sacs, which in turn ventilated the lungs. This respiratory grade was retained by winged *Archaeopteryx*. Dromaeosaurs, troodonts, and oviraptorosaurs were even more avian in having much longer sternal plates with ossified sternal ribs, as well as ossified uncinate processes. Note that the sterna of winged dromaeosaurs were much longer than those of tachyenergetic kiwis. The air sac driven respiratory complexes of these dinosaurs should have been as well developed as those of some living birds. It is not yet clear whether any theropods with gastralia used the latter to ventilate abdominal air sacs that extended all the way to the pubes (Claessans 2004a; O'Connor and Claessens 2005), or if the ventilatory air sacs were limited to the ribcage as in some ratites which may retain the basal condition (Fig. 37.9; Paul 2002; Sereno et al. 2008). The absence of the extreme exercise levels associated with powered flight tends to favor the latter possibility. Nor is it certain to what if any degree theropods had unidirectional lung air flow, although its basic presence is suggested by air sac ventilation.

Ruben et al. (2003) questioned whether the lateral edges of the dinosaurs' sternal plates were thick enough to form adequate hinge joints with the ossified sternal ribs. The lateral edge of the dromaeosaur sternum is considerably thicker than the main body of the plate, and well-developed concavities formed hinge joints with the sternal ribs. Sternal edges are thicker in modern birds, but the entire plate is thicker, too, and this may represent strengthening for advanced flight retained even in flightless examples. In any case, the sternal-rib joints are not even ossified in active bird chicks. For that matter, the statement by Ruben et al. (2003) that "there is no compelling evidence that any dinosaur possessed thoracic skeletomuscular modifications to ventilate a birdlike lung air sac system" is contrary to the fact that no other extinct forms have ribcages and sterna so similar to those birds. If their criteria were applied to many living birds, it must be concluded that they, too, lack a birdlike respiratory system.

It has been argued that soft tissue evidence instead favors a crocodilian pelvovisceral pump in predatory dinosaurs (see Chapter 36, this volume; Ruben et al. 2003; Chinsamy and Hillenius 2004). This idea started with

37.9. Phylogenetic chart of thoraces, pelves, and respiratory tracts conservatively restored or known in archosaurs: (A) Basal archosaur *Euparkeria* with basal tetrapod system. (B) Modern crocodilian with pelvovisceral pump. Avepods with increasingly well-developed air sac complexes: (C) *Coelophysis;* (D) *Allosaurus;* (E) dromaeosaur composite; in D And E abdominal air sacs may have extended more posteriorly than shown. (F) Kiwi with short sternum and air sacs limited to ribcage; (G) duck with extremely large sternum attached to all ribs and extensive abdominal air sacs. Sauropod with air sacs: (H) *Apatosaurus.* Ornithopod with diaphragm: (I), hadrosaur. Articulated pubes in anterior view in B, D and E and (J), unidentified late Cretaceous avepod dinosaur; (K) *Archaeopteryx;* and (L) basal crocodilian *Hesperosuchus* (medially incomplete and possibly broader than shown). Lungs irregular lined, pulmonary diverticula stippled, pelvodiaphragmatic muscles fine-lined, ossified gastralia, uncinate processes, ribs, and sterna solid black; restored cartilagenous sternal elements and sternal ribs heavy lined; E-G hinge articulation of sternum with coracoid (only partly included) indicated by arrows.

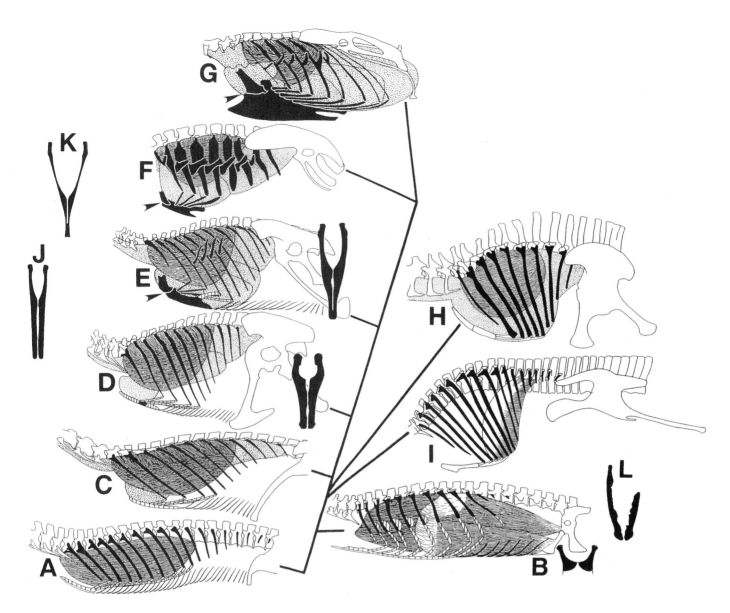

the identification of a deep liver behind an arched septum in one specimen of small *Sinosauropteryx*, but the arc may be an illusion due to breakage and repair work. The position of the poorly preserved soft tissue is more compatible with remnants of the digestive tract (Paul 2002). The counter effort by Chinsamy and Hillenius (2004) to demonstrate the presence of the arc only serves to illustrate the ambiguity of this inadequately preserved fossil. In the juvenile *Scipionyx*, the liver appears to be deep. In any case, livers are highly variable in size in vertebrates, tend to be large in growing juveniles and in flesh eating carnivores, and span the entire depth of the body in some birds (Paul 2002). The young predatory dinosaur's big liver is not, therefore, compelling evidence for the presence of a crocodilian type septum. Nor is the high position of its posterior intestine, which, if it represents the true life arrangement, represents a basal condition that may have been retained as preavian abdominal air sacs developed. The unusually low position of the colon in birds does not have any obvious functional link to the presence of abdominal air sacs. Instead it appears related to the splitting of the pubes associated with posterior migration of the abdomen, a process well under way in some of the most birdlike theropods. Ruben et al. (2003) persist in arguing that basal birds had strongly retroverted pubes bearing hypopubic cups that indicate the presence of air sacs and were absent in dinosaurs, despite the absence of the cup in any early birds and the presence of greater pubic retroversion in some dinosaurs than was present in some basal avians (Paul 2002).

The hypothesis that dromaeosaurs, which not only were very similar in form to *Archaeopteryx*, but also had avian respiratory adaptations not seen in the latter, had a crocodilian system is not logical. Theropod pubes were not well suited for anchoring the diaphragmatic muscles that ventilate lungs by pulling on the organs because they were generally more slender than those of even early crocodilians, sometimes extremely so (Fig. 37.9). Nor did the long-posterior-ribbed theropods have the crocodilian-style lumbar region that facilitates dramatic fore-and-aft motion of the viscera.

The inferior soft tissue evidence does not overturn the overwhelming osteological evidence that has created a broad consensus that the group ancestral to birds evolved increasingly birdlike, tachyaerobic-capacity, air sac–ventilated respiratory complexes, and that crocodilian-like respiration was limited to that group (Perry 1989; Paul 1998, 2002; Claessans 2004a, b; O'Connor and Claessens 2005; O'Conner 2006; Sereno et al. 2008).

Sauropods are interesting in that, next to theropods, they had the most birdlike ribcages. Not only was the ceiling of the ribcage deeply corrugated, but highly pneumatic axial skeletons indicate well-developed air sacs (Daniels and Pratt 1992; Paul and Leahy 1994; Paul 1998, 2002; Perry and Reuter 1999; Wedel 2003, 2005, 2009; O'Conner 2006; Lehman and Woodward 2008; Sander and Clauss 2008; Perry et al. 2009); sauropods had a means of ventilating the latter. Instead of being firmly anchored and/or closely spaced in order to support the massive belly, the long posterior ribs were widely separated and had birdlike hinge joints, giving them the mobility needed to operate the posterior respiratory tract, which was probably limited to the ribcage in view of the absence of gastralia. If so, a boosted aerobic exercise capacity is likely, but it was probably modest in the manner of similarly slow elephants. The skeletal anatomy of the prosauropod thorax is not readily restorable because it is not sufficiently similar to that of living analogs.

Ornithischians lack evidence of the pneumatic and other adaptations associated with air sacs. Lacking any living descendants or obvious analogs, it is difficult to restore the nature and operation of nonornithopod ornithischian lungs. Large anterior pubic processes may have played a respiratory role, but its exact function is difficult to discern due to the lack of an analog. Perry (1989) suggested that the retroverted pubes supported abdominal muscles that acted as a pseudodiaphragm to ventilate the lungs. This is most applicable to nonornithopods, which lacked a lumbar region. Ornithopods are interesting in that they had a well-developed lumbar region; its topography was more like that of mammals than the lumbar regions seen in some other archosaurs. They further mimicked mammals in lacking gastralia or a long procumbent pubis. It is therefore possible that a vertical muscular diaphragm was supported by the steeply sloped ribs that defined the anterior border of the lumbar region (Paul and Leahy 1994). If so, this is most compatible with elevated metabolic rates. Although the ornithischian lung probably dead-ended, some degree of unidirectional airflow cannot be ruled out considering its existence in crocodilians.

There is widespread agreement that dinosaurs had four-chambered hearts, probably complete double pumps in the avian-mammalian manner. This is in accord with the fact that many dinosaurs, unlike the consistently horizontal postured reptiles, habitually carried their heads well above heart level. It is doubtful that incomplete four-chambered hearts can generate the blood pressures over 100 mmHg needed to pump blood well above the organ (Paul 1998; Seymour et al. 2008). It is also doubtful that complete four-chambered hearts are compatible with low metabolic rate ectothermy, because it probably requires the ability to shunt blood between sides of the heart. Producing the high blood pressures that complete four-chambered hearts can generate also requires higher levels of work (Seymour et al. 2008), which raises the resting metabolic rate along with the greater work that has to be conducted by supporting organs.

The long-necked sauropods are especially significant in this regard. Paul (1998) explains that it would have been maladaptive for sauropods to evolve extremely long necks primarily to browse near ground level because the trivial amount of energy to be saved from not needing to walk a few more meters to reach a plant were counterbalanced or negated by the energy costs of growing, maintaining, and operating such a large structure, as well as breathing through a long trachea (Daniels and Pratt 1992; Sander and Clauss 2008) and swallowing through a long esophagus. Sauropods also possess adaptations for holding the neck highly erect, either at the base of the neck or via rearing (Paul 1998; 2006; Dzemski and Christian 2007; Mallison 2009; Taylor et al. 2009; Christian 2010). The only pair of co-fused sauropod neck vertebrae are flexed upwards ~10 degrees and, if the rest of the neck base vertebrae are flexed about the same amount, the brain could be regularly elevated many meters above heart level in longer-necked sauropods, especially those with high shoulders (Fig. 37.8); this is true whether the necks were used for high browsing and/or display as in giraffes. Because even brief brain hypoxia could lead to dangerous falls, it is presumed that blood pressures well above those generated by living animals had to be produced. That would have required oversized cardiac organs regularly producing exceptional power. The energy flow of such large, hardworking pumps should have boosted resting metabolic rates far above reptilian levels. The work of other organs to support the super heart should have further boosted the rate of metabolism (Paul and Leahy 1994; Paul 1998). In this view, oversized hearts paid for themselves by allowing sauropods to access floral resources beyond the reach of shorter herbivores, much as the enlarged and expensive flight muscles of flying tetrapods (which increase metabolic rates by a third) or the big brains of apes and humans (which increase metabolic rates by as much as a quarter) provide benefits that more than make up for the energy drain (Paul and Leahy 1994; Paul 1998; Taylor et al. 2011). As juvenile sauropods grew taller, the work that needed to be done by the heart and other organs would have had to increase. In this scenario, instead of decreasing with maturity as is usual, the metabolic rate may have increased with growth in sauropod dinosaurs (Paul 1998; see sections on cruising speeds and bone isotopes).

Seymour et al. (2011) found that the bone foramina of most bradyaerobic reptiles are narrower than those of more aerobically capable varanids and trachyaerobic mammals because the oxygen delivering blood vessels have to be larger in the latter. The exceptionally large bone foramina of all dinosaurs measured (their mass estimates are appropriate, including the immature brachiosaur) indicate they were strongly tachyaerobic.

Incubating

Eggs that are not incubated by adults must be deposited in warm places that offer adequate protection from the elements. Because it is maladaptive to leave eggs exposed to strongly fluctuating temperatures and precipitation, nests that leave eggs exposed to the weather constitute evidence of incubation via brooding. Brooding to protect exposed eggs from excessive heat or rain or from predators is a secondary action that only occurs because the eggs had to be exposed in order to allow body heating in the first place. A few reptiles and many birds brood eggs, but do so in very different ways. Snake brooding involves coiling their sinuous bodies around a pile of eggs and using low frequency contractions of the long thoracic muscles to generate heat that is transferred to the eggs. Reptilian brooding therefore is a form of aerobic exercise that raises heat production to a level similar to the resting metabolic rate of tachymetabolic animals. Bird eggs are not laid in a pile and the bird sits among or atop the egg/s, covering it or them with its body, partly spread wings, and tail. Body heat generated by an elevated metabolic rate, plus boosted thermogenesis when conditions are especially cold, is transferred to the eggs. Avian brooding behavior involves the application of resting energy production sometimes augmented by metabolic cold response (in this case shivering).

The brooding posture of at least some maniraptor dinosaurs—body set amidst a ring of partially exposed eggs, and the arms and tail draped over the latter (see Chapter 30 this volume)—is not utilized by any reptiles. This was an unambiguously preavian arrangement that allowed both soil and body heat to be used for incubation (Clark et al. 1999; Varricchio et al. 1999; Paul 2002). The best-preserved oviraptorid nest eggs were in at least partial contact with the body of the brooding adult. Coverage of the eggs was probably via a combination of the body and long feathers, as with ratites. The size of eggs, ring nests, and skeletons indicate that the biggest incubating dinosaurs reached two tonnes. The short, rigid, lightly muscled trunks of the dinosaurs were poorly suited for using low frequency muscle contractions to generate body heat in the manner of brooding snakes. It is therefore concluded that general body heat via an elevated resting metabolism warmed the eggs. The nesting behavior of birdlike dinosaurs is not compatible with classic reptilian energetics and firmly supports the presence of a tachyenergetic preavian reproductive thermal physiology.

Open Nests

There is good evidence that baby hadrosaurs dwelled in open nests which they were not able to depart (see Chapter 30 this volume; Horner 2000; Horner and Dobb 1997), leaving them exposed to extremes of temperature and rain (Paul 1994c). Juvenile reptiles do not live in open nests because they are not able to thermoregulate well enough to survive the harsh conditions. Only tachymetabolic bird nestlings with well-developed heat production and thermal controls are left exposed to the elements. The same metabolic requirements probably applied to hadrosaur and any similar dinosaur nestlings.

For quantitative comparisons of rates of growth to be meaningful, they should compare the equivalent, fastest sustained portion of rapid juvenile growth (Case 1978; Erickson et al. 2001; Padian et al. 2001; Paul 2002). Ruben (1995) and Ruben and Jones (2000) used mass at sexual maturity when comparing growth rates. This is inappropriate because, at one extreme, body mass can be lost before sexual maturity (some birds and chameleons) which may occur long after growth has ceased (Erickson et al. 2007; Karsten et al. 2008); at the other extreme, many animals are still in rapid juvenile growth and at a minority of adult body mass when they begin reproducing. Between these extremes, many animals experience a period of slow, subadult growth before fecundity. Using sexual maturity as the critical end of growth point arbitrarily lowers the growth values of animals that become sexually mature late in the growth period. Nor is it always possible to determine the onset of sexual maturity in extinct forms. For example, the comparison by Ruben (1995) and Ruben and Jones (2000) of the growth of large, 50+ kg *Troodon* specimens (that were probably adults that had completed most growth) to 30 kg subadult crocodilians (that are still growing rapidly) understated the overall estimated growth rate of the dinosaur compared to that of the crocodilian whose adult mass is hundreds of kilograms.

Comparisons of growth rates as a potential indicator of thermophysiology need to be limited to animals living in the same energetic context. This includes natural conditions to the exclusion of agricultural captivity. The tachyenergetic, homeothermic human caretakers of captive animals can supply the latter with amounts of food that they could not gather on their own at so little cost, greatly boosting growth rates (Grenard 1991; Wintner 2000; Walsh et al. 2002). Since dinosaurs did not live on farms or in zoos, citing the boosted growth of a farm raised alligator as a model for the Mesozoic archosaurs is no more logical than citing a well-slopped farm hog for the same purpose.

Nor is it appropriate to directly compare the growth rates of primarily terrestrial vertebrates with those of hydrodynamically streamlined swimmers because the latter are able to mature much faster than land-based counterparts with similar energetics (Fig. 37.10). The growth of great whales can be astonishingly swift, being an order of magnitude higher than seen in giant land mammals. Giant bradyenergetic and poikilothermic fish and leatherback turtles grow much more rapidly than big terrestrial reptiles and match land mammals of similar size. If marginal data using growth rings in their armor is correct (see Schwimmer 2002 for cautions) the growth of freshwater super crocodilians of 8 tonnes (Erickson and Brochu 1999; Sereno et al. 2001; Schwimmer 2002; the much higher mass estimate in Lehman 2007 appears excessive) may have been well above that of terrestrial reptiles, although not as fast as giant land mammals. Modern crocodilians grow

37.10. Actual and estimated growth rates, mostly terrestrial amniotes, plus some large aquatic forms, faster growing sex plotted unless otherwise indicated. Extant taxa either wild, or nonagricultural captives, terrestrial examples enclosed by least area polygons; re, perennial reptiles; ba, altricial birds; bp, precocial birds; me, monotremes; mm, marsupials; mp, most placentals; pr, primates. Multiple taxa listed under same label in order of ascending mass; W, basking and whale shark; F, Leedsichthys; L, leatherback turtle; T, tortoise; Q, annual chameleon; X, ora; C, alligator, estuarine crocodile; c, Deinosuchus and Sarcosuchus are coincident; y, Coelophysis; q, Sinornithomimus; o, Oviraptor, Citipatis; d, dromaeosaur, Sauronitholestes, "Deinonychus"; t, tiny troodont, Troodon; u, Shuvuuia; a, Allosaurus; r, Albertosaurus, Tyrannosaurus; p, Massospondylus, Plateosaurus; s, British sauropod, Janenschia, Alamosaurus, Apatosaurus; e, Scutellosaurus; i, Psittacosaurus; c, Pachyrhinosaurus, Centrosaurus (provisional); ?, small ornithopods without deep growth rings, Australian subpolar taxon, Texan taxon, Dryosaurus; h, Hypacrosaurus; K, kiwi (female, male); m, Megalapteryx, Dinornis; U, emu; O, ostrich; M, red kangaroo (female, male); H, human; P, S, F and B, sperm, sei, fin and blue whales; G, giraffe; I, hippo; R, white rhino; E, African elephant, cow, bull, projected "world record" bull. Data and estimate sources include Parker and Stott (1965), Reid (1990), Paul (1994b, 2002), Zug and Barham (1996), Wintner (2000), Sereno et al. (2001), Walsh et al. (2002), Padian et al. (2004), Sandar and Klein (2005), Turvey et al. (2005), Bybee (2006), Erickson et al. (2007), Lehman (2007), Liston (2007), Cooper et al. (2008), Karsten et al. (2008), Lehman and Woodward (2008), Varricchio et al. (2008), Erickson and Druckenmiller (2011), with adult masses of extinct taxa corrected (usually reduced) as necessary. Inserts are same scale adult 60 tonne sauropod and 5 tonne elephant with new babies to compare adult/juvenile growth distances, bar equals 4 m.

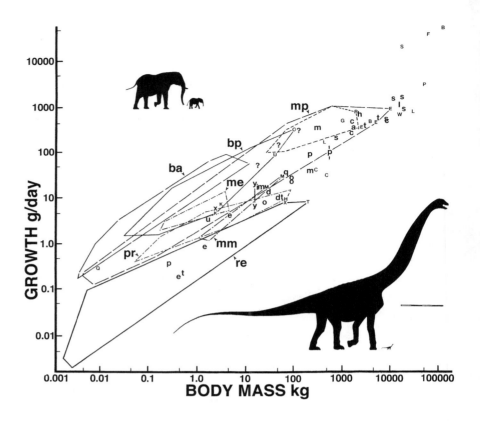

rather slowly compared to other aquatic vertebrates, but appear to outpace more terrestrial large reptiles (the claim by Ruben [1995] that the growth of alligators exceeds that of marsupials manyfold is exaggerated – male red kangaroos grow faster than wild male alligators).

Because dinosaurs were not streamlined water lovers, their growth is best compared to those of similarly land-based living amniotes. Bradyenergetic ectotherms grow slowly compared to many tachyenergetic endotherms (Fig. 37.10; Case 1978; Paul 2002). Typical lizards grow at a small fraction of the rate of shrews, and horses reach a quarter tonne in a couple of years, while giant tortoises reach just a few kilograms in the same period. The most rapidly growing small reptiles are chameleons that die well within a year (Karsten et al. 2008). Growth rates of wild oras are not known, but captives indicate that they do not grow faster than humans; Bengal monitors apparently grow even more slowly (Auffenberg 1981, 1994; Walsh et al. 2002). The fastest growing perennial land reptiles come close to matching and in a few cases exceed the slowest growing tachyaerobes. Among wild birds and mammals, those with the lowest metabolic rates (smaller moas if their rings mark true age, kiwis, monotremes, marsupials, edentates, and tenrecs) often grow more slowly than more energetic examples, but numerous exceptions show that growth rates can vary widely even among land tetrapods with broadly similar levels of energy production. For example, precocial, large ratite chicks are less energetic than nursing mammals and semi-altricial children that share their savanna habitat, yet the birds grow as rapidly as the nonprimate mammals and much more rapidly than humans. There is a similarly large variation in rates of growth of reptiles of the same adult mass.

The reasons that energy inefficient tachyenergetic terrestrial tetrapods can grow fast and energy efficient bradyenergetic ones cannot are not fully understood. Relatively stable body temperatures that optimize the internal conditions for growth apparently facilitate swift growth. The ability to deliver nutrients with sufficient rapidity to quickly convert them into large amounts of body tissues may be the most critical factor. If so, gathering large amounts of food per unit time normally necessitates high levels of sustained activity, which on land requires high capacity digestive, circulatory, and respiratory systems, which in turn involve an elevated rate of energy expenditure to operate. This does not apply in water where the low cost of streamlined swimming reduces the cost of gathering food so dramatically that growth rates are greatly boosted compared to land animals with similar energetics. The relative thermal stability of water also accelerates the maturation of aquatic bradyenergetic ectotherms. But the growth disparity between energy levels still applies among swimmers, and the superior growth rate potential of tachyenergetic marine animals may explain their size superiority compared to bradyenergetic swimmers. The ability of annual chameleons to gather large amounts of food with little effort via an ambush projectile system probably explains why they can dedicate so much of their juvenile energy budget to growing faster than other nonaquatic reptiles, rather than to quicker locomotion or to reproduction, which is delayed until growth is completed.

Otherwise rapid juvenile growth on land in excess of the conventional terrestrial reptilian maximum rate requires, and is diagnostic of, the presence of tachyenergetic homeothermy in either the juveniles themselves or their caretakers. This becomes increasingly true as the gap increases. It is also increasingly true the more that harsh winter conditions inhibit the annual growth period. Aerobically capable, homeothermic, self-feeding juveniles have the high activity levels (see section on cruising) needed to seek out and consume large amounts of food and keep themselves warm. The parents of poikilothermic bird chicks are high-metabolic-rate endotherms that use their body heat to keep their charges warm and strenuously work to bring the latter enormous quantities of food at no energy cost to the juveniles. The onset of reproductive ability before adult mass is reached slows the growth of humans and other adolescently reproductive animals (Erickson et al. 2007). Land reptiles, whether juveniles or their parents, cannot produce enough body heat to keep warm enough to grow fast when the environment is cool, nor can they sustain the high activity levels needed to find and consume enough food to grow as fast as birds and many land mammals. This also explains why adult land reptiles do not try to gather and bring food to their young (see section on cruising and socialization), so even parented lizards grow slowly (Duffield and Bull 2002).

Adult growth is not metabolically diagnostic. Growth does not entirely cease with adulthood in most but not all reptiles and is determinate in most but not all mammals and birds. But some male kangaroos never stop growing. And cessation of growth and typical age of death are essentially coincident in bull African elephants.

Although the presence of growth rings or a particular bone type may not be diagnostic of a particular metabolism, the rates of growth they record may be more indicative. Once thought to be limited to bradyenergetic vertebrates, and absent in dinosaurs, growth rings set deep in skeletal elements have been identified in mammals, modern birds, and dinosaurs (Reid 1990 and Chapter 31 this volume). If lines formed by severe slowing or cessation of growth are deposited annually, as in living reptiles, then they can be used to estimate growth rates when issues such as loss of inner rings due to bone reabsorption are taken into account. However, the risk of misestimating age when using bone rings may be under appreciated. Castanet et al. (1993), Bjorndal et al. (1998), Klomp and Furness (1992), and Broughton et al. (2002) warn that bone zones do not record age in every taxon, and their accuracy needs to be verified with field observations for each species, even when relatives lay down annuli. In a number of birds, rings often do not record actual age (Lewis 1979; Manikowski and Walasz 1980; Nelson and Brookhout 1980; Klomp and Furness 1992; Broughton et al. 2002). Most of the discrepancies indicate that the rings are multi-annular, their number often exceeding true age in younger adults multifold. Potential nonannual causes of growth lines include multiple climatic or food-based growth stressors in a given year, reproduction before adult mass is reached, and illness. Because is not possible to entirely verify the reliability of a given dinosaur's bone rings as age markers, growth rates based on growth zones are potential minimum values. Another serious, but underappreciated, problem vexes the use of growth rings to age fossil vertebrates–only those that are aged with rings can be assessed and graphically plotted. This problem is acute because the lack of deep rings is evidence that the subject grew more rapidly than land reptiles, so relying on growth rings to estimate growth rates risks missing the higher end of a group's maximum growth performance, and statistical analysis of growth within such a group covers only those constituents that possess rings rather than the group as a whole. Since a slow growing small animal should lay down at least one growth ring if it takes at minimum a year to reach adult size, the absence of any rings indicates that a small animal reached or approached final mass in about a year. Because slow growing, bradyenergetic terrestrial reptiles consistently deposit annuli throughout life in the wild (assuming they live at least one year), the absence of zonal bone in a terrestrial tetrapod is indicative of a boosted energy budget.

Peak protodinosaur and dinosaur growth rates have been calculated by a number of researchers (Chinsamy 1993; Paul 1994b, 2002; Varricchio 1997; Curry 1998; Erickson et al. 2001, 2004, 2007; Padian et al. 2001, 2004; Horner and Padian 2004; Sander and Klein 2005; Bybee 2006; Lehman 2007; Cooper et al. 2008; Lehman and Woodward 2008; Fostowicz-Frelik and Sulej 2010; Stein et al. 2010; Erickson and Druckenmiller 2011). The very low end of the growth rate estimated for a prosauropod by Lehman (2007) is excluded because it is not the maximum rate. The ring-based growth rate of a very small troodont is low even for a reptile, much less for a feathered near-avian that apparently incubated its eggs. It is also much slower than near relatives and other maniraptors, so this is a candidate for multiannular rings, perhaps due to early onset of reproduction. Otherwise, estimated maximal rates are never below those seen in slower growing birds and mammals and are above those of large monitors and other land reptiles. Because the growth of some large dinosaurs matches that of similarly massive land mammals it is unlikely that their growth bands are multiannular. The maturation of the former is not intermediate to that of similar sized mammals and those expected for nonaquatic megareptiles. Sander and Klein

(2005) conclude that *Plateosaurus* had a reptilelike variable growth pattern; alternately the rings did not consistently record genuine age. The large difference in growth rates estimated for *Coelophysis* also may or may not be real. A number of dinosaur taxa, including small and large ornithopods and nonpolar ceratopsids, lack growth rings at least until they approached maturity (Winkler 1994; Chinsamy et al. 1998; Padian et al. 2004; Lehman 2007; Cooper et al. 2008; Erickson and Druckenmiller 2011). Because the bone matrix is of the zoneless fibro-lamellar type associated with continuous rapid growth, it is probable that growth rates of these ornithischians were well above the reptilian maximum. Lehman's (2007) speculation that ceratopsids grew slowly was correspondingly improbable, and has been refuted by a polar specimen (Erickson and Druckenmiller 2011). In order to avoid excluding the small ornithopods from the graphic comparison, growth rates calculated on the tentative assumption they reached mass at death within a year are provisionally plotted. The conclusion by Erickson et al. (2001) that larger dinosaurs had higher proportional growth rates than smaller dinosaur taxa cannot be verified, their method being untestable because it automatically excludes potentially fast growing small taxa that lack zonal bone, may be contaminated by multiannular rings in some taxa, and includes apparently exaggerated growth rates for some large dinosaurs (nor are Erickson et al.'s extremely sigmoidal dinosaur growth curves showing nearly no initial growth supported by Lehman [2007] and Cooper et al. [2008]). The rapid initial growth indicated by avian style growth plates in dinosaur embryos and by the apparently fast size expansion of hadrosaur nestlings is most compatible with tachyenergy (Paul 1994b; Weishampel et al. 2008).

The growth of juvenile sauropods was particularly remarkable. They had to increase mass tens of thousands fold over a few decades without the benefit of mother's milk while moving on energy-demanding land (Fig. 37.10). Milk-fed elephants and whales need to expand mass only a dozen fold, and the latter enjoy very low costs of transport. The proposal that continent-dwelling sauropods growing from the size of geese to whales in decades rather than centuries achieved their extraordinary growth performance with reptilian energetics is not credible (Paul 1998; Sander and Clauss 2008). The somewhat slower growth of dwarfed island sauropods remains in the tachyenergetic zone (Stein et al. 2010).

Because the growth of lesser sized dinosaurs and even protodinosaurs also appears to normally exceed the terrestrial reptile maximum, the general diagnosis is at least some degree of endothermic tachyenergy starting very early in the clade, if not before (de Ricqlès et al. 2008). The ability of dinosaurs to grow rapidly even in habitats marked by the severe winters described later enhances this conclusion. The modest growth rates of many small dinosaurs indicate they diverted resources from growth to reproduction while immature in a manner atypical for most modern mammals and all extant birds (Erickson et al. 2007). The possibly variable growth rates of prosauropods are compatible with an extinct, intermediate thermophysiology. The unusually low metabolic rates calculated on the basis of observed dinosaur growth rates by Gillooly et al. (2006) are suspect because their obscure system seems to be biased towards producing reptilian results, as exposed by their refusal to restore the metabolism of a feathered theropod because it "is fundamentally different from the eight more *reptile-like* species" of dinosaurs they examined (emphasis added). In other words, they did not wish to acknowledge that their method would restore an insulated, birdlike dinosaur as a bradyenergetic ectotherm (also see Escarguel 2009; Eagle et al. 2011).

Attempts to calculate growth rates of basal birds, such as Padian et al. (2001), Chinsamy (2002), de Ricqlès et al. (2003a), Chiappe et al. (2008), and Erickson et al. (2009) are debatable in part because the record of juvenile growth has been wiped out by resabsorbtion of all but the thin outer wall of the bones in many cases (Paul 2002). Erickson et al. (2009) have concluded that their growth was sufficiently rapid to fall into the lower zone of the tachyenergetic, probably due to reproduction before maturity.

Modest declines in resting metabolic rates relative to mass (up to a third) are observed in birds and mammals as they mature. It has been suggested that dinosaurs achieved rapid juvenile growth with high metabolic rates and then experienced a dramatic decline to or near reptilian energy levels upon maturation (Farlow 1990; Lambert 1991; Sander and Clauss 2008; perhaps Perry et al. 2009). Not documented in tetrapods, such extreme metabolic shifts are improbable because they would have required a massive and implausible reconfiguration of cellular structure and physiology (Ruben 1995) and left adult dinosaurs with an aerobic exercise capacity grossly inferior to that of their young. As noted above and below, sauropods may have experienced an unusual, cardiac-forced increase in relative metabolism as they grew up.

Teeth, Beaks, and Cheeks, Lazy Flesh Eating Dinosaurs, Food Passage Rates, and Shrunken Heads

It has long been contended that the mouths of sauropods were too small for them to have gathered enough food to stoke a high energy budget, a view reinforced by the badly flawed analysis of Weaver (1983). Paul (1991, 1998) and Christiansen (1999) falsified this myth. Due to the scaling of metabolic rates and the relatively slow digestion of giant herbivores, tachyenergetic sauropods would have had the relatively moderate energy and food needs seen in big mammals (Christiansen 2004). There is no correlation between mouth size and metabolic level in living herbivores—lizards are much broader mouthed than birds and mammals. The portion of the mouths of sauropods used to crop food were as large as those of herbivorous mammals and birds relative to body mass (Fig. 37.11) and had the capacity to gather enough fodder to feed tachyenergetic-level metabolic rates on a bite by bite basis even if energy budgets scale to the ¾s power. For example, the mouth of the largest brachiosaurs was up to three times broader than that of an indricothere, six times broader than a giraffe's, and could swallow a small child whole (Paul 1998). Assuming that the tachyenergetic sauropod fed for 14 hours a day, at an average rate of three bites per minute, then each bite would have to be only a third of a kilogram, just a dozen ounces (less if food consumption scales to the 2/3s power). The delicate-skulled, square-mouthed *Nigersaurus* had an unusually broad mouth for a large herbivore. The dimensions of small headed prosauropod mouths were also normal. Some stegosaurs had very narrow mouths, but even these should have been able to attain a per-bite rate of consumption to achieve a tachyenergetic energy budget, especially if they were mesoaerobic (Paul

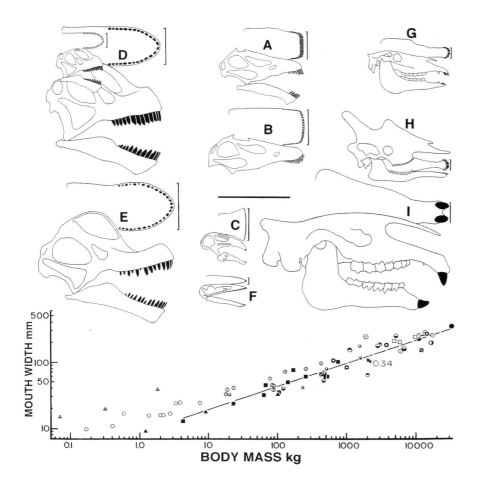

37.11. (Top) Skulls and cropping tooth arcades and beaks drawn to the same scale. A, Diplodocus (scaled to 12 tonnes); B, Apatosaurus (11 or 18); C, Nigersaurus(1.9); D, Camarasaurus adult (14.2) and juvenile (0.64); E, Giraffatitan (33); F, elephant bird (0.4); G, giraffe (1.1); H, zebra (0.4); I, Indricotherium (11). Cropping teeth are solid black, mouth widths are indicated by vertical bars; bar equals 500 m. (Bottom) Mouth width as a function of body mass in herbivorous lizards (triangle with white dot), tortoises (solid triangle); sauropods, Giraffatitan (solid circle), Camarasaurus (hollow circle), Jobaria (circle with solid right half); in ascending mass Euhelopus, Mamenchisaurus youngi, Omeisaurus (circle with solid lower half), Shunosaurus (circle solid upper half); diplodocoids (circle with central dot) in ascending order of mass Nigersaurus, Diplodocus and Apatosaurus; prosauropods (circle with solid lower right half) in ascending order of mass Anchisaurus, Massospondylus, Yunnanosaurus, Plateosaurus, Riojasaurus, Jingshanosaurus, Lufengosaurus; Stegosaurus (smaller circle with solid upper half); herbivorous and omnivorous flying birds (open circle), living (circle with interrupted vertical line) and extinct (circle with vertical line) ratites; African bovids (solid square), giraffe (square with an open dot), equid (square with diagonal bar), elephants (open square), indricothere (square with opposite diagonal bar). Line is a linear regression for living ungulates, r = 0.86, it shows that mouth width is essentially isometric as size changes. Doubling the estimated masses of the dinosaurs would not significantly alter the results.

1998). At the opposite head size extreme, ceratopsids and iguanodontoids had well-developed dental batteries that suggest a rate of food intake and processing above that seen in reptiles; this is especially true of the refined grinding complexes of hadrosaurs. The rate of food processing should have been boosted by the presence of cheek tissues that retained food while being chewed. Cheeks are partly ossified in some ankylosaur specimens. Condors possess elastic cheek tissues (Hoyos et al. 1994, 37) that may parallel those of ornithschians, therizinosaurs, and some prosauropods and sauropods (Upchurch et al. 2007).

Feduccia (1996) implied that because predatory dinosaurs lacked the precisely occluding, heterodont teeth of mammalian carnivores, their food consumption should have been more similar to that of reptiles. This view is falsified by predatory birds, which establish that occluding, heterodont teeth are not necessary to eat enough flesh to stoke a high energy budget. Feduccia also implied that predatory dinosaurs were ambush predators that spent so little time hunting that they had low food budgets. Big cats are ambush predators that spend most of their time sleeping and resting (Schaller 1972), so this criterion does not meet the qualifications of a good paleometabolic character. Besides, various predatory dinosaurs may have been pursuit predators (Paul 1988).

Feduccia further argued that predatory dinosaurs digested their food slowly, to the point that bone was completely destroyed and is absent in their fossil coprolites in the manner of bradymetabolic reptiles, rather than birds and mammals whose high energy demands force food intake, passage, and assimilation to be so rapid that intact bone fragments are passed. Actually, theropod coprolites do contain the numerous, often nearly pristine, bone fragments that show food passage was rapid in the tachyenergetic manner (Chin et al. 1998). Even more extraordinary evidence for very rapid rates of food passage derives from the presence of undigested muscle cells (probably those of a herbivorous dinosaur) in another large coprolite assignable to a tyrannosaur (Chin et al. 1999). At the other end of the avepod dinosaur size spectrum, the intestines of a small avepod dinosaur are short and deep in section, suggesting to Sasso and Signore (1998) that rates of food absorption were high. The fecal and soft tissue evidence favors high rates of food intake and passage in dinosaurs.

Brains

Large brained sharks are bradyenergetic, and some small brained tuna are tachyenergetic (Fig. 37.12), so there is no clear link between brain size and metabolism and the variable brain sizes of dinosaurs provide no particular information about their thermoenergetics (Paul 1991, 2002).

Bone Isotopes

Barrick and Showers (1994), Barrick et al. (1996, 1997), Fricke and Rogers (2000), and Amiot et al. (2006) have used oxygen isotope ratios in fossil bones to discriminate between ectothermic heterotherms and endothermic homeotherms. Their success in diagnosing fossil reptiles as heterothermic, while other skeletons in the same sediments exhibited homeothermic patterns, supports the viability of their methodology because post depositional alteration of bone isotopes should produce uniform results. The presence of homeothermy is diagnostic of elevated metabolic rates in creatures too small to be mass homeotherms in essentially all terrestrial habitats. It is theoretically possible that a homeothermic gigantotherm had a low metabolic rate in a region that did not experience a cool season, but this is not true in more seasonal high latitudes. Marked heterothermy can reflect either a short or long term thermal cycle, with the proviso that sufficiently large animals are subject only to the latter. Heterothermy at all sizes is compatible with all basic metabolic levels including standard bradyenergetic ectothermy and tachyenergy associated with strong torpor on a daily and/or seasonal basis.

So far, almost all of a wide variety of predaceous and herbivorous dinosaurs examined – including some small adults and juveniles – have been restored as homeothermic. The degree of homeothermy appears to have been too high to be fully compatible with low rates of metabolism in even the larger examples, especially because some of the large dinosaurs dwelled at high latitudes with a pronounced cool season. Because the Mesozoic dinosaurs' degree of homeothermy was similar to that of birds and mammals living in more seasonal Cenozoic climates, it has been concluded that their level of homeothermy was somewhat less than the mammalian standard. These views are compatible with either a metabolic rate somewhat lower than in placentals or a less developed metabolic response to cold, or both. One large armored dinosaur tested as strongly heterothermic (Barrick et al. 1997). It is not clear whether this reflects a low metabolic rate or hibernation during a high latitude winter (see below).

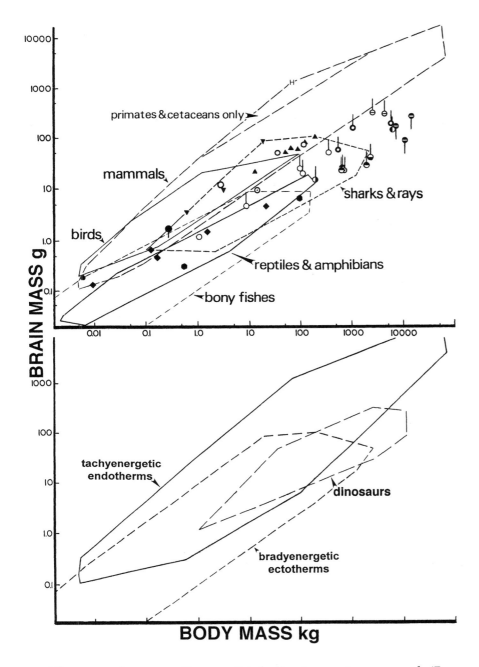

37.12. (Top) Brain mass as a function of body mass in vertebrates: human (H); bradyenergetic myliobatiform rays (inverted triangles), reef sharks (triangles), and elephant nose electric fish (small hexagon); tachyenergetic tuna (large hexagons); dinosaurs (circles, vertical projections cover variation depending on how much of brain cavity is filled with neural mass). Multiple taxa listed under same label in order of ascending mass: sauropodomorphs (solid with horizontal white line) *Plateosaurus, Camarasaurus* juvenile, *Diplodocus, Giraffatitan*; ceratopsians, *Protoceratops* (solid left half); *Triceratops* (solid right half); *Euplocephalus* (solid lower half); stegosaurs (solid upper halves) *Kentrosaurus, Stegosaurus*; large ornithopods (horizontal line) *Kritosaurus, Edmontosaurus*; smaller ornithopods (thin rings) *Leaellynasaura* juvenile, *Hypsilophodon, Dryosaurus, Thescelosaurus, Tenontosaurus, Camptosaurus*; large theropods (thick rings) *Ceratosaurus, Allosaurus, Tyrannosaurus*; small cerato-sauran (central dot) *Coelophysis*; small coelurosaurs (thin ringed) *Sinornithoides* juvenile?, *Troodon, Dromiceiomimus; Archaeopteryx* (solid). (Bottom) Demonstration with least area polygons of the extensive overlap of vertebrates with high and low metabolic rates. Data is from Paul (2002) and Palmeri et al. (2008).

The unusual pattern of isotope distribution in growing sauropods (Barrick and Russell 2001) is compatible with an increase in relative metabolic rate forced by increasing height (see sections on circulation and cruising speeds).

Barrick et al. (1998) used oxygen isotope ratios in fossil bones to estimate mass specific metabolic rates. They concluded that the energy budgets of the dinosaurs they examined were intermediate to those of bradyenergetic reptiles and supraenergetic placentals. However, they underestimated the mass specific metabolism of the dinosaurs by a factor of two, in that they overstated their body masses by the same value, so the disparity between dinosaurs and the placental average is not so great. Amiot et al. (2006) and Eagle et al. (2011) restored the body temperatures of a variety of Jurassic and Cretaceous dinosaurs whose sizes range from small to gigantic sauropods, 33–38°C, straddling the marsupial-placental zones; that enormous sauropods were at the higher end of the range supports their being especially energetic.

Feathers and Other Fluff

Both ectotherms and endotherms may lack insulation in the form of feathers or fur. This is true of all reptiles and some mammals. Among the former are multi-tonne equatorial examples, in which bulk insulation is operative. Suids and humans are more modest sized endotherms that lack integumentary insulation in warm climates. The nearly hairless southern Asian naked bat shows that even flying endotherms of less than 200 g do not require insulation in a sufficiently benign thermal environment.

That some birds and mammals use direct solar radiation to raise body temperatures (Ruben 1995; Geiser 2002; Ruben and Jones 2000; Ruben et al. 2003) does not constitute evidence that ectotherms can be insulated. After all, basking birds and mammals are tachymetabolic endotherms. Ectotherms lack insulation because it thermally decouples them from the environmental heat they critically depend upon. Nor, as demonstrated by classic experiments, does insulation confer homeothermy when the metabolic rate is low. This is probably why well over ten thousand insulated living tetrapod species are tachyenergetic endotherms and none are bradyenergetic ectotherms. Some immobile altricial bird chicks that cannot thermoregulate are insulated, but this situation is not applicable to active juveniles and adults. Some ectothermic, bradyenergetic insects are insulated with furlike structures, but these are so small that gaining heat from the environment is not significantly hindered.

The presence of extensive feather insulation may come as close as any single characteristic to being a paleometabolic Rosetta stone that marks its owner as a tachyenergetic endotherm. Until recently, the most serious problem with feathers as a paleometabolic character has been a general absence of preservation, but this situation is rapidly changing. *Archaeopteryx* was feathered not only aerodynamically, but probably all over its body (Paul 2002; Christiansen and Bonde 2004). The general consensus that the early bird was a high metabolic rate endotherm is superior to the speculation of a few that its energetics were reptilian. Much the same applies to the numerous avepods and ornithischians, most small but some fairly large (Zheng et al 2009, Xu et al. 2012), which sported fibers on the head, neck, body, tail, arms, and legs that are probably hollow (Paul 2002). The hypothesis that the fibers are external rather than internal collagen (Ruben et al., Chapter 36 this volume) is supported by their absence on other tetrapods in the same sediments; the nearly consistent posterior streamlined orientation in the manner of fur and feathers; their presence on the skull, metatarsus, and distal tail where the skin is too thin to produce extensive collagen; the bushy, tuftlike structure preserved extending well distal to the tip of the tail (Ji et al. 2007); and the presence of color melanosomes in the fibers (Li et al. 2010; Zhang et al. 2010). Indeed, the obvious color banding of the tail fibers of the original *Sinosauropteryx* specimen should have made their external origin similarly obvious from the beginning (Fig. 37.13).

The restriction of external fibers to a halo is a common result of preservational bias (as shown by many fossil bird and mammal specimens). The extent of fiber from the front of the head to the tip of the tail favors

37.13. Fossils are showing that a variety of small predaceous and herbivorous dinosaurs bore possible insulation: top, ornithischian heterodontosaur; bottom, theropod Sinosauropteryx.

a complete body covering in many examples, and occasional fibers seen on the flanks of some avepod trunks and legs (personal observation) support the presence of insulating body coverings in at least some dinosaurs. The possible combination of lateral scales and dorso-ventral fibers along a short tail segment in a small avepod (Chiappe and Gohlich 2010), if not a preservational illusion, is no more indicative of a general absence of body insulation among small theropods than are areas of naked integument on the necks, tails, and limbs of some terrestrial mammals and birds. It is not yet known when insulation appeared in dinosaurs or their ancestors. Its existence in early pterosaurs, basal ornithischians, and theropods suggests it was widely distributed among dinosaurs, if not archosaurs (Paul 1988). Archosaur fibers may first have served other purposes, such as display, and developed a thermoregulatory function as they covered enough of the body to provide insulation (as supported by the presence of both display and insulation fibers in some dinosaurs; Xu et al. 2009; Zheng et al. 2009). The absence of insulation in some small dinosaurs (Mayr et al. 2002) is not energetically definitive because they weighed over a kilogram, lived in relatively warm locales or could have burrowed in colder regions, and fat may have provided insulation. Insulation on large dinosaurs may have helped them retain body heat in habitats with chilly winters (Xu et al. 2012).

Polar Dinosaurs

Dinosaurs of a wide variety of types and sizes have been found at very high paleolatitudes over the temporal span of the dinosaurs' existence (Paul 1988, 1994a; Clemens and Nelms 1993; Molnar and Wiffen 1994; Chinsamy et al. 1998; Constantine et al. 1998; Tarduno et al. 1998; Rich and Vickers-Rich

2001; Rich et al. 2002; Fiorillo 2004; Bell and Snively 2008; Godefroit et al. 2008). Winter conditions included a long lack of the solar and ambient heat ectotherms depend on. Some polar habitats have been restored as relatively mild even in winter, with reptiles present. Other locations experienced winter frosts and freezes so harsh, and summer conditions so chilly and sunless, that low energy reptiles are rare or entirely absent from nonmarine sediments that produce substantial dinosaurs and mammals (Clemens and Nelms 1993; Rich et al. 2002; Godefroit et al. 2008; Spicer and Herman 2010). Dinosaurs of the reptile-free Alaskosiberian Arctic coastal plain toward the end of the Cretaceous were subjected to over four months of dark with midwinter mean temperatures of −2°C, implying cold dips as low as −10°C and frequent blizzards. Even midsummer mean temperatures were a chilly ~14°C, with occasional peaks of barely warm ~23°C and persistent cloud cover. Southern Australian dinosaurs apparently experienced even harsher winters that produced permafrosts. The stark differences in tetrapod presence and absence indicate that the mammals, birds, and other dinosaurs were dramatically superior to reptiles in coping with the harsh polar conditions: sharp winter cold and months of darkness, followed by a paucity of basking opportunities during cool summers when clouds usually blocked the warming rays of the midnight sun, combined to exclude solar powered, bradyenergetic turtles, lizards, and crocodilians.

Chinsamy et al. (1998) and Constantine et al. (1998) suggested that the presence of growth rings in a polar ornithomimosaur's bones is compatible with its having hibernated during the winter, in contrast to the continuous activity implied by the absence of the same in a small ornithopod. But bone ring formation does not require hibernation; it is quite possible that the little ornithopod completed growth before any winter hibernation could be recorded (see above). Small predaceous and herbivorous dinosaurs could evade winter conditions by burrowing (Varricchio et al. 2007; Simpson et al. 2010), but it is has not been documented that any denning polar dinosaurs had the well-developed zonal bone that results from hibernation, especially in bradyenergetic vertebrates. Most large dinosaurs were ill suited for torpor because they were too large to seek shelter from predators and the elements; the plausible exceptions being shallow bodied, well-armored forms whose thicker integument would have provided protection against predation and perhaps frostbite. The ability of at least some large, unarmored polar dinosaurs to grow rapidly (Erickson and Druckenmiller 2011) is therefore further evidence of their being markedly more energetic than reptiles.

The possibility that dinosaurs migrated well away from the poles to reach warm latitudes in the winter is implausible. The great distances required are impractical for terrestrial bradyaerobes to cover and excessive even for large tachyaerobes much less the juveniles (Paul 1988, 1994a; Fiorillo 2004). Bone microsturcture supports that polar dinosaurs were permanent residents (Chinsamy et al. 2012). In some locations apparently subject to extensive winter freezes and snows, marine barriers prevented resident large dinosaurs from migrating (Molnar and Wiffen 1994; Bell and Snively 2008), so large theropods and sauropods were able to generate the internal heat needed to overcome the steep temperature gradient well enough to keep their bare skins above freezing. A danger never faced by marine organisms, frostbite cannot be prevented by bradyenergetic inertial homeothermy over an extended period. Nor can low-metabolic-rate gigantothermy defend core body temperatures in a chronically cold, sunless climate (Spotilla et al. 1991).

If bradyenergetic dinosaurs had been able to migrate or hibernate, doing so would not have done them any good in the harsher polar locales. Even in the summer the combination of chilling temperatures, breezes, and rain with mostly cloudy skies would have barred big, fat insulated dinosaurs from operating as inertial homeotherms—no simulation has shown otherwise—so bradyenergetic high-latitude dinosaurs were not viable at any time of the year. The importance of dinosaurs living in cold polar habitats to the great metabolic debate cannot therefore be exaggerated; one way or another their energy output must have been above reptilian levels, and because the Alaskan dinosaurs were at least the same genera as those much further south, it is hardly likely that the latter were dramatically less energetic, perhaps less so than today's elephants are vis-à-vis the woolly mammoth that was a highly specialized polar taxon. (The once common use of dinosaurs as indicators of warm climates is obsolete as realized by Ostrom [1970].) The smaller polar dinosaurs probably also needed insulation. Sufficient winter forage should have been available to supplement the fat deposits of overwintering dinosaurs (Bell and Snively 2008).

Keeping Cool

A common argument against big dinosaurs having high levels of heat production is the danger of overheating. This is a myth on multiple levels (Paul 1991, 1998). The problem is moot if internal heat production in normally active animals does or can scale in conformity to surface area as some research noted earlier implies—the current inability to firmly determine this basic scaling patterns renders an overheating modeling that assumes a particular value unduly speculative (as per Spotilla et al. 1991; O'Conner and Dodson 1999; Seebacher et al. 1999; Seebacher 2003). Even if internal heat production does scale to the ¾s power, the 0.08 difference over the 2/3s scaling of surface area is not great. Nor is overheating just a problem of the large; if anything, the opposite is more true (Paul 1998). Expose a dog to the sun on a hot day without shade or water and it will be dead in a few hours. Do the same to an elephant and it will suffer no permanent harm. There is no evidence that elephants ever die from heat under natural conditions; in hot droughts the big bulls exhibit the lowest mortality (smaller females and juveniles die from starvation). Some of the biggest elephants dwell in deserts, where they have been observed crossing barren lands at midday under cloudless skies with air temperatures at or exceeding 40°C (Paul 1998). O'Conner and Dodson (1999) propose that tachyenergtic African elephants can tolerate high temperatures only because their large ear pinnae provide sufficient cooling, but the ears are only effective as long as environmental temperatures are not so high that outward radiation is blocked (Hiley 1975). Elephants actually minimize ear flapping as ambient temperatures approach body temperature (Buss and Estes 1971). That even larger extinct proboscideans and indricotheres proved able to tolerate high temperatures without the aid of super ears directly falsifies suggestions that similarly energetic, pinnaeless dinosaurs up to 20 tonnes would have had serious problems with overheating (contra spurious calculations by Spotilla et al. [1991] and O'Conner and Dodson [1999] that even 1-tonne tachyenergetic, low latitiude tetrapods suffer serious heating issues despite the many mammal species that fit that bill). Nor does the problem dramatically change over 20 tonnes (contra O'Conner and Dodson 1999).

Assuming internal heat production scales to ¾s, the overheating problem is unavoidable only if environmental temperatures remain sufficiently high for 24 hours each day. That can happen only in extremely hot equatorial deserts unlikely to contain giant animals in the first place. Otherwise, the day's heat is always interrupted by a much cooler night that provides a thermal escape for large animals. Because giants have low mass specific metabolic rates, even if tachyenergetic metabolic rates scale to the ¾s power it takes many hours for body temperatures to rise a few degrees if all internal heat is stored (Fig. 37.13). In this case, having a low surface area is an advantage because giants can shut down most heat gain from the environment. Large tachyenergetic tetrapods with flexible body temperatures as described in the section on modern avian–mammalian energetics can allow internal temperature to rise as high as 46.5°C (in some mammals and birds) during the day; the excess heat is then unloaded at night into the dark sky (Paul 1998; Ostrowski et al. 2003; Ostrowski and Williams 2006; Weissenböck et al. 2011). This strategy is especially useful if water for evaporative cooling is in short supply. The bigger an animal is, the better this system works, so tachyenergetic sauropods were in no particular danger of losing their cool. Those who model dinosaur thermoregulation have failed to integrate this basic strategy in their simulations.

The errancy of the belief that dinogiants should not be tachyenergetic lest they cook themselves is ironically exposed by the correct observation by Seebacher et al. (1999) and Seebacher (2003) that because bradyenergetic leviathans have lower mean body temperatures than tachyenergetic ones, the first are actually more vulnerable to overheating than their high energy counterparts: high body temperatures allow elephants to remain exposed to the noon desert sun, while low energy dinosaurs would have had to find shade or water or die. It is time to put this illogical fable to final rest.

Battle of the Titans: Gigantothermy versus Terramegathermy

There are two competing hypotheses concerning the energetics of mega-animals (those that exceed one tonne): gigantothermy and terramegathermy (Spotila et al. [1991] and Paladino et al. [1997] versus Paul and Leahy [1994], Paul [1998]; Amiot et al. [2006] and Pontzer et al. [2009] reject the first hypothesis and tacitly support the second). The former hypothesis predicts that the metabolics of giants converge to produce a consistent uniformity. Plagued by inconsistent definitions, use, and predictions, it is not always clear whether gigantothermy postulates a convergence of metabolic rates as well as thermoregulatory performance, or whether the supposed metabolic convergence is toward the reptile level, the mammal level, or in between. Terramegathermy is limited to land giants and starts with the observation that among fully terrestrial tetrapods none that are clearly bradyaerobic ectotherms have exceeded about one tonne (the largest being extinct monitors and tortoises). All land tetrapods that exceed a tonne–mammals and dinosaurs – are, or always exhibit anatomical evidence of being, tachyaerobic endotherms (Fig. 37.14). The same is true of all tall land animals. Terramegathermy argues that in order to be a land giant, animals must have high power aerobic power systems to dwell in 1 G. Mesoaerobic edentates reach ~7 tonnes; above that supraanerobiosis may be required. Proposals that the most massive and tallest land animals should or must be bradyaerobic are correspondingly illogical since the largest living land animals are tachyaerobic.

As explained earlier, the convergence of metabolic rates predicted by at least some versions of gigantothermy has not been verified. Regardless of the metabolic condition of giants, gigantothermy is highly dependent upon the homeothermy provided by mass which supposedly provides consistently high activity levels. But if aerobic exercise capacity can differ by a factor of ten in giants, so should their sustainable locomotory potential. As discussed earlier, the more stable body temperatures that can be enjoyed by giant reptiles do not translate into 24 hours worth of high level aerobic exercise capacity. Nor could energy-inefficient anaerobic power production be, as Coulson (1979) believed, used to sustain high levels of activity in giants, because it would result in intense and sustained fatigue, illness, and death. Pontzer et al. (2009) agree that giant vertebrates benefit more from high levels of aerobic power production than smaller ones due to the problems of moving large masses at higher speeds on land. It is doubtful whether the low levels of aerobic power production characteristic of bradyaerobes is sufficient to run the high pressure circulatory system needed in giants, especially tall giants, living in 1 G. The slow growth inherent to low metabolic rates in fully terrestrial land animals almost certainly prevents them from growing fast enough to become gigantic.

High aerobic power produces the high locomotory power, elevated circulatory pressures and rapid growth that easily and fully explain the anatomy and gigantism of megadinosaurs. The extreme height of sauropods especially required high level energetics at the supraaerobic level. Conversely, the anatomy of large dinosaurs supports the presence of tachyaerobiosis. A basic contradiction is inherent to the proposal that because sauropods were able to exceed the 15–20 tonne maximum of the largest proboscideans and indricotheres (Paul 1997; Christiansen 2004) they and other megadinosaurs should have been bradyaerobic land animals, when all known examples of the latter fail to match the size of the only tetrapods that have proven able to equal most megadinosaurs in bulk, the tachyaerobic megamammals. The great bulk of megadinosaurs is therefore fully compatible with and helps verify terramegathermy, while refuting gigantothermy.

37.14. Estimate of how long tachyaerobic terrestrial tetrapods can store internal body heat before overheating assuming a 6–8° C rise in body temperature, and total active energy budgets either 2 times mammalian resting metabolic rate assuming it scales to 0.75 power, or 1.3 times latter due to suppression of activity and or metabolism during peak heat hours; external heat largely excluded by a combination of low surface areas and high and rising body temperatures. Lined zone covers living examples, solid black extinct giants with indricothere indicated by vertical line; circles indicate observed heat storage times for ungulates (Paul 1993).

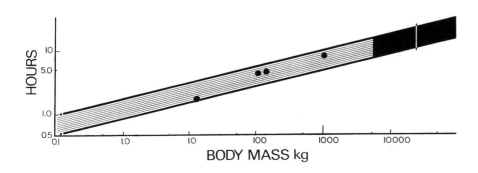

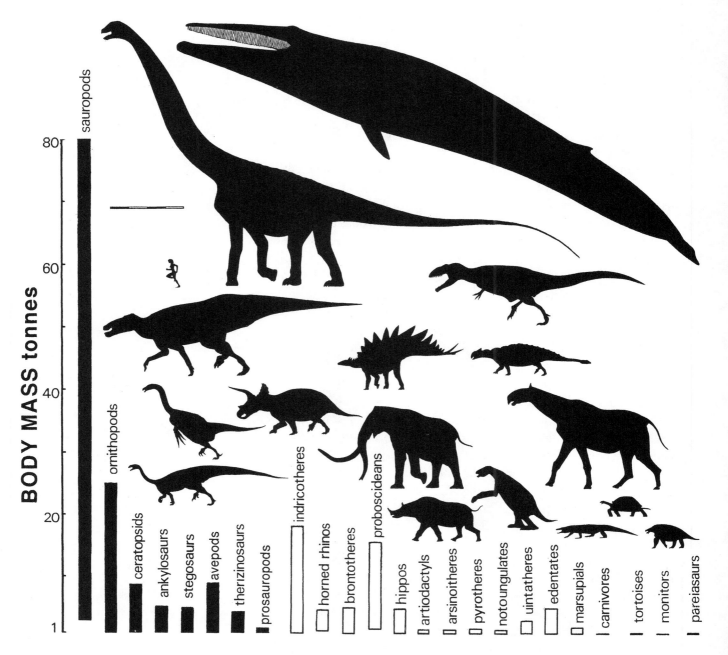

BODY MASS tonnes

sauropods · ornithopods · ceratopsids · ankylosaurs · stegosaurs · avepods · therizinosaurs · prosauropods · indricotheres · horned rhinos · brontotheres · proboscideans · hippos · artiodactyls · arsinoitheres · pyrotheres · notoungulates · uintatheres · edentates · marsupials · carnivores · tortoises · monitors · pareiasaurs

37.15. Approximate mass ranges over 1 tonne in continental tetrapod groups; it is not certain whether monitors reached a tonne. Very large representatives of many groups and blue whale shown to same scale; bar equals 4 m.

Lebensraum

Dinosaurian gigantothermy is directly refuted by the evidence of their bones, which preserve the 36–37°C body temperatures predicted by tachy-energetic terramegathermy (Amiot 2006), not the ~30°C temperatures predicted by inertial homeothermy (Spotilla et al. 1991; O'Conner and Dodson 1999; Seebacher et al. 1999; Seebacher 2003).

The extreme size of some dinosaurs, especially the bigger sauropods and the giant flesh eaters, has led some researchers to argue that they could not have sustained viable populations and species diversity over time and in a given habitat if they had avian–mammalian-level energy budgets (Farlow et al. 1995; Paladino et al. 1997; Van Valkenburgh and Molnar 2000; Burness et al. 2001; Farlow and Pianka 2003; McNab 2009). In this view, proboscideans and indricotheres approaching twenty tonnes are the largest tachyenergetic herbivores that can live on a large continent without overtaxing the resource base, and a tonne is the largest that similarly energetic predators can be. If giants have lower, reptilian energy budgets, then individuals can be much larger on the same resource base.

The general Lebensraum hypothesis assumes that tachyenergetic animals must have populations similar to those of living elephants or big predators in order to be viable in the long term. The population of adult African elephants prior to modern hunting pressures was in the millions (Nowak 1999), but such large numbers are by no means necessary. African rhino populations were historically much lower. The moas of New Zealand were successful and diverse for millions of years even though adult populations of the most populous species were probably only in the tens of thousands (Holdaway and Jacomb 2000). An Africa-sized area should have been able to support hundreds of thousands of herbivorous sauropods. Big cats, canids, and hyenids numbered in the hundreds of thousands in India in the early twentieth century (Nowak 1999) and presumably were more numerous before humans populated the subcontinent. A similar Mesozoic land area should have been able to support tens of thousands of gigantic tachyenergetic predaceous theropods. Farlow and Pianka (2003) assert that giant tachyaerobic theropods needed implausibly large territories. Although nomadic lions can have home ranges up to 4000 sq km, entire prides of nonmigratory lions can have home ranges as small as 20 sq km (Nowak 1999), so a giant tachyenergetic theropod with a total energy budget equivalent to one or a few lion prides could have territory of a few dozen to a few hundred square kilometers—quite reasonable for such a large, fast cruising tachyaerobe.

Big cats use their heads and powerfully muscled arms and legs to dispatch large prey. Lacking weapon bearing limbs, canids and hyaenids need to increase their total mass via large numbers to overwhelm large prey. Great theropods did not have arms as powerful as those of felids; their feet lacked substantial weaponry and it is doubtful that they were able to hunt in organized packs (Paul 2008), so their enormous bulk may be an adaptation for attacking prey items as big as elephants and whales largely with their jaws and teeth. The abundance of poorly protected juveniles produced by large r-strategist dinosaurs may have provided large theropods an abundance of body-mass-building food denied to modern land carnivores by more protective K-strategist megaherbivore parents (Sander and Clauss 2008). Because sauropod heads were relatively small they could evolve the extremely long necks that in turn encouraged the development of exceptionally large bodies to anchor them (Paul 1998, Sander and Clauss 2008). Dental batteries and large brains force the heads of herbivorous mammals to be large, limiting the length of the neck and the necessity for oversized bodies.

Comparing the population dynamics of large dinosaurs to mammals is complicated by their dramatically differing reproductive potentials. Moas were and giant mammals are K-strategists that reproduce at low rates. Also, mammal young are dependent upon a lactating parent for survival. Giant dinosaurs were r-strategist "weed species" that laid large numbers of eggs, which hatched into young that had at least the potential to mature without adult care. Some chameleons have an annual life cycle in which only their eggs and then juveniles are extant for part of the year (Karsten et al. 2008). It is possible that the entire adult population of a given dinosaur species could accidentally be lost and the juveniles could grow up keep the species going with the ability to expand into the available space via rapid reproduction. In general, dinosaurs could have remained viable over the long term with small adult populations, allowing the population of the latter to be much smaller than seen in mammalian giants. If so, then individual adults could have grown much larger (Janis and Carrano 1992; Paul 1994b, 1998, 2008; Sander and Clauss 2008).

Another complication is the possibility that strongly elevated CO_2 levels boosted the productivity of Mesozoic plants well above modern levels, allowing a given area to support a much larger herbivore and predator biomass. The Lebensraum hypothesis cannot refute high energy budgets in dinosaurian giants or island archosaurs because Mesozoic floral productivity is unknown, giant tachyenergetic dinosaurs could have had reasonably large populations, and the minimum adult population that giant, nonnursing r-strategists need to be viable over geological time is unknown. The use of another aspect of fossil population dynamics, predator/prey ratios, to restore dinosaur energetics was criticized as unverifiable and no longer receives significant attention, so it is disturbing to see some who oppose P/P-ratio evidence for tachyenergetic dinosaurs using the seriously defective living space hypothesis as evidence for bradyenergetic dinosaurs.

In a miniature Lebensraum argument, Elzanowski (2002) suggested that the small, arid Solnhofen islands were not able to support a viable population of tachyenergetic *Archaeopteryx*, a point that would also apply to the habitat's small theropod dinosaurs *Compsognthus* and *Juravenator*. But the large dodo as well as other flightless birds lived on an island only 50 km in diameter, and the dinosaurs and the urvogel may have had a much greater geographical range than the fossils imply.

Size Squeezes

McNab (1978) proposed that high-metabolic-rate endothermy evolves during a size squeeze in which the homeothermy evolved in large animals can only be retained by increasing mass-specific heat production as size decreases. The downsizing that marked the evolution of mammals from therapsids is cited as an example of this effect. Because many dinosaurs were large, it is has been argued that they did not experience the initial size squeeze necessary to induce endothermy. The size squeeze hypothesis is open to question on multiple grounds. Although not falsified, it is only an unverified hypothesis, so the absence of a size squeeze cannot be used to falsify elevated metabolic rates. The last point is also true in that, even if the size squeeze did drive the evolution of tachyenergetics in mammals, it may not have been the key factor in other tetrapods. While the size squeeze hypothesis assumes that the primary motive for evolving elevated metabolic rates is improved thermoregulation, the aerobic capacity hypotheses do not require a size squeeze since the advantages of increased aerobic capacity accrue at any dimension.

As for dinosaurs, the simple fact is that they did experience an initial size squeeze. The protodinosaurs from which they all evolved were quite small, at only a couple of hundred grams.

Citing various lines of evidence, Paul (1988), Schweitzer and Marshall (2001), de Ricqlès et al. (2003a, 2008), Seymour et al. (2004), Farmer and Sanders (2010), and Fostowicz-Frelik and Sulej (2010) suggest that elevated metabolic rates and aerobic capacity, perhaps aided by some degree of unidirectional airflow in the lung, and external insulation (Xu et al. 2009) began to develop in Triassic basal archosaurs that included the common ancestor of crocodilians, pterosaurs, and dinosaurs. Better resolution of this question may await the finding of integument preserving shallow-water lagerstatten from the early Mesozoic. Because potentially tachyaerobic early archosaurs were brevischian, their energetics should have been most similar to similarly short-hipped basal dinosaurs. The evidence for some level of endothermic tachyaerobiosis in the ancestors of dinosaurs reinforces the case that the latter were not bradyaerobic ectotherms and that they inherited and further developed tachyenergy rather than evolving it independently. Derived crocodilians, and perhaps some other crurotarsans, subsequently lost high metabolic rate endothermy in this scenario (Carrier 1987; Carrier and Farmer 2000a; Paul 2002; de Ricqlès et al. 2003a, 2008; Seymour et al. 2004; Farmer and Sanders 2010).

Dinosaur Ancestry

Energetic Synthesis of Dinosaurs Early and Late, Small and Great

It is often said that we can never really know the thermoenergetics of dinosaurs, but this is no more or less true than whether we can tell whether or not evolution occurred. Many details will always be hazy, but broader patterns can be discerned. The preponderance of the evidence on the issue of dinosaur metabolics is clear. No dinosaur or early bird exhibits the anatomical features, slow walking pace, or very slow growth associated with reptilian bradyenergy. Indicators that have been offered as evidence of a reptilelike metabolism such as growth rings, problems with living space, overheating, limited food intake, and processing capacity, or inadequate flight muscles and respiratory conchae have proved errant or open to serious question. In energetic terms, dinosaurs were not reptiles.

But not all dinosaurs had uniform avian-mammalian metabolisms (contra Paul 1991). Although erect legged, fast growing, and showing signs of moving in organized groups, brevischian protodinosaurs, basal tetradactyl theropods, and prosauropods lacked the large leg muscles typical of birds and mammals. This conflicted combination is compatible with no extant metabolic and locomotory combination and is indicative of an extinct mesoenergetic organization that was unstable in that the sustained walking speed potential of the long striding legs could not be fully exploited by the limited musculature and mesoaerobic exercise capacity. The anatomy of brevischian dinosaurs is the least incompatible with heteroenergy, but this hypothesis is inferior to their being mesometabolic because compelling evidence that even early dinosaurs were bradymetabolic animals strongly dependent on external heat is lacking, and because heteroenergy remains speculative in view of the absence of any living examples of terrestrial bradymetabolic tachyaerobes. It is possible that basal small dinosaurs and even protodinosaurs were insulated, a trait perhaps inherited from earlier archosaurs (and other more derived tachyaerobic archosaurs, including gracile terrestrial crocodilians).

It is not surprising that the brevischian, mesoenergetic dinosaurs of the Triassic and early Jurassic were quickly and totally replaced by the longoschian avepod, sauropod, and ornithischian dinosaurs that eliminated the disparity between leg and pelvis proportions, a condition inherited by birds. Their expanded, tachyaerobic musculature was able to take full advantage of the sustainable locomotory and migratory potential of the erect legs recorded in their long stride trackways. Various longoschian dinosaurs also exhibit a host of other attributes indicative of elevated aerobic capacity including sophisticated, bird- or mammal-like thoraxes that probably contained high capacity respiratory tracts, rapid food passage rates, insulation, egg incubation, parental care, reptile-free polar habitats that could not sustain inertial homeotherms any time of the year, and the ability to grow rapidly, to sizes never achieved by land reptiles, and to extreme heights. All of these features are fully compatible with and easily explained by tachymetabolism, tachyaerobiosis, and endothermy. They are not so readily explainable with lower level metabolics, and the presence of typical dinosaurs in lands where even the summers were too chilly and cloudy for bradyenergetic creatures of any size is especially damning for the obsolete hypothesis of reptilelike dinosaur energetics. All longoschian dinosaurs were probably tachyenergetic as per the basic vertebrate manner seen in birds and mammals.

Because the living longoschian dinosaurs closest to the nonavian members of the group, ratites, have resting metabolisms in the lower supraaerobic zone, it is likely that no or few dinosaurs had higher basal metabolisms. In that case, most or all dinosaurs did not reach the Kleiber level. This is in accord with the 33–38°C body temperatures recorded in their bones. Ironically the most plausible exception was the tallest sauropods with the hardest working hearts, which may have driven their resting metabolisms up to the standard avian-mammalian range. Assuming longoschian dinosaurs were tachymetabolic, some or all may have had the higher maximum/minimum aerobic ratios seen in some tetrapods, leaving some of the dinosaurs mesometabolic and supraaerobic. Nor is there good reason to think that any dinosaur grew up fast with high metabolic rates and then experienced a dramatic metabolic decline upon adulthood; if anything, maturing giant sauropods may have experienced a moderate form of the opposite as indicated by their exceptional combination of ever taller heads demanding increasingly oversized hearts, apparently rising cruising speeds, and unusual bone isotope patterns as they matured. None of the above features is compatible with the bradyaerobic capacity of reptiles. There is no evidence that any nonavian dinosaur was hypermetabolic like the most energetic mammals and birds, or had their exceptionally high body temperatures.

Far from being uniform, the energetics of the anatomically diverse longoschian dinosaurs were probably correspondingly diverse. Even if sauropods had exceptionally high resting metabolisms, their exercise capacity and energy budgets should have been modest like those of similarly slow elephants. The respiratory complexes of many ornithischians do not appear to have the capacity seen in ornithopods and saurischians, and the feeding complexes of the same ornithischians also appear to have been modest in capacity. These factors imply that at least some nonornithopod ornithischians, the armored forms particularly, may have been mesoenergetic in the manner of edentates. Protected against weather and predation by armor, high latitude ankylosaurs may have sought out deep woods within which to hibernate through long winters. Ceratopsids are especially interesting. On one hand, their respiratory complex, dominated by an exceptionally rigid ribcage, seems to lack the sustained capacity of the diaphragm driven breathing of rhinos. On the other hand, the great horned dinosaurs had much larger areas for muscle attachments than seen in rhinos, indicating a much more powerful musculature. Such oversized leg muscles could not have been tachyanaerobic in the manner of reptiles because their burst power would have been extraordinary, yet would have soon left them too fatigued to fight off attackers. Instead of running long distances, ceratopsids may have used their enormous leg muscles to aerobically power brief but intense high speed charges at tyrannosaurs, which could be repeated indefinitely. The sophisticated respiratory apparatus of ornithopods and avepods are evidence for energetics at or at least very near the supraaerobic level. So are the very high walking speeds of both avepods and ornithopods and the grinding dental batteries of the latter. The fast moving ornithopods and avepods should have had excellent sustainable aerobic capacity at all sizes. The aerobic power of the vagile avepod dinosaurs may have tended to increase during the Mesozoic as pelves became larger and legs longer, culminating in the ostrich-mimics and tyrannosaurs. An exception to the theropod pattern was the strange therizinosaurs. Their enormous bellies and oversized toe claws suggest they were slow animals with reduced aerobic exercise potential. They may have been avepodian equivalents of ground sloths, with similarly mesometabolic and aerobic energetics.

It is very unlikely that any dinosaur had heat producing brown fat (Mezentseva et al. 2008). It is possible that many or all dinosaurs were semiheterothermic, in the manner common to some mammals and many birds. Brevischian dinosaurs should have been especially prone to partial heterothermy. Basking was probably common in the brevischian dinosaurs, smaller examples of which may have been mesotherms dependent upon both external and internal heat. Less intense basking may have been practiced by some small endothermic longoschian examples in the manner of birds. Metabolic and insulatory adaptations for cold should have been better developed in higher latitude and altitude dinosaurs than their warmer climate relatives.

The view that for 150 million years dinosaurs somehow used reptilelike energetics and warm Mesozoic climes to become highly active, bird- and mammal-like animals, while failing to evolve the boosted aerobic metabolisms actually needed to do so, is not logical and is a dwindling minority opinion at a time when a growing and spectacular array of insulated and winged dinosaurs are being discovered. Most researchers now agree with the once seemingly radical proposition (Ostrom 1970, Bakker 1972) that dinosaurs are the acme fulfillment of the selective pressures outlined by the aerobic exercise hypothesis to evolve elevated oxygen powered exercise capacity in order to achieve high levels of sustained activity on land and in the air. With long striding legs from their earliest iteration, they did so in a manner more clear-cut than seen in therapsids and mammals, the latter remaining small and short-legged until dinosaurs outside Aves disappeared at the κ/ρ boundary. Had dinosaurs not evolved boosted exercise capacity, their evolution and that of other land tetrapods would have been dramatically different. More energetic mammals could have evolved to fill the sustained activity, large animal roles starting in the Jurassic. Dinosaurs would not have been able to become as big as rhinos, elephants, and whales, as tall as trees, move in herds and take care of their young, or dwell in the colder polar habitats. Nor would dinosaurs have evolved powered flight in the form of elegant birds and the daylight skies would be dominated by membrane winged bats, some perhaps larger than any in actual existence.

To see living animals with dinosaur-like energetics look not to sprawling oras and gators or flippered leatherbacks. For the gigantic end of the size spectrum, look to giraffes, rhinos, and elephants. Most of all consider our feathered friends, especially the flightless ratites of the southern continents and islands, which are living dinosaurs whose metabolics are probably not very different from their Mesozoic ancestors.

References

Alexander R. 1992. *Exploring Biomechanics.* New York: Scientific American Library.

Amiot, R., et al. 2006. Oxygen isotopes from biogenic apatites suggest widespread endothermy in Cretaceous dinosaurs. *Earth and Planetary Science Letters* 246: 41–54.

Askew, G., R. Marsh, and C. Ellington. 2001. The mechanical power output of the flight muscles of blue-breasted quail during take-off. *Journal of Experimental Biology* 204: 3601–3619.

Auffenberg, W. 1981. *The Behavioral Ecology of the Komodo Monitor.* Gainesville: University Press of Florida.

Auggenberg, W. 1994. *The Bengal Monitor.* Gainesville: University Press of Florida.

Bakker, R. 1972. Anatomical and ecological evidence of endothermy in dinosaurs. *Nature* 238: 81–85.

Barrick, R., W. Showers, and A. Fischer. 1996. Comparison of thermoregulation of four ornithischian dinosaurs and a varanid lizard from the Cretaceous Two Medicine Formation: Evidence from oxygen isotopes. *Palaios* 11: 295–305.

Barrick, R., and W. Showers. 1994. Thermophysiology of *Tyrannosaurus rex*: evidence from oxygen isotopes. *Science* 265: 222–224.

Barrick, R., M. Stoskopf, and W. Showers. 1997. Oxygen isotopes in dinosaur bone. In J. Farlow and M. Brett-Surman (eds.), *The Complete Dinosaur*, 474–490. Bloomington: Indiana University Press.

Barrick, R., and D. Russell. 2001. Physiological implications of ontogenetic variability in oxygen isotope distribution in sauropods. *Journal of Vertebrate Paleontology* 21: 32A.

Barrick, R., D. Russell, and W. Showers. 1998. How much did dinosaurs eat: metabolic evidence from oxygen isotopes. *Journal of Vertebrate Paleontology* 18(3): 26A.

Bell, P., and E. Snively. 2008. Polar dinosaurs on parade: a review of dinosaur migration. *Alcheringa* 32: 271–284.

Bennett, A. 1983, Ecological consequences of activity metabolism. In R. Huey, E. Painka, and T. Schoener (eds.), *Lizard Ecology*, 11–23. Cambridge: Harvard University Press.

———. 1991. The evolution of aerobic capacity. *Journal of Experimental Biology* 160: 1–23.

Bennett, A., and J. Ruben. 1979. Endothermy and activity in vertebrates. *Science* 206: 649–654.

Bennett, A., R. Seymour, D. Bradford, and G. Webb. 1985. Mass-dependence of anaerobic metabolism and acid-base disturbance during activity in the saltwater crocodile, *Crocodylus porosus. Journal of Experimental Biology* 118: 161–171.

Bernal, D., et al. 2005. Mammal-like muscles power swimming in a cold-water shark. *Nature* 437: 1349–1352.

Bickler, P., and R. Anderson. 1986. Ventilation, gas exchange and aerobic scope in a small monitor lizard, *Varanus gilleni. Physiological Zoology* 59: 76–83.

Bjorndal, K., et al. 1998. Age and growth in sea turtles: limitations of skeletochronology for demographic studies. *Copeia* 1998: 23–30.

Block, B. et al. 1993. Evolution of endothermy in fish: mapping physiological traits on a molecular phylogeny. *Science* 260: 210–214.

Brill, R. 1986. On the standard metabolic rates of tropical tunas, including the effect of body size and acute temperature change. *Fishery Bulletin* 85: 25–35.

Brill, R., and P. Bushnell. 1991. Metabolic and cardiac scope of high energy demand teleosts, the tunas. *Canadian Journal of Zoology* 69: 2002–2009.

———. 2001. The cardiovascular system of tunas. *Fish Physiology* 19: 79–120.

Broughton, J., D. Rampton, and K. Holanda. 2002. A test of an osteologically based age determination technique in the double-crested cormorant *Phalacrocorax auritus*. *Ibis* 144: 143–146.

Burness, G., J. Diamond, and T. Flannery. 2001. Dinosaurs, dragons, and dwarfs: The evolution of maximal body size. *Proceedings of the National Academy of Sciences* 98: 14518–14523.

Buss, I., and J. Estes. 1971. The functional significance of movements and positions of the pinnae of the African elephant. *Journal of Mammalogy* 52: 21–27.

Bybee, P., A. Lee, and E. Lamm. 2006. Sizing the Jurassic theropod dinosaur *Allosaurus*: assessing growth strategy and evolution of ontogenetic scaling of limbs. *Journal of Morphology* 267: 347–359.

Carey, F., J. Teal, J. Kanwisher, and K. Lawson. 1971. Warm-bodied fish. *American Zoologist* 11: 137–145.

Carpenter, K. 1999. *Eggs, Nests, and Baby Dinosaurs.* Bloomington: Indiana University Press.

Carrier, D. 1987. The evolution of locomotor stamina in tetrapods: circumventing a mechanical restraint. *Paleobiology* 13: 326–341.

Carrier, D., and C. Farmer. 2000a. The evolution of pelvic aspiration in archosaurs. *Paleobiology* 26: 271–293.

———. 2000b. The intergration of ventilation and locomotion in archosaurs. *American Zoologist* 40: 87–100.

Case, T. 1978. On the evolution and adaptive significance of postnatal growth in the terrestrial vertebrates. *Quarterly Review of Biology* 53: 243–282.

Castanet, J., H. Francillon-Vieillot, F. Meunier, and A. de Ricqlès. 1993. Bone and individual aging. In B. Hall (ed.), *Bone* 7, 245–283. Boca Raton, Fla.: CRC Press.

Chiappe, L., and U. Gohlich. 2010. Anatomy of *Juravenator starki* from the Late Jurassic of Germany. *Neues Jahrbach fur Geologie und Palaontologie Abhandlungen* 258: 257–296.

Chiappe, L., J. Marugan-Lobon, S. Ji, and Z. Zhou. 2008. Life history of a basal bird: morphometrics of the Early Cretaceous *Confuciusornis*. *Biology Letters* doi:1098/rsbl.2008.0409.

Chin, K., T. Tokaryk, G. Erickson, and L. Calk. 1998. A king-sized theropod coprolite. *Nature* 393: 680–682.

Chin, K., D. Eberth, and W. Sloboda. 1999. Exceptional soft-tissue preservation in a theropod coprolite from the Upper Cretaceous Dinosaur Park Formation of Alberta. *Journal of Vertebrate Paleontology* 19(3): 37A.

Chinsamy, A. 1993. Bone histology and growth trajectory of the prosauropod dinosaur *Massospondylus carinatus* Owen. *Modern Geology* 18: 319–329.

Chinsamy, A. 2002. Bone microstructure of early birds. In L. Chiappe and L. Witmer (eds.), *Mesozoic Birds*, 421–431. Berkeley: University of California Press.

Chinsamy, A., and W. Hillenius. 2004. Physiology of nonavian dinosaurs. In D. Weishampel, P. Dodson, H. Osmólska (eds.), *The Dinosauria*, 643–659. Berkeley: University of California Press.

Chinsamy, A., T. Rich, and P. Vickers-Rich. 1998. Polar dinosaur bone histology. *Journal of Vertebrate Paleontology* 18: 385–390.

Chinsamy, A., D. B. Thomas, A. R. Tumarkin-Deratzian, and A. R. Fiorillo. 2012. Hadrosaurs Were Perennial Polar Residents. *The Anatomical Record: Advances in Integrative Anatomy and Evolutionary Biology* 295(4): 610–614.

Christian, A. 2010. Some sauropods raised their necks – evidence for high browsing in *Euhelopus zdanskyi*. *Biology Letters* 6 (6): 823–825

Christiansen, P. 1999. On the head size of sauropodomorph dinosaurs: implications for ecology and physiology. *Historical Biology* 13: 269–297.

———. 2004. Body size in proboscideans, with notes on elephant metabolism. *Zoological Journal of the Linnean Society* 140: 523–549.

Christiansen, P., and N. Bonde. 2004. Body plumage in *Archaeopteryx*: a review, and new evidence from the Berlin specimen. *Comptes Rendus Palevol* 3: 99–118.

Claessens, L. 2004a. Dinosaur gastralia; origin, morphology and function. *Journal of Vertebrate Paleontology* 24: 89–106.

———. 2004b. Archosaurian respiration and the pelvic girdle aspiration breathing of crocodyliforms. *Proceedings of the Royal Society of London B* 271: 1461–1465.

———. 2009a. A cineradiographic study of lung ventilation in Alligator missisippiensis. *Journal of Experimental Zoology* 311A: 563–585.

———. 2009b. The skeletal kinematics of lung ventilation in three basal bird taxa. *Journal of Experimental Zoology* 311A: 586–599.

Clark, J., M. Norell, and L. Chiappe. 1999. An oviraptorid skeleton from the Late Cretaceous of Ukhaa Tolgod, Mongolia, preserved in an avian-like brooding position over an oviraptorid nest. *American Museum Novitates* 3265: 1–36.

Clemens, W., and L. Nelms. 1993. Paleoecological implications of Alaskan terrestrial vertebrate fauna in latest Cretaceous time at high paleolatitudes. *Geology* 21: 503–506.

Clemente, C., P. Withers, and G. Thompson. 2009. Metabolic rate and endurance capacity in Australian lizards. *Biological Journal of the Linnean Society* 97: 664–676.

Constantine, A., A. Chinsamy, P. Vickers-Rich, and T. Rich. 1998. Periglacial environments and polar dinosaurs. *South African Journal of Science* 94: 137–141.

Cooper, L., A. Lee, M. Taper, and J. Horner. 2008. Relative growth rates of predator and prey dinosaurs reflect effects of predation. *Proceedings of the Royal Society B*: doi.1098/rspb.2008.0912.

Coulson, R. 1979. Anaerobic glycolysis: the Smith and Wesson of the heterotherms. *Perspectives in Biology and Medicine* 22: 465–479.

Curry, K. 1998. Histological quantification of growth rates in *Apatosaurus*. *Journal of Vertebrate Paleontology* 18(3): 37A.

Daniels, C., and J. Pratt. 1992. Breathing in long necked dinosaurs; did the sauropod have bird lungs? *Comparative Biochemistry and Physiology* 101A: 43–46.

Dausmann, K., J. Glos, J. Ganzhorn, and G. Heldmaier. 2004. Hibernation in a tropical primate. *Nature* 429: 825–826.

de Ricqlès, A., K. Padian, and J. Horner. 2003a. On the bone histology of some Triassic pseudosuchian archosaurs and related taxa. *Annales de Paleontologie* 64: 85–111.

de Ricqlès, A., K. Padian, J. Horner, E. Lamm, and N. Myhrvold. 2003b. Osteohistology of *Confuciusornis sanctus*. *Journal of Vertebrate Paleontology* 23: 373–386.

de Ricqlès, A., K. Padian, F. Knoll, and J. Horner. 2008. On the origin of high growth rates in archosaurs and their ancient relatives: Complementary histological studies on Triassic archosauriforms and the problem of a "phylogenetic signal" in bone histology. *Annales de Paleontologie* 94: 57–76.

Dodds, P. 2010. Optimal form of branching supply and collection networks. *Physical Review Letters* 104: 04872.

Dodds, P., D. Rothman, and J. Weitz. 2001. Re-examination of the "¾-law" of metabolism. *Journal of Theoretical Biology* 209: 9–27.

Dodson, P. 1991. Lifestyles of the huge and famous. *Natural History* 100: 30–34.

Duffield, G., and C. Bull. 2002. Stable social organization in an Australian lizard, *Egernia stokesii*. *Naturwissenschaften* 89: 424–427.

Dzemski, G., and A. Christian. 2007. Flexibility along the neck of the ostrich, and consequences for the reconstruction of dinosaurs with extreme neck length. *Journal of Morphology* 268: 701–714.

Eagle, R., et al. 2011. Dinosaur body temperatures determined from isotopic ($^{13}C-^{18}O$) order in fossil biominerals. *Science* doi 10.1126/science.1206196.

Else, L., and A. Hulbert. 1985. An allometric comparison of mammalian and reptilian tissues: the implications for the evolution of endothermy. *Journal of Comparative Physiology* 156: 3–11.

Elzanowski, A. 2002. Archaeopterygidae. In L. Chiappe and L. Witmer (eds.), *Mesozoic Birds*, 129–159. Berkeley: University of California Press.

Erickson, G., and C. Brochu. 1999. How the 'terror crocodile' grew so big. *Nature* 398: 205–206.

Erickson, G., and P. Druckenmiller. 2011. Longevity and growth rate estimates for a polar dinosaur: a *Pachyrhinosaurus* specimen from the North Slope of Alaska showing a complete developmental record. *Historical Biology* 23(4): 327–334.

Erickson, G., K. Rogers, and S. Yerby. 2001. Dinosaurian growth patterns and rapid avian growth rates. *Nature* 412: 429–433.

Erickson, G., et al. 2004. Gigantism and comparative life-history parameters of tyrannosaurid dinosaurs. *Nature* 430: 772–775.

Erickson, G., et al. 2007. Growth patterns in brooding dinosaurs reveals the timing of sexual maturity in nonavian dinosaurs and genesis of the avian condition. *Biology Letters* 3: 558–561.

Erickson, G., et al. 2009. Was dinosaurian physiology inherited by birds? Reconciling slow growth in *Archaeopteryx*. *PLoS One* 4: e7390.

Escarguel, G. 2006. On dinosaur thermophysiology: Comment on "Dinosaur fossils predict body temperature." *PLoS Biology* 4(8): doi:10.1371/journal.pbio.0040248.

Farlow, J. O. 1990. Dinosaur energetics and thermal biology. In D. Weishampel, P.

Dodson and H. Osmólska (eds.), *The Dinosauria*, 43–62. Berkeley: University of California Press.

Farlow, J. O., and E. Pianka. 2003. Body size overlap, habitat partitioning and living space requirements of terrestrial vertebrate predators: Implications for the paleoecology of large theropod dinosaurs. *Historical Biology* 16: 21–40.

Farlow, J. O., P. Dodson, and A. Chinsamy. 1995. Dinosaur biology. *Annual Review of Ecological Systematics* 26: 445–471.

Farmer, C., and K. Sander. 2010. Unidirectional airflow in the lungs of alligators. *Science* 327: 338–340.

Fedak, M., and H. Seeherman. 1979. Reappraisal of energetics of locomotion shows identical cost in bipeds and quadrupeds including ostrich and horse. *Nature* 282: 713–716.

Feduccia, A. 1996. *The Origin and Evolution of Birds*. New York: Yale University Press.

Fieler, C., and B. Jayne. 1998. Effects of speed on the hindlimb kinematics of the lizard *Dipsosaurus dorsalis*. *The Journal of Experimental Biology* 201: 609–622.

Fiorillo, A. 2004. The Dinosaurs of Arctic Alaska. *Scientific American* 291: 84–91.

Fostowicz-Frelik, L., and T. Sulej. 2010. Bone histology of *Silesaurus opolensis* from the Late Triassic of Poland. *Lethaia* 43: 137–148.

Fricke, H., and R. Rogers. 2000. Multiple taxon-multiple locality approach to providing evidence for warm-blooded theropod dinosaurs. *Geology* 28: 799–802.

Geiser, F., N. Goodship, and C. Parey. 2002. Was basking important in the evolution of mammalian endothermy? *Naturwissenschaften* 89: 412–414.

Geist, N. 2000. Nasal respiratory turbinate function in birds. *Physiological and Biochemical Zoology* 73: 581–589.

Gillooly, J., A. Allen, and E. Charnov. 2006. Dinosaur fossils predict body temperatures. *PLoS Biology* 4: 1467–1469.

Glazier, D. 2009. A unifying explanation for diverse metabolic scaling in animals and plants. *Biological Reviews* 85: 111–138.

Godefroit, P., L. Golovneva, S. Shchepetov, G. Garcia, and P. Alekseev. 2008. The last polar dinosaurs: high diversity of latest Cretaceous dinosaurs in Russia. *Naturwissenshaften* 96(4): 459–501. doi: 10.1007/s00114-008-0499-0

Graham, J., and K. Dickson. 2004. Tuna comparative physiology. *Journal of Experimental Biology* 207: 4015–4024.

Grenard, S. 1991. *Handbook of Alligators and Crocodiles*. Malabar: Krieger Publishing.

Groscolas R. 1990. Metabolic adaptations to fasting in emperor and king penguins. In L. Davis and J. Darby (eds.), *Penguin Biology*, 269–296. London: Academic Press.

Hammer, W., and W. Hickerson. 1994. A crested theropod dinosaur from Antarctica. *Science* 264: 828–830.

Hayes, J. 2010. Metabolic rates, genetic constraints, and the evolution of endothermy. *Evolutionary Biology* 22: 1868–1877.

Hayes, J., and T. Garland. 1995. The evolution of endothermy: testing the aerobic capacity model. *Evolution* 49: 836–847.

Heinrich, B. 1993. *The Hot Blooded Insects*. Cambridge: Harvard University Press.

Hengst, R., et al. 1996. Biological consequences of Mesozoic atmospheres: respiratory adaptations and functional range of *Apatosaurus*. In N. Macleod and G. Keller (eds.), *Cretaceous-Tertiary Mass Extinctions: Biotic and Environmental Changes.*, 327–347. New York: W. W. Norton and Co.

Hiley, P. 1975. How the elephant keeps it cool. *Natural History* 84: 34–41.

Hill, R., and G. Wise. 1989. *Animal Physiology*. New York: Harper Row.

Holdaway, R., and C. Jacomb. 2000. Rapid extinction of the moas: model, test and implications. *Science* 287: 2550–2254.

Horner, J. 2000. Dinosaur reproduction and parenting. *Annual Review of Earth and Planetary Sciences* 28: 19–45.

Horner, J., and E. Dobb. 1997. *Dinosaur Lives*. New York: HarperCollins.

Horner, J., and K. Padian. 2004. Age and growth dynamics of *Tyrannosaurus rex*. *Proceedings of the Royal Society of London B* 271: 1875–1880.

Hoyos, J., A. Elliot, and J. Sargatal. 1994. *Handbook of the Birds* 2. Barcelona: Lynx Edicons.

Hoyt, D., and G. Kenagy. 1988. Energy costs of walking and running gaits and their aerobic limits in golden-mantled ground squirrels. *Physiological Zoology* 61: 34–40.

Janis, C., and M. Carrano. 1992. Scaling of reproductive turnover in archosaurs and mammals: why are large terrestrial mammals so rare? *Acta Zoologica Fennica* 28: 201–206.

Ji S., Q. Ji, J. Lu, and C. Yuan. 2007. A new giant compsognathid dinosaur with long filamentous integuments from Lower Cretaceous of Northeastern China. *Acta Geologica Sinica* 81: 8–15.

Karsten, K., L. Andriamandimboarisoa, S. Fox, and C. Raxworthy. 2008. A unique life history among tetrapods: an annual chameleon living mostly as an egg.

Proceedings of the National Academy of Sciences 105: 8980–8984.

Kenagy, G., and D. Hoyt. 1989. Speed and time-energy budget for locomotion on golden-mantled ground squirrels. *Ecology* 70: 1834–1839.

Klomp, N., and R. Furness. 1992. A technique which may allow accurate determination of the age of adult birds. *Ibis* 134: 245–249.

Kolokotrones, T., V. Savage, E. Deeds, and W. Fontana. 2010. Curvature in metabolic scaling. *Nature* 464: 753–756.

Kruuk, H. 1972. *The Spotted Hyena.* Chicago: University of Chicago Press.

Lambert, W. 1991. Altriciality and its implications for dinosaur thermoenergetic physiology. *Neues Jahrbuch für Geologie und Paläontologie* 182: 73–84.

Langman, V., et al. 1995. Moving cheaply: energetics of walking in the African elephant. *Journal of Experimental Biology* 198: 629–632.

Langman, V., and M. Maloiy. 1989. Passive obligatory heterothermy of the giraffe. *Journal of Physiology* 415: 89P.

Lehman, T. 2007. Growth and population age structure in the horned dinosaur *Chasmosaurus.* In K. Carpenter (ed.), *Horns and Beaks: Ceratopsian and Ornithopod Dinosaurs,* 259–317. Bloomington: Indiana University Press.

Lehman, T., and H. Woodward. 2008. Modeling growth rates for sauropod dinosaurs. *Paleobiology* 34: 264–281.

Lewis, J. 1979. Periosteal layers do not indicate ages of sandhill cranes. *Journal of Wildlife Management* 43: 269–271.

Li, Q., et al. 2010. Plumage color patterns of an extinct dinosaur. *Science* 327: 1369–1372.

Liston, J. 2007. *A Fish Fit For Ozymandias?: The Ecology, Growth and Osteology of* Leedsichthys. Unpublished PhD Thesis. Faculty of BioMedical & Life Sciences, University of Glasgow, Scotland.

Lutcavage, M., P. Bushnell, and D. Jones. 1990. Oxygen transport in the leatherback sea turtle *Dermochelys coriacea. Physiological Zoology* 63: 1012–1024.

Mallison H. 2009. Rearing for food? Kinetic/dynamic modeling of bipedal/tripodal poses in sauropod dinosaurs. *Tribute to Charles Darwin and Bernissart Iguanodons: New Perspectives on Vertebrate Evolution and Early Cretaceous Ecosystems,* www.naturalsciences.be.science/colloquia.

Manning, P. 2008. *T. rex* speed trap. In P. Larson and K. Carpenter (eds.), *Tyrannosaurus rex, the Tyrant King,* 205–231. Bloomington: Indiana University Press.

Manikowski, S., and K. Walasz. 1980. An attempt to age pigeons using layered structure of tibia. *Ornis Scandinavia* 11: 73–74.

Mayr, G., D. Peters, G. Plodowski, and O. Vogel. 2002. Bristle-like integumentary structures at the tail of the horned dinosaur *Psittacosaurus. Naturwissenschaften* 89: 361–365.

McKechnie, A., and B. Lovegrove. 2002. Avian facultative hypothermic responses: a review. *Condor* 104: 705–724.

McNab, B. 1978. The evolution of endothermy in the phylogeny of mammals. *American Naturalist* 112: 1–21.

———. 2009. Resources and energetics determined dinosaur body size. *Proceedings of the National Academy of Sciences* 106: 12189–12188.

Meier, A., and A. Fivizzari. 1980. Physiology of migration. In S. Gauthreaux, (ed.), *Animal Migration, Orientation and Navigation,* 225–282. London: Academic Press.

Meng, Q., et al. 2004. Parental care in an ornithischian dinosaur. *Nature* 431: 145–146.

Mezentseva, N., J. Kumaratilake, and S. Newman. 2008. The brown adipocyte differentiation pathway in birds: an evolutionary road not taken. *BMC Biology* 6: 17.

Molnar, R., and J. Wiffen. 1994. A Late Cretaceous polar dinosaur fauna from New Zealand. *Cretaceous Research* 15: 689–706.

Murrish D. 1973. Respiratory heat and water exchange in penguins. *Respiration Physiology* 19: 262–270.

Nelson, R., and T. Brookhout. 1980. Counts of periosteal layers invalid for aging Canada Geese. *Journal of Wildlife Management* 44: 518–521.

Norris, S. 1998. Of mice and mammoths. *Bioscience* 48: 887–892.

Nowak, R. 1999. *Walker's Mammals of the World,* 6th ed. Baltimore: The Johns Hopkins University Press.

O'Conner, M., and P. Dodson. 1999. Biophysical constraints on the thermal ecology of dinosaurs. *Paleobiology* 25: 341–368.

O'Connor, P. 2006. Postcranial pneumaticity: an evaluation of soft-tissue influences on the postcranial skeleton and the reconstruction of pulmonary anatomy in archosaurs. *Journal of Morphology* 10: 1199–1226.

O'Connor, P., and P. Claessens. 2005. Basic avian pulmonary design and flow-through ventilation in non-avian theropod dinosaurs. *Nature* 436: 253–256.

Ostrom, J. 1970. Terrestrial vertebrates as

indicators of Mesozoic climates. *North American Paleontologocical Convention Proceedings* D: 347–376.

Ostrowski, S., and J. Williams. 2006. Heterothermy of free-living Arabian sand gazelles in a desert environment. *The Journal of Experimental Biology* 209: 1421–1429.

Ostrowski, S., J. Williams, and K. Ismael. 2003. Heterothermy and the water economy of free-living Arabian oryx. *The Journal of Experimental Biology* 206: 1471–1478.

Owerkowicz, T., and A. Crompton. 2001. Allometric scaling of respiratory turbinates and trachea in mammals and birds. *Journal of Vertebrate Paleontology* 21(3): 86A.

Packard, G., and G. Birchard. 2008. Traditional allometric analysis fails to provide a valid predictive model for mammalian rates. *The Journal of Experimental Biology* 211: 3581–3587.

Padian, K., A., de Ricqlès, and J. Horner. 2001. Dinosaurian growth rates and bird origins. *Nature* 412: 405–408.

Padian, K., J. Horner, and A. de Ricqlès. 2004. Growth in small dinosaurs and pterosaurs: the evolution of archosaurian growth strategies. *Journal of Vertebrate Paleontology* 24: 555–571.

Paladino, F., M. P. O'Connor, and J. Spotila. 1990. Metabolism of leatherback turtles, gigantothermy, and thermoregulation of dinosaurs. *Nature* 344: 858–860.

Paladino, F., J. Spotila, and P. Dodson. 1997. A blueprint for giants: Modeling the physiology of large dinosaurs. In J. Farlow and M. Brett-Surman (eds.), *The Complete Dinosaur,* 491–504. Bloomington: Indiana University Press.

Palmeri, G., et al. 2008. Brain morphology and immunohistochemical localization of the gonadotropin-releasing hormones in the bluefin tuna. *European Journal of Histochemistry* 52: 19–28.

Parker, H., and P. Stott. 1965. Age, size and vertebral calcification in the basking shark, *Cetorhinus maximus. Zoologischie Mededelingen* 40: 305–319.

Parsons, K. 2001. *Drawing out Leviathan.* Bloomington: Indiana University Press.

Paul, G. 1988. Physiological, migratorial, climatological, geophysical, survival and evolutionary implications of Cretaceous polar dinosaurs. *Journal of Palaeontology* 62: 640–652.

———. 1991. The many myths, some old, some new, of dinosaurology. *Modern Geology* 16: 69–99.

———. 1994a. Physiology and migration of North Slope dinosaurs. In D. Thurston and K. Fujita (eds.), 1992 *Proceedings International Conference on Arctic Margins*, 405–408. Anchorage: U.S. Department of the Interior.

———. 1994b. Dinosaur reproduction in the fast lane: implications for size, success, and extinction. In K. Carpenter, K. Hirsch, and J. Horner (eds.), *Dinosaur Eggs and Babies*, 244–255. Cambridge: Cambridge University Press.

———. 1994c. Thermal environments of dinosaur nestlings: implications for endothermy and insulation. In K. Carpenter, K. Hirsch, and J. Horner (eds.), *Dinosaur Eggs and Babies*, 279–287. Cambridge: Cambridge University Press.

———. 1996. The art of Charles R. Knight. *Scientific American* 274(6): 74–79.

———. 1997. Dinosaur models: the good, the bad, and using them to estimate the mass of dinosaurs. In D. Wolberg, E. Stump, and G. Rosenberg (eds.). *Dinofest International Symposium Proceedings*, 129–154. Philadelphia Academy of Sciences, Philadelphia.

———. 1998. Terramegathermy and Cope's Rule in the land of titans. *Modern Geology* 23: 179–217.

———. 2000. Restoring the life appearance of dinosaurs. In G. Paul (ed.), *The Scientific American Book of the Dinosaur*. New York: Byron Preiss.

———. 2002. *Dinosaurs of the Air*. Baltimore: The Johns Hopkins University Press.

———. 2006. Fused *Camarasaurus* cervicals preserve an erect, not horizontal neck. *Journal of Vertebrate Paleontology* 26(3): 109A.

———. 2008. The extreme lifestyles and habits of the gigantic tyrannosaurid superpredators of the Late Cretaceous of North America and Asia. In P. Larson and K. Carpenter (eds.), Tyrannosaurus rex, *the Tyrant King*, 307–352. Bloomington: Indiana University Press.

Paul, G., and G. Leahy. 1994. Terramegathermy in the time of the titans: Restoring the metabolics of colossal dinosaurs. *The Paleontological Society Special Publication* 7: 177–198.

Persons, W., and P. Currie. 2011. The tail of *Tyrannosaurus*: reassessing the size and locomotive importance of the M. caudofemoralis in non-avian theropods. *The Anatomical Record* 294: 119–131.

Perry S. 1989. Mainstreams in the Evolution of Vertebrate Respiratory Structures. In A. King and J. McLelland (eds), *Form and Function in Birds* 4: 1–67. London: Academic Press.

Perry, S., A. Christian, T. Breuer, N. Pajor, and J. Codd. 2009. Implications of an avian-style respiratory system for gigantism in sauropod dinosaurs. *Journal of Experimental Zoology* 311A: 600–610.

Pontzer, H., V. Allen, and J. Hutchinson. 2009. Biomechanics of running indicates endothermy in bipedal dinosaurs. *PLoS One* 4: e7783.

Priede, I. 1985. Metabolic scope in fishes. In P. Tyler and P. Calow (eds), *Fish Energetics: New Perspectives*, 33–64. Baltimore: The Johns Hopkins University Press.

Prough, F., C. Janis, and J. Heiser. 2005. *Vertebrate Life*. San Francisco: Benjamin Cummings.

Quick, D., and J. Ruben. 2009. Cardio-pulmonary anatomy in theropod dinosaurs: implications from extant archosaurs. *Journal of Morphology* 270: 1232–1246.

Reid, R. 1990. Zonal "growth rings" in dinosaurs. *Modern Geology* 15: 19–48.

Rich, T., and P. Vickers-Rich. 2001. *Dinosaurs of Darkness*. Bloomington: Indiana University Press.

Rich, T., P. Vickers-Rich, and R. Gangloff. Polar dinosaurs. *Science* 295: 979–980.

Rinehart, L., A. Heckert, S. Lucas, and M. Celeskey. 2008. Growth, allometry, and age/size distribution of the Late Triassic theropod dinosaur Coelophysis bauri. *Journal of Vertebrate Paleontology* 28: 132A.

Ruben, J. 1991. Reptilian physiology and the flight capacity of *Archaeopteryx*. *Evolution* 45: 1–17.

———. 1995. The evolution of endothermy in mammals and birds. *Annual Review of Physiology* 57: 69–95.

Ruben, J. and T. Jones. 2000. Selective factors associated with the origin of fur and feathers. *American Zoologist* 40: 585–596.

Ruben, J., T. Jones, and N. Geist. 2003. Respiratory and reproductive paleophysiology of dinosaurs and early birds. *Physiology and Biochemical Zoology* 76: 141–164.

Ruben J., et al. 1997. Lung structure and ventilation in theropod dinosaurs and birds. *Science* 278: 1267–1270.

Russell, D. 1989. *An Odyssey in Time: The Dinosaurs of North America*. Ottawa: National Museum of Natural Sciences.

Sander, P., and N. Klein. 2005. Developmental plasticity in the life history of a prosauropod dinosaur. *Science* 310: 1800–1802.

Sander, P., and M. Clauss 2008. Sauropod gigantism. *Science* 322: 200–201.

Sasso, C., and M. Signore. 1998. Exceptional soft-tissue preservation in a theropod dinosaur from Italy. *Nature* 392: 383–387.

Savage, V., et al. 2004. The prominence of quarter-power scaling in biology. *Functional Ecology* 18: 257–282.

Schaller, G. 1972. *The Serengeti Lion*. Chicago: University of Chicago Press.

Schwarz, D., O. Wings, and C. Meyer. 2007a. Peneumaticity and soft-tissue reconstruction in the neck of diplodocid and dicraeosaurid sauropods. *Acta Palaeontologica Polonica* 52: 167–188.

Schweitzer, M., and C. Marshall. 2001. A molecular model for the evolution of endothermy in the theropod-bird lineage. *Journal of Experimental Zoology* 291: 317–338.

Schwimmer, D. 2002. *King of the Crocodylians – the Paleobiology of* Deinosuchus. Bloomington: Indiana University Press.

Seebacher, F., G. Grigg, and L. Beard. 1999. Crocodiles as dinosaurs: behavioral thermoregulation in very large ectotherms leads to high and stable body temperatures. *The Journal of Experimental Biology* 202: 77–86.

Seebacher, F. 2003. Dinosaur body temperatures: the occurrence of endothermy and ectothermy. *Paleobiology* 29: 105–122.

Sereno, P., H. Larsson, C. Sidor, and B. Gado. 2001. The giant crocodyliform *Sarcosuchus* from the Cretaceous of Africa. *Science* 294: 1516–1519.

Sereno, P., et al. 2008. Evidence for avian intrathoracic air sacs in a new predatory dinosaur from Argentina. *PLoS ONE* 3(9): e3303.

Seymour, R. et al. 2008. Evidence for endothermic ancestors of crocodiles at the stem of archosaur evolution. *Physiology and Biochemical Zoology* 77: 1051–1067.

Seymour, R., S. Smith, C. White, D. Henderson, and D. Schwarz-Wings. 2011. Blood flow to long bones indicates activity metabolism in mammals, reptiles and dinosaurs. *Proceedings of the Royal Society B: Biological Sciences* doi.10.1098/rspb.2011.0968.

Shine, R. 1988. Parental care in reptiles. In C. Gans (ed.), *Biology of the Reptilia*, 275–329. New York: Alan Press.

Simpson, E. et al. 2010. Predatory digging behavior by dinosaurs. *Geology* 38: 699–702.

Spicer, R., and A. Herman. 2010. The Late Cretaceous environment of the Arctic: A quantitative reassessment based on plant fossils. *Paleogeography, Palaeoclimatology, Palaeoecology* 295: 423–442.

Spotila, J., M. O'Connor, P. Dodson, and F. Paladino. 1991. Hot and cold running dinosaurs: body size, metabolism and migration. *Modern Geology* 16: 203–227.

Stein, K. et al. 2010. Small body size and extreme cortical bone remodeling indicate phyletic dwarfism in *Magyarosaurus dacus*. *Proceedings of the National Academy of Sciences* 107: 9258–9263.

Tarduno, J., et al. 1998. Evidence for extreme climatic warmth from Late Cretaceous arctic vertebrates. *Science* 282: 2241–2244.

Taylor, M., M. Wedel, and D. Naish. 2009. Head and neck posture in sauropod dinosaurs inferred from extant animals. *Acta Palaeontologica Polonica* 54: 213–220.

Taylor, M., D. Hone, M. Wedel, and D. Naish. 2011. The long necks of sauropods did not evolve primarily through sexual selection. *Journal of Zoology* doi: 10.1111/j.1469-7998.2011.00824.x.

Thomas, R., and E. Olson. 1980. *A Cold Look at the Hot Blooded Dinosaurs.* Washington D.C.: AAAS.

Thompson, G., and P. Withers. 2000. Standard and maximal metabolic rates of goannas. *Physiological Zoology* 70: 307–323.

Thulborn, T., 1992. The demise of the dancing dinosaurs? *The Beagle, Records of the Northern Territory Museum of Arts and Sciences* 9: 29–34.

Turvey, S., O. Green, and R. Holdaway. 2005. Cortical growth marks reveal extended juvenile development in New Zealand moa. *Nature* 435: 940–943.

Upchurch, P., P. Barrett, X. Zhao, and X. Xu. 2007. A re-evaluation of *Chinshakiangosaurus chunghoensis*: implications for cranial evolution in basal sauropod dinosaurs. *Geological Magazine* 144: 247–262.

Uriona, T., and C. Farmer. 2008. Recruitment of the diaphragmaticus, ischiopubis and other respiratory muscles to control pitch and roll in the American alligator. *Journal of Experimental Biology* 211: 1141–1147.

Van Valkenburgh, B., and R. Molnar. 2000. Dinosaurian and mammalian predator guilds compared. *Journal of Vertebrate Paleontology* 20: 75A.

Varricchio, D. 1997. Growth and embryology. In P. Currie and K. Padian (eds.), *Encyclopedia of Dinosaurs*, 282–288. San Diego: Academic Press.

Varricchio, D., F. Jackson and C. Trueman. 1999. A nesting trace with eggs for the Cretaceous theropod dinosaur *Troodon formosus*. *Journal of Vertebrate Paleontology* 19: 91–100.

Varricchio, D., et al. 2008. Mud-trapped herd captures evidence of distinctive dinosaur sociality. *Acta Palaeontologica Polonica* 53: 567–578.

Varricchio, D. A. Martin, and Y. Katsura. 2007. First trace and body fossil evidence of a burrowing, denning dinosaur. *Proceedings of the Royal Society* B doi:10.1098/rspb. 2006.0443.

Walsh, T., D. Chiszar, G. Birchard, and K. Tirtodinigrat. 2002. Captive management and growth, 178–195. In J. Murphy, C. Ciofi, C. Panouse, and T. Walsh (eds.), *Biology and Conservation of Komodo Dragons.* Washington D.C.: Smithsonian Institution Press.

Weaver, J. 1983. The improbable endotherm: the energetics of the sauropod dinosaur *Brachiosaurus*. *Paleobiology* 9: 173–182.

Wedel, M. 2003. Vertebral pneumaticity, air sacs, and the physiology of sauropod dinosaurs. *Paleobiology* 29: 243–255.

——. 2005. Postcranial pneumatization in sauropods and its implications for mass estimates. In J. Wilson and K. Curry-Rogers (eds.). *The Sauropods: Evolution and Paleobiology*, 201–228. Berkeley: University of California Press.

——. 2009. Evidence for Bird-like air sacs in saurischian dinosaurs. *Journal of Experimental Zoology* 311A: 611–628.

Weishampel, D. et al. 2008. New oviraptorid embryos from Bugin-Tsav, Nemegt Formation, Mongolia, with insights into their habitat and growth. *Journal of Vertebrate Paleontology* 28: 1110–1119.

White, C., T. Blackburn, and R. Seymour. 2009. Phylogenetically informed analysis of the allometry of mammalian basal metabolic rates supports neither geometric nor quarter-power scaling. *Evolution* 63: 2658–2667.

Winkler, D. 1994. Aspects of growth in the Early Cretaceous Proctor Lake ornithopod. *Journal of Vertebrate Paleontology* 14: 53A.

Withers, P., and J. Williams. 1990. Metabolic and respiratory physiology of an arid-adapted Australian bird, the spinifex pigeon. *Condor* 82: 99–100.

Witmer, L. 1995. Homology of facial structures in extant archosaurs, with special references to paranasal pneumaticity and nasal conchae. *Journal of Morphology* 225: 269–327.

Witmer, L., and S. Sampson. 1999. Nasal conchae and blood supply in some dinosaurs: physiological implications. *Journal of Vertebrate Paleontology* 19(3): 85A.

Wintner, S. 2000. Preliminary study of vertebral growth rings in the whale shark *Rhinocodon typus*, from the east coast of South Africa. *Envionmental Biology of Fishes* 59: 441–451.

Xu, X., K. Wang, K Zhang, Q. Ma, L Xing, C. Sullivan, D. Hu, S. Cheng, and S. Wang. 2012. A gigantic feathered dinosaur from the Lower Cretaceous of China. *Nature* 484: 92–95. Zhang, F. et al. 2010. Fossilized melanosomes and the colour of Cretaceous dinosaurs and birds. *Nature* 463: 1075–1078.

Zug, G., and J. Parham. 1996. Age and growth in leatherback turtles, *Dermochelys coriacea*: a skeletochronological analysis. *Chelonian Conservation Biology* 2: 173–183.

38.1. The heart and associated arteries in (A) a lizard, (B) a crocodile, and (C) a dinosaur, assumed to have a complete double-pump circulation. Figures A and B after Bellairs and Attridge (1975). Abbreviations: a = aorta (dorsal aorta); ca = carotids; LA = left auricle (or atrium); FP = Foramen of Panizza. LP = left pulmonary artery; LS = left systemic arch; LV = left ventricle; RA = right auricle (or atrium); RP = right pulmonary artery; RS = right systemic arch; RV = right ventricle; V = ventricle. The auricles (atria) should be pictured as lying in part behind the ventricles, and opening into them in the areas shaded black. In the lizard (A), deoxygenated blood and oxygenated blood are pumped by the same single ventricle, with only some separation by incomplete ventricular septa. In the crocodile (B), there are separate right and left ventricles, but the right and left systemic arches communicate at their bases through an aperture (the Foramen of Panizza), which allows oxygenated blood to pass into both systemic arches when the lungs are in use. It closes during diving, and some deoxygenated blood then passes into the left systemic arch without having to be pumped through the lungs. The pattern suggested for dinosaurs (C) is supposed to be derived from the crocodilian one by loss of the left systemic arch, as suggested by L. S. Russell (1965), and corresponds with what is now seen in birds. As an adaptation to diving, the Foramen of Panizza is not a likely feature of terrestrial archosaurs ancestral to crocodiles and dinosaurs.

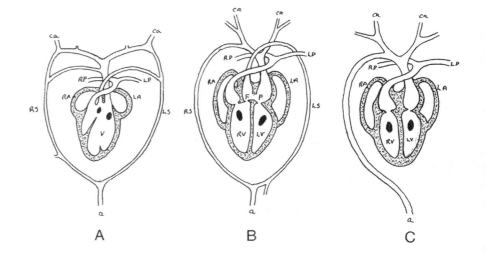

A   B   C

"Intermediate" Dinosaurs: The Case Updated

38

†R. E. H. Reid (1924–2007)

In the previous edition of this book, I explained why I see most nonavian dinosaurs as probably having been intermediate animals (Reid 1997). In this updated chapter, as then, I first note three (not just two) main possibilities:

1. They were cold-blooded like modern reptiles, as used to be assumed.
2. They were warm-blooded, as claimed by Bakker (e.g., 1972) and de Ricqlès (e.g., 1974).
3. They had no modern physiological counterparts, as Regal and Gans (1980) thought possible.

As before, I reject the first possibility, and see the third as best fitting the evidence. I accept the arguments of Seymour (1976) and Ostrom (1980) for double-pump circulation (Fig. 38.1C), but do not see this or fast growth as meaning that they must have been warm-blooded. The early background material is largely unchanged, but the last two sections have been rewritten extensively. "Evidence from Bone" now begins with accounts of the histology of bone and bone tissues thought significant by de Ricqlès (1974), as background to discussing what they tell us. He and I mostly use the same terminology, but I reject his way of dividing periosteal bone tissues into two contrasting types and his use of fibrolamellar bone as evidence of endothermy. I also note cases in which different growth patterns were apparently controlled by genetics, not thermal physiology, and suspect that a potential for fast growth has been common to all tetrapods. "Matching the Pieces" now starts with a summary of the principles I try to follow, adds some new points, and asks two so far unanswered questions.

Heights and Hearts

Modern reptiles, other than crocodiles, have three-chambered hearts (Fig. 38.1A) with a single ventricle and no complete separation of the pulmonary circulation to the lungs from the systemic circulation to the body. This primitive arrangement is adequate because the vertical distance through which blood has to circulate is not great, even in the largest reptiles. Dinosaurs, however, not only reached much larger sizes, they also stood upright and sometimes held their heads high above their backs on long necks. The head of the giant sauropod *Brachiosaurus*, for instance, if its pose has been interpreted correctly, was carried as much as 11 meters (36 feet) above the

"Warm" Blood, "Cold" Blood, and Other Notions

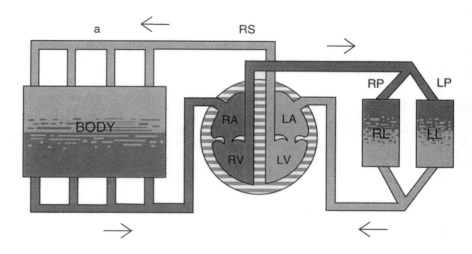

38.2. How blood would circulate in a dinosaur with complete double-pump circulation. LL = left lung; RL = right lung; other abbreviations as in Figure 38.1. Oxygenated and deoxygenated blood shown light and dark gray respectively.

ground and 7.5 meters (24.6 feet) above its heart. The pressure required to circulate blood through such vertical distances would be greater than in any modern reptile and almost certainly too great to be applied to blood vessels in the lungs without causing a fatal leakage of blood fluid. This points to dinosaurs having some form of double-pump circulation, as in mammals and birds, with four-chambered hearts and with complete separation of the pulmonary and systemic circulations. Crocodiles (Fig. 38.1B) have a primitive form of double-pump circulation, with four-chambered hearts, but they do not stand upright, and the right- and left-hand systemic arches communicate (note that in contrast to the opinion held here, Seymour et al. [2004] take the structure of the crocodilian heart as evidence that they are derived from endothermic ancestors). Large dinosaurs, walking upright, would have needed a fully evolved double-pump system; and this would probably have been of the type now seen in birds (Figs. 38.1C, 38.2), which appears to have arisen from the crocodilian type by suppression of the left systemic arch (or, left aorta: compare Figs. 38.1B and C), as Russell (1965) saw.

If this conclusion is correct, as I think, there was a major anatomical difference between dinosaurs and modern reptiles, which could well have opened the way to further physiological differences. Because this type of circulatory system is now seen only in warm-blooded animals, its postulated presence implies that dinosaurs could have been warm-blooded. The question then is not whether they differed from modern reptiles physiologically, but how much they differed.

What Is "Warm" Blood?

We need to look at the technical terminology used in defining "cold" and "warm" blood scientifically as well as some associated concepts relevant to discussion of dinosaurs. First, it needs to be understood that the terms cold and warm blood involve more than just temperature. We call amphibians and reptiles cold-blooded because many, which have lower temperatures than our own, feel cold when they are handled; but some reptiles can have temperatures approaching our own or even higher. In optimum conditions, the body temperatures of American alligators can reach 33° to 34° C (Coulson and Hernandez 1983), close to the human level (36° C), while the desert iguana *Dipsosaurus* has a temperature of 40° to 42° C when active.

Scientifically, three pairs of contrasting terms are used to describe animals. They are called: (1) homoiothermic (or homeothermic) if they are able to maintain a steady temperature in their normal environments, irrespective of daily variations in atmospheric (ambient) temperature, or poikilothermic if they cannot maintain a steady temperature, (2) endothermic if they maintain an activity temperature by means of internally generated heat, or ectothermic if they depend on external heat sources, and, (3) tachymetabolic if their metabolism runs at a high rate, or bradymetabolic if it runs at a low one. Ideally, warm-blooded animals are homoiothermic, endothermic, and tachymetabolic, with rapid metabolism providing an internal heat source, while cold-blooded animals are poikilothermic and bradymetabolic. The conditions they show are described as, for example, homoiothermy, endothermy, and tachymetabolism in the warm-blooded case. In practice, however, not all warm-blooded forms conform to the ideal pattern; some birds (e.g., hummingbirds, chickadees) save energy by lowering their temperatures overnight. Such forms, in which this habit appears to be secondary, are distinguished from typical endotherms as heterotherms. The naked mole rat *Heterocephalus* of East Africa, which spends its life in burrows, is almost completely poikilothermic. The temperatures maintained by modern endotherms also vary, being generally between 28° to 30° C in sloths and monotremes, between 33° and 36° C in marsupials, between 36° and 38° C in most placentals, and between 40° and 41° C in birds.

A mistake which needs to be avoided is the notion that poikilotherms have no control over their temperatures, which will simply correspond with those of their environments. Ectothermic lizards, which cannot maintain an optimum temperature overnight, can often maintain it quite precisely during daytime, at levels above or below the ambient temperature, by various behavioral and physiological means. Even if dinosaurs were simply normal reptiles, they would probably have similar abilities.

"Warm" Blood through Bulk

In living tetrapods, homoiothermy is restricted to endotherms with high metabolic rates; but some authors have argued that an ectotherm can become homoiothermic if it grows large enough, without needing a high metabolic rate. This supposed condition has been called inertial homoiothermy (McNab and Auffenberg 1976) or simply mass homoiothermy (de Ricqlès 1974). The basic idea was derived from experiments on alligators made by Colbert et al. (1946), who found that large individuals take longer to change temperature than small ones. This is explicable as due to body volume increasing by cubes of linear dimensions, whereas surface area increases by squares only. For any given temperature change, more heat must pass through a given surface area in a large form than in a small one. Extrapolating upward from their alligator data, they concluded that a 10-ton cold-blooded dinosaur would take 86 hours to change its temperature by 1° C. Because there are only 24 hours in a day, such an animal would be able to maintain an essentially steady temperature despite normal daily temperature variations.

In its modern form, however, the concept dates from a mathematical study by Spotila et al. (1973). Using a model dinosaur to have body diameter

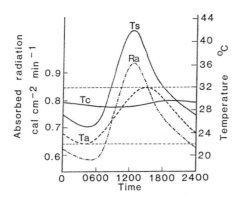

38.3. Inertial (or ectothermic) homoiothermy in a 1 meter diameter model dinosaur devised by Spotila et al. (1973), whose Figures 2a and 2b are here combined. In Florida-like conditions, with air temperatures cycling daily between 22° and 32° C, the internal (core) temperature (Tc) of a 1 meter diameter animal, resembling an American alligator physiologically, would vary only between 28.5° and 29.6° C, and thus by only 1.1° C daily despite a 10° C daily variation in air temperature (Ta). Surface temperature (Ts) would, however, vary more widely, with a pattern paralleling that of absorbed radiation (Ra).

From Reid 1987, Figure 1, by permission; recaptioned.

of 1 meter, a 5-centimeter layer of insulating fat beneath its skin, and the energy budget of an American alligator, they found: (1) in a climate like that of Florida, with air temperatures cycling between 22° and 32° C daily, its core temperature (Fig. 38.3, Tc) would remain between 28.5° and 29.6°, and (2) the time taken for 63.7% of a change between two equilibrium temperatures, termed the "time constant," would be 48 hours. Many dinosaurs had larger bodies, and some were much larger. This general picture has been confirmed by various later studies. For instance, using Spotila et al.'s method, Pough et al. (1989) obtained a time constant of over five days (127 hours) for a 2 m diameter dinosaur. A recent study by Seebacher et al. (1999) is instead based on body temperature data from the largest modern crocodile, *Crocodylus porosus*. By scaling up data from specimens ranging from 32–1010 kg in weight, they conclude that the body (core) temperature of a 10 tonne (10,000 kg) dinosaur would vary by less than 0.1° in conditions in which the ambient temperature cycled daily between 20° and 38°C. Such studies show clearly that large dinosaurs could have been homoiothermic without being endotherms. Small forms and juveniles, however, would still be poikilothermic and would have to use other means of regulating temperature. Halstead (1975), for instance, took the minimum size for homoiothermy through bulk as that of the prosauropod *Plateosaurus*, and pictured the small ornithopod *Fabrosaurus*, only 1 meter long, as moving in and out of sunlight to regulate its temperature, in the manner described as shuttling heliothermy when practiced by modern lizards.

Dinosaurs pictured as mass homoiotherms have been called warm-blooded by some authors, including Halstead (1975), but this equates warm-bloodedness with homoiothermy only. As was seen by de Ricqlès (1974), raised temperatures resulting from mass effects would presumably cause some upward shift in resting metabolic rates; but a high metabolic rate maintained by a high steady temperature depending on bulk is clearly a different matter from a high steady temperature maintained by a high metabolic rate independently of bulk. So, dinosaurs envisaged in this manner should not be called warm-blooded if we want the term to mean the same as in mammals and birds. The terms *endothermy* and *endotherm* also need to be used with the same meanings. Some authors have treated endotherms as simply animals in which most of the heat in the body is of internal origin; but, because all metabolic activity produces heat, this could apply to a large enough ectotherm with a low metabolic rate, again as a result of the "cube and square effect" (de Ricqlès 1980, 139). Such endothermy, depending on bulk and termed "mass endothermy" by de Ricqlès (1983, 234), is clearly different from the tachymetabolic endothermy of mammals and birds.

Two other features of modern reptiles can be pictured as likely to have supplemented mass effects in dinosaurs. First, McNab and Auffenberg (1976) found that heat exchange through the surface occurs more quickly in lizards than in mammals below about 100 kilograms in weight, but less quickly above that weight (Fig. 38.4). If dinosaurs resembled lizards, this would reinforce the mass effect of large size; and again most reached much larger sizes. Second, Regal and Gans (1980) pictured mass effects as supplemented by vascular control over heat exchanges, through contraction or dilation of superficial blood vessels as in some modern reptiles. This combination of mass effects with physiological control has been called gigantothermy by Spotila et al. (1991).

"High" Metabolic Rates

Although it is commonly said that endotherms have high metabolic rates while ectotherms have low ones, the true picture is much more complex. First, the metabolic rates of animals are not constant but vary with their level of activity. Every animal has a basic metabolic rate, termed its resting or standard metabolic rate (SMR), which is defined by the minimum level of oxygen consumption required to sustain life at a standard temperature and pressure. Any form of activity, even digestion, requires a higher rate. An active ectotherm can have a higher metabolic rate than a resting endotherm of the same size. When endotherms are said to have high metabolic rates, this means that their SMRs are generally six or more times higher than those of comparable ectotherms. At 28° C, for instance, the SMR of a 70-kilogram American alligator is said to be less than 4% that of a 70-kilogram human (Coulson and Hernandez 1983). Thus, if high metabolic rates are claimed as grounds for seeing dinosaurs as endotherms, this must refer to their basic metabolic rates and not to activity metabolism. Some small dinosaurs were probably highly active, with high levels of activity metabolism and resultant heat production; but the endothermy of mammals and birds is not based on activity metabolism, but on high basic metabolic rates which do not depend on activity. This is why they can maintain high steady temperatures while sleeping, even at small sizes, which a dinosaur maintaining a high temperature through activity could not do.

Second, the statement that endotherms have higher basic metabolic rates than ectotherms applies only to animals of similar size, because of the two ways in which SMRs change with size. In terms of total oxygen consumption, they are inevitably highest in large forms and lowest in small ones; but minimum oxygen consumption per unit mass, termed the mass-specific SMR, is instead highest in small forms (see Box 38.1) to the extent that small ectothermic lizards can have higher mass-specific SMRs than large endothermic mammals (Fig. 38.5). These changes in SMRs with size have two implications for dinosaurs. One is that small forms would automatically have higher mass-specific SMRs than large ones, without this

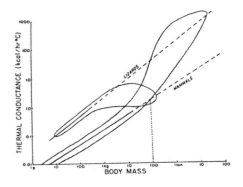

38.4. How thermal conductance changes with size in lizards and mammals according to McNab (1978), whose Figure 1 is here simplified. At sizes under 1 kilogram, the thermal conductance of the skin of a lizard can be as much as ten times that of the skin of a mammal of similar weight. Above 1 kilogram, however, the conductance values converge, and they reverse their relationship above 100 kilograms. If the conductance of dinosaurian skin was similar to that of large lizards, or even lower at larger sizes, this would contribute substantially to the maintenance of inertial homoiothermy.

From Reid 1987, Figure 2, by permission; recaptioned.

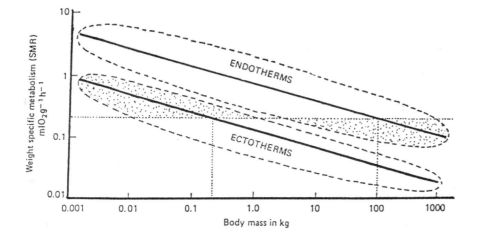

38.5. How mass-specific standard metabolic rates change with size in ectotherms and endotherms. In both, weight-specific SMR declines from the smallest to the largest, with the averaged regression lines for their scatters following the slopes of about −0.25. As a result, small ectotherms in the shaded area at left have *higher* metabolic rates than large endotherms in the shaded area at right. This is why the terms *low* and *high* metabolic rates, when used comparatively, must always be understood as referring to animals of similar size.

Modified from McFarland and Heiser 1979, Figure 7.21, lower part; this version from Reid 1987, Figure 4, by permission; recaptioned.

Box 38.1. Metabolic Rates and Sizes

A characteristic feature of vertebrates (and many invertebrates) is that while their total oxygen consumption increases with size, their consumption per unit of body mass instead decreases. If the mass-specific standard metabolic rates (SMRs) of modern vertebrates are plotted logarithmically, those of ectotherms and endotherms cluster around parallel regression lines, with −0.25 slopes, which fall at levels corresponding with their average difference in metabolic rates (Fig. 38.5). We have hence to expect a similar pattern in dinosaurs of different sizes, irrespective of whether they were ectotherms, endotherms, or somehow intermediate between the two. This, of course, assumes that one basic type of physiology was common to all dinosaurs; but this assumption is justified in showing what would then be expected.

In the following tables, data from Coulson and Hernandez (1983) for SMRs in mammals and alligators are analyzed in terms of percentages, for ease of comparison. First, figures for representative mammals from a very small shrew to the largest mammal known (the blue whale) provide a picture of the range of variation among members of one class. (See Table 1.)

Table 1. Mammals

	Mass in kilograms	O₂ used in liters/kg/day	Comparative SMRS by mass			
Shrew	0.002	322.8	100%			
Mouse	0.02	88.0	26%	100%		
Rat	0.2	27.7	8.6%	32.9%	100%	
Dog	10.0	8.0	2.5%	9.5%	28.9%	100%
Human	70.0	5.38	1.7%	6.4%	19.4%	67.3%
Blue whale	100,000.0	0.49	0.15%	0.58%	1.8%	6.1%

If you multiply the figures in the first two columns, you will find that the total minimum oxygen consumption (= total SMR) for the whale is more than 75,000 times that for the shrew; but at the mass-specific level, that of the shrew is nearly 660 times that of the whale, and 60 times that of an average (70-kg) human.[1] This is why the difference between the total figures is roughly two-thirds that of the 10-kilogram dog, one-fifth that of the 200-gram rat, one-sixteenth that of a 20-gram mouse, and one-sixtieth of a 2-gram shrew. If we assume the same pattern for dinosaurs, the mass-specific SMR of a 100-tonne sauropod would be under one-tenth (9.1%) that of a 70-kilogram deinonychid, and about one-sixteenth that of a 10-kilogram hatchling dinosaur.

The same authors' figures for American alligators give a striking picture of how much SMR can change during growth, when adults are much larger than juveniles. (See Table 2.)

From these figures, a 20,000-times mass increase leads SMR to fall by a factor of nearly 28. How far dinosaurs would parallel alligators is conjectural, but the figures give an indication of the scale of change to be expected in large forms up to 700 kilograms in mass, assuming that hatchling masses were between 0.5 and 7 kg.

Table 2. Alligators

Mass in kilograms	O₂ in liters/ kg/day	Relative change in SMR during growth				
0.035	2.35	100%				
0.050*	2.07*	88.1%				
0.12	1.56	66.4%				
0.5*	1.05*	44.7%	100%			
1.0	0.77	32.8%	73.3%	100%		
3.5*	0.56*	23.8%	53.3%	72.7%	100%	
7.0	0.40	17.0%	38.1%	51.9%	71.4%	100%
70.0	0.18	7.66%	17.1%	23.4%	32.1%	45%
700.0	0.084	3.6%	8.0%	10.9%	15.0%	21%

*From Coulson and Hernandez 1983, fig 2.1, table 2.1.

Note: 1. Read the comparative SMR percentage vertically by column. For example, in the first column, the shrew = 100%; thus, the SMR of the blue whale compared to the shrew is 0.15%. In the next column, the mouse = 100%; the SMR of the blue whale compared to the mouse is 0.58%.

implying any movement toward endothermy—although, lacking the temperature stabilization of mass effects, they could have been under greater selection pressure to move toward it. Assuming parallels, for instance, with Eckert et al.'s cat and elephant figures (1988, table 16.2), the total SMR of a 3,833 kilogram (3.833 tonne) tyrannosaur would be about 158 times that of a 2.5-kilogram compsognathid; but the latter's mass-specific SMR would be 9.7 times that of the tyrannosaur thus almost a whole order higher. The contrast between it and a 30-tonne sauropod, for example, would have been even greater.

In addition, mass-specific SMRs can fall considerably during growth if adults are much larger than juveniles, as appears to have been generally the case in large dinosaurs. Walter Coombs (1980), for instance, judged *Psittacosaurus* hatchlings to weigh only 0.7% of adult weight, and Ted Case's (1978) estimates of 2.9 kilograms and 5,300 kilogram (5.3 tonnes) for hatchlings and adults of *Hypselosaurus* give the still lower figure of only 0.055%. For comparison, values obtained by Coulson and Hernandez (1983) show the mass-specific SMR of a 7-kilogram alligator as only 17% of that of a 35-gram hatchling, falling further to 7.66% at 70 kilograms and only 3.6% at 700 kilograms (see "Metabolic Rates and Sizes" above). A considerable fall in mass-specific SMR during growth is hence a predictable feature of all large dinosaurs, irrespective of whether they were ectotherms, endotherms, or intermediates.

A Digression: Why Regulate Temperature?

Regulation of temperature (thermoregulation) by reptiles, mammals, and birds has been noted above without comment on why this should happen. Explained simply, regulation of temperature within narrow limits in terrestrial species is usually interpreted (e.g., Pough et.al. 1989) as allowing maximum efficiency in the coordination of the thousands of enzyme-catalyzed and temperature-sensitive biochemical processes on which life depends. High optimum temperatures are said to allow faster nervous reactions than lower ones, and a permanently stabilized temperature allows activity to be independent of external heat sources. An ectothermic dinosaur growing large enough for full mass homoiothermy would gain the same independence in activity, plus a raised and stabilized metabolic rate, without the energetic cost of true endothermy, in which 90% of energy expenditure is devoted to heat production.

In addition, some thermoregulation is related to a need to keep temperature within limits outside which life processes will break down irreversibly. Because of their dependence on external heat sources, many ectothermic reptiles can tolerate cooling to well below their activity temperatures; but most endotherms are killed by hypothermia, if heat losses outstrip their ability to compensate by increased heat production. The difference is particularly striking in some semiaquatic reptiles. American alligators, for instance, have an optimum body temperature of approximately 31.5° C, but during winter cold spells they can survive in water as cold as 4–5° C (Brisbin et al. 1982), in which unprotected humans can die in less than 15 minutes. At the other extreme, both ectotherms and endotherms have little tolerance to core temperatures above their normal ranges and adopt various methods of avoiding this problem. Panting, for instance, is a physiological method of dispersing excess heat, by increasing respiratory cooling; and many forms living in warm deserts take refuge in shade or underground during the hottest hours. How tolerant small dinosaurs would have been to lowered temperatures would depend on what type of physiology they had, but overheating would potentially be a problem for all large dinosaurs, especially in warm conditions and strong sunlight, irrespective of their thermal physiology.

A Model for Dinosaurs

Although de Ricqlès (1974, 1980) hinted at a physiological difference between dinosaurs and modern tetrapods, the details of how they could have differed first emerged in a model proposed by Philip Regal and Carl Gans (1980) and designed to allow large animals to make optimum use of a limited food supply. Their ideal dinosaur was an animal with complete double-pump circulation, a temperature stabilized by bulk, and low a metabolic rate (SMR), with further features including aerobic activity metabolism, vascular control over heat exchanges, and tolerance of temperature instability. This combination, they thought, could give such animals "extraordinary abilities to be active and grow rapidly on a limited food supply," as well as explain how giants, such as the sauropods, could avoid overheating and feed themselves with only small heads.

Besides having double-pump circulation, such dinosaurs would differ from modern reptiles in a second major way if their activity metabolism were aerobic. In modern forms, reliance on anaerobic glycolysis as an energy source restricts high activity to short periods, producing oxygen debts which then have to be repaid; whereas energy release by direct oxidation, as in mammals and birds, would give dinosaurs a similar capacity for continuous activity. Although monitor lizards (also known as varanids or goannas) show partial aerobic activity, complete double-pump circulation would presumably allow its full development in dinosaurs. A low basic metabolic rate was specified because the higher this is, the more food animals of similar size need to sustain it—modern mammals, for instance, need ten to thirteen times as much as modern reptiles (Pough 1979; Nagy 2001). Whether dinosaurs were really adapted to limited food supplies is debatable (see Farlow et al. 1995), but a low SMR would also let more of the energy derived from their food be devoted to growth. Vascular control over heat exchange, by dilation or contraction of superficial blood vessels, is known from various modern reptiles and would add an active control over heat loss to the passive insulation of Spotila et al.'s (1973) model. Some attempts to dismiss the concept of mass homoiothermy (e.g., Desmond 1975) have assumed that dinosaurs had no active means of controlling heat exchanges; but, because they were live animals and not inanimate objects, this is unlikely.

Because full double-pump circulation is now restricted to endotherms, it needs to be asked whether its possession would have converted dinosaurs into endotherms automatically, thus not allowing the existence of animals of the type which Regal and Gans envisaged. This is possible but does not seem likely. Crocodiles have a form of double-pump circulation, in which systemic arteries carry only oxygenated blood except during diving, when the lungs are out of use, yet they show no sign of moving toward becoming endotherms. On the other hand, I think it unlikely that the three-way combination of full double-pump circulation, full aerobic activity, and a capacity for fast continuous growth, could be evolved without at least some upward movement of basic metabolic rates resulting. Hence, I have pictured Regal-Gans dinosaurs as likely to show at least some movement toward endothermy (Reid 1987, 148–152). The question then becomes how far they would move toward it.

This leads to the question of how endotherms evolve. Nobody knows; and while comparing modern examples can give us an insight into what changes must happen, it cannot tell us how or why they happen. It seems fairly likely, however, that the process can only begin in forms more advanced physiologically than any modern reptiles. We can only guess what this would require, but two things now seen only in mammals and birds—not in modern reptiles—are complete double-pump circulation and fully aerobic activity. This suggests that both must be evolved before endotherms can start to evolve, with the aerobic activity seen in endotherms also probably first made possible by this kind of circulation (Reid 1987, 146). But, if this is only the point at which endotherms can start to evolve, how and why do they?

One possible answer to these questions is provided by two kinds of progressive changes seen in therapsids—mammal-like reptiles from which mammals were evolved. First, beginning as animals that show no signs of having greater food needs or higher respiration rates than typical endotherms, they change later into forms showing evidence of both. Second, as some grew more mammal-like, they also grew smaller—starting as large animals and ending as small ones by the time the first mammals appeared. This led McNab (1978) to see endotherms arising through animals first achieving homoiothemy inertially (i.e., as mass homoiothermy) and then maintaining it, while following a trend to small sizes, by raising their basic metabolic rates and developing external insulation. The background to such a trend would be adaptation to adult exploitation of what were originally juvenile food resources, down to insectivorous level. On its own, such a trend would only cause metabolic rates to rise to those usual in small ectotherms (Fig. 38.5, lower field, right to left); but, for animals committed to homoiothermy, to maintain it they would have to rise to the levels seen in small endotherms (Fig. 38.5, upper field, top left) and would probably need to acquire external insulation. This picture is, of course, hypothetical; but the changes on which it is based are factual and striking in the case of the reduction in size. Early therapsids, called dinocephalians, were massive animals with skulls between 40–80 cm (15¾–31½ inches) long; but the late nonmammalian therapsids closest to mammals (tritylodonts and trithelodonts) were rat-sized, and the earliest known mammals were mouse sized. Of the latter, the skull of *Morganucodon* is only about 22.6 mm (⅞ inch) long, and the lower jaw of *Kuehneotherium* only approximately 13.4 mm (½ inch) long. So, whether or not McNab's explanation is correct, it seems clear that mammalian endothermy was evolved at small sizes. This has the further implication that large modern mammals are endotherms because they had small ancestors, and not because they need to be.

Although this picture is hypothetical, intermediate animals must once have existed for endotherms to be evolved from ectotherms. As emphasized by Kemp (1982) in his study of the origin of mammals, the switch from ectothermy to endothermy cannot be simply a matter of a single mutational jump. On the contrary, it must involve a large complex of correlated and cumulative mutations affecting all biological systems, all of which will need to be kept in balance for the animals concerned to stay viable. Especially, this would apply to the enzyme systems on which all life processes depend. Evolving intermediates would also have to be viable at every stage of the process and potentially capable of freezing at some stage if this were somehow advantageous. In the warm Late Triassic conditions in which they first appeared, this could have happened to dinosaurs and I think it likely that it did.

In contrast to mammals, nothing that we know of the history of archosaurs or of prearchosaurian diapsids suggests that dinosaurs had ancestors that followed a trend to smaller sizes; and, after first appearing, many followed an opposite trend to larger sizes. Early forms were small by dinosaurian standards, but were still much larger animals than the earliest mammals. Whether some early forms (e.g., *Eoraptor*, *Staurikosaurus*) were true dinosaurs or only close relatives (dinosauromorphs) is disputed; but they, and the accepted ones (e.g., *Fabrosaurus*), are generally animals in the 1–2 meters length range, which could have weighed up to 30 kilograms (66 pounds). Judged from skull size, the early mammal *Morganucodon* would have weighed about as much as a large house mouse, which can weigh up to 35 grams or one ounce. So, if we accept McNab's picture of how endotherms are evolved, there is nothing to suggest that dinosaurs began their history as endotherms. While the trend to large sizes, which many then followed, would have led them away from endothermy, not toward it, and instead toward homoiothermy based on bulk (mass or inertial homoiothermy). This led to my seeing most nonavian dinosaurs as probably no more advanced physiologically than deinocephalian therapsids and, except in a single stock leading to birds, as what I described as failed endotherms (Reid 1984b, 596; 1987, 148–151)—meaning animals sufficiently advanced to have been able to evolve into endotherms, but which took a wrong turn, blocking them from doing so. So, in calling them intermediate animals, I do not mean animals in process of evolving into endotherms, but animals of the kind that Regal and Gans envisaged—neither ectotherms nor endotherms like modern ones, but "a third kind of animal, between the two and different from both" (Reid 1990, 46).

Second Digression: What Do Birds Tell Us?

Birds are of interest for several reasons. Living birds are endotherms, they had dinosaur ancestors, and some authors call them dinosaurs. So, if they are both endotherms and dinosaurs, does this mean that all dinosaurs were endotherms?

It does not, for two separate reasons. First, the practice of calling birds dinosaurs is a result of the rules of cladistic (phylogenetic) classification, under which all taxa (groups) distinguished have to be monophyletic in Hennig's sense. This term originally only meant "having a single common ancestor;" but in cladistic usage, taxa are only monophyletic if they include all descendants of the group's earliest common ancestor—groupings which do not are rejected as paraphyletic. So a taxon Dinosauria has to include birds to be monophyletic sensu Hennig, and birds are, in that sense, dinosaurs. They are, however, only dinosaurs in the same sense that mammals are therapsids, and their being called dinosaurs has no physiological implication.

Second, while modern birds are endotherms, they all belong to the order Neornithes, whose radiation into modern orders began in the Cretaceous and early Tertiary (Feduccia 1995; Clarke et al. 2005). We can assume that neornithine birds were endotherms by the time their evolutionary radiation began; but that tells us nothing about other Cretaceous Ornithurae (e.g., *Ambiortus, Hesperornis, Ichthyornis*) or the then dominant Enantiornithes (or opposite birds), let alone *Archaeopteryx* or forms close to it. Because we call *Archaeopteryx* a bird, it is tempting to see it as warm-blooded and even to extend this to the small feathered theropods now known to have existed; but Ruben (1991) showed *Archaeopteryx* could have flown with only the energy resources available to modern lizards and saw avian endothermy as probably evolved in birds themselves. If so, the original function of feathers in nonavian theropods could have been the retention of heat produced by activity, not endothermy; and, if feathers evolved as is currently thought likely (Sues 2001, Fig. 1), their efficiency as thermal insulation could have been limited initially. Some physiological difference between Enantiornithes and modern birds is also implied by Chinsamy et al.'s (1994, 1995) demonstration of two cases in which their bone was formed slowly—the birds seemingly took several years to reach full size. Variations in the size of *Archaeopteryx* specimens (cf. Benton 1987, Fig. 2; Walker 1985, 131) could be seen as suggesting that they grew throughout life, like modern reptiles.

Thus, modern birds being endotherms tells us nothing certain about the thermal physiology of other kinds of birds or of nonavian dinosaurs. We do know that a single stock of small feathered theropods was able to give rise to both birds and endotherms; but we do not know whether endothermy preceded flight, or flight endothermy. Feathers do imply a capacity for retaining endogenous heat, but they cannot tell us whether the heat source was resting metabolism or activity and so they cannot be taken as evidence of endothermy. However, we can say that, if the endothermy of birds was evolved in birds themselves, their nonavian ancestors cannot have had more than some intermediate type of physiology.

Evidence from Bone

Bone has been claimed to throw light on dinosaurian physiology since early in the present controversy (e.g., Bakker 1972; de Ricqlès 1974, 1976), but its use as evidence involves problems. First, even if some bone tissues are thought to reflect metabolic rates, they can still provide no direct evidence of how temperature was regulated. Second, the only modern tetrapods with which dinosaurs can be compared histologically are ectotherms and fully evolved endotherms, there now being none of the intermediate forms that must once have existed. Third, the same types of bone are now found in both ectotherms and endotherms, which only differ in the extent to which different types are developed. And fourth, in dinosaurs large enough for mass effects, there is no way of telling whether endotherm-like conditions reflect endothermy or simply homoiothermy. A further artificial problem is that nomenclature introduced by de Ricqlès (1974) makes ectotherms and endotherms seem more different histologically than they are (see "thermal physiology: periosteal bone" below).

Histological Background

Bone is a mineralized tissue, formed mainly from collagen and hydrated calcium phosphate (hydroxyapatite) and is normally alive throughout life. It is typically produced by cells termed osteoblasts, some of which are enclosed in it as it grows and are then known as osteocytes. The collagen forms fine fibrils, grouped in bundles on which the phosphate is deposited in microcrystal form. Resulting tissues are mostly divisible into coarsely and finely bundled types, with bundle diameters reaching up to approximately 30 µ in the first case but only 2–4 µ in the second. Minor amounts of bone, described as metaplastic, may also be formed by local mineralization of tendons, ligaments, or cartilage without osteoblasts being present. Texturally, bone forms a continuum between dense and spongy tissues, described technically as compact and cancellous respectively. Compact bone is vascular if it encloses blood capillaries, which occupy spaces called vascular canals, or is avascular if these are absent. Bone-depositing tissues, which coat external and internal bone surfaces, are called the periosteum and endosteum respectively, and bone that they deposit is periosteal or endosteal correspondingly. Bone tissues formed during growth are classed as primary, and others which may replace them later are classed as secondary.

The bone tissues relevant here are periosteal bone, which is primary, and Haversian bone, which is secondary. Periosteal bone is bone laid down under the soft periosteum during growth and is typically compact, although finely cancellous in some marine forms and sometimes initially during growth. It has simple variants of wholly periosteal origin (periosteal bone sensu stricto) and compound ones made partly from bone formed in internal vascular spaces (Fig. 38.6a). Simple periosteal bone can be a texturally uniform tissue (e.g., Fig. 38.7c) or show fine lamellation resulting from

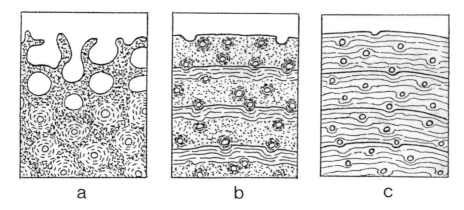

38.6. Fibrolamellar and zonal bone. (a) Fibrolamellar bone, showing its initial formation as finely cancellous bone (top), and subsequent compaction with the formation of primary osteons (lower half). (b) Zonal bone, with lamellated annuli between zones with small primary osteons in a non-lamellated matrix. (c) Zonal bone formed wholly from lamellated tissue, with zones defined by resting lines.

small scale cyclical variations in its microstructure (the fine lines in Fig. 38.6c). Bone formed internally is described as osteonic and builds structures called primary osteons, whose shapes depend on those of the spaces they are formed in. These can form irregular networks, parallel tunnels, radial fissures, or circumferential clefts, and the tissues which result are called reticulate, parallel osteoned, radiate, or laminar correspondingly. In different compound tissues, the content of osteonic bone can be anything from barely detectable to over 70% of all bone present. Bone formed periosteally is typically a coarsely bundled tissue; but osteonic bone is finely bundled and usually of a finely layered type called lamellar bone. The fiber bundles in bone formed periosteally are commonly stratified parallel to the external surface; but in the fastest growing cases they are interwoven without order, and the tissue then formed is called woven bone. This is seen chiefly in embryonic periosteal bone and a postnatal composite tissue called fibrolamellar bone. Growth can be continuous, periodic, or show cyclical alternation of faster and slower growing tissues (Figs. 31.7 and 38.6). In simple periodic growth, successive increments called zones are separated by features termed resting lines (a.k.a., lines of arrested growth [LAGs]), marking pauses in growth (the strong lines in Fig. 38.6c); while alternating growth rates are marked by the zones having features termed annuli built from lamellated tissue between them (Fig. 38.6b). These zones mark the periods of faster growth and the annuli the slower ones. Some such tissues can show both resting lines and annuli. Bone with zones is termed zonal, while continuously formed bone is azonal. In modern forms, cyclical zonation is typically annual, although irregular interruptions of growth can also occur.

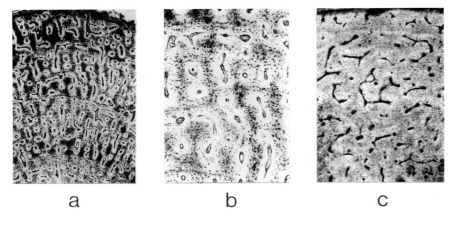

38.7. Bone from two modern reptiles. (a) Bone from a young American alligator, showing obvious compound structure and four weakly defined zones. (b) The same tissue enlarged, showing the unstratified character of the periosteal framework (the dark trabeculae). Similar tissues can form zones between typical annuli in larger examples, as in some *Crocodylus* species (e.g., *C. porosus, C. johnstoni.* (c) Vascular bone without growth rings, from the femur of a half-grown Galapagos tortoise.

38.8. Haversian bone. (a) The start of Haversian bone formation as seen in an *Allosaurus* lachrymal, showing unreplaced periosteal bone at the top of the figure, and resorption spaces and secondary osteons in its middle and lower parts. Three of the secondary osteons only thinly line resorption spaces, being at a stage at which bone redeposition had barely started, but others (e.g., at bottom centre) form typical compact Haversian systems. A resorption space at lower right was encroaching on the central Haversian system below it, illustrating what leads to the formation of dense Haversian bone. For an example of that tissue see Figure 38.9b . (b) Haversian bone in its cancellous form, as seen in an *Allosaurus* ischium. Large, thin-walled, secondary osteons enclose spaces which in life would have contained marrow tissues and their blood supply. A few compact Haversian systems are also seen in places.

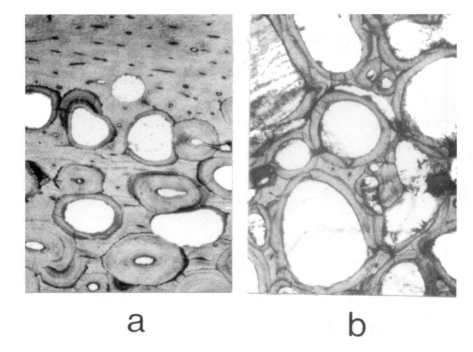

a b

Haversian bone (Figs. 38.8–38.9; Plate 6), in contrast, is a tissue which replaces bone formed during growth and is formed at rates generally unrelated to growth rates. The extent to which it replaces primary tissues has a range from almost total to not at all in different cases. In vascular compact bone, the replacement process begins with some vascular canals being enlarged, from several times to many times their original diameters, by bone resorbing cells called osteoclasts. Resorption then ceases and osteoblasts deposit new bone centripetally to form structures called secondary osteons. These are typically cylindrical with a single central vascular canal and with a sharply defined periphery which marks the outward limit of bone resorption. Their cross sections are the Haversian systems of traditional medical

38.9. How the formation of Haversian bone can reduce vascularity. These figures show fibrolamellar periosteal bone (a) and dense Haversian bone (b) from a dorsal rib of *Tyrannosaurus* at the same magnification. The large number of small primary vascular canals in the periosteal bone contrasts strongly with the small number of secondary ones in the Haversian bone. The figures are from different parts of a single transverse section.

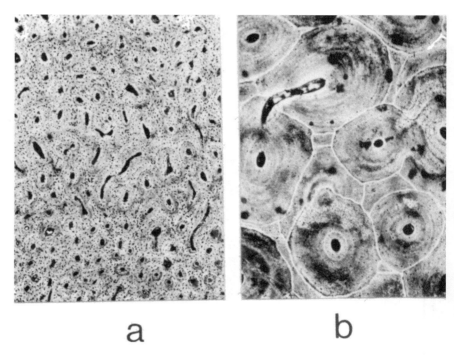

a b

nomenclature, so called from their vascular canals having been first reported (though not as vascular) by the pioneer anatomist Clopton Havers (1691). If the process continues, the remaining primary tissue is progressively replaced and resorption spaces start to encroach on the earliest secondary osteons. The ultimate result is a tissue formed entirely from the last formed secondary osteons and portions of partly resorbed earlier ones. This is called dense Haversian bone or simply Haversian bone. Replacement of periosteal by Haversian bone begins in the interior and spreads outwards, hence becoming progressively more complete inwards.

There are also two variant developments. In one, periosteal or dense Haversian bone can be replaced by a secondarily cancellous tissue, built from thin-walled tubular osteons formed in large resorption spaces. Continued reconstruction can again produce a tissue with a mixture of complete and partly resorbed osteons. In the second, the cancellous endochondral bone produced as bones grow in length (Figs. 31.1 and 31.5) can be resorbed and replaced by a new cancellous tissue with its bony struts (trabeculae) aligned differently, in some cases in ways which reflect local stress fields. These tissues are not usually called Haversian, despite their similar origins.

That is most of what you need to know here, but one warning is also needed – in the source literature, the term laminar is used in four different ways. When Foote introduced it in 1916, he applied it to both zonal bone and a fibrolamellar tissue with cyclically formed circumferential vascular networks, which he saw as showing most clearly what he meant by laminar. When Gross (1934) introduced the term zonal, he restricted the term laminar to the fibrolamellar type. Enlow and Brown (1956) and Enlow (1969) instead called zonal bone laminar and described the fibrolamellar type as plexiform. De Ricqlès (1974 etc.) recognized this confusion, but used both plexiform and laminar for forms of the fibrolamellar type, according to whether radial vascular canals were conspicuous or inconspicuous. I follow Gross because the two types distinguished by de Ricqlès are completely intergrading. There are also two further points worth noting. First, Enlow and Brown (1956) misinterpreted primary osteon as meaning primary vascular canal and called bone with primary osteons protohaversian. Enlow corrected this later (1969). Second, the term fibrolamellar dates only from 1972 and so does not appear in earlier source papers (e.g., Gross 1934; Enlow 1969) even when that kind of bone is described.

Circulatory Implications

Although dinosaurian bone histology has been claimed to imply high metabolic rates, its only certain implication is that they had highly efficient circulatory systems. This is implied by many having been able to form azonal fibrolamellar bone for as long as they grew, even when they reached larger sizes than any known terrestrial mammal (e.g., *Brachiosaurus*: Gross 1934, Fig. 17–19). Two features of this composite tissue make it the fastest growing form of periosteal bone. Its non-osteonic parts are built from woven bone, which grows faster than other bone tissues, and are also formed as finely cancellous bone (Fig. 38.6a), allowing a given volume of new bone to occupy more space than it would if deposited solidly. This gives a maximized radial growth rate, with the tissue becoming compact bone as its primary osteons are formed. And, when formed azonally and mainly from osteonic bone (e.g., Fig. 31.9), it points to growth at rates now found only in large, fast-growing mammals and birds. It then follows that dinosaurs must have had cardiovascular and haemal systems able to support such growth, which would not have been possible otherwise. Tetrapods can be programmed genetically to grow less quickly than they could, as we are ourselves; but none can grow faster than a rate which depends on how fast their circulatory systems can supply growing tissues with the necessary substrates and energy. Hence, the hearts of such dinosaurs must have been able to maintain the required capillary flow rates for as long as fast growth lasted, even when this was to sizes which would cause a massive increase in vascular resistance to blood flow as major vessels lengthened and capillary networks multiplied. Taken with comparable growth now being seen only in forms with complete double-pump circulation (e.g., birds and mammals), this can be seen as collateral evidence for this type of circulation in dinosaurs, reinforcing the haemodynamic argruments of Seymour (1976) and Ostrom (1980).

Three comparisons between dinosaurs and their crocodilian cousins are also of interest. First, the ability of dinosaurs to grow at the high rates implied by their bone suggests that their blood's capacities for the carriage and delivery of tissue-building substrates and oxygen could have been higher than in crocodilians. The blood of alligators, for instance, carries only 40% of the oxygen carried by mammalian blood (Coulson 1984). Higher capacities would reduce the flow rates required to sustain rapid growth, and the work that their hearts would have to do to maintain them. Second, in a study of alligators Coulson and Hernandez (1983) found their blood flow to slow down progressively during growth, in response to increasing vascular resistance and to proportional changes in heart size and aortic diameters. Comparing specimens weighing 1.9 kg and 99 kg, they found that, relative to body weight, the heart of the larger individual weighed less than half of that of the smaller one, while its aorta had relatively less than a sixth of the diameter. For dinosaurs to maintain rapid growth to large sizes, the growth of their hearts and major blood vessels would have had to keep pace with that of their bodies much more closely (Reid 1996). Third, however large crocodilians grow, their hearts do not have to pump blood through large vertical distances. Those of large dinosaurs did and so will also have needed to keep pace with large increases in the distances involved during growth from small hatchlings to large adults. The various demands of fast growth to large sizes on dinosaurian hearts, especially on those of sauropods, point to their having been large with powerful left ventricles. Moreover, the pressures these would have to be able to generate would not be just a matter of the vertical distances through which blood had to be pumped, because vascular resistance to blood flow would also increase progressively and do so in all directions from vertical to horizontal.

Thermal Physiology: Haversian Bone

Haversian bone was the first bone tissue to be claimed to show that dinosaurs had endothermic metabolic rates. It is now seen mainly in mammals, which are endotherms, but was known to be widespread in dinosaurs from a pioneer study by Seitz (1907), who figured examples from genera as diverse as *Plateosaurus, Diplodocus, Allosaurus, Iguanodon, Stegosaurus,* and *Triceratops*. This led Bakker (1972) and de Ricqlès (1974, 1976) to see its presence as evidence that dinosaurs were endotherms. Following medical opinion (Amprino, 1967), both saw it as involved in calcium and phosphate exchanges between bone and body fluids, with its rapid and extensive development implying high exchange rates, and, by inference, endothermic metabolic rates. For Bakker, the key factors were the high vascularity of its dense form and Amprino's view that such exchanges occur in the last formed parts of the latest Haversian systems. De Ricqlès put more emphasis on exchanges in the formative process itself (e.g., 1976, 140–141), in both its dense and cancellous forms, but otherwise argued on essentially the same lines as Bakker.

These arguments are reasonable, but the overall picture is much more complex than these authors made it seem. The primary problems are that Haversian bone occurs in some ectotherms (Plate 6) as well as in endotherms, and also that it does not occur in some mammals and many birds (Bouvier, 1977). Especially, it does not occur in the smallest mammals and birds, which have the highest mass-specific metabolic rates (Fig. 38.5), ruling out any causal connection between it and high metabolic levels (McNab 1978). When seen at all in modern reptiles it is typically restricted to occurrences of scattered Haversian systems, or some replacement of primary compact bone by cancellous bone; but it can be extensively developed in the bony scutes of alligators after sexual maturity (Ferguson 1984). Its dense form also occurs in various fossils which are not likely endotherms, as seen by Bouvier. Seitz (1907), for instance, figured its dense form from the crocodiles *Mystriosaurus, Metriorhynchus,* and *Goniopholis,* and from the sea lizard *Mosasaurus*; and I found it developed to human levels in the tortoise *Geochelone triserrata* (Reid 1987, Fig. 3b). Incompletely developed, with periosteal bone persisting between the osteons, it was found by Enlow and Brown (1957) in the early amphibians *Eryops, Edops, Trimerorhachis,* and *Archeria,* and in the primitive reptile *Limnoscelis*. The process of bone resorption and redeposition, on which replacement of primary bone depends, is even known from the cellular bone of cephalaspid agnathans and the acellular bone (aspidin) of pteraspids (e.g., Moy-Thomas and Miles 1971, 19, 42). Thus, this process dates from almost the start of known vertebrate history, and has not even a historical connection with the origins of endothermy.

A second group of problems then arises from various nonthermal factors being known or thought to trigger the formation of Haversian bone, and by its level of development sometimes varying in ways with no evident explanation. Because of its absence in small endotherms, it seems to be somehow dependent on size and so presumably adult weight. Some authors (e.g., Ostrom 1980) have seen its function as mechanical strengthening, but the evidence is ambiguous. In the head of the human femur, secondary trabeculae replacing cancellous endochondral bone are arranged in ways reflecting both compressional and tensional stresses (e.g., Pritchard 1979, Fig. 2, 25, 26); but, in the limb bones of cattle, Currey (1959) found the tensile strength of secondary osteons to be less than that of their primary bone. In mammals, dense Haversian bone tends to be most extensive when primary vascularity is low, as it is in humans after infancy and less developed when it is high; but this is not always so. Enlow (1962a) found it sometimes developed to reinforce muscular attachment when muscles are attached in areas affected by metaphyseal remodelling (cf. Fig. 31.3) and as a replacement for dead bone in areas where osteocytes have died. Ruth's (1957) study of mammals convinced him that its general extent and rate of development in them are both controlled genetically.

Dinosaurs themselves then add further elements to the puzzle. The papers of Bakker (1972) and de Ricqlès (1974, 1976) cited above can give an impression that dense Haversian bone is always extensively developed in dinosaurs; but this is not so in genera I have sampled, in which its level of development in major limb bones varies from the highest seen in mammals to as low as is now usual in crocodiles. In examples with medullary cavities, from forms as big as *Allosaurus* or *Iguanodon*, midshaft sections of femora or tibiae can show it limited to occasional isolated osteons seen near the medullary margin, or even altogether absent. It can, however, also form radial blocks of dense Haversian bone under areas where major muscles were attached, especially when bone formed at the surface was metaplastic bone ossified from fibrous tissue attaching the muscles (e.g., Reid 1984, Fig. 19, 22, 36, 37). In such cases its development was clearly not related to thermal physiology. Other bones can instead show it generally and extensively developed, sometimes leaving only superficial remnants of periosteal bone; but all major limb bones in which I have found it extensive have been cored by secondary cancellous bone, grading outwards into dense Haversian bone, instead of by a simple medullary cavity. As a final twist, sex may have sometimes been involved. In American alligators, Ferguson (1984) found that Haversian replacement of primary bone in scutes was more extensive in females than in males. He thought this could be due to their use as a source of the material used to build eggshells. In a recent study of the Tendaguru sauropods *Barosaurus*, *Brachiosaurus*, *Dicraeosaurus*, and *Janenschia*, Sander (2000) found major limb bones of *Barosaurus* to be histologically dimorphic, showing two different growth styles and two correlated levels of Haversian bone development, with the higher one higher than in any of the other three genera. Like Ferguson, he saw this as possibly reflecting egg formation in females.

Haversian bone thus provides no sure evidence of dinosaurs having the high metabolic rates of modern endotherms. If it were restricted to endotherms the matter would be different, despite its absence in the smallest forms; but it is not, and, judged from the fossil record, never has been. Indeed, the processes of bone resorption and replacement which lead to its formation are even known from early jawless fishes, whose modern relatives are lampreys and hagfish. Add that we now have no living intermediate forms for comparison, and what controlled the production of Haversian bone in dinosaurs is currently best treated as unknown. I can, however, offer a suggestion. During the growth of limb bones, the marrow can only expand radially at the expense of preexisting periosteal bone, which has to be resorbed to provide extra space; and, in some of my samples, low and high levels of Haversian bone development seem to correlate with whether this was provided by formation of a simple medullary cavity or a core of secondary cancellous bone, grading outward into Haversian bone. Which way the needed extra space was provided was presumably determined genetically and not by thermal physiology. This suggests that the same could have been true of all Haversian bone development in dinosaurs, except when other special factors triggered it.

Thermal Physiology: Periosteal Bone

Fibrolamellar periosteal bone in its laminar form was mentioned briefly by Bakker (1972), but only in the context of Currey (1960) having found it still more vascular than dense Haversian bone. The corollary, not mentioned by Bakker, is that replacement of fibrolamellar bone by Haversian bone reduces vascularity (Fig. 38.9). This is because the secondary osteons of Haversian bone are typically larger than comparable (i.e., cylindrical) primary osteons, and sometimes much larger. As a result, the vascular canals become fewer and more widely spaced—the more so, the larger the osteons. This suggests that the high vascularity of dinosaurian fibrolamellar bone was more concerned with sustaining fast growth than with subsequent ionic exchanges.

In contrast, de Ricqlès (1974, 1976) saw periosteal bone as providing a way of distinguishing between fossil ectotherms and endotherms, because it records the ways in which tetrapods grow (cf. Chapter 31, this volume, last part). As he saw, its histology varies according to growth rate. When growing most slowly, periosteal bone forms simple, sparsely vascular or avascular bone, while its fastest growth produces highly vascular fibrolamellar bone. Modern reptiles grow only at slow to moderate rates and often show periodic growth, producing zonal bone; but most mammals grow continuously for as long as growth lasts, unless events in their lives interrupt growth. Many can grow quickly to large sizes, forming azonal fibrolamellar bone. So could dinosaurs, and this led de Ricqlès to see their fibrolamellar bone as evidence of their having been endotherms. He was careful to point out that such bone is strictly only evidence of growth at rates now seen in large, fast-growing mammals and birds; but, because such growth is now seen only in endotherms, it can be thought to be only possible for endotherms.

This argument can seem reasonable at first sight and has been followed by some authors, but it does not stand up when viewed critically. Its basic flaw is that fibrolamellar bone can have no causal connection with endothermy. It does not form the bones of the smallest kinds of mammals and birds, despite their having the highest mass-specific metabolic rates (Fig. 38.5); and we ourselves do not form it after infancy despite being endotherms. Our rate of growth is controlled by genetics, not our thermal physiology.

The main snag, however, is in there being two critical errors in de Ricqlès' picture of the bone tissues concerned. Those he called lamellar-zonal were depicted as formed entirely from lamellated simple bone, with no content of compound bone (de Ricqlès 1974, Fig. 1), and fibrolamellar bone was said to be formed quickly and *continuously* (1974, 53, my italics). But Enlow (1969, 55) recognized the zonal bone of some turtles and crocodiles as partly built from fibrolamellar bone, which was formed periodically and not continuously. He did not call it fibrolamellar because that term was not then in use; but his description of how it forms (1969, 55–57) is unmistakable, and he figured examples showing simple primary osteons from a fresh water turtle (or terrapin), a crocodile, and an alligator (1969, Figs. 12, 13, and 15). De Ricqlès's picture of two sharply distinct tissues was thus misleading and made modern ectotherms and endotherms seem more different histologically than they are. While histology and depositional continuity do interact, they are properly separate matters and do not have the fixed correlation he implied. More importantly, these two papers give very different pictures of what can be inferred from fibrolamellar bone. By treating it as always formed continuously, which now only happens in endotherms, de Ricqlès made its presence seem to be evidence of high metabolic rates. But Enlow showed that it can form at rates no higher than those experienced by turtles and crocodiles when they are growing most quickly—thus at any rates between theirs and those of endotherms. The perspective provided by Enlow seems to be the one with which we need to view it in dinosaurs.

There are two further complications. First, use of the term lamellar-zonal can give the impression that tissues called by this name are built from lamellar bone sensu stricto (i.e., lamellar finely bundled bone: Pritchard 1950, 12); but grossly lamellated tissues of the type shown in de Ricqlès (1974, Fig. 1) are usually built from stratified (parallel-fibered) coarsely bundled bone, as recognized by Enlow (1969, 49–50, 57–58) in the type he called pseudolamellar. This is why I call such tissues lamellated, not lamellar, and do not use the term lamellar-zonal (Reid 1990, 24). Second, when he wrote his 1974 paper, de Ricqlès thought that bone with zonal growth rings was absent in dinosaurs, claiming that it had long been known that their compact bone was either fibrolamellar or Haversian (1974, 63) and describing their periosteal bone in all investigated dinosaurs, including carnosaurs, prosauropods, sauropods, and ornithopods, as showing very high rates of continuous (not cyclical) growth (1974, 65). In his 1980 paper, he still held that zonal bone does not occur in them, but I showed that it does, first in a sauropod (Reid 1981) and then in the other groups (1987, Fig. 5; 1990, Figs. 1–28). This changed the perspective in which his argument should be viewed.

While such new data show dinosaurian periosteal bone histology to be more complex than he thought, they are essentially matters of detail, not of principle. In the latter field, I see the main question as whether histological data from modern endotherms can really be used to identify fossil ones. I strongly suspect that they cannot be. De Ricqlès (1974) thought that they could, and he and his followers still treat fast growth as pointing to high metabolic rates; but in this they seem to me to rely on the 19th Century notion that the present is the key to the past, which dates back to the time of Charles Lyell and other early geologists. This concept is of undoubted value as applied to purely physical phenomena, such as erosional cycles or plate tectonics, which have been operating since somewhere in Precambrian times; but is it equally applicable to data from modern evolutionary end forms, with long histories of progressive anatomical change behind them? The snag in assuming this is that all modern tetrapods are either ectotherms or fully evolved endotherms, with none of the intermediate forms which must once have existed for endotherms to have evolved from ectotherms. Therefore it seems unlikely that none of these intermediates is represented in the known fossil record. Fast continuous growth to large sizes is now seen only in endotherms, but is it only possible for endotherms or is it an ability inherited from preendothermic ancestors who now have only endothermic descendants? We do not know and currently have no way of finding out. It could be either, and all we can do is to try to assess which seems more likely in terms of the evidence we have. For instance, in the lineage leading from the earliest theropods to neornithine birds, it could be that progressive anatomical changes were paralleled by progressive physiological ones, rather than endothermy dating from the earliest forms to grow quickly. So, in dealing with a clade with its roots in the early Triassic, and no known endothermic members but neornithines, I prefer to remember that the past was the father of the present and to limit assumptions based on modern forms to features shared by ectotherms and endotherms. Fibrolamellar bone, for instance, reflects fast growth in both ectotherms and endotherms, but it does not reflect high metabolic rates in ectotherms.

Turning to the details of periosteal bone histology in dinosaurs, we can first note that periosteal bone is not restricted to the two types described by de Ricqlès (1974). These were simply the tissues that he saw as especially distinctive of ectotherms and endotherms respectively. Bone tissues that tetrapods can form periosteally include lamellar bone sensu stricto as well as other simple and compound tissues whose periosteal content is non-lamellar (Table 38.1). In dinosaurs whose bone I have sampled (list: Reid 1996, 25), lamellar bone sensu stricto occurs mainly in primary osteons, although also sometimes in annuli, and in tissues in which bone formed periosteally was non-lamellar form—a complex of intergrading simple and compound types, of which any could be formed periodically (zonally) or continuously (azonally). These tissues are of three main histological types:

1. Uniform or lamellated simple tissues, built from stratified (parallel-fibered) bone.
2. Compound bone with primary osteons in a stratified periosteal tissue.
3. Typical fibrolamellar bone, in which the periosteally formed groundmass is unstratified woven bone, except in laminar variants in which cyclical formation of closely spaced circumferential vascular networks could impose a secondary stratification.

Judged from similar modern examples, these three types reflect low, moderate, and high growth rates respectively. Vascularity is typically high, but simple tissues can also be sparsely vascular or avascular. Azonal formation of fibrolamellar bone predominates in limb bones; but they can also show zonal formation of this tissue or one of the slower growing ones, or start growth azonally but switch to a zonal pattern later. Species whose limb bones grew azonally can show zonal patterns in other bones; and some bones which grew asymmetrically can show zonal and azonal growth on opposite surfaces, with these conditions intergrading between them. Occasionally, asymmetrical growth led to different types of tissues intergrading laterally. The type of tissue formed could be the same throughout growth, or could change with age from a faster to a slower growing type through part or all of the range noted above. One can also find cases in which active growth stopped abruptly, to be followed by accretion of avascular lamellated simple bone which presumably formed slowly. In azonal tissues, slowing growth could cause progressively reduced vascularity; but identical vascular patterns can also occur in tissues formed at different speeds. In zonal tissues, slowing growth could instead lead to the width of zones decreasing progressively without the type of bone formed changing, implying that the periods of active growth became progressively shorter.

Table 38.1. How periosteal bone tissues seem related. For me, different combinations of the varying conditions shown here produce a complex of intergrading tissue types, of which any can be formed either zonally or azonally. Lamellation here means the type seen in Fig. 38.8c, called pseudolamellar by Enlow (1969). Bone-forming primary osteons is typically lamellar bone sensu Weidenreich, but can also be a non-lamellar tissue, or of both types with the lamellar bone innermost.

Description	Simple	Compound
Vascularity	Present or absent	Always present
Primary osteons	Absent	Present
Lamwellation	Present or absent	Absent
Periosteal bone sensu stricto	Stratified ("parallel-fibered")	Woven
Zonation	Present or absent	

The periosteal bone histology of nonavian dinosaurs was thus not as simple as de Ricqlès (1974) supposed, and I prefer to assess it in terms of what we now know about it. Summed up briefly, we are dealing with animals which could grow in both mammal-like and typically reptilian manners, sometimes doing both in different bones, and which formed a range of periosteal bone tissues roughly equivalent to those now seen in crocodiles, ratite birds, and large, fast-growing mammals together. This combination is not typical of any group of animals now living and is my primary reason for seeing dinosaurs as somehow unique physiologically. But, just what do these tissues tell us about their physiology, and, equally important, what can they not tell us?

Taking fibrolamellar bone first, there is no doubt that this tissue implies fast growth, which it shows in all known modern cases; but the notion that its formation implies high (endothermic) metabolic rates does not fit Enlow's finding (1969, 55–57) that some turtles (Plate 5) and crocodiles form it during periods of rapid growth. Some authors (e.g., Paul 1991) have claimed that this only happens in farmed examples raised in ideal conditions; but the young wild American alligator whose fibrolamellar bone was figured in this book's first edition (Reid 1997, Figs.32.7a,b) here (Figs. 38.7a, b) was from North Carolina, where the pools in which these alligators live can freeze over in winter. Padian et al. (2004, 565–566) called it anomalous because it did not fit their phylogenetic views; but they thought it a single isolated specimen, collected by me, when it was one of the Ferguson specimens noted in this context in the first of my two papers which they cited (Reid 1996, 46). But, more to the point, it is also now 25 years since Ferguson showed that wild alligators form fibrolamellar bone periodically, as an incidental result of his work on alligator palates.

Starting in 1959, workers researching growth rates in wild Louisiana alligators, for which their ages needed to be known, began marking young examples of known ages (mostly hatchlings) and releasing them into the wild for recapture and study in later years. Twenty years later, Chabreck and Joanen (1979) showed a rough correlation between age and length; but this approach worked best with young examples and less well with older ones. Because Ferguson needed accurate ages for examples of any age, he instead used counts of periosteal growth rings seen in their bones, first comparing ring counts with known ages (1–25) in 20 recaptured specimens, to confirm the rings as annual and to find a means of allowing for rings lost as medullary cavities expanded (Ferguson 1982; 1984, 252–253; Ferguson et al. 1982, 275–276). In doing this, he found the rings to result from a cyclical alternation of thin annuli and thicker vascular zones, in which the canals were enclosed within primary osteons. By injecting live specimens with marker dyes, he showed that bone in the zones was formed during summer growth periods. He did not call this bone fibrolamellar, because his interest was in alligator ages, not the thermal physiology of dinosaurs; but he showed that it was histologically, and I called it by that name after figuring one of his specimens to show this (Reid 1984b, Fig. 1d, 594). In 2004, the number of wild alligators from which fibrolamellar bone was known was not one but 21, including 20 recorded by Ferguson (1984, table I); and Tumarkin-Deratzian 2007) has recently added three more, this time from Florida. Other species in which similar bone is formed include *Crocodylus porosus* and *C. johnsoni* (Ferguson personal communication 1982). So, I think we can be sure that the specimen I figured in 1997 and here (Fig. 38.7a, b) was not anomalous and that fibrolamellar bone can be formed at any metabolic rate between those of modern (neornithine) birds and those of American alligators.

Dinosaurs, admittedly, could form fibrolamellar bone continuously, in a manner now seen only in large, fast-growing endotherms; but we do not know what progress, if any, toward endothermy must be made before such growth becomes possible. And if the first forms for which it was possible were not endotherms, as evidence from therapsids suggests, it need not be evidence of high metabolic rates in dinosaurs. Therapsids were advanced mammal-like reptiles (Synapsida), which differed histologically from their pelycosaur precursors in typically having fibrolamellar periosteal bone, but in primitive genera show no other signs of being endotherms. Especially, forms called dinocephalians, which were dominant in the first (early Late Permian) faunas. They were as advanced histologically as modern mammals (de Ricqlès 1974, 63), but had none of the skeletal features implying increased food requirements and respiratory rates as seen in the most anatomically advanced later forms (bauriid therocephalians and cynodonts). As seen in the cynodont precursors of mammals (Brink 1956; Jenkins 1971; Kemp 1982; Hillenius 1994), these included:

1. Multicusped teeth, allowing food to be chewed before swallowing, and showing mammal-like occlusion in the most advanced forms (e.g., *Cynognathus, Diademodon*).
2. Related patterns of jaw suspension and musculature.
3. Complete false palates, allowing them to breath while feeding – implying this was needed.
4. Special nasal turbinals (maxilloturbinals), concerned with controlling respiratory water losses caused by raised metabolic and respiratory rates.
5. Shortened posterior lumbar ribs, suggesting the presence of a diaphragm.

None of these features is present in dinocephalians despite their mammalian bone histology, throwing serious doubt on whether the latter can be evidence of high metabolic rates. After comparing rival arguments for their having been endotherms (Bakker 1975) or only inertial (mass) homoiotherms (McNab 1978), Kemp (1982, 96–98) judged that the truth could lie somewhere between the two, assessing them as probably having "a metabolic rate above that of typical modern reptiles, but still well below that of mammals," and thought mammalian thermoregulation "most improbable." If this judgment is correct, as I think likely, they can be seen as animals which had taken a first step along the road to endothermy but were still a long way from attaining it. This is the most that fast mammal-like growth need mean in dinosaurs. What made this first step possible can now only be guessed; but it would probably be something which must happen before endotherms can start to evolve, and could be the evolution of double-pump circulation. It does not seem to have been the evolution of hair, because the skin of *Estemmenosuchus* shows no sign of it (Chudinov 1968).

To dinocephalians we can also add a number of still older forms, which again had fibrolamellar bone but whose having been endotherms is doubtful. Enlow and Brown (1957, 204 and plate 25, Fig. 3) found an obvious fibrolamellar tissue (their protohaversian type) in long bones of the pelycosaur *Ophiacodon*, which seems to have been a semiaquatic fish eater. Enlow (1969, 71) noted similar tissues as occurring in the cotylosaurs *Labidosaurus* and *Captorhinus* and the rhachitome amphibian *Parioxys*, specifically pointing out their being of a type found in mammals. Enlow and Brown's figure of bone from a *Labidosaurus* rib (1957, plate 15, Fig. 1) shows a tissue with obvious primary osteons, indistinguishable histologically from fibrolamellar bone figured by de Ricqlès et al. (2003, Fig. 6a) from the Early Cretaceous bird *Confuciusornis*. Their figure of bone from a *Captorhinus* jawbone (Enlow and Brown 1957, plate 15, Fig. 2) shows a similar tissue, also figured more clearly by de Ricqlès and Bolt (1983, e.g., Fig. 9b); and while their *Parioxys* figure (1956, plate 11, Fig. 3) shows only an azonal vascular tissue, a zoned fibrolamellar one was figured by de Ricqlès (1978, plate 19, Fig. 3) from the stereospondyl amphibian *Stenotosaurus* and closely matches one I have figured from a sauropod (Reid 1981, Fig. 1c). None of these genera shows any other sign of being endotherms, which in any case seems unlikely in amphibians. That a primitive reptile (*Labidosaurus*) and a feathered, avian-grade archosaur (*Confuciusornis*) could have had similar growth rates is believable; but whether they had similar types of thermal physiology could be doubted.

This leads to the question of whether the diversity of groups in which fibrolamellar bone occurs reflects its independent evolution in each of them, or reflects iterative manifestations of a common potential for forming it. There is no certain answer to this question, but their diversity seems to make the latter more likely. Independent evolution would have had to have occurred in at least temnospondyl amphibians (rhachitomes and stereospondyls), captorhinid cotylosaurs, ophiacodont pelycosaurs, basal therapsids, chelonians, and archosauromorphs. In the latter it would also appear to have been evolved twice, first in the prearchosaurian erythrosuchids and later in forms leading to dinosaurs. Because nearly all known Mesozoic mammals are well below the minimum size above which modern ones form it, post-Cretaceous ones could have had to reevolve it. It seems to me more likely that a potential for forming fibrolamellar bone has been inherent (or plesiomorphic) in tetrapods since early in amphibian history, before the divergence of anthracosaurs and temnospondyls, and has since become activated whenever some factor or factors have made rapid growth beyond some critical size possible. In mammals, its postnatal occurrences seem to mark extensions of an embryonic mode of bone formation into later life (Reid 1990, 43–44) and this could also apply to other cases.

The second major question is what dinosaurian zonal bone can tell us, but we first need to note a new, recently discovered complication. In a study of Tendaguru (Tanzanian) examples of the sauropods *Barosaurus*, *Brachiosaurus*, *Dicraeosaurus*, and *Janenschia*, Sander (2000) found that fibrolamellar bone they formed continuously can have a hidden zonation, in the form of a new kind of cyclically formed growth lines which he calls polish lines. These are only seen when polished sections are viewed by incident light and with bright field illumination, and mark thin bands of bone whose hardness and related reflectivity are slightly less than in the rest of the tissues (Sander 2000, Fig. 1, 2). Their formation appears to have been annual, as zonation typically is in modern reptiles, and to mark periodic disturbances of the process of bone mineralization without tissue formation being affected. How widely they occur in other dinosaurs is currently unknown, but they add a new factor to our puzzle. Until 2000, dinosaurs which formed fibrolamellar bone continuously appeared to be unaffected by climatic cycles, but now it seems that even large sauropods were affected.

Turning to typical zonation, it needs to be emphasized that this is not just a matter of bones showing growth rings when sectioned. These can occur in any individual tetrapod whose growth was interrupted by adverse events or conditions, such as injury, illness, starvation, or climatic events, and then be marked by resting lines or annuli identical with those seen in zonal bone. But, whereas the latter are typically cyclical and constantly present, examples which only mark events in individual lives are neither and can be spotted in fossils by comparing bones from different individuals. In dinosaurs, examples are known from the Cleveland-Lloyd allosaurs (Reid 1996, 34). Bones can also show noncyclical growth rings marked only by slight color changes, due to different levels of diagenetic staining. These presumably reflect slight depositional changes in the bone's mineral microstructure, but what caused such changes is unknown. In small dinosaurs, such features can be difficult to distinguish from true zonal ones; but in larger forms they are typically few in number (e.g., 1–3) and widely spaced, unlike those seen in tissues figured here as zonal (Figs. 38.10–13).

This leads to the question of what causes zonation in periosteal bone. Because its cyclicity is typically annual in modern reptiles, apart from minor aberrations due to zones sometimes being duplicated, it has often been assumed to be simply a matter of seasonally changing conditions affecting the growth of animals dependent on external heat sources, as ectotherms are (e.g., Peabody 1961); but this may not be the whole story. Farming methods can double the natural growth rates of American alligators (Joanen and McNease 1976); but examples kept at constant temperatures and with a constant food supply are said still to form bone with zones and annuli, which only become less distinct instead of being eliminated (M. W. J. Ferguson

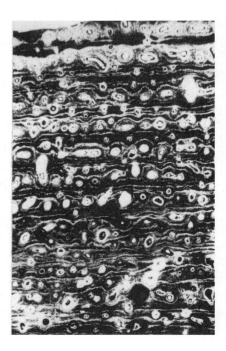

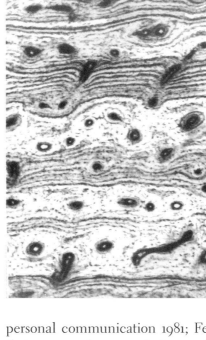

38.10. Zonal bone formed from alternating layers of fibrolamellar compound bone and lamellated simple bone, implying a cyclical alternation of periods of faster and slower growth. The tissue figured is part of a small patch of zonal bone found in an *Allosaurus* radius, in which it seems to have marked a local manifestation of an endogenous (genetically controlled) cyclicity which was normally suppressed.

38.11. Zonal bone from the pubis of a sauropod dinosaur, from the Lower Bajocian of Northamptonshire, England. This tissue is strikingly similar to that figured by Gross (1934, Figure 4) from *Nothosaurus* as typical of reptilian zonal bone.

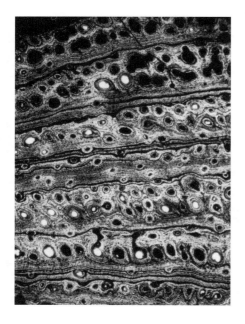

38.12. Bone with zonal growth rings from a rib of an allosaur-sized carnosaur, from the Wessex Formation (Lower Cretaceous, Wealden, ?Barremian) of the Isle of Wight, England.

personal communication 1981; Ferguson et al. 1982, 276). This suggests that some endogenous factor could also be involved in their production, perhaps in the form of a genetically inbuilt biological clock, timed to run in time with seasons but not directly controlled by them (Reid 1983, 401). This is speculation only, but is of interest for two reasons. First, if it were so, transitions from zonal to azonal growth could occur through a gene or a group of genes switching off, as those for tooth production have in birds. Second, some occurrences of zonal bone in dinosaurs do not fit its reflecting climatic cycles.

If one looks only at long bones and relies on comparisons with modern species, it is possible to see dinosaurs as having included both ectotherms and endotherms, distinguished by growth of their periosteal bone being zonal and azonal respectively. Contrasting forms showing both types of growth occur at all latitudes, and can also be found together in some localities, then presumably representing forms which lived in the same conditions (Chinsamy et al. 1998, 388). The theropod *Baryonyx*, for instance, which formed zonal bone (Reid 1990, Fig. 15, 16) was found at a locality (Smokejacks Pit in Surrey, England) which also yielded bones of an *Iguanodon* which grew azonally. Near Grange Chine in the Isle of Wight, the bones of a similar *Iguanodon* occur with those of the large carnosaur whose zonal bone is shown here (Fig. 38.12). But such tissues cannot reflect different types of thermal physiology when both occur in different bones of one animal, as in *Megalosaurus* from Woodstock near Oxford (Fig. 38.14), or on opposite sides of an *Allosaurus* ischium (Reid 1990, 29), when what controlled each of them when they were formed must have been different. The best clue we have to what this was comes from bones which grew asymmetrically. When their periosteal bone was zonal, the zones are thickest on the side that grew fastest. Some also show resting lines (LAGs) between zones becoming less distinct as zones thicken, until a zonal tissue passes laterally into an azonal one. What then controlled whether bone formation was zonal or azonal was clearly not thermal physiology but the genetically controlled local growth rates responsible for asymmetric growth; and, while this does not prove that contrasting occurrences in different bones were also controlled genetically, it does not seem unreasonable to think so when all their other differences reflect differences in genetic blueprints. We then seem to be dealing with animals genetically programmed to form periosteal bone azonally when or where its rate of accretion was above some critical level, but also able to form it zonally when or where the rate was lower. If so, this was a notable difference between dinosaurs and endothermic mammals, in which slow growth does not lead to a switch to zonal growth patterns.

On this basis, I see the zonal bone of nonavian dinosaurs as one of two principal keys to understanding nonavian dinosaur physiology, of equal importance to their azonal fibrolamellar bone and pointing to their having been animals with no true modern physiological counterparts. Horner et al. (1999) have tried to discount it as evidence by claiming resting lines (their LAGs) to be plesiomorphic features, on which no reliable conclusions can be based; but my argument is based on its cyclicity, not the simple occurrence of LAGs, on its wide distribution in such dinosaurs, and on their ability to form zonal and azonal at the same time during growth, and presumably at the same metabolic rate. To be plesiomorphic, moreover, features need to be genetically heritable; and I doubt that the general liability of tetrapods to having growth interrupted by adverse events or conditions has any specific genetic basis. It seems more likely to be a matter of physio-biochemical parameters being breached, as in the cases of death from either hypothermia or hyperthermia but with less serious consequences. Interrupted growth in an elk from Montana (Horner et al. 1999) is likely to have been of this type, and, if it were annually cyclical, so are cold winters in Montana. Chinsamy (1998, Fig. 4) has figured zoned bone from a polar bear but these animals spend much of their lives in freezing conditions in a region of annual polar darkness and related fluctuations in food supplies. Sander and Andrássy (2006) have recorded examples from large mammals from the Pleistocene of Germany, including reindeer, wooly rhinoceros, and mammoths; but these lived during the Weichselian (last) glaciation and so again in cold climates. Other examples listed by Klevezal and Kleinenberg (1969) are almost all from small mammals from cold regions, many of which hibernate, and marine forms which live in or visit cold seas. In contrast, none of the 41 dinosaurs from which I (Reid 1990) figured zonal growth rings lived in comparable conditions, so far as can be judged from other relevant evidence. The sauropod, whose zonal bone is seen here (Fig. 38.11; Reid 1990, Fig. 1), lived in what is now Northamptonshire, England, in a period when reef corals could build patch reefs in the nearby Cotswold Basin. *Baryonyx* (Reid 1990, Fig. 15, 16) and an Isle of Wight carnosaur (Fig. 38.12) shared their Wealden (Early Cretaceous) environment with the crocodiles *Goniopholis* and *Bernissartia*; while the Late Cretaceous ornithopods *Orthomerus* (Reid 1990: Figure 23) and *Rhabdodon* (Fig. 38.13) lived in lands lapped by the tropical ocean Tethys and so presumably in a climate like that of modern southeast Asia and Indonesia. In North America, Late Cretaceous dinosaurs from the Judith River Formation of Montana shared their world with the giant crocodile *Deinosuchus*, as did those of the Dinosaur Park Formation of Alberta with *Leidysuchus*. Thus, Horner et al. (1999) and Sander and Andrássy (2006) use mammals that lived in cold temperate, subarctic, or

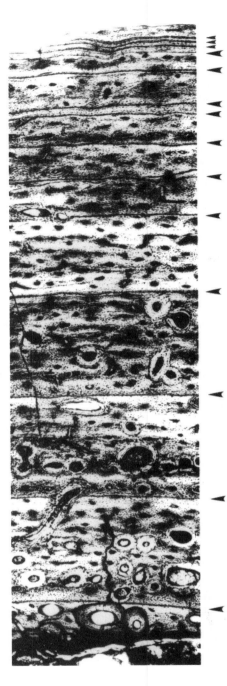

38.13. Bone with zonal growth rings from the femur of the ornithopod *Rhabdodon,* from the Maastrichtian of Szenpeterfalva, Romania. Progressive thinning of the rings toward the external surface (at top) implies progressively slowing growth, like that seen in modern crocodiles, although the thickness of the rings shows growth as faster than in any known crocodile. The specimen was collected by Baron Ferenc Nopcsa, before World War I.

38.14. Contrasting styles of periosteal growth in bones of *Megalosaurus*. (a) zonal (periodic) growth in a pubis. (b) azonal (continuous) growth in a femur. When found in different bones from one animal these conditions cannot reflect different types of thermal physiology, as supposed by de Ricqlès in his early work (1974–1980), and point to which pattern was followed having been controlled genetically, assuming that interruption of growth by some exogenous factor would have affected both bones. Figure a also illustrates zonal bone formed almost wholly from fibrolamellar bone, and progressive reduction in the thickness of zones late in life, showing growth and then slowing progressively. This change occurring without change in the type of bone formed points to reduction in the length of the periods in which bone was formed actively, and not a change in the rate of its formation during them. Large resorption spaces in the lower part of the figure show Haversian bone as apparently still spreading when the animal died.

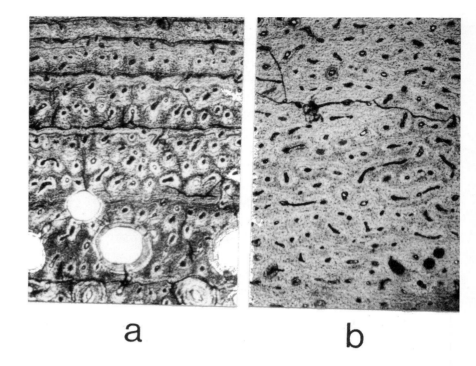

a   b

arctic conditions as a basis for passing judgment on dinosaurs that lived in warm temperate, subtropical, or tropical ones; I think they have misled themselves by doing so.

Last, it remains to compare what periosteal bone can and cannot tell us. In brief, it tells us: (1) nonavian dinosaurs could form fibrolamellar bone continuously, as now only happens in endotherms, (2) they could also form periosteal bone periodically, like typical reptiles, (3) they could show both styles of growth in different bones or even in different parts of the same bone. But that is all. What it cannot tell us includes:

1. Whether nonavian dinosaurs had the high metabolic rates some claim.
2. Whether fast growth to large sizes required a high metabolic rate or only cardiovascular, haemal, digestive, and respiratory systems able to sustain it.
3. What their standard metabolic rate was—in terms of liters of O_2 used per kg body weight per hour.
4. What their core temperature was—in degrees Celsius.
5. How they controlled body temperature.
6. Whether small species could maintain daytime temperatures overnight.
7. How much of the heat in their bodies, at any given size, was endogenous or exogenous in origin.
8. What proportions of endogenous heat were derived from basal metabolism, activity, or other functions, including growth.
9. How much of the body heat in large herbivores was generated by gut floras—without expenditure of energy.
10. What proportions of energy available from their food were devoted to fueling basal metabolism, activity, and other functions, including growth.

11. What their average daily food intake was—in terms of kg of food per kg body weight.
12. What the calorific value of that food was—per kg.
13. What proportion of the energy potentially available from their food was recovered in digestion.
14. How tolerant they were to hypothermia.

All these things need to be known before we can build a full picture of nonavian dinosaurian physiology, and periosteal bone cannot tell us any of them. It does show that they must have differed somehow from all modern reptiles and, in my view, probably also from any living tetrapods; but, in terms of all we need to know about them, this is barely the tip of the iceberg.

Other Evidence

In mammalian limb bones, large parts of the surfaces of their shafts can be formed by bone of endosteal origin, instead of by periosteal bone. This is the result of an external remodelling process called lateral drifting, in which periosteal bone is resorbed on one side of the shaft while being added on the other (Enlow 1962b, Figs. 7–9). In dinosaurs, in contrast, exposed endosteal tissues are usually seen only in metaphyseal parts, if at all, as in modern reptiles. In this respect, dinosaurian bone histology is typically reptilian and not mammal-like. Occurrences of different growth rings counts in different bones (Horner et al. 1999) were also caused in the same way as in crocodiles.

Isotopic data can potentially be used to infer the temperatures at which bone tissues were formed; but they cannot show whether the heat sources involved were external or internal, and they may not record how much temperature varied if bone deposition was not continuous (Reid 1997, box 2). Results can also be affected by diagenetic alteration or by the isotopic content of meteoric water during life (e.g., Thomas and Carlson 2004).

Cavernous Bones

Cavernous bones have no direct bearing on thermal physiology, but have features which point to nonavian saurischians having birdlike respiratory systems. This in turn suggests that they also had aerobic activity metabolism.

In sauropodomorphs and nonavian theropods, cavernosity occurs in vertebrae and ribs, but is currently not known from limb bones—the hollow ones of theropods having seemingly been marrow-filled (Reid 1996). In vertebrae, it is seen mainly in those of the neck and trunk regions, but can also occur further back. It can occur in both external and internal forms, or be purely internal apart from surfaces being pierced by small foramina. External cavernosity is best known from sauropod vertebrae in which deep indentations called pleurocoels in the sides of the centra can leave only a thin partition between those of the opposite sides in the middle parts. From these spaces smaller cavities run into the condylar parts (e.g., Janensch 1947, Figs. 1, 4: *Brachiosaurus*). In a contrasting condition common in nonavian theropods (e.g., *Allosaurus*: Fig. 38.15), there is little or no external indentation, but each side of the centrum has a small foramen leading into an internal complex of cavernous spaces, aligned more or less longitudinally. These foramina are also called pleurocoels, but are more comparable with

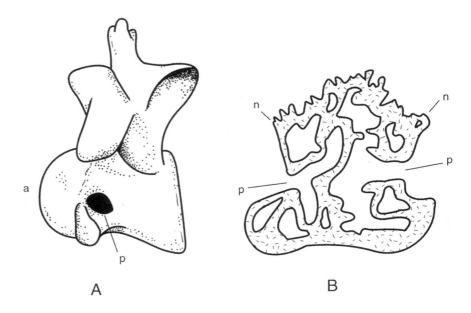

38.15. (A) Cervical vertebra of the theropod *Allosaurus,* showing a pleurocoel (p) in the form of a small lateral perforation; a = anterior. (B) Transverse section of a detached centrum through the pleurocoels, showing how they open into internal cavities in the manner of pneumatopores. The neural arch has been detached along the line of the neurocentral suture (n). A, X ⅔. B, X 1.

apertures called pneumatopores through which branches of the avian air-sac system enter cavernous bones in birds. Various combinations of these conditions can also occur, and some internally cavernous centra have completely hollow middle parts. Internal cavernosity can also occur in neural arches, extending into the articular processes (zygapophyses) near which further small foramina are then seen. In ribs, internal cavernosity occurs in the proximal parts and mainly in the tuberculum, from which a tubular cavity may run a short way down the shaft (e.g., Reid 1996, Figs. 74, 75). Again small external foramina lead into the internal cavities (e.g., Janensch 1947, Fig. 13).

The resemblance of such bones to the pneumatic ones seen in modern birds has led authors from Richard Owen onwards to see them as also pneumatic (Owen 1856 and, e.g., Seeley 1870; Nopcsa 1917; Romer 1945; Swinton 1934; Janensch 1947). This view is supported by evidence from their bone histology, which follows the same general pattern as in modern pneumatic bones apart from differences related to size. As in modern cases, the internal cavities were produced by bone resorption during growth, after which bone partitions remaining between them were coated on both sides with endosteal lining bone. When the bone resorbed was cancellous, any not resorbed was typically compacted, which does not happen when medullary cavities expand by resorption in marrow-filled bones (cf. Reid 1996, Figs. 50, 52). And, in some cases, complete resorption of bone between expanding internal cavities was followed by formation of intercavernous partitions built entirely from endosteal lining bone and indistinguishable histologically from similar features seen in birds and in the pneumatic parts of elephant skulls (Reid 1996, 39–40). Gross morphology and bone histology thus both point to saurischians having had an air-sac system like that of birds, diverticula from which will have occupied the pleurocoels and internal spaces seen in their bones, and also a similar specialized style of respiration (cf. Fig. 38.16).

Birds are unique among modern tetrapods in having lungs through which air flows in one direction only, via open-ended tubes called parabronchi, instead of being pumped in and out alternately (or tidally) as in humans. When they breathe in, air inhaled through the trachea bypasses the anterior

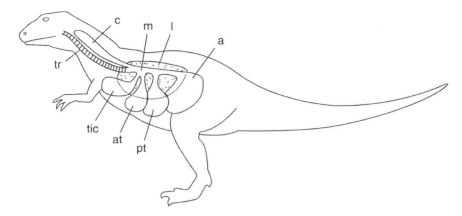

38.16. A theropod dinosaur envisaged as having an air-sac system like that of a bird. Figure shows structures of the right side as seen from the left. tr = trachea; l = lung; c = cervical sac; tic = interclavicular sac; at = anterior thoracic sac; pt = posterior thoracic sac; a = abdominal sac; m = mesobronchus. For a similar recent picture, see O'Connor and Claessens (2005, Figure 4).

part of each lung through a passageway called a mesobronchus and flows into their posterior parts and the posterior air sacs. As this happens, air in the lungs already is forced into the anterior air sacs. When they breathe out, air in the posterior sacs is pumped into the lungs and air in the anterior ones is expelled through the trachea. Air thus flows forward through the lungs during both inhalation and exhalation. This specialized style of respiration, termed parabronchial, is more efficient than the normal tidal sort, allowing some birds to live or fly at heights not accessible to mammals and can be seen as implying that nonavian saurischians would have had full aerobic activity if they breathed in that manner. Without strict proof it seems likely that they did, although two caveats need noting. First, the only forms from which parabronchial lungs are known are the modern birds (Neornithes), so that they, like modern avian endothermy, could have been evolved in birds themselves. Second, the specialized thoracic structure seen in modern bird skeletons is not known from any nonavian saurischian (cf. Ruben et al. 1997, Fig. 35.10), nor seen in *Archaeopteryx* despite its being classed as a bird. Hence, it could be thought that they could only have had normal tidal lungs, with some muscular operating mechanism like that seen in crocodiles. But crocodiles lack an air-sac system, whereas two lines of evidence point firmly to nonavian saurischians having had one sufficiently developed for their axial skeletons to be extensively pneumaticized. Moreover, an air-sac system must presumably be evolved at first in the soft parts, where its only likely function would be enhancement of respiratory efficiency. Thus it seems likely that nonavian saurischians did have parabronchial lungs, although probably more primitive ones than those now found in birds. Modern avian respiration and its links with thoracic structure can be seen as a specialized flight-related version of something which existed in saurischians before birds started flying. Indeed, as seen by Britt (1997, 593), such lungs could first have been evolved in some predinosaurian archosaur, ancestral to both dinosaurs and pterosaurs–the latter also having had cavernous bones.

This leads to the question of whether picturing nonavian dinosaurs as having both complete double-pump circulation and full aerobic activity metabolism involves implying that they would have had high (endothermic) metabolic rates. For those who think rigidly that the present is the key to the past, the answer is inevitably yes, because the two are now only found together in endotherms (birds and mammals). But this is not so if one thinks that both need to be evolved before endotherms can start to evolve (Reid 1987, 146), or if one thinks it possible that avian endothermy was evolved by birds themselves (cf. Ruben 1991), for example as an energetic by-product of their evolving a capacity for sustained flapping flight. Some upward shift in basic rates, compared with those of modern reptiles which rely on anaerobic activity, would then be expected; but all that need be implied otherwise is that nonavian dinosaurs would have had a potential for evolving into endotherms – which need only have been realized in a single line leading to modern birds.

Much in contrast, ornithischians lacked cavernous postcranial bones and so could have lacked an air-sac system or had one without this being evident. Their shared common origin with saurischians can be thought to make their having one more likely; while, in phylogeny, pneumaticization of the vertebral column could have been prevented or lost secondarily as a consequence of its lateral reinforcement by their ossified tendons. But, whatever the answer, they do not seem to have been disadvantaged compared with saurischians, as might be expected if they lacked parabronchial lungs and saurischians had them. Those we know did not reach the sizes of the largest known sauropods (e.g., *Seismosaurus*, *Argentinosaurus*); but that could have been due to mechanical limitations, not physiological ones, such as the greater weight of non-pneumatic vertebrae compared with cavernous ones.

Matching the Pieces

To start this section, it seems useful to state the principles I have always tried to follow in assessing the evidence and arguments involved. First and foremost, critical assessment of any hypothesis or argument must be based primarily on searching for evidence which does not fit it, and not simply on how much appears to. This is because the core discipline of science is test, not conviction, and because we need to rule out the untenable to determine what is possible. Second, if even one piece of evidence cannot fit some hypothesis or argument, it must be taken as outweighing any number of others which seem to. Third, in trying to reach conclusions, one must always work from evidence to answer, and not try to fit evidence to some preconceived conviction. Fourth, the most that any piece of evidence should be taken as meaning is the most that it must mean, however much more it might mean. Fifth, if two kinds of evidence from the same source appear to point to different conclusions, one or both must be being misinterpreted. And sixth, no statement which is simply a conclusion should ever be represented or accepted as factual or final. This is a minimalist approach, but I think it the most realistic one.

Correspondingly, in assessing other studies I do not accept arguments or assertions based on actualistic assumptions that cannot be verified, like the notions that erect limbs and fast growth imply high metabolic rates. This puts me at odds with others who do rely on them, but I think that this is justified. For me, both these notions are flawed in depending on the uniformitarian view that the present is the key to the past, which, while of proven value in interpreting the physical world, is potentially misleading if applied to anatomical features of modern evolutionary end forms. Fully erect limbs and fast growth to large sizes are now seen only in endotherms; but we do not know whether they are only possible for endotherms or first appeared in preendothermic ancestors that now have only endothermic descendants. To find out which is the case, we would have to be able to determine the actual mass-specific standard metabolic rate (SMR) of every archosauromorph which ever had erect limbs or grew quickly—in the latter case, back to the prearchosaurian erythrosuchids, whose periosteal bone (Gross 1934, Fig. 9) was of the type figured by de Ricqlès (1974, Fig. 3) as typical fibrolamellar bone; and there is currently no way of doing this. Without this information, neither of these notions can be verified; and, with it unavailable, they are untestable hypotheses of the kind that the philosopher Karl Popper (1956) rejected as unscientific. This is reasonable grounds for treating assertions which depend on them with caution. Arguments based on such hypotheses can, of course, suggest possibilities, but they remain only arguments and cannot establish facts.

Turning to dinosaurs, we can first rule out the notion that their being either ectotherms or endotherms like modern ones are the only possibilities. Despite various convergences, modern terrestrial tetrapods are clearly divisible into ectotherms and endotherms, with widely separated mean basic metabolic rates (Fig. 38.5); but this is because we no longer have the intermediate animals which must once have existed for endotherms to be evolved from ectotherms. If we could travel through time and study successive members of the lineages which led to mammals and birds, we would find them to show continuous spectra of progressively advancing conditions and not divisible into ectotherms and endotherms—or, ectotherms, intermediates, and endotherms—except on some arbitrary basis, such as chosen mean basic metabolic rates at some specified size. So this is a case in which the present need not be the key to the past—it could be misleading if treated as such.

We can also dismiss the possibility that the mammal-like bone histology of many dinosaurs was simply a consequence of homiothermy based on bulk. First, if this were so, we would not find mammal-like bone in dinosaurs like the greyhound-sized *Hypsilophodon*, much too small to show appreciable mass effects. Second, assuming fibrolamellar bone to mark rapid growth, as it always does at present, it appears that dinosaurs could grow rapidly to large and very large sizes and so must have had digestive, respiratory, cardiovascular, and haemal systems able to support such growth. It would not have been possible without them and so cannot have resulted simply from mass homoiothermy or mass endothermy. And, taken with the haemodynamic arguments of Seymour (1976) and Ostrom (1980), this makes it highly probable that they had the complete double-pump circulation envisaged by Regal and Gans (1980). If so, this is a major difference between them and all modern reptiles and a probable major factor in their evidently unreptilian lifestyles.

Complete double-pump circulation is, however, now found only in endotherms, so that, if we ascribe it to dinosaurs, we need to ask whether this would imply that they were endotherms. There is no certain answer to this question, because we do not know whether this type of circulation and endothermy are evolved concurrently or consecutively. If endothermy depends on complete double-pump circulation, as is possible, it cannot evolve first; but, while they could evolve together, a complete double-pump system may have to exist before endotherms can start to evolve. So, even if a dinosaur were found with a birdlike heart preserved, this would only mean that it could have been an endotherm—not that it was one.

A fossil heart being simply four-chambered also need not suggest endothermy. Modern (eusuchian) crocodiles have four-chambered hearts, but they retain the left systemic arch lost in endothermic birds and show no sign of moving toward endothermy, despite neotenous retention of a modified embryonic blood shunt (the Foramen of Panizza) which allows them to fill both systemics with oxygenated blood when not submerged (Fig. 38.1b). If we rule out the "good reptiles" picture of dinosaurs, the remaining possibilities are their having been endotherms, intermediates, or a mixture of both. In each of these cases, a factor involved is aerobic activity metabolism, which is a characteristic of modern endotherms and a feature of the Regal-Gans intermediate model. Here the evidence is more limited than that for complete double-pump circulation and only direct in the case of the seeming pneumaticity of saurischian vertebrae. This is reasonable evidence for the presence of a system of pulmonary air sacs, enhancing respiratory efficiency and allowing at least some progress toward the level seen in birds. No similar evidence is seen in ornithischians, but they do not appear to have been disadvantaged compared with saurischians, as they probably would have been with only anaerobic activity. But that is the limit of evidence from dinosaurs themselves. That various small bipeds had limbs, which imply that they could run at high speeds, is not helpful because, while we can estimate the speeds they could have reached, we have no way of knowing how long they could sustain them. We can guess that fully aerobic activity is made possible by complete double-pump circulation, because it now is seen only in forms with that type of circulatory system; but, if we ascribe it to dinosaurs on that basis, we are basing an inference on an inference. Purely anaerobic activity in dinosaurs would, however, have resulted in oxygen debts, as it does in modern reptiles, and this could have been a serious problem for large forms. According to Coulson (1984), for instance, a 100-ton dinosaur which used up all the energy available to it anaerobically would have needed at least three weeks to replace it; and, with predators looking for animals in trouble, it would not have much chance of survival despite its huge size. So dinosaurs having aerobic activity metabolism can be seen as at least more likely than their lacking it.

Next, it needs to be emphasized that claims that nonavian dinosaurs were endotherms have been based entirely on evidence which only shows that they could have been, without showing that they were. For that to be done, they would have to be shown to have resembled modern mammals and birds in all physiological essentials, especially:

1. In having similar mass-specific SMRs which would cluster around the upper regression line in Figure 38.5 or one not significantly below it.
2. In using the same average proportion of the energy derived from their food for non-active heat production.
3. Within the same size range, in being able to maintain similar high steady temperatures independently of bulk or activity, and in balancing heat gains and heat losses by physiological means.
4. In having a similar low tolerance to hypothermia as compared with that of ectothermic reptiles.

Currently there is no way in which any of these things can be ascribed to them with certainty, for two main reasons. First, without living examples, there is no way in which any of them being so can be verified directly. Second, while dinosaurs have some anatomical features which are now seen only in endotherms, they have none that can be identified as only possible for endotherms. This is because we do not know the thermal physiology of the oldest form in which they occur and so cannot tell how much progress toward endothermy they require. Together these two problems rule out any certain demonstration that nonavian dinosaurs were endotherms by any means currently available. So, they leave the Regal-Gans picture as at least a possible alternative. There is also the problem that we do not know whether endotherms can evolve without the presence of external insulation, which in archosaurs is known only in pterosaurs, birds, and some nonavian theropods. If required, and absent in other dinosaurs, the potential for evolving into endotherms could have been limited to theropods.

In this light, all current arguments for dinosaurs in general being endotherms can be seen as combining a basic reliance on the present distinctness of ectotherms and endotherms with a fundamental dependence on uniformitarian assumptions that cannot be verified; and some can be faulted on logical or factual grounds. Because nonavian dinosaurs had fully erect limbs, which are now seen only in endotherms, some authors (e.g., Ostrom 1970; Bakker 1972) have assumed that only endotherms can have enough energy to walk upright; but standing and walking on erect limbs takes less energy than doing so on non-erect ones. A preendothermic shift to aerobic activity may have first made erect limbs possible, as Regal and Gans implied (1980, 185). In the case of small, lightly-built dinosaurs it could have been simply a matter of evolving a suitable hip joint. The same assumption has recently led Horner and Padian (2004, 665) to assert that, if nonavian dinosaur were ectotherms, we would have to assume that they could not have had fully erect limbs; but this is false logic. Because their skeletons show that they did, we have to assume that ectotherms can—if they first evolve suitable limb and girdle structures—and that the reason no modern reptiles have them is not their not being endotherms but their not being dinosaurs. De Ricqlès (1974) rightly saw fibrolamellar periosteal bone as implying fast growth; but he wrongly supposed that it is only formed continuously (1974, 53), which now only happens in endotherms, and this was also the case in dinosaurs (1974, 65). Their fibrolamellar bone does show nonavian dinosaurs as able to grow quickly to large sizes; but, as Regal and Gans saw, all this implies is the ability to supply growing tissues with the necessary substrates and energy at the rate such growth requires. In other words, while it does imply high circulatory efficiency, it need not imply high metabolic rates.

It is, of course, not in doubt that nonavian dinosaurs must have differed physiologically from modern ectothermic reptiles; but the two main lines of argument that led to some seeing them as endotherms only tell us this and not how much they differed. Several other kinds of evidence suggest the same thing, but all are equally inconclusive. Bakker's predator-prey ratios (1972) appeared to show that theropods at least were endotherms, although throwing no light on their prey; but his data have not been published and others (e.g., Farlow 1976) have not been able to confirm his claims. Late Cretaceous dinosaurs known from the Alaskan North Slope (Brouwers et al. 1987) could be thought to have needed to be endotherms to live there; but botanical evidence (Herman and Spicer 1976) points to a climate in which the Arctic Ocean did not freeze even in winter and these dinosaurs could have been summer migrants rather than residents. This could have been due to winter darkness limiting adequate food for large herbivores in summer months, especially in the case of low browsers relying on herbaceous plants which died down when summer ended, or simply to the general problem of finding food in darkness. The bone beds in which some of them occur also suggest winter kills, perhaps of forms that only entered the area during abnormally warm summers. The presence of feathers in some small nonavian theropods (e.g., *Caudipteryx*) could be thought to imply that they were endotherms, as modern birds are; but feathers only imply a capacity for retaining bodily heat and cannot tell us how it was generated. Conserving heat produced by activity could have been their original function. Birds themselves are frustratingly unhelpful in this context. They could have inherited endothermy from some small feathered nonavian theropod, as the size of *Archaeopteryx* suggests; but they could have evolved it themselves instead, as thought by Ruben (1991)—we have no way of knowing which happened. Ostrom's (1974) much copied misinterpretation of *Archaeopteryx* as a flightless runner also made it seem much closer functionally to nonavian maniraptors than it really was. As seen by Feduccia and Tordoff (1979), the asymmetry of the principal (primary) feathers on its forelimbs is an aerodynamic adaptation to their use in flight as individual aerofoils, as by modern birds, and implies that birds had already been flying for long enough to evolve both this way of using them and the neuromuscular mechanisms needed for doing so. But the main problem is that all modern birds belong to the Neornithes, which alone survived the Late Cretaceous (K/T) extinctions. Because all are endotherms, we can assume that their ancestors were endotherms when they started their Paleocene radiation; but this tells us nothing about the other kinds of birds that did not survive. Indeed, if we follow the principle that features of modern forms should not be ascribed to fossil relatives "outside the group of extant taxa" in which they occur (Horner and Padian 2004, 665), modern avian endothermy should not be ascribed to any nonneornithine birds, let alone to nonavian theropods.

Adding to these problems, there is evidence suggesting that some nonavian dinosaurs were not endotherms and even some which could mean that none were. Some authors have thought that sauropods could not have eaten enough to be endotherms because of the small size of their heads in relation to their bodies. Halstead (1975) saw *Apatosaurus* as having to eat for more than 24 hours a day to be one. And a detailed study of *Brachiosaurus* and its food resources led Weaver (1983) to see endothermy as impossible for sauropods weighing more than 55 metric tons and "improbable to impossible" below that weight (173, abstract). If the lifestyles and energy sources of these very different animals are contrasted, her views can also be seen as supported by Erickson et al.'s finding (2001, 430) that sauropods grew at high rates now seen only in whales among mammals. First, whales are medium supported animals, which do not have to meet the energy costs of terrestrial life at large sizes and heavy weights. Second, their fast growth is fueled until weaning by the richest milk known in modern mammals, after which they feed as carnivores with access to rich food supplies. How the energy available to growing sauropods is assessed will, of course, depend on what assumptions are made; but, with head size limiting how much food could be eaten in 24 hours, it seems at least doubtful whether growing sauropods dependent on the diet depicted by Weaver (1983, tables 2–4) could have supported endothermy as well as whalelike growth rates.

In other developments, Reid (1987b, 272) saw the occurrence of serrated, steak-knife teeth—like those of theropods in two kinds of extinct terrestrial crocodiles (sebecid mesosuchians and pristichampsid eusuchians) and the living Komodo monitor—as suggesting that theropods had a crocodile-like digestive system, geared only to low energy requirements. Ruben and his colleagues (1997) argued that lack of nasal maxilloturbinals in nonavian dinosaurs would have caused unsustainable respiratory water losses if they had had the high metabolic and respiratory rates of endotherms, also noting genera in which the dimensions of the nasal passages suggest an ectothermic respiration rate. An isotopic study of bones of *Tyrannosaurus* and *Giganotosaurus* by Barrick and Showers (1999) suggests that both had intermediate metabolic rates, while a haemodynamic one by Seymour and Lillywhite (2000) led them to see erect necks as only possible for sauropods "if they had metabolic rates considerably lower than expected for endotherms" (1886, conclusions) and impossible if they were endotherms (1883, abstract). Van Valkenburgh and Molnar (2002) found the diversity of dinosaurian carnivores in various faunas to be greater than is usual in mammals and saw this as more readily explicable if theropods had lower metabolic requirements than mammals; while Farlow and Pianka (2003) see the presence of two large tyrannosaurs (*Gorgosaurus* and *Daspletosaurus*) in the Dinosaur Park Formation of northwestern Canada as made possible by their being "more like ectotherms than endotherms in their metabolic rates." Horner and Padian (2004) disparage ecological approaches on the grounds of the assumptions involved, but they rely on the assumption that fast growth implies high metabolic rates. Amiot et al. (2006) list isotopic data from turtles, crocodiles, and dinosaurs, which they see as showing that high metabolic rates were widespread in Cretaceous dinosaurs; but they make the flawed assumption that their dinosaurs must have been either simple mass homoiotherms or endotherms (2006, 52) and ignore the intermediate alternative.

This leads to the question of whether bone can throw light on these problems, as some authors suppose. It potentially could, if we knew more about the factors which control its formation; but it currently has limitations which prevent any certain conclusions being based on it. First, no type of bone is found only in ectotherms or only in endotherms. At fibrillar level, both can form tissues including lamellar finely bundled bone to woven coarsely bundled bone. At histological level, both can form periosteal bone including uniform simple bone to fibrolamellar compound bone; and both can form it periodically (zonally) or continuously (azonally). Both can also remodel bones externally or reconstruct them internally; and both can form metaplastic bone as well as normal osteoblastic tissues. Modern ectotherms and endotherms do differ in the kinds of bone which are commonest in them, but even here overlaps occur. Haversian and fibrolamellar bone are seen mainly in endotherms, but can also occur in turtles and crocodiles. Continuous growth is seen mainly in mammals and birds, but can also occur in reptiles (e.g., Fig. 38.7c); and, while cyclically periodic growth is most typical of reptiles, it can also occur in mammals and flightless birds. The largely avascular character of periosteal bone in lepidosaurs (tuataras, lizards, snakes; e.g., Enlow [1969, Figs. 20, 21]) is a secondary ("derived") feature, not typical of ectotherms in general; and the smallest mammals and birds can have similar bone, despite having the highest metabolic rates. So, no type of bone, style of growth, or vascularity level can be used to tell whether extinct forms were ectotherms or endotherms with certainty.

Second, growth fast enough for fibrolamellar periosteal bone to form need not imply a high metabolic rate as de Ricqlès (1974) supposed. When he introduced that notion, he pictured zonal tissues (his lamellar-zonal type) as built solely from lamellated simple bone (1974, Fig. 1), and thought that fibrolamellar bone is formed not just quickly but continuously (1974, 53), as now only happens in endotherms; but Enlow (1969, 55–57) had earlier described its periodic formation in zonal tissues from turtles and crocodiles and did not call it fibrolamellar only because that term had not then been introduced. It follows that the simple ability to form fibrolamellar bone cannot require a higher metabolic rate than these ectotherms have when they are growing fastest. So, it can presumably be formed at any metabolic rates between theirs and those of modern endotherms. In these reptiles, osteonic bone is admittedly often restricted to thin and sparsely cellular linings to the vascular canals; but such linings can grade into typical cylindrical osteons with several circlets of osteocytes around the central canal, and equally simple osteons can occur in some dinosaurs (e.g., Reid 1990, Fig. 7, prosauropod *Euskelosaurus*).

This, however, only shows that fast growth as such need not imply high metabolic rates, and still leaves their bone as showing that dinosaurs could grow to large sizes much more quickly than any modern reptiles. The largest modern reptile is the saltwater crocodile (*Crocodylus porosus*), of which a 5.5 meter (18 foot) long example examined by Ferguson (in Reid 1984b, 584) had taken up to 70 years to reach a weight of 823.7 kg (1816 pounds, or approximately ⅘ of a ton); but the young Bajocian sauropod in whose pubis and ischium I found the first certain examples of dinosaurian zonal bone (Reid 1981) seems to have reached at least the size of a large (5–6 ton) modern elephant in only 28 or 29 years (Reid 1987, 144; 1990, 26). From recent estimates (Erickson et al. 2004, Horner and Padian 2004a), the large theropod *Tyrannosaurus rex* grew even faster, reaching weights between 5000–6000 kg (approximately 5–6 tons) in only 18–22 years. Growth rates change with age, but, in terms of age and final weight only, a *T. rex* reaching 5000 kg in 18 years would have grown approximately 24 times as fast as Ferguson's *C. porosus*. Such contrasts indicate some major physiological differences between nonavian dinosaurs and modern reptiles; but they still do not tell us that such dinosaurs had high metabolic rates, because the only things we know to be needed for such growth are the right genetic makeup (genome), an adequate food intake, and digestive, respiratory, and cardiovascular systems able to supply growing tissues with the necessary substrates and energy at the rate that it requires. A high basal metabolic rate could be a further requirement; but there is currently no proof that it is, because actual metabolic rates can only be measured in live animals. Further, if there were a causal connection between fast growth and endothermy, all endotherms would have to grow quickly, which we humans do not, and the periodic fast growth of some turtles and crocodiles would not be possible. Admittedly, the formation of new tissue is one of the sources of endogenous heat, and how much is produced is presumably greatest when growth is fastest; but, if it were the main heat source in endotherms, they would have to grow for all their lives, or cease to be endotherms if they stopped growing. Neither happens.

The critical question then emerges as whether fast growth to large sizes requires high metabolic rates, and so implies them in extinct forms as de Ricqlès and his followers infer, or whether instead its present restriction to endotherms is an artefact of tetrapod history, due to forms which would give a different answer now being extinct, as Regal and Gans thought (1980, 187). Either is possible and neither can be confirmed or refuted by any means currently available. Which answer is thought the more likely then depends on how we look at the evidence. An uniformitarian approach leads automatically to nonavian dinosaurs being pictured as animals with high metabolic rates; but it depends on the assumption that fast growth to large sizes requires a high metabolic rate, which Regal and Gans (1980) rejected. An alternative then is to argue (Reid 1984b, 595) that what we need to know is the thermal physiology of the oldest forms to grow in this manner, and then to cite evidence that the mammal-like reptiles called therapsids had this ability well before they evolved into endotherms. If so, such growth need not imply endothermic metabolic rates.

The most striking event in the history of synapsids earlier than mammals was the abrupt replacement of early forms called pelycosaurs by therapsids at the start of Late Permian times. The bone histology of pelycosaurs is like that now typical of ectothermic reptiles and nothing about them suggests that they were anything else; but that of therapsids had mammal-like features, in even the most primitive forms, and was already completely mammal-like in large early carnivores called dinocephalians (de Ricqlès 1974, 63) despite their showing no other signs of being endotherms (Kemp 1982, 98). These were also lacking in their gorgonopsid successors and first appeared near the end of the Late Permian in advanced therocephalians and early examples of the mainly Triassic cynodonts from which mammals came. Dinocephalians also had teeth suited only to tear and bolt feeding and show no signs of needing more food than pelycosaurs the same size; but these late forms had teeth showing that they chewed their food, as we do, as well as false palates which let them breathe while chewing, implying that they needed to, and nasal turbinal bones of a type which in mammals bear tissues used to limit respiratory water losses (Hillenius 1994). Then, if they are accepted as the first therapsid endotherms, the ability of dinocephalians to grow quickly to large sizes many millions of years earlier is opposite to what would be expected if such growth were only possible for endotherms and must have some different explanation. But, if so, what was it?

There is no certain answer to this question, but one obvious possibility is increased circulatory efficiency produced by the earliest appearance of complete double-pump circulation. Judged from embryology and adult anatomy in modern forms, this condition arises after four-chambered hearts have evolved, by loss of whichever of the two systemic arches seen in modern reptiles (Fig. 38.1) originally carried venous blood from the right ventricle to the aorta. This has happened in both mammals and birds, which are the only known tetrapod endotherms, despite their different ancestries and the involvement of different systemics (the right in mammals, the left in birds); and I see it as the factor which first makes possible both fully aerobic activity and the start of a trend toward endothermy (Reid 1987, 146). If this view is correct, what led to the bone histological differences between pelycosaurs and therapsids was not the evolution of endothermy but of this type of circulatory system in some ancestral therapsid, some of whose descendants were then able to move progressively toward endothermy as they evolved the means to sustain and control it. This fits Kemp's assessment of dinocephalians as having "a metabolic rate above that of typical modern reptiles, but still well below that of mammals where additional adaptations are needed to increase food and oxygen uptake" (1982, 98), and points to this being the most that should be inferred in other forms with mammal-like bone histologies. In this view, all that fast growth to large sizes need imply in nonavian dinosaurs is their having the complete double-pump circulation that haemodynamics imply (Seymour 1976; Ostrom 1980) as well as the other features cited above as known to be needed, without their having the high (endothermic) metabolic rates that de Ricqlès and his followers see as implied. These rival views are both untestable by any available direct method; but, with a gap of most of Late Permian time between therapsids first growing in this manner and their first showing anatomical signs of endothermy, I see taking account of this as more realistic than relying on single time plane data from modern evolutionary end forms as de Ricqlès (1974) did. It would, of course, still be possible that the metabolic rates of such dinosaurs, or some of them, in time rose higher than the minimum levels implied; but such growth does not show that they did, and should not be claimed to have done.

This leaves the question of what light occurrences of bone with zonal growth rings can throw on nonavian dinosaurs. De Ricqlès' early view that they were endotherms depended partly on his wrongly supposing that such bone does not occur in them (1974, 65); and he changed his views when I showed that they did (Reid 1981), instead interpreting sauropod physiology in terms of a combination of mass homoiothermy and mass endothermy, and ending "it is thus not necessary to assume endothermic (i.e., at avian or mammalian levels) energetics to account for the growth pattern in sauropods" (1983, 230). In recent years, however, he and his colleagues (e.g., Horner et al. 1999) have tried to discount its significance, by claiming that the resting lines (LAGS) between zones are plesiomorphic features which also occur in many mammals; but this is misleading. That primitive (plesiomorphic) characters are not admissible in phylogenetic systematics (cladistics) has no implication that they cannot be significant physiologically or that only inherited characters can be plesiomorphic. Any individual tetrapod can potentially have its growth slowed or halted by adverse events or conditions which temporarily upset its normal physiological balance non-lethally; but, while the annuli or resting lines which can result do not differ from those seen in zonal bone, they are unique to the affected individuals and only show that an event occurred. Because they are neither inherited nor inheritable they cannot be plesiomorphic. But what may be, if Castenet et al. (1977) and de Buffrénil and Buffetaut (1981) are correct, is the annually cyclical zonation seen in many modern reptiles, which these authors see as endogenous in origin and hence inbuilt genetically. If so, it has two implications for interpreting occurrences in dinosaurs. First, as a feature seen first in some early amphibians, its only likely origin is as something advantageous to ectotherms. Second, if zonation is inbuilt genetically, its loss in phylogeny is basically a matter of genes switching off, as in the loss of teeth or digits (Reid 1993, 401).

Then, whatever may be true of modern reptiles, some occurrences of zonal and azonal tissues in dinosaurs are not explicable as related to either climate or thermal physiology. Neither can account for occurrences of both zonal and azonal bone in different bones from one animal, especially if the former occurs in a core bone and the latter in a limb bone (Fig. 38.14); and lateral intergradations of zonal and azonal tissues, as in some bones which grow asymmetrically, point plainly to which was then formed depending on genetically determined local growth rates. If this was so in all cases in which both types were produced, we seem to be dealing with animals which could form zonal tissues when or where the overall local growth rate was below some critical figure, but could also form it azonally when or where the rate was higher. So far as I know, this is not usual in any non-dinosaurian tetrapods, although it is also reflected in some early birds (Chinsamy et al. 1994, 1995, e.g., Figs. 7a, b) and the slowly growing moas (Turvey et al. 2005, Fig. 2). Two further comparisons with modern forms are also relevant. First, in modern mammals, forms showing annually interrupted growth are nearly all from cold regions and are also mainly small forms, e.g., hamster-size; but nonavian dinosaurs showing zonation also include large forms from warm regions (e.g., *Baryonyx*, *Rhabdodon*), some having bone more reptilian than any tissues known from comparable mammals (e.g., Fig. 38.13; compare with relevant figures in Enlow and Brown [1958]). Second, even when they grew periodically in a manner resembling that of crocodiles, nonavian dinosaurs could still grow at rates now seen only in large, fast-growing mammals. The sauropod whose bone is seen in Fig. 38.11 (see also Reid 1981, Figures 1c–e) grew at roughly the same speed as modern elephants (Reid 1987, 144). Horner and Padian (2004a, 1879) found the same to have been true of *Tyrannosaurus*. Add to this the finding that they uniquely grew faster the larger they were (Erickson et al. 2001). This mixture of reptilian, mammal-like, and unique features gives a strong impression that these were animals with no true modern counterparts.

In short, there is no certain evidence that nonavian dinosaurs were endotherms, and some evidence suggests that most were not or that none were. The remaining possibility is that some or all of them had an intermediate type of physiology, of the kind envisaged by Regal and Gans or some variant of it. This answer seems to fit the evidence reasonably. Because modern birds (Neornithes) are endotherms, some part of the lineage leading to them must have consisted of intermediate forms; and these could have been the nonavian theropods from which they were derived, especially if avian endothermy was only evolved in birds themselves as suggested by Ruben

(1991). Haemodynamics and bone implying fast growth to large sizes point strongly to dinosaurs having complete double-pump circulation, which seems as certain as it can be without preserved soft parts; and this could have allowed them the aerobic activity that Regal and Gans postulated. High circulatory efficiency would probably have led to some rise in mean basic metabolic rates, to above the mean level found in modern reptiles; but this need not have been to endothermic levels and cannot have been unless the arguments of Ruben et al. (1997) are mistaken. It would, however, have increased their capacity for achieving mass homoiothermy, and reduced the size at which they could achieve it. Vascular control of heat exchanges through the skin is not demonstrable, but is a mechanism known from modern reptiles (e.g., Weathers and Mogareidge 1971). So, the Regal-Gans model provides at least a feasible alternative to dinosaurs having been endotherms. But, is it likely that dinosaurs had a type of physiology which is not seen in any modern animals?

It seems that it is and that this is the key to understanding both their post-Triassic predominance in terrestrial faunas and the ways in which different kinds of evidence can otherwise seem contradictory. First, if the origin of endothermy requires a trend to small sizes, as argued by McNab (1978), the trend to large sizes shown by dinosaurs from early in their history would have led them away from endothermy, not toward it, but toward the exploitation of bulk as a means of achieving homoiothermy. Second, the terrestrial mammals which now fill roles once filled by dinosaurs are fully evolved endotherms, in which a high proportion of the energy derived from their food is expended in passive heat production. But, as descendants of ancestors which spent their Mesozoic years in the size range of modern rats and mice, they can be seen as being endotherms more by inheritance than necessity and as having a style of physiology more appropriate to, rat-sized animals than to elephant-sized ones. In contrast, with no prior commitment to endothermy, being animals of the kind that Regal and Gans envisaged would have been possible for nonavian dinosaurs, and, in warm Mesozoic conditions, would have given them a less energetically wasteful way of functioning as large active animals. With basic metabolic rates well below those of comparable endotherms, they would have needed correspondingly less food, to the advantage of herbivores in times in which plant foods were mostly of low calorific value (cf. Weaver 1983). Also, the less available energy was wasted on needless heat production, the more could be used to fuel growth. This could be how sauropods grew to huge sizes on such diets at the high rates implied by their bone. They would also not have needed the high respiration rates of endotherms (cf. Ruben et al. 1997). Low rates of passive heat production could be why giants like *Seismosaurus* and *Argentinosaurus* could reach sizes which Spotila (1980) saw overheating as making impossible for endotherms. So the Regal-Gans picture of dinosaurs can be seen as not just feasible, but likely to be basically the correct one.

This then leaves the question: could some nonavian dinosaurs being more mammal-like histologically than others mean they were endotherms although others were not? This is possible, but unlikely for three reasons. First, we have to expect that different dinosaurs differed as much physiologically as do different modern reptiles or different mammals, even if they all shared the same general type of physiology. Second, apart from the feathers now known from some theropods, nothing shows dinosaurs as having any means of regulating temperature but those employed by modern reptiles; and even feathers were probably only used originally for simple heat retention, not endothermic thermoregulation. And third, on reviewing the material used for my general survey of dinosaurian bone histology (Reid 1996), I have found that reptilelike and mammal-like features occur in seemingly random combinations, with no correlation between occurrences of different mammal-like features. Haversian bone can be scarcely to extensively developed, whether growth was periodic or continuous, and the only bone in which I found evidence of determinate growth (Fig. 31.11) had grown periodically. Taken with the intergradation of zonal and azonal tissues in some bones which grew asymmetrically, this points plainly to such variations having been controlled genetically and not implying any major differences in thermal physiology.

If these arguments are accepted, the most appropriate picture of dinosaurs is as animals which differed from all living reptiles in the nature of their circulatory systems and in related ways that affected activity and growth, but were still at least basically ectothermic and remained reptilian in the ways in which they regulated temperature. If they could be studied as living animals, I would expect the metabolic rates of different ones to cluster about a regression line lying somewhere between those for modern ectotherms and endotherms (cf. Fig. 38.5) and closer to the former than the latter. Any overlap between scatter fields would probably be mainly or entirely with that for modern ectotherms. Many large forms were big enough to be mass homoiotherms as adults, while their juveniles and forms too small for mass effects would regulate activity temperatures in the same ways as modern reptiles. In large herbivores, especially sauropods, homoiothermy may have been partly based on heat produced by their gut floras (cf. Farlow 1987). In warm Mesozoic conditions, the only forms likely to have needed to move toward endothermy would be small forms living at high altitudes or high latitudes, where diurnal or seasonal temperatures varied more than in warm lowlands; but there is currently no sure evidence that any ever achieved it.

Is there any actual evidence that nonavian dinosaurs were animals of the kind that Regal and Gans envisaged? There may be. One notable difference between their picture and the endothermic one is that one of its implications is potentially testable and may have been tested already. If homoiothermy in such dinosaurs depended more on mass effects than in fully evolved endotherms, there should have been a minimum size below which their temperatures would have fallen appreciably overnight – the more so the smaller they were. This would have affected the growth of their bones in a way which should be detectable if a suitable technique could be devised. This could be why Erickson et al. (2001, Figure 3) found small dinosaurs to grow more slowly than larger ones – which grew more quickly the larger they were. Their sample (six genera: see Chapter 31, this volume) is too small for any certain conclusion to be drawn; but, while their result is unexpected if such dinosaurs are pictured as endotherms, it is what would be expected if they are pictured as by Regal and Gans. And Gillooly et al.

(2006) see such differences as meaning that large forms had higher temperatures than small ones and take this as actual evidence of dependence on mass (intertial) homoiothermy.

Finally, there are two things which suggest that nonavian dinosaurs did differ physiologically from anything now living. First, there is their seeming failure to radiate extensively as mouse to rat-sized animals—as ectothermic lizards and endothermic mammals and birds have done. Judged from their skeletons, few of them weighed less than 20 kg as adults and very few (e.g., *Microraptor*) less than one kilogram. Yet many small mammals weigh less than a kilogram, with weights ranging down to as little as 2–3 grams in pygmy shrews. Similarly, many birds weigh less than a kilogram, with weights ranging down to 1.6 grams in male bee hummingbirds. We tend to think of dinosaurs in terms of the large ones, of which many were in the 0.5–5.0 tonne range or near it, with sauropod giants reaching 30 tonnes or more. Unless the fossil record is misleading, a rarity of small forms is also one of their features and, if genuine, is potentially more important physiologically. Probably there were more small forms than we know, due simply to small bones being less often fossilized than large ones, but nothing we know currently suggests that any were ever as small as the smallest modern mammals and birds. Second, there is their striking failure to radiate as marine forms—as turtles, crocodiles, lizards, birds, and mammals have all done at various times. Remains of some genera (e.g., *Cetiosaurus, Megalosaurus, Scelidosaurus*) have been found in marine deposits, but none shows any signs of marine adaptations—they seem to be terrestrial forms whose carcasses were washed out to sea. Most notably, there are no known dinosaurian equivalents of sea otters, seals and their allies, or dolphins and whales, despite the sea then teeming with life forms on which swimming theropods could have preyed. Nor are there any known marine herbivores like modern sea cows (manatees and dugongs) or the extinct desmostylians. Even the fish eating theropod *Baryonyx* was a terrestrial swamp dweller. So, if nonavian dinosaurs were really the endothermic counterparts of today's mammals and birds, as some think, why should they not have done as mammals, birds, and, in one case, turtles, crocodiles and lizards have done? The obvious and simple answer is that something stopped them from doing so; but, if so, what was it? As noted above, the seeming rarity of very small forms could be due to differential preservation, but, if not, it does not seem likely to be due to competition with lizards, which some (e.g., *Compsognathus*) ate or with Mesozoic mammals, if these were nocturnal as has been supposed. Jurassic seas were broadly dominated by plesiosaurs and ichthyosaurs, of which some plesiosaurs (pliosaurs) probably ate other reptiles, but this did not deny a marine life to teleosaurid and metriorhynchid crocodiles, of which the latter had aquatic adaptations which paralleled those seen in ichthyosaurs. In the Early Cretaceous, two kinds of lizards (dolichosaurs and aigialosaurs) related to monitors moved into the sea as small semiaquatic forms, while the related Late Cretaceous mosasaurs were fully marine forms, some of which reached large sizes (e.g., *Tylosaurus*). In the Late Cretaceous, there were also seagoing turtles (e.g., *Archelon, Protostega*) and dyrosaurid crocodiles, but still no marine theropods, unless one counts the seagoing birds which appeared at that time and were theropods in the cladistic sense. All this points to those two dinosaurian failures not having ecological origins and so having some different cause. An alternative possibility is that their own physiology gave them only a limited capacity for adapting to life as small animals (Reid 1987, 151), except in a single stock leading to birds. Their physiology also barred them from adapting to marine life despite their avian descendants being able to. That Cretaceous seagoing birds were not just land-based visitors, but included forms as highly adapted to marine life as the wingless diver *Hesperornis*, suggests they differed physiologically from their nonavian ancestors and even hints that no nonavian dinosaurs ever had endothermic metabolic rates.

My case for most or all nonavian dinosaurs having been intermediate animals ends there. It remains inconclusive and will probably be rejected or ignored by authors committed to treating them as endotherms; but my aim is to show you that a case can be made and can be seen as fitting the available evidence. My own preference for the intermediate picture depends chiefly on my views on their fibrolamellar bone, which I think have been misinterpreted. Here I argue as follows.

1. The ability of nonavian dinosaurs to grow quickly to large sizes, forming large amounts of fibrolamellar bone, is genuine and implies some major difference between them and all modern reptiles. What we need to ask is thus not whether they differed from modern forms, but why and how much they differed.

2. Why nonavian dinosaurs could grow quickly to large sizes cannot be determined; but I think that the most likely answer is their having had complete double-pump circulation, produced from the type still seen in crocodiles by loss of the left systemic arch (cf. Fig. 38.1b, cb) as suggested by Russell (1965). This seems to have happened at least twice in archosaurs, because bone like that of dinosaurs is also seen in the earlier erythrosuchids (e.g., Gross 1934, Figs. 9–12). The crocodilian (eusuchian) Foramen of Panizza (Fig. 38.1b) I take to be a secondary adaptive feature, related to their semiaquatic lifestyle and evolved from an embryonic blood shunt allowing blood to be diverted from the lungs until hatching.

3. All arguments for endothermy in forms precursory to mammals and birds are fundamentally flawed if they ignore the former existence of the intermediate forms that must have existed for endotherms to be evolved from ectotherms. Modern tetrapods are clearly divisible into ectotherms and endotherms, despite the secondary heterothermy of some endotherms; but they are evolutionary end forms, and must have had precursors whose metabolic rates spanned the whole of the gap between those now typical of ectotherms and endotherms respectively (cf. Fig. 38.5), including halfway forms not classifiable as either. So, in dealing with forms ancestral to known endotherms, I hold that they need to be seen in this perspective and not treated as though they must have been either ectotherms or endotherms in the modern sense.

4. The critical question is whether fast growth to large sizes requires high metabolic rates, or only the ability to feed growing tissues with the necessary substrates and energy at a rate which makes it possible. Today, only endotherms grow in this manner; but, to

know whether only forms with high metabolic rates can do so, we need to know those of the earliest forms that did, and not just those of their latest descendants. We do not, and, so long as this is so, we cannot be sure that such growth implies high metabolic rates, especially in view of its appearance in therapsids before they show other signs of endothermy. It shows that nonavian dinosaurs could have been endotherms; but it does not show that they were.

5. As relevant historically, I note: (1) Enlow (1969) recorded the occurrence of simple fibrolamellar bone in two kinds of ectothermic reptiles (turtles; alligators, and crocodiles) five years before de Ricqlès' early study (1974) appeared, and (2) Ferguson (1982) showed its formation in wild American alligators nine years before Paul (1991) called this a myth, and 22 years before a specimen I figured (Reid 1997, Figs. 32.7; Fig. 38.7 here) was called anomalous by Padian et al. (2004). As evidence which has largely been ignored, I note the failure of nonavian dinosaurs to radiate extensively as small (mouse to rat-sized) animals or to adapt to marine life. Both suggest physiological limitations which modern birds and mammals do not have.

In this light, I think it clearly possible that most or all nonavian dinosaurs were essentially animals of the kind envisaged by Regal and Gans (1980, 185)—unlike anything now living physiologically. They probably had higher metabolic rates than modern reptiles and differed from them in their circulatory anatomy but not in their thermal physiology. This could account for their striking success as large animals in Mesozoic (not modern) conditions, for the huge sizes reached by sauropods, for their failure to radiate extensively as small animals, and for the largest growing the fastest. It could also have been a factor in their failure to adapt to the sea as well as in their final extinctions. Movement toward endothermy would probably have been greatest in small forms, especially if they had feathers; but actual avian endothermy could have evolved in birds in the course of their acquiring a capacity for sustained flapping flight. In its modern form, coupled to perfected parabronchial respiration, it could also have been limited to Ornithurae, or even to the Neornithes which alone survived the Cretaceous extinctions. They are, after all, the only dinosaurs we know to have evolved into endotherms.

References

Amiot, R., C. Lécuyer, E. Buffetaut, G. Escarguel, F. Fluteau, and F. Martineau. 2006. Oxygen isotopes from biogenetic apatites suggest widespread endothermy in Cretaceous dinosaurs. *Earth and Planetary Science Letters* 246: 41–54.

Amprino, R. 1967. Bone histophysiology. *Guy's Hospital Medical Report* 116(2): 51–69.

Bakker, R. T. 1972. Anatomical and ecological evidence of endothermy in dinosaurs. *Nature* 238: 81–85.

———. 1975. Dinosaur renaissance. *Scientific American* 232 (4): 58–78.

———. 1986. *The Dinosaur Heresies: New Theories Unlocking the Mystery of the Dinosaurs and Their Extinction.* New York: William Morrow.

Barrick, R. E., and W. J. Showers. 1994. Thermophysiology of *Tyrannosaurus rex*: Evidence from bone isotopes. *Science* 265: 222–224.

——. 1999. Thermophysiology and biology of *Giganotosaurus*: comparison with *Tyrannosaurus*. *Palaeontologica Electronica* 2(2).

Bellairs, A. d'A., and J. Attridge. 1975. *Reptiles* (Fourth Edition). Essex, U. K.: Anchor Press.

Benton, M. J. 1987. Why *Archaeopteryx* is not a fake but suffers from too much publicity. *Geology Today* (July-August 1987): 118–121.

Bouvier, M. 1977. Dinosaur Haversian bone and endothermy. *Evolution* 31: 449–450.

Brink, A. S. 1956. Speculations on some advanced mammalian characteristics in the higher mammal-like reptiles. *Palaeontologia Africana* 4: 77–96.

Brisbin, I. L. Jr., E. A. Standora, and M. J. Vargo. 1982. Body temperature and behavior of American alligators during cold winter weather. *American Midland Naturalist* 107: 209–218.

Britt, B. B. 1997. Postcranial pneumaticity. In P. J. Currie and K. Padian (eds.), *Encyclopedia of Dinosaurs*, 590–595. San Diego: Academic Press.

Brouwers, E. M., W. A. Clemens, R. A. Spicer, T.A. Ager, L. D. Carter, and H. V. Sliter. 1987. Dinosaurs on the North Slope, Alaska: high latitude Late Cretaceous environments. *Science* 237: 1608–1610.

Case, T. J. 1978. Speculations on the growth rate and reproduction of some dinosaurs. *Paleobiology* 4: 320–328.

Castanet, J., F. Meunier, and A. J. de Ricqlès. 1977. L'enregistrement de la croissance cyclique par le tissue osseux chez les Vertébrés poikilothermes: donnés comparative et essai de synthèse. *Bulletin Biologique de la France et de la Belgique* 3: 183–202.

Chabreck, R. H., and T. Joanen. 1979. Growth rates of American alligators in Louisiana. *Herpetologica* 35: 51–57.

Chinsamy, A. 1990. Physiological implications of the bone histology of *Syntarsus rhodesiensis* (Saurischia: Theropoda). *Palaeontologia africana* 27: 77–82.

——. 1993a. Bone histology and growth trajectory of the prosauropod dinosaur *Massospondylus carinatus*. *Modern Geology* 118: 319–329.

——. 1993b. Image analysis and the physiological implications of the vascularisation of femora in archosaurs. *Modern Geology* 19: 101–108.

——. 1995. Ontogenetic changes in the bone histology of the Late Jurassic ornithopod *Drysaurus lettowvorbecki*. *Journal of Vertebrate Paleontology* 15: 96–104.

Chinsamy, A., L. M. Chiappe, and P.

Dodson. 1994. Growth rings in Mesozoic birds. *Nature* 368: 196–197.

——. 1995. Mesozoic avian bone microstructure: physiological implications. *Paleobiology* 21: 561–574.

Chinsamy, A., T. Rich, and P. Vickers-Rich. 1998. Polar dinosaur bone histology. *Journal of Vertebrate Paleontology* 18: 385–390.

Chudinov, P. 1968. On the structure of the skin in theromorph reptiles. Akad. Nauk. USSR 179: 207–210 (in Russian).

Clarke, J. A., C. P. Tambussi, J. I. Noriega, G. M. Erickson, and R. A. Ketcham. 2005. Definitive fossil evidence for the extant avian radiation in the Cretaceous. Nature 433: 305–308.

Colbert, E. H., R. B. Cowles, and C. M. Bogert. 1946. Temperature tolerances in the American alligator, and their bearing on the habits, evolution and extinction of the dinosaurs. *Bulletin of the American Museum of Natural History* 86: 327–373.

Coombs, W. 1980. Juvenile ceratopsians from Mongolia: The smallest known dinosaur specimens. *Nature* 283: 380–381.

Coulson, R. A. 1984. How metabolic rate and anaerobic glycolysis determine the habits of reptiles. In M. W. J. Ferguson (ed.), *The Structure Development and Evolution of Reptiles*, 425–441. Orlando, Fla., and London: Academic Press.

Coulson, R. A., and T. Hernandez. 1983. Alligator metabolism: studies on chemical reactions *in vivo*. *Comparative Biochemistry and Physiology* 74: i–iii, 1–182.

Currey, J. D. 1959. Differences in the tensile strength of bone of different histological types. *Journal of Anatomy*, London 93: 87–105.

——. 1960. Differences in the blood-supply of bone of different histological types. *Quarterly Journal of Microscopical Science* 101: 351–370.

——. 1962. The histology of the bone of a prosauropod dinosaur. *Palaeontology* 5: 238–246.

de Buffrénil, V. 1982. Données preéliminaire sur la présence de lignes de'arrêt de croissance périostiques dans la mandibule du marsouin commun, *Phocoena phocoena* (L.), et leur utilisation comme indicateur de l'âge. *Journal Canadien de Zoologie* 60: 2557–2567.

de Buffrenil, V., and E. Buffetaut. 1981. Skeletal growth lines in an Eocene crocodilian skull from Wyoming as an indicator of ontogenetic age and paleoclimatic conditions. *Journal of Vertebrate Paleontology* 1: 57–66.

de Ricqlès, A. J. 1968. Recherces palèohistologiques sur les os longs des tètrapods.

I. Origine du tissue osseux plexiforme des dinosauriens sauropodes. *Annales de Palèontologie (Vertèbres)* 54: 133–145.

——. 1974. Evolution of endothermy: Histological evidence. *Evolutionary Theory* 1: 51–80.

——. 1976. On bone histology of fossil and living reptiles, with comments on its functional and evolutionary significance. In A. d'A. Bellairs and C. B. Cox (eds.), *Morphology and Biology of Reptiles*, 123–150. London: Academic Press.

——. 1980. Tissue structure of dinosaur bone: Functional significance and possible relation to dinosaur phiysiology. In R. D. K. Thomas and E. C. Olson (eds.), *A Cold Look at the Warm-Blooded Dinosaurs*, 103–139. American Association for the Advance of Science Selected Symposium 28. Boulder, Colo.: Westview Press.

——. 1983. Cyclical growth in the long limb bones of a sauropod dinosaur. *Acta Palaeontological Polonica* 28: 225–232.

de Ricqlès, A. J., and J. R. Bolt. 1983. Jaw growth and tooth replacement in *Captorhinus aguti* (Reptilia: Captorhinomorpha): a morphological and histological analysis. *Journal of Vertebrate Paleontology* 3: 7–24.

de Ricqlès, A. J., K. Padian, J. R. Horner, E.-T. Lamm, and N. Myhrvold. 2003. Osteohistology of *Confuciusornis sanctus* (Theropoda: Aves). *Journal of Vertebrate Paleontology* 23: 373–386.

Desmond, A. J. 1975. *The Hot-Blooded Dinosaurs*. London: Blond and Briggs.

Eckert, R., D. Randall, and G. Augustine. 1988. *Animal Physiology: Mechanisms and Adaptations*. 3rd ed. New York: W. H. Freeman.

Enlow, D. H. 1962a. Functions of the Haversian system. *American Journal of Anatomy* 110: 268–306.

——. 1962b. A study of the post-natal growth and remodeling of bone. *American Journal of Anatomy* 110: 79–102.

——. 1969. The bone of reptiles. In C. Gans and A. d'A. Bellairs (eds.), *Biology of the Reptilia*, vol. 1, 45–80. London and New York: Academic Press.

Enlow, D. H., and S. O. Brown. 1956. A comparative histological study of fossil and recent bone tissue. Part I. *Texas Journal of Science* 8: 405–443.

——. 1957. A comparative histological study of fossil and recent bone tissue. Part II. *Texas Journal of Science* 9: 186–214.

——. 1958. A comparative histological study of fossil and recent bone tissues. Part III. *Texas Journal of Science* 10: 187–230.

Erickson, G. M., K. Curry Rogers, and S. A.

Yerby. 2001. Dinosaurian growth patterns and rapid avian growth rates. *Nature* 412: 429–433.

Erickson, G. M., P. J. Makovicky, P. J. Currie, M. A. Norell, S. A. Yerby, and C. A. Brochu. 2004. Gigantism and comparative life-history parameters of tyrannosaurid dinosaurs. *Nature* 430: 772–775.

Farlow, J. O. 1976. A consideration of the trophic dynamics of a Late Cretaceous large-dinosaur community (Oldman Formation). *Ecology* 57: 841–857.

——. 1987. Speculations about the diet and digestive physiology of herbivorous dinosaurs. *Paleobiology* 13: 60–72.

Farlow, J. O., and E. R. Pianka. 2003. Body size overlap, habitat partitioning and living space requirements of terrestrial predators: implications for the paleoecology of large theropod dinosaurs. *Historical Biology* 16: 21–40.

Feduccia, A. 1995. Explosive evolution in Tertiary birds and mammals. *Science* 267: 637–638.

Feduccia, A., and H. B. Tordoff. 1979. Feathers of *Archaeopteryx*: asymmetric vanes indicate aerodynamic function. *Science* 203: 1021–1022.

Ferguson, M. W. J. 1982. The structure and development of the palate in *Alligator mississippiensis*. Ph.D. Thesis, Queen's University of Belfast, Ireland.

——. 1984. Craniofacial development in *Alligator mississippiensis*. In M. W. J. Ferguson (ed.), *Zoological Society of London Symposia* 52: 223–273. London: Academic Press.

Ferguson, M. W. J., L. S. Honig, P. Bringas, Jr., and H. C. Slavkin. 1982. In vivo and in vitro development of first branchial arch derivatives in *Alligator mississippiensis*. In A. D. Dixon and B. Sarnat (eds.), *Factors and Mechanisms Influencing Bone Growth*, 275–286. Alan R. Liss.

Gillooly, J. F., A. P. Allen, and E. L. Charnov. 2006. Dinosaur fossils predict body temperatures. *Public Library of Science Biology* 4: 1467–1469.

Gross, W. 19334. Die Typen des mikroskopischen Knochenbaues bei fossilen Stegocephalen und Reptilien. *Zeitschrift für Anatomie* 103: 731–764.

Halstead, L. B. 1975. *The Evolution and Ecology of the Dinosaurs*. London: Peter Lowe.

——. 1979. *The Evolution of the Mammals*. London: Book Club Associates.

Havers, C. 1691. *Osteologia Nova*. London: Samuel Smith.

Herman, A. B., and R. A. Spicer. 1996. Palaeobotanical evidence for a warm Cretaceous Arctic Ocean. *Nature* 380: 330–333.

Hillenius, W. J. 1994. Turbinates in therapsids: evidence for Late Permian origins of mammalian endothermy. *Evolution* 48: 207–229.

Hohnke, L. A. 1973. Haemodynamics in the Sauropoda. *Nature* 244: 309–310.

Horner, J. R., and K. Padian. 2004a. Age and growth dynamics of *Tyrannosaurus rex*. *Proceedings of the Royal Society of London B* 271: 1875–1880.

——. 2004b. Dinosaur physiology. In D. B. Weishampel, P. Dodson, and H. Osmólska (eds.), *The Dinosauria* (second edition), 660–671. Berkeley: University of California Press.

Horner, J. R., A. de Ricqlès, and K. Padian. 1999. Variation in dinosaur skeletochronology indicators: implications for age assessment and physiology. *Paleobiology* 25: 295–304.

Hotton, N. III. 1980. An alternative to dinosaur endothermy: The happy wanderers. In R. D. K. Thomas and E. C. Olson (eds.), *A Cold Look at the Warm-Blooded Dinosaurs*, 311–350. American Association for the Advancement of Science Selected Symposium 28. Boulder, Colo.: Westview Press.

Janensch, W. 1947. Pneumatizität bei Wirbeln von Sauropoden und anderen Saurischiern. *Palaeontographica*, Suppl. VII: 1–25.

Jenkins, F. A. 1971. The postcranial skeleton of African cynodonts. *Bulletin of the Peabody Museum of Natural History* 36: 1–216.

Joanen, T., and L. McNeese. 1976. Culture of immature American alligators in controlled environmental chambers. *Proceedings of the Annual Meeting of the World Maricultural Society* 7: 201–211.

——. 1980. The effects of a severe winter freeze on wild alligators in Louisiana. In *Crocodiles: Proceedings of the 9th Working Meeting of the Crocodile Specialist Group*, 21–32. Gland, Switzerland: World Conservation Union.

Kemp, T. S. 1982. *Mammal-Like Reptiles and the Origin of Mammals*. London and New York: Academic Press.

Klevezal, G. A., and S. E. Kleinenberg. 1969. *Age Determination of Mammals from Layered Structures in Teeth and Bone*. Jerusalem: Israel Program for Scientific Translations.

Lehman, T. M. 1997. Late Campanian dinosaur biogeography in the western interior of North America, 223–240. In D. L. Wolberg, E. Stump, and G. D. Rosenberg (eds), *Dinofest International*. Philadelphia: Academy of Natural Sciences.

McFarland, W. N.; F. H. Pough; T. J. Cade; and J. B. Heiser (eds.). 1979. *Vertebrate Life*. London and New York: Collier Macmillan International.

McNab, B. K. 1978. The evolution of endothermy in the phylogeny of mammals. *American Naturalist* 112: 1–21.

McNab, B. K., and W. Auffenberg. 1976. The effect of large body size on the temperature regulation of the Komodo dragon, *Varanus komodoensis*. *Comparative Biochemistry and Physiology* 55A: 345–350.

Moy-Thomas, J. A., and R. S. Miles. 1971. *Paleozoic Fishes*. Chapman and Hall, London.

Nopcsa, F. 1917. Über Dinosaurier 2: die Riesenformen unter den Dinosauriern. *Zentralblatt über Mineralogie* 1917: 332–348.

O'Connor, P. M., and L. P. A. M. Claessens. 2005. Basic avian pulmonary design and flow-through ventilation in nonavian theropod dinosaurs. *Nature* 436: 253–256.

Ostrom, J. H. 1970. Terrestrial vertebrates as indicators of Mesozoic climates. *Proceedings of the North American Paleontological Convention* D: 347–376.

——. 1974. *Archaeopteryx* and the origin of flight. Quarterly Review of Biology 49: 27–47.

——. 1980. The evidence for endothermy in dinosaurs. In R. D. K. Thomas and E. C. Olson (eds.), *A Cold Look at the Warm-Blooded Dinosaurs*, 15–54. American Association for the Advancement of Science Selected Symposium 28. Boulder, Colo.: Westview Press.

Owen, R. 1856. *Monograph of the Fossil Reptilia of the Purbeck and Wealden Formations, Part III, Dinosauria* (Megalosaurus): 1–26. Palaeontological Society, London.

Padian, K. 1997. Physiology. In P. J. Currie and K. Padian (eds), *Encyclopedia of Dinosaurs*, 552–557. San Diego: Academic Press.

Padian, K., J. R. Horner, and A. de Ricqlès. 2004. Growth in small dinosaurs and pterosaurs: the evolution of archosaurian growth strategies. *Journal of Vertebrate Paleontology* 24: 555–571.

Paul, G. S. 1991. The many myths, some old, some new, of dinosaurology. *Modern Geology* 16: 69–99.

Popper, K. 1956. *The Logic of Scientific Discovery*. London: Hutchinson.

Pough, F. H. 1979. Modern reptiles. In W. N. McFarland, F. H. Pough, T. J. Cade, and J. B. Heiser (eds.), *Vertebrate Life*, 455–513. London and New York: Collier Macmillan International.

Pough, F. H., J. B. Heiser, and W. N. Mc-Farland. 1989. *Vertebrate Life*. 3rd ed. London and New York: Collier Macmillan and Macmillan Publishing.

Pritchard, J. J. 1979. *Bones*. Burlington, North Carolina: Carolina Biological Supply Company.

Regal, P. J., and C. Gans. 1980. The revolution in thermal physiology. In R. D. K. Thomas and E. C. Olson (eds.), *A Cold Look at the Warm-Blooded Dinosaurs*, 167–188. American Association for the Advancement of Science Selected Symposium 28. Boulder, Colo.: Westview Press.

Reid, R. E. H. 1981. Lamellar-zonal bone with zones and annuli in the pelvis of a sauropod dinosaur. *Nature* 292: 49–51.

———. 1984a. The histology of dinosaurian bone, and its possible bearing on dinosaurian physiology. In M. W. J. Ferguson (ed.), *The Structure, Development, and Evolution of Reptiles*, 629–633. Orlando, Fla., and London: Academic Press.

———. 1984b. Primary bone and dinosaurian physiology. *Geological Magazine* 121: 589–598.

———. 1987a. Bone and dinosaurian "endothermy." *Modern Geology* 11: 133–154.

———. 1987b. Review of *The Dinosaur Heresies* by R. T. Bakker. *Modern Geology* 11: 271–280.

———. 1990. Zonal "growth rings" in dinosaurs. *Modern Geology* 15: 19–48.

———. 1993. Apparent zonation and slowed late growth in a small Cretaceous theropod. *Modern Geology* 18: 391–406.

———. 1996. Bone histology of the Cleveland-Lloyd dinosaurs and of dinosaurs in general. Part I: Introduction: Introduction to bone tissues. *Brigham Young University Geology Studies* 41: 25–71.

———. 1997. Dinosaurian physiology: the case for "intermediate" dinosaurs. In J. O. Farlow and M. K. Brett-Surman (eds.), *The Complete Dinosaur*, 449–473. Bloomington: Indiana University Press.

Romer, A. S. 1945. *Vertebrate Paleontology*. Chicago: University of Chicago Press.

Ruben, J. A. 1991. Reptilian physiology and the flight capacity of *Archaeopteryx*. *Evolution* 45: 1–17.

———. 1995. The evolution of endothermy in mammals and birds: From physiology to fossils. *Annual Review of Physiology* 57: 69–95.

Ruben, J. A., A. Leitch, W. Hillenius, N. Geist, and T. Jones. 1997. New insights into the metabolic physiology of dinosaurs. In J. O. Farlow and M. K. Brett-Surman (eds.), *The Complete Dinosaur*, 505–518. Bloomington: Indiana University Press.

Russell, L. S. 1965. Body temperature of dinosaurs and its relationship to their extinction. *Journal of Paleontology* 39: 497–501.

Ruth, E. B. 1953. Bone studies II: an experimental study of the Haversian-type vascular channels. *American Journal of Anatomy* 93: 429–455.

Sander, M. 2000. Longbone histology of the Tendaguru sauropods: implications for growth and biology. *Paleobiology* 26: 466–488.

Sander, P. M., and P. Andrássy. 2006. Lines of arrested growth and long bone histology in Pleistocene large mammals from Germany: what do they tell us about dinosaur physiology? *Palaeontographica* A 277: 143–159.

Seebacher, F., G. C. Grigg, and L. A. Beard. 1999. Crocodiles as dinosaurs: behavioural thermoregulation in very large ectotherms leads to high and stable body temperatures. *Journal of Experimental Biology* 202: 77–86.

Seeley, H. G. 1870. On *Ornithopsis*, a gigantic animal of the pterodactyle kind from the Wealden. *Annals and Magazine of Natural History (Series 4)*: 5: 279–283.

Seitz, A. L. L. 1907. Vergleichender Studien über den mikroskopischen Knochenbau fossiler und rezenter Reptilien und dessen Bedeutung für das Wachstum und Umbildung des Knochengewebes in allgemeinen. *Nova Acta, Abhandlungen der kaiserlichen Leopold-Carolingischen deutschen Akademie der Naturforscher* 87: 230–370.

Seymour, R. S. 1976. Dinosaurs, endothermy and blood pressure. *Nature* 262: 207–208.

Seymour, R. S., C. L. Bennett-Stamper, S. D. Johnston, D. R. Carrier, and G. C. Grigg. 2004. Evidence for endothermic ancestors of crocodiles at the stem of archosaur evolution. *Physiological and Biochemical Zoology* 77: 1051–1067.

Seymour, R. S., and H. B. Lillywhite. 2000. Hearts, neck posture and metabolic intensity of sauropod dinosaurs. *Proceedings of the Royal Society London B* 267: 1883–1887.

Spotila, J. R. 1980. Constraints of body size and environment on the temperature regulation of dinosaurs. In R. D. K. Thomas and E. C. Olson (eds.), *A Cold Look at the Warm-Blooded Dinosaurs*, 233–252. American Association for the Advancement of Science Selected Symposium 28. Boulder, CO.: Westview Press.

Spotila, J. R., P. W. Lommen, G. S. Bakken, and D. M. Gates. 1973. A mathematical model for body temperature of large reptiles: Implications for dinosaur ecology. *American Naturalist* 107: 391–404.

Spotila, J. R., M. P. O'Connor, P. Dodson, and F. V. Paladino. 1991. Hot and cold running dinosaurs: Body size, metabolism and migration. *Modern Geology* 16: 203–227.

Sues, H.-D. 2001. Ruffling feathers. *Nature* 410: 1036–1037.

Swinton, W. E., 1934. *The Dinosaurs*. London: Thomas Murby.

Thomas, K. J. S., and S. Clarkson. 2004. Microscale δO and δC isotopic analysis of an ontogenetic series of the hadrosaurid dinosaur *Edmontosaurus*: implications for physiology and ecology. *Palaeogeography, Palaeoclimatology, Palaeoecology* 206: 257–287.

Tumarkin-Deratzian, A. 2007. Fibrolamellar bone in wild adult *Alligator mississippiensis*. Journal of Herpetology 41: 341–345..

Turvey, S.T., O. R. Green, and R. N. Holdaway. 2005. Cortical growth marks reveal extended juvenile development in New Zealand moa. *Nature* 435: 940–943.

Van Valkenburgh, B., and R. E. Molnar. 2002. Dinosaurian and mammalian predators compared. *Paleobiology* 28: 527–543.

Varricchio, D. J. 1993. Bone microstructure of the Upper Cretaceous theropod dinosaur *Troodon Formosus*. *Journal of Vertebrate Paleontology* 13: 99–104.

Walker, A. 1985. The braincase of *Archaeopteryx*. In M. K. Hecht, J. H. Ostrom, G. Viohl, and P. Wellnhofer (eds), *The Beginnings of Birds: Proceedings of the International Archaeopteryx Conference*, Eichstatt 1984, 123–134.

Weathers, W. W., and K. R. Mogareidge. 1971. Cutaneous vascular response to temperature changes in the spiny-tailed iguana, *Ctenosaura hemilopha*. *Copeia* 1971: 548–551.

Weaver, J. C. 1983. The improbable endotherm: the energetics of the sauropod dinosaur *Brachiosaurus*. *Paleobiology* 9: 173–182.

Yao, J.-X., Y. Zhang, and Z. L. Tang. 2002. Histological study on the Late Cretaceous ornithomimid and hadrosaurid. *Acta Palaeontologica Sinica* 14: 241–250.

39.1. Magnetic "stripes" of the sea floor surrounding the Mid-Atlantic Ridge southwest of Iceland. New oceanic crust forms from volcanic eruptions at the crest of the ridge. As the lava solidifies, iron-bearing minerals align with the prevailing magnetic field of the earth. As the sea floor splits and moves to either side away from the ridge, a given magnetic band is torn in two to form a mirror-image pattern across the Mid-Atlantic Ridge. New molten rock rises from below to replace the solid crust that has been carried away. Should the earth's magnetic field reverse while this is happening, the newly formed oceanic crust will have a polarity opposite from that prior to the reversal. In this diagram, dark-colored bands correspond to regions of oceanic crust that are magnetized with "normal" polarity, like that which prevails now. Light-colored bands correspond to "reversed" polarity. Redrawn from Press and Siever 1994.

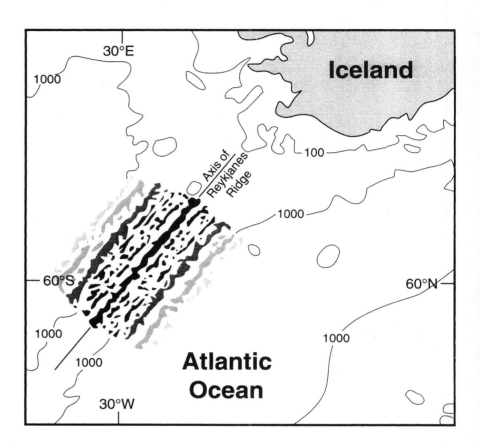

Principles of Paleobiogeography in the Mesozoic

39

Ralph E. Molnar

The past was different; otherwise there could be no history. It has even been said that the past is a different country. But a derivative quotation (Stross 2006) is more to the point: "not only is the past a foreign country, it's one that doesn't issue visas." We don't take this seriously, but simply imagine that the world of the past was very much like the present. But our imaginations are weak – the past was a foreign world and a trip into the Mesozoic would take us to a place unrecognizable except to specialists in the evolution and history of the earth.

We are raised on the notion that life has evolved, and so we realize that the creatures and plants of the Mesozoic were different from those now alive. But the climate and geography of the earth have also changed. To understand the distribution and ecology of dinosaurs, we need to understand how these aspects of the environment differed from those familiar today and, more important, how we can discern this from the raggle-taggle remains of ancient organisms and the sedimentary detritus deposited when they lived and died.

Given the "doesn't issue visas" provision, we also need to examine how we know what we think we know about the past. Let's start with the histories of continents and oceans.

Continental Drift and Plate Tectonics: A Brief History

The great voyages of European discovery of the sixteenth century – or rather the maps derived from them – revealed that the east and west coastlines of the Atlantic Ocean more or less matched. Sir Francis Bacon remarked in 1610 that this match could be no coincidence, and M. François Placet suggested that the ocean had been created by the biblical deluge, separating what previously had been a single landmass. In the early twentieth century, the idea arose that the continents on both sides of the ocean had slipped apart, creating the Atlantic in their wake.

Although not the first to propose such a theory, the German meteorologist Alfred Wegener (1915) stated the hypothesis in the most detail and was first vilified, then hailed as the father of continental drift. Although the idea was supported by some geologists, mostly from the Southern Hemisphere, it did not gain general acceptance in Europe or North America. Much has been made of the fact that Wegener was not a geologist but a meteorologist – an "outsider." However, there were two good reasons for rejecting continental drift. One was that Wegener didn't propose any mechanism that could propel the massive continents around the surface of the earth, through solid rock; and the other was that geological evidence for drift, although clear in South America and southern Africa, was harder to find in the northern continents. Most geologists of the time were not wealthy enough to travel about the world simply to examine the foreign geology in detail. So they regarded continental drift as a novel, but unlikely, hypothesis.

The change in attitude occurred in the 1960s, and it came about precisely from the discovery of convincing answers to these two objections.

The Evidence for Plate Tectonics

The exploration of the Atlantic floor, particularly near the Mid-Atlantic Ridge, revealed that it was largely composed of solidified lavas. These lavas retained a record, a signature, of the direction of the earth's magnetic field when they solidified. These signatures formed clear bands parallel to the ridge (Fig. 39.1). The earth's magnetic field has repeatedly reversed polarity in the past, interchanging North and South (magnetic, of course) poles, and this banding showed that the ages of the lavas varied from new (adjacent to the ridge) to older (away from the ridge). It was as if the sea floor had formed at the ridge and slowly spread away, and this process—appropriately termed "sea-floor spreading"—is just what happened. As the sea floor spread, so the continents also moved, sometimes driving over the floors of other oceans. North America, moving with the spreading Atlantic floor, was driven over the Pacific floor to the west, elevating the Cordilleran Mountains (including the Rockies, Sierra Nevada, and others) in the process.

This evidence for sea-floor spreading was far from all the evidence for drift, which included geographical features, geological formations, and the ranges of fossil animals and plants that could be matched across oceans, most notably across the South Atlantic in both South Africa and southern Brazil (Colbert 1973). In both places, local geologists had long been convinced of the reality of continental drift. Of course, North America had drifted west from Europe, but the evidence for this was less apparent because this separation had occurred at least seventy million years before South America separated from Africa. Most dramatically, the development of satellites and of methods for precisely measuring small distances and very short periods of time allowed measurement of the movement of the continents at several centimeters per year—roughly the rate at which your fingernails grow.

The Mechanism of Continental Movement

At first, the mechanism was not so convincingly obvious as the evidence. Geophysicists had long known that although small pieces of rock (small as compared to the masses that make up the continents) are rigid and brittle on the surface of the earth, at depth, under great pressure and heat, they become plastic and flow. In the earth's mantle (the thick region of the earth's interior separating the crust outside from the outer core within), they move in great slow convection currents, rising from below, spreading along the surface of the mantle (beneath the crust), and then sinking back when they have cooled sufficiently. It is their slow spread along the surface that seemingly drags the lighter, floating continents (and sea floors), enabling the continents to override the ocean floors. When the currents of the mantle

39.2. Diagram of the mechanism of plate tectonics. The earth's crust is shown in dark gray, the rigid uppermost part of the mantle in black, and the hotter, less rigid mantle below in light gray. Molten rock from the mantle rises at a mid-ocean ridge (or its equivalent on land, a rift valley), solidifies, and then moves away to either side of the ridge. Eventually the lithospheric plate on which oceanic crust rides subsides into the mantle at a subduction zone, which is expressed at the earth's surface as a deep-ocean trench. The sinking plate melts, and the lighter of the molten material rises to the surface to form a volcanic island arc or a chain of volcanic mountains near the edge of a continent. Isolated plumes of molten material that erupt onto the ocean floor create chains of volcanic islands, like the Hawaiian chain, as the plate moves over them.

Ralph E. Molnar

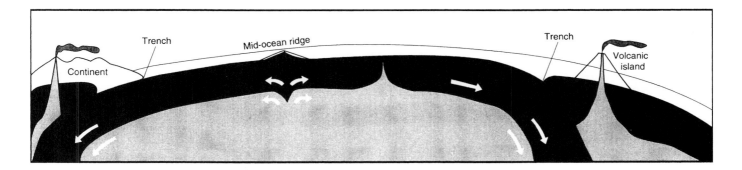

have cooled, they descend again into the depths of the planet (Fig. 39.2). So to geophysicists, the salient features of geography are not the continents and oceans but the crustal plates—including both continents and sea floors that float on the mantle (Fig. 39.3), presumably borne along by its currents. These plates give this theory its name—plate tectonics.

This implies that continents are permanent features (although they may merge or divide), but oceans and oceanic islands (which rise from sea floor, not from submerged continental crust) and most other such features are transient. And further, if continents have histories, these can be represented as a kind of cladogram or cladogram analog. We shall return to these realizations in the next section (Theories of Biogeography). Note that not only continents can be included in these ersatz cladograms, but also lakes

39.3. From the plate tectonic viewpoint, the primary features of the earth's surface are the lithospheric plates, not the more obvious but superficial continents and oceans. This figure illustrates the modern pattern of lithospheric plates. Redrawn from Duxbury and Duxbury 1994.

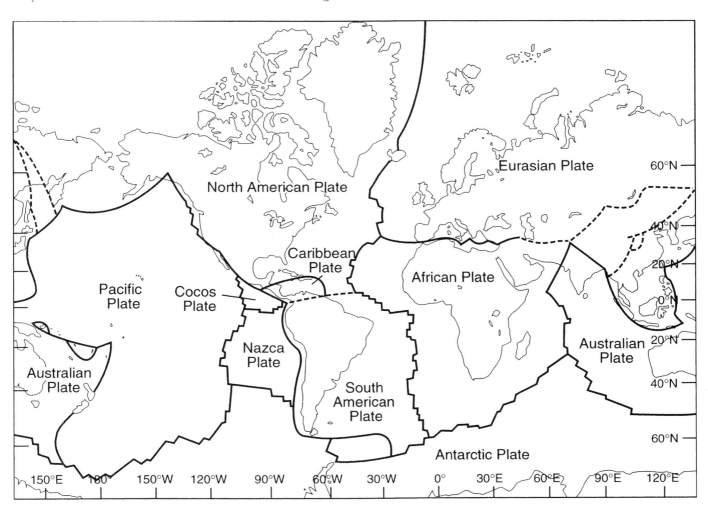

and continental islands, which rise from submerged continental crust, but probably not oceanic islands.

Dissent from Plate Tectonics and Continental Drift

Continental drift and the theory of its mechanism, plate tectonics, are widely, but not universally, accepted. Chatterjee and Hotton (1992) provide a selection of the dissenting views, and there is also a website—run by Australian geologists—devoted to this issue: http://www.ncgt.org. The dissent is of two kinds—that from continental drift and that from plate tectonics. The later, although interesting, is not yet relevant to paleobiogeography (no creatures live in the earth's mantle—we think). On the other hand, if continental drift is not as we suppose it to be, then our interpretations of how the distributions of animals and plants came about are likely incorrect. We remember that many biogeographers prior to the general acceptance of drift regarded the distributions of organisms to be entirely consistent with their notion of fixed continents.

There was, of course, a unified theory of continental form and placement before plate tectonics, which was that mountains and ocean basins resulted from wrinkling of the crust following from the contraction of the earth as it cooled. By the mid-twentieth century, little attention was paid to this notion, but it is now generally accepted as an explanation of major features on both Venus and Mars. Neither, so far, shows convincing evidence for plate tectonics.

All (or almost all) of the alternative theories were put forward by scientists to resolve what they saw as difficulties in the theory of plate tectonics. All have some evidence to support them. The adherents of plate tectonics, who are the majority, believe that the past success of this concept indicates that further research will resolve these difficulties, real or apparent, in a way supporting plate tectonics. And so far they have been right—but we cannot be sure that new discoveries will not generate another Wegener. Several of these alternative views, however, have influential supporters in places such as Australia, New Zealand, India, and Russia, and this must be remembered when reading paleobiogeographical papers from these countries.

Theories that the continents have moved laterally, from place to place, give different interpretations of biogeography from those that hypothesize that the continents have remained in place. Some views, especially that the continents moved vertically—sometimes partly submerged, sometimes not—may be consistent with plate tectonics and have no special paleozoogeographical implications. The sinking of the Lord Howe Rise east of Australia seems to be associated with the fragmenting of the eastern part of the Australasian plate on its northward drift. The notion of sunken lands in the western Pacific (e.g., Pacifica or the Darwin Rise) may be relevant to the distribution of Asiamerican dinosaurs, such as tyrannosaurs or ceratopsians. Unlike some modern zoogeographers, no dinosaurian zoogeographers yet take the vertical motions concept seriously.

Plate tectonics is the most popular interpretation because of its success in linking together seemingly disparate kinds of geological observations. This theory relates paleobiogeography, the position and structure of mountains and other topographic features, and the location of mineral deposits

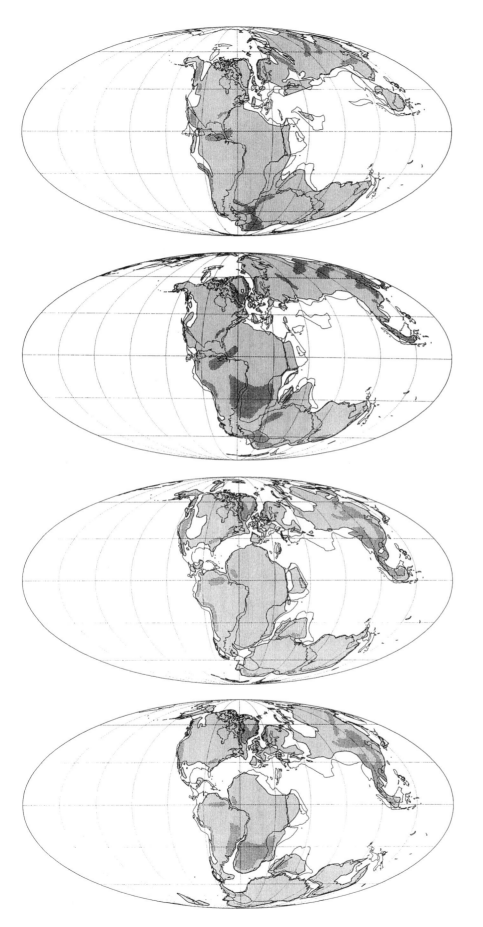

39.4. Paleogeographic reconstruction of continental positions during the Late Triassic (Carnian through Rhaetian Ages). Light stipple shows landmasses, highlands in heavier stipple. The modern outlines of the continents are indicated by light lines. The major landmasses of the world have come together to form the supercontinent Pangaea (or Pangea), although a shallow, narrow seaway separates North America from Europe. Western Europe consists of a series of small islands, a configuration that will remain through the remainder of the Mesozoic Era. Map from Smith et al. 1994.

39.5. Paleogeographic reconstruction for the Early Jurassic (Pliensbachian Age). From Smith et al. 1994.

39.6. Paleogeographic reconstruction for the Middle Jurassic (Callovian Age). The Atlantic Ocean is in its infancy and a shallow seaway has invaded western North America. From Smith et al. 1994.

39.7. Paleogeographic reconstruction for the Late Jurassic (Tithonian Age). India, Antarctica, and Australia (East Gondwanaland) are separating from Africa and South America (West Gondwanaland). From Smith et al. 1994.

39.8. Paleogeographic reconstruction for the Early Cretaceous (Hauterivian-Barremian Ages). The North Atlantic continues to grow, and the South Atlantic is starting to become a significant body of water. From Smith et al. 1994.

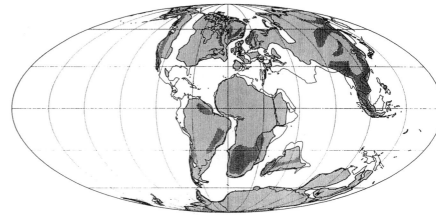

39.9. Paleogeographic reconstruction for the later Early Cretaceous (Albian Age). Antarctica and India are widely separated from Africa, India has lost contact with Australia and become an island, and shallow seas flood much of Australia and North America. From Smith et al. 1994.

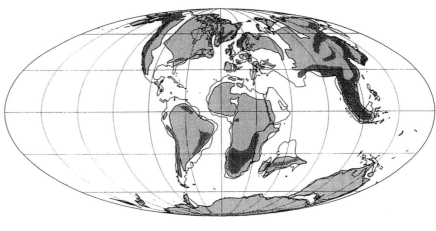

39.10. Paleogeographic reconstruction for the early Late Cretaceous (Turonian Age). Shallow seaways cover many of the continents, and Madagascar has separated from India. From Smith et al. 1994.

39.11. Paleogeographic reconstruction for the Late Cretaceous (Campanian Age). Continental flooding continues, and the continents are starting to move into positions reminiscent of modern geography. By the end of the Cretaceous Period (Maastrichtian Age), North America will have made a transient contact with South America, and the seas will have largely withdrawn from the continents. From Smith et al. 1994.

(among other things) in a way not yet done by the alternatives and certainly not by any previous geological theory. In other words, so far it has more explanatory power than other theories. It has proved quite influential in dinosaurian studies. Under the previous paradigm of static continents, no one seriously considered *Dryosaurus* from the Late Jurassic of the Rocky Mountain region and *Dysalotosaurus* from the Late Jurassic of southern Tanzania to be closely related. They lived too far apart. But we now know that in the Late Jurassic, Africa was adjacent to South America and not far from North America, so considering them as two species of a single genus (*Dryosaurus*) no longer seems unlikely on geographical grounds.

Figures 39.4 through 39.11 summarize current thinking about the changing shape of world geography during the Mesozoic Era. The position of the continents had a profound effect on the distribution of dinosaurian faunas.

Theories of Biogeography

Having some notion of its geographical background, let's now consider the aims of biogeography. Biogeography attempts to explain the distributions of organisms in terms of four factors: climate, the geographical distribution of resources and underlying soils or rock types, barriers to dispersal (such as mountains, deserts, and seas), and the evolutionary history of the organisms in the areas of interest. There was a feeling among some biogeographers that biogeographical processes occur faster than evolutionary ones, and certainly faster than large-scale geological ones—that, for example, if the animals of an island are exterminated by a Krakatau-like volcanic eruption, those that reappear when the island cools are drawn from the same populations of the same species as were the original inhabitants.

Here we are interested in much longer times. Paleobiogeography attempts to make sense of the distributions of plants and animals both in terms of their evolution and the evolution of the earth's surface, i.e., the positions and past movements of continents, islands, and oceans, as well as past climates.

Historical Biogeography versus Ecological Biogeography

Paleontologists assume that evolutionary (and geographical) history is the major factor in determining distributions of organisms. Not all biogeographers agree. Some argue that it is solely or predominantly ecological considerations—specifically climate, competition, predation, and availability of food—that determine distributions, just as some ecologists argue that it is solely ecological factors that determine how biotic communities are put together. These biogeographers saw that the number of species in a region is governed by the size (area) of that region. That being so, what role can historical processes play? If, on an island for example, there were only opportunity (i.e., area) for a small number of species to coexist, then an invading organism had a given probability of becoming established there, regardless of its history and that of the island. And this seems to be the case for islands, and close analogs of islands such as high mountain ranges isolated by deep valleys, tropical valleys isolated by intervening mountains, and the remnants of forests separated by cultivated land. This process of assembling such an insular community is assumed to be rapid in terms of geological time. It should be noted that these biogeographers did not entirely neglect historical factors; they just regarded them as subordinate to ecological ones. And then, about 25 years ago, shortly after continental drift became widely accepted, the situation changed and ecological biogeography was pretty much forgotten, at least among paleobiogeographers. After all, the phenomena and issues of concern for ecological biogeography are not easy to see in the incomplete fossil record.

Vicariance Biogeography

Before the acceptance of continental drift, no one realized that continents had histories. Mountains had risen and eroded, and epicontinental seas flooded the lowlands, but there were only two kinds of important changes, biogeographically speaking. These were the establishment and loss of the land bridges linking different continents and changes in the climate—and that was climatic, not continental, history. Thus, paleobiogeography was simply biogeography on a grander, or at least longer, scale. With the recognition of continental drift, it became clear that continents had histories. For example, during the Paleozoic Era, North America was adjacent to Australia, but early in the Mesozoic it was adjoined to Europe, and it is now connected to South America. Such different geographical relationships obviously had a great effect on the potential paths of migration and dispersal.

Furthermore, the movement of the continents suggested that, like a ship, they had carried with them complements of (evolving) plants and animals. Where continents had long been in contact, such as Europe and Asia, there would have been time for extensive intermingling of faunas, but where continents had been isolated, such as Australia and South America during the Paleogene Period of the Cenozoic Era, unique faunas would be expected. Vicariance biogeography (Fig. 39.12) considers the history and distributions of organisms in relation to the history and movements of continents and islands. It also explicitly concerns itself with the evolutionary history of organisms, because understanding the phylogenetic relationships

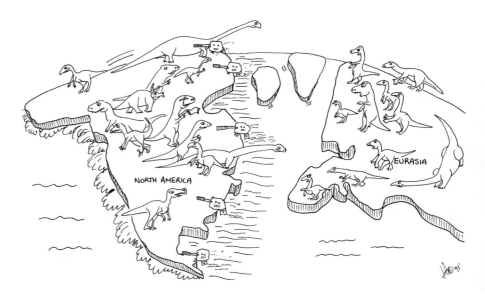

39.12. A cartoon illustrated the vicariance model of animal distribution—whole continents move, carrying along their faunas (and floras).

Drawing by S. Hocknull.

of organisms is essential to understanding their distribution and how that distribution came about. Vicariance biogeography accepts that the main mechanism for distributing animals and plants across the globe is the movement of continents and smaller landmasses (such as microcontinents).

Note that vicariance biogeography applies in only a limited way to marine organisms. It is relevant to shallow-water, benthic (bottom-living) organisms, and fossils show that benthic invertebrates have crossed the Pacific on microcontinents, but is irrelevant to the swimmers and drifters of the high seas. Although the oceans and their basins may, in a sense, be said to move (and sea floors clearly do), they cover so much of the earth's surface that it is mainly currents, temperature, salinity, and physical barriers, such as continents, that determine the distributions of pelagic organisms.

Physical barriers are largely established by continental drift. For example, mountains are created by colliding plates, seas by separating plates, and deserts by both the movement of continents into regions of little rainfall and the creation of such regions by the rain shadows of mountain ranges.

Dispersalist Biogeography

The school of biogeography (as well as paleobiogeography) that prevailed before the acceptance of continental drift—and therefore didn't realize it was just a school rather than the whole science—is now termed dispersalist (or Wallacean) biogeography (Fig. 39.13). The name follows from the preferred method of explaining the distributions of organisms by the movement of individuals (or populations) across continents and seas—in other words, the dispersal of organisms. Dispersalist biogeography accepts that the main mechanisms for distributing animals and plants is their own dispersal, whether by flying, swimming, walking, drifting, or even being blown away by a hurricane. (Its alternative name, Wallacean, honors Alfred Russel Wallace, who pioneered the science of biogeography.) Dispersal was aided by land bridges for large terrestrial animals and by winds and ocean currents for small buoyant animals and seeds. It was inhibited or prevented altogether by oceans, mountains, and deserts, as well as by regions with unsuitable climates.

Dispersalist biogeographers took to heart the concept of allopatric speciation—that species usually or always arose in small, geographically isolated

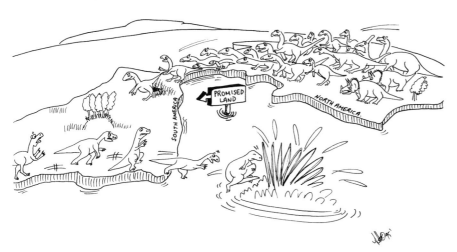

39.13. The dispersalist model of animal distributions. Movement of animals is significant, not that of continents. This may be because the continents don't move (an idea no longer generally accepted), or because animals move from place to place much faster than continents drift.

Cartoon by S. Hocknull.

populations. To acquire the broad geographical ranges often seen, animals and plants would have to emigrate outward from the small areas, or centers of origin, where they originated. To get to islands, or different continents, they would have to migrate or disperse, by intent or by accident, across oceans and land bridges—unless, of course, they could fly. But even then, because most animals stick to some home territory, they would still disperse only when seeking a new territory or as the result of some unusual event, such as a hurricane.

Dispersalist paleobiogeography thus considered it important to discover where species or evolving lineages of organisms originated and how they arrived (both literally and figuratively) at their present distributions. Although there were some notable exceptions, such as the American paleontologist William Diller Matthew, some dispersalist biogeographers thought that dispersal occurred almost instantaneously compared with evolution. Thus, distributions were the results of evolution, not components of it. Although vicariance biogeographers have criticized and played down the role of this school in understanding the changing distributions of plants and animals, some explanations by dispersalists have stood the test of time. The lineages of horses, for example, are widely accepted to have first appeared in North America, and then dispersed to Eurasia and Africa, eventually becoming extinct on their homeland—North America. Similarly, we think that ceratopsians, tyrannosaurs, and hadrosaurs originated in Asia before dispersing into North America sometime in the Cretaceous.

Dispersalist and vicariance biogeography share some conceptual elements—the role of geography in speciation (i.e., allopatric speciation) and the land bridges of dispersalists theoretically (although not geographically) correspond to the continental collisions of vicariance. They differed in that vicariance treated the movements of whole faunas and floras rather than species as populations or individuals, and vicariance movements were geologically slow as opposed to the relatively faster movements of populations and individuals.

Recent phylogenies based on molecular data lead some biogeographers to claim that dispersal is a better explanation of biogeographical patterns than vicariance, which is an unnecessary hypothesis (e.g. Waters and Craw 2006), and that critical biogeographical processes occur much more rapidly than continental drift and evolution (Queiroz 2005).

Panbiogeography

There is a third school popular in some parts of the world (e.g., New Zealand). This is panbiogeography (Fig 39.14), founded by Leon Croizat. This school is significant in that it is often taken to have been the forerunner of vicariance biogeography—but not by Croizat himself. Croizat, originally a botanist, was a prolific and engaging writer, but one who preferred painting a picture in words of his ideas to stating them explicitly. Basically he took the view that biogeography was intimately related to evolution and to time in geologically long periods; Croizat was not an ecological biogeographer. Organisms—that is, lineages of organisms—were older than the geography of today, and so modern biogeography is the result of paleobiogeography (and evolution). Croizat did not accept that speciation necessarily took place at

39.14. The panbiogeographic model of animal distributions – animals and continents both evolve together.

Cartoon by S. Hocknull.

one point, geographically speaking, and then organisms spread from that point of origin. Apparently, he regarded gene flow as more rapid than speciation. Instead, many species originated over large regions, including where they now live, and their ranges subsequently changed, either by extinction or through the emergence or submergence of lands; he did not accept plate tectonics. Croizat used the image of a shattered pane of glass in a frame. The original pane was the ancestral species, the fragments the later, descendant species. The fragments did not move – they are constrained within the frame – and he thought that the descendant species did not necessarily disperse (or be transported by moving land masses) to their present ranges. He did not reject dispersal out of hand, but instead felt that dispersal "fine-tuned" the distributions we see today derived from those of the ancient past.

Panbiogeography is basically a way of looking at and studying the patterns of distributions of animals and plants, without – at least according to its proponents – presupposing how they came to be. Unlike the more widespread schools that presuppose mechanisms for moving animals and plants across the globe and explain modern distributions in terms of these, panbiogeography endeavors to explain past distributions in terms of modern ones. This is a different way of thinking, but it is sound in that we can observe processes of the present but only infer those of the past.

The "Modern Synthesis" of Biogeography

We can characterize, almost parody, dispersalist zoogeography as assuming that animals move, but not continents, and vicariance zoogeography as assuming that continents move, but not animals. Although there was a time when vicariance biogeographers doubted the role of dispersal, as Brooks and McLennan (2002) point out, if there were no dispersal, all organisms would still live at the spot where life originated; so most biogeographers now accept that both occur. There is an important difference in scale here. Dispersalist biogeography treats the movements of organisms, exemplified by individuals, which may be adults or young. Vicariance biogeography treats the movements of entire faunas and floras. Dispersalist biogeography is concerned with relatively short times, geologically speaking, and vicariance with long periods. In this light, we would expect most of the distributions at large scales to be explained by vicariance biogeography, while dispersalist biogeography accounts for the details of and the exceptions to, mostly at small scales, vicariance theory. No one would use vicariance biogeography to explain why lions, zebras, hyenas, and hippos, but not giraffes, crocodiles, and okapis live within Ngorongoro Crater, although all of them inhabit the nearby Serengeti Plain (or other parts of Africa). Likewise, we no longer appeal to dispersalist theory to explain why (until recently, geologically speaking) the large tetrapods in Australia were marsupials and birds (and one great goanna), and in New Zealand large ground-dwelling birds, rather than eutherian mammals as in most other lands.

The biogeographical application of continental drift unifies many aspects of biogeography into a single geological/biological concept. Dispersalist biogeography took physical barriers as givens inserted into the theory from geology. In vicariance biogeography, these are results, even deductions, of continental drift. Dispersal is, obviously, governed by geography. Creatures can disperse from Alaska to Siberia because they are nearby, but not from Alaska (or Siberia) to New Zealand—animals cannot teleport. Geography in turn is governed by continental drift. So, dispersal is dependent on and is constrained by, vicariance; to put it metaphorically, dispersal is an actor in the theater of vicariance. This is not a necessary situation. Were there creatures on Mars, where there is no continental drift, all biogeography would be dispersalist, as it will be here when drift ceases. It just so happens that drift occurs on earth and therefore, it governs dispersal.

DISTINGUISHING VICARIANCE FROM DISPERSAL

The creation of phylogenetic systematic methodology by Willi Hennig gave us a method of determining the genealogy, basically the history, of living organisms. This was easily extended to fossil organisms, but the methods are applicable to anything that has a history (although, as far as I know, they have not yet been applied in astronomy or the social sciences). So a dendrogram representing the history of continents, for example, can be drawn. One can go further and apply such methods to faunas or floras, as well (e.g., Gates et al. 2010). These give powerful tools for understanding what happened in the past.

Both dispersal and vicariance can explain biogeographical patterns, so how do we tell which is true in a given situation? We are assuming that the phylogenies of the creatures of interest are known or, at least, can be determined. If vicariance is true, then whole suites of organisms, whole floras and faunas, move. And, when they move, they create the opportunity for allopatric speciation to take place in these floras and faunas, which is sometimes termed vicariance speciation. In this case, the cladograms of unrelated lineages of organisms (unrelated clades) should all show the same basic pattern, and this pattern, in turn, should match that of the history of the continents involved, based on geological data (as mentioned in the passage on the mechanism of continental drift).

But one can't just match the cladogram for geographical regions with that for lineages of organisms—however much this superficially appears to be done—for this would be comparing the proverbial apples and oranges. Instead, one must match area cladograms—those in which the terminal branches are regions of the world and which come in two kinds. They may be based on geographical history or on the usual kind of cladograms in which the terminal clades have been replaced with the regions in which those taxa live. In comparing area cladograms based on lineages of organisms with those based on paleogeography, one is comparing what are, after a fashion, genealogies. But there are differences to be aware of. As often illustrated, the genealogies of the continents and the cladograms of organisms look the same. But lineages of plants and animals diverge, giving the familiar dichotomously branching tree. Their area cladograms also dichotomously diverge. This is not the general case for continents, which not only separate and drift apart, but also merge—remember Pangaea. Thus, their genealogies sometimes branch dichotomously and diverge. At other times, continents converge and the branches of their cladograms merge. So, only selected, appropriate parts of the area cladograms of continents are compared with those of lineages of organisms.

To return to distinguishing vicariance from dispersal, if one looks only at a single clade, the match between its area cladogram and that of the continents just might be fortuitous. Therefore, one matches the genealogy of the continents, their area cladogram, with those of several clades, all clades of dinosaurs for example, or all of crocodyliforms (topics of Upchurch et al. 2002 and Turner 2004, respectively) or even of less closely related forms. Taking mythological creatures as examples, mythological because this should minimize preconceptions as to their (mythological) phylogeny and how they came to their (mythological) distributions, we can demonstrate this in a suitably simplified fashion. In Fig. 39.15, we see distributions (Fig. 39.15B and 39.15C) that match that of the area cladogram of the continents (Fig. 39.15A). (Antarctica, with no mythological creatures known to me, is excluded.) If these area cladograms all match up (as in Fig 39.15), and especially if the organisms have rather different dispersal abilities, then one may conclude that their distributions were determined by those of their continents. In other words, they were due to vicariance, not to dispersal. Upchurch et al. (2002) and Turner (2004) include further refinements of this method. (The mythological data, incidentally, comes from Nigg (2002), supplemented by Borges (1969) with a little pedagogical modification.)

What if the diagrams do not match? Well, we know that the geographical area cladogram must be the standard to which the other is compared, for that is—insofar as the geologists have got it right—the history of the regions. If the diagrams match, then the organisms "stayed at home" on their continents, and their distribution is due to vicariance. And if all the lineages duly speciated, and none died out or emigrated (i.e., dispersed), the diagrams will match. So it is the discrepancies of the organisms' area cladograms from the geographical area cladogram (here, that of the continents) that give us the means to determine which distributions are due to vicariance and which are not.

In cases where the area cladograms of organisms don't match those of continents, too many terminal branches indicates that the extra branches are due to dispersal, and too few terminal branches implies that some of the branches are missing, possibly due to extinction of some lineages. Returning

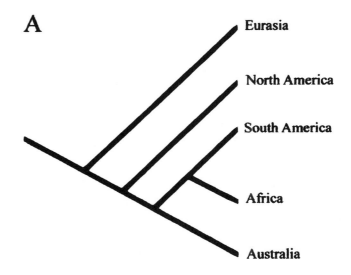

39.15. Area cladograms for the continents (A) and two clades of mythological creatures (B and C). The topological congruence of the area cladograms indicates that vicariance is the preferred explanation for the distributions of the creatures (more details in text). Because there are no Antarctic mythologies, Antarctica is not included here. It should also be noted that morphology is no guide to phylogeny for mythological animals.

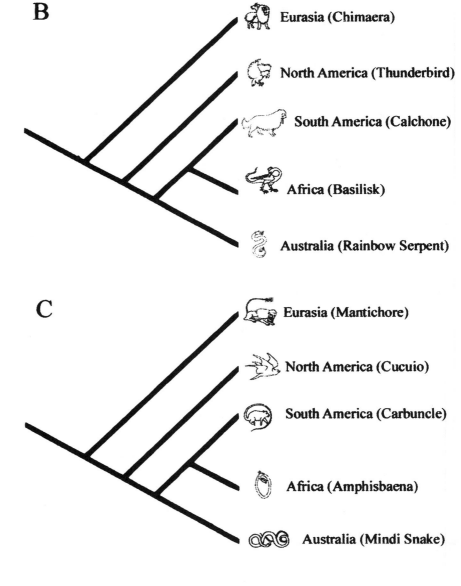

Ralph E. Molnar

to our mythological examples, consider the situation of Fig. 39.16. We see a general match to the area cladograms of the continents (Fig. 39.16A), except for the crocotta, that appears in Africa. Such a supernumerary appearance is likely due to dispersal. If it were the chimera (instead of the crocotta) that appeared in both Eurasia and Africa, further phylogenetic analysis would be necessary to determine if the chimera were the sister group of, or were ancestral to, the axehandle hound, carbuncle, and catoblephas. If ancestral, then the chimera simply survived in Africa without vicariant speciation; if the sister group, then it dispersed to Africa.

And what if there are too few terminal branches, as in Figure 39.17? There is almost a match between the area cladograms (Figs. 39.17A and 39.17B), but with the Australian clade missing (in Fig. 39.17B). Here the situation is less straightforward. The clade could be missing because its members never got to Australia, or because they did, but became extinct at some time before the present. (If we are looking at fossils, there is a third possibility—they were there, but their fossils have not yet been found.) Thus, we need to consult another area cladogram, e.g., that of Figure 39.17C. Actually we should consult several, to establish a general pattern of area cladograms, but we'll settle for one that represents the general pattern. Because

A

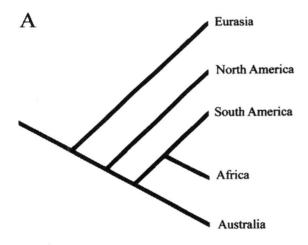

Eurasia

North America

South America

Africa

Australia

B

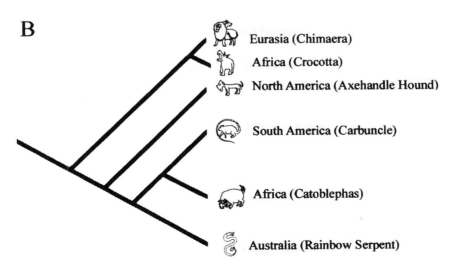

Eurasia (Chimaera)

Africa (Crocotta)

North America (Axehandle Hound)

South America (Carbuncle)

Africa (Catoblephas)

Australia (Rainbow Serpent)

39.16. Area cladogram (B) with more branches (for the chimera clade) than expected from the geographical area cladogram (A). These area cladograms are not topologically congruent, and the extra African branch suggests dispersal of the crocotta into Africa.

39.17. Area cladogram (B) with fewer branches than expected from geographical area cladogram (A). This might indicate either extinction of the Australian clade or that it never occurred in Australia. If, as here, the cladogram (B) does not conform to the general pattern of comparable cladograms (exemplified by C), we may suggest that there once was a sister group to the minhocao + flying serpent clade in Australia, but its members have become extinct.

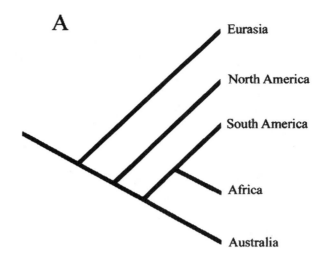

A

- Eurasia
- North America
- South America
- Africa
- Australia

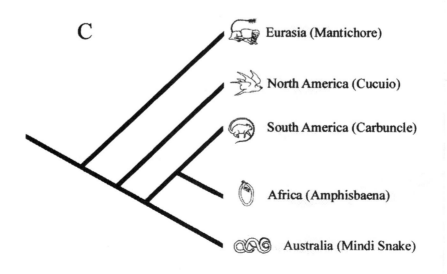

B

- Eurasia (Barometz)
- North America (Pinnacle grouse)
- South America (Minhocao)
- Africa (Flying Serpent of the Western Desert)
- Australia - none

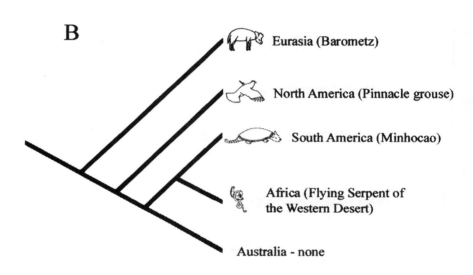

C

- Eurasia (Mantichore)
- North America (Cucuio)
- South America (Carbuncle)
- Africa (Amphisbaena)
- Australia (Mindi Snake)

this does have an Aussie sister group of the carbuncle + amphisbaena clade (and matches the geographical area cladogram of Fig. 39.17A) we can suggest that local extinction of members of the minhocao + flying serpent lineage is most likely the explanation. If the South American-African clades did not generally have Australian sister groups, then we could conclude that relatives of the minhocao + flying serpent clade never reached Australia.

Be forewarned, things are not as simple as in the examples here, and in actual research complications often appear. For further information, one may consult the classic work of Gareth Nelson and Norman Platnik (1981). I also found chapter 4 of Brooks and McLennan (2002) illuminating (but it is not for beginners). More advanced applications are found in Hunn and Upchurch (2001).

The complications arise because, metaphorically speaking, the past does not issue visas. We (except for some quantum physicists) believe there was a unique past, not a myriad of them that have converged to give the present. But the older the past we study, the less information survives to understand this past. There may not be multiple pasts, but clearly there are multiple pictures or models of the past that seem consistent with what we know now. To find out which is most consistent with our knowledge, statistical tests must be applied (as in Gates et al. 2010). Bayesian statistics is a useful method here, and, although hardly material for an already long introductory chapter on paleobiogeography, it is worth mentioning that chapter 9 of Baeyer (2003) is a readable introduction to Bayesian analysis.

Paleobiogeography explicitly requires several kinds of information. On the geographical side, these include the past positions and configurations of the continents and of smaller terrains and microcontinents (like New Zealand) as they move across the oceans. Whether these lands have been submerged (like New Zealand in the Oligocene) and when and if they later reappeared (like New Zealand in the Miocene) is also important. One must know the phylogeny of the group of interest, and often of other groups as well, and when the clades of interest diverged from their sister clades. For some applications faunal lists are important, where the fauna has been adequately sampled, as are the analytical techniques used. With such data, and the appropriate analytical techniques, more can be learned than just what lived where and when. Discerning when dispersal versus vicariance was important is but the first result. Did the region of interest have a uniform fauna, or was it divvied up into faunal provinces? Did the breakup or merger of landmasses affect diversification (cladogenesis) or extinction? Did climatic changes? Are there preferred regions where taxa tend to originate? Even whether clades lived in a region where they are not represented in the fossil record may be investigated.

TWO OTHER FACTORS

So far, we have considered vicariance and dispersal, and these certainly have attracted much attention. But there are two other factors that influenced the composition of communities in the past and even now. Consider the makeup of land-dwelling tetrapod communities in, say, the Rocky Mountains region of North America 60 million years ago, in the Paleocene Epoch. There were various kinds of mammals trotting about—condylarths, marsupials, multituberculates, plesiomorphic ungulates, even primates—along with turtles, crocodilians, champsosaurs, frogs, salamanders and such. But who would have predicted such a fauna 10 million years earlier? Then the dinosaurs were dominant; now—that is, in the Paleocene—they're gone, except for birds. So extinction also plays a role in the makeup of communities—if your clade has become extinct, you don't contribute to the community structure. This extinction was global, as far as we can tell, and so had more-or-less uniform biogeographical effects. But extinction can occur one place and not another, and that clearly does affect the geographical distribution of organisms.

For example, in Jurassic North America the large carnivorous dinosaurs included both tetanurans (*Allosaurus* and *Torvosaurus*) and ceratosaurs (*Ceratosaurus*). At some time during the Cretaceous, ceratosaurs disappeared and allosauroid tetanurans (e.g., *Acrocanthosaurus*) became the predominant large predators. Later in the Cretaceous, they too disappeared and tyrannosaurs were prominent. In other lands, however, other things happened. In South America, for example, Jurassic tetanurans (e.g., *Condorraptor* and *Piatnitzkysaurus*) apparently gave way to allosauroids (*Giganotosaurus*) and ceratosaurs (abelisaurs) in the later Cretaceous. In North America, the extinction of the Jurassic ceratosaur and allosaur clades resulted in, or permitted, coelurosaurs (tyrannosaurs) to be the prominent predators. But in South America—where tyrannosaurs apparently never arrived, although other North American clades did—allosaurs and ceratosaurs retained their prominence. A more recent, and more widely known, example, mentioned previously, is the distribution of the equid clade, horses, zebras, and such. Although they had a long history in North America, and well as Eurasia, after the Pleistocene, they had disappeared from the New World and were reintroduced by humans. With an incomplete fossil record, a paleontologist of the future might be confused by the appearance of an unusual convergence of North American horses to those of Eurasia, not realizing that there had been extinction followed by dispersal. So extinction has a clear effect on biogeographic patterns, whether local or mass extinction.

The other factor is one that might have been thought of as dispersal, if dispersalists had known about continental drift. They didn't, of course, and so this aspect was never considered. After widespread acceptance of the role of plate tectonics, dispersal came to be retroactively redefined as the movement of individuals across some kind of geographic barrier (e.g., Nelson and Platnik 1981). This is the mechanism that accounts for what kinds of animals (and plants) are found on the Galapagos Islands, for example. But there is another kind of dispersal that comes into play when geographical barriers cease to exist, as when climatic change brings moisture to deserts, when high mountains erode down, or—if you're a marine creature—when land bridges become submerged. Then, not just lone individuals can disperse, but suites of individuals from different species can all proceed across the defunct barrier. This mechanism is known as geodispersal (Lieberman 2005), to distinguish it from the dispersal of individuals. Geodispersal came into play when a dry land route opened between North and South America in the Late Cretaceous, allowing members of the hadrosaur and ceratopsian clades to emigrate into South America. It had previously also allowed those of the ceratopsian, hadrosaur, and tyrannosaur clades to move into North America when the western North American mini-continent was connected to Asia.

Extinction removes players from the stage, so to speak, and allows others to take their place. It may operate quickly or slowly, over the entire planet or just a continent (or less) or an archipelago. Only one clade may be affected or many. Geodispersal probably operates over time periods intermediate between those characteristic of vicariance and dispersal. It can operate equally well for land-dwelling and marine creatures. It transfers organisms from the landmass on which (or the ocean in which) they evolved to others, or from one part of the landmass to others, and may involve individuals, clades, or faunas (and floras). Vicariance explains how separate landmasses come to have their complements of animals and plants. Geodispersal shows how they may intermix when landmasses come together. These four mechanisms—dispersal, vicariance, extinction, and geodispersal—seem to account for both biogeographic patterns at a given time and the changes in these patterns with time.

Predictions in Biogeography

From our knowledge of continental positions throughout the history of the earth, we should be able to work out what kinds of animals inhabited areas and times for which the fossil record is unknown. For example, Antarctica is covered by a massive ice cap that makes collecting fossils difficult, although not impossible. If we wish to get some idea of the Cretaceous dinosaurs of Antarctica, we can look to the distributions of dinosaurs of other southern continents, which are reasonably well known. From these distributions we can work out what dinosaurs lived in Cretaceous Antarctica. In fact, this was done in 1989 (Molnar 1989) when a prediction that small ornithopods lived in Antarctica was later verified (Hooker et al. 1991).

The discovery of dinosaurian fossils in Antarctica, together with the finding of remains of forests, shows that Antarctica was not always as cold as it is now. Because Antarctica has been in the vicinity of the South Pole since at least the beginning of the Cretaceous, this is an impressive example of climatic change over geological time, a topic to which we now turn.

Changes in the weather are more than well known; in fact, in some parts of the world weather is almost synonymous with swift and unexpected change. But the weather changes more than from day to day; it changes from month to month, with the seasons, and even over historical time, not to mention geological time. Long-term changes—those which take longer than a few years—are considered climate, rather than weather, so in understanding dinosaurian zoogeography, it is the Mesozoic climate, rather than the Mesozoic weather, that is important. Climate is controlled by well-known factors such as the input of heat from the sun and the position of the earth in its orbit, among others. Both of these have changed, but the former so slowly that we needn't worry about it here, even for climate, and the latter—at the moment—seems not to have had as great an influence on Mesozoic climate as it has since, especially during the ice ages.

Paleoclimatology

The composition of the atmosphere—not which gases are present, but in what concentrations—also affects climate, as we should all now be well aware. Carbon dioxide levels during the Mesozoic apparently were some four (or more) times what they are presently (cf. Berner and Kothavala 2001; Bergman et al. 2004). However, because this gas was presumably uniformly spread through the atmosphere, it would seem to have had no effect on the zoogeography of land-dwelling animals. But climate is also linked to geography and therefore to continental drift. As continents move north or south, they stray into regions of different climates. A prime example of this is Australia. Early in the Cenozoic Era, this continent was situated in the southern temperate zone, with its southern coast near the Antarctic Circle. The climate was cool and moist. Today Australia is farther north, almost straddling the Tropic of Capricorn, and it is basically a desert continent. It has crept north from the region where ascending air masses dropped their moisture as rain, into the region where descending air masses heat as they fall and so absorb any water available on the ground. So a cool, moist climate came to be replaced by a warm, dry one. This instance shows how the concept of drift can explain some apparent changes in climate seen in the geological record. In fact, here the global climate has not changed much—what changed was the location of Australia.

Some geologists believe that continents, and their drift, can influence climate in a more profound way. Continental collisions can create mountain ranges, and mountains influence both how the air flows and how it is heated. It has been suggested, for example, that the rise of the Himalayas profoundly affected the climate during the Cenozoic (Raymo and Ruddiman 1992).

Evidence of Past Climates in Sedimentary Rocks

Mud cracks and raindrop impressions are obvious indicators of the weather of the past. But they do indicate weather rather than climate, unless there is layer after layer of them. Geochemical changes that occur in sediments during and just after their deposition are better indicators of climate. Several different kinds of continental sediments are laid down only, or usually, under certain climatic conditions. Coals, for example, form when the climate is moist enough to support abundant plant growth, often in swamps or similarly moist environments; there are no desert coals. Clay minerals may indicate the degree of weathering and therefore the amount of precipitation. Red beds are oxidized sediments and were once thought to indicate arid conditions. But this view was unfounded, and they are now thought to form under several different kinds of climates. When the air is dry, water evaporates, leaving behind any dissolved salts it may contain. Seawater carries lots of dissolved substances, so when it evaporates, specific minerals known as evaporites are left behind, often in specific sequences. These, when found in thick beds, indicate a dry climate—if not really hot, at least warm enough to evaporate large amounts of seawater. Similarly some minerals may be deposited by evaporating groundwater. Such occurrences, especially the caliches formed by deposition of carbonate, also indicate aridity.

Of the various sedimentary paleoclimatic indicators, the most controversial are tillites. Tillites are rocks formed from tills—unsorted deposits of the rocks, gravel, sand, and silt that result from glacial action. The controversy concerns how rocks can be conclusively identified as tillites. Marine sediments, although not discussed here, can also provide paleoclimatic evidence – coral reef deposits, for example, indicate warm climates.

None of these methods gives specific measurements, such as temperatures, of past climatic conditions, but temperatures can be calculated from the ratio of oxygen isotopes (16 to 18). The rate at which these isotopes are taken up by various compounds, being a chemical process, depends on the temperature, so, with appropriate mathematical treatment, the ratio can indicate a temperature. These are often sea or groundwater temperatures, rather than air temperatures, but they usually indicate at least roughly what the air temperatures were. However, one must be wary. In lowland New Guinea, groundwater temperatures near rivers flowing down from the mountains reflect the cold air temperatures high in the mountains rather than the tropical conditions of the lowland. Furthermore, oxygen-isotope ratios may be altered by diagenesis long after the sediments bearing them were deposited. Even so, with careful interpretation, they provide important information.

Paleobotanical Evidence for Past Climate

Fossil plants provide another way to gain some idea of past climates. Animals can shelter from the climate, but plants usually can't (unless they are annuals, or die back to roots or tubers). Thus, the kinds of plants that lived together in a place indicate what the climate of that place was like, at least in a general way. There is, however, always some uncertainty about whether the fossil plants had the same taste in climate as do their modern descendants.

The weather is such an important factor to plants in some places that they develop specific adaptations to deal with it. Thus, climates can be deduced from plant morphology (Dilcher 1973). Leaf size and the nature of leaf margins are often used. Large leaves and leaves with entire (i.e., smooth) margins (Fig. 39.18) indicate warm or moist conditions, and small leaves and leaves with notched ("nonentire") edges suggest cool or dry conditions. It is the proportion of the large leaves or those with entire margins among those fossilized at a site that is used as the indicator, not simply the occurrence of these kinds of leaves. And the inferences are necessarily a bit vague – conditions are interpreted as warm or moist (or both), but temperatures or rainfall amounts are not given as such. Drip tips, elongate tips to leaves (Fig. 39.18), are taken to imply that there was so much rain – at least seasonally – that specific structures were needed to facilitate the shedding of the water. These correlations between climate and leaves were recognized in angiosperms, but some paleobotanists have applied them to other plants as well. Seasonality may be indicated by growth rings ("tree rings"). And Eocene (early Cenozoic) *Banksia* seeds, similar to those of today that require fire to germinate, indicate seasonal dry conditions, conducive to fires.

Climatic conditions can be inferred both from features of the plant communities and from morphological features of individual plants. As mentioned above, the proportion of leaves (from all of the plants) with serrate margins is related to mean annual temperature. However, as with the

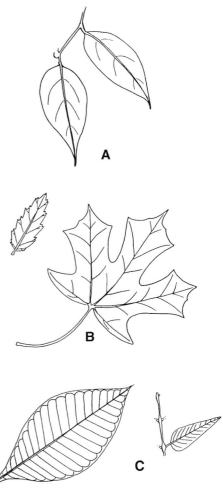

39.18. Features of leaves useful for inferring past climates. (A) Drip tips at the ends of leaves indicate high rainfall. (B) Notched leaf margins suggest cool or dry climates. (C) Unnotched, entire margins suggest wet or warm conditions. See Dilcher 1973 for details.

oxygen-isotope methods, caution must be used in inferring the temperature. The proportions of leaves preserved may be altered during the processes of fossilization, and soil types and amounts of rainfall also affect this proportion (Wing and Greenwood 1993).

Simulations of Climate

The basic factors that determine the climate are well known – the amount of sunlight, the position of the earth in its orbit, the amount of solar radiation reflected off clouds, seas, and ice, among others (Fig. 39.19) – so it should be possible to use a computer to simulate the weather and the climate as well as to predict their changes. This use of computers, like many others, began in the 1940s with the Hungarian mathematician John von Neumann. He quickly discovered, however, that it wasn't a simple problem and that many calculations were needed to simulate, and then to forecast, the weather. In fact, calculating the weather twenty-four hours in advance took nearly twenty-four hours to do. As computers became more sophisticated and faster, simulating the global climate became feasible. Simulations in the 1970s showed the potential importance of the greenhouse effect and so encouraged the development of better models. Run backward, these models have been used to simulate climates of the past.

The basics of the simulations for predicting the weather (and climate) are straightforward. Both can be expressed in numbers. These numbers are the values of certain properties of the weather, such as wind speed, barometric pressure, amount of rainfall per year or per season, etc. All are interrelated (see Fig. 39.19). Pressure changes cause winds, for example. The values of these at any one time can be calculated from their values at some previous time – in other words, tomorrow's weather depends on the weather today. This is simply saying that there is causation even in the weather.

The interrelations among these variables are complex, but well understood, so they can be expressed as a series of equations that can be handled by a computer. The equations describe the dynamics of a fluid, the gases

39.19. Simple model of the major features that influence weather and climate. These parameters and their interrelationships can be expressed as equations, which in turn can be used to simulate weather and climate.

Redrawn from Casti 1991.

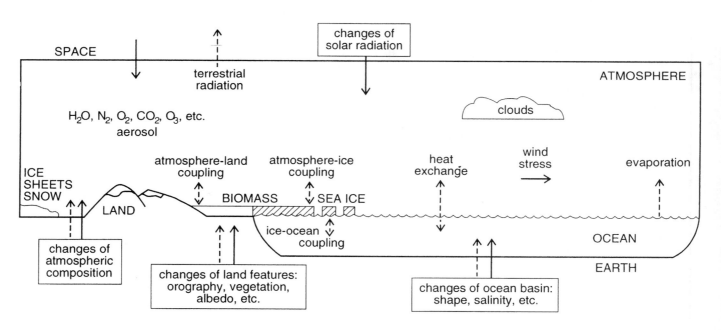

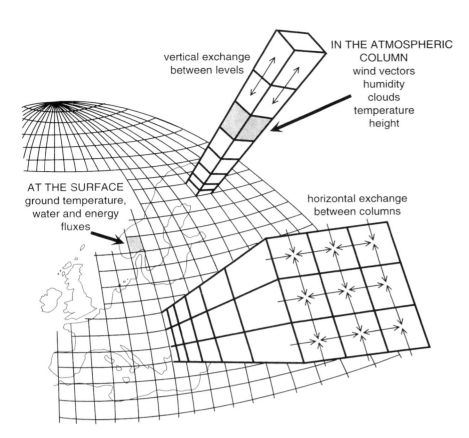

vertical exchange
between levels

IN THE ATMOSPHERIC
COLUMN
wind vectors
humidity
clouds
temperature
height

AT THE SURFACE
ground temperature,
water and energy
fluxes

horizontal exchange
between columns

39.20. In quantitative models of climate (and weather), the atmosphere if imagined as a three-dimensional grid. Calculations are then carried out for each cell in this grid. The large size of these cells necessary to keep the calculations from being prohibitively complex is one source of error in such simulations.

Redrawn from Casti 1991.

of the atmosphere, in four dimensions, including the familiar three dimensions of space and the fourth of time. They relate the input of heat from the sun to evaporation, changes in pressure, and air movement, both horizontally (winds) and vertically (updrafts and downdrafts). Vertical movements may result in precipitation, rain- or snowfall, and winds can transport both moisture and heat. Specifically, five equations describe (1) horizontal and (2) vertical movements of the air, (3) conservation of mass in going from one to the other (so that, for example, no simulated air vanishes when a downdraft hits the simulated ground but instead transforms into a wind), (4) heating of the air, and (5) conservation of mass for the moisture in the air (as either humidity or precipitation). Each equation is then calculated for a given part of the atmosphere, imagined (i.e., simulated) as a three-dimensional grid (Fig. 39.20; cf. Casti 1991 for details).

But climate cannot be simulated merely by simulating the weather for a long time. Although the one stems from the other, they involve distinct processes—such as changes in carbon dioxide concentration—with obviously different time frames, and their simulations require different information. For example, in simulating the weather, we wish to know—among other things—the temperature at various times during the day. For climatic processes, it is the average temperature that is important; for these processes, the differences in temperature at different times of the day are just trivial fluctuations. Variations in the amount of heat arriving at the earth that occur over long periods of time have little (direct) effect on the weather, but they are influential in determining the climate.

There are two basic kinds of numerical models. Those starting with values of the variables—temperature, pressure, etc.—for some specific time are used for forecasting. Thus, they can calculate (in theory, anyway) tomorrow's weather from information about today's. The second type, which calculates weather from general, or average, values of variables such as the amount of heat arriving at the earth, are used in reconstructing the climate of the past (and future). The most widely used of these are the GCMs, general circulation models.

It was generally hoped that such climatic models would allow us to accurately re-create, mathematically, the climates that prevailed at various times of geological history. This proved easier to hope than to do. Paleoclimatic models have encountered several problems and difficulties and even simulations of the modern climate are not exact. First, there are problems with simulating the interactions among the atmosphere and other components of the planet. The oceans, ice caps, topography of the land, and even clouds all affect the atmosphere and weather but have proved difficult to simulate in detail. Second, there are problems with the actual computation of the models. Heated air moves much more quickly than heated ocean water, and the difference in rates of movement creates problems for the models because the rate of computation, and hence simulated movement is the same whether simulating air or water. The simulations are also rather coarse-grained, which in recent models means that the weather is assumed to be uniform over patches about 500 kilometres across (Fig. 39.20). This is like assuming that the weather is always the same from New York City to North Carolina, or in both San Francisco and Los Angeles. The simulations represent only large-scale features.

However, the most extreme computational problems are those related to dynamic chaos, lightly referred to as the "butterfly effect." This occurs when imperceptibly small changes in the values of the variables can lead to substantially different results—in other words, when small differences at first are greatly amplified later on into large or even very large ones—the proverbial butterfly who flaps her wings in Rio one week, which leads to a hurricane in London (or wherever) the next. The situation is exaggerated in paleoclimatic models because there are no real (i.e., measured) values to work with in the first place. All the values—temperatures, pressures, and such—are inferred, and so doubtless less reliable than those we can measure today.

Viewed in this light, the last problem is perhaps no surprise. Simulated past climates do not always agree with climatic implications drawn from fossils. Simulations of the Permian (late Paleozoic) climate of central Gondwanaland (the part that is now South Africa) indicate extreme temperatures, to 60°C in the summer and −40°C in the winter. Yet a wide variety of fossil plants and vertebrates (largely therapsids) suggest a much more temperate climate. Perhaps the plants and therapsids were adapted to such extremes? But the sedimentary indicators and even oxygen-isotope paleotemperatures don't support these extremes either. It seems the problem is one of resolution and topography (Yemane 1993). There is evidence for large lakes in the region, with the water acting to cool the climate during the summers and warm it in winter. There is a similar lack of agreement between simulation and evidence for the Eocene (early Cenozoic) climate of central North America (Wing and Greenwood 1993). When considering claims for the ability of dinosaurs to survive where simulations indicate seasonally freezing temperatures, we must ask if there may have been similar ameliorating factors (such as the presence of high-latitude plant cover; Otto-Bliesner and Upchurch 1997).

But these problems all seem to concern the details of the simulations rather than the basics. Unless there is something important that we don't know about the climate, climatic simulation looks to be a promising tool for interpreting paleobiogeography. It may be, however, that there are important things we don't know about the climate, for there have been two such surprises. Clouds have turned out to absorb from three to five times as much sunlight as modelers had thought (Kerr 1995). And the climate (and sea ice) around Antarctica has a tendency to run in cycles of eight to ten years in duration (Yuan et al. 1996). Nonetheless, the use of simulations helps point up the importance of such discoveries, so that they can be incorporated into later generations of simulations.

Geography, Climate, and Evolution

Geography and climate have certainly had some effect on the course of evolution, even if by and large they have attracted little attention from many evolutionary biologists. But they are also related to the processes of evolution, if only because they are factors in natural selection. Animals and plants survive, mate, and reproduce in an environment, and the conditions of that environment play a role in determining their success at these activities. Climate and geography are major features of the environment and so may be expected to have a significant effect on survival, mating, and reproduction. The effect of natural selection on any organism is its success or failure in these three activities, so geography and climate are factors in this. Geography and climate may enter into selection in four basic ways—by the effects of land area, by the degree of separation or isolation between regions, by regions where species tend to originate ("cradles"), and by the effects of climatic change.

The basic, and still most widely accepted, mechanism for the origin of new species involves geography. This is the process of allopatric speciation. Briefly, it involves the separation of a species into two parts, with no possibility of mating between any members of one part and those of the other. Allopatric speciation comes about by the emplacement of some barrier, usually geographical, such as a sea, between the two parts. And if the barrier is in place long enough, the two populations are thought to diverge genetically to the point that reproduction between them ceases and new species are formed. Allopatric speciation was an integral part of dispersalist biogeography and remains so in vicariance biogeography.

Another component of the dispersalist school was the importance of land bridges. These were proposed in pre-drift times as a major means of dispersal of terrestrial animals from one continent to another. Although they were more a means of getting creatures around than a factor in evolution, they did play this role. When land bridges appeared, contact was possible between previously isolated faunas, and this "end of isolation" resulted in the evolution of new taxa and the extinction of old (the dispersalist analog of geodispersal). But this leads us to another relationship, one between area and faunal diversity. Simply stated, it can be shown that, other things being equal, the landmass with the greater area will have the greater number of animal species (cf. Brown 1995). Islands have less diverse faunas than continents not only because they are difficult to reach, but also because their areas are smaller than those of continents.

This effect suggested to paleontologists, such as Henry Fairfield Osborn, early last century, that large landmasses were "cradles" where new species and higher taxa originated. And it was this hypothesis that led him to fund the American Museum Mongolian expeditions of the 1920s under Roy Chapman Andrews (see Chapter 5, this volume). Now that we accept continental drift, it seems that Osborn was right—many dinosaurian and mammalian clades did originate in Asia—but for the wrong reasons. If Osborn's idea were correct, we would expect the greatest rates of diversification and appearance of new forms in land animals to have occurred when all the continents were together in the supercontinent of Pangaea; but, as far as we can tell, this didn't happen.

The idea of great landmasses being cradles for speciation was never controversial, but it led to a notion that was—that the animals that evolved on the largest landmasses, including modern Eurasia, were exposed to the most intense selection. In consequence, it was thought that they would outcompete corresponding types of animals that had evolved on smaller landmasses. Thus, it was no surprise to discover that North American/ Eurasian mammals drove many of the native South America mammals to extinction when the former reached South America in the late Cenozoic. Well, maybe. The situation isn't all that clear, because some of the southerners, such as glyptodonts and ground sloths, prevailed against the tide of northern mammals to the extent that ground sloths made it all the way to Alaska (almost to Asia). In addition, armadillos have settled down and are surviving well in North America even against that most competitive of mammals—humans.

The real problem in working out just what happened during the great American faunal interchange of the Late Cenozoic, and why, is twofold. First, you can't run experiments on intercontinental faunal exchanges—the scale is too big, the process takes too long, and the cost is too high. And second, we have a good record of only one of the invasions—that of South America, as just mentioned.

The last of the effects noted in dispersalist biogeography, and again recognized by the vicariance school, is the effect of geographic isolation. This is not the kind of small-scale isolation involved in allopatric speciation, but isolation on the grand scale—the kind now enjoyed by Australia, New Zealand, and Antarctica. This grand isolation allowed the evolution over the Cenozoic Era of faunas and floras unlike those found anywhere else. The difficulty of reaching these places is shown by their biotas—prior to the coming of humans, especially Europeans, of course. In Australia, the major terrestrial herbivores were marsupials, and the predators, reptiles. In New Zealand, the plant-eaters were flightless birds (moas), and the predators flying birds. On little Cocos island (in the Indian Ocean), the main land animals were (and still are) crabs. Isolation also permitted the survival of animals and plants all of whose close relatives elsewhere had become extinct, the relicts or so-called living fossils. Australia, isolated throughout much of the Cenozoic Era, still retains a strong complement of such creatures.

The coming of variance biogeography did little to change our appreciation of these relationships between geography and the course of evolution—they are, after all, fairly obvious. This school did add some concepts, though, such as the carrying of whole faunas and floras across oceans on large or small fragments of continents, semifacetiously termed "Noah's arks." And it did sharpen our understanding of the interaction of geography and evolution. But its major contribution came from reconsidering isolation. During the course of the Mesozoic Era, a single large, contiguous landmass broke into several smaller (though still large) isolated or semi-isolated continents. This necessarily split up the populations of the land-dwelling and freshwater animals. In effect, it imposed allopatric speciation on a large scale.

Bjorn Kurtén (1969) proposed that the breaking up of the supercontinent Pangaea increased the diversity of land-dwelling animals, and this seems to be the explanation for the origin of those dinosaurian and mammalian groups in central Asia that Osborn thought originated as a result of the large area of that continent. It has been suggested, largely on what might be called circumstantial evidence, that the origins not only of some major groups of (nonavian) dinosaurs, but also of major groups of mammals, birds, and frogs, were related to the breakup of Pangaea (Hedges et al. 1996). If correct—and the agreement between the dates of the origin of the groups and of the breakup of Pangaea is a bit vague—this indicates that continental drift can have evolutionary effects that reverberate up the taxonomic hierarchy and produce new families and orders as well as new species.

Recently more novel relationships have been proposed. First, and probably most important, is the role of climate. Elisabeth Vrba has proposed that climatic changes tend to occur more or less in synchrony around the world (cf. Tudge 1993). These changes, not surprisingly, drive changes in evolution. Vrba has analyzed the fossil records of late Cenozoic African mammals in detail and claims that such data provide support for her concept. It is not simply that the climate changes and the animals respond; her idea is more subtle. With changes in climate, small animals or specialist feeders may become extinct and be replaced by new species, while those with a more eclectic appetite simply change their diet. In Africa, the record shows extinctions and replacements in the specialist bovids (bovids include cattle, sheep, and antelope) around 2.25 million years ago, but no change in generalist feeders such as impala (which are also bovids) and pigs. This was a period of climatic change, at least in Africa, northern South America, and China. Vrba's idea seems plausible, and much of the criticism of it is actually over whether the climatic and faunal changes are worldwide, or even Africa-wide, and not over their evolutionary impact (Tudge 1993).

Down under, the influence of climate on evolution has been taken up by paleomammalogist Tim Flannery (1994), who has argued that Aussie ecosystems are significantly different from those of the northern continents because of climatic conditions. The Australian climate is unpredictable—there may be rain, lots or little, in a given wet season, or it may be dry for a decade. Meteorologists can't tell and neither can the local flora and fauna. Flannery uses the example that animals such as North America's grizzly bears, which hibernate during the winter knowing that they can find sufficient food come spring, did not evolve in Australia. They could hibernate, all right, but depressingly often the food would come during the hibernation or not at all. Instead, most Australian animals (and plants) are geared to the rains—where the rains come, estivation breaks. For those at the top of the food chain, it helps to able to wait out the dry periods, which effectively means surviving famine. This implies a lower metabolic rate, and as mentioned, many of the Australian top predators are (or were) lizards, snakes, and terrestrial crocodilians. Flannery claims more extensive evolutionary effects than Vrba, but for a less speculative reason—continental drift. He believes that the climatic change that occurred when Australia drifted north into the drier, and less predictable, subtropical climate has had a profound effect on evolution on the "island continent."

Since 1980, several rather speculative ideas have been proposed about interactions between evolution and geography. It is worthwhile to briefly consider these to appreciate recent thought in this area, although they are not yet established beyond doubt.

It has been proposed that several Cenozoic North American plants and vertebrates originated in the Arctic and later spread south over the continent (Hickey et al. 1983). No mechanism for this was given, but the notion of a geographic cradle that preferentially produces new species is in the air again. David Jablonski (1993) argued that since the Paleozoic, marine invertebrate species have tended to originate in tropical waters. Jablonski and Bottjer (1991) also proposed that marine invertebrates tend to originate in near-shore regions of the continental shelf. It has also been claimed that the Antarctic region was a cradle for marine invertebrates during the early Paleogene (Zinsmeister and Feldman 1984). Although seemingly contradictory, this and the claim for tropical origins are not really in conflict. There is nothing mystical about these regions that churn out new species. Antarctica became separated from the other continents in the Eocene (early Paleogene) and therefore allopatric speciation came into play. The tropics have a diverse biota and, if nothing else, provide lots of opportunities for new niches and lots of lineages to speciate.

Geerat Vermeij, who also works on marine invertebrates, maintains that the intensity of predation is reduced in polar regions (Vermeij 1987). These then are safe places where selection—at least selection due to predation—is reduced and relict forms can persist.

So far, such notions have been applied (it now appears, mistakenly) to dinosaurs only in Australasia, which was near the South Pole through much of the Mesozoic and which housed both relict and precocious non-dinosaurian tetrapods.

Probably the most significant new trend in biogeography is exemplified in the work of the expatriate Ugandan Jonathan Kingdon (1990). He has produced a detailed study of African biogeography, synthesizing modern ecology and climates with their changes over the past several million years. Kingdon's work integrates almost all of the themes mentioned here. In contrast to the grand sweep of Vrba's research, Kingdon's looks at the interactions of geography, topography, climate, and evolution at the fine scale, with an eye toward conservation. He has traced the changing fortunes of deserts and forests, as well as cradles where a great diversity of habitat in small areas has stimulated speciation, and regions that have remained stable and acted as refuges. The concept of refuges, regions where animals and plants survive times when most of their ranges are not suitable for them, has become important. It has obvious relevance for conservation, but also for our understanding of biogeography. The rainforests of Central America, for example, and possibly even those of the Amazon Basin, may be only about 10,000 years old (Lewin 1984). They may have regenerated from older rainforest species that survived drier and otherwise adverse times in refuges. An encouraging aspect of Kingdon's approach is that it allows prediction of where endemic and relict species may be found. This is not only useful to conservation efforts, but also shows that geographical understanding isn't restricted to any one or a few instances, but is applicable to all places over geological history.

Studies on the dinosaurs of what is now the Rocky Mountains region of western Canada, the United States, and Mexico, that in the Late Cretaceous was the long, narrow island of Laramidia, are now revealing its paleozoogeography in greater detail than was previously possible. It appears that, contrary to expectation for such large creatures, dinosaurs made up several distinct north-to-south faunas (Sampson et al. 2010). Although not carried out in the detail of Kingdon's work, this is an approach to such fine-grained analysis.

How All This Applies to Dinosaurs

So, finally we get around to dinosaurs. Like lions, pandas, wombats, and elk, dinosaurs were animals whose distributions were subject to climate and the locations of seas, mountains, and deserts. Looking at paleogeography and paleoclimates indicates that some forms, such as *Ammosaurus* in North America and *Protoceratops* in Asia, were desert dwellers, and others—especially fin-bearing taxa, such as *Spinosaurus*, *Ouranosaurus*, and *Amargasaurus*—seem to have been adapted to tropical climates. Dinosaurs in Antarctica, southern Australia, Alaska, and New Zealand lived in polar or near-polar regions.

This tells us more than just what environments were inhabited by dinosaurs, interesting though that may be. It also illuminates dinosaurian evolution and features of tetrapod evolution in general. The existence of dinosaurs in the tropics, near the poles, and in deserts shows that climate was not a limiting factor in dinosaurian distribution and evolution—for the group as a whole, if not for individual species. The absence of oceanic, aquatic dinosaurs, however, suggests that some factor in their evolution, be it genetic or due to competition with the already existing marine saurians (or to something else), kept them from exploiting this environment. Near-polar dinosaurs lived in climates not suitable for large modern lepidosaurs or tortoises (Molnar and Wiffen 1994), or even, for that matter, contemporaneous ones (cf. Clemens and Nelms 1993). These dinosaurs had some physiological or behavioral method for coping with cool climates not possessed by modern large ectothermic tetrapods. The evidence that dinosaurs were endothermic in the same sense as are modern large mammals (being tachymetabolic endotherms) is not convincingly clear-cut. However, the evidence is clear that they could and did survive in climates and places that modern large lizards, snakes, and tortoises do not (Rich 1996; Vickers-Rich 1996; Wiffen 1996).

Interpretations such as these are both enlightening and fun. But they are not, or at least not always, simple and straightforward. Not every species or even every genus that ever lived is represented by fossils. In other words, the fossil record isn't complete. Look at dinosaurs – 44% are known from approximately the final 20% of Mesozoic time. Over a third (365 out of 864) of the dinosaurian body fossil localities listed by Weishampel (2004) are from the Late Cretaceous, which represents 21% (about one-fifth) of the period during which (nonavian) dinosaurs lived. Although the numbers of localities are not as biased toward the more recent as are the numbers of dinosaur specimens, both show that the more recent dinosaurs are better known than the more ancient forms. Any interpretations from the fossil record, of dinosaurs or anything else, have to contend with this loss of information, and it gets worse as the fossils get older.

This situation shows up in dinosaurian biogeography with fossils that seem to be "out of place." The distribution of ceratopsians has seemed odd since early last century. Ceratopsids, such as *Triceratops*, *Styracosaurus*, and kin inhabited North America, and protoceratopsians inhabited Asia (with a few in North America). If the protoceratopsians could get to North America from Asia, why did the ceratopsids not go in the opposing direction? Indeed, ceratopsid fossils had been reported from Asia, but they proved mistaken. Thus, the latest of these, the Uzbek *Turanoceratops*, was often disregarded. But wrongly as it turns out, for *Turanoceratops* seems to be the sister group of *Triceratops* (Sues and Averianov, 2009). Then another ceratopsid, *Sinoceratops*, was described from China (Xu et al. 2010). Likewise, a single specimen reportedly of a pachycephalosaur (*Majungatholus*) from Madagascar was the only representative of pachycephalosaurs from the Southern Hemisphere (but see below). And *Timimus* is the only ornithomimosaur from the Southern Hemisphere (*Elaphrosaurus*, previously thought to be an ornithomimosaur, now is recognized as a ceratosaur).

More problematic is *Serendipaceratops*, a neoceratopsian also reported from the Early Cretaceous of southeastern Australia. Not only is it from Australia, where no other neoceratopsian material has been found, but it is also one of the few (three) Early Cretaceous neoceratopsians, so it is, in a sense, both out of place and out of time (Fig. 39.21). Furthermore, if the animal in question truly is a neoceratopsian, this implies that our understanding of ceratopsian evolution is based on only part of the lineage and perhaps not even a representative part. Compare this with another recent report of an out of place neoceratopsian (Godefroit and Lambert 2007). *Craspedodon*, from the Coniacian-Santonian (83.5–89.3 million years ago) of Belgium, has long been known and interpreted as an iguanodontian ornithopod. Known only from isolated teeth, these have recently been reinterpreted as having derived from a neoceratopsian. Neoceratopsians are known from the Santonian (and earlier) of Asia and could have, at some previous time, emigrated to Europe. Cranial remains of the Hungarian *Ajkaceratops* confirm that they did. Comparing these cases, *Serendipaceratops* dates from a time when neoceratopsians seem to have been rare and would have taken a long and tortuous route to Australia (leaving no indications behind in the lands that must have been traversed), while *Craspedodon* dates from a time when neoceratopsians were almost certainly abundant and could have had a (relatively) short and direct route of emigration west to Europe.

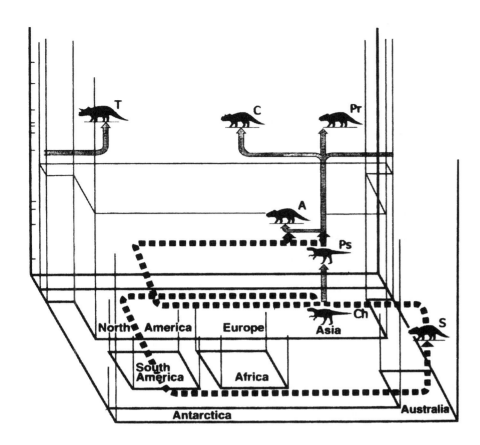

39.21. A fossil that is "out of place"–the reported Early Cretaceous neoceratopsian, Serendipaceratops (S), from Australia. This figure illustrates just how far Serendipaceratops is from the generally accepted picture of ceratopsian evolution. Cretaceous time is indicated as a dark vertical line along the left margin of the diagram (later is toward the top). Landmasses are represented by boxes along the base of the diagram, with arrows projected upward from the bases indicating evolution of ceratopsians in those areas. As usually understood, ceratopsian evolution took place in Asia, beginning with a group of Late Jurassic dinosaurs (chaoyangosaurs, Ch; Zhao et al. 1999), and proceeded through the Early Cretaceous psittacosaurs (Ps) to neoceratopsians. The earliest neoceratopsians (e.g., Asiaceratops, A) are reported from the end of the Early Cretaceous in central Asia, and protoceratopsids (Pr) and ceratopsids (T) appear in the Late Cretaceous. The path of ceratopsian evolution as usually understood is shown here with solid arrows, the extensive detours necessary to accommodate Serendipaceratops with dashed arrows. The detour to accommodate the Belgian Craspedodon © is much shorter.

These examples emphasize how incomplete our knowledge of dinosaurian biogeography is. All of the animals described above are from regions that are poorly known, "dinosaurily" speaking, and all are known from fragmentary specimens. Perhaps they indicate that some groups of dinosaurs were much more widespread than we realize. But they may also indicate convergence, in that the animals in question–more exactly, the parts found–look like those of pachycephalosaurs, ornithomimosaurs, and neoceratopsians, but only because of similar factors in natural selection, not because they are closely related. For example, we now know that *Majungatholus* is not a pachycephalosaur at all, but the theropod *Majungasaurus* (Krause et al. 2007). Like pachycephalosaurs, it too evolved thickened dome-like structures on the skull roof. This suggests that we should be skeptical about the identities of fragmentary, out of place fossils. We can be sure about them only when further, more complete fossils have been found. Indeed, other interpretations of both *Timimus* and *Serendipaceratops* have already been suggested (Agnolin et al. 2010). But *Turanoceratops* and *Sinoceratops* remain–the notion that ceratopsids inhabited Asia as well as North America has proved correct, even if all the evidence for it for some 80 years prior to 2009 was mistaken. The lesson is that one must be skeptical, both in accepting such out of place claims and in rejecting them; it is, after all, alright to simply say "we don't know."

Although we must be cautious about the details, the basic outlines–discussed in the following chapters–of dinosaurian biogeography seem to be pretty well known. Still, there is plenty of room for exciting new discoveries to expand, and maybe even substantially change, our understanding of dinosaurian biogeography and evolution.

References

Agnolin, F. L., M. D. Ezcurra, D. F. Pais, and S. W. Salisbury. 2010. A reappraisal of the Cretaceous non-avian dinosaur faunas from Australia and New Zealand: evidence for their Gondwanan affinities. *Journal of Systematic Palaeontology*, 8: 257–300.

Baeyer, H. C. von, 2003. *Information, The New Language of Science*. London: Weidenfeld & Nicolson.

Berner, R. A., and Z. Kothavala, 2001. GEOCARB III: A revised model of atmospheric CO_2 over Phanerozoic time. *American Journal of Science*, 301: 182–204.

Bergman, N. M.; T. M. Lenton, and A. J. Watson. 2004. COPSE: A new model of biogeochemical cycling over Phanerozoic time. *American Journal of Science*, 304: 397–437.

Borges, J. L. 1969. *The Book of Imaginary Beings*. New York: Avon Books.

Brooks, D. R., and D. A. McLennan, 2002. *The Nature of Diversity*. Chicago: University of Chicago Press.

Brown, J. H. 1995. *Macroecology*. Chicago: University of Chicago Press.

Casti, J. L. 1991. *Searching for Certainty: What Scientists Can Know about the Future*. New York: William Morrow.

Chatterjee, S., and N. Hotton III (eds.). 1992. *New Concepts in Global Tectonics*. Lubbock: Texas Tech University Press.

Clemens, W. A., and L. G. Nelms. 1993. Paleoecological implications of Alaskan terrestrial vertebrate fauna in latest Cretaceous time at high paleolatitudes. *Geology* 21: 503–506.

Colbert, E. H. 1973. *Wandering Lands and Animals*. London: Hutchinson.

Dilcher, D. L. 1973. A paleoclimatic interpretation of the Eocene floras of southeastern North America. In A. Graham (ed.), *Vegetation and Vegetational History of Northern Latin America*, 39–59. Amsterdam: Elsevier.

Duxbury, A. C., and A. B. Duxbury. 1994. *An Introduction to the World's Oceans*. 4th ed. Dubuque, Iowa: Wm. C. Brown.

Flannery, T. F. 1994. *The Future Eaters*. Sydney: Reed Books.

Gates, T. A., S. D. Sampson, L. E. Zanno, E. M. Roberts, J. G. Eaton, R. L. Nydam, J. H. Hutchison, J. A. Smith, M. A. Loewen, and M.A. Getty. 2010. Biogeography of terrestrial and freshwater vertebrates from the late Cretaceous (Campanian) western interior of North America. *Palaeogeography, Palaeoclimatology, Palaeoecology* 291: 371–387.

Godefroit, P., and O. Lambert. 2007. A re-appraisal of *Craspedodon lonzeensis* Dollo, 1883, from the Upper Cretaceous of Belgium: the first record of a neoceratopsian dinosaur in Europe? *Bulletin de l'Institut Royal des Sciences Naturelles de Belgique, Science de la Terre* 77: 83–93.

Hedges, S. B.; P. H. Parker; C. G. Sibley; and S. Kumar. 1996. Continental breakup and the ordinal diversification of birds and mammals. *Science* 381: 226–229.

Hickey, L. J.; R. M. West; M. R. Dawson; and D. K. Choi. 1983. Arctic terrestrial biota: Paleomagnetic evidence of age disparity with mid-northern latitudes during Late Cretaceous and Early Tertiary. *Science* 221: 1153–1156.

Hooker, J. J.; A. C. Milner; and S. E. K. Sequeira. 1991. An ornithopod dinosaur from the Late Cretaceous of west Antarctica. *Antarctic Science* 3: 331–332.

Jablonski, D. 1993. The tropics as a source of evolutionary novelty through geological time. *Nature* 364: 142–144.

Hunn, C. A., and P. Upchurch. 2001. The importance of time/space in diagnosing the causality of phylogenetic events: towards a "chronobiogeographical" paradigm? *Systematic Biology* 509: 391–407.

Jablonski, D., and D. J. Bottjer. 1991. Environmental patterns in the origins of higher taxa: the post-Paleozoic fossil record. *Science* 252: 1831–1833.

Kerr, R. A. 1995. Darker clouds promise brighter future for climate models. *Science* 267: 454.

Kingdon, J. 1990. The genesis archipelago. *BBC Wildlife* 8 (5): 296–302.

Krause, D. W., S. D. Sampson, C. A. Forster, and P. Dodson. 2007. Overview of the history of discovery, taxonomy, phylogeny, and biogeography of *Majungasaurus crenatissimus* (Theropoda: Abelisauridae) from the Late Cretaceous of Madagascar. *Society of Vertebrate Paleontology Memoir* 8: 1–20.

Kurtén, B. 1969. Continental drift and evolution. *Scientific American* 220 (3): 54–64.

Lewin, R. 1984. Fragile forests implied by Pleistocene pollen. *Science* 226: 36–37.

Lieberman, B. S. 2005. Geobiology and paleobiogeography: tracking the co-evolution of the Earth and its biota. *Palaeogeography, Palaeoclimatology, Palaeoecology* 219: 23–33.

Molnar, R. E. 1989. Terrestrial tetrapods in Cretaceous Antarctica. In J. A. Crame (ed.), *Origins and Evolution of the Antarctic Biota*, 131–140. Special Publication 47. London: Geological Society.

Molnar, R. E., and J. Wiffen. 1994. A Late Cretaceous polar dinosaur fauna from New Zealand. *Cretaceous Research* 15: 689–706.

Nelson, G., and N. Platnik. 1981. *Systematics and Biogeography*. New York: Columbia University Press.

Nigg, J. 2002. *The Book of Dragons & Other Mythical Beasts*. New York: Barron's Educational Series.

Otto-Bliesner, B. L., and G. R. Upchurch, Jr. 1997. Vegetation-induced warming of high-latitude regions during the Late Cretaceous period. *Nature* 385: 804–807.

Press, F., and R. Siever. 1994. *Understanding Earth*. New York: W. H. Freeman.

Queiroz, A. de. 2005. The resurrection of oceanic dispersal in historical biogeography. *Trends in Ecology and Evolution* 20: 68–73.

Raymo, M. E., and W. F. Ruddiman. 1992. Tectonic forcing of late Cenozoic climate. *Nature* 359: 117–122.

Rich, T. 1996. The significance of polar dinosaurs in Gondwana. *Memoirs of the Queensland Museum* 39: 711–717.

Rich, T. H., and P. Vickers-Rich. 1994. Neoceratopsians and ornithomimosaurs: Dinosaurs of Gondwana origin? *National Geographic Research and Exploration* 10: 129–131.

Sampson, S. D., M. A. Loewen, A. A. Farke, E. M. Roberts, C. A. Forster, J. A. Smith, and A. L.. Titus. 2010. New horned dinosaurs from Utah provide evidence for intracontinental dinosaur endemism. *Public Library of Science One* 5(9): 1–12.

Stross, C. 2006. *The Jennifer Morgue*. Urbana: The Golden Gryphon Press.

Sues, H-D. and Averianov, A., 2009. *Turanoceratops tardabilis*—the first ceratopsid dinosaur from Asia. *Naturwissenschaften* 96: 645–652.

Tudge, C. 1993. Taking the pulse of evolution. *New Scientist* 139 (1883): 32–36.

Turner, A. H. 2004. Crocodyliform biogeography during the Cretaceous: evidence of Gondwanan vicariance from biogeographical analysis. *Proceedings of the Royal Society of London B* 271: 2003–2009.

Upchurch, P; C. A. Hunn; and D. B. Norman 2002. An analysis of dinosaurian biogeography: evidence for the existence of vicariance and dispersal patterns caused by geological events. *Proceedings of the Royal Society of London B* 269: 613–621.

Vermeij, G. 1987. *Evolution and Escalation*. Princeton, N.J.: Princeton University Press.

Vickers-Rich, P. 1996. Early Cretaceous polar tetrapods from the Great Southern Rift valley, southeastern Australia. *Memoirs of the Queensland Museum* 39: 719–723.

Waters, J. M.., and D. Craw. 2006. Goodbye Gondwana? New Zealand biogeography, geology, and the problem of circularity. *Systematic Biology* 55: 351–356.

Wegener, A. 1915. *The Origin of Continents and Oceans*. Braunschweig: Fried. Vieweg and Sohn. (Reprinted by Dover in 1966.)

Weishampel, D. B., P. M. Barrett, R. A. Coria, J. le Loeuff, Xu X., Zhao X., A. Sahni, E. M. P. Gomani, and C. R. Noto. 2004. Dinosaurian distribution. In D. B. Weishampel, P. Dodson, and H. Osmolska (eds.), *The Dinosauria*, (2nd ed.), 517–606. Berkeley: University of California Press.

Wiffen, J. 1996. Dinosaurian paleobiology: A New Zealand perspective. *Memoirs of the Queensland Museum* 39: 725–731.

Wing, S. L., and D. R. Greenwood. 1993. Fossils and fossil climate: The case for equable continental interiors in the Eocene. *Philosophical Transactions of the Royal Society of London B* 341: 243–252.

Xu X., Wang K., Zhao X., and Li D. 2010. First ceratopsid dinosaur from China and its biogeographical implications. *Chinese Science Bulletin* 55: 1631–1635.

Yemane, K. 1993. Contribution of late Permian palaeogeography in maintaining a temperate climate in Gondwana. *Nature* 361: 51–54.

Yuan X., M. A. Cane, and D. G. Martinson. 1996. Cycling around the South Pole. *Nature* 380: 673–674.

Zhao X., Cheng Z., and Xu X. 1999. The earliest ceratopsian from the Tuchengzi Formation of Liaoning, China. *Journal of Vertebrate Paleontology* 19: 681–691.

Zinsmeister, W. J., and R. M. Feldman. 1984. Cenozoic high latitude heterochroneity of Southern Hemisphere marine faunas. *Science* 224: 281–283.

40.1. Example of the early "Chondrastean" *Dictyopyge*.

Non-Dinosaurian Vertebrates

40

Nicholas C. Fraser

It is undoubtedly appropriate to refer to the Mesozoic Era as the Age of Dinosaurs, but it was much more than that. Swimming in the Mesozoic seas were giant crocodiles, dolphinlike ichthyosaurs and enormous plesiosaurs. The air was alive with pterosaurs ranging in size from the sparrow-sized *Pterodactylus* to the truly monstrous *Quetzalcoatlus*, which was the size of a small airplane. Consequently, the Mesozoic is also sometimes referred to as the Age of Reptiles. But we can go beyond even this basic statement, for it was essentially in the first part of the Mesozoic (the Triassic) that the seeds of today's terrestrial ecosystems were sown.

In addition to dinosaurs, the Triassic period witnessed the first lissamphibians (frogs and salamanders), crocodiles, turtles, sphenodontians (and, by extension, their sister group, the squamates), and mammals. It is not even unreasonable to say that we can also trace the origin of birds back to the Triassic. It is now quite universally accepted among paleontologists that the birds arose from within the theropod dinosaur lineage and that Aves merely represent derived theropod dinosaurs. There are even some who argue that true bird remains have been found in the Triassic. On the basis of fragmentary vertebrate remains from the Dockum, Chatterjee (1990) erected the genus *Protoavis* which he considered to be a bird. However, this assessment remains highly controversial.

Whereas large plesiosaurs, such as *Elasmosaurus*, or gigantic azhdarchid pterosaurs, such as *Quetzalcoatlus*, have attracted as much attention as their dinosaur contemporaries, quite a few of the taxa discussed in this chapter have been previously overlooked. Small, 1-cm fragments of the sphenodontian *Opisthias* jaw bones have seemed insignificant alongside the 2-m femur of a sauropod dinosaur from the same formation, and tiny partial mammal teeth are readily overlooked in a matrix containing 30-cm *Tyrannosaurus rex* teeth! Yet these smaller vertebrates are tremendously important in their own right, and their study is necessary for a more complete understanding of the Mesozoic world. Indeed, it is typically from the smaller, less conspicuous fossils that we have documentation of the early evolution of modern vertebrate groups. These "modern" groups were present throughout Mesozoic and Tertiary times, although admittedly there were times when they were rare or even absent from the fossil record.

This chapter examines most of the major groups of vertebrates that were contemporaries of the dinosaurs.

Fish

When using the term *fish*, people can be referring to any of a number of quite different animals. The earliest forms were the agnathan (jawless) fishes that extend back to the Cambrian. The first agnathans (the ostraco-derms) are characterized by an extensive exoskeleton comprising a solid carapace or bony shield that covered the head together with large bony plates or scales over the rest of the body. Most of these had died out by the end of the Devonian, but two lineages, the lampreys and the hagfish, survive today, indicating that agnathans lived during the dinosaurs' reign.

The first jawed fish date back to the Silurian. Many of the Paleozoic jawed fish, the placoderms, were, like the jawless ostracoderms, heavily armored and were likely poor swimmers and largely benthonic. They are not closely related to either of today's major divisions of fishes: the cartilaginous fishes (Chondrichthyes) or bony fishes (Osteichthyes).

The Chondrichthyes can be divided into two groups: the sharks, skates, and rays (Elasmobranchii) on the one hand, and the chimaeras or ratfish (Holocephali) on the other. Throughout their history, sharks have been predominantly marine predators. Although the elasmobranchs date back to the latest Silurian, the evolutionary history of modern sharks (Neoselachii) only goes back to the Triassic. Some of the Triassic sharks were actually freshwater forms. The neoselachians radiated in the Jurassic and Cretaceous, and a wide variety of essentially modern-looking sharks and rays were denizens of the Mesozoic seas. Likewise, Mesozoic holocephalians such as *Ischyrodus* from the Upper Jurassic were very like the modern chimaeras.

Almost from their beginnings, sometime in the Devonian or possibly even the Silurian, the bony fish fall neatly into two groups. Those with fleshy lobes forming the bases of their fins are the Sarcopterygians, meaning "lobe fins," and include the lungfish and coelacanths. The second group is the Actinopterygia, which means "ray fins." These, naturally enough, have fins supported by series of cartilaginous or bony rods.

The sarcopterygians are particularly important to our understanding of the origin of tetrapods sometime in the Devonian. But with respect to the Mesozoic, remnants of sarcopterygians are also frequently encountered in freshwater and terrestrial assemblages in the form of lungfish plates. Although lungfish declined in diversity after the Carboniferous, two lineages persisted right through the age of dinosaurs and their tooth plates are very characteristic elements in many dinosaur-producing types of sediment (Fig. 40.1). The coelacanths also arose in Devonian times and lived both in marine and freshwater environments. They are sometimes referred to as tassel-tails because of the shape of their caudal fin. Of course, the coelacanths are a particularly celebrated group of fishes because they were long thought to have died out in the Cretaceous. That concept quickly changed in 1938 when the fishing vessel, *The Nerine*, netted a specimen off the coast of Africa. Since then a number of specimens have been caught off the Comores Islands in the Indian Ocean.

Many of the early ray-finned fishes are often placed in a chondrostean ("cartilage-bone") group. In reality, they do not constitute a natural assemblage of organisms, but they do represent an early evolutionary stage in the radiation of actinopterygians as a whole. Although they survive to this day in the form of the birchirs, paddlefishes and sturgeons, they began to wane during Triassic times and do not appear to have been particularly abundant during the Mesozoic as a whole. Holosteans are another grade of organization of actinopterygians. The modern gars and bowfin are surviving examples of holosteans. Extinct holosteans include the pycnodonts, deep bodied fishes with elongate dorsal and anal fins that possessed relatively small mouths equipped with a pavement of crushing teeth. Widespread in the Jurassic and Cretaceous, pycnodonts likely inhabited quiet waters of reefs where they could have fed on mollusks and echinoderms.

Some authors include two other groups of Jurassic and Cretaceous fishes, the aspidorhynchids and the pachycormids, within the holostean grade, but others regard them as early teleosts: fishes in both groups are characterized by long bodies. Many also have an elongated rostrum, which in some forms is akin to that of the swordfish. Some were large-bodied suspension-feeding planktonivores, ecologically equivalent to today's whale sharks and basking sharks (Friedman et al. 2010)

As they are today, the teleosts were by far the most common fish in Mesozoic waters. These actinopterygian fishes first appear at the end of the Triassic. The large teleost, *Xiphactinus*, from the Cretaceous of North America, is a typical member of the Icthyodectiformes, a fairly primitive group of extinct teleosts. Fossils of other large fish are occasionally found in the body cavity of specimens of this 4-m giant, vividly illustrating its predatory habit! By the Early Cretaceous, radiation of the elopomorphs (eels and their allies) and clupeomorphs (herrings and their kin) had begun; by Middle and Late Cretaceous times, groups such as the acanthomorphs (cod, haddock) and percomorphs (flatfishes, tunas, seahorses) had appeared. It should be stressed, however, that most modern members of these groups only appeared after the end of the Mesozoic.

Some of the higher order taxa mentioned in this chapter had subsets with temporal ranges that only partially overlapped those of the dinosaurs. In these instances, the discussion centers mostly on the dinosaur contemporaries. The so-called Amphibia is one such group.

Amphibian

Amphibians today fall into a clearly defined group of animals that in all probability form a monophyletic assemblage, the Lissamphibia (Parsons and Williams 1963; Milner 1988). But when we delve into Paleozoic and Mesozoic times, the term *Amphibia* really only applies to a grade of organization—tetrapods that are not amniotes. These range from the polydactylous Devonian fully aquatic forms like *Ichthyostega* and *Acanthostega* (Clack 2000) to many of the anthracosaurs that may have been almost completely emancipated from water and approached the reptilian grade of organization. With respect to the Mesozoic and the contemporaries of the dinosaurs, only the temnospondyls and lissamphibians need concern us.

The temnospondyl grade of organization included a diverse array of forms (Milner 1990), most of which were around long before the dinosaurs. In the Permian, for example, they included forms like *Eryops* that were very much adapted to a terrestrial existence. However, many temnospondyl families disappeared at the close of the Permian, and most of those that survived into the Mesozoic show adaptations to a fully aquatic existence. They continued to decline throughout the Triassic, so that towards the end of the period, when the dinosaurs began their rise to prominence,

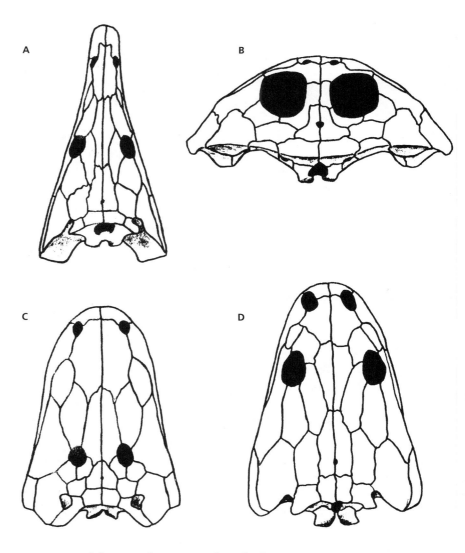

40.2. Skulls of a variety of Triassic temnospondyls showing the variability in form. A, the trematosaur, *Tertrema;* B, the plagiosaur, *Gerrothorax;* C, the metoposaur, *Metoposaurus;* D, the capitosaur, *Cyclotosaurus.*

temnospondyls are rather rare with only the capitosaurs, metoposaurs, and plagiosaurs still occurring in decent numbers. By the close of the Triassic, practically all temnospondyl groups had died out, leaving just two lineages extending into the Jurassic (Warren and Hutchinson 1983) and only one into the Cretaceous (Warren et al. 1991; Warren et al., 1997).

Of the Mesozoic temnospondyls (Fig. 40.2), the trematosaurids formed a particularly diverse family in the early part of the Triassic and had a global distribution, but after the Ladinian records are very scarce. The slender-snouted, ghariallike *Tertrema* is one of the better-known forms (Fig. 40.2a). The plagiosaurs were probably bottom dwellers employing a suction-gulping technique to snare their food. They had exceptionally short, but very wide skulls with very large dorsally facing orbits. *Gerrothorax* and *Plagiosaurus* were two typical members (Fig. 40.2B). *Plagiosuchus* apparently retained external gills as adults.

The chigutisaurids, which were characterized by short deep skulls, were another group of suction gulpers, and it has been suggested that they lived in the open water. Chigutisaurids ranged throughout the Triassic and include forms like *Pelorocephalus* from the Ischigualasto Formation. Remarkably, one chigutisaurid has been described from the Cretaceous of

Australia (Warren et al. 1991; Warren et al., 1997), so they also must have been rare components of certain Jurassic assemblages.

A second group of temnospondyls—the brachyopoids—extended into the Jurassic. These relatively small aquatic forms were undoubted suction gulpers and are characterized by short and deep skulls. Although not a particularly diverse group, they are largely restricted to Late Permian and Early Triassic sediments. However, three taxa are known from the Jurassic of Asia and brachyopoids are now known to range at least to the Middle Jurassic and probably the Upper Jurassic (Shiskin 1991).

The metoposaurids were larger animals, many exceeding 2 m in length. Their skulls were rather like those of today's alligators, with a relatively short snout and an elongate posterior region (Fig. 40.2c). However their eyes were positioned much further anteriorly than those of an alligator. It is likely that, like alligators, metoposaurs were semiaquatic, spending their days lurking around lake margins and entering the water to catch fish. *Metoposaurus* and *Buettneria* are just two genera within this family.

The capitosaurids and mastodonsaurids (Fig. 40.2d) were also rather alligator-like and somewhat similar to each other. They can easily be distinguished from the metoposaurids by their eyes, which are set well back on the skull. The large mastondsaurids would have been particularly fearsome, with some attaining lengths in excess of 6 m. Although not quite reaching the monstrous proportions of mastodonsaurids, capitosaurids still included 3-m beasts and are probably the most widespread and best-known of the Triassic temnospondyls. *Cyclotosaurus* is a common genus in the Upper Triassic of Germany and is known from a number of different species.

Aside from the Triassic temnospondyls, the amphibian faunas of the Mesozoic were probably not too dissimilar to the modern lissamphibians. Thus, from the Jurassic onward, frogs, salamanders, and caecilians are known. Indeed, a plausible sister group for the anurans (frogs and toads) in the form of *Triadobatrachus* (Fig. 40.3) dates back to the Lower Triassic of Madagascar, but the oldest known undoubted anuran is *Prosalirus* from the Early Jurassic Kayenta Formation (Shubin and Jenkins 1995, Jenkins and Shubin 1998). In general, frogs do not become really widespread until the Cretaceous. Likewise, at least three modern salamander families extend back at least as far as the Early Cretaceous. Today, the legless caecilians are fossorial forms that are known from about 160 species, but their fossil record is very sparse. However, one form, *Eocaecilia*, also from the Kayenta Formation, has been described on the basis of a number of specimens, but this early form even possessed limbs (Jenkins and Walsh 1993).

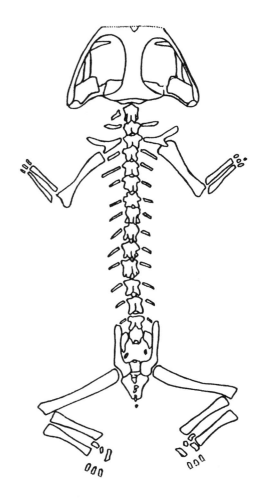

40.3. *Triadobatrachus,* a possible early frog from the Triassic of Madagascar (after Estes and Reig, 1973).

Reptiles

Reptiles, like amphibians, are a paraphyletic group: not only mammals, but also birds are derived from within their ranks. Traditional classifications of reptiles relied heavily on the nature of the temporal region and the presence and numbers of openings, or fenestrations, in the side of the skull. The absence of any such openings was typically seen as the primitive condition for reptiles, whereas those groups with openings were considered more advanced. This view was underscored by the fact that the oldest reptiles lacked any lateral temporal openings. Such a configuration of the temporal

40.4. Fenestration patterns in the skull of tetrapods. A, anapsid; B, euryapsid; C, synapsid; D, diapsid; E, diapsid with an antorbital fenestration.

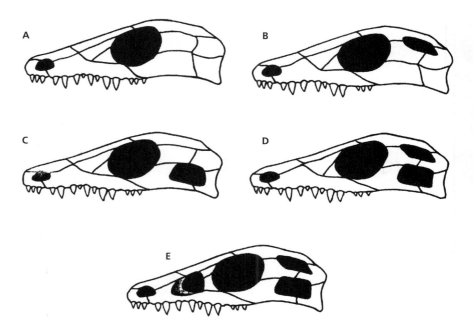

region (Fig. 40.4A) is referred to as *anapsid* (literally, "no arches"). Because chelonians are anapsid, they were widely viewed as the most primitive group of living reptiles.

Historically, three fundamentally different arrangements of the fenestrations were recognized (Fig. 40.4). Skulls with a single fenestra positioned high on the side of the skull and largely bounded by the parietal, postorbital, postfrontal, and squamosal bones, were euryapsid (Fig. 40.4B). These typically included forms such as the plesiosaurs. Skulls characterized by a single fenestra in the lower half of the temporal were synapsid (Fig. 40.4C). This configuration is typical of mammal-like reptiles. Finally, skulls possessing two lateral temporal fenestrae were diapsid ("double arch") (Fig. 40.4D, E), and, among others, this typifies the crocodiles and dinosaurs. In addition, it was thought that a host of different forms, including the lizards and snakes, were derived from this double-arched condition, and they too were considered diapsid although at least one of the two arches was missing.

This very basic division of reptiles has changed radically in recent years. Much more importantly, with the cladistic revolution, a number of major phylogenetic studies have been undertaken in which large numbers of characters and character states have been analyzed. It now seems that the position of the temporal openings is just one of a great number of variable osteological characters, and is not singularly critical to phylogenetic analyses. For example, many other features of the skull and postcranium indicate that traditional euryapsids, such as the plesiosaurs and ichthyosaurs, are part of the diapsid radiation. Thus, in this instance, it would seem that the presence of an opening in the upper part of the temporal region is of more significance. The absence of a lower opening may simply be a secondary closure or representative of an earlier stage in the development of the true diapsid condition. But more surprising is the fact that a number of varied studies indicate that chelonians may also be part of that same diapsid radiation. In this case, it is even conceivable that the lack of temporal openings might be a derived condition within that radiation. In addition to studies based on traditional osteological data that indicate diapsid affinities for chelonians (DeBraga and Rieppel 1997; Rieppel and Reisz 1999), a study based on mitochondrial DNA (Zardoya and Meyer 1998) concluded that the turtle's closest living relatives are birds and crocodiles, with lizards and snakes being more distantly related. However, this remains a very contentious issue and a recent study once again found Chelonia to lie outside the Diapsida (Lyson et al. 2010).

Anapsids

Omitting the chelonians from consideration leaves a single group of anapsids, the Procolophonoidea, that were contemporaries of the dinosaurs. Procolophonoids first appear in the Late Permian and became widespread and relatively common throughout the Triassic. They were rather small, squat herbivores that may have looked similar to the "horned toad," *Phrynosoma.*

In most procolophonoids, the orbit is extensively developed posteriorly and, in many Upper Triassic forms, it takes on a very characteristic keyhole shape (Fig. 40.3). It is clear that the extension of the orbital margin into the temporal region achieves the same development of bony margins for attachment of jaw musculature that discrete fenestrations provide in diapsids and synapsids.

In dorsal view, the skulls of Late Triassic procolophonoids were broad and triangular. The mandible was typically deep and robust, and marginal tooth numbers were reduced. The bulbous teeth may also be transversely broadened with rather complex facets. Many of the Late Triassic members display extensive spiky outgrowths of the quadratojugal region of the skull. The cheeks of *Hypsognathus* from the Newark Supergroup bristle with such spikes (Fig. 40.5). Although *Hypsognathus* has not been found together with dinosaur remains, dinosaur trackways and footprints are well documented from contemporaneous horizons in the Newark Supergroup. Thus, the Wolfville Formation in Nova Scotia yields trackways of, among other dinosaur ichnotaxa, *Atreipus*; it is also known for *Hypsognathus* skeletal material. Procolophonoids similar to *Hypsognathus* also occur in the Chinle Group (home to *Coelophysis* among others), the Norian fissure deposits of England (alongside the basal sauropodomorphs *Thecodontosaurus* and

40.5. Restoration of the Late Triassic procolophonoid, *Hypsognathus.*

Pantydraco and the dinosauromorph or basal theropod *Agnosphitys*) and the Lossiemouth Sandstone of Scotland (together with the enigmatic *Saltopus*).

The phylogenetic relationships of procolophonoids have been the subject of much debate. Some have argued a sister group relationship with the pareiasaurs, a group of large heavy-built Permian herbivores, while others strongly favor a sister-group relationship with the turtles (Reisz and Laurin 1991). Should the latter ultimately prove to be the best-supported hypothesis, and if chelonians do indeed fall within Diapsida, then the procolophonoids may represent one of the early offshoots of the diapsid radiation.

Diapsids

Two main diapsid lineages are generally recognized: the Lepidosauromorpha and the Archosauromorpha. Setting aside the question of the Chelonia for the moment, of the living reptiles, the lizards, snakes, and tuatara fall within the Lepidosauromorpha and the crocodiles together with birds are the sole surviving members of the Archosauromorpha. It has been postulated that the principal differences between these two major groups can be attributed to basic differences in posture and locomotion (Carroll 1988). By and large, lepidosauromorphs might be considered sprawlers moving by means of mediolateral excursions of the limbs. At the same time, the backbone is strongly flexed from side to side as the animal moves. In all likelihood, one of the characteristics of lepidosauropmorphs that can be linked with this type of gait is the presence of a large sternum. Jenkins and Goslaw (1983) showed that the sternum prevents the shoulder from moving backward when the forelimb is brought backward. By contrast, the archosauromorphs might be considered more upright citizens of the past! In the ultimate development of the archosauromorph locomotor pattern (dinosaurs), the backbone was kept rigid and the hind legs moved backward and forward in a parasagittal plane.

Unfortunately, these supposed distinctions become a little fuzzy when one tries to compare the stance and posture of some of the early archosauromorphs (such as the protorosaurs) with that of contemporary lepidosauromorphs. Equally, the living archosauromorphs—the crocodiles and alligators—could hardly be exemplified as having an upright stance: unless exercising their "high walk," they adopt the classic sprawling pose on the river bank. Finally, the sternum is not a structure that ossifies very well and, as a consequence, is generally poorly preserved in the fossil record. Therefore, its absence in a fossil is not meaningful. One is forced to rely on more technical details, such as the presence of a specialized opening, the ectepicondylar foramen, in the distal end of the humerus or perhaps the shape of the quadrate, to determine whether a diapsid is a lepidosauromorph or an archosauromorph

CHELONIA (TURTLES)

The oldest known turtle is *Odontochelys* from the Triassic of southern China (Li et al. 2009). This ancestral form lacks the dorsal bony carapace covering the dorsal surface of the body; in contrast to all other turtles, however, it bore teeth in both the upper and lower jaws.

A number of beautifully preserved specimens of additional Triassic turtles are known, particularly from Germany, although they do not form a major component of Triassic terrestrial assemblages. Since that time, however, they diversified greatly, although the basic turtle body has remained relatively consistent: toothless skull, ribs lying external to the girdle elements, and, of course, the trademark shell.

Modern turtles can be divided into two basic turtle groups—the Pleurodira (so-called side necked turtles) and the Cryptodira. Very few pleurodires survive to this day, but they do date back to the Jurassic. The cryptodires also arose during Jurassic times and they have since become the more diverse of the two groups. They radiated on land, in freshwater, and in the world's Mesozoic oceans. One particular family, the Protostegidae, had members reaching astronomical proportions. *Archelon*, for example, attained almost 4 m in length. Although turtles are quite prevalent in Jurassic and Cretaceous sediments, they are frequently added to faunal lists simply on the basis of isolated plates that are readily recognizable as chelonian but cannot be identified beyond that basic level.

The Euryapsids

There is nothing conclusive regarding the specific position of the euryapsids within Diapsida. Some authorities (DeBraga and Rieppel 1997; Rieppel and Reisz 1999) consider sauropterygians (plesiosaurs and their allies) to be basal relatives of the lepidosauromorphs, whereas another view supports sauropterygians and ichthyosaurs as basal archosauromorph relatives (Merck 1997).

ICHTHYOSAURS (EARLY TRIASSIC TO MID-CRETACEOUS)

The ichthyosaurs, or "fish reptiles," superficially resembled toothed-whales (odontocetes) (Fig. 40.6). Indeed, they can be viewed as the Mesozoic ecological equivalents of certain toothed-whales or open-water sharks. There is certainly no disputing the remarkable similarity in the body outline of

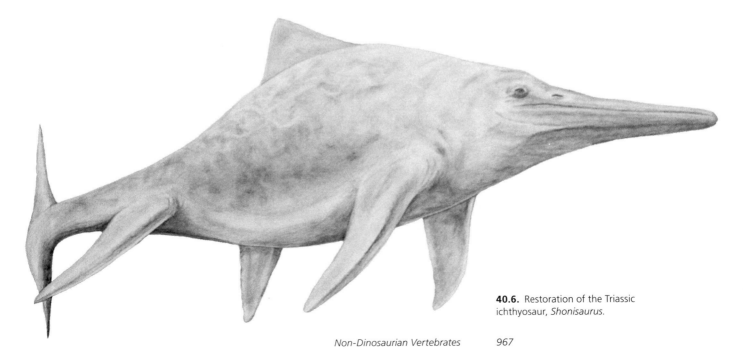

40.6. Restoration of the Triassic ichthyosaur, *Shonisaurus*.

classic Jurassic ichthyosaurs like *Ichthyosaurus* and *Ophthalmosaurus* to that of a bottle-nose dolphin. All ichthyosaurs had essentially the same body plan with both front and hind limbs modified to form paddle-shaped organs. In addition, restorations of the animal (Fig. 40.6) typically show a large dorsal fin and a large lunate tail fin like that of a modern-day tuna. This is based on the fact that some specimens preserve carbonized images of the body outline showing prominent dorsal and tail fins. This does not automatically mean that all ichthyosaurs possessed such fins, but it does seem highly probable; in the absence of such a large dorsal fin, they would have lacked stability in the water column. Many may have swum by holding the body rigid and generating a strong propulsive force simply by moving the tail quickly back and forth.

It is widely accepted that ichthyosaurs were completely unable to come out on dry land, and there is substantial evidence to indicate that they gave birth to live young at sea. A number of specimens have been found with embryos preserved within the mother's body, and there is at least one individual that apparently died during childbirth – the baby apparently stuck in the birth canal as it was born tail first.

The earliest ichthyosaurs appeared in the Lower Triassic – approximately 15–20 million years before the first dinosaurs walked on the earth. Interestingly, they also died out long before the dinosaurs disappeared, the last ones living approximately 90 million years ago. Perhaps they failed to compete successfully with the advanced sharks that began to radiate at that time.

The oldest ichthyosaur is *Chaohusaurus* from southern China (Young and Dong 1972). It did not have the dolphinlike form of later members of the group and is one of the smallest forms of ichthyosaur. Even so, some of the largest ichthyosaurs were among the earlier members of the group, although recent finds indicate one later species attaining a length of at least 23 m. *Shonisaurus*, from the Upper Triassic, attained a length of 15 m. Such giants would undoubtedly have been formidable predators of the oceans. Ichthyosaurs seem to have reached their peak diversity in the Jurassic, and the numerous fabulously preserved specimens from Germany and England give us a glimpse of fast-moving predators like tunas and some sharks. Their main prey would have been fish and cephalopods

SAUROPTERYGIANS

The sauropterygians comprise three distinct groups, the placodonts, nothosaurs and plesiosaurs. The plesiosaurs are best familiarized as the Loch Ness Monster of the Mesozoic seas, and they are widespread throughout the Jurassic and Cretaceous. The placodonts and nothosaurs were entirely Triassic.

Nothosaurs

I like to think of nothosaurs as mini Loch Ness Monsters – miniature forerunners of the long-necked plesiosaurs. They are characterized by long necks and paddle-shaped limbs. At least some gave birth to their young live rather than laying eggs (Cheng et al. 2004). They are particularly abundant and diverse in the Middle Triassic of Europe and China, but did not overlap with the dinosaurs.

Plesiosaurs

Except for a few primitive Early Jurassic species (Ketchum and Benson 2010), plesiosaurs are readily split into two distinct groups, the plesiosauroids which typically have rather small heads at the end of sometimes extraordinarily long necks, and the pliosauroids with a much more conspicuous and robust head atop a relatively much smaller neck (Fig. 40.7). *Elasmosaurus* from the Late Cretaceous was the longest of all plesiosaurs. It is particularly notable for being the central character in initiating the great rivalry between Marsh and Cope, for Cope put the skull on the wrong end of the body in his initial restoration of *Elasmosaurus*, and it was Marsh that pointed out the error of his soon-to-be rival. Conybeare actually described plesiosaurs as "snakes threaded through the bodies of turtles" which is a very apt descriptor!

In both plesiosauroids and pliosauroids the limbs were modified to form paddles or flippers. Over the years, there has been some debate concerning the way these modified limbs were used. It was initially suggested that plesiosaurs used their limbs as oars to row through the water. This was certainly a very attractive idea given the shape of the flippers. But although the paddles could be feathered, they could not be removed from the water, and therefore the force of the backstroke would create a counter thrust that would cancel out at least part of the forward motion. An alternative method would be a kind of underwater flight that is employed today by penguins and sea turtles. In this case, the up-and-down movement of the paddles would provide lift, with the flippers describing a figure eight. The shape of the flipper with an aerofoil cross section certainly supported this hypothesis (Robinson 1975). However, an examination of the range of movements possible at the shoulder girdle indicated that plesiosaurs were incapable of significant up-and-down movement of the flipper. So another explanation was sought. The answer to the problem seems to be a version that incorporates elements of the flying model and the rowing theory. Godfrey (1984) suggested that the tip of the flipper described a crescent-shaped path. In this case, the primary propulsive force would come from the retraction of the flipper—similar to the rowing model. But in the recovery stroke there would have been a component of upward lift that would have minimized the counterthrust as the limb was brought back into its original position again. In fact, this is precisely what is seen in living sea lions.

An interesting corollary of the locomotor patterns concerns the reproductive habits of plesiosaurs. Although no plesiosaur specimens have been found with associated embryos, neither have any plesiosaur eggs been described, so we have no way of telling what their reproductive strategy was. Nevertheless, detailed examinations of the skeletons of plesiosaurs indicate that they had very rigid trunk regions with broad, platelike limb girdles and little movement possible between the individual vertebrae. As a result, if they did lay eggs, it would have been a very strenuous trip up the beach; they would have been incapable of arching and bouncing themselves up the beach in the manner of sea lions. Instead, they would have been forced

40.7. Restorations of the Cretaceous plesiosauroid, *Elasmosaurus* (above), and the Late Jurassic pliosauroid, *Liopleurodon* (below; reproduced courtesy of Pat Gulley).

to drag their enormous masses across the beach in the way that sea turtles do today.

Placodonts

The third lineage of sauropterygians was restricted entirely to the Triassic. Placodontia comprises a group of somewhat superficially turtlelike animals, and include *Henodus*, *Paraplacodus* and *Placodus* itself. Like the turtles, some placodonts developed a series of bony plates over the body that would have afforded these rather slow-moving marine animals some measure of protection from predator attacks. Placodonts were specialized mollusk feeders with procumbent front teeth that were employed to pry their food off the seafloor. Large molariform teeth positioned toward the back of the jaws crushed the shells.

CHORISTODERES

One group of diapsid reptiles whose relationships remain obscure is the Choristodera. For many years this group was thought to comprise solely of superficially gaviallike aquatic animals as exemplified by *Champsosaurus* and *Simoedosaurus* from the Upper Cretaceous and early Tertiary of North America and Europe. However, it has now been recognized that several smaller taxa, including *Pachystropheus* from the Late Triassic (Storrs and Gower 1993) and *Cteniogenys* from the Jurassic (Evans 1989), can also be assigned to the choristoderes. They are characterized by a flattened skull with a flared temporal region that extends back to either side of the first neck vertebra; some members have pachostotic ribs and gastralia, and all have three sacral vertebrae. Spectacular choristodere remains have also come from the famed Yixian Formation of China. Gao and colleagues (2000) have shown that *Monjurosuchus* is also a choristodere. Remains from Liaoning Province show that this particular genus had webbed feet and a rather soft skin bearing small scales that are particularly tiny on the ventral surface. *Monjurosuchus* also had a closed lower temporal fenestra. Additional choristodere material continues to come to light, and we now know that they occupied a diverse array of Mesozoic aquatic niches. For example, two long-necked forms from China and Japan were apparently convergent with nothosaurs.

LEPIDOSAUROMORPHS

Living lepidosauromorphs are almost exclusively lizards or snakes (Squamata). However, there are two extant nonsquamate lepidosauromorphs, both referred to the genus *Sphenodon*. Called sphenodontians (or sometimes rhynchocephalians), *Sphenodon punctatus* and *Sphenodon guentherii* are today strictly confined to a handful of isolated islands off New Zealand. However, in the Late Triassic, the sphenodontians were very widespread, and apparently the most ubiquitous of the lepidosauromorphs.

The sphenodontians have acrodont teeth, that is, the teeth are fused to the summit of the jaw, rather than set in sockets. They typically possess a robust lower jaw with a characteristic process on the dentary that extends back well beyond the level of the coronoid process (Fig. 40.8a). As a consequence, they can be easily recognized, even on the basis of quite fragmentary material. That the modern-day *Sphenodon* is often described as a "living fossil" is something of a misconception. Admittedly, the fully diapsid skull of *Sphenodon*, with a completely intact lower temporal arcade, is somewhat reminiscent of early diapsids such as *Petrolacosaurus* and *Youngina*. However, this is almost certainly a secondarily derived condition in *Sphenodon*. Moreover, Mesozoic sphenodontians were rather diverse and widespread. So that in the Triassic there are a number of relatively small forms including the widespread genus *Clevosaurus* (Fig. 40.8). During Jurassic times, they become much more diverse, and we have the aquatic *Palaeopleurosaurus* and pleurosaurs from Germany, *Opisthias* and *Eilenodon* from the famed Morrison Formation, and some interesting tiny individuals from Mexico. Cretaceous forms are rarer but still include taxa like *Toxolophosaurus*. Some sphenodontians, including *Eilenodon* and *Toxolophosaurus*, are characterized by transversely broadened cheek teeth and an edentulous anterior region of the jaw; indicative of an adaptation toward herbivory.

More recently, two unusual sphenodontians have been described from the Albian of Mexico. The aptly named *Ankylosphenodon pachyostosus* has pachyostotic ribs and vertebrae and was apparently an aquatic herbivore (Reynoso 2000). Since it occurs in an essentially marine depositional environment, it may have had a similar lifestyle to the modern-day marine iguana. The second form, *Pamizinsaurus*, was covered in small rounded osteoscutes, giving the animal a "beaded" appearance. Reynoso (1997) noted the similarity of the dermal armor to that of the Gila monster, *Heloderma*, and suggested that in *Pamizinsaurus* it is also associated with protection in an open, dry, terrestrial environment.

Other Cretaceous sphenodontians include *Derasmosaurus*, from the famed Pietraroja deposits (Albian) of southern Italy, and a variety of forms from the Kota Formation of India. In addition, Kellner describes one Upper Cretaceous sphenodontian (possibly Cenomanian) from Patagonia (author personal communication) that bears a dentition resembling that of *Eilenodon*.

There is a good fossil record for the Squamata (lizards and snakes) extending back to the Middle Jurassic. However, for a long time there were no older records, and this became something of an enigma given that, as we have just seen, their sister group appears in the Carnian. It is also worth pointing out that the earliest known sphenodontian, *Brachyrhinodon*, is actually nested within the crown group Sphenodontia (Rhynchocephalia), thereby implying an even longer record for both sphenodontians and

40.8. A, Skull of the sphenodontian *Clevosaurus* in lateral view; B, restoration of the Triassic sphenodontian, *Clevosaurus hudsoni* (after Fraser 1988).

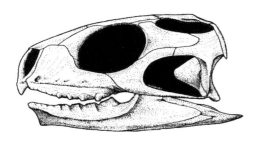

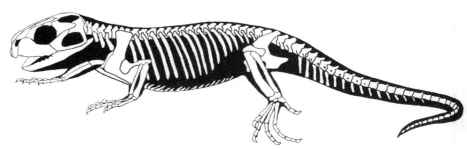

squamates—probably extending back at least to sometime in the Middle Triassic, if not before. It further suggests that the seeds of other modern-day groups might also have been around earlier in the Triassic, but that we have failed to recognize them. Perhaps their most diagnostic characters were not fully developed early in their history.

It was, perhaps, inevitable then that Triassic records would show up; the discovery of *Tikiguania* from the Late Triassic of India (Datta and Ray 2006) came as no surprise. Even so, Triassic records remain rare; there are numerous unequivocal records of squamates (lizards at least) from the Middle Jurassic onward, however. Squamates can be broadly split into Iguania (iguanas and chamaeleons) and Scleroglossans (all other lizards, snakes and amphisbaenians), with the Scleroglossans normally further subdivided into anguimorphs, scincomorphs, and gekkotans. While rather spotty, Jurassic records of the group include iguanians, scincomorphs, anguimorphs, and possibly even gekkotans.

By the Late Jurassic and Early Cretaceous, the squamate record improves considerably, but many forms such as *Bavarisaurus* and *Eichstaettisaurus* from the famed Solnhofen limestones are difficult to place satisfactorily, although they are typically considered to be basal scleroglossans. Also within the Mesozoic there are some rather unusual lizards that apparently filled very different niches from modern-day lizards. For instance, the Polyglyphanodontines (members of the teiids), have broad cheek teeth exhibiting a certain resemblance to those of mammals and procolophonoids. The upper and lower teeth are not identical but form complementary occlusal surfaces. *Bicuspidon* has teeth that exhibit a chisellike, transversely broadened distal end (Nydam and Cifelli 2002). Nydam and colleagues (2000) also showed that teeth of *Peneteius* from the Upper Cretaceous were very like those of mammals in their overall shape. It appears to have processed its food by cutting and crushing in a manner similar to that of mammals.

The oldest undisputed snake remains come from the Upper Cretaceous of North America, South America, and North Africa, and Upper Cretaceous rocks have yielded at least seven families of snakes. The origin of snakes from within Squamata is something of a contentious issue. A series of aquatic snakes from the Late Cretaceous of the Middle East bear rudimentary hindlimbs. Some authors have argued that these limbed snakes are among the most basal and that they show a close relationship to the mosasaurs (a fully marine group of Cretaceous varanids) (e.g., Caldwell and Lee 1997). Others contend that these limbed snakes are actually macrostomatans (Zaher 1998)—in other words, quite advanced snakes, and this now seems to be the more widely accepted hypothesis (e.g., Head 2002). The genus *Najash* from the Late Creataceous of Argentina shows adaptations for burrowing (Apesteguía and Zaher 2006) and also suggests that snakes had a terrestrial origin.

Of course, the mosasaurs themselves are rather aberrant by today's lizard standards. Of the 3000 or so species of living lizard, only one, the marine iguana *Amblyrhynchus*, can be considered marine, and it spends most of its time on the rocky shores of the Galápagos Islands. However, in the Cretaceous there were three families of fully marine lizards: the dolichosaurs, aigialosaurs, and mosasaurs. All three families are considered to

be varanoids, the radiation of lizards that is notable today for the monitor lizards, including the monstrous Komodo dragon. Whereas the dolichosaurs and aigialosaurs are rather poorly known groups, the mosasaurs include quite a diverse array of complete skeletons from the Upper Cretaceous. These include *Clidastes*, *Tylosaurus*, *Platecarpus*, and the enormous *Plotosaurus* that attained lengths of 10 m. With long tails and slender bodies, they probably moved through the water with an anguilliform swimming motion (Fig. 40.9). The legs were modified to form flippers. They typically occur in sediments deposited in near-shore coastal environments. They probably had a good appetite for mollusks, in particular the ammonites that are frequently found in the same deposits.

The final group of living squamates is the amphisbaenians. These are very specialized burrowing reptiles that, with exception of one genus, *Bipes*, lack limbs and girdle elements. The bones of the skull are very solid, which is in contrast to the pattern of lizards and snakes. The skull is actually used as the digging tool. Only one form, *Sineoamphisbaenia*, from the Upper Cretaceous of China, can be unequivocally said to have lived at the same time as the dinosaurs. There are, however, some fragmentary remains from Asia that have been tentatively assigned to Amphisbaenia.

ARCHOSAUROMORPHS

The recognition of a discrete archosauromorph division of the diapsids has resulted in some Permian and Triassic taxa, traditionally viewed as precursors of the squamates (lizards and snakes), being more closely allied with the archosaurs. Thus *Prolacerta*, or the "prelizard," with its elongated neck is apparently a misnomer. *Prolacerta*, together with *Macrocnemus* and *Tanystropheus* sporting an absurdly elongated neck (Fig. 40.10), form a group of their own, the protorosaurs. Most of these were aquatic, or semiaquatic, animals and were especially widespread during Middle Triassic times. Yet one or two, such as *Tanytrachelos*, range into Late Triassic sediments that are also known for dinosaur trackways.

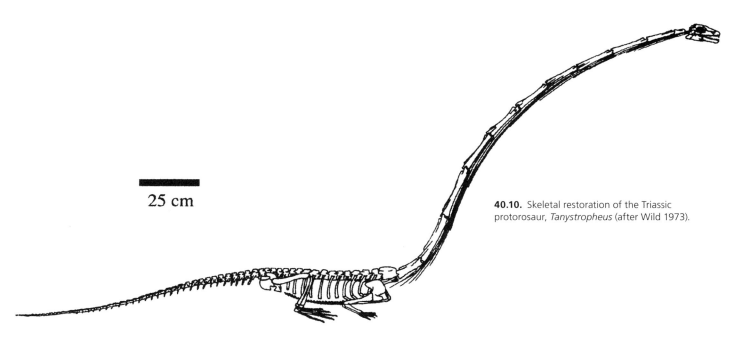

25 cm

40.10. Skeletal restoration of the Triassic protorosaur, *Tanystropheus* (after Wild 1973).

The position of many other Late Triassic lizardlike taxa, including the gliding kuehneosaurs, is still unclear. The remarkable *Longisquama*, with its array of spinelike structures along its back, and *Sharovipteryx*, with a gliding membrane stretched across its hindlimbs, are difficult to assess because of a combination of highly modified features overprinting certain characteristics and the poor preservation of critical parts of the skeleton.

Many of the early archosauromorph lineages failed to survive beyond the Triassic and therefore were at best only briefly (on a geological scale at least) contemporaneous with the dinosaurs.

The rhynchosaurs, once aligned with the sphenodontians, were widespread in the Middle Triassic, but also a major component of many Late Triassic terrestrial assemblages. These large quadrupeds with barrel-shaped bodies (Fig. 40.11) used their beaklike jaws to snip off plant material. Some have suggested that rhynchosaurs and many basal archosuars were unable to compete with the early dinosaurs (e.g. Charig 1980, 1984); others have suggested that the dinosaurs simply radiated into the niches that were being vacated as a result of the extinction of the rhynchosaurs and other major groups (e.g. Benton 1987, 1991). Much more has been written on this subject in recent times. For example, Brusatte et al. (2008) suggested that historical contingency rather than competitive superiority was instrumental in the rise of the dinosaurs. Whatever the mechanisms involved, rhynchosaurs apparently died out sometime before the close of the Triassic.

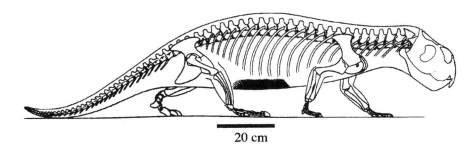

20 cm

40.11. The Triassic rhynchosaur, *Hyperodapedon* (after Benton 1983).

One particularly strange group of Triassic animals that has recently come to light is the Drepanosauria. Their position within Diapsida is unclear, but they appear to be more akin to the archosauromorphs than the lepidosauromorphs (Dilkes 1998) and presently are most closely linked with the protorosaurs (Renesto et al. 2010). They are characterized by a very thin, rodlike scapula, well-developed neural spines in the region of the shoulder girdle that typically fuse, a barrel-shaped trunk, pronounced neural and hemal spines in the tail vertebrae giving the tail a compressed leaflike appearance, and sometimes a clawlike bone at the end of the tail. Some may have had a prehensile tail. Although some, like *Megalancosaurus*, appear to be highly adapted to life in the trees, others, like the tiny *Hypuronector* (Colbert and Olsen 2001) known informally as the deep-tailed swimmer, were almost certainly fully aquatic forms. *Hypuronector* occurs in the Lockatong Formation of the Newark Supergroup, a unit that is also known to include facies containing numerous dinosaur footprints.

For the small school that continues to argue that the origin of birds falls outside Theropoda, *Megalancosaurus* has attracted some attention as a possible alternative. However, any resemblance to birds is purely superficial; like *Longisquama*, *Megalancosaurus* does not offer a viable alternative to Theropoda.

ARCHOSAURIA

A more exclusive grouping of the archosauromorphs is the Archosauria, or "Ruling Reptiles," which of course also include the dinosaurs themselves. A number of other major archosaurian lineages can be recognized, notably the crocodiles and pterosaurs, but there is no current consensus on all the details of their phylogeny.

Originally, the key-defining feature of the Archosauria was considered to be the presence of an antorbital fenestra (Fig. 40.4E). The function of this additional opening in the skull has been the subject of much debate. Most recently, in a series of elegant studies, Witmer (1997) has argued a strong case that it is associated with the extension of an air sinus system. With one or two exceptions, most authors still accept that the antorbital fenestra is a valid character that probably evolved just once and helps to define a monophyletic clade of organisms (but see the discussion on pterosaurs later in this chapter). However, most authors do not equate this node in the archosauromorph clade with Archosauria. Instead, Gauthier defined Archosauria to include all descendents of the most recent common ancestor of birds and crocodiles, and, as such, excluded the Proterochampsidae, Erythrosuchidae, and Proterosuchidae, despite the fact that these three families have an antorbital fenestra. This view was also adopted by Parrish (1993). He called the more inclusive group *Archosauriformes*.

Sereno (1991) provides one basic starting point for an introduction to the divisions of the basal archosaurs (see also Parrish, this volume). As he points out, it has long been recognized that early in archosaur history there was a split into two major clades (Fig. 40.12). These were traditionally based on the understanding that a rotary-style ankle joint, that is a definitive ball-and-socket joint between the two principal bones of the ankle, evolved

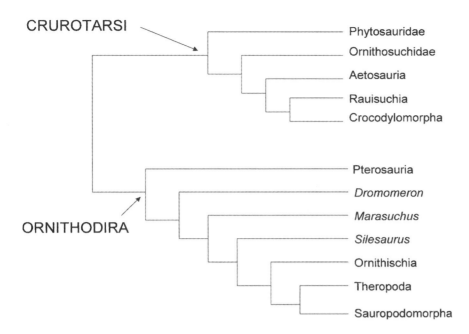

CRUROTARSI

- Phytosauridae
- Ornithosuchidae
- Aetosauria
- Rauisuchia
- Crocodylomorpha

ORNITHODIRA

- Pterosauria
- *Dromomeron*
- *Marasuchus*
- *Silesaurus*
- Ornithischia
- Theropoda
- Sauropodomorpha

40.12. The phylogeny of the Archosauriformes after Sereno (1991). An antorbital fenestra is present in all the named taxa.

independently in two apparently separate lineages. These are the so-called crocodile-normal and the crocodile-reversed types of ankle joint; the two lineages have been termed the *Pseudosuchia* (suchians and parasuchians) and the *Ornithosuchia* (Ornithosuchidae, pterosaurs, and dinosaurs), respectively. However, in an extensive analysis of a wide range of cranial and postcranial characters, Sereno (1991) suggested that the rotary ankle joint evolved just once and that the Ornithosuchidae should be grouped together with suchians and parasuchians. For this group, Sereno used the term *Crurotarsi*, and this has generally been accepted in recent studies. With the exclusion of the Ornithosuchidae, the second lineage, known as the Avemetatarsalia (or Ornithidira), essentially comprises just the Pterosauria and Dinosauromorpha. Neither of these two major lineages exhibits a rotary joint in the ankle.

CRUROTARSI

The interrelationships within Crurotarsi continue to be a subject of intense debate with no agreement on which group is most basal and which is most closely related to the crocodiles (see Brusatte et al. 2010). Apart from the crocodiles, four principal crurotarsan groups briefly lived alongside, or at least were contemporaries of, the dinosaurs. These are the phytosaurs (the parasuchians of some authors), rauisuchians, stagonolepids (or aetosaurs) and ornithosuchids. The superficially crocodile-like phytosaurs (Fig. 40.13) are usually considered the most basal members of the Archosauria and are major components of many northern hemisphere assemblages.

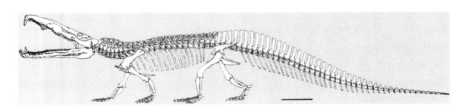

40.13. Restoration of a large rutiodontine phytosaur (after Long and Murry 1995).

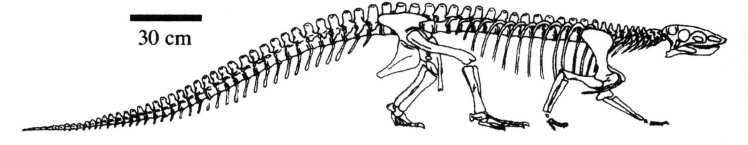

30 cm

40.14. The aetosaur, *Stagonolepis* (after Walker 1961).

However, like today's Nile crocodile or American alligator, phytosaurs were semiaquatic animals that typically possessed elongate rostra and jaws bristling with acutely conical, serrate teeth. They appear to have filled a very similar niche to living crocodiles and alligators. For much of their day they probably lazed on the edges of rivers and lakes, but took to the water to lie mostly submerged, waiting for unsuspecting prey to come to the water's edge. Like crocodiles, they possessed bony scutes in the skin. Unlike crocodiles, the external nares were not positioned at the end of the snout but had shifted back to a point on top of the skull, so that in many instances the nostrils were close to the eye sockets. As a consequence, when the animal was lazing at the water surface, only the eyes and the outline of the nostrils just in front were above the water.

Of the crurotarsans, only the aetosaurs were obligate herbivores (Fig. 40.14). These armored quadrupedal forms are restricted to Upper Triassic sequences. They bore leaf-shaped teeth, but both the upper and lower jaws were edentulous anteriorly. They may have rooted around in the dirt a bit like pigs in their search for roots and stems. At approximately 3 m long, *Stagonolepis* and *Longosuchus* were typical in size for aetosaurs.

The rauisuchians were mostly large quadrupedal carnivores that assumed an erect posture, with the limbs drawn under the body in a manner similar to dinosaurs. Some of these large carnivores, such as *Postosuchus* from the American Southwest, may have been bipedal. Indeed, a remarkable convergence in body form and anatomical details can be seen between some crurotarsan archosaurs such as *Effigia* (Nesbitt, 2007) and contemporary theropod dinosaurs. In dinosaurs, the erect posture is achieved through an offset proximal head on the femur that turns into a laterally facing socket (acetabulum) in the hip girdle. In the rauischians on the other hand, the femur does not have an offset head, but the acetabulum is angled downward. But other anatomical details such as the very prominent bootlike foot on the pubis is present both in *Effigia* and theropod dinosaurs.

The ornithosuchids were large predatory archosaurs that may have been habitually bipedal. It is likely that forms such as *Ornithosuchus* (Fig. 40.15) were very like theropod dinosaurs, assuming the role of top predator in the paleocommunities they inhabited, although they lacked most of the subtle osteological characters that are diagnostic of dinosaurs. It is surprising, therefore, to find one of the best-known ornithosuchids, *Ornithosuchus*, from the Lossiemouth Sandstone, alongside one of the earliest putative dinosaurs, *Saltopus* (Huene 1910). *Saltopus*, however, is based on a single rather indistinct fossil and its dinosaurian status has been questioned and Benton and Walker (2011) now consider it more likely to be a dinosauriform rather than a true dinosaur.

0.5 m

CROCODILES

40.15. The ornithosuchid, *Ornithosuchus* (after Walker 1964).

Obviously, the final group of crocodile-normal archosaurs to be discussed here is the crocodiles themselves. Crocodiles must rank alongside the dinosaurs and pterosaurs as major success stories. Although often cited as a conservative group that has changed little over their 230 or so million year history, crocodiles form a very diverse group. The first crocodiles are the sphenosuchians, relatively small animals such as *Terrestrisuchus* (Fig. 40.16) from the Upper Triassic (Crush 1984). A far cry from the sluggish semiaquatic crocodiles and alligators that we see today, the sphenosuchians were typically agile terrestrial animals, highly adapted to a cursorial predatory existence. It is even possible that some of these early crocodiles were habitually bipedal. Despite being fleet-footed creatures, they still possess most of the key features that characterize crocodiles today, including the elongate wrist bones (radiale and ulnare), an anteriorly inclined quadrate contacting the prootic, some degree of pneumatization of the cranial bones, and an attenuated posterior process on the coracoid bone.

The protosuchids of the Late Triassic and Early Jurassic were also fully terrestrial forms, but their skull was somewhat more dorsoventrally compressed giving them more of a typical crocodilian shape. Fairly early in their history the evolution of an aquatic habit took place. Thus, in the

40.16. Restoration of the Triassic sphenosuchid crocodilian, *Terrestrisuchus* (after Sereno and Wild 1992).

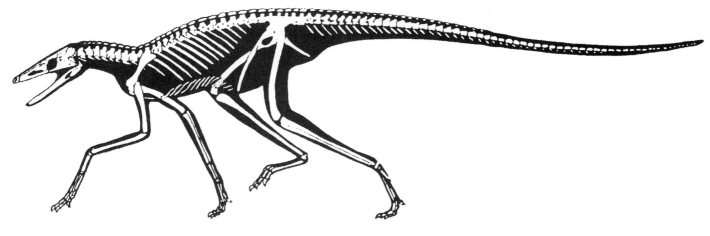

Lower Jurassic deposits of Germany and England we see fully marine forms such as *Steneosaurus*. But the most specialized aquatic crocodiles were the Jurassic metriorhynchids. *Metriorhynchus* for example had lost the heavy dermal armor and its limbs took the form of flippers. Even more striking was the sharp down turning of the vertebral column towards the end of the tail that, in exactly the same way as the ichthyosaurs, supported a tail fin.

Later on in the Cretaceous, some crocodiles attained exceptional size with *Deinosuchus* and *Sarcosuchus* being of proportions rivaling even the largest meat-eating dinosaurs. It is even possible that these large crocodiles were capable of ambushing large theropods that came down to the water's edge to drink.

Non-Dinosaurian Ornithodirans

The Pterosauria

Pterosaurs and dinosaurs have been grouped together for a long time, and today the prevalent view among vertebrate paleontologists is still that they do indeed form a natural group of animals. There is some skepticism, however, among one or two workers. These views have been most strongly articulated by Rupert Wild, and most recently Peters (2000). Both authors have argued that the pterosaurs were more closely related to the protorosaurs. Certainly, as pointed out by Bennett (1996), some of the principal features shared by dinosaurs and pterosaurs are associated with the hindlimb, and as such could simply be a result of convergence toward a similar stance and locomotor pattern. Admittedly, it would still be necessary to invoke the origin of the antorbital fenestra on at least two separate occasions.

Pterosaurs are a unique group of ancient vertebrates that are characterized by an enormously elongate fourth finger that supported a membranous wing. Pterosaurs are often broadly separated into two groups: the rhamphorhynchoids and the pterodactyloids. Although the former is something of a ragbag that is not strictly monophyletic, they are loosely characterized by an elongate tail, a relatively short neck, and small head. The pterodactyloids appeared in the Late Jurassic and form a more natural grouping of animals. They are characterized by a much-reduced tail and elongate wrist bones. However, a recent discovery suggests that it will become even harder to maintain these generalized distinctions in future. *Darwinopterus* (Lu et al. 2010) is a remarkable animal that appears to be a rhamphorhynchoid/pterodactyloid hybrid with a pterodactyloid skull (complete with a combined naso-antorbital opening) on a rhamphorhynchoid body (with a long tail and a long fifth toe).

The earliest pterosaurs originate from the Late Triassic of Europe and Greenland. *Eudimorphodon*, *Peteinosaurus*, and *Preondactylus* have all been described from northern Italy; *Eudimorphodon* has also been documented from Greenland (Jenkins et al. 2001). A fourth genus, *Austriodactylus*, has been recorded from the Norian of Austria (Dalla Vecchia et al. 2002). This particular genus is notable for the presence of a sagittal cranial crest. Such crests are widespread in Late Jurassic and Cretaceous pterodactyloids, but clearly they appeared early in the evolutionary history of pterosaurs. Both *Eudimorphodon* and *Austriodactylus* have distinct heterodont dentitions with long, peglike teeth anteriorly and multicusped teeth more posteriorly. In addition, *Eudimorphodon* has two caniniform teeth in the middle of the maxilla, and *Austriodactylus* has a number of large bladelike teeth positioned at intervals along the maxilla.

Typically, pterodactyloids have much larger heads relative to the length of their bodies than rhamphorhynchoids do. During the Cretaceous, some giant pterosaurs, such as *Pteranodon*, evolved, and the truly monstrous *Quetzalcoatlus* with a wingspan possibly approaching 12 m was equivalent in size to a small airplane (Fig. 40.17). The traditional view of pterosaurs was one of wonderfully efficient flyers and gliders, but clumsy and comical ground waddlers. The reason for this image was that the wing membrane was presumed to have attached to the hind limb in a fashion not unlike that of a bat wing. Certainly, specimens of pterosaurs that preserve remnants of the membrane suggest that it was attached near the ankle. However, some workers have disputed this idea, and the examination of other specimens indicate that perhaps the wing was attached nearer to the trunk (Padian 1983). If this were the case, there was at least a possibility that pterosaurs were capable of a much brisker, upright walk. Indeed, evidence from three-dimensionally preserved specimens of certain pterosaur hip bones and hindlimbs suggests that the limb might have been held in an upright position like that of the dinosaurs. Some of the most recent studies argue that pterosaurs adopted a stance and gait that was somewhat intermediate between the two outlined above. Based on trackways known as *Pteraichnus*, widely thought to have been made by pterosaurs (Lockley et al. 2001), Bennett (1997) made the case that the makers were neither sprawling quadrupedal forms nor birdlike bipedal animals. Instead, they appear to have been walking on all fours, but with an erect posture.

These are clearly opposing theories, but perhaps they don't have to be mutually exclusive. Although the modification of the forelimb into a wing certainly poses some restrictions, this still doesn't rule out the possibility that different groups of pterosaur evolved different locomotor patterns. After all, even among a closely related group of animals such as the birds, we see a great disparity of locomotor types. The waddling of penguins is a far cry from the high-speed run of an ostrich. It is at least conceivable that there were somewhat different locomotor patterns and postures within pterosaurs. So, the question remains unresolved. A study of the shape of the claws in pterosaurs adds a further twist to the debate. Krauss and colleagues (2002) reported that the manus claws were significantly more curved than those of the pes and within the range of curvature recorded for raptors and perching birds. This difference in curvature was particularly marked in rhamphorhynchoids. They noted that in bats, which primarily hang from their feet but also use their manus claws for crawling along branches of trees, both manus and pes claws exhibited high degrees of curvature. Krauss and colleagues suggested that the low degree of curvature in pterosaur pes claws spoke against an arboreal existence and instead proposed that pterosaurs used their manus to manipulate food while adopting a bipedal stance on the ground.

The tiny *Anurognathus* from the Jurassic of Germany was a curious rhamphorhynchoid with a much-reduced tail. It had a very short snout and

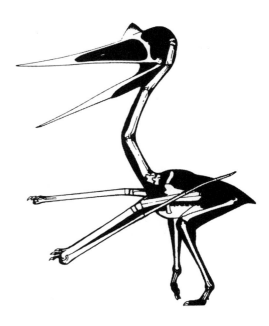

40.17. Restoration of the giant pterosaur, *Quetzalcoatlus,* with an erect, bipedal posture (after Peters 1989).

enormous eyes, and seems to have been well adapted to catching small insects on the wing.

Although some of the smaller pterosaurs were insectivorous, many pterosaurs were undoubtedly fish eaters. The Jurassic *Rhamphorhynchus* probably trawled its lower jaw through the water and pierced fish with its long, wide-spaced teeth. One of the best known of all the pterosaurs is *Pteranodon*, first discovered by Williston while working the Niobrara Chalk for O. C. Marsh. It is likely that the toothless *Pteranodon* also trawled the sea, but swallowed its prey whole.

The diversity of head and beak crests that evolved among pterosaurs probably served principally for display and recognition functions (Bennett 2002a). *Thalassodromeus* from the Cretaceous of Brazil had an enormous, albeit very thin, crest that is at least ¾ the length of the skull. Radiating across the surface of the crest is a series of grooves that are thought to have accommodated a network of blood vessels. Thus, like the plates in *Stegosaurus*, it has been postulated that the crest was used in thermoregulation as well as serving a display function (Kellner and Campos 2002). Kellner and Campos also noted that *Thalassodromeus* had a rather unique scissor-like jaw action. They argued that the only living animal with such a jaw is the skimmer, *Rynchops*. In order to catch fish and crustaceans *Rynchops* skims its beak through seawater; it is conceivable that *Thalassodromeus* fed in a similar manner. The beak of *Thalassodromeus* does not exhibit the same blunt cross section seen in the beak of skimmers, however, so there remains some doubt. One Cretaceous form, *Pterodaustro*, bore multitudes of incredibly long and thin teeth on the lower jaw, and together they formed a sieve somewhat akin to the baleen of mysticete whales. It could probably force mouthfuls of water through the mesh of teeth and trap small marine organisms that it could then lick off and swallow. The pterosaurs died out alongside the last non-avian dinosaurs (excluding birds) at the end of the Cretaceous.

DINOSAUROMORPHS

The discovery of the dinosauromorph *Dromomeron* in deposits alongside some of the earliest theropods indicates that they did coexist with true dinosaurs. Other dinosauromorphs, including *Silesaurus* from the Carnian of Poland (Dzik 2003), may also have lived alongside early dinosaurs. The earliest record of a dinosauromorph is *Asilisaurus* from the Middle Triassic (Anisian) of Tanzania (Nesbitt et al. 2010) which predates the oldest accepted dinosaur body fossils. It is interesting to note that both *Asilisaurus* and *Silesaurus* were almost certainly herbivores and that herbivory evolved independently from a carnivorous ancestor in silesaurids, ornithischians, and sauropodomorphs.

Mammals

Although mammals are not typically considered alongside dinosaurs, they arose at approximately the same time, and as Lillegraven and colleagues (1979) so succinctly put it in the title of their book, the Mesozoic (the Age of Dinosaurs) accounts for two-thirds of the evolutionary history of mammals!

Mammals are descended from the synapsid ("mammal-like reptiles") level of reptile organization (see Fig. 40.4C). True mammals can be characterized as warm-blooded tetrapods that suckle their young, possess hair or fur, have three bones in the middle ear, and have a single bone (the dentary) forming the lower jaw. However, there are many other features, such as the presence of multicusped and multirooted teeth, that can be used to identify mammals, and these are often of great significance in analyzing the sometimes poor and fragmentary fossil record. Of course, all the characters that we use to define mammals were not simply acquired together overnight, and as a consequence there are a variety of synapsid reptiles exhibiting varying degrees of "mammalness." The almost complete absence of completely reliable indicators of physiology causes much discussion surrounding which group of synapsids form the sister group to mammals. In recent years, a variety of different groups have been touted as closest to the ancestry of mammals. These include the Thrinaxodontidae, Probainognathidae, Tritheledontidae, and Tritylodontidae. Current support is strongest for the carnivorous tritheledonts or the herbivorous tritylodonts. Both range from the Late Triassic to the Early Jurassic, were contemporaries of the early dinosaurs, and were rather rodentlike in appearance.

The earliest undisputed record of a mammal is *Adelobasileus* from the Dockum Formation of Texas, which is about 225 million years old (late Carnian) and practically contemporaneous with the first dinosaurs. Other Triassic mammals include the morganucodontids, haramyids, and kuehneotheriids, many of which have been found in the famous fissure deposits of southwest Britain. Although for the most part the fissure assemblages comprise completely disassociated elements (rather than articulated material), a widely accepted restoration of a generalized morganucodontid has been produced based on these disarticulated elements together with an almost complete skeleton of *Megazostrodon* from southern Africa (Fig. 40.18).

In addition to the members of the extant infraorder Eutheria (placentals), and the orders Marsupialia and Monotremata, a plethora of extinct

40.18. Rendition of an early mammal such as *Megazostrodon*.

orders have been recognized from the Mesozoic. These include the Triconodonta, Docodonta, Multituberculata, and Eupantotheria.

The docodonts are a small group of mammals known just from the Middle and Upper Jurassic of North America and Europe. They were about the size of a small mouse and are characterized by their rectangular molars with high buccal cusps. *Haldanodon* from the Guimarota mines of Portugal is perhaps the best known. It has noticeably robust and stocky limb elements and it seems likely that it was adapted to a burrowing lifestyle and was perhaps semiaquatic, somewhat like the modern-day desmanes. Other docodonts include *Docodon* from the Morrison Formation and *Simpsonodon* from Kirtlington, England.

The most diverse group of Mesozoic mammals is the Multituberculata. They appear in the Upper Jurassic (but they may have close affinities with the Upper Triassic haramyids) and range through the Cretaceous. The skull is superficially like that of a rodent, largely because of the presence of procumbent incisors separated by a diastema from a series of broad grinding molars. It also seems likely that multituberculates were largely herbivorous.

The so-called Eupantotheria include forms such as the Dryolestids and henkelotherids that lack tribosphenic molars (see discussion later in this section). As such, they are still basal in their evolutionary position among mammals. The dryolestids have very characteristic molars that make them easy to identify, even within fragmentary material. They were probably insectivores and are distributed throughout Europe and North and South America. They survived into the Late Cretaceous. *Henkelotherium* from Guimarota is represented by one superb skeleton that indicates a small insectivorous arboreal animal. But complete mammal skeletons such as the *Henkelotherium* specimen are generally rare in the Mesozoic and are mostly restricted to one or two unique horizons and localities such as the Yixian Formation of Liaoning Province, China, or the late Cretaceous Ukhaa Tolgod locality in Mongolia.

The famed Yixian Formation has yielded a plethora of exquisite fossils aside from feathered dinosaurs and birds. Fish, mollusks, insects, and plants are well represented, but one of the most significant fossils is a symmetrodont mammal (Hu et al. 1997). *Zhangheotherium* is probably part of the early therian radiation that preceded the separation of living marsupials and placentals. Interestingly, it also indicates that multituberculates are more closely related to therians than any other mammalian lineage.

The majority of mammals are represented in the Mesozoic fossil record on the basis of teeth; consequently, dental characters have played a very important role in determining phylogenetic relationships and distribution of the various Mesozoic mammalian taxa.

The so-called tribosphenic molar has been central to many of the discussions. This very distinctive molar is capable of both grinding (*tribein*) and shearing *(sphen)*, and it characterizes placentals and Marsupials (living therians) and their close fossil relatives. It is not present in living monotremes, and the other Mesozoic mammals described to this point (such as the symmetrodonts and eupantotheres). In the past, the tribosphenic molar was thought to have arisen in the northern hemisphere sometime in the early Cretaceous. However, *Ambondra*, a mammal from the Middle Jurassic of Madagascar (Flynn et al. 1999) has fully tribosphenic molars and, moreover, apparently groups together with the earliest known monotreme (*Steropodon* from the early Cretaceous of Australia) and *Austriktribosphenos*, a placental mammal (possibly even a member of the living hedgehog family). This, therefore, raises the distinct possibility that the tribosphenic condition arose in two separate lineages—one in the northern hemisphere and the other in the southern.

Certainly the records for Mesozoic mammals are exceptionally abundant and varied, but it must still be said that they remained relatively small and, by their Tertiary standards, relatively inconspicuous during the Age of Dinosaurs.

Summary

It is easy to become blinkered when looking at the Mesozoic world, to see only the dinosaurs. It is not surprising that we are totally enthralled by their apparent dominance of the terrestrial realm, but without an even more complex and diverse array of contemporaneous life forms they could not have existed at all. For vertebrates alone, they ranged from the giant azhdarchid pterosaurs of the Cretaceous to tiny mouse-sized mammals, sometimes known only on the basis of isolated teeth. The seas, too, contained a plethora of non-dinosaurian vertebrates, including the monstrous plesiosaurs and ichthyosaurs. And, aside from these long-extinct groups, the template for the modern world was also laid during the Mesozoic, so that the first turtles, frogs, crocodiles, and mammals also walked alongside the world's first dinosaurs.

When we look at all components of terrestrial assemblages, it is immediately apparent that the Mesozoic was quite simply an age of great vertebrate diversity and the dawn of modern terrestrial ecosystems.

References

Apesteguía, S., and H. Zaher. 2006. A Cretaceous terrestrial snake with robust hindlimbs and a sacrum. *Nature* 440: 1037–1040.

Bennett, S. C. 1996. The phylogenetic position of the Pterosauria within the Archosauromorpha. *Zoological Journal of the Linnean Society* 118: 261–308.

——. 1997. Terrestrial locomotion of pterosaurs: a reconstruction based on *Pteraichnus* trackways. *Journal of Vertebrate Paleontology* 17: 104–113.

——. 2002a. Soft tissue preservation of the cranial crest of the pterosaur *Germanodactylus* from Solnhofen. *Journal of Vertebrate Paleontology* 22: 43–48.

——. 2002b. A second specimen of *Anurognathus* from the Solnhofen Limestone of southern Germany. *Journal of Vertebrate Paleontology* 22 (suppl. to 3): 36A.

Benton, M. J. 1983. The Triassic reptile *Hyperodopedon* from Elgin: Functional morphology and relationships. *Philosophical Transactions of the Royal Society of London B* 302: 605–720.

Benton, M. J. 1987. Mass extinctions among families of non-marine tetrapods: The data. *Mémoires de la Société Géologique de France, n.s.* 150: 21–32.

Benton, M. J. 1991. What really happened in the Late Triassic? *Historical Biology* 5: 263–278.

Benton, M. J., and A. D. Walker. 2011 Saltopus, a dinosauriform from the Upper Triassic of Scotland. *Earth and Environmental Science Transactions of the Royal Society of Edinburgh* 101: 285–299

Brusatte, S., M. J. Benton, M. Ruta, and G. T. Lloyd. 2008. Superiority, competition, and opportunism in the evolutionary radiation of dinosaurs. *Science* 321: 1485–1488.

Brusatte, S. L., M. J. Benton, J. B. Desojo, and M. C. Langer. 2010. The higher-level phylogeny of Archosauria (Tetrapoda: Diapsida). *Journal of Systematic Palaeontology* 8: 1–47.

Caldwell, M. W., and M. S. Y. Lee 1997. A snake with legs from the marine Cretaceous of the Middle East. *Nature* 386: 705–709.

Carroll, R. L. 1988. *Vertebrate Paleontology and Evolution*. New York: W. H. Freeman.

Charig, A. J. 1980. Differentiation of lineages among Mesozoic tetrapods. *Mémoires de la Société Géologique de France, n. s.* 139: 207–210.

Charig, A. J. 1984. Competition between therapsids and archosaurs during the Triassic period: A review and synthesis of current theories. In Ferguson, M. W. J. (ed.), *The structure, development and evolution of reptiles*, 597–628. London: Academic Press.

Chatterjee, S. 1990. *Protoavis* and the early evolution of birds. *Palaeontographica Abteilung A* 254: 1–100.

Cheng, Y., X. Wu and Q. Ji. 2004. Triassic marine reptiles gave birth to live young. *Nature* 432: 383–386.

Clack, J. A. 2000. The origin of tetrapods. In H. Heatwole and R. L. Carroll (eds.), *Amphibian Biology 4, Palaeontology: The Evolutionary History of Amphibians*, 979–1029. Chipping Norton, NSW, Australia: Surrey Beatty and Sons.

Colbert, E. H., and P. E. Olsen. 2001. A new and unusual aquatic reptile from the Lockatong Formation of New Jersey (Late Triassic, Newark Supergroup). *American Museum Novitates* 3334: 1–24.

Crush, P. J. 1984. A Late Triassic sphenosuchid from Wales. *Palaeontology* 27: 131–157.

Dalla Vecchia, F. M., R. Wild, H. Hope, and J. Reitner. 2002. A crested rhamphorynchoid pterosaur from the late Triassic of Austria. *Journal of Vertebrate Paleontology* 22: 196–199.

Datta, P. M., and S. Ray. 2006. Earliest lizard from the Late Triassic (Carnian) of India. *Journal of Vertebrate Paleontology* 26: 795–800.

DeBraga, M. and O. Rieppel. 1997. Reptile phylogeny and the interrelationships of turtles. *Zoological Journal of the Linnean Society* 120: 281–354.

Dilkes, D. W. 1998. The Early Triassic rhynchosaur *Mesosuchus browni* and the interrelationships of basal archosauromorph reptiles. *Philosophical Transactions of the Royal Society of London B* 353: 501–541.

Dzik, J. 2003. A beaked herbivorous archosaur with dinosaur affinities from the early Late Triassic of Poland. *Journal of Vertebrate Paleontology* 23: 556–574.

Estes, R., and O. A. Reig. 1973. The early fossil record of frogs: A review of the evidence. In J. L. Vial (ed.), *Evolutionary Biology of the Anurans: Contemporary Research on Major Problems*, 11–63. Columbia: University of Missouri Press.

Evans, S. E. 1989. New material of *Cteniogenys* (Reptilia: Diapsida) and a reassessment of the systematic position of the genus. *Neues Jahrbuch für Geologie und Paläontologie, Monatshefte* 1989: 577–589.

Flynn, J. J., J. M. Parrish, B. Rakotosamimanana, W. F. Simpson, and A.E. Wyss. 1999. A Middle Jurassic mammal from Madagascar. *Nature* 401: 57–60.

Fraser, N. C. 1988. The Osteology and relationship of *Clevosaurus* (Reptilia: Sphenodontida). *Philosophical Transactions of the Royal Society of London B* 321: 125–178.

Friedman, M., K. Shimada, L. D. Martin, M. J. Everhart, J. Liston, A. Maltese, and M. Triebold. 2010. 100-million-year dynasty of giant planktonivorous bony fishes in the Mesozoic seas. *Science* 327: 990–993.

Gao, K., S. E. Evans, Q. Ji, M. Norell, and S. Ji. 2000. Exceptional fossil material of a semi-aquatic reptile from China: the resolution of an enigma. *Journal of Vertebrate Paleontology* 20: 417–421.

Godfrey, S. 1984. Plesiosaur subaqueous locomotion: a reappraisal. *Neues Jahrbuch für Geologie und Paläontologie, Monatshefte* 11: 661–672.

Head, J. J. 2002. Phylogenetic significance of vertebral morphology in snakes: implications for interpreting the fossil record. *Journal of Vertebrate Paleontology* 22 (suppl. to 3): 63A.

Huene, F. von 1910. Ein primitiver dinosaurier aus der mittleren Trias von Elgin. *Geologische und Palaontologische Abhandlungen* (N.F.) 8: 315–322

Hu, Y., Y. Wang, Z. Luo, and C. Li. 1997. A new symmetrodont mammal from China and its implications for mammalian evolution. *Nature* 390: 137–142.

Jenkins, F. A. Jr., and D. M. Walsh. 1993. An Early Jurassic caecilian with limbs. *Nature* 365: 246–250.

Jenkins, F. A., Jr., and G. E. Goslaw, Jr. 1983. The functional anatomy of the shoulder of the Savannah monitor lizard (*Varanus exanthematicus*). *Journal of Morphology* 175: 195–216.

Jenkins, F. A., Jr., and N. H. Shubin. 1998. *Prosalirus bitis* and the anuran caudopelvic mechanism. *Journal of Vertebrate Paleontology* 18: 495–510.

Jenkins, F. A., Jr., N. H. Shubin, S. Gatesy, and K. Padian. 2001. A diminutive pterosaur (Pterosauria: Eudimorphodontidae) from the Greenlandic Triassic. *Bulletin of the Museum of Comparative Zoology* 156: 151–170.

Kellner, A. W. A., and D. de A. Campos. 2002. A large skimming pterosaur (Pterodactyloidea, Tapejaridae) from the Early Cretaceous of Brazil. *Journal of Vertebrate Paleontology* 22 (suppl. to 3): 73A.

Ketchum, H. F., and R. B. J. Benson. 2010. Global interrelationships of Plesiosauria (Reptilia, Sauropterygia) and the pivotal role of taxon sampling in determining the outcome of phylogenetic analyses. *Biological Review* 85: 361–392.

Krauss, D. A., A. Gupta, J. Santarosa, and N. Scivoletti. 2002. Claw geometry is an indicator of the terrestrial habits of pterosaurs. *Journal of Vertebrate Paleontology* 22 (suppl. to 3): 76A.

Li, C., X.-C. Wu, O. Rieppel, L.-T. Wang, and J. Zhao. 2009. Ancestral turtle from the late Triassic of southwestern China. *Nature* 456: 497–501.

Lillegraven, J. A., Z. Kielan-Jaworowska, and W.A. Clemens. 1979. *Mesozoic Mammal. The First Two-thirds of Mammalian History*. Berkeley: University of California Press.

Lockley, M. G., J. L. Wright, W. Langston, and E. S. West. 2001. New pterosaur track specimens and track sites in the Late Jurassic of Oklahoma and Colorado, their paleobiological significance and regional ichnological context. *Modern Geology* 24: 179–203.

Long, R. A. and P. A. Murry. 1995. Late Triassic (Carnian and Norian) tetrapods from the southwestern United States. *New Mexico Museum of Natural History & Science Bulletin* 4: 1–254.

Lu, J., D. M. Unwin, X. Jin, Y. Liu, and J. Qi. 2010. Evidence for modular evolution in a long-tailed pterosaur with a pterodactyloid skull. *Proceedings of the Royal Society B* 277: 383–389.

Lyson, T. R., G. S. Bever, B.-A. S. Bhullar, W. G. Joyce, and J. A. Gauthier. 2010. Transitional fossils and the origin of turtles. *Biology Letters* 6: 830–833.

Merck, J. W. 1997. A phylogenetic analysis of the euryapsid reptiles. Unpublished Ph.D. dissertation. University of Texas at Austin.

Milner, A. R. 1988. The relationships and origins of living amphibians. In M. J. Benton (ed.), *The Phylogeny and Classification of the Tetrapods, Vol. 1*, 59–102, Oxford: Clarendon Press.

———. 1990. The radiations of temnospondyl amphibians. In P. D. Taylor and G. P. Larwood (eds.), *Major Evolutionary*

Radiations, 321–349. Oxford: Clarendon Press.

Nesbitt, S. J. 2007. The anatomy of *Effigia okeeffeae* (Archosauria, Suchia), theropod-like convergance, and the distribution of related taxa. *Bulletin of the American Museum of Natural History* 302: 1–84.

Nesbitt, S. J., C. A. Sidor, K. D. Angielczyk, R. M. H. Smith, and L. M. A. Tsuji. 2010. Ecologically distinct dinosaurian sister groups shows early diversification of Ornithodira. *Nature* 464: 95–98.

Nydam, R. L., and R. L. Cifelli 2002. A new teiid lizard from the Cedar Mountain Formation (Albian-Cenomanian Boundary) of Utah. *Journal of Vertebrate Paleontology* 22: 276–285.

Nydam, R. L., J. A. Gauthier, and J. J. Chiment. 2000. The mammal-like teeth of the Late Cretaceous lizard *Peneteius aquilonias* Estes 1969 (Squamata, Teiidae). *Journal of Vertebrate Paleontology* 20: 628–631.

Padian, K. 1983. Functional analysis of flying and walking in pterosaurs. *Paleobiology* 9: 218–239.

Parrish, J. M. 1993. Phylogeny of the Crocodylotarsi, with reference to archosaurian and crurotarsan monophyly. *Journal of Vertebrate Paleontology* 13: 287–308.

Parsons, T., and E. Williams 1963. The relationship of modern Amphibia. *Quarterly Review of Biology* 38: 26–53.

Peters, D. 1989. *A Gallery of Dinosaurs & Other Early Reptiles*. New York: A. Knopf.

———. 2000. A re-examination of four prolacertiforms with implications for pterosaur phylogenesis. *Rivista Italiana di Paleontologia e Stratigrafia* 106: 293–336.

Reisz, R. R., and M. Laurin. 1991. *Owenetta* and the origin of turtles. *Nature* 349: 324–326.

Reynoso, V. 1997. A beaded sphenodontian (Diapsida: Lepidosauria) from the Early Cretaceous of central Mexico. *Journal of Vertebrate Paleontology* 17: 52–59.

———. 2000. An unusual aquatic sphenodontian (Reptilia: Diapsida) from the Tlayua Formation (Albian), Central Mexico. *Journal of Paleontology* 74: 133–148.

Rieppel, O., and R. R. Reisz. 1999. The origin and early evolution of turtles. *Annual Review of Ecology and Systematics* 30: 1–22.

Renesto, S., J. A. Spielmann, S. G. Lucas, and G. T. Spagnoli. 2010. The taxonomy and paleobiology of the Late Triassic (Carnian-Norian: Adamanian-Apachean) drepanosaurs (Diapsida: Archosauromorpha: Drepanosauromorpha). *New Mexico Museum of Natural History and Science Bulletin* 46: 1–81.

Robinson, J. A. 1975. The locomotion of plesiosaurs. *Neues Jahrbuch für Geologie Paläontologie Abhandlungen* 149: 286–332.

Sereno, P. C. 1991. Basal archosaurs: phylogenetic relationships and functional implications. *Society of Vertebrate Paleontology Memoir* 2: 1–53; suppl. to *Journal of Vertebrate Paleontology* 11 (4).

Sereno, P. C., and R. Wild. 1992. *Procompsognathus*: theropod, "thecodont" or both? *Journal of Vertebrate Paleontology* 12: 435–458.

Shiskin, M. A. 1991. [A labyrinthodont from the late Jurassic of Mongolia: Evolution of early amphibians (Plagiosauroidea).] *Paleontologichesiky* 1991 (1): 81–95. [In Russian]

Shubin, N. H., and F. A. Jenkins, Jr. 1995. An Early Jurassic jumping frog. *Nature* 377: 49–52.

Storrs, G. W., and D. J. Gower. 1993. The earliest possible choristodere (Diapsida) and gaps in the fossil record of semi-aquatic reptiles. *Journal of the Geological Society, London* 150: 1103–1107.

Walker, A. D. 1961 Triassic reptiles from the Elgin area: *Stagonolepis, Dasygnathus* and their allies. *Philosophical Transactions of the Royal Society of London B.* 244: 103–204.

Walker, A. D. 1964. Triassic reptiles from the Elgin area: *Ornithosuchus* and the origin of carnosaurs. *Philosophical Transactions of the Royal Society of London B* 248: 53–134.

Warren, A. A., and M. N. Hutchinson. 1983. The last labyrinthodont? A new brachyopoid (Amphibia, Temnospondyli) from the Early Jurassic Evergreen Formation of Queensland, Australia. *Philosophical Transactions of the Royal Society of London B* 303: 1–62.

Warren, A. A., L. Kool, M. Cleeland, T. H. Rich, and P. V. Rich. 1991. Early Cretaceous Labyrinthodont. *Alcheringa* 15: 327–332.

Warren, A. A., P. V. Rich, and T. H Rich. 1997. The last labyrinthodonts? *Palaeontographica A* 247: 1–24.

Wild, R. 1973. Die Triasfauna der Tessiner Kalkalpen. XXIII. *Tanystropheus longobardicus* (Bassani) (Neue Ergebnisse). Abhandlungen der Schweizerischen Paläontologischen Gesellschaft 95: 1–162.

Witmer, L. M. 1997. The evolution of the antorbital cavity of archosaurs: a study in soft-tissue reconstruction in the fossil record with an analysis of the function of pneumaticity. *Society of Vertebrate Paleontology Memoir* 3: 1–73; supplement to *Journal of Vertebrate Paleontology* 17(1).

Young, C. C., and Z. M. Dong. 1972. [*Chaohusaurus geishanensis* from Anhui Province]. In [*Aquatic reptiles from the Triassic of China*] C. C.Yong and Z. M. Dong (eds.), 11–14. Academia Sinica, Institute of vertebrate Palaeontology and Palaeoanthropology, Memoir 9 Peking. [Chinese]

Zaher, H. 1998. The phylogenetic position of *Pachyrhachis* within snakes (Squamata, Serpentes). *Journal of Vertebrate Paleontology* 18: 1–3.

Zardoya, R., and A. Meyer 1998. Complete mitochondrial genome suggests diapsid affinities of turtles. *Proceedings of the National Academy of Sciences of the United States* 95: 14226–14231.

41.1. Reconstructions of Pangaea at 220 Ma (above) and 240 Ma (below). Grey areas indicate land. Note northward drift of Pangaea during the Triassic.

Modified from base maps provided by PaleoMap Project © Christopher R. Scotese (University of Texas at Arlington).

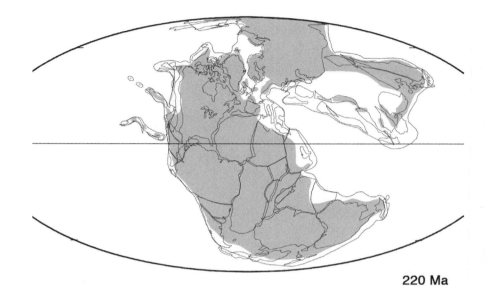

220 Ma

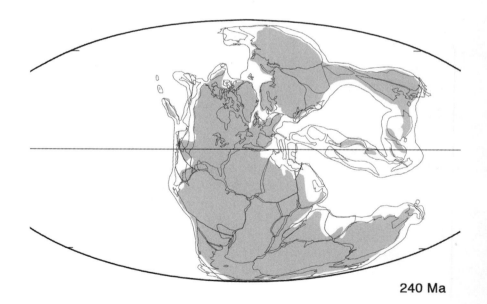

240 Ma

Early Mesozoic Continental Tetrapods and Faunal Changes

41

Hans-Dieter Sues

Most of the principal groups of present-day continental vertebrates, or their closest relatives, first appeared in the fossil record during the Triassic Period—mammals, turtles, archosaurian reptiles (crocodylians and the precursors of birds, dinosaurs), lepidosaurian reptiles (rhynchocephalians and squamates), frogs, and salamanders (because they are the sister-group of frogs). This chapter briefly reviews the distribution of the principal assemblages of Triassic and Early Jurassic continental tetrapods in time and space and discusses the profound changes that occurred among tetrapod communities during the early Mesozoic. For a comprehensive recent survey of Triassic life on land and the biotic changes during this period the reader is referred to Sues and Fraser (2010).

During the Late Permian, the most diverse and abundant tetrapods on land were therapsids, but many members of this group vanished during the Late Permian or at the end of the Permian Period in what is generally considered the greatest episode of extinction in the history of life (Erwin 2006). At the beginning of the Triassic Period, a few remaining therapsid lineages (especially dicynodonts) still constituted the vast majority of terrestrial tetrapods although their taxonomic diversity was low. By the beginning of the Jurassic Period, dinosaurs had become the dominant group of large continental tetrapods (Olsen et al. 2002; Brusatte et al. 2010; Langer et al. 2010). Archosaurian reptiles rapidly diversified during the Triassic, while only a few groups of therapsids (including the precursors of mammals) persisted in moderate diversities. The possible causes for this major change in continental vertebrate communities are not yet fully understood, but the basic structure of land ecosystems established during the early Mesozoic has persisted to the present day. Subsequently, the principal change in continental vertebrate communities was the replacement of dinosaurs other than birds by mammals in many ecological roles at the beginning of the Cenozoic Era. However, one lineage of dinosaurs—birds—vastly outnumbers mammals in terms of species numbers even today.

For the entire length of the Triassic, the continents formed a single vast landmass—the supercontinent Pangaea (Fig. 41.1). Apparently few, if any, significant barriers to dispersal of land-living animals across Pangaea existed. The breakup of the supercontinent began during the Triassic, but significant separation of landmasses did not occur until the Jurassic. Late Triassic and Early Jurassic assemblages of continental vertebrates share some taxa across Pangaea, but there are also differences in faunal composition.

During the Triassic, there was a worldwide climatic trend toward warmer and increasingly drier conditions (Sellwood and Valdes 2006), which was, at least in part, driven by a northward drift of Pangaea through some 10 degrees of latitude (Golonka 2007; Fig. 41.1). The sole exception to this trend appears to have been an interval of increased humidity during the Carnian (Simms and Ruffell 1989; Prochnow et al. 2006). The slow northward drift of Pangaea continued into the Jurassic Period so that, by the Early Jurassic, paleogeographic reconstructions place more than 50% of the exposed land surface of the supercontinent north of the paleoequator.

Triassic Faunal Succession

Paleontologists have long attempted to correlate the various early Mesozoic tetrapod assemblages with the stages of the Triassic Period on the Standard Global Chronostratigraphic Scale (SGCS)–Induan and Olenekian (Early Triassic); Anisian and Ladinian (Middle Triassic); and Carnian, Norian, and Rhaetian (Late Triassic). Such efforts are fraught with difficulties because most of the SGCS stages are formally based on sequences of marine strata in the European Alps. Direct correlation between continental and marine sedimentary rocks has proven challenging because only in rare instances have fossils of continental tetrapods and plants been preserved in Triassic marine strata. Radiometric dates for volcanic rocks interbedded with sedimentary deposits and paleomagnetic data increasingly provide ways for reliable correlation of marine and continental strata.

Lucas (1998) proposed a comprehensive scheme for the global correlation of Triassic continental strata based on their vertebrate assemblages. He defined a succession of eight "Land Vertebrate Faunachrons" (LVFs) during the Triassic Period. Each LVF is characterized by the first appearance in the fossil record of a particular tetrapod taxon. For example, Lucas defines the first (oldest) LVF, the Lootsbergian, based on the first appearance of the dicynodont therapsid *Lystrosaurus*. He augmented the definition of each LVF by adding other characteristic tetrapod taxa. Thus, the Lootsbergian LVF is also characterized by the presence of the procolophonid parareptile *Procolophon*, the cynodont therapsid *Thrinaxodon*, and the archosauriform reptile *Proterosuchus*. Lucas's LVFs provide a temporal sequence independent from the SGCS. However, many details of this scheme are still in dispute (Rayfield et al. 2005, 2009; Lucas et al. 2007), and considerable refinement is needed before a succession of LVFs can be employed for the correlation of early Mesozoic continental strata across Pangaea.

The following section briefly outlines the temporal succession of major early Mesozoic assemblages of continental vertebrates.

Early Triassic

Earliest Triassic (Induan) assemblages of continental tetrapods are best documented from the *Lystrosaurus* Assemblage Zone (Beaufort Group) of the Karoo Basin in South Africa (Rubidge 2005) . The *Lystrosaurus* Assemblage Zone is dominated by the eponymous dicynodont therapsid *Lystrosaurus*, which at many localities comprises more than 90% of the identifiable vertebrate fossils. In general, however, the diversity of therapsids was much reduced compared to that of the preceding Late Permian. *Lystrosaurus* is found together with a few other tetrapods such as the superficially crocodile-like archosauriform *Proterosuchus*. *Lystrosaurus* and *Proterosuchus* also occur together in India (Bandyopadhyay 1999) and China (Li et al.

2008). *Lystrosaurus* (Kalandadze 1975) and archosauriform reptiles closely related to *Proterosuchus* (Sennikov 1995) are known from the Early Triassic of European Russia. Early Triassic communities of continental tetrapods from European Russia and Australia are characterized by a great diversity of temnospondyls but include few therapsids (Cosgriff 1984; Ivakhnenko et al. 1997).

Stratigraphically slightly younger, late Early Triassic (Olenekian) to early Middle Triassic (Anisian) tetrapod assemblages are known from the *Cynognathus* Assemblage Zone of South Africa (Rubidge 2005). The zone is characterized by the presence of the large carnivorous cynodont *Cynognathus* (with a skull length of up to 40 cm) and the herbivorous or omnivorous gomphodont cynodont *Diademodon*. These cynodont therapsids occur together with the large dicynodont *Kannemeyeria* (with a skull length of up to 50 cm) and archosauriform reptiles such as the large *Erythrosuchus* (with a skull length of up to 1 m) and the small (about 50 cm long) *Euparkeria*. *Cynognathus* is also known from the Early Triassic of Argentina (Bonaparte

41.2. (Above) Reconstructed skull of the Late Triassic rhynchosaur *Hyperodapedon huxleyi* in dorsal and left lateral views. Modified from Chatterjee and Roy Chowdhury (1974). Scale bar equals 10 cm. (Below) Skull of the Middle Triassic dicynodont *Rhinodicynodon gracile* (holotype) in left lateral view. Scale bar equals 5 cm. After Kalandadze (1970).

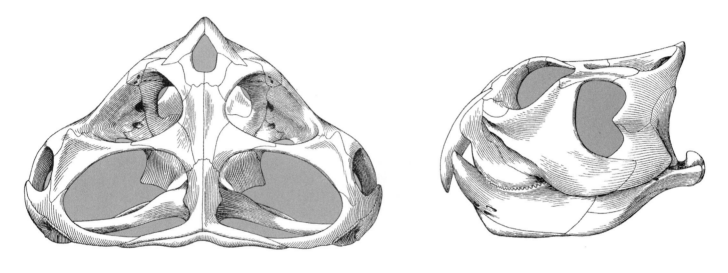

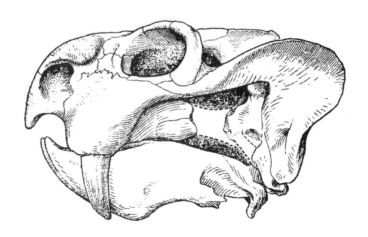

1978), and relatives of *Kannemeyeria* (Fig. 41.2) and *Erythrosuchus* occur in China (Li et al. 2008) and Russia (Sennikov 1995; Ivakhnenko et al. 1997).

Late Early to early Middle Triassic assemblages of continental tetrapods from the Middle and Upper Buntsandstein in Central Europe and the Moenkopi Formation of the American Southwest are characterized by the predominance of temnospondyls. Abundant tracks indicate the presence of the two major lineages of archosaurian reptiles—crurotarsans (crocodylomorphs and their relatives) and ornithodirans (dinosaurs and their relatives)—already in the late Early Triassic (Brusatte et al. 2011).

Middle to Early Late Triassic

The Middle to early Late Triassic tetrapod assemblages of Gondwana are best documented from northwestern Argentina (Romer 1966, 1973; Bonaparte 1978, 1997) and southern Brazil (Huene 1935–42; Langer et al. 2007). They are characterized by a great abundance of rhynchosaurs (Fig. 41.2) and traversodont cynodonts. Traversodont cynodonts are characterized by molarlike postcanine teeth that met in occlusion in a manner similar to that in mammals. Large dicynodonts, such as *Stahleckeria* (with a skull length of up to 60 cm and an estimated weight of one metric ton or more), are an additional important element of these assemblages. Most notable is the diversity of archosaurian reptiles, which include the oldest known dinosaurs and various ornithodirans closely related to them (Bonaparte 1978, 1997; Sereno et al. 1993; Sereno 1997; Brusatte et al. 2010; Langer et al. 2010).

The best-known Middle Triassic tetrapod assemblage comes from the Chañares Formation of northwestern Argentina (Romer 1966, 1973; Bonaparte 1978, 1997; Rogers et al. 2001). It includes the very common traversodont cynodont *Massetognathus*, but, unlike other more or less coeval assemblages from Gondwana, lacks rhynchosaurs. Additional faunal elements comprise carnivorous cynodonts (*Chiniquodon, Probainognathus*), a small archosaur related to crocodylomorphs (*Gracilisuchus*), the small, slender-limbed ornithodirans *Marasuchus* and *Lagerpeton*, which were closely related to dinosaurs (Sereno and Arcucci 1994a,b), large dicynodonts (*Dinodontosaurus*), and the large "rauisuchian" *Luperosuchus* (with a skull length of 60 cm). (The name "rauisuchian" is used in quotation marks throughout this chapter because the monophyly of this group of crurotarsan archosaurs is still questionable.) *Chiniquodon, Dinodontosaurus,* and *Massetognathus* are also known from the lower part (*Dinodontosaurus* Assemblage Zone) of the Santa Maria Formation of Rio Grande do Sul (Brazil). The Chañares Formation and the *Dinodontosaurus* Assemblage Zone of the Santa Maria Formation are considered Middle Triassic (Ladinian) in age (Lucas 1998; Rogers et al. 2001).

Late Triassic

Rhynchosaurs (*Hyperodapedon*, often referred to as *Scaphonyx*) and large traversodont cynodonts (especially *Exaeretodon*) are the most abundant elements in the late Carnian to early Norian tetrapod communities from the Ischigualasto Formation of northwestern Argentina (Bonaparte 1978, 1997; Martinez et al. 2011) as well as from the *Hyperodapedon* Assemblage Zone of the Santa Maria Formation and the overlying lower part of the Caturrita Formation in Rio Grande do Sul, Brazil (Langer et al. 2007). Large (up to 6 m long) "rauisuchian" archosaurs such as *Prestosuchus* and *Saurosuchus* were the top predators in these assemblages. The lower part of the Ischigualasto Formation, the age of which has recently been radiometrically bracketed between 231.4 and 225.9 million years ago (Martinez et al. 2011), has yielded a diversity of early dinosaurs including *Eodromaeus*, *Eoraptor*, *Herrerasaurus*, *Panphagia*, and *Pisanosaurus* (Bonaparte 1978, 1997; Sereno et al. 1993; Martínez and Alcober 2009; Brusatte et al. 2010; Langer et al. 2010; Nesbitt 2011; Martinez et al. 2011). The presence of the ornithischian *Pisanosaurus* indicates that the evolutionary divergence of the two principal dinosaurian lineages predated the late Carnian. As medium-sized to large carnivores and herbivores, dinosaurs quickly became an important element of the tetrapod assemblages during the Norian and Rhaetian stages of the Triassic (Brusatte et al. 2010; Langer et al. 2010; Martinez et al. 2011).

Two groups of amphibious or fully aquatic carnivores, metoposaurid temnospondyls and the superficially crocodile-like phytosaurs (Fig. 41.3), were particularly common in the classical Late Triassic (Carnian through Rhaetian) assemblages of continental tetrapods from Laurasia, specifically the formations of the Keuper in Germany and adjoining regions (Schoch and Wild 1999; Sues and Fraser 2010) and the Chinle Formation and Dockum Group of the American Southwest (Long and Murry 1995; Lucas 1998; Parker and Martz 2011). A diverse Carnian tetrapod assemblage including metoposaurs and phytosaurs also occurs in the Timezgadiouine Formation of Morocco (Dutuit 1976). Metoposaurs and phytosaurs are rare or absent in most of the known Late Triassic communities of continental tetrapods from Gondwana. A notable exception is the lower faunal assemblage of the Maleri Formation of India, which includes metoposaurs and phytosaurs

41.3. (Above) Reconstructed skeleton of the Late Triassic "rauisuchian" *Postosuchus kirkpatricki* in lateral view. (Below) Composite reconstruction of the skeleton of a Late Triassic large phytosaur in lateral view, based on the skull of *Smilosuchus* and postcranial bones of *Leptosuchus* and related taxa. Scale bars each equal 50 cm. Modified from Long and Murry (1995).

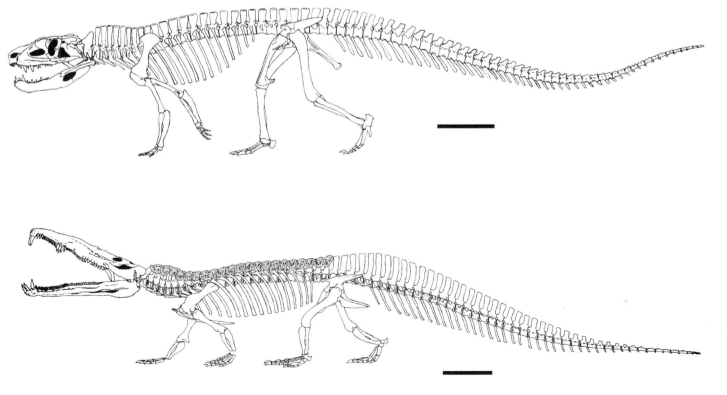

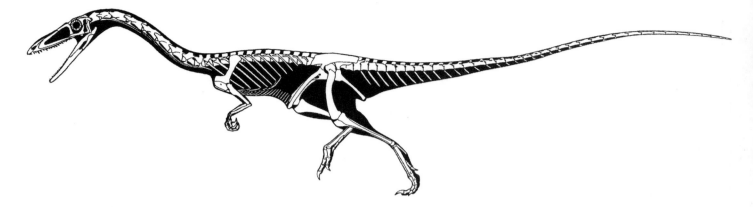

41.4. Reconstructed skeleton of the Late Triassic theropod dinosaur *Coelophysis bauri* in lateral view. From Paul (1993).

along with the traversodont cynodont *Exaeretodon* and the rhynchosaur *Hyperodapedon* (Bandyopadhyay 1999). *Hyperodapedon* is also present in the Lossiemouth Sandstone Formation of Scotland (Benton and Walker 1985).

Other important Late Triassic faunal elements from both Gondwana and Laurasia include the heavily armored aetosaurs, "rauisuchians" (Fig. 41.3), and early dinosaurs (Fig. 41.4) (Long and Murry 1995; Brusatte et al. 2010; Langer et al. 2010; Nesbitt 2011).

The middle to late Norian tetrapod assemblages from the Arnstadt Formation and its equivalents in Germany and adjoining countries are famous for the occurrence of the large sauropodomorph dinosaur *Plateosaurus* (Schoch and Wild 1999). A virtually identical assemblage of reptiles, temnospondyls, and synapsids (including mammaliaforms) occurs in the Fleming Fjord Formation of East Greenland (Jenkins et al. 1994).

In the northern part of Pangaea, rhynchosaurs and traversodont cynodonts are found together with metoposaurs in the Wolfville Formation (Carnian) of Nova Scotia, Canada, and a traversodont (*Boreogomphodon*) is the most common tetrapod in the Tomahawk Member of the Vinita Formation (Carnian) in Virginia (Sues and Olsen 1990).

To date, the best-known latest Triassic tetrapod assemblage from Gondwana has been recovered from the upper third of the Los Colorados Formation (late Norian–Rhaetian) of northwestern Argentina (Bonaparte 1972, 1997; Arcucci et al. 2004). It includes the quadrupedal sauropodomorph *Riojasaurus* (up to 11 m long), the coelophysoid theropod *Zupaysaurus*, the large "rauisuchian" *Fasolasuchus* (with a skull length of up to 95 cm), an ornithosuchid (*Riojasuchus*), diverse crocodylomorphs (*Hemiprotosuchus* and *Pseudhesperosuchus*), a tritheledontid cynodont (*Chalimia*), and the basal turtle *Palaeochersis*. In its faunal composition, the Los Colorados assemblage is intermediate between typically Late Triassic and Early Jurassic communities of continental tetrapods.

Early Jurassic Tetrapod Assemblages

Reassessment of the age of many allegedly Late Triassic continental strata worldwide as Early Jurassic by Olsen and Galton (1977, 1984) invalidated the traditional view that there were few Early Jurassic records of continental tetrapods. For many years, our knowledge of Early Jurassic terrestrial reptiles appeared to be restricted to stray finds of dinosaurs and pterosaurs in marine strata from northwestern Europe. Strata in southern Germany and southwestern Britain, and fissure-fillings in southwestern Britain, some of which straddle the Triassic–Jurassic boundary, have yielded countless disarticulated bones and teeth of small tetrapods and helped fill in the picture to some extent (Clemens 1980; Benton and Spencer 1995). The Early Jurassic fissure-fillings of southern Wales are famous for the occurrence of the mammaliaform *Morganucodon* and apparently represent communities on islands that later were inundated by the advancing sea (Benton and Spencer 1995).

Some of the most diverse assemblages of Early Jurassic continental tetrapods are known from the Lower Lufeng Formation (Dark Red Beds and Dull Purplish Beds) of Yunnan, China (Young 1951; Luo and Wu 1995; Li et al. 2008). They include diverse sauropodomorph dinosaurs, especially *Lufengosaurus*. However the assemblages also have a variety of small to medium-sized tetrapods – sphenodontian lepidosaurs (*Clevosaurus*), a basal crocodylomorph (*Dibothrosuchus*), protosuchid crocodyliforms (e.g., *Platyognathus*), abundant tritylodontid cynodonts (e.g., *Bienotherium*), and mammaliaforms including *Morganucodon*.

Morganucodon is now known from the Lower Jurassic fissure-fillings of southern Wales, the Kayenta Formation of Arizona, and the Lower Lufeng Formation of Yunnan (Kielan-Jaworowska et al. 2004). The closely related (possibly synonymous) *Erythrotherium* occurs in the Upper Elliot Formation of southern Africa. The Upper Elliot and Clarens formations of southern Africa and neighboring countries share with the McCoy Brook Formation of Nova Scotia the presence of the tritheledontid cynodont *Pachygenelus*, the crocodyliform *Protosuchus*, and the sphenodontian *Clevosaurus* (Shubin et al. 1991; Sues et al. 1994, 1996; Rubidge 2005). The small tritylodontid cynodont *Oligokyphus*, first reported from the Lower Jurassic of southern Germany and southwest Britain (Benton and Spencer 1995), is also found in the Kayenta Formation (Sues 1985) and the Lower Lufeng Formation (Luo and Sun 1994). Such cosmopolitanism among Early Jurassic continental vertebrates is surprising because the breakup of Pangaea was already well under way. However, recent work indicates that there is less taxonomic homogeneity in other groups of continental tetrapods (e.g., dinosaurs) than previously thought (Rowe et al. 2011).

Based on the currently available fossil record, Early Jurassic assemblages of continental tetrapods are primarily characterized by the absence of such typically Late Triassic groups as metoposaurid temnospondyls, most crurotarsan archosaurs, and procolophonid parareptiles. Apparently few new groups of continental vertebrates appeared during the Early Jurassic. At the same time, dinosaurs rapidly diversified and became abundant. This change is particularly well illustrated by the record of tracks across the Triassic–Jurassic boundary in the Newark Supergroup of eastern North America (Olsen et al. 2002).

The Rise of Dinosaurs: Competition or Ecological Opportunism?

Paleontologists have proposed a number of scenarios to explain the diversification of archosaurian reptiles during the Triassic Period and their replacement of therapsids as the dominant group of continental tetrapods. Most of these scenarios posit that certain morphological and/or inferred physiological innovations present in archosaurs provided these reptiles with selective advantages over potential competitors, especially therapsids. Many center on the differences in limb posture and gait between therapsids and other tetrapods on the one hand and archosaurian reptiles on the other.

Therapsids were obligatory quadrupeds with more or less sprawling, wide-track gaits (Hotton 1980). Archosaurs had more erect gaits with limbs held and moved closer to the body. This may have facilitated more rapid and, even in many early forms, at least partially bipedal locomotion.

Robinson (1971) and Hotton (1980) suggested that archosaurian reptiles replaced therapsids in part because the former were able to excrete nitrogen as urea with little loss of water. Present-day reptiles and birds dispose of nitrogen in the form of uric acid, either as slurry (in birds) or as a nearly dry pellet (in lizards), whereas extant mammals (and presumably their therapsid precursors) almost exclusively excrete nitrogen as urea, which requires considerable amounts of water to be flushed out of the body. Given increasingly drier climatic conditions during the early Mesozoic, Hotton argued that the water-saving nitrogen metabolism of archosaurs, along with their locomotor specializations, gave them a competitive advantage over therapsids.

In Charig's (1984) scenario, the fully erect limb posture of dinosaurs resulted in superior locomotor abilities and supposedly accounted for the tremendous evolutionary success of this group. Thus, the carnivorous dinosaurs became more efficient predators that, in due course, would have outcompeted other carnivorous archosaurs and cynodonts as well as eliminated non-dinosaurian herbivores (dicynodonts, gomphodont cynodonts, and rhynchosaurs) of the time. The disappearance of these plant-eaters, in turn, could have accelerated the evolutionary diversification of ornithischian and sauropodomorph dinosaurs to exploit the vegetation resources (Charig 1984). However, the assumed competitive superiority of dinosaurs is questionable. Several groups of crurotarsan archosaurs independently evolved erect posture during the Late Triassic, yet, with the exception of crocodylomorphs, none survived the extinction event at the end of the Triassic.

Benton (1983) first questioned whether large-scale competition between dinosaurs with erect limb posture and other tetrapods with sprawling limb posture over a period of several millions of years could have resulted in the former outcompeting the latter. Indeed, there exists no evidence to build a compelling case for a competition scenario. Sereno (1997) and Brusatte et al. (2008, 2010) also argued that early dinosaurs were opportunists that merely took over ecological niches vacated by the extinction of presumed competitors, such as most crurotarsan archosaurs.

Early Mesozoic Extinctions among Continental Tetrapods

One or two major episodes of extinction among continental tetrapods occurred during the early Mesozoic. It has long been known that numerous groups of marine invertebrates, especially among ammonoid cephalopods, bivalves, and brachiopods, vanished at the end of the Triassic Period (Hallam and Wignall 1997). In fact, this event is considered one of the five great extinctions during the Phanerozoic Eon. Colbert (1958) first noted the disappearance of many lineages of tetrapods at the Triassic–Jurassic boundary. More recent surveys of early Mesozoic tetrapod biodiversity have repeatedly confirmed this observation (e.g., Benton 1986, 1994; Olsen and Sues 1986).

The stratigraphically well-constrained fossil record from the Triassic–Jurassic sedimentary strata of the Newark Supergroup in eastern North

America supports the hypothesis of an extinction event in continental ecosystems at the end of the Triassic (Olsen et al. 2002). The Newark Supergroup represents the remnants of thousands of meters of sedimentary and volcanic rocks deposited in a chain of rift basins that formed during a lengthy episode of crustal extension and thinning preceding the Jurassic breakup of Pangaea and the opening of the northern Atlantic Ocean (Olsen 1997). It has yielded a series of stratigraphically well-constrained assemblages of continental tetrapods that range in time from possibly the Middle Triassic to the Early Jurassic. The occurrence of abundant tetrapod remains of earliest Jurassic age in the lower part of the McCoy Brook Formation in Nova Scotia is particularly relevant in the present context (Olsen et al. 1987; Shubin et al. 1991; Sues et al. 1994, 1996). Although the localities of this unit represent a variety of paleoenvironments, they all share the absence of certain groups of tetrapods, such as metoposaurs and phytosaurs, which characterized Late Triassic communities of continental tetrapods in North America and elsewhere. The Early Jurassic tetrapod assemblages from the Lower Lufeng Formation of China and from the Upper Elliot and Clarens formations of southern Africa show similar changes in faunal composition.

The precise number of tetrapod lineages disappearing at the end of the Triassic remains uncertain, but the extinctions do not appear to have been taxonomically or ecologically selective (Olsen and Sues 1986; Benton 1986, 1994). On the other hand, the tetrapods found in the McCoy Brook Formation and elsewhere all represent groups that were already present in the Late Triassic, and apparently few new tetrapod taxa originated at the beginning of the Jurassic Period (Olsen and Sues 1986; Olsen et al. 1987; Benton 1994). It is noteworthy that dinosaurs appear to have been largely unaffected by the end-Triassic extinction event.

The extensive fossil record of pollen and spores from the Newark Supergroup indicates a major floral change at or near the end of the Triassic (Olsen et al. 2002). Taxonomically diverse Late Triassic assemblages of pollen and spores gave way to greatly impoverished Early Jurassic assemblages that are almost entirely composed of pollen of cheirolepidaceous conifers (form genus *Classopollis* or *Corollina*). However, no clear signs of major floral changes have been detected in other regions of the world (e.g., Achilles 1981).

The causes for the biotic changes at the Triassic–Jurassic boundary are still not understood. Olsen et al. (1987) linked the end-Triassic extinction event to the impact of a large extraterrestrial object that created the Manicouagan crater in Québec. This vast feature, now a ring-shaped lake with a diameter of some 70 km (Fig. 41.5), may have originally had a diameter of about 100 km. However, Hodych and Dunning (1992) and Ramezani et al. (2005) radiometrically redated the age of the impact event as 214 ± 1 and ca. 215.5 million years, respectively. These dates place the impact well within the Norian stage, ruling it out as the proximate cause for the end-Triassic extinctions. Parker and Martz (2011) have recently noted that changes in the composition of continental tetrapods and palynomorph assemblages in the Chinle Formation during the middle Norian are consistent with the age of the Manicouagan impact event. More work is needed to establish whether this biotic turnover is a regional or global phenomenon.

41.5. Satellite image of the Late Triassic Manicouagan impact feature, now visible as a large annular lake, in northern Québec, Canada.

© *NASA (image # PIA-03434).*

Other indirect evidence for an end-Triassic impact event has been published; Bice et al. (1992) reported on what they interpret as shocked quartz in a section of Triassic–Jurassic boundary strata in Italy, but Hallam and Wignall (1997) subsequently questioned this report. Walkden et al. (2002) found a Late Triassic layer of glass spherules and shocked quartz grains in southwest Britain, but these ejecta were likely generated by the Manicouagan impact (Thackrey et al. 2009). Olsen et al. (2002) observed a small increase in iridium in the latest Triassic continental strata of the Newark Supergroup. At present, the evidence for an end-Triassic impact event is slim compared to that for the catastrophe caused by the Chicxulub impact at the end of the Cretaceous Period (Hallam and Wignall 1997), and, furthermore, no impact site of the appropriate age has been identified.

Hallam and Wignall (1997) favored terrestrial causes for the end-Triassic extinctions, especially an episode of significant volcanic activity along the margins of the future Atlantic Ocean and a major drop in sea level at the end of the Triassic. The flood basalts of the vast Central Atlantic Magmatic Province (known by its acronym CAMP; Marzoli et al. 1999), which extended from eastern North America (Fig. 41.6) to southern Brazil and West Africa, formed over a period of a few million years around the Triassic–Jurassic boundary and were related to the rifting leading to the breakup

41.6. Triassic-Jurassic boundary section along the shores of the Bay of Fundy at Five Islands, Nova Scotia. The massive North Mountain Basalt, which is part of the Central Atlantic Magmatic Province (CAMP) and is earliest Jurassic in age, overlies cyclically bedded Late Triassic lacustrine strata of the Blomidon Formation.

Courtesy of Paul E. Olsen.

of Pangaea (McHone and Puffer 2003). Based on evidence from plant fossils at the Triassic–Jurassic boundary in Greenland, McElwain et al. (1999) hypothesized a substantial increase in the atmospheric concentration of the greenhouse gas CO_2 at the Triassic–Jurassic boundary. Such an increase could have been linked to the significant gaseous and aerosol emissions generated by the CAMP volcanic activity. Whiteside et al. (2010) recently presented geochemical evidence that the end-Triassic extinction event began slightly before the eruption of the earliest CAMP basalts in eastern North America but simultaneous with the oldest CAMP basalts in Morocco. This is the best evidence to date for a volcanic causation of this extinction event.

Benton (1991) argued for a second, even larger extinction among continental tetrapods during the Late Triassic, at the end of the Carnian stage. In his view, dinosaurs subsequently rapidly diversified during the Norian and continued to do so after the end-Triassic extinction event. However, Rogers et al. (1993) and other authors (e.g., Dzik et al. 2008) have challenged the validity of the numbers of last appearances of taxa in the late Carnian, which would greatly diminish the magnitude of the proposed end-Carnian extinction event. Furthermore, the chronostratigraphic placement of the boundary between the Carnian and Norian stages has also changed considerably in recent years. Muttoni et al. (2004) have set this boundary at 227–228 million years, which places many faunal assemblages long considered Carnian in age in the Norian stage and thus requires reconsideration of a possible second extinction event during the Late Triassic.

The Triassic Period was the time during which most of the principal groups of extant continental tetrapods, or their closest relatives, made their first appearance in the fossil record, along with the first dinosaurs and pterosaurs. All major lineages of dinosaurs were already present during the late Carnian, and close relatives of Dinosauria extended back to the Early Triassic (Brusatte et al. 2011). Sauropodomorph and theropod dinosaurs became abundant, at least in some regions, during the Norian and Rhaetian, but ornithischians appear to have to been uncommon until the Early Jurassic (Benton 2006; Brusatte et al. 2010; Langer et al. 2010; Nesbitt 2011).

References

Achilles, H. 1981. Die rätische und liassische Mikroflora Frankens. *Palaeontographica B* 179: 1–86.

Arcucci, A. B., C. A. Marsicano, and A. T. Caselli. 2004. Tetrapod association and palaeoenvironment of the Los Colorados Formation (Argentina): a significant sample from Western Gondwana at the end of the Triassic. *Geobios* 37: 557–568.

Bandyopadhyay, S. 1999. Gondwana vertebrate faunas of India. *Proceedings of the Indian National Science Academy A* 65: 285–313.

Benton, M. J. 1983. Dinosaur success in the Triassic: a noncompetitive ecological model. *Quarterly Review of Biology* 58: 29–55.

———. 1986. The Late Triassic tetrapod extinction events. In K. Padian (ed.), *The Beginning of the Age of Dinosaurs: Faunal Change across the Triassic-Jurassic Boundary,* 303–320. New York: Cambridge University Press.

———. 1994. Late Triassic to Middle Jurassic extinctions among continental tetrapods: testing the pattern. In N. C. Fraser and H.-D. Sues (eds.), *In the Shadow of the Dinosaurs: Early Mesozoic Tetrapods,* 366–397. New York: Cambridge University Press.

Benton, M. J., and P. S. Spencer. 1995. *Fossil Reptiles of Great Britain.* London: Chapman & Hall.

Benton, M. J., and A. D. Walker. 1985. Palaeoecology, taphonomy, and dating of Permo-Triassic reptiles from Elgin, north-east Scotland. *Palaeontology* 28: 207–234.

Bice, D., C. R. Newton, S. McCauley, P. W. Reiners, and C. A. McRoberts. 1992. Shocked quartz at the Triassic-Jurassic boundary in Italy. *Science* 255: 443–446.

Bonaparte, J. F. 1972. Los tetrápodos del sector superior de la formación Los Colorados, La Rioja, Argentina (Triásico superior). I Parte. *Opera Lilloana* 22: 1–183.

———. 1978. El Mesozoico de América del Sur y sus tetrápodos. *Opera Lilloana* 26: 5–596.

———. 1997. *El Triásico de San Juan-La Rioja, Argentina, y sus Dinosaurios.* Buenos Aires: Museo Argentino de Ciencias Naturales "Bernardino Rivadavia."

Brusatte, S. L., M. J. Benton, M. Ruta, and G. T. Lloyd. 2008. Superiority, competition, and opportunism in the evolutionary radiation of dinosaurs. *Science* 321: 1485–1488.

Brusatte, S. L., S. J. Nesbitt, R. B. Irmis, R. J. Butler, M. J. Benton, and M. A. Norell. 2010. The origin and early radiation of dinosaurs. *Earth-Science Reviews* 101: 68–100.

Brusatte, S. L., G. Niedźwiedzki, and R. J. Butler. 2011. Footprints pull origin and diversification of dinosaur stem-lineage deep into Early Triassic. *Proceedings of the Royal Society B* 278: 1107–1113.

Charig, A. J. 1984. Competition between therapsids and archosaurs during the Triassic period: a review and synthesis of current theories. In M. W. J. Ferguson (ed.), *The Structure, Development and Evolution of Reptiles,* 597–628. London: Academic Press.

Chatterjee, S., and T. Roy Chowdhury. 1974. Triassic Gondwana vertebrates from India. *Indian Journal of Earth Sciences* 1: 96–112.

Clemens, W. A. 1980. Rhaeto-Liassic mammals from Switzerland and West Germany. *Zitteliana* 5: 51–92.

Colbert, E. H. 1958. Tetrapod extinctions at the end of the Triassic. *Proceedings of the National Academy of Sciences* 44: 973–977.

Cosgriff, J. W. 1984. The temnospondyl labyrinthodonts of the earliest Triassic. *Journal of Vertebrate Paleontology* 4: 30–46.

Dutuit, J.-M. 1976. Introduction à l'étude paléontologique du Trias continental marocain. Description des premiers Stégocéphales recueillis dans le couloir d'Argana (Atlas occidental). *Mémoires du Muséum National d'Histoire Naturelle C* 36: 1–253.

Dzik, J., T. Sulej, and G. Niedźwiedzki. 2008. A dicynodont-theropod association in the latest Triassic of Poland. *Acta Palaeontologica Polonica* 53: 733–738.

Erwin, D. H. 2006. *Extinction: How Life on Earth Nearly Ended 250 Million Years Ago.* Princeton: Princeton University Press.

Fraser, N. C. 2006. *Dawn of the Dinosaurs: Life in the Triassic*. Bloomington: Indiana University Press.

Golonka, J. 2007. Late Triassic and Early Jurassic palaeogeography of the world. *Palaeogeography, Palaeoclimatology, Palaeoecology* 244: 297–307.

Gradstein, F. M., J. G. Ogg, and A. G. Smith (eds.) 2004. *A Geologic Time Scale 2004*. Cambridge: Cambridge University Press.

Hallam, A., and P. B. Wignall. 1997. *Mass Extinctions and Their Aftermath*. Oxford: Oxford University Press.

Hodych, J. P., and G. R. Dunning. 1992. Did the Manicouagan impact trigger end-of-Triassic mass extinction? *Geology* 20: 51–54.

Hotton, N., III. 1980. An alternative to dinosaur endothermy: the happy wanderers. In R. D. K. Thomas and E. C. Olson (eds.), *A Cold Look at the Warm-Blooded Dinosaurs*, 311–350. AAAS Selected Symposia Series, Vol. 28. Boulder: Westview Press.

Huene, F. von. 1935–42. *Die fossilen Reptilien des südamerikanischen Gondwanalandes. Ergebnisse der Sauriergrabungen in Südbrasilien 1928/29*. Munich: C. H. Becksche Verlagsbuchhandlung.

Ivakhnenko, M. F., V. K. Golubev, Y. M. Gubin, N. N. Kalandadze, I. V. Novikov, A. G. Sennikov, and A. S. Rautian. 1997. [*Permian and Triassic Tetrapods of Eastern Europe*.] Moscow: GEOS. [In Russian.]

Jenkins, F. A., Jr., N. H. Shubin, W. W. Amaral, S. M. Gatesy, C. R. Schaff, L. B. Clemmensen, W. R. Downs, A. R. Davidson, N. Bonde, and F. Osbæck. 1994. Late Triassic vertebrates and depositional environments of the Fleming Fjord Formation, Jameson Land, East Greenland. *Meddelelser om Grønland* 32: 1–25.

Kalandadze, N. N. 1970. [New Triassic kannemeyeriids from southern Cisuralia.] In K. K. Flerov (ed.), [*Materials on the Evolution of Terrestrial Vertebrates*], 51–57. Moscow: Nauka. [In Russian.]

———. 1975. [First find of a lystrosaur on the territory of the European part of the USSR.] *Paleontologicheskii Zhurnal* 1975(4):140–142. [In Russian.]

Kielan-Jaworowska, Z., R. L. Cifelli, and Z.-X. Luo. 2004. *Mammals from the Age of Dinosaurs: Origins, Evolution, and Structure*. New York: Columbia University Press.

Langer, M. C., A. M. Ribeiro, C. L. Schultz, and J. Ferigolo. 2007. The continental tetrapod-bearing Triassic of south Brazil. In S. G. Lucas and J. A.

Spielmann (eds.), The Global Triassic. *New Mexico Museum of Natural History and Science Bulletin* 41: 201–218

Langer, M. C., M. D. Ezcurra, J. S. Bittencourt, and F. E. Novas. 2010. The origin and early evolution of dinosaurs. *Biological Reviews* 85: 55–110.

Li, J., X. Wu, and F. Zhang (eds.). 2008. *The Chinese Fossil Reptiles and Their Kin. Second Edition*. Beijing: Science Press.

Long, R. A., and P. A. Murry. 1995. Late Triassic (Carnian and Norian) tetrapods from the southwestern United States. *New Mexico Museum of Natural History and Science Bulletin* 4: 1–254.

Lucas, S. G. 1998. Global Triassic tetrapod biostratigraphy and biochronology. *Palaeogeography, Palaeoclimatology, Palaeoecology* 143: 347–384.

Lucas, S. G., A. P. Hunt, A. B. Heckert, and J. A. Spielmann. 2007. Global Triassic tetrapod biostratigraphy and biochronology: 2007 status. *New Mexico Museum of Natural History & Science Bulletin* 41: 229–240.

Luo, Z., and A. Sun. 1994. *Oligokyphus* (Cynodontia: Tritylodontidae) from the Lower Lufeng Formation (Lower Jurassic) of Yunnan, China. *Journal of Vertebrate Paleontology* 13: 477–482.

Martínez, R. N., and O. A. Alcober. 2009. A basal sauropodomorph (Dinosauria: Saurischia) from the Ischigualasto Formation (Triassic, Carnian) and the early evolution of Sauropodomorpha. *PLoS One* 4(2): e4397.

Martinez, R. N., P. C. Sereno, O. A. Alcober, C. E. Colombi, P. R. Renne, I. P. Montañez, and B. S. Currie. 2011. A basal dinosaur from the dawn of the dinosaur era in southwestern Pangaea. *Science* 331: 206–210.

Marzoli, A., P. R. Renne, E. M. Piccirillo, M. Ernesto, G. Belliena, and A. De Min. 1999. Extensive 200-million-year-old continental flood basalts of the Central Atlantic Magmatic Province. *Science* 284: 616–618.

McElwain, J. C., D. J. Beerling, and F. I. Woodward. 1999. Fossil plants and global warming at the Triassic-Jurassic boundary. *Science* 285: 1386–1390.

McHone, J. G. and J. H. Puffer. 2003. Flood basalt provinces of the Pangean Atlantic Rift: regional extent and environmental significance. In P. M. LeTourneau and P. E. Olsen (eds.), *The Great Rift Valleys of Pangea in Eastern North America. Volume One: Tectonics, Structure, and Volcanism*, 141–154. New York: Columbia University Press.

Muttoni, G., D. V. Kent, P. E. Olsen, P.

DiStefano, W. Lowrie, S. M. Bernasconi, and F. M. Hernández. 2004. Tethyan magnetostratigraphy from Pizzo Mondello (Sicily) and correlation to the Late Triassic Newark astrochronological polarity timescale. *Geological Society of America Bulletin* 116: 1043–1058.

Nesbitt, S. J. 2011. The early evolution of archosaurs: relationships and the origin of major clades. *Bulletin of the American Museum of Natural History* 352: 1–292.

Nesbitt, S. J., R. B. Irmis, and W. G. Parker. 2007. A critical re-evaluation of the Late Triassic dinosaur taxa of North America. *Journal of Systematic Palaeontology* 5: 209–243.

Olsen, P. E. 1997. Stratigraphic record of the early Mesozoic breakup of Pangea in the Laurasia-Gondwana rift system. *Annual Review of Earth and Planetary Sciences* 25: 337–401.

Olsen, P. E., and P. M. Galton. 1977. Triassic-Jurassic extinctions: are they real? *Science* 197: 983–986.

———. 1984. A review of the reptile and amphibian assemblages from the Stormberg of southern Africa, with special emphasis on the footprints and the age of the Stormberg. *Palaeontologia Africana* 25: 87–110.

Olsen, P. E., D. V. Kent, H.-D. Sues, C. Koeberl, H. Huber, A. Montanari, E. C. Rainforth, S. J. Fowell, M. J. Szajna, and B. W. Hartline. 2002. Ascent of dinosaurs linked to an iridium anomaly at the Triassic-Jurassic boundary. *Science* 296: 1305–1307.

Olsen, P. E., N. H. Shubin, and M. H. Anders. 1987. New Early Jurassic tetrapod assemblages constrain Triassic-Jurassic tetrapod extinction event. *Science* 237: 1025–1029.

Olsen, P. E., and H.-D. Sues. 1986. Correlation of continental Late Triassic and Early Jurassic sediments, and the Triassic-Jurassic tetrapod transition. In K. Padian (ed.), *The Beginning of the Age of Dinosaurs: Faunal Change across the Triassic-Jurassic Boundary*, 321–351. New York: Cambridge University Press.

Parker, W. G., and J. W. Martz. 2011. The Late Triassic (Norian) Adamanian–Revueltian tetrapod faunal transition in the Chinle Formation of Petrified Forest National Park, Arizona. *Earth and Environmental Science Transactions of the Royal Society of Edinburgh* 101: 231–260.

Paul, G. S. 1993. Are *Syntarsus* and the Whitaker Quarry theropod the same genus? *New Mexico Museum of Natural History & Science Bulletin* 3: 397–402.

Prochnow, S. J., L. C. Nordt, S. C. Atchley,

and M. R. Hudec. 2006. Multi-proxy paleosol evidence for Middle and Late Triassic climate trends in eastern Utah. *Palaeogeography, Palaeoclimatology, Palaeoecology* 232: 53–72.

Ramezani, J., S. A. Bowring, M. S. Pringle, F. D. Winslow III, and E. T. Rasbury. 2005. The Manicouagan impact melt rock: a proposed standard for the intercalibration of U-Pb and ^{40}Ar/^{39}Ar isotopic systems. *Goldschmidt Conference Abstracts* 2005: A321.

Rayfield, E. J., P. M. Barrett, R. McDonnell, and K. J. Willis. 2005. A Geographical Information System (GIS) study of Triassic vertebrate biochronology. *Geological Magazine* 142: 327–354.

Rayfield, E. J., P. M. Barrett, and A. R. Milner. 2009. Utility and validity of Middle and Late Triassic "land vertebrate faunachrons." *Journal of Vertebrate Paleontology* 29: 80–87.

Robinson, P. L. 1971. A problem of faunal replacement on Permo-Triassic continents. *Palaeontology* 14: 131–153.

Rogers, R. R., C. C. Swisher III, P. C. Sereno, A. M. Monetta, C. A. Forster, and R. N. Martinez. 1993. The Ischigualasto tetrapod assemblage (Late Triassic, Argentina) and 40Ar/39Ar dating of dinosaur origins. *Science* 260: 794–797.

Rogers, R. R., A. B. Arcucci, F. Abdala, P. C. Sereno, C. A. Forster, and C. L. May. 2001. Paleoenvironment and taphonomy of the Chañares Formation tetrapod assemblage (Middle Triassic), northwestern Argentina: spectacular preservation in volcanigenic concretions. *Palaios* 16: 461–481.

Romer, A. S. 1966. The Chañares (Argentina) Triassic reptile fauna. I. Introduction. *Breviora* 247: 1–14.

——. 1973. The Chañares (Argentina) Triassic reptile fauna. XX. Summary. *Breviora* 413: 1–20.

Rougier, G. W., M. S. de la Fuente, and A. B. Arcucci. 1995. Late Triassic turtles from South America. *Science* 268: 855–858.

Rowe, T. B., H.-D. Sues, and R. R. Reisz. 2011. Dispersal and diversity in the earliest North American sauropodomorph dinosaurs, with a description of a new taxon. *Proceedings of the Royal Society B* 278: 1044–1053.

Rubidge, B. S. 2005. Re-uniting lost continents – fossil reptiles from the ancient Karoo and their wanderlust. *South African Journal of Geology* 108: 135–172.

Schoch, R., and R. Wild. 1999. Die Wirbeltier-Fauna im Keuper von Süddeutschland. In N. Hauschke and V. Wilde (eds.), *Trias – Eine ganz andere Welt*, 395–408. Munich: Verlag Dr. Friedrich Pfeil.

Sellwood, B. W., and P. J. Valdes. 2006. Mesozoic climates: general circulation models and the rock record. *Sedimentary Geology* 190: 269–287.

Sennikov, A. G. 1995. [*Early Thecodonts of Eastern Europe.*] Moscow: Nauka. [In Russian.]

Sereno, P. C., and A. B. Arcucci. 1994a. Dinosaurian precursors from the Middle Triassic of Argentina: *Lagerpeton chanarensis. Journal of Vertebrate Paleontology* 13: 385–399.

——. 1994b. Dinosaurian precursors from the Middle Triassic of Argentina: *Marasuchus lilloensis*, gen. nov. *Journal of Vertebrate Paleontology* 14: 53–73.

Sereno, P. C., C. A. Forster, R. R. Rogers, and A. M. Monetta. 1993. Primitive dinosaur skeleton from Argentina and the early evolution of Dinosauria. *Nature* 361: 64–66.

Shubin, N. H., A. W. Crompton, H.-D. Sues, and P. E. Olsen. 1991. New fossil evidence on the sister-group of mammals and early Mesozoic faunal distributions. *Science* 251: 1063–1065.

Simms, M. J., and A. H. Ruffell. 1989. Synchroneity of climatic change and extinctions in the Late Triassic. *Geology* 17: 265–268.

Sues, H.-D. 1985. First record of the tritylodontid *Oligokyphus* (Synapsida) from the Lower Jurassic of western North America. *Journal of Vertebrate Paleontology* 5: 328–335.

Sues, H.-D., and N. C. Fraser. 2010. *Triassic Life on Land: The Great Transition.* New York: Columbia University Press.

Sues, H.-D., and P. E. Olsen. 1990. Triassic vertebrates of Gondwanan aspect from the Richmond basin of Virginia. *Science* 249: 1020–1023.

Sues, H.-D., N. H. Shubin, and P. E. Olsen. 1994. A new sphenodontian (Lepidosauria: Rhynchocephalia) from the McCoy Brook Formation (Lower Jurassic) of Nova Scotia, Canada. *Journal of Vertebrate Paleontology* 14: 327–340.

Sues, H.-D., N. H. Shubin, P. E. Olsen, and W. W. Amaral. 1996. On the cranial structure of a new protosuchid (Archosauria: Crocodyliformes) from the McCoy Brook Formation (Lower Jurassic) of Nova Scotia, Canada. *Journal of Vertebrate Paleontology* 16: 34–41.

Thackray, S., G. Walkden, A. Indares, M. Horstwood, S. Kelley, and R. Parrish. 2009. The use of heavy mineral correlation for determining the source of impact ejecta: a Manicouagan distal ejecta case study. *Earth and Planetary Science Letters* 285: 163–172.

Walkden, G., J. Parker, and S. Kelley. 2002. A Late Triassic impact ejecta layer in southwestern Britain. *Science* 298: 2185–2188.

Whiteside, J. H., P. E. Olsen, T. Eglinton, M. E. Brookfield, and R. N. Sambrotto. 2010. Compound-specific carbon isotopes from Earth's largest flood basalt eruptions directly linked to the end-Triassic mass extinction. *Proceedings of the National Academy of Sciences* 107: 6721–6725.

Young, C. C. 1951. The Lufeng saurischian fauna in China. *Palaeontologia Sinica*, n. s., C, 13: 1–96.

Dinosaurian Faunas of the Later Mesozoic

42

Matthew T. Carrano

This chapter discusses the evolution of dinosaurian faunas during the last two-thirds of the Mesozoic Era, from the beginning of the Middle Jurassic Period through the end of the Late Cretaceous (i.e., from 176 to 65.5 mya). Three related concepts are addressed: (1) the biogeography of dinosaur groups, (2) the composition of dinosaur faunas on the different continents, and (3) the appearance, extinction, and diversity of different dinosaur clades. The discussion will rely primarily on counts of genera and species for each time interval, examined in light of geographic location, ancient environment (geology and sedimentology), and paleoecology (diet, habit, contemporaneous species, and similar topics). Taken together, these data allow us to connect the biological patterns of dinosaur evolution with the geological patterns of continental tectonics, and thus to understand some of the major changes that occurred in dinosaur faunas throughout the later Mesozoic Era.

Several problems must be considered in any study of this type, especially when examining the available data and interpreting the conclusions drawn from them. Perhaps the most important consideration for the reader is that our knowledge of the dinosaur fossil record is incomplete but rapidly improving. Because of this, dramatic changes in our understanding of dinosaur evolution can occur over a relatively short period of scientific study. For example, thanks to numerous discoveries made in the Yixian Formation of Liaoning Province, China since 1995 (e.g., Ji et al. 1998, 2001; Xu et al. 1999), feathers are now known to have first evolved in basal coelurosaurs, rather than basal birds, and thus their origin is unlikely to have been directly linked to the evolution of flight. These new discoveries have radically altered a scientific idea held since the mid-nineteenth century in a matter of just a few years.

Therefore the first, and most significant, problem concerns the quality of the dinosaur fossil record. Paleontologists have known for decades that the record of terrestrial faunas throughout the Mesozoic is patchy. The better-sampled Mesozoic time intervals (e.g., the Late Jurassic and Late Cretaceous) represent less than 20% of the total time in which dinosaurs existed, and even these intervals are unevenly sampled. As a result, we can only view dinosaur evolution through a series of windows that may not provide a complete or accurate picture of the entire process. Using another analogy, we are attempting to understand an entire story by reading only a few phrases and pages excerpted from a very long book.

Problems and Pitfalls

The quality of the fossil record not only relates to time intervals, but also to the geographic locations where dinosaur fossils have been found. Whole continents are lacking in dinosaur fossils from particular time intervals (e.g., the Late Cretaceous of Australia, or the Late Jurassic of Madagascar). This means that we cannot know anything about dinosaur evolution in these places at these times. If important groups of dinosaurs originated there and then, our studies cannot detect this, and as a result our view of the biogeographic history of dinosaurs will be flawed.

Biogeography and Dinosaurs

There are many ways for terrestrial organisms to move from place to place. The study of these processes, and the resulting geographic patterns of animals and plants, is called biogeography. Several different biogeographic processes combine to produce most patterns we see during the evolution of any group (Fig. 42.1).

The first process, dispersal, happens when organisms expand their range or cross a barrier (e.g., a mountain range or a stretch of ocean) to inhabit a new place. The likelihood of dispersal depends on the biological qualities of the organisms as well as the physical nature of any barriers. For example, some vertebrates have a much easier time crossing large bodies of water because they are small enough to raft on floating vegetation (e.g., Zink et al. 2000; Roxworth et al. 2002).

A second process, vicariance, happens when organisms exist across a large landmass that later splits apart due to tectonics. In this scenario, the organisms travel with the different landmasses and end up widely separated millions of years later. Vicariance is often viewed as an important process for dinosaur biogeography, because most major dinosaur groups evolved in the early Mesozoic, when Pangaea was still intact, but continued to evolve as the continents broke apart.

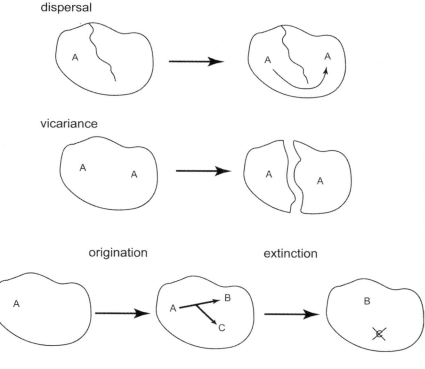

42.1. Basic biogeographic processes. In dispersal, species A migrates to a new region, around a water barrier (river). In vicariance, species A exists on two different landmasses because it was present before the landmass physically split apart. Origination produces two new species (B, C) from a single ancestor (A), while extinction causes the loss of one of them (C).

Matthew T. Carrano

Two final and inevitable factors are origination and extinction of species. Origination is crucial because it pinpoints both the time and place of origin for each dinosaur species and each dinosaur group. In doing so, it provides a starting point for all future evolution of that species or group. In contrast, extinction eliminates species from particular regions and time periods. Unfortunately, the effect of extinction – that is, the absence of a species – can sometimes be indistinguishable from other processes that also result in species absence (e.g., the presence of a barrier to dispersal).

Together, dispersal, vicariance, origination, and extinction should explain the fundamental history of dinosaur evolution through space and time. How each of these processes interacts with the others, and the probability of each occurring, can be estimated for present-day organisms but are virtually unknown for the Mesozoic. For example, the major types of dinosaurs should have been able to reach most regions of the world simply by dispersal, but different types of barriers may have prevented this for different groups. Regardless, once dinosaur groups were widely dispersed, they would also have been able to "ride" the continents across the globe. The story is undoubtedly complicated, and research is only beginning to address its more challenging aspects. Nonetheless, we have a great deal of information about dinosaur faunas through time and space, and the changes and differences between them provide some hints at the larger picture of dinosaur evolution.

Later Mesozoic Plate Tectonics and Climate

A brief outline of plate tectonics and broad climatic changes during the later Mesozoic provides a necessary physical backdrop for the story of dinosaur faunas and their evolution.

The Middle Jurassic world was still very similar to that of the Late Triassic and Early Jurassic (Sues, this volume; Fig. 42.2A, B). Most of the major landmasses retained connections, resulting in few significant geographic barriers to the dispersal of terrestrial vertebrates. During the Middle Jurassic (Fig. 42.2C), the Americas began to separate from Europe and Africa as the Atlantic Ocean opened from north to south, but this ocean remained narrow. Africa was connected to Europe near its western end, but a wide seaway – a precursor to the Tethys Sea – separated them to the east. Sea levels were high at this time, so much of the low-lying land was under water. Europe was reduced to a series of large islands.

As during the earlier Mesozoic, both northern and southern land areas were largely temperate, separated from the wet-summer equatorial region by bands of arid habitat (Hallam 1993; Rees et al. 2000). This situation continued into the Late Jurassic (Fig. 42.2D). Although the Atlantic Ocean widened, it was still insufficient to fully separate the Americas (to the west) from Europe and Africa (to the east). Furthermore, the wide eastern separation between Africa and Europe by an arm of the Tethys Sea slowly began to close during this time. Eastern Asia was a very long way by land from South America, Antarctica, and Australia, but it is possible that some terrestrial species could have dispersed this great distance during the Late Jurassic.

Continental separation finally began to show in earnest during the long interval of the Early Cretaceous (Fig. 42.2E). Not only did the Atlantic Ocean finally create a formidable barrier between North America and

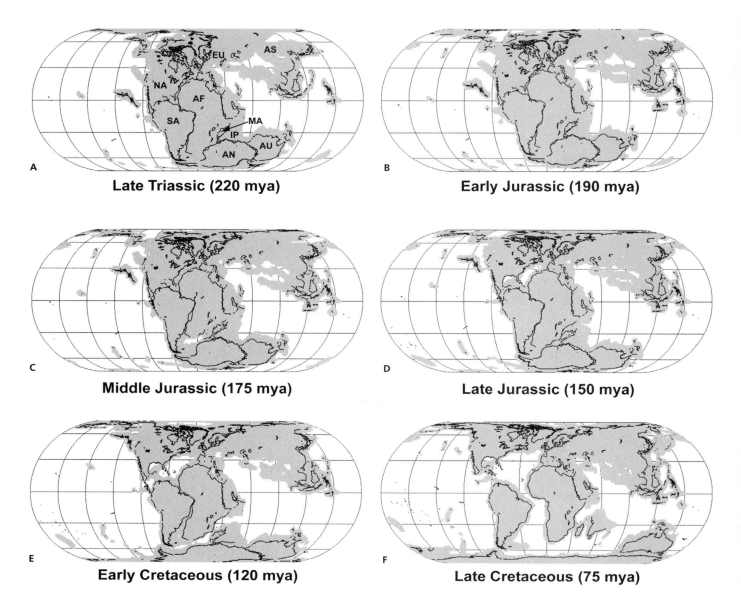

A Late Triassic (220 mya)

B Early Jurassic (190 mya)

C Middle Jurassic (175 mya)

D Late Jurassic (150 mya)

E Early Cretaceous (120 mya)

F Late Cretaceous (75 mya)

42.2. Tectonic history of major landmasses through the Mesozoic. Continental crust is in gray, oceanic crust is in white.

Maps generated courtesy of the Paleobiology Database (www.paleodb.org), using plotting software © John Alroy (2002) and tectonic reconstructions © Chris Scotese (2002). Abbreviations: AF = Africa, AN = Antarctica, AS = Asia, AU = Australia, EU = Europe, IP = Indo-Pakistan; MA = Madagascar; NA = North America, SA = South America.

Europe, but the northern and southern continents also began to separate. As Africa continued to swing northward, it approached Europe. In the south, South America began to pull away from Africa and Australia-Antarctica. These changes brought an increase in coastline, and therefore more coastal habitats, but also may have increased habitat diversity by altering weather and ocean current patterns (Hallam 1985; Rees et al. 2000).

The Late Cretaceous world saw greater continental fragmentation than any other time in the Mesozoic (Fig. 42.2F). Laurasia (in the north) was nearly completely separated from Gondwana (in the south), and even these landmasses had begun to fragment. Africa closed on Europe considerably, with (possibly) some intermittent connections along its northern coast. Eastern Asia and North America shared a land bridge across the modern Bering Strait. In the south, a sequence of continental splits broke away Africa, followed by South America, Australia-Antarctica, and finally India-Pakistan and Madagascar. These last were the final landmasses to separate, some 80 mya. In addition, high sea levels created epicontinental seas that flooded large areas of western Africa, central North America, and Europe, although these had greatly receded by latest Cretaceous times (Ziegler and Rowley 1998).

The result of all this tectonic activity was that the the last of the dinosaurs' worlds was the opposite of the first. Instead of a (Late Triassic) world of global connections, the last dinosaurs lived in a (Late Cretaceous) world of barriers and separation.

Late Triassic and Early Jurassic (Carnian–Toarcian; 228–176 mya)

These faunas have been addressed in the previous chapter (Sues, this volume), and are only briefly summarized here for comparison. The most important qualities of earlier Mesozoic dinosaur faunas can still be seen in both the Late Triassic and Early Jurassic Periods (Fraser and Sues 1994). Three important points serve as useful benchmarks for comparison with later faunas and highlight the changing ecological roles of dinosaurs.

First, although all major dinosaur groups had appeared by the Early Jurassic (and most by the Late Triassic; Sereno 1999; Carrano and Wilson 2001), total dinosaur diversity was lower during these times. This is especially true within the broadest clades, such as Ornithischia and Saurischia, each of which was represented only by some of their constituent groups. It is also the case for more restricted clades, such as Theropoda and Sauropoda, each of which became much more diverse during the later Mesozoic (Fig. 42.3).

Second, Late Triassic and Early Jurassic faunas were cosmopolitan (Russell and Bonaparte 1997; Holtz et al. 2004a). Dinosaur types were spread widely across the globe, thanks to the interconnectedness of the major landmasses. As a result, each major group was probably present in each major geographic region. Indeed, some closely related species were

Later Mesozoic Dinosaur Faunas

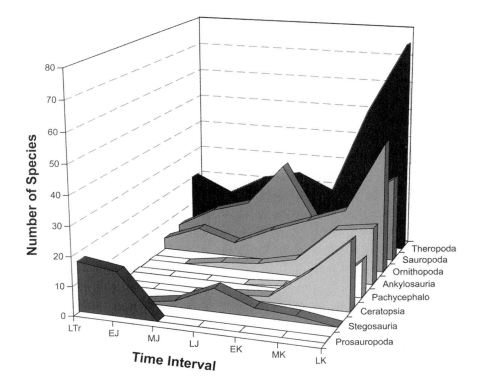

42.3. Diversity of dinosaur groups through the Mesozoic. Abbreviations: LTr = Late Triassic, EJ = Early Jurassic, MJ = Middle Jurassic, LJ = Late Jurassic, EK – earliest Cretaceous, MK = medial Cretaceous, LK = latest Cretaceous; Pachycephalo = Pachycephalosauria.

Table 42.1. Major Dinosaur-Producing Geologic Formations and Groups (Italicized) on Each Landmass, Arranged by Time Interval

	Middle Jurassic	Late Jurassic	Earliest Cretaceous	Medial Cretaceous	Latest Cretaceous
North America		Morrison	Cedar Mountain	Cedar Mountain Cloverly *Trinity*	*Montana* *Belly River* Edmonton Kaiparowits Fruitland
South America	Cañadon Asfalto		La Amarga Santana	*Neuquén*	*Neuquén* *Salta* *Chubut* *Bauru* *Malargüe*
Europe	*Great Oolite*	Louriñha Villar del Arzobispo	*Wealden* *Purbeck Limestone* Villar del Arzobispo		Arén Gosau Sânpetru
Africa	El Mers	Tendaguru Tiouraren(?)	*Irhazer*	*Tegama* Kem Kem Baharîje	
Indo-Pakistan					Lameta
Madagascar	Isalo III				Maevarano
Antarctica					
Australia			*Otway* *Strzelecki*		
Asia	Balabansai Shishugou Xiashaximiao	*Khorat* Shangshaximiao Shishugou	*Jehol* *Khorat*	Bayn Shire Dzharakuduk Khodzhakul *Jehol* *Tugulu* *Hayang*	Nemegt Djadokhta Bayan Mandahu Iren Dabasu Bostobe *Wangshi*

widely dispersed; for example, the theropod *Coelophysis* and its relatives are known from South Africa (Raath 1969), western North America (Rowe 1989), Europe (Rauhut and Hungerbühler 2000), and China (Irmis 2004). Conversely, there are few obvious faunal differences between these broad regions. Environmental variations were also less pronounced, with much of the Pangaean interior subjected to a warm, seasonally arid climate (Hallam 1993; Rees et al. 2000).

Third, these earlier faunas document a shift in abundance from the Late Triassic to the Early Jurassic, during which dinosaurs went from being a significant minority in many environments to near dominance (Benton 1991; Irmis et al. 2007; Brusatte et al. 2008; Sues, this volume). By the Early Jurassic, dinosaurs were the dominant amniote group in nearly every terrestrial fauna worldwide. Along with this shift came a significant decline in the diversity and abundance of many synapsids and basal archosaurs. Most Late Triassic faunas contained many other types of archosaurian predators and herbivores (e.g., rauisuchians, phytosaurs, rhynchosaurs, poposaurids, aetosaurs, and ornithosuchids), but these taxa dwindled or became extinct by and during the Early Jurassic, while dinosaurs diversified. Whether this transition was gradual or abrupt, or the result of competition or catastrophe, is currently under debate (e.g., Benton 1991, 2004; Olsen et al. 2002, 2003; Thulborn 2003; Irmis et al. 2007; Brusatte et al. 2008).

Middle Jurassic (Aalenian–Callovian; 175–161 mya)

In general, Middle Jurassic dinosaur faunas are far too poorly understood to allow for general characterizations. They are best known from Argentina,

Table 42.2. Dinosaur genera and their geographic locations, arranged by time interval. The Late Triassic and Early Jurassic are included for comparison (see Sues, this volume, for details). The first number indicates well-defined species; the number in parentheses reflects the addition of dubious or undescribed species; some of these are known only from footprints. Abbreviation: Pachycephalo = Pachycephalosauria. Data from Weishampel et al. (2004), emended with information from the Paleobiology Database (http://www.paleodb.org/).

	Late Triassic 45(72)	Early Jurassic 53(70)	Middle Jurassic 35(59)	Late Jurassic 74(101)	Earliest Cretaceous 47(79)	Medial Cretaceous 111(182)	Latest Cretaceous 202(281)
North America 136(185)	5(11) Theropoda: 5(8) Prosauropoda: 0(1) Ornithischia: 0(2)	9(11) Theropoda: 4(5) Prosauropoda: 3(3) Ornithopoda: 0(1) Thyreophora: 1(2)	0(4) Theropoda: 0(2) Sauropoda: 0(1) Ornithopoda: 0(1)	26(33) Theropoda: 8(10) Sauropoda: 10(13) Ornithopoda: 5(7) Stegosauria: 2(2) Ankylosauria 1(1)	11(15) Theropoda: 3(5) Sauropoda: 2(2) Ornithopoda: 4(6) Ankylosauria: 2(2)	18(27) Theropoda: 4(8) Sauropoda: 2(4) Ornithopoda: 5(7) Ankylosauria: 6(6) Ceratopsia: 1(2)	71(94) Theropoda: 15(22) Sauropoda: 1(1) Ornithopoda: 19(27) Ankylosauria: 10(12) Pachycephalo: 9(9) Ceratopsia: 17(23)
South America 67(76)	13(14) Saurischia: 1(1) Theropoda: 6(6) Prosauropoda: 4(5) Sauropoda: 1(1) Ornithopoda: 1(1)	2(2) Prosauropoda: 1(1) Ornithopoda: 1(1)	4(5) Theropoda: 2(2) Sauropoda: 2(3)	2(6) Theropoda: 0(2) Sauropoda: 2(3) Ornithopoda: 0(1)	1(6) Theropoda: 0(2) Sauropoda: 1(2) Ornithopoda: 0(1) Ankylosauria: 0(1)	10(26) Theropoda: 11(15) Sauropoda: 9(9) Ornithopoda: 0(2)	33(44) Theropoda: 13(17) Sauropoda: 15(19) Ornithopoda: 5(7) Ankylosauria: 0(1)
Europe 95(138)	11(22) Theropoda: 4(13) Prosauropoda: 7(9)	6(9) Theropoda: 3(3) Sauropoda: 1(2) Ornithischia: 0(1) Thyreophora: 2(3)	11(13) Theropoda: 8(8) Sauropoda: 1(3) Ornithopoda: 1(1) Stegosauria: 1(1) Ankylosauria 1(1)	19(26) Theropoda: 8(9) Sauropoda: 6(9) Ornithopoda: 3(4) Stegosauria: 2(2) Ankylosauria: 2(3)	22(33) Theropoda: 7(11) Sauropoda: 3(6) Ornithopoda: 8(11) Stegosauria: 0(1) Ankylosauria: 3(3) Pachycephalo: 1(1)	13(25) Theropoda: 3(7) Sauropoda: 2(6) Ornithopoda: 4(6) Stegosauria: 1(1) Ankylosauria: 3(5)	17(33) Theropoda: 3(10) Sauropoda: 4(6) Ornithopoda: 7(12) Ankylosauria: 3(4) Ceratopsia: 0(1)
Africa 57(68)	12(16) Theropoda: 1(1) Prosauropoda: 5(9) Sauropoda: 2(2) Ornithischia: 2(2) Ornithopoda: 2(2)	13(16) Theropoda: 2(2) Prosauropoda: 3(4) Sauropoda: 3(3) Ornithopoda: 5(6) Thyreophora: 0(1)	3(4) Theropoda: 1(2) Sauropoda: 4(4)	11(11) Theropoda: 3(3) Sauropoda: 6(6) Stegosauria: 1(1) Ornithopoda: 1(1)	6(8) Theropoda: 2(4) Sauropoda: 1(2) Stegosauria: 1(1)	12(20) Theropoda: 8(13) Sauropoda: 2(4) Ornithopoda: 2(3)	0(2) Theropoda: 0(1) Sauropoda: 0(1)
Indo-Pakistan 10(22)	1(1) Theropoda: 1(1)	5(6) Theropoda: 1(2) Sauropoda: 4(4)	0(3) Theropoda: 0(1) Sauropoda: 0(2)	1(1) Sauropoda: 1(1)	0(0)	0(1) Sauropoda: 0(1)	9(22) Theropoda: 5(12) Sauropoda: 4(9) Ankylosauria: 0(1)
Madagascar 4(15)	0(3) Theropoda: 0(1) Prosauropoda: 0(2)	0(0)	2(6) Theropoda: 0(4) Sauropoda: 2(2)	0(0)	0(0)	0(0)	3(5) Theropoda: 2(2) Sauropoda: 1(2) Ornithischia?: 0(1)
Antarctica 3(6)	0(0)	2(2) Theropoda: 1(1) Prosauropoda: 1(1)	0(0)	0(0)	0(0)	0(0)	4(4) Theropoda: 1(1) Ornithopoda: 2(2) Ankylosauria: 1(1)
Australia 8(14)	0(1) Theropoda: 0(1)	0(2) Ornithopoda: 0(1)	0(3) Theropoda: 0(1) Sauropoda: 1(1) Ornithopoda: 0(1)	0(0)	0(2) Theropoda: 0(1) Ornithopoda: 0(1)	11(19) Theropoda: 4(6) Sauropoda: 1(2) Ornithopoda: 5(5) Ankylosauria: 1(5) Ceratopsia: 0(1)	0(1) Theropoda: 0(1)
Asia 167(217)	3(4) Theropoda: 1(2) Sauropoda: 2(2)	16(22) Theropoda: 2(4) Prosauropoda: 7(10) Sauropoda: 4(5) Ornithopoda: 1(1) Thyreophora: 2(2)	15(21) Theropoda: 6(7) Sauropoda: 5(9) Ornithopoda: 2(2) Stegosauria: 1(2) Ankylosauria: 1(1)	15(24) Theropoda: 4(7) Sauropoda: 6(8) Ornithopoda: 1(3) Stegosauria: 3(5) Ceratopsia: 1(1)	7(15) Theropoda: 3(5) Sauropoda: 0(3) Ornithopoda: 2(4) Stegosauria: 0(1) Ceratopsia: 2(2)	47(64) Theropoda: 20(25) Sauropoda: 4(8) Ornithopoda: 4(8) Stegosauria: 1(1) Ankylosauria: 8(10) Ceratopsia: 10(12)	65(76) Theropoda: 34(37) Sauropoda: 4(4) Ornithopoda: 13(15) Ankylosauria: 4(5) Pachycephalo: 4(7) Ceratopsia: 6(8)

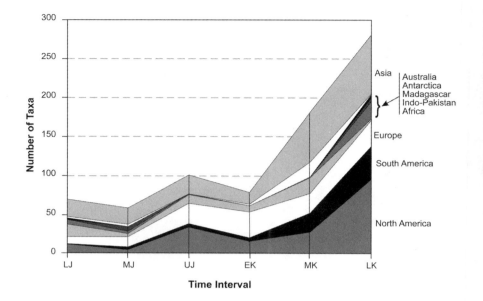

42.4. Overall dinosaur diversity through the later Mesozoic, divided by continent. Abbreviations as in Figure 42.3.

England, France, and China. Fragmentary Middle Jurassic dinosaur remains are also known from Australia, Madagascar, and isolated sites across Russia (Weishampel et al. 2004; see Tables 42.1, 42.2, and Figs. 42.4, 42.5).

Most Middle Jurassic theropods were primitive tetanurans (primarily megalosaurs and allosaurs), but some small coelurosaurs are known (e.g., *Proceratosaurus*; Woodward 1910; Holtz et al. 2004). Sauropods were represented by basal forms, such as *Cetiosaurus* (Upchurch and Martin 2003), *Patagosaurus* (Bonaparte 1986), and the omeisaurids (Martin-Rolland 1999), although representatives of the two main neosauropod lineages were probably present as well (Wilson 2002; Upchurch and Martin 2003). Prosauropods, the characteristic herbivores of earlier Mesozoic dinosaur faunas, were extinct.

In their place, ornithischians had diversified significantly. Among them, stegosaurs were common but ankylosaurs were rare (known from China and England; Galton 1983; Dong 1993a). Marginocephalians are known only by the poorly-dated ceratopsian *Chaoyangsaurus* (Zhao et al. 2003) and seem either to have been rare or to have preferred upland and other nondepositional environments. Ornithopods were the most common ornithischians. Most were small-bodied species with relatively simple dentitions (e.g., *Agilisaurus*; Peng 1992), but iguanodontians were also present (e.g., *Callovosaurus*, Ruiz-Omeñaca et al. 2006). Phylogenetically, Middle Jurassic dinosaurs were closer to Late Jurassic than to Early Jurassic forms (Holtz et al. 2004a).

The dinosaurs in these faunas seem to have remained rather cosmopolitan, if one considers the broader groups to which the individual species belong. Unlike Early Jurassic and Late Triassic faunas, however, they were not truly Pangaean – there are no examples in which the same dinosaur genus is believed to have existed in distant geographic regions. However, with better samples this view may yet change. Indeed, little can be said about these dinosaur faunas in general other than that they tended to be dominated by saurischians. Ornithischians are unknown from South America and Madagascar at this time and remain rare in Australia (known from a single track site; Rich and Vickers-Rich 2003) and the European beds. However, most European deposits are marine, with dinosaurs occurring

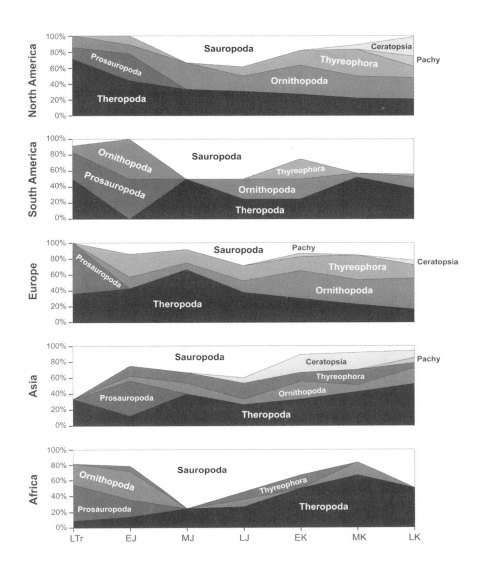

42.5. Percentages of major dinosaur groups on five major landmasses through the Mesozoic. Abbreviations as in Figure 42.3; Pachy = Pachycephalosauria.

as isolated carcasses that floated out to sea. Thus, even these records may poorly represent the true compositions of the nearby terrestrial faunas.

Other Middle Jurassic archosaurs included only crocodylians and pterosaurs, although both groups were diverse; all other archosaur clades had gone extinct. Nonarchosaurian reptiles were almost entirely small-bodied (<10 kg), with the exception of temnospondyl amphibians, and never dominated in abundance over dinosaurs. Mammals were represented by several archaic groups, including docodonts, symmetrodonts, and multituberculates, although nonmammalian synapsids (e.g., traversodontids) persisted.

Ecologically, the terrestrial Middle Jurassic was truly dinosaur-dominated. Large and small theropods were among the major predators, a role they shared with crocodylians. Smaller predators and insectivores included lizardlike diapsids, sphenosuchian crocodylians, amphibians, and mammals (e.g., Evans and Milner 1994; Clark et al. 2004). Sauropods were the largest herbivores, cropping and simply masticating low and high plants; they would fill this ecological role until the end of the Cretaceous (Upchurch and Barrett 2000). Alongside, many ornithischian herbivores (especially ornithopods) became increasingly refined oral processors (Norman and Weishampel 1991). Most other herbivorous animals were small and specialized, such as sphenodontians and late-surviving traversodontids.

Late Jurassic (Oxfordian–Tithonian; 161–145.5 mya)

The Late Jurassic world is much better known thanks to extensive outcrops in the western United States, Tanzania, China, and Portugal. In particular, the Morrison Formation deposits in the United States are some of the richest dinosaur-bearing deposits in the world (Dodson et al. 1980). Certain areas of China also have productive Late Jurassic outcrops (Shangshaximiao and Shishugou Formations), and spottier deposits occur in England, France, Germany, Mongolia, and Thailand (Weishampel et al. 2004; Buffetaut et al. 2005; Buffetaut and Suteethorn 2007; see Tables 42.1, 42.2 and Figs. 42.4, 42.5).

Among theropods, allosaurs dominated the large predator niche and were often the most common large predators (Madsen 1976), with ceratosaurs and megalosaurs as rarer faunal elements. Small theropods included the first diverse assemblage of coelurosaurs, among them early tyrannosauroids and dromaeosaurs (Rauhut 2000; Xu and Zhang 2005). Sauropods had fully diversified into diplodocoid and macronarian lineages, with members of each often inhabiting the same environments. In fact, sauropods were the most abundant large herbivores in almost every Late Jurassic fauna (Dodson et al. 1980), and their frequently high local diversity implies some degree of niche partitioning (Fiorillo 1991; Upchurch and Barrett 2000).

Among thyreophoran ornithischians, stegosaurs remained diverse, alongside a smaller number of ankylosaur species. Both groups were composed of low-browsing, large-bodied herbivores. Marginocephalians remained rare and were probably not diverse at this time. Minimally, the pachycephalosaur and ceratopsian lineages had diverged at this time, as evidenced by the primitive ceratopsian *Yinlong* (Xu et al. 2006). In contrast, ornithopod diversity had significantly increased and now included larger-bodied iguanodontians, many of which exhibited increasingly derived dental adaptations for mastication (Norman and Weishampel 1991). Smaller, cursorial forms (hypsilophodonts) were also common worldwide.

Late Jurassic dinosaur faunas still show some cosmopolitan characteristics, as shown by the great similarities between North American, African, and European species (Holtz et al. 2004a). For example, *Allosaurus*, *Ceratosaurus*, *Torvosaurus*, and *Stegosaurus* are present in both Portugal and the United States (Pérez-Moreno et al. 1999; Mateus et al. 2006; Escaso et al. 2007), and theropods similar to *Allosaurus* and *Ceratosaurus* are known from Tanzania (Janensch 1925). Related species are present in other places, such as China, although not the same genera (*Sinraptor* instead of *Allosaurus*, for example). Certain ornithopods seem to show a similar cosmopolitanism (Galton 1980), whereas sauropod groups, such as omeisaurids and dicraeosaurids, may have been more geographically restricted.

Late Jurassic mammals included archaic groups, such as triconodonts, docodonts, and symmetrodonts (Kielan-Jaworowska et al. 2004). The rodentlike multituberculates were some of the first mammalian herbivores; some of these weighed up to several kilograms. Reptilian and amphibian diversity was similar to the Middle Jurassic, but members of extant groups (lissamphibians and lacertilians; Evans and Chure 1998, 1999; Evans et al. 2005) were more diverse in comparison to archaic forms (temnospondyls and eolacertilians). Crocodylians were very similar in outward appearance to modern forms, with a decline in sphenosuchian diversity. *Archaeopteryx* provides evidence that birds had appeared, but they were not yet common. Pterosaurs dominated the skies and exhibited a wide range of body sizes.

Earliest Cretaceous (Berriasian–Barremian; 145.5–125 mya)

The first part of the Early Cretaceous is discussed here separately from the last third, which is grouped with the first part of the Late Cretaceous under the informal term "medial Cretaceous." This deviates from common practice, which is to use only formal geological epochs, but it provides three time intervals that are closer in duration. More importantly, these three Cretaceous intervals have their own distinct characteristics.

The earliest Cretaceous is not as well known as the Late Jurassic, but ongoing exploration is increasing our knowledge of this important time period. The best-known deposits are in western Europe, from England to Spain, but significant finds have also been made in western North America, South America, western Africa, and northeastern China. Other sites are known from Australia, Russia, and Thailand (Weishampel et al. 2004; see Tables 42.1, 42.2, and Figs. 42.4, 42.5). Together, they provide a broad geographic sample of dinosaur faunas, the first in which nearly every continent is represented.

By this time, dinosaur diversity had grown considerably over previous time intervals. For example, theropods were represented by nearly every type of large-bodied (allosaurs, megalosaurs, tyrannosaurs) and small-bodied (dromaeosaurs, troodontids, oviraptorosaurs, ornithomimids) form. Importantly, several theropod lineages had begun to diversify into noncarnivorous diets (therizinosaurs, oviraptorosaurs, and ornithomimosaurs). Although primitive sauropods had become rare, both the diplodocoid and macronarian lineages were quite diverse, with several types of each often coexisting. Ornithopods continued to diversify, with iguanodontians becoming large and developing a more sophisticated dental apparatus. Both ankylosaurid and nodosaurid ankylosaurs were present as well as stegosaurs. Marginocephalians were not common, with rare records of primitive ceratopsians.

With the onset of the Laurasia–Gondwana split, dinosaur faunas began to acquire more distinctive characteristics in each geographic region (Bonaparte and Kielan-Jaworowska 1987; Russell and Bonaparte 1997). The broadest similarities are between the faunas within the northern (Laurasian) and southern (Gondwanan) regions, but in fact important smaller regional differences can also be seen (Holtz et al. 2004a). Within continents, however, most faunas seem relatively close to one another.

For example, the many European sites in the Wealden Group are quite similar, but show differences with contemporary North American faunas. In both, brachiosaur and titanosaurian sauropods were common herbivores along with diverse iguanodontians, hypsilophodontians, and nodosaurids (Galton 1981; Martin and Buffetaut 1992; Blows 1998; Norman and Barrett 2002). However, coelurosaurs were the most common small predators in the Wealden, alongside carcharodontosaurids (*Concavenator* and *Neovenator*; Hutt et al. 1996; Ortega et al. 2010) and a spinosaurid (*Baryonyx*; Charig and

Milner 1997). Stegosaurs persisted alongside marginocephalians (*Stenopelix*; Sues and Galton 1982). In contrast, a large dromaeosaur (*Utahraptor*; Kirkland et al. 1993) and an early therizinosaur (*Falcarius*; Kirkland et al. 2005) are known from North America, but no stegosaurs or spinosaurs.

Recent discoveries from the Yixian Formation in Liaoning, China (which may span Barremian to Aptian time) supplement those from across eastern and southeastern Asia and document the presence of diverse small coelurosaur predators, including troodontids, dromaeosaurids, compsognathids, and tyrannosauroids (e.g., Zhou et al. 2003; Xu and Zhang 2005), along with herbivorous theropods (e.g., Xu et al., 1999). The record of larger predators is more poorly known, but includes allosauroids (*Fukuiraptor*, *Siamotyrannus*) elsewhere in Asia (Holtz et al. 2004b; Benson et al. 2010a). Titanosaur sauropods were the largest herbivores, often living alongside iguanodontian ornithopods (*Fukuisaurus*) and psittacosaurid ceratopsians.

The South American record is poor, but has remains of abelisaurs, titanosaurs (Rauhut et al. 2003), dicraeosaurids (*Amargasaurus*), and fragmentary ornithischians. In Africa, the Irhazer beds may pertain to this time, and include a carcharodontosaurid, a small ceratosaur, and rebbachisaurid and titanosaur sauropods (Lapparent 1960; Sereno et al. 1999). In South Africa, fossils include a stegosaur, an ornithopod, and the small coelurosaur *Nqwebasaurus* (de Klerk et al. 2000). None of these faunas are known well enough to adequately compare them to those of other regions.

By the earliest Cretaceous, many modern types of living tetrapod groups were evident, including frogs, salamanders, numerous types of lizards and turtles, and crocodylians; many of them held ecological roles similar to their modern relatives (Evans and Sigogneau-Russell 1997; Evans et al. 2004). Although Early Cretaceous bird fossils are generally rare, the rich localities of Liaoning Province (Zhou et al. 2003) demonstrate that birds could be locally diverse and common. Still, pterosaurs remained the most prominent flying vertebrates globally, although ecological partitioning is evident (in both feeding and body size) between these two volant groups (Unwin 1999; Wang et al. 2005). Modern mammal groups (especially eutherians and marsupials) were present, alongside more archaic forms, and these small endotherms seem to have occupied specialized small herbivore, insectivore, and carnivore roles (Kielan-Jaworowska et al. 2004; Hu et al. 2005).

Medial Cretaceous (Aptian–Turonian, 125–89.3 mya)

The medial Cretaceous is represented by deposits across North America, as well as South America, northern and western Africa, western Europe, central Asia, and China. Other sites are known from Australia, Japan, Italy, and Mongolia (Weishampel et al. 2004; see Tables 42.1, 42.2 and Figs. 42.4, 42.5) and provide an increasingly dense global sample.

Among saurischians, theropods reached gigantic sizes, both among carcharodontosaurids and spinosaurids, although tyrannosaurs had begun to reach large sizes as well. Smaller theropods included some ceratosaurs and an increasingly diverse set of coelurosaurs, especially the herbivorous and omnivorous lineages (ornithomimosaurs, therizinosaurs, and oviraptorosaurs). Sauropod diversity was also high, particularly among titanosaurians.

Diplodocoids were represented almost entirely by specialized, grazing rebbachisaurids (Sereno and Wilson 2005).

Ornithischian diversity included many types of ornithopods, including numerous iguanodontians and some early hadrosaurs, along with a variety of primitive forms such as *Tenontosaurus* and hypsilophodontians. Nodosaurids and ankylosaurids were diverse and widespread, but stegosaurs were restricted to Asia. The most common marginocephalians were psittacosaurs, but early derived ceratopsians were also present (e.g., *Yamaceratops*, *Zuniceratops*; Makovicky and Norell 2006). Pachycephalosaur fossils are not known.

As in the earliest Cretaceous, the greatest faunal similarities are within continents. For example, similar dinosaurs are present in many of the medial Cretaceous faunas of North America. In the Cloverly and Antlers Formations, a large allosauroid (*Acrocanthosaurus*) was the dominant predator, with small coelurosaurs (e.g., *Deinonychus*, *Microvenator*) occupying more specialized niches (Ostrom 1970; Harris 1998). In the Cedar Mountain Formation faunas (Cifelli et al. 1999; Kirkland et al. 1999), herbivorous and omnivorous theropods such as therizinosaurs and oviraptorosaurs were present. In all these faunas, titanosauriform sauropods were common, but so were ornithopods, typically represented by *Tenontosaurus* and large iguanodontians. Stegosaurs had vanished, replaced by ankylosaurids and nodosaurids.

More broadly, northern African and South American faunas were alike and reminiscent of the earlier Wealden in Europe in having two types of large predatory dinosaurs, carcharodontosaurids and spinosaurs (Stromer 1915, 1931, 1932; Sereno et al. 1996, 1998; Taquet and Russell 1998). These similarities have led to suggestions of earlier interchange between Europe and Africa (Russell and Bonaparte 1997). Some small coelurosaurs were present, but these cohabited with a more ancient lineage, the ceratosaurs. Ornithischians of all kinds were generally rare (an exception being the ornithopods of Gadoufaoua, Niger; Taquet 1976; Taquet and Russell 1999), and in their place sauropods were the dominant herbivores, including titanosaurians and the highly specialized rebbachisaurids (Lapparent 1960; Sereno et al. 1999; Wilson 2002).

Allosauroids remained the dominant larger predators in Asia (*Chilantaisaurus*, *Shaochilong*) (Benson and Xu 2008; Brusatte et al. 2009), along with less common tyrannosaurs. Both titanosaurs and iguanodontian ornithopods (Norman 1996, 2002) were diverse. Psittacosaurs were common and widespread across eastern Asia (Zhen et al. 1985; Dong 1993b; Sereno 2000), while stegosaurs and ankylosaurs were rarer. Herbivorous theropods included ornithomimosaurs, oviraptorosaurs, and therizinosaurs, accompanied by small predators such as dromaeosaurs and troodontids.

The Australian faunas of this time include ankylosaurs, coelurosaurs, small ornithopods, and sauropods, but they are still fragmentarily known and do not show clear affinities to other continents (Rich and Vickers-Rich 1989; Molnar and Wiffen 1994). Indeed, some elements appear to be relictual, possibly indicating some degree of isolation (Vickers-Rich 1996; Russell and Bonaparte 1997; Thulborn and Turner 2003) but others suggest exchange with other Cretaceous faunas (Benson et al. 2010b; Barrett et al. 2011). These high-latitude environments show a predominance of smaller-bodied taxa, but it is not clear whether this is genuine or a depositional and facies artifact.

Many modern lizards groups were present (Nydam 2002; Nydam and Cifelli 2002), along with marine and terrestrial turtles, lissamphibians, and a significant radiation of teleost fishes. The first snakes appeared in the Cenomanian, although these were primarily marine (e.g., Zaher and Rieppel 2004). Crocodylians of modern aspect lived alongside exotic forms with complex teeth and specialized dietary adaptations (e.g. O'Connor et al. 2010). Pterosaurs remained common and attained gigantic sizes, while birds remained small but were widespread. Cretaceous mammals included diverse multituberculates, marsupials, and eutherians (Kielan-Jaworowska et al. 2004), as well as early monotremes (e.g., Flannery et al. 1995) and surviving members of more ancient lineages.

Latest Cretaceous (Coniacian–Maastrichtian, 89.3–65.5 mya)

The dinosaur faunas of the latest Cretaceous are the best known of all, although they come from a somewhat more restricted geographic range than those of the earlier parts of this period. Because the latest Cretaceous is the most recent interval of the Mesozoic Era, rocks from this time have produced the most dinosaur fossils and many of the best preserved. The richest beds are in western North America, eastern Asia, and Argentina, but other productive Late Cretaceous sites are present in France, Brazil, Uzbekistan, Madagascar, and India-Pakistan (Weishampel et al. 2004; see Tables 42.1, 42.2, and Figs. 42.4, 42.5).

Saurischian dinosaurs were at their most diverse in terms of species numbers, but this may reflect the higher level of detail in the available record. In fact, numerous saurischian groups were reduced in overall diversity or absent. For example, theropods were represented almost exclusively by ceratosaurs and coelurosaurs; spinosaurs did not survive past the Cenomanian, and only a single latest Cretaceous allosaur is known (*Orkoraptor*; Benson et al. 2010a). The largest tyrannosaurids were the top Late Cretaceous predators in North America and eastern Asia; abelisaurids filled these roles on many southern landmasses. All known sauropods were macronarians, and most of these were derived titanosaurians (Sereno et al. 1999; Wilson 2002).

Among ornithischians, both hadrosaurids and ceratopsids achieved high levels of species diversity; often multiple species of each group were present in the same environment. A few primitive, smaller-bodied ornithopods and basal iguanodontians (e.g., *Gasparinisaura*, *Thescelosaurus*) persisted into the Late Cretaceous, as well as small, primitive ceratopsians (e.g., *Protoceratops*, *Leptoceratops*). Stegosaurs were probably extinct by this time (the identification of the Indian *Dravidosaurus* as a stegosaur remains questionable), and in their stead lived a diverse array of nodosaurid and ankylosaurid ankylosaurs. Even pachycephalosaurs were at their peak, although they were never common (Sullivan 2006).

The north–south continental split is most apparent in latest Cretaceous dinosaur faunas (Bonaparte and Kielan-Jaworowska 1987; Russell and Bonaparte 1997; Holtz et al. 2004a). Laurasian dinosaur faunas – particularly those from eastern Asia and western North America – are very much alike, often sharing the same genera or closely related forms (e.g., Sereno 1999, 2000; Currie 2003; Godefroit et al. 2003). All large predatory dinosaurs in Laurasia were coelurosaurs (i.e., tyrannosaurids), while smaller coelurosaurs included carnivores, omnivores, and herbivores. Sauropods were uncommon, represented only by titanosaurians (Lucas and Hunt 1989; Wilson 2002). The most abundant herbivores were hadrosaurs and ceratopsians, often found in bone beds with hundreds of individuals (e.g., Rogers 1990; Ryan et al. 2001). Both groups were highly advanced and specialized chewers, although quite divergent in their jaw mechanics (Weishampel 1984; Ostrom 1964; Norman and Weishampel 1991; Dodson et al. 2004). Ankylosaurs and pachycephalosaurs were less common, but many different species of each were present.

Likewise, similarities are seen between various Gondwanan faunas, particularly South America, Madagascar, and India (Krause et al. 1999). In contrast to the Laurasian faunas, ceratosaurs were the dominant large predators, with both coelurosaurs and other ceratosaurs occupying small-predator niches (Bonaparte 1996). Sauropods were the most common herbivores, including both small (saltasaurids) and enormous (e.g., *Argentinosaurus*) species (Bonaparte and Coria 1993; Bonaparte 1996; Powell 2003). Ornithopods and ankylosaurs are known only from South America during this time (e.g., Salgado and Coria 1996; Coria 1999), and both were rarer than sauropods but probably more common than is currently appreciated. No marginocephalians are known from these faunas.

Yet the overall picture is complex. Europe at this time was a chain of large islands, and its fauna shows similarities to those of both Laurasia and Gondwana (Le Loeuff 1991; Le Loeuff and Buffetaut 1995; Holtz et al. 2004a). Here ceratosaurs and coelurosaurs lived alongside titanosaurs, derived iguanodontians, and nodosaurs, with rare marginocephalians (Lindgren et al. 2007; Osi et al. 2010). Africa is not yet studied well enough to determine whether its faunas are more like those from other southern continents, similar to Europe, or distinct from other Gondwanan landmasses (Carrano and Sampson 2008). Fragmentary remains indicate the presence of titanosaurs and ceratosaurs (Kennedy et al. 1987; Smith and Lamanna 2006; Carrano and Sampson 2008). The Antarctic Peninsula has produced remains of ankylosaurs, small theropods, hadrosaurids, and primitive ornithopods (Weishampel et al. 2004; Case et al. 2007).

Among other vertebrates, modern groups continued to diversify into the latest Cretaceous, following trends established earlier in the period (e.g., Bonaparte 1996; Russell and Bonaparte 1997). Overall, faunas acquired an increasingly modern aspect, especially with regard to the presence of modern-type lepidosaurs, crocodylians, birds, and mammals, although in many cases more primitive lineages remained common (Gao and Norell 2000; Kielan-Jaworowska et al. 2004; Candeiro et al. 2006). Gondwanan crocodylians show an especially broad morphological diversity (e.g., Buckley et al. 2000). Considerable debate exists as to whether stem members of the modern orders of mammals (Foote et al. 1999; Archibald et al. 2001) and birds (Clarke et al. 2005) were present in any significant numbers.

42.6. Diversity of herbivorous dinosaurs through time, indicating different feeding types and major groups of contemporaneous plants. Modified from Fastovsky and Smith (2004).

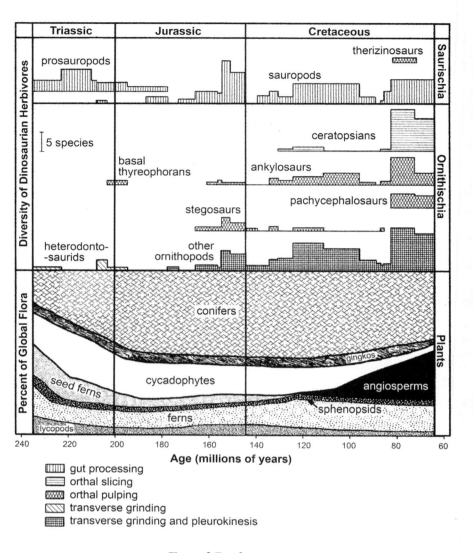

Patterns of Evolution

Faunal Replacement

Several broad patterns of dinosaur faunal evolution can be highlighted from the preceding discussion. First, a gradual and ongoing replacement occurred throughout the Mesozoic, as earlier groups of dinosaurs went extinct and new groups arose (Figs. 42.3, 42.6, 42.7). Among carnivorous dinosaurs, several evolutionary waves of theropods rose to dominance and receded (e.g., Bakker et al. 1992). Primitive forms such as herrerasaurids formed the first wave, followed by coelophysoids. Both groups shared the top carnivore niche with other archosaurs in the Late Triassic, but by the Early Jurassic only theropods remained (e.g., Benton 1991; Sues this volume). Coelophysoids remained dominant into the Early Jurassic (Rauhut and Hungerbühler 2000; Carrano and Sampson 2004), and were replaced by a succession of non-coelurosaur tetanurans – first megalosaurs (Middle Jurassic), then allosaurs (Late Jurassic) and spinosaurs (medial Cretaceous) (Bakker et al. 1992). These shifts were primarily shifts in dominance, with new groups becoming more common (and occasionally more diverse), while older groups rarely went extinct but persisted in supporting roles. By the end of the Late Cretaceous, however, only two groups remained – ceratosaurs dominated in Gondwana and coelurosaurs dominated in Laurasia.

Matthew T. Carrano

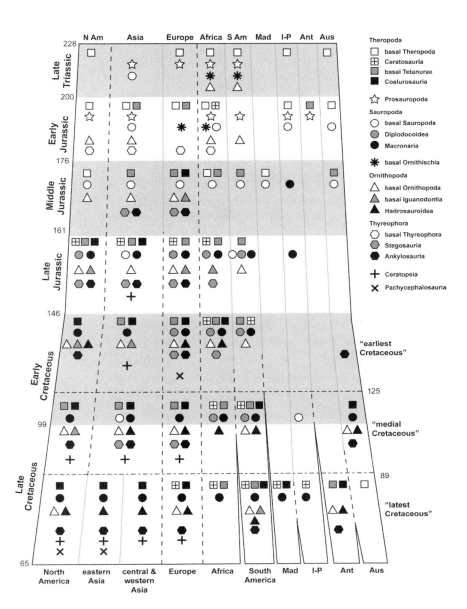

42.7. Summary of faunal changes through time. Symbols indicate the presence of different dinosaur groups on each landmass during each time interval. Continents are shown as columns, separated by gray lines for convenience. The onset of continental separation is shown by a dashed black line, when interchange may still have been possible. Splitting events are shown as physical divisions between the columns. Abbreviations: Ant = Antarctica, Aus = Australia, I-P = Indo-Pakistan, Mad = Madagascar, N Am = North America, S Am = South America. Numbers indicate millions of years ago.

Among herbivores (Fig. 42.3), the prosauropods were the first large-bodied forms, initially evolving alongside other herbivorous archosaurs (e.g., dicynodonts, aetosaurs). By the Early Jurassic, prosauropods and early sauropods constituted the only large herbivores, and afterward sauropods remained in this role until the end of the Mesozoic. Both groups appear to have been specialized as croppers and gizzard processors, with relatively limited oral mastication (Calvo 1994; Barrett 2000; Upchurch and Barrett 2000; Barrett and Upchurch 2005). Both high- and low-feeding taxa were present. Alone among sauropods, rebbachisaurids seem to have developed as specialized grazers with an unusual dentition (Sereno and Wilson 2005). The Cretaceous also saw the rise of herbivorous and omnivorous theropods, specifically the ornithomimosaurs, oviraptorosaurs, and therizinosaurs (Barrett 2005).

A series of smaller-bodied ornithischian herbivores arose in parallel to the sauropodomorphs, eventually specializing as chewers, slicers, and grinders of Mesozoic plants. Many forms may have had nonmuscular cheeks as an aid to oral processing of food, unlike sauropodomorphs (Galton 1973).

Ornithopods dominated the earlier faunas, with feeding mechanisms not strikingly distinct from those of prosauropods (Norman and Weishampel 1991). Subsequently, ornithopods embarked on a steady size increase and became more ungulate-like in the Cretaceous, culminating with hadrosaurs (Weishampel 1984; Carrano et al. 1999). The earliest thyreophorans and marginocephalians were grossly similar to early ornithopods in body form and size, but remained rare. By the Middle Jurassic, the thyreophoran lineage became common (mostly stegosaurs, replaced by ankylosaurs in the Cretaceous) as low-foliage feeders. Marginocephalians split into pachycephalosaurs, which were smaller and more ornithopod-like in form, and ceratopsians, which became much larger and developed their own specialized system of slicing and chewing (Fig. 42.6; Ostrom 1964; Dodson et al. 2004).

Thus, among both carnivorous and herbivorous dinosaurs, there were apparent general trends toward a greater diversity of feeding types throughout the later Mesozoic. The diminishment or loss of certain earlier groups (e.g., prosauropods, stegosaurs) was at least partly accommodated by similar adaptations in later groups (e.g., therizinosaurs, ankylosaurs), so that there was little obvious loss of ecological diversity toward the Late Cretaceous. Indeed, the opposite seems to be true, with a number of specialized ecological types present in medial and latest Cretaceous communities that had been entirely absent during prior time intervals, particularly those involving extensive oral mastication of plant matter.

Diversification

This pattern of replacement was accompanied by an increase in overall diversity as well. The number of different dinosaur species increased from one in the Middle or early Late Triassic (the original dinosaurian ancestor) to hundreds by the Late Cretaceous (Fig. 42.4).

This is partly explained by the simple diversification of lineages – as time goes on, there are more opportunities for innovation, which leads to the evolution of new species. But extinction should also happen frequently, balancing out this effect to some degree. Nonetheless, only two major dinosaur groups – prosauropods and stegosaurs – appear to have gone extinct before the end of the Cretaceous (Figs. 42.3, 42.6). Although other groups (e.g., sauropods) were once thought to have experienced a marked decline from the Jurassic to the Cretaceous, this is no longer clearly the case. Instead, sauropods may have occupied increasingly specific habitats in Laurasia as hadrosaurs and ceratopsians diversified (Lehman 1987, 2001; Lucas and Hunt 1989) while remaining both diverse and abundant throughout much of Gondwana. Thus, the increase in different habitat types throughout the Mesozoic may have encouraged the evolution of new dinosaur species and permitted the coexistence of a greater number of species at any one time (see below).

Regardless of the cause, Cretaceous dinosaur faunas often contained more species than Jurassic ones. Perhaps more significantly, these Cretaceous dinosaurs also exhibited a greater diversity of sizes and shapes than those of the Jurassic. Both increases – in numbers of species and in ecological/morphological types – were the result of continued evolutionary innovation in the absence of a major selective global extinction.

Tectonics and Climate

It is difficult to identify clear connections between climate change and dinosaur evolution during the Mesozoic Era at anything other than a very broad scale. Certainly large-scale changes in climate are apparent, tied predominately to the huge tectonic changes and associated shifts in continental connections and positions.

For example, during the Late Triassic and Early Jurassic, much of Pangaea had a relatively uniform, arid climate across the interior. This was at least partly due to the difficulties associated with transporting atmospheric moisture across the great landlocked stretches of the supercontinent (Rees et al. 2000; Loope et al. 2004). As a result, dinosaurs (and their faunas) in one region were often similar to those in another, not only because of geographic connections, but also thanks to broad environmental similarities.

As the continents separated during the Jurassic (Figs. 42.2, 42.7), new habitats and climate regimes were created, especially as the Pangaean interior fragmented and became less arid and uniform. Increased coastlines afforded still more habitat diversity along all the edges of the continents. During these changes, we see more differentiation between dinosaur faunas in different places at any one time. Although most major dinosaur groups were still widespread, there were fewer instances of the same genus or species present in distant locations. By the Late Cretaceous, it was common to find entirely different groups of species living in different environments but in close geographic proximity. Such differences are clear between the upland and nearshore Campanian formations of the Western Interior (Horner et al. 1992; Brinkman et al. 1998; Sampson et al. 2010) or between the different Maastrichtian habitats of the same region (Lucas 1981; Lehman 1987, 2001).

The Late Cretaceous of North America provides one of the few clear examples of contemporaneous segregation of dinosaur habitats based on environment. Similar smaller-scale climate effects are difficult to observe in other places and time periods, largely because we have less data (both climatic and dinosaurian) to examine. There are few places where a finely resolved dinosaur record is available from a single place for several million years, so it is difficult to know how, and to what degree, dinosaurs evolved in response to local climate changes.

References

Archibald, J. D., A. O. Averianov, and E. G. Ekdale. 2001. Late Cretaceous relatives of rabbits, rodents, and other extant eutherian mammals. *Nature* 414: 62–65.

Bakker, R. T., J. Siegwarth, D. Kralis, and J. Filla. 1992. *Edmarka rex*, a new, gigantic theropod dinosaur from the middle Morrison Formation, Late Jurassic of the Como Bluff outcrop region. *Hunteria* 2(9): 1–24.

Barrett, P. M. 2000. Prosauropod dinosaurs and iguanas: speculations on the diets of extinct reptiles. In H.-D. Sues (ed.), *Evolution of Herbivory in Terrestrial Vertebrates: Perspectives from the Fossil Record*, 42–78. Cambridge: Cambridge University Press.

———. 2005. The diet of ostrich dinosaurs (Theropoda: Ornithomimosauria). *Palaeontology* 48(2): 347–358.

Barrett, P. M., and P. Upchurch. 2005. Sauropodomorph diversity through time: paleoecological and macroevolutionary implications. In K. A. Curry Rogers, and J. A. Wilson (eds.) *The Sauropods: Evolution and Paleobiology*, 125–156. Berkeley: University of California Press.

Barrett, P. M., R. B. J. Benson, T. H. Rich, and P. Vickers-Rich. 2011. *Biology Letters*

Benson, R. B. J., and X. Xu. 2008. The anatomy and systematic position of the theropod dinosaur *Chilantaisaurus tashuikouensis* Hu, 1964 from the Early Cretaceous of Alanshan, People's Republic of China. *Geological Magazine* 145: 778–789.

Benson, R. B. J., M. T. Carrano, and S. L. Brusatte. 2010a. A new clade of archaic large-bodied predatory dinosaurs Theropoda: Allosauroidea that survived to the latest Mesozoic. *Naturwissenschaften* 97: 71–78.

Benson, R. B. J., P. M.Barrett, T. H. Rich, and P. Vickers-Rich. 2010b. A southern tyrant reptile. *Science* 327: 1613.

Benton, M. J. 1991. What really happened in the Late Triassic? *Historical Biology* 5: 263–278.

——. 2004. Origin and relationships of Dinosauria. In Weishampel 2004, 7–19.

Blows, W. T. 1998. A review of Lower and Middle Cretaceous dinosaurs of England. In Lucas 1998, 29–38.

Bonaparte, J. F. 1986. Les dinosaures (Carnosaures, Allosauridés, Sauropodes, Cétosauridés) du Jurassique Moyen de Cerro Cóndor (Chubut, Argentina). *Annales de Paléontologie (Vert.-Invert.)* 72(3): 325–386.

——. 1996. Cretaceous tetrapods of Argentina. *Münchner Geowissenschaften Abhandlungen* (A) 30: 73–130.

Bonaparte, J. F., and R. A. Coria. 1993. Un nuevo y gigantesco saurópodo titanosaurio de la Formación Rio Limay (Albanio-Cenomaniano) de la Provincia del Neuquén, Argentina. *Ameghiniana* 30: 271–282.

Bonaparte, J. F., and Z. Kielan-Jaworowska. 1987. Late Cretaceous dinosaur and mammal faunas of Laurasia and Gondwana. In P. J. Currie, and E. H. Koster (eds.), *Fourth Symposium on Mesozoic Terrestrial Ecosystems*, 24–29. Tyrrell Museum of Paleontology, Drumheller, Alberta.

Brinkman, D. B., M. J. Ryan, and D. A. Eberth. 1998. The paleogeographic and stratigraphic distribution of ceratopsids (Ornithischia) in the upper Judith River Group of western Canada. *Palaios* 13: 160–169.

Brusatte, S., R. B. J. Benson, D. J. Chure, X. Xu, C. Sullivanand D. W. E. Hone. 2009. The first definitive carcharodontosaurid (Dinosauria: Allosauroidea) from Asia, the evolution of Asian Cretaceous theropods and the delayed ascent of tyrannosauroids. *Naturwissenschaften* 96: 1051–1058.

Buckley, G. A., C. A. Brochu, D. W.

Krause, and D. Pol. 2000. A pug-nosed crocodyliform from the Late Cretaceous of Madagascar. *Nature* 405: 941–944.

Buffetaut, E., and V. Suteethorn. 1998. Early Cretaceous dinosaurs from Thailand and their bearing on the early evolution and biostratigraphical history of some groups of Cretaceous dinosaurs. In Lucas 1998, 205–210.

Buffetaut, E., and V. Suteethorn. 1999. The dinosaur fauna of the Sao Khua Formation of Thailand and the beginning of the Cretaceous radiation of dinosaurs in Asia. *Palaeogeography, Palaeoclimatology, Palaeoecology* 150: 13–23.

——. 2007. A sinraptorid theropod (Dinosauria: Saurischia) from the Phu Kradung Formation of northeastern Thailand. *Bulletin de la Société Géologique de France* 178(6): 497–502.

Buffetaut, E., V. Suteethorn, H. Tong, and A. Kosir. 2005. First dinosaur from the Shan–Thai Block of SE Asia: a Jurassic sauropod from the southern peninsula of Thailand. *Journal of the Geological Society, London* 162(3): 481–484.

Calvo, J. O. 1994. Jaw mechanics in sauropod dinosaurs. *GAIA* 10: 183–193.

Candeiro, C. R. A., A. G. Martinelli, L. d. S. Avilla, and T. H. Rich. 2006. Tetrapods from the Upper Cretaceous (Turonian–Maastrichtian) Bauru Group of Brazil: a reappraisal. *Cretaceous Research* 27: 923–946.

Carrano, M. T., and S. D. Sampson. 2004. A review of coelophysoids (Dinosauria: Theropoda) from the Early Jurassic of Europe, with comments on the late history of the Coelophysoidea. *Neues Jahrbuch für Geologie und Paläontologie Monatshefte* 2004(9): 537–558.

——. 2008. The phylogeny of Ceratosauria (Dinosauria: Theropoda). *Journal of Systematic Palaeontology* 6(2): 183–236.

Carrano, M. T., and J. A. Wilson. 2001. Taxon distributions and the tetrapod track record. *Paleobiology* 27(3): 564–582.

Carrano, M. T., C. M. Janis, and J. J. Sepkoski, Jr. 1999. Hadrosaurs as ungulate parallels: lost lifestyles and deficient data. *Acta Palaeontologica Polonica* 44(3): 237–261.

Case, J. A., J. E. Martin, and M. A. Reguero. 2007. A dromaeosaur from the Maastrichtian of James Ross Island and the Late Cretaceous Antarctic dinosaur fauna. In A. K. Cooper, C. R. Raymond, and the ISAES Editorial Team (eds.), *Antarctica: A Keystone in a Changing World. Online Proceedings of the 10th ISAES*, 1–4. United States Geological Survey and the National Academies,

USGS Open-File Report 2007–1047, Short Research Paper 083.

Charig, A. J., and A. C. Milner. 1997. *Baryonyx walkeri*, a fish-eating dinosaur from the Wealden of Surrey. *Bulletin of the Natural History Museum, Geology Series* 53(1): 11–70.

Cifelli, R. L., R. L. Nydam, J. D. Gardner, A. Weil, J. G. Eaton, J. I. Kirkland, and S. K. Madsen. 1999. Medial Cretaceous vertebrates from the Cedar Mountain Formation, Emery County, Utah: the Mussentuchit local fauna. In Gillette 1999, 219–242.

Clark, J. M., X. Xu, C. A. Forster, and Y. Wang. 2004. A Middle Jurassic sphenosuchian from China and the origin of the crocodylian skull. *Nature* 430: 1021–1024.

Clarke, J. A., C. P. Tambussi, J. I. Noriega, G. M. Erickson, and R. A. Ketcham. 2005. Definitive fossil evidence for the extant avian radiation in the Cretaceous. *Nature* 433: 305–308.

Coria, R. A. 1999. Ornithopod dinosaurs from the Neuquén Group, Patagonia, Argentina: phylogeny and biostratigraphy. In Y. Tomida, T. H. Rich, and P. Vickers-Rich (eds.), *Proceedings of the Second Gondwanan Dinosaur Symposium*, 47–60. Tokyo: National Science Museum Monographs.

Currie, P. J. 2003. Allometric growth in tyrannosaurids (Dinosauria: Theropoda) from the Upper Cretaceous of North America and Asia. *Canadian Journal of Earth Sciences* 40: 651–665.

de Klerk, W. J., C. A. Forster, S. D. Sampson, A. Chinsamy, and C. F. Ross. 2000. A new coelurosaurian dinosaur from the Early Cretaceous of South Africa. *Journal of Vertebrate Paleontology* 20(2): 324–332

Dodson, P., A. K. Behrensmeyer, R. T. Bakker, and J. S. McIntosh. 1980. Taphonomy and paleoecology of the dinosaur beds of the Jurassic Morrison Formation. *Paleobiology* 6(2): 208–232.

Dodson, P., C. A. Forster, and S. D. Sampson. 2004. Ceratopsidae. In Weishampel 2004, 494–513.

Dong, Z. 1993a. An ankylosaur (ornithischian dinosaur) from the Middle Jurassic of the Junggar Basin, China. *Vertebrata PalAsiatica* 31(4): 257–266.

Dong, Z. 1993b. Early Cretaceous dinosaur faunas in China: an introduction. *Canadian Journal of Earth Sciences* 30(10–11): 2096–2100.

Escaso, F., F. Ortega, P. Dantas, E. Malafaia, N. L. Pimentel, X. Pereda-Suberbiola, J. L. Sanz, J. C. Kullberg, M. C. Kullberg, and F. Barriga. 2007. New evidence of shared dinosaur across Upper

Jurassic proto-North Atlantic: *Stegosaurus* from Portugal. *Naturwissenschaften* 94: 367–374.

Evans, S. E., and D. J. Chure. 1998. Morrison lizards: structure, relationships and biogeography. *Modern Geology* 23(1–4): 35–48.

Evans, S. E., and D. J. Chure. 1999. Upper Jurassic lizards from the Morrison Formation of Dinosaur National Monument, Utah. In Gillette 1999, 151–159.

Evans, S. E., and A. R. Milner. 1994. Middle Jurassic microvertebrate assemblages from the British Isles. In N. C. Fraser and H.-D. Sues (eds.), *In the Shadow of the Dinosaurs: Early Mesozoic Tetrapods*, 303–321. Cambridge: Cambridge University Press.

Evans, S. E., P. M. Barrett, and D. J. Ward. 2004. The first record of lizards and amphibians from the Wessex Formation (Lower Cretaceous: Barremian) of the Isle of Wight, England. *Proceedings of the Geologists' Association* 115: 239–247.

Evans, S. E., C. Lally, D. J. Chure, A. Elder, and J. A. Maisano. 2005. A Late Jurassic salamander (Amphibia: Caudata) from the Morrison Formation of North America. *Zoological Journal of the Linnean Society* 143: 599–616.

Fastovsky, D. E., and J. B. Smith. 2004. Dinosaur paleoecology. In Weishampel 2004, 614–626.

Fiorillo, A. R. 1991. Dental microwear on the teeth of *Camarasaurus* and *Diplodocus*: implications for sauropod paleoecology. In Z. Kielan-Jaworowska, N. Heintz, and H. A. Nakrem (eds.), *Fifth Symposium on Mesozoic Terrestrial Ecosystems and Biota, Extended Abstracts*, 23–24. Contributions from the Paleontological Museum, University of Oslo.

Flannery, T. F., M. Archer, T. H. Rich, and R. Jones. 1995. A new family of monotremes from the Cretaceous of Australia. *Nature* 377: 418–420.

Foote, M., J. P. Hunter, C. M. Janis, and J. J. Sepkoski, Jr. 1999. Evolutionary and preservational constraints on origins of biologic groups: divergence times of eutherian mammals. *Science* 283: 1310–1314.

Fraser, N. C. and H.-D. Sues. 1994. Comments on Benton's "Late Triassic to Middle Jurassic extinctions among continental tetrapods." In N. C. Fraser and H.-D. Sues (eds.), *In the Shadow of the Dinosaurs: Early Mesozoic Tetrapods*, 398–400. Cambridge: Cambridge University Press.

Galton, P. M. 1973. The cheeks of ornithischian dinosaurs. *Lethaia* 6: 67–89.

Galton, P. M. 1980. *Dryosaurus* and *Camptosaurus*, intercontinental genera of Upper Jurassic ornithopod dinosaurs. *Mémoires de la Société géologique de France, nouvelle série* 139: 103–108.

Galton, P. M. 1981. *Craterosaurus pottonensis* Seeley, a stegosaurian dinosaur from the Lower Cretaceous of England, and a review of Cretaceous stegosaurs. *Neues Jahrbuch für Geologie und Paläontologie, Abhandlungen* 161(1): 28–46.

Galton, P. M. 1983. *Sarcolestes leedsi* Lydekker, an ankylosaurian dinosaur from the Middle Jurassic of England. *Neues Jahrbuch für Geologie und Paläontologie Monatshefte* 1983(3): 141–155.

Gao, K., and M. A. Norell. 2000. Taxonomic composition and systematics of Late Cretaceous lizard assemblages from Ukhaa Tolgod and adjacent localities, Mongolian Gobi Desert. *Bulletin of the American Museum of Natural History* 249: 1–118.

Gillette, D. D. (ed.). 1999. *Vertebrate Paleontology in Utah*. Salt Lake City: Utah Geological Survey.

Godefroit, P., Y. L. Bolotsky, and V. R. Alifanov. 2003. A remarkable hollow-crested hadrosaur from Russia: an Asian origin for lambeosaurines. *Comptes Rendus Palevol* 2: 143–151.

Hallam, A. 1985. A review of Mesozoic climates. *Journal of the Geological Society, London* 142: 433–445.

———. 1993. Jurassic climates as inferred from the sedimentary and fossil record. *Philosophical Transactions: Biological Sciences* 341(1297):387–396.

Harris, J. D. 1998. Large, Early Cretaceous theropods in North America. In Lucas 1998, 225–228.

Holtz, T. R., Jr., R. E. Chapman, and M. C. Lamanna. 2004a. Mesozoic biogeography of Dinosauria. In Weishampel 2004, 627–642.

Holtz, T. R., Jr., R. E. Molnar, and P. J. Currie. 2004b. Basal Tetanurae. In Weishampel 2004, 71–110.

Horner, J. R., D. J. Varricchio, and M. B. Goodwin. 1992. Marine transgressions and the evolution of Cretaceous dinosaurs. *Nature* 358: 59–61.

Hu, Y., J. Meng, Y. Wang, and C. Li. 2005. Large Mesozoic mammals fed on young dinosaurs. *Nature* 433: 149–152.

Hutt, S., D. M. Martill, and M. J. Barker. 1996. The first European allosaurid dinosaur (Lower Cretaceous, Wealden Group, England). *Neues Jahrbuch für Geologie und Paläontologie Monatshefte* 1996(10): 635–644.

Irmis, R. B. 2004. First report of *Megapnosaurus* (Theropoda: Coelophysoidea) from China. *PaleoBios* 24(3): 11–18.

Irmis, R. B., S. J. Nesbitt, K. Padian, N. D. Smith, A. H. Turner, D. T. Woody, and A. Downs. 2007. A Late Triassic dinosauromorph assemblage from New Mexico and the rise of dinosaurs. *Science* 317: 358–361.

Janensch, W. 1925. Die Coelurosaurier und Theropoden der Tendaguru-Schichten Deutsch-Ostafrikas. *Palaeontographica Supplement VII* (1): 1–100.

Ji, Q., P. J. Currie, M. A. Norell, and S.-A. Ji. 1998. Two feathered dinosaurs from northeastern China. *Nature* 393: 753–761.

Ji, Q., M. A. Norell, K.-Q. Gao, S.-A. Ji, and D. Ren. 2001. The distribution of integumentary structures in a feathered dinosaur. *Nature* 410: 1084–1088.

Kennedy, W. J., H. C. Klinger, and N. J. Mateer. 1987. First record of an Upper Cretaceous sauropod dinosaur from Zululand, South Africa. *South African Journal of Science* 83:173–174

Kielan-Jaworowska, K., R. L. Cifelli, and Z.-X. Luo. 2004. *Mammals from the Age of Dinosaurs: Origins, Evolution, and Structure*. New York: Columbia University Press.

Kirkland, J. I., D. Burge, and R. Gaston. 1993. A large dromaeosaur (Theropoda) from the Lower Cretaceous of eastern Utah. *Hunteria* 2(10): 1–16

Kirkland, J. I., R. L. Cifelli, B. B. Britt, D. L. Burge, F. L. DeCourten, J. G. Eaton, and J. M. Parrish. 1999. Distribution of vertebrate faunas in the Cedar Mountain Formation, east-central Utah. In Gillette 1999, 201–217.

Kirkland, J. I., L. E. Zanno, S. D. Sampson, J. M. Clark, and D. D. DeBlieux. 2005. A primitive therizinosauroid dinosaur from the Early Cretaceous of Utah. *Nature* 435(3468): 84–87

Krause, D. W., R. R. Rogers, C. A. Forster, J. H. Hartman, G. A. Buckley, and S. D. Sampson. 1999. The Late Cretaceous vertebrate fauna of Madagascar: implications for Gondwanan paleobiogeography. *GSA Today* 9(8): 1–7.

Lapparent, A. F. 1960. Les dinosauriens du "Continental Intercalaire" du Sahara central. *Mémoires de la Société géologique de France, nouvelle série* 88A: 1–57.

Lehman, T. M. 1987. Late Maastrichtian paleoenvironments and dinosaur biogeography in the Western Interior of North America. *Palaeogeography, Palaeoclimatology, Palaeoecology* 60: 189–217.

———. 2001. Late Cretaceous dinosaur

provinciality. In D. H. Tanke and K. Carpenter (eds.), *Mesozoic Vertebrate Life: New Research Inspired by the Paleontology of Philip J. Currie*, 310–328. Bloomington: Indiana University Press.

Le Loeuff, J. 1991. The Campano-Maastrichtian vertebrate faunas from southern Europe and their relationships with other faunas in the world: palaeobiogeographical implications. *Cretaceous Research* 12(2): 93–114.

Le Loeuff, J., and E. Buffetaut. 1995. The evolution of Late Cretaceous non-marine vertebrate fauna in Europe. In A. Sun, and Y. Wang (eds.). *Sixth Symposium on Mesozoic Terrestrial Ecosystems and Biota, Short Papers*, 181–184. Beijing: China Ocean Press.

Lindgren, J., P. J. Currie, M. Siverson, J. Rees, P. Cederström and F. Lindgren. 2007. The first neoceratopsian dinosaur remains from Europe. *Palaeontology* 50(4): 929–937

Loope, D. B., M. B. Steiner, C. M. Rowe, and N. Lancaster. 2004. Tropical westerlies over Pangaean sand seas. *Sedimentology* 51: 315–322.

Lucas, S. G. 1981. Dinosaur communities of the San Juan Basin: a case for lateral variations in the composition of Late Cretaceous dinosaur communities. In S. G. Lucas, J. K. Rigby, Jr., and B. S. Kues (eds.), *Advances in San Juan Basin Paleontology*, 337–393. Albuquerque: University of New Mexico Press.

Lucas, S. G., and A. P. Hunt. 1989. *Alamosaurus* and the sauropod hiatus in the Cretaceous of the North American Western Interior. In J. O. Farlow (ed.), *Paleobiology of the Dinosaurs*, 75–85. Special Paper 238. Boulder: Geological Society of America.

Lucas, S. G., J. I. Kirkland, and J. W. Estep (eds.). 1998. *Lower and Middle Cretaceous Terrestrial Ecosystems*. New Mexico Museum of Natural History and Science Bulletin 14, Albuquerque.

Madsen, J. H., Jr. 1976. *Allosaurus fragilis*: a revised osteology. *Utah Geological and Mineral Survey Bulletin* 109: 1–163.

Makovicky, P. J., and M. A. Norell. 2006. *Yamaceratops dorngobiensis*, a new primitive ceratopsian (Dinosauria: Ornithischia) from the Cretaceous of Mongolia. *American Museum Novitates* 3530: 1–42

Martin, V., and E. Buffetaut. 1992. Les *Iguanodon*s (Ornithischia – Ornithopoda) du Crétacé inférieur de la region de Saint-Dizier (Haute-Marne). *Revue de Paléobiologie* 11(1): 67–96.

Martin-Rolland, V. 1999. Les sauropodes chinois. *Revue Paléobiologie, Genève* 18(1): 287–315.

Mateus, O., A. Walen, and M. T. Antunes. 2006. The large theropod fauna of the Lourinhã Formation (Portugal) and its similarity to that of the Morrison Formation, with a description of a new species of *Allosaurus*. In J. R. Foster and S. G. Lucas (eds.), *Paleontology and Geology of the Upper Jurassic Morrison Formation*, 123–129. Bulletin 36. Albuquerque: New Mexico Museum of Natural History and Science.

Molnar, R. E., and J. Wiffen. 1994. A Late Cretaceous polar dinosaur fauna from New Zealand. *Cretaceous Research* 15: 689–706.

Norman, D. B. 1996. On Mongolian ornithopods (Dinosauria: Ornithischia). 1. *Iguanodon orientalis* Rozhdestvensky 1952. *Zoological Journal of the Linnean Society* 116: 303–315.

Norman, D. B. 2002. On Asian ornithopods (Dinosauria: Ornithischia). 4. *Probactrosaurus* Rozhdestvensky, 1966. *Zoological Journal of the Linnean Society* 136: 113–144.

Norman, D. B., and P. M. Barrett. 2002. Ornithischian dinosaurs from the Lower Cretaceous (Berriasian) of England. *Palaeontology* 68: 161–189.

Norman, D. B., and D. B. Weishampel. 1991. Feeding mechanisms in some small herbivorous dinosaurs: processes and patterns. In J. M. V. Rayner and R. J. Wootton (eds.), *Biomechanics in Evolution*, 161–181. Cambridge: Cambridge University Press.

Nydam, R. L. 2002. Lizards of the Mussentuchit Local Fauna (Albian-Cenomanian boundary) and comments on the evolution of the Cretaceous lizard fauna of North America. *Journal of Vertebrate Paleontology* 22(3): 645–660.

Nydam, R. L., and R. L. Cifelli. 2002. Lizards from the Lower Cretaceous (Aptian-Albian) Antlers and Cloverly Formations. *Journal of Vertebrate Paleontology* 22(2): 268–298.

O'Connor, P.M., J. W. Sertich, N. J. Stevens, E. M. Roberts, M. D. Gottfried, T. L. Hieronymus, Z. A. Jinnah, R. Ridgely, S. E. Ngasala, and J. Temba. 2010. The evolution of mammal-like crocodyliforms in the Cretaceous Period of Gondwana. *Nature* 466: 748–751.

Olsen, P. E., D. V. Kent, H.-D. Sues, C. Koeberl, H. Huber, A. Montanari, E. C. Rainforth, S. J. Fowell, M. J. Szajna, and B. W. Hartline. 2002. Ascent of dinosaurs linked to an iridium anomaly at the Triassic-Jurassic boundary. *Science* 296: 1305–1307.

Olsen, P. E., H.-D. Sues, E. C. Rainforth, D. V. Kent, C. Koebert, H. Huber, A. Montanari, S. J. Fowell, M. J. Szajna, and B. W. Hartline. 2003. Response to comment on "Ascent of dinosaurs linked to an iridium anomaly at the Triassic-Jurassic boundary." *Science* 301: 169c-d.

Ortega, F., F. Escaso, and J. L. Sanz. 2010. A bizarre, humped Carcharodontosauria Theropoda from the Lower Cretaceous of Spain. *Nature* 467: 203–206

Ostrom, J. H. 1964. A functional analysis of jaw mechanics in the dinosaur *Triceratops*. *Postilla* 88: 1–35.

Ostrom, J. H. 1970. Stratigraphy and paleontology of the Cloverly Formation (Lower Cretaceous) of the Bighorn Basin area, Wyoming and Montana. *Peabody Museum Bulletin* 35: 1–234.

Peng, G. 1992. Jurassic ornithopod *Agilisaurus louderbacki* (Ornithopoda: Fabrosauridae) from Zigong, Sichuan, China. *Vertebrata PalAsiatica* 30(1): 39–53.

Pérez-Moreno, B. P., D. J. Chure, C. Pires, C. M. Da Silva, V. Dos Santos, P. Dantas, L. Póvoas, M. Cachão, J. L. Sanz, and M. G. de Carvalho. 1999. On the presence of *Allosaurus fragilis* (Theropoda: Carnosauria) in the Upper Jurassic of Portugal: first evidence of an intercontinental dinosaur species. *Journal of the Geological Society, London* 156: 449–452.

Powell, J. E. 2003. Revision of South American titanosaurid dinosaurs: palaeobiological, palaeobiogeographical and phylogenetic aspects. *Records of the Queen Victoria Museum Launceston* 111: 1–173.

Raath, M. A. 1969. A new coelurosaurian dinosaur from the Forest Sandstone of Rhodesia. *Arnoldia (Rhodesia)* 4(28): 1–25.

Rauhut, O. W. M. 2000. The dinosaur fauna from the Guimarota mina. In T. Martin and B. Krebs (eds.), *Guimarota – A Jurassic Ecosystem*, 75–82. Munich: Verlag Dr. Friedrich Pfeil.

Rauhut, O. W. M. and A. Hungerbühler. 2000. A review of European Triassic theropods. *GAIA* 15: 75–88.

Rauhut, O. W. M., G. Cladera, P. A. Vickers-Rich, and T. H. Rich. 2003. Dinosaur remains from the Lower Cretaceous of the Chubut Group, Argentina. *Cretaceous Research* 24: 487–497.

Raxworthy, C. J., M. R. J. Forstner, and R. J. Nussbaum. 2002. Chameleon radiation by oceanic dispersal. *Nature* 415: 784–787.

Rees, P. A., A. M. Ziegler, and P. J. Valdes.

2000. Jurassic phytogeography and climates: new data and model comparisons. In B. T. Huber, K. G. Macleod and S. L. Wing (eds.), *Warm Climates in Earth History*, 297–318. Cambridge: Cambridge University Press.

Rich, T. H. V., and P. Vickers-Rich. 1989. Polar dinosaurs and biotas of the Early Cretaceous of southeastern Australia. *National Geographic Research* 5(1): 15–53.

———. 2003. *A Century of Australian Dinosaurs*. Queen Victoria Museum and Art Gallery and Monash Science Centre, Monash University, Launceston.

Rogers, R. R. 1990. Taphonomy of three dinosaur bone beds in the Upper Cretaceous Two Medicine Formation of northwestern Montana: evidence for drought-related mortality. *Palaios* 5: 394–431.

Rowe, T. B. 1989. A new species of the theropod dinosaur *Syntarsus* from the Early Jurassic Kayenta Formation of Arizona. *Journal of Vertebrate Paleontology* 9(2): 125–136.

Ruiz-Omeñaca, J. I., X. Pereda Suberbiola, and P. M. Galton. 2007. *Callovosaurus leedsi*, the earliest dryosaurid dinosaur (Ornithischia: Euornithopoda) from the Middle Jurassic of England. In K. Carpenter (ed.), *Horns and Beaks: Ceratopsian and Ornithischian Dinosaurs*, 3–16. Bloomington: Indiana University Press.

Russell, D. A., and J. F. Bonaparte. 1997. Dinosaurian faunas of the later Mesozoic. In J. O. Farlow, and M. K. Brett-Surman (eds.), *The Complete Dinosaur*, 644–661. Bloomington: Indiana University Press.

Ryan, M. J., A. P. Russell, D. A. Eberth, and P. J. Currie. 2001. The taphonomy of a *Centrosaurus* (Ornithischia: Certopsidae) bone bed from the Dinosaur Park Formation (Upper Campanian), Alberta, Canada, with comments on cranial ontogeny. *Palaios* 16: 482–506.

Salgado, L., and R. A. Coria. 1996. First evidence of an ankylosaur (Dinosauria, Ornithischia) in South America. *Ameghiniana* 33(4): 367–371.

Sampson, S. D., and M. A. Loewen. 2010. Unraveling a radiation: a review of the diversity, stratigraphic distribution, biogeography, and evolution of horned dinosaurs Ornithischia: Ceratopsidae. In M. J. Ryan, B. J. Chinnery-Allgeier, and D. A. Eberth eds., New Perspectives on Horned Dinosaurs: The Royal Tyrrell Museum Ceratopsian Symposium, 405–427. Indiana University Press, Bloomington.

Sereno, P. C. 1999. The evolution of dinosaurs. *Science* 284: 2137–2147.

———. 2000. The fossil record, systematics, and evolution of pachycephalosaurs and ceratopsians of Asia. In M. J. Benton, M. A. Shishkin, D. M. Unwin, and E. N. Kurochkin (eds.), *The Age of Dinosaurs in Russia and Mongolia*, 480–516. Cambridge: Cambridge University Press.

Sereno, P. C., and J. A. Wilson. 2005. Structure and evolution of a sauropod tooth battery. In K. A. Curry Rogers, and J. A. Wilson (eds.) *The Sauropods: Evolution and Paleobiology*, 157–177. Berkeley: University of California Press.

Sereno, P. C., D. B. Dutheil, M. Iarochene, H. C. E. Larsson, G. H. Lyon, P. M. Magwene, C. A. Sidor, D. J. Varricchio, and J. A. Wilson. 1996. Predatory dinosaurs from the Sahara and Late Cretaceous faunal differentiation. *Science* 272(5264): 986–991.

Sereno, P. C., A. L. Beck, D. B. Dutheil, B. Gado, H. C. E. Larsson, O. W. M. Rauhut, R. W. Sadleir, C. A. Sidor, D. J. Varricchio, G. P. Wilson, and J. A. Wilson. 1998. A long-snouted predatory dinosaur from Africa and the evolution of spinosaurids. *Science* 282: 1298–1302.

Sereno, P. C., A. L. Beck, D. B. Dutheil, H. C. E. Larsson, G. H. Lyon, B. Moussa, R. W. Sadleir, C. A. Sidor, D. J. Varricchio, G. P. Wilson, and J. A. Wilson. 1999. Cretaceous sauropods from the Sahara and the uneven rate of skeletal evolution among dinosaurs. *Science* 286: 1342–1347.

Smith, J. B., and M. C. Lamanna. 2006. An abelisaurid from the Late Cretaceous of Egypt: implications for theropod biogeography. *Naturwissenschaften* 93/11: 242–245

Stromer, E. 1915. Ergebnisse der Forschungsreisen Prof. E. Stromers in den Wüsten Ägyptens. II. Wirbeltier-Reste der Baharîje-Stufe (unterstes Cenoman). 3. Das Original des Theropoden *Spinosaurus aegyptiacus* nov. gen., nov. spec. *Abhandlungen der Königlich Bayerischen Akademie der Wissenschaften Mathematisch-physikalische Klasse Abhandlung* 28(3): 1–31.

———. Ergebnisse der Forschungsreisen Prof. E. Stromers in den Wüsten Ägyptens. II. Wirbeltier-Reste der Baharîjestufe (unterstes Cenoman). 10. Ein Skelett-Rest von *Carcharodontosaurus* nov. gen. *Abhandlungen der Bayerischen Akademie der Wissenschaften Mathematisch-naturwissenschaftliche Abteilung, Neue Folge* 9: 1–23.

———. 1932. Ergebnisse der Forschungsreisen Prof. E. Stromers in den Wüsten Ägyptens. II. Wirbeltierreste der Baharîje-Stufe (unterstes Cenoman). 11.

Sauropoda. *Abhandlungen der Bayerischen Akademie der Wissenschaften Mathematisch-naturwissenschaftliche Abteilung, Neue Folge* 10: 1–21.

Sues, H.-D. 2012. Early Mesozoic continental tetrapods and faunal changes. In M. K. Brett-Surman and J. O. Farlow (eds.), *The Complete Dinosaur, 2nd ed.*, 988–1002. Bloomington: Indiana University Press.

Sues, H.-D., and P. M. Galton. 1982. The systematic position of *Stenopelix valdensis* (Reptilia: Ornithischia) from the Wealden of north-western Germany. *Palaeontographica Abteilung A* 178 (4–6): 183–190.

Sullivan, R. M. 2006. A taxonomic review of the Pachycephalosauridae (Dinosauria: Ornithischia). In S. G. Lucas and R. M. Sullivan (eds.), *Late Cretaceous Vertebrates from the Western Interior*, 347–365. Bulletin 35. Albuquerque: New Mexico Museum of Natural History and Science.

Taquet, P. 1976. Géologie et paléontologie du gisement de Gadoufaoua (Aptien du Niger). *Cahiers de Paléontologie* 1976: 1–191.

Taquet, P., and D. A. Russell. 1998. New data on spinosaurid dinosaurs from the Early Cretaceous of the Sahara. *Comptes Rendus de l'Académie des Sciences à Paris, Sciences de la Terre et des Planètes* 327: 347–353.

———. 1999. A massively-constructed iguanodont from Gadoufaoua, Lower Cretaceous of Niger. *Annales de Paléontologie* 85(1): 85–96.

Thulborn, R. A. 2003. Comment on "Ascent of dinosaurs linked to an iridium anomaly at the Triassic-Jurassic boundary." *Science* 301: 169b.

Thulborn, T., and S. Turner. 2003. The last dicynodont: an Australian Cretaceous relict. *Proceedings of the Royal Society of London B* 270: 985–993.

Unwin, D. M. 1999. Pterosaurs: back to the traditional model? *Trends in Ecology and Evolution* 14 (7): 263–268.

Upchurch, P. and P. M. Barrett. 2000. The evolution of sauropod feeding mechanisms. In H.-D. Sues (ed.), *Evolution of Herbivory in Terrestrial Vertebrates: Perspectives from the Fossil Record*, 79–122. Cambridge: Cambridge University Press.

Upchurch, P., and J. Martin. 2003. The anatomy and taxonomy of *Cetiosaurus* (Saurischia, Sauropoda) from the Middle Jurassic of England. *Journal of Vertebrate Paleontology* 23 (1): 208–231.

Vickers-Rich, P. 1996. Early Cretaceous polar tetrapods from the Great Southern

Rift Valley, southeastern Australia. *Memoirs of the Queensland Museum* 39 (3): 719–723.

Wang, X., A. W. A. Kellner, Z. Zhou, and D. d. A. Campos. 2005. Pterosaur diversity and faunal turnover in Cretaceous terrestrial ecosystems in China. *Nature* 437: 875–879.

Weishampel, D. B. 1984. Evolution of jaw mechanics in ornithopod dinosaurs. *Advances in Anatomy, Embryology and Cell Biology* 87: 1–110.

Weishampel, D. B., P. Dodson, and H. Osmólska (eds.). 2004. *The Dinosauria* (2nd ed.). Berkeley: University of California Press.

Weishampel, D. B., P. M. Barrett, R. A. Coria, J. Le Loeuff, X. Xu, X.-J. Zhao, A. Sahni, E. Gomani, and C. R. Noto. 2004. Dinosaur distribution. In D. B. Weishampel, P. Dondson, and H. Osmólska (eds.), *The Dinosauria*, 517–606. Berkeley: University of California Press.

Wilson, J. A. 2002. Sauropod dinosaur phylogeny: critique and cladistic analysis. *Zoological Journal of the Linnean Society* 136: 217–276.

Woodward, A. S. 1910. On a skull of *Megalosaurus* from the Great Oolite of Minchinhampton (Gloucestershire). *Quarterly Journal of the Geological Society of London* 66 (262): 111–115.

Xu, X., and F. Zhang. 2005. A new maniraptoran dinosaur from China with long feathers on the metatarsus. *Naturwissenschaften* 92: 173–177.

Xu, X., Z.-l. Tang, and X.-l. Wang. 1999. A therizinosauroid dinosaur with integumentary structures from China. *Nature* 399: 350–354.

Xu, X., C. A. Forster, J. M. Clark, and J. Mo. 2006. A basal ceratopsian with transitional features from the Late Jurassic of northwestern China. *Proceedings of the Royal Society of London B* 273: 2135–2140.

Zaher, H., and O. C. Rieppel. 2002. On the phylogenetic relationships of the Cretaceous snakes with legs, with special reference to *Pachyrhachis problematicus* (Squamata, Serpentes). *Journal of Vertebrate Paleontology* 22 (1): 104–109.

Zhao, X., Z. Cheng, and X. Xu. 1999. The earliest ceratopsian from the Tuchengzi Formation of Liaoning, China. *Journal of Vertebrate Paleontology* 19 (4): 681–691.

Zhen, S., B. Zhen, N. J. Mateer, and S. G. Lucas. 1985. The Mesozoic reptiles of China. *Bulletin of the Geological Institute of the University of Uppsala, new series* 11: 133–150.

Zhou, Z., P. M. Barrett, and J. Hilton. 2003. An exceptionally preserved Lower Cretaceous ecosystem. *Nature* 421: 807–814.

Ziegler, A. M. and D. B. Rowley. 1998. The vanishing record of epeiric seas, with emphasis on the Late Cretaceous "Hudson Seaway." In T. J. Crowley and K. Burke (eds.), *Tectonic Boundary Conditions for Climate Reconstructions*, 147–165. Oxford: Oxford University Press.

Zink, R. M., R. C. Blackwell-Rago, and F. Ronquist. 2000. The shifting roles of dispersal and vicariance in biogeography. *Proceedings of the Royal Society of London B* 267: 497–503.

Dinosaur Extinction: Past and Present Perceptions

43

J. David Archibald

When I ask most schoolchildren what happened to the dinosaurs, they tell me that dinosaurs became extinct because a giant asteroid hit the Earth and instantly wiped them all out. When I then ask them if birds are related to dinosaurs, they most often answer yes, many saying that birds are dinosaurs. If they are then confronted with the quandary that if birds are dinosaurs and birds are alive today, how can dinosaurs be extinct, I am most often greeted with a stare or a grin. A very few respond that it was only the big dinosaurs on the ground that became extinct but birds did not. Although inelegant, these student answers are mostly correct.

Many adults do not grasp this distinction, mostly because they were raised to believe that dinosaurs and birds are not closely related, and that such a diverse group as birds that is unique in its mode of flight surely must be its own group. Most research and researchers have shown that birds and other dinosaurs are close kin; still, putting birds in a group with other dinosaurs is too much to swallow. What has happened in this apparent dilemma is that biological distinctiveness and evolutionary relationship have been confounded. Yes, birds are very distinct and in a number of ways unique, but we do not classify species on unique characteristics that they alone possess, but rather on derived characteristics they share uniquely with other species. For example, flight appears to be a character uniquely derived in birds, but feathers appears to be uniquely derived in birds as well as some of their nonflighted relatives that we call dinosaurs. This means that there is a group for birds, Aves; a more inclusive group, Theropoda, including Aves and a variety of both feathered and nonfeathered dinosaurs; and an even more inclusive group, Dinosauria, including theropods and other groups, such as sauropods, hadrosaurs, and ceratopsians.

This question of group, or more correctly, clade or lineage identification is an important issue for the study of extinction because we must precisely define a particular group of species, what biologists call a *taxon*, before we can study such issues as its demise. Thus, for Dinosauria we can say with considerable certainty that all onithischians (bird-hipped dinosaurs), all sauropods, and all theropods except for some uncertain number of bird species perished at or shortly following the Cretaceous/Tertiary (K/T) or Cretaceous/Paleogene (K/PG) boundary. The rate at which this occurred, whether it was a global phenomenon, how many species were involved, and how other species were affected are all less well known and are the subjects of this chapter.

A final misconception that requires discussion is the quality of the fossil record immediately preceding and following the extinction of dinosaurs. One may believe that we possess a good global record of dinosaur

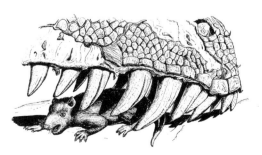

extinction. This is unfortunately not the case. New studies in other parts of the world show promise, but in none of these do we have a good vertebrate fossil record up to and including the K/T boundary. The only region in the world where we do have good record is in the Western Interior of North America. Localities in New Mexico and Texas preserve some aspects of the K/T boundary fossil record, but by far the best records are from Alberta, Saskatchewan, Montana, and the Dakotas (Archibald, 1996, 2011).

Past Theories

Theories of the extinction of nonavian dinosaurs started appearing soon after they were discovered and named in the first half of the nineteenth century. Three of the better-known compilers of extinction theories were Glenn Jepsen in the 1960s, Alan Charig in the 1970s and 1980s, and Michael Benton in the 1990s. Depending on how one parses their lists, Jepsen (e.g., 1963) listed something like 48 theories in one long, paragraph-length sentence. The highest number of extinction theories claimed to have been complied were the 80 or so reported to Archibald (1996) by Alan Charig. Although Charig (1983, 1989) discussed various theories, an extensive list was not published. In 1990, Benton provided a much more methodically organized and referenced list of possible causes. He also provided opinions as to whether theories were "deliberate jokes," "speculative ideas," or "supported by some evidence." Table 43.1 is a slightly modified and updated version of Benton's tabulation. Again, exact numbers are somewhat difficult to compute, but Benton tabulates about 66 theories, not including 10 or so variations.

With so many theories, it is no wonder that dinosaur extinction remains one of the perennially hot topics concerning dinosaurs. I will not detail all the theories, but will provide a short synopsis of three of the more widely discussed. Even these three include a combination of factors listed in Table 43.1. Before turning to these current theories, it is worthwhile to examine what we know and what we think we know concerning the pace of dinosaur extinction—was it gradual, instantaneous, or something in between? The question of the pace of the extinctions has engendered debate second only to that surrounding the cause or causes of the extinctions.

Past and Present Evidence for and against Gradual Dinosaur Extinction

With the record of the extinction of dinosaurs currently limited to western North America, it is not surprising this is where researchers have most frequently turned to assess the speed at which dinosaurs became extinct. Two major questions are posed. The first question is whether the number of dinosaur species declined toward the later part of the Cretaceous, usually meaning the last 10 million years of the Cretaceous. The second is whether the number of species declined gradually, very rapidly, or somewhat in between during the time leading up to the end of the Cretaceous, usually the last days or months up to the last hundreds or tens of thousands of years of the Cretaceous.

In this section, I address what can be called the gradual arguments for these changes. Opinions vary on what happened to the number of dinosaur species during the last 10 million years of the Cretaceous (in the Western Interior). A widely held view is that the number of genera and

Table 43.1. Various theories of dinosaur extinction, mostly after Benton (1990) with modifications. Many of the sources for these theories can be found in Benton (1990).

I. Biotic causes
  1. Medical problems
    A. Metabolic disorders
      a. Slipped vertebral discs
      b. Malfunction or imbalance of hormone systems
        i. Overactivity of pituitary gland and excessive (acromegalous?) growth of
          bones and cartilage
        ii. Malfunction of pituitary gland, leading to excess growth of unnecessary and
          debilitating horns, spines, and frills
        iii. Imbalances of vasotocin and estrogen levels, leading to pathological
          thinning of egg shells
    B. Diminution of sexual activity
    C. Cataract blindness
    D. Disease: caries, arthritis, fractures, and infections at maximum in Late Cretaceous reptiles
    E. Epidemics
    F. Parasites
    G. AIDS caused by increasing promiscuity
    H. Change in ratio of DNA to cell nucleus
    I. Mental disorders
      a. Dwindling brain and consequent stupidity
      b. Absence of consciousness and ability to modify behavior
      c. Development of psychotic suicidal factors
      d. Paleo-weltschmerz
    K. Genetic disorders: excessive mutation rate induced by high levels of cosmic rays or ultraviolet
      light, leading to a small population size burdened by a high genetic load, and consequent
      vulnerability to environmental shock
  2. Racial senility (phylogeronty)
    A. Evolutionary drift into senescent overspecialization, as evinced in gigantism, spinescence
      (e.g., loss of teeth, and degenerate form)
    B. Racial old age (Will Cuppy [1964]: "the Age of Reptiles ended because it had gone on long
      enough and it was all a mistake in the first place")
    C. Increasing levels of hormone imbalance leading to ever-increasing growth of unnecessary
      horns and frills
  3. Biotic interactions
    A. Competition with other animals
      a. Competition with the Asian mammals that invaded North America
      b. Competition with caterpillars, which ate all of the plants
    B. Predation
      a. Overkill capacity by predators (carnosaurs ate themselves out of existence)
      b. Egg eating by mammals, which reduced hatching success of the young and
        drained gene pools
    C. Floral changes
      a. Spread of angiosperms and reduction in availability of gymnosperms, ferns, etc., leading
        to a reduction of fern oils in dinosaur diets and lingering death by terminal constipation
      b. Floral change and loss of marsh vegetation
      c. Floral change and increase in forestation, leading to a loss of habitat
      d. Reduction in availability of plant food as a whole
      e. Presence of poisonous tannins and alkaloids in the angiosperms
      f. Presence of other poisons in plants
      g. Lack of calcium and other necessary minerals in plants
      h. Rise of angiosperms and their pollen, leading to extinction of dinosaurs by
        terminal hay fever
II. Abiotic (physical) causes
  1. Terrestrial explanations
    A. Climatic change
      a. Too hot because of high levels of carbon dioxide in the atmosphere and the
        "greenhouse effect"
        i. High temperature and increased aridity
        ii. Spermatogenesis inhibited
        iii. Unbalanced male:female ratio of hatchlings
        iv. Juveniles killed
        v. Overheating in summer, especially if dinosaurs were endothermic
      b. Too cold
        i. Embryonic development inhibited
        ii. Constant body temperature impossible to maintain by endothermic dinosaurs
          lacking insulation
        iii. Hibernation during cold impossible due to size of dinosaurs
        iv. Cold winter temperatures fatal even to inertial homeotherms (i.e., not endotherms)

Table 43.1. *continued*

    c. Too dry
    d. Too wet
    e. Reduction in climatic equability and increase in seasonality
  B. Atmospheric change
    a. Changes in the pressure or composition of the atmosphere (e.g., excessive amounts of oxygen from photosynthesis)
    b. High levels of atmospheric oxygen, leading to fires following an impact
    c. Low levels of carbon dioxide, removing the "breathing stimulus" of endothermic dinosaurs
    d. Excessively high levels of carbon dioxide, causing asphyxiation of dinosaur embryos in the eggs
    e. Extensive vulcanism and the production of volcanic dust
    f. Poisoning by selenium from volcanic lava and dust
    g. Toxic substances in the air, possibly produced from volcanoes, causing thinning of dinosaur egg shells
  C. Oceanic and topographic change
    a. Marine regression
    b. Lowering of global sea level, leading to dinosaur extinction, assuming they were underwater organisms
    c. Floods
    d. Mountain building, for example, the Laramide Revolution
    e. Drainage of swamp and lake habitats
    f. Stagnant oceans caused by high levels of carbon dioxide
    g. Bottom-water anoxia at start of transgression
    h. Spillover of Arctic water (fresh) from its formerly enclosed condition into the oceans, leading to reduced temperatures worldwide, reduced precipitation, and a 10-year drought
    i. Reduced topographic relief
    j. Reduction in number of terrestrial habitats
    k. Fragmentation of terrestrial habitats
  D. Other terrestrial catastrophes
    a. Sudden vulcanism
    b. Fluctuation of gravitational constants
    c. Shift of the earth's rotational poles
    d. Extraction of the moon from the Pacific Basin
    e. Poisoning by uranium sucked up from the soil
2. Extraterrestrial explanations
  A. Entropy; increasing chaos in the Universe and hence loss of large organized life forms
  B. Sunspots
  C. Cosmic radiation and high levels of ultraviolet radiation
  D. Destruction by solar flares of the ozone layer, letting in ultraviolet radiation
  E. Ionizing radiation
  F. Electromagnetic radiation and cosmic rays from the explosion of a nearby supernova
  G. Interstellar dust cloud
  H. Flash heating of atmosphere by entry of meteorite
  I. Oscillations about the galactic plane
  J. Impact of an asteroid, comet, or comet showers, causing extinction by a number of postulated mechanisms
    a. Acid rain
    b. Months or years of freezing temperatures
    c. Global wildfires
    d. Regional wildfires
    e. Super hurricanes
    f. Sudden increase in temperature
    g. Airborne debris
    h. Debris in freshwater streams
    i. Months or years of darkness
    k. Tsunamis
3. Multiple extraterrestrial or terrestrial factors
  A. Bolide impact triggering or increasing volcanic activity with concomitant results
  B. Volcanic activity, followed by marine regression and habitat fragmentation, and bolide impact with concomitant results.

species declined considerably over this 10 million years (e.g., Sullivan 1998; Archibald, 2011; Archibald and MacLeod, 2007). Usually, this comparison is based on the two best-studied Late Cretaceous dinosaur faunas in the Western Interior: the approximately 75-million-year-old (middle to late Campanian) fauna from Dinosaur Provincial Park ("Judith River" in Archibald

1996) of southern Alberta and the approximately 66-million-year-old (late Maastrichtian) Hell Creek fauna of eastern Montana and western Dakotas. In 1996, Archibald indicated that the absolute generic diversity of dinosaurs decreased by 40% (dropping from 32 to 19) from the Judith River to the Hell Creek dinosaur fauna. Weishampel and colleagues (2004) provided a more recent taxonomic compilation, which included an intermediately aged fauna (early Maastrichtian) from the Horseshoe Canyon Formation, southern Alberta, which can be added to these computations. According to Weishampel et al. (2004), going from oldest to youngest, the Dinosaur Park Formation has yielded 32 named genera, the Horseshoe Canyon Formation has yielded 25 named genera, and the Hell Creek Formation has yielded 18. In each successive fauna there is turnover, but there is also a net loss of seven genera between each successive fauna. This translates to a 20% decline from the Dinosaur Park Formation fauna to the Horseshoe Canyon Formation fauna, and a 28% decline from the Horseshoe Canyon Formation fauna to the Hell Creek Formation fauna, for a total decline of almost 44% during the last 10 million years of the Cretaceous in at least the northern part of the Western Interior of North America. The often rarer nonavian saurischians (reptile-hipped dinosaurs), which at this time and place were represented by only theropods, declined by almost half during this interval (from eleven to six genera). Of particular interest was the dramatic reduction in numbers of species of the large, probably herding ornithischian ceratopsids (from five to two) and hadrosaurids (from seven to two) (Archibald 1996; Weishampel et al., 2004). Even most recent studies still continue to support the clear evidence for the decline of nonavian dinosaurs in the Late Cretaceous of North America (Archibald, 2011).

As noted, the second issue related to the rate of dinosaur extinction is whether dinosaurs declined gradually, very rapidly, or not at all as one approaches the к/т boundary. Unlike the previous discussion that dealt with the last 10 million years of the Cretaceous, this interval of time is the last one million or so years of the Cretaceous. The last study to argue that there was gradual decline in the number of species of dinosaurs in the last few million years of the Cretaceous was that of Sloan and colleagues in 1986 in the area of Bug Creek in eastern Montana. They based this on an examination of the number and variety of isolated dinosaur teeth, which they argued decrease as one approaches the к/т boundary in eastern Montana. Later it was shown that much of the section they used in their analysis includes Paleocene channel deposits that had cut into underlying Cretaceous sediments (Lofgren 1995). Thus, dinosaur teeth in the Bug Creek sequence are usually regarded as having been reworked, and any decrease at subsequently higher localities would simply be the result of fewer and fewer reworked teeth. These were called *Zombie taxa* by Archibald (1996).

Another view for the same 10-million-year interval suggests dinosaur diversity remained the same or even increased. Russell (1984) has been one of the major advocates of this idea. He argued that the Dinosaur Park Formation ("Judith River Formation" in Russell 1984) is much better sampled than the Hell Creek Formation. Thus, according to him, any apparent decline is the result of a discrepancy in sampling effort. Moreover, a comparison of faunas

Past and Present Evidence for and against Catastrophic Dinosaur Extinction

that preceded the Dinosaur Park fauna as well as those that succeeded it suggests that the Dinosaur Park fauna possesses an anomalously diverse dinosaur fauna. According to this reading of the fossil record by Russell, one would need to normalize for the number of localities in the Dinosaur Park Formation as well as the Hell Creek Formation to obtain meaningful comparative data between these fossil-bearing units.

More recently, Russell and Manabe (2002) provided evidence supporting the view that there is a decline in dinosaur taxa going from the approximately 75-million-year-old Dinosaur Park Formation to the 66-million-year-old Hell Creek Formation. Their numbers of named dinosaur genera differ somewhat from those cited in the previous section by Weishampel and colleagues (2004). Russell and Manabe (2002) recognized 30 named dinosaur genera from the Dinosaur Park Formation and 21 from the Hell Creek Formation. They indicated that the diversity of small dinosaurs was similar in the two faunas, but that the Hell Creek assemblage had less than half of the large dinosaurs found in the Dinosaur Park assemblage. These results indicate that Russell has modified his earlier views that there was no evidence for decline of dinosaurs in the last 10 million years of the Cretaceous in North America.

In 1990, Dodson quantified data on nonavian dinosaurs throughout their 160-million-year existence. He concluded that during the last 10 or so million years of the Cretaceous there was no evidence to indicate that dinosaurs were in decline, but also that the data do not support either gradual or abrupt extinction.

In addition to those arguing for a gradual decline of dinosaurs during the very last years of the Cretaceous, there are also those who argue that there was no such decline of dinosaurs during this interval. Thus, dinosaur extinction is commensurate with catastrophic events. In 1991, Sheehan and colleagues published a study purporting to show no decline in either kinds of dinosaurs or numbers of individuals as one approaches the K/T boundary. They wrote that they were tracking the diversity of eight families of dinosaurs vertically through the uppermost Cretaceous Hell Creek Formation. They argued that they were testing whether they could detect any change in the relative abundance of individuals or families of dinosaurs as one approaches the K/T boundary. They reasoned that if there were no change, the diversity of dinosaurs did not decrease when approaching the K/T boundary. If, however, change were detectable in the relative abundances, there was a diversity change approaching the K/T boundary. The study concluded that there was no discernible change, which they argued is compatible with theories of catastrophic extinction.

Unfortunately, their methodology precluded them from addressing the question they wished to ask—whether there was discernible change. According to these authors, the eight families of dinosaurs that they included in their study represent 14 genera. They used familial-level data in their analysis rather than the 14 genera because they felt that generic-level data could be misleading and that many fossils must be excluded because they were not identifiable to generic level. This means genera could become extinct within the Hell Creek without being detected by the authors. The distribution of genera in families used in this study shows that six of the 14 genera could have become extinct without any drop in familial diversity. This is a 43% generic extinction without the loss of a single family. Thus, this study used a taxonomic level (family) that is far too coarse to detect any decline in dinosaur diversity as one approaches the K/T boundary, if there were such a decline. There were other problems with this study. As shown by Hurlbert and Archibald (1995), the statistical approaches used in the Sheehan et al. (1991) study were not capable of detecting whether there was a decline, an increase, or no change in dinosaur diversity. Also, most of the specimens identified in the field in the Sheehan et al. (1991) study were not collected, so it is impossible to verify identifications (Pearson et al. 2002).

A more recent study published by Pearson and colleagues (2002) collected over 10,000 vertebrate specimens from microsites, and parts of 41 dinosaur skulls from most of the stratigraphic extent of the Hell Creek Formation, western Dakotas. As in most other studies (e.g., Archibald 1996), they found that the uppermost two meters is essentially devoid of fossils. Using a rarefaction analysis, they found no evidence for "a decline in vertebrate diversity through the formation or dinosaur diversity in the three meters below the K/T boundary" (145). They argued that their results are not commensurate with gradual vertebrate extinctions at the end of the Cretaceous. In this study, they recovered some 61 vertebrate taxa, only about two-thirds of which were identified to at least the generic level. This contrasts with the next most recent study that identified at least 107 vertebrate taxa to at least the generic level (Archibald 1996). This earlier study, however, lacked the stratigraphic completeness of the Pearson et al. (2002) study. When all 61 vertebrates are considered, the Pearson et al. (2002) study is quite convincing as to the relative stability of the fauna throughout the Hell Creek Formation. When dinosaurs alone are considered, however, the picture is somewhat different. Of the 13 nonavian dinosaurs recognized in this study, only eight are identified below the family level. Thus, the same taxonomic coarseness found with the Sheehan et al. (1991) study might also pertain here. Second, the actual pattern of the disappearance of these 13 taxa below the K/T boundary is as follows (Pearson et al. 2002, fig. 4): seven disappear between 5 and 0 m below (one is in fact found above 0 m, but still in Cretaceous aged rocks), three disappear between 10 and 5 m, and two disappear between 15 and 10 m, while the final two are gone between 40 and 35 m below. Thus, the disappearances of nonavian dinosaurs show a more stepped pattern than the overall more static pattern for all vertebrates combined. Whether this should be read literally as a gradual or stepwise extinction or can be treated statistically as indistinguishable from a simultaneous extinction remains an open question. The vertebrate fossil record, especially of large, rare taxa such as dinosaurs, is certainly poorer than the marine record of invertebrates. Yet even for invertebrates at the K/T boundary, Payne (2003) found "that the data are not always sufficient to rule out either stratigraphically simultaneous extinctions or a literal reading of extinction rate as representing gradual or episodic extinction" (37). The Pearson et al. (2002) study shows that the vertebrate fauna of the Hell Creek Formation appears to have been stable throughout most of its deposition. It does not, however, address the question of whether the disappearances of nonavian dinosaurs and other vertebrates were gradual or sudden.

Coda on the Rates of Dinosaur Extinction

The previous two sections examined past and present evidence concerning the diversity of dinosaurs during (1) the last 10 million years of the Cretaceous and (2) the waning one million years or so before and at the K/T boundary. For the longer-term pattern, the evidence continues to support a decline of 30 to 45% generic diversity of dinosaurs. For the shorter-term pattern, better support is emerging that vertebrate diversity held steady in the waning days of the Cretaceous, but the evidence for gradual versus catastrophic dinosaur extinction remains unclear. Both of these patterns are biased in that they are from the northern Western Interior of North America.

Current Theories

Here I examine three of the major causes proposed for the extinction of dinosaurs—asteroid (or more generally, bolide) impact, marine regression and habitat regression, and massive eruptions of flood basalts. None of these are mutually exclusive, and a number of scenarios have been proposed that combined parts of each of the three. In general, the three causes, in the order listed above, range from very short to very long term in both duration and effect.

The most catastrophic of the three causes is the asteroid impact theory (Alvarez et al. 1980) that argues that a 10-km (6-mi) asteroid struck the Earth 66 mya, producing ejecta and a plume reaching far enough into the atmosphere to spread around the globe, blocking the Sun. The cessation of photosynthesis resulted in death and extinction of many plants, the herbivores that fed on them, and the carnivores that fed on the herbivores (but see Pope 2002). The probable crater, named Chicxulub, has been located near the tip of the Yucatan Peninsula (e. g., Hildebrand 1993). At 180 km (110 mi) across, it is one of the larger such structures on Earth. In addition to the crater, two other important pieces of physical evidence supporting an impact are an increase in the element iridium at the K/T boundary, and minerals, especially quartz grains, showing shocked lamellae in two directions. A high level of iridium at the Earth's surface and double lamellae are both more indicative of an impact than volcanism. Some of the more proximate effects of an asteroid impact include acid rain, globe wildfire, sudden temperature increases or decreases, infrared radiation, tsunamis, and super-hurricanes (e.g., Archibald, 2011; Archibald and Fastovsky, 2004).

The next cause, marine regression and habitat fragmentation, occurred over a longer interval of time, from tens to hundreds of thousands of years. *Marine regression* refers to draining of epicontinental seas. One of the greatest such regressions is recorded in rocks near the end of the Cretaceous Period some 66 mya. Estimates suggest that 29 million square km (11.2 million square mi) of land were exposed during this interval (Smith et al. 1994), more than twice the next largest such addition of land during the past 250 million years. This land mass is approximately the size of Africa. There were marked proximate effects of this regression—major loss of low coastal plain habitats, fragmentation of the remaining coastal plains, establishment of land bridges, extension of freshwater systems, and climatic change with a general trend towards cooling on the newly emerged landmasses.

Massive eruptions of flood basalts, the Deccan Traps, on the Indian subcontinent occurred over a much longer interval than marine regression, perhaps millions of years surrounding the к/т boundary (Courtillot 1999). The volume of material estimated to have been erupted over this 4-million-year interval would cover both Alaska and Texas to a depth of 610 m (2000 ft). Proximate causes resulting from such massive volcanism have not been as well studied as those for marine regression or asteroid impact, but newer studies are underway (Keller et al. 2008). Proximate causes, however, have been argued to be similar for both impact and volcanism. Climatic changes caused by massive eruptions would have been longer term.

To test these three causes, one must examine what happened not only to dinosaurs, but also to other species that lived with these creatures. Such a study (Archibald and Bryant, 1990; Archibald, 1996, 2011) examined the best-known к/т sections located in eastern Montana. The assemblage included 107 well-documented species of vertebrates belonging to 12 major monophyletic lineages. Of these species, only 19 are nonavian dinosaurs. Because of a poor fossil record, pterosaurs and birds (a clade of saurischian dinosaurs) could not be included in this sample. Although there has been debate as to whether a number of major clades of modern birds appeared before the к/т boundary, Chiappe (2007) argued that unquestioned members of modern bird clades are not known in the fossil record until the early Tertiary, whereas Lindlow (2011) supports the appearance of some modern clades before the к/т boundary. Pterosaurs, however, had dwindled to at most one family (Archibald 2011). Of the 107 species of vertebrates, 52 (49%) survived the к/т boundary. This level of survival is only about 10% lower than that for similar intervals before and after the к/т boundary. The amount of survival at the к/т boundary for vertebrates in the Western Interior appears to have been higher than that for a number of other major groups of plants and animals. Nevertheless, the effects on vertebrates were profound. Whether the extinction of nonavian dinosaurs was gradual or catastrophic, nonavian dinosaurs, the large land vertebrates for much of the Mesozoic Era, were replaced by mammals in probably only a few thousands or tens of thousands of years. Mammals, including our own clade, Primates, rapidly diversified, filling niches left by dinosaurs and creating new ones.

The most obvious pattern of extinction among these 12 major vertebrate clades is that extinctions were concentrated in five of the clades: sharks and relatives, lizards, marsupials, ornithischians, and saurischians. Species in these five clades account for 75% of the extinctions. This demonstrates that the к/т extinctions were highly selective; any theory of extinction must account for this selectivity.

Using the vertebrate fossil record, we cannot test directly whether the seas underwent a major regression, whether there were major volcanic eruptions, or whether an asteroid struck Earth. We can instead test the various proximate causes that have been proposed as the result of these ultimate causes. Starting with the asteroid impact, we can test the suggestions of acid rain, sharp temperature decrease, and global wildfire. (Because the biotic effects of volcanic eruption have not been explored as extensively and

The Pattern of K/T Vertebrate Extinctions

Comparing the K/T Vertebrate Record to the Possible Causes

are considered similar to the effects of impact, this will not be discussed further.) We know from work on extant species and habitats that among vertebrates acid rain hurts aquatic organisms most. Except for sharks and relatives, however, aquatic species did very well through the K/T transition. If a sharp, short temperature spike occurred, we know from modern vertebrates that wholly or partially terrestrial ectotherms should have been most affected. Once again, most of these, except lizards, did well through the K/T boundary.

Whether dinosaurs should be considered endotherms, ectotherms, or another kind of physiology remains controversial. Finally, global wildfire, which is argued to have consumed 25% of all above ground biomass, should have been a nearly equal opportunity killer, outright burning many creatures and suffocating others in aquatic systems with a hyper-influx of detritus. In addition to the lack of substantial amounts of charcoal at the K/T boundary (Belcher et al. 2003), the fossil record is highly selective, rendering global wildfire a very unlikely scenario. A global fire had been suggested as a result of a thermal pulse, which also would have outright killed all unprotected plants and animals (Robertson et al. 2004) until it was realized that much of such a pulse would have been blocked by any incoming material (Goldin and Melosh 2009).

Of all of these proximate causes of an asteroid impact, what seems most important is the blocking of sunlight, causing a cessation of photosynthesis. This had been considered enough to kill and cause the extinction of up to 80% of the plants species in at least some areas of North America. By 2004, it was argued (Wilf and Johnson 2004) that the plant extinction rate in at least the middle portion of North America would have been only 57%. Even this lower level would have had a devastating effect on large herbivores, especially if already stressed by other events, such as marine regression and habitat fragmentation.

As global marine regression began in the last few million years of the Cretaceous, tremendous new tracts of dry land were added. But all known dinosaur-bearing vertebrate localities near the K/T boundary are from coastal plain habitats that were being drastically reduced during marine regression. This reduced dramatically the size of these habitats, stranding dinosaurs in ever-smaller areas, much as humans are doing to the habitats of large mammals in Africa today.

This stressed dinosaur populations, setting them up for the final biotic insult caused by even a smaller asteroid impact and massive volcanism. At the same time, the coastlines were retreating away from the Western Interior, taking the sharks and relatives with them. They could traverse freshwater courses up to a few hundreds of kilometers, but not thousands of kilometers. The marine connection was severed. Although low coastal streams disappeared, the total freshwater systems held their own and, in many cases, increased in length as the coastline retreated. Thus, freshwater species did very well, with descendants such as paddlefish, sturgeon, and gar still plying the Missouri–Mississippi river systems.

The lowering of sea level also reconnected once separated landmasses, such as eastern Asia and western North America. The earliest ungulate relatives, first known from North America, appear at this time. A very rare occurrence of the early archaic ungulate, *Protungulatum*, is definitely known from the Hell Creek Formation in Montana some 300,000 years before the K/T boundary (Archibald et al. 2011). Possibly latest Cretaceous archaic ungulates are also reported from sites in Canada (Fox 1989). These new ungulates have a dental morphology that resembles that of the opossum-like marsupials, which arose in North America some 100 mya and had been very common for at least the 20 million years leading up to the K/T boundary, 66 mya. It seems likely that the appearance of these ungulates in North America spelled competitive doom for the marsupials. Interestingly, when they both appear in South America a few million years after the K/T boundary, they have spilt the difference, with the ungulates becoming more strictly herbivores while the marsupials became omnivores and carnivores, including large saber-toothed marsupial cats. The one group that cannot be explained by this globally testable hypothesis is the drastic reduction of lizards in the Western Interior. A more local, ad hoc explanation is that the suggested climatic shift to a wetter, rainier climate in areas such as eastern Montana following the K/T boundary may have driven out the more dry-adapted lizards.

Summary

Clearly, among the public, asteroid impact is the most widely recognized cause of dinosaur extinction. It is also popular among scientists, but there remain many unanswered questions regarding its killing mechanisms. For some time after the impact theory was purposed in 1980, impacts by various bolides were argued to have caused cyclic extinctions, including the five mass extinctions in the past 550 million years. More analysis, however, has caused the idea of cyclical episodes of extinction to fade. Although impacts on Earth are quite common, the evidence now indicates that such impacts do not correlate well with episodes of mass extinction, while both marine regression and massive volcanism do correlate well with such episodes (MacLeod 2004). All three of these events occurring at the K/T boundary provided a unique set of events in the history of planet Earth.

References

Alvarez, L. W., W. Alvarez, F. Asaro, and H. Michel. 1980. Extraterrestrial cause for the Cretaceous-Tertiary extinction. *Science* 208: 1095–1108.

Archibald, J. D. 1996. *Dinosaur Extinction and the End of an Era: What the Fossils Say.* New York: Columbia University Press.

———. 2011. *Extinction and Radiation: How the Fall of the Dinosaurs Led to the Rise of the Mammals.* Baltimore: The Johns Hopkins University Press.

Archibald, J. D. and L. Bryant. 1990. Differential Cretaceous-Tertiary extinctions of non-marine vertebrates: Evidence from northeastern Montana. In V. L. Sharpton and P. Ward (eds.), *Global Catastrophes in Earth History: An Interdisciplinary Conference on Impacts,* *Volcanism, and Mass Mortality.* Geological Society of America, Special Paper 247: 549–562.

Archibald, J. D. and D. E. Fastovsky. 2004. Dinosaur extinction. In D. B. Weishampel, P. Dodson, and H. Osmólska (eds.), *The Dinosauria,* 2nd ed., 672–684. Berkeley: University of California Press.

Archibald, J. D. and N. MacLeod. 2007. Dinosaurs, extinction theories for. *Encyclopedia of Biodiversity,* Elsevier, 1–9.

Archibald, J. D., Y. Zhang, T. Harper, and R. L. Cifelli. 2011. *Protungulatum,* confirmed Cretaceous occurrence of an otherwise Paleocene eutherian (placental?) mammal. *Journal of Mammalian Evolution* 18 (3): 153–161, doi: 10.1007/s10914-011-9162-1.

Belcher, C. M., M. E. Collinson, A. R. Sweet, A. R. Hildebrand, and A. C. Scott. 2003. Fireball passes and nothing burns—the role of thermal radiation in the Cretaceous-Tertiary event: Evidence from the charcoal record of North America. *Geology* 31: 1061–1064.

Benton, M. J. 1990. Scientific methodologies in collision: the history of the study of the extinction of the dinosaurs. *Evolutionary Biology* 24: 371–400.

Charig, A. J. 1983. *A New Look at Dinosaurs*. New York: Facts on File.

———. 1989. The Cretaceous-Tertiary boundary and the last of the dinosaurs. In W. G. Chaloner and A. Hallam (eds.), *Evolution and Extinction*, 147–158. Cambridge: Cambridge University Press.

Chiappe, L. M. 2007. *Glorified Dinosaurs: The Origin and Early Radiation of Birds*. Hoboken, N.J.: Wiley.

Courtillot, V. E. 1999. *Evolutionary Catastrophes: The Science of Mass Extinction*. Cambridge: Cambridge University Press.

Dodson, P. 1990. Counting dinosaurs: How many kinds were there? *Proceedings of the National Academy of Science*. 87: 7608–7612.

Fox, R. C. 1989. The Wounded Knee local fauna and mammalian evolution near the K/T boundary, Saskatchewan, Canada. *Palaeontographica Abteilung A Palaeozoologie-Stratigraphie* 208: 11–59.

Goldin, T. A., and H. J. Melosh. 2009. Self-shielding of thermal radiation by Chicxulub impact ejecta: Firestorm or fizzle? *Geology* 37: 1135–1138.

Hildebrand, A. R. 1993. The Cretaceous/Tertiary impact (or the dinosaurs didn't have a chance). *Journal of the Royal Astronomical Society of Canada* 87: 77–118.

Hurlbert, S. and J. D. Archibald. 1995. No evidence of sudden (or gradual) dinosaur extinction at the K/T boundary. *Geology* 23: 881–884.

Jepsen, G. L. 1963. Terrible lizards revisited. *Princeton Alumni Weekly* Nov. 26: 6–19.

Keller, G., T. Adatte, S. Gardin, A. Bartolini, and S. Bajpai. 2008. Main Deccan volcanism phase ends near the K–T boundary: Evidence from the Krishna–Godavari Basin, SE India. *Earth and Planetary Science Letters* 268: 293–311.

Lindow, B. E. K. 2011. Bird evolution across the K-Pg boundary and the basal neornithine diversification. In G. Dyke and G. Kaiser (eds.) *Living Dinosaurs: The Evolutionary History of Modern Birds*. Chichester, Wiley-Blackwell, 338–354.

Lofgren, D. L. 1995. The Bug Creek Problem and the Cretaceous-Tertiary Transition at McGuire Creek, Montana. *University of California Publications in the Geological Sciences* 140: 1–185.

MacLeod, N. 2004. Identifying Phanerozoic extinction controls: statistical considerations and preliminary results. In A. B. Beaudoin and M. J. Head (eds.) *The Palynology and Micropaleontology of Boundaries*, 11–33. London: Geological Society, Special Publications.

Payne, J. L. 2003. Applicability and resolving power of statistical tests for simultaneous extinction events in the fossil record. *Paleobiology* 29: 37–51.

Pearson, D. A., T. Schaefer, K. R. Johnson, D. J. Nichols, and J. P. Hunter. 2002. Vertebrate biostratigraphy of the Hell Creek Formation in southwestern North Dakota and northwestern south Dakota. In J. H. Hartman, K. R. Johnson. and D. J. Nichols (eds.) *The Hell Creek Formation and the Cretaceous-Tertiary Boundary in the Northern Great Plains*. Geological Society of America, Special Paper 361: 1145–168.

Pope, K. O. 2002. Impact dust not the cause of the Cretaceous-Tertiary mass extinction. Geology 30: 99–102.

Robertson, D. S., M. C. McKenna, O. B. Toon, S. Hope, and J. A. Lillegraven.

2004. Survival in the first hours of the Cenozoic. *Geological Society of America Bulletin* 116: 760–768.

Russell, D. A. 1984. The gradual decline of the dinosaurs: Fact or fallacy? *Nature* 307: 360–361.

Russell, D. A. and M. Manabe. 2002. Synopsis of the Hell Creek (uppermost Cretaceous) dinosaur assemblage. In J. H. Hartman, K. R. Johnson. and D. J. Nichols (eds.) *The Hell Creek Formation and the Cretaceous-Tertiary Boundary in the Northern Great Plains*. Geological Society of America, Special Paper 361: 169–176.

Sheehan, P. M., D. E. Fastovsky, R. G. Hoffman, C. B. Berghaus, and D. L. Gabriel. 1991. Sudden extinction of the dinosaurs: Latest Cretaceous, upper Great Plains, U.S.A. *Science* 254: 835–839.

Smith, A. G., D. G. Smith, and B. M. Funnell. 1994. *Atlas of Mesozoic and Cenozoic Coastlines*. Cambridge: Cambridge University Press.

Sloan, R. E., J. K. Rigby, Jr., L. M. Van Valen, and D. L. Gabriel. 1986. Gradual dinosaur extinction and simultaneous ungulate radiation in the Hell Creek Formation. *Science* 234: 1173–1175.

Sullivan, R. M., 1998. The many myths of dinosaur extinction: Decoupling dinosaur extinction from the asteroid impact. In D. L., Wolberg, K. Gittis, L. Carey, and A. Raynor (eds), *The Dinofest Symposium*, April 17–19, 1998, Academy of Natural Sciences, Philadelphia: 58–59.

Weishampel, D. B., et al. 2004. Dinosaur Distribution. In D. B. Weishampel, P. Dodson, and H. Osmólska (eds.), *The Dinosauria*, 2nd ed., 517–613. Berkeley: University of California Press.

Wilf, P., and K. R. Johnson. 2004. Land plant extinction at the end of the Cretaceous: A quantitative analysis of the North Dakota megafloral record. *Paleobiology* 30: 347–368.

Life after Death: Dinosaur Fossils in Human Hands

44

Daniel J. Chure

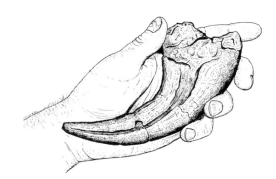

Of all the extinct denizens to have walked our plant, none have captured the imagination of scientists and the public as have dinosaurs. Their (generally) huge size, their (sometimes) ferocious nature, and their high coefficients of weirdness have made them popular cultural icons. Dinosaurs have taken on many diverse roles, including the uniformed, gas-pumping sauropod of the Sinclair Oil Company; the polite, purple Barney, purveyor of proper manners for rug rats; and the land version of the great white shark in the Jurassic Park movies (Glut 1980, 2002; Glut and Brett-Surman 1997; Lazendorf 2000; Sarjeant 2001). Wildly painted multiple casts of both their bones and footprints have even been used as the subject of exhibits by the postmodernist artist Allan McCollum (Mitchell 1998).

More than half of all known dinosaur species have been described since 1970, and every week brings more publications about new discoveries. No longer are we finding just the fossilized bones of these amazing creatures. Fossilized eggs, nests, embryos, skin impressions, feathers and feathery integuments, internal organs, footprints and trackways, and even osteocytes, capillaries, and red blood cells are all providing remarkable insight into the evolution, biology, and lifestyles of the Dinosauria. Truly, we are living in the Golden Age of Dinosaur Research and there is no end in sight to this extraordinary and exciting time.

Certainly the discovery of a dinosaur fossil is an exciting event. Be it a tooth, a bone, a footprint, pieces of an eggshell, or impressions of skin, each of these specimens is a tangible link to the long-vanished world of the dinosaurs. Making such a discovery is not something limited to members of scientific expeditions. Many important paleontological discoveries have been made by nonscientists. Given the abundant and global distribution of dinosaur fossils (see Weishampel et al. 2004 for an exhaustive review) it is not unexpected that someone who is not a paleontologist might come across a new dinosaur fossil sticking out of the ground. Granted, the chances of making such a find is higher while hiking and exploring in arid regions, such as the Inter Mountain West, where vegetation is sparse and cities and roads are uncommon. But even more mundane activities can lead to a chance discovery. For example, Mr. Charlie Fickle found a partial skeleton of the great predatory dinosaur *Tyrannosaurus rex* while walking his dog at a housing construction site in Littleton, CO (Carpenter and Young 2002).

From Curios to Cash Cows

Fossils have long captured the human imagination. There was a widespread trade in fossils as far back as the Paleolithic age in Europe (Oakley 1965). They were well known to ancient civilizations in both the Old and New

World, although their interpretation was radically different from ours (Buffetaut 1987; Edwards 1976; Mayor 2000, 2005). Fossils have even played a central role in a nasty incident of academic and political intrigue and subsequent judicial proceedings in the early eighteenth century of Würzburg (Jahn and Woolf 1963). And fossils have a long, if ultimately ineffective, history of use in medicine (Kennedy 1976). Most of the medical use of fossils involves invertebrates, but New Age devotees have used dinosaur bone as a "grounding medium," as part of healing pouches and healing patches used in "earth preservation ceremonies," and in magical wands (Anonymous 1992).

However, for most of the history of the science of paleontology, fossils have been natural curiosities attracting the attention of scientists and the occasional interested amateur (Brett-Surman 1997). Private collections amassed by these individuals often became the core of the emerging museums in the sixteenth and seventeenth centuries. The earliest scientific publication on a dinosaur bone is that of Plot (1677). This specimen, the distal end of a femur, was well illustrated by Plot and Brookes (1763) and appears to belong to a predatory dinosaur known as a megalosaur. Unfortunately, this specimen is lost. Other early dinosaur discoveries, all also now lost, are reported by Lhuyd (1699) and Platt (1758). Sarjeant (1997a, 1997b) identifies a femoral shaft in the collections of the Woodwardian Museum of Cambridge as being collected in 1728; it is the earliest discovered dinosaur fossil that is still in existence. During the latter part of the eighteenth and early part of the nineteenth centuries, dinosaur fossils slowly continued to be gathered, although their true nature went unrecognized. However, as vertebrate paleontology expanded in the nineteenth century, the significance of dinosaurs became apparent and the study of their remains exploded, leading to what has been called the First Golden Period of dinosaur discovery (Brett-Surman 1997).

These discoveries were often fortuitous. However, commercial operations, such as stone quarries, proved to be a consistent source of vertebrate fossils, including dinosaurs. Among the most famous of these are the quarries at Stonesfield, England, where many of the early discoveries previously mentioned, as well as that of *Megalosaurus*, were found (Benton and Spencer 1995; Delair and Sarjeant 1975). It was not until the latter half of the nineteenth century that the first crews hired specifically to collect vertebrate fossils undertook sustained field operations (Howard 1975; Lanham 1992). With the explosive growth in vertebrate paleontology in general and dinosaur paleontology in particular, there was an increased international demand for specimens both for research and museum exhibits. In response to this need, a small group of professional vertebrate fossil collectors arose. These individuals found, collected, prepared, and mounted specimens that they then sold to museums. The most famous of these is the Sternberg family, which worked primarily in North America (Rogers 1999). It is worth noting that the Sternbergs sold to institutions and museums. There was no significant market of private collectors for large vertebrate skeletons at that time.

Seven-Figure Dinosaurs!

A watershed event in the commercial value of fossil vertebrates in general and dinosaurs in particular was the discovery of the *Tyrannosaurus rex* skeleton popularly known as Sue. The celebration of its completeness, the recognition of its great scientific importance, and the ensuing legal battles, trial, and prison sentences generated immense interest on both sides of the issue of the sale of Sue. An objective history of this event has yet to be written. Pfeiffer (2001) provides a popular account, but one not sympathetic to the U.S. government's prosecution of the case. Larson and Donnan (2002) is an insider's account of the story.

When the dust settled, Sue was purchased at auction by the Field Museum of Natural History (Chicago) for nearly $8,000,000. Somewhat later, the North Carolina State Museum of Natural Science paid $3,000,000 for another large predatory dinosaur *Acrocanthosaurus atokensis* (Monastersky 1998). With prices such as these, everyone saw dollar signs when it came to dinosaur bones and skeletons. Another *T. rex* specimen, colloquially known as Mr. Z-rex, was advertised with a price in excess of $10,000,000, although the asking price was apparently never met. Certainly the financially lucrative nature of fossil vertebrates has contributed to the problems of theft. Even isolated, shed tooth crowns of tyrannosaurids have been marketed with prices in excess of $1,000.

The Second Extinction of the Dinosaurs

As land-living vertebrates, the dinosaurs already had two strikes against them when it came to becoming fossils. Most continental environments are erosional rather than depositional, so the chances of remains being buried in sediment (the first step to becoming a fossil) are not great. To end up in a museum, a dinosaur's bones had to become buried and then survive tens or even hundreds of millions of years in the earth's crust, avoid destruction by metamorphism and erosion, and appear at our planet's surface sometime during just the last two centuries in a place where inquisitive humans might find, collect, and preserve them (Colbert 1968). Once collected, prepared, and curated, dinosaur fossils should finally be safely preserved for present and future generations. However, such is not always the case.

Warfare and violent social revolution wracked the twentieth century, with nearly fifty significant conflicts across the globe (Miller 1997). The increasing destructive capability of modern weapons and political ideologies that require the physical destruction of the social order they replace have both taken a serious toll on dinosaurs and other fossils. This area has not received as much attention and documentation as the impacts on cultural resources (for example, Akinsha et al. 1995; Nicholas 1995). Even so, it is clear that major collections and many significant specimens have been destroyed (Bigot 1945; Chure 2003; Nothdurft et al. 2003; Psihoyos 1994, 252–254). Although this dinosaur destruction is best documented for collections in Western Europe, even that loss is poorly understood. This is an area of dinosaur history (and vertebrate paleontology in general) that is ripe for further research. An outstanding starting point for such work is the Society of Vertebrate Paleontology News Bulletin. Reading the reports of foreign correspondents following the end of World War II gives chilling and poignant firsthand accounts of the destruction (for example, Anonymous 1945, 1947).

Destruction during World War I was spotty or at least poorly reported. There seems to have been little looting, although Heintz (1933, 11) reports that fossil vertebrate collections were taken from Tartu University (Estonia) by Russian forces and only returned after the end of the conflict. The most famous loss during this conflict is that of a shipment of dinosaurs collected by the Sternbergs for the British Museum of Natural History. This lot, including a hadrosaur mummy, was on the SS *Mount Temple* when it was sunk by the German raider *Moewe* in the Atlantic Ocean, some 600 miles west of England on December 6, 1916 (Spalding 2001; Tanke et al. 2002).

World War II was an entirely different matter. The scope, ferocity, and duration of this conflict resulted in the loss of more dinosaur specimens (and fossil vertebrates) than all other conflicts in history combined. Virtually any bad thing that can happen to a fossil collection occurred during this conflict. The losses came in many forms. The most obvious was physical

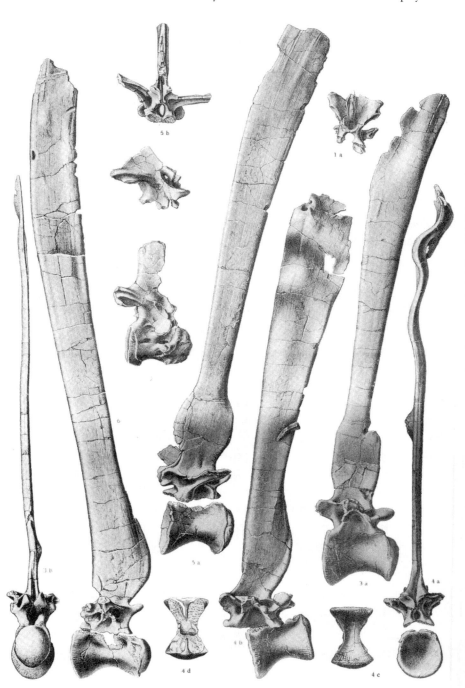

44.1. The vertebrae of the theropod dinosaur *Spinosaurus*. The elongated neural spines are the origin of the name. This is part of Stromer's Egyptian dinosaur collection, which was destroyed in Munich during World War II.

Daniel J. Chure

loss. Bombing destroyed entire fossil reptile collections in both Allied and Axis countries (Bigot 1945; Nothdurft et al. 2003; Psihoyos 1994). The museums were not the target of the bombing, but their locations in cities near military targets and the poor accuracy of aerial bombardment, which often required repeated bombing raids, ensured that such damage was inevitable. Given the international nature of vertebrate paleontology, it is not surprising that some specimens were destroyed far from their native lands. A number of dinosaur fossils from the United States went to the *Senckenbergische Naturforschende Gesellschaft* via exchange with the American Museum of Natural History and were destroyed (Crumly 1984). Figure 44.1 shows a small part of Ernst Stromer's Egyptian dinosaur collection, which was lost in Munich (Nothdurft et al. 2003). Augee and colleagues (1986) provide an account of the bizarre and tragicomic history of fossil marsupials collected in Australia and also lost in Munich.

The paleontologist Karl Beurlen was the director of the *Zoologische Staatssammlung München*, which contained what was probably the largest fossil vertebrate collection in Central Europe (Fig. 44.2). He was an ardent Nazi who considered the evacuation of any of the museum's collections to be an expression of a defeatist attitude that the Allies could strike deep into Bavaria. In his opinion, if Germany lost the war, the survival of the collections was of no importance. As a result, there was no formal evacuation of the collections. However curators of several departments did move and subsequently saved important neontological collections (Crumly 1984). Unfortunately, the same cannot be said for the fossil vertebrates, and most of them, including Stromer's collection of Cretaceous vertebrates from Egypt, were nearly completely destroyed during bombing raids (Crumly 1984; Nothdurft et al. 2003).

The brutal nature of fighting in cities also had disastrous consequences. The decision of Nazi Germany to fight to the very end resulted in intense,

44.2. The Alte Academie in Munich destroyed by Allied bombing in 1944. This museum housed the Bavarian State Collection of Paleontology and Geology, which included Stromer's Egyptian dinosaur collection, which was destroyed during the raid.

house-to-house combat in cities across Europe all the way to the *Fürher-bunker* in central Berlin. Apparently, fighting went on for several days between German and Russian forces inside the *Museum of Naturkunde* in Berlin in April of 1945. Fighting was often room-to-room. When resistance was encountered, the room was cleared by the use of hand grenades.

Fossil vertebrate collections that did survive were often subject to poor conditions long after the end of hostilities. In some museums, collections were moved and sealed in secure places within the museum (*Museum für Naturkunde*), or dispersed elsewhere in the country (British Museum of Natural History). Crumly (1984) reports that 85% of the paleontology collections of the *Stuttgarter Museum für Naturkunde*, Ludwigsburg, survived by being evacuated and stored in a salt mine. However, given the fragile and brittle nature of fossil bone, damage to specimens was inevitable, even when moves occurred in the best of conditions (see, for example, Young 1947).

The theft of scientific equipment and collections was part of a broader plan of Nazi cultural looting in Europe during World War II. In his 21 February 1946 Nuremberg Trial testimony, Alfred Rosenberg, the German Reich minister for the occupied territories, stated, "I have entrusted the *Einsatzstab Rosenberg* [Rosenberg Battalions] for the Occupied Territories with the listing and detailed handling of all cultural valuables, research materials, and scientific work in libraries, archives, research institutions, museums, et cetera, found in public and religious establishments, as well as in private houses." The "detailed handling" meant bringing the contents of those institutions back to Germany. That the looting and destruction of such scientific institutions was more extensive in Eastern Europe and the former Soviet Union than in Western Europe was no accident. This reflected different policies of the Nazi state in each theater of conflict. In the west, the intent was to establish a series of client states for Nazi Germany. In the east, the objective was the destruction of the native populations and their wholesale replacement by a new German landed gentry (see, for example pages 932–934 concerning the memorandum "Some Thoughts on the Treatment of the Alien Population in the East" in Noakes and Pridham 1990). Information on the activities of the Einsatzstab Rosenberg and other details of the destruction and looting of scientific institution is available through the Avalon Project at the Yale Law School, Nuremberg Trial Proceedings. Of particular interest for the reader is the February 21, 1946, testimony available online at http://www.yale.edu/lawweb/avalon/imt/proc/02-21-46.htm.

Such looting was not restricted to one side. After World War II, the Russian authorities seized most of the library, as well as all microscopes, photographic equipment, and lamps in Berlin's *Museum für Naturkunde*. This was viewed as reparations. The material went to Russia and was never returned (Crumly 1984).

Another important loss was that of data associated with specimens. Accession and catalog data, field notes, maps, photographs, and other materials were destroyed (Nothdurft et al. 2003). If the specimen survived, there may have no longer been any information associated with it. If the specimen had been described, photos, drawings, and measurements may still exist in the scientific literature even if the specimen did not. However, many specimens had not been described, and the loss of the specimen and data make it as though the fossil never existed.

Coupled with the loss of specimens and facilities was the loss of scientists and support staff. Some scientists were conscripted for long periods. Some servicemen were killed in combat, whereas others were taken as prisoners of war and served long stints in captivity or were never returned to their native country (see, for example, Woerdeman 1948). Civilians were also killed by fighting and bombing. Curator Walter Arndt of the *Zoologisches Museum für Naturkunde*, Berlin, was executed by the Nazis in June 1944 for expressing a doubt about Germany's ultimate victory (Crumly 1984). After years of harassment, the trilobite researcher Rudolf Kaufman was killed because he was Jewish (Fortey 2000, 166–171). There was a lengthy postwar denazification process that disrupted the careers of even those who were opponents of the Nazi regime, whereas scientists who supported or sympathized with the Nazis were relieved of their positions in museums and universities. In the lands directly ravaged by combat, postwar conditions precluded any scientific work—mere survival was struggle enough (Anonymous 1945, 1947; Crumly 1984).

As codified by Chure (2000a), the three principles of collections management are:

> If it exists, people will collect it.
> If people collect it, they will exhibit it.
> If people exhibit it, someone will try to steal it.

The strongest evidence in support of these principles is that a bustier worn by Madonna on one of her tours was stolen from the Frederick's of Hollywood Bra Museum in California during a break-in during the 1990s. It was never recovered.

The impact of the international trade of antiquities has long been recognized and documented (Meyer 1973). Documentation of the impacts of the international trade in fossil vertebrates has not received the same detailed attention. Chure (2000a) provides a summary of the global theft of dinosaurs and other fossil vertebrates from museum collections and exhibits. Long (2003) is a longer first-person account of tracking fossils stolen from the field and the lab around the world. This book is an excellent starting point for anyone interested in exploring this issue.

The impact of specimens illegally collected from the field is often hard to evaluate. In many cases, the stolen fossil was unknown to scientists and there is no way to evaluate the significance of the specimen when the only thing remaining is a hole in the ground. However, fossils stolen from the field have later been recovered on other continents, showing that the movement of these objects is international (Stokstad 2002).

There is a large and lucrative international trade in fossils, especially vertebrates. Even a brief perusal of catalogs and websites shows that among vertebrates, the highest prices are demanded by dinosaur fossils. The high prices of these specimens put them out of the reach of researchers and most museums and institutions. Once sold into private ownership, specimens

Going, Going, Gone!

are often lost to science (Padian 2000). Furthermore, the origin of many of these specimens may be uncertain. However, a number of countries, such as Argentina and the People's Republic of China, ban outright the export of fossil vertebrate remains, so questions of legality arise over merely possessing such fossils (Dalton 2004; Steghaus-Kovac 2002; Schmidt 2000).

Given the high prices fossils demand, it is not surprising that they have been stolen from virtually every type of institutional setting. Chure (2000a) and Long (2003) provide many examples, so only a few will be given here.

1. *Private collections:* One of the few known specimens of the earliest bird, *Archaeopteryx*, was in the private collection of Mr. Eduard Opitsch in Pappenheim, Germany. Upon Mr. Opitsch's death, an inventory of his estate failed to turn up the specimen. There is no evidence that the owner sold the specimen, and it is presumed stolen. It has not been recovered (Wellnhofer 1992).

2. *Commercial fossil concerns:* There have been several instances where commercial fossil dealers have reported the theft of specimens from their exhibit tables at trade shows.

3. *Public museums:* A break-in at the *Museum of Paleontology* at the University of La Rioja, Argentina, resulted in the loss of cast and original material of a prosauropod dinosaur and the type specimens of several mammal-like reptiles (Arcucci 1994).

4. *Government facilities:* During the winter season, when the visitor center was closed, thieves kicked in the doors at the Cleveland-Lloyd Dinosaur Quarry and stole approximately 30 dinosaur bones (Anonymous 1996). Thieves stole parts of a sauropod bone off the quarry face within the Quarry Visitor Center at Dinosaur National Monument during a busy summer day with many visitors in the building (Chure 2000a).

5. *Museum collection storage areas:* The best-known example of this category involves the Paleontological Museum of the Russian Academy of Sciences in Moscow. Several thefts in the middle 1990s resulted in the loss of numerous vertebrate fossils, including type specimens of ceratopsians and tyrannosaurids (Abbott 1996, 1998).

6. *Museum collection exhibit areas:* The thefts at the Paleontological Museum in Moscow included a series of Triassic amphibian skulls, including several type specimens, from an exhibit case (Shiskin 1992). The fact that the lock on the exhibit case was not broken led to the suspicion that institutional staff might have been involved (Feder and Abbott 1994). However, a similar theft of rhinoceros skulls occurred at the University of Michigan. In this case, the thieves dismantled the exhibit case, took what they wanted, and reassembled the case. As a result, the missing specimens were not noticed for some time.

When it comes to theft, it is clear that fossils have achieved the same status as cultural objects such as vases, sculptures, and paintings. There is enough of a market for such items that it is unlikely the threat will dissipate any time soon. As a result, increased security policies and access restrictions have been implemented. However, better communication and public awareness is critical. Many databases exist for stolen or missing cultural items. See, for example, the FBI National Stolen Art File at http://www.fbi.gov/majcases/arttheft/art.htm. Such databases have been useful in the recovery of missing items. The time has arrived for a similar database for stolen paleontological specimens. Some thoughts on such a database are given in Chure (2000a).

Finding a dinosaur fossil is pretty exciting. But what do you do next? Exhibit impulse control—don't start digging up the specimen! That's a difficult process that requires skill, patience, and often significant resources (Chure 2000b, 2002; Gillette 1997; Leiggi et al. 1994). You don't want to carelessly wreck the bones of a dinosaur that died millions and millions of years ago. But there are even more fundamental issues that need to be addressed before anyone considers collecting the dinosaur. The first is documenting the site location. The other is land ownership and, by extension, who owns the fossil. Just because you found the fossil doesn't mean it's yours.

Documenting the Site

You might have found the most important dinosaur ever discovered, but that will be of little import if the site cannot be found again. Ideally, you would be available to guide someone back to the fossils. But if not, someone who had never been there before should be able to find it.

Photographs are of prime importance. Photos should be general, showing the terrain and horizon and giving a general setting for the site. The fossil site should be well marked on the general photos. In addition, closer photos should be taken giving more detail about the location of the site, the rock type, and how the exposed fossils look. Are there bone chips and pieces of bone on the surface or can the bones only be seen in cross-section in the rock? Many digital cameras also have the ability to record digital video. If possible, you should shoot some video of the site and narrate it.

Of course, the advent of inexpensive GPS has made site documentation easier and often more accurate. If you carry a GPS system on your travels, get the coordinates of the site. GPS data should complement photographic documentation, however, not substitute for it. Both data together provide the best chance that the site can be found again. Additional information, such as a written description of the route to the site, sketch maps, and the site plotted on a topographic map are also useful.

Land Ownership

A review of the global mosaic of laws concerning ownership of fossils is far beyond the scope of this chapter. The laws and regulations governing the ownership of fossils in the United States are addressed here. It is impossible to tease out those regulations that apply just to dinosaurs, so regulations applying to all fossils are included. The reader should be aware that in other countries the legal situation may be quite different. For example, the concept of "Crown Property" in Canada, England, and Australia creates a quite different philosophical framework than one sees in the United States. These differences are reflected in the recommendations about fossil ownership and trade suggested by the Australian paleontologist John Long (2003).

Land ownership is a crucial issue in the ownership of fossils; it is your responsibility to know whose land you are on. Land ownership falls into a few broad categories.

1. PRIVATE PROPERTY

Unless you are the landowner, you need the landowner's permission to be on his or her land. In some cases, such as the more heavily populated East Coast, private land is often fenced, marked, or signed. This is not always the case in the more open areas of the central and western United States. Nevertheless, being on the land without permission is trespassing.

Fossils of any sort (dinosaurs, pollen, footprints, etc.) on private land are the property of the landowner. What happens to them is at the landowner's discretion. They can be left to erode to dust. They can be ground up and used to line a driveway. They can be dug up and sold. Many landowners, however, are willing to work with scientists and others interested in the fossils on their land. Over the last 150 years, private landowners have made many significant contributions to vertebrate paleontology.

If you find a fossil on private land, approach the landowner and talk about what you have discovered. Offer to take the landowner back to the site. If you can, explain the potential scientific significance of what you have found and suggest contacting a paleontologist to get a professional opinion. However, observe the landowner's wishes. Don't contact a scientist or the press and don't advertise the find without the landowner's agreement. If there are loose fossils on the surface, you might collect just a few to show the owner the kind of fossils discovered. However, digging fossils up and disturbing the surface of the land could upset the owner and close off a site.

2. FEDERAL PUBLIC LANDS

There are many types of public land. In all cases, the landowner is a federal, state, or local government entity. As such, the fossils are often considered the property of the citizenry and part of our nation's heritage, and their collection is often regulated or restricted.

Federal lands are those owned and managed by the U.S. national government. Lands owned by Native American tribes are not federal lands and are considered separately (see category 4). Fossils are considered to be a nonrenewable resource. With the exception of some types of fossils (such as pollen and spores) and energy minerals (such as coal), most fossils are considered rare by land managers. As a general principle, fossils on federal land are managed "for their scientific, educational, and, where appropriate, recreational values" (Anonymous 2000).

There are many federal agencies, only a subset of which has land management responsibilities. However, these agencies have control over large

Table 44.1. Requirements for obtaining a collecting permit for fossils on lands managed by the U.S. federal government. From Anonymous 2000.

Agency	Qualifications	Permit Types	Other	Repository
Bureau of Land Management (BLM)	Graduate degree in paleontology or related topics, or equivalent experience with one who meets that standard.	Survey/limited surface collection (<1 sq m disturbance), or excavation (1 sq m surface disturbance or more).	Reports required annually and at the end of the project. Work in Special Management Areas requires additional reviews.	Designated by permit applicant; must meet Department of the Interior (DOI)/BLM standards.
Bureau of Reclamation (BOR)	Similar to BLM.	Scientific collecting permit.	None.	Designated by BOR or permit applicant; must have letter from repository showing intent to accept specimens.
US Forest Service (FS)	Same as BLM.	Varies with forest unit from survey and inventory to excavation and collection.	Reports required annually and at the end of the project. Work in Wilderness Areas may be restricted.	Designated in application for Special Use Permit; must meet FS standards. Standards added to permit.
US Fish and Wildlife Service (FWS)	Related to nature of work.	Special Use permit required for survey or collection.	Reports required at the end of the project.	Similar to BLM.
National Park Service (NPS)	In revision; qualifications and experience to conduct scientific study or representative of reputable scientific or educational institution or state/federal agencies.	Scientific research and collection.	Reports required annually.	At NPS units or in an approved repository designated by permit applicant; must meet DOI/NPS standards.

tracts of land, primarily in the western United States. The National Park Service (NPS), the Bureau of Land Management (BLM), and the U.S. Forest Service are familiar agencies that act as land managers. However, the Bureau of Reclamation, the U.S. Fish and Wildlife Service, and the Department of Defense also control land, much of which has fossils of some sort. Each of these federal land management agencies has a mission given to it by Congress. As a result, regulations governing the permitting and collecting of fossils varies between them (Table 44.1). Anonymous (2000) provides a readable and useful summary of the management of fossil resources on federal lands. Kuizon (2006) provides a short but very useful summary of the legislative and regulatory history of fossil resources on federal lands.

In 2009, a major piece of U. S. federal legislation was passed that specifically concerned fossil resources. The Paleontological Resources Preservation Act (P.L. 111-011 Title VI Subtitle D) was signed into law on March 30, 2009, as part of the Omnibus Public Land Management Act of 2009. This law and the details of its legislative history can be viewed at http://www.blm.gov/wo/st/en/prog/more/CRM/paleontology/paleo_legislation.html. The law includes new authorities and specific mandates related to the preservation, management, and protection of fossil resources on lands administered by federal agencies. Development of policy and implementation of this act will take some time, but those efforts are already under way.

All federal agencies recognize the rarity of vertebrate fossils; such fossils are often managed much more stringently than other types of fossils. These greater restrictions apply to dinosaurs. All agencies require that permits be obtained before a fossil vertebrate can be collected. These permits are issued for scientific and educational purposes only. The permit stipulates that the repository for all collected specimens will be a reputable scientific or educational institution. If you find a dinosaur on federal land, you will not be able to legally take any of it home.

As the leading conservation agency of the federal government, the National Park Service has the greatest restrictions on the fossil resources on its lands. It is illegal to disturb, remove, or destroy any natural or cultural resources on NPS lands. If you find any kind of fossil, vertebrate, invertebrate, plant, footprint, trail, etc., you cannot remove it. That is pretty straightforward. Other agencies manage their fossil resources in a more diverse manner, which reflects their differing missions (Table 44.2). In contrast to most other agencies, the BLM allows for the collection of common fossil invertebrates and petrified wood without permit. In some areas, the BLM has specifically designated sites where nonvertebrate fossils can be collected without permit. In other areas, such collection might be prohibited in some areas because of special conditions. Furthermore, surface disturbance while collecting fossils can impact other resources. These resources may be cultural (such as an arrowhead chipping site) or biological (like endangered plants). Damaging them, even unintentionally, could result in fines or prosecution.

The Presidential Executive Order on Sacred Sites (E.O. 13007 May 24, 1996) requires federal land managing agencies to avoid adversely affecting the physical integrity of Native American Sacred Sites on their land. Additional restrictions apply to any fossil collecting at or near such sites.

Table 44.2. Different agencies of the U.S. Department of the Interior (DOI) management approaches to collecting fossils. From Anonymous 2000.

Agency	Invertebrates	Vertebrates	Petrified Wood	Other Fossil Plants
BLM	Reasonable amounts for personal use; no permit required.	Must have a permit.	Up to 25 lbs/day/person + 1 piece, not to exceed 250 lbs/year for non-commercial use. BLM treats petrified wood as a mineral material.	Reasonable amounts for personal use; no permit required.
BOR	Permit required; scientific purposes only.	Permit required; scientific purposes only.	Permit required; scientific purposes only.	Permit required; scientific purposes only.
FWS	Special Use permit required; scientific or educational purposes only.	Special Use permit required; scientific or educational purposes only.	Special Use permit required; scientific or educational purposes only.	Special Use permit required; scientific or educational purposes only.
NPS	Permit required; scientific or educational purposes only.	Permit required; scientific or educational purposes only.	Permit required; scientific or educational purposes only.	Permit required; scientific or educational purposes only.

If this seems confusing and complicated, then you have achieved enlightenment. As previously stated, you need to know whose land you are on when you find a fossil. There are a variety of businesses that sell land ownership maps and some of these maps may be available free online. However, these may be at too coarse a scale or simply inaccurate. Land boundaries can change as lands are exchanged or sold. To get the most current information, you should go to the local office of the federal agency whose land you will be on to find out where boundaries are, if there are specially designated or prohibited areas for collecting fossils, and any other management concerns. Ignorance is not an excuse, and you have the responsibility of knowing who owns the land your fossil discovery is on.

3. STATE, MUNICIPAL, AND OTHER LOCAL LANDS

Each state has its own set of regulations governing the collection of fossils on its land. West (1989, 1991) provides a summary of such state regulations. Although somewhat dated, these publications give a good feel for the kind of permitting required and how this varies across the country. Regulations are likely more variable at the county and city level. For example, even if there are no specific regulations concerning fossils, digging in a city park is likely to be viewed as damaging public property. Once again, you are responsible for finding out what regulations and restrictions apply to your situation.

4. NATIVE AMERICAN LANDS

Indian lands are held in trust by the federal government and managed for the benefit of a tribe or an individual. The government serves in the role of a trustee. Thus, the tribe or individual may use fossil resources on their land for economic benefit. Such was the reason for the auctioning of the *T. rex* known as Sue (Pfeiffer 2001). The Bureau of Indian Affairs role as trustee is limited to ensuring that any commercial collecting of fossil is of economic benefit to the tribe or landowner. Anyone wishing to work on such lands needs permission from the tribe or landowner for access. Any conditions stipulated are set by the tribe or landowner and not by the Bureau of Indian Affairs (Anonymous 2000). As with other lands, you are responsible for knowing where you are.

44.3. It should always be this good! Jensen resident "Pop" Goodrich stands before the eight tail vertebrae of *Apatosaurus,* the original discovery at the Carnegie Quarry, one of the greatest dinosaur quarries ever found. This site eventually became Dinosaur National Monument. Some 1500 dinosaur bones are still visible as a permanent in situ exhibit within the Quarry Visitor Center at the monument.

Contact a Professional

Understanding the significance of a discovery is not often easy. Granted, the initial discovery at the site that would become Dinosaur National Monument was obviously important. Eight articulated tail vertebrae of a sauropod dinosaur projecting out of the ground is a pretty good clue that something important may lie just beneath the surface (Fig. 44.3). However, in most cases the initial find is not so spectacular. In fact, it may be quite mundane. However, to the trained and experienced eye, even small fragments may show critical morphological features that reveal the importance of the find. A find does not have to be a new species to be valuable. Most dinosaur species are known from less than one complete specimen, so even a partial skeleton or a few bones can provide new data that can help improve our understanding of the biology and evolutionary position of that dinosaur (Fig. 44.4). In fact, there is likely more to be gained scientifically from a good skeleton of a known species than from a few bones of a new species. In addition, a specimen can be valuable because it provides a significant temporal or geographic range extension. Or it might provide insights into ancient environments.

Not all of this may be discernible from the initial discovery, but a paleontologist may be able to read much from the few scraps available and evaluate whether an excavation is warranted. That's why it is so important to contact a professional. After all, you wouldn't do your own crime scene investigation just because you like the TV program CSI. You'd call the experts. It's no different with dinosaurs.

There are many ways that you can help with the process of scientific discovery. Paleontologists are often glad to have help in excavating and preparing fossils, especially those of large vertebrates. Such excavations may go on for several years depending on the hardness of the fossil-bearing rock and the extensiveness of the bone deposit. Once back in the lab, the dinosaur still needs to be removed from the rock and volunteer help may be

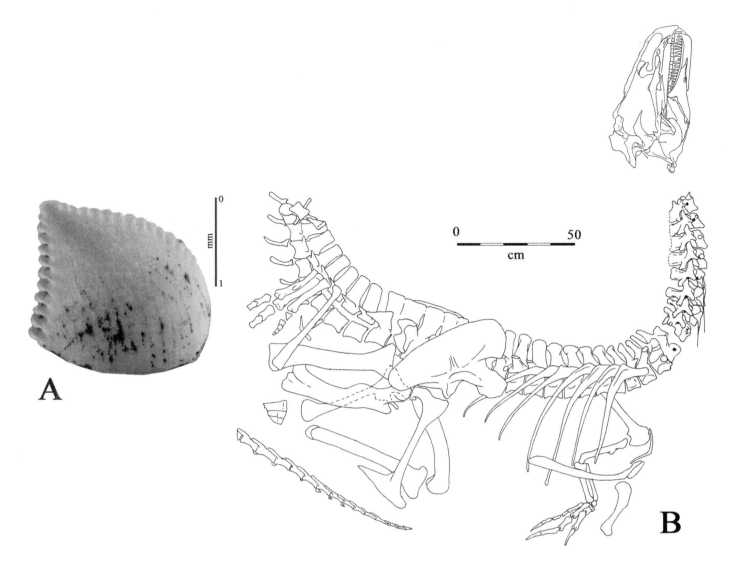

needed. Developing the ability to prepare a fossil takes time, however, and the process is complicated when the matrix is hard. It may take considerable training to develop the requisite skill to extract a fossil from its rocky blanket. Developing a good relationship with the staff of a scientific institution can lead to additional fieldwork and excavation. Gillette (1994) offers a fine example of how a discovery made by two observant hikers led to a long and arduous excavation and recovery of one of the largest dinosaurs ever found.

The contributions made by fossil discoverers are often recognized in a number of ways. Acknowledgment in a scientific publication is common. If the specimen is a new species, it is sometimes named after the discoverer. Sometimes a cast duplicate of some part of the specimen is presented to the discoverer, allowing that person to have a tangible reminder of when they found a piece of the distant past. However, none of this should be demanded. Regardless of the recognition accorded, you can take satisfaction that as the discoverer of a dinosaur fossil, whether a bone, a footprint, or something else, you have helped opened the door to a long vanished world of dinosaurs—undoubtedly the most fantastic period in the history of life on Earth.

44.4. From the ridiculous to the sublime. A, The type and only known specimen of *Koparion douglassi,* a small theropod from the Late Jurassic Morrison Formation of Utah. Although just an isolated tooth crown, the specimen is distinctive and shows a morphology reminiscent of troodontids. This is an example of how even a small fossil can show important features. Modified from Chure 1994. B, Quarry map for a new specimen of *Allosaurus* from Dinosaur National Monument, showing how remarkably complete some fossils can be. The initial indication of this skeleton was just a few toe bones on the surface of a sandstone layer. Modified from Chure 2000c.

References

Abbott, A. 1996. Missing dinosaur skulls raise new fears of smuggling in Moscow. *Nature* 384: 499.

———. 1998. Moscow's missing fossils come under new scrutiny. *Nature* 391: 724.

Akinsha, K., G. Kozlov, and S. Hochfield. 1995. *Beautiful Loot: The Soviet Plunder of Europe's Art Treasures*. New York: Random House.

Anonymous. 1945. Paleontology in Central Europe. *Society of Vertebrate Paleontology News Bulletin* 15: 6–8.

———. 1947. English translation of Professor Ernst Stromer's "Report on Fossil Vertebrate Collections in Germany." *Society of Vertebrate Paleontology News Bulletin* 22: 23–27.

———. 1992. Who is investing in fossils? *Fossil Index: The Newsletter for Collectors* 2 (2): 1–2, 8, 10.

———. 1996. Cleveland-Lloyd Dinosaur Quarry suffers break-in. *Al's Archives, Newsletter for Paleontology, Archeology, and Natural History Enthusiasts*. College of Eastern Utah's Prehistoric Museum, Price, Utah (Winter 1996) 13.

———. 2000. Assessment of Fossil Management on Federal and Indian Lands. *Report of the Secretary of the Interior to the U.S. Congress.*

Arcucci, A. 1994. Stolen fossils. *Society of Vertebrate Paleontology News Bulletin* 161: 62.

Augee, M., R. Dehm, and L. Dawson. 1986. The Munich collection of Wellington Cave fossil marsupials. *Australian Zoologist* 22 (4): 3–6.

Benton, M.J., and P. S. Spencer. 1995. *Fossil Reptiles of Great Britain*. London: Chapman-Hall.

Bigot, A. 1945. La destruction des collections et des bibliothèques scientifiques de Caen. *Bulletin de la Sociétè linnéenne de Normandie, Caen* (suppl.): 1–75.

Brett-Surman, M.K. 1997. A chronological history of dinosaur paleontology. In J. O. Farlow and M. K. Brett-Surman (eds.), *The Complete Dinosaur*, 707–720. Bloomington: Indiana University Press.

Brookes, R. 1763. *The Natural History of Waters, Earths, and Minerals, with their virtues, properties, and mineral uses; to which is added the method in which Linnaeus has treated these subjects*. Volume V. London: Newberry.

Buffetaut, E. 1987. *A Short History of Vertebrate Paleontology*. London: Croom Helm.

Carpenter, K., and D. B. Young. 2002. Late Cretaceous dinosaurs from the Denver Basin, Colorado. *Rocky Mountain Geology* 37 (2): 237–254.

Chure, D. J. 1994. *Koparion douglassi*, a new dinosaur from the Morrison Formation (Upper Jurassic) of Dinosaur National Monument; the oldest troodontid (Theropoda: Maniraptora). *Brigham Young University Geology Studies* 40 (1): 11–15.

———. 2000a. New threats to old bones: The theft of fossil vertebrates from museum collections. *Cultural Resource Management*. National Park Service, U.S. Department of the Interior, Cultural Resources 23 (5): 18–22.

———. 2000b. Digging them up. In G. S. Paul (ed.) *The Scientific American Book of Dinosaurs*, 46–51. New York: St. Martin's Press.

———. 2000c. A New Species of *Allosaurus* from the Morrison Formation, Dinosaur National Monument (UT-CO) and a Revision of the Theropod Family Allosauridae. Ph.D. dissertation, Department of Earth and Environmental Sciences, Columbia University, N.Y.

———. 2002. Raising the Dead. In J. G. Scotchmoor, B. H. Breithaupt, D. A. Springer, and A. R. Fiorillo (eds.), *Dinosaurs: The Science behind the Stories*, 127–136. American Geological Institute.

———. 2003. Bullets, bombs, and bones: the impact of 20th century warfare on fossil vertebrate collections. *Journal of Vertebrate Paleontology* 23 (suppl. to 3): 40A.

Colbert, E. H. 1968. *Men and Dinosaurs*. New York: E.P. Dutton. Reprint, 1984, *The Great Dinosaur Hunters and Their Discoveries*, New York: Dover Press.

Crumly, C. R. 1984. Saving a legacy: natural history collections in Germany before and after World War II. *Curator* 27 (3): 205–219.

Dalton, R. 2004. Dinosaur eggs escape sale as smuggling claims unearthed. *Nature* 430: 493.

Delair, J. B., and W. A. S. Sarjeant. 1975. The earliest discoveries of dinosaurs. *Isis* 66: 5–25.

Edwards, W. N. 1976. *The Early History of Paleontology*. Publication no. 658. London: British Museum (Natural History).

Feder, T., and A. Abbott. 1994. Concern grows over "trade" in Russian fossils. *Nature* 371: 729.

Fortey, R. 2000. *Trilobite! Eyewitness to Evolution*. New York: Alfred A. Knopf.

Gillette, D. D. 1994. *Seismosaurus: The Earth Shaker*. New York: Columbia University Press.

——. 1997. Hunting for dinosaur bones. In J. O. Farlow and M. K. Brett-Surman (eds.), *The Complete Dinosaur*, 64–77. Bloomington: Indiana University Press.

Glut, D. F. 1980. *The Dinosaur Scrapbook.* Secaucus, N. J.: Citadel Press.

——. 2002. Dinosaurs in the movies: Fiction and fact. In J. G. Scotchmoor, B. H. Breithaupt, D. A. Springer, and A. R. Fiorillo (eds.) *Dinosaurs: The Science behind the Stories*, 171–178. American Geological Institute.

Glut, D.F., and M. K. Brett-Surman. 1997. Dinosaurs and the media. In J. O. Farlow and M. K. Brett-Surman (eds.), *The Complete Dinosaur*, 675–706. Bloomington: Indiana University Press.

Heintz, A. 1933. *Revision of the Estonian Arthrodira. Part I. Family Homosteiidae Jaekel.* Publications of the Geological Institution of the University of Tartu, no. 38.

Howard, R.W. 1975. *The Dawnseekers: The First History of American Paleontology.* New York: Harcourt Brace Jovanovich.

Jahn, M. E., and D. J. Woolf. 1963. *The Lying Stones of Dr. Johann Bartholomew Adam Beringer, being his Lithographiae Wirceburgensis.* [Translated and annotated by the authors]. Berkeley: University of California Press.

Kennedy, C. B. 1976. A fossil for what ails you: The remarkable history of fossil medicine. *Fossil Magazine*, May 1976: 42–57, 108.

Kuizon, L. 2006. Legislative and regulatory history of paleontological resources. In S. G. Lucas, J. A. Spielmann, P. M. Hester, P. J. Kenworthy, and V. L. Santucci (eds.), *Fossils from Federal Lands*, 102–108. New Mexico Museum of Natural History and Science Bulletin 34.

Lanham, U. 1992. *The Bone Hunters: The Heroic Age of Paleontology in the American West.* New York: Dover Press.

Larson, P.L., and K. Donnan. 2002. *Rex Appeal: The Amazing Story of Sue, the Dinosaur That Changed Science, the Law, and My Life.* Montpelier, Vt.: Invisible Cities Press.

Lazendorf, J. 2000. *Dinosaur Imagery.* New York: Academic Press.

Leiggi, P., C. R. Schaff, and P. May. 1994. Macrovertebrate collecting: field organization and specimen collecting. In P. Leiggi and P. May (eds.), *Vertebrate Paleontological Techniques*, 1: 59–77. Cambridge: Cambridge University Press.

Lhuyd, E. 1699. *Lithophylactii Brittanici Ichnographia.* London.

Long, J. 2003. *The Dinosaur Dealers: Mission–To Uncover International Fossil Smuggling.* London: Allen and Unwin.

Mayor, A. 2000. *The First Fossil Hunters: Paleontology in Greek and Roman Times.* Princeton: Princeton University Press.

——. 2005. *Fossil Legends of the First Americans.* Princeton: Princeton University Press.

Meyer, K. E. 1973. *The Plundered Past.* New York: Atheneum.

Miller, D. 1997. *A Century of Warfare: The History of Conflict in the 20th Century.* New York: Crescent Books.

Mitchell, W. J. T. 1998. *The Last Dinosaur: The Life and Times of a Cultural Icon.* Chicago: University of Chicago Press.

Monastersky, R. 1998. Another dinosaur sells for millions. *Science News* 153: 95.

Nicholas, L. H. 1995. *The Rape of Europa: The Fate of Europe's Treasures in the Third Reich and the Second World War.* New York: Random House Vintage Books.

Noakes, J., and G. Pridham, eds. 1990. *Nazism: A History in Documents and Eyewitness Accounts 1919–1945. Volume II: Foreign Policy, War, and Racial Extermination.* New York: Schocken Books.

Nothdurft, W., J. Smith, M. Lamann, K. Lacovara, J. Poole, and J. Smith. 2003. *The Lost Dinosaurs of Egypt: The Astonishing and Unlikely True Story of One of the Twentieth Century's Greatest Paleontological Discoveries.* New York: Random House.

Oakley, K. 1965. Folklore of Fossils. *Antiquity* 39: 9–16, 117–125.

Padian, K. 2000. Feathers, fakes, and fossil dealers: how the commercial sale of fossils erodes science and education. *Palaeontologia Electronica* 3(2), editorial 2, http://palaeo-electronica.org/2000_2/editor/padian.htm.

Pfeiffer, S. 2001. *Tyrannosaurus Sue: The Extraordinary Saga of the Largest, Most Fought Over T. Rex Ever Found.* New York: W. H. Freeman.

Platt, J. 1758. An account of a fossile thigh-bone of a large animal, dug up at Stonesfield, near Woodstock, in Oxfordshire. *Philosophical Transactions of the Royal Society of London* 50 (II): 542–527.

Plot, R. 1677. *Natural History of Oxfordshire. Being an essay towards the natural history of England.* Oxford.

Psihoyos, L. 1994. *Hunting Dinosaurs.* New York: Random House.

Rogers, K. 1999. *The Sternberg Fossil Hunters: A Dinosaur Dynasty.* Missoula, Mont.: Mountain Press.

Sarjeant, W. A. S. 1997a. History of dinosaur discoveries. In P. J. Currie and K. Padian (eds.), *Encyclopedia of Dinosaurs*, 340–347. New York: Academic Press.

——. 1997b. The earliest discoveries. In J. O. Farlow and M. K. Brett-Surman (eds.), *The Complete Dinosaur*, 3–11. Bloomington: Indiana University Press.

——. 2001. Dinosaurs in Fiction. In D. H. Tanke and K. Carpenter (eds.), *Mesozoic Vertebrate Life*, 504–529. Bloomington: Indiana University Press.

Schmidt, A. C. 2000. The *Confuciusornis sanctus*: An examination of Chinese cultural property law and policy in action. *Boston College International and Comparative Law Review.* Student Publications. 23 (2): 185–228.

Shiskin, M. A. 1992. Russian Triassic amphibians stolen. *Lethaia* 25: 360.

Spalding, D. A. E. 2001. Bones of contention: Charles H. Sternberg's Lost Dinosaurs. In D. H. Tanke and K. Carpenter (eds.), *Mesozoic Vertebrate Life*, 481–503. Bloomington: Indiana University Press.

Steghaus-Kovac, S. 2002. Tug-of-war over mystery fossil. *Nature* 295: 121–1213.

Stokstad, E. 2002. China regains fossils seized in California. *Science* 296: 2311.

Tanke, D. H., N. L. Hernes, and T. E. Guldberg. 2002. The 1916 sinking of the SS *Mount Temple*: historical perspectives on a unique aspect of Alberta's paleontological heritage. *Canadian Palaeobiology* 7: 5–26.

Weishampel, D. B., P. M. Barrett, R. A. Coria, J. Le Loeuff, X. Xing, Z. Xijin, A. Sahni, E. M. P. Gomani, and C. R. Noto. 2004. Dinosaur Distribution. In D. B. Weishampel, P. Dodson, and H. Osmólska (eds.), *The Dinosauria*, 2nd ed., 517–606. Berkeley: University of California Press.

Wellnhofer, P. 1992. Missing *Archaeopteryx*. *Society of Vertebrate Paleontology News Bulletin* 155: 53–54.

West, R. M. 1989. State regulations of geological, paleontological, and archaeological collecting. *Curator* 32: 281–319.

——. 1991. Survey: state regulation of geological, paleontological, and archaeological collecting. *Curator* 34: 199–209.

Woerdeman, M. W. 1948. Morphology in the Netherlands during the Second World War. *Acta Neerlandica Morphologiae* 6: 1–7.

Young, C. C. 1947. Report from China. *Society of Vertebrate Paleontology News Bulletin* 22: 27–30.

Dinosaurs and Evolutionary Theory

Kevin Padian and Elisabeth K. Burton

45

For many of us who work on dinosaurs, they are the prime examples that we consider when we think of how amazing the process of evolution on Earth has been. Where they came from, how they evolved, how all their bizarre and terrifying forms diversified–these were the questions that set our minds spinning from the first time we saw their representations in pictures, books, and movies. This is why we read the books by Ned Colbert and Roy Chapman Andrews under our sheets at night until our flashlight bulbs burned out. A distant world populated by giant reptiles–*reptiles?*–seemed incredible; one might as well have thought of giant, intelligent frogs ruling an alternate past universe, were there no fossils to rein in our imaginations. But it wasn't just the animals themselves. It was what they meant to our understanding of nature that rocked our worlds. Their existence meant that life had changed through time, that animals had evolved to replace those that had become extinct, that perhaps our own species itself was the result of evolution. And we were by no means the first generation to confront these questions.

In fact, this is how the ancient landscape of mosasaurs, ichthyosaurs, plesiosaurs, pterosaurs, and finally dinosaurs revealed itself to European scholars of the late eighteenth and nineteenth centuries, who had to make sense of it all. Put yourself in the place of a common person in their world, knowing nothing at all of what you know today, and think of how you might have reacted to the news of the discovery of *Iguanodon*. A fifty-foot giant lizard with teeth like a hippo? Well, you might not laugh today, but that's only because you know we have records of hundreds of kinds of these things from the Mesozoic Era.

But step back a bit to think about the world in which dinosaurs were first conceived, and it's a different story. Consider that Richard Owen named the Dinosauria in 1842, six years after Darwin returned from the voyage of HMS *Beagle*, five years after the young Queen Victoria ascended the throne (which she occupied until 1901, two years after O. C. Marsh died), a decade after the death of Cuvier, two decades before the start of the American Civil War. In some ways, 1842 seems so close to us, and in other ways so far–like the dinosaurs themselves, really.

And yet, if you looked up "evolution" or "ecology" in the *Encyclopaedia Britannica* (1853–1860) when Darwin was writing *On the Origin of Species by Means of Natural Selection* (1859), you would find no entries for those terms. "Transmutation" gets a very short explanation and is regarded as no more than speculation. But transmutation is what we would call "speciation" today, and it became the main word that Darwin used in the *Origin of Species* to refer to the eventual effects of natural selection and other

45.1. Three pictorial calling cards of Richard Owen, likely from the 1850s.

Collection of the Kevin Padian.

processes. The section on "Palaeontology" in that encyclopedia was written by none other than Richard Owen (Fig. 45.1), who accepted the progression of life through time, but unlike many pre-Darwinian scientists, did not regard it as a process of improvement. Nor, on the other hand, did he accept transmutation as a viable scientific hypothesis.

In the first edition of this book, Hugh Torrens (1997) described how he established (1992) that Owen did not name and define the Dinosauria in his 1841 speech to the British Association for the Advancement of Science, as had been thought, but in the printed version of that talk, which did not appear until 1842 (despite the 1841 publication date on the title page, which refers to the date of the meeting). Between the December 1841 meeting and the publication of its proceedings in spring 1842, Owen's thought matured. His naming of the Dinosauria, and what he made of them, had a great deal to do with his constantly developing biological philosophy.

In this chapter we discuss how dinosaurs have influenced the development of evolutionary theory. At times, we stray as far as to include other Mesozoic reptiles that have contributed to the subject. Our thesis, simply put, is that in roughly pre-*Origin* days, dinosaurs and other fossil vertebrates were instrumental to the formation and testing of evolutionary theory; that from approximately the publication of the *Origin of Species*, the evidence of paleontology—despite spectacular discoveries ranging from *Archaeopteryx* to *Apatosaurus*—had little effect on the development of evolutionary theory; and that in the past century dinosaurs and most other fossil vertebrates have not spurred the development of evolutionary theory, but have been used when possible as test cases for evolutionary hypotheses. (Theories, in scientific parlance, purport to explain substantial bodies of fact and observation; hypotheses test more specific, limited propositions.) However, one can also make the case that recent evidence about the growth rates and trajectories of dinosaurs has affected at least one area of evolutionary theory of vertebrates: the supplanting of typological thinking (see following discussion) by evolutionary thinking.

Designing the Dinosaur

Few papers have provided a more useful perspective to the origin of their field than Adrian Desmond's 1979 essay, "Designing the Dinosaur: Richard Owen's Response to Robert Edmond Grant." Desmond contrasted the ambitious connivance of Owen, a defender of the status quo who would seemingly go to any rhetorical and political lengths to gain ground over his enemies (Desmond 1982), with the fortunes of the hapless Grant, an expert

Table of Strata and Order of Appearance of Vertebrate Life upon the Earth.

Era	Strata	Period	Order of Appearance of Life	Class
TERTIARY or NEOZOIC	Turbary. / Shell Marl. / Glacial Drift. / Brick Earth. (Bone-Caves.)	Pleistocene	**MAN** by Remains. / ... by Weapons.	Birds and Mammals.
	Norwich / Red / Coralline } Crag.	Pliocene	Mammals under present geographical distribution.	
	Faluns. / Molasse.	Miocene	Ruminantia. Quadrumana. / Proboscidia. (Birds, Orders of.) (Mammals, Orders of.)	
	Gyps. / London / Plastic } Clays.	Eocene	Rodentia. / Ungulata. Carnivora.	
SECONDARY or MESOZOIC	Maestricht. / Upper Chalk. / Lower Chalk. / Upper Greensand. / Lower Greensand.	Cretaceous	Cycloid. / Ctenoid. } FISHES. Mosasaurus. Polyptychodon. / BIRDS, by Bones. / Procœlian Crocodilia. / Pterodactyles.	Reptiles.
	Weald Clay. / Hastings Sand. / Purbeck Beds. / Kimmeridgian. / Oxfordian. / Kellovian. / Forest Marble. / Bath-Stone. / Stonesfield Slate. / Great Oolite. / Lias. / Bone Bed. / U. New Red Sandstone. / Muschelkalk. / Bunter.	Wealden / Oolite (U. M. L.) / Trias	Iguanodon. / Marsupials.—Chelonia by Bones. / Pliosaurus. / Birds by Bones and Feathers. / Marsupials. / Ichthyopterygia. (Amphicœlian Crocodilia. Pterosauria. Homocercal Fishes.) / **MAMMALIA** / **AVES,** by Foot-prints. / Sauropterygia. / Labyrinthodontia.	
PRIMARY or PALÆOZOIC	Marl-Sand. / Magnesian Limestone. / L. New Red Sandstone.	Permian	Sauria. / Chelonia, by Foot-prints.	Fishes.
	Coal-Measures. / Mountain Limestone. / Carboniferous Slate.	Carboniferous	**REPTILIA** ganocephala.	Vertebrates.
	U. Old Red Sandstone. / Caithness Flags. / L. Old Red Sandstone. / Ludlow.	Devonian	ganoid. / placo-ganoid. / placoid. (Heterocercal.) / **PISCES**	
	Wenlock. / Caradoc. / Llandeilo. / Lingula Flags. / Cambrian.	Silurian	Echinoderms. Annelides. Bivalves. Trilobites. Pteropods. Brachiopods. Gastropods. Cephalopods, 4-gilled. / Fucoids. Zoophytes.	Invertebrates.

45.2. Richard Owen was no simple creationist, as this chart of geology and life from his *On the Anatomy of Vertebrates* (1866) shows. Like everyone else of his time, he did not know how old the Earth or its geological formations were, but he accepted the progression of life through time. He simply could not accept that its foundational mechanism was entirely materialistic.

invertebrate zoologist and theorist of morphology whose abrasive, naïve personality got the better of him in struggles for status in the scientific community of their day (Desmond 1979; Thomson 2009). Grant was a Lamarckian (see following discussion) in the sense that he accepted what we would today call evolution (transmutation of species) and that life, by becoming more complex through time, showed progressivist tendencies. (*Progressivism* in those days was a neutral term differentiated from progress, which had a more meliorist and teleological slant, suggesting continual improvement and preordained goals.) Also like Lamarck, Grant was a thoroughgoing philosophical materialist: he had no religious sentiments. Owen was not, but he was not a traditional creationist either.

Owen rejected materialism because it denied the hand of the First Cause, the Designer, the Creator, who did not interfere with His work but created the mysterious laws on which all of nature is based. The presence of that First Cause was essential to the upper-class Tory interests that backed Owen, and it was equally essential that he protect their interests by harnessing science to their worldview. Owen's classically trained audience learned that the ideal Platonic entities cannot change, so biological entities also could not change, and so species and other taxa were imanent, eternal forms (Dewey 1910). So, although Owen could accept the progression of life through time (and wrote the first comprehensive British text on *Palaeontology* in 1860; see Fig. 45.2), he could not accept the transmutation of one species into another, which Grant (like Lamarck before him, Chambers contemporaneously, and Darwin after him) was frankly advocating.

It seems strange to us today to think that someone could accept the fossil record but not be able to conceive of a materialist mechanism for the transmutation of one form (species) into another. Such was the Victorian world, full of contradictions (McGowan 2001; Wilson 2002; Young 2007). Owen's answer—as far as we can understand it, because his prose is so obscure—was predicated on the acceptance that all vertebrates are built on a basic body plan, which scholars for generations had called the Archetype (Russell 1916; Desmond 1982; Padian 1997, 2007). The archetype looked like a stylized amphioxus (*Branchiostoma*), but was mostly meant to be an abstract Aristotelian concept, rather than a Platonic ideal form (Rupke 1993), and not an actual, fossilizable animal.

But there the distinction began to blur. Owen seemed to be saying that the production (it would be misleading in today's parlance to use the word *evolution*) of new forms involved a kind of modification of the archetypal plan that was called forth by the demands of ecological habitat, which modified the form of the archetypal plan through developmental change. And, of course, that was only the material basis of the change; the "First Cause" behind it was a matter for theology. Thus, for Owen, the diversity of life was not the result of existing species splitting into new species, but rather of new forms appearing through the independent modifications of a basal archetypal plan, as called forth by environmental exigencies. If you're confused, Owen's contemporaries such as Darwin and Huxley were nonplussed: phrases such as "the continuous operation of the ordained becoming of living things" made their heads spin.

Desmond's (1979) central thesis is that Owen "invented" the Dinosauria as a concept not to show that "reptiles" could approach the condition of birds and mammals, but to show that *all* they could do was approach it. They differed from other reptiles, certainly, because they were so large, but were not aquatic like the ichthyosaur and plesiosaur; they had five sacral vertebrae instead of only two (this at least was true of the forms available to Owen), and their hips and legs showed that they stood upright like birds and mammals (Fig. 45.3). However, Owen claimed, the similarities were only

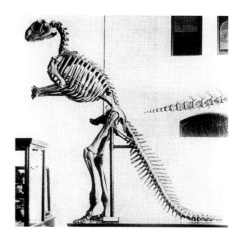

45.3. The original mount of *Hadrosaurus* in the Academy of Natural Sciences in Philadelphia. The skull was based on *Iguana;* the kangaroo-pose was constructed by the famous artist Waterhouse Hawkins, after studies by Joseph Leidy.

superficial. Dinosaurs were reptiles, not birds or mammals, as their morphology clearly showed. So all the presumably "advanced" features were, as Owen could claim, "purely adaptive." That is, they were just reflections of the animal's function, not of its physiology (Padian 1995). In this way, Desmond argued, Owen could neutralize the arguments of his progressivist opponents by showing that ancient reptiles were more sophisticated in structure than their present-day counterparts. Still, they were only reptiles. So much for progressivism. Furthermore, by showing how different dinosaurs were from other reptiles, he could frustrate the allegation that they must have evolved from some other ancient species.

This kind of argument is often called *typological* or *essentialist*, although the terms are not interchangeable. *Typology* has the odor of "stereotype," and it refers to characterizing a whole class of things by the possession of certain features, thereby emphasizing these constancies rather than the possibilities afforded by the variation among individuals in the class. *Essentialism* is similar in emphasizing constancies, but it explains these by reference to an ideal form. Ernst Mayr (e.g., 1963, 1982) railed against typology in twentieth-century biology, but he meant something rather different than Owen did. Mayr was contrasting the "type" specimen approach of taxonomists at the time, an approach that seemed to generalize the characteristics of the "type" standard-bearer of the species to all its individuals, with the "populational" approach that he favored, in which natural variation in populations would not be discounted but seen for what it really was, the raw material of evolution.

So it is strange that Chinsamy and Hillenius (2004) would advance the argument that because ectothermy is the primitive condition for amniotes, dinosaurs were likely ectothermic and that it is not even necessary to substantiate this position with evidence; it is in effect a default conclusion. Other biologists (e.g., Paladino et al. 1997) have asserted that dinosaurs were basically ectotherms, and that large dinosaurs were no more than "inertial homeotherms," maintaining a constant body temperature because they were so large. It is true, but trivial, to note that large animals have more stable body temperatures than smaller ones; they are more resistant to the vicissitudes of ambient temperatures. However, the phylogenetic record shows that dinosaurs clearly departed from other reptiles in their growth rates and underlying metabolisms (Padian et al. 2001), and actual growth data show that models of dinosaurian growth based on living animals are far off the mark (Lee and Werning 2008). Furthermore, small dinosaurs could not have been inertial homeotherms, and constancy of temperature says nothing either way about metabolic or growth rates. Paladino and his colleagues, like Chinsamy and Hillenius, are unwittingly repeating the same typological arguments used by Richard Owen (Padian and Horner 2002, 2004).

If the very taxonomic creation of the Dinosauria was amid swirls of evolutionary controversy, how have the dinosaurs stimulated evolutionary thinking since the 1840s?

How Can Fossils Influence Evolutionary Theory?

E. C. Olson (1966) asked this question and determined that with respect to the two main pillars of evolutionary theory, paleontology could do a great deal to illuminate the history and phylogeny of life, but could say little directly about natural selection. We agree with his first point but have reservations about the second. It is true that at the populational level the fossil record cannot practicably measure natural selection in any testable way (adaptive advantage of phenotypes, heritability of advantages, or differential reproduction of phenotypes from generation to generation: Brandon 1996). However, within a given lineage or clade, the effects of natural selection certainly can be tested in the measurable improvement or elaboration of structures that can be linked to particular functions or behaviors. One can see in the lineage of ornithopods, for example, how simple leaf-shaped teeth give way to a more heterodont dentition with a diastema and cheek teeth more adapted to crushing plants, a trend taken to extremes by the great grinding batteries of the iguanodontids and hadrosaurs. This is the macroevolutionary dimension of adaptation, which by definition, is the result of natural selection (Williams 1992). But Olson was quite right that in the populational sense of this question, it is difficult for fossils to contribute.

Olson (1966) allowed that many ideas about how evolution generally works have come from studies of the record of past life; but he noted too that authors from Chambers (anonymously, 1844) to "Cope, as a neo-Lamarckian, Osborn with his concept of aristogenesis, Schindewolf and catastrophism or neocatastrophism, and Teilhard de Chardin with the Omega point" stood as evidence that speculation on that basis has not always been unconstrained, and its hypotheses not always enduring. Olson was writing just before the renaissance of paleobiology and macroevolution that began in the 1970s, notably with the publication of T. J. M. Schopf's seminal edited volume *Models in Paleobiology* (1972), which featured among other landmark articles the first full statement of Eldredge and Gould's hypothesis of punctuated equilibria, and the founding soon after by Ralph Johnson and his colleagues of the journal *Paleobiology*. But Olson predicted that new insights would continue to come from paleontological induction. Of course, he was talking about paleontology in general, not just about dinosaurs or other Mesozoic vertebrates.

Yet Mesozoic reptiles played a large part in the understanding of two very important axioms of natural history that were established in the eighteenth century. The first is that faunas and floras had changed through time, and the second is that the animals and plants of the past were largely extinct. The first axiom resulted from the understanding of stratigraphy that culminated in the first major geologic map of England and Scotland, produced by William Smith in 1801 (Winchester 2001), and later in the mapping of the horizons of the Paris Basin by Brongniart and Cuvier (Rudwick 1997). The second is commonly attributed to Cuvier with his demonstration that the mosasaur was extinct. However, the concept of extinction has a longer history. It was stated by Steno in the mid-1600s (but ignored), reiterated by Buffon in his mid-eighteenth-century *Encyclopedia of Natural History*, and by Lamarck. But by the late 1700s, it was understood among European savants that the creatures found as fossils did not represent living forms (and vice versa), and that the appearance of fossil plants and animals in places where they cannot live today indicates that the climate of the Earth has changed through time. This was quite obvious, for example, to Collini, who described in 1784 the first known remains of what Cuvier later called

45.4. One of two sketches of a reconstruction of the first known pterosaur specimen, a small *Pterodactylus* in the Natural History Cabinet at Mannheim, sent by Professor Jean Hermann of Strasbourg to Cuvier in 1800. These are the earliest known restorations of a pterosaur. For why Hermann reconstructed it as he did, see Taquet and Padian (2004).

the "ptero-dactyle" (Taquet and Padian 2004; Fig. 45.4). These facts were also perfectly well known to Thomas Jefferson and Benjamin Franklin (Thomson 2008a, 2008b).

Thus, despite any classical ideas about the Great Chain of Being, the Natural System, and the impossibility of extinction (showing the Creator's work was not perfect), the fossil record clearly showed that floras and faunas, as well as climates, had changed continually through time, and that extinction was a ubiquitous fact of life (Bowler 2003). Among the great extinct reptiles that drove this point home, mosasaurs, ichthyosaurs, plesiosaurs, and pterosaurs were the most prominent (Desmond 1976; McGowan 2001; Cadbury 2000; Taquet and Padian 2004). Dinosaurs did not play a role here because their remains were not recognized for what they were; before Owen named them, they were generally thought to be some sort of giant crocodile, lizard, or mammal.

Darwin and Dinosaurs

Darwin does not mention dinosaurs in his published work, the watershed of evolutionary theory in Victorian times. There are several reasons for this. First, he never studied them. Second, they were the subjects of huge fights and arguments, first between Owen and Mantell (Rupke 1994; Dean 1999, Cadbury 2000), and later between Owen and Huxley (Desmond 1994), and Darwin had no wish to annoy Owen any more than he had to. Third, whereas Darwin clearly thought about the macroevolutionary scales of time, space, and change, he wrote mostly about phenomena that took place within species. He was trying to get his audience to accept that there was natural variation within species (he did not really think at the population level) and that some of this variation would be useful to its bearers in some circumstance, enabling them to survive better and leave more offspring with these traits to the next generation. His clear message was that of extrapolation (Gould 2002): given enough time and opportunity (or pressure), these changes would cause adaptation of some segments of a species to branch off and form a new species. However, he had no concrete evidence for this. So the great extinct beasts that held everyone's interest in thrall, and that proved a great success and spectacle in the Crystal Palace Exhibition of 1851, did not figure in his theory of evolution. On the other hand, Darwin's strong supporter, T. H. Huxley (1868, 1870a, b), was the first to propose the ancestry of birds from dinosaurs, thus crossing traditional Linnean boundaries in a daring and dramatic way.

Similarly, when neo-Darwinism, the Modern Synthesis of Evolution, arose in the early decades of the twentieth century, it also had little to do with the lessons of paleontology (Gould 1980); it focused on population biology and models, not on macroevolutionary theory, which was thought to take care of itself by extrapolation from populational processes—just as Darwin had assumed. So the Synthesis, by leaving out paleontology from making any theoretical predictions of note, was "unfinished" (Eldredge 1985), and so it has remained (Gould 2002).

This does not mean, however, that dinosaurs and their ilk could not or did not contribute to the formation of evolutionary theory; it means simply that the part of evolutionary theory encompassed by the Modern Synthesis had little use for them. But the Modern Synthesis also paid very little attention to morphological theory (Davis 1949; Ghiselin 1980; Hamburger 1980), functional morphology and biomechanics, or development; in fact, in Ernst Mayr's (1963) 800-page magnum opus *Animal Species and Evolution*, the bible of late-century evolution students, only one page is devoted to the topic of evolution and development, and it is limited to stating that little is understood on this topic—despite decades of experiments and observations by Bateson, deVries, Waddington, Lerner, Schmalhausen, Riedl, and many others. For most of two decades now, evolutionary developmental biology ("evo-devo"), the fusion of paleontology, morphology, phylogenetics, developmental biology, and developmental genetics, has been the hottest area in biology (e.g., Carroll et al. 2005). The questions were always there; we just needed the tools to solve them and the interest in framing evolutionary questions that addressed more the origin of natural variations than their deployment in populations under the agents of selection, drift, and so on. It remains to be seen what relevance, if any, the populational thinking of the Modern Synthesis will have for the "post-Modern" synthesis that evo-devo and related fields represent. But dinosaurs are already in play.

Dinosaurs as Exemplars of Evolutionary Principles

Let us turn, then, to a few historical examples of how dinosaurs have been used either to stimulate or to test hypotheses in evolutionary biology. The foremost student of these problems has been David B. Weishampel, who both singly and in collaboration with colleagues such as Cora Jianu and Wolf Reif, has plumbed the evolutionary thinking of important historical figures in dinosaur paleontology.

The nineteenth century did not lack colorful figures in paleontology (Wallace 1999), nor notions about how evolution worked that often invoked the mystical, the teleological, and the vitalistic (see Olson's [1966] remarks in previous discussion). The Law of Irreversibility was among many supposed laws and rules that applied to evolution. It denied the possibility that structures once lost could be regained. This, at least, was the view of Belgian paleobiologist Louis B. Dollo, who was one of the first scientists to do a comprehensive study of the skeletons of the dinosaur *Iguanodon bernissartensis* that had been excavated from a coal mine in Belgium (Dollo 1883). The hands of *Iguanodon* had a couple of strange features, including a blocky set of wrist bones, a conical thumb spike formed of the coalesced metacarpal and two phalanges of digit I, and – although Dollo never seems to have mentioned it – a fifth digit that bore four phalanges instead of the usual two. Basal reptiles and other basal amniotes had four phalanges, but dinosaurs did not. Here, it would seem, is an example of the Law of Irreversibility that Dollo himself may have overlooked.

An explicit argument for irreversibility was made by Gerhard Heilmann, an artist who wrote the most influential book in the twentieth century on the origin of birds (Heilmann 1926). After an exhaustive survey of the skeleton and soft part systems of living birds and other reptiles, plus their fossil relatives, Heilmann was all but convinced that birds must have evolved from small carnivorous dinosaurs. He stopped short of this conclusion because no known theropod was found with clavicles. Clavicles were known as the wishbone (furcula) of birds; if theropods had lost them, they could not be regained, according to Dollo's Law, so, for Heilmann, theropods could not be the ancestors of birds. Heilmann threw avian origins back to a much more remote and shadowy group of reptiles, the "thecodonts," and there the problem essentially remained until John Ostrom successfully argued for the theropod ancestry in the 1970s.

The origins of birds and their flight were popular topics in the late nineteenth century, and this subject has often been reviewed (e.g., Ostrom 1974, 1975; Gauthier 1986; Shipman 1998; Dingus and Rowe 1998; Padian and Chiappe 1998; Padian 2001a, b). They are separate problems in the sense that bird origins is a straightforward phylogenetic question, whereas the origin of flight involves many aspects of functional morphology, aerodynamics, paleoecology, and physiology, against which the phylogeny serves as an independent line of evidence like any other (Padian 2001a). But it has not always been so; in fact, for most of the history of these questions, they have been inseparable for some workers (e.g., Bock 1986).

Perhaps the strongest contributions to evolutionary theory of the "bird origins" problem are that it forced biologists to consider the origin of one traditional Linnean "Class" from another (Aves from "Reptilia"), and that it showed that single lines of evidence were insufficient to solve major evolutionary questions. Famously, T. H. Huxley urged Darwin to make use of the newly discovered specimens of *Archaeopteryx* in later editions of the *Origin of Species*, but Darwin demurred (again, he was talking about changes within species, not those that could be seen to connect major groups; and Owen and others were still arguing about what exactly *Archaeopteryx* was; Desmond 1976). O. C. Marsh was greatly heartened by a congratulatory letter he received from Darwin on receipt of Marsh's (1880) monograph on the Odontornithes, or toothed Cretaceous birds, including *Hesperornis* and *Ichthyornis* (Schuchert and LeVene 1940) – although Darwin never mentioned these animals in his work either.

The discovery of clavicles in several nonavian theropods removed the obstacle of Heilmann's classic objection to the theropod origin of birds; the demonstration that the fingers of the bird hand are morphologically I-II-III rather than II-III-IV, a problem that had split ornithologists for over a century, could only have been solved by extinct theropods that bore the intermediate stages (Padian and Chiappe 1998). However, it is difficult to say that these are really substantial contributions to general evolutionary theory, although the latter discovery is clearly a counterexample to Morse's "law" that digits always reduce from both lateral and medial sides. However, these problems could not have been solved with the tools of population biology, and not even with those of developmental biology alone.

The origin of birds and their flight was only one of many paleobiological subjects treated by the colorful Hungarian paleontologist Baron Franz von Nopcsa in the early decades of the twentieth century. Weishampel and Reif (1984) reviewed Nopcsa's impressive contributions to dinosaurian paleobiology and macroevolution. Although, as they noted, Nopcsa's ideas were sometimes peculiar or poorly informed, there is no doubt of his creative genius or of the utility of many of his contributions. For our present purposes, a few examples will suffice. In the early 1900s, Nopcsa published a series of papers that described several new species of dinosaurs from the Siebenbürgen region of Romania; their small size and relatively primitive characteristics led him to the inference that they were dwarf species (see Jianu and Weishampel 1999), and that Siebenbürgen represented an island area during the Cretaceous. Weishampel and Reif (1984) commented that Nopcsa's work was often overlooked in later considerations of European Cretaceous faunas. The island arc concept has recently been resuscitated with interesting new dinosaur evidence (Benton et al. 1997, 2006; Sander et al. 2006).

Taking a stand against other neo-Lamarckians of his day and using "several cases of reversible evolution in fossil reptiles" (Weishampel and Reif 1984), Nopcsa proposed a significant reinterpretation of Dollo's Law of Irreversibility. In doing so, he suggested that an organism in immediate threat of extinction can only dodge this fate through "reversal to an embryonic stage," a concept familiar to us today as heterochrony. However, rather than reject Dollo's Law outright, he simply modulated it with his observations, astutely noting that evolutionary adaptation "is only possible to the degree that is allowed by its physiology." He further emphasized the role of physiology in the phenomenon of parallel evolution; he posits that the mastodon, for example, could speciate from its ancestor in several different places simultaneously "because the same environmental stimulus hits the same physiological system" in many individuals at once, and therefore parallel evolution is a common consequence of the dominating factor of physiology. There is nothing at all wrong with this reasoning, even though it creates conniptions for simple cladistics; Osborn used his own similar observations to argue for a mystical process called aristogenesis that allegedly directed these patterns.

Nopcsa's other interests included dinosaur taxonomy, jaw mechanics, sexual dimorphism, paleohistology, and some theoretical ideas about evolution and tectonics that were rather dated or poorly informed even in his time (Weishampel and Reif 1984). Yet his reputation as an original and synthetic thinker endures. Nopcsa amassed a large collection of histological slides of dinosaurs and other vertebrates, the whereabouts of which are currently unknown (Weishampel and Reif 1984). He seems to have started this work, which has a tradition in paleontology going back to the mid-nineteenth century, as a way to try to identify isolated bone fragments taxonomically. However, he gave up when he realized that histology varies among elements of skeletons and through growth, which shows an insight not always appreciated by later workers. Nopcsa also recognized that it was possible to tell adults from juveniles by their histological profiles, a perception still useful today (Nopcsa and Heidsieck 1933).

For some decades after Nopcsa, dinosaurs had very little to do with evolutionary theory. These decades correlated with quiet times that produced very little new research either in the field or the laboratory. The Dinosaur Renaissance of the 1970s began with John Ostrom's fairly innocuous article (1970) that suggested that living reptiles may not be good models for dinosaurs in the "typical" way often pictured, because living reptiles live in a great variety of habitats and climates, and dinosaur distributions suggest an even more varied regime of habitats and climates. Claims that dinosaurs were therefore "warm-blooded" (whatever that meant to individual authors) became the biggest question in vertebrate paleontology of the 1970s, marshaling a variety of lines of evidence including social behavior, stance and gait, bone histology, parental care, biogeography, and functional morphology (summaries in Desmond 1976; Thomas and Olson 1980). Just as vociferous were the opponents of these claims; and although little was settled in those years, the effect on paleontology may have been to begin thinking of extinct animals more as once-living, and hence tractable to biological analysis, than as long-dead and intractable to biological analysis.

A Frontier for Dinosaurs in Evolutionary Theory?

Many kinds of studies, from systematics (Gauthier 1986; Sereno 1986) to biogeography (Upchurch et al., 2002) to molecular biology (Schweitzer and Marshall 2001), have used dinosaurs to apply methods, techniques, and questions that are used to study virtually every other group of organisms, past and present. In the main, however, for more than a century, dinosaurs and other Mesozoic reptiles have not been the basis on which new evolutionary insights and hypotheses have been proposed—at least not those that affect more than the individual taxa in question. But this situation applies to nearly all taxa, from snails to whales, from *Drosophila* to *Anolis*. The iconic organisms of evolutionary textbooks—fruit flies, flour beetles, Caribbean lizards, stickleback fishes, Galapagos finches—all have their evolutionary stories to tell, but in any of them do we find real generalizations about evolutionary theory? Or is it more accurate to say that they embody certain kinds of evolutionary processes and phenomena as good examples of the expectations of theory?

If the latter is correct, as we think it is—because a phenomenon found in one organism can best stimulate research to see if it applies to others—then dinosaurs have their role to play in evolutionary theory, provided that the right questions are asked. As noted previously, the paleontological record best addresses macroevolutionary questions, rather than population- or genetic-level questions; this particularly applies to organisms such as fossil vertebrates that are generally less numerous and less represented stratigraphically than invertebrates.

So, what can "dinosaurs as dinosaurs," to use Peter Dodson's (1974) phrase, tell us about evolutionary theory as it relates to themselves? We submit that new dinosaur research has given us a synthetic and integrative picture of what dinosaurs were like as living animals. They began as small components of the Late Triassic fauna alongside their very similar precursors (Irmis et al. 2007), yet they were different. They were almost all bipedal; their stance was erect, and their gait parasagittal. They grew much more quickly than the other reptiles that had dominated their ecosystems but mostly died out by the end of the Triassic (Padian et al. 2001). Dinosaurs diversified quickly in the Jurassic, evolving several herbivorous lineages, some of which attained large size and quadrupedality. They produced large clutches of eggs, but not as large as those of other reptiles (Padian and Horner 2004). At least some of them appear to have invested substantial parental care (Horner and Makela 1979; Norell et al., 1995). They evolved a great array of horns, crests, frills, spikes, plates, and armor that have suggested several functions, but were almost certainly used in species recognition (Main et al. 2005).

These and many other new discoveries about dinosaurian biology lead us to the very problem that began this essay, the one that occasioned the baptism of dinosaurs themselves: their separation from other reptiles. Dinosaurs were not very much like reptiles of today, and their biology cannot simply be explained by enlarged models of lizards or crocodiles (Padian and Horner 2002, 2004; Lee and Werning 2008); the actual data don't fit the models. Richard Owen was right and wrong: dinosaurs were certainly different from other reptiles, as he perceived, but they were not simply glorified reptiles that were "adaptively modified" in some ways while still retaining the basic physiology of turtles or lizards. Many of the changes that separate birds from other reptiles first appeared in nonavian dinosaurs (Padian and Chiappe 1998; Padian 2001a). The acceptance that the evolution of birdlike features was a long road that began in the Triassic, well before birds were even a glint in an ancestral maniraptoran's eye, is perhaps the principal insight that dinosaurs have made to our understanding of their own biology in recent years.

Probably the greatest contribution that dinosaurs have made to evolutionary theory is in the theory of their own evolution. And because their evolutionary trajectories were so different in so many fundamental ways from those of living vertebrates, they provide great lessons for the possibilities of vertebrate evolution that are not afforded by our living menagerie. This, in the end, is probably what attracts so many of us to them.

Using Dinosaurs in Teaching Evolutionary Theory

How dinosaurs have been involved in the development of evolutionary theory is an academic, professional exercise that is mainly the interest of scholars of these arcane matters. It is a different matter to consider the transmission of what scholars know about dinosaurs to the classroom. Unless the insights that have been gained about the biology of dinosaurs, and what this has to say about the understanding of biology in general, are transmitted to the next generation, they will be available only to scholars. And the often-repeated but misinformed ideas in K–12 textbooks will continue.

Dinosaurs and other fossil vertebrates can be of great use in illustrating evolutionary principles and concepts, particularly macroevolutionary ones. The evolution of major groups and major adaptations–the emergence of tetrapods onto land, the rise to dominance of dinosaurs, the origin of birds, the origin of mammals, the return of whales to the sea, and many more major developments in evolution–can only be demonstrated by recourse to the history of life. Microevolution, the change that occurs within populations, is nonproblematic for antievolutionists (because they can regard it simply as variation within "created kinds"), and is covered copiously in most U.S. high school and college texts. Over time, however, creationists have successfully blocked full explanation of major changes in evolution from K–12 curricula, in some cases even arranging for the term *evolution* to be replaced by wording such as "change over time."

The explanation of major evolutionary change needs to be presented in more than a cartoonish account of a procession of groups appearing and disappearing through time, as it is in K–12 and even most college-level books (Padian 2008). Instead, students should be taught the specific methods by which evolutionists understand and study how major evolutionary changes occur. If concepts such as major evolutionary change, which involve illustrations with dinosaurs and other extinct animals, are taught in university curricula and well represented in university textbooks, they can be brought to the K–12 level; with no representation in university curricula, however, there is no reason to teach these concepts below the college level (Padian 2008).

For example, Fig. 45.5 shows some salient features in the emergence of birds from other theropod dinosaurs. Following the nodes along the backbone of the tree, students can learn the sequence of evolution of features related to the skeleton, flight apparatus, feathering, and many other features. These figures can be supplemented with illustrations that teach students

45.5. Cladogram showing the sequence of acquisition of some major adaptive features in the origin of birds and their flight. From Padian 2001a.

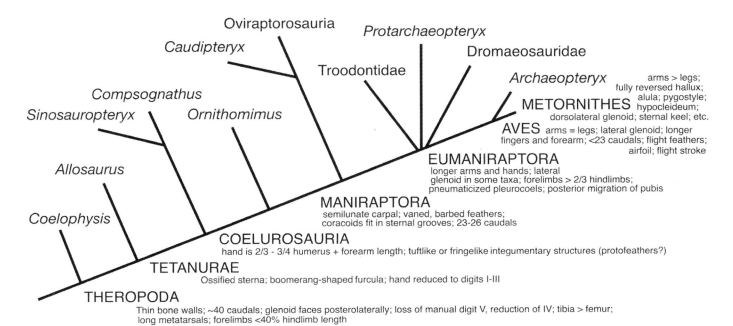

that macroevolution is not simply speculation and arm-waving; rather, there is a variety of independent sources of evidence, including phylogenies corroborated by many sets of characters in the skeletons and by fossil evidence of integumentary coverings, including several kinds of fossils no longer seen in living birds. In this way, the dinosaurs will continue to teach us all about evolutionary theory—and to provide an even more fascinating dimension of science to future generations of students reading about them under the covers at night.

Acknowledgments

We thank Randall Irmis and Sarah Werning for comments on the manuscript, and D. B. Weishampel for help with historical questions. This is University of California Museum of Paleontology Contribution No. 2002.

References

Benton M. J., E. Cook, D. Grigorescu, E. Popa, and E. Tallodi. 1997. Dinosaurs and other tetrapods in an Early Cretaceous bauxite filled fissure, northwestern Romania. *Palaeogeography, Palaeoclimatology, Palaeoecology* 130: 275–292.

Benton, M. J., N. J. Minter, and E. Posmosanu. 2006. Dwarfing in ornithopod dinosaurs from the Early Cretaceous of Romania. In Z. Csiki (ed.), *Mesozoic and Cenozoic Vertebrates and Paleoenvironments; Tributes to the Career of Prof. Dan Grigorescu*, 79–87. Bucharest: Ars Docendi.

Bock, W. J. 1986. The arboreal origin of avian flight. *Memoirs of the California Academy of Sciences* 8: 57–72.

Bowler, P. J. 2003. *Evolution : The History of an Idea* (2nd ed.). Berkeley: University of California Press.

Brandon, R. N. 1996. *Concepts and Methods in Evolutionary Biology*. Cambridge: Cambridge University Press.

Cadbury, D. 2000. *Terrible Lizard: The First Dinosaur Hunters and the Birth of a New Science*. New York: Henry Holt & Co.

Carroll, S. B., J. K. Grenier, and S. D. Weatherbee. 2005. *From DNA to Diversity: Molecular Genetics and the Evolution of Animal Design*. Malden, Mass.: Blackwell Publishing.

[Chambers, R.] 1844. *Vestiges of the Natural History of Creation*. London: John Churchill.

Chinsamy, A., and W. J. Hillenius. 2004. Physiology of non-avian dinosaurs. In D. B. Weishampel, P. Dodson, and H. Osmólska (eds.), *The Dinosauria*, 643–659. Berkeley: University of California Press.

Collini, C. 1784. Sur quelques zoolithes du Cabinet d'Histoire naturelle de S.A.S.E. Palatine et de Bavière, à Mannheim. *Acta Academia Theodoro-Palatinae, Pars Physica* 5 (1784): 58–103.

Darwin, C. 1859. *On the Origin of Species by Means of Natural Selection*. London: John Murray.

Davis, D. D. 1949. Comparative anatomy and the evolution of vertebrates. In G. L. Jepson, E. Mayr, and G. G. Simpson (eds.), *Genetics, Paleontology, and Evolution*, 64–89. Princeton, N.J.: Princeton University Press.

Dean, D. R. 1999. *Gideon Mantell and the Discovery of Dinosaurs*. Cambridge: Cambridge University Press.

Desmond, A. J. 1976. *The Hot-Blooded Dinosaurs: a Revolution in Palaeontology*. New York: Dial.

———. 1979. Designing the dinosaur: Richard Owen's response to Robert Edmund Grant. *Isis* 70: 224–234.

———. 1982. *Archetypes and Ancestors: Palaeontology in Victorian London 1850–1875*. London: Blond & Briggs.

———. 1994. *Huxley: The Devil's Disciple*. London: Michael Joseph.

Dewey, J. 1910. The influence of Darwinism on philosophy. In J. Dewey, *The Influence of Darwinism on Philosophy and Other Essays in Contemporary Thought*, 1–19. New York: Henry Holt & Co.

Dingus, L., and T. Rowe. 1998. *The Mistaken Extinction: Dinosaur Evolution and the Origin of Birds*. New York: W.H. Freeman.

Dodson, P. 1974. Dinosaurs as dinosaurs. *Evolution* 28: 494–497.

Dollo, L.1883. Note sur les restes de dinosauriens rencontrés dans le Crétacé Supérieur de la Belgique. Bulletin du Musée Royal d'Histoire Naturelle de Belgique 2: 205–221

Eldredge, N. 1985. *Unfinished Synthesis: Biological Hierarchies and Modern Evolutionary Thought.* New York: Oxford University Press.

Encyclopedia Britannica (1853–1860), 8th ed. Boston: Little, Brown & Co.

Gauthier, J. A. 1986. Saurischian monophyly and the origin of birds. *Memoirs of the California Academy of Sciences* 8: 1–55.

Ghiselin, M. T. 1980. The failure of morphology to assimilate Darwinism. In E. Mayr and W. B. Provine (eds.), *The Evolutionary Synthesis: Perspectives on the Unification of Biology,* 180–192. Cambridge, Mass.: Harvard University Press.

Gould, S. J. 1980. Is a new and general theory of evolution emerging? *Paleobiology* 6: 119–130.

———. 2002. *The Structure of Evolutionary Theory.* Cambridge, Mass.: Belknap Press.

Hamburger, V. 1980. Evolutionary theory in Germany: a comment. In E. Mayr and W. B. Provine (eds.), *The Evolutionary Synthesis: Perspectives on the Unification of Biology,* 180–192. Cambridge, Mass.: Harvard University Press.

Heilmann, G. 1926. *The Origin of Birds.* New York: Appleton & Co.

Horner, J. R., and R. Makela. 1979. Nest of juveniles provides evidence of family structure among dinosaurs. *Nature* 282: 296–298.

Huxley, T. H. 1868. On the animals which are most nearly intermediate between birds and reptiles. *Geological Magazine* 5: 357–365.

———. 1870a. Further evidence of the affinities between the dinosaurian reptiles and birds. *Quarterly Journal of the Geological Society of London* 26: 12–31.

———. 1870b. On the classification of the Dinosauria, with observations on the Dinosauria of the Trias. *Quarterly Journal of the Geological Society of London* 26: 31–50.

Irmis, R. B., S. J. Nesbitt, K. Padian, N. D. Smith, A. H. Turner, D. Woody, and A. Downs. 2007. A Late Triassic dinosauromorph assemblage from New Mexico and the rise of dinosaurs. *Science* 317: 358–361.

Jianu, C. M. and D. B. Weishampel. 1999. The smallest of the largest: a new look at possible dwarfing in sauropod dinosaurs. *Geologisches Mijnbouw* 78: 335–343.

Lee, A. H., and S. Werning. 2008. Sexual maturity in growing dinosaurs does not fit reptilian growth models. *Proceedings of the National Academy of Sciences* (USA) 105: 582–587.

Main, R. P., A. de Ricqlès, J. R. Horner, and K. Padian. 2005. The evolution and function of thyreophoran scutes: implications for plate function in stegosaurs. *Paleobiology* 31: 293–316.

Marsh, O. C. 1880. Odontornithes: A monograph on the extinct toothed birds of North America. *US Geological Exploration of the 40th Parallel.* Washington, D.C.: U.S. Government Printing Office.

Mayr, E. 1963. *Animal Species and Evolution.* Cambridge, Mass.: Harvard University Press.

———. 1982. *The Growth of Biological Thought.* Cambridge, Mass.: Belknap Press.

McGowan, C. 2001. *The Dragon Seekers: How an Extraordinary Circle of Fossilists Discovered the Dinosaurs and Paved the Way for Darwin.* Cambridge, Mass.: Perseus.

Norell, M. A., J. M. Clark, L. M. Chiappe, and D. Dashzeveg. 1995. A nesting dinosaur. *Nature* 378: 774–776.

Nopcsa, F., and E. Heidsieck. 1933. On the bone histology of the ribs in immature and half-grown trachodont dinosaurs. *Proceedings of the Royal Zoological Society* 1933: 221–223.

Olson, E. C. 1966. The role of paleontology in the formulation of evolutionary thought. *BioScience* 16: 37–40.

Ostrom, J. H. 1970. Terrestrial vertebrates as indicators of Mesozoic climates. *North American Paleontological Convention (Chicago, IL, 1969), Proceedings D:* 347–376.

———. 1974. *Archaeopteryx* and the origin of flight. *Quarterly Review of Biology* 49: 27–47.

———. 1975. The origin of birds. *Annual Reviews of Earth and Planetary Sciences* 3: 55–77.

Owen, R. 1842. Report on British fossil reptiles, Part II. *Reports of the British Association for the Advancement of Science* 1841: 60–204.

———. 1866. On the Anatomy of Vertebrates. London: Longmans, Green & Co.

Padian, K. 1995. Pterosaurs and typology: archetypal physiology in the Owen-Seeley dispute of 1870. In W.A.S. Sarjeant (ed.), *Vertebrate Fossils and the Evolution of Scientific Concepts,* 285–298. Yverdon, Switzerland: Gordon and Breach.

———. 1997. The rehabilitation of Richard Owen. *BioSience* 47(7): 446–452.

———. 2001a. Cross-testing adaptive hypotheses: phylogenetic analysis and the origin of bird flight. *American Zoologist* 41: 598–607.

———. 2001b. The false issues of bird origins: an historiographic perspective. In J. A. Gauthier and L. M. Gall (eds.), *New Perspectives on the Origin and Early Evolution of Birds,* 485–499. New Haven: Yale University Press.

———. 2007. Richard Owen's Quadrophenia: The pull of opposing forces in Victorian cosmogony. Introductory essay in R. Amundson (ed.), *The Nature of Limbs* by Richard Owen, 53–91. Chicago: University of Chicago Press.

———. 2008. Trickle-down evolution: an approach to getting major evolutionary adaptive changes into textbooks and curricula. *Integrative and Comparative Biology* 48: 175–188.

Padian, K., and L. M. Chiappe. 1998. The origin and early evolution of birds. *Biological Reviews* (Cambridge) 73: 1–42.

Padian, K., and J. R. Horner. 2002. Typology versus transformation in the origin of birds. *Trends in Ecology and Evolution* 17: 120–124.

———. 2004. Dinosaur physiology. In D. B. Weishampel, P. Dodson, and H. Osmólska (eds.), *The Dinosauria,* 660–671. Berkeley: University of California Press.

Padian, K., A. J. de Ricqles, and J. R. Horner. 2001. Dinosaurian growth rates and bird origins. *Nature* 412: 405–408.

Paladino, F. V., J. R. Spotila, and P. Dodson. 1997. A blueprint for giants: modeling the physiology of large dinosaurs. In J. O. Farlow and M. K. Brett-Surman (eds.), *The Complete Dinosaur,* 491–504. Bloomington: Indiana University Press.

Rudwick, M. J. S. 1997. *Georges Cuvier, Fossil Bones, and Geological Catastrophes.* Chicago: University of Chicago Press.

Rupke, N. A. 1993. Richard Owen's vertebrate archetype. *Isis* 84: 231–251.

———. 1994. *Richard Owen: Victorian Naturalist.* New Haven: Yale University Press.

Russell, E. S. 1916. *Form and Function: A Contribution to the History of Animal Morphology.* London: John Murray.

Sander, M., O. Mateus, T. Laven, and N. Knötschke. 2006. Bone histology indicates insular dwarfism in a new Late Jurassic sauropod dinosaur. *Nature* 441: 739–741.

Schopf, T. J. M. (ed.) 1972. *Models in Paleobiology.* San Francisco: Freeman, Cooper & Co.

Schuchert, C. S., and C. M. LeVene. 1940. *O.C. Marsh: Pioneer in Paleontology.* New Haven: Yale University Press.

Schweitzer, M. H., and C. L. Marshall. 2001. A molecular model for the evolution of endothermy in the theropod-bird lineage. *Journal of Experimental Zoology*

(*Molecular Development and Evolution*) 291: 317–338.

Sereno, P. C. 1986. Phylogeny of the bird-hipped dinosaurs (Order Ornithischia). *National Geographic Research* 2: 234–256.

Shipman, P. 1998. *Taking Wing:* Archaeopteryx *and the Evolution of Bird Flight.* New York: Simon & Schuster.

Taquet, P., and K. Padian. 2004. The earliest known restoration of a pterosaur and the philosophical origins of Cuvier's *Ossemens Fossiles. PalEvol (Comptes Rendus à l'Academie des Sciences, Paris)* 3: 157–175.

Thomas, R. D. K., and E. C. Olson (eds.). 1980. *A Cold Look at the Warm-Blooded Dinosaurs.* AAAS Selected Symposium 28. Boulder, Colo.: Westview Press.

Thomson, K. S. 2008a. Jefferson, Buffon, and the moose. *American Scientist* 96: 200–202.

——. 2008b. *The Legacy of the Mastodon: The Golden Age of Fossils in America.* New Haven: Yale University Press.

——. 2009. *The Young Darwin.* New Haven: Yale University Press.

Torrens, H. S. 1992. When did the dinosaur get its name? *New Scientist*, April 4: 40–44.

——. 1997. Politics and Paleontology: Richard Owen and the invention of dinosaurs. In J. O. Farlow and M. K. Brett-Surman (eds.), *The Complete Dinosaur*, 175–190. Bloomington: Indiana University Press.

Upchurch, P., C. Hunn, and D. Norman. 2002. An analysis of dinosaurian biogeography: evidence for the existence of vicariance and dispersal patterns caused by geological events. *Proceedings of the Royal Society of London*, Series B, 269: 613–621.

Wallace, D. R. 1999. *The Bonehunters' Revenge: Dinosaurs, Greed, and the Greatest Scientific Feud of the Gilded Age.* Boston: Houghton Mifflin.

Weishampel, D. B., and W. E. Reif. 1984. The work of Franz Baron Nopcsa (1877–1933): dinosaurs, evolution and theoretical tectonics. *Jahrbuch der Geologischen BundesanstaltWien* 127: 187–203.

Williams, G. C. 1992. *Natural Selection: Domains, Levels, and Challenges.* Oxford: Oxford University Press.

Wilson, A. N. 2002. *The Victorians.* London: Hutchinson.

Winchester, S. 2001. *The Map That Changed the World: William Smith and the Birth of Modern Geology.* New York: HarperCollins.

Young, D. 2007. *The Discovery of Evolution* (2nd ed.). Cambridge: Cambridge University Press.

Appendix: Dinosaur-Related Web Sites

DINOBASE
   http://dinobase.gly.bris.ac.uk

DINODATA
   http://www.dinodata.info (requires subscription)

THE DINOSAURIA: MUSEUM OF PALEONTOLOGY,
UNIVERSITY OF CALIFORNIA, BERKELEY
   http://www.ucmp.berkeley.edu/diapsids/dinosaur.html

I WANT TO BE A PALEONTOLOGIST! –
ADVICE FOR STUDENTS AND PARENTS
   http://www.priweb.org/ed/lol/careers.html

DINOSAURNEWS – DINOSAURS IN THE POPULAR PRESS
   http://www.dinosaurnews.org

EARTH SCIENCE LITERACY INITIATIVE
   http://www.earthscienceliteracy.org

THE PALEOBIOLOGY DATABASE
   http://www.paleodb.org/cgi-bin/bridge.pl

NATIONAL CENTER FOR SCIENCE EDUCATION
   http://ncse.com
   http://www.facebook.com/evolution.ncse

PALEONTOLOGICAL MAPS
   http://www.scotese.com
   http://cpgeosystems.com/paleomaps.html

THE PALEONTOLOGY PORTAL
   http://www.paleoportal.org

SOCIETY OF VERTEBRATE PALEONTOLOGY
   http://www.vertpaleo.org

TETRAPOD ZOOLOGY
   http://blogs.scientificamerican.com/tetrapod-zoology/

TREE OF LIFE WEB PROJECT
   http://tolweb.org

UNDERSTANDING EVOLUTION
   http://evolution.berkeley.edu

Glossary

Abduction to spread away from the body or axis of a limb, as with limbs and fingers.

Abductor muscles muscles that enable the moving away or spreading of parts of the skeleton, such as limbs and fingers.

Acceleration increase in the growth rate of a descendent species compared with its ancestor, producing a peramorphic descendant.

Acetabulum the hip socket, where the thighbone (femur) articulates with the pelvis.

Adenosine triphosphate (ATP) the biochemical fuel for cellular metabolism.

Adduction to draw toward the body (as in a limb) or to bring together (as in fingers).

Adductor muscles muscles that pull parts of the skeleton toward the body (as in a limb) or bring them together (as in fingers).

Age a subdivision of a geologic epoch.

Aggradational deposits sedimentary layers that accumulate during the filling in of a depositional basin.

Agnathans jawless fishes.

Agonistic display behaviors used to threaten or drive off rivals.

Akinetic without movement.

Allometry the growth of part of an organism in relation to the growth of another part or of the whole organism; the measure of such growth. An increase in size relative to the whole organism is called "positive allometry." A relative decrease in size is known as "negative allometry." If the relative size and shape remains the same, growth is said to be "isometric."

Allopatric speciation the formation of new species after the geographic isolation of different populations of what was formerly a single species. Over time, the separated populations become genetically different enough that interbreeding is no longer possible. The species of finches on the Galapagos Islands provide a good example.

Altriciality born or giving birth to young that are helpless and require care for some length of time.

Alula the small, feathered digit ("thumb") on the anterior edge of a bird's wing.

Amnion one of the internal membranes that characterize the amniote egg.

Amniote egg the shelled egg of more derived (advanced) land vertebrates, characterized by a series of membranes (amnion, chorion, etc.) that surround, protect, and nourish the developing embryo.

Amniotes tetrapods that produce by means of the amniote egg or its derivatives; they include reptiles, birds, and mammals.

Amphiplatyan vertebrae with centra that are flat, both anteriorly and posteriorly.

Anagenesis evolutionary change in which one taxon replaces another without branching.

Analogous used to describe different structures in organisms that serve the same function but are not derived from the same ancestral structure. The wings of birds, bats, and pterosaurs are good examples.

Anamniotes vertebrates that do not reproduce by means of the amniote egg; fishes and amphibians.

Anapsids amniotes with a solid, unperforated skull in the temple region behind the eye such as turtles.

Angiosperms a group of seed plants in which the seed is surrounded by a fruit; the flowering plants.

Anterior toward the front end (sometimes referred to as "cranial")

Antibodies proteins produced by the body that bind to foreign molecules, tagging them for attack by the immune system.

Antiserum a serum containing antibodies.

Antorbital cavity a space within the skull between the nares and the orbit

Antorbital fenestra an opening in the skull in front of the orbit and behind the external nares. This is one of the characters that is used to define the Archosauria.

Antorbital fossa a depression that surrounds the antorbital fenestra.

Apatite a calcium phosphate mineral found in bones and teeth.

Apomorphy a derived character.

Appendicular skeleton bones of the limbs and limb girdles.

Archetype Richard Owen's concept of a metaphysical, generalized body plan or blueprint, of which the observed body forms of living animals are a physical manifestation.

Archosauriforms a subgroup of archosauromorphs that includes archosaurs and some earlier, closely related groups.

Archosauromorphs a group of diapsids that includes rhynchosaurs, protorosaurs, trilophosaurs, and archosauriforms.

Archosaurs derived archosauriforms, including dinosaurs, birds, pterosaurs, crocodilians, and their close relatives.

Articular a bone toward the rear of the mandible, by which the dinosaurian lower jaw articulates with the quadrate bone of the cranium.

Astragalus a large, proximal anklebone that articulates with the tibia.

Atlas the first neck vertebra.

Autotrophs organisms that manufacture their own food.

Axial skeleton bones of the body midline: skull, vertebral column, and tail.

Binomial system the practice of using two names (generic and specific) in the formal scientific name of a species. Invented by Caspar Bauhin and John Ray.

Biochron a short interval of geologic time defined on the basis of its fossil content.

Biomass the amount of living material represented by the combined individual masses of all the animals of a population.

Biomineralization the process by which living organisms produce minerals, often to harden or stiffen existing tissues to form, for example, shell or bone.

Bone spur a bony projection associated with arthritis, aging, or infection (osteomyelitis) that forms along joints (osteophytes) or at the attachment of a tendon or ligament (enthesophytes).

Brachial enlargement an enlargement of the spinal cord that reflects enhanced neural activity at the level of the forelimb.

Bradymetabolic used to describe organisms with slow metabolic rates.

Braincase a group of small, tightly sutured bones that surround and protect the brain.

Branch on a cladogram, a line connecting a taxon to the node joining it to another taxon. The branch represents the divergence of a taxon from its closest relatives.

Bryophytes mosses and their relatives.

Buccinator a thin quadrilateral muscle between the maxilla and the mandible at the side of the face.

Calcaneum a proximal ankle bone that articulates with the fibula; forms the heel of mammals.

Calcareous layer the hard outer portion of an eggshell.

Caliche a hardened lime deposit formed during the evaporation of pore water from soils.

Callus the bony tissue that develops around a fracture site, which begins as soft callus and is remodeled into mineralized hard callus as healing progresses.

Cancelli spaces within the spongy bone underlying the outer solid cortical bone, lined with endosteum and filled with marrow.

Capitulum the ventral projection by which a rib articulates with a vertebra.

Carotid artery the artery that supplies the head and neck with oxygenated blood.

Carpals bones of the wrist.

Carpometacarpus a fused mass of hand bones (incorporating carpals and metacarpals). A carpometacarpus is typical for modern birds and their close relatives but also evolved independently in alvarezsaurids (theropods).

Caudal toward the tail.

Caudals tail vertebrae.

Centrum the spool-shaped ventral portion of a vertebra.

Cephalic of or relating to the head.

Ceratobranchial each of the paired ventral cartilaginous sections of the branchial arch that supports the gills in fishes.

Ceratohyal one of bony or cartilaginous elements that make up the hyoid.

Cerebellum the part of the brain between the brain stem and the cerebrum associated with muscle coordination and equilibrium.

Cerebral cortex the surface layer of gray matter of the cerebrum associated with coordinating sensory and motor function.

Cerebrum the enlarged front or upper part of the brain consisting of the cerebral hemispheres, considered the seat of conscious mental processes.

Cervicals neck vertebrae.

Chevrons V-shaped bones beneath the caudal vertebrae.

Chitin substance forming the hard outer shell of insects, arachnids, and crustaceans.

Choanae openings of the nasal tract in the roof of the mouth.

Chondrichthyes fishes with a cartilaginous skeleton; sharks, skates, and rays.

Chondroblasts immature cartilage-producing cells.

Chondroclasts cartilage-resorbing cells.

Cingulum a shelf near the base of a tooth crown, just above the tooth root.

Clade a genetically related group of organisms; also known as a monophyletic group.

Cladistics phylogenetic systematics. This system replaces the older Linnaean System of ranked taxa.

Cladogenesis evolutionary change characterized by the branching of new taxa.

Clavicles collarbones, which attach the shoulder girdle to the sternum.

Cleidoic egg an egg enclosed in membranes and a shell; the amniote egg.

Condyle a rounded, knobby joint.

Conifers a group of gymnosperms that includes pines, larches, spruces, firs, and their relatives.

Convergences characters shared by groups but acquired independently and therefore not useful for assigning groups to the same clade.

Coprolites fossilized feces.

Coracoid a bone of the pectoral girdle, located ventral to the scapula.

Coronoid process a projection of bone in the upper surface of the lower jaw, behind the tooth row, to which jaw-closing muscles attach.

Cranial toward the head.

Cranial nerves the paired nerves that enter or emerge directly from the brain and pass through openings in the skull to connect with the periphery of the body.

Cranium the skeleton of the head minus the lower jaws.

Crocodylotarsians the archosaurian group that includes crocodilians and their close relatives.

Crurotarsians in one classification scheme of archosaurs, a lineage including crocodylotarsians and ornithosuchids.

Cycadophytes a group of gymnosperms; cycads and cycadeoids.

Definition the meaning of a taxon name. It is defined by its member taxa or by a statement of ancestry.

Deltopectoral crest a bony flange on the upper arm bone (humerus) that served for muscle attachment.

Dentary the tooth-bearing bone of the mandible.

Dentine a hard tissue forming the core of teeth.

Diagenesis physical and chemical changes affecting sediments (and any contained bones) after burial.

Diagnosis the way in which a taxon is recognized. Fossil vertebrate taxa are diagnosed on the basis of skeletal features.

Diaphragm a sheet of muscle separating the chest cavity from the abdomen in synapsids.

Diaphysis the shaft of a long bone.

Diapophyses articulation sites of ribs to the vertebrae.

Diapsids amniotes with two temporal openings on each side of the skull.

Diastema a gap in the tooth row.

Diencephalon the posterior part of the forebrain, containing the thalamus, hypothalamus, and part of the pituitary gland.

Diffuse idiopathic skeletal hyperostosis (DISH) a disease characterized by calcification and ossification of soft tissues, mainly ligaments and sites where ligament and tendon connect to the bone.

Digit a toe or finger.

Digitigrade used to describe animals that walk only on their toes, with the main hand and foot bones off the ground.

Dinosauria the common ancestor of *Corvus* (the crow) and *Triceratops* (a ceratopsian dinosaur) and all of its descendents. The two main member clades are the Saurischia and the Ornithischia. As originally defined by Owen in 1842, the "deinos" part means "fearfully great," not "terrible."

Diploid used to describe cells or organisms that have two of each kind of chromosome in the cell nucleus.

Dispersalist biogeography a school of thought that interprets the geographic distribution of organisms in terms of movements of the organisms themselves across the earth's surface.

Distal away from the central portion of the animal.

DNA [deoxyribonucleic acid] the basis of heredity; the double-helix-shaped molecule that contains the genetic instructions used in the development and function of all known living organisms with the exception of some viruses. DNA coding segments are called genes. DNA serves as a template for making RNA, a molecule that carries the blueprint for making the proteins that do the actual work in cells and bodies.

Dorsal toward the top (literally back) of an animal.

Dorsals vertebrae between the shoulder and pelvic girdles.

Dura mater the membrane that envelopes the brain and spinal cord.

Durham's Law even in the best of circumstances, only about 10% of the actual original biota is preserved as fossils.

Durophagy the consumption of hard-shelled food items, such as nuts, shelled mollusks, or crabs.

Dyschondroplasia a disease that affects the development of bone and cartilage in the ends of tubular bones and causes distorted growth in length and tumorlike cartilage masses.

Eburnation a grooving of bone articular surfaces caused by bone-on-bone rubbing due to the absence of the normally intervening cartilage, associated with severe osteoarthritis.

Ectotherm an organism that derives most of its body heat from sources outside itself.

Edentulous toothless.

Enamel a hard tissue forming the outer covering of teeth.

Encephalization quotient the ratio of the observed brain size of an animal divided by the brain size predicted on the basis of body mass.

Endocast an internal cast of a hollow object, for example, a braincase.

Endochondral bone bone formed by progressive ossification of a cartilage mass, constituting the majority of bones in the body.

Endocrine the system of glands that secrete hormones into the bloodstream.

Endosacral enlargement an enlargement of the opening in the sacrum for the spinal cord that reflects enhanced neural supply to the hind limbs, and also possibly a structure known as the glycogen body.

Endosteum a bone-forming tissue located in a bone's interior.

Endotherm an organism that obtains most of its body heat from its own metabolism.

Eon a very long interval of geologic time; the Phanerozoic Eon, for example, includes the Paleozoic, Mesozoic, and Cenozoic Eras.

Epaxial musculature the muscles on the dorsal side of an animal.

Epaxial tendons ossified tendons located above the vertebral centra, running across the neutral spines.

Epiphyses the terminal ends of endochondral bones, adjacent to the site of the cartilaginous growth plate.

Epiphysis (pineal gland) a small endocrine gland located near the center of the brain between the two hemispheres that produces melatonin, a hormone that affects wake/sleep patterns and seasonal functions.

Epipophyses bony processes at the rear end of vertebrae.

Epithalamus a dorsal posterior segment of the diencephalon that connects the limbic system to other parts of the brain.

Epitopes complex folded regions of molecules that allow antibodies to determine whether the molecules are from the organism or are intruders.

Epoch a subdivision of a geologic period.

Era a long interval of geologic time made up of geologic periods. The Mesozoic Era, during which dinosaurs lived, is subdivided into the Triassic, Jurassic, and Cretaceous periods. The Mesozoic Era is a subdivision of an even longer interval of time, the Phanerozoic Eon.

Euryapsids reptiles with a single temporal opening placed high on each side of the skull; probably a derived group of diapsids. Example = the plesiosaurs.

Evagination creation of a pouch.

Evaporites sedimentary rocks formed by the evaporation of seawater and the crystallization of minerals previously dissolved in water.

Evolutionary systematics (evolutionary taxonomy, gradistics) an eclectic approach to classification based on the Linnean system and the overall morphological similarities among organisms; attempts to recognize ancestor–descendant relationships.

Exostosis a noncancerous bony growth projecting from a bone surface.

Extension an unbending movement around a joint in a limb, such as the knee or elbow.

Facial nerve the seventh cranial nerve, which supplies motor fibers to the muscles of the face and jaw, and both sensory and parasympathetic fibers to the tongue and palate.

Femur the thigh bone.

Fenestra an opening (or window) into a bone.

Fibrolamellar bone a rather open (filled with blood vessels) hard tissue characteristic of fast-growing bone.

Fibula the smaller, outer bone of the lower hind leg.

Flexion a bending movement inward around a joint in a limb, such as the knee or elbow; or a forward raising of the arm or leg at the shoulder or hip joint.

Foramen (pl. foramina) an opening through which muscles, nerves, arteries, veins, and other structures pass.

Foramen magnum a large opening in the back of the skull, through which the spinal cord passes.

Formation a formally defined, mapable sedimentary rock unit.

Fossa a pit, groove, or depression, in bone.

Fossil evidence of life in the geologic past. This includes the actual remains of an organism or trace fossils left by an organism.

Frontal a bone of the skull roof, located just behind the nasal.

Furcula the wishbone, a structure that connects the pectoral girdle with the sternum. Found in birds and some dinosaurs, it takes the place of, and may be derived from, the clavicles.

Gametophyte a haploid plant that produces eggs and sperm.

Gastralia belly ribs that help support the viscera.

Gastroliths "stomach stones" found within the gut regions of herbivorous dinosaurs, presumably used to process food.

Gene a unit of heredity in a living organism residing on a stretch of DNA. Genes contain the codes (specifications) for every protein or functional RNA chain. Genes hold the information to build and maintain an organism's cells and pass genetic traits to offspring.

Genetic sequence phrase used to refer to the primary structure of a nucleic acid, such as biological DNA, which is the order of nucleotides that comprise the whole molecule and that has the capacity to carry information.

Genome the entire DNA sequence of living things; each species has its characteristic genome.

Ghost lineage a hypothetical extension of the geologic range of a taxon, earlier than when the taxon is first seen in the geologic record, predicted on the basis of the earliest geologic occurrence of the taxon's sister taxon.

Gigantothermy use of large body size, circulatory adjustments, and body insulation to maintain a high, constant body temperature with low rates of metabolism.

Glenoid the socket in the shoulder girdle to which the humerus attaches.

Glycogen body in the spinal cord of birds, an oval structure that is made of specialized cells containing large amounts of glycogen and that is probably related to energy storage. Glycogen bodies may have been present in some dinosaurs and likely are the structure once (erroneously) thought to be a "second brain" in animals such as *Stegosaurus*.

Gnathostomes jawed vertebrates.

Gymnosperms a paraphyletic group of seed plants, including conifers, seed ferns, cycadophytes, and their relatives, in which the seed is not surrounded by fruit.

Hallux the toe on the inside edge of the foot. In humans the hallux is normally termed the big toe.

Haploid used to describe cells or organisms that have a single chromosome of each type in the cell nucleus.

Heterochronoclines the collective name for evolutionary trends from ancestors to descendants that shows increasingly more juvenile or more adult traits.

Heterochrony a change in the timing or rate of developmental events, relative to the same events in the ancestor, leading to changes in size and shape. The concept was introduced by Ernst Haeckel in 1875.

Heterocoelous vertebrae vertebrae with centra that are saddle shaped at both ends as in birds and turtles. Centra that are flat at both ends, as in mammals, are acelous.

Heterotrophs organisms that cannot manufacture their own food, but must feed, directly or indirectly, on other organisms.

Homeostasis internal regulation of a system to maintain a stable, constant condition.

Homeotherm (Homoiotherm) an organism that maintains a fairly constant body temperature.

Homologous a word used to describe anatomical structures in different organisms derived from the same structure in their common ancestor.

Homoplasy a shared similarity between two taxa that is explained by convergence, character reversal, or chance.

Humerus the upper arm bone.

Hydrophobicity physical property of a molecule that is repelled from water.

Hyoids throat bones located at the base of the tongue.

Hyolingual the musculature system that moves the tongue for catching and consuming food.

Hypantrum a small, anterior projection at the base of the neural spine that articulates with the hyposphene of the preceding vertebra.

Hypaxial musculature the muscles on the ventral side of an animal.

Hypaxial tendons ossified tendons that run across the chevrons between caudal vertebrae.

Hypermorphosis a type of heterochrony in which peramorphic morphologies will evolve if the period of growth in the descendant is extended and may affect the whole organism if sexual maturity is delayed, because fast juvenile growth rates will persist.

Hyposphene a small, posterior projection at the base of the neural spine that articulates with the hypantrum of the following vertebra.

Hypothalamus part of the diencephalon located just anterior to the brain stem in all vertebrates; links the nervous and endocrine systems via the pituitary gland; controls body temperature, hunger, thirst, fatigue, sleep, and circadian cycles.

Ichnocoenosis a footprint assemblage.

Ichnofabric the manner in which trace fossils affect the texture of a sedimentary deposit.

Ichnofacies sedimentary deposits of a particular kind that repeatedly have the same distinctive track assemblages.

Ichnology the study of footprints and other trace fossils.

Ichnotaxonomy the naming and classification of trace fossils.

Ilium the anterior-most bone of the pelvis; attaches the pelvic girdle to the sacrum.

Insertion (of muscle) the end of the muscle attaching to the freely moving bone of its joint.

Integument covering or enclosing layer of an organism, such as skin.

Interorbital septum bony partition between the large orbital cavities of birds.

Intrinsic muscles muscles fully contained within a structure, such as the "hand" or "foot" of quadrupeds.

Intromittent organ external organ of a male specialized to deliver sperm during copulation.

Involucrum a layer of new bone growth deposited by a rapidly expanding periosteum, frequently seen in bone infected by bacteria.

Ischial boot a structure on the distal end of the ischium.

Ischium the more posterior of the lower bones of the pelvis.

Isometry see Allometry

Jugal the cheek bone, located posterior of the maxilla and below the orbit.

Keratin a protein associated with hard epidermal tissue such as nails.

Kinesis a movement of an organism in response to a stimulus.

Labyrinthodonts the first clade of amphibians (in the Devonian) whose teeth have complex infoldings of enamel.

Lacrimal (lachrymal) a bone positioned between the antorbital fenestra and the orbit.

Lamellar-zonal bone a layered, rather dense hard tissue characteristic of slowly growing bone; often shows growth rings.

Lateral away from the midline of an animal.

Lateral temporal fenestra an opening in the side of the skull, behind the orbit. It is very prominent in theropods.

Laterosphenoid an ossification of neurocranial cartilage in the anterior sidewall of the braincase of crocodilians and birds.

Lepidosauromorphs lizardlike diapsids.

Lepidosaurs a subgroup of lepidosauromorphs that includes lizards, snakes, and the tuatara.

Levators a muscle that raises a body part.

Ligament a fibrous tissue that connects bones to other bones.

Lingual located near or next to the tongue.

Lipid one of the basic structural components of living cells, along with proteins and carbohydrates. Fat is a lipid.

Lissamphibians the modern amphibian groups; frogs, toads, salamanders, and caecilians.

Lumbar vertebrae vertebrae of the lower back in mammals.

Lytic lesions an area of bone that has been destroyed due to various diseases.

Mandible the lower jaw.

Mandibular fossa the depression in the temporal bone that articulates with the mandibular condyle.

Mantle a thick region of the earth's interior, located beneath the planet's crust but external to the core.

Manus collective term for the bones of the forefoot (or hand).

Maxilla the posterior tooth-bearing bone of the upper jaw.

Maxillary fenestra an opening in the skull in front of the antorbital fenestra of theropod dinosaurs.

Medial toward the midline of an animal.

Medulla (of brain) the posterior portion of the brainstem.

Megatracksites single surfaces or thin packages of sedimentary beds that are rich in fossilized footprints over a large geographic area.

Melatonin a chemical secreted by the pituitary gland and associated with regulating the circadian cycle; also related to the mechanism by which some amphibians and reptiles change the color of their skin.

Membranous labyrinth a collection of fluid-filled tubes and chambers within the bony labyrinth of the inner ear that houses the receptors for the senses of equilibrium and hearing.

Mesencephalon the midbrain.

Metacarpals bones of the forefoot or hand (excluding the fingers); bones of the "palm" of the hand.

Metaphysis the region of a bone located between the diaphysis and the epiphysis. In growing animals, it contains the cartilaginous growth plate.

Metatarsals bones of the foot (excluding the toes).

Monophyletic groups taxa composed of a single taxon and all of its descendants.

Motor nerves neurons that control the muscles.

Multicameral lungs lungs subdivided into many small chambers.

Myology the study of muscles and muscle tissue.

Nares openings in the skull for the nostrils.

Nasal a bone on the top of the skull, to the rear of the premaxilla.

Neognaths one of two clades of neornithine birds, including virtually all living birds except for tinamous and the flightless ratites; (neognath means "new jaws," although neognath palate structure may be more primitive than that of their sister clade, the palaeognaths).

Neontology the part of biology that deals with now-living (recent) organisms.

Neoteny decrease in the growth rate of a descendent compared with its ancestor, producing a peramorphic descendant.

Nested hierarchy the arrangement of taxa into a series of larger and more inclusive groups.

Neural relating to the nerves or nervous system.

Neural arch the dorsal portion of a vertebra, located above the centrum, and surrounding the spinal cord.

Neural canal an opening in a vertebra, located above the centrum, through which the spinal cord passes.

Neural spine process projecting dorsally from a vertebral neural arch.

Node the point where two or more lines in a cladogram meet; in cladistics, a node constitutes a taxon that contains all of the descendant taxa that ultimately meet at that node.

Node-based definitions taxon definitions that take the form of "the most recent common ancestor for taxon X and taxon Y, and all descendants of that common ancestor." Dinosauria is a node-based taxon.

Nomenclature the official naming of taxa.

Notochord a rod of stiff tissue running along the back of chordate animals at some point in their lives. The notochord is replaced by the vertebral column in many vertebrates.

Obturator foramen an opening in the pubis, located near the acetabulum.

Obturator process a bony projection from the ischium.

Occipital condyle a rounded, knobby joint by which the skull articulates with the vertebral column.

Occiput the area in the back part of the skull where the neck attaches.

Octaval (VIII) nerve one of the cranial nerves.

Ocular relating to the eye.

Olecranon a process on the ulna for muscle attachment.

Olfactory bulb a structure anterior to the cerebral hemispheres that receives olfactory (smell) information from the nose and transmits it to the olfactory lobe of the brain.

Olfactory turbinates thin bones lined with sensory (olfactory) epithelia, located in the nasal passages.

Opisthocoelous used to describe vertebral centra with convex anterior faces and concave posterior faces.

Opisthopubic used to describe a pubis that is directly rearward.

Optic chiasma the x-shaped intersection of the left and right optic nerves on the undersurface of the brain.

Optic lobes either of the two prominences of the midbrain concerned with vision.

Optic tectum a paired structure that forms the roof of the vertebrate midbrain.

Orbit the opening in the skull for the eye.

Origin (of muscle) the point at which a muscle attaches to a bone (usually) or another muscle.

Ornithischia the clade of dinosaurs that includes the stegosaurs, anklyosaurs, ornithopods, ceratopsians, pachycephalosaurs, and their common ancestors.

Ornithodirans the archosaurian lineage that includes dinosaurs and birds.

Ornithosuchians in one classification scheme of archosaurs, a group that includes ornithosuchids and ornithodirans.

Ossified tendons tendons that have undergone bony transformation; can be pathological, but in hadrosaurs ossified tendons are a normal feature connecting across vertebrae to strengthen the backbone.

Osteichthyes bony fishes.

Osteoarthritis joint disease caused by degeneration of the cartilage cushion on the ends of bones and characterized by the formation of osteophytes at joint margins, bone-on-bone rubbing, and changes in bone density.

Osteoblasts bone-forming cells.

Osteochondroma an abnormal bony protuberance with a cap of cartilage.

Osteoclasts bone-resorbing cells.

Osteocytes cells involved in the maintenance of bony tissue.

Osteoderms bones that form in the skin (e.g., stegosaur plates and anklyosaur scutes).

Osteoma benign tumor composed of bone tissue.

Osteomyelitis an infectious disease of the bone that causes inflammation and, frequently, pus formation.

Osteophytes overgrowths of bone that form adjacent to articular surfaces.

Oviparity able to produce eggs that develop and hatch outside the mother's body.

Pace the distance between two successive footprints of the opposite feet (right to left, or left to right).

Paedomorphocline an evolutionary trend of a sequence of progressively more paedomorphic species.

Paedomorphosis a phenotypic or genotypic change in which the adults of a descendant species retain traits previously occurring only in juveniles of the ancestral species, either by growing at a slower rate (neoteny), growing for a shorter period of time (progenesis), or relative delay in offset of growth of a particular trait.

Palaeognaths one of two clades of neornithine birds; (palaeognath means "old jaws," a reference to the birds' palate anatomy, once thought to be more primitive than that of their sister clade the neognaths). The clade includes tinamous and flightless ratites—kiwis, cassowaries, emus, rheas, and ostriches.

Palate the bony roof of the mouth.

Palatine one of the bones of the palate, positioned toward the front of the skull and lateral to the vomer.

Palatoquadrate in some fishes, the dorsal portion of the mandibular arch.

Palichnostratigraphy subdivision of the time intervals represented by sedimentary rocks on the basis of fossilized footprint assemblages.

Palpebral a small bone in the eyelid.

Pangaea (Pangea) the supercontinent composed of the modern, presently separated continents of the world that existed during the late Paleozoic and early Mesozoic Eras.

Parabronchial lung the complex lung of birds, in which air flows in the same direction across the lung whether the bird inhales or exhales. The operation of the lung involves the use of extensive air sacs external to the lung itself.

Paramesonephric duct one of two ducts in an embryo that, in a female, develop into the oviducts, uterus, cervix, and upper vagina.

Paranasal sinuses air-filled spaces, communicating with the nasal cavity, within the bones of the skull and face.

Paraphyletic groups taxa consisting of a single ancestor and some, but not all, of its descendants. Example = the pelycosaurs.

Parasagittal parallel to an animal's midline.

Parataxonomy a classification that is parallel to the Linnean taxonomic system. Parataxonomy does not reflect the actual taxonomic relationships of the organism themselves, but rather classifies objects made by the organisms, such as footprints or eggs.

Parietal a bone of the rear portion of the skull roof, located behind the frontal.

Parietal organ a part of the diencephalon present in some vertebrates that may include a photoreceptive "eye" and is associated with the pineal gland, which regulates circadian rhythm and hormone production for thermoregulation. Parietal organs are found in tuatara, lizards, frogs and lampreys, and some species of fish, such as tuna and pelagic sharks, where it is visible as a light-sensitive spot on top of the head.

Parsimony the scientific principle that the simplest explanation is the best for any phenomenon. In systematics, parsimony is the principle that the evolutionary tree that requires the smallest number of evolutionary changes is the most likely approximation of the true historical pattern of phylogeny.

Pelvic girdle (pelvis) the complex of bones by which the hind limb attaches to the body; includes the ilium, ischium, and pubis.

Pelycosaurs a paraphyletic group of basal synapsids (once called the "mammal-like reptiles"). An example is *Dimetrodon*.

Peramorphocline an evolutionary trend of a sequence of progressively more peramorphic species.

Peramorphosis a phylogenetic change in which individuals of a descendant species undergo a greater extent of growth than their ancestors by delaying offset of growth (hypermorphosis), growing faster (acceleration), or beginning growth of a trait relatively earlier (predisplacement).

Perichondrium a coating of tissue that lines the periphery of the growing cartilage precursor of a bone and deposits the cartilage.

Perimortem around the time of death.

Period one of the major intervals of geologic time. The Triassic, Jurassic, and Cretaceous periods were the time intervals during which dinosaurs dominated terrestrial faunas. Periods are subdivided into epochs, and periods are grouped together into eras.

Periosteal bone bone formed by the periosteum.

Periosteum bone-forming tissue at the periphery of a growing bone.

Pes collective term for the bones of the hind foot.

Phalanx (pl. phalanges) bones of the fingers or toes.

Phosphatic fossilization when phosphates replace the organic structures during fossilization cell structure and soft tissue can be preserved.

Phylogenetics the study of evolutionary relatedness among various groups of organisms.

Phylogenetic systematics (cladistics) an approach of classification based strictly on the interrelationships among clades of organisms.

Phylogeny an evolutionary tree depicting ancestor–descendant relationship.

Pinnation angle angle between the line of orientation of muscle fibers and the line of muscle action.

Pituitary gland an endocrine gland below the hypothalamus at the base of the brain that secretes hormones regulating homeostasis.

Plantigrade a word used to describe animals (including humans) that walk flat-footed, with the metatarsals against the ground.

Plate tectonics the unifying theory of the earth sciences. The surface of the earth is shaped by the interaction of large tectonic plates composed of the crust and outer mantle.

Pleurocoels openings along the lateral surfaces of vertebrae into chambers inside the centrum or neural arch.

Pleurokinesis the development of a hinge between the maxilla and the remainder of the skull, such that the maxillae swing outward when the jaws are closed.

Pneumatic system delivers air to organism for breathing and buoyancy.

Poikilotherms animals whose body temperatures fluctuate in response to changing environmental temperatures.

Polymerase chain reaction a technique that allows large-scale replication of selected gene segments.

Polyphyletic groups groups with multiple ancestors; considered invalid by taxonomists of whatever persuasion.

Pontine flexure a bend in the axis of the embryological central nervous system that marks the junction where, in the mature brain, the cerebellum and medulla meet.

Postacetabular process the portion of the ilium posterior to the acetabulum.

Postcrania collective term for all the bones of the skeleton other than the skull.

Postdisplacement an alteration in development in which a developmental process begins later in a descendant than in its ancestor and may not have been completed by the time maturity is reached, producing a paedomorphic morphology.

Posterior toward the rear end.

Postorbital process the rear, upper edge of the eye socket which is a projection from the frontal bone.

Postzygapophyses bony projections at the rear ends of vertebral neural arches that articulate with prezygapophyses of the following vertebrae.

Preacetabular process the portion of the ilium anterior to the acetabulum.

Precociality refers to species in which the young are relatively mature and mobile from the moment of birth or hatching (as opposed to altricial, where the young are born or hatched helpless).

Predentary a bone at the front end of the lower jaw in ornithischians.

Predisplacement an alteration in development in which a developmental process begins earlier in a descendant than in its ancestor resulting in a greater degree of growth of that structure in the descendant, producing a peramorphic morphology.

Premaxilla the anterior tooth-bearing bone of the upper jaw.

Prepubis an anteriorly directed process of the pubis.

Presacrals vertebrae anterior to the sacrum.

Prezygapophyses bony projections at the front ends of vertebral neural arches that articulate with postzygapophyses of the preceding vertebrae.

Primary osteons structures formed by layers of bone that are deposited inward from the walls of tunnels surrounding blood vessels in newly formed bone.

Primitive characters characters found in all members of a group under study and possibly in taxa outside that group. It is a relative term, not an absolute term.

Procoelous used to describe vertebral centra with concave anterior faces and convex posterior faces.

Progenesis a precocious offset of growth, usually earlier onset of sexual maturity in the descendant compared with the ancestor, producing a paedomorphic morphology.

Promaxillary fenestra an opening in the skull in front of the maxillary fenestra and the antorbital fenestra in theropod dinosaurs.

Prosencephalon the forebrain.

Protein organic compounds made of amino acids. The sequence of amino acids in a protein is defined by a gene. Proteins are the means by which genetic instructions are carried out.

Proteome the set of proteins expressed by a genome, cell, tissue, or organism.

Protraction the act of extending forward or outward.

Protractors muscles that extend a part.

Proximal toward the central portion of an animal.

Pseudo-acromion process a spur of bone on the shoulder blade (scapula) for muscle attachment.

Pteridophytes a paraphyletic group of vascular plants that includes ferns, horsetails, and club mosses.

Pteridosperms fernlike plants that reproduce by seeds.

Pterygoid a large bone in the rear part of the roof of the mouth, located behind the vomer and lateral to the braincase.

Pubic boot structure at the end of the pubic bone where locomotor and support muscles attach.

Pubis the more anterior of the lower bones of the pelvis.

Pygostyle a fused mass of caudal vertebrae, typical for pygostylian birds where (in most members of the clade) it supports the rectrices and their associated musculature.

Quadrate a large bone at the rear of the skull, to which the dinosaurian lower jaw articulates.

Radius the smaller, more anteriorly placed bone of the forearm.

Rectrices (sing. rectrix) the long, stiff feathers that form a bird's tail fan. Rectrices aid in the generation of both thrust and lift to assist flight (compare Remiges).

Red beds reddish sedimentary rocks. The red color comes from the abundance of iron in the rocks.

Regional heterothermy having a different temperature in different parts of the body.

Remiges (sing. remex) the long, stiff feathers on the wings of a bird. Their primary function is to aid in the generation of both thrust and lift, thereby enabling flight (compare Rectrices).

Respiratory turbinates thin, complex structures of bone or cartilage, lined with respiratory epithelia and located in the nasal airway.

Retraction act or ability to draw back.

Retroarticular process the bone projection above the mandibular fossa.

Reversal a transformation of an advanced character back to the ancestral state.

Rhamphotheca a horny covering over the anterior tips of the upper and lower jaws.

Rhombencephalon the hindbrain.

Rostral in ceratopsians, a bone located in front of the premaxilla in the upper jaw.

Rostrum a long snout region.

Sacculus the smaller chamber of the membranous labyrinth of the ear.

Sacral enlargement an enlargement of the spinal cord that reflects enhanced neural activity at the level of the hind limb.

Sacrals the vertebrae that articulate with the pelvis.

Sacrum a structure formed from the fusion of the sacral vertebrae.

Saurischia the clade of dinosaurs that includes the theropods, the sauropodomorphs, and their common ancestor.

Scapula shoulder blade.

Sea-floor spreading the creation of new oceanic crust at mid-ocean ridges; crust that then moves laterally away from the ridge.

Semicircular canals the loop-shaped parts of the inner ear forming a sense organ associated with equilibrium.

Sensory nerves nerves that convey impulses from the sensory organs to the brain.

Sequestrum a fragment of dead bone surrounded by living bone tissue as a consequence of circulatory disruption due to rapid lifting of the bone membrane by swelling caused by infection.

Serotonin a neurotransmitter that narrows blood vessels.

Serum (pl. sera) blood from which cells and fibrin (a fibrous protein associated with blood clotting) have been removed.

Sesamoid a bone that is embedded within a tendon where the tendon passes over a joint (such as the hand, knee, and foot) and that protects the tendon and increases its mechanical effect.

Sexual display behaviors used to attract a mate.

Shared derived characters (synapomorphies) characters shared by two or more descendant taxa that depart from the primitive configuration of the characters.

Shell membrane the inner, organic layer of an eggshell.

Shell units abutting and interlocking components of the hard, calcareous layer of an eggshell.

Sister taxon (sister group) a taxon that shares a splitting event with another taxon is the sister taxon of the latter. Two sister taxa share a common node on a cladogram.

Somatosensory the sensory system associated with sensing touch, temperature, body position, and pain.

Spondyloarthropathy a general term for many joint diseases such as arthritis.

Spondylosis deformans a degenerative condition of the annulus fibrosis of the intervertebral disc, characterized by the growth of bony spurs from the margins of vertebral centra.

Squamosal a bone on the posterior surface of the skull.

Standard metabolic rate (SMR) the minimal rate of energy expenditure by a resting, fasting ectotherm under specified temperature conditions.

Stem-based definitions taxon definitions that take the form "taxon X and all organisms sharing a more recent common ancestor with taxon X than with taxon Y." Example = ornithopods.

Sternum the breastbone, formed by the fusion of a series of bones along the ventral midline of the trunk.

Streptostylic the manner in which the quadrate bone freely articulates with the skull thus allowing for movement.

Stride the distance between two successive footprints from the same foot.

Subnarial foramen a small opening in the skull below the nose.

Supraorbitals small bones along the upper rim of the eye opening (orbit) of the skull.

Supratemporal fenestra an opening in the top of the skull, behind the orbit.

Sutures immovable joints between bones.

Symphysis a joint between two bones, connected by fibrous tissues, that allows limited movement between bones.

Synapomorphies shared derived characters.

Synapsids amniotes with a single, laterally placed opening low on either side of the skull, behind the orbit; includes pelycosaurs and mammals.

Syndesmophytes pathological, bony projecting vertebral growths due to ossification of outer fibers of annulus fibrosis of intervertebral discs.

Synonymy different names assigned to the same taxon.

Synovial membrane membrane that secretes a lubricating fluid (as by a tendon sheath).

Synsacrum a structure formed by fusion of dorsal or tail vertebrae with sacral vertebrae to form a single unit.

Syrinx a specialized vocal organ unique to birds and located deep within the chest.

Systematics the scientific study of the diversity of organisms within and among clades. This is not the same as taxonomy.

Tachymetabolic used to describe animals with rapid metabolic rates.

Taphonomy the study of how organisms decay over time and become fossilized.

Tarsals ankle bones.

Tarsometatarsus a bone in the lower leg, formed from the fusion of tarsal (ankle) and metatarsal (foot) bones. A tarsometatarsus is typical for birds.

Taxon a named group of organisms.

Taxonomy the scientific practice and study of labeling and ordering organisms into groups.

Temporal bone at the temple region of the skull.

Tendon white fibrous connective tissue that ties a muscle with some other part, such as a bone, and transmits the force that a muscle exerts.

Tetrapods the four-footed, land-living vertebrates (including secondarily aquatic forms).

Thagomizer collective term for the tail spikes of stegosaurs, coined by Gary Larson.

Therapsids a group of nonmammalian synapsids, known in noncladistic parlance as premammals, or mammal-like reptiles.

Thoracic vertebrae vertebrae of the chest region in mammals.

Tibia the larger, more medial bone of the lower hind leg.

Tillites sedimentary rocks formed from the consolidation of glacial deposits.

Trabeculae bony struts forming the framework of endochondral bone.

Trigeminal nerve the fifth cranial nerve, which carries sensory nerves from the face, and motor nerves associated with biting, chewing, and swallowing.

Triosseal canal in birds the opening formed where the scapula combines with the dorsal ends of the coracoid and furcula (wishbone), through which the tendon of the muscle used to raise the wing passes from the sternal keel to the humerus.

Trochanter refers to a number of rough raised areas on the upper part of the femur.

Tuberculum the dorsal projection by which a rib articulates with a vertebra.

Type specimen the actual individual specimen first used to name a new taxon.

Ulna one of the two bones of the forearm; it is larger and more posteriorly located than the radius, and forms the lower elbow joint.

Uncinate process hook-shaped process on the sides of the top surface of a spinal vertebra which prevents the vertebra from sliding off the vertebra below it.

Undertracks footprints formed by the transmission of the trackmaker's weight into buried sediment layers, deforming those layers. The animal makes a true print on the surface it walks across, but also makes a series of undertracks of underlying sediment surfaces.

Unguals the terminal bones of fingers or toes; they support, and lie underneath, horny claws or nails.

Ureotelic used to describe animals (synapsids) whose nitrogenous wastes are released in the form of urea.

Uricotelic used to describe animals (reptiles) whose nitrogenous wastes are released in the form of uric acid.

Valleculae valleys, crevices, or depressions in the body.

Vascular plants plants with a skeleton of conducting tissues that distribute water and food products throughout the body of the plant.

Vascular system the body's system of vessels that transport fluids such as blood.

Ventral toward the bottom (literally belly) of an animal.

Vertebra one of the units of the vertebral column or backbone.

Vestibular system part of the labyrinth of the inner ear which contributes to balance and sense of spatial orientation.

Vicariance biogeography a school of thought that interprets the history and geographic distribution of organisms in relation to the history and movements of continents and islands.

Viscera the digestive tract and other internal organs; guts.

Viviparity able to produce living young.

Volant capable of flight.

Vomer one of the bones of the palate, located at the anterior end of the skull.

Vomeronasal bulb a chemosensory bulb that receives input from the vomeronasal organ, often associated with detection of pheromones.

Zygote a fertilized egg.

Index

Ameghino, Carlos, 112
Ameghino, Florentino, 112
amino acids, 275
Amiot, R., 852, 853, 909
Amitabha, 405
ammonites, 234–35
ammonoids, 340
Ammosaurus, 424, 432, 953
amniotes: classification of, 221, 317; locomotion of, 319; metabolic and thermoregulatory characteristics of, 786–98; musculature of, 155; necks of, 657; skulls of, 316
amphibians, 961–63; aerobic capacity of, 822; bones of, 627; brains of, 198, 853; classification of, 215; cold-blooded, 874; energetics of, 820–21; and keratins, 277; locomotion of, 827; lungs of, 804; and respiration, 794; skull type of, 317; trackways of, 746
amphisbaenians, 974
anaerobic capacity and power, 820, 823, 825, 906
analogous anatomical features, 134, 135–36
anapsids, 317, 317, 964, 965–66
Anas platalea, 398
Anatalavis, 398
anatids, 398–99
anatomical directional terms, 136–38
Anatosaurus, 199
Anatotitan, 210
Anchiceratops, 536
Anchiornis, CP 18, 97, 366, 382
anchisaurians, 432–34, 437, 439, 440
Anchisauridae, 426
Anchisaurus, 13, 424, 427, 432, 440, 655, 851
Andalgalornis steuletti, 406
Anderson, J. F., 198, 630
Andersson, Johann G., 77, 78
Andrássy, P., 899
Andrews, Roy Chapman, 4, 77–78, 78, 79, 83, 85, 950, 1057
angiosperms, CP 27, 568, 945
anhingas (Anhingidae), 400, 401
Animantarx, 517, 519
animatronic dinosaurs, 300–301
ankles, 150
ankylosaurids: armor of, 516, 522; classification of, 505–506; distribution of, 517; features that characterize, 520; feeding of, 523–24; limbs of, 513; nasal airways of, 837; origins of, 521, 1015; paedomorphosis in, 775; pelvis of, 514; shoulder blade of, 512; skeletal reconstruction of, 506; skull of, 507, 507, 508; tail of, 511; teeth of, 509–10, 511, 775; vertebrae of, 512
ankylosaurines, 506
ankylosaurs, 505–24; armor of, 148, 485, 505, 505, 512, 514–16, 515, 516, 520, 521–23, 803; and Asian discoveries, 87, 90, 92, 99; biology and behavior of, 520, 521–22,

522; body mass of, 860; cheeks of, 851; classification of, 484, 486; distribution of, 517, 519–21; and early dinosaur discoveries, 18; early theories on, 332; eggs and nests of, 614; and evolutionary theory, 1007; feeding of, 523–24, 594; footprints and trackways of, 725, 726; limbs of, 494, 645; locomotion of, 827; musculature of, 155, 158, 161, 164, 168, 172, 177; neuroanatomy of, 201–202, 509; ontogenetic changes in, 768; origin and evolution of, 517–21, 517, 518, 519, 1010, 1012, 1013, 1015, 1016, 1017, 1018; postcranial skeleton of, 511–12, 511, 512, 513, 514; quadrupedality of, 726; skeletal features, 506, 518; skull of, 506–11, 507, 508, 510, 511; tails of, 511, 511, 522, 642–43; taxonomy of, 505–506; teeth of, 164, 509–10; trunks of, 650
Ankylosaurus, CP 28; classification of, 519; distribution of, 517, 520; limbs of, 644; skull of, 507, 508; teeth of, 510, 511; trunk of, 650
Ankylosphenodon pachyostosus, 972
Anomalogonatae, 395
Anomoepus, 723, 724, 737, 740, 742
Anoplosaurus, 517
Anseranas semipalmata, 398
anseriforms, 393, 394, 398, 398, 399
Antarctica: and biogeography, 943; and evolution of dinosaurs, 1008, 1009, 1010; and faunal succession, 1019; and geographic isolation, 950; and paleoclimatology, 943, 948, 953; and plate tectonics, 929, 1006; and southern continent dinosaur hunters, 114–15
Antarctic sheathbills (Chionididae), 407
Antarctopelta, 517
Antarctosaurus, 84, 447
anterior (term), 137, 138
Antetonitrus, 424, 432
anthracosaurs, 961
antibodies, 273–74, 274, 275, 276–79
antorbital fenestra, 139–40, 977
anurans, 963
Anurognathus, 981–82
Apatosaurus: bone growth of, 630; buoyancy of, 651; center of mass of, 636; classification of, 460, 461; and evolutionary theory, 1058; food requirements of, 908; footprints and trackways of, 725; and fossil collecting, 1052; growth rate of, 466–67, 467, 468, 769, 772, 846; limbs of, 145, 147, 645; locomotion of, 645; mounting of, 293; mouth of, 851; nasal airways of, 837; neck of, 657; pathologies of, 669, 671, 682, 682, 693, 694, 694, 696, 697; pelvic anatomy of, 146; peramorphosis in, 771–72; size of, 464; skeletal reconstruction of, 451; skull of, 661; sternal ribs of, 803; tail of, 640; tooth-marked

bones of, 593; tripodal stances of, 648–49; vertebrae of, 143, 448, 454
Aphanapteryx, 405
Apodidae, 411
apodiforms, 410
apodimorphs, 410
Appalachiosaurus, 360
Apsaravis, 392
Apterygidae, 396
Apteryx, 396
Aptornis (adzebills), 405
Aquila chrysaetos, 409
Aramidae (limpkins), 405
Araucaria (tree), 572, 572
Araucarioxylon, 310, 312–13
Arbour, V. M., 168
Archaeoceratops, 527, 536, 538
archaeocete whales, 286, 288
archaeology, 4, 7–8
archaeopterygids, 389
Archaeopteryx, CP 4; anatomy of, 388–89; arboreal existence of, 811; and Asian dinosaur hunters, 97; classification of, 366, 379; early theories on, 380; ectothermy of, 833; endocasts of, 192; and European dinosaur hunters, 57; and evolutionary theory, 70, 1013, 1058, 1065; feathers of, 281, 388, 854, 883, 908; footprints and trackways of, 724; growth of, 883; and Lebensraum hypothesis, 862; lungs of, 810–11, 903; musculature of, 174; nasal airways of, 796; neuroanatomy of, 201; Ostrom on, 381–82, 908; pelvic anatomy of, 364, 810–11, 810, 811; peramorphosis in, 777; and private fossil collections, 1046; relationship to birds, 240, 379, 379; respiration of, 840; skeletal reconstruction of, 281, 367; specimen, 388; thermoregulation of, 791
Archaeoraptor, 262–63
Archaeornis, 367
archaeornithine birds, 796
Archaeornithomimus, 81
archaeotrogonids, 411
Archaeovolans repatriatus, 253
Archean Eon, CP 1, 226
Archelon, 967
Archetype body plan, 1060
Archibald, J. David, 1028, 1031, 1033, 1035
Archosauria, 221, 322, 322, 323, 976–77
archosauriforms: and Archosauria, 976; classification of, 977; decline of, 327; early history of, 320–27, 320, 321, 322, 323, 324, 325, 326; features that characterize, 318, 318; locomotion of, 319–20; skulls of, 321; and succession of vertebrates, 319; teeth of, 348
archosauromorphs, 318, 318, 341, 966, 974–76
archosaurs, 316–27; appearance of, 318; and archosauromorphs, 975; brain of, 194, 200; and cladistics, 221; classification

of, 322, 977; and Crurotarsi, 978; early history of, 320–27; and endothermy, 791–92; evolutionary patterns, 319–20, 1011; extinction of, 341; and feathers, 791–92, 791; footprints and trackways of, 736–37; gait of, 746; genitalia of, 606; insulation of, 855; lungs of, 810; and Mesozoic faunal changes, 989; musculature of, 153, 155, 156, 158, 159, 160, 161, 164, 167, 169, 180; neuroanatomy of, 197, 201, 203; and North American discoveries, 69; pathologies of, 705; posture of, 319–20, 996; pulmonary system of, 917; reproduction of, 605, 606, 607–608; and succession of vertebrates, 319; teeth of, 660
Arcucci, A. B., 112, 322
Ardeidae (herons), 394, 400, 403
Argentavis, 410
Argentina, 111–13, 337, 1046
Argentinosaurus, 107, 464, 466, 904
Argusianus argus, 401
Argyrosaurus, 464
aristogenesis, 1066
armadillos, 950
armature, 290, 290, 294, 296–97, 296, 297, 298, 302
Arndt, Walter, 1045
Arrhinoceratops, 536
arsinoitheres, 860
arthritis, 689–96
artiodactyls, 860
Asia: and biogeography, 239, 240–41, 241, 954, 955; and early dinosaur discoveries, 4–5, 73–102, 108–109; and evolution of dinosaurs, 1008, 1009, 1010, 1011, 1015, 1016; and faunal succession, 1019; and geology, 238; key sites in, 95; and paleoclimatology, 953; and plate tectonics, 1006. *See also specific countries*
Asilisaurus, 336, 982
asities (Philepittidae), 418
asteroid impacts, 342, 997–98, 1027, 1030, 1034–36, 1037
Athenaeum, 30
Atlantic Ocean, 926, 930, 997, 1005
Atlantosauridae, 459, 460
Atlasaurus, 461
Atreipus, 734, 965
Aucasaurus, 354, 355, 655
Aufderheide, A. C., 668
Auffenberg, W., 827, 876
Augee, M. R., 1043
auks (Alcidae), 407, 407
Auroraceratops, 536, 538, 544
Austinornis, 392, 393, 400
Australia: and biogeography, 936, 954; and early dinosaur discoveries, 8–9; and evolution of dinosaurs, 1008, 1009, 1010, 1015; and faunal succession, 1019; and geographic isolation, 950–51; and paleoclimatology, 944, 951, 953; and plate

tectonics, 928, 929, 1006; and southern continent dinosaur hunters, 113–14
Australian treecreepers, 418
Australovenator, 359
Austriktribosphenos, 985
Austriodactylus, 980–81
Austroraptor, 368
Austrosaurus, 461
Avaceratops, 536
Avemetatarsalia, 977
avepods, 860, 864, 865
averostrans, 352, 353, 373
Aves, 156, 215–16, 221, 221, 379, 959, 1027
avetheropods, 221, 355, 356, 357
avialians: and *Archaeopteryx*, 379, 389; classification of, 366, 367, 379; locomotion of, 372; secondarily flightless members of, 383; size of, 373; term, 379; trends in, 379. *See also* birds
avimimids, 647–48
Avimimus, 365–366
axial skeleton, 142–44
Ayyasami, K., 489
Azendohsaurus, 336, 439

Bachelet, Abbé, 47
Bacon, Francis, 925
Bactrosaurus, 81, 552
Badiostes, 416
Bagaceratops, 92, 240–41, 536, 539–40, 766, 778–79, 779
Bahariasaurus ingens, 56
Baird, Donald, 13
Bakewell, Robert, 28
Bakker, R. T., 70–71, 184, 309, 332, 463, 499, 653, 873, 889, 890, 891, 907
Balaeniceps rex, 404
balance, 196, 648, 650, 652
Balearica, 406
Bambiraptor, 803
Banks, Joseph, 26, 27
Barapasaurus, 91, 109, 460, 461
Bara Simla, 73, 83–84
barbets (Capitonidae), 415
Barden, Holly, 492
Barosaurus: bones of, 890, 897; classification of, 461; fossils of, 447; neuroanatomy of, 204; respiration of, 658; tripodal stances of, 648–49
Barrett, P. M., 114, 434
Barrick, R. E., 790, 852, 853, 909
Barsbold, Rinchen, 89
Baryonyx: and air sacs, 804; bones of, 898, 899; classification of, 357; diet of, 594; and European dinosaur hunters, 57; size of, 360
basal sauropodomorpha. *See* prosauropods
Basilosaurus, 286
Bates, K. T., 639
bathornithids, 406
Bauhin, Caspar, 209
Bavarisaurus, 973

Bayesian analysis, 941
bee-eaters (Meropidae), 414
Beipiaosaurus, 97
Belgirallus, 405
Bellusaurus, 461
Bennett, S. C., 323, 980, 981
Benson, R. B. J., 355, 356, 357, 359
benthic organisms, 441, 738–39, 933
Benton, M. J., 42, 319, 332–33, 427, 487, 1028, 1029–30
Berberosaurus, 352, 353
Berkey, Charles P., 74, 79
Berman, D. S., 695
Bernissartia, 899
Berruornis, 410
beta keratins, 276–81, 276, 277
Beurlen, Karl, 1043
Bhattacharji, Durgasankar, 83–84
Bian Meinian (Edward M. Bien), 86
Bice, D. M., 342
Bicuspidon, 973
"Big Al": death posture of, 676–77; fracture repair in, 678–80, 679, 680, 686; infection in, 688–89, 689; pseudopathologies in, 377; size of, 655
biochrons, 233–34
biogeography: basic processes, 1004; historical vs. ecological, 931–32; and Late Mesozoic dinosaurian fauna, 1004–1005; and Modern Synthesis, 935–36; and predictions, 943, 953; theories of, 931–43
biomechanics, 636–63; bipedal stance (rearing), 648–50, 650; body mass, 637–39; center of mass, 636, 648–49, 650; and evolutionary theory, 1064; head, 659–62, 661; limbs, 643–48, 646, 651–55; neck, 655–59; posture, 743; tails, 639–43, 643; trunk, 650–51
biostratigraphy, 233–35, 236
bipedality: and aerobic capacity, 830; bipedality-quadrupedality transition, 651–52; and Crurotarsi, 978; and early dinosaur discoveries, 62; and evolutionary theory, 1068; and footprints and trackways, 114, 724, 725–26, 725, 728; mounting skeletons of, 295; and musculature, 171, 176, 179; museum exhibits of, 286; posture of, 203, 293; and pterosaurs, 981–82; and size of dinosaurs, 651–52; and southern continent discoveries, 113, 114; and tails, 642; of theropods, 348, 726; and velocity, 830. *See also* bipedal stance (rearing)
bipedal stance (rearing): for feeding, 579; and footprints and trackways, 471; and manus claws, 653; mechanics of, 648–50; for reproduction, 473, 648
Bipes, 974
Bird, R. T., 71, 470, 589, 718–19, 735, 744, 745
birds, 379–420; abdominal cavity of, 807; air sacs of, 685–86, 800, 802–803, 811–12,

839–42; anatomy of, 379–80, 384–88, 385, 386, 387, 388; arboreal existence of, 811; archosaur predecessors of, 322; and Asian discoveries, 97, 98–99; biomechanics of, 650–51; bird-hipped dinosaurs, 111, 113; birds of prey (raptors), 409–10; body temperatures of, 790; bones of, 889; brains of, 194, 195, *195*, 198, 199, 853; classification of, 379, 382–83, 883; Coraciiformes, 395, 414–15; cuckoos, 413; digits of, 777; and dinosaur extinctions, 1027, 1035; dorsal-ventral separation in, 171; enantiornithines, 390–91; and energetics, 821, 833; evolution of, 777, 1013, 1016, 1017, 1060–61, 1065, 1068, 1069–70, *1069*; falcons, 416, 417; feathers of, 383, 384, 387–88, 854; flamingos, 408; and flight, 97, 383–84, 822, 833, 1065; flightless birds, 383, 386, 389, 392, 397, 399, 405, 406, 417; footprints and trackways of, 725, 740–41, *741*, 743; fossil record of, 379, 380; future of, 419–20; gait of, 830; Galloanserae, 394, 398–400; genetic alteration of, expressing dinosaurian characteristics, 302; genitalia of, 605–606; and geologic time, 224, 240; grebes, 408–409; growth rates of, 846, *846*, 847, 850, 910; hands of, 382–83; healing response in, 679; hearts of, 656, 790, 839, 872, 874; Heilmann's *Origin of Birds*, 380–81; and heterochrony, 777; insulation of, 854–55, *854*; juveniles, 384; keratin proteins of, 276, 277; limb growth in, 628; locomotion of, 830; lungs of, 799, 800, 801, 803–12, 902–903; and Mesozoic faunal changes, 989, 996; metabolic rates of, 786, 850, 866, 877; metabolic type of, 786, 793, 797–98, 812, 821, 882, 883–84, 908, 918; mousebirds, 413–14, *413*; mouths of, 851; musculature of, 151–52, 155, 159, 160, 164, 165, 169, 177, 180, 181–82, 183; nasal airways of, 792, 793, 834, *834*, 835, 836, 837; necks of, 657; Neoaves, 394, 395, 400–407; and nomenclature, 154; ornithurines, 392; ornithuromorphans, 391–92; ossified tendons of, 693; and Owen's progression of life, 1059; paleognaths, 395–98; and paleopathology, 674–75, 685–86, 687, 688, 692–93, 702–703, 705; parrots, 416–17; passerine radiation, 417–19; pathologies of, 698–99; pelvic anatomy of, 810–11, *810*, *811*, 831, 841–42; Piciformes, 415–16; pigeons, 412–13; premaxillae and maxilla of, 435; radiation of, 388–90, 584; raptors (birds of prey), 821; reproduction of, 604, 605, 608; respiratory conchae of, 835–36; respiratory turbinates of, 792–94, *792*, 794; shorebirds and kin, 407–408; strisores, 410–12; thermoregulation of, 791, 793–95, 852, 854; and theropods, 380–83; of Triassic

period, 959; vertebrae of, 452, 695; wing-assisted incline running, 372–73, 384
Bishtahieversor, 360
Bissektipelta, 517
bite forces, 371, 662
Bjorndal, K., 848
Black, Davidson, 77
black-necked screamers, 398
black-throated huet-huet, 418
Blikanasauridae, 426
Blikanasaurus, 424, 432, 458
blood pressure, 656–57, 843
Boas, J. E. V., 381
body mass, 637–39, 786–87, 824, 831–33, *831*, 845–46, *846*, 860
body temperatures, 787, 790, 821, 824, 853–54, 859. *See also* thermoregulation
Bohaiornis, 391
Bonaparte, J. F., 112, 332, 351, 460, 488–89
bone paleopathology: broken bones, 677, *678*, 678, 680–86, 684; diagnosis, 672–74, *672*, *673*; disease models, 674–76; fracture repair, 678–80, *679*, 680, 685; healthy bones, 704–705; infections, 672, 679, 686–89, *688*, 690, 692, 700–703, *701*; pseudopathologies and postmortem damage, 677, *677*, 678; tumors, 696–98, *697*, 698
bones and skeletons, 135–48; and anatomical directional terms, 136–38; appendicular skeleton, 142, 144–48, *145*, *146*, *147*; and armature, 290, *290*, 294, 296–97, *296*, 297, 298, 302; axial skeleton, 142–44; bone accretion rates, 630, 631; bone isotopes, 852–54; bone spurs, 698; and cancellous tissue, 887; and cartilage, 623–24, *624*, 625, 626, 697–99; casts of, 288–89, 294; cavernous bones, 900–904, 902, 903; and circulatory systems, 887–88; circumference of, 198; and comparative anatomy, 290; composition of, 138; compression of, 265; dermal bones, 138; development of, 623–25, *624*; and ectothermy, 788–89; endochondral bone, 138; endochondral bone and epiphyses, 624, *624*, 625–28, *625*, *626*, 628, 633; of forelimbs, 652–53; and growth rates, 770, 789, 893–94; growth rings, 629–30, 629, 631, *631*, 897, 898, 899, 912–13; and heterochrony, 769; histology, 101, 788–89, 884–87, 888–90, 891–901, 905, 909–10; homology and analogy in, 135–36; and intermediate dinosaurs, 884–901; and limb motion, 291–93, *291*, 292; and locomotion, 644–45, 648; missing bones, 265–66, 296, 297; mounting of, 289–90, 290, 293–99, 294, 295, 296, 297, 298, 299; names of, 136; in necks, 657–58; oxygen isotope (O^{16}:O^{18}) ratios in, 790, 853; and rearing postures, 648; and size of dinosaurs, 646–47, 646; strength of, 644–45, 648; tooth-marked bones, 591–93, *592*,

662, 700; weight of, 290, 296, *296*. *See also* bone paleopathology; fibrolamellar bone; fossils; Haversian bone; periosteal bone; skulls; teeth; zonal bone
bony fishes (Osteichthyes), 960
bony-toothed birds (Pelagornithidae), 399, 402, *402*
boobies and gannets (Sulidae), 400, 401–402
Borges, J. L., 937
Borogovia, 727
Borsuk-Bialynicka, M., 183
botany: and ecological changes, 341; and illustration of dinosaurs, 305, 307, 308, 309, 310, 312–13; and paleoclimatology, 945–46; and sauropods, 473–74. *See also* plants
Bothrosauropodidae, 459
Bottjer, D. J., 952
Bourdon, E., 402
Bouvier, M., 889
bovids, 951
bowerbirds, 418
brachiosaurs, 111, 460, 658, 796, 850
Brachiosaurus: anatomy of, 457, 458; balance of, 653; and bipedal stance, 649, 650; body mass estimates of, 639; body-to-brain weight relationship, 199, 200; bones of, 633, 890, 897; buoyancy of, 651, 745; center of mass of, 636; classification of, 460, 461; feeding of, 473, 908; fossils of, 447; limbs of, 456, 643, 646, 652, 653; neck of, 656, 658; neuroanatomy of, 204, 498; peramorphosis in, 773; pulmonary system of, 873–74; size of, 464; skeletal reconstruction of, 451; skull of, 448–49, 450; tail of, 640; vertebrae of, 454; weight of, 630
Brachyceratops, 541
brachylophosaurids, 559, 593–94
brachyopoids, 963
Brachypteraciidae (ground-rollers), 414
Brachyrhinodon, 972
bradyaerobic organisms, 820, 823
bradyenergetic organisms, 820, 821
bradymetabolic organisms, 819–20, 875
brain: and metabolic rates, 852; reconstruction methods for, 262; and respiratory turbinates, 794; size of, 197–99, 852, 853. *See also* paleoneurology
braincase, 191; components of, 141; and endocasts, 191, 192–93, 199; function of, 192; and paleoneurology field, 191; and size of brain, 195, 198, 200; and skull components, 138
Branca, Wilhelm von, 51
branch-based taxon, 220–21, *221*
Branchiostoma, 1060
Brazil, 113
Breithaupt, B., 250, 261
Brett-Surman, M. K., 155
Breviceratops, 778–79, *779*
Britt, B. B., 903

footprints and trackways of, 728, 729, 730, 731, 732, 733; fossils of, 70; growth rates of, 849; hands of, 564; limbs of, 645; locomotion of, 563; musculature of, 151, 157, 164, 172, 174, 177; nasal airways of, 795–96, 837; ontogenetic changes in, 765; paedomorphosis in, 774; pathologies of, 681, 684, 685, 697, 698; peramorphosis in, 773; pleurokinetic system of, 562; as prey, 595, 699; radiation of, 583; skulls of, 139; social nature of, 563; tooth-marked bones of, 591; and wartime loss of fossils, 1042

Hadrosaurus: bipedality of, 62; early scholarship on, 551, 724; and evolutionary theory, *1060*; pathologies of, 683, 685; skeletal reconstruction of, 287, 288

Haematopodidae (oystercatchers), 407

Hagryphus, 366

Haines, R. W., 628

Halcyonidae, 414

Halcyornis, 416

halcyornithids, 416

Haldanodon, 984

Hall, J., 248

Hallam, A., 998

Hallet, Mark, 309

Halstead, L. B., 876, 908

Hammer, William, 114

hammerkops, 404

hands (manus), *134*, *135–36*, *145*, *176*, *348*, *776*

Hanna, R. R., 677, 679, 689

Hanssuesia, 531, 532, 533

Haplocanthosaurus, 461

Haplocheirus, 95, 364, 365

Harberetzer, J., 253

Harpymimus, 363

harriers, 409

Harris, G., 253

Hatcher, John Bell, 266, 473, 669, *669*

Haughton, S. H., 110–11, 212

Haversian bone, CP 6, 888–91; characteristics, 886–87, *886*; definition of, 674; development of, 678–79, 683–84, *684*, 900, 915; and endothermy, 909–10; vascularity, 886, 889–90, *891*

Hawkins, Benjamin Waterhouse, 40, 287, 288–90, *289*, 299, 300, *1060*

hawks, 409

Hayashi, Shoji, 497, 499

Hayden, Ferdinand, 61, 723

heads: mechanics of, 659–62, *661*; and scanning technology, 262. *See also specific components, including* teeth *and* skulls

hearing, 196

hearts: double-pump circulation, 872, 873, *873–74*, 880–81, 888, 896, 903, 905, 906, 911–12, 914, 917; and endothermy, 905–906; and growth rates, 888; and metabolic rates, 790–91; and neck posture, 656–57, 843–44

Hedin, Sven, 84

height of dinosaurs, 638

Heilmann, Gerhard, 380–81, 1065, 1066

Heinrich, R. E., 771

Heintz, A., 1041–42

Heliornis fulica, 405

Heliornithidae (finfoots), 405

Hell Creek Mural, CP 14

Heloderma, 972

hemipodes, 407

Hemiprocnidae, 411

Henderson, D. M., 639, 745, 747

Henkelotherium, 984

Hennig, E., 487–88

Hennig, Willi, 70, 216, 936

Henodus, 971

herbivores: coprolites of, 595, 596–98, *597*; digestive systems of, 651; and dinosaur extinctions, 1034, 1036; and evolutionary theory, *1018*, *1019*, *1020*, *1068*; fossilized stomach contents of, 593–94; and gastroliths, 594; habitats of, 584; jaws of, 434–35; limbs of, 651; ornithopods, 551; and plants, 569, 575, 577, 578 (*see also* plants); quadrupedality of, 651; sauropods, 449, 472–74; trunks of, 650

herd sizes, 581–82

Hermann, Jean, *1063*

Hernandez, T., 878, 879, 888

Herodotus, 6

herons (Ardeidae), 400, 403

Herpetotheres cachinnans, 416

herrerasaurs, 330, 337, 339, 341, 348, 350–51

Herrerasaurus: classification of, 330, 337, 339; features that characterize, 337–38; footprints and trackways of, 731, 733; fossils of, 336; limbs of, *172*, *181*, 654; musculature of, *167*, *172*, *176*, *181*; neck of, 655; and origins of theropods, 348, 350; pelvis of, 802, 809; skeletal reconstruction of, 333, 349; skull of, 660; and southern continent discoveries, 112

Hesperonychus, 347, 368

Hesperornis, 392, *392*, 393, 1065

Hesperornithes, 379

hesperornithiforms, 379, 392

hesperornithines, 392, *392*, 393

Hesperosaurus: armor of, 495, *496*, 498, 499; classification of, *486*; distribution of, 488; limbs of, *494*; neck of, 497; ossified tendons of, 493

heterochrony, 760–81; classifying heterochrony, 762–64; definition of, 761; and developmental tradeoffs, 775–76; dissociated heterochrony, 775–76; and evolutionary novelties, 776–78; and evolutionary theory, 1066; and evolutionary trends, 778–79; in the fossil record, 768–69; and life history strategies, 776, 779–80; ontogenetic growth, 764–68, 776; paedomorphosis, 760, 762–63, 768–69,

770, 774–75; peramorphosis, 760, 762–63, 768–73, 773; six processes of, 760, 763; and size of dinosaurs, 769–74

heterodontosaurs: classification of, 550, 558–60; and diversity of herbivores, *1018*; and feathered dinosaurs, 98; hands of, 563; insulation of, 855; locomotion of, 563; teeth of, 562

Heterodontosaurus: classification of, 559; footprints and trackways of, 728, 731, 733; hands of, 564; limbs of, 644, 647; skeletal reconstruction of, 560; and southern continent discoveries, 111

heteroenergetic organisms, 820

Hexinlusaurus, 550, 560, 561

hibernation, 856–57

Hieraaetus moorei, 409, 410

high-field magnetic resonance imaging, 193

Hillenius, W., 840, 841, 1061

Himalayas, 944

Hintze, Ferdinand F., 69

hippopotamus, 860

hips, *150*

histology, 1066–67

Hitchcock, Edward, 9, 61, 713, 722–25, *723*, 732, 737, 739, 740

hoatzin (*Hoazinoides*), 413

Hodych, J. P., 997

Hoffet, Josué Heilmann, 5, 85

Hoffstetter, R., 487

Holland, W. J., 65, 669, 743–44

Hollanda, 392

Holliday, C. M., 154, 156, 157, 159, 160, 164

Holmes, George Bax, 29, 30–31, 34, 46

Holmes, R., 544, 745

Holmgren, Nils, 381

holocephalians, 960

Holosteans, 960–61

Holt, E. C., 110

Holtz, T. R., Jr., 71, 364, 370

Homalocephale, 526, 530, 531, 533, 533

Homalosauropodidae, 459

homeothermy, 876; and body mass, 824; and bone isotopes, 852; definition of, 875; of ectotherms, 787; food requirements of, 582; of mammals, 821; mass homeothermy, 875–76, 912, 914; and Mesozoic Era climates, 787, 798; and size of dinosaurs, 875–76

hominids, 110, 775

Homo erectus ("Peking Man"), 77, 86

homology, *134*, *135–36*

Homo sapiens, *134*, *135–36*, 221

Hone, D. W. E., 371, 465

honeyeaters, 418

honeyguides (Indicatoridae), 415

Hongshanornis, 392

Hongshanosaurus, 536–37

Hooke, Robert, 9–10

Hooper, K., 252

hoopoes (Upupidae), 414, 415, 418

Hopkins, William, 39

museums, 285–302; attendance of, 286; and demand for fossil collection, 1040; early exhibits in, 286–90, 286, 287; education and outreach of, 265–68; and field documentation, 128; and illustration of dinosaurs, 307; mounting of skeletons, 289–90, 290, 293–99, 294, 295, 296, 297, 298,299; online access to, 257; plastic casts in, 294; renovation of existing exhibits, 294–95; and restorations of dinosaurs, 299–301; skeletons on exhibit in, 265, 292; technological advances in, 255–57; and theft of dinosaur fossils, 1046; virtualizing specimens, 255–56; virtual museums, 301; and wartime loss of fossils, 1042–45

Musophagidae (turacos), 413
Mussaurus, 424, 427, 432, 438, 768
Muttaburrasaurus, 550, 554, 554, 556
Mycteria, 403
Myhrvold, N. P., 695
Mymoorapelta, 514, 517, 518
Mynde, J., 11
mythological creatures, 4, 6–7, 74, 75, 937–39, 938, 939

Naish, D., 531
Najash, 973
naming of dinosaur specimens, 1053
Nanotyrannus, 774, 795, 797
nasal airways, 795–98, 797, 833–38, 834, 836, 837, 864, 909
Nasmyth, Alexander, 41
Native Americans, 7, 8, 61, 1048, 1050–51
Natunaornis, 412
natural selection: and allopatric speciation, 949, 950; and evolutionary theory, 1057–58, 1062; and heterochrony, 778; and life history strategies, 776; and North American dinosaur hunters, 68
Necker, R., 205
necks, 166, 170–71, 655–59
Nedoceratops, 536
Nelson, Gareth, 941
Nelson, Nels C., 74, 82–83
nemegtosaurids, 462
Nemegtosaurus, 90, 461, 462, 768, 773
Nemegt Valley and Basin, 87–88, 90, 95
Neoaves, 394, 395, 400–407
neoceratopsians, 472, 537–38, 545, 778–79, 954–55
neo-Darwinism, 1064
Neodrepanis coruscans, 418
Neogaeornis, 404
Neogene Period, CP 1
Neognathae, 394, 394
neognaths, 387, 394
neornithines: anatomy of, 384–88; classification of, 379, 393–95, 394; diversity of, 419–20; enantiornithines, 390–91; endothermy of, 883, 908;

hands of, 382–83; Ornithurae, 392; rise of, 392–93. *See also* birds
Neosauropoda, 460, 461, 462, 464, 465
neotetrapods, 340, 341, 350–51, 353, 371
Neovenator, 359, 360
Neovenatoridae, 359
Neozoic Era, 1059
Nesbitt, S. J., 351, 353
Nestor notabilis, 416
nests. *See* eggs and nests
Nettapterornis, 398
Neumann, John von, 946
Neuquén Basin, 112–13
Neuquenornis, 391
Neuquensaurus, 454, 461, 465
neutron tomography, 260
Newberry, John Strong, 62
Newman, B. H., 655
New World vultures (Cathartidae), 409
New World warblers, 419
New Zealand: and biogeography, 936; dissenting opinions on plate tectonics, 928; and early dinosaur discoveries, 9; and geographic isolation, 950; and paleoclimatology, 953; and southern continent dinosaur hunters, 113–14
Ngorongoro Crater, 936
Nicrosaurus, 324
Nigersaurus, CP 3, 449, 461, 472, 850, 851; mouth of, 850
nightbirds, 410
nightjars (Caprimulgidae), 410
Niobrarasaurus, 517, 519, 519
Nipponosaurus, 85
Noah's flood, 4
"Noah's Raven," 4, 9, 61
noasaurids, 353, 372, 373
Noasaurus, 107, 354
node-based taxon, 220–21, 221
Nodocephalosaurus, 517
nodosaurids: armor of, 516, 516, 521; cheeks of, 163; classification of, 505; distribution of, 517, 521; evolution of, 519, 1015, 1017; feeding of, 524; limbs of, 513, 514; ontogenetic changes in, 768; pelvis of, 514; shoulder blade of, 512; skeletal reconstruction of, 506; skull of, 507, 507, 508, 509; tail of, 511; teeth of, 509, 510, 511; vertebrae of, 512
Nodosaurus, 517, 519
nomenclature, 136, 154
Nomina Anatomica Avium (NAA) nomenclature system, 136, 137
Nopcsa, F., 52–54, 53, 204–205, 380, 484, 487, 488, 774, 1066–67
Norell, Mark, 96
Norian Age: boundaries of, 229, 341; and earliest dinosaurs, 336, 337, 340, 342, 343, 343; ecological change in, 340–41; major events of, 342; and mass extinctions, 999; theropods of, 348
Norman, D. B., 55, 484, 550

North Africa, 55–57
North America: and biogeography, 942, 954; and dinosaur extinctions, 1028; and dispersalist biogeography, 934; and early dinosaur discoveries, 61–72; and evolution of dinosaurs, 1008, 1008, 1009, 1010, 1011, 1012, 1013, 1015, 1016; and faunal succession, 1019; and paleoclimatology, 948, 953; and plate tectonics, 926, 930, 1006
Northcutt, R. G., 194
northern goshawk, 409
Nothdurft, W., 56
nothosaurs, 968–69
Nothronychus mckinleyi, 363
notoungulates, 860
Novas, F. E., 112, 333, 339, 489
Nqwebasaurus, 111, 1014
Nupharanassa, 408
Nyctibiidae (potoos), 410, 411

Obruchev, Vladimir A., 77, 80
occipital condyle, 141–42
O'Conner, M., 857–58
Odontoanserae, 399
Odontochelys, 966
Odontopterygiformes, 402
Odontornithes, 1065
Ogygoptynx, 410
Oheim, K., 248
oilbirds (Steatornithidae), 410, 412
Ojoceratops, 536
Olduvai Gorge, 110
olfactory system, 193, 793, 794
Oligokyphus, 995
Olsen, George, 64, 66, 80, 82
Olsen, K. H., 699
Olsen, P. E., 342, 740, 994, 997, 998
Olson, E. C., 1061–62
Omeisaurus, 86, 460, 461, 851
omnivores, 434–36
omnivoropterygids, 379
Omnivoropteryx, 253, 390
Omosaurus, 483, 487
ontogeny, 68–70, 101–102, 212
Ophthalmosaurus, 968
Opisthias, 959, 972
Opisthocoelia, 445
Opisthocoelicaudia, 90, 451, 457, 461, 471, 650
Opisthocoelinae, 461
Opitsch, Eduard, 1046
optic lobes, 195
optic nerves, 195
orders, 214–15
Ordovician invertebrate shells, 275
Organ, C. L., 167, 693
origination biogeography, 1004, 1004–1005
origins of dinosaurs, 331–32.
   See also evolution
Orkoraptor, 359
Ornatotholus, 533

643; mechanics of, 639–43, 643, 657; mobility of, 148; musculature of, 167–68, 181; and orientation of body, 642; and trackways, 470, 639–40, 723, 723, 747–48; and tripodal stances, 473, 579, 648–49, 650; vertebral fusion and joint disease in, 693–96; as weapons, 642–43, 695

takahe, 405

Talarurus, 517, 519, 520, 725

Talpanas, 399

Tanius, 552

Tanke, D. H., 20, 681, 689, 698, 699, 700, 703

Tanycolagreus, 359

Tanystropheus, 318, 974, 975

Tanytrachelos, 974

Tanzania, 50–52

tapaculos, 418

Taquet, P., 5, 56–57, 76, 110

Tarbosaurus, 88, 90, 193, 362, 727, 772

Tarchia, 517, 520

Tasmaniosaurus, 321

Taubatornis, 410

Tawa, 349, 350–51

taxa, 209–10, 220–22

Taxodiaceae (bald cypress family), 572

taxonomy, 25–42, 209–14, 1027. *See also* clades and cladistics

Taylor, Bert L., 493

Taylor, E. L., 570

Taylor, T. N., 570

Tazoudasaurus, 454

technology, 246–68; in body mass estimates, 639; and collections, 255–57; digitization of resources, 248; in education and outreach, 265–68; in excavations, 128–31; in the field, 247–52, 714, 718; in footprint documentation, 714, 718, 719, 720; importance of, 247; in the laboratory, 252–55; in paleoneurology, 192–93; in paleopathology, 671, 673, 673; in prospecting for potential sites, 128–31, 248–49; in research settings, 257–64; virtual museums, 301

Technosaurus, 558

Tecovasaurus, 336

teeth: and Asian discoveries, 76, 92; *Bicuspidon*, 973; and digestive system, 909; and dinosaur extinctions, 1031; and early dinosaur discoveries, 10–11, 12; and early mammals, 984–85; and evolutionary theory, 1062; and feeding behaviors, 160, 589, 590; and market for fossil discoveries, 1041; mechanics of, 659–60; and Mesozoic faunal changes, 992, 995; and musculature of dinosaurs, 166; and plants as food source, 661; and predation, 659, 699–700; premaxillary teeth, 510, 561, 562, 775; and pterosaurs, 981, 982; and rate of food intake, 850–51; and skull anatomy, 139–40; sphenodontians, 971–72; of theropods,

1053; tooth-marked bones, 591–93, 592, 662, 700; ziphodont teeth, 348, 350, 362

Tehuelchesaurus, 461

Teilhard de Chardin, Pierre, 83, 84, 86, 1062

Teleosauridae, 326

teleosts, 961

Telmatosaurus, 53, 557, 774

temnospondyls, 340, 341, 961–63, 962, 993–94, 995, 1011

Tendaguru, 50–52, 110, 447, 897

Tenontosaurus: classification of, 550, 554, 555; developmental tradeoffs in, 776; and evolution of dinosaurs, 1015; footprints and trackways of, 728, 729, 730, 731, 732, 733; ontogenetic changes in, 765; paedomorphosis in, 774, 775; peramorphosis in, 771, 773; as prey, 369, 555, 588, 590, 593; sexual maturity in, 765; skeletal reconstruction of, 555

teratorns, 409–10

teratosaurids, 426

terns, 407

terramegathermy, 858–60

terrestrial ecosystems, 959

Terrestrisuchus, 326, 801–802, 802, 809, 979, 979

Tertiary boundary, 342, 1027–28, 1031, 1032–35, 1036–37

Tertrema, 962, 962

tetanurans, 796, 942

tetanurines, 351, 353, 354–56, 371, 372, 660

tetrapods: and biogeography, 941; classification of, 215–16; and competition vs. ecological opportunism, 995–96; and derived characters, 219; and evolutionary theory, 1069; features that characterize, 135; food of, 579, 880; and German discoveries, 50; growth rates of, 628, 846–47, 888; and Mesozoic extinctions, 996–1000; musculature of, 152, 171, 174, 175; nomenclature for, 136; ontogenetic changes in, 776–77; thermoregulation of, 857–58; and Triassic faunal succession, 990–95

Texascephale, 533

Texasetes, 517

texture mapping, 260–61

Teyuwasu barberenai, 336, 338

Thailand, 76, 91, 93, 99, 102

Thalassodromeus, 982

thecodont hypothesis, 381

thecodontians, 216, 340, 381, 1065

thecodontosaurids, 426, 427–28

Thecodontosaurus, 18, 428–29, 440; classification of, 330, 424, 427; evolution of, 440; limbs of, 651

Thecospondylus, 205

theft of dinosaur fossils, 1044, 1045–47

therapsids: evolution of, 317; food of, 579; growth rate of, 911; and Mesozoic faunal changes, 996; metabolic rates of, 895;

nasal airways of, 795; size of, 882; and succession of vertebrates, 319, 911

therizinosaurs: and Asian discoveries, 90, 97, 98–99; and biogeography, 239–40; body mass of, 860; classification of, 363; eggs and nests of, 615–16; and evolution of dinosaurs, 1015, 1018, 1019; and feathered dinosaurs, 97; features that characterize, 364–65; footprints and trackways of, 727; locomotion of, 372; musculature of, 175; pubis orientation, 364; size of, 373

Therizinosaurus, 90, 365

thermoregulation: and body mass, 824–25; characteristics of, 786–92; and colder climates, 855–57; and energetics, 820, 821, 854, 857–58, 865; and feathers, 791–92, 854, 883, 915; function of, 879–80; in hot conditions, 857–58; and pterosaurs, 982; and respiratory turbinates, 792–98, 792, 793, 794, 796, 797, 798; and size of organisms, 824–25, 875–77, 877; and vascular sinuses, 197

theropods, 347–73; and air sacs, 651, 804–805, 808; and amateur discoveries, 57; and Asian discoveries, 76, 81, 84–85, 87, 89–95, 97–98; and birds, 380–83; bones of, 624, 627, 898, 901; brains of, 194, 195; characteristics of, 348, 362; classification of, 330, 347; coprolites of, 598, 598, 852; deinonychosaurs (raptors), 366–69, 367; diet of, 594; digestive systems of, 909; digits of, 777; and early dinosaur discoveries, 4, 12, 47; early theories on, 332; early theropods, 351–59, 355–56, 358; ectothermy of, 787, 812; eggs and nests of, 613–16, 619; endocasts of, 192; and endothermy, 791; and evolutionary theory, 1007, 1007, 1010, 1011, 1011, 1012, 1014, 1015, 1017, 1018–20, 1065, 1066, 1069, 1069–70; and extinctions, 1000, 1027, 1031; and faunal succession, 1019; and feathered dinosaurs, 97, 98; feeding of, 371–72, 591, 593; footprints and trackways of, 71, 589–90, 712, 721, 722, 727, 728, 729, 729, 732, 733–34, 733, 735, 736, 746–47, 747; fossils of, 327, 1053; gait of, 830; and geologic time, 240; growth rates of, 764–65, 770; hunting behavior, 589–90; insulation of, 855; limbs of, 372, 647–48, 651, 654, 655; locomotion of, 348, 372–73, 726, 746–48, 747, 750–51, 751, 829; lung morphology and ventilatory mechanisms, 803–12; metabolic capacities of, 787; musculature of, 161, 164, 168, 169, 169, 172, 173, 174, 181; nasal airways of, 795, 796, 797, 835–37; necks of, 655; neotheropods, 350–51, 353, 371; neuroanatomy of, 200–201; noncarnivorous theropods, 362–66, 371; and North American discoveries, 69, 70; origins of, 348–51, 349; paedomorphosis in, 774; pathologies of, 688–89, 700–701; pelvic anatomy

and trackways of, 730, 731, 732, 733, 736; growth rates of, 770, 770; jaws of, 660; limbs of, 647–48, 654–55; locomotion of, 362, 372; olfactory abilities of, 370; paedomorphosis in, 774; pathologies of, 681, 688; peramorphosis in, 770–71; predation of, 362, 699; sizes of, 373, 770–71; skulls of, 770–71; stereopsis of, 370; tooth-marked bones of, 591

tyrannosauroids: and Asian discoveries, 97; basal tyrannosauroids, 359–60; classification of, 360; eggs and nests of, 614; and evolution of dinosaurs, 1012, 1014; and feathered dinosaurs, 97; features that characterize, 360–61, 371; feeding of, 362, 371–72, 591; forelimbs of, 775; hands of, 362; locomotion of, 361, 372; size of, 360, 362, 373; skeletal reconstruction of, 361; teeth of, 591, 593

tyrannosaurs: and Asian discoveries, 88, 93, 94, 99; and biogeography, 942–43; and dispersalist biogeography, 934; and evolution of dinosaurs, 1014; and feathered specimens, 102; footprints and trackways of, 727; juveniles, 705; mass-specific SMR of, 879; musculature of, 157, 161, 164, 166, 179, 181; paedomorphosis in, 769; pathologies of, 692, 704–705; pelvic anatomy of, 832

Tyrannosaurus: bite forces, 662; body mass estimates of, 639; body-to-brain weight relationship, 199; bone growth of, 630, 631, 886; brain of, 194, 199; coprolites of, 598; developmental tradeoffs in, 775; digitization of, 258, 260; feeding of, 593; footprints and trackways of, 736, 747; and fossil collection, 1039, 1040–41, 1051; growth rate of, 629, 764, 765, 770, 770, 781, 846, 910, 913; and heterochrony, 781; limbs of, 145, 147, 292, 646, 647, 654–55, 770, 775, 776; locomotion of, 184, 647; metabolic rate of, 909; mounted skeleton of, 298; musculature of, 150, 151, 152, 165, 173, 174, 175, 176, 184; as museum exhibit, 285, 286; nasal airways of, 837; neuroanatomy of, 200–201, 205; ontogenetic changes in, 765, 772; ossified tendons of, 641; paedomorphosis in, 775; pathologies of, 672, 673, 686–89, 686, 687, 696, 700–701, 701, 705; pelvic anatomy of, 146, 832; peramorphosis in, 771, 775–76; posture of, 285, 286; predation of, 699, 700, 781; restoration of, 301; and r–K continuum, 781; sexual maturity in, 765; size of, 347, 359, 362; skeletal reconstruction of, 361; skull of, 141, 660, 661, 662; speed estimates for, 749–50; tail of, 640, 641; and taxonomy, 210, 211; teeth of, 659; used as model saurischian, 183; vertebrae of, 492. *See also* "Sue" (*Tyrannosaurus rex* specimen)

Tyrannotitan, 358

Tyrrell, Joseph Burr, 65–66
Tytthostonyx, 403

Udanoceratops, 536, 539
uintatheres, 860
Unaysaurus, 424, 427, 440
Unenlagia, 802
ungulates, 1036–37
uniformitarianism: definition of, 225; in geologic time, 225; in geology, 236; in metabolic rates, 911; in paleopathology, 668, 671, 674, 702; in physiology, 622, 904
United Kingdom, 11–13, 18–19, 76, 337
university curricula, 1069–70
Upchurch, G. R., Jr., 584
Upchurch, P., 427, 460, 462, 653, 937, 941
Upupidae (hoopoes), 414, 418
Upupiformes, 415
urogenital systems, 602
Utahceratops, 536
Utahraptor, 368
Uzbekistan, 954

Vagaceratops, 536
Valdosaurus, 554
Vannier, M. W., 253, 262
Van Straelen, Victor, 55
Van Valkenburgh, B., 909
Varanus (Megalania) prisca, 776, 831
Vargas, A. O., 777
variation within species, 212, 1063–64
Varricchio, D. J., 630, 701
Vasconcelos, C. C., 350–51
vascular tissues, 197
Vastanavis, 417
Vegavis, 393
Velociraptor, CP 7; and Asian discoveries, 81, 90; classification of, 368; and feathered dinosaurs, 97–98; fossils of, 250; musculature of, 175; nasal airways of, 837; and paleobehavior, 99; pelvic anatomy of, 811; predation of, 539, 590–91; skeletal reconstruction of, 367
Velocisaurus, 354
ventral (term), 137, 138
Vermeij, Geerat, 952
Versluys, J., 164
vertebra and vertebral column, 142–43, 143, 144, 148, 693–96, 694
vestibular structures, 196
vicariance biogeography: and allopatric speciation, 949; and continental drift, 932; contrasted with dispersal, 935–41; described, 932–33; and geodispersal, 943; and geographic isolation, 950–51; and Late Mesozoic dinosaurian fauna, 1004–1005; and other biogeographic processes, 1004
Vickaryous, M. K., 518
Vickers-Rich, Patricia, 113
Vila, B., 744–45
Virchow, Rudolf, 667

vision, 195
Vjushkovia triplocostata, 321, 321
vocalization, 154
volcanism: and dinosaur extinctions, 1034–36, 1037; and Mesozoic faunal changes, 998–1000; volcanic ash falls, 237; volcanic islands, 926–27; volcanic stratiform rocks, 228
Vrba, Elisabeth, 951, 952
Vulcanodon, 460, 461, 693
vultures, 409

Waimanu, 404
Walkden, G., 998
Walker, A. D., 174–75, 381
Walker, Cyril, 390
Walker, William, 57
Wallace, Alfred Russel, 933
Wang, S. C., 581
Wannanosaurus, 529, 530, 531, 533
warblers, 419
Warburton, Henry, 15, 25
warm-blooded organisms, 875
waterfowl or wildfowl (Anseriformes), 398–99
Weaver, J., 850, 908, 909
Webb, Philip Barker, 25–26
Webster, Thomas, 12
Wedel, M. J., 804
Weems, R. E., 728
Wegener, Alfred, 925
weights of dinosaurs, 199, 630–31, 638
Weishampel, D. B., 157, 164, 771, 954, 1031, 1032, 1064, 1066
Wellnhoferia, 367
Wennerbom, John, 39
Werner, Abraham Gottlob, 12
Werning, S., 765
West Africa, 110
whales, 466–67, 838, 846, 908–909, 1069
Wheeler, E. A., 583, 584
Wheeler, P., 775
Whitehurst, John, 12
white-light scanners, 250, 261, 261
Whiteside, J. H., 999
White's thrush, 419
whooper swans, 398
Whyte, M. A., 489
Wichmann, R., 447
Wied, Wilhelm zu, 54
Wiffen, Joan, 114
Wignall, P. B., 998
Wild, R., 323, 980
wildfire, global, 1036
Wilhite, R., 258
Williamson, T. E., 531, 533
Williamsonia, 573
Willis, K. J., 570
Williston, S. W., 63, 383, 668, 670, 671, 982
Wilson, J. A., 136, 144, 460, 462, 472, 645, 727
Wing, S. L., 584

This book was designed by Jamison Cockerham and set in type by Tony Brewer at Indiana University Press and printed by Sheridan Books, Inc.

The text type is Electra, designed by William A. Dwiggins (circa 1935), and the headings and captions are set in Frutiger, designed by Adrian Frutiger (circa 1975), both issued by Adobe Systems. The display type is Futura, designed by Paul Renner (in 1927), also issued by Adobe.